Calculus Formulas

In the following, a, b, c, and n are constants; u and v are functions of x; and x and y are functions of t. Logarithmic expressions are to the base $e = 2.718\,28\ldots$ All angles are measured in radians.

DERIVATIVES

1. $\frac{d}{dx}[cu] = c\frac{du}{dx}$

2. $\frac{d}{dx}[u + v] = \frac{du}{dx} + \frac{dv}{dx}$

3. $\frac{d}{dx}\left[\frac{u}{v}\right] = \frac{v\frac{du}{dx} - u\frac{dv}{dx}}{v^2}$

4. $\frac{d}{dx}[uv] = u\frac{dv}{dx} + \frac{du}{dx}v$

5. $\frac{d}{dx}[u]^n = nu^{n-1}\frac{du}{dx}$

6. $\frac{d}{dx}[a^u] = (a^u \ln a)\frac{du}{dx}$

7. $\frac{d}{dx}[\sin ax] = a\cos ax$

8. $\frac{d}{dx}[\cos ax] = -a\sin ax$

9. $\frac{d}{dx}[\tan ax] = a\sec^2 ax$

10. $\frac{d}{dx}[\ln u] = \frac{1}{u}\frac{du}{dx}$

11. $\frac{du}{dt} = \frac{du}{dx}\frac{dx}{dt}$ (the "chain rule")

12. $\frac{du}{dv} = \frac{\left(\frac{du}{dx}\right)}{\left(\frac{dv}{dx}\right)}$

INTEGRALS

1. $\int a\,dx = ax + c$

2. $\int [u + v]\,dx = \int u\,dx + \int v\,dx + c$

3. $\int_a^b u\,dv = (uv)\Big|_a^b - \int_a^b v\,du$

4. $\int \frac{dx}{ax + b} = \frac{1}{a}\ln(ax + b) + c$

5. $\int x^n\,dx = \frac{x^{n+1}}{n+1} + c \qquad (n \neq -1)$

6. $\int \sin ax\,dx = -\frac{1}{a}\cos ax + c$

7. $\int \cos ax\,dx = \frac{1}{a}\sin ax + c$

8. $\int \tan ax\,dx = -\frac{1}{a}\ln(\cos ax) + c$

9. $\int \sin^2 ax\,dx = \frac{x}{2} - \frac{\sin 2ax}{4a} + c$

10. $\int \cos^2 ax\,dx = \frac{x}{2} + \frac{\sin 2ax}{4a} + c$

11. $\int (\sin ax)(\cos ax)\,dx = \frac{\sin^2 ax}{2a} + c$

12. $\int \frac{dx}{\sqrt{a^2 - x^2}} = \begin{cases} \sin^{-1}\left(\frac{x}{|a|}\right) + c \\ -\cos^{-1}\left(\frac{x}{|a|}\right) + c \end{cases}$

13. $\int \frac{dx}{a^2 + x^2} = \frac{1}{a}\tan^{-1}\left(\frac{x}{a}\right) + c$

14. $\int \frac{dx}{(a^2 + x^2)^{3/2}} = \frac{x}{a^2\sqrt{a^2 + x^2}} + c$

University Physics

University Physics

Alvin Hudson
Occidental College

Rex Nelson
Occidental College

Harcourt Brace Jovanovich, Inc.
New York San Diego Chicago San Francisco Atlanta
London Sydney Toronto

COVER: White light reflected from a soap film produces interference colors. Photograph by Karol J. Mysels.

Photograph and illustration sources appear on pages 961–964.

Printed in the United States of America

Library of Congress Catalog Card Number: 81-80607

ISBN: 0-15-592960-7

Preface

This text is designed for the usual two- or three-semester introductory physics course for scientists and engineers. We have written it for students who share a desire to learn physics but who may have a wide range of talent or prior preparation. New mathematical concepts are introduced as needed, consistent with the progress of topics in an introductory calculus course. Physics concepts are presented initially in broad terms, then formulated mathematically, and subsequently sharpened and illustrated with practical examples. During the first few weeks of instruction, it is important to give special attention to establishing good habits in approaching problems and to be explicit in elucidating the steps of reasoning that trained physicists are accustomed to using. Consequently, the first few chapters emphasize the importance of a systematic approach to problem solving. With some careful assistance at this stage, many students who would otherwise fall by the wayside can be sustained and supported until they feel more comfortable with a new (for many of them) way of analytic thinking.

The sophistication with which material is presented becomes greater in subsequent chapters as the student becomes more familiar with the analytic procedures and linear thinking required in physics. Our topic sequence follows the traditional pattern of mechanics (including special relativity), wave motion, electricity and magnetism, optics, and a brief discussion of quantum ideas. Some rearrangement of the order or omission of certain topics is possible.

We believe our emphasis on methods of approaching the subject is unique. We try to dispel the widespread notion that learning physics requires memorizing "all those formulas." Instead, we emphasize that the physicist's approach is to employ a few general principles that have immensely wide applicability. In mechanics, for example, we present just three ways of solving mechanics problems: using Newton's laws, using ideas of energy and momentum, and using the relation between a conservative force and a potential (a relation that is developed more fully in electricity and magnetism). We encourage students to examine a problem for the clues that suggest which method is likely to be the most effective in solving it. Each example therefore focuses on the crucial first steps in analyzing phenomena by stating the general principle in equation form before specific symbols are substituted. Furthermore, just as students are trained to use good English in their writing, we are careful to write "good mathematics," making the physical reasoning clear and precise in each step of a mathematical presentation.

Recognizing that many users of this text may not be physics majors, we set forth a wide range of examples and problems. Brief special topics and comments also are included from time to time to illustrate physics principles in a variety of applications, though we adhere to the physicist's way of analyzing phenomena.

Yet our coverage of subject matter fully prepares students to continue to the next level of physics courses as currently taught in major institutions. An accompanying instructor's manual presents suggested syllabi for courses of varying lengths and emphases. We also indicate sections that can be omitted without affecting later continuity, thereby offering some leeway for instructors to tailor the content to match their own preferences or to meet the needs of the class at hand.

The metric system (SI) is used throughout. However, since the transition period to exclusive metric usage will continue for several years, the British engineering system of measurement is introduced in mechanics. We have also retained a few non-SI units, such as the *atmosphere* and the *electron-volt*, because of their great convenience and widespread usage.

The extensive problems and questions (over 1680) are the result of much careful thought. To avoid page-turning, answers to odd-numbered problems are given *in situ*. Problems are arranged in three groups. Group A problems are the easiest, primarily testing basic concepts and equations. Group B problems are of average difficulty, involving some insight into the physics of real situations. Group C problems are more difficult, usually requiring intermediate calculations before the final answer is obtained. For the convenience of instructors whose students are just beginning their study of calculus, problems in mechanics (Chapter 2 through Chapter 14) are placed in Group C if they *require* calculus; of course, problems in other groups also may be solved using calculus methods if desired. In later chapters, problems requiring calculus will be found in all three groups. Within each group, problems are arranged in the order of presentation in the text.

Each chapter has a summary to aid in reviewing important concepts and a collection of questions that extends the use of the main ideas of the chapter. Many of the questions offer good subjects for discussion; some are open-ended and quite challenging, having no unique answers.

Chapter length has been dictated by the logic of the subject matter and by our desire to reveal the coherence of related ideas, a pattern that enables students to grasp the unity of various parts of the subject. From time to time, we provide short Perspectives, which stand back from the subject to point out the truly significant ideas from the maze of details. (See, for example, the Perspective following Chapter 4.)

Chapter 11, Accelerated Frames and Inertial Forces, is presented with greater rigor than is customary. In this day of manned space ventures, students have a high interest in the subject. Chapter 13, Special Relativity, also deserves comment. Instead of injecting bits of relativity at various points throughout the text while students are first encountering classical ideas, we present the conclusions of relativity as a unified topic in a single chapter toward the end of the classical mechanics section. The discussion at this point is mainly limited to space and time and to mass-energy considerations. (For instructors who prefer to postpone the subject, Chapter 13 may be incorporated with Chapter 28, where we give a modest treatment of the relativistic transformation of electric and magnetic fields.) A supplementary discussion of relativity at the end of the book presents derivations of the basic conclusions and philosophical implications in greater detail. The well-known "paradoxes" of relativity are included here for their value in dramatizing the new ways of thinking about space and time.

In Chapter 27, *AC* Circuit Theory, we recognize the growth of interest in this field and its applications to modern technology. We have also given somewhat more emphasis than is usual to the behavior of light, Chapters 28–32, because of the growing impact of lasers and other optical instruments in applied physics and engineering. Finally, we believe an introduction to the basic wave-particle duality of both matter and radiation is important in a first look at physics, so the final two chapters discuss the quantum nature of radiation and the wave characteristics of particles.

The writing of this text and the use of a preliminary version in the classroom have extended over a decade. We are indebted to the many persons who gave valuable criticism and advice. Among these are our colleagues, Professors Tim Sanders and Stuart Elliott, as well as our students who continually provided the critical responses every author seeks. The manuscript has been reviewed at various stages by many individuals, who offered numerous helpful suggestions and comments. Among the reviewers we wish to thank are Raymond Adams, California State University; David Boulware, University of Washington; Ricardo Gomez, California Institute of Technology; Vivian Johnson, Purdue University; Jerome Pine, California Institute of Technology; Stanley Shepherd, Pennsylvania State University; and Gabriel Weinreich, University of Michigan. We also are grateful to the staff of Harcourt Brace Jovanovich, who provided admirable help and encouragement at crucial moments. In particular, we would like to recognize the contributions of our manuscript editor, Susan Harter Collette, and our art editor, Sue Lasbury. Finally, to our valuable secretary, Julia Loewe, we extend our deep appreciation and high praise for her uncommon good sense and patience as she assisted us in typing the evolving versions of the manuscript.

Alvin Hudson
Rex Nelson

Contents

Preface v

Introduction 1 Chapter 1

1.1 Physics 1
1.2 The Domain of Physics 1
1.3 Theory and Fact 3
1.4 Why Mathematics? 4
1.5 How to Use This Book 4
Serendipity in Science *5*

Kinematics: A Description of Motion 6 Chapter 2

2.1 Space and Time 6
2.2 Standard Units 9
2.3 Metric Prefixes 9
2.4 Coordinate Systems and Frames of Reference 10
2.5 Rectangular Cartesian Coordinates for Two Dimensions 10
2.6 Position Vector **r** 11
2.7 Displacement Vector Δ**r** 12
2.8 Vector Addition and Subtraction 14
2.9 Vectors in Three Dimensions 15
2.10 Velocity and the Concept of a Limit 16
2.11 Acceleration 19
2.12 Kinematic Equations for Linear Motion with Constant Acceleration 21
2.13 Derivation of the Kinematic Equations Using Calculus 25
2.14 Graphical Relations Between x, v, and a 26

2.15 Dimensions and Conversion of Units 28
2.16 Significant Figures 29
2.17 Numerical Calculations 29
2.18 Motion in Two and Three Dimensions 35
SUMMARY 40
QUESTIONS 41
PROBLEMS 42

Chapter 3 **Circular Motion** **47**

3.1 Introduction 47
3.2 Polar Coordinates 47
3.3 Velocity and Acceleration in Circular Motion 48
3.4 Tangential Acceleration 48
3.5 Centripetal Acceleration 48
3.6 Derivation of Acceleration for Circular Motion Using Unit Vectors 51
3.7 General Curvilinear Motion 53
SUMMARY 55
QUESTIONS 55
PROBLEMS 56

Chapter 4 **Dynamics—The Physics of Motion** **59**

4.1 Introduction 59
4.2 Observations and Experiments in Particle Motion 60
4.3 Analysis of the Experimental Data 62
4.4 Standard Mass 63
4.5 Momentum 64
4.6 Newton's Second Law 64
4.7 Mass and Weight 66
4.8 Units for Mass and Weight 67
4.9 Applying Newton's Second Law 69
4.10 Tension 71
4.11 Friction 74
4.12 Newton's Third Law 77
SUMMARY 84
QUESTIONS 84
PROBLEMS 86

PERSPECTIVE 94

Chapter 5 **Work and Energy** **95**

5.1 Introduction 95
5.2 Work 96
5.3 Work and the Scalar Product of Vectors 97
5.4 Work Done by a Varying Force 98
5.5 Spring Forces 101
5.6 Kinetic Energy 103

5.7 Kinetic Energy Change with a Variable Force 106
5.8 Gravitational Potential Energy 106
5.9 Work and Mechanical Energy 108
Lilliputians and Brobdingnagians 110
5.10 Potential Energy Stored in a Spring 112
5.11 Thermal Energy Associated with Friction 113
5.12 The Work-Energy Relation 115
5.13 Internal Work 118
5.14 Power 119
Energy for the Future 120
5.15 Efficiency 123
5.16 Mechanical Advantage 125
SUMMARY 126
QUESTIONS 128
PROBLEMS 129

Conservative Forces and Conservation of Energy 133 Chapter 6

6.1 Introduction 133
6.2 Conservative Forces 133
6.3 Conservative Forces and Potential Energy 136
6.4 Nonconservative Forces 137
6.5 The Energy Theorem 138
6.6 The Simple Harmonic Oscillator and Energy Graphs 140
6.7 Energy Conservation with Nonconservative Forces 145
SUMMARY 147
QUESTIONS 148
PROBLEMS 149

Conservation of Linear Momentum 154 Chapter 7

7.1 Introduction 154
7.2 The Conservation of Linear Momentum 154
7.3 Impulse 159
7.4 A Conveyor-Belt Problem 162
7.5 The Rocket 164
SUMMARY 166
QUESTIONS 166
PROBLEMS 167

Collisions 171 Chapter 8

8.1 Introduction 171
8.2 Elastic and Inelastic Collisions 172
8.3 The Analysis of Collisions 172

8.4 Center of Mass 177
8.5 Relative Velocities Using a Geometrical Method 180
8.6 The Center-of-Mass Frame of Reference 182
8.7 Collisions in the Center-of-Mass Frame 184
8.8 Momentum Vector Diagrams for Elastic Collisions 186
SUMMARY 189
QUESTIONS 190
PROBLEMS 191

PERSPECTIVE 196

Chapter 9 Dynamics of Rotational Motion for a Point Mass 197

9.1 Introduction 197
9.2 Notation for Angular Motion 197
9.3 Torque 200
9.4 Moment Arm and Moment of a Force 201
9.5 Moment of Inertia 202
9.6 The Vector Product of Two Vectors 203
9.7 Angular Momentum and the Torque on a Particle 204
9.8 Conservation of Angular Momentum 207
9.9 Notation for the General Case 209
9.10 A Note About Vectors Representing Rotational Quantities 209
SUMMARY 210
QUESTIONS 211
PROBLEMS 212

Chapter 10 The Dynamics of Rigid Bodies 215

10.1 Introduction 215
10.2 The Center of Gravity 215
10.3 Center of Mass for an Extended Object 217
10.4 Equilibrium Positions of a Rigid Body 220
10.5 Static and Dynamic Equilibrium of Rigid Bodies 221
10.6 Translational Motion of Large Objects 226
10.7 Rotational Motion of Large Objects 228
10.8 Rotation of a Rigid Body About a Fixed Axis of Symmetry 230
10.9 Moment of Inertia of a Solid Object 231
10.10 Calculation of Moments of Inertia 232
10.11 Radius of Gyration 233
10.12 Parallel-Axis Theorem 234
10.13 Kinematic Equations for Rotational Motion 234
10.14 Rotational Dynamics of Rigid Bodies 238
10.15 The Axial Vectors ω and α 240
10.16 Moving Axes of Rotation 243
10.17 Conservation of Angular Momentum 245
10.18 Conservation of Energy in Rotational Motion 247
10.19 The Gyroscope 250

10.20 Elastic Properties of Matter 253
SUMMARY 256
QUESTIONS 258
PROBLEMS 259

Accelerated Frames and Inertial Forces 271 Chapter 11

11.1 Introduction 271
11.2 Inertial Forces 271
11.3 Linearly Accelerated Frames of Reference 272
11.4 Rotating Frames of Reference 278
11.5 Mathematical Operators 279
11.6 Coriolis and Centrifugal Forces 280
11.7 Comments 285
11.8 Terminology 288
SUMMARY 289
QUESTIONS 290
PROBLEMS 290

Harmonic Motion 295 Chapter 12

12.1 Introduction 295
12.2 Simple Harmonic Motion 296
12.3 Some Characteristics of SHM 299
12.4 Circle-of-Reference Analogy for SHM 300
12.5 The Simple Pendulum 302
12.6 The Torsional Pendulum 303
12.7 The Physical Pendulum 304
12.8 Oscillations of a Two-Mass System 306
12.9 Resonance 309
Damped Oscillations *310*
Forced Oscillations *311*
SUMMARY 315
QUESTIONS 316
PROBLEMS 316

Special Relativity 321 Chapter 13

13.1 Introduction 321
13.2 The Fundamental Postulates of Special Relativity 322
13.3 How to Make Measurements 323
13.4 Comparison of Clock Rates 323
Albert Einstein *324*
13.5 Comparison of Length Measurements Along the Direction of Motion 326
13.6 Relativistic Momentum 326
13.7 Relativistic Velocity Addition 329

13.8 Relativistic Energy 330
13.9 Binding Energy 332
13.10 The Twin Paradox 333
SUMMARY 334
QUESTIONS 336
PROBLEMS 336

Chapter 14 **Gravitation** 340

14.1 Introduction 340
14.2 Kepler's Laws 341
14.3 Newton's Law of Universal Gravitation 342
14.4 A Note About Newton 342
14.5 The Gravitational Force Between a Particle and an Extended Mass 343
14.6 The Gravitational Field 347
Variations in g *348*
14.7 Gravitational Field Lines 349
14.8 The Cavendish Experiment 350
14.9 Gravitational Potential Energy 350
14.10 Escape Velocity and Binding Energy 351
14.11 Satellite and Planetary Motions 353
14.12 Inertial and Gravitational Mass 354
14.13 General Relativity 355
SUMMARY 356
QUESTIONS 357
PROBLEMS 358

Chapter 15 **Fluids** 363

15.1 Introduction 363
15.2 Density 364
15.3 Pressure 365
15.4 Pascal's and Archimedes' Principles 368
15.5 Surface Tension 370
15.6 Fluids in Motion 372
15.7 Bernoulli's Principle 374
15.8 Examples of Bernoulli Effects 376
SUMMARY 378
QUESTIONS 379
PROBLEMS 380

Chapter 16 **Wave Motion** 384

16.1 Introduction 384
16.2 The Wave Equation 385
16.3 A General Solution to the Wave Equation 387
16.4 A Particular Solution to the Wave Equation 389
16.5 Wave Speeds 391

16.6 Waves in Two and Three Dimensions 393
16.7 Energy Considerations in Wave Motion 396
16.8 Reflection of Waves 397
16.9 The Superposition Principle and Standing Waves 399
16.10 Doppler Shift 402
16.11 Shock Waves 403
16.12 Beats 404
SUMMARY 406
QUESTIONS 407
PROBLEMS 408

PERSPECTIVE 413

Heat and Temperature 414 Chapter 17

17.1 Introduction 414
17.2 Temperature 414
17.3 The Celsius and Fahrenheit Temperature Scales 415
17.4 Thermal Expansion 416
17.5 Volume and Area Thermal Expansion 418
17.6 Heat 419
17.7 Absorption of Heat 421
17.8 Heat Conduction 423
Keep Cool 425
17.9 The Constant-Volume Gas Thermometer 426
17.10 An Improved Definition for the Ideal-Gas Thermometer 429
SUMMARY 433
QUESTIONS 433
PROBLEMS 434

The Ideal Gas and Kinetic Theory 438 Chapter 18

18.1 Introduction 438
18.2 The Ideal Gas 438
18.3 Model of an Ideal Gas 443
SUMMARY 449
QUESTIONS 450
PROBLEMS 451

The First Law of Thermodynamics 455 Chapter 19

19.1 Basic Concepts 455
19.2 Heat, Energy, Work, and the First Law 457
19.3 Reversible and Irreversible Processes 458
19.4 Analysis of Specific Processes 459
Isothermal Expansion 460
Specific Heats of an Ideal Gas 462
Isovolumic Process 462
Isobaric Expansion 463
Adiabatic Process 464

19.5 Energy Variables and the Equipartition Theorem 469
The Motion of a Mass Point *469*
The Rigid Dumbbell *470*
The Vibrating Dumbbell *470*
19.6 Clues to a More Correct Theory 473
SUMMARY 474
QUESTIONS 475
PROBLEMS 475

Chapter 20 The Second Law of Thermodynamics 479

20.1 The Second Law 480
20.2 The Carnot Cycle 481
20.3 Efficiency of Engines 484
20.4 Other Types of Engines 487
20.5 Proof That No Engine Can Exceed the Carnot Efficiency 489
20.6 The Kelvin Absolute Temperature Scale 490
20.7 The Third Law of Thermodynamics 491
SUMMARY 492
QUESTIONS 493
PROBLEMS 493

Chapter 21 Entropy 497

21.1 Entropy from a Macroscopic Viewpoint 497
21.2 Entropy from a Microscopic Viewpoint 501
21.3 Entropy and the Second Law 503
21.4 Entropy and Unavailable Energy 506
21.5 Thermodynamic Heat Death 507
Bits of Information *508*
21.6 Entropy and Information 510
21.7 Perpetual Motion Devices 510
SUMMARY 512
QUESTIONS 512
PROBLEMS 512

PERSPECTIVE 516

Chapter 22 Electrostatics 517

22.1 Electrostatic Forces 518
22.2 Coulomb's Law 519
22.3 Electrostatic Fields 524
22.4 Electric Field Lines 526
22.5 The Electric Dipole 528
22.6 Gauss's Law 532
22.7 Electric Potential 536

22.8 Relationships Between Fields and Potentials 546
22.9 The Gradient of V 547
22.10 Equipotential Surfaces 550
SUMMARY 554
QUESTIONS 555
PROBLEMS 556

Capacitance and the Energy in Electric Fields 560 Chapter 23

23.1 Capacitance 560
23.2 Combinations of Capacitors 565
23.3 Dielectrics 567
23.4 Potential Energy of Charged Capacitors 571
23.5 Energy Stored in an Electric Field 574
SUMMARY 576
QUESTIONS 577
PROBLEMS 577

Electrodynamics 581 Chapter 24

24.1 Electromotive Force 581
24.2 Conductors and Resistors 583
24.3 Ohm's Law 587
24.4 Joule's Law 588
Superconductivity 590
24.5 Resistors in Series and in Parallel 591
24.6 Multiloop Circuits and Kirchhoff's Rules 595
24.7 The Superposition Principle 596
24.8 Applications 599
The Voltmeter 599
The Ammeter 600
The Wheatstone Bridge 602
The Potentiometer 602
24.9 *RC* Circuits 604
SUMMARY 607
QUESTIONS 608
PROBLEMS 609

Magnetostatics 615 Chapter 25

25.1 Magnetic Fields 615
25.2 Motion of a Charged Particle in a Magnetic Field 617
The Cyclotron 620
25.3 The Lorentz Force Law 621
25.4 Magnetic Force on a Current-Carrying Conductor 623

25.5 Magnetic Dipoles 625
25.6 Applications 628
Galvanometer *628*
Hall Effect *630*
Linear Mass Spectrometer *632*
25.7 Magnetic Flux 633
25.8 Comments About Units 634
SUMMARY 635
QUESTIONS 635
PROBLEMS 636

Chapter 26 Magnetodynamics 640

26.1 The Biot-Savart Law 640
26.2 Ampère's Law 644
26.3 Faraday's Law 649
26.4 Lenz's Law 654
26.5 Self-Inductance 656
26.6 Mutual Inductance 658
26.7 *RL* Circuits 658
26.8 Energy in Inductors 661
26.9 Magnetic Properties of Materials 664
Paramagnetism *664*
Diamagnetism *664*
Ferromagnetism *665*
26.10 Transformers 669
SUMMARY 671
QUESTIONS 673
PROBLEMS 674

Chapter 27 *AC* Circuit Theory 680

27.1 Circuits with Resistance Only 681
27.2 Circuits with Capacitance Only 681
27.3 Circuits with Inductance Only 683
27.4 Series *RLC* Circuits 684
27.5 Impedance in Series *RLC* Circuits 687
27.6 Impedance in Parallel *RLC* Circuits 692
27.7 Resonance 694
Series Resonance *695*
Parallel Resonance *698*
27.8 Power in *AC* Circuits 699
SUMMARY 703
QUESTIONS 704
PROBLEMS 705

Chapter 28 Electromagnetic Radiation 710

28.1 Maxwell's Equations 712
28.2 Electromagnetic Waves 717

28.3 Radiation 725
28.4 Energy in Electromagnetic Waves 727
28.5 Momentum of Electromagnetic Waves 730
SUMMARY 734
QUESTIONS 735
PROBLEMS 736

739 Geometrical Optics Chapter 29

29.1 Wavefronts and Rays 739
29.2 Huygens' Principle 741
29.3 Reflection by a Plane Mirror 742
Using Huygens' Principle 742
Using Fermat's Principle 743
29.4 Reflection by Spherical Mirrors 744
Case 1. Concave Mirror: Real Image 746
Case 2. Concave Mirror: Virtual Image 747
Case 3. Convex Mirror: Virtual Image 747
29.5 Ray Diagrams: Linear Magnification 751
Case 1. Concave Mirror: Real Image 752
Case 2. Concave Mirror: Virtual Image 753
Case 3. Convex Mirror: Virtual Image 753
29.6 Refraction at a Plane Surface 755
29.7 Total Internal Reflection 759
29.8 Refraction at a Spherical Surface 761
29.9 Thin Lenses 763
The First Surface 765
The Second Surface 765
The Combined Result 766
29.10 Diopter Power 767
29.11 Thin Lens Ray-Tracing and Image Size 768
Case 1. Converging Lens: Real Image 769
Case 2. Converging Lens: Virtual Image 770
Case 3. Divergent Lens: Virtual Image 770
29.12 Combinations of Lenses 771
29.13 Optical Instruments 773
The Simple Magnifier 773
The Eye 776
Eyeglasses 778
The Astronomical Telescope 780
The Simple Microscope 783
The Slide Projector 783
29.14 Aberrations 784
Spherical Aberrations 785
Astigmatism 785
Curvature of Field 786
Distortion 786
Chromatic Aberration of Lenses 786
SUMMARY 787
QUESTIONS 789
PROBLEMS 790

Chapter 30 Interference 794

30.1 Double-Slit Interference 794
30.2 Multiple-Slit Interference 805
30.3 Interference by Thin Films 806
30.4 Interference Produced by Thin Wedges 808
30.5 The Michelson Interferometer 810
SUMMARY 813
QUESTIONS 814
PROBLEMS 815

Chapter 31 Diffraction 819

31.1 Single-Slit Diffraction 822
Half-Wave Zones *822*
Phasors *823*
31.2 The Diffraction Grating 827
31.3 Fraunhofer Diffraction by a Circular Aperture 836
31.4 X-Ray Diffraction 839
31.5 Fresnel Diffraction 842
Circular Aperture *842*
Circular Obstacle *844*
31.6 The Fresnel Zone Plate 845
31.7 Holography 846
SUMMARY 849
QUESTIONS 850
PROBLEMS 851

Chapter 32 Polarization 854

32.1 Polaroid 856
32.2 Birefringence 858
32.3 Polarization by Reflection and Scattering 859
32.4 Wave Plates and Circular Polarization 861
32.5 Optical Activity 865
32.6 Interference Colors and Photoelasticity 865
SUMMARY 866
QUESTIONS 867
PROBLEMS 868

Chapter 33 The Quantum Nature of Radiation 870

33.1 The Spectrum of Cavity Radiation 871
33.2 Attempts to Explain Cavity Radiation 872
Wien's Theory *873*
The Rayleigh-Jeans Theory *874*
33.3 Planck's Theory 875
33.4 The Photoelectric Effect 878
33.5 The Compton Effect 884
33.6 The Dual Nature of Electromagnetic Radiation 887

SUMMARY 889
QUESTIONS 890
PROBLEMS 891

Wave Nature of Particles 894 Chapter 34

34.1 Models of an Atom 894
The Thomson Model *895*
The Rutherford Model *895*
The Bohr Model *897*
34.2 The Correspondence Principle 900
34.3 De Broglie Waves 901
34.4 The Davisson-Germer Experiments 903
34.5 Wave Mechanics 906
34.6 The Uncertainty Principle 909
34.7 The Complementarity Principle 913
34.8 A Brief Chronology of Quantum Theory Development 914
SUMMARY 915
QUESTIONS 916
PROBLEMS 916

Derivations of Special Relativity Conclusions 919 Supplemental Topic

ST.1 Setting Clocks in Synchronism 919
ST.2 The Galilean Transformation 920
ST.3 The Lorentz Transformation 921
ST.4 Comparison of Clock Rates 922
ST.5 Comparison of Length Measurements Parallel to the Direction of Motion 923
ST.6 Relativistic Momentum 924
ST.7 Proper Measurements 926
ST.8 Relativistic Velocity Addition 928
ST.9 The Nonsynchronism of Moving Clocks 929
ST.10 The Twin Paradox 932
PROBLEMS 934

Appendices 937

A. Prefixes for Decimal Multiples and Submultiples of Ten 938
B. Mathematical Symbols 939
C. Conversion Factors 940
D. Trigonometric Formulas 943
E. Mathematical Approximations, Formulas, and Conversions 945
F. Fourier Analysis 947
G. Calculus Formulas 949

H. Finite Rotations 952
I. Trigonometric Functions 953
J. Periodic Table of the Elements 954
K. Constants and Standards 955
L. Terrestrial and Astronomical Data 956
M. SI Units 957

Photograph and Illustration Sources 961

Index 965

University Physics

There is no field of science that speaks more intimately than physics of man's speculative inquiry into the natural world. At its most basic, physics dwells on the ordinary miracles which we see about us every day—the changes associated with the rising and setting of the sun; the movements of the atmosphere, the earth, and its waters; the colors and sounds which provide the pathways to nature; and the forms of balance and motion which we witness and experience as we move through the physical world. At its most profound, the field of physics deals with the subtlest of scientific concepts—the quantization of matter and energy, the interweaving of time and space, and the spreading and condensation of matter in the vast, undulating universe. All of the other natural sciences, being in some way reflections of the physical world, depend upon the field of physics for some understanding of natural phenomena and for some of the most effective methods of analyzing them.*

Frederick Seitz, President
Rockefeller University

* Reprinted with permission from Dr. Frederick Seitz's review of the book *Physics in Perspective*, National Academy of Sciences, 1972, in *American Scientist*, vol. 61, May–June 1973, page 352.

chapter 1 Introduction

The chess-board is the world; the pieces are the phenomena of the universe; the rules of the game are what we call the laws of Nature. The player on the other side is hidden from us. We know that his play is always fair, and patient. But also we know, to our cost, that he never overlooks a mistake, or makes the smallest allowance for ignorance.

THOMAS HUXLEY

The eternal mystery of the world is its comprehensibility.

ALBERT EINSTEIN

1.1 Physics

Physics is beautiful—to a physicist. The physicist finds it gratifying and pleasing that so wide a range of phenomena can be understood in terms of just a few simple ideas. As a beginning student, on the other hand, you may view physics as a mathematical maze of an overwhelming number of definitions and equations—a study you may begin with considerable apprehension. One of the purposes of this book is to persuade you that physics is fundamentally a collection of just a few simple concepts, but very powerful concepts that bring insight and understanding to a surprisingly large array of phenomena.

Of course, we are oversimplifying the situation. There *are* a wealth of details to learn in this book, and many different procedures to master. And, naturally, there is the "growing edge" of physics as a discipline, where the attempts to formulate theories to explain new phenomena are most exciting, frustrating, and puzzling. If any simplicity is present, it sometimes is certainly well concealed. Nevertheless, most of the material in this book does illustrate the striking characteristic of physics as a science: the explanation of an incredible variety of happenings in terms of just a few fundamental ideas. These basic ideas are worth knowing.

1.2 The Domain of Physics

Nothing known to humans is too large or too small to arouse the interests of physicists. Table 1–1 illustrates the extraordinary range of sizes, time intervals, and energies that form the domain of physics. The astrophysicist is concerned not only with the enormous values associated with astronomical sizes, but also with stellar energy sources, which involve the minuscule sizes and times associated with atoms, nuclei, and fundamental particles. The extent to which both ends of the scales in Table 1–1 may be pushed even further is perhaps limited only by the apparatus and imagination of experimental and theoretical physicists. The range of quantities we encounter in everyday life, though important, is a very small fraction of the total. The first sections of this book are devoted primarily to this fraction.

Even though the magnitudes of quantities within physics are essentially unbounded, the physicist employs a clever strategy to do his work. He chooses his own battleground by studying phenomena that he can more or less isolate

TABLE 1–1

The Sizes, Time Intervals, and Energies Involved in the Study of Physics

Magnitude[1]	Size (meters)	Time (seconds)	Energy[2] (joules)
			Energy equivalent of the sun's mass ($E = mc^2$)
10^{40}			Daily energy output of all the stars in our galaxy
			Kinetic energy of the earth's orbital motion
			Daily output of solar energy
10^{30}			
	Known part of the universe		
			Earth's daily receipt of solar energy
10^{20}	Diameter of our galaxy		
			Energy of a strong earthquake
	Distance to the nearest star	Estimated age of the universe	
		Rotational period of our galaxy	Kinetic energy of a cyclone
			Footnote 3
	Earth's orbital diameter	Light transit time of our galaxy	
10^{10}	Sun's diameter	Century	
	Earth's diameter	Year	
		Day	Kinetic energy of a fast car
	Height of Mt. Everest		Kinetic energy of a pitched baseball
	Largest plants		
10^{0}	Humans	Time between heartbeats	See Footnote 2
		Period of an audible sound wave	
	Thickness of this page	Microsecond	Kinetic energy of a housefly
	Giant molecules	Lifetime of the atomic excited state	
		Period of an atomic clock	
10^{-10}	Atoms		Mass energy of a proton
		Period of molecular rotation	
	Nuclei	Period of a visible light wave	Mass energy of an electron
	Elementary particles		
		Period of an x-ray wave	Energy required to produce an ion pair
10^{-20}			
		Time for light to cross an atomic nucleus	Kinetic energy per molecule per degree Kelvin

[1] The notation 10^{26}, for example, is a shorthand method of writing 1 followed by 26 zeros; the notation 10^{-15} is equal to 1 divided by 10^{15}.
[2] One joule of energy is approximately the kinetic energy acquired by a small apple in falling from an apple tree to the head of a person meditating under the tree.
[3] Daily energy output of Hoover Dam; solar energy per day on 5 sq km; burning of 7000 tons of coal; energy released in complete fission of 1 g of ^{235}U; energy equivalent ($E = mc^2$) of 1 g of matter.

experimentally from the surroundings. This is a tremendous advantage over other sciences such as biology, which must grapple with systems of vastly greater complication, dealing with entities such as the heart or an individual cell, which can only be studied in the context of a larger living organism.

1.3 Theory and Fact

Among the great achievements of the human mind is its perception of the physical world. Within the last 400 years, we have become careful, systematic observers of our physical environment. Observation through experimentation gave rise to a structure of statements, called theories, describing a self-consistent view of natural phenomena.

A theory is not a statement of fact. It is a conjecture about the behavior of nature that distills the results of a vast amount of experimental evidence and intellectual effort. Any theory may require revision if new experimental evidence shows disagreement. Thus it can never be known with certainty that a theory is a "law of nature" expressing some absolute universal truth; an experiment that contradicts the theory may be done tomorrow. Often, one may begin to suspect a theory is wrong only after discovering a variety of very small discrepancies, none of which is individually large enough to cause real concern. In some cases the accumulation of several persistent disagreements with theory in the seventh or eighth decimal place in a measurement can be a clue to a radically different approach that completely overthrows the former theory. Chapter 13 traces one example of the questioning and downfall of a respected theory (classical mechanics) by the much more successful theory of special relativity.

We develop great faith in some theories because they have repeatedly withstood the tests of experiments. Indeed, we have little confidence in a theory unless we can subject it to a variety of experiments that approach the phenomenon from numerous different angles. Because the central ideas in physical theories fit in such a wide variety of contexts, we gain great confidence in their validity. This building up of a vast interconnecting network of concepts gives physics an extremely firm foundation to further extend our understanding of the universe. Unfortunately, this aspect also makes it difficult to give up some cherished ideas, and the history of physics shows that new theories are often accepted only grudgingly, after much fervent rear-guard action to preserve the status quo. After all, physics is a *human* endeavor.

Ultimately, the appeal of physics rests on the faith that all of nature can be described in a simple way by a few fundamental theories that are yet to be discovered. It may be an elusive goal, but our curiosity drives us toward it. How are new theories devised? Physicists do *not* use that widely-touted "scientific method," which gives a list of instructions on how one should go about constructing a new theory. The real situation is much more complex. Generally, it involves a wilder, no-holds-barred approach, sparked by inexplicable flashes of insight, and every bit as mysterious as the creativity in other fields.

One fact should be recognized. Our philosophical "frame of reference" undoubtedly affects the manner in which we approach new phenomena and forms our general picture of the universe. In ancient times, it was perfectly natural for astronomers to look for the explanation of the motions of planets and stars in terms of *circles*—in their view, the only "perfect" paths suitable for heavenly objects to travel. Our modern views are no doubt similarly colored by preconceptions and ways of thinking that are so deeply ingrained we are hardly conscious of them.

Another intriguing question is the extent to which our basic patterns of thinking limit our ability to perceive the universe. The models of reality we construct are undoubtedly linked with the way our brains are "wired up." Of course, this wiring, in turn, must be closely related to evolutionary pressures from the environment; that is, the wiring must somehow be intimately linked with the constraints and demands of physical reality. So perhaps there is a basic underlying similarity in the functioning of our brain and the physical and mechanistic constraints of what we call "reality."

1.4 Why Mathematics?

Much of the beauty of physics lies not only in what is said but also in how it is said. The statements of physics are precise in content. It is neither precise nor economical to say, "The force of the earth's gravity becomes less at greater distances from the earth." Newton's law of universal gravitation states precisely and succinctly: $F = GMm/r^2$, where M and m are two different masses, r is the distance between their centers, and G is the universal gravitational constant. There is no room for misinterpretation of this mathematical statement. More importantly, mathematical statements may be manipulated according to those logical processes contained in the structure of mathematics. By combining various equations, new relationships may be revealed. For example, if $F = GMm/r^2$ and $F = mv^2/r$, then their combination leads to $v = \sqrt{GM/r}$, a useful relation for the velocities (v) of spacecraft circling the earth at various orbital radii.

To truly understand physics, you must read it in its natural mode of expression—the language of mathematics. Although translating physics at the level of this text into a nonmathematical description would be possible, the result would be an extremely lengthy discourse, extremely difficult to follow. The mathematical techniques used in this book will be described briefly in context as they arise. If you cannot follow a particular development, determine whether your difficulty is in the understanding of the physical concepts, or in your ability to follow the math procedures. You may have to review some mathematical techniques. For example, although the amount of calculus used in this volume is quite small, you may find it helpful to review the equations summarized in Appendix G. Above all, do not let the mathematics obscure the physics. Always be conscious of the separation of a physical concept from the purely mathematical procedures used to deal with the concept.

1.5 How to Use This Book

This text will help you learn physics in the easiest way we know. What makes physics hard for some beginning students is usually not the mathematics. Instead, difficulty stems from not catching on soon enough to the "physicist's approach"—namely, viewing phenomena in terms of just a *few* general principles. For example, much of this volume deals with the mechanics of linear and angular motion, yet basically we use only three methods to solve mechanics problems:

(1) Newton's laws relating force, **F**, mass, m, and acceleration, **a**: $\mathbf{F} = m\mathbf{a}$.
(2) Relations between work, energy, and momentum.
(3) The relation between a conservative force, F, and potential energy, $U(x)$: $F = -dU/dx$.

Serendipity in Science

In "The Three Princes of Serendip," a Persian fairy tale, the heroes had the fortunate knack of making valuable discoveries by accident. This trait, called serendipity, is also a notable characteristic of scientific investigation—provided the investigator keeps his or her eyes open:

Roentgen was not looking for x-rays.

Becquerel was not looking for radioactivity.

Kamerlingh Onnes was not looking for superconductivity.

Geophysicists were not trying to verify continental drift when they set out to study magnetic anomalies in the ocean floor.

Hahn and Strassman were not trying to find nuclear fission when they were attempting to create transuranic elements.

Fleming was not looking for penicillin.

Faraday was not looking for the basis of an industrial revolution (the electric generator).

Bell Telephone scientists were not looking for the remnant of the Big Bang when they found the microwave background radiation.

Davisson and Germer were not looking for evidence of de Broglie waves.

Herz was not looking for the photoelectric effect.

When Einstein worked out the theory of stimulated emission from atoms (which led to the laser), he was not trying to invent a remarkable new technique for eye surgery, or a new tool for the microbiologist, or a new surveying instrument, or a new communications industry, or . . . (The list of potential laser applications can be extended almost without limit.)

. . . and on, and on.

The secret to learning mechanics easily is to learn how to use these three principles. Then, in approaching a problem, look for those clues that reveal which of the three methods is applicable. From then on the solution is almost always straightforward. In addition to the physics itself, therefore, we will be emphasizing *the methods of using physics.* If you skim rapidly through the text, you will be awed at the large amount of material. But keep your eye on the few giants in the forest—in the case of mechanics, the three central ideas just listed. What most of the text deals with are just various examples and applications of a very few basic principles. Concentrate on these fundamental ideas.

A final word of advice. The first few chapters may seem a bit tedious. But early in the game we present certain "standard" approaches to problems, which are the simplest procedures in the long run. *Read them carefully and adopt these suggested thought patterns from the beginning.* It is much easier than trying to unlearn awkward ways of approaching the subject later on. You will find that such care in the beginning will pay off handsomely.

chapter 2 Kinematics: A Description of Motion

"What's the good of Mercator's North Poles and Equators,
Tropics, Zones, and Meridian Lines?"
So the Bellman would cry: and the crew would reply
"They are merely conventional signs!"

LEWIS CARROLL
(*The Hunting of the Snark*)

Much of our understanding of nature comes from observing the motion of objects. In this chapter we will develop a description for the motion of a single point as it moves through space. Although a point is a geometrical concept quite different from everyday objects such as baseballs and automobiles, we shall see that the actual motion of many objects is most easily described as the motion of a single point (the "center of mass"), plus the rotation of the object about that point. Postponing a discussion of rotation, let us begin here with a description of the motion of a single point as it moves through space.

2.1 Space and Time

Kinematics is concerned with two basic questions: "Where?" and "When?" Though the questions are simple, the answers are potentially quite complicated if we inquire about phenomena outside our ordinary daily experiences. For example, the physics of very high speeds, or of events involving intergalactic distances or submicroscopic dimensions, is quite different from our common-sense ideas. We will discuss these interesting subjects in later chapters. For the present we shall adopt the space and time of Newton—those concepts we gradually developed as a result of our everyday experiences.

Space is assumed to be continuously **uniform** and **isotropic.** These two terms mean that space has no "graininess," and that whatever its properties may be, they are independent of any particular direction or location. In the words of Isaac Newton, "Absolute space, in its own nature, without relation to anything external, remains always similar and unmovable." Every object in the universe exists at a particular location in space, and an object may change its location by moving through space as time goes on. We specify the location of a particular point in space by its relation to a frame of reference.

Time, according to Newton, is also absolute in the sense that it "flows on" at a uniform rate. We cannot speed it up or slow it down in any way. In Newton's words, "Absolute, true, and mathematical time, of itself, and from its own nature, flows equably without relation to anything external, and by another name is called duration." Time is assumed to be continuous and ever advancing, as might be indicated by a clock.

Space and time are wholly independent of each other, though it is recognized that all physical objects must exist simultaneously in both space *and* time. That is, it seems impossible to imagine an object existing somewhere in space for no time at all, or existing for a finite length of time but being nowhere in space.

Remarkably, many of these traditional ideas turn out to be naive and inconsistent with experimental evidence. The world is just different from the picture we form from our common-sense, intuitive ideas. Space and time, by themselves, are concepts that are difficult (or perhaps impossible) to define in terms of anything simpler. However, we can *measure* space and time in unambiguous ways. We define certain operations by which we obtain numerical measurements of these quantities using rulers and clocks, based upon standard units of space and time.

For many years, our standard of time was based on astronomical observations of the earth's rotation. However, with the development of very precise **atomic clocks** (which are based on the frequencies of radio waves emitted by certain atoms), it was found that the rotation of the earth varies in a complicated fashion by about 1 part in 10^8. Figure 2–1 shows these variations in daylength when measured by atomic clocks that are consistent with one another to a few parts in 10^{11}. The earth's rotation rate is affected by such factors as the seasonal variations due to polar ice melting, tidal friction, and even sunspot activity (which affects the circulation of the earth's atmosphere). Earthquake activity also may be a factor. Because of these variations in the earth's rotation, in 1967 the 13th General Conference on Weights and Measures, attended by 38 nations, adopted an atomic standard for time.

Similarly, our former standard of length was the distance between two marks on a platinum–iridium bar kept at Sèvres, France. However, such a standard is subject to the risks of war, theft, or natural catastrophes, as well as to the difficulties associated with producing secondary standards for use in other laboratories. Therefore, in 1960, the fundamental length standard was redefined in terms of the wavelength of light emitted during a transition between two atomic energy levels. Thus the standard can be reproduced from just a description of the apparatus needed to generate the wavelength of light; it is not necessary to make copies of some physical object.

Some methods for measuring various objects and the sizes of these objects are given in Figure 2–2.

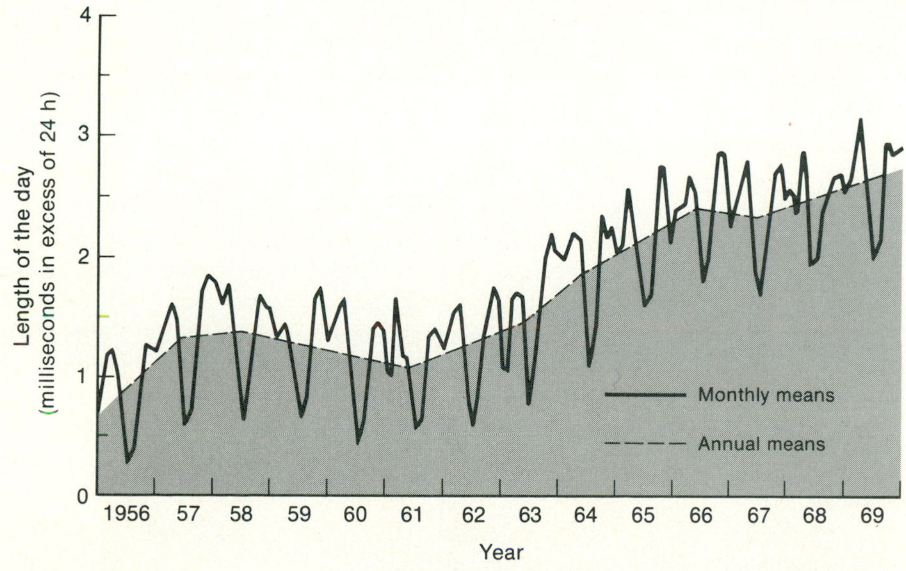

Figure 2–1
Variations in the length of the day, obtained by comparing atomic time with a time scale based on the earth's rotation.

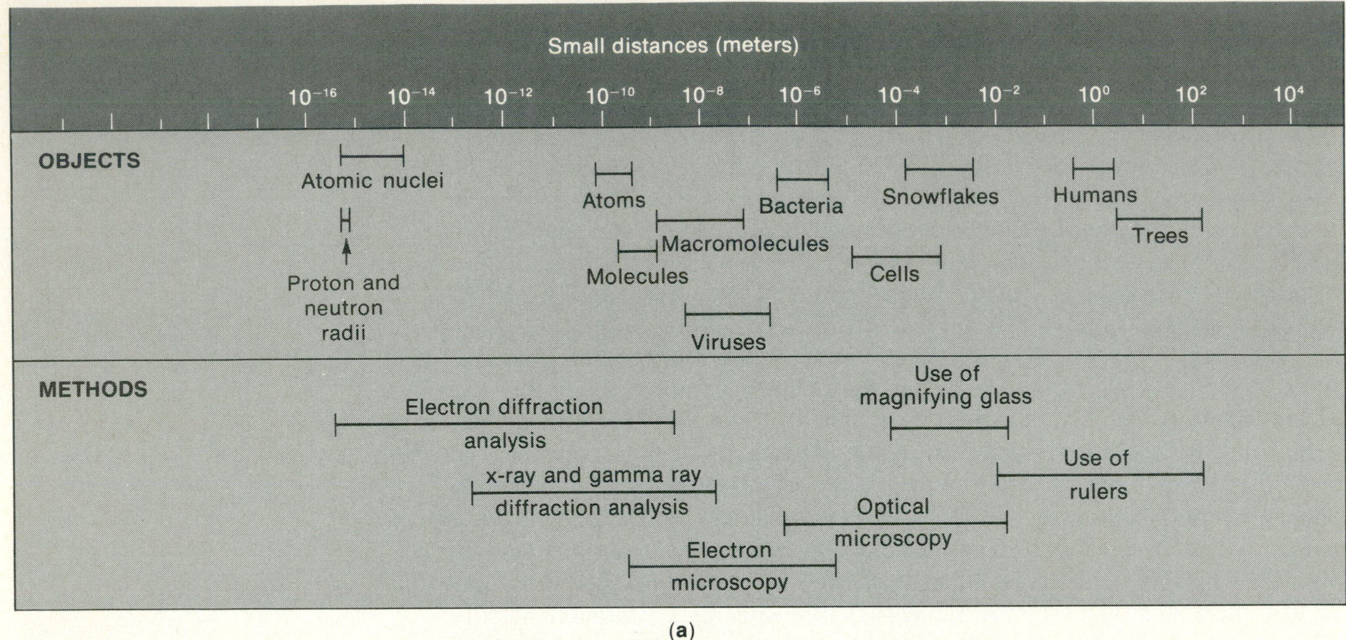

(a)

Large distances (meters)

10^{2} 10^{4} 10^{6} 10^{8} 10^{10} 10^{12} 10^{14} 10^{16} 10^{18} 10^{20} 10^{22} 10^{24} 10^{26}

OBJECTS

Mountains

Neutron stars

Diameters of planets

Radii of planetary orbits

Stellar diameters (including white dwarfs)

Earth-moon distance

Earth-sun distance

10^{0} 10^{2} 10^{4} 10^{6} 10^{8} 10^{10} (light-years)

Distance to the nearest star

Globular cluster diameters

Galactic diameters

Distances to galaxies

Distance to the nearest galaxy

METHODS

Radar reflection

Earth-based triangulation procedures

Use of odometers

(stellar diameters) Stellar interferometry and (primarily) evaluation of luminosity-distance relations

Measurement of stellar parallaxes

Evaluation of cepheid variables

Study of main-sequence-star characteristics

Study of blue-supergiant characteristics

Study of supernova characteristics

Evaluation of the Doppler shift of spectral lines (Hubble's law)

Study of galaxy characteristics

(b)

Figure 2–2

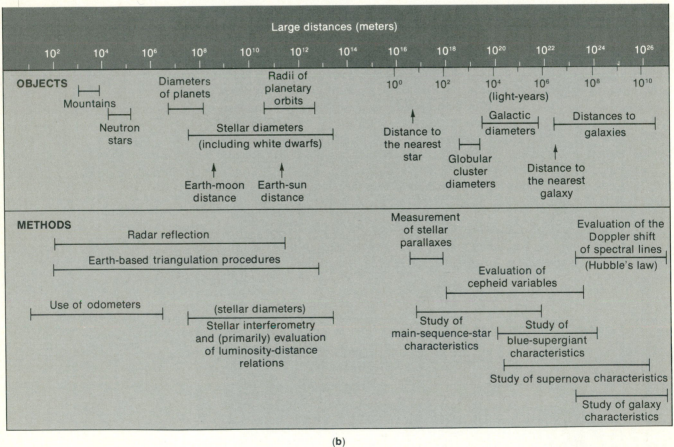

(a) Because of overlapping ranges in measuring techniques, one has more confidence in measuring extremely small distances than in measuring very large distances.

(b) Determinations of astronomical distances become less accurate as the distance increases. Furthermore, the calibration of methods for very large distances depends on the validity of methods for lesser distances.

2.2 Standard Units

The standard units of time and length may be described as follows:

An interval of time. The fundamental unit is the **second** (s), which by international agreement[1] is defined as the duration of 9 192 631 770 periods of radiation corresponding to the transition between two specific energy levels in the atomic isotope cesium 133.

An interval of length. The fundamental unit is the **meter** (m), which is defined independently of the time interval. By international agreement the meter is defined as exactly 1 650 763.73 wavelengths of the orange light emitted from the isotope krypton 86.

Other length intervals are in popular use. The most common (in English-speaking countries) are the familiar units of inches (in), feet (ft), yards (yd), and miles (mi). In 1959 the yard was formally defined to be

$$1 \text{ yd} = 0.9144 \text{ m} \qquad \text{(exactly)}$$

which is equivalent to

$$1 \text{ in} = 2.54 \text{ cm} \qquad \text{(exactly)}$$

English units have long been employed in engineering and in everyday usage, so they will be with us for some time in spite of the long-overdue decision in the United States to adopt the convenient metric measure, which the majority of other countries in the world use. Most examples and problems in this text employ metric units, with a few using English units to establish familiarity with the older usage.

2.3 Metric Prefixes

In the metric system, multiples of a standard unit are designated by a prefix that indicates positive or negative powers of 10. Some common prefixes are:

Multiplying Factor	Prefix	Symbol	Pronunciation
10^{-9}	nano-	n	na′nō
10^{-6}	micro-	μ	mī′krō
10^{-3}	milli-	m	mĭl′ĭ
10^{3}	kilo-	k	kĭl′ō
10^{6}	mega-	M	mĕg′ȧ
10^{9}	giga	G	jĭ′gȧ

Certain older units of length are still occasionally used, although they are gradually being replaced by SI units. They include:

$$1 \text{ angstrom (Å)} = 10^{-10} \text{ m}$$
$$1 \text{ micron } (\mu \text{ or } \mu\text{m}) = 10^{-6} \text{ m}$$

[1] The General Conference on Weights and Measures periodically holds international meetings to recommend units suitable for adoption by the various countries. In this text, we generally adopt their recommendation of metric units and abbreviations, known as SI units (*Systeme Internationale d'Unités*), which are increasingly used throughout the world. Abbreviations are the same for the singular and plural, and no periods are used.

The angstrom is named after the nineteenth century Swedish physicist, A. J. Ångstrom, and the micron is from the Greek word *mikron*, meaning "very small."

Additional information is given in Appendix A.

2.4 Coordinate Systems and Frames of Reference

If we wish to locate a point in space, the simplest way is to describe how far the point is from some other known point, and in what direction. Both pieces of information are necessary: the *distance* and the *direction*. This information is given by first designating a **coordinate system**—for example, the familiar x, y, and z axes called Cartesian coordinates. Cartesian coordinates will be discussed in the next section; various other coordinate systems, such as those with a particular cylindrical or spherical symmetry, will be introduced later.

Every coordinate system involves a particular point, the **origin** (O), and specified directions or **axes** that enable points in space to be described relative to the origin and the axes. It turns out that all coordinate systems *at rest with respect to one another* are related in that certain important characteristics of motion are identical as measured in the various systems. Namely, the *velocity* and *acceleration* of a point have the same values in each system. Taken together, various coordinate systems at rest with respect to one another form a **frame of reference.** A frame of reference is always associated with some physical object. For example, it may be the nucleus of an atom, the laboratory bench, a rotating space station, the sun, or the sum total of all the so-called "fixed" stars—stars so distant their motions with respect to one another are undetectable.

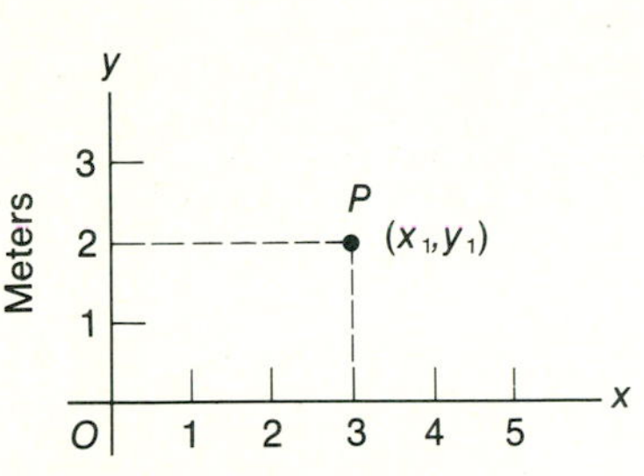

Figure 2–3
Cartesian coordinates in two dimensions designating the location of a point P in a plane at $(x_1, y_1) = (3\text{ m}, 2\text{ m})$.

2.5 Rectangular Cartesian Coordinates for Two Dimensions

The location of a point in a plane may be described by reference to two mutually perpendicular lines in the plane whose intersection defines the origin O (Figure 2–3). As you know, one line is labeled x and is called the x axis, and the other is the y axis. Each axis has a unique direction away from the origin indicated by the letters x and y. Distances from O along these directions are taken to be positive; in the opposite direction, points are labeled as the negative of the distance from the origin. In order to specify the location of a point, equal intervals of length are established along each axis using a standard ruler, and the basic unit is stated in parentheses near each axis. For the example shown, the point P is located 3 meters along the x axis and 2 meters along the y axis. A convenient notation is (3 m,2 m), with the first number indicating the x coordinate and the second number indicating the y coordinate.

An extension to three dimensions is provided by a third axis, the z axis, which is perpendicular to both the x and y axes. The positive direction of the z axis may be chosen in either of the two directions perpendicular to the x-y plane. However, by convention it has become customary always to use a right-handed coordinate system, defined in the following way. If the direction of the positive x axis is rotated toward the direction of the positive y axis, this establishes a sense of rotation that would cause a right-handed screw to advance along the positive z direction. An alternative criterion is that if the fingers of the right hand are coiled around to point in the direction of rotation, then the extended thumb points along the positive z axis. Figure 2–4 illustrates perspective drawings of right-handed coordinate systems. The point P in Figure 2–4(a) has x, y, and z coordinates, respectively, of (3 m,2 m,1 m).

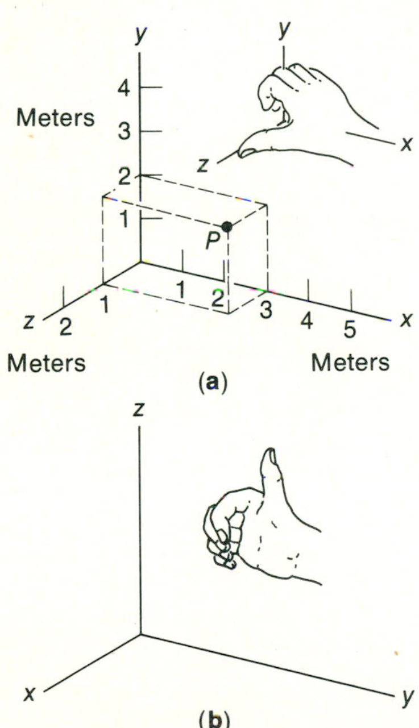

Figure 2–4
Right-handed Cartesian coordinate systems for designating a point in three dimensions. Though (a) and (b) show the coordinate axes with different orientations, they are both right-handed coordinate systems.

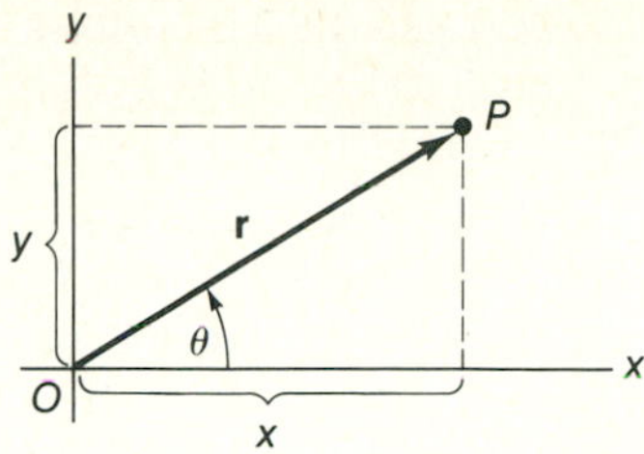

The projection of **r** onto the x and y axes marks off the coordinates (x,y) of the point P.

(**a**)

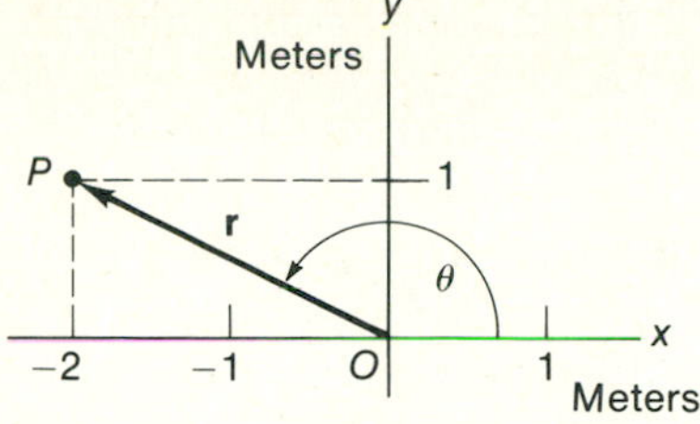

This position vector has the components $x = -2$ m and $y = +1$ m.

(**b**)

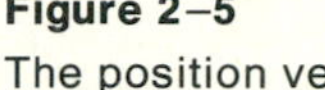

Figure 2–5
The position vector **r** designates the point P with respect to the coordinate axes.

2.6 Position Vector r

The location of a point P with respect to a given coordinate system may also be designated by a position vector **r**, as shown in Figure 2–5(a). A **vector** is a quantity that has both *magnitude* (with units) and *direction*. In this instance, the **position vector r** is a directed line segment from the origin O to the point P. The projection of **r** onto the x and y axes involves dropping perpendicular lines from the point P to the axes, thus marking off the coordinates (x,y) of the point. The magnitudes of these projections along the axes, called the **scalar components** of the vector **r**, are designated r_x and r_y. Since the position vector is always drawn from the origin, an alternative and simpler method of notation is to label the components simply as x and y. Finding the components of a vector is known as **resolving** the vector into its components.

Vectors are distinguished in printing by using bold-faced type (**r**) and in handwriting by any of several different methods (commonly $\vec{r}$, $\bar{r}$, or $\underset{\sim}{r}$). The **magnitude** of a position vector is a scalar quantity; that is, it has a numerical length value (with units) but does not contain the direction information. It is printed as $|\mathbf{r}|$ or r (ordinary type).

From trigonometry, we obtain the following relations for two-dimensional vectors:

$$r = \sqrt{r_x^2 + r_y^2} \qquad \left.\begin{aligned} x &= r\cos\theta \\ y &= r\sin\theta \end{aligned}\right\} \qquad \textbf{(2–1)}$$

$$\tan\theta = \frac{y}{x}$$

where θ is the angle measured *counterclockwise* from the $+x$ axis.

EXAMPLE 2–1

A position vector **r** has a magnitude of 0.10 m and is at an angle of 150° with respect to the $+x$ axis, as shown in Figure 2–6. Find the scalar components of the vector.

SOLUTION
As shown in the figure, the components are

x component: $x = -r\cos 30° = -(0.10\text{ m})(0.866) =$ $\boxed{-0.0866\text{ m}}$

y component: $y = r\sin 30° = (0.10\text{ m})(0.500) =$ $\boxed{0.0500\text{ m}}$

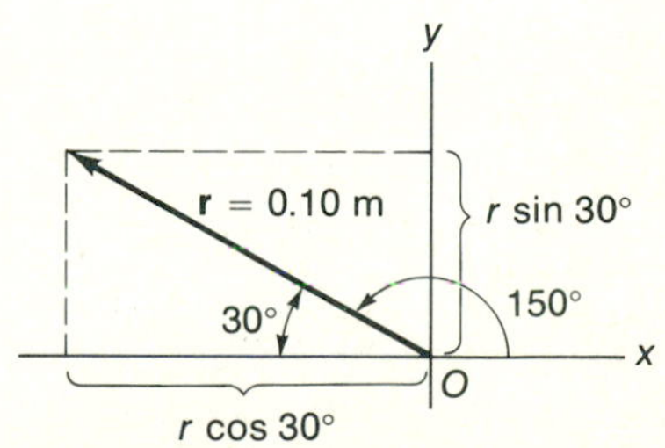

Figure 2–6
Example 2–1

A vector may be multiplied by an ordinary number (a scalar) to increase or decrease its length, and it may be multiplied by minus one to reverse its direction. These operations are illustrated in Figure 2–7.

There is a very important reason for using vectors: If an equation is expressed in vector form, *it is a valid equation in every coordinate system in the frame of reference*, whether Cartesian, cylindrical, spherical, or other. Vector notation makes the mathematical form very simple. In a sense, it strips away the cumbersome mathematical details associated with a particular coordinate system so that the fundamental relationships of physics stand out clearly. Many modern theories in physics would involve almost insurmountable mathematical difficulties if it were not for the neatness of vector notation.

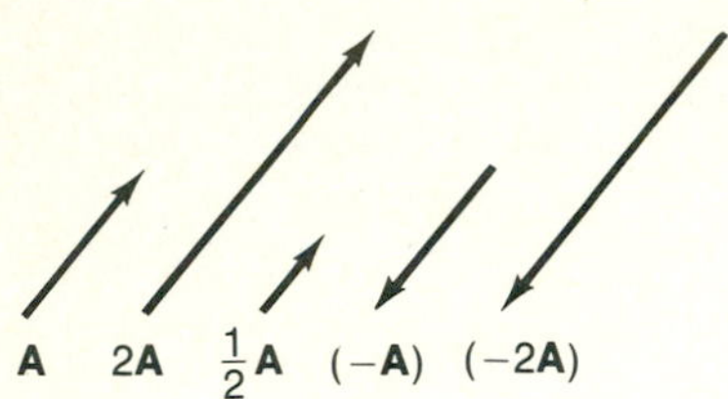

Figure 2–7
Multiplication of the vector **A** by various scalar quantities.

Furthermore, the laws of physics have an important characteristic, called **invariance,** under translation and rotation of coordinate axes. The property of invariance means that the laws preserve exactly the same mathematical form even though the coordinate system is **translated** (the origin is moved to a new location while the axes are kept parallel) and **rotated** (the axes are rotated to a new orientation in space). Since vectors also have this invariance property, the mathematical techniques for vector analysis are ideal for expressing certain laws of physics.

2.7 Displacement Vector $\Delta\mathbf{r}$

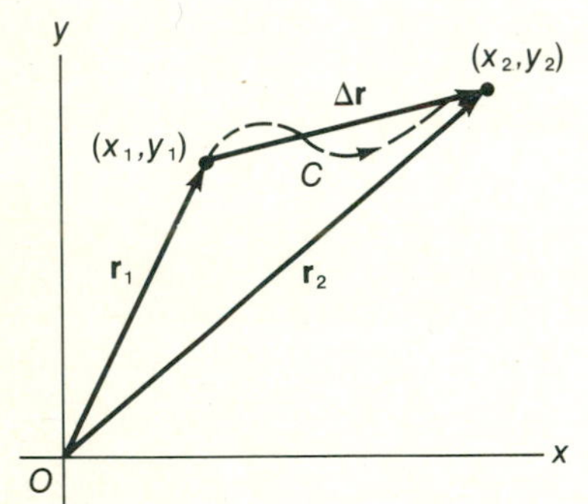

Figure 2–8
The displacement vector $\Delta\mathbf{r}$. Although the object moves along the curved path *C*, its net displacement is $\Delta\mathbf{r}$.

Kinematics provides a description of every detail of the motion of an object from one point in space to another. We will begin our study of kinematics by considering this motion without regard to how rapidly it takes place. Consider the two-dimensional case shown in Figure 2–8. Suppose the actual motion from (x_1,y_1) to (x_2,y_2) is along the curved path *C* (the dashed line in the figure). Regardless of the particular path followed, the net **displacement** from (x_1,y_1) to (x_2,y_2) is characterized by the vector $\Delta\mathbf{r}$ in Figure 2–8. Note that the magnitude of the displacement is not the total distance along the path traveled, but is instead the shortest straight line from the starting point to the end point.

We may express any vector displacement in terms of the corresponding displacements along the *x* and *y* directions. To do this, we define a **unit vector** $\hat{\mathbf{x}}$, in the $+x$ direction, with a magnitude equal to unity. Similarly, $\hat{\mathbf{y}}$ is a unit vector in the $+y$ direction, also with a magnitude of unity (see Figure 2–9). Unit vectors are always distinguished by the symbol "^" placed over them.[2]

[2] An alternative notation for unit vectors is $\hat{\mathbf{i}}$ and $\hat{\mathbf{j}}$, in place of the unit vectors $\hat{\mathbf{x}}$ and $\hat{\mathbf{y}}$ defined here. However, in physics there seems to be a trend toward defining unit vectors in terms of the symbols for their respective coordinate axes.

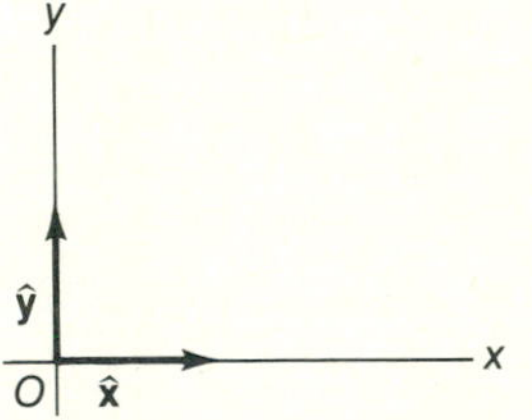

The unit vectors $\hat{\mathbf{x}}$ and $\hat{\mathbf{y}}$ have magnitudes of unity and directions along the respective positive axes.

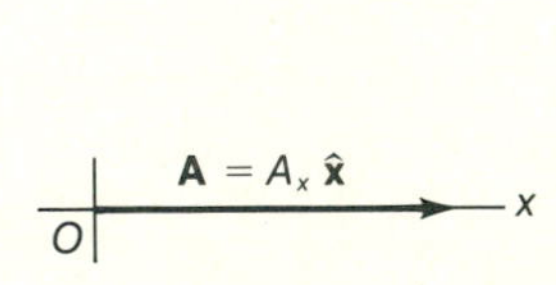

In one dimension, a vector **A** may be expressed as the magnitude A_x times the unit vector $\hat{\mathbf{x}}$.

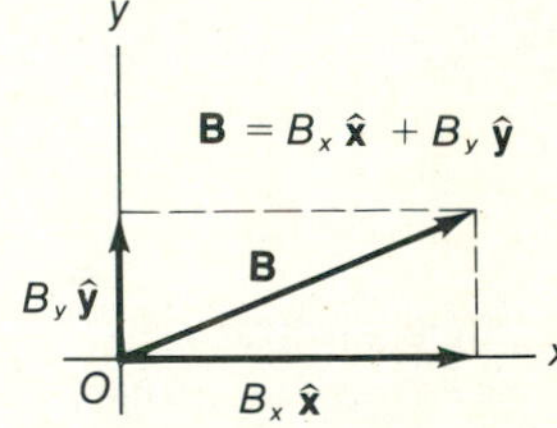

In two dimensions, a vector **B** that has scalar components B_x and B_y may be expressed as $\mathbf{B} = B_x\hat{\mathbf{x}} + B_y\hat{\mathbf{y}}$.

Figure 2–9
Use of the unit vectors $\hat{\mathbf{x}}$ and $\hat{\mathbf{y}}$.

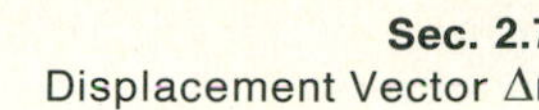

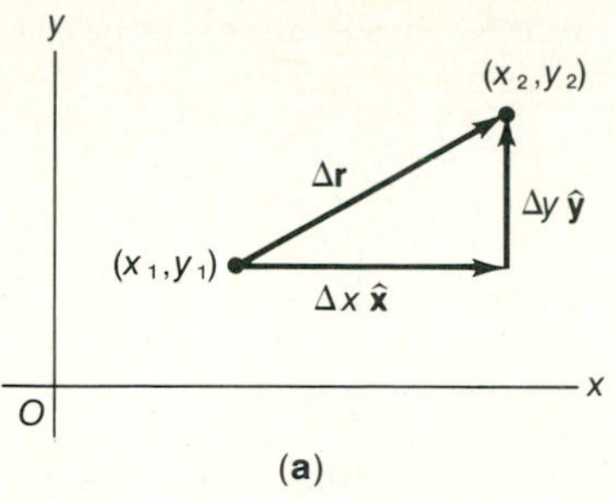

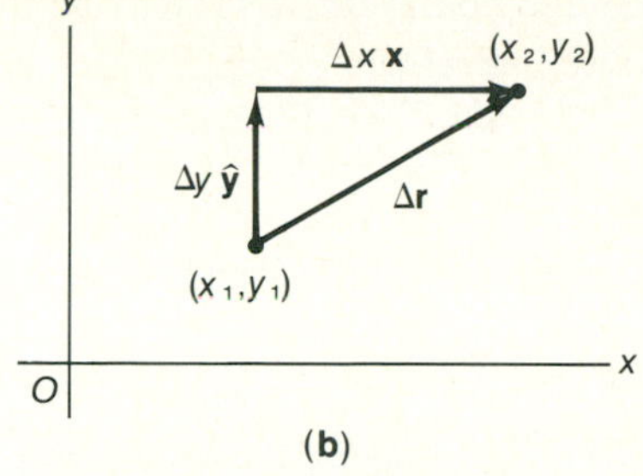

Figure 2–10
The displacement vector $\Delta\mathbf{r}$ has the vector components $\Delta x\,\hat{\mathbf{x}}$ and $\Delta y\,\hat{\mathbf{y}}$. Thus, $\Delta\mathbf{r} = \Delta x\,\hat{\mathbf{x}} + \Delta y\,\hat{\mathbf{y}}$.

As an example of the use of unit vectors, consider the displacement vector $\Delta\mathbf{r}$, whose scalar components are $\Delta x = (x_2 - x_1)$ and $\Delta y = (y_2 - y_1)$. As shown in Figure 2–10, the vector $\Delta\mathbf{r}$ may be written as

DISPLACEMENT VECTOR

$$\Delta\mathbf{r} = \Delta x\,\hat{\mathbf{x}} + \Delta y\,\hat{\mathbf{y}} \tag{2–2}$$

whose magnitude Δr is (from the Pythagorean Theorem)

$$(\Delta r)^2 = (\Delta x)^2 + (\Delta y)^2$$

Here, $\Delta x\,\hat{\mathbf{x}}$ and $\Delta y\,\hat{\mathbf{y}}$ are called **vector components** of $\Delta\mathbf{r}$ in the respective x and y directions.

EXAMPLE 2–2

Consider the displacement from the point (2 m,1 m) to the point (5 m,5 m). Find the magnitude and direction of the displacement vector $\Delta\mathbf{r}$.

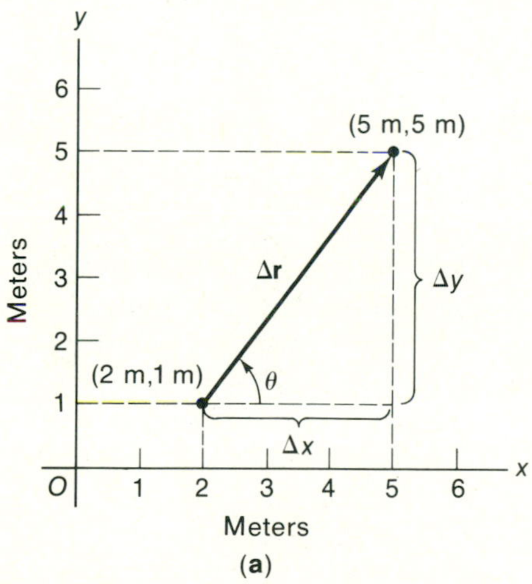

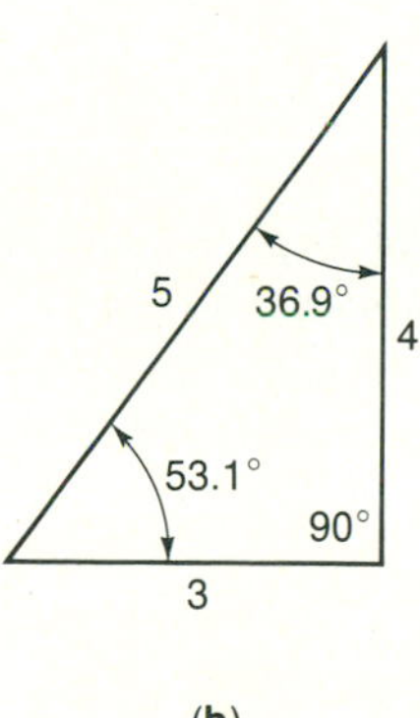

Figure 2–11
Example 2–2

SOLUTION

The displacement is depicted graphically in Figure 2–11(a). From the Pythagorean Theorem we obtain the magnitude of the vector.

$$\begin{aligned}(\Delta r)^2 &= (\Delta x)^2 + (\Delta y)^2 \\ &= (5\text{ m} - 2\text{ m})^2 + (5\text{ m} - 1\text{ m})^2 \\ &= (3\text{ m})^2 + (4\text{ m})^2 = (9 + 16)\text{ m}^2 = 25\text{ m}^2\end{aligned}$$

Therefore $\quad \Delta r = \sqrt{25\text{ m}^2} = \boxed{5.00\text{ m}}$

The direction of the vector may be specified by the angle θ it makes with the positive x direction. Thus

$$\tan \theta = \frac{\Delta y}{\Delta x} = \frac{4 \text{ m}}{3 \text{ m}} = 1.33$$

In other words, θ is the angle whose tangent is 1.33. This is written as

$$\theta = \tan^{-1}(1.33)$$

where the -1 does *not* indicate a reciprocal, but is the usual shorthand notation for the previous sentence. Carrying out the operation on a pocket calculator (or from a table of trigonometric functions), we find

$$\boxed{\theta = 53.1^\circ}$$

As shown in Figure 2–11(b), the acute angles for this 3-4-5 right triangle are 53.1° and 36.9°.

2.8 Vector Addition and Subtraction

In comparing vectors and performing other mathematical operations such as addition and subtraction, we may translate vectors anywhere in the coordinate space for convenience. We must be careful, however, to preserve their magnitudes and directions with respect to the axes.

Two vectors are added graphically by drawing the first vector and then placing the tail of the second vector at the head of the first, preserving its magnitude and direction. The single vector that represents the sum or addition of the two vectors is another vector that starts at the tail of the first and extends to the head of the second. Thus, in Figure 2–12(a), the displacement vector $\Delta\mathbf{r}$ is equal to the addition of the two component displacements $\Delta x\,\hat{\mathbf{x}}$ and $\Delta y\,\hat{\mathbf{y}}$. We may reverse the order of adding the two vectors and still arrive at the correct net displacement, as shown in Figure 2–12(b). Because the order in which the vectors are added is immaterial, the vectors are said to have the mathematical property of "obeying the *commutative law* in addition," or simply, "*commuting* in addition."

Several displacement vectors may be added together to find the **resultant** (or **net**) displacement vector. The graphical construction shown in Figure 2–12 is called the *polygon* method of vector addition. Here, three displacement vectors, **A**, **B**, and **C**, are added together to obtain the resultant displacement vector **D**. Again, the particular order in which the vectors are added is not significant.

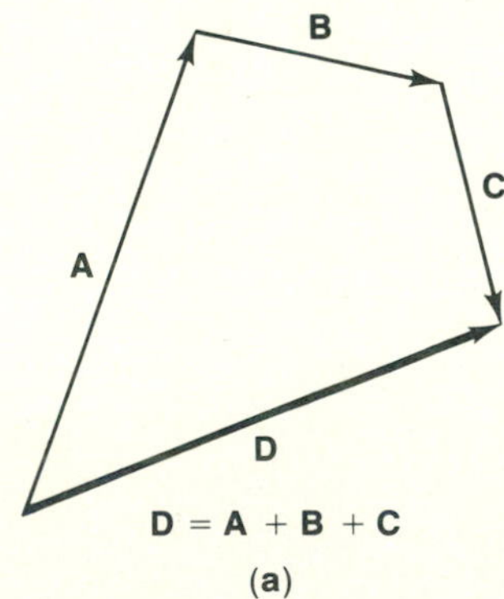

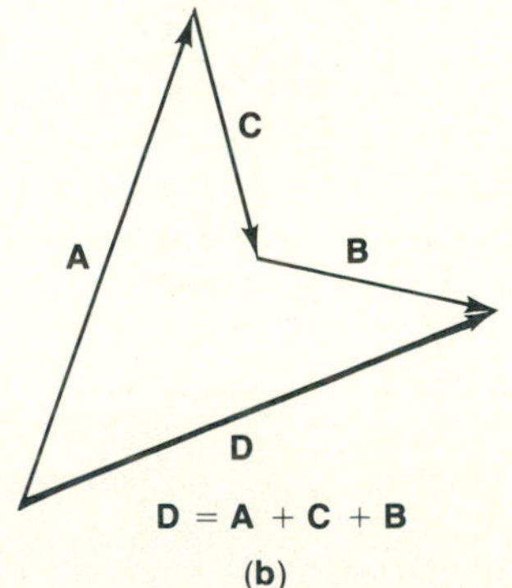

Figure 2–12
The order in which vectors are added is immaterial; in each case, the resultant vector **D** is the same.

When analytical solutions are desired, the x components of each individual vector are added together to obtain the x component of the resultant, and the same is done for the y component. Figure 2–13 makes this procedure easily verifiable for the case $\mathbf{C} = \mathbf{A} + \mathbf{B}$.

$$\mathbf{C} = C_x\hat{\mathbf{x}} + C_y\hat{\mathbf{y}} \qquad \text{(where } C_x = A_x + B_x \text{ and } C_y = A_y + B_y\text{)} \tag{2–3}$$

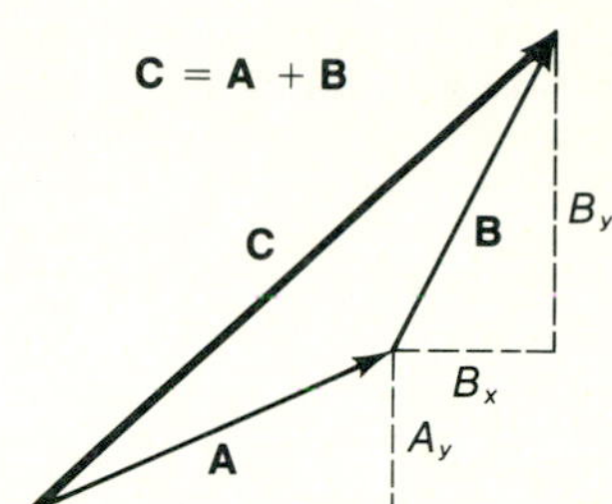

Figure 2–13
The sum of the x components of **A** and **B** equals the x component of **C**; the same is true for the y components.

To *subtract* one vector from another, remember that subtraction is really the addition of a negative quantity. Thus

$$\left.\begin{aligned} &\mathbf{C} = \mathbf{A} - \mathbf{B} && \text{(vector subtraction)} \\ \text{becomes} \quad &\mathbf{C} = \mathbf{A} + (-\mathbf{B}) && \text{(a process of vector addition)} \end{aligned}\right\} \tag{2–4}$$

and we follow the usual procedure for addition, as shown in Figure 2–14. Note carefully that in vector subtraction, the vectors do *not* commute, since $(\mathbf{A} - \mathbf{B}) = -(\mathbf{B} - \mathbf{A})$. In fact, in some cases the difference between two vectors may have a larger magnitude than the sum (Figure 2–15).

A word of caution: Use a negative sign in a vector diagram only to designate the negative of another vector. Never add a minus sign to a vector just because it happens to point along the negative direction of a coordinate axis. (For example, labeling the downward vector for the acceleration of gravity $-\mathbf{g}$ is incorrect; it should just be labeled $\mathbf{g}$.) Adding a minus sign to such a vector adds an ambiguity, since it might be reasoned that the "true" positive vector really points in the opposite direction. Vector diagrams are correct as they stand, independent of what particular direction in space is chosen to be negative. Ignoring this distinction sometimes leads to confusion, so we emphasize again: *never use a negative sign in a vector diagram unless it specifically designates the negative of another vector.*

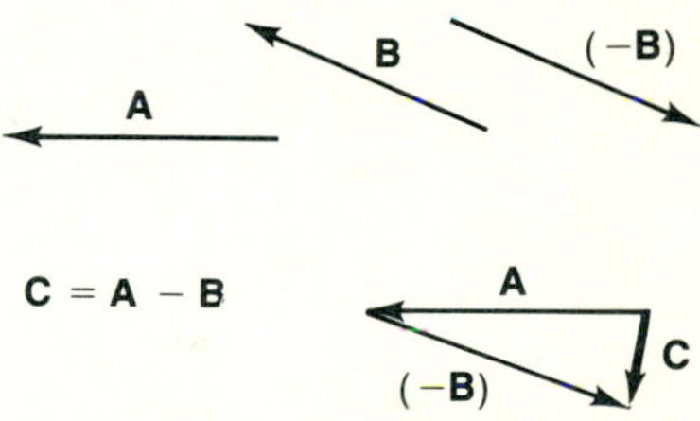

Figure 2–14
The vector subtraction $\mathbf{A} - \mathbf{B} = \mathbf{C}$ is found by *adding* to the vector **A** the vector $(-\mathbf{B})$.

2.9 Vectors in Three Dimensions

We may easily extend the concept of vector displacements to three dimensions by defining the unit vector $\hat{\mathbf{z}}$ in the $+z$ direction, with a magnitude of unity. Figure 2–16 illustrates the three-dimensional case.

Figure 2–17 depicts the coordinate systems most commonly used to represent three dimensions. At this time, we will use primarily Cartesian coordinates for three-dimensional situations. The other two coordinate schemes are displayed here for comparison.

Given:

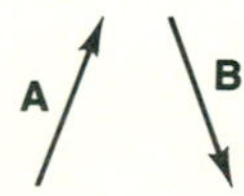

Addition: $\mathbf{A} + \mathbf{B} = \mathbf{C}$

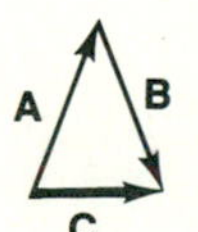

Subtraction: $\mathbf{A} - \mathbf{B} = \mathbf{D}$

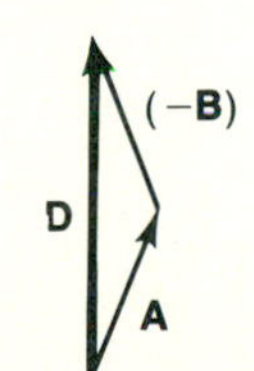

Figure 2–15
In certain cases, subtracting one vector from another may produce a vector that has *larger* magnitude than does the sum of the same two vectors.

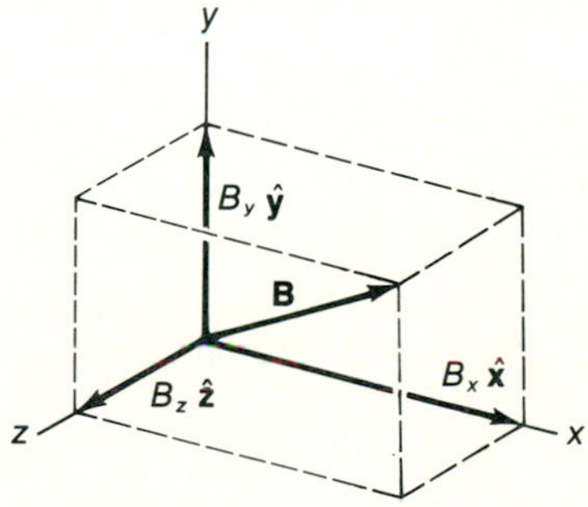

Figure 2–16
In three dimensions, a vector **B** has the scalar components B_x, B_y, and B_z, and the vector is written as $\mathbf{B} = B_x\hat{\mathbf{x}} + B_y\hat{\mathbf{y}} + B_z\hat{\mathbf{z}}$.

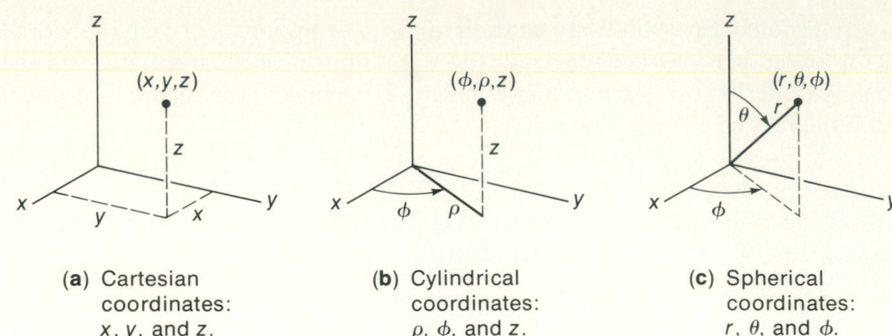

Figure 2–17
Common coordinate systems for locating a point in three spatial dimensions. Each involves *three* parameters: (x,y,z), (ρ,ϕ,z), and (r,θ,ϕ).

2.10 Velocity and the Concept of a Limit

To fully describe the motion of an object as it moves through space, we would like to specify how fast and in which direction it is moving at every instant of time. The idea that an object could be at a definite location in space at a certain moment yet simultaneously be moving through that point involves some subtleties that puzzled philosophers for 2000 years. The Greek philosopher Zeno (495–435 B.C.), for example, described several famous paradoxes that hinged on his bafflement over the concept of velocity—an idea that could not be precisely defined until Newton (1642–1727) and Gottfried Welhelm von Leibnitz (1646–1716) developed differential calculus in the seventeenth century. We will discuss the concept of velocity carefully because it is one of the fundamental ideas in mechanics.

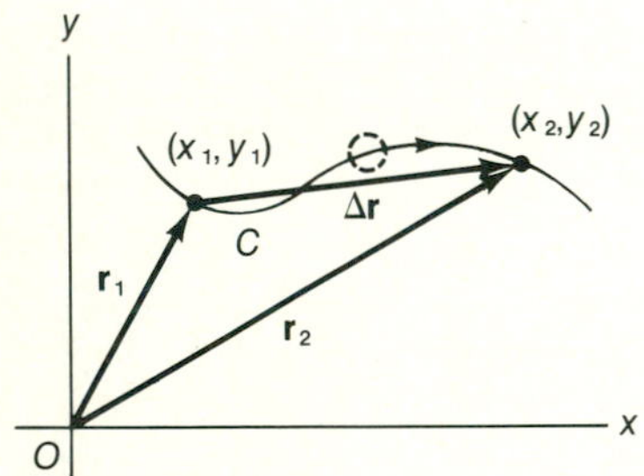

Figure 2–18
The displacement vector $\Delta\mathbf{r}$ extends from the point (x_1,y_1) to the point (x_2,y_2).

Consider an automobile trip between two cities located 100 km apart "as the crow flies." Figure 2–18 illustrates the situation with a displacement $\Delta\mathbf{r}$, whose length represents 100 km from point (x_1,y_1) to point (x_2,y_2). Of course, the actual path traveled by the automobile may be over some curved path such as C, a *longer* distance than the net displacement $\Delta\mathbf{r}$. In physics we define the **average velocity** $\mathbf{v}_{\text{ave}}$ to be

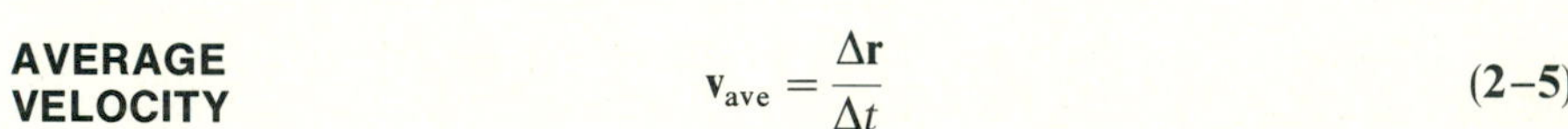

AVERAGE VELOCITY

$$\mathbf{v}_{\text{ave}} = \frac{\Delta\mathbf{r}}{\Delta t} \tag{2–5}$$

where Δt is the time elapsed as we move along C from (x_1,y_1) at t_1 to (x_2,y_2) at t_2. Thus, $\Delta t = (t_2 - t_1)$. Note that if the road were along the curved path, we would have to average a bit more than 50 km per hour to cover the distance in 2 h. Nevertheless, we define the average velocity to be the net displacement divided by the time involved, regardless of how convoluted or lengthy the actual path traveled between the two points. For a round trip, the net displacement is zero.

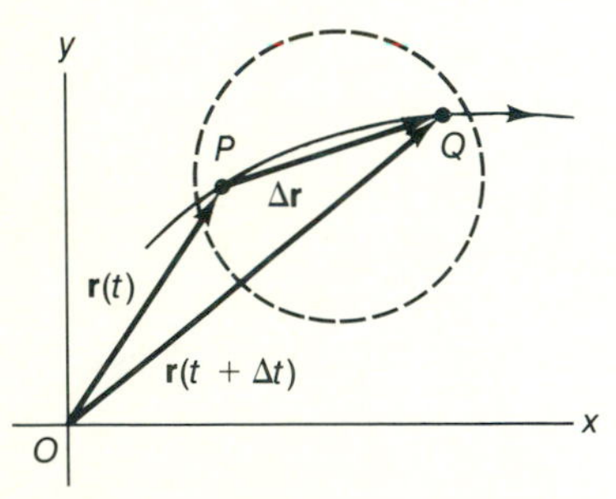

Figure 2–19
During a time Δt, the moving object undergoes a change in displacement $\Delta\mathbf{r}$.

In such a journey along some curved path C, it is clear that part of the time we may travel faster (or slower) than 50 km per hour and that we do not always travel in the direction of $\Delta\mathbf{r}$. To deal with these detailed changes of motion, we need to be able to describe the exact velocity at any particular point along the actual path. Consider a small segment of the curve C, indicated by the dashed circle in Figures 2–18 and 2–19 (enlarged). What is the precise velocity at the point P as represented by $\mathbf{r}(t)$? We obtain the answer by first noting that the average velocity between point P and a nearby point Q involves the net displacement $\Delta\mathbf{r}$ and the time Δt required to travel this displacement. Thus the average velocity from P to Q is

$$\mathbf{v}_{\text{ave}} = \frac{\Delta \mathbf{r}}{\Delta t} \qquad (2\text{–}6)$$

But now suppose the point Q is chosen closer and closer to P, as in Figure 2–20. Although both $\Delta \mathbf{r}$ and Δt each become smaller and smaller, their *ratio* does not necessarily become small. Indeed, as we successively compute the average velocities for these smaller and smaller time intervals, the ratio $\Delta \mathbf{r}/\Delta t$ approaches a limiting value called the **instantaneous velocity,** $\mathbf{v}(t)$, at the point $\mathbf{r}(t)$. Although you may not yet have discussed this limiting process in your calculus class, you will learn about it soon; we present it here for the benefit of those acquainted with the idea. In mathematical notation, this limiting process is written

$$\mathbf{v}(t) = \lim_{\Delta t \to 0} \frac{\Delta \mathbf{r}}{\Delta t} = \lim_{\Delta t \to 0} \left[\frac{\mathbf{r}(t + \Delta t) - \mathbf{r}(t)}{\Delta t} \right]$$

which, in the limit, becomes

INSTANTANEOUS VELOCITY

$$\mathbf{v}(t) = \frac{d\mathbf{r}}{dt} \qquad (2\text{–}7)$$

where the symbol $d\mathbf{r}/dt$ is called "the **derivative** of $\mathbf{r}$ with respect to t" and represents the *time rate of change of position.* Common units are m/s, ft/s, and mi/h. Note that in the limiting process, the slope of the chord PQ approaches the slope of the tangent to the curve at P. *Thus the direction of* $\mathbf{v}(t)$ *is tangent to the curve and points in the direction the object travels along the path.*

If you are not familiar with derivatives, the concept may seem mysterious. However, the procedure for finding $d\mathbf{r}/dt$ is quite straightforward. Knowing the mathematical expression for $\mathbf{r}(t)$, its derivative can usually be found from just a few rules for differentiation. Appendix G is a table of all derivatives used in this text. Until you become more familiar with derivatives in a calculus course, this table will be a convenient source for finding the derivative of any mathematical expression we use.

In physics, it is customary to use the term **velocity** to represent the vector quantity, both magnitude and direction. The word **speed** denotes just the magnitude. Thus, *instantaneous velocity* is a vector, and *speed* is its magnitude only.

We should mention one aspect of the word "speed" in connection with average values. *Average speed* refers to the total length of the path traveled divided by the total elapsed time. Thus the winner of the Indianapolis 500 might have an average speed (including pit stops) of 160 mi/h. However, since he ended where he started, his *average velocity* would be zero. In dealing with

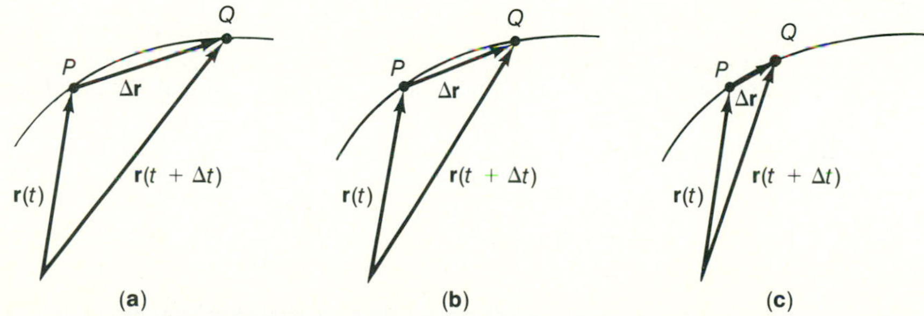

Figure 2–20
As the point Q is chosen closer and closer to P, the ratio $\Delta \mathbf{r}/\Delta t$ approaches the limiting value $d\mathbf{r}/dt$, and the direction of $\Delta \mathbf{r}$ approaches the tangent to the curve at P.

average quantities, then, average speed is not necessarily the magnitude of the average velocity. This peculiar distinction will not arise often later, since we usually will be dealing only with instantaneous quantities.

The instantaneous vector velocity is defined without reference to any particular coordinate system; it is valid in all. In rectangular coordinates, it is expressed as follows:

$$\mathbf{v}(t) = v_x(t)\hat{\mathbf{x}} + v_y(t)\hat{\mathbf{y}} + v_z(t)\hat{\mathbf{z}} \tag{2-8}$$

where the scalar components of the velocity vector are the respective derivatives of the displacement components:

$$v_x = \frac{dr_x}{dt}, \qquad v_y = \frac{dr_y}{dt}, \qquad \text{and} \qquad v_z = \frac{dr_z}{dt} \tag{2-9}$$

EXAMPLE 2–3

An object moves away from the origin of a coordinate system along the $+x$ direction. Its position from the origin as a function of time is expressed by the following equation:

$$x(t) = 4t^3 \qquad [\text{where } x(t) \text{ is in meters if } t \text{ is in seconds}]$$

Find the instantaneous speed at the time $t = 2$ s.

SOLUTION

If we differentiate the expression for the displacement $x(t)$ with respect to time, we obtain the general expression for the instantaneous speed $v(t)$:

$$x(t) = 4t^3$$

$$v(t) = \frac{dx}{dt}$$

Since the expression for $x(t)$ involves a constant times a power of the variable t, we use the following two rules from Appendix G:

$$\frac{d}{dt}(cu) = c\frac{du}{dt} \qquad \text{(where } c \text{ is any constant and variable } u \text{ is a function of the variable } t\text{)}$$

and

$$\frac{d}{dt}(u^n) = n^{n-1}\frac{du}{dt} \qquad \text{(where } n \text{ is a number)}$$

In this example, $c = 4$ and $u = t^3$. Thus:

$$v(t) = \frac{d}{dt}(4t^3) = 4\left[\frac{d}{dt}(t^3)\right] = (4)\left(3t^2\frac{dt}{dt}\right)$$

The quantity dt/dt is unity, so we have

$$v(t) = 12t^2 \qquad \text{(where } v \text{ is expressed in meters per second if } t \text{ is in seconds)}$$

Note that the factor 12 must have units of m/s^3 associated with it to make the equation dimensionally correct.

Finally, to solve the problem, we determine the numerical value of $v(t)$ at $t = 2$ s:

$$v(2\text{ s}) = \left(12\,\frac{\text{m}}{\text{s}^3}\right)(2\text{ s})^2 = \boxed{48.0\,\frac{\text{m}}{\text{s}}}$$

2.11 Acceleration

In the previous section we defined velocity as the *time rate of change of position.* The most intriguing types of motion are those for which the velocity is changing. Everyone who has ever driven a car and stepped on the "accelerator" knows that in everyday usage, acceleration means an increase in the speed of the automobile. In physics, the term is given a more precise definition: **acceleration** is the *time rate of change of velocity.*

The mathematical development of the definition for acceleration is very similar to the previous discussion concerning velocity. Figure 2–21(a) shows the velocity of a point at two different locations as it moves along a curved path. (In the most general case, the point may speed up or slow down as it travels along the path; here we depict a case where the point speeds up.) The vectors $\mathbf{v}(t)$ and $\mathbf{v}(t + \Delta t)$ are drawn with their tails positioned at the respective locations. To investigate how these vectors change with time, we construct the diagram shown in Figure 2–21(b). In the figure, the vector $\Delta\mathbf{v}$ represents the *change of velocity* during the time Δt. That is:

$$\Delta\mathbf{v} = \mathbf{v}(t + \Delta t) - \mathbf{v}(t) \qquad \textbf{(2–10)}$$

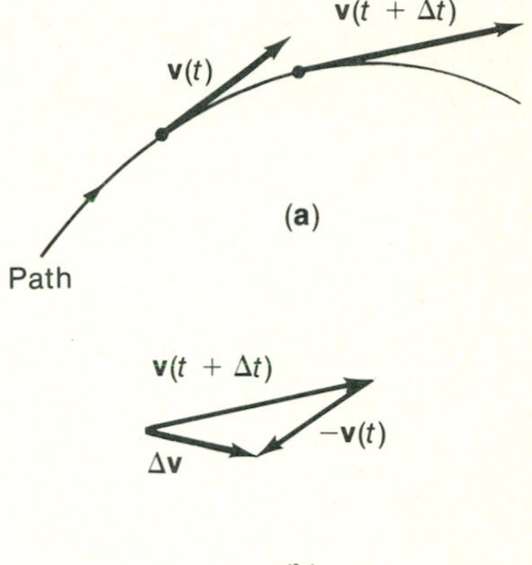

Figure 2–21
During the time Δt, the velocity vector **v** may change both its magnitude and direction by an amount $\Delta\mathbf{v}$.

The **average acceleration $\mathbf{a}_{ave}$**, is defined to be

AVERAGE ACCELERATION

$$\mathbf{a}_{ave} = \frac{\Delta\mathbf{v}}{\Delta t} \qquad \textbf{(2–11)}$$

Then, analogously with the definition of instantaneous velocity, we define the **instantaneous acceleration, a**(t), to be the time rate of change of velocity at time t.

INSTANTANEOUS ACCELERATION

$$\mathbf{a}(t) = \lim_{\Delta t \to 0} \frac{\Delta\mathbf{v}}{\Delta t} = \frac{d\mathbf{v}}{dt} \qquad \textbf{(2–12)}$$

Since the units for acceleration describe the ratio of velocity to time, they may be expressed as (meters per second) per second. This is written as m/s^2 in metric units; in English units (feet per second) per second is written ft/s^2.

In rectangular coordinates, the acceleration is expressed as:

$$\mathbf{a}(t) = a_x(t)\hat{\mathbf{x}} + a_y(t)\hat{\mathbf{y}} + a_z(t)\hat{\mathbf{z}} \qquad \textbf{(2–13)}$$

where the scalar components of the acceleration vector are the respective derivatives of the velocity components:

$$a_x = \frac{dv_x}{dt}, \qquad a_y = \frac{dv_y}{dt}, \qquad \text{and} \qquad a_z = \frac{dv_z}{dt}$$

□ □ □

These three concepts—*position*, *velocity*, and *acceleration*—are the key ideas of kinematics. With them, we can conveniently describe the complicated motion of a point, no matter how arbitrary. The power and versatility of the concepts are impressive.

One possible source of confusion needs to be mentioned. The directions of the position, velocity, and acceleration vectors are not necessarily the same. For example, a thrown object falling freely under gravity follows a parabolic path. Though the velocity vector is always tangent to the path, in this case the acceleration vector (the acceleration due to gravity) is straight down (Figure 2–22).

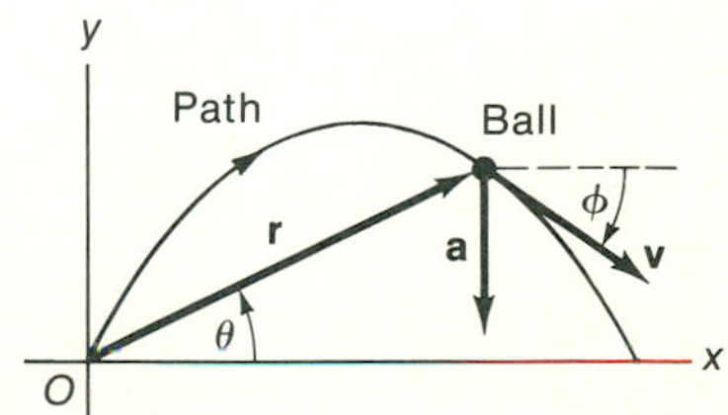

Figure 2–22
A thrown ball follows a parabolic path. At the instant shown, vectors representing the ball's position (**r**), velocity (**v**), and acceleration (**a**) point in three different directions.

Another source of confusion is the fact that an object is not necessarily moving in the direction it is accelerating. Indeed, an object may be instantaneously at rest, though accelerating, as is the situation at the top of the trajectory of an object thrown vertically upward (Figure 2–23). As you work with these concepts, guard against these easily misunderstood subtleties—they are some of the same misconceptions that prevented the early Greek philosophers from comprehending the nature of motion.

At the top of the trajectory, the instantaneous velocity is zero, and the acceleration is 9.8 m/s² downward.

Figure 2–23
The trajectory of an object thrown vertically upward. Throughout the object's free-fall motion (moving both upward and downward), the acceleration is the constant value of 9.8 m/s^2 (or 32 ft/s^2) *downward*. Thus an object may be instantaneously at rest, yet still be accelerating.

EXAMPLE 2–4

The trajectory of a ball thrown from the origin of a coordinate system is shown in Figure 2–22 (page 19). Under the action of gravity (if we ignore air friction and other small effects), the ball moves along a parabolic path. Its position $\mathbf{r}(t)$ from the starting point at the origin is expressed in analytic form as

$$\mathbf{r}(t) = (32\,t)\hat{\mathbf{x}} + (64t - 16t^2)\hat{\mathbf{y}} \qquad \text{(where } r \text{ is in feet and } t \text{ is in seconds)}$$

Find the position, velocity, and acceleration at $t = 3$ s.

SOLUTION

Position: Substituting the value $t = 3$ s into the equation for $\mathbf{r}(t)$, we obtain

$$\mathbf{r}(3) = 96\hat{\mathbf{x}}\text{ ft} + 48\hat{\mathbf{y}}\text{ ft}$$

The magnitude of the displacement is found from the Pythagorean Theorem:

$$|\mathbf{r}(3)| = \sqrt{96^2 + 48^2} = \boxed{107.2\text{ ft}}$$

Note that this is *not* the distance along the trajectory the ball travels.

The angle that $\mathbf{r}$ makes with the horizontal direction is found from

$$\tan\theta = \frac{r_y}{r_x} = \frac{48\text{ ft}}{96\text{ ft}} = 0.500$$

$$\theta = \tan^{-1}(0.500) = \boxed{26.5^\circ}$$

Velocity: Recalling that the velocity, $\mathbf{v}(t)$, is defined as the time rate of change of the position $\mathbf{r}$(t), we follow the rules for differentiation given in Example 2–3 to obtain the general expression

$$\mathbf{v}(t) = \frac{d\mathbf{r}}{dt} = 32\hat{\mathbf{x}} + (64 - 32t)\hat{\mathbf{y}} \qquad \text{(where } \mathbf{v} \text{ is in feet per second and } t \text{ is in seconds)}$$

At $t = 3$ s, we obtain the numerical value

$$\mathbf{v}(3) = 32\hat{\mathbf{x}}\,\frac{\text{ft}}{\text{s}} - 32\hat{\mathbf{y}}\,\frac{\text{ft}}{\text{s}}$$

The magnitude of $\mathbf{v}(3)$ is found from the Pythagorean Theorem:

$$|\mathbf{v}(3)| = \sqrt{\left(32\,\frac{\text{ft}}{\text{s}}\right)^2 + \left(-32\,\frac{\text{ft}}{\text{s}}\right)^2} = \boxed{45.3\,\frac{\text{ft}}{\text{s}}}$$

The angle $\mathbf{v}$ makes with the horizontal direction is found from

$$\tan\phi = \frac{v_y}{v_x} = \frac{-32\text{ ft}}{32\text{ ft}} = -1.00$$

The angle ϕ is therefore

45° below the $+x$ direction

Acceleration: Since $\mathbf{a}(t) = d\mathbf{v}/dt$, by differentiating

$$\mathbf{a}(t) = \frac{d\mathbf{v}}{dt} = \frac{d}{dt}[32\hat{\mathbf{x}} + (64 - 32t)\hat{\mathbf{y}}]$$

we obtain

$$\mathbf{a}(t) = \boxed{-32.0\hat{\mathbf{y}}\ \frac{\text{ft}}{\text{s}^2}}$$

The acceleration is thus in the negative y direction and constant in value.

2.12 Kinematic Equations for Linear Motion with Constant Acceleration

Now that vector definitions for position, velocity, and acceleration have been expressed in general terms, let us turn to the special case of *motion along a straight line with constant acceleration.* In analyzing such motion, we will derive three **kinematic equations** that are extremely useful in solving problems. These equations do not contain any new information that is not inherently in the fundamental definitions of **r**, **v**, and **a**. But they make explicit certain relationships between these concepts, and they provide the most useful starting point in the analysis of motion.

For motion along a straight line, we choose a rectangular coordinate system that is oriented so that one of the axes (for example, the x axis) is along the line. Then, components of position, velocity, and acceleration lie along this direction, and the y and z components are zero. The *vector* equations then become *scalar* equations, as illustrated here for the x direction:

Vectors	Scalars	
$\mathbf{r} = r_x\hat{\mathbf{x}} + r_y\hat{\mathbf{y}} + r_z\hat{\mathbf{z}}$	$r = x$	**(2–14)**
$\mathbf{v} = \frac{dr_x}{dt}\hat{\mathbf{x}} + \frac{dr_y}{dt}\hat{\mathbf{y}} + \frac{dr_z}{dt}\hat{\mathbf{z}}$	$v = \frac{dx}{dt}$	**(2–15)**
$\mathbf{a} = \frac{dv_x}{dt}\hat{\mathbf{x}} + \frac{dv_y}{dt}\hat{\mathbf{y}} + \frac{dv_z}{dt}\hat{\mathbf{z}}$	$a = \frac{dv}{dt}$	**(2–16)**

By using these *scalar* symbols with appropriate plus and minus signs, we correctly express both the magnitude and direction of the vectors **r**, **v**, and **a**. Scalar equations often are simpler to handle mathematically than vector equations, and they are useful for analyzing problems that involve curvilinear motion in two or three dimensions, since the component directions can be considered separately along the path and at right angles to the path.

Let us begin our discussion of the *straight-line* motion of a point moving with *constant* acceleration along the $+x$ axis. At $t = 0$, the point may have some initial velocity v_0; after this, the velocity changes uniformly until it reaches some different value v at $t = t$ (see Figure 2–24).

The algebraic relationship between velocity and time indicated by the graph in Figure 2–21 follows from the definition of acceleration. In the case of *constant* acceleration, the acceleration a is the same as the average acceleration a_{ave}. Choosing $\Delta v = (v - v_0)$ and $\Delta t = (t - t_0)$, we have

$$a = \frac{\Delta v}{\Delta t} = \frac{(v - v_0)}{(t - t_0)} = \frac{(v - v_0)}{(t - 0)}$$

or, rearranging,

$$\boxed{v = v_0 + at \qquad \text{(only for constant } a\text{)}} \tag{2–17}$$

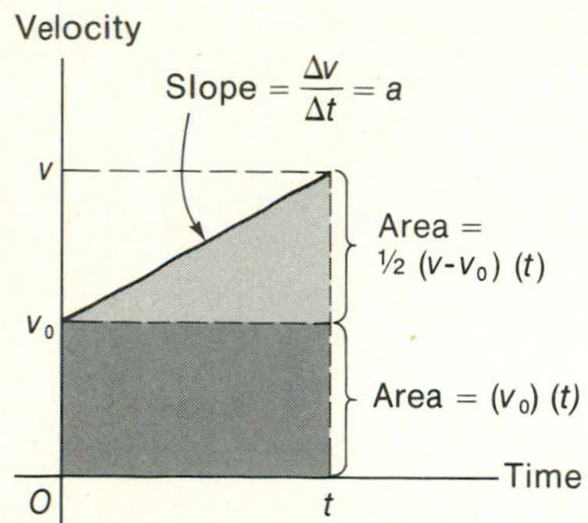

Figure 2–24
A graph of velocity versus time for straight-line motion with constant acceleration. The area under the curve represents the *net* displacement during the time $t = 0$ to $t = t$.

This is the first of three kinematic equations that will be useful in solving problems.

The relationship between position (x) and time (t) can be found simply from geometric considerations of the straight-line graph (Figure 2–24). We proceed as follows. The total area A under the straight line is made up of two parts:

the area of the rectangle: $v_0 t$

the area of the triangle: $\frac{1}{2}(v - v_0)t$

Adding these together, we obtain

$$A = \left(\frac{v + v_0}{2}\right)t \tag{2–18}$$

The average velocity is $v_{ave} = \Delta x/\Delta t$, or, rearranging,

$$\Delta x = v_{ave}\,\Delta t$$

But since the acceleration is constant, we may also express the average velocity as the arithmetic mean of the initial and final velocities:

$$v_{ave} = \left(\frac{v + v_0}{2}\right)$$

Substituting this value in the previous equation, along with $\Delta x = (x - x_0)$ and $\Delta t = (t - t_0)$, we have (for $t_0 = 0$):

$$(x - x_0) = \left(\frac{v + v_0}{2}\right)t \tag{2–19}$$

When we compare this with Equation (2–18), we see that the net displacement $(x - x_0)$ corresponds to the total area (A) under the graph. This identification of the net displacement with the area[3] under the velocity-versus-time curve is a very general relationship, true also for more complicated cases in which velocity varies nonlinearly with time.

---

[3] In calculating the area under a mathematical curve representing physical quantities, we always multiply together the relevant numerical values *with their units*. Thus, the "area" will not necessarily have dimensions of a distance squared. For example, in Figure 2–25 we multiply a *velocity* and a *time* to obtain an "area" that has units of *distance:*

$$[\text{velocity}][\text{time}] = \left[\frac{\text{distance}}{\text{time}}\right][\text{time}] = [\text{distance}]$$

We obtain a second kinematic equation by substituting $v = (v_0 + at)$ into Equation (2–19), and rearranging:

$$x = x_0 + v_0 t + \tfrac{1}{2}at^2 \qquad \text{(only for constant } a\text{)} \tag{2–20}$$

A third[4] useful kinematic equation is obtained by eliminating t from Equations (2–17) and (2–20):

$$v^2 = {v_0}^2 + 2a(x - x_0) \qquad \text{(only for constant } a\text{)} \tag{2–21}$$

This expression is employed when the time is of no interest in a problem.

A procedure for simplifying the use of the kinematic equations is to place the origin of the coordinate system for the object at $t = 0$. This causes x_0 to be equal to zero, thus eliminating one term from the equations. Of course, it may not always be possible to shift the origin in this way, but if it is, the procedure has the advantage of eliminating one parameter right from the start.

The three kinematic equations are so useful that they are collected together here. *Every problem involving straight-line motion with constant acceleration can be solved using these equations only.*

Kinematic Equations for Straight-Line Motion with Constant Acceleration

$$v = v_0 + at \tag{2–22}$$

$$x = x_0 + v_0 t + \tfrac{1}{2}at^2 \qquad \text{(only for constant } a\text{)} \tag{2–23}$$

$$v^2 = {v_0}^2 + 2a(x - x_0) \tag{2–24}$$

Similar equations could also be written for the y and z directions.

A word of caution about signs: The kinematic equations must always be written exactly as shown in Equations (2–22), (2–23), and (2–24). If you know in advance that an unknown quantity has a negative numerical value, do not change the general equation by inserting a negative sign with the symbol. Negative signs should be used only when substituting specific numerical values for the general symbols (or other symbols that represent specific values).

The reason for this procedure is a fundamental one. The kinematic equations are valid in *any* coordinate system *regardless of the directions chosen as positive or negative*. That is, the choice of a particular coordinate system (and thus the designation of positive and negative directions) does not alter the basic relationships between the vector parameters as expressed in the general

---

[4] Sometimes Equation (2–19) is given as a fourth kinematic equation:

$$x = x_0 + \left(\frac{v + v_0}{2}\right)t \qquad \text{(only for constant } a\text{)}$$

However, every solvable kinematics problem can be worked out using only the three kinematic equations listed. So in a sense this fourth equation is extraneous, and it therefore is not included in the list.

kinematic equations. It only affects the sign of numerical values when measured with respect to the chosen coordinate system. So always write the equations as shown, letting negative signs arise only when substituting numerical values.

The same procedure is followed when problems are solved in algebraic form (rather than with specific numbers). Here, *secondary symbols* are substituted into the kinematic equations. If they represent values in the directions chosen as negative, they will have a negative sign. For example, in this text the symbol g (the acceleration due to gravity) always represents a *positive* number (32.0 ft/s^2 or 9.80 m/s^2 at the earth's surface).[5] If the downward direction has been chosen as negative, the acceleration a of a freely falling object becomes $-g$. However, if the downward direction were chosen as positive, then a would equal $+g$.

EXAMPLE 2–5

A ball is dropped from rest at a location where the acceleration due to gravity is $g = 32$ ft/s^2. Where is the ball when its speed is 16 ft/s? In solving this problem, compare the method of solution when the upward direction is assumed positive with that when the downward direction is assumed positive.

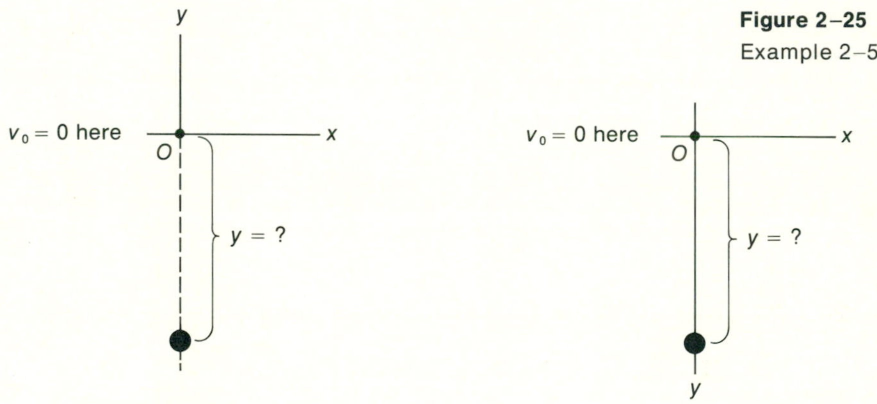

Figure 2–25
Example 2–5

SOLUTION

We place the origin O of the coordinate system where the object is located at $t = 0$. Thus the initial displacement y_0 equals 0. We indicate the direction assumed positive by sketching a set of axes oriented appropriately.

Assuming the *upward* direction is positive, the ball has an acceleration of $a_y = -32$ ft/s^2.

Given: $a_y = -32$ ft/s^2
$v_0 = 0$
$v = -16$ ft/s
$y = ?$

Assuming the *downward* direction is positive, the ball has an acceleration of $a_y = 32$ ft/s^2.

Given: $a_y = 32$ ft/s^2
$v_0 = 0$
$v = 16$ ft/s
$y = ?$

[5] Unless otherwise specified, throughout this text we will assume a and g to have the above values. Actually, because of centrifugal effects that will be discussed in Chapter 11, the value at sea level varies with the latitude. There is also an additional altitude variation as the distance from the center of the earth increases.

Comparing the given quantities with the kinematic equations, we see that Equation (2–24) is useful. It is valid regardless of which direction is chosen to be positive. Since displacement is only along the vertical direction, we write the equation in terms of y:

General equation

$$v^2 = v_0^{\,2} + 2a_y(y - y_0)$$

General equation

$$v^2 = v_0^{\,2} + 2a_y(y - y_0)$$

(Note: Even though we can guess correctly that y will come out negative in the left-hand solution, we do *not* anticipate this fact by writing a minus sign with y in the general equation. The solution itself will produce a negative sign in the final answer.)

Substituting numerical values

$$\left(-16\,\frac{\text{ft}}{\text{s}}\right)^2 = (0)^2 + 2\left(-32\,\frac{\text{ft}}{\text{s}^2}\right)(y - 0)$$

$$256\,\frac{\text{ft}}{\text{s}} = \left(-64\,\frac{\text{ft}}{\text{s}^2}\right)y$$

Solving for *y*

$$y = \frac{256\,\dfrac{\text{ft}^2}{\text{s}^2}}{-64\,\dfrac{\text{ft}}{\text{s}^2}} = -4\text{ ft}$$

Since y was chosen to be positive in the *upward* direction, the minus sign signifies that the ball is 4 ft *below* its starting point.

Substituting numerical values

$$\left(16\,\frac{\text{ft}}{\text{s}}\right)^2 = (0)^2 + 2\left(32\,\frac{\text{ft}}{\text{s}^2}\right)(y - 0)$$

$$256\,\frac{\text{ft}^2}{\text{s}^2} = \left(64\,\frac{\text{ft}}{\text{s}^2}\right)y$$

Solving for *y*

$$y = \frac{256\,\dfrac{\text{ft}^2}{\text{s}^2}}{64\,\dfrac{\text{ft}}{\text{s}}} = 4\text{ ft}$$

Since y was chosen to be positive in the *downward* direction, the positive answer signifies that the ball is 4 ft *below* its starting point.

2.13 Derivation of the Kinematic Equations Using Calculus

The kinematic equations may be obtained quite easily in a direct fashion using integral calculus. We will show the derivation here for students who already know the basic meaning of an integral. (Appendix G includes a summary of integration formulas used in this text.)

The defining equations for

$$v = \frac{dx}{dt} \qquad \text{and} \qquad a = \frac{dv}{dt}$$

may also be written in terms of **integrals** (or **antiderivatives**) as

$$x = \int v\,dt \qquad \text{and} \qquad v = \int a\,dt$$

For the case of *constant* acceleration a, the equation for the velocity v integrates to give

$$v = \int a\,dt = a\int dt = at + C_1$$

where C_1 is a constant of integration whose value is determined by noting the "initial conditions," namely, that when $t = 0$, the velocity v is equal to v_0.

Substituting these values in the equation, we have

$$v = at + C_1$$

$$v_0 = (a)(0) + C_1 \qquad \text{(at } t = 0\text{)}$$

Therefore: $$C_1 = v_0$$

Thus we arrive at the first kinematic equation:

$$v = v_0 + at \qquad \text{(only for constant } a\text{)}$$

We can then carry the derivation one more step by substituting this expression for velocity into the equation for position (x) and integrating.

$$x = \int v\,dt$$

$$= \int (v_0 + at)\,dt$$

$$x = v_0 t + \tfrac{1}{2}at^2 + C_2$$

The constant of integration C_2 is again determined from the initial conditions: at $t = 0$, the position x is equal to x_0. This leads to the second kinematic equation:

$$x = x_0 + v_0 t + \tfrac{1}{2}at^2 \qquad \text{(only for constant } a\text{)}$$

The first term (x_0) is the initial position of the particle; that is, the position at $t = 0$. The last two terms are just the net displacement that occurs during the time interval 0 to t, in agreement with Equation (2–23) derived from geometric considerations. Of course, a geometric interpretation of the process of integration is the summing up of the area under a curve, so the two different ways we used—graphically and analytically—are essentially equivalent. (The third kinematic equation is merely an algebraic combination of the first two, so calculus is not needed for its derivation.)

2.14 Graphical Relations Between *x*, *v*, and *a*

There are a number of interesting graphical relationships between x, v, and a. Suppose we know a particular x versus t function. The derivative of x with respect to time gives the velocity function. Since velocity is the rate at which displacement changes with time, it is the *slope of the tangent* to the graph at that instant (see Figure 2–26). By a similar reasoning, the acceleration at any instant is the corresponding slope of the tangent to the v versus t curve. Because of these relationships, it is possible to graphically derive curves for v and a as functions of time from the displacement curve.

One may also play the game in the opposite direction. As just mentioned, integration may be interpreted as a summing up of the area beneath the graph of a function. Thus, given the expression for v versus t, a graph for the manner in which the displacement *changes* may be obtained by plotting the value of the accumulated area under the velocity curve as one progresses along in time from $t = 0$. (This process does not, however, reveal a possible initial displacement x_0, a number corresponding to the constant of integration. Physically this constant is related to where we place the origin of the coordinate system.) Similarly, the velocity v may be obtained from the a versus t graph.

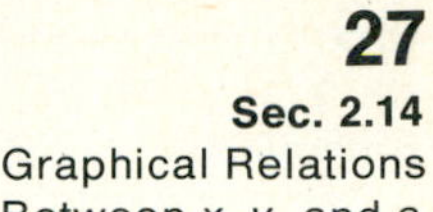

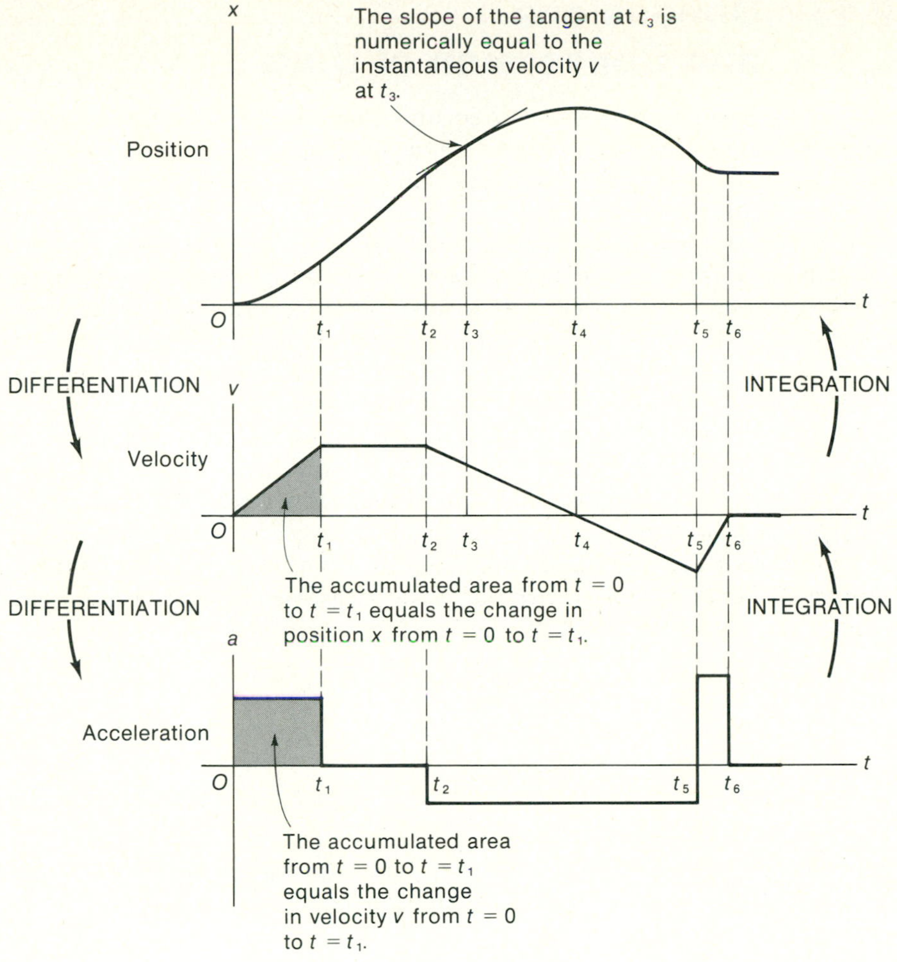

Figure 2–26

Graphical relations between x, v, and a for an object starting at rest at the origin and moving along the $+x$ axis. Between $t = 0$ and $t = t_1$, the slope of the position curve increases uniformly, so the velocity increases correspondingly. Between t_1 and t_2, the slope of the position curve is constant, so the velocity remains constant. At t_4, the slope of the position curve is zero, so at that instant the velocity has fallen to zero. From t_4 to t_5, the position slope is negative; thus, the velocity is negative there. Finally, after t_6, the object is at rest (zero slope), so the velocity returns to zero at t_6. Thus, as the graphs illustrate, the slope of the position curve is the velocity ($\Delta x/\Delta t = v$), and the slope of the velocity curve is the acceleration ($\Delta v/\Delta t = a$).

Integration is involved in going in the reverse direction. Initially, the acceleration is constant. As time goes by, the total accumulated area under the acceleration curve equals the change in velocity (which starts at zero). Since the area accumulates at a uniform rate, the velocity increases as a straight line. Correspondingly, the accumulated area under the velocity curve equals the change in position (which also starts at zero). Here, the area accumulates at an ever increasing rate, so the position changes more rapidly as time goes by. (Since the velocity curve is a straight line, it can be shown that the position curve increases parabolically.) From t_4 to t_6, the *negative* area under the velocity curve causes the position to decrease. In this example, the net area under the complete velocity curve is positive, so the final position has a positive (nonzero) value. The net area under the complete acceleration curve is zero, so the final velocity returns to its initial value of zero.

2.15 Dimensions and Conversion of Units

In substituting numerical values into equations one must use the same system of units throughout. For example, if a term in an equation represents velocity, it could be expressed as meters per second, miles per hour, or any other combination of a length interval divided by a time interval. But to be logically consistent, we must use the *same* units throughout any given equation. For example, every length interval could be expressed in terms of meters, and all time intervals in seconds, but we cannot mix meters with kilometers or feet, nor seconds with hours or days.

At a fundamental level, we can check the internal consistency of an equation by examining the **dimensions** of each term. We say that velocity has dimensions of length divided by time, or $[L/T]$. Note the distinction between dimensions and units. Length and time are *dimensions*, while meters, feet, seconds, and hours are units. Just as apples and automobiles cannot be added together, to be logically consistent *each term in an equation must have the same dimensions.* As an illustration, let us investigate Equation (2–23) in terms of these fundamental dimensions, recalling that acceleration has dimensions of [velocity/time]. We have

$$x = x_0 + v_0 t + \tfrac{1}{2}at^2$$

$$\text{Dimensions:}\quad [L] = [L] + \left[\frac{L}{T}\right][T] + \frac{\left[\frac{L}{T}\right]}{[T]}[T^2]$$

Pure numerical factors such as the $\frac{1}{2}$ in the last term have dimensions of unity (1) and are of no consequence in the analysis of dimensional consistency, so we omit them.

The dimensions cancel just like algebraic quantities to verify that each term in this equation does indeed involve the same dimensions of length.

$$[L] = [L] + \left[\frac{L}{\cancel{T}}\right][\cancel{T}] + \left[\frac{L}{\cancel{T^2}}\right][\cancel{T^2}]$$

$$[L] = [L] + \quad [L] \quad + \quad [L]$$

Examining an equation for dimensional consistency is an easy way to check for errors in problem solutions. Of course, dimensional consistency alone does not guarantee that an equation is correct, since it does not account for numerical factors such as $\frac{1}{2}$ or π. But it is a powerful method of checking internal consistency, and it will readily reveal an error in the dimensions of any term.

Since we cannot add lengths of inches, feet, meters, etc., without first converting to the same units, we need to develop a simple method for the conversion of units. A foolproof scheme is to *multiply by a ratio equaling unity.* Some examples of such conversion ratios are

$$\left(\frac{12\ \text{in}}{1\ \text{ft}}\right) = 1 \qquad \text{and} \qquad \left(\frac{60\ \text{s}}{1\ \text{min}}\right) = 1$$

Thus, to express the snail's-pace of six inches per hour in terms of feet per second, we use appropriate conversion factors formed by these ratios, multiplied so that unwanted units cancel:

$$6\,\frac{\not{\text{in}}}{\not{\text{h}}}\underbrace{\left(\frac{1\,\not{\text{h}}}{60\,\not{\text{min}}}\right)\left(\frac{1\,\not{\text{min}}}{60\text{ s}}\right)\left(\frac{1\text{ ft}}{12\,\not{\text{in}}}\right)}_{\text{Conversion ratios}} = (6)\left(\frac{1}{60}\right)\left(\frac{1}{60}\right)\left(\frac{1}{12}\right)\frac{\text{ft}}{\text{s}}$$

$$= 0.000\,139\,\frac{\text{ft}}{\text{s}} = 1.39 \times 10^{-4}\,\frac{\text{ft}}{\text{s}}$$

A particularly useful factor for converting speeds comes from the relationship:

$$30\,\frac{\text{mi}}{\text{h}} = 44\,\frac{\text{ft}}{\text{s}} \qquad \text{or} \qquad \frac{30\,\frac{\text{mi}}{\text{h}}}{44\,\frac{\text{ft}}{\text{s}}} = 1$$

Other conversion ratios are given in the conversion tables in Appendix C.

A suggestion: When substituting values into equations, it is easier to keep track of which quantities are in the numerator and denominator if you avoid the use of the solidus (/), as in ft/s, and instead use a horizontal line, as in $\frac{\text{ft}}{\text{s}}$.

2.16 Significant Figures

In quoting experimental data, it is customary to make distinctions between numbers such as 4 and 4.00. The first is accurate to only one significant figure, while the second implies an accuracy to three significant figures. That is, the true value for the final digit is assumed to lie within a range of about one adjacent integer: say, from 3 to 5 in the first instance and from 3.99 to 4.01 in the second. A number such as 0.000 05 is accurate to only one significant figure; the zeros immediately following the decimal point are not "significant," as becomes obvious when the number is written in scientific notation (5×10^{-5}).

Today most students use pocket calculators, which commonly carry out computations to eight or ten figures. However, in this text we will adopt the following general procedure for examples and problems: *Numerical calculations will be carried out to three significant figures in the final answer, assuming that numbers as given in a problem are precise.* Thus, if data are given as $v_{\text{ave}} = 3.44$ m/s and $t = 2$ s, the displacement would be calculated as (3.44 m/s)(2 s) = 6.88 m, not 6.8800 (which would appear on the calculator), nor 7 m (which would equal the number of significant figures in $t = 2$ s). This general rule will be adequate for our purposes in this text.

You should be aware, however, that in the laboratory the number of significant figures involved in a measurement may be of great importance. When information is known about the uncertainty associated with a particular experimental measurement, it is customary to quote the range explicitly: for example, as 7.75 ± 0.04 m. Here, the range of values (unless specified otherwise) is taken to mean there is a 66% probability that the true value lies within this range and a 34% probability that the true value lies either above or below the indicated limits.

2.17 Numerical Calculations

Many examples in physics texts appear to be quite trivial, and indeed they are, at least at first glance. Although it may be difficult to become excited over where a thrown object will be two seconds later or how long it takes a chunk of ice to slide down an incline, the ideas behind these problems can be fascinating. Newtonian mechanics represents one of the crowning achievements of human creative intellect. In beginning a study of mechanics, it is necessary to focus on simple and trivial situations so that the physics will "stand out" without being swamped by a mass of nonessential details. So in working these problems always look beyond the example itself to the *methods of analysis*. These same methods landed a man on the moon.

The following example illustrates highly important procedures for solving all types of kinematic problems. The solution is exceptionally long and detailed in order to clarify those steps in thinking that are standard procedures for most problems in mechanics. If you approach all problems using these procedures, you will find that most solutions become quite simple and straightforward.

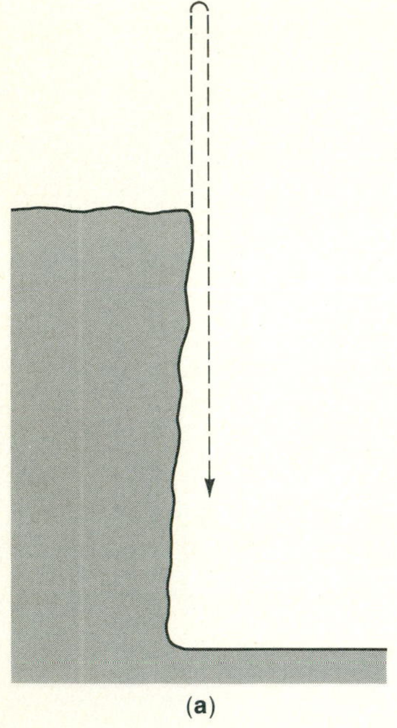

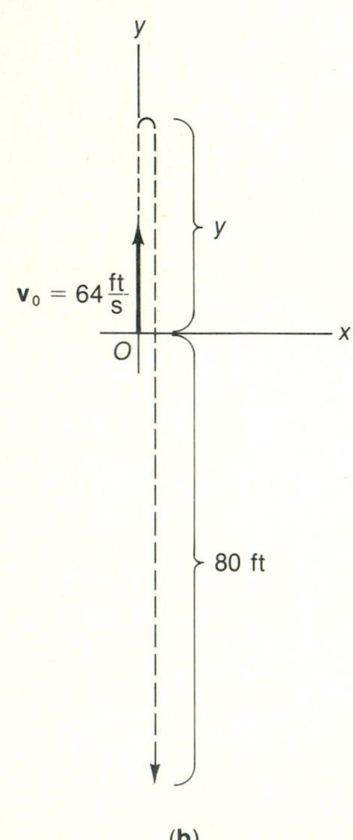

Figure 2–27
Example 2–6

EXAMPLE 2–6

A ball is thrown vertically upward from the edge of a cliff with an initial speed of 64 ft/s, as shown in Figure 2–27(a). The force of gravity produces a constant acceleration of 32 ft/s^2 downward. (a) What maximum height above its launching point does the ball travel, and (b) how long is the ball in flight before falling to a point 80 ft below its starting point?

SOLUTION

(1) We first try to spot the *type* of problem. In this case, the problem involves linear motion (vertical) with constant acceleration. Therefore the kinematic equations apply.

(2) We then sketch the motion, establishing a coordinate system with the y axis vertical (upward direction positive) and with the origin at the initial location of the ball ($t = 0$), as shown in Figure 2–27(b).

(3) We next list the known and unknown parameters of interest and compare these with the kinematic equations. All numerical values must be in a consistent system of units—in this case, the English system. For part (a), we seek the maximum height the ball travels. So we first consider the "subproblem" of the ball traveling upward, with a "final" velocity of zero at the top.

$$y_0 = 0$$
$$v_0 = 64\,\frac{\text{ft}}{\text{s}}$$
$$a = -32\,\frac{\text{ft}}{\text{s}^2}$$
$$v = 0$$
$$y = ?$$

Note that a must be negative because it is directed downward (in the negative y direction).

Comparing the known and unknown parameters with the kinematic equations, we see that when y is substituted for x, Equation (2–24) involves the single unknown we seek.

General equation $\quad v^2 = v_0{}^2 + 2a(y - y_0)$

Let us first solve for y in terms of symbols and then substitute the numerical values:

$$\frac{(v^2 - v_0{}^2)}{2a} = (y - y_0)$$

$$y = \frac{(v^2 - v_0{}^2)}{2a} + y_0 = \frac{0 - (64)^2\,\frac{\text{ft}^{\not 2}}{\not s^2}}{2\left(-32\,\frac{\not \text{ft}}{\not s^2}\right)} + 0 = \boxed{64.0\text{ ft}}$$

It is usually safest to do the necessary algebraic manipulation in *symbol* form before substituting numerical values: one is apt to make far fewer errors juggling symbols than manipulating numbers. Furthermore, since some symbols may disappear in the process, one avoids substituting numerical values in an equation at one stage and then canceling them out later in the analysis.

To solve part (b) of the problem (that is, to find the time required to reach a point 80 ft below the origin), we again list all known and unknown values. Note that in this part of the problem, v is no longer equal to zero, y (the displacement) is listed as -80 ft, and t is the new unknown.

$$y_0 = 0$$
$$v_0 = 64\,\frac{\text{ft}}{\text{s}}$$
$$a = -32\,\frac{\text{ft}}{\text{s}^2}$$
$$y = -80\text{ ft}$$
$$t = ?$$

Comparing this new list with the kinematic equations, we see that Equation (2–23) involves the single unknown parameter t, with all other symbols being known (assuming that y is substituted for x).

General equation $\qquad y = y_0 + v_0 t + \frac{1}{2}at^2$

Since this is a quadratic equation, we rearrange the terms in descending orders of the variable (t):

$$\tfrac{1}{2}at^2 + v_0 t + (y_0 - y) = 0$$

We then substitute the numerical values:

$$\frac{1}{2}\left(-32\,\frac{\text{ft}}{\text{s}^2}\right)(t^2) + \left(64\,\frac{\text{ft}}{\text{s}}\right)(t) + [0 - (-80\text{ ft})] = 0$$

After simplifying and factoring, we obtain

$$t^2 - 4t - 5 = 0 \qquad \text{(where } t \text{ is in seconds)}$$
$$(t - 5)(t + 1) = 0$$
$$t = 5\text{ s} \qquad \text{and} \qquad t = -1\text{ s}$$

Since we are only interested in times that occur later than $t = 0$, we discard the negative answer as not meaningful for this problem. The correct answer is thus

$$\boxed{t = 5.00\text{ s}}$$

A note on extraneous roots: In solving kinematics problems, we will sometimes arrive at "incorrect" answers such as $t = -1$ s in the previous example. These answers most often crop up as solutions to quadratic equations (in kinematics, particularly those involving t). When we begin a problem at $t = 0$, we specify the initial and future conditions, which the kinematic equations carry forward in a correct fashion. But these equations are also correct for times *before* $t = 0$, *provided the same conditions apply*. In the example, these conditions did not apply (that is, the ball was not moving freely with $a = -32$ ft/s^2 prior to $t = 0$), and we therefore had to reject the negative answer.

Sometimes, however, both roots *are* correct. For example, had we asked for the time(s) when the ball was 30 ft above its starting point, we would obtain two values, both of which would be correct: the ball is at that location at two different times, going up and going down. *In every problem involving multiple answers, reasoning about the physical situation must be used to reject or to retain such values.*

Rejecting the extraneous root in this example was obviously the proper procedure. But the history of physics shows that the most "logical" procedure is not always the prudent one. For example, in formulating a relativistic theory for quantum mechanics, the German physicist Erwin Schrödinger obtained a quadratic equation for the kinetic energy of an electron that gave two solutions having positive and negative values. But since kinetic energy ($\frac{1}{2}mv^2$) is proportional to the velocity squared, what physical meaning could one give to values less than zero? A velocity in either direction, when squared, gives a positive value. How could one possibly obtain negative values? This reasoning seemed justification for rejecting the negative roots as physically meaningless. Yet, much to everyone's surprise, the English physicist Paul Dirac later pursued the consequences of retaining the negative roots and in the process developed an exciting new theory that predicted the existence of positrons and other antimatter particles—the building blocks of modern nuclear theory. So one must always be wary of "intuitively obvious" conclusions.

EXAMPLE 2–7

A motorist traveling 45 mi/h due east applies the brakes so that his (uniform) deceleration is 4 ft/s^2.

(a) How far does he travel during the 2 s interval immediately after applying the brakes?
(b) What is his velocity at the end of this time interval?
(c) How long after applying the brakes does he come to rest?
(d) What total distance does the car travel while decelerating?

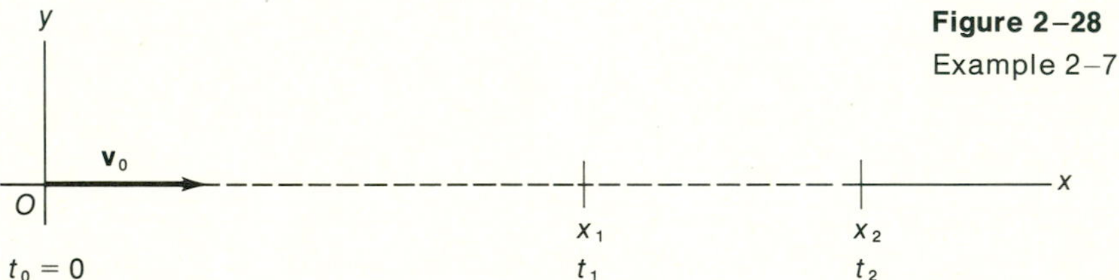

Figure 2–28
Example 2–7

SOLUTION

Since this is a problem in straight-line motion with constant acceleration, our kinematic equations apply. We first make a sketch with a coordinate system whose origin we choose to be the location of the object at $t = 0$ (the moment the brakes are applied).

We note that two different distances are involved: parts (a) and (b) involve the first two seconds of travel, and parts (c) and (d) involve a different distance. To keep matters straight, we will assign the subscript "1" to the first distance and "2" to the second.

Part (a): We let the displacement $x = x_1$ represent the distance traveled in time $t_1 = 2$ s. To express all parameters in a consistent system of units, we must convert the initial velocity of 45 mi/h into units of ft/s. We therefore multiply the velocity by the conversion ratio (44 ft/s)/(30 mi/h) = 1:

$$v_0 = 45\,\frac{\cancel{\text{mi}}}{\cancel{\text{h}}}\underbrace{\left(\frac{44\,\dfrac{\text{ft}}{\text{s}}}{30\,\dfrac{\cancel{\text{mi}}}{\cancel{\text{h}}}}\right)}_{\substack{\text{Conversion}\\ \text{ratio}}} = 66\,\frac{\text{ft}}{\text{s}}$$

In addition, we have

$$a = -4\,\frac{\text{ft}}{\text{s}^2} \qquad \text{(Note the minus sign, which indicates that acceleration is in the negative } x \text{ direction.)}$$

$$t_1 = 2\text{ s}$$

$$x_0 = 0$$

$$x_1 = ?$$

Comparing the above values with the kinematic equations, we see that Equation (2–23) leads to the desired value:

$$x_1 = x_0 + v_0 t_1 + \tfrac{1}{2} a t_1^{\,2}$$

Substituting numerical values, we obtain

$$x_1 = 0 + \left(66\,\frac{\text{ft}}{\text{s}}\right)(2\text{ s}) + \frac{1}{2}\left(-4\,\frac{\text{ft}}{\text{s}^2}\right)(4\text{ s}^2) = 132\text{ ft} - 8\text{ ft} = \boxed{124\text{ ft}}$$

Part (*b*): The velocity v_1 after 2 s may be found from Equation (2–22):

$$v_1 = v_0 + at_1$$

Substituting numerical values, we have

$$v_1 = 66\,\frac{\text{ft}}{\text{s}} + \left(-4\,\frac{\text{ft}}{\text{s}^2}\right)(2\text{ s}) = 66\,\frac{\text{ft}}{\text{s}} - 8\,\frac{\text{ft}}{\text{s}}$$

$$v_1 = \boxed{58.0\,\frac{\text{ft}}{\text{s}}} \qquad \text{(The positive value signifies that the car travels toward the east.)}$$

Part (*c*): At the instant the car stops (parts c and d), we have different values of v, t, and x, which we indicate with the subscript "2." Thus:

$$v_0 = 66\,\frac{\text{ft}}{\text{s}}$$

$$v_2 = 0$$

$$a = -4\,\frac{\text{ft}}{\text{s}^2}$$

$$t_2 = ?$$

$$x_0 = 0$$

$$x_2 = ?$$

Comparing this information with the kinematic equations, we see that Equation (2–22) will lead to the desired answer:

$$v_2 = v_0 + at_2$$

Solving for t_2, we have

$$t_2 = \frac{v_2 - v_0}{a}$$

Substituting numerical values yields

$$t_2 = \frac{0 - 66\,\frac{\text{ft}}{\text{s}}}{-4\,\frac{\text{ft}}{\text{s}^2}} = \boxed{16.5\text{ s}}$$

Part (*d*): Since we now know the time for stopping, we may use either Equation (2–23) or (2–24) to obtain the total distance the car travels while stopping. Let us use Equation (2–24):

$$v_2^{\,2} = v_0^{\,2} + 2a(x_2 - x_0)$$

Solving for x_2, we obtain

$$x_2 = \frac{v_2^{\,2} - v_0^{\,2}}{2a} - x_0$$

Substituting numerical values gives us

$$x_2 = \frac{0 - \left(66 \frac{\text{ft}}{\text{s}}\right)^2}{2\left(-4 \frac{\text{ft}}{\text{s}^2}\right)} - 0 = \boxed{545 \text{ ft}}$$

EXAMPLE 2–8

A highway motorist travels at a constant velocity of 45 mi/h in a 30 mi/h zone. Unfortunately, a motorcycle police officer has been watching from behind a billboard, and at the moment the speeding motorist passes the billboard, the police officer accelerates uniformly from rest to overtake her. If the acceleration of the police officer is 10 ft/s^2, how long does it take to reach the motorist?

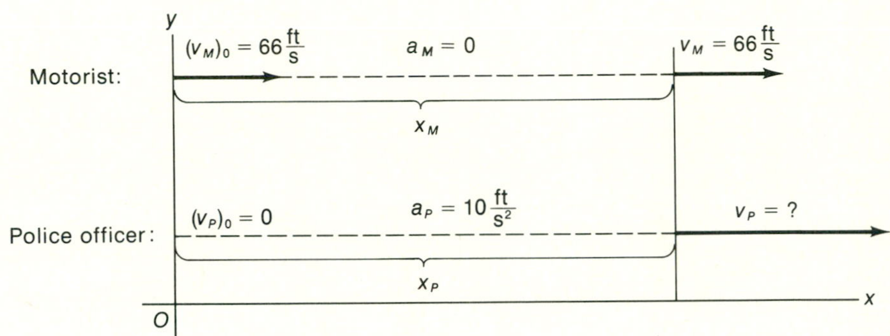

Figure 2–29
Example 2–8

SOLUTION

Our first step in solving the problem is to establish a coordinate system with the origin at the billboard and with the positive x axis oriented in the direction of motion. We shall use the subscript M to indicate values for the motorist and the subscript P to indicate values for the police officer.

The next step is to convert the velocity units to units of feet per second:

$$(v_M)_0 = \left(45 \frac{\text{mi}}{\text{h}}\right) \underbrace{\left(\frac{44 \frac{\text{ft}}{\text{s}}}{30 \frac{\text{mi}}{\text{h}}}\right)}_{\text{Conversion ratio}} = 66 \frac{\text{ft}}{\text{s}}$$

We also have:

Motorist	**Police officer**
$(v_M)_0 = 66 \frac{\text{ft}}{\text{s}}$	$(v_P)_0 = 0$
$a_M = 0$	$a_P = 10 \frac{\text{ft}}{\text{s}^2}$
$v_M = 66 \frac{\text{ft}}{\text{s}}$	$v_P = ?$
$x_M = ?$	$x_P = ?$
$t_M = ?$	$t_P = ?$

We note from the sketch that the distances traveled by the motorist and the police officer are equal. Also, the time interval to travel this distance is the same for both, so we need not carry the subscripts on t. We find that Equation (2–23) applies, and we set the two distances equal to each other:

$$\overbrace{(v_M)_0 t + \tfrac{1}{2}a_M t^2}^{x_M} = \overbrace{(v_P)_0 t + \tfrac{1}{2}a_P t^2}^{x_P}$$

Rearranging and solving for t, we have

$$\tfrac{1}{2}(a_M - a_P)t^2 = [(v_P)_0 - (v_M)_0]t$$

$$t = \frac{[(v_P)_0 - (v_M)_0]}{\tfrac{1}{2}(a_M - a_P)}$$

Substituting numerical values gives us

$$t = \frac{0 - 66\,\frac{\cancel{\text{ft}}}{\cancel{\text{s}}}}{\frac{1}{2}\left(0 - 10\,\frac{\cancel{\text{ft}}}{\cancel{\text{s}^2}}\right)} = \boxed{13.2\ \text{s}}$$

Note in the above example how helpful the sketch was in revealing the clue to the method of solution. The sketch emphasized the "rendezvous" aspect of the problem, in which two different people arrived at the same place at the same time. In this case, they also started at the same location simultaneously. These facts, which determined our method of approach, might have been overlooked without the diagram. In general, then, *the most important step in solving physics problems is first to make a sketch of the physical situation, showing notation and as much other information as possible on the diagram.*

2.18 Motion in Two and Three Dimensions

When Galileo and Newton were making tremendous advances in the science of mechanics, a common problem studied was the motion of objects near the earth's surface moving under the effect of gravity. Fortunately, if the objects were made small enough and massive enough, the complications of air friction could be ignored. And for local motions involving only minor differences in elevation, the earth's gravitational effects could be assumed constant, giving the same acceleration (about 9.8 m/s^2 or 32 ft/s^2 downward) to all objects. Consequences of the earth's rotation could also be ignored. With these approximations, Galileo discovered that *the horizontal and vertical components of motion are completely independent of each other.*[6] In other words, these components can be analyzed separately, thereby simplifying problems that involve motion in more than one dimension.

From now on, then, we will solve trajectory problems by analyzing their horizontal and vertical component motions separately. And we will continue to assume that the acceleration due to gravity (designated by the symbol g) has the standard value: 32 ft/s^2 or 9.8 m/s^2. Note that g is always a positive number representing just the magnitude (not the direction) of the acceleration due to gravity.

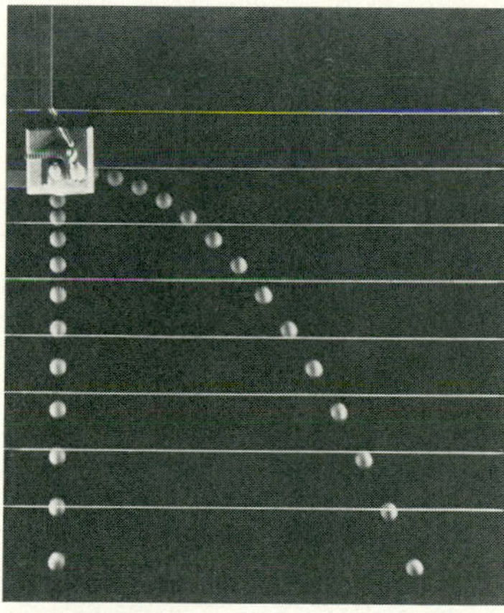

Figure 2–30
A multiflash photograph of two golf balls. At the instant one ball was dropped from rest, the other was projected horizontally. Their vertical motions are identical, demonstrating that horizontal and vertical components of free-fall motion are independent. The light flashes occurred 1/30 s apart, and the white lines are horizontal strings 15 cm apart.

[6] This independence of component motions is not always valid. In the next chapter we will examine the motion of an object whirling about on the end of a string, where there is a close relationship between the x and y component motions, and in Chapter 11 we will discuss the coriolis force, in which the force in one direction depends on the object's velocity in another direction. But for trajectory motion where g is constant and where there is no air friction, the x and y component motions are independent.

We previously developed three kinematic equations for motion in a straight line, given by Equations (2–22), (2–23), and (2–24). For more general motion that involves all three coordinate directions, we use corresponding equations for the three Cartesian components of the motion that make explicit the coordinate directions to which they refer. Here is the complete list:

Cartesian Components of Kinematic Equations for Motion with Constant Acceleration

$v_x = (v_x)_0 + a_x t$	$v_y = (v_y)_0 + a_y t$	$v_z = (v_z)_0 + a_z t$
$x = x_0 + (v_x)_0 t + \frac{1}{2}a_x t^2$	$y = y_0 + (v_y)_0 t + \frac{1}{2}a_y t^2$	$z = z_0 + (v_z)_0 t + \frac{1}{2}a_z t^2$
$v_x^2 = (v_x)_0^2 + 2a_x(x - x_0)$	$v_y^2 = (v_y)_0^2 + 2a_y(y - y_0)$	$v_z^2 = (v_z)_0^2 + 2a_z(z - z_0)$

At any instant, the parameter t in the above equations has the same value in all equations.

EXAMPLE 2–9

A projectile is fired with a velocity of $v_0 = 50$ m/s at an angle of $\theta_0 = 55°$ with respect to the horizontal. The projectile strikes a hillside 60 m higher than the level of firing, as shown in Figure 2–31(a). (a) How long is the projectile in flight? (b) How far horizontally does it travel? (c) Find its velocity at the instant it strikes the hillside.

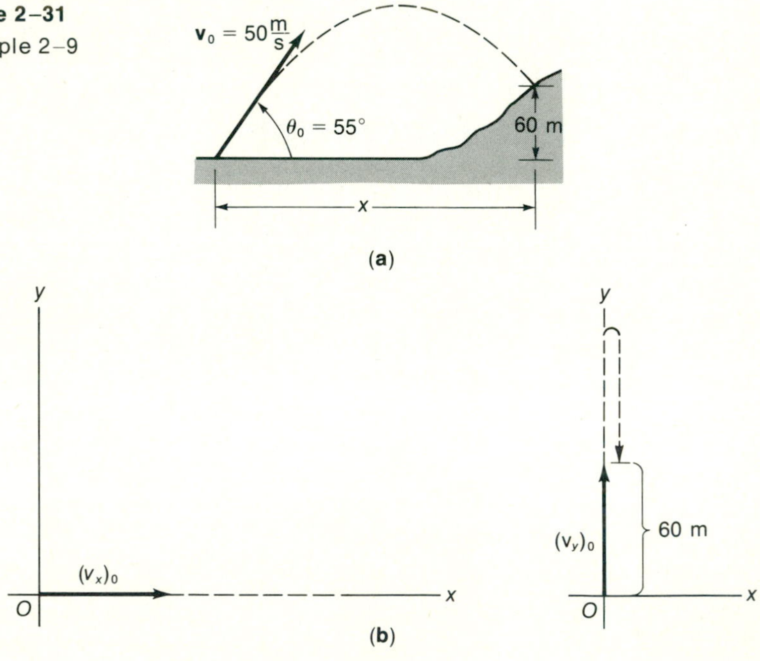

Figure 2–31
Example 2–9

SOLUTION

We first choose a coordinate system whose origin is the firing point, with positive directions along the horizontal x axis and the vertical y axis. The trajectory lies wholly in the x-y plane. Since we will analyze horizontal and vertical components of the motion separately, we make separate sketches of the component motions and tabulate initial and final values of the parameters, keeping careful account of the plus and minus directions.

Horizontal component	**Vertical component**
$v_0 = 50\,\frac{\text{m}}{\text{s}}$	$v_0 = 50\,\frac{\text{m}}{\text{s}}$
$\theta_0 = 55^\circ$	$\theta_0 = 55^\circ$
$(v_x)_0 = v_0 \cos\theta_0$	$(v_y)_0 = v_0 \sin\theta_0$
$a_x = 0$	$a_y = -g$
$x_0 = 0$	$y_0 = 0$
$x = ?$	$y = 60\text{ m}$
$t = ?$	$t = ?$

Parts (a) and (b): Comparing the known and unknown values with the kinematic equations, we choose Equation (2–23) as leading directly to the value of t in the y component equation (which, of course, is the same time t in the x component equation).

$$x = x_0 + (v_x)_0 t + \tfrac{1}{2}a_x t^2$$
$$x = 0 + (v_0 \cos\theta_0)t + 0$$

Because this equation has the two unknowns x and t, we cannot proceed further until we obtain the value of t from the y component analysis.

$$y = y_0 + (v_y)_0 t + \tfrac{1}{2}a_y t^2$$
$$y = 0 + (v_0 \sin\theta_0)t - \tfrac{1}{2}gt^2$$

Rearranging in the usual form for a quadratic, we have

$$\tfrac{1}{2}gt^2 - (v_0 \sin\theta_0)t + y = 0$$

We then use the formula for the roots of a quadratic equation (see Appendix E):

Given: $at^2 + bt + c = 0$

Then: $t = \dfrac{-b \pm \sqrt{b^2 - 4ac}}{2a}$

So for our case:

$$t = \frac{v_0 \sin\theta_0 \pm \sqrt{(v_0 \sin\theta_0)^2 - 2gy}}{g}$$

Substitution of numerical values yields

$$t = \frac{\left(50\,\frac{\text{m}}{\text{s}}\right)(\sin 55^\circ) \pm \sqrt{\left[\left(50\,\frac{\text{m}}{\text{s}}\right)(\sin 55^\circ)\right]^2 - 2\left(9.8\,\frac{\text{m}}{\text{s}^2}\right)(60\text{ m})}}{9.8\,\frac{\text{m}}{\text{s}^2}}$$

$$t = 6.46\text{ s} \quad \text{and} \quad 1.89\text{ s}$$

We reject the 1.89 s answer, since this earlier time represents the vertical displacement of 60 m when the projectile was traveling *upward*. Therefore, the desired value is 6.46 s, when the projectile is 60 m above the starting point but traveling downward. We substitute this value in the equation for the x component of the displacement.

Substituting the value of t from the y component equations, we have

$$x = \left(50\,\frac{\text{m}}{\text{s}}\right)(\cos 55^\circ)(6.46\text{ s})$$

$$\boxed{x = 185\text{ m}}$$

The answers are thus: (a) The projectile is in flight for $\boxed{6.46\text{ s.}}$

(b) It travels horizontally a distance of $\boxed{185\text{ m.}}$

Part (c): To determine the velocity of the projectile just as it strikes the hillside, we use the horizontal component for the initial velocity, since this component has not changed:

$$v_x = v_0 \cos\theta_0 = \left(50\,\frac{\text{m}}{\text{s}}\right)(\cos 55^\circ) = 28.7\,\frac{\text{m}}{\text{s}}$$

To find the vertical component at $t = 6.46$ s, we use the following kinematic equation:

$$v_y = (v_y)_0 + a_y t = \left(50\,\frac{\text{m}}{\text{s}}\right)(\sin 55°) - \left(9.8\,\frac{\text{m}}{\text{s}^2}\right)(6.46\text{ s}) = -22.4\,\frac{\text{m}}{\text{s}}$$

The minus sign signifies that the vertical component is in the *negative* y direction. The velocity vector thus has the components shown below:

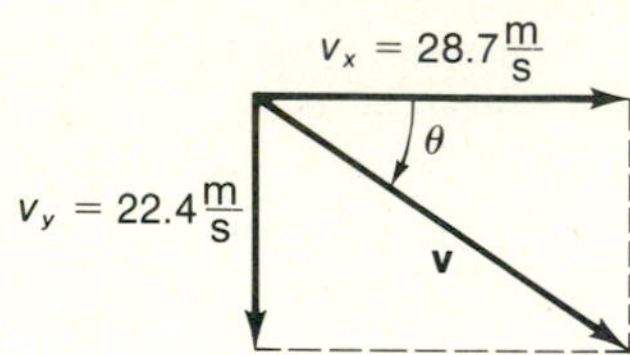

The magnitude and direction of the velocity vector at $t = 6.46$ s are:

$$v = \sqrt{v_x{}^2 + v_y{}^2} = \sqrt{\left(28.7\,\frac{\text{m}}{\text{s}}\right)^2 + \left(-22.4\,\frac{\text{m}}{\text{s}}\right)^2} = \boxed{36.4\,\frac{\text{m}}{\text{s}}}$$

$$\theta = \tan^{-1}\left(\frac{v_y}{v_x}\right) = \tan^{-1}\left(\frac{-22.4\,\frac{\text{m}}{\text{s}}}{28.7\,\frac{\text{m}}{\text{s}}}\right) = \boxed{-38.0°}$$

Here, the minus sign indicates that the angle is *below* the horizontal.

EXAMPLE 2–10

A ball is tossed from an upper-story window in a building. The ball is given an initial velocity of 8 m/s at an angle of 20° below the horizontal, as shown in Figure 2–32(a).

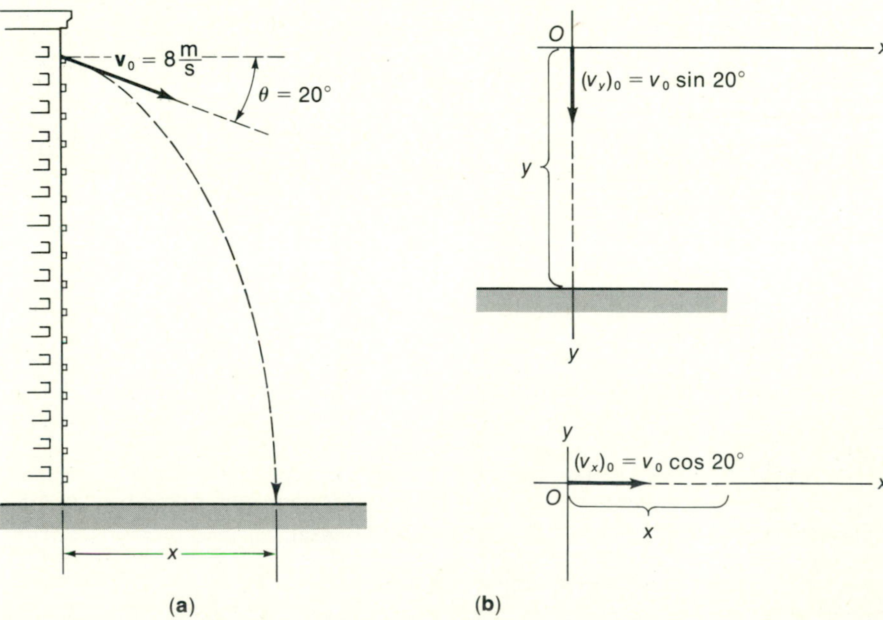

Figure 2–32
Example 2–10

The ball strikes the ground 3 s later. (a) How far horizontally from the base of the building does the ball strike the ground? (b) How high is the window? (c) How long does it take the ball to reach a point 10 m below the level of launching? (Ignore air friction.)

SOLUTION

Since all motion is downward, it is convenient to choose the downward direction as positive. We make separate sketches for the x and y components of the motion. The x-y axes indicate our choice of the launching point to be the origin of the coordinate system, with positive directions downward and to the right, as shown.

Horizontal component

Data for part (a)

$$(v_x)_0 = v_0 \cos 20^\circ$$

$$(v_x)_0 = \left(8\,\frac{\text{m}}{\text{s}}\right)(0.940) = 7.52\,\frac{\text{m}}{\text{s}}$$

$$a_x = 0$$
$$t = 3\text{ s}$$
$$x_0 = 0$$
$$x = ?$$

Vertical component

Data for part (b)

$$(v_y)_0 = v_0 \sin 20^\circ$$

$$(v_y)_0 = \left(8\,\frac{\text{m}}{\text{s}}\right)(0.342) = 2.74\,\frac{\text{m}}{\text{s}}$$

$$a_y = 9.8\,\frac{\text{m}}{\text{s}^2}$$ (g is positive because the positive direction is downward)

$$t = 3\text{ s}$$
$$y_0 = 0$$
$$y = ?$$

By comparing the known data with the kinematic equations, we see that Equation (2–23) is appropriate in both cases:

$$x = x_0 + (v_x)_0 t + \tfrac{1}{2}a_x t^2$$

Substituting numerical values, we have

$$x = 0 + \left(7.52\,\frac{\text{m}}{\text{s}}\right)(3\text{ s}) + \frac{1}{2}(0)(3\text{ s})^2$$

$$\boxed{x = 22.6\text{ m}}$$

$$y = y_0 + (v_y)_0 t + \tfrac{1}{2}a_y t^2$$

Substituting numerical values, we have

$$y = 0 + \left(2.74\,\frac{\text{m}}{\text{s}}\right)(3\text{ s}) + \frac{1}{2}\left(9.8\,\frac{\text{m}}{\text{s}^2}\right)(3\text{ s})^2$$

$$\boxed{y = 52.3\text{ m}}$$

Part (c): Here we are concerned only with the vertical component of motion. The given data are:

$$(v_y)_0 = 2.74\,\frac{\text{m}}{\text{s}}$$ (from part b)

$$a_y = 9.8\,\frac{\text{m}}{\text{s}^2}$$
$$y_0 = 0$$
$$y = 10\text{ m}$$
$$t = ?$$

By comparing these parameters with the kinematic equations, we choose Equation (2–23):

$$y = y_0 + (v_y)_0 t + \tfrac{1}{2}a_y t^2$$

$$10\text{ m} = 0 + \left(2.74\,\frac{\text{m}}{\text{s}}\right)(t) + \frac{1}{2}\left(9.8\,\frac{\text{m}}{\text{s}^2}\right)(t^2)$$

Rearranging (and omitting units for clarity), we have

$$4.90t^2 + 2.74t - 10 = 0$$

This equation cannot be factored by inspection, so we use the formula for the roots of a quadratic equation (see Appendix E):

$$t = \frac{-b \pm \sqrt{b^2 - 4ac}}{2a} = \frac{-2.74 \pm \sqrt{(2.74)^2 - (4)(4.90)(-10)}}{2(4.90)} = +1.18\text{ s},\ -1.74\text{ s}$$

Since we started at time $t = 0$ when the ball was thrown, we reject the negative root as not applicable to our problem. Therefore:

$$\boxed{t = 1.18 \text{ s}}$$

Note that the negative time represents the time prior to the start of our problem when a freely falling ball could have been located 10 m below the window, traveling upward so that after it reached its maximum upward motion and was falling downward, it "latched on" to our problem at $t = 0$. In some problems, *both* roots are correct, so you should always consider the context of the situation to justify the rejection (or retention) of each root that the mathematical manipulation provides.

Summary

The basic definitions for describing the motion of a particle are:

$$\left.\begin{aligned}
&\text{Displacement:} && \mathbf{r} = r_x\hat{\mathbf{x}} + r_y\hat{\mathbf{y}} + r_z\hat{\mathbf{z}} \\
&\text{Velocity:} && \mathbf{v} = \frac{d\mathbf{r}}{dt} = \frac{dr_x}{dt}\hat{\mathbf{x}} + \frac{dr_y}{dt}\hat{\mathbf{y}} + \frac{dr_z}{dt}\hat{\mathbf{z}} \\
&\text{Acceleration:} && \mathbf{a} = \frac{d\mathbf{v}}{dt} = \frac{dv_x}{dt}\hat{\mathbf{x}} + \frac{dv_y}{dt}\hat{\mathbf{y}} + \frac{dv_z}{dt}\hat{\mathbf{z}}
\end{aligned}\right\}\ \text{Three-dimensional motion}$$

These vector quantities have both magnitude and direction.

For the one-dimensional case, we use scalars to represent the magnitudes, and plus or minus signs to indicate the directions.

$$\left.\begin{aligned}
r &= x \\
v &= \frac{dx}{dt} \\
a &= \frac{dv}{dt}
\end{aligned}\right\}\ \text{One-dimensional motion}$$

Certain kinematic equations, which can be derived from the above definitions, are useful in solving problems. The symbol x represents a displacement along any straight-line direction.

$$\left.\begin{aligned}
v &= v_0 + at \\
x &= x_0 + v_0t + \tfrac{1}{2}at^2 \\
v^2 &= v_0{}^2 + 2a(x - x_0)
\end{aligned}\right\}\ \text{(only for constant } a\text{)}$$

These equations are simplified if the origin of the coordinate system is chosen to be the location of the object at $t = 0$, since in that case x_0 becomes equal to zero.

To avoid confusion in using the kinematic relations, associate plus or minus signs only with *numerical values* substituted into the equations (or with *symbols that represent numerical values*).

Displacement, velocity, and acceleration vectors need not be in the same direction. For example, a particle moving in the $+x$ direction and *slowing down* will have an acceleration in the $-x$ direction. [This is because the final speed v is smaller numerically than the initial speed v_0, making $\Delta v = (v - v_0)$ a negative number.]

In solving problems involving motion, the following standard procedure is helpful:

(1) Identify the *type* of problem. If the problem involves linear motion with constant acceleration, the kinematic equations apply.
(2) Make a sketch of the motion, drawing x-y axes to indicate the origin of the coordinate system and the directions that are positive. Include the notation you adopt and as much other information as possible to make evident the relations between various elements of the problem. If feasible, put the origin at the object's location at $t = 0$ to eliminate the x_0 term from the kinematic equations.
(3) List known and unknown parameters (in a consistent system of units) and compare with the kinematic equations. The main goal will be to find as many independent equations as there are unknowns. In some instances, these equations will suggest themselves as you examine the sketch of the physical situation.

Conversion of units is easily accomplished by multiplying by a ratio equaling unity.

For two- and three-dimensional motion, each component of the motion is solved separately, using scalar equations for the component directions. A common application of this method is the analysis of the trajectory of objects falling freely under the action of gravity. The symbol g represents the *magnitude* of the acceleration due to gravity: 9.8 m/s^2 or 32 ft/s^2.

Questions

1. Which of the following quantities are scalars, which are vectors, and which are neither: speed, velocity, volume, temperature, time, area, and color?
2. Can you combine two vectors of different magnitudes so that the resultant is zero? How about three vectors of different magnitudes? State the criterion that would be necessary for three vectors to add to zero.
3. Suppose you had a ruler with which (by interpolation) you could measure lengths to the nearest tenths of a millimeter. What would be the percentage of error in determining the length of this page? In determining the thickness of a page? (Think about the latter question a moment; there is a way to reduce the error considerably.)
4. Discuss methods by which the distance between the earth and the moon could be determined.
5. Name several phenomena that are repetitive that might serve as a time standard.
6. A solar day is the time it takes the earth to rotate once with respect to the sun. A sidereal day is the time it takes the earth to rotate once with respect to the fixed stars. The earth rotates about its axis in the same sense as it rotates about the sun. Which is longer, and why: a solar day or a sidereal day?
7. A lunar month is the time it takes the moon to return to the same phase. A sidereal month is the time it takes the moon to return to a given position with respect to the background of fixed stars. Which month is longer, and why? The moon rotates about the earth in the same sense as the earth moves about the sun.
8. Without any standard rulers, watches, or masses, describe how Robinson Crusoe could have set up a system of units for numerical measurements.
9. Can an object be moving north yet have acceleration toward the south?
10. Can an object reverse the direction of its motion even though it has constant acceleration?

11. Can an object reverse the direction of its acceleration even though it continues to move in the same direction?

12. Since time seems to proceed only in one direction, is it a vector?

13. Suppose a movie film of various freely falling objects (no air friction) were projected backward. Would the resultant motions be in agreement with the kinematic equations? (This procedure is equivalent to replacing t with $-t$ in the equations.)

14. Describe several phenomena whose motion, when filmed, would appear normal if the film were projected backward for viewing.

15. A motion picture film of a freely falling object shows the object moving downward and accelerating downward. If the film were run backward, would it show the object accelerating upward or downward?*

Problems

2A–1 Atoms are roughly 10^{-10} m in diameter and nuclei are about 10^{-15} m in diameter. If an atom were enlarged to 2 m in diameter, determine the diameter of the nucleus in millimeters. *Answer:* 0.020 mm

2A–2 The diameter of our disk-shaped galaxy is about 10^5 light-years. Andromeda, our nearest galactic neighbor, is about 2 million light-years away. If we represent our galaxy by a dinner plate 25 cm in diameter, determine the distance to the next dinner plate.

2A–3 The nearest star is about 4×10^{13} km away. If our sun (diameter = 1.4×19^9 m) were represented by a cherry pit 7 mm in diameter, determine the distance to the next cherry pit. *Answer:* 200 km

2A–4 A ship sails due north for 40 km, northwest for 100 km, then 30° south of west for 50 km. Find the net displacement from the starting point.

2A–5 Walking along streets laid out in a square grid pattern, a child walks 2 blocks west, 3 blocks north, then 2 blocks west.

(a) Find the total distance traveled.

(b) Find the net displacement (magnitude and direction) from the starting point.

Answers: (a) 7 blocks (b) 5 blocks; 36.9° north of west

2A–6 Given the vectors:

$$\mathbf{A} = 4\hat{\mathbf{x}} + 3\hat{\mathbf{y}}$$

$$\mathbf{B} = 2\hat{\mathbf{x}} - 2\hat{\mathbf{y}}$$

(a) Make freehand sketches for the vector sum $\mathbf{C} = \mathbf{A} + \mathbf{B}$ and for the vector subtraction $\mathbf{D} = \mathbf{B} - 2\mathbf{A}$.

(b) Express **C** and **D** in terms of polar coordinates (a magnitude and an angle with respect to the $+x$ axis).

2A–7 Given the vectors:

$$\mathbf{A} = 2\hat{\mathbf{x}} + 6\hat{\mathbf{y}}$$

$$\mathbf{B} = 4\hat{\mathbf{x}} - 1\hat{\mathbf{y}}$$

(a) Make freehand sketches (reasonably to scale) for the vector sum $\mathbf{C} = \mathbf{A} + \mathbf{B}$ and for the vector subtraction $\mathbf{D} = \mathbf{A} - \mathbf{B}$.

(b) Find analytic solutions in terms of unit vectors for **C** and **D**. Express **C** and **D** in terms of polar coordinates (a magnitude and an angle with respect to the $+x$ axis).

Answers: (a) $\mathbf{C} = 6\hat{\mathbf{x}} + 5\hat{\mathbf{y}}$, $\mathbf{D} = -2\hat{\mathbf{x}} + 7\hat{\mathbf{y}}$
(b) $\mathbf{C} = 7.81$ at 39.8°; $\mathbf{D} = 7.28$ at 106°

2A–8 The speed of light is 3×10^8 m/s. Calculate the time it takes light to travel across a nucleus whose diameter is 2×10^{-15} m.

---

* Reprinted with permission from *Thinking Physics*, Volume 1, by L. Epstein and P. Hewitt (1979), page 14. Published by Insight Press, 614 Vermont Street, San Francisco, CA 94107.

2A-9 Astronomical distances are so large that convenient units of distance in astronomy are:

1 **light-year:** The distance light travels in a vacuum in one year at a speed of 3.00×10^8 m/s.

1 **astronomical unit (AU):** The average radius of the earth's orbit about the sun. By definition of the International Astronomical Union, 1 AU $\equiv 149\,600 \times 10^6$ m.

1 **parsec (pc):** The distance at which 1 AU subtends an angle of 1 second of arc.

Express the light-year and parsec in meters.

Answers: 1 light-year $= 9.46 \times 10^{15}$ m; 1 parsec $= 3.08 \times 10^{16}$ m

2A-10 A truck traveling at 10 m/s doubles its speed by accelerating uniformly for 10 s.

(a) Find the acceleration.

(b) Determine the distance traveled by the truck while accelerating.

2A-11 A baseball is given a speed of 10 m/s by a player whose throwing hand moves in a straight line for 0.80 m. Find the average acceleration the ball experienced.

Answer: 62.5 m/s^2

2A-12 The circumference of the earth is 24 902 mi. Using conversion factors you know without looking them up, convert this distance to inches. Express this answer in scientific notation.

2A-13 A motorist travels from Los Angeles to San Francisco (425 mi) in 8 h. Calculate the average speed for the journey. Express the answer in both miles per hour and feet per second. *Answer:* 53.1 mi/h; 77.9 ft/s

2A-14 A ball is thrown vertically upward with a speed of 20 m/s.

(a) Determine the time it takes for the ball to reach the top of its trajectory.

(b) Find the height of the trajectory.

(c) Find the speed of the ball when it returns to its starting point.

2A-15 A stone is dropped from rest 2.0 m above the ground. Calculate the time that has elapsed when it strikes the ground. *Answer:* 0.639 s

2A-16 Jumping off a diving board 3 m above the water, a diver launches himself with a speed of 2 m/s at an angle of 60° above the horizontal. Determine the amount of time he is in the air.

2A-17 A ball is thrown straight upward with an initial speed of 12 m/s. Find the displacement and velocity (including direction) after (a) 1 s and (b) 2 s.

Answers: (a) 7.10 m above the launching point; 2.20 m/s (upward)
(b) 4.40 m above the launching point; -7.60 m/s (downward)

2A-18 A ball is tossed vertically upward from the edge of a cliff, as shown in Figure 2-33. It rises above the cliff 5 m and then falls vertically downward, striking the ground 15 m below the point launched.

(a) Find the initial speed of the ball.

(b) Determine the amount of time the ball was in the air.

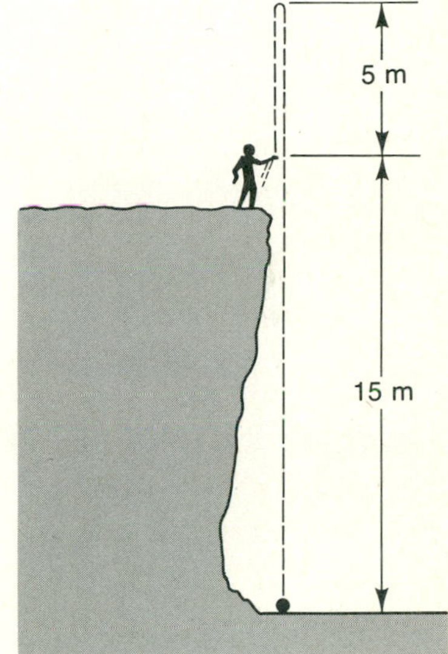

Figure 2-33
Problem 2A-18

2A-19 A projectile is launched at an angle of 53.1° with respect to the horizontal, as shown in Figure 2-34. It strikes a target that is 120 m away horizontally and 160 m below the launching point. Find the initial speed of the projectile. Assume $g = 10$ m/s^2.

Answer: 25 m/s

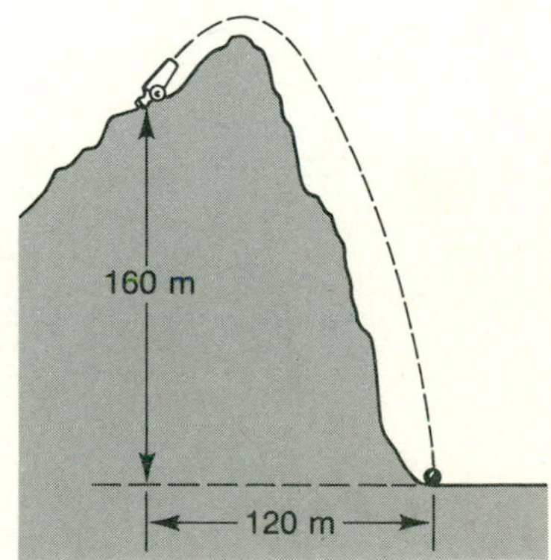

Figure 2-34
Problem 2A-19

2B-1 Given the following vectors:

$$\mathbf{A} = 3\hat{\mathbf{x}} + 4\hat{\mathbf{y}}$$
$$\mathbf{B} = 2\hat{\mathbf{x}} - 2\hat{\mathbf{y}}$$

Sketch freehand graphical solutions and also solve analytically to obtain the magnitude and direction of these vectors:

$$\mathbf{C} = \mathbf{A} + \mathbf{B}$$
$$\mathbf{D} = \mathbf{A} - \mathbf{B}$$
$$\mathbf{E} = \mathbf{B} - 2\mathbf{A}$$

Answers: $\mathbf{C} = 5.39$ at 21.8°
$\mathbf{D} = 6.08$ at 80.5°
$\mathbf{E} = 10.8$ at 259.2°

2B–2 A novice golfer on the green takes three strokes to sink the ball. The successive displacements are 4 m due north, 2 m northeast, and 1 m 30° west of south. Starting at the same initial point, an expert golfer could make the hole in what single displacement?

2B–3 An automobile traveling at 20 m/s slows uniformly to one-half its speed at a rate of 2 m/s^2.

(a) Determine the amount of time this takes.

(b) Determine the distance covered during this time. *Answers:* (a) 5.00 s (b) 75.0 m

2B–4 A coin is tossed vertically upward with a speed of 4 m/s. Determine the time when the coin is 0.50 m above the starting point. Explain why there are two answers to this problem.

2B–5 Water drips at a steady rate from a faucet that is 30 cm above a sink. Just as one drop hits the sink, another is falling through the air and a third is leaving the faucet. Assuming $g = 10$ m/s^2, determine the number of drops per minute that emerge from the faucet. *Answer:* 485 drops/min

2B–6 A subway train is designed so that the maximum acceleration the passengers experience is 1.5 m/s^2.

(a) If two subway stations are 800 m apart, find the minimum time required between stops at these stations.

(b) Find the maximum speed the subway train acquires between these stops.

2B–7 A stone is dropped from a bridge 45 m above a river. One second later, another stone is thrown vertically downward so that both stones strike the water simultaneously. Find the initial speed the second stone had. *Answer:* 12.7 m/s

2B–8 One runner covered the 100-m dash in 10.3 s. Another runner came in second at a time of 10.8 s. Assuming the runners traveled at their average speeds for the entire distance, determine how far behind the second runner was when the winner crossed the finish line.

2B–9 When the sun is directly overhead, a hawk dives toward the ground at a speed of 5 m/s. If the direction of his motion is at an angle of 60° below the horizontal, calculate the speed of his shadow moving along the ground. *Answer:* 2.50 m/s

2B–10 The bore of a rifle is directed horizontally toward the center of a target 100 m away, but the bullet strikes 10 cm below the center. Calculate the muzzle velocity of the bullet.

2B–11 A ball is thrown horizontally with a speed of 6 m/s from a balcony, and it is caught by a person standing 10 m from the base of the building. Determine the height of the balcony. *Answer:* 13.6 m

2B–12 A motorist drives along a straight road at 15 m/s. Just as she passes a parked motorcycle police officer, the officer starts to accelerate at 2 m/s^2 to overtake her. Maintaining this constant value of acceleration, (a) determine the amount of time it will take the police officer to reach the motorist. Find (b) the speed and (c) the total displacement of the police officer as he overtakes the motorist.

2B–13 A stone is thrown with an initial velocity of 8 m/s at an angle of 50° above the horizontal. The launching point is the top of a cliff 20 m high.

(a) Find the maximum height above the base of the cliff the stone reaches.

(b) Determine the amount of time it is in the air before hitting the ground, which extends horizontally from the cliff base.

(c) Determine the horizontal distance the stone travels during its flight.

(d) Determine the speed and angle at which it strikes the ground.

Answers: (a) 21.9 m (b) 2.74 s (c) 14.1 m (d) 21.4 m/s; 13.9° from the vertical

2B–14 A ball is launched at an angle of 48° above the horizontal with a speed of 12 m/s. Find the x and y coordinates of the ball (relative to the starting point) when its velocity makes an angle of 35° below the horizontal.

2B–15 A ball is thrown with initial speed v_0 at an angle θ with respect to the horizontal, where $0 < \theta < 90°$. Show that the ball's trajectory is a parabola. (Hint: Eliminate t from the equations and express y as a function of x. Recall that the general equation of a parabola has the form $y = ax + bx^2$.)

Answer: $y = (\tan\theta)x + \left[\dfrac{g}{2(v_0\cos\theta)^2}\right]x^2$

2B–16 A mountain climber tosses a piton to his girlfriend, who is climbing above him. The initial velocity of the piton is 12 m/s at 55° with respect to the horizontal. At the instant the piton is caught, it is traveling horizontally. Calculate the distance between the climbers.

Figure 2–35
Problem 2B–16

2B–17 A stone is dropped from rest into a well 50 m deep. If the speed of sound is 330 m/s, determine the time that has elapsed when the splash is heard. *Answer:* 3.34 s

2B–18 A ball is tossed vertically to a catcher 5 m above the point of release. If the catcher is unable to catch a ball traveling faster than 6.0 m/s, find the minimum and maximum time of flight of the ball.

2B–19 A stone is thrown horizontally from a bridge 25 m high. If the tosser sees the splash along a line of sight 45° below the horizontal, (a) determine the initial speed of the stone. (b) Determine the velocity (magnitude and direction) with which the stone strikes the water. *Answers:* (a) 11.1 m/s (b) 24.7 m/s; 26.5° from the vertical

2B–20 The velocity of an object in one-dimensional motion is shown in Figure 2–36. Sketch graphs of the displacement versus time and the acceleration versus time, showing numerical values for points at which the quantity undergoes a change in slope.

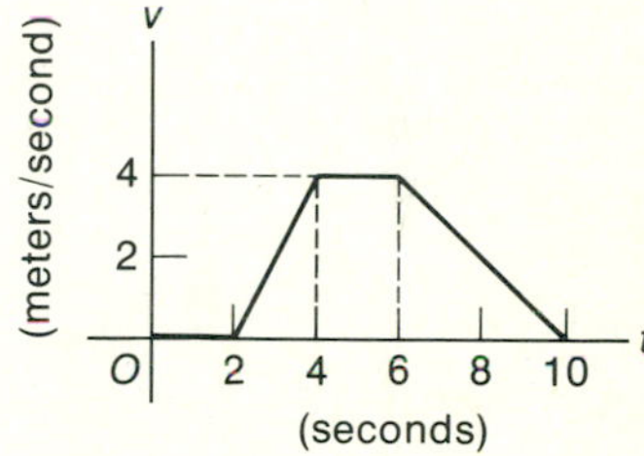

Figure 2–36
Problem 2B–20

2B–21 A projectile is launched at an angle of 50° with respect to the horizontal, striking a target as shown in Figure 2–37. Determine the initial speed of the projectile.
Answer: 55.4 m/s

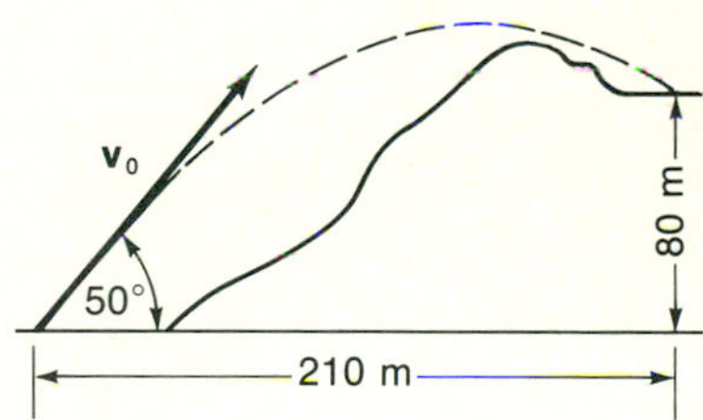

Figure 2–37
Problem 2B–21

2B–22 Calculate the difference in minutes between an hour and a microcentury. Which is longer?

2B–23 Starting with one year, use conversion factors you know (without looking them up) to determine the number of seconds in a year. Express the answer in scientific notation. (Note that the significant figures in the answer are, within 0.5%, equal to π. This fact will help you to remember this useful conversion factor.) *Answer:* 3.15×10^7 s

2B–24 A motorist travels one mile at 15 mi/h. Determine the speed at which he should travel the next mile in order to average 30 mi/h for the entire two-mile trip. (Caution: The answer is *not* 45 mi/h.)

2C–1 A hot air balloon is descending at 3 m/s when a passenger releases a sandbag.

(a) Calculate the speed of the sandbag (relative to the earth) one second later.

(b) If the balloon's descent is immediately slowed to 2 m/s upon release of the sandbag, determine how far below the balloon the sandbag is one second after its release.

Answers: (a) 12.8 m/s (b) 5.90 m

2C–2 Consider the following two vectors:

$$\mathbf{E} = 2\hat{\mathbf{x}} - 3\hat{\mathbf{y}} + 4\hat{\mathbf{z}}$$

$$\mathbf{F} = 4\hat{\mathbf{x}} + 1\hat{\mathbf{y}} - 2\hat{\mathbf{z}}$$

Find, analytically, (a) the vector $\mathbf{G} = \mathbf{E} + \mathbf{F}$, (b) the vector $\mathbf{H} = \mathbf{F} - \mathbf{E}$, and (c) the vector $\mathbf{J}$ such that $\mathbf{E} - 2\mathbf{F} + \mathbf{J} = 0$.

2C–3 A brick falls from a building and passes from the top to the bottom of a window 2 m tall in 0.20 s. Determine how high above the top of the window the brick fell, assuming the brick had no initial velocity. *Answer:* 4.15 m

2C–4 An elevator accelerates and decelerates uniformly at 2 ft/s² while moving between adjacent floors 12 ft apart. Find the minimum travel time between stops.

2C–5 A projectile is launched at an angle θ above the horizontal. Derive an expression for the line-of-sight angle ϕ from the launching point to the projectile at its maximum elevation.

Answer: $\phi = \tan^{-1}\left(\dfrac{\tan\theta}{2}\right)$

2C–6 A traffic light is located at an intersection where the speed limit is 30 mph. Suppose it takes 0.5 s, on the average, for a driver to apply the brakes after seeing the light turn yellow.

(a) If the average rate of deceleration is 10 ft/s² (assumed uniform), what should be the minimum distance from the stop line for a driver traveling at the speed limit when the light turns yellow?

(b) If the light turns red just as this driver stops at the stop line, how long was the yellow light displayed?

2C–7 Starting with the kinematic equations, derive the following expression for the maximum height y_m reached by a projectile launched at an angle θ above the horizontal with an initial speed v_0.

Answer: $y_m = \dfrac{v_0^2 \sin^2 \theta}{2g}$

2C–8 Two adjacent subway stops are 1400 ft apart. If the subway train accelerates uniformly at 3 ft/s^2 and decelerates uniformly at 4 ft/s^2, find (a) the minimum travel time between station stops, (b) the maximum speed attained, and (c) the average speed between stops. Include freehand graphs for the displacement, velocity, and acceleration as functions of time. Indicate on the graphs the numerical coordinates for the instant the motorman switches from the accelerator to the brake.

2C–9 Starting with the kinematic equations, derive an expression for the horizontal range R of a projectile fired from ground level on a horizontal plane with an initial velocity v_0 at an angle θ above the horizontal.

Answer: $R = \dfrac{v_0^2 \sin 2\theta}{g}$

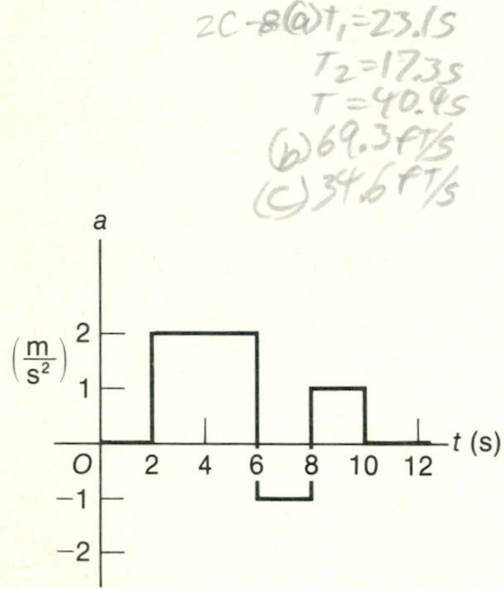

Figure 2–38
Problem 2C–12

2C–10 Starting from rest, an automatic subway train is controlled by a computer that causes its motion for the first 5 s to follow the relation $x = At^3$ for the distance x covered. Find expressions for (a) the speed and (b) the acceleration as functions of time. (c) If the train acquires a speed of 4 m/s after 5 s, what is the value of A (including units)?

2C–11 By differentiating the equation for the horizontal range R in Problem 2C–9, show that the initial elevation angle θ for the maximum range is 45°.

2C–12 An object starts from rest and accelerates along a straight line, as shown in Figure 2–38. Sketch graphs of the velocity versus time and the displacement versus time, showing numerical values for points at which t = 2, 6, 8, and 10 s.

2C–13 If the horizontal range R and the maximum height H of a projectile are equal in magnitude, determine the initial launching angle θ. (For your analysis, start with the kinematic equations.) *Answer:* 76.0°

2C–14 Copy the graphs of velocity versus time shown in Figure 2–39. Directly below each graph, sketch separate freehand graphs of the displacement versus time and the acceleration versus time. Use the same time scale in all graphs.

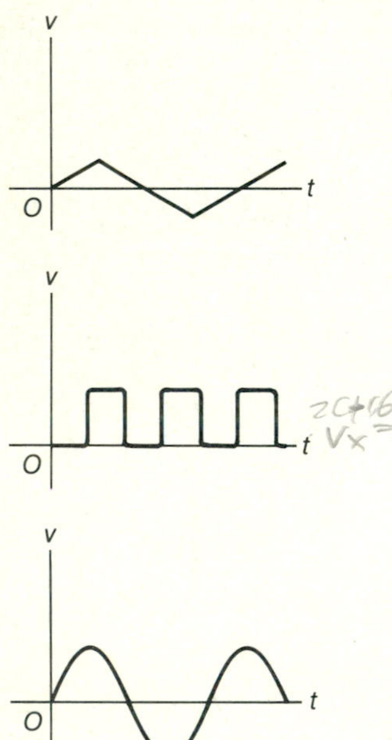

Figure 2–39
Problem 2C–14

2C–15 A ball is launched with an initial speed and launching angle such that its horizontal range is R.

(a) Show that R equals $x^2/(x - y)$, where x and y are the coordinates of the ball at any point after it passes the maximum height of the trajectory (and before it strikes the ground).

(b) Does the fact that g does not appear in the formula mean that the range would be the same on the moon, where g is smaller? Explain.

2C–16 A grasshopper can jump a maximum horizontal distance of 1 m. If it spends negligible time on the ground, find the fastest speed with which it can travel down the road.

2C–17 In his *Two New Sciences*, Galileo states that "for elevations (angles of projection) which exceed or fall short of 45° by equal amounts, the ranges are equal." Prove this statement.

2C–18 A tardy commuter runs at a constant speed of 3 m/s in an attempt to catch a train that is about to pull away from the station. The commuter runs parallel to the tracks and is still a distance x_0 from the door at the end of the train when the train starts to accelerate from rest at 0.5 m/s^2.

(a) Sketch a freehand (qualitative) graph for the position $x_C(t)$ of the commuter, letting $x_C = 0$ at $t_0 = 0$. On the same graph, sketch a family of curves for positions $x_T(t)$ of the train for various values of x_0. Include cases where the commuter catches the train, as well as cases where the train is missed.

(b) Find the largest value of x_0 for which the commuter catches the train.

Answer: (b) 9 m

One had to be a Newton to notice that the moon is falling, when everyone sees that it doesn't fall.

PAUL VALÉRY

chapter 3 Circular Motion

3.1 Introduction

A very important case of planar motion is that of traveling in a circle. Examples include the various parts of rotating machinery, an automobile moving along a curved road, electrons in the Bohr model of the hydrogen atom, and an earth satellite in a circular orbit. Because circular motion is so common, we shall now develop a convenient notation for describing the position, velocity, and acceleration of a particle traveling in a circle. One interesting feature is the fact that a particle moving in a circle at constant speed has acceleration. Although this characteristic may seem a bit puzzling at first, it follows directly from the basic definition of acceleration: $\mathbf{a} = d\mathbf{v}/dt$. The velocity vector $\mathbf{v}$ is constant in *magnitude*, but its *direction* is constantly changing, and it is this change in direction that is related to the acceleration.

3.2 Polar Coordinates

Polar coordinates are the most convenient for describing circular motion. Consider a particle moving along a circular path $a \rightarrow b \rightarrow c$, as shown in Figure 3–1. The center of the circle is the origin O of the polar coordinate system. For convenience, a rectangular coordinate system with x and y axes is also shown.

The position of the particle along the path is described by two quantities: r, the magnitude of the position vector $\mathbf{r}$ from O to the particle, and θ, the angle the vector makes relative to an arbitrary direction. (Here, we chose the $+x$ axis as the reference direction.) The angle θ is called the **angular position.**

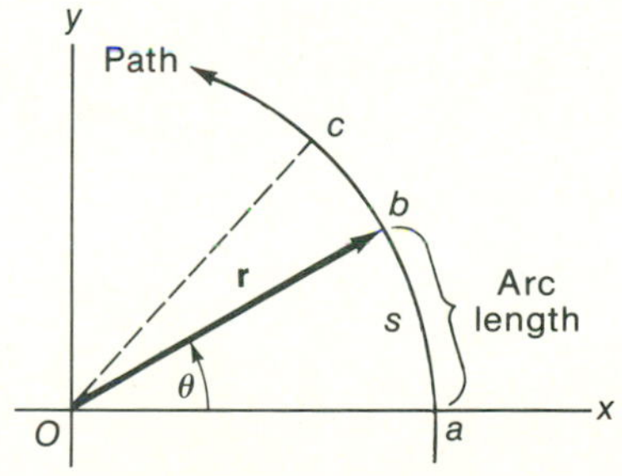

Figure 3–1
The polar coordinates r and θ describe the motion of a particle moving along the circular path $a \rightarrow b \rightarrow c$.

In kinematic equations dealing with circular motion, the angular position is always expressed in units of radians, abbreviated rad and defined as the ratio of the arc length s to the radius r.

ANGULAR POSITION $$\theta = \frac{s}{r} \qquad (3\text{–}1)$$

Since the radian is a ratio of lengths, it has dimensions of unity. The angle subtended by a complete circle (that is, the angle made by one complete revolution about the origin) is 2π rad. Therefore:

$$2\pi \text{ rad} = 360^\circ$$

or

$$1 \text{ rad} = \frac{360^\circ}{2\pi} \approx 57.3^\circ$$

3.3 Velocity and Acceleration in Circular Motion

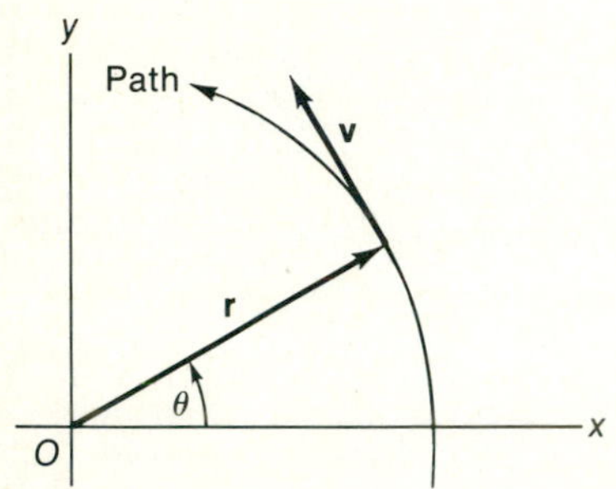

Figure 3–2
The velocity **v** for a particle in circular motion is always tangent to the path.

The velocity vector for a particle moving along a circular path is always *tangent to the path* (see Figure 3–2). This vector *may* change its magnitude (speeding up or slowing down), and it is *always* changing its direction. In general, the result is two acceleration components, *tangential* and *radially inward*, which are quite simple to use in solving problems.

3.4 Tangential Acceleration

In the previous chapter, we analyzed straight-line motion and found that speeding up or slowing down produced on acceleration *along* the direction of motion. For circular motion, you may think of the **tangential acceleration** component as a straight-line path bent around the circular arc, resulting in an acceleration component that is tangent to the path whenever the speed changes.

TANGENTIAL ACCELERATION $$a_t = \frac{dv}{dt} \quad \text{(tangent to the path)} \qquad (3\text{–}2)$$

Note that Equation (3–2) is not a vector equation even though the acceleration component is always tangent to the path. If the object is speeding up, a_t is *in* the direction of motion; if it is slowing down, a_t is still tangent to the curve, but *opposite* to the direction of motion. If the speed is constant, the tangential acceleration is zero.

3.5 Centripetal Acceleration

The other acceleration component is perpendicular to the path. It is always *radially inward*, toward the center of the circle, and is called **centripetal acceleration,** a_{cp} (from the Latin *centrum*, meaning "center," and *petere*, meaning "to seek"). Be careful not to confuse this with centri*fugal* acceleration, a_{cf} (from *fugere*, meaning "to flee"), which is radially outward. (Centrifugal acceleration will be discussed in Chapter 11.)

To investigate the centripetal acceleration component, we look first at the special case of a particle that has *uniform circular motion* (that is, a particle moving *at constant speed* around a circle). Here, the direction of the velocity vector is constantly changing, though its magnitude remains constant. Figure 3–3(a) shows two velocity vectors, $\mathbf{v}(t)$ and $\mathbf{v}(t + \Delta t)$, for two points along the

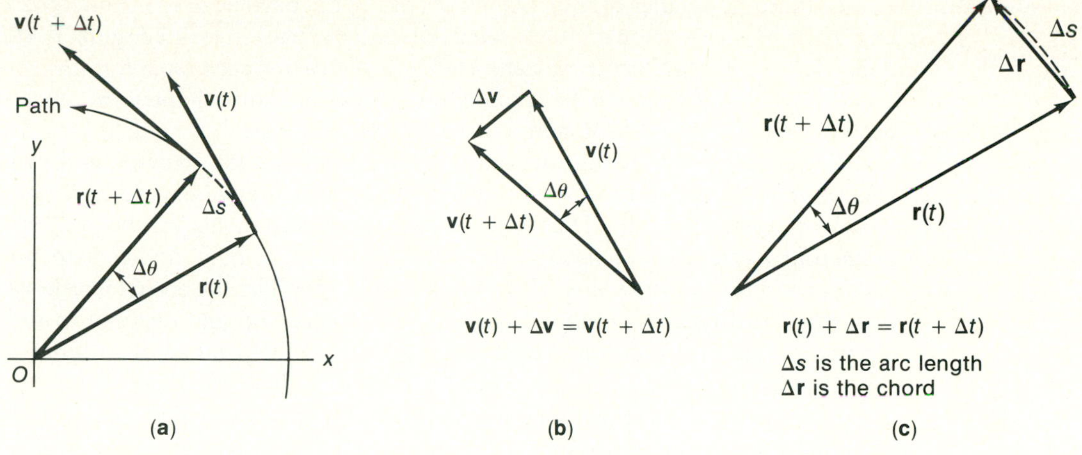

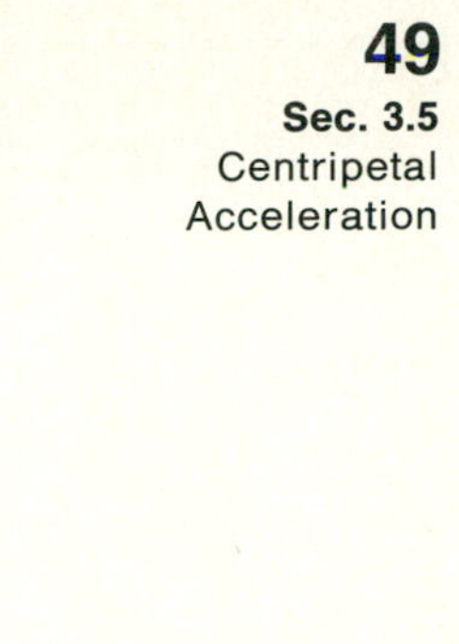

Figure 3–3
In uniform circular motion, a particle moves along the curved path Δs (shown dashed) with constant speed. Both the velocity vectors (b) and the position vectors (c) change direction through the angle $\Delta\theta$, though their magnitudes remain constant.

path. To investigate the change in velocity, we construct the vector diagram of Figure 3–3(b), sketching the velocity vectors from a common point while preserving their magnitudes and directions. We see that

$$\mathbf{v}(t) + \Delta\mathbf{v} = \mathbf{v}(t + \Delta t) \tag{3–3}$$

or

$$\Delta\mathbf{v} = \mathbf{v}(t + \Delta t) - \mathbf{v}(t) \tag{3–4}$$

Note that the angle $\Delta\theta$ between the velocity vectors is the same as the angle $\Delta\theta$ between the position vectors $\mathbf{r}(t)$ and $\mathbf{r}(t + \Delta t)$ in Figure 3–3(c).[1] Furthermore, if $\Delta\theta$ is very small, the arc length Δs approaches the chord length Δr. In fact, in the limit as $\Delta\theta \to 0$ (and $\Delta t \to 0$), we may assume that Δs equals Δr with negligible error.

The acceleration is defined as $\lim_{\Delta t \to 0} \Delta\mathbf{v}/\Delta t$. Note carefully the direction of $\Delta\mathbf{v}$. As $\Delta\theta \to 0$, the direction of $\Delta\mathbf{v}$ becomes *radially inward*, at right angles to $\mathbf{v}$ itself. From similar triangles, we have

$$\frac{\Delta v}{v} = \frac{\Delta r}{r} \approx \frac{\Delta s}{r} \tag{3–5}$$

or

$$\Delta v \approx \left(\frac{v}{r}\right)\Delta s \tag{3–6}$$

We can substitute this value for Δv into the defining equation for acceleration, Equation (2–12), noting that v/r is constant:

$$a_{cp} = \lim_{\Delta t \to 0} \frac{\Delta v}{\Delta t} \approx \left(\frac{v}{r}\right) \lim_{\Delta t \to 0} \frac{\Delta s}{\Delta t} \tag{3–7}$$

In the limit, the approximation becomes an equality, and since $\lim_{\Delta t \to 0} \Delta s/\Delta t$ is defined as the instantaneous speed v along the path, we have

CENTRIPETAL ACCELERATION

$$a_{cp} = \frac{v^2}{r} \quad \text{(radially inward)} \tag{3–8}$$

We can verify that v^2/r does, indeed, have dimensions of acceleration:

$$\textit{Dimensions of } \frac{v^2}{r}\text{:} \quad \frac{\left[\frac{L}{T}\right]^2}{[L]} = \left[\frac{L}{T^2}\right]$$

[1] The reason is obvious if you remember that $\mathbf{v}$ is always at right angles to $\mathbf{r}$. Thus, as the radius moves through an angle $\Delta\theta$, the velocity vector also moves through this angle.

In uniform circular motion (constant speed), the particle *always* undergoes a centripetal acceleration of v^2/r toward the center of the circle (see Figure 3–4). *Note that the particle never gets any closer to the center even though it always accelerates in that direction.* (As pointed out in the previous chapter, an object need not move in the direction of its acceleration vector.)

If the motion is nonuniform (that is, if the particle speeds up or slows down while it travels around the circle), it will also undergo a tangential acceleration of dv/dt [Equation (3–2)]. Thus both acceleration components will be present: tangential acceleration *along the path of the particle*, $a_t = dv/dt$; and centripetal acceleration *toward the center of the circle*, $a_{cp} = v^2/r$. The tangential acceleration is *in* the direction of motion if the particle speeds up and *opposite* to the direction of motion if the particle slows down. The centripetal acceleration, on the other hand, is always *radially inward*. Figure 3–5 illustrates these two right-angle components of the total acceleration for $a_t = 0.600\ \text{m/s}^2$ and $a_{cp} = 0.800\ \text{m/s}^2$. From the Pythagorean Theorem, we find that the net acceleration **a** has a magnitude $a = \sqrt{a_t^2 + a_{cp}^2}$, and its direction may be specified as shown in Figure 3–5.

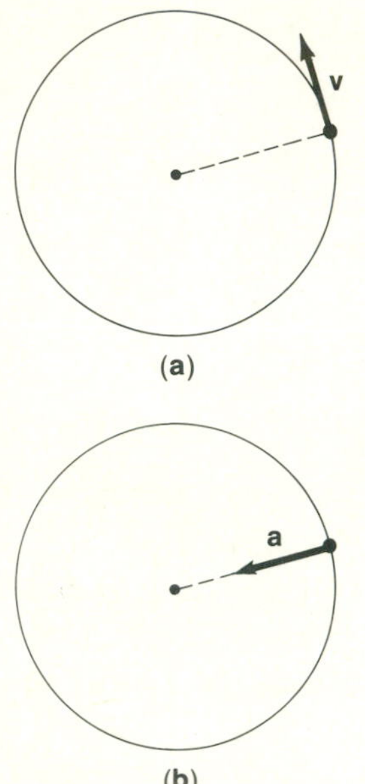

Figure 3–4
For motion in a circle *at constant speed*, the velocity vector (a) and the acceleration vector (b) are at right angles.

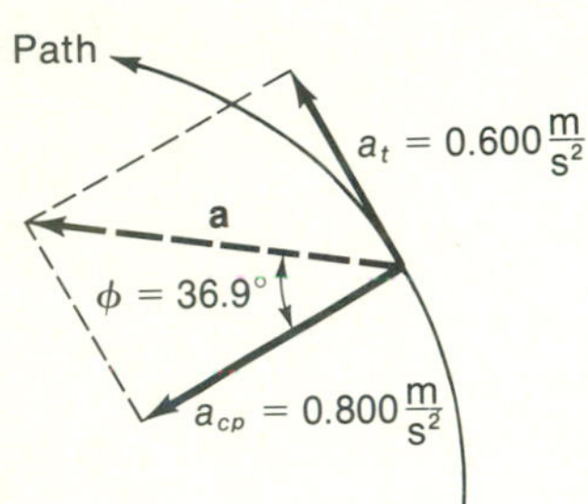

Figure 3–5
Acceleration components a_t and a_{cp} and the corresponding total acceleration **a** for a particle undergoing nonuniform circular motion.

EXAMPLE 3–1

What is the acceleration of a particle that is 12 cm from the center of a phonograph record turning at a constant speed of 33.3 revolutions per minute?

SOLUTION

Since the record is turning at a constant speed, the acceleration is entirely centripetal. Therefore the magnitude of the acceleration is given by

$$a_{cp} = \frac{v^2}{r}$$

The speed may be found from

$$v = \frac{\text{Distance traveled in one revolution}}{\text{Time for one revolution}}$$

The time for one revolution is called the **period,** T, of the motion.

$$T = \left(\frac{1}{33.3}\right)\cancel{\text{min}}\underbrace{\left(\frac{60\ \text{s}}{1\ \cancel{\text{min}}}\right)}_{\substack{\text{Conversion}\\\text{ratio}}} = 1.80\ \text{s}$$

Thus:

$$v = \frac{2\pi r}{T}$$

and therefore:

$$a_{cp} = \frac{v^2}{r} = \frac{\left(\frac{2\pi r}{T}\right)^2}{r} = \frac{4\pi^2 r}{T^2}$$

Substitution of numbers gives us:

$$a_{cp} = \frac{4\pi^2(0.12\ \text{m})}{(1.80\ \text{s})^2} = \boxed{1.46\ \frac{\text{m}}{\text{s}^2}}$$

Note that the particle accelerates toward the center of the circle even though its velocity is tangent to the circle, as was shown in Figure 3–4. This illustrates the fact that the velocity and acceleration vectors are never in the same direction for motion along a curved path. The direction of **a** depends on the *change* in **v**, which in this case is not in the direction of **v** itself.

EXAMPLE 3–2

An automobile whose speed is increasing at a rate of 0.6 m/s^2 travels along a circular road of radius $r = 20$ m. When the instantaneous speed of the automobile is 4 m/s, find (a) the tangential acceleration component, (b) the centripetal acceleration component, and (c) the magnitude and direction of the total acceleration.

SOLUTION

(a) The tangential acceleration component a_t depends only on the rate of change of the speed: $a_t = dv/dt$. As stated in the problem, this is

$$\boxed{a_t = 0.600\,\frac{m}{s^2}}$$

(b) The centripetal acceleration component a_{cp} depends on the instantaneous speed v and the radius r.

$$a_{cp} = \frac{v^2}{r} = \frac{\left(4\,\dfrac{\text{m}}{\text{s}}\right)^2}{(20\text{ m})} = \boxed{0.800\,\frac{\text{m}}{\text{s}^2}}$$

(c) The acceleration components were illustrated in Figure 3–5. From the Pythagorean Theorem, we find that the magnitude of the total acceleration is

$$a = \sqrt{a_t^2 + a_{cp}^2} = \sqrt{\left(0.600\,\frac{\text{m}}{\text{s}^2}\right)^2 + \left(0.800\,\frac{\text{m}}{\text{s}^2}\right)^2} = \boxed{1.00\,\frac{\text{m}}{\text{s}^2}}$$

The direction of **a** can be specified as making a certain angle with respect to some known direction. For example, if we use the inward radial direction as a reference, we have

$$\tan\phi = \frac{\left(0.600\,\dfrac{\text{m}}{\text{s}^2}\right)}{\left(0.800\,\dfrac{\text{m}}{\text{s}^2}\right)} = 0.750$$

$$\phi = \tan^{-1}(0.750) = \boxed{36.9^\circ}$$

Note that again the velocity and acceleration vectors are not in the same direction.

3.6 Derivation of Acceleration for Circular Motion Using Unit Vectors

If you know how to differentiate sines and cosines, you will appreciate a more direct derivation of the acceleration components.[2] It involves the unit vectors $\hat{\mathbf{r}}$ and $\hat{\boldsymbol{\theta}}$, which are used in polar coordinates. Figures 3–6 and 3–7 illustrate these unit vectors.

[2] The rules for differentiating $\sin\theta$ and $\cos\theta$ when θ varies with respect to time are:

$$\frac{d}{dt}(\sin\theta) = \cos\theta\,\frac{d\theta}{dt}$$

$$\frac{d}{dt}(\cos\theta) = -\sin\theta\,\frac{d\theta}{dt}$$

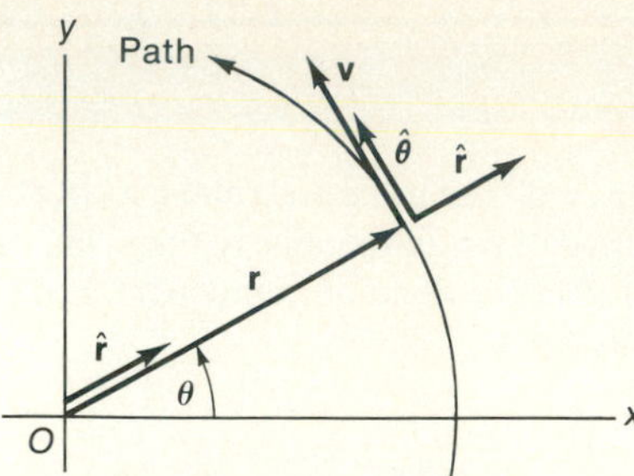

(**a**) The unit vector $\hat{\mathbf{r}}$ is in the direction of **r**, and $\hat{\theta}$ is perpendicular to **r** in the direction of increasing θ. As θ changes with time, the unit vectors rotate in direction along with **r**.

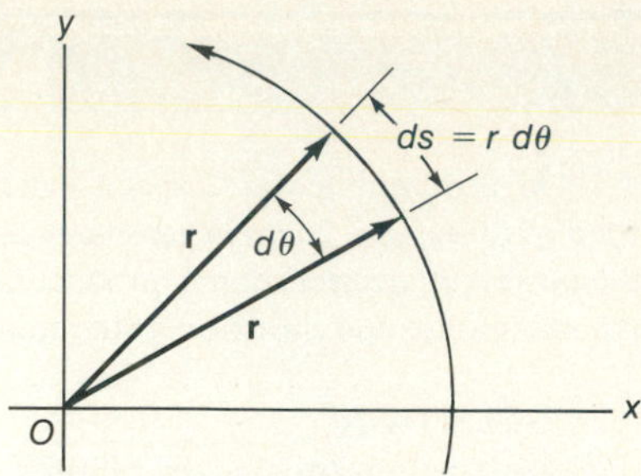

(**b**) As **r** moves through the angle $d\theta$, the particle moves along the arc a distance $ds = r\,d\theta$.

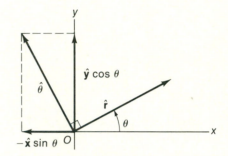

Figure 3–6
The unit vectors $\hat{\mathbf{r}}$ and $\hat{\theta}$ for polar coordinates.

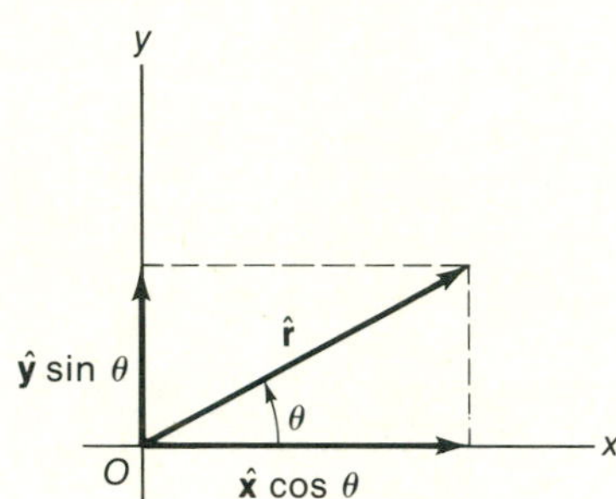

(**a**) The unit vector $\hat{\mathbf{r}}$ has vector components in terms of the unit vectors $\hat{\mathbf{x}}$ and $\hat{\mathbf{y}}$.

(**b**) The unit vector $\hat{\theta}$ is at right angles to the unit vector $\hat{\mathbf{r}}$. In terms of $\hat{\mathbf{x}}$ and $\hat{\mathbf{y}}$, its vector components are as shown.

Figure 3–7
Relations between the polar unit vectors $\hat{\mathbf{r}}$ and $\hat{\theta}$ and the rectangular unit vectors $\hat{\mathbf{x}}$ and $\hat{\mathbf{y}}$.

$\hat{\mathbf{r}} = \hat{\mathbf{x}} \cos\theta + \hat{\mathbf{y}} \sin\theta$ This is a unit vector in the radially outward direction; thus it lies along the position vector **r** in the direction of increasing **r**. **(3–9)**

$\hat{\boldsymbol{\theta}} = -\hat{\mathbf{x}} \sin\theta + \hat{\mathbf{y}} \cos\theta$ This is a unit vector at right angles to $\hat{\mathbf{r}}$; it lies along the direction of increasing θ. **(3–10)**

In contrast with the unit vectors $\hat{\mathbf{x}}$, $\hat{\mathbf{y}}$, and $\hat{\mathbf{z}}$, which remain fixed in space, the unit vectors $\hat{\mathbf{r}}$ and $\hat{\boldsymbol{\theta}}$ are functions of the angle θ, and thus they change direction as the position vector rotates.

Consider the general motion of a particle moving along a circular path with instantaneous values of **r** and **v**, as shown in Figure 3–6. The position vector **r**, which designates the location of the particle, is represented as

$$\begin{aligned} \mathbf{r} &= r\hat{\mathbf{r}} \\ \mathbf{r} &= r(\hat{\mathbf{x}} \cos\theta + \hat{\mathbf{y}} \sin\theta) \end{aligned} \qquad \textbf{(3–11)}$$

We find the velocity **v** of the particle by the usual differentiation $\mathbf{v} = d\mathbf{r}/dt$, noting that the scalar r and the unit vectors $\hat{\mathbf{x}}$ and $\hat{\mathbf{y}}$ are constants. The angle θ, however, does change with time. Finding $d\mathbf{r}/dt$, we have

$$\begin{aligned} \mathbf{v} &= r\left(-\hat{\mathbf{x}} \sin\theta \frac{d\theta}{dt} + \hat{\mathbf{y}} \cos\theta \frac{d\theta}{dt}\right) \\ \mathbf{v} &= r\frac{d\theta}{dt}(-\hat{\mathbf{x}} \sin\theta + \hat{\mathbf{y}} \cos\theta) \end{aligned} \qquad \textbf{(3–12)}$$

From Equation (3–1), we note that the arc length distance s is related to r and θ by

$$s = r\theta$$

The speed of the particle as it moves along the circular path is $v = ds/dt$, or (since r is constant):

$$v = r\frac{d\theta}{dt} \tag{3–13}$$

Thus, Equation (3–12) becomes

$$\mathbf{v} = v\underbrace{(-\hat{\mathbf{x}}\sin\theta + \hat{\mathbf{y}}\cos\theta)}_{\hat{\boldsymbol{\theta}}}$$

$$\mathbf{v} = v\hat{\boldsymbol{\theta}} \tag{3–14}$$

The acceleration **a** is found in the usual way for differentiating **v**. However, in the general case, the speed v will increase or decrease, so we have the derivative of the product[3] of two factors: v and the quantity in parentheses.

$$\mathbf{a} = \frac{d}{dt}[v(-\hat{\mathbf{x}}\sin\theta + \hat{\mathbf{y}}\cos\theta)] \tag{3–15}$$

$$= v\frac{d}{dt}(-\hat{\mathbf{x}}\sin\theta + \hat{\mathbf{y}}\cos\theta) + \frac{dv}{dt}(-\hat{\mathbf{x}}\sin\theta + \hat{\mathbf{y}}\cos\theta)$$

$$= v\left(-\hat{\mathbf{x}}\cos\theta\frac{d\theta}{dt} - \hat{\mathbf{y}}\sin\theta\frac{d\theta}{dt}\right) + \frac{dv}{dt}(-\hat{\mathbf{x}}\sin\theta + \hat{\mathbf{y}}\cos\theta)$$

$$\mathbf{a} = v\frac{d\theta}{dt}\underbrace{(-\hat{\mathbf{x}}\cos\theta - \hat{\mathbf{y}}\sin\theta)}_{-\hat{\mathbf{r}}} + \frac{dv}{dt}\underbrace{(-\hat{\mathbf{x}}\sin\theta + \hat{\mathbf{y}}\cos\theta)}_{\hat{\boldsymbol{\theta}}} \tag{3–16}$$

However, from Equation (3–13) we note that $d\theta/dt = v/r$, so the above equation becomes

$$\mathbf{a} = \underbrace{-\left(\frac{v^2}{r}\right)\hat{\mathbf{r}}}_{\text{Radially inward}} + \underbrace{\left(\frac{dv}{dt}\right)\hat{\boldsymbol{\theta}}}_{\text{Tangential}} \tag{3–17}$$

The first term is the *centripetal acceleration*, $a_{cp} = v^2/r$, in the $-\hat{\mathbf{r}}$ direction (and thus radially inward), and the second term is the *tangential acceleration*, $a_t = dv/dt$, in the $\hat{\boldsymbol{\theta}}$ direction.

3.7 General Curvilinear Motion

With only a slight extension of the ideas just developed, we may describe the motion of a particle traveling with changing speed along *any* curved path. At any given point along such a path, the curve may be approximated by a segment of a circle. The radius r of the circle that "best fits" the curve at that location is

[3] The rule for differentiation of a product is

$$\frac{d}{dt}(uv) = u\frac{dv}{dt} + \frac{du}{dt}v$$ (where u and v are functions of the variable t)

called the **radius of curvature** at that point, and the center of the circle is called the instantaneous **center of curvature.** For an arbitrary curve, the value of r will change continuously along the path. If we know the values of r, v, and dv/dt at a given point in the path, we can calculate the acceleration components in the usual way. Thus, as in Figure 3–8, if a particle travels along the curved path with instantaneous speed v, it always has a centripetal acceleration component perpendicular to the path (toward the instantaneous center of curvature). Whether or not it also has a tangential component of acceleration depends on whether or not there is a rate of change of speed (dv/dt) of the particle at that instant.

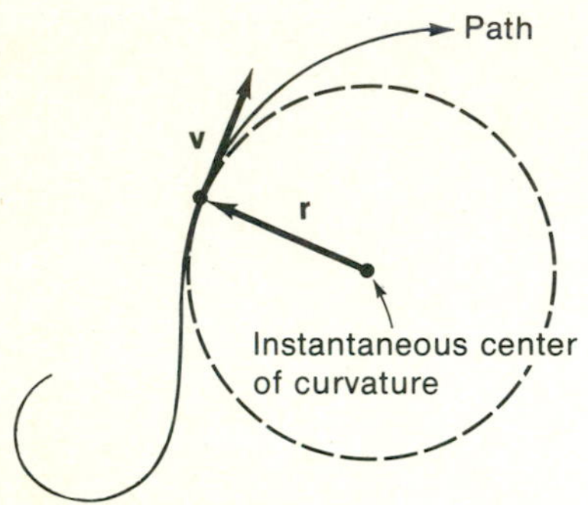

Figure 3–8
Motion of a particle with changing speed along an arbitrarily curved path.

EXAMPLE 3–3

An automobile travels along a curved road. As it traverses a section whose radius of curvature is 100 m, the driver slows down by reducing her speed 5 km/h each second (see Figure 3–9). At the instant she is traveling 60 km/h, what is the acceleration of the automobile?

SOLUTION

As usual, we analyze the tangential and centripetal acceleration components separately. First, let us calculate the tangential component:

$$a_t = \left(\frac{5\,\frac{\text{km}}{\text{h}}}{1\text{ s}}\right)\underbrace{\left(\frac{1000\text{ m}}{1\text{ km}}\right)\left(\frac{1\text{ h}}{60\text{ min}}\right)\left(\frac{1\text{ min}}{60\text{ s}}\right)}_{\text{Conversion ratios}} = 1.39\,\frac{\text{m}}{\text{s}^2}$$

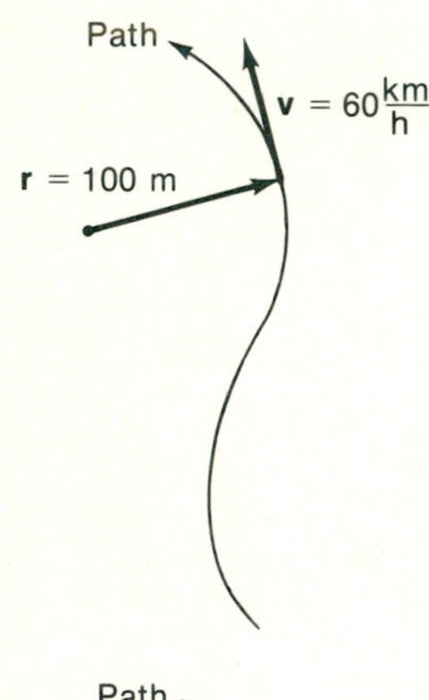

Because the car is slowing down, this tangential component of acceleration is *opposite* to the direction of motion of the car.

To calculate the centripetal acceleration component, we first need to convert the speed to units of meters per second.

$$v = \left(60\,\frac{\text{km}}{\text{h}}\right)\underbrace{\left(\frac{1000\text{ m}}{1\text{ km}}\right)\left(\frac{1\text{ h}}{60\text{ min}}\right)\left(\frac{1\text{ min}}{60\text{ s}}\right)}_{\text{Conversion ratios}} = 16.7\,\frac{\text{m}}{\text{s}}$$

Then:

$$a_{cp} = \frac{v^2}{r} = \frac{\left(16.7\,\frac{\text{m}}{\text{s}}\right)^2}{100\text{ m}} = 2.78\,\frac{\text{m}}{\text{s}^2}$$

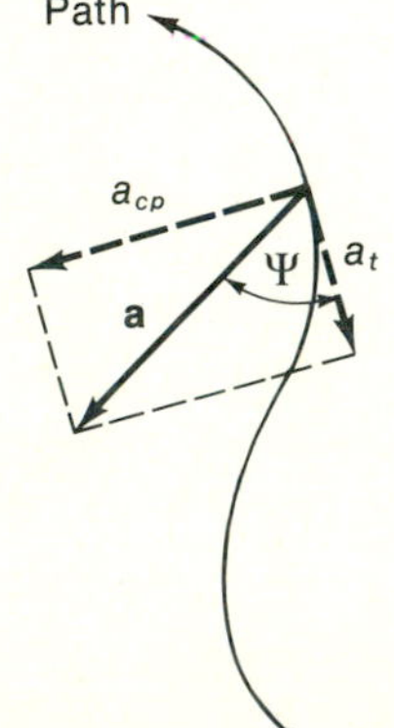

This component is at right angles to the path, toward the instantaneous center of curvature.

The magnitude of the resultant acceleration is

$$a = \sqrt{a_t^{\,2} + a_{cp}^{\,2}} = \sqrt{\left(1.39\,\frac{\text{m}}{\text{s}^2}\right)^2 + \left(2.78\,\frac{\text{m}}{\text{s}^2}\right)^2} = \boxed{3.11\,\frac{\text{m}}{\text{s}^2}}$$

The direction of the acceleration is specified here by the angle Ψ that it makes with the tangent to the path.

$$\tan\Psi = \frac{a_{cp}}{a_t} = \frac{\left(2.78\,\frac{\text{m}}{\text{s}^2}\right)}{\left(1.39\,\frac{\text{m}}{\text{s}^2}\right)} = 2.00$$

$$\Psi = \tan^{-1}(2.00) = \boxed{63.4^\circ}$$

Figure 3–9
Example 3–3

Summary

The position of a particle undergoing circular motion is designated by the vector **r** from the center of the circle.

$$\textit{Position:}\quad \mathbf{r}$$

$$\textit{Velocity:}\quad \mathbf{v} = \frac{d\mathbf{r}}{dt}$$

$$\textit{Acceleration:}\quad \mathbf{a} = \frac{d\mathbf{v}}{dt}$$

A particle exhibiting *uniform* circular motion (motion at a constant speed v) has centripetal acceleration.

CENTRIPETAL ACCELERATION (toward the center of the circle)

$$a_{cp} = \frac{v^2}{r}$$

(due to the *change in direction* of **v**)

Note that the particle never gets any closer to the center of the circle even though it always accelerates in that direction.

A particle undergoing *nonuniform* circular motion (speeding up or slowing down) also has a tangential acceleration component.

TANGENTIAL ACCELERATION (tangent to the path)

$$a_t = \frac{dv}{dt}$$

(due to the *change in magnitude* of **v**)

If the particle is speeding up, a_t is *in* the direction of motion; if it is slowing down, a_t is *opposite* to the direction of motion.

In general, the magnitude of the total (net) acceleration **a** is

$$a = \sqrt{{a_t}^2 + {a_{cp}}^2}$$

The direction of **a** may be specified as making a certain angle with respect to some known direction (such as the inward radial direction or the tangent to the path). This angle can be found from the acceleration components using trigonometry.

In the more general case of motion along an arbitrarily curved path, at any given instant the path has a *radius of curvature* and a *center of curvature*, both of which may change continuously as the particle moves along. Knowing r, v, and dv/dt at any instant, the acceleration components may be found in the usual way.

Questions

1. Under what conditions is an automobile accelerating, yet neither speeding up nor slowing down?
2. Considering instantaneous values, can an object have zero velocity and nonzero acceleration? Can it have zero acceleration and nonzero velocity?
3. Can an object have constant speed, yet be changing its velocity? Can it have constant velocity, yet be changing its speed?
4. An airplane pilot pulls out of a dive along a path that lies in a vertical plane. What does it mean to say he experienced an acceleration of a certain number of g's?

5. The next time you see an automatic clothes washer in action, estimate the acceleration (in terms of g) of the clothes during the spin-dry cycle.
6. Due to the earth's rotation, a point on the equator has a speed of about 1000 mi/h. Describe precisely the frame of reference in which this statement is true.
7. The orbital motion of the moon about the earth lies in approximately the same plane as the earth's orbital motion about the sun. However, the path of the moon about the sun is always concave toward the sun. Explain how these seemingly contradictory statements can be true.

Problems

3A–1 Viewed from the earth, the moon and sun have about the same angular size: approximately one-half degree (0.5°).
(a) Calculate this angle in radians.
(b) Calculate the angular size, in radians, of a dime (diameter = 1.8 cm) held at arm's length away from your eye (say, 60 cm).
Answers: (a) 8.73×10^{-3} rad (b) 0.030 rad

3A–2 In the Bohr model of the hydrogen atom, an electron travels with a speed of 2.19×10^{6} m/s around a proton in a circular orbit of radius 5.29×10^{-11} m. Find the acceleration of the electron.

3A–3 The moon moves in a nearly circular orbit about the earth. Determine the centripetal acceleration of the moon as it moves in orbit. *Answer:* 2.72×10^{-3} m/s²

3A–4 Chicago is located at 41.9° north latitude. Find the city's centripetal acceleration due to the earth's rotation.

3A–5 During the warm-up of a helicopter motor prior to take off, the motor shaft turns at 300 rpm. If a rotor blade is 4 m long, determine the speed with which the tip of each rotor blade is moving through the air. *Answer:* 126 m/s

3A–6 An ant is sitting 10 cm from the axis of a phonograph record rotating at 33.3 revolutions per minute. Calculate the ant's acceleration, expressed in meters per second squared.

3A–7 An amusement park ride carries riders around in a horizontal circle of radius 5 m. If the centripetal acceleration the passengers experience must be limited to 0.4 g, determine the maximum speed of the passengers. *Answer:* 4.43 m/s

3A–8 A modern ultracentrifuge uses magnetic suspension to replace the usual mechanical bearings. Devices have been built that produce centripetal accelerations of $10^{9}g$. Such instruments are used to determine within 1% the atomic masses of molecules in the range of about 50 atomic mass units (amu) to giant molecules such as the tobacco mosaic virus, whose atomic mass is roughly 100 million amu. The radius of the rotating object is small: $r = 10^{-2}$ mm. Calculate (a) the rotational frequency in revolutions per second and (b) the speed of a particle at that radius. (Note: Small steel balls of this radius will explode from centrifugal effects when the peripheral speed reaches about 1000 m/s.)

3B–1 At the earth's surface, the angular diameter of the sun is about 0.5°. Calculate how far from your eye a dime (diameter = 18 mm) must be in order to just block out the sun's disk. *Answer:* 2.06 m

3B–2 Find the speed of a satellite traveling in a circular orbit about the earth at an altitude of 300 km above the earth's surface. The acceleration due to gravity at that location is 8.9 m/s². The mean radius of the earth is 6371 km.

3B–3 A typical pulsar is believed to be an extremely dense neutron star about 40 km in diameter, rotating about 1 rev/s.
(a) Find the acceleration of a particle on the equator of such a star.
(b) Find the acceleration of a particle on the surface at 45° north latitude (that is, halfway from the equator toward the polar axis).
Answers: (a) 7.90×10^{5} m/s² (b) 5.58×10^{5} m/s²

3B–4

(a) Calculate the centripetal acceleration of a point on our equator due to the earth's rotation.

(b) Calculate the centripetal acceleration of the earth in its orbit about the sun.

3B–5 A high-speed dentist's drill driven by an air turbine achieves a rotational speed of 350 000 rpm. The diameter of the cutting burr is 1 mm.

(a) Find the linear speed of the outer edge of the burr.

(b) Determine the acceleration of the outer edge. Express this acceleration as a multiple of the standard gravitational acceleration g. *Answers:* (a) 18.3 m/s (b) $6.85 \times 10^4\ g$

3B–6 A particle moves along a circle at a constant speed of 10 m/s. Determine the magnitude and direction of the change in velocity corresponding to a movement one-third the way around the circle.

3B–7 An automobile travels around a curve of radius 80 m at a constant speed of 10 m/s.

(a) Calculate the acceleration of the car.

(b) If the car now slows down uniformly to rest in 6 s, calculate the tangential acceleration component a_t during this process.

(c) At the instant the car is traveling 8 m/s (while slowing down), find the magnitude and direction of the total acceleration of the car. Include a sketch to clarify the direction of the acceleration vector.

Answers: (a) 1.25 m/s^2 toward the center of curvature of the road
(b) 1.67 m/s^2
(c) 1.85 m/s^2; 64.4° back from the inward radial direction

3B–8 A projectile is launched with an initial speed v_0 at an angle θ_0 above the horizontal. For a short distance near the top of its trajectory, the parabolic path may be approximated as a circular arc. Find the equation for radius r of this arc in terms of v_0, θ_0, and g.

3C–1 An accelerometer is an instrument for measuring the acceleration of moving objects. While driving an automobile equipped with an accelerometer, a driver notices that when traveling at a constant speed of 45 mph, the accelerometer indicates an acceleration of 0.15 g to the right. (Accelerometers often indicate measurements in terms of multiples of gravitational acceleration.)

(a) Find the radius of the turn being executed by the car.

(b) Is the car turning right or left? *Answers:* (a) 908 ft (b) right

3C–2 A racetrack has a circular turn whose radius is 500 m. The curve is 800 m long and joins two straightaways. Determine the angle between the directions of the two straightaways.

3C–3 A car travels along a curved road whose radius of curvature is 300 m. The driver applies the brakes, slowing the car at the rate of 1.2 m/s^2. At the instant the car's speed is 15 m/s, calculate the car's acceleration (magnitude and direction). Include a sketch showing the acceleration vector. *Answer:* 1.42 m/s^2; 32° from $-\mathbf{v}$

3C–4 A ball on the end of a string is whirled around in a horizontal circle of radius 0.3 m. The plane of the circle is 1.2 m above the ground. The string breaks and the ball lands 2 m (horizontally) away from the point on the ground directly beneath the ball's location when the string breaks. Find the centripetal acceleration of the ball during its circular motion.

3C–5 For a satellite to move in a stable circular orbit at a constant speed, its centripetal acceleration must be inversely proportional to the square of the radius of the orbit.

(a) Show that the tangential speed of a satellite is inversely proportional to the square root of the radius of the orbit.

(b) Show that the time required for one orbit is proportional to the three-halves power of the orbital radius.

3C–6 An x-y coordinate system is used to describe locations on a circular roadway of radius 10 m. At $t = 0$, a motorcyclist passes the origin of the coordinate system, traveling in the $+x$ direction with constant speed of 3 m/s.

(a) Determine the time it takes the motorcyclist to travel one-fourth the way around the circle.

(b) Find the x and y coordinates of the motorcyclist 8 s after passing the origin.

(c) Find the magnitude and direction of the displacement of the motorcyclist that occurs between $t = 8$ s and $t = 12$ s.

(d) Find the magnitude and direction of the instantaneous velocity vectors at the beginning and end of the time interval of part (c).

(e) Repeat part (d) for the instantaneous acceleration vectors.

3C–7 A race car is initially traveling at 180 mph around a circular track whose circumference is 1 mi. The car slows uniformly to a stop in one turn of the track.

(a) Find the tangential acceleration of the car.

(b) Find the centripetal acceleration when the car is $\frac{1}{2}$ mi from the stop.

(c) Find the total acceleration of the car in part (b) (magnitude and direction). Include a sketch to specify the direction clearly.

Answers: (a) -6.60 ft/s^2 (b) 20.7 ft/s^2
(c) 21.7 ft/s^2; 17.7° back from the inward radial direction

Nature and Nature's laws lay hid in night,
God said: "Let Newton be!", and all was light.

ALEXANDER POPE

chapter 4 Dynamics — The Physics of Motion

4.1 Introduction

In Chapters 2 and 3 we developed the kinematics of a particle; that is, given the acceleration of a particle we can describe all its possible motions relative to a frame of reference. We now take up a more interesting question: "*Why* do objects accelerate?"

Isaac Newton[1] gave an answer, saying that an object accelerates because *forces act on it.* You probably have already seen a mathematical form of this statement known as Newton's second law:

$$\mathbf{F} = m\mathbf{a} \tag{4-1}$$

where m is the so-called inertial mass of the object. Although this statement may seem simple, its implications are profound and we shall spend considerable time discussing them. In addition to its power in explaining many natural phenomena, Newton's theory implies several assumptions about the nature of space and time. Interestingly, although Newton's second law withstood the test of experimental verification for over 200 years, it has been shown to be only a very good approximation, casting doubt on these assumptions about space and time. The law failed only when our experimental techniques were sufficiently refined to examine masses moving near the speed of light.

Newton's theory of mechanics incorporated three fundamental laws, which along with his ideas about gravitation formed a way of viewing the universe that was truly astonishing. It explained such diverse phenomena as the motions of planets, the tides of the oceans, the falling of an apple, and the precession of the earth's equinox (the slow wobbling of the earth's axis that is completed once every 26,000 years). It was a strictly causal and mechanistic theory, one that provoked reverberations in many areas of human concern for two centuries. Philosophy, politics, theology, and numerous other disciplines all felt its impact.

---

[1] Isaac Newton, born on Christmas, 1642 (about the time Galileo died), was clearly one of the geniuses in human history. His *Mathematical Principles of Natural Philosophy* ("The Principia") stirred such excitement in the seventeenth century that many areas of human thought were changed as a result. In addition to being the "father" of classical mechanics, by the age of 24 Newton had invented calculus, discovered the binomial theorem, developed a complete theory of gravitation, and gained considerable insight into the nature of white light.

4.2 Observations and Experiments in Particle Motion

We have seen that the motion of a particle may be classified as either one of *constant velocity* (no acceleration) or of *accelerated* motion. Before Newton's time, Galileo gained much understanding of motion with constant velocity. By experimenting with balls rolling in troughs inclined at smaller and smaller angles, Galileo reasoned that if friction and other outside influences were reduced to zero, a moving object ought to continue moving *forever* with the same velocity. This was considered a radical statement at the time, one that clearly violated the Aristotelian view of motion, which had prevailed for centuries. According to Aristotle, the natural state of an object was one of *rest*. For example, if we stop pushing a cart, it eventually comes to rest; if a ball rolls down a hill and comes to a horizontal surface, it gradually slows down and stops. But Galileo demonstrated that this widely held "common sense" explanation was wrong.

Newton adopted Galileo's new insight as the first of his three laws. In modern language, it is:

NEWTON'S FIRST LAW **An object continues in a state of rest, or in motion with constant velocity, unless acted on by unbalanced forces.**

Note that this first law does not distinguish between an object at rest and one in motion with constant velocity. Both objects have zero net force on them, and therefore both have accelerations of zero. Furthermore, there is no distinction between an object with no forces and one with several forces that balance one another to produce zero *net* force. An asteroid moving in interstellar space keeps moving with constant velocity because there is no force acting on it. If we keep pushing on a moving cart to balance the retarding force of friction, the cart maintains its uniform velocity because the net force acting on the cart is zero.

Although Newton's first law contains the word "force," it does not constitute a good definition of the concept; it only tells us how to recognize the presence or absence of a net force. If an object accelerates, something called a force is causing it. If, instead, an object does not accelerate (that is, it either remains at rest or continues moving in a straight line with constant velocity), there is no net force on it. This unaccelerated motion is explained by attributing a property called **inertia** to matter. Because of inertia, objects resist acceleration and thus they behave according to the first law, often called the "law of inertia."

If the first law does not define a force, what is it and how does it arise? Clearly other nearby objects are the source of the disturbing influence. In particular, forces arise from some type of *interaction between two objects.*

Newton's first law also tells us how to identify what is called an **inertial frame of reference,** that is, one in which Newton's law of inertia holds true. It is clear that our evaluation of an object's motion depends on the frame of reference we use to measure the motion. If an object is observed to be at rest as measured in one frame of reference, then in a second frame that is accelerating with respect to the first, the object would be accelerating.

Recognition of an inertial frame is an important first step in applying Newtonian mechanics. In an inertial frame, objects with no net forces on them remain at rest or move with constant velocity in a straight line. Sometimes such a frame is said to be "at rest with respect to the fixed stars." In practice, the earth is often taken to be an inertial frame in spite of its axial rotation and orbital motion around the sun, since these accelerations are so small that effects due to them usually can be ignored.

Finally, once we find one inertial frame, all other frames of reference that have uniform (unaccelerated) motion with respect to that one are also inertial frames. Thus, if an object is at rest in one frame, observers in all the other inertial frames would see the object moving with uniform velocity. In other words, the object moves without acceleration in all inertial frames, just as an object with no net force on it should behave according to Newton's first law. The fact that Newton's laws apply equally well in all inertial frames is called the *principle of relativity;* we will have much more to say about this in Chapter 13, "Special Relativity."

□ □ □

Newton's second law will seem a more plausible statement if we first describe some experiments involving accelerated motions. We observe that accelerated motion results from the interaction of at least two objects. For example, if a ball is dropped from rest, it accelerates toward the earth because of an interaction between the earth and the ball which we call *gravity.* A cart accelerates if it is pulled or pushed by another object. An automobile is accelerated through its interaction with the road. These examples, however, are needlessly complicated for our initial investigation.

In order to simplify the investigation of forces and accelerations, we devise an experiment that involves just two objects and their mutual interaction in a particularly simple way. The apparatus is a horizontal "air track" that supports "carts" on a film of air forced through small holes on the top surfaces of the track (see Figure 4–1). This apparatus has two desirable features:

(1) The motion of the carts is limited to one dimension, thus simplifying the analysis.
(2) A cart on the track has essentially zero net force on it, since the track produces a vertical force that balances the downward force of gravity and the film of air allows the cart to move horizontally with negligible friction.

Two carts on an air track may be made to interact with each other by various means: for example, a compressed spring situated between them, an explosive cap that blows them apart, or a pair of magnets attached to the carts so that they repel each other.

When a single cart is placed on the track, it maintains its state of motion. That is, if the cart is initially at rest it remains at rest, and if it is moving it continues to move with essentially constant velocity. These facts are in agreement with Newton's first law: In the absence of an interaction with another cart, the cart maintains its state of motion.

Suppose we next allow two carts to interact through a coiled spring that pushes them apart. We start the carts at rest with a compressed spring between

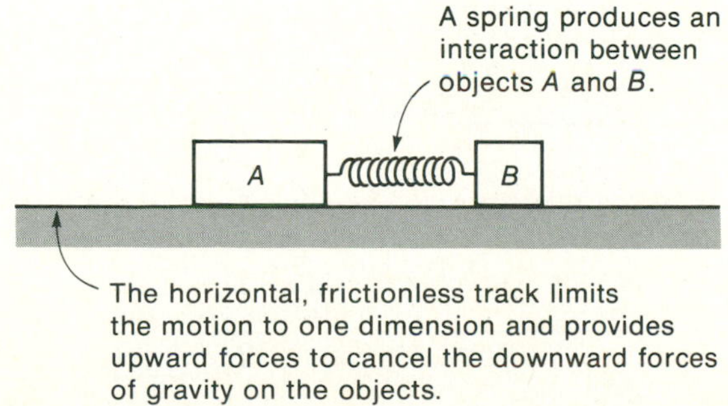
A spring produces an interaction between objects *A* and *B*.
A
B
The horizontal, frictionless track limits the motion to one dimension and provides upward forces to cancel the downward forces of gravity on the objects.

Figure 4–1
A hypothetical experiment to investigate the interaction of two objects *A* and *B*.

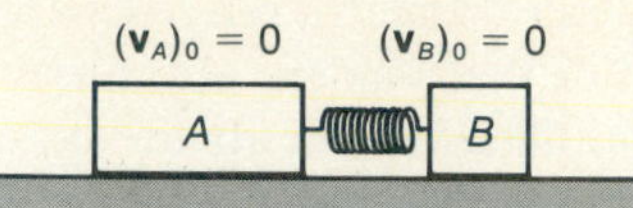

(a) Before the interaction, the carts are at rest.

(b) After the interaction, the carts move apart, with each maintaining a constant velocity.

Figure 4–2
An experiment to investigate the interaction of two carts on an air track.

them. Figure 4–2(a) is a sketch of the carts before the spring is allowed to expand, and Figure 4–2(b) shows the state of motion after the spring has propelled the carts apart (and the spring is no longer in contact with the carts). The carts now move in opposite directions with constant speeds v_A and v_B.

Suppose we repeat the experiment numerous times, varying the compression of the spring. In each case, if we calculate the ratio v_A/v_B, we find that it always has the same value regardless of how strong or weak the interaction is. Furthermore, if we repeat the experiment using other kinds of interactions such as magnets and explosive caps, we find that the ratio is still the same. We therefore conclude that v_A/v_B is *a constant number, independent of the nature of the interaction.* The discovery of a quantity that remains constant even though other aspects of an experiment are changed frequently leads to new understanding and insight. Watch for such quantities. They are often very powerful clues in analyzing physical phenomena.

4.3 Analysis of the Experimental Data

The next step in our experimental procedure is to develop a more precise mathematical statement that describes the experimental results. Here we assume that the mathematical symbols have a close correspondence to the physical quantities they represent. In particular, we assume that carrying out the logical rules of mathematical operations does not destroy this correspondence between the symbols and the physical concepts.

Figure 4–3 is a graphical representation of the speed of the carts as a function of time for a particular experiment. From the graph we see that the speed of cart A increases during the interaction and reaches a value of $v_A = 3$ m/s at the time the interaction ceases. Cart B attains a speed of $v_B = 5$ m/s during the same time period. The speeds of the carts during the interaction are shown by dashed curves to indicate the lack of experimental data during the acceleration.

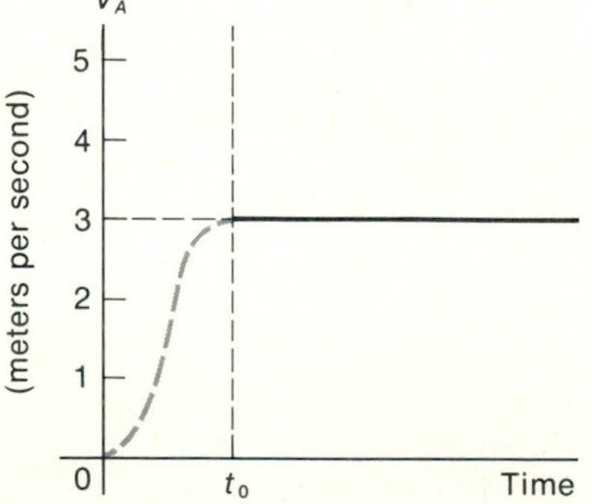

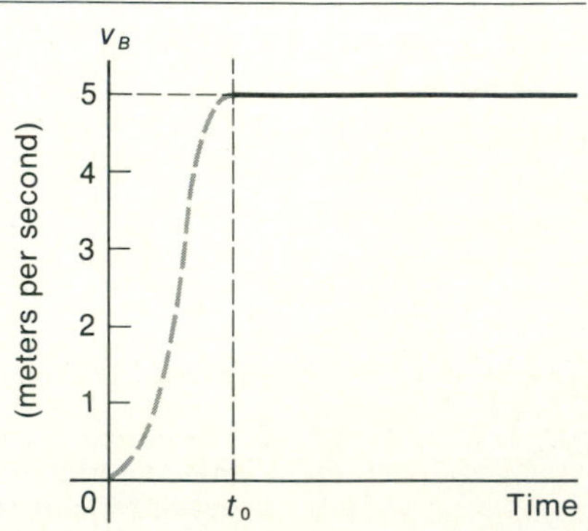

Figure 4–3
Plots of the velocities of carts A and B versus time.

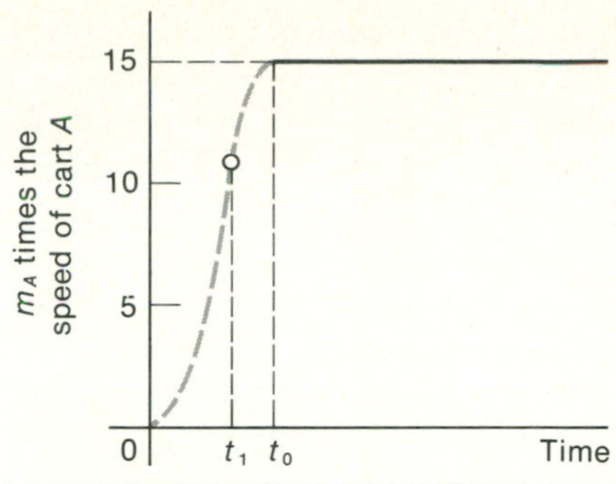

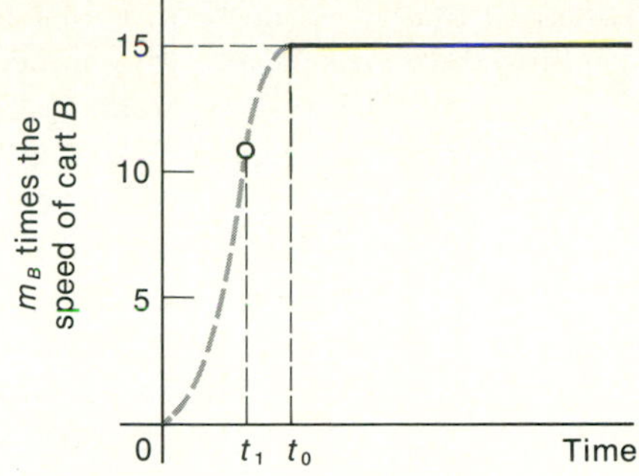

Figure 4–4
Plots of $m_A v_A$ and $m_B v_B$ versus time.

The graphs in Figure 4–3 may be represented in another way. If the values of the speed of cart A were multiplied by a number $m_A = 5$, and the values of the speed of cart B were multiplied by a number $m_B = 3$, the graphs would become those shown in Figure 4–4. Clearly, the values $m_A v_A$ and $m_B v_B$ have the same value (15 units).

Of greater interest is the shape of the two curves for time values less than t_0. We will show that these two curves must have exactly the same shape. The reasoning is as follows: Suppose the interaction considered in this example had ceased at time t_1 (Figure 4–4), where t_1 is less than t_0. Since experiments have indicated that the ratio v_A/v_B is *independent of the nature of the interaction*, m_A times v_A would still be equal to m_B times v_B for this "modified" interaction, which is turned off at t_1. Similarly, for *every* instant of time less than t_0, corresponding values of $m_A v_A$ would still be equal to $m_B v_B$. Therefore, the two curves have the same shape.

Furthermore, since the shapes are identical, the *slopes* of the two curves at corresponding times must also be equal. This may be written as

$$\frac{d}{dt}(m_A v_A) = \frac{d}{dt}(m_B v_B) \tag{4-2}$$

At this point certain quantities have revealed themselves as being particularly significant and therefore seem to warrant a general definition. The numbers m_A and m_B used in this example were arbitrary; any pair of numbers that satisfies the equation

$$\frac{m_B}{m_A} = \frac{3}{5} \tag{4-3}$$

would have been satisfactory. Clearly, if the value of m_A were set equal to 1, then m_B would equal $\frac{3}{5}$. Indeed, a definite number could be ascribed to *any other cart* when compared with cart A on an air track. Interestingly, this number appears to be associated with that property of an object by which the object resists acceleration. The larger the number, the smaller is the acceleration. The property is that of inertia, and we define the number to be a measure of the **inertial mass** (or simply the **mass**) of the object.

4.4 Standard Mass

A standard of mass—the *International Prototype Kilogram*—has been established in the form of a platinum–iridium cylinder kept at Sèvres, France. It is defined to have a mass of 1 kilogram (kg). Thus we could obtain an empirical definition of the mass m of any object by allowing the object to interact with the

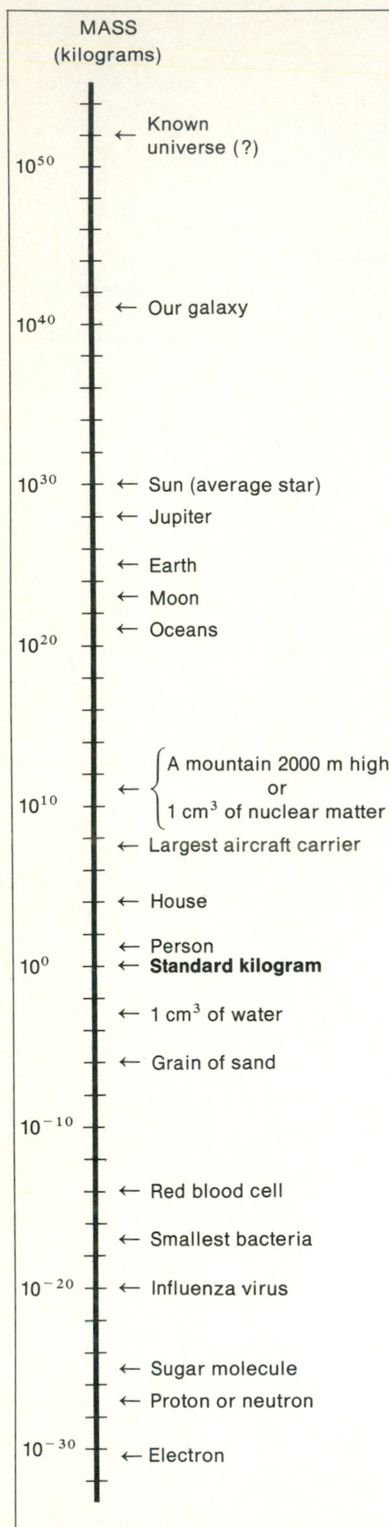

TABLE 4–1

The masses of various objects in relation to the International Prototype Kilogram

"standard mass." (A more practical method of mass determination involves weighing the object on a beam balance.)

Since the scientific community has adopted atomic standards for length and time, it has been suggested that an atomic standard would also be desirable for mass in case some calamity befell the platinum cylinder. Furthermore, through the action of cosmic rays, atoms are continually being knocked off (or added to) the cylinder. However, present techniques of measurement do not permit us to determine the number of atoms in a macroscopic object to the same degree of accuracy as their masses can be compared using an equal-arm beam balance. Therefore, the standard of mass for objects of everyday size remains the International Prototype Kilogram. Table 4–1 gives the masses of a variety of objects in relation to this standard.

An atomic standard is in use in atomic and nuclear physics. Though the masses of individual atoms cannot be compared directly with the standard mass, atoms and molecules can be compared with one another to even greater precision than that in ordinary mass determinations. The atomic mass standard is defined as follows[2]:

1 unified atomic mass unit (u) = $\frac{1}{12}$ of the mass of an atom of carbon-12 (the nucleus plus six electrons)

$$1\ u = 1.66 \times 10^{-27}\ \text{kg}$$

4.5 Momentum

The quantity *mass times velocity*, which appeared in Equation (4–2), is perhaps the most important single quantity in the study of mechanics, appearing in numerous different contexts. This quantity is called **momentum,** and is denoted by the symbol $\mathbf{p}$. In cases involving motion in two dimensions (such as flat-bottomed objects interacting on a horizontal air table) or three dimensions (for example, colliding nuclear particles with trajectories in three dimensions), we can verify experimentally that the momentum is a vector quantity having both magnitude and direction. Thus:

MOMENTUM $$\mathbf{p} = m\mathbf{v} \qquad \textbf{(4–4)}$$

The analysis of the interaction between two objects A and B can be written in terms of the vector momentum $\mathbf{p}$ of each object. Note that regardless of the type of interaction, if the objects start at rest they acquire velocities in *opposite* directions. Hence, in *vector* form, Equation (4–2) must include a minus sign as follows:

$$\frac{d\mathbf{p}_A}{dt} = -\frac{d\mathbf{p}_B}{dt} \qquad \textbf{(4–5)}$$

where $\mathbf{p}_A = m_A\mathbf{v}_A$ and $\mathbf{p}_B = m_B\mathbf{v}_B$.

4.6 Newton's Second Law

Newton summarized all the experimental behavior of interacting masses in his second law. In essence, this law defines a net force $\mathbf{F}$ in terms of the change it produces in the momentum of the mass m. To emphasize that it is the *net*

[2] The standard formerly was based on oxygen-16. But because many more carbon compounds than oxygen compounds have been found useful in these studies, in 1960 the standard was changed.

(resultant) force that brings about the change in momentum, we shall use the summation sign[3] as shown:

NEWTON'S SECOND LAW

$$\Sigma \mathbf{F} = \frac{d\mathbf{p}}{dt} \qquad \textbf{(4–6)}$$

The unit of force is called a **newton** (N). From Equation (4–6) it can be seen that a force of 1 N produces a change of momentum equal to 1 kg·m/s². To gain a feeling for this force unit, we may say that one newton is about the force you exert when you hold a small apple in your hand.

If we write out the derivative $d(m\mathbf{v})/dt$ explicitly (see Appendix G, rule 4, for the derivative of the product $m\mathbf{v}$), we obtain the following equation for Newton's second law:

$$\Sigma \mathbf{F} = m\frac{d\mathbf{v}}{dt} + \frac{dm}{dt}\mathbf{v} \qquad \textbf{(4–7)}$$

Because the mass of the particle does not change, $dm/dt = 0$. And since $d\mathbf{v}/dt = \mathbf{a}$, we may write Newton's second law (when m is constant) as

NEWTON'S SECOND LAW

$$\Sigma \mathbf{F} = m\mathbf{a} \quad \text{(for constant mass)} \qquad \textbf{(4–8)}$$

Newton's second law contains a wealth of implications. First, the concept **F** does agree with our everyday idea of a force as a push or a pull applied to an object. Note that a force cannot exist by itself; it must be applied to some object.

All the forces in mechanics fall into two broad categories: contact forces and noncontact forces. **Contact forces** are simply those that involve physical contact between two or more objects. For example, the force that a rope under tension or a stretched or compressed spring exerts on an attached object is a contact force. Another example is the frictional force that occurs when one object slides across another (or when one object *tends* to slide without actually moving). Frictional forces are always parallel to the surfaces involved. (Analogous frictional forces are also present when an object moves through the air or through a fluid.)

An example of a **noncontact force** is the gravitational force of attraction between nearby objects that are not touching each other. Every particle in one of the objects attracts every particle in the other object with a gravitational force given by Newton's law of gravitation:

$$F = G\frac{m_1 m_2}{r^2}$$

where r is the distance between the particles of masses m_1 and m_2 and G is the universal gravitational constant. We will discuss gravitational forces in detail in Chapter 14.

Other noncontact forces in nature include electric forces and magnetic forces, which will be treated later. Actually, since the so-called forces of contact—spring forces, friction, tensions in ropes, and so forth—are ultimately traceable to electrical forces between atoms, we recognize that forces can be grouped into the following basic categories: *weak electromagnetic* forces, *gravitational* forces, and *nuclear* forces. However, at this stage we will be dealing primarily

[3] The Greek capital letter sigma, Σ, is used in mathematics to denote summation. Here, it designates the *vector sum* of all the forces acting on an object.

with certain contact forces, for which we will continue to use the common names of friction, spring force, and so on, and the noncontact force of gravity. In each case, we must identify the external object that is exerting the force on the object under investigation.

Are forces true vector quantities? Formally, they are defined to be vectors by Newton's second law. But defining them to be vectors does not necessarily make them so. We must perform experiments to verify that the physical concept of a force does, indeed, follow the mathematical rules of vector addition and subtraction. We discover experimentally that forces *do* behave as vectors.

In Newton's view, the second law is a *causal* statement: a net force *causes* a mass to accelerate. Of course, we may use the expression in either of two ways. First, if we know the forces on an object, we may calculate the object's acceleration. Alternatively, if we observe an object accelerating, we may *infer* that there is a net force on the object and calculate its magnitude and direction. However, note that *acceleration does not cause a force to act on the object.* Rather, in Newton's view, it is the other way around. In the traditional meaning of "cause and effect," a net force on an object causes that object to accelerate.

Finally, although Newton's laws are an accurate description of the mechanical behavior of masses ranging from baseballs to galaxies, they do not give the whole story. As you will see in Chapter 13, Einstein's theory of relativity indicates that a modification is needed when masses move with speeds that are significant compared with the speed of light: 3×10^8 m/s. Strictly speaking, this means that Newton's laws are "wrong." However, for ordinary needs the discrepancy is negligible. Thus $\Sigma \mathbf{F} = m\mathbf{a}$ is adequate for objects of everyday size moving with velocities much less than the speed of light. Indeed, the relativistic equations reduce to Newtonian expressions as the speed becomes slower.

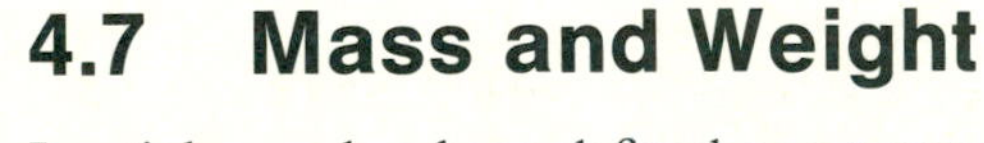

4.7 Mass and Weight

Inertial mass has been defined as a measure of the inertial property of matter. In principle, we may determine the mass of an object by observing its interaction with another object of known mass. But from a practical point of view, techniques such as the use of air tracks are not always feasible, particularly when dealing with large objects or when great precision is desired.

In practice, the mass of an object is more easily determined by measuring the force of gravity on the object. The validity of this method arises from an important fact: in the absence of the buoyant and resistive effects of the air, *all objects fall with the same acceleration.*[4] The magnitude of this acceleration is denoted by the symbol g.

$$g \approx 9.80\ \text{m/s}^2 \qquad \text{or} \qquad 32\ \text{ft/s}^2$$

If all objects fall with the same acceleration at a given location, it follows that *the force of gravity upon an object must be directly proportional to the mass of the object.* Suppose we let g represent the acceleration of a freely falling object and W denote the force of gravity on the object. Then we apply Newton's second law

$$\Sigma \mathbf{F} = m\mathbf{a}$$

to obtain

$$W = mg \tag{4–9}$$

Figure 4–5
In Galileo's day, one of the theories of planetary motion was that planets moved because invisible angels, beating their wings, pushed the planets forward along their orbits about the sun. We are smarter now. Instead of angels providing a *tangential* force, the angels must apply the force *at right angles* to the motion. And we replace the word "angels" by the word "gravity." (After Richard Feynman, *The Character of Physical Law*, MIT Press, 1973.)

[4] The acceleration due to gravity depends somewhat upon geographic location and elevation. For sea level, at 0° latitude $g = 9.780\ 39\ \text{m/s}^2 = 32.087\ 8\ \text{ft/s}^2$ and at 90° latitude $g = 9.832\ 17\ \text{m/s}^2 = 32.257\ 7\ \text{ft/s}^2$.

This experiment to determine the force of gravity mg is not limited to free-fall situations. The relation $W = mg$ always holds, independent of the velocity or acceleration of an object.

The force of gravity on an object can be found by using a weighing device, such as the spring scale illustrated in Figure 4–6. There are two forces on the mass: the upward force **F**, which the stretched spring exerts, and the downward force of gravitational attraction **W**, which the earth exerts. These forces add vectorially to zero when the object is at rest. The scale may be calibrated in any appropriate force units. Since $|\mathbf{F}| = |\mathbf{W}|$, the scale indicates the magnitude of the force of gravity W on the object. Though later we will introduce a slightly more sophisticated definition, for the present we may define the term **weight** to be the force of gravity on an object.

The weight of an object (in newtons) is found by multiplying its mass (in kilograms) by the local value of g (in meters per second squared). This means that a newton has units of

$$\mathrm{N} = \frac{\mathrm{kg \cdot m}}{\mathrm{s}^2} \tag{4–10}$$

The dimensions of force (and weight) are thus $[\mathrm{M \cdot L/T^2}]$.

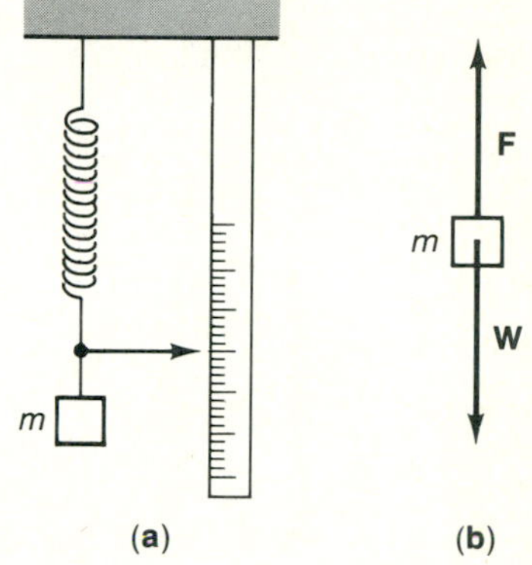

Figure 4–6
(a) A spring scale for determining the weight of the mass m. The pointer on the spring moves along a stationary scale. (b) Forces on the mass m. The vector **F** is the force that the stretched spring exerts on the mass, and the vector **W** represents the force due to gravity.

4.8 Units for Mass and Weight

In scientific work today, the preferred metric system of units is the *Système International* (abbreviated SI), a refined version of the older MKS system (*m*eter, *k*ilogram, *s*econd). The SI system was developed over a period of years and given official international status in 1971 by the General Conference on Weights and Measures. It is constructed from the seven base units and two supplementary units listed in Table 4–2. All other units are derived from these

TABLE 4–2

SI Base and Supplementary Units

	Quantity	Unit name	Unit symbol
Base units	Length	meter	m
	Mass	kilogram	kg
	Time	second	s
	Electric current	ampere	A
	Thermodynamic temperature	kelvin	K
	Amount of substance	mole	mol
	Luminous intensity	candela	cd
Supplementary units	Plane angle	radian	rad
	Solid angle	steradian	sr

TABLE 4–3

A Comparison of Units for F = *ma*

	SI	**British engineering**
Force	newton (N)	pound (lb)
Mass	kilogram (kg)	slug (—)
Acceleration	$\frac{\text{meters}}{\text{second}^2}\left(\frac{\text{m}}{\text{s}^2}\right)$	$\frac{\text{feet}}{\text{second}^2}\left(\frac{\text{ft}}{\text{s}^2}\right)$

nine. Note that in the SI system, force is a derived unit; that is, it is defined as a combination of the base units of *mass*, *length*, and *time*.

In contrast, most engineering work in the United States has been in the **British engineering system,** in which force is fundamental and the unit for mass is the derived unit. Here the force unit **pound** (lb) is basically defined in terms of the force of gravity on a mass of 0.453 592 37 kg at a location (45° north latitude at sea level) where $g = 9.806\ 65\ \text{m/s}^2$. This means that[5]

$$1\ \text{lb} \approx 4.45\ \text{N}$$

The mass unit is derived from the force unit through the equation $m = W/g$. The unit for mass is the **slug** (from "sluggish"), and there is no official abbreviation. As an example, an object that weighs 32 lb has a mass of 1 slug.

$$m = \frac{W}{g} = \frac{32\ \text{lb}}{32\ \frac{\text{ft}}{\text{s}^2}} = 1\ \text{slug}$$

Figure 4–7 may be helpful in comparing various units for mass and weight. Remember that weight is a *force* and not a mass. Table 4–3 summarizes the relationships between the units in the systems.

[5] It is also true that a 1 kg mass weighs about 2.205 lb. However, one cannot form a true conversion ratio from the relation 1 kg ≈ 2.205 lb since it involves a *mass* on one side and a *weight* on the other—two different concepts. Furthermore, the exact numbers depend on the local value of g. With care, however, this "equivalence" can be used to convert mass in SI units into the approximate weight of the object in the British engineering system.

This object has a mass of 2 kg. Its weight is 19.6 N.

$$W = mg = (2\ \text{kg})\left(9.8\frac{\text{m}}{\text{s}^2}\right) = 19.6\ \text{N}$$

(**a**) The SI system.

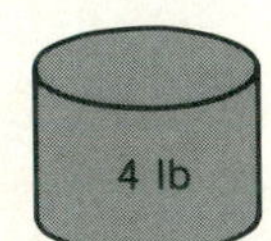

This object weighs 4 lb. Its mass is $\frac{1}{8}$ slug.

$$m = \frac{W}{g} = \frac{(4\ \text{lb})}{\left(32\frac{\text{ft}}{\text{s}^2}\right)} = \frac{1}{8}\ \text{slug}$$

(**b**) The British engineering system.

Figure 4–7
Two common systems of units. In each, weight is a *force*, not a mass.

EXAMPLE 4–1

Object A is labeled 3 kg and object B is labeled 3 lb. What are the mass and weight of each object? Give answers in the SI system or the British engineering system, whichever is appropriate.

SOLUTION

Object A is labeled 3 kg. Since kilogram is a unit of mass in the SI system, we conclude:

$$\text{Object } A \text{ has a mass of } \boxed{3\ \text{kg}}$$

$$\text{Its weight is } W = mg = (3\ \text{kg})\left(9.8\ \frac{\text{m}}{\text{s}^2}\right) = \boxed{29.4\ \text{N}}$$

Note that the combination of units kg·m/s^2 is equivalent to the unit newton.

Object B is labeled 3 lb. Since pound is a unit of force in the British engineering system, we conclude:

$$\text{Object } B \text{ weighs } \boxed{3\ \text{lb}}$$

$$\text{Its mass is } m = \frac{W}{g} = \frac{(3\ \text{lb})}{\left(32\ \frac{\text{ft}}{\text{s}^2}\right)} = \boxed{0.093\,8\ \text{slug}}$$

Note that the combination of units lb·s^2/ft is equivalent to the unit slug.

4.9 Applying Newton's Second Law

If we know the forces that act on an object, we can find the object's acceleration by applying Newton's second law, $\Sigma\mathbf{F} = m\mathbf{a}$. The first step is to find the net force $\Sigma\mathbf{F}$ acting on the object, which is accomplished most easily by "isolating" the object from its surroundings. We imagine a surface that completely surrounds the object, then make a sketch showing (as vectors) all the forces that act *through* this surface *on* the object. Such a sketch is called an **isolation diagram** or **free-body diagram.** Addition of the force vectors gives the net force. Then, once the net force has been found, solving $\Sigma\mathbf{F} = m\mathbf{a}$ is essentially an algebraic procedure.

In each of the examples that follow, note the importance of the free-body diagram. It is here that most of the physical analysis takes place. In addition, be aware of the new symbols and procedures that are introduced; these will become standard notation in this text.

A block pulled along a frictionless plane.

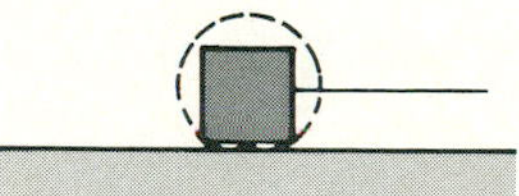

An imaginary surface that isolates the block from its surroundings.

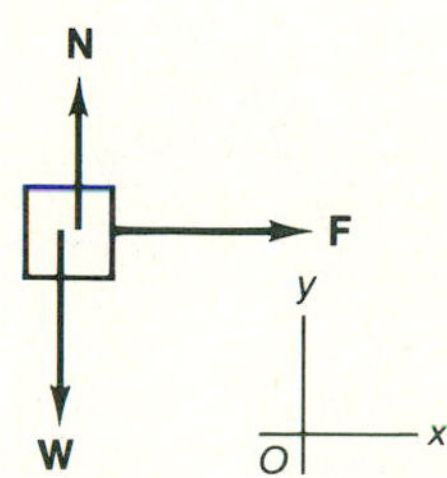

A free-body diagram.

Figure 4–8
Example 4–2

EXAMPLE 4–2

Consider a block of mass $m = 2$ kg on a horizontal, frictionless surface. A string is attached to the block and pulled horizontally with a force of 4 N (see Figure 4–8). Starting at rest, how far will the block move in 3 s?

SOLUTION

If we find the acceleration, we can use the kinematic equations to calculate the distance the block moves in 3 s. We find the acceleration by applying Newton's second law. The first step is the creation of a free-body diagram. We imagine the block to be encased in an imaginary surface and then sketch all forces that act through the surface on the block. When possible, it is helpful to sketch force vectors with the tail of the arrow located at the point where the force is applied. In the case of gravity, we place the tail at the center of mass, which is also the geometrical center for symmetrical objects.

The forces on the block are:

W = downward force due to gravity
N = upward force the surface exerts on the block
F = horizontal force on the block

The notation **N** stands for the *n*ormal force (perpendicular to the surface) that any surface exerts on an object. Although this force generally is spread out over the entire surface of contact, for the present we may show it as a single force concentrated near the center of the surface.

Our next step is to apply $\Sigma \mathbf{F} = m\mathbf{a}$ in two component directions at right angles. Because the component equations are one-dimensional, we use *scalar* equations: $\Sigma F_x = ma_x$ and $\Sigma F_y = ma_y$, with plus and minus signs to keep track of directions. Since we know in advance that the object accelerates horizontally toward the right, we choose this as one of the component directions. This choice and the corresponding positive directions are indicated by the tiny set of x and y axes shown in the free-body diagram of Figure 4–8.

Vertical component

Choosing the upward direction as positive, we have

$$\Sigma F_y = ma_y$$

Horizontal component

Choosing the direction toward the right as positive, we have

$$\Sigma F_x = ma_x$$

We may now substitute the symbols representing the numerical values. If, as in this case for gravity, the force is in the direction chosen as negative, then we use a minus sign with the letter: $(-W)$. This does not violate our rule of never changing the signs in the *general* equation, $\Sigma \mathbf{F} = m\mathbf{a}$. But when we substitute numerical values (or specific letters representing numerical values), then the appropriate signs are used with the letter. In this one-dimensional case, we thus use scalars with plus and minus signs for the vector forces.

$$N + (-W) = m(0)$$
$$N = W$$

This information is not needed in the solution.

$$F = ma_x$$
$$a_x = \frac{F}{m} = \frac{(4 \text{ N})}{(2 \text{ kg})} = 2.00 \frac{\text{m}}{\text{s}^2}$$

Finally, using the kinematic equation, we obtain

$$x = x_0 + v_{x0}t + \tfrac{1}{2}a_x t^2$$
$$x = 0 + 0 + \frac{1}{2}\left(2 \frac{\text{m}}{\text{s}^2}\right)(3 \text{ s})^2$$
$$x = \boxed{9.00 \text{ m}}$$

EXAMPLE 4–3

A 2-kg block starts from rest and slides down a frictionless surface at 30° with respect to the horizontal [see Figure 4–9(a)]. How long does it take to travel 4 m along the surface?

SOLUTION

If we find the acceleration, we can use the kinematic equations to calculate the time it takes to travel 4 m. As always, we first sketch a free-body diagram [Figure 4–9(b)]. Note that there are only two forces acting on the block:

W = downward force due to gravity
N = normal force exerted by the surface on the block

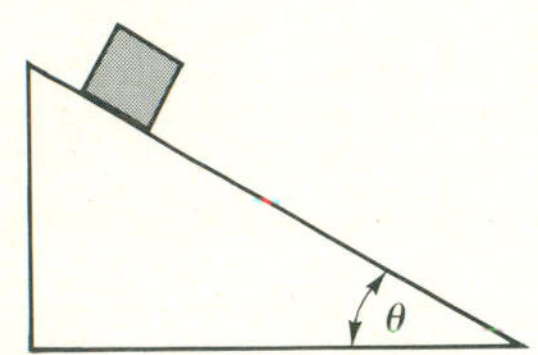

(a) A block sliding down a frictionless incline.

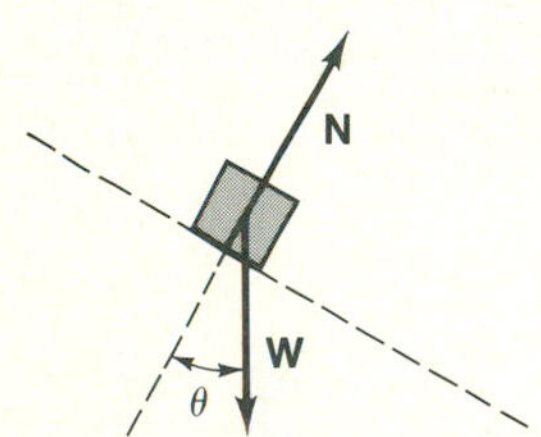

(b) A free-body diagram.

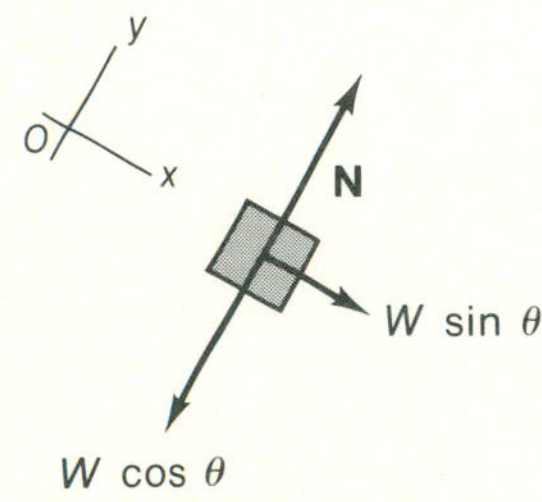

(c) A free-body diagram with forces resolved into components in two mutually perpendicular directions.

Figure 4–9
Example 4–3

Since we wish to apply $\Sigma\mathbf{F} = m\mathbf{a}$ along two mutually perpendicular directions, we look ahead and note that the block accelerates downward parallel to the surface. So we choose this as one of the two component directions for resolving the forces. We then draw a second free-body diagram to show the components, indicating the positive directions by a small set of axes [Figure 4–9(c)].

In sketching free-body diagrams, we avoid drawing a force vector and its components on the same diagram. If both are shown on the same sketch, it in effect puts the force in twice. Instead, two different diagrams should be drawn, as in Figure 4–9(b) and (c).

Parallel to the surface

$$\Sigma\mathbf{F}_x = m\mathbf{a}_x$$

$$(W \sin\theta) = ma_x$$

Solving for a_x and substituting mg for W, we have

$$a_x = \frac{mg \sin\theta}{m} = g \sin\theta$$

$$a_x = \left(9.8\ \frac{\text{m}}{\text{s}^2}\right)(\sin 30^\circ)$$

$$a_x = 4.90\ \frac{\text{m}}{\text{s}^2}$$

The time it takes to travel 4 m from rest is found from the kinematic equation:

$$x = x_0 + (v_x)_0 t + \tfrac{1}{2}a_x t^2$$

Solving for t^2, we have

$$t^2 = (x - x_0 - v_{x0}t)\left(\frac{2}{a_x}\right)$$

$$t^2 = (4\ \text{m} - 0 - 0)\left(\frac{2}{4.9\ \frac{\text{m}}{\text{s}^2}}\right)$$

$$t^2 = 1.63\ \text{s}^2$$

$$t = \pm 1.28\ \text{s}$$

Discarding the negative root as not significant in the context of this problem, we have

$$\boxed{t = 1.28\ \text{s}}$$

Perpendicular to the surface

$$\Sigma\mathbf{F} = m\mathbf{a}_y$$

$$(N - W\cos\theta) = ma_y$$

Solving for N and substituting mg for W, we have

$$N = ma_y + mg\cos\theta$$

$$N = 0 + (2\ \text{kg})\left(9.8\ \frac{\text{m}}{\text{s}^2}\right)(\cos 30^\circ)$$

$$N = 0 + (2\ \text{kg})\left(9.8\ \frac{\text{m}}{\text{s}^2}\right)(0.867)$$

$$N = 16.97\ \text{N}$$

This answer is not needed, but note that the normal force is less than the weight of the block ($W = 19.6$ N).

4.10 Tension

We often use strings, ropes, or wires to transmit a force from one point to another. Unless stated otherwise, we will assume that such strings and ropes have negligible mass, so we can ignore the force of gravity on them. Consider a string being pulled at opposite ends by forces of equal magnitude, as shown in Figure 4–10(a). If each force has the magnitude T, we say that the string is under a **tension** T that exists throughout the length of the string. This occurs because, on a microscopic scale, distances between all the atoms in the string have been

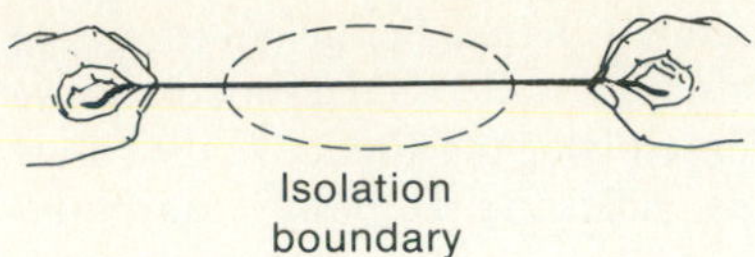

(a) Pulling on both ends of a string creates a tension T at all points in the string.

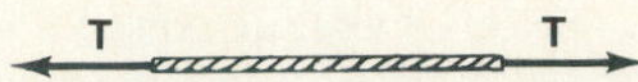

(b) A free-body diagram for the segment of the string inside the isolation boundary.

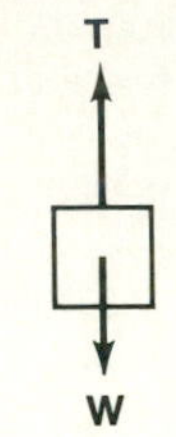

(d) A free-body diagram for the block.

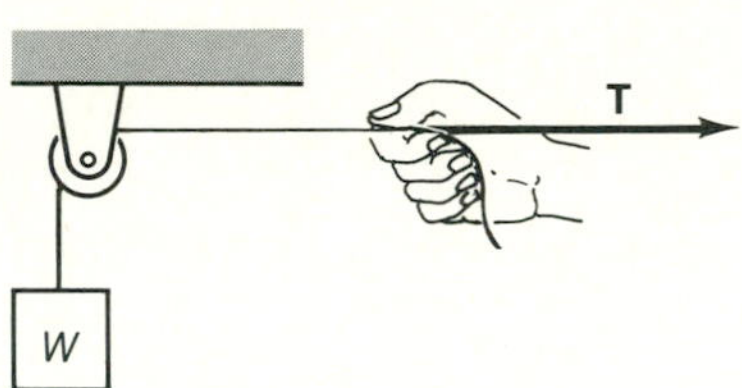

(c) The rope passing over the pulley transmits the tension force **T** to the block.

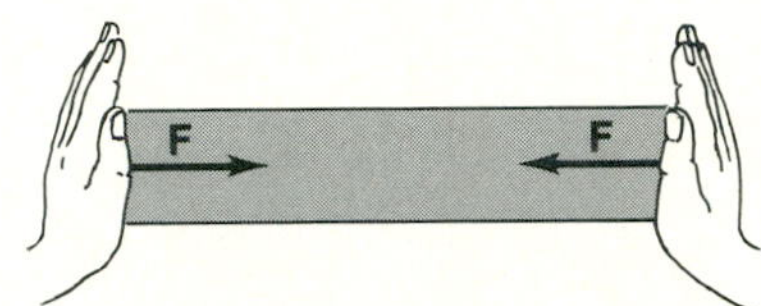

(e) A rigid beam under compression forces **F**.

Figure 4–10
Forces of tension and compression on various objects.

slightly increased, causing forces of attraction between adjacent atoms. Figure 4–10(b) shows a free-body diagram for the segment of the string enclosed in the "isolation boundary." At the point where this boundary cuts across the string, the atoms on one side exert attractive forces on the atoms on the other side. Thus, the segment we have isolated experiences tension forces as shown. Note that the tension is present uniformly throughout the entire length of the string. No matter where we cut across it with an isolation boundary, the tension force **T** arises on *each* of the cut ends, since this is the force that would be required to hold each cut end in place.

If a rope is passed around a pulley, the same tension forces exist, but they are redirected. Similarly, a rigid rod or beam may transmit forces of **compression** (in the opposite directions) as well as those of tension when the rod is subjected to forces that tend to shorten the length of the rod [see Figure 4–10(e)].

EXAMPLE 4–4

A 4-lb ball swings with constant speed in a horizontal circle of radius $r = 2$ ft. The ball is suspended from a 5-ft rope, as shown in Figure 4–11(a). (a) Find the tension in the rope. (b) How long does it take the ball to make one complete revolution?

SOLUTION

(a) As always, the first step is to draw a free-body diagram. For convenience, we choose a perspective as viewed in the plane of motion when the ball is located farthest toward the right. The distances involved indicate that $\theta = \sin^{-1} \frac{2}{5} = 23.6^\circ$.

There are only two forces acting on the ball, as shown in Figure 4–11(b).

W = downward force on the ball due to gravity

T = tension in the rope at an angle of 23.6° with respect to the vertical

If the lack of a radially outward force on the ball seems puzzling, try to find some other object that touches the ball and pushes outward; there is none. Therefore, since each force on an object must have its origin in some other object outside the isolation boundary, in this example there *is* no outward force. *Trying to add such a (nonexistent) outward force is the most common error in analyzing circular motion.*

We next choose two coordinate directions and resolve all forces into components along these directions. Looking ahead, we know that the ball is accelerating toward the center of the circle (motion in a circle at constant speed). Therefore, we choose this direction as one of the two component directions, as shown in Figure 4–11(c).

Horizontal component

$$\Sigma F_x = ma_x$$

$$T \sin\theta = ma_x$$

Solving for a_x, we have

$$a_x = \frac{T \sin\theta}{m}$$

Substituting the value for T found from the vertical component analysis at the right, we have

$$a_x = \frac{(4.37 \text{ lb})(\sin 23.6°)}{\left(\dfrac{4 \text{ lb}}{32 \dfrac{\text{ft}}{\text{s}^2}}\right)} = 14.0 \frac{\text{ft}}{\text{s}^2}$$

Vertical component

$$\Sigma F_y = ma_y$$

$$T \cos\theta - W = ma_y$$

Solving for T, we have

$$T = \frac{ma_y + W}{\cos\theta} = \frac{m(0) + (4 \text{ lb})}{(\cos 23.6°)}$$

$$\boxed{T = 4.37 \text{ lb}}$$

(b) The centripetal acceleration a_x toward the center of the circle may also be pressed in terms of the tangential speed v of the ball and the radius r.

$$a_x = \frac{v^2}{r}$$

Solving for v^2, we have

$$v^2 = a_x r$$

$$v^2 = \left(14.0 \frac{\text{ft}}{\text{s}^2}\right)(2 \text{ ft}) = 28.0 \frac{\text{ft}^2}{\text{s}^2}$$

$$\boxed{v = 5.29 \frac{\text{ft}}{\text{s}}}$$

The time to make one revolution is

$$\text{time} = \frac{\text{distance}}{\text{velocity}} = \frac{2\pi r}{v} = \frac{2\pi(2 \text{ ft})}{5.29 \dfrac{\text{ft}}{\text{s}}} = \boxed{2.38 \text{ s}}$$

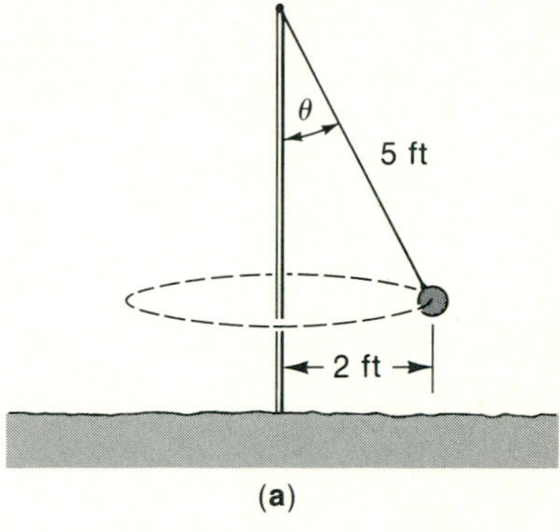

(a)

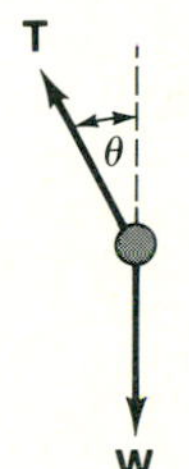

(b) The free-body diagram.

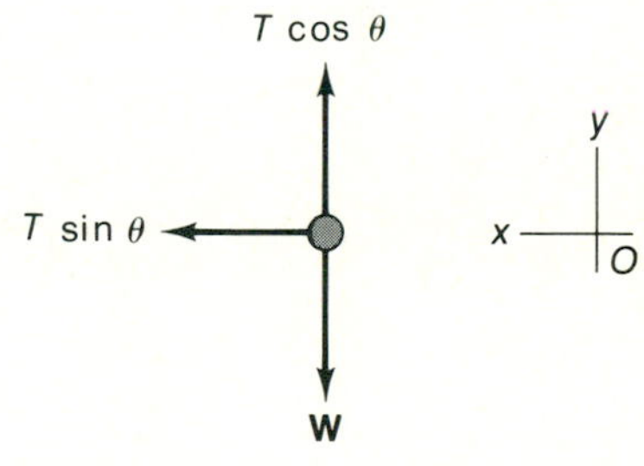

(c) A free-body diagram with forces resolved into two mutually perpendicular directions.

Figure 4–11
Example 4–4

We may summarize the general procedure for solving $\Sigma \mathbf{F} = m\mathbf{a}$ as follows:

(1) Sketch a free-body diagram, showing all forces that act *through* an isolation boundary (from outside) *on* the object inside the isolation boundary.

(2) Choose an x-y coordinate system suitable to the problem. If you know that the object accelerates in a particular direction, it will be convenient to choose one of the (positive) axes to lie in this direction.
(3) Redraw the diagram, resolving all forces into the two component directions.
(4) Apply $\Sigma F_x = ma_x$ and $\Sigma F_y = ma_y$ and solve for the required quantity.

4.11 Friction

When one object slides across another, each one exerts a force of friction on the other. These forces are always *tangent* to the surfaces in contact, and the direction of the frictional force on each object is opposite to the direction of motion relative to the other object. We use the symbol **f** to denote frictional forces. Note that **f** and **N** are the tangential and normal components of the *same* force, namely, the force that the surface exerts on the object. When added together, these two components give the single net force that a surface may exert on an object. The components are convenient because of their empirical relationship, as explained below.

Friction is a very complicated phenomenon that is not well understood.[6] A careful analysis must consider the interaction between the more or less jagged bumps on a microscopic level that occur even on the most highly polished surfaces (refer to Figure 4–12 in Example 4–5, page 76). Contamination by oxides, surface films of other substances, the degree of polish, and other variables may greatly alter the amount of friction that is present. If two metals are used, very high temperatures occur at the microscopic points of contact. It is thought that momentary welds may occur, and thus friction in this case might be the force necessary to break these welds. For two dissimilar substances, the situation is more complex. But for all cases, including those in which there is no sliding, a frictional force will develop if attempts are made to cause sliding.

In spite of some poorly understood aspects of friction, three interesting facts emerge from experimental studies:

(1) The friction force developed is approximately *proportional to the normal force* **N** that one surface exerts on the other.
(2) The friction force is approximately *independent of the area of contact* between the objects.
(3) The magnitude of the friction force is approximately *constant over a range of speeds* between about one centimeter per second and a few meters per second.

It is not surprising that friction would be proportional to the normal force squeezing the two surfaces together. The *microscopic* areas of contact on an atomic scale may be only about 1/10,000 or less of the *macroscopic* area of the surfaces in contact, so one would expect that a greater normal force would crush together more of these microscopic contacts. On the other hand, the fact that the friction force is independent of the macroscopic area of contact (for a given normal force) is a bit unexpected. This means, for example, that regardless of which face of a rectangular brick is in contact with the other object, the friction force is the same. This behavior becomes plausible when we realize that the total area of microscopic contact depends on the *pressure* (force per

[6] See, for example, E. Rabinowicz, "Stick and Slip," *Scientific American*, May 1956, pages 109–18.

unit area) and not just the *normal force*. The normal force is the same no matter which face is down, but the pressure varies with the size of the face. The greater pressure with the smallest face down probably increases the total area of microscopic contact up to about the same value as when one of the larger faces is down.

Friction may be reduced, of course, by lubrication. An even more effective way to reduce the amount of friction is to interpose a layer of air or gas between the surfaces, as an air track so vividly illustrates. Perhaps the best display of friction reduction is in the case of a rotating object suspended in a vacuum by magnetic forces, so that there is no physical contact with the rotating object except by the magnetic field itself.

In this text we will idealize friction forces and pretend the word "approximately" is absent from the three experimental characteristics just listed. So henceforth we will assume that *the friction force on a sliding object is exactly proportional to N and does not vary with the area of contact or with the speed of sliding*. For any two given materials, the constant of proportionality between f and N, however, does depend on whether the surfaces are actually sliding across each other. Friction forces thus fall into two classes:

KINETIC FRICTION
(when the surfaces slide across each other)

$$f_k = \mu_k N \tag{4–11}$$

STATIC FRICTION
(when no sliding occurs)

$$f_s \leq \mu_s N$$

where μ is the corresponding constant of proportionality between f and N. The constant μ is called the **coefficient of friction.**

Suppose an external force is applied horizontally to an object in an attempt to make it slide across a horizontal surface. If friction prevents sliding, we have the *static* case; the force of static friction f_s will be whatever value (and whatever direction) is necessary to keep the object in equilibrium. As the applied force is made larger and larger, the static friction also increases up to some maximum value given by the equal sign in Equation (4–12). When the object finally "breaks away" from the surface and begins sliding, the force of kinetic friction is usually less than the maximum static friction force; that is, usually $\mu_k < \mu_s$.

Keep in mind that the kinetic friction force $\mathbf{f}_k$ is always directed *opposite* to the object's velocity $\mathbf{v}$ (relative to the surface). The static friction force $\mathbf{f}_s$ may be in either direction parallel to the surface, depending upon the direction of the force necessary to prevent sliding. In solving problems, the direction of $\mathbf{f}_k$ must be properly chosen on the free-body diagram to obtain the correct answer.

Equations (4–11) and (4–12) are empirical, rule-of-thumb approximations for dry, unlubricated surfaces. The real situation is much more complex. Nevertheless, these relations are obeyed closely enough by many common substances that they may be used in many engineering applications. We will assume that they always apply.[7]

---

[7] Some comment should be made regarding *rolling friction* associated with the use of wheels to move objects across a level surface. The mechanism involved in rolling friction is quite different from that of sliding friction. Because no substance is perfectly rigid, all objects will deform somewhat when pressed against another object. It is believed that rolling friction arises primarily from energy lost when elastically deformed objects fail to regain their original shapes immediately upon release of the stress. As an example of the difference in magnitude between rolling friction and sliding friction, the coefficient of friction for steel *sliding* on steel is about 0.20, whereas a steel ball *rolling* on steel has an equivalent coefficient of friction of only about 0.002.

EXAMPLE 4–5

A 16-lb block rests on a horizontal surface. A horizontal force of 5 lb is required to cause the block to break away and start moving, and a force of 3 lb is sufficient to keep the block sliding at constant speed once it is set into motion. Calculate the static and kinetic coefficients of friction.

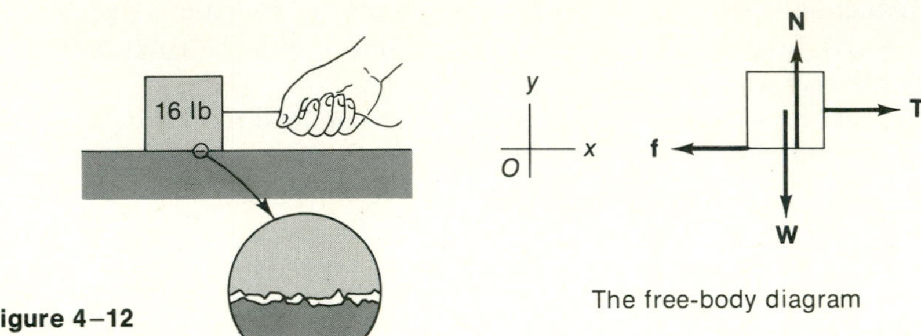

The free-body diagram

Figure 4–12
Example 4–5

SOLUTION

As always, we sketch a free-body diagram, as shown in Figure 4–12. This diagram applies for both cases if

$$f = f_s = \mu_s N \quad \text{(static case, maximum friction)}$$

and

$$f = f_k = \mu_k N$$

Vertical component

$$\Sigma F_y = ma_y$$
$$(N - W) = ma_y$$
$$N = ma_y + W = m(0) + 16 \text{ lb}$$
$$N = 16 \text{ lb}$$

This value for N is substituted into the equation at right.

Horizontal component

Static case (maximum friction)

$$\Sigma F_x = ma_x$$
$$(T - \mu_s N) = ma_x$$
$$\mu_s N = T - ma_x$$
$$\mu_s = \frac{T - ma_x}{N} = \frac{5 \text{ lb} - m(0)}{16 \text{ lb}}$$
$$\mu_s = \boxed{0.313}$$

Kinetic case (constant velocity)

$$\Sigma F_x = ma_x$$
$$(T - \mu_k N) = ma_x$$
$$\mu_k = \frac{(T - ma_x)}{N} = \frac{3 \text{ lb} - m(0)}{16 \text{ lb}}$$
$$\mu_k = \boxed{0.188}$$

EXAMPLE 4–6

A 4-kg box is dragged across a horizontal floor by pulling with a 20-N force on a rope that makes an angle of 35° with the horizontal. The coefficient of kinetic friction is 0.20. Find the acceleration of the block. (See Figure 4–13.)

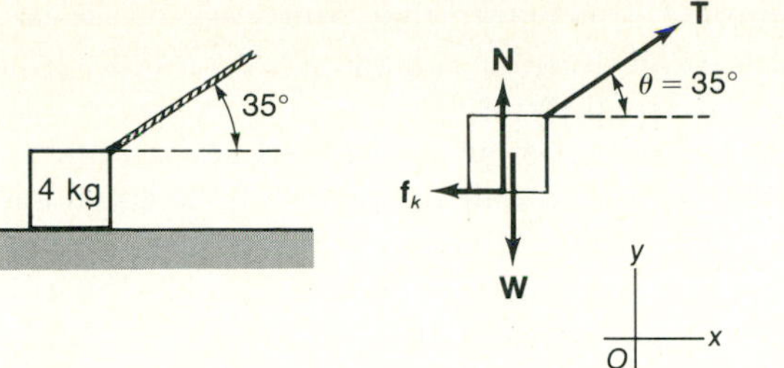

A free-body diagram.

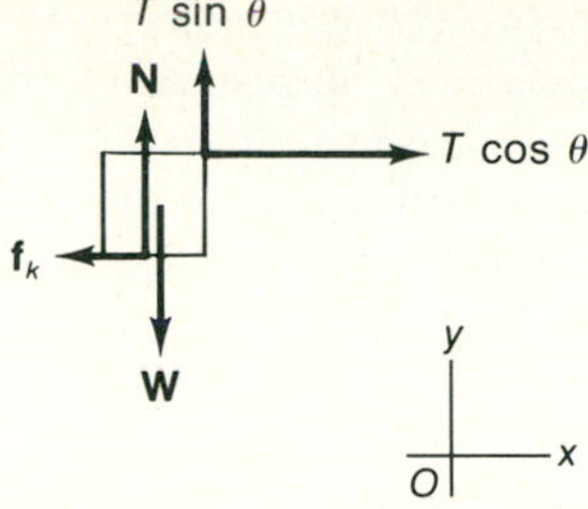

A free-body diagram with the forces resolved into horizontal and vertical directions.

Figure 4–13
Example 4–6

SOLUTION

We sketch the usual free-body diagram and resolve all forces into two mutually perpendicular directions. Since the box has acceleration in the horizontal direction, we choose that as one of the component directions.

Vertical component

$$\Sigma F_y = ma_y$$

There is no vertical acceleration, so

$$N + T \sin \theta - W = ma_y$$

$$N = ma_y - T \sin \theta + W$$

$$= m(0) - (20 \text{ N})(0.574) + (4 \text{ kg})\left(9.8 \frac{\text{m}}{\text{s}^2}\right)$$

$$\boxed{N = 27.7 \text{ N}}$$

This value is substituted into the equation for the horizontal component.

Horizontal component

$$\Sigma F_x = ma_x$$

$$T \cos \theta - f_k = ma_x$$

$$T \cos \theta - \mu_k N = ma_x$$

$$a_x = \frac{(T \cos \theta - \mu_k N)}{m}$$

$$= \frac{(20 \text{ N})(0.819) - (0.20)(27.7 \text{ N})}{4 \text{ kg}}$$

$$\boxed{a_x = 2.71 \frac{\text{m}}{\text{s}^2}}$$

4.12 Newton's Third Law

Before we can analyze more interesting physical situations, we need to introduce Newton's third law. It is a very simple law to state, yet a curiously puzzling and subtle one when you first come upon it. In Newton's words: "To every action there is always opposed an equal reaction; or, the mutual actions of two bodies on each other are always equal, and directed to contrary parts."

In modern English, this principle may be stated as follows:

NEWTON'S THIRD LAW **When two objects exert forces on each other, the force that object *A* exerts on *B* is always equal in magnitude and opposite in direction to the force that *B* exerts on *A*.**

It is clear that Newton's use of the words "action" and "reaction" was a semantic convenience, since there is no implication of cause and effect in the third law. Undoubtedly Newton wished to emphasize that forces always come in pairs and that the word "action" has a counterpart in "reaction," while the word "force" does not commonly call to mind any complementary twin. But keep in mind that forces always do come in *pairs*, acting in *opposite directions*, on *two different objects.* It is important to realize that either one of the two forces may be designated the action and the other one the reaction; that is, the forces arise simultaneously and have equal status.

Confusion over this point sometimes arises when one considers a situation such as a person pushing against a wall. The wall will exert a force equal and opposite to that of the hand, and it may seem as though the person has "caused" the wall to "respond." Of course the person made the decision to push the wall, and in that sense he or she "caused" the forces to arise. But in a more fundamental sense, the interaction between the wall and the hand is one of atoms repelling each other. If two atoms are brought close together, *mutual* forces of repulsion arise *simultaneously*. This is the basic interaction.

Figure 4–14
Some pairs of forces that can be described by Newton's third law. Note that the diagrams in (b) and (c) do not include all the forces involved; therefore they are not complete free-body diagrams.

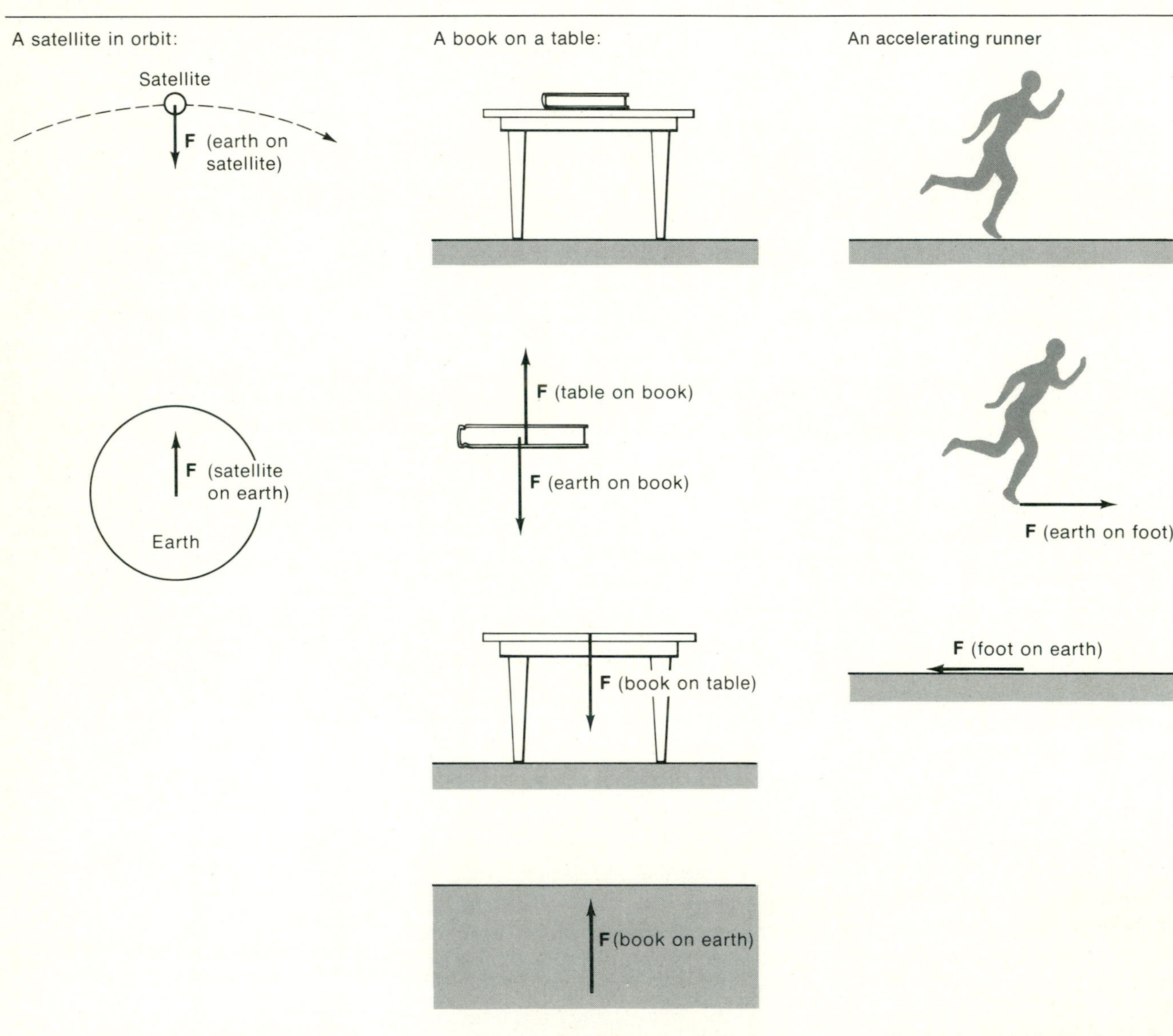

(a) The gravitational force the *earth* exerts on the *satellite* is equal in magnitude and opposite in direction to the gravitational force of the *satellite* on the *earth*.

(b) Here we illustrate two pairs of forces: the pair of gravitational forces between the book and the earth, and the pair of contact forces between the table and the book.

(c) The horizontal friction force the *earth* exerts on the runner's *foot* is equal in magnitude and opposite in direction to the friction force the *foot* exerts on the *earth*.

Another important feature of Newton's third law is that *the two forces always act on two different objects, never on the same object.* Forgetting this fact is the most common source of difficulty in using the third law. Finally, keep in mind that the third law says something about forces, but nothing about the motion of objects or how they accelerate. (We must use the second law, $\Sigma\mathbf{F} = m\mathbf{a}$, to acquire this latter information.)

Before we analyze the third law in specific cases, let us identify in a few simple instances just which forces constitute third-law pairs. Figure 4–14 shows a few examples.

□ □ □

We have already had a preview of Newton's third law. Recall the discussion of the experiments using two masses, A and B, on an air track. It was found experimentally that

$$\frac{d\mathbf{p}_A}{dt} = -\frac{d\mathbf{p}_B}{dt} \tag{4-13}$$

Using Newton's second law, $\mathbf{F} = d\mathbf{p}/dt$, we may substitute the appropriate force for each term. Thus the relation can also be written as

$$\mathbf{F}_{B\text{ on }A} = -\mathbf{F}_{A\text{ on }B} \tag{4-14}$$

This mathematical statement of the third law looks simple enough, yet many students find it hard to grasp. One difficulty is that "common-sense" explanations for many situations are wrong. For example, consider the following explanation: "The winning team in a tug of war wins because it pulls harder on the rope than the losing team." According to Newton's third law, this statement is false. Team A, at all times, pulls *exactly* as hard on the rope as does Team B, regardless of who wins. (This assumes that the rope has negligible mass.)

Thus Newton's third law really says it is *impossible* for any object A to exert a greater (or lesser) force on another object B than B exerts on A. In other words, the two forces that comprise the mutual interaction between any two objects are in all cases, at all times, *exactly equal in magnitude.* This is true whether one or both of the objects remains at rest, is moving with constant speed, or is accelerating.

What is the correct explanation in the example just cited? How can one team win the tug of war if each team pulls *exactly* as hard as the other team *at all times*? We need to approach Newton's third law in easy steps. Let us begin by considering the situation depicted in Figure 4–15(a).

Three objects interact in this example: the woman, the sled, and the earth. Because no acceleration in the vertical direction occurs, each object has zero *net* vertical force acting on it. Therefore, we concentrate our attention only on the horizontal force components. Figure 4–15(b) shows the appropriate isolation diagrams.

The horizontal force components are:

$\mathbf{F}_1$ = force woman exerts on sled
$\mathbf{f}_1$ = friction force earth exerts on sled
$\mathbf{F}_2$ = force sled exerts on woman
$\mathbf{f}_2$ = friction force earth exerts on woman
$\mathbf{f}_3$ = friction force woman exerts on earth
$\mathbf{f}_4$ = friction force sled exerts on earth

(a)

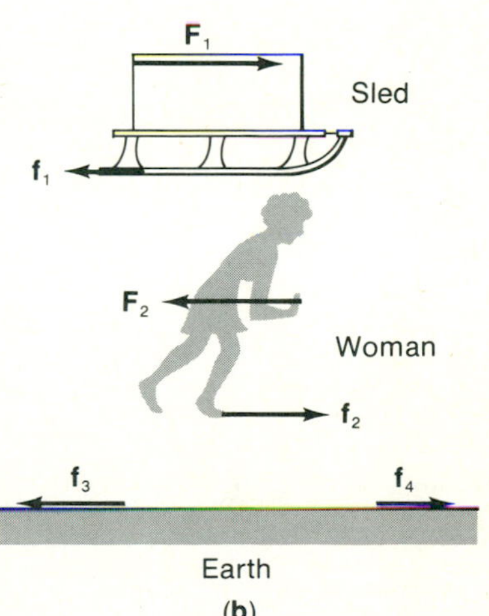

(b)

Figure 4–15
(a) A woman pushes a sled along the ground. (b) The horizontal force components involved in the action.

Newton's third-law pairs of forces (magnitude only) are:

$$F_1 = F_2 \qquad f_1 = f_4 \qquad f_2 = f_3$$

These third-law pairs of forces are always equal in magnitude and opposite in direction, regardless of whether the sled is at rest, moving with constant velocity, or accelerating.

What determines whether or not the sled accelerates? Newton's *second* law gives the answer: *the net force on the sled.* It is true that the sled always pushes back on the woman just as hard as the woman pushes on the sled. But this fact has nothing to do with whether or not the sled accelerates. Only the forces on the sled determine the sled's acceleration. Thus we have the clue to understanding motion: *the only factor that determines whether or not an object will accelerate is the sum of the forces on that object.* That is the meaning of $\Sigma\mathbf{F} = m\mathbf{a}$.

This example clarifies the old puzzle: "If the cart pulls as hard on the horse as the horse pulls back on the cart, how can the cart ever start moving?" The answer is that only the forces *on the cart* determine its motion. The fact that two different forces (on two different objects) happen to be equal has no influence on whether the cart (or the horse) accelerates. True, these two forces add up to zero. But why should they be added together? They act on *different* objects. Only the net force *on an object* determines that object's acceleration.

Why should we study Newton's third law? It is not a relation that, by itself, tells how objects will move. We should learn its implications, however, because it is often helpful in identifying unknown forces (we may know *one* of a third-law pair). Additionally, of course, it is part of the complete Newtonian explanation of the role of forces, their relationships, and how—as $\Sigma\mathbf{F} = m\mathbf{a}$ states—forces cause objects to accelerate.

EXAMPLE 4–7

A 70-kg man, initially at rest, starts to run with an acceleration of 2 m/s^2. Analyze the forces involved.

SOLUTION

Note the *direction* of the force of static friction $\mathbf{f}_s$ in the free-body diagram in Figure 4–16. The man, knowing Newton's third law, makes this forward force come into existence by pushing *backward* on the earth with his foot. The equal and opposite static friction force the earth exerts on the man is the net horizontal force producing the acceleration. Note carefully the distinction that it is *not* the backward force the man's foot exerts *on the earth* that accelerates the man. Instead, it is the forward force the earth exerts *on the man.* Only the forces *on* an object determine that object's motion.

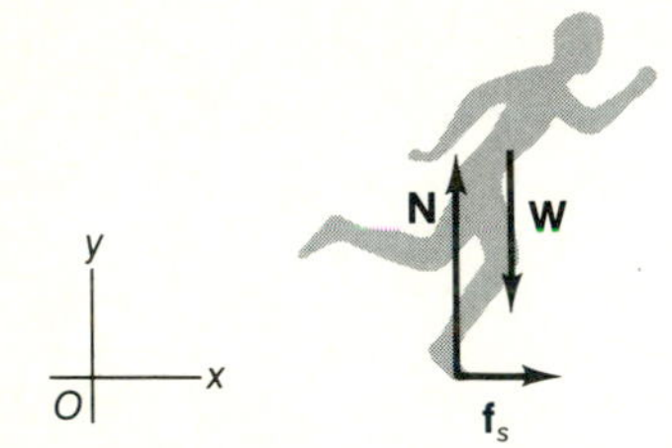

Figure 4–16
Example 4–7

$$\Sigma F_x = ma_x$$
$$f_s = ma$$
$$f_s = (70\ \text{kg})\left(2\ \frac{\text{m}}{\text{s}^2}\right) = \boxed{140\ \text{N}}$$

The previous example should demolish the common misconception that "friction always opposes motion." The relationship is more subtle than this.

The direction of the friction force that a surface exerts on an object is opposite to the direction of the motion (or tendency to move) of the object *relative to the surface*. Here, the man's foot tends to slip *backward* relative to the earth though the man himself moves forward. So instead of "opposing motion," it is actually the *forward* force of static friction that the earth exerts on the man, and it is this force that causes him to accelerate. Similarly, a highway exerts a forward force on automobile tires, causing a car to speed up. And it is the steel rails that push a train forward. These and numerous other examples illustrate $\Sigma\mathbf{F} = m\mathbf{a}$ in action.

EXAMPLE 4–8

Find the external force **F** that will give the lower block in Figure 4–17(a) an acceleration of 2 ft/s² toward the right. Assume that the pulley is frictionless and that the coefficient of kinetic friction μ_k between all flat sliding surfaces is 0.20.

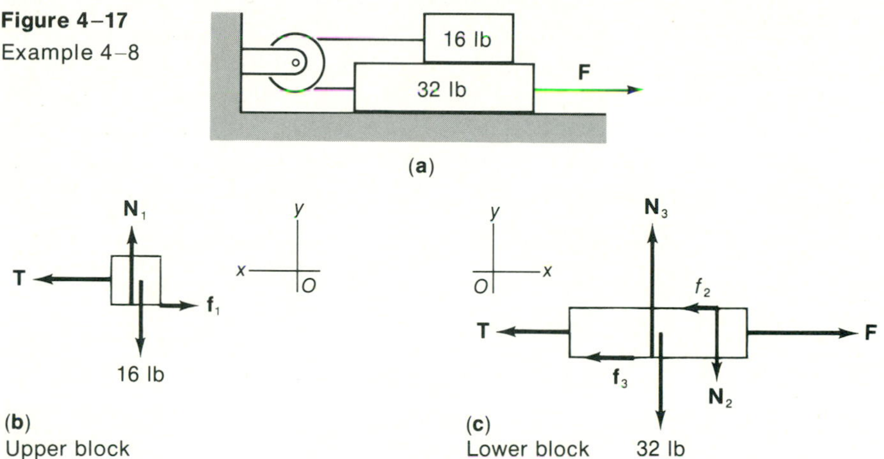

Figure 4–17
Example 4–8

(b)
Upper block

(c)
Lower block

SOLUTION

We first make the usual free-body diagrams for the two blocks [Figure 4–17(a) and (b)]. Since each block is in vertical equilibrium, from Newton's second law we have:

$$\textit{Upper block:}\quad \Sigma F_y = 0$$
$$N_1 = 16 \text{ lb}$$
$$\textit{Lower block:}\quad \Sigma F_y = 0$$
$$N_3 = N_2 + 32 \text{ lb}$$

From Newton's third law, we know that the following forces are equal in magnitude:

$$N_1 = N_2 \qquad \text{and} \qquad f_1 = f_2$$

Therefore:

$$N_3 = 16 \text{ lb} + 32 \text{ lb} = 48 \text{ lb}$$

Also:

$$f_1 = \mu_k N_1 = (0.20)(16 \text{ lb}) = 3.20 \text{ lb}$$
$$f_3 = \mu_k N_3 = (0.20)(48 \text{ lb}) = 9.60 \text{ lb}$$

Thus, the forces on the two blocks are:

Upper block

Assuming the positive direction is toward the left (since this is the direction of acceleration), we have

$$\Sigma F_x = ma_x$$
$$T - f_1 = ma_x$$
$$T = ma_x + f_1$$
$$T = \left(\frac{16\ \text{lb}}{32\ \frac{\text{ft}}{\text{s}^2}}\right)\left(2\ \frac{\text{ft}}{\text{s}^2}\right) + (3.20\ \text{lb})$$
$$\boxed{T = 4.20\ \text{lb}}$$

This value is substituted into the equation at right.

Lower block

Assuming the positive direction is toward the right (in the direction of acceleration of the block), we have

$$\Sigma F_x = ma_x$$
$$(F - T - f_2 - f_3) = ma_x$$
$$F = ma_x + T + f_2 + f_3$$
$$F = \left(\frac{32\ \text{lb}}{32\ \frac{\text{ft}}{\text{s}^2}}\right)\left(2\ \frac{\text{ft}}{\text{s}^2}\right) + 4.20\ \text{lb} + 3.20\ \text{lb} + 9.60\ \text{lb}$$
$$\boxed{F = 19.0\ \text{lb}}$$

For an important case of Newtonian motion, consider the photograph in Figure 4–18(a), which shows a car traveling at constant speed on a banked roadway. There is a certain critical speed such that *no* friction forces between the road and tires are involved—only the normal force is present [see Figure 4–18(b) and (c)]. At this speed, even if the roadway were slick with ice, the car could successfully negotiate the curve.

If the car travels faster than this particular speed, the road then exerts a friction force *tangent to the road surface* toward the inside of the circle. If the car travels slower than this critical speed, the friction force is also *tangent to the road surface* but toward the outer edge. (In both cases, the normal force **N** is also altered in magnitude.) But for *every* case of motion in a circle at constant speed, the sum of the force components in the horizontal direction gives the car its acceleration ($a_x = v^2/r$) toward the center of the circle. Therefore, for the

Figure 4–18
Forces on a car traveling a banked roadway at a certain constant speed such that there are no friction forces between the road and tires. The angle of banking is θ.

(**a**) A banked roadway on an automobile testing grounds.

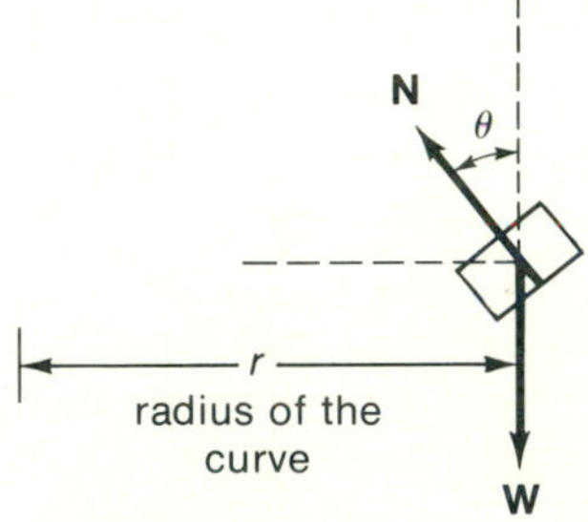

(**b**) Forces on the car. At a certain critical (constant) speed, the road exerts only a normal force **N** on the car. No friction force tangent to the road surface is present at this speed.

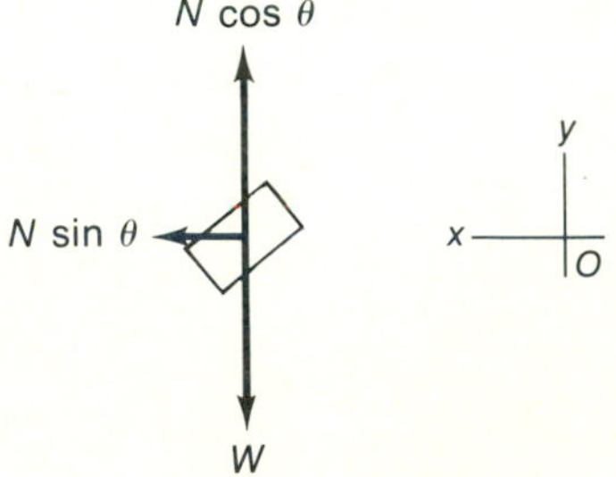

(**c**) Even though traveling at constant speed, the car is accelerating toward the center of the curved path. Therefore we choose that direction (the *x* axis) as one of the two mutually perpendicular directions for resolving force components.

horizontal direction, $\Sigma \mathbf{F}_x = m\mathbf{a}_x$. For the vertical direction, $\Sigma \mathbf{F}_y = 0$, because the car has no acceleration in the y direction.

☐ ☐ ☐

The next example illustrates a useful method of approaching problems that ask you to compare two similar situations.

EXAMPLE 4–9

While approaching a planet circling a distant star, a space traveler determines the planet's radius to be half that of the earth. After landing on the surface, the traveler finds the acceleration due to gravity to be twice that on the earth's surface. Find the mass of the planet M_p in terms of the mass of the earth M_e. (Newton's law of gravitation for the gravitational force between two masses, m_1 and m_2, separated by a distance r, is $F_{\text{grav}} = G(m_1 m_2/r^2)$ where G is the universal gravitational constant.)

SOLUTION

Whenever we wish to compare two similar situations, it is easiest to set up a ratio by dividing an equation for one case (in symbol form) by the same equation for the other case. This way, all the common terms cancel, and the calculations are greatly simplified.

In solving the problem, we use the subscript p for symbols referring to the planet and the subscript e for the earth symbols. When we apply $\Sigma \mathbf{F} = m\mathbf{a}$ for a freely falling object of mass m at the surfaces of the planet and of the earth, we have

For the earth	**For the planet**
$\Sigma \mathbf{F} = m\mathbf{a}$	$\Sigma \mathbf{F} = m\mathbf{a}$
$G\dfrac{M_e m}{r_e^2} = mg_e$	$G\dfrac{M_p m}{r_p^2} = mg_p$

Dividing one equation by the other yields

$$\frac{\left(G\dfrac{M_p m}{r_p^2}\right)}{\left(G\dfrac{M_e m}{r_e^2}\right)} = \frac{mg_p}{mg_e}$$

Simplifying, we obtain

$$\left(\frac{M_p}{M_e}\right)\left(\frac{r_e^2}{r_p^2}\right) = \frac{g_p}{g_e}$$

Expressing the parameters for the planet in terms of symbols for the earth gives us

$$g_p = 2g_e \qquad \text{and} \qquad r_p = \tfrac{1}{2}r_e$$

Substituting these parameters into our equation, we have

$$\left(\frac{M_p}{M_e}\right)\left(\frac{r_e^2}{(\frac{1}{2}r_e)^2}\right) = \frac{2g_e}{g_e}$$

Many of the unknown symbols cancel, leading to the simple expression

$$\frac{M_p}{M_e}(4) = 2$$

Rearranging, we obtain

$$\boxed{M_p = \tfrac{1}{2}M_e}$$

Summary

Newton's three laws of motion are the following:

(1) An object continues in a state of rest, or of uniform motion in a straight line, unless compelled to change that state by a net force acting on it.
(2) $\Sigma\mathbf{F} = d(m\mathbf{v})/dt$ or $\Sigma\mathbf{F} = m\mathbf{a}$
(3) When two objects exert forces on each other, the force object A exerts on B is always equal in magnitude and opposite in direction to the force B exerts on A.

In this text, we use the following units for $\Sigma\mathbf{F} = m\mathbf{a}$:

	SI units	British engineering units
F	newton (N)	pound (lb)
m	kilogram (kg)	slug (—)
a	$\frac{\text{meters}}{\text{second}^2}\left(\frac{\text{m}}{\text{s}^2}\right)$	$\frac{\text{feet}}{\text{second}^2}\left(\frac{\text{ft}}{\text{s}^2}\right)$

We use the word "weight" to designate the force of gravity on an object ($W = mg$).

The clue to using Newton's laws is a free-body diagram that shows all the forces that act *through* an isolation boundary (from outside) *on* the object inside the isolation boundary. If necessary, the diagram is drawn a second time with forces resolved into components along two mutually perpendicular directions appropriate for the problem. *The importance of good free-body diagrams cannot be emphasized too strongly.*

The force that a surface exerts on an object is most conveniently analyzed in terms of two components: **N** is the component perpendicular (normal) to the surface, and **f** is the tangential (friction) component. Friction forces are designated by two classes, where the coefficients of friction for the two classes are almost always unequal ($\mu_k < \mu_s$):

Kinetic friction: $f_k = \mu_k N$ (when surfaces are sliding across each other)

Static friction: $f_s \leq \mu_s N$ (when no sliding occurs)

The equal sign in the static friction relationship applies only when the surfaces are on the verge of slipping and the maximum static friction force is thus developed.

Questions

1. Must an object move in the direction of the force on it?

2. A string is placed under tension by means of two spring balances, one at each end, each indicating T newtons. If the string is cut and a single spring inserted, will it read T or $2T$?

3. It is possible to set up a system of measurement with force, length, and time as the fundamental units. In such a system, what are the dimensions of mass?

4. If the mass of atoms could be determined with sufficient precision, describe the advantages of establishing an atomic standard for mass.
5. A ball is tossed straight upward. Taking into account the effect of air resistance, is the time going up longer or shorter than the time falling down?
6. Consider a ping-pong ball launched at an arbitrary angle with the horizontal. Because of air friction, the path is not exactly parabolic. At what point (or points) along the path is the speed a maximum? A minimum?
7. An object of mass m is supported from the ceiling by a cord A, and a similar cord B is attached to the bottom of the mass and hangs freely below. If cord B is pulled down with a gradually increasing force, cord A breaks. But if cord B is jerked down suddenly, cord B breaks. Explain.
8. Two objects of equal mass are on opposite pans of a balance scale. If the apparatus is in an elevator that is accelerating upward, does the scale remain balanced?
9. Two raindrops have different radii but the same density. If the frictional force of the air on the drops is directly proportional to the speed, which (if either) will have the greater terminal speed?
10. Explain the distinction between *mass* and *weight*.
11. Use Newton's laws of motion to explain (a) how a salt shaker works, (b) how a helicopter can hover at a fixed location, and (c) how a jet plane accelerates.
12. The vertical loop in *The Great American Revolution* roller coaster (Figure 4–19) is not a perfect circle; instead, it has a smaller radius of curvature near the top of the loop than for other portions. Discuss the advantages of designing the loop in this fashion.

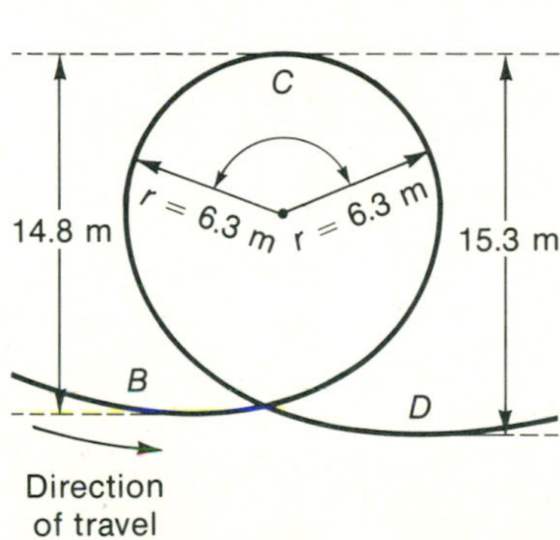

Figure 4–19

The Great American Revolution. The world's first vertical-loop roller coaster began operation at Six Flags Magic Mountain amusement park in Valencia, California, in 1976. The vertical loop is not a perfect circle, but has the configuration shown above. The upper portion of the loop, however, is exactly circular. Following are some technical details regarding the motion of the center-of-mass of the loaded cars:

Radius of upper (circular) portion of loop:	6.3 m
Height above point B at start of last free-coasting descent before reaching loop:	21.6 m
Initial speed of cars when they begin the free-coasting descent before loop:	5.07 m/s
Speed of cars at point B:	20.2 m/s
Speed of cars at point C:	9.7 m/s
Speed of cars at point D:	19.2 m/s
Average mass of train of cars with passengers:	5440 kg
Maximum g force during ride (not on loop):	4.94

Problems

4A–1 A 2-kg object has (only) two forces acting on it in the $+x$ and $+y$ directions, as shown in Figure 4–20. Find the acceleration of the object (magnitude and direction).
Answer: 5.00 m/s^2, 36.9° above the $+x$ direction

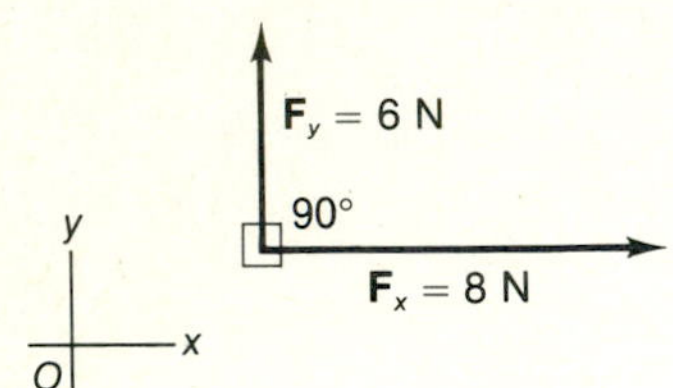

Figure 4–20
Problem 4A–1

4A–2 For each of the following cases, sketch a free-body diagram for the underlined object. Then redraw the diagram, resolving all forces into two mutually perpendicular directions appropriate for the solution. Write equations for $\mathbf{F} = m\mathbf{a}$ for the directions that have force components.

(a) A falling feather that has reached terminal velocity (and therefore is falling at constant speed).
(b) A football at the highest point in its trajectory.
(c) A truck driver in a truck that is traveling along an unbanked horizontal curve of radius r at constant speed v.
(d) A child on a moving swing at the highest point in its arc.
(e) A child on a moving swing at the lowest point in its arc.
(f) A roller coaster car traveling over the top of a curved section of track in a vertical plane.

4A–3 A net force of 20 N acts on a 5-kg object.
(a) What acceleration does the object have?
(b) Determine how far the object travels starting from rest to acquire a speed of 8 m/s.
Answers: (a) 4.00 m/s^2 (b) 8.00 m

4A–4
(a) Determine the force of gravity on a 160-lb man.
(b) Determine the mass of a 160-lb man.
(c) Determine the force of gravity on a 20-kg object.
(d) Determine the mass of a 20-kg object.
(e) Find the net force that will give a 160-lb man an acceleration of 32 ft/s^2.
(f) Find the net force that will give a 20-kg object an acceleration of 9.8 m/s^2.

4A–5 An object weighing 30 N is acted on by the force of gravity and a constant horizontal force of 40 N. Find the net displacement of the object after 6 s (magnitude and direction), assuming that the object starts from rest. *Answer:* 294 m, 36.9° below the horizontal

4A–6 A 100-lb boy hangs on a rope tied to two trees, as shown in Figure 4–21. The two segments of the rope are labeled A and B. Find the tension in each segment and specify which segment has the greater tension.

Figure 4–21
Problem 4A–6

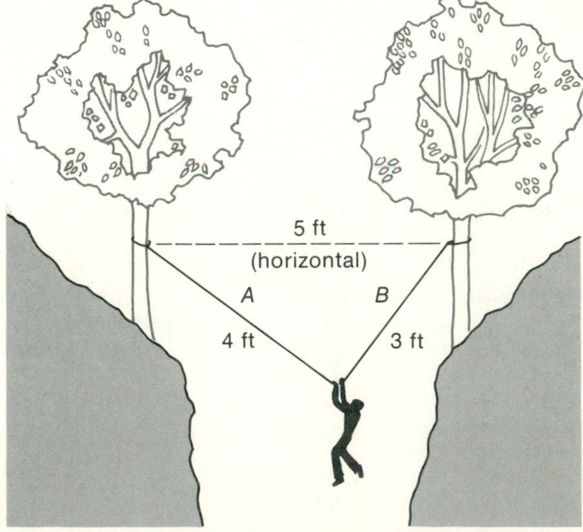

4A–7 A 160-lb man has an acceleration of 4 ft/s^2.

(a) Find the net force acting on the man.

(b) Determine how far he has traveled when his speed is 12 ft/s^2. Assume he starts from rest. *Answers:* (a) 20.0 lb (b) 18.0 ft

4A–8 A person is using a twisted sheet to escape from an upper story of a burning building by hanging the sheet vertically between the window and the ground. Explain how a 160-lb man could safely use it to descend to the ground even though the sheet has a breaking strength of only 140 lb.

4A–9 A 64-lb wagon is pulled along a horizontal road by a horizontal force of 4 lb. Friction forces are negligible.

(a) Calculate how far the wagon travels in 5 s, starting at rest.

(b) Calculate its speed at the end of this 5-s interval. *Answers:* (a) 25.0 ft (b) 10.0 ft/s

4A–10 In driving over a hill that has the curvature of a vertical circle of radius 18 m, a driver notices that she barely remains on the seat. Find the speed of the vehicle in kilometers per hour.

4A–11 Two blocks, connected by a string, are on a horizontal frictionless surface as shown in Figure 4–22.

(a) What horizontal force **F** will cause them to accelerate at 0.3 m/s^2?

(b) Find the tension T in the string between the blocks while they are accelerating. *Answers:* (a) 2.40 N (b) 0.900 N

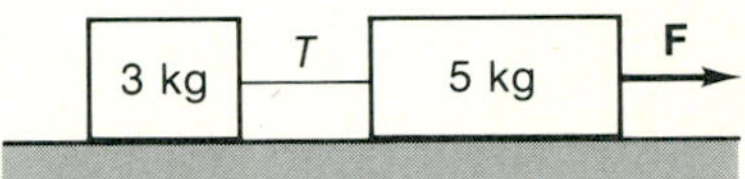

Figure 4–22
Problem 4A–11

4A–12 Two students suspend a 20-lb sign from the midpoint of a cable fastened at the same horizontal level on two buildings 100 ft apart. The point of suspension of the sign sags 1 ft below a line joining the two ends of the cable. Find the tension in the cable.

4A–13 One end of a 30-ft rope is tied to a tree; the other end is tied to an automobile stuck in the mud. The motorist pulls sideways on the midpoint of the rope, displacing it a distance of 2 ft. If the motorist exerts a force of 80 lb under these conditions, determine the force exerted on the automobile. *Answer:* 300 lb

4A–14 In a game of tetherball, the rope supporting a 10-lb ball is 5 ft long, as indicated in Figure 4–23.

(a) Find the tension in the rope as the ball swings in a horizontal circle of radius 3 ft. Include a free-body diagram for the ball, with forces resolved along two directions at right angles.

(b) Determine how long it takes the ball to make one complete revolution.

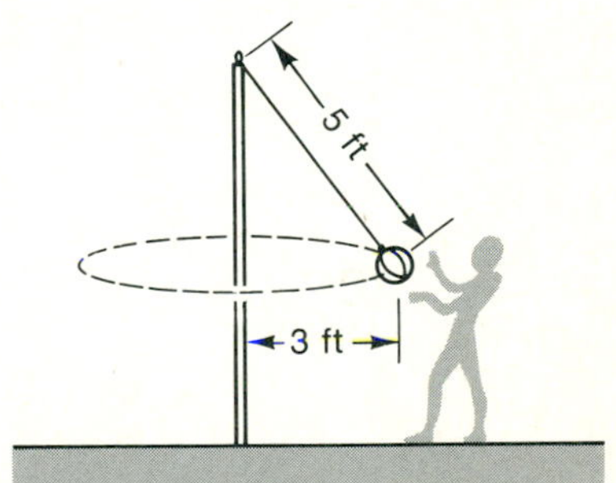

Figure 4–23
Problem 4A–14

4A–15 A 30-kg boy in a swing is pushed to one side and held at rest by a horizontal force **F** so that the swing ropes are 30° with respect to the vertical.

(a) Calculate the magnitude of **F**.

(b) Calculate the tension in each of the two ropes supporting the swing under these conditions. *Answers:* (a) 170 N (b) 170 N

4A–16 A box is set into motion sliding along a horizontal floor with an initial speed of 6 ft/s. It comes to rest 2 s later. Determine the coefficient of kinetic friction between the box and floor.

4A–17 A sled and its occupant weigh 120 N. The coefficient of kinetic friction between the sled and the ice is 0.070.

(a) Determine the horizontal force applied to the sled that will keep it moving with constant velocity.

(b) Determine the horizontal force that must be applied to give the sled an acceleration of 0.60 m/s^2. *Answers:* (a) 8.40 N (b) 15.8 N

4A–18 In the previous problem, if it requires a horizontal force of 10 N to start the sled moving from rest, calculate the coefficient of static friction between the sled and the ice.

4A–19 A block is placed on a plane inclined at 60° with respect to the horizontal. If the block slides down the plane with an acceleration of $g/2$, determine the coefficient of kinetic friction between the block and plane. *Answer:* 0.732

4A–20 In Figure 4–24, two boxes are in contact with each other on a horizontal frictionless surface.

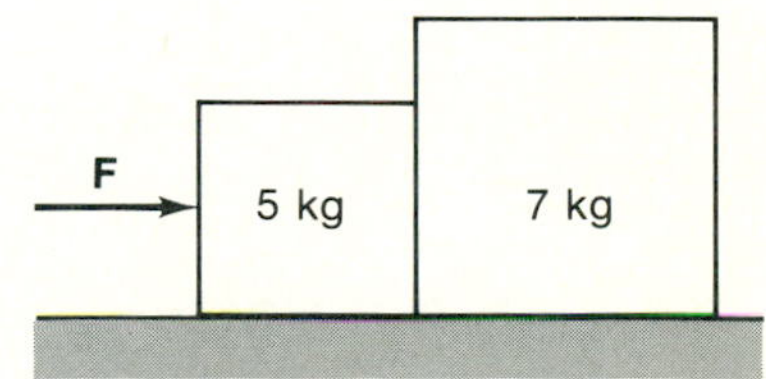

Figure 4–24
Problem 4A–20

(a) Determine the horizontal force F applied to the 5-kg box that will give the system an acceleration of 2 m/s^2.

(b) Under the conditions of part (a), determine the force the 7-kg box exerts on the 5-kg box.

4B–1 Two forces act on an 8-lb object: the downward force of gravity and a constant horizontal force. No other forces are present. Starting at rest, the object is observed to have an acceleration of 40 ft/s^2. Find (a) the magnitude of the horizontal force and (b) the direction of the acceleration. (c) Is the motion of the object along a straight line or a parabola?

Answers: (a) 10 lb
(b) 53.1° below the horizontal
(c) A straight line

4B–2 A rope has a breaking tension of 250 N. Find the minimum acceleration at which a boy weighing 300 N can safely slide down the rope.

4B–3 A child stands on the surface of a frozen pond, 12 m from the shore. If the coefficient of static friction between her boots and the ice is 0.05, determine the minimum time required for the child to reach the shore without slipping. *Answer:* 7.00 s

4B–4 A roller coaster car travels over the top of a circular section of track in a vertical plane. (The curve is concave downward.) The radius of curvature is 20 ft, and the total weight of the car plus passengers is 3000 lb. At the top of the curve, the speed of the car is 20 ft/s.

(a) What is the acceleration of the car (including direction) just as it travels over the topmost point?

(b) At this instant, what net force acts on the car plus passengers?

(c) What force does the track exert against the car at the top of the curve?

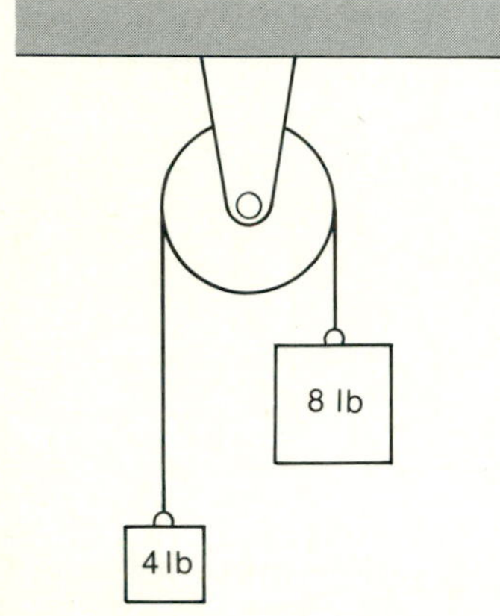

Figure 4–25
Problem 4B–5

4B–5 Figure 4–25 depicts two blocks connected by a string of negligible mass that passes over a frictionless pulley. The blocks are released from rest.

(a) By applying Newton's laws, find the acceleration of the blocks.

(b) Determine the tension in the string while the blocks are accelerating.

(c) Determine how fast the 8-lb block is going after it travels from rest a distance of 6 in.

Answers: (a) 10.7 ft/s^2 (b) 5.33 lb (c) 3.27 ft/s

4B–6 Assume $g = 10$ m/s^2 and sketch free-body diagrams for each of the masses in Figure 4–26. Find the upward acceleration of the 4-kg block for each case.

4B–7 A mass is suspended from the ceiling by a string of length ℓ. It is set into motion so that the string makes a constant angle θ with the vertical while the mass moves at constant speed in a horizontal circle, as illustrated in Figure 4–27. Find the time t for the mass to make one complete revolution in terms of ℓ, θ, and g.

Answer: $t = 2\pi\sqrt{\dfrac{\ell \cos\theta}{g}}$

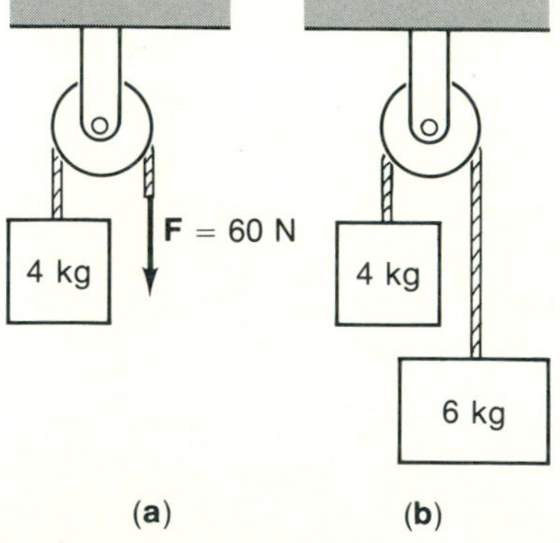

Figure 4–26
Problem 4B–6

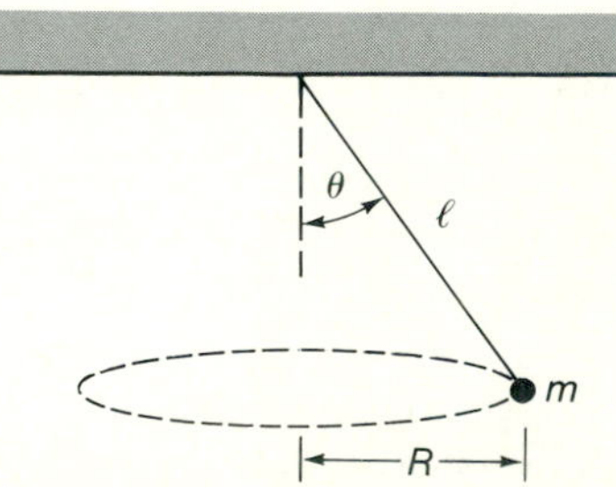

Figure 4–27
Problem 4B–7

4B–8 In Figure 4–28, calculate the weight W that will give the 3200-lb elevator an upward acceleration of 4 ft/s^2 if the constant friction force between the elevator and rails is 200 lb.

4B–9 For the situation illustrated in Figure 4–29, find the force **F** that will pull the box at constant speed along the floor. The coefficient of kinetic friction between the box and floor is 0.50, as indicated. *Answer:* 7.54 lb

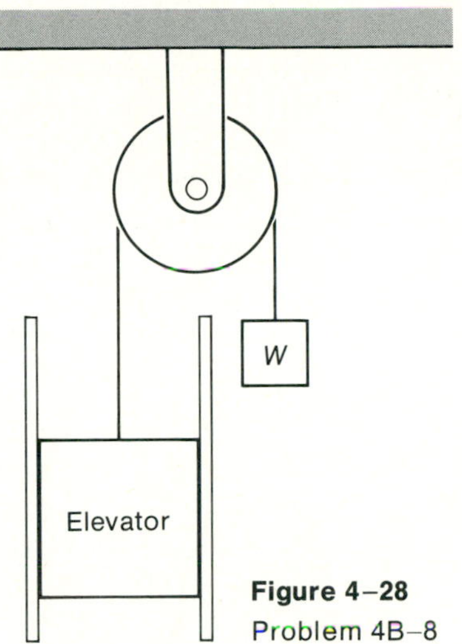

Figure 4–28
Problem 4B–8

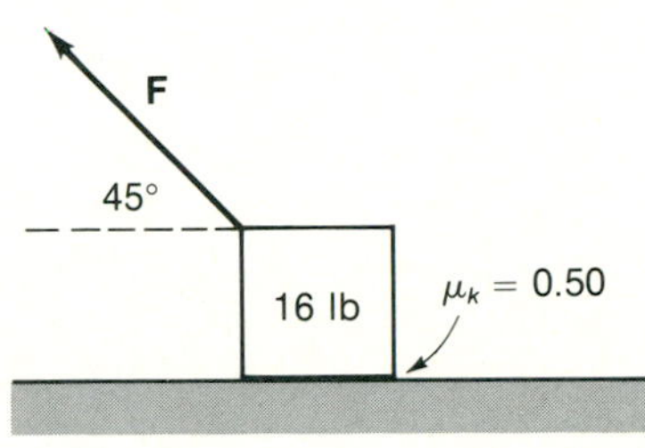

Figure 4–29
Problem 4B–9

4B–10 A 4-kg block is pulled along a horizontal surface by a force **F** of 20 N, as shown in Figure 4–30. Determine the acceleration of the block if the coefficient of kinetic friction between the block and the surface is 0.20.

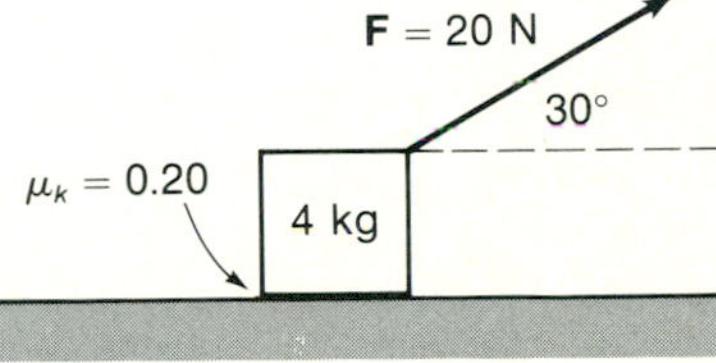

Figure 4–30
Problem 4B–10

4B–11 A 500-kg crate is being carried in a truck traveling horizontally at 15 m/s. If the coefficient of static friction between the crate and the truck is 0.40, determine the minimum stopping distance for the truck such that the object will not slide. *Answer:* 287 m

4B–12 In Figure 4–31, a 500-kg horse pulls a sledge of mass 100 kg, as shown. The combination has a forward acceleration of 1 m/s^2 when the friction force on the sledge is 500 N. Find (a) the tension in the connecting rope and (b) the force of friction on the horse. (c) Verify that the total force of friction the earth exerts will give to the total system an acceleration of 1 m/s^2.

Figure 4–31
Problem 4B–12

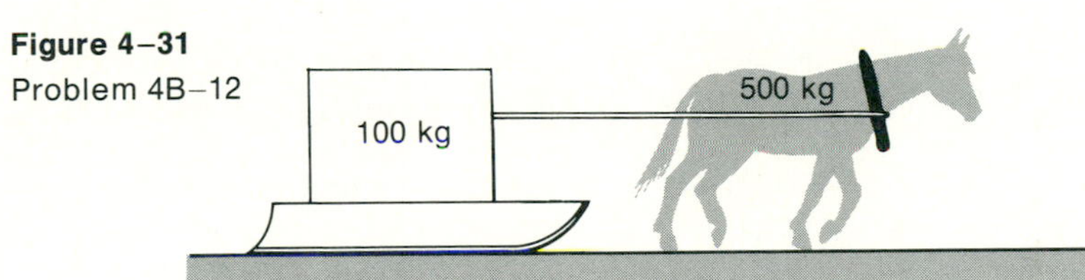

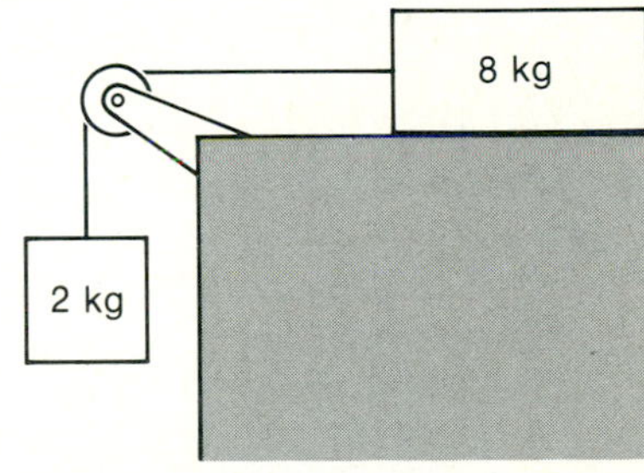

Figure 4–32
Problem 4B–13

4B–13 Two blocks are connected by a light cord, as illustrated in Figure 4–32. Assume that the pulley has negligible friction and that $g = 10$ m/s^2. If the blocks are released from rest, (a) determine the acceleration of the blocks and (b) find the tension in the connecting cord. *Answers:* (a) 2.00 m/s^2 (b) 16.0 N

4B–14 Two blocks are connected together by a cord, as shown in Figure 4–33. The coefficient of kinetic friction between the blocks and the level surface is 0.50.

(a) Find the force **F** that will give the system an acceleration of 2 m/s^2.

(b) Find the tension T in the connecting cord between the blocks. Assume $g = 10$ m/s^2.

Figure 4–33
Problem 4B–14

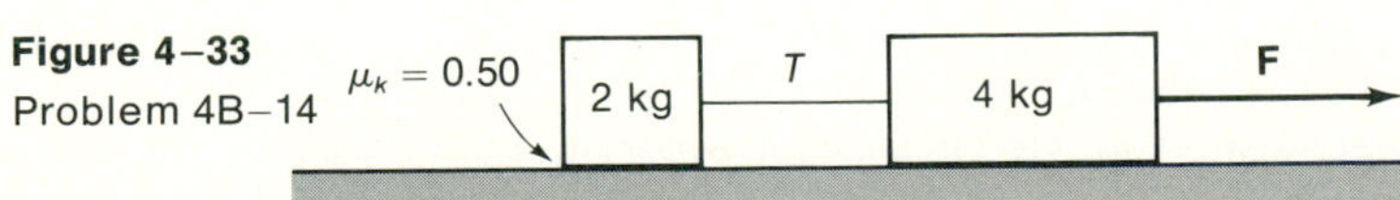

4B–15 Once it is set into motion, a block slides at constant speed down an incline at 20° with respect to the horizontal. Find the coefficient of kinetic friction between the block and the incline. *Answer:* 0.364

4B–16 A 5-kg box slides down a plane inclined at 41° with respect to the horizontal.
(a) Determine the force of friction on the box if the coefficient of kinetic friction is 0.3.
(b) Determine the acceleration of the box down the incline.

4B–17 This problem asks you to compare the maximum friction force required for two different situations. In the first instance, a car is traveling along a straight road with initial speed v_0 when the brakes are applied, bringing the car uniformly to rest with an acceleration of magnitude a. In the second instance, the car is traveling around a curve of radius r on a level road with the same initial speed v_0 when the brakes are applied, bringing the car uniformly to rest in the *same distance* traveled along the road as in the first case. Show that the maximum friction force between the road and the tires is *greater* for the curved road by a factor of

$$\sqrt{1+\frac{1}{r}\left(\frac{v_0^2}{a}\right)^2}$$

(This explains why applying the brakes while traveling around a curve must be done with extra caution.)

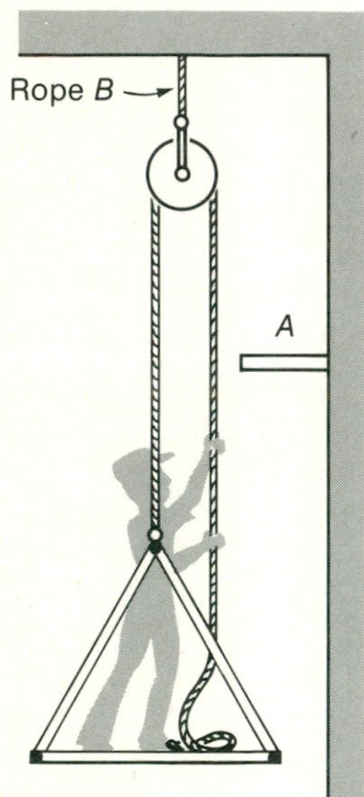

Figure 4–34
Problem 4B–18

4B–18 To paint the side of a building, a painter normally hoists himself up by pulling on the rope as shown in Figure 4–34, fastening the free end of the rope to his scaffold. The painter and scaffold together weigh 200 lb, and the rope B supporting the pulley can withstand up to 300 lb without breaking. On one occasion, however, the painter decides to fasten the free end of the rope to a very firm and secure pipe, which projects from the side of the building at A. Describe and explain the immediate disaster that befalls the painter.

4B–19 An automobile travels with speed v_0. If the coefficient of static friction between the road and tires is μ_s, show that the shortest distance in which the automobile can be stopped without skidding is $v_0^2/2g\mu_s$. (Note: Since generally $\mu_k < \mu_s$, if skidding occurs the distance increases. For this reason, in emergency stops it is wise to apply the brakes as strongly as possible without locking the wheels.)

4B–20 In Figure 4–35, the man and platform together weigh 180 lb. Determine how hard the man would have to pull to hold himself off the ground. (Or is it impossible? If so, explain why.)

4B–21 An amusement-park ride is in the form of a large cylinder with its central axis vertical. Riders position themselves against the inside curved wall of the cylinder, and the cylinder starts to rotate about its axis. After it acquires a certain rotational speed, the floor is lowered, leaving the riders "pinned" against the wall; frictional forces prevent the riders from sliding downward.
(a) Sketch a free-body diagram for a rider under these conditions.
(b) If the radius of the cylinder is R, derive an expression for the minimum value of μ_s, the coefficient of static friction between the riders and the wall, when the speed of the riders is v. *Answer:* (b) gR/v^2

Figure 4–35
Problem 4B–20

4B–22 A 2000-lb car travels 60 mi/h on an unbanked circular road with a radius of 400 yards.
(a) Determine the total force of friction the road exerts on the car.
(b) Determine the angle θ at which the road should be banked in order to make the friction force zero at this speed.
(c) If a car traveled 30 mi/h on the road banked at the angle in part (b), find the magnitude of the friction force on the car. (Hint: In what direction is this friction force?)

4B–23 This question refers to *The Great American Revolution* roller coaster (see Figure 4–19 for technical data).
(a) Is it necessary to wear a seat belt to prevent falling out of the car as it travels around the top of the loop? Explain.
(b) If the answer to part (a) is "yes," find the force of the seat belt on a 70-kg passenger at the top of the loop. If the answer is "no," find the force of the seat on the passenger. *Answer:* (b) 359 N

4B–24 Again referring to Figure 4–19, assume that the speed of the car at one side of the loop (traveling down) is 15 m/s and that the loop is circular at that point with a radius of 6.3 m.

(a) Determine the force, if any, that the seat exerts on a 70-kg passenger there.

(b) Determine the direction of the net acceleration of the passenger in part (a).

4C–1 An amusement-park "fun house" contains a rotating cylinder of radius 2 m. A child is sitting partway up one side so that she remains at rest a distance h above the bottom of the rotating cylinder (see Figure 4–36). If the coefficient of kinetic friction between the child and cylinder is 0.40, find the distance h. *Answer:* 0.143 m

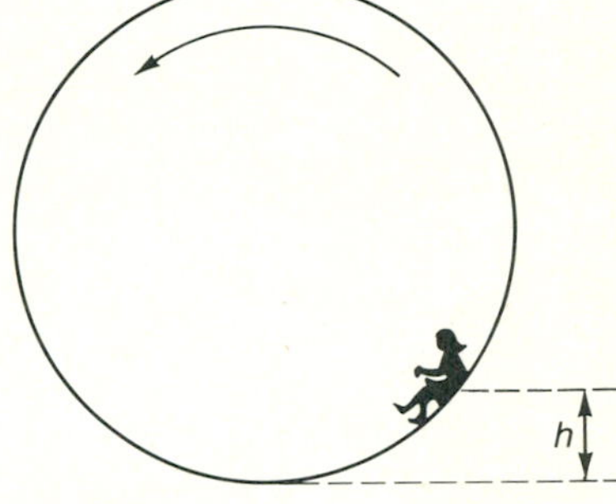

Figure 4–36
Problem 4C–1

4C–2 The following series of events is from a classic comic sequence in a motion picture of the 1920s. A farmer attempts to lower a small barrel of bricks from the upper-story window of a barn by means of a pulley arrangement, as shown in Figure 4–37. By suddenly applying his full weight to the end of the rope, the farmer dislodges the barrel from the windowsill. He hangs on tightly to the rope, but unfortunately the loaded barrel weighs more than the farmer, so it pulls him up in the air while it descends. Upon striking the ground, the bottom of the barrel breaks, spilling the bricks on the ground. The farmer descends rapidly while the broken barrel is pulled into the air. When the barrel strikes the pulley, it breaks apart and the barrel staves fall freely down upon the prostrate farmer below. Using the data listed, calculate the total elapsed time for the sequence of events. The windowsill is 2 m below the pulley and 10 m above the ground. Assume $g = 10 \text{ m/s}^2$, and ignore the weight of the rope.

	Mass
Farmer	80 kg
Loaded barrel	100 kg
Empty (broken) barrel	40 kg

Note: Some clarifications are necessary. When the loaded barrel strikes the ground, the farmer continues to move upward, slowed only by the force of gravity. He misses the pulley both in his upward flight and subsequent downward flight. Still holding the rope while moving downward, he launches the broken barrel upward. Assume that the sudden tension on the rope slows his speed to two-thirds of its value. Finally, when the broken barrel (moving upward) hits the pulley, assume that the pulley brings it instantaneously to rest, allowing the fragments to fall freely.

(Though this should not be included with the problem, the final indignity occurred when the farmer finally let go of the rope: The metal hook on the other end of the rope had sufficient weight to descend, pulling the entire rope through the pulley so it also tumbled down upon the stunned farmer.)

Figure 4–37
Problem 4C–2

4C–3 Standing on tiptoe as illustrated in Figure 4–38 produces considerable tension in the Achilles tendon, attached near the heel. The three forces on the foot (**T**, **A**, and **W**)

Figure 4–38
Problem 4C–3

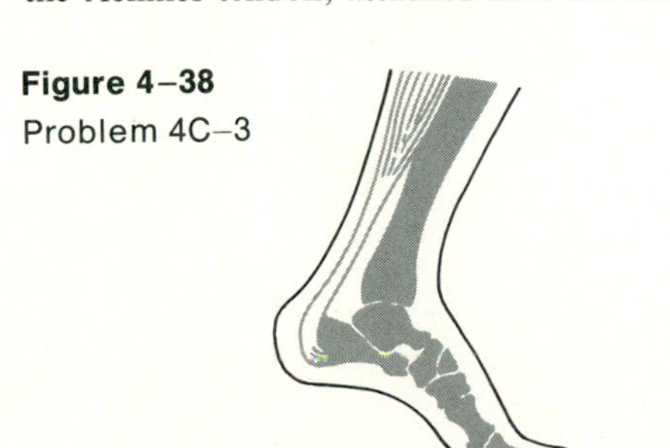

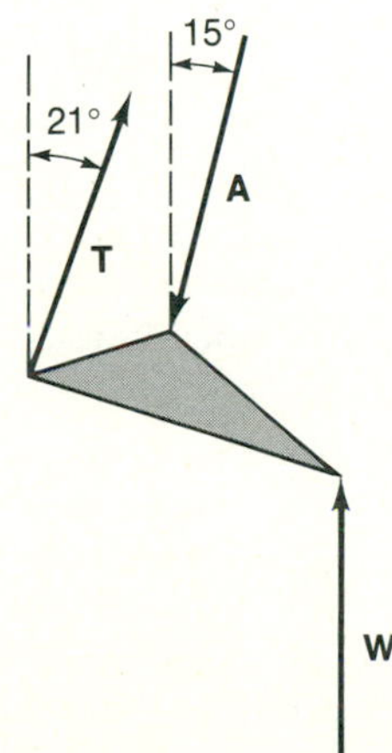

are in equilibrium; hence a graphical addition of the forces forms a closed triangle. Find the tension in this tendon when a 180-lb man stands on one toe in this manner. (Hint: See Appendix D for the solution of a triangle with arbitrary angles.) *Answer:* 446 lb

4C–4 In a drag race, a motor car is driven from a standing start over a quarter-mile distance (1320 ft) in the shortest possible elapsed time. The terminal speed at the end of the course is (approximately) measured by photocells placed symmetrically on either side of the finish line, spaced 132 ft apart. Suppose a 3600-lb car covers the distance in 15 s, with a terminal speed of 120 mph as measured by the photocells. Assuming constant acceleration, find the following:

(a) The friction force that was present.
(b) The average coefficient of friction between the tires and the road. (The word "average" is used because some slipping occurs during the initial acceleration period.)
(c) The average horsepower the motor put out. (The estimate will be too low because some of the energy went into rotational motion of the flywheel, crankshaft, wheels, and so on, which may be ignored here.)
(d) If the terminal speed is calculated as (132 ft)/(time to travel between the photocells), determine the difference in speeds between the instants of passing the first and second photocells.

(For an interesting analysis of drag racing, see G. T. Fox, "On the Physics of Drag Racing," *American Journal of Physics* **41**, 311, March 1973.)

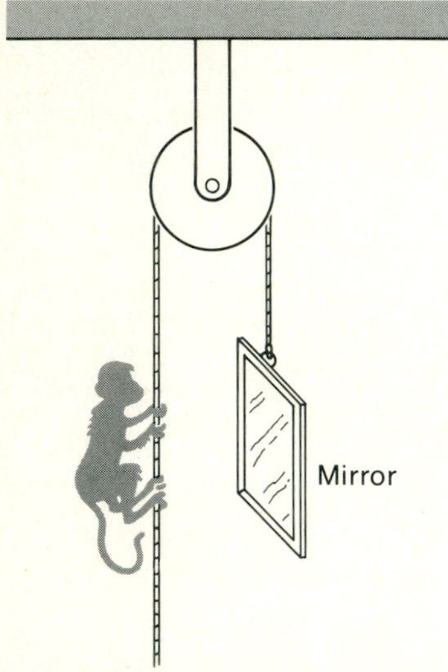

Figure 4–39
Problem 4C–6

4C–5 A block of mass m starts at rest and slides without friction down an incline at 30° with respect to the horizontal.

(a) Determine how long it takes to acquire a speed of 50 cm/s.
(b) Determine how far it travels during this time. Assume $g = 10\ \text{m/s}^2$.

Answers: (a) 0.100 s (b) 0.025 m

4C–6 A humorous illustration of Newton's laws involves a frictionless pulley with a rope of negligible mass supporting a monkey on one side of the pulley and a mirror of equal mass on the other side (see Figure 4–39). Explain why the monkey, starting at rest and then climbing up or down the rope, always remains opposite to the mirror so that he continually sees his own reflection.

4C–7 In Figure 4–40, determine the mass m that will allow the 10-kg block to accelerate down the frictionless incline at $0.2\ \text{m/s}^2$. Assume that the pulley has no friction and that $g = 10\ \text{m/s}^2$. *Answer:* 4.71 kg

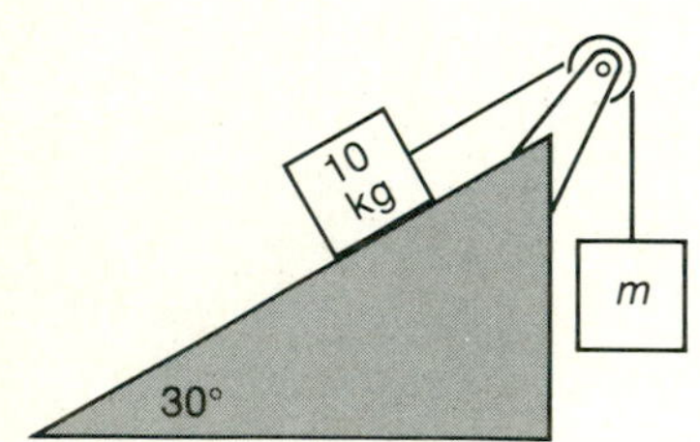

Figure 4–40
Problem 4C–7

4C–8 An airplane can be flown along a parabolic trajectory (sometimes called a "free-fall" flight path) to simulate a gravity-free environment for passengers in the plane's frame of reference. Using both the up and down portions of the parabola, flights under these conditions have been sustained for longer than a minute. With some oversimplification, the forces on the moving plane may be separated into:

(1) the gravitational force
(2) the engine thrust (forward along the plane's axis)
(3) air drag (a frictional force opposite to the direction of motion)
(4) airlift (perpendicular to the plane's axis)

(a) Determine the condition placed upon the horizontal component of velocity throughout the free-fall trajectory.
(b) Make free-body diagrams and explain why the plane's axis cannot be pointed along the flight path while climbing or descending.

4C–9 As indicated in Figure 4–41, the coefficient of static friction between the top and bottom blocks is 0.40. Find the maximum horizontal force **F** that will accelerate the blocks toward the right without causing slipping between the blocks. The horizontal surface is frictionless. *Answer:* 31.4 N

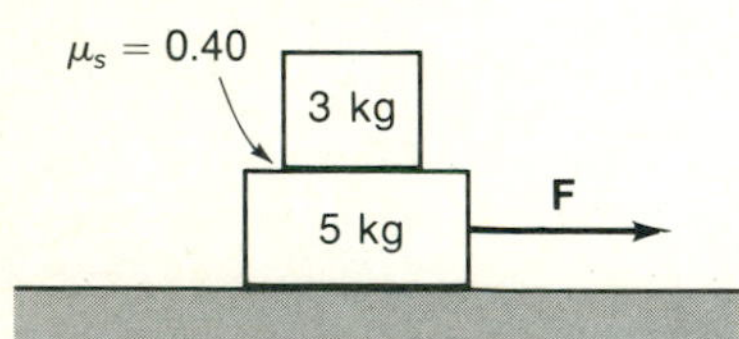

Figure 4–41
Problem 4C–9

4C–10 The following expression gives the force (in newtons) on a sphere of radius r (in meters) exerted by a stream of air with speed v (in meters per second):

$$F = arv + br^2v^2$$

where a and b are constants with appropriate SI units. Their numerical values are $a = 3.1 \times 10^{-4}$ and $b = 0.87$. Using this formula, find the terminal speed for water droplets falling under their own weight in air, taking these values for the drop radii:

(a) 10 μm
(b) 100 μm
(c) 1 mm

Note that for (a) and (c) it is possible to obtain accurate answers, without solving a quadratic equation, by considering which of the two contributions to the air resistance is dominant and ignoring the lesser contribution.

(d) Find the radius of the falling drop such that the two contributions are equal.
(e) Find the terminal speed of a drop having the radius given in answer to part (d).

4C–11 In Figure 4–42, the external force **F** causes the 4-kg block to accelerate at 0.50 m/s^2.

(a) Taking into account the coefficients of kinetic friction indicated, find the tension in the cord connecting the two blocks during this motion.
(b) Find the magnitude of **F**. To simplify the calculation, assume $g = 10$ m/s^2.

Answers: (a) 5.00 N (b) 17.0 N

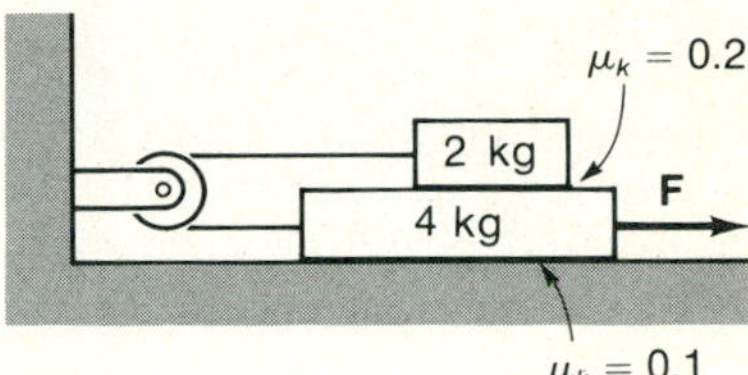

Figure 4–42
Problem 4C–11

4C–12 An amusement-park ride consists of a circular array of small cubicles in which riders stand, leaning against the outer wall as the apparatus rotates in a horizontal plane (see Figure 4–43). While rotating at constant angular speed, the apparatus is slowly tilted to an angle of 60° with respect to the horizontal. If the riders are carried around in a circle of radius 16 ft, determine the minimum rotational speed in revolutions per second that will prevent riders from falling out, even without restraining straps and with no friction between the riders and the floor.

Figure 4–43
Problem 4C–12

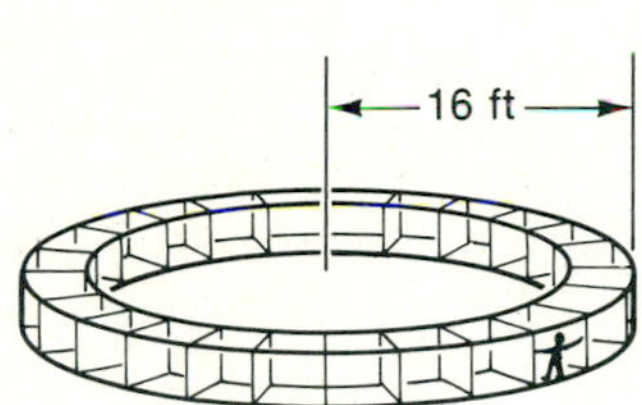

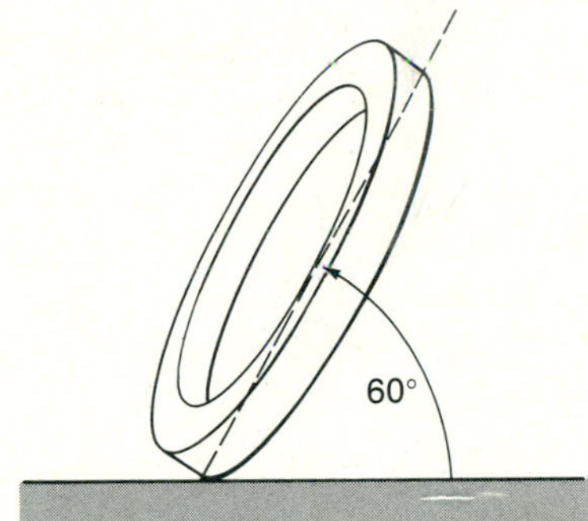

Perspective

Sometimes it is easy to become overwhelmed by the many trees in the forest and fail to see the structure of a subject as a whole. In particular, physics has a few giant trees that rise above the surrounding landscape—central ideas that unify and simplify the subject. It will be helpful in our progress to occasionally stand back from the subject and emphasize these important ideas. This is the first of four such commentaries.

In trying to understand nature's behavior, we first established a precise and convenient way of describing *motion*:

(1) the definitions of position, velocity, and acceleration

(2) the kinematic equations

Then, thanks to Newton's genius, we have a classical description of how forces produce a change in the motion of an object:

(3) Newton's laws (with particular attention to $\Sigma\mathbf{F} = m\mathbf{a}$)

It must be emphasized that $\Sigma\mathbf{F} = m\mathbf{a}$ can be applied only if we understand the meaning of $\Sigma\mathbf{F}$.

A free-body diagram of all the forces acting *on* an object is the first essential step in all problems we have discussed so far. This preliminary analysis of forces is where the physics lies. Knowing the forces, the remaining procedures are generally just algebraic. So concentrate on always approaching problems from this viewpoint: look first at the forces acting on an object.

But sometimes we do not know the forces involved. Or they are so complicated (or they act for such a short time, as in a collision) that we cannot deal with them directly. The next few chapters will introduce two different approaches that are useful in these cases. These new ideas are *the conservation of energy and momentum* and *the relation between a conservative force and a potential.* In this introductory course, we will learn just three basic methods of analyzing problems in mechanics:

(1) Newton's laws
(2) Ideas of energy and momentum
(3) Relations between a conservative force and a potential

Keep your eye on these important themes, which appear again and again as we discuss a variety of topics. Learn to spot the clues in a physical situation that tip you off as to which method may be the most effective—it will greatly simplify the solving of problems. This is the physicist's approach: analyzing a wide variety of phenomena in terms of just a few powerful concepts.

I love work. I can sit and watch it for hours!

ANONYMOUS

chapter 5 Work and Energy

5.1 Introduction

Energy is an extremely important concept in physics. Although it appears in many different forms, it is not a physical substance, but a calculated quantity. It is everywhere, cropping up in various forms in all branches of physics. Because energy can appear in such different guises, we need to discuss it from numerous different viewpoints.

Though we often speak of an object as "possessing" a certain amount of energy, and calculate how energy is transferred from one object to another, the amount of energy an object has depends crucially on the frame of reference we use. The energy content is different in different frames. For example, if we observe an object moving in our frame of reference, we calculate a finite kinetic energy for the object. But in a frame of reference moving along with the object (and thus the object is at rest in that frame), its kinetic energy is zero. So energy is not a kind of substance that an object may possess more or less of. Instead, it is a calculated quantity based on measurements made in a given frame of reference.

The reason energy plays such an important role in physics is that it is conserved. Once we determine the energy content of an isolated system, regardless of how that system changes in the future *the total energy remains the same.* Thus, energy can only be changed from one form to another; it cannot be created or destroyed. This conclusion, based on experiment, is known as the **law of conservation of energy.**

Conserved quantities—those that remain the same even though various physical changes take place—are important concepts for the physicist. They are powerful because of their generality. A conservation statement says nothing about *how* a specific process may take place. But it guarantees that in every possible instance, if we calculate the quantity before a process occurs, it will have the same value afterward. This fact alone is often a great aid in the analysis of physical phenomena. Using energy concepts, we frequently can solve problems in dynamics more easily than by a direct application of Newton's laws. Basically, the energy concept is a useful accounting scheme that helps keep track of what happens when objects interact.

5.2 Work

The first example of energy we shall take up is work. The concept of work has a more restricted meaning in physics than it does in common usage. As you will see, it represents the *transfer* of energy from one system to another. The simplest case is for a constant force **F** moving through a straight-line displacement D, as illustrated in Figure 5–1(a). Here, the **work** W is defined as follows:

The work *W* done by a constant force F is the product of two factors: the component of force parallel to the direction of the displacement *D*, and the magnitude of the displacement.

$$\text{Work} = (\text{Force})(\text{Displacement}) \qquad W = (F)(D) \qquad \text{(for } \mathbf{F} \text{ constant and in the direction of a straight-line displacement } D\text{)} \tag{5-1}$$

If the force **F** is at an angle θ with respect to the displacement, as in Figure 5–1(b), we have

$$W = (F\cos\theta)(D)$$

or

$$W = (F_{\parallel})(D) \qquad \text{(for } \mathbf{F} \text{ constant and at an angle } \theta \text{ with respect to a straight-line displacement } D\text{)} \tag{5-2}$$

where $F_{\parallel}$ is the component of **F** parallel to the direction of the displacement.

In the SI metric system, the unit of work is the **newton-meter** (N·m), which is also given the name **joule** (J).[1] The **foot-pound** (ft·lb) is the customary unit in the British engineering system. The dimensions of work are

$$(\text{Force})(\text{Distance}) \Rightarrow \left[\frac{M\cdot L}{T^2}\right][L] = \left[\frac{ML^2}{T^2}\right]$$

Using the symbol W for work usually causes no confusion with our previous use of the same symbol for weight, the force of gravity. The context of a problem makes the distinction clear. It is also helpful to use the symbol F_{grav} or mg for the force of gravity.

Note that if D is equal to zero, the work done by **F** is also zero. A great deal of physical exertion may occur in pushing against a building, but unless

[1] This unit is named in honor of the son of a wealthy English brewer, an amateur physicist, Sir James Joule (1818–1889). His name is pronounced "jowl," though in the United States the unit is most frequently pronounced "jool." Perhaps motivated by a desire to obtain cheaper power for his father's breweries as well as by a love of experimentation, at the age of 19 Joule began to investigate the relation between the work needed to drive an electric generator and the amount of heat produced by the electric current generated. Five years later his results were presented to the British Association, where they met with general silence. The following year, a paper on the same subject was rejected by the Royal Society. It was an additional five years until Joule's important treatise "On the Mechanical Equivalent of Heat" was accepted in the *Philosophical Transactions*. Even then, the committee to whom the paper had been referred deleted Joule's (correct) claim that friction involved the conversion of mechanical work into thermal energy.

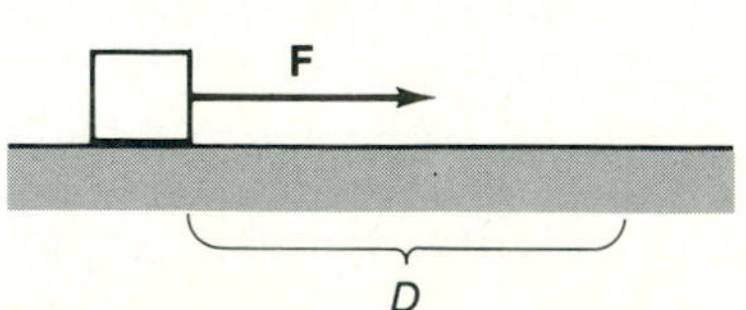

(**a**) **F** parallel to the displacement

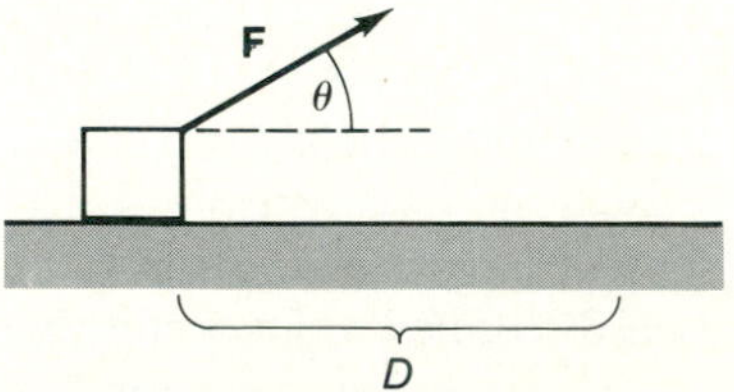

(**b**) **F** at a constant angle θ with respect to the displacement

Figure 5–1
A constant force **F** moves through a straight-line displacement D.

the building moves, no work is done in the physics sense. True, our tensed muscles involve energy transfers in a biological sense (between microscopic muscle units and their surroundings), where chemical energy is transformed into mechanical tensing of the muscle tissue. As a result of the waste products of these chemical processes, our muscles get tired and we may begin to get warm and perspire. However, unless we exert *a force through a distance*, the work, as defined in physics, is zero.

Table 1–1 (page 2) provides interesting information regarding the energy transfers involved in various processes.

EXAMPLE 5–1

A broom is pushed across the floor by a constant force **F** of 50 N directed downward along the handle of the broom, making an angle of 60° with the floor. Find the work done by the force in moving the broom a distance of 3 m along the floor.

SOLUTION

Because the force and the angle remain constant, we use Equation (5–2).

$$W = (F_{||})(D) = (F)(\cos\theta)(D)$$

$$W = (50\text{ N})(0.500)(3\text{ m}) = 75\text{ N}\cdot\text{m}$$

Since the newton-meter is called the joule:

$$\boxed{W = 75.0\text{ J}}$$

5.3 Work and the Scalar Product of Vectors

As mentioned previously, work is defined in terms of the product of a force and a displacement. Although both of these quantities are vectors, work is the product of the *component* of the force in the direction of the displacement and the *magnitude* of the displacement. This is simply the product of two scalar quantities, and it therefore yields a scalar result.

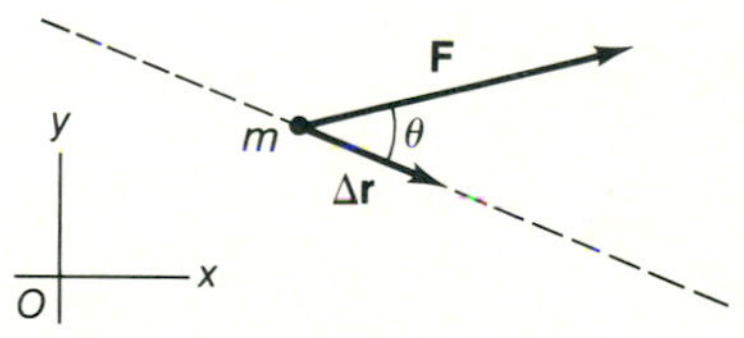

Figure 5–2
A force **F** moves a particle of mass *m* a distance **Δr**.

For discussing work, it is convenient to define a vector multiplication that produces a scalar quantity. Suppose a constant force **F** moves a particle through a displacement **Δr** as shown in Figure 5–2. Only the *component* of **F** in the direction of **Δr** does work on the particle. One way of describing this would be as follows:

$$\Delta W = F\,\Delta r\cos\theta \qquad (5\text{–}3)$$

where θ is the angle between the forward directions of **F** and **Δr**.

Another convenient notation expresses the scalar product of the two vectors (**F** and **Δr**), as defined in vector algebra. This product is denoted by $\mathbf{F}\cdot\Delta\mathbf{r}$ and is defined as

$$\mathbf{F}\cdot\Delta\mathbf{r} \equiv |\mathbf{F}|\,|\Delta\mathbf{r}|\cos\theta$$
$$= F\,\Delta r\cos\theta$$

This is precisely the definition of work.

WORK $$\Delta W = \mathbf{F}\cdot\Delta\mathbf{r} \qquad \text{(for constant }\mathbf{F}) \qquad (5\text{–}4)$$

Note that this equation expresses work, ΔW, without reference to any particular coordinate system. This is one of the advantages of equations in vector form. They are true for *all* coordinate systems and are usually much more concise and easy to write than when written in a given coordinate system. For example, if $\mathbf{F}$ and $\Delta\mathbf{r}$ are expressed in rectangular coordinates, $\mathbf{F} = F_x\hat{\mathbf{x}} + F_y\hat{\mathbf{y}}$ and $\Delta\mathbf{r} = \Delta x\,\hat{\mathbf{x}} + \Delta y\,\hat{\mathbf{y}}$, then the work $\Delta W = \mathbf{F}\cdot\Delta\mathbf{r}$ becomes

$$\Delta W = F_x\,\Delta x + F_y\,\Delta y \tag{5-5}$$

In three dimensions, the corresponding expression is

$$\Delta W = F_x\,\Delta x + F_y\,\Delta y + F_z\,\Delta z \tag{5-6}$$

EXAMPLE 5-2

A boy pulls a sled by exerting a constant force of 20 N at an angle of 40° with the horizontal. What work does the boy do in pulling the sled a distance of 3 m?

SOLUTION

The force and the angle remain constant. Establishing the $+x$ direction in the direction of the displacement (*not* in the direction of the force), we apply

$$\Delta W = \mathbf{F}\cdot\Delta\mathbf{r} = F_x\Delta x = (F\cos\theta)(\Delta x)$$

$$\Delta W = (20\text{ N})(0.766)(3\text{ m}) = 46.0\text{ N·m} = \boxed{46.0\text{ J}}$$

Work may be either positive or negative. For example, hold a book in your hand. If you now raise the book a vertical distance h at constant speed, the force you exert on the book during its ascent is equal in magnitude to the weight of the book. Thus the work done is positive because the force is in the direction of the displacement. In lowering the book, the force you exert is still upward but the displacement is downward. In this case, the cosine of the angle θ in the dot product $\mathbf{F}\cdot\Delta\mathbf{y}$ is -1, and the work done by the force you exert is negative. If the book is moved along a horizontal path at a constant speed, the work done is zero because $\theta = 90°$ and $\cos 90° = 0$.

Of course, in raising the book, it is necessary to momentarily exert an upward force F slightly greater than the book's weight (mg) to initially set it in motion. And to bring it to rest at the top, the force, for a moment, must be slightly less than the weight of the book. A graph of $F(y)$ as a function of y might be as shown in Figure 5–3. The area under the curve—the product of $F(y)$ and y—represents the work done. But since the time to perform the task does not affect the work done, the slight changes in F at the beginning and end of the path may be made negligibly small. Therefore, in this case we may ignore the slight accelerations involved in starting and stopping the book. (Later we will show that regardless of the magnitude of the accelerations involved, the extra area under the curve at the start just balances the deficiency of the area at the end.) Thus, the total work is simply the rectangular area: $W = (mg)(h)$.

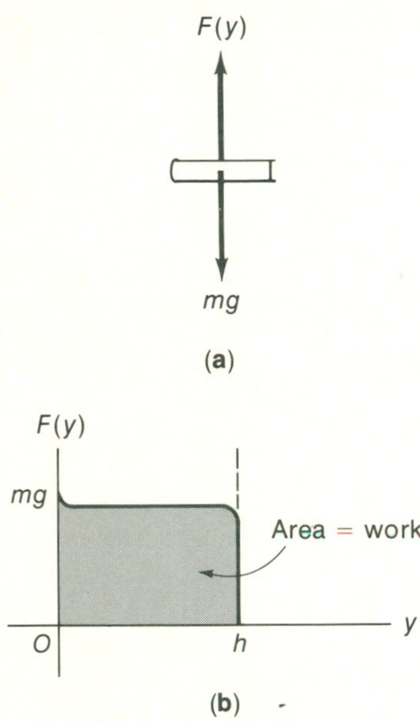

Figure 5–3
A book of mass m is raised vertically a distance h. The area under the curve is essentially the rectangular area of side lengths mg and h, representing the work W done by the external force $F(y)$. Thus, $W = (mg)(h)$.

5.4 Work Done by a Varying Force

In most natural phenomena, forces vary with time, with position, or with both. The concept of work developed in the previous sections for a constant force may be extended to include such varying forces.

We will first consider the case in which the force varies with the displacement along a straight line in the $+x$ direction. Here, we need to express the force $\mathbf{F}(x)$ as a function of position and investigate the incremental work ΔW done by the force as it moves along each incremental distance $\Delta\mathbf{x}$ of the path. The force $\mathbf{F}(x)$ on a particle depends upon the displacement x, as shown in Figure 5–4. The incremental work (ΔW) done by $\mathbf{F}(x)$ in moving the particle through an incremental distance Δx is given by

$$\Delta W = \mathbf{F}(x) \cdot \Delta\mathbf{x}$$

Since $\mathbf{F}(x)$ and $\Delta\mathbf{x}$ are always in the same direction, the dot product gives $\cos 0° = 1$, so we may write the equation as

$$\Delta W = F(x)\Delta x$$

In order to calculate the total work W required to move the particle from $x = a$ to $x = b$, it is necessary to sum the work done for each incremental change in displacement as the force moves from a to b.

Let us divide the total distance into k intervals: $\Delta x_1, \Delta x_2, \ldots, \Delta x_k$. Initially, the force moves the particle from $x = a$ to $x = a + \Delta x_1$. Although the force $F(x)$ varies slightly during this motion, we may approximate it closely by assuming that it is constant with the value $F(x_1)$, which it has at some intermediate position x_1 within this increment. Thus the work ΔW_1 for the first increment is approximately the area of the rectangle $F(x_1)\Delta x_1$. (Its value becomes more precise as Δx_1 becomes smaller.) Similarly, the work ΔW_2 for the second increment Δx_2 is $F(x_2)\,\Delta x_2$, and so on.

An approximation of the total work W done in moving from $x = a$ to $x = b$ is the sum of all the incremental areas, written analytically as

$$W \approx \sum_{n=1}^{n=k} F(x_n)\Delta x_n$$

where the interval from $x = a$ to $x = b$ is divided into k intervals, each Δx wide, and $F(x_n)$ is the value of the force assigned to the nth interval. This value of the total work becomes exact as the incremental distances become infinite in number while their lengths become infinitely small. That is:

$$W = \lim_{k\to\infty} \sum_{n=1}^{n=k} F(x_n)\Delta x_n \qquad \text{(from } x = a \text{ to } x = b\text{)} \tag{5–7}$$

where $F(x_n)$ is the value of the force at x_n. In the limit, the above equation is merely the *total area under the curve* of $F(x)$ versus x between $x = a$ and $x = b$.

$$\text{Work} = \text{Area under the curve of } F(x) \text{ versus } x \tag{5–8}$$

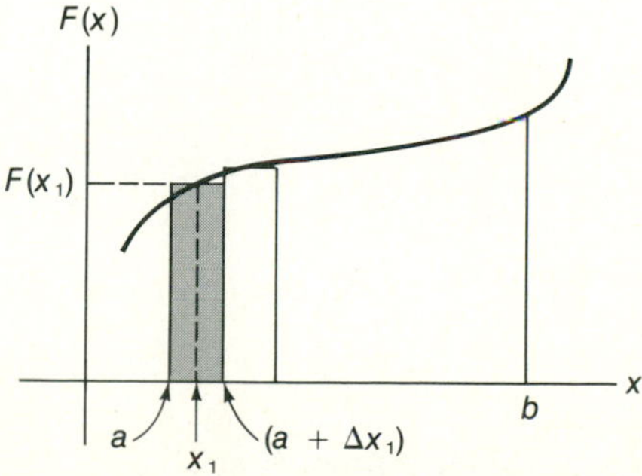

Figure 5–4
As the particle moves along the x axis from a to b, the force $F(x)$ changes its value.

The area under a curve may be calculated easily if the force $F(x)$ varies *linearly* with x. For more complicated cases, integration[2] is used.

A special point should be mentioned with regard to associating the area under a curve with a quantity such as work. In the calculation, the units that are assigned to each axis of the graph must be included. In this case, the area has units of *force* times distance, or (newtons)(meters) instead of the usual area units of $(\text{meters})^2$.

EXAMPLE 5–3

A flexible chain of length ℓ and mass m is slowly pulled at constant speed up over the edge of a table by a force $F(x)$ parallel to the edge of the table top as shown in Figure 5–5. What work is done by $F(x)$? Assume there is no friction between the table and the chain.

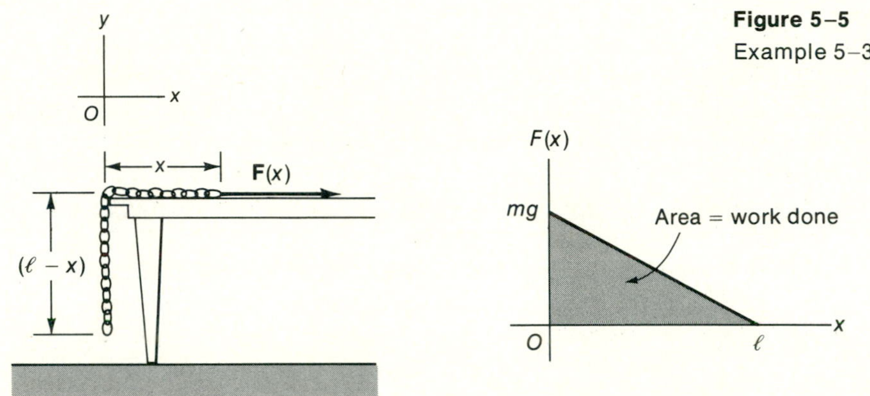

Figure 5–5
Example 5–3

SOLUTION

The force required to pull the chain varies with distance. It is equal to the weight of that fraction of the chain $[(\ell - x)/\ell]$ still hanging over the edge. (The table supports the

---

[2] For those students who already have studied integration, the calculus notation for Equation (5–7) is shown below:

$$W = \int_a^b F(x)\,dx$$

Graphically, this is depicted as:

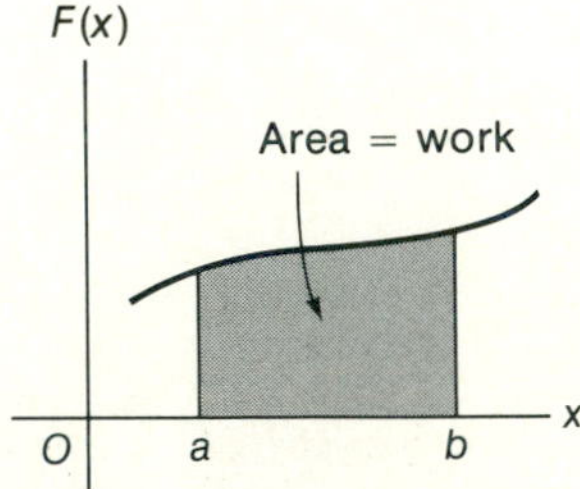

This means that in the limit the summation $\Sigma F(x_n)(\Delta x_n)$ from a to b becomes the integral $\int_a^b F(x)\,dx$. The integral represents the area under the curve and equals the work done by $F(x)$ in moving from a to b.

In more general form, this result may be expressed using vector notation for a straight-line integration along the x direction from x_1 to x_2:

$$W = \int_{x_1}^{x_2} \mathbf{F} \cdot d\mathbf{x} \tag{5–9}$$

Appendix G-III explains how to evaluate this integral.

other part of the chain.) Choosing the edge of the table to be $x = 0$, the force $F(x)$ is

$$F(x) = \left(\frac{\ell - x}{\ell}\right)mg$$

$$F(x) = \frac{mg}{\ell}(\ell - x) \qquad \textbf{(5–10)}$$

The total work done in exerting this force over the displacement from $x = 0$ to $x = \ell$ may be calculated by finding the area under the graph of $F(x)$ versus x. The graph is a straight line, as shown in Figure 5–5(b), and the triangular area under the graph is $\frac{1}{2}(mg)(\ell)$. So the total work done in pulling the chain onto the table is[3]

$$\boxed{W = \frac{mg\ell}{2}}$$

5.5 Spring Forces

To a good approximation, the force a spring exerts is proportional to the distance it is stretched or compressed from its relaxed length. Since a spring exerts a force always in the direction *opposite* to the displacement, it is called a **restoring force.** This behavior is known as **Hooke's Law,**[4] expressed

[3] An alternative solution for students who are familiar with integration is as follows:

$$W = \int \mathbf{F} \cdot d\mathbf{x}$$

We first account for the dot product by noting that **F** and $d\mathbf{x}$ are always in the same direction. So $\mathbf{F} \cdot d\mathbf{x} = |\mathbf{F}|\,|d\mathbf{x}| \cos 0°$. Since $\cos 0° = 1$, the expression under the integral is simply $F(x)\,dx$. [We write it as $F(x)$ rather than simply as F to emphasize that the force is a function of x.] Now, since we wish to integrate only over the range $x = 0$ to $x = \ell$, we show these limits on the integral sign in the usual notation.

$$W = \int_0^\ell F(x)\,dx$$

We next substitute the value for $F(x)$ and bring out from under the integral sign any factors that remain constant during the integration.

$$W = \int_0^\ell \left(\frac{mg}{\ell}\right)(\ell - x)\,dx = \left(\frac{mg}{\ell}\right)\int_0^\ell (\ell - x)\,dx$$

The first term under the integral sign involves the integral of the constant factor ℓ, and the second term is the integral of x itself. Referring to the table of integrals (Appendix G-II), we have $\int \ell\,dx = \ell x$ and $\int x\,dx = x^2/2$. Thus:

$$W = \left(\frac{mg}{\ell}\right)\left(\ell x - \frac{x^2}{2}\right)\Bigg|_0^\ell$$

The symbol $|_0^\ell$ means that we are to substitute the upper limit $(x = \ell)$ for every x in the expression, and subtract from this the value of the expression when we substitute the lower limit $(x = 0)$ for every x. This procedure completes the evaluation of the integral.

$$W = \left(\frac{mg}{\ell}\right)\left(\ell^2 - \frac{\ell^2}{2}\right) - \left(\frac{mg}{\ell}\right)\left(\ell(0) - \frac{0^2}{2}\right)$$

$$W = \left(\frac{mg}{\ell}\right)\left(\frac{\ell^2}{2} - 0\right) = \boxed{\frac{mg\ell}{2}}$$

For this problem, the integration method may seem more laborious than merely calculating (by inspection) the area under a straight-line curve. However, for more complicated functions of $F(x)$, it usually is not possible to calculate the area by inspection, so the integration method is the simplest and most straightforward.

[4] This law is named after the English physicist Robert Hooke (1635–1703), who maintained priority on his discovery regarding spring behavior by publishing his law in anagram form: *ceiiinosssttuv*. A few years later he divulged the solution as an alphabetical listing of the letters in the Latin sentence *Ut tensio, sic vis* ("As the extension, so the force").

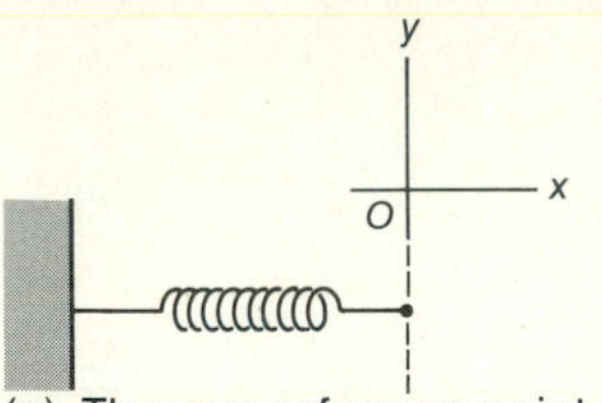

(**a**) The zero reference point is chosen at the end of the unstressed spring.

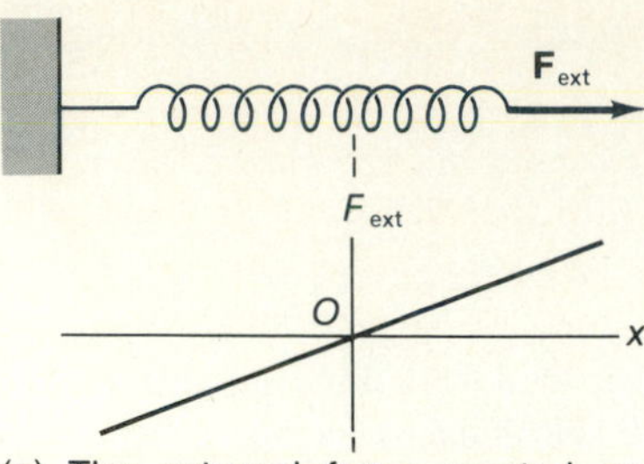

(**c**) The external force exerted on the spring is $F_{ext} = kx$

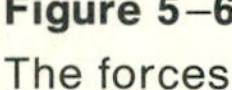

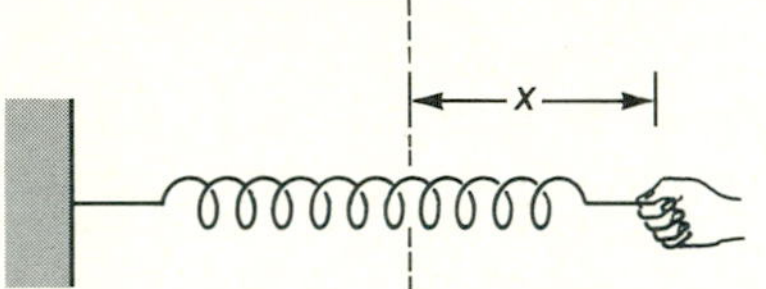

(**b**) The end of the spring is stretched a distance *x* from its unstressed position.

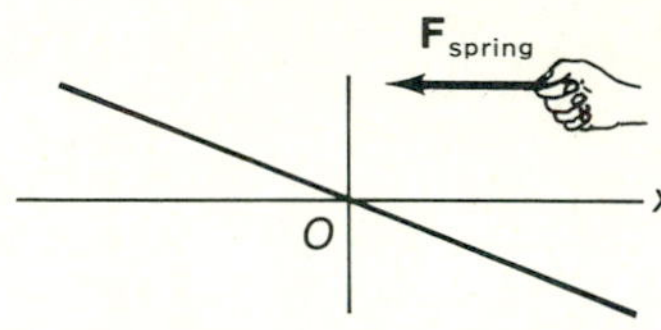

(**d**) The force exerted by the spring on the hand is $F_{spring} = -kx$

Figure 5–6
The forces involved when the end of a spring is stretched (or compressed) a distance *x* from its unstressed location. The slope of the plotted line is equal (in magnitude) to the spring constant *k*.

analytically as

HOOKE'S LAW $$F = -kx \qquad (5\text{–}11)$$

where k is the **spring constant,** a characteristic of the spring expressed in units of force/distance, and x is the distance the spring is stretched or compressed from its relaxed length.

If we apply an external force F_{ext} to change the length of a spring, this force is in the opposite direction to the spring force, namely, $F_{ext} = kx$. (Here, we assume the spring's mass is negligible and therefore may be ignored.) In solving problems involving springs, it is important to distinguish between forces exerted *on* the spring and forces exerted *by* the spring on some other object. Figure 5–6 clarifies this distinction.

EXAMPLE 5–4

The relaxed length of a certain Hooke's Law spring is 0.30 m. Suppose we apply an external force F_{ext} to stretch the spring. How much work is done by this force as the spring is stretched from 0.10 m to 0.20 m? The spring constant k is 30 N/m.

SOLUTION

We establish the zero reference for distance as the location of the free end when the spring is relaxed (see Figure 5–7). (If the zero is placed at some other location, the equation

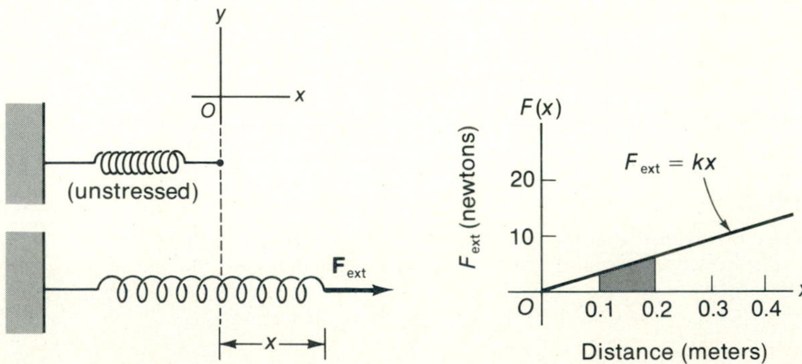

Figure 5–7
Example 5–4

for the force would not have the simple Hooke's Law form.) Thus, according to Hooke's Law, the equation for the force the *spring* exerts is $F = -kx = -(30\ \text{N/m})(x)$. The external force applied to stretch the spring has the same magnitude but is opposite in direction: $F_{ext} = kx = +(30\ \text{N/m})(x)$. *Considering the zero reference level at the free end of the relaxed spring*, we calculate the work done by F_{ext} in stretching the end of the spring from the position $x = 0.10$ m to $x = 0.20$ m by finding the area under this portion of the curve for $F(x)$ versus x. This shaded area in Figure 5–7 is found by subtracting the area of the smaller triangle ($x = 0$ to $x = 0.10$ m) from the area of the larger triangle ($x = 0$ to $x = 0.20$ m).

$$\begin{aligned} \text{Larger triangle:}\quad & \tfrac{1}{2}(0.20)k(0.20) = 0.020k \\ \text{Smaller triangle:}\quad & \tfrac{1}{2}(0.10)k(0.10) = 0.005k \\ & \text{Difference} = 0.015k \end{aligned}$$

Since $k = 30$ N/m, the area is $(0.015)(30) = 0.450$ in units of force times distance. Thus[5]:

$$W = \boxed{0.450\ \text{J}}$$

5.6 Kinetic Energy

Let us now consider the effect of work done on an object when the acceleration is not zero and the object's speed therefore changes. Figure 5–8 illustrates a mass m on a frictionless surface such as an air track. A horizontal force **F** is applied which, for our first case, we make constant in magnitude and direction. The downward force of gravity is exactly balanced by the upward force of the surface. Since friction is negligible, the *resultant net force* on the mass is simply the force **F**.

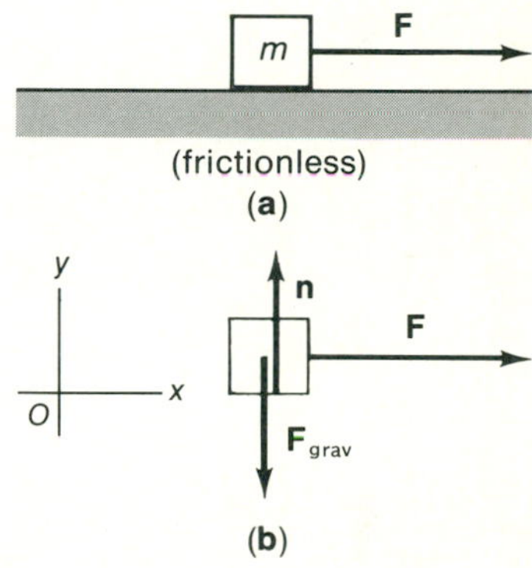

Figure 5–8
A mass *m* on an air track is subjected to a constant horizontal force **F**.

[5] An alternative solution using calculus is as follows:

$$W = \int \mathbf{F} \cdot d\mathbf{x}$$

Since **F** and $d\mathbf{x}$ are always in the same direction, the dot product gives $\cos 0^\circ = 1$. Therefore:

$$W = \int F(x)\,dx$$

Putting in the limits and substituting kx for $F(x)$, we have

$$W = \int_{0.10\ \text{m}}^{0.20\ \text{m}} kx\,dx = k\left(\frac{x^2}{2}\right)\Bigg|_{0.10\ \text{m}}^{0.20\ \text{m}}$$

We next substitute the upper limit ($x = 0.20$ m) for x and then subtract the value of the expression when we substitute the lower limit ($x = 0.10$ m) for x:

$$W = \left(\frac{k}{2}\right)(0.20\ \text{m})^2 - \left(\frac{k}{2}\right)(0.10\ \text{m})^2$$

$$W = \left(\frac{30\,\frac{\text{N}}{\text{m}}}{2}\right)(0.040\ \text{m}^2 - 0.010\ \text{m}^2)$$

$$W = (\tfrac{1}{2})(30)(0.030)\ \text{N·m}$$

Since 1 N·m is equivalent to 1 J, we have

$$W = \boxed{0.450\ \text{J}}$$

Because this net external force **F** is constant, we have constant acceleration a, and the mass will thus move according to the kinematic equation

$$v = v_0 + at$$

or, rearranging:
$$a = \left(\frac{v - v_0}{t}\right) \tag{5–12}$$

The distance x moved while changing from v_0 to v is given by the kinematic equation

$$x = v_0 t + \tfrac{1}{2}at^2$$

When we substitute the above value for a into this expression and simplify, we obtain

$$x = v_0 t + \frac{1}{2}\left(\frac{v - v_0}{t}\right)t^2$$

$$x = \tfrac{1}{2}(v + v_0)t$$

We can now evaluate the work W done on the mass by the net force F in moving the distance x.

$$W = (F)(x)$$

For F we may write $ma = m(v - v_0)/t$, and for the distance x we have $x = \frac{1}{2}(v + v_0)t$. Substituting these values into the expression for work, we have

$$W = (ma)(x) = m\left[\frac{v - v_0}{t}\right]\left[\frac{1}{2}(v + v_0)t\right]$$

or, simplifying:
$$W = \tfrac{1}{2}mv^2 - \tfrac{1}{2}m{v_0}^2 \tag{5–13}$$

The term $\frac{1}{2}mv^2$ is called the **kinetic energy** K of the mass m; this is energy the object possesses by virtue of its motion.

KINETIC ENERGY
$$K = \tfrac{1}{2}mv^2 \tag{5–14}$$

As Equation (5–13) states, *the work done by the net resultant force acting on a mass produces a change in the kinetic energy of the mass.* Although our derivation considered a *constant* net force, in the next section we will show that the result is a general one, true also when the net force varies with distance.

Kinetic energy (K) and work (W) have the same units: joules or foot-pounds. Starting at rest, the work done on a mass by a net external force **F** is energy transferred *from* the system that produces the force to the system of the mass. The mass gains energy in the amount $\frac{1}{2}mv^2 - \frac{1}{2}m{v_0}^2$; this gain in kinetic energy is a result of the increased speed of the mass.

We may now make the general statement that *any moving mass has the capacity for doing work on some other object.* To see this, consider another object (your hand, for example) that brings the moving mass to rest by exerting a force F on it. By Newton's third law, an equal-magnitude force F is thus exerted on your hand. This force moves through some distance as the mass is brought to rest, doing work on your hand. The amount of work done is just equal to the original kinetic energy of the mass.

This process is an example of the interchange of energy between systems. The work W done on a mass by a net external force in accelerating the mass from rest gives kinetic energy to the mass according to $W = \frac{1}{2}mv^2$. Conversely, the moving mass, in being brought to rest, performs work on some other system in the same amount: $\frac{1}{2}mv^2 = W$. As will be shown later, this is part of an overall conservation of energy principle in which energy is interchanged between systems by the performance of work.

□ □ □

The case of circular motion at constant speed is an interesting example that deserves special comment. Here, the net resultant force on the mass (the centripetal force) is always radially inward at right angles to the direction of motion. Since work is done on a mass only when there is a component of force in the direction of motion, the centripetal force does no work on the mass, and the speed of the mass remains constant. Thus its kinetic energy also remains constant.

EXAMPLE 5–5

(a) How much work is required to double the speed of a 4-kg object initially moving at 2 m/s? (b) What constant net force will accomplish this change within 6 m?

SOLUTION

(a) We start with the relation between work and kinetic energy:

$$W = \tfrac{1}{2}mv^2 - \tfrac{1}{2}mv_0^{\,2}$$

Substituting the numerical values, we have

$$W = \tfrac{1}{2}m(v^2 - v_0^{\,2}) = \tfrac{1}{2}(4\text{ kg})\left[\left(4\frac{\text{m}}{\text{s}}\right)^2 - \left(2\frac{\text{m}}{\text{s}}\right)^2\right]$$

$$\boxed{W = 24.0\text{ N}\cdot\text{m}}$$

(b) One method for determining the net force is as follows:

$$W = (F)(x)$$

$$(24\text{ N}\cdot\text{m}) = (F)(6\text{ m})$$

$$\boxed{F = 4.00\text{ N}}$$

An alternative method utilizes the kinematic equation (2–24) to find the required acceleration.

$$v^2 = v_0^{\,2} + 2a(x - x_0)$$

Solving for a, we have

$$a = \frac{v^2 - v_0^{\,2}}{2(x - x_0)}$$

Substitution of numerical values gives

$$a = \frac{\left(4\frac{\text{m}}{\text{s}}\right)^2 - \left(2\frac{\text{m}}{\text{s}}\right)^2}{2(6\text{ m} - 0)} = 1\frac{\text{m}}{\text{s}^2}$$

Then, from $\Sigma F = ma$, we obtain

$$\Sigma F = ma$$

$$F = (4\text{ kg})\left(1\frac{\text{m}}{\text{s}^2}\right) = \boxed{4.00\text{ N}}$$

5.7 Kinetic Energy Change with a Variable Force

Again, consider the one-dimensional case of Figure 5–8, but now suppose we allow $F(x)$ to vary with the distance x. However, as the force moves the mass through a very small incremental distance Δx, the force F during that movement may be considered constant. The corresponding incremental work ΔW done by the force is then

$$\Delta W = F\,\Delta x$$

Since

$$F = m\frac{\Delta v}{\Delta t}$$

where Δv is the change in speed during the movement, we have

$$\Delta W = m\frac{\Delta v}{\Delta t}\Delta x$$

or

$$\Delta W = m\,\Delta v\frac{\Delta x}{\Delta t} \qquad \textbf{(5–15)}$$

The quantity $\Delta x/\Delta t$ is the average speed of the mass during the time Δt. If the speed changes from an initial value v_i to a final value v_f, the average value is

$$\frac{\Delta x}{\Delta t} = \frac{(v_f + v_i)}{2}$$

If we substitute this value and $\Delta v = v_f - v_i$ into Equation (5–15), the incremental work becomes

$$\Delta W = \tfrac{1}{2}m(v_f - v_i)(v_f + v_i) = \tfrac{1}{2}mv_f^2 - \tfrac{1}{2}mv_i^2 = \tfrac{1}{2}m\,\Delta v^2$$

The total work W done in moving the mass from position a to position b is the sum of the work increments:

$$W = \sum_a^b \Delta W = \frac{1}{2}m\sum_a^b \Delta v^2$$

When we expand the right side of this equation, we have

$$W = \tfrac{1}{2}m[(v_1^2 - v_a^2) + (v_2^2 - v_1^2) + \cdots + (v_b^2 - v_{b-1}^2)]$$

All but the initial and final values of v^2 cancel, so we obtain

$$W = \tfrac{1}{2}mv_b^2 - \tfrac{1}{2}mv_a^2$$

This result is a general expression, analogous to Equation (5–13) for a constant force. Thus the work done on a mass by a net external force, *whether variable or not*, equals the change in kinetic energy of the mass.

5.8 Gravitational Potential Energy

Usually, more than just one force acts on an object. There may be several forces—gravity, friction forces, tensions in springs, and so on—that act simultaneously on the object. Let us investigate the relationship between work and energy for each one of these individual forces.

Consider a constant upward force **F** applied to a mass in the presence of gravity (Figure 5–9). There are two forces on the mass: the upward force **F** and the downward force of gravity $\mathbf{F}_{grav}$. (Here we use the symbol $\mathbf{F}_{grav}$ instead of **W** to avoid confusion with work.) In this example, we assume $\mathbf{F} > \mathbf{F}_{grav}$.

What work is done by **F** as it moves the object vertically from $y = a$ to $y = b$? We will assume the distance moved is small enough so that we may ignore the variation of the earth's gravitational force with distance from the earth's center. Therefore, the force of gravity $\mathbf{F}_{grav}$ on the object remains constant throughout. We are considering the general case in which **F** is larger than $\mathbf{F}_{grav}$, so the object accelerates in its upward motion. Choosing the upward direction as positive, we start with Newton's second law.

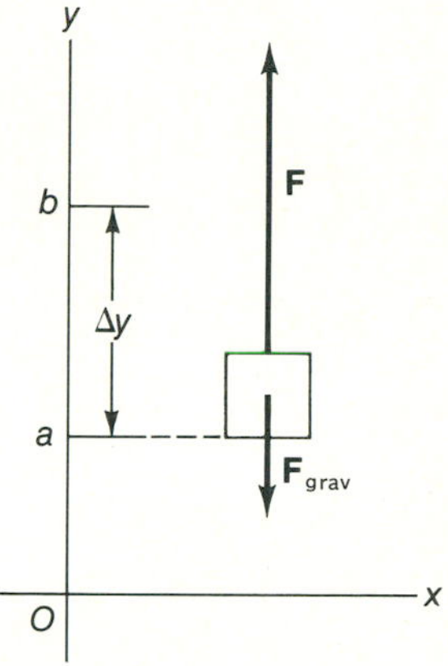

Figure 5–9
Raising an object vertically against gravity.

$$\Sigma\mathbf{F} = m\mathbf{a}$$
$$\mathbf{F} + \mathbf{F}_{grav} = m\mathbf{a}$$
$$\mathbf{F} = m\mathbf{a} - \mathbf{F}_{grav} \tag{5–16}$$

The work ΔW_F done by **F** in moving a mass a distance $\Delta\mathbf{y}$ is

$$\Delta W_F = \mathbf{F} \cdot \Delta\mathbf{y}$$

Substituting $\mathbf{F} = m\mathbf{a} - \mathbf{F}_{grav}$, we have

$$\Delta W_F = m\mathbf{a} \cdot \Delta\mathbf{y} - \mathbf{F}_{grav} \cdot \Delta\mathbf{y} \tag{5–17}$$

The first term on the right is simply $(ma)(\Delta y)$. (This is true because **a** and $\Delta\mathbf{y}$ are always in the same direction, so the dot product involves $\cos 0° = 1$.) Referring to the derivation of Equation (5–13), we see that $(ma)(\Delta y)$ equals the change in kinetic energy ΔK of the mass. Hence:

$$\Delta W_F = \Delta K - (\mathbf{F}_{grav} \cdot \Delta\mathbf{y}) \tag{5–18}$$

The second term on the right, including the minus sign, is called the *change in gravitational potential energy* ΔU_{grav} of the mass m. To evaluate this term, we first consider the dot product. Since $\mathbf{F}_{grav}$ is in the downward direction and $\Delta\mathbf{y}$ is upward, the cosine of 180° is -1. We therefore have another minus sign as a result of the dot product. That is:

$$\Delta U_{grav} = -(-mg)(\Delta y)$$

or

$$\Delta U_{grav} = (mg)(\Delta y) \tag{5–19}$$

Note that this is only the *change* in gravitational potential energy, not an absolute value for U_{grav}. For convenience, it is possible to assign specific values to the gravitational potential energy everywhere *if we define the value to be zero at some particular point*. The location chosen as a reference point is purely arbitrary; it depends on the circumstances of the problem. Often it is simplest to define the zero reference point $U_{grav} \equiv 0$ at the lowest location that the mass m has in a problem, setting the origin of the coordinate system $y = 0$ at that location (positive direction upward). Then, for any vertical distance y above this reference, the **gravitational potential energy** U_{grav} is

$$\Delta U_{grav} = (mg)(\Delta y)$$
$$U_{(y=y)} - \underbrace{U_{(y=0)}}_{\substack{\text{Defined} \\ \text{to be zero}}} = (mg)(y) - (mg)(0) \tag{5–20}$$

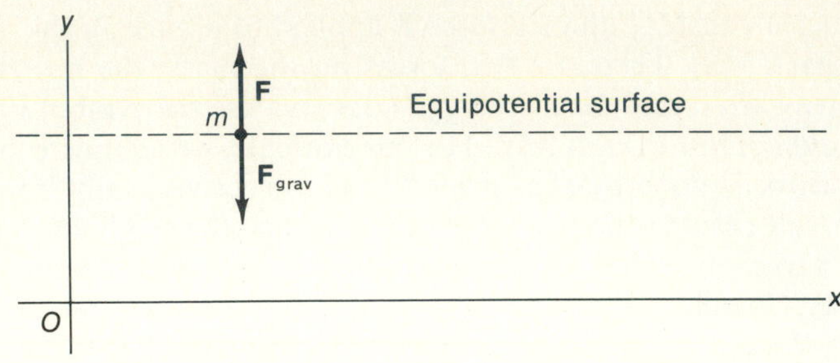

Figure 5–10
If a mass is moved horizontally, the force of gravity $\mathbf{F}_{grav}$ does no work (since $\Delta W = \mathbf{F}_{grav} \cdot d\mathbf{x} = 0$). Thus, all points on a given horizontal surface have the same gravitational potential energy.

Thus:

GRAVITATIONAL POTENTIAL ENERGY of a mass *m* near the earth's surface

$$U_{grav} = mgy \tag{5–21}$$

where $U_{grav} \equiv 0$ at $y = 0$

(assumes constant g and positive direction upward)

Because the value of U_{grav} depends on the location of the reference point $U_{grav} \equiv 0$, this location should always be specified in solving problems.

Note that changes in gravitational potential energy occur only for displacements in the *vertical* direction. Moving the mass *parallel* to the earth's surface (in the x direction) does not change the value of U_{grav} since, in this case, the force $\mathbf{F}_{grav}$ is at right angles to $\mathbf{x}$ and therefore does no work. It is helpful to think in terms of **equipotential surfaces**—surfaces on which the gravitational potential energy has a constant value (see Figure 5–10). Equipotential surfaces are parallel to the surface of the earth.

5.9 Work and Mechanical Energy

In the previous section, we discussed the case where the only forces acting on the object were a given force **F** plus the force of gravity. (Friction was notably absent.) We found that the work ΔW_F done by the given force **F** was

WORK AND MECHANICAL ENERGY

$$\Delta W_F = \Delta K + \Delta U_{grav} \tag{5–22}$$

The symbol ΔK represents the kinetic energy added to the object (by altering its speed) and ΔU_{grav} represents the gravitational potential energy added (by changing the elevation of the object). Together, ΔK and ΔU_{grav} are called **mechanical energy.**

Using this relationship between work and energy, we may redefine the gravitational potential energy of a mass at any location, $U_{grav}(x,y)$:

GRAVITATIONAL POTENTIAL ENERGY of a mass *m* near the earth's surface (Alternative definition)

The work required to move the mass from a reference point y_0 (where $U_{grav} \equiv 0$) to the point y *without producing any net change in the kinetic energy.*

EXAMPLE 5–6

A 4-kg mass, initially at rest, is lifted against the force of gravity by a constant force of 60 N. (a) How much work is done in raising the mass a vertical distance of 2 m? (b) What changes in energy of the mass does the work produce? (c) What is the final speed of the mass?

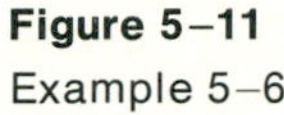
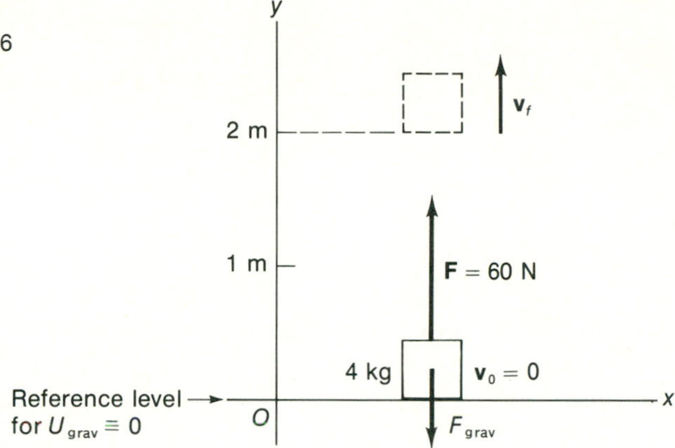

Figure 5–11
Example 5–6

SOLUTION

(a) Because the force is constant (Figure 5–11), the work is

$$\Delta W = \mathbf{F} \cdot \Delta \mathbf{y}$$

The dot product gives cos 0° = 1.

$$\Delta W = (60 \text{ N})(2 \text{ m}) = 120 \text{ N·m} = \boxed{120 \text{ J}}$$

This work appears as a change in the kinetic energy of the mass and a change in the gravitational potential energy of the mass.

(b) The change in gravitational potential energy ΔU_{grav} is (defining $U_{\text{grav}} \equiv 0$ at $y = 0$):

$$\Delta U_{\text{grav}} = mg\,\Delta y$$

$$\Delta U_{\text{grav}} = (4 \text{ kg})(9.8 \text{ m/s}^2)(2 \text{ m}) = 78.4 \text{ N·m}$$

$$\Delta U_{\text{grav}} = \boxed{78.4 \text{ J}}$$

The rest of the energy given to the mass appears as a change in kinetic energy ΔK:

$$\Delta W = \Delta U_{\text{grav}} + \Delta K$$

$$(120 \text{ J}) = (78.4 \text{ J}) + \Delta K$$

$$\Delta K = \boxed{41.6 \text{ J}}$$

(c) We calculate the final speed v_f from

$$\Delta K = \tfrac{1}{2}mv_f^{\,2} - \tfrac{1}{2}mv_0^{\,2}$$

$$(41.6 \text{ J}) = \tfrac{1}{2}(4 \text{ kg})(v_f^{\,2}) - \tfrac{1}{2}(4 \text{ kg})(0)$$

$$v_f = \sqrt{20.8\,\frac{\text{J}}{\text{kg}}} = \sqrt{20.8\,\frac{\text{m}^2}{\text{s}^2}} = \boxed{4.56\,\frac{\text{m}}{\text{s}}}$$

Lilliputians and Brobdingnagians

In the novel Gulliver's Travels *by Jonathan Swift* (c. *1726*), *Gulliver visits some rather strange countries. In Lilliput, the inhabitants are only one-twelfth as large in linear dimensions as Gulliver, while in Brobdingnag the local residents are twelve times larger than he is. This change of scale has fascinating consequences regarding physical abilities. Swift correctly figured out some of the consequences, but in some cases he was wrong, or did not think of some of the startling implications. Let us consider just a few.*

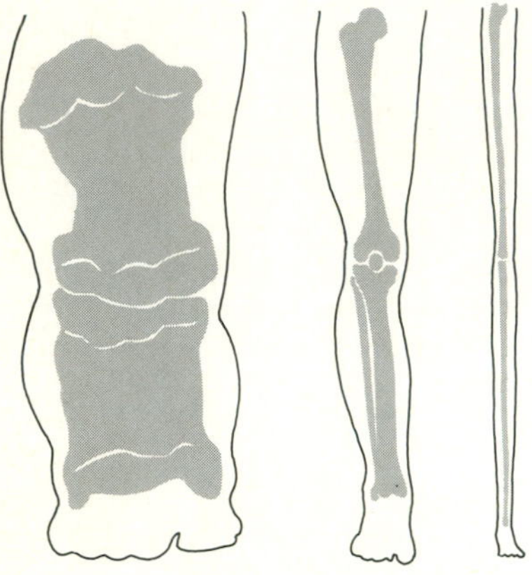

Body weight increases as the cube of the linear distance, but bone strength increases only as the square of the distance. Therefore, as linear dimensions increase, the size of bones must be relatively larger in proportion to body size to support the increased weight. Though the linear dimensions of the persons differ by factors of 12 in adjacent drawings, they are drawn here all to the same length for comparison.

The function of our skeletal bones is to support our weight, and they must also be able to resist breakage if we try to bend or twist them. How does the size of a bone alter its ability to resist compressional forces? If a piece of wood of cross-section 5 cm × 5 cm *can support a weight W, then a piece twice as large could support a weight of 4W (since four pieces of the original size, side by side, would have a cross-section of* 10 cm × 10 cm. *What about the strength of the muscles required to move our limbs and perform work? Since muscle fibers act a bit like stretched rubber bands, which can be contracted at will, the number of such bands is roughly proportional to the cross-sectional area of the muscle. So doubling the linear dimension makes the muscles four times as strong, the same factor by which our bones increase their compressional strength. This is fortunate, since as we grow larger the force of our muscles keeps right in step with the strength of our bones.*

But note that a person's weight goes up with the volume, which is proportional to the cube of the linear dimension. So someone twice as large would weigh eight times as much, requiring the bones to be relatively much larger in size to support this extra weight. This is borne out in nature. For example, the bones of a mouse are only about 5% to 8% of its weight, while those of a person are about 17%. Bones of larger animals such as horses or elephants are a still larger fraction of the animal's body weight. It is very unlikely that a Brobdingnagian would have a sufficiently strong skeleton with enough body weight left over for the enormous muscles that would be necessary to move the huge mass of bones and muscle. In the case of animals and fish that live in water, the buoyant force of the water supports most of the weight, so the skeleton and muscles do not need to be as large as these calculations imply. Unfortunately, a whale out of water is usually crushed to death because its bones and muscles themselves cannot support the weight of its enormous mass.

Another interesting consequence of size is the effect on walking speed. The natural rate of swinging the leg back and forth varies inversely with the square root of the length. Let us compare the speed of walking of an adult with that of a child whose leg is half as long. For the same angle of swing, the distance covered by one step of the adult is twice as long as the child's. But the frequency at which the child takes steps is only $\sqrt{2}$ *times larger than that of an adult. Thus, if an adult takes one step of length* ℓ, in the same time *the child will take* $\sqrt{2}$ *steps, each only half as long. Thus the speed of walking of the child—the length of each step times the number of steps per second—is only* $\frac{1}{2}(\sqrt{2}) = 1/\sqrt{2}$ *that of the adult.*

If a child is forced to walk at a normal adult walking speed, his or her leg muscles very soon become extremely tired and painful. So when a parent who is unaware of this difference walks with a small child, it is not surprising that the child quickly becomes tired trying to keep up.

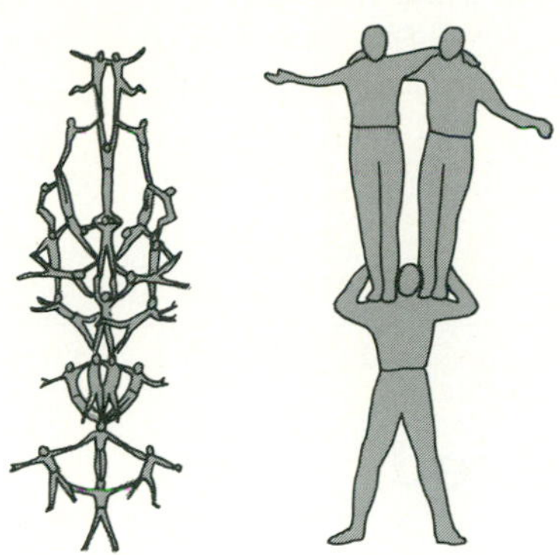

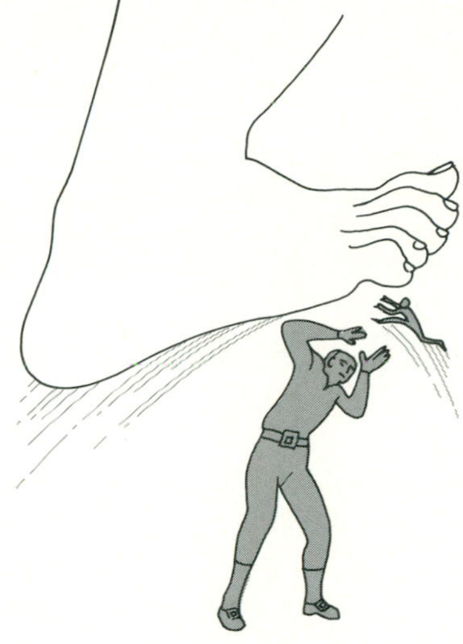

A man can support two persons his own weight. By change-of-scale reasoning, we may conclude that a Lilliputian could perform the impressive feat of supporting 24 other Lilliputians, but it would take six Brobdingnagians to lift one Brobdingnagian.

In jumping, our muscles exert a force through the distance our center of gravity is raised while our feet are in contact with the ground. This work is transformed into gravitational potential energy. Using change-of-scale reasoning, we may conclude that the height of jumping should be *independent* of the size of the person. Since a man can jump about his own height, a Lilliputian could also jump this same distance, or 12 times his height. On the other hand, a Brobdingnagian could jump vertically only about half the length of his foot—a rather unimpressive feat. In nature, one sees this same general behavior. A flea can jump 200 times its own length; a grasshopper can hop about 30 times its length; and a dog can jump just a few times its length. An elephant cannot jump at all.

References: P.K. Weyl, Men, Ants, and Elephants, *Viking Press, 1959, and F.W. Went, "The Size of Man,"* American Scientist, *400, 56, 1968.*

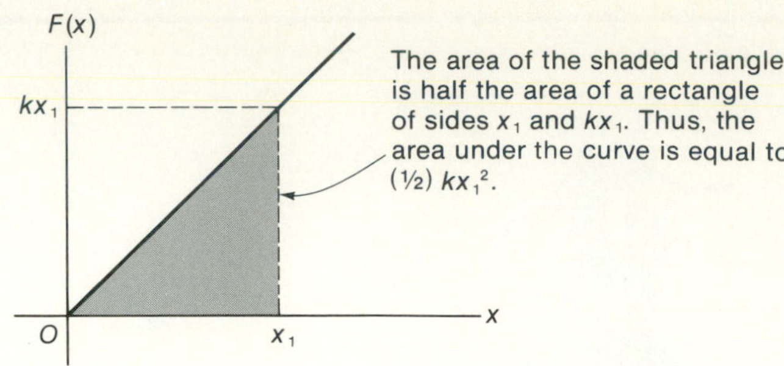
F(x)
kx₁
O
x₁
x
The area of the shaded triangle is half the area of a rectangle of sides x₁ and kx₁. Thus, the area under the curve is equal to (½) kx₁².

Figure 5–12
The shaded area under the curve represents the work required to stretch a Hooke's Law spring from 0 to x_1. The force varies with distance according to $F(x) = kx$.

5.10 Potential Energy Stored in a Spring

A system that contains a spring involves another type of potential energy—energy stored in the spring when it is stretched or compressed. Recall from Example 5–4, in which a force F stretched a "Hooke's Law" spring, that the work done in stretching the spring from $x = 0$ to $x = x_1$ was equal to the area under the $F(x)$ versus x graph. As shown in Figure 5–12, this area equals $\frac{1}{2}kx_1^2$.

The stretched spring is said to contain **potential energy** U_{spring}, which is stored in the elongated spring. The spring is capable of doing work on some external system if allowed to move back to its original, relaxed position, thus exerting a force through a distance as it moves.

POTENTIAL ENERGY STORED IN A STRESSED SPRING

$$U_{\text{spring}} = \tfrac{1}{2}kx^2 \qquad (5\text{–}23)$$

(where k is the force constant of the spring and x is the distance the spring is stretched or compressed from its relaxed position)

Of course, ultimately this energy resides in the distorted atomic bonds between atoms of the spring. When the interatomic distances are altered by stretching the spring, the electrical potential energy between pairs of atoms is changed. For our purposes, however, we may adopt the large-scale view that treats the energy stored in a stressed spring, $\frac{1}{2}kx^2$, as one form of potential energy that a mechanical system may have.

EXAMPLE 5–7

(a) What work is required to stretch the end of a Hooke's Law spring ($k = 200$ N/m) from its relaxed position, $x = 0$, to $x = 0.040$ m? (b) What additional work would be required to stretch the spring from $x_1 = 0.040$ m to $x_2 = 0.060$ m?

SOLUTION

(a) The work ΔW done in stretching the spring from its relaxed position equals the change in potential energy ΔU_{spring}.

$$\Delta W = \Delta U_{\text{spring}}$$

$$\Delta W = \tfrac{1}{2}kx_1^{\,2} - \tfrac{1}{2}kx_0^{\,2}$$

$$\Delta W = \frac{1}{2}\left(200\,\frac{\text{N}}{\text{m}}\right)(0.040\text{ m})^2 - \frac{1}{2}\left(200\,\frac{\text{N}}{\text{m}}\right)(0)^2 = \boxed{0.160\text{ J}}$$

(b) The additional work ΔW done in stretching the spring from its new position, $x_1 = 0.040$ m, to a position 2 cm away, where $x_2 = 0.060$ m, is:

$$\Delta W = \Delta U_{\text{spring}}$$

$$\Delta W = \tfrac{1}{2}kx_2^{\,2} - \tfrac{1}{2}kx_1^{\,2} = \tfrac{1}{2}k(x_2^{\,2} - x_1^{\,2})$$

$$\Delta W = \frac{1}{2}\left[200\ \frac{\text{N}}{\text{m}}\right][(0.060\ \text{m})^2 - (0.040\ \text{m})^2] = \boxed{0.200\ \text{J}}$$

5.11 Thermal Energy Associated with Friction

When friction is present between sliding surfaces, some **thermal energy** always appears, raising the temperature of the objects. Basically, thermal energy is the sum of the kinetic and potential energies associated with the random motions of atoms and molecules. Friction is one way of increasing these motions.

In physics, the meaning of the term "thermal energy" differs from that of "heat," as the following definitions illustrate:

THERMAL ENERGY **The kinetic and potential energies associated with the random motions of atoms and molecules; also called *internal energy*.**

HEAT **The thermal energy transferred from one system to another as a result of temperature differences.**

Thus, the term "thermal energy" describes *energy residing in a body*, while "heat" applies only to the *transfer* of thermal energy from one system to another.

Although kinetic energy of random motions is involved in thermal energy, these are *internal* energies, distinct from the kinetic energy of a macroscopic object traveling with a speed v. We usually use the term "(mechanical) kinetic energy" to refer to energy associated with the large-scale motion of a mass as a whole, while the term "thermal energy" (or "internal energy") applies only to energies of the random motions at the submicroscopic level. Thus, in one case it is an *organized motion*, with all the atoms traveling in the same direction, and in the other case it is the *random motion* of atoms.

□ □ □

Consider Figure 5–13, which depicts a mass m being accelerated along a frictional surface by a given force **F**. The coefficient of kinetic friction is μ_k, and the kinetic friction force is f_k. Clearly, the vertical forces balance, leading to the following force relationship in the horizontal direction:

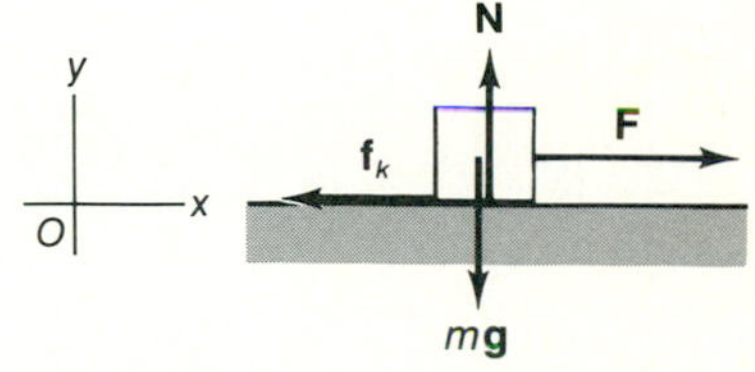

Figure 5–13
An external force F accelerates a mass horizontally.

$$\Sigma \mathbf{F}_x = m\mathbf{a}_x$$

$$(\mathbf{F} + \mathbf{f}_k) = m\mathbf{a}$$

$$\mathbf{F} = (m\mathbf{a} - \mathbf{f}_k)$$

The work ΔW_F done by the given force **F** in moving a distance Δx in the presence of friction is

$$\Delta W_F = \mathbf{F} \cdot \Delta\mathbf{x}$$

Substituting $m\mathbf{a} - \mathbf{f}_k$ for $\mathbf{F}$, we have

$$\Delta W_F = m\mathbf{a} \cdot \Delta\mathbf{x} - \mathbf{f}_k \cdot \Delta\mathbf{x}$$

As shown previously in the derivation of Equation (5–13), the first term $m\mathbf{a} \cdot \Delta\mathbf{x}$ on the right is just the change in kinetic energy ΔK. The second term, including its minus sign, is called the *change in thermal energy* $\Delta U_{\text{thermal}}$ developed by frictional forces between two sliding surfaces. Since f_k and x are in opposite directions, the dot product gives $\cos 180° = -1$, adding another minus sign. Therefore $\Delta U_{\text{thermal}} = -(-f_k)(\Delta x)$, or

THERMAL ENERGY DEVELOPED BY FRICTION FORCES

$$\Delta U_{\text{thermal}} = (f_k)(\Delta x) \tag{5–24}$$

Thus, whenever an object slides along a surface with friction present, there is an increase in the thermal energy of the system by an amount $U_{\text{thermal}} = (f_k)(\Delta x)$. This thermal energy is shared between the object and the surface. (Note that it is incorrect to calculate *two* such quantities, one for the friction on the block and the other for the equal and opposite friction force on the surface.)

EXAMPLE 5–8

A block of mass m is dragged 3 m across the floor at constant speed by an external horizontal force $\mathbf{F} = 6$ N, as shown in Figure 5–14. The force $\mathbf{F}$ has the same magnitude as the kinetic friction force f_k, so no acceleration occurs. Discuss the energy changes that occur.

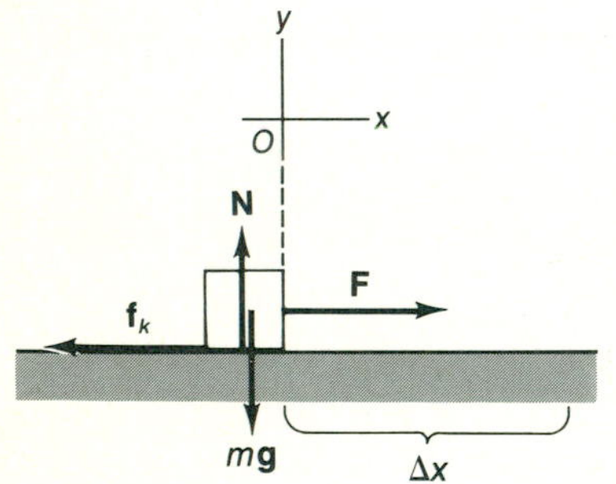

Figure 5–14
Example 5–8

SOLUTION

The work ΔW done by the external force F on the block is

$$\Delta W_F = (F)(\Delta x) = (6 \text{ N})(3 \text{ m}) = \boxed{18.0 \text{ J}}$$

The thermal energy developed between the block and surface is

$$\Delta U_{\text{thermal}} = (f_k)(\Delta x) = (6 \text{ N})(3 \text{ m}) = \boxed{18.0 \text{ J}}$$

The external force F adds energy to the block by doing work on it. Where does this energy go? It does not change the gravitational potential energy (because there is no change in vertical elevation), and it does not change the kinetic energy of the block (because the velocity is constant). It also does not involve spring potential energy. The only energy change that occurs is the increase in thermal energy $\Delta U_{\text{thermal}}$. We conclude that, in this case, the work done by the external force $\mathbf{F}$ ultimately appears as thermal energy through the action of friction.

It will be helpful at this point to summarize the relations between the work ΔW_F done on a mass by a given force $\mathbf{F}$ and the various energy changes that occur as a result. We will include all the forms of energy we have discussed so far. The energy ΔW_F added to the system may appear in the following ways:

(a) The *kinetic energy* of the mass may change from $\frac{1}{2}mv_0^2$ to $\frac{1}{2}mv^2$.
(b) The *gravitational potential energy* may change from mgy_0 to mgy.

(c) The *spring potential energy* may change from $\frac{1}{2}kx_0^2$ to $\frac{1}{2}kx^2$.
(d) The *thermal energy* of the system may increase by $(f_k)(\Delta x)$.

This careful accountability of where the energy goes may be extended to any mechanical system involving any type of force, spring, mass, incline, and so forth. In general:

$$\left.\begin{aligned}\begin{bmatrix}\text{The energy we add to a}\\ \text{system by doing work on}\\ \text{it with a given force } \mathbf{F}\end{bmatrix} = {} & [\text{change in kinetic energy}]\\ & + \begin{bmatrix}\text{change in gravitational}\\ \text{potential energy}\end{bmatrix}\\ & + \begin{bmatrix}\text{change in spring}\\ \text{potential energy}\end{bmatrix}\\ & + [\text{change in thermal energy}]\end{aligned}\right\} \quad (5\text{–}25)$$

$$\Delta W_F = \Delta K + \Delta U_{\text{grav}} + \Delta U_{\text{spring}} + \Delta U_{\text{thermal}}$$

Since the energy changes may be either gains or losses, it is important to keep careful account of plus and minus signs.

5.12 The Work-Energy Relation

There is another convenient way to express the relation between work and energy. It considers the work done by *each individual force* that acts on a system: the gravity force, the spring force, friction forces, and so on, in addition to other forces produced by other means. Referring to the derivation of the energy terms on the right-hand side of Equation (5–25), we may make the following observations:

(a) The term ΔU_{grav} may be expressed as the *negative* of the work done by $\mathbf{F}_{\text{grav}}$; that is, $\Delta U_{\text{grav}} = -(\mathbf{F}_{\text{grav}} \cdot \Delta \mathbf{y})$.
(b) The term ΔU_{spring} may be expressed as the *negative* of the work done by $\mathbf{F}_{\text{spring}}$; that is, $\Delta U_{\text{spring}} = -\int \mathbf{F}_{\text{spring}} \cdot d\mathbf{x}$.
(c) The term $\Delta U_{\text{thermal}}$ may be expressed as the *negative* of the work done by $\mathbf{f}_k$; that is, $\Delta U_{\text{thermal}} = -(\mathbf{f}_k \cdot \Delta \mathbf{x})$.

Consequently, we may move these energy terms to the left-hand side of the equal sign in Equation (5–25), leaving only ΔK on the right. Since each of these energy terms equals the negative of the work done by the corresponding force, when moved to the left-hand side they can be replaced by the *positive* work done by those forces. Thus:

$$\underbrace{\Delta W_{\text{grav}}}_{\substack{\text{Work done}\\ \text{by gravity}}} + \underbrace{\Delta W_{\text{spring}}}_{\substack{\text{Work done by a}\\ \text{spring force}}} + \underbrace{\Delta W_{\text{friction}}}_{\substack{\text{Work done by}\\ \text{kinetic friction}\\ \text{(always negative}\\ \text{because the}\\ \text{direction of the}\\ \text{kinetic friction}\\ \text{force is opposite}\\ \text{to that of the}\\ \text{displacement)}}} + \underbrace{\Delta W_F}_{\substack{\text{Work done by}\\ \text{all the other}\\ \text{external forces}\\ \text{on the system}}} = \underbrace{\Delta K}_{\substack{\text{Change in}\\ \text{kinetic energy}\\ \text{of the system}}} \quad (5\text{–}26)$$

In using Equations (5–25) and (5–26), be very careful of the signs involved in each separate calculation of the work done.

THE WORK-ENERGY RELATION

$$\begin{bmatrix}\text{Work done by the}\\ \textbf{net resultant force}\end{bmatrix} = \begin{bmatrix}\text{Change in kinetic}\\ \text{energy}\end{bmatrix} \quad \textbf{(5–27)}$$

The above expression is known as the **work-energy relation.** Note that the left-hand side represents the work done by the *net resultant force* on the object, that is, the sum of *all* the forces present. Thus, if the net force acting on an object is known, it is a simple matter to calculate the change in kinetic energy of the object.

EXAMPLE 5–9

Starting from rest, a 2-kg block is pulled across the floor by a horizontal force of 6 N. The friction force is 4 N. Find the speed of the block after moving 3 m.

SOLUTION

The net horizontal force on the block is

$$\Sigma F_x = (6\text{ N} - 4\text{ N}) = 2\text{ N}$$

Applying the work-energy relation, Equation (5–27), we obtain

$$\text{Work done by } F_{\text{net}} = \Delta K$$

$$(F_{\text{net}})(x) = \tfrac{1}{2}mv^2 - \tfrac{1}{2}mv_0^{\,2}$$

where $\frac{1}{2}mv_0^{\,2} = 0$. Solving for v gives

$$v = \sqrt{\frac{2(F_{\text{net}})(x)}{m}} = \sqrt{\frac{(2)(2\text{ N})(3\text{ m})}{(2\text{ kg})}} = \boxed{2.45\,\frac{\text{m}}{\text{s}}}$$

EXAMPLE 5–10

This example includes all the forms of energy we have discussed so far. As shown in Figure 5–15, a given force **F** = 30 N acts parallel to the incline as it accelerates a block of mass $m = 2$ kg up the 30° incline, with a coefficient of kinetic friction μ_k equal to 0.30. A spring whose force constant is $k = 40$ N/m is attached to the block, which starts from rest at a position ($x_0 = 0$) where the spring is unstressed. Find the speed of the block after traveling 0.20 m up the incline.

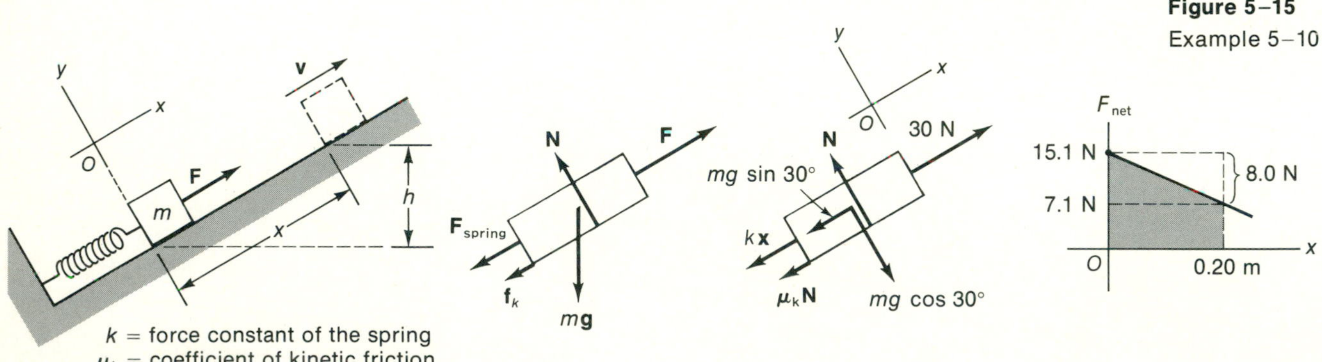

Figure 5–15
Example 5–10

SOLUTION

We first sketch a free-body diagram to determine the forces parallel and perpendicular to the incline. Since forces perpendicular to the incline do not do work, we consider only those forces parallel to the direction of motion. Their magnitudes are

$$F = 30\ \text{N}$$

$$F_{\text{spring}} = kx = \left(40\ \frac{\text{N}}{\text{m}}\right)(x)$$

$$f_k = \mu_k n = (\mu_k)(mg\cos 30^\circ) = (0.30)(2\ \text{kg})\left(9.8\ \frac{\text{m}}{\text{s}^2}\right)(0.867) = 5.10\ \text{N}$$

$$F_{\text{grav}}\ (\text{parallel to incline}) = mg\sin 30^\circ = (2\ \text{kg})\left(9.8\ \frac{\text{m}}{\text{s}^2}\right)(0.500) = 9.80\ \text{N}$$

We will solve the problem by two different methods: Equation (5–25) and Equation (5–27).

Method 1:

We calculate the energy the force F adds to the system by doing work on it, and equate this to the changes that occur in various forms of energy.

$$\begin{bmatrix}\text{Work done on the system} \\ \text{by the given force } \mathbf{F}\end{bmatrix} = [\Delta K + \Delta U_{\text{grav}} + \Delta U_{\text{spring}} + \Delta U_{\text{thermal}}]$$

The vertical change in elevation is $h = x\sin 30^\circ = (0.20\ \text{m})(0.500) = 0.10\ \text{m}$. Therefore:

$$(F)(x) = \tfrac{1}{2}mv^2 + mgh + \tfrac{1}{2}kx^2 + (f_k)(x)$$

$$(30\ \text{N})(0.20\ \text{m}) = \tfrac{1}{2}(2\ \text{kg})(v^2) + (2\ \text{kg})\left(9.8\ \frac{\text{m}}{\text{s}^2}\right)(0.10\ \text{m}) + \frac{1}{2}\left(40\ \frac{\text{N}}{\text{m}}\right)(0.04\ \text{m}^2) + (5.1\ \text{N})(0.20\ \text{m})$$

$$6.0\ \text{N·m} = (1\ \text{kg})(v^2) + 1.96\ \text{N·m} + 0.80\ \text{N·m} + 1.02\ \text{N·m}$$

$$v^2 = \frac{2.22\ \text{N·m}}{1\ \text{kg}} = 2.22\left(\frac{\text{m}}{\text{s}}\right)^2$$

$$\boxed{v = 1.49\ \frac{\text{m}}{\text{s}}}$$

Method 2:

Alternatively, we may use the work-energy relation, Equation (5–27).

$$\begin{bmatrix}\text{Work done by the} \\ \textbf{net resultant force}\end{bmatrix} = [\text{Change in kinetic energy}]$$

The net resultant force is parallel to the incline in the direction of motion.

$$F_{\text{net}} = 30\ \text{N} - \left(40\ \frac{\text{N}}{\text{m}}\right)(x) - 5.1\ \text{N} - 9.8\ \text{N}$$

$$F_{\text{net}} = (15.1 - 40x)\ \text{N} \qquad (\text{where } x \text{ is in meters})$$

We see that F_{net} varies with the distance x (because of the spring). Being linear in x, it is a straight-line function that varies from its initial value (at $x = 0$):

$$F_{\text{net}} = [15.1 - (40)(0)]\ \text{N} = 15.1\ \text{N}$$

to its final value (at $x = 0.20$ m):

$$F_{\text{net}} = [15.1 - (40)(0.20)]\ \text{N} = 7.1\ \text{N}$$

In the graph of this expression, the area under the curve is equal to the work done by F_{net} as it moves between these limits.[6]

$$\begin{aligned} &\text{Area of rectangle:} \quad (7.1\ \text{N})(0.20\ \text{m}) = 1.42\ \text{N}\cdot\text{m} \\ &\text{Area of triangle:} \quad \tfrac{1}{2}(8.0\ \text{N})(0.20\ \text{m}) = 0.80\ \text{N}\cdot\text{m} \\ &\qquad\qquad\qquad\qquad\qquad \text{Total} = 2.22\ \text{N}\cdot\text{m} \end{aligned}$$

Substituting this value into the work-energy relation, we have

$$\begin{bmatrix}\text{Work done by the} \\ \text{net resultant force}\end{bmatrix} = \Delta K$$

$$2.22\ \text{N}\cdot\text{m} = (\tfrac{1}{2}mv^2 - \tfrac{1}{2}mv_0{}^2)$$

$$2.22\text{N}\cdot\text{m} = [\tfrac{1}{2}(2\ \text{kg})v^2 - 0]$$

$$v^2 = \frac{2.22\ \text{N}\cdot\text{m}}{1\ \text{kg}} = 2.22\left(\frac{\text{m}}{\text{s}}\right)^2$$

$$\boxed{v = 1.49\,\frac{\text{m}}{\text{s}}}$$

5.13 Internal Work

(**a**) The girl, initially at rest, sets herself in motion by pushing on the fixed wall.

(**b**) A free-body diagram of the external forces on the girl-plus-cart system.

Figure 5–16
A special case in which the external force $\mathbf{F}_1$ that accelerates the system does no work on the system. The kinetic energy the system acquires comes from an *internal* source of energy, namely, the muscular activity of the girl's arm.

In certain special cases, the source of energy for the work done on a system is not obvious. Consider, for example, a girl in a cart that rolls on a horizontal surface without friction. If the girl sets herself in motion by pushing on a fixed wall, the forces shown in Figure 5–16(b) act on the girl-plus-cart system. The force $\mathbf{F}_1$ that the wall exerts on the girl's hand is the net external force that accelerates the system. But even though the system thereby gains kinetic energy, *the force* $\mathbf{F}_1$ *does no work* since it does not move through any distance. Where does the energy come from? In this instance, the energy originates in the muscular activity of the girl's arm. It is thus an *internal* source of energy (somewhat analogous to energy stored in a compressed spring which is then allowed to expand).

Another example would be raising oneself from a squatting position. The gain in gravitational potential energy comes about as a result of the muscular work done *internally* within the system and not from the external upward force that the ground exerts. Thus, sometimes the external forces alone are not sufficient to identify all the sources of energy involved in a process; internal

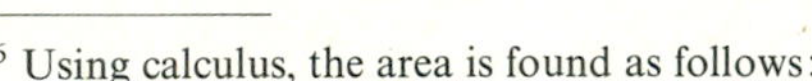

[6] Using calculus, the area is found as follows:

$$W = \int \mathbf{F}(x)\cdot d\mathbf{x}$$

The dot product gives $\cos 0° = 1$, so the integral becomes $W = \int F(x)\,dx$. Substituting for $F(x)$ and indicating the limits on x gives us

$$W = \int_0^{0.20\ \text{m}} (15.1 - 40x)\,dx$$

Carrying out the integration (see Appendix G), we obtain

$$W = \left(15.1x - 40\frac{x^2}{2}\right)\Bigg|_0^{0.20\ \text{m}}$$

Finally, we evaluate the expression at the upper limit ($x = 0.20$ m) and subtract its value at the lower limit ($x = 0$).

$$W = (3.02 - 0.80)\ \text{N}\cdot\text{m} - (0) = \boxed{2.22\ \text{N}\cdot\text{m}}$$

energy sources must also be included in the analysis.

$$[\Delta W_{\text{ext}} + \Delta W_{\text{int}}] = \Delta K$$

5.14 Power

Power is defined as the *time rate at which work is done.* In many situations, it is not sufficient to know how fast an object is moving after work is done on the object. How rapidly the object acquires the speed may also be of great importance. We wish not only to know the total amount of energy transfer involved, but are often willing to pay a premium price for a rapid accomplishment of the task. A chipmunk running in a small treadmill could eventually raise an automobile 20 ft vertically. But we usually want to accomplish the task more rapidly with an engine, which has a greater "power rating." Thus, power is an important concept for our industrialized society.

We begin with the definition of work:

$$\Delta W = \mathbf{F} \cdot \Delta \mathbf{x}$$

where ΔW is the increment of work done by the force $\mathbf{F}$ in moving the object a distance $\Delta \mathbf{x}$. If the object moves $\Delta \mathbf{x}$ in a time Δt, then the time rate at which work is done is

$$\frac{\Delta W}{\Delta t} = \mathbf{F} \cdot \frac{\Delta \mathbf{x}}{\Delta t}$$

In the limit, as Δt becomes infinitesimal, the time rate at which work is done is called the **power** P.

$$P = \lim_{\Delta t \to 0} \frac{\Delta W}{\Delta t} = \lim_{\Delta t \to 0} \left(\mathbf{F} \cdot \frac{d\mathbf{x}}{dt} \right) = \mathbf{F} \cdot \frac{d\mathbf{x}}{dt} = \mathbf{F} \cdot \mathbf{v}$$

POWER

$$P = \frac{dW}{dt} \tag{5–28}$$

$$P = \mathbf{F} \cdot \mathbf{v} \tag{5–29}$$

Another useful quantity is the **average power,** which is defined as the work done over an extended interval of time divided by that interval of time.

$$P_{\text{ave}} = \frac{\text{Work}}{\text{Time}} \tag{5–30}$$

Power is measured in the following units:

SI system: The units are (N·m/s) and (1 J/s), also called the **watt** (W)[7]; a common multiple is the *kilowatt* (kW), which is equal to 1000 watts.

British engineering system: The unit of power is (ft·lb/s).

Another common unit in engineering usage is the **horsepower** (hp), originally based on the maximum work an average horse can produce over an extended period of time. (Actually, most horses fall short of this value, whereas a human, for a very short period of time, can do work at a rate somewhat above one

[7] This unit honors James Watt (1736–1819), a Scottish inventor who developed the automatic steam engine.

Energy for the Future

The problem of energy for the future is one of the most complex questions facing our civilization. The subject is an intricate mix of technical knowledge, economics, politics, and personal feelings regarding how much damage to the environment one is willing to tolerate. This latter factor alone is such a volatile issue, with widely varying opinions, that any discussion of energy problems is bound to be a complicated matter.

The United States, with only 6.4% of the world's population, now consumes about 35% of the total energy produced. And we are demanding energy at a rapidly increasing rate, not only because of population increases, but also because the per capita *increase in energy usage has climbed drastically in the past few decades. All predictions show that such increases will continue. When this situation is combined with the fact that the major fraction of the world's peoples—found in the underdeveloped countries—are anxious to obtain more energy, too, the long-range view of the clamor for energy becomes a dramatic and sobering one.*

Let us look at just one aspect of this complex subject: the geometric growth in demand for energy. It is fascinating that the growth pattern of almost any measurable quantity initially shows a geometric characteristic (also called an exponential growth). This is true for automobile sales, rabbit populations, the number of articles in technical publications, bacteria growth, the numbers of visitors to national parks, and so on.[1] *But exponential growth patterns cannot continue forever. They* must *come to a halt eventually. And whenever exponentials are involved, "eventually" can be unexpectedly soon, as the following example shows.*

All studies of possible energy sources for the future conclude that there are only two sources that are truly abundant enough to meet long-range demands: solar energy and fusion of deuterium from the ocean. Solar radiation is essentially an inexhaustible source, presumably lasting about 5 billion years, the estimated remaining lifetime of our sun. On the other hand, only about 1 in 6500 hydrogen atoms in sea water is the isotope deuterium ($^2_1$H). *This is a finite source, but an impressively large one. When certain light nuclei such as deuterium combine (undergo* fusion*), some mass m disappears in the reaction. From Einstein's famous relation* $E = mc^2$, *we know that an equivalent energy* E *is thereby released.*

In 1980 it cost less than a dime to extract the deuterium from one gallon of sea water. If this much deuterium were utilized in fusion reactions to the maximum possible advantage, the resultant energy produced would be equivalent to burning 300 gallons of gasoline. Another way of saying it is that 1 km^3 *of sea water has greater energy content than all the world's known fossil fuels combined!*

Of course, the big "if" is whether or not fusion can be achieved economically. The technical problems remaining to be solved seem enormous. But let us be

[1] *Interesting discussions of exponential growth patterns are found in Ralph Lapp,* The Logarithmic Century *(Englewood Cliffs, N.J: Prentice-Hall,* 1973). *See also Albert A. Bartlett,* Forgotten Fundamentals of the Energy Crisis, American Journal of Physics, 46, *876 (1978).* Bartlett *has also written an entertaining series of articles called "The Exponential Function," in* The Physics Teacher, 14, *393 (1976) through* 17, *23 (1979).*

optimistic and assume that fusion power is just around the corner. What will be its impact in meeting the world's long-range needs? By their nature, fusion reactors will most likely be used to produce electricity.

In 1974, the world consumption of electricity was about 5×10^{12} kW·h/yr. *If we obtained this energy from the fusion of deuterium in the ocean, how long would the deuterium last? Estimates vary widely. Depending on the assumed efficiencies of the extraction and fusion processes, as well as the efficiency of the distribution system, the numbers quoted range from several hundred million years to many billions of years. Suppose we adopt the optimistic estimate and assume that ocean deuterium could provide the world's electrical needs* at the current rate *for 5 billion years—the estimated lifetime of our sun—a total of* 25×10^{21} kW·h. *This sounds impressive. But now let us recalculate the situation if we assume a 3% per year increase*[2] *in the demand for electricity. Letting* a *be the initial yearly consumption,* r *the yearly rate of growth, and* n *the total number of years, we have*

$$a + a(1 + r) + a(1 + r)^2 + \cdots + a(1 + r)^{n-1} = \text{Total available energy}$$

The sum of the geometric series on the left can be written as

$$\sum_{k=0}^{k=n-1} a(1 + r)^k = a\left[\frac{(1 + r)^n - 1}{r}\right]$$

Substituting numbers for a 3% per year increase and solving for n, *the number of years required to use up the total available energy, shows that the deuterium will last only 637 yr.*

A limit of 5 billion years versus 637 years! *This is, indeed, a striking example of the effects of a steady rate of growth of only 3% per year. Every geometric increase eventually "blows up." Of course, the calculation is facetious in the sense that long before 637 years have elapsed, other limits will have imposed themselves. For example, if the world population growth remains constant (at about 2% per year),*[3] *in about 330 years there will be one person for each square meter of the earth's land surface. Obviously, things will come to a halt long before that happens. But regardless of the direction from which the problem is approached, a constant growth of any parameter results in a geometric (or exponential) increase. Such an unchecked increase ultimately means disaster, and there is no alternative. Furthermore, such problems as food, energy, and pollution ultimately hinge on the most basic consideration of all: the number of people on earth. At some point, world population growth must stop. If it is not curtailed by direct choice, then eventually it will be cut back by war, famine, pestilence, or pollution.*

[2] *This is a conservative estimate; current worldwide figures are higher. And considering the eagerness of underdeveloped countries to move into the modern age, once abundant energy becomes available the rate of increase will probably be even greater.*

[3] *As of 1980,* the daily *net gain was roughly 200,000 persons. Consider, for example, the extra farm acreage for food, the generation of additional energy, the new manufacturing plants for goods, or the new educational facilities that must be provided* every 24 hours *to cope with this many extra people!*

horsepower.) The horsepower is an English unit defined as

$$1 \text{ hp} \equiv 550 \text{ ft}\cdot\text{lb/s}$$

A convenient form for remembering conversion factors for power is (to three significant figures)

$$\underset{\text{(exact)}}{1 \text{ hp} = 550 \text{ ft}\cdot\text{lb/s}} \approx \underset{\text{(approximate)}}{746 \text{ watts}} \tag{5–31}$$

A convenient rule-of-thumb approximation is 1 hp $\approx \frac{3}{4}$ kW.

Although the kilowatt is normally thought of as an electrical unit, it is, nevertheless, similar to all other units of power. If desired, we could state the power of automobile engines in kilowatts rather than horsepower. The difference in usage stems from a historical reason; electrical devices were developed at a later time, after the metric system came into general use.

A convenient unit of *energy* (not power) is the **kilowatt-hour** (kW·h), which is defined as the amount of work done by a device working at a constant rate of one kilowatt for one hour.

EXAMPLE 5–11

An outboard motor propels a boat through the water at a constant speed of 10 mi/h. The water resists the forward motion of the boat with a force of 15 lb. How much power is produced by the outboard motor?

SOLUTION

Since the boat is moving at constant speed, the forward force of the boat produced by the action of the motor just equals the magnitude of the resistive force of the water on the boat. Thus, the power produced by the motor is

$$P = \mathbf{F} \cdot \mathbf{v}$$

or

$$P = (15 \text{ lb})\left(10 \frac{\text{mi}}{\text{h}}\right)\underbrace{\left(\frac{44 \frac{\text{ft}}{\text{s}}}{30 \frac{\text{mi}}{\text{h}}}\right)}_{\substack{\text{Conversion}\\ \text{ratio}}} = 220 \frac{\text{ft}\cdot\text{lb}}{\text{s}}$$

Converting to units of horsepower gives

$$P = \left(220 \frac{\text{ft}\cdot\text{lb}}{\text{s}}\right)\underbrace{\left(\frac{1 \text{ hp}}{550 \frac{\text{ft}\cdot\text{lb}}{\text{s}}}\right)}_{\substack{\text{Conversion}\\ \text{ratio}}} = \boxed{0.40 \text{ hp}}$$

EXAMPLE 5–12

A boy with a mass of 60 kg runs up a staircase in 2.5 s. If the height of the staircase is 3 m, how much average power has the boy expended in going up it?

SOLUTION

Assuming the boy moves at constant speed, his kinetic energy does not change. However, the muscles in his legs do work, which increases his gravitational potential energy by *mgh*. The average power is

$$P_{\text{ave}} = \left(\frac{\text{Work}}{\text{Time}}\right) = \frac{mgh}{t}$$

$$P_{ave} = \frac{(60\text{ kg})\left(9.8\dfrac{\text{m}}{\text{s}^2}\right)(3\text{ m})}{(2.5\text{ s})} = 706\,\frac{\text{N}\cdot\text{m}}{\text{s}}$$

Since N·m/s is defined as a watt, we have

$$\boxed{P_{ave} = 706\text{ W}}$$

In units of horsepower, the boy does work at the rate of

$$P_{ave} = 706\text{ W}\underbrace{\left(\frac{1\text{ hp}}{746\text{ W}}\right)}_{\text{Conversion ratio}} = \boxed{0.946\text{ hp}}$$

5.15 Efficiency

Machines are designed to convert energy into some form of useful work. However, because of frictional effects, the work performed by the machine is always less than the energy put into the machine. The **efficiency** e of a machine is defined as:

EFFICIENCY e OF A MACHINE

$$e = \frac{\text{Useful energy output}}{\text{Energy input}} \qquad \textbf{(5–32)}$$

The efficiency is often expressed as a percentage: percent efficiency $= e \times 100\%$.

EXAMPLE 5–13

A certain motor uses 3 kW of electrical power while operating an inclined conveyor system that raises crushed rock a vertical distance of 10 m at the rate of 40 000 kg/h. What is the efficiency of the system?

SOLUTION

In one second, the work done is

$$\text{Work done in 1 s} = \frac{\Delta mgh}{\Delta t}$$

$$\text{Work done in 1 s} = \frac{(4 \times 10^4\text{ kg})\left(9.8\dfrac{\text{m}}{\text{s}^2}\right)(10\text{ m})}{(1\text{ h})\left(\dfrac{3600\text{ s}}{1\text{ h}}\right)} = 1089\,\frac{\text{J}}{\text{s}}$$

The energy input per second is 3 kW = 3000 J/s. Therefore, the efficiency e is

$$e = \frac{\text{Work done each second}}{\text{Energy input each second}}$$

$$e = \frac{1089\,\dfrac{\text{J}}{\text{s}}}{3000\,\dfrac{\text{J}}{\text{s}}} = \boxed{0.363} \quad \text{or} \quad \boxed{36.3\%}$$

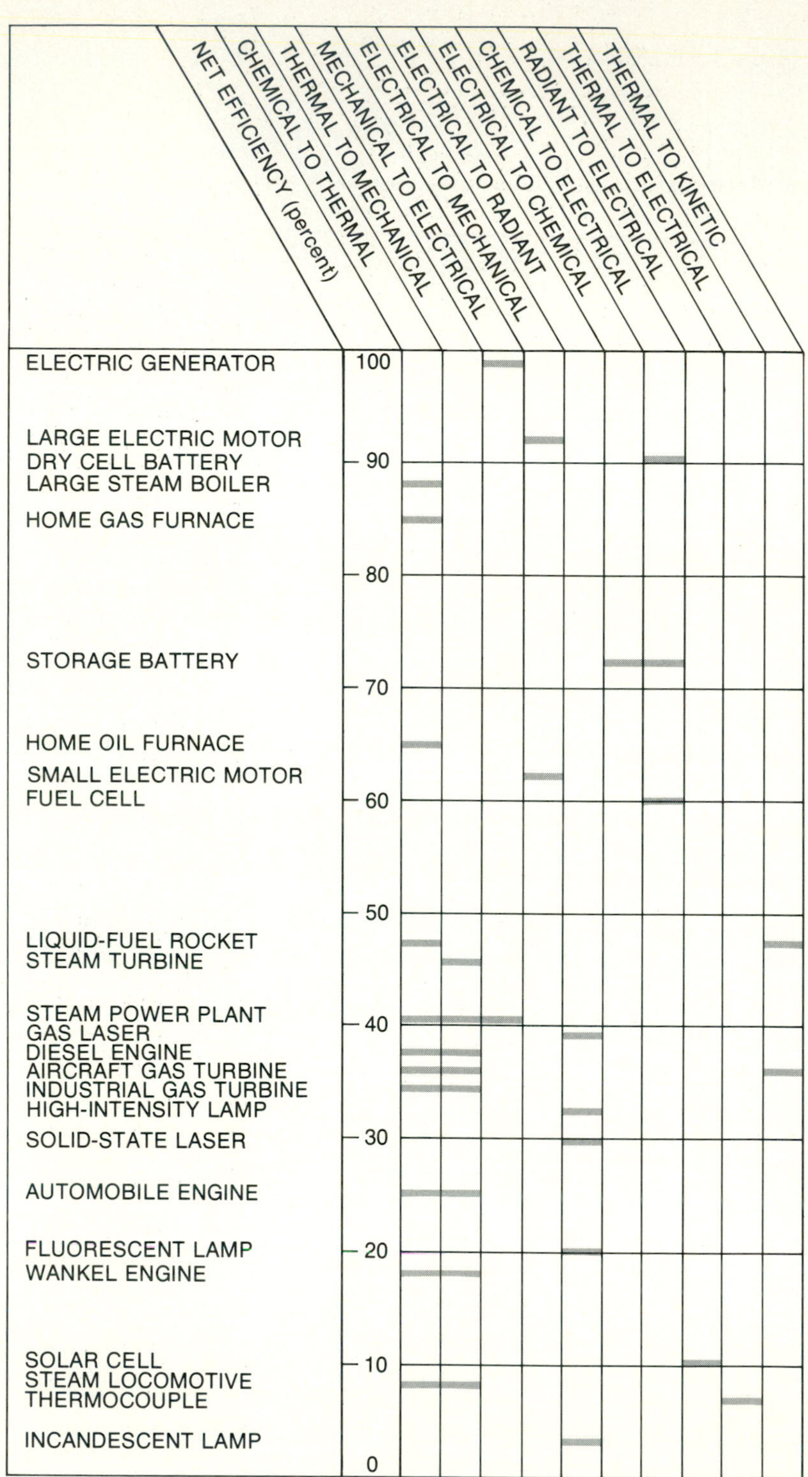

Figure 5–17

Typical efficiencies of various energy converters range from a high of 99% for large electric generators to less than 5% for incandescent lamps.

5.16 Mechanical Advantage

A useful characteristic of some devices and machines is that they convert a small force into a very large one. Of course, we cannot get something for nothing; if friction is negligible, the work input must equal the work output. That is, when a small force moves through a large distance, the large force must move through a small distance. Another class of devices changes the speed with which a force moves. A small force moving rapidly is converted into a large force moving slowly, again with conservation of energy (power input equals the power output, ignoring friction). Examples of devices with mechanical advantages of greater than one are illustrated in Figure 5–18.

A simple way to express the force relationship for a given machine is to define its **actual mechanical advantage.**

ACTUAL MECHANICAL ADVANTAGE

$$\text{A.M.A.} = \frac{\text{Output force}}{\text{Input force}} \tag{5–33}$$

Figure 5–18
Some devices that provide mechanical advantages larger than unity.

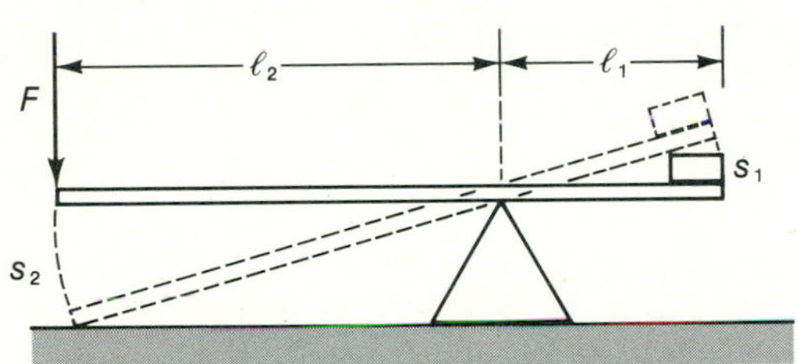

(a) The lever has an ideal mechanical advantage of ℓ_2/ℓ_1. (Archimedes discovered the relationship of levers in the second century B.C. He declared, "Give me a place to stand on and I will move the earth!")

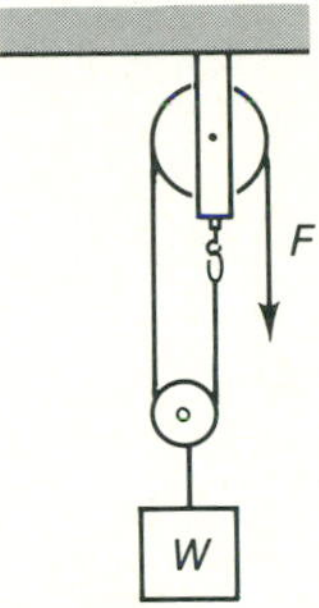

(b) This pulley system has an ideal mechanical advantage of 2 to 1. (The topmost pulley merely changes the direction of the force for convenience. If it were eliminated and the left-hand rope pulled upward, the ideal mechanical advantage still would be 2 to 1.)

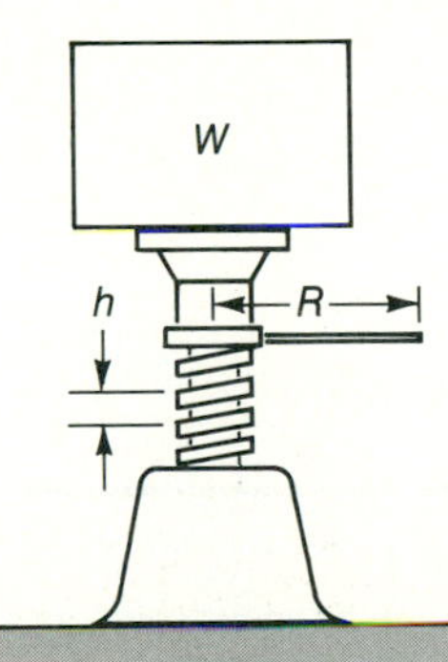

(c) In operating a jackscrew, a horizontal force is applied perpendicular to the handle at a distance R from the center of the screw. One complete turn raises the load a vertical distance h, with an ideal mechanical advantage of $2\pi R/h$.

(d) A differential chain hoist has two pulleys of radii R_1 and R_2 rigidly connected together. (Sprockets in the pulleys engage the chain links so that no slippage occurs.) Pulling down on one chain as shown will raise the weight W, with an ideal mechanical advantage of $2R_2/(R_2 - R_1)$. If the difference in the radii $(R_2 - R_1)$ is small, the mechanical advantage can be made very large.

In the ideal case where friction is negligible (and the device itself has negligible weight), we have

$$(\text{Work output}) = (\text{Work input})$$
$$(F_o)(x_o) = (F_i)(x_i)$$

or

$$\frac{F_o}{F_i} = \frac{x_i}{x_o}$$

The ratio of the *distances* the forces move is called the **ideal mechanical advantage.**

IDEAL MECHANICAL ADVANTAGE

$$\text{I.M.A.} = \frac{\text{Distance the input force moves}}{\text{Distance the output force moves}} \tag{5-34}$$

The efficiency e of any machine is

$$e = \frac{\text{A.M.A.}}{\text{I.M.A.}} \tag{5-35}$$

EXAMPLE 5–14

Consider the block-and-tackle pulley system shown in Figure 5–19. (a) Find the ideal mechanical advantage of the system. (b) If a 40-N force F will raise a 100-N load W, what is the actual mechanical advantage? (c) Find the efficiency.

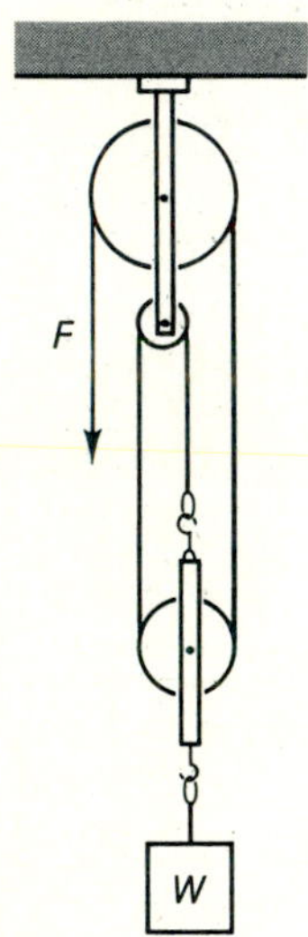

Figure 5–19
Example 5–14

SOLUTION

(a) Assuming no friction and negligible weight for the rope and pulleys, the tension T in the rope would be the same throughout the system. Because three ropes support the weight, we have $W = 3T$. When the free end of the rope moves a distance x, the weight is raised a distance h. From conservation of energy (for the ideal case), we have

$$\text{Work input} = \text{Work output}$$
$$(T)(x) = (3T)(h)$$

or

$$x = 3h$$

Thus

$$\text{I.M.A.} = \frac{\text{Distance the input force moves}}{\text{Distance the output force moves}} = \frac{3h}{h} = \boxed{3}$$

(b) The actual mechanical advantage is

$$\text{A.M.A.} = \frac{\text{Output force}}{\text{Input force}} = \frac{100\text{ N}}{40\text{ N}} = \boxed{2.50}$$

(c) The efficiency is

$$e = \frac{\text{A.M.A.}}{\text{I.M.A.}} = \frac{2.5}{3} = \boxed{0.833 \quad \text{or} \quad 83.3\%}$$

Summary

Work is the transfer of energy from one system to another when a force acts through a displacement. The work ΔW done by a constant force $\mathbf{F}$ moving a distance $\Delta\mathbf{r}$ is

$$\Delta W = \mathbf{F} \cdot \Delta\mathbf{r}$$
$$\Delta W = |\mathbf{F}|\,|\Delta\mathbf{r}| \cos\theta$$

In rectangular coordinates this is

$$\Delta W = F_x \Delta x + F_y \Delta y + F_z \Delta z$$

In the metric system (SI), work is measured in units of *newton-meter* (N·m), also called the *joule* (J). In the British engineering system, the unit is the *foot-pound* (ft·lb).

The work may be found by calculating the area under the $F(x)$ versus x graph, a straightforward procedure if the force varies *linearly* with distance. (In more complicated cases, the integral $W = \int_a^b \mathbf{F}(r) \cdot d\mathbf{r}$ represents the area under the curve from a to b.)

Energy is a calculated quantity with the important characteristic of conservation. Any energy added to a system must be accounted for as changes in the various forms of energy within the system. Work and energy are measured in the same units.

The work done on a system by a given external force $\mathbf{F}$ may result in several different energy changes:

ΔK The change in *kinetic energy* due to motion ($K = \frac{1}{2}mv^2$).

ΔU_{grav} The change in *gravitational potential energy* ($U_{\text{grav}} = mgy$ where $U_{\text{grav}} \equiv 0$ at $y = 0$, assuming a constant value for g at all locations).

ΔU_{spring} The change in *potential energy stored in a stressed spring* ($U_{\text{spring}} = \frac{1}{2}kx^2$, where k is the force constant and x is the distance the spring is stretched or compressed from its relaxed position).

$\Delta U_{\text{thermal}}$ The change in *internal energy* due to kinetic friction forces ($U_{\text{thermal}} = (f_k)(x)$, always a positive quantity).

Letting ΔW_F represent the work done by this given external force $\mathbf{F}$, we have

$$\Delta W_F = [\Delta K + \Delta U_{\text{grav}} + \Delta U_{\text{spring}} + \Delta U_{\text{thermal}}]$$

An alternative form of this expression is known as the *work-energy relation*. It points out that the total work done by *all* the forces acting on a particle equals the change in kinetic energy of the particle. That is:

$$[\Delta W_F + \Delta W_{\text{grav}} + \Delta W_{\text{spring}} + \Delta W_{\text{thermal}}] = \Delta K$$

(Caution: In calculating the work done by each individual force, give careful attention to plus and minus signs for the various forms of work.) The total work done by all the forces acting on a particle is the same as the work done by the *resultant* force on the particle. Therefore:

WORK-ENERGY RELATION

$$\begin{bmatrix}\text{Work done by the} \\ \text{resultant force} \\ \text{acting on a particle}\end{bmatrix} = \Delta K$$

Power P is the time rate of doing work:

$$P = \frac{dW}{dt}$$

which may also be written as

$$\frac{dW}{dt} = \frac{\mathbf{F} \cdot d\mathbf{x}}{dt} = \mathbf{F} \cdot \mathbf{v}$$

Units of power are (N·m/s) = (J/s) = *watt* (W) in the metric system (SI) and (ft·lb/s) in the British engineering system. A common unit of power in engineering is the *horsepower* (hp). Relations between the various power units are:

$$1 \text{ hp} = 550 \frac{\text{ft}\cdot\text{lb}}{\text{s}} \approx 746 \text{ watts}$$

(exact) (approximate)

A useful rule-of-thumb approximation is 1 hp $\approx \frac{3}{4}$ kW.

The various characteristics of machines may be summarized as follows:

Efficiency e is the ratio

$$e = \frac{\text{Work output}}{\text{Energy input}}$$

(often expressed as the *percent efficiency*, $e \times 100\%$)

Actual Mechanical Advantage:

$$\text{A.M.A.} = \frac{\text{Output force}}{\text{Input force}}$$

Ideal Mechanical Advantage:

$$\text{I.M.A.} = \frac{\text{Distance the input force moves}}{\text{Distance the output force moves}}$$

Note that $$e = \frac{\text{A.M.A.}}{\text{I.M.A.}}$$

Questions

1. Give an example of a net force acting on a moving object for a finite length of time without doing any work on the object.
2. Give at least five examples of energy-conversion processes occurring in the room you now occupy.
3. A ball on the end of a cord is swung in a horizontal circle at constant speed. What net work is done on the ball in one revolution?
4. How many different meanings can you find for the word "*power*," other than its definition as used in physics?
5. Describe a system that has kinetic energy but no net momentum. Can you think of a system having momentum but no kinetic energy?
6. Check the consistency of the data claimed by a car manufacturer regarding the acceleration capability, the weight, and the maximum horsepower ratings for a car.
7. If you ride an elevator from the top of a tall building to the ground floor, what happened to the loss of potential energy you experienced?
8. A spring is compressed tightly and fastened in this position. It is then dissolved in acid. What happened to the potential energy stored in the compressed spring?
9. Two equal masses are connected by a strong spring. They are placed on the floor with one mass above the other. The upper mass is pushed down, compressing the spring, and then released. Is it possible to make the top mass pull the bottom mass off the floor by this technique? What is the net external force on the system? Does this net force do any work (that is, does it move through a distance)?
10. The block and tackle illustrated in Figure 5–20 will not work properly. Explain why, and what would happen if the free end of the rope were held fixed.
11. Two ropes hang down from a box, which conceals some sort of mechanical device. Experimentally, one finds that a force of 10 N on rope A will support a load of 30 N

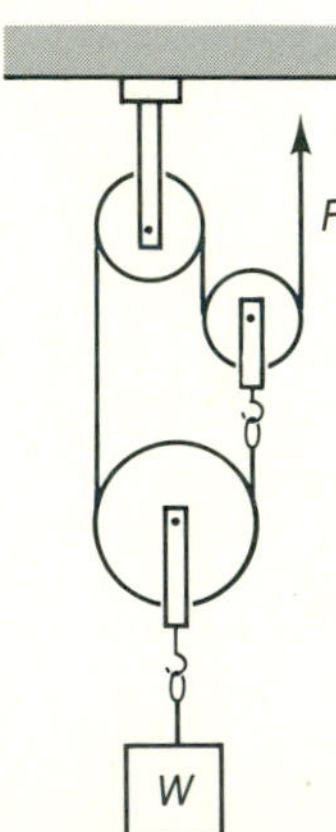

Figure 5–20
Question 10

on rope *B*. Assuming that friction is negligible, describe several different kinds of machines that might be in the box.

12. In the previous question, suppose doing 20 J of work on rope *A* results in the machine performing 18 J of work. Determine the efficiency of the machine and the actual mechanical advantage of the machine.

Problems

5A–1 Find the amount of work (in foot·pounds) it takes to raise 2 tons of roofing tiles from the ground to a roof 30 ft higher. *Answer:* 1.2×10^5 ft·lb

5A–2 Starting at rest, a man pushes with a constant force on a 4-kg box, accelerating it up an incline 6 m long at 30° with respect to the horizontal. Friction is negligible. At the top of the incline, the box has acquired a speed of 2 m/s.
- **(a)** Calculate the kinetic energy the box gained.
- **(b)** Calculate the gain in potential energy of the box.
- **(c)** Calculate the work the man did.
- **(d)** Calculate the constant force (parallel to the incline) the man exerted.

5A–3 A child weighing 200 N sits in a swing at rest. The ropes supporting the swing are 3 m long. A friend pushes horizontally on the swing seat, moving the child to a new (stationary) position such that the ropes are at 36.0° with respect to the vertical. Calculate the work done by the friend in moving the child to this new position. *Answer:* 120 J

5A–4 The force required to stretch a Hooke's Law spring varies from zero to 50 N as the end of the spring is stretched 12 cm from its unstressed position.
- **(a)** Find the force constant k of the spring.
- **(b)** Find the work done in stretching the spring.

5A–5 A woman does 1300 J of work in lifting a 12-kg bucket of water from rest 10 m up a well. Find the kinetic energy the bucket has at the top. *Answer:* 124 J

5A–6 A wad of paper with a mass of 2 g is thrown upward with a speed of 15 m/s. The wad reaches a height of 10 m above the point of release. Calculate the work done by the force of air resistance.

5A–7 Calculate how far an automobile would have to fall freely from rest in order to acquire the same kinetic energy it has going 60 mi/h. *Answer:* 121 ft

5A–8 A man lifts a 30-kg box from the ground to a location 1.5 m higher.
- **(a)** Find the amount of work the man has done against the force of gravity.
- **(b)** Find the work done by the force of gravity.
- **(c)** Find the total work done by the man and the force of gravity.

5A–9 A 40-kg child slides from rest down an amusement-park slide, descending a vertical distance of 4 m. Calculate the frictional thermal energy developed along the way if the child's speed at the bottom is 3 m/s. *Answer:* 1390 J

5A–10 At 30 mi/h, the force of wind resistance on an automobile is 400 lb. Determine the amount of horsepower required to overcome the air resistance.

5A–11 A tugboat pulls a raft of logs through the water at 3 m/s. The tension in the towing cable is 10^4 N. Find the rate at which the tugboat is doing work. *Answer:* 30 kW

5A–12 A 4-kg block is given an initial speed of 10 m/s on a horizontal surface. The coefficient of kinetic friction between the block and the surface is 0.30. Using energy principles, find out how far the block will slide.

5A–13 A 130-lb student runs up a flight of stairs (14-ft vertical change in elevation) in 2.6 s. Find the average horsepower that she developed. *Answer:* 1.27 hp

5A–14 Assuming an electrically operated hoist is 45% efficient, calculate the amount of mass that could be lifted vertically a distance of 3 m with an energy input of 2 kW·h.

5A–15 Derive the expression for the ideal mechanical advantage of the lever in Figure 5–18(a). (Assume forces move parallel to their motions.)

5A–16 Find the ideal mechanical advantage of the pulley system in Figure 5–21.

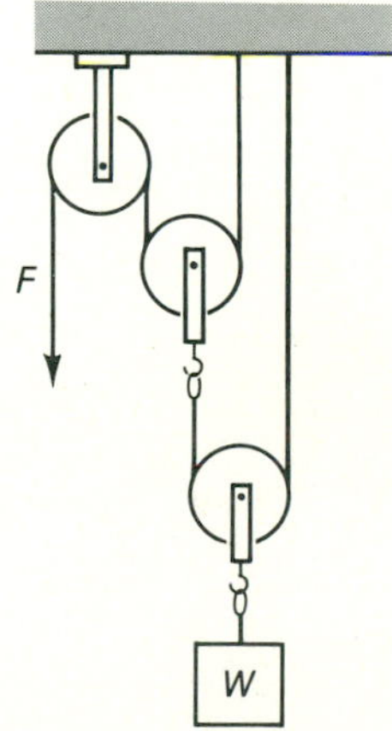

Figure 5–21
Problem 5A–16

5A–17 Derive the expression for the ideal mechanical advantage of the jackscrew in Figure 5–18(c).

5A–18 Show that the efficiency of a machine is equal to A.M.A./I.M.A.

5B–1 A man raises a 60-lb crate of potatoes from the ground, carries it 40 ft across level ground, and sets the crate on the bed of a truck 3 ft above the ground. In the physics sense, calculate the work he did. *Answer:* 180 ft·lb

5B–2 A 1.5-kg brick falls from rest from the top of a tall building. Calculate the work done by gravity during the first 2 s.

5B–3 A 50-kg crate slides from rest down a ramp inclined at 30° with respect to the horizontal.

(a) If the crate moves 4 m along the ramp, find the work done by the force of gravity.

(b) If the crate acquires a speed of 5 m/s, calculate the amount of thermal energy developed. *Answers:* (a) 980 J (b) 355 J

5B–4 A box rests on a rough loading ramp inclined at 27° with the horizontal. The box is shoved down the ramp with an initial speed of 2 m/s, and it comes to rest after sliding 3 m along the ramp. Find the coefficient of kinetic friction between the box and the ramp.

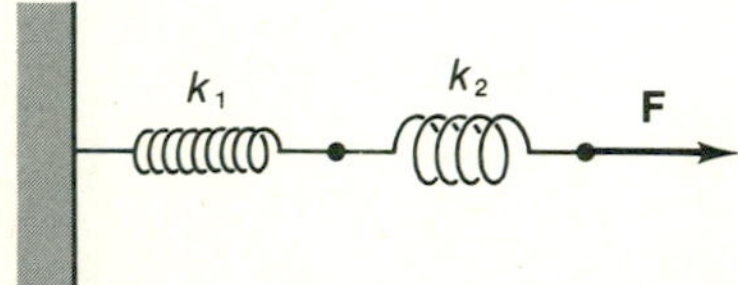

Figure 5–22
Problem 5B–8

5B–5 A bullet with a mass of 5 g and a speed of 600 m/s strikes a tree and penetrates the tree to a depth of 4 cm.

(a) Find the average frictional force stopping the bullet.

(b) Assuming the frictional force to be constant, determine how much time elapsed between the moment the bullet entered the tree and the moment it stopped. *Answers:* (a) 2.25×10^4 N (b) 1.33×10^{-4} s

5B–6 A children's pastime is to walk around with springs fastened on the bottom of each shoe. A 30-kg child wears identical Hooke's Law springs on each foot. When she stands on one foot only, the spring compresses a distance of 4 cm from its unstressed length. If she now leaps straight up and falls from rest a distance of 20 cm before the springs touch the ground, determine the maximum compression the springs will undergo if she lands with her weight distributed equally on both feet.

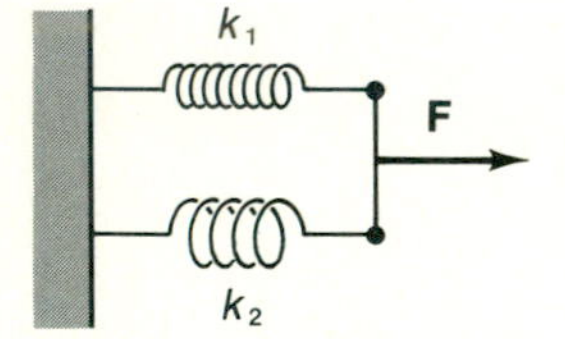

Figure 5–23
Problem 5B–9

5B–7 The ball launcher in a pinball machine has a spring whose force constant is 1.20 N/cm. The surface on which the ball moves is inclined at 10° with respect to the horizontal. If the spring is compressed a distance of 5 cm, find the launching speed of a 100-g ball when the spring is released. *Answer:* 1.68 m/s

5B–8 Two Hooke's Law springs with different spring constants, k_1 and k_2, are fastened end-to-end, as illustrated in Figure 5–22.

(a) Show that the combination acts like an equivalent single spring whose spring constant is $k_1k_2/(k_1 + k_2)$.

(b) What fraction of the total energy stored in the combination resides in the spring that has the spring constant k_1?

5B–9 Two Hooke's Law springs with different spring constants, k_1 and k_2, are fastened together as shown in Figure 5–23.

(a) Demonstrate that the combination acts like an equivalent single spring with a spring constant of $k_1 + k_2$.

(b) What fraction of the total energy stored in the combination resides in the spring that has the spring constant k_1?

Answer: (b) $\dfrac{k_1}{k_1 + k_2}$

5B–10 If it takes 4 J of work to stretch a Hooke's Law spring 10 cm from its unstressed length, determine the extra work required to stretch it an additional 10 cm.

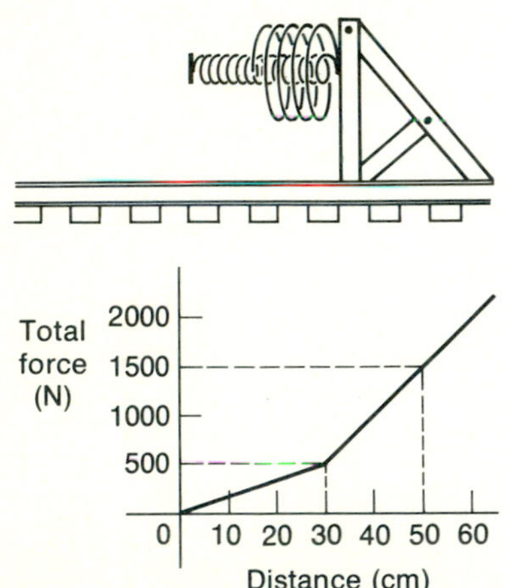

Figure 5–24
Problem 5B–11

5B–11 A 6000-kg freight car rolls along rails with negligible friction. The car is brought to rest by a combination of two coiled springs, as illustrated in Figure 5–24. Both springs obey Hooke's Law. After the first spring compresses a distance of 30 cm, the second spring (acting with the first) increases the force for additional compression, as shown in the graph. If the moving freight car is brought to rest 50 cm after first contacting the spring system, find its initial speed v_0. *Answer:* 0.303 m/s

5B–12 In the previous problem, determine the force constant of the second spring.

5B–13 A rope tow pulling skiers up a 30° slope 600 m long moves at 3 m/s and carries a maximum of 120 passengers at any one time. The average mass of each passenger is 80 kg. Neglecting friction, determine the power a motor must have in order to operate the tow under maximum load conditions. *Answer:* 141 kW

5B–14 A freight elevator has a mass of 4×10^4 kg. It is raised vertically 120 m in 20 s. Find the average rate at which the cable does work on the elevator.

5B–15 Electrical energy is usually sold by the kilowatt-hour (kW·h). At 5¢ per kW·h, calculate the cost per month (30 days) to burn a 100-W light bulb continuously. *Answer:* \$3.60

5B–16 A 3000-lb automobile traveling at 60 mi/h is braked to a stop in 10 s. Assuming a constant stopping force, find (a) the amount of work done by the brakes in stopping the automobile and (b) the average power expended during the braking.

5B–17 A racing car weighing 1000 lb accelerates uniformly to 100 mi/h in a distance of a quarter-mile. Calculate the average horsepower expended during the acceleration, assuming 30% of the power output of the engine goes toward overcoming friction. *Answer:* 44.2 hp

5B–18 The force required to tow a barge through the water at a constant speed is directly proportional to the speed. If it takes a 3 kW motor to tow the barge at 2 m/s, determine the motor power required to tow the barge at 4 m/s.

5B–19 If 40% of the energy output of a motor goes into frictional losses, find the power (in kilowatts) a motor must have in order to raise a 1200-kg elevator 60 m in $\frac{1}{2}$ min. *Answer:* 39.2 kW

5B–20 A typical washing machine motor is rated at $\frac{1}{2}$ hp. If solar cells trap sunlight and convert it to electrical energy with 15% efficiency, calculate the area of solar cells perpendicular to the incoming sunlight required to operate the washing machine. Solar energy per unit of time at normal incidence at the top of the earth's atmosphere is 1340 W/m^2. Atmospheric absorption reduces this to about 840 W/m^2 at sea level (where the washing machine is located).

5B–21 Suppose the block and tackle of Example 5–14 is used to pull a car from a mudhole. A nearby tree provides an anchor point. Decide which end of the block-and-tackle system should be fastened to the tree: the single-pulley end or the double-pulley end. Why?

5B–22 Determine the efficiency of the block-and-tackle system from Example 5–14.

5B–23 Find the weight of a load that can be raised with the jackscrew shown in Figure 5–18(c) if the applied force is 80 N, the lever arm distance is 40 cm, the screw threads are 8 mm apart, and the efficiency of the device as a machine is 70%. *Answer:* 1.76×10^4 N

5B–24 Explain the change-of-scale reasoning that concludes that a Lilliputian can jump vertically 12 times his own height, while a Brobdingnagian can jump only $\frac{1}{12}$ his height. Start with the assumption that a man can jump vertically a distance equal to his height.

5C–1 A railroad flatcar of mass 2×10^4 kg is coasting at 2 m/s without friction along a level track. The car collides with a fixed bumper spring at the end of the track, causing a maximum compression of the spring by 0.50 m. If the spring is constructed so that when compressed a distance x it exerts the force $F = -kx^2$, determine the force constant k (including units). *Answer:* 9.6×10^5 N/m^2

5C–2 A certain spring force varies according to $F = -kx^3$ (where $k = 200$ N/m^3) rather than according to the Hooke's Law relationship. Find the work done in stretching the spring from $x = 0.1$ m to $x = 0.2$ m.

5C–3 A varying force acting on a 4-kg object causes it to have a displacement given by $x = 2t - 3t^2 + t^3$ (where x is in meters and t is in seconds). The object starts at rest. Find the work this force does on the object in the first 3 s of motion. (Hint: What is the velocity as a function of time?) *Answer:* 242 J

5C–4 A particle of mass m initially at rest is subjected to a net force of the form $F = F_0 \sin 2\pi(t/a)$ (where F_0 and a are constants with appropriate units).

(a) Derive an expression for the maximum speed acquired by the mass.
(b) Determine how far the mass moves in acquiring the maximum speed.

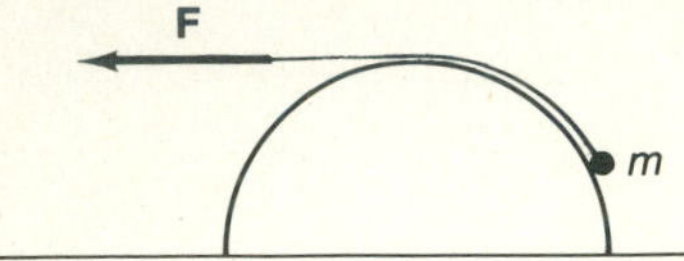

Figure 5–25
Problem 5C–5

5C–5 A small mass m is pulled to the top of a frictionless half-cylinder by a cord passing over the top of the cylinder, as illustrated in Figure 5–25. By directly integrating $W = \int \mathbf{F} \cdot d\mathbf{s}$, find the work done in moving the mass from the bottom to the top of the half-cylinder at constant speed. The radius of the cylinder is R. *Answer:* mgR

5C–6 Two spheres, each of mass m and radius R, are initially at rest with their centers separated a distance D ($D > 2R$). The spheres are attracted toward each other by a force inversely proportional to the square of the distance x between their centers. That is, $F = k/x^2$, where k is a constant with appropriate units. If one sphere is held fixed and the other is released, find the speed of the moving sphere when it collides with the other sphere.

5C–7 One of the kinematic equations is $v^2 = v_0{}^2 + 2a(s - s_0)$. Derive this equation from work-energy relationships and state any assumptions that are necessary.

5C–8 Starting at rest, a 5-kg object is acted upon by the variable force indicated in Figure 5–26. Find the total work done by the force. (Hint: If you can find the acceleration versus time, you can determine the velocity versus time, and hence the value of the velocity at $t = 3$ s.)

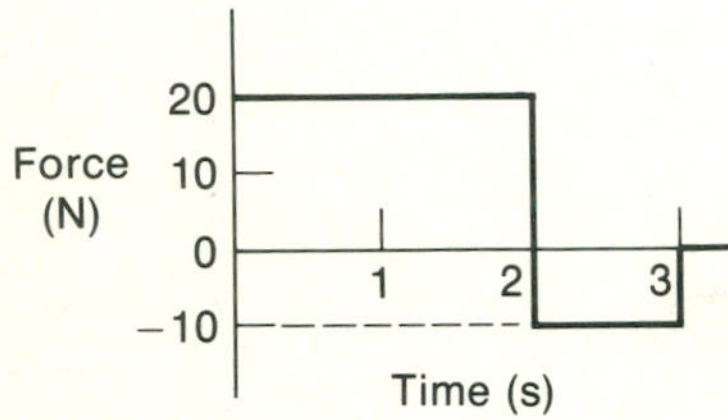

Figure 5–26
Problem 5C–8

5C–9 A mass m falls from rest under the influence of gravity. Show that the average power supplied to the mass by the force of gravity is $P_{\text{ave}} = m\sqrt{g^3h/2}$ where h is the distance the mass falls.

5C–10 While a 1500-kg automobile is moving at a constant speed of 15 m/s, the motor supplies 15 kW of power to overcome friction, wind resistance, and so forth.
(a) Find the effective retarding force due to all friction effects combined.
(b) Find the power the motor would need to supply to drive the automobile up an 8% grade (8 m vertically for every 100 m horizontally) at 15 m/s.
(c) Determine the downgrade that the car would coast at 15 m/s.

5C–11 Two different Hooke's Law springs (with force constants k_1 and k_2) are fastened together and stretched a total distance L, as illustrated in Figure 5–27. The unstressed lengths of the springs are ℓ_1 and ℓ_2, where $L > (\ell_1 + \ell_2)$. Find the equilibrium position x of the point P joining the two springs.

Answer: $\dfrac{k_1\ell_1 + k_2(L - \ell_2)}{k_1 + k_2}$

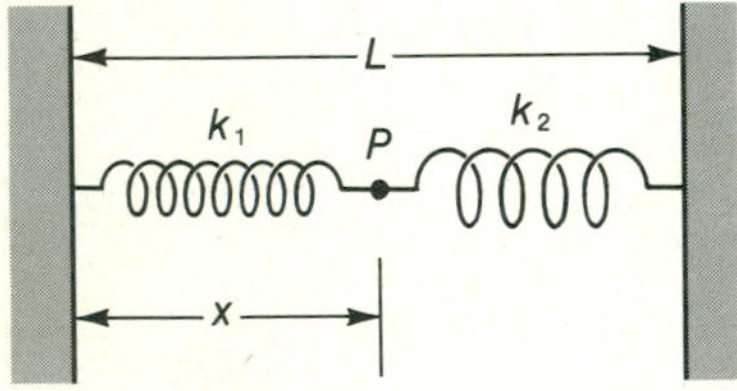

Figure 5–27
Problem 5C–11

5C–12 Derive the expression for the ideal mechanical advantage of the differential chain hoist in Figure 5–18(d). (Hint: Consider the amount of chain that moves when one revolution of the upper pulley occurs. Remember that the work input equals the work output.)

5C–13 Explain the change-of-scale reasoning that concludes that a Lilliputian can support 24 other Lilliputians, while it takes six Brobdingnagians to support just one other. As a starting point, assume that a man can support twice his weight.

Energy is eternal delight.

WILLIAM BLAKE

chapter 6

Conservative Forces and Conservation of Energy

6.1 Introduction

The previous chapter described the different forms of energy involved when various types of forces act through a distance. We also stated the important fact that the total energy in an isolated system remains the same throughout any process—the conservation of energy principle. Let us now discuss the conservation of energy from another viewpoint.

6.2 Conservative Forces

In analyzing mechanical systems, it is convenient to distinguish between two classes of forces: *conservative* and *nonconservative*. To introduce this idea, it will be simpler to first discuss a few specific examples of conservative forces, then later give a more general definition. The word "conservative" is used for these forces because they are closely linked with the conservation of energy principle.

Consider a particle of mass m acted upon by the force of gravity and some other force **F**. Figure 6–1 shows the particle as it moves under the action of these forces along an arbitrary path in the x-y plane. The (constant) force of

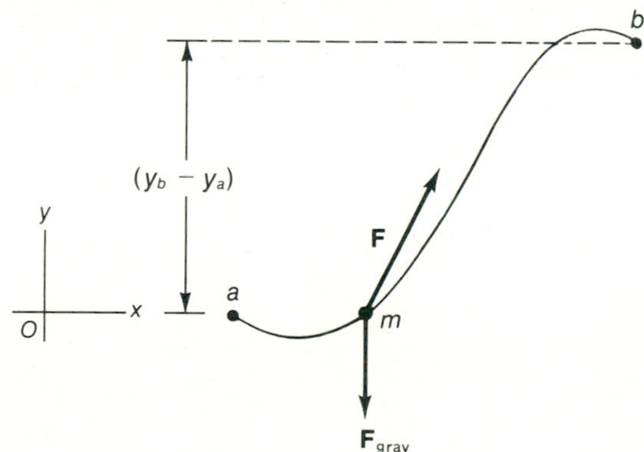

Figure 6–1

As the particle is moved along the curved path from a to b, the work done by the gravitational force $\mathbf{F}_{grav}$ depends only on the difference in elevation $(y_b - y_a)$, and not on the particular path followed.

gravity $\mathbf{F}_{grav}$ acts always in the downward direction, while the other force **F** may vary in any way whatsoever. Let us focus our attention on the gravitational force.

In Chapter 5 it was shown that in moving from a to b, the *negative* of the work done by the force of gravity[1] is equal to the change in gravitational potential energy mgy of the particle–earth system. That is (assuming the $+y$ axis upward):

$$-(\mathbf{F}_{grav} \cdot \Delta\mathbf{y}) = mgy_b - mgy_a \qquad \textbf{(6–1)}$$

Note that no work is done by gravity for *horizontal* motion (because of the 90° angle in the dot product: $\mathbf{F}_{grav} \cdot \Delta\mathbf{x}$). The force of gravity does work only for *vertical displacements*. Every curved path may be thought of as a series of infinitesimal horizontal and vertical steps in which work is done by the gravitational force only along the vertical steps. Thus the work done depends only on the difference in elevation $(y_b - y_a)$, and not on the particular path followed. In Figure 6–2, all paths between a and b involve the same net change in gravitational potential energy and thus the same net work done by the gravitational force. These characteristics of $\mathbf{F}_{grav}$ constitute one approach to defining a conservative force.

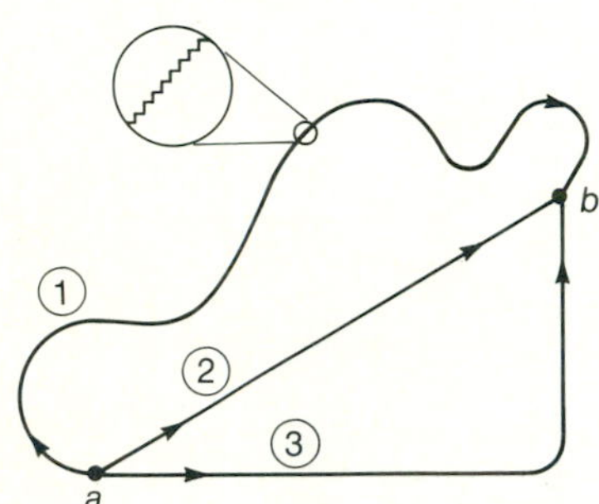

Figure 6–2
All paths between a and b involve the same change in gravitational potential energy. As shown in the enlarged segment, a curved path may be represented by a series of infinitesimal horizontal and vertical steps; the force of gravity does work only along the vertical steps.

Suppose we contrast the situation just described with one in which friction is involved. If we apply a force **F** to move an object along a path where friction is present, the amount of thermal energy produced by friction will depend on the length of the path. Thus the work done in moving the object from point a to point b will vary for different paths. The work done is *not* independent of the path. Furthermore, this thermal energy is not "recoverable" in the same sense that mechanical energy—kinetic and potential—is recoverable; that is, it cannot be easily transformed from one form to another.

As a concrete example, suppose an object is thrown straight upward. The only force acting is gravity. While ascending, the negative work done by gravity is related to the increase in the potential energy of the object. Upon descending, the reverse is true: the positive work done by gravity is related to the decrease in potential energy. The exact transformation of the work done into a change in the mechanical energy of the system (and the complete reverse change if the process is reversed) is the central idea of a conservative force. Thus gravity is a conservative force.

Referring again to Figure 6–2, note that if we follow any path backward from b to a, the change in potential energy (and the work done by gravity) is just the *negative* of the value for the forward paths. Therefore, for any round trip that forms a complete, closed loop, the net work done by the force of gravity is zero. For example, in Figure 6–2, if we traveled from a to b along path ① and returned via path ③, the total work done by the gravitational force in this round trip would be zero. This fact furnishes a second way to define a conservative force, namely, that the net work done in moving a mass around any closed loop is zero. In mathematical notation,[2] this is written as

FOR ALL CONSERVATIVE FORCES

$$\underbrace{\Sigma(\mathbf{F} \cdot \Delta\mathbf{r})}_{\text{Closed path}} = 0 \qquad \textbf{(6–2)}$$

---

[1] In this chapter we discuss gravitational forces for motion near the earth's surface, where g is essentially constant. Even for motion far from the earth's surface where the gravitational acceleration varies as $1/r^2$, the gravitational force is still a conservative one.

[2] In integral form, the notation is

$$\oint \mathbf{F} \cdot d\mathbf{r} = 0$$

where the circle on the integral sign means that the path of integration must form a closed loop.

In a similar way, ideal spring forces are conservative. The negative work done by the spring force when it is stretched is related to the increase in the potential energy of the stretched spring. Upon returning to its unstressed position, the spring force does positive work related to the decrease in potential energy of the spring. Thus spring forces are conservative forces.

To summarize, a **conservative force** may be characterized in either of two ways. A force is conservative if:

CONSERVATIVE FORCE

(a) the net work done in moving a mass between two points depends only on the location of the points and not on the particular path followed

or (b) the net work done in moving a mass through any round-trip path is zero

Either of these ideas may be used to identify a conservative force; the two criteria are equivalent. If an isolated system involves only conservative forces, it is said to be a **conservative system.**

Some common examples of conservative forces include all those forces that depend only upon the distance from a point and that have only a radial component. The *gravitational* force, the *electrostatic* force, and the *ideal spring* force are of this type. Examples of *non*conservative forces include friction and spring forces that exceed their limits of elasticity.

A subtle point regarding friction should be clarified. On a *microscopic* scale, forces of friction are just those between atoms and molecules. Strictly speaking, these, too, are conservative forces. The random motions of the molecules represent continuous exchanges between the kinetic energies of motion and the potential energies due to the varying distances between atoms. On this level, the system as a whole is a conservative one.

But for many physical situations, we concentrate our attention on the *large-scale* motions of objects. We concern ourselves with the motions on an atomic scale only to include them as "thermal energy" when it is present. In fact, kinetic friction is sometimes called a *dissipative* force, since it causes some mechanical energy to "disappear" into the form of thermal energy. In the absence of frictional effects, mechanical energy associated with the large-scale motions of objects is conserved if conservative forces are the only forces involved. For these cases, no mechanical energy "disappears" as heat.

Thus the distinction regarding conservative and nonconservative forces is a bit artificial when we include both the microscopic and macroscopic views. But it is a convenient distinction, because so frequently we keep track only of the large-scale mechanical motions of objects.

EXAMPLE 6–1

Show, for the one-dimensional case, that *all* forces that are constant in magnitude but opposite to the direction of motion are nonconservative. An example of such a force would be the frictional force on an object launched up a rough inclined plane which then returns to its starting point.

SOLUTION

Consider a round-trip motion along the x axis between two points a and b. From a to b, $\mathbf{F}_x$ is negative while the displacement $\mathbf{x}$ is positive. On the return trip, the reverse is true. Therefore:

$$\text{Work}_{a\to b} = \mathbf{F}_x \cdot \mathbf{x} = (-F_x)(b-a)$$

and

$$\text{Work}_{b\to a} = \mathbf{F}_x \cdot \mathbf{x} = (F_x)(a-b) = -(F_x)(b-a)$$

The total work for the round trip is

$$\text{Work}_{a\to b} + \text{Work}_{b\to a} = -2F_x(b-a)$$

If the force were conservative, the net work done in a round trip would be zero. Therefore, the force F is nonconservative. Friction is an example of this type of force.

6.3 Conservative Forces and Potential Energy

For every conservative force, one can define an associated potential energy. In fact, it is *only* for conservative forces that this can be done. In each case, the potential energy expression is a function only of the position. We have previously discussed two examples:

Conservative force	Associated potential energy function
1. Gravitational force:	
$F_{\text{grav}} = -mg$ (positive direction is vertically upward)	$U_{\text{grav}} = mgy$ where $U_{\text{grav}} \equiv 0$ at $y = 0$
2. Spring force:	
$F_{\text{spring}} = -kx$ (x is the distance the spring is stretched or compressed from its relaxed length)	$U_{\text{spring}} = \frac{1}{2}kx^2$ where $U_{\text{spring}} \equiv 0$ at $x = 0$

For one dimension, the relation between a conservative force $F(x)$ and the associated potential energy $U(x)$ is expressed as follows:

RELATION BETWEEN A CONSERVATIVE FORCE AND ITS RELATED POTENTIAL ENERGY FUNCTION

In summation[3] form:

$$U_b - U_a = -\sum_{a\to b} [F(x)\,\Delta x] \tag{6-3}$$

In derivative form:

$$F(x) = -\frac{d}{dx}[U(x)] \tag{6-4}$$

An important advantage of analyzing problems in terms of energy rather than forces is that energy is a *scalar* quantity. When several forces are present

[3] In calculus notation, the integral form is

$$U_b - U_a = -\int_a^b F(x)\,dx \qquad \text{[6-3(a)]}$$

at the same time, acting in various directions, the mathematics of *scalar* quantities are simpler than those of *vector* relations for forces. Furthermore, in many cases (particularly those dealing with small systems such as atoms and nuclei), the energy state of a system can be measured fairly easily, whereas the forces themselves cannot.

Note that the energy a system possesses is unique to a particular frame of reference. For example, a mass with finite kinetic energy in one frame of reference will have zero kinetic energy in a coordinate frame moving with the mass.

Also note that potential energy is always a property of a system, not just a single particle. It depends solely on the relative separation of the components of the system. True, we sometimes speak of the gravitational potential energy mgy of a particle of mass m located a distance y above some zero reference level. But, strictly speaking, it is the particle–earth system that shares the potential energy as a whole. If we drop a ball, it accelerates toward the earth; simultaneously, the earth accelerates toward the ball. But because of the great difference between the masses of the two objects ($m_{\text{ball}} \ll m_{\text{earth}}$), the motion of the earth is negligible and we therefore ignore the kinetic energy due to the earth's motion. Of course, for astronomical systems in which both masses are of comparable size (the earth–moon system, for example), we must treat the system as a whole and keep track of the motion of each component.

One interesting characteristic of potential energy is that although we have a universal formula for kinetic energy, $K = \frac{1}{2}mv^2$, we cannot write a single formula that is valid for all forms of potential energy. In each case, the potential energy a system possesses depends on the positions of all the parts of the system relative to one another, whether it is a system of masses connected by springs, a system of stars attracted to each other by gravity, or a system of electrically charged particles. Furthermore, the changes in potential energy that occur as the system moves from one configuration to another must be calculated for each particular system as a whole.

6.4 Nonconservative Forces

Forces we call **nonconservative** are those that do not meet the stated criteria for conservative forces—for example, friction. Suppose the mass shown previously in Figure 6–2 were being moved through a viscous medium so that a force of friction acted on it. Because frictional forces are opposite to the direction of motion, the work done by a frictional force in a round-trip journey cannot be zero. Also, the work done in traveling from a to b would not be independent of the path, since friction would develop more thermal energy for the longer paths than for the shortest one.

Another example of a nonconservative force occurs when a spring is stretched beyond its elastic limit, so that the material of the spring becomes permanently deformed. This is a nonconservative situation because some energy remains trapped thereafter in the deformed material. Thus the work done in stretching the spring from a to b is not equal to the negative of the work done when the spring moves back from b to a.

Some forces, such as aerodynamic drag on high-speed aircraft, are velocity-dependent. These forces also are nonconservative, since the work done depends on the *speed* of travel along the path and thus violates the criterion that in moving between two points, the work done depends only on the location of the two points. Similarly, if a force depends explicitly on *time*, it is a nonconservative force. In addition, there are also certain electromagnetic forces of induction that are nonconservative. It should be emphasized, however, that even in the presence of nonconservative forces, the *overall* conservation of energy for an isolated system still holds, provided we consider *all* forms of energy—not just mechanical energy but also such internal forms as thermal energy and energies associated with the permanent deformation of materials. In the broad sense, nature always seems to obey the conservation of energy principle.

This principle may seem straightforward and almost self-evident when it is presented, as we have done, in an abbreviated and oversimplified fashion. Yet it took many decades of struggle and heated argument during the nineteenth century to arrive at the clear understanding we have today. Many investigators approached the central ideas independently from different directions, leading to some bitter controversies regarding priority and rightful claims to fame.[4]

6.5 The Energy Theorem

In the previous chapter we described the work-energy relation, which states that the total work done by all the forces acting on a mass equals the change in kinetic energy of the mass. This relation is valid for all types of forces, conservative or not. But because many physical systems involve only conservative forces (or negligible nonconservative forces), it will be useful to derive a statement of the conservation of energy for these special cases.

Consider an object acted upon by several forces, all of which are conservative. The net force $\mathbf{F}_{\text{net}}$ is therefore also conservative. Then, the work done by these conservative forces may be written as the negative of the change in the associated potential energies.

$$\Sigma(\mathbf{F}_{\text{net}} \cdot \Delta\mathbf{x}) = -(U_b - U_a) \qquad \textbf{(6–5)}$$

By the work-energy relation

$$\Sigma(\mathbf{F}_{\text{net}} \cdot \Delta\mathbf{x}) = K_b - K_a$$

we have

$$-(U_b - U_a) = K_b - K_a$$

$$U_a + K_a = U_b + K_b \qquad \textbf{(6–6)}$$

This implies that the *total mechanical energy* E (equal to U plus K) of a conservative system is the same at point a as at point b. Since these points are arbitrary, we may express the idea in more general terms. If E_0 represents the initial energy, the energy E at any later time will be the same. This principle is the **conservation of mechanical energy** when conservative forces are the only significant forces present.

CONSERVATION OF MECHANICAL ENERGY

$$E_0 = E \qquad \textbf{(6–7)}$$

where $E_0 = U_0 + K_0$ and $E = U + K$.

Recall that the term "mechanical energy" refers to potential and kinetic energies combined; it specifically excludes thermal energy arising from friction or other forms of energy associated with nonconservative forces. The potential energy U includes all forms of potential energy: gravitational energy, the energy stored in a stressed spring, electrical potential energy, and so on. In solving problems involving conservation of mechanical energy, it is important to always specify a zero reference location where $U \equiv 0$. For motion under the action of gravity, usually the lowest point of interest is the most convenient.

[4] For an interesting discussion, see "Where Credit Is Due—The Energy Conservation Principle," V. V. Raman, *The Physics Teacher*, February 1975.

In the analysis of complicated mechanical systems it is particularly valuable to discover quantities that remain constant throughout a process. These quantities are called *constants of the motion*; the total mechanical energy E is one of these quantities (provided that the system is isolated and that only conservative forces act). Thus, regardless of the particular processes that occur in the system, the total energy at the start equals the total energy at the end, and we need not concern ourselves with the details of the interactions.

EXAMPLE 6–2

Suppose a mass m, initially at rest, is dropped from a height h above the floor (see Figure 6–3). Using the appropriate energy equations, derive an expression for the speed v of the mass just before it strikes the floor.

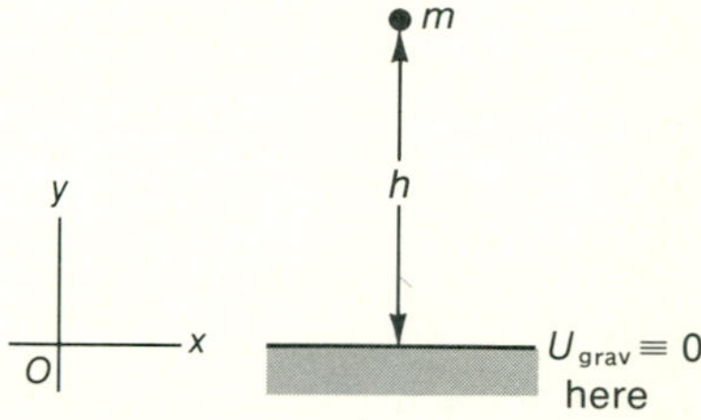

Figure 6–3
Example 6–2

SOLUTION

Since the only force present is the *conservative* force of gravity, the total mechanical energy $E = U + K$ is constant. For convenience, we choose the zero reference level for potential energy to be the lowest point of interest in the problem: the floor. That is, $U_{grav} \equiv 0$ for $y = 0$. Then, equating initial and final energies, we have

$$E_0 = E$$
$$U_0 + K_0 = U + K$$
$$mgy + 0 = 0 + \tfrac{1}{2}mv^2$$
$$\boxed{v = \pm\sqrt{2gh}}$$

The minus sign is the appropriate one by our choice of coordinate systems. The plus sign indicates that the mass, moving solely under the action of gravity, could have been traveling upward at $y = 0$ with the indicated speed and "latched on" to our problem with zero kinetic energy at $y = h$. In other words, the plus sign refers to the possible motion of the mass before the beginning of our problem and is therefore not a relevant answer.

We could have obtained the same answer using Newton's laws of motion.

$$\Sigma \mathbf{F} = m\mathbf{a}$$
$$-mg = ma$$
$$a = -g$$

Substituting the appropriate values into the kinematic equation

$$v^2 - v_0{}^2 = 2a(y - y_0)$$

we have

$$v^2 - 0 = 2(-g)(0 - h)$$
$$v^2 = 2gh$$
$$\boxed{v = \pm\sqrt{2gh}}$$

Note that in this problem, application of the energy theorem leads to the solution more directly than an application of Newton's laws of motion. This is very often the case when conservative forces are the only ones present.

EXAMPLE 6–3

Starting with an initial speed v_0, a mass m slides down a curved frictionless track, arriving at the bottom with a speed v (see Figure 6–4). From what vertical elevation h did it start?

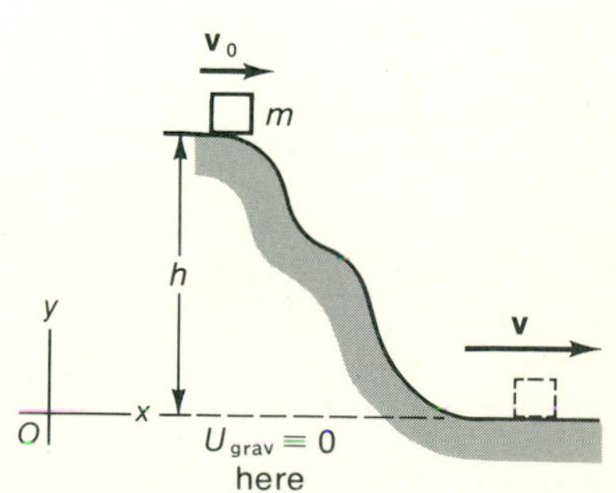

Figure 6–4
Example 6–3

SOLUTION

The force the frictionless track exerts on the mass is the normal force, which is always at right angles to the motion of m. Therefore, this force does no work and is not involved in any energy transfer considerations. The conservative force of gravity is the only other force present, so we have conservation of mechanical energy. Establishing the zero reference level for potential energy at the bottom of the track, we have

$$E_0 = E$$
$$U_0 + K_0 = U + K$$
$$mgh + \tfrac{1}{2}mv_0^2 = 0 + \tfrac{1}{2}mv^2$$

Solving for h gives us

$$\boxed{h = \frac{(v^2 - v_0^2)}{2g}}$$

(Note: A common error in this type of problem is to apply the kinematic equation $v^2 = v_0^2 + 2ay$, obtaining an answer in the above form. However, this reasoning is incorrect. The displacement of the mass as it moves along the curved track is not h, nor does the mass have constant acceleration a. Remember that the kinematic equations apply only for *straight-line motion* with *constant acceleration*.)

6.6 The Simple Harmonic Oscillator and Energy Graphs

Figure 6–5 depicts a Hooke's Law spring attached to a mass m which rests on a horizontal frictionless surface. At the *equilibrium position* $x = 0$, the spring is unstressed and the net force on the mass is zero. If we displace the mass from the equilbrium position and release it, the mass will oscillate back and forth in a manner called **simple harmonic motion**[5] (SHM). Vibrating or oscillating motions such as this are important in many areas of physics, and all are analyzed in terms of SHM. Some examples are pendulum motion, vibrations of air molecules as a sound wave passes by, and oscillations of the balance wheel in watches.

[5] The adjective "simple" means that it involves just a *single* frequency of oscillation, and the word "harmonic" refers to the fact that the equations for describing the motion are the *harmonic* functions in mathematics (sines and cosines).

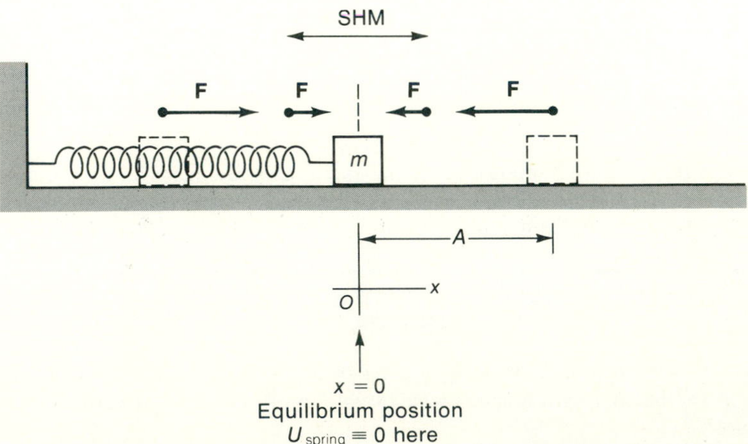

Figure 6–5
Simple harmonic motion. The force $F = -kx$ on the mass is shown for several different positions. It is a *restoring* force, acting always toward the equilibrium position at $x = 0$.

In SHM, the net force on the object is the Hooke's Law relation $F = -kx$. It has two characteristics:

(a) It is a *restoring* force. That is, if the object is displaced from its equilibrium position toward the right, the force is toward the left; if the object is displaced toward the left, the force is toward the right.
(b) The force is *proportional to the displacement* away from the equilibrium position. The constant of proportionality is k.

In the figure, the SHM is along the $\pm x$ direction under the action of the Hooke's Law force

$$F = -kx \tag{6-8}$$

The potential energy function associated with this force is

$$U = \tfrac{1}{2}kx^2 \tag{6-9}$$

We set the zero reference point for potential energy $U \equiv 0$ at the equilibrium position $x = 0$.

The maximum displacement away from the origin is called the **amplitude** A of the motion. At the locations $x = \pm A$, the mass momentarily has zero velocity as it changes direction, so the kinetic energy is zero at these extreme positions. With friction absent the system is conservative, and for the total mechanical energy (at $x = \pm A$), we have

$$E = U + K$$
$$E = \tfrac{1}{2}kA^2 + 0$$
$$E = \tfrac{1}{2}kA^2 \tag{6-10}$$

The total mechanical energy E thus depends only upon the value of the *force constant* k and the *amplitude* A. Since the kinetic energy is always $\frac{1}{2}mv^2$, we may write a general expression for the mechanical energy at all points of the motion:

$$E = U + K$$
$$\tfrac{1}{2}kA^2 = \tfrac{1}{2}kx^2 + \tfrac{1}{2}mv^2 \tag{6-11}$$

The relationships between these energies are illustrated graphically in Figure 6–6. The potential energy U is shown by the dashed curve, the kinetic

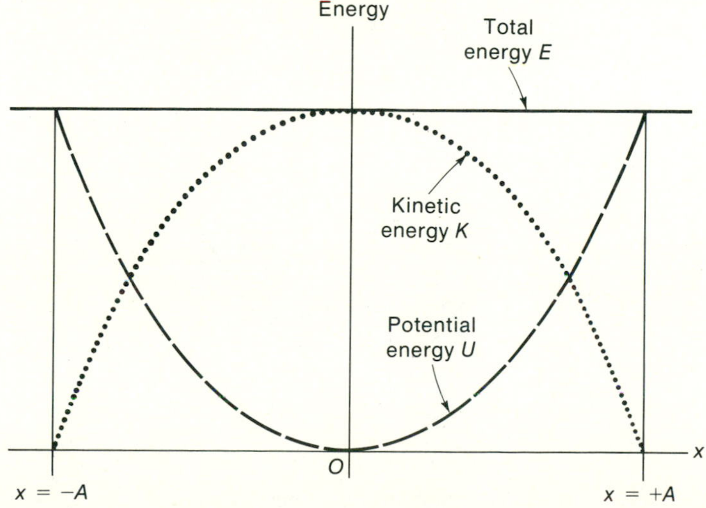

Figure 6–6
An energy diagram for simple harmonic motion along the x axis. Curves are shown for the kinetic energy K, the potential energy U, and the total energy E. For a given total energy E, the mass moves between limits $x = \pm A$ defined by the points of intersection of the potential energy curve U and the total energy E. At all times, $U + K = E$.

energy K by the dotted curve, and the total energy E by the solid line. Note that since $\frac{1}{2}mv^2$ is always positive, the only meaningful part of the graph is the range of x where the kinetic energy curve is above the axis. Thus the mass cannot travel beyond the limits $x = \pm A$, sometimes called the "turning points" of the motion. At these points the potential energy rises to the value of E, the total energy. In SHM, the curves for both U and K are parabolic in shape, symmetrical about the equilibrium position.

From the graph we see that when the kinetic energy K is a maximum the potential energy U is zero, and, conversely, when the potential energy U is a maximum the kinetic energy K is zero. Thus we may express the (constant) total energy $E = U + K$ in either of two ways:

$$\text{Total energy:} \quad E = \tfrac{1}{2}mv_{max}^2 \quad \text{(the maximum kinetic energy)} \tag{6–12}$$

or

$$\text{Total energy:} \quad E = \tfrac{1}{2}kA^2 \quad \text{(the maximum potential energy)} \tag{6–13}$$

Note that the total energy is proportional to the *square* of the amplitude A. As will be shown later, half the energy in SHM is in kinetic form and half is in potential form, on the average. This equal sharing of the total energy in these two forms is a notable characteristic of simple harmonic motion.

For systems involving conservative forces only, a sketch of the potential energy curve often aids greatly in understanding the motion. Figure 6–7 shows a potential energy curve versus displacement for one-dimensional motion that is more complicated than SHM. The total energy E is represented by the horizontal line, since it has the same constant value throughout the motion. At an arbitrary location such as $x = a$, the value of U and K add together to give E. Furthermore, a particle with the value of E is confined to move between the limits marked off by the turning points. The particle is said to be "trapped within a potential well." As the particle moves back and forth between these limits, the motion is more complicated than SHM, yet the mechanical energy relation still holds. The energy shifts from one form to another, with the total energy E remaining constant in time.

We can extend these ideas a bit further to imagine a frictionless roller coaster constructed in the shape of the potential energy curve. Then, if a particle of mass m is allowed to slide along the roller coaster under the action of gravity, it will travel forever, moving back and forth between the turning-point limits of the motion. At any particular *horizontal* displacement, the particle has kinetic and potential energies proportional to those of the actual mass moving in *its* one-dimensional motion. This analogy also provides a convenient way to remember the minus sign in $F_x = -dU/dx$. In regions where the slope dU/dx is positive ("uphill" as x increases), the force is in the $-x$ ("downhill") direction.

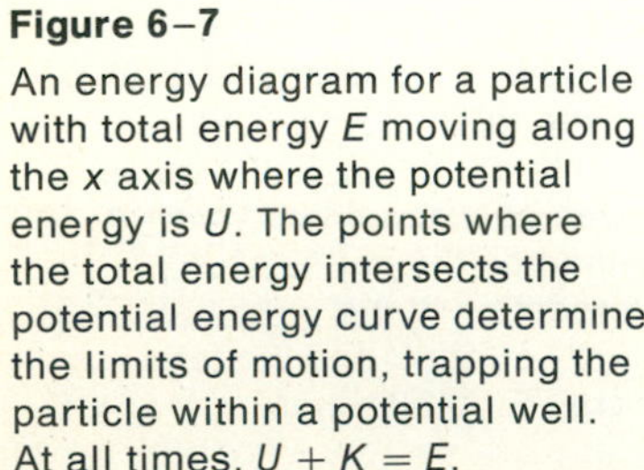

Figure 6–7
An energy diagram for a particle with total energy E moving along the x axis where the potential energy is U. The points where the total energy intersects the potential energy curve determine the limits of motion, trapping the particle within a potential well. At all times, $U + K = E$.

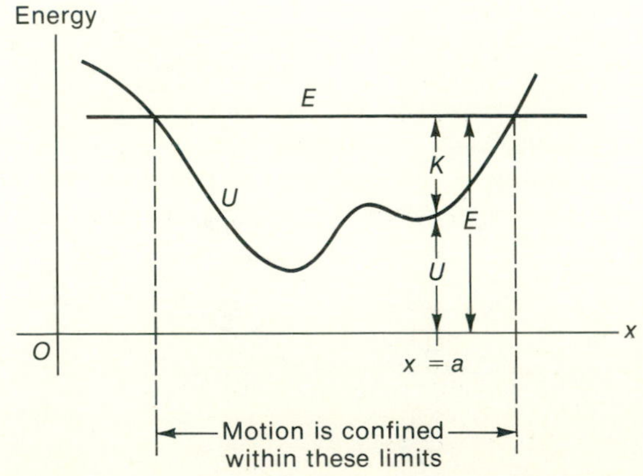

The roller-coaster analogy works because the gravitational potential energy for the particle on the roller coaster is proportional to the vertical displacement y, just as in the mathematical graph the potential energy is proportional to the vertical (energy) axis. In using the analogy, however, we must remember that the actual motion of the mass being analyzed is in *one* dimension, rather than the two physical dimensions of the roller coaster. Furthermore, the acceleration of the mass along the roller coaster is not the same as the acceleration of the actual mass in one-dimensional motion. So the analogy, while useful, must be used with care.

EXAMPLE 6–4

A 2-kg mass moves along a line under the influence of a force that is described by the potential energy

$$U = \left(6.5\,\frac{\text{J}}{\text{m}}\right)|x|$$

where x is the distance (in meters) along the line from the reference point (see Figure 6–8). The total energy E of the mass is 10 J. (a) Determine the distance the mass will travel from the position of equilibrium before reversing direction. (b) Find the maximum speed acquired by the mass.

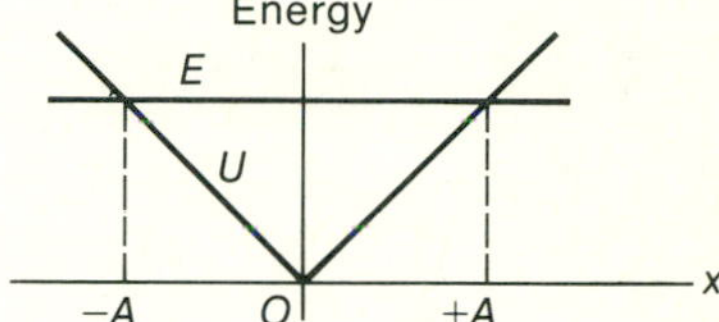

Figure 6–8
Example 6–4

SOLUTION

(a) At the turning point, the kinetic energy is zero. From the expression for the mechanical energy, we have

$$U + K = E$$

$$\left(6.5\,\frac{\text{J}}{\text{m}}\right)|A| + 0 = 10\text{ J}$$

$$\boxed{A = 1.54\text{ m}}$$

(b) The speed is a maximum when the potential energy is a minimum (at $x = 0$). Again, from

$$U + K = E$$

we have

$$0 + \tfrac{1}{2}mv_{max}{}^2 = E$$

$$\tfrac{1}{2}(2\text{ kg})v_{max}{}^2 = 10\text{ J}$$

$$v_{max} = \pm\sqrt{10}\,\frac{\text{m}}{\text{s}} = \boxed{\pm 3.16\,\frac{\text{m}}{\text{s}}}$$

The plus or minus sign in the result means that the mass may be moving in either direction. Both values are correct.

EXAMPLE 6–5

A particle of mass m is moving in a vertical circular path under the action of gravity, as shown in Figure 6–9. The particle is held in this path by a string of length ℓ between the particle and the center of the circle. The speed v_t at the top of the path is sufficient to keep the string taut at all times. Find the speed of the particle at the bottom. The only forces acting on the particle are the force of gravity and the tension in the string.[6]

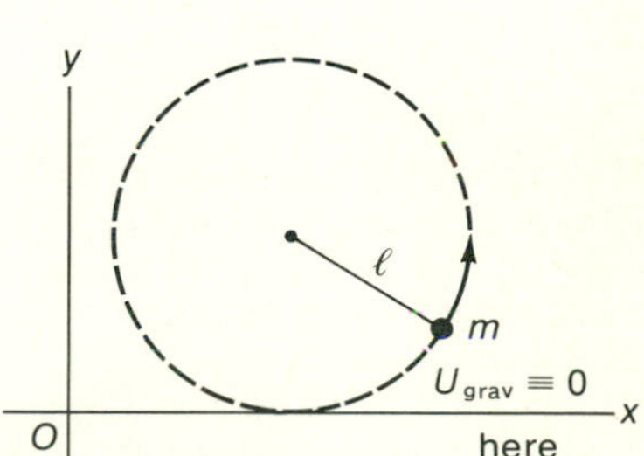

Figure 6–9
Example 6–5

[6] Regarding motion in a vertical circle, you should convince yourself that if the string is taut at the topmost point, it remains taut everywhere. Furthermore, if the speed of the particle is gradually reduced, the location where the string first becomes slack is at the topmost point. (Similar remarks apply to the contact force that the track exerts on an object sliding around a vertical loop-the-loop track.)

SOLUTION

Ideally, we assume that flexing the string does not produce any energy in the form of frictional heating. Furthermore, the particle always moves at right angles to the direction of the tension **T**, so the tension force does no work. Since only conservative forces act, we may apply the conservation of energy relation. We establish the zero reference level for $U_{\text{grav}} \equiv 0$ at the bottom of the circle. Thus:

$$E_{\text{top}} = E_{\text{bottom}}$$

$$U_t + K_t = U_b + K_b$$

$$mg(2\ell) + \tfrac{1}{2}mv_t^2 = 0 + \tfrac{1}{2}mv_b^2$$

Canceling the common factor m and solving for v_b, we obtain

$$4g\ell + v_t^2 = v_b^2$$

$$\boxed{v_b = \pm\sqrt{4g\ell + v_t^2}}$$

EXAMPLE 6–6

The potential energy function associated with the force between two atoms in a diatomic molecule may be approximated as follows:

$$U(x) = \frac{c}{x^{12}} - \frac{d}{x^6} \tag{6–14}$$

where c and d are positive constants (with units) and x is the distance between the atoms. A qualitative sketch of the $U(x)$ versus x curve is shown in Figure 6–10(a). At temperatures above absolute zero, the molecule will undergo oscillations along a line joining the two

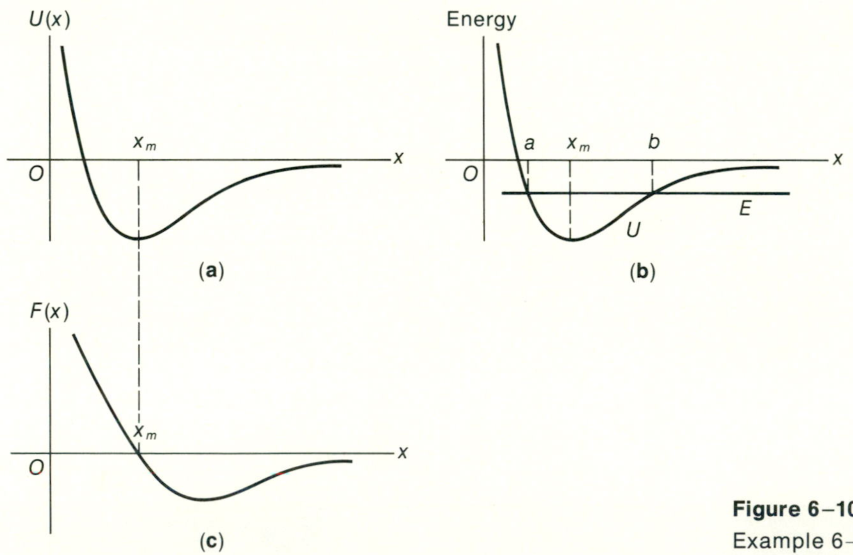

Figure 6–10
Example 6–6

atoms.[7] For a given total energy E, this motion will change from a minimum distance a to a maximum separation b, as shown in Figure 6–10(b). As the temperature is lowered, the oscillations become smaller and the separation approaches the distance x_m, at which the potential energy U is a minimum. (a) Find the force $F(x)$ between the atoms and sketch a curve of $F(x)$ versus x. (b) Find the distance x_m.

[7] The nonsymmetry of the shape of the potential energy curve explains the fact that most substances expand upon heating. For higher temperatures E increases and the average separation between the atoms becomes greater.

SOLUTION

(a) Since the force is conservative, we have

$$F(x) = -\frac{dU}{dx} = -\frac{d}{dx}\left[\frac{c}{x^{12}} - \frac{d}{x^6}\right]$$

Taking the derivative (as explained in Appendix G) gives us

$$\boxed{F(x) = \frac{12c}{x^{13}} - \frac{6d}{x^7}}$$

This force is sketched in Figure 6–10(c). In drawing graphs such as this, it is helpful to remember that the force is proportional to the *negative slope* of the $U(x)$ versus x curve.

(b) We find the distance x_m by noting that the slope of the potential energy curve is zero at this point. So we set the derivative of U equal to zero and solve for x_m.

$$\frac{dU}{dx} = 0$$

$$\left(-\frac{12c}{x_m^{13}} + \frac{6d}{x_m^{7}}\right) = 0$$

$$x_m^{6} = \frac{12c}{6d}$$

$$x_m = \boxed{\sqrt[6]{\frac{2c}{d}}}$$

6.7 Energy Conservation with Nonconservative Forces

The ideas of energy conservation may be used even when nonconservative forces such as friction are present. To see how these nonconservative forces fit into the overall scheme of work and energy, we write the work-energy relation [Equation (5–27)], separating the work done by nonconservative forces ΔW_{nc} from that done by conservative forces ΔW_c.

$$\Delta W_{nc} + \Delta W_c = \Delta K$$

From Equation (6–5), we know that the work done by conservative forces is equal to the negative of the change in potential energy, or $-\Delta U$. Thus:

$$\Delta W_{nc} - \Delta U = \Delta K$$

$$\Delta W_{nc} = \Delta K + \Delta U \qquad \textbf{(6–15)}$$

Substituting $U - U_0$ for ΔU and $K - K_0$ for ΔK, and rearranging, we have

$$\Delta W_{nc} + U_0 + K_0 = U + K \qquad \textbf{(6–16)}$$

Friction is a common nonconservative force. When friction is the only nonconservative force present in an isolated system, we can write Equation (6–16) in a way that makes the idea of conservation of energy more apparent. Because the work that friction does is always negative, we have $\Delta W_{nc} = -(f_k)(\Delta x) = -\Delta U_{\text{thermal}}$. Recognizing that thermal energy is just another form of energy, we include this, too, in the overall accounting of energy transfers that occur in an isolated system, equating the initial energy E_0 with the final energy E.

$$E_0 = E$$

$$\underbrace{(U_{\text{thermal}})_0}_{\substack{\text{always zero}\\ \text{at the start}\\ \text{of a problem}}} + U_0 + K_0 = \underbrace{U_{\text{thermal}}}_{\substack{\|\\ (f_k)(\Delta x)}} + U + K \qquad \textbf{(6–17)}$$

For the situations we discuss here, thermal energy is not transferred back into mechanical energy, so any thermal energy that might be present at the start remains as thermal energy throughout the problem. Thus it would be an added quantity on both sides of the equation and can be ignored. Writing the initial thermal energy as zero, the final value is then the result of the nonconservative force of friction acting during the process.

EXAMPLE 6–7

Referring to the sketch in Example 6–3, suppose that a portion of a roller coaster has the following data:

Mass of car plus passengers:	$m = 2000$ kg
Initial speed at top:	$v_0 = 4$ m/s
Height of drop:	$h = 10$ m

(a) Assuming friction is negligible, find the speed at the bottom. (b) Suppose that because of friction the actual speed at the bottom is 13.8 m/s. Find the thermal energy developed during the descent.

SOLUTION

(a) We establish the zero reference for potential energy $U_{\text{grav}} \equiv 0$ at the bottom of the track, where $y = 0$, with the positive direction up. Applying the conservation of energy principle for conservative forces only, we have

$$E_0 = E$$

$$K_0 + U_0 = K + U$$

$$\tfrac{1}{2}mv_0^2 + mgh_0 = \tfrac{1}{2}mv^2 + 0$$

Solving for v gives us

$$v = \sqrt{v_0^2 + 2gh_0} = \sqrt{\left(4\,\frac{\text{m}}{\text{s}}\right)^2 + (2)\left(9.8\,\frac{\text{m}}{\text{s}^2}\right)(10\text{ m})} = \boxed{14.6\,\frac{\text{m}}{\text{s}}}$$

(b) Note that the actual speed at the bottom is less than this; the "missing" mechanical energy appears as thermal energy due to friction. Applying the overall conservation of energy principle and including thermal energy due to friction effects, we have

$$E_0 = E$$

$$(U_{\text{thermal}})_0 + K_0 + U_0 = U_{\text{thermal}} + K + U$$

$$0 + \tfrac{1}{2}mv_0^2 + mgh_0 = U_{\text{thermal}} + \tfrac{1}{2}mv^2 + mgh$$

Solving for U_{thermal}, we have

$$U_{\text{thermal}} = m\left[\frac{v_0^2 - v^2}{2} + g(h_0 - h)\right]$$

$$U_{\text{thermal}} = [2000\text{ kg}]\left[\frac{\left(4\,\frac{\text{m}}{\text{s}}\right)^2 - \left(13.8\,\frac{\text{m}}{\text{s}}\right)^2}{2} + \left(9.8\,\frac{\text{m}}{\text{s}^2}\right)(10\text{ m} - 0)\right]$$

$$U_{\text{thermal}} = \boxed{2.16 \times 10^4\text{ J}}$$

The conservation of mechanical energy principle was an important signpost along the road toward a comprehensive conservation principle. The discovery of exceptions to $E_0 = E$, such as the "disappearance" of some mechanical energy while parts of the system became warmer, spurred efforts to understand the effects of nonconservative forces and to include thermal energy and other previously unknown forces. The entire science of thermodynamics was born when thermal energy was interpreted as just kinetic and potential energies on an atomic level. The extension of energy conservation to all branches of physics—sound energy, electromagnetic energy, nuclear energy, and so on—provided one of the great unifying ideas in physics. Although progress in this direction was extremely difficult and generated much controversy over a period of two centuries as the best thinkers of the time wrestled with the concept of energy, today there is no question that energy seems to be conserved in nature. It can be neither created nor destroyed, only changed from one form to another. Such an all-inclusive general principle is possible because at the most elementary level, everything boils down to just the kinetic and potential energies of the basic constituents of matter. And at a submicroscopic level, all the basic forces between the fundamental particles in the universe seem to be *conservative forces.*

Summary

A force is *conservative* if

(a) the work done in moving a mass between two points depends only on the location of the end points and not on the particular path

or **(b)** the work done in moving a mass through any round-trip path is zero.

(The two criteria are equivalent.) Nonconservative forces are generally those that involve friction or that depend on velocity or time.

When all the forces present in an isolated system are conservative, the *system* is conservative and we can introduce the concept of the potential energy of the system U. We then have the *conservation of mechanical energy* E:

$$E_0 = E \qquad (\text{where } E = U + K)$$

It is important always to specify the zero reference location for $U \equiv 0$. For motions under the action of gravity, usually the lowest point of interest is the most convenient zero reference.

The relation between a conservative force $\mathbf{F}$ and its associated potential energy U may be expressed in two forms (*note the minus signs*):

As a summation[8]

$$U_b - U_a = -\sum_{a \to b} [F(x)\Delta x]$$

As a derivative

$$F_x = -\frac{d}{dx}[U(x)]$$

[8] In integral form:

$$U_b - U_a = -\int_a^b \mathbf{F} \cdot d\mathbf{x}$$

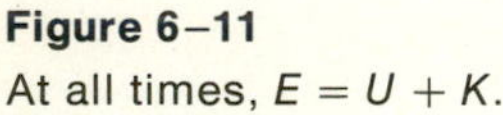

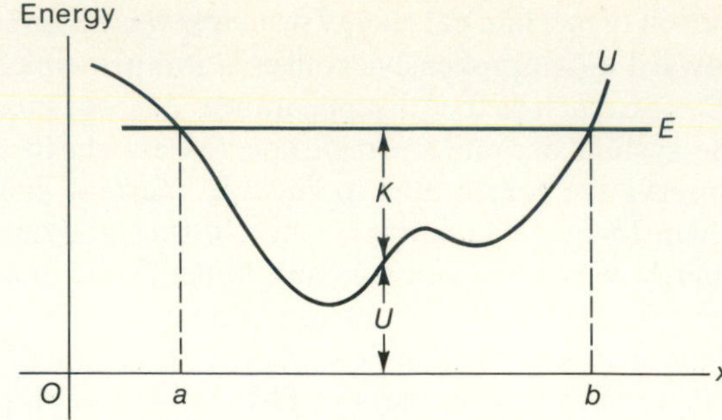

Figure 6–11
At all times, $E = U + K$.

A *potential well* is formed by a potential energy function that (for a given E) confines the motion between limits $a \leq x \leq b$, as shown in Figure 6–11. At every point of the motion, $E = U + K$.

If the net force on a mass is a *Hooke's Law force* ($F = -kx$), the mass will undergo *simple harmonic motion* (SHM). The maximum displacement from the equilibrium position is called the *amplitude A*. For SHM:

$$E = U + K$$

$$E = \tfrac{1}{2}kx^2 + \tfrac{1}{2}mv^2$$

Also: $$E = \tfrac{1}{2}kA^2 \quad \text{or} \quad \tfrac{1}{2}mv_{max}^2$$

which is a constant value. In SHM, the total energy E is shared equally, on the average, between kinetic and potential forms.

A roller-coaster analogy for one-dimensional potential energy curves is often helpful in understanding the motion. However, remember that although the roller coaster is two dimensional, the motion being analyzed is in one dimension only.

When nonconservative forces are also present, the work W_{nc} done by the nonconservative forces (usually negative) is added to the initial energy as follows:

$$W_{nc} + E_0 = E$$

If friction is the only nonconservative force, the thermal energy produced by kinetic friction may be included in the overall accounting of energy transfers that occur:

$$E_0 = E$$

$$\underbrace{(U_{\text{thermal}})_0}_{\substack{\text{always zero}\\ \text{at the start}\\ \text{of a problem}}} + U_0 + K_0 = \underbrace{U_{\text{thermal}}}_{\substack{\|\\ (f_k)(x)}} + U + K$$

Questions

1. Determine the most obvious form of energy in each of the following systems: a moving airplane, a flashlight battery, a flowing river, a stretched rubber band, a piece of TNT, a candy bar, a rock on top of Mt. Everest.
2. Name a conservative force not mentioned in this chapter.

3. Considering that the main forces between the earth, moon, and sun are gravitational (a *conservative* force), what causes the kinetic energy of the earth's rotation to change? (Refer to Figure 2–1.)
4. Name a nonconservative force other than friction.
5. The normal temperature of the human body is 98.6°F (37°C), usually higher than that of the surroundings. Trace the energy transfer processes involved in maintaining this temperature, starting with your body's metabolism and going back to the ultimate source of energy, the sun.
6. Sketch the potential energy well that limits the motion of a ball bouncing vertically as it makes elastic collisions with a horizontal surface. (Plot the potential energy as a function of the vertical distance y above the surface.)
7. Sketch the potential energy well that limits the motion of a pendulum bob when it swings to the 90° positions on either side of the vertical. (Plot the potential energy as a function of the angle the string makes with the vertical.)

Problems

6A–1 A ball is thrown vertically upward with a speed of 15 m/s. Use conservation of energy principles to find the speed of the ball 8 m above its starting point.

Answer: 8.26 m/s

6A–2 A ball of mass m is released from rest above a horizontal surface. It bounces elastically, returning after each bounce to a height h above the surface.

(a) Make a graph of energy versus distance above the surface, showing separate curves for the total energy E, the kinetic energy K, and the gravitational potential energy U_{grav}.

(b) Find the time interval between bounces.

6A–3 A bead of mass m slides without friction along a wire bent in a vertical circle. If the bead starts from rest at the top, what force does the wire exert on it when it has slid halfway down?

Answer: $2mg$

6A–4 A ball of mass m is suspended from the ceiling by a string of length ℓ, as shown in Figure 6–12. It is held to one side, at position A, by a horizontal thread, so that the string makes an angle of θ with the vertical. The horizontal thread is then burned, allowing the ball to swing freely.

(a) Using conservation of energy principles, find the speed of the ball as it passes through the lowest point B in terms of m, ℓ, θ, and g.

(b) Find the tension in the string at point A before the thread is burned.

(c) Find the tension in the string at point C.

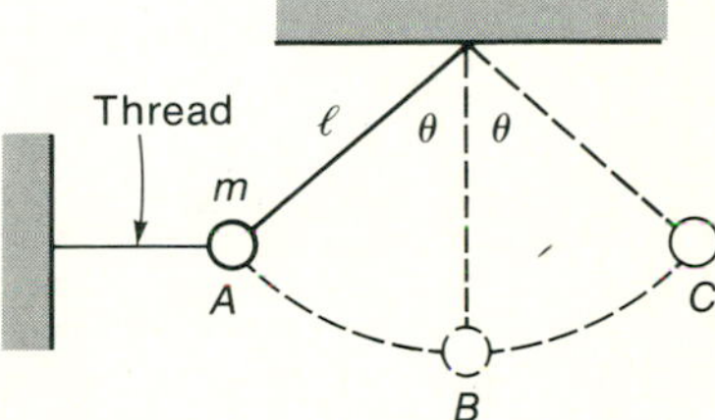

Figure 6–12
Problem 6A–4

6A–5 A pendulum bob of mass m swings on a string of length ℓ. When the bob is at its highest point, the string makes an angle θ ($<90°$) with the vertical. Using energy principles, derive an expression for the speed of the bob at its lowest point.

Answer: $\sqrt{2g\ell(1 - \cos\theta)}$

6A–6 A carnival stuntman coasts down a frictionless track in the form of a loop-the-loop, as shown in Figure 6–13. Find the minimum height H necessary for the bicycle to remain in contact with the track at all times. The loop is a circle of radius R. (Hint: The point at the top of the loop is the critical situation.)

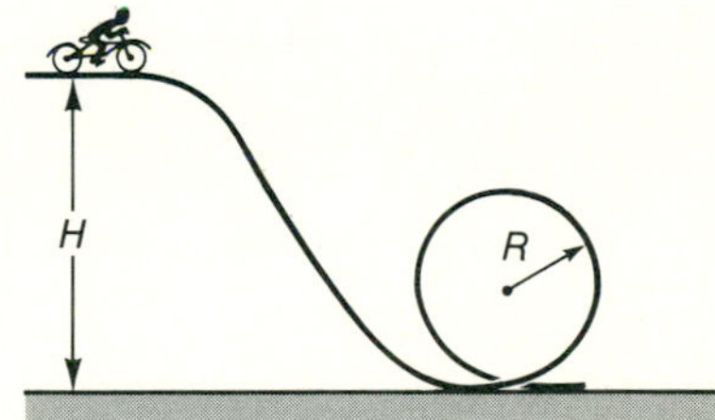

Figure 6–13
Problems 6A–6, 6B–6, and 6C–6

6A–7 A ball of mass m on the end of a string is set into circular motion in a vertical plane. The length of the string is 0.60 m.

(a) Find the minimum speed the ball must have at the bottom of the loop to keep the string barely taut at the top.

(b) Find the tension in the string under these conditions when the ball is halfway between top and bottom.

Answers: (a) 5.42 m/s (b) $3mg$

6A–8 A mass m on the end of a string is set into circular motion in a vertical plane. At the top of the loop, the string has a tension of $2mg$. Find the tension in the string at the bottom.

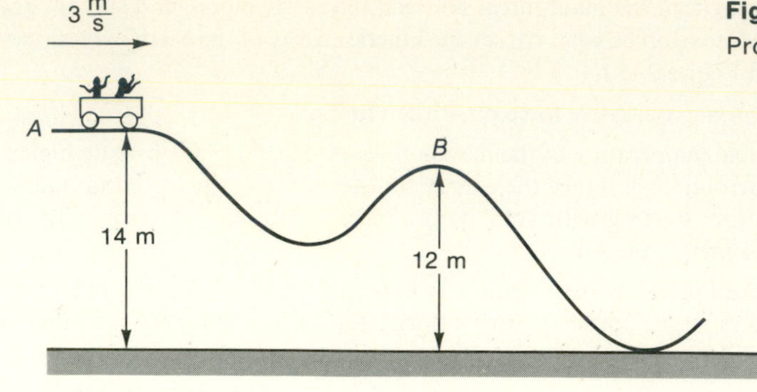

Figure 6–14
Problem 6A–9

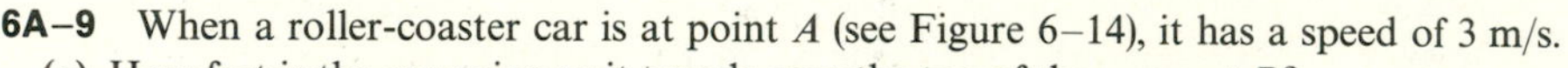

6A–9 When a roller-coaster car is at point A (see Figure 6–14), it has a speed of 3 m/s.

(a) How fast is the car going as it travels over the top of the curve at B?

(b) Find the smallest radius of curvature of the track at B such that the car will not lose contact with the rails. *Answers:* (a) 6.94 m/s (b) 4.92 m

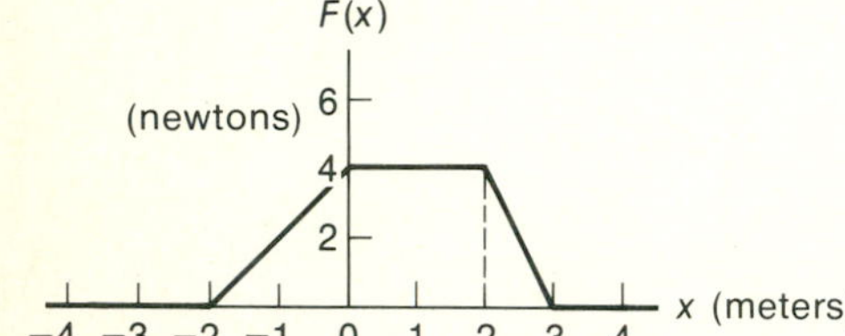

Figure 6–15
Problems 6A–10 and 6C–2

6A–10 A conservative force $F(x)$ acts on a particle that moves in the $\pm x$ direction, as shown in Figure 6–15. Make a graph of the potential energy $U(x)$ versus x for $-4 < x < +4$, with the potential energy at $x = -4$ defined to be zero. Indicate numerical values of U at $x = 0$ m, 2 m, and 3 m.

6B–1 A particle of mass m starts at rest and slides down a frictionless track as shown in Figure 6–16. It leaves the track horizontally, striking the ground as indicated in the sketch. At what height H did it start above the ground? *Answer:* 0.20 m

6B–2 A small mass m rests at the top of a smooth hemisphere of radius r. It is displaced slightly, so that it slides from rest down the hemisphere, without friction. It remains in contact with the surface until it reaches the location described by the radius vector at the angle θ, as shown in Figure 6–17. From this point on, it loses contact with the surface and follows a free-fall parabolic path. Find the angle θ. (Hint: Just before the mass reaches the position θ, the surface is still exerting a normal force on it as it travels in a circle with instantaneous speed v. But just as the mass leaves the surface, this contact force drops to zero. Investigate this dividing-line situation.)

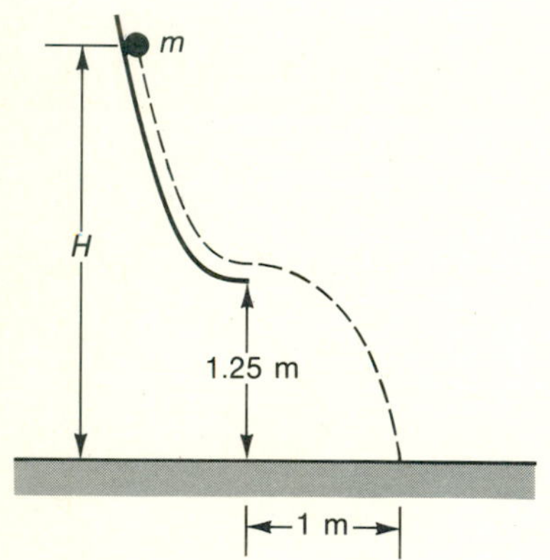

Figure 6–16
Problem 6B–1

6B–3 A Hooke's Law spring has a relaxed length of 15 cm and a force constant of 8 N/m. It is placed on the floor in a vertical position, then compressed to half its length. A 10-g mass is placed on top of the spring. The system is released, allowing the mass to be propelled upward. What maximum height above the floor does the mass reach? *Answer:* 0.305 m

6B–4 In Figure 6–18(a), a mass m rests on a spring, compressing it a distance d from its relaxed length of s_0. Suppose that instead the mass is released from rest when it barely touches the unstressed spring.

(a) Find the maximum compression distance d_{max} of the spring as the mass moves downward.

(b) During this process, what maximum speed does the mass attain?

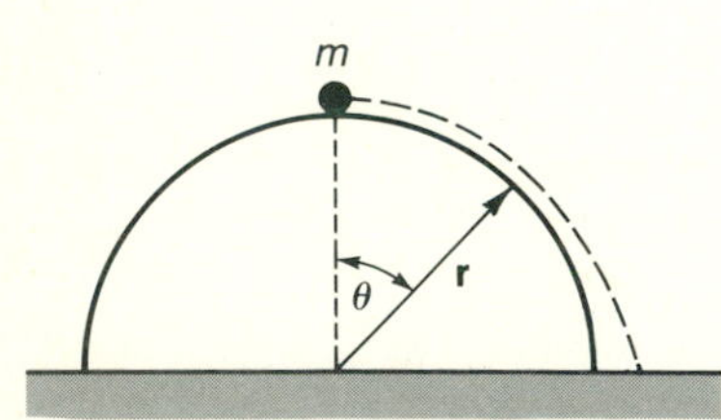

Figure 6–17
Problem 6B–2

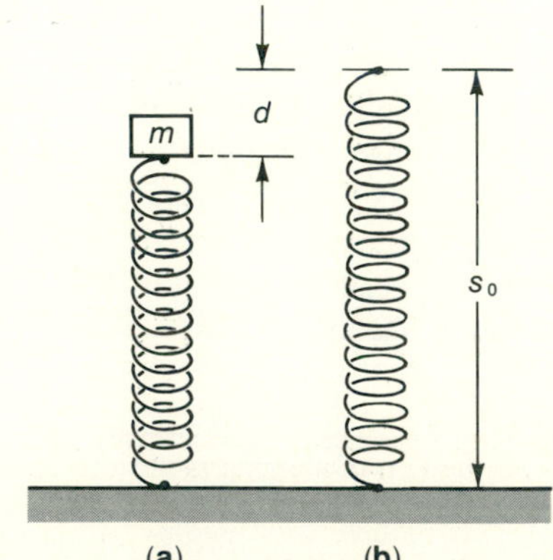

Figure 6–18
Problem 6B–4

6B–5 A ball of mass m is fastened on the end of a rigid rod of length ℓ and negligible mass. The rod rotates freely about a fixed horizontal axis, as shown in Figure 6–19. The ball is released from rest when the rod makes an angle of 53.1° with the vertical. In terms of m, g, and ℓ, find the tension T in the rod as the ball swings through the lowest point.

Answer: $4.20mg$

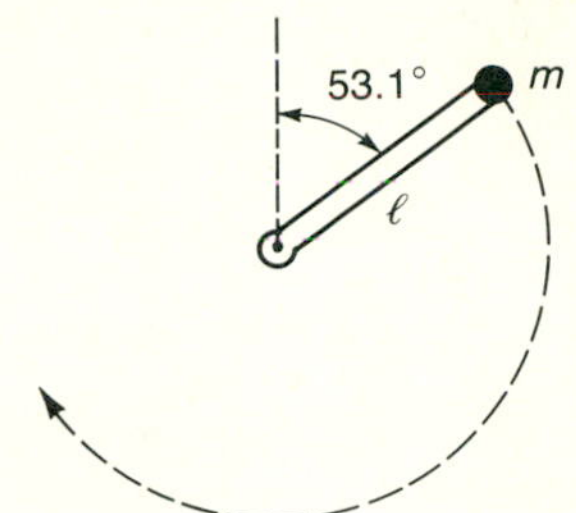

Figure 6–19
Problem 6B–5

6B–6 In Figure 6–13, the bicyclist coasts without friction from rest at the top of a track. Suppose as he passes the topmost point of the loop, the track exerts a force on the bicycle of 3 times the weight of the rider-plus-bicycle. Find the height H.

6B–7 A particle of mass m slides without friction down a curved track of radius R as shown in Figure 6–20. It starts at rest at point A, where the track is vertical.

(a) Show that the speed of the particle as it passes point B (a vertical distance $R/2$ below the starting point) is $\sqrt{gR}$.

(b) Find the tangential and perpendicular acceleration components of the particle as it passes point B.

(c) Find the magnitude and direction of the force the track exerts on the particle at B.

Answers: (b) g (c) $3mg/2$; radially inward

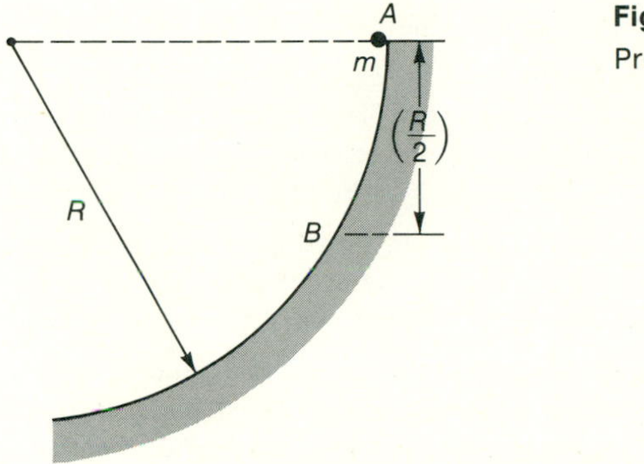

Figure 6–20
Problem 6B–7

6B–8 A particle moves along the $\pm x$ axis. The potential energy as a function of position is shown in Figure 6–21. (The curved segments, b units long each, are parabolic.) Make a graph of the force $F(x)$ on the particle as a function of x.

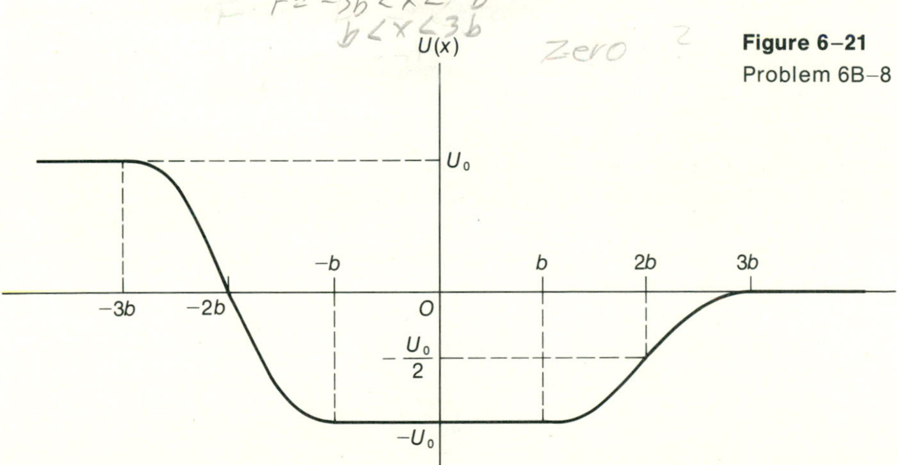

Figure 6–21
Problem 6B–8

6B–9 A mass m is suspended by a flexible cord of length ℓ so that it swings as a simple pendulum. The mass is held with the cord horizontal and released from rest. Suppose there is a small peg fixed a distance $\frac{2}{3}\ell$ below the point of suspension so that it interrupts the motion of the cord. After passing the lowest point, the mass thus swings in a vertical circle of radius $\frac{1}{3}\ell$ as shown in Figure 6–22. (Assume the mass misses the supporting cord at the top of its trajectory.) Find the tension in the cord at the instant the mass reaches point A, the highest point of its motion after the cord contacts the peg.

Answer: mg

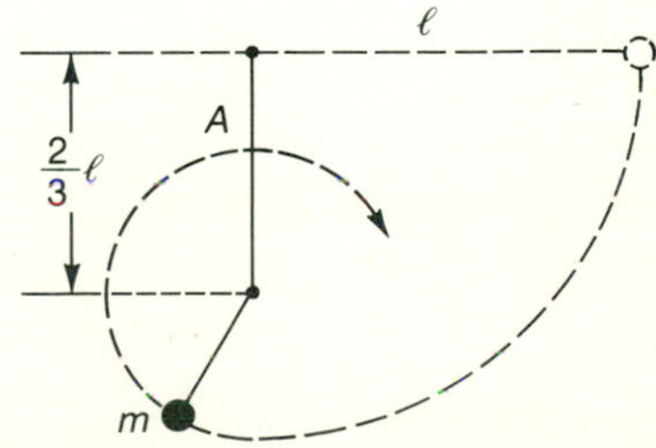

Figure 6–22
Problem 6B–9

6B–10 In the previous problem, at what distance y below the point of suspension should the peg be placed so that the cord barely remains taut at the topmost point as the mass swings in the small circle about the peg?

6B–11 This problem refers to The Great American Revolution roller coaster (see Figure 4–19 for technical data). How much energy is "lost" to frictional effects as a loaded train of cars travels the last free-coasting descent before reaching the loop? Note that the cars have a finite speed when they begin this descent. *Answer:* 1.12×10^5 J

6B–12 In the previous problem, how much mechanical energy is "lost" to frictional effects as a loaded train travels around the loop B–C–D with the speeds listed in the figure?

6C–1 Assume that the motion of an object is governed solely by the following force:

$$F = -4x + 3x^2$$

where F is in newtons and x is in meters. Find the change in potential energy when the object moves from $x = 1$ m to $x = 2$ m. *Answer:* -1.00 J

6C–2 In Figure 6–18(a), a mass m rests on a spring compressing it a distance d from its relaxed length of s_0. Suppose that the mass is now pushed down on the spring an *additional* distance of $3d$. When released from rest at this point, how high above the point of release will the mass rise, assuming the mass is not attached to the spring?

6C–3 A ball of mass m on the end of a string is set into motion in a vertical circle. The string remains taut at all times. (The exact speed at the top of the circle is not specified; it may be any arbitrary value v, provided the conditions are met.) Prove that the tension in the string when the ball is at the bottom exceeds the tension when it is at the top by $6mg$.

6C–4 For values of x between $-\pi/2$ and $+\pi/2$, a conservative force has the following behavior: $F(x) = -k \sin x$. Make sketches of $F(x)$ and the associated potential energy function $U(x)$. Choose the zero reference level for potential energy at $x = 0$.

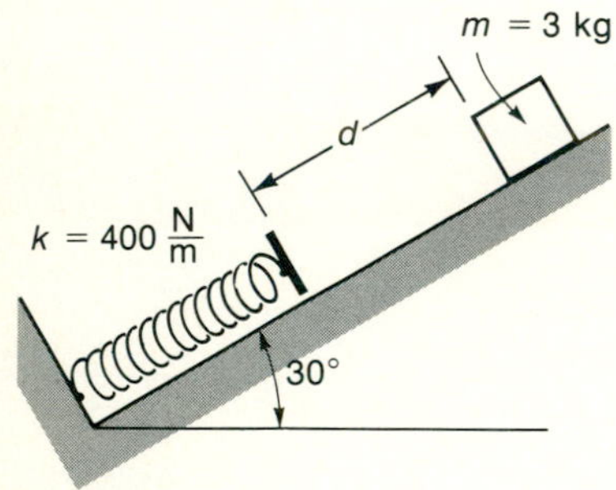

Figure 6–23
Problem 6C–5

6C–5 A 3-kg mass starts at rest and slides a distance d down a smooth 30°-incline, where it contacts a spring of negligible mass as shown in Figure 6–23. It slides an additional 0.2 m as it is brought momentarily to rest by compressing the spring (force constant $k = 400$ N/m). Find the initial separation d between the mass and the spring. *Answer:* 0.344 m

6C–6 As shown in Problem 6A–6, a bicyclist coasts from rest without friction, starting from an initial height $H = 3R$. As he coasts upward in the loop-the-loop, find his acceleration (magnitude and direction) when he is a distance R above the ground.

6C–7 A mass moves under the influence of a potential energy function of the form $U = a + b(x - x_0)^2$.

(a) For what value of x is the force on the mass zero?

(b) Find the magnitude and direction of the force on the mass at $x = 0$.

Answers: (a) x_0 (b) $2bx_0$; $+x$ direction

6C–8 Consider a path in the x-y plane described by the straight lines (0,0) to (3,0) to (3,2) to (0,2) to (0,0), where the points (x, y) are in units of meters.

(a) Show that the force $\mathbf{F} = 6xy\hat{\mathbf{x}} + 3x^2\hat{\mathbf{y}}$ (in newtons) is conservative by calculating the work done by this force in traveling each individual line and showing that the sum for the complete round trip is zero.

(b) Verify that the force $\mathbf{F} = x^2y\hat{\mathbf{x}} + x^2\hat{\mathbf{y}}$ is nonconservative by calculating the total work done by the force around the same path as in part (a).

6C–9 The total energy E of a mass moving in one dimension without friction is

$$E = \tfrac{1}{2}mv^2 + U$$

where U is the potential energy. The total energy is constant; that is, $dE/dt = 0$. Show that these relationships lead to Newton's second law ($F = m\,dv/dt$).

6C–10 A particle moves in a region where its potential energy varies inversely with the distance r from the origin; that is, $U(r) = A/r$, where A is a constant.

(a) Give the SI units of A.

(b) Find the x and y components of the force on the particle.

(c) Verify that the force is conservative.

6C–11 A mass moves in a region where the potential energy is $U(x) = A(1 - \cos ax)$. Show that for small values of x, the mass is pulled back toward $x = 0$ by the force $F = Aa^2x$. (Hint: Use the small-angle approximation for $\cos ax$.)

6C–12 A particle of mass m moves in one-dimensional motion in a region where its potential energy is given by:

$$U(x) = \frac{A}{x^3} - \frac{B}{x} \qquad (A \text{ and } B \text{ are constants})$$

The general shape of this function is shown in Figure 6–24.

(a) Find the static equilibrium position x_0 of the particle in terms of m, A, and B.
(b) Determine the depth U_0 of this potential well.
(c) In moving along the x axis, what maximum force toward the *negative* x direction does the particle experience?

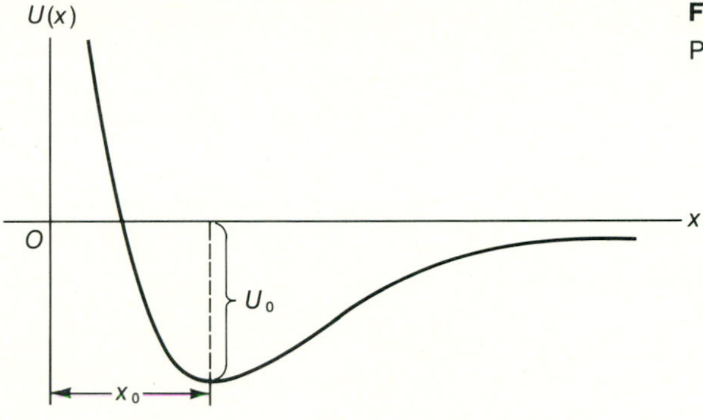

Figure 6–24
Problem 6C–12

6C–13 A particle moves along the $+x$ axis in a region where the potential energy $U(x)$ of the particle is given by:

$$U(x) = 6x^2 - 2x^3 \qquad (U \text{ is in joules and } x \text{ in meters})$$

A qualitative curve for $U(x)$ is shown in Figure 6–25.

(a) For what range of values of x is the force on the particle in the negative x direction?
(b) What is the largest value of the total mechanical energy for which oscillatory motion is possible?

Answers: (a) $0 \leq x \leq 2$ m (b) 8 J

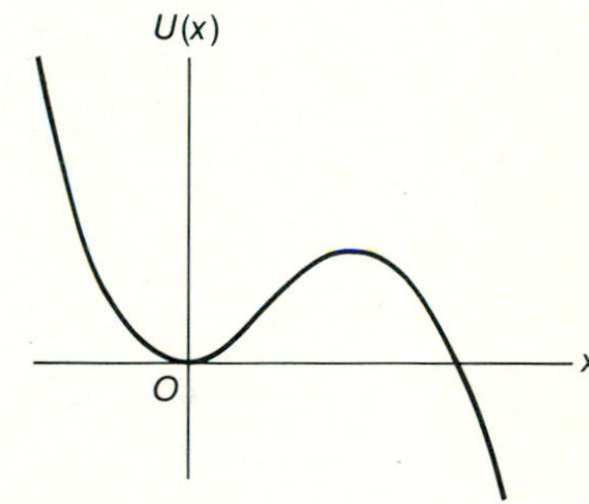

Figure 6–25
Problem 6C–13

chapter 7

Conservation of Linear Momentum

At 90 miles drove Eddie Shawn
The motor stopped, but Ed kept on.
ANONYMOUS

7.1 Introduction

In the previous chapter, we introduced the conservation of mechanical energy, which applies to every isolated conservative system. This principle provides a very powerful method of analyzing mechanical processes. Its effectiveness comes from the great generality of the statement. One need not specify the particular details regarding the forces of an interaction—the conservation of mechanical energy holds for *all* processes in a conservative system, even though we do not explicitly know the forces involved.

We shall now discuss a similar conservation relation which is equally powerful: the conservation of linear momentum. It is even more general than the conservation of mechanical energy since it is valid in *every* process, conservative or nonconservative. In fact, it is one of the most fundamental and important statements one can make in mechanics.

The theorem does not state any facts about the behavior of interacting masses that are not already contained in Newton's laws. In fact, in Chapter 4 we actually obtained a statement of the conservation of momentum by experimenting with masses on an air track.

When applicable, the two conservation relations for energy and momentum permit an analysis of processes in terms of the *positions* and *velocities* of masses. In contrast, Newton's laws involve *forces* and *accelerations*. These are useful distinctions to remember since the type of information available in a problem is the clue to which method will be most effective in its solution. Of course, some problems may require both Newton's laws and the conservation relations for a complete analysis.

7.2 The Conservation of Linear Momentum

Consider the interaction of two particles, m_A and m_B. For simplicity, let us imagine the particles sliding on a frictionless, horizontal surface, so that the only forces altering their horizontal motions are those between the particles themselves. In Figure 7–1, the particles are shown approaching each other, colliding, and then separating. We shall use the subscript "0" to designate *initial*

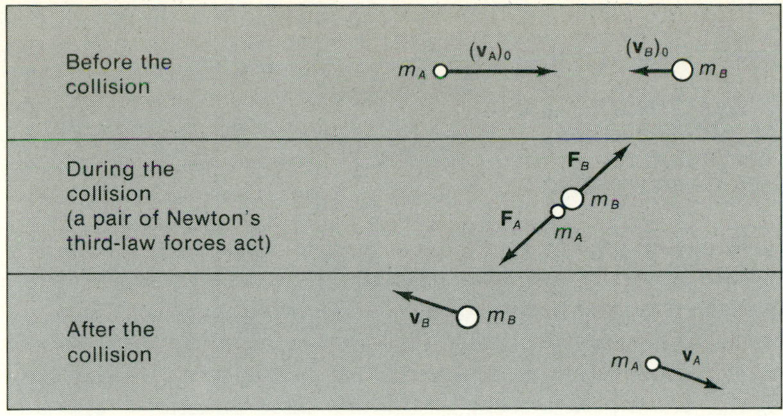

Figure 7–1
A view from above of the collision between two masses sliding on a horizontal frictionless surface.

values before the collision and no subscript for the *final* values after the interaction is completed.

If the interaction involves only forces that obey Newton's third law, during the collision we have

$$\mathbf{F}_A = -\mathbf{F}_B \tag{7–1}$$

where $\mathbf{F}_A$ is the force on m_A and $\mathbf{F}_B$ is the force on m_B.

Newton's *second* law states that a force $\mathbf{F}$ may be written in terms of the time rate of change of momentum ($m\mathbf{v}$):

$$\frac{d}{dt}(m_A\mathbf{v}_A) = -\frac{d}{dt}(m_B\mathbf{v}_B)$$

where $\mathbf{v}_A$ is the velocity of m_A and $\mathbf{v}_B$ is the velocity of m_B. Since the momentum terms $m_A\mathbf{v}_A$ and $m_B\mathbf{v}_B$ occur repeatedly throughout this derivation, it is convenient to make the substitution $\mathbf{p} = m\mathbf{v}$. Thus:

$$\frac{d}{dt}(\mathbf{p}_A) = -\frac{d}{dt}(\mathbf{p}_B) \tag{7–2}$$

$$\frac{d\mathbf{p}_A}{dt} + \frac{d\mathbf{p}_B}{dt} = 0 \tag{7–3}$$

$$\frac{d}{dt}(\mathbf{p}_A + \mathbf{p}_B) = 0 \tag{7–4}$$

This expression implies that the total momentum ($\mathbf{p}_A + \mathbf{p}_B$) of the two interacting particles does not change during the interaction.

There is another way of expressing this fact. Since the total momentum has the same value before and after the interaction, it remains *constant in time*, as long as the system is isolated (that is, as long as no outside forces act to change the overall momentum). Thus we arrive at the **conservation of linear momentum:**

$$\mathbf{p}_A + \mathbf{p}_B \qquad \text{remains constant in time} \tag{7–5}$$

Letting the subscript "0" refer to initial values and no subscript refer to final values, we have

$$(\mathbf{p}_A + \mathbf{p}_B)_0 = \mathbf{p}_A + \mathbf{p}_B \tag{7–6}$$

Generalizing this expression to include any number of particles that make up an isolated system, we obtain

CONSERVATION OF LINEAR MOMENTUM FOR AN ISOLATED SYSTEM OF PARTICLES

$$\Sigma \mathbf{p}_0 = \Sigma \mathbf{p} \tag{7–7}$$

This principle is true for systems that involve *all* types of forces, including friction. The only strict qualification is that the system must be isolated; that is, all external forces must be zero, or must add to produce zero net force.

Since the equation is a *vector* equation, it must also hold true for the *components* of the vectors, as measured in any inertial frame of reference. For a rectangular coordinate system, we obtain three equations, one for each coordinate direction.

$$\begin{aligned} \text{For } x\text{:} \quad & (p_{Ax})_0 + (p_{Bx})_0 = p_{Ax} + p_{Bx} \\ \text{For } y\text{:} \quad & (p_{Ay})_0 + (p_{By})_0 = p_{Ay} + p_{By} \\ \text{For } z\text{:} \quad & (p_{Az})_0 + (p_{Bz})_0 = p_{Az} + p_{Bz} \end{aligned} \tag{7–8}$$

Here, $(p_{Ax})_0$ means the x component of $\mathbf{p}_A$ before the collision, and so forth. The usual method of solution in two-dimensional problems is to consider each component direction separately using Equations (7–8).

The conservation of linear momentum is valid in any inertial frame of reference. True, observers in different moving frames of reference looking at the same two particles interacting would measure different values for the momentum of each particle. However, all would agree on the *conservation* of linear momentum for the interaction. Moreover, as will be shown in Chapter 10, the principle is true not only for two-particle interaction, but also for *an isolated system of any number of particles*, from a triatomic molecule to a galaxy containing 10^{12} stars.

EXAMPLE 7–1

A 4-kg mass moves at an initial speed of 3 m/s on a horizontal frictionless surface. The mass collides and sticks to a 2-kg mass, which was initially at rest (see Figure 7–2). Find the speed of the combined masses after the collision.

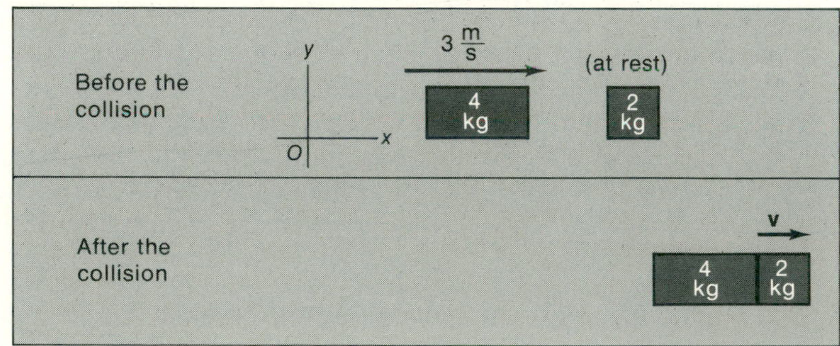

Figure 7–2
Example 7–1

SOLUTION

Since the initial momentum is just that of the 4-kg mass, the final momentum after the collision is along the same direction and has the same magnitude. Therefore, the collision is one-dimensional. We choose the $+x$ coordinate to be in the direction of the initial momentum and adopt the following notation:

	Before the collision	After the collision
$m_A = 4$ kg	$(v_A)_0 = 3$ m/s	The combined masses move with the same common velocity.
$m_B = 2$ kg	$(v_B)_0 = 0$	$v_A = v_B = v$

Applying the conservation of linear momentum principle, we have

$$\Sigma \mathbf{p}_0 = \Sigma \mathbf{p}$$

$$(m_A v_A)_0 + (m_B v_B)_0 = (m_A + m_B)v$$

Solving for v, we obtain

$$v = \frac{(m_A v_A)_0 + (m_B v_B)_0}{(m_A + m_B)} = \frac{(4\ \text{kg})\left(3\ \frac{\text{m}}{\text{s}}\right) + (2\ \text{kg})(0)}{(4\ \text{kg} + 2\ \text{kg})} = \boxed{2.00\ \frac{\text{m}}{\text{s}}}$$

EXAMPLE 7–2

A boy at the earth's surface tosses a ball of mass m straight upward. Considering that the total momentum of the ball–earth system must remain constant, find the ratio of the two kinetic energies:

$$\frac{K_{\text{earth}}}{K_{\text{ball}}} = \frac{\frac{1}{2}m_e v_e^2}{\frac{1}{2}m_b v_b^2}$$

at the instant just after the ball is released.

SOLUTION

We consider the mass of the boy-plus-earth to be the mass of the earth m_e. Since the contact force and the force of gravity between the ball and the boy-plus-earth are *internal* forces in the system, the conservation of linear momentum holds for the ball–earth system as a whole. We choose an inertial frame of reference in which the system is initially at rest ($\Sigma \mathbf{p}_0 = 0$). We then apply the conservation of linear momentum principle and solve for $\mathbf{v}_e$:

$$\Sigma \mathbf{p}_0 = \Sigma \mathbf{p}$$

$$(m_b \mathbf{v}_b)_0 + (m_e \mathbf{v}_e)_0 = m_b \mathbf{v}_b + m_e \mathbf{v}_e$$

$$0 + 0 = m_b \mathbf{v}_b + m_e \mathbf{v}_e$$

$$\mathbf{v}_e = -\frac{m_b}{m_e}\mathbf{v}_b$$

The minus sign verifies that the velocities of the ball and earth are in opposite directions.

Substituting the expression for $\mathbf{v}_e$ into the kinetic energy ratio, we have

$$\frac{K_e}{K_b} = \frac{\frac{1}{2}m_e\left(\frac{m_b}{m_e}v_b\right)^2}{\frac{1}{2}m_b v_b^2} = \boxed{\frac{m_b}{m_e}}$$

For a numerical comparison in the above example, suppose the mass of the ball were 0.6 kg. Since the mass of the earth is about 6×10^{24} kg, the ratio of kinetic energies would be $\sim 10^{-25}$. Thus the earth acquires negligible kinetic energy. Because of this fact, in Chapter 6 when we discussed the potential and kinetic energies of a ball thrown upward, we were justified in ignoring the earth's kinetic energy. As mentioned in Chapter 6, potential energy is always a property of a *system* rather than a property of a single particle, and in many cases we cannot ignore the significant sharing of energies between all parts of a system. But for tossing a ball upward, where $m_e \gg m_b$, essentially all of the kinetic energy in the system resides in the ball (because the velocity the earth acquires is extremely small), so the simplified approach used in Chapter 6 was justified. Note, however, that after the ball is tossed, it has an upward momentum *exactly equal in magnitude* to the downward momentum of the boy-plus-earth.

EXAMPLE 7–3

Two ice skaters, A and B, approach each other at right angles. Skater A has a mass of $m_A = 50$ kg, and she is traveling in the $+x$ direction at 2 m/s. Skater B has a mass of $m_B = 70$ kg, and he is moving in the $+y$ direction at 1.5 m/s. They collide and cling together. (a) Find the final velocity of the couple. (b) Find the fractional change in kinetic energy ($\Delta K/K_0$) during the collision.

Figure 7–3
Example 7–3

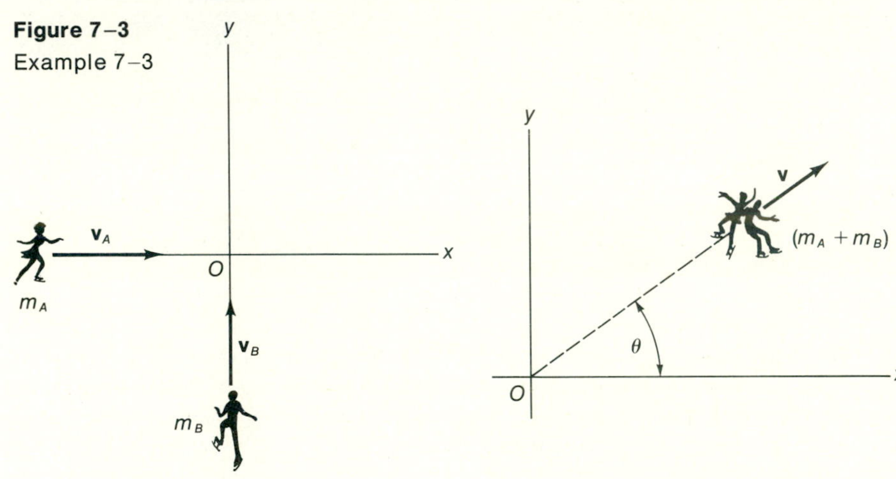

SOLUTION

(a) Using the notation shown in Figure 7–3, we apply the conservation of linear momentum principle for the two component directions [Equations (7–8)].

$$x \text{ direction:} \quad (p_x)_0 = p_x$$

$$m_A v_A = (m_A + m_B)v \cos\theta \qquad \textbf{(7–9)}$$

$$y \text{ direction:} \quad (p_y)_0 = p_y$$

$$m_B v_B = (m_A + m_B)v \sin\theta \qquad \textbf{(7–10)}$$

Dividing Equation (7–10) by Equation (7–9), we obtain

$$\left(\frac{m_B v_B}{m_A v_A}\right) = \tan\theta$$

$$\theta = \tan^{-1}\left[\frac{(70 \text{ kg})\left(1.5\,\dfrac{\text{m}}{\text{s}}\right)}{(50 \text{ kg})\left(2\,\dfrac{\text{m}}{\text{s}}\right)}\right] = \tan^{-1}(1.05) = 46.4^\circ$$

The final speed v is found by substituting this angle in Equation (7–9). Solving that equation for v gives us

$$v = \frac{m_A v_A}{(m_A + m_B)\cos\theta} = \frac{(50\text{ kg})\left(2\,\frac{\text{m}}{\text{s}}\right)}{(50\text{ kg} + 70\text{ kg})(\cos 46.4°)} = \boxed{1.21\,\frac{\text{m}}{\text{s}}}$$

(b) The fractional change in the kinetic energy during the collision is

$$\left(\frac{\Delta K}{K_0}\right) = \left(\frac{K - K_0}{K_0}\right) = \left(\frac{K}{K_0} - 1\right)$$

where $$K_0 = \tfrac{1}{2}m_A v_A^2 + \tfrac{1}{2}m_B v_B^2$$

$$K_0 = \tfrac{1}{2}(50\text{ kg})\left(2\,\frac{\text{m}}{\text{s}}\right)^2 + \tfrac{1}{2}(70\text{ kg})\left(1.5\,\frac{\text{m}}{\text{s}}\right)^2 = 179\text{ J}$$

and $$K = \tfrac{1}{2}(m_A + m_B)v^2$$

$$K = \tfrac{1}{2}(50\text{ kg} + 70\text{ kg})\left(1.21\,\frac{\text{m}}{\text{s}}\right)^2 = 87.6\text{ J}$$

Thus, the fractional change in the kinetic energy is

$$\left(\frac{K}{K_0} - 1\right) = \left(\frac{87.6\text{ J}}{179\text{ J}} - 1\right) = \boxed{-0.511}$$

The negative sign indicates that some of the initial kinetic energy is "lost" during the collision process.

7.3 Impulse

Newton's second law states that at every instant, the net force on a mass produces a corresponding time-rate-of-change of momentum. In many cases, the forces involved in collisions and other interactions that last a very short time are not known explicitly. Yet it is often relatively easy to determine the resultant change of momentum that occurs. In this section we will examine the concept of impulse, which has been found useful in these instances.

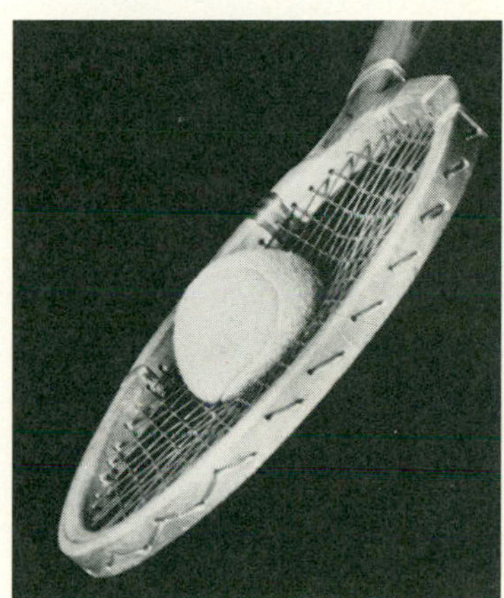

Figure 7–4
A short-duration flash photograph illustrates conditions during an impulsive force. The large amount of distortion indicates a large force between the racket and ball.

We write Newton's second law as

$$\mathbf{F} = \frac{d\mathbf{p}}{dt}$$

Integrating[1] both sides of this equation with respect to time, we have

$$\int_{t_0}^{t} \mathbf{F}\,dt = \int_{\mathbf{p}_0}^{\mathbf{p}} d\mathbf{p} \qquad \textbf{(7–11)}$$

where $\mathbf{p}_0$ is the initial momentum at time t_0, and $\mathbf{p}$ is the final momentum at time t. We integrate the right-hand side to obtain the change in momentum, $\mathbf{p} - \mathbf{p}_0$, written $\Delta\mathbf{p}$.

$$\int_{t_0}^{t} \mathbf{F}\,dt = \Delta\mathbf{p} \qquad \textbf{(7–12)}$$

The left-hand side is called the **impulse.** It represents the area under the curve of $F(t)$ versus t during the time interval of the interaction (see Figure 7–5).

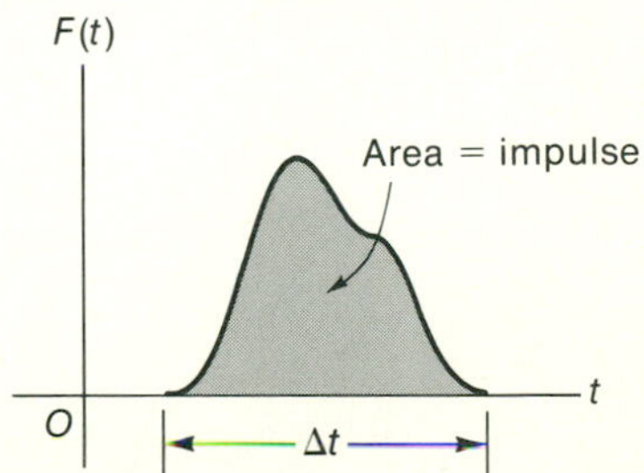

Figure 7–5
The force $F(t)$ involved in the collision between a ball and bat. The interaction occurs during the time Δt. The impulse is equal in magnitude to the shaded area under the curve.

[1] See Appendix G-III for a discussion of definite integrals.

Note that the impulse is a *vector* quantity, having the same direction as the *change* of momentum $\Delta\mathbf{p}$.

IMPULSE

$$\underbrace{\int_{t_0}^{t} \mathbf{F}\,dt}_{\text{Impulse}} = \underbrace{\Delta\mathbf{p}}_{\text{Change in momentum}} \tag{7–13a}$$

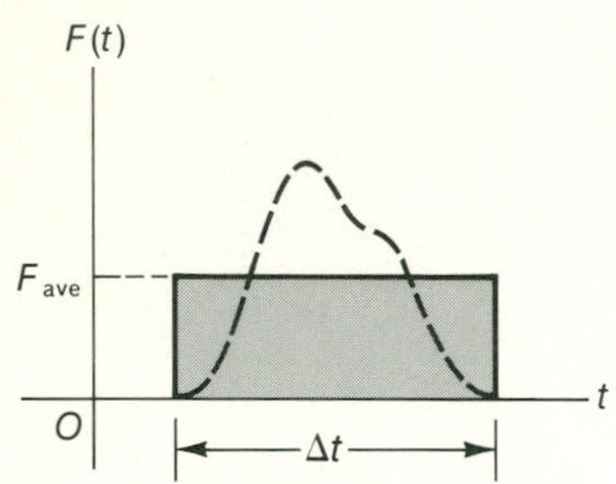

Figure 7–6
The constant force F_{ave} acting for the same time interval Δt produces the same impulse as in Figure 7–5 since the areas under the two graphs are the same.

As an example, consider a baseball as a bat strikes it. The time of the collision Δt is very small, perhaps a few hundredths of a second. The actual force function $F(t)$ may be similar to that illustrated in Figure 7–5. Although it would be difficult to determine exactly how the force $F(t)$ varies with time, it is useful to estimate the (constant) *time-average force*[2] F_{ave} acting for the time interval Δt that will produce the same change of momentum in the ball. The corresponding areas under the two curves in Figures 7–5 and 7–6 are the same. (Of course, for a short time the actual force may be much larger than F_{ave}.)

IMPULSE

$$\underbrace{\mathbf{F}_{ave}\,\Delta t}_{\text{Impulse}} = \underbrace{\Delta\mathbf{p}}_{\text{Change in momentum}} \tag{7–13b}$$

EXAMPLE 7–4

A 1-lb ball is traveling horizontally with a speed of 32 ft/s when it is struck with a bat, giving the ball a speed of 96 ft/s in the opposite direction. Assuming the bat was in contact with the ball for 0.040 s, find the average force F_{ave} the bat exerted against the ball.

SOLUTION

The time of interaction is so short that the force of gravity causes negligible motion to the ball during the actual collision. So we assume the only force acting is that caused by the bat. Choosing a coordinate system such that the ball is traveling initially in the $+x$ direction, we have

$$\mathbf{p}_0 = (m)(\mathbf{v}_0) = \left(\frac{1}{32}\,\text{slug}\right)\left(32\,\frac{\text{ft}}{\text{s}}\right) = 1\,\frac{\text{slug}\cdot\text{ft}}{\text{s}} \qquad \text{(toward } +x \text{ direction)}$$

$$\mathbf{p} = (m)(\mathbf{v}) = \left(\frac{1}{32}\,\text{slug}\right)\left(-96\,\frac{\text{ft}}{\text{s}}\right) = -3\,\frac{\text{slug}\cdot\text{ft}}{\text{s}} \qquad \text{(toward } -x \text{ direction)}$$

We then apply Equation (7–13b) and solve for $\mathbf{F}_{ave}$:

$$\mathbf{F}_{ave}\Delta t = \mathbf{p} - \mathbf{p}_0$$

$$\mathbf{F}_{ave} = \frac{\mathbf{p} - \mathbf{p}_0}{\Delta t} = \frac{-3\,\dfrac{\text{slug}\cdot\text{ft}}{\text{s}} - 1\,\dfrac{\text{slug}\cdot\text{ft}}{\text{s}}}{0.040\ \text{s}} = \boxed{-100\ \text{lb}} \quad \text{(toward the } -x \text{ direction)}$$

Note that one pound is equal to (1/32) slug under standard conditions of gravity and that the units [slug·ft/s^2] are the same as [pound].

---

[2] The *time-average* force $\mathbf{F}_{ave}$ over the interval of time $\Delta t = t - t_0$ is defined mathematically as

$$\mathbf{F}_{ave} = \frac{1}{\Delta t}\left[\int_{t_0}^{t} \mathbf{F}\,dt\right] = \frac{\text{Impulse}}{\Delta t}$$

Thus:

$$\mathbf{F}_{ave}\,\Delta t = \text{Impulse}$$

EXAMPLE 7–5

A golf ball with a mass of 0.045 kg is struck by a golf club. Assume that during the time the club is in contact with the ball, the force exerted by the club on the ball is given by the relationship shown in Figure 7–7. What is the speed of the golf ball as it leaves the head of the golf club?

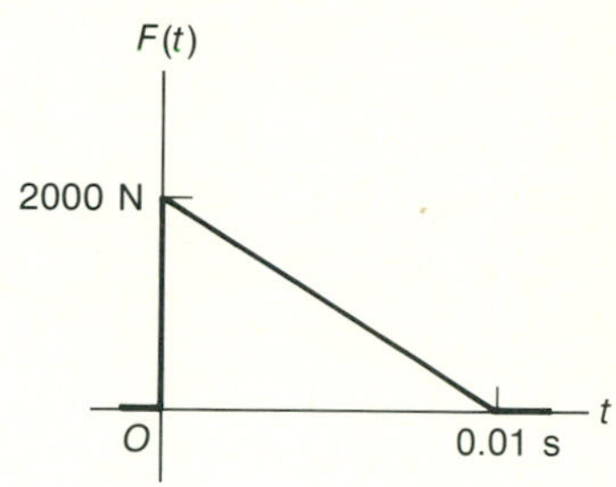

Figure 7–7
Example 7–5

SOLUTION

We use Equation (7–13a) for this problem:

$$\int_{t_0}^{t} \mathbf{F}\,dt = m\mathbf{v} - m\mathbf{v}_0$$

The impulse is numerically equal to the area under the $F(t)$ versus t curve. Since mathematically the force is a straight-line relationship, the area is one-half that of a rectangle with sides 2000 N and 0.01 s.

$$\text{Impulse} = \tfrac{1}{2}(2000\ \text{N})(0.01\ \text{s}) = \boxed{10.0\ \text{N}\cdot\text{s}}$$

Because the newton has units of [kg·m/s²], this impulse is the same as 10 kg·m/s. Thus:

$$\text{Impulse} = \text{Change in momentum}$$

$$10\,\frac{\text{kg}\cdot\text{m}}{\text{s}} = mv - mv_0$$

$$v = \frac{10\,\dfrac{\text{kg}\cdot\text{m}}{\text{s}}}{m} + mv_0$$

$$v = \frac{10\,\dfrac{\text{kg}\cdot\text{m}}{\text{s}}}{0.045\ \text{kg}} + (0.045\ \text{kg})(0) = \boxed{222\,\frac{\text{m}}{\text{s}}}$$

EXAMPLE 7–6

A stream of water is directed at an angle θ (with respect to the normal) against a flat surface, as shown in Figure 7–8. The initial speed of the stream is v, its cross-sectional area is A, and the density of the water is ρ. After striking the surface, the fluid spreads out parallel to the surface. Essentially, the component of water motion parallel to the surface does not change, but the component normal to the surface is reduced to zero. Find the average force (perpendicular to the surface) that the stream exerts against the surface.

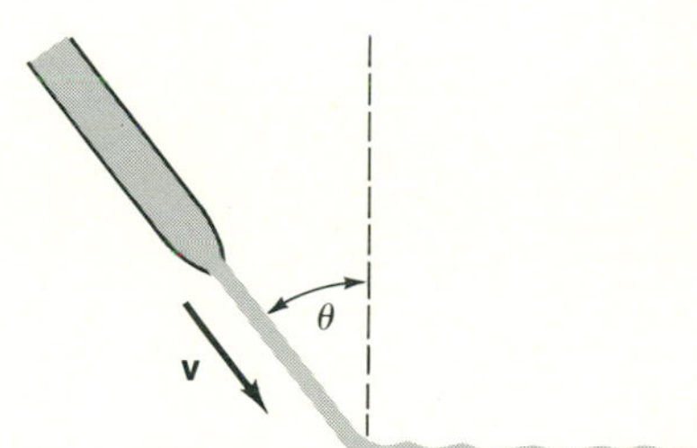

Figure 7–8
Example 7–6

SOLUTION

In a time Δt:

The volume of water striking the surface is:	$Av\,\Delta t$
The mass of this amount of water is:	$\rho Av\,\Delta t$
The momentum Δp of this amount is:	$\rho Av^2\,\Delta t$

Only the component perpendicular to the surface will exert a normal force against it.

The component of momentum perpendicular to the surface is: $\rho Av^2\,\Delta t\cos\theta$

The normal force the surface exerts against the stream to bring this component of momentum to zero is equal in magnitude (by Newton's third law) to the normal force the stream exerts against the surface.

$$F = \frac{\Delta p}{\Delta t} = \frac{\rho Av^2\,\Delta t\cos\theta}{\Delta t} = \boxed{\rho Av^2\cos\theta}$$

7.4 A Conveyor-Belt Problem

Suppose that sand from a stationary hopper falls on a moving belt at the rate of 8 slugs/s, as shown in Figure 7–9(a). The belt is supported by frictionless rollers and moves at a constant speed of 2 ft/s under the action of a constant horizontal force **F** applied to the belt.

Find the rate of change of the sand's momentum in the horizontal direction. (The change of momentum in the vertical direction is taken care of by the supporting rollers and the belt; it is therefore of no concern in this problem.)

Each second, a mass of 8 slugs acquires a horizontal velocity of 2 ft/s (from rest). Therefore:

$$\frac{\Delta p}{\Delta t} = \frac{m\,\Delta v}{\Delta t} = \frac{(8\text{ slugs})\left(2\,\frac{\text{ft}}{\text{s}}\right)}{(1\text{ s})} = \boxed{16.0\,\frac{\text{slug}\cdot\text{ft}}{\text{s}^2}}$$

Find the friction force f which the belt exerts on the sand to produce this change in momentum.

From Newton's second law, we have

$$\mathbf{f} = \frac{\Delta \mathbf{p}}{\Delta t} = \boxed{16.0\text{ lb}\qquad\text{(toward the right)}}$$

Find the external force $\mathbf{F}_{ext}$ on the belt that is required to keep it moving at constant speed v. (This force is supplied by the motor that drives the belt.)

From Newton's third law, if the belt exerts a horizontal force **f** on the sand toward the right, the sand exerts an equal force $-\mathbf{f}$ on the belt toward the left. Therefore, from $\Sigma\mathbf{F} = 0$, to keep the belt moving at constant speed the motor must exert a force of

$$\mathbf{F}_{ext} = \boxed{16.0\text{ lb}\qquad\text{(toward the right)}}$$

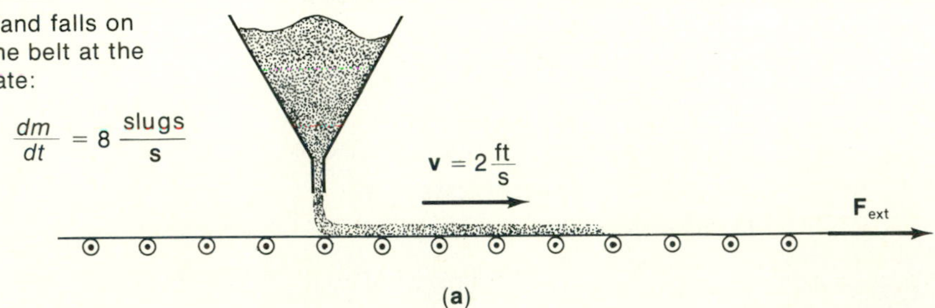

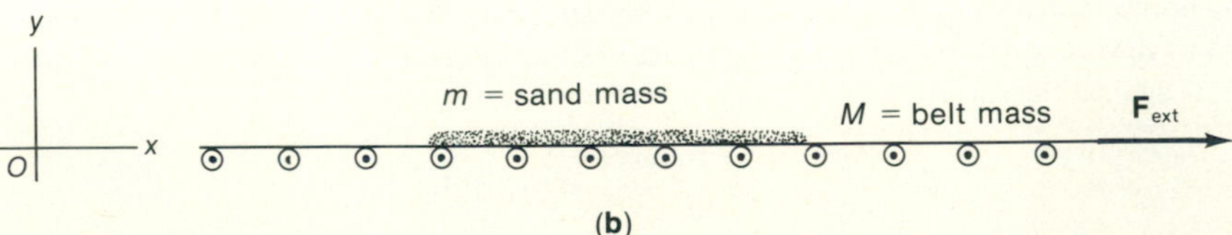

Figure 7–9
(a) The external force $\mathbf{F}_{ext}$ pulls the conveyor belt at constant speed. (b) The "system" consists of the belt plus the sand.

What is the work done by $\mathbf{F}_{ext}$ in one second?

The force $\mathbf{F}_{ext}$ moves through a distance of 2 ft per second, so

$$\Delta W = \mathbf{F} \cdot \Delta \mathbf{x} = (16 \text{ lb})(2 \text{ ft}) = \boxed{32.0 \text{ ft·lb}}$$

What kinetic energy does the incoming sand acquire each second?

In one second:

$$\Delta K = \tfrac{1}{2}\Delta m v^2 = \tfrac{1}{2}(8 \text{ slugs})\left(2\,\frac{\text{ft}}{\text{s}}\right)^2 = \boxed{16.0 \text{ ft·lb}}$$

Explain why the answers to the previous two questions are different. (That is, explain why only *half* the work done by $\mathbf{F}_{ext}$ appears as kinetic energy gained by the sand. Where does the other half go?)

This problem is an example of a continuous inelastic impact in which a constant *rate* of momentum change occurs. As the sand hits the belt and is accelerating horizontally from rest to the belt speed, there must be some sliding that occurs between the belt and the sand. It turns out that regardless of whether we assume 1 s or 1/100 s for the acceleration time, the thermal energy developed by frictional forces between the belt and the sand exactly accounts for the "missing" energy. (Note: For this calculation, the distance the friction force acts is the *relative* sliding distance between the sand and belt while acceleration occurs.)

We also may account for the "missing" energy by applying Newton's second law. We use a stationary frame of reference attached to the earth and choose our "system" to be the *moving belt plus the sand on the belt*, as illustrated in Figure 7–9(b). The belt has a mass M (which is constant), while the sand has an instantaneous mass m that increases with time at the rate dm/dt. The total momentum $\mathbf{P}$ of the system is $(m + M)\mathbf{v}$, which changes with time because m changes. Application of Newton's second law to the system gives

$$\Sigma \mathbf{F} = \frac{d\mathbf{P}}{dt}$$

$$F_{ext} = \frac{d}{dt}[(m + M)v] = \frac{d}{dt}(mv + Mv)$$

$$F_{ext} = m\frac{dv}{dt} + \frac{dm}{dt}v + M\frac{dv}{dt} + \frac{dM}{dt}v$$

The first term on the right represents possible acceleration (dv/dt) of the mass m of the sand already moving with the belt. In this case, there is no such acceleration. Because dv/dt and dM/dt are both equal to zero, we have

$$F_{ext} = \frac{dm}{dt}v$$

The power supplied by the external force F_{ext} is

$$\text{Power} = \mathbf{F}_{ext} \cdot \mathbf{v} = \left(\frac{dm}{dt}v\right)v = \frac{d}{dt}(mv^2)$$

which may be written as

$$\text{Power} = 2\frac{d}{dt}(\tfrac{1}{2}mv^2) = 2\frac{dK}{dt}$$

Thus, the power supplied by the external motor is exactly *twice* the rate at which the kinetic energy of the sand is increasing. Note that this result is true for all cases, since in this formal analysis we did not specify a particular time interval for accelerating the sand.

7.5 The Rocket

Initially, a rocket carries along all the material it will eject during its motion. Through a combustion process that usually involves a fuel and an oxidizer, the gaseous products of combustion are ejected from the rocket tail with a high speed v_r *relative to the rocket.* The ejected gases represent a large amount of momentum in the backward direction. Since the total momentum of the system as a whole (rocket plus exhaust gases) remains constant, the rocket itself gains forward momentum. In most instances, the ejection speed v_r and the rate of mass expulsion $\Delta m/\Delta t$ are essentially constant, so by Newton's third law the forward thrust that the ejected gases exert on the rocket is similarly constant.

Consider the rocket shown in Figure 7–10, aimed vertically upward at the earth's surface. We will analyze the total system (rocket plus exhaust gases) from an inertial frame attached to the earth, assuming constant g and no air friction.[3] At the instant depicted in (a), the rocket's upward speed is v, and the total mass of the rocket plus the fuel it carries is m. During a time Δt, the rocket loses mass by an amount Δm_R. The mass loss by the rocket appears as a mass gain by the exhaust Δm_E, which has been ejected downward with a speed v_r *relative to the rocket*. This results in a gain of speed Δv by the rocket and a change in the rocket's mass from m to $m + \Delta m_R$. Note that since Δm_R represents a mass *loss*, it is intrinsically negative. The speed v_E of the ejected gases relative to the earth is the difference between the speed of the rocket v relative to the earth and the speed of the ejected gases v_r relative to the rocket:

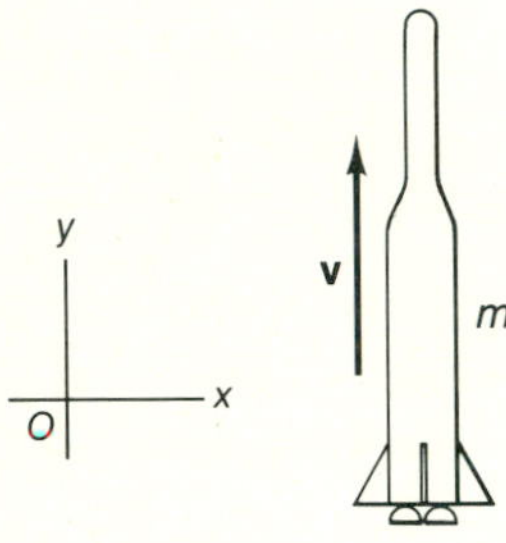

(**a**) at time t

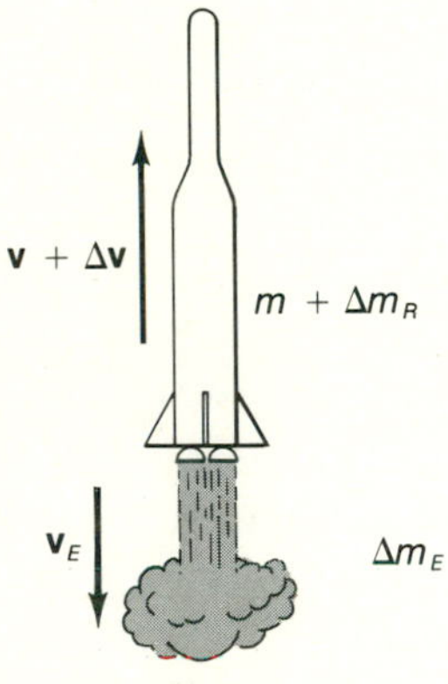

(**b**) at time $t + \Delta t$

Figure 7–10
The operation of a rocket as analyzed in an inertial frame attached to the earth.

$$v_E = v - v_r \qquad \textbf{(7–14)}$$

We now apply to the system as a whole the relation between the impulse and the change of momentum. The only external force on the system is that of gravity, $m\mathbf{g}$, acting downward. The motion is one-dimensional, and we assume that the upward direction is positive.

$$\text{Impulse} = \text{Change in momentum}$$

$$F_{\text{ext}}\,\Delta t = \begin{bmatrix}\text{Final}\\ \text{momentum}\end{bmatrix} - \begin{bmatrix}\text{Initial}\\ \text{momentum}\end{bmatrix}$$

$$-mg\,\Delta t = [\underbrace{(m + \Delta m_R)(v + \Delta v)}_{\text{Rocket}} + \underbrace{\Delta m_E(v - v_r)}_{\text{Exhaust}}] - [mv] \qquad \textbf{(7–15)}$$

Since $\Delta m_E = -\Delta m_R$, Equation (7–15) reduces to

$$\frac{\Delta v}{\Delta t} = -\frac{1}{m}(v_r + \Delta v)\frac{\Delta m_R}{\Delta t} - g$$

As Δt approaches zero, $\Delta v/\Delta t$ becomes the acceleration, Δv approaches zero, and $\Delta m_R/\Delta t$ becomes dm/dt. In the limit, the acceleration $a = dv/dt$ of the

[3] Actually, air friction near the earth's surface, particularly at high speeds, becomes a very important factor. Of course, the presence of air is not essential to developing the forward thrust on the rocket: The engine also works in the vacuum of outer space.

rocket is

$$\frac{dv}{dt} = -\frac{v_r}{m}\left(\frac{dm}{dt}\right) - g \qquad \textbf{(7–16)}$$

where dm/dt is the rate of change of mass of the rocket (also intrinsically negative). The rocket will accelerate upward if

$$\frac{v_r}{m}\left|\frac{dm}{dt}\right| > g$$

or if $v_r|dm/dt|$, called the *effective thrust*, is greater than the weight mg of the rocket.

We can easily obtain an expression for the rocket's velocity v as a function of time by integrating Equation (7–16) with respect to t. Assuming the rocket starts at rest at $t = 0$, we have

$$\int_0^v dv = -v_r \int_{m_0}^{m} \frac{dm}{m} - g\int_0^t dt \qquad \textbf{(7–17)}$$

Carrying out the integration (see Appendix G-III), we obtain

$$v = v_r \ln\left(\frac{m_0}{m}\right) - gt \qquad \text{(assumes constant } g\text{)} \qquad \textbf{(7–18)}$$

In a region where g is negligible (or in the operation of a "rocket sled," which runs on horizontal rails so that the effects of gravity are not a factor), the equation has a particularly simple form:

$$v = v_r \ln\left(\frac{m_0}{m}\right) \qquad \text{(no gravity)} \qquad \textbf{(7–19)}$$

The final velocity when all propellant has been used is thus determined by the *mass ratio at burnout* (m_0/m_{burnout}). Equation (7–19) shows two important facts about rocket propulsion: (a) the exhaust velocity v_r should be as high as possible and (b) the mass ratio at burnout must be as large as is feasible. Unfortunately, since the payload is contained in m_{burnout} (as well as the empty fuel tanks and rocket engine), there is an upper limit to this ratio. (For this reason, most high-performance rockets are multistage vehicles with smaller rockets as the payload of the first rocket. When one rocket has burned its fuel, it is discarded and the next rocket is ignited.)

A *jet engine*, like a rocket, ejects a burned-up material from its exhaust port. However, in contrast to a rocket, a jet engine takes in large quantities of air to burn the jet fuel. So in analyzing the forces on a jet engine, one must take into account the additional reaction force on the engine due to the amount of air scooped up in flight, and also add this mass of air to the fuel exhaust.

EXAMPLE 7–7

Consider a rocket launched vertically with the following characteristics: initial mass $m_0 = 15\,000$ kg (including a propellant mass of 10 000 kg), exhaust speed $v_r = 2.5 \times 10^3$ m/s, and burning time $t = 60$ s. Assuming a constant value for g, find (a) the burnout velocity and (b) the acceleration of the rocket just before burnout.

SOLUTION

(a) Equation (7–18) is appropriate:

$$v_{\text{burnout}} = v_r \ln\left(\frac{m_0}{m_{\text{burnout}}}\right) - gt$$

Substituting numerical values gives

$$v_{\text{burnout}} = \left(2.5 + 10^3\ \frac{\text{m}}{\text{s}}\right)\ln\left(\frac{15\ 000\ \text{kg}}{5000\ \text{kg}}\right) - \left(9.8\ \frac{\text{m}}{\text{s}^2}\right)(60\ \text{s}) = \boxed{2.16 \times 10^3\ \frac{\text{m}}{\text{s}}}$$

(b) The acceleration a is equal to dv/dt, so we use Equation (7–16):

$$a = -\frac{v_r}{m}\left(\frac{dm}{dt}\right) - g$$

Noting that dm/dt is a negative number (because dm is a *loss* of mass from the system), we have

$$a = -\left(\frac{2.5 \times 10^3\ \frac{\text{m}}{\text{s}}}{5 \times 10^3\ \text{kg}}\right)\left(\frac{-10^4\ \text{kg}}{60\ \text{s}}\right) - 9.8\ \frac{\text{m}}{\text{s}^2}$$

$$a = \boxed{73.5\ \frac{\text{m}}{\text{s}^2}} \qquad \text{(about 7.5 } g\text{)}$$

Summary

For an isolated system of particles that remains free from external forces, the total linear momentum $\Sigma\mathbf{p}$ is constant in time.

CONSERVATION OF LINEAR MOMENTUM

$$\Sigma\mathbf{p}_0 = \Sigma\mathbf{p}$$

In the absence of external forces, this principle *always* holds, regardless of whether mechanical energy is also conserved. Note that momentum, $\mathbf{p} = m\mathbf{v}$, is a vector quantity. In two-dimensional problems, the conservation of momentum is applied to the x and y component directions separately.

The *impulse* is equal to the net change in momentum:

$$\underbrace{\int_{t_0}^{t} \mathbf{F}\,dt}_{\text{Impulse}} = \underbrace{\Delta\mathbf{p}}_{\text{Change in momentum}}$$

In an impulsive collision between two particles, the exact function $\mathbf{F}(t)$ is usually unknown. However, assuming a time interval Δt for the interaction, one can calculate a constant force $\mathbf{F}_{\text{ave}}$ acting for the time Δt that will produce the same change in momentum.

$$\underbrace{\mathbf{F}_{\text{ave}}\,\Delta t}_{\text{Impulse}} = \underbrace{\Delta\mathbf{p}}_{\text{Change in momentum}}$$

Questions

1. If you were at rest in the middle of a perfectly smooth pond of ice, how could you get off? Also, how could you get in the middle (at rest) in the first place?
2. Explain how a rocket can accelerate in outer space, where there is no air to push against.

3. Before vacuum cleaners were invented, rugs were often cleaned by suspending them from a clothesline and beating them with a rug beater. Would this process best be described as beating the dust out of the rug, or beating the rug out of the dust?
4. Analyze how momentum is conserved when a ball is bounced off a wall.
5. Can a sailboat be moved by blowing a stream of air against the sail using an electric fan carried aboard the boat?
6. Two people stand on a railroad flatcar initially at rest. If they jump off the rear end one after the other, each with the same velocity relative to the car, the car will acquire a greater speed than if they both jump off simultaneously. Explain.
7. An hourglass with all the sand in the upper section is balanced at rest by an equal mass, as shown in Figure 7–11. (The pulley is frictionless.) The sand now is allowed to fall into the lower section of the hourglass.
   (a) When the sand first starts to flow (but before it strikes the bottom of the hourglass), is the hourglass in equilibrium?
   (b) While the sand is flowing and building up in the bottom section, is the hourglass in equilibrium? If not, what is its acceleration? If so, does the hourglass have any velocity?
   (c) After all the sand has come to rest in the bottom section, what is the state of motion of the hourglass? If moving, describe the motion; if at rest, describe the location with respect to the initial position.
8. A thread holds a cork completely submerged near the bottom of a beaker of water. The beaker rests on one pan of a balance scale, with an equal mass m on the other pan. The thread breaks. *While the cork is rising upward through the water*, does the balance read zero, or does it indicate that the beaker weighs less or more than mg?
9. If a can of compressed air is punctured and the escaping air blows toward the right, the can will accelerate to the left in a rocket-like fashion. Now consider a vacuum can (originally at rest) that is punctured. As shown in Figure 7–12, the air blows in toward the left as it enters the can. After the air has entered the can, will the can be moving toward the left, toward the right, or not be moving?*

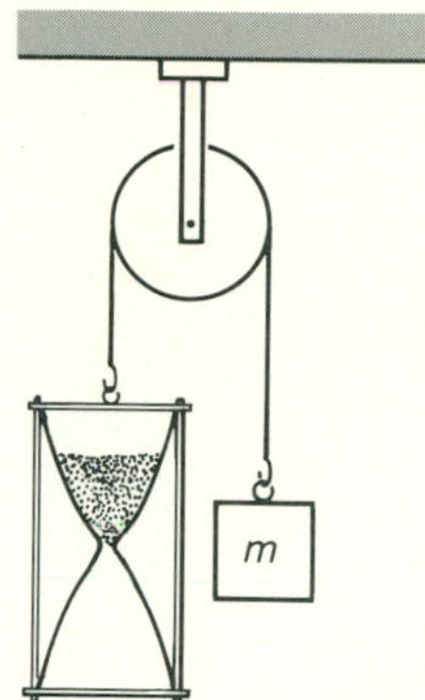

Figure 7–11
Question 7

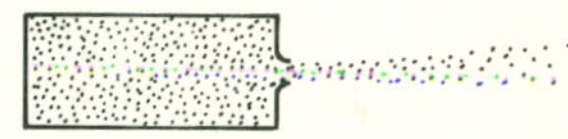

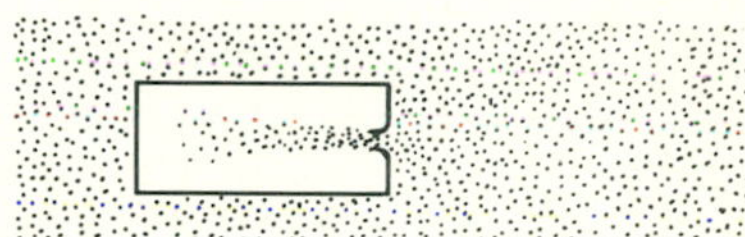

Figure 7–12
Question 9

Problems

7A–1
(a) Find the momentum of a 1000-kg car traveling 12 m/s.
(b) If the car collides with a brick wall and comes to rest in 0.5 s, what average force did the wall exert on the car? *Answers:* (a) 1.20×10^4 kg·m/s (b) 2.40×10^4 N

7A–2 A 0.4-kg ball is thrown with a speed of 10 m/s. How fast would a 2-g ball bearing have to move to have the same momentum as the ball? 2000 m/s

7A–3 A 200-ton train of freight cars moving at 5 mi/h coasts into a tunnel, collides, and connects with a 25-ton freight car moving inside the tunnel. The train emerges from the other end of the tunnel traveling at 3.6 mi/h. Determine the speed and direction of the freight car in the tunnel prior to the collision.
Answer: 7.60 mi/h, approaching the train

7A–4 A mass m moving with a speed v strikes and sticks to a similar mass initially at rest. Find the fractional change in the kinetic energy, $(K - K_0)/K_0$.

7A–5 A rifle weighing 40 N fires a 5-g bullet with a velocity of 1000 m/s. With what speed does the rifle recoil? *Answer:* 1.23 m/s

7A–6 A 3.5-kg mass moving at a speed of 15 m/s collides with a 5-kg object initially at rest. They stick together. Find the velocity of the combination after the collision. 6.18 m/s

---

* Reprinted with permission from *Thinking Physics*, Volume 1, by L. Epstein and P. Hewitt (1979), page 37. Published by Insight Press, 614 Vermont Street, San Francisco, CA 94107.

7A–7 A 7.5-kg mass slides on a horizontal frictionless surface with a speed of 4 m/s. It collides head-on and sticks to a 10-kg mass initially at rest. Find the kinetic energy "lost" in the collision. *Answer:* 34.3 J

7A–8 Two gliders at rest on an air track have masses of 3 kg and 4 kg. They compress a spring of negligible mass between them but are not connected to the spring. The system is now released. If the 3-kg mass acquires a speed of 4 m/s, how fast does the 4-kg mass move?

7A–9 A 3-kg mass moving at 7.5 m/s has its direction reversed without any change in speed by a constant force acting for 0.05 s. Find the direction and magnitude of the force.
Answer: 900 N, opposite to the particle's original velocity

7A–10 Rework the conveyor-belt example of Section 7.4 using a time interval of 0.01 s instead of 1.0 s.

7A–11 A stream of 0.50-g pellets is directed against a wall perpendicularly. The pellets strike at a rate of 15 pellets/s, and each pellet moves with a speed of 250 m/s.
(a) If the pellets stick to the wall, what average force do they exert against the wall?
(b) If instead they rebound elastically, what is the force?
Answers: (a) 1.88 N (b) 3.75 N

7B–1 A 150-lb man is sliding without friction on ice due north at 3 ft/s. A girl weighing 60 lb is moving east at 7 ft/s. They collide and hang on to each other. Find the magnitude and direction of the velocity of the man–girl combination.
Answer: 2.93 ft/s, 47° N of E

7B–2 A 1600-kg truck traveling 10 m/s approaches an intersection at the same time as a 1000-kg car traveling 25 m/s approaches at a right angle from the cross street. The vehicles collide and remain locked together. Find the direction of the wreckage just after the collision, specifying the angle with respect to the initial direction of the truck.

7B–3 A 0.75-kg ball is dropped from rest 1 m above a horizontal floor. The ball rebounds to a height of 85 cm. What impulse did the floor exert on the ball?
Answer: 6.38 N·s upward

7B–4 A 2.5-lb weight is suspended by a 30-inch string. The weight is struck a sharp horizontal blow, causing it to swing upward until the string makes an angle of 30° below the horizontal. Find the magnitude of the impulse imparted by the blow.

7B–5 A 5-kg object initially at rest is subjected to a constant force of 6 N for 5 s, then the force decreases linearly to zero in 3 s. Determine the final speed of the mass.
Answer: 7.80 m/s

7B–6 A 4-lb object moves toward the $+x$ direction with a speed of 10 ft/s. It collides and sticks to an 8-lb object initially at rest. The combination then collides and sticks to a 3-lb object moving toward the $-x$ direction. If the combination of the three masses comes to rest after the second collision, what was the speed of the 3-lb object?

7B–7 A mass slides without friction on a horizontal surface with a velocity of 7.5 m/s in the $+x$ direction. It is blown apart into two fragments of masses 3 kg and 5 kg. After the explosion, the fragments slide along the surface, with the 3-kg mass having a velocity of 15 m/s at an angle of 37° with the x axis. Find the magnitude and direction of the velocity of the 5-kg mass. *Answer:* 7.22 m/s, −48.4°

7B–8 A particle of mass m moves with a speed v along the $+x$ direction. What impulse (magnitude and direction) will cause it to move along the $+y$ direction with the same speed?

7B–9 A 5-g bullet moving at 700 m/s is stopped by a fixed wooden block. Assuming the block exerts a constant force of 8×10^3 N in stopping the bullet, find (a) the time it takes to stop the bullet, (b) the depth of penetration of the bullet in the block, (c) the work done by the block in stopping the bullet, and (d) the change in kinetic energy of the bullet.
Answers: (a) 4.37×10^{-4} s (b) 0.153 m
(c) 1.22×10^{-3} J (d) 1.23×10^3 J

7B–10 A surprising demonstration involves dropping an egg from a third-floor window to land on a foam rubber pad 2 in thick without breaking. If the pad compresses to a minimum thickness of 0.20 in, find the average force (assumed constant) during the deceleration of a 2-oz egg that has fallen 36 ft from rest. (If you wish to try the experiment, station someone nearby to catch the egg on its first bounce to a height of about a yard. If the egg strikes bare ground after falling from this 3-ft height, it invariably breaks.)

7B–11 Two masses, m and M, initially at rest, are forced apart by a spring of negligible mass. If the total work done by the spring is W, show that the speed of m is given by:

$$v = \left[\frac{2MW}{m(m + M)}\right]^{1/2}$$

7B–12 In an investigation of the tolerance of the human body in crash-type impacts, Dr. John Stapp decelerated from a speed of 632 mi/h to rest in 1.4 s. What net force did his special restraining straps exert on him during this abrupt stop? Express the force as a multiple of the force of gravity on Dr. Stapp. (He survived without permanent injury!)

7B–13 A 0.100-kg baseball traveling horizontally at 20 m/s is hit so that it now travels three times faster in a direction 30° above the horizontal. It passes directly above the pitcher.

(a) Sketch a vector diagram to find the change in momentum of the ball.

(b) If the bat was in contact with the baseball for 0.002 s, what average force (magnitude and direction) did it exert on the baseball?

Answers: (a) 7.80 kg·m/s, 22.6° above the horizontal
(b) 3900 N, 22.6° above the horizontal

7B–14 Two masses, m and km (where k is a constant), have equal initial speeds v_0 along the $+x$ and $+y$ directions, respectively. The two masses collide and stick together.

(a) In terms of the given symbols, find the final velocity $\mathbf{v}$ (magnitude and direction) of the combined masses.

(b) Calculate the numerical ratio v/v_0 for $k = 2$.

7B–15 A mass, initially at rest, splits explosively into three fragments. A 3-kg fragment moves along the $+x$ direction with a speed of 4 m/s, and a 2-kg fragment moves along the $+y$ direction with a speed of 2.5 m/s. Find the momentum (magnitude and direction) of the third fragment. *Answer:* 13.0 kg·m/s, 22.6° relative to the $+x$ direction

7B–16 A 9-kg mass initially at rest splits explosively into three fragments. Two of the fragments, each of mass 2 kg, separate at right angles along the $+x$ and $+y$ directions with speeds, respectively, of 1.5 m/s and 2 m/s. Find the speed and direction of the remaining fragment.

7B–17 A stream of water traveling at 20 m/s at the rate of 600 L/min is directed horizontally against a vertical wall. The water spreads out along the wall without splashing back appreciably. What force does the stream exert against the wall? (The density of water is 10^3 kg/m^3, and 1 m^3 = 10^3 L.) *Answer:* 200 N

7B–18 In Figure 7–13, the flexible chain has a length l and a mass m. It is lowered onto the table top with constant velocity v. Make a graph of the force the chain exerts on the table as a function of time, setting t equal to zero when the first link touches the table. Include values of the force in terms of ℓ, m, and v for all inflection points of the graph. (Hint: In addition to the weight of the chain at rest on the table, what additional force is required to bring the moving chain to rest?)

Figure 7–13
Problem 7B–18

7B–19 A stream of water is directed against a stationary turbine blade, which reverses the direction of the stream, as shown in Figure 7–14. The speed of the water is v both before and after the impact. If the mass of water per unit of time is μ, find the force that the stream exerts on the blade. *Answer:* $2.00\mu v$

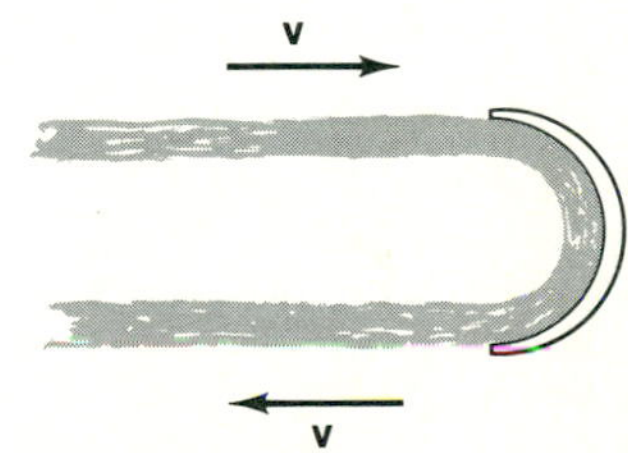

Figure 7–14
Problem 7B–19

7B–20 A 5000-kg empty freight car rolls without friction under a stationary hopper filled with sand, as illustrated in Figure 7–15. The car approaches the hopper with a speed of 2 m/s. The hopper trapdoor is opened, and a steady stream of sand falls into the car as it passes beneath the hopper. If a total of 10 000 kg of sand is loaded, find the final speed of the filled freight car.

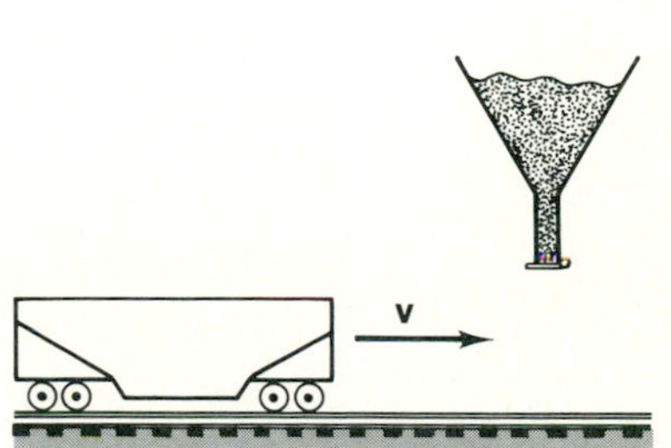

Figure 7–15
Problem 7B–20

7B–21 In the previous problem, suppose the freight car is 10 m long and the 10 000 kg of sand is spread uniformly over the full length of the car. Find the horizontal force required to keep the car moving at a constant speed of 2 m/s while being filled.

Answer: 4.00×10^3 N

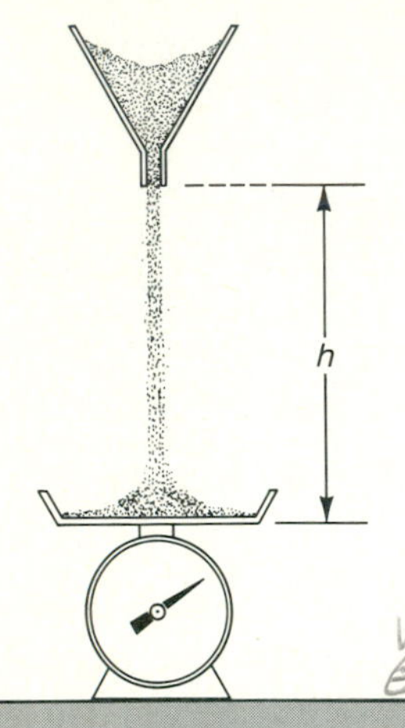

Figure 7–16
Problem 7B–22

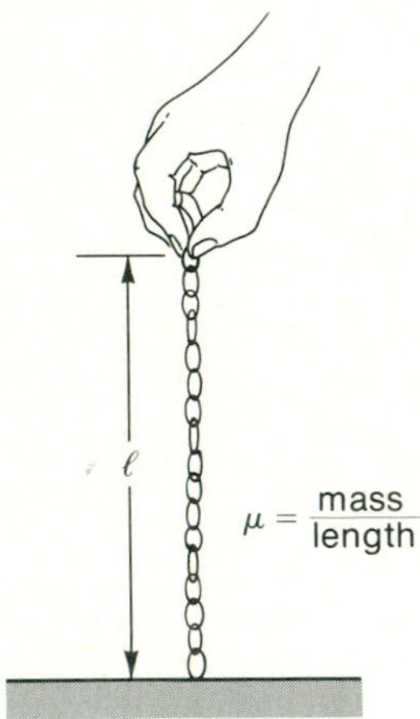

Figure 7–17
Problem 7C–5

7B–22 During a time interval of 2 s, a steady stream of lead shot falls onto the pan of a scale, as shown in Figure 7–16. The rate of flow of the shot is 2 lb/s, and the stream strikes the pan (without rebounding) with a speed of 16 ft/s. Make a graph of the scale reading (in pounds) versus time, beginning at $t = -1$ s (one second before the shot first touches the pan) to $t = 3$ s (one second after the last shot reaches the pan). Indicate on the graph the numerical values for significant points of the curve.

7B–23 A steady stream of lead shot is released from rest at a height h above a container on a scale. The shot flows at the rate of n particles per second, and each particle has a mass m. If the scale is set to read zero before the shot is released, what is the reading on the scale t seconds after the stream first begins to strike the container?

Answer: $nmg\left(t + \sqrt{\frac{2h}{g}}\right)$

7C–1 An isolated mass M, initially at rest, splits explosively into two unequal parts.

(a) What fraction of the total energy of the explosion is given to the smaller portion of mass m?

(b) Make a graph of this fraction versus m. *Answer:* (a) $(M - m)/M$

7C–2 A 15-kg mass slides without friction on a horizontal surface. It moves due north at a speed of 10 m/s. A 5-kg mass moving from the northeast at a speed of 25 m/s collides with and sticks to the 15-kg mass. With what velocity must a 3-kg mass be ejected from the combination so that the remaining mass moves with a final velocity of 10 m/s due north?

7C–3 A 2.5-kg mass is accelerated from rest by a force $F = At^2$, where $A = 0.75\ \text{N/s}^2$ in SI units.

(a) Determine the speed of the mass 15 s after application of the force.

(b) Find the constant force that would achieve this speed in the same time.

Answers: (a) 22.5 m/s (b) 3.75 N

7C–4 An 8-kg mass is accelerated from rest by a force $F = At - Bt^2$, where $A = 24$ N/s and $B = 1.2\ \text{N/s}^2$ in SI units.

(a) Find the maximum speed achieved by the mass before coming once again to rest.

(b) When will the mass come to rest?

7C–5 A flexible chain of length ℓ and mass per unit length μ is held at rest above a table with the bottom link barely touching the surface, as shown in Figure 7–17. If the chain is allowed to fall, the force it exerts on the table starts at zero and increases to some maximum value just as the final portion of the chain reaches the table. Show that the maximum force the chain exerts against the table is three times the weight of the chain. (Hint: In addition to the weight of the chain already on the table, what additional force is required to bring the topmost segment Δx of the chain to rest? Note that the time to bring the topmost segment to rest is essentially $\Delta x/v$, where v is the speed acquired after falling a distance h.)

7C–6 A flexible chain of length ℓ and mass m is lifted vertically with constant velocity v from a pile on a tabletop (Figure 7–13). Make a graph of the force F the hand exerts as a function of time t, indicating specific values of F at all inflection points of the graph.

7C–7 In tests to determine how much acceleration the human body can stand, Air Force volunteers ride rocket-powered sleds that run on rails. The sled can be brought to rest very suddenly by a "water brake." The rapid braking action is achieved by scooping up water into the moving sled from a water trough beneath the rails. Depending on whether the passenger is facing backward or forward, the deceleration is called, picturesquely, "eyeballs-in" or "eyeballs-out." Captain Eli Beeding, Jr., has undergone a deceleration of 83 g for a few hundredths of a second without suffering permanent ill effects. Consider a case where the initial speed of the sled is 175 mi/h and a deceleration of 83 g occurred for 0.005 s.

(a) Find the average force applied to Captain Beeding, who weighs 180 lb, during this interval.

(b) How fast was the sled going after the deceleration?

(c) If the sled and passenger weighed 2400 lb before braking, find the amount of water (in pounds) that was scooped into the sled in the braking procedure.

Answers: (a) 14 940 lb (b) 243 ft/s (c) 132 lb

Nature, it seems, is the popular name
for milliards and milliards and milliards
of particles playing their infinite game
of billiards and billiards and billiards.

PIET HEIN
["Atomyriades," *Grooks* (1966)]

chapter 8 Collisions

8.1 Introduction

Collisions play a large part in our everyday lives. Included in the term "collisions" are all those frequent instances when one object impacts against another in a short, but violent interaction. With each footstep our heels collide with the floor; when a teacup is set on its saucer (even gently) it involves a sudden, large acceleration; games of sport involve violent collisions of balls with a variety of bats, rackets, clubs, cue sticks, and so forth. Coupling spacecraft components in flight is a collision phenomenon. Physicists make detailed examinations of collisions between fundamental particles, nuclei, and atoms, since knowledge of these types of interaction provides an insight into the nature of the colliding particles. Although we shall restrict our discussion to the collision of only two objects, the method of study is applicable to the simultaneous collision of more than two.

A remarkable feature in the analysis of collisions is the fact that for many purposes we need not know anything about the specific forces that come into play. The only ideas we utilize are the conservation of energy and momentum. Thus, the method we present here is particularly effective in those cases where we are ignorant of the exact force relationships and even of the precise motions of the objects during the collision.

What is a "collision"? The central ideas are:

(1) The interaction between the objects is sufficiently abrupt so that we can identify a relatively short time interval Δt during which the interaction takes place. Before and after Δt, the objects do not interact.
(2) The collision forces are so large that external forces from outside the system of colliding particles can be ignored. The only forces that come into consideration are *internal* forces between the two objects.

The time interval Δt is a relative matter. In the case of colliding nuclei it may be 10^{-23} seconds, while for colliding galaxies it could be a billion years. The criteria are that no interaction occurs outside the time interval and that Δt is short compared with other significant time intervals in the physical situation.

For some collision problems, we may obtain the desired unknowns by using just the principle *Impulse = change of momentum*, developed in the previous chapter. However, here we delve more fully into two- and three-dimensional situations, and also introduce the simplifying procedure of viewing the collision in the *center-of-mass* frame of reference.

8.2 Elastic and Inelastic Collisions

Collisions are generally classified as elastic or inelastic, depending on whether or not kinetic energy is conserved in the collision. In an **elastic collision,** the total kinetic energy of the particles beforehand equals the total kinetic energy after the collision. (This type of collision is also sometimes called *perfectly elastic.*) In contrast, if some energy is "lost" to other forms of energy, such as to thermal energy due to frictional heating or to the energy of permanent deformation of the objects, then the collision is called **inelastic.** Interactions in which kinetic energy is gained—for example, in an explosion—are also called inelastic. Finally, if one moving object "captures" another so that the two masses move off together, then the collision is called *completely inelastic.*

The conservation of linear momentum holds for all types of collisions. In these collisions, we assume that external forces from outside the colliding particles are not significant. For example, in the collision of a ball and bat, or of atomic nuclei, the forces the objects exert on each other are much larger than the external forces of gravity, so during the collision time Δt we may ignore gravitational forces in the analysis. In other cases, outside forces add to zero, so they may be ignored. The only important forces in collisions are the Newton's third-law forces between the colliding particles themselves.

Of course, from a fundamental point of view, the conservation of energy principle also holds for all types of collisions, provided we include all forms of energy (frictional heating, sound energy, the energy to deform materials, and so on). But in inelastic collisions, it is often difficult to calculate directly the internal energy changes involved. For example, we must deduce the amount of energy that goes into the permanent deformation of an object by keeping track of the mechanical energy "lost" to these forms during the collision. Since this amount cannot be calculated directly, it cannot be included in the accounting of interchanges between various forms of mechanical energy. Thus when we say that energy is not conserved in inelastic collisions, we mean that *mechanical* energy is not conserved. The conservation of linear momentum, however, is true for all collisions, elastic or not.

8.3 The Analysis of Collisions

Figure 8–1 illustrates a collision of two masses, m_A and m_B, sliding on a horizontal frictionless surface. The system of two colliding masses has zero net external force, so the conservation of momentum applies. *Almost without exception the first step in every collision analysis is the application of the conservation of momentum principle.*

$$\Sigma \mathbf{p}_0 = \Sigma \mathbf{p}$$

or

$$(\mathbf{p}_A)_0 + (\mathbf{p}_B)_0 = \mathbf{p}_A + \mathbf{p}_B \tag{8–1}$$

which represents the following three independent relationships:

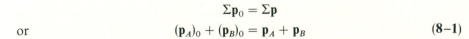

$$\left.\begin{aligned} x \text{ component:} \quad (p_{Ax})_0 + (p_{Bx})_0 &= p_{Ax} + p_{Bx} \\ y \text{ component:} \quad (p_{Ay})_0 + (p_{By})_0 &= p_{Ay} + p_{By} \\ z \text{ component:} \quad (p_{Az})_0 + (p_{Bz})_0 &= p_{Az} + p_{Bz} \end{aligned}\right\} \tag{8–2}$$

where $(p_{Ax})_0$ is the x component of the initial momentum of particle A, and so forth.

If the collision is an elastic one, we also have the conservation of kinetic energy.

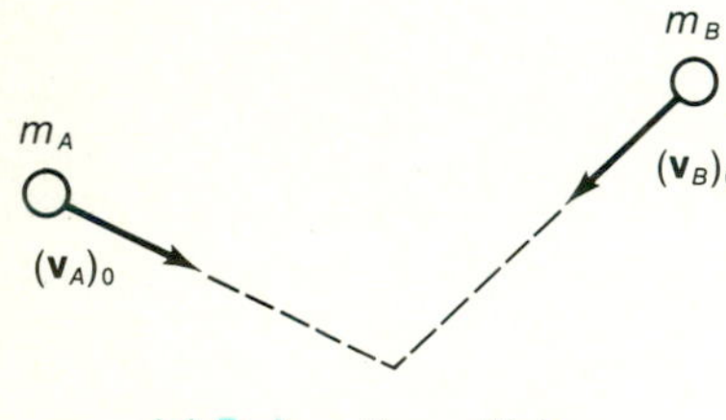

(a) Before the collision

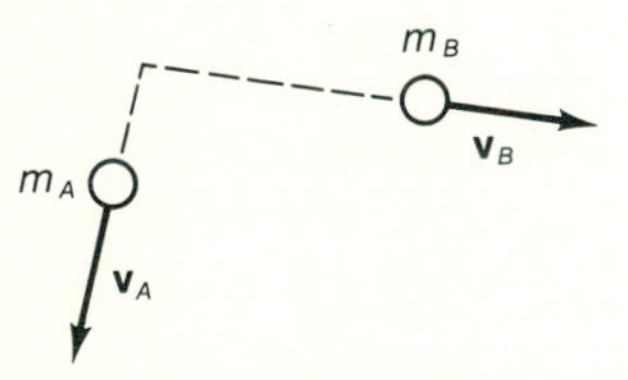

(b) After the collision

Figure 8–1
The collision of two particles sliding on a horizontal frictionless surface.

$$E_0 = E$$

$$\tfrac{1}{2}m_A(v_A)_0{}^2 + \tfrac{1}{2}m_B(v_B)_0{}^2 = \tfrac{1}{2}m_Av_A{}^2 + \tfrac{1}{2}m_Bv_B{}^2 \tag{8-3}$$

where $(v_A)_0{}^2 = (v_{Ax})_0{}^2 + (v_{Ay})_0{}^2 + (v_{Az})_0{}^2$, and so on. The masses m_A and m_B are usually known along with the initial values of their velocities; if so, the desired unknowns are the final velocities. For a given pair of masses, we thus search for six unknowns: the three Cartesian components for each of the two final velocities. Yet Equations (8–2) and (8–3) contain only four independent statements. So, in general, we need additional information regarding the outcome in order to solve the problem. This might be, for example, the final velocity of one of the particles.

In all cases, the resolution of the vector momenta into three component directions [Equation (8–2)] is a straightforward procedure that always works. But when possible, use the shortcut method of sketching simple vector diagrams for the conservation of momentum [Equation (8–1)]; it will save tedious algebra.

EXAMPLE 8–1

Consider an elastic collision between two masses, m and $3m$, which are moving toward each other with the same initial speed v_0. (a) If the masses separate after the collision along the same line of their approach (that is, it is a one-dimensional collision), what are the final speeds of the masses? (b) If instead the smaller mass moves away at right angles to its initial direction, what are the final speeds of the masses and what is the angle of scattering of the larger mass?

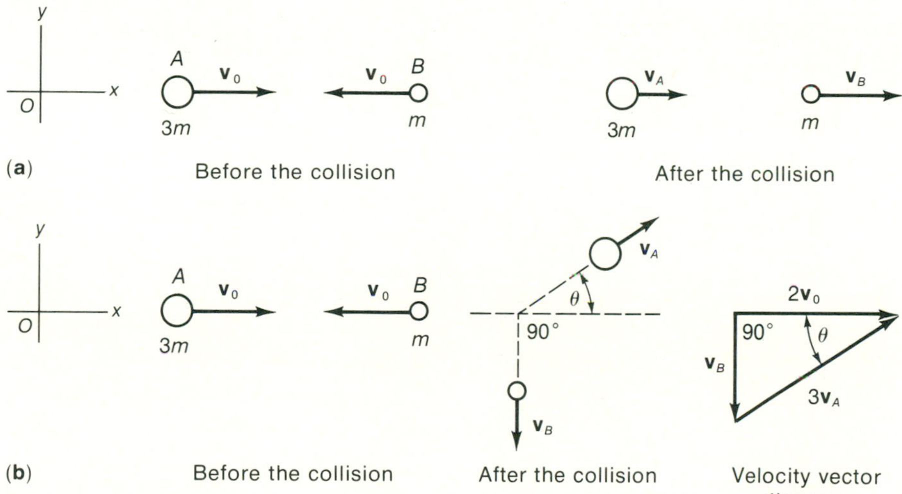

Figure 8–2
Example 8–1

SOLUTION

(a) We sketch the problem as shown in Figure 8–2(a), labeling the masses A and B and labeling their respective final speeds v_A and v_B. Although we may guess intuitively that the smaller mass moves toward the right after the collision, the motion of the larger mass is a bit more uncertain. We will *assume* that it also travels toward the right; if it actually moves toward the left, our answer will have a minus sign. We choose the positive direction toward the right and indicate this choice by the small set of coordinate axes.

Our next step is to apply the conservation of momentum principle. Since it is a one-dimensional situation, we write the following *scalar* equation, using plus and minus signs to designate directions:

$$\Sigma(p_x)_0 = \Sigma p_x$$

$$(3m)(v_0) + (m)(-v_0) = (3m)(v_A) + (m)(v_B)$$

or

$$2v_0 = 3v_A + v_B \tag{8-4}$$

Because the collision is *elastic*, the conservation of kinetic energy also applies:

$$\Sigma K_0 = \Sigma K$$

$$\tfrac{1}{2}(3m)v_0^2 + \tfrac{1}{2}(m)v_0^2 = \tfrac{1}{2}(3m)v_A^2 + \tfrac{1}{2}(m)v_B^2$$

or

$$4v_0^2 = 3v_A^2 + v_B^2 \qquad \textbf{(8–5)}$$

If we solve for v_B in Equation (8–4) and square the result, we have

$$v_B^2 = 4v_0^2 - 12v_0v_A + 9v_A^2$$

From Equation (8–5) we obtain a second equation for v_B^2.

$$v_B^2 = 4v_0^2 - 3v_A^2$$

Subtracting the above equation from the previous equation gives

$$0 = -12v_0v_A + 12v_A^2$$

which leads to two possible answers: $v_A = v_0$ or $v_A = 0$. To choose between them, we reason as follows. If $v_A = v_0$, then Equation (8–4) is

$$2v_0 = 3v_0 + v_B$$

or

$$v_B = -v_0$$

This result is physically meaningless since it implies that the two masses must pass through each other during the collision (and, from the conservation of momentum, that the $3m$ mass has a velocity of v_0—a result indistinguishable from "no collision"). Therefore, we accept the alternative answer as correct:

$$\boxed{v_A = 0}$$

and, from Equation (8–4):

$$\boxed{v_B = 2v_0}$$

Thus the $3m$ mass comes to rest and the mass m reverses its direction and acquires a speed twice its original value.

(b) For the second case, we have a two-dimensional situation, as illustrated in Figure 8–2(b). Although this part could be solved by a simultaneous solution of the conservation of energy and momentum equations in the x and y directions, a graphical solution using the momentum vectors is simpler. We write the conservation of momentum for this two-dimensional case as a single *vector* equation (without resolving the vectors into components).

$$\Sigma \mathbf{p}_0 = \Sigma \mathbf{p}$$

$$3m\mathbf{v}_0 - m\mathbf{v}_0 = 3m\mathbf{v}_A + m\mathbf{v}_B$$

or

$$2\mathbf{v}_0 = 3\mathbf{v}_A + \mathbf{v}_B$$

This vector equation may be diagramed as a closed right triangle, as shown.

The Pythagorean Theorem yields

$$(3v_A)^2 = (2v_0)^2 + v_B^2$$

or

$$4v_0^2 = 9v_A^2 - v_B^2 \qquad \textbf{(8–6)}$$

This equation has two unknowns: v_A and v_B. To obtain another relation between the unknowns, we apply the conservation of energy principle:

$$E_0 = E$$

$$\tfrac{1}{2}(3m)v_0^2 + \tfrac{1}{2}(m)v_0^2 = \tfrac{1}{2}(3m)v_A^2 + \tfrac{1}{2}(m)v_B^2$$

or

$$4v_0^2 = 3v_A^2 + v_B^2 \qquad \textbf{(8–7)}$$

Solving Equations (8–6) and (8–7) simultaneously, we obtain

$$\boxed{v_A = \sqrt{\tfrac{2}{3}}v_0} \quad \text{and} \quad \boxed{v_B = \sqrt{2}v_0}$$

The angle θ may be found from the vector diagram:

$$\sin\theta = \left(\frac{v_B}{3v_A}\right)$$

Substituting the above values for v_A and v_B in terms of v_0, we have

$$\sin\theta = \left(\frac{\sqrt{2}v_0}{3\sqrt{\frac{2}{3}}v_0}\right) = 0.577$$

$$\theta = \sin^{-1}(0.577) = \boxed{35.3^\circ}$$

As is well known by billiard players, when one particle collides elastically with an equal-mass particle at rest, the recoil trajectories have an interesting relationship. If spin and friction can be neglected, the two particles move away at *right angles* with respect to each other (see Figure 8–3).

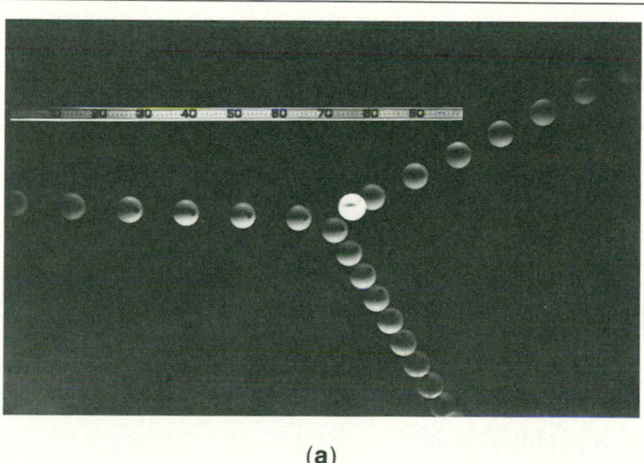

(a)

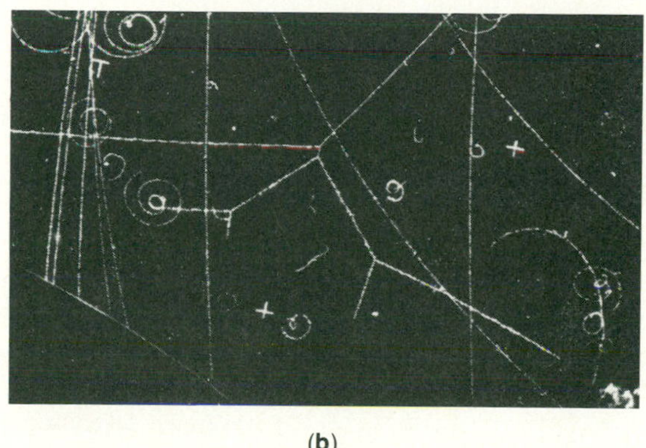

(b)

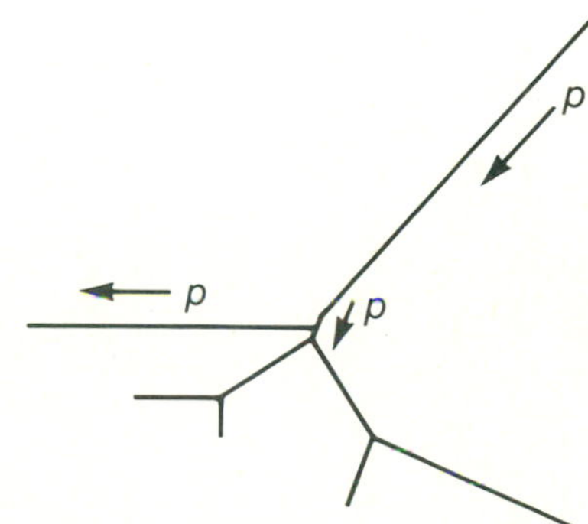

Figure 8–3

(a) A multiple-flash photograph of a collision between two billiard balls having equal masses. One ball was initially at rest, so it appears whiter where it was stationary during several flashes. Assuming the collision is perfectly elastic and frictionless, the angle between the recoiling balls is theoretically 90° for this situation. However, if spinning motion occurs or if friction is appreciable, the angle may be slightly different, as in this photograph, where it is about 87°.

(b) Elastic collisions between protons in a liquid hydrogen bubble chamber. The right-angle relationship is not obvious from this single photograph because all the trajectories do not lie in the plane of the figure. However, stereoscopic views verify that the angle between scattered particles is always 90°. (For very high-energy collisions, relativistic effects cause the angle to become $<90°$.)

A bubble chamber contains a superheated liquid (often hydrogen) under high pressure. The passage of a charged particle leaves a trail of ionization in the liquid, which creates nucleation centers for the formation of vapor bubbles when the pressure on the liquid is suddenly reduced. The bubbles grow for a few hundredths of a second to a size convenient for photographing. Then, after photographing the tracks, the pressure is restored, the bubbles dissolve, and the chamber is ready to record another event—all within a fraction of a second.

EXAMPLE 8–2

In a *ballistic pendulum* system, a rectangular wooden block of mass M is suspended by four cords which permit it to swing as a pendulum without rotating (see Figure 8–4). A bullet of mass m traveling with speed v strikes the block and becomes embedded in it before the block has time to move appreciably. As a result of the impact, the bullet-plus-block swings upward, rising to a maximum vertical distance h. (a) Find the initial speed of the bullet. (b) Find the fraction of the initial kinetic energy that was converted into other forms of energy (mainly permanent deformations of the block and bullet).

Figure 8–4
Example 8–2

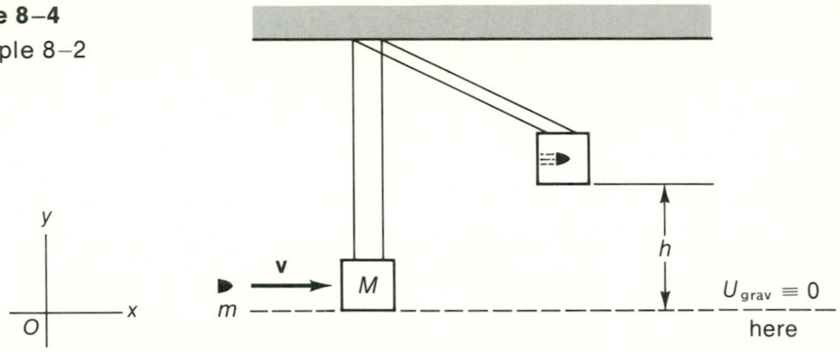

SOLUTION

(a) The collision between the bullet and the block is *completely inelastic*, because the two objects stick together after colliding. We apply the conservation of momentum principle for the one-dimensional collision itself:

$$(P_x)_0 = P_x$$

$$mv + M(0) = (m + M)V$$

where V is the speed of the bullet-plus-block combination *the instant just after the collision*, before the block has had time to move appreciably. Solving for V, we have

$$V = \left(\frac{m}{m + M}\right)v$$

The block-plus-bullet now swings up to the maximum elevation h. This latter part of the problem involves only conservative forces in which the kinetic energy of motion is converted to gravitational potential energy. We thus can apply the conservation of mechanical energy for the swinging motion of the pendulum, choosing the lowest point as the zero reference level for gravitational potential energy.

$$E_0 = E$$

$$(U_{\text{grav}})_0 + K_0 = U_{\text{grav}} + K$$

$$0 + \tfrac{1}{2}(m + M)V^2 = (m + M)gh + 0$$

Substituting the value for V found above and solving for v, we have

$$\boxed{v = \left(\frac{m + M}{m}\right)\sqrt{2gh}}$$

(b) During the collision, the fractional change in mechanical energy ($\Delta E/E_0$) of the bullet-plus-block system is

$$\frac{K - K_0}{K_0} = \frac{K}{K_0} - 1 = \frac{\frac{1}{2}(m + M)V^2}{\frac{1}{2}mv^2} - 1$$

Substituting $(m/(m + M))v$ for V gives

$$\frac{\frac{1}{2}(m + M)\left[\left(\frac{m}{m + M}\right)v\right]^2}{\frac{1}{2}mv^2} - 1 = \frac{m}{m + M} - 1 = \frac{m}{m + M} - \frac{m + M}{m + M} = \boxed{-\left(\frac{M}{m + M}\right)}$$

The minus sign indicates that ΔE is a *decrease* in energy. For a typical case, $m = 4$ g and $M = 2$ kg, which results in about 99.8% of the initial kinetic energy being transformed into nonmechanical forms such as the energy used in frictional heating and in forcing apart the wood fibers and deforming the bullet. Also, a small portion appears as sound energy of the impact.

8.4 Center of Mass

So far in our discussions we have treated masses as if they were ideal *point* particles. This is a useful simplification and one that is entirely valid in a great many cases. However, all real objects are extended bodies, not mathematical points. We often are interested in the exact motion of an extended object as it simultaneously *rotates* and *translates* through space, perhaps even vibrating internally as well (see Figure 8–5). How do we deal with these more complicated situations?

Fortunately, there is one point associated with every object, called the *center of mass*, that greatly simplifies the analysis. It moves exactly as if the entire mass of the object were concentrated at this point and as if it were acted upon by the same *external* forces as the actual object. In figuring out the motion of the center of mass, we can ignore all the *internal* forces between various parts of the object, thereby simplifying the calculation.

The procedure we will introduce is very useful not only for an extended object but also for a system of many separate bodies that interact—all the way from a vibrating molecule to a galaxy of 10^{12} stars. In this chapter we will consider the center of mass for a system of just two particles. It is a simple matter to extend this idea to three or more particles and to solid objects.

In a system of two particles, m_1 and m_2, with velocities $\mathbf{v}_1$ and $\mathbf{v}_2$, respectively, the **center-of-mass** point (abbreviated *CM*) has the following properties:

(a) The total mass $M = m_1 + m_2$ is assumed to be concentrated at the center-of-mass location x_{CM}.

(b) This single mass point M has the same momentum $\mathbf{p}$ as the entire system; that is, $\mathbf{p} = m_1\mathbf{v}_1 + m_2\mathbf{v}_2$. We may also express the momentum as $\mathbf{p} = M\mathbf{v}_{CM}$, where $\mathbf{v}_{CM}$ is the velocity of the center-of-mass point.

Thus, $\mathbf{p} = m_1\mathbf{v}_1 + m_2\mathbf{v}_2$ and $\mathbf{p} = (m_1 + m_2)\mathbf{v}_{CM}$. Equating these expressions, we have

$$(m_1 + m_2)\mathbf{v}_{CM} = m_1\mathbf{v}_1 + m_2\mathbf{v}_2 \tag{8–8}$$

VELOCITY OF THE CENTER OF MASS FOR TWO PARTICLES

$$\mathbf{v}_{CM} = \frac{m_1\mathbf{v}_1 + m_2\mathbf{v}_2}{m_1 + m_2} \tag{8–9}$$

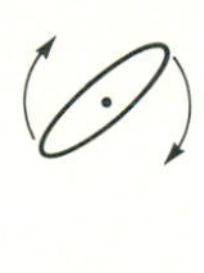

(a) Pure rotation about a fixed point.

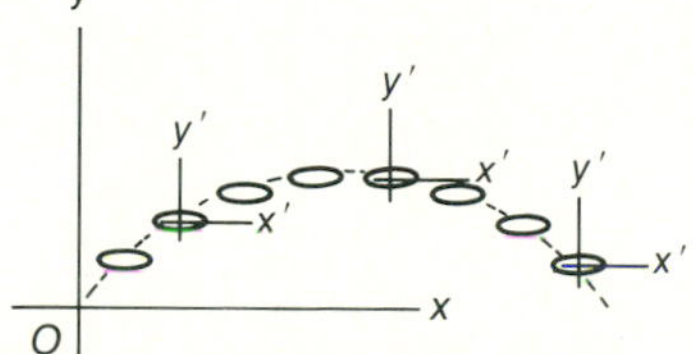

(b) Pure translational motion.

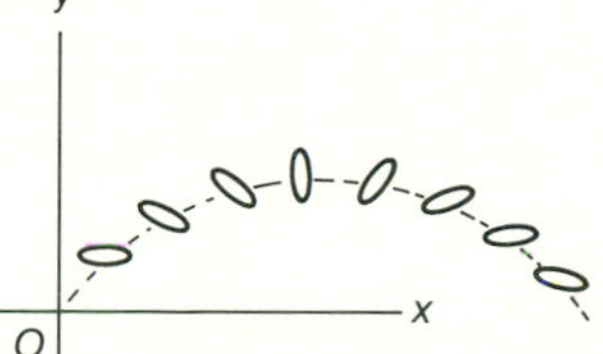

(c) Combined rotational and translational motion.

Figure 8–5
Some possible motions of a rigid body moving in two dimensions. Such motions can also occur in three dimensions.

The location x_{CM} of the center-of-mass point[1] is given by

$$x_{CM} = \frac{m_1 x_1 + m_2 x_2}{m_1 + m_2} \tag{8-10}$$

Note that taking the derivative of Equation (8–10) with respect to time yields Equation (8–9).

To extend this definition of location to three dimensions, we use similar expressions for the additional coordinates of the CM:

$$y_{CM} = \frac{m_1 y_1 + m_2 y_2}{m_1 + m_2} \quad \text{and} \quad z_{CM} = \frac{m_1 z_1 + m_2 z_2}{m_1 + m_2} \tag{8-11}$$

These component relations are all contained within the general expression for the location of the center of mass in terms of the position vector $\mathbf{r}_{CM}$:

LOCATION OF THE CENTER OF MASS FOR TWO PARTICLES

$$\mathbf{r}_{CM} = \frac{m_1 \mathbf{r}_1 + m_2 \mathbf{r}_2}{m_1 + m_2} \tag{8-12}$$

Figure 8–6 sketches these relationships. For two particles, the center of mass always lies on a line joining the particles, closer to the more massive particle. In sketches, we will indicate the center-of-mass location by a small cross (**X**) labeled CM.

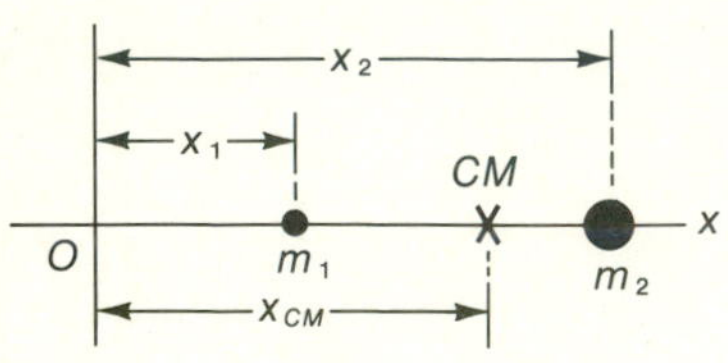

One dimension

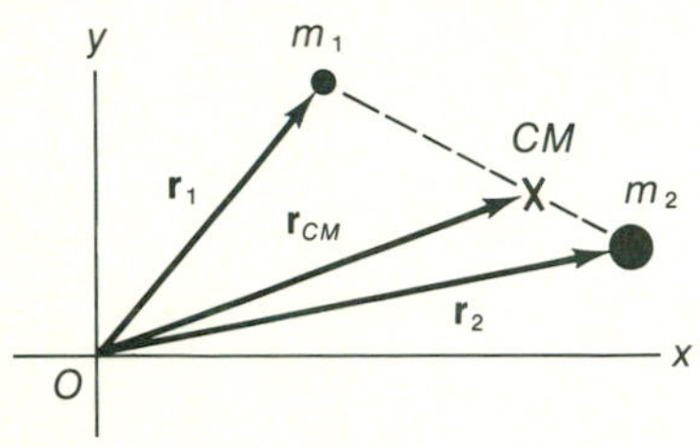

Two dimensions

Figure 8–6
The location of center of mass CM for two particles.

EXAMPLE 8–3

Given: $m_1 = 2$ kg at $x_1 = 1$ m, and $m_2 = 3$ kg at $x_2 = 3$ m (see Figure 8–7). Find the center of mass.

SOLUTION

$$x_{CM} = \frac{m_1 x_1 + m_2 x_2}{m_1 + m_2} = \frac{(2\text{ kg})(1\text{ m}) + (3\text{ kg})(3\text{ m})}{2\text{ kg} + 3\text{ kg}} = \frac{11\text{ kg·m}}{5\text{ kg}} = \boxed{2.20\text{ m}}$$

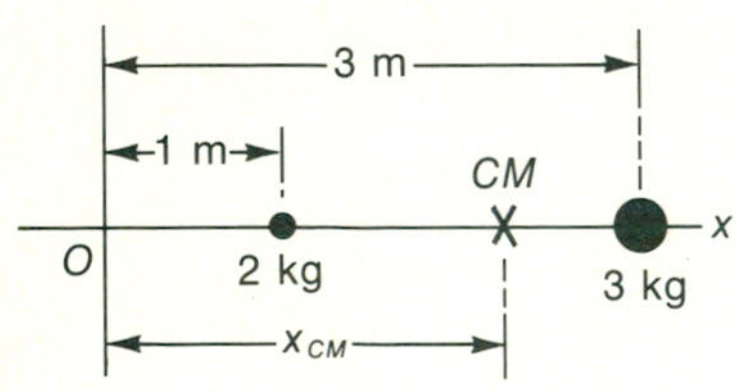

Figure 8–7
Example 8–3

EXAMPLE 8–4

Find the position of the center of mass for the following mass points:

$m_1 = 2$ kg at (1 m,2 m)
$m_2 = 3$ kg at (3 m,2 m)
$m_3 = 4$ kg at (–1 m,–4 m)

SOLUTION

The x component: $$x_{CM} = \frac{m_1 x_1 + m_2 x_2 + m_3 x_3}{m_1 + m_2 + m_3}$$

$$x_{CM} = \frac{(2\text{ kg})(1\text{ m}) + (3\text{ kg})(3\text{ m}) + (4\text{ kg})(-1\text{ m})}{2\text{ kg} + 3\text{ kg} + 4\text{ kg}} = 0.778\text{ m}$$

The y component: $$y_{CM} = \frac{m_1 y_1 + m_2 y_2 + m_3 y_3}{m_1 + m_2 + m_3}$$

$$y_{CM} = \frac{(2\text{ kg})(2\text{ m}) + (3\text{ kg})(2\text{ m}) + (4\text{ kg})(-4\text{ m})}{2\text{ kg} + 3\text{ kg} + 4\text{ kg}} = -0.667\text{ m}$$

Therefore, the CM is located at the point $\boxed{(0.778\text{ m}, -0.667\text{ m}).}$

[1] Equation (8–10) may be considered the *mass-weighted mean* of x_1 and x_2. Similarly, Equation (8–9) is the *mass-weighted mean* of $\mathbf{v}_1$ and $\mathbf{v}_2$.

EXAMPLE 8–5

Find the position and velocity (magnitude and direction) of the center of mass for the following two particles, which are in motion (see Figure 8–8):

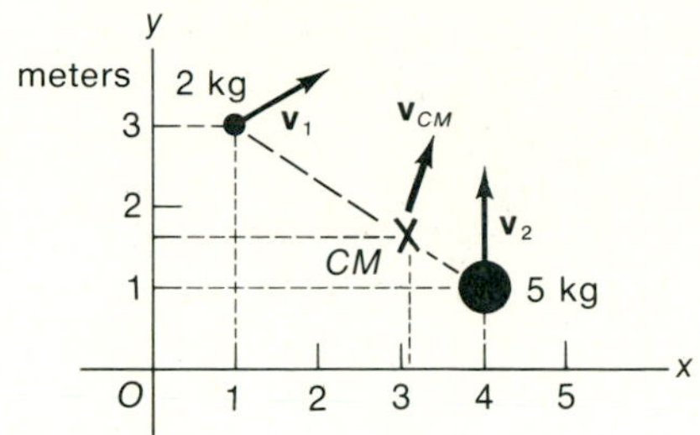

(a) The position and velocity of the two particles

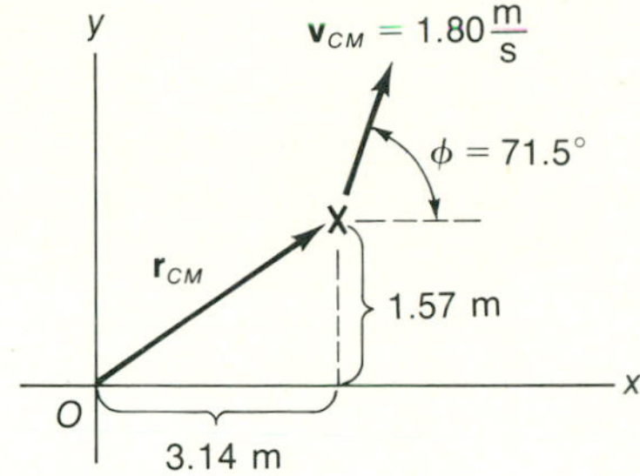

(b) The position and velocity of the center of mass

Figure 8–8
Example 8–5

Mass	Position	Velocity
$m_1 = 2$ kg	(1 m,3 m)	$v_1 = \left(2\frac{\text{m}}{\text{s}}\right)\hat{\mathbf{x}} + \left(1\frac{\text{m}}{\text{s}}\right)\hat{\mathbf{y}}$
$m_2 = 5$ kg	(4 m,1 m)	$v_2 = (0)\hat{\mathbf{x}} + \left(2\frac{\text{m}}{\text{s}}\right)\hat{\mathbf{y}}$

SOLUTION

$$\mathbf{r}_{CM} = \frac{m_1\mathbf{r}_1 + m_2\mathbf{r}_2}{m_1 + m_2} = \frac{[2\text{ kg}][(1\text{ m})\hat{\mathbf{x}} + (3\text{ m})\hat{\mathbf{y}}] + [5\text{ kg}][(4\text{ m})\hat{\mathbf{x}} + (1\text{ m})\hat{\mathbf{y}}]}{2\text{ kg} + 5\text{ kg}}$$

$$= \frac{(2\hat{\mathbf{x}} + 6\hat{\mathbf{y}} + 20\hat{\mathbf{x}} + 5\hat{\mathbf{y}})\text{ kg}\cdot\text{m}}{7\text{ kg}} = \frac{(22\hat{\mathbf{x}} + 11\hat{\mathbf{y}})\text{ kg}\cdot\text{m}}{7\text{ kg}} = \boxed{(3.14\text{ m})\hat{\mathbf{x}} + (1.57\text{ m})\hat{\mathbf{y}}}$$

$$\mathbf{v}_{CM} = \frac{m_1\mathbf{v}_1 + m_2\mathbf{v}_2}{m_1 + m_2} = \frac{[2\text{ kg}]\left[\left(2\frac{\text{m}}{\text{s}}\right)\hat{\mathbf{x}} + \left(1\frac{\text{m}}{\text{s}}\right)\hat{\mathbf{y}}\right] + [5\text{ kg}]\left[(0)\hat{\mathbf{x}} + \left(2\frac{\text{m}}{\text{s}}\right)\hat{\mathbf{y}}\right]}{2\text{ kg} + 5\text{ kg}}$$

$$= \frac{(4\hat{\mathbf{x}} + 2\hat{\mathbf{y}} + 10\hat{\mathbf{y}})\dfrac{\text{kg}\cdot\text{m}}{\text{s}}}{7\text{ kg}} = (0.571\hat{\mathbf{x}} + 1.71\hat{\mathbf{y}})\frac{\text{m}}{\text{s}}$$

$$= \sqrt{(0.571)^2 + (1.71)^2}\,\frac{\text{m}}{\text{s}} = \boxed{1.80\frac{\text{m}}{\text{s}}}$$

$$\tan\phi = \left(\frac{v_y}{v_x}\right) = \left(\frac{1.71\dfrac{\text{m}}{\text{s}}}{0.571\dfrac{\text{m}}{\text{s}}}\right) = 2.99$$

$$\phi = \tan^{-1}(2.99) = \boxed{71.5^\circ}$$

8.5 Relative Velocities Using a Geometrical Method

Suppose a mass m has a location $\mathbf{r}$ as measured in an inertial frame of reference that we designate with the symbol S. Then consider a second inertial frame of reference S' moving with respect to S with constant velocity $\mathbf{V}$, as shown in Figure 8–9. (For convenience, we align the x' and y' axes parallel to the corresponding axes of S, and consider the velocity $\mathbf{V}$ to be along the $+x$ direction.) At the instant shown, the moving origin O' is located at $\mathbf{h}$ with respect to O, and $\mathbf{V} = d\mathbf{h}/dt$. From the vector diagram we see that

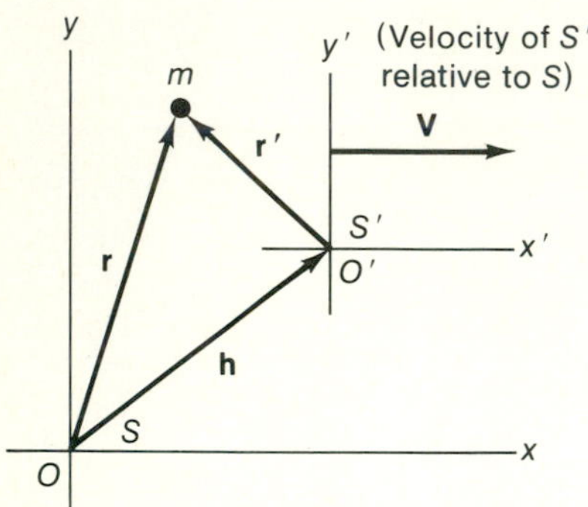

Figure 8–9
The S' frame of reference moves with constant velocity $\mathbf{V}$ (along the $+x$ direction) relative to the S frame. At the instant shown, the origin O' of the moving frame is located at $\mathbf{h}$ with respect to O.

$$\mathbf{r} = \mathbf{r}' + \mathbf{h} \tag{8–13}$$

where $\mathbf{r}'$ is the location of m with respect to O'. Now if the mass m has a velocity $\mathbf{v} = d\mathbf{r}/dt$ as measured in S, we take the time derivative of the above equation to determine the relation between $\mathbf{v}$ and $\mathbf{V}$.

$$\frac{d\mathbf{r}}{dt} = \frac{d\mathbf{r}'}{dt} + \frac{d\mathbf{h}}{dt}$$

or

GALILEAN VELOCITY ADDITION

$$\mathbf{v} = \mathbf{v}' + \mathbf{V} \tag{8–14}$$

where $\mathbf{v}$ is the velocity of m as measured in S, $\mathbf{v}'$ is the velocity of m as measured in S', and $\mathbf{V}$ is the velocity of S' relative to S.

Obviously, Galilean velocity addition is useful when only *two* relative velocities are involved (the velocity of S' relative to S, and the velocity of m relative to S'). But what about more complicated situations involving many different relative velocities in the same problem? Fortunately, there is a very straightforward method that applies in every case of relative velocities. It involves three steps:

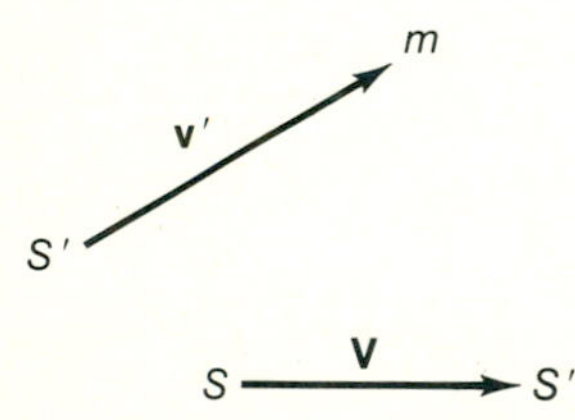

(**a**) Velocity vectors with labels on the heads and tails.

Step 1. First, make a sketch of each velocity vector, labeling the *point* of the arrow with the name of the *object*, and the *tail* of the arrow with the name of the *reference frame* in which that velocity is measured. Figure 8–10(a) illustrates the procedure.

Step 2. Redraw the diagram, moving the vectors around so that *the same labels join together*, regardless of whether the vectors join head-to-tail, head-to-head, or tail-to-tail. One must be careful, of course, to retain the magnitudes and directions of all vectors. (This geometrical procedure is sometimes called the "domino" method after the game of the same name.)

Step 3. On this diagram, an arrow drawn between any two labels represents the vector velocity of the *label at the point* relative to the *label at the tail* [Figure 8–10(b)].

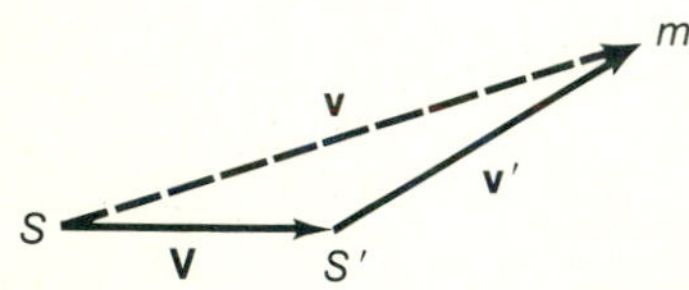

(**b**) The diagram drawn a second time with the S' labels joined together. The dashed arrow represents the velocity $\mathbf{v}$ of the mass m relative to the S frame. Note that $\mathbf{v} = \mathbf{V} + \mathbf{v}'$.

Figure 8–10
The geometrical method of analyzing relative velocities.

A little thought will show that this labeling scheme and the geometrical method of constructing the vector diagram are based on good physical reasoning. It clearly shows the relationships between both the magnitudes and directions of the various velocities involved.

EXAMPLE 8–6

A train moves due east at 4 m/s along a level, straight track. A girl on the train rolls a ball along the floor with a speed of 2 m/s relative to the train. The ball is launched toward a point directly across the aisle from south to north. Find the velocity of the ball in the earth's frame of reference.

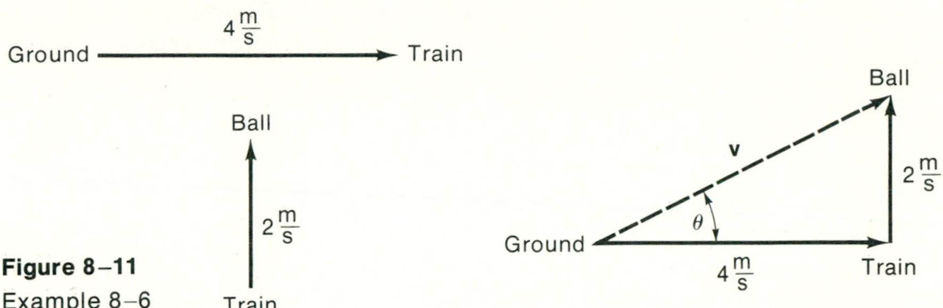

Figure 8–11
Example 8–6

SOLUTION

We draw the velocity vectors with appropriate labels and then redraw the vectors with similar labels joined (see Figure 8–11). From the Pythagorean Theorem, we see that the velocity **v** has the following magnitude:

$$v = \sqrt{\left(4\frac{\text{m}}{\text{s}}\right)^2 + \left(2\frac{\text{m}}{\text{s}}\right)^2} = \sqrt{20\left(\frac{\text{m}}{\text{s}}\right)^2} = \boxed{4.47\frac{\text{m}}{\text{s}}}$$

The direction of **v** is at an angle θ given by $\tan\theta = v_y/v_x$.

$$\tan\theta = \left(\frac{2\frac{\text{m}}{\text{s}}}{4\frac{\text{m}}{\text{s}}}\right) = 0.500$$

$$\theta = \tan^{-1}(0.500) = \boxed{26.6^\circ \text{ N of E}}$$

EXAMPLE 8–7

Suppose a man on a ship walks due west at the speed of 2 km/h relative to the ship. The ship is traveling due north at 11 km/h relative to the water, while the water flows at a

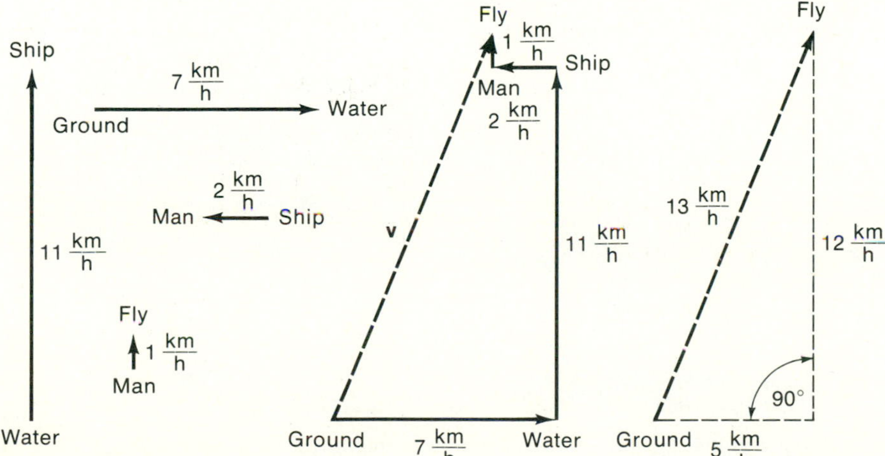

Figure 8–12
Example 8–7

rate of 7 km/h due east with respect to the ground. A fly on the shoulder of the man crawls due north at 1 km/h relative to the man. Find the fly's speed relative to the ground.

SOLUTION

We first sketch each vector separately, showing the correct magnitude and direction and labeling the point and tail of each. Next, we redraw the vectors, moving them around so that the same labels join together (see Figure 8–12). We then identify the velocity vector we seek, whose label at the point moves relative to the label at the tail.

Because of the right-angle directions in the statement of the problem, we can simplify the diagram to a right triangle, as shown. Applying the Pythagorean Theorem, we have

$$v = \sqrt{\left(5\,\frac{\text{km}}{\text{h}}\right)^2 + \left(12\,\frac{\text{km}}{\text{h}}\right)^2} = \sqrt{169\left(\frac{\text{km}}{\text{h}}\right)^2} = \boxed{13.0\,\frac{\text{km}}{\text{h}}}$$

Vectors representing other relative velocities—for example, the ship relative to the ground, or the fly with respect to the water—could be obtained easily from the same diagram.

8.6 The Center-of-Mass Frame of Reference

We are now in a position to make a simplification in the analysis of two-particle collisions. We will show that the total momentum in the center-of-mass frame is zero. Because of this fact, the analysis of collisions is much easier if viewed in a frame of reference that is moving with the center of mass. In fact, the nuclear physicist, who obtains a great deal of information about fundamental particles from collision processes, always shifts from the laboratory (*LAB*) frame to the center-of-mass (*CM*) frame because the analysis is much more straightforward and symmetrical in the *CM* frame.

Figure 8–13(a) shows successive "snapshots" during the elastic collision between a particle with mass m_A and initial velocity $(\mathbf{v}_A)_0$ and a target particle

Figure 8–13
Successive snapshots of a one-dimensional elastic collision as viewed in the laboratory (*LAB*) frame of reference and in the center-of-mass (*CM*) frame. In each diagram, the location of the center of mass is shown by a cross.

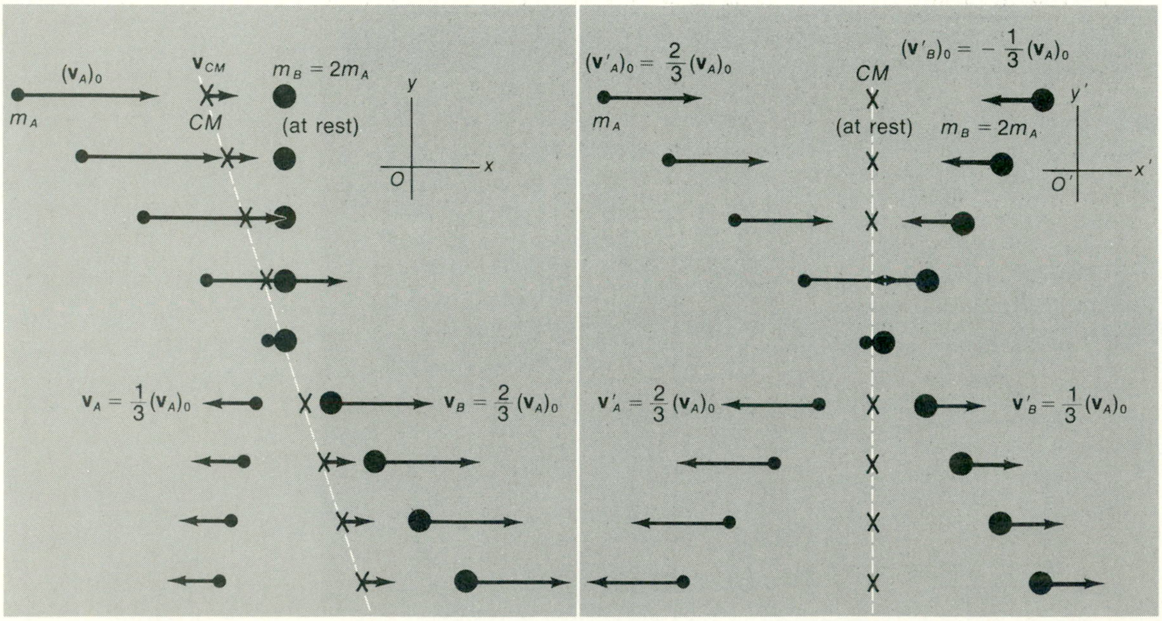

(**a**) In the *LAB* frame of reference, the velocity of the center of mass $\mathbf{v}_{CM}$ remains constant.

(**b**) In the *CM* frame of reference, the center of mass remains at rest.

Fig. 8-13

with $m_B = 2m_A$, initially at rest. The collision is viewed in the *LAB* frame of reference. From Equation (8–9), we find that the velocity of the center of mass $\mathbf{v}_{CM}$ is

$$\mathbf{v}_{CM} = \frac{m_A(\mathbf{v}_A)_0 + m_B(\mathbf{v}_B)_0}{m_A + m_B} = \frac{m_A(\mathbf{v}_A)_0 + m_B(0)}{m_A + 2m_A} = \tfrac{1}{3}(\mathbf{v}_A)_0$$

Now imagine a frame of reference S' moving with the center of mass. We will use prime signs (′) to distinguish measurements made with respect to this frame from those made in the laboratory frame. To determine relative velocities in the *CM* frame, we use the geometrical method. As shown in Figure 8–14, particle m_A moves with the following initial velocity in the *CM* frame:

$$(\mathbf{v}'_A)_0 = (\mathbf{v}_A)_0 - \mathbf{v}_{CM} = (\mathbf{v}_A)_0 - \tfrac{1}{3}(\mathbf{v}_A)_0 = \tfrac{2}{3}(\mathbf{v}_A)_0$$

Particle m_B approaches with a speed equal to the motion of the *CM* frame itself:

$$(\mathbf{v}'_B)_0 = (\mathbf{v}_B)_0 - \mathbf{v}_{CM} = 0 - \mathbf{v}_{CM} = -\mathbf{v}_{CM}$$

Thus, in the *CM* frame the total momentum of the two particles is

$$(m_A\mathbf{v}'_A)_0 + (m_B\mathbf{v}'_B)_0 = m_A[(\mathbf{v}_A)_0 - \mathbf{v}_{CM}] + m_B[(\mathbf{v}_B)_0 - \mathbf{v}_{CM}] \qquad \textbf{(8–15)}$$

Multiplying out the right-hand side, we have

$$(m_A\mathbf{v}'_A)_0 + (m_B\mathbf{v}'_B)_0 = m_A(\mathbf{v}_A)_0 + m_B(\mathbf{v}_B)_0 - (m_A + m_B)\mathbf{v}_{CM}$$

But from Equation (8–8) we find that the right-hand side of this equation always equals zero. Therefore, *in the CM frame the total momentum of the system is zero.* A common name for this frame is the **zero-momentum frame.**

The interesting simplicity and symmetry of collisions in the *CM* frame is evident in Figure 8–13(b). Here the incident particle approaches from the left with a velocity $(\mathbf{v}'_A)_0 = \tfrac{2}{3}(\mathbf{v}_A)_0$, while the target particle approaches from the right with a velocity $(\mathbf{v}'_B)_0 = -\mathbf{v}_{CM} = -\tfrac{1}{3}(\mathbf{v}_A)_0$. After the collision, they separate with the *same* respective speeds. In fact, for all *elastic* collisions between two particles as viewed in the *CM* frame, the speeds of approach before the collision are exactly equal to the speeds of the particles after the collision. This

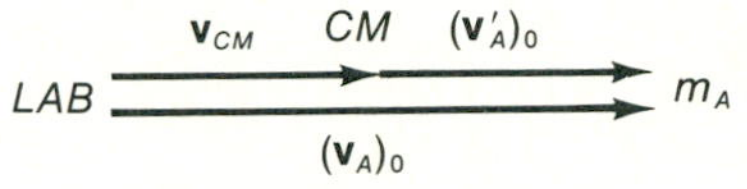

Figure 8–14
The vector diagram $(\mathbf{v}_A)_0 = \mathbf{v}_{CM} + (\mathbf{v}'_A)_0$ expresses the relationships between the velocities of m_A in the *LAB* and *CM* frames of reference.

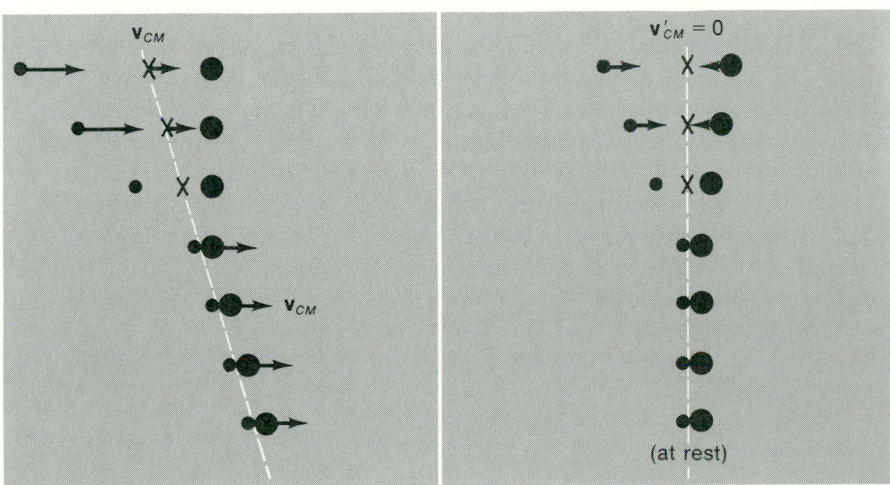

(a) In the *LAB* frame, the center of mass has a velocity $\mathbf{v}_{CM}$ before and after the collision.

(b) In the *CM* frame, the center of mass remains at rest at all times.

Figure 8–15
A *completely inelastic* collision (one-dimensional) in which the two particles stick together after impact.

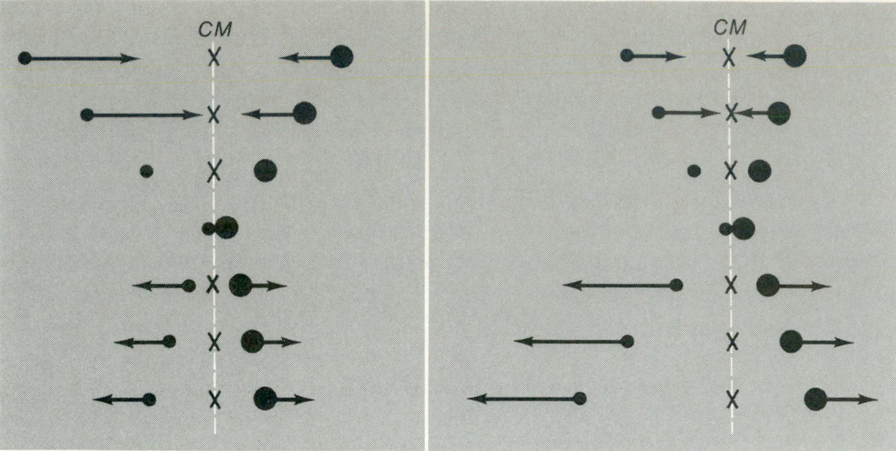

(a) A partially inelastic collision in which some of the initial kinetic energy is transformed into internal potential energy of the particles.

(b) An explosive collision in which potential energy (perhaps stored in a compressed spring or a chemical explosive) is transformed into additional kinetic energy.

Figure 8–16
One-dimensional *inelastic* collisions in the *CM* frame of reference.

is one of the advantages of using the *CM* frame. For a general two-dimensional collision viewed in the *LAB* frame, we have four velocities to keep track of (the initial and final speeds of both particles) and two scattering angles. On the other hand, in the *CM* frame we have only the two initial velocities (the final velocities are just the reverse of the initial values) and one scattering angle.

There are also advantages to using the *CM* frame for *in*elastic collisions. In a completely inelastic collision, for example, the particles stick together and remain at rest in the *CM* frame, a particularly simple situation. For partially inelastic or explosive collisions, the symmetries in the *CM* frame are still helpful, as Figures 8–15 and 8–16 illustrate.

8.7 Collisions in the Center-of-Mass Frame

Much of our knowledge about fundamental particles comes from investigating collisions between atomic and nuclear particles. However, there is a feature of these collisions that prevents a portion of the initial kinetic energy (as measured in the laboratory frame of reference) from partaking in the collision process. This has important implications in the design of collision experiments in nuclear physics. Fortunately, as we will now demonstrate, the energy analysis in the *CM* frame automatically takes into account this additional complication.

Imagine a one-dimensional collision in which the incoming particle travels along the x axis and strikes the target particle "head-on," so that subsequent motions of both particles are only along the x axis. The particle masses are labeled m_A and m_B, with initial velocities $(v_A)_0$ and $(v_B)_0$. (Because this is a one-dimensional case, we may use *scalar* symbols with plus and minus signs for the velocities.) What is the energy involved in such a collision? Because it is an *elastic* collision, the total kinetic energy must be the same before and after the collision. However, since the velocity of a particle depends on the frame of reference used, the kinetic energy as measured in the *LAB* frame will differ from that in the *CM* frame. Let us investigate the relationship.

In the *LAB* frame, the initial kinetic energy is

$$(K_{LAB})_0 = \tfrac{1}{2}m_A(v_A)_0^2 + \tfrac{1}{2}m_B(v_B)_0^2 \qquad \textbf{(8–16)}$$

Referring back to Figure 8–14, we see that the relations between the velocities in the *LAB* and *CM* frames are

$$(v_A)_0 = (v'_A)_0 + v_{CM}$$
$$(v_B)_0 = (v'_B)_0 + v_{CM}$$

Substituting these values into Equation (8–16) gives us

$$(K_{LAB})_0 = \tfrac{1}{2}m_A[(v'_A)_0^2 + 2(v'_A)_0 v_{CM} + v_{CM}^2] + \tfrac{1}{2}m_B[(v'_B)_0^2 + 2(v'_B)_0 v_{CM} + v_{CM}^2]$$

$$(K_{LAB})_0 = [\tfrac{1}{2}m_A(v'_A)_0^2 + \tfrac{1}{2}m_B(v'_B)_0^2] + \{[m_A(v'_A)_0 + m_B(v'_B)_0]v_{CM}\} + [\tfrac{1}{2}(m_A + m_B)v_{CM}^2] \qquad \textbf{(8–17)}$$

These three terms are

$$(K_{LAB})_0 = (K'_{CM})_0 + \begin{bmatrix}\text{always zero because the} \\ CM \text{ frame is the zero-} \\ \text{momentum frame (in which} \\ m_A v'_A + m_B v'_B = 0)\end{bmatrix} + \begin{bmatrix}\text{kinetic energy of} \\ \text{the motion of the} \\ \text{center of mass of} \\ \text{the system}\end{bmatrix}$$

Thus:

$$(K_{LAB})_0 = (K'_{CM})_0 + \tfrac{1}{2}Mv_{CM}^2 \qquad \text{(where } M = m_A + m_B) \qquad \textbf{(8–18)}$$

This equation is valid in all cases, not just in the one-dimensional situation we used for the derivation.

From the conservation of momentum principle we know that in the absence of external forces on the two-particle system, the velocity of the center of mass remains constant. Therefore, the term $\tfrac{1}{2}Mv_{CM}^2$ represents an amount of kinetic energy "trapped" in the motion of the center of mass itself. This means that a certain portion of the initial kinetic energy (as measured in the *LAB* frame of reference) is unavailable for such functions as changing the internal potential energies of the particles or altering their kinetic energies. Thus the only energy that can partake in the energetics of the collision is the kinetic energy $(K'_{CM})_0$, measured in the *CM* frame.

As an example, when moving protons strike stationary helium nuclei ($m_{He} = 4m_p$), one-fifth of the kinetic energy of the incoming protons as measured in the *LAB* frame is tied up in the motion of the center of mass. Therefore this fraction of the initial energy cannot take part in the reaction. If instead the more massive helium nuclei were moving toward stationary protons, the situation would be even worse: four-fifths of the initial kinetic energy in the *LAB* frame would be unavailable for the reaction. However, when analyzed in the *CM* frame, all the initial energy is available, and so the analysis becomes simpler in that frame.

Recently, scattering experiments have been designed so that two beams of high-energy particles collide head-on with each other. If the particles have the same momenta (in opposite directions), the *LAB* frame is also the *CM* frame, and thus all the initial kinetic energies produced by the particle accelerators in the laboratory are utilized in the reactions. Even if the momenta are not exactly equal, such "colliding-beam" experiments have distinct energy advantages compared with just one beam incident upon a stationary target.

EXAMPLE 8–8

A proton of mass m and speed v collides with a deuterium nucleus (mass $2m$) in a one-dimensional head-on collision. (a) Find the fraction of the initial kinetic energy measured in the *LAB* system that is trapped in the motion of the center of mass. (b) Find the initial energy in the *CM* system. (c) Verify that the sum of parts (a) and (b) equals the initial kinetic energy in the *LAB* system. That is, $(K_{LAB})_0 = (K'_0)_{CM} + \frac{1}{2}Mv_{CM}^2$.

SOLUTION

(a) In the *LAB* frame, the initial kinetic energy is

$$(K_{LAB})_0 = \tfrac{1}{2}m_A v_A{}^2 + \tfrac{1}{2}m_B v_B{}^2 = \tfrac{1}{2}mv^2 + 0$$

Referring back to Equation (8–9), we find that the velocity of the center of mass in the *LAB* frame is

$$v_{CM} = \left(\frac{m_A}{m_A + m_B}\right)v = \tfrac{1}{3}v$$

Therefore, in the *LAB* frame, the kinetic energy of the center-of-mass motion is

$$\tfrac{1}{2}Mv_{CM}{}^2 = \tfrac{1}{2}(m + 2m)(\tfrac{1}{3}v)^2 = \tfrac{1}{2}(3m)(\tfrac{1}{9}v^2) = \boxed{\tfrac{1}{3}(\tfrac{1}{2}mv^2)}$$

Thus, one-third of the initial kinetic energy in the *LAB* frame is locked up in the center-of-mass motion.

(b) In the *CM* frame, the two particles approach each other with the following velocities:

$$v'_A = v_A - v_{CM} = v - \tfrac{1}{3}v = \tfrac{2}{3}v$$
$$v'_B = v_B - v_{CM} = 0 - \tfrac{1}{3}v = -\tfrac{1}{3}v$$

Thus, in the *CM* frame the total kinetic energy is

$$(K'_{CM})_0 = \tfrac{1}{2}m_A(v'_A)^2 + \tfrac{1}{2}m_B(v'_B)^2 = \tfrac{1}{2}(m)(\tfrac{2}{3}v)^2 + \tfrac{1}{2}(2m)(-\tfrac{1}{3}v)^2$$

$$(K'_{CM})_0 = \tfrac{2}{9}mv^2 + \tfrac{1}{9}mv^2 = \boxed{\tfrac{1}{3}mv^2}$$

(c) We wish to verify that

$$(K_{LAB})_0 = (K'_{CM})_0 + \tfrac{1}{2}Mv_{CM}{}^2$$
$$\tfrac{1}{2}mv^2 \stackrel{?}{=} \tfrac{1}{3}mv^2 + \tfrac{1}{6}mv^2$$
$$\tfrac{1}{2}mv^2 = \tfrac{1}{2}mv^2$$

Thus, $(K_{LAB})_0$ is indeed the sum of parts (a) and (b).

(**a**) Before the collision.

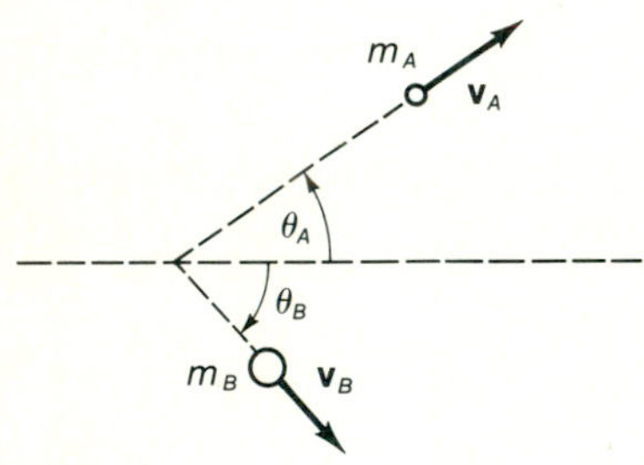

(**b**) After the collision.

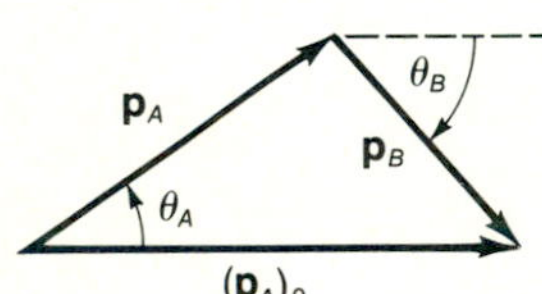

(**c**) Momentum diagram in the *LAB* frame. The three momenta have unequal magnitudes, and the two scattering angles are different. From $\Sigma \mathbf{p}_0 = \Sigma \mathbf{p}$, we have $(\mathbf{p}_A)_0 = \mathbf{p}_A + \mathbf{p}_B$.

Figure 8–17

A collision between an incident particle m_A and a target particle m_B initially at rest in the *LAB* frame. The incident particle is scattered at an angle θ_A, and the target particle recoils at an angle θ_B.

8.8 Momentum Vector Diagrams for Elastic Collisions

Frequently a vector diagram illustrating the conservation of momentum ($\Sigma \mathbf{p}_0 = \Sigma \mathbf{p}$) helps greatly in the analysis of the collision of two particles. Consider a nuclear scattering experiment in which a moving particle collides with a target nucleus initially at rest. Figures 8–17 and 8–18 show the features of such a collision. In the *LAB* frame, *two* scattering angles are involved. However, in the *CM* frame, only *one* scattering angle is present, since in that frame the two

All the initial and final momenta have the same magnitudes.

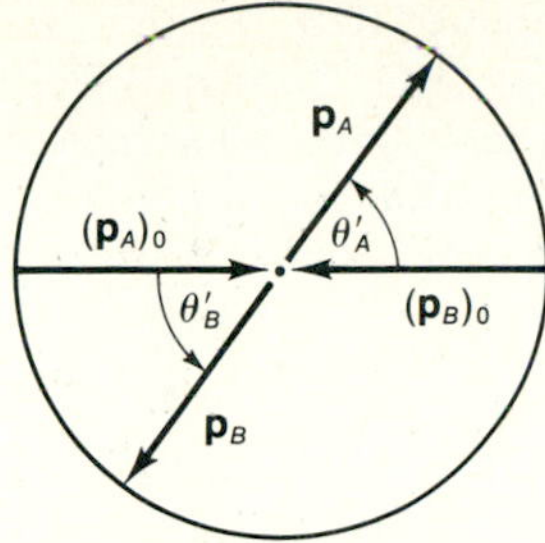

(**a**) An *elastic* collision.

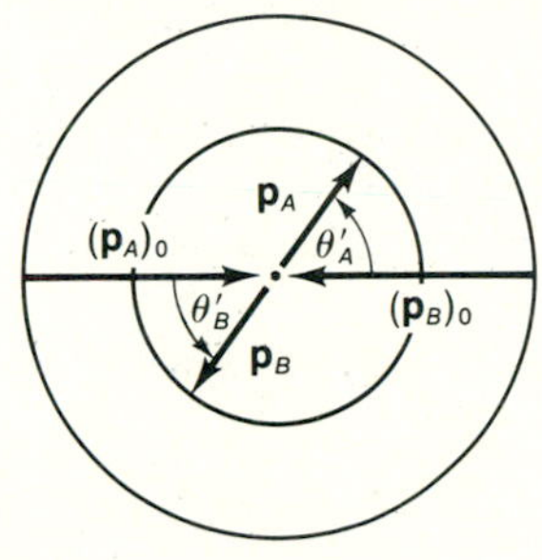

(**b**) An *inelastic* collision.

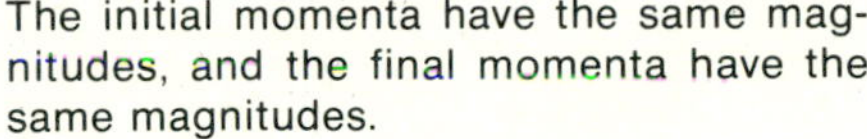
The initial momenta have the same magnitudes, and the final momenta have the same magnitudes.

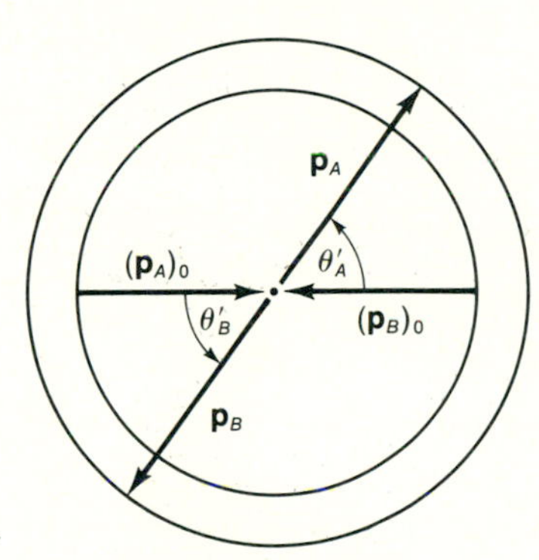

(**c**) An *explosive* collision.

Figure 8–18
Momentum vectors for collisions in the *CM* frame of reference. Note that the scattering angles θ'_A and θ'_B are always equal and that other symmetries simplify the analyses.

particles approach along the same line with equal and opposite momenta, and after the collision they separate at some angle along the same line with equal and opposite momenta. The relations between the scattering angles in the *LAB* and *CM* frames can be found by making a relative velocity diagram with labels on the heads and tails of the vectors (see Figure 8–19).

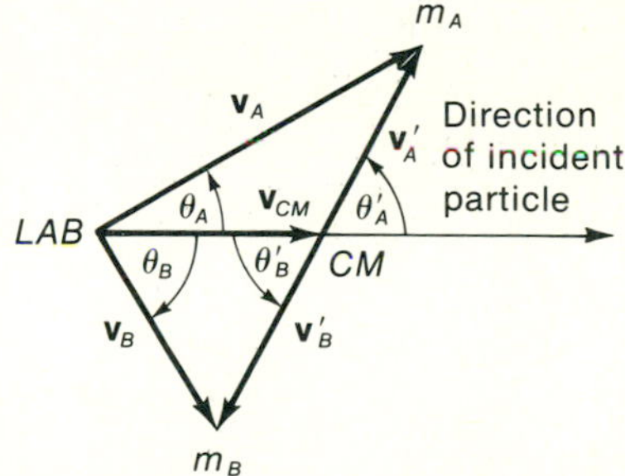

Figure 8–19
In the *LAB* frame of reference, m_A is incident on a target particle m_B initially at rest. After an *elastic* collision, the particle velocities are related as shown. Measurements in the *LAB* frame are unprimed; those in the *CM* frame are primed. Note that $|\mathbf{v}_{CM}| = |\mathbf{v}'_B|$ and that $\theta'_A = \theta'_B$.

EXAMPLE 8–9

A proton of mass m and velocity $\mathbf{v}$ approaches and strikes a helium nucleus (mass $4m$) initially at rest. As measured in the *LAB* frame of reference, the proton is scattered 30° with a speed of $v/2$. (a) By investigating the momentum vector diagram for the *LAB* frame, find the scattering angle (*LAB* frame) and the speed (in terms of v) of the helium nucleus after the collision. (b) Find the speed of the scattered proton in the *CM* frame and the scattering angles of the particles in the *CM* frame. (c) Is kinetic energy gained, lost, or unchanged in the collision?

SOLUTION

(a) Figure 8–20(a) shows a momentum vector diagram for the *LAB* frame. The scattering angle for the helium nucleus is θ and its speed is V. Appendix D gives useful information for solving triangles with various unknown quantities. In this case, V is easily obtained by using the *law of cosines*:

$$a^2 = b^2 + c^2 - 2bc \cos \alpha$$

$$(4mV)^2 = (mv)^2 + \left(m\frac{v}{2}\right)^2 - 2(mv)\left(m\frac{v}{2}\right)(\cos 30^\circ)$$

$$16V^2 = v^2 + \frac{v^2}{4} - v^2(0.866)$$

$$V^2 = \frac{0.384}{16}v^2 = 0.024v^2$$

$$V = \boxed{0.155v}$$

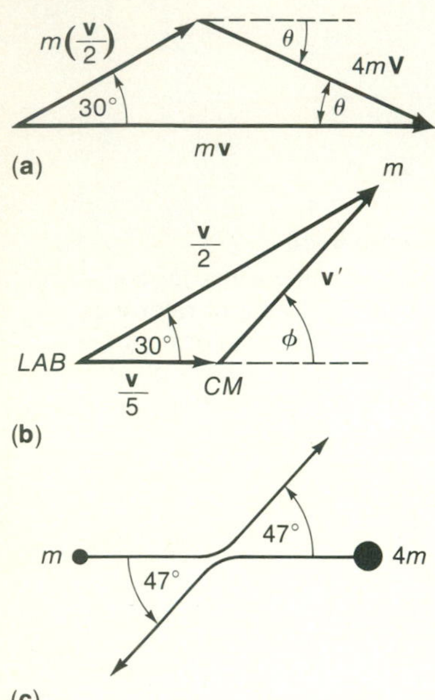

Figure 8–20
Example 8–9

The *law of sines* then yields a value for θ:

$$\frac{\sin \gamma}{c} = \frac{\sin \alpha}{a}$$

Substituting the notation from our problem and solving for $\sin \theta$, we have

$$\frac{\sin \theta}{m\left(\frac{v}{2}\right)} = \frac{\sin 30^\circ}{4mV}$$

$$\sin \theta = \frac{m\left(\frac{v}{2}\right)\sin 30^\circ}{4mV}$$

$$\sin \theta = \frac{\left(\frac{v}{2}\right)(0.500)}{4(0.155v)} = (0.403)$$

$$\theta = \sin^{-1}(0.403) = \boxed{23.8^\circ}$$

(b) We now draw a velocity vector diagram for the relative motions of the scattered proton [Figure 8–20(b)]. The motion of the *CM* is given by:

$$v_{CM} = \left(\frac{m_p}{m_p + m_{\text{He}}}\right)v = \left(\frac{m}{m + 4m}\right)v = \frac{v}{5}$$

The speed v' of the scattered proton in the *CM* frame is obtained using the law of cosines:

$$a^2 = b^2 + c^2 - 2bc \cos \alpha$$

$$(v')^2 = \left(\frac{v}{2}\right)^2 + \left(\frac{v}{5}\right)^2 - 2\left(\frac{v}{2}\right)\left(\frac{v}{5}\right)\cos 30^\circ$$

$$v' = \boxed{0.342v}$$

Then, we use the law of sines to find the scattering angle ϕ of the proton in the *CM* frame:

$$\frac{\sin \beta}{b} = \frac{\sin \alpha}{a}$$

$$\frac{\sin(180^\circ - \phi)}{\left(\frac{v}{2}\right)} = \frac{\sin 30^\circ}{v'}$$

$$\frac{\sin(180^\circ - \phi)}{\left(\frac{v}{2}\right)} = \frac{\sin 30^\circ}{0.342v}$$

$$\phi = \boxed{47.0^\circ}$$

This is the scattering angle for *both* particles in the *CM* frame, as shown in Figure 8–20(c).

(c) We compare the kinetic energy before (K_0) and after (K) the collision. Although the *change* would be the same whether we use the *LAB*- or the *CM*-frame velocities of the particles, in this problem it is simpler to use the *LAB* values.

Before: $K_0 = \frac{1}{2}mv^2$

After: $K = \frac{1}{2}m\left(\frac{v}{2}\right)^2 + \frac{1}{2}(4m)(0.155v)^2 = 0.346(\frac{1}{2}mv^2)$

Since the final energy is less than the initial energy, kinetic energy is lost in the collision. (This "lost" energy goes into internal excitation energy of the helium nucleus.) The collision is thus an inelastic one.

Summary

Collisions between particles are classified as:

elastic: if the kinetic energy is conserved.

inelastic: if some kinetic energy "disappears" into other forms such as thermal energy, energy of permanent deformation, or internal potential energy. Explosive collisions, which add kinetic energy to the system, are also inelastic.

completely inelastic: when one mass "captures" another and the two masses move off together.

For isolated systems, the conservation of linear momentum *always* holds, whether or not mechanical energy is also conserved.

The location of the *center of mass* (*CM*) of two particles is

$$x_{CM} = \frac{m_1x_1 + m_2x_2}{m_1 + m_2}$$

or, in three dimensions:

$$\mathbf{r}_{CM} = \frac{m_1\mathbf{r}_1 + m_2\mathbf{r}_2}{m_1 + m_2}$$

The velocity of the center of mass of two particles is

$$\mathbf{v}_{CM} = \frac{m_1\mathbf{v}_1 + m_2\mathbf{v}_2}{m_1 + m_2}$$

The center-of-mass point for a system moves as if the entire mass ($M = m_1 + m_2$) were concentrated at that point and acted upon by the same *external* forces as the actual system. These ideas may be extended to any number of particles.

Scattering experiments in which an incoming particle is scattered by a target particle (initially at rest) are important methods for studying atomic and nuclear structure as well as nuclear forces. Though measurements are made in the *LAB* frame of reference, the analysis is usually easier in the *CM* frame. In the *LAB* frame, the velocity of the center of mass must be the same after the collision as before the collision. Therefore, the energy $\frac{1}{2}Mv_{CM}^2$ represents kinetic energy "trapped" in the motion of the center of mass, which is thus unavailable to partake in the energetics of the collision. It is for this reason (and because of certain symmetries) that the *CM* frame of reference is usually the most convenient for analyzing collisions.

Relative velocity considerations (and transfer between the *LAB* and *CM* frames) are most easily handled using a geometrical method that clarifies their relationships:

(1) Label the *point* of each velocity vector with the name of the moving object and the *tail* with the name of the reference frame in which that velocity is measured.
(2) Redraw the vectors, *fitting identical labels together* (keeping magnitudes and directions correct).
(3) An arrow drawn between any two labels represents the vector velocity of the *label at the point* relative to the *label at the tail.*

Comparisons between the velocity of a particle in two inertial frames of reference, the *S* and *S'* frames, are made via the *Galilean velocity addition* relation:

$$\mathbf{v} = \mathbf{v}' + \mathbf{V}$$

where $\mathbf{v}$ is the velocity of the particle as measured in S, $\mathbf{v}'$ is the velocity of the particle relative to *S'*, and $\mathbf{V}$ is the velocity of *S'* relative to *S*. A prime sign (′) is used to identify all measurements made in the *S'* frame; unprimed quantities refer to the *S* frame.

Questions

1. A rock is allowed to fall to the ground (where it remains). What happened to the momentum the rock had just before it struck the ground?
2. What is meant by saying that two objects "touch" during a collision? What about collisions between two protons, which may be repelled by their electric charges when they are relatively far apart?
3. Object *A* undergoes a one-dimensional elastic collision with object *B*, which is initially at rest. Find the mass ratio m_A/m_B such that object *B* acquires (a) the largest possible speed, (b) the largest possible momentum, (c) the largest possible kinetic energy, (d) the smallest possible speed, and (e) the same speed as object *A* before the collision.
4. Consider a nuclear collision when the target particle is initially at rest. Explain why the trajectories of the particles before and after the collision must all lie in a plane. Is this true if both particles are moving before, as well as after, the collision?
5. On a "bumpmobile," it is claimed people can move themselves along the floor by repeatedly hitting a supporting block with a sledgehammer (see Figure 8–21). Is this possible? If so, what should be the relative coefficients of static friction between the person and the block, compared with the block and the floor? If it works, why isn't it a case of shifting the center of mass of a system (the block-person-sledgehammer) by *internal* forces alone?
6. Suppose a mouse is inside a closed cardboard box at rest on the floor. Is there some way the mouse could make the box move across the floor? To simplify the job, should the coefficient of friction between the box and floor be rather large, rather small, or negligible?

Figure 8–21
Question 5

Problems

8A–1 Two point masses, 1 kg and 4 kg, are 1 m apart. How far from the 4-kg mass is the center of mass?
Answer: 0.200 m

8A–2 Two point masses are 2 m apart. Their center of mass is located 0.40 m from one of the masses, which is equal to 2 kg. What is the other mass?

8A–3 Two masses, m and $2m$, approach each other along paths at right angles. They collide and stick together, moving off with speed 2 m/s and at an angle of 37° with respect to the original direction of m. Find the initial speed of each mass.
Answers: m: 4.80 m/s; $2m$: 1.80 m/s

8A–4 A 7.5-kg mass moving eastward at a speed of 20 m/s is followed by a 5-kg mass moving at 15 m/s.

(a) Find the speed of the center of mass.
(b) Find the velocity of each mass relative to the center of mass.
(c) Determine the momentum of each mass relative to the center of mass.

8A–5 A 4-kg mass moving along the $+x$ direction with a speed of 5 m/s strikes a 10-kg mass initially at rest. The 4-kg mass rebounds in the $-x$ direction with a speed of 2 m/s. Is this collision elastic? If not, how much kinetic energy was lost or gained?
Answer: No; 34.8 J lost

8A–6 A billiard ball moving at 5 m/s strikes a stationary ball of the same mass. After the collision, the ball moves at 4.33 m/s at an angle of 30° with respect to the original line of motion. Assuming an elastic collision (and ignoring friction and rotational motion), find the magnitude and direction of the struck ball's velocity.

8A–7 A 10-kg mass initially at rest explodes into three pieces. A 4.5-kg piece goes north at 20 m/s, and another 2-kg piece moves eastward at 60 m/s.

(a) Determine the magnitude and direction of the velocity of the third piece.
(b) Find the energy of the explosion.

Answers: (a) 42.9 m/s, 37° S of W (b) 7720 J

8A–8 A mass m moving with speed v_0 strikes and sticks to a mass M initially at rest. The combined masses move off with speed v. In terms of m and M, what fraction of the initial kinetic energy is lost in the collision?

8A–9 If a speedboat is aimed directly across a river, its motion is downstream at 53.1° with respect to the bank, as shown in Figure 8–22. The water flows at 6 ft/s, and the river is 100 ft wide. Suppose instead that the boat were aimed somewhat upstream so that it traveled (with the same speed through the water) perpendicular to the bank. How long would it take to cross the river?
Answer: 18.9 s

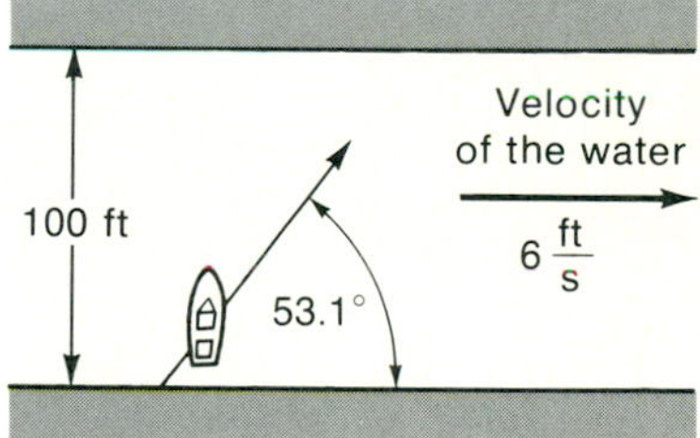

Figure 8–22
Problem 8A–9

8A–10 A 900-kg cannon fires a 100-kg ball with a speed of 75 m/s *relative to the cannon.* What is the recoil speed of the cannon relative to the ground?

8A–11 A 20-g lead pellet is fired into an 800-g ballistic pendulum bob made of wood. The subsequent swinging motion of the bob causes it to rise a maximum vertical distance of 10 cm.

(a) Find the speed of the bob-plus-pellet at the instant just after the pellet embeds itself in the bob.
(b) Find the initial speed v_0 of the pellet.
(c) Find the fraction of the initial kinetic energy of the pellet that went into penetrating the wood block, deforming the lead pellet, and so on. Express this as $(\Delta K/K_0) \times 100\%$.

Answers: (a) $\sqrt{2}$ m/s (b) 58.0 m/s (c) 97.6%

8B–1 Find the coordinates of the center of mass for the following system of three masses: a 2-kg mass at (1 m,2 m), a 4-kg mass at (3 m,1 m), and a 7-kg mass at (−1 m,−1 m).
Answer: ($\frac{7}{13}$ m,$\frac{1}{13}$ m)

8B–2 For the following three point masses, find the distance from the center of mass to the 5-kg mass: a 5-kg mass at (2 m, −5 m), a 3-kg mass at (0,6 m), and a 2-kg mass at (−3 m,4 m).

8B–3 Two objects, each of mass m, are approaching each other along the same straight line with speeds v_1 and v_2. They collide and stick together. Show that the loss in total kinetic energy is $\frac{1}{4}m(v_1 - v_2)^2$.

8B–4 A 0.25-kg mass moving with a speed of 2 m/s collides elastically with a 0.75-kg mass initially at rest. The incoming mass is deflected at 37°.

(a) Find the speed of each mass after the collision.

(b) Determine the final direction of motion of the 0.75-kg mass.

8B–5 A 3-kg mass initially at rest is struck head-on by a 1-kg mass moving at a speed of 5 m/s. After the collision, the two masses move separately along the line of motion of the 1-kg mass. If one-fourth of the initial kinetic energy is lost in the collision, what are the final velocities of the masses? *Answers:* −1.81 m/s, 2.27 m/s

8B–6 A 3-kg mass moves with a speed of 16 m/s toward a 5-kg mass at rest. In a coordinate system in which the 5-kg mass is at rest, (a) what is the total momentum of the masses and (b) what is the kinetic energy of the masses? In the *CM* frame of reference, (c) what is the total momentum of the masses and (d) what is the kinetic energy of the masses?

8B–7 A 50-kg object is initially northbound at a speed of 16 m/s. A 20-kg object traveling eastward collides elastically with the 50-kg object in such a way that the 20-kg object is deflected northward, and the 50-kg object is deflected eastward. Find the final speed of each object. *Answers:* 20-kg object: 40 m/s
50-kg object: 16 m/s

8B–8 At the top of its trajectory, a mortar shell explodes into two fragments: a 2.5-kg piece moving southwest with a horizontal speed of 100 m/s and a 3.5-kg piece moving north with a horizontal speed of 70 m/s. Find the horizontal direction of motion of the shell just before the explosion.

8B–9 An old-fashioned coal-burning train travels due north at 20 m/s when the wind is known to be blowing due west. The engineer notes that the trail of smoke from the smokestack makes an angle of 20° with respect to the train's length. Assuming the smoke acquires the speed of the wind as soon as it leaves the smokestack, what is the wind speed? *Answer:* 7.28 m/s

8B–10 An airplane moves at 100 mi/h relative to the air. Although the plane is headed due west, the pilot notes that her velocity relative to the ground is 120 mi/h, 20° south of west.

(a) Find the magnitude of the wind velocity relative to the ground.

(b) From what direction is the wind blowing?

8B–11 A riverboat travels east at 8 ft/s relative to the water, while the water flows south at 3 ft/s relative to the ground. A bird flies horizontally in a northeast direction at 5 ft/s relative to the air, and the air moves at 4 ft/s due west relative to the ground. Find the magnitude and direction of the bird's motion relative to a passenger sitting in a deck chair on the boat. *Answer:* 10.69 ft/s, 37.7° N of W

8B–12 A motorboat travels through the water at a constant speed of 2 m/s.

(a) Starting from the south bank of a river flowing 1 m/s west, in what direction should the boat be aimed in order to arrive at a point on the opposite bank directly across from the starting point? (Hint: The boat must be aimed somewhat upstream so that its velocity relative to the earth is due north.)

(b) Find the magnitude of the boat's velocity relative to the earth in part (a).

(c) If the river is 50 m wide, how long does it take the boat to reach its destination?

(d) If instead the boat traveled to a point 50 m downstream (on the south shore), then returned to the starting point, how much time would be required for the round trip?

(e) How much longer or shorter is this time than the round trip across the river and back in part (c)?

(f) Suppose the boat were aimed directly across the river, so that it landed on the opposite shore somewhat downstream. How much time is required to cross the river this way?

(g) Is there some other direction the boat could be aimed so that it could cross the river in a shorter time than in part (f)? Explain.

8B–13 An airplane flies a daily round trip along a straight line between two cities, A and B, located a distance L apart. The speed of the plane is V relative to the air. Calculate the round trip times if a wind of speed v relative to the ground is blowing (a) *along* the direction from A to B and (b) *perpendicular* to the line from A to B. (c) Show that the round trip parallel to the wind direction takes longer than the time for the crosswind journey by a factor of:

$$\frac{1}{\sqrt{1-\frac{v^2}{V^2}}}$$

8B–14 In a ballistic pendulum experiment, the experimenter wants the bob to swing so that the suspending strings make a maximum angle of 20° with the vertical. If 8-g bullets traveling 400 m/s are used with a 3-kg bob, how long should the strings be?

8B–15 For elastic collisions between two particles of equal mass in which the target particle is initially at rest, show that the recoiling particles always move off at right angles to each other. (Ignore the case of a head-on collision in which the incident particle comes to rest; for this unique case, the angle between the recoiling particles is undefined. Hint: Draw a vector diagram for the conservation of momentum. Then consider the expression for the conservation of energy. This latter equation implies something about the properties of the triangle in the momentum diagram.)

8B–16 A 3-kg mass slides in the $+x$ direction on a horizontal frictionless surface with a speed of 5 m/s. It collides with a 2-kg mass initially at rest at the origin. Two seconds after the collision, the 3-kg mass is at the point (4 m,2 m).

(a) What is the speed of the center of mass?
(b) Where is the center of mass 2 s after the collision?
(c) Where is the 2-kg mass 2 s after the collision?
(d) Was the collision elastic?

8C–1 A 2-kg mass moves uniformly from (3 m,4.5 m) to (2 m,−6 m) while a 5-kg mass moves uniformly from (1.5 m,−4 m) to (−2 m,5 m). If the process takes 2 s, what is the speed of the center of mass? *Answer:* 2.21 m/s

8C–2 A particle of mass m traveling with initial speed v_0 collides with another particle of mass $2m$ initially at rest. The incident particle is scattered at an angle of 37° with respect to its original direction and has a final speed of $v_0/2$.

(a) Find the angle θ and the speed v of the target particle after the collision.
(b) Is kinetic energy conserved in this collision?

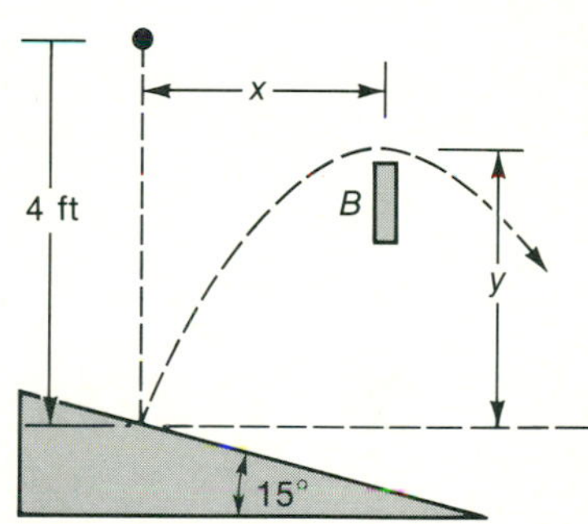

Figure 8–23
Problem 8C–3

8C–3 A ball bearing manufacturer uses an impact test to reject defective balls. Specimens are dropped from rest and fall 4 ft onto a massive steel plate inclined at 15°, as shown in Figure 8–23. To pass the test, the balls must clear a barrier B at the top of their rebound trajectories. It is desired that any ball whose collision with the impact plate is less than perfectly elastic should be rejected. Assuming the balls have negligible size, find the position of the barrier by determining x and y. (Note: In an elastic collision, the angles of *incidence* and *rebound*, as measured with respect to the normal to the surface, are equal.) *Answer:* (3.46 ft,3.00 ft)

8C–4 A ball bearing is dropped from rest at a height h_0 above a horizontal steel slab. It bounces repeatedly against the slab, rising to smaller and smaller heights after successive bounces because the impacts are not perfectly elastic (see Figure 8–24). For each bounce, the ratio

$$\frac{\text{Velocity just after impact}}{\text{Velocity just before impact}}$$

has the constant magnitude ε. In terms of ε, h_0, and n, devise a formula for the height h_n to which the ball bearing rises after the nth bounce.

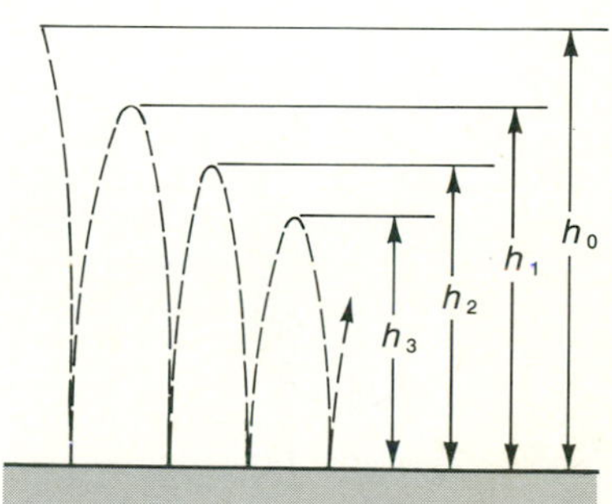

Figure 8–24
Problem 8C–4

8C–5 A mass m, moving with initial speed v_0, undergoes a head-on elastic collision with a mass M initially at rest ($M > m$). The collision is one-dimensional, so that after the interaction both masses are moving along the original line of motion. (a) Find the ratio

$$\frac{\text{Final kinetic energy}}{\text{Initial kinetic energy}}$$

for m. (b) Suppose, instead, M were incident on m (initially at rest). Find the same ratio for M.

Answers: (a) $\left(\dfrac{M-m}{M+m}\right)^2$ (b) the same as (a)

8C–6 A moving mass m_1 collides elastically with a mass m_2 initially at rest. After the collision, m_1 moves at 90° relative to its original direction.

(a) Find the scattering angle ϕ (relative to the incident particle direction) for the motion of m_2 after the collision.

(b) Show that m_2 must be greater tham m_1.

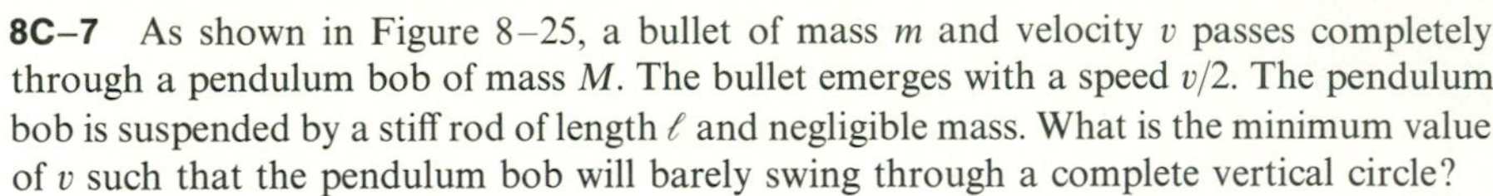

8C–7 As shown in Figure 8–25, a bullet of mass m and velocity v passes completely through a pendulum bob of mass M. The bullet emerges with a speed $v/2$. The pendulum bob is suspended by a stiff rod of length ℓ and negligible mass. What is the minimum value of v such that the pendulum bob will barely swing through a complete vertical circle?

Answer: $4M\sqrt{g\ell}/m$

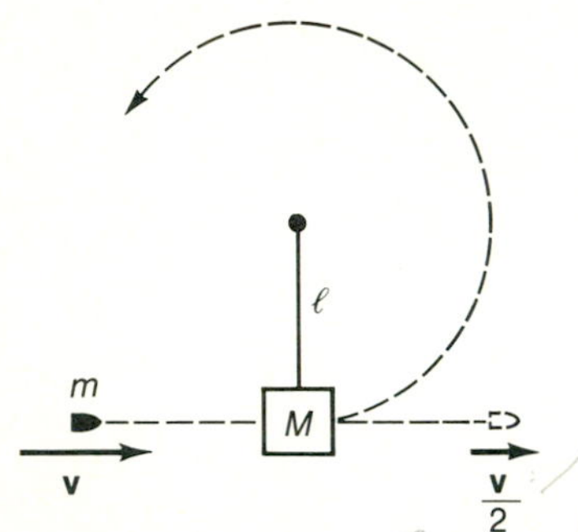

Figure 8–25
Problem 8C–7

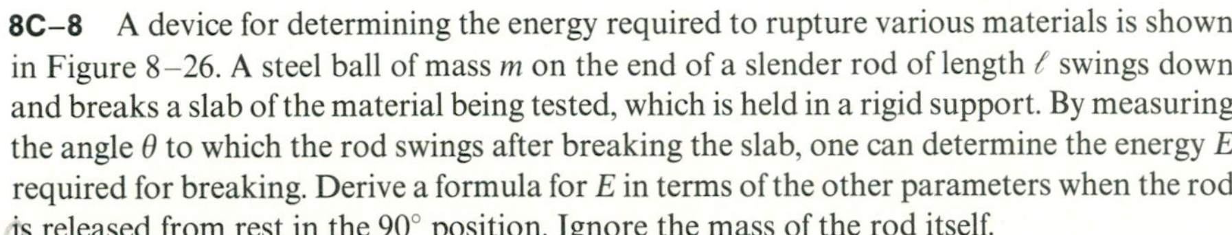

8C–8 A device for determining the energy required to rupture various materials is shown in Figure 8–26. A steel ball of mass m on the end of a slender rod of length ℓ swings down and breaks a slab of the material being tested, which is held in a rigid support. By measuring the angle θ to which the rod swings after breaking the slab, one can determine the energy E required for breaking. Derive a formula for E in terms of the other parameters when the rod is released from rest in the 90° position. Ignore the mass of the rod itself.

8C–9 Two men, each of mass m, stand on a cart which has frictionless wheels. The cart has a mass M and is initially at rest. The two men now jump off the rear of the cart with a horizontal velocity v relative to the cart. Show that the final speed of the cart is greater if the men jump one after the other rather than in unison. (An extension of this problem shows that a rocket exhaust becomes more efficient as the exhaust material becomes more finely divided.)

8C–10 A boat aims due east, as measured by the ship's compass, with a speed of 6 km/h relative to the water. The water is moving north relative to the ground at 7 km/h, and a wind is blowing relative to the ground from south to north. A passenger in a deck chair on the boat notes that the smoke from the ship's funnel makes an angle of 30° with respect to the wake of the ship. Assuming the smoke remains at rest with respect to the air as soon as it leaves the funnel, find the speed of the wind relative to the ground. (This problem has two possible answers; find both of them.)

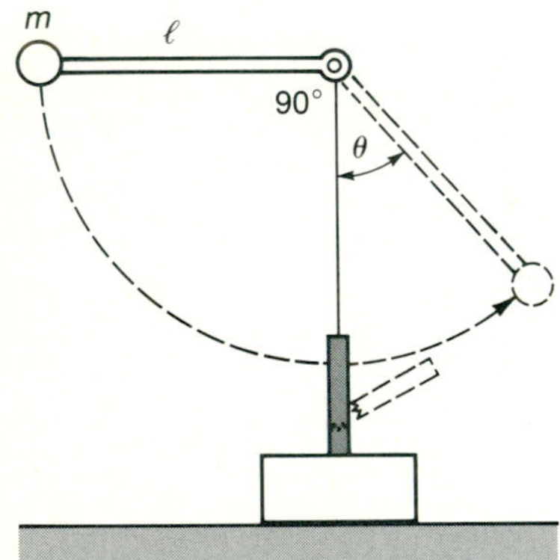

Figure 8–26
Problem 8C–8

8C–11 As shown in Figure 8–27, a stream of water with speed v and mass per unit time μ (both relative to the ground) impacts against a turbine blade. The blade has constant motion in a straight line such that the water stream has a speed $v/4$ relative to the ground after impact in a direction opposite to the incoming stream.

(a) How fast is the blade moving relative to the ground?

(b) In terms of μ and v, what force does the stream exert on the moving blade?

Answers: (a) $(3/8)v$ (b) $(25/32)\mu v$

8C–12 In a nuclear scattering experiment, a proton of mass m and initial speed v_0 collides elastically with another proton at rest. After the collision, one of the protons is observed to have a velocity v_1 at 37° with respect to the motion of the incident proton. (The angle between the final velocities of the protons is, of course, 90°.)

(a) Sketch velocity vector diagrams (*before* and *after* the collision) for both the *LAB* and *CM* frames of reference.

(b) Sketch a vector diagram illustrating the conservation of momentum (*LAB* frame) and determine the final velocity v_2 of the other proton in terms of v_0.

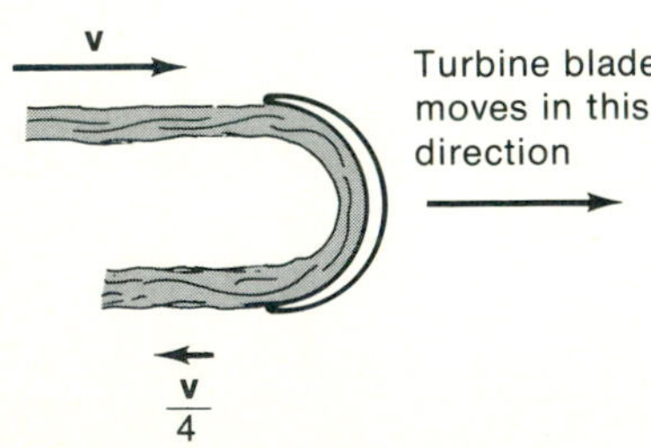

Figure 8–27
Problem 8C–11

8C–13 A 2-kg mass moves with a speed of 32 m/s toward a 6-kg mass initially at rest in the laboratory frame of reference. When viewed from the center of mass system, both masses are deflected 90° in an elastic collision.

(a) Through what angle is the 2-kg mass deflected in the *LAB* frame?

(b) Relative to the incident direction of the 2-kg mass, what is the direction of motion of the 6-kg mass in the *LAB* frame?

Answers: (a) 71.6° (b) 45°

8C–14 A particle of mass m and kinetic energy K collides elastically with a similar particle of mass m initially at rest. After the collision, their velocities (in the *LAB* frame) are at 90°

with respect to each other (because they have equal masses). Also, after the collision the incident particle has 4 times the kinetic energy of the other particle in the *LAB* frame.

(a) In terms of m and K, find the final speed and direction of the more energetic particle in the *LAB* frame. (Hint: Start with a velocity vector diagram for the final velocities.)

(b) At what angle is the incident particle scattered in the *CM* frame?

Answers: (a) $\sqrt{\frac{8K}{5m}}$, at 26.6° relative to incident direction

(b) 53.1°

8C–15 A bead of mass m slides without friction along a wire circle oriented rigidly in a vertical plane, as shown in Figure 8–28. The bead is released from rest at one side of the circle, sliding down under gravity to collide elastically with another bead of mass $3m$, which is initially at rest at the bottom.

(a) Find the height to which each bead rises above the bottom after the collision. The radius of the circle is R.

(b) The beads continue to slide without friction until they collide a second time. Find the speed of each bead immediately after this second elastic collision.

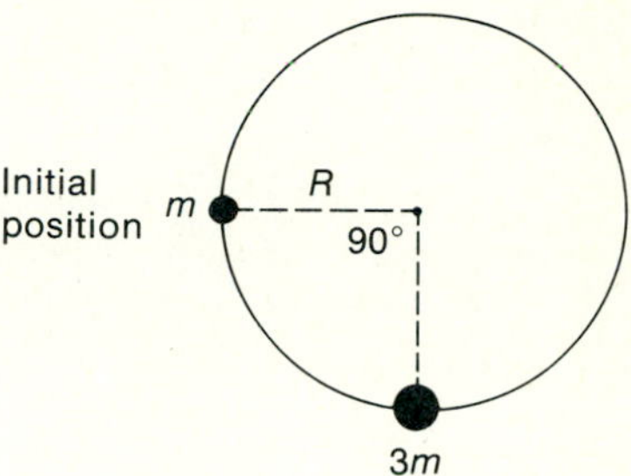

Figure 8–28
Problem 8C–15

Perspective

We pause again to point out the few giant pinnacles in the mountainous terrain we have traveled. In the last few chapters we introduced (1) work and energy, (2) momentum, (3) the conservation of energy and momentum, and (4) the relationship between a conservative force and its associated potential energy function. In addition to $\Sigma \mathbf{F} = m\mathbf{a}$, these concepts of *energy* and *momentum* are extremely useful in solving problems. The *conservation relations* are most helpful when we do not know the forces explicitly, or when the forces are complicated or act for an extremely short time (as in collisions). The clue to their applicability is whether or not a system is *isolated* from its surroundings—that is, whether forces act between the system and the external world. If there is no transfer of energy to or from a system, the total energy remains constant. And if there is no net external force acting on the system, the total momentum of the system remains constant. Of course, there may be interchanges of energy and momentum between various internal parts of the system. But the total remains constant for an isolated system.

One point deserves comment. In *inelastic* collisions, some energy goes into forms that often cannot be calculated directly: deformations, thermal energy, and so on (though, of course, energy is still conserved in the strict sense). In these cases, it may not be possible to set up an energy conservation equation directly. However, the conservation of momentum always applies to isolated systems, regardless of the type of interactions.

Ideas of conservation are helpful even if the system is not isolated. We then recognize that the energy (or momentum) added to a system increases the energy (or momentum) content of the system by just that amount. Therefore, keep in mind the boundary that separates a system from its surroundings, and watch for energy or momentum transfers across this boundary. For these cases, the *work-energy relation* is helpful: The work done by *all* the forces acting on a system (that is, the *net* force) equals the change in kinetic energy of the system.

The final important tool we have for solving problems—the connection between a conservative force and its associated potential energy function—will be discussed more fully later in this text. Often, theories in physics are written in terms of energy rather than forces, since energies, being *scalars*, are easier to cope with mathematically than *vector* forces. Analyzing a problem via either route is possible because of the close connection between conservative forces and potential energy.

To summarize, you have learned four powerful methods for solving problems: *Newton's laws, the conservation relations*, the *work-energy relation*, and the *relationship between a conservative force and its associated potential energy*. Become well acquainted with the earmarks that characterize each method. Do we know the forces? Is the system isolated? Are only conservative foces involved? Once the problem is classified, the same initial steps are taken for every problem of that type. If we know the forces, we draw a free-body diagram. If the system is isolated, we write expressions for $E_0 = E$ and $\mathbf{p}_0 = \mathbf{p}$. And if conservative forces are the only ones involved, we consider $F = -dU/dx$.

When we use this approach, the large amount of material we have covered sorts itself into just a few broad categories that are easy to remember. This simplification is characteristic of the physicist's way of looking at things—identifying the few central ideas that crop up again and again in all phenomena. In terms of these ideas, the vast array of happenings in the universe fit together into a basically simple conceptual scheme.

"I find," said 'e, "things very much as 'ow I've always found,
For mostly they goes up and down or else goes round and round."

P.R. CHALMERS
(*Roundabouts and Swings*)

chapter 9 Dynamics of Rotational Motion for a Point Mass

9.1 Introduction

Have you ever watched a small child learn how to push open a swinging door? Very soon the child discovers two facts. A given force **F** is most effective in causing the door to rotate about its hinges if (a) the force is applied *perpendicular* to the door and (b) the force is applied *as far from the axis* of the hinges as possible. So more than just the *magnitude* of the force is significant. In order to have a true measure of the effectiveness of giving the door some angular motion, we must include both of the above factors as well as the size of the force. We will introduce these concepts by investigating the rotation of a *point mass* about a fixed axis. Fortunately, the equations for rotational motion are closely analogous to the equations for straight-line motion, which we already know.

9.2 Notation for Angular Motion

It will be useful to develop a special notation for describing rotational motion. Although the symbols defined here are normally used to describe the most common type of such motion—the rotation of a rigid body about a fixed axis—they can equally well describe the motion of a particle that moves in a circle. We shall limit our attention to that type of motion in this chapter.

Consider a point mass m attached to a rigid rod of negligible mass which rotates about a fixed axis, as shown in Figure 9–1. On diagrams, we use a star (☆) to indicate a fixed axis of rotation perpendicular to the plane of the paper. The position of the particle is specified by the radius r and the angle θ that the radius vector makes with respect to some fixed reference line, often the $+x$ axis. The angle θ is the angular position of the particle, defined as the ratio of the arc length s to the radius r.

ANGULAR POSITION

$$\theta = \frac{\text{arc length } s}{r} \qquad \textbf{(9–1)}$$

The angle is measured in **radians** (rad). Since the unit is a ratio of two lengths, its dimensions are unity [1]. One sense of rotation is chosen as *positive*; the opposite sense is then *negative*.

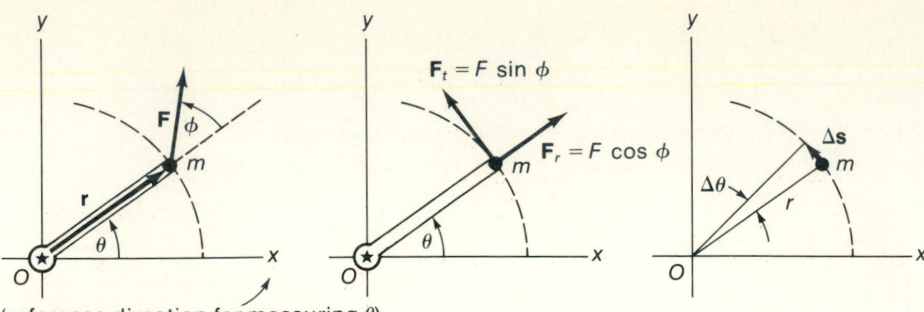

Figure 9–1
The point mass m is attached to a rigid rod of negligible mass free to rotate about the fixed axis ☆ perpendicular to the plane of the paper. The positive sense of rotation is chosen to be counterclockwise.

(**a**) As the mass moves around the circle, the angle ϕ is held constant.

(**b**) The tangential component of **F** is $\mathbf{F}_t = F \sin \phi$.

(**c**) The mass moves along the arc of the circle a distance $\Delta s = r\,\Delta\theta$.

If the rod rotates counterclockwise, during a time Δt the angular position will increase by an amount $\Delta\theta$. The **average angular speed** ω_{ave} is defined as

$$\omega_{ave} = \frac{\Delta\theta}{\Delta t} \tag{9–2}$$

where ω is the Greek letter *omega.* In the limit as $\Delta t \to 0$, we define the **instantaneous angular speed** ω:

$$\omega = \lim_{\Delta t \to 0} \frac{\Delta\theta}{\Delta t}$$

INSTANTANEOUS ANGULAR SPEED

$$\omega = \frac{d\theta}{dt} \tag{9–3}$$

The units of angular speed are radians per second (rad/s). Because the radian has dimensions of unity, the dimensions of angular speed are $[T^{-1}]$.

If the angular speed increases by $\Delta\omega$ in a time Δt, the **average acceleration** α_{ave} is defined as

$$\alpha_{ave} = \frac{\Delta\omega}{\Delta t} \tag{9–4}$$

where α is the Greek letter *alpha.* In the limit as $\Delta t \to 0$, this becomes the **instantaneous angular acceleration** α:

$$\alpha = \lim_{\Delta t \to 0} \frac{\Delta\omega}{\Delta t}$$

INSTANTANEOUS ANGULAR ACCELERATION

$$\alpha = \frac{d\omega}{dt} \tag{9–5}$$

Angular acceleration is measured in radians per second squared (rad/s^2). The dimensions of α are $[T^{-2}]$.

The angle θ involved in angular motion may also be expressed in terms of degrees or revolutions, with angular speed correspondingly measured in degrees per second, revolutions per second, or even revolutions per minute (rpm). Similarly, angular acceleration may be expressed in degrees per second squared or revolutions per second squared. The relations between these angular measurements are

$$1 \text{ rev} = 2\pi \text{ rad} = 360°$$

and

$$1 \text{ rad} \cong 57.3°$$

Conversion ratios are easily constructed from these relationships.

There is a close relationship between these angular quantities and the tangential motion of the particle along the arc of the circle. As the mass point travels the arc length s, the angular position increases by the angle θ, Equation (9–1):

$$s = r\theta$$

Differentiating both sides with respect to time (with r constant), we obtain the relation between the speed v and the angular speed ω.

$$\frac{ds}{dt} = r\frac{d\theta}{dt}$$

or

$$v = r\omega \tag{9–6}$$

Similarly, the *tangential acceleration* a_t and the *angular acceleration* α are related. Again taking the derivative with respect to time, we have

TANGENTIAL ACCELERATION

$$\frac{dv}{dt} = r\frac{d\omega}{dt}$$

$$a_t = r\alpha \tag{9–7}$$

The tangential acceleration (expressed in meters per second squared) is present only if the angular speed ω is changing—or, in other terms, if the speed v of the particle is changing. If v is increasing, a_t is positive; if v is decreasing, a_t is negative. However, recall from Chapter 3 that even if there is no tangential acceleration, the moving particle always has centripetal acceleration toward the center of the circle: $a_{cp} = v^2/r$. Since $v = r\omega$, the centripetal acceleration component may also be expressed as

CENTRIPETAL ACCELERATION (in terms of ω)

$$a_{cp} = r\omega^2 \tag{9–8}$$

The relations between the tangential quantities and the angular quantities are summarized below. Since the defining equation for θ is in radian units, all angular quantities in the equations below also must be in radian measure.

$$\left.\begin{aligned} s &= r\theta \\ v &= r\omega \\ a_t &= r\alpha \end{aligned}\right\} \text{Angular quantities } \textit{must} \text{ be in radian measure.} \tag{9–9}$$

It is helpful to remember that English symbols are used for *tangential* quantities (s,v,a_t), while Greek symbols are always used for *rotational* quantities (θ,ω,α).

Acceleration component	In terms of r and v	In angular measure	
Tangential Acceleration (tangent to the path) Present only if the speed v is changing	$a_t = \frac{dv}{dt}$	$a_t = r\alpha$	(9–10)
Centripetal Acceleration (radially inward) Always present, even for constant speed	$a_{cp} = \frac{v^2}{r}$	$a_{cp} = r\omega^2$	

The total acceleration a is found by combining the two right-angle components: $a = \sqrt{a_t^2 + a_{cp}^2}$, with the direction of **a** specified by some trigonometric function, as shown previously in Chapter 3 (Figure 3–5).

9.3 Torque

Consider a point mass m attached to a rigid rod (of negligible mass) that is free to rotate about a fixed axis (☆) at the origin O in Figure 9–1. Suppose we apply a force **F** to the mass at an arbitrary angle ϕ with respect to the rod, and investigate the effectiveness of this force in giving the mass some rotational acceleration about the origin. In other words, just as a force **F** gives a mass *linear* acceleration, there is an angular analogue that gives the mass in Figure 9–1 an *angular* acceleration.

First, we note that only the *tangential* force component contributes toward the tangential acceleration along the circular path. (The radial force component, and the rigid rod itself, may contribute radial forces, but these cannot change the speed of the particle along the path.) The tangential component is $F_t = F \sin \phi$, as shown in Figure 9–1(b). If we keep ϕ constant as the mass moves tangentially around the circle, the tangential component of the force $F_t = F \sin \phi$ will be constant in magnitude.

We now calculate the work done by F_t in moving a short distance $\Delta s\,(= r\,\Delta\theta)$ along the arc. This work will increase the kinetic energy of the mass by increasing its tangential speed from v to $v + \Delta v$. Applying the work-energy relation, we have

$$\text{Work done} = \text{Change in kinetic energy}$$
$$(F_t)(\Delta s) = \tfrac{1}{2}m(v + \Delta v)^2 - \tfrac{1}{2}mv^2$$
$$(F \sin \phi)(r\,\Delta\theta) = \tfrac{1}{2}m[v^2 + 2v\,\Delta v + (\Delta v)^2 - v^2]$$

To obtain the time rate at which work is done on the mass, we divide both sides by Δt.

$$(F \sin \phi)\left(r\frac{\Delta\theta}{\Delta t}\right) = \tfrac{1}{2}m\left[\frac{2v\,\Delta v + (\Delta v)^2}{\Delta t}\right]$$

Taking the limit as $\Delta t \to 0$, we note that $d\theta/dt = \omega$ and $dv/dt = a_t$.

$$(F \sin \phi)(r\omega) = \tfrac{1}{2}m2va_t + \lim_{\Delta t \to 0} \tfrac{1}{2}m\frac{(\Delta v)^2}{\Delta t}$$

The second term on the right equals zero in the limit because $\Delta v \to 0$ as $\Delta t \to 0$. Making the substitution $v = r\omega$, we now have

$$(F \sin \phi)(r\omega) = mr\omega a_t$$

Dividing by ω gives us

$$rF \sin \phi = mra_t$$

Finally, to obtain the expression in terms of the angular acceleration α, we substitute $a_t = r\alpha$.

$$\underbrace{rF \sin \phi}_{\text{torque}} = (mr^2)(\alpha) \qquad \textbf{(9–11)}$$

The left-hand side of this equation is defined as the **torque** τ (Greek letter *tau*) about the axis of rotation at the origin O; it produces the angular acceleration α of the point mass m.

TORQUE τ ABOUT THE ORIGIN O

$$\underset{(\text{about } O)}{\tau} = rF \sin \phi \qquad \textbf{(9–12)}$$

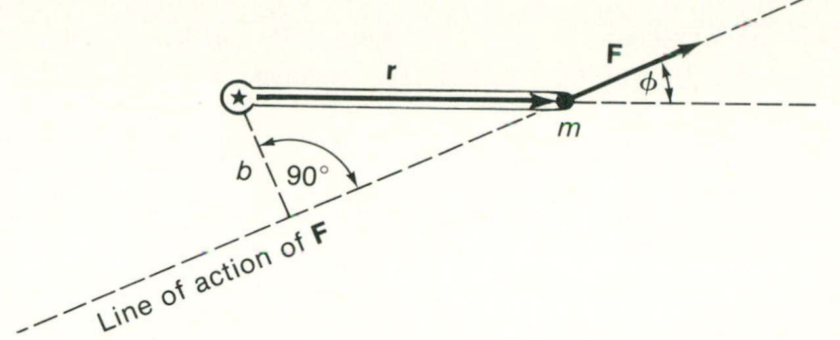

Figure 9–2
The distance b is the *moment arm* (or *lever arm*) about the axis of rotation of the force **F**.

where r is the distance from the origin O (the axis of rotation) to the point of application of the force and ϕ is the smaller angle between the *forward* directions of the radius vector **r** and the force vector **F**. Torque is measured in units of newton-meters (N·m) or pound-feet (lb·ft).

Note the presence of the following two significant factors: the distance r from the axis and the tangential component of the force $F \sin \phi$. Also recognize the close analogy between linear and angular concepts: a *force* gives a mass *linear* acceleration, while a *torque* gives a mass *angular* acceleration about a given axis.

9.4 Moment Arm and Moment of a Force

There is an alternative way of thinking about torque. It makes use of the concept of the *moment arm* (or *lever arm*) of the force **F** about the axis of rotation. We define the **line of action** of a force as the line drawn along the force vector, extended in both directions. The **moment arm** is defined as the perpendicular distance between the line of action of the force **F** and the axis of rotation. In Figure 9–2 the moment arm is $b = r\cos(90° - \phi) = r \sin \phi$.

$$\text{(Moment arm)} \qquad b = r \sin \phi \qquad \textbf{(9–13)}$$

Hence, we may also express the torque τ as

TORQUE τ ABOUT THE ORIGIN O

$$\underset{\text{(about } O)}{\tau} = \text{(force)(moment arm)} \qquad \textbf{(9–14)}$$

Although Equations (9–12) and (9–14) are equivalent, in many instances it is simpler to calculate the torque as the force times the moment arm. Torque is also called the *moment of the force* **F** *about* O, or more simply, the *turning moment*.

Another point should be mentioned: Regardless of the choice of plus and minus directions for coordinate axes, all products of (force)(moment arm) are considered to be positive magnitudes. The plus and minus signs for torques are then assigned in accordance with which sense of rotation is chosen as positive. We indicate this choice by the symbols (+↷ and ↶+).

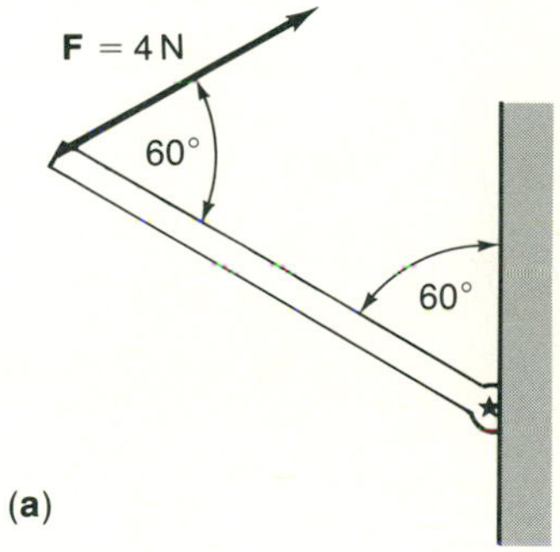

(a)

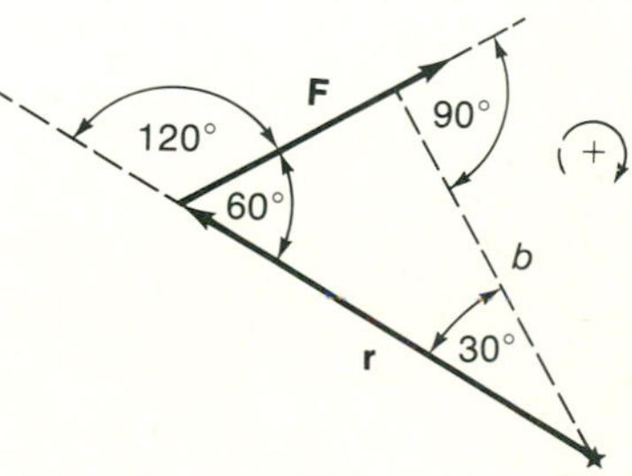

(b)

Figure 9–3
Example 9–1

EXAMPLE 9–1

A meter stick is free to rotate about a fixed axis located at one end and oriented perpendicular to its length. A force $F = 4$ N is applied to the other end, as shown in Figure 9–3(a). Find the magnitude of the torque this force produces about the axis.

SOLUTION

We sketch the significant vectors and choose the clockwise sense of rotation as positive [Figure 9–3(b)]. From Equation (9–12), we have

$$\underset{(\text{about }☆)}{\tau} = rF\sin\phi$$

Here, $\phi = 120°$, the angle between the *forward* directions of **r** and **F**.

$$\underset{(\text{about }☆)}{\tau} = (1\text{ m})(4\text{ N})(\sin 120°) = (1\text{ m})(4\text{ N})(0.866) = \boxed{3.46\text{ N}\cdot\text{m}}$$

We also could have solved the problem using the concept of the moment arm.

$$\underset{(\text{about }☆)}{\tau} = (F)(\text{moment arm}) = (F)(r\cos 30°) = (4\text{ N})(1\text{ m})(0.866) = \boxed{3.46\text{ N}\cdot\text{m}}$$

9.5 Moment of Inertia

We have obtained an expression [Equation (9–11)] relating the torque on a particle and its angular acceleration α:

$$\underbrace{rF\sin\phi}_{\text{torque}} = (mr^2)(\alpha)$$

We already have examined the left-hand side of the equation; let us now look at the right-hand side. As pointed out previously, *force* is the agent that gives a particle *linear* acceleration, and *torque* is the agent that gives a particle *angular* acceleration. In Newton's second law, $F = ma$, we interpret m as the inertial property by which a mass resists being given *linear* acceleration a. Similarly, we now interpret the factor in parentheses (mr^2) as the inertial property by which a particle resists being given *angular* acceleration α about the origin O of the coordinate system. For angular motion, this inertial property is called the **moment of inertia** I.

MOMENT OF INERTIA I ABOUT THE ORIGIN O FOR A SINGLE MASS POINT m

$$\underset{(\text{about }O)}{I} = mr^2 \qquad \textbf{(9–15)}$$

Moment of inertia is measured in units of $\text{kg}\cdot\text{m}^2$ or $\text{slug}\cdot\text{ft}^2$. Note that these units are very different from the corresponding inertia units of a mass for the linear case. All moments of inertia have units of a *mass* times a *distance squared*.

If we combine Equation (9–15) with Equation (9–11), torque $= (mr^2)(\alpha)$, we arrive at an important equation for rotational motion:

NEWTON'S SECOND LAW FOR ANGULAR MOTION (scalar form)

$$\underset{(\text{about }O)}{\tau} = I\alpha \qquad \textbf{(9–16)}$$

It is helpful to keep in mind the close analogy between the linear and angular expressions:

$$\textit{Linear:}\quad F = ma$$

$$\textit{Angular:}\quad \tau = I\alpha$$

It is also important to remember that torque and moment of inertia always refer to a particular line: the axis of rotation (☆). If we were to choose a different axis of rotation, the same force **F** would produce a different torque because the moment arm would be different. Similarly, the moment of inertia is different for different axes. For this reason, the particular axis chosen should always be specified when using $\tau = I\alpha$.

EXAMPLE 9–2

A 64-lb boy sits on the outer rim of a playground merry-go-round 10 ft in diameter (see Figure 9–4). (a) Find the moment of inertia of the boy about the central axis. (b) If a net tangential force of 6 lb is applied to the boy, find the rotational acceleration he will experience.

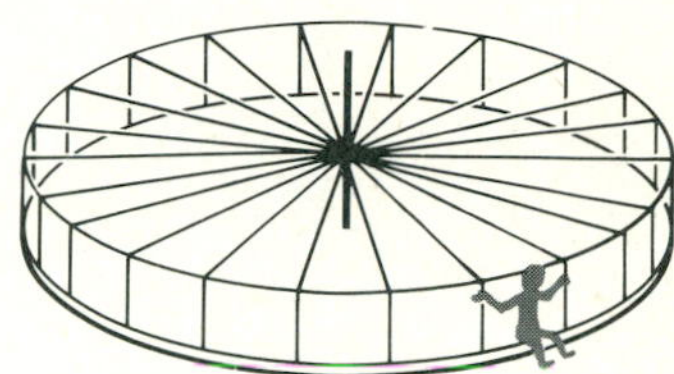

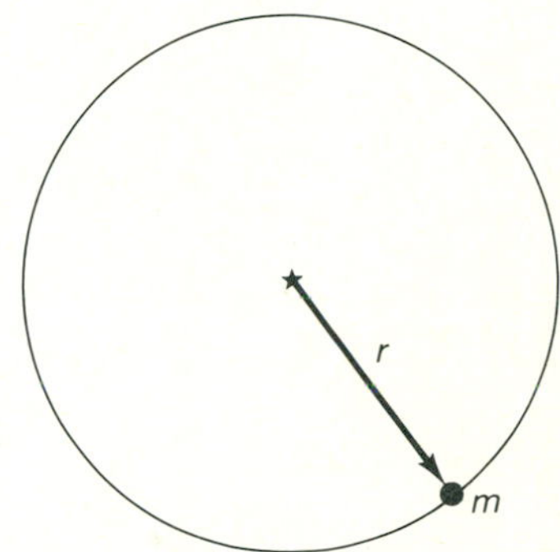

A view from above.

Figure 9–4
Example 9–2

SOLUTION

(a) We designate the axis by the symbol ☆. From Equation (9–15), we have

$$\underset{(\text{about } ☆)}{I} = mr^2 = \left(\frac{64 \text{ lb}}{32 \dfrac{\text{ft}}{\text{s}^2}}\right)(5 \text{ ft})^2 = \boxed{50.0 \text{ slug}\cdot\text{ft}^2}$$

(b) To find the rotational acceleration, we use Equation (9–16).

$$\underset{(\text{about } ☆)}{\tau} = I\alpha$$

$$\alpha = \frac{\tau}{I} = \frac{(F)(r)\sin\phi}{I}$$

Noting that the units of a *slug* are $\text{lb}\cdot\text{s}^2\cdot\text{ft}^{-1}$, we have

$$\alpha = \frac{(6 \text{ lb})(5 \text{ ft})(\sin 90^\circ)}{\left(50 \dfrac{\text{lb}\cdot\text{s}^2}{\text{ft}}\right)\text{ft}^2} = \boxed{0.600 \frac{\text{rad}}{\text{s}^2}}$$

Note that we must add "radians" to the answer to designate the proper units for rotational motion (to distinguish from "revolutions").

9.6 The Vector Product of Two Vectors

There is a convenient mathematical notation that enables us to write $\tau = rF\sin\phi$ in a particularly concise and useful form. It is the **vector product** of two vectors, written as **A X B** (read as "*A* cross *B*"). This product defines a third vector whose magnitude is

$$|\mathbf{A} \times \mathbf{B}| \equiv AB\sin\phi \tag{9–17}$$

where ϕ is the smaller angle between the forward directions of **A** and **B**. The direction of (**A X B**) is defined to be perpendicular to the plane containing **A** and **B**. Of course, there are two directions perpendicular to the plane. However, by convention we choose the direction according to a "right-hand rule," as illustrated in Figure 9–5.[1] Rotating the first vector **A** through the angle ϕ into the direction of the second vector **B** establishes a sense of rotation. If the fingers of the *right* hand are curled around in this rotational sense, the extended thumb

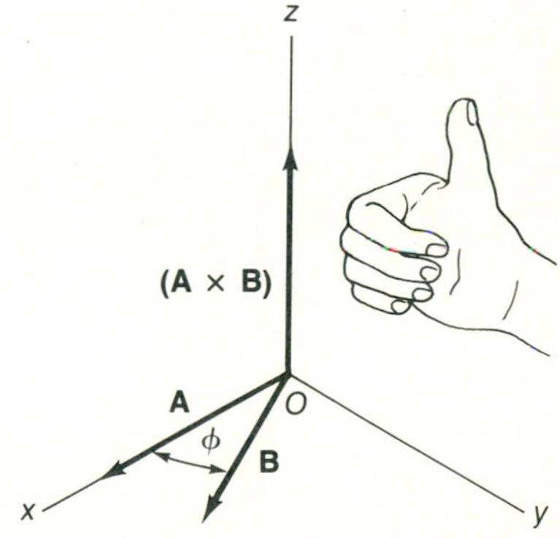

Figure 9–5
The "right-hand rule" for cross products. The vectors **A** and **B** lie in the *x-y* plane. Rotating **A** through the angle ϕ into the direction of **B** establishes a sense of rotation. If the fingers of the *right* hand are curled around in this rotational sense, the extended thumb points in the direction of the vector (**A X B**).

[1] Our choice of a *right-hand rule* is consistent with our convention of always choosing a *right-hand coordinate system*—that is, one in which a 90° right-hand rotation about the *z* axis carries the *x* axis into the *y* axis. Thus, $\hat{\mathbf{x}} \times \hat{\mathbf{y}} = \hat{\mathbf{z}}$.

points in the direction of the vector (**A X B**). Note that a "left-hand rule" would produce a vector in the opposite direction. Also, note that if the vectors are multiplied in the opposite order, the resultant vector is in the opposite direction. Hence:

$$(\mathbf{A} \times \mathbf{B}) = -(\mathbf{B} \times \mathbf{A})$$

So the *order* of writing the vectors is important. A final point: If a scalar multiplies one of the two vectors, the product may also be written as the scalar times the cross product. Thus:

$$(m\mathbf{A} \times \mathbf{B}) = m(\mathbf{A} \times \mathbf{B})$$

Expressing physical concepts in vector form is a convenient notation. The reason is that equations written in vector form are correct in *any* coordinate system: rectangular, polar, spherical, and so on. And an even more fundamental reason is that vector equations *retain the same form* even if the coordinate system is translated or rotated to a new orientation. (The laws of physics are believed to remain the same under coordinate translation or rotation.)[2] Furthermore, vector notation is often much simpler then writing out all the component equations separately. So vector notation is, indeed, the best way to express the fundamental relations of physics.

Let us now make use of the cross-product notation in expressing the vector equation for torque.

TORQUE τ ABOUT THE ORIGIN O

$$\underset{(\text{about } O)}{\boldsymbol{\tau}} = \mathbf{r} \times \mathbf{F} \qquad \textbf{(9–18)}$$

$$\text{Magnitude } \tau = rF \sin \phi$$

The torque $\boldsymbol{\tau}$ is thus a vector defined according to the right-hand rule of Figure 9–5. It is drawn through the point O, along the axis of rotation. Note that this vector is a bit unusual: No physical object is traveling along, or pointing in, the (right-hand-thumb) direction we choose. Instead, this vector describes a rotational concept associated with the plane that is perpendicular to the vector.

9.7 Angular Momentum and the Torque on a Particle

A net force **F** applied to a point mass m will give it a *linear* acceleration **a**, causing the mass to acquire *linear* momentum $\mathbf{p} = m\mathbf{v}$. Or, more concisely, a force **F** produces a time rate of change of *linear* momentum. A similar concept in angular motion is that a torque $\boldsymbol{\tau}$ applied to a particle will produce a time rate of change of *angular* momentum. The vector concept of angular momentum is a bit more complex than that of linear momentum, since angular momentum (like torque and moment of inertia) must always be expressed with respect to a given axis. For the present, we shall restrict our discussion to the case of motion of the particle in the x-y plane. The fixed axis, about which we will calculate torques and angular momenta, passes through the origin O perpendicular to the x-y plane.

We define the **angular momentum L** about the origin O of a particle of mass

---

[2] Hidden behind this statement is a fundamental assumption that space is homogeneous and isotropic. (*Homogeneous* is defined as having the same properties throughout, regardless of *position*, while *isotropic* refers to the same properties throughout, regardless of *direction*.) If this assumption about space is correct, it should not make any difference *where* we locate the origin of a coordinate system, or *in what direction* we orient the system of axes.

m moving with velocity **v** (and thus having linear momentum **p**) as the vector:

ANGULAR MOMENTUM L ABOUT THE ORIGIN *O*

$$\underset{\text{(about } O)}{\mathbf{L}} = \mathbf{r} \times \mathbf{p} \tag{9–19}$$

$$\text{Magnitude } L = rp \sin \phi$$

where ϕ is the smaller angle between the forward directions of **r** and **p**. The direction of **L** is perpendicular to the plane containing **r** and **p**, according to the right-hand rule, and the magnitude of **L** is the product of the *moment arm* times the *momentum*. Sometimes the angular momentum is called the *moment about O of the linear momentum* **p**.

Figure 9–6 illustrates this concept. Note that if we moved the point about which we calculate the angular momentum (the origin O) to some other location, the angular momentum about the new location would be different (because the vector **r** would be different). Thus it is important always to specify the axis (or point) about which the angular momentum is to be considered.

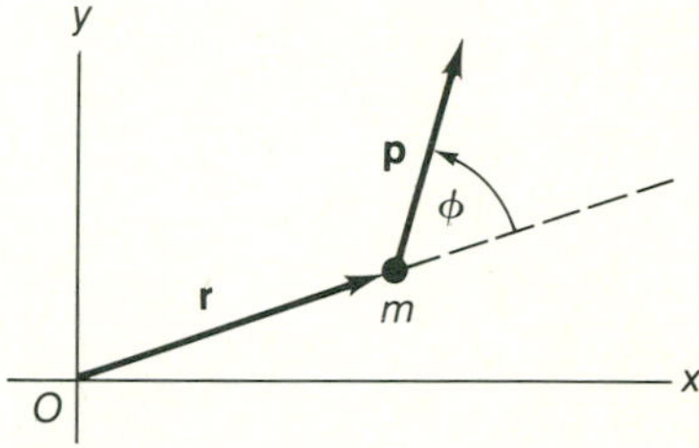

Figure 9–6
A particle of mass m moves with linear momentum **p**. At the instant shown, the particle is at the distance **r** from the origin. (The vectors **r** and **p** lie in the plane of the diagram.) According to the right-hand rule, the angular momentum $\mathbf{L} = \mathbf{r} \times \mathbf{p}$ about the origin is perpendicular to the x-y plane, *out* of the paper at the point O.

For the special case of circular motion, the radius vector **r** and the momentum **p** are always at right angles. The sin ϕ factor in the cross product gives $\sin 90° = 1$, so $L = rp \sin \phi$ becomes:

ANGULAR MOMENTUM L (for circular motion)

$$\underset{\text{(about } O)}{L} = mvr = mr^2\omega \tag{9–20}$$

Angular momentum is measured in units of kg·m²/s or slug·ft²/s. Be careful not to confuse these with the units of *linear* momentum: kg·m/s or slug·ft/s.

EXAMPLE 9–3

Suppose the playground merry-go-round in Example 9–2 were rotating at 12 rpm. If the boy sitting on the rim weighs 64 lb, what would be the magnitude of his angular momentum about the axis? The radius of the merry-go-round is 5 ft.

SOLUTION

Since the boy is traveling in a circle, the angular momentum L about the axis is

$$\underset{\text{(about ☆)}}{L} = mr^2\omega$$

where

$$m = \frac{w}{g} = \left(\frac{64\ lb}{32\ \frac{\text{ft}}{\text{s}^2}}\right) = 2 \text{ slugs}$$

$$r = 5 \text{ ft}$$

$$\omega = \left(12\ \frac{\text{rev}}{\text{min}}\right)\underbrace{\left(\frac{2\pi \text{ rad}}{1 \text{ rev}}\right)\left(\frac{1 \text{ min}}{60 \text{ s}}\right)}_{\text{Conversion ratios}} = 0.40\pi\ \frac{\text{rad}}{\text{s}}$$

Therefore:

$$L = (2 \text{ slugs})(5 \text{ ft})^2\left(0.40\pi\ \frac{\text{rad}}{\text{s}}\right) = \boxed{62.8\ \frac{\text{slug·ft}^2}{\text{s}}}$$

EXAMPLE 9–4

Suppose the boy in the previous example runs along the ground in a straight line tangent to the rim of the merry-go-round and jumps on board. If his speed is constant (6 ft/s), find his angular momentum **L** about the axis of the merry-go-round at the instant he jumps on.

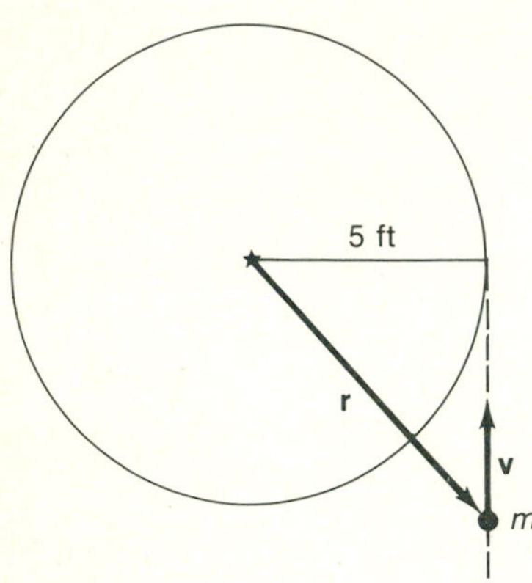

Figure 9–7
Example 9–4

SOLUTION

In Figure 9–7, the vector **r** from the axis to the boy specifies his instantaneous location as he runs along. His linear momentum is $p = mv = (2 \text{ slugs})(6 \text{ ft/s}) = 12 \text{ slug·ft/s}$. His angular momentum is $\mathbf{L} = \mathbf{r} \times \mathbf{p}$ (with magnitude $L = rp \sin \phi$). At the instant he jumps on board, the vectors **r** and **p** are at 90°, with the length of **r** being 5 ft. So the magnitude of **L** is:

$$L = rp \sin \phi = (5 \text{ ft})\left(12 \frac{\text{slug·ft}}{\text{s}}\right)(\sin 90°) = \boxed{60.0 \frac{\text{slug·ft}^2}{\text{s}}}$$

In the sketch, the direction of **L** (right-hand rule) is *out of the paper*, located at the axis ☆.

EXAMPLE 9–5

Consider a particle of mass m moving with constant speed v in a straight line. Show that the angular momentum **L** about any given point has the same (constant) value for all points along the line.

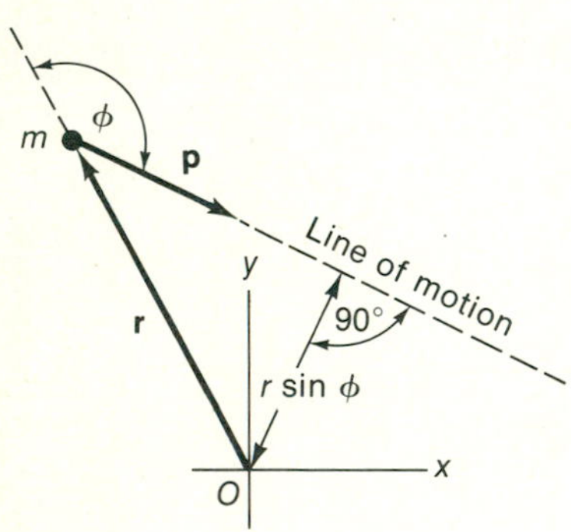

Figure 9–8
Example 9–5

SOLUTION

Figure 9–8 shows the mass m moving with a constant linear momentum $\mathbf{p} = m\mathbf{v}$ past the arbitrary point O. The problem is to show that even though ϕ and **r** change as the mass moves, **L** about the origin remains constant. By definition:

$$\mathbf{L} = \mathbf{r} \times \mathbf{p}$$

or

$$L = rp \sin \phi$$

where L is the magnitude of **L** directed into the plane of the sketch. We regroup the factors as $L = (r \sin \phi)p$, so that $r \sin \phi$ is identified with the moment-arm distance between the line of motion and the point O, which is *constant*. Because both p and $r \sin \phi$ remain constant as the mass moves along, the angular momentum **L** about O has the same constant value as the mass moves.

$$\boxed{\mathbf{L} = \text{constant in time}}$$ (directed into the plane of the sketch at the point O)

Let us now derive the important relationship between the torque τ and the angular momentum **L**. The first step is to take the time derivative of the angular momentum, $\mathbf{L} = \mathbf{r} \times \mathbf{p}$. When taking the derivative of a cross product, we must carefully maintain the *order* of writing the vectors and their derivatives. (If we reverse the order, we reverse the direction of the vector cross product.) The rule for the derivative of a cross product is similar to the usual rule for ordinary products (Appendix G-I):

$$\frac{d}{dt}(\mathbf{r} \times \mathbf{p}) = \left(\frac{d\mathbf{r}}{dt} \times \mathbf{p}\right) + \left(\mathbf{r} \times \frac{d\mathbf{p}}{dt}\right) \tag{9–21}$$

where we have been careful not to change the order of the terms.

We next note two facts: $d\mathbf{r}/dt = \mathbf{v}$ and $\mathbf{p} = m\mathbf{v}$. Substituting these relations into the above equation, we have

$$\frac{d\mathbf{L}}{dt} = (\mathbf{v} \times m\mathbf{v}) + \left(\mathbf{r} \times \frac{d\mathbf{p}}{dt}\right) \tag{9–22}$$

But since the vector $\mathbf{v}$ is in the same direction as the vector $m\mathbf{v}$, the cross product of these two vectors is zero. (Recall that the definition of the cross product involves the sine of the angle between the two vectors. If they are collinear, the angle is 0°, and $\sin 0° = 0$.) Also, we recognize that $d\mathbf{p}/dt = \mathbf{F}$, so Equation (9–22) may be written as

$$\frac{d\mathbf{L}}{dt} = \mathbf{r} \times \frac{d\mathbf{p}}{dt} = \mathbf{r} \times \mathbf{F} \tag{9–23}$$

Finally, we recall that the torque about the origin O is $\boldsymbol{\tau} = \mathbf{r} \times \mathbf{F}$. Thus, we have

THE ROTATIONAL ANALOGUE FOR NEWTON'S SECOND LAW

$$\underset{\text{(about } O)}{\Sigma\boldsymbol{\tau}} = \frac{d\mathbf{L}}{dt} \tag{9–24}$$

Of course, both $\boldsymbol{\tau}$ and $\mathbf{L}$ must be measured relative to the same point O.

It is helpful to remember that Equation (9–24) is the rotational analogue of $\mathbf{F} = d\mathbf{p}/dt$. Just as a force $\mathbf{F}$ applied to a particle of mass m produces a time rate of change of *linear* momentum, the torque $\boldsymbol{\tau}$ applied to a particle produces a time rate of change of *angular* momentum.

9.8 Conservation of Angular Momentum

Suppose we have a system in which no external torques act; that is, the system is *isolated* as far as torques are concerned. If the net torque $\Sigma\boldsymbol{\tau}$ on a particle about some axis O is zero, then it follows directly from Equation (9–24) that

$$\frac{d\mathbf{L}}{dt} = 0$$

or

$$\underset{\text{(about } O)}{\mathbf{L}} = \text{constant in time}$$

This is a statement of the **conservation of angular momentum.** Using the subscript "0" for the initial value, and no subscript for the final value, we have

CONSERVATION OF ANGULAR MOMENTUM

$$\underset{\text{(about } O)}{\mathbf{L}_0} = \mathbf{L} \quad \text{(if the net torque about } O \text{ is zero)} \tag{9–25}$$

This conservation relation is just as useful in solving *angular* motion problems as the conservation of linear momentum ($\mathbf{p}_0 = \mathbf{p}$) is in analyzing *linear* motion problems.

Many examples can be found in nature to illustrate motion in which the net torque is zero. A force whose direction always passes through a fixed point is called a **central force.** If we choose that point as the point O about which to consider torques and angular momentum, then the torque about O is always zero (because the force passes through O, making the moment arm zero). The angular momentum about O is thus constant in time. As an example, the gravitational attraction that the sun exerts on a planet is a central force, because the force on the planet is always toward the center of the sun. Hence, at all points along the planetary orbit, the planet's angular momentum about the sun is constant in time. The motion of an electron in the Bohr model of the hydrogen atom is also a case of a central force—the force with which the proton attracts the electron. Thus the angular momentum of the electron about the proton is constant.

EXAMPLE 9–6

Consider a satellite of mass m traveling in an elliptical orbit about the earth such that its distance from the center of the earth at *apogee* (farthest from the earth) is four times its distance from the earth's center at *perigee* (closest approach), as illustrated in Figure 9–9. Find the ratio of the satellite's speed at apogee, v_a, to its speed at perigee, v_p.

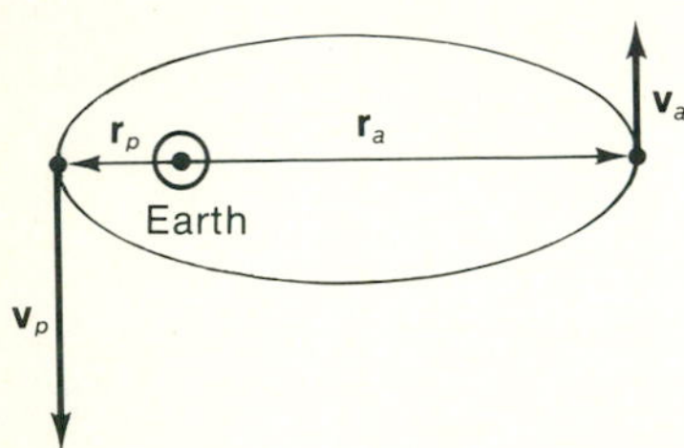

Figure 9–9
Example 9–6

SOLUTION

The force of gravity on the satellite is a *central* force, always passing through the center of the earth as the satellite moves along its elliptical orbit. Therefore, the angular momentum L about the earth's center is constant in time.

$$L_a = L_p$$

From Equation (9–20), these angular momenta are

$$mv_a r_a = mv_p r_p$$

Thus we obtain the ratio

$$\frac{v_a}{v_p} = \frac{mr_p}{mr_a} = \boxed{\frac{1}{4}}$$

EXAMPLE 9–7

Show that as a planet moves around the sun, its radius vector (from the sun to the planet) sweeps out equal areas in equal times.

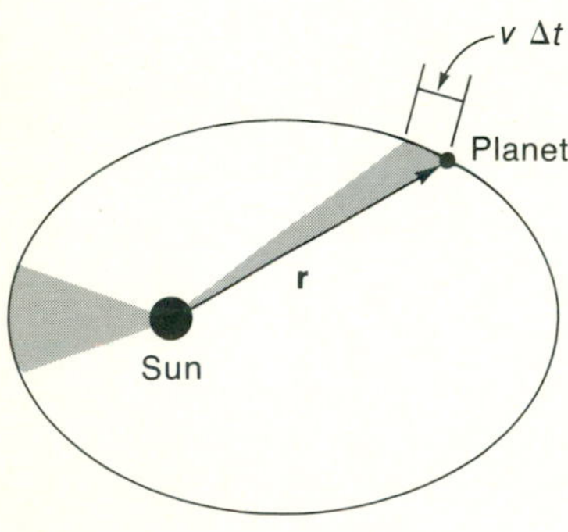

Figure 9–10
Example 9–7

SOLUTION

Figure 9–10 shows the elliptical orbit of a planet circling the sun (with the ellipticity greatly exaggerated). Because the gravitational force on the planet is always toward the center of the sun, there is no net torque on the planet about the sun. Hence the angular momentum of the planet is constant in time. (We assume that the sun is at rest.)

In a time Δt, the planet moves a distance $v\,\Delta t$. The triangular area ΔA swept out by the motion of the radius vector in the time Δt is approximated by

$$\Delta A \approx \tfrac{1}{2}(r)(v\,\Delta t)(\sin\theta)$$

where θ is the angle between the radius vector $\mathbf{r}$ and the velocity vector $\mathbf{v}$. As the value of Δt becomes smaller, the value of ΔA approaches the above value. Multiplying numerator and denominator by m, dividing by Δt, and letting $\Delta t \to 0$, we have

$$\lim_{\Delta t \to 0} \frac{\Delta A}{\Delta t} = \frac{dA}{dt} = \frac{1}{2m}(mvr\sin\theta)$$

The term $(mvr \sin\theta)$ is the angular momentum L of the planet about the sun, which is constant in time. Thus, the rate dA/dt at which the radius vector sweeps out area is also constant in time, for all portions of the orbit. That is, *the radius sweeps out equal areas in equal times.* Note that this conclusion depends on the fact that the force of gravity is a *central* force, producing no torque on the planet.

9.9 Notation for the General Case

Up to this point, all of our examples have been for the special case in which both the radius vector **r** and the force **F** were in the *x*-*y* plane. In the general case, these vectors may lie in *any arbitrary direction.* Figure 9–11 illustrates this situation. The fixed point about which the torque is calculated is the origin *O*. Note how the vector equation for torque ($\tau = \mathbf{r} \times \mathbf{F}$) can be applied in three dimensions to locate the vector τ. This illustrates one of the advantages of *vector* equations—namely, that they remain correct even though the coordinate system is rotated to a new orientation in space.

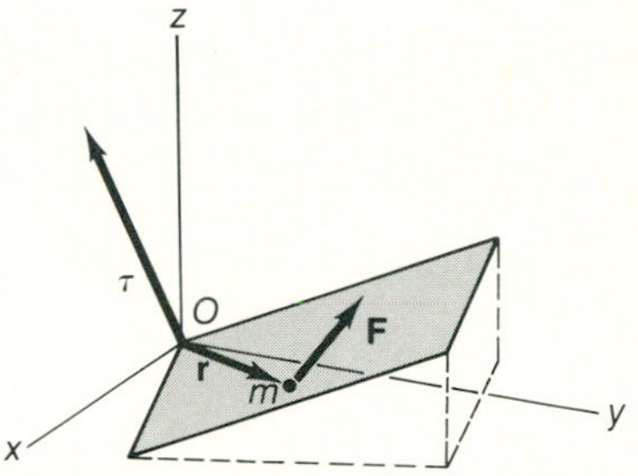

Figure 9–11
In the general case, **r** and **F** may be at any arbitrary angles. Here they lie in the tilted plane. The torque τ about *O* due to the force **F** on the mass point *m* is a vector perpendicular to the tilted plane, according to the right-hand rule.

9.10 A Note About Vectors Representing Rotational Quantities

Thus far we have introduced two angular quantities that are represented by vectors: τ and **L**. Although we defined the direction of these vectors by using the right-hand rule, we could equally well have chosen a left-hand rule, with the vectors pointing in the opposite direction. No physical object is traveling along (or pointing in) the direction we choose for these vectors. Instead, they describe rotational concepts associated with the plane perpendicular to the vectors.

An unusual subtlety comes to light if we investigate how these vectors behave when the physical situation they represent is viewed in a mirror. Why do we consider a mirror view? Primarily because it provides one test of the fundamental nature of space itself. The argument runs as follows. If we actually construct any given piece of apparatus as it appears in a mirror, would this mirror-image apparatus behave according to the same laws of physics we have developed for the "original" world? Our intuition says yes. For example, a mirror-image automobile constructed with right-hand drive and with all the motor parts reversed in position would still run satisfactorily. Indeed, prior to 1956, all physical phenomena were thought to have this mirror-symmetry property—namely, that the mirror-image of any occurrence represents a possible actual situation in the real world, obeying the same laws of physics. The discovery of an exception in 1956 (the nonconservation of parity)[3] emphasized that it is not a trivial question to ask how physical concepts behave when reflected in a mirror. Rather, the question probes something very basic about space itself: whether or not there is a preferred right or left handedness to empty space. The answer is not clear, at least for the concept of parity. Physicists and philosophers are currently attempting to understand more fully the implications of this situation.

To illustrate the unusual behavior of vectors representing rotational concepts, consider a displacement $\Delta\mathbf{r}$ and an angular momentum **L**. We will investigate two cases: when the physical situation is such that these vectors are (a) parallel and (b) perpendicular to the mirror. In Figure 9–12, imagine that the mirror-world physical situation is *actually constructed* and the appropriate vectors are then added to represent the rotational properties of the mirror-world apparatus.

In case (a), the displacement vector $\Delta\mathbf{r}$ retains its original direction. But the mirror-image disk rotates in the opposite sense; thus the angular momentum vector **L** reverses its direction. In case (b), the displacement vector reverses its

[3] A good introductory article on this is "The Overthrow of Parity," Phillip Morrison, *Scientific American*, April 1957. Also see the very interesting discussion in *The Feynman Lectures on Physics*, Leighton and Sands, Addison Wesley (1963), Volume I, Chapter 52.

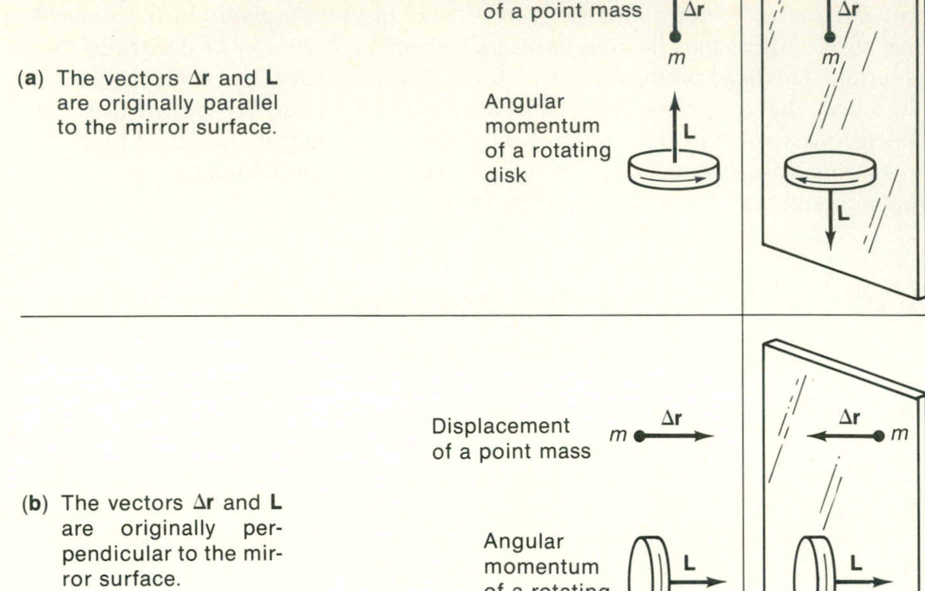

(**a**) The vectors **Δr** and **L** are originally parallel to the mirror surface.

(**b**) The vectors **Δr** and **L** are originally perpendicular to the mirror surface.

Figure 9–12
An experiment to test the behavior of different types of vectors when the physical situation is reflected in a mirror. Imagine that the mirror view represents real physical occurrences, with the appropriate vectors added later. (The sketches do *not* depict reflections of the vectors themselves, but just reflections of the physical displacement or apparatus.)

direction, while the angular momentum vector retains its direction. If we similarly investigate other vectors representing angular quantities, we conclude that rotational vectors behave differently from "ordinary" vectors when the physical situation is reflected in a mirror. For this reason, two different names are used to classify vectors. Vectors associated with angular concepts are called *axial vectors* (or *pseudo-vectors*), while "ordinary" vectors such as force and displacement are called *polar vectors.*

Summary

In diagrams, a star symbol (☆) may be used to represent a fixed axis of rotation perpendicular to the plane of the drawing. It is often most convenient to place the origin O at the axis of rotation.

The *torque* τ about the origin O produced by the force **F** applied at a point **r** from the origin is

TORQUE τ
$$\underset{(\text{about } O)}{\tau} = \mathbf{r} \times \mathbf{F}$$

or
$$\underset{(\text{about } O)}{\tau} = rF \sin \phi$$

where ϕ is the smaller angle between the forward directions of **r** and **F**. The direction of τ is obtained by applying the right-hand rule for cross-product vector multiplication.

The torque may also be expressed as

$$\underset{\text{(about } O)}{\tau} = \text{(force)(moment arm)}$$

where the *moment arm* is the distance between the line of action of the force and the axis of rotation.

The *moment of inertia I* of a point mass *m* located at a distance *r* from the center of rotation is given by

MOMENT OF INERTIA *I*

$$\underset{\text{(about } O)}{I} = mr^2$$

The *angular momentum*, **L**, of a point mass with linear momentum **p** at a distance **r** from the axis of rotation is defined by

ANGULAR MOMENTUM L

$$\underset{\text{(about } O)}{\mathbf{L}} = \mathbf{r} \times \mathbf{p}$$

This definition of **L** leads to Newton's second law for angular motion of a particle:

NEWTON'S SECOND LAW FOR ANGULAR MOTION

$$\underset{\text{(about } O)}{\Sigma\tau} = \frac{d\mathbf{L}}{dt}$$

For cases where *I* remains constant in magnitude, this reduces to

$$\underset{\text{(about } O)}{\Sigma\tau} = I\alpha$$

From Newton's second law for rotation, it follows that if the net torque on the particle is zero, the angular momentum of a particle is conserved.

CONSERVATION OF ANGULAR MOMENTUM

$$\underset{\text{(about } O)}{\mathbf{L}_0} = \mathbf{L} \quad \text{(if the net torque about } O \text{ is zero)}$$

Questions

1. Are units of *radian* measure satisfactory for the analysis of kinematic problems regardless of which systems of units are used for measuring other quantities?
2. In Europe, some surveying is done using the *grad* as a unit of angular measure. There are 100 grads in a quarter-circle arc. Compare the advantages and disadvantages of this unit with *degrees*.
3. Explain why an object traveling in a *straight* line can have angular momentum about some point.
4. An object travels with constant speed in a straight line. If it has an angular momentum *L* about a certain point *O* (not on the line), what is its angular momentum about a point *O* that is three times farther from the line? What is its angular momentum about a point *on* the line?
5. A particle moves only in the *x*-*y* plane. Find the possible directions of its angular momentum about the origin of the coordinate system.

6. Two identical satellites, A and B, are put into circular orbits around the earth with radii R and $2R$, respectively. Which (if either) has the greater linear speed? Angular speed? Angular momentum? Rotational kinetic energy?

7. Two ice skaters, each of mass m and speed v, are approaching each other along parallel paths separated by a distance d. Just as they are about to pass, one holds a light bamboo stick toward the other, who grasps it firmly. Describe the subsequent motion of the skaters as they cling to the rod. What would be the motion if one skater had more mass than the other?

8. Consider a particle moving along a straight line with constant velocity. Does the radius vector from a fixed point (not on the line) to the particle sweep out equal areas in equal time?

Problems

9A–1 A 0.40-kg bead slides without friction on a wire bent into a horizontal circle of diameter 0.60 m. What force tangent to the circle will give the bead an angular acceleration of 4 rad/s^2?
Answer: 0.480 N

9A–2 A small 2-kg mass moves in a circle 3 m in diameter with a constant speed of 4 m/s.
(a) Find the moment of inertia of the mass about the center of the circle.
(b) Find the angular momentum about the center of the circle.
(c) Determine the force, tangent to the circle, that will give the mass an angular acceleration about the center of 5 rad/s^2.

9A–3 Two point masses, each 4 kg, are joined by a light rigid rod 0.40 m long. This "dumbbell" system is placed on a horizontal frictionless surface and set into rotation about a vertical axis through the center of mass with an angular speed of 3 rad/s.
(a) Find the moment of inertia of the system about the center of mass.
(b) Find the angular momentum about the center of mass.
(c) What tangential force, applied to each mass, will produce an angular acceleration of 6 rad/s^2?
Answers: (a) 0.320 kg·m^2 (b) 0.960 kg·m^2/s (c) 4.80 N

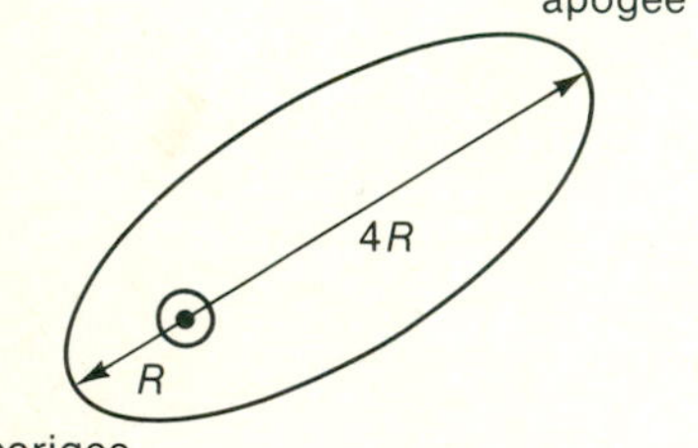

Figure 9–13
Problem 9A–4

9A–4 As shown in Figure 9–13, a satellite moves in an elliptical orbit about the earth such that at perigee and apogee positions, the distances from the earth's center are, respectively, R and $4R$. Find the ratio of the orbit speeds at the two positions: $v_{\text{apogee}}/v_{\text{perigee}}$.

9A–5 A proton of mass 1.7×10^{-27} kg moves in a straight line with a constant speed of 5×10^6 m/s as it approaches a stationary neutron. The path of the proton misses the neutron by 1×10^{-11} m. What is the angular momentum of the proton about the neutron?
Answer: 8.50×10^{-32} kg·m^2/s

9A–6 A mass m on the end of a string is set into motion in a vertical circle so that it travels around the circle continually thereafter. When the mass is at the top, its angular momentum about the center of the circle is less than when it is at the bottom. Explain the source of the torque that increases and decreases the angular momentum during the motion.

9A–7 Four point masses m are situated at the corners of a square formed by four (massless) rods of length ℓ connected together. Find the moment of inertia of this object when it is rotating about (a) an axis perpendicular to the plane of the square, passing through its center, (b) an axis in the plane of the square, passing through the center of one edge and the center of the square, (c) an axis through one edge, and (d) an axis running diagonally from one corner through the opposite corner.
Answers: (a) $2m\ell^2$ (b) $m\ell^2$ (c) $2m\ell^2$ (d) $m\ell^2$

9B–1 A force $\mathbf{F} = 4\hat{\mathbf{x}} + 3\hat{\mathbf{y}} + 0\hat{\mathbf{z}}$ (in newtons) is applied to the point (5 m,12 m,0). Find the magnitude and direction of the torque about the origin of the coordinate system.
Answer: 32.9 N·m in the $-\hat{\mathbf{z}}$ direction

9B–2 A 5-kg mass is at the point (-4 m,3 m) traveling at 4 m/s in the $+x$ direction.
(a) What is its angular momentum about the origin?
(b) What will its angular momentum about the origin be one second later?

9B–3 A model airplane whose mass is 0.75 kg is tethered by a wire so that it flies in a circle 30 m in radius. The airplane engine provides a thrust of 0.80 N perpendicular to the tethering wire.

(a) Find the torque the engine thrust produces about the center of the circle.
(b) Find the angular acceleration of the airplane when in level flight.
(c) Find the linear acceleration of the airplane tangent to its flight path.

Answers: (a) 24.0 N·m (b) 3.56×10^{-2} rad/s² (c) 1.07 m/s²

9B–4 A satellite travels about the earth in an elliptical orbit whose perigee distance is two earth radii above the earth's surface. At apogee, it travels with one-fourth the speed it has at perigee. In terms of the earth's radius R, what is the maximum distance of the satellite from the earth's surface?

9B–5 The earth–sun distance varies from 9.14×10^7 mi (perihelion) to 9.45×10^7 mi (aphelion). The minimum orbital speed of the earth is 18.2 mi/s. Find the maximum orbital speed.

Answer: 18.8 mi/s

9B–6 Figure 9–14 depicts a mass m moving with initial speed v_0 in a circle of radius r_0 on a horizontal frictionless surface. A string fastened to the mass and passing through a small hole supplies the centripetal force. The string is now pulled slowly to a new position so that the mass travels in a circle of radius $r_0/2$. In terms of m, r_0, and v_0, find (a) the final speed of the mass and (b) the total work done in pulling the string to its new position. (c) Show that the work done equals the change in kinetic energy of the mass. (Hint: Is angular momentum about the hole conserved? Is energy conserved?)

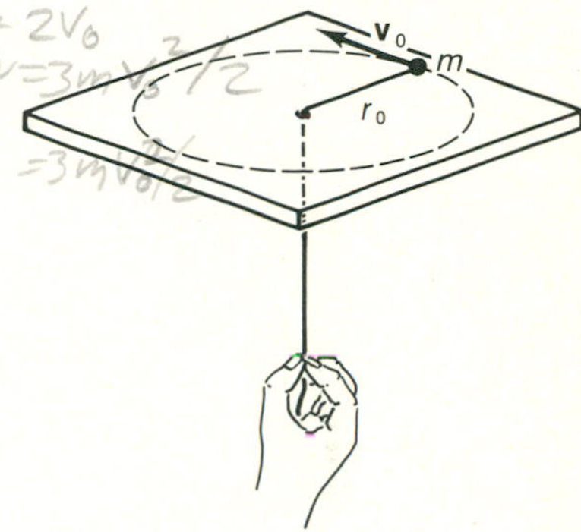

Figure 9–14
Problems 9B–6 and 9C–6

9B–7 Two unequal (point) masses, M and m, a distance D apart, rotate with the same angular speed about their common center of mass. In terms of the given symbols, find the ratio of their angular momenta about the center of mass: L_M/L_m. *Answer:* m/M

9B–8 A satellite of mass m moves in a circular orbit of radius R around the earth. Its angular momentum is L. In terms of m, L, and R, what is its kinetic energy?

9C–1 A simple pendulum with bob mass m and string length ℓ is released from rest when the string makes an angle of θ ($<90°$) with the downward vertical. Using the concept of torque, calculate the work done by gravity in lowering the pendulum bob to its lowest point. Show this result is the same as the bob weight times the vertical distance it moves.

9C–2 An earth satellite weighing 5 tons is in circular orbit about the earth. It encounters a small air resistance that causes the satellite to slowly spiral toward the earth. At a height of 500 mi above the earth's surface, the period of revolution is 101.5 min. After spiraling to 400 mi above the earth, the period is 98.1 min.

(a) What are the initial and final angular momenta of the satellite about the center of the earth? Assume that the earth's radius is 4000 mi.
(b) Assuming the spiral is so gradual that every orbit may be assumed circular, what are the initial and final orbital speeds?

(Note that even though the air resistance is a force opposite to the orbital velocity, the velocity of the satellite increases.)

9C–3 The satellite in the previous problem changed its orbit radius as described in 500 days. Estimate the magnitude of the frictional force in pounds.

Answer: 1.97×10^{-3} lb

9C–4 Figure 9–15 shows a mass m moving with speed v_0 on a horizontal frictionless surface. A string fastened to the mass winds up on a fixed peg as the mass moves, causing the mass to follow a spiral path inward. If the speed is v_0 when the string length is r_0, what is the speed when the string length has been reduced to $r_0/2$? Explain your reasoning. (Hint: The instantaneous velocity of the mass is always at right angles to the string. Is angular momentum about the center of the peg conserved? Is energy conserved?)

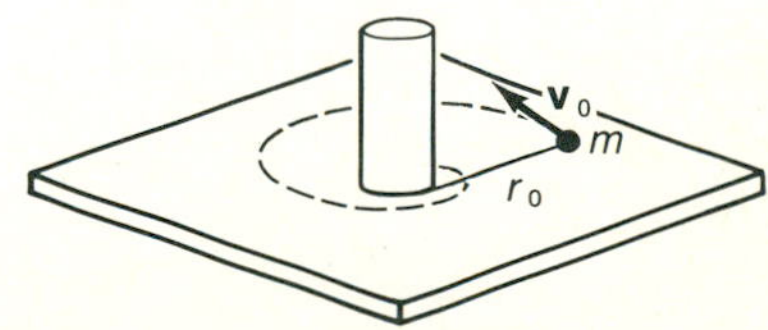

Figure 9–15
Problems 9C–4 and 9C–6

9C–5 Two identical masses are at opposite sides of a circle centered at the origin. Each moves with the same (constant) angular velocity about the origin in the x-y plane. Show that the angular momentum of the masses is the same about every point in the x-y plane.

9C–6 In this problem, you are asked to compare two situations. In Figure 9–15, the initial conditions are a mass m moving with speed v_0 when the string length is r_0. At a later time, the mass will have moved to some point P (not shown) on the spiral path, where it will have a speed v. Now look at the situation depicted in Figure 9–14. Suppose, in this case, we start the mass at the *same point* with the *same initial speed* v_0 as in the first case. But now we cleverly pull the string so that the mass follows the identical spiral path as in the first case. In this new situation, when the mass arrives at the point P will it have the same speed as in the first case? If so, justify your reasoning. If not, explain why not.

9C–7 Kepler's Second Law of Planetary Motion may be stated as follows: The line joining any planet to the sun sweeps out equal areas in equal times. Show that $dA/dt = L/(2m)$, where dA/dt is the rate at which the line sweeps out area, L is the angular momentum of the planet about the sun, and m is the mass of the planet. (Hint: Consider the essentially triangular area swept out by the radius when it swings through an angle $d\theta$.)

9C–8 A particle of mass m is released from rest at the point $(x_1,0,0)$. It thereafter falls freely under the action of gravity in the $+y$ direction (assumed positive downward).

(a) Find the torque ($\mathbf{r} \times \mathbf{F}$), about the origin O, that acts on m and show that it is constant in time.

(b) Derive an expression for the angular momentum ($\mathbf{r} \times \mathbf{p}$) about the origin O as a function of time.

9C -9 A 2-kg mass moves in a circle of radius 5 m. Starting at rest, the angular momentum L of the mass about the center of the circle varies with time as $L = 3t^2$, where L is in units of $\text{kg}\cdot\text{m}^2/\text{s}$ and t is in seconds.

(a) What torque (about the center of the circle) acts on the mass?

(b) Derive an expression for the angular velocity $\omega(t)$ as a function of time.

Answers: (a) $6t$ (in units of Newton·meters if t is in seconds)
(b) $0.060t^2$ (in units of radians per second if t is in seconds)

But in physics I soon learned to scent out the paths that led to the depths, and to disregard everything else, all the many things that clutter up the mind, and divert it from the essential. The hitch in this was, of course, the fact that one had to cram all this stuff into one's mind for the examination, whether one liked it or not.

ALBERT EINSTEIN

chapter 10 The Dynamics of Rigid Bodies

10.1 Introduction

Up to this point, we have limited our discussion to the motion of single particles. The extension of Newton's laws to the more general cases of solid objects is a notable achievement of the Newtonian viewpoint. The complete theory is quite complicated and requires a multivolume set of books to treat the subject thoroughly. Nevertheless, it is an elegant and satisfying field of physics.

In a sense, this chapter contains the "heart" of Newtonian physics: the analysis of the general motion of an extended object moving through space. It is a fairly long chapter with many details. But we will only be using just the same few methods for solving problems that we discussed previously:

(1) Newton's second law: $\Sigma \mathbf{F} = m\mathbf{a}$ (or $\Sigma \boldsymbol{\tau} = I\boldsymbol{\alpha}$)
(2) Conservation of energy and momentum
(3) Work-energy relation

All the seemingly large number of topics in this chapter aim toward the common goal of applying these three methods to the general motion of extended objects.

Fortunately, many interesting problems in engineering and physics involve a relatively simple case: *the rotation of a symmetrical object about an axis of symmetry through its center of mass.* We will treat this situation in some detail. If you always watch for the close analogies between the equations of *linear* and *angular* motions, the extension of our methods of analysis to the motion of rigid bodies will be straightforward and easy to learn. One new concept, the *center of gravity*, greatly simplifies the treatment, so we will start with that topic.

10.2 The Center of Gravity

Every object near the earth experiences a force of gravity which is distributed throughout the material, acting alike on each atom of the object. This fact raises a perplexing question: If an object is subjected to a torque about some axis because of gravitational forces, how do we go about summing up the enormous number of terms involved, each the product of an incremental mass times its own particular moment-arm distance?

Fortunately, we can reduce the problem to just a single term by defining the **center of gravity** (*CG*) of an object: *that point at which we may consider the total force of gravity* $\mathbf{W}$ *to be concentrated for the purpose of calculating torque.*

It is by this strategem that we are able to extend Newton's laws for a particle to the case of an extended object moving (and rotating) in a gravitational field.

Consider a two-dimensional object that we approximate by an arbitrary figure cut from a thin sheet of metal. Suppose we orient the object in a vertical plane and fasten it at some point to a horizontal axis of rotation (☆) at right angles to the plane of the sheet (see Figure 10–1). The object is free to rotate about this fixed axis.

The total mass M of the object may be thought of as being made up of N incremental mass particles Δm_i.

$$M = \Delta m_1 + \Delta m_2 + \Delta m_3 + \cdots \Delta m_N \tag{10–1}$$

You may imagine several hundred increments, or, if you wish to recognize the gravitational force on each atom present, a number of the order of 10^{23}. In Figure 10–1(a) we show the forces of gravity on a few of the mass elements and also the origin O of a coordinate system located at the axis of rotation (☆).

What is the total net torque $\Sigma\tau$ about O due to the individual forces of gravity on all the separate mass elements? Each force of gravity $\Delta \mathbf{W}_i$ has its own moment-arm distance x_i, so the gravitational torque about O on each element is $x_i \Delta W_i$ (where $\Delta W_i = g\,\Delta m_i$). The total torque is therefore

$$\underset{\text{(about } O)}{\Sigma\tau} = x_1 \Delta W_1 + x_2 \Delta W_2 + x_3 \Delta W_3 + \cdots$$

$$\underset{\text{(about } O)}{\Sigma\tau} = \sum_{i=1}^{i=N} x_i \Delta W_i \tag{10–2}$$

If x_{CG} represents the moment-arm distance of the center of gravity, CG, then

$$\underset{\text{(about } O)}{\Sigma\tau} = x_{CG} W \tag{10–3}$$

Equating the above two expressions for torque, we solve for x_{CG} to obtain

CENTER OF GRAVITY (one dimension)

$$x_{CG} = \frac{\sum_{i=1}^{i=N} (x_i \Delta W_i)}{\sum_{i=1}^{i=N} \Delta W_i} \tag{10–4}$$

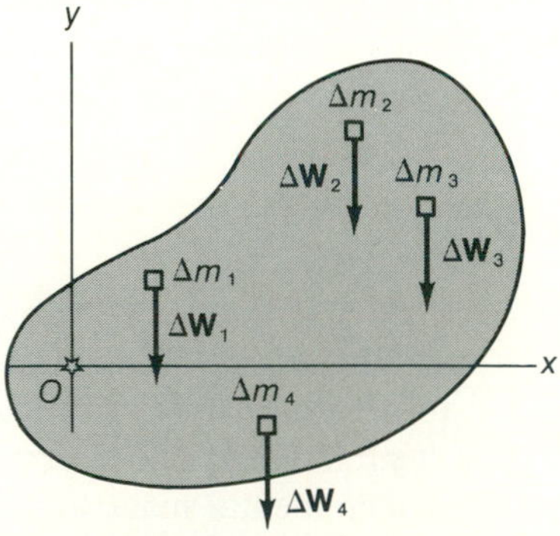

(**a**) Each element of mass Δm_i has a gravitational force $\Delta \mathbf{W}_i$ on it (only a few are shown).

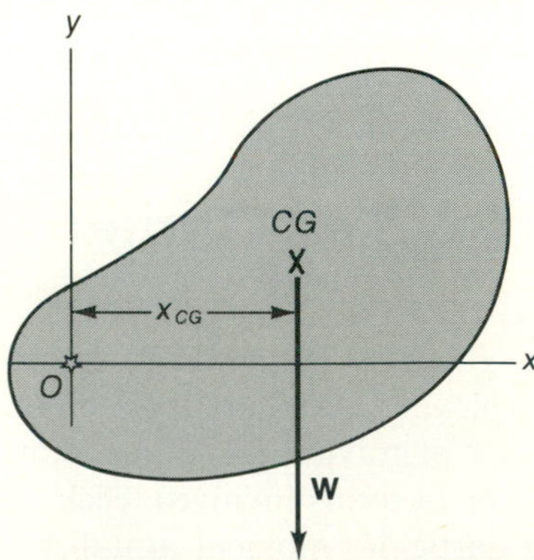

(**b**) If the total force of gravity **W** is imagined to act at the point CG, the torque about O is the same as in (**a**).

Figure 10–1
A two-dimensional object free to rotate about a fixed horizontal axis (☆) under the action of gravity.

Note that the denominator is just $\Delta W_1 + \Delta W_2 + \Delta W_3 + \cdots = W$, the total weight of the object.

Rotating the object by 90° so that gravity acts along the original x direction of the body, a similar argument leads to the y coordinate y_{CG}. For a three-dimensional object, the z coordinate z_{CG} could also be found. The sums may be carried out directly if the system under consideration involves just a few mass elements. For extended objects that have a continuous distribution of matter, we use ideas of the calculus to change the *finite sum* to an *integral* involving the gravity force dW on infinitesimal mass elements dm. In equation form, we have:

Delta notation for finite mass elements	**Calculus notation for uniform mass distributions**	
$x_{CG} = \dfrac{\sum_{i=1}^{i=N} x_i \, \Delta W_i}{\sum_{i=1}^{i=N} \Delta W_i}$	$x_{CG} = \dfrac{\int x \, dW}{\int dW}$ (integrated over the entire body)	**(10–5)**

We use similar expressions for y_{CG} and z_{CG}. The equations for the three components combine into just a single *vector* equation:

CENTER OF GRAVITY (CG)

$$\mathbf{R}_{CG} = \frac{\sum_{i=1}^{i=N} \mathbf{r}_i \, \Delta W_i}{\sum_{i=1}^{i=N} \Delta W_i} \qquad \mathbf{R}_{CG} = \frac{\int \mathbf{r} \, dW}{\int dW} \tag{10–6}$$

(integrated over the entire body)

where $\mathbf{r}_i$ (or $\mathbf{r}$) is the position vector of the mass element Δm_i (or dm) and $\mathbf{R}_{CG}$ is the position vector of the center of gravity. Note that the denominator in each expression is simply the total weight W of the object.

The *CG* need not lie within the actual material of the object itself. For example, a doughnut's *CG* is at the geometric center of the hole. For geometrically symmetrical objects, the center of gravity is at the *geometric center* of the object if the density (mass/volume) is uniform throughout (or if it varies symmetrically about the axis).

10.3 Center of Mass for an Extended Object

There is a close connection between the *CG* and the center of mass *CM* of an object. Note that $\Delta W_i = g \, \Delta m_i$ and that $dW = g \, dm$. Thus, when the acceleration due to gravity is a constant value throughout a region under consideration, we have a factor g in both the numerator and the denominator of Equation (10–6).

$$\mathbf{R}_{CG} = \frac{\sum_{i=1}^{i=N} \mathbf{r}_i g \, \Delta m_i}{\sum_{i=1}^{i=N} g \, \Delta m_i} \qquad \mathbf{R}_{CG} = \frac{\int \mathbf{r} g \, dm}{\int g \, dm}$$

Canceling g, we obtain the definition of the **center of mass** (CM) for an object of finite size:

CENTER OF MASS (*CM*)

$$\mathbf{R}_{CM} = \frac{\sum_{i=1}^{i=N} \mathbf{r}_i \, \Delta m_i}{\sum_{i=1}^{i=N} \Delta m_i} \qquad \mathbf{R}_{CM} = \frac{\int \mathbf{r}\, dm}{\int dm} \qquad (10\text{–}7)$$

(integrated over the entire body)

where $\mathbf{R}_{CM}$ is the position vector of the center of mass. The location of this vector is independent of whether any gravity forces are present or not.[1] It is solely a property of the object itself, dependent only on the way the mass is distributed.

EXAMPLE 10–1

Three mass points are located at rest as follows (Figure 10–3):

Mass	Coordinates (x, y)
$m_1 = 1$ kg	(2 m,0 m)
$m_2 = 2$ kg	(1m,2 m)
$m_3 = 3$ kg	(4 m,3 m)

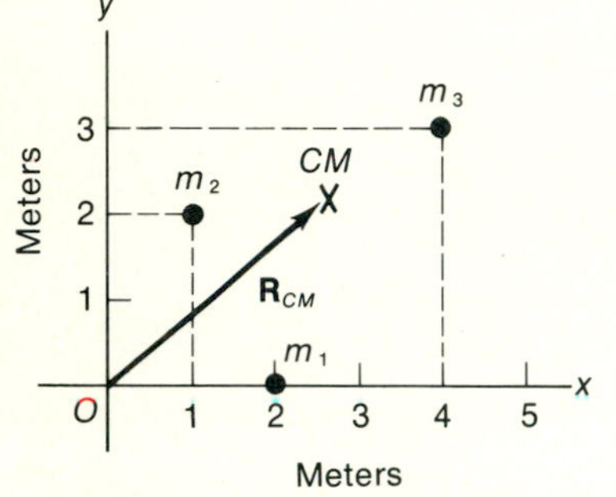

Figure 10–3
Example 10–1

Locate the center of mass.

SOLUTION

Since this problem involves point masses, we use Equation (10–7) in the delta notation form:

$$x_{CM} = \frac{\sum_{i=1}^{i=N} x_i \, \Delta m_i}{\sum_{i=1}^{i=N} \Delta m_i} = \frac{(x_1)(m_1) + (x_2)(m_2) + (x_3)(m_3)}{m_1 + m_2 + m_3}$$

[1] When g is uniform throughout the region, the CG and CM coincide. However, in a *nonuniform* gravitational field, they are not necessarily the same point for an extended object. As an example, consider a huge, uniform, "one-dimensional" rod that is supported at a single point at an angle in the earth's field (Figure 10–2). Since the acceleration due to gravity becomes less at greater distances from the earth, the forces of gravity on the upper half of the object are less than those on the lower half. Hence the CG will be located slightly downward from the geometrical center of the object. Since the location of the CG depends on the particular position and orientation of the object in the nonuniform field, the CG is not a fixed point in the object. However, for a uniform object the CM is the geometrical center and remains so under all conditions. Practically, this difference is negligible except in bodies of astronomical size. For problems in this text, you may assume the CG and CM coincide unless stated otherwise.

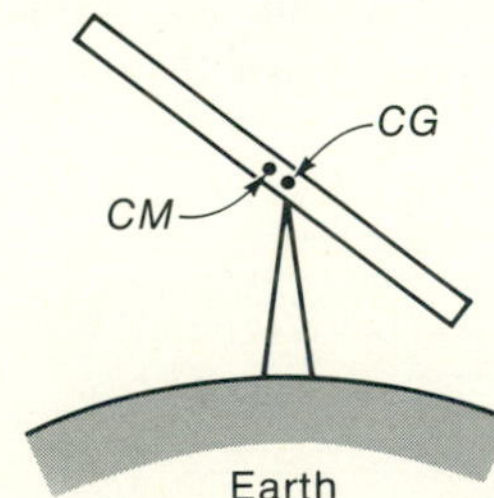

Figure 10–2
In a nonuniform gravitational field, the *CG* and *CM* do not necessarily coincide. The *CM* is at the geometrical center; the *CG* is slightly toward the lower end.

$$x_{CM} = \frac{(2\ \text{m})(1\ \text{kg}) + (1\ \text{m})(2\ \text{kg}) + (4\ \text{m})(3\ \text{kg})}{(1 + 2 + 3)\ \text{kg}} = 2.67\ \text{m}$$

and similarly;

$$y_{CM} = \frac{\sum_{i=1}^{i=N} y_i\,\Delta m_i}{\sum_{i=1}^{i=N} \Delta m_i} = \frac{(0\ \text{m})(1\ \text{kg}) + (2\ \text{m})(2\ \text{kg}) + (3\ \text{m})(3\ \text{kg})}{(1 + 2 + 3)\ \text{kg}} = 2.17\ \text{m}$$

So the center of the mass is located at the point (2.67 m,2.17 m)

EXAMPLE 10–2

Find the *CM* of the L-shaped figure cut from a thin, uniform sheet of material, shown in Figure 10–4(a).

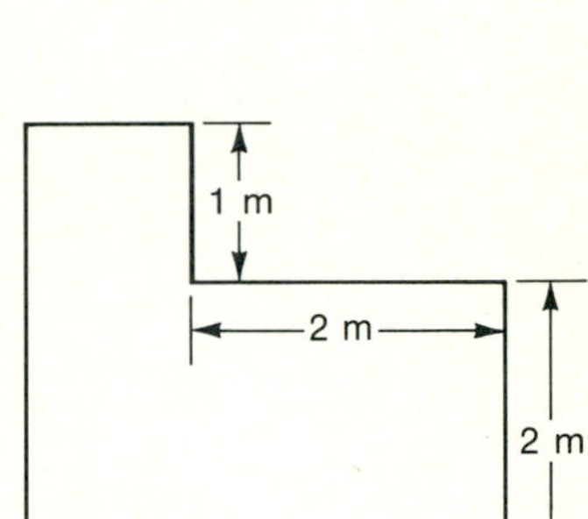

(a)

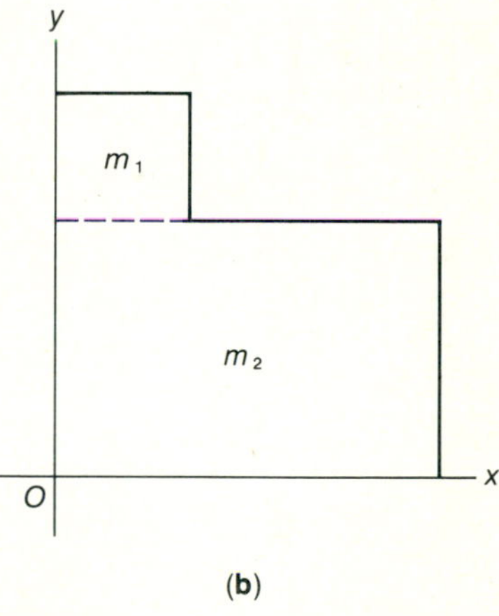

(b)

Figure 10–4
Example 10–2

SOLUTION

We simplify the analysis by imagining the object is divided into two symmetrical pieces, as shown in Figure 10–4(b). We know that the *CM* for each segment is at its geometrical center. The mass m of an area A of the sheet is: $m = (A)(\text{density})(\text{thickness})$. Thus:

	Mass of segment (area)(ρt)	Coordinates of *CM* (x, y)
m_1:	(1 m)(1 m)(ρt)	(0.50 m,2.5 m)
m_2:	(3 m)(2 m)(ρt)	(1.5 m,1 m)

Therefore:

$$x_{CM} = \frac{x_1 m_1 + x_2 m_2}{m_1 + m_2} = \frac{(0.50\ \text{m})(1\ \text{m})(1\ \text{m})(\rho t) + (1.5\ \text{m})(3\ \text{m})(2\ \text{m})(\rho t)}{(1\ \text{m})(1\ \text{m})(\rho t) + (3\ \text{m})(2\ \text{m})(\rho t)} = \frac{9.5\ \text{m}}{7.0} = 1.36\ \text{m}$$

$$y_{CM} = \frac{y_1 m_1 + y_2 m_2}{m_1 + m_2} = \frac{(2.5\ \text{m})(1\ \text{m})(1\ \text{m})(\rho t) + (1\ \text{m})(3\ \text{m})(2\ \text{m})(\rho t)}{(1\ \text{m})(1\ \text{m})(\rho t) + (3\ \text{m})(2\ \text{m})(\rho t)} = \frac{8.5\ \text{m}}{7.0} = 1.21\ \text{m}$$

The coordinates of the *CM* are thus (1.36 m,1.21 m)

EXAMPLE 10–3

The triangular figure illustrated in Figure 10–5 is cut from a uniform slab of material. Find its center of mass.

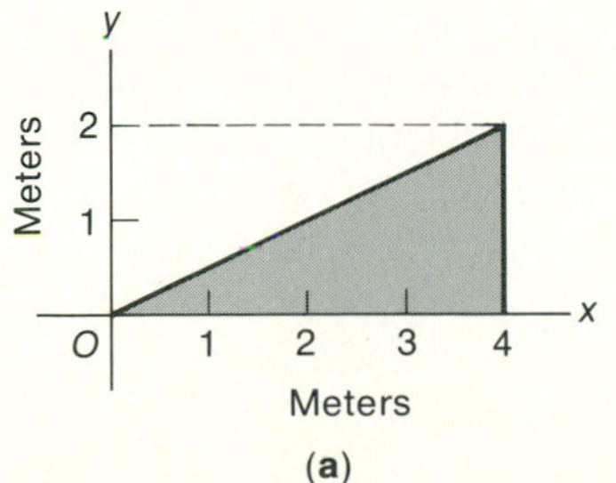

(a)

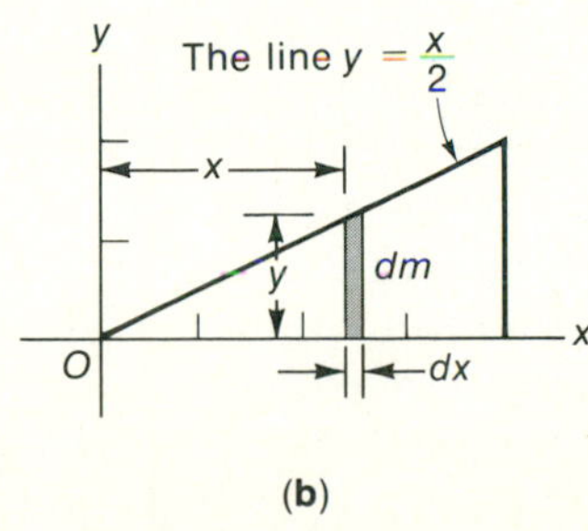

(b)

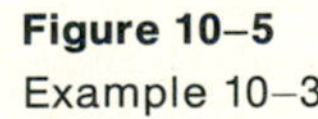
Figure 10–5
Example 10–3

SOLUTION

Because the material is uniform, its density (mass/volume) is a constant value ρ. We represent the thickness of the slab perpendicular to the plane of the figure by t. (Both parameters, ρ and t, will cancel in the analysis.) For this case, the mass is distributed throughout the object, so we must use integration to sum all the elements in the form $r\,dm$. So, for the coordinate x_{CM} we choose mass elements dm, *all of which are the same distance* x *from the origin*, as shown in Figure 10–5(b). Note that the height of the mass element dm is limited by the line $y = x/2$.

We must now express dm in a form involving a variable that will permit us to carry out the integration. *We always do this by noting that*

$$dm = \rho\,dV$$

where dV is the volume of dm in Figure 10–5(b).

$$dV = \text{(height)(thickness)(width)} = yt\,dx = \left(\frac{x}{2}\right)t\,dx$$

Therefore: $$dm = \rho t\left(\frac{x}{2}\right)dx$$

and we thus have expressed dm in terms of the variable dx.

$$x_{CM} = \frac{\int x\,dm}{\int dm} = \frac{\int_0^{4\text{ m}} x\rho t\left(\frac{x}{2}\right)dx}{\int_0^{4\text{ m}} \rho t\left(\frac{x}{2}\right)dx}$$

$$x_{CM} = \frac{\left(\frac{\rho t}{2}\right)\int_0^{4\text{ m}} x^2\,dx}{\left(\frac{\rho t}{2}\right)\int_0^{4\text{ m}} x\,dx} = \frac{\left(\frac{x^3}{3}\right)\Big|_0^{4\text{ m}}}{\left(\frac{x^2}{2}\right)\Big|_0^{4\text{ m}}} = \frac{2}{3}\,x\Big|_0^{4\text{ m}} = \frac{8}{3}\text{ m} = 2.67\text{ m}$$

The x coordinate is thus $\frac{2}{3}$ the distance along the x axis from the point of a right triangle toward the opposite side. In a similar fashion, we could find the y coordinate of the CM by choosing an element of mass dm and width dy, parallel to the x axis. A calculation similar to the one just performed leads to the result $y_{CM} = 0.667$ m. The x and y coordinates of the center of mass are thus

(2.67 m,0.667 m)

In general, the center of mass of any triangle of uniform mass density is one-third the height of the triangle from any side.

10.4 Equilibrium Positions of a Rigid Body

The location of the *CG* of an object is of crucial importance to the stability of the object. In general, a body suspended at a point will hang so that its *CG* comes to rest directly below the point of suspension. This provides a simple experimental method for locating the *CG* of an irregularly shaped object for which the mathematical calculation of Equation (10–7) would be difficult to perform. As illustrated in Figure 10–6, when suspended from an arbitrary point *A* the *CG* lies on a vertical line through *A*. Repeating the procedure from some other point *B* establishes a second line. The *CG* is thus the point of intersection of the two lines.

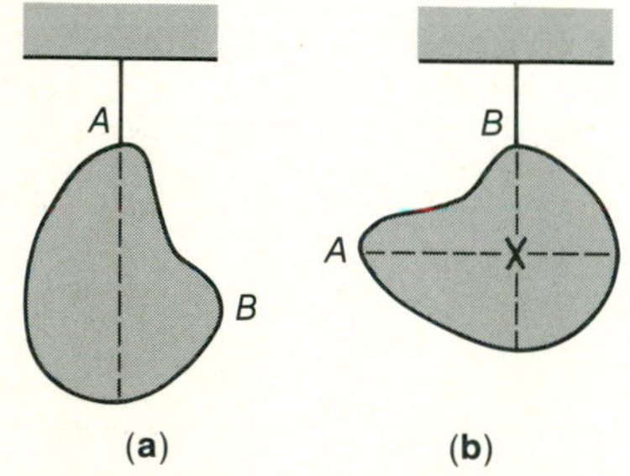

Figure 10–6
The *CG* of an irregularly shaped object may be located by suspending it from two arbitrary points. Vertical lines from the two points of suspension intersect at the *CG*.

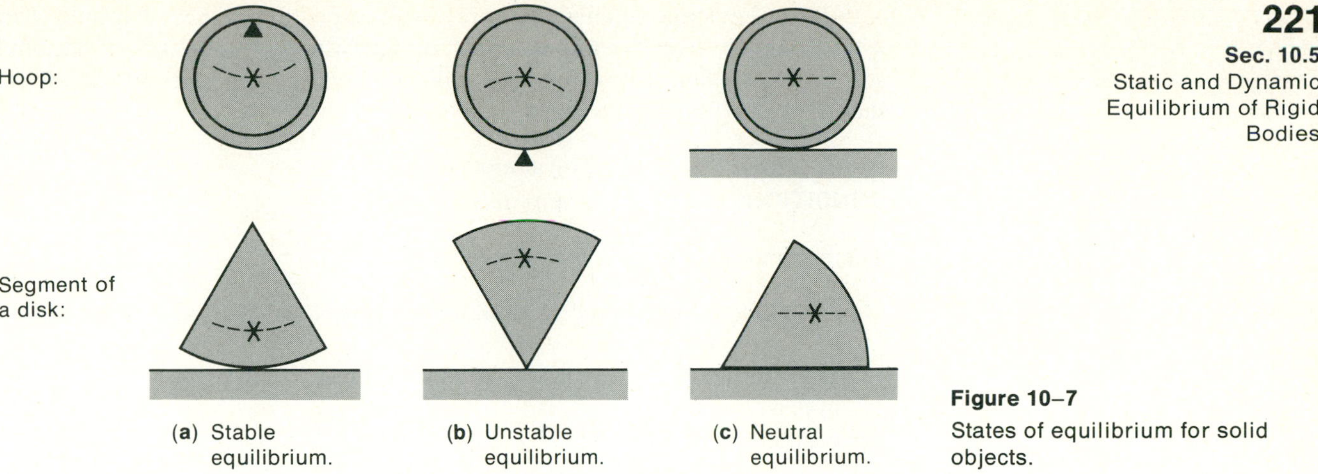

Figure 10–7
States of equilibrium for solid objects.

If an object rests on a knife edge or similar surface so that a slight displacement (or rotation) *raises* the *CG*, the object is said to be in **stable equilibrium.** If the *CG* is *lowered* by a small displacement or rotation, the object is then in **unstable equilibrium.** Finally, if the object may move without changing the height of its *CG*, the object is in **neutral equilibrium.** Figure 10–7 illustrates these three cases.

In all cases of equilibrium, the object has no tendency to rotate (though unstable equilibrium is admittedly a rather precarious state). This is because the lever arm is zero for the torque about the point of suspension due to the force *W* through the *CG*. Hence no torque is present to cause rotation.

10.5 Static and Dynamic Equilibrium of Rigid Bodies

A large area of concern to the engineer is the stability and equilibrium of structures. Engineers seek to understand the forces and torques of all structural members so that bridges do not fall, nor buildings collapse, nor a chair crumple under an overweight person. This branch of mechanics is called *statics*. It deals

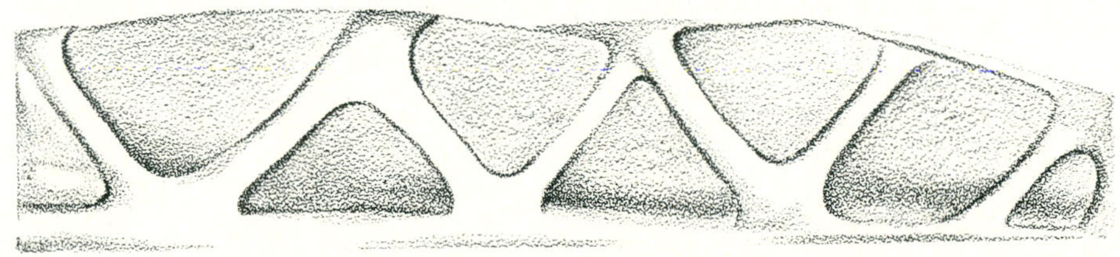

(a) The internal structure of a bone in a vulture's wing.

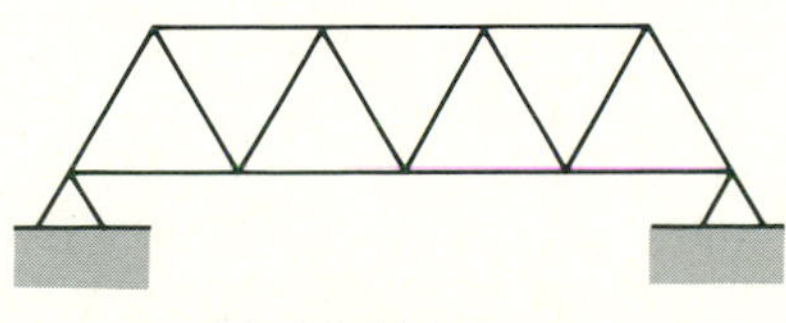

(b) A bridge truss.

Figure 10–8
Nature has evolved efficient designs for combining strength and lightness. All birds have many hollow bones with internal struts similar to those inside airplane wings or to the struts of a bridge truss.

with objects in **static equilibrium** (at rest with respect to an inertial frame) or in **dynamic equilibrium** (moving with constant velocity without linear or rotational accelerations). The central idea of equilibrium can be summarized succinctly: there is *no net force* and *no net torque.*

EQUILIBRIUM CONDITIONS

$$\left.\begin{array}{ll}\text{Translational:} & \Sigma\mathbf{F}_{\text{ext}} = 0 \\ \text{Rotational:} & \Sigma\tau_{\text{ext}} = 0\end{array}\right\} \qquad \textbf{(10–8)}$$

The method of analysis is a familiar one:

(a) Sketch a free-body diagram for the object showing all *external* forces acting on it.
(b) Apply the two conditions of equilibrium:

$$\text{First condition:} \quad \Sigma\mathbf{F}_{\text{ext}} = 0$$

$$\text{Second condition:} \quad \Sigma\tau_{\text{ext}} = 0 \qquad \text{(about } any \text{ point)}$$

Care must be taken to sketch all forces acting at their precise points of application, so that the correct moment-arm distances for calculating torques will be evident in the diagram.

If an object is in translational equilibrium ($\Sigma\mathbf{F} = 0$), the torques about all parallel axes have the same value. Thus we need to investigate the torque about only one axis to test for rotational equilibrium ($\Sigma\tau = 0$). Because, *in equilibrium situations*, we are free to choose *any* axis within (or outside) the object for calculating torques, we can often simplify the analysis by purposely choosing a point through which one or more *unknown* forces pass. Since their moment arms are then zero, these forces do not appear in the equation for the torques.

EXAMPLE 10–4

Two boys weighing 150 N and 200 N sit at opposite ends of a uniform 3-m plank weighing 250 N. (a) Find the fulcrum point at which the plank will balance. (b) Find the force the fulcrum exerts on the plank.

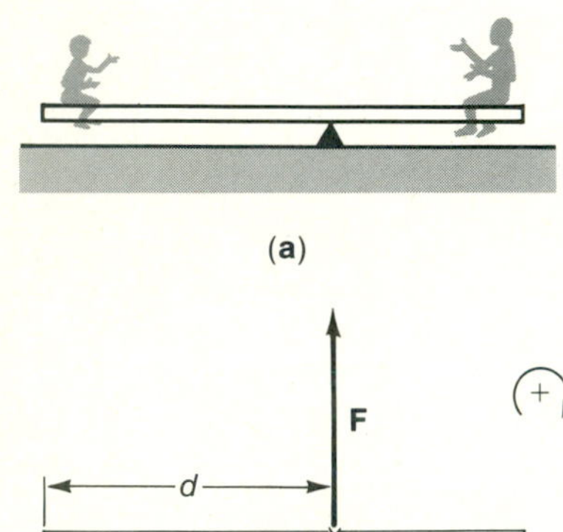

Figure 10–9
Example 10–4

SOLUTION

In Figure 10–9(b), we let **F** be the upward force of the fulcrum applied at the unknown distance d from the left end. We choose the point (☆) about which to calculate torques at the fulcrum point, through which the unknown force **F** acts. *Thus we do not need to include this unknown force in calculating torques about* ☆. Because the plank is uniform, we consider the force of gravity to be concentrated at its geometric center. All forces are vertical, and we indicate positive directions for x and y (with the origin at ☆).

(a) The second condition of equilibrium is

$$\underset{\text{(about ☆)}}{\Sigma\tau} = 0$$

Assuming *clockwise* torques are positive (and indicating this choice by (+↻)), we have

$$(200\text{ N})(3\text{ m} - d) - (250\text{ N})(d - 1.5\text{ m}) - (150\text{ N})(d) = 0$$

$$(975\text{ N·m}) - (600\text{ N})(d) = 0$$

$$\boxed{d = 1.63\text{ m}}$$

Note that in calculating torques, all moment-arm *distances* are taken as *positive* numbers. Then appropriate plus or minus signs are assigned to the torques to signify clockwise or counterclockwise directions, depending on which rotational sense was chosen as positive.

(b) To find the unknown force F, we apply the first condition of equilibrium.

$$\Sigma F_y = 0$$

$$(F - 150\text{ N} - 250\text{ N} - 200\text{ N}) = 0$$

$$F = \boxed{600\text{ N}}$$

EXAMPLE 10–5

In Figure 10–10, a horizontal beam supports a 200-N weight. The beam is attached to the wall with a *pin hinge*, which may exert a *force* on the beam in *any* direction but which cannot exert a *torque* about its own axis. If the weight of the beam is negligible, find (a) the tension **T** in the cable and (b) the force **F** the hinge exerts on the beam.

Figure 10–10
Example 10-5

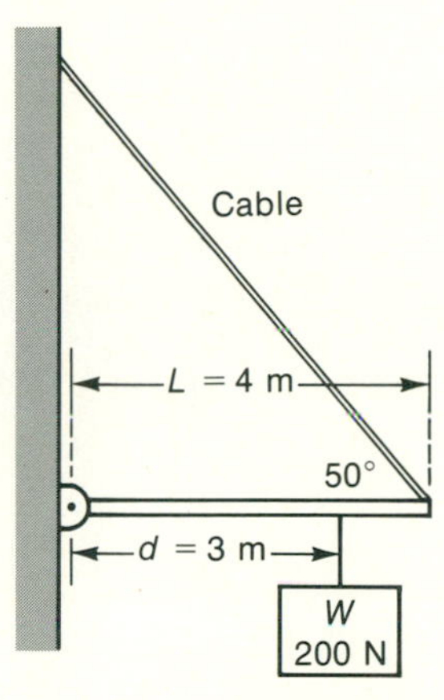

(**a**) A horizontal beam supports a 200-N load.

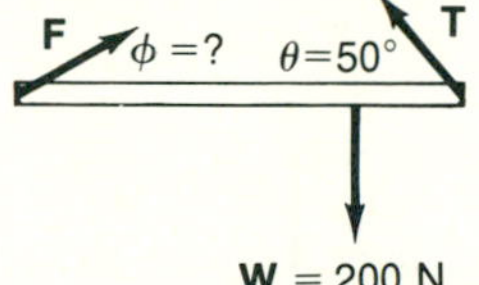

(**b**) A free-body diagram for the beam.

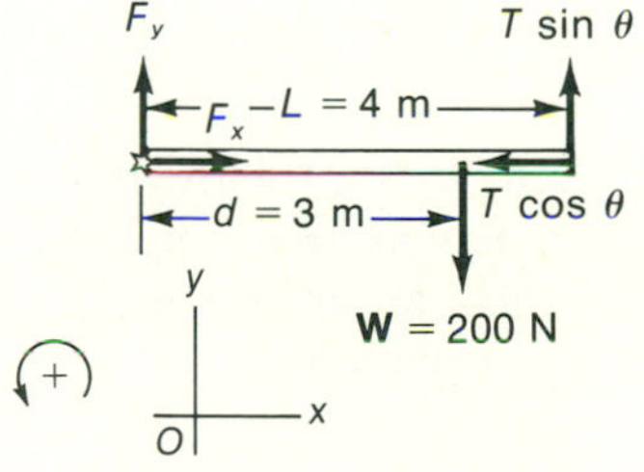

(**c**) A free-body diagram, with all forces resolved into two perpendicular directions.

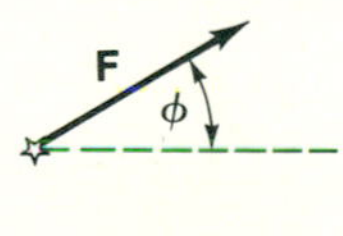

(**d**) The force the hinge exerts on the beam makes an angle ϕ with the horizontal.

SOLUTION

(a) We first sketch free-body diagrams for the beam, as shown in Figures 10–10(b) and (c). Since we know neither the magnitude nor the direction of the hinge force, it is easiest to solve for its horizontal and vertical components, F_x and F_y, separately. We can guess their directions: to counteract the horizontal component of **T**, F_x must be toward the *right*; to prevent the left end of the beam from falling, F_y must be *up*. (It is always best to make the diagram as correct as possible by doing such reasoning in advance. However, if an incorrect choice is made, the answer will come out negative for that force, signaling that the actual force is opposite to the assumed direction.)

Three unknown forces pass through the hinge, so we choose the hinge as the point for calculating torques. (Because the moment arms for these forces will be zero, they will not produce any torques about the hinge.) We show a star ☆ on the diagram to indicate this choice. We then choose counterclockwise rotations as positive, indicated on the diagram with the symbol ↺+. Also, positive directions for the x and y axes are shown. We now apply the second condition of equilibrium:

$$\underset{\text{(about the hinge)}}{\Sigma\tau} = 0$$

$$(T \sin\theta)(L) - (W)(d) = 0$$

$$T = \frac{(W)(d)}{(\sin\theta)(L)} = \frac{(200\ \text{N})(3\ \text{m})}{(\sin 50^\circ)(4\ \text{m})} = \boxed{196\ \text{N}}$$

(b) We next apply the first condition of equilibrium.

For the x *direction:*

$$\Sigma F_x = 0$$

$$F_x - T\cos\theta = 0$$

$$F_x = T\cos\theta = (196\ \text{N})(\cos 50^\circ) = 126\ \text{N}$$

For the y *direction:*

$$\Sigma F_y = 0$$

$$F_y - W + T\sin\theta = 0$$

$$F_y = W - T\sin\theta$$

$$F_y = 200\ \text{N} - (196\ \text{N})(\sin 50^\circ) = 49.9\ \text{N}$$

The magnitude of the hinge force is

$$F = \sqrt{F_x^2 + F_y^2} = \sqrt{(126)^2 + (49.9)^2}\ \text{N} = \boxed{136\ \text{N}}$$

The direction of the hinge force [Figure 10–10(d)] is:

$$\tan\phi = \frac{F_y}{F_x} = \frac{49.9\ \text{N}}{126\ \text{N}} = 0.396$$

$$\phi = \tan^{-1} = \boxed{21.6^\circ}$$

In these examples, it has been advantageous to start with $\Sigma\tau = 0$ before applying $\Sigma\mathbf{F} = 0$. Often this is the best approach in analyzing equilibrium situations, since the torque condition may furnish an answer directly rather than give an equation with two or more unknown symbols.

EXAMPLE 10–6

A ladder of length ℓ and negligible weight leans against a smooth wall at an angle of 60° with the horizontal. The coefficient of static friction at the bottom is 0.50. How far up the ladder (in terms of ℓ) can a person climb before the ladder slips?

SOLUTION

Figure 10–11(b) shows a free-body diagram with the force of the ground on the ladder shown as two components, F_x and F_y. (Note that the force the ground exerts is *not* necessarily along the direction of the ladder.) Since the wall is smooth, we may assume that friction is negligible; that is, $F_y = 0$ at the wall. Positive directions for rotations and for the x and y directions, as well as the point about which torques are to be calculated, are indicated by the appropriate symbols.

Figure 10–11
Example 10–6

Smooth wall

$\mu_s = 0.50$

(a)

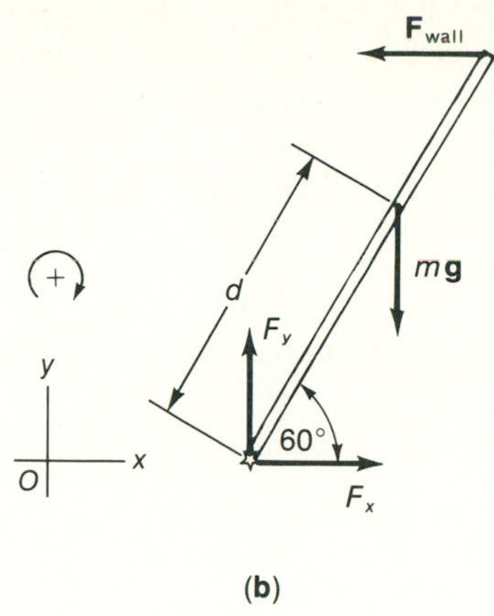

(b)

Applying the second condition of equilibrium, we have

$$\underset{\text{(about the base)}}{\Sigma\tau} = 0$$

$$[(mg)(d \cos \theta) - (F_{\text{wall}})(\ell \sin \theta)] = 0$$

$$d = \frac{(F_{\text{wall}})(\ell)(\tan \theta)}{mg}$$

Applying the first condition of equilibrium gives us

For the x *direction:*

$$\Sigma F_x = 0$$
$$(F_x - F_{\text{wall}}) = 0$$
$$F_x = F_{\text{wall}}$$

For the y *direction:*

$$\Sigma F_y = 0$$
$$(F_y - mg) = 0$$
$$F_y = mg$$

We also know that the friction force F_x on the ladder is

$$F_x = (\mu_s)(F_y) \qquad \text{(for } \textit{maximum} \text{ static friction)}$$

Combining the various equations we have obtained, we arrive at the following expression for d:

$$d = \frac{(\mu_s)(mg)(\ell \tan \theta)}{mg}$$

$$d = (\mu_s)(\ell \tan \theta) = (0.50)(\ell)(\tan 60°) = \boxed{0.866\,\ell}$$

EXAMPLE 10–7

Two uniform 100-N planks are joined with a hinge and placed in a vertical plane on a frictionless surface, as shown in Figure 10–12. A chain of negligible weight is fastened at the center of each plank so that it prevents the assembly from collapsing. Find (a) the tension in the chain and (b) the force of the hinge on each plank.

SOLUTION

We can obtain the upward force of the ground on the planks by drawing a free-body diagram for the object as a whole, as in Figure 10–12(b). In this case, the hinge and chain forces are wholly internal (inside the isolation boundary), so we may ignore them. From

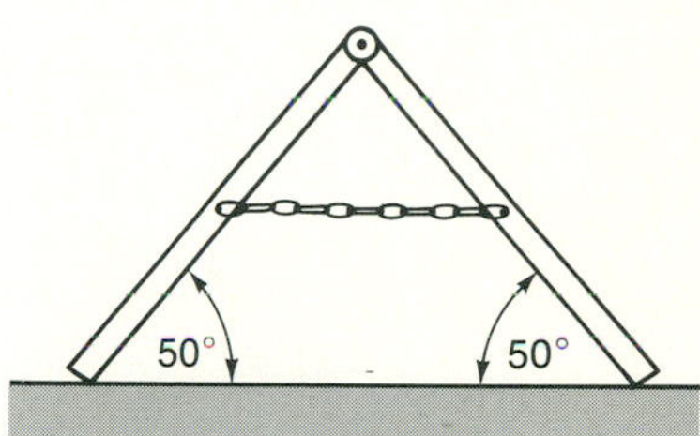

(a)

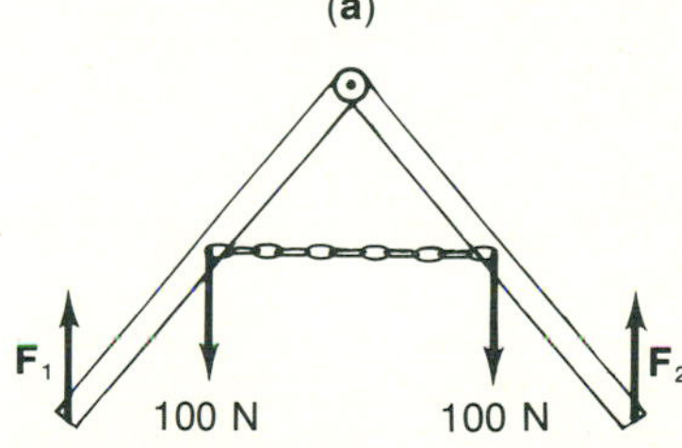

(b) A free-body diagram for the system as a whole.

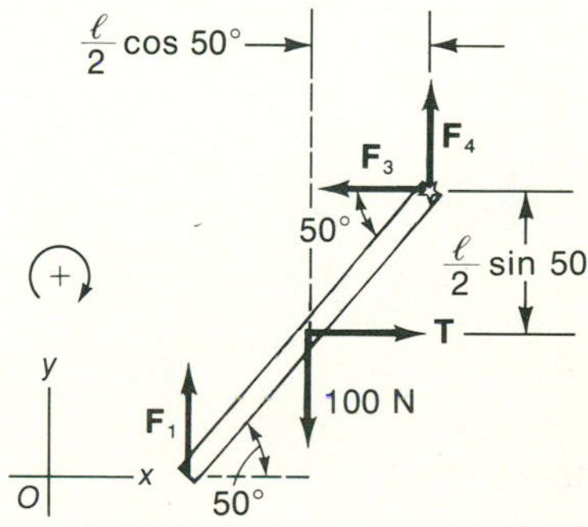

(c) A free-body diagram for the left plank.

Figure 10–12
Example 10-7

symmetry considerations and $\Sigma F_y = 0$, we conclude that

$$F_1 = F_2 = 100\text{ N}$$

In order to determine the chain force, we need to bisect the chain with an isolation boundary. (In objects with several parts, it is often necessary to isolate one or more subunits in order to make the desired forces appear in a force diagram.) So we next sketch a free-body diagram, Figure 10–12(c), for just one plank, designating the length of the plank by ℓ. Since we do not know the direction of the hinge force, it is shown with both x and y components, labeled F_3 and F_4. Positive directions for rotations and for the x and y axes are indicated. Applying the conditions of equilibrium, we have

$$\Sigma F_y = 0$$
$$100\text{ N} - 100\text{ N} + F_4 = 0$$
$$F_4 = 0$$

Similarly:

$$\Sigma F_x = 0$$
$$F_3 = T$$

Because unknown forces pass through the hinge, we choose that point (☆) for calculating torques. Moment-arm distances are shown on the sketch.

$$\underset{\text{(about the top)}}{\Sigma\tau} = 0$$

$$(100\text{ N})(\ell \cos 50^\circ) - (100\text{ N})\left(\frac{\ell}{2}\cos 50^\circ\right) - T\left(\frac{\ell}{2}\sin 50^\circ\right) = 0$$

$$100\text{ N} - 50\text{ N} - \frac{T}{2}\tan 50^\circ = 0$$

or

$$T = \frac{100\text{ N}}{\tan 50^\circ} = \frac{100\text{ N}}{1.1918} = \boxed{83.9\text{ N}}$$

From $F_3 = T$, we have for the hinge force:

$$F_3 = \boxed{83.9\text{ N (horizontal, toward the left)}}$$

10.6 Translational Motion of Large Objects

When a large object is thrown into the air, it may tumble end-over-end as it moves through space in a manner that seems quite complicated. However, as we will show, this complex motion can be described as a combination of *translational motion of the center of mass* and *rotational motion about the center of mass.* In solving problems, we always analyze these two types of motion separately.

First, let us consider the translational motion of the center of mass. A large, rigid object may be thought of as an assembly of particles, tied together by internal forces, perhaps with external forces on them as well. We may imagine these incremental mass particles to be as small as we wish. The only criterion is that they be small enough so that each obeys Newton's laws of motion for point masses. We shall let N represent the total number of such particles that comprise the entire object.

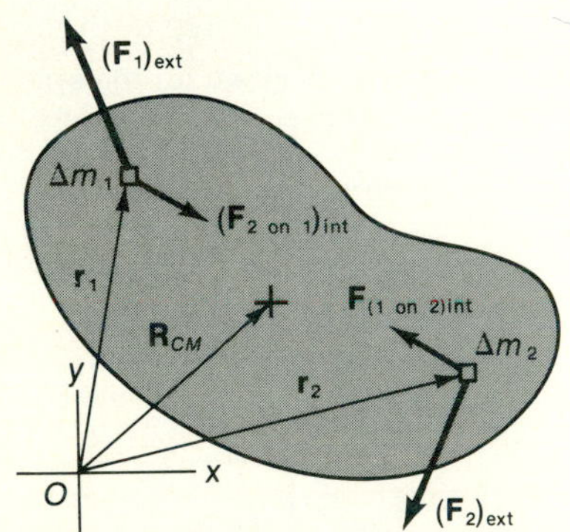

Figure 10–13
The two incremental masses Δm_1 and Δm_2 are part of the large object. They have both *internal* ($\mathbf{F}_{\text{int}}$) and *external* ($\mathbf{F}_{\text{ext}}$) forces acting on them.

Figure 10–13 shows a large object and two of the incremental masses, Δm_1 and Δm_2. The corresponding position vectors are $\mathbf{r}_1$ and $\mathbf{r}_2$, drawn from the origin O. Each individual mass obeys Newton's second law:

$$\left.\begin{aligned}\Sigma\mathbf{F}_1 &= \Delta m_1 \frac{d^2\mathbf{r}_1}{dt^2} \\ \Sigma\mathbf{F}_2 &= \Delta m_2 \frac{d^2\mathbf{r}_2}{dt^2}\end{aligned}\right\} \qquad \textbf{(10–9)}$$

where $\Sigma\mathbf{F}_1$ and $\Sigma\mathbf{F}_2$ are the net forces acting, respectively, on Δm_1 and Δm_2. Each of these forces arises from two possible sources:

(a) *Internal* forces $\Sigma\mathbf{F}_{\text{int}}$ due to mutual interactions between the particles making up the object.
(b) *External* forces $\Sigma\mathbf{F}_{\text{ext}}$ that arise from sources external to the object.

Thus:

$$\Sigma\mathbf{F}_1 = \Sigma(\mathbf{F}_1)_{\text{int}} + \Sigma(\mathbf{F}_1)_{\text{ext}} \qquad \text{and} \qquad \Sigma\mathbf{F}_2 = \Sigma(\mathbf{F}_2)_{\text{int}} + \Sigma(\mathbf{F}_2)_{\text{ext}} \qquad \textbf{(10–10)}$$

with similar equations for all other particles in the object.

Summing up all the terms of Equation (10–10) for the object as a whole, we have:

$$\Sigma\mathbf{F}_{\text{ext}} + \Sigma\mathbf{F}_{\text{int}} = \Sigma\left(\Delta m_i \frac{d^2\mathbf{r}_i}{dt^2}\right) \qquad \textbf{(10–11)}$$

The sum of the *external* forces on all the particles, $\Sigma\mathbf{F}_{\text{ext}}$, is just the net external force on the object as a whole. This is made up of the familiar forces of springs, strings, friction, gravity, and so on. All the *internal* forces are of the Newton's third law type, so they form *pairs* of forces, *each pair being equal and opposite force vectors.* Hence their total sum is zero:

$$\Sigma\mathbf{F}_{\text{int}} = 0$$

The summation on the right is to be carried out from $i = 1$ to $i = N$, over the entire object.

To simplify still further, recall that we have defined the center of mass $\mathbf{R}_{CM}$ of a system of particles to be:

$$\mathbf{R}_{CM} = \frac{\Sigma(\Delta m_i \mathbf{r}_i)}{\Sigma\Delta m_i}$$

The second derivative of this equation with respect to time is:

$$\frac{d^2\mathbf{R}_{CM}}{dt^2} = \frac{\Sigma\left(\Delta m_i \dfrac{d^2\mathbf{r}_i}{dt^2}\right)}{\Sigma\Delta m_i} \qquad \textbf{(10–12)}$$

Since $\Sigma\Delta m_i = M$, the total mass of the object, and $d^2\mathbf{R}_{CM}/dt^2 = \mathbf{a}_{CM}$, we combine Equations (10–11) and (10–12) to obtain:

NEWTON'S SECOND LAW FOR THE TRANSLATIONAL MOTION OF A LARGE OBJECT

$$\Sigma\mathbf{F}_{\text{ext}} = M\mathbf{a}_{CM} \qquad \textbf{(10–13)}$$

This equation has the same form as Newton's second law for a particle. It describes the important conclusion:

The translational motion of the center of mass of a large object of mass *M* is just that of a single particle of mass *M* moving in response to the external forces.

We can express this conclusion in the neat, Newton's second law form only because we made use of the concept of the center of mass of an object. So the idea of a *CM* (and a *CG*) is essential for applying the methods of Newton to the general motion of extended objects.

10.7 Rotational Motion of Large Objects

In the last section we showed that all *internal forces* of objects add to zero and therefore that the rate of change of *linear* momentum $d\mathbf{p}/dt$ of the *CM* of a solid object depends only on the net *external force.* By a similar argument, we will now show that all *torques due to internal forces* about any axis add to zero and therefore that the rate of change of *angular* momentum $d\mathbf{L}/dt$ depends only on the net *external torque.*

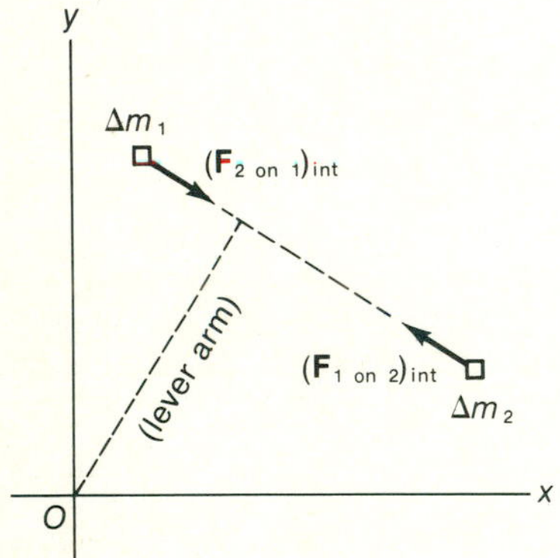

Figure 10-14
The two mass elements Δm_1 and Δm_2 have a mutual interaction that produces the Newton's third law pair of forces $(\mathbf{F}_{1 \text{ on } 2})_{\text{int}}$ and $(\mathbf{F}_{2 \text{ on } 1})_{\text{int}}$. These forces are equal in magnitude and act along the line joining the two particles. Thus they have the same lever-arm distance from the origin *O* of an inertial frame. Because they are in opposite directions, their net torque about *O* is zero.

Figure 10–14 shows two elemental masses Δm_1 and Δm_2, which are part of a larger object. The internal forces between these masses form a Newton's third law pair: equal in magnitude, opposite in direction, and acting along the same line (hence they have the same lever arm about *O*). We thus find that their net torque about *O* is zero. Choosing counterclockwise torques as positive, we have:

$$\underset{(\text{about } O)}{\Sigma\tau} = (\mathbf{F}_{2 \text{ on } 1})_{\text{int}}(\text{lever arm}) - (\mathbf{F}_{1 \text{ on } 2})_{\text{int}}(\text{lever arm})$$

$$\underset{(\text{about } O)}{\Sigma\tau} = (\text{F}_{2 \text{ on } 1} - \text{F}_{1 \text{ on } 2})_{\text{int}}(\text{lever arm}) = 0$$

The same reasoning may be extended to all particles in the system. *All the internal forces come in pairs whose torques about* O *add to zero.* Thus, we need only be concerned with *external* torques on the object. If the axis *O* about which the torques are taken is the center of mass of the object, the net external torque is related to the time rate of change in angular momentum in a very simple way. If the vector sum of all the individual angular momenta for all the elemental masses about the center of mass is defined by

$$\mathbf{L} \equiv \sum_{i=1}^{i=n} \mathbf{L}_i \tag{10–14}$$

then we state without proof

NEWTON'S SECOND LAW FOR THE ROTATIONAL MOTION OF A LARGE OBJECT

$$\underset{(\text{about the } CM)}{\Sigma\tau_{\text{ext}}} = \frac{d\mathbf{L}}{dt} \tag{10–15}$$

This result may be stated in words as

The net external torque on an object (about an axis through the *CM*) equals the time rate of change of the total angular momentum L about the *CM*.

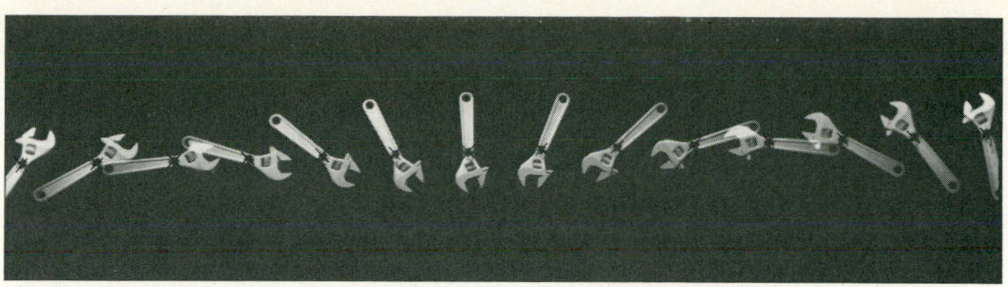

(a) A wrench sliding across a horizontal frictionless surface. Since there is zero net force on the wrench, its *CM* (indicated by the cross) travels in a straight line at constant speed as the wrench spins around.

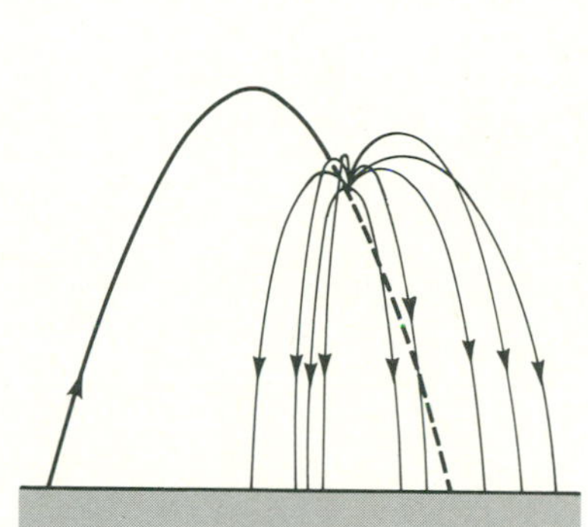

(b) If a mortar shell explodes in flight, the *CM* of the falling fragments continues in the same parabolic trajectory it would have had if the explosion had not occurred.

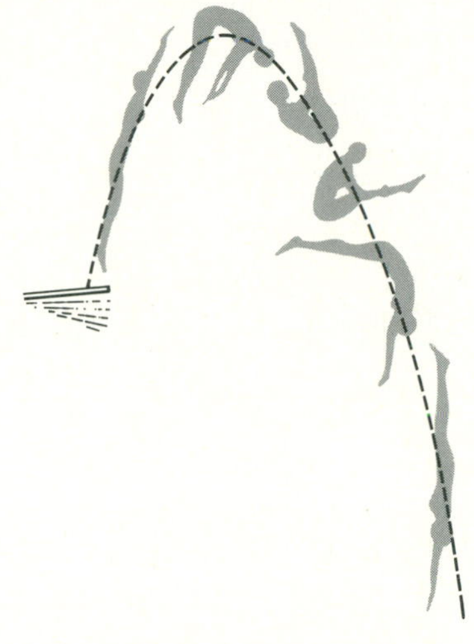

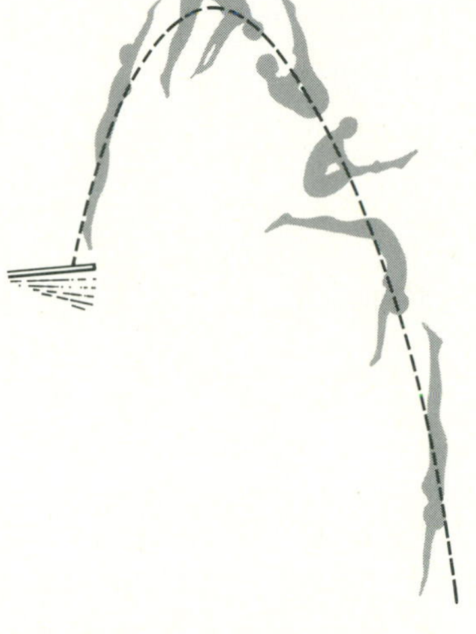

(c) After leaving the diving board, a diver cannot alter the parabolic path of her *CM*.

Figure 10–15
Internal forces cannot alter the motion of the *CM* of a system of particles. The *CM* always moves under the action of the net *external* force only. Thus the *CM* of a freely falling system always has a parabolic trajectory (assuming uniform g and ignoring air friction).

Equations (10–14) and (10–15) are valid for any object or isolated system of particles. Nowhere in the argument was the *rigidity* of the object an essential feature; the important point was that Newton's third law forces were the only ones involved. Therefore, the conclusions also apply to nonrigid systems, such as a globular cluster of stars or a wriggling cat falling through the air, trying to land on its feet. For example, the explosion of a mortar shell in flight [Figure 10–15(b)] does not alter the parabolic path of the *CM* of the entire system of particles as they fall to the ground along various trajectories. Because the forces that blow apart the fragments of the shell are entirely *internal* forces, the only external force on the system is gravity. Hence the *CM* follows the same free-fall trajectory it would have taken had the explosion not occurred. Similarly, once a diver leaves the diving board she cannot change the parabolic path of her *CM*, regardless of how she moves her arms and legs [Figure 10–15(c)].

EXAMPLE 10–8

During the construction of a space platform in a region where gravity is negligible, two astronauts exert equal forces **F** in opposite directions on a long uniform beam, as shown in Figure 10–16(a). Find the net force and the net torque about the center of mass of the beam.

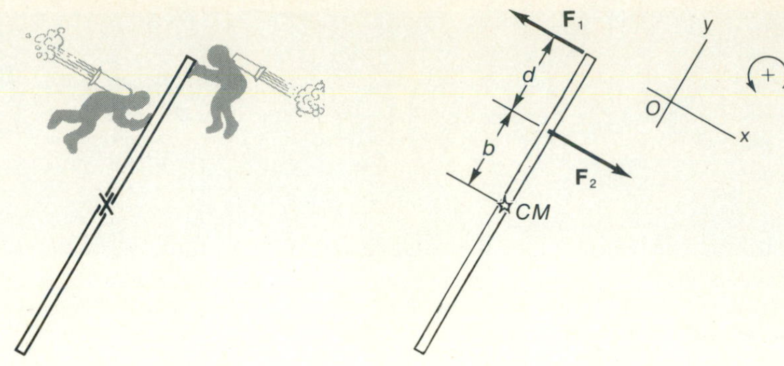

Figure 10–16
Example 10–8

(**a**) The astronauts exert forces perpendicular to the beam.

(**b**) A free-body diagram for the beam.

SOLUTION

Choosing the $+x$ direction as in Figure 10–16(b), the net force is:

$$\Sigma \mathbf{F}_x = \mathbf{F}_2 - \mathbf{F}_1 = 0 \qquad \boxed{\text{zero net force}}$$

Taking torques about the *CM* (assuming counterclockwise torques are positive), we have

$$\underset{\text{(about the } CM)}{\Sigma \tau} = F(b + d) - F(b) = \boxed{Fd}$$

Since there is no net force, there is no translational motion. But because of the net torque *Fd*, the beam will increase its angular momentum (counterclockwise) about the *CM*. If the forces continue to be applied at right angles to the beam as it moves, the beam will continue to have rotational acceleration with its *CM* remaining at rest in an inertial frame.

The forces $\mathbf{F}_1$ and $\mathbf{F}_2$ illustrate what is called a **couple**: *two forces equal in magnitude, opposite in direction, and not collinear.* The quantity *Fd* is the value of the couple. Note that it does not contain any reference to the distance *b* from the *CM*. Hence it follows that the torque of the couple is the same for any arbitrary point about which the torque may be calculated. Because the sum of the two forces is zero in every direction, they cannot be combined into a single force. They exert only a *torque Fd* on an object, not a net force.

10.8 Rotation of a Rigid Body About a Fixed Axis of Symmetry

A common situation in physics and engineering involves a symmetrical rigid body rotating about a fixed axis that passes symmetrically through its center of mass. Common examples include all types of rotating machinery. We will analyze several examples of this special case, confining our discussion to rigid objects that have *physical symmetry* about the axis of rotation. The criterion for a physical symmetry axis is that for every mass element Δm on one side of the axis, there is a similarly located equal mass element Δm on the diametrically opposite side of the axis.

Although there are no truly rigid substances in nature, for most cases we may assume that any distortions or deformations (which might alter the relative positions of the various atoms in the object) are negligibly small. So all parts of the body maintain fixed relative positions with respect to one another, even though we apply torques and forces to the system.

10.9 Moment of Inertia of a Solid Object

Let us now take up the calculation of the kinetic energy associated with rotational motion. As usual, we imagine the entire mass of the body to be made up of a collection of particles so small that we consider the mass of a given particle Δm_i to be concentrated at a single point r_i radially away from the axis of rotation (see Figure 10–17). Note that r_i is the *radius of the circular path*, not the position vector from the origin. The kinetic energy of such a particle is

$$K_i = \tfrac{1}{2}\Delta m_i v_i^2 \qquad \textbf{(10–16)}$$

Since the particle travels in a circle of radius r_i and has a linear speed $v_i = r_i\omega$ (where ω is the angular speed of rotation), we may also express the kinetic energy of the particle as

$$K_i = \tfrac{1}{2}\Delta m_i r_i^2 \omega^2 \qquad \textbf{(10–17)}$$

The total rotational kinetic energy K for the body as a whole is the sum of the kinetic energy of all the particles:

$$\Sigma K_i = \tfrac{1}{2}(\Delta m_1 r_1^2 + \Delta m_2 r_2^2 + \cdots)\omega^2$$

$$K = \tfrac{1}{2}(\Sigma \Delta m_i r_i^2)\omega^2 \qquad \textbf{(10–18)}$$

The quantity in parenthesis is given a special name. It is called the **moment of inertia** I of the body with respect to the given axis of rotation.

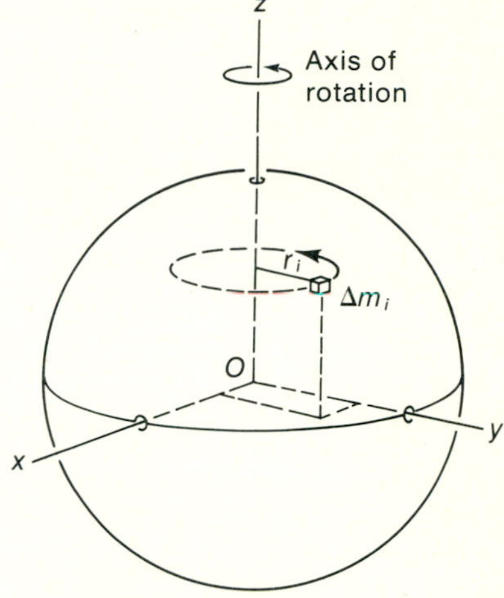

Figure 10–17
The element Δm_i located somewhere within the solid sphere. As the sphere rotates, the mass element moves in a circle of radius r_i.

MOMENT OF INERTIA I FOR A COLLECTION OF MASS POINTS

$$I = \Sigma \Delta m_i r_i^2 \qquad \textbf{(10–19)}$$

Note that I depends on the particular axis of rotation and on the way the mass of the body is distributed about the axis. Moment of inertia is measured in units of kilogram-meters^2 or slug-feet^2.

Using the symbol I, we may now express the rotational kinetic energy as:

$$K = \tfrac{1}{2}I\omega^2 \qquad \textbf{(10–20)}$$

As with all angular quantities, there is a close analogy with the linear case: $K = \tfrac{1}{2}mv^2$. The *moment of inertia* in angular motion corresponds to *mass* in linear motion; similarly, the *angular velocity* ω corresponds to *linear velocity* v.

EXAMPLE 10–9

Four equal point masses, each of mass m, are held at the corners of a square of side ℓ by rigid rods of negligible mass, as sketched in Figure 10–18. Find the moment of inertia of the object for rotation about an axis through the CM and perpendicular to the plane of the object.

SOLUTION

By symmetry, the CM is located at the center of the square. Thus, each mass is the same distance r from the CM. From the Pythagorean Theorem, we have

$$r^2 + r^2 = \ell^2$$

$$r^2 = \frac{\ell^2}{2}$$

Thus, the moment of inertia is

$$I = \Sigma \Delta m_i r_i^2 = 4m\frac{\ell^2}{2} = \boxed{2m\ell^2}$$

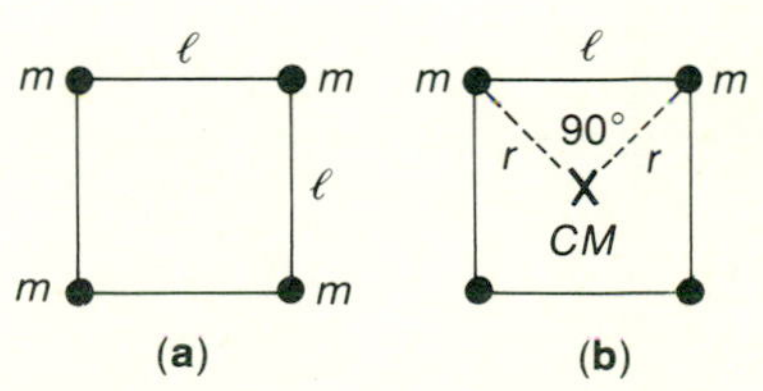

Figure 10–18
Example 10–9

10.10 Calculation of Moments of Inertia

We shall next take up the calculation of the moment of inertia I for bodies that are made up of a continuous distribution of matter of uniform density. For these cases we must use the ideas of calculus. The clue to each calculation is a judicious choice of an element of mass dm, such that *all of it resides at the same distance* r *from the axis of rotation.* Then Equation (10–19) may be written in integral form:

MOMENT OF INERTIA I FOR A SOLID BODY

$$I = \int r^2\, dm \qquad \text{(10–21)}$$

(integration to be taken over the entire body)

For the cases we shall consider, the body will have some feature of symmetry about the axis of rotation, which we will utilize in choosing the mass element dm. The following example illustrates the overall procedure for calculating the moment of inertia.

EXAMPLE 10–10

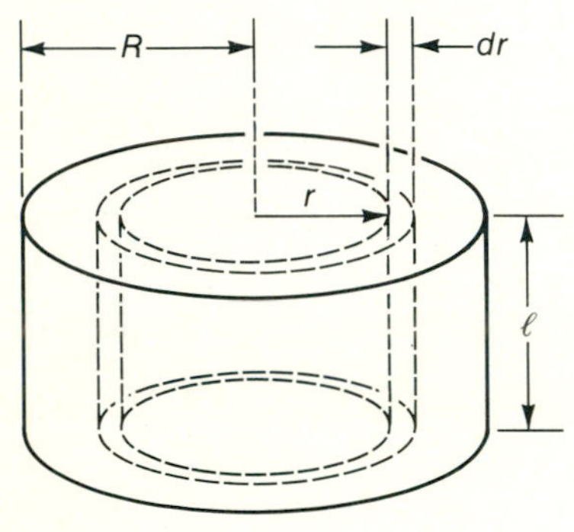

Figure 10–19
Example 10–10

Find the moment of inertia about the axis for a solid cylinder of uniform density ρ, length ℓ, radius R, and total mass M (see Figure 10–19).

SOLUTION

We choose the element of mass dm to be the cylindrical shell of radius r, thickness dr (assumed to be negligible), and length ℓ. Since dr has negligible thickness, all the mass of the shell is essentially at the same distance r from the axis of rotation. *This is the necessary criterion for the mass element we choose in each case.*

The mass dm of this shell is found by recognizing that

$$dm = \rho\, dV \qquad \text{(10–22)}$$

where ρ is the density of the substance (mass/volume) and dV is the volume element. From geometrical considerations, we have

$$dV = (2\pi r\, dr)\ell \qquad \text{(10–23)}$$

Combining these equations, we obtain

$$dm = 2\pi r \rho \ell\, dr \qquad \text{(10–24)}$$

Thus, by these steps we convert the variable of integration from dm (which is not defined specifically at first) into dr, which is a familiar parameter that varies from zero to the radius R of the cylinder. The moment of inertia I is thus:

$$I = \int r^2\, dm = 2\pi\ell \int_0^R \rho r^3\, dr \qquad \text{(10–25)}$$

Since, for this case, the density ρ is uniform throughout the cylinder, we may remove it from the integral sign. Integrating the expression (see Appendix G-II and G-III) gives

$$I = 2\pi\ell\rho \int_0^R r^3\, dr = \frac{2\pi\ell\rho r^4}{4}\bigg|_0^R = \frac{2\pi\ell\rho R^4}{4} \qquad \text{(10–26)}$$

We make one further simplification in the answer by noting that the total mass M of the cylinder may be expressed as:

$$M = \text{(density)(volume)} = (\rho)(\pi R^2 \ell)$$

Factoring out this value for M, we have

$$\boxed{I = \tfrac{1}{2}MR^2} \quad \text{(uniform cylinder about its axis)} \qquad \text{(10–27)}$$

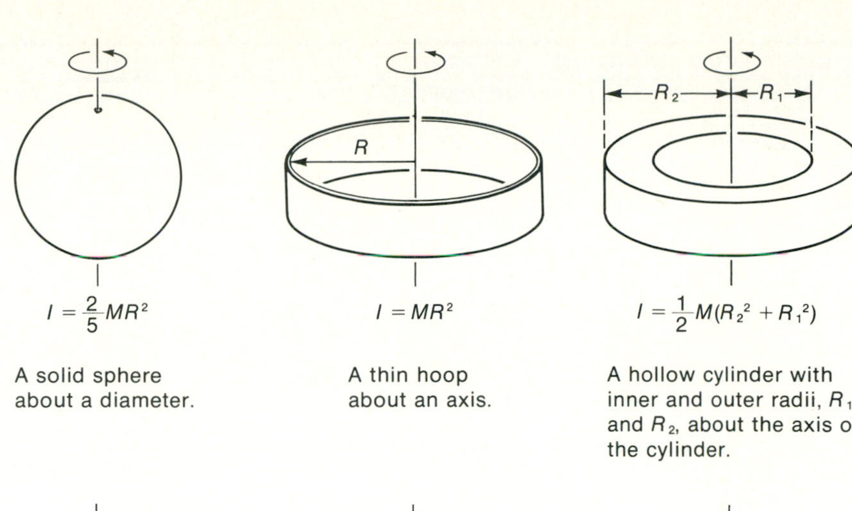

$I = \frac{2}{5}MR^2$

A solid sphere about a diameter.

$I = MR^2$

A thin hoop about an axis.

$I = \frac{1}{2}M(R_2^2 + R_1^2)$

A hollow cylinder with inner and outer radii, R_1 and R_2, about the axis of the cylinder.

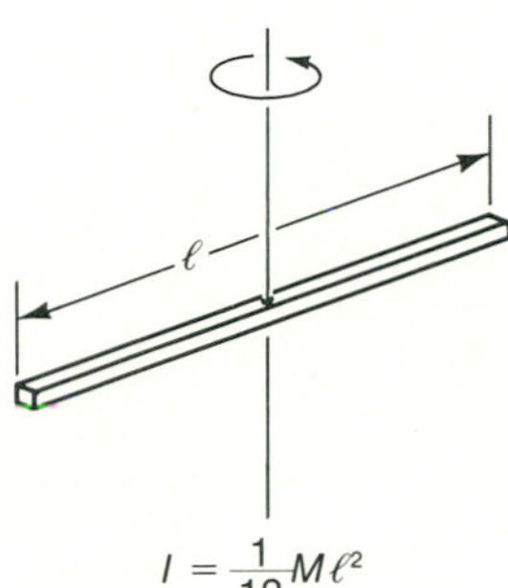

$I = \frac{1}{12}M\ell^2$

A slender rod of total length ℓ. The axis is through the center and perpendicular to the rod.

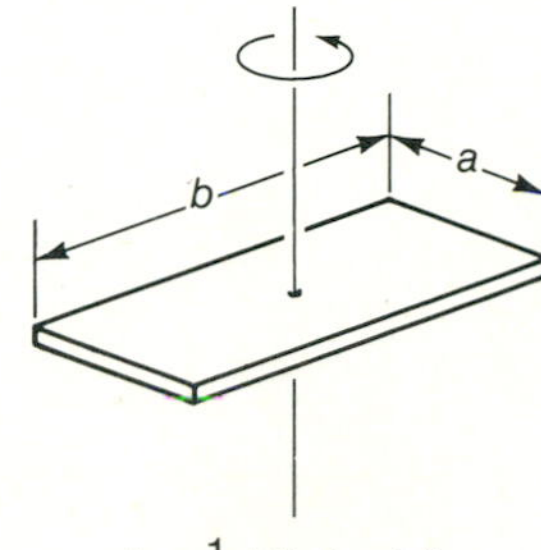

$I = \frac{1}{12}M(a^2 + b^2)$

A thin rectangular plate. The axis is through the center and perpendicular to the plate.

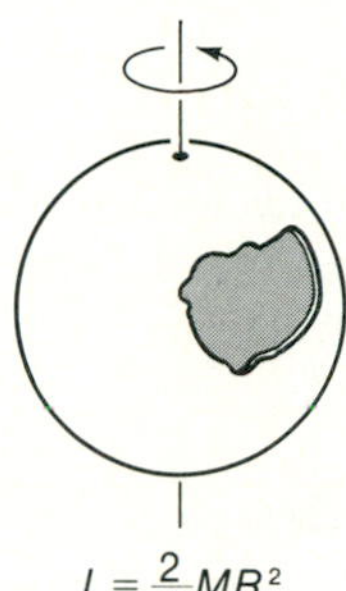

$I = \frac{2}{3}MR^2$

A thin spherical shell about a diameter.

Figure 10–20
Moments of inertia for various symmetrical bodies.

Some common moments of inertia for various symmetrical objects are given in Figure 10–20. In each case, the axis of rotation passes symmetrically through the center of the mass and the object is made of a material of uniform density. Note that each result is in the form of the product of a *mass* times the *square of a distance.*

10.11 Radius of Gyration

Many situations of practical interest involve the rotation of objects that are physically too complicated to permit a calculation of the moment of inertia. A space satellite provides a good example. Nevertheless, we do need to define a moment of inertia for use in calculations. The procedure is as follows. The mass of the object and the value of I are *experimentally* determined. (For example, a known torque is applied and the resulting angular acceleration α is measured.) The value of I thus determined can then be expressed as the product of the mass of the object times the square of a distance called the **radius of gyration** k. The radius of gyration may be interpreted as the distance from the axis of rotation at which the entire mass may be concentrated for the purpose of calculating the moment of inertia. (Note that the location where we consider the mass to be concentrated is *not* the *CM* of the object.)

$$I = Mk^2 \qquad \textbf{(10–28)}$$

where k is the radius of gyration.

EXAMPLE 10–11

Find the radius of gyration k for a solid sphere of mass M and radius R about an axis through its CM.

SOLUTION

The moment of inertia of a solid sphere about an axis through its CM is $I = \frac{2}{5}MR^2$. Therefore:

$$Mk^2 = \frac{2}{5}MR^2$$

$$\boxed{k = \sqrt{\tfrac{2}{5}}R}$$

10.12 Parallel-Axis Theorem

Another useful relation, the **parallel-axis theorem,** gives the moment of inertia *about any axis parallel to an axis through the* CM if its moment of inertia about the *CM* is known. The theorem is:

$$I = I_{CM} + Md^2 \tag{10–29}$$

where I is the moment of inertia about an axis parallel to the axis through the center of mass and I_{CM} is the moment of inertia about the axis that passes symmetrically through the center of mass. The perpendicular distance between the two parallel axes is d. The relation is perhaps better understood if you think of the motion as *rotation about the* CM plus *translation of the* CM *point in a circle about the new axis.* The extra term is thus the added contribution of treating the object as a *point* mass concentrated at the *CM* and moving in a circle about the new axis.

EXAMPLE 10–12

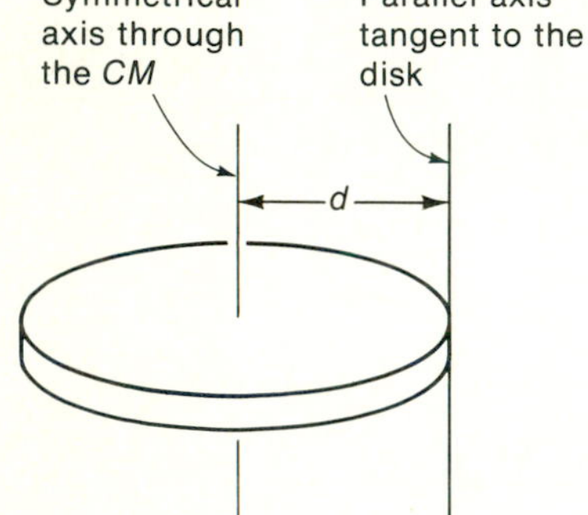

Figure 10–21
Example 10–12

Find the moment of inertia of a circular disk about an axis through its edge and perpendicular to the plane of the disk (see Figure 10–21).

SOLUTION

The moment of inertia about an axis through the CM and perpendicular to the plane of the disk is $\frac{1}{2}MR^2$. We thus apply the parallel-axis theorem to obtain:

$$I = I_{CM} + Md^2 = \tfrac{1}{2}MR^2 + MR^2 = \boxed{\tfrac{3}{2}MR^2}$$

10.13 Kinematic Equations for Rotational Motion

Before discussing the dynamics of rigid bodies we need to obtain a convenient set of *kinematic equations for rotational motion.* Recall that in describing one-dimensional motion, we used the linear parameters of *position x, velocity v,* and *acceleration a.* And we developed the convenient kinematic equations [(2–22),

(2–23), and (2–24)] for describing linear motion with constant acceleration. We shall now develop a similar set of kinematic equations for rotational motion using the rotational parameters of *angular position* θ, *angular velocity* ω, and *angular acceleration* α.

We start with the definition of angular acceleration, $\alpha = d\omega/dt$, and rearrange the terms:

$$\alpha\, dt = d\omega$$

Keeping the angular acceleration α constant, we integrate both sides of the equation, bringing the constant α out from under the integral sign.

$$\alpha \int dt = \int d\omega$$

To set limits on the integration, we start at time $t_0 = 0$ and allow for the possibility of some initial velocity ω_0. As time proceeds to some later value t, the angular velocity changes to the later value ω.

$$\alpha \int_0^t dt = \int_{\omega_0}^{\omega} d\omega$$

Carrying out the integration (see Appendix G-II and G-III), we evaluate the result at the upper limit and subtract the value at the lower limit:

$$\alpha t \Big|_0^t = \omega \Big|_{\omega_0}^{\omega}$$

$$\alpha(t - 0) = (\omega - \omega_0)$$

Rearranging, we obtain the *first kinematic equation* for rotational motion.

$$\omega = \omega_0 + \alpha t \qquad \text{(only for constant } \alpha\text{)} \qquad \textbf{(10–30)}$$

To obtain another equation, we substitute $\omega = d\theta/dt$, rearrange the terms, and integrate. Note that ω_0 is a constant.

$$\frac{d\theta}{dt} = \omega_0 + \alpha t$$

$$\int_{\theta_0}^{\theta} d\theta = \int_0^t (\omega_0 + \alpha t)\, dt$$

where θ_0 is an initial angular position. When we evaluate the result at the upper limit and subtract the value at the lower limit, we obtain

$$\theta \Big|_{\theta_0}^{\theta} = \left(\omega_0 t + \alpha \frac{t^2}{2}\right)\Bigg|_0^t$$

$$(\theta - \theta_0) = \omega_0(t - 0) + \alpha \frac{(t^2 - 0^2)}{2}$$

Rearranging, we obtain the *second kinematic equation* for rotational motion.

$$\theta = \theta_0 + \omega_0 t + \tfrac{1}{2}\alpha t^2 \qquad \text{(only for constant } \alpha\text{)} \qquad \textbf{(10–31)}$$

Just as we did with the linear kinematic equations, we eliminate t between Equations (10–30) and (10–31) to obtain the *third kinematic equation* for rotational motion.

$$\omega^2 = \omega_0{}^2 + 2\alpha(\theta - \theta_0) \qquad \text{(only for constant } \alpha\text{)} \qquad \textbf{(10–32)}$$

These three[2] equations and the analogies between linear motion and rotational motion are so important that they are summarized below. Particularly note the close ties between the tangential motion of a particle as described by the tangential parameters of *arc length s*, *velocity v* (which is always tangent to the path), and the *tangential acceleration component* a_t, and the corresponding angular quantities of θ, ω, and α.

Tangential quantities		Angular quantities		Relations between tangential and angular quantities
Tangential displacement	s	Angular displacement	θ	$s = r\theta$
(Tangential) velocity	$v = \dfrac{ds}{dt}$	Angular velocity	$\omega = \dfrac{d\theta}{dt}$	$v = r\omega$
Tangential acceleration	$a_t = \dfrac{dv}{dt}$	Angular acceleration	$\alpha = \dfrac{d\omega}{dt}$	$a_t = r\alpha$
				NOTE: θ, ω, and α must be in *radian* measure

If you keep these relationships in mind, the angular kinematic equations are easier to remember, since they closely resemble the form of the linear kinematic equations.

The Kinematic Equations for Rotational Motion

Tangential motion along the circumference		Rotational motion		
$v = v_0 + a_t t$	(only for constant a_t)	$\omega = \omega_0 + \alpha t$	(only for constant α)	**(10–33)**
$s = s_0 + v_0 t + \frac{1}{2} a_t t^2$		$\theta = \theta_0 + \omega_0 t + \frac{1}{2}\alpha t^2$		**(10–34)**
$v^2 = v_0^2 + 2a_t(s - s_0)$		$\omega^2 = \omega_0^2 + 2\alpha(\theta - \theta_0)$		**(10–35)**

EXAMPLE 10–13

A wheel, 4 ft in diameter, initially at rest, begins to turn with a constant angular acceleration of 0.6 rad/s^2. Consider a point on the rim with an initial angular displacement of $\theta_0 = 0°$ at $t_0 = 0$.

---

[2] As in the linear case, sometimes a fourth kinematic equation is listed:

$$\theta = \theta_0 + \left(\frac{\omega + \omega_0}{2}\right)t \qquad \text{(only for constant } \alpha\text{)}$$

However, every solvable kinematics problem can be worked out using just the three kinematic equations given.

(a) Find the angular speed of the wheel 2 s after it begins to turn.
(b) Calculate the total distance s the point travels during the 2-s interval after the wheel begins to turn.
(c) Determine the location of the point at $t = 2$ s.
(d) Find the acceleration of the point at $t = 2$ s.

SOLUTION

Tabulating known values in consistent units, we have

$$r = 2 \text{ ft}$$

$$\alpha = 0.6 \frac{\text{rad}}{\text{s}^2}$$

$$t = 2 \text{ s}$$

$$\omega_0 = 0$$

$$\omega = ?$$

(a) Examining the kinematic equations, we choose Equation (10–33) as appropriate for this question.

$$\omega = \omega_0 + \alpha t$$

$$\omega = 0 + \left(0.6 \frac{\text{rad}}{\text{s}^2}\right)(2 \text{ s}) = \boxed{1.20 \frac{\text{rad}}{\text{s}}}$$

(b) If we knew the angular displacement θ during the first 2 s, we could find the tangential displacement s along the circumference from $s = r\theta$ [see Figure 10–22(a)].

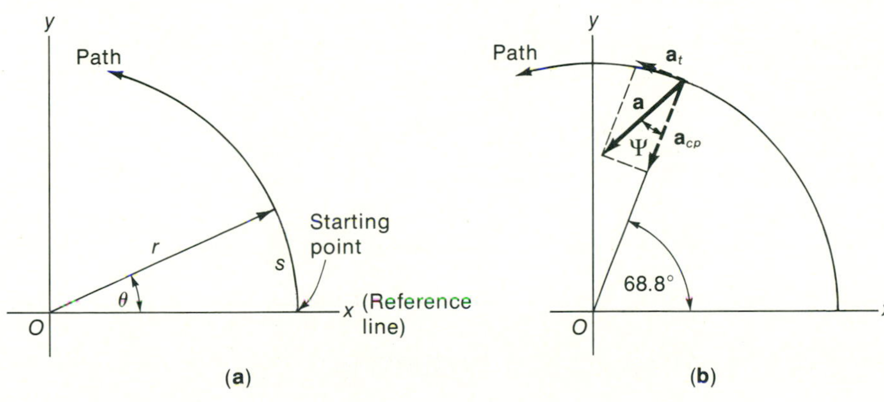

Figure 10–22
Example 10–13

Therefore, we start with Equation (10–33).[3]

$$\theta = \theta_0 + \omega_0 t + \tfrac{1}{2}\alpha t^2$$

$$\theta = 0 + 0 + \frac{1}{2}\left(0.6 \frac{\text{rad}}{\text{s}^2}\right)(2 \text{ s})^2 = 1.2 \text{ rad}$$

Thus:

$$s = r\theta$$

$$s = (2 \text{ ft})(1.2 \text{ rad}) = \boxed{2.40 \text{ ft}}$$

Note that the unit "radian," having dimensions of unity, is omitted from the final answer.

(c) We found in part (b) that at $t = 2$ s, the point is located at an angle $\theta = 1.2$ rad from the reference line ($\theta = 0°$). For convenience, we convert this angle to degrees by

[3] Alternatively, we could have used the tangential form of Equation (10–33), $s = s_0 + v_0 t + \frac{1}{2}a_t t^2$, instead of the angular form.

multiplying a conversion factor equal to unity:

$$\theta = 1.2\ \text{rad}\underbrace{\left[\frac{360^\circ}{2\pi\ \text{rad}}\right]}_{\text{Conversion factor}} = \boxed{68.8^\circ}$$

(d) The acceleration has two components, which we calculate separately.

Tangential component (constant in time):

$$a_t = r\alpha$$

$$a_t = (2\ \text{ft})\left(0.6\ \frac{\text{rad}}{\text{s}^2}\right) = 1.20\ \frac{\text{ft}}{\text{s}^2}$$

Centripetal component:

$$a_{cp} = \frac{v^2}{r}$$

Since $v = r\omega$, we have

$$a_{cp} = r\omega^2$$

$$a_{cp} = (2\ \text{ft})\left(1.20\ \frac{\text{rad}}{\text{s}}\right)^2 = 2.88\ \frac{\text{ft}}{\text{s}^2}$$

The total acceleration **a** at $t = 2$ s may be found from the Pythagorean Theorem.

$$a = \sqrt{a_{cp}^2 + a_t^2} = \sqrt{\left(2.88\ \frac{\text{ft}}{\text{s}^2}\right)^2 + \left(1.20\ \frac{\text{ft}}{\text{s}^2}\right)^2} = \boxed{3.12\ \frac{\text{ft}}{\text{s}^2}}$$

The direction of **a** is shown in Figure 10–22(b). The angle Ψ that **a** makes with the inward radial direction is given by

$$\tan\Psi = \frac{a_t}{a_{cp}} = \frac{1.20\ \frac{\text{ft}}{\text{s}^2}}{2.88\ \frac{\text{ft}}{\text{s}^2}} = 0.417$$

$$\Psi = \tan^{-1}(0.417) = \boxed{22.6^\circ}$$

10.14 Rotational Dynamics of Rigid Bodies

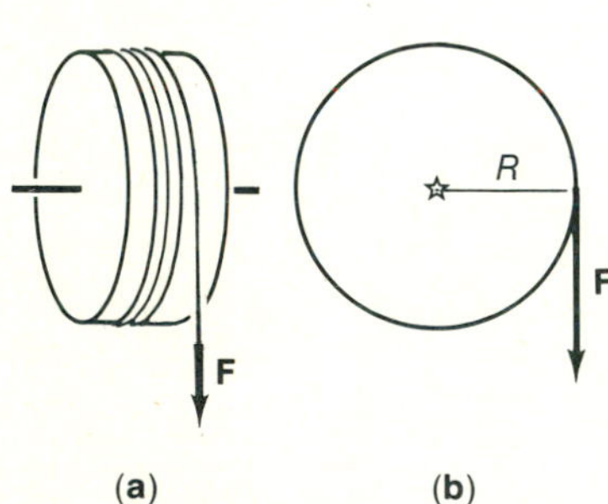

Figure 10–23
A solid, uniform cylinder rotates about a fixed axis under the action of the force **F**.

Figure 10–23 illustrates a solid, uniform cylinder that may rotate about a fixed axis (along the cylinder axis) on frictionless bearings. A string of negligible mass is wound around the cylinder. By pulling on the string with a force **F**, we apply a torque τ = (force)(moment-arm) to the cylinder.

The work dW done to the cylinder during a small rotation $d\theta$ (which causes the string to move a linear distance $ds = R\,d\theta$) is:

$$dW = \mathbf{F}\cdot d\mathbf{s} = |\mathbf{F}|\,|d\mathbf{s}|\cos\phi \qquad \textbf{(10–36)}$$

Here, the distance $d\mathbf{s}$ is always parallel to the force **F**, so the $(\cos\phi)$ factor always equals 1. Hence:

$$dW = F\,ds = FR\,d\theta$$

$$dW = \tau\,d\theta \qquad \textbf{(10–37)}$$

where τ is the torque about the axis produced by the force F.

The rate at which work is being done on the cylinder is found by dividing Equation (10–37) by dt:

$$\frac{dW}{dt} = \tau \frac{d\theta}{dt}$$

$$\frac{dW}{dt} = \tau\omega \qquad (10\text{–}38)$$

where ω is the instantaneous angular speed (in radians per second) of the rotating object.

From the work-energy relation we conclude that as a result of this work, the rotational kinetic energy of the cylinder increases at the rate

$$\frac{dK}{dt} = \frac{d}{dt}(\tfrac{1}{2}I\omega^2) = \tfrac{1}{2}I\frac{d}{dt}(\omega^2) = I\omega\frac{d\omega}{dt}$$

$$\frac{dK}{dt} = I\omega\alpha \qquad (10\text{–}39)$$

where α is the instantaneous angular acceleration (in radians/second^2) of the rotating body. Comparing Equations (10–38) and (10–39), we conclude

NEWTON'S SECOND LAW FOR ROTATIONAL MOTION (if *I* remains constant)

$$\underset{\text{(about the } CM)}{\tau} = I\alpha \qquad (10\text{–}40)$$

It might be helpful to pause a moment to see where we are. We now have the *kinematic equations* for both linear and rotational motions, and we have *Newton's second law* for both linear and rotational motions. Armed with these, we can describe the general motion of a rigid object through space and also analyze why it moves as it does.

EXAMPLE 10–14

The catapult shown in Figure 10–24 is an ancient weapon used to hurl large stones. The bent wooden beam acts like a spring to exert a force (via the cable) on the catapult arm. The cable is attached to the arm at a distance of $r_1 = 2$ m from the arm pivot. The stone has a mass of $m = 100$ kg and is located at a distance of $r_2 = 7$ m from the pivot. The rope, which holds the catapult ready to fire, is cut. At the instant when $\phi = 30°$, the cable tension $T = 2 \times 10^4$ N. Neglecting the mass of the catapult arm, determine the instantaneous angular acceleration of the stone about the pivot.

Figure 10–24
Catapult analyzed in Example 10–14

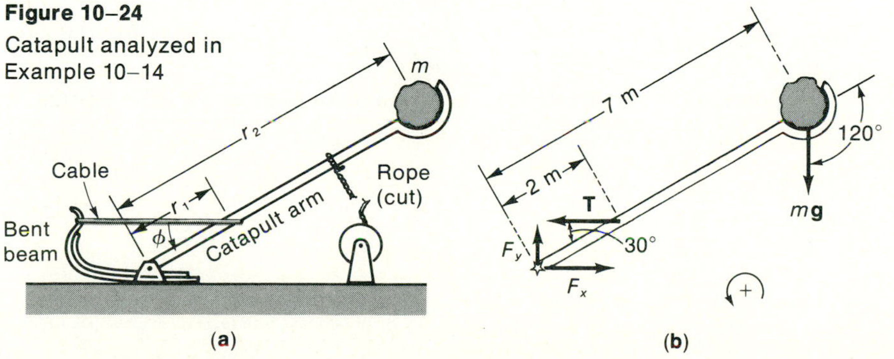

SOLUTION

Assuming counterclockwise torques are positive, the torque about the pivot applied to the mass m is (see Figure 10–24):

$$\underset{\text{(about the pivot)}}{\Sigma\tau} = \mathbf{r}_1 \times \mathbf{T} - \mathbf{r}_2 \times m\mathbf{g}$$

$$\begin{aligned}\Sigma\tau &= Tr_1 \sin(\pi - 30^\circ) - mgr_2 \sin 120^\circ \\ &= (2 \times 10^4\ \text{N})(2\ \text{m})(0.500) - (100\ \text{kg})\left(9.8\ \frac{\text{m}}{\text{s}^2}\right)(7\ \text{m})(0.866) \\ &= 2 \times 10^4\ \text{N·m} - 5.94 \times 10^3\ \text{N·m} = 1.41 \times 10^4\ \text{N·m}\end{aligned}$$

The moment of inertia of the mass m a distance $r_2 = 7$ m from the axis of rotation is

$$I = mr^2$$

$$I = (100\ \text{kg})(7\ \text{m})^2 = 4.90 \times 10^3\ \text{kg·m}^2$$

The angular acceleration is obtained from

$$\underset{\text{(about the pivot)}}{\Sigma\tau} = I\alpha$$

$$(1.41 \times 10^4\ \text{N·m}) = (4.90 \times 10^3\ \text{kg·m}^2)\alpha$$

$$\alpha = \frac{1.41 \times 10^4\ \text{N·m}}{4.90 \times 10^3\ \text{kg·m}^2} = \boxed{2.88\ \frac{\text{rad}}{\text{s}^2}}$$

Note: We can estimate the speed acquired by the stone by assuming this angular acceleration is essentially constant while the arm moves through an angle of 90° before the stone leaves the catapult. With this approximation, the stone acquires a speed of over 20 m/s ($\gtrsim$ 45 mi/h).

10.15 The Axial Vectors ω and α

We can obtain a *vector* form of $\tau = I\alpha$ by defining the *axial* vectors $\boldsymbol{\omega}$ and $\boldsymbol{\alpha}$.[4] An angular velocity is represented by a vector $\boldsymbol{\omega}$ directed *along the axis of rotation* according to the right-hand-rule convention illustrated in Figure 10–25. Angular acceleration $\boldsymbol{\alpha}$ ($d\omega/dt$) is similarly defined. Thus we may write Equation (10–40) in *vector* form:

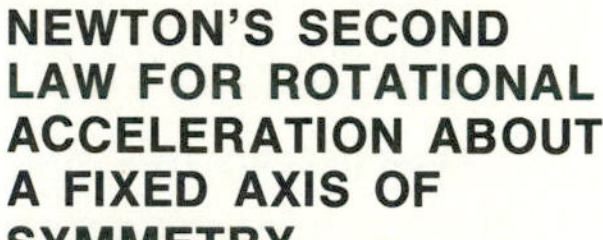

NEWTON'S SECOND LAW FOR ROTATIONAL ACCELERATION ABOUT A FIXED AXIS OF SYMMETRY

$$\underset{\text{(about the CM)}}{\Sigma\boldsymbol{\tau}} = I\boldsymbol{\alpha} \qquad \textbf{(10–41)}$$

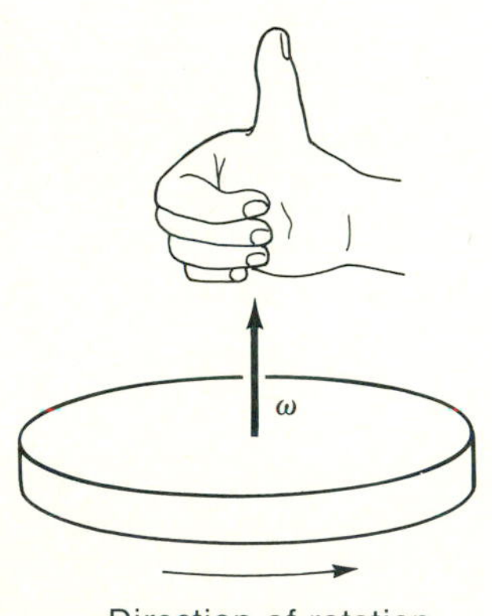

Figure 10–25
The axial vector ω according to the right-hand rule. Grasping the axle with the right hand so that the fingers are curled in the sense of rotation, the extended thumb points in the direction of ω.

Although $\boldsymbol{\tau}$ and $\boldsymbol{\alpha}$ always point in the same direction (as do $\mathbf{F}$ and $\mathbf{a}$ in the linear case), the angular velocity $\boldsymbol{\omega}$ will point in the *opposite* direction for the case of deceleration (as will $\mathbf{v}$ in the linear case).

Defining ω as a vector permits the **angular momentum L** to be written in this convenient vector form:

ANGULAR MOMENTUM ABOUT A FIXED AXIS OF SYMMETRY

$$\underset{\text{(about the CM)}}{\mathbf{L}} = I\boldsymbol{\omega} \qquad \textbf{(10–42)}$$

[4] By analogy with the vectors for the linear case (**s**,**v**,**a**,), it seems plausible that an angular displacement θ might similarly be an axial vector to complete the set ($\boldsymbol{\theta}$,$\boldsymbol{\omega}$,$\boldsymbol{\alpha}$). However, this is *not* the case, as explained in Appendix H.

where, of course, both **L** and I must be calculated about the same axis. (Note the analogy with the linear momentum: $\mathbf{p} = m\mathbf{v}$.)

EXAMPLE 10–15

A uniform, solid cylinder of mass M and radius R is supported by a frictionless axle, as shown in Figure 10–26(a). A light cord is wrapped around the cylinder, and a tension T is maintained to accelerate the cylinder from rest. Find (a) the angular acceleration α of the cylinder and (b) the angular velocity ω after a length d of the cord is unwound. (c) Find numerical values for the above when $M = 2$ kg, $R = 0.10$ m, $T = 5$ N, and $d = 0.40$ m.

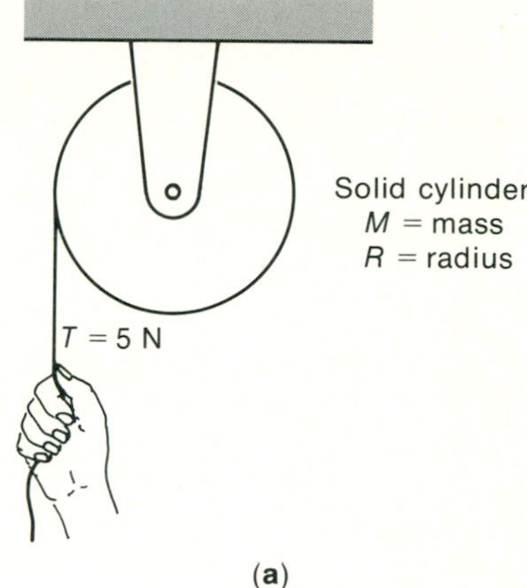

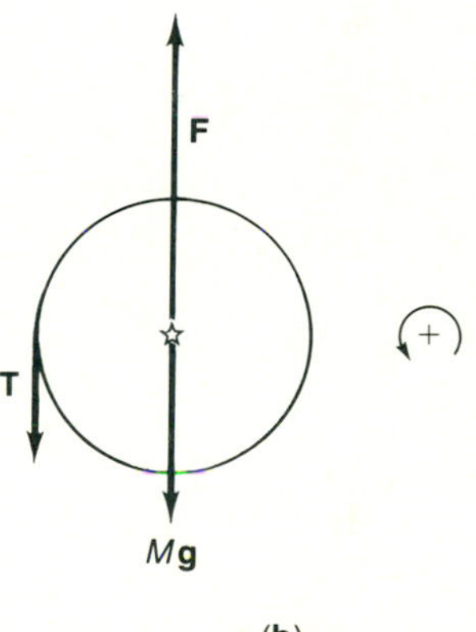

Figure 10–26
Example 10–15

SOLUTION

(a) A free-body diagram is shown in Figure 10–26(b), with counterclockwise rotations designated as positive. We apply

$$\underset{\text{(about the axis)}}{\Sigma\tau} = I\alpha$$

$$(T)(R) = (\tfrac{1}{2}MR^2)\alpha$$

$$\alpha = \boxed{\frac{2T}{MR}}$$

The direction of $\boldsymbol{\alpha}$ is the same as the direction of $\boldsymbol{\tau}$ (out of the plane of the figure through the axis of rotation).

(b) As a length d of the cord unwinds, the angle through which the cylinder rotates from rest is $\theta = d/R$. We find the angular speed of the cylinder at this instant from Equation (10–35).

$$\omega^2 = \omega_0{}^2 + 2\alpha(\theta - \theta_0)$$

$$\omega^2 = (0)^2 + 2\left(\frac{2T}{MR}\right)\left(\frac{d}{R} - 0\right)$$

$$\omega = \boxed{\frac{2}{R}\sqrt{\frac{Td}{M}}}$$

Because the cylinder *accelerates*, the direction of $\boldsymbol{\omega}$ is the *same* as the direction of $\boldsymbol{\alpha}$ and $\boldsymbol{\tau}$.

(c) Substituting numerical values yields

$$\alpha = \frac{2T}{MR} = \frac{(2)(5\text{ N})}{(2\text{ kg})(0.10\text{ m})} = \boxed{50.0\ \frac{\text{rad}}{\text{s}^2}}$$

$$\omega = \frac{2}{R}\sqrt{\frac{Td}{M}} = \left(\frac{2}{0.10\text{ m}}\right)\sqrt{\frac{(5\text{ N})(0.40\text{ m})}{(2\text{ kg})}} = \boxed{20.0\ \frac{\text{rad}}{\text{s}}}$$

EXAMPLE 10–16

Suppose that in the previous example, we hung a mass of weight $W = 5$ N on the cord instead of pulling on it with a force of $T = 5$ N [see Figure 10–27(a)]. (a) Will the angular acceleration of the cylinder be the same as in Example 10–15? (b) If not, calculate its numerical value (using $M = 2$ kg, $R = 0.10$ m, and $W = mg = 5$ N). (c) Find the tension in the cord.

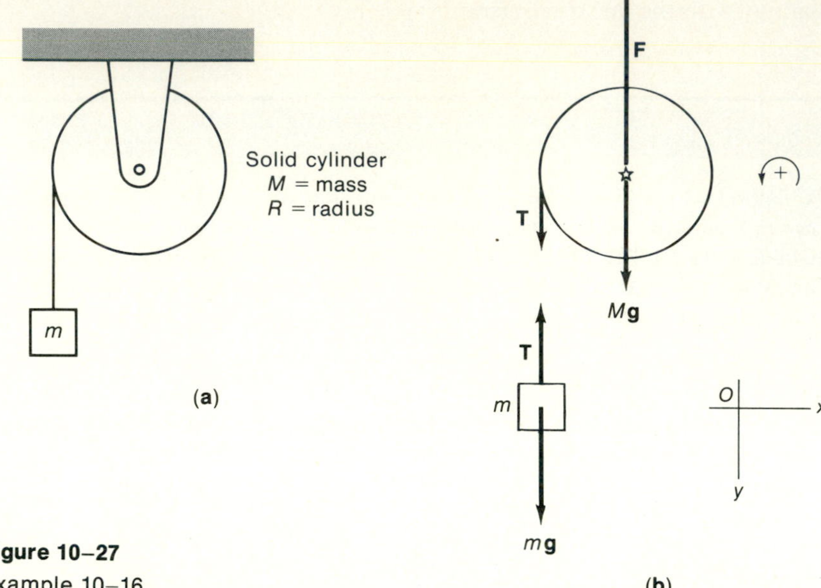

Figure 10–27
Example 10–16

SOLUTION

(a) Figure 10–27(b) shows free-body diagrams of the cylinder and mass. We can infer by inspection that the tension T must be *less* than that in the previous example. If it were equal to mg, the mass would have zero net force and would therefore not accelerate. So T must be *less* than mg, and thus the angular acceleration of the cylinder must also be *less*.

(b) We now apply Newton's second law to the mass m and to the cylinder. Assuming the cord does not slip, the linear and angular accelerations are related through $a = R\alpha$. Because of this relation, we must be consistent in our choices of positive directions for a and α. Choosing a downward direction for a as positive, we must choose a counterclockwise rotation for α as positive.

Translation of *m*

$$\Sigma F = ma$$
$$(mg - T) = ma$$
$$m(g - a) = T$$

Rotation of the cylinder

$$\underset{\text{(about the } CM)}{\Sigma\tau} = I\alpha$$
$$TR = (\tfrac{1}{2}MR^2)\alpha$$
$$T = \tfrac{1}{2}MR\alpha$$

Substituting $a = R\alpha$ and eliminating T by equating the two expressions, we have

$$m(g - R\alpha) = \tfrac{1}{2}MR\alpha$$
$$mg = R\alpha\left(\frac{M}{2} + m\right)$$
$$\alpha = \frac{2mg}{R(M + 2m)}$$

Since the value of m is

$$m = \frac{W}{g} = \frac{5\ \text{N}}{\left(9.8\ \dfrac{\text{m}}{\text{s}^2}\right)} = 0.510\ \text{kg}$$

we obtain

$$\alpha = \frac{2mg}{R(M + 2m)} = \frac{2(5\ \text{N})}{(0.10\ \text{m})[(2\ \text{kg}) + (2)(0.51\ \text{kg})]} = \boxed{33.1\ \frac{\text{rad}}{\text{s}^2}}$$

(c) The tension in the cord may be found from

$$T = \tfrac{1}{2}MR\alpha = \tfrac{1}{2}(2\text{ kg})(0.10\text{ m})\left(33.1\ \frac{\text{rad}}{\text{s}^2}\right) = \boxed{3.31\text{ N}}$$

Note that this is *less* than the 5-N tension in Example 10–15.

10.16 Moving Axes of Rotation

The previous discussion regarding rotation about a *fixed* axis of symmetry also applies to one other special case. Consider a cylinder rolling down an incline. Here, the axis of the cylinder is not fixed, but rather *accelerates* down the incline, moving along with the body as its *CM* accelerates linearly. But note that this acceleration of the axis is in a *parallel* fashion, *always maintaining the axis in the same direction in space.* We may always analyze this type of problem by choosing the *CM* as the point about which the torques and angular momenta are calculated, and then applying $\Sigma \mathbf{F} = m\mathbf{a}$ and $\Sigma\tau = I\alpha$. The analysis is therefore the usual procedure of breaking down the motion into two parts: *translational acceleration* of the *CM* and *rotational acceleration* about the *CM*.

This procedure is summarized below. It applies equally well to the accelerated motion of a ball rolling down an incline or to the tumbling motion of a satellite in its orbit about the earth.

Translation	Rotation	
$\Sigma \mathbf{F}_{\text{ext}} = M\mathbf{a}_{CM}$ $\mathbf{P} = M\mathbf{v}_{CM}$	$\Sigma\tau_{\text{ext}} = I\alpha$ (only about the *CM*) $\mathbf{L} = I\omega$ (only about the *CM*)	These apply only in the special cases of rotation about a stationary axis of symmetry through the *CM*, or when the object accelerates in such a way that its axis always keeps the same (parallel) orientation in space. Both *M* and *I* must remain constant.
$\Sigma \mathbf{F}_{\text{ext}} = \dfrac{d\mathbf{P}}{dt}$	$\Sigma\tau = \dfrac{d\mathbf{L}}{dt}$ (about any axis)	While the translation equation applies in all cases, the rotation equation applies only if the point about which the torques are calculated is the center of mass, or if the point is *not* accelerating.

EXAMPLE 10–17

A solid sphere of mass *M* and radius *R* starts from rest and rolls without slipping down an incline, as shown in Figure 10–28. Find the translational speed of the *CM* of the sphere at the bottom of the incline.

SOLUTION

Because there is no slipping, the force of friction (which provides the torque fR about the *CM*) is a *static* friction force. However, since we do not know whether or not the sphere is *on the verge of slipping*, we cannot assume that the *maximum* static friction force is being

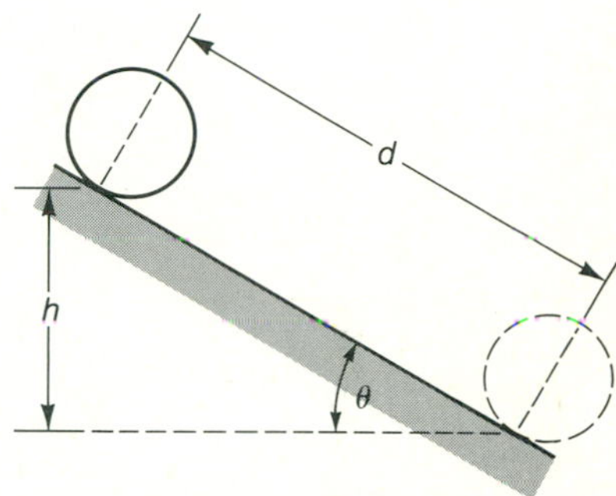

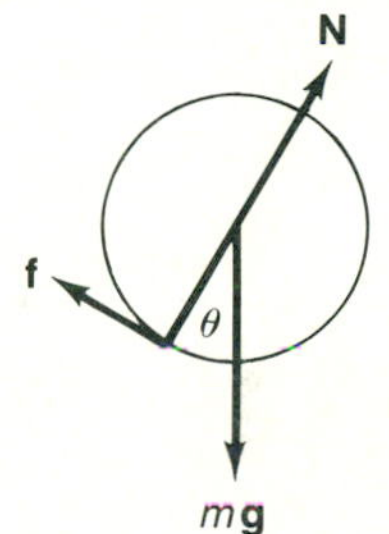

A free-body diagram.

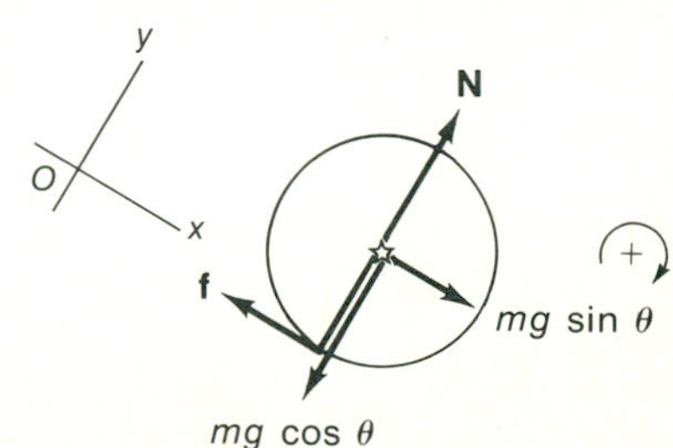

A free-body diagram with forces resolved into two component directions at right angles. Because the *CM* accelerates down the incline, we choose this direction as positive.

Figure 10–28
Example 10–17

developed (that is, $f \neq \mu_s N$). We break the motion into two parts:

Rotational acceleration about the CM

$$\Sigma\tau = I\alpha$$
(about the CM)

$$fR = \tfrac{2}{5}MR^2\alpha$$
$$f = \tfrac{2}{5}MR\alpha$$

Since $a_{CM} = R\alpha$, we have:

$$f = \tfrac{2}{5}Ma_{CM}$$

Translational acceleration of the CM

$$\Sigma F = ma$$
$$(Mg \sin\theta - f) = Ma_{CM}$$

We substitute the value for f from the rotational analysis and solve for a_{CM}:

$$\left(Mg \sin\theta - \frac{2Ma_{CM}}{5}\right) = Ma_{CM}$$

$$\tfrac{5}{7}(g \sin\theta) = a_{CM}$$

To obtain the translational speed of the CM after the sphere has moved from rest a distance d, we note that the acceleration a_{CM} is constant. Therefore, we may use the linear kinematic equation:

$$v^2 = v_0{}^2 + 2a(x - x_0)$$
$$v^2 = (0)^2 + 2(\tfrac{5}{7}\, g \sin\theta)(d - 0)$$
$$v = \boxed{\sqrt{\frac{10gd \sin\theta}{7}}}$$

The previous example illustrates an important point regarding a rolling object. When a spherical object of radius R rolls on a surface *without slipping*, the arc length s subtending the angle through which it turns is, in effect, transferred to the horizontal surface as the linear displacement x_{CM} of the center-of-mass point (see Figure 10–29). That is:

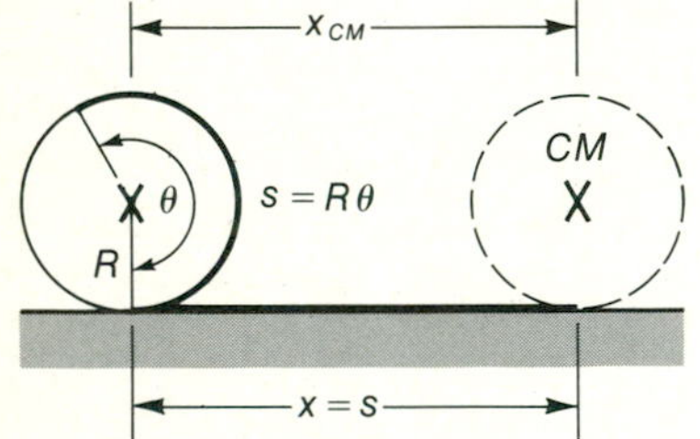

Figure 10–29
A spherical object that rolls without slipping.

$$s_{CM} = R\theta$$

By successive differentiation, we also have:

$$v_{CM} = R\omega \qquad \text{and} \qquad a_{CM} = R\alpha$$

Because of this connection between the *linear* and *rotational* motions, it is necessary to choose the positive x direction and the positive rotational direction consistently. That is, a positive linear displacement of the CM must correspond to a positive angular displacement about the CM.

EXAMPLE 10–18

A spinning cylinder of mass M and radius R is placed on a horizontal surface and released such that initially its CM is at rest. Because of the spinning motion, the cylinder will begin to translate along the surface, acquiring linear speed, while it slows down its rotational speed [see Figure 10–30(a)]. The coefficient of kinetic friction between the cylinder and surface is μ_k. Consider the condition when the cylinder is rotating and sliding against the surface, while accelerating linearly. Find (a) the translational acceleration a_{CM} of the CM and (b) the rotational acceleration α of the cylinder about its CM.

SOLUTION

(a) Figure 10–30(b) shows a free-body diagram. Since sliding occurs, the force of kinetic friction is $f = \mu_k N$. This force provides the net horizontal force for accelerating the CM and also the (negative) torque about the CM which slows down the angular speed. There is a connection between the linear and rotational motions, so we must choose the

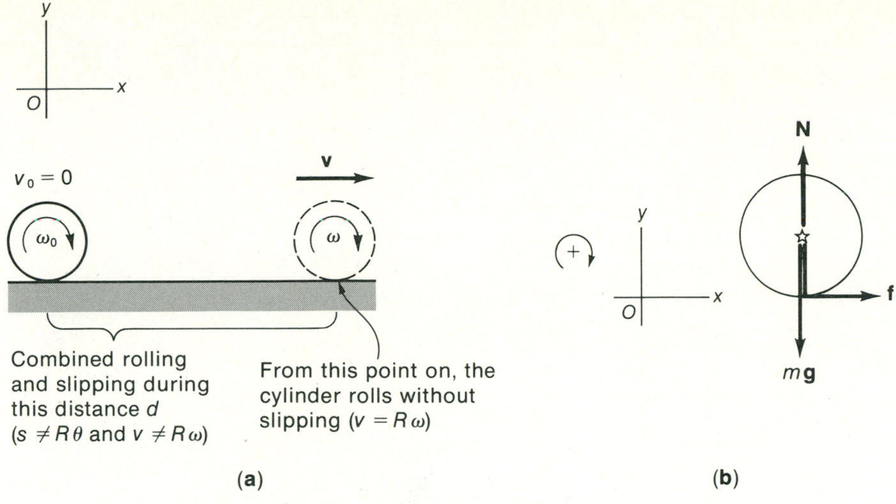

Figure 10–30
Example 10–18

positive directions for translation and rotation consistently. That is, the $+x$ direction is toward the right, and the clockwise rotations are chosen as positive.

Because there is no acceleration in the vertical direction, the vertical forces must add to zero:

$$\Sigma F_y = 0$$
$$(N - Mg) = 0$$
$$N = Mg$$

Investigating the horizontal motion, we see that the force of friction f is the net horizontal force. Thus:

$$\Sigma F_x = ma_x$$
$$f = Ma_{CM}$$
$$\mu_k N = Ma_{CM}$$

In the analysis of the vertical component, we found $N = Mg$. Therefore:

$$\mu_k Mg = Ma_{CM}$$

or

$$a_{CM} = \boxed{\mu_k g}$$

(b) Since we have chosen the clockwise sense of rotation as positive, the torque about the CM is negative, making the angular acceleration α also negative. We apply Newton's second law for rotation:

$$\underset{\text{(about the } CM)}{\Sigma\tau} = I\alpha$$

$$-(fR) = (\tfrac{1}{2}MR^2)\alpha$$

$$\alpha = -\frac{\mu_k Mg}{\frac{1}{2}MR} = \boxed{-\frac{2\mu g}{R}}$$

Note that because the cylinder is sliding, the linear acceleration a does *not* equal $R\alpha$. Also, $s \neq R\theta$ and $v \neq R\omega$. These relations are true only for the case of rolling without slipping.

10.17 Conservation of Angular Momentum

We have seen that the torque $\boldsymbol{\tau}$ is equal to the rate of change of angular momentum $\mathbf{L}$: $\boldsymbol{\tau} = d\mathbf{L}/dt$. Thus, *whenever the external torque on a system is zero, the angular momentum* $\mathbf{L}$ *about a fixed axis remains constant in time.* This is expressed

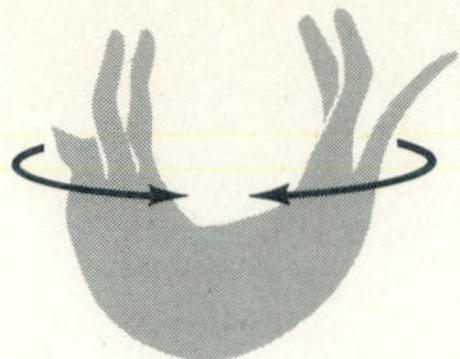

(**a**) The basic maneuver turns the U-shaped position inside-out, rotating the cat's back from the outside to the inside of the U. The front and rear portions rotate in *opposite* directions, *conserving angular momentum.*

(**b**) Generally, the fore part of the body begins to rotate first, so the cat may at least land on its front paws if the drop is a short one. The front legs are drawn in toward the stomach, reducing the moment of inertia while turning. The hind legs are left extended until the front half has been partly turned, then the front paws are extended as the hind legs are drawn in and the rear rotated. During the entire process, the tail whips rapidly around to help conserve the overall angular momentum.

Figure 10–31
Since there is no external torque on a falling cat, how does the cat perform the legendary feat of always landing right-side-up even though dropped upside-down? The intricate motions of the cat conserve angular momentum.

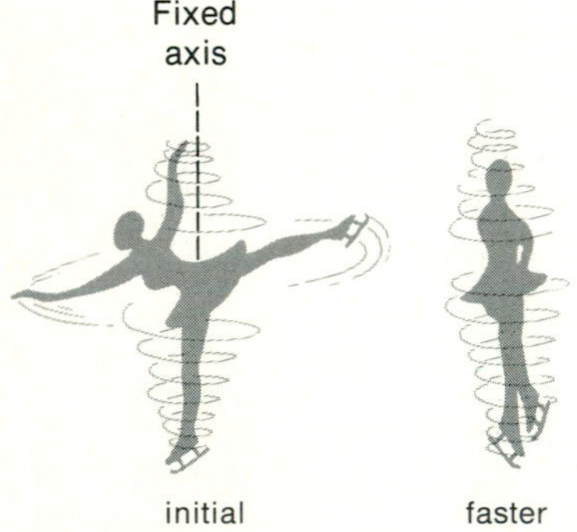

Figure 10–32
An ice skater makes use of the conservation of angular momentum by starting to rotate with arms and legs extended. Pulling all parts of the body as close to the central axis as possible increases the rotational speed greatly.

by the **conservation of angular momentum** relation, which equates the initial and final values:

CONSERVATION OF ANGULAR MOMENTUM (about some axis)

$$\mathbf{L}_0 = \mathbf{L} \qquad \text{or} \qquad I_0\omega_0 = I\omega \tag{10–43}$$

This principle is just as useful a method for solving problems as the analogous statement of the conservation of linear momentum: $\mathbf{p}_0 = \mathbf{p}$. Watch for the clue to its applicability: *whenever the net external torque is zero.* Examples of the conservation of angular momentum are illustrated in Figures 10–31 and 10–32.

EXAMPLE 10–19

A horizontal rod supports two small spheres, each of mass $m = 0.20$ kg, in symmetrical positions 0.10 m from the axis, as shown in Figure 10–33. The spheres are small enough to be considered point masses, and the moment of inertia of just the rod alone about the vertical axis through its center is 0.070 kg·m². While spinning freely about the vertical axis at 80 rpm, latches holding the spheres are released. The spheres slide outward and come to rest against the stops at the ends of the rod. Find the final angular speed of the system.

SOLUTION

Because there is no external torque about the axis, conservation of angular momentum applies. We first convert the rotational speed units:

$$\omega_i = 80\,\frac{\cancel{\text{rev}}}{\cancel{\text{min}}}\underbrace{\left[\frac{1\,\cancel{\text{min}}}{60\text{ s}}\right]\left[\frac{2\pi\text{ rad}}{1\,\cancel{\text{rev}}}\right]}_{\substack{\text{Conversion}\\ \text{ratios}}} = 8.37\,\frac{\text{rad}}{\text{s}}$$

We then have:

$$L_0 = L$$

$$[(I_{\text{rod}})_0 + (I_{\text{spheres}})_0]\omega_0 = [(I_{\text{rod}}) + (I_{\text{spheres}})]\omega$$

$$\omega = \omega_0\,\frac{(I_{\text{rod}})_0 + (I_{\text{spheres}})_0}{(I_{\text{rod}}) + (I_{\text{spheres}})}$$

$$\omega = \left(8.37\,\frac{\text{rad}}{\text{s}}\right)\frac{(0.070\text{ kg}\cdot\text{m}^2) + (2)(0.20\text{ kg})(0.10\text{ m})^2}{(0.070\text{ kg}\cdot\text{m}^2) + (2)(0.20\text{ kg})(0.40\text{ m})^2}$$

$$\omega = \left(8.37\,\frac{\text{rad}}{\text{s}}\right)\frac{(0.074\text{ kg}\cdot\text{m}^2)}{(0.13\text{ kg}\cdot\text{m}^2)} = \boxed{4.76\,\frac{\text{rad}}{\text{s}}}$$

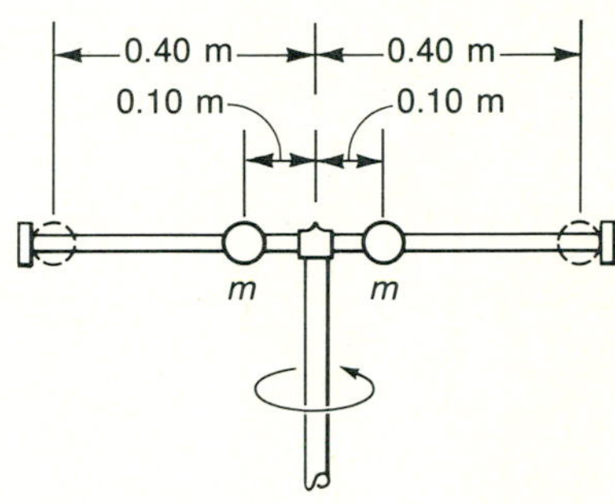

Figure 10–33
Example 10-19

10.18 Conservation of Energy in Rotational Motion

The conservation of energy principle is true, of course, for the rotational motion of a solid object. The analogies with the linear case are so close that we will just state the expressions for the various forms of energy we have treated before. If you note the similarities between the rotational and linear expressions, it will make them easier to remember. The expressions for energy in rotational motion, Equations 10–44 to 10–50, are compared with their linear analogues in Table 10–1.

For an isolated (conservative) system, we have the conservation of mechanical energy:

CONSERVATION OF MECHANICAL ENERGY

$$E_0 = E$$

$$U_0 + K_0 = U + K \qquad \textbf{(10–51)}$$

The *work-energy theorem* (Chapter 5) also applies in the rotational case:

For *translational* motion of the *CM*: [Work done by the net resultant *force*] = [Change in *translational* kinetic energy of the *CM*]

For *rotational* motion about the *CM*: [Work done by the net resultant *torque*] = [Change in *rotational* kinetic energy about the *CM*]

TABLE 10–1

Energy in rotational motion	Linear analogy
(1) The work W done by an external torque in moving through an angle θ: $$W = \int \tau\, d\theta \qquad \textbf{(10–44)}$$	$W = \int F\, dx$
(2) The gravitational potential energy depends only on the elevation h_{CM} of the *center of mass* above the reference level: $$U_{\text{grav}} = Mgh_{CM} \qquad \textbf{(10–45)}$$	$U_{\text{grav}} = mgh$
(3) Spiral springs (such as attached to a balance wheel in a watch) have a torque constant κ (Greek letter *kappa*) describing their "Hooke's Law behavior." *The restoring torque is proportional to the angular displacement.* $$\tau = -\kappa\theta \qquad \textbf{(10–46)}$$ where κ is measured in units of torque/radian. The potential energy stored in a spiral spring that has been rotated through an angle θ from its equilibrium position is: $$U_{\text{spring}} = \tfrac{1}{2}\kappa\theta^2 \qquad \textbf{(10–47)}$$	Linear spring Hooke's Law $F = -kx$ $U_{\text{spring}} = \tfrac{1}{2}kx^2$
(4) The kinetic energy of an extended object is handled in two parts: (a) the *translational* kinetic energy of the motion of the CM: $$K_{\text{trans}} = \tfrac{1}{2}Mv_{CM}{}^2 \qquad \textbf{(10–48)}$$ (b) the *rotational* kinetic energy associated with rotations about the CM: $$K_{\text{rot}} = \tfrac{1}{2}I\omega^2 \qquad \textbf{(10–49)}$$ The total kinetic energy K for general motion is: $$K = K_{\text{trans}} + K_{\text{rot}} \qquad \textbf{(10–50)}$$	Kinetic energy for a mass point: $K = \tfrac{1}{2}mv^2$

EXAMPLE 10–20

A uniform solid sphere rolls without slipping along a horizontal surface at constant linear speed. What fraction of its total kinetic energy is in the form of rotational kinetic energy about the CM?

SOLUTION

The fraction we seek is:

$$\frac{K_{\text{rot}}}{K_{\text{total}}} = \frac{K_{\text{rot}}}{K_{\text{trans}} + K_{\text{rot}}} = \frac{\frac{1}{2}I\omega^2}{\frac{1}{2}Mv_{CM}{}^2 + \frac{1}{2}I\omega^2} = \frac{\frac{1}{2}(\frac{2}{5}MR^2)\omega^2}{\frac{1}{2}Mv_{CM}{}^2 + \frac{1}{2}(\frac{2}{5}MR^2)\omega^2}$$

Since the sphere rolls without slipping, $v_{CM} = R\omega$:

$$\frac{K_{\text{rot}}}{K_{\text{total}}} = \frac{\frac{2}{5}v_{CM}{}^2}{v_{CM}{}^2 + \frac{2}{5}v_{CM}{}^2} = \boxed{\frac{2}{7}}$$

EXAMPLE 10–21

A mass m is attached to a cord that is wound around a uniform solid cylinder, as shown earlier in Figure 10–27(a). The cylinder has a mass $M = 4m$ and a radius R, and is supported on frictionless bearings. Assuming the system is moving freely, use energy conservation principles to find the linear acceleration a of the mass m. (This is an alternative solution to the type of problem analyzed in Example 10–16. Usually the conservation-of-energy approach illustrated here is more direct than using Newton's laws.)

SOLUTION

It is convenient to assume the mass m starts at rest and moves down a distance h. (This distance will cancel out in the calculation.) It is a conservative system and we easily find the speed of the mass after falling this distance by applying the conservation of energy principle. We choose the final position of m as the reference level for gravitational potential energy.

$$E_0 = E$$
$$(U_{grav})_0 + (K_{trans})_0 + (K_{rot})_0 = U_{grav} + K_{trans} + K_{rot}$$
$$(mgh)_0 + (\tfrac{1}{2}mv^2)_0 + (\tfrac{1}{2}I\omega^2)_0 = mgh + \tfrac{1}{2}mv^2 + \tfrac{1}{2}I\omega^2$$
$$mgh + (0) + (0) = (0) + \tfrac{1}{2}mv^2 + \tfrac{1}{2}[\tfrac{1}{2}(4m)R^2]\omega^2$$

Since the cord unwinds from the radius R of the cylinder without slipping, $v = R\omega$:

$$2gh = v^2 + \tfrac{1}{2}(4)v^2 = 3v^2$$
$$v = \sqrt{\tfrac{2}{3}gh}$$

The linear acceleration of m is found from the kinematic equation:

$$v^2 = {v_0}^2 + 2a(y - y_0)$$
$$\tfrac{2}{3}gh = 0 + 2ah$$
$$\boxed{a = \tfrac{1}{3}g}$$

EXAMPLE 10–22

A solid sphere of mass M and radius R starts from rest and rolls without slipping down an incline of height h. From energy considerations, find the translational speed of the CM of the sphere at the bottom of the incline. (This example illustrates an alternative approach to the problem discussed in Example 10–17.)

SOLUTION

It may seem that the static friction force on the sphere moves along as the object rolls, and we might (mistakenly) conclude that it does some work, or that some thermal energy is developed. This is not true; the rolling process merely *transfers* the static friction force from one point of contact to another as the object rolls. Furthermore, in rolling without slipping, one surface does *not* slide across another, so no thermal energy is developed. (Only *kinetic* friction forces produce thermal energy according to $f_k x$.)

The system is conservative and we thus apply the conservation of mechanical energy, establishing the zero reference level for gravitational potential energy at the lowest point of interest: the bottom of the incline.

$$E_0 = E$$
$$(U_{grav})_0 + (K_{trans})_0 + (K_{rot})_0 = U_{grav} + K_{trans} + K_{rot}$$
$$(Mgh_{CM})_0 + (\tfrac{1}{2}M{v_{CM}}^2)_0 + (\tfrac{1}{2}I\omega^2)_0 = Mgh_{CM} + \tfrac{1}{2}M{v_{CM}}^2 + \tfrac{1}{2}I\omega^2$$
$$Mgh_{CM} + (0) + (0) = (0) + \tfrac{1}{2}M{v_{CM}}^2 + \tfrac{1}{2}(\tfrac{2}{5}MR^2)\omega^2$$

Because the sphere rolls without slipping, $v_{CM} = R\omega$:

$$2gh_{CM} = v_{CM}^2 + \tfrac{2}{5}v_{CM}^2$$

Referring back to Figure 10–28(a), we see that $h = d \sin\theta$:

$$\boxed{v_{CM} = \sqrt{\frac{10gd\sin\theta}{7}}}$$

EXAMPLE 10–23

After turning a spiral spring through an angular displacement of 60° from its equilibrium position, it requires a torque of 0.30 N·m to hold the spring in the wound position. Assuming a Hooke's Law behavior in which the restoring torque is proportional to the angular displacement, (a) find the torque constant κ of the spring and (b) find the potential energy stored in the wound spring.

SOLUTION

(a) The angle θ is

$$\theta = 60^\circ\left(\frac{2\pi\text{ rad}}{360^\circ}\right) = \frac{\pi}{3}\text{ rad}$$

From $\tau = \kappa\theta$, we have:

$$\kappa = \frac{\tau}{\theta} = \frac{(0.30\text{ N·m})}{\left(\frac{\pi}{3}\text{ rad}\right)} = \boxed{0.287\,\frac{\text{N·m}}{\text{rad}}}$$

(b) The potential energy in the stressed spring is

$$U_{\text{spring}} = \tfrac{1}{2}\kappa\theta^2 = \frac{1}{2}\left(0.287\,\frac{\text{N·m}}{\text{rad}}\right)\left(\frac{\pi}{3}\text{ rad}\right)^2 = 0.157\text{ N·m} = \boxed{0.157\text{ J}}$$

Note that the unit "radian" is omitted from the final answer, since energy is appropriately expressed as newton-meters or joules.

EXAMPLE 10–24

In bringing an automobile to a sudden stop, the brakes are applied so that one wheel partially skids and partially turns while being brought to rest. If the wheel of radius R turns through a total angle of θ while skidding a linear distance D along the road ($D > R\theta$), how much thermal energy is developed by the kinetic friction force f_k?

SOLUTION

Since the wheel does roll somewhat, the rolling motion potentially transfers the friction force from one point of contact to another without that force doing work or developing thermal energy. Thus we need to find the *net* sliding distance between the tire surface and the roadway; this distance is $D - R\theta$. The thermal energy developed is therefore

$$U_{\text{thermal}} = \boxed{f_k(D - R\theta)}$$

10.19 The Gyroscope

The general motion of a gyroscope is so complex that entire books have been devoted solely to explaining its characteristics. The most familiar of a gyroscope's peculiarities is illustrated in Figure 10–34(a). When a spinning gyroscope is supported at one end we expect its axis to tilt downward. Instead, the axis swings around in a horizontal plane in a motion called **precession,** seeming to "defy gravity." Of course, no violation of the "law" of gravity occurs—the gyroscope is only obediently complying with $\boldsymbol{\tau} = d\mathbf{L}/dt$.

The gyroscope is an example of rotational motion in which the axis of rotation does not move parallel to itself. To analyze this motion, we will apply $\boldsymbol{\tau} = d\mathbf{L}/dt$ by calculating the torque about a *fixed point* rather than about the center of mass. Consider the vector diagram of Figure 10–34(b). If the spinning wheel were supported at both ends, its angular momentum would be $\mathbf{L}(t) = I\boldsymbol{\omega}$, along the $+y$ axis. However, when supported at only one end as shown, the force of gravity $m\mathbf{g}$ on the *CM* of the wheel produces a torque about the fixed point of support: $\boldsymbol{\tau} = \mathbf{r} \times m\mathbf{g}$, as shown in Figure 10–34(c). By the right-hand rule, the torque $\boldsymbol{\tau}$ is perpendicular to both $\mathbf{r}$ and $m\mathbf{g}$, in the direction shown in part (b) of the figure. Because $\boldsymbol{\tau} = \Delta\mathbf{L}/\Delta t$, this torque, acting for a time Δt, produces a change in angular momentum $\Delta\mathbf{L}$ that is perpendicular to $\mathbf{L}$. The vector sum of $\mathbf{L}$ and $\Delta\mathbf{L}$ lies in the x-y plane, displaced an angle $\Delta\phi$ from $\mathbf{L}$. As $\Delta t \to 0$, the new angular momentum $\mathbf{L}(t + \Delta t)$ has the same magnitude as the original value, but in a different direction. The angular momentum vector lies along the axis of the spinning gyroscope, so the gyroscope axis turns in a horizontal plane, rotating about the pivot point with the precessional angular speed

$$\omega_p = \frac{\Delta\phi}{\Delta t}$$

To calculate this precessional speed we note from the geometry that $\Delta L = L\,\Delta\phi$. Division by Δt gives us

$$\frac{\Delta L}{\Delta t} = L\frac{\Delta\phi}{\Delta t}$$

But $|\tau| = \Delta L/\Delta t$, and $\omega_p = \Delta\phi/\Delta t$, so

$$\tau = L\omega_p \tag{10–52}$$

For Figure 10–34(a), the precessional speed is thus

$$\omega_p = \frac{mgr}{L} \tag{10–53}$$

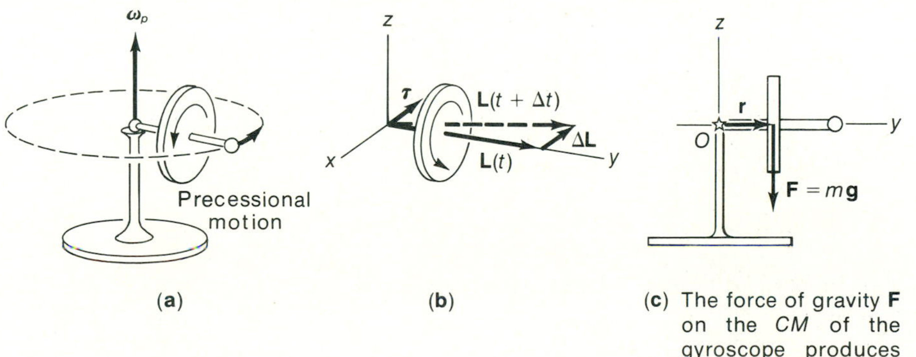

(c) The force of gravity **F** on the *CM* of the gyroscope produces a torque $\boldsymbol{\tau} = (\mathbf{r} \times \mathbf{F})$ about the pivot point *O into* the plane of the diagram.

Figure 10–34
When a spinning gyroscope is supported at one end only, it undergoes precessional motion ω_p, as shown in (a).

with the direction of ω_p along the $+z$ axis. In vector form, Equation (10–51) is

$$\tau = \omega_p \mathbf{X} \mathbf{L} \qquad \textbf{(10–54)}$$

or (since $\mathbf{L} = I\omega$):

$$\tau = \omega_p \mathbf{X}\, I\omega \qquad \textbf{(10–55)}$$

In general, the direction of precessional motion may be determined by the following statement: *The angular momentum vector tends to seek the direction of the applied torque vector.* This accounts for the startling behavior of a gyroscope when a torque is applied to its axis. For example, when a rotating gyroscope is held with its axis horizontal, attempting to tilt the axis upward causes it to swing toward the right or left, depending on the sense of rotation of the gyroscope.

EXAMPLE 10–25

The rotor of a small gyroscope has a mass of 0.220 kg, and its moment of inertia about its axis is 2.5×10^{-4} kg·m^2. (The mass of the supporting frame is negligible.) The rotating gyroscope is supported by a pivot point at one end, with its axis horizontal and with the *CM* of the rotor 5 cm from the pivot. The gyroscope precesses in a horizontal plane, performing one revolution every 6 s.

(a) Find the angular velocity ω (in revolutions per minute) of the spinning rotor.

(b) Suppose, looking from the pivot point along the axis, the rotor spins counterclockwise. What would be the rotational sense of precessional motion looking down on the gyroscope from above?

SOLUTION

(a) Figure 10–35 illustrates the physical situation. Since $\mathbf{F}_{grav}$ ($=m\mathbf{g}$) and $\mathbf{r}$ are at right angles, the torque τ about the pivot point has a magnitude of $(r)(mg) \sin 90° = (r)(mg)$. Because the gyroscope precesses in a horizontal plane, we know ω_p must be vertical (though at this point we do not know whether it is up or down). Therefore, ω_p and $\mathbf{L}$ are at right angles, so the cross product $|\omega_p \mathbf{X} \mathbf{L}| = (\omega_p)(I\omega) \sin 90° = (\omega_p)(I\omega)$. Equation (10–54) expresses a relation between the significant parameters:

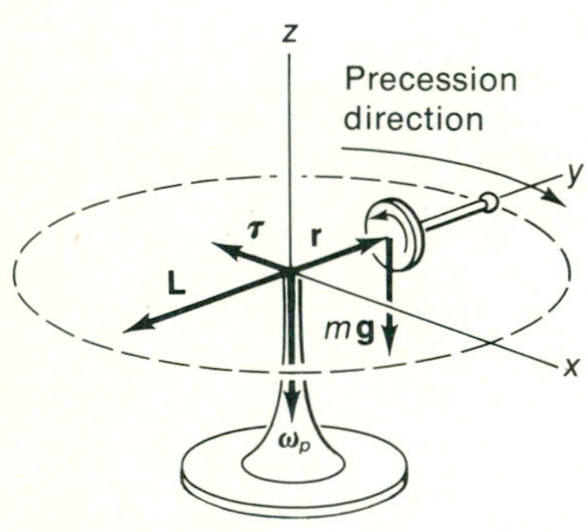

Figure 10–35
Example 10–25

$$|\tau| = |\omega_p \mathbf{X}\, I\omega|$$

$$(r)(mg) = (\omega_p)(I\omega)$$

$$\omega = \frac{rmg}{\omega_p I} \qquad \textbf{(10–56)}$$

Determining the value of ω_p in radian measure, we have

$$\omega_p = \left(\frac{1\ \text{rev}}{6\ \text{s}}\right)\underbrace{\left(\frac{2\pi\ \text{rad}}{1\ \text{rev}}\right)}_{\text{Conversion ratio}} = \frac{\pi}{3}\frac{\text{rad}}{\text{s}}$$

Substituting numerical values in Equation (10–55) and converting to units of revolutions per minute yields

$$\omega = \frac{(0.050\ \text{m})(0.220\ \text{kg})\left(9.8\ \frac{\text{m}}{\text{s}^2}\right)}{\left(\frac{\pi}{3}\frac{\text{rad}}{\text{s}}\right)(2.5 \times 10^{-4}\ \text{kg·m}^2)} = \frac{1294}{\pi}\frac{\text{rad}}{\text{s}}$$

$$\omega = \left(\frac{1294}{\pi}\frac{\text{rad}}{\text{s}}\right)\underbrace{\left(\frac{60\ \text{s}}{1\ \text{min}}\right)\left(\frac{1\ \text{rev}}{2\pi\ \text{rad}}\right)}_{\text{Conversion ratio}} = \boxed{3932\ \text{rpm}}$$

(b) Referring to Figure 10–35, we see that a *clockwise* precession direction (that is, ω_p down), when viewed from above, agrees with $\tau = \omega_p \times \mathbf{L}$.

10.20 Elastic Properties of Matter

In a chapter dealing with rigid bodies, we should again point out that actually there is no perfectly rigid substance and that the concept of a rigid body is merely a convenient mathematical abstraction. All matter becomes deformed if external forces are applied. For homogeneous and isotropic materials, elastic properties are conveniently described by specifying three different constants, known as the **elastic moduli,** that characterize how the physical dimensions of an object change when we apply force to it.

We can deform a piece of matter in a variety of ways. We can apply pressure to it in an attempt to squeeze it into a smaller volume. We can stretch or compress it along one direction. We can "shear" it in a manner similar to pushing horizontally on the top cover of a book while the book rests on a table. In each case, the *applied force per unit area* at the surface (or within the material) is called the **stress.** The resultant deformation of the material is the **strain,** a dimensionless ratio that measures the *fractional change in physical dimensions.*

As a first case, consider stretching (or compressing) a rod along its length, as illustrated in Figure 10–36(a). If the force changes from F to $F + \Delta F$, the length will change from L_0 to $L_0 + \Delta L$. The *stress* is thus $\Delta F/A$, where A is the cross-sectional area of the rod; the fractional change in length, or *strain,* is $\Delta L/L_0$.

LONGITUDINAL STRETCHING OR COMPRESSION

$$\text{Strain} \equiv \frac{\Delta L}{L_0} \tag{10–57}$$

Similarly, if an object is submerged in water [Figure 10–36(b)] and subjected to an increase in pressure from P to $P + \Delta P$, its volume will change from V_0 to $V_0 + \Delta V$, leading to a *strain* of $\Delta V/V_0$.

VOLUME COMPRESSION

$$\text{Strain} \equiv \frac{\Delta V}{V_0} \tag{10–58}$$

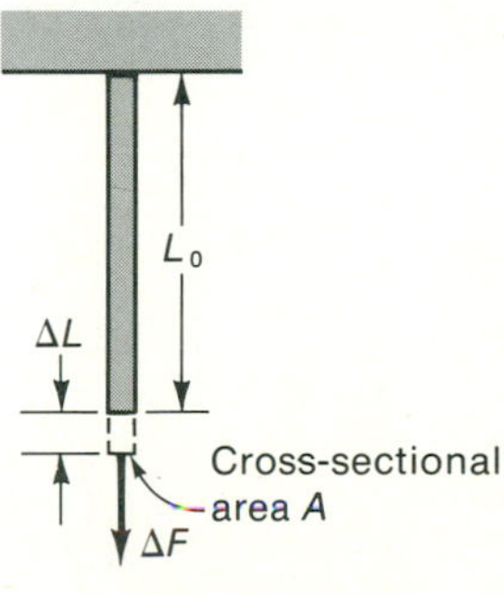

(a) A longitudinal *stress* $\Delta F/A$ is caused by the force ΔF normal to the cross-sectional area A. The resultant *strain* is $\Delta L/L_0$.

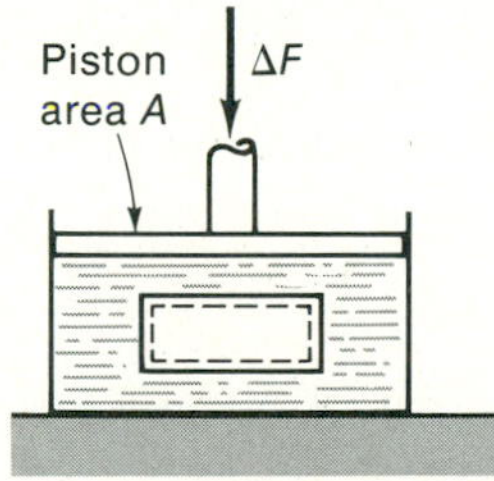

(b) A block is subjected to a compression *stress* of the pressure $\Delta P = \Delta F/A$. The resultant compressional *strain* is $\Delta V/V_0$.

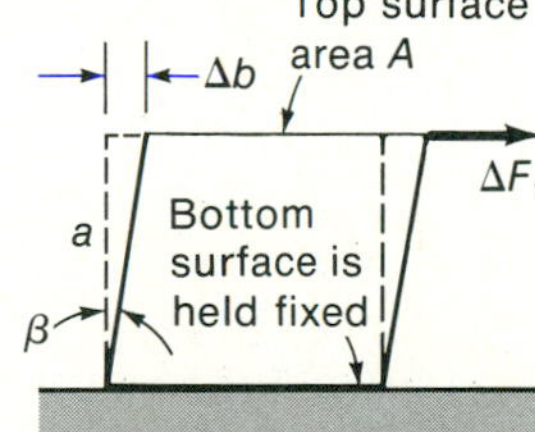

(c) A rectangular block undergoes a shearing *stress* of $\Delta F_t/A$ due to the force ΔF_t *tangent* to the top surface area A. Tan $\beta = \Delta b/a$ is a measure of the shearing *strain*.

Figure 10–36
Three different ways of deforming matter.

If an object is subjected to shear, as in Figure 10–36(c), the *strain* is

SHEARING

$$\text{Strain} \equiv \tan\beta = \frac{\Delta b}{a} \tag{10–59}$$

In each case, the elastic modulus is the proportionality constant between stress and strain; that is, (stress) = (modulus)(strain). If the stress is closely *proportional* to the strain, the modulus is essentially constant and the material is said to obey Hooke's Law. The following expressions for the elastic moduli are valid as long as the fractional change in the physical dimension is very small (that is, as long as $\Delta L \ll L_0$, $\Delta V \ll V_0$, and the angle β is very small).

YOUNG'S MODULUS (Longitudinal stretching or compression)

$$Y = \frac{\text{stress}}{\text{strain}} = \frac{\left(\dfrac{\Delta F}{A}\right)}{\left(\dfrac{\Delta L}{L_0}\right)} \tag{10–60}$$

BULK MODULUS (Volume compression)

$$B = \frac{\text{stress}}{\text{strain}} = -\frac{\left(\dfrac{\Delta F}{A}\right)}{\left(\dfrac{\Delta V}{V_0}\right)} = -\frac{\Delta P}{\left(\dfrac{\Delta V}{V_0}\right)} \tag{10–61}$$

where ΔP is the change in pressure. (A minus sign in the definition gives a positive value to B, since an *increase* in pressure corresponds to a *decrease* in volume.)

SHEAR MODULUS (Shearing)

$$S = \frac{\text{stress}}{\text{strain}} = \frac{\left(\dfrac{\Delta F_t}{A}\right)}{\left(\dfrac{\Delta b}{a}\right)} = \frac{\left(\dfrac{\Delta F_t}{A}\right)}{\tan\beta} \tag{10–62}$$

Representative values are listed in Table 10–2.

All substances have an *elastic limit*, beyond which recovery from the deformation is not complete. When this happens, the object is said to acquire a *permanent set.*

TABLE 10–2

Approximate Values of Elastic Moduli

Material	Young's Modulus Y $\left(\text{in } \frac{\text{N}}{\text{m}^2}\right)$	Bulk Modulus $\left(\text{in } \frac{\text{N}}{\text{m}^2}\right)$	Shear Modulus $\left(\text{in } \frac{\text{N}}{\text{m}^2}\right)$
Glass (crown)	7×10^{10}	5×10^{10}	3×10^{10}
Aluminum	7×10^{10}	7.5×10^{10}	2.4×10^{10}
Cast bronze	8.1×10^{10}	9.6×10^{10}	3.4×10^{10}
Copper	12.3×10^{10}	13.1×10^{10}	4.5×10^{10}
Steel	20.6×10^{10}	18.1×10^{10}	8.9×10^{10}
Tungsten	35×10^{10}	31×10^{10}	14×10^{10}
Water	—	0.21×10^{10}	—
Mercury	—	2.8×10^{10}	—

Anisotropic materials have physical properties that vary with direction. For example, in a piece of wood, the grain structure causes the moduli to have different values along different axes. Single crystals are anisotropic if the atomic arrangement along one direction is different from that along another direction. Most metallic solids are *polycrystalline* (that is, they are made up of microscopic crystals oriented randomly), so their individual anisotropic properties average out and the material as a whole behaves as an isotropic substance.

EXAMPLE 10–26

A steel piano wire 1.12 m long has a cross-sectional area of 6×10^{-3} cm^2. When under a tension of 115 N, how much does it stretch?

SOLUTION

Young's modulus for steel (Table 10–1) is $Y = 20.6 \times 10^{10}$ N/m^2. From Equation (10–59), we obtain

$$Y = \frac{\left(\dfrac{\Delta F}{A}\right)}{\left(\dfrac{\Delta L}{L_0}\right)}$$

$$\Delta L = \frac{L_0 \Delta F}{YA} = \frac{(1.12\text{ m})(115\text{ N})}{\left(20.6 \times 10^{10}\,\dfrac{\text{N}}{\text{m}^2}\right)(6 \times 10^{-7}\text{ m}^2)}$$

$$\Delta L = 1.04 \times 10^{-3}\text{ m} = \boxed{1.04\text{ mm}}$$

EXAMPLE 10–27

If a cubic meter of ocean water on the surface sinks to a depth of about 3 km, where the pressure is 300 atm, what change in volume occurs? Assume that sea water has the same compressibility as fresh water. (*Note:* One atmosphere is equal to 1.013 × 105 N/m^2.)

SOLUTION

Conversion of pressure units gives

$$(300\text{ atm})\underbrace{\left(\frac{1.013 \times 10^5\,\dfrac{\text{N}}{\text{m}^2}}{1\text{ atm}}\right)}_{\text{Conversion ratio}} = 3.04 \times 10^7\,\frac{\text{N}}{\text{m}^2}$$

We use the bulk modulus for compression:

$$B = -\frac{\Delta P}{\dfrac{\Delta V}{V_0}}$$

Solving for ΔV and substituting numerical values (including $B = 0.21 \times 10^{10}$ N/m^2 from Table 10–1), we have

$$\Delta V = -\frac{V_0 \Delta P}{B} = -\frac{(1\text{ m}^3)\left(3.04 \times 10^7\,\dfrac{\text{N}}{\text{m}^2}\right)}{\left(0.21 \times 10^{10}\,\dfrac{\text{N}}{\text{m}^2}\right)} = \boxed{-1.45 \times 10^{-2}\text{ m}^3}$$

Thus, the new volume of the cubic meter of sea water is $(1 - 0.0145)\ \mathrm{m}^3$, or $0.9855\ \mathrm{m}^3$. In spite of this small compressibility, it has been estimated that the mean sea level would be about 30 m higher than it is if water were truly incompressible.

Summary

The location of the *center of mass* (*CM*) is defined as follows:

For a system of several mass points Δm_i	**For a solid object with distributed mass**
$x_{CM} = \dfrac{\Sigma x_i \Delta m_i}{\Sigma \Delta m_i}$	$x_{CM} = \dfrac{\int x\,dm}{\int dm}$

with similar equations for y and z.

The *center of gravity* (*CG*) of a solid object or system of particles is the single point at which we may consider the total force of gravity $\mathbf{W}$ to be concentrated for the purpose of calculating the torque due to gravity on the object. The *CG* is found from equations similar to the x_{CM} calculations, except that the summation (or integration) is over the variable $\Delta W_i = g\,\Delta m_i$ (or $dW = g\,dm$). If g is uniform over the entire body, the *CG* and *CM* coincide.

The *moment of inertia* I about a given axis is:

$$I = \int r^2\,dm$$

(integration over the entire body)

The element of mass dm is equal to $\rho\,dV$ (where ρ is the density and dV is the volume element). Note that dV must be chosen so that the entire element is the *same distance* r from the axis of rotation. Figure 10–20 gives the moments of inertia for several common symmetrical objects.

The *radius of gyration* k is the distance from the axis of rotation at which we may assure the entire mass of an object is concentrated for the purpose of calculating the moment of inertia: $I = Mk^2$.

The *parallel-axis theorem* gives the moment of inertia I about any axis parallel to an axis through the *CM* of an object. If the distance between the two parallel axes is d, and if I_{CM} is the moment of inertia about an axis through the *CM*, then:

$$I = I_{CM} + Md^2$$

Defining axial vectors $\boldsymbol{\omega}$ and $\boldsymbol{\alpha}$ according to the right-hand rule, we express the *angular momentum* $\mathbf{L}$ as $\mathbf{L} = I\boldsymbol{\omega}$. (Both $\mathbf{L}$ and I must refer to the *same* axis of rotation.)

When symmetrical objects undergo *rotation* about an axis of symmetry, or *translation in which the axis maintains its direction in space*, or both, then the motion is most easily analyzed by applying Newton's second laws for translation of the *CM* and rotation about the *CM*:

	Accelerated motion keeping the axis parallel	**Static or dynamic equilibrium**
Translation:	$\Sigma\mathbf{F}_{\text{ext}} = M\mathbf{a}_{CM}$	$\Sigma\mathbf{F}_{\text{ext}} = 0$

Rotation:

$$\underset{\text{(only about the } CM)}{\Sigma\tau_{\text{ext}}} = I\alpha \qquad \underset{\text{(about any axis)}}{\Sigma\tau_{\text{ext}}} = 0$$

In analogy with Newton's second law for the linear case, $F = d\mathbf{p}/dt$, the most general rotational expression is

$$\underset{\text{(about any axis)}}{\tau} = \frac{d\mathbf{L}}{dt}$$

which is valid in all cases.

A particle undergoing rotational motion (constant r) may be described using the tangential parameters of *arc length* s, *speed* v (always tangent to the path), and the *tangential acceleration component* a_t. An alternative description in terms of the rotational parameters θ, ω, and α is also possible. They are related as follows:

$$\left.\begin{aligned} s &= r\theta \\ v &= r\omega \\ a_t &= r\alpha \end{aligned}\right\} \text{ The angular parameters } \theta, \omega, \text{ and } \alpha \text{ must be in } radian \text{ measure}$$

Keeping these relations in mind, the *kinematic equations for rotational motion* of an extended object (constant angular acceleration α) closely resemble the linear kinematic equations.

The Kinematic Equations for Circular Motion

Tangential motion along the circumference (where s is the arc length and a_t is the tangential acceleration component)		Angular motion	
$v = v_0 + a_t t$	(only for constant a_t)	$\omega = \omega_0 + \alpha t$	(only for constant α)
$s = s_0 + v_0 t + \frac{1}{2}a_t t^2$		$\theta = \theta_0 + \omega_0 t + \frac{1}{2}\alpha t^2$	
$v^2 = v_0{}^2 + 2a_t(s - s_0)$		$\omega^2 = \omega_0{}^2 + 2\alpha(\theta - \theta_0)$	

In the absence of external torques, the angular momentum $\mathbf{L}$ (about O) for an isolated system remains constant in time. Thus we have the *conservation of angular momentum*:

$$\mathbf{L}_0 = \mathbf{L}$$
$$I_0\omega_0 = I\omega$$

For an isolated conservative system, the *conservation of mechanical energy* holds:

$$E_0 = E$$

where

$$\underbrace{E}_{\substack{\text{Total}\\ \text{energy}}} = \underbrace{\tfrac{1}{2}\kappa\theta^2}_{\substack{\text{Potential}\\ \text{energy stored}\\ \text{in a spiral}\\ \text{spring obeying}\\ \tau = -\kappa\theta}} + \underbrace{Mgh_{CM}}_{\substack{\text{Gravitational}\\ \text{potential}\\ \text{energy of}\\ \text{the } CM}} + \underbrace{\tfrac{1}{2}Mv_{CM}{}^2}_{\substack{\text{Translational}\\ \text{kinetic energy}\\ \text{of the } CM}} + \underbrace{\tfrac{1}{2}I\omega^2}_{\substack{\text{Rotational}\\ \text{kinetic energy}\\ \text{about the } CM}}$$

If the system is not isolated, the *work-energy theorem* holds:

For *translational* motion of the *CM*: [Work done by the net resultant *force*] = [Change in *translational* kinetic energy of the *CM*]

For *rotational* motion about the *CM*: [Work done by the net resultant *torque*] = [Change in *rotational* kinetic energy about the *CM*]

When a torque is applied to change the direction of the axis of a rotating gyroscope, the axis will change its orientation with an angular velocity $\boldsymbol{\omega}_p$ (called *precession*) according to:

$$\tau = \omega_p \times \mathbf{L}$$

The direction of precessional motion may be found from the following: *The angular momentum vector* **L** *tends to seek the direction of the applied torque vector* $\boldsymbol{\tau}$.

For small deformations that obey Hooke's Law, the elastic properties of a homogeneous and isotropic material are expressed by three *elastic moduli*, each of which is the ratio of a *stress* to a *strain:*

YOUNG'S MODULUS (Longitudinal stretching or compression)

$$Y = \frac{\text{stress}}{\text{strain}} = \frac{\left(\dfrac{\Delta F}{A}\right)}{\left(\dfrac{\Delta L}{L_0}\right)}$$

BULK MODULUS (Volume compression)

$$B = \frac{\text{stress}}{\text{strain}} = -\frac{\left(\dfrac{\Delta F}{A}\right)}{\left(\dfrac{\Delta V}{V_0}\right)} = -\frac{\Delta P}{\left(\dfrac{\Delta V}{V_0}\right)}$$

SHEAR MODULUS (Shearing)

$$S = \frac{\text{stress}}{\text{strain}} = \frac{\left(\dfrac{\Delta F_t}{A}\right)}{\left(\dfrac{\Delta b}{a}\right)} = \frac{\left(\dfrac{\Delta F_t}{A}\right)}{\tan\beta}$$

Questions

1. Consider a circus aerialist flying through the air, turning somersaults as he goes. Can he change the path of his center of mass from that of free-fall motion?
2. A ladder leans against a wall. As a woman climbs to the top, is the ladder more likely to slip when she is near the top or near the bottom? Why?
3. How can a man weighing more than 200 lb weigh himself on a bathroom scale that has a capacity of only 190 lb?
4. Support a yardstick horizontally from below using one finger of each hand. Gradually draw your fingers closer together. No matter where your fingers start from, they always meet at the center of the yardstick. Explain.

5. Explain how a wind blowing uniformly in a straight line can exert a torque on windmill vanes to set them into rotation.

6. Suppose you have a sphere made of a material whose density you do not know. Can you determine whether the sphere is solid or a hollow, spherical shell without drilling a hole in it?

7. A woman on a rotating stool holds two equal masses at arm's length. While the stool is rotating, she drops the masses. Does her own angular momentum change? Can you apply conservation of angular momentum to this situation?

8. When a moving automobile is brought to rest, what happens to the angular momentum of the wheels? Is conservation of angular momentum violated in this instance?

9. Why are all satellites launched from Cape Canaveral in an eastward rather than a westward direction?

10. Most helicopters have a small propeller mounted in the tail with the axis of rotation in a horizontal position perpendicular to the longitudinal axis of the helicopter. Why?

11. Two uniform, solid disks are in a race down an incline. They have equal masses, but the radius of one is twice the radius of the other. Which wins? If they had the same radius but unequal masses, which would win?

12. A uniform, solid sphere and a uniform, solid disk are in a race down an incline. If they have equal masses and radii, which wins? If they have equal masses, is it possible for either to win by changing their relative radii? If their radii are equal, is it possible for either to win by changing their relative masses?

13. A disk is mounted on frictionless bearings with the axis in a vertical position. A mouse sits on the outer rim with the disk initially rotating. The mouse now walks in toward the axis along a line painted radially on the disk. What happens to the angular speed of the disk? Considering just the forces on the disk alone, what is the source of the torque that changes the disk's angular speed?

14. Suppose a million motorists at the equator, each with a 2000-kg car, started from rest and began to drive eastward with a speed of 100 m/s. Could one experimentally verify within 24 h that the length of the day had been altered? The best atomic clocks are accurate to about 1 part in 10^{11}.

15. A satellite revolves about the earth in a circular orbit. Explain whether the following quantities remain constant or change during the motion: kinetic energy, gravitational potential energy, linear momentum, and angular momentum. Which are constant and which are variable for an elliptical orbit?

16. Suppose the power output of a motor is directly proportional to the angular speed ω of the shaft. Find a relation between the torque exerted by the shaft and the angular speed.

Problems

10A–1 A man sits on one end of a uniform plank of length ℓ. The plank is at rest, supported on small frictionless wheels on horizontal rails. The man and the plank have equal masses m. The man now walks to the other end of the plank and sits down. Find the total distance (relative to the ground) the plank moves during the process (Hint: Does the *CM* of the man-plus-plank change its location during the process?) *Answer:* $\ell/2$

10A–2 A circular hole of radius 8 cm is cut out of a thin rectangular metal sheet with outside dimensions of 20 cm × 30 cm. The center of the hole is along the center line of the rectangle, 20 cm from one end. Find the center of mass. (Hint: Treat the missing portion as "negative mass" to be combined with the original uncut object.)

10A–3 A uniform rod 3 m long is pivoted at one end about a vertical axis so that the rod can rotate freely in the x-y plane. A force of 16 N is applied to the free end at an angle of $\theta = 30°$, as shown in Figure 10–37.

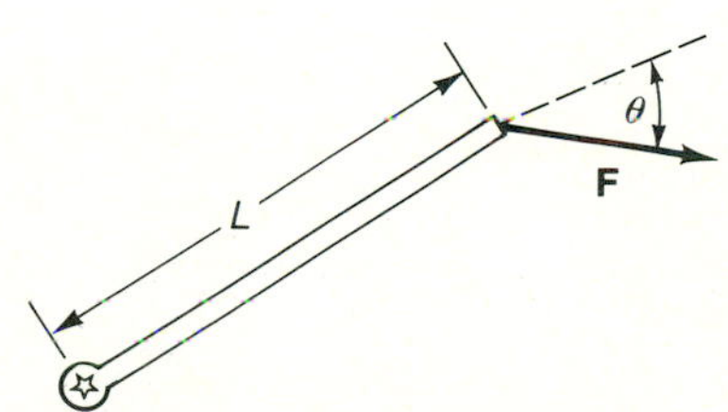

Figure 10–37
Problem 10A–3

(a) Find the moment arm of the force about the axis.
(b) What torque about the axis is applied to the rod?

Answers: (a) 1.50 m (b) 24.0 N·m

10A–4 A uniform chain of mass m and length ℓ hangs from a hook, as shown in Figure 10–38(a). Determine the amount of work required to raise the middle link to the hook, as shown in (b). (Hint: Consider what happens to the center of mass of various portions of the chain.)

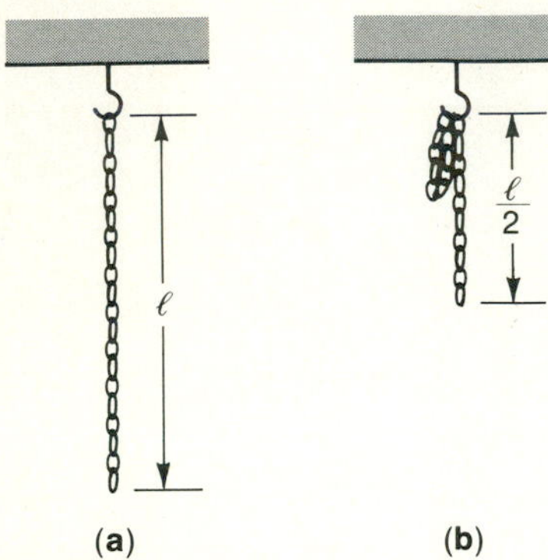

Figure 10–38
Problems 10A–4 and 10C–6

10A–5 Consider a square of edge length b. At each corner, a force is applied in the plane of the square, parallel to an edge, so that each force produces a torque about the center in the same direction. Each force has the same magnitude.
(a) Find the net torque about the center of the square.
(b) Compute the torque about one corner.

Answers: (a) $2bF$ (b) $2bF$

10A–6 In Figure 10–39, the load weighs 200 N and the beam has negligible weight. Find (a) the tension T in the cable as indicated and (b) the horizontal and vertical components of the force that the pivot hinge exerts on the beam.

10A–7 A uniform rod of length ℓ weighing 300 N is hung from the ceiling by a hinge at one end.
(a) What horizontal force applied to the free end will hold the rod at 37° with respect to the horizontal?
(b) Find the magnitude and direction of the force the hinge at B exerts against the lever.

Answers: (a) 200 N (b) 361 N at 56.3° up from the horizontal

10A–8 A uniform 12-ft ladder weighing 40 lb leans against the corner of the wall shown in Figure 10–40. The contact at the wall corner is essentially frictionless. Find the force the wall exerts against the ladder.

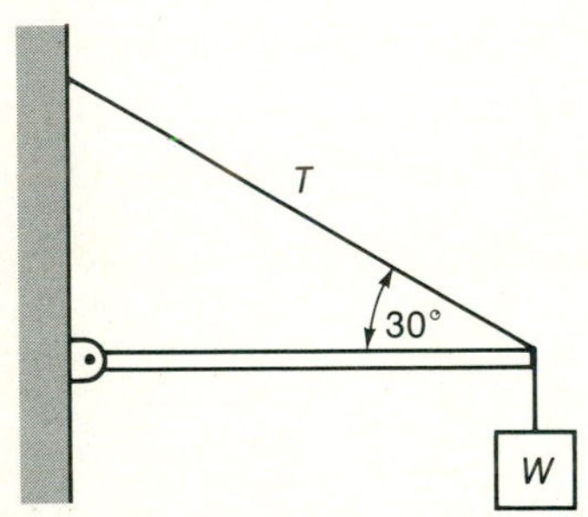

Figure 10–39
Problems 10A–6 and 10B–9

10A–9 In Figure 10–41, the load weighs 200 N and the beam has negligible weight. Find (a) the tension T in the cable as indicated and (b) the horizontal and vertical components of the force the pivot hinge exerts on the beam.

Answers: (a) 200 N (b) 173 N toward right, 100 N up

10A–10 A sphere of mass m rests in the corner formed by a vertical wall and a plane inclined at an angle θ with respect to the horizontal. All surfaces are smooth. Find the magnitude of the forces the wall and the plane exert on the sphere.

10A–11 A uniform cylinder weighing 200 N is suspended by cables so that it leans against a smooth wall, as shown in Figure 10–42. What force does the cylinder exert against the wall?

Answer: 115 N

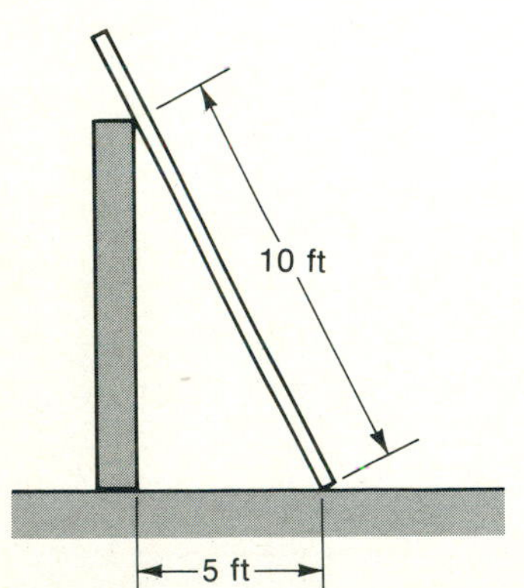

Figure 10–40
Problems 10A–8 and 10B–6

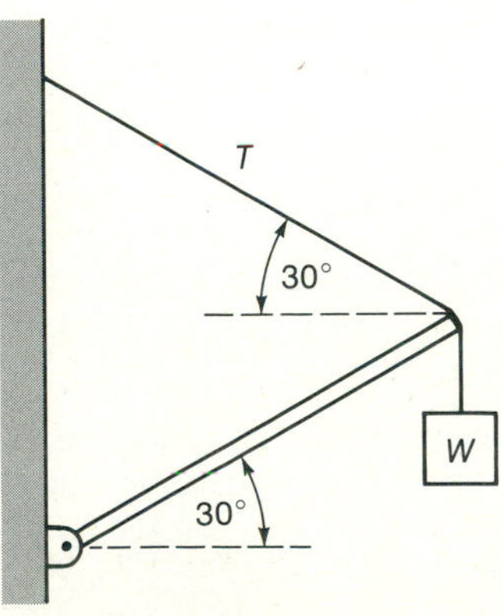

Figure 10–41
Problems 10A–9 and 10B–10

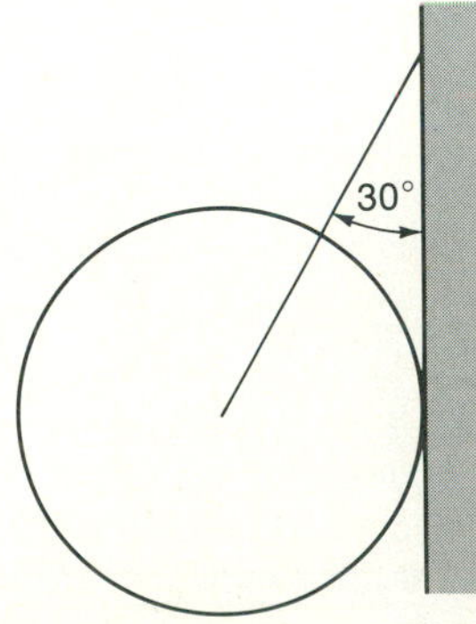

Figure 10–42
Problem 10A–11

10A–12 A man exerts a horizontal force of 180 N in the test apparatus shown in Figure 10–43. Find the horizontal force his biceps muscle exerts on his forearm.

10A–13 List the following in order of increasing angular speed and express their values in radians per second: the minute hand of a clock, the hour hand of a clock, an astronaut in a circular orbit about the earth during a 100-min period, and a grindstone turning at 6000 rpm.

Answers: astronaut: 1.05×10^{-3} rad/s, hour hand: 1.75×10^{-3} rad/s, minute hand: 0.105 rad/s, grindstone: 628 rad/s

10A–14 Devise a conversion ratio for converting revolutions per minute to radians per second (and vice versa).

10A–15 A phonograph turntable is turning initially at $33\frac{1}{3}$ rpm. When the power to the turntable is turned off, the turntable slows down at a constant rate of 0.20 rad/s^2.

(a) How many seconds elapse before the turntable stops?

(b) How many revolutions will the turntable make before stopping?

Answers: (a) 17.4 s (b) 4.85 rev

10A–16 A flywheel has a mass of 500 kg and a radius of gyration of 60 cm.

(a) What constant torque will give the flywheel an angular speed of 600 rpm in 20 s, starting from rest?

(b) If the radius of the flywheel is 80 cm, what constant tangential force would produce this torque?

10A–17 A uniform solid sphere (mass M, radius R) rolls without slipping down an incline at angle θ with respect to the horizontal. It starts from rest and rolls a distance s along the incline. Find the linear speed v of the CM of the sphere at the bottom by two methods: (a) by using conservation of energy principles and (b) by applying Newton's laws.

Answer: $\sqrt{\dfrac{10gs \sin \theta}{7}}$

10A–18 A regulation basketball has a 10 in diameter and weighs 20 oz. It may be approximated as a thin, spherical shell of negligible wall thickness. Starting at rest, how long will a basketball take to roll without slipping 12 ft down an incline at 30° with respect to the horizontal? (Hint: Use conservation of energy to find the speed at the bottom, then apply the kinematic equations.)

10A–19 A ball of diameter D rolls without slipping along a horizontal table top with constant speed v. The ball rolls off the edge, falling to the floor a vertical distance h below. While in the air, how many revolutions does the ball make?

Answer: $\dfrac{v}{\pi D}\sqrt{\dfrac{2h}{g}}$

10A–20 A uniform solid sphere and a uniform solid cylinder, each with mass M and radius R, have a race (from rest) rolling without slipping down an inclined plane. Which (if either) wins? Explain your reasoning.

10A–21 A cube of gelatin 7 cm on an edge is resting in a dish. Pushing horizontally on the top face with a force of 0.2 N causes the top to undergo a displacement of 3 mm. Calculate the shear modulus for the gelatin. *Answer:* 952 N/m^2

10A–22

(a) Find the minimum diameter of a steel wire 18 m long that will elongate no more than 9 mm when a load of 380 kg is hung on the lower end.

(b) If the elastic limit for this steel is 3×10^8 N/m^2, will permanent deformation occur?

10A–23 At a depth of 400 km below the earth's surface, the pressure is 1.36×10^{10} Pa. If a cubic centimeter of copper were brought to the surface from this depth, what would be its volume? *Answer:* 1.104 cm^3

10B–1 In Figure 10–44, the scales read $W_1 = 95$ lb and $W_2 = 85$ lb. Neglecting the weight of the supporting plank, how far from the woman's feet is her center of mass?

Answer: 3.17 ft

10B–2 A circular area of radius R is cut from a uniform sheet of metal. A circular hole is now cut out of this area, as shown in Figure 10–45. The radius of the hole is $R/3$, and

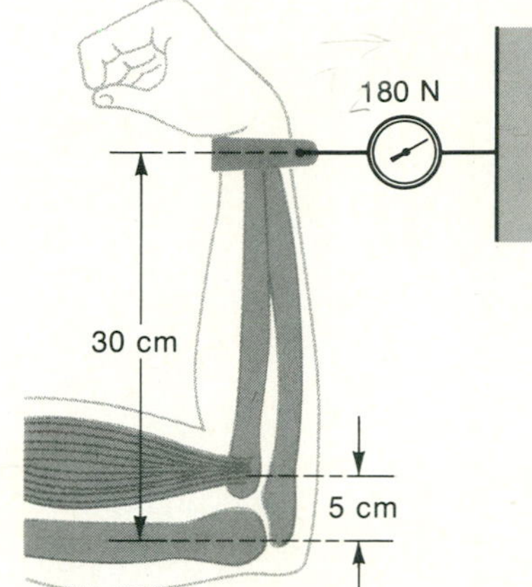

Figure 10–43
Problem 10A–12

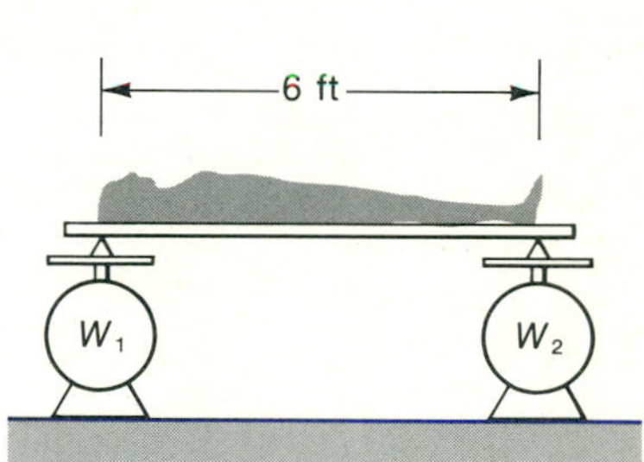

Figure 10–44
Problem 10B–1

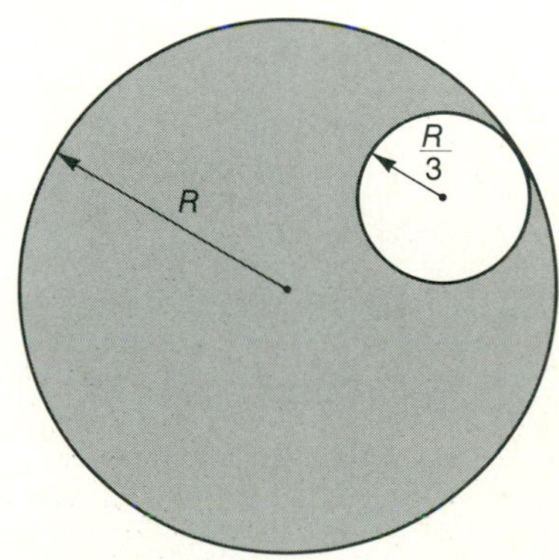

Figure 10–45
Problem 10B–2

the hole is tangent to the outer rim of the area. How far from the center of the original circle is the center of mass? (Hint: Treat the missing material as "negative mass" to be combined with the original uncut area.)

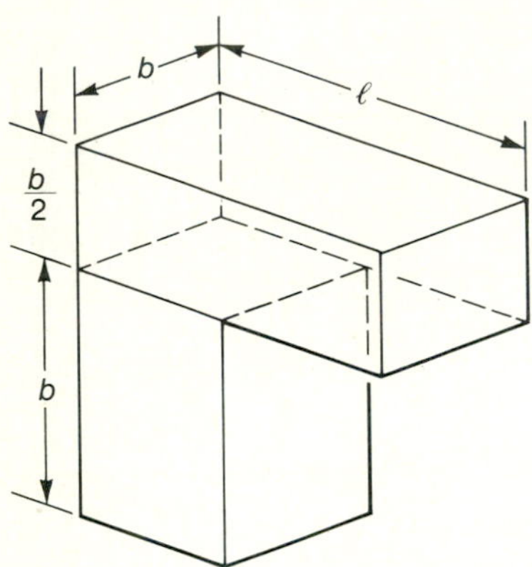

Figure 10–46
Problem 10B–3

10B–3 Two solid blocks made of the same uniform material are glued together as shown in Figure 10–46. One block is a cube of edge b, and the other is a rectangular block of dimensions $b \times b/2 \times \ell$. Find the maximum length ℓ in terms of b such that the combination will not tip over. *Answer:* $b(1 + \sqrt{3})$

10B–4 Two identical uniform bricks of length L are placed in a stack over the edge of a horizontal surface with the maximum overhang possible without falling (see Figure 10–47). Find the distance x.

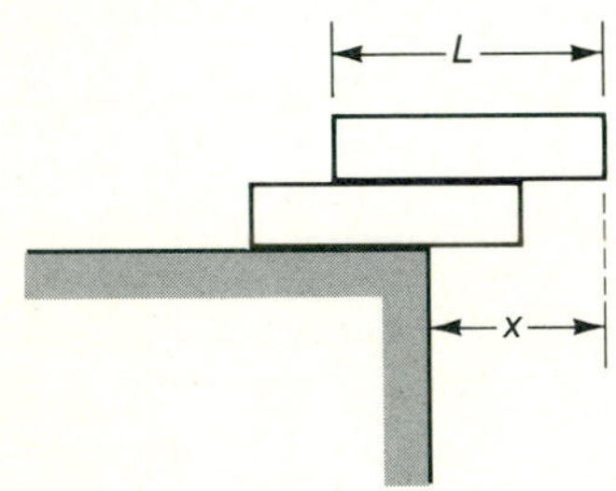

Figure 10–47
Problems 10B–4 and 10C–3

10B–5 A ladder of length ℓ leans against a smooth wall at an angle θ with respect to the horizontal. A man climbs to within $\ell/4$ from the top, at which point the ladder begins to slip along the floor. Show that the coefficient of static friction between the ladder and floor is $3/(4 \tan \theta)$. Neglect the weight of the ladder in comparison with the man's weight.

10B–6 A uniform 12-ft ladder weighing 20 lb leans against the wall, as shown in Figure 10–40. The contact at the wall corner is frictionless. Find (a) the force the wall exerts on the ladder and (b) the force the ground exerts on the ladder. (Include the magnitude and direction of each.)

10B–7 Figure 10–48 shows a claw hammer as it pulls a nail out of a horizontal surface. If 40 lb of force is exerted horizontally as shown, find (a) the force exerted by the hammer claws on the nail and (b) the force exerted by the surface on the point of contact with the hammer head. Assume the force the hammer exerts on the nail is parallel to the nail.

Answers: (a) 277 lb (b) 260° at 67.7° with respect to horizontal

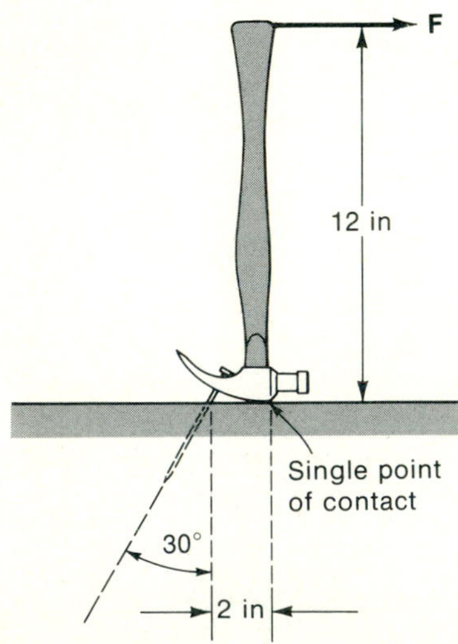

Figure 10–48
Problem 10B–7

10B–8 A flexible chain weighing 40 N hangs between two hooks located at the same height. At each hook, the chain makes an angle of 42° with the horizontal.

(a) Find the magnitude of the force each hook exerts on the chain.
(b) Find the tension in the chain at its midpoint.

(Hint: Make a free-body diagram for half the chain.)

10B–9 In Figure 10–39, the load weighs 200 N and the uniform beam weighs 100 N. Find (a) the tension T in the cable as indicated and (b) the horizontal and vertical components of the force the pivot hinge exerts on the beam.

Answers: (a) 500 N (b) $F_x = 433$ N, $F_y = 50$ N

10B–10 In Figure 10–41, the load weighs 200 N and the uniform beam weighs 100 N. Find (a) the tension T in the cable as indicated and (b) the horizontal and vertical components of the force the pivot hinge exerts on the beam.

10B–11 A man pushes a broom along the floor by exerting a force downward along the handle of the broom. If the coefficient of kinetic friction between the floor and broom is μ_k, find the minimum angle between the broom handle and the vertical in order to move the broom at a constant speed. Ignore the mass of the broom. *Answer:* $\theta = \tan^{-1} \mu_k$

10B–12 An object weighing 40 N is held at arm's length with the forearm horizontal. Referring to Figure 10–49, calculate the force **F** that the brachialis muscle exerts, as shown. Assume that the mass of the hand and forearm together is 1.5 kg, with the center of mass located midway between the elbow joint and the object.

10B–13 In Figure 10–50 the boy tries to roll a lawn roller (mass = 70 kg) over a curb 6 cm high. The radius of the roller is 30 cm. What minimum horizontal force is required?

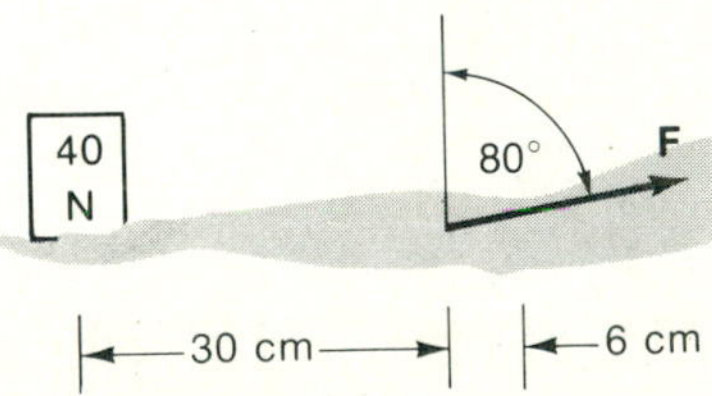

Figure 10–49
Problem 10B–12

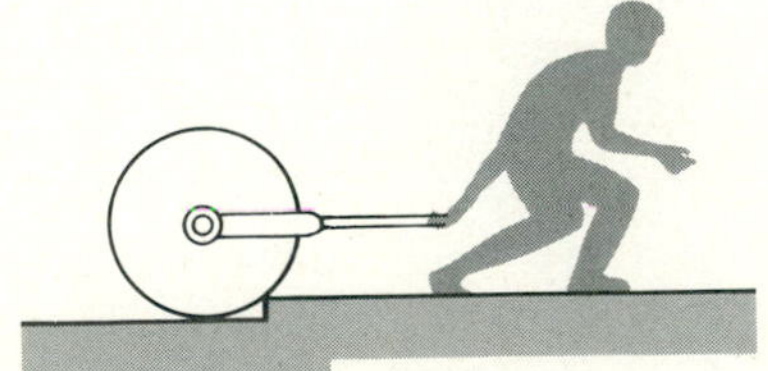

Figure 10–50
Problem 10B–13

(Hint: Consider the situation where the roller is just barely lifted off the ground and is held at rest against the curb corner. The horizontal force necessary to hold the roller in this position is the "dividing-line" value—any greater force will cause the roller to rotate on over the curb.) *Answer:* 525 N

10B–14 Figure 10–51 shows an L-shaped control lever, hinged at B.

(a) What force **F** applied at right angles to the handle (as shown) will cause the lever to exert a horizontal force of 30 lb against an object (not shown) at point A?

(b) Find the magnitude and direction of the force the hinge at B exerts against the lever.

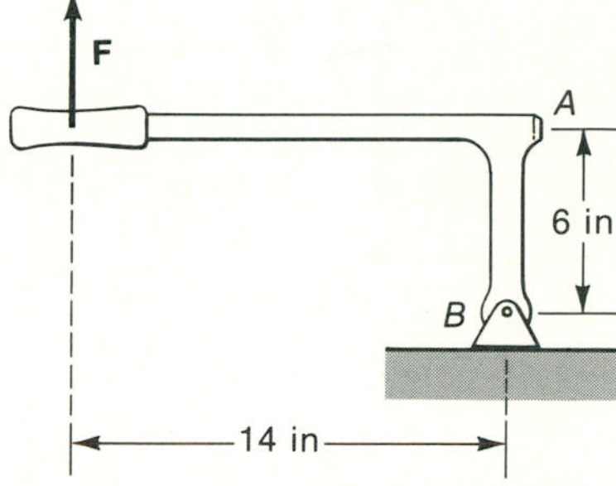

Figure 10–51
Problem 10B–14

10B–15 A hollow cylinder of mass 40 kg has inner and outer radii of 2 cm and 4 cm, respectively. Find the radius of gyration about the axis of the cylinder. *Answer:* 3.16 cm

10B–16 The string of a Yo-yo is wound around an inner shaft of radius 2 mm. The mass of the Yo-yo is 200 g and its radius of gyration is 2 cm. (Ignore the thickness of the string itself.)

(a) When released from rest, how long does it take to unwind 1 m of string?

(b) While descending, what is the ratio of the rotational kinetic energy of the Yo-yo to its translational kinetic energy?

10B–17 A cloth tape is wound around the outside of a uniform solid cylinder (mass M, radius R) and fastened to the ceiling, as shown in Figure 10–52. The cylinder is held with the tape vertical and then released from rest.

(a) While descending, does the CM of the cylinder move to the left, the right, or straight down? Explain your reasoning.

(b) Show that the acceleration of the CM is $2g/3$.

10B–18 A boy rolls a thin hoop (of mass M and radius R) along the ground by pushing horizontally on it with a small, frictionless wheel mounted on the end of a stick, as shown in Figure 10–53. He thus exerts a horizontal force through the center of mass of the hoop without causing any tangential force on the rim.

(a) Sketch a free-body diagram for the hoop.

(b) If the coefficient of static friction between the hoop and the ground is μ, what maximum force F can be applied to the rolling hoop without causing it to slide?

(Note: The maximum force is *not* equal to the maximum static friction force.)

10B–19 A uniform solid cylinder rolls without slipping down an incline at an angle θ with respect to the horizontal. Show that the minimum coefficient of static friction to prevent slipping is $(\tan \theta)/3$.

10B–20 A carnival stuntman rides a motorcycle on the inside vertical wall of a cylinder, as shown in Figure 10–54. His CM travels in a horizontal circle of radius 4 m. The coefficient of static friction between the tires and wall is 0.70. (a) Find the minimum speed v and (b) the corresponding angle θ that will enable him to maintain this motion.

10B–21 As a motorcyclist travels around a curve on a horizontal roadbed with a speed of 30 km/h, her center of mass travels in a circle of radius 10 m. Find the angle to the vertical at which she leans. *Answer:* 35.3°

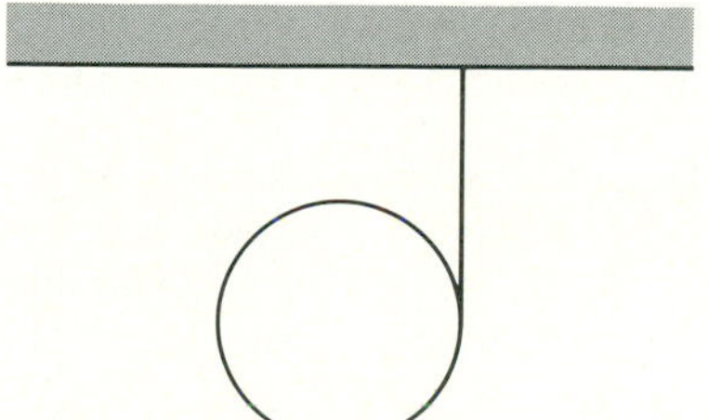

Figure 10–52
Problem 10B–17

Figure 10–53
Problem 10B–18

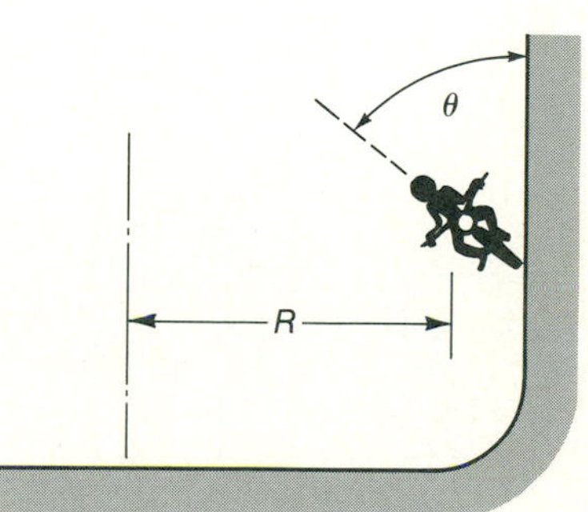

Figure 10–54
Problem 10B–20

10B–22 Two uniform solid cylinders, one with mass M and radius R, the other with mass $2M$ and radius $R/2$, have a race (from rest) rolling without slipping down an inclined plane. Which (if either) wins? Explain your reasoning.

10B–23 This problem describes a method of determining the moment of inertia of an irregularly shaped object such as the payload for a satellite. Sometimes satellites are set into rotation to provide scanning for cameras, so it is important to know the moment of inertia of all components. Figure 10–55 shows one method of determining them experimentally. A mass m is suspended by a cord wound around the inner shaft (radius r) of a turntable supporting the object. When the mass is released from rest, it descends uniformly a distance h, acquiring a speed v. Show that the moment of inertia I of the equipment (plus turntable) is $mr^2\ (2gh/v^2 - 1)$.

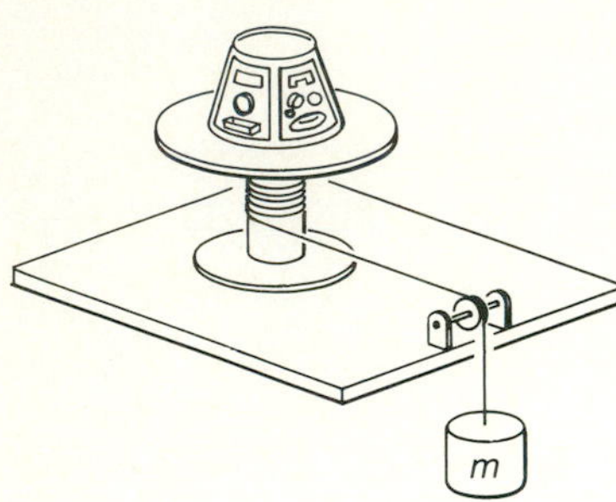

Figure 10–55
Problem 10B–23

10B–24 A 60-kg grindstone in the form of a uniform solid cylinder 50 cm in diameter is brought from rest to an angular speed of 800 rpm in 20 s.

(a) What constant torque was applied to the grindstone?

(b) A metal tool is now pressed against the grindstone with a radially inward force of 20 N. If the coefficient of kinetic friction between the tool and grindstone is 0.50, what power must the motor deliver to the grindstone to maintain its rotational speed of 800 rpm? Assume the bearing friction is negligible.

10B–25 In Figure 10–56, the uniform solid cylinder, mass 5 kg and radius 10 cm, is supported on a frictionless horizontal axis. The blocks, masses 2 kg and 4 kg, are released from rest. There is no slipping between the rope and cylinder. By applying Newton's laws for each block and the cylinder, find (a) the tensions T_1 and T_2 in the rope while the blocks are accelerating and (b) the speed of the 4-kg block after descending 4 m from rest.

Answers: (a) $T_1 = 24.2$ N, $T_2 = 30.0$ N (b) 4.30 m/s

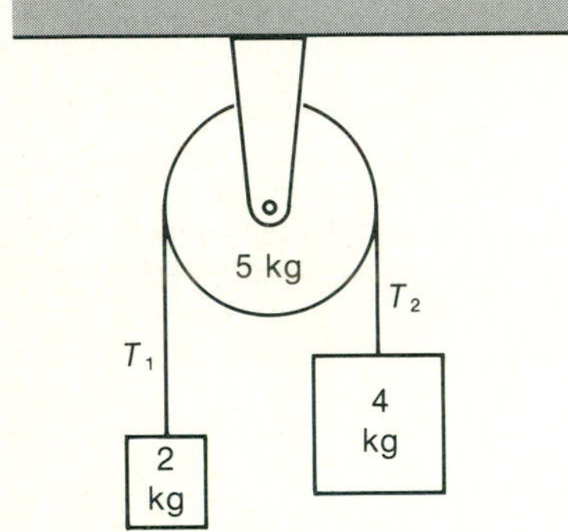

Figure 10–56
Problems 10B–25 and 10B–26

10B–26 Solve question (b) of the previous problem by applying conservation of energy principles.

10B–27 Figure 10–57 shows an elevator and a counterbalance mass m. When loaded, the elevator and passengers have a total mass of 2000 kg, and the counterbalance mass is 2800 kg. Suppose the brake on the driving mechanism fails and the elevator begins to descend from rest, raising the counterbalance. Neglect the mass of the supporting pulleys and sources of friction.

(a) Find the downward acceleration of the elevator.

(b) What is the tension T in the cable while the elevator is descending?

Answers: (a) $0.222g$ (b) 1.52×10^4 N

10B–28 An automobile whose wheels are 90 cm in diameter is traveling at 30 m/s.

(a) Find the angular speed of a wheel about its axle.

(b) The wheels are brought uniformly to rest as they make 40 rev. Find the angular acceleration.

(c) How far does the car travel while slowing down?

10B–29 A race car accelerates at a constant rate while going around a circular track one mile in circumference. The speed of the car increases uniformly from 40 mi/h to 80 mi/h while traversing three-quarters of a mile.

(a) Find the tangential acceleration of the car.

(b) Find the centripetal acceleration when the car is traveling 80 mi/h.

Answers: (a) 3.91 ft/s^2 (b) 16.4 ft/s^2

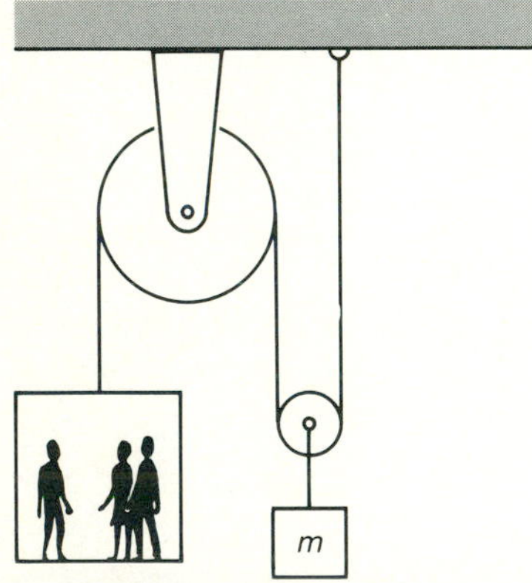

Figure 10–57
Problem 10B–27

10B–30 A rigid body starts at rest and rotates about a fixed axis with constant angular acceleration. Every point not on the axis undergoes a tangential acceleration a_t and a centripetal acceleration a_{cp}. Consider an arbitrary point off the axis. Show that after turning from rest through an angle of one radian, the centripetal acceleration is twice the tangential acceleration.

10B–31 A space station in the form of a large wheel of mass M and radius of gyration k is constructed in a region where gravity is negligible. To provide an artificial gravity for astronauts near the outer rim a distance R from the axis of rotation, the wheel is set into rotation by turning on four rocket motors situated symmetrically as in Figure 10–58.

(a) Find the angular speed ω that will provide an artificial gravity at the rim equal to g, the value at the earth's surface.

(b) If each rocket exerts a tangential force F, for how long t must they be turned on to give the wheel the necessary angular speed (starting from rest)?

Answers: (a) $\sqrt{g/R}$ (b) $\dfrac{Mk^2}{4F}\sqrt{\dfrac{g}{R^3}}$

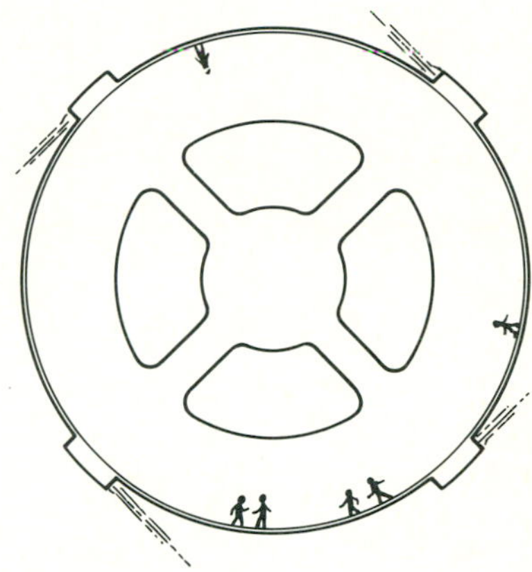

Figure 10–58
Problems 10B–31 and 10B–32

10B–32 Suppose, through an unbelievable error, one of the rockets in the previous problem was mounted backward. All four rockets are turned on for a very short time interval Δt (much shorter than in the previous problem), then turned off. Thus, the rockets are turned off before the wheel moves appreciably. However, the wheel does receive an angular *impulse*. Assuming the wheel starts at rest, derive formulas in terms of the given symbols for the translational and rotational speeds of the wheel relative to an inertial frame of reference after the motors are turned off.

10B–33 A rigid rod 1 m long of negligible mass is pivoted at one end about a vertical axis so that it rotates freely in a horizontal plane. A 0.4-kg mass is attached to the free end. A horizontal force of 5 N is applied perpendicularly to the rod at its midpoint.

(a) Find the angular acceleration of the rod about the pivot.

(b) Starting at rest, through how many radians will it turn in 4 s?

(The force is always perpendicular to the rod.) *Answers:* (a) 6.25 rad/s^2 (b) 50 rad

10B–34 A thin rod of length D and negligible mass has small equal masses m fastened at each end. It is placed at rest on a horizontal, frictionless surface. A ball of putty, also of mass m, slides with speed v toward the rod at right angles to the rod's length, striking and sticking to one of the masses.

(a) Find the angular speed ω of the system just after the collision.

(b) What is the ratio of the initial kinetic energy of the putty to the final kinetic energy of the moving system?

(c) Will the rod subsequently make a 360° rotation? If not, through what maximum angle will the rod rotate?

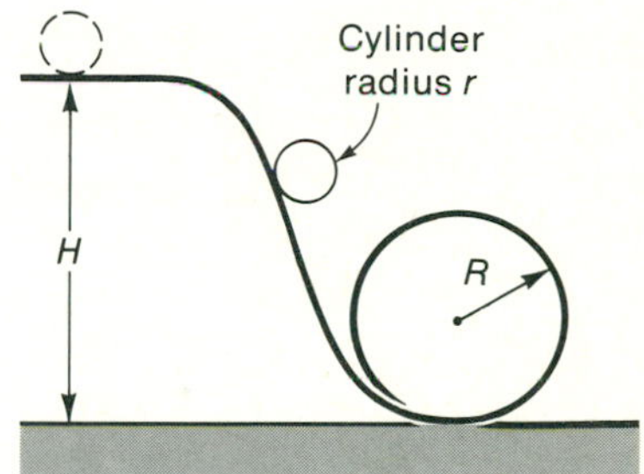

Figure 10–59
Problem 10B–37

10B–35 An amusement park ride is in the form of a horizontal turntable mounted on bearings that have negligible friction. The turntable's radius is 3 m, its mass is 1000 kg, and its radius of gyration is 2 m. The turntable is held stationary while an 80-kg man runs around the outer rim with a speed of 2 m/s. The turntable brakes are now released and the man then stops running and sits down at the outer edge. Find the final rotational speed of the turntable. (Hint: Is the total angular momentum of the system conserved?)

Answer: 0.102 rad/s

10B–36 In the previous problem, the 80-kg man is at the center of the turntable while it rotates freely at 1 rev every 3 s. The man now crawls radially outward on the moving turntable and sits at the outer rim. Find the final speed of the turntable.

10B–37 In Figure 10–59, a uniform solid cylinder of radius r rolls without slipping down the loop-the-loop track, starting from rest on a track a distance H above the bottom of the loop. If the radius of the loop is R (where $R > r$), find the minimum distance H that will ensure that the cylinder maintains contact with the track at all times.

Answer: $[11(R-r)]/4$

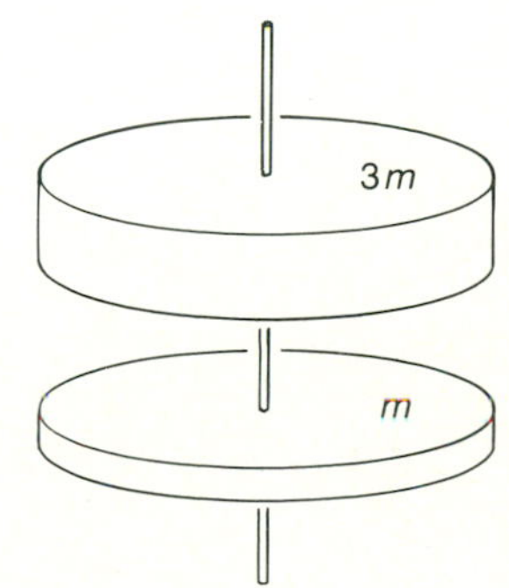

Figure 10–60
Problem 10B–38

10B–38 In Figure 10–60, two uniform disks have masses m and $3m$ and equal radii R. They are mounted as shown on the same vertical axis with frictionless bearings. The upper disk is given an initial angular velocity ω_0 and then allowed to fall onto the lower disk, which is initially at rest. Friction between the disk surfaces causes them to rotate together with a common rotational speed ω. In terms of the given symbols, find (a) the final angular speed ω and (b) the total amount of frictional thermal energy generated while the disks were sliding against each other. (Assume that the vertical distance the top disk falls is very small, so changes of gravitational potential energy are not important here.) (c) What would be a *linear* analogue to this rotational "collision"?

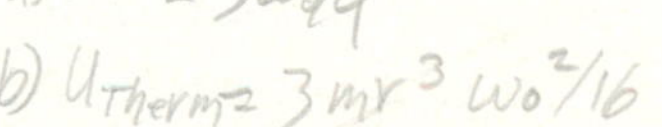

10B–39 It has been proposed that a bus be powered by a single large flywheel mounted on a vertical axis in the lower portion of the bus. The flywheel would be brought up to speed at every bus stop. Investigate the feasibility of such a system, assuming that the bus must develop an average of 100 horsepower during the 15 min between stops, that the flywheel must weigh no more than 2 tons, and that the maximum angular velocity of the flywheel is 100 rpm.

(a) How much rotational energy must the flywheel have before leaving each bus stop?
(b) What must the radius of gyration of the flywheel be?
(c) Do you think this proposal is feasible? *Answer:* (b) 8.50 ft

10B–40 A satellite has a moment of inertia of 21.7 kg·m^2 about its axis. It is desired to rotate the satellite through an angle of 24° by activating a small electric motor whose axis of rotation is parallel to that of the satellite.

(a) If the moment of inertia of the rotating part of the motor is 8×10^{-4} kg·m^2, what total number of revolutions should the motor undergo?
(b) Must the axis of the motor coincide with the axis of the satellite, or could it equally well be placed off-axis (keeping both axes parallel)?

10B–41 A bicyclist is coasting straight ahead. If the bicycle starts to fall toward the right, will the gyroscopic action of the rotating front wheel tend to turn the wheel toward the right, or toward the left? Justify your answer. (Note: The gyroscopic effect, though real, is very small compared with other effects that govern the motion of a bicycle.)

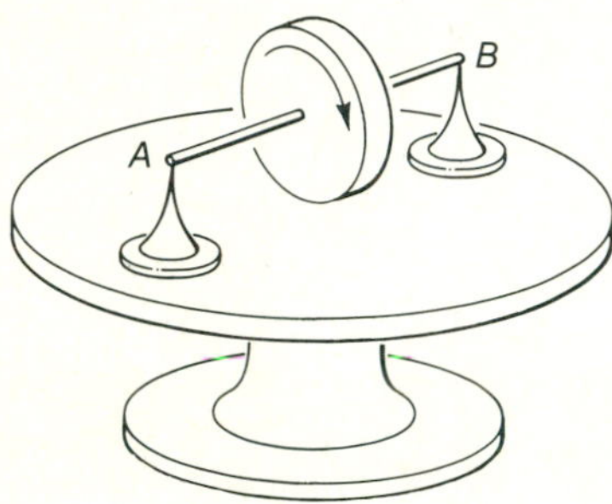

Figure 10–61
Problem 10B–42

10B–42 The axle of a spinning gyroscope is supported symmetrically on two pedestals with the axis horizontal, as shown in Figure 10–61. The system rests on a horizontal turntable. If the turntable is now rotated clockwise as viewed from above, which pedestal, A or B, will exert the greatest upward force? Explain your reasoning.

10B–43 When a wire is stretched, the elongation x is proportional to the stretching force F. Thus, the material has a Hooke's Law behavior: $F = kx$. Express the force constant k in terms of Young's modulus Y, the length L_0, and the cross-sectional area A.
Answer: AY/L_0

10B–44 Two sheets of metal are joined together by four steel rivets, each 7 mm in diameter. Find the maximum force that can be applied to pull the sheets apart tangentially if the shearing stress on the rivets must not exceed 8×10^7 N/m^2. Assume the stress is shared equally by the rivets.

10C–1 A thin sheet of metal of uniform thickness is cut into the shape bounded by the lines $x = a$ and $y = \pm kx^2$ (where distances are in meters), as shown in Figure 10–62. Find the coordinates of the center of mass. *Answer:* $(0, 3a/4)$

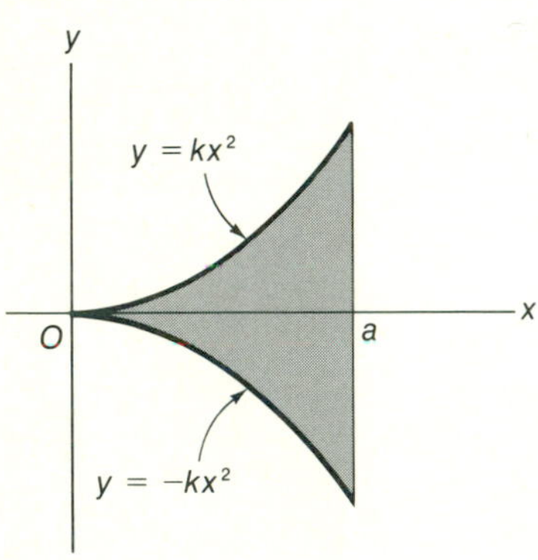

Figure 10–62
Problems 10C–1 and 10C–18

10C–2 A thin, uniform sheet of metal is cut into a circular area of radius R. This area is then cut in half along a diameter. Find the center of mass of one of the halves. (Hint: For a mass element dm, choose a thin strip parallel to the straight edge.)

10C–3 Consider an indefinite number of identical bricks stacked to produce the maximum overhang without falling similar to the situation shown earlier in Figure 10–47. Show that for n bricks, the distance x is given by $x_n = (L/2)\,\Sigma_{k=1}^{n}(1/k)$. This series slowly diverges; that is, $\lim_{n\to\infty} x_n = \infty$. Thus, the overhang distance approaches infinity as the number of bricks approaches infinity! (Hint: Start with the topmost brick and work down the stack, deriving a mathematical expression for the distance x.)

10C–4 A thin rod, 3 m long, rests on a horizontal surface. The *linear mass density* (mass/length) varies linearly with distance along the rod, being zero at one end and 50 kg/m at the other.

(a) Find the total mass of the rod.
(b) How much work is done in raising the heavier end 1.5 m above the surface?

10C–5 Prove the following statement: *If an object is in translational equilibrium, then the net torque on the object is the same value for every point about which the torque is calculated.*

10C–6 A uniform chain of mass m and length ℓ hangs from a hook, as shown earlier in Figure 10–38.

(a) Derive an expression for the force $F(y)$ as a function of the distance y required to raise the middle link to the hook.

(b) Verify by calculating $\int \mathbf{F} \cdot d\mathbf{y}$ that the work done equals the gain in gravitational potential energy of the centers of mass for various portions of the chain.

10C–7 A power line follows a 30° bend in the road, where it is supported by a vertical pole AC. The lines are horizontal where they are fastened to the pole at A. To prevent the pole from bending over, a guy wire AB is installed symmetrically as shown in Figure 10–63, making an angle of 30° with respect to the pole (that is, the line BC bisects the 150° angle). If the tension in the lines is T, find the tension in the guy wire. *Answer:* $1.04T$

10C–8 The sides of the stepladder in Figure 10–64 are each 4 m long, hinged together at point B. The ladder rests on a horizontal frictionless surface and is prevented from sliding apart by the horizontal guy wire DE, fastened as shown. The center of mass of each side is at the center of the side. Side AB weighs 80 N, and side BC weighs 40 N. Find the force of the ground on the base of the ladder (a) at point A and (b) at point C. (c) Find the tension T in the guy wire. a) 50 N b) 70N c) 23.1N

10C–9 A cloth tape is wound around the inner diameter of a spool (inner and outer radii: r and R). As shown in Figure 10–65 the spool is placed on a level surface and the tape is pulled up at an angle θ with the vertical so that the spool rotates at constant angular speed but does not translate. There is some friction between the spool and surface. (a) Make a free-body diagram and (b) find the angle θ. *Answer:* $\sin^{-1}(r/R)$

10C–10 Two uniform rods of equal length are hinged together and hung vertically as shown in Figure 10–66(a). The upper rod weighs 100 N, and the lower rod weighs 200 N. What single force applied at the lower end will hold them in the position shown in (b), with the lower end directly below the point of suspension?

10C–11 Figure 10–67 shows a rotating vertical shaft AB with a centrifugal governor designed to shut off the power if the angular speed exceeds 300 rpm. The governor has four connecting links (with frictionless pivots), each 20 cm long and of negligible mass. The collar slides without friction on the shaft and has a mass $M = 2$ kg. A control mechanism shuts off the power when the center of mass of the collar is raised to within 10 cm of the top. For convenience, assume $g = 10 \text{ m/s}^2$.

(a) What are the values of the two equal masses m?

(b) At maximum speed, what is the tension in each of the upper links? Include appropriate free-body diagrams. *Answers:* (a) 0.506 kg (b) 60.0 N

10C–12 A tricycle handlebar is locked so that it will not turn. A forward horizontal force is applied to one pedal in the "down" position. Assuming no skidding occurs, will the tricycle go forward or backward? Explain your reasoning.

10C–13 During a docking maneuver in a region where gravity is negligible, it is desired to give a 1100-kg space capsule a linear acceleration of 0.06 m/s^2 and a rotational acceleration of 0.08 rad/s^2. One control jet applies a force $\mathbf{F}_1$ to the CM of the capsule, and a second control jet 3 m from the CM applies a force $\mathbf{F}_2$ in a direction opposite to $\mathbf{F}_1$. The moment of inertia of the capsule about its CM is $4900 \text{ kg}\cdot\text{m}^2$.

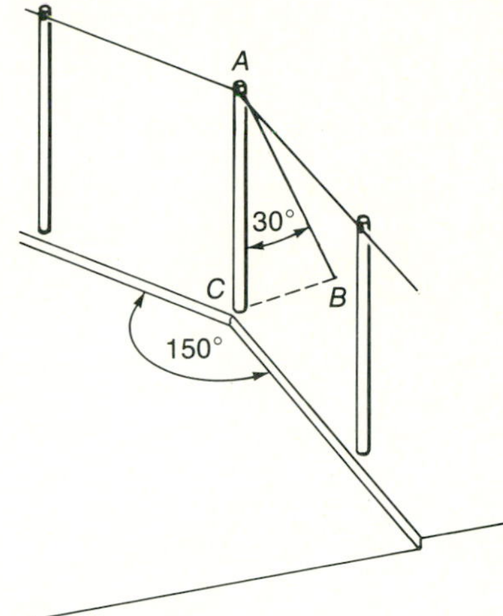

Figure 10–63
Problem 10C–7

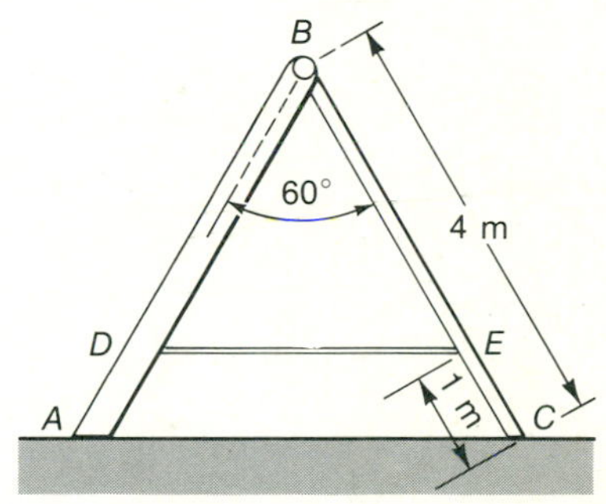

Figure 10–64
Problem 10C–8

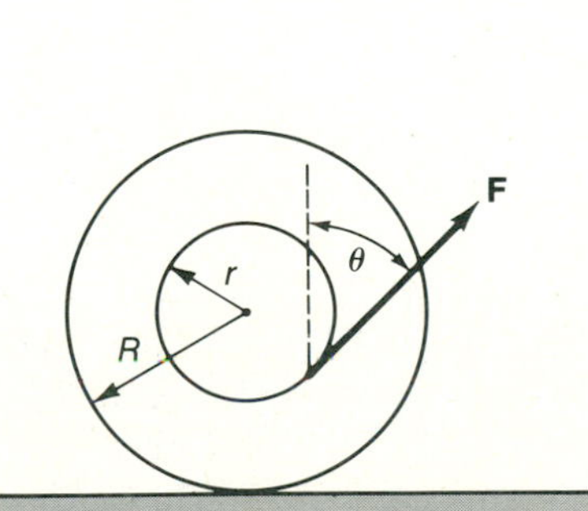

Figure 10–65
Problem 10C–9

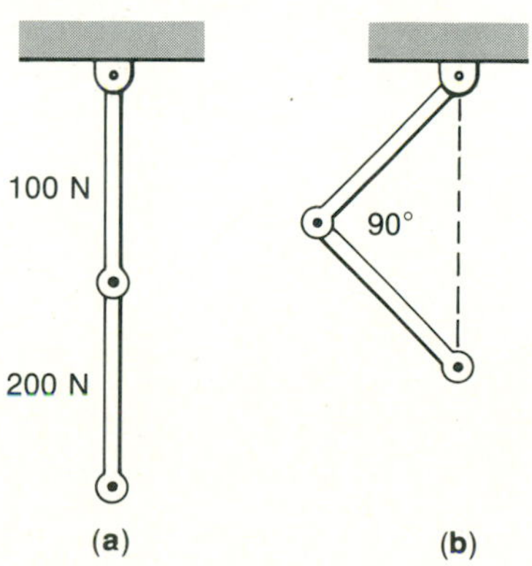

Figure 10–66
Problem 10C–10

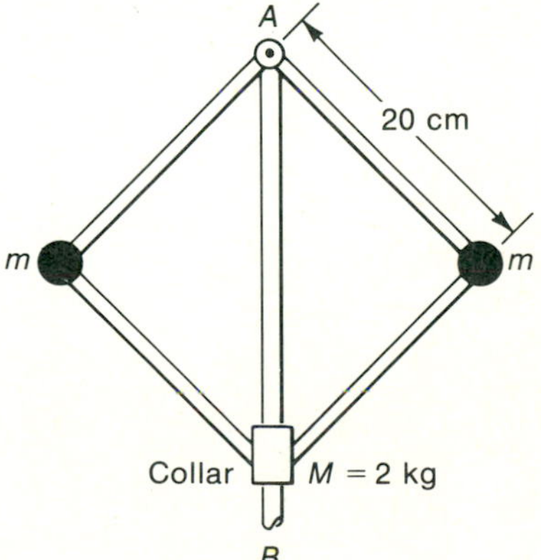

Figure 10–67
Problem 10C–11

(a) Find the magnitudes of $\mathbf{F}_1$ and $\mathbf{F}_2$ if the capsule is to accelerate linearly in the direction of $\mathbf{F}_1$.
(b) Would the *magnitudes* of the accelerations be different if the two forces were interchanged?
(c) Would the *directions* of the accelerations be different? Explain.

Answer: (a) $F_1 = 197$ N, $F_2 = 13.0$ N

10C–14 Using the parallel-axis theorem, find the moment of inertia of a uniform solid sphere about an axis tangent to the surface of the sphere.

10C–15 Consider a thin slab of material in the x-y plane with its center of mass at the origin. If I_x is the moment of inertia about the x axis and I_y the moment of inertia about the y axis, show that the moment of inertia of the slab about the z axis is $I_z = I_x + I_y$ This is called the *perpendicular-axis theorem.*

10C–16
(a) Derive the expression for the moment of inertia I of a uniform, slender rod of length ℓ and mass m about an axis through the center of mass perpendicular to the rod.
(b) Extend this result to find the moment of inertia of a uniform, thin, rectangular slab about an axis in the plane of the slab through its center of mass parallel to one edge.
(c) Using the perpendicular-axis theorem (see previous problem), derive an expression for the moment of inertia of a uniform rectangular slab about an axis perpendicular to the plane of the slab through its center of mass.

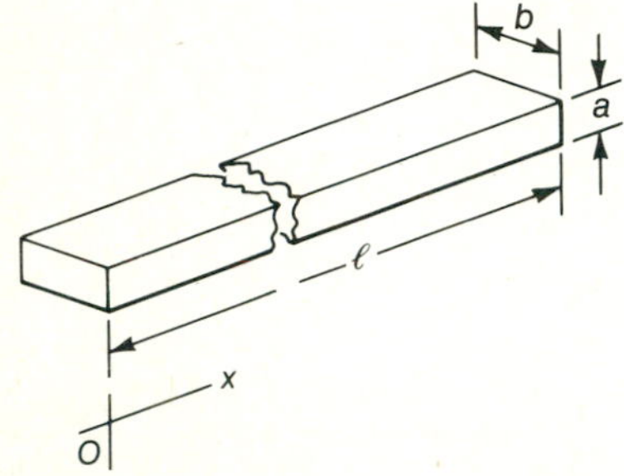

Figure 10–68
Problem 10C–17

10C–17 As shown in Figure 10–68, a long, thin stick of dimensions $a \times b \times \ell$ is made of nonuniform material whose density ρ varies linearly with distance x along the stick according to $\rho = kx^3$. (Note: Both a and b are negligible compared with ℓ.)
(a) What are the units of k in the metric system?
(b) Show that the total mass M is $abk\ell^4/4$.
(c) In terms of M and ℓ, find the moment of inertia I about an axis through the end $x = 0$, perpendicular to the length of the stick.

Answer: (c) $2M\ell^2/3$

10C–18 Find the moment of inertia about the y axis of the thin metal sheet cut as in Figure 10–62. Assume the mass per unit area of the sheet is σ.

10C–19 A wheel undergoes a constant angular acceleration of 2 rad/s^2. During a 3-s interval, the wheel turns through a total angular displacement of 90 rad. If the wheel started from rest, how long had it been turning before the beginning of the 3-s interval?

Answer: 13.5 s

10C–20 A wheel rolls without slipping along level ground with a constant speed of 6 m/s. Relative to the ground, determine the instantaneous velocity of a particle on the rim of the wheel when the particle is on the front edge of the wheel.

10C–21 A woman sits on the outer rim of an amusement park ride in the form of a giant horizontal turntable 20 ft in diameter. Starting from rest, it accelerates uniformly and reaches a rotational speed of 0.50 rev/s in 6 s.
(a) Find the angular acceleration in rad/s^2.
(b) Find the tangential and radial acceleration components of the woman 3 s after the turntable starts to accelerate.
(c) Determine the direction of the woman's (total) acceleration **a** for part (b), and include a sketch to indicate this direction.

Answers: (a) $(\pi/6)$ rad/s^2 (b) $(5/3)\pi$ ft/s^2, $(5/2)\pi^2$ ft/s^2
(c) 25.3 ft/s^2, 12° forward from the inward radial direction

10C–22 Suppose a space station is being assembled in outer space, where gravity is negligible. A construction piece is in the form of a long, thin I-beam of mass 100 kg and length 8 m. A single force of 20 N is applied to one end of the isolated beam. The beam starts "at rest" with respect to the fixed stars, and the force is initially applied at right angles to the beam. This force is maintained constant in magnitude and in direction with respect to the fixed stars.
(a) Describe *quantitatively* the motion of the center of mass of the beam.
(b) Describe *qualitatively* the motion of the beam about its center of mass.

10C–23 Toppling chimneys often break apart in mid-fall because the mortar between the bricks cannot withstand much tension force. This tension supplies the centripetal forces on the topmost segments which they need to keep them traveling in a circle. Consider a long uniform rod of length ℓ pivoted at the lower end. The rod starts at rest in a vertical position and falls over under the influence of gravity. What fraction of the length of the rod will have a tangential acceleration greater than the component of gravitational acceleration in the tangential direction? *Answer:* $(2/3)\ell$

10C–24 A uniform cylinder of mass M and radius R rolls without slipping down a plane inclined at an angle θ with respect to the horizontal. While rolling, the cylinder accelerates under the influence of three forces: the force of gravity, the friction force of the plane, and the normal force of the plane. Suppose we mentally abolish gravity and the plane. There exists a *single* force that will produce the same translational and rotational accelerations through space. Find (a) the magnitude and direction of this force and (b) its point of application.

10C–25 A billiard ball of radius R rests on a frictionless, horizontal surface. The ball is struck a horizontal blow a distance h above the surface. For what value of h will there be no tendency for the ball to slip as it begins to roll? (Hint: The linear impulse $F\,\Delta t$ gives to the CM a change in momentum $m\,\Delta v$; correspondingly, the angular impulse about the CM produces a change in the angular momentum about the CM. What is the relation between these effects so that there will be no slipping?) *Answer:* $h = (7/5)R$

10C–26 A wheel of mass m and radius of gyration k is supported by a fixed, horizontal axle of radius r (which is slightly smaller than the hole in the wheel). The wheel is given an initial angular velocity ω_0. Because of friction the wheel slows down.

(a) Find the point of application of the friction force f on the wheel, and explain why it is *not* at the top of the hole.

(b) In terms of μ_k, r, g, and k, find the magnitude of the angular acceleration α.

10C–27 One end of a long, thin rod of length D is fastened to the floor with a hinge. The free end is raised so the rod makes an angle θ with the horizontal, then released from rest. Find the linear speed v of the free end just before it strikes the floor. (Hint: Use conservation of energy principles.) *Answer:* $\sqrt{3gD \sin\theta}$

10C–28 Suppose the cylinder of Example 10–18 has an initial angular speed ω_0 when it is placed on the horizontal surface with its CM initially at rest. Find the distance d the cylinder rolls (and slides) along the surface until it begins to roll without slipping.

10C–29 A uniform solid bowling ball (mass M, radius R) is launched in pure translation (without rotation) along the bowling alley floor with an initial speed v_0. During an initial distance D, it partially slides while gaining rotational speed, after which it rolls without slipping. The constant force of friction is f.

(a) In terms of the given symbols, find the linear speed v of the ball when it begins to roll without slipping.

(b) Find the distance D.

(Note: While traveling the distance D, $a \neq R\alpha$.) *Answers:* (a) $5v_0/7$ (b) $12mv_0^2/49f$

10C–30 *Velikovsky Problem.* In 1950, Macmillan Publishers released a book by Immanuel Velikovsky titled "Worlds in Collision," in which the author maintained that in about 1500 B.C. the planet Jupiter ejected material which, in the form of a comet, passed close to the earth. (It was claimed that this comet later turned into the planet Venus.) The resultant shower of meteorites falling on the earth was supposed to have stopped the earth's rotation, at least for a short time. Velikovsky also maintained that later, in 747 B.C., and again in 687 B.C., "Mars caused a repetition of the earlier catastrophes on a smaller scale."

Let us analyze this theory from a physical viewpoint to see if it is reasonable. We will make rough calculations using only one significant figure for each parameter. For this problem, assume the following data in SI units:

Mass of the earth:	6×10^{24} kg	Approximate values
Radius of the earth:	6×10^{6} m	
Angular speed:	7×10^{-5} rad/s	

(a) Find, in SI units, the angular momentum of the earth that has to be counteracted to stop its rotation (and, incidentally, that must be supplied from somewhere to start up its rotation afterwards).

(b) In the most favorable case, the meteorites would have fallen on the equator tangentially at grazing incidence. Let us suppose they traveled parallel to the earth's orbit about the sun, but opposite in direction to the earth's orbital motion. At impact, assume their speed relative to the earth's center of mass is 60 km/s (about twice the speed of the earth in its orbit about the sun). What is the total mass of these meteorites if, in a tangent collision at the equator, they are to stop the earth's angular rotation about its axis?

(c) What would be the total volume of these meteorites, assuming they were of the stony type whose density is essentially that of the earth's crust: 2700 kg/m^2? (About 94% of all meteorites are of the stony type. For comparison, the moon's volume is about 22×10^{18} m^3.)

(d) Assume, for the moment, that after the impact of part (b), the center of mass of the earth continued in its orbital motion about the sun. What was the total kinetic energy that was transformed into other forms of energy by the stopping of the earth's rotation about its axis and the change in velocity of the captured meteorites?

(e) What would happen to the motion of the earth's center of mass as a result of the meteorite collision of part (d)?

(f) The solar energy falling per unit of time at normal incidence on a unit area of the earth's atmosphere is 1340 W/m^2. For comparison with the answer in part (d), find the total energy received by the earth from the sun in a 24-h period. What might be the consequences of this energy if it were concentrated in a relatively small region of the earth's surface where the meteorite collision supposedly occurred?

(g) Daily growth rings in fossil corals indicate that the earth's rotation has been remarkably constant for the past 600 million yr except for a gradual lengthening of the day, possibly due to tidal friction effects. Thus Velikovsky's theory is faced with the problem of starting the earth rotating again at precisely its former angular speed. Barring another fortuitous collision with meteorites, could gravitational forces (which are *central* forces between spherical bodies) change the state of rotation of the earth?

(h) Critics of Velikovsky point out what might happen to the oceans, atmosphere, people, and other loose objects at the earth's surface if the earth were to stop rotating within a short time interval. Try to get around this objection by assuming some rather long time interval (say, an hour or two) for the slowing-down process. Estimate the sideways acceleration that would occur at the earth's equator and discuss briefly whether or not such an event would be survivable.

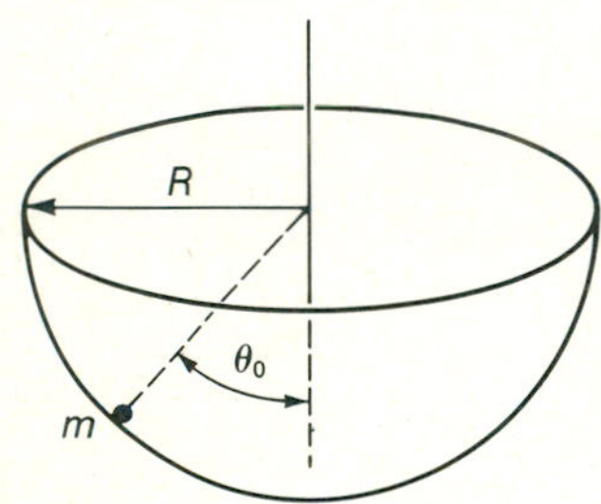

Figure 10–69
Problem 10C–32

10C–31 A wire of length L and cross-sectional area A is stretched a distance ΔL.

(a) Show that work W done in stretching the wire is $W = [YA(\Delta L)^2]/2L$.

(b) Find the energy per unit volume stored in the stretched wire (ignoring that in the elongation ΔL).

10C–32 As shown in Figure 10–69, a particle of mass m moves without friction on the inside of a fixed, hemispherical bowl of radius R. At an angular position θ_0, it is given an initial horizontal speed v_0. In terms of θ_0 and R, find the minimum speed v_0 that will enable the particle to barely escape from the bowl. (Hint: Use conservation of energy principles.)

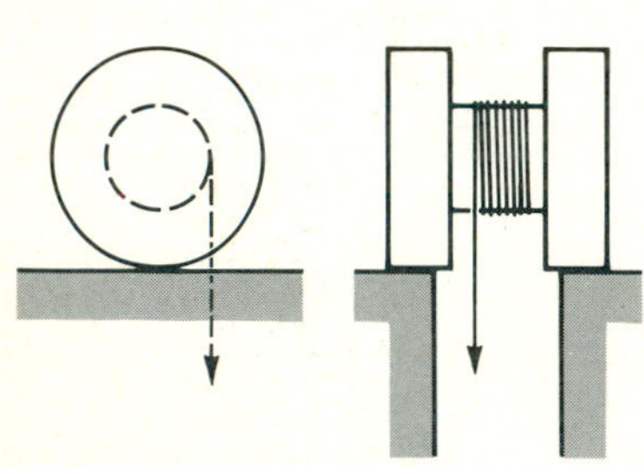

Figure 10–70
Problem 10C–33

10C–33 Figure 10–70 shows a spool in the form of two uniform disks, each of mass M and radius R, with an inner solid cylinder of mass M and radius $R/2$. A cord wound around the inner cylinder extends downward and applies a constant vertically downward force T to the spool. As a consequence, the spool accelerates horizontally as it rolls along the table without slipping. (A slot in the table permits the cord to remain vertical as the spool moves.) In terms of M and T, find (a) the linear acceleration a of the center of mass and (b) the horizontal force exerted on the spool.

Answer: (b) $\dfrac{4T}{11}$

10C–34 A gyroscope rotor is in the shape of a uniform solid disk 12 cm in radius. It is mounted in the middle of a 23-cm axle which is supported at one end with the axle horizontal. If the rotor is spinning at 1200 rpm, find the precession ω_p (in revolutions per minute).

Happy is he who gets to know the reasons for things.

VIRGIL (70–19 B.C.)

chapter 11 Accelerated Frames and Inertial Forces[1]

11.1 Introduction

Almost everyone has had the experience of riding in an automobile that turns a corner rapidly, and of being "thrown" toward the outer edge of the curve. To the passengers, this force seems very real—in fact, the passengers may even be "thrown" against the side of the car with a fairly large impact. Similarly, passengers in an elevator accelerating upward experience a force that apparently pushes them harder against the floor. And riders in a moving automobile that is braked rapidly believe there is a force propelling them toward the front of the car so violently that, in the absence of seatbelts, they brace themselves against the dashboard to counteract this force. As a final example, it has been proposed that spaceships on long voyages be given a rotation to provide astronauts with an "artificial gravity," or an outward "centrifugal force," as it is sometimes called. Experiments have shown that such an artificial gravity force is important for reasons of health as well as for comfort, since muscle tissue and bones tend to deteriorate if they are not exercised by working against this artificial gravity, or by doing some other exercise.

Each of these examples describes the same special type of force: a force that has no physical source or external object causing it. Its origin lies in the fact that *the observer is in an accelerated frame of reference*. To understand these forces requires a reorientation in our usual way of thinking. Although this may seem hard at first, if you follow the discussion step by step it will not be difficult.

11.2 Inertial Forces

In previous chapters, all the forces discussed may be traced to a "source"; that is, some other object or physical system causes the force. We will call these forces "real" forces, signifying that in each case the origin of the force may be traced to some other object. A few examples are listed below:

(a) Tensions in springs, strings, and so on.
(b) Forces of contact in which one object pushes or pulls on another because of physical contact between the objects.

[1] A portion of this chapter is adapted from one of the authors' (AMH) discussion in *Physics—A New Introductory Course*, Science Teaching Center, M.I.T. 1965.

(c) Gravitational forces of attraction between masses.
(d) Electrical forces between objects that have a net electric charge.
(e) Magnetic forces (ultimately traceable to forces between electric charges in motion).
(f) Nuclear forces.

As we delve deeper into the nature of forces, we discover that what we conveniently call forces of contact, tensions in strings, cohesion forces between atoms and molecules, and so on, are ultimately electromagnetic in origin; in other words, they are forces between charges on electrons and protons. So our list of real forces reduces to the following basic classifications:

Gravitational forces	(between masses)
Electromagnetic forces	(between charged particles)
Nuclear forces	(between certain fundamental particles, only significant at distances comparable to nuclear dimensions)

The kinds of forces listed above are involved in the analysis of phenomena viewed from an inertial frame of reference. Analysis by an observer in an *accelerated* frame of reference involves an additional class of forces that are evident only to such an accelerated observer. Because these forces are associated with the inertial property of matter (rather than being "caused" by another object or physical system), they are called **inertial forces.**[2] A phenomenon such as a ball bouncing in an accelerating freight car may be analyzed either by an observer in the inertial frame of the railway station or by an observer in the accelerated frame of the freight car. To the observer in the station, no inertial forces are present, but to the observer in the car, inertial forces must be considered in the analysis of motion. The choice of a frame of reference for the analysis is entirely a matter of convenience.

11.3 Linearly Accelerated Frames of Reference

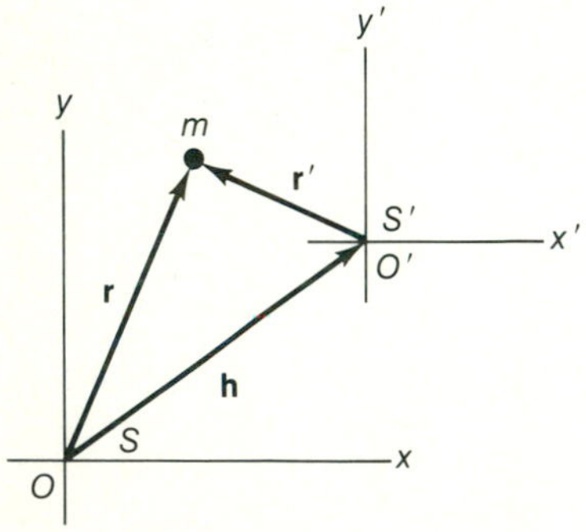

Let us establish a notation to describe the motion of a mass point m as seen[3] in two different frames of reference. Consider a coordinate system S that is at rest with respect to an inertial frame of reference. Another coordinate system S' has arbitrary translational motion with respect to the S system (see Figure 11–1). Measurements made in the S' system will be designated by primed symbols. (For convenience, we position the x', y', and z' axes of S' so that they are parallel with the corresponding axes of S.) The position vector $\mathbf{h}$ locates the origin O' with respect to O.

Observers in the S and S' frames describe the location of the same mass point m by the position vectors $\mathbf{r}$ and $\mathbf{r}'$, respectively. These vectors are related by the usual rule of vector addition:

$$\mathbf{r} = \mathbf{r}' + \mathbf{h} \tag{11–1}$$

Figure 11–1
Observers in the S and S' frames of reference designate the location of the mass m by the position vectors $\mathbf{r}$ and $\mathbf{r}'$, respectively. The vector $\mathbf{h}$ locates the origin of the S' frame relative to the origin of the S frame.

[2] Other common names for inertial forces are *fictitious* or *pseudo* forces. We prefer to avoid such terms since they may imply that the forces are somewhat imaginary. As we shall see, inertial forces are just as verifiable as the force of gravity.

[3] The phrase "as *seen* in a given frame of reference" is not to be interpreted literally. "Viewing" or "observing" the motion of an object in a frame of reference here means "*making space and time measurements of the object* relative to that particular frame of reference."

If both m and S' move, then the velocity of m as measured in the two frames of reference is found by taking derivatives of the position vectors with respect to time:

$$\frac{d\mathbf{r}}{dt} = \frac{d\mathbf{r}'}{dt} + \frac{d\mathbf{h}}{dt}$$

or

$$\mathbf{v} = \mathbf{v}' + \mathbf{v}_h \tag{11–2}$$

where $\mathbf{v}_h$ denotes the velocity of O' relative to O. Similarly, if m has acceleration, we differentiate again to obtain

$$\frac{d\mathbf{v}}{dt} = \frac{d\mathbf{v}'}{dt} + \frac{d\mathbf{v}_h}{dt}$$

or

$$\mathbf{a} = \mathbf{a}' + \mathbf{a}_h \tag{11–3}$$

where $\mathbf{a}_h$ represents any accelerations of O' relative to O. Equation (11–3) is a general statement applicable in all cases.

We shall first examine the special case of S' moving *uniformly* (no acceleration or rotation) with respect to S, where the acceleration $\mathbf{a}_h$ is therefore equal to zero. Since Newton's laws are valid in an inertial frame (S), if m is subject to a (real) external force $\mathbf{F}$, we have

$$\textit{As seen in S:}\quad \Sigma\mathbf{F}_{\text{real}} = m\mathbf{a} \tag{11–4}$$

Multiplying Equation (11–3) by m, we obtain

$$m\mathbf{a} = m\mathbf{a}' + m\mathbf{a}_h \tag{11–5}$$

But since $\mathbf{a}_h = 0$, this reduces to

$$m\mathbf{a} = m\mathbf{a}'$$

or, by Equation (11–4):

$$\textit{As seen in S':}\quad \Sigma\mathbf{F}_{\text{real}} = m\mathbf{a}' \qquad \text{(when } S' \text{ has uniform translation and no acceleration)} \tag{11–6}$$

Thus Newton's first and second laws are valid for a uniformly moving frame of reference, that is, a frame with no acceleration. We call all such frames **inertial frames** because Newton's first law—the law of inertia—is correct in these frames.

Let us now consider the more interesting case where S' accelerates uniformly relative to S, with constant linear acceleration $\mathbf{a}_h$. Rearranging Equation (11–5), we have

$$m\mathbf{a} - m\mathbf{a}_h = m\mathbf{a}' \tag{11–7}$$

Since $\Sigma\mathbf{F}_{\text{real}} = m\mathbf{a}$, we may write:

$$\Sigma\mathbf{F}_{\text{real}} - m\mathbf{a}_h = m\mathbf{a}' \tag{11–8}$$

Here lies a crucial step in the derivation. In the S' frame, the mass m is observed to have the acceleration $\mathbf{a}'$. In terms of Newtonian mechanics, an object accelerates because a net force acts on it. Newton's way of thinking is deeply ingrained in our common sense, and in order to retain the form of Newton's law, we consider the quantity $\mathbf{F} - m\mathbf{a}_h$ as the net force $\Sigma\mathbf{F}'$ acting on m to give it the acceleration $\mathbf{a}'$ as observed in S'. This net force $\Sigma\mathbf{F}'$ is made up of two parts: a real force $\Sigma\mathbf{F}_{\text{real}}$ and an inertial force $(-m\mathbf{a}_h)$. Thus, in the accelerated frame

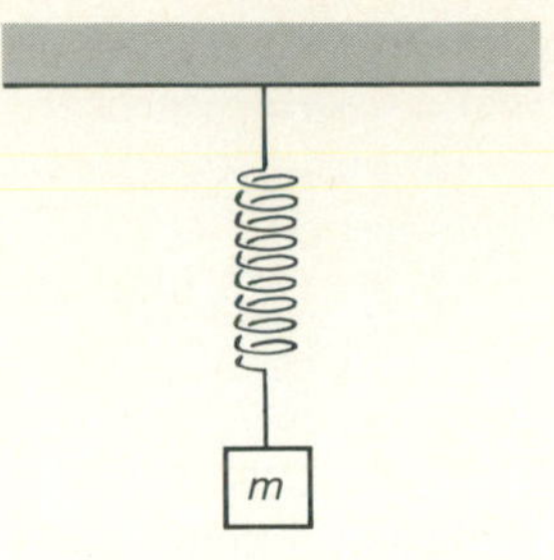

(a) A mass on a spring.

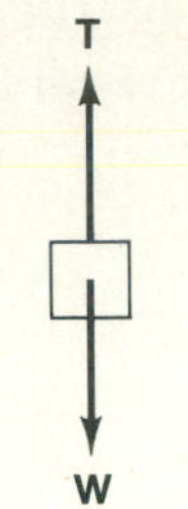

(b) A free-body diagram for a mass hanging on a spring.

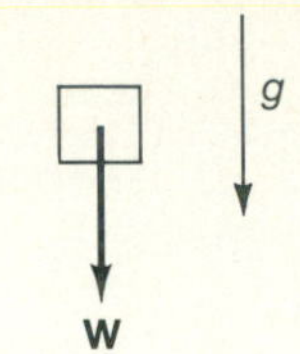

(c) A freely falling mass has downward acceleration equal to g.

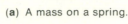

Figure 11–2
These analyses are in an inertial frame of reference at rest in the presence of a gravitational field **g**.

we write Newton's second law as follows:

NEWTON'S SECOND LAW (in a frame with linear acceleration $\mathbf{a}_h$)

$$\underbrace{\Sigma\mathbf{F}_{\text{real}}}_{\substack{\text{Sum of all the}\\ \text{real forces}}} + \underbrace{(-m\mathbf{a}_h)}_{\substack{\text{Inertial}\\ \text{force}}} = \Sigma\mathbf{F}' = m\mathbf{a}' \qquad (11\text{–}9)$$

This is not merely a mathematical trick. As you will see, observers in the accelerated frame can experimentally verify the presence of the force $(-m\mathbf{a}_h)$ just as they can verify the presence of a gravitational force.

We need to make a special comment regarding signs. In equations, we will always associate the minus sign with this inertial force: $\mathbf{F}_{\text{inertial}} = (-m\mathbf{a}_h)$. Note that this is *opposite* to the direction of the acceleration of the frame $\mathbf{a}_h$. This procedure is consistent with our interpretation of the net force on an object as the *sum* of all the forces acting on it. Thus, in the accelerated frame the net force $\Sigma\mathbf{F}'$ is the *sum* of the real plus the inertial force:

$$\Sigma\mathbf{F}' = \Sigma\mathbf{F}_{\text{real}} + \mathbf{F}_{\text{inertial}}$$

where $\mathbf{F}_{\text{inertial}} = -m\mathbf{a}_h$.

A note of caution: In drawing vector diagrams, we follow the custom of never using a minus sign with the symbol on the arrow. The diagram itself displays the proper directions for the forces, independent of the particular direction designated "minus" by the choice of coordinate system. Thus, having determined the correct direction of the inertial force to be opposite to $\mathbf{a}_h$, in sketching the vector we label it simply with its magnitude ma_h. We use a minus sign with a vector arrow only when it specifically represents the negative of a vector.

We cannot trace the origin of the term $(-m\mathbf{a}_h)$ to any other physical system or object. In this sense it is "fictitious." But to observers in the S' frame, it is really "there." Let us discuss this point a bit further. Consider how we detect the force of gravity in an inertial frame of reference. If we hang a mass m on the end of a spring [see Figure 11–2(a)], the spring elongates to exert an upward spring force **T**, exactly balancing the downward force of gravity **W**. In fact, we infer the downward force precisely because the spring elongates. When we combine **T** with the inferred force **W**, there is zero net force on m, and the mass remains at rest. If the spring breaks, the mass accelerates downward, which provides further evidence of a gravitational force. Applying $\Sigma\mathbf{F} = m\mathbf{a}$, we say that the single gravitational force **W** produces the downward acceleration **g**: $\mathbf{W} = m\mathbf{g}$.

Suppose we perform the same experiment in a gravity-free region but in a frame of reference S' which has linear acceleration $\mathbf{a}_h$ as shown in Figure 11–3. Observers in that frame note an elongation of the spring, signifying an upward spring force on m. Since the mass is at rest in this frame, however, they infer

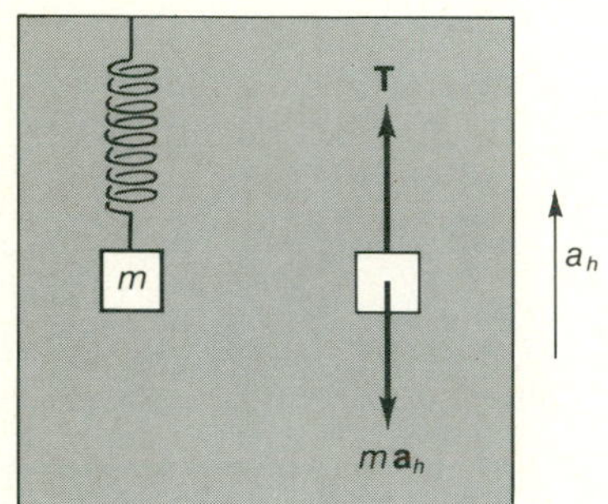

(a) Observers in the accelerated box believe there are two forces acting on m to produce equilibrium conditions. Thus the observers explain why m remains at rest in their frame of reference.

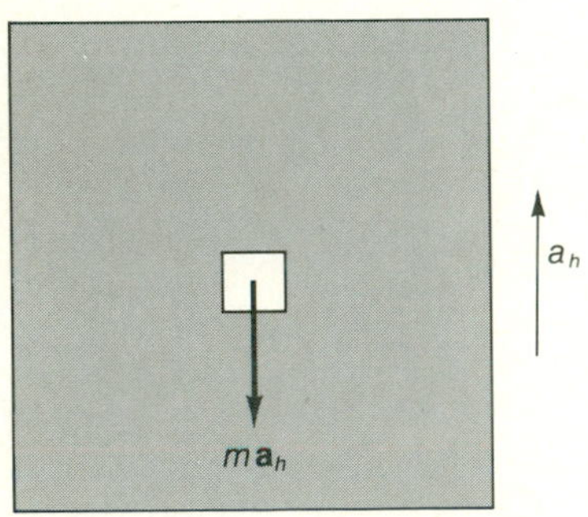

(b) If the spring breaks, observers see the mass m accelerate toward the floor of the box in response to the unbalanced force $m\mathbf{a}_h$ downward.

Figure 11–3
These analyses are in a frame of reference attached to the box, which has upward acceleration a_h in a region where there is no gravitational field. It is therefore an accelerated frame of reference

the presence of a second "downward" force $-m\mathbf{a}_h$ on the mass to make it remain at rest in S'. Furthermore, if the spring breaks, they observe the mass to move "downward" with an acceleration $-\mathbf{a}_h$ in response to the unbalanced force $-m\mathbf{a}_h$.

Thus, observers in the accelerated frame S' infer the presence of the inertial force $-m\mathbf{a}_h$ in exactly the same way we infer the force of gravity, and the inertial force is just as real to them as the force of gravity is to us. There are some novel features, however. Though we assign the cause of gravity to be the presence of a nearby mass (the earth), the origin of the inertial force cannot be traced to any other object or system. Since no *agent* exerts it, it does not obey Newton's third law. *The inertial force is a consequence of analyzing the situation from an accelerated frame of reference and represents a desire to preserve the form of Newton's second law.* We cling strongly to the idea that an object accelerates because there is an unbalanced force acting on it.

Is gravity an inertial force? It has much in common with inertial forces. For example, in a stationary frame in the presence of gravity, all freely falling masses have the same acceleration g. Similarly, in an accelerated frame (with no gravity), all free masses undergo the acceleration $-\mathbf{a}_h$. So, like gravity, the inertial force is proportional to the mass m. Furthermore, just as we may consider all gravitational forces to be acting on a single point (the center of mass of an object), inertial forces may be analyzed as if they acted on the center of mass of the object.

To pursue this connection further, hold your arm out horizontally. Can you sense the downward force of gravity acting on your arm? Most people will answer, "Yes, of course I can feel the force of gravity pulling my arm downward." But think again. You actually do *not* have any sensation of a gravitational force acting downward on each atom of your arm. You "feel" only the muscular tension you must exert to apply an *upward* force on your arm to hold it in equilibrium. *The downward force of gravity itself creates no physical sensation at all in your arm.* You are aware only of forces you exert to *oppose* the force of gravity.

Another example may emphasize the fact that we do not experience directly the force of gravity. An astronaut circling the earth in a satellite will note that any object left to itself will float freely inside the space capsule. Indeed, if not strapped down, he himself will float. The astronaut feels no net force on himself, yet an appreciable gravitational force is acting on him. (For a space station 300 mi above the earth's surface, the force of gravity is about 86% of its value at the earth's surface.) Surprising as it may seem, we cannot "feel" the force of gravity.

□ □ □

Let us consider a few examples of motion as observed in a linearly accelerated frame of reference. To clarify the analysis, we shall present parallel treatments as seen in the S and S' frames. In all the examples of this chapter, we will assume that observers in S and S' know which way is "up" and that both use the same gravitational force W on objects. The only difference between the two descriptions that the observers give for the same occurrence is the presence or absence of inertial forces.

EXAMPLE 11–1

Consider a mass m suspended by a string from the ceiling of a railroad car. The car has constant acceleration as shown in Figure 11–4, causing the mass to hang at a steady angle θ with the vertical. Find the angle θ in terms of the other symbols given.

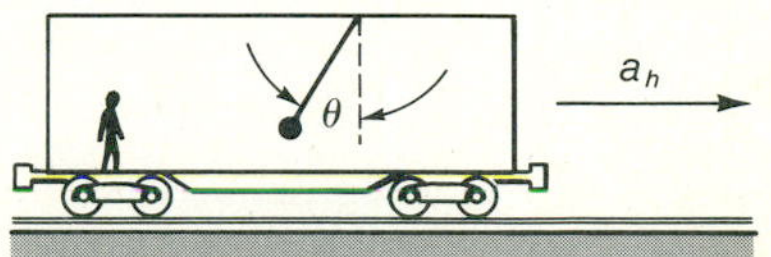

Figure 11–4
Example 11–1

SOLUTION

We will analyze this problem as seen in S, an inertial frame of reference attached to the ground, and S', an accelerated frame attached to the car.

As seen in S

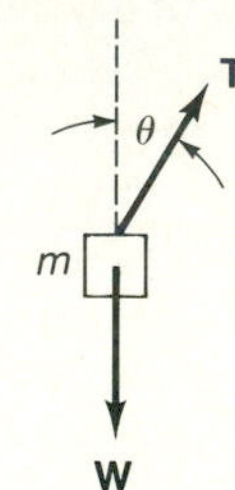

Because the object accelerates toward the right, we resolve forces into two mutually perpendicular directions, one of which is the direction of the acceleration.

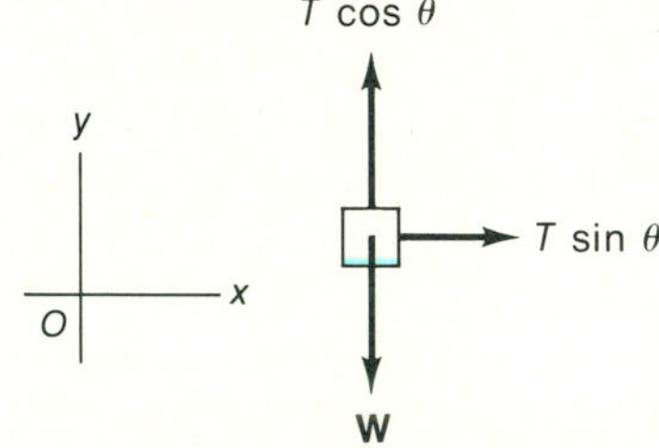

Horizontal component

$$\Sigma F_x = ma_x$$

$$[T \sin \theta] = ma \qquad [1]$$

Vertical component

$$\Sigma F_y = 0 \qquad \text{(Equilibrium)}$$

$$[T \cos \theta] = mg \qquad [2]$$

Dividing [1] by [2] gives

$$\frac{T \sin \theta}{T \cos \theta} = \frac{ma}{mg}$$

$$\boxed{\tan \theta = \frac{a}{g}}$$

As seen in S'

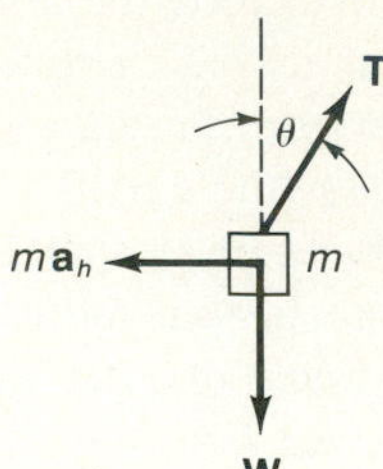

We resolve forces into two mutually perpendicular directions. Note the presence of the inertial force $m\mathbf{a}_h$ toward the *left* (because the car accelerates toward the *right*).

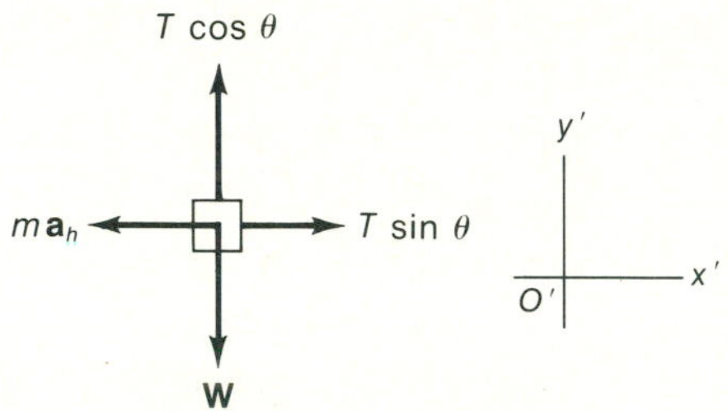

Horizontal component

$$\Sigma F'_x = 0 \qquad \text{(Equilibrium)}$$

$$[T \sin \theta - ma_h] = 0 \qquad [1]$$

Vertical component

$$\Sigma F'_y = 0 \qquad \text{(Equilibrium)}$$

$$[T \cos \theta] = mg \qquad [2]$$

Dividing [1] by [2] gives

$$\frac{T \sin \theta}{T \cos \theta} = \frac{ma_h}{mg}$$

$$\boxed{\tan \theta = \frac{a_h}{g}}$$

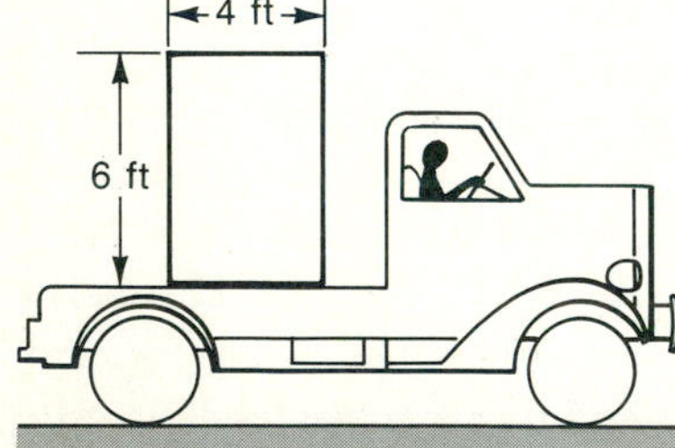

Figure 11–5
Example 11–2

EXAMPLE 11–2

A 200-lb packing case, 3 ft × 4 ft × 6 ft, is on a truck traveling 30 mi/h. The driver applies the brakes, coming to rest as soon as possible without overturning the box. The center of mass of the box is at its geometric center, and it is placed on the truck as shown in Figure 11–5. Assuming the box does not skid, find the minimum stopping distance for the truck without overturning the box.

SOLUTION

In braking, the box will tend to tip over (assuming no skidding) by rotating about the edge closest to the front of the truck. We will solve for the acceleration **a** that will barely tip over the box. This will be the dividing-line situation where the box is just barely rotated so that all its weight is supported by the front edge, but where it is still in rotational equilibrium. Therefore, the normal and friction forces, **N** and **f**, are drawn through this front edge. Note that since there is no sliding, we *cannot* assume that the static friction is its maximum value. Therefore, $f_s \neq \mu_s N$.

As seen in *S*

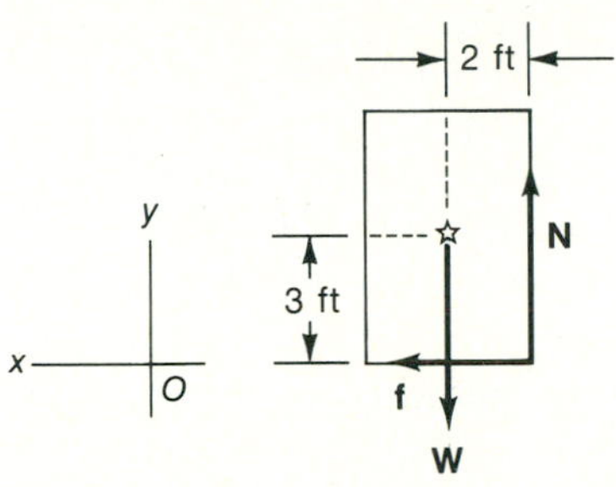

The box accelerates toward the *left* with acceleration **a**. It is in rotational equilibrium.

Horizontal component

$$\Sigma F_x = ma_x$$

$$f = ma \qquad [1]$$

Vertical component

$$\Sigma F_y = 0 \qquad \text{(Equilibrium)}$$

$$N = mg \qquad [2]$$

Rotational equilibrium

Because acceleration is involved (linear in this case), we must calculate torques *only about the CM*.

$$\underset{\text{(about the } CM)}{\Sigma\tau} = 0 \qquad \text{(Rotational equilibrium)}$$

Clockwise torques = Counterclockwise torques

$$f\,(3\text{ ft}) = N\,(2\text{ ft}) \qquad [3]$$

Substituting [1] and [2] in [3] gives

$$ma\,(3\text{ ft}) = mg\,(2\text{ ft})$$

$$\boxed{a = \tfrac{2}{3}g}$$

As seen in *S′*

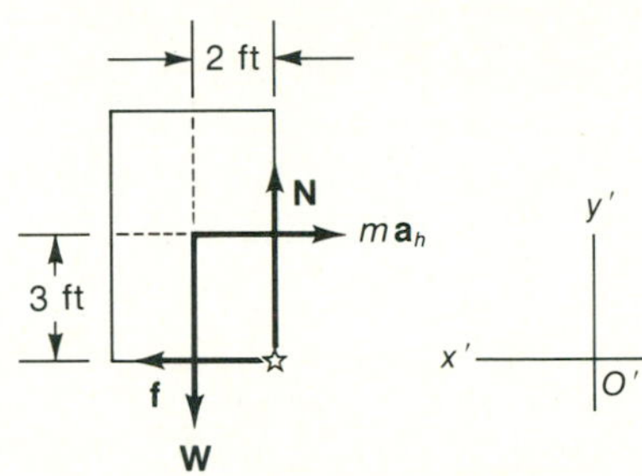

The box is in translational and rotational equilibrium. Since *S′* accelerates toward the left, we include the inertial force $(-m\mathbf{a}_h)$ toward the right.

Horizontal component

$$\Sigma F'_x = 0 \qquad \text{(Equilibrium)}$$

$$f = ma_h \qquad [1]$$

Vertical component

$$\Sigma F'_y = 0 \qquad \text{(Equilibrium)}$$

$$N = mg \qquad [2]$$

Rotational equilibrium

Because in this frame the box is in both translational and rotational equilibrium, we are free to investigate torques about *any axis*. To eliminate one unknown, we choose the front edge as the axis.

$$\underset{\text{(front edge)}}{\Sigma\tau'} = 0 \qquad \text{(Rotational equilibrium)}$$

Clockwise torques = Counterclockwise torques

$$ma\,(3\text{ ft}) = mg\,(2\text{ ft}) \qquad [3]$$

$$\boxed{a = \tfrac{2}{3}g}$$

Note that the mass of the box is superfluous information. The same deceleration of the truck would apply regardless of the mass of the box. However, the fact that the *CM* is located in the geometric center *is* important.

We will next take up the case of a *rotating* frame of reference. Our discussion will be limited to the case of a frame with *constant* angular velocity (that is, the frame has no angular acceleration). However, even though it rotates with constant angular velocity, it is an *accelerated* frame of reference because of centripetal acceleration.

11.4 Rotating Frames of Reference

When a moving object is viewed from an inertial frame and also a rotating frame, differences in the two descriptions can be demonstrated using a simple apparatus. A small cart of mass m is made to move on a straight track with constant velocity $\mathbf{v}$. Viewed from an inertial frame, the cart has *zero net force* exerted on it, since its acceleration is zero. But suppose that a turntable rotating with constant angular speed is placed below the cart and a felt pen attached to the cart. As the cart moves, the pen traces out its path on the rotating turntable (see Figure 11–6).

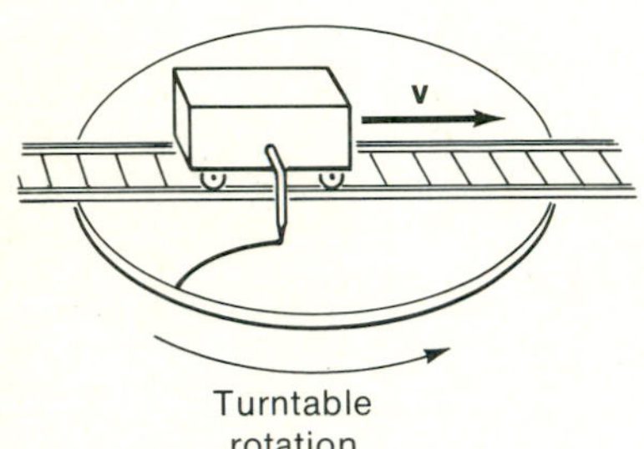

Figure 11–6
A small cart moves at constant velocity **v** in a straight line. A felt pen attached to the cart traces out its path on the rotating turntable.

As viewed from a rotating frame of reference attached to the turntable, the path of object m (as recorded by the pen trace) will be similar to one of the general shapes shown in Figure 11–7. Because of the curved trajectory, observers in the rotating frame of reference would conclude—if they believed in $\Sigma\mathbf{F} = m\mathbf{a}$—that *the object must have a net unbalanced force on it to make it follow a curved path.* The object clearly accelerated. Yet, as viewed from an inertial frame of reference, the object has *zero* net force on it. Here is another case of inertial forces arising in an accelerated frame of reference—in this instance, a rotating frame.

We shall now develop additional useful notation. Consider two frames:

S Frame	A stationary (or inertial) frame at rest with respect to the "fixed" stars.
S' Frame	A frame attached to a merry-go-round rotating with constant angular velocity $\boldsymbol{\omega}$ relative to S. As usual, measurements with respect to the rotating frame will be *primed* (′) to distinguish them from *unprimed* measurements made with respect to the inertial frame S.

To take advantage of the symmetry, we adopt the spherical coordinates r, θ, and ϕ, as shown in Figure 11–8(a). We place the origins of the two frames, O and O', coincident at the center of the merry-go-round so that the same identical position vector $\mathbf{r}$ (which equals $\mathbf{r}'$) describes the location of a point in either frame of reference. The z and z' axes are also made coincident; the merry-go-round rotates about this axis with constant angular velocity $\boldsymbol{\omega}$, as shown in Figure 11–8(c).

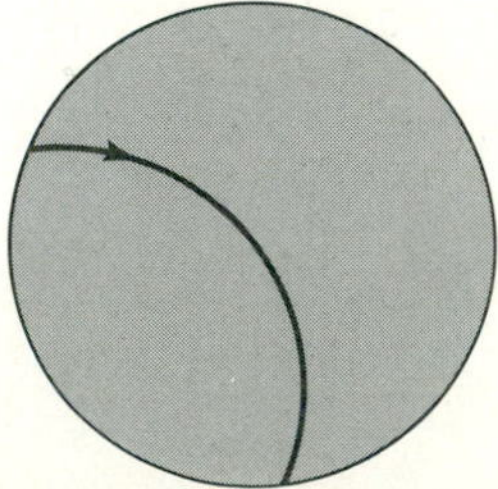
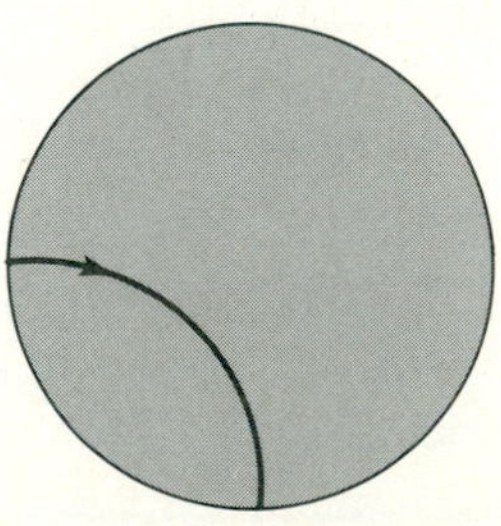
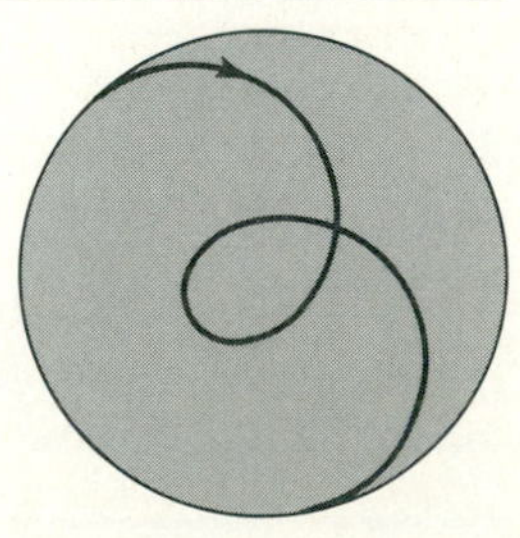

Figure 11–7
In the rotating frame of reference, the path of the moving cart of Figure 11–6 will be similar to one of these general curves (or more complex), depending on the angular speed of the turntable.

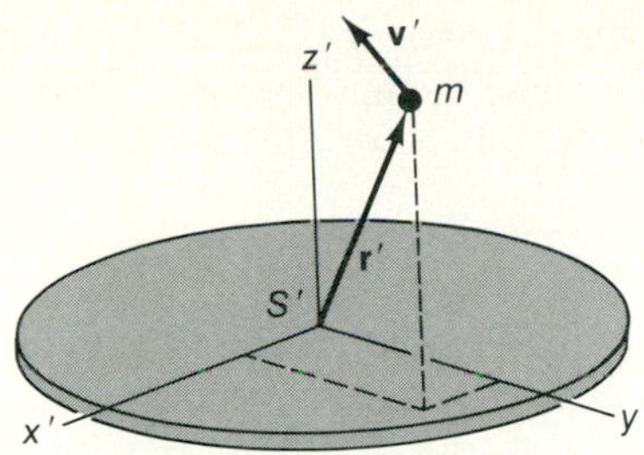

Figure 11–9
The mass m is given an arbitrary velocity $\mathbf{v}'$ with respect to the rotating S' frame.

Consider a mass *m at rest in the rotating frame* at some arbitrary location $\mathbf{r}'$. As seen in the S frame, however, this mass will move in a circle. In terms of spherical coordinates, this motion is along the ϕ direction. Thus, in S, the mass has a velocity $\mathbf{v}_\phi$, where

$$|\mathbf{v}_\phi| = (r \sin \theta)\omega$$

or

$$\mathbf{v}_\phi = \boldsymbol{\omega} \times \mathbf{r} \tag{11–10}$$

Now suppose we allow the mass m to have *any arbitrary velocity* $\mathbf{v}'$ with respect to the rotating S' frame, as shown in Figure 11–9. In the usual manner in which we compare relative velocities, the velocity $\mathbf{v}$ relative to the inertial system S is

$$\mathbf{v} = \mathbf{v}' + \mathbf{v}_\phi$$

or

$$\mathbf{v} = \mathbf{v}' + \boldsymbol{\omega} \times \mathbf{r} \tag{11–11}$$

Note that $\mathbf{v}'$, as an arbitrary velocity, can be in any direction whatsoever. Since $\mathbf{v} = d\mathbf{r}/dt$, we may write Equation (11–11) as

$$\left(\frac{d\mathbf{r}}{dt}\right)_S = \left(\frac{d\mathbf{r}}{dt}\right)_{S'} + \boldsymbol{\omega} \times \mathbf{r} \tag{11–12}$$

where the subscript S signifies that the measurements are made with respect to the stationary frame and the subscript S' means that they are made with respect to the rotating frame. Because the displacement vector $\mathbf{r}$ is the same in either coordinate system, we also make the substitution $\mathbf{r} = \mathbf{r}'$; therefore the same symbol $\mathbf{r}$ appears in each term. Setting up an equation so that each term represents a mathematical operation on the *same quantity* (in this case, $\mathbf{r}$) leads to a useful generalization known as a "transformation equation." The transformation equation obtained in the next section is a simple way to relate the motion as viewed in one frame to the motion as viewed in another frame.

11.5 Mathematical Operators

Every mathematical operation such as "add 6," "extract the square root," or "take the first derivative with respect to t" can be represented by a characteristic symbol called a **mathematical operator:**

$$+6, \qquad \sqrt{}, \qquad \text{or} \quad \frac{d}{dt}$$

In a sense, operators by themselves are just the mathematical instructions in compact form. They acquire significance only when applied to some mathematical function. As we shall use them here, the operator indicates a certain mathematical operation to be performed on whatever function is written immediately after the operators.

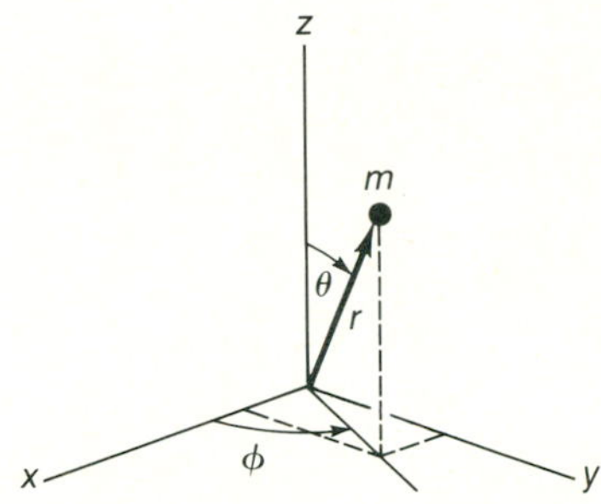

(a) The spherical coordinates r, θ, and ϕ.

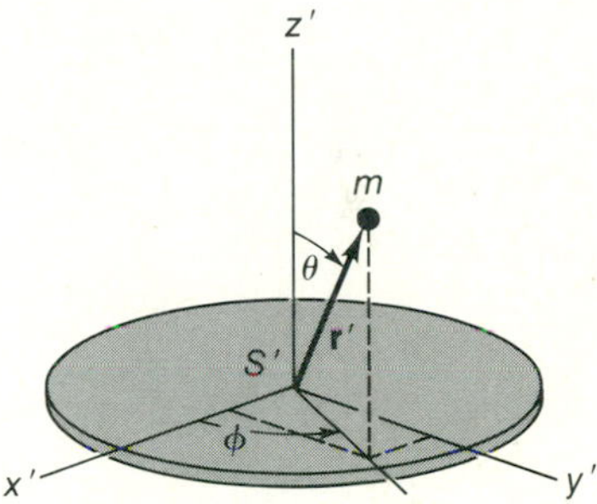

(b) The S' frame is attached to a rotating merry-go-round whose axis of rotation is the z' axis. The location of the mass m at rest with respect to the merry-go-round is the position vector $\mathbf{r}'$.

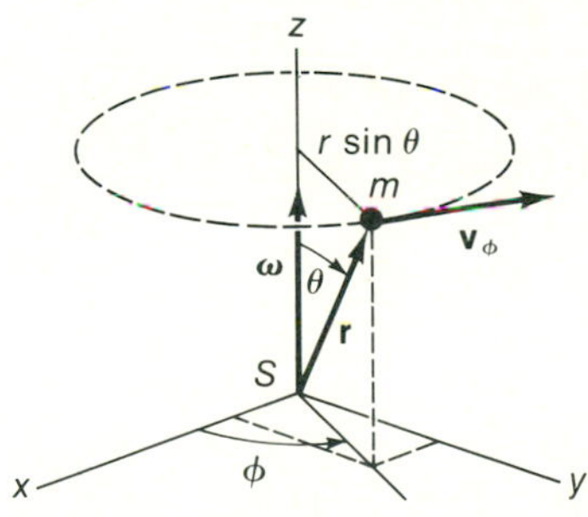

(c) The S frame is an inertial (stationary) frame of reference. In it, the mass m (which is at rest in S') moves in a circle of radius $r \sin \theta$ along the ϕ direction with a velocity $\mathbf{v}_\phi$. (The S' frame rotates with constant angular velocity $\boldsymbol{\omega}$ relative to S.)

Figure 11–8
Relations between the S and S' frames.

We may interpret Equation (11–12) as a certain mathematical operation on the displacement vector **r**. We can thus write the mathematical instructions alone as

$$\left(\frac{d}{dt}\right)_S = \left(\frac{d}{dt}\right)_{S'} + \boldsymbol{\omega} \times \qquad \textbf{(11–13)}$$

This equation is an **operator equation** (or **transformation**) that relates the motion of an object as seen in a stationary frame of reference to the motion as seen in a rotating frame. Although we will not prove it, this equation may be applied to *any* vector describing the particle's motion. Furthermore, it is valid even when the origins and the z axes are not coincident.

The first term on the right of the equal sign is the motion as seen in the S' frame, and the second term is the additional effect of the rotation of S' relative to S. It can be shown that this equation correctly converts motions as viewed in the S' frame to the same motions as viewed in the S frame (and vice versa). It is similar to a foreign language dictionary that forms a link between two different languages.

11.6 Coriolis and Centrifugal Forces

Having obtained a transformation equation, we will use it to investigate the case of an object that has an arbitrary velocity **v** with respect to S. How is the motion described in the S and S' frames? If we apply the operator equation to the velocity **v**, we obtain

$$\left(\frac{d\mathbf{v}}{dt}\right)_S = \left(\frac{d\mathbf{v}}{dt}\right)_{S'} + \boldsymbol{\omega} \times \mathbf{v} \qquad \textbf{(11–14)}$$

The term on the left is the acceleration as measured in the stationary frame; we designate this acceleration by the standard symbol **a**. Substituting for **v** from Equation (11–11), the right-hand side becomes

$$\mathbf{a} = \left[\frac{d}{dt}(\mathbf{v}' + \boldsymbol{\omega} \times \mathbf{r})\right]_{S'} + \boldsymbol{\omega} \times (\mathbf{v}' + \boldsymbol{\omega} \times \mathbf{r})$$

Because ω is constant and hence $d\omega/dt = 0$, the equation expands to

$$\mathbf{a} = \left(\frac{d\mathbf{v}'}{dt}\right)_{S'} + \boldsymbol{\omega} \times \left(\frac{d\mathbf{r}}{dt}\right)_{S'} + \boldsymbol{\omega} \times \mathbf{v}' + \boldsymbol{\omega} \times (\boldsymbol{\omega} \times \mathbf{r}) \qquad \textbf{(11–15)}$$

Since $(d\mathbf{r}/dt)_{S'} = \mathbf{v}'$ and $(d\mathbf{v}'/dt)_{S'} = \mathbf{a}'$, we may write Equation (11–15) as[4]

$$\mathbf{a} = \mathbf{a}' + 2(\boldsymbol{\omega} \times \mathbf{v}') + \boldsymbol{\omega} \times (\boldsymbol{\omega} \times \mathbf{r}') \qquad \textbf{(11–16)}$$

Multiplying Equation (11–16) by the mass m of the particle, we recognize the left-hand side as the net (real) force on the mass as seen in the S frame.

$$m\mathbf{a} = \Sigma\mathbf{F}_{\text{real}}$$

Thus

$$\Sigma\mathbf{F}_{\text{real}} = m\mathbf{a}' + 2m(\boldsymbol{\omega} \times \mathbf{v}') + m[\boldsymbol{\omega} \times (\boldsymbol{\omega} \times \mathbf{r}')] \qquad \textbf{(11–17)}$$

Now, in the rotating S' frame, the particle m is observed to have an acceleration $\mathbf{a}'$. We preserve the form of Newton's second law in the rotating frame by re-

[4] We should make a remark about the last term of this equation, $\boldsymbol{\omega} \times (\boldsymbol{\omega} \times \mathbf{r})$. (In vector algebra, it is called a "triple product.") It is essential to perform the cross product inside the parentheses *before* performing the other cross product. If the operations are performed in the opposite order, the cross product $(\boldsymbol{\omega} \times \boldsymbol{\omega})$ will be zero, because $\sin 0^\circ$ always equals zero. Note also that because $\mathbf{r} = \mathbf{r}'$, we here write $\mathbf{r}'$ to help emphasize that these terms are significant only as measured in the primed frame of reference.

arranging the equation in the following manner:

$$\underbrace{\Sigma \mathbf{F}_{\text{real}} - 2m(\boldsymbol{\omega} \times \mathbf{v}') - m[\boldsymbol{\omega} \times (\boldsymbol{\omega} \times \mathbf{r}')]}_{\Sigma \mathbf{F}'} = m\mathbf{a}' \quad (11\text{–}18)$$

Thus, the net force $\Sigma \mathbf{F}'$ on the object as seen in the rotating frame is made up of the real forces seen in the stationary frame (which combine to give $\Sigma \mathbf{F}_{\text{real}}$) and two inertial forces, called the **coriolis force** and the **centrifugal force.** These latter two forces are attributable to the fact that we are viewing the motion from an accelerated frame of reference; they are never present in the stationary frame of reference.

Following is a summary of the three forces as seen in the rotating frame:

REAL FORCE
$\Sigma \mathbf{F}_{\text{real}}$

This is the sum of all the real forces on the object, such as forces of contact, tensions in springs, the force of gravity, electrical forces, and magnetic forces. Only these real forces are seen in an inertial frame of reference.

CORIOLIS FORCE
$\mathbf{F}_{\text{cor}} = -2m(\boldsymbol{\omega} \times \mathbf{v}')$

The coriolis force is a *deflecting* force, always at right angles to the velocity *v′* of the mass *m*. If the object has no velocity in the rotating frame of reference, there is no coriolis force. It is an inertial force not seen in a stationary frame of reference. Note the minus sign.

CENTRIFUGAL FORCE
$\mathbf{F}_{\text{cf}} = -m[\boldsymbol{\omega} \times (\boldsymbol{\omega} \times \mathbf{r}')]$

The centrifugal force depends on position only and is always radially outward. Like the coriolis force, it is an inertial force not seen in a stationary frame of reference. Again, note the minus sign.

The centrifugal force term, $-m[\boldsymbol{\omega} \times (\boldsymbol{\omega} \times \mathbf{r}')]$, may be simplified. Referring to Figure 11–10, $\boldsymbol{\omega} \times \mathbf{r}'$ is directed into the plane of the figure with a magnitude $\omega r' \sin\theta$. Performing the other cross product, $-m\boldsymbol{\omega} \times$ [this vector], we obtain a final vector of magnitude $-m\omega^2 r' \sin\theta$ directed radially away from the axis of rotation. This direction is along the direction of $\mathbf{r}'_{\perp}$, which itself has the magnitude $r' \sin\theta$. Thus:

$$-m[\boldsymbol{\omega} \times (\boldsymbol{\omega} \times \mathbf{r}')] = m\omega^2 \mathbf{r}'_{\perp} \quad (11\text{–}19)$$

It is easy to remember that the centrifugal force is always *perpendicularly outward from the axis of rotation.*

To summarize, we write *Newton's second law* $\Sigma \mathbf{F}' = m\mathbf{a}'$ *for a rotating frame of reference* as

NEWTON'S SECOND LAW (in a rotating frame)

$$\underbrace{\Sigma \mathbf{F}_{\text{real}} + \mathbf{F}_{\text{cor}} + \mathbf{F}_{\text{cf}}}_{\Sigma \mathbf{F}'} = m\mathbf{a}' \quad (11\text{–}20)$$

where the following terms are the *inertial forces* seen only in a rotating frame:

$$\mathbf{F}_{\text{cor}} = -2m(\boldsymbol{\omega} \times \mathbf{v}') \quad \text{and} \quad \mathbf{F}_{\text{cf}} = -m[\boldsymbol{\omega} \times (\boldsymbol{\omega} \times \mathbf{r}')]$$

The minus signs are important in the definitions of the inertial force terms. Not only do they identify the correct directions in space, but they are consistent with our interpretation of the net force on the object as the *sum* of all the forces on it, as shown in Equation (11–20). In sketching vector diagrams, we do not use a minus sign with the label on the vector; the diagram itself gives the spatial

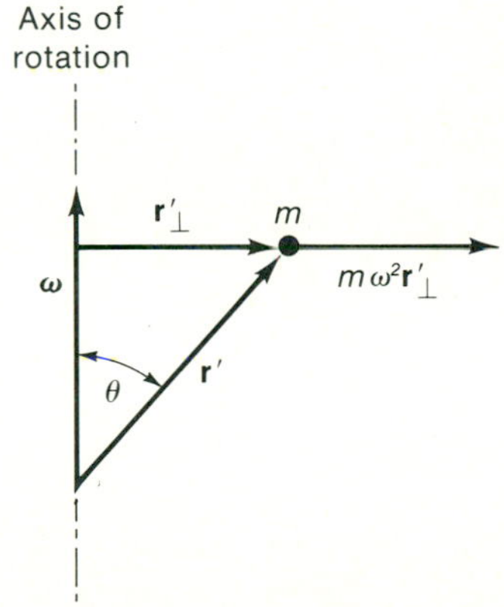

Figure 11–10
The vector $-m[\boldsymbol{\omega} \times (\boldsymbol{\omega} \times \mathbf{r}')]$ is a vector of magnitude $m\omega^2 r'_{\perp}$ and is directed perpendicularly away from the axis of rotation.

directions, and is independent of the particular coordinate direction chosen as minus.

EXAMPLE 11–3

A ball on the end of a cord swings in a horizontal circle of radius $r = 0.60$ m while the cord makes an angle of 35° with respect to the vertical. Find the time required for the ball to make one revolution.

SOLUTION

We will analyze the situation from two frames of reference: the S frame at rest and the S' frame rotating at the same speed as the ball, as shown in Figure 11–11.

(a) As seen in S, the ball travels in a horizontal circle.

Observer sits on the turntable

(b) As seen in S', the ball is at rest.

Figure 11–11
Example 11–3

As seen in S

The ball moves in a circle at a constant speed and has an acceleration toward the center of the circle. We therefore choose this horizontal direction as one of the component directions.

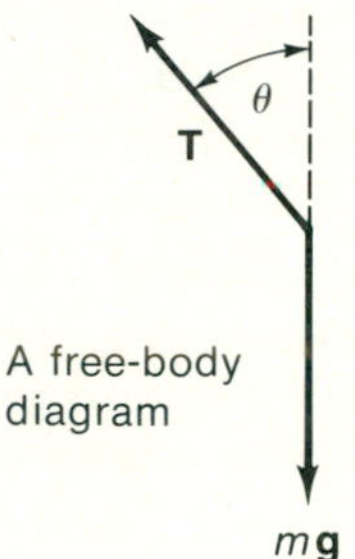

A free-body diagram

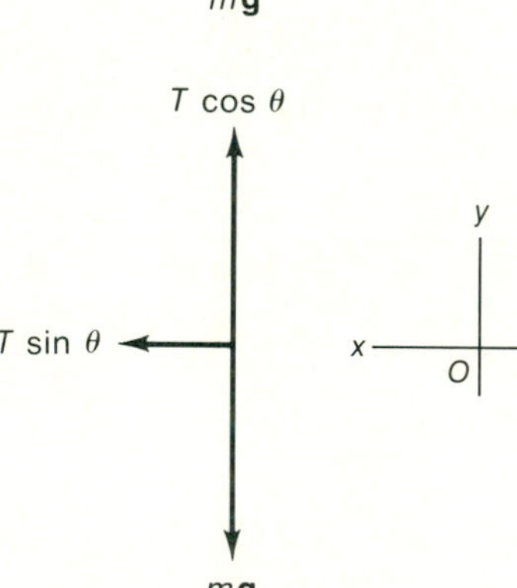

A free-body diagram with forces resolved into two mutually perpendicular directions.

Horizontal component

$$\Sigma F_x = ma_x$$

$$T \sin\theta = m\left(\frac{v^2}{r}\right) \qquad [1]$$

$$\Sigma F_y = 0 \qquad \text{(Equilibrium)}$$

$$T\cos\theta = mg \qquad [2]$$

As seen in S'

The ball is at rest in this rotating frame. Therefore, there is no coriolis force, but there is an outward centrifugal force:

$$-m[\boldsymbol{\omega} \times (\boldsymbol{\omega} \times \mathbf{r}')]$$

equal in magnitude to $m(v^2/r)$.

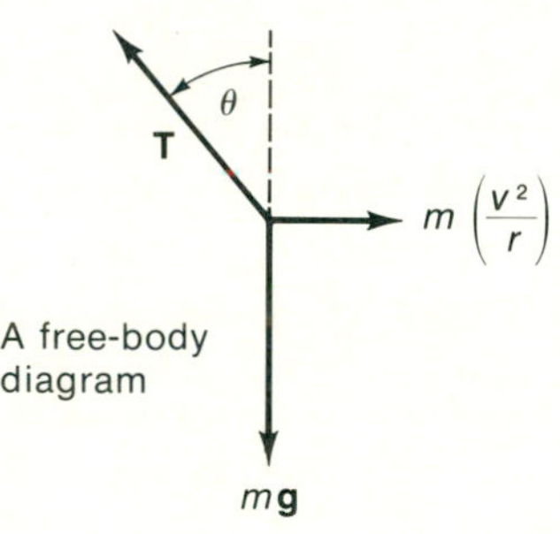

A free-body diagram

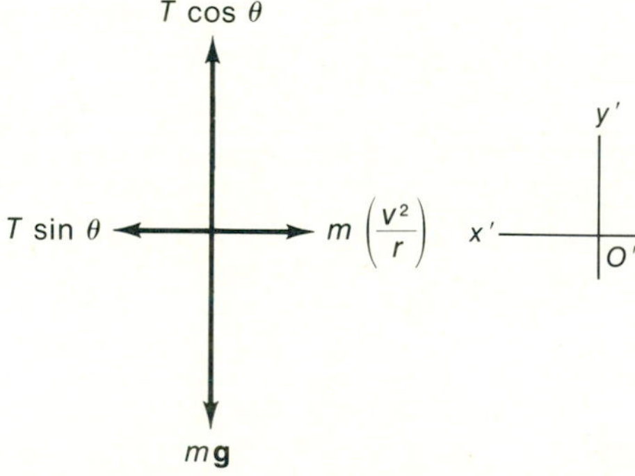

A free-body diagram with forces resolved into two mutually perpendicular directions.

Horizontal component

$$\Sigma F_x = 0 \qquad \text{(Equilibrium)}$$

$$T \sin\theta = m\left(\frac{v^2}{r}\right) \qquad [1]$$

$$\Sigma F_y = 0 \qquad \text{(Equilibrium)}$$

$$T\cos\theta = mg \qquad [2]$$

Dividing [1] by [2], we have

$$\frac{T \sin\theta}{T \cos\theta} = \frac{m\left(\dfrac{v^2}{r}\right)}{mg}$$

$$\tan\theta = \frac{v^2}{rg}$$

Dividing [1] by [2], we have

$$\frac{T \sin\theta}{T \cos\theta} = \frac{m\left(\dfrac{v^2}{r}\right)}{mg}$$

$$\tan\theta = \frac{v^2}{rg}$$

Both analyses lead to the same equation. Solving for v and substituting the numbers, we obtain the following numerical result:

$$v = \sqrt{rg \tan\theta} = \sqrt{(0.60\text{ m})\left(9.8\,\frac{\text{m}}{\text{s}^2}\right)(\tan 35^\circ)} = 2.03\,\frac{\text{m}}{\text{s}}$$

The time for one revolution is

$$t = \frac{2\pi r}{v} = \frac{2\pi(0.60\text{ m})}{2.03\,\dfrac{\text{m}}{\text{s}}} = \boxed{1.86\text{ s}}$$

EXAMPLE 11–4

A 200-g mouse runs radially outward on a merry-go-round, which is turning at an angular speed of 10 rpm. The speed of the mouse is constant at 0.50 m/s relative to the merry-go-round. (a) Draw a free-body diagram for the mouse as seen in a frame attached to the merry-go-round. (b) Find the force of friction that the surface exerts on the mouse when it is 2 m from the axis of rotation.

SOLUTION

In this problem, the analysis is far simpler in the rotating frame because in that frame the mouse moves at constant speed in a straight line. It is *unaccelerated* motion. By contrast, in an inertial frame the mouse moves in a complicated spiral path with constantly changing speed and acceleration. Thus we will choose the rotating frame for our analysis. But even in the rotating frame, there are several forces involved. If we take them one at a time, the analysis is straightforward; it is a case of applying $\Sigma\mathbf{F}' = m\mathbf{a}'$.

(a) The forces are not coplanar, so we will make a sketch in three-dimensional perspective, assuming a sense of rotation as indicated in Figure 11–12(a). In addition to the real forces (such as gravity and friction) acting on the mouse, we also have two inertial

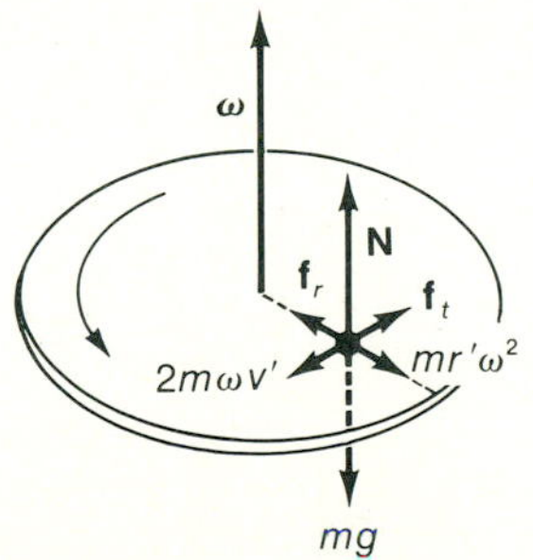

(a) Forces on the mouse as seen in the rotating frame. The sketch is in a three-dimensional perspective; all forces are mutually perpendicular.

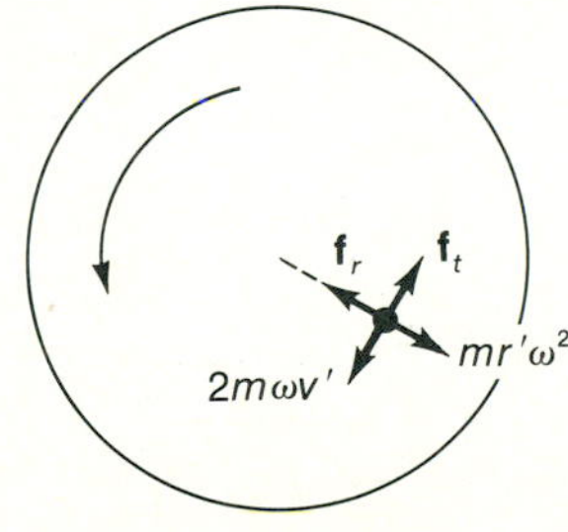

(b) Looking vertically down from above on the mouse (as seen in the rotating frame of reference).

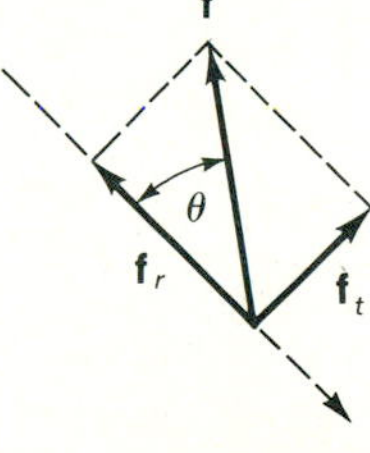

(c) The friction force **f** has the components $\mathbf{f}_t$ and $\mathbf{f}_r$.

Figure 11–12
Example 11–4

forces, and the coriolis force and the centrifugal force. Thus the following forces are involved:

$W = mg$, the downward force of gravity

$N =$ the upward normal force exerted by the merry-go-round

$2m\omega v' = |-2m(\omega \times \mathbf{v}')|$, the coriolis force, whose direction is obtained from the right-hand rule for finding the cross product

$mr'\omega^2 = |-m[\boldsymbol{\omega} \times (\boldsymbol{\omega} \times \mathbf{r}')]|$, the centrifugal force, which is radially outward

Since the mouse moves in a straight line at constant speed, it is in dynamic equilibrium in this frame (zero net force). Hence the two horizontal forces (the coriolis and the centrifugal) must be balanced by another horizontal force, that of friction. For convenience, we represent the friction force as two perpendicular force components, radial and tangential:

$f_r =$ the radial (inward) component of friction, which balances the outward centrifugal force

$f_t =$ the tangential component of friction, which balances the coriolis force

Note that this is a case of static friction, since the mouse is not sliding. We do not know whether or not the mouse is on the verge of slipping; therefore, $f_s \neq \mu_s N$.

(b) To calculate the friction force, we sketch a force diagram showing the horizontal components only (see Figure 11–12b).

Radial component	**Tangential component**
$\Sigma F_r' = 0$ (Equilibrium)	$\Sigma F_t' = 0$ (Equilibrium)
$f_r = mr'\omega^2$	$f_t = 2m\omega v'$

Converting numerical values to SI units, we have:

$$m = 0.20 \text{ kg}$$
$$r = 2 \text{ m}$$
$$v' = 0.50 \text{ m/s}$$
$$\omega = \left(10 \frac{\text{rev}}{\text{min}}\right)\underbrace{\left(\frac{1 \text{ min}}{60 \text{ s}}\right)\left(\frac{2\pi \text{ rad}}{1 \text{ rev}}\right)}_{\text{Conversion ratios}} = \frac{\pi}{3.0} \frac{\text{rad}}{\text{s}}$$

Substituting the numbers into the above equations, we obtain

$$f_r = mr'\omega^2 = (0.20 \text{ kg})(2 \text{ m})\left(\frac{\pi \text{ rad}}{3.0 \text{ s}}\right)^2 = 0.438 \text{ N}$$

$$f_t = 2m\omega v' = (2)(0.20 \text{ kg})\left(\frac{\pi \text{ rad}}{3.0 \text{ s}}\right)\left(0.50 \frac{\text{m}}{\text{s}}\right) = 0.209 \text{ N}$$

The force of friction has the following magnitude:

$$f = \sqrt{f_r^2 + f_t^2} = \boxed{0.485 \text{ N}}$$

The direction of the friction force is given by:

$$\tan\theta = \frac{f_t}{f_r} = \frac{0.209 \text{ N}}{0.438 \text{ N}} = 0.477$$

$$\theta = \tan^{-1}(0.477) = \boxed{25.5^\circ}$$

See Figure 11–12(c) for the designation of angle θ.

Note that if the merry-go-round were not rotating, there would be no horizontal forces at all on the moving mouse. However, with rotation at constant speed, there is a radially outward force that increases with distance (the centrifugal force) as well as a constant sideways force (the coriolis force). To maintain a constant speed radially outward, the mouse must "dig in its heels" to cause a friction force that counteracts these inertial forces.

EXAMPLE 11–5

A marksman follows a rapidly moving target by swinging his rifle from *left to right* about a vertical axis while holding the barrel horizontal. Just as the bullet is about to leave the barrel, it has a muzzle velocity of 800 m/s. (a) If, at this instant, the rifle is rotating at 1 rad/s, what is the magnitude of the coriolis force on the 20-g bullet? (b) Find the direction of force the rifle barrel exerts on the bullet.

SOLUTION

(a) The rotating frame turns with the rifle barrel. The angular velocity ω of this frame is vertically downward. The magnitude of the coriolis force on the bullet is equal to

$$F_{\text{cor}} = |-2m(\boldsymbol{\omega} \times \mathbf{v}')|$$

Figure 11–13(a) shows the direction of this force *on the bullet*. Because of the 90° relationship between $\boldsymbol{\omega}$ and $\mathbf{v}$, we have the magnitude:

$$F_{\text{cor}} = 2m\omega v'$$

$$F_{\text{cor}} = (2)(0.020\ \text{kg})\left(1\ \frac{\text{rad}}{\text{s}}\right)\left(800\ \frac{\text{m}}{\text{s}}\right) = \boxed{32\ \text{N}}$$

(b) The force the bullet exerts on the rifle barrel is in a direction opposite to that which the rifle exerts on the bullet. In the rotating frame, the bullet has no sideways acceleration. Its only acceleration is in the radial direction. Figure 11–13(b) shows the forces on the bullet just before it leaves the rifle barrel.

F_1 = force the expanding gas exerts on the bullet

F_{cor} = coriolis force on the bullet

F_{cf} = centrifugal force on the bullet

F_2 = force of the rifle barrel to balance F_{cor} and thus make the bullet in equilibrium for the θ component of motion

Therefore the force the barrel exerts on the bullet is F_2. It is equal in magnitude to the coriolis force (32 N) and is present only if the rifle maintains its constant angular speed ω during the firing of the bullet. In practice, this is essentially impossible, since the marksman cannot exert such an extremely brief sideways force. He will feel the jolt, however!

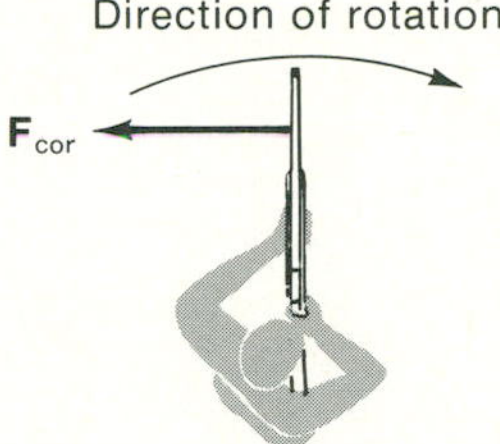

(a) The view from above in the rotating frame, showing just the coriolis force on the moving bullet. The vector $\boldsymbol{\omega}$ is into the page.

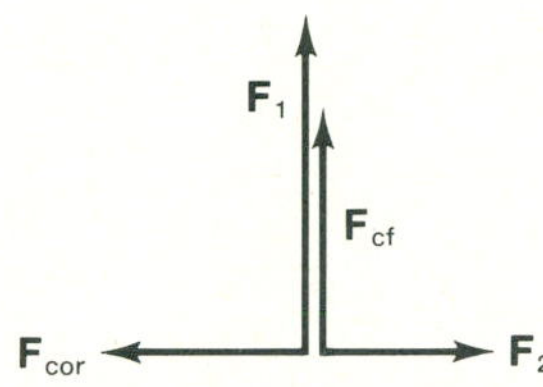

(b) The horizontal forces on the bullet just before it leaves the barrel, as viewed from above in the rotating frame.

Figure 11–13
Example 11–5

11.7 Comments

As we have seen, inertial forces arise solely from our choice of an accelerated (here, rotating) frame of reference and from our desire to preserve the format of Newton's second law, $\Sigma\mathbf{F} = m\mathbf{a}$. We have a strong tendency to interpret observed accelerations as a result of some net force on the object. But there may be a deeper puzzle here: Inertial forces are not all that imaginary. In an accelerated frame of reference, one is easily convinced that these inertial forces really exist; they produce real, physically measurable effects. Furthermore, we verify the presence of an inertial force in exactly the same manner as we verify the presence of a gravitational force. The only difference is that we cannot trace the origin of inertial forces to some other physical object as we customarily do in the case of gravity and other real forces. Inertial forces thus do not obey Newton's third law.

In many problems, the accelerated frame is the most convenient one. For example, the motion of a Foucault pendulum can be explained in terms of the inertial forces that arise when the pendulum is viewed from the rotating frame (see Figure 11–14). And in studying the action of high-speed centrifuges, it is simpler to analyze the situation from a frame of reference in which the liquid is at rest rather than as a fixed observer watching the whirling container.

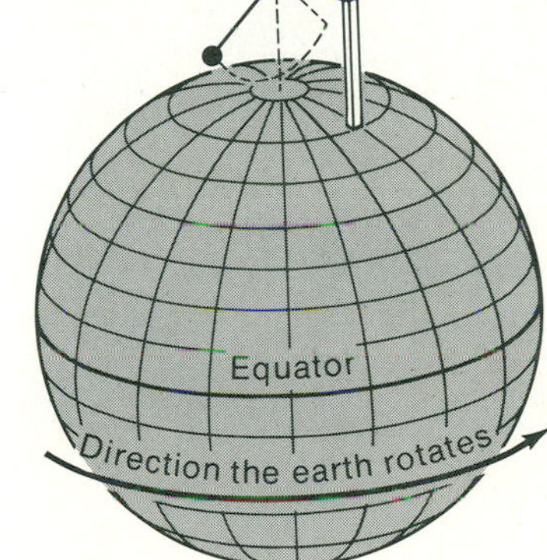

Figure 11–14
The Foucault pendulum. A swinging plumb bob, suspended from a point along the polar axis of the earth, is the simplest case that illustrates the earth's rotation. The plane of the swing maintains its direction in inertial space while the earth rotates beneath it. Viewed in the rotating earth frame, the plane of the swing rotates once a day. (At other latitudes, the situation is more complex. The rate of rotation equals $\omega \sin\theta$, where ω is the earth's rotational speed and θ is the latitude.)

Coriolis forces are evident in many phenomena. Because of coriolis forces, most cyclones are in a counterclockwise direction in the Northern Hemisphere and a clockwise direction in the Southern Hemisphere (see Figure 11–15). Also, the coriolis forces on the circulation of winds around an area of high atmospheric pressure tend to maintain the high pressure area as a sharply defined weather system. Since the horizontal coriolis forces are least near the equator, weather systems with their associated winds often dissipate near the equator, creating the doldrums. Interesting effects in atomic physics are produced by the coriolis force on vibrating atoms in a polyatomic molecule that is rotating as a whole.

Figure 11–15

Over 40% of the earth's surface is visible in this photograph taken from synchronous orbit 22,300 mi above equatorial Brazil by NASA's Applications Technology Satellite in 1968. Each time the satellite revolves, at 94 rpm, the camera records a hair-thin strip of the earth's image, 0.004 in wide, from horizon to horizon. In a 24-min period the station records on film a complete picture, detailing cloud patterns and movement. Note that because of coriolis forces, the predominant spiral patterns rotate in opposite senses in the Northern and Southern Hemispheres.

(**a**) The common housefly and its relatives have *halteres*, small knobs on the ends of rods that vibrate in a plane during flight. This photograph shows halteres of a crane fly. When the fly turns, the plane of oscillation tends to maintain its orientation by utilizing an effect similar to that of a Foucault pendulum, giving the fly information about the rate of turn, etc. When the fly accelerates linearly, the masses at the end of the rods lag behind due to their inertia, thus providing a signal that is interpreted as linear acceleration. If the halteres are immobilized or removed, the insect cannot fly in a straight line and instead flies erratically in small, random circles.

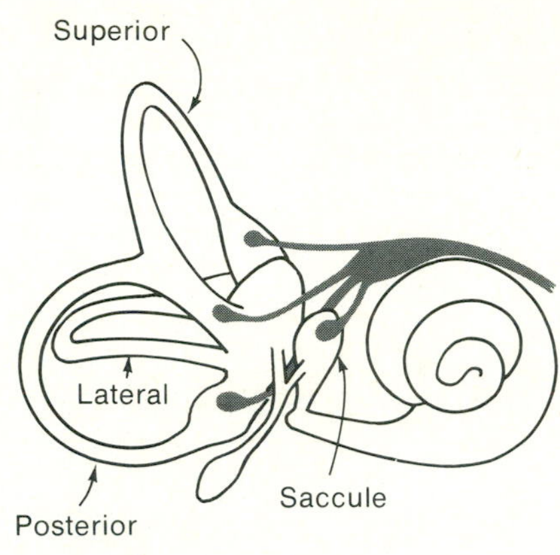

(**b**) The semicircular canals of the human ear are three circular tubes that contain fluid. When the head is turned, the fluid lags behind due to its inertia. Relative motion between the fluid and the canals is detected by nerve fibers in the tube walls; these nerve fibers inform the brain that rotational motion is occurring. The canals are situated in approximately three mutually perpendicular planes, so they can detect rotary motion about any given axis. Translational acceleration is detected by the motion of solid particles of calcium carbonate suspended in a gelatinous substance, which is contained in a small sac called the saccule. When the head accelerates linearly, the solid particles lag behind. Nerve fibers projecting from the walls of the saccule detect the resultant motion of the particles relative to the walls. If a person in a rotating frame of reference turns his or her head, coriolis effects on the moving fluid produce unusual signals to the brain, incorrectly signifying rotation about some other axis. The resulting confusion produces feelings of vertigo and nausea. Such effects may be serious in proposed space stations, which are given appreciable rotation for producing artificial gravity.

Figure 11–16

Biological organisms have inertial guidance systems that detect motions with respect to an inertial frame of reference. Two examples are illustrated here.

The coriolis force is also of crucial significance in long-range projectile motion.[5] Biological systems also utilize inertial forces for detecting their own linear and rotational motion relative to an inertial frame of reference (see Figure 11–16).

Sometimes it is claimed that the water draining out of a washbowl circulates in a preferred direction due to the coriolis force. In most of these cases, the coriolis force is negligible compared with other, larger effects—particularly those due to residual angular momentum resulting from the manner in which the container was filled. These effects persist sometimes for days, so the effect of the coriolis force itself is difficult to demonstrate. However, if very precise and careful experiments are performed, the so-called "bathtub vortex" indeed exhibits a preferred direction due to the coriolis force, opposite in Northern and Southern Hemispheres. You can investigate coriolis effects yourself by walking a radial line inward or outward on a rotating merry-go-round.

EXAMPLE 11–6

Suppose that at some future date, high-speed mail service is established by missile from a town in central Africa on the equator to Helsinki, Finland (on the same longitude line). If the coriolis force is neglected in programming the missile flight, will the missile land to the east or west of Helsinki? Analyze the problem from the (rotating) earth's frame of reference.

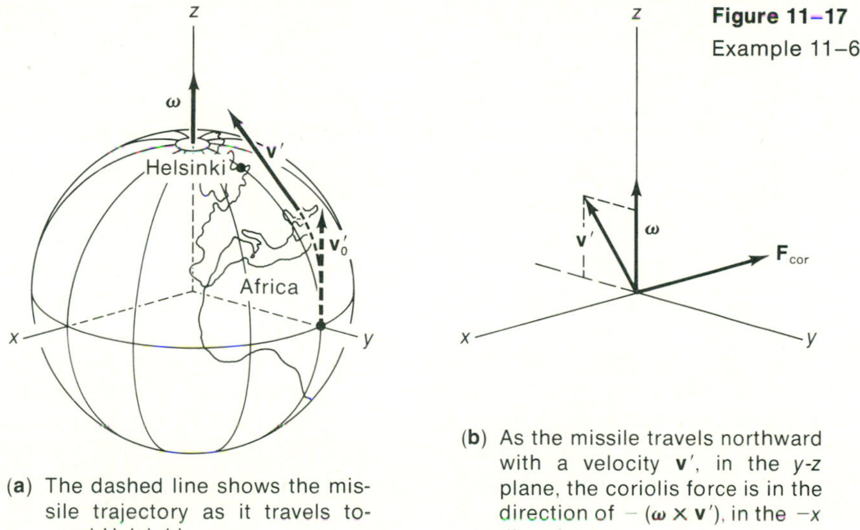

Figure 11–17
Example 11–6

(**a**) The dashed line shows the missile trajectory as it travels toward Helsinki.

(**b**) As the missile travels northward with a velocity $\mathbf{v}'$, in the y-z plane, the coriolis force is in the direction of $-(\boldsymbol{\omega} \times \mathbf{v}')$, in the $-x$ direction.

SOLUTION

Figure 11–17 shows that the coriolis force is zero at the initial launching (because the earth's rotation $\boldsymbol{\omega}$ and the initial velocity $\mathbf{v}'_0$ are parallel). As the flight progresses, however, an angle develops between these vectors. The coriolis force $\mathbf{F}_{cor} = -2m(\boldsymbol{\omega} \times \mathbf{v}')$ is thus eastward, causing the missile to land east of Helsinki.

[5] During a naval engagement near the Falkland Islands (off the tip of South America) that occurred early in World War I, British gunners were surprised to see their accurately aimed salvos falling 100 yd to the left of the German ships. The designers of the sighting mechanisms were well aware of the coriolis deflection and had carefully taken this into account, but they apparently were under the impression that all sea battles took place near 50° N latitude and never near 50° S latitude. The British shots fell, therefore, at a distance from the targets equal to *twice* the coriolis deflection. (A footnote in Jerry B. Marion, *Classical Dynamics of Particles and Systems*, 2nd ed., New York: Academic Press, 1965, p. 346.)

11.8 Terminology

Among various authors there are differences in the usage of the term "centrifugal force," and there are also several sources of possible confusion in other concepts associated with accelerated frames. For example, a common source of confusion is in the use of the word "inertial" for forces viewed in an accelerated frame of reference. A stationary frame of reference is an inertial frame—that is, one in which Newton's first law (the law of inertia) is valid *without* the addition of inertial forces. Thus, inertial forces do *not* occur in inertial frames.

Confusion in terminology also has arisen from the fact that for historical reasons, the last two terms of Equation (11–16) are usually called, respectively, the "coriolis acceleration" and the "centrifugal acceleration." But since these vector accelerations have directions *opposite* to the direction of the corresponding force terms of Equation (11–20), we have the peculiar situation that, defined this way, the coriolis force does *not* produce the coriolis acceleration and the centrifugal force does *not* produce the centrifugal acceleration.

A further widespread source of confusion is the fact that some engineering texts define the centrifugal force differently from the usage in physics. One particular definition treats centrifugal and centripetal forces as a Newton's third-law pair of forces: opposite in direction and acting on two different objects. For example, if a stone on the end of a string is whirled around in a circular path, the string exerts an inward centripetal force on the stone, while the stone (according to some authors) is said to exert an outward centrifugal force on the string. Thus, according to this particular definition, the centrifugal force does not act on the object going in a circle. Defining the word "centrifugal" in this manner seems unfortunate, since it then prevents applying this descriptive adjective to the outward inertial force on objects when viewed in a rotating frame of reference. Of course, we may adopt any definition that proves to be helpful. In this text, we define the terms as used currently by most authors.

Popular discussions in newspapers and other sources for the nontechnical reader are frequently incorrect, since they often attribute outward centrifugal forces as acting on an object going in a circle when clearly the phenomenon is being viewed from a stationary frame of reference, thus agreeing with *neither* the engineering nor the physics definition. As an example, consider the following statement: "The moon does not fall as it circles the earth because the (outward) centrifugal force just balances the inward force of gravity; thus there is no net force on the moon to make it fall." This statement is incorrect in the frame of reference attached to the earth, and also in an inertial, nonrotating frame. If the moon had *no* net force on it toward the earth, it would not move in a circle around the earth but *in a straight line according to Newton's first law* (see Figure 11–18). Of course, in a frame of reference in which the moon is *at rest* relative to the earth's *CM* (that is, a frame centered on the earth that rotates once every 27.3 days with respect to the fixed stars), the inward force of gravity *is* exactly balanced by the outward inertial force, so that in such a rotating frame the moon would indeed have no net force on it. (These comments ignore the earth's orbital motion about the sun.)

In making free-body diagrams for circular motion, students sometimes make the common mistake of adding a "centripetal force vector" radially inward, equal to mv^2/r. *There is no such extra force.* Of course, from $\Sigma\mathbf{F} = m\mathbf{a}$, there must be a net inward force acting on the object to make it travel in a circle. But this inward force $\Sigma\mathbf{F}$ always arises in the usual ways: from gravity, a tension in a string, an electrical force, an inertial force, or any other conventional force. The phrase "centripetal force" is merely the *name* we give to this net radially-inward force. It is *not* an extra force that mysteriously arises because an object goes in a circle.

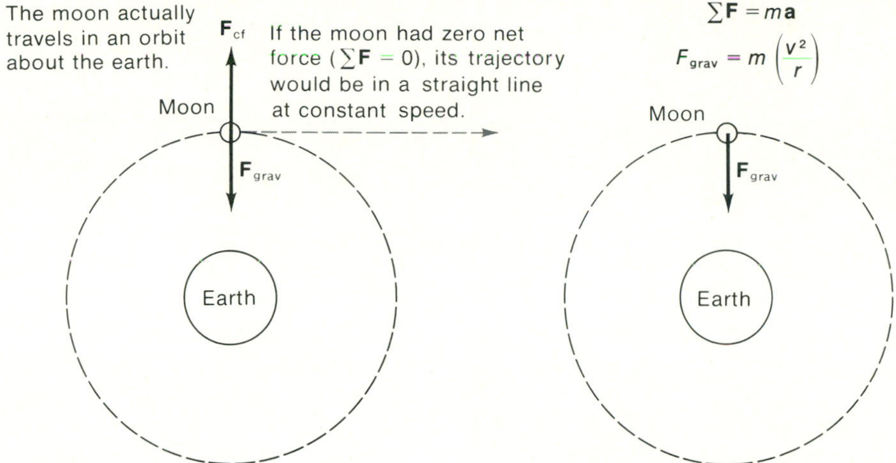
The force of gravity is the net force that causes the moon to have the centripetal acceleration necessary for motion in a circle.
∑F = ma
F_grav = m (v²/r)
The moon actually travels in an orbit about the earth.
F_cf
If the moon had zero net force (∑F = 0), its trajectory would be in a straight line at constant speed.
Moon
F_grav
Earth
Moon
F_grav
Earth

(**a**) An incorrect free-body diagram for the forces on the moon as seen in an inertial frame of reference.

(**b**) The correct free-body diagram for forces on the moon as seen in an inertial frame of reference.

Figure 11–18

Sometimes it is erroneously claimed that "the outward centrifugal force on the moon just balances the inward force of gravity." If this were true, the moon would have no net inward force to deflect its motion; it would travel in a straight line at constant speed, as illustrated in (a). Instead, the unbalanced force of gravity causes the moon to continually "fall" toward the earth as it travels in a circle at constant speed, as shown in (b).

Summary

In a frame of reference that has *constant linear acceleration* $\mathbf{a}_h$, we preserve the form of Newton's second law $\Sigma\mathbf{F}' = m\mathbf{a}'$ by adding the inertial force $-m\mathbf{a}_h$ to the real forces that might be present on a point mass m.

In a rotating frame of reference that has *constant angular velocity* ω, we preserve $\Sigma\mathbf{F}' = m\mathbf{a}'$ by adding two inertial forces, the coriolis and centrifugal forces. The *coriolis* force is present only if there is a velocity $\mathbf{v}'$ as seen in the rotating frame:

CORIOLIS FORCE

$$\mathbf{F}_{\text{cor}} = -2m(\boldsymbol{\omega} \times \mathbf{v}')$$

It is a sideways deflecting force at right angles to $\mathbf{v}'$ and to ω. *In a nonrotating frame of reference, there is no such thing as a coriolis force.* The *centrifugal* force is always present (except on the axis) and is radially outward from the axis of rotation:

CENTRIFUGAL FORCE

$$\mathbf{F}_{\text{cf}} = -m[\boldsymbol{\omega} \times (\boldsymbol{\omega} \times \mathbf{r}')]$$

It is equal in magnitude to $mv^2/r_\perp = m\omega^2 r_\perp$, where $r_\perp$ is the distance from the axis of rotation to the mass m. In *a nonrotating frame of reference, there is no such thing as a centrifugal force.*

In an accelerated frame, inertial forces are just as evident as forces of gravity—in fact, we infer their presence in exactly the same way we infer the presence of gravitational forces. However, the origin of inertial forces is not due to other physical objects (as is the case for real forces). Instead, they arise because we wish to preserve the form of Newton's second law, $\mathbf{F}' = m\mathbf{a}'$, in an accelerated frame of reference. Inertial forces are never present in an inertial frame of reference.

Questions

1. A pickup truck with a vertical windshield moves with a speed v_w. Raindrops are falling straight down with a constant terminal speed v_r. What is the angle at which the drops strike the windshield?
2. A girl riding on one side of a train tosses a ball across the aisle. The ball is caught by her friend sitting on the other side. If the train moves with constant velocity, describe the ball's motion in the train's frame and also in the earth's frame. Repeat for the case when the train is accelerating forward in a straight line. What is the situation if the train is traveling around a curve at constant speed?
3. What (precise) measurements in a very wide elevator could distinguish between the elevator at rest at the earth's surface and the elevator accelerating upward in the absence of gravity?
4. Suppose you were in a windowless room without any standard measuring instruments. How could you tell whether or not the room was rotating about a vertical axis? How could you determine whether the axis of rotation was inside or outside the room? How could you determine the direction of rotation?
5. In the previous question, what instruments would you need to determine the numerical value of the angular velocity of the room?
6. At the equator, suppose you drop a rock down a vertical (unoccupied) mine shaft. Considering the earth's rotation, toward which wall of the shaft will the rock veer?
7. Automobile seats with headrests help prevent "whiplash" injuries to passengers if the car is hit from behind. For a whiplash incident, explain what happens as viewed from the reference frame attached to the automobile seat.
8. In the frame of reference of the (rotating) earth, describe all the forces that act on a book resting on a table at your particular location on the earth's surface.
9. A truck is transporting a tub full of water down a straight hill, coasting freely down the incline without friction. Ignore the sloshing that might occur as the motion is started and consider the "steady-state" situation after transient motions have subsided. Is the water surface parallel to the earth's surface or to the incline? Explain.

Problems

11A–1 A woman in an elevator weighs a mass labeled "2.50 kg" on a spring balance. If the spring balance reads 30 N, find the acceleration of the elevator. Use the elevator's frame of reference for your analysis and assume that the local acceleration due to gravity is 9.8 m/s^2. *Answer:* 2.2 m/s^2 upward

11A–2 A 160-lb man stands on a bathroom scale while riding in an elevator that is accelerating downward at 4 ft/s^2. What does the scale indicate as the man's weight? Analyze the problem (a) from an inertial frame of reference and (b) from the elevator's frame.

11A–3 A 10-lb weight is lifted by a cord with a breaking strength of 15 lb. By analyzing the problem in the accelerated frame of reference, determine the maximum acceleration with which the weight may be lifted. *Answer:* 8 ft/s^2

11A–4 A ball is dropped from the ceiling of a glass-walled elevator that is accelerating downward at $\frac{3}{4}$g. Describe the acceleration of the ball as viewed by an observer (a) at rest with respect to the elevator and (b) riding in the elevator.

11A–5 An acrobatic pilot experiences weightlessness as she passes over the top of a loop-the-loop maneuver. If her speed is 100 mph at the time, find the radius of the loop. Solve the problem both in the airplane's frame of reference and in an inertial frame.

Answer: 672 ft

11A–6 A 1000-lb uniform steel beam 25 ft long rests on the ground. A 180-lb man sits on one end.

(a) How far from the man should a lifting cable be fastened to the beam, so that the beam plus the man will not tip when lifted at constant speed?

(b) Explain why no disaster happens even when they are lifted with acceleration.

11A–7 Find the ideal banking angle of a race track curve with a radius of 200 ft if the speed of the car making the curve is 60 mph. Solve the problem from the car's frame of reference.

Answer: 50.4°

11A–8 In a popular amusement-park ride, the riders position themselves on the inner wall of a rotating cylinder whose axis of rotation is vertical. The floor is then lowered, leaving the riders "pinned" to the wall by centrifugal effects and prevented from sliding downward by friction forces. In terms of the cylinder radius R, the angular speed ω, and g, find the minimum coefficient of static friction that will prevent sliding. Analyze the problem in the rotating frame of reference.

11A–9 A small box rests 12 ft from the center of a merry-go-round. If the coefficient of static friction between the floor and box is 0.30, find the merry-go-round's maximum angular velocity in revolutions per minute that still will not cause the box to slip. Solve the problem twice: (a) from the merry-go-round's frame and (b) from an inertial frame.

Answer: 8.54 rpm

11A–10 In a lecture demonstration, a smoothly sliding puck of mass 0.5 kg is launched radially inward from the outer rim of a rotating circular, horizontal platform 1.2 m in diameter. The angular velocity of the platform is 2 rad/s, and the rotation is clockwise as viewed from above. If the launching speed is 2 m/s as seen in the rotating frame, find, at the instant just after launching, (a) the centrifugal force on the puck and (b) the coriolis force on the puck. Include the directions these forces act.

11B–1 A 2.5-kg mass falls under the influence of gravity with an acceleration of 9.8 m/s^2. With what additional force must the object be pushed downward to accelerate the mass at 12 m/s^2? Analyze the problem (a) from an inertial frame of reference and (b) in a frame of reference at rest with respect to the mass. (Assume the earth is not rotating.)

Answer: 5.5 N

11B–2 An elevator moves upward with constant acceleration $a = 2$ m/s^2. An object is dropped from rest (relative to the elevator) at a height of 2 m above the floor of the elevator. How long will it be before the object strikes the floor? Solve the problem in terms of the elevator's frame of reference.

11B–3 An *accelerometer* is a device that measures acceleration. The basic principle on which a simple accelerometer works is shown in Figure 11–19. A mass m, supported on frictionless wheels, is constrained by two identical springs. A pointer fastened to the mass indicates its position on a scale. If the device undergoes constant linear acceleration along the direction of the springs, the mass is displaced from its equilibrium mid-position due to inertial effects. (A damping arrangement, not shown, eliminates oscillations due to starting and stopping.) If $m = 0.1$ kg, what is the force constant k of each spring that will result in a deflection of 5 cm for an acceleration of 5 m/s^2? Solve the problem twice: in an inertial frame and in the frame of the accelerometer.

Answer: 5 N/m

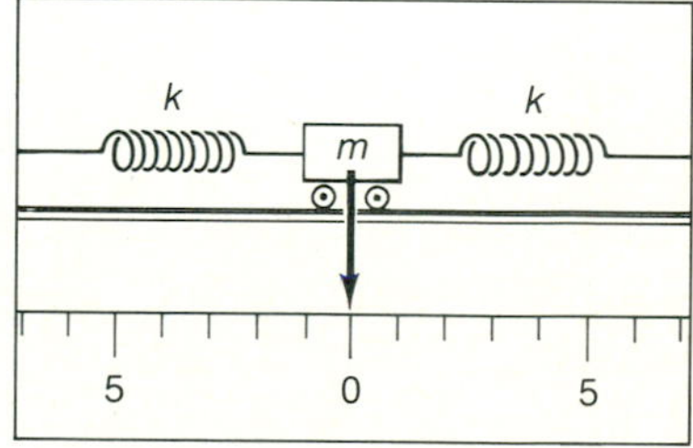

Figure 11–19
Problem 11B–3

11B–4 A 600-N person is standing on a cart that is accelerating horizontally. In order to maintain equilibrium, the person must lean at 15° from the vertical.

(a) In which direction does the person lean?

(b) Find the magnitude of the acceleration.

(c) Find the magnitude of the force the person exerts on the floor of the cart.

11B–5 A 3-lb weight is suspended by a string from the ceiling of a dragster that is accelerating forward at 12 ft/s^2. If the weight is at rest relative to the car, find (a) the angle the string makes with the vertical and (b) the tension in the string.
Analyze the problem from the accelerating frame of reference.

Answers: (a) 20.6° (b) 3.20 lb

11B–6 A 1-kg mass hangs from the ceiling of an enclosed truck. An observer notes that the string supporting the mass makes a steady angle of 10° with the vertical toward the rear of the truck. Explain how the observer, by measuring the tension in the string, can determine whether the truck is going uphill at a constant velocity or accelerating forward on level ground.

11B–7 A train runs due east in New Orleans (30° north latitude). Find the direction of the total coriolis force on the train. *Answer:* South, 60° above the horizon

11B–8 If you run due west on your college campus, what is the precise direction of the total coriolis force on you?

11B–9 The center of mass of a small cubical box is at its geometrical center. The box is placed on a rotating horizontal turntable so that its *CM* is at a distance *R* from the axis of rotation. The box is orientated so that one face is perpendicular to the radial direction. If frictional forces are sufficient to keep the box from sliding, what minimum rotational speed ω of the turntable will cause the box to tip over about its outer edge? Analyze the problem in the rotating frame of reference. *Answer:* $\sqrt{g/R}$

11B–10 A girl stands near the outer edge of a merry-go-round as it rotates at 10 rpm. She pitches a ball horizontally with an initial speed of 20 m/s, and observes that the ball's trajectory swerves toward her right.

(a) Find the initial horizontal radius of curvature of the ball's path.

(b) As the pitcher looks past the center of the merry-go-round, does the distant landscape move to her right or to her left?

11B–11 A model train runs around a small horizontal circular track with speed v at the equator. A small weighing scale is on one of the flatcars, and a mass m rests on the scale.

(a) Calculate the total range in "apparent weight" as measured by this scale due to the coriolis force in the rotating earth's frame of reference. Express the answer in terms of m, v, the earth's rotational speed ω, the earth's equatorial radius R, and the acceleration due to gravity g.

(b) In which direction is the flatcar moving when the "weight" is the greatest?

Answers: (a) $4m\omega v$ (b) westward

11B–12 Assume the earth is a rotating perfect sphere (rather than its actual oblate spheroid shape). An Eskimo steps on a bathroom scale at the North Pole and finds he weighs exactly 1000 N.

(a) If he steps on the same scale at the equator, by how much will the scale reading change?

(b) If we now take into account the differing polar and equatorial radii of the earth, will the change in the Eskimo's weight be greater or less than in part (a)?

11B–13 A small bug of mass 2 g is located 50 cm from the center of a horizontal turntable rotating at 2 rad/s in the counterclockwise direction as viewed from above. The bug remains in a fixed position as seen in the rotating frame. Sketch free-body diagrams (as viewed from above the turntable) that show only the horizontal components of the forces on the bug, including their numerical values. Do this for (a) an inertial frame of reference and (b) the turntable's frame. Suppose the bug now is walking radially outward at a constant speed of 10 cm/s relative to the turntable along a radial line painted on the surface. (c) Repeat parts (a) and (b) when the bug is 50 cm from the center.

Answers: (a) friction: 4×10^{-3} N, radially inward
(b) The above plus an outward centrifugal force of 4×10^{-3} N
(c) The forces in (a) and (b) plus a coriolis force 8×10^{-4} N toward the bug's right and an equal and opposite friction force

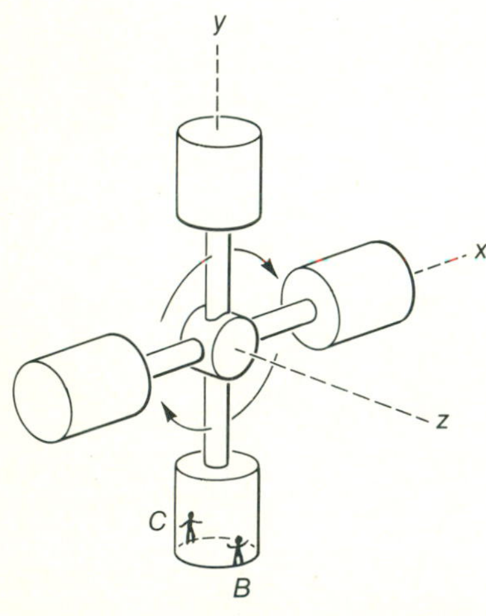

Figure 11–20
Problem 11B–14

11B–14 A proposed rotating space station has four cylindrical sections connected to a central axis by shafts (see Figure 11–20). Astronauts would move along these shafts by facing and grasping an endless moving ladder. The cylinders are 33 ft in diameter and 40 ft long; the shafts are 60 ft from the cylinders to the central axis. The station rotates in the x-y plane, as shown, at 4 rpm.

(a) In moving radially inward from cylinder A toward the central axis, in what direction should the astronaut be facing so that the coriolis force will tend to press her closer to the ladder, rather than push her sideways or away from the ladder? (If it cannot be done, explain why not.)

(b) Astronauts at the outermost (flat) end of the cylinders will experience what fraction of earth gravity g?

(c) Two astronauts in cylinder A are located at C and B (parallel to the z axis). Astronaut B throws an orange toward astronaut C. In the thrower's frame of reference, in which direction is the trajectory curved (or is it straight)?

11C–1 An inclined plane is fastened to a cart that is accelerating horizontally at 6 m/s^2. A mass is placed on the plane, where it remains at rest (in the cart's frame) even though the inclined plane is frictionless.

(a) Make a sketch showing the position of the inclined plane relative to the direction of acceleration.

(b) What angle α does the plane make with the horizontal? Analyze part (b) from the frame of reference of the cart and also from an inertial frame. *Answer:* 31.5°

11C–2 An inclined plane makes an angle θ with the horizontal. A block on the plane is on the verge of slipping when the plane is not moving. Find the horizontal acceleration of the plane that will barely cause the block to begin to slip *up* the incline. Analyze the problem in the accelerated frame of reference.

11C–3 A marksman aims his rifle at a target moving from left to right. The rifle barrel is held in a horizontal position during this motion. The rifle rotates at 1.5 rad/s, and the speed of the 5-g bullet just as it emerges from the rifle is 500 m/s. In the rotating frame, what is the coriolis force on the bullet just as it leaves the barrel? In which direction is this force?

Answer: 7.5 N, toward the left

11C–4 A ball is dropped from the ceiling of a freight car that is accelerating horizontally. The ball strikes the floor and bounces several times.

(a) Make a sketch of the trajectory of the ball during its first few bounces as seen in the freight car's frame of reference.

(b) Describe a situation not involving acceleration that would result in a similar trajectory.

11C–5 A set of rails is installed at the outer edge of a merry-go-round of radius R. A self-propelled cart of mass m moves on these rails at a speed so that the cart is at rest in an inertial frame while the merry-go-round turns with constant angular speed ω. Find the net force on the cart as analyzed (a) in a frame of reference attached to the ground and (b) in a frame of reference attached to the merry-go-round.

Answers: (a) $m\omega^2 R$, inward (b) zero

11C–6 A horizontal turntable has a smooth tube fastened along a radial line (length R). A small mass m is held inside the tube at the point $r = R/2$. While the turntable is rotating at constant angular speed ω, the mass is released, allowing the mass to slide outward freely.

(a) As calculated in the rotating frame, how much work is done by the centrifugal force on the mass while m slides outward from $r = R/2$ to $r = R$?

(b) In the rotating frame, what outward radial speed v' does the mass have just as it emerges from the tube?

11C–7 A merry-go-round revolves at a constant angular speed of 0.50 rad/s in a counterclockwise direction when viewed from above. A 40-kg rider walks tangentially (in the same direction the merry-go-round is moving) at a constant speed of 2 m/s relative to the merry-go-round, maintaining a constant radius of 2 m from the axis. Analyze the situation in the rotating frame to find the magnitude and direction of the following: (a) the centrifugal force on the rider, (b) the coriolis force on the rider, and (c) the total force of friction on the rider. (d) Show that in an inertial frame, the single friction force of part (c), acting alone, explains the motion as observed in that frame. *Answers:* (a) 20 N, radially outward
(b) 80 N, radially outward
(c) 180 N, radially inward

11C–8 An object at the earth's equator has a velocity (relative to the earth) such that the coriolis and centrifugal forces on the object are equal in magnitude and in the same direction.

(a) Write an expression for the magnitude of this velocity, defining each symbol that appears in your expression.

(b) Describe the direction of the velocity, including a pictorial sketch for clarity.

11C–9 At a radial distance r from the center of a merry-go-round rotating at an angular speed ω, an object is dropped from a height h above the floor. By analyzing the motion in the merry-go-round frame, show that the approximate horizontal distance between the point of impact on the floor and the point directly below the point of release is $\omega^2 rh/g$. What assumptions are made in your analysis?

We know very little, and yet it is astonishing that we know so much, and still more astonishing that so little knowledge can give us so much power.

BERTRAND RUSSELL

chapter 12 Harmonic Motion

12.1 Introduction

Many objects undergo a *periodic* motion, which repeats itself regularly in time. Pendulums swing to and fro, pistons in a gasoline engine oscillate back and forth, and all atoms and molecules vibrate about their equilibrium positions. Vibrations of air columns in organ pipes, horns, and strings of musical instruments have similar oscillatory motion. And alternating-current flow in electrical circuits is described by the same equations that describe the oscillatory motion of mechanical systems. Periodic motion is thus a widespread phenomenon occurring in almost all branches of physics.

We will analyze the common features among all vibrating systems by investigating a single mass m that moves in one dimension on a frictionless horizontal surface. This may seem a trivial example. But its simplicity enables us to derive easily the general equations that apply in a great many cases. A spring is attached to the mass as shown in Figure 12–1(a). When the mass is

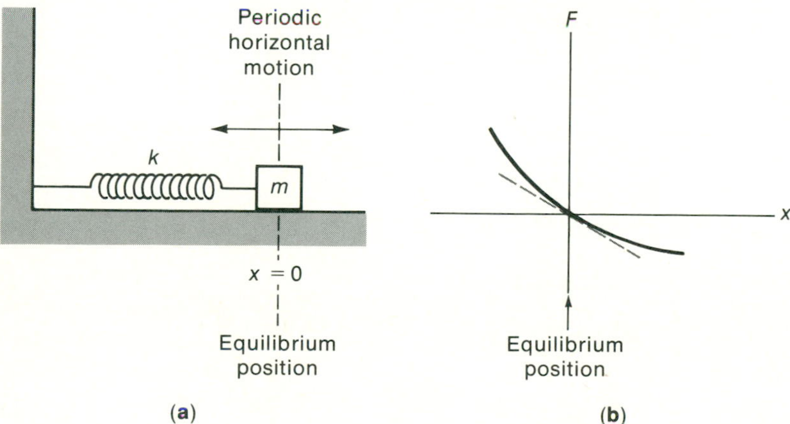

Figure 12–1
Many restoring forces in nature closely approximate the ideal linear restoring force exerted by a spring (shown dashed): $F = -kx$.

displaced a distance x away from its equilibrium position, the spring provides a *restoring* force on the mass that is proportional to the displacement. This force is called a Hooke's Law force:

HOOKE'S LAW FORCE

$$F = -kx \tag{12–1}$$

where k is the **force constant,** expressed in newtons/meter.

Many physical systems have a force that closely obeys this law. And even if the force behavior is not a linear one, for small displacements a straight-line relation is often a good approximation (see Figure 12–1b).

12.2 Simple Harmonic Motion

We shall now develop mathematical expressions for analyzing this oscillatory motion. We start, as usual, with Newton's second law:

$$F_x = ma_x$$

$$F_x = m\frac{d^2x}{dt^2} \tag{12–2}$$

Substituting the Hooke's Law relation for the force, we obtain

$$-kx = m\frac{d^2x}{dt^2} \tag{12–3}$$

or

$$\frac{d^2x}{dt^2} = -\left(\frac{k}{m}\right)x \tag{12–4}$$

To "solve" this equation, we must find a function $x(t)$ which, when we take its second derivative with respect to time, equals some constant times the original function itself. Sine and cosine functions have this property. For our purpose here, we need to write them as a function of time. The angle θ may be thought of as the angle ωt, where ω is the **angular frequency,** measured in radians/second, and t is the time. (Note that here the value of ω is always a constant, in contrast to its usage in rotational motion, where the rotational frequency ω may vary.) So we try the general expression:

$$x(t) = A\cos(\omega t + \phi) \tag{12–5}$$

where ϕ = an arbitrary constant called the **initial phase angle** in radians.

ω = the **angular frequency,** measured in radians/second. It is equal to $2\pi f$, where f represents the *frequency* of the oscillation, measured in cycles/second, or in hertz (Hz).[1] The time for one complete cycle of the function is the *period* $T = 1/f$, measured in seconds.

A = the **amplitude,** or maximum displacement, of the periodic motion; it is measured in units of length.

Because of the trigonometric relation $\cos(\omega t + \pi/2) = -\sin(\omega t)$, the use of the phase angle ϕ makes Equation (12–5) a general expression that includes sine as well as cosine functions. Mathematically, these are included in the *harmonic functions.* The equations we are developing describe a type of periodic

[1] This frequency unit is named in honor of Heinrich Hertz (1857–1894) who first experimentally verified the existence of electromagnetic waves. It has units of s^{-1}, though it is often expressed in cycles/ second or vibrations/second. Of course, "cycles" or "vibrations" are merely words describing the physical phenomenon and have no inherent units.

motion known as **simple harmonic motion** (SHM): "simple" because it involves just a single frequency rather than several simultaneous frequencies. The name "periodic motion" might be more apt, but the usage of "simple harmonic motion" is so widespread that we continue the practice.

Differentiating Equation (12–5) twice with respect to time (see Appendix G-I), we obtain:

$$x = A\cos(\omega t + \phi)$$

$$\frac{dx}{dt} = -A\omega\sin(\omega t + \phi)$$

$$\frac{d^2x}{dt^2} = -A\omega^2\cos(\omega t + \phi)$$

$$\frac{d^2x}{dt^2} = -\omega^2 x \qquad \textbf{(12–6)}$$

Comparing Equations (12–4) and (12–6), we note that our trial solution does agree with Equation (12–4) if

$$\frac{k}{m} = \omega^2 \qquad \textbf{(12–7)}$$

Substituting $\omega = 2\pi f$ and solving for f, we obtain the *frequency* in SHM:

$$\frac{k}{m} = (4\pi^2 f^2)$$

SHM FREQUENCY

$$f = \frac{1}{2\pi}\sqrt{\frac{k}{m}} \qquad \textbf{(12–8)}$$

The *period* T is the time for one complete cycle of the motion. That is:

$$T = \frac{1}{f} \qquad \textbf{(12–9)}$$

SHM PERIOD

$$T = 2\pi\sqrt{\frac{m}{k}} \qquad \textbf{(12–10)}$$

Equation (12–5) represents the *displacement* x away from the equilibrium position for a particle undergoing SHM. A typical graph of displacement versus time is shown in Figure 12–2(a). The initial phase angle ϕ enables us to start measuring time from any arbitrary initial displacement of the mass. The period T need not be measured from peak to peak as shown, but is the time interval from any point on the graph to the corresponding time after the object traverses one complete cycle of its motion.

The mathematical expressions for the velocity and acceleration of the mass are obtained by taking successive derivatives with respect to time, as summarized in the following equations:

SIMPLE HARMONIC MOTION

Displacement: $$x = A\cos(\omega t + \phi) \qquad \textbf{(12–11)}$$

Velocity: $$\frac{dx}{dt} = v = -A\omega\sin(\omega t + \phi) \qquad \textbf{(12–12)}$$

Acceleration: $$\frac{dv}{dt} = a = -A\omega^2\cos(\omega t + \phi)$$

$$= -\omega^2 x \qquad \textbf{(12–13)}$$

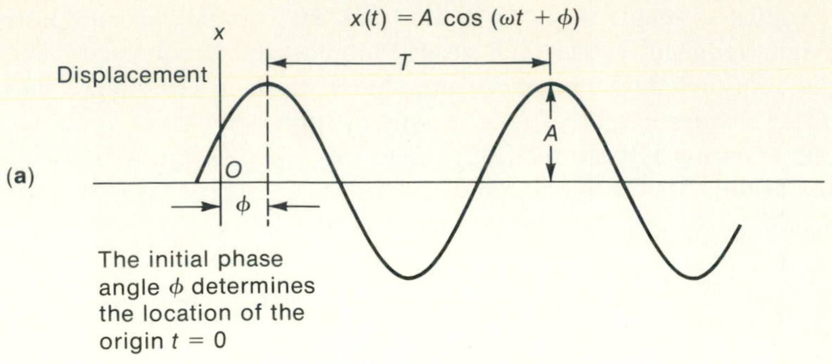

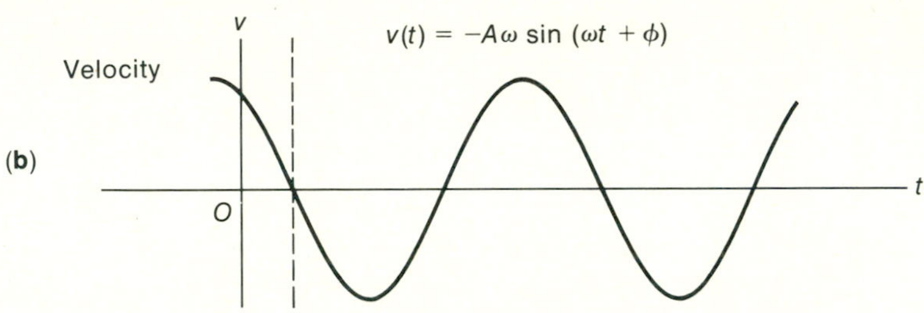

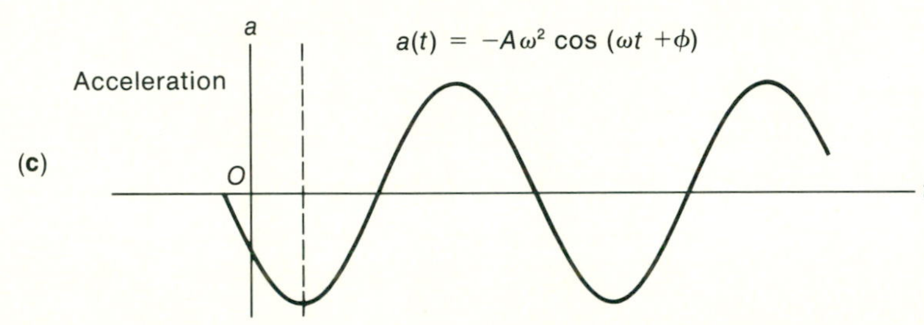

Figure 12–2
Simple harmonic motion for a particle with displacement $x(t) = A \cos(\omega t + \phi)$. If the initial phase angle ϕ is zero, the vertical coordinate is shifted to the dotted-line position.

Note that:

$$\omega = 2\pi f = \frac{2\pi}{T} \tag{12–14}$$

EXAMPLE 12–1

A 200-g mass that is attached to a Hooke's Law spring as in Figure 12–1(a) oscillates in SHM with an amplitude of 4 cm. The force constant of the spring is 25 N/m. Find (a) the frequency of oscillation and (b) the time for one complete cycle of the oscillation. Consider the instant when the mass is displaced toward the right 2 cm from the equilibrium position, and traveling toward the right. Find, at this instant, (c) the velocity and (d) the acceleration of the mass.

SOLUTION

(a) The frequency of oscillation is given by Equation (12–8):

$$f = \frac{1}{2\pi}\sqrt{\frac{k}{m}}$$

Substituting the numerical values (after conversion to SI units) gives

$$f = \frac{1}{2\pi}\sqrt{\frac{25 \text{ N/m}}{0.200 \text{ kg}}} = \boxed{1.78 \frac{\text{cycles}}{\text{s}}} = \boxed{1.78 \text{ Hz}}$$

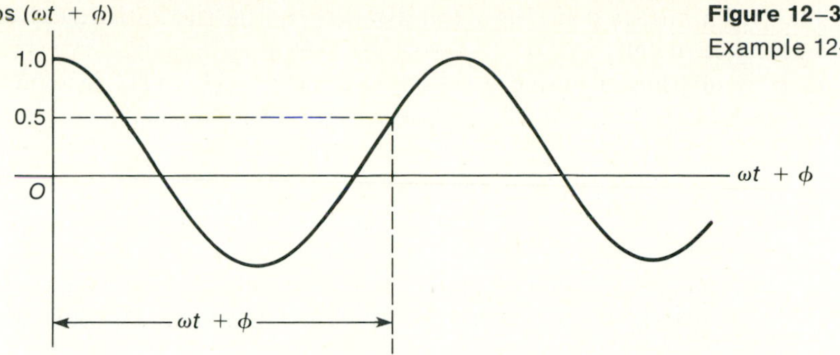

Figure 12–3
Example 12–1

(b) The period T is given by Equation (12–9):

$$T = \frac{1}{f} = \frac{1}{\left(1.78\ \dfrac{\text{cycles}}{\text{s}}\right)} = \boxed{0.562\ \text{s}}$$

(c) To determine the velocity, we begin by substituting numerical values into the equation for displacement and solving for the angle $\omega t + \theta$. We then can take the first derivative to obtain v.

$$x = A\cos(\omega t + \phi)$$
$$(0.02\ \text{m}) = (0.04\ \text{m})\cos(\omega t + \phi)$$
$$0.50 = \cos(\omega t + \phi)$$

Although $(\omega t + \phi) = \cos^{-1}(0.50) = 60^\circ$, if we graph the motion (as in Figure 12–3), we see that the angle we seek is instead

$$(\omega t + \phi) = 300^\circ$$

since the function is increasing at this value. The velocity is thus:

$$v = -A\omega \sin(\omega t + \phi)$$
$$v = -(0.04\ \text{m})(2\pi)\left(1.78\ \frac{\text{cycles}}{\text{s}}\right)\sin 300^\circ$$
$$v = \boxed{0.387\ \frac{\text{m}}{\text{s}}} \quad \text{(toward the right)}$$

(d) The acceleration is given by

$$a = -\omega^2 x$$
$$a = -\left[(2\pi)\left(1.78\ \frac{\text{cycles}}{\text{s}}\right)\right]^2 (0.02\ \text{m})$$
$$a = \boxed{-2.50\ \frac{\text{m}}{\text{s}^2}} \quad \text{(toward the left)}$$

Note that the velocity and the acceleration are in opposite directions because the mass is slowing down.

12.3 Some Characteristics of SHM

The characteristics of SHM may be summarized as follows:

(1) The frequency f (and thus the period $T = 1/f$) is independent of the amplitude of the motion. All possible oscillations of the system, regardless of their various amplitudes, vibrate with the same frequency.

(2) The two arbitrary constants, A and ϕ, depend on the initial displacement velocity of the mass.
(3) Instead of using a cosine function in Equation (12–11), we could have expressed it as a sine (by changing the initial phase angle ϕ):

$$x = A \sin(\omega t + \phi)$$

Sines and cosines are included in both expressions.
(4) Whether or not a system exhibits SHM depends on whether its restoring force is strictly proportional to the displacement. Although this proportionality is true for many systems in nature (provided the displacement is small), all physical systems deviate from this strict proportionality relationship if the displacements are made large enough. Their motions then are not described by a single sine or cosine function, but by a more complicated expression. Nevertheless, no matter how irregular the motion is, if it is a *periodic* one (that is, if it repeats itself exactly in successive periods of time) it can always be described by a combination of pure sine and cosine functions of various frequencies. This mathematical method is called *Fourier analysis* and is described in Appendix F.

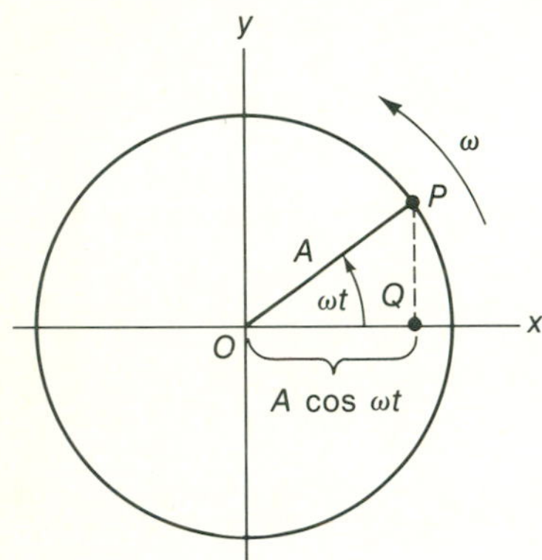

Figure 12–4
A *circle of reference* for simple harmonic motion. The actual motion of the object undergoing SHM is the point Q as it moves back and forth along the $\pm x$ axis.

12.4 Circle-of-Reference Analogy for SHM

The previous discussion may seem to involve a large number of different equations. However, there is a convenient geometrical model that is of considerable aid in remembering the various relations. It is called a **circle of reference.**

Consider an imaginary point P moving in a circle of radius A with constant angular speed ω. At any instant, the angle with respect to the x axis is ωt, as shown in Figure 12–4. The projection of point P on the x axis is the point Q. As P travels around the circle, its projection Q oscillates back and forth along the x axis in SHM. The displacement of Q from the origin is given by $x = A \cos \omega t$, identical to Equation (12–11) when the initial phase angle ϕ is zero.

By the proper choice of the initial phase angle ϕ, we may start counting time ($t = 0$) when the point is at any arbitrary position on the circle. This implies that the actual object undergoing SHM has the corresponding initial displacement x at that instant. (In general, SHM may be described as the projection of uniform circular motion along *any* diameter, not just the x axis).

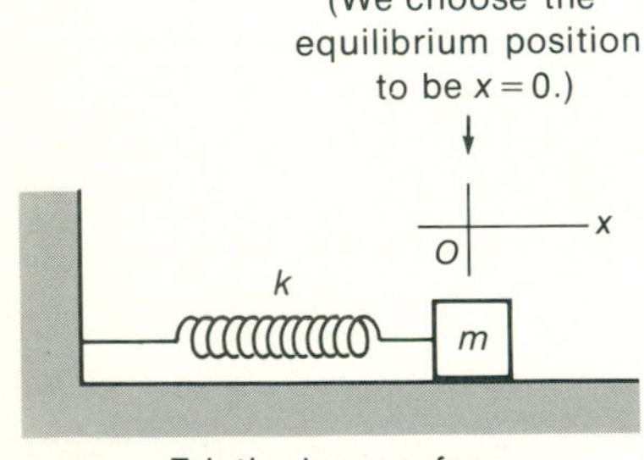

Figure 12–5
Example 12–2

EXAMPLE 12–2

A 2-kg mass is attached to a spring and placed on a horizontal, frictionless surface as in Figure 12–5. For the spring chosen, a force of 20 N is required to hold the mass at rest when it is displaced 0.2 m from its equilibrium position, the origin of the x axis. The mass is now released from rest (with an initial displacement of $x = 0.2$ m), and it subsequently undergoes SHM oscillations. (a) Find the force constant k of the spring. (b) Find the frequency of the SHM oscillations. (c) Determine the time it takes to make one complete cycle of the oscillatory motion. (d) Find the maximum speed of the mass. (e) Determine the maximum acceleration. Where does it occur? (f) Find the total energy of the oscillating system. (g) Calculate the velocity and acceleration when the displacement is one-third the maximum value.

SOLUTION

(a) When displaced at $x = +0.2$ m, the spring is exerting a force $F = -20$ N (minus because it is in the negative x direction). Hence, from Equation (12–1), the constant k is

$$k = -\frac{F}{x} = -\frac{(-20\text{ N})}{0.2\text{ m}} = \boxed{100\,\frac{\text{N}}{\text{m}}}$$

(b) From Equation (12–8), we have

$$f = \frac{1}{2\pi}\sqrt{\frac{k}{m}} = \frac{1}{2\pi}\sqrt{\frac{100\,\frac{\text{N}}{\text{m}}}{2\text{ kg}}} = \boxed{1.13\,\frac{\text{cycles}}{\text{s}} \quad (\text{or } 1.13\text{ s}^{-1}\text{, or } 1.13\text{ Hz})}$$

(c) We use Equation (12–9) to obtain

$$T = \frac{1}{f} = \frac{1}{1.13\text{ s}^{-1}} = \boxed{0.885\text{ s}}$$

(d) From Equation (12–12), we have

$$v = -A\omega \sin(\omega t + \phi)$$

where $\sin(\omega t + \phi)$ is equal to ± 1 for the maximum speed. Thus:

$$v_{\text{max}} = \pm A\omega = \pm(0.2\text{ m})(2\pi)(1.13\text{ s}^{-1}) = \boxed{\pm 1.41\,\frac{\text{m}}{\text{s}}}$$

The maximum speed occurs when passing through the equilibrium position, $x = 0$. The $\pm$ symbol indicates the mass travels in either direction at this point.

(e) From Equation (12–13), we have:

$$a = -\omega^2 x$$

The minus sign indicates that the acceleration a is opposite in direction from the displacement x. The acceleration has its maximum magnitude when the displacement equals its maximum value A. Hence:

$$|a_{\text{max}}| = \omega^2 A = (2\pi f)^2 A = 4\pi^2(1.13\text{ s}^{-1})^2(0.2\text{ m}) = \boxed{10.1\text{ m/s}^2}$$

The maximum acceleration thus occurs at the extremities of the motion.

(f) From Chapter 6, we have $E = (U_{\text{sp}})_{\text{max}}$:

$$E = \tfrac{1}{2}k{x_{\text{max}}}^2 = \frac{1}{2}\left(100\,\frac{\text{N}}{\text{m}}\right)(0.2\text{ m})^2 = \boxed{2.00\text{ J}}$$

We could also calculate the total energy from:

$$E = \tfrac{1}{2}m{v_{\text{max}}}^2 = \tfrac{1}{2}(2\text{ kg})\left(1.41\,\frac{\text{m}}{\text{s}}\right)^2 = \boxed{2.00\text{ J}}$$

(g) For intermediate values of these quantities (neither maximum nor minimum), we must identify the argument of the sine or cosine. This is easily found from Equation (12–11):

$$x = A\cos(\omega t + \phi)$$

We solve for the argument of the cosine function as follows:

$$\frac{A}{3} = A\cos(\omega t + \phi)$$

or

$$\cos(\omega t + \phi) = \tfrac{1}{3}$$

$$(\omega t + \phi) = \arccos\tfrac{1}{3} = 70.5^\circ\ (= 1.23\text{ rad})$$

To obtain the instantaneous velocity and acceleration at this location, we use Equations (12–12) and (12–13):

$$v = -A\omega \sin(70.5^\circ) = -(0.2\text{ m})(2\pi)(1.13\text{ s}^{-1})(0.943) = \boxed{-1.34\,\frac{\text{m}}{\text{s}}}$$

$$a = -\omega^2 x = 4\pi^2(1.13\text{ s}^{-1})^2\left(\frac{0.2\text{ m}}{3}\right) = \boxed{-3.36\,\frac{\text{m}}{\text{s}^2}}$$

12.5 The Simple Pendulum

Consider an idealized pendulum consisting of a mass point m suspended from an inextensible string of negligible weight, as shown in Figure 12–6(a). The pendulum swings in a vertical plane under the influence of gravity. As we shall demonstrate, for cases limited to small angles the motion is approximately SHM.

The forces are as shown in Figure 12–6(b). We now must decide which directions are appropriate for resolving forces. Since the mass always moves perpendicular to the string, we choose to resolve force components *perpendicular* and *parallel* to the string, as shown in Figure 12–6(c). Perpendicular to the string, there is a net restoring force component of $mg \sin\theta$, which accelerates the bob back toward the center position. So we choose this direction for the $+x$ axis.

For small angles, we may replace $\sin\theta$ by its approximate value $\theta(= s/\ell)$ to obtain:

$$F_{\text{restoring}} \approx -mg\theta = -\left(\frac{mg}{\ell}\right)s \qquad \textbf{(12–15)}$$

The restoring force F is thus (approximately) proportional to the displacement s, and the periodic motion is approximately SHM with an effective force constant $k = mg/\ell$. From Equation (12–10) we obtain the period T of a simple pendulum (for small-angle motions)·

SIMPLE PENDULUM PERIOD

$$T \approx 2\pi\sqrt{\frac{\ell}{g}} \qquad \textbf{(12–16)}$$

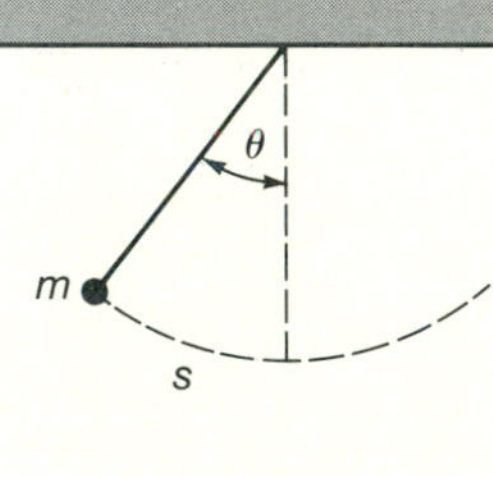

(a) A simple pendulum. The displacement s of the mass point is measured along the curved trajectory.

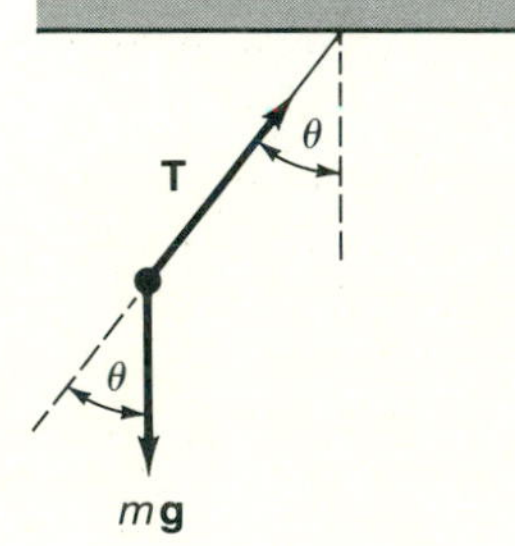

(b) A free-body diagram.

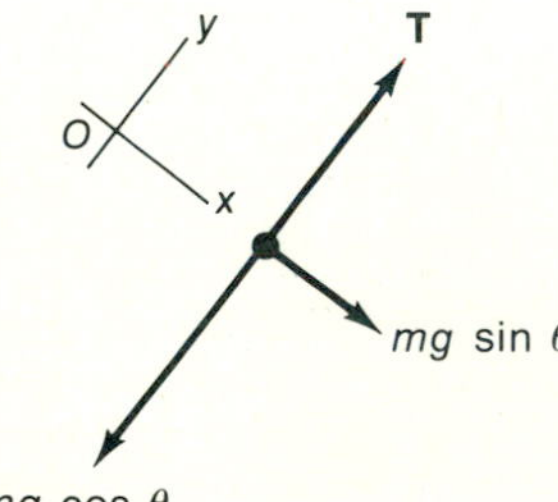

(c) A free-body diagram with force components in two mutually perpendicular directions.

Figure 12–6
Forces on the bob of a simple pendulum.

The exact equation for the period T is:

$$T = 2\pi\sqrt{\frac{\ell}{g}}\left[1 + \frac{1^2}{2^2}\sin^2\left(\frac{\theta}{2}\right) + \frac{(1^2)(3^2)}{(2^2)(4^2)}\sin^4\left(\frac{\theta}{2}\right) + \cdots\right] \qquad \textbf{(12–17)}$$

For deflections smaller than $\theta = 15°$, Equations (12–16) and (12–17) differ by less than 0.5%; in these cases, then, Equation (12–16) is satisfactory.

EXAMPLE 12–3

Find the length, in meters, of a simple pendulum that makes 12 oscillations per minute.

SOLUTION

Since $T = 1/f$, we may write the equation for simple pendulum motion as

$$\frac{1}{f} = 2\pi\sqrt{\frac{\ell}{g}}$$

Solving for the length ℓ, we obtain

$$\ell = \frac{g}{4\pi^2 f^2}$$

We have already

$$f = \left(12\,\frac{\text{oscillations}}{\text{min}}\right)\left(\frac{1\text{ min}}{60\text{ s}}\right) = 0.20\text{ s}^{-1}$$

Substitution of numerical values gives

$$\ell = \frac{\left(9.8\,\frac{\text{m}}{\text{s}^2}\right)}{4\pi^2(0.20\text{ s}^{-1})^2} = \boxed{6.20\text{ m}}$$

12.6 The Torsional Pendulum

Any physical object suspended by a wire attached to its center of mass can undergo periodic *rotations* about an axis coincident with the wire (see Figure 12–7). The restoring torque is exerted by the twisted wire and, for small angles, is often directly proportional to θ, the angular displacement. Thus the system meets the criterion for SHM in the angle θ, where the **torsion constant** is the Greek letter kappa κ:

$$\tau = -\kappa\theta \qquad \textbf{(12–18)}$$

The constant κ is measured in units of newton-meters/radian or pound-feet/radian.

Newton's second law for the case of angular motion is $\tau = I\alpha$, or:

$$\tau = I\,\frac{d^2\theta}{dt^2} \qquad \textbf{(12–19)}$$

which, when combined with Equation (12–18), leads to

$$\frac{d^2\theta}{dt^2} = -\left(\frac{\kappa}{I}\right)\theta \qquad \textbf{(12–20)}$$

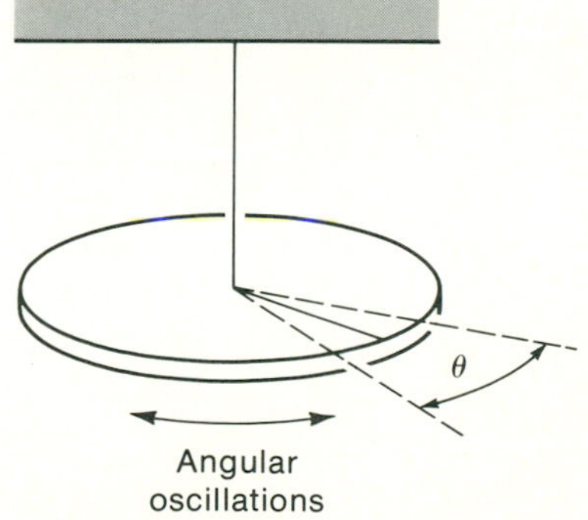

Figure 12–7
A uniform disk, suspended as shown, is one example of a torsional pendulum. A radial line drawn on the disk will undergo angular oscillations that are simple harmonic motion in the coordinate θ.

By analogy with the solution of the similar equation for the linear case (Equation 12–4), a solution to this differential equation is

$$\theta = \theta_0 \cos(\omega t + \phi)$$

where θ_0 is the *angular amplitude* of the motion, measured in radians. Comparing the notation to Equation (12–4) we may, by analogy, write the expression for the period T of the oscillation:

TORSIONAL-PENDULUM PERIOD

$$T = 2\pi\sqrt{\frac{I}{\kappa}} \tag{12–21}$$

(As long as the restoring torque follows a Hooke's Law relationship, we need not use the approximation sign that appears in the simple pendulum equation.)

EXAMPLE 12–4

Two identical small masses, each with $m = 0.002$ kg, are fastened to the ends of a bar of negligible mass and of length $\ell = 0.10$ m. The bar is suspended horizontally from a thin wire fastened at its midpoint, forming a torsional pendulum, as shown in Figure 12–8. When set into rotational oscillations about a vertical axis passing through the wire, the period of the oscillations is 10 min. Find the torsion constant κ for the wire.

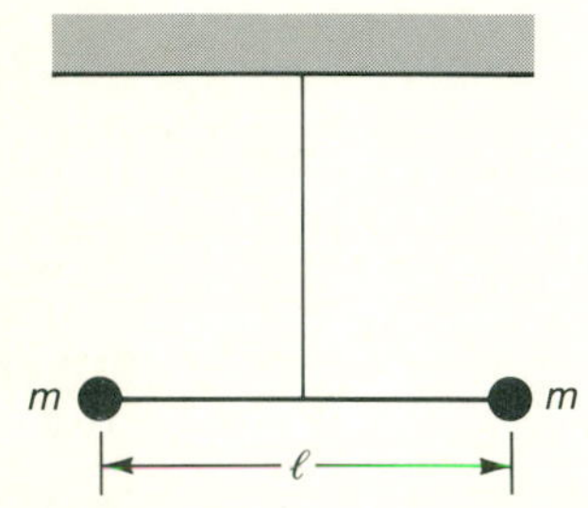

Figure 12–8
Example 12–4

SOLUTION

From the equation for torsional-oscillation periods, we solve for the torsion constant κ:

$$T = 2\pi\sqrt{\frac{I}{\kappa}}$$

$$\kappa = \frac{4\pi^2 I}{T^2}$$

The moment of inertia I (about the point of suspension) for the masses on the ends of the rod is

$$I_1 = mr^2 = m\left(\frac{\ell}{2}\right)^2$$

for each mass individually. For the two-mass system, this becomes

$$I = 2m\left(\frac{\ell}{2}\right)^2 = \tfrac{1}{2}m\ell^2$$

Thus:

$$\kappa = \frac{4\pi^2}{T^2}\left(\tfrac{1}{2}m\ell^2\right) = \frac{2m\pi^2\ell^2}{T^2}$$

where $T = 10 \text{ min}\left(\frac{60 \text{ s}}{1 \text{ min}}\right) = 600$ s. Substitution of numerical values yields

$$\kappa = \frac{2(2.0 \times 10^{-3} \text{ kg})(\pi^2)(0.10 \text{ m})^2}{(600 \text{ s})^2} = \boxed{1.10 \times 10^{-9} \frac{\text{N·m}}{\text{rad}}}$$

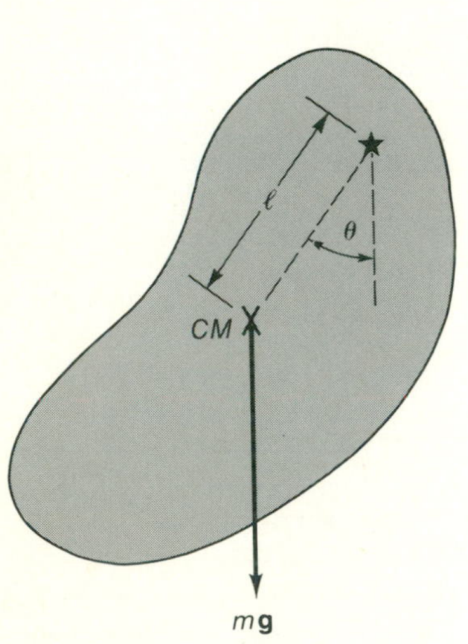

Figure 12–9
A physical pendulum (an extended body) oscillates about its equilibrium position when pivoted about a fixed horizontal axis

12.7 The Physical Pendulum

A rigid body of any shape, when pivoted about a fixed horizontal axis through any point a distance ℓ from its center of mass, will undergo oscillations under the influence of gravity (see Figure 12–9). In contrast to the simple pendulum whose (ideal) point mass is suspended from a weightless cord, a pendulum made from a real, physical body is called a **physical pendulum.**

In this case the restoring torque τ is provided by the force of gravity acting through the center of mass to give

$$\tau = -mg\ell \sin\theta \tag{12–22}$$

For small angular displacements, we may approximate $\sin\theta$ by θ:

$$\tau = -mg\ell\theta \tag{12–23}$$

Hence, the restoring torque constant κ

$$\kappa = mg\ell \tag{12–24}$$

leads to a period

$$T = 2\pi\sqrt{\frac{I}{\kappa}} = 2\pi\sqrt{\frac{I}{mg\ell}} \tag{12–25}$$

where I is the moment of inertia of the object about the fixed axis. (The parallel-axis theorem, $I = I_{CM} + Md^2$, explained in Section 10.12, is useful in determining the moment of inertia about axes that do not pass through the CM.)

The length L of a simple pendulum having the same period as a given physical pendulum locates a point called the **center of oscillation** CO on the physical pendulum (see Figure 12–10). The center of oscillation and the pivot axis are **conjugate points:** If a physical pendulum is pivoted at the center of oscillation, the former pivot point becomes the new center of oscillation and the period is the same as before.

The point CO is also called the **center of percussion** relative to the pivot axis. If a short, impulsive force is applied (at right angles to L) at the center of percussion, the initial motion will involve only the rotation about the pivot axis. Baseball players are familiar with the phenomenon. For example, if a baseball hits a bat very far from the center of percussion, a noticeable "sting" will be experienced by the batter. However, a ball impacting against the exact center of percussion will produce no impulsive reaction against the batter's hands—that is, the angular rotation of the bat about the pivot axis (the hands) is changed abruptly, but without appreciable change in the translational motion of the axis.

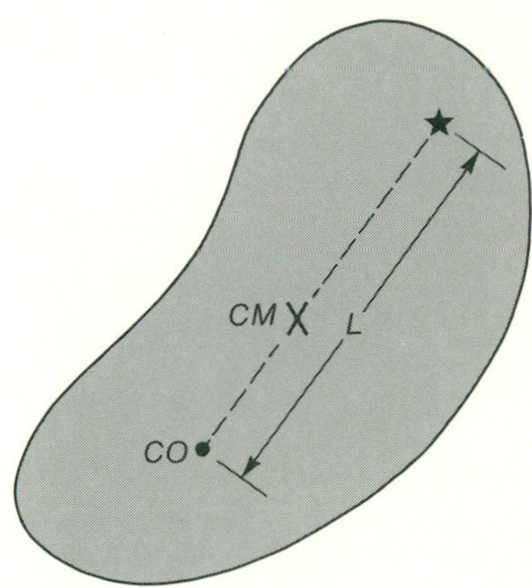

Figure 12–10
The length L locates the center of oscillation CO with respect to the pivot axis.

EXAMPLE 12–5

One end of a thin, uniform rod of length $L = 1.1$ m and mass $m = 2.5$ kg is fastened to a fixed horizontal axis so that the rod can swing back and forth in a vertical plane. For small amplitude oscillations, find (a) the approximate period T of the motion and (b) the length L' of a simple pendulum having the same period.

SOLUTION

(a) The period of a physical pendulum (for small-amplitude motion) is given by Equation (12–25): $T = 2\pi\sqrt{I/mg\ell}$, where ℓ is the distance from the axis to the CM. In Figure 12–11, the moment of inertia of a thin rod about an axis through one end and perpendicular to the rod may be found by using the parallel-axis theorem or by direct integration.

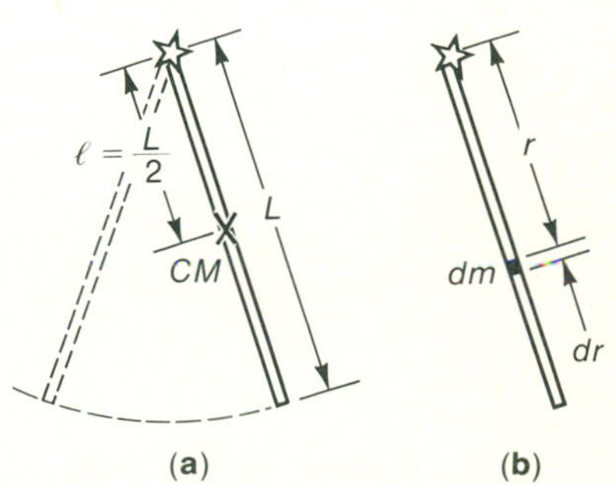

Figure 12–11
Example 12–5

By the parallel-axis theorem

From Chapter 10, the moment of inertia of a thin rod about an axis through the CM and perpendicular to the rod's length is:

$$I = \tfrac{1}{12}mL^2$$

By the parallel-axis theorem (Equation 10–29), the moment of inertia for the parallel axis through the end is

$$I = I_{CM} + Md^2$$

$$I = \tfrac{1}{12}mL^2 + m\left(\frac{L}{2}\right)^2 = \tfrac{1}{3}mL^2$$

By direct integration

The general expression for the moment of inertia (Equation 10–21) is

$$I = \int r^2\,dm$$

We choose the mass element to be

$$dm = \rho_\ell dr$$

where ρ_ℓ is the mass/unit length (in this case, m/L). Therefore:

$$I = \int_0^L r^2 \rho_\ell dr = \rho_\ell \int_0^L r^2\,dr$$

$$I = \rho_\ell \frac{r^3}{3}\bigg|_0^L = \tfrac{1}{3}\rho_\ell L^3$$

But since $\rho_\ell = m/L$, we have

$$I = \tfrac{1}{3}mL^2$$

Thus, the period is

$$T = 2\pi\sqrt{\frac{I}{mg\ell}} = 2\pi\sqrt{\frac{(\frac{1}{3}mL^2)}{mg\left(\frac{L}{2}\right)}} = 2\pi\sqrt{\frac{2L}{3g}} = 2\pi\sqrt{\frac{2(1.1\text{ m})}{3\left(9.8\frac{\text{m}}{\text{s}^2}\right)}} = \boxed{1.72\text{ s}}$$

(b) The period of a simple pendulum of length L' is $T = 2\pi\sqrt{L'/g}$. Equating this expression with Equation (12–25) and solving for L', we obtain

$$2\pi\sqrt{\frac{L'}{g}} = 2\pi\sqrt{\frac{I}{mg\left(\frac{L}{2}\right)}}$$

$$L' = \frac{2Ig}{mgL} = \frac{2(\frac{1}{3}mL^2)g}{mgL} = \tfrac{2}{3}L = \tfrac{2}{3}(1.1\text{ m}) = \boxed{0.733\text{ m}}$$

12.8 Oscillations of a Two-Mass System

In many instances, we do not have a situation as in Figure 12–1 wherein a single mass m moves under the influence of a spring attached to an essentially infinite mass such as the earth. Rather, a more common case is that of two masses of comparable size, moving under the influence of mutual forces that arise between them. An example is a diatomic molecule, which vibrates along the line joining the two atoms. Figure 12–12 illustrates a simple system with these features. Fortunately, if we introduce just one new concept, the *reduced mass* μ, the equations we have already obtained for single-mass oscillations will also describe the behavior of the two masses coupled together.

After establishing a coordinate system to measure distance, the two masses are located by the coordinates x_1 and x_2. The force that the spring exerts on each mass will depend on the amount by which the spring is compressed or stretched from its unstressed length ℓ. We follow the same procedure as in the case of a single mass, letting the symbol x designate the amount by which the spring is compressed or stretched from its unstressed length. For the two-mass system, we therefore have

$$x = (x_2 - x_1) - \ell \tag{12–26}$$

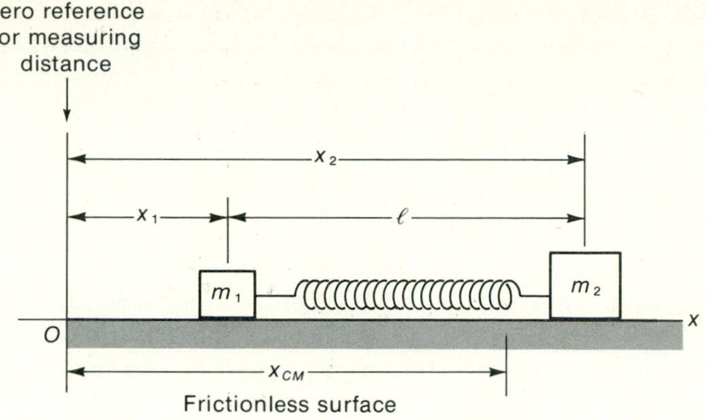

Figure 12–12
Two masses, m_1 and m_2, coupled together by a spring of negligible mass. The unstressed length of the spring is ℓ.

Thus, if $x = 0$, the spring is unstressed; if x is positive, the spring is stretched; and if x is negative, the spring is compressed.

Application of Newton's second law of motion to each of the masses, m_1 and m_2, gives

$$m_1 \frac{d^2x_1}{dt^2} = F_1 \tag{12–27}$$

and

$$m_2 \frac{d^2x_2}{dt^2} = F_2 \tag{12–28}$$

where the force $\mathbf{F}_1$ (which the spring exerts on m_1) is equal and opposite to the force $\mathbf{F}_2$ (which the spring exerts on m_2):

$$\mathbf{F}_1 = -\mathbf{F}_2$$

Multiplying Equation (12–27) by m_2 and Equation (12–28) by m_1, and then subtracting, we obtain

$$m_1 m_2 \frac{d^2}{dt^2}(x_2 - x_1) = -F_1(m_1 + m_2) \tag{12–29}$$

Since, by Hooke's Law:

$$F_1 = -kx \tag{12–30}$$

and since

$$\frac{d^2}{dt^2} x = \frac{d^2}{dt^2}(x_2 - x_1)$$

Equation (12–29) may be written

$$\left(\frac{m_1 m_2}{m_1 + m_2}\right) \frac{d^2x}{dt^2} = -kx \tag{12–31}$$

Comparing this equation with Equation (12–3), we can simplify the notation by defining a new mass μ, called the **reduced mass.**

REDUCED MASS

$$\mu = \left(\frac{m_1 m_2}{m_1 + m_2}\right) \tag{12–32}$$

The reduced mass gains its name from the fact that it is always less than either m_1 or m_2. It may be easily remembered by the phrase, "the product over the sum."

We now interpret the equation

$$\frac{d^2x}{dt^2} = -\frac{k}{\mu}x \tag{12–33}$$

as describing the oscillations of a single (reduced) mass μ, which we imagine to be under the influence of the *same* spring attached to a rigid wall (see Figure 12–13). Note the important distinction that x is the displacement away from the equilibrium position of the (hypothetical) reduced mass μ. It does not indicate the displacement of either of the actual masses involved in the real motion. However, each mass, m_1 and m_2, oscillates with the same frequency as μ, whose period is given by

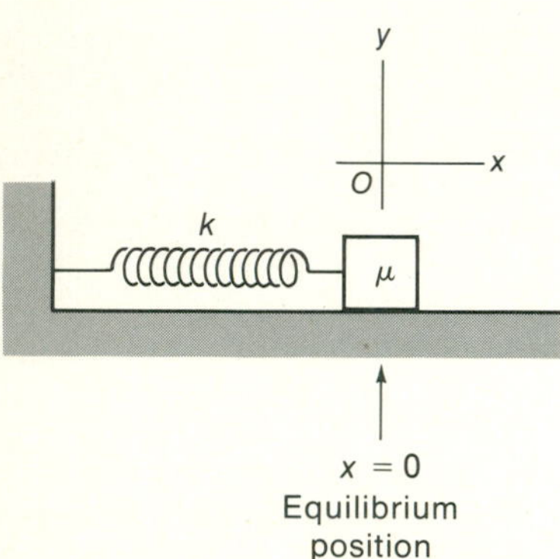

Figure 12–13
The reduced-mass analogy is helpful in analyzing a system of two masses coupled together by a spring (Figure 12–12). If a reduced μ were connected to a rigid wall by the same spring, it would oscillate at the mass frequency as the two coupled masses.

$$T = 2\pi\sqrt{\frac{\mu}{k}} \tag{12–34}$$

Adopting a reduced-mass analogy for two-body systems is widely used in physics because it greatly simplifies the mathematics for problems of this type.

Carrying through the analogy to utilize the other equations for SHM, we let $A = A_1 + A_2$:

$$v = -A\omega \sin(\omega t + \phi)$$
$$a = -\omega^2 x$$

The equations for the energy are (the derivation will be left for a problem):

$$\left.\begin{aligned} K &= \tfrac{1}{2}\mu v^2 \\ U &= \tfrac{1}{2}kx^2 \\ E &= U + K \end{aligned}\right\} \quad \text{(as seen in the } CM \text{ frame}^2\text{)} \qquad \begin{aligned}(12\text{–}35)\\(12\text{–}36)\\(12\text{–}37)\end{aligned}$$

Here, v and a denote the *relative* velocity and the *relative* acceleration of one mass as referred to the other. This follows directly from the definition of x:

$$x = (x_2 - x_1) - \ell$$

Taking successive derivatives with respect to time, we obtain

$$\left.\begin{aligned} v &= v_2 - v_1 \\ a &= a_2 - a_1 \end{aligned}\right\} \tag{12–38}$$

The analogy even extends to rotations of the two masses about their common center of mass, such as in the rotational motions of double-star systems. Here, the moment of inertia I is given by

$$I = \mu r^2 \tag{12–39}$$

where r is the separation of the two masses. Equations for the angular momentum and rotational kinetic energy are also valid:

[2] These equations represent the energy of the *relative motion* of m_1 and m_2. It is the energy that would be seen in a frame of reference attached to the center of mass of the system. (It is often called the *internal* energy of the system.) In some other inertial frame of reference, the center of mass itself would be moving, contributing an additional term to the expression for the total energy in that frame of reference. For this reason, it is often simplest to analyze systems from the center-of-mass frame.

$$\left.\begin{aligned} L &= \mu r^2 \omega \\ K &= \tfrac{1}{2}\mu r^2 \omega^2 \end{aligned}\right\} \quad \text{(in the } CM \text{ frame)} \qquad \begin{aligned} &(12\text{–}40) \\ &(12\text{–}41) \end{aligned}$$

where ω is the angular velocity of one mass about the other.

EXAMPLE 12–6

A 5-kg mass and a 1-kg mass rest on a horizontal frictionless surface and are joined by a spring with a spring constant of 50 N/m. Suppose this spring is compressed and tied in that state by a string. If the string is burned when the system is at rest, the two masses will separate until the spring tension pulls them together again, continuing to oscillate in this fashion toward and away from each other. (a) Find the period of the oscillation. (b) If the 5-kg mass were held fixed and the 1-kg mass were allowed to oscillate, what would the period of oscillation be?

SOLUTION

(a) From Equation (12–34), we have

$$T = 2\pi\sqrt{\frac{\mu}{k}}$$

where

$$\mu = \frac{(5\text{ kg})(1\text{ kg})}{(5\text{ kg} + 1\text{ kg})} = \tfrac{5}{6}\text{ kg}$$

Then

$$T = 2\pi\sqrt{\frac{(\frac{5}{6}\text{ kg})}{\left(50\,\frac{\text{N}}{\text{m}}\right)}} = \boxed{0.811\text{ s}}$$

(b) If the 5-kg mass were held fixed, the period of oscillation of the 1-kg mass would be (from Equation 12–10):

$$T = 2\pi\sqrt{\frac{m}{k}} = 2\pi\sqrt{\frac{1\text{ kg}}{\left(50\,\frac{\text{N}}{\text{m}}\right)}} = \boxed{0.889\text{ s}}$$

12.9 Resonance

It is easy to cause a swing to move with a large amplitude. We merely give it small, repeated pushes, timed to coincide with the natural frequency of the swing (or a submultiple, such as every other swing). Furthermore, we must apply the impulses at the proper phase of the motion, so that we do not oppose the motion and thereby reduce rather than increase the amplitude. These two factors—applying the impulses at the *natural frequency* and at the *proper phase*—are important considerations in the type of motion we will discuss next.

The phenomenon whereby a system responds selectively to certain natural frequencies of applied impulses is called **resonance.** It occurs in many branches of physics—for example, in mechanics, optics, and electrical circuits, as well as in atomic and nuclear physics. A mechanical pendulum has just one natural frequency at which it resonates. Systems such as a vibrating string or a drumhead have many natural frequencies, corresponding to the variety of standing waves that may be generated. In this introductory treatment, we will limit our attention to one-dimensional systems. The conclusions, however, may easily be extended to the more complicated two- and three-dimensional situations.

Damped Oscillations

Consider a mass on a Hooke's Law spring (Figure 12–14). If there were no friction, the system would oscillate with the resonant frequency ω_0 (from Equations 12–7 and 12–8):

RESONANT FREQUENCY (no friction)

$$\omega_0 = 2\pi f_0 = \sqrt{\frac{k}{m}} \qquad \textbf{(12–42)}$$

where ω_0 is expressed in radians/second.

What effect does friction have? In many physical situations, it is found that friction is closely proportional to the *velocity* of the motion.[3] That is:

FRICTION FORCE

$$f_k = -bv = -b\frac{dx}{dt} \qquad \textbf{(12–43)}$$

where the constant of proportionality b is called the **damping coefficient.**

Applying Newton's second law, we see there are *two* forces acting on the mass:

$$\Sigma F = ma$$

$$-kx - b\frac{dx}{dt} = m\frac{d^2x}{dt^2}$$

Rearranging, we have

$$m\frac{d^2x}{dt^2} + b\frac{dx}{dt} + kx = 0 \qquad \textbf{(12–44)}$$

This differential equation is called the **equation of motion** for the mass. A possible solution (that is, x as a function of t that satisfies the equation) is

$$x = Ae^{-\left(\frac{b}{2m}\right)t}\cos(\omega' t - \alpha) \qquad \textbf{(12–45)}$$

where the amplitude, $Ae^{-(b/2m)t}$, decreases exponentially with time because frictional effects are absorbing energy from the vibrating mass, which oscillates with an angular frequency ω'. When a relatively small amount of friction is present, the system oscillates with a resonant frequency somewhat lower than ω_0:

RESONANT FREQUENCY (with a small amount of friction)

$$\omega' = \sqrt{\omega_0{}^2 - \left(\frac{b}{2m}\right)^2} \qquad \textbf{(12–46)}$$

The *phase angle* α and the *amplitude A* depend on the initial conditions at $t = 0$. In Figure 12–15, the mass is released from rest with an initial displacement x_0 (here, the amplitude A). The various curves show the qualitative effect of increasing the friction force. Such motion is called **damped harmonic motion,** since the amplitude of vibration gradually dies out exponentially as the energy of oscillation is transformed into frictional heating.

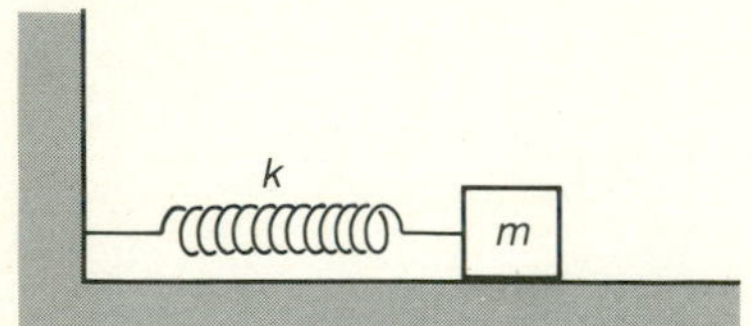

(a) The horizontal surface has friction.

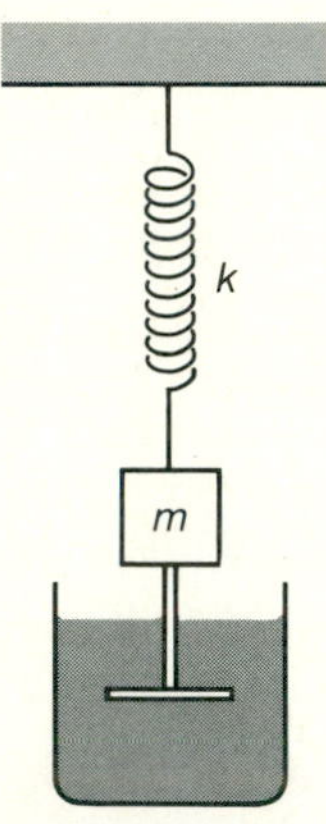

(b) A disk fastened to the mass is submerged in a fluid, providing a frictional damping force that opposes the motion.

Figure 12–14
Examples of damped harmonic motion. In each case, frictional forces oppose the motion of the mass.

[3] At moderate speeds, friction between sliding surfaces is fairly constant. But at faster speeds, the friction force does vary, usually becoming less. For motion through fluids, however, the friction force generally becomes larger for faster speeds. In certain cases of *laminar* flow (to be discussed in Chapter 15), the friction force is often quite closely proportional to the velocity.

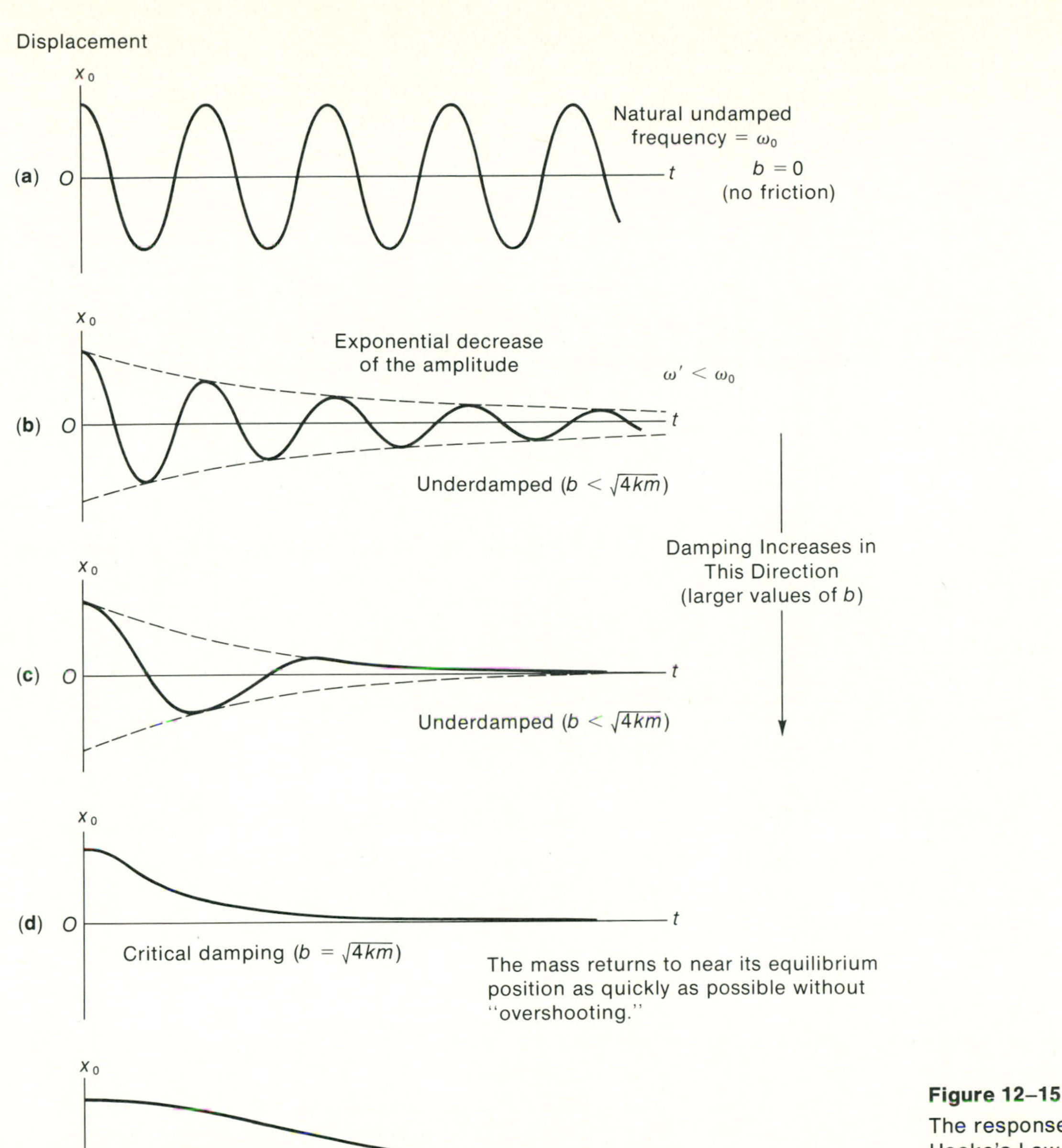

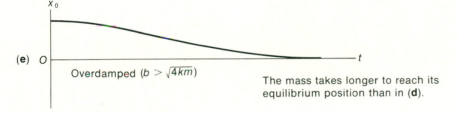

Figure 12–15
The response of a mass on a Hooke's Law spring with increasing amounts of friction. In each case, the mass is released from rest with an initial displacement x_0.

Forced Oscillations

Let us examine what happens when we apply an external oscillatory force on the system. We will assume the added force has the form:

$$\text{Driving force} \qquad F = F_0 \sin \omega t$$

To avoid confusion in the notation, we will first define ω as the driving frequency, ω' as the resonant frequency with slight damping, and ω_0 as the resonant frequency with no damping. The driving frequency ω may have any value.

The mass now has *three* different forces on it: a Hooke's Law force, $-kx$, a resistance force, $-b(dx/dt)$, and the driving force, $F_0 \sin \omega t$. The motion of the mass is, of course, governed by Newton's second law:

$$\Sigma F = ma$$

$$\left[F_0 \sin \omega t - kx - b\frac{dx}{dt}\right] = m\frac{d^2x}{dt^2} \tag{12–47}$$

Rearranging, we write it in the following form:

$$m\frac{d^2x}{dt^2} + b\frac{dx}{dt} + kx = F_0 \sin \omega t \tag{12–48}$$

The solutions of this differential equation describe the motion of the mass. What are the general characteristics of the solutions?

If we suddenly apply the sinusoidal force, the resultant motion of the mass depends on two factors in the solution: a *transient* term and a *steady-state* term. The beginning stages of the motion may involve very large displacements (see Figure 12–16). The transient motion depends on the initial displacement and velocity of the mass and on the particular phase of the sinusoidal force at the instant it is applied. The transient curve, shown dashed in Figure 12–16(b), is just Equation (12–45) with $\alpha = \pi/2$, one of the solutions for the (undriven) damped harmonic motion.

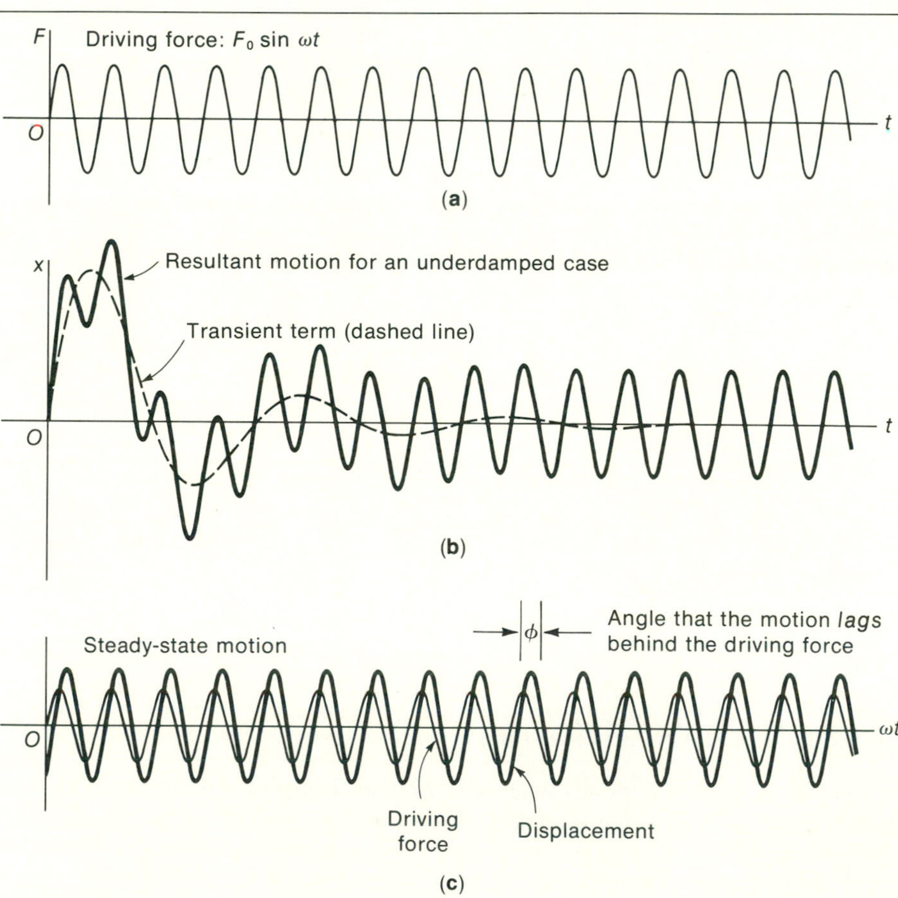

Figure 12–16

Forced oscillations with damping. Curve (a): The harmonic driving force. Curve (b): The resultant motion for an underdamped case, including the transient response, which decreases exponentially at a rate dependent upon the amount of friction (the damping coefficient b). Curve (c): After the transient effects die down, the mass oscillates in steady-state motion with the frequency ω of the driving force, but its motion lags behind the phase of the driving force by an angle ϕ.

During the transient period, the driving force is increasing the kinetic and potential energies of the system until the average rate at which energy is fed into the system just equals the average rate of loss of energy through frictional effects. If we wait awhile, the transient motion dies out exponentially, and the system settles down to a pure sinusoidal oscillation, called **steady-state** motion. The mass then vibrates in *forced* harmonic motion at the driving frequency ω. Its motion lags behind the phase of the driving force by an angle ϕ.

We present the equation of motion for the steady-state case (a solution to Equation 12–48) without proof:

FORCED HARMONIC MOTION (steady-state)

$$x = A \sin(\omega t - \phi) \tag{12–49}$$

where the amplitude A is:

$$A = \frac{\left(\dfrac{F_0}{m}\right)}{\left[(\omega_0{}^2 - \omega^2)^2 + \left(\dfrac{\omega b}{m}\right)^2\right]^{1/2}} \tag{12–50}$$

The phase angle ϕ by which the displacement lags behind the driving force is given by:

$$\tan\phi = \frac{\left(\dfrac{\omega b}{m}\right)}{\omega_0{}^2 - \omega^2} \tag{12–51}$$

An interesting feature of forced harmonic motion is the way the amplitude changes as the driving frequency varies. Figure 12–17 shows that with no friction, the amplitude goes to infinity at the resonant frequency ω_0. (Of course, no real systems behave this way. Either some friction is present, or the restoring force deviates from a Hooke's Law force as the amplitude increases, and thus a different differential equation would apply.) As the damping increases, the resonance curve becomes relatively broad, with the *peak of the curve*—the resonant

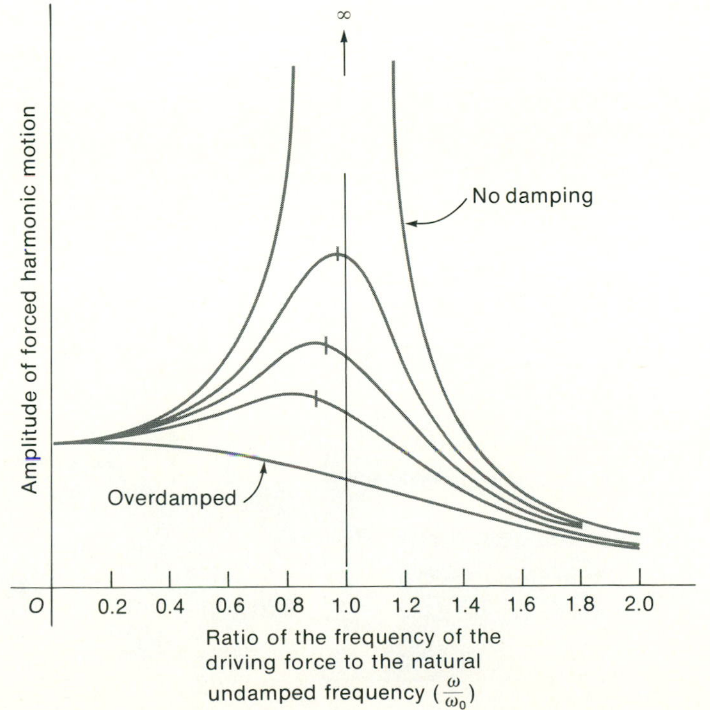

Figure 12–17
The amplitude of forced harmonic motion as a function of the frequency ω of the driving force. The various curves are for different amounts of damping. The small vertical lines are the damped resonant frequencies ω' (see Equation 12–46) with no driving force.

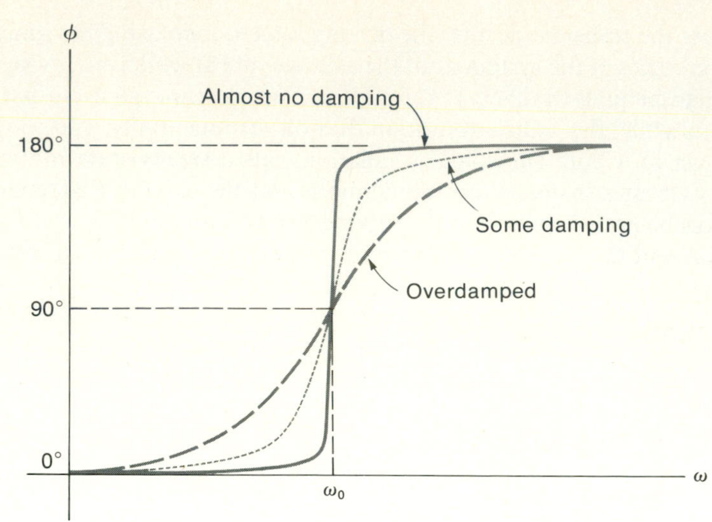

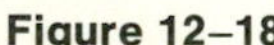
Figure 12–18
The phase angle ϕ (by which the displacement lags behind the driving force) versus the driving frequency ω. Curves for three different amounts of damping are shown.

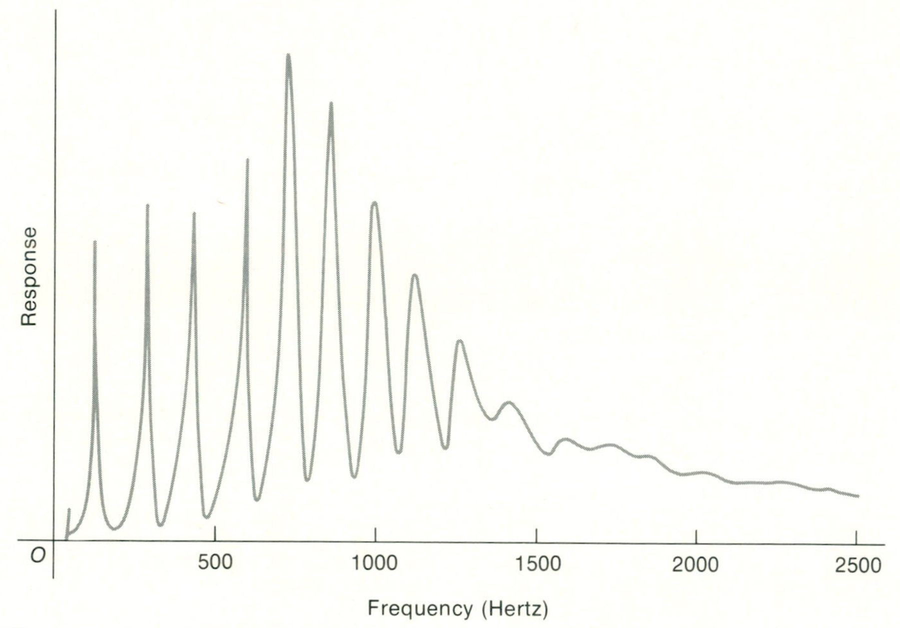

Figure 12–19
Resonant frequencies in a nineteenth-century cornet. The cornet's response to sound vibrations at the mouthpiece is plotted versus the frequency of the vibration. Peaks in the response curve indicate resonances. The number of nodes in the standing wave pattern within the cornet increases by one at each successive peak.

Figure 12–20
The Tacoma Narrows Bridge near Puget Sound, Washington, was known locally as "Galloping Gertie." With moderate wind velocities it oscillated vertically in standing-wave patterns with 0–8 nodes between the two towers. On July 1, 1940, a 40- to 45-mi/h wind produced vertical vibrations with peak-to-peak motions of 5 ft, closing the bridge to traffic. After a few hours, the main span suddenly developed torsional oscillations at 14 vibrations/min, causing the bridge to break and fall 190 ft into the water.

frequency—occurring at lower and lower frequencies.[4] For the *overdamped* case, there is no peak. Figure 12–18 shows how the phase angle ϕ changes with frequency. If a system can oscillate in several different standing wave patterns, there will be a series of resonant peaks, each corresponding to a particular resonant frequency (see Figure 12–19).

Resonance phenomena play an important part in many practical situations. For example, in the design of loudspeakers, unavoidable resonances should be broad so as not to overemphasize certain frequencies. Sometimes in machines such as automobiles and rockets, components may be insufficiently braced, so that they have natural frequencies low enough to coincide with vibrations inherent in the operation of the machine. These components then undergo troublesome large-amplitude oscillations when the motor runs at speeds that generate vibrations at the corresponding resonant frequencies. As a final example, when a group of soldiers crosses a bridge, they should "break step" in case the frequency of their footsteps in unison happens to match one of the resonant frequencies of the structure. A spectacular example of resonance vibrations occurred in the collapse of the Tacoma Narrows Bridge in 1940 (see Figure 12–20).

Summary

Many natural phenomena within the scope of classical mechanics involve motions of masses under the influence of "restoring" forces, which direct the masses toward positions of static equilibrium. The Hooke's Law relation $F = -kx$ is applicable in a great many cases, particularly for small-amplitude motions. Phenomena described by Hooke's Law are summarized in Table 12–1.

[4] The "resonant frequency" for damped systems may be defined in several different ways. It is usually defined as we have stated: *the frequency at maximum amplitude*. However, in certain applications, it becomes more useful to define it as the frequency at which the mass has its maximum velocity. This latter is the frequency of *maximum power transfer* to the system; it occurs at $\omega = \omega_0$, the resonant frequency with no damping.

TABLE 12–1

	Single-body linear motion	Angular motion	Two-body motion in the *CM* frame
Hooke's Law	$F = -kx$	$\tau = -\kappa\theta$	$F = -kx$ where x is the amount of stretching ($x > 0$) or compression ($x < 0$) of the spring
Displacement	$x = A\cos(\omega t + \phi)$	$\theta = \theta_0 \cos(\omega t + \phi)$	$x = A\cos(\omega t + \phi)$
Angular frequency	$\omega = \sqrt{\dfrac{k}{m}}$	$\omega = \sqrt{\dfrac{\kappa}{I}}$	$\omega = \sqrt{\dfrac{k}{\mu}}$ where $\mu = \dfrac{m_1 m_2}{m_1 + m_2}$
Velocity	$v = -A\omega\sin(\omega t + \phi)$	$\dfrac{d\theta}{dt} = -A\omega\sin(\omega t + \phi)$	$v = -A\omega\sin(\omega t + \phi)$
Acceleration	$a = -A\omega^2\cos(\omega t + \phi)$ or $a = -\omega^2 x$	$\alpha = -A\omega^2\cos(\omega t + \phi)$ or $\alpha = -\omega^2\theta$	$a = -A\omega^2\cos(\omega t + \phi)$ or $a = -\omega^2 x$
Kinetic energy	$K = \frac{1}{2}mv^2$	$K = \frac{1}{2}I\left(\dfrac{d\theta}{dt}\right)^2$	$K = \frac{1}{2}\mu v^2$ (in the *CM* frame)
Potential energy	$U = \frac{1}{2}kx^2$	$U = \frac{1}{2}\kappa\theta^2$	$U = \frac{1}{2}kx^2$ (in the *CM* frame)

For small oscillations, the motion of simple and physical pendulums may be described as simple harmonic motion, that is, motion under the influence of Hooke's Law. This may be summarized as follows:

	Simple pendulum	Physical pendulum	
Displacement	$\theta = \theta_0 \cos(\omega t + \phi)$	$\theta = \theta_0 \cos(\omega t + \phi)$	
Angular frequency	$\omega = \sqrt{\frac{g}{\ell}}$	$\omega = \sqrt{\frac{mg\ell}{I}}$	where I is the moment of inertia about the axis and ℓ is the distance from the axis to the *CM*
Period	$T = 2\pi\sqrt{\frac{\ell}{g}}$	$T = 2\pi\sqrt{\frac{I}{mg\ell}}$	

Damped harmonic motion occurs when friction is present in oscillating systems. When a SHM driving force is applied to such systems, *forced harmonic motion* takes place. After an initial period of *transient* effects that die out exponentially, the system oscillates in *steady-state motion* at the frequency of the driving force, but its displacement lags behind the phase of the driving force. *Resonance* is the establishment of especially large amplitudes when the driving impulses occur close to one of the natural oscillation frequencies of the system.

Questions

1. A pendulum clock is carried in an elevator. Does the clock run slow or fast (a) when the elevator accelerates upward and (b) when it accelerates downward?
2. Will a pendulum swinging with a large amplitude have a longer or shorter period than when swinging with a small amplitude?
3. The period of vibration of a mass suspended from an ideal (massless) spring is $T = 2\pi\sqrt{m/k}$. If we now take into account the mass of the spring, the value of m should be increased by one-third the mass of the spring. Explain qualitatively why this value (rather than the entire mass of the spring) is reasonable. Why is it not just *half* the mass of the spring?
4. One technique of geological exploration for ore deposits or oil is to employ a pendulum whose period can be precisely determined. How does this method work?
5. A sphere filled with water is hung by a cord and set swinging as a pendulum. Because of a small hole in the bottom of the sphere, the water slowly leaks out. Explain why the period of the pendulum first increases, then slightly decreases.
6. What combination of simple harmonic motions will trace out an infinity sign: ∞?

Problems

12A–1 A 2-kg mass on the end of an ideal spring is pulled vertically downward from its equilibrium position a distance of 5 cm and released from rest. The mass then oscillates in SHM with a period of 8 s.

(a) Find the velocity of the mass 3 s after being released.
(b) Determine the force constant on the spring.

Answers: (a) 2.79 cm/s (b) 1.23 N/m

12A–2 A 3-kg mass is suspended by a spring of force constant 48 N/m. The mass is pulled down from its equilibrium position a distance of 10 cm and released at $t = 0$. Write an equation (in SI units) for the displacement as a function of time.

12A–3 A 100-g mass hangs on the end of a Hooke's Law spring suspended vertically. When 40 g are added, the spring stretches an additional 5 cm. With this extra mass, the spring is now set into vertical oscillation with an amplitude of 10 cm.

(a) Find the frequency of the motion.

(b) How long does it take for the mass to travel from the mid-position to a point of maximum displacement?

(c) Find the net force on the total mass when it is at a point of maximum displacement upward. *Answers:* (a) 1.20 s^{-1} (b) 0.208 s (c) 0.80 N downward

12A–4 An object undergoes SHM vibrations according to $x = 0.04 \cos(2t)$, where all quantities are in SI units. Find (a) the maximum acceleration, (b) the period of the motion, and (c) the displacement 0.5 s after it passes the midpoint traveling in the negative x direction.

12A–5 A 2-kg mass hangs on the end of a spring in the position of rest. It requires a force of 2 N to stretch the spring an additional 4 cm. If the force is then suddenly removed, the mass executes SHM vibrations. Find (a) the total energy of the motion and (b) the frequency. *Answers:* (a) 0.04 J (b) 0.796 s^{-1}

12A–6 The balance wheel of a wristwatch undergoes torsional oscillations with an amplitude of π rad and a frequency of 2 oscillations/s. Find the maximum angular speed of the wheel.

12A–7 A 16-lb object hangs from a spring. It undergoes SHM oscillations in a vertical direction with an amplitude of 6 in and a frequency of 2 cycles/s. Find (a) the period of the oscillations, (b) the force constant of the spring, (c) the maximum speed of the object, (d) the maximum acceleration of the object, (e) the total energy of the oscillating system, (f) the speed when the object is at half its maximum displacement, and (g) the acceleration when it is at half its maximum displacement.

Answers: (a) 0.50 s (b) 79.0 lb/ft (c) 6.28 ft/s (d) 79.0 ft/s^2 (e) 9.88 ft · lb (f) 5.45 ft/s (g) 39.5 ft/s^2

12A–8 A uniform meter stick is pivoted at the 70-cm mark so that it swings in small-amplitude oscillations about a horizontal axis perpendicular to the length of the stick. What is the period of these oscillations?

12A–9 Two masses on an air track have weights of 2 lb and 6 lb. They are connected by a spring of negligible mass and set into SHM vibrations with a resultant period of 2 s. (The *CM* of the system remains at rest.)

(a) Find the reduced mass μ for this system.

(b) Find the force constant of the spring.

(c) In terms of ℓ, the separation of the two masses when in equilibrium, how far is the *CM* from the more massive object?

Answers: (a) 0.0468 slug (b) 0.462 lb/ft (c) $\ell/4$

12A–10 Solve the previous problem for two masses with weights of 2 kg and 5 kg, vibrating in SHM with a period of 3 s.

12B–1 A simple pendulum 40 cm long undergoes small-amplitude oscillations where the acceleration of gravity is 10 m/s^2.

(a) Find the period of the oscillations.

(b) If the pendulum is located in an elevator accelerating upward at 1 m/s^2, what is the period?

(c) Repeat for downward acceleration at 1 m/s^2.

Answers: (a) 1.25 s (b) 1.20 s (c) 1.33 s

12B–2 A 1200-kg car oscillates vertically with a period of 0.60 s when empty. If six persons, each of mass 80 kg, get into the car, how far down will the supporting springs be depressed?

12B–3 A 2-kg mass hangs at rest on a spring whose force constant is 240 N/m. An additional 1-kg mass is now added, and the combination is released from rest at the initial position of the original mass.

(a) To what maximum distance below this initial position do the masses descend after being released?

(b) What is the frequency of the oscillations? *Answers:* (a) 8.17 cm (b) 1.42 s^{-1}

12B–4 A mass, fastened to a spring, rests on a horizontal frictionless surface. It undergoes SHM oscillations of amplitude A.

(a) When the displacement is $A/2$, what fraction of the total energy is in the form of kinetic energy?

(b) What is the displacement when half the total energy is in the form of kinetic energy?

12B–5 A penny is placed on a horizontal surface that is undergoing SHM oscillations in a horizontal direction with frequency f and amplitude A. If the penny does not slip on the surface, find, in terms of f and A, the minimum coefficient of static friction μ_s between the penny and the surface. *Answer:* $4\pi^2 f^2 A/g$

12B–6 It is claimed that the time shown on pendulum clocks is different when the moon is overhead from when it is diametrically on the opposite side of the earth. Explain in which situation the clock would run faster, and why.

12B–7 A 100-g block is placed on top of a 200-g block. The coefficient of static friction between the blocks is 0.20. The lower block is now moved back and forth horizontally in SHM oscillations with an amplitude of 6 cm.

(a) Keeping the amplitude constant, what is the highest frequency for which the upper block will not slip relative to the lower block?

(b) What is the maximum external force on the combination as they move in part (a)? *Answers:* (a) 0.910 s^{-1} (b) 0.588 N

12B–8 In the previous problem, suppose the lower block (mass = 200 g) is moved vertically in SHM, rather than horizontally. The frequency is held constant at 2 oscillations/s while the amplitude is gradually increased.

(a) Find the amplitude at which the upper block (mass = 100 g) will no longer maintain contact with the lower block.

(b) Determine the maximum driving force on the combination when moving as in part (a).

12B–9 Show that the change in period ΔT of a simple pendulum when the acceleration due to gravity changes by Δg is $-T\,\Delta g/2g$.

12B–10 A uniform, solid disk of mass 7 kg and radius 6 cm is suspended by a wire through its center of mass, with the plane of the disk horizontal. It undergoes angular oscillations as a torsional pendulum with a period of 12 s.

(a) Find the torsion constant for the supporting wire.

(b) Suppose the disk is now replaced by an object with unknown rotational inertia I. If the rotational period is 17 s, what is the value of I?

12B–11 A uniform, solid cylinder of mass 5 kg and radius 4 cm is suspended as a torsion pendulum by a steel wire so that it undergoes rotational oscillations about its axis of symmetry. The torque constant for the wire is 4×10^{-4} N·m/rad. Find the period of the torsional oscillations. *Answer:* 19.9 s

12B–12 A thin hoop of radius 20 cm is supported on a horizontal knife-edge so that it swings as a physical pendulum in the plane of the hoop.

(a) Find the period for small-amplitude oscillations.

(b) What is the length of a simple pendulum having the same period?

12B–13 The diatomic molecule O_2 vibrates along the line joining the two atoms with a frequency of 4.74×10^{11} s^{-1}. The mass of each atom is 2.66×10^{-26} kg. Assuming the atoms are effectively bound together by a Hooke's Law spring, find the force constant of the spring in SI units. *Answer:* 0.118 N/m

12B–14 The normal pace of walking is determined by the natural period of oscillation of the leg as a pendulum. Approximate the adult leg as a uniform rod 1 m long, pivoted at one end.

(a) What is the natural frequency of small-amplitude oscillations?

(b) If the leg swings $\pm 20°$ from the vertical in taking each step, what is the normal walking speed?

(c) For the same angle of swing, what would be the normal speed of walking for a child whose leg length is 60 cm?

12B–15 A 0.20-kg mass moves in simple harmonic motion according to the equation $y = A \cos \omega t$, where $A = 10$ cm, and $\omega = 4$ rad/s. Find the displacement (a) when $t = 0$ s and (b) when $t = 1$ s. (c) How long does it take to move from the equilibrium position to a displacement equal to half the amplitude? (d) Find the maximum kinetic energy of the mass. (e) Determine the total energy of the oscillating system.

Answers: (a) 0.10 m (b) -0.0654 m (c) 1.57 s (d) 0.04 J (e) 0.04 J

12B–16 A 2-kg mass is suspended by a spring whose force constant is 200 N/m. Because of friction, the mass moves in underdamped harmonic motion. The mass is displaced 0.20 m from its equilibrium position and released from rest. Six seconds later, its amplitude has been reduced to 0.16 m.

(a) Find the damping coefficient associated with the friction force.
(b) Determine the resonant frequency of the system.

12B–17 A 4-kg mass is suspended by a spring whose force constant is 100 N/m. A resistive force (damping coefficient = 5 N·s/m) is supplied by a dashpot arrangement. If the mass is given an initial displacement and released from rest, what is the frequency of the resultant oscillations? *Answer:* 0.790 Hz

12C–1 Show that the formula for the period of oscillation of a physical pendulum contains, as a special case, the period of a simple pendulum.

12C–2 A pendulum clock keeps perfect time at a location where the acceleration due to gravity is exactly 9.80 m/s^2. When the clock is moved to a higher altitude, it loses 8 s per day. Find the value of g at the new location.

12C–3 A uniform, solid cylinder of radius R is supported on a horizontal axis perpendicular to the plane of the disk so that the disk oscillates as a physical pendulum. The axis is adjustable, so it may pass through various points a distance r from the center of the disk, as shown in Figure 12–21. Consider small-amplitude oscillations only.

(a) If T_0 is the period when $r = R$ and T is the period for other values of r, make a graph of T/T_0 versus r/R, choosing points along the radius that are spaced $R/4$ apart.
(b) By differentiating, show that the minimum period occurs at $r = R/\sqrt{2}$.

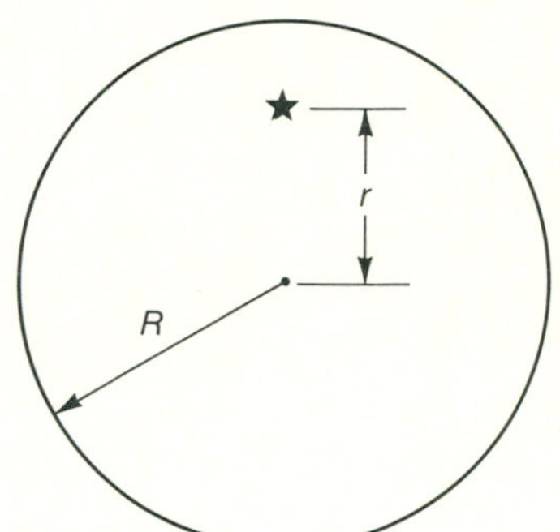

Figure 12–21
Problem 12C–3

12C–4 Figure 12–22 shows the pendulum of a grandfather clock in the form of a uniform, solid disk of mass 2 kg and radius 10 cm, fastened at one edge to a slender rod of mass 1 kg and length ℓ. Find the value of ℓ so that for small-amplitude oscillations, the pendulum ticks once each second (that is, the period is 2 s).

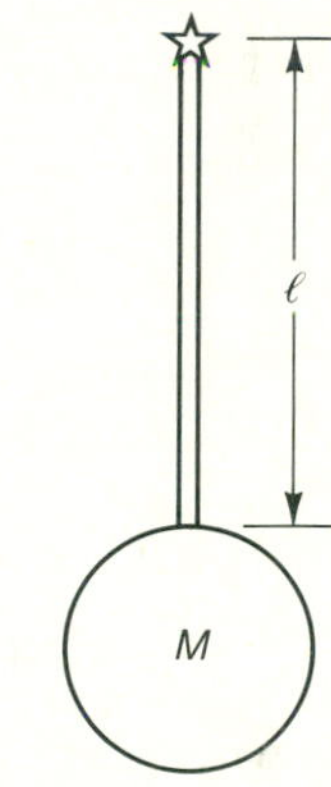

Figure 12–22
Problem 12C–4

12C–5 A uniform spring of unstressed length ℓ and force constant k is cut in two so that one piece is twice as long as the other.

(a) In terms of k, find the force constant of each of the two pieces.
(b) If identical masses were hung on one end of each piece, what would be the ratio of the frequencies? *Answers:* (a) $3k$, $1.5k$ (b) $\sqrt{2}:1$

12C–6 A uniform spring has an unstressed length ℓ and a force constant k. It is cut into two pieces whose lengths are related according to $\ell_1 = n\ell_2$, where n is an integer. In terms of n and k, derive formulas for the force constants k_1 and k_2 of the two pieces.

12C–7 A mass m on the end of a spring of force constant k stretches the spring to a length ℓ when at rest. The mass is now set into motion so it executes up-and-down vibrations while swinging back and forth as a pendulum. The mass moves in a horizontal figure-eight pattern, as shown in Figure 12–23(b). Express the force constant k in terms of m, ℓ, and g. *Answer:* $4mg/\ell$

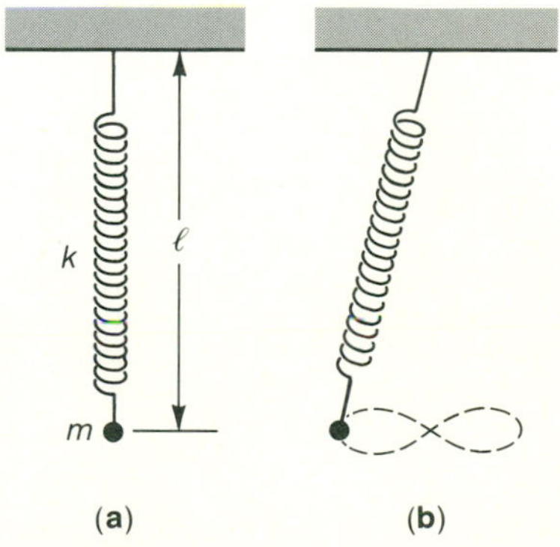

Figure 12–23
Problem 12C–7

12C–8 Two springs each have unstressed lengths of 20 cm and force constants, of $k_1 = 40$ N/m and $k_2 = 80$ N/m. They are fastened to a small mass, $m = 0.60$ kg, resting on a horizontal, frictionless surface. The springs are stretched in opposite directions and fastened to two hooks 60 cm apart (see Figure 12–24). Assuming the mass has negligible size, (a) how far from the left hook is the equilibrium position of the mass? (b) What is the frequency of SHM oscillations of the mass along the direction of the springs?

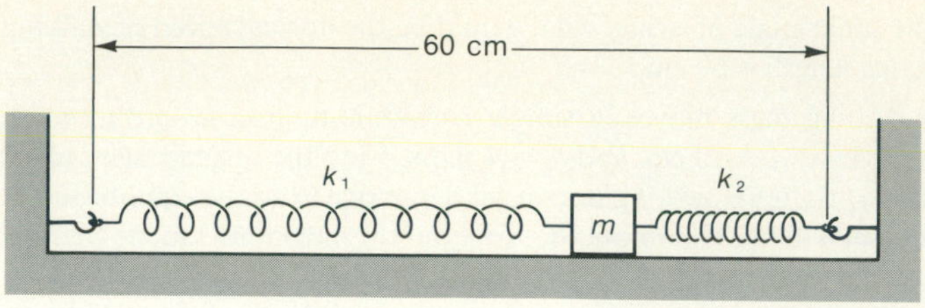

Figure 12–24
Problem 12C–8

12C–9 Starting with appropriate equations for the laboratory frame of reference, derive Equations (12–35) and (12–36) for the kinetic and potential energies of a two-mass system in the *CM* frame.

12C–10 A damped oscillating system has a mass of 40 kg and a force constant of 6000 N/m. The damping coefficient associated with the friction force is 200 N·s/m. A sinusoidal driving force, $F = (20\text{ N}) \sin \omega t$, is applied to the mass, and the frequency of the driving force is varied across the resonance peak of the system.

(a) Find the frequency for maximum-amplitude motion.

(b) Find the amplitude of the motion in part (a).

12C–11 A driving force causes a damped oscillator to undergo the following (steady-state) displacement as a function of time: $x = A \sin \omega t$. If the resistive force is $-bv$, find how much work is done against the resistive force during one complete cycle of the motion. (Hint: How are force and velocity related to work?) *Answer:* $\pi b A^2 \omega$

12C–12 A 0.40-kg mass is hung on a spring whose force constant is 160 N/m. A resistive force, $f = -bv$, impedes the motion of the mass, where $b = 4$ N·s/m. A driving force of (20 N) sin $(25t)$ is applied to the mass, and steady-state oscillations are attained after the transient response dies out.

(a) Find the amplitude of the oscillations.

(b) By how many degrees does the motion lag behind the applied sinusoidal force?

Oh, that Einstein, always skipping lectures. . . .
I certainly never would have thought he could do it.

HERMAN MINKOWSKI

Newton, forgive me.

ALBERT EINSTEIN

chapter 13 Special Relativity

13.1 Introduction

Two revolutions in physics occurred in the early part of the twentieth century, radically changing our concepts of the universe. One was the work of several people over a period of decades: the development of quantum mechanics. The other was the theory of special relativity,[1] published by Albert Einstein in 1905. Einstein's theory not only led to apparent paradoxes that seemed to violate common sense in the most radical way, but it completely changed our basic understanding of space and time. As far as is known today, special relativity unquestionably describes the way the world "is."

The main difficulties in understanding special relativity are not mathematical ones. Rather, the challenge comes from our reluctance to discard deeply ingrained ideas about space and time. We grow up using Newtonian concepts to explain physical phenomena, and it is disturbing to have cherished beliefs overthrown. Furthermore, the structure of our language reflects these common-sense classical notions, so this adds to the difficulty of gaining a new perspective. Of course, the classical way of thinking cannot be completely wrong, since it does serve admirably to explain everyday experiences. But scientists exploring the fine details of natural phenomena must abandon classical concepts and deal with a more modern theory.

The basic question relativity asks is, "If a given phenomenon is viewed from two different frames of reference that have uniform relative motion with respect to each other, how do the two measurements of the phenomenon compare?"

Einstein points out that making a measurement involves determining *where* and *when* something happens in *space* and *time*. In particular, we seek four quantities about an *event* that happens at a given point in space and at a given time:

A point event (x,y,z,t)

These four quantities are, of course, measured in some inertial frame of reference that we will call the S frame, assumed to be "at rest." Another frame of reference, the S' frame, moves with constant velocity $\mathbf{V}$ along the $+x$ direction of S. For

[1] The *special* theory deals with frames of reference that have constant motion in a straight line relative to each other. The *general* theory, published in 1916, treats accelerated frames of reference (see Section 14.13).

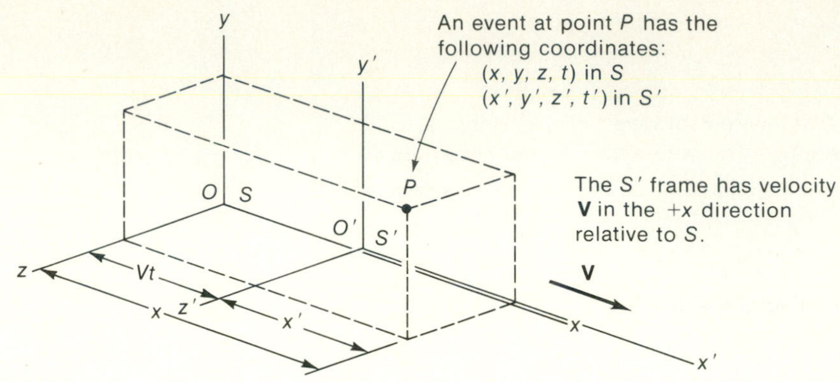
y
y′
An event at point P has the following coordinates:
(x, y, z, t) in S
(x′, y′, z′, t′) in S′
O
S
P
O′
S′
The S′ frame has velocity
V in the +x direction
relative to S.
V
z
Vt
x
z′
x′
x
x′

Figure 13–1
The coordinate systems S and S'. The frames are coincident at the time $t = t' = 0$. At a later time, the origins are a distance Vt apart.

convenience, we align the two frames so that their origins and respective axes are coincident at the time $t = 0$ (see Figure 13–1). As before, unprimed quantities designate measurements made in the S frame, while primed (′) quantities are for the S' frame. Each frame is equally valid, and measurements made in either frame correctly measure space and time for that frame. Both frames of reference are *inertial frames*, since neither has acceleration. Relativity shows that it is only the *relative* velocity that is important, not which system is imagined to be "at rest." We could equally well assume S' is at rest and S moves with a velocity $-\mathbf{V}$ (in the negative x' direction). The basic conclusions of relativity would be exactly the same.

For relatively slow velocities, Newtonian mechanics is valid in all inertial frames of reference (as anyone who has ridden in a smoothly moving airplane will testify). Stated another way, there is no *mechanical* effect by which observers in the S and S' frames could determine which frame is "truly" moving and which is at rest. This fact is known as the Galilean Relativity Principle: *The laws of Newtonian mechanics are the same in all inertial frames.* Einstein felt that this relativity principle should apply to *all* the laws of physics, not just mechanics. But a basic contradiction arose, because in classical relativity the laws of electromagnetism (Maxwell's equations, discussed in Chapter 27) were not the same for all frames of reference. However, Einstein found that if one assumes that the speed of light in a vacuum ($c = 3 \times 10^8$ m/s) has the same value in all frames of reference, independent of the motion of the light source or of the observer, then Maxwell's equations have the same form in all frames.[2]

13.2 The Fundamental Postulates of Special Relativity

Einstein based his theory of relativity on two assumptions:

BASIC POSTULATES OF SPECIAL RELATIVITY

(1) All the laws of physics have the same form in all inertial frames (the Principle of Relativity).

(2) The speed of light in a vacuum has the same value c in all inertial frames (the Principle of the Constancy of the Speed of Light).

The entire theory of special relativity is derived from just these two postulates. As a consequence, Einstein showed that Newtonian mechanics is only approxi-

[2] The currently accepted value for c is $(2.997\,925 \pm 0.000\,003) \times 10^8$ m/s.

mately correct, usable in cases where velocities are small compared with the speed of light. It is comforting, however, that Einstein's relativistic mechanics reduce to Newtonian mechanics when $v \ll c$.

We shall examine some important conclusions of special relativity. By themselves, they seem paradoxical and contrary to common sense. But if we consider all the conclusions of special relativity together, and manage to give up our Newtonian concepts of absolute space and time, they form a coherent and satisfying theory, one that has been amply verified experimentally. We will state the results of relativity here without proof. The corresponding derivations and a more complete discussion of the implications will be found as a Supplementary Topic at the end of the book.

13.3 How to Make Measurements

Basically, all measurements reduce to determining the four quantities associated with a point event: (x,y,z,t). Einstein suggests that, in principle, a meter-stick framework be extended throughout the frame of reference and an observer be stationed at every location within the frame. Each observer has a clock that has been synchronized with all other clocks in the frame. Every event is to be measured by a "local" observer, situated where the event occurs. The spatial coordinates (x,y,z) of the event are found by reference to the meter-stick framework in that vicinity, and the time (t) is given by the observer's clock. Because all measurements are of *local* events, one does not have to take into account the transit times that would be involved for light signals to travel from some distant event to the observer. Even if such cases were allowed, however, the conclusions of relativity would be the same.

13.4 Comparison of Clock Rates

Suppose that observers in the S frame of reference decide to measure the rate at which a clock in the moving S' frame keeps time. According to relativity, during a time interval in which a single clock in S' indicates an elapsed time T_0, clocks in the S frame will indicate a longer time interval T according to

TIME DILATION

$$T = \frac{T_0}{\sqrt{1-\beta^2}} \quad \text{(where } T_0 \text{ must be a time interval measured by a } \textit{single} \text{ clock)} \tag{13–1}$$

Here we introduce the notation:

DEFINITION OF β

$$\beta \equiv \frac{V}{c} \tag{13–2}$$

because the ratio V/c will appear frequently hereafter in equations. Since the factor $\sqrt{1-\beta^2}$ is always less than unity, $T > T_0$, and we conclude that *moving clocks run slower than clocks at rest*. This effect is called **time dilation.** The moving clocks run slower not because the motion somehow deforms them so that they show an incorrect time; rather, it is *time itself* that is different for a moving frame of reference compared with the time scale in a stationary frame. All clocks show the correct time for their local frames of reference.

An even more startling feature of time dilation is that since either frame may be considered "at rest," observers in S' would find that clocks in S run

Albert Einstein

Between 1900 and 1927, there were two great revolutions in physics: quantum mechanics and relativity. The former grew from contributions by many physicists (including Einstein), but relativity was the creation of Einstein alone, a stunning accomplishment ranking easily with the achievements of Newton.

Albert Einstein was born in Ulm, Germany, in 1879. His father owned a small electrochemical shop. In his early schooling, Einstein did so poorly his parents were afraid he might be retarded. He did not speak at all before the age of three, nor fluently until he was almost nine. He particularly disliked the rigid discipline and authoritarian teaching methods common in German schools. His relatives predicted he would never amount to much, and his high school teachers considered him a "disruptive influence," asking him to leave school, which he did at age 15. Yet during this time he was intensely interested in geometry, algebra, and calculus; these he studied diligently on his own. After a year of roaming about in Northern Italy, at age 16 (two years younger than most applicants) he took the entrance examination for admission to the Federal Institute of Technology in Zurich, a renowned engineering school. He failed the test because of deficiencies in modern languages, zoology, and biology. After returning to high school to earn his diploma and doing some extra studying with the help of a friend, he took the exam again and was admitted. He seemed an indifferent student, uninspired by the old-fashioned nature of the curriculum, attending classes sporadically, and spending considerable time in the local cafes. But he also thought a great deal about physics and during this time taught himself Maxwell's theory of electromagnetism. He graduated in 1900 with no particular distinction.

Perhaps it was Einstein's middling academic record that prevented him from obtaining the immediate teaching position he desired. After an unsatisfactory interval of trying to earn a living by tutoring poor students, he obtained a job in the Swiss patent office in Bern through the aid of a friend. It was an undemanding position with modest pay, but it left a great deal of spare time for his absorbing intellectual pursuits. During the next eight years, Einstein made remarkable contributions to physics. Though isolated from the ferment and stimulation of an academic environment, he completed his doctoral thesis and published several papers on statistical mechanics and molecular motions. The year 1905 was truly a banner period, in which he published four short papers on the photoelectric effect, Brownian motion, and the special theory of relativity. In spite of Einstein's questionable background, the scientific community began to recognize the value of his accomplishments. He was offered numerous professorial positions in various universities; he accepted those in Zurich, Prague, and finally a prestigious

more slowly than their own S' clocks. The effect is entirely symmetrical: *Observers in each frame find that the other "moving" clocks run slower than clocks at rest in their own frame.* All measurements depend on the frame of reference of the observer, and each frame has its own scale of time, which does not necessarily agree with the time scale in other frames. The question "Do moving clocks *really* run slower?" is properly answered by pointing out that according to all measurements made on moving clocks, yes, they certainly *do* run slower than clocks in our own frame of reference. It is not an illusion. All clocks show the correct time in their own frame of reference. There is simply no absolute time scale, valid in all frames. By itself, this conclusion may seem paradoxical. But

appointment at the University of Berlin which left him entirely free from specified duties.

In 1916, Einstein published his general theory of relativity. Its abstract, mathematical nature made acceptance slow until one of its predictions—the bending of starlight in the strong gravitational field of the sun—was experimentally verified by a group of English physicists in 1919. After that, Einstein's reputation soared in academic circles and with the general public (for whom he became the perfect symbol of the absent-minded brilliant professor, whose theories, it was reputed, "only seven people in the world could understand"). In 1921, he was awarded the Nobel Prize in physics—not for relativity (!), but for his explanation of the photoelectric effect.

Einstein was noted for his warm, generous personality and his sly sense of humor. He was an accomplished musician, playing his violin or piano frequently. Mozart and Bach were his favorites. He had a dogged persistence in intellectual pursuits, repeatedly seeking simplicity and unity in describing nature. This fondness for simplicity, for eliminating all but essentials, was also evident in his personal life: in his clothes and in his behavior.

Unfortunately, political events—World War I, increasing nationalism, and the rise of the Nazis—had considerable impact on Einstein's life. His invited lectures in France and England were occasionally boycotted by some professors whose nationalistic feelings apparently overwhelmed their scientific interests. Being a Jew and a confirmed pacifist who refused to support the German war effort, Einstein became the target of Nazi anti-Semitism. His prestige protected him for a time, but in 1933 he decided to emigrate to the United States, settling after a few years at the Institute for Advanced Study at Princeton. He continued to work on a unified field theory in which he attempted (unsuccessfully) to combine gravitation and electromagnetism into a single theoretical structure.

In his later years, Einstein devoted much attention to pacifist ideas, the Zionist movement, world government, and similar social and political issues. He became a passionate and fearless spokesman for causes of human freedom. Some considered him naive, but all believed in his sincerity. He frequently was perplexed and saddened by the contradictions of people and politics. In 1939, concerned about the rising fury in Europe and aware of German research in uranium fission, he lent his name to a letter to President Roosevelt urging immediate investigation into the possibility of a nuclear bomb. After the war, in response to criticism in a Japanese journal reproaching him for this involvement, he wrote: "There are circumstances in which I believe the use of force is appropriate—namely, in the face of an enemy bent on destroying me and my people."

Einstein died in 1955. His most famous legacy—the truly brilliant insight of relativity—gives a new unity and clarity to our understanding of the universe.

when all aspects of special relativity are taken together, they form a most logical and impressive structure that agrees completely with experimental evidence.

EXAMPLE 13–1

A clock at rest in the S' frame gives a "tick" once each second. Thus, as measured in the S' frame, the time interval between ticks is $T_0 = 1$ s. If the S' frame has a velocity of $0.80c$ relative to the S frame, what is the time between ticks as determined in the S frame?

SOLUTION

Since $T_0 = 1$ s and $\beta = 0.80$, we have

$$T = \frac{T_0}{\sqrt{1-\beta^2}} = \frac{(1\text{ s})}{\sqrt{1-0.64}} = \frac{(1\text{ s})}{\sqrt{0.36}} = \boxed{1.67\text{ s}}$$

The moving clock thus runs slower than our own clocks at rest.

13.5 Comparison of Length Measurements Along the Direction of Motion

Consider a meter stick at rest in the S' frame, aligned along the direction of relative motion of the two frames, the x and x' axes. As measured in the S' frame, its length is L_0. But when measured from the S frame, the length of the moving stick is the distance between the ends of the stick when they are located *simultaneously* in the S frame. Relativity theory predicts that as measured in the S frame, the length L of the moving meter stick is

LENGTH CONTRACTION
$$L = L_0\sqrt{1-\beta^2} \qquad \text{(13–3)}$$
(where L_0 must be a measurement made in a frame in which the object is at rest)

Again, $\beta = V/c$. Because the factor $\sqrt{1-\beta^2}$ is always less than unity, $L < L_0$, and we conclude that *the length of a moving object along the direction of its motion is less than the length when measured at rest.* (Distances perpendicular to the direction of motion are unchanged.) The effect is called **length contraction.** As with time dilation, this is a symmetrical effect. Observers in S' would measure precisely the same contraction for lengths in the S frame that are parallel to the direction of motion. No contradiction is involved, since the measurements are made in two different frames. Each frame has its own scale of space, which does not necessarily agree with the space scales in other frames.

EXAMPLE 13–2

A meter stick moving with speed $0.60c$ is oriented parallel to the direction of motion. Find the length of this meter stick as measured by an observer at rest.

SOLUTION

Since $L_0 = 1$ m and $\beta = 0.60$, we have

$$L = L_0\sqrt{1-\beta^2} = (1\text{ m})\sqrt{1-0.36} = (1\text{ m})\sqrt{0.64} = \boxed{0.800\text{ m}}$$

Thus the moving meter stick is shorter than a meter stick at rest.

13.6 Relativistic Momentum

Relativity theory predicts that the momentum of a moving particle is greater than the classical value $\mathbf{p} = m\mathbf{v}$. In particular, the momentum does not increase linearly with velocity, but instead approaches infinity as the speed approaches c.

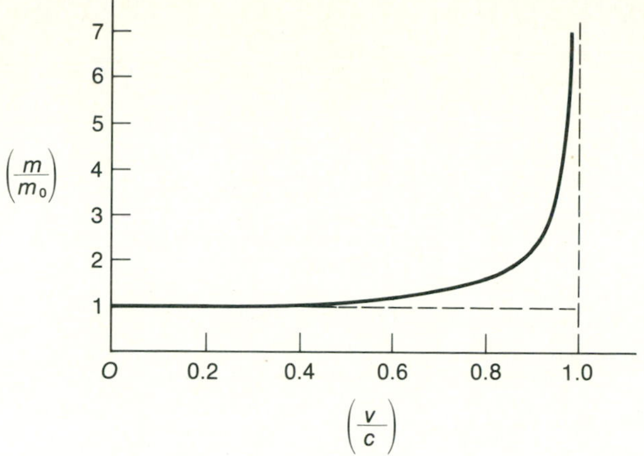

Figure 13–2
The relativistic mass m (in terms of m_0) as a function of the ratio of the speed v to the speed of light c.

RELATIVISTIC MOMENTUM

$$\mathbf{p} = \frac{m_0 \mathbf{v}}{\sqrt{1 - \beta^2}} \tag{13–4}$$

Here we have used the symbol m_0 to represent the **rest mass** of the particle, that is, its mass when measured at rest. For convenience, it was customary in the past to define a **relativistic mass** as follows:

RELATIVISTIC MASS *m*

$$m = \frac{m_0}{\sqrt{1 - \beta^2}} \tag{13–5}$$

With this notation, relativistic momentum is written simply as $\mathbf{p} = m\mathbf{v}$. Though convenient, this notation has led to some misunderstanding. It is a misconception to think there is an increase in the intrinsic mass property of the particle itself; rather, the factor $\sqrt{1 - \beta^2}$ arises because of the nature of space and time, which affects velocity measurements. It makes sense to consider the mass of a particle as a fundamental property that does not change. All of relativity can be expressed without defining a relativistic mass, and among modern authors there is a trend away from its use. However, it was widely used in the past and in some situations it simplifies the discussion, so it is included here.

Because of the relativistic mass increase, c is the upper limit to the velocity attainable by any particle that has a rest mass. Figure 13–2, a graph of Equation (13–5), shows the relationship for the increase as the speed approaches the velocity of light. As this occurs, the relativistic mass approaches infinity. Consequently, an increasingly larger force is required to further accelerate the particle. It would take an infinite amount of energy to achieve the speed c; thus it is impossible for a material particle to be accelerated to the speed c or to surpass it. The speed of light is truly an upper limit.[3]

There is convincing experimental evidence for this limiting velocity. The results of one experiment are given in Figure 13–3, showing a plot of the square of the speed of electrons versus the kinetic energy of the electrons. On the scale

[3] It has been proposed that particles called *tachyons*, which always travel faster than c, might exist. For them, the speed of light would be a lower limiting velocity. The existence of such particles is consistent with special relativity; approached from either side, c remains an impenetrable barrier. So far, experiments to detect them have been unsuccessful and they may not exist. For more information, see: G. Feinberg, "Particles That Go Faster Than Light," *Scientific American*, 223, No. 2 (Feb. 1970), p. 69.

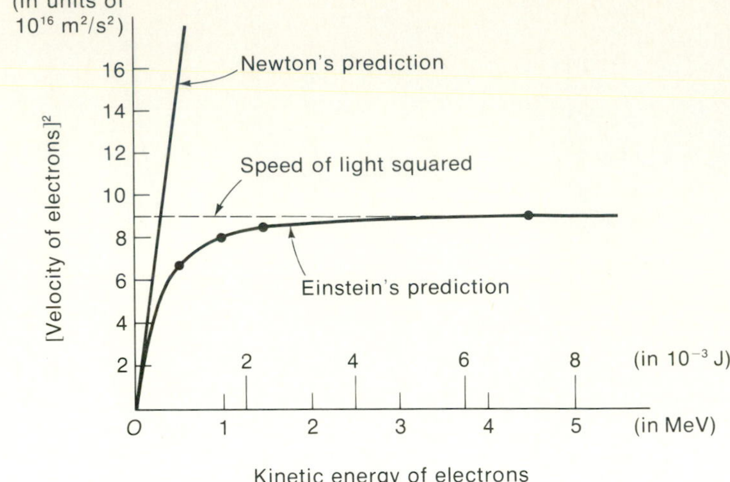

Figure 13–3
Experimental evidence for the speed of light as a limiting velocity for any material particle.

Figure 13–4
The Stanford two-mile linear accelerator for electrons (an interstate highway passes over the accelerator). The operation of the accelerator verifies special relativity. Electrons emerging from the accelerator differ from the speed of light by only about 5 parts in 10^{11}. If classical (Galilean) relativity were correct and the relativistic mass increase did not occur, the accelerator would need to be only a few inches long to achieve this speed.

of this graph, electrons emerging from Stanford's two-mile accelerator (Figure 13–4) would be plotted about 236 m to the right, but they still would be traveling at a velocity less than c.

EXAMPLE 13–3

Electrons emerging from Stanford's two-mile linear accelerator are traveling at 99.999 999 97% the speed of light. Find their momentum in terms of m_0c.

SOLUTION

Their momentum is not $m_0v \approx m_0c$ as classical theory predicts, but instead is given by Equation (13–4). Because β is extremely close to unity, we may use the approximation (see Appendix E):

$$1 - \beta^2 = (1 + \beta)(1 - \beta) \approx 2(1 - \beta)$$

For $\beta = 0.999\,999\,999\,7$, the factor $(1 - \beta)$ is equal to 3×10^{-10}. Hence, $\sqrt{1 - \beta^2} \approx \sqrt{2(1 - \beta)} = \sqrt{6 \times 10^{-10}}$, and we have

$$p = \frac{m_0v}{\sqrt{1 - \beta^2}} \approx \frac{m_0v}{\sqrt{6 \times 10^{-10}}} \approx \boxed{4.08 \times 10^4 m_0c}$$

This agrees with the experimentally measured momentum for the electrons when they are deflected by a magnetic field as they emerge from the accelerator. Although it is common to speak of these electrons as having a relativistic mass 4×10^4 greater than their rest mass, we emphasize again that this change occurs because of the unusual properties of space and time, not because of any peculiar changes in the mass itself.

EXAMPLE 13–4

A baseball moves at 30 m/s. By what fraction does its true relativistic momentum differ from the classical value of m_0v?

SOLUTION

Here the velocity is very small compared with c, so we use the following approximation, valid for $\beta^2 \ll 1$ (see Appendix E):

$$(1 \pm \beta^2)^n \approx 1 \pm n\beta^2 \qquad (\text{for } \beta \ll 1)$$

Therefore: $$\frac{1}{\sqrt{1-\beta^2}} = (1-\beta^2)^{-1/2} \approx 1 + \frac{\beta^2}{2}$$

The fraction we seek is:

$$\frac{\text{difference}}{m_0 v} = \frac{p - m_0 v}{m_0 v} = \frac{p}{m_0 v} - 1 = \frac{m_0 v(1-\beta^2)^{-1/2}}{m_0 v} - 1 \approx 1 + \frac{\beta^2}{2} - 1 \approx \frac{\beta^2}{2}$$

The numerical value of β is:

$$\beta = \frac{v}{c} = \frac{30 \text{ m/s}}{3 \times 10^8 \text{ m/s}} = 1 \times 10^{-7}$$

So: $$\frac{\beta^2}{2} = \frac{1 \times 10^{-14}}{2} = \boxed{5 \times 10^{-15}}$$

Thus the relativistic correction is negligible for speeds we usually encounter in everyday experience.

Note that for problems like Examples 13–3 and 13–4, which involve speeds of $v \ll c$ and $v \approx c$, the approximation formulas help to avoid awkward procedures such as directly calculating $\sqrt{1 - (0.999\,999\,999\,7)^2}$, an operation beyond the capability of most pocket calculators. If you find yourself tangled in such unwieldy operations, you have not made the appropriate approximations before substituting numerical values.

13.7 Relativistic Velocity Addition

Suppose an object has a speed v' along the $+x$ direction in the S' frame of reference. The S' frame itself has the speed V along the $+x$ direction relative to the S frame. Relativity theory predicts that the speed v of the object as measured in the S frame is

RELATIVISTIC VELOCITY ADDITION (for velocities along the $\pm x$ direction)

$$v = \frac{v' + V}{1 + \left(\frac{v'V}{c^2}\right)} \qquad \textbf{(13–6)}$$

For speeds much less than c, this expression reduces to the classical velocity addition relation $v = v' + V$. If any velocities are in the $-x$ (or $-x'$) direction, minus signs are used with the corresponding numerical values.

What happens if both of the velocities, v' and V, are close to the speed of light? Can this result in a velocity greater than c? No. The successive addition of *any number* of such velocities less than c, all in the same direction, still results in a final velocity less than c.

EXAMPLE 13–5

Suppose two stars, A and B, recede from the earth in opposite directions, with speeds as shown in Figure 13–5(a). Find the speed star B would have for observers on star A.

SOLUTION

In terms of the notation we have developed, star A is the S frame, while the earth (S' frame) is the moving frame ($V = 0.7c$), in which star B is observed to have the speed $v' = 0.8c$ relative to the earth (Figure 13–15b). Using the relativistic velocity addition

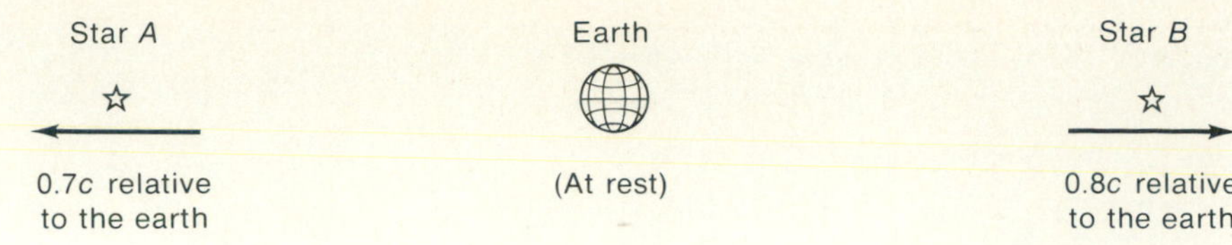

(a) As seen in the earth's frame of reference (the S' frame)

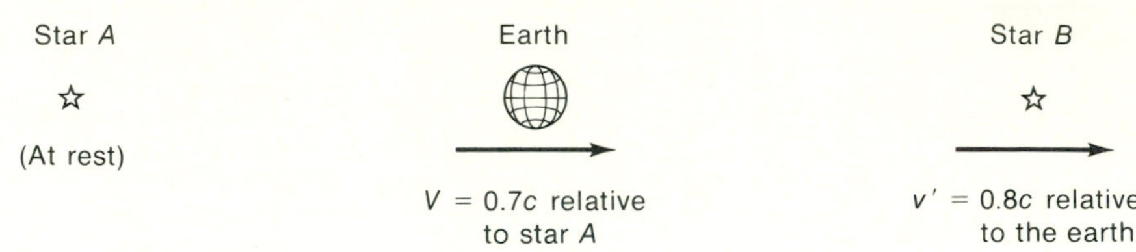

(b) As seen in star A's frame of reference (the S frame)

Figure 13–5
Example 13–5

formula, Equation (13–6), we have

$$v = \frac{v' + V}{1 + \left(\frac{v'V}{c^2}\right)} = \frac{(0.8c + 0.7c)}{1 + \left[\frac{(0.8)(0.7)c^2}{c^2}\right]} = \boxed{0.962c}$$

Note that this is less than the speed of light.

13.8 Relativistic Energy

In classical physics, we say a particle gains kinetic energy equal to the work done in increasing its speed from rest. In relativity, the same idea holds true. For work performed in a straight-line displacement of a particle of rest mass m_0, the kinetic energy K is

RELATIVISTIC KINETIC ENERGY

$$K = mc^2 - m_0c^2 \tag{13–7}$$

where m_0 is the rest mass and m is the relativistic mass (equal to $m_0/\sqrt{1-\beta^2}$). Thus kinetic energy is an increase in the term mc^2. Einstein took a bold leap and suggested that *all* relativistic mass represents energy, obtaining the famous expression that gave birth to the nuclear age:

TOTAL ENERGY

$$E = mc^2 \tag{13–8}$$

The **total energy** E of a particle is therefore the sum of its **rest energy** m_0c^2 and its **kinetic energy** K.

$$\underset{\text{Total energy}}{E} = \underset{\text{Rest energy}}{m_0c^2} + \underset{\text{Kinetic energy}}{K} \tag{13–9}$$

Equation (13–8) has been amply verified experimentally. The use of nuclear reactors to generate electrical power is one example. Because of the large numerical value of c, the energy equivalent to even a small mass is impressive.

EXAMPLE 13–6

A penny has a mass of about 3 g. Compute the energy that would be released if its rest mass were entirely converted into energy.

SOLUTION

$$E = mc^2 = (0.003 \text{ kg})(3 \times 10^8 \text{ m/s})^2 = \boxed{2.70 \times 10^{14} \text{ J}}$$

This is about equal to the maximum energy output of Hoover Dam for $2\frac{1}{2}$ days.

If we combine the definitions of E, K, and p, we may obtain the following useful relations:

$$E^2 = (m_0c^2)^2 + (pc)^2 \tag{13–10}$$

$$p = \frac{1}{c}\sqrt{K^2 + 2m_0c^2K} \tag{13–11}$$

$$v = \frac{pc^2}{E} \tag{13–12}$$

If the total energy E is much greater than the rest energy m_0c^2, the first term of Equation (13–10) may be neglected, giving the useful approximation:

$$E \approx pc \qquad (\text{for } E \gg m_0c^2) \tag{13–13}$$

In calculations involving fundamental particles, it is often convenient to express energies in terms of electron volts (eV) or mega-electron-volts (MeV), where 1 MeV = 1.602×10^{-13} J (see Table 13–1). Momentum is conveniently expressed in terms of MeV/c. *A semantics note:* When we say "a particle has an energy of 2 MeV," we mean its *kinetic* energy is 2 MeV.

TABLE 13–1

Rest-Mass Energies of Fundamental Particles

Particle	Symbol	m_0c^2 (in MeV)
Electron (or positron)	e or e^- (e^+)	0.511
Muon	$\mu^{\pm}$	105.659
Pi meson (neutral)	π^0	134.963
Pi meson (charged)	$\pi^{\pm}$	139.567
Atomic mass unit	u	931.502
Proton	p	938.280
Neutron	n	939.573
Deuteron	d or ^{2}H	1875.628
Alpha particle	α or ^{4}He	3727.411

EXAMPLE 13–7

Find (a) the momentum and (b) the speed of a proton whose kinetic energy equals its rest energy.

SOLUTION

(a) From Equation (13–11) we have

$$p = \frac{1}{c}\sqrt{K^2 + 2m_0c^2K} = \frac{K}{c}\sqrt{1 + 2\frac{m_0c^2}{K}}$$

For the case where $K = m_0c^2$, we obtain

$$p = \frac{m_0c^2}{c}\sqrt{1+2} = \frac{(938\ \text{MeV})}{c}\sqrt{3} = \boxed{1625\ \frac{\text{MeV}}{c}}$$

(b) For $K = m_0c^2$, we have $E = m_0c^2 + K = 2m_0c^2$. Thus, from Equation (13–12):

$$v = \frac{pc^2}{E} = \frac{\left(1625\ \frac{\text{MeV}}{c}\right)(c^2)}{(2)(938\ \text{MeV})} = \boxed{0.866c} \quad \text{or} \quad \boxed{2.60 \times 10^8\ \frac{\text{m}}{\text{s}}}$$

EXAMPLE 13–8

Find (a) the total energy E, (b) the kinetic energy K, and (c) the momentum p of an electron moving with speed $v = 0.6c$.

SOLUTION

(a)
$$E = mc^2 = \frac{m_0c^2}{\sqrt{1-\beta^2}}$$

Since $\sqrt{1-\beta^2} = \sqrt{1-(0.6)^2} = 0.8$, and $m_0c^2 = 0.511$ MeV, we have

$$E = \frac{0.511\ \text{MeV}}{0.8} = \boxed{0.639\ \text{MeV}}$$

(b) The kinetic energy is

$$K = E - m_0c^2 = 0.639\ \text{MeV} - 0.511\ \text{MeV} = \boxed{0.128\ \text{MeV}}$$

(c) The momentum is $p = m_0v/\sqrt{1-\beta^2}$. Multiplying numerator and denominator by c^2, we have

$$p = \frac{m_0c^2v}{\sqrt{1-\beta^2}c^2} = \frac{(0.511\ \text{MeV})(0.6c)}{(0.8)(c^2)} = \boxed{0.383\ \frac{\text{MeV}}{c}}$$

13.9 Binding Energy

By comparing the rest mass of a nucleus with the rest masses of the individual particles that make up the nucleus, we find that in general the mass of a nucleus is less than the sum of the masses of its parts. If an amount of rest mass Δm_0 disappears when particles combine to form a system such as a nucleus, the equivalent amount of energy, $\Delta E = \Delta m_0c^2$, is called the **binding energy** of the particles in the nucleus. To break the nucleus into its individual parts, this amount of energy must be supplied from some external source, such as by bombarding the nucleus with another particle.

EXAMPLE 13–9

A deuteron is composed of a neutron and a proton bound together. Referring to Table 13–1, calculate how much energy would be required to break up the deuteron into a proton and a neutron.

SOLUTION

The combined rest energies of a proton and a neutron are 938.280 MeV + 939.573 MeV = 1877.853 MeV. The rest energy of a deuteron, 1875.628 MeV, is subtracted from this to yield 2.22 MeV, the binding energy of the deuteron.

13.10 The Twin Paradox

The so-called *twin paradox* has generated more violent controversy than any other topic in relativity.[4] Briefly stated, the paradox is as follows. Two twins live on the earth. One decides to take a relativistic trip to a distant star and return. According to relativity, upon his return the traveling twin will be younger than the brother who remained on earth. The paradox arises when one asks why the traveling twin cannot claim that in his frame of reference, his earth brother moved away from him and returned, and thus the earth twin (not the traveling twin) would be younger upon their reunion. After all, does not relativity tell us that absolute motion is a fiction? Cannot either twin be considered the stationary one and thus the situation be symmetrical? No. Because the traveling twin must accelerate in some fashion to change his velocity for the return trip, acceleration is involved only with the traveling twin's frame of reference. Acceleration is an absolute, not a relative, matter, so the situation is not a symmetrical one. The consequences are laborious to straighten out, but the conclusion is inescapable: The traveling twin really would be younger upon his return compared with the twin who stayed home.

Several experiments using radioactive nuclei confirm this result. The first direct experiment using macroscopic clocks was made in 1971, when four precise cesium clocks were flown on commerical jet flights around the world, once eastward and once westward.[5] The results confirm the twin paradox effect. It does seem odd that two clocks, initially synchronized, *both of which always show the proper time*, will disagree after being separated and then brought together in this manner. Nevertheless, this is the essence of the twin paradox. It is merely a consequence of the fact that there is no absolute time.

As a final comment, a startling example of the twin paradox is a hypothetical straight-line trip in which travelers on a spaceship undergo constant acceleration g throughout, accelerating the first half of the outward journey, decelerating the second half, and coming to rest at the destination. The return

---

[4] An excellent source of information about relativity is *Resource Letter SRT*-1 (*Selected Reprints: Special Relativity Theory*), published by the American Institute of Physics, 335 East 45th St., New York, NY 10017. For an interesting discussion of the historical origins of relativity, see G. Holton, *American Journal of Physics* 28, 627 (1960). A comprehensive discussion of the twin paradox is L. Marder, *Time and the Space Traveller*, University of Pennsylvania Press, 1971.

[5] See two consecutive articles: J. C. Hafele and R. E. Keating, "Around-the-World Atomic Clocks," *Science*, 177, 14 July 1972, pp. 166–70.

trip is made in a similar fashion. Such constant acceleration of g would be comfortable for the travelers, since it simulates earth-gravity conditions. For a round trip to Andromeda galaxy, 2 million light-years away, the elapsed time in the spaceship would be only 59 years. Yet the earth would be more than 4 million years older upon the travelers' return. For a similar round trip lasting 78 years in the spaceship's frame, it would be possible to reach a destination 500 million light-years away, returning to find the earth more than one billion years older. Such trips are essentially impossible, however, because of practical engineering difficulties (not because of any limitations in the laws of nature).[6]

EXAMPLE 13–10

A space traveler wishes to journey to a star 20 light-years away and return. Assuming she travels at a constant speed $v = 0.80c$ (with negligible times for acceleration and turn-around), how much younger is she upon her return than her twin sister, who stayed at home? (Note: By expressing the unit light-years as $c\cdot$yr, the factor c will often cancel in equations.)

SOLUTION

In the *earth's* frame, the time for a round trip at constant speed is

$$t_{\text{earth}} = \frac{\text{distance}}{\text{speed}} = \frac{40\ c\cdot\text{yr}}{0.80c} = 50\ \text{yr}$$

The proper round-trip distance L_0 (measured in the earth's frame) is 40 $c\cdot$yr. In the traveler's frame, this distance is shorter because of length contraction:

$$L = L_0\sqrt{1-\beta^2} = (40\ c\cdot\text{yr})\sqrt{1-(0.80)^2} = 24\ c\cdot\text{yr}$$

Therefore, in the space traveler's frame the time it takes to cover the round-trip distance is

$$t_{\text{traveler}} = \frac{\text{distance}}{\text{speed}} = \frac{24\ c\cdot\text{yr}}{0.80c} = 30\ \text{yr}$$

So, upon her return, the twin is (50 − 30) yr = **20 years younger** than her sister.

In this chapter we have focused on those aspects of special relativity that involve space and time, energy and momentum. The major significance of relativity lies in its application to atomic and nuclear physics, electric and magnetic fields, as well as astrophysics and cosmology. This brief introduction should whet your appetite for further study of this fascinating subject. Relativity theory is surely one of the towering achievements of the human mind.

Summary

Special relativity compares measurements of a given phenomenon made in two different frames of reference (S and S'), which have uniform relative velocity $\mathbf{V}$ with respect to each other. The frames are aligned so that their origins and

[6] For an interesting discussion of space travel, see E. M. McMillan, "The 'Clock Paradox' and Space Travel," *Science* 126, pp. 381–384 (1957), and S. von Hoerner, "The General Limits of Space Travel," *Science* 137, pp. 18–23 (1962).

respective axes are coincident at $t = t' = 0$, and S' moves in the $+x$ direction relative to S. For convenience, we define $\beta = V/c$.

A point event (x,y,z,t) in the S frame
(x',y',z',t') in the S' frame

Each event is to be measured by a *local* observer, situated where the event occurs and equipped with a clock that has been synchronized with other clocks in the frame.

Postulates of special relativity

(1) *All the laws of physics have the same form in all inertial frames* (the Principle of Relativity).

(2) *The speed of light in a vacuum has the same value* c *in all inertial frames* (the Principle of the Constancy of the Speed of Light).

Time dilation $$T = \frac{T_0}{\sqrt{1-\beta^2}}$$ (where T_0 must be a time interval measured by a *single* clock)

Length contraction $$L = L_0\sqrt{1-\beta^2}$$ (where L_0 must be a measurement made in a frame in which the object is at rest)

Relativistic momentum $$\mathbf{p} = \frac{m_0\mathbf{v}}{\sqrt{1-\beta^2}}$$

Relativistic mass $$m = \frac{m_0}{\sqrt{1-\beta^2}}$$ (where m_0 is the rest mass)

Relativistic velocity addition (for velocities along the $\pm x$ direction) $$v = \frac{v' + V}{1 + \left(\dfrac{v'V}{c^2}\right)}$$

Relativistic energy

Total energy $E = mc^2$ or $E = m_0c^2 + K$

Kinetic energy $K = mc^2 - m_0c^2$

Other energy relations $E^2 = (m_0c^2)^2 + (pc)^2$

$$p = \frac{1}{c}\sqrt{K^2 + 2m_0c^2K}$$

$$v = \frac{pc^2}{E}$$

If $E \gg m_0c^2$, then $E \approx pc$.

Mathematical approximations

When $\beta^2 \ll 1$ $$\frac{1}{\sqrt{1-\beta^2}} \approx 1 + \frac{\beta^2}{2}$$

$$\sqrt{1-\beta^2} \approx 1 - \frac{\beta^2}{2}$$

When $\beta^2 \approx 1$ $1 - \beta^2 = (1+\beta)(1-\beta) \approx 2(1-\beta)$
In these cases, it is often useful to solve for the quantity $1 - \beta$, the amount by which β differs from 1.

Binding energy

If an amount of rest mass Δm_0 disappears when particles combine, the equivalent energy $\Delta E = \Delta m_0 c^2$ is called the binding energy of the particles in the system.

Twin paradox

If one twin goes on a relativistic round-trip journey, that twin will be younger upon returning than the twin who remained at home.

Questions

1. What were Galileo's contributions to special relativity?
2. Explain how it is possible for the moving spot on an oscilloscope screen to move across the screen faster than the speed of light without violating relativity.
3. Discuss what life would be like if the speed of light were, say, 100 km/h.
4. List several quantities whose measured values would be different in two inertial frames in relative motion. Other than the speed of light, what quantities would have the same values in these two frames?
5. Under what circumstances would you be older than your parents?
6. Interestingly, there is nothing in special relativity that forbids speeds faster than c as long as such particles *always* travel faster than c. Approaching the speed of light from either side, the speed c seems to be an effective barrier that cannot be "penetrated" from either direction. It is proposed that particles that always travel faster than c be called "tachyons," after the Greek word *tachos*, meaning "speed." Experiments have been performed to detect them, without success. What might be some properties of tachyons? Could they have a rest mass? What would be some consequences for fundamental ideas about causality? (See Bilaniuk and Sudarshan, "Particles Beyond the Light Barrier," *Physics Today*, May 1969, p. 43.)
7. Explain why it has been suggested that the "theory of relativity" could equally well be called the "theory of absolutism."
8. In a famous science fiction story, aliens kidnap several people and take them away in a spaceship. One person remarks, "We are traveling at the speed of light—look at your watches." Someone does and exclaims, "My God! My watch has stopped!" What is the blunder the author made in writing this incident?

Problems

13A–1 According to his wristwatch, an astronaut takes 2 min to eat a chocolate bar.
(a) If the astronaut is traveling with a speed of $0.5c$ relative to the earth, determine the amount of time that elapses in the earth's frame of reference during this time interval.
(b) Find the distance the spaceship travels during this time.

Answers: (a) 2.31 min (b) 1.16 $c\cdot$min

13A–2 A meter stick, oriented parallel to the direction of motion, and a 1-kg object are on board a spaceship that has a speed $v = 0.6c$ relative to the earth. Find (a) the length of the meter stick and (b) the mass of the object as measured in the earth's frame of reference. (c) If it takes an astronaut 6 h to do her physics homework, calculate the time it takes her as measured in the earth's frame of reference. (d) According to observers on earth, how far (in $c\cdot$hr) does the spaceship travel during this time?

13A–3 Alpha Centauri is a star about 4 light-years away. A rocketship travels at constant speed from the earth to this star in one day as measured by the rocketship's occupants.

(a) Find the speed of the rocketship relative to the earth. Express your answer as the amount by which β differs from 1. [Hint: Because β is so nearly equal to 1, use the convenient approximation $1 - \beta^2 = (1 + \beta)(1 - \beta) \approx 2(1 - \beta)$.]

(b) In the rocketship's frame, how far away is the star at the beginning of the trip?

Answers: (a) $1 - \beta \approx 2.35 \times 10^{-7}$ (b) One c·day

13A–4 A certain quasar recedes from the earth with a speed $v = 0.87c$. A jet of material is ejected from the quasar toward the earth with a speed of $0.55c$ relative to the quasar. Find the speed of the ejected material relative to the earth.

13A–5 An astronomer observes that two distant galaxies are traveling away from the earth in opposite directions, each with speed $v = 0.9c$. What would an observer on one galaxy measure for the speed of the other galaxy? *Answer:* $v_x = 0.994c$

13A–6 It is estimated that the total energy input to the U.S. economy in the year 1975 was about 8×10^{19} J. Assuming all this energy came from nuclear reactions in which mass is converted to energy according to $E = mc^2$, determine the total mass annihilation that would be involved.

13A–7 Determine an object's speed if its kinetic energy equals its rest energy.

Answer: $v = 0.866c$

13A–8 The rest energy of a tritium nucleus, ^{3}He (two protons and a neutron), is 2808.413 MeV. Find the energy required to remove one proton, resulting in a deuteron, ^{2}H, plus the proton.

13B–1 Electrons emerging from Stanford's linear electron accelerator differ from the speed of light by about 5 parts in 10^{11}. Find this difference in centimeters/second.

Answer: 1.5 cm/s

13B–2 In 1849, H. L. Fizeau experimentally determined the speed of light by sending a light beam through the slots of a rotating toothed wheel to a distant mirror 8633 m away. Upon return of the reflected light pulses, if the rotation speed of the wheel was just right the light pulses could again pass through the slot openings between the teeth and thus be seen by the experimenter. On the other hand, with a different rotation speed the teeth interrupted the return light pulses, so no light was observed (see Figure 13–6). Thus, as the wheel was speeded up, the observer would see a gradual progression from brightness to darkness to brightness, and so on, depending on whether the return pulses met a tooth or a slot on the rotating wheel. Fizeau reported that as the wheel was speeded up, the first "eclipse" of the return pulses occurred when the speed of rotation was 12.6 rev/s. The wheel had 720 teeth and 720 slots, all of the same width. Using these data, find the speed for light that Fizeau must have calculated. (His value was somewhat larger than more accurate determinations made later.)

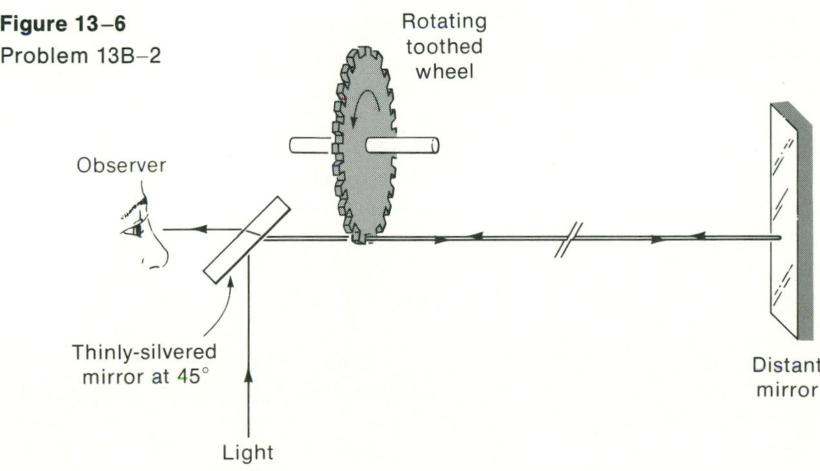

Figure 13–6
Problem 13B–2

13B–3 A beam of π^+ pions has a speed of $0.7c$. When at rest, the pions have an average lifetime of 2.6×10^{-8} s before disintegrating.

(a) In the laboratory frame of reference, how long, on the average, will the moving pions live before disintegrating?

(b) On the average, how far will they travel through the laboratory in this time?

Answers: (a) 3.64×10^{-8} s (b) 7.65 m

13B–4 An astronaut wishes to visit the Andromeda galaxy (2 million light-years away) in a one-way trip that will take 30 yr in the spaceship's frame of reference. Assuming his speed is constant, how fast must he travel relative to the earth? Express your answer as the amount by which β differs from 1.

13B–5 The half-life of a given sample of radioactive particles is the time it takes for half the initial number of particles to undergo a disintegration. A group of radioactive particles moving at a speed of $0.8c$ travels through the laboratory a distance of 30 m. Half the particles survive the trip. Find the half-life of the particles in their own frame of reference.

Answer: 22.5 m/c or 7.5×10^{-8} s

13B–6 A particle of rest mass M_0 moving at $v_1 = 0.6c$ collides head-on and sticks to another particle of rest mass m_0 moving at $v_2 = 0.8c$ in the opposite direction. After the collision, the combined mass is at rest with respect to the laboratory. Find the ratio of the rest masses: M_0/m_0.

13B–7 A certain type of meson decays at rest into two equal-mass particles, which are ejected in opposite directions with speeds of $0.8c$. Suppose the meson is traveling through the laboratory with a speed $v = 0.6c$ when the decay particles are emitted along the line of motion (in opposite directions). Find the speeds of the two decay particles as measured in the laboratory frame. *Answers:* $0.946c$ and $-0.385c$

13B–8 At normal incidence at the top of the earth's atmosphere, the energy from the sun per unit time is about 1340 W/m^2. From this information, estimate the mass loss per second ($E = mc^2$) of the sun.

13B–9 A free neutron will decay into a proton, an electron, and a massless particle called an antineutrino. From the difference between the mass of the neutron and the masses of the decay particles, calculate the total kinetic energy (in joules) the decay particles would have if the neutron were initially at rest. *Answer:* 1.25×10^{-13} J

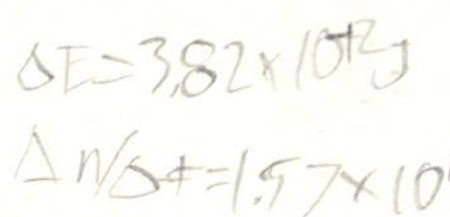

13B–10 The fusion of two deuterons (^{2}H) will form an alpha particle (^{4}He).

(a) Calculate the energy released in this reaction due to the decrease in rest mass.

(b) How many such reactions must occur each second to light a 60-W lightbulb?

13C–1 If you travel on a jet plane from Los Angeles to New York (4000 km air distance), at an average speed of 1000 km/h, how much younger are you on arrival than you would have been had you remained in Los Angeles during the time it took the plane to make the journey? (Hint: Since $v \ll c$ and thus $\beta^2 \ll 1$, use an approximation for $\sqrt{1 - \beta^2}$. Also note that T, the time that would have been spent in Los Angeles, is extremely close to T_0, the time spent on the plane.) *Answer:* 6.17 ns

13C–2 A spaceship of proper length L travels past the earth with a speed $v = (4/5)c$. When a clock at the tail of the spaceship reads $t' = 0$ (and when earth clocks also read $t = 0$), a light signal is sent from the tail to the front of the spaceship. Determine the time at which the signal reaches the front end of the ship (a) according to spaceship clocks and (b) according to earth clocks. (c) The answers to parts (a) and (b) are *not* related according to the time dilation formula. Why not? (d) The light signal is reflected by a mirror at the front end back toward the rear. Find the time at which it reaches the rear according to rocket clocks. (e) Find the time in (d) according to earth clocks. (f) Are the answers to parts (d) and (e) related according to the time dilation formula? Explain why or why not.

13C–3 At exactly noon in our frame of reference, a clock moving with speed $v = 0.8c$ reads 12:00 (noon) as it passes the origin of our frame.

(a) How far away will it be when its hands indicate 1 s after 12:00? (Leave the symbol c in the answer.)

(b) When the clock face reads 1 s after 12:00, a light signal is sent from the clock back toward the origin of our frame of reference. At what time (in our frame) does this signal arrive at our origin? *Answers:* (a) 1.33 c·s (b) 3.00 s

13C–4 Electrons in Stanford's 10 000-ft linear accelerator attain a final velocity of $(0.999\,999\,999\,7)c$.

(a) In a frame of reference moving at this speed, how long is the accelerator? (Use the appropriate mathematical approximation.)

(b) Traveling at this (constant) speed, how long would it take to travel this distance in the frame of reference of an electron?

(c) How long would the journey take as measured by a Stanford physicist?

13C–5 Imagine that a runner carries a mirror 1 m (in the runner's frame of reference) in front of her face to observe her own reflection as she runs (see Figure 13–7). Her speed is $0.6c$ relative to the earth. She blinks.

(a) In the runner's frame of reference, how much time after she blinks will she see the blink of her mirror image?

(b) In the earth's frame of reference, what is this time interval? Leave the symbol c in the answers. *Answers:* (a) $2.00 \text{ m}/c$ (b) $2.50 \text{ m}/c$

Figure 13–7
Problem 13C–5

13C–6 A golf ball travels with a speed of 90 m/s. By what fraction does its relativistic mass differ from its rest mass m_0? That is, find the ratio $(m - m_0)/m_0$.

13C–7 The total energy E of a proton from a high-energy accelerator is 5 times its rest energy E_0 (equal to m_0c^2). In terms of its rest energy E_0, find (a) its kinetic energy K and (b) its momentum p. (c) Find the value of β (equal to v/c). When appropriate, leave the symbol c in the answer. *Answers:* (a) $K = 4E_0$ (b) $p = \sqrt{24}E_0/c$ (c) $\beta = \sqrt{24/25}$

13C–8

(a) According to Newtonian mechanics, determine the amount of work required to accelerate an electron from rest to $0.80c$.

(b) Find the work required according to special relativity. Express your answers in terms of m_0c^2.

13C–9 A high-energy proton has a speed of approach v relative to a proton at rest on the earth. Find the speed V relative to the earth of a frame of reference in which the two protons have equal speeds.

Answer: $V = v\left(\dfrac{1 - \sqrt{1 - \beta^2}}{\beta^2}\right)$ (where $\beta \equiv v/c$)

13C–10 One way of expressing the relativistic mass increase is the fraction f by which the relativistic mass exceeds its rest mass: that is, $f \equiv (m - m_0)/m_0$. Derive the following expression for the speed ratio $\beta = v/c$ in terms of f: $\beta = \sqrt{f(f + 2)}/(f + 1)$.

chapter 14 Gravitation

Gravitation, n. The tendency of all bodies to approach one another with a strength proportioned to the quantity of matter they contain—the quantity of matter they contain being ascertained by the strength of their tendency to approach one another.

AMBROSE BIERCE
(*The Enlarged Devil's Dictionary*)

I am sitting here 93 million miles from the sun on a rounded rock which is spinning at the rate of 1000 miles an hour. . . and my head pointing down into space with nothing between me and infinity but something called gravity which I can't even understand, and which you can't even buy any place so as to have some stored away for a gravityless day. . . .

RUSSELL BAKER
[*American Journal of Physics 43,* 704 (1975)]

14.1 Introduction

What is gravity? This question has puzzled scientists from the time of the early Greeks. Controversy still exists. The most accepted theory about gravity—Einstein's general relativity—agrees with the few experimental tests that have been made so far, though it would be reassuring to have further evidence. Several additional experiments to test general relativity are currently in preparation.[1]

Until the seventeenth century, the most widespread opinion as to why objects fall toward the earth was that this was the "natural" tendency of such objects. However, since the tendency was obviously not followed by the sun, "heavenly" bodies were thought to obey different laws of motion. This double standard in thinking about celestial and terrestrial motions persisted for about 2000 years, from the time of Aristotle ($\sim$340 B.C.). These ideas gave to the earth a unique position in the universe. We were *the* center, and our physical laws were different from those of the rest of the universe.

Such ideas die hard. Although now we accept easily the insignificant place of the earth in the grand scheme of things, giving up the abode of humans as the center of the universe was a painful experience. In 1600, the Italian philosopher Bruno was burned at the stake for his belief (among other heretical ideas) in the Copernican view that the sun really was the center of a "solar system." Later, in 1633, Galileo was tried and imprisoned for similar thoughts.

Newton's theory of gravitation grew out of his attempts to understand the motion of the moon and planets. The groundwork had been laid by Copernicus, Kepler, Tycho Brahe, and others, who described celestial motions with great precision and fitted them into a relatively simple model wherein the planets revolved around the sun in elliptical orbits. But why did planets move in this fashion? What force would produce Kepler's elliptical paths? Newton's great insight took the bold step of searching for an explanation of heavenly motions

[1] A few alternative theories of gravitation have recently been proposed. While agreeing with the main features of Einstein's theory, these new proposals do differ in certain details. Unfortunately, experiments sensitive enough to distinguish between these newer theories and Einstein's general relativity are extremely difficult to devise and carry out. So far, all astronomical data agree (within experimental limits of error) with Einstein's theory.

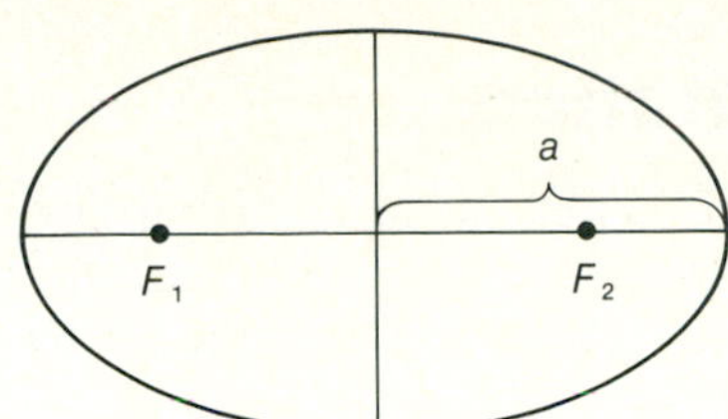

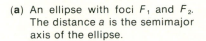

(a) An ellipse with foci F_1 and F_2. The distance a is the semimajor axis of the ellipse.

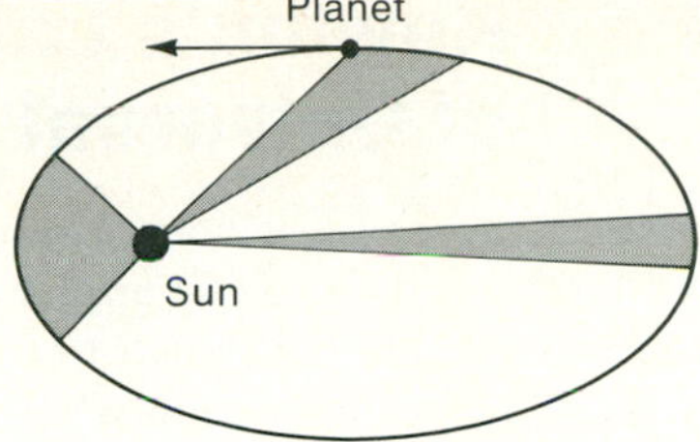

(b) Kepler's second law: As a planet moves about the sun, at all parts of its orbit the radius vector from the sun to the planet sweeps out equal areas in equal times. A planet thus travels faster when it is closer to the sun.

Figure 14–1
Planets move about the sun in elliptical orbits with the sun at one focus of the ellipse. In these sketches, the elongation of the ellipse is greatly exaggerated to clarify the equal-area relationship; actual planetary orbits are much closer to circles.

in terms of physical laws that had been derived from observations on the earth. Let us briefly review this story.

14.2 Kepler's Laws

By the end of the sixteenth century, the observation of astronomical bodies had progressed to a remarkable degree. Tycho Brahe, a Danish nobleman, built the first large observatory—a truly phenomenal installation—on an island near Copenhagen. Although the telescope had not yet been invented, Tycho painstakingly recorded precise observations of the motions of the planets over a period of several years, using mechanical gadgetry and the unaided eye. After Tycho's death, his assistant, Johann Kepler, studied these data and devised three simple laws that described the motions of the planets with much greater accuracy than any previous explanations had done. Instead of traveling in circles about the sun as Copernicus believed (following the Greeks), the planets, according to Kepler, travel in ellipses (see Figure 14–1). Kepler's three laws are as follows:

KEPLER'S LAWS OF PLANETARY MOTIONS

(1) Each planet moves along an elliptical path with the sun at one focus.

(2) The radius vector from the sun to a planet sweeps out equal areas in equal intervals of time.

(3) The square of the period T of a planet's orbital motion is proportional to the cube of its mean distance[2] from the sun R. That is, T^2/R^3 is a constant for all planets.

Kepler also verified that these laws could be applied to another planetary system, the moons of Jupiter. Thus his theory seemed to be a general one.

But the reason behind these regularities was unknown. The concept of "force" was still not well established, although Galileo provided a first, important step toward a clear idea of force when he proposed it as the cause of changes in motion. Unquestionably, Newton made the significant giant strides when he published the *Principia*, which formulated his laws of motion and theory of gravitation.

[2] The derivation of Kepler's third law from Newton's laws of mechanics and gravitation reveals the "mean distance" as the length of the semimajor axis of elliptical orbit (see Figure 14–1a).

14.3 Newton's Law of Universal Gravitation

Newton's impressive accomplishment was his statement of a general law of gravitation, valid for falling apples as well as for holding the moon and planets in their orbits. He concluded that between any two particles in the universe, there is a force of attraction whose magnitude is

$$F = G\frac{m_1 m_2}{r^2} \tag{14–1}$$

where m_1 and m_2 are the respective masses of the two particles, r is the distance between the particles, and G is the universal gravitational constant, experimentally determined to be $G = 6.672 \times 10^{-11}\ \mathrm{N \cdot m^2/kg^2}$.

In vector form, the force that m_1 exerts on m_2 is

$$\mathbf{F}_{1\text{ on }2} = -G\frac{m_1 m_2}{r^2}\hat{\mathbf{r}}_{1\text{ to }2} \tag{14–2}$$

where $\hat{\mathbf{r}}_{1\text{ to }2}$ is the unit vector from m_1 to m_2. The minus sign makes it a force of attraction.

Some features of Newton's law of gravitation should be emphasized. First of all, it does not answer the fundamental questions of what gravity *really* is or why such forces exist. But it does identify forces that explain the way certain objects move and express concisely the characteristics of these forces. They always come in pairs; that is, two particles mutually exert gravitational forces of attraction on each other, equal in magnitude, opposite in direction, and acting on two different objects—a third-law pair of forces. Such pairs of forces occur between every two particles in the universe, independent of the presence of other objects. If several masses are present, the superposition principle applies: The total gravitational force on any one particle is found by adding all the individual forces on it as vectors.

Unlike the forces between two electrical charges, which may be either attractive or repulsive, forces of gravitation are always attractive. Another difference between gravitational forces and electric or magnetic forces is that there is no known shield or screen effective against gravitational forces, whereas a suitable metal plate or screen will inhibit the electrical force between two charges or a soft iron plate of sufficient thickness will cut off magnetic forces between two magnets.

It is interesting, however, that the equation describing the electric force between two point charges q_1 and q_2:

$$F = k\frac{q_1 q_2}{r^2} \qquad \begin{pmatrix}\text{In vector form:} \\ \mathbf{F}_{1\text{ on }2} = k\dfrac{q_1 q_2}{r^2}\hat{\mathbf{r}}_{1\text{ to }2}\end{pmatrix} \tag{14–3}$$

has the same inverse-square-law form as the law of gravitation (though the electric force may be either attractive or repulsive, depending on the sign of the charges). So it is tempting to hunt for a single "unified field theory" that will describe both gravitation and electric forces. But so far all such attempts have failed.

14.4 A Note About Newton

Newton was 23 years old when he developed his theory of gravitation. A student at Cambridge University, he left during the plague of 1665–1666 to live and study in isolation in his home in the country. Although his health was somewhat frail, this was a very productive time for Newton. A falling apple in his garden reputedly caused Newton to speculate that perhaps the same force that accelerated the apple also governed the moon's motion.

But a major obstacle still remained. If, indeed, there was a force of attraction between two *particles*, what was the force between the apple and an *extended* mass such as the earth? After all, nearby mass presumably would attract the apple with more force than equal mass on the other side of the earth. To solve this preliminary problem, Newton invented the calculus,[3] another major achievement of his unquestioned genius.

In 1684, a friend, Edmund Halley (of Halley's comet fame), was trying to determine what type of force would cause elliptical orbits according to Kepler's Laws. Newton remarked that he had solved the problem years ago; it was an **inverse-square-law** force, $F \propto 1/r^2$. Halley persuaded Newton to publish these and other results, and after two years of intense work, Newton completed his *Philosophiae Naturalis Principia Mathematica.* Its publication in 1687 established Newton as one of the greatest scientific geniuses of all time. In the *Principia*, Newton swept away 2000 years of the two separate theories—celestial and terrestrial—of Aristotelian teaching. Newton's theory explained Kepler's three laws as just another example of the universal mechanics, valid everywhere. His other striking accomplishments in the *Principia*—in mathematics, optics, and astronomy—reinforced great faith in the human ability to understand and control natural phenomena.

14.5 The Gravitational Force Between a Particle and an Extended Mass

In developing the calculus to determine the force between a particle and an extended mass, Newton came to two important conclusions. Provided the large mass has spherical symmetry,[4] then:

(a) For external points, the large mass may be considered as if its entire mass were concentrated at its center.

(b) If the particle is anywhere inside a uniform, hollow, spherical shell of mass, there is no net gravitational force on the particle due to the shell.

We shall now derive these two results. The integration is somewhat involved, but if you follow it a step at a time it will provide an instructive example of how integrals are set up.

Consider first the gravitational attraction between a slender ring of total mass m' and a particle of mass m located on the ring's axis (see Figure 14–2). An element of the ring's mass dm, located a distance s from the particle, exerts a force $d\mathbf{F}$ on m of magnitude

$$dF = G\frac{m\,dm}{s^2} \qquad \textbf{(14–4)}$$

Now, consider all the elements of the ring. Because of symmetry, the force components that are *perpendicular* to the x axis add up to zero. Thus, we need to consider only the components parallel to the axis. As we add the contributions

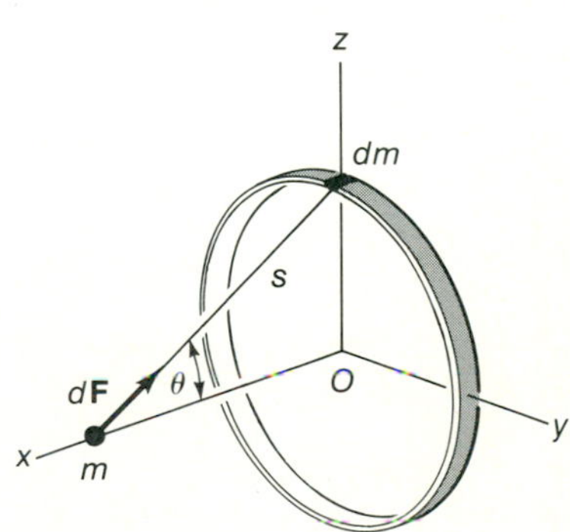

Figure 14–2
The thin ring of total mass m' lies in the y-z plane. The particle is on the axis of the ring.

[3] At the time, there were other ideas of calculus "in the air." Newton's mathematics teacher, Isaac Barrow, had already proved a theorem equivalent to the fundamental theorem of calculus. Gottfried Wilhelm von Leibnitz, a German philosopher and mathematician, developed the calculus independently nine years later, with a notation closer to today's usage.

[4] Spherical symmetry means that although the density may vary with the radius, it does not depend on the particular angular direction away from the origin. The density thus is a function of r alone: $\rho = f(r)$.

of all the elements around the ring, the $\cos\theta$ remains constant during the summation and thus comes out from under the integral sign. Hence, the resultant force along the x axis F_x is the sum:

$$F_x = \int \cos\theta\, dF = \int G\frac{m\cos\theta\, dm}{s^2} = G\frac{m\cos\theta}{s^2}\underbrace{\int dm}_{\substack{\parallel \\ m'}} = G\frac{m'm\cos\theta}{s^2} \quad \textbf{(14–5)}$$

where the integration is over the entire ring. (Note that we *cannot* assume all the mass of the ring to be concentrated at its center for calculating the gravitational force on the particle. The simplification works only for masses that have spherical symmetry.)

Now consider the gravitational attraction between a particle and a thin-walled spherical shell, shown in Figure 14–3. The total mass of the shell is M, and its wall thickness is negligible compared with the radius R. We choose an element of mass dm on the sphere such that all the mass of the element is the same distance from the particle.

The element dm (shown shaded) is a circular strip cut from the spherical shell. By analogy with Equation (14–5), the force of dF on m is:

$$dF = G\frac{m\cos\theta\, dm}{s^2} \quad \textbf{(14–6)}$$

However, we must express the element dm in terms of another parameter that is more convenient for integrating over the entire sphere.

The radius of the circular-strip element is $R\sin\phi$, and the area of the shell's surface that makes up the strip is

$$\text{Area of circular strip} = (2\pi R\sin\phi)(R\,d\phi) \quad \textbf{(14–7)}$$

Thus, if we designate the *mass per unit area* of the shell as σ, the mass dm of the strip is

$$dm = (\text{area})\left(\frac{\text{mass}}{\text{unit area}}\right) = (2\pi R\sin\phi)(R\,d\phi)(\sigma) = 2\pi\sigma R^2\sin\phi\, d\phi \quad \textbf{(14–8)}$$

To set up the integral for the entire sphere, we express convenient relations between the variables s and ϕ using the law of cosines:

$$s^2 = R^2 + x^2 - 2Rx\cos\phi \quad \textbf{(14–9)}$$

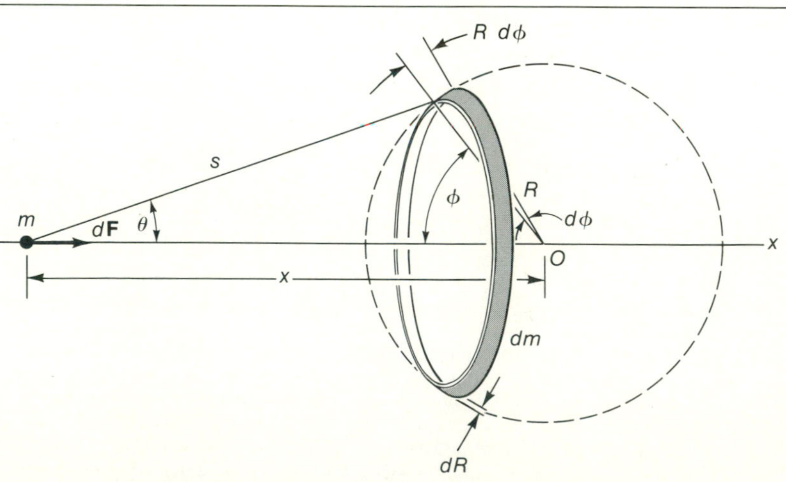

Figure 14–3
The force $d\mathbf{F}$ of gravitational attraction on a particle of mass m by a circular strip dm, which is cut from a spherical shell of radius R and thickness dR. The particle is outside the shell a distance x from the center of the shell.

Taking differentials of both sides and rearranging, we obtain (since R and x are constants here):

$$2s\,ds = 2Rx \sin \phi \, d\phi \tag{14–10}$$

$$R \sin \phi \, d\phi = \frac{s\,ds}{x} \tag{14–11}$$

We also note, from Figure 14–3, that

$$\cos \theta = \frac{x - R \cos \phi}{s}$$

which, by using the law of cosines, may be written:

$$\cos \theta = \frac{x^2 - R^2 + s^2}{2xs} \tag{14–12}$$

Substituting Equations (14–8, 14–11, and 14–12) in Equation (14–6), we obtain

$$dF = \frac{Gm\pi\sigma R}{x^2}\left(\frac{x^2 - R^2}{s^2} + 1\right)ds \tag{14–13}$$

This is the force exerted on the particle m by the circular-strip element. As we add the forces due to similar strip elements over the entire shell, we note that the only variable is s. Its limits extend from $(x - R)$ to $(x + R)$.

Conveniently, the integration has a simple answer. Setting aside the constant for the moment, we have for the integral itself:

$$\int_{(x-R)}^{(x+R)} \left(\frac{x^2 - R^2}{s^2} + 1\right) ds = 4R \tag{14–14}$$

Giving, for the total force of the particle:

$$F = \int_{\substack{\text{Entire}\\ \text{shell}}} dF = \frac{G\,m\sigma 4\pi R^2}{x^2} \tag{14–15}$$

Since the total mass of the shell is $M = (\sigma)(4\pi R^2)$, we have

$$F = G\frac{mM}{x^2} \tag{14–16}$$

This is the same answer we would obtain if all the mass of the spherical shell were concentrated at its center.

A solid sphere may be thought of as made up of a large number of concentric shells, similar to the layers of an onion. Even though the individual shells may have different densities, the mass of each may be considered concentrated at its center. Thus, *provided the solid sphere has spherical symmetry* (that is, the density is a function of r alone), *the gravitational force the sphere exerts on an external particle is the same as if all the sphere's mass were concentrated at its center.*

What about a particle located *inside* a spherical shell? We may calculate the force directly by noting that Equation (14–14) now has limits of integration from $(R - x)$ to $(R + x)$. The result is zero: *A particle located anywhere inside a uniform, spherical shell has zero net force on it due to gravitational attraction by the shell.*

We may also obtain this last result by an interesting chain of reasoning that does not involve complicated integrals. Consider the particle to be located at an arbitrary point P inside the shell. Imagine, now, that a narrow cone is constructed with its apex at the point, extending out in an arbitrary fashion. The

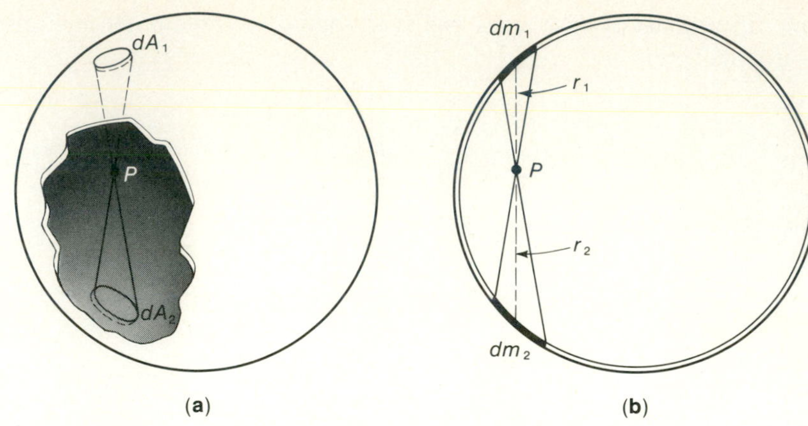

Figure 14–4
The point P is arbitrarily located inside a spherical shell. Equal elementary solid angles $d\Omega$ extend in opposite directions from P to intercept, respectively, the areas dA_1 and dA_2 on the shell. The segments of the shell within the solid angles have masses dm_1 and dm_2.

cone will intercept, on the shell, an element of area dA_1 (see Figure 14–4). If we construct a similar cone with an equal solid angle in the opposite direction, it will intercept an area dA_2.

In general, a solid angle Ω subtends an area A on any sphere of radius R, centered at the apex of a solid angle, according to:

$$\text{Solid angle } \Omega \equiv \frac{A}{R^2} \qquad \text{(measured in } \textit{steradians}\text{)} \tag{14–17}$$

In this case, the area A is perpendicular to R. But in Figure 14–4, the areas dA_1 and dA_2 are not necessarily perpendicular to r_1 and r_2 because P is not necessarily at the center of the shell. But the *angle of obliquity* θ between the direction of r and the area dA is the same in both cases. So from this fact and Equation (14–17), we conclude that the intercepted areas on the shell in Figure 14–4 are directly proportional to the square of the distance from P. And because the shell has uniform thickness and density, the masses dm_1 and dm_2 are, respectively, also *directly* proportional to ${r_1}^2$ and ${r_2}^2$. But gravity is an inverse-square-law relation. Hence, for a particle m located at P, the gravitational attractions of the elemental masses on m are *inversely* proportional to ${r_1}^2$ and ${r_2}^2$. When these two factors of proportionality are combined, the factors r^2 and $1/r^2$ cancel. *Thus each element attracts* m *with the same force, equal in magnitude, but opposite in direction.* The net force on m is zero.

We may divide the entire shell in this way, accounting for all its mass. Thus we conclude that the resultant force on the particle is zero for any location within the shell.

We can now discuss an interesting problem, that of the net force on a particle moving within a tunnel bored diametrically through the earth (see Figure 14–5). At any arbitrary location, the net gravitational force on the particle is due solely to that fraction of the earth's mass contained within a sphere of radius r. (As we have seen, the spherical shell external to this radius produces no net force on m.) *Assuming a uniform density ρ for the earth* (contrary to the actual situation), the mass M' inside a sphere of radius r (shown shaded in the figure) is

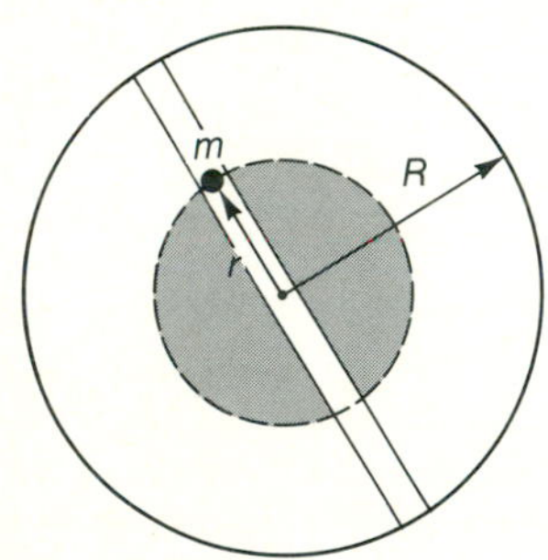

Figure 14–5
A particle moving in a tunnel along the diameter of the earth is acted upon by a force due to the portion of the earth's mass shown shaded here.

$$M' = \rho V' = \rho \tfrac{4}{3}\pi r^3$$

The mass M' exerts a gravitational attraction equal to

$$F = -G\frac{mM'}{r^2} = -G\frac{m}{r^2}(\rho \tfrac{4}{3}\pi r^3) = -\underbrace{\left(\frac{Gm\rho\, 4\pi}{3}\right)}_{k} r = -kr$$

where the minus sign indicates a direction toward the center of the earth and k is a constant. This is just the criterion for SHM. If we ignore the rotation of the earth, a particle dropped into such a hole would oscillate back and forth in SHM. Similarly, it can also be proved that a particle sliding through a straight, smooth tunnel joining *any* two arbitrary points on the surface of the earth (not necessarily a diameter) will undergo SHM with the same period of oscillation as in the diametric hole.

Figure 14–6 graphs the gravitational force on a particle at various distances from the center of a homogeneous solid sphere.

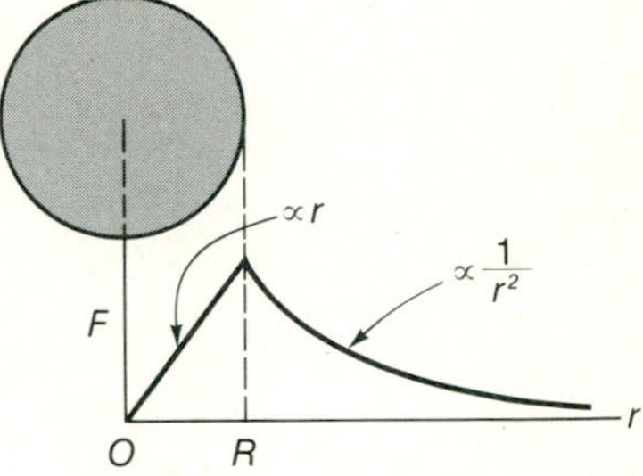

Figure 14–6
The magnitude of the gravitational force **F** on a particle as a function of the distance r from the center of a solid sphere of uniform density and radius R.

EXAMPLE 14–1

Find the gravitational force of attraction between a uniform sphere of mass M and a uniform, thin rod of length ℓ and mass m oriented as shown in Figure 14–7.

SOLUTION

Since the sphere is uniform, its entire mass may be considered to be concentrated at its center.

By Newton's law of universal gravitation, the force dF on the mass element dm is

$$dF = G\frac{M}{x^2}\,dm \tag{14–18}$$

where

$$dm = \frac{m}{\ell}\,dx$$

Substituting for dm and integrating over the length of the rod, we obtain

$$F = \int_r^{r+\ell} G\frac{mM}{Lx^2}\,dx = G\frac{mM}{\ell}\left(\frac{1}{x}\right)\Bigg|_r^{r+\ell} = G\frac{mM}{\ell}\left[\frac{\ell}{r(r+\ell)}\right] = G\frac{mM}{r(r+\ell)}$$

As expected, if $r \gg \ell$, then the rod behaves almost like a point mass a distance r from the center of the sphere.

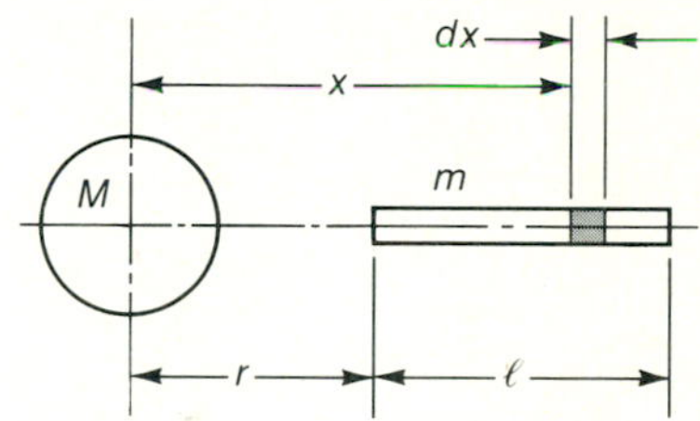

Figure 14–7
Example 14–1

14.6 The Gravitational Field

When Newton's theory of gravitation was first published, perhaps the most serious criticism was that it was difficult to conceive of a force that could act at a distance. In Newton's view, the force of gravity somehow reached across completely empty space to affect the motion of distant objects. Although he could not specify how the force arose, Newton felt that because the theory accounted satisfactorily for the observed behavior of bodies, it was a success.

About a century and a half later, Michael Faraday, in developing ideas about electromagnetism, suggested a new way of thinking about forces of interaction between distant objects. Although his idea explained the forces between electrically charged bodies, the same concept—a field—may be applied equally to gravitational attraction. This concept greatly simplifies the mathematical discussions of all action-at-a-distance forces.

In the case of gravity, the idea is introduced as follows. Consider a mass A that exerts a gravitational force of attraction on mass B. We say that object A modifies the space surrounding it by establishing a gravitational field. This field extends outward in all directions, falling to zero at infinity. A mass B, located within this field, experiences a force due to the field *at that point*. It

Variations in g

Very precise measurements of g, *the acceleration due to gravity, have revealed some interesting variations. With the development of an extremely stable and sensitive instrument—the superconducting gravimeter—changes in the value of* g *due to deformation of the solid earth are clearly revealed. If the earth's shape is altered so that the surface is moved closer to or farther from the center of the earth, this shift can be detected by the resultant changes in* g.

The superconducting gravimeter has remarkable sensitivity: $\Delta g/g \approx 10^{-11}$. *Its absolute accuracy is 0.2%. The largest changes in* g *are due to the two daily "tides" in the solid earth produced by the sun and moon. Note in the graph in Figure 14–8 (obtained near La Jolla, California) that as the weeks go by, the maxima in these effects shift in phase due to changes in the relative positions of the sun and moon. These tidal effects produce a vertical change of about 20 cm/day in distance of the earth's surface from its center. Other variations are produced by the water tides near La Jolla that load the earth's surface with additional compressional forces somewhat out-of-phase with the earth tides. Similarly, increases in barometric pressure squeeze the earth's surface closer to the center of the earth. All these effects correlate clearly with certain variations in the data.*

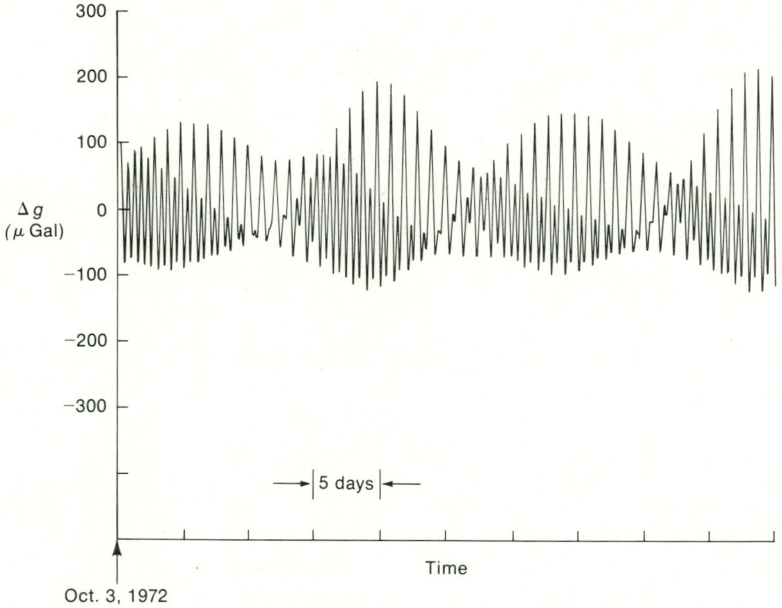

Figure 14–8
Variation of g versus time at La Jolla, California. Note that 1 Gal $\equiv 10^{-2}$ m/s^2. (Research by R. J. Warburton and J. M. Goodkind, University of California at San Diego, La Jolla, California, and C. Beaumont, Department of Energy, Mines, and Resources, Ottawa, Canada.)

Such graphs are analyzed by the mathematical procedure known as Fourier analysis *(Appendix F), which separates the individual frequency components that contribute to the overall oscillations and thereby permits their identification with various natural causes. The apparatus is being improved in sensitivity and stability, so that eventually one might be able to detect possible change in* G, *the universal gravitational constant. If such changes actually occur, it would have considerable implications for physics, astronomy, and cosmology. This technique also has exciting possibilities for earthquake prediction. There is evidence that prior to some earthquakes, the surface of the earth rises dramatically a few centimeters in a geologically sudden time.*

This is another example in which developing new techniques for precise measurements has an unpredictable "fallout" in other areas. It is a pattern repeated over and over again in the history of science.

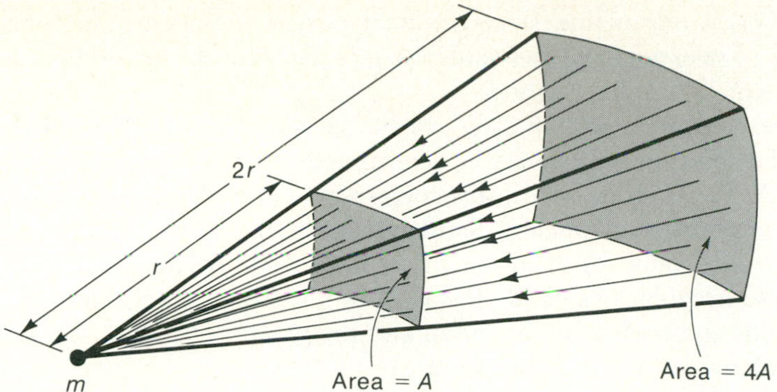

Figure 14–9

The inverse-square-law relationship for the gravitational field surrounding a point mass m. In the sketch, the solid angle extending from the point cuts two areas from spherical surfaces of radii r and $2r$. Since the sides of the squares are proportional to the distance r, the *areas* are proportional to the square of the distance. Field lines are radial, so the same number of lines that pass through area A also pass through the area $4A$. Because the field strength g is proportional to the number of lines per unit area, when the distance is doubled the field is only one-fourth as strong. Thus, $g \propto 1/r^2$.

is thus the *local* presence of a gravitational field that produces the force, not some distant object. (Of course, the situation is symmetrical: Mass A experiences a gravitational force because of the field set up by mass B.)

The gravitational field is a *vector field*. At each point in space, it has both a magnitude and direction, defined by:

$$\mathbf{g} = \frac{\mathbf{F}}{m} \tag{14–19}$$

The **gravitational field g** at any point is thus the force per unit mass at that location. If several masses are present, the total field at a point is the vector sum of the individual fields.

14.7 Gravitational Field Lines

One way of thinking about gravitational fields is to visualize an array of **field lines** such that at every point on a line, the tangent points in the direction of the gravitational force. For a spherical mass, the lines would thus be radially inward. The **strength** of the gravitational field at any location is proportional to the *number of field lines passing through a unit area oriented perpendicular to the lines.* Figure 14–9 shows that such an array of radial lines does, indeed, lead to an inverse-square-law relation for the force If a unit area is moved twice as far away, the number of lines penetrating the area is only one-fourth as great.

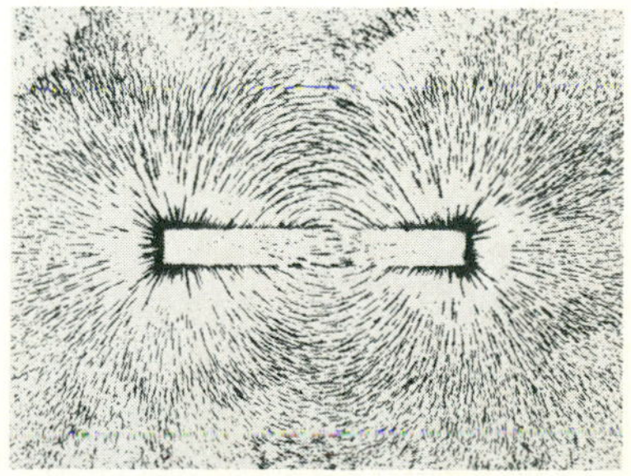

Figure 14–10

In the space surrounding a magnet, iron filings arrange themselves in lines parallel to the direction of the magnetic force. The resultant pattern depicts the *magnetic field.*

Do field lines exist in reality? Our familiarity with a *magnetic field*, depicted by iron filings in Figure 14–10, does lure one into believing that such magnetic lines of force actually do exist. But they are merely a visual aid to our thinking; the helpfulness of the concept justifies our use of it. Because of this artificiality, the number of field lines we imagine for any given situation is purely arbitrary, and any proportionality we find convenient is equally valid.

Numerical values for the magnitude of **g** listed in handbooks for various locations on the earth include the centrifugal effects due to the rotation of the earth. This is reasonable, since for most purposes one considers the earth as a frame of reference and the value for **g** is determined locally by a pendulum experiment or similar method.

The accuracy with which g can be measured is surprising. *Variations* in g (not the absolute values) may be determined within 1 part in 10^{11}, by a *gravity meter*. (At the earth's surface, this corresponds to a change in elevation of less than a tenth of a millimeter.) Similar instruments carried in an airplane can detect underground oil fields or mineral deposits by the small changes in g that these irregularities cause. Installed in a satellite, the instrument can map mountains and craters on the moon and the planets or detect "mascons" (concentrated masses beneath the surface).

14.8 The Cavendish Experiment

The numerical value for the universal gravitational constant G (not to be confused with the gravitational field g) was not accurately known until 1798, when the English physicist Henry Cavendish used a torsion balance to determine it. The apparatus consists of two small spheres of mass m mounted on the ends of a horizontal rod, which is suspended by a very thin metal or quartz fiber (see Figure 14–11). Two larger lead spheres, each of mass M, are placed so that the gravitational attraction between pairs of large and small masses tends to twist the rod about the axis of the supporting fiber. A mirror attached to the rod reflects a beam of light to a distant scale to record the small deflections that occur. The gravitational forces involved are quite small, typically of the order of 10^{-10} N (roughly 10 000 times smaller than the weight of a human hair).

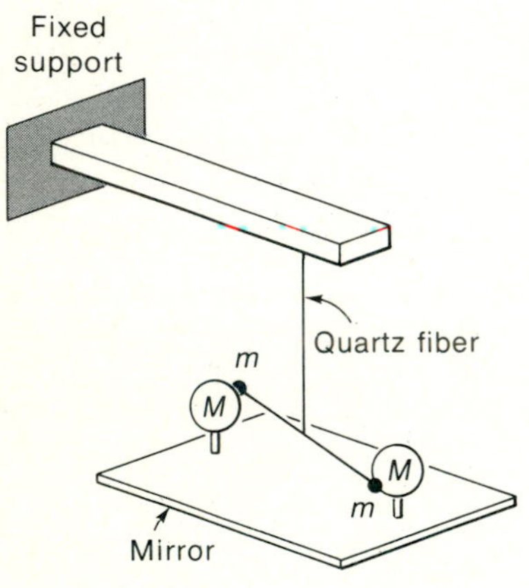

Figure 14–11
A gravitational torsion balance used by Cavendish to determine the universal gravitational constant G.

Because of experimental difficulties, G is perhaps the least known of all fundamental constants. As mentioned earlier, the currently accepted value is:

$$G = 6.672 \times 10^{-11}\ \frac{\text{N}\cdot\text{m}^2}{\text{kg}^2} \qquad \textbf{(14–20)}$$

The force of gravity is of negligible importance in determining the motion of objects unless at least one of them is massive.

14.9 Gravitational Potential Energy

We mentioned in Chapter 6 that forces of gravity are conservative forces. We may therefore define a potential function $U(r)$ that is related to the conservative force F by

$$U_b - U_a = -\int_a^b \mathbf{F}\cdot d\mathbf{r} \qquad \textbf{(14–21)}$$

Previously, we considered only those cases in which the mass remained close to the earth's surface, where the value of g is essentially constant. We will now extend the discussion to include motions of satellites and rockets, whose trajectories may involve large distances above the earth. The *gravitational potential energy* function we will derive is associated with the system as a whole: the satellite-plus-earth. It will have only one variable, the separation r between the satellite and the earth's center.

$$U_b - U_a = -\int_a^b \left(-G\frac{Mm}{r^2}\right)dr = GMm\int_a^b \frac{dr}{r^2} \qquad \textbf{(14–22)}$$

(The expression for F includes a minus sign to indicate a force of attraction, that is, in the $-r$ direction.) For motions of m that may be far from the earth's surface, it is convenient to define the *zero reference* of potential energy at $r = \infty$, when the particle is infinitely far away. The path of integration from the zero-reference location to any given point is perfectly arbitrary, since for conservative forces, the value of the integral is independent of the path. To make the integration easy, we choose a radial path. Thus, as the particle moves from infinity $[U(\infty) \equiv 0]$ to the arbitrary location r external to the earth, we obtain a general expression for the **gravitational potential energy** $U(r)$:

$$U(r) - 0 = GMm \int_{\infty}^{r} \frac{dr}{r^2} = -G\frac{Mm}{r}\Bigg|_{\infty}^{r}$$

GRAVITATIONAL POTENTIAL ENERGY

$$U(r) = -G\frac{Mm}{r} \quad \text{(outside the earth where } U \equiv 0 \text{ at } r = \infty) \qquad \textbf{(14–23)}$$

Figure 14–12 shows the gravitational potential energy for points both inside and outside the earth (see Problem 14C–7).

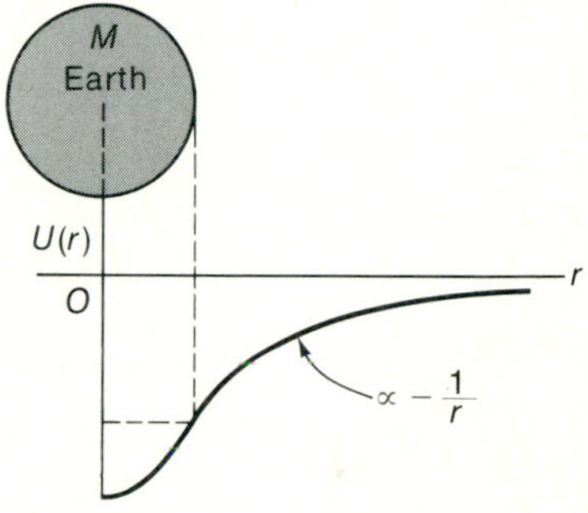

Figure 14–12
The gravitational potential energy $U(r)$ of the earth-plus-a-particle system.

14.10 Escape Velocity and Binding Energy

Energy considerations provide a useful way to calculate the minimum initial velocity of an object at the earth's surface such that it would escape from the earth, never to return. (In our calculations, we shall ignore the frictional resistance effects of the earth's atmosphere and other effects due to such factors as the earth's rotation and the presence of other astronomical bodies.)

The problem is an example of the conservation of energy. If, at the earth's surface, an object has a speed v, its kinetic energy K at that point is $\frac{1}{2}mv^2$. Thus, the total energy is

$$E = U + K$$

To *barely* escape without falling back to the earth requires that the object reach infinity with zero kinetic energy. If v_e is the minimum **escape speed** at the earth's surface, then

$$E_0 = E$$
$$U_0 + K_0 = U + K$$
$$-G\frac{Mm}{R} + \tfrac{1}{2}mv_e^{\,2} = 0 + 0$$

Solving for v_e, we have

ESCAPE SPEED FROM THE EARTH

$$v_e = \sqrt{\frac{2GM}{R}} \qquad \textbf{(14–24)}$$

which is equal to 11.2 km/s at the earth's surface (about 7 mi/s, or 25 000 mi/h). The *direction* of launching the particle is unimportant (within the assumptions of ignoring atmospheric friction, the earth's rotation,[5] gravitational effects of the moon, and so on).

[5] We have calculated the escape velocity for a nonrotating earth. That is, the speeds would be for a frame of reference *at rest* relative to the "fixed" stars. (Classically, a satellite outside the earth's atmosphere does not "know" that the earth is rotating beneath it.) If we wanted to find the escape velocity relative to the launching site, we would need to make a correction for the rotational velocity of the earth's surface and for the direction of launch relative to that velocity.

It is interesting that escape speeds are independent of the masses of the escaping particles. If we look into the situation for the earth's atmosphere, however, the lighter components (hydrogen and helium) escape more easily than the heavier components (oxygen and nitrogen). The reason is that for a gas at a uniform temperature, each atom or molecule has, on the average, the same kinetic energy $K = \frac{1}{2}mv^2$. Thus light atoms and molecules have faster average speeds than heavy ones. Of course, each particle does not move with exactly the same average speed; some have faster-than-average and some have slower-than-average speeds. Both hydrogen and helium in our atmosphere have sufficient speeds in their upper range of velocities to escape the earth's gravitational potential. In fact, it is believed that the earth and solar system were formed from interstellar gas and dust containing mostly hydrogen and helium, with only one or two percent of other elements. Today, however, hydrogen and helium comprise only a few parts per million in the earth's atmosphere. On the moon, the escape velocity is so low that the moon cannot keep an appreciable atmosphere of any sort on its surface. In contrast, the planet Jupiter, whose mass is 318 times that of the earth, has an atmosphere of about 84% hydrogen atoms, 15% helium, and only 1% heavier molecules, a chemical composition believed comparable to that of the interstellar gas from which the solar system was formed.

Molecules escape from the atmosphere only at high altitudes, where the air density is so low that a molecule with a speed $>v_e$ has an appreciable chance of escaping without colliding with another molecule. For the earth, this escape altitude is roughly above 600 km, in a region called the *exosphere*, where temperatures are of the order of 1500°C. (Such a high temperature does not mean that an astronaut or satellite would be burned up at this location. The air is so rarefied that air molecules cannot transfer their energies to an object faster than the object radiates energy away.)

One convenience in solving orbital trajectory problems from an energy viewpoint is that potential energy is a *scalar*, and scalars are easier to deal with mathematically than vectors. For example, the mutual gravitational potential energy of the three masses shown in Figure 14–13 may be found by adding the potential energy for each pair of masses:

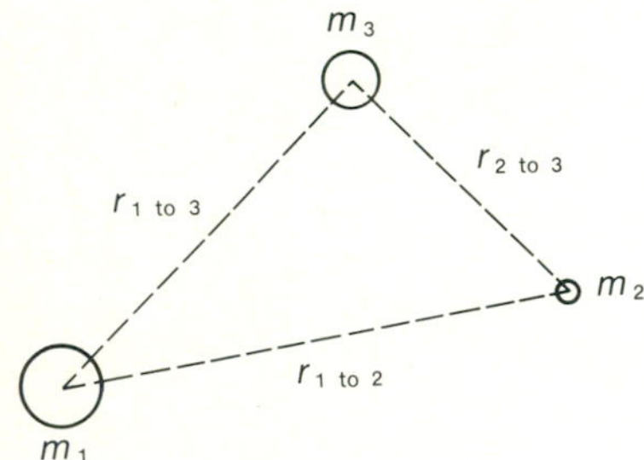

Figure 14–13
A system of three masses.

$$U = -G\frac{m_1m_2}{r_{1\text{ to }2}} - G\frac{m_2m_3}{r_{2\text{ to }3}} - G\frac{m_1m_3}{r_{1\text{ to }3}} \qquad \textbf{(14–25)}$$

Equation (14–25) gives the gravitational potential energy of the system *with respect to the zero potential energy when all the masses are separated infinitely far from one another*. If we wish to separate the masses, we must supply this amount of energy in the form of work done in pulling them apart against the mutual forces of gravitational attraction. This (negative) potential energy is called the *binding energy* holding the system together.

EXAMPLE 14–2

How much work is done by the force of gravity in assembling a uniform sphere of radius R and density ρ? Assume that the sphere is assembled of elemental masses that are originally so far apart that the gravitational forces among them are essentially zero.

SOLUTION

We approach this problem by considering the work dW done by gravity in bringing a mass element dm from infinity (where $U \equiv 0$) to the spherical mass of radius r already assembled. Since gravity is a conservative force, this work is the negative of the change in gravitational potential energy of the mass element. That is:

$$dW = -dU$$

From Equation (14–23), the change in gravitational potential energy is

$$dU = -G\frac{M}{r}\,dm$$

where M is the mass already assembled, given by

$$M = \tfrac{4}{3}\pi r^3 \rho$$

Substituting for M, we have

$$dW = G\frac{4\pi r^2 \rho}{3}\,dm$$

To keep the assembled mass spherical in shape, we smooth out dm uniformly over the surface. It thus adds a spherical shell of mass dm and thickness dr:

$$dm = 4\pi r^2 \rho\, dr$$

Thus:

$$dW = G\frac{16\pi^2 \rho^2}{3} r^4\, dr$$

Integrating, we obtain

$$W = \int_0^R G\frac{16\pi^2\rho^2}{3} r^4\, dr = \boxed{G\frac{16}{15}\pi^2\rho^2 R^5 = G\frac{3M^2}{5R}}$$

14.11 Satellite and Planetary Motions

An interesting contemporary problem involves the motion of a satellite m about the earth M. Because the earth is so much more massive than the satellite, we may consider the earth to be at rest in an inertial frame and the satellite m to have various possible orbits about the earth. Figure 14–14 sketches trajectories for various values of the total energy $E = U + K$, when the initial velocity $\mathbf{v}_0$ is at right angles to a line joining the satellite and the center of the

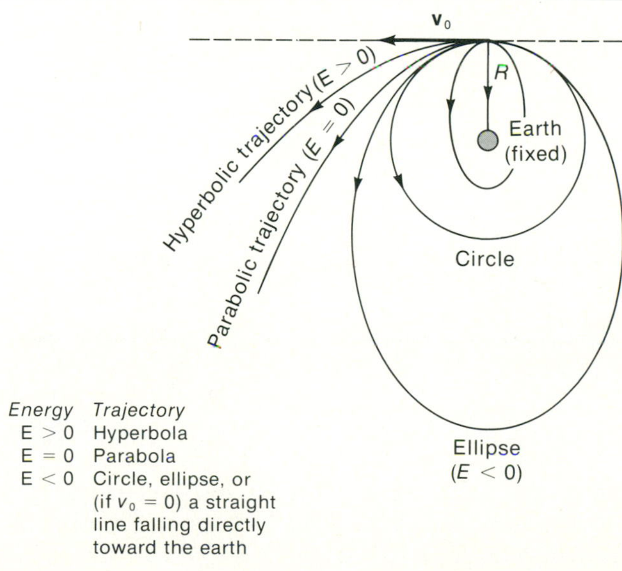

Figure 14–14
Possible trajectories of a satellite projected with various initial velocities v_0, as shown at a distance R from the center of the earth. The type of trajectory depends on the total energy $E = U + K$.

earth. It is useful to remember not only that gravitational forces are conservative, but that for spherical bodies they are central forces that can exert no torque on the satellite. Consequently, the *total energy* and the *angular momentum* are conserved. Applying these two principles to elliptical orbits, we may easily calculate various characteristics: for example, the speed at any point or the *perigee* (the closest approach to the earth's center) and the *apogee* (the farthest distance from the earth's center).

EXAMPLE 14–3

A satellite moves in an elliptical trajectory about the earth. The minimum and maximum distances from the surface of the earth are 400 km and 3000 km. Find the speeds of the satellite at apogee and perigee.

SOLUTION

Since the mass of the satellite is negligible compared with the earth's mass, we take the *CM* of the earth to be at rest. Gravity is a *central* force (that is, it acts through the *CM* of the earth), so the angular momentum of the satellite about the earth's *CM* remains constant in time. Therefore, the conservation of angular momentum holds true.

Using subscripts a and p for the apogee and perigee positions, we have (assuming the earth's radius is 6.36×10^6 m):

$$r_a = 9.36 \times 10^6 \text{ m}$$
$$r_p = 6.76 \times 10^6 \text{ m}$$

The conservation of angular momentum gives us

$$L_p = L_a$$
$$mv_pr_p = mv_ar_a$$
$$v_pr_p = v_ar_a \tag{14–26}$$

Applying the conservation of energy, we obtain

$$E_p = E_a$$
$$U_p + K_p = U_a + K_a$$
$$-G\frac{Mm}{r_p} + \tfrac{1}{2}mv_p^2 = -G\frac{Mm}{r_a} + \tfrac{1}{2}mv_a^2$$
$$2GM\left(\frac{1}{r_a} - \frac{1}{r_p}\right) = \left(v_a^2 - v_p^2\right) \tag{14–27}$$

Since we know the numerical values of G, M, r_p, and r_a, we have two equations (14–26 and 14–27) to determine the two unknowns (v_p and v_a). Solving the equations simultaneously, we obtain

$$\boxed{v_p = 8250\ \frac{\text{m}}{\text{s}}} \quad \left(\text{or } 18\,500\ \frac{\text{mi}}{\text{h}}\right) \quad \text{and} \quad \boxed{v_a = 5960\ \frac{\text{m}}{\text{s}}} \quad \left(\text{or } 13\,300\ \frac{\text{mi}}{\text{h}}\right)$$

14.12 Inertial and Gravitational Mass

Up to this point, we have sidestepped a curious puzzle. There are two, seemingly different, properties of mass: a *gravitational* attraction for other masses and an *inertial* property that resists acceleration. These two attributes are apparently distinct. To designate them, we will use the subscripts g and i and write:

$$\text{Gravitational property} \qquad W = m_g g \qquad \textbf{(14–28)}$$

$$\text{Inertial property} \qquad F = m_i a \qquad \textbf{(14–29)}$$

The numerical value for the gravitational constant G was chosen to make the magnitudes of m_g and m_i numerically equal. But regardless of how G is chosen, the strict *proportionality* of m_g and m_i has been established experimentally to an extremely high degree: a few parts in 10^{12}. Thus it appears that gravitational and inertial mass are, indeed, exactly proportional.

But why? They seem to involve two entirely different attributes: a force of mutual gravitational attraction between masses and the resistance a single mass has regarding acceleration. This puzzled Newton and other physicists until Einstein published his theory of gravitation known as *general relativity* in 1916. It is a mathematically complex theory, and thus we will be able only to hint at the elegance and insight Einstein achieved.

(**a**) An observer at rest in a uniform gravitational field where the acceleration due to gravity is g.

14.13 General Relativity

In Einstein's view, the remarkable coincidence that m_i and m_g seem to be exactly proportional was evidence for a very intimate and basic connection between the two concepts. He pointed out that no *mechanical* experiment (such as dropping a mass) could distinguish between the two different situations sketched in Figure 14–15(a and b). In each case, if the observer released a mass from his hand, it would undergo a downward acceleration of g relative to the floor of the box.

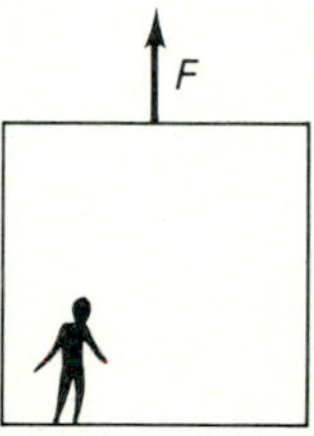

(**b**) An observer in a region where gravity is negligible, but whose frame of reference is accelerated through space (by the force F) with an acceleration equal to g.

Einstein carried this idea further to propose, as one of two fundamental postulates in his general theory of relativity, that no experiment whatsoever, mechanical *or otherwise*, could distinguish the difference between the two cases. This extension to include all phenomena (not just mechanical ones) has interesting consequences. For example, a ray of light sent horizontally across the box would appear, in case (b), to bend downward toward the floor as the box accelerated upward to meet it. Therefore, proposed Einstein, in case (a), a ray of light should be bent downward by the presence of the gravitational field. No such bending is predicted in Newton's theory of gravitation.

The two postulates of Einstein's **general relativity** theory may be stated as follows:

(1) *All the laws of nature may be stated so that they have the same form for observers in any space-time frame of reference, whether accelerated or not.* (This is the *principle of covariance.*[6])

(2) *In the neighborhood of any given point, a gravitational field is equivalent in every respect to an accelerated frame of reference in the absence of gravitational effects.* (This is the *principle of equivalence.*)

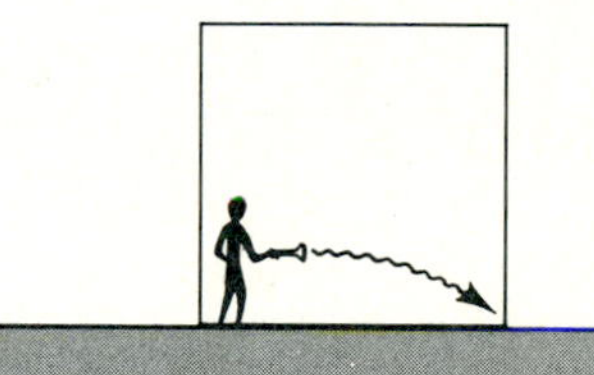

(**c**) If (**a**) and (**b**) are truly equivalent as Einstein proposed, then a ray of light would be bent in a gravitational field. Such an effect has been experimentally verified by light and radio signals that pass close to the strong gravitational field of the sun.

Figure 14–15

According to Einstein, these (a) and (b) frames of reference are equivalent in every way. No experiment of any sort could distinguish any difference.

The second postulate implies that gravitational mass and inertial mass are completely *equivalent*, not just proportional. What were thought to be two different types of mass are actually, in a basic sense, identical.

One interesting effect predicted by general relativity is that time scales are altered by gravity. In the presence of gravity, a clock runs more slowly than

[6] An equation that has the same form after transformation to another frame of reference is *covariant* with respect to the transformation.

one situated where gravity is negligible. Another predicted effect is that spectral lines emitted by atoms in the presence of a strong gravitational field are *red-shifted* to lower frequencies when compared with the same spectral emissions in a weak field. This gravitational red shift has been detected in spectral lines emitted by atoms in massive stars. It has also been verified on the earth by comparing the frequency of gamma rays emitted from nuclei separated vertically by about 20 m.

The second postulate suggests that a gravitational field may be "transformed away" at any point by choosing an appropriately accelerated frame of reference—a freely falling one. Einstein developed an ingenious way of describing the exact amount of acceleration necessary. He specifies a certain quantity, the *curvature of space-time*, that describes the gravitational effect at every point. In fact, the curvature of space-time completely replaces Newton's gravitational theory. According to Einstein, there is no such thing as a gravitational force. Rather, the presence of a mass such as the sun causes a curvature of space-time in its vicinity, and this curvature dictates the path that all freely moving objects follow. As a consequence, the earth (with no gravitational force on it) travels in a curved path around the sun.

If the concentration of mass becomes very great, as is believed to occur when a large star exhausts its nuclear fuel and collapses to a very small volume, a **black hole** may be formed. Here, the curvature is so extreme that within a certain distance from the center, all matter and light become trapped. Theoretically, it is possible for one star of a double-star system to evolve to the black hole stage. Possible experimental verification of such a combination is the observation of unusual bursts of x-ray radiation when ionized matter from the ordinary star spirals into the black hole.[7]

Summary

Newton's law of universal gravitation between two particles is

$$F_{\text{grav}} = G\frac{m_1 m_2}{r^2}$$

where m_1 and m_2 are the gravitational masses of the two particles, r is the distance between the particles, and G is the universal gravitational constant, equal to $6.672 \times 10^{-11}\ \text{N}\cdot\text{m}^2/\text{kg}^2$.

For a sphere of matter whose density is a function of r alone (that is, the sphere has spherical symmetry), the gravitational force it exerts on a mass point located externally is the same as if all the sphere's mass were concentrated at its center. A spherical shell exerts no net force on a mass point located inside. Therefore, at an arbitrary point inside a solid sphere (located a distance r from the center), a mass point experiences gravitational attraction only for that portion of the sphere's mass contained within a sphere of radius r.

A *gravitational field* $\mathbf{g} = \mathbf{F}_{\text{grav}}/m$ is a useful aid in thinking about gravity.

Gravitational *potential energy* U is defined by

$$U_2 - U_1 = -\int_1^2 \mathbf{F}\cdot d\mathbf{r}$$

[7] For an introduction to curved space-time, see J. J. Callahan, "The Curvature of Space in a Finite Universe," *Scientific American*, August 1976, pp. 90–100. For a discussion of the central role that relativity theory plays in cosmology, see William Kaufmann III, "The Cosmic Frontiers of General Relativity" (Boston: Little, Brown and Co., 1977). As a further reference, see Martin Gardner, "The Relativity Explosion" (New York: Vintage Books, Random House, 1977).

If we choose $U(\infty) \equiv 0$, then

$$U(r) = -G\frac{Mm}{r}$$

In classical theory, two different attributes of mass are exactly proportional:

m_g = *gravitational mass* (the property of attraction for other masses)
m_i = *inertial mass* (the property of resisting acceleration)

The value of G is chosen so that there is numerical equivalence of m_g and m_i.

In *general relativity*, Einstein proposed that in the neighborhood of any given point, a gravitational field is equivalent in every respect to an accelerated frame of reference in the absence of gravity. This is the *principle of equivalence*, which implies that gravitational and inertial masses are equal, not just proportional. The *principle of covariance* states that all the laws of nature are the same for observers in any frame of reference, accelerated or not. These two postulates provide the basis for general relativity.

Questions

1. Can a satellite be placed in orbit about the earth so that it travels in a plane not passing through the earth's center?
2. Is it possible to determine the mass of another planet if we know only the radius of its orbit and the mass of the earth? If not, what minimum additional information would allow us to determine it?
3. Neglecting air friction and the earth's rotation, can you place a satellite in orbit about the earth by firing it from a gigantic land-based cannon?
4. The average density of matter in interstellar space is roughly one hydrogen atom per cubic centimeter. Our sun has a mass of about 2×10^{30} kg. What was the initial volume of the gas cloud that formed our sun if it had 5 times greater density than the average?
5. Suppose that during the next week, the value of G gradually became half of its present value. Describe some changes that would occur during the week.
6. If the gravitational force on objects is directly proportional to their masses, why don't large masses fall with greater acceleration than small ones?
7. A pendulum clock and a wristwatch are carried to the moon. Which will run faster on the moon?
8. The mass of the moon was known before any human traveled there and could measure the acceleration of falling objects on its surface. How was this mass determined?
9. Suppose you are weighing an object on a spring scale at the equator. Consider two situations: weighing it at noon when the sun is directly overhead and at midnight when the sun is directly "beneath" the earth. Ignoring the earth's daily rotation but considering its orbital motion about the sun, will the two weighings differ? If the earth's daily rotation is considered, will the answer change?
10. Disscuss what happens to a small space station when an astronaut inside jumps from one wall to the opposite wall. What would occur if the astronaut started running around the inside (circular) wall?
11. The gravitational force the sun exerts on the moon is about twice as great as the gravitational force the earth exerts on the moon. Why doesn't the sun pull the moon away from the earth during a total eclipse?
12. If all space were empty except for two nearby masses, say, two drops of water, the drops would, according to Newton's law of gravity, be attracted together. Now suppose all space were full of water except for two bubbles somewhat close to each

other. Would the bubbles be attracted together, repelled from each other, or would there be zero net force on each? (Ignore possible surface tension effects.)*

Problems

14A–1 Which exerts a greater force of gravitational attraction on objects on the earth: (a) the moon or (b) the sun? Calculate these forces on a 1-kg mass.
Answers: (a) 3.32×10^{-5} N (b) 5.92×10^{-3} N

14A–2 When a falling meteor is at a distance above the earth's surface of 3 times the earth's radius, what is its acceleration due to the earth's gravity?

14A–3 In terms of g, the acceleration at the earth's surface, find the gravitational acceleration at a height above the earth's surface equal to the earth's diameter.
Answer: $g/9$

14A–4 Astrologers claim that the changing locations of the planets affect our lives. Estimate the ratio of the gravitational force exerted on you by a 2000-kg automobile when you are 2 m from its center of mass to the gravitational force exerted on you by the planet Mars (mass $\approx 6.5 \times 10^{23}$ kg) at its distance of closest approach (about 56 million km). For ease of calculation, assume you and the automobile are both point masses.

14A–5 An astronaut weighs 140 N on the moon's surface. When he is in a circular orbit about the moon at an altitude equal to the moon's radius, what gravitational force does the moon exert on him?
Answer: 35.0 N

14A–6 In introductory physics laboratories, a typical Cavendish balance for measuring the gravitational constant G uses lead spheres of masses 1.5 kg and 15 g, whose centers are separated about 4.5 cm. Calculate the gravitational force between these spheres at that distance.

14A–7 Consider a satellite that is orbiting the earth very near the earth's surface. Using only $g = 9.8\ \text{m/s}^2$ and the earth's radius $R = 6.37 \times 10^6$ m, (a) find the period of revolution (in minutes) and (b) the orbital speed.
Answers: (a) 84.4 min (b) 7.90 km/s

14A–8 A rocket is fired vertically, moving upward with a constant acceleration $2g$. After 40 s, the rocket motors are turned off and the rocket subsequently moves under the action of gravity alone, with negligible air resistance. Ignoring the variation of g with altitude, find (a) the maximum height the rocket reaches and (b) the total flight time from launch until the rocket returns to earth. (c) Sketch a freehand (qualitative) graph of velocity versus time for the entire flight.

14A–9 A nonrotating spherical planet has mass M and radius R. A particle is fired off radially from the surface of the planet with speed $v = \sqrt{GM/2R}$. (This is one-half the escape speed from the planet.) Calculate the farthest distance from the center of the planet that the particle reaches.
Answer: $4R/3$

14A–10 The planet Uranus has a mass of about 14 times the earth's mass, and its radius is equal to about 3.7 earth radii.

- **(a)** By setting up ratios with the corresponding earth values, find the acceleration due to gravity on the surface of Uranus.
- **(b)** Ignoring the rotation of the planet, find the minimum escape velocity from the surface of Uranus.

14A–11 The force of gravity on the surface of the moon is about one-sixth that on the surface of the earth. If the radius of the moon is about one-quarter that of the earth, find the ratio of the average mass density of the moon relative to the average mass density of the earth.
Answer: 2/3

14A–12 Calculate the escape velocity from the moon's surface.

14A–13 Show that the magnitude of potential energy of a satellite in circular orbit is exactly twice its kinetic energy.

---

* Reprinted with permission from *Thinking Physics*, Vol. 1, by L. Epstein and P. Hewitt (1979), p. 125. Published by Insight Press, 614 Vermont Street, San Francisco, CA 94107.

14A–14 At what point between the earth and the moon is the net gravitational force on a small mass zero?

14A–15 In the hydrogen atom, the electron and proton are separated by 5×10^{-11} m. If the only force between them were gravitational, what would be the period of revolution of the electron orbiting around the proton? Their masses are $m_e = 9.11 \times 10^{-31}$ kg and $m_p = 1.67 \times 10^{-27}$ kg. *Answer:* 1.40 yr

14A–16 A satellite moves in an elliptical orbit about the earth such that at perigee and apogee positions, the distances from the earth's center are, respectively, D and $4D$.

(a) Find the ratio of the speeds at the two positions: $v_{\text{perigee}}/v_{\text{apogee}}$.

(b) Find the ratio of the *total* energy (kinetic and potential) at the same positions: $E_{\text{perigee}}/E_{\text{apogee}}$.

14B–1 Kepler's third law states that the square of the period of a planet's orbital motion is proportional to the cube of its mean distance from the sun. Derive an expression in symbol form for this constant of proportionality, assuming circular orbits. Define each symbol in your expression, and compute the numerical value of the constant.
Answer: $4\pi^2/GM$, 3.00×10^{-19} s^2/m^3

14B–2 A "synchronous" earth satellite revolves in a circular orbit so that it appears stationary to an observer on the earth.

(a) Explain why the satellite's orbit can lie only in the equatorial plane.

(b) Determine the radius of this orbit from the earth's center.

(c) Find the farthest north latitude on the earth from which this satellite would be visible.

14B–3 A "synchronous" satellite, which always remains above the same point on a planet's equator, is put in orbit around Jupiter to study the famous red spot. Jupiter rotates once every 9.9 h. Use the data of Appendix L to find the distance above Jupiter's equator that such an orbit would be. *Answer:* 8.74×10^7 m

14B–4 Two stars of masses M and $2M$, separated a distance D, rotate in circular orbits about their common center of mass.

(a) Locate the center of mass.

(b) Find the period of their orbital motions in terms of M, D, and G, the universal gravitational constant.

14B–5 In a certain double-star system, the two stars rotate in circular orbits about their common center of mass. The stars are spherical, they have the same density ρ, and their radii are R and $2R$. Their centers are a distance $5R$ apart. Find the period T of the stars' orbital motion in terms of ρ, R, and G. *Answer:* $\sqrt{125\pi/3G\rho}$

14B–6 X-ray pulses from Cygnus X-1, a celestial x-ray source, have been recorded during high-altitude rocket flights. The signals can be interpreted as originating when a blob of ionized matter orbited a black hole with a period of 5 milliseconds. If the blob were in a circular orbit about a black hole whose mass is 10 times the mass of the sun, what would be the radius of the orbit?

14B–7 Consider two solid, uniform, spherical objects of the same density ρ. One has a radius R, the other a radius $2R$. They are in outer space where gravitational fields from other objects are negligible. If they are at rest with their surfaces touching, what is the contact force between the objects due to their gravitational attraction?
Answer: $(128/81)G\pi^2R^4\rho^2$

14B–8

(a) Determine the amount of work (in joules) that must be done on a 100-kg payload to to reach a height of 1000 km above the earth's surface.

(b) Determine the amount of additional work that is required to put the payload into orbit at this elevation.

14B–9 A rocket is given an initial speed vertically upward of $v_0 = 2\sqrt{Rg}$ at the surface of the earth (radius R). The rocket motors are then cut off, and the rocket coasts thereafter under the action of gravitational forces only. (Ignore atmospheric friction and the earth's rotation.) Derive an expression for the subsequent speed v as a function of the distance r from the center of the earth in terms of g, R, and r. *Answer:* $\sqrt{2Rg(1 + R/r)}$

14B–10 Let M and R represent the earth's mass and radius.

(a) With what minimum speed v_0 must an object be projected vertically from the earth's equator to barely reach a distance of 2 earth radii above the earth's surface? Ignore the earth's rotation and atmospheric friction.

(b) If the earth's rotation is now taken into account, will this increase, decrease, or not affect the answer for (a)? Explain.

14B–11 Three particles, each of mass m, are arranged to form an equilateral triangle of length ℓ on each side. How much work would be required to separate the masses infinitely far apart? *Answer:* $3Gm^2/\ell$

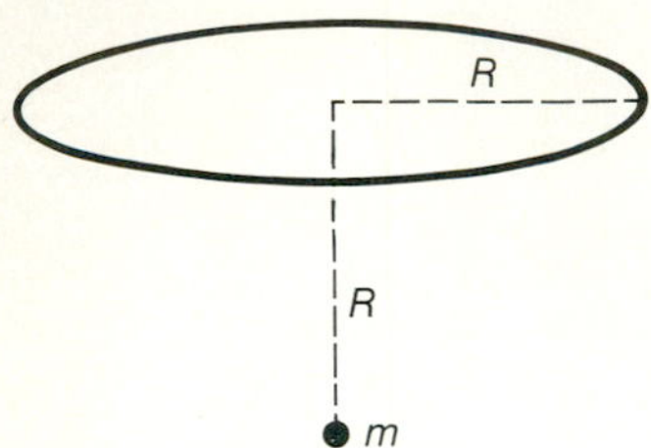

Figure 14–16
Problem 14B–13

14B–12 Consider a star of mass m located at the center of a thin, ring-shaped nebula of total mass M and radius R. In terms of m, M, R, and G, find the gravitational potential energy of the star. Assume that the potential energy is defined to be zero when the star is infinitely far from the nebula.

14B–13 Consider the ring-shaped nebula in the previous problem. With what speed v would the star pass through the center of the ring if it approached along the axis of the ring, starting at rest a distance R from the center of the ring (see Figure 14–16)? *Answer:* $\sqrt{(GM/R)(2-\sqrt{2})}$

14B–14 Ignoring the presence of the other planets, find the escape velocity of a rocket launched at Cape Canaveral that will enable it to leave the solar system. (Note: Take into account both the sun and the earth.) Is the assumption regarding the other planets a plausible one? Justify your answer.

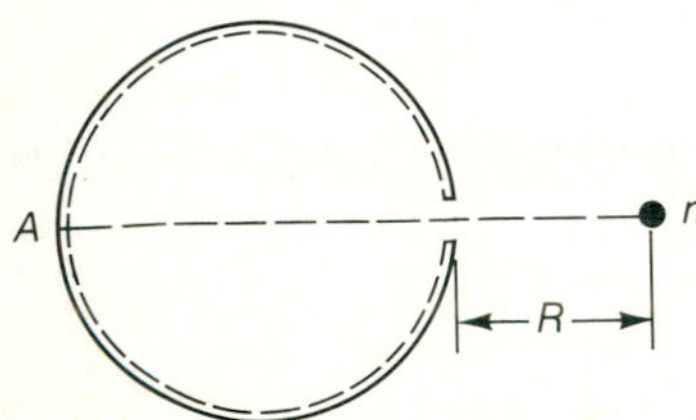

Figure 14–17
Problem 14B–15

14B–15 A thin, spherical shell of total mass M and radius R is held fixed. There is a small hole in the shell, as shown in Figure 14–17. A mass m is released from rest a distance R from the hole along a line that passes through the hole and also through the center of the shell. The mass subsequently moves under the gravitational force of the shell (only). How long does it take the mass to travel from the *hole* to the point A diametrically opposite? *Answer:* $2\sqrt{R^3/GM}$

14B–16 After exhausting its nuclear fuel, the ultimate fate of our sun is possibly to collapse to a *white dwarf* state, having approximately the mass of the sun, but the radius of the earth. Calculate (a) the average density of the white dwarf, (b) the acceleration due to gravity at its surface, and (c) the gravitational potential energy of a 1-kg object at its surface ($U_{\text{grav}} = 0$ at infinity).

14B–17 One of Jupiter's moons has an orbital radius of about 10^6 km and a period of about 7 earth-days. Use these data to estimate the mass of Jupiter. *Answer:* 1.62×10^{27} kg

14C–1 Figure 14–18 shows a small mass and a thin rod, each of mass m, positioned along the same straight line as shown. The small mass is a distance L from the end of the rod, whose length is $2L$. Find the force of gravitational attraction the rod exerts on the small mass. (Hint: Consider the force exerted by the mass element $dm =$ (mass/length) dx and integrate to find the total force.) *Answer:* $Gm^2/3L^2$

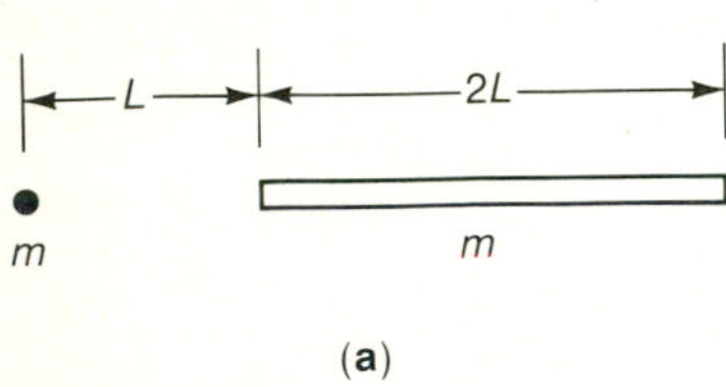

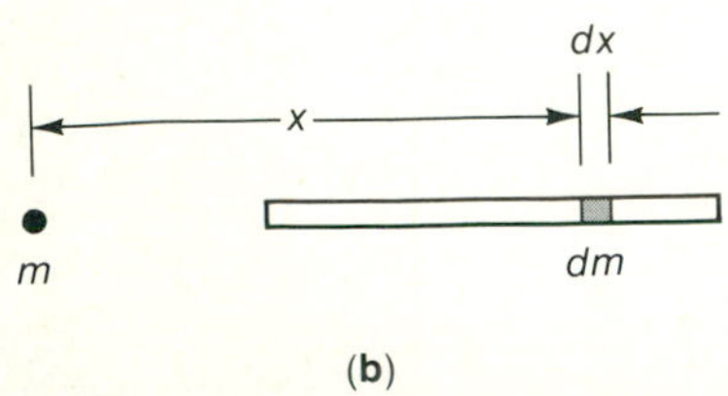

Figure 14–18
Problem 14C–1

14C–2 The density of a spherical distribution of mass is given by $\rho = kr$ where $r \leq R$ and $\rho = 0$ where $r > R$.

(a) Derive an expression for the total mass.

(b) Derive an expression for the gravitational field for all values of r.

14C–3 An object slides without friction on a *straight* track (that is, the track does not conform to the curvature of the earth). If the center of the track is horizontal, the object will slide back and forth along the track. Find the period of small-amplitude oscillations. *Answer:* 1.41 h

14C–4 The equatorial and polar radii of the earth are $r_e = 6.378 \times 10^6$ m and $r_p = 6.357 \times 10^6$ m. They differ by an amount $\Delta r = 22 \times 10^3$ m (about 13.6 mi).

(a) Derive a general equation for the fractional change in the *gravitational force* on an object of mass m when it is moved from the north pole to the equator. (The fractional change is defined as the ratio of change in the force to the original force.)

(b) Calculate its numerical value.

14C–5 A small cylindrical hole is drilled along the diameter of a uniform, solid sphere whose original mass was M and whose radius is R. The mass of the material excavated is m. Find the gravitational force on a small mass m located along the hole's axis a distance $2R$ from the center of the sphere. (Hint: Drilling the hole essentially subtracts the force due to the material excavated; see Problem 14C–1.)

Answer: $\dfrac{Gm}{R^2}\left(\dfrac{M}{4} - \dfrac{m}{3}\right)$

14C–6 Imagine that the entire sun collapsed to a sphere of radius R_g such that the work required to remove a small mass m from the surface would be equal to the rest energy m_0c^2 of the mass. This radius is called the *gravitational radius* for the sun. Find R_g. (It is believed that the ultimate fate of many stars is to collapse to their gravitational radii or smaller.)

14C–7 Imagine that a small hole is drilled to the center of the earth (mass M and radius R), so that a mass point m can move radially inside the earth. Derive an expression for the gravitational potential energy $U(r)$ of the mass point as a function of r (for $r < R$). (Hint: Since $U(R) = -GMm/R$, work in from this point using $\Delta U = -\int \mathbf{F} \cdot d\mathbf{r}$. Then, at the point r, find the fraction of the earth's mass that exerts a force on m.)

Answer: $-\dfrac{GMm}{2R}\left[3 - \left(\dfrac{r}{R}\right)^2\right]$

14C–8 Imagine a hole drilled straight down to the center of the earth. A small mass m is dropped into the hole. Ignoring the earth's rotation and all sources of friction, find the speed of the mass just as it reaches the earth's center. (You may use the answer to the previous problem if you wish.)

14C–9 Assume the earth to be a homogeneous sphere of radius R and mass M. In the following cases, show that the restoring force on the mass m is proportional to the displacement away from the equilibrium position and that therefore the motion will be simple harmonic. Also show that in each case, the period T of the SHM oscillation has the same value.

(a) A smooth, straight hole is bored along a diameter of the earth and a small mass m is dropped into the hole. (The mass will subsequently undergo SHM from one side of the earth to the other.)

(b) A smooth, straight hole is bored along a chord from a point P_1 on the earth's surface to any other arbitrary point P_2 on the surface (not along a diameter). A mass m is dropped into the hole at one end. *Answer:* $T = 2\pi\sqrt{R^3/GM} = 84.5$ min

14C–10 Assume the earth to be a homogeneous sphere of radius R and mass M. Show that the following three situations lead to a period T, which is the same as that found in the previous problem: (a) a simple pendulum of a length equal to the earth's radius, $\ell = R$, oscillating with small amplitude in a *uniform* gravitational field equal to that at the earth's surface, (b) a simple pendulum of essentially infinite length (Figure 14–19) whose mass oscillates with small amplitude in the true gravitational field at the earth's surface (that is, in a field proportional to $1/r^2$), and (c) a satellite orbiting the earth in a circular trajectory that just skims the earth's surface (assume the radius of the orbit is the radius of the earth).

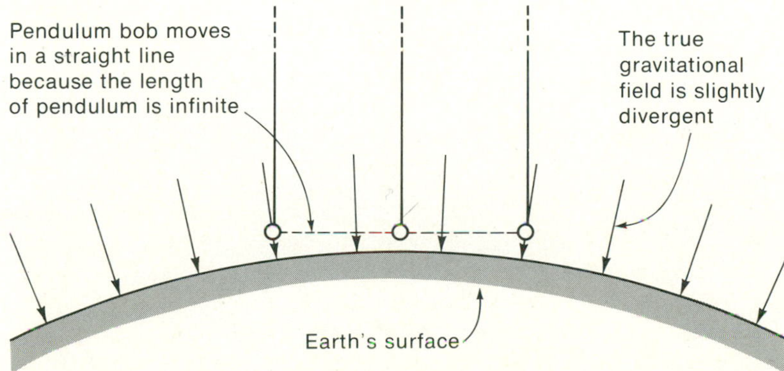
Pendulum bob moves
in a straight line
because the length
of pendulum is infinite
The true
gravitational
field is slightly
divergent
Earth's surface

Figure 14–19
Problem 14C–10

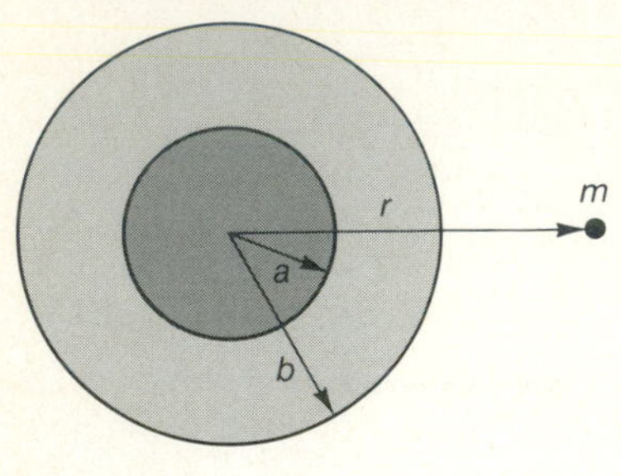

Mass of spherical shell = M

Figure 14–20
Problem 14C–12

14C–11 Show that the two expressions we have developed for gravitational potential energy, $U = -GMm/r$ and $U = mgh$, are consistent. (Hint: Assume r changes to $r + h$, where $h \ll r$. Over this small distance, g may be assumed constant.)

14C–12 A sphere of uniform matter has a spherical hole at its center, as shown in Figure 14–20. The mass of the spherical shell is M, and the inner and outer radii are a and b, respectively. Consider the gravitational force $F(r)$ exerted by this spherical shell on a mass point m at various distances r from the center for $0 < r < \infty$.

(a) Sketch a graph of $F(r)$ versus r, indicating the exact values (in terms of the given parameters and G) at the points $r = 0$, a, and b.

(b) Sketch a similar graph for the gravitational potential energy, $U(r)$ versus r, where $U(\infty) \equiv 0$.

14C–13 On the earth's surface, a woman can jump straight up, raising her center of mass a distance h. Find the radius of the largest asteroid (of density equal to the earth's average density) from which the woman could completely escape by jumping equally fast.

Answer: $\sqrt{R_{\text{earth}}h}$

14C–14 After a supernova explosion, a star may undergo a gravitational collapse to an extremely dense state known as a neutron star, in which all the electrons and protons are squeezed together to form neutrons. A neutron star with a mass about equal to that of the sun would have a radius of about 10 km. Find (a) the acceleration due to gravity at its surface, (b) the weight of a 70-kg man at its surface, and (c) the energy required to remove a neutron from its surface to infinity.

14C–15 It is claimed that a commercially available, portable gravity meter is sensitive enough to detect changes in g to 1 part in 10^{11}. At the earth's surface, what change in elevation would produce this variation? Assume the radius of the earth is 6×10^6 m. (Hint: See Appendix E for a useful approximation for $1/(1 + x)$ when $x \ll 1$.)

Big whirls have little whirls,
That feed on their velocity;
And little whirls have lesser whirls,
And so on to viscosity.

LEWIS RICHARDSON

[summarizing his classic paper,
The Supply of Energy From and To Atmospheric Eddies (1920)]

chapter 15
Fluids

15.1 Introduction

This chapter has two main themes: the pressure within a fluid at rest and the variations of pressure in a fluid in motion. *Hydrodynamics* is a field of engineering that deals with fluids in motion, and *aerodynamics* deals with gas flow. Our discussion will be limited to the very simplest cases. There will be little new physics introduced in this chapter, but it will be an interesting example of applying Newton's laws and the conservation of energy to a new physical situation.

Matter is commonly said to exist as a *solid* or a *fluid*. The word **fluid** denotes an ability to flow and thus includes both liquids and gases. Another state of matter, a highly ionized gas with equal amounts of positive and negative particles, is called a **plasma.** (Sometimes a plasma is classified as a special type of fluid.) Recently, the properties of plasmas have been intensively studied as part of the problem of achieving controlled nuclear fusion reactions.

These broad classifications, however, oversimplify the situation. In some instances, the dividing line between a liquid and a gas (or a liquid and a solid) is not at all clear-cut. Some familiar "solids," such as glass, are more properly classified as an incompletely solidified liquid. No solid is completely rigid. And many solid substances exist in several distinct forms with different melting points, densities, specific heats, and so on. There are, for example, three different forms of solid carbon, and seven different kinds of ice.

We shall not concern ourselves with these complications here. Instead, we will discuss the behavior of an imaginary **ideal liquid**[1]—one that is incompressible and completely nonviscous. **Viscosity** is the property by which fluids and gases offer resistance to objects moving through them. Such resistive forces are not constant in magnitude, but depend on the velocity. For example, if a small metal sphere is dropped into thick oil, the sphere soon acquires a constant **terminal velocity**—the speed at which the viscous resistive force just equals the force of gravity on the sphere. Thus there is zero net force and the object falls without acceleration. Viscosity is also associated with resistive forces present when layers of fluid or gas slide over one another.

[1] Of all real liquids, water is among those that approximate an ideal liquid fairly closely, at least for pressures and flow velocities commonly encountered.

15.2 Density

As mentioned previously, the **density** ρ of matter is defined as

DENSITY
$$\rho \equiv \frac{\Delta m}{\Delta V} \tag{15-1}$$

where Δm is the mass of matter occupying a volume ΔV. Assuming an ideal liquid, we note that the density remains constant for all pressures. In the case of water, this is a reasonable assumption, since it requires a pressure increase of about 200 times atmospheric pressure to compress a given volume by only 1%. In the English system, it is customary to use the *weight density* in units of lb/ft^3 (see Table 15–1).

The **specific gravity** s of a substance is defined as the ratio of the density of the substance to the density of water. Being a ratio of two densities, it is a number without units.

TABLE 15–1

The Densities of Some Materials

		Mass density $\equiv \rho$ (kg/m^3)	Weight density $\equiv \rho g$ (lb/ft^3)
Gases			
Interstellar space		10^{-18} to 10^{-21}	
Vacuum pump		$\sim 10^{-14}$	
Hydrogen (H_2)	At 0°C and 1 atm	0.0899	0.00560
Air	At 0°C and 1 atm	1.293	0.08071
Oxygen (O_2)	At 0°C and 1 atm	1.429	0.08921
Radon (Rn)	At 0°C and 1 atm	9.73	0.607
Tungsten fluoride (WF_6)	At 0°C and 1 atm	12.9	0.805
Solids and Liquids			
Styrofoam		~ 30	
Walnut wood		$0.64–0.70 \times 10^3$	40–43
Ice		0.917×10^3	
Water (at 0°C and 1 atm)		1.000×10^3	62.4
(at 0°C and 50 atm)		1.002×10^3	
(at 100°C and 1 atm)		0.958×10^3	
Sea water		1.025×10^3	64.0
Beryllium	Metallic elements	1.84×10^3	114.9
Aluminum	Metallic elements	2.7×10^3	168.5
Mercury	Metallic elements	1.36×10^4	848.8
Uranium	Metallic elements	1.87×10^4	1167.
Osmium	Metallic elements	2.25×10^4	1400.
Earth, core		9.5×10^3	
crust		2.8×10^3	
average		5.52×10^3	344.7
Sun, center		1.6×10^5	
average		1.4×10^3	
White dwarf stars (core)		$10^8 - 10^{15}$	
Nuclear matter		$\sim 10^{17}$	

SPECIFIC GRAVITY

$$s \equiv \frac{\rho}{\rho_{\text{water}}} \tag{15-2}$$

The term "specific gravity" is, perhaps, a misnomer, since it has nothing to do with gravity.

15.3 Pressure

Let us now consider the area of contact between a fluid and its container. The container can only exert a force *normal* to the fluid interface. This conclusion follows from the assumption that the fluid is nonviscous. That is, if the fluid is subjected to a tangential force, the "layers" of the fluid would merely slide over one another without friction. Thus an ideal fluid cannot sustain a tangential force, so it in turn cannot exert a tangential force on any surface with which it is in contact. It can exert forces only perpendicularly to the surface.

We may extend these comments to apply to regions within the fluid itself. Imagine an arbitrarily oriented elementary area situated somewhere within the fluid. The surrounding fluid can only exert forces that are perpendicular to the surface of this area. Thus it becomes convenient to define the **pressure** p acting on a fluid in terms of the force ΔF exerted perpendicularly on an elementary area ΔA:

PRESSURE

$$p \equiv \lim_{\Delta A \to 0} \frac{\Delta F}{\Delta A} \tag{15-3}$$

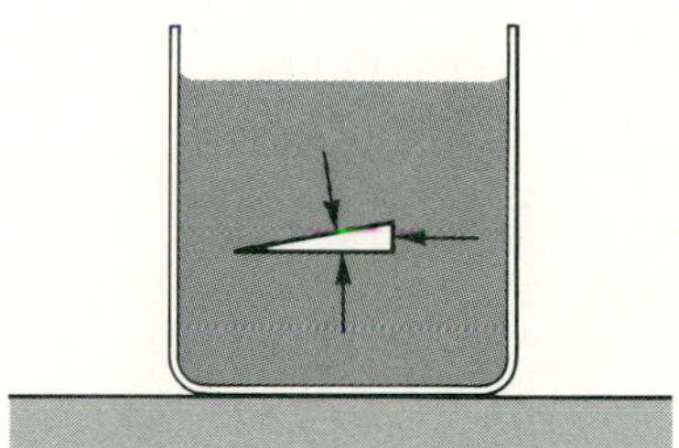

Figure 15–1
A liquid exerts forces on a submerged object that are always normal to the object's surface area.

Even though force is a vector quantity, the pressure is defined to be a *scalar* quantity, having no particular direction. It is measured in units of force per unit area or in a variety of other units indicated by Equations (15–7).

The lack of a particular direction for pressure in a fluid is illustrated by the following example. Consider a small, solid, triangular wedge submerged in a liquid (Figure 15–1). If the density of the liquid and the wedge are equal, there is no tendency for the wedge to sink or float upward: it is in equilibrium. The liquid exerts forces on the wedge that are always normal to the surface, *regardless of the orientation of the wedge.* Thus, pressure at a given location in a fluid does not have a specific direction.

As we shall now demonstrate, the pressure within a liquid at rest depends on two factors: the pressure on the surface of the liquid and the depth below the surface. Consider a small volume element of liquid at a depth h below the surface, as in Figure 15–2a. The element is in the form of a thin rectangular

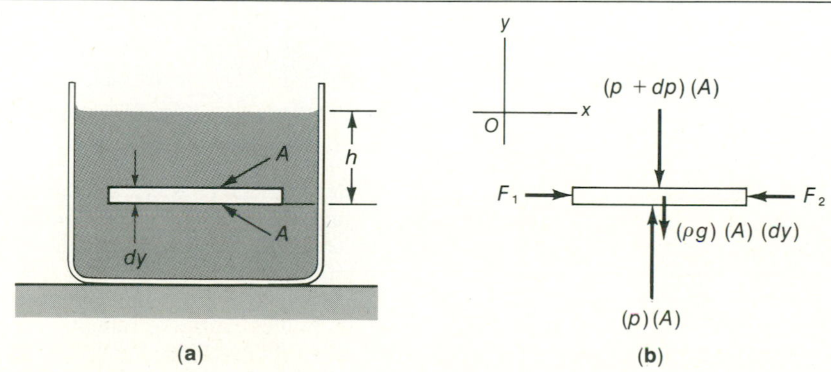

Figure 15–2
(a) An elementary volume of liquid in the form of a thin, horizontal, rectangular slab of area A and thickness dy. (b) A free-body diagram showing the forces on the element of volume in (a).

slab of vertical thickness dy and mass dm. Each horizontal face has an area A. The volume is at rest, so the sum of the forces on it must add to zero. In every instance, the forces are perpendicular to the surfaces and equal to (pressure)(area).

A free-body diagram for the slab is shown in Figure 15–2b. By symmetry, the horizontal force components add to zero. The vertical forces also add to zero, but we must include the force of gravity on the slab itself. This force is $(dm)g = (\rho\, dV)g = (\rho A\, dy)g$. The force of the fluid on the lower face of the slab is (pA) and that on the upper face is $(p + dp)(A)$, where dp is the infinitesimal pressure difference between the upper and lower sides of the slab. Hence, summing all the vertical forces on the slab, we have

$$\Sigma F_y = 0$$

$$(p)(A) - (p + dp)(A) - (\rho g)(A)(dy) = 0$$

$$dp = -\rho g\, dy \qquad \textbf{(15–4)}$$

Let us establish a zero reference of vertical distance at the top surface of the liquid, with upward displacements designated positive. Then as we go down into the liquid, dy changes negatively, and the minus sign of Equation (15–4) ensures that the pressure change dp will be positive. The pressure p at an arbitrary depth h below the surface of a liquid thus becomes

$$\int_{p_0}^{p} dp = \int_{y=0}^{y=-h} -\rho g\, dy \qquad \textbf{(15–5)}$$

$$p - p_0 = \rho g h$$

PRESSURE IN A FLUID AT REST

$$p = p_0 + \rho g h \qquad \textbf{(15–6)}$$

Figure 15–3

In 1650, a German engineer named Otto von Guericke devised the first vacuum pump that could create near-zero pressures. In Magdeburg, he demonstrated the surprisingly large pressure of the atmosphere by showing that 16 horses could not pull apart two metal hemispheres, 51 cm in diameter, that had been held together and then evacuated. He no doubt knew the dramatic advantage of using 16 horses pulling in opposite directions, rather than creating the same force by attaching one hemisphere to a fixed support and pulling on the other half with just 8 horses.

where p_0 is the pressure at the surface and h is the depth below the surface, expressed as a *positive* number.

In a given situation, the pressure p_0 on the surface may be one standard atmosphere of pressure (1 atm), or any other value.

STANDARD ATMOSPHERIC PRESSURE

$$
\begin{aligned}
1 \text{ atm} &= 1.013 \times 10^5 \text{ N/m}^2 = 1.013 \times 10^5 \text{ Pa} \\
&\quad [\text{a } \textit{pascal} \text{ (Pa) is the SI unit for pressure}] \\
&= 1.013 \times 10^6 \text{ dynes/cm}^2 \\
&= 1013 \text{ millibars} \\
&= 14.7 \text{ lb/in}^2 = 2116 \text{ lb/ft}^2 \\
&= 76.0 \text{ cm Hg}
\end{aligned}
\qquad \textbf{(15–7)}
$$

Meteorologists often express local values of atmospheric pressure in the unit *bar*, defined to be 10^6 dynes/cm^2 (equal to 10^5 N/m^2). Thus one standard atmosphere is 1013 millibars. Another expression in Equation (15–7) uses the shorthand notation "cm Hg" for the phrase "centimeters of mercury." This usage arises from the fact that a pressure of one standard atmosphere supports a column of mercury 76.0 cm high in a device called a *barometer*. Note carefully, however, that pressure always has fundamental units of a *force per unit area*, not a length in centimeters of mercury.

As a practical convenience in engineering, pressure gauges sometimes are set to read zero when the pressure is at 1 atm. Thus, two ways of specifying pressure are in common usage: **absolute pressure** (or, in physics, just *pressure*) and **gauge pressure** (the pressure minus one atmosphere). Gauge pressures of less than 1 atm have negative numerical values. Hence, a gauge pressure of -4.0 lb/in^2 represents an absolute pressure of $+10.7$ lb/in^2.

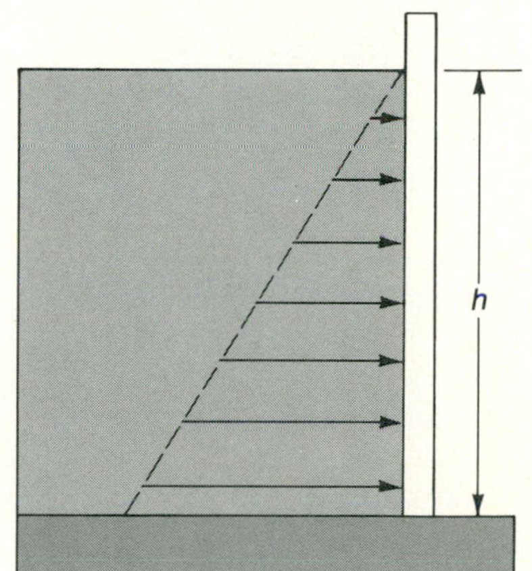

EXAMPLE 15–1

A certain diving pool has the dimensions 15 ft × 30 ft. When filled with water to a constant depth of $h = 10$ ft, (a) what total force does the water exert on the bottom? (b) What is the pressure at the bottom? (c) What total force does the water exert on a vertical end wall (15 ft wide, 10 ft deep)? In this example, consider the water alone, ignoring air pressure. (Note: A cubic foot of water weighs 62.4 lb.)

SOLUTION

(a) The total force the water exerts on the bottom is its weight:

$$\left(\frac{\text{weight}}{\text{volume}}\right)(\text{volume}) = \left(62.4 \frac{\text{lb}}{\text{ft}^3}\right)(15 \text{ ft})(30 \text{ ft})(10 \text{ ft}) = \boxed{2.81 \times 10^5 \text{ lb}}$$

(b) The pressure at the bottom (ignoring air pressure) is:

$$p = \rho g h = \left(62.4 \frac{\text{lb}}{\text{ft}^3}\right)(10 \text{ ft}) = \boxed{624 \frac{\text{lb}}{\text{ft}^2}}$$

(c) The pressure on the end wall varies uniformly from zero at the top to $\rho g h$ at the bottom (see Figure 15–4). The pressure p at a depth y due to the water alone is:

$$p = \rho g y$$

The force dF on the element of area dA is

$$dF = p\,dA = \rho g y \ell\,dy$$

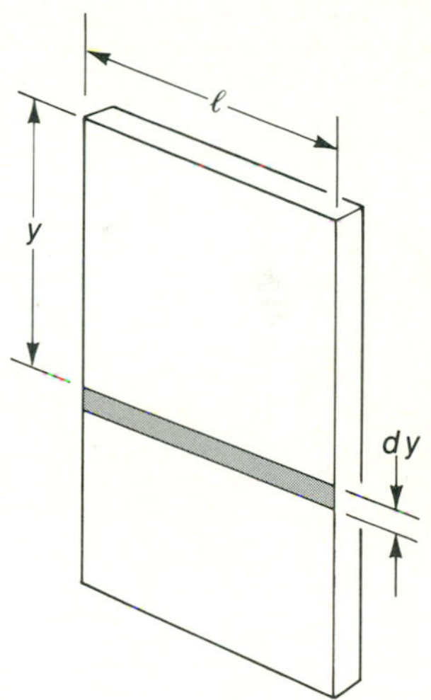

Figure 15–4
The pressure on a vertical wall due to a liquid (not counting air pressure on the liquid surface) varies uniformly from zero at the surface to $\rho g h$ at the bottom. At a depth y below the surface where the horizontal element of area $\ell\,dy$ is located, the pressure is $\rho g y$.

The total force F is thus:

$$F = \int dF = \int_0^h \rho g y \ell \, dy = \frac{\rho g h^2 \ell}{2} = \frac{\left(62.4 \frac{\text{lb}}{\text{ft}^3}\right)(100 \text{ ft}^2)(15 \text{ ft})}{2} = \boxed{4.68 \times 10^4 \text{ lb}}$$

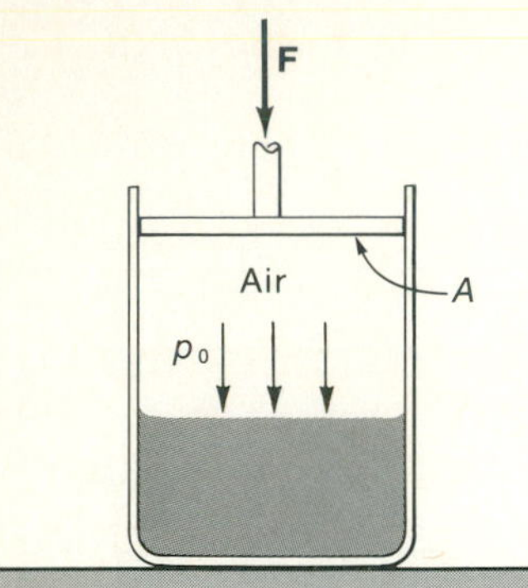

Figure 15–5
A liquid subjected to various pressures by a movable piston. (The air space above the liquid does not need to be present.)

Alternative method for (c): Because the side wall of the pool has the same width at all depths, the pressure p varies *uniformly* with the depth y. Thus the average pressure p_{ave} is that halfway down, or $\rho gh/2$. Multiplying this average pressure by the total area of the end wall gives

$$F = (p_{ave})(\text{area}) = \left(\frac{\rho g h}{2}\right)(h\ell) = \boxed{\frac{\rho g h^2 \ell}{2}}$$

(Note that this method does not work if the width of the side wall varies with depth.)

15.4 Pascal's and Archimedes' Principles

If a liquid is confined in an enclosed vessel, any desired external pressure may be applied to the liquid, as illustrated schematically by the piston in Figure 15–5. Equation (15–6) still applies. If we ignore the weight of the enclosed air itself, the air serves merely to communicate the force of the piston to the liquid. Obviously, the air does not need to be present for Equation (15-6) to be valid. The fact that pressure at any depth in a liquid also includes the constant amount p_0 applied externally to the liquid was discovered by Blaise Pascal (1623–1662):

PASCAL'S PRINCIPLE **Pressure applied to an enclosed fluid is transmitted undiminished to all parts of the fluid and to the walls of the container.**

Another important principle regarding fluids was discovered by Archimedes (287–212 B.C.) in his attempt to determine whether the king's crown was of solid gold or had been alloyed with silver. Reputedly, Archimedes discovered this principle after noting the apparent partial loss of weight on his arms and legs while submerged in his bath. Presumably he weighed equal amounts of gold and silver in air and also when submerged in water to verify that, indeed, the crown was made of pure gold. His conclusion:

ARCHIMEDES' PRINCIPLE **Any body, wholly or partially submerged in a fluid, is buoyed up by a force equal to the weight of the displaced fluid.**

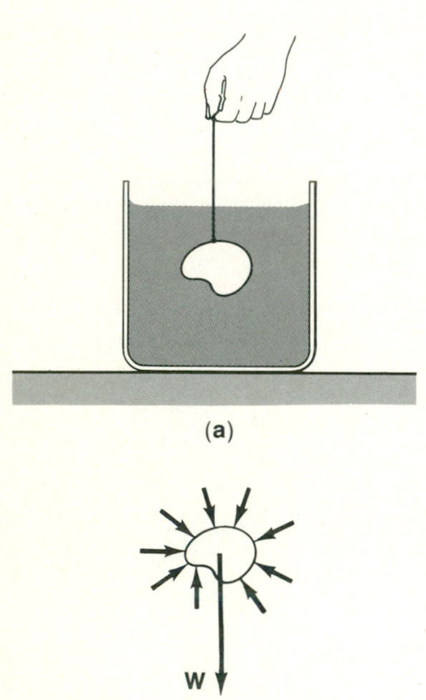

Figure 15–6
A submerged object displaces a portion of the fluid equal to its own volume. Before being displaced, the fluid portion had forces on it due to the surrounding fluid, as shown in (b). The sum of these forces is equal and opposite to **W**, the weight of the displaced fluid.

This principle becomes obvious when we make a free-body diagram for the portion of fluid destined to be displaced when an object is submerged (see Figure 15–6b). Before being displaced, the net buoyant force on this portion exerted by the surrounding fluid must equal the weight W of the portion itself. Furthermore, the buoyant force may be considered to be concentrated at the center of gravity of the displaced fluid. This point is called the **center of buoyancy** (or the **center of buoyant force**).

Archimedes' principle has a practical application in the design of boats. The center of gravity of a boat must be located relative to the center of the buoyant force so as to produce a torque that tends to right the boat if tipped (**a**) In a properly designed boat (with ballast), the torque tends to right the boat.

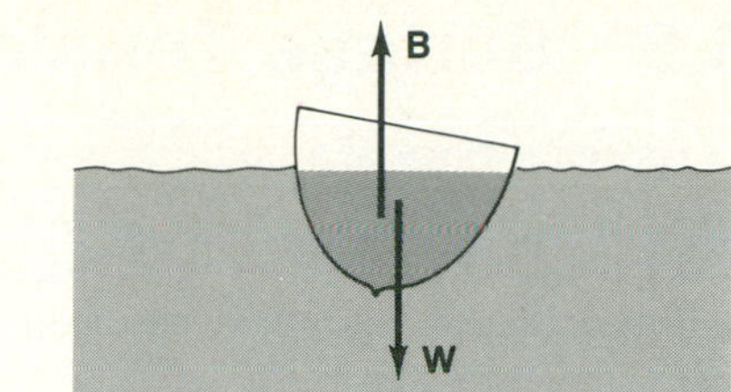

(**b**) If designed improperly, the torque tends to tip the boat even more.

Figure 15–7
The buoyant force **B** and the weight **W** produce a torque on a boat tipped at an angle. In (a), the point of application of **B** is at the center of gravity of the displaced fluid and changes for different angular positions; **W** is at the boat's center of gravity.

Figure 15–8
When the 480-ton "Piero Riero Gambini" was launched in Teduccio, Italy, on January 26, 1952, someone forgot to install the ballast necessary to maintain the upright position of the ship. As the ship turned over, more than 50 people were thrown overboard, all escaping serious injury.

(see Figure 15–7a). As the boat rotates to various angles, the center of buoyancy usually shifts its position relative to the boat because the shape of the displaced fluid changes. Standing up in a canoe is particularly hazardous because the center of gravity of the canoe plus the passengers is raised so high that the condition of Figure 15–7(b) is very easily achieved. Figure 15–8 is a further illustration.

In the laboratory, Archimedes' principle is useful for determining the density of an unknown liquid. If a liquid less dense than water is poured into one arm of a U-tube containing some water, the fluid levels will adjust themselves similar to Figure 15–9. Point a is the interface between the water and the less dense liquid. Points a and b in the water are the same depth beneath the water surface c. Thus they will have the same pressure. This pressure, due to the liquid only, is

$$p_a = p_b = \rho_w g h_1 \tag{15–8}$$

where ρ_w is the density of water. (We ignore here the pressure of the atmosphere; that is, all expressions are gauge pressures.)

But the pressure at a is also determined by the depth h_2 and the density of the unknown liquid ρ_x. Thus:

$$p_a = \rho_x g h_2 \tag{15–9}$$

Solving the above two equations for ρ_x leads directly to

$$\rho_x = \rho_w \left(\frac{h_1}{h_2} \right) \tag{15–10}$$

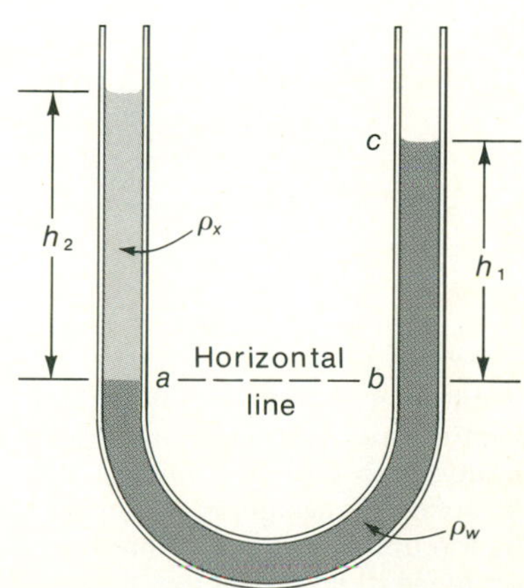

Figure 15–9
A U-tube is filled with water; a less dense liquid is then poured into the left arm of the tube.

15.5 Surface Tension

An exposed surface of a liquid acts as a stretched membrane that exerts forces on the liquid enclosed. The reason is that molecules at the surface are attracted to their neighbors with a finite net force that tends to pull the molecule into the liquid. A molecule wholly submerged, on the other hand, feels an average net force of zero (see Figure 15–10).

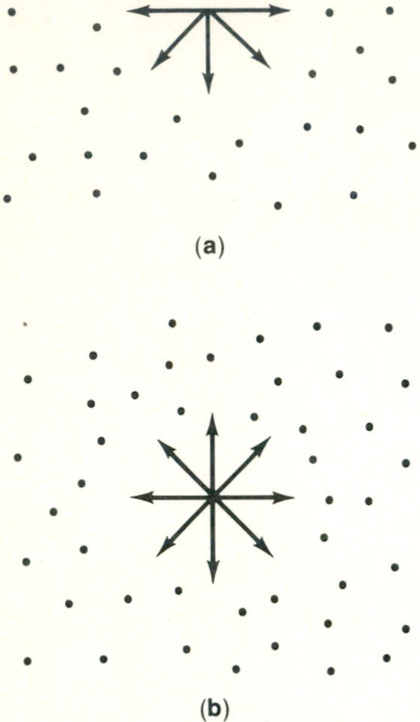

Figure 15–10
Forces due to neighboring molecules on a molecule of a liquid (a) at the surface and (b) inside the liquid.

To create a new surface of a liquid, we in effect must do work on the liquid to pull some molecules away from their neighbors. Let the Greek symbol gamma, Γ, represent the *work done per unit area.* Figure 15–11 illustrates a simple apparatus to determine Γ. If we pull the horizontal wire downward a short distance y, the film of liquid is stretched and exerts an upward force on the wire (which we work against). This indicates that the surface energy increases when the area is increased. If the width of the wire frame is ℓ, the work we do in pulling the wire a distance y equals the increase in surface energy. Thus:

$$\Gamma = \frac{\text{work}}{\text{area created}} = \frac{Fy}{2\ell y} = \frac{F}{2\ell} \qquad \textbf{(15–11)}$$

Note the factor of 2, which accounts for the *two* surfaces of the film (front and back).

Thus we may think of surface tension Γ not only as the work done per unit area, but also as a *force per unit length* of any line in the surface. Unlike the force associated with a stretched string ($F = -ky$), the force involved here is essentially constant with distance as the film expands.

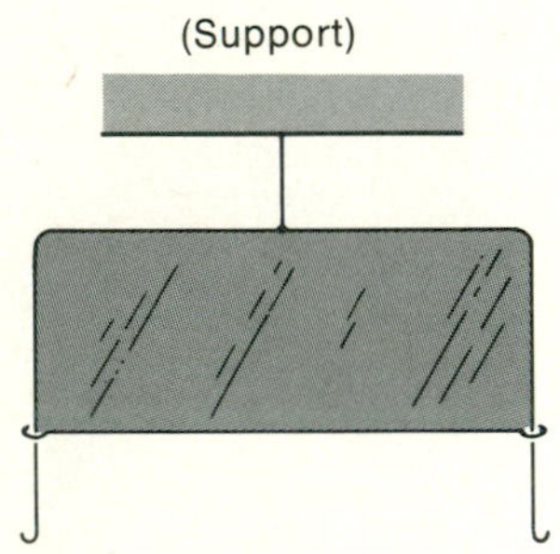

Figure 15–11
A thin film of liquid in a wire frame. The horizontal wire at the bottom can slide in a vertical direction. Its weight is supported by the upward forces of the *two* surfaces of the liquid film.

EXAMPLE 15–2

How much work is done by an atomizer in converting 1 liter of water entirely into droplets, each 10 μm in diameter? Consider only the work done in forming the surface area of the droplets and assume that the surface tension of water is 0.075 N/m.

SOLUTION

The work done in creating a surface area A is given by

$$W = \Gamma A$$

The volume of n droplets each with a radius r is

$$V = \tfrac{4}{3}n\pi r^3$$

and the total surface area of the droplets is

$$A = 4n\pi r^2$$

Eliminating n between these equations for the total volume and total area, we have

$$A = \frac{3V}{r}$$

The work required is then the surface tension Γ times the area A:

$$W = \frac{3\Gamma V}{r}$$

Substitution of appropriate values in SI units, where 1 liter $= 10^{-3}$ m^3 and 1 μm $= 10^{-6}$ m yields

$$W = \left(0.075\ \frac{\text{N}}{\text{m}}\right)\frac{(3)(10^{-3}\ \text{m}^3)}{(5 \times 10^{-6}\ \text{m})} = \boxed{45.0\ \text{J}}$$

Surface tension creates very large pressures within small droplets of liquid. Consider a typical cloud droplet of diameter D. If we mentally slice the droplet in two, the forces acting on half the droplet are as shown in Figure 15–12. In the diagram, one of the surface tension force vectors is labeled $\mathbf{\Gamma}$. Remember, however, that $\mathbf{\Gamma}$ is a force *per unit length*: The total force due to surface tension is found by summing over the entire length of the boundary line of the film. Also use care to note whether just *one* surface is involved (as in a solid drop) or *two* surfaces (as in a hollow bubble). Equating horizontal forces on the half-droplet, we have

$$(p)(\text{area}) = (\Gamma)(\text{circumference of cut edge})$$
$$p\pi R^2 = \Gamma 2\pi R$$

Solving for p yields

$$p = \frac{2\Gamma}{R}$$

For a cloud droplet of radius $\sim 10^{-6}$ m, the internal pressure due to surface tension effects may be many times atmospheric pressure.

Tubes with very small bores are called **capillary tubes.** When they are partially immersed in a liquid, the liquid will rise or be depressed in the capillary, depending on whether the liquid "wets" the tube wall or not (see Figure 15–13). That is, whether there is a rise or fall depends on whether **cohesion** between like molecules of the liquid is greater than the **adhesion** between unlike molecules of the liquid and wall. In the latter case, the liquid wets the tube wall and a concave surface (called a **meniscus**) develops at the top of the column, causing the liquid to rise. On the other hand, if the liquid does not wet the tube wall, the meniscus is convex and the column is depressed below the level of the liquid surface outside the capillary. For reasons described below, this change in height of the column of water in a capillary tube depends on the angle of contact the meniscus makes with the wall (see Figure 15–14).

The pressure change Δp across a liquid surface of radius R can be shown (by the analysis of Figure 15–12) to equal

$$\Delta p = \frac{2\Gamma}{R} \qquad \textbf{(15–12)}$$

Hence, for a meniscus of radius R in a tube of radius r, we have:

$$\cos\theta = \frac{r}{R} \qquad \textbf{(15–13)}$$

Combining Equations (15–12) and (15–13), along with $\Delta p = \rho g h$ (where h is the rise of the liquid), we have

$$h = \frac{2\Gamma\cos\theta}{\rho g r} \qquad \textbf{(15–14)}$$

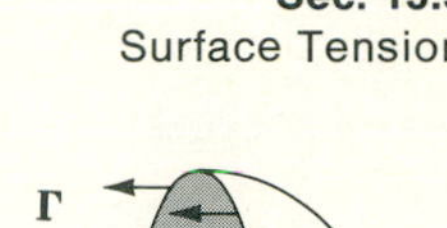
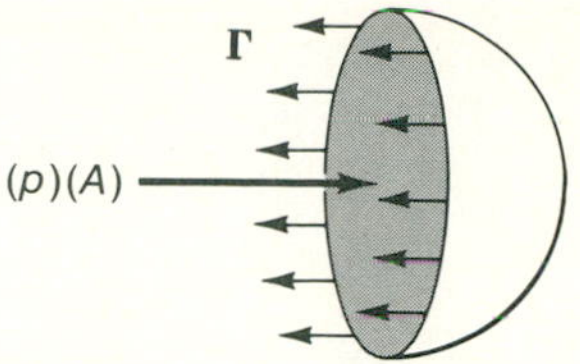

Figure 15–12
A free-body diagram for one-half of a liquid drop. The internal pressure of the missing half produces a force of $(p)(A)$ toward the right. Surface tension produces a force of $(\Gamma)(2\pi R)$ toward the left.

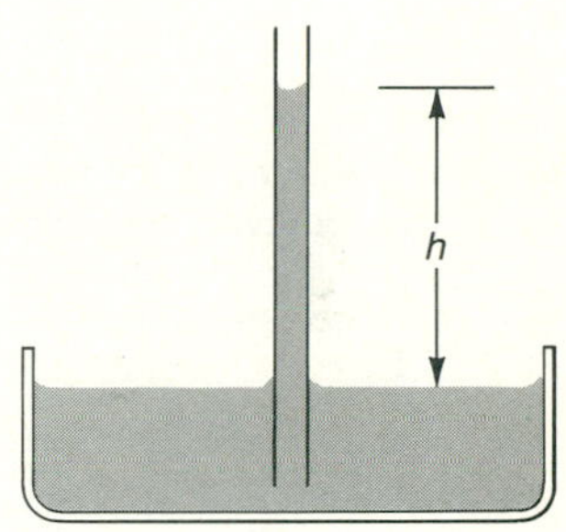

Figure 15–13
The rise of a liquid in a capillary tube (shown enlarged).

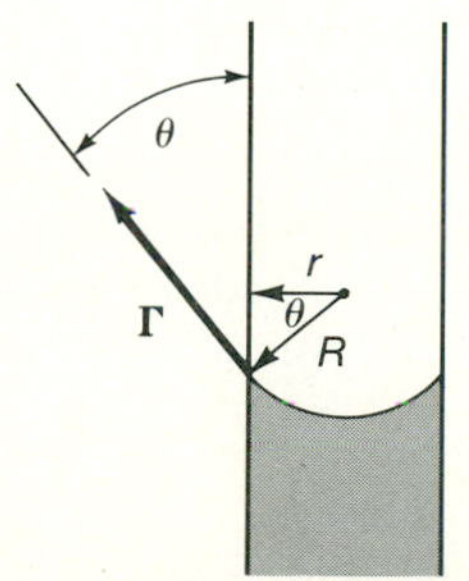

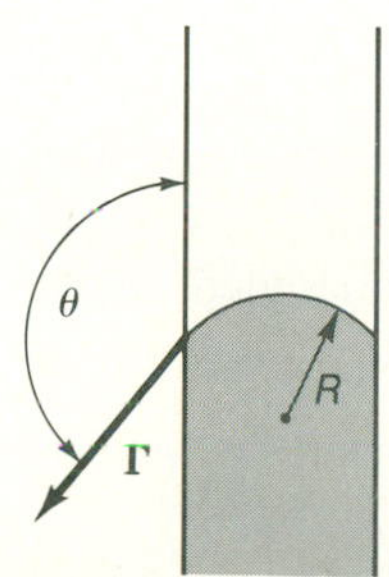

Figure 15–14
The angle of contact θ between the meniscus and the wall of a capillary tube. In each case, $\mathbf{\Gamma}$ represents the surface tension force per unit length of the liquid surface along the line where it joins the tube wall.

For $\theta < 90°$, h is positive and the liquid rises. If $\theta > 90°$, h is negative and the liquid is depressed.

Surface tension effects may support small objects that are more dense than the liquid. Provided the liquid does not wet the object, the deformed surface exerts a net upward force to balance the weight of the object. Certain small water bugs have a waxy coating on their feet, which prevents wetting and enables them to "walk" easily on the water surface. Some even secrete a substance from the abdomen that lowers the surface tension behind them; the resultant imbalance in surface tension forces propels them forward through the water at a lively speed.

Water droplets form on rinsed objects because of surface tension. In hard-water areas, the residue left behind after evaporation is often unsightly. For this reason, a "wetting" agent that greatly lowers the surface tension is sometimes used in the rinse cycle of automatic dishwashers to prevent spot formation on glassware. Photographers also use a wetting agent in rinsing film negatives to prevent similar spot formation.

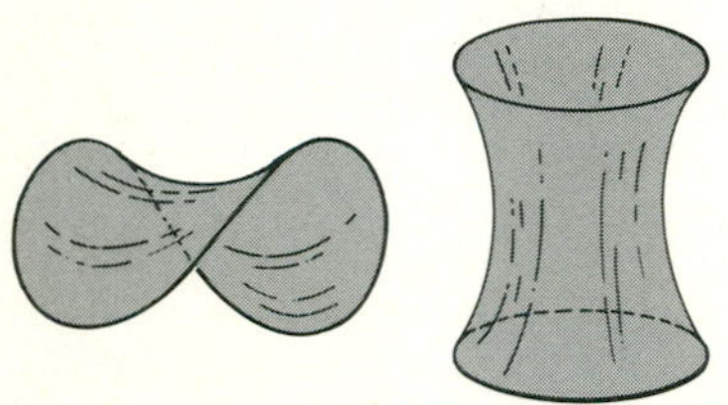

Figure 15–15
Wire frames, dipped in a soap solution and removed, will form soap films that illustrate minimum-area problems in mathematics.

Surface tension is temperature-dependent, generally weakening as the temperature is increased. A small piece of floating ice moves around in an erratic fashion because unequal melting rates cause the surface tension to vary on different sides, thus pulling the ice in one direction or another.

A soap film seeks a shape with the lowest energy, which (because of surface tension) is a shape with the minimum surface area. This property provides a simple means of demonstrating the solutions to minimum-area problems in mathematics. Wire frames are constructed to determine the "boundary conditions" for the edges of the film. When dipped in a soap solution, the resulting film automatically forms the minimum-area surface attached to the boundaries (see Figure 15–15).

15.6 Fluids in Motion

The general subject of fluid dynamics is a complex one. The following descriptive terms are used to characterize particular types of fluid flow:

nonsteady	or	steady
compressible	or	incompressible
viscous	or	nonviscous
rotational	or	irrotational

We shall primarily discuss the simpler cases that involve only the last-named terms in each of these pairs.

Because the fluid mass is nonrigid, it may undergo very turbulent motion, where the trajectories of various particles of the fluid are so complicated that a precise mathematical analysis is not feasible. For nonturbulent motion, called **steady flow** or **smooth flow,** if the rate of flow is moderate, at any fixed point in space the velocity is constant with time and can easily be measured.

A convenient way to think of fluid flow is by imagining **streamlines.** These are lines that, for steady flow, trace out the path of any given particle of the fluid. At every point, the streamline is tangent to the fluid velocity.[2] A number of clever techniques have been developed to make streamlines visible; a few are illustrated in Figure 15–16. When streamlines are well defined and do not become entangled due to turbulence, the situation is called **streamline flow** or **laminar flow** (from the Latin *lamina*, meaning a "thin plate" or "layer").

[2] In nonsteady flow, the velocities change with time and the streamlines do not necessarily coincide with the trajectories of fluid particles.

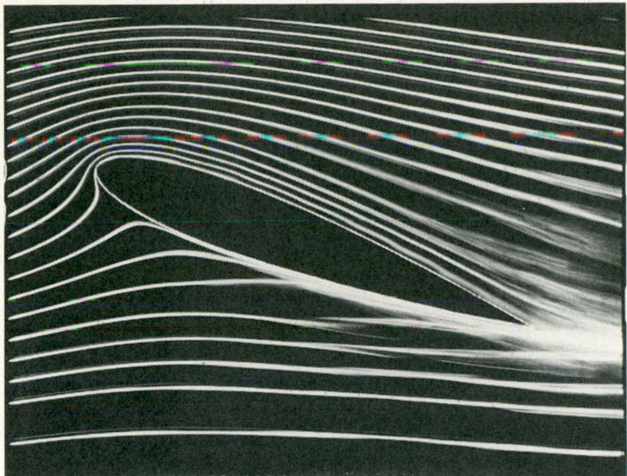

(a) Smoke lines around an air-foil model in a wind tunnel illustrate streamline flow.

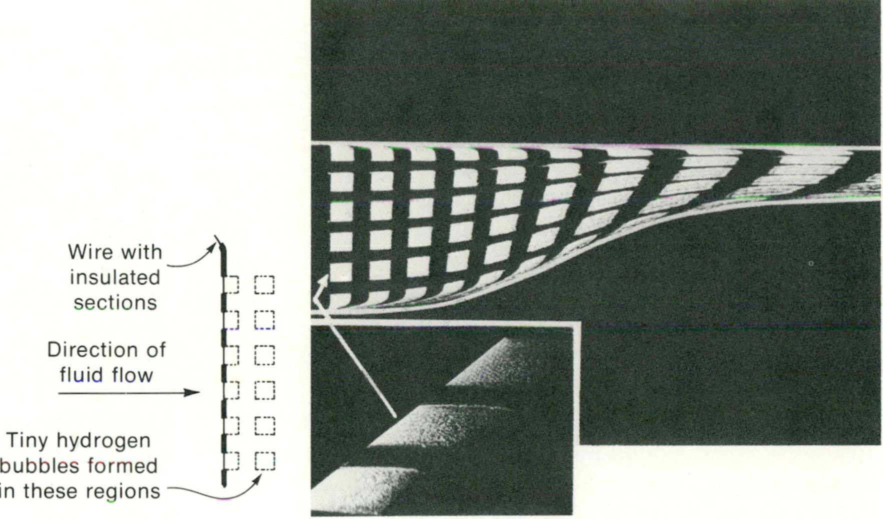

Figure 15–16
Some of the methods used in flow visualization.

(b) A thin wire, stretched normal to the flow direction in a water tank, has short sections coated with insulation. When a negative voltage is applied to the wire, electrolysis occurs at the uninsulated portions, producing myriads of tiny hydrogen bubbles in the water. By turning the voltage on and off, a series of time-streak markers is generated. As these markers are carried along in the stream, they give much information about the flow of the water as, in this example, it passes into a narrower region.

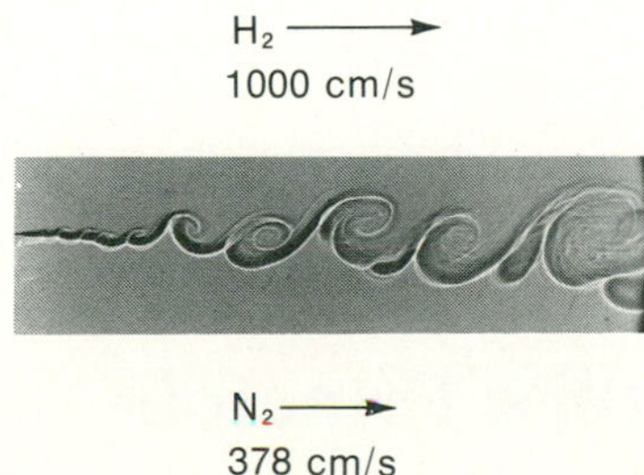

(c) Two different gases flowing side by side at different speeds cause eddies to form at the boundary. This illustrates a shadow technique useful with gases that have different indices of refraction.

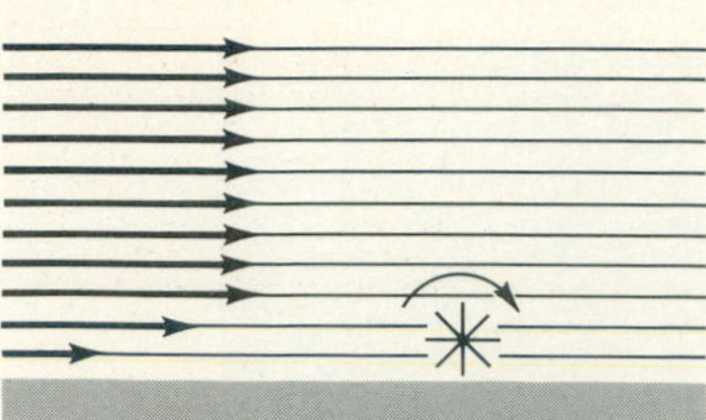

(a) Layers of fluid near the bottom of a river flow more slowly because of viscosity. If a small paddle wheel were placed near the bottom, it would be turned by the flowing stream, indicating *rotational* flow.

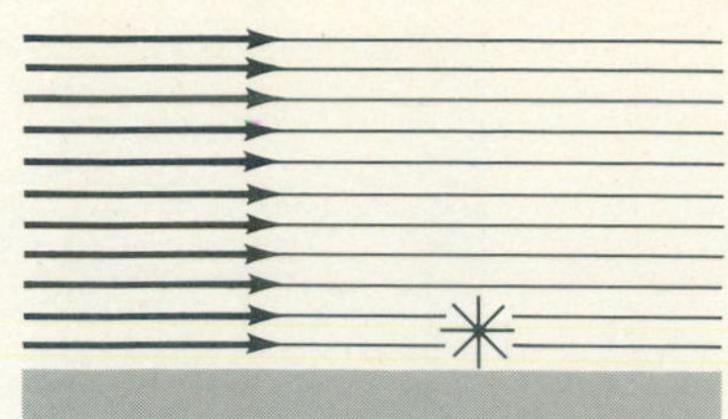

(b) An ideal fluid is nonviscous. All portions flow with the same velocity. A paddle wheel would not be turned, indicating *irrotational* flow.

Figure 15–17
Examples of rotational and irrotational flow in a river whose streamlines are parallel. Vectors indicate the flow velocities at various depths.

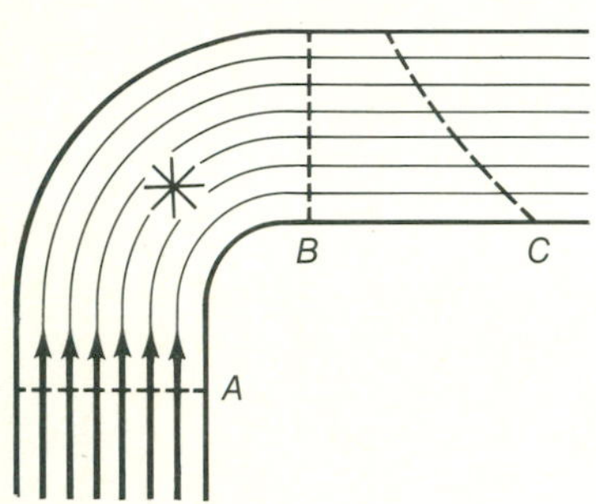

Figure 15–18
If particles at the dotted line *A* reach *B* simultaneously, the faster speeds along the outside of the curve will cause rotational flow. On the other hand, flow in the curved portion will be irrotational if streamlines along the outer region travel slower, so that the particles will reach the dotted line at *C* simultaneously.

A flow is *irrotational* when any small element of fluid has zero angular momentum about any axis through its center of mass. To test whether a flow is rotational or not, it is helpful to imagine a tiny paddle wheel immersed at arbitrary locations (and orientations) within the flowing fluid. If the paddle wheel does not rotate, the flow is irrotational.

Rotational flow does not necessarily mean the streamlines are curved. Even with parallel streamlines, the flow may be rotational. For example, all parts of a river do not flow at the same rate, even though each particle of fluid moves parallel to all others. Near the bottom and sides, the velocity is slower due to friction of the various "layers" of water sliding over one another (see Figure 15–17a). As mentioned earlier, viscosity is a measure of this frictional effect. An *ideal* fluid is nonviscous. That is, there are no tangential forces between layers of fluid in relative motion. Figure 15–17b illustrates nonviscous, irrotational flow. When streamlines are curved, the flow may or may not be rotational. Figure 15–18 illustrates this point.

A bundle of streamlines makes up what is known as a *tube of flow*, or *stream tube* (see Figure 15–19). The streamlines on the outer boundary of the tube form a sort of "pipe" for all streamlines inside, since in laminar flow no lines can cross the boundary. All streamlines entering one end at area A_1 must emerge from the other end through area A_2.

If the fluid is incompressible, its density ρ is constant throughout. Thus the volume of fluid entering the tube at one end per unit time must equal the volume of fluid leaving the other end per unit time. In a time Δt, the entering fluid causes the area A_1 to move a distance $v_1\,\Delta t$, sweeping out a volume $A_1 v_1\,\Delta t$. Similarly, the volume leaving the other end of the tube in a time Δt is $A_2 v_2\,\Delta t$. Equating these volumes, $A_1 v_1\,\Delta t = A_2 v_2\,\Delta t$, and dividing by Δt, we obtain the *equation of continuity*:

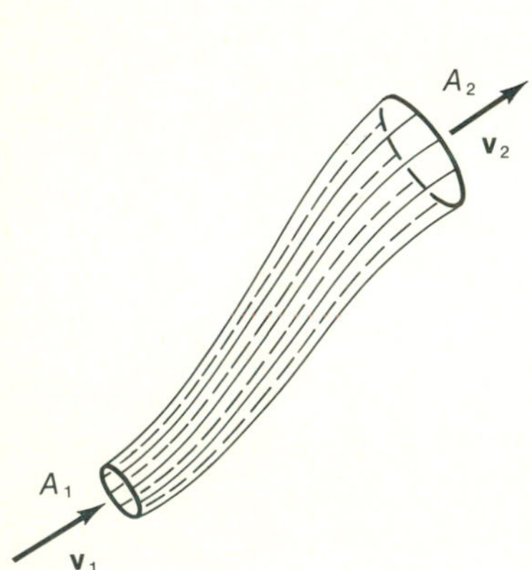

Figure 15–19
A tube of flow within a moving fluid.

EQUATION OF CONTINUITY

$$A_1 v_1 = A_2 v_2 \tag{15–15}$$

This equation shows that the velocity is greatest where the cross-sectional area is smallest.

15.7 Bernoulli's Principle

Armed with these basic concepts, we are in a position to discuss the fundamental relation of fluid flow first stated by Daniel Bernoulli, a Swiss mathematician and scientist. This relation does not involve new principles, but merely applies *conservation of energy* to the case of an ideal fluid in steady flow. Bernoulli's principle expresses very useful relations between the *pressure*, *speed*, and *elevation* in fluid motion.

Consider the irrotational flow of an incompressible, nonviscous fluid through a pipe of varying cross-section and elevation, as in Figure 15–20. We shall compare the total energy situation for two regions, designated ① and ②. If we consider the fluid between these two regions as an isolated system, then any work done *on* the system from the outside will change the kinetic and potential energy content of the system. Let us calculate these quantities.

At region ①, there is a force toward the right on the fluid due to the pressure of the fluid outside the system. This force is p_1A_1. Similarly, at ② there is a force of p_2A_2 toward the left. Now, as the fluid flows, the cross-sectional area A_1 moves a distance Δx_1 at ①, while the area A_2 moves a distance Δx_2 at ②. The equation of continuity [Equation (15–15)] applies here: Because the fluid is incompressible, the shaded volumes swept out at ① and ② are equal and may be written as m/ρ, where m is the mass of that volume of fluid.

The force at ① does work *on* the system equal to:

$$\Delta W_1 = \mathbf{F} \cdot \Delta \mathbf{x} = (p_1A_1)(\Delta x_1) = p_1 m/\rho \qquad \textbf{(15–16)}$$

At region ②, the work done on the system is negative, because the force is directed opposite to the displacement:

$$\Delta W_2 = \mathbf{F} \cdot \Delta \mathbf{x} = -(p_2A_2)(\Delta x_2) = -p_2 m/\rho \qquad \textbf{(15–17)}$$

Forces of gravity also do work as the shaded volume at ① is (in effect) raised in elevation to ②. (The heights y_1 and y_2 are measured to the respective centers of the pipe.) Again, because the force and displacement are opposite, the work done on the system by gravity is negative:

$$\Delta W_g = \mathbf{F} \cdot \Delta y = -(mg)(y_2 - y_1) \qquad \textbf{(15–18)}$$

Thus, the total *net work* ΔW done *on* the system by the external fluid and gravity is:

$$\Delta W = \frac{m}{\rho}(p_1 - p_2) - mg(y_2 - y_1) \qquad \textbf{(15–19)}$$

The change in *kinetic energy* that results is given by

$$\Delta K = \tfrac{1}{2}m(v_2{}^2 - v_1{}^2) \qquad \textbf{(15–20)}$$

Applying the work-energy theorem (work done = change in kinetic energy), we equate the two previous equations, and rearrange, to obtain

$$\overbrace{p_1 + \tfrac{1}{2}\rho v_1{}^2 + \rho g y_1}^{①} = \overbrace{p_2 + \tfrac{1}{2}\rho v_2{}^2 + \rho g y_2}^{②} \qquad \textbf{(15–21)}$$

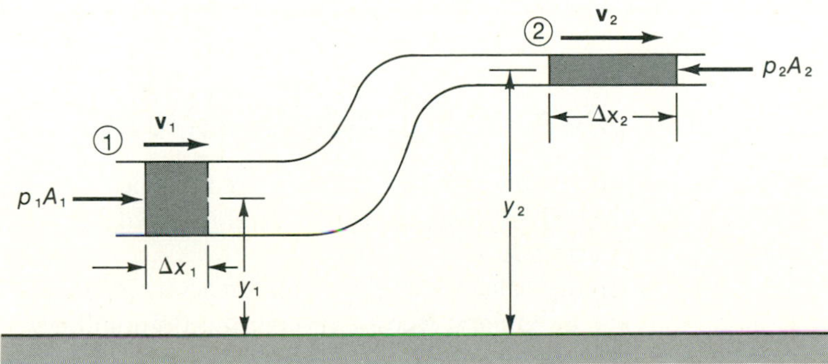

Figure 15–20
The sketch illustrates the general case of irrotational flow of an incompressible, nonviscous fluid through a pipe of varying cross-section and elevation. Bernoulli's principle is a statement of the conservation of energy for this situation.

Note that the term $\rho g y$ is basically a gravitational potential energy term, with y the average distance above some arbitrary zero reference level.

Finally, since ① and ② may refer to *any* two locations, we drop the subscripts and state:

BERNOULLI'S PRINCIPLE **For steady, nonviscous, incompressible flow:**

$$p + \tfrac{1}{2}\rho v^2 + \rho g y = \text{a constant} \tag{15–22}$$

for all points along the same streamline.

Bernoulli's principle is easy to remember when the three terms are recognized, respectively, as the work done per unit volume (by the surrounding fluid), the kinetic energy per unit volume, and the gravitational potential energy per unit volume. Thus this principle is just a convenient way of expressing the conservation of energy for fluid flow.

EXAMPLE 15–3

Using Bernoulli's principle, determine the **velocity of efflux** v from a horizontal orifice located at a depth h below the water level in a large tank.

SOLUTION

Figure 15–21 illustrates the situation. We investigate the points ① at the top surface and ② just outside the orifice; the points are connected by streamlines. Because the tank is large, as the water flows out we assume that the distance h does not change appreciably and that the velocity at ① is therefore essentially zero. The pressure at both points is atmospheric pressure, so we use *gauge pressure* to simplify the algebra.

$$\overbrace{p_1 + \tfrac{1}{2}\rho v_1{}^2 + \rho g h_1}^{①} = \overbrace{p_2 + \tfrac{1}{2}\rho v_2{}^2 + \rho g h_2}^{②}$$
$$0 + \quad 0 \quad + \rho g h = 0 + \tfrac{1}{2}\rho v_2{}^2 + \quad 0$$

$$\boxed{v_2 = \sqrt{2gh}}$$

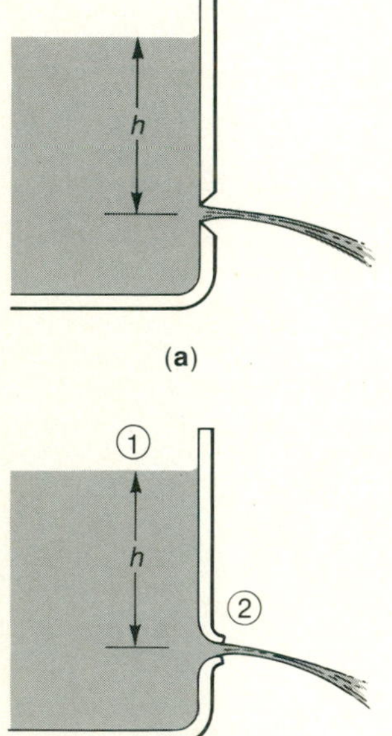

Figure 15–21
Example 15–3

This result is known as **Torricelli's law.** An interesting effect occurs if the opening is circular and sharp as in Figure 15–21(a). The fluid continues to accelerate a short distance beyond the opening, thus contracting the cross-sectional area of the jet (recall the equation of continuity). The emerging jet stream has a characteristic shape known as the *vena contracta*. The emerging jet thus appears as though it were emanating from a smooth nozzle of smaller diameter, as in Figure 15–21(b). Torricelli's law applies to the stream after the contraction is complete.

15.8 Examples of Bernoulli Effects

The fact that the pressure is lowest where the velocity is highest is commonly called the Bernoulli effect. It may be amusingly demonstrated in a variety of ways, as shown in Figure 15–22.

The motion of spinning balls moving through a viscous medium (such as air) is dependent on several different factors, though the Bernoulli effect is the

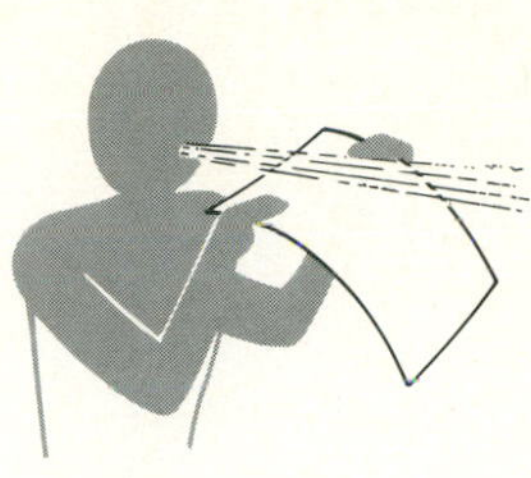

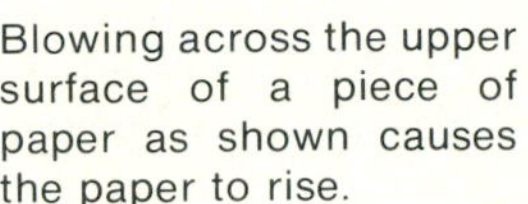

Blowing across the upper surface of a piece of paper as shown causes the paper to rise.

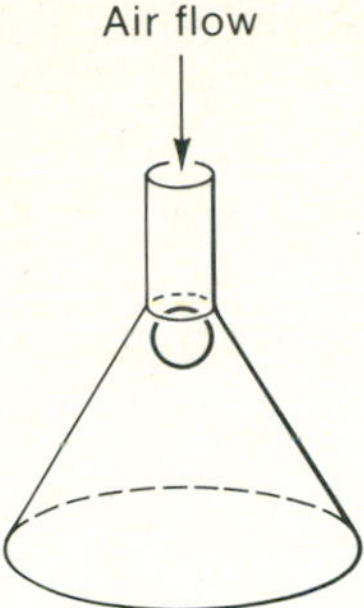

A ping pong ball is supported in an inverted funnel when air is blown downward into the funnel.

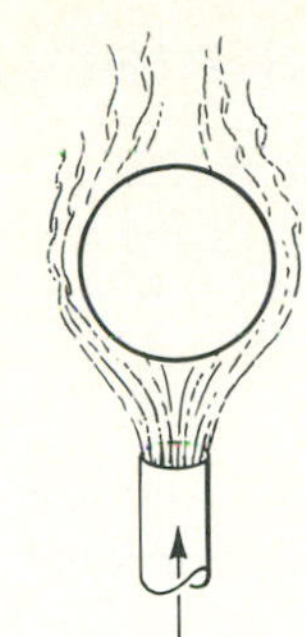

A large beach ball is supported by a vertical jet of air. If the ball moves to one side, the Bernoulli effect pulls it back into the stream of air.

Figure 15–22 Examples of the Bernoulli effect

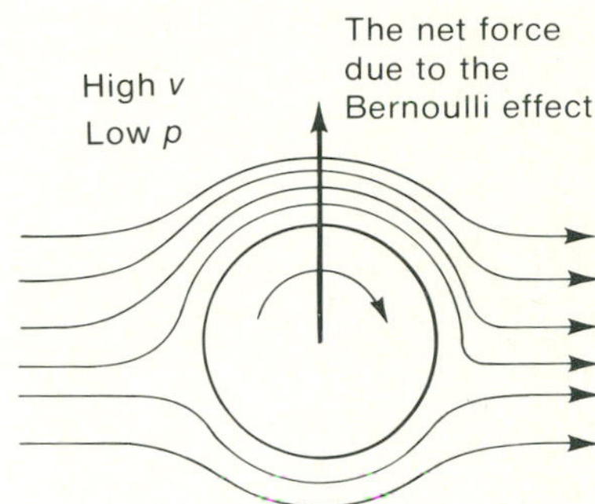

Figure 15–23
Streamlines around a thrown ball as viewed in a frame of reference in which the *CM* of the ball is at rest. For a nonrotating ball, the streamlines divide symmetrically, producing equal pressures above and below the ball. However, if the ball spins as shown, the viscosity of the air results in more air flow above the ball than below. Correspondingly, the air velocity is faster above the ball, producing a lower pressure at the top, and the net force is up.

most pronounced. If given a spinning motion, a thrown ball will follow a curved trajectory (different from the familiar parabolic path). It is convenient to analyze the situation from a frame of reference in which the center of mass of the ball is at rest (see Figure 15–23). By roughening the surface of the ball, the ball is more effective in dragging air around one side while spinning. The dimples on a golf ball similarly ensure that a boundary layer of air adheres to the surface and thus enhances the "lift" that can be produced by the Bernoulli effect.

The dynamic lift on airplane wings is achieved by shaping the wing to force more air over the upper surface (refer to Figure 15–16a). Air travels faster over the top surface than the bottom surface, causing a lower pressure on the top than the bottom. As a result, the net force due to the Bernoulli effect is upward. (The impact of the air against the lower surface of a wing in this position also provides an upward force component.)

The **Venturi effect** is the name given to the fact that the pressure in a pipeline reduces when the cross-sectional area is made smaller. (The pipe must be tapered in a certain way to avoid turbulence.) A practical application of this effect is the manner in which gasoline vapor is drawn into the manifold of an automobile engine; a Venturi throat connected to the carburetor produces a region of low pressure. Similarly, atomizers and aspirators utilize the Venturi effect. A Venturi meter (Figure 15–24) is a convenient device for measuring flow rates. The rate of flow through the pipe is faster at the narrowed portion, thus producing lower pressures at that point. The pressure difference between the narrow region and the normal-diameter section may conveniently be measured with a U-tube, as shown. By applying Bernoulli's principle, the flow velocity v can be shown to be equal to

$$v = \left[\frac{2gh\left(\dfrac{\rho'}{\rho} - 1\right)}{\left(\dfrac{A_1}{A_2}\right)^2 - 1} \right]^{1/2} \qquad \textbf{(15–23)}$$

where the symbols are as defined in Figure 15–24.

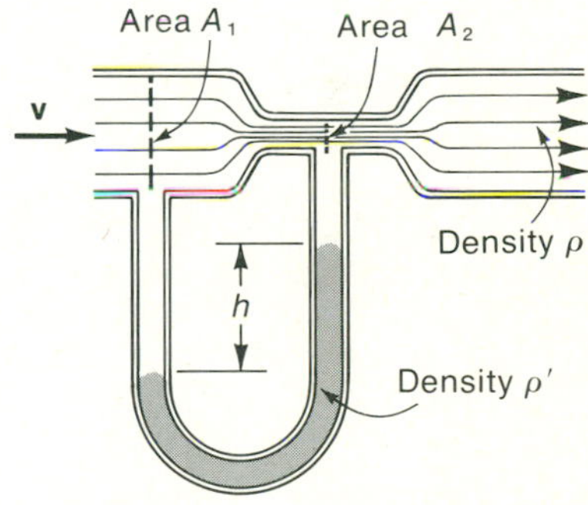

Figure 15–24
A Venturi meter for measuring the flow rate of a fluid through a pipe. The difference in heights of the mercury (or other liquid) in the U-tube is related to the rate of flow through the pipe. Where the pipe is smaller, the flow is faster, causing a reduced pressure.

Summary

Fluids at Rest

Pressure is a *scalar* quantity, defined as

$$p = \lim_{\Delta A \to 0} \frac{\Delta F}{\Delta A}$$

Fluids at rest always exert forces normal to any surface the fluid contacts.

The pressure p at a depth h below the surface of the fluid is

$$p = p_0 + \rho g h$$

where p_0 is the pressure on the fluid surface, ρ is the density of the fluid (mass/unit volume), and h is the depth below the surface, expressed as a positive number.

A pressure of *one standard atmosphere* is equal to

$$\begin{aligned}(1\text{ atm}) &= 1.013 \times 10^5\text{ Pa}\\ &= 1.013 \times 10^5\text{ N/m}^2\\ &= 1.013 \times 10^6\text{ dynes/cm}^2\\ &= 1013\text{ millibars}\\ &= 14.7\text{ lb/in}^2\\ &= 2116\text{ lb/ft}^2\\ &= 76.0\text{ cm Hg}\end{aligned}$$

The following two principles apply to fluids at rest:

PASCAL'S PRINCIPLE **Pressure applied to an enclosed fluid is transmitted undiminished to all parts of the fluid and to the walls of the container.**

ARCHIMEDES' PRINCIPLE **Any body, wholly or partially submerged in a fluid, is buoyed up by a force equal to the weight of the displaced fluid.**

The surface of a fluid acts as a stretched membrane, exerting a surface tension force Γ (force/length) along an edge of the surface. Surface tension may also be expressed as the "work per unit area created" when the surface area is changed. Surface tension forces explain the rise (or fall) of fluid levels in capillary tubes, the pressure inside a bubble or droplet, and other phenomena.

Fluids in Motion

In this text, fluids in motion are analyzed only for the case of *ideal fluids*—that is, incompressible and nonviscous fluids. We also limit our considerations to irrotational, steady flow.

Bernoulli's principle is a statement of the conservation of energy for liquid flow:

BERNOULLI'S PRINCIPLE $$p + \tfrac{1}{2}\rho v^2 + \rho g y = \text{a constant}$$

For some situations, it is convenient to use *gauge pressure*, which is the reading on a pressure gauge set to indicate zero when an absolute pressure of 1 atm is present.

Questions

1. A Roman aqueduct supports the water in a canal as it crosses a ravine. Suppose a loaded barge moves slowly across the aqueduct. Will the load on the aqueduct increase, decrease, or remain the same?

2. Consider the three containers (of negligible mass) in Figure 15–25, each filled to the same depth of water. The bottom areas are equal; therefore the force the fluid exerts on the bottom area of each is the same. But if they were placed on scales, they would not weigh the same. Explain this fact, known as the "hydrostatic paradox."

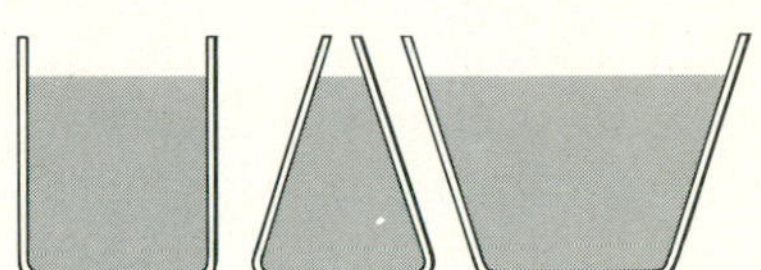

Figure 15–25
Question 2

3. An object floats on the surface of a container of mercury. If a bell jar is now placed over the container and the air pumped out, will the object sink a bit, rise slightly, or remain at the same depth?

4. Two objects with different densities are placed on opposite pans of an equal-arm balance, which indicates they have the same mass. The balance is now placed in an airtight chamber and the air is evacuated. Now the balance indicates their masses are not equal. Which object has the greater mass?

5. Fill a water tumbler completely full of water. Place a piece of stiff paper or cardboard over the glass and, holding it in place, invert the tumbler. Remove your hand from the paper. The water does not fall out! Explain.

6. Explain why a helium-filled balloon, rising in the earth's atmosphere, eventually reaches a maximum height. In which case would the maximum height be greater: if the balloon maintained a constant volume or if its volume increased with altitude (keeping the amount of gas inside the same)?

7. Occasionally fishermen use a line whose breaking strength is less than the weight of the fish they catch. How can they reel in such a fish without breaking the line?

8. An ice cube is floating in a glass of water. It melts. Does the water level rise, fall, or remain the same? What would happen if the ice cube had a small air bubble in it? If the ice cube had a small pebble in it? What would happen if the water and glass were initially at room temperature (rather than 0°C)?

9. Alcohol has a slightly lower density than water. Do ice cubes float higher or lower in a mixed drink than in pure water?

10. A barge loaded with scrap iron is floating in a lake. If the scrap iron is tossed overboard, will the water level in the lake rise, fall, or remain the same?

11. Paper towel manufacturers brag about how quickly their product can soak up a spilled liquid. What causes the liquid to be absorbed? If two different brands have different speeds for this process, what is the physical difference in the towels?

12. Most cloth tents are waterproofed so that the fibers of the tent material do not absorb water, but air can pass through the tiny openings of the woven material. Campers in rainy areas are familiar with the fact that touching the inside of a cloth tent will often cause it to leak. Explain why the tent does not leak before it is touched, and why it does afterward.

13. Following is a famous puzzle. If two small objects are floating close together in a liquid that wets the surfaces of both, they will be drawn together. If the liquid does not wet the surface of either, they will also be drawn together. However, if the liquid wets one object but not the other, they will move apart. Explain.

14. Two sheets of paper are suspended vertically side by side a few centimeters apart. If you blow between them, they move together. Why?

15. Two automobile drivers traveling parallel to each other will experience a force pushing them toward each other. Use Bernoulli's principle to explain this situation. (If they do not use caution, two large ships moving side by side in the ocean can get into trouble because of this effect.)

16. As mentioned in the chapter, certain water bugs that "walk" on water secrete a substance behind them that lowers the surface tension of the water; the resultant

imbalance of surface tension forces propels them forward. Where does the energy come from that increases the bug's kinetic energy?

17. Hold a sugar cube so that it barely touches the surface of coffee in a cup. Where does the energy come from that raises the coffee vertically against gravity as it is drawn up into the sugar cube?

18. A boat is floating in a swimming pool. A cork is pulled from a hole in the bottom of the boat. Explain any changes of the water level in the pool that occur (a) as the water first starts to squirt up into the boat and (b) after the boat has sunk to the bottom.

19. Occasionally, large icebergs turn over with dangerous implications for nearby vessels. Since 89% of an iceberg is submerged, how can this happen, considering that so much of an iceberg's mass is below sea level?

20. Because of the equatorial bulge of the earth, the source of the Mississippi River is closer to the center of the earth than its mouth. Explain how the river flows "uphill."

Problems

15A–1 In air, an object weighs 15 N. When immersed in water, the same object weighs 12 N. When immersed in another liquid, it weighs 13 N. Find (a) the density of the object and (b) the density of the other liquid. *Answers:* (a) 5000 kg/m^3 (b) 667 kg/m^3

15A–2 A fisherman rows to the center of a pond and throws an anchor overboard; the anchor goes to the bottom of the pond. As a consequence, will the level of the pond be raised, stay the same, or be lowered? Explain.

15A–3 One hundred milliliters of water are poured into a U-tube which has a cross-sectional area of 1.0 cm^2. One hundred milliliters of oil, which has a density 0.80 that of water, are then poured down one side of the U-tube so that the oil floats on the water. Find the difference in height of the liquid surfaces in the two sides of the U-tube.
Answer: 20 cm

15A–4 An ice cube floats in a glass of water. As the ice cube melts, will the level of the water in the glass rise, fall, or remain unchanged? Explain.

15A–5 In Figure 15–26, the pistons in the hydraulic lift are circular in shape and positioned at the same horizontal level. Find the force **F**, applied as shown, that will support the 2400-lb load. *Answer:* 50 lb

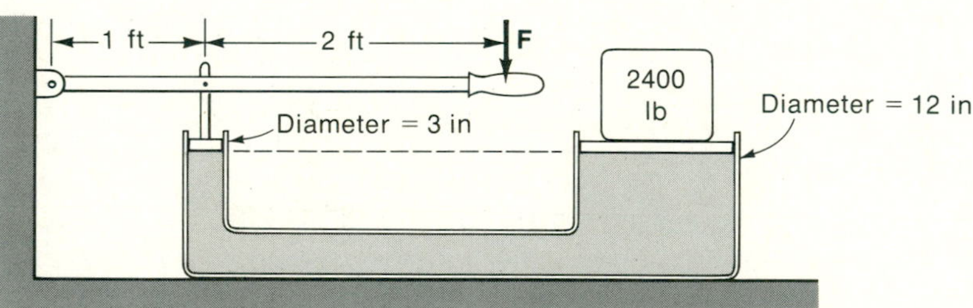

Figure 15–26
Problem 15A–5

15A–6 Find the pressure (in atmospheres) inside a water droplet 10^{-6} m in diameter. The surface tension of water is 0.075 N/m.

15A–7 Ethyl alcohol (specific gravity = 0.80) rises 6.12 cm in a glass capillary tube 0.20 mm in diameter. The liquid surface contacts the glass tangentially at an angle of 0°. Find the surface tension of ethyl alcohol. *Answer:* 0.024 N/m

15A–8 Find the pressure due to surface tension effects (a) inside a water droplet 4 μ in diameter and (b) inside a thin-walled water bubble of the same diameter.

15A–9 Water flows through a pipe with an internal diameter of 4 cm at a speed of 10 m/s. If the pipe joins a pipe with an internal diameter of 2 cm, what is the speed of flow in the smaller pipe? *Answer:* 40 m/s

15A–10 A small airplane has a wing area of 35 m^2. The speed of the air over the top surface of the wing is 60 m/s, and it is 55 m/s under the bottom surface. If the density of air is 1.30 kg/m^3, find the net lifting force on the wing.

15A–11 A horizontal pipe 10 cm in diameter has a smooth reduction to a pipe 5 cm in diameter. If the pressure of the water in the larger pipe is 8×10^4 N/m^2 and the pressure in the smaller pipe is 6×10^4 N/m^2, at what rate does water flow through the pipes?
Answer: 12.8 kg/s

15B–1 A tank of horizontal dimensions 6 ft × 12 ft contains water to a depth of 8 ft. Neglecting air pressure, find (a) the force on the bottom area and (b) the force on the 6 ft × 8 ft side area due to the water alone. *Answers:* (a) 35 940 lb (b) 11 980 lb

15B–2 A cube with an edge length ℓ is completely submerged in a liquid of density ρ. If the top face of the cube is horizontal and a distance D below the surface of the liquid, find the force on each surface of the cube. Find the net force on the cube.

15B–3 A rectangular box rests on the bottom of a deep tank of water. Show that the total force on a vertical face of the box is the area of the face times the pressure at the center of the face.

15B–4 Two solid objects, one made of steel and one of aluminum, have the same apparent weight when submerged in water. If they are weighed in air, which (if either) is heavier? Justify your answer.

15B–5 A layer of oil (specific gravity = 0.80) is resting on top of a quantity of water (specific gravity = 1.00). A uniform rectangular block floats at the interface between the oil and water such that one-third of the block's volume extends into the oil. Find the specific gravity of the block. *Answer:* 0.933

15B–6 What fraction of the total volume of an ice cube floating in water is above the surface of the water?

15B–7 A block of wood (specific gravity = 0.70) is 40 cm long, 20 cm wide, and 5 cm thick. What volume of lead (specific gravity = 11.35), fastened to the bottom, will cause the wood to sink in water so that its top is barely even with the water surface?
Answer: 116 cm^3

15B–8 To hold a block of wood completely under water requires a downward force of one-eighth the block s weight in air. Find the weight density of the wood in pounds per cubic feet.

15B–9 A uniform rod whose specific gravity is 0.50 is supported at one end by a rope. The other end floats freely in water, as shown in Figure 15–27.
(a) Sketch a free-body diagram for the rod.
(b) What fraction of the rod's length is submerged? *Answer:* $(1 - 1/\sqrt{2})$

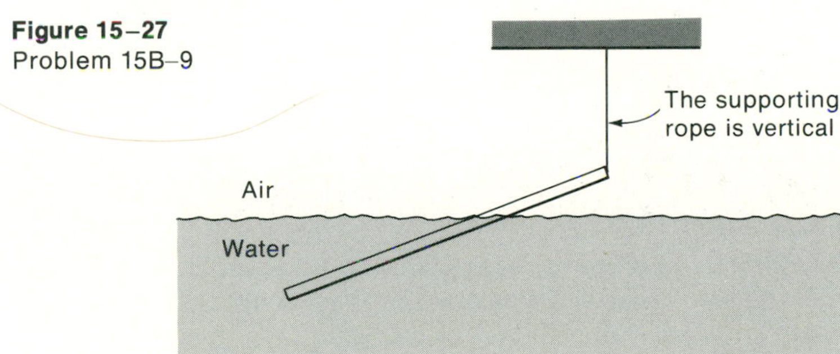
The supporting rope is vertical
Air
Water

Figure 15–27
Problem 15B–9

15B–10 In the previous problem, suppose the water level changes gradually so that the angle of the rod with respect to the horizontal varies between 10° and 80°. Does the tension in the rope change or remain constant as the angle of the rod changes? Explain your reasoning.

15B–11 A long, uniform, wooden spar buoy is anchored in fresh water so that it is at an angle with the surface, as shown in Figure 15–28. If the specific gravity of the wood is 0.50, what fraction of the buoy's length extends above the water? *Answer:* $(1 - 1/\sqrt{2})$

15B–12 Derive Equation (15–23).

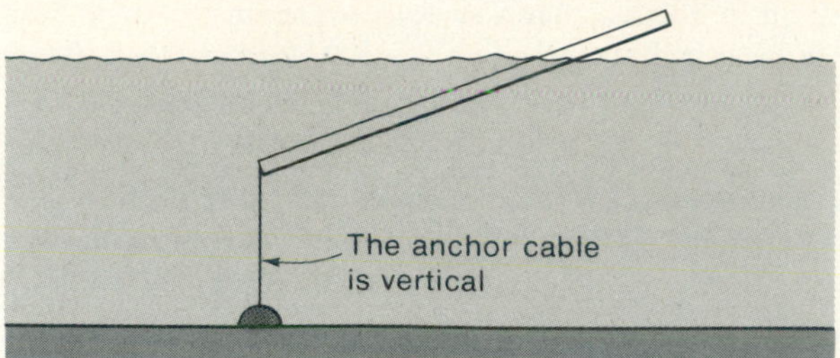

Figure 15–28
Problem 15B–11

15B–13 A pipe with cross-sectional area A has a bend in a horizontal plane, as shown in Figure 15–29. Adjacent connecting pipes (not shown) exert equal horizontal forces F applied to the flanges. If the speed of the liquid is v and its density is ρ, determine F and show that it is independent of the angle θ. *Answer:* $\rho A v^2$

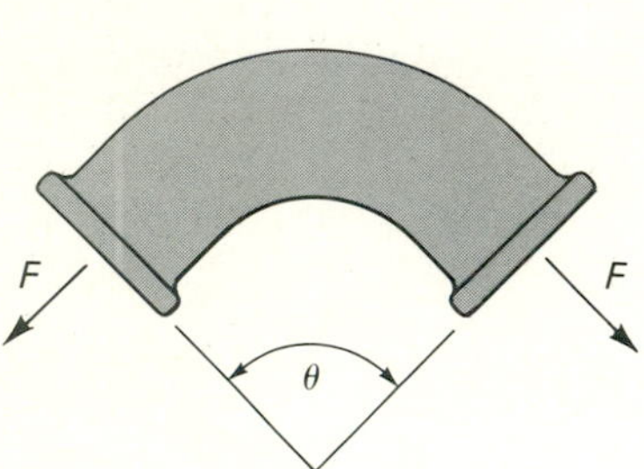

Figure 15–29
Problem 15B–13

15B–14 Torricelli's law, $v = \sqrt{2gh}$, gives the speed v of a liquid emerging from a hole a distance h below the surface of a liquid in a container. It assumes the liquid level has zero velocity. Using Bernoulli's principle, derive a version of Torricelli's law that includes the speed v_0 of the liquid level.

15B–15 What power is required to pump 50 L/s of water through a horizontal pipe with a pressure difference of 1 atm? *Answer:* 5065 W

15B–16 Water is forced out of a fire extinguisher by air pressure, as shown in Figure 15–30. How much gauge air pressure (in atmospheres) is required to have a water jet with a speed of 30 m/s when the water level is 0.50 m below the nozzle?

15B–17 In an oil pipeline that carries oil of (weight) density 50 lb/ft^3, the gauge pressure is 5 lb/in^2 above atmospheric pressure and the velocity is 4 ft/s. The cross-sectional area of the pipe at this point is 1.3 in^2. At a second point in the pipeline 8 ft below, the cross-sectional area is 1.0 in^2. Neglecting viscosity effects, find the gauge pressure at the second point. *Answer:* 7.71 lb/in^2

15C–1 Show that by ignoring the buoyant effect of the atmosphere, the fractional error $\Delta W/W$ in weighing an object is the ratio of the density of air to the density of the object.

15C–2 A rectangular tank contains water to a depth of 2.5 m. If the width of one end of the tank is 8 m, (a) find the total force exerted on the end of the tank by the water. (Ignore atmospheric pressure.) (b) Find the torque about the bottom edge of the end of the tank due to this force.

15C–3 Consider a soap film on a wire hoop with a thread loop embedded in the film, as shown in Figure 15–31a. When the film inside the loop is broken, the loop forms a circular hole of radius R, as shown in Figure 15–31b. Show that the tension in the thread loop is $T = 2R\Gamma$, where Γ is the surface tension of the film.

15C–4 The density ρ of air is proportional to the pressure p (that is, $\rho = kp$, where k is a constant). Show that the pressure in the atmosphere varies with height according to $p = p_0 e^{-kgh}$, where h is the height corresponding to the pressure p. (Hint: Investigate the forces on a horizontal slab of air of height dy and area A. As the pressure changes from p to $p + dp$, going in the negative y direction, the y coordinate changes by $-dy$.)

15C–5 A cylindrical can is weighted to float vertically as in Figure 15–32. The cross-sectional area of the cylinder is A and its mass is m. Find the period T of small-amplitude

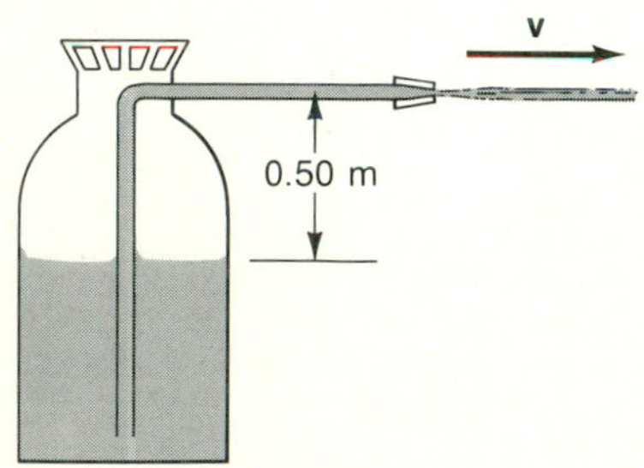

Figure 15–30
Problem 15B–16

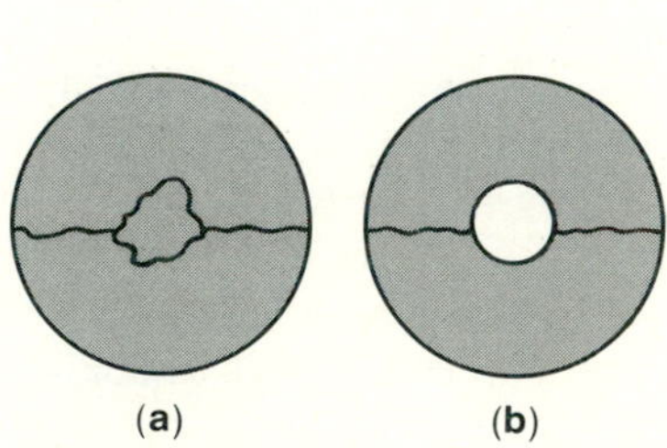

Figure 15–31
Problem 15C–3

Figure 15–32
Problem 15C–5

vertical oscillations in a fluid of density ρ. (Hint: If the cylinder is depressed a small distance y, what is the restoring force on the cylinder?) *Answer:* $T = 2\pi\sqrt{m/\rho A g}$

15C–6 One hundred cubic centimeters of mercury are poured into a U-tube with a cross-sectional area of 1 cm^2. The mercury will oscillate back and forth in the bottom of the tube until static equilibrium is reached. Find the period of oscillation.

15C–7 A vertical tank filled with water to a depth H has a small hole drilled in its side a distance h from the bottom. The water jet emerges horizontally.

(a) At what value of h will the jet of water strike the ground at the maximum horizontal distance from the tank?

(b) Prove that this maximum horizontal distance is H. *Answer:* $H/2$

15C–8 A bucket of water spins with constant speed about a vertical axis through the center of the bucket. Show that in static equilibrium, the surface of the water assumes a parabolic shape. (Hint: In the rotating frame of reference, investigate the forces on a small mass of water dm at the surface of the liquid. The effective weight of dm, composed of the force of gravity plus the radially outward centrifugal force, must be perpendicular to the liquid surface. From this, show that for the surface, dy/dx is equal to kx, where k is a constant. This leads to $y = k'x^2$, the equation of a parabola.)

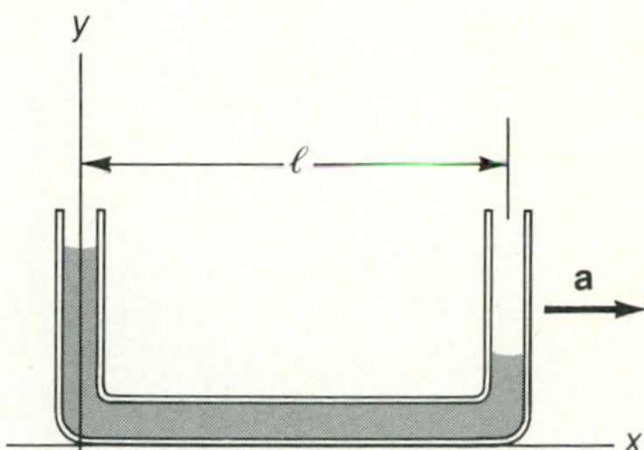

Figure 15–33
Problem 15C–9

15C–9 A glass tube, partially filled with a liquid, is in the form of a rectangular U-tube with horizontal and vertical sections, as shown in Figure 15–33. Find the difference in heights between the liquid surfaces in the vertical columns when the tube is moved in the $+x$ direction with constant acceleration $\mathbf{a}$. *Answer:* $a\ell/g$

15C–10 A cork is held by a thread near the bottom of a beaker filled with water. The beaker is placed on one pan of an equal-arm balance, and an equal mass is placed on the other pan so that the pointer reads at the zero-balance position. The thread breaks and the cork rises. While the cork is rising through the water (and before it reaches the top), will the pan supporting the beaker rise, fall, or remain at rest? Explain.

chapter 16
Wave Motion

An ocean traveler has even more vividly the impression that the ocean is made of waves than that it is made of water.

A.S. EDDINGTON

[*The Nature of the Physical World*, Cambridge (1929)]

16.1 Introduction

Wave motion is a phenomenon common to almost all branches of physics. Waves seem to be everywhere in nature: There are water waves, sound waves, radio waves, and light waves. All are described by a mathematical expression called the *wave equation*, which has the same basic form for all types of waves. In this chapter we will limit our discussion to *mechanical* waves, which are generated in any elastic medium by rapidly displacing a small portion of the medium. As you will see, because the medium has properties of mass and elasticity, the disturbance is propagated with a speed that is characteristic of each particular medium.

Basically, *all wave motion transmits energy and momentum from one point in space to another without any matter itself actually moving from one point to the other*. That is, particles of the medium do not themselves travel along with the wave. Instead, they oscillate back and forth about some equilibrium position as the wave passes by. Only the *disturbance* is propagated. When the disturbance arrives at the distant point, it sets into motion the particles at that location, giving them kinetic energy and momentum. As an example, when water waves originating in a South Pacific storm travel at about 50–70 km/h and transmit extra energy and momentum to ocean waves on the California shore, 10 000 km away, none of the water in the South Pacific actually travels to California. Only the wave disturbance makes the journey.

There are two general classes of traveling waves, as illustrated in Figure 16–1. In **transverse waves,** the particles of the medium oscillate *perpendicular* to the direction of propagation of the wave. In **longitudinal waves,** the particles of the medium oscillate *parallel* to the direction of propagation of the wave. Traveling waves on a rope are transverse since the rope vibrates perpendicular to the direction of propagation. Sound waves in air are longitudinal waves since gas molecules in the air oscillate parallel to the direction of propagation. Some types of waves, such as those on the surface of water, involve both transverse and longitudinal motions simultaneously.

If the medium is given just a one-time "bump" or "jolt," the disturbance that propagates outward from the source is called a **wave pulse.** Examples would be the explosive sound of a handclap as it propagates through the air, or the disturbance moving along a stretched rope after the rope is struck sharply near one end. If instead the disturbance is repetitive, a series of pulses called a

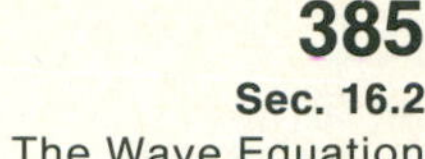

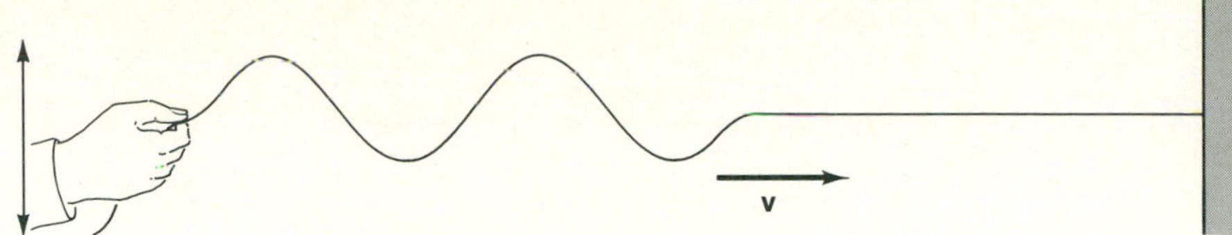

(a) *Transverse waves* on a stretched rope. The particles of rope are displaced perpendicularly to the direction of propagation of the wave.

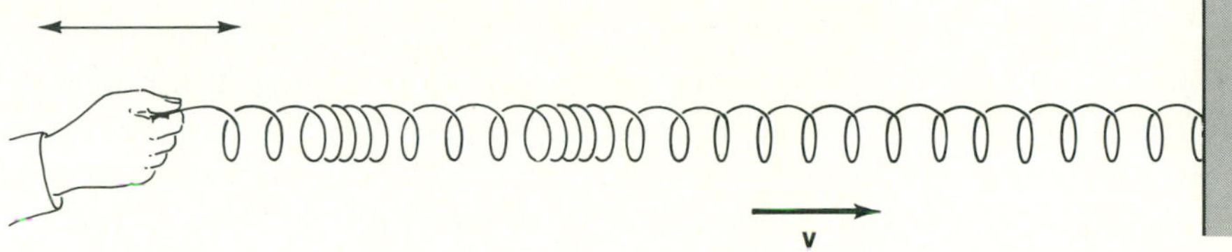

(b) *Longitudinal waves* in a stretched spring. The coils of the spring are displaced parallel to the direction of propagation of the waves.

Figure 16–1
When a portion of a medium is rapidly displaced back and forth, a periodic wave disturbance is generated. This disturbance propagates through the medium with a speed v whose numerical value is characteristic of that particular medium.

wave train is generated. A common form of wave train occurs when the disturbance is simple harmonic motion, so that each particle of the medium undergoes SHM oscillations as the wave train passes by. This type of disturbance is known as a **sinusoidal wave train.** If a repetitive disturbance has a more complex motion, it is called a **periodic wave train.**

16.2 The Wave Equation

For simplicity, we will start with just a one-dimensional medium, later extending the results to two and three dimensions. Consider a rope stretched along the x axis with a force F. (To avoid confusion with the notation for the period T of simple harmonic waves, here we will use the symbol F for the tension in the rope.) If we rapidly move one end up and down just once, as in Figure 16–2, the disturbance will initiate a single transverse wave pulse that travels along the rope. In the simplest (ideal) case, the wave pulse retains its shape,

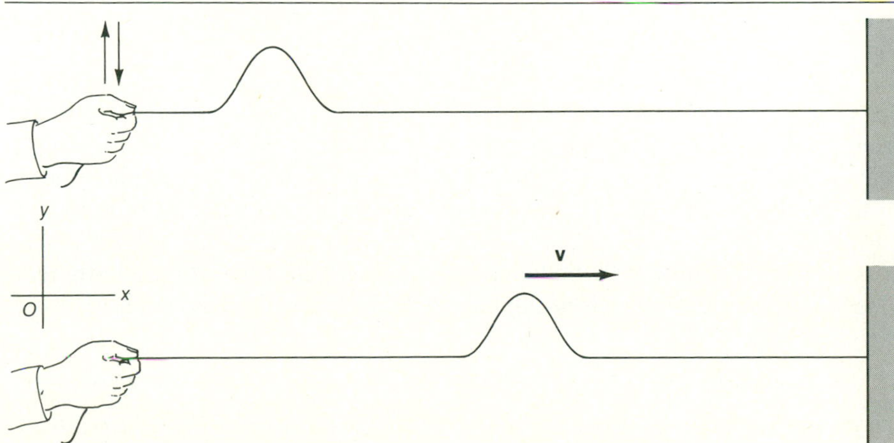

Figure 16–2
A single, rapid, up-and-down motion in the y direction generates a transverse pulse which travels along the x direction on a stretched rope. (The amplitude of the pulse is greatly exaggerated in this sketch.)

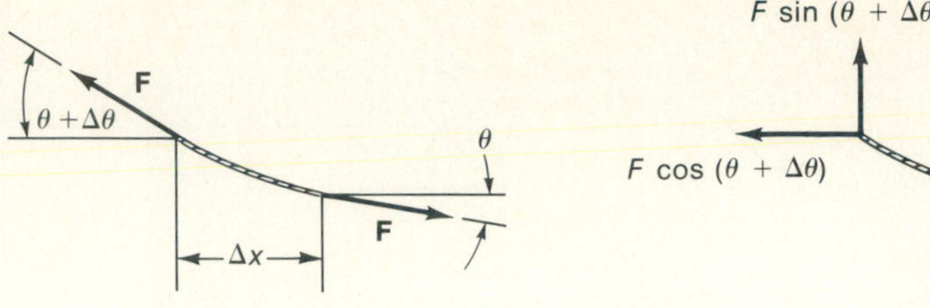

(a) Isolation diagram for the small string segment Δx.

(b) Forces on the segment Δx resolved into horizontal and vertical components.

Figure 16–3
As the pulse passes by, the forces on the string segment Δx cause it to accelerate upward. The mass of the segment is $\mu\,\Delta x$.

or **profile,** as it moves along.[1] At any instant, the traveling disturbance is limited to just a small portion of the rope as each particle, in turn, undergoes a momentary transverse motion and then returns to its original position at rest.

Several parameters are useful for describing this motion:

y = the displacement of the particles of the rope away from their equilibrium positions (in this case, the displacement is perpendicular to the direction of propagation)

x = the distance along the direction of propagation of the pulse

t = the time

The rope has a *mass per unit length* (linear mass density) of μ and the transverse pulse travels along the rope with speed v. We assume the displacement of the transverse motion is so small that the tension F remains essentially constant as the pulse passes.

Let us focus our attention on a small segment Δx of the rope while it undergoes the transverse displacement due to the traveling pulse. By applying $\Sigma \mathbf{F} = m\mathbf{a}$ to the *transverse* motion, we will derive an equation for the *longitudinal* speed v of the pulse. A free-body diagram for this segment is shown in Figure 16–3. We make two assumptions, valid for small-pulse amplitudes: (1) the angle θ is always small and (2) the displacement y is always small enough that the tension F is the same throughout. Because θ is small, we may write:

$$\sin\theta \approx \tan\theta \approx \frac{\partial y}{\partial x} \qquad \textbf{(16–1)}$$

where $\partial y/\partial x$ is called the partial derivative of y with respect to x. This just means that the derivative with respect to x is to be taken in the usual way *with all other variables* (in this case, t) *treated as constants.* That is, the derivative is taken only with respect to x. As an example, for the function $y = x^2 + t^3$, the partial derivatives are $\partial y/\partial x = 2x$ and $\partial y/\partial t = 3t^2$.

It is always helpful to keep in mind the physical meaning of mathematical symbols. Here, $\partial y/\partial x$ is the *slope* of the rope (at each point x) and $\partial y/\partial t$ is the *transverse velocity* of the rope particles (at each point x). Note the distinction between the two velocities associated with wave motion: the *transverse* velocity of the rope particles and the *longitudinal* velocity of propagation of the wave along the rope.

For small values of θ, the functions $\cos\theta$ and $\cos(\theta + \Delta\theta)$ are both very close to 1. Their difference is negligible, so there is essentially no net horizontal

[1] In many physical situations, the elastic medium does not exactly obey Hooke's Law, particularly for large-amplitude motions. As a consequence, waves of different frequencies have different speeds. This causes the pulse to spread out as it travels along (a property known as *dispersion*), so that its profile gradually becomes lower and wider. Also, if the medium has internal frictional effects, causing it to heat up as it absorbs energy from the wave, the profile will become smaller in amplitude, or *attenuated.*

force on the segment. However, the difference between the *sines* of these two small angles is relatively large. Thus there *is* an appreciable net vertical (transverse) force $(F_{net})_{trans}$ on the segment. This force is:

$$(F_{net})_{trans} = F\sin(\theta + \Delta\theta) - F\sin\theta \qquad \textbf{(16–2)}$$

$$(F_{net})_{trans} = F\left[\left(\frac{\partial y}{\partial x}\right)_{x+\Delta x} - \left(\frac{\partial y}{\partial x}\right)_x\right] \qquad \textbf{(16–3)}$$

where for small θ, $\sin\theta \approx \tan\theta \approx \partial y/\partial x$, the slope of the segment at x. The quantity $\partial y/\partial x$ is some function $f(x,t)$. The notation $(\partial y/\partial x)_{x+\Delta x}$ means that the function $\partial y/\partial x$ is to be evaluated at $x + \Delta x$; similarly, $(\partial y/\partial x)_x$ is to be evaluated at x. As shown in any calculus book, the quantity in brackets is part of the definition of the *second derivative* of y with respect to x. That is:

$$\lim_{\Delta x \to 0} \frac{\left[\left(\frac{\partial y}{\partial x}\right)_{x+\Delta x} - \left(\frac{\partial y}{\partial x}\right)_x\right]}{\Delta x} = \frac{\partial^2 y}{\partial x^2} \qquad \textbf{(16–4)}$$

The partial second derivative symbol is an instruction to take the partial derivative twice:

$$\frac{\partial^2 y}{\partial x^2} \equiv \frac{\partial}{\partial x}\left(\frac{\partial y}{\partial x}\right)$$

holding the other variable t constant.

By combining Equations (16–3) and (16–4), we obtain for the net force:

$$(F_{net})_{trans} = F\left(\frac{\partial^2 y}{\partial x^2}\right)\Delta x \qquad \textbf{(16–5)}$$

Noting that the mass m of the segment is $\mu\,\Delta x$, we apply Newton's second law for the transverse motion:

$$\Sigma F_y = ma_y$$

Since $a_y = \partial^2 y/\partial t^2$, we have:

$$F\left(\frac{\partial^2 y}{\partial x^2}\right)\Delta x = \mu\,\Delta x\left(\frac{\partial^2 y}{\partial t^2}\right) \qquad \textbf{(16–6)}$$

Dividing both sides by Δx and rearranging, we obtain

$$\frac{\partial^2 y}{\partial t^2} = \left(\frac{F}{\mu}\right)\frac{\partial^2 y}{\partial x^2} \qquad \textbf{(16–7)}$$

The dimensions of (F/μ) are $(ML/T^2) \div (ML) = L^2/T^2$, or a velocity squared. So we simplify the notation by substituting $(F/\mu) = v^2$ to obtain the **wave equation:**

THE WAVE EQUATION

$$\frac{\partial^2 y}{\partial t^2} = v^2\frac{\partial^2 y}{\partial x^2} \qquad \textbf{(16–8)}$$

Referring back to Equations (16–5) and (16–6), note that the wave equation was obtained simply by applying to $\Sigma F = ma$ to the transverse motion of the rope.

16.3 A General Solution to the Wave Equation

Having an equation that tells us how traveling waves behave, we next seek *solutions* to the wave equation—that is, particular functions of the two variables x and t that describe particular cases. Whatever functions we obtain must

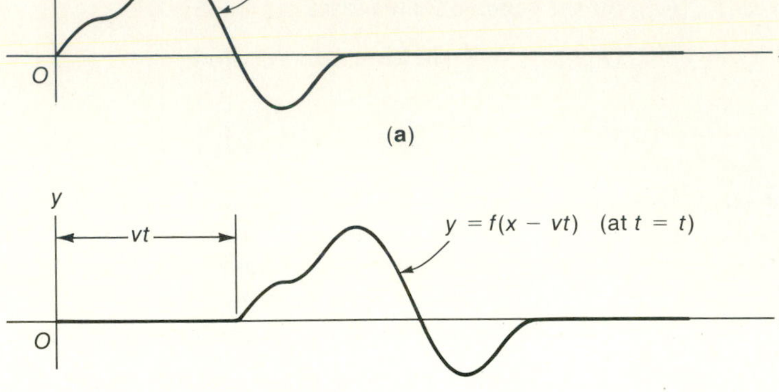

Figure 16–4
An arbitrary wave shape traveling along a stretched string in the $+x$ direction with velocity v.

agree with Equation (16–8) in that the second derivative with respect to time must equal v^2 times the second derivative with respect to distance.

As a concrete example, let us consider waves traveling on a stretched string aligned along the x axis. Figure 16–4 shows an arbitrary wave shape moving in the $+x$ direction. In (a), the wave disturbance is shown at time $t = 0$, when the shape at that instant is described by some function $f(x)$:

$$y = f(x) \qquad (\text{at } t = 0) \tag{16–9}$$

At a time t later, the wave will have moved along the string with velocity v, preserving its exact shape (if friction and dispersion effects are negligible). So the function that now describes the wave has the form:

$$y = f(x - vt) \qquad (\text{at } t = t) \tag{16–10}$$

At first glance you may be puzzled that the above equation with its minus sign does, indeed, represent a wave traveling in the $+x$ direction. It is easy to demonstrate. Let us focus our attention on a specific point x in the wave, say, the highest peak in Figure 16–4. Suppose at $t = 0$ the displacement of the peak from the equilibrium position is 2 mm:

$$y_{(\text{at } t=0)} = 2 \text{ mm}$$

At a later time t this peak will be found at the location $x = 2 \text{ mm} + vt$. Solving for the displacement $y = 2$ mm, we have:

$$y_{(\text{at } t=t)} = 2 \text{ mm} = (x - vt)$$

As t increases, x also must increase if $(x - vt)$ is to maintain the same value. We can see that this is what occurs, since the peak (and every other part of the wave) moves along in the $+x$ direction with the same speed v. So we have discovered an important fact about solutions to the wave equation for motion in the $+x$ direction: *They all must be functions of the quantity* $(x - vt)$. *Any* function[2] written in general as $y = f(x - vt)$ is such a traveling wave. (This can be shown by substituting the value of y into the wave equation and verifying that, indeed, the partial second derivative with respect to time equals v^2 times the partial second derivative with respect to distance.) The *particular* function will depend on the shape of the wave we wish to represent.

Let us verify that the speed of the wave is actually v. If we follow a particular point of the wave shape (say, the peak) as it moves along the x axis, we must have

$$(x - vt) = \text{a constant} \tag{16–11}$$

[2] The phrase "a function of the quantity $(x - vt)$" means that x and t must always occur in the form $(x - vt)$. Thus $\sin(x - vt)$ and $(x - vt)^2$ are satisfactory, but $x^2 - (vt)^2$ is not.

Differentiating with respect to time gives

$$\frac{dx}{dt} - v = 0$$

or

$$\frac{dx}{dt} = v \qquad (16\text{–}12)$$

Thus, the speed with which a point on the wave (and therefore the wave as a whole) moves along the x axis is, indeed, v. This is called the **phase velocity** of the wave, because specifying a particular point on the wave is called specifying the **phase** of the wave.

By similar reasoning, we conclude that waves traveling in the *negative* x direction will be functions of $(x + vt)$. Thus:

$y = f(x - vt)$ (waves traveling in the *positive* x direction)

$y = f(x + vt)$ (waves traveling in the *negative* x direction)

16.4 A Particular Solution to the Wave Equation

Let us now take up a particular case—that of a *sinusoidal* wave train traveling along a stretched string. This is one of the easiest waves to deal with mathematically. It is also an apt choice because *any periodic wave*, no matter how complex, can be represented as a sum of sine and cosine waves of various frequencies. This type of sum, discussed in Appendix F, is called a Fourier series. If we learn to handle the case of a single sinuosidal wave, we will have the basic knowledge to handle periodic waves of any complexity by using the Fourier series analysis.

Figure 16–5 illustrates a sinusoidal wave. It is generated when one end of the string is displaced transversely in simple harmonic motion. There are numerous ways of writing the mathematical expression for a traveling sinusoidal

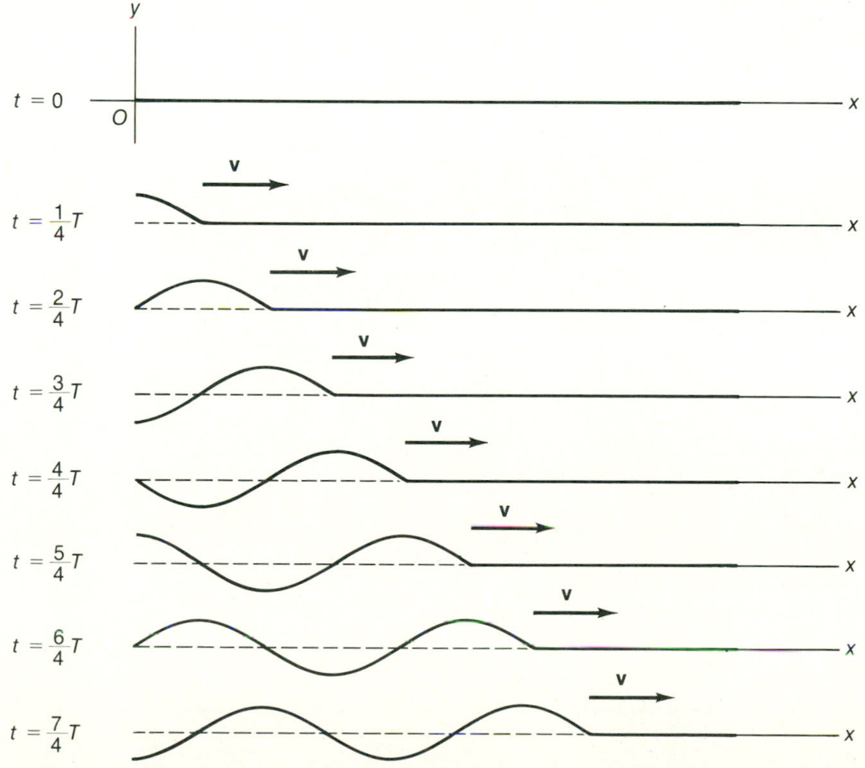

Figure 16–5
Successive stop-motion views of the traveling sinusoidal wave generated when one end of a stretched string is displaced transversely in simple harmonic motion. The time T is one period of the SHM.

wave. One way is:

$$y = A \sin \frac{2\pi}{\lambda}(x - vt) \tag{16–13}$$

This equation brings out the fact that the wave is a function of $(x - vt)$ and thus is truly a wave traveling in the $+x$ direction. It may be more useful, however, to introduce the notation we used in SHM (see Figure 16–6):

- $\lambda =$ **wavelength,** the distance between two adjacent points having the same phase
- $T =$ **period,** the time for one complete vibration to pass a given point
- $f =$ **frequency,** $(f = 1/T)$ the number of vibrations per second occurring at a given point
- $A =$ **amplitude,** the magnitude of the maximum y displacement

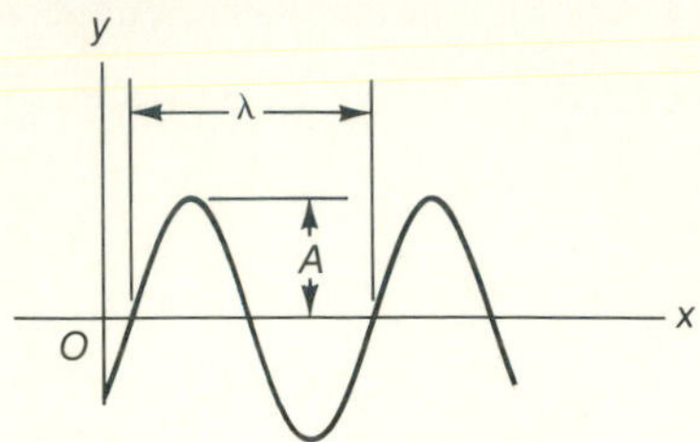

(a) By holding t fixed, we focus attention on the (instantaneous) transverse displacement of the string along its entire length. The wavelength λ is the distance between two adjacent points that are in the same phase.

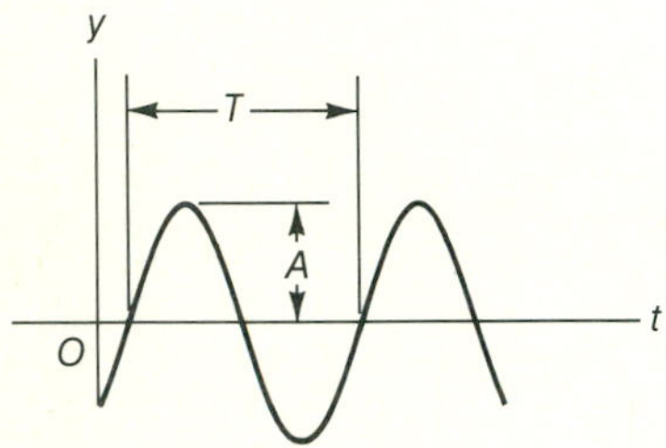

(b) By holding x fixed, we look at the transverse motion in time of the part of the string located at the point x. It undergoes SHM in the y direction. The period T is the time required for one complete vibration.

Figure 16–6
A sinusoidal traveling wave of the form $y = A \sin(2\pi x/\lambda - 2\pi t/T)$. The magnitude of the maximum displacement is called the *amplitude.*

Physically, the **source** is responsible for the *frequency*, and the **medium** (here, the stretched string) determines the *speed.* Combining these two leads to the *wavelength.* Since a time T is required for the wave to travel a distance of one wavelength λ at the speed v, we have:

$$\lambda = vT \qquad \text{or} \qquad f\lambda = v \tag{16–14}$$

Substituting these symbols into Equation (16–13), we obtain another convenient form of the solution:

$$y = A \sin 2\pi\left(\frac{x}{\lambda} - \frac{t}{T}\right) \tag{16–15}$$

A still more concise and useful notation makes use of the **wave number** k and the **angular frequency** ω (equal to $2\pi f$).

$$\text{Wave number} \qquad k \equiv \frac{2\pi}{\lambda}$$

$$\text{Angular frequency} \qquad \omega \equiv \frac{2\pi}{T}$$

(Be careful not to confuse this k with the same symbol used for the radius of gyration and the spring constant.) Since kx and ωt appear in the argument of a sine, these two terms must be dimensionless (expressed in *radian* measure).

Making the substitutions in Equation (16–15), we obtain:

TRAVELING SINUSOIDAL WAVES

$$\left.\begin{aligned} y &= A \sin(kx - \omega t) \quad &&\text{(sinusoidal waves traveling in the } +x \text{ direction)} \\ y &= A \sin(kx + \omega t) \quad &&\text{(sinusoidal waves traveling in the } -x \text{ direction)} \end{aligned}\right\} \tag{16–16}$$

Before we present numerical examples, let us verify that the solution we have obtained does agree with Equation (16–8), the general wave equation. We need the partial derivatives with respect to both time and space.

Time derivatives	Space derivatives
$y = A \sin(kx - \omega t)$	$y = A \sin(kx - \omega t)$
$\dfrac{\partial y}{\partial t} = -\omega A \cos(kx - \omega t)$	$\dfrac{\partial y}{\partial x} = kA \cos(kx - \omega t)$
$\dfrac{\partial^2 y}{\partial t^2} = -\omega^2 A \sin(kx - \omega t)$	$\dfrac{\partial^2 y}{\partial x^2} = -k^2 A \sin(kx - \omega t)$

We see that the second derivatives have a relationship that agrees with the wave equation:

$$\frac{\partial^2 y}{\partial t^2} = \left(\frac{\omega}{k}\right)^2 \frac{\partial^2 y}{\partial x^2} \tag{16–17}$$

But $\omega/k = (2\pi/T) \div (2\pi/\lambda) = \lambda/T = f\lambda = v$. In comparison with Equations (16–7) and (16–8), we see that:

THE SPEED *v* OF TRANSVERSE WAVES ON A STRETCHED STRING

$$v = f\lambda = \sqrt{\frac{F}{\mu}} \tag{16–18}$$

recalling that F is the tension in the string and μ is the mass per unit length. So our solution does agree with the wave equation. We also found that the constant v really is the speed of the waves. Note that the speed increases if the tension F increases or if the mass per unit length μ decreases. In either case, the *restoring acceleration* increases, causing the string to return to its equilibrium more quickly. Thus the relation agrees with our physical intuition regarding the behavior of the string.

EXAMPLE 16–1

A sinusoidal wave moving along a string under tension is described by the equation:

$$y(x,t) = 0.0020 \sin(10x - 120t) \qquad \text{(in SI units)}$$

where y is the transverse displacement of the string in meters, x is the distance along the string in meters, and t is the time in seconds. Find (a) the amplitude of the transverse displacement of the string, (b) the wavelength of the traveling wave, (c) its frequency of oscillation, and (d) the speed of propagation of the wave.

SOLUTION

Comparing $y(x,t) = 0.0020 \sin(10x - 120t)$ with the general form of a wave moving along a taut string, $y = A \sin(kx - \omega t)$, we see that

(a) $A = \boxed{0.0020 \text{ m}}$

(b) $k = \dfrac{2\pi}{\lambda} = 10$, or $\lambda = \dfrac{2\pi}{10} = \boxed{0.628 \text{ m}}$

(c) $\omega = 2\pi f = 120$, or $f = \dfrac{120}{2\pi} = \boxed{19.1 \text{ Hz}}$

(d) $v = f\lambda = (19.1 \text{ s}^{-1})(0.628 \text{ m}) = \boxed{12.0 \dfrac{\text{m}}{\text{s}}}$

16.5 Wave Speeds

We have discussed the speed of transverse waves on a stretched string. Other types of waves travel with speeds that depend on the particular elastic properties of the medium. For example, sound waves are longitudinal waves of compression and rarefaction that are transmitted with different speeds through solids, liquids, and gases. The audible range that a young human ear can sense is from about 20 Hz to 20 000 Hz. Longitudinal mechanical waves below audible frequencies are called **infrasonic** waves, while those above audible frequencies are **ultrasonic** waves.[3]

We now state (without proof) expressions for the speeds of various types of waves. Note that in each case the speed has the form

$$v = \sqrt{\frac{\text{Elastic modulus}}{\text{Density}}} \tag{16–19}$$

(Elastic moduli, which characterize the properties of deformed materials, were discussed in Section 10.20.)

SPEED OF COMPRESSION-RAREFACTION WAVES IN A LIQUID OR SOLID ROD

$$v = \sqrt{\frac{B}{\rho}} \tag{16–20}$$

where B is the bulk modulus for the medium and ρ is the density. The *primary waves*, or *P-waves*, generated by earthquakes are similar to this type (though a precise expression for the speed of earthquake P-waves is a bit more complicated than this equation). Primary waves travel faster than other earthquake waves, so they are the first to arrive at a distant sensing station. The *secondary waves*, or *S-waves*, are transverse waves that cause a shear strain on the solid medium as they pass.

SPEED OF TRANSVERSE (SHEAR) WAVES IN A SOLID

$$v = \sqrt{\frac{S}{\rho}} \tag{16–21}$$

where S is the shear modulus for the solid. Since liquids and gases cannot support a shear stress, these waves exist only in solids. No shear waves travel deeper than about 2900 km below the surface, implying that the core of the earth becomes liquid at this depth. (Seismic-wave evidence also implies a solid inner core at a depth of about 5100 km.) A third type of earthquake wave is *long waves*, or *L-waves*, which travel on the surface at slightly lower speeds speeds than the S-waves. Long waves are similar to water waves, involving both transverse and longitudinal motions, and are by far the most damaging.

The speed of sound in air is somewhat more complicated because it depends on several factors. From the previous discussion, one might guess that the relation should be $v = \sqrt{B/\rho}$. Indeed, Newton, using good physical reasoning, derived a formula similar to this. However, it is incorrect by about 20%.[4] Due to an effect unknown in Newton's time, the bulk modulus for gases can have two different values, depending on whether the *compressions* and *rarefactions* (regions of pressure *above* and *below* the equilibrium value for pressure) occur at constant temperature (*isothermal*) or without heat transfer (*adiabatic*). As shown by Pierre Simon de Laplace about a century later, for sound the process is adiabatic.

SPEED OF SOUND IN GASES

$$v = \sqrt{\frac{B_{\text{adiabatic}}}{\rho}} = \sqrt{\frac{\gamma B_{\text{isothermal}}}{\rho}} \tag{16–22}$$

[3] Many animal species can hear ultrasonic frequencies. For example, porpoises can detect frequencies to about 120 kHz, though at the lower range they are limited to those above 10 000 Hz. Thus a porpoise cannot hear musical notes on a piano (whose highest frequency is 4186 Hz), nor most human speech, nor most sounds emitted by certain underwater loudspeakers whose upper frequency range is limited. As a consequence, some experiments on communicating with porpoises were inherently doomed to failure because the experimenter was unaware of the lower limit of the porpoise's hearing ability. Few animals have the low-frequency sensitivity of humans.

[4] In order to convince skeptics of the correctness of his theories, Newton occasionally went to rather extreme lengths to bring his theories into better agreement with experimental data. As related by Richard W. Westfall in "Newton and the Fudge Factor," *Science*, Vol. 179 (Feb. 23, 1973), p. 751, a rather obvious example was his attempt to account for the 20% disagreement between theory and experimental values by introducing "correction factors" that had little scientific basis. Such a human reaction on Newton's part should not obscure the immense magnitude of his accomplishments.

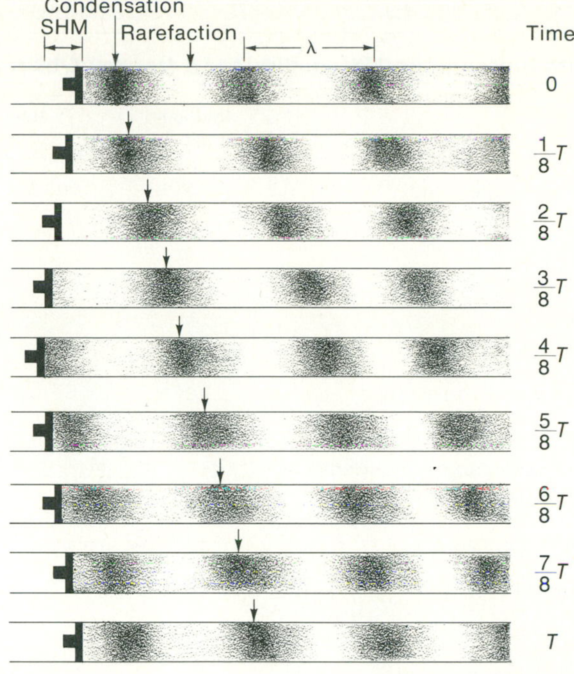

Figure 16–7
Successive "stop-movement" illustrations of a longitudinal sinusoidal wave generated in an air-filled tube by a piston at one end moving in SHM. The pattern of condensations and rarefactions moves along the tube with constant velocity as each particle of the medium undergoes SHM along the direction of propagation.

where $\gamma = C_p/C_v$ (equal to 1.4 for air), the ratio of the molar specific heat of the gas at constant pressure to that at constant volume.[5] This is equivalent to

$$v = \sqrt{\gamma P/\rho} \qquad \textbf{(16–23)}$$

where P is the gas pressure. At 20°C, the experimental value for the speed of sound in air is

$$v = 344 \text{ m/s} \qquad \text{or} \qquad 1129 \text{ ft/s} \qquad \textbf{(16–24)}$$

which is in good agreement with Equation (16–23). A more complete analysis shows that the speed of sound varies with the square root of the absolute temperature K (equal to 273° + °C).

16.6 Waves in Two and Three Dimensions

The equations we have developed for the transverse waves on a stretched string also describe other types of wave motion. We need only interpret the displacement y to be *the motion away from the equilibrium position of the particles of the medium.* The equations apply equally well to transverse waves and to longitudinal waves.

If a piston at one end of a tube of air is made to vibrate with SHM, longitudinal sinusoidal waves are generated (see Figure 16–7). As the gas molecules in the air vibrate about their equilibrium position, condensations are formed, followed by adjacent rarefactions. The pattern of pressure differences thus generated propagates along the tube with constant velocity, as shown by the arrow in the figure. These, of course, are sound waves.[6] Some characteristics of the response of the human ear are shown in Table 16–1 and Figure 16–8.

[5] The *molar specific heat* is defined as the amount of heat required to raise the temperature of one mole of a substance 1°C. In the case of a gas, it can be done at constant pressure (C_P) or at constant volume (C_V). This concept will be discussed in detail in Chapter 19.

[6] The human ear is so sensitive it can detect pressure variations as small as about 3 parts in 10^{10}. This corresponds to a sound intensity of 10^{-12} watts/meter^2 (W/m^2), for which the amplitudes of the motions of air molecules are only about one-tenth the radius of an atom.

TABLE 16–1

Approximate Sound Intensities for a Nearby Observer

	Intensity (W/m^2)	Intensity level[a] (dB)
Large rocket engine	10^6	180
Jet airplane takeoff	10^3	150
Pain-producing sound	1	120
Rock-and-roll band	10^{-1}	110
Jackhammer or riveter	10^{-2}	100
Level above which prolonged exposure can permanently damage hearing	10^{-3}	90
Busy street traffic	10^{-5}	70
Ordinary conversation	10^{-6}	60
Average whisper	10^{-10}	20
Barely audible sound	10^{-12}	0

[a] The *intensity level* β in *decibels* of sound is defined as

$$\beta \equiv 10 \log \frac{I}{I_0} \tag{16–25}$$

where the reference level for intensity I_0 is, by international agreement, 10^{-12} W/m^2. The unit *decibel* (dB) is named after Alexander Graham Bell.

Human hearing is sensitive over a very large range. Thus an intensity level based on the *logarithm* of intensity ratios is convenient for two reasons: first (and more important) is the fact that human perception of sound intensity is linear on such a scale, where for every increase of 10 dB, the "loudness" of a sound is approximately doubled. Moreover, the least perceptible *change* in intensity is of the order of one decibel, independent of the intensity. Second, the range of perceptible intensity is about 12 orders of magnitude, making a *logarithmic* scale mathematically more convenient than a linear scale.

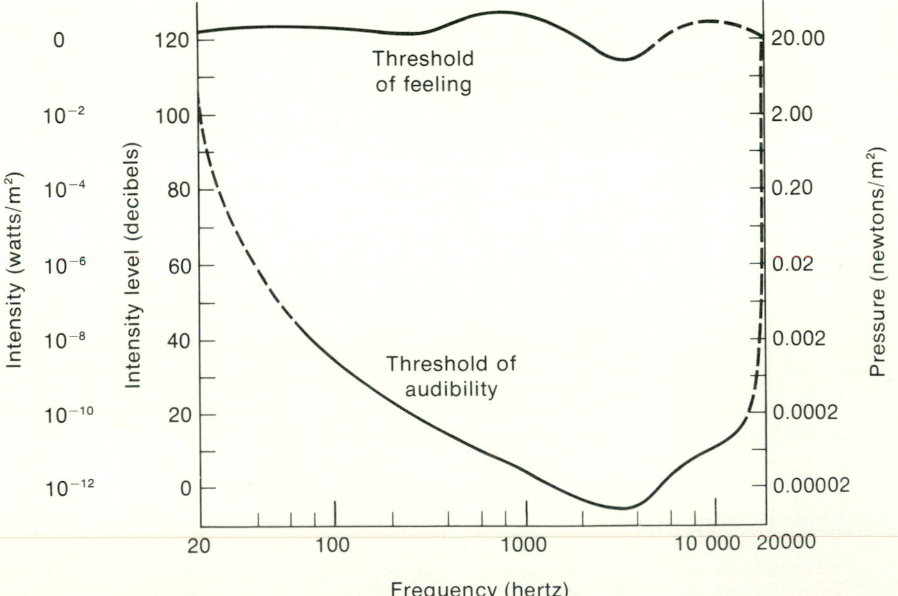

Figure 16–8

The ranges of audible frequencies and intensities for a sensitive human ear. For about half the population, the minimum intensity for a sound to be heard is roughly 10 to 100 times greater than the lower curve. Sustained exposure to sound level above about 90 dB can result in permanent impairment of hearing.

We may even write similar equations of the form $\theta = \theta_0 \sin(kx - \omega t)$ to designate *torsional* waves in which a "twisting" displacement is propagated along a taut wire or rod (Figure 16–9).

Transverse waves propagating along a surface (such as those on a drumhead or metal sheet) require two spatial variables, say x and y (with the transverse motion in the z direction), as well as the time variable t. Waves in three dimensions propagating from a point (such as sound waves emanating from an explosion) or from a line source (such as a vibrating wire) require three spatial variables.

In dealing with waves in two or three dimensions, it is useful to define a **wavefront.** For periodic waves, *each particle on a wavefront has the same phase.* A two-dimensional example would be water waves expanding out from the point where a rock was tossed into a pond. Here the wavefronts are concentric circles. All three-dimensional wavefronts, when observed farther and farther from the source, approach a plane configuration known as **plane waves** (Figure 16–10b). Plane waves are particularly easy to deal with mathematically, so they are often discussed in the introductory treatments of various types of wave motion. One notable feature of plane waves is that the entire wave profile does not change its amplitude as the wave moves along, assuming there is no energy dissipation in the medium itself. If some energy is "lost" as the plane wave progresses (usually by heating the medium), then the amplitude of the wave motion gradually becomes smaller. This is reasonable, since (as in SHM) the energy of the wave is proportional to the square of the amplitude. Of course, in the case of cylindrical or spherical wave propagation, the energy associated with each wavefront is spread over an increasingly larger area. Therefore, in these cases there will be a decrease in amplitude, even in the absence of dissipative conditions.

Water waves are an interesting example that involves longitudinal and transverse motions simultaneously. As the wave passes by, particles on the surface can have circular trajectories in a vertical plane, as shown in Figure 16–11. The speed of propagation of water waves is not the same for all wavelengths, a property known as *dispersion.* It depends, among other factors, on the depth of the water, becoming slower in shallower water. This explains the tendency of ocean waves to curl over and "break" as they approach the shore.

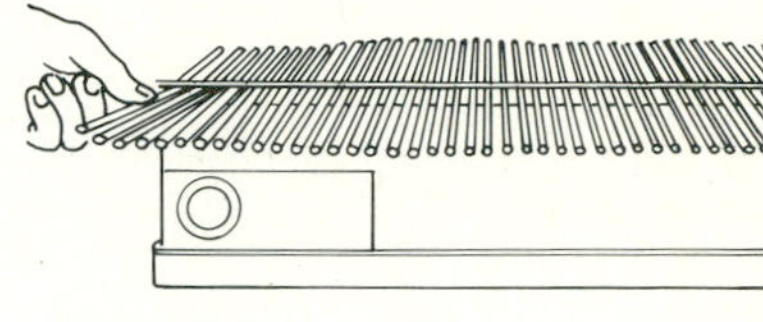

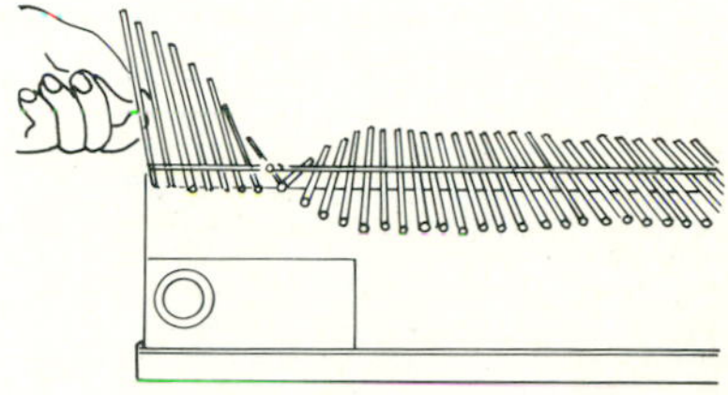

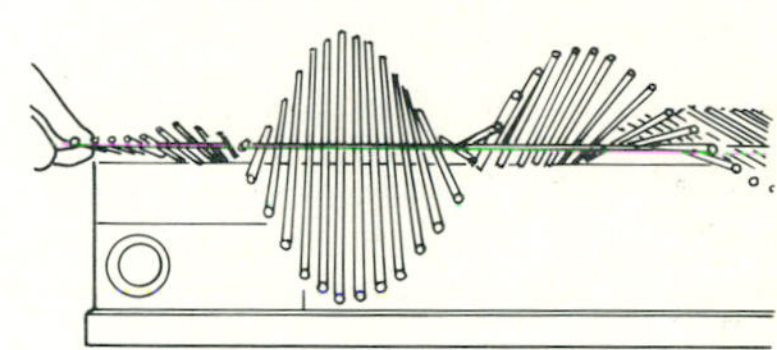

Figure 16–9
An apparatus developed by the Bell Telephone Laboratories for demonstrating wave motion utilizes torsional waves. A series of metal rods are fastened at their midpoints to a stiff "backbone" wire supported horizontally. When the end rod is rotated in angular SHM displacements, a sinusoidal torsion wave is propagated along the wire.

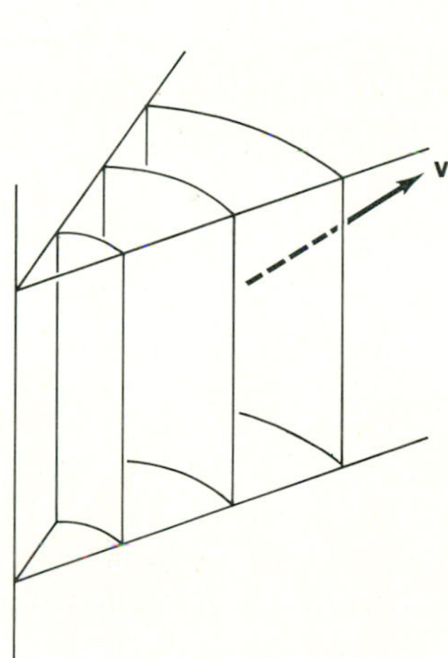

(**a**) A portion of the cylindrical wavefronts moving radially away from a line source.

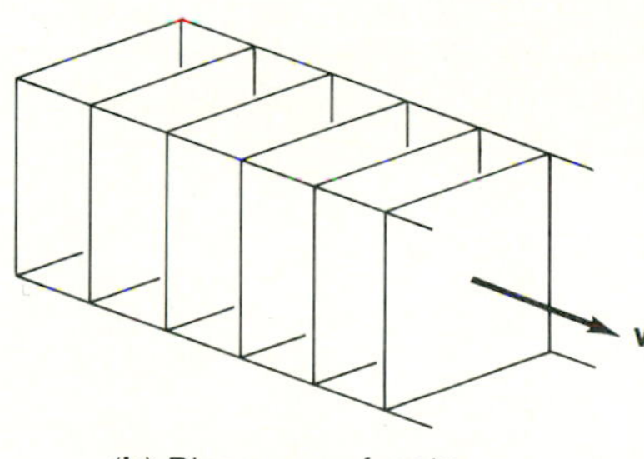

(**b**) Plane wavefronts.

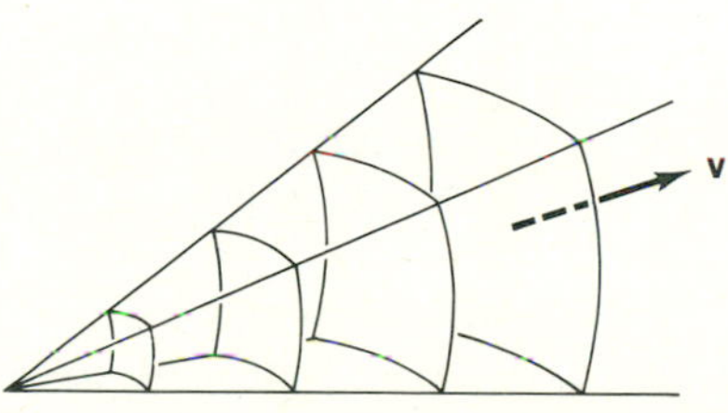

(**c**) A portion of the spherical wavefronts moving away from a point source.

Figure 16–10
Wavefronts are imaginary surfaces containing points that have the same phase. At very great distances from a point or line source, a small portion of the curved wavefront becomes essentially a plane wave.

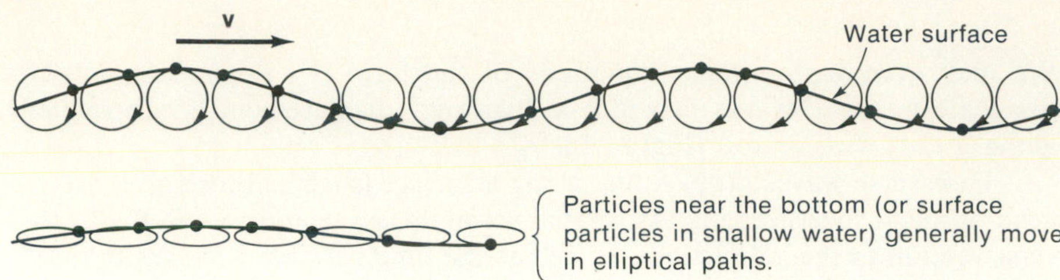
v
Water surface
Particles near the bottom (or surface particles in shallow water) generally move in elliptical paths.

Figure 16–11
Water waves involve both longitudinal and transverse motions of the medium.

The tops of the waves have sufficient forward momentum to outrun the main part of the wave, whose speed slows down as it progresses into shallower water.

16.7 Energy Considerations in Wave Motion

We have pointed out that wave propagation transports energy and momentum through a distance without particles of the medium actually traversing the distance. Energy is injected at the source of the waves, and the wave disturbance then carries it to a distant location, where energy may be extracted. We will now calculate the rate at which energy is transported by sinusoidal waves past a given point on a stretched string. The mass per unit length of the string is μ, so a small mass segment dm is equal to $\mu\,dx$. Each mass element dm undergoes SHM with the same frequency $\omega = 2\pi f$ and the same amplitude A. Recall from Chapter 6 that the total energy E for simple harmonic motion is $\frac{1}{2}kA^2$, where k is the equivalent force constant associated with the Hooke's Law force. From Chapter 12 we know that $k = m\omega^2$, so the total energy dE of the mass element dm is

$$dE = \tfrac{1}{2}\,dm\,\omega^2 A^2$$

$$dE = \tfrac{1}{2}\mu\,dx\,\omega^2 A^2 \qquad \textbf{(16–26)}$$

For a wave traveling to the right, this energy comes from the work done by the force that the string segment at the left exerts on the mass element dm. In turn, the mass element itself does work on the adjacent mass element to its right, passing the energy along. The rate at which energy is passed along the string, dE/dt, is the *power* transmitted past a given point by the waves. Dividing Equation (16–26) by dt, we have

$$\text{Power} = \frac{dE}{dt} = \tfrac{1}{2}\mu\,\frac{dx}{dt}\,\omega^2 A^2 \qquad \textbf{(16–27)}$$

Since $dx/dt = v$, the wave speed, we have

POWER TRANSMITTED BY WAVES ON A STRING

$$\text{Power} = \underbrace{(\tfrac{1}{2}\mu\omega^2 A^2)}_{\text{Energy per unit length}}(v) \qquad \textbf{(16–28)}$$

The rate of energy flow is proportional to the velocity v of propagation, as expected. Also, the power is proportional to the *square* of the frequency f (equal to $\omega/2\pi$) and the *square* of the amplitude A. If you possess high fidelity equipment you will now understand why the loudspeaker cone that generates low frequencies (the "woofer") must be so much larger and undergo greater amplitude oscillations than the loudspeaker cone that generates high frequencies (the "tweeter"). *For the same power radiated,* the lower the frequency, the greater must be the amplitude of the sound waves.

For sound waves in three dimensions, the rate of energy flow per unit area normal to the direction of propagation is defined to be the **intensity** I of the sound wave. A derivation analogous to Equation (16–28) leads to

$$I = \frac{\text{Power}}{\text{Unit area}}$$

$$I = \left(\frac{\text{Energy}}{\text{Unit volume}}\right)(\text{Velocity})$$

INTENSITY I OF PLANE SOUND WAVES (power transmitted per unit area)

$$I = \underbrace{(\tfrac{1}{2}\rho\omega^2 A^2)}_{\substack{\text{Energy per}\\ \text{unit volume}}}(v) \qquad \textbf{(16–29)}$$

where ρ is the mass density (mass/volume) of the air. The intensity I is measured in watts/meter^2. Note the similarity of the form of this equation with the corresponding equation for the power transmitted by waves on a string [Equation (16–28)].

EXAMPLE 16–2

Find the amplitude of air-molecule oscillation in a plane sound wave that has a frequency of 3000 Hz and intensity level of 60 dB. Use 343 m/s as the speed of sound and 1.21 kg/m^3 as the density of air.

SOLUTION

The decibel scale of sound intensity is useful in comparing the physiological effects of different sound intensities, but physically the intensity itself is more significant. We may convert a 60 dB intensity level to intensity by application of Equation (16–25):

$$\beta = 10 \log \frac{I}{I_0}$$

where $I_0 = 10^{-12}$ W/m^2. Solving for I, we have

$$I = I_0\, 10^{\beta/10}$$

Substituting numerical values, we have

$$I = 10^{-12}\ \text{W/m}^2\ 10^{60/10} = 10^{-6}\ \text{W/m}^2$$

We use this value for the intensity in Equation (16–29):

$$I = (\tfrac{1}{2}\rho\omega^2 A^2)(v)$$

Solving for A and substituting numerical values yields

$$A = \left[\frac{2I}{\rho\omega^2 v}\right]^{1/2} = \left[\frac{(2)\left(10^{-6}\ \frac{\text{W}}{\text{m}^2}\right)}{\left(1.21\ \frac{\text{kg}}{\text{m}^3}\right)\left(2\pi\left[3000\ \frac{\text{rad}}{\text{s}}\right]\right)^2\left(343\ \frac{\text{m}}{\text{s}}\right)}\right]^{1/2} = \boxed{3.68 \times 10^{-9}\ \text{m}}$$

16.8 Reflection of Waves

Consider a single pulse that travels down a string fastened to a rigid wall at the far end, as in Figure 16–12. If the incident pulse has a vertically upward displacement, it exerts an upward force on the wall when it arrives. By Newton's

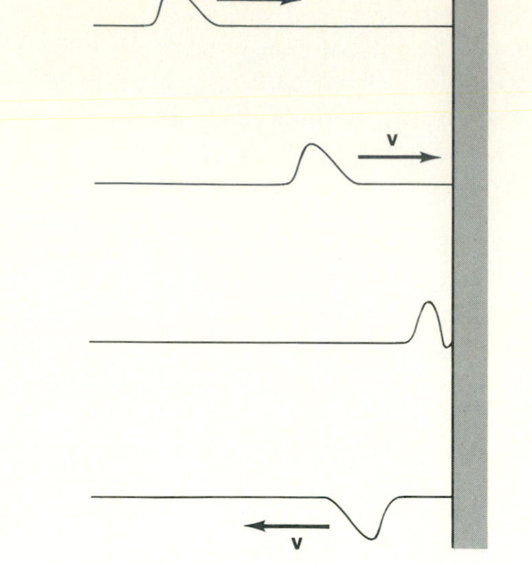

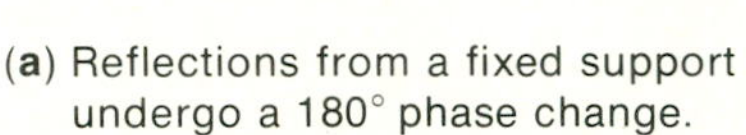

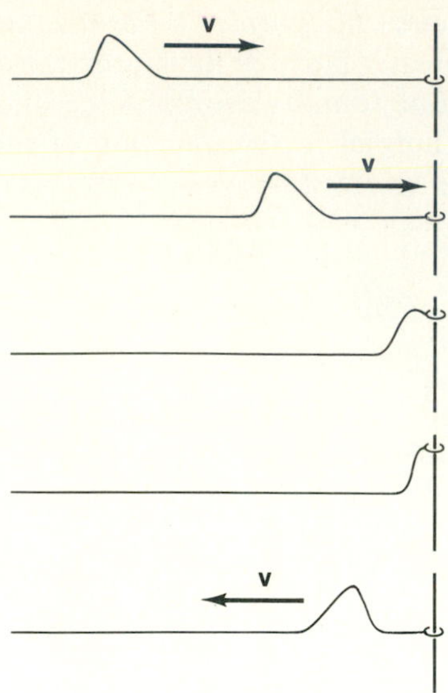

Figure 16–12
Reflections of a traveling pulse on a string. The pulse has an asymmetrical wave shape to aid in understanding the reflection process. Whether or not there is a phase change depends on how the end of the string is terminated.

(**a**) Reflections from a fixed support undergo a 180° phase change.

(**b**) The end of the string is free to move vertically if fastened to a frictionless, massless ring on a vertical rod. The pulse is reflected with no phase change.

third law, the wall therefore exerts an equal force in the downward direction on the end of the string. (This reaction force would be absent if the wall were not present.) The wall's downward force on the string generates a downward pulse that subsequently travels along the string in the opposite direction. The pulse has been "reflected." It changed from a positive amplitude in the incident pulse to a negative amplitude in the reflected pulse, thus undergoing a phase change of 180°.

If instead the far end of the string is free to move vertically, as in Figure 16–12(b), the arrival of the pulse causes the free end to overshoot the normal amplitude of the pulse. (It actually reaches twice the normal amplitude.) This process generates a reflected pulse that is positive; hence no phase change occurs in this case.

A string may be terminated so that its end is free to move in the vertical direction by tying it to another string of very long length whose mass per unit length is negligible. The *discontinuity* in the mass per unit length (from μ to essentially zero) gives rise to the reflected pulse. If μ becomes greater across the junction, the reflected pulse has a 180° phase change; if μ becomes less, the reflected pulse has no phase change. In general, any discontinuity generates both reflected and transmitted pulses whose amplitudes depend on the type of discontinuity. The energy of the incident pulse is split between the reflected and transmitted pulses, so their amplitudes are less than the incoming wave amplitude.

It is possible to achieve 100% transmission of pulses across a discontinuity so that no reflected pulse occurs. This will happen if μ changes negligibly over a distance that is large compared with the pulse length, so that the discontinuity is a sufficiently gradual one. For traveling waves (not single pulses) that involve just one frequency, a discontinuity in μ may be "smoothed over" in a short distance by the following method. A short segment of string called a "matching stub," one-quarter wavelength long, is inserted between the two strings. If the mass per unit length for the matching stub has the geometric mean value $\sqrt{\mu_1\mu_2}$, which is intermediate between the values of μ for the two strings, then no reflection occurs. This antireflection technique has analogies in other branches of physics: an example is the antireflection coating on photographic lenses. The coating forms a quarter-wavelength matching stub for electromagnetic waves, preventing reflection of certain frequencies. (Here, the parameters that must be matched are the "index of refraction" of the glass and of the air, factors that determine the speed of electromagnetic waves.)

16.9 The Superposition Principle and Standing Waves

If two waves of the same frequency are traveling in opposite directions on a stretched string, they create a **standing-wave** pattern of displacements. These standing waves are the basis for all musical instruments. They arise in the following manner.

Consider Figure 16–13, which shows the combination of two waves of equal amplitude and wavelength, traveling in opposite directions on the same string. The two displacements combine, or *superpose*, to produce a stationary vibration pattern in which certain points remain fixed, while all other points oscillate in time with the same frequency. The fact that two or more wave displacements add together in this fashion is known as the **superposition principle.**[7] Because the wave equation is *linear*, combinations of solutions such as $ay_1 + by_2$ are also a solution. The superposition principle applies (at least approximately) to all forms of small-amplitude wave motion: water waves, sound waves, electromagnetic waves, and so on. Depending on the phase of the two displacements, their combination produces increased amplitude, called **constructive interference,** or diminished amplitude, called **destructive interference.**

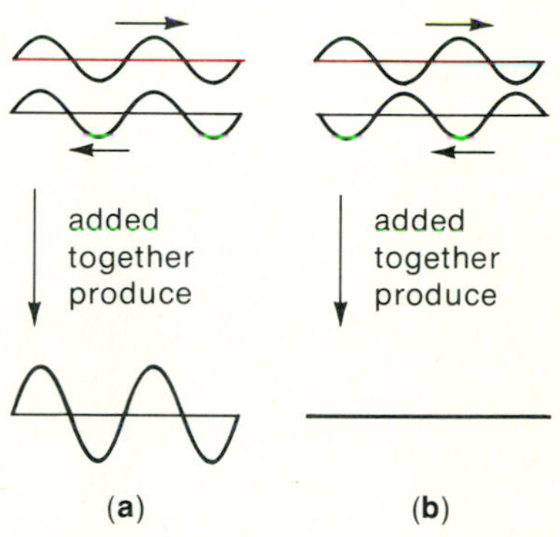

Figure 16–13
When waves traveling in opposite directions are superposed, they produce (a) a double-amplitude displacement, (b) zero displacement, or any intermediate amplitude between these extremes.

The characteristics of standing waves can easily be demonstrated mathematically. Consider two waves of equal amplitude, frequency, and wavelength, traveling in opposite directions:

$$y_1 = A \sin (kx - \omega t) \qquad (+x \text{ direction})$$
$$y_2 = A \sin (kx + \omega t) \qquad (-x \text{ direction})$$

Combining them, we have another possible solution to the wave equation:

$$y = y_1 + y_2 = A\left[\sin (kx - \omega t) + \sin (kx + \omega t)\right] \qquad \textbf{(16–30)}$$

We now use the trigonometric relation for the sum of the sines of two angles:

$$\sin x + \sin y = 2 \sin \tfrac{1}{2}(x + y) \cos \tfrac{1}{2}(x - y) \qquad \textbf{(16–31)}$$

Making use of this relation, the combined expression may be written:

STANDING WAVES

$$y = [2A \sin (kx)] \cos (\omega t) \qquad \textbf{(16–32)}$$

The first factor of the equation, $2A \sin (kx)$, is the amplitude at various positions x along the string, and the second factor, $\cos (\omega t)$, makes all points along the string vibrate in SHM with the same frequency ω.

[7] The superposition principle is used frequently in physics, because it is usually valid for small amplitudes, resulting in mathematical simplifications. However, it is worth pointing out that often it does not hold. For example, whenever Hooke's Law is violated (as in large-amplitude motions), the effect of combining two waves is not merely the sum of the individual waves. Hi-fi enthusiasts recognize this as *intermodulation distortion*, which occurs when two frequencies sounded simultaneously generate additional frequencies because of nonlinearities in the system.

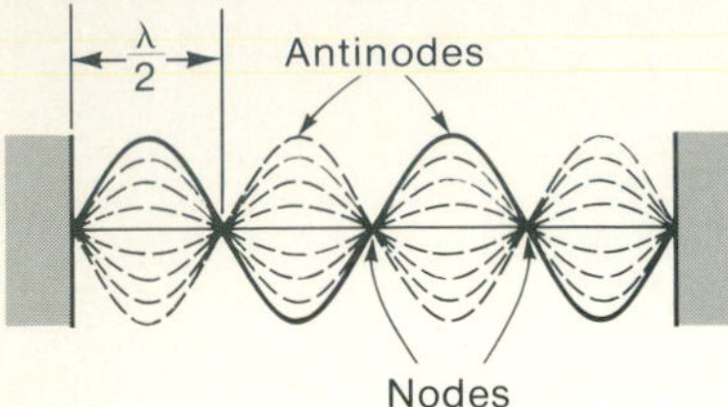

Figure 16–14
The pattern of oscillations associated with a standing wave. The distance between adjacent nodes is one-half wavelength.

The fixed points of *minimum* amplitude are called **nodes,** as sketched in Figure 16–14. The amplitude is zero at the locations where

$$kx = 0, \pi, 2\pi, 3\pi, \ldots$$

or, since $k = 2\pi/\lambda$:

$$x = 0, \frac{\lambda}{2}, 2\frac{\lambda}{2}, 3\frac{\lambda}{2}, \ldots \qquad \text{(location of the nodes)}$$

Thus:

$$\text{Distance between adjacent nodes} = \frac{\lambda}{2}$$

The positions of *maximum* amplitude are called **antinodes,** spaced halfway between the nodes, where:

$$kx = \frac{\pi}{2}, 3\frac{\pi}{2}, 5\frac{\pi}{2}, \ldots$$

$$x = \frac{\lambda}{4}, 3\frac{\lambda}{4}, 5\frac{\lambda}{4}, \ldots \qquad \text{(location of the antinodes)}$$

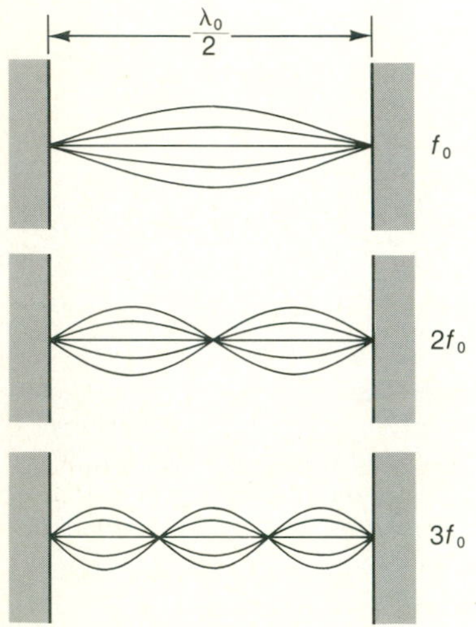

Figure 16–15
Various standing waves on a string of fixed length have integral frequency relationships.

If the string is fastened to rigid supports located at the nodal positions, then the standing wave oscillations continue indefinitely in time (provided, of course, there is no energy dissipation through friction, radiated sound energy, or vibration of the support). This occurs because no energy can flow past a node if the string is stationary. Instead, each point on the string is undergoing SHM oscillations (in the transverse direction), with all points between two adjacent nodes being in phase with each other. Any two points separated by a single node are 180° out of phase with each other. Since nodes must occur at the ends which are held rigid, the length of the string contains *integral numbers of half wavelengths.* A string can simultaneously vibrate with different frequencies (Figure 16–15). Thus, for a given stretched string there is a family of standing waves whose frequencies are known as **harmonic frequencies:** f_0, $2f_0$, $3f_0$, and so on. Each "allowed" frequency in this series is consistent with the existence of a node at the fixed ends of the string. The lowest frequency f_0 is called the **fundamental frequency,** and successively higher frequencies are the **second harmonic,** the **third harmonic,** and so forth.

Similar harmonic series are associated with standing-wave sound vibrations in open-end and closed-end organ pipes (Figure 16–16). In the "resonant" standing-wave patterns in these cases, a node must always occur at a closed end; antinodes generally occur at open ends and also at the air-intake end

Figure 16–16
A jet of air blown into a pipe at one end and directed against the edge of a side opening will set up standing-wave oscillations. The edge against which the air is blown always forms an antinode. The other end of the pipe may be either open or closed. Because the fundamental wavelength λ_0 is different in cases (a) and (b), pipes of equal length will have different fundamental frequencies. In general, more than one of these modes exist simultaneously in the pipe.

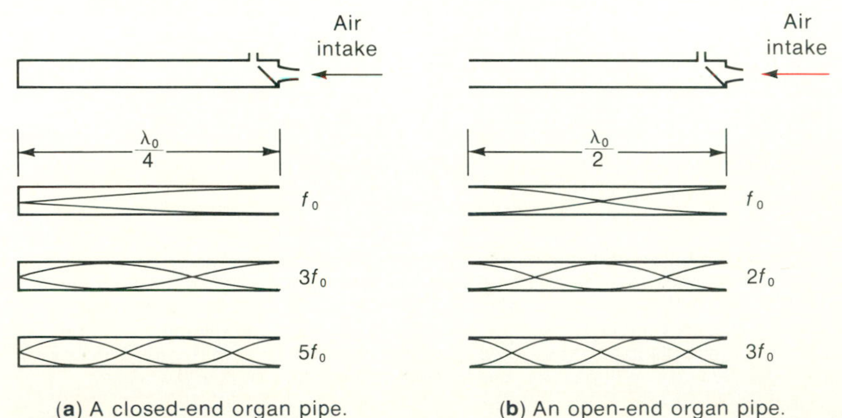

(**a**) A closed-end organ pipe.

(**b**) An open-end organ pipe.

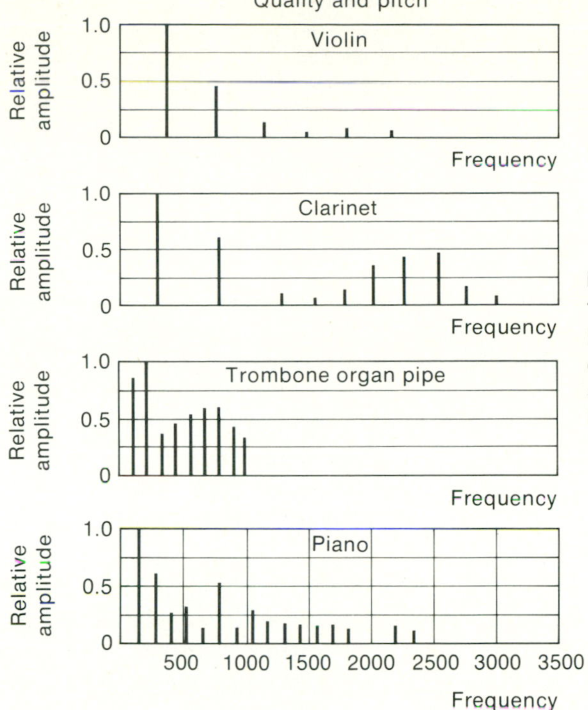

Figure 16–17
The amplitudes of various harmonics provide the distinguishing sound when notes are played on various musical instruments.

where vibrations are initiated. In a closed-end organ pipe, all even harmonics are missing. The number and amplitudes of the harmonics resent in a musical instrument determine the quality, or *timbre*, of a note and distinguish the note from those of other instruments (see Figure 16–17).

Standing-wave vibrations occur for two-dimensional membranes as well as for three-dimensional objects, as illustrated in Figure 16–18. In these cases, the natural frequencies of the various modes of vibration usually do *not* form harmonic series that are integral multiples of some fundamental frequency. Instead, the frequencies are related (in the simpler cases) to Bessel's functions and other mathematical functions. Generally, these standing-wave patterns involve nodal *lines* or nodal *surfaces*.

The study of three-dimensional standing-wave oscillations is extremely important for many areas of physics and engineering. The design of loudspeakers is an obvious practical example. Perhaps of greater significance are the standing-wave patterns associated with models of atoms. These studies provide helpful visual and mathematical models that explain much of the data of atomic physics.

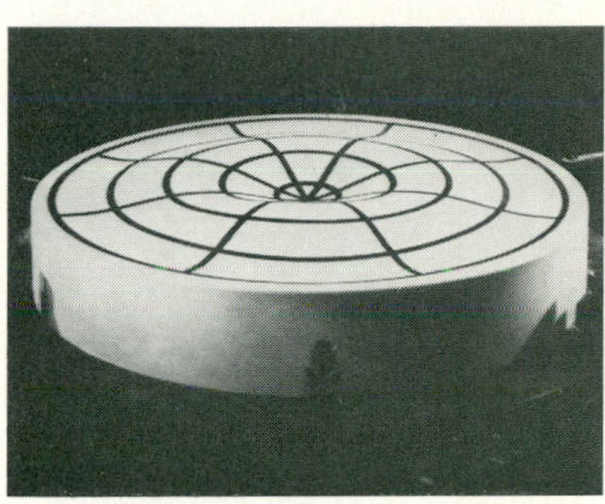

(a) A circular drumhead may be set into standing-wave vibrations by a loudspeaker beneath the drumhead.

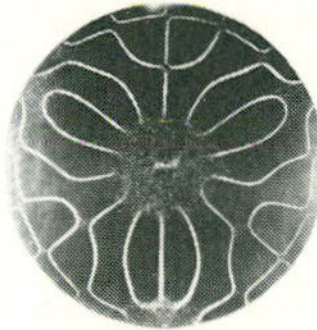

(b) Standing-wave patterns on a horizontal metal plate may be visualized by sprinkling fine sand on the vibrating plate. The sand collects along nodal lines where the motion of the plate is negligible. The patterns are called Chladni figures.

Figure 16–18
Standing-wave vibrations of two-dimensional surfaces.

EXAMPLE 16–3

The E-string on a violin is 33 cm long and its mass per unit length is 5.46×10^{-4} kg/m. What tension F is required to make it vibrate in its fundamental mode at 660 Hz?

SOLUTION

The fundamental-mode standing wave fits one-half wavelength within the total length ℓ of the string. Therefore:

$$\lambda = 2(\ell) = 2(33\ \text{cm}) = 66\ \text{cm}\underbrace{\left(\frac{1\ \text{m}}{100\ \text{cm}}\right)}_{\substack{\text{Conversion}\\ \text{factor}}} = 0.660\ \text{m}$$

The velocity of the waves on the string can be obtained from

$$v = f\lambda = (660\ \text{s}^{-1})(0.660\ \text{m}) = 436\ \frac{\text{m}}{\text{s}}$$

The tension F is then found from

$$\frac{F}{\mu} = v^2$$

or

$$F = (\mu)(v)^2 = \left(5.46 \times 10^{-4}\ \frac{\text{kg}}{\text{m}}\right)\left(436\ \frac{\text{m}}{\text{s}}\right)^2 = \boxed{104\ \text{N}}$$

16.10 Doppler Shift

The frequency of sound perceived by an observer depends not only upon the relative motion of the observer and the source of the sound, but also on whether the observer or source is moving relative to the medium through which the sound is propagated. This phenomenon was first explained by Christian Doppler, an Austrian physicist, in 1842.

We will begin by discussing the two cases: that in which the *observer* moves relative to the medium and that in which the *source* moves relative to the medium. Consider a source that emits sound of frequency f. If the source is at rest with respect to still air, the wavelength in air is λ, and the sound waves travel with a speed v. A stationary observer would receive, in one second, a number of vibrations equal to $f = v/\lambda$. However, if the observer is moving *toward* the source with a speed v_{ob} relative to the medium (air), he receives more vibrations per second. As well as the number v/λ he would receive if at rest, he also receives an additional number of vibrations equal to v_{ob}/λ, the number he "overtakes" in one second (Figure 16–19a). The *observed frequency* f' is thus

$$f' = \left(\frac{v}{\lambda} + \frac{v_{ob}}{\lambda}\right) = f\left(\frac{v + v_{ob}}{v}\right) \qquad \text{(observer moving)} \qquad \textbf{(16–33)}$$

If, on the other hand, the source moves toward the observer with a speed v_s relative to the medium (Figure 16–19b), then the wavelengths of the sound in air are shorter. In one second, f vibrations are squeezed together in the distance $(v - v_s)t$. Thus the wavelength in air is the shorter value $\lambda' = (v - v_s)/f$, and the stationary observer receives the frequency:

$$f' = \frac{v}{\lambda'} = f\left(\frac{v}{v - v_s}\right) \qquad \text{(source moving)} \qquad \textbf{(16–34)}$$

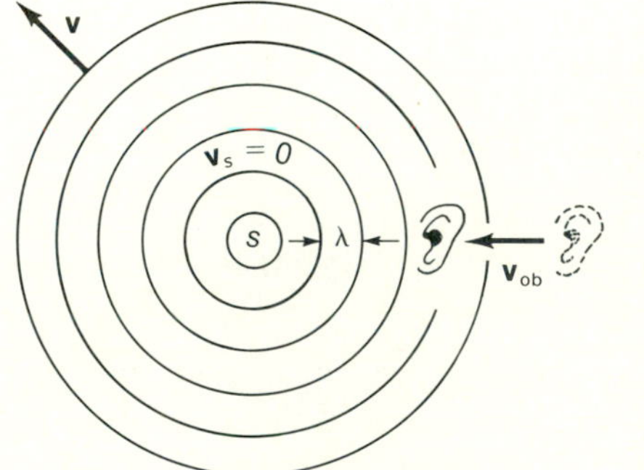

(a) Source at rest, observer in motion.

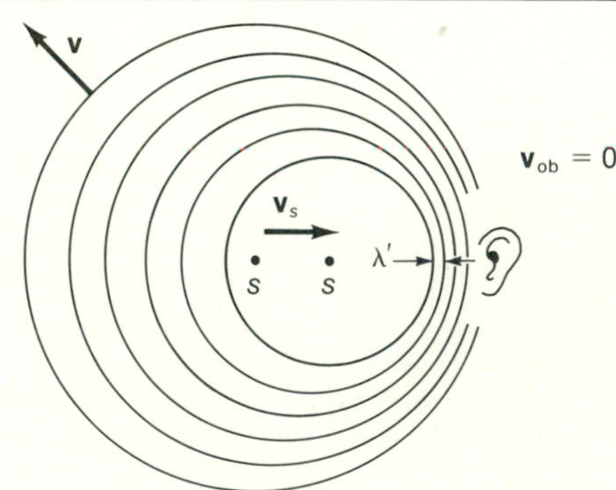

(b) Observer at rest, source in motion.

Figure 16–19
The Doppler effect due to motions of the source or the observer. All speeds are relative to the medium, air.

If motions are in the opposite direction to those discussed above, the signs of v_{ob} and v_s are reversed.

We may combine all effects into the single formula

DOPPLER SHIFT FOR SOUND

$$f' = f\left(\frac{v \pm v_{ob}}{v \mp v_s}\right) \tag{16–35}$$

The speeds v_{ob} and v_s are measured *relative to the medium*, namely the air. The signs are to be chosen so that the observed frequency is *higher* for relative *approach* of source and observer, and *lower* for relative *separation.*

EXAMPLE 16–4

A low-flying airplane skims the ground at a speed of 200 m/s as it approaches a stationary observer. A loud horn whose fundamental frequency is 400 Hz is carried on the airplane. (a) What frequency does the ground observer hear? (b) If instead the horn were on the ground, what frequency would the airplane pilot hear as she approached? The air is at rest with respect to the ground. Though the speed of sound in gases is independent of the pressure, it does vary with temperature changes. Assume for this problem that the speed of sound in air is 340 m/s.

SOLUTION

In this case the source and observer are approaching, so the signs in the Doppler Shift formula are chosen to be:

$$f' = f\left(\frac{v + v_{ob}}{v - v_s}\right)$$

(a) Since $v_{ob} = 0$, we have

$$f' = f\left(\frac{v}{v - v_s}\right) = f\left(\frac{1}{1 - \frac{v_s}{v}}\right) = (400\ \mathrm{s}^{-1})\left(\frac{1}{1 - \frac{200\ \mathrm{m/s}}{340\ \mathrm{m/s}}}\right) = \boxed{971\ \mathrm{Hz}}$$

(b) Here, $v_s = 0$, so

$$f' = f\left(\frac{v + v_{ob}}{v}\right) = f\left(1 + \frac{v_{ob}}{v}\right) = (400\ \mathrm{s}^{-1})\left(1 + \frac{200\ \mathrm{m/s}}{340\ \mathrm{m/s}}\right) = \boxed{635\ \mathrm{Hz}}$$

Note that although, the source and the observer are approaching each other in both cases, the observed frequency is different, depending on which is moving relative to the medium.

16.11 Shock Waves

Equation (16–35) is not valid if v_s or v_{ob} exceeds v. Yet these situations can occur, such as when a supersonic aircraft moves faster than the speed of sound in air. In these cases, a wave front with a large and sudden increase in pressure develops, called a *shock wave*, which causes a "sonic boom" when it reaches the ground. Figure 16–20 indicates the origin of the shock wavefront. The moving source generates spherical wavefronts that expand outward from various points along the path. The outer edges of these wavefronts combine to form a conical envelope that has a half-angle θ. From the geometry, we see that

$$\sin\theta = \frac{v}{v_s} \tag{16–36}$$

Analogous effects[8] occur when a charged particle moves through a transparent medium faster than the speed of light in the medium. The moving particle generates a conical wavefront of light called *Cerenkov radiation*, after its discovery in 1934 by the Russian physicist P. A. Cerenkov. Although the particle travels faster than the speed of light *in the medium*, it still does not exceed the relativistic limit of c, the speed of light *in a vacuum*.

16.12 Beats

If two sound waves pass through the same point, their displacements add according to the superposition principle. If they have the same amplitude, but only slightly different frequencies, they will combine as shown in Figure 16–21. The resultant variations in amplitude are known as **beats.**

$$y_1 = A \sin \omega_1 t \qquad (\text{where } \omega_1 = 2\pi f_1)$$

$$y_2 = A \sin \omega_2 t \qquad (\text{where } \omega_2 = 2\pi f_2)$$

$$y = y_1 + y_2 = A(\sin \omega_1 t + \sin \omega_2 t) \tag{16–37}$$

Using the trigonometric relation:

$$\sin x + \sin y = 2 \sin \left(\frac{x+y}{2}\right) \cos \left(\frac{x-y}{2}\right) \tag{16–38}$$

we may write Equation (16–37) as:

$$y = \underbrace{\left[\sin \left(\frac{\omega_1 + \omega_2}{2}\right) t\right]}_{\text{Vibrational frequency}} \underbrace{\left[2A \cos \left(\frac{\omega_1 - \omega_2}{2}\right) t\right]}_{\text{Amplitude variation}} \tag{16–39}$$

The resultant displacement oscillates with the average frequency $(f_1 + f_2)/2$, and the amplitude varies with the frequency[9] $(f_1 - f_2)$.

[8] In the case of water waves, the situation is more complex. A moving ship leaves a wake pattern that always has the half-angle $\theta = \sin^{-1}(1/3)$, *regardless of the speed of the ship.* See H. D. Keith, *American Journal of Physics*, Vol. 25 (1957), p. 466.

[9] The amplitude is a maximum whenever $\cos 2\pi[(f_1 - f_2)/2]t$ equals ± 1. *Each* of the two values occurs once in every cycle. Hence the number of maximum amplitude situations (beats per second) is given by $2(f_1 - f_2)/2 = (f_1 - f_2)$, the *difference* in frequency.

Figure 16–20
The development of a shock wave.

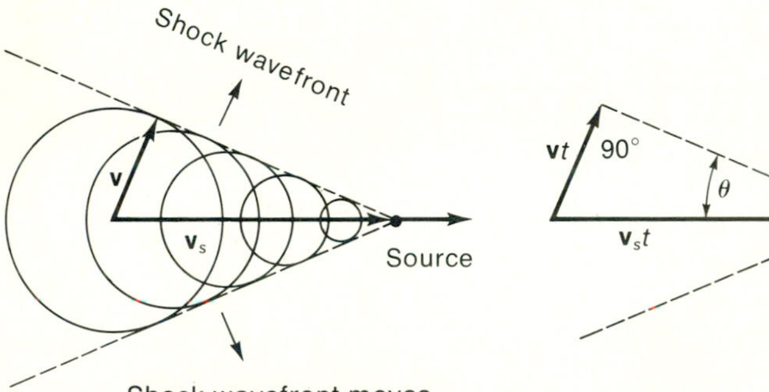

(a) A sound source moves with a speed $\mathbf{v}_s$ that is faster than the speed **v** of sound waves in the medium. At any given instant, the outer edges of the spherical wavefronts combine to produce a conical envelope called a shock wave.

(b) In time t, the sound waves move a distance $\mathbf{v}t$ while the source moves a distance $\mathbf{v}_s t$.

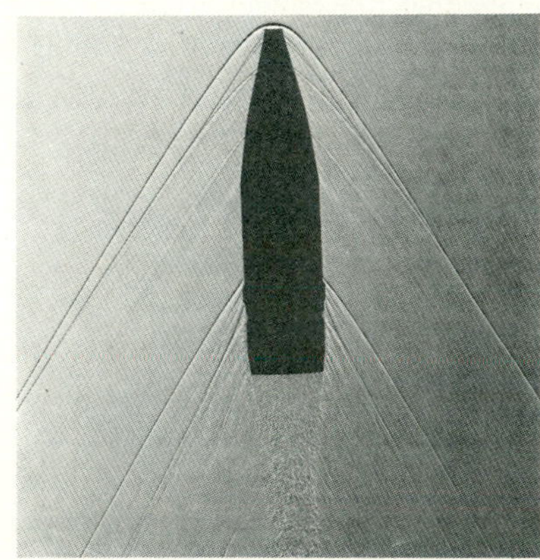

(c) The shock waves from a supersonic projectile are shown in this photograph.

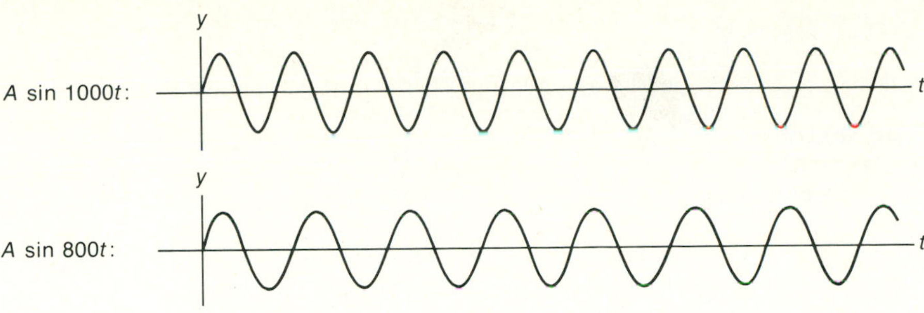

(**a**) Two sound waves of different frequencies: $\omega_1 = 1000$ rad/s and $\omega_2 = 800$ rad/s.

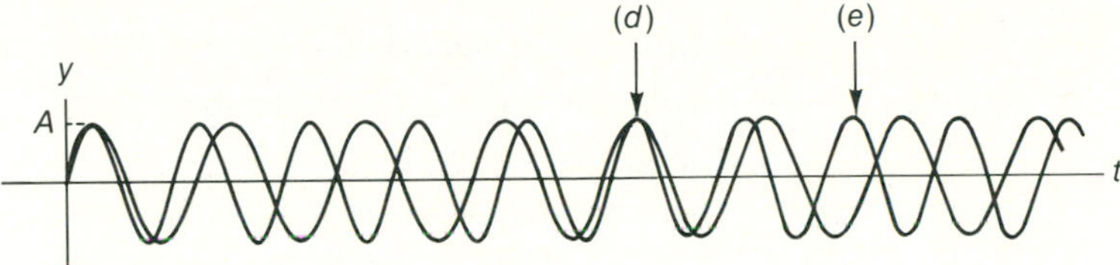

(**b**) The two waves are plotted on the same graph. At (*d*), the waves are in phase and they add together to produce an amplitude of 2*A*. At (*e*), the waves are 180° out of phase, combining to produce zero amplitude at that instant.

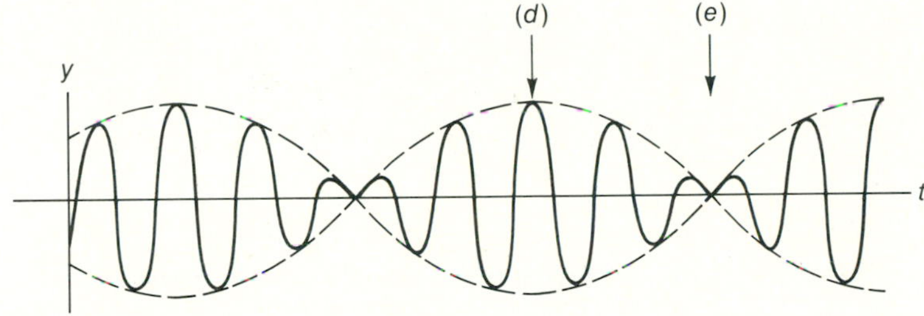

(**c**) The sum of the two waves has a resultant displacement whose amplitude varies with the difference of the two frequencies: $\omega_1 - \omega_2 = 200$ rad/s.

Figure 16–21
The superposition of two waves of different frequencies.

The human eardrum has a *nonlinear* response; that is, it would not undergo equal positive and negative displacements if, for example, the wave of Figure 16–21(b) were incident. Instead, the eardrum moves more easily in one direction than the other. Thus the eardrum has an average nonzero displacement, as shown dashed in Figure 16–22. As a result, if the sound is intense enough we hear the beat frequency $(f_1 - f_2)$ as well as f_1 and f_2. If the eardrum had a linear response (that is, moved equally easily for positive and negative displacements), then a symmetric wave such as Figure 16–21(c) would produce *zero* average displacement. The beat frequency would then not be heard, at least for the higher frequencies. (At sufficiently low frequencies, the variations in amplitude themselves caused a sound sensation of a definite frequency.)

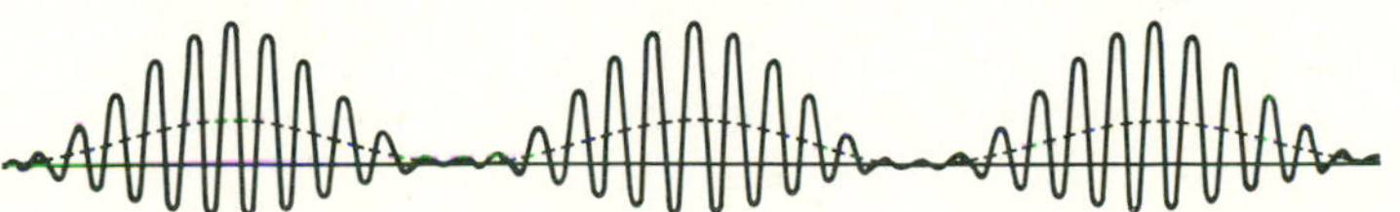

Figure 16–22
The nonlinear response of a human eardrum results in an average displacement (shown dashed) that fluctuates with the beat frequency.

Summary

All waves are described by an equation of the same general form:

THE WAVE EQUATION (for one dimension)

$$\frac{\partial^2 y}{\partial t^2} = v^2 \frac{\partial^2 y}{\partial x^2}$$

There are two general classes of waves: *transverse* and *longitudinal.* Transverse sinusoidal waves of small amplitude propagated along a string under tension in the $+x$ direction are described by the following solution to the wave equation:

$$y(x,t) = A \sin(kx - \omega t)$$

where:

y = transverse displacement of the string
A = amplitude (the maximum displacement)
x = distance along the string
t = time
k = wave number $= 2\pi/\lambda$
ω = angular frequency $= 2\pi/T = 2\pi f$

For a wave propagated in the *negative* x direction, the argument of the sine is $(kx + \omega t)$.

The *wavelength* λ, *frequency* f, and *speed of propagation* v are related:

$$\lambda f = v$$

The time for one complete vibration is the *period* $T = 1/f$; thus

$$\frac{\lambda}{T} = v$$

For waves on a string of mass per unit length μ and tension F:

$$v = \sqrt{\frac{F}{\mu}}$$

The *intensity level* β in *decibels* (dB) is

$$\beta = 10 \log \frac{I}{I_0}$$

where the reference intensity I_0 is equal to 10^{-12} W/m^2.

The *power* P transmitted by waves on a string is

Waves on a string $\qquad$ $\text{Power} = \underbrace{(\tfrac{1}{2}\mu\omega^2 A^2)}_{\text{Energy per unit length}}(v)$

where μ is the mass per unit length. The power per unit area, or the intensity I, transmitted by plane sound waves is

Sound waves $\qquad$ $I = \underbrace{(\tfrac{1}{2}\rho\omega^2 A^2)}_{\text{Energy per unit volume}}(v)$

where ρ is the mass per unit volume, or density, of the medium.

Waves moving along a string are reflected from a fixed end 180° out of phase and reflected from a free end in phase.

The general equation $y = A \sin(kx - \omega t)$ also describes *longitudinal* sinusoidal waves if we interpret the displacement y to mean the distance (*parallel* to the direction of propagation) that the particles move away from their equilibrium locations.

Waves may combine (*superpose*) to produce interference phenomena. A wave moving in one direction may interfere with its reflection to produce standing waves. The points where the amplitudes are a minimum in a standing-wave pattern are called *nodes*; the distance between adjacent nodes is $\lambda/2$. *Antinodes* are the locations of maximum amplitude, located midway between the nodes. For a stretched string, the fixed ends are nodal points. For an organ pipe, nodes occur at closed ends of the pipe, and antinodes are (essentially) at the open ends.

Standing-wave patterns in two and three dimensions involve *nodal lines* or *nodal surfaces*. Associated with each vibrating system is a family of "allowed" frequencies corresponding to all possible standing-wave patterns that may exist. Mechanical systems may simultaneously vibrate in more than one allowed frequency.

Waves of two different frequencies moving in the same direction may combine together. The amplitude of the combined wave varies with the *difference frequency* $(f_1 - f_2)$. In the case of sound waves, the difference frequency may be detected by the human ear in addition to the two original frequencies f_1 and f_2.

Complex period waves involving several different frequencies may be analyzed as a sum of sine and cosine waves by a method known as Fourier analysis (Appendix F).

If the source of sound or the observer moves relative to the propagating medium, the frequency f' of the observed sound is changed:

DOPPLER SHIFT (for sound)

$$f' = f\left(\frac{v \pm v_{ob}}{v \mp v_s}\right)$$

where f is the frequency of the sound at the source, v is the speed of propagation through the medium, v_s is the speed of the source relative to the medium, and v_{ob} is the speed of the observer relative to the medium. The signs are chosen to make the observed frequency higher for a relative *approach* of source and observer and lower for relative *separation.*

If a source of vibrations moves through air faster than the speed of sound, a conical *shock wave* develops. The half-angle θ of the cone is given by

$$\sin\theta = \frac{v}{v_s}$$

Analogous effects occur with light when a charged particle moves through a transparent medium faster than the speed of light in the medium (Cerenkov radiation).

Questions

1. Musicians often characterize a sound by its pitch, loudness, and timbre (or tone quality). Find the physical characteristics of a sound wave that correspond to these attributes.
2. Both transverse and longitudinal waves in solid objects involve displacements of the medium that are *linear* harmonic motions. Could you generate a wave motion by *angular* harmonic displacements? What would the wave be like?
3. The next time you make a cup of instant coffee in a ceramic or glass cup (it does not work in a plastic foam cup), try this demonstration. Fill the cup containing the instant

coffee with water that has just been raised to the boiling point. Immediately tap the rim of the cup repeatedly with your spoon. The pitch of the tapping will change dramatically. Explain.

4. Suppose the speed of sound in air varied with the frequency, with higher frequencies traveling faster. What would a musical selection sound like from your stereo as you gradually walked away from the loudspeaker? The speed of light in a transparent medium such as water *is* frequency dependent. What would be a visual analogy to the audio effect you described?
5. Suppose a source of sound of a definite frequency is at a fixed distance from you. If there is a wind blowing steadily from the source toward you, will the observed pitch change? What if the wind is blowing away from you toward the source? What will happen in each of the cases if, while you listen to the sound, the wind starts to blow, changing its speed from zero to some final, steady value?
6. If you have lived in an apartment house, you probably noticed that when your neighbors played their stereo, the bass notes sounded to you much more pronounced than the higher frequencies. What does this imply about the absorption of various frequencies of the walls between you and your neighbors?
7. The next time you are in an enclosed shower stall, hum sounds of various frequencies and note the pronounced resonance effects of the enclosure. Explain why it is more pronounced with massive marble walls than with lightweight wooden partitions.
8. Sing a note of fixed pitch while you successively mouth the vowels: a, e, i, o, u. Describe how the frequency components of the sound change for the different vowels.
9. Hold a vibrating tuning fork near your ear with the two prongs vertical. Why does the loudness of the sound vary noticeably as you rotate the fork about a vertical axis?

Problems

16A–1 Compressional waves in steel travel at about 5000 m/s. Find the wavelength of an ultrasonic compression wave of frequency 60 000 Hz. *Answer:* 8.33 cm

16A–2 The human eardrum responds to sound waves in a frequency range of about 20 to 20 000 Hz. Find the corresponding wavelengths (in centimeters) in air.

16A–3 Verify that the right-hand side of Equation (16–23) has the dimensions of *speed*.

16A–4 Find the period and wavelength in air of the musician's standard "A" note, which vibrates at 440 Hz.

16A–5 An organ pipe closed at one end has a fundamental frequency of 110 Hz.

(a) If the speed of sound in air is 340 m/s, what is the length of the pipe?
(b) What would be the length of a pipe open at both ends?
(c) What is the next higher frequency of standing waves for the pipe in part (a)?
(d) In part (b)?

Answers: (a) 0.773 m (b) 1.55 m (c) 330 Hz (d) 220 Hz

16A–6 Find the three lowest natural resonant frequencies of an organ pipe 1 m long when the speed of sound is 340 m/s if (a) both ends of the pipe are open and (b) one end is closed and the other end open.

16A–7 The overall length of a piccolo is 32 cm. The resonating air column vibrates as a pipe open at both ends.

(a) Find the frequency of the lowest note a piccolo can play, assuming the speed of sound in air is 330 m/s.
(b) Opening holes in the side effectively shortens the length of the resonant column. If the highest note a piccolo can sound is 4000 cycles/s, find the distance between adjacent nodes for this mode of vibration.

Answers: (a) 515 Hz (b) 4.13 cm

16A–8 A 40-ft rope weighing 8 lb is stretched between two supports. A sharp blow at one end of the rope causes a pulse to travel to the other end and return in a total transit time of 1 s.

(a) Determine the tension in the rope.
(b) How many pulses per second must be initiated in order to cause two pulses to cross paths at the midpoint at the moment a pulse is being reflected from each end?

16A–9 A string 2 m long has a mass of 1.8 g. One end is attached to a vibrator oscillating at 60 Hz. What tension in the string will result in transverse standing waves with five segments along the string's length? *Answer:* 2.07 N

16A–10 A string of linear density 10^{-3} kg/m and length 3 m is stretched between two points. One end is vibrated transversely at 200 Hz. What tension in the string will establish a standing-wave pattern with 3 loops along the string's length?

16A–11 The speed of longitudinal waves in a thin brass rod is 3480 m/s. Find the two lowest standing-wave frequencies for longitudinal vibrations in a brass rod 1 m long clamped at one end. Include sketches of the standing-wave patterns.
Answer: 870 Hz, 2610 Hz

16A–12 The following equation represents a standing wave on a stretched string: $y = A \sin(2\pi x/\lambda) \sin(2\pi t/T)$. One end of the string is at $x = 0$, and the other end is at $x = \ell$.
(a) Find the relation between λ and ℓ for standing waves.
(b) Rewrite the equation (in SI units) for the case where the amplitude is 2 cm, the distance between nodes is 20 cm, and the frequency is 80 cycles/s.

16A–13 A warning gong on an emergency vehicle has a fundamental frequency of 500 vibrations/s. On a day when the speed of sound is 340 m/s, a man on the street hears a pitch of 531 vibrations/s for the gong as the vehicle approaches. How fast is the vehicle traveling? *Answer:* 19.9 m/s

16A–14 Passengers on a train traveling at 20 m/s approach a railroad crossing that has a stationary warning gong whose basic pitch is 400 cycles/s. The speed of sound in air is 330 m/s on that day.
(a) What is the frequency heard by the passengers as they approach the crossing?
(b) What frequency do passengers hear after they pass the crossing and are traveling away from the gong?
(c) If the train carried a gong of frequency 400 cycles/s, what frequency would an observer on the ground hear as the train approached at 20 m/s?

16A–15 A jet fighter plane travels in horizontal flight at Mach 1.2 (that is, 1.2 times the speed of sound in air). At the instant an observer on the ground hears the shock wave, what is the angle her line-of-sight makes with the horizontal as she looks at the plane?
Answer: 56.4°

16A–16 A wire of density ρ and diameter d is stretched with a force F between stationary supports a distance ℓ apart. Devise (a) a formula for the speed v of transverse waves and (b) a formula for the frequency f_n of the nth harmonic for standing-wave vibrations.

16A–17 In certain ranges of a piano keyboard, more than one string is tuned to the same note to provide extra loudness. For example, the note at 110 Hz has two strings at this pitch. If one string slips from its normal tension of 600 N to 540 N, what beat frequency will be heard when the two strings are struck simultaneously? *Answer:* 5.64 Hz

16B–1 A construction worker hammers a nail at a steady pace so that there is 0.8 s between successive hits. A child observing the worker from some distance away notes that the sound of each hammer strike arrives equally spaced in time between the observed hits. Find the distance between the child and the construction worker. *Answer:* 860 m

16B–2 A small boat loaded with explosives blows up. An underwater swimmer hears the sound of the explosion through the water, then raises his head out of the water and hears another sound of the same explosion that traveled through the air at a slower speed.
(a) From the bulk modulus for water, calculate the speed of sound in the water.
(b) If the time delay between the arrival of the two sounds is 5 s, determine the distance from the swimmer that the explosion occurred.

16B–3 An explosion occurs on a railroad track 1.7 km from an observer sitting on the track.
(a) Find the velocity of compressional sound waves traveling through the steel rails.

(b) Find the time delay between the two sounds of the explosion heard by the observer: the sound as it traveled through the rails and the sound as it traveled through the air. *Answer:* 4.61 s

16B–4 A person stands 100 ft in front of a flight of concrete steps, each step of which is 1 ft wide from the edge to the next step. She claps her hands once sharply and listens to the echo from the flight of steps.

(a) What will be the fundamental frequency of the series of echo pulses if the speed of sound in air is 1100 ft/s?

(b) Suppose, instead, she stands a distance ℓ (at right angles) from a picket fence whose vertical pickets are equally spaced a distance d apart ($\ell \gg d$). She claps her hands sharply once. Describe qualitatively the sound she hears by reflection from the fence.

16B–5 The smallest intensity difference detectable by the human ear is about 0.6 dB. Show that this represents an increase of about 15% in the energy emitted per second by the source.

16B–6 A sawtooth pulse has the shape shown in Figure 16–23 as it travels at 100 m/s along a stretched wire. (The transverse amplitude is greatly enlarged in the sketch.) At $t = 0$, the leading edge of the pulse is 6 m from a fixed end of the wire. Make a sketch to show the shape of the wire 0.10 s later, with significant dimensions of the pulse shape indicated in your sketch.

Figure 16–23
Problem 16B–6

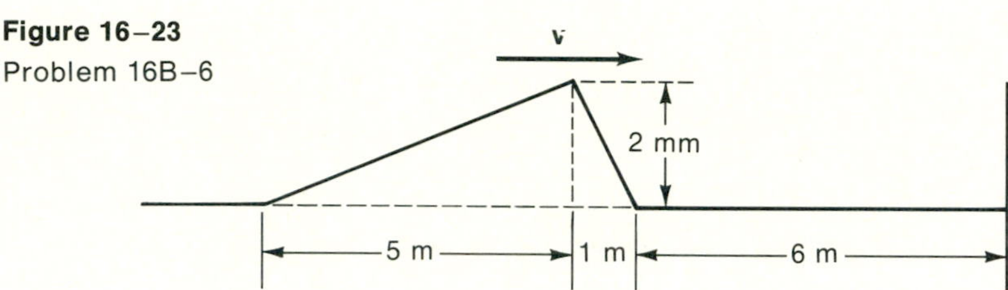

16B–7 A flexible chain has a mass per unit length of μ. It is formed into a loop which is then set into rotation with angular speed ω, so that it becomes quite rigid, like a hoop (see Figure 16–24).

(a) Neglecting gravity, show that the tension in the chain is $\mu R^2\omega^2$. (Hint: Draw a free-body diagram for a small segment of the chain $\Delta\ell$ long that subtends an angle $\Delta\theta$. Apply $\mathbf{F} = m\mathbf{a}$ for a mass $\mu\,\Delta\ell$ moving in a circle.)

(b) Show that if a small kink is formed in the chain, it will remain stationary in space as the chain rotates.

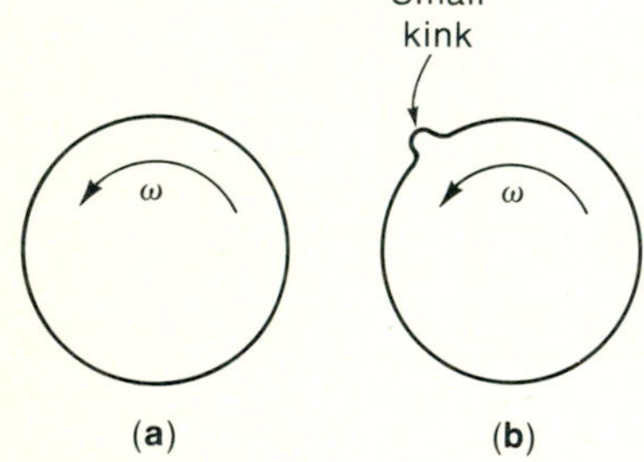

Figure 16–24
Problem 16B–7

16B–8 A cello A-string vibrates in its fundamental mode with a frequency of 220 vibrations/s. The vibrating segment is 70 cm long and has a mass of 1.2 g.

(a) Find the tension in the string.

(b) Determine the frequency of the harmonic that causes the string to vibrate in 3 segments.

16B–9 On a marimba, the wooden bar that sounds a tone when struck vibrates as a free bar in transverse standing waves with antinodes at each end. The lowest frequency note is 87 Hz, produced by a bar 40 cm long.

(a) Find the speed of transverse waves on the bar.

(b) The loudness and duration of the emitted sound are enhanced by a resonant pipe suspended vertically below the center of the bar. If the pipe is open at the top end only and the speed of sound in air is 340 m/s, what is the length of the pipe to resonate with the bar in part (a)?

Answers: (a) 69.6 m/s (b) 0.977 m

16B–10 A stretched wire vibrates in its fundamental mode at a frequency of 400 vibrations/s. What would be the fundamental frequency if the wire were half as long, with twice the diameter, and with 4 times the tension?

16B–11 A rope is stretched so that transverse waves have a speed of 16 m/s. A standing-wave pattern of transverse vibrations is established with 2 m between adjacent nodes. If the maximum amplitude of these vibrations is 2 cm, what maximum transverse acceleration does the rope experience at an antinode? *Answer:* 12.6 m/s^2

16B–12 A vertical glass tube, 1.2 m long, is filled with water, which can be adjusted to any desired height. A scale for measuring the distance from the top (open) end of the tube to the water level is provided. A tuning fork vibrating at 528 Hz is held at the upper end of the tube. List all the scale readings for positions of the water level that cause resonance in air at 20°C.

16B–13 On a day when the speed of sound in air is 330 m/s (and there is no wind), a source of sound emits a sound of frequency 1000 Hz. What frequency will a listener hear under the following conditions?

(a) The listener moves toward the source with a speed of 30 m/s.
(b) The listener is at rest, and the source moves toward the listener at 30 m/s.
(c) Both source and listener are at rest, but now there is a 30 m/s wind blowing from the source toward the listener.

Answers: (a) 1091 Hz (b) 1100 Hz (c) 1000 Hz

16B–14 A source of sound and an observer are stationed a fixed distance from each other. When the air is still, the observer determines the frequency, velocity, and wavelength of the received sound. How will these parameters change when (a) a steady wind blows from the source to the observer and (b) the wind blows from the observer to the source?

16B–15 The *Mach number* used by aerodynamicists is the ratio of the speed of an aircraft to the speed of sound in the air through which the plane is flying. If the supersonic jet Concorde flies at Mach 2.1, where the local speed of sound is 320 m/s, find the half-angle of the resultant conical shock wave. *Answer:* 28.4°

16B–16 The pitch of a pipe organ and a piano are identical for the note at 440 Hz when the speed of sound in air is 340 m/s. The temperature rises so that the speed increases to 346 m/s. What beat frequency will be heard when this note is simultaneously sounded by both instruments? (Assume the piano pitch does not change.)

16C–1 Show that the equation $y = A(x - ct)^n$ is a possible solution to the wave equation (Equation 16–8), where A, c, and n are constants.

16C–2 Show that the general equation $y = f(x - vt)$ is a solution to the wave equation.

16C–3 Show that $y = Ae^{i(kx - \omega t)}$ is a possible solution to the wave equation, where $i = \sqrt{-1}$, and A, k, and ω are constants.

16C–4 A rule of thumb for estimating the distance from an observed lightning flash is to count the seconds that elapse between seeing the flash and hearing the thunder. This number divided by five is approximately the distance in miles.

(a) Find the percentage error in this calculation for air at 20°C.
(b) Devise a similar (simple) rule for the distance in kilometers, and give the error in your rule.

16C–5 A stone is dropped from rest into a well. The sound of the splash is heard exactly 2 s later. Find the depth of the well. *Answer:* 18.56 m

16C–6 The equation for transverse waves on a rope is $y = 0.02 \sin (5x - 80t)$ where x and y are in meters and t is in seconds. Find (a) the amplitude, (b) the frequency, (c) the wavelength, and (d) the speed of waves on the rope. (e) Determine the displacement 40 cm from the origin at the time $t = 0.10$ s. (f) Find the transverse velocity of the point of the rope in part (e).

16C–7 At a depth of 600 km in the earth, the velocities of the P and S earthquake waves are, respectively, 10.3 km/s and 5.66 km/s. Because the medium extends for long distances in three dimensions, Equation (16–20) is modified for P-waves by replacing B with $(B + 4/3\ S)$. Assuming an average density of 3.9 g/cm³ for earth rocks at that location, calculate the bulk modulus B and the shear modulus S. *Answer:* $B = 2.47 \times 10^{11}$ N/m²
$S = 1.25 \times 10^{11}$ N/m²

16C–8 As a sound wave travels through the air, it produces pressure variations (above and below atmospheric pressure) given by

$$p = 1.27 \sin \pi (x - 340t)$$

where p is in pascals, x is in meters, and t is in seconds. Find (a) the amplitude of the pressure variations, (b) the frequency, (c) the wavelength in air, and (d) the speed of the sound wave.

16C–9 Continuous transverse sine waves with a frequency of 100 vibrations/s travel down a string stretched with tension 0.40 N, whose mass per unit length is 2×10^{-3} kg/m.

Design a matching stub that will permit these waves to pass unreflected onto a string under the same tension but whose mass per unit length is 8×10^{-3} kg/m.

Answer: $\mu = 4.00 \times 10^{-3}$ kg/m, 2.50 cm long

16C–10 The energy per unit area in a plane wave front remains constant as the wave propagates through space (if there are no dissipative forces present).

(a) How does the energy per unit area of a spherical wave front emitted from a point source vary with distance from the source?
(b) For a cylindrical wave front emitted from a line source?
(c) How does the *amplitude* of the wave motion vary with distance for part (a)?
(d) For part (b)?

16C–11 The amplitude of the vibrations of a certain sound wave is increased by 50%. What is the increase in the decibel level of the sound? *Answer:* 3.52 dB

16C–12

(a) If the intensity of a sound changes by a factor of 5, by how much does the decibel level change?
(b) What is the ratio of the amplitudes that corresponds to this decibel change?

16C–13 A point on a violin string is vibrating transversely at 500 cycles/s with an amplitude of 1 mm. Find the maximum speed and acceleration of this point.

Answer: 3.14 m/s, 9.87×10^3 m/s^2

16C–14 A steel piano wire whose mass per unit length is 0.1 g/cm is under a tension of 500 N.

(a) If its fundamental frequency of vibration is 200 Hz, find the length of the wire.
(b) If the amplitude of vibration at the center of the wire is 2 mm, find the maximum transverse speed of the center.

16C–15 One end of a 2-g cord, 2 m long, is fastened to the end of a cord 1 m long whose mass is $\frac{1}{4}$ g. The combination is stretched with a tension of 3.6 N and tied down at the free ends (making a total length of 3 m). Find the lowest frequency of a standing-wave pattern with a node at the junction of the two cords. *Answer:* 60.0 Hz

16C–16 Suppose a stretched rubber band behaved as a Hooke's Law spring with a force constant k. (It doesn't!) Consider a segment of rubber band stretched to twice its unstressed length. If the band is now "twanged," it will emit a sound of frequency f_0.

(a) If the band is now stretched to four times its unstressed length, what would be the frequency when twanged? (Don't forget to account for the changing mass-to-length ratio.)
(b) Now actually try the experiment. The frequency response will be different from that predicted in part (a). Describe (qualitatively) in what way the rubber band differs from a strict Hooke's Law behavior in order to explain your experimental results.

16C–17 A violinist attempts to sound a note of 523 Hz but her finger is not quite at the correct distance from the bridge and the actual note is at 530 Hz. If her finger is 25 cm from the bridge, find the distance and the direction she should move her finger to obtain the desired pitch. *Answer:* 0.335 cm

16C–18 A train is traveling at 10 m/s as it approaches a tunnel in the smooth, vertical face of a rock cliff. The speed of sound is 340 m/s. The engineer sounds a whistle with a fundamental frequency of 300 Hz. A rider in the caboose simultaneously hears the train whistle and also the echo from the wall.

(a) How many beats/s does the rider hear?
(b) What would be the beat frequency heard by an observer on the ground near the passing caboose? (Hint: Treat reflections as waves emitted from a (moving) mirror image of the source.)

16C–19 Various vowel sounds, such as "ee," "ah," and "oo," can all be spoken using the same fundamental frequency of the vocal cords. The differing harmonic content is achieved by rearrangements of the resonant cavities in the throat and mouth. By filling the lungs with helium, for which the speed of sound is about three times that of air, the characteristics of the speaking voice can be altered considerably. Discuss what changes might (or might not) occur when speaking with helium-filled lungs.

Perspective

We have now completed our introduction to mechanics and it is worthwhile again to step back a bit from the subject to point out the prominent features. Basically, we have learned just three approaches for solving problems in mechanics:

(1) *Newton's laws*
(2) *Energy and momentum* (the conservation relations and the work-energy relation)
(3) *The connection between a conservative force and its associated potential energy*

We showed how these ideas could explain a variety of phenomena, and we discussed three special topics: inertial forces in accelerated frames, Newton's great insight into gravitation, and Einstein's brilliant theory of special relativity. Of course, we have had to carry along a rather large bagful of facts, definitions, and other information. But those few giant ideas are the heart of mechanics, providing a solid foundation and framework upon which to build as we investigate other phenomena.

The next five chapters show that heat effects are basically Newtonian mechanical principles operating on a microscopic scale. (Heat was formerly thought to be a mysterious, weightless substance that flowed from one object to another.) Particularly impressive are the elegant statements concerning energy and its availability: *the laws of thermodynamics*. The second of these laws is probably the most pervasive and profound law in physics.

chapter 17 Heat and Temperature

To engage in experiments on heat was always one of my most agreeable employments.

BENJAMIN THOMPSON
(COUNT RUMFORD)

17.1 Introduction

When a rubber ball is dropped to the floor, it begins to bounce up and down with a continual interchange of gravitational potential energy and kinetic energy. As the bouncing continues, the ball gradually loses both kinetic and potential energies with each successive bounce. Where does the energy go? Is it irretrievably lost? When we observe mechanical systems left to themselves, we conclude that they tend to run down. Is the universe also doomed to run down with a loss of mechanical energy? These questions are within the scope of *thermodynamics*, a study of thermal energy and its conversion to other forms of energy, such as mechanical, chemical, or electrical energies.

17.2 Temperature

Until now, all the concepts we have studied were ultimately based on the fundamental physical quantities of *mass*, *length*, and *time*. To measure them, we chose the arbitrary units of the *kilogram*, *meter*, and *second*, and carefully defined the operations to be performed for their measurement. All other concepts in mechanics were derived from these three basic entities. In describing *thermal* phenomena, we will find that one additional fundamental concept must be introduced; it cannot be derived directly from the other three. This new concept is *temperature*. In common with other basic ideas in physics, the definition of temperature was originally based on our everyday senses; then the concept evolved through a gradual process of refinement and change. Its full and precise meaning is intimately connected with the *operations* we carry out to measure it.

As a reliable way of measuring hotness and coldness, various instruments, called *thermometers*, have been devised; these furnish accurate and reproducible measurements. All use a change in some physical property of a certain substance to indicate the degree of hotness. Successful thermometers have involved the following types of changes in specific materials: (a) the expansion of solids and liquids, (b) pressure and volume changes of a gas, (c) changes in electrical resistance, and (d) color changes in heated filaments and in certain chemical reactions. Each thermometer is usable only over a limited range of temperatures that avoids phase changes of the substance. The various scales are calibrated against each other in overlapping regions.

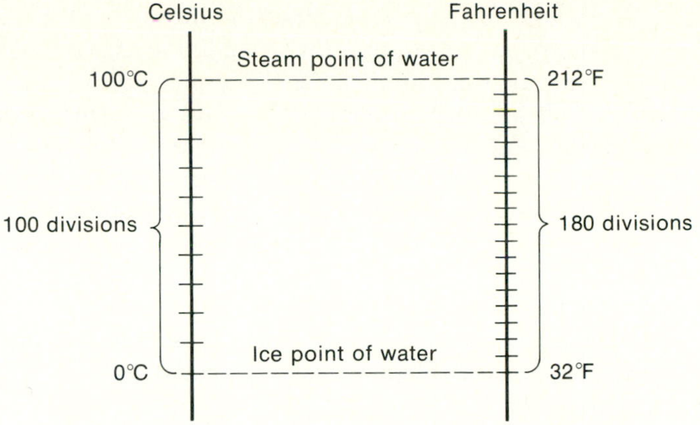

Figure 17–1
The Celsius and Fahrenheit temperature scales, which are based on the expansion of a column of liquid mercury

17.3 The Celsius and Fahrenheit Temperature Scales

No doubt you are familiar with the liquid-column thermometer containing mercury or colored alcohol. This thermometer is useful over a range of temperatures for which the liquid does not boil or freeze, nor the glass soften. To establish a numerical scale of temperature T, a linear scale is laid out adjacent to the liquid column and two clearly defined temperature points are chosen. These two points are the temperatures at which (a) water exists in equilibrium as a liquid and solid (the *ice point*) and (b) water exists in equilibrium as a liquid and gas (the *steam point*). Both points are determined at one standard atmosphere of pressure. A unit of temperature change ΔT is then chosen by dividing the scale between these points into a specified number of equal length intervals.

Two scales in common use are the Celsius and Fahrenheit scales[1] invented in the early 1700s by physicists of those names. They designated the fixed points and the number of intervals as shown in Figure 17–1. Points on these scales are measured in *degrees Celsius* (°C) or *degrees Fahrenheit* (°F). To convert from one system to the other, a variety of formulas may be used. One that has a certain symmetry (and therefore may be easier to remember) is

TEMPERATURE CONVERSION FORMULA

$$5(\mathrm{F} + 40°) = 9(\mathrm{C} + 40°) \qquad \textbf{(17–1)}$$

where the symbols C and F represent the temperature readings on the corresponding scales. Note that because the zeros differ in the two scales, we must *not* write equations linking corresponding readings such as 0°C = 32°F or 10°C = 50°F. (To see why, subtract the first equation from the second one.) However, *intervals of temperature difference* measured in units of *Celsius degrees* (C°) or *Fahrenheit degrees* (F°) may correctly be expressed as equations. For example:

$$5\ \mathrm{C}° = 9\ \mathrm{F}° \quad \text{(temperature } \textit{intervals}\text{)} \qquad \textbf{(17–2)}$$

[1] Formerly, the Celsius scale was called the *centigrade* scale. In 1948, the name was formally changed in recognition of the scale's originator, Anders Celsius, a Swedish astronomer (1701–1744). The German physicist Gabriel Fahrenheit (1686–1736) invented the mercury thermometer.

To avoid later confusion, be sure to understand this distinction between specific *temperatures* on the scales (measured, for example, in *degrees Celsius*: °C) and *temperature intervals* (measured in *Celsius degrees*: C°).

If different liquids are used as the working substances in various liquid-column thermometers, and a linear scale is laid out as just described, no two thermometers will read exactly alike over the range of the scale. This is because no two substances expand exactly alike as they become hotter. Nor will such temperature scales agree with those based on other physical changes, such as variations in electrical resistance or changes in color. Which thermometer measures the "true" temperature? *It is just a matter of definition. We may choose any particular thermometer scale as our standard and calibrate all others against it.* In this sense, we define **temperature** as the reading on the standard temperature scale.

But which temperature scale shall we choose? There is an "absolute" scale, the *Kelvin thermodynamic temperature scale*, which by international agreement is the standard scale for all basic scientific work. It is truly a fundamental scale in that it can be defined independently of the properties of any particular substance. But to describe it, we must first learn a few principles of thermodynamics, so a discussion of the Kelvin scale will be postponed until Chapter 21. For the time being, we will use the familiar mercury-column thermometer marked in Celsius degrees, which will be satisfactory for discussing the phenomena we take up in this chapter.

EXAMPLE 17–1

What Celsius temperature corresponds to 98.6°F, the "normal" temperature of the human body?

SOLUTION

Substituting the numerical values into Equation (17–1), we have

$$5(\mathrm{F} + 40^\circ) = 9(\mathrm{C} + 40^\circ)$$
$$5(98.6^\circ\mathrm{F} + 40^\circ) = 9(\mathrm{C} + 40^\circ)$$

Solving for C, the temperature in °C, yields

$$\mathrm{C} = \tfrac{5}{9}(98.6 + 40^\circ) - 40^\circ = \boxed{37.0^\circ\mathrm{C}}$$

17.4 Thermal Expansion

Nearly all materials expand as the temperature increases.[2] We can understand this behavior qualitatively in terms of increased molecular motions as substances become hotter. In a simplified view, forces between adjacent atoms in a

[2] An important exception is water in the temperature range 0°C to 4°C. Over this small range, water *contracts* with increasing temperature. As a consequence, ice formation takes place first on the *surface* of a body of water. To illustrate, consider a pond of water at, say, 10°C, with air temperatures below the freezing point. The surface layer cools and becomes more dense, sinking to the bottom and forcing lighter water to the top. (The process of mixing because of different densities is called *convection*.) The mixing proceeds until essentially all of the water has been cooled to 4°C. From this stage on, further cooling of the surface makes the top layer *less dense*, so that it floats on the surface, continuing to cool toward the freezing point without further mixing. And when ice forms, it remains on the top, since its density is less than that of the water below. The lower depths of the pond freeze only as heat is further extracted by *conduction* through the intervening layers of ice. The consequences of this unusual behavior of water for marine life, and for the evolution of life itself, are quite profound.

solid may be represented in terms of the associated potential energy function sketched in Figure 17–3. At very low temperatures, the average separation distance between atoms is close to r_0. As the temperature increases, the vibrating atoms gain energy and oscillate between wider limits of motion. *Because the potential energy curve is asymmetric*, the *restoring force* (proportional to the negative slope of the curve) is less at separations greater than r_0 than at separations less than r_0, and the *accelerations* of the atoms are correspondingly less. The atoms thus spend relatively more time at larger separations than they do at distances less than r_0. Therefore, the *time average* of the motion leads to greater average separation distance between atoms, and the material expands upon heating. If the potential energy function were perfectly symmetrical, no thermal expansion effects would occur. (In a few instances, *transverse*, or shearing, modes of vibration predominate over the longitudinal vibrations discussed here, causing the object to contract with increasing temperature. Such substances are fairly rare, however.)

Let us investigate the amount of thermal expansion ΔL that occurs for an object whose original length is L_0. Experimentally, over a limited temperature range, the *fractional* change in length $\Delta L/L_0$ is closely proportional to the temperature change ΔT. For most engineering applications, the relationship may be assumed to be linear without serious error. Thus, we may write:

$$\frac{\Delta L}{L_0} = \alpha \Delta T \tag{17–3}$$

LINEAR THERMAL EXPANSION

$$\Delta L = L_0 \alpha \Delta T \tag{17–4}$$

where α, the constant of proportionality, is called the **coefficient of linear thermal expansion.** Since $\Delta L/L_0$ is dimensionless, the units of α are *per Celsius degree*, written (per C°) or $(\text{C}°)^{-1}$. Because the Celsius and Kelvin temperature intervals are identical, numerical values of α are the same for either scale. However, since a Fahrenheit degree is only $\frac{5}{9}$ as large as a Celsius degree, values of the coefficients for the Fahrenheit scale are only $\frac{5}{9}$ as large. Table 17–1 lists some representative coefficients.

Since the expanded length L equals $L_0 + \Delta L$, we may also write:

$$L = L_0(1 + \alpha \Delta T) \tag{17–5}$$

Figure 17–2
Inadequate allowance for thermal expansion caused these rails to buckle in an unusual heat wave in New Jersey, derailing a train car.

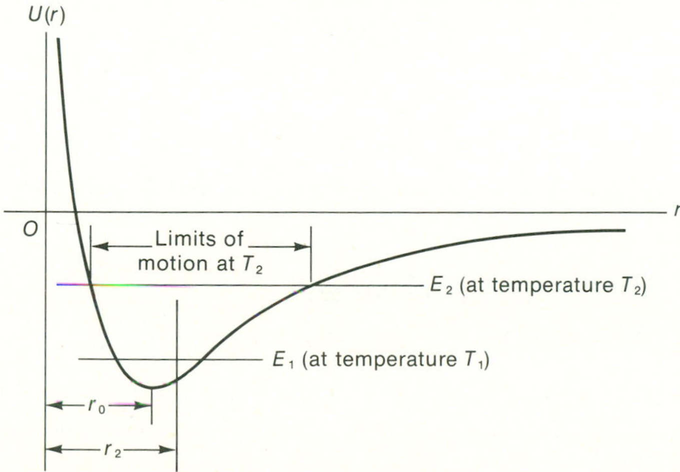

Figure 17–3
The potential energy $U(r)$ for two atoms in a crystalline solid as a function of the separation distance r between their nuclei. At a low temperature T_1, the average separation is close to r_0, the distance corresponding to the bottom of the curve. At a higher temperature T_2, the average separation increases to r_2 because of the asymmetry of the curve, causing the material to expand upon heating.

TABLE 17–1

Average Coefficients of Linear Thermal Expansion

Material	Average α $(C^\circ)^{-1}$
Aluminum	24×10^{-6}
Brass	19×10^{-6}
Glass (various types of soft glass)	$4\text{–}9 \times 10^{-6}$
Glass (Pyrex)	3×10^{-6}
Ice	51×10^{-6}
Invar (36% Ni, 64% Fe)	0.9×10^{-6}
Oak (parallel to fiber)	5×10^{-6}
Oak (across fiber)	34×10^{-6}
Quartz (fused)	0.4×10^{-6}
Steel	12×10^{-6}

EXAMPLE 17–2

The length of the standard meter is exactly 1 meter at 0°C. This standard, kept at the International Bureau of Weights and Measures near Paris, France, is made of a platinum-iridium alloy that has an average coefficient of linear thermal expansion of $8.9 \times 10^{-6}\ (C^\circ)^{-1}$. Assuming all the numerical values are exact (and α remains constant), find the length of the bar at 37°C.

SOLUTION

The change in length ΔL is given by

$$\Delta L = L_0 \alpha \Delta T$$

Substituting values, we have

$$\Delta L = (1.000\ldots\text{ m})[8.9 \times 10^{-6}\ (C^\circ)^{-1}](37\ C^\circ) = 3.293 \times 10^{-4}\text{ m}$$

The expanded length is thus:

$$L = L_0 + \Delta L = \boxed{1.000\ 329\ 3\text{ m}}$$

17.5 Volume and Area Thermal Expansion

Experiments on thermal expansion show that the fractional change in volume $\Delta V/V_0$ is (over a limited temperature range) proportional to the temperature change ΔT.

$$\frac{\Delta V}{V_0} = \beta \Delta T \tag{17-6}$$

VOLUME THERMAL EXPANSION

$$\Delta V = V_0 \beta \Delta T \tag{17-7}$$

where the constant of proportionality β is called the **coefficient of volume thermal expansion.** It is related to the coefficient of linear thermal expansion α in the following way. Consider a cubical volume[3] V_0 of material with edge length L_0. As the material is heated, each edge will expand according to $L = L_0 + \Delta L$. The new volume V is thus:

$$V = (L)^3 = (L_0 + \Delta L)^3 = L_0{}^3 + 3L_0{}^2(\Delta L) + 3L_0(\Delta L)^2 + (\Delta L)^3$$

$$V = L_0{}^3\left[1 + 3\left(\frac{\Delta L}{L_0}\right) + 3\left(\frac{\Delta L}{L_0}\right)^2 + \left(\frac{\Delta L}{L_0}\right)^3\right] \tag{17-8}$$

In the usual cases, $\Delta L/L_0 \ll 1$, so we may drop the higher-order terms with negligible error. Therefore, making use of Equation (17–3), we have

$$V = V_0(1 + 3\alpha \Delta T) \tag{17-9}$$

Since $V = V_0 + \Delta V$, we conclude that

$$\Delta V = V_0 3\alpha \Delta T \tag{17-10}$$

Comparison with Equation (17–7), $\Delta V = V_0 \beta \Delta T$, reveals that

$$\beta = 3\alpha \tag{17-11}$$

A hole in a solid object expands exactly as the missing material would expand. Thus the volume of the hollow glass bulb of a thermometer changes its volume exactly as an equal volume of solid glass would change its volume. Precision instruments must make corrections for such effects.

By a similar analysis, it can be shown that the thermal expansion of an area A is given by

AREA THERMAL EXPANSION

$$\Delta A = A_0 2\alpha \Delta T \tag{17-12}$$

and:

$$A = A_0(1 + 2\alpha \Delta T) \tag{17-13}$$

The area of a hole expands just as the missing material would expand. Thus area changes under thermal expansion just as in a photographic enlargement.

17.6 Heat

Why does an object get hotter or colder? Until the early part of the nineteenth century, the generally accepted answer was: "Because it gains or loses *caloric*, an invisible, weightless fluid that can flow from one object to another." The caloric theory explained a wide range of phenomena that involved a change in temperature. For example, when a hot object (supposedly containing much

[3] Any arbitrarily shaped volume of homogeneous material may be thought of as an assembly of elementary cubes. As the size of the cubes becomes infinitesimal, the volume of the object approaches the volume of the cubes. Thus the result we are deriving for a cube applies also to a volume of any shape whatsoever.

caloric) was brought into contact with a cold body, the excess caloric in the hot object flowed into the cold one, until they both reached the same intermediate temperature. Furthermore, according to the caloric theory hot bodies expanded because the extra caloric repelled itself and pushed the atoms a bit farther apart.

In 1798 the idea was refuted. An American,[4] who later became Count Rumford of Bavaria, was hired to bore cannons for the Bavarian Army. According to the accepted caloric theory, the drill and cannon became hot during the boring process because the caloric was released as the metal was shaved into chips. However, Rumford noted that the drill and cannon became *even hotter* using dull drills that cut no metal at all. The source of caloric apparently was inexhaustible, a fact that was hard to reconcile with the caloric theory. He suggested, rather, that the source of heat seemed to be the work done by the horses turning the drill. Later, several other investigations concluded that heat was really just another form of energy and that it was conserved.

□ □ □

In everyday conversation, the words heat and temperature are often used interchangeably. Other phrases, such as heat energy, heat content, heat of the day, or the heat in a body, add to the confusion. The following definitions, used in scientific discussions, may help sort out the various concepts.

TEMPERATURE **The degree of "hotness" as measured on some definite scale; "what a thermometer measures."**

HEAT **Energy transferred between a system and its surroundings due to temperature differences only.**

This definition of heat has some subtleties. As stated, heat is energy *transferred* from one system to another system (when they are in *thermal contact*) because of a temperature difference. Before the energy is transferred, and after the energy arrives in the system, it is not called heat (and *never* heat energy). More properly, it is called

THERMAL ENERGY (OR INTERNAL ENERGY) **Kinetic and potential energies due to the random motions of atoms and molecules.**

The *temperature* of a system is thus the parameter that describes whether or not the system will transfer thermal energy to another system when brought into thermal contact with it.

Energy is a rather peculiar and elusive concept in spite of its extreme importance in science. It is everywhere in various forms: work (mechanical energy in transit), gravitational potential energy, kinetic energy, chemical energy, heat (thermal energy in transit), and so on. Yet in no instance is energy ever measured directly. Instead, it is always deduced indirectly from other measurements. For example, the energy associated with mechanical work is found by measuring a force and a distance, and thermal energy is found by

---

[4] Benjamin Thompson (Count Rumford) was sort of a political knave—an opportunist who took kickbacks on military contracts and on several occasions spied for various foreign governments. Though a major in the Colonial Army, during the War of Independence his sympathies were with the Tories, so he abandoned his wife and child and fled to Europe. Despite his deviousness in political and business dealings, he was an astute engineer-scientist whose scientific fame is justly deserved.

measuring a mass and a temperature. But the concept of energy and, particularly, its conservation is perhaps the most important and firmly established concept in all of science. In the words of one author: "Energy explains everything."

Although the literature of science and engineering is filled with references to the "calorie" as a unit of heat, it is, after all, just another form of energy. A *calorie* (cal) was originally defined as the amount of heat required to raise the temperature of one gram of water (at one atmosphere pressure) one Celsius degree, from 14.5°C to 15.5°C.[5] In 1948, the Ninth General Conference on Weights and Measures adopted the mechanical energy unit, the *joule* (J), as the official unit of heat. The relation between heat units and mechanical units may be obtained from experiments in which a definite amount of mechanical energy is converted (through frictional effects) into thermal energy. The first accurate measurements were made by James Joule, a British brewer, who tried a variety of schemes, including stirring water vigorously with slotted paddle wheels driven by heavy weights that descended from pulleys. Modern experiments obtain more precise values by electrically heating a resistor immersed in water, giving the value:

MECHANICAL EQUIVALENT OF HEAT

$$1 \text{ cal} = 4.184 \text{ J} \quad \text{(defined to be exact)} \qquad \textbf{(17–14)}$$

This equation now defines the calorie unit (SI).

17.7 Absorption of Heat

The absorption of heat by a substance always produces some change. In addition to thermal expansion, its temperature may rise or it may undergo a **phase change** in which its physical characteristics are altered without a change in temperature. As an example, suppose some ice initially below the freezing point is heated. The temperature of the ice gradually rises until it begins to melt.[6] From this point on, the temperature does not increase further until all the ice has changed into water. Further heating raises the temperature of the water to the boiling point, at which another phase change to the vapor state occurs, while the water-vapor mixture remains at 100°C. Additional heating then increases the temperature of the water vapor above 100°C. So there are two distinct types of changes: a temperature increase and a change in the physical characteristics of the material.

The number of calories required to raise the temperature of one gram of a substance one degree Celsius is called the **specific heat** of the substance. Since specific heat varies slightly with temperature, precise determinations should specify the temperature range involved. Specific heat is designated by the symbol c, defined as follows:

SPECIFIC HEAT c

$$c = \left(\frac{1}{m}\right)\left(\frac{\Delta Q}{\Delta T}\right) \qquad \textbf{(17–15)}$$

[5] Nutritionists and weight watchers also make use of a unit called the (food) calorie, one thousand times larger than the calorie defined above. To say that a candy bar "contains 150 calories" means that after removing the water from it by dehydration, the candy bar will produce (150 × 1000) calories (as defined in physics) when burned in an atmosphere of pure oxygen.

[6] At any given temperature, some ice may also be *subliming*, that is, changing directly into the vapor state without first melting.

where m is the mass of the substance, ΔQ is the quantity of heat added, and ΔT is the change in temperature. This equation may also be expressed in integral form:

$$Q = m \int_{T}^{T+\Delta T} c\,dT \tag{17–16}$$

where the specific heat is included within the integral because it may depend on temperature. Table 17–2 shows some representative specific heat values.

The amount of heat transferred during changes of phase (which include condensation and freezing as well as melting and boiling) is called **latent heat.** For water, the amount of heat that must be extracted to change water at 0°C to ice at 0°C is 80 cal/g (3.34×10^5 J/kg). The latent heat of vaporization[7] for converting water to steam, both at 100°C, is 540 cal/g (2.26×10^6 J/kg). The heat of vaporization for water is the largest for any liquid listed in the Handbook of Chemistry and Physics, because of the strong attraction between water molecules and the consequently large amount of energy that must be expended to separate them into the vapor state. In general terms, the quantity of heat Q required to bring about a phase change is given by:

HEAT TO PRODUCE A PHASE CHANGE

$$Q = mL \tag{17–17}$$

where m is the mass of the substance and L is the latent heat.

[7] The latent heat of evaporation depends on the temperature. For example, for water at 0°C, the latent heat of evaporation is 595 cal/g (2.49×10^6 J/kg).

TABLE 17–2

Specific Heats of Some Common Substances

Substance	Specific heat[a] [cal/(g·C°)]	Temperature (Celsius)
Water	1 (exactly, by definition)	14.5–15.5
	1.00	0–100
Ice	0.487	0
Water vapor	0.421 (constant pressure)	100
Aluminum	0.218	0–100
Iron	0.110	0–100
Copper	0.093	0–100
Gold	0.031	0–100
Glass, common (crown)	0.20	19–100
Wood	0.33	20
Air	0.238 (constant pressure)	0–200

[a] A unit of heat commonly used in engineering is the *British thermal unit* (Btu), defined as the amount of heat required to raise the temperature of one pound of water from 63°F to 64°F. This definition makes specific heats in cal/(g·C°) numerically the same as those in units of Btu/(lb·F°).

EXAMPLE 17–3

Suppose 300 g of water initially at 30°C is poured over a 50-g ice cube that is initially at -5°C. Find the final temperature of the water if all the ice melts or the temperature if the ice does not entirely melt.

SOLUTION

The heat lost by the water is transferred to the ice cube. The ice cube may or may not completely melt. As a first trial, we will assume that it does melt. The heat lost by the water $(\Delta Q)_w$ equals the heat gained by the ice $(\Delta Q)_i$:

$$(\Delta Q)_w = (\Delta Q)_i$$
$$m_w c_w (\Delta T)_w = m_i c_i (\Delta T)'_i + m_i L_F + m_i c_w (\Delta T)_i$$

where $(\Delta T)'_i$ = the temperature rise of the ice
L_F = the latent heat of freezing
$(\Delta T)_i$ = the temperature increase of the melted ice

The numerical substitutions yield:

$$(300\text{ g})\left(1\ \frac{\text{cal}}{\text{g}}\right)(30°\text{C} - T) = (50\text{ g})\left(0.487\ \frac{\text{cal}}{\text{g}}\right)(5°\text{C}) + (50\text{ g})\left(80\ \frac{\text{cal}}{\text{g}}\right)$$
$$+ (50\text{ g})\left(1\ \frac{\text{cal}}{\text{g}}\right)(T - 0°\text{C})$$

$$T = \boxed{13.9°\text{C}}$$

If our assumption about all of the ice melting were incorrect, no positive value of T would have satisfied the heat-balance equation.

17.8 Heat Conduction

If one object is placed in physical contact with another at a different temperature so that heat flows across the area of contact, the process is called the transfer of heat by **conduction.** The ability of a substance to conduct heat is specified in a characteristic constant k, the **thermal conductivity** for the substance. It is defined from the following experimental setup. Consider a uniform slab of material of length Δx and cross-sectional area A whose ends are kept at a temperature difference ΔT (equal to $T_2 - T_1$), as illustrated in Figure 17–4. (It is assumed that no heat leaves the sides of the slab.)

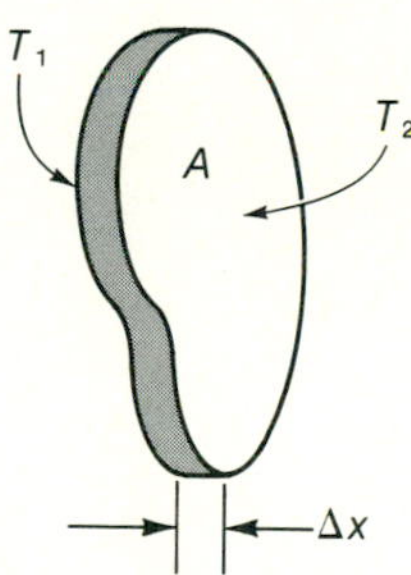

Figure 17–4
A slab of material of thickness Δx and cross-sectional area A. Opposite faces are at temperatures T_1 and T_2.

The amount of heat ΔQ transferred in a time Δt is found experimentally to be:

HEAT CONDUCTION

$$\frac{\Delta Q}{\Delta t} = -kA\frac{\Delta T}{\Delta x} \qquad \textbf{(17–18)}$$

where ΔQ is the quantity of heat transferred and Δt is the time interval. Because heat always flows from a higher to a lower temperature, ΔQ is positive if $\Delta T/\Delta x$ is negative: hence the minus sign in Equation (17–18). The quantity

$\Delta T/\Delta x$ is often called the temperature *gradient*, a mathematical term for the rate of change of a quantity with respect to distance. Table 17–3 lists thermal conductivities for several substances.

EXAMPLE 17–4

A copper sheet 1 mm thick is bonded to a steel sheet 2 mm thick. The outer surface of the copper is maintained at 200°C and the outer surface of the steel is kept at 20°C, as shown in Figure 17–5. (a) Find the temperature of the steel–copper interface. (b) How many calories of heat are conducted through the combination per square meter per second?

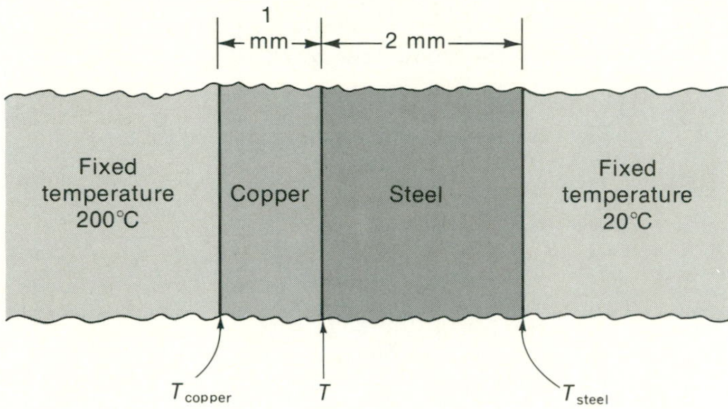

Figure 17–5
Heat conduction through a compound wall of two different materials.

SOLUTION

(a) The heat flow through the copper must be the same as through the steel (otherwise there would be a steady increase of thermal energy at the interface between the two metals). Applying Equation (17–18), we have

$$\left(\frac{\Delta Q}{\Delta t}\right)_{\text{copper}} = \left(\frac{\Delta Q}{\Delta t}\right)_{\text{steel}}$$

$$\left(-kA\frac{\Delta T}{\Delta x}\right)_{\text{copper}} = \left(-kA\frac{\Delta T}{\Delta x}\right)_{\text{steel}}$$

If T is the temperature of the interface, then

$$k_{\text{copper}}\left(\frac{T_{\text{copper}} - T}{(\Delta x)_{\text{copper}}}\right) = k_{\text{steel}}\left(\frac{T - T_{\text{steel}}}{(\Delta x)_{\text{steel}}}\right)$$

Substituting the appropriate values, we obtain

$$\left(0.918\,\frac{\text{cal}}{\text{cm}\cdot\text{C}^\circ\cdot\text{s}}\right)\left(\frac{200^\circ\text{C} - T}{0.1\text{ cm}}\right) = \left(0.13\,\frac{\text{cal}}{\text{cm}\cdot\text{C}^\circ\cdot\text{s}}\right)\left(\frac{T - 20^\circ\text{C}}{0.2\text{ cm}}\right)$$

$$T = \boxed{188^\circ\text{C}}$$

(b) The heat flow may be obtained by using the value of the interface temperature in Equation (17–18) for the steel sheet (or the copper sheet).

$$\frac{\Delta Q}{\Delta t} = -kA\frac{\Delta T}{\Delta x}$$

$$\frac{\Delta Q}{\Delta t} = \left(0.13\,\frac{\text{cal}}{\text{cm}\cdot\text{C}^\circ\cdot\text{s}}\right)\left(100\text{ cm}\right)^2\left(\frac{188^\circ\text{C} - 20^\circ\text{C}}{0.2\text{ cm}}\right) = \boxed{1.09 \times 10^6\text{ cal/s}}$$

Keep Cool*

If the human body did not have sufficient cooling mechanisms, we would soon "boil over." The efficiency of muscles in performing work is quite poor—less than 20%. So about 80% of the energy derived from food is converted to thermal energy deep inside the body. In a resting adult, this thermal energy production is about 1800 kcal/day, *or* 90 watts. *If this internal energy were not dissipated, it would cause a temperature rise of about* 3 C°/h, *resulting in death in just a few hours.*

Interestingly, thermal conduction through human tissue material is very low, the thermal conductivity being approximately that of cork, a good heat insulator. Much of the heat formed inside the body would be trapped if it were not for one mechanism: the circulation of the blood. Blood absorbs heat easily and transports it rapidly to the capillaries near the surface of the skin. There, the thermal energy must be conducted through only a thin layer of tissue to reach the skin's surface.

At the same time, increased blood temperatures are measured by the body's thermostat, the hypothalamus, *located deep within the brain. It sends information that triggers various heat-dissipating mechanisms, including the sweat glands. Perspiration absorbs about* 540 cal/g, *as it evaporates, changing phase from a liquid to the vapor state. This is such an effective cooling mechanism that under experimental conditions, humans have survived without injury an air temperature of* 125°C (257°F) *for a length of time long enough to cook a steak.*

In cold weather, on the other hand, a low skin temperature may cause the capillaries to shrink, thus restricting blood flow near the surface of the skin. This thickens the layer of tissue without blood circulation, thereby becoming a good heat insulator that protects the inner body from excessive heat loss.

* Source: Paul Davidovits, *Physics in Biology and Medicine*, Prentice-Hall, 1975, pp. 142–44.

TABLE 17–3

Thermal Conductivities of Some Common Substances

Substance	Thermal conductivity [cal/(cm·C°·s)]
Silver	1.006
Copper	0.918
Aluminum	0.480
Steel	0.13
Glass	0.002 4
Water	0.001 43
Rubber	0.000 45
Wood (pine)	0.000 2
Cork	0.000 1
Air	0.000 06

17.9 The Constant-Volume Gas Thermometer

In examining various substances and any physical properties that might be used in devising a thermometer, one effect stands out from all others in its sensitivity and reproducibility. Though cumbersome to use, it is far more precise than the familiar mercury-column thermometer. This instrument is the constant-volume gas thermometer illustrated in Figure 17–6. A gas, usually hydrogen or helium, is contained in a bulb with some provision for keeping the volume constant. The *pressure of the gas* is the thermometric property used to define various temperatures.

We set up the temperature scale in the following way. For convenience, we choose a *linear* relationship between the temperature and the particular property utilized (here, the pressure of the gas). Thus, the temperature T on this scale is related to the pressure P by a constant proportionality factor b:

$$T = bP \tag{17–19}$$

and the ratio of any two temperatures equals the ratio of the two pressures:

$$\frac{T_1}{T_2} = \frac{P_1}{P_2} \tag{17–20}$$

Suppose we now experiment with a variety of real gases at different pressures. A temperature scale is set up for each, using the *ice point* and the *steam point* to establish a linear scale, as shown in Figure 17–7. (We will avoid high

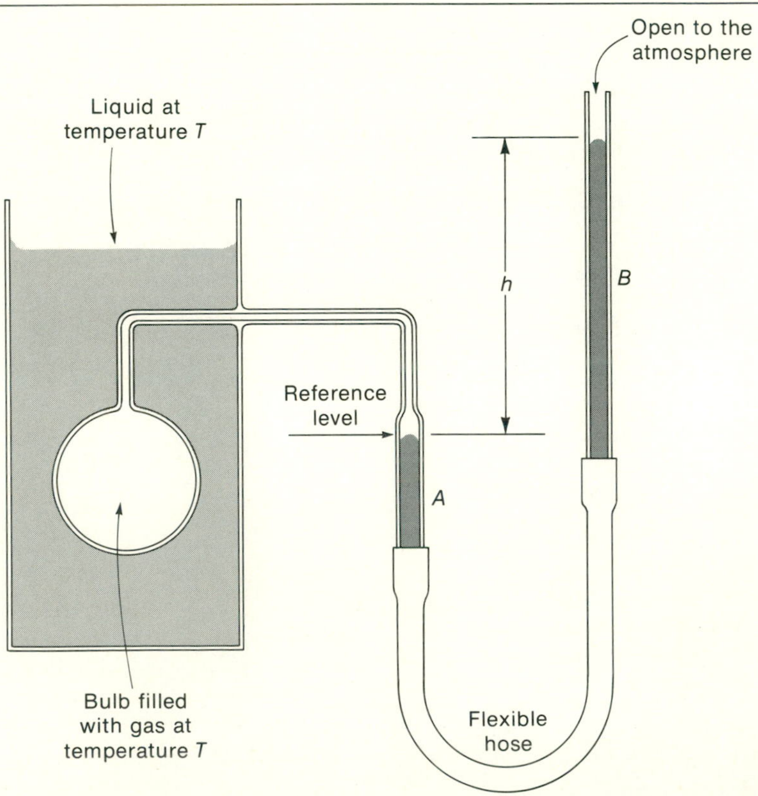

Figure 17–6
A constant-volume gas thermometer. The gas-filled bulb is immersed in a liquid whose temperature is to be determined. By raising or lowering the tube *B*, the height of the mercury column in tube *A* is adjusted so the mercury level is kept at the reference position. The gas is thus maintained in a constant volume. The pressure is then determined by the difference *h* in the heights of the two mercury columns. In practice, many refinements and corrections are incorporated.

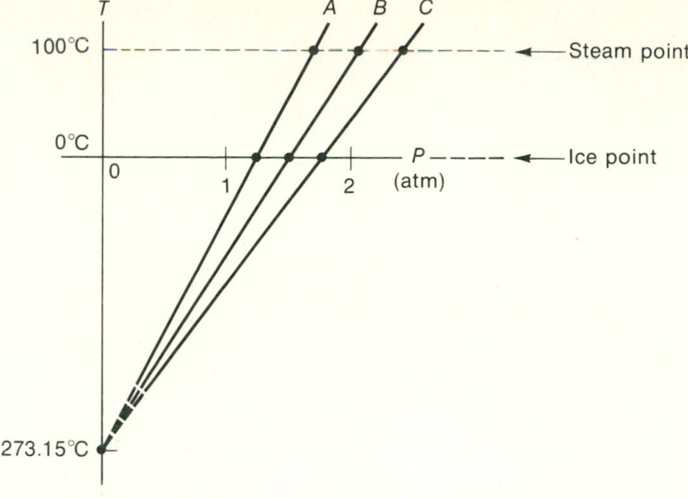

Figure 17–7
Three constant-volume gas thermometer scales (*A*, *B*, and *C*) using different amounts of the same gas. Although in general the slopes of the straight lines differ for various amounts of gas, all lines extrapolate to the same lowest temperature, −273.15°C, at which point the pressure would be zero.

pressures, where certain complications arise due to the finite size of the molecules, and instead use fairly low pressures.) The temperature scales established for different types of gases (such as H_2, He, O_2, and air) are not identical even though the amounts of gas are adjusted to give the same pressure at a given temperature (say, the ice point). However, the differences are very slight: *All real gases tend toward the same behavior if the pressure is made sufficiently low.* Because the gas thermometer we are describing is based on this "ideal" behavior, it is often called the **ideal-gas thermometer scale.**

But now a new fact emerges. Although the straight lines will have different slopes for different pressures (at a given temperature), *all lines approach the same lowest temperature, at which the pressure falls to zero.* So it becomes useful to define a gas temperature scale whose zero is at this lowest point. No thermometer can actually be carried down to this "absolute" zero because all gases liquefy before reaching it. Nevertheless, by extrapolation, its relation to the Celsius scale can be accurately determined: −273.15°C. Thus we have defined a constant-volume ideal-gas temperature scale whose zero has a more fundamental meaning (in terms of a limiting value) than the zero of the Celsius scale. Strictly speaking, however, we have not defined the temperature $T = 0$ itself.

A thermometer scale that does precisely define absolute zero (and whose definition does not depend on the properties of any substance) was devised by Lord Kelvin in the late 1800s. The Kelvin scale and the ideal-gas scale are identical over the range of usable temperatures. If we adopt the temperature unit kelvin (K) for use with the gas thermometer, the relationship between the Kelvin scale and the Celsius scale may be given by (see Figure 17–8):

RELATION BETWEEN THE CELSIUS AND KELVIN TEMPERATURE SCALES

$$\mathrm{K} = \mathrm{C} + 273.15^\circ \qquad \textbf{(17–21)}$$

where the symbols K and C are the corresponding temperature readings. Since the scales differ by just an added constant number, it follows that the temperature *intervals* are the same on the two scales.

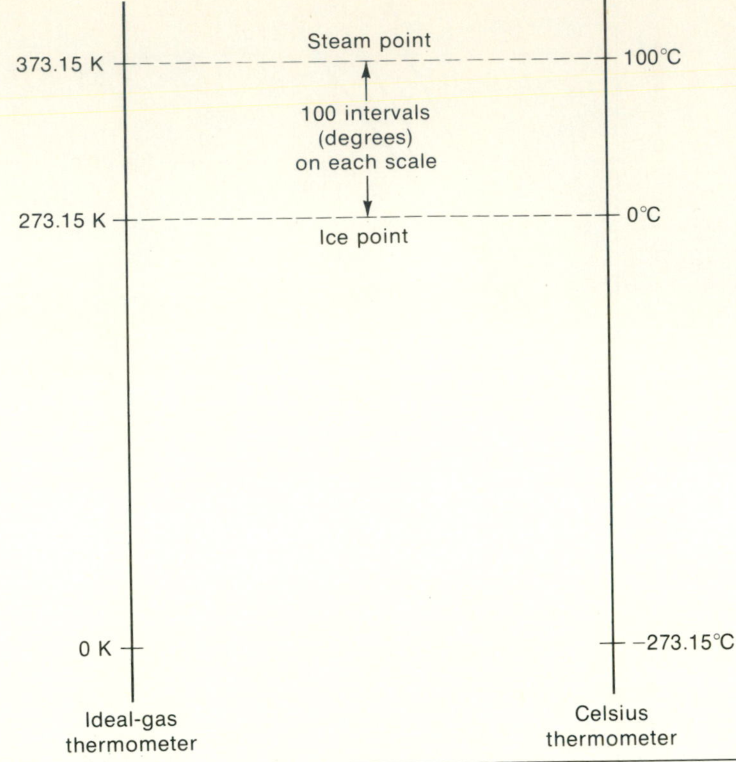

Figure 17–8
The relation between the Celsius scale and the ideal-gas thermometer scale (as defined prior to 1954), labeled in kelvins (K).

In working numerical problems, we will use 273° as the difference, rather than the more exact value 273.15°, in order to simplify the arithmetic. By international agreement, the symbol K stands for "degrees Kelvin," known as "kelvins," so the symbol (°) is omitted for Kelvin temperatures. When indicating a temperature *interval*, it is customary to write explicitly "K degrees" or "C degrees." Because the degree symbol (°) is used for the Celsius scale, 5 C° represents an *interval* of five degrees anywhere on the Celsius scale, while 5°C represents one particular point on the scale.

The Kelvin temperature scale is intimately connected with the theory of thermodynamics (to be discussed in detail in Chapter 19). In fact, all thermodynamic equations are derived on the basis of the Kelvin scale. Therefore, *in all thermodynamic equations in the remaining chapters, the temperature T will always be expressed in kelvins* (K).

EXAMPLE 17–5

Find the Kelvin temperature for the freezing point of mercury: −39°C.

SOLUTION

To three significant figures, the relation between the Kelvin and Celsius scales is:

$$K = C + 273^\circ$$

$$K = (-39^\circ C) + 273^\circ = \boxed{234\ K}$$

17.10 An Improved Definition for the Ideal-Gas Thermometer

The definition of the ideal-gas scale was accepted until 1954, when a more workable procedure was adopted by international agreement. The new definition does not appreciably change its relation to the Celsius scale, but experimentally the scale is more precisely defined. In establishing the new scale, the problems were the difficulties of maintaining the fixed points. For example, the exact ice point (pure ice in equilibrium with air-saturated water at one atmosphere) was easily upset as ice melted and formed *pure* water next to the ice, preventing contact with the *air-saturated* water. There were other experimental difficulties with the steam point. In the new scheme, just a single fixed point is defined—the **triple point** of water—at which ice, water, and water vapor coexist in equilibrium (see Figure 17–9). It is a unique point on the temperature scale, more accurately achieved experimentally than the ice and steam points. Since the triple point occurs at 0.01°C, agreement between the old and new temperature scales was obtained by defining this fixed point to be $T_{tr} = 273.16$ K.

With just one fixed point, how are the other points on the scale determined? They are obtained by a method of successive approximations that eliminates the slight differences between scales based on different gases and different pressures. The method is as follows. We will designate the successive measurements by the subscripts 1, 2, 3, ... Consider a given gas thermometer using a certain type of gas, for example, helium. The gas bulb is surrounded by

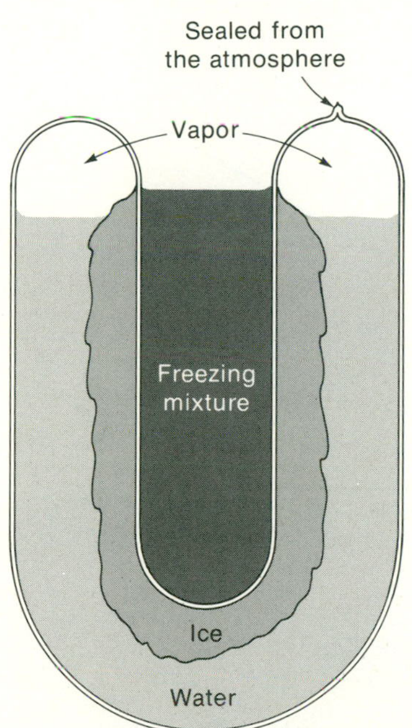

(**a**) A freezing mixture is put into the central well to form a layer of ice.

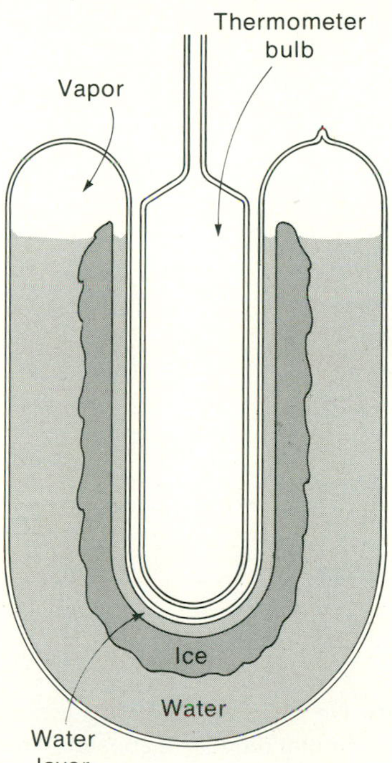

(**b**) The thermometer bulb replaces the freezing mixture in the well and melts a thin layer of ice, allowing the ice, water, and vapor phases to coexist in equilibrium.

Figure 17–9
Simplified features of an apparatus for determining the triple point temperature of water.

water at its triple point temperature T_{tr}. Suppose, for the amount of gas used in this first trial, the pressure in the gas bulb at the triple point temperature is 100 mm of Hg. We label this pressure $(P_{tr})_1$. Using this thermometer, the pressure P_1 at some other temperature T_1 (for example, that of condensing steam) can be found using Equation (17–20):

$$\frac{T_1}{T_{tr}} = \frac{P_1}{(P_{tr})_1}$$

Substituting $T_{tr} = 273.16$ K, $(P_{tr})_1 = 100$ mm of Hg, and rearranging, we obtain a first approximation for the steam temperature T_1:

$$T_1 = (273.16 \text{ K})\left(\frac{P_1}{100 \text{ mm of Hg}}\right)$$

We then perform the experiment again using a *smaller amount of gas in the bulb*. Suppose the value of $(P_{tr})_2$ is 80 mm of Hg. The value we will obtain for the condensing steam temperature T_2 will be slightly different from the value for T_1:

$$T_2 = (273.16 \text{ K})\left(\frac{P_2}{80 \text{ mm of Hg}}\right)$$

We continue making additional measurements as the amount of gas in the bulb is reduced to lower and lower pressures. These data are plotted on a graph, forming one of the lines in Figure 17–10. Finally, in the limit as the triple point pressure approaches zero, we obtain the temperature we seek, that of condensing steam.

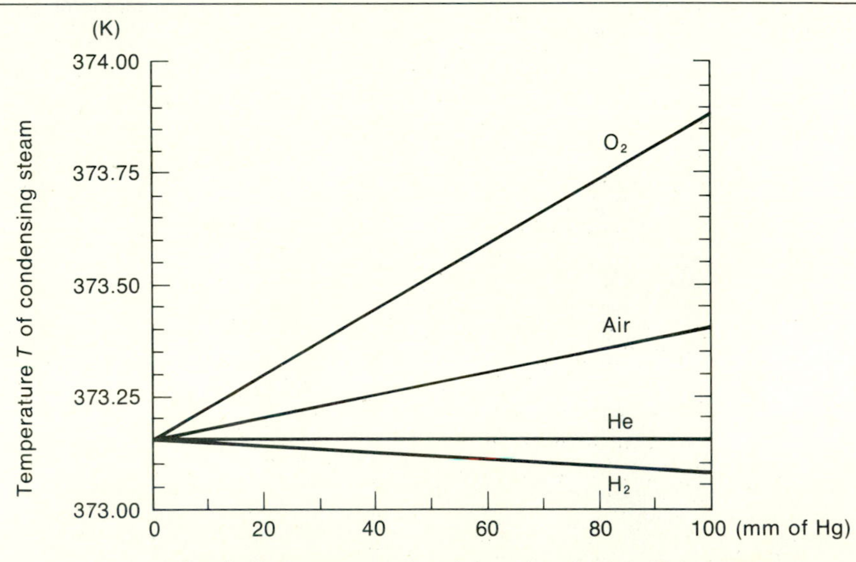

Figure 17–10
Experimental data for a constant-volume gas thermometer using various gases to determine the temperature of condensing steam. The curves are obtained by making successive readings as the gas pressure at the triple point is made lower and lower. At any given value of P_{tr}, various gases give slightly different readings. (The discrepancies are very small—the scale of the ordinate is greatly amplified.) In the limit as the pressure approaches zero, all readings converge on the same numerical value.

Note that other gases will give other straight lines. (Because the ordinates in Figure 17–10 are stretched considerably, these differences are slight, though real.) But in the limit, *they all "zero in" on the same temperature reading for condensing steam*: 373.15 K. So it really makes no difference which gas we use. Thus, the general equation that defines temperatures on the Kelvin constant-volume ideal-gas thermometer is

KELVIN TEMPERATURE FOR A CONSTANT-VOLUME IDEAL-GAS THERMOMETER

$$T = (273.16\ \text{K}) \lim_{P_{tr} \to 0} \left(\frac{P}{P_{tr}}\right) \qquad (17\text{–}22)$$

Figure 17–11 will clarify the relation between this improved definition of the constant-volume ideal-gas thermometer scale and the Celsius scale. (Try not to confuse the two numbers 273.15 and 273.16.) A few significant temperatures are shown in Figure 17–12.

□ □ □

As you no doubt realize by now, the definition of a temperature scale has unexpected subtleties. Let us review a few of the facts regarding the establishment of an ideal-gas scale.

(1) The scale is independent of the properties of any particular gas, though it does depend on "ideal" behavior, toward which all gases tend as the pressure is reduced.
(2) We have not defined the absolute zero temperature $T = 0$ (except as an extrapolation which, at this stage of our discussion, may or may not hold true).
(3) We did not base our definition on any principle that the pressure of a gas at constant volume is inherently proportional to the temperature. Making such a statement before we establish a temperature scale is logically inconsistent. We did *assume* a linear relationship between P and T. But we could just as well have set up a scale based on a logarithmic or any other mathematical function.

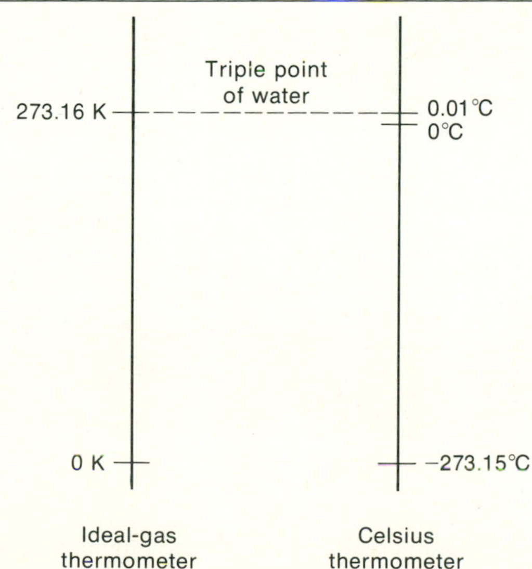

Figure 17–11
The relation between the Celsius scale and the ideal-gas scale (as defined after 1954), labeled in kelvins.

Finally, one remark concerning the temperature $T = 0$. It is incorrect to assume that at this temperature all molecular motion ceases. On the contrary, modern theory based on quantum mechanics shows that as a substance approaches absolute zero, its molecules retain a finite amount of kinetic energy known as the *zero-point energy*. It is the minimum possible, but it is not zero.

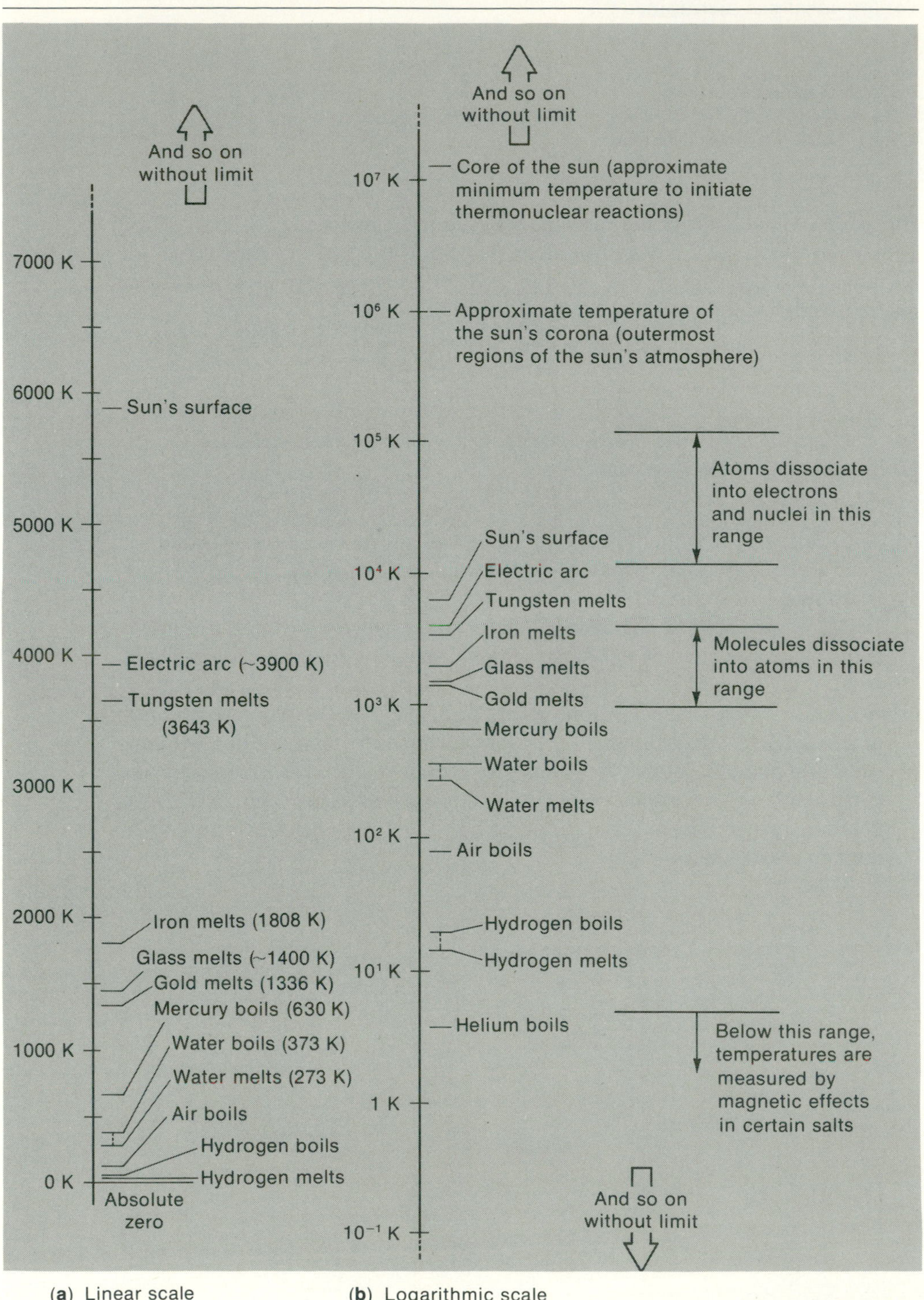

Figure 17–12

Some significant points on the Kelvin temperature scale. The logarithmic scale is useful for presenting information at extremely high and low values.

Summary

Temperature scales

Both the Fahrenheit and Celsius scales are based on the freezing and boiling points of water (at standard atmospheric pressure):

Freezing	32°F	or	0°C
Boiling	212°F	or	100°C

The conversion formula is $5(F + 40°) = 9(C + 40°)$, where C and F are the corresponding temperature readings.

Thermal expansion

Linear expansion $\quad L = L_0(1 + \alpha \Delta T)$

where α is the *coefficient of linear expansion.*

Area expansion $\quad A = A_0(1 + 2\alpha \Delta T)$

Volume expansion $\quad V = V_0(1 + 3\alpha \Delta T)$

The last expression is sometimes written with the substitution $3\alpha = \beta$, the *coefficient of volume expansion.*

Mechanical equivalent of heat
(the definition of the calorie, an SI unit):

$$1 \text{ cal} = 4.184 \text{ J}$$

Specific heat c $\quad \Delta Q = mc\Delta T$

Latent heat L $\quad Q = mL$

Heat conduction $\quad \dfrac{\Delta Q}{\Delta t} = -kA\dfrac{\Delta T}{\Delta x}$

where k is the *thermal conductivity* of the material and Δt is the time interval.

It is possible to devise a constant-volume ideal-gas temperature scale, which has many advantages as a standard. The relation between the Celsius scale and the ideal-gas scale (in kelvins) is:

$$K = C + 273.15°$$

where K and C are the corresponding temperature readings. In numerical problems, for ease of calculation we usually use 273° as the difference between these scales.

Questions

1. Discuss some of the errors inherent in a simple mercury-in-glass thermometer that has a uniform bore printed with a linear scale.
2. When the bulb of a mercury-in-glass thermometer is plunged into boiling water, the indicated temperature first goes down slightly before it begins to rise. Explain.
3. A common thermostat employs a bimetallic strip consisting of two thin strips of different metals bolted together. (The metals have different thermal expansion coefficients.) Why is this useful as a temperature sensor?
4. A brass ring is heated and slipped over a (cold) aluminum cylinder, fitting snugly. After the combination comes to a common temperature, is it possible to separate the two objects again without cutting them apart?

5. Suppose we set up a temperature scale that was reversed, with the hotter temperatures having smaller numerical values. Discuss the pros and cons of such a scale.

6. Both wave propagation and heat conduction involve the transfer of energy. What are the similarities and differences between these two processes?

7. Give a detailed explanation of why specific heats measured in units of cal/g·C° have the *same* numerical values as those measured in units of Btu/lb·F°.

8. Look at Table 17–1, the coefficients of thermal expansion. If the values are multiplied by $\frac{5}{9}$, they are correct for use with a Fahrenheit thermometer. Explain.

9. Suppose you are served coffee several minutes before you are ready to drink it. In which case will the coffee be warmer later: if you put the cream in when you are served, or just before you drink it? (Note: Cooling is largely due to evaporation from the surface, since the cup sides provide fairly good insulation.)

Problems

17A–1 Find the Celsius and Kelvin temperatures corresponding to (a) −19°F, (b) 22°F, and (c) 98.6°F (the "normal" body temperature for adults).

Answers: (a) −28.3°C, 244.7 K (b) −5.56°C, 267.4 K (c) 37.0°C, 310.0 K

17A–2 What are the Celsius and Kelvin temperatures corresponding to (a) −38°F, (b) 17°F, and (c) 49°F?

17A–3 Express the following temperatures in Fahrenheit degrees: (a) −36°C, (b) 89°C, (c) 156 K, and (d) 348 K.

Answers: (a) −32.8°F (b) 192.2°F (c) −178.6°F (d) 167°F

17A–4 The world's longest ship in 1971 was the *Europoort* oil tanker, made of steel, whose length was given as 1141 ft and 15/16 in. What temperature change would make the ship's length exactly 1142 ft?

17A–5 When heated from −20°C to +30°C, a 50-ft lead pipe lengthens by 0.90 in. Find the coefficient of expansion of the lead. *Answer:* 3×10^{-5}/C°

17A–6 Steel railroad rails 50 ft long are placed with small gaps between their ends to allow for thermal expansion. If the rails are laid when the temperature is 10°C, what minimum gap between rails will prevent them from touching when the temperature rises to 50°C? The coefficient of linear expansion for the rails is 1.1×10^{-5} (per C°).

17A–7 A steel measuring tape is calibrated to read correctly at 20°C. If the temperature falls to −10°C, what correction should be applied to a reading of exactly 20.000 m?

Answer: Add 7.2 mm

17A–8 How many calories of heat must be added to a 40-g ice cube at 0°C to change it to water at 27°C?

17A–9 How many calories of heat are required to heat 200 g of iron from 20°C to 90°C?

Answer: 1540 cal

17A–10 How much heat is required to change 2 g of ice at −20°C to steam at 130°C?

17A–11 Adding 1400 cal to a 600-g metal object raises its temperature from 27°C to 36°C. Find the specific heat of the metal. *Answer:* 0.259 cal/g·C°

17A–12 To heat 100 g of a substance from 20°C to 28°C requires 690 cal. Find the specific heat of the substance.

17A–13 A 2-kg bronze object at 90°C is submerged in 1 L of water at 20°C. The combination finally comes to an equilibrium temperature of 32°C. Find the specific heat of the bronze. *Answer:* 0.103 cal/g·C°

17A–14 The bottom of an aluminum pan has a diameter of 20 cm and a thickness of 1 mm. When placed on a stove burner, a constant temperature difference of 4.2 C° is maintained between the two surfaces of the bottom of the pan. How many calories will flow through the pan bottom in 1 min?

17A–15 One end of a copper bar 10 cm long, with a cross-sectional area of 4 cm^2, is kept in ice at 0°C, and the other end is in contact with steam at 100°C. Neglecting heat losses from the sides of the bar, find (a) the amount of ice (in grams) that is melted in 10 min and (b) the amount of steam (in grams) that condenses to water at 100°C in 10 min.

Answer: (a) 275 g (b) 40.8 g

17B–1 At what Fahrenheit temperature are the Kelvin and Fahrenheit temperatures numerically equal? *Answer:* 574°F

17B–2 Steel railroad rails 60 ft long are laid down with 0.10-in spaces between them on a day when the temperature is 95°F. What space will be between the rails when the temperature drops to −20°F? The coefficient of linear expansion for the steel is 12×10^{-6} $(°C)^{-1}$.

17B–3 A 5-gal aluminum container is filled to the brim with turpentine when the temperature is 15°C. When the temperature rises to 27°C, 6.36 fluid oz spills over from the container. Determine the volume coefficient of expansion for turpentine.

Answer: 9×10^{-4} (per C°)

17B–4 The volume coefficient of expansion for carbon tetrachloride is $5.81 \times 10^{-4} (C°)^{-1}$. If a 50-gal steel container is filled completely with carbon tetrachloride when the temperature is 10°C, how much will spill over when the temperature rises to 30°C?

17B–5 Suppose 500 g of BB shot at 100°C are added to 300 g of water in a 200-g glass beaker initially at 20°C. The combination eventually comes to an equilibrium temperature of 23.4°C. The specific heat of the glass is 0.20 cal/g·C°. Assuming that heat loss to the surroundings is negligible, find the specific heat of the shot. *Answer:* 0.0302 cal/g·C°

17B–6 A 50-g ice cube at −20°C is dropped into a container of water at 0°C. How much water will freeze onto the ice?

17B–7 Suppose 100 g of water in a 200-g glass beaker (specific heat = 0.20 cal/g·C°) is at 27°C. How many grams of copper at 100°C must be added to raise the temperature of the mixture to 31°C? *Answer:* 87.3 g

17B–8 A 2-g lead bullet traveling at 300 m/s strikes a 1-kg wooden block in a ballistic pendulum experiment. If half the mechanical energy that "disappears" goes into heating the bullet, calculate the temperature rise of the bullet. For lead, $c = 0.0305$ cal/g·C°.

17B–9 The brick wall of a building has dimensions of 4 m × 10 m and is 15 cm thick. How much heat (in calories) flows through the wall in a 12-h period when the average inside and outside temperatures are, respectively, 20°C and 5°C?

Answer: 2.59×10^7 cal

17B–10 Figure 17–13 shows three round copper bars. Their upper ends are kept at the same temperature, and their lower ends are kept at a (lower) common temperature. Rank them as heat conductors, listing the best first.

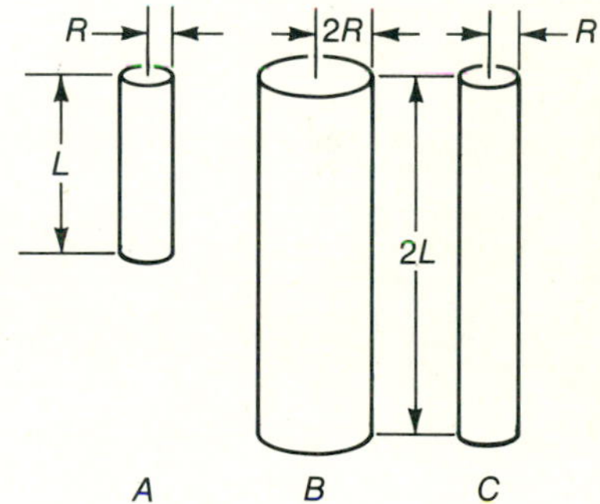

Figure 17–13
Problem 17B–10

17B–11 A commercially available "heat-transfer pipe" consists of a sealed, evacuated tube, 11 in long, 3/8 in in diameter, and lined with a capillary wick that contains a working fluid. Heating one end causes the fluid to vaporize, filling the tube quickly with vapor. Upon contact with an unheated area, the vapor condenses and thus gives up its heat of vaporization to the wall. It is claimed that the pipe is "hundreds of times" better as a heat conductor than the best metal conductor. Calculate the equivalent thermal conductivity from these data: maximum temperature gradient = 0.25°C/in and thermal power transfer = 60 W. Compare with the values given in Table 17–3.

17C–1 The glass bulb of a mercury thermometer contains 0.40 cm³ of mercury. The diameter of the stem bore is 0.10 mm.

(a) Ignoring the expansion of the glass, predict the distance the column of mercury would move along the stem when the temperature changes from 25°C to 40°C. The volume coefficient of expansion for mercury is $1.82 \times 10^{-4} (C°)^{-1}$.

(b) If the column actually moves 11.9 cm, what is the volume coefficient of expansion for the glass? *Answers:* (a) 13.9 cm (b) $2.6 \times 10^{-5} (C°)^{-1}$

17C–2

(a) On a day when the outside temperature is constant at 32°F, how much energy is lost per 24-h period through a plate glass window whose dimensions are 1 m × 2 m and which is 4 mm thick, from a room whose air temperature is 70°F? The thermal conductivity for the glass is 2×10^{-3} cal/(cm·s·C°).

(b) If the house is heated electrically at a cost of 8¢/(kW·h), and there are 10 windows of that size in the house, how much does the homeowner pay each 24 h to heat up the outdoors by heat escaping through the windows?

17C–3 The procedure of "shrink fitting" a brass disk inside an aluminum cylinder is to machine the disk slightly larger than the inside diameter of the cylinder when both are at room temperature (27°C). When both are heated to 100°C, the cylinder expands more than the disk, allowing the disk to be inserted. Upon cooling to room temperature, the cylinder shrinks more than the disk, to hold the disk snugly in place. If the cylinder diameter is exactly 8.0000 cm at 27°C, what should be the corresponding diameter of the disk so that when both are heated to 100°C, the cylinder will be 0.001 cm larger than the disk?

Answer: 8.0039 cm

17C–4 A box with a total surface area of 1.2 m^2 and a wall thickness of 4 cm is made of insulating material. A 10-W electric heater inside the box maintains the inside temperature at a steady 15°C above the outside temperature. Find the thermal conductivity of the insulating material.

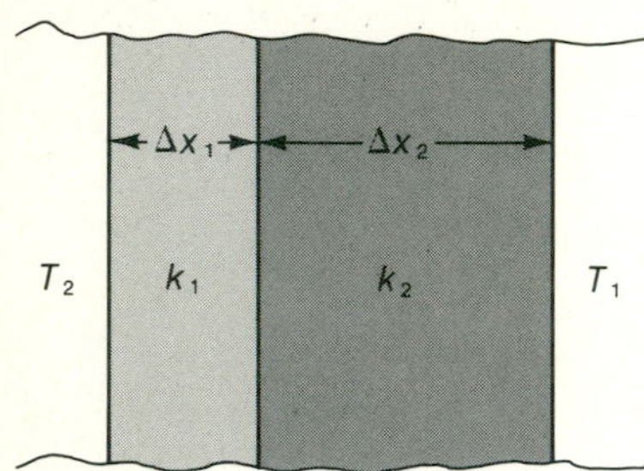

Figure 17–14
Problem 17C–5

17C–5 Heat is conducted through a compound wall formed of two different substances in contact, as shown in Figure 17–14. The part with thickness Δx_1 has thermal conductivity k_1, and the other part has thickness Δx_2 and thermal conductivity k_2. The outer walls are maintained at temperatures T_2 and T_1 as shown.

(a) Find the steady state temperature T_x at the interface between the substances.

(b) Show that the heat transfer per unit time through an area A of the wall is given by

$$\frac{\Delta Q}{\Delta t} = -\frac{A(T_2 - T_1)}{\left(\dfrac{\Delta x_1}{k_1} + \dfrac{\Delta x_2}{k_2}\right)}$$

Note: This result may be generalized for n sections in contact to

$$\frac{\Delta Q}{\Delta t} = -\frac{A(T_2 - T_1)}{\sum \dfrac{\Delta x_n}{k_n}}$$

Answer: (a) $\dfrac{T_2 k_1 \, \Delta x_2 + T_1 k_2 \, \Delta x_1}{k_2 \, \Delta x_1 + k_1 \, \Delta x_2}$

17C–6 A layer of air sandwiched between two window panes greatly improves the heat insulation characteristics compared with just the two glass panes in contact without the air layer. For panes of glass 3 mm thick (each) and an air layer 1 cm thick, find the ratio of the heat conducted per unit time: (without an air layer)/(with an air layer). Assume a unit area A. You may use the results of the previous problem and the numerical data in Table 17–3.

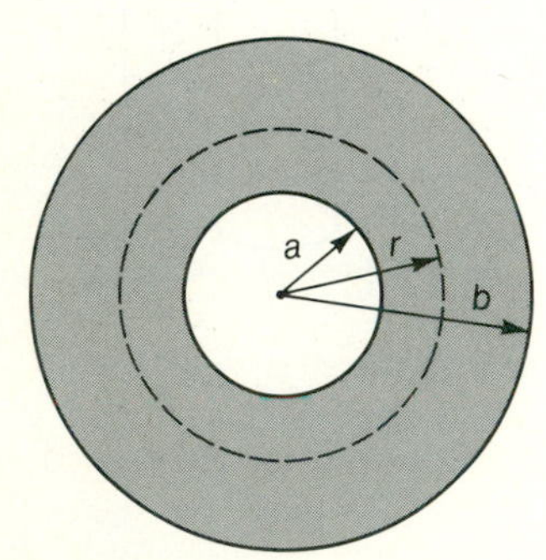

Figure 17–15
Problem 17C–7

17C–7 A steam pipe is surrounded by a layer of insulating material, as shown in Figure 17–15. The inner and outer radii of the insulation are a and b, and the inner and outer surfaces are at temperatures T_2 and T_1, respectively. Show that the radial heat flow per unit time $\Delta Q/\Delta t$ for a length of pipe L is given by

$$\frac{\Delta Q}{\Delta t} = -\frac{2\pi k(T_2 - T_1)}{[\ln(b/a)]/L}$$

where k is the thermal conductivity of the insulation. (Hint: If A is the area of the surface at radius r, and dT/dr is the temperature gradient at that radius, then $\Delta Q/\Delta t = -kA \, dT/dr =$ constant. Rearrange and integrate this expression over the appropriate ranges of the variables.)

17C–8 A steam pipe 8 cm in diameter is surrounded by insulating material 3 cm thick. The temperature of the pipe is 100°C, and the outer surface of the insulation is at 30°C. The thermal conductivity of the insulation is 2×10^{-4} cal/s·cm·C°. Find the temperature gradient dT/dr at the inner and outer surfaces of the insulation.

17C–9 A spherical ball is surrounded by a layer of insulating material. The inner and outer radii of the insulation are a and b, and the inner and outer surfaces are at temperatures

T_2 and T_1, respectively. Show that the heat flow per unit time $\Delta Q/\Delta t$ radially outward is given by

$$\frac{\Delta Q}{\Delta t} = -\frac{4\pi k(T_2 - T_1)}{\left(\frac{b-a}{ab}\right)}$$

where k is the thermal conductivity of the insulation. (See the hint in Problem 17C–7.)

17C–10 Two steam pipes, A and B, are surrounded by jackets of insulating material. The diameter of pipe B is twice that of A, and the total volume of insulating material per unit length surrounding B is twice that of A. The thickness of the insulation of A equals the radius of pipe A. Find the ratio of the heat loss per unit time: (pipe B)/(pipe A).

17C–11 A pond of water at 0°C is covered with a layer of ice 4 cm thick. If the air temperature stays constant at -10°C, how long will it be before the ice thickness is 8 cm? (Hint: The incremental heat dQ extracted from the water through the thickness x of ice is the amount required to freeze a thickness dx of ice. That is, $dQ = L\rho A\,dx$, where ρ is the density of the ice, A is the area, and L is the latent heat of freezing.)

Answer: 4.17×10^4 s = 11.6 h

17C–12 Assuming the specific heat of the human body is the same as that of water, how much water would have to evaporate from the skin of a 70-kg man to cool his body by one degree Celsius? The heat of vaporization of water at 37°C (the "normal" temperature of the human adult) is 575 cal/g.

chapter 18 The Ideal Gas and Kinetic Theory

Not too small
Not too inhomogeneous
Not too homogeneous
CORNELIUS LANCZOS
(on the criteria for systems which may be analyzed statistically)

18.1 Introduction

In this chapter, we take our first step to probe beneath the gross characteristics of matter to see if we can understand what is going on in terms of the motions and interactions of atoms and molecules. To comprehend the microscopic "machinery" underlying matter in all its forms is an extremely formidable task. For one thing, atoms and molecules do not follow the rules of Newtonian mechanics. Instead, the radically different approach of *quantum mechanics* is necessary. Although we will postpone a discussion of quantum mechanics, as a preliminary introduction to this approach we can carve out a simple problem to tackle: matter in its *gaseous* phase. In order to limit our discussion to situations in which quantum effects are negligible, we will avoid extremely low temperatures, where the gas approaches its phase change to a liquid, and very high pressures, where the gas molecules are squeezed so close together that the finite size of the molecules themselves must be taken into account.

In the next four chapters, you will note that we approach the subject from two different perspectives. Thermodynamics deals with the *macroscopic* view of matter, that is, with measurements we carry out in the laboratory, such as determining the volume and temperature changes of a substance when we heat it or compress it. (What is seen in a microscope is also classified as a macroscopic phenomenon.) The other approach to understanding the behavior of matter is called *statistical mechanics;* it deals with the *microscopic* view, that is, with the motions of atoms and molecules themselves. Because we are unable to follow explicitly the individual motions of each atom or molecule separately, mathematical methods are used to deal with the *average* behavior of these entities.

These two views of matter, the macroscopic and the microscopic, together form a unified picture of the world, since the macroscopic features are really just averages over time of a great number of microscopic interactions. For example, the pressure of a gas is basically just the average of extremely large numbers of molecular impacts against the walls of the container.

18.2 The Ideal Gas

When we devise a physical theory in science, we usually begin by defining certain concepts and assigning a convenient shorthand symbol to them. Then the relations between the concepts are expressed in a mathematical equation involving the symbols. In describing the physical characteristics of matter, such mathematical relations are known as *equations of state*, because the equation describes the physical state of the matter under all possible conditions. For most substances, the equation of state is so complicated that it is not known. However, one example can be imagined—that of the so-called *ideal gas*—that does have a very simple equation of state. It is an important equation, since real gases at moderate pressures and temperatures closely follow this behavior.

In early experiments with gases, it became apparent that for a given amount of gas, the significant parameters were the pressure P, the volume V, and the temperature T. The equation of state is thus some mathematical function of the form $f(P,V,T)$. The first successes in discovering the proper relations were those of Robert Boyle (1660), J. A. C. Charles, and Joseph Louis Gay-Lussac (1802). Performing a large number of experiments on given masses of various gases, they found the following relations:

BOYLE'S LAW

$$PV = [\text{constant}] \quad \text{(at constant } T\text{)} \tag{18–1}$$

CHARLES' AND GAY-LUSSAC'S LAW

$$V = [\text{constant}][T] \quad \text{(at constant } P\text{)} \tag{18–2}$$

Both of these relations may be combined together into a single equation:

$$PV = [\text{constant}][T] \tag{18–3}$$

From the following reasoning, we conclude that the constant in Equation (18–3) must be *proportional to the mass of gas under consideration.* Suppose we have a given volume of gas at certain fixed values of P and T. If we consider just a segment of this gas, say, if we partition off one-half the original volume, we have a second sample at the same values of P and T but only half the volume. To keep Equation (18–3) valid for both samples of gas, we see that the constant must contain the information that the smaller volume has only one-half the amount of gas as the original volume.

A convenient way to express the amount of a gas is by a dimensionless quantity called the number of moles. The **mole** (mol) is defined as:

ONE MOLE (mol) **The amount of any substance that contains as many elementary entities as there are atoms in 0.012 kg of carbon-12. This number is called Avogadro's number: $N_A = 6.022\,17 \times 10^{23}$ particles/mole.**

When the mole is used, the elementary entities must be specified. They may be atoms, molecules, ions, electrons, other particles, or specified groups of such particles. Thus, for example, one may have a mole of electrons, a mole of atoms, or a mole of molecules.

Chemists have discovered some interesting facts about the mole which we will review here. **Standard conditions** of temperature and pressure (abbreviated STP) are 0°C and 1 atm. *At STP, one mole of any gas occupies the same volume:*

VOLUME OF ONE MOLE OF ANY GAS AT 0°C AND 1 ATM 22.4 liters (or 0.0224 m^3)

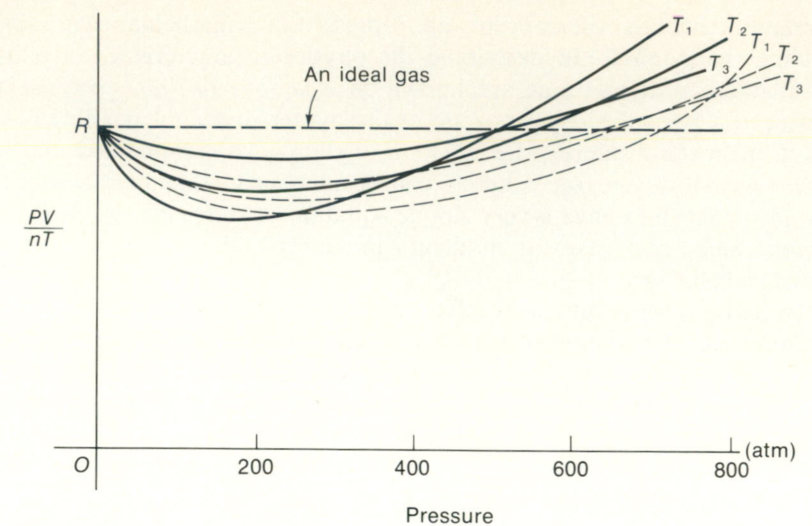

Figure 18–1
The qualitative behavior of two real gases. (One set of curves is shown dashed.) The values of PV/nT are shown for three different temperatures. Note that as the pressure is reduced, all real gases approach the same value of PV/nT. This value is R, the universal gas constant.

Another helpful fact is that the mass of one mole, expressed in grams, is numerically equal to its **molecular mass**[1] in **unified atomic mass units.**

UNIFIED ATOMIC MASS UNIT (u) **The unified atomic mass unit (u) is defined as $\frac{1}{12}$ the mass of an atom of carbon-12 (including its electrons): 1 u $\approx$ 1.660 53 $\times$ 10^{-27} kg.**

Incorporating these relations, the constant in Equation (18–3) is chosen to be nR, where n is the number of moles present and R is the experimentally determined universal gas constant in units that will make the equation consistent:

$$\begin{aligned} R &= 8.314 \text{ J/mol·K} \\ &= 0.0821 \text{ L·atm/mol·K} \\ &= 1.986 \text{ cal/mol·K} \end{aligned}$$

We thus arrive at the **equation of state for an ideal gas:**

IDEAL GAS LAW $$PV = nRT \qquad \textbf{(18–4)}$$

where T must be expressed *in kelvins.* (Recall from the last chapter that *all temperatures in thermodynamic equations must be expressed in kelvins.*)

Equation (18–4) is called the *ideal* gas law because it is the behavior toward which all real gases tend as the pressure becomes very low. This is sketched qualitatively in Figure 18–1. As the pressure is reduced, the experimental values of PV/nT for all gases approach the value for R. An **ideal gas,** then, can be conveniently defined as a (hypothetical) gas for which the value of PV/nT has the constant value R for all values of the pressure. In the range of a few atmospheres, this behavior is closely approximated by most real gases,

[1] The molecular mass is closely related to what is called the **molecular weight.** The molecular weight is found by adding the **atomic weights** of the atoms making up a given molecule. For example, since the atomic weights of hydrogen (H) and oxygen (O) are, respectively, 1 and 16, the molecular weight of water (H_2O) is 18. Its molecular mass is 18 u. Also, 18 g of water is one mole of water, containing Avogadro's number of H_2O molecules.

particularly monatomic gases. The extrapolation is quite a small step. In fact, helium at low pressures behaves essentially like an ideal gas until we approach within a few degrees of absolute zero on the Kelvin scale. So the concept of an "ideal gas" is not so far from what is actually realizable. *Unless otherwise specified, we will assume in numerical problems that all real gases behave as an ideal gas.*

The equation of state for an ideal gas may be visualized as a three-dimensional surface, as in Figure 18–2. Lines of constant temperature are called **isothermals,** or **isotherms** (from the Greek *isos*, meaning "equal" or "the same"). When projected on the pressure-volume plane, the isotherms form a family of hyperbolas. (Recall that the mathematical form of a hyperbola is $PV =$ constant.)

For real substances, the three-dimensional surface representing the equation of state that covers all phases—solid, liquid, and gas—is generally very complicated. In fact, the exact equation of state is often unknown. However, useful diagrams that represent the qualitative features can be drawn, as illustrated in Figures 18–3 and 18–4. Two features should be noted. In Figure 18–3, the gas cannot be liquefied above the *critical temperature*, no matter how great a pressure is applied. On the pressure-temperature diagram of Figure 18–4(b), the *triple point* indicates the pressure and temperature at which the substance may coexist as a solid, liquid, and vapor.

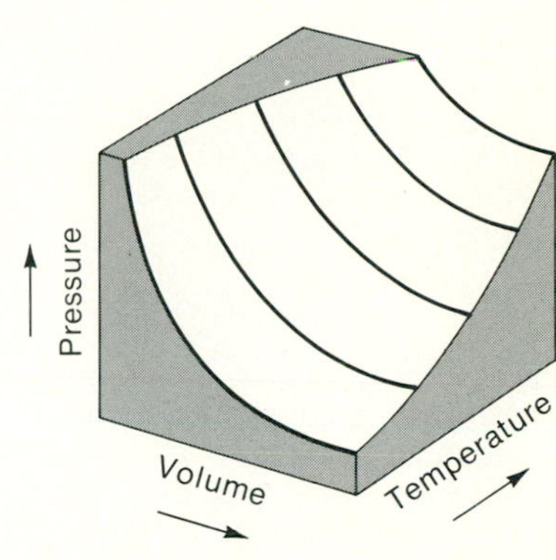

(**a**) The pressure-volume-temperature surface for an ideal gas. Lines of constant temperature are drawn on the surface.

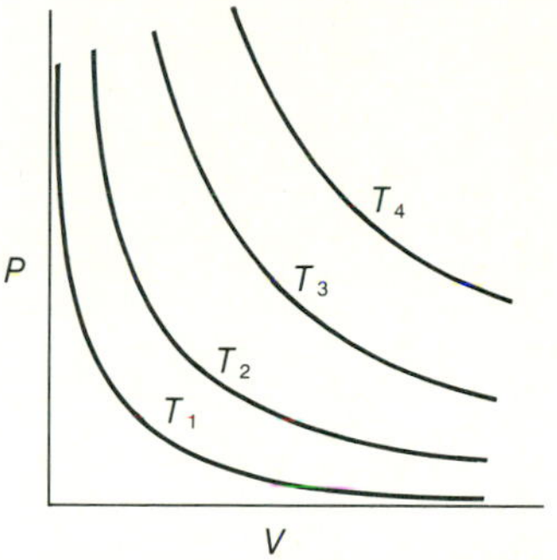

(**b**) In a projection on the pressure-volume plane, the lines of constant temperature (isotherms) form a family of hyperbolas.

Figure 18–2

The functional relationship of pressure, volume, and temperature, represented in two ways.

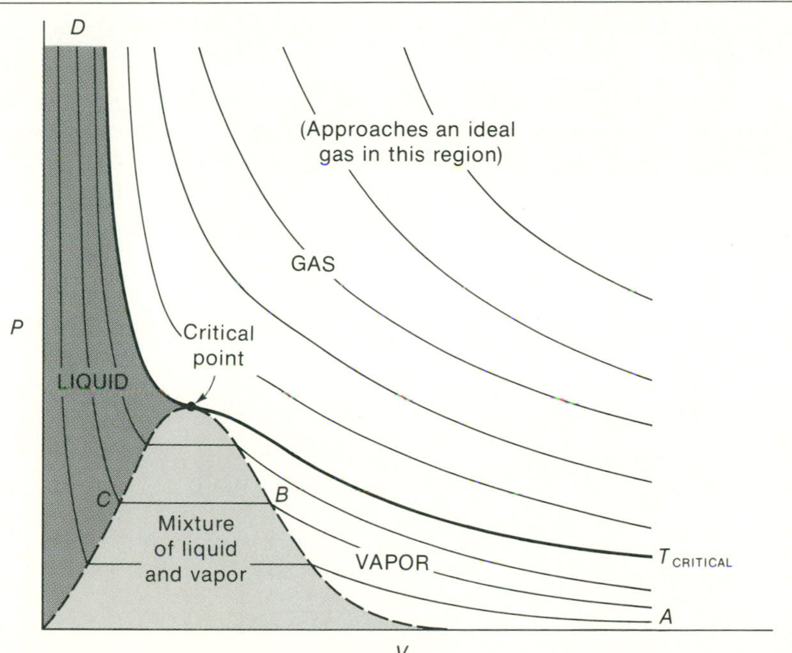

Figure 18–3

A qualitative diagram showing isotherms for a gas that liquefies. For temperatures above $T_{critical}$, the gas cannot be liquefied at any pressure because the average energy of collisions is more than enough to break apart the binding forces that hold molecules together as a liquid. Along the isotherm from *A* to *B*, the substance exists in its vapor phase. (It is customary to use the term "vapor" for gaseous states below $T_{critical}$.) Condensation occurs from *B* to *C*, creating a mixture of liquid and vapor at the constant *vapor pressure* for that temperature. For the pure liquid phase from *C* to *D*, the isotherm is almost vertical, signifying that a great increase in pressure is required to compress the liquid to a smaller volume.

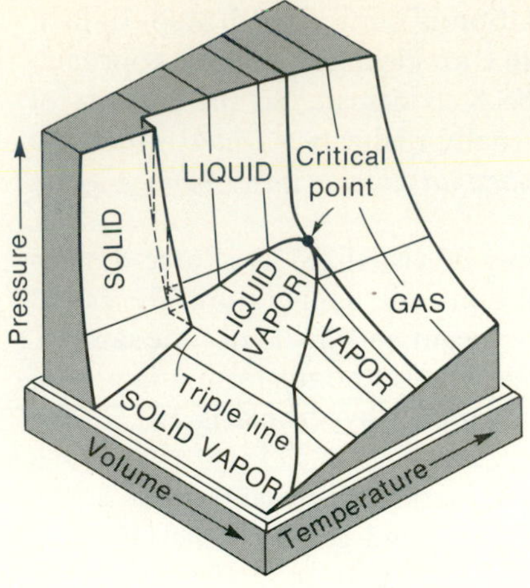

(a) The pressure-volume-temperature surface for a substance that expands on freezing.

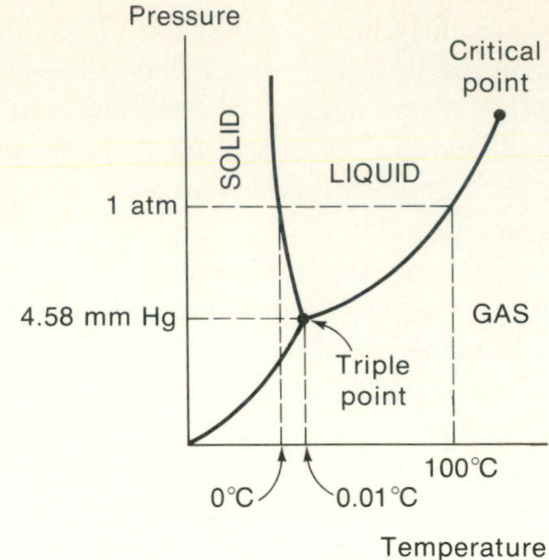

(b) The projection of (a) on the pressure-temperature plane. The intersection of the three lines that separate different phases is called the triple point. (Numerical values given are for water.)

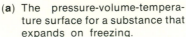

Figure 18–4
A three-dimensional surface for the equation of state for a substance that expands on freezing (such as water). The scales have been distorted to clarify certain features.

EXAMPLE 18–1

A high-altitude research balloon contains helium gas. At its maximum altitude of 20 km, the outside temperature is $-50°C$ and the pressure has dropped to 40 mm of Hg. The volume of the balloon at this location is 800 m^3. Assuming the helium has the same temperature and pressure as the surrounding atmosphere, find the number of moles of helium in the balloon.

SOLUTION

Since we assume that all real gases behave as an ideal gas, the equation of state of an ideal gas, $PV = nRT$, relates the parameters for any given conditions. The first step is to convert all data into a consistent system of units. We choose SI units and $R = 8.31$ J/mol·K. As mentioned, temperature must be expressed in kelvins. Therefore, from Equation (17–21):

$$\text{K} = \text{C} + 273° = (-50°\text{C}) + 273° = 223\ \text{K}$$

The units of pressure must be changed to newtons/meter^2. At standard pressure, 760 mm of Hg corresponds to 1.013×10^5 N/m^2; we therefore have

$$(40\ \text{mm Hg})\underbrace{\left(\frac{1.013 \times 10^5\ \dfrac{\text{N}}{\text{m}^2}}{760\ \text{mm of Hg}}\right)}_{\text{Conversion ratio}} = 5.33 \times 10^3\ \frac{\text{N}}{\text{m}^2}$$

Finally, from $PV = nRT$ we obtain

$$n = \frac{PV}{RT} = \frac{\left(5.33 \times 10^3\ \dfrac{\text{N}}{\text{m}^2}\right)(800\ \text{m}^3)}{\left(8.31\ \dfrac{\text{J}}{\text{mol}\cdot\text{K}}\right)(223\ \text{K})} = \boxed{2300\ \text{mol}}$$

EXAMPLE 18-2

Find the mass of the helium in the previous example.

SOLUTION

The "molecular weight" of helium is 4.00, so one mole of helium has a mass of 4.00 g. Therefore, 2300 mol have a mass of

$$m = \left(4.00\,\frac{\text{g}}{\text{mol}}\right)(2300\text{ mol}) = 9200\text{ g} = \boxed{9.20\text{ kg}}$$

EXAMPLE 18-3

Consider the amount of helium in the previous two examples. (a) What volume of tank will contain this much helium at standard conditions? (b) At 27°C and 170 atm?

SOLUTION

(a) Since one mole at standard conditions (0°C and 1 atm) occupies 22.4 L (equal to 0.0224 m^3), 2300 mol at standard conditions will occupy a volume 2300 times larger:

$$V = (2300\text{ mol})\left(\frac{0.0224\text{ m}^3}{\text{mol}}\right) = \boxed{51.5\text{ m}^3}$$

(b) Although we could find the volume by solving $PV = nRT$, there is another method that is sometimes more convenient. This is to recognize that for a given type of gas, any two states can be equated through the relation:

$$\frac{P_1V_1}{nT_1} = \frac{P_2V_2}{nT_2} \qquad \textbf{(18–5)}$$

Because symbols such as P and V appear as common factors on both sides, conversion of units may sometimes be avoided (since the same conversion factors would appear on each side). The temperature, however, must be expressed in kelvins, since conversion from °C to K involves an *additive* number rather than a *multiplicative* constant. Here, $T_2 = 27°\text{C} + 273° = 300\text{ K}$. Solving for the volume V_2 gives

$$V_2 = \frac{P_1V_1\not{n}T_2}{P_2\not{n}T_1} = \frac{(1.0\text{ atm})(51.5\text{ m}^3)(300\text{ K})}{(170\text{ atm})(273\text{ K})} = \boxed{0.333\text{ m}^3}$$

18.3 Model of an Ideal Gas

We shall now describe a simple model for an ideal gas that initially involves only two attributes of the molecules: *mass* and *velocity*. It is interesting that we can explain the general behavior of an ideal gas with such a simple model. Although we will not carry out all the additional steps, the theory can be improved by adding a finite size and shape for each molecule (instead of considering them point masses) and thereby accounting for such properties as viscosity, the coefficient of diffusion, and thermal conductivity. By adding the feature of long-range forces between the molecules, we can explain the additional phenomena of liquefaction and solidification. And if we assume some substructure to each molecule and atom, we can predict the chemical, electrical, and magnetic properties of matter, the spectral emission and absorption of light, and so forth. This is the pattern of development for all physical theories. We start with the simplest possible model that will lead to some correct predictions regarding experimental results. Later modifications extend and improve the agreement.

Sometimes the process continues until all but a seemingly slight disagreement remains, perhaps in the fifth or sixth decimal place of the numerical data. At this point, a dramatic breakthrough might be made by someone who proposes a radically different model that not only explains all that the old theory did, but also correctly predicts the slight discrepancy remaining in the old theory. These are the exciting developments of physics that bring justifiable honor to such men as Newton, Einstein, Bohr, Planck, Schrödinger, and many others.

Our first model for an ideal gas contains the following assumptions:

(1) The gas consists of a large number of identical point masses. (These are, of course, approximations to atoms or stable molecules, and we will occasionally designate them by these names even though, in this first approximation, the only attributes we give them are mass and velocity.) The mass of each particle is m.
(2) The particles have random motions with various velocities, and they undergo elastic collisions with the wall,[2] conserving momentum and kinetic energy.
(3) There are no forces between the particles and the walls except during collisions, which are assumed to occur in negligible time compared with the times between collisions. This implies that the particle-wall forces are *short range* only, comparable to dimensions of real molecules. (The kinetic energy that is momentarily changed to potential energy during the collisions is converted back to kinetic energy so quickly that we may ignore this effect and assume that the energy of the particles is entirely kinetic.)

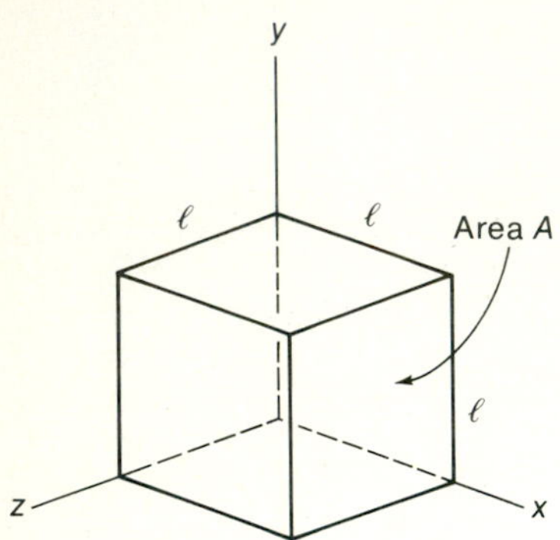

Figure 18–5
A cubical enclosure containing $\mathcal{N}$ particles per unit volume, each of mass m, in random motions with various velocities.

For simplicity's sake, we assume the gas contains $\mathcal{N}$ particles per unit volume and is enclosed in a cubical container of edge ℓ, as in Figure 18–5. (The volume of the enclosure will cancel out later in the derivation, so we are justified in choosing this convenient shape.)

Consider a particle whose velocity is $\mathbf{v}$. Its components are v_x, v_y, and v_z. Now let us concentrate on the number of particles with a velocity component of magnitude $|v_{x1}|$ in the $\pm x$ direction. Of course, no particle has *precisely* the component value $|v_{x1}|$. A more accurate statement would be: "components in the range from $|v_{x1}|$ to $|v_{x1} + \Delta v|$, where Δv is a small but finite interval" or, using a numerical illustration, "in the range from 1000 m/s to 1001 m/s." For brevity, we will not carry along the Δv notation. but it should be kept in mind in the following discussion.

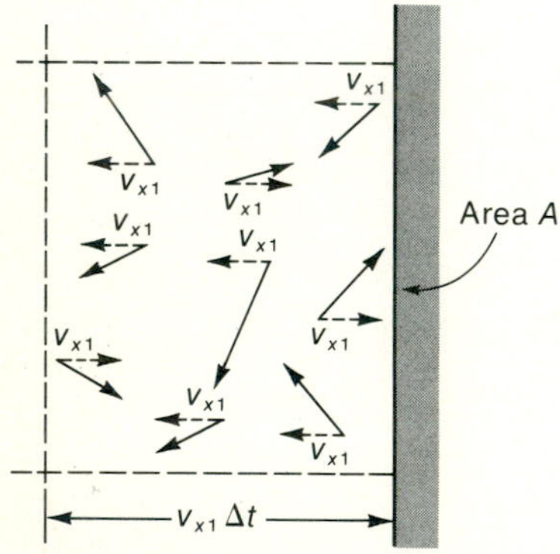

Figure 18–6
This sketch shows particles that have a $\pm x$ velocity component of magnitude v_{x1} and that are located in a volume $(v_{x1}\,\Delta t)(A)$ next to the wall. Since on the average, half are traveling in the $-x$ direction, only half of them will strike the wall in a time Δt. (Molecules that travel out of the chosen volume before striking the wall area are compensated for by essentially the same number of molecules outside that travel into the volume.)

Let us use the designation n_1 to represent the number of particles per unit volume that have the velocity component of magnitude $|v_{x1}|$. Of these particular particles, a certain number will strike the end wall of area A at $x = \ell$ within the time interval Δt. They obviously must be within a distance $v_{x1}\Delta t$ from the wall if they are to reach the wall in time Δt. Hence, they are located in a volume $(v_{x1}\,\Delta t)(A)$, as shown in Figure 18–6. Since, on the average, one-half of the particles will be traveling away from the wall, the number of particles striking the wall area A in time Δt is:

$$\left(\frac{1}{2}\right)\begin{bmatrix}\text{number of particles per}\\ \text{unit volume with}\\ \text{velocity component } |v_{x1}|\end{bmatrix}\begin{bmatrix}\text{total volume situated close}\\ \text{enough for particles to}\\ \text{strike the wall in time } \Delta t\end{bmatrix}$$

$$= (\tfrac{1}{2})[n_1][(v_{x1}\,\Delta t)(A)] = \frac{n_1 v_{x1} A\,\Delta t}{2} \qquad \textbf{(18–6)}$$

[2] If the particles are true point masses, they cannot make contact with one another (because they are of infinitesimal size). Perhaps a more realistic model would be to assume tiny hard spheres of negligible, though finite, size. It does not make any difference which model we choose for this initial discussion, since elastic collisions of molecules with one another conserve both momentum and energy. Hence, this type of collision in no way alters the overall behavior of the gas.

Each particle will deliver an impulse $F_1\,\Delta t = \Delta p_1$ to the wall equal to $2mv_{x1}$. (The factor of 2 occurs because the velocity component v_{x1} is exactly reversed in the elastic collision process.) Thus the total impulse on the wall due to all particles with this velocity component is:

$$\text{Impulse} = \begin{bmatrix}\text{number striking}\\ \text{the wall in time } \Delta t\end{bmatrix}\begin{bmatrix}\text{impulse delivered by}\\ \text{each individual impact}\end{bmatrix}$$

$$F_1\,\Delta t = \left[\frac{n_1 v_{x1} A\,\Delta t}{2}\right][2mv_{x1}]$$

$$F_1 = n_1 m A v_{x1}^2 \tag{18–7}$$

Since pressure $P_1 = F_1/A$, we have:

$$P_1 = n_1 m v_{x1}^2 \tag{18–8}$$

All the other velocity components are then:

$$P_2 = n_2 m v_{x2}^2$$
$$P_3 = n_3 m v_{x3}^2$$
$$\vdots$$

and the total pressure P due to all velocity components in the x direction is their sum:

$$P = P_1 + P_2 + P_3 + P_4 \cdots \tag{18–9}$$

which, from Equation (18–8), may be written:

$$P = m(n_1 v_{x1}^2 + n_2 v_{x2}^2 + n_3 v_{x3}^2 + \cdots) \tag{18–10}$$

At first glance, Equation (18–10) looks impossible to work with, since at this stage we really do not know any of the terms in the parentheses. But, by making a slight digression, we will show that the quantity in parentheses can be written in a very meaningful and simple way.

How do we calculate the *average value* of a quantity that has a range of different values? For example, what is the average age (to the nearest year) of students in your class? If n_1 students have age a_1, n_2 have age a_2, and so forth, the average age $\bar{a}$ for students in the class is:

$$\text{Average age } \bar{a} = \frac{(n_1 a_1 + n_2 a_2 + n_3 a_3 + \cdots)}{(n_1 + n_2 + n_3 + \cdots)} \tag{18–11}$$

The *bar* over a symbol is used to signify the average value of that parameter.[3]

Equation (18–11) is valid for any quantity that has a range of values, so we apply it to the *x-component of velocity squared* for our ideal gas model:

$$\overline{v_x^2} = \frac{(n_1 v_{x1}^2 + n_2 v_{x2}^2 + n_3 v_{x3}^2 + \cdots)}{(n_1 + n_2 + n_3 + \cdots)} \tag{18–12}$$

The denominator is just the total number of particles per unit volume. Using the symbol $\mathcal{N}$ for this quantity, we have:

$$\mathcal{N}\overline{v_x^2} = (n_1 v_{x1}^2 + n_2 v_{x2}^2 + n_3 v_{x3}^2 + \cdots) \tag{18–13}$$

We may use this expression to rewrite Equation (18–10):

$$P = m\mathcal{N}\overline{v_x^2} \qquad \text{(due to the } x\text{-component only)} \tag{18–14}$$

[3] Equation (18–11) is an example of a *weighted average.* Instead of the simple arithmetic mean of the ages, each age value is multiplied by a *weighting factor* n_i (here it is the number of students who have that particular age value).

From the Pythagorean Theorem in three dimensions, for a single particle:

$$v^2 = v_x{}^2 + v_y{}^2 + v_z{}^2 \qquad \textbf{(18–15)}$$

and, averaging all other particles:

$$\overline{v^2} = \overline{v_x{}^2} + \overline{v_y{}^2} + \overline{v_z{}^2} \qquad \textbf{(18–16)}$$

Because the particle motions are completely random (there is no wind or drift of particles in a particular direction), we also know that

$$\overline{v_x{}^2} = \overline{v_y{}^2} = \overline{v_z{}^2} \qquad \textbf{(18–17)}$$

Therefore, from the above two equations, we conclude:

$$\overline{v_x{}^2} = \tfrac{1}{3}\overline{v^2} \qquad \textbf{(18–18)}$$

We may now write Equation (18–14) as:

$$P = \tfrac{1}{3}\mathcal{N} m\overline{v^2} \qquad \textbf{(18–19)}$$

or

$$P = \tfrac{2}{3}\mathcal{N}\tfrac{1}{2}m\overline{v^2} \qquad \textbf{(18–20)}$$

where the term $\frac{1}{2}m\overline{v^2}$ is the *average translational kinetic energy per particle.* Since $\mathcal{N}$ is the number of particles per unit volume, the total number N in a volume V is:

$$N = \mathcal{N} V \qquad \textbf{(18–21)}$$

Combining the last two equations, we have

$$PV = \tfrac{2}{3}N(\tfrac{1}{2}m\overline{v^2}) \qquad \textbf{(18–22)}$$

Comparing this equation with the equation of state for an ideal gas:

$$PV = nRT$$

we obtain:

$$nRT = \tfrac{2}{3}N(\tfrac{1}{2}m\overline{v^2}) \qquad \textbf{(18–23)}$$

A mole of any substance contains Avogadro's number of molecules, denoted by N_A. If both sides of Equation (18–23) are divided by N_A, we have:

$$\frac{nRT}{N_A} = \frac{2}{3}\frac{N}{N_A}(\tfrac{1}{2}m\overline{v^2}) \qquad \textbf{(18–24)}$$

The factor R/N_A is the universal gas constant expressed for one molecule, known as **Boltzmann's constant** k:

BOLTZMANN'S CONSTANT

$$k = \frac{R}{N_A} = 1.3805 \times 10^{-23} \frac{\text{J}}{\text{K·molecule}}$$

The factor N/N_A is the number of moles, n. Making these substitutions in Equation (18–24) and rearranging,[4] we obtain:

For one molecule $\quad \tfrac{1}{2}m\overline{v^2} = \tfrac{3}{2}kT \quad$ (In terms of microscopic parameters) **(18–25)**

[4] It might be concluded from Equation (18–25) that as $T \to 0$, all molecular motion ceases. This is not true. At very low temperatures, the behavior the particles exhibit is described by quantum mechanics, which reveals that the lowest possible energy state (the *ground state*) is *not* one in which the particles are completely at rest. Instead, as we approach absolute zero, the kinetic energy of the molecules approaches a finite value called the *zero-point energy*.

This equation expresses the important insight that kinetic theory provides. It identifies the *micro*scopic concept of the translational kinetic energy of molecules into the *macro*scopic variable of temperature T. It is interesting that with just a thermometer, one can infer the average translational speeds of ideal gas molecules. Because the factor $\frac{3}{2}k$ is a constant, the mean kinetic energies of various molecules are equal at the same temperature. That is, gases such as H_2, He, and O_2 (which approximate an ideal gas at moderate temperatures and pressures) all have the same average kinetic energy per molecule at the same temperature, even though their masses differ.

Thus the kinetic theory model "explains" the equation of state for an ideal gas with the following correlations. First, the *absolute temperature* (in kelvins) is related to the *translational kinetic energy* of the molecules. Second, the *pressure* exerted by the gas on the walls of its container is the average force per unit area due to the elastic collisions of the molecules against the walls.

EXAMPLE 18–4

Refer to Figure 18–7 and calculate the *root-mean-square* speed, $v_{\text{rms}} = \sqrt{\overline{v^2}}$, of hydrogen molecules (H_2) at room temperature, 20°C. (Note that the average of the squared speed $\overline{v^2}$ is *not* the same as the square of the average speed $\bar{v}^2$.)

SOLUTION

The expression that relates temperature to the average squared speed of molecules is

$$\tfrac{1}{2}m\overline{v^2} = \tfrac{3}{2}kT$$

or

$$v_{\text{rms}} = \sqrt{\overline{v^2}} = \sqrt{\frac{3kT}{m}} \qquad \textbf{(18–26)}$$

The mass of an individual molecule may be found from its molecular weight and Avogadro's number:

$$m = \frac{\text{molecular weight}}{N_A} = \frac{2\,\dfrac{\text{g}}{\text{mol}}}{6.02 \times 10^{23}\,\dfrac{\text{molecules}}{\text{mol}}} = 3.32 \times 10^{-24}\,\frac{\text{g}}{\text{molecule}}$$

$$m = \left(3.32 \times 10^{-24}\,\frac{\text{g}}{\text{molecule}}\right)\left(\frac{1\text{ kg}}{10^3\text{ g}}\right) = 3.32 \times 10^{-27}\,\frac{\text{kg}}{\text{molecule}}$$

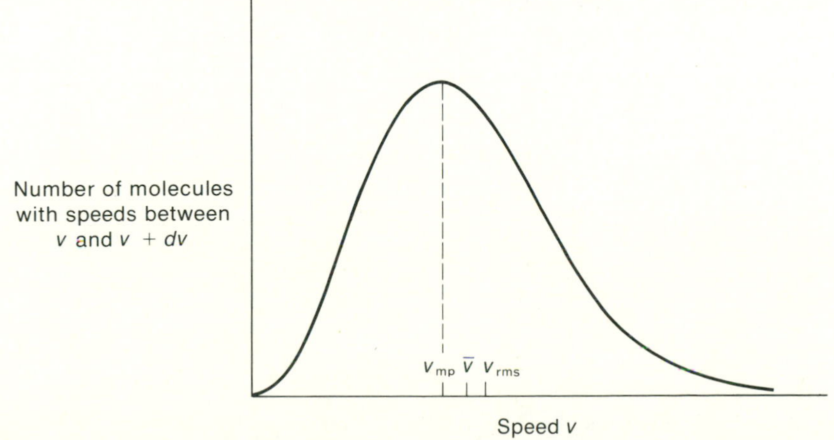

Figure 18–7
The speed distribution for molecules of a gas in thermal equilibrium. The *most probable speed* v_{mp} is the peak of the curve. Note that the average speed $\bar{v}$ is at a slightly higher value than the peak because of the asymmetry of the distribution and that v_{rms} is at a still higher value because the root-mean-square value gives greater weight to the faster speeds. This curve is known as the *Maxwell-Boltzmann distribution,* after the physicists who first devised it.

The other parameters are:

$$k = 1.38 \times 10^{-23} \frac{\text{J}}{\text{K·molecule}}$$

$$T = (20°\text{C} + 273) = 293 \text{ K}$$

Therefore: $$v_{\text{rms}} = \sqrt{\frac{(3)\left(1.38 \times 10^{-23} \dfrac{\text{J}}{\text{K·molecule}}\right)(293 \text{ K})}{3.32 \times 10^{-27} \dfrac{\text{kg}}{\text{molecule}}}} = \sqrt{3.65 \times 10^{6} \frac{\text{J}}{\text{kg}}}$$

Recalling that 1 J = 1 N·m and that the units of newtons are kilograms·meters/second², we have

$$v_{\text{rms}} = \boxed{1910 \frac{\text{m}}{\text{s}}}$$

This velocity is about 50% greater than the speed of sound in hydrogen gas, and somewhat faster than a high-speed rifle bullet.

EXAMPLE 18–5

In terms of *macroscopic* parameters, derive an equation for v_{rms} analogous to Equation (18–26) in the previous example.

SOLUTION

We begin by multiplying both sides of Equation (18–26) by N_A:

$$\tfrac{1}{2}(mN_A)\overline{v^2} = \tfrac{3}{2}(kN_A)T$$

Since $kN_A = R$ and $mN_A = M$ (the mass of one mole), we have

For one mole $$\tfrac{1}{2}Mv^2 = \tfrac{3}{2}RT \qquad \text{(in terms of macroscopic parameters)} \qquad \textbf{(18–27)}$$

or $$v_{\text{rms}} = \boxed{\sqrt{\overline{v^2}} = \sqrt{\frac{3RT}{M}}} \qquad \textbf{(18–28)}$$

EXAMPLE 18–6

The individual speeds of a collection of 10 particles are (in meters/second) 1, 2, 2, 3, 3, 3, 4, 4, 5, and 7. Find (a) the average speed $\overline{v}$ of the particles and (b) the root-mean-square speed $v_{\text{rms}} = \sqrt{\overline{v^2}}$.

SOLUTION

(a) The average speed is:

$$\overline{v} = \frac{\Sigma n_i v_i}{\Sigma n_i} = \frac{[1 + 2(2) + 3(3) + 2(4) + 5 + 7]\dfrac{\text{m}}{\text{s}}}{10} = \boxed{3.40 \frac{\text{m}}{\text{s}}}$$

(b) The root-mean-square speed is:

$$v_{\text{rms}} = \sqrt{\overline{v^2}} = \sqrt{\frac{\Sigma n_i v_i^2}{\Sigma n_i}} = \sqrt{\frac{(1)^2 + 2(2)^2 + 3(3)^2 + 2(4)^2 + (5)^2 + (7)^2}{10}} = \boxed{3.77 \frac{\text{m}}{\text{s}}}$$

Note that v_{rms} is greater than $\overline{v}$. This is always the case, because the average of squared quantities gives greater weight to the larger quantities.

EXAMPLE 18–7

Molecules in the rarefied atmosphere above 600 km (where the temperature is roughly 1500 K) have an appreciable chance of escaping the earth's gravitational attraction without colliding with other molecules provided they exceed the *escape speed*, Equation (14–24): $v_e = \sqrt{2GM/R}$, where G is the universal gravitational constant (6.67×10^{-11} N·m²/kg²), M is the mass of the earth (5.98×10^{24} kg), and R is the distance from the center of the earth ($R_{\text{earth}} = 6.370 \times 10^3$ km). (a) Calculate the escape speed at this altitude. (b) Compare with the root-mean-square speed for hydrogen molecules at 1500 K.

SOLUTION

(a) At an altitude of 600 km, the radius $R = R_{\text{earth}} + 600 \text{ km} = 6370 \text{ km} + 600 \text{ km} = 6970 \text{ km}$.

$$v_e = \sqrt{\frac{2GM}{R}} = \sqrt{\frac{(2)\left(6.67 \times 10^{-11}\,\dfrac{\text{N}\cdot\text{m}^2}{\text{kg}^2}\right)(5.98 \times 10^{24}\text{ kg})}{6.97 \times 10^6\text{ m}}} = \boxed{1.07 \times 10^4\,\frac{\text{m}}{\text{s}}}$$

(b) Since the mass of one mole of hydrogen molecules is $M = 2\,g$, from Equation (18–28) we have

$$v_{\text{rms}} = \sqrt{\frac{3RT}{M}} = \sqrt{\frac{(3)\left(8.31\,\dfrac{\text{J}}{\text{mol}\cdot\text{K}}\right)(1500\text{ K})}{(0.002\text{ kg})}} = \boxed{4.32 \times 10^3\,\frac{\text{m}}{\text{s}}}$$

Although the root-mean-square speed is only about 0.4 of the escape speed, some molecules will have speeds many times the root-mean-square value and therefore will have sufficient energy to escape. In the 4.6 billion years since the formation of the earth, there has been ample time for essentially all of the hydrogen and helium in the earth's primordial atmosphere to diffuse outward from lower altitudes and escape.

Summary

A convenient unit for expressing the amount of a substance is the *mole.*

> *One mole* (mol) of any substance contains as many elementary entities as there are atoms in 0.012 kg of carbon-12. This number is called *Avogadro's number:* $N_A = 6.022\,71 \times 10^{23}$ particles/mole. The entities may be atoms, molecules, ions, electrons, other particles, or specified groups of such particles.

One mole of any gas at 0°C and 1 atm (STP) occupies the volume 22.4 L (0.0224 m³). The mass of one mole expressed in grams is numerically equal to its *molecular weight.* (It is also equal to its *molecular mass*, expressed in *unified atomic mass units* (u), where $1\text{ u} \approx 1.66 \times 10^{-27}$ kg.)

As the pressure is lowered, the behavior of all real gases tends toward the same *equation of state*, expressed by the *ideal gas law:*

$$PV = nRT$$

where P = pressure
V = volume
T = temperature *in kelvins*
n = number of moles
R = universal gas constant (= 8.314J/mol·K = 0.0821 L·atm/mol·K = 1.986 cal/mol·K)

A *kinetic theory* model for an ideal gas (in which molecules are considered point masses that undergo perfectly elastic collisions) leads to useful connections between macroscopic and microscopic concepts and their relations to the kelvin temperature scale:

For one molecule $\frac{3}{2}kT = \frac{1}{2}m\overline{v^2}$

where k is the *Boltzmann constant* ($= R/N_A = 1.3805 \times 10^{-23}$ J/K·molecule).

For one mole $\frac{3}{2}RT = \frac{1}{2}M\overline{v^2}$

where M is the mass of one mole.

In all thermodynamic equations, the temperature T must be expressed in kelvins.

Questions

1. By reference to Figure 18–4(b), give a qualitative explanation for the phrase used by ice skaters: "The ice is slick today."
2. What physical evidence can you offer that atoms exist?
3. What evidence can you give to show that the kinetic theory model for a gas (gas molecules as mass points undergoing elastic collisions) is better than a model that treats gas molecules as fluffy, springy globs that expand to fill the available space?
4. Explain why the extremely large numbers of molecules that are in a container of gas make the kinetic theory analysis much simpler.
5. Even though the average velocity of gas molecules in a container is zero, the average speed is not. Why not?
6. Can you add heat to a monatomic gas without changing the temperature of the gas? What about a diatomic gas? What about a liquid?
7. Refer to Figure 18–7 for the *speed* distribution of molecules of a gas in thermal equilibrium. The *velocity* distribution is usually drawn for one component direction (say, the x direction) and has a much different shape. It is symmetrical about the origin, extending to plus and minus values, and has the familiar bell-shaped, or *Gaussian*, distribution (also known in mathematics as the *error function*). Explain the qualitative differences between these two distributions. Why is one symmetrical about zero and the other asymmetrical about some positive peak value?
8. In Figure 18–7, which speed of a gas molecule, v_{mp}, $\bar{v}$, or v_{rms}, has the average kinetic energy?
9. A few hundred kilometers above the earth's surface, the temperature of the air is over 1000 K. Yet astronauts at that altitude must wear clothes heated by an external power source in order to keep warm. Explain this apparent paradox.
10. Consider a mixture of two different gases (with two different atomic weights). Justify the statement that at thermal equilibrium each kind of gas has the same speed distribution it would have if the other gas were absent (refer to Figure 18–7).
11. If a gas is quickly compressed without adding any heat to it, the temperature of the gas will increase. On the basis of the kinetic theory, explain how the gas molecules acquired their increased speeds.
12. Since the Kelvin temperature scale is related to the mean kinetic energy of gas molecules, does the theory of relativity, with its limit on the fastest possible speed c, imply that there is a maximum highest temperature?
13. Suppose that opposite walls of a container of gas are maintained at different temperatures. Describe the process of heat conduction through the gas.

Problems

In the following problems, assume all gases are ideal.

18A–1

(a) How many electrons are required to make a mass of 1 kg?

(b) How many moles of electrons are there in 1 kg of electrons?

Answers: (a) 1.10×10^{30} electrons (b) 1.82×10^{6} mol

18A–2 The world population in the year 2000 is estimated to be 6.3 billion persons. How many moles of human beings is this?

18A–3

(a) How many moles of water (H_2O) are there in a glass containing 200 g of water (about 7 oz)?

(b) How many water molecules does the glass contain?

Answers: (a) 11.1 mol (b) 6.68×10^{24} molecules

18A–4 Consider a gas under standard conditions. Assume each molecule has a diameter of 10^{-10} m. Imagine that the molecules are in a cubical array, equally spaced from one another along the three coordinate directions. Find the ratio of the distance between adjacent molecules to the size of a molecule.

18A–5 Assume the earth's atmosphere has the same density at all elevations that it has at sea level (1.29 kg/m^3). Find the height of the atmosphere that would give the observed pressure of 1 atm at sea level. (For comparison, Mt. Everest is 8.85 km high.)

Answer: 8.01 km

18A–6 By volume, air is composed of approximately 78% nitrogen (N_2), 21% oxygen (O_2), and 1% other gases. Ignoring the 1% other gases, use these facts to find the mass of a cubic meter of air at STP.

18A–7

(a) Find the number of moles of oxygen gas (O_2) that occupy 244 ft^3 at STP.

(b) Determine the mass of this amount of O_2 gas.

Answers: (a) 308 mol (b) 9.86 kg

18A–8 The pressure in a tank of nitrogen is 2700 lb/in^2 at 70°F. If the nitrogen would occupy 300 ft^3 at 1 atm and 70°F, find the volume of the tank.

18A–9

(a) Find the volume of 200 mol of nitrogen gas (N_2) at STP.

(b) Determine its mass.

Answers: (a) 4.48 m^3 (b) 5.60 kg

18A–10 An automobile tire is inflated to a gauge pressure of 24 lb/in^2 at a temperature of 27°C. After driving awhile, the pressure increases to 34 lb/in^2. Find the temperature of the air inside the tire.

18A–11 If 6 m^3 of hydrogen at STP are compressed in a tank at a pressure of 136 atm and a temperature of 27°C, find the volume of the tank. *Answer:* 48.5 L

18A–12 A large balloon for high-altitude research has a volume of 500 m^3. If helium is available in 40-L tanks at 150 atm pressure, find the number of tanks required to fill the balloon at 1 atm pressure.

18A–13 The temperature of the sun's interior is about 2×10^{7} K.

(a) Find the average translational kinetic energy of a proton in the sun's interior.

(b) Find its root-mean-square speed.

Answers: (a) 4.14×10^{-16} J (b) 7.04×10^{5} m/s

18A–14 Find the kinetic energy of a mole of ideal monatomic gas at 0°C.

18A–15 A fusion reaction with deuterium (molecular mass = 2) occurs only if the kinetic energy of the deuterium is greater than about 1.2×10^{-13} J. Find the temperature required to initiate fusion reactions with deuterium. *Answer:* 5.80×10^{9} K

18B–1 There are approximately 10^{21} kg of water in the earth's surface. Suppose a 12-oz soda-pop can of dye (volume = 32 cm^3, specific gravity = 1.00, and molecular mass = 200)

were thoroughly mixed into all the oceans. How many molecules of dye would a canful of ocean water then contain? *Answer:* on the average, 3.48 molecules

18B–2 The smallest object that can be seen in a microscope using visible light is on the order of one micrometer ($1 \mu m = 10^{-6}$ m). How many atoms of carbon are contained in a cube whose edge is 1 μm long? The atomic mass of carbon is 12 and its density (in graphite form) is 2550 kg/m^3.

18B–3 The smallest object that can be distinguished using an electron microscope is on the order of one nanometer (1 nm = 10^{-9} m). How many atoms of gold are contained in a cube whose edge is 1 nm long? The atomic mass of gold is 197 and its specific gravity is 19.3. *Answer:* about 59 atoms

18B–4 About the best vacuum that can be obtained in the laboratory is 10^{-10} mm of Hg. How many gas molecules per cubic centimeter does this vacuum contain at room temperature (27°C)?

18B–5 The dimensions of a desk top are 3 ft × 6 ft. Find the force (in pounds) on the desk top due to the atmosphere. *Answer:* 38 100 lb, or about 19 tons

18B–6 Recall from Problem 18A-6 that by volume, air is composed of approximately 78% nitrogen (N_2), 21% oxygen (O_2), and 1% other gases. Ignoring the 1% other gases, use these facts to calculate the lifting force on a helium-filled balloon with a volume of 1 m^3 at a pressure of 1 atm.

18B–7

(a) Find the pressure of 10 mol of carbon dioxide gas (CO_2) stored in a 500-L tank at 25°C.

(b) Find the density of the gas in the tank.

Answers: (a) 0.489 atm (b) 0.888 kg/m^3

18B–8 There are roughly 10^{59} nuclear particles (neutrons and protons) in an average star and about 10^{11} stars in a typical galaxy. Galaxies tend to form in clusters of (on the average) about 10^3 galaxies, and there are about 10^9 clusters in the known part of the universe.

(a) Approximately how many neutrons and protons are there in the known universe?

(b) Suppose all this matter were compressed into a sphere of nuclear matter such that each nuclear particle occupied a volume of 10^{-45} m^3 (about the volume of a neutron or proton). What would be the radius of this sphere of nuclear matter?

(c) How many moles of nuclear particles are there in the observable universe?

18B–9 An automobile tire contains air at 26 lb/in^2 (gauge pressure) at 27°C. What would be the gauge pressure at 60°C? *Answer:* 27.8 lb/in^2

18B–10 Make a freehand sketch of a pressure-temperature diagram (similar to Figure 18–4b) for a substance that contracts upon freezing. Do not include numerical values, but show the correct qualitative behavior of the lines that separate the different phases.

18B–11 On a day when the atmospheric pressure is 1 atm and the temperature is 20°C, a diving bell in the shape of a cylinder 4 m tall, closed at the upper end, is lowered into water to aid in the construction of an underground foundation for a bridge tower. The water inside the diving bell rises to within 1.5 m of the top and the temperature drops to 8°C.

(a) Find the air pressure inside the bell.

(b) How far below the surface of the water is the bell located? (In actual use, additional air is pumped in, forcing the water out to provide working space for the construction workers.) *Answers:* (a) 2.56 atm (b) 15.7 m

18B–12 Using the principles of mechanics (not thermodynamics) derive a conversion factor relating the two energy units *joule* and *liter-atmosphere*. Verify your relation by converting the numerical value of the gas constant R using one of the units to its value using the other unit.

18B–13 Oxygen is sold commercially in a tank that contains 1.63 ft^3 of the gas under 2200 lb/in^2 pressure when the temperature is 70°F. What would be the volume of oxygen at 1 atm pressure and 70°F? *Answer:* 244 ft^3

18B–14 A bubble of air, 0.20 cm in diameter, is formed at the bottom of a lake where the temperature is 4°C. It rises 25 m to the surface, where the water temperature is 24°C.

Assuming the air in the bubble always has the temperature of the surrounding water, find the size of the bubble just as it reaches the surface. The atmospheric pressure is 1 atm.

18B–15 Consider an ideal gas of density ρ and pressure P.

(a) Show that the root-mean-square speed v_{rms} of the gas molecules is $\sqrt{3P/\rho}$.

(b) Air under standard conditions has a density of 1.29 kg/m^3. Treating air as a hypothetical, ideal gas of molecular mass M, show that the equivalent molecular mass of air is 28.9. (Hint: Remember that 1 mol of an ideal gas occupies 22.4 L under standard conditions.)

18B–16 At what temperature is the root-mean-square speed of oxygen molecules (O_2) equal to the escape velocity from the earth?

18B–17 The rectangular box sketched in Figure 18–8 contains a single atom of mass m and speed v traveling only in the $\pm x$ direction. The atom makes perfectly elastic collisions with the end faces (area = ℓ^2). Starting with fundamental principles, derive an expression for the average pressure P_{ave} on the end faces in terms of m, v, and ℓ.

Answer: $mv^2/3\ell^3$

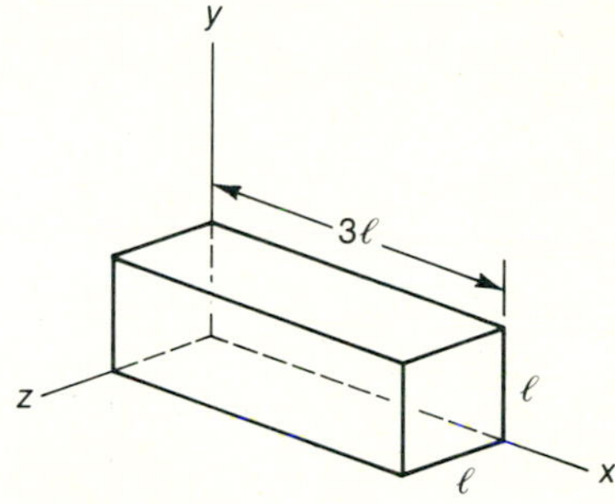

Figure 18–8
Problems 18B–17 and 18B–20

18B–18 A group of particles has the following speeds (in m/s): 3, 4, 6, 6, 7, 8, 8, 8, 10, and 13. Find (a) the average speed and (b) the root-mean-square speed.

18B–19

(a) Find the root-mean-square speed of hydrogen molecules (H_2) under standard conditions.

(b) Assume that all molecules have the root-mean-square speed and that one-third travel along the $\pm x$ direction, one-third along the $\pm y$ direction, and one-third along the $\pm z$ direction. How many collisions per second are there on a wall area of 1 cm^2?

Answers: (a) 1.84×10^3 m/s (b) 8.20×10^{23} collisions/second

18B–20 The box in Figure 18–8 contains 6×10^{12} hydrogen molecules (H_2). Make the simplifying assumption that one-third of the atoms travel only along the $\pm x$ direction, one-third along the $\pm y$ direction, and one-third along the $\pm z$ direction. Also assume that all atoms travel with the root-mean-square speed associated with 27°C. Find the pressure on the wall in terms of ℓ (SI units).

18B–21 It has been estimated that if the root-mean-square speed of an air molecule is greater than about 20% of the escape speed, there has been sufficient time since the formation of the earth for those molecules to have escaped from our atmosphere.

(a) Calculate the root-mean-square speed for nitrogen (N_2) molecules at a temperature 1500 K, the approximate temperature at an altitude of 600 km where a molecule has an appreciable chance of escaping without colliding with other molecules in the air.

(b) Compare the root-mean-square value with the minimum escape speed of 10.7 km/s at that altitude.

Answer: (b) 9.3% of the escape speed

18C–1 A 40-lb piston (frictionless) is placed on top of a circular cylinder containing an ideal monatomic gas at 27°C and 1 atm (see Figure 18–9). The piston compresses the gas somewhat as it sinks down to an equilibrium position below the top of the cylinder.

(a) How far down from the top does the piston sink if the gas is maintained at 27°C?

(b) To what temperature must the gas be raised in order to restore the piston to the top of the cylinder?

Answers: (a) 1.04 in (b) 329 K = 56°C

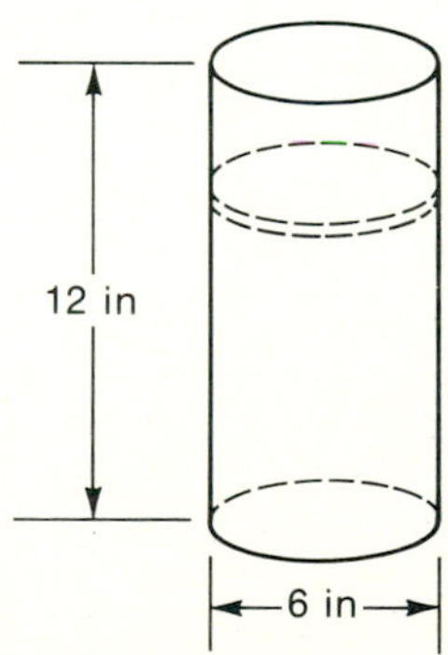

Figure 18–9
Problem 18C–1

18C–2 In the previous problem, suppose the gas is compressed even more by pouring mercury on top of the piston. Find the maximum depth of mercury before it starts to spill over the top. The gas is maintained at 27°C during the process. (Assume the piston has negligible thickness.)

18C–3 A beam of gas molecules can be formed by vaporizing a substance in a heated oven and allowing the gas molecules to emerge from a small hole into an evacuated region (see Figure 18–10). The speed of the molecules in the beam can be measured by two rotating disks, separated a distance x, that have slits in them to allow molecules of a certain speed to pass through both slits to a detector. If the offset angle between the two slits is θ, find the angular speed ω of the disks that will detect molecules of speed v.

Answer: $\omega = v\theta/x$

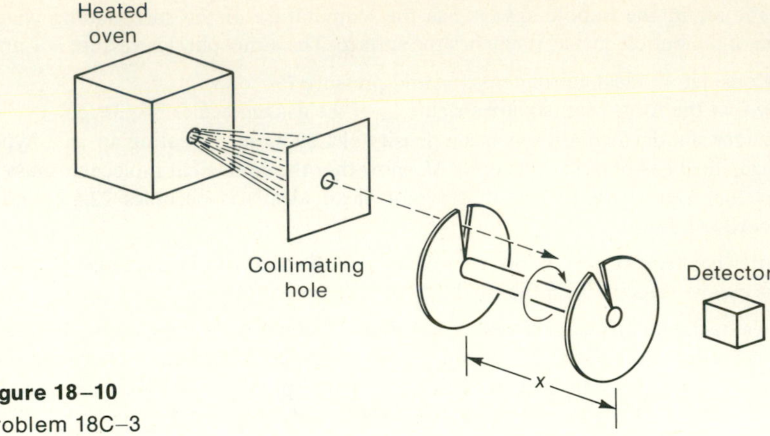

Figure 18–10
Problem 18C–3

18C–4 In the previous problem, consider a beam of mercury atoms (molecular mass = 200). If the disks are separated 20 cm and their rotation speed is 100 rev/s, what offset angle θ will detect mercury atoms with the root-mean-square speed for a temperature of 800 K? *Answer:* 0.399 rad

18C–5 Consider the apparatus in Problem 18C–3. The disks are separated a distance of 40 cm and they rotate at 100 rev/s. Suppose the offset angle between the centers of the slits is set to detect a molecular speed of 400 m/s. The beam is 8 cm from the disks' axis of rotation, and the slit through which the beam passes is 2 mm wide. If a range of molecular speeds is present, what are the minimum and maximum speeds of molecules that can reach the detector? *Answer:* 385 m/s, 417 m/s

The energy produced by breaking down the atom is a very poor kind of thing. Anyone who expects a source of power from the transformation of these atoms is talking moonshine.

ERNEST RUTHERFORD, 1933
(five years before fission was accidentally discovered by Hahn and Strassman)

chapter 19 The First Law of Thermodynamics

We are at a transition point in our study of heat. Having introduced the concept of temperature and discussed the changes that occur in objects when heat is added or taken away, we now approach the "heart" of the matter. The most important fact of all is that *heat is just another form of energy*. The mechanical equivalent of heat (in the form of work) opens up the possibility of getting useful work in exchange for a given quantity of thermal energy. This conversion is what makes our present-day civilization possible. The industrial revolution of the late eighteenth and nineteenth centuries, with its proliferation of engines and machines, stimulated numerous studies in that branch of physics known as *thermodynamics*. (The name itself links together the two concepts, heat and mechanical work.) Among the interesting conclusions that came out of these studies were:

(1) It is impossible to construct a perpetual motion device.
(2) It is impossible to build an engine operating in a cycle that is 100% efficient in converting heat to work.

Negative statements such as these—expressing the impossibility of doing something—are among the most powerful principles in science. We will have more to say about them later.

19.1 Basic Concepts

Physics students are always being told to "choose a system." The choice is crucial, since it defines the portion of the universe we wish to study and enables us to learn how the "outside world" affects that system. The word **system** is used in a very general sense: It may designate any type of engine or machine, a combination of gas and liquid enclosed in a container, or a quantity of matter in the center of the sun. We always imagine a well-defined boundary enclosing the system, because we must keep account of energy transfers across the boundary between the system and its surroundings, whether by the flow of heat or by the performance of work.

To simplify the discussion, we shall pick one particular system as a concrete example: n moles of an ideal gas enclosed in a cylinder with a movable, frictionless piston, as illustrated in Figure 19–1. Our system is the gas itself. Because the piston is frictionless, it may move without generating heat in the walls of the cylinder.

The cylinder is constructed to be a good heat insulator, so no heat leaks in or out of the gas unnoticed. However, we may purposely cause some heat transfer by placing the end of the cylinder in **thermal contact** with an external hot (or cold) object called a *heat reservoir*, shown in Figure 19–2. Heat flows in or out of the gas because of temperature differences between the system and the heat reservoir. In addition, the gas may do work on the "outside world" by pushing the piston outward against atmospheric pressure or some other external force. Conversely, work may be done on the system by an outside force if it compresses the gas by moving the piston inward. The direction of the work "flow" will depend on whether the piston moves inward or outward.

The physical state of a thermodynamic system may be described by the parameters in an **equation of state** (discussed in the previous chapter). For our specific case of n moles of an ideal gas in a cylinder, the parameters are the pressure P, the volume V, and the absolute temperature T in kelvins. The equation of state is $PV = nRT$. Other systems will have different parameters depending on their properties. For example, a rubber band may be described in terms of temperature, tension, and length.

An equation of state exists for our system only if all the measurable properties (such as P and T) remain static throughout the system and do not change with time. Such a condition is known as **thermodynamic equilibrium.** The system will approach this equilibrium condition by itself through heat conduction and gas flow within the system, provided it is isolated from all outside influences. We will exclude effects of gravity, since this would cause both the pressure and the density within the gas to change with elevation, as illustrated by the earth's atmosphere. We purposely limit our considerations to small systems for which such gravitational effects are negligible.

One aspect of thermodynamic equilibrium concerns the conditions under which no heat transfer will occur. If two bodies with different temperatures are placed in thermal contact, heat will flow until they are in **thermal equilibrium** at the same (intermediate) temperature, after which no more heat transfer takes place. Carrying this one step further forms the basis for the **zeroth law of thermodynamics.**[1]

[1] The first and second laws of thermodynamics were developed before it was realized that to be logically consistent, a prior statement regarding thermal equilibrium and temperature was needed. Thus it became necessary to backtrack one step in numerical order to give the statement its proper priority.

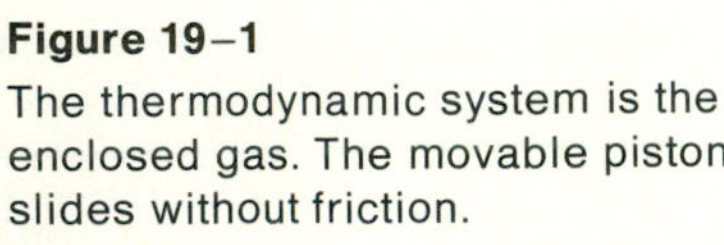

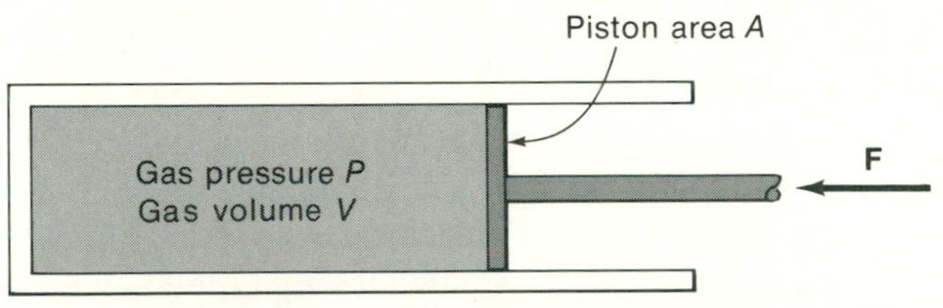

Figure 19–1
The thermodynamic system is the enclosed gas. The movable piston slides without friction.

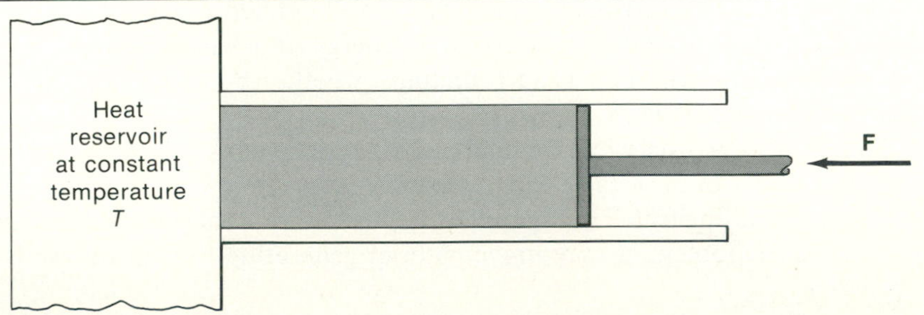

Figure 19–2
The end of the cylinder has a heat-conducting wall through which heat can be transferred to or from the gas by placing it in contact with a heat reservoir.

ZEROTH LAW OF THERMODYNAMICS **If each of two systems is in thermal equilibrium with a third system (which may be a thermometer), then the two systems are in thermal equilibrium with each other.**

The important concept of the zeroth law is that the *temperature* of a system is the parameter by which one measures the conditions of thermal equilibrium with another system, thus ensuring that no heat transfer will subsequently occur.

19.2 Heat, Energy, Work, and the First Law

We are now ready to discuss the first law of thermodynamics. It may seem to you an obvious relation, one that hardly merits the honor of being named "the first law." Yet it is truly an important foundation stone in the structure of physics.

The conservation of mechanical energy is a similarly important principle that you may now feel seems obvious. Yet if you read the history of the development of this principle you will discover that for centuries it eluded some of the greatest thinkers in science. Energy is a subtle concept in that it is never measured directly. It resides in numerous forms: There is kinetic energy of motion of a mass, potential energy due to the relative positions of two or more masses that interact with each other, energy transfers in the form of work, and so forth. These are forms that involve the motion or position of the center of mass of objects. But we are also familiar with other types of energy: chemical energy residing in a battery or in a firecracker and thermal energy, which is the random motion of atoms and molecules. These latter examples are called the **internal energy** of a system, since they are due to the random motions or configurations of atoms and molecules on a microscopic scale.

What the first law does is to incorporate these internal energy forms into a general conservation relation with other forms of energy. The particular type of internal energy is not important. It may be thermal energy, or energy stored in a tightly wound spring (which, of course, is traceable to the increased potential energies of atoms and molecules when their mutual distances are altered). All these forms are lumped together under the symbol U, which represents the internal energy of the system.

The first law is a generalization of the results of all the experiments done where a strict accounting is made of the energy interchanges that occur. It cannot be proved or derived, but it expresses our observations of nature's behavior when we deal with energies in all forms: That is, energy is conserved. This first law is:

$$\begin{bmatrix}\text{Heat added to}\\ \text{the system}\end{bmatrix} = \begin{bmatrix}\text{Change in the internal}\\ \text{energy of the system}\end{bmatrix} + \begin{bmatrix}\text{Work done on the external}\\ \text{world by the system}\end{bmatrix}$$

Using the notation *i* and *f* for *initial* and *final* values, we write the equation symbolically as:

$$Q = U_f - U_i + W$$

or

FIRST LAW OF THERMODYNAMICS

$$Q = \Delta U + W \qquad \textbf{(19–1)}$$

where Q is the heat added to the system, $\Delta U = (U_f - U_i)$ is the change in the internal energy of the system, and W is the work done by the system on the outside world. From the equation, we can see that the heat Q added to a system must appear as some other form of energy, either as increased internal energy or as work the system does on the outside world, or both. No heat is lost.

Note carefully that only *changes* in internal energy are specified; nothing is said about the total internal energy residing in a system. Also keep in mind the direction of the energy flow. As defined, Q is positive when heat is *added to* the system, W is positive when the system does work *on* the external world, and ΔU is positive (and thus the internal energy *increases*) if the final value U_f is *greater* than the initial value U_i. If any of these changes occur in the opposite direction, the corresponding *numerical* value will be negative. (Of course, we never change the sign in the general statement of the first law itself.)

EXAMPLE 19–1

A solar panel with an area of 1.5 m^2 generates electrical power to drive an electric hoist. When the panel is directed toward the sun, it absorbs 40% of the solar power of 0.84×10^3 W/m^2 falling on it. (The remaining 60% is reflected or radiated away.) If the hoist lifts a 20-kg mass with a speed of 1.7 m/s, find the rate at which the internal energy of the panel-hoist system is changing.

SOLUTION

Energy is added to the system at the rate of dQ/dt; the internal energy of the system changes at the rate dU/dt; and the system does work at the rate dW/dt. Applying the first law, $\Delta U = Q - W$, we have

$$\frac{dU}{dt} = \frac{dQ}{dt} - \frac{dW}{dt}$$

Substitution of the appropriate numerical values yields

$$\frac{dU}{dt} = (0.40)\left(0.84 \times 10^3\ \frac{\text{W}}{\text{m}^2}\right)(1.5\ \text{m}^2) - (20\ \text{kg})\left(9.8\ \frac{\text{m}}{\text{s}^2}\right)\left(1.7\ \frac{\text{m}}{\text{s}}\right)$$

$$= 504\ \text{W} - 333\ \text{W} = \boxed{171\ \text{W}}$$

This rate of internal energy change results in an increase of temperature of both the solar panel and the hoist motor, which further results in a decreased efficiency. In practical application, a means of cooling the panel and motor would be provided.

19.3 Reversible and Irreversible Processes

We bring about changes of state in a system by adding or subtracting heat from it, or by doing work on it, or by causing the system to do work on the outside world. Let us consider these processes in detail, keeping in mind our ideal gas system as a concrete example.

Heat will flow if there is a temperature difference between the system and its surroundings, which is accomplished by placing the system in thermal contact with an external heat reservoir. (An auxiliary refrigerator or heater will maintain the desired temperature in the reservoir.) If the temperature difference between the system and the reservoir is *nonzero*, heat spontaneously flows from the hotter region to the cooler one until thermal equilibrium is established between the two bodies. From that point on, the system remains in equilibrium with the reservoir. Such a process is called **irreversible** because the change occurs only in one direction, which drives it toward the equilibrium condition. All natural processes are irreversible.

If the temperature differences are *infinitesimal* (very small, approaching zero), the direction of heat flow could be reversed by an infinitesimal change in the temperature of the reservoir. This more idealized process is called **reversible.** In it, the system never deviates away from equilibrium conditions except by infinitesimal amounts, which may be treated using the methods of calculus. The advantage of this type of process is that because the system remains essentially in equilibrium, the *equation of state applies at all times during the reversible process.* The system is, in effect, carried through a sequence of equilibrium states, in what is sometimes called a *quasi-static process.* So we can manipulate the equation of state mathematically to predict how the system ought to behave, thus comparing theory with experiments.

All processes we bring about with real physical equipment are irreversible because they involve *finite* differences of temperature and pressure. But with care, reversible processes can be approximated with a real apparatus if temperature and pressure differences are made arbitrarily small. This generally means the changes proceed at a very slow rate (though not all slow processes are reversible).

A reversible process does not necessarily mean the system is changed from state 1 to state 2, and then the reverse path is taken back to state 1. But such reversibility *could* be brought about, thus restoring the system and the external world to exactly their original conditions, provided there is no friction of any sort. In our ideal gas system, if the moving piston had friction its motion would transform some energy into thermal energy. This thermal energy would be in the form of random kinetic and potential energies of molecular motions in the walls and piston. As we shall see, we cannot convert these random motions *completely* back into the "organized" motion of the piston.

□ □ □

The concepts we applied to our example of a gas in a cylinder could be applied to *any* system: a battery that causes a resistor to heat because of an electric current, the surface film on a bubble that contracts or expands, a block of lead that strikes the floor after falling under the action of gravity, a mixture of various compounds undergoing a chemical reaction, a magnetic substance in the presence of a varying magnetic field, and so forth. Although the parameters of the appropriate equation of state will vary from case to case and may involve many more than just the three in our example, the thermodynamic considerations are equally valid for all well-defined systems.

19.4 Analysis of Specific Processes

We shall now examine some specific processes that will tie together all the concepts just discussed. We will carry our system of an ideal gas through a variety of processes, changing P, V, and T as heat is added (or taken away) and as work is done on the system (or by the system on the outside world).

Most of the processes we will consider are *reversible*, so the equation of state applies at all times. In effect, we travel along a path on the pressure-volume-temperature surface, as sketched in the example of Figure 19–3(a). For all cases, we will diagram the journey on the convenient pressure-volume projection of Figure 19–3(b).

It is helpful to give characteristic names to certain common processes:

iso<u>thermal</u> change	The temperature is held constant. (Greek: *isos*, "equal" or "identical," and *therma*, "heat.")
iso<u>volumic</u> change	The volume is held constant. (This occasionally is called an iso*choric* change. Greek: *choras*, "a place" or "space.")
iso<u>baric</u> change	The pressure is held constant. (Greek: *baros*, "weight" [of the air].)
<u>adiabatic</u> change	No heat transfer occurs. (Greek: *adiabatos*, "not passable.") All "sudden" processes are adiabatic ones, since heat flow usually requires a finite amount of time to occur.

Of course, there are an infinite number of possible processes that do not fall into any of these categories. However, these four are singled out as common (and mathematically easy) processes to investigate. For each, we will perform the process *reversibly*, so the equation of state applies at all times. Let us take them up in turn.

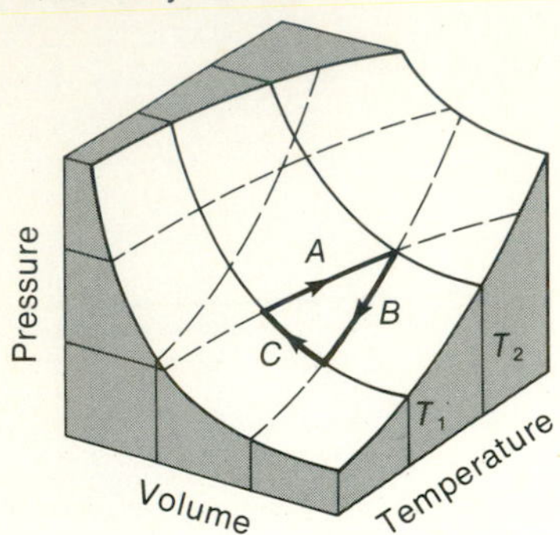

(**a**) The pressure-volume-temperature surface for an ideal gas. The surface is defined by the equation PV = (constant)(T). Lines of constant P, constant V, and constant T are drawn on the surface.

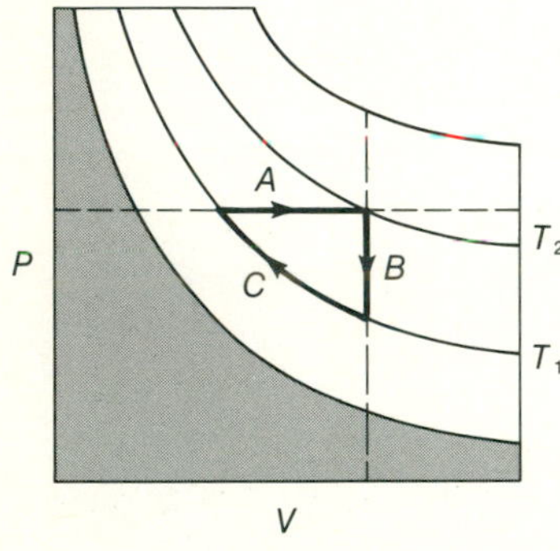

(**b**) A projection of the surface on the pressure-volume plane. The same three processes sketched in (**a**) are shown. The lines of constant temperature (isotherms) form a family of hyperbolas, as is evident from the equation PV = (constant).

Figure 19–3

A series of processes on the pressure-volume-temperature surface. Process A is an expansion of the gas at constant pressure. Process B reduces the pressure of the gas at constant volume (by cooling). Process C compresses the gas, keeping the temperature constant.

Isothermal Expansion

For this case, we hold the temperature constant as the parameters change from the initial values P_i and V_i to the final values P_f and V_f. The system is thus carried along an isothermal line as shown in Figure 19–4(a). Since the internal energy U is associated with the random motions of atoms and molecules, for an ideal gas U depends only on temperature. (Recall the result from kinetic theory analysis for the average translational kinetic energy of one mole of gas: $(1/2)M\overline{v^2} = (3/2)RT$.) Therefore, for an isothermal process, $\Delta U = 0$ because the temperature does not change.[2]

As the gas expands, it performs work on the external world. This work energy must come from some place. Unless the internal thermal energy of the gas is replaced, the gas will cool down. (On a microscopic scale, the molecules rebound from the retreating piston with decreased velocities.) So, to maintain constant temperature, we add heat to the gas during the process by keeping the cylinder in thermal contact with a heat reservoir. As the piston moves outward, the force the gas exerts on the piston is $F = PA$. Hence the gas does work:

$$\Delta W = \mathbf{F} \cdot \mathbf{\Delta x} = PA\,\Delta x = P\,\Delta V \qquad \textbf{(19–2)}$$

Using calculus notation:

$$W = \int_{V_i}^{V_f} P\,dV \qquad \textbf{(19–3)}$$

Note that this work is the area under the path on a pressure-volume diagram.

[2] These statements are true for *monatomic* gases, whose atoms approximate the mass points of the kinetic theory model. *Diatomic* gases, whose molecules may have internal vibrations, are a slightly different story, as we will see in the next section.

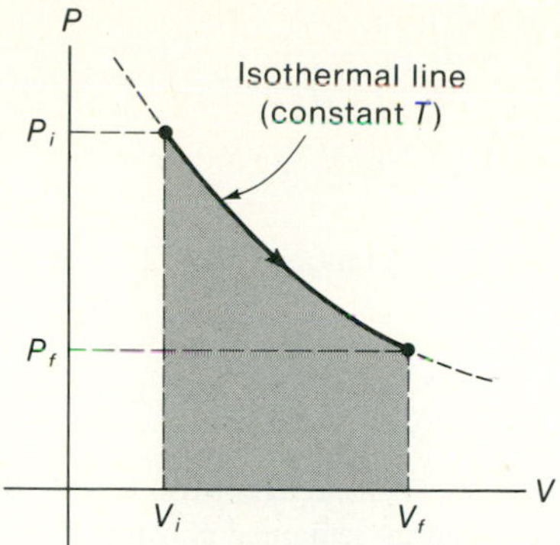

(**a**) The shaded area under the curve represents the work *W* done *by* the system *on* the external world. (If the process occurred in the opposite direction, *W* would by a negative number representing the work done *by* the external world *on* the system.)

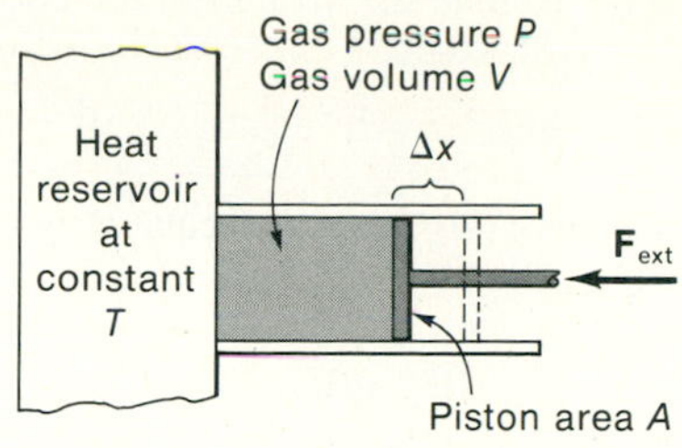

(**b**) To maintain constant temperature, the cylinder is kept in thermal contact with a heat reservoir. In moving the piston a distance Δx against the external force, the gas does work: $\Delta W = F\,\Delta x = PA\,\Delta x = P\,\Delta V$.

Figure 19–4
An isothermal expansion

The ideal gas law furnishes the mathematical relationship between P and the variable V, namely, $P = nRT/V$. Hence:

$$W = \int_{V_i}^{V_f} \frac{nRT}{V}\,dV = nRT \int_{V_i}^{V_f} \frac{dV}{V} \tag{19–4}$$

$$W = nRT \ln\left(\frac{V_f}{V_i}\right)$$

Keeping in mind that $\Delta U = 0$ for an isothermal change, we apply the first law of thermodynamics:

$$Q = \Delta U + W$$

$$Q = 0 + \int P\,dV$$

ISOTHERMAL PROCESS

$$Q = nRT \ln\left(\frac{V_f}{V_i}\right) = W \tag{19–5}$$

Thus, the heat Q added to the gas is converted entirely to work W done by the system on the external world.

If the process were carried out in the reverse direction, the external world would do work on the system (to compress the gas) and heat would have to be transferred to the reservoir to keep the gas temperature constant. In this case, the numerical values of both Q and W would be negative, signifying energy transfers in the directions opposite to those in the fundamental definitions of these quantities. This sign convention may seem trivial in this example. However, it is extremely important to observe it in every case, particularly in the more complex examples we discuss in later sections.

EXAMPLE 19–2

How much work is required to compress isothermally 2 g of oxygen, initially at STP, to one-half its original volume? (Assume the oxygen behaves as an ideal gas.)

SOLUTION

We can find the work from Equation (19–5):

$$W = nRT \ln\left(\frac{V_f}{V_i}\right)$$

Since oxygen (O_2) has a molecular weight of 32, the number of moles is $n = 2/32 = 0.0625$. Thus:

$$W = (0.0625 \text{ mol})\left(8.31 \frac{\text{J}}{\text{mol}\cdot\text{K}}\right)(273 \text{ K}) \ln\left(\frac{1}{2}\right) = \boxed{-98.3 \text{ J}}$$

The minus sign verifies that the work was done by some external agent on the gas. Since the temperature of the gas did not change, its internal energy remained constant. To keep the process isothermal, 98.3 J of heat had to be transferred from the gas to the heat reservoir.

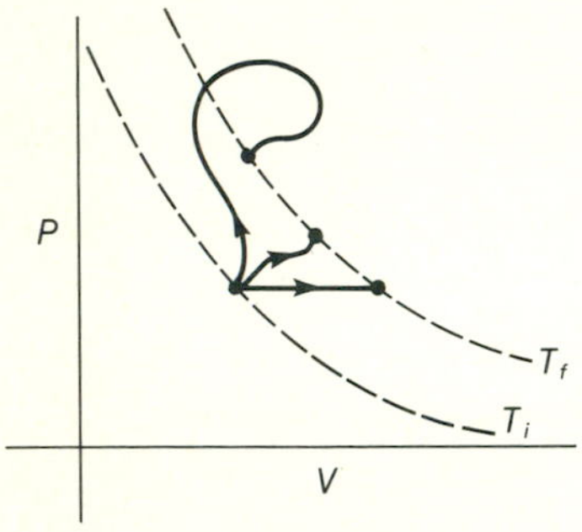

Figure 19–5
In changing the temperature from T_i to T_f, we may travel along various paths. Since the area under the path represents the work done, each path involves a different amount of work. However, ΔU is the same for each, so (from the first law) Q must be different for each path.

Specific Heats of an Ideal Gas

How much heat is required to raise the temperature of a given amount of gas from T_i to T_f? One mole is a convenient quantity to consider. We will do this reversibly, meaning that the process involves only infinitesimal differences of temperature and pressure at all times. To bring about a temperature difference, we can travel along a variety of paths from one isotherm to another, as shown in Figure 19–5. In fact, we have an infinite number of paths to choose from. Since the change of temperature ΔT is the same for each path, *the internal energy change ΔU is the same for all cases.* But the first law, $Q = \Delta U + W$, implies that the *amount of heat Q* required for each path will be different, because W (the area under the various curves) is different for each path. Suppose we want to define a meaningful molar specific heat C.[3] How can we do this when the amount of heat Q required to bring about a given temperature change is not a single, unique amount? In general, each path requires a different amount of heat to bring about the same temperature change.

To resolve the difficulty, we pick two particular paths that occur frequently in various processes, namely, changes at *constant volume* and at *constant pressure.* This results in two distinct molar specific heats for gases (instead of an infinite number of values). Their designations are as follows:

C_V = molar specific heat at *constant volume*

C_P = molar specific heat at *constant pressure*

The defining equations for the **molar specfic heats** are:

MOLAR SPECIFIC HEATS C_V AND C_P

$$Q = nC_V\,\Delta T \quad \text{(constant volume)} \tag{19–6}$$

$$Q = nC_P\,\Delta T \quad \text{(constant pressure)} \tag{19–7}$$

where n is the number of moles.

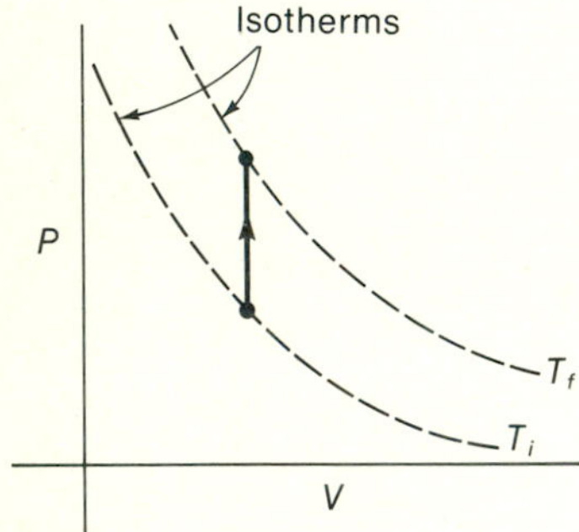

Figure 19–6
An isovolumic change involves C_V. No work is done by the system on the external world because the piston does not move.

Isovolumic Process

Let us next discuss the constant-volume process, as illustrated in Figure 19–6. The volume does not change, so $W = 0$ and (from the first law):

$$Q = \Delta U + W$$

$$nC_V\,\Delta T = \Delta U + 0 \tag{19–8}$$

[3] Recall that the specific heat c for solids and liquids is defined from $Q = mc\,\Delta T$. In these cases, one needs extremely high pressures to keep the volume from increasing as T increases. So experimental values are usually at constant pressure (1 atm).

This equation expresses the internal energy for a temperature change ΔT. But because the internal energy for an ideal gas depends *only* on the temperature, ΔU has the same value for the same ΔT *regardless of the process.* It may be at constant volume, constant pressure, or any other process that carries the gas through the same temperature difference. Hence, for a volume and temperature change at constant *pressure*, where the work done is $P\,\Delta V$, we have:

$$Q = \Delta U + W$$

$$nC_P\,\Delta T = nC_V\,\Delta T + P\,\Delta V \qquad \textbf{(19–9)}$$

From the ideal gas law, we have (for constant pressure):

$$P\,\Delta V = nR\,\Delta T$$

Combining the above two equations, we obtain the useful relation:

$$C_P - C_V = R \qquad \textbf{(19–10)}$$

It is reasonable that C_P is larger than C_V. As heat is added to a gas at constant pressure, the gas must do work in expanding that requires a larger amount of energy to bring about a given change in internal energy. In contrast, as heat is added at constant volume, no work is done by the gas, so less heat energy is required to produce the same change in internal energy. Equation (19–10) has been experimentally confirmed for real gases.

We emphasize again that regardless of what type of process the ideal gas is carried through, the change in internal energy is $\Delta U = nC_V\,\Delta T$, even though the process may not be at constant volume. Thus the molar specific heat at constant volume is

$$C_V = \frac{1}{n}\frac{dU}{dT} \qquad \textbf{(19–11)}$$

Isobaric Expansion

This process is one in which the pressure remains constant (see Figure 19–7). In our example, if we wish to keep the pressure constant as the piston moves outward, we find that the temperature of the gas must be increased. This is obvious from the equation of state: $PV = nRT$. At constant pressure, the volume of a given quantity of gas is directly proportional to its temperature. On a microscopic scale, we can see that as a gas becomes more diluted in a larger volume, molecular impacts on the walls would become less frequent, causing a drop in pressure. However, by increasing the temperature we increase the speed of the molecules, thus compensating for this effect.

The work W done by the system on the external world is:

$$W = \int_{V_i}^{V_f} P\,dV = P\int_{V_i}^{V_f} dV$$

$$W = P(V_f - V_i) \qquad \textbf{(19–12)}$$

Since for any process involving a temperature change ΔT, the change in internal energy ΔU is:

$$\Delta U = nC_V\,\Delta T \qquad \textbf{(19–13)}$$

the first law gives the following:

$$Q = \Delta U + W$$

ISOBARIC PROCESS

$$nC_P\,\Delta T = nC_V\,\Delta T + P(V_f - V_i) \qquad \textbf{(19–14)}$$

EXAMPLE 19–3

Suppose the gas of Example 19–2 (2 g of oxygen initially at STP) were compressed isobarically to one-half its original volume. (a) How much work is required for the compression? (b) What temperature change occurs in this process?

SOLUTION

(a) The work is the shaded area of Figure 19–7. Since P is constant, it equals:

$$W = \int_{V_i}^{V_f} P\,dV = P(V_f - V_i)$$

Because $V_f = V_i/2$, we have

$$W = P\left(\frac{V_i}{2} - V_i\right) = -\frac{PV_i}{2}$$

We thus need to find the initial volume of 2 g of O_2 at STP. From the equation of state:

$$P_iV_i = nRT_i$$

$$V_i = \frac{nRT_i}{P}$$

Therefore:

$$W = -\frac{PV_i}{2} = -\frac{\cancel{P}nRT_i}{2\cancel{P}} = -\frac{(\frac{2}{32}\text{ mol})\left(8.31\ \frac{\text{J}}{\text{mol·K}}\right)(273\text{ K})}{2} = \boxed{-70.9\text{ J}}$$

The minus sign signifies work was done by the external world on the gas.

(b) From Equations (19–10) and (19–14), in terms of temperature differences the work done on the gas is:

$$W = nR\,\Delta T = nR(T_f - T_i)$$

Solving for T_f and substituting numerical values gives

$$T_f = \frac{W}{nR} + T_i = \frac{(-70.9\text{ J})}{(\frac{2}{32}\text{ mol})\left(8.31\ \frac{\text{J}}{\text{mol·K}}\right)} + 273\text{ K} = -137\text{ K} + 273\text{ K} = \boxed{136\text{ K}}$$

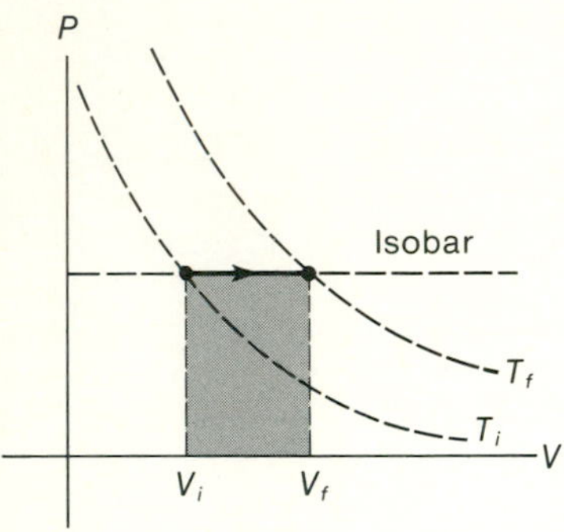

Figure 19–7
An isobaric expansion involves C_P. The shaded area under the path represents the work done by the system on the external world. (An isobaric compression would occur in the reverse direction, with the area then representing the work done by the external world on the gas.)

Adiabatic Process

In this process, no heat is transferred in or out of the system. For our ideal gas, we thus remove the heat reservoir from contact with the cylinder. (In actual practice, the apparatus may be surrounded with some insulating material, such as cork, asbestos, or firebrick.) Also, since heat flow is a relatively slow process, if we carry out the process rapidly, it will be adiabatic. For example, in the operation of a gasoline engine, the compression stroke is an adiabatic compression of the mixture of gasoline vapor and air. Similarly, during the power stroke the combustion products are exhausted to the atmosphere in an adiabatic expansion. Other cases of adiabatic processes occur naturally in our atmosphere. Through meteorologic processes, expansion and compression of the air often happen so quickly that appreciable heat flow does not have time to occur. In addition, the volume of air involved is effectively insulated from its surroundings.

We will now derive some useful relations for adiabatic processes. Applying the first law when $Q = 0$:

$$Q = \Delta U + W$$
$$0 = (U_f - U_i) + W$$

By rearranging, we see that the change in internal energy is the negative of the work done on the external world:

ADIABATIC PROCESS

$$U_f - U_i = -W \tag{19–15}$$

In a compression, the work W is a *negative* numerical quantity, so the internal energy increases ($U_f > U_i$); that is, the gas heats up. In an expansion, the work W is a *positive* quantity, so the internal energy decreases and the gas cools. All this seems reasonable when we consider the kinetic theory model. On a microscopic scale, when gas atoms bounce elastically off the approaching piston, their velocities increase, corresponding to an increase in temperature and internal energy. If they bounce off a piston that is moving away, they lose some velocity and thus the temperature falls and the internal energy decreases.

To obtain another useful expression, consider an infinitesimal, reversible adiabatic process. The first law, in differential form, is:

$$dQ = dU + dW$$

Substituting $dQ = 0$, $dU = nC_V\,dT$, and $dW = P\,dV$:

$$0 = nC_V\,dT + P\,dV \tag{19–16}$$

From the equation of state $PV = nRT$, we may write:

$$P\,dV + V\,dP = nR\,dT \tag{19–17}$$

Eliminating dT from the above two equations and making the substitution $C_P - C_V = R$, we have (after a bit of algebra):

$$\frac{dP}{P} + \left(\frac{C_P}{C_V}\right)\frac{dV}{V} = 0 \tag{19–18}$$

Defining the ratio of specific heats C_P/C_V by the symbol γ (Greek *gamma*) and integrating this expression, we obtain:

$$\ln P + \gamma \ln V = [\text{constant}] \tag{19–19}$$

ADIABATIC PROCESS

$$PV^\gamma = [\text{constant}] \tag{19–20}$$

Note that only V, not PV, is raised to the γ power.

Since $C_P > C_V$, the value of γ is always larger than 1. This means that adiabatic lines on the pressure-volume diagram are "steeper" than isothermal lines (see Figure 19–8). This is reasonable when we remember that in an adiabatic compression the gas heats up and thus we move to a higher isothermal line.

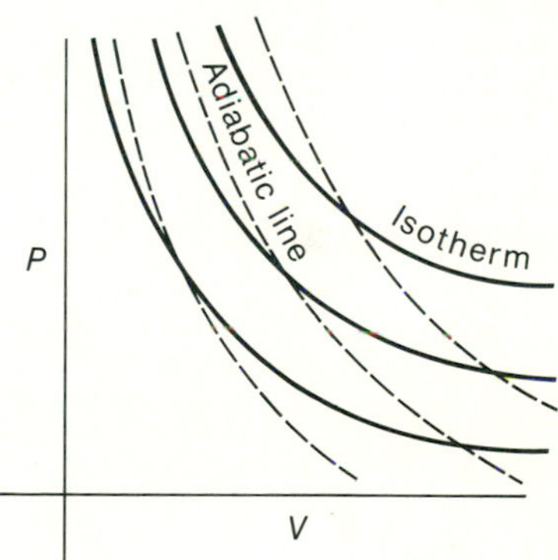

Figure 19–8
Adiabatic lines (dashed) are "steeper" than isotherms (solid).

EXAMPLE 19–4

Again, consider the gas of Example 19–2 (2 g of oxygen initially at STP). If the sample of gas is adiabatically compressed to one-half its original volume, find the final values of (a) the pressure (in atmospheres) and (b) the temperature. The value of γ for O_2 is 1.40.

SOLUTION

(a) For adiabatic processes, $P_i(V_i)^\gamma = P_f(V_f)^\gamma$. Solving for P_f and substituting numerical values[4] gives:

$$P_f = P_i\left(\frac{V_i}{V_f}\right)^\gamma = (1\ \text{atm})\left(\frac{2}{1}\right)^{1.40}$$

We calculate this by taking the logarithm of both sides:

$$\log P_f = 1.40 \log 2 = 0.421$$

The antilog is:

$$P_f = \boxed{2.64\ \text{atm}}$$

(b) The temperature may be found from:

$$\frac{P_iV_i}{n_iT_i} = \frac{P_fV_f}{n_fT_f}$$

Since $n_i = n_f$ and $V_f/V_i = 1/2$, we obtain:

$$T_f = \frac{P_fV_fT_i}{P_iV_i} = \frac{(2.64\ \text{atm})(1)(273\ \text{K})}{(1.00\ \text{atm})(2)} = \boxed{360\ \text{K}}$$

EXAMPLE 19–5

In this example, we carry a sample of an ideal gas around a *closed cycle*, returning to the original state. Consider one mole of an ideal gas undergoing a reversible process consisting of ① an isothermal expansion from (P_0, V_0, T_0) to double its volume. Then the gas ② is compressed isobarically to its original volume and ③ is heated at constant volume to its original pressure. Figure 19–9 diagrams the cycle. For each of the three processes, calculate (a) the change in internal energy, (b) the heat added to the gas, and (c) the work done by the gas. For the entire cycle, find (d) the net change in internal energy, (e) the work done by the gas, and (f) the total heat added to the gas.

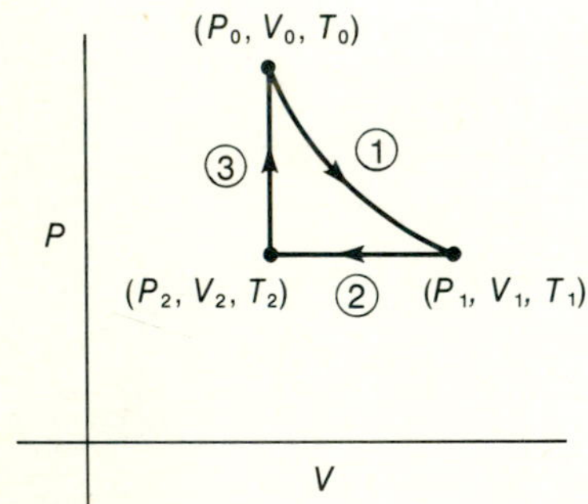

Figure 19–9
Example 19–5

SOLUTION

By examining Figure 19–9 and applying the ideal gas law, we know that $P_1 = P_2 = P_0/2$, $V_2 = V_0$, $V_1 = 2V_0$, and $T_1 = T_0$.

① **Isothermal expansion:**
Since the internal energy U depends only on the temperature and $\Delta T = 0$:

$$\boxed{\Delta U_1 = 0}$$

The work W_1 done by the gas is given by Equation (19–5):

$$W_1 = nRT_0 \ln\left(\frac{V_1}{V_0}\right)$$

$$W_1 = \boxed{RT_0 \ln 2}$$

By the first law:

$$Q = \Delta U + W$$

[4] Since the pressure is a common factor on both sides of the equation, the same conversion factor applies to each. Thus we may use "mixed" units such as atmospheres and cubic meters instead of only SI units throughout.

Since $\Delta U = 0$, the added heat Q_1 is:

$$Q_1 = W_1 = \boxed{RT_0 \ln 2}$$

② **Isobaric Compression:**
The work W_2 done by the gas is given by:

$$W_2 = \int P\,dV$$

$$W_2 = \int_{2V_0}^{V_0} \frac{P_0}{2}\,dV = \frac{P_0}{2}(V_0 - 2V_0) = -\frac{P_0V_0}{2} = \boxed{-\frac{RT_0}{2}}$$

From Equation (19–7), the heat Q_2 added to the gas during this compression is given by:

$$Q_2 = nC_P\Delta T = (1)C_P\Delta T = C_P(T_2 - T_0)$$

where the value of T_2 can be calculated from:

$$\frac{P_0V_0}{n_0T_0} = \frac{P_2V_2}{n_2T_2}$$

Since $n_0 = n_2$:

$$T_2 = T_0\left[\frac{\left(\frac{P_0}{2}\right)V_0}{P_0V_0}\right] = \frac{T_0}{2}$$

Thus:
$$Q_2 = \boxed{-\frac{C_PT_0}{2}}$$

By the first law, we find ΔU_2, the change in internal energy:

$$\Delta U_2 = Q_2 - W_2$$

$$\Delta U_2 = -\frac{C_PT_0}{2} + \frac{RT_0}{2} = \frac{T_0}{2}(R - C_P) = \boxed{-\frac{T_0}{2}(C_V)}$$

③ **Isovolumic Heating:**
Since the volume does not change:

$$\boxed{W_3 = 0}$$

The heat Q_3 added during this change to the gas is given by:

$$Q_3 = nC_V\,\Delta T = (1)C_V\,\Delta T = C_V\left(T_0 - \frac{T_0}{2}\right) = \boxed{\frac{T_0}{2}C_V}$$

We find the change in internal energy ΔU_3 from the first law:

$$Q = \Delta U + W$$

$$\Delta U_3 = \boxed{\frac{T_0C_V}{2}}$$

(d) The net change in internal energy ΔU for the entire cycle is found by adding the changes for each portion of the cycle.

$$\Delta U = \Delta U_1 + \Delta U_2 + \Delta U_3$$

$$\Delta U = 0 + \frac{T_0}{2}(R - C_P) + \frac{T_0}{2}C_V = \frac{T_0}{2}[R - (C_P - C_V)]$$

$$\Delta U = \boxed{0}$$

We could have predicted there was no net change in the internal energy from the fact that the initial and final temperatures are the same for this cyclic process, since the internal energy of an ideal gas depends only on temperature.

(e) The work W done by the gas for the entire cycle is found by adding the work for each portion of the cycle.

$$W = W_1 + W_2 + W_3$$

$$W = RT_0 \ln 2 + \left(-\frac{RT_0}{2}\right) + 0 = RT_0(\ln 2 - \tfrac{1}{2})$$

$$W = \boxed{0.193RT_0}$$

Note that W is a positive quantity, verifying that work was done by the gas on the outside world.

(f) The total heat added to the gas for the entire cycle is:

$$Q = Q_1 + Q_2 + Q_3$$

$$Q = RT_0 \ln 2 - \frac{T_0}{2}(C_P) + \frac{T_0}{2}(C_V) = RT_0(\ln 2 - \tfrac{1}{2})$$

$$Q = \boxed{0.193RT_0}$$

This example verifies the first law, $Q = \Delta U + W$, for a closed cycle where $\Delta U = 0$.

TABLE 19–1

Summary of Thermodynamic Relations for an Ideal Gas

Equation of state for an ideal gas: $PV = nRT$

For any two equilibrium states, 1 and 2, of an ideal gas: $\frac{P_1V_1}{T_1} = \frac{P_2V_2}{T_2}$

$$Q = \Delta U + W$$

Process	Q	ΔU	W
Isothermal $\Delta T = 0$ $(\Delta U = 0)$	$nRT \ln \frac{V_2}{V_1} = P_1V_1 \ln \frac{P_1}{P_2}$	0	(equal to Q)
Isovolumic $\Delta V = 0$ $(W = 0)$	$nC_V\,\Delta T$	$nC_V\,\Delta T$	0
Isobaric $\Delta P = 0$	$nC_P\,\Delta T$	$nC_V\,\Delta T$	$P\,\Delta V$
Adiabatic $Q = 0$ $P_1(V_1)^\gamma = P_2(V_2)^\gamma$	0	$nC_V\,\Delta T$	$\int_{V_1}^{V_2} P\,dV = \frac{P_2V_2 - P_1V_1}{1 - \gamma}$

Note: In these expressions, the specific heats C_V and C_P are assumed constant: $R = C_P - C_V$.

Table 19–1 summarizes the application of the first law to the reversible processes we have discussed. Note that if a temperature change occurs by *any* process, the change in internal energy is always $\Delta U = nC_V \Delta T$.

19.5 Energy Variables and the Equipartition Theorem

A significant concept that emerges from the kinetic theory is the connection between the internal energy of a system and the kinetic energy of the particles that make up the system. We shall now investigate systems made up not just of mass-point particles, but of more complicated particles: diatomic and triatomic molecules. We will see that the internal energy now includes not only kinetic energies of motion, but also potential energies associated with molecular distortions. We must first focus our attention on the number of variables required to describe the total energy a particle has.

The Motion of a Mass Point

Consider one of the atoms in our kinetic theory model of an ideal gas. It is a single point mass, free to move in three dimensions, as shown in Figure 19–10. The only kind of energy it may have is kinetic energy due to its translational motion.[5] In terms of velocity components, the total energy is

$$E = \tfrac{1}{2}mv_x^{\,2} + \tfrac{1}{2}mv_y^{\,2} + \tfrac{1}{2}mv_z^{\,2} \qquad \textbf{(19–21)}$$

Here, we need three variables to describe the energy: v_x, v_y, and v_z, associated with the three coordinate directions. If the system contained N mass points, all moving independently, we would need $3N$ variables to describe the total energy.

[5] As usual, we are not considering gravitational potential energy *mgy* for these examples.

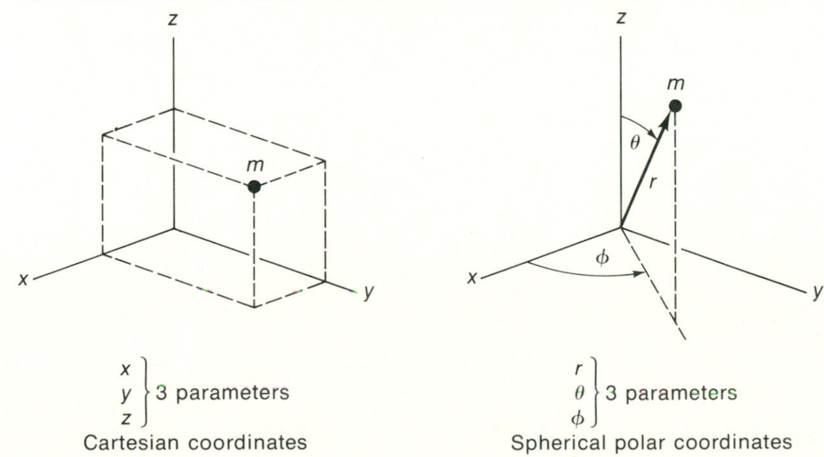

Figure 19–10
A single mass point *m* free to move in three dimensions. Regardless of the type of coordinate system used, *three* parameters are required to describe its position in space.

The Rigid Dumbbell

For a diatomic gas, we can imagine a "dumbbell" model for the molecule, that is, two mass points joined together by a *rigid* rod as in Figure 19–11. How many variables are required to describe the energy this rigid dumbbell has? First of all, the molecule as a whole may translate through space to give three kinetic energy components. Letting $M = m + m$, the mass of the molecule, these translational kinetic energy components are $\frac{1}{2}Mv_x^2$, $\frac{1}{2}Mv_y^2$, and $\frac{1}{2}Mv_z^2$. But in addition the molecule may rotate about its center of mass. How many additional quantities are required to specify this rotational energy? To describe the direction of a line in space, we need two additional parameters: the angles θ and ϕ. A possible rotational motion is associated with each of these two angular directions. If the respective moments of inertia are I_θ and I_ϕ, the dumbbell may have rotational kinetic energy of $\frac{1}{2}I_\theta\omega_\theta^2$ associated with the θ coordinate and $\frac{1}{2}I_\phi\omega_\phi^2$ associated with the ϕ coordinate. (Since we are assuming true *point* masses, the moment of inertia about the line joining the masses is zero, so we need not consider that type of rotation.) Thus, the total energy of a rigid dumbbell is:

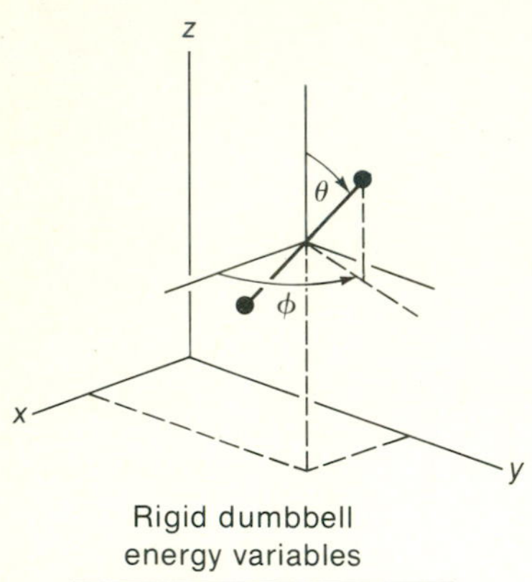

Rigid dumbbell energy variables

3	for *CM* translation
2	for angle orientations
5	energy variables

Figure 19–11
A diatomic molecule composed of two mass points joined as a rigid dumbbell. Rotation about the axis joining the mass points cannot be defined if the masses are true points with negligible dimensions.

$$E = \underbrace{\tfrac{1}{2}Mv_x^2 + \tfrac{1}{2}Mv_y^2 + \tfrac{1}{2}Mv_z^2}_{\text{Translations}} + \underbrace{\tfrac{1}{2}I_\theta\omega_\theta^2 + \tfrac{1}{2}I_\phi\omega_\phi^2}_{\text{Rotations}}$$

To specify the total energy, we need five variables: v_x, v_y, v_z, ω_θ, and ω_ϕ. If a system contained N such molecules, all moving independently, we would need $5N$ variables to describe the total energy.

The Vibrating Dumbbell

Now let us modify the rigid dumbbell to make a more realistic model, recognizing that actual diatomic molecules can vibrate along the lines joining the two atoms (see Figure 19–12). We assume these vibrations are SHM. Thus there will be some effective force constant k for the vibrations. The total energy for SHM oscillations of this type is:

$$E = \tfrac{1}{2}\mu v_{x'}^2 + \tfrac{1}{2}kx'^2 \quad \text{(for SHM vibrations only)} \tag{19–22}$$

where μ is the reduced mass of the two-mass system and x' is the distance between the atoms. Thus two more parameters are needed: $v_{x'}$ and x', associated, respectively, with the kinetic and potential energies of SHM vibrations. Thus the total energy expression is:

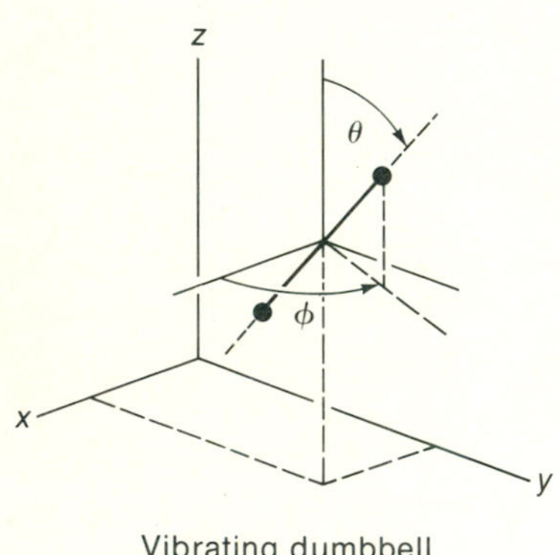

Vibrating dumbbell energy variables

3	for *CM* translation
2	for angle orientations
2	for SHM vibrations
7	energy variables

Figure 19–12
A diatomic molecule that can undergo SHM vibrations along the line joining the two mass points has *seven* energy variables.

$$E = \underbrace{\tfrac{1}{2}Mv_x^2 + \tfrac{1}{2}Mv_y^2 + \tfrac{1}{2}Mv_z^2}_{\text{Translations}} + \underbrace{\tfrac{1}{2}I_\theta\omega_\theta^2 + \tfrac{1}{2}I_\phi\omega_\phi^2}_{\text{Rotations}} + \underbrace{\tfrac{1}{2}\mu v_{x'}^2 + \tfrac{1}{2}kx'^2}_{\text{SHM vibrations}}$$

To express the total energy, then, seven variables must be specified: v_x, v_y, v_z, ω_θ, ω_ϕ, $v_{x'}$, and x'. If the system contained N such molecules, there would be $7N$ variables.

In summary, if a system contains N molecules, $3N$ variables would be required to describe the total energy of a *monatomic* system, $5N$ variables would be required for a system of *rigid diatomic molecules*, and $7N$ variables would be needed for a system of *vibrating diatomic molecules*.

Why is this concept useful? It is because the number of variables required to describe the energy is essential to a theoretical statement known as the **equipartition of energy.** This principle was first derived by James Clerk Maxwell

$\theta_i = \theta_r = \theta$

$\beta = \gamma + \theta = \alpha + 2\theta \therefore \alpha + \beta = 2\gamma$

$\alpha = \frac{h}{OM} \quad \beta = \frac{h}{IM} \quad \gamma = \frac{h}{CM}$

$$\frac{1}{OM} + \frac{1}{IM} = \frac{2}{CM}$$

$\alpha = \gamma + \theta_i$ $\quad \alpha - \beta = 2\gamma$

$\theta_r = \gamma + \beta$

$$\frac{1}{OM} - \frac{1}{IM} = \frac{2}{CM}$$

$\theta_i = \gamma + \alpha$

$\beta = \theta_r + \gamma \quad \theta_r = \beta - \gamma$

$\alpha - \beta = -2\gamma$

$$\frac{1}{OM} - \frac{1}{IM} = -\frac{2}{CM}$$

$\frac{1}{p} + \frac{1}{q} = \frac{2}{R_{od}}$ | $f = \frac{R}{2}$ | $1/p + 1/q = 1/f$

$M = \frac{S_i}{S_o} = \frac{-q}{p} \qquad M_T = M_1 \times M_2 + M_3 \cdots$

$n = \frac{c}{v_{med}} \qquad n_1 \sin\theta_1 = n_2 \sin\theta_2 \qquad \sin\theta_c = \left(\frac{n_2}{n_1}\right)$

$1/f = (n-1)\left(\frac{1}{R_1} - \frac{1}{R_2}\right)$

ang. Mag. $m = \frac{\angle \text{ subtended w/magnifier}}{\angle \text{ unaided @ 25cm}}$

Aberrations:

- Spherical aberration
- Astigmatism
- Curvation of field
- Distortion
- chromatic aberration (lenses)

$$\frac{\partial^2 E_y}{\partial x^2} = \mu_0 \epsilon_0 \frac{\partial^2 E_y}{\partial t^2} \qquad \frac{\partial^2 B_z}{\partial x^2} = \mu_0 \epsilon_0 \frac{\partial^2 B_z}{\partial t^2}$$

for sinusoidal EM wave

$$E_y = E_{y0} \sin \omega t \qquad B_z = B_{z0} \sin \omega t$$

$$c = \frac{1}{\sqrt{\mu_0 \epsilon_0}} \qquad E_y = c B_z$$

Energy Density $u_E = \frac{1}{2} \epsilon_0 E^2 \qquad u_B = \frac{1}{2\mu_0} B^2$

Energy flow $S = \frac{1}{\mu_0} (\vec{E} \times \vec{B}) \qquad S_{ave} = \frac{1}{2\mu_0} E_{y0} B_{z0} = \frac{\frac{dU}{dt}}{A}$

S is Poynting

Momentum: $U = cp$ p = momentum

Light pressure: $P = \frac{S_{ave}}{c}$ (total absorbtion)

$P = \frac{2 S_{ave}}{c}$ (total reflection)

Visible: $\lambda = 400 - 700\,nm$ $1\,\text{Å} \equiv 10^{-8}\,cm$

Radio & TV – Microwave – Infrared – Vis. – U.V. – X ray – Gamma

$A = A_0 \sin(kx - \omega t)$ $k = \frac{2\pi}{\lambda}$ $\omega = \frac{2\pi}{T} = 2\pi f$

$v = \frac{\omega}{k}$ = velocity (speed of prop.)

using statistical arguments. It applies only to *classical* particles, that is, those that behave according to Newtonian mechanics. The theorem states:

EQUIPARTITION OF ENERGY **An average energy of $\frac{1}{2}kT$ is associated with each of the variables required to specify the total energy of a system of particles.**

This means that the energy a system possesses is shared *equally* among all the possible ways of storing energy in the system.

How well does this theorem agree with experimental findings? A straightforward test is to examine experimental values for the molar specific heat capacity of gases. The theoretical values are computed as follows. For a monatomic gas, we have seen that $3N_A$ variables are required to specify the total energy for one mole (N_A = Avogadro's number). Therefore the internal energy is $U = (3N_A)[\frac{1}{2}kT]$. Since $N_A k = R$ (Chapter 18), we have:

$$U = \tfrac{3}{2}RT \qquad \text{(per mole for a monatomic gas)} \qquad \textbf{(19–23)}$$

Equation (19–13) in differential form relates this quantity to specific heat:

$$dU = nC_V\, dT$$

$$C_V = \frac{1}{n}\frac{dU}{dT}$$

For one mole, this is:

$$C_V = \left(\frac{1}{1}\right)\frac{d}{dT}(\tfrac{3}{2}RT) = \tfrac{3}{2}R$$

Handbook values for specific heats are usually given for constant pressure. Thus:

$$C_P = C_V + R = \tfrac{5}{2}R = 20.8\,\frac{\text{J}}{\text{mol·K}} \qquad \text{(Monatomic gas: point masses)} \qquad \textbf{(19–24)}$$

For a diatomic molecule without vibrations, we add two more rotational energy variables for a total of seven, each sharing $\frac{1}{2}RT$ energy per mole:

$$C_P = \tfrac{7}{2}R = 29.1\,\frac{\text{J}}{\text{mol·K}} \qquad \text{(Diatomic molecule: no vibrations)} \qquad \textbf{(19–25)}$$

If SHM vibrations are present, we add two more energy variables:

$$C_P = \tfrac{9}{2}R = 37.4\,\frac{\text{J}}{\text{mol·K}} \qquad \text{(Diatomic molecule: with vibrations)} \qquad \textbf{(19–26)}$$

Rigid polyatomic molecules (without vibration) would add one more energy variable to those in the corresponding diatomic case, to account for rotations about the third axis we formerly ignored. Thus:

$$C_P = \tfrac{8}{2}R = 33.2\,\frac{\text{J}}{\text{mol·K}} \qquad \text{(Polyatomic molecule: no vibrations)} \qquad \textbf{(19–27)}$$

The various modes of vibration for different polyatomic molecules are quite complex; since agreement between theory and experiment is poor for these cases, we will not calculate numerical values. Nevertheless, it is clear that if the many possible modes of vibration are included, C_P should range to higher values.

We can also predict what the value of γ should be. For a monatomic gas:

$$\gamma = \frac{C_P}{C_V} = \frac{(\frac{5}{2}R)}{(\frac{3}{2}R)} = \frac{5}{3} = 1.67 \qquad \text{(Monatomic gas)}$$

For a diatomic gas without vibrations, we have:

$$\gamma = \frac{C_P}{C_V} = \frac{(\frac{7}{2}R)}{(\frac{5}{2}R)} = \frac{7}{5} = 1.40 \qquad \text{(Diatomic gas: no vibrations)}$$

A diatomic gas with vibrations would give:

$$\gamma = \frac{C_P}{C_V} = \frac{(\frac{9}{2}R)}{(\frac{7}{2}R)} = \frac{9}{7} = 1.29 \qquad \text{(Diatomic gas: with SHM vibrations)}$$

For a polyatomic gas with no vibrations:

$$\gamma = \frac{C_P}{C_V} = \frac{(\frac{8}{2}R)}{(\frac{6}{2}R)} = \frac{8}{6} = 1.33 \qquad \text{(Polyatomic gas: no vibrations)}$$

with somewhat lower values if vibrations are included.

Table 19–2 compares these predictions with some experimental data. We see the agreement is excellent for monatomic gases. However, diatomic gases present a puzzle, since the data seem to imply that vibrational energy storage does not occur appreciably. And this lack of agreement becomes even more noticeable for polyatomic gases. Perhaps we oversimplified the model by considering point masses and pure SHM oscillations. Or have we pushed the classical theory beyond the limits of its applicability? As the next section shows, these data represent revealing effects that have their origin in quantum phenomena that Newtonian mechanics cannot explain.

TABLE 19–2

Theoretical and Experimental Values of Molar Specific Heats for Gases at 1 atm and ~300 K[a]

Type of gas	Gas	$C_P \frac{\text{J}}{\text{mol}\cdot\text{K}}$ Experimental	Theoretical	$\gamma = \frac{C_P}{C_V}$ Experimental	Theoretical
Monatomic	He	20.8	20.8	1.66	1.67
	Ar	20.7		1.67	
Diatomic	H_2	28.8	29.1 (no vibration) 37.4 (with vibrations)	1.41	1.40 (no vibration) 1.29 (with vibrations)
	O_2	29.3		1.40	
	NO	29.3		1.40	
	Cl_2	34.1		1.36	
Polyatomic	CO_2	36.6	33.2 (no vibration) (higher with vibrations)	1.32	1.33 (no vibration) (lower with vibrations)
	SO_2	40.7		1.29	
	CCl_4	74.0		1.13	
	C_2H_6	48.5		1.20	

[a] The data are for various temperatures, all within 15°C to 25°C.

19.6 Clues to a More Correct Theory

If we investigate specific heats of gases over a wide range of temperatures, an interesting behavioral trend is observed. Figure 19–13 shows C_V for hydrogen. From the graph, it appears that as the temperature is reduced to lower and lower values, certain energy variables no longer partake in the equal sharing of energy as the equipartition theorem demands. Classical statistical mechanics can offer no reason for this "freezing out" of some of the energy variables. According to classical theory, the specific heat should be independent of temperature.

This behavior remained mysterious until about 1910 when the development of quantum mechanics predicted just such a phenomenon. Although a discussion of quantum mechanics must be postponed, the reason that certain energy variables become ineffective in sharing energy can be given now. The heart of the argument lies in an important assumption made by Max Planck in 1899 (later modified by Niels Bohr and others):

PLANCK'S QUANTUM HYPOTHESIS

An oscillator cannot have any amount of energy, but only integral multiples of a basic energy unit given by

$$E = nhf$$

where n is an integer, h is Planck's constant, and f is the frequency. The energy is *quantized*.

The diatomic molecule H_2 has definite quantized energy states for rotational and vibrational motions. At temperatures of about 293 K (room temperature), the average translational kinetic energy is about 6×10^{-21} J. Many rotational energy states are much lower than this, so as a result of random collisions the molecule is set into rotation in one or another of these states and thus rotational modes of energy storage are significant. But the first "quantum jump" in energy associated with *vibratory* motion is about 10 times larger than the average translational energy. Hence, only a negligible number of molecules can gain, through random collisions, enough energy to reach a vibrational energy state. For this reason, the two energy variables associated with SHM vibrations are effectively "frozen out" below a few thousand kelvins, and below about 300 K the rotational modes also are frozen out. This certainly is an odd state of affairs. Thinking classically, it seems reasonable that diatomic molecules

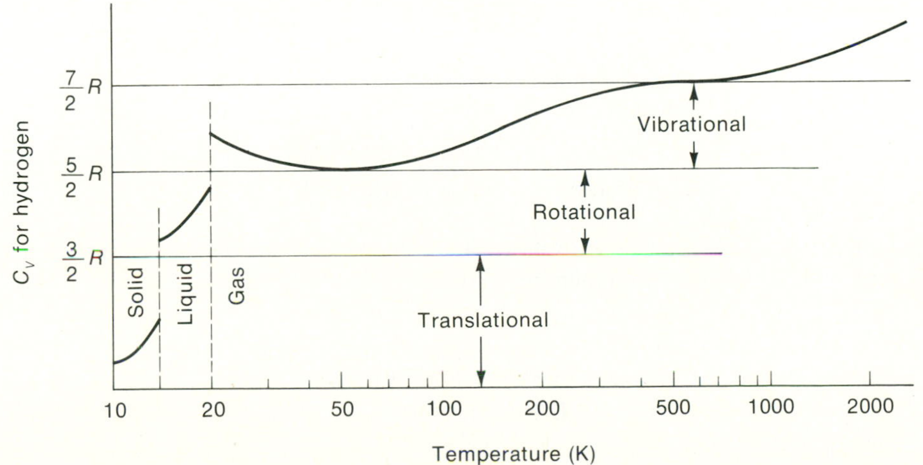

Figure 19–13
The variation of molar specific heat C_V for hydrogen as a function of temperature. As the temperature increases, the average energy per energy variable becomes large enough to raise molecules to the first excited (quantized) energy states for, successively, the rotational and the vibrational modes. At still higher temperatures, internal energy states within atoms bring additional energy variables into consideration.

of a gas, traveling through empty space, ought to be able to rotate freely. Yet the theory of quantum mechanics implies that if the temperature is low enough, this is not so. The molecules translate but do not rotate.

In general, as the temperature rises, the average energy per energy variable becomes large enough to excite many molecules into higher rotational and vibrational energy states. Hence, a smaller fraction of the total internal energy resides in the *translational* kinetic energy form associated with the temperature. Therefore, the specific heat of most substances increases with increasing temperature as more energy variables become "unfrozen." (Remember that temperature is proportional only to the *translational* kinetic energy.)

Summary

The basic concepts covered in this chapter are as follows:

System Any object that has a well-defined boundary across which we keep track of heat flow and work done. (Our specific example is an ideal gas enclosed in a cylinder with a movable, frictionless piston.)

Equation of state The mathematical equation relating variables that fully describe the physical condition of the system. (The equation of state for our example is $PV = nRT$.)

Thermodynamic equilibrium The condition that exists when all the measurable properties of a system (such as P, V, and T) are the same throughout the system and do not change with time.

Zeroth law of thermodynamics: A law stating that if two systems are each in thermal equilibrium with a third system (which may be a thermometer), then the two systems are in thermal equilibrium with each other.

Process The changes that occur when a system undergoes a transition from one state to another. If the direction of energy flow can be reversed by an infinitesimal change in a state parameter, then the process is *reversible*. If a finite difference in state variables is involved, the process is *irreversible*.

First law of thermodynamics: $Q = \Delta U + W$, where Q is the heat added to the system, ΔU (equal to $U_f - U_i$) is the change in the internal energy of the system, and W is the work done by the system on the external world. (If the direction of energy flow is reversed, the numerical value is negative. The signs in the general statement of the first law itself should never be changed.)

We applied the first law of thermodynamics to several reversible processes for our ideal gas system:

isothermal	(constant temperature)
isovolumic	(constant volume)
isobaric	(constant pressure)
adiabatic	(no heat transfer)

Table 19–1 summarizes the application of the first law for these reversible processes. It is important to remember that the internal energy U is a function only of the absolute temperature T. If an ideal gas undergoes a temperature change ΔT, the internal energy of the gas changes by $\Delta U = nC_V\,\Delta T$ for *all* processes, whether or not the process was at constant volume.

"Sudden" processes are generally adiabatic, since heat transfer requires time to take place.

The *molar specific heats* of an ideal gas, C_P and C_V, are related to the *universal gas constant R*:

$$C_P - C_V = R$$

For *all* processes:
$$\frac{P_f V_f}{nT_f} = \frac{P_i V_i}{nT_i}$$

For *adiabatic* processes:
$$P_f V_f^{\gamma} = P_i V_i^{\gamma} \qquad \text{(where } \gamma = C_P/C_V\text{)}$$

The *equipartition theorem* states that an average energy of $(\frac{1}{2})kT$ is associated with each of the variables required to specify the total energy of a system of particles.

Planck's quantum hypothesis states that an oscillator can have only discrete energy values given by $E = nhf$, where n is an integer, h is Planck's constant, and f is the frequency of the oscillator.

Questions

1. What ultimately happens to the energy in the sound of a handclap? Where did the energy come from?
2. Is it possible to tell whether the increase in the internal energy of a system came from work being done on the system or from heat?
3. Canvas water bags are often used by desert travelers. Explain why this type of container (rather than, say, a metal container) cools the water somewhat.
4. Can a given amount of thermal energy be entirely converted into work? Can a given amount of work be entirely converted into thermal energy?
5. When a gas expands adiabatically, it does work on the external world because the volume changes. What is the source of energy for this work?
6. A Thermos bottle contains cold lemonade. It is shaken for a few moments. Does the temperature of the lemonade rise? Was heat added to the lemonade? Has the internal energy of the lemonade increased?
7. A laboratory source of "dry ice" (solidified CO_2) is sometimes provided by a small cylinder of CO_2 gas. When the tank valve is opened, the escaping gas turns into "snowflakes," which are immediately squeezed together into solid cakes of dry ice. Explain the gas-expansion process in terms of the first law.
8. Hold your mouth wide open and blow gently on your wrist from a distance of about 10 cm. Your breath will feel warm. Now purse your lips and repeat. Why does your breath feel cool in the second case?
9. Suppose a box contains just eight molecules in random motion. Occasionally, by chance, all eight molecules will find themselves in the right half of the box. This is the reverse of a "free expansion," a process that is considered *irreversible*. How do you account for this?

Problems

In the following problems, assume all gases are ideal.

19A–1 Assume that 2 mol of an ideal monatomic gas at STP are heated at constant pressure until the volume doubles.

(a) Find the final temperature.
(b) Determine the amount of work done by the gas.
(c) Determine the amount of heat added to the gas.
(d) Find the change in internal energy of the gas.

Answers: (a) 546 K (b) 4538 J (c) 1.13×10^4 J (d) 6806 J

19A–2 Assume that 0.40 mol of an ideal monatomic gas at 6 atm and 27°C are heated at constant pressure until the volume doubles.

(a) Find the amount of work (in joules) done by the gas.
(b) How much heat (in joules) was added to the gas?
(c) Find the change in internal energy of the gas.

19A–3

(a) How much work (in joules) is required to compress isothermally 0.16 g of helium at STP to 1/10 the original volume?
(b) Find the amount of heat (in joules) extracted from the gas during the process.
(c) What was the change in internal energy of the gas?
(d) Determine the final volume of the helium.

Answers: (a) 209 J (b) 209 J (c) 0 (d) 0.0896 L

19A–4 Suppose 0.20 mol of an ideal monatomic gas, initially at 20 atm and 27°C, is expanded isothermally to twice the initial volume.

(a) Find the amount of work (in joules) done by the gas.
(b) What heat (in joules) is added to the gas?
(c) What is the change in internal energy of the gas?
(d) Find the final volume of the gas.

19A–5 A liter of argon at STP is expanded suddenly to 3 L. Find (a) the final pressure and (b) the final temperature. *Answers:* (a) 0.160 atm (b) 131 K

19A–6 An ideal diatomic gas ($\gamma = 1.40$) at STP is expanded suddenly to 4 times its original volume. Find (a) the final pressure and (b) the final temperature.

19A–7 The roof of a house built to absorb the solar radiation incident upon it has an area of 7 m × 10 m. The solar radiation at the earth's surface is 840 W/m^2. On the average, the sun's rays make an angle of 60° with the plane of the roof.

(a) If 15% of the incident energy is converted into useful electrical power, how many kilowatt-hours per day of useful energy does this source provide? Assume the sun shines for an average of 8 h/day.
(b) If the average household user pays about 6.0¢/kW·h, what is the monetary saving of this energy source per day? *Answers:* (a) 61.1 kW·h (b) $3.67

19B–1 Suppose 3 mol of an ideal monatomic gas are compressed isothermally from an initial pressure of 20 atm to 60 atm at 27°C.

(a) Find the amount of work (in joules) done on the gas.
(b) How much heat (in joules) is extracted?
(c) What is the change in internal energy of the gas?
(d) Find the final volume of the gas.

Answers: (a) 8220 J (b) 8220 J (c) 0 (d) 1.23 L

19B–2 Assume that 4 mol of an ideal monatomic gas undergo an adiabatic expansion from an initial volume of 1 m^3 and a temperature of 300 K to a final volume of 10 m^3.

(a) How much work was done by the gas?
(b) What is the final temperature?
(c) If, instead, the gas expanded isothermally, how much work was done?
(d) In part (c), where did the energy come from to perform the work?

19B–3 An ideal monatomic gas at 27°C is suddenly compressed to a smaller volume. It is next cooled at constant volume to its original temperature and then allowed to expand isothermally to its original volume.

(a) Sketch the process of a pressure-volume diagram.
(b) If 5000 cal of heat were removed from the gas during the entire process, how much net work was done on the gas? *Answer:* (b) 3.35×10^4 J

19B–4 Consider a mixture of monatomic and diatomic gases in equilibrium at the same temperature. Find the quantitative ratios of monatomic to diatomic gases for (a) the average energy and (b) the average translational energy. Assume the diatomic molecules undergo rotational motion, but that vibration along the line joining the atoms does not occur.

19B–5 Suppose 1600 J of heat are added to 8 g of nitric oxide gas (NO) initially at 10°C and 60 atm pressure. Assume the gas behaves as an ideal diatomic gas with rotational motions, but no vibrations.

(a) Keeping the pressure constant, find the final temperature of the gas.
(b) Find the final volume of the gas. *Answers:* (a) 216°C (b) 0.178 L

19B–6 Consider a model of hypothetical one-dimensional "gas" made of beads free to slide without friction along a straight wire and to rotate about the wire.

(a) How many energy variables does each bead have?
(b) A model for a two-dimensional "gas" might be flat, disk-shaped "molecules," free to rotate as they slide around on a horizontal surface without friction. How many energy variables does each disk have?

19B–7 Starting with the expression for the work done by the gas in an adiabatic change, $W = (P_2V_2 - P_1V_1)/(1 - \gamma)$, show that this is equal to $nC_v(T_1 - T_2)$.

19B–8 Consider a room in a house whose air is at a (cool) temperature T_1 and a pressure P_1. The furnace is now turned on, heating the room air to a higher temperature T_2. Because of cracks around windows and doors, some air escapes to the outside so the pressure remains at P_1.

(a) Show that the total thermal energy of the air in the room after heating is the same as before heating.
(b) Explain where the energy from the furnace went. (Ignore the energy that goes into heating the walls, ceiling, and so on.)

19B–9 Show that the isothermal bulk modulus $B = -V\,dP/dV$ for an ideal gas equals the pressure P.

19C–1 A liter of an ideal monatomic gas at standard conditions expands isothermally to a volume of 2 L.

(a) Find the amount of work (in joules) done by the gas.
(b) If the gas is next allowed to expand adiabatically to a volume of 3 L, find the amount of work (in joules) done by the gas during this expansion.
(c) Determine the final temperature of the gas.
(d) If the gas is then compressed isothermally and adiabatically back to its original state, how much work is done on the gas during the isothermal process?
(e) During the adiabatic process?
(f) During the complete cycle, find the net work done by the gas on the external world. Include a pressure-volume diagram for the cycle.

Answers: (a) 70.4 J (b) 35.8 J (c) 208.7 K (d) −53.8 J (e) −35.8 J (f) 16.6 J

19C–2 An ideal monatomic gas, initially at P_1, V_1, and T_1, is carried through a three-step cycle: an isothermal expansion to V_2, an isobaric compression to the original volume, and an isovolumic heating to the initial pressure and temperature.

(a) Diagram the cycle on a pressure-volume diagram.
(b) In terms of the given parameters and the gas constants, R and C_v (equal to $\frac{3}{2}R$), find the number of moles of gas there are.
(c) Referring to part (b), find the temperature T_2 after the isobaric compression.
(d) For each of the three processes, write an equation for the temperature as a function of the appropriate variable.

19C–3 Starting at P_0, V_0, T_0, an ideal monatomic gas is carried through the following three processes: an isothermal expansion to $2V_0$, an isobaric compression to V_0, and an isovolumic change to the original conditions. Show that the work done by the gas on the external world for this complete cycle is $P_0V_0(\ln 2 - 1/2)$. Sketch the process on a pressure-volume graph.

19C–4 Suppose 0.4 mol of an ideal diatomic gas is confined at 300 K and an initial volume of 2 L. Assume the molecules may undergo rotational motions, but no vibrations along the line joining the atoms. Calculate the work done by the gas in expanding to twice its original volume by the following processes: (a) isobaric, (b) isothermal, and (c) adiabatic. Include a freehand sketch showing the three processes on a pressure-volume diagram. (d) Find the heat transfers that occur in processes (a) and (b).

19C–5 Assume oxygen behaves as an ideal diatomic gas with rotational motion but no vibrations.

(a) How many joules of heat will raise 4 g of oxygen gas (O_2) from 27°C to 40°C if the pressure is maintained constant at 20 atm?

(b) What is the final volume of the gas?
(c) How much work is done by the gas during this process?
(d) What is the change in internal energy of the gas?

Answers: (a) 47.3 J (b) 1.61×10^{-4} m^3 (c) 13.5 J (d) 33.8 J

19C–6 An insulated container has a partition separating 1 mol of helium at 0°C from 2 mol of oxygen (O_2) at 100°C. The partition is removed and the gases mix and come to thermal equilibrium.

(a) Find the final temperature of the mixed gases assuming the oxygen molecules undergo rotations but no vibrations.
(b) Determine the effective molar specific heat at constant volume for this mixture.

19C–7 The carbon dioxide molecule (CO_2) is a linear configuration with the carbon atom in the center and the oxygen atoms spaced equally on opposite sides.

(a) Describe the four modes of oscillation of the molecule (two are linear, two are transverse).
(b) Assuming each mode is SHM, what is the value of γ at a temperature when all modes of oscillation occur?

Answer: (b) 13/11

(Comment on claims of a perpetual motion machine using magnets) O that the gods would bring to a miserable end such fictitious, crazy, deformed labours, with which the minds of the studious are blinded!

WILLIAM GILBERT
De Magnete (1600)
[*American Journal of Physics, 44* (1976)]

chapter 20
The Second Law of Thermodynamics

If a book is held above the floor and dropped, what energy interchanges occur? We start with some initial potential energy, $U_{grav} = mgh$, in the book-plus-earth system. After the book comes to rest on the floor, that energy resides as thermal energy in the slightly warmer temperature of the book, floor, air, and walls of the room, which absorbs the sound waves generated by the impact. The initial potential energy is entirely converted into thermal energy,[1] and the net result is to increase random motions on a microscopic scale. Energy is conserved.

But what about the reverse process? Can we restore everything back to its original state? Could we convert the *disordered* motions associated with thermal energy back into the *ordered* motion of mechanical energy to raise the book to its original height, cooling the walls and floor to their initial temperatures? Although such a process would not violate the *first* law, which states that heat is a form of energy and is conserved, the second law claims that it is virtually impossible.

Consider another example: Can a steamship be propelled across the ocean by taking in sea water, extracting thermal energy from it to run the engines that drive the ship, and dumping ice and salt overboard as it goes? This sounds like a clever proposal to utilize the thermal energy in the ocean water itself rather than having to carry along diesel oil or some other source of energy for the ship's engines. The suggestion certainly does not violate the first law. But again, according to the second law it is *virtually impossible.* What about the adverb "virtually"? Doesn't it imply a loophole for an inventor to devise an engine that does accomplish the complete conversion of thermal energy to work? Not at all. As you will see, the impossibility of this is one of the most certain statements that can be made. The second law of thermodynamics has the widest generality of any principle in science. It is perhaps the most "unrepealable" of all the so-called laws of nature, having less chance of being changed or discarded at some time in the future than any other principle you will learn. When its implications are understood, the second law reveals such power and generality that many scientists confess a sense of awe.

The first law states that energy exists in various forms and that it can be converted from one form to another with complete accounting of where the

[1] If the floor or book were permanently distorted from the impact, some energy would also reside "internally" in the increased potential energy of the altered intermolecular distances. However, this would not change the overall argument of complete energy conservation.

TABLE 20–1

Estimated Energy Sources for Goods and Services in the United States

Year	Humans	Animals	Machines
1800	14%	80%	6%
1900	10%	52%	38%
1974	~1%	~0%	99%

Adapted from Putnam, *Energy in the Future*, Van Nostrand, 1953.

energy goes in every case. Energy cannot be created out of nothing, nor can it be destroyed. The second law, however, prohibits certain types of energy changes. Some energy conversion processes we would like to accomplish just cannot be done.

20.1 The Second Law

The second law of thermodynamics may be stated in various forms, many of which seem far removed from other ways of describing it.

We will start our discussion by focusing attention on certain energy-conversion processes: those that are to be repeated over and over. That is, we will discuss *heat engines*, which go through repeated cycles of operation, converting heat to work. In essence, most engines (exceptions include those using electric batteries or solar cells) involve this basic transformation of *thermal energy* to work. This is true whether their source of energy is gasoline, coal, oil, natural gas, burning wood, or nuclear reactions. Obviously such processes are of prime importance to our present-day way of life.

The industrial revolution of the nineteenth century brought about a radical change in the sources of the energy, as Table 20–1 shows. The changes in the way we live resulting from improved ways of converting energy are dramatically stated by Gerald Piel.[2] He points out that the average man can do manual labor at the rate of about 1.7×10^8 J/yr. On this basis, the U.S. per capita usage (in 1974) of electrical energy alone equivalently provided each man, woman, and child with 185 servants. Any comparison of life today with the "good old days" ought to include such considerations.

In 1824, Nicolas Léonard Sadi Carnot, a 28-year-old French military engineer, gained great insight into the fundamentals of energy conversion processes. He wrote a treatise on the efficiency of steam engines that revolutionized all thinking about the problem. His accomplishment was remarkable in that it occurred before the microscopic view of thermal energy was known and before the first law of thermodynamics was formulated. In fact, the incorrect "caloric fluid" explanation was still the predominant theory of heat. Nevertheless, Carnot was able to arrive at what later became the second law. Following are three ways of stating this law.[3]

[2] See his book: *Science in the Cause of Man*, Knopf, 1962.

[3] In the next chapter, we will state the second law in still other ways in terms of the concept *entropy*. This different approach will further clarify why we believe so strongly in the second law.

The Kelvin-Planck statement:

It is impossible to build a cyclic engine that converts thermal energy into work with 100% efficiency.

or

The Clausius statement:

It is impossible to construct a cyclic device that transfers thermal energy from a cold object to a hot object without work being performed on the device by the external world.

or

Thermal energy will not, of its own accord, flow from a cooler object to a warmer object.

We will show that these statements are entirely equivalent. Any hypothetical engine that violates one statement also would violate the others. No exceptions have ever been found. The first statement dashes our hopes of ever accomplishing a 100% efficient heat engine whose work output equals the amount of thermal energy input. The other two statements recognize the fact that heat seems only to flow "downhill," from higher to lower temperatures. As we will see in the next chapter, the second law of thermodynamics can be justified in terms of statistical arguments that are overwhelming in their logic. For the present, we will accept the second law and use it as a valid principle in the discussions that follow.

20.2 The Carnot Cycle

Carnot achieved a very worthy goal: he invented a series of processes that represent the maximum possible efficiency attainable by *any* engine operating between the same two temperatures. He disregarded the details of the operation of specific engines and instead concentrated on the fact that basically, all engines do the following:

(a) All engines absorb heat at a relatively high temperature.
(b) All engines perform some work on the external world.
(c) All engines exhaust some heat at a lower temperature.

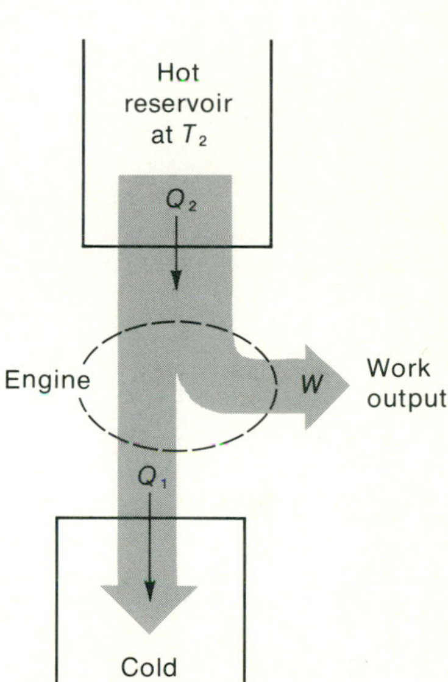

Figure 20–1
A schematic diagram of a heat engine. The engine absorbs a quantity of heat Q_2, performs an amount of work W on the external world, and rejects a quantity of heat Q_1. The engine operates between the two constant-temperature heat reservoirs at T_2 and T_1. From the first law (the conservation of energy), the net work done by the engine is $W = Q_2 - Q_1$.

Figure 20–1 illustrates these characteristics. A **heat reservoir** is a source of heat for which the temperature is maintained constant regardless of how much heat is added to or taken from the source. Thus, in the figure, the heat Q_2 is supplied from the (higher) temperature reservoir at T_2, while Q_1 is heat rejected to the (lower) temperature reservoir at T_1. *In designating temperatures with subscripts, we will follow the custom of using higher numbers for higher temperatures.* Thus, the sequence $T_1, T_2, T_3, \ldots$ represents increasingly higher temperatures (always in kelvins).

The Carnot cycle is a series of reversible changes—that is, it may be run backwards if desired. In principle, the cycle may operate using any type of material as the *working substance*. During the cycle there is no net change in the working substance; it merely acts as a carrier of the thermal energy. A Carnot-cycle engine may be devised using a non-ideal gas, a liquid, a magnetic substance, or other substance. For simplicity, we will illustrate the cycle using

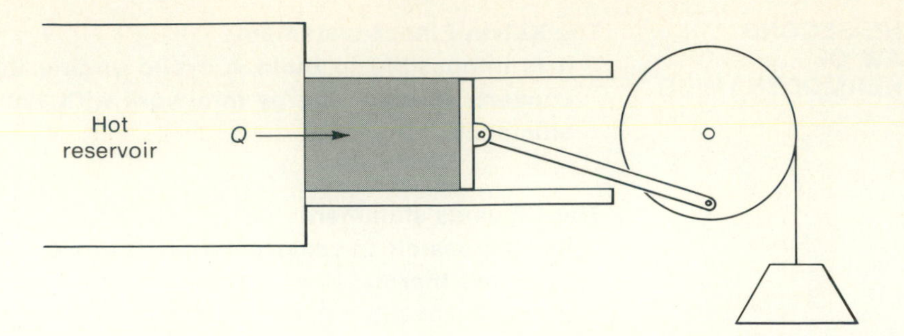

Figure 20–2
One method of converting heat into mechanical work. As heat is added to the gas, the volume of the gas expands and thereby raises the weight.

an ideal gas in a cylinder with a frictionless piston. Figure 20–2 suggests a way in which the piston might be hooked to a wheel so that as the wheel turns, a weight is lifted against gravity, representing useful work done on the external world.

The **Carnot cycle** involves *isothermal* and *adiabatic* processes, as diagramed in Figure 20–3.

THE CARNOT CYCLE

Step 1 (*A* to *B*)
Heat Q_2 is absorbed from the high-temperature reservoir (T_2) as the gas expands isothermally from *A* to *B*.

Step 2 (*B* to *C*)
The gas expands adiabatically from *B* to *C*, cooling to a lower temperature T_1.

Step 3 (*C* to *D*)
The gas is compressed isothermally from *C* to *D*, rejecting heat Q_1 to the low-temperature reservoir.

Step 4 (*D* to *A*)
The gas is restored to its original state by an adiabatic compression from *D* to *A*.

Recall that the work done while moving along a path on a pressure-volume diagram is equal to the area under the curve. For $A \rightarrow B \rightarrow C$, the

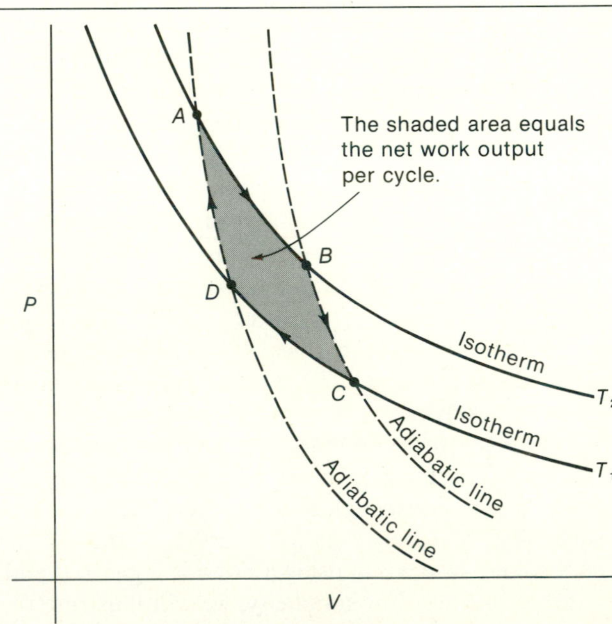

Figure 20–3
The Carnot cycle traverses along two isothermal and two adiabatic paths on a pressure-volume diagram in the direction $A \rightarrow B \rightarrow C \rightarrow D \rightarrow A$. The difference in slopes between the isothermal and adiabatic lines has been exaggerated to clarify the diagram.

engine does work *on the external world.* For $C \to D \to A$, the external world does work *on the engine.* Therefore, the net useful work per cycle done by the engine on the external world is the shaded area within the closed path, as shown in the figure. The work W is *positive* for *clockwise* traversal of the closed-loop path and *negative* for *counterclockwise* traversal.

Table 20–2 summarizes the various energy changes, which we derived in Chapter 19. Note that the *net* heat added to the gas $(Q_2 - Q_1)$ equals the work W done by the gas on the external world. The net change in internal energy is zero, so the gas returns to its original state, and we are ready to repeat the cycle. The sequence is performed repeatedly as the engine continues to run.

EXAMPLE 20–1

Verify that for the Carnot cycle, the heat rejected in the isothermal process $C \to D$:

$$Q_1 = nRT_1 \ln\left(\frac{V_D}{V_C}\right)$$

can also be written as:

$$Q_1 = (-)nRT_1 \ln\left(\frac{V_B}{V_A}\right)$$

TABLE 20–2

The Carnot Engine

Process	Work done by the gas (W)	Heat added to the gas (Q)	Change in the thermal energy of the gas (ΔU)
$A \to B$ Isothermal expansion $P_A V_A = P_B V_B$	$nRT_2 \ln\left(\frac{V_B}{V_A}\right)$	$Q_2 = nRT_2 \ln\left(\frac{V_B}{V_A}\right)$	0
$B \to C$ Adiabatic expansion $P_B(V_B)^\gamma = P_C(V_C)^\gamma$	$\frac{nR}{\gamma - 1}(T_2 - T_1)$	0	$nC_V(T_1 - T_2)$
$C \to D$ Isothermal compression $P_C V_C = P_D V_D$	$nRT_1 \ln\left(\frac{V_D}{V_C}\right)$ or $-nRT_1 \ln\left(\frac{V_B}{V_A}\right)$	$Q_1 = nRT_1 \ln\left(\frac{V_D}{V_C}\right)$ or $-nRT_1 \ln\left(\frac{V_B}{V_A}\right)$	0
$D \to A$ Adiabatic compression $P_D(V_D)^\gamma = P_A(V_A)^\gamma$	$\frac{nR}{\gamma - 1}(T_1 - T_2)$	0	$nC_V(T_2 - T_1)$
Net change for one cycle	$nR(T_2 - T_1) \ln\left(\frac{V_B}{V_A}\right)$	$nR(T_2 - T_1) \ln\left(\frac{V_B}{V_A}\right)$	0

SOLUTION

For the two isothermal processes in the Carnot cycle:

$$P_A V_A = P_B V_B$$

and

$$P_C V_C = P_D V_D$$

Also, for the two adiabatic processes:

$$P_B(V_B)^\gamma = P_C(V_C)^\gamma$$

and

$$P_D(V_D)^\gamma = P_A(V_A)^\gamma$$

Multiplying these four equations together and canceling the factor $P_A P_B P_C P_D$ common to both sides gives

$$V_A V_C (V_B V_D)^\gamma = V_B V_D (V_A V_C)^\gamma$$

Rearranging:

$$(V_A V_C)^{1-\gamma} = (V_B V_D)^{1-\gamma}$$

Extracting the $(1 - \gamma)$ root of both sides leads to

$$V_A V_C = V_B V_D$$

or

$$\frac{V_A}{V_B} = \frac{V_D}{V_C}$$

Substituting this value in the expression $Q_1 = nRT_1 \ln(V_D/V_C)$ leads to

$$Q_1 = nRT_1 \ln\left(\frac{V_D}{V_C}\right) = nRT_1 \ln\left(\frac{V_A}{V_B}\right) = \boxed{-nRT_1 \ln\left(\frac{V_B}{V_A}\right)}$$

20.3 Efficiency of Engines

Of prime significance to our industrialized society is the cost of running an engine. To make the engine perform useful work, we must pay for a given quantity of heat input Q_2 to the engine. (The second law guarantees that the heat input Q_2 is always greater than the work W we get out.) A useful measure of the **efficiency** e of an engine is the ratio of the useful work the engine does to the thermal energy we supply to make the engine operate.

ENGINE EFFICIENCY

$$e = \frac{\text{Work done per cycle}}{\text{Heat input per cycle}} \qquad \textbf{(20–1)}$$

$$e = \frac{Q_2 - Q_1}{Q_2}$$

$$e = 1 - \frac{Q_1}{Q_2} \qquad \textbf{(20–2)}$$

where the values Q_1 and Q_2 are the *magnitudes* of the heat interchanges that occur.[4] To convert efficiencies to percent values, we simply multiply by 100%.

The value Q_2 must be greater than Q_1 (the difference is the net work done). As we will demonstrate later in this chapter, the efficiency of every engine is necessarily *less* than unity because *it is impossible to restore the gas to its original condition without rejecting some heat to a lower temperature reservoir.* Thus, Q_1 can never be zero to make the efficiency 100%, even in the theoretically ideal case in which there are no losses due to friction.

[4] By defining efficiency as we have, we do not here carry along possible plus or minus signs to indicate the direction of heat flow in or out of the system. Such signs must, however, always be used in the *first* law, $Q = \Delta U + W$, which expresses the conservation of energy.

It should be noted that we are discussing *cyclic* devices. In contrast, for the *once-only* isothermal expansion of an ideal gas (which occurs without reference to an exhaust heat reservoir), heat is completely converted into work with 100% efficiency. But what we analyze in this chapter are *cyclic* devices, in which *the working substance is restored to its original condition after each cycle*, ready to repeat the process.

As we will show later, the Carnot cycle is the most efficient engine that can be devised for operating between two temperature reservoirs. *No engine can possibly outperform the Carnot engine.* The Carnot efficiency thus sets the theoretical upper limit achievable by any engine that has ever been invented or can ever be devised. The Carnot cycle and the second law of thermodynamics are clearly of importance for maintaining our technological civilization.

The efficiency for a Carnot engine may also be expressed in a form that is often more useful than Equation (20–2). From Table 20–2 we see that the amount of heat absorbed and rejected during the isothermal expansion and compression processes are:

$$\left.\begin{array}{ll}\text{Heat absorbed} & Q_2 = nRT_2 \ln\left(\dfrac{V_B}{V_A}\right) \\[2ex] \text{Heat rejected} & Q_1 = (-)nRT_1 \ln\left(\dfrac{V_B}{V_A}\right)\end{array}\right\} \text{(for an ideal gas)}$$

The ratio of these magnitudes is:

$$\frac{Q_1}{Q_2} = \frac{T_1}{T_2} \tag{20–3}$$

Thus the Carnot efficiency may also be expressed in terms of the temperatures:

CARNOT EFFICIENCY

$$\underset{\text{(Carnot)}}{e} = 1 - \frac{Q_1}{Q_2} = 1 - \frac{T_1}{T_2} \tag{20–4}$$

This equation applies to all Carnot engines, regardless of the working substance. *The ultimate efficiency depends only on the temperatures.*

EXAMPLE 20–2

A steam engine is operated in a cold climate where the exhaust temperature is at 0°C. (a) Calculate the theoretical maximum efficiency of the engine using an intake steam temperature of 100°C. (b) If, instead, superheated steam at 200°C is used, find the maximum possible efficiency.

SOLUTION

(a) Since the theoretical maximum efficiency of any heat engine is the Carnot efficiency, for steam at 100°C (373 K) we have

$$e = 1 - \frac{T_1}{T_2}$$

$$e = 1 - \frac{273\ \text{K}}{373\ \text{K}} = \boxed{26.8\%}$$

(b) Using superheated steam at 200°C (473 K), the theoretical maximum efficiency is

$$e = 1 - \frac{273\ \text{K}}{473\ \text{K}} = \boxed{42.3\%}$$

Because of this increase in efficiency at higher temperatures, most steam engines use superheated steam.

The overall efficiency of actual steam engines is of the order of 15% to 20%. Gasoline engines have efficiencies of roughly 20% to 25%, while that of diesel engines may be as high as about 40%.

Since a Carnot cycle is reversible, it may be operated in the reverse direction. A device that does this, called a *Carnot refrigerator*, extracts heat from a low-temperature reservoir and rejects it to a high-temperature reservoir (see Figure 20–4). A measure of the effectiveness of this energy transfer is the **performance coefficient** μ, defined as:

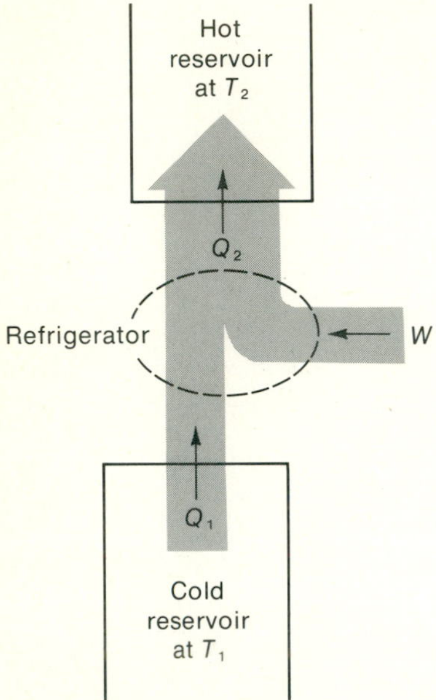

Figure 20–4
A Carnot refrigerator. Heat Q_1 is extracted from the cold reservoir, W is the work required to run the engine, and Q_2 is the heat rejected to the hot reservoir. From the conservation of energy, $W + Q_1 = Q_2$.

PERFORMANCE COEFFICIENT OF A REFRIGERATOR

$$\mu = \frac{\text{Heat extracted from the cold reservoir}}{\text{Work input}} \qquad \textbf{(20–5)}$$

$$\underset{\text{(Carnot)}}{\mu} = \frac{Q_1}{Q_2 - Q_1} = \frac{T_1}{T_2 - T_1} \qquad \textbf{(20–6)}$$

EXAMPLE 20–3

(a) Derive the expression for μ [Equation (20–6)]. (b) Calculate the performance coefficient for a Carnot refrigerator operating between 0°C (273 K) and 30°C (303 K).

SOLUTION

(a) Refer to Figure 20–4, the conservation of energy requires that the sum of the work done on the Carnot refrigerator plus the heat extracted from the cold reservoir equals the heat rejected to the hot reservoir:

$$W + Q_1 = Q_2$$

Therefore, $W = Q_2 - Q_1$, so:

$$\mu = \frac{Q_1}{Q_2 - Q_1}$$

From Equation (20–3), we have:

$$Q_2 = Q_1\left(\frac{T_2}{T_1}\right)$$

Substituting this value for Q_2 gives

$$\mu = \frac{Q_1}{Q_1\left(\frac{T_2}{T_1}\right) - Q_1} = \boxed{\frac{T_1}{T_2 - T_1}}$$

(b) Substituting numerical values into the equation for μ, we have

$$\mu = \frac{273\ \text{K}}{303\ \text{K} - 273\ \text{K}} = \boxed{9.1}$$

This is the theoretical maximum value for a Carnot refrigerator operating between the given temperatures. For actual refrigerators, the performance coefficient rarely exceeds 6 or 7. As a practical example, the performance of a small home refrigerator or air-conditioning unit is often described in terms of the rate of heat transfer, usually in terms of British thermal units per hour (Btu/h), divided by the electrical power input, in watts (W). A typical small air-conditioning unit transfers about 25 000 Btu/h, with a power consumption of 4000 W.

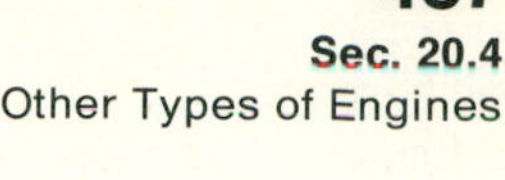

For this case, the performance coefficient [Equation (20–5)] would be:

$$\frac{\text{Heat transfer}}{\text{Work input}} = \frac{25\,000\ \dfrac{\text{Btu}}{\text{h}}}{4000\ \text{W}} = 6.25\ \frac{\text{Btu}}{\text{W}\cdot\text{h}} \underbrace{\left(\frac{0.293\ \text{W}\cdot\text{h}}{1\ \text{Btu}}\right)}_{\substack{\text{Conversion}\\ \text{ratio}}} = \boxed{1.83}$$

The performance coefficient is disappointingly low, but it is typical of small air conditioners.

An interesting application of the refrigerator principle is the *heat pump*, which warms a house in winter by extracting heat from the ground outside. Such a method is particularly effective where underground water provides a relatively constant temperature source. A further feature is that some heat pumps can reverse the flow of refrigerant; thus is may also be used for cooling the house in summer.

EXAMPLE 20–4

Consider a heat pump whose performance coefficient is 6. Compare the amount of heat supplied for heating a house with the amount of work expended in operating the heat pump.

SOLUTION

The performance coefficient (Equation 20–5) is

$$\mu = \frac{\text{Heat extracted from cold reservoir}}{\text{Work input}}$$

In terms of the symbols in Figure 20–4:

$$\mu = \frac{Q_1}{W}$$

$$6 = \frac{Q_1}{W}$$

But the heat delivered to the house is $Q_2 = Q_1 + W$, or $Q_2 = 6W + W = 7W$. Thus:

$$\frac{\text{Heat delivered to the house}}{\text{Work input}} = \frac{7W}{W} = \boxed{7}$$

The advantage is obvious. If we pay for 1 unit of work to operate the device, we obtain 7 units of heat delivered to the house. This is in contrast to heating the house with electric heaters, wherein the process of converting electrical energy to heat is essentially 100% efficient. In that case we obtain only 1 unit of heat for 1 unit of electrical energy we pay for.

20.4 Other Types of Engines

Of course, actual engines utilize numerous types of cycles other than the Carnot cycle. While no real engine can eliminate frictional effects completely and thus be classified as a reversible engine, it is instructive to analyze various cycles as if they were reversible and thus gain insight into the theoretical maximum efficiency possible for any given engine. In principle, any arbitrary reversible cycle may be thought of as a combination of a large number of Carnot engines, with the heat exhausted by one engine supplying the heat intake for the adjoining engine (see Figure 20–5). Traveling around each Carnot cycle

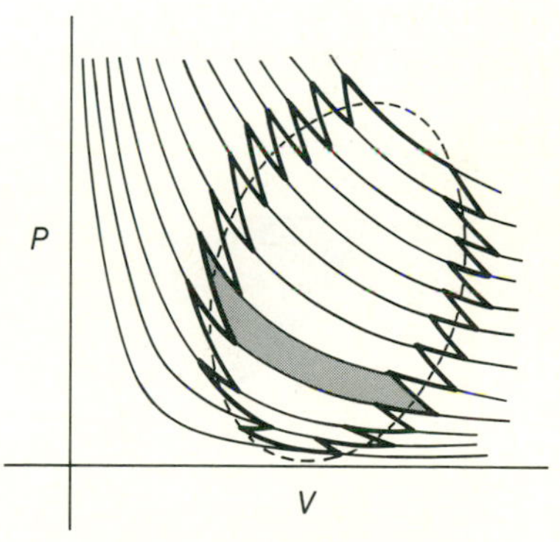

Figure 20–5
Any arbitrary reversible cycle (shown dashed) may be approximated by a family of Carnot cycles operating along isotherms that are arbitrarily close to one another. (One of the Carnot cycles is shaded for ease of visualization.) The closer together the isotherms are chosen to be, the more accurate the approximation.

in sequence is equivalent to going once around the outer zigzag perimeter. The common isotherms between adjacent cycles will be traversed twice, in opposite directions, thereby canceling these contributions to the heat transfer and work done. The closer together we choose the isotherms to be, the more closely we approximate the dotted path. In the limit, we thus arrive at the Carnot efficiency for the arbitrary reversible cycle. All *ir*reversible cycles (and all real engines) have *less* efficiency than the Carnot cycle. The Carnot efficiency truly represents an upper limit to the efficiency with which we can convert thermal energy to work.

One example of an engine that utilizes a different process from that of the Carnot cycle is the *internal combustion engine* (the gasoline motor in automobiles). A simplified approximation to the cycle that occurs in this engine is shown in Figure 20–6. Known as the *Otto cycle*, after the German engineer who first analyzed its behavior, it is assumed to be frictionless and reversible. The net work per cycle is the shaded area. Air is the working substance, assumed to behave as an ideal gas in the following processes:

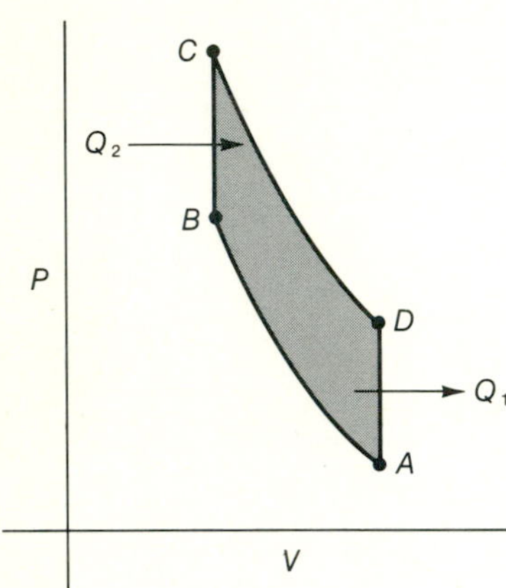

Figure 20–6
The Otto Cycle, an idealized approximation for the processes that occur in an internal combustion engine.

(1) Air at atmospheric pressure is compressed adiabatically from A to B. (This corresponds to the *compression stroke* in the real engine in which a mixture of air and gasoline vapor is compressed. The ratio V_A/V_B, called the compression ratio, is typically about 10.)
(2) The air is then heated at constant volume from B to C. (This approximates the rapid explosion of the gasoline vapor, ignited by a spark timed to occur near the maximum compression.[5])
(3) The air expands adiabatically from C to D, performing work as the air temperature drops. (This approximates the *working stroke* of the engine.)
(4) The air is cooled at constant volume from D to A, thus permitting the cycle to start again. (This approximates the *exhaust stroke* of the engine.)

Only the two isovolumic processes involve heat exchanges. Assuming the air to behave as an ideal gas, the theoretical efficiency of the engine (that is, the ratio of the work output to the heat input) is calculated to be:

OTTO CYCLE EFFICIENCY

$$\frac{\text{Work output}}{\text{Heat input}} = 1 - \frac{1}{\left(\dfrac{V_A}{V_B}\right)^{\gamma-1}} \tag{20-7}$$

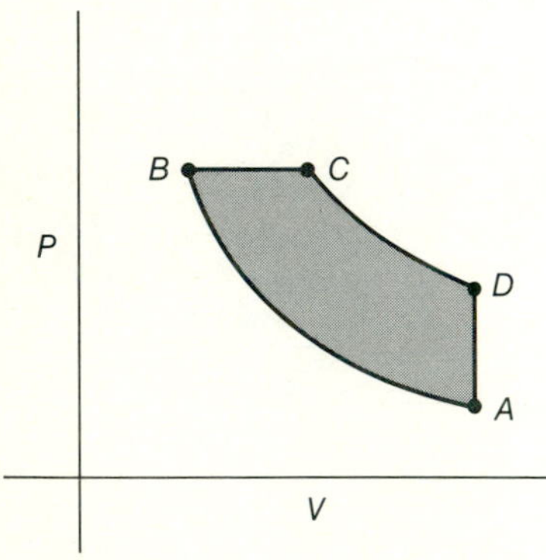

Figure 20–7
An idealized approximation for *the diesel cycle.*

For a compression ratio of 8 and $\gamma = 1.4$, for example, the theoretical efficiency is 56%. (Of course, the actual efficiency of the engine is much lower than this theoretical value.) Note that the efficiency improves as the compression ratio is increased.

Another type of engine that does not use the Carnot cycle is the *diesel engine*. An idealized diesel cycle is shown in Figure 20–7. Initially, air at atmospheric pressure (point A) is compressed adiabatically, increasing its temperature enough so that fuel oil injected at the end of the compression stroke (point B) ignites spontaneously. (No spark plugs are necessary.) The explosion of the vapor is slower than in a gasoline engine, and this process is approximated in this highly simplified version as an expansion first at constant pressure, then adiabatically, along the line $B \to C \to D$. Finally, the exhaust stroke restores the engine to its original condition. Since no oil vapor is present during

[5] If the compression is too high, the temperature may rise enough to ignite the gasoline vapor spontaneously before the proper time, an undesirable condition known as pre-ignition.

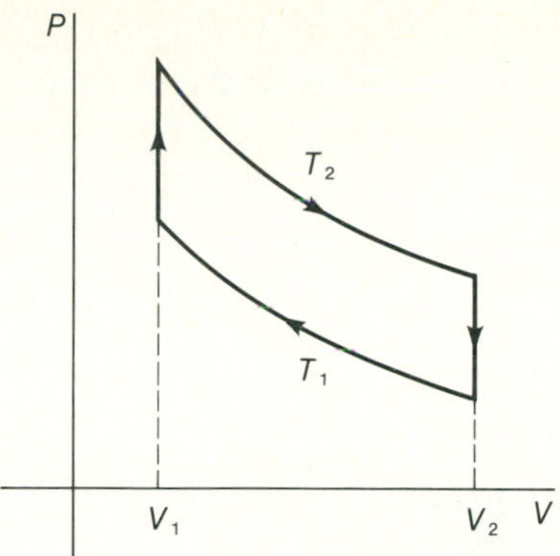

(a) An idealized cycle of a Stirling heat engine, patented in 1827.

(b) A kangaroo rat.

Figure 20–8

The Stirling heat engine utilizes heat exchanges in a way similar to that in the nose of the kangaroo rat, which lives in the Arizona desert. (See Knut Schmidt-Nielsen, *How Animals Work,* Cambridge University Press, 1972, pages 22–25.) These heat-exchange processes are the single most important factor in the water balance of small desert rodents. The idealized Stirling engine employs two isotherms and two contant-volume processes. A recent application of the Stirling engine provides a method of attaining very low temperatures (below liquid helium) that approach absolute zero.

compression, pre-ignition cannot occur; thus the compression ratio may be higher than that in an internal combustion engine. Typical ideal efficiencies are 50% to 60%.

20.5 Proof That No Engine Can Exceed the Carnot Efficiency

Using the second law of thermodynamics, Carnot was able to draw several interesting conclusions. His logic remains valid today even though at that time the first law was not known, nor was the molecular-motion basis for heat understood. The first conclusion we will discuss is known as Carnot's theorem:

CARNOT'S THEOREM **The Carnot cycle is the most efficient of all possible engines operating between the same two temperatures.**

We will not present the proof of this theorem in the complicated (but correct) fashion that Carnot did without the first law. Instead, we will give a simpler version using the first law. The argument is as follows. Suppose an inventor claims he has devised an irreversible engine that is more efficient than a Carnot engine. To prove his claim false, we hook up his gadget to a Carnot engine, as shown in Figure 20–9. Now, a Carnot engine is reversible, which means we may operate it in the reverse direction. The size of the Carnot engine is chosen so that its operation requires an amount of work input W exactly equal to the work output W' of the proposed engine.

The inventor's claim that the efficiency of his engine exceeds that of the Carnot engine translates mathematically into:

$$\left(\frac{W'}{Q'_2}\right)_{\substack{\text{proposed}\\ \text{engine}}} > \left(\frac{W}{Q^2}\right)_{\substack{\text{Carnot}\\ \text{engine}}} \qquad \text{(assumed true)} \qquad \textbf{(20–8)}$$

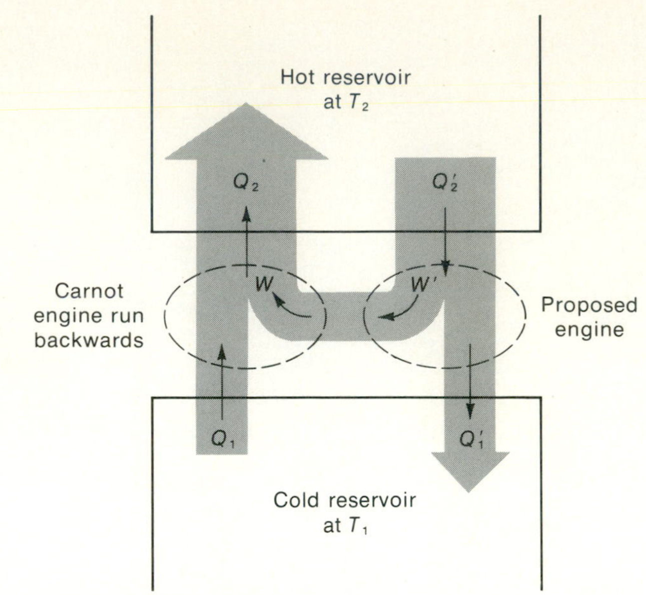

Figure 20–9
One way of proving that a Carnot engine is the most efficient of all possible engines. The work output of the proposed engine is arranged to drive the Carnot engine backwards (as a refrigerator). Primed symbols represent values for the proposed engine.

Since $W' = W$, this implies that:

$$Q_2 > Q_2' \tag{20–9}$$

From the first law, we have:

$$W = Q_2 - Q_1 \qquad \text{and} \qquad W' = Q_2' - Q_1'$$

Again, since $W' = W$, we have, using the inequality of Equation (20–9):

$$Q_1 > Q_1' \tag{20–10}$$

Thus, the overall effect of the two engines operating together as shown is to transfer some thermal energy from the cold reservoir to the hot reservoir *without any net work being done on the combined system by the external world.* This violates the second law of thermodynamics: It is impossible to construct a cyclic device that transfers thermal energy from a cold reservoir to a hot reservoir without work being performed on the device by the external world. Therefore, the assumed efficiency of the proposed engine cannot be greater than the Carnot efficiency.

In a similar fashion, if the proposed engine is a reversible one, its efficiency also cannot exceed that of a Carnot cycle. For this case, either engine may be run backwards. Therefore, the same reasoning leads to the conclusion that *all reversible engines operating between the same two temperatures have the same efficiency—the Carnot efficiency.* Furthermore, since no mention of the particular type of working substance had to be made during the proof, it follows that *the efficiency of a reversible Carnot engine is independent of the working substance used in the engine.*

20.6 The Kelvin Absolute Temperature Scale

We now can describe the absolute temperature scale proposed by Kelvin in 1848. As you will see, it is defined independently of the physical properties of any particular substance. Instead, it is based on the efficiency of a reversible engine:

$$e = 1 - \frac{Q_1}{Q_2} = 1 - \frac{T_1}{T_2} \tag{20–11}$$

Kelvin suggested that this equation can form the basis for defining a new temperature scale:

$$\frac{T_1}{T_2} = \frac{Q_1}{Q_2} \tag{20–12}$$

where T_1 and T_2 are two different points on the Kelvin scale. Since only their ratio is defined, we need to assign a numerical value for one specific temperature on the scale. To make the temperature *intervals* correspond to the Celsius scale, the unique point is defined to be the triple point of water (see Chapter 18), assigned a value of 273.16 K. Since the triple point is 0.01°C, the zero in the Kelvin scale corresponds to −273.15°C.

Note that the Kelvin scale is defined in terms of an energy *ratio rather than in terms of the physical properties of a particular substance.* The triple point of water comes into the definition only as a means of making temperature intervals correspond to the Celsius scale.

The defining equation for Kelvin temperatures may be written in terms of the heat transfers that occur when a Carnot engine operates with one temperature reservoir at the triple point of water. Thus, if Q heat transfer occurs at temperature T, and $Q_{tr\ pt}$ occurs at $T_{tr\ pt} \equiv 273.16$ K, then:

THE KELVIN ABSOLUTE TEMPERATURE SCALE

$$T \equiv (273.16\ \text{K})\left(\frac{Q}{Q_{tr\ pt}}\right) \tag{20–13}$$

Because this temperature scale is defined without reference to any particular substance, it is truly an absolute scale. Its temperature intervals can be made to correspond exactly to those of the ideal-gas constant-volume scale (see Chapter 17) by defining the triple point temperature for water to be 273.16 K.

20.7 The Third Law of Thermodynamics

If we could obtain a cold-temperature reservoir at $T_1 = 0$ K, the Carnot efficiency would be $1 - (T_1/T_2) = 1 - 0 = 100\%$. Unfortunately, it is an experimental fact that as we try to reach colder and colder temperatures, each succeeding step becomes more difficult to accomplish. In fact, there are also good theoretical reasons to accept a statement known as the third law of thermodynamics:

THIRD LAW OF THERMODYNAMICS **It is impossible to reduce the temperature of a body to absolute zero in a finite number of steps.**

If a Carnot cycle were operated with the cold reservoir at absolute zero ($T_1 = 0$), Equation (20–3) implies that no heat would be rejected to the cold reservoir ($Q_1 = 0$). Since the work W is equal to $Q_2 - Q_1$, such an engine would transform a quantity of thermal energy Q_2 entirely into work, in violation of the second law. Thus the third law forms a consistent and logical structure together with the other three laws: the zeroth, first, and second.

Though we will not deal with the third law further, it is included here to complete the basic foundation upon which the science of thermodynamics rests.

Summary

The *second law of thermodynamics* may be stated in several equivalent ways. Three such statements are:

The Kelvin-Planck statement:
It is impossible to build a 100% efficient cyclic engine that converts thermal energy into work.

The Clausius statements:
It is impossible to construct a cyclic device that transfers thermal energy from a cold reservoir to a hot reservoir without work being performed on the device by the external world.

Thermal energy will not, of its own accord, flow from a cooler object to a warmer object.

The *Carnot cycle* is a series of reversible processes that operate along isothermal and adiabatic lines. This cycle has the maximum possible efficiency attainable by any heat engine operating between two given temperature reservoirs (*Carnot's theorem*). The efficiency of heat engines is defined as:

Efficiency
$$e = \frac{\text{Work output}}{\text{Heat input}} = 1 - \frac{Q_1}{Q_2}$$

$$\underset{\text{(Carnot)}}{e} = 1 - \frac{Q_1}{Q_2} = 1 - \frac{T_1}{T_2}$$

Cycles for various other types of heat engines may be approximated by idealized reversible processes; each permits a calculation of the theoretical maximum efficiency.

The *Kelvin absolute temperature scale* (K) is based on the heat exchanges involved in a Carnot cycle. It is defined without reference to any particular substance. Its zero corresponds to −273.15°C, and the temperature intervals are identical to those on the Celsius scale.

A Carnot cycle, being reversible, may be operated backwards as a *refrigerator*, transferring heat from a cold object to a hot object and requiring work W to operate the device. A measure of the effectiveness of this process is the *performance coefficient* μ:

Performance coefficient of a refrigerator
$$\mu = \frac{\text{Heat extracted from the cold reservoir}}{\text{Work input}} = \frac{Q_1}{Q_2 - Q_1}$$

$$\underset{\text{(Carnot)}}{\mu} = \frac{Q_1}{Q_2 - Q_1} = \frac{T_1}{T_2 - T_1}$$

The *third law* (along with the other laws of thermodynamics) completes the basic foundation upon which the science of thermodynamics rests.

Third law of thermodynamics — *It is impossible to reduce the temperature of a body to absolute zero in a finite number of steps.*

Questions

1. Is it possible to cool a kitchen by leaving the refrigerator door wide open?
2. Explain the contradiction involved in a specific "perpetual motion" machine that violates the first law.
3. Analyze a specific "perpetual motion" machine that violates the *second* law, and explain where the contradiction lies.
4. An inventor announces he has devised a machine that takes in air, converts some of the thermal energy of the air into work, and rejects the cooler air out of an exhaust nozzle. He claims the device is not violating the first law, since energy is conserved. He also claims the device is not violating the second law because not *all* the thermal energy of the air was converted to work. Prove that his device cannot work because it *does* violate one (or both) of the thermodynamic laws.
5. Using the second law as a central idea in your discussion, explain why it is impossible for a refrigerator to cool an object to absolute zero.
6. In 1977, three businesspeople invested $17 000 with a British inventor who claimed to have developed an automobile that runs on water alone. The process uses current from the battery to decompose water into hydrogen and oxygen. The engine runs on this hydrogen gas, driving the car and recharging the battery in the meantime. According to the inventor, "all you do is pull up to a water faucet every 100 miles or so." Discuss this invention.

Problems

In the following problems, assume all gases are ideal.

20A–1 A proposal for an energy-conserving heat engine calls for utilizing the surface ocean water in the tropics as the high-temperature reservoir (22°C) and water at a depth of 700 m as the low-temperature reservoir (5°C). Determine the maximum thermal efficiency of such an engine. *Answer:* 5.76%

20A–2 Find the maximum efficiency of a heat engine operating between a bubbling hot spring at 100°C and an air temperature of 5°C.

20A–3 A Carnot engine operates between temperatures of 400 K and 300 K. If it absorbs 600 J of heat from the higher temperature reservoir, determine the amount of work the the engine performs. *Answer:* 150 J

20A–4 A heat engine whose efficiency is 15% performs 1000 J of work. Find the amount of heat it absorbs from the high-temperature reservoir.

20A–5 A heat engine whose Carnot efficiency is 30% absorbs heat from a reservoir at 400 K. Determine the temperature of the cooler reservoir. *Answer:* 280 K

20A–6 An ideal heat engine performs work at the rate of 1000 kW. If it absorbs heat from a reservoir at 800 K, how much heat is discarded to the low-temperature reservoir (300 K) each second?

20A–7 A heat engine performs 1000 J of work while rejecting 6000 J of heat to the cold-temperature reservoir. Find the efficiency of the engine. *Answer:* 14.2%

20A–8 Suppose a Carnot engine is run backwards as a refrigerator. It extracts heat from a refrigerator at 250 K and ejects it at a temperature of 350 K. For each joule of heat removed from the refrigerator, how many joules are ejected to the high-temperature reservoir?

20A–9 A refrigerator with a performance coefficient of 5 extracts heat from a reservoir at −50°C. Find the temperature of the hotter reservoir. *Answer:* −5.40°C

20B–1 A heat engine operating between 200°C and 80°C achieves 20% of the maximum possible efficiency. What energy input will enable the engine to perform 10^4 J of work?
Answer: 1.97×10^5 J

20B–2 Sketch a Carnot cycle on a temperature-volume diagram.

20B–3 An ideal monatomic gas is carried around the cycle $a \rightarrow b \rightarrow c \rightarrow d$ in Figure 20–10. Find, in terms of P_0 and V_0, the net work done by the gas. *Answer:* $(4/3)P_0V_0$

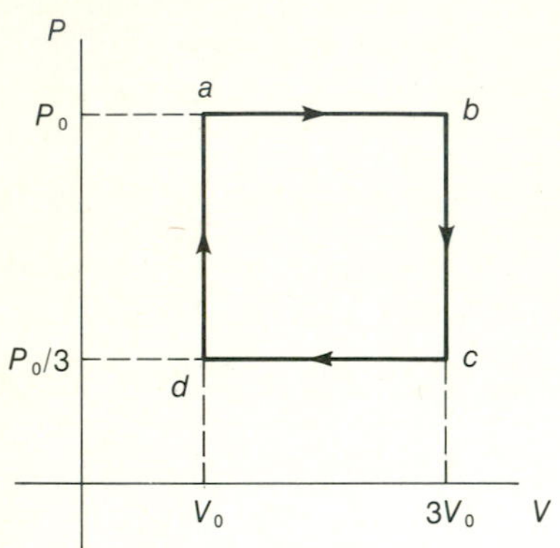

Figure 20–10
Problems 20B–3 and 20B–4

20B–4 In the previous problem, suppose the gas had been carried through the processes $a \rightarrow b \rightarrow c$ and then taken by a *straight-line* path from point c to point a. What would be the net work done per cycle by the gas?

20B–5 A Carnot refrigerator operates between a room temperature of 27°C and the temperature of the freezing compartment of a refrigerator. Find the work input that will extract 10 J of heat (a) from a compartment at 0°C and (b) from a compartment at −50°C.
Answers: (a) 0.99 J (b) 3.45 J

20B–6 A 0.4-kW motor operates a Carnot refrigerator between −10°C and 27°C. Find (a) the amount of heat per second extracted from the low-temperature reservoir and (b) the amount per second ejected to the high-temperature reservoir.

20B–7 How much work (in joules) is required to pump 1000 cal of thermal energy from frozen ground at 0°C into a house at 27°C using a Carnot refrigerator engine? (b) What total amount of heat is added to the house? *Answers:* (a) 414 J (b) 4600 J

20B–8 A house is heated using a heat pump that has a performance coefficient of 3. If a monthly heating bill is $45 for the electricity to operate the heat pump, what would have been the cost if ordinary electrical heaters were used instead of the heat pump?

20B–9 Working steadily, the average adult can perform manual labor of about 0.13 kW·h in an 8-h day.

- **(a)** Estimate the number of persons it would take to crank electric generators (assumed 100% efficient) to light a lecture hall using 6000 W (typical for a well-lighted classroom of 100 students).
- **(b)** Find the cost per 8-h day if these persons were paid the legal minimum wage.
- **(c)** Find the cost per day at a power company rate of 7¢ per kW·h.

(Note: The above estimate is for fluorescent illumination. Incandescent bulbs require three to four times the power for the same illumination. Power for heating and air conditioning has not been considered.)
Answers: (a) 370 persons (b) $9620.00 (at $3.25/h) (c) $3.36

20B–10 Solve the previous problem using the power consumption of your own classroom.

20B–11 Starting with Equation (20–8), derive the inequality expressed by Equation (20–10).

20C–1 A reversible engine operates in the cycle shown in Figure 20–11(a), which consists of two adiabatic and two isovolumic processes. Find the efficiency in terms of V_1 and V_2.
Answer: $1 - \left(\frac{V_1}{V_2}\right)^{(\gamma-1)}$

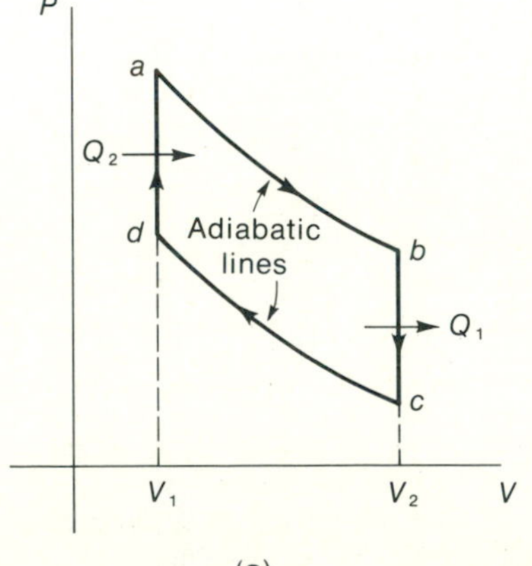

(a)

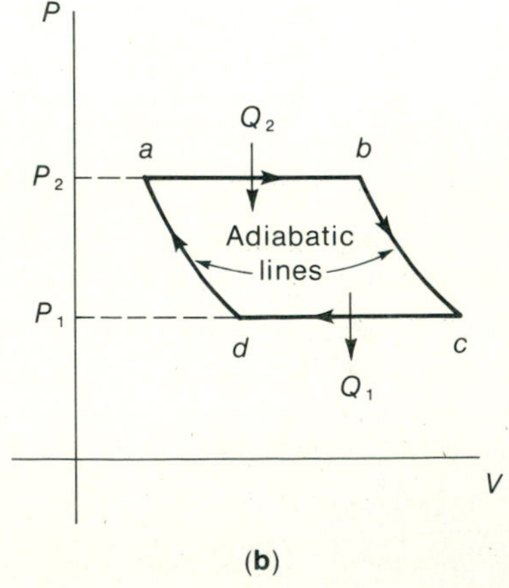

(b)

Figure 20–11
Problems 20C–1 and 20C–2

20C–2 A reversible engine operates in the cycle shown in Figure 20–11(b), which consists of two adiabatic lines and two isobars. Find the efficiency in terms of P_1 and P_2.

20C–3 An ideal heat engine carries 0.2 mol of an ideal gas ($\gamma = 5/3$) around the cycle shown in Figure 20–12. Process $a \to b$ is isovolumic, $b \to c$ is adiabatic, and $c \to a$ is isobaric. The temperatures at a and b are shown in the diagram.

(a) Find the unknown values of P, V, and T at points a, b, and c.

(b) Find the net work done per cycle.

Answers: (a) a: 4.92 L; b: 1.67 atm; c: 6.69 L, $T_c = 408$ K
(b) 52.7 J

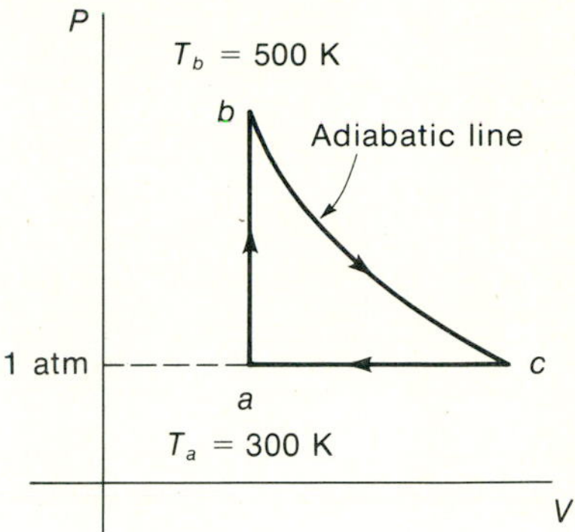

Figure 20–12
Problem 20C–3

20C–4 Two Carnot engines operate between the same two isothermal reservoirs. One uses an ideal monatomic gas, the other an ideal diatomic gas (with rotations but no vibrations). Each starts initially at the same values of P_0 and V_0, and each undergoes the same isothermal expansion to P_1 and V_1.

(a) Do both engines have the same efficiency?

(b) Is the heat absorbed during the isothermal expansion the same for both? (If not, which is greater?)

(c) Is the work per cycle the same for both? (If not, which is greater?) In each case, explain your reasoning.

20C–5 A reversible engine operates in the cycle shown in Figure 20–13, which consists of two isothermal and two isovolumic changes. Calculate the efficiency of the engine and show that it is the same as that of a Carnot cycle operating between the same two temperatures.

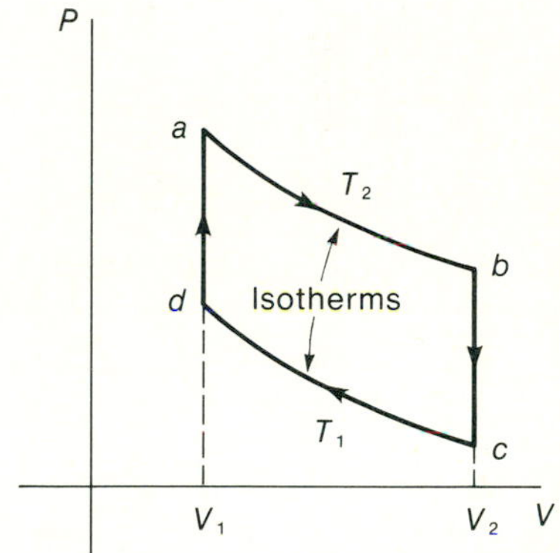

Figure 20–13
Problem 20C–5

20C–6 Two Carnot engines operate between the same two temperatures. One uses an ideal monatomic gas, the other an ideal diatomic gas. The minimum and maximum pressures are the same for both engines, and the minimum and maximum volumes are the same for both. Is the work per cycle the same for both? If not, which is greater and why? (Note: You do not need to do mathematical calculations; thermodynamic reasoning alone should be sufficient.)

20C–7 A refrigerator extracts heat from its freezing section at $-15°C$ and exhausts heat to the room at 27°C. It achieves 30% of the maximum theoretical thermal efficiency. What is the input requirement (in watts) to the motor if the refrigerator can freeze 40 g of water at 27°C to ice at $-15°C$ each minute? *Answer:* 173 W

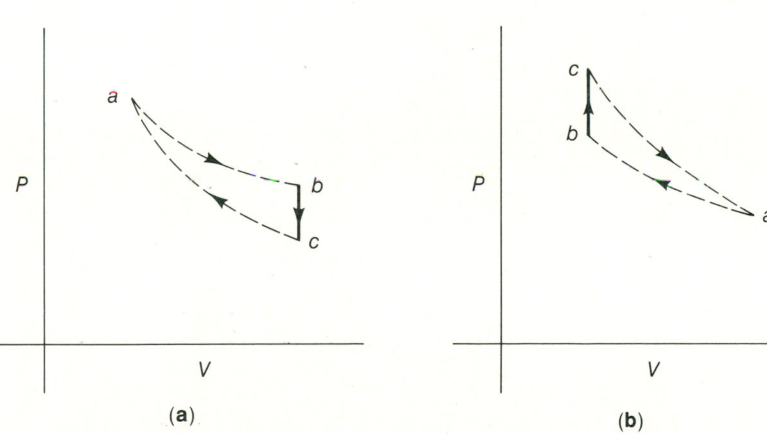

Figure 20–14
Problem 20C–8

20C–8

(a) Show that if two adiabatic lines intersected as in Figure 20–14(a) (the dashed lines), it is possible to violate the first law of thermodynamics by going around the cycle $a \to b \to c \to a$.

(b) Show that the cycle $a \to b \to c \to a$ in Figure 20–14(b) violates the *second* law of thermodynamics. In both figures, processes $b \to c$ are at constant volume.

20C–9 A new absolute temperature scale called "degrees Newtie" (for "new T") is based on the melting point and boiling point of a special compound found only on the moon. A Carnot engine operating between these temperatures absorbs 800 J at the higher temperature and rejects 600 J at the lower temperature. If these two temperatures are defined to be 100 Newtie degrees apart, find the temperatures (in degrees Newtie) at which this compound melts and boils. *Answer:* 400° N, 300° N

20C–10 Suppose two adiabatic lines on a pressure-volume diagram intersected. By considering a closed cycle formed by these lines and an isothermal expansion line, show that the second law of thermodynamics would be violated.

chapter 21 Entropy

The future belongs to those who can manipulate entropy; those who understand but energy will be only the accountants. . . . The early industrial revolution involved energy, but the automatic factory of the future is an entropy revolution.

FREDERIC KEFFER

Laws of Thermodynamics:
1. You cannot win.
2. You cannot break even.
3. You cannot stop playing the game.

In this chapter we will take another look at the second law, approaching it from a completely different perspective than we did in the previous chapter. The approach we will now take involves a new concept: *entropy*, which not only is interesting in itself, but which gives us a much deeper understanding of the second law. The laws of thermodynamics are among the most general in all of physics. The first law tells us that heat is a form of energy and it is conserved. The second law prohibits certain energy-conversion processes, *even though the overall energy is conserved*. So the second law is an important statement about what *cannot* be done.

Processes that occur in nature—so-called "natural" processes—often involve heat flow from hot to cold objects. In fact, there is no perfect heat insulator, so heat-leakage processes occur all the time. On the other hand, the reverse process, heat flow from a cold object to a hot one, never occurs naturally. You cannot get warm by snuggling up to an icicle, even though energy would be conserved in the transfer process. Thermal energy seems only to "run downhill" from higher temperatures to lower ones.

This is one meaning of the phrase "an irreversible process." The "uphill" flow of heat never occurs spontaneously on a macroscopic level. True, one can force thermal energy from a colder object to a hotter one. All refrigerators do this. But it is accomplished only at the expense of additional changes elsewhere: either the performance of work on the system by the external world or the rejection of still more thermal energy to some still colder object. *Thus, all natural processes involving heat flow are irreversible.*

The common theme in all such natural processes is that they involve an increase in randomization on a microscopic level. These dual ideas, macroscopic heat flow and microscopic randomization, are two ways of dealing with the same basic phenomenon. We will begin with the macroscopic viewpoint.

21.1 Entropy from a Macroscopic Viewpoint

Let us investigate certain aspects of the thermodynamic processes we discussed in the last chapter. From Equation (20–12) we have the following relation for a Carnot cycle:

$$\frac{Q_2}{T_2} = \frac{Q_1}{T_1} \tag{21–1}$$

We shall now consider heat transfers Q as algebraic quantities that are positive when heat enters a system and negative when heat is removed from the system. (Although this usage differs from that used in discussing Carnot cycles, it is the same sign notation used in the first law and the accepted usage in discussing entropy.) Adopting this sign convention, we conclude from Equation (21–1) that if we sum the quantity Q/T around one complete (reversible) Carnot cycle, the result is zero:

$$\frac{Q_2}{T_2} + \frac{Q_1}{T_1} = 0$$

Or, in more compact notation:

For a Carnot cycle $$\sum_{\substack{\text{(around the}\\ \text{entire cycle)}}} \left(\frac{Q}{T}\right) = 0 \qquad \textbf{(21–2)}$$

We recall that a Carnot cycle is, by definition, a *reversible* cycle. But what about other reversible processes that do not follow a Carnot path? What is the sum of (Q/T) for these cases? As we showed in the last chapter, any arbitrary, reversible cycle may be approximated by a series of isothermal and adiabatic steps that form a family of Carnot cycles (refer to Figure 20–5). Therefore, since we can approximate any reversible cycle by a series of Carnot cycles, we may conclude that the summation $\Sigma(Q/T)$ around any reversible cycle is also zero.

For any reversible cycle $$\sum_{\substack{\text{(around the}\\ \text{entire cycle)}}} \left(\frac{Q}{T}\right) = 0 \qquad \textbf{(21–3)}$$

In 1865, the German physicist Rudolf Clausius christened the quantity (Q/T) as "the **entropy** S" after the Greek word *trope*, meaning "change" or "transformation." (Clausius thought of the concept as measuring the potentiality of a system for bringing about a transformation.) For the change in entropy, this becomes

CHANGE IN ENTROPY ΔS $$S_2 - S_1 = \int_1^2 \frac{dQ}{T} \qquad \textbf{(21–4)}$$

where the entropy S is measured in units of joules/kelvin. The integral may be taken along any reversible path between the two states because, by Equation (21–3), the two states lie on some closed path describing a reversible cycle.

Note that we have not assigned an absolute numerical value to an entropy scale; we have obtained only the *change* in entropy. In this respect, entropy is similar to gravitational potential energy: *Only the changes in it are significant.* (However, just as we can for gravitational potential energy, we could, if desirable, assign a reference value to the entropy of a system in any particular state and then follow subsequent changes as the system proceeds away from that reference state.)

The concept of entropy is particularly useful because it is a true *variable of state.* That is, it behaves mathematically so that the change in the value of the parameter from one state to another *depends only on the two states themselves, not on the path that connects the states.* All state variables (including the internal energy U, the pressure P, and the temperature T) behave in this fashion. (Note that Q and W are *not* state variables.) An *equation of state* expresses the relation between state variables that a system in equilibrium must always obey.

We obtained the *temperature* T and the *internal energy* U from the zeroth and first laws, respectively. The *entropy* S is related to the second law. Entropy is an important concept because, as we will show later, it can be interpreted as a measure of the "quality" of the energy a system possesses, that is, the "availability" of the energy for doing work. Just knowing the total energy a system has is not enough for determining how much work the system can do. Only its entropy gives that information.

□ □ □

Up to this point we have discussed only complete, reversible *cycles*. But what about a process that carries a system just from one state to another? Even though such a process may not be reversible, we can always calculate the change in entropy between the initial and final states by considering *any* convenient reversible process connecting the two states. *This is possible because entropy is a state variable.* No matter which path we follow from state 1 to state 2, the change in entropy is the same.

As an example, let us consider the entropy change that occurs when an ideal gas is initially confined to a small volume by a partition that is later removed, permitting the gas to undergo a *free expansion* from V_i to V_f (see Figure 21–1). This is an *ir*reversible process. That is, the pressure, density, and temperature throughout the gas are not essentially uniform at all times; they deviate more than the *incremental* amounts allowed in reversible changes. For example, the faster, "hotter" molecules will move into the empty space more quickly than the slower molecules. Thus, since the gas is not in thermal equilibrium during the expansion, we cannot apply the equation of state in calculating the entropy change. In a sense, we lose mathematical control of the situation. However, *since the entropy of a system depends only on the state of the system*, the entropy change during this irreversible expansion must equal the entropy change for *any reversible path* that connects the same two states. This is the method we use when seeking entropy changes for an irreversible process: We merely calculate the entropy change for some other, convenient, reversible path connecting the same initial and final states.

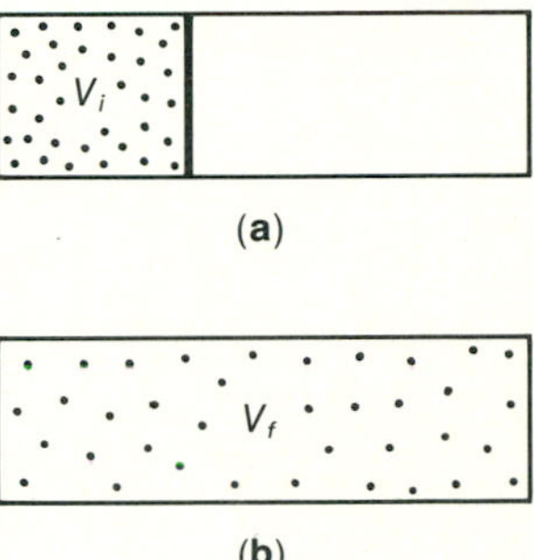

Figure 21–1
In a free expansion, the gas is allowed to expand into a larger volume that previously was a vacuum.

In a free expansion, the total energy of the gas does not change, but there is a very significant difference: the entropy increases. To pursue this statement further, consider the following line of reasoning. Since the expansion of the gas is into a vacuum, no work is done by the gas; therefore $W = 0$. Furthermore, the chamber is insulated, so no thermal energy is added to the gas; thus $Q = 0$. From the first law:

$$Q = \Delta U + W$$

we have:

$$0 = \Delta U + 0$$

Since $\Delta U = 0$, the temperature remains constant. Therefore, an isothermal expansion is one type of reversible process that connects the two states. Note that even though we are investigating a free expansion, *because entropy depends only on the state of the system* we can calculate the entropy change for a *different* process we already know about (an isothermal expansion), recognizing that the answer we obtain is also the entropy change for the free expansion we are investigating.

According to the analysis in Chapter 19, Equation (19–5), an isothermal expansion from V_i to V_f requires the addition of an amount of heat Q:

$$Q = nRT \ln\left(\frac{V_f}{V_i}\right) \qquad \textbf{(21–5)}$$

The entropy change (Q/T) for this reversible expansion (and also for the free expansion we seek) is thus an increase by the amount:

ENTROPY CHANGE IN A FREE EXPANSION

$$\Delta S = nR \ln\left(\frac{V_f}{V_i}\right) \qquad \textbf{(21–6)}$$

In addition to the increase in entropy for a *free* expansion, there is an entropy increase every time we pop a balloon, exhale, expel gases from an automobile exhaust, and so forth.

Macroscopic processes always proceed so as to increase entropy. This one-way direction in the change of entropy has been called "nature's arrow," because it indicates the direction in which all natural processes proceed. Mathematically, this is expressed as

$$\Delta S > 0 \qquad \text{(21–7)}$$

Natural
(irreversible)
processes

EXAMPLE 21–1

What is the entropy change when 2 kg of ice at 0°C melt to water at 0°C?

SOLUTION

Here, the temperature remains constant throughout the process: $T = 0°\text{C} = 273$ K. The heat transferred is due to the phase change from ice to water. We will assume the heat transfer occurs only as a result of *incremental* differences in temperature. Thus the process is *reversible*, and we may use Equation (21–4). The latent heat of freezing of ice is $L = 79.6$ cal/g $= 3.34 \times 10^5$ J/kg. Thus:

$$Q = mL = (2\text{ kg})(3.34 \times 10^5\text{ J/kg}) = 6.68 \times 10^5\text{ J}$$

and the entropy change is

$$\Delta S = \int \frac{dQ}{T} = \frac{\Delta Q}{T} = \frac{6.68 \times 10^5\text{ J}}{273\text{ K}} = \boxed{2.45 \times 10^3\ \frac{\text{J}}{\text{K}}}$$

EXAMPLE 21–2

A hot stone of mass m_2, specific heat c_2, and temperature T_2 is dropped into a container of cool water of mass m_1, specific heat c_1, and temperature T_1 (where $T_1 < T_2$). The final temperature of the combination is T_f. (Ignore the effects of the container.) What entropy change occurs in this process?

SOLUTION

In actuality, the process is an irreversible one, since finite temperature differences are involved. However, if we imagine a reversible process connecting the initial and final states, the entropy change would be the value we seek (since entropy changes do not depend on the path). So we imagine that the stone is gradually (reversibly, that is) cooled to T_f, with only incremental temperature differences and incremental heat transfers. Similar restrictions apply to the warming of the water. On this basis, the entropy change ΔS is

$$\Delta S = \underset{\text{water}}{\int \frac{dQ}{T}} + \underset{\text{stone}}{\int \frac{dQ}{T}}$$

The amount of heat transfer will be $dQ = mc\,dT$. Thus:

$$\Delta S = m_1c_1\int_{T_1}^{T_f}\frac{dT}{T} + m_2c_2\int_{T_2}^{T_f}\frac{dT}{T}$$

$$= m_1c_1(\ln T_f - \ln T_1) + m_2c_2(\ln T_f - \ln T_2)$$

$$\Delta S = \boxed{m_1c_1\ln\left(\frac{T_f}{T_1}\right) + m_2c_2\ln\left(\frac{T_f}{T_2}\right)}$$

The first term is positive because $T_f > T_1$, and the second term is negative because $T_f < T_2$. Moreover, it can be shown that since $T_1 < T_f < T_2$, the positive term is always larger in magnitude, resulting in an entropy increase in every case.

EXAMPLE 21–3

One end of a copper rod is in thermal contact with a reservoir at $T_2 = 500$ K and the other end is in thermal contact with a reservoir at $T_1 = 300$ K. If 2000 cal are conducted from one end to the other, with no change in the temperature distribution along the rod, find (a) the entropy change of each reservoir, (b) the entropy change of the rod, and (c) the entropy change of the universe

SOLUTION

(a) The heat involved is

$$Q = (2000\text{ cal})\underbrace{\left(\frac{4.184\text{ J}}{1\text{ cal}}\right)}_{\substack{\text{Conversion}\\ \text{factor}}} = 8368\text{ J}$$

The entropy change of the hot reservoir is therefore

$$\Delta S_2 = \frac{Q}{T_2} = \frac{-8368\text{ J}}{500\text{ K}} = \boxed{-16.7\,\frac{\text{J}}{\text{K}}}$$

The entropy change in the cold reservoir is

$$\Delta S_1 = \frac{Q}{T_1} = \frac{+8368\text{ J}}{300\text{ K}} = \boxed{+27.9\,\frac{\text{J}}{\text{K}}}$$

(b) The rod has not undergone a net change of thermal energy or temperature. Therefore:

$$\boxed{\Delta S_{\text{rod}} = 0}$$

(c) The net change of entropy for the universe is

$$\Delta S_{\text{universe}} = \Delta S_1 + \Delta S_2 + \Delta S_{\text{rod}} = 27.9\,\frac{\text{J}}{\text{K}} + \left(-16.7\,\frac{\text{J}}{\text{K}}\right) + 0$$

$$= \boxed{11.2\,\frac{\text{J}}{\text{K}}}$$

21.2 Entropy from a Microscopic Viewpoint

As with other topics in thermodynamics, the concept of entropy can be approached from microscopic considerations (that is, from a statistical analysis of molecular motions) as well as from macroscopic concepts (from pressure, temperature, and other parameters measured with laboratory instruments). Let us now look at the process just discussed—the free expansion of an ideal gas—from a microscopic viewpoint.

According to kinetic theory, we imagine the gas molecules to be mass points in random motions. Initially all the gas is confined to the volume V_i [Figure 21–1(a)]. When the partition is removed, the molecules eventually become more or less evenly distributed throughout the larger volume V_f. The exact distribution is a matter of probability.

To determine the probabilities for the distributions, we begin by finding the probability for the various molecular locations involved in the free expansion process. Just at the instant after the partition is removed (and before the molecules have a chance to move into the other half of the chamber), we find all the molecules in the initial volume V_i. We can easily estimate the probability of this particular configuration occurring "naturally" through random motions of molecules in a larger volume V. In that case, since each molecule would travel freely throughout the total volume, the probability W_i of finding one particular molecule in the region V_i, which is a fraction (V_i/V) of the total volume V, is just the ratio of the volumes:

Probability $$W_i = \frac{V_i}{V} \qquad \text{(for one molecule)} \tag{21–8}$$

As we add more molecules to the system, if they move independently of one another the probabilities multiply together. Therefore, the probability that N molecules will simultaneously be found in the volume V_i is just

$$W_i = \left(\frac{V_i}{V}\right)^N \qquad \text{(for } N \text{ molecules)} \tag{21–9}$$

Similarly, the probability of finding all N molecules in some larger volume V_f is

$$W_f = \left(\frac{V_f}{V}\right)^N \qquad \text{(for } N \text{ molecules)} \tag{21–10}$$

The ratio for these two probabilities is

$$\frac{W_f}{W_i} = \frac{\left(\frac{V_f}{V}\right)^N}{\left(\frac{V_i}{V}\right)^N} = \left(\frac{V_f}{V_i}\right)^N \tag{21–11}$$

If we now take the natural logarithm of this equation and multiply by Boltzmann's constant k, we can draw a very interesting conclusion.

$$k \ln\left(\frac{W_f}{W_i}\right) = nN_A k \ln\left(\frac{V_f}{V_i}\right) \tag{21–12}$$

where we write the number of molecules N as nN_A, the number of moles times Avogadro's number. From Chapter 18 we know that $N_A k$ is the universal gas constant R, so this equation may be written as:

$$k \ln W_f - k \ln W_i = nR \ln\left(\frac{V_f}{V_i}\right) \tag{21–13}$$

But we have found from thermodynamic considerations [specifically, Equation (21–6)] that when n moles of a gas undergo a free expansion from V_i to V_f, the change in entropy ΔS is

$$S_f - S_i = nR \ln\left(\frac{V_f}{V_i}\right) \tag{21–14}$$

Note that the right-hand sides of Equations (21–13) and (21–14) are identical. We thus make the following connection between *entropy* and *probability*:

ENTROPY *S* AND PROBABILITY *W*

$$S = k \ln W \tag{21–15}$$

Although our discussion used the specific example of the free expansion of an ideal gas, a more rigorous development of the statistical interpretation of entropy leads to the same conclusion.

EXAMPLE 21–4

Again considering a free expansion, let us verify that the macroscopic and microscopic approaches lead to the same conclusion. Suppose one mole of an ideal gas undergoes a free expansion to four times its initial volume V_i. The initial and final temperatures are, of course, the same. (a) Using a macroscopic approach, calculate the entropy change. (b) Find the probability W_f that all the molecules will, through random motions, be found simultaneously in the original volume V_i. (c) Using the probability considerations of part (b), calculate the change in entropy ΔS for the free expansion and show that it agrees with part (a):

$$\Delta S = S_f - S_i = k \ln W_f - k \ln W_i = k \ln\left(\frac{W_f}{W_i}\right)$$

SOLUTION

(a) From Equation (21–6), we obtain

$$\Delta S = nR \ln\left(\frac{V_f}{V_i}\right) = (1)R \ln\left(\frac{4V_i}{V_i}\right) = \boxed{R \ln 4}$$

(b) The probability W_i that a single molecule will be in V_i is just the ratio of the volumes:

$$W_i = \frac{V_i}{4V_i} = \frac{1}{4} \qquad \text{(for one molecule)}$$

For one mole (N_A) of molecules (which move independently of one another), the probability that all of them will be in V_i is the extremely small number:

$$W_i = \boxed{\left(\frac{1}{4}\right)^{N_A}} \qquad \text{(for } N_A \text{ molecules)}$$

(c) The probability W_f of all N_A molecules being in the volume $4V_i$ is 1 (since they cannot be anywhere else).

$$\Delta S = k \ln\left(\frac{W_f}{W_i}\right) = k \ln\left[\frac{1}{(1/4)^{N_A}}\right] = N_A k \ln 4 = \boxed{R \ln 4}$$

This is the same answer as in part (a), which dealt with macroscopic parameters.

21.3 Entropy and the Second Law

The connection between entropy and probability gives us a new insight into the second law. Entropy tends to increase for a very obvious reason: Every natural process tends toward the state of maximum probability. That is, systems always change toward the most probable configuration, the state of maximum randomization. This increase cannot be "proved" any more than the downhill direction of heat flow can be "proved." It is just a statistical tendency. But because macroscopic systems involve such an incredibly large number of molecules, this tendency is overwhelmingly what happens.

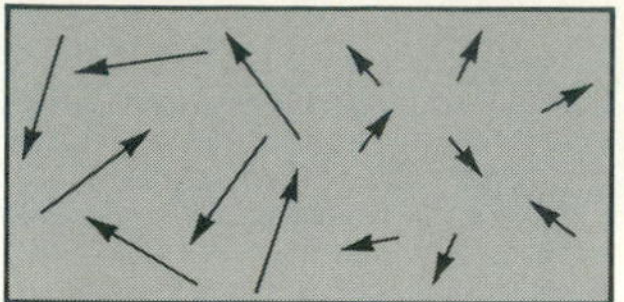

(a) An *ordered* arrangement and a very *unlikely* distribution of molecular speeds.

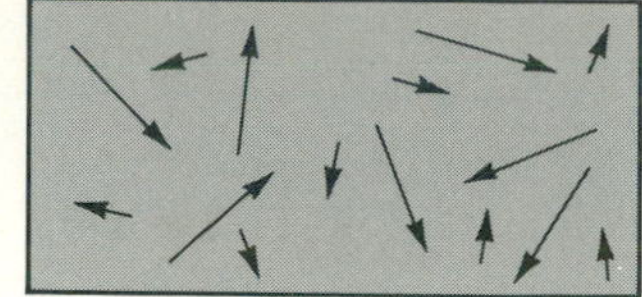

(b) A *disordered* arrangement and a *likely* distribution of molecular speeds.

Figure 21–2
A box of gas in two of the possible states of molecular motions that may exist. Note that each state contains the same total energy.

From considerations such as these, we can express the second law in terms of entropy. The following statement is entirely equivalent to other forms of the second law.

SECOND LAW OF THERMODYNAMICS (Entropy form)	$\Delta S > 0$	**For all natural (that is, irreversible) processes**
	$\Delta S = 0$	**Only for reversible processes**

A specific example will be helpful. Consider a box of gas in which molecules with speeds above the median value are gathered together in the left-hand side and molecules with slower speeds are gathered in the other side, as shown in Figure 21–2(a). This particular arrangement is, of course, very unlikely, since random molecular motions tend to bring about a more or less uniform mixing of the fast and slow groups, as in Figure 21–2(b). In the first case, the phrases "an *ordered* arrangement" or "*unlikely* probability" describe the situation. In the second case, the phrases "a *disordered* arrangement" or "*likely* probability" are more applicable. If there were only 10 molecules in the box, occasionally we might find, at some instant, the five faster molecules in one half and the five slower ones in the other half. But as we increase the number of molecules, the chances of random fluctuations producing a fast-slow separation become truly negligible: not zero, but extremely close to zero.

Let us estimate this probability. The chance of any one molecule being in its "correct" half of the box as a result of random motion is 1/2. As we add more molecules to the system, assuming they move independently, the probabilities multiply together. Thus, if we have, say, 100 molecules, the probability W of achieving a fast-slow separation through random motion is:

$$W = \begin{bmatrix} \text{Probability of the 50} \\ \text{faster molecules being} \\ \text{in the left half} \end{bmatrix} \begin{bmatrix} \text{Probability of the 50} \\ \text{slower molecules being} \\ \text{in the right half} \end{bmatrix}$$

$$W = \left(\frac{1}{2}\right)^{50}\left(\frac{1}{2}\right)^{50} = \left(\frac{1}{2}\right)^{100} = \frac{1}{1.27 \times 10^{30}}$$

or about 1 chance in 10^{30}. As the number of molecules in the system increases, the probability rapidly becomes even smaller.[1] Thus the ordered arrangement

[1] Large numbers, written in scientific notation, are sometimes deceptive. For example, 10^{30} is about 10 000 000 000 000 *times* the total number of seconds since the earth was formed about 4.5 billion years ago.

of Figure 21–2(a), when extrapolated from 100 molecules to a mole of gas ($\sim 10^{23}$ molecules), has an extremely low probability of occurring.

The second law is really "known" by everyone. Its consequences are everywhere about us. Every time you see something which, if a motion picture film of the event were run backwards, would seem silly, then you have seen the second law at work. Handwriting is not "scooped up" into a ballpoint pen; automobile exhausts do not take in smoggy air and convert it to gasoline and clear air; an athlete never becomes refreshed by running backwards around a track as she absorbs her perspiration; hair clippings never leap off the barbershop floor and reunite themselves with their respective mates; a diver does not leap backwards out of the swimming pool onto the diving board with the waves and splashes suddenly turning to a smooth water surface; and a pan of water on the stove flame does not freeze (heat does not flow from cold objects to hot ones.) In fact, *almost every macroscopic happening you observe clearly illustrates the second law.*

When an ice cube melts, entropy increases. The second law does not imply, of course, that the melted ice cube could not be refrozen. But to do so, the freezer would necessarily bring about a *still greater* increase in entropy by its own operation. Once the ice cube melts, it is impossible to have everything exactly as it was.

Of course we create order, for example, every time we fit a jigsaw puzzle together or build an automobile. Living organisms themselves become increasingly ordered as they grow, producing a local decrease in entropy as long as they continue to assemble molecules in a highly organized pattern. The human brain is a masterpiece of low-entropy concentration. But the creation of each of these ordered systems, when analyzed in terms of the overall entropy changes in the system *and in its surroundings*, always results in a net entropy increase. An isolated system, by itself, cannot decrease its entropy. The entropy can only remain the same or increase.

It is difficult to conceive of the universe ever operating the other way around, of all natural processes ever tending toward greater order. For example, if we place 100 white marbles on top of 100 similar black marbles in a box and shake them, they inevitably get mixed up. The opposite effect, starting with a random mixture and finding that upon shaking they *almost always* sort themselves into separate black and white groups, seems so unusual as to be inconceivable. For this reason, the second law is perhaps the most unrepealable of all the laws of physics.

Occasionally the question is posed: "But since the 'downhill' flow of heat is only a matter of probability, isn't there a definite chance that a pan of water on the fire will freeze instead of heat up?" Or perhaps: "Couldn't the random motions of molecules, by chance, combine to cause a small object to jump up from the table? Thus, *random* motions of atoms would be partially converted to the *ordered* motion of the kinetic energy of an object as a whole." True, the probabilities are not absolutely *zero*. However, the statistics of the situation reveal what is involved. As an example, consider a 1-g mass at rest on a table. It is in thermal equilibrium, continually interchanging energy with its surroundings. On a microscopic scale, however, there are random fluctuations in these energy interchanges. Some atoms momentarily have larger-than-average energy, and at other times they have lower-than-average energy. In principle, these statistical variations could result in a slight excess of momentum in one direction. This excess might be large enough to cause the mass to spontaneously jump up from the table a small distance, converting the kinetic energy of random molecular motion into gravitational potential energy. But the chance that we will observe this happening is truly insignificant. It can be shown that the fraction of the time the mass will gain, through statistical fluctuations, enough energy to jump at least 1 mm vertically against gravity is only about:

$$1 \text{ part in } 10^{1\,000\,000\,000\,000\,000} \quad \text{or} \quad 10^{10^{15}}$$

To gain some perspective on this number, the total number of electrons, protons, and neutrons in all the galaxies of the known universe is perhaps on the order of 10^{82}.

Of course, the size of the mass involved is an important factor. We can actually see such statistical fluctuations of position if we observe (with a microscope) very small particles suspended in a fluid. The unequal molecular impacts on various sides of the particle cause it to dance around in a haphazard, zigzag fashion known as *Brownian motion.* Even for larger masses, statistical fluctuations have noticeable consequences. The human eardrum fluctuates with thermal motion at an average amplitude of about 10^{-10} m. If an incoming sound is so faint that it does not cause the eardrum to vibrate with amplitudes greater than this threshold value, we hear the thermal "noise" of our eardrums instead of the incoming sound.

The one-way behavior of entropy changes for large-scale systems is further dramatized by one author[2] in a colorful manner. He writes: "At room temperature, conversion of a single calorie of thermal energy completely into potential energy is a less likely event than the production of Shakespeare's complete works fifteen quadrillion times *in succession without error* by a tribe of wild monkeys punching randomly on a set of typewriters." (Emphasis added.) Because the second law is based ultimately on statistical considerations, its validity is overwhelming. There seems to be no way to get around it.

21.4 Entropy and Unavailable Energy

The two configurations of Figure 21–2 have an important difference other than the obvious one of order and disorder. The ordered arrangement of (a) could do some work because, in a sense, the two groups of molecules constitute high- and low-temperature reservoirs between which we could operate an engine. But the more disordered arrangement of (b) could not do this, since the entire system is at the same temperature. *Note that each configuration contains exactly the same total amount of thermal energy.* But in Figure 21–2(b) this energy is unavailable for doing work.[3] This fact is at the heart of the second law. Just having some energy does not tell the whole story. It is also significant whether or not the energy is in a form that is available for doing work. Of course, any source of thermal energy at a uniform temperature can run an engine provided we also have another source at a still lower temperature to act as a low-temperature reservoir. But this does not change the basic argument. We merely enlarge our considerations to include *both* sources and the same general conclusion follows: Whenever heat flows from a warmer to a colder body, this energy becomes less available for doing work. Although the energy is conserved, it has become "degraded." *In this sense, entropy is a measure of the unavailability of energy for doing work.*

---

[2] *The Second Law*, H.A. Bent, Oxford University Press, 1965. Also, if you wish to read a particularly clear and enjoyable discussion of thermodynamics, see the short book by H.C. Van Ness, *Understanding Thermodynamics*, McGraw-Hill Paperbacks, 1969.

[3] Some humorist has pointed out that the second law proves Hell must be a place that has the same temperature throughout. Otherwise, the resident thermodynamicists would use temperature differences to run a refrigerator to cool off a local region.

EXAMPLE 21-5

Suppose a quantity of heat Q is conducted from a hot reservoir at T_2 to a cold reservoir at T_1. (a) Find the entropy change of the universe. (b) What maximum work could have been done using this heat Q?

SOLUTION

(a) The entropy change involved in extracting heat Q from the hot reservoir is $-Q/T_2$. The entropy change in adding heat Q to the cold reservoir is $+Q/T_1$. The total entropy change of the universe is the sum of these changes:

$$\boxed{\Delta S_{\text{universe}} = \frac{Q}{T_1} - \frac{Q}{T_2}}$$

which is a positive quantity because $T_2 > T_1$.

(b) The amount of work we could have done is *not* equal to Q, the heat transferred. Instead, the maximum amount of work is limited by the Carnot efficiency, $e = 1 - T_1/T_2$, operating between these two temperatures.

$$W = (e)(Q) = \left(1 - \frac{T_1}{T_2}\right)(Q) = (T_1)\left(\frac{Q}{T_1} - \frac{Q}{T_2}\right) = \boxed{T_1\,\Delta S_{\text{universe}}}$$

Thus, the maximum work that could have been done is the lower temperature T_1 times the entropy change $\Delta S_{\text{universe}}$ of the universe.

Let us now stand back from the details of our discussion of entropy and the second law and summarize some conclusions. We can express the second law in various forms. The following statements are not equally precise, but each does contain the central feature of the second law.

THE SECOND LAW

(a) The entropy of the universe tends to increase.
(b) Heat tends to flow from hot to cold bodies.
(c) The disorder in the universe tends to increase.
(d) Natural processes tend to degrade thermal energy.
(e) The existence of any cyclic process whose *net* effect is the transfer of heat from a cold body to a hot body is impossible.
(f) The existence of any cyclic process whose *net* effect is to convert thermal energy into an equivalent amount of work is impossible.
(g) Things generally tend to get mixed up.

A convenient way to state the two laws of thermodynamics is:

(1) *The energy in the universe remains constant.*
(2) *The entropy in the universe tends to increase.*

21.5 Thermodynamic Heat Death

The second law seems to imply an ultimate "heat death" of the universe where, through the "downhill" flow of heat, all objects eventually come to thermal equilibrium at the same temperature (about 4 K)—a state of maximum entropy

Bits of Information

As part of an experiment to discover the least amount of visual information a picture may contain and still be recognizable, scientists at Case Western Reserve University use a computer to reproduce pictures with varying amounts of information content. In Figure 21–3, the picture is divided into more than 700 hexagons, with the average density in each hexagon printed as an even tone of one of 16 levels of gray. Each hexagon contains 4 bits of information, since we could determine which of the 16 possible levels of grayness a particular hexagon has by asking 4 yes-or-no questions. The possibilities are successively narrowed down by dividing the remaining choices into two equal groups and inquiring if one of the two groups contains the correct choice. Thus, each answer in the following illustration represents one bit of information.*

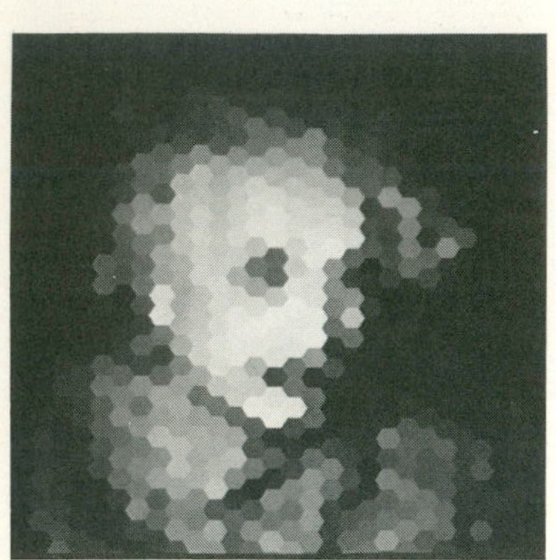

Figure 21–3
The information content of this picture has been deliberately reduced by a computer program to the minimum number of bits that will still allow recognition.

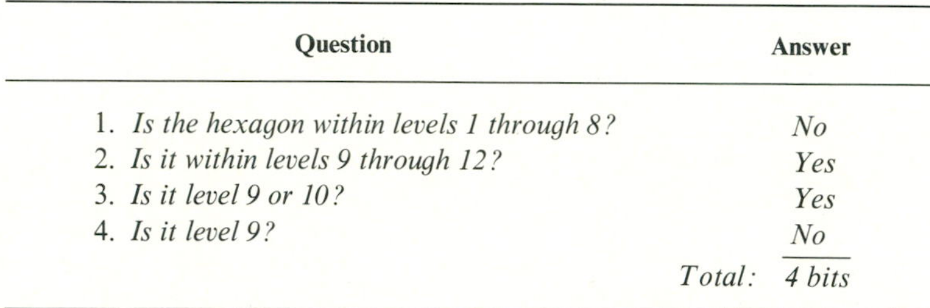
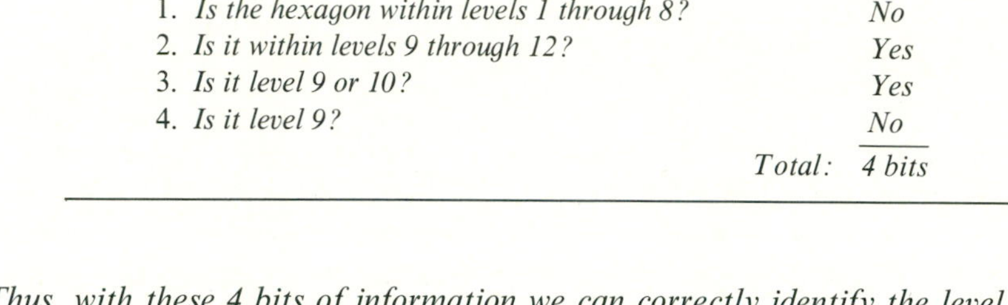

Question	Answer
1. *Is the hexagon within levels 1 through 8?*	*No*
2. *Is it within levels 9 through 12?*	*Yes*
3. *Is it level 9 or 10?*	*Yes*
4. *Is it level 9?*	*No*
Total:	*4 bits*

Thus, with these 4 bits of information we can correctly identify the level as 10.

The original picture, of course, contained smooth variations of density. The process of reduction to discrete hexagons introduces a noise *into the reproduction because of the sharp edges. The way our brain interprets images is particurlaly sensitive to straight lines and geometrical shapes, so this noise introduced by the sampling technique obscures and confuses our interpretation of the information. If you squint or defocus your eyes, jiggle the picture, and view it from a distance to minimize the noise, recognition improves considerably and you should easily identify the portrait.*

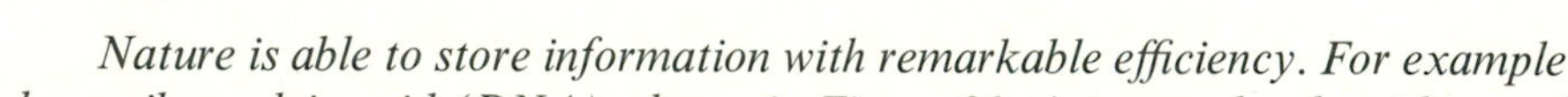

□ □ □

Nature is able to store information with remarkable efficiency. For example, deoxyribonucleic acid (DNA), shown in Figure 21–4, is a molecule within each

* Barry Isenstein processed the photograph shown here; the original photograph, "Albert Einstein 1938," is courtesy of Lotte Jacobi.

for our universe as a whole. True, the second law does point in this direction unless, of course, fresh amounts of energy at low entropy were injected into our universe.[4] But we need not worry about this heat death. Another sort of fate may overtake us before that happens. It involves the fact that a few percent of the energy output of stars is believed to be in the form of neutrino emission. Neutrinos are particles without electric charge, whose rest mass is zero. They always travel at the speed of light, and have the remarkable property of hardly

[4] Alternatively, maybe everything will "wind itself up" again. There is a cosmological model that proposes a "big bang" expansion for our entire universe, followed later by a contraction to a very small region, which initiates another big bang, and so on indefinitely.

living cell of every living organism, containing perhaps as many as 10 billion atoms in advanced organisms such as humans. If all the DNA in an average adult were uncoiled and laid out in a line, it would reach a length of more than 10 billion kilometers—the diameter of our solar system. The DNA molecule carries the master plan for reproducing the cell, including all information about how the organism will develop and function. It is estimated that nature uses about 50 to 100 atoms in DNA to code one bit of information.

Professor Richard Feynman of Caltech has an interesting way of emphasizing how remarkable a feat of information storage this process is.[†] *He considers the problem of storing all the information contained in three large libraries of the world: the* Library of Congress *in Washington, D.C., the* Bibliotheque Nationale *in Paris, and the* British Museum Library. *Allowing for some duplication, there were (in 1959) perhaps 24 million volumes of interest in these three libraries. So let's consider the information content of 24 million volumes, each the size of an average* Encyclopaedia Brittanica *volume. We will include all photographs and illustrations by coding the density of each halftone dot individually. If we allow 100 atoms to store one bit of information, the total information content of all these 24 million volumes would occupy a cube only* 1/8 mm *on each edge! Of course, the readout process might be a bit troublesome. . . . But nature seems to accomplish an equivalent readout task in the living cell.*

[†] These remarks were part of Feynman's after-dinner talk, "There's Plenty of Room at the Bottom," given at a banquet of the American Physical Society meeting at Caltech, December, 1959. For interesting theories of the discovery of the DNA structure, see James D. Watson, *The Double Helix*, Atheneum, 1968, and Anne Sayre, *Rosalind Franklin and DNA*, Norton, 1975.

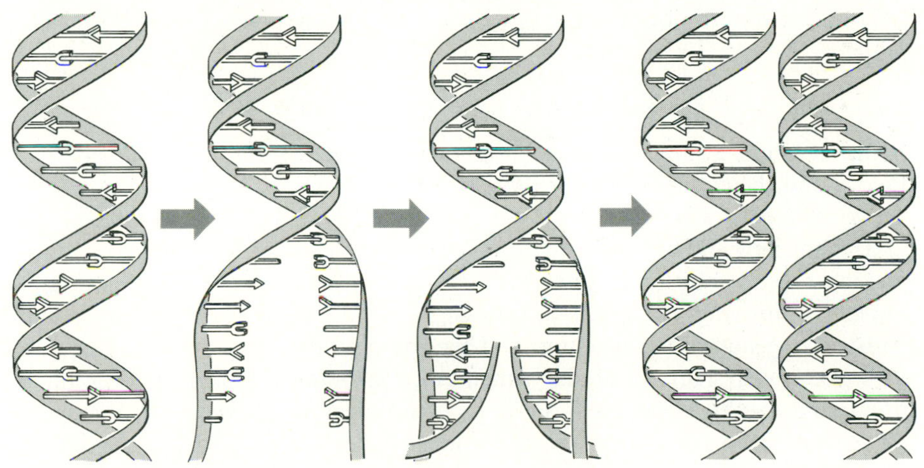

Figure 21–4
A schematic representation of a DNA molecule. All the genetic information for the reproduction of a living cell is contained within the spiral structure of the DNA molecule, which resembles a twisted ladder, or a *double helix.* Each rung of the ladder is a linked pair of molecules chosen from four types of molecules called *nucleotides.* During cell division, the ladder untwists and comes apart like a zipper, each rung breaking at the midpoint. Each nucleotide can fit together with only one other type of nucleotide. So each broken link automatically selects the particular matching molecule from the surrounding cell fluid to rebuild the rung of the ladder, reconstructing the missing half of the helix and thus making two replicas of the original DNA.

interacting with matter at all. Each square centimeter at the earth's surface supposedly receives about 10^{11} neutrinos per second from our sun. It has been estimated that a neutrino emitted from nuclear processes in stars could travel, on the average, through perhaps hundreds of light-years of solid matter before being absorbed. A neutrino moving at the speed of light through the known universe since the "big bang" would have only one chance in 10^{25} of being absorbed.[5] So practically all of the neutrinos ever emitted are still traveling.

[5] In spite of their incredible elusiveness, neutrinos have been detected experimentally. For the story of this accomplishment, see "Neutrinos from the Sun" by J. Bahcall, *Scientific American*, p. 28, July 1969.

Thus, apparently a sizable fraction of the energy of the universe is continually being converted into a flux of neutrinos traveling at the speed of light, with negligible interactions, through matter and empty space. In comparison with the heat death of the second law, this "neutrino death" will occur a lot sooner.

But long before either of these "running-down" fates occurs, a different sort of catastrophe for human life seems certain. Astronomers feel reasonably confident in their understanding of the evolution of stars. According to accepted theory, when our sun reaches the end of its hydrogen-burning phase it will brighten and expand, becoming a *red giant* that engulfs Mercury and Venus, extending perhaps to the size of the earth's orbit. (The star Antares is currently in this red-giant stage of evolution.) The resulting fiery destruction of all life on earth will occur about 5 billion years from now, far sooner than either the hypothetical heat death of the universe or the neutrino death. In the meantime, as newcomers who just recently arrived on the scene, we have considerable time yet to watch and enjoy the universe.

21.6 Entropy and Information

Ideas of entropy have also found application in *information theory*, a discipline developed primarily in the past few decades. In information theory, the basic definition of **information** I is:

$$I = -\ln W \tag{21-16}$$

where W is the probability of guessing a particular message before it is received (or, in some contexts, the probability of sending that particular message).

The probability W may refer to a simple yes-or-no unit of information, known as a *bit* (of information), or to a more complex message. The similarity between

$$S = k \ln W \qquad \text{and} \qquad I = -\ln W$$

points to a close *mathematical* analogy between the equations of information theory and the equations of entropy. In particular, *information* corresponds to *negative entropy*. One author[6] points out that this analogy ". . .helps scientists in devising new and ingenious methods of coding information, efficiently multiplexing several telephone conversations into a single channel, designing amplifiers, improving the music-to-scratch ratio for phonograph records, and constructing automatic machinery. This analogy also helps in studying our own nervous system, vocabulary, memory, reasoning patterns of the human mind, and maybe 'mind' itself. Perhaps it can even be applied to problems in the social sciences." Obviously, these ideas have played a large part in the explosive growth and development of computers and communication systems.

21.7 Perpetual Motion Devices

Occasionally an inventor will claim to have developed a perpetual motion machine (Figure 21–5). Invariably, these inventions fall into two classes: *those that violate the first law* ("perpetual motion machines of the first kind," which create their own energy sources) and *those that violate the second law* ("perpetual motion machines of the second kind," which involve zero or negative changes in entropy). Most physicists have such a deep and firm belief in the principles of thermodynamics that the mere fact the invention violates one or

[6] See *Physics for the Inquiring Mind*, E.M. Rogers, Princeton University Press, 1960.

the other law is often sufficient reason to dismiss the claim without further analysis. Such cavalier treatment usually fails to satisfy the inventor, who demands to know at what particular point in his or her proposed mechanism the fault lies.

During the nineteenth century, working models of perpetual motion machines were exhibited at fairs and other public places. As might be expected, each was found to involve some measure of deceit, much to the chagrin of those people who had invested in stock the inventor offered for sale. (A few were based on changes in atmospheric pressure or similar "obscure" energy sources and thus were not true perpetual motion devices.) The fact remains that so far, no machine has ever operated in a way that violates the thermodynamic laws. In 1775, the French Academy of Sciences decided to no longer consider any claims for perpetual motion devices. The British Patent Office will not accept applications for any form of perpetual motion machine. The U.S. Patent Office will accept such applications only if accompanied by a *working* model of the device, a stipulation which, understandably, no one has been able to meet.

Figure 21–5
Can you explain the fallacy in these proposals for perpetual motion machines?

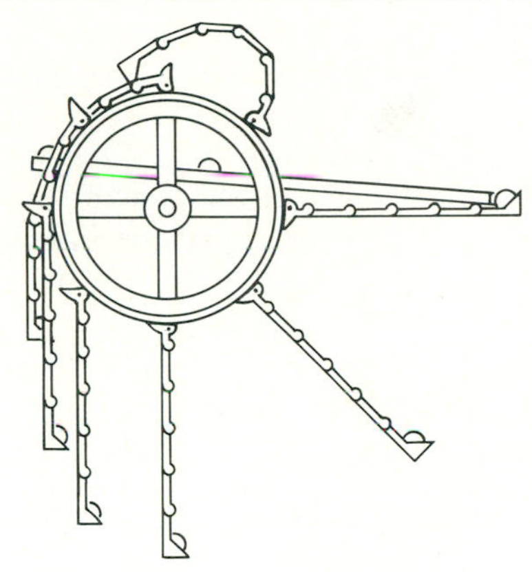
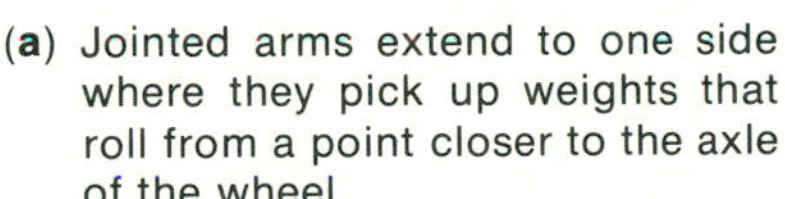

(a) Jointed arms extend to one side where they pick up weights that roll from a point closer to the axle of the wheel.

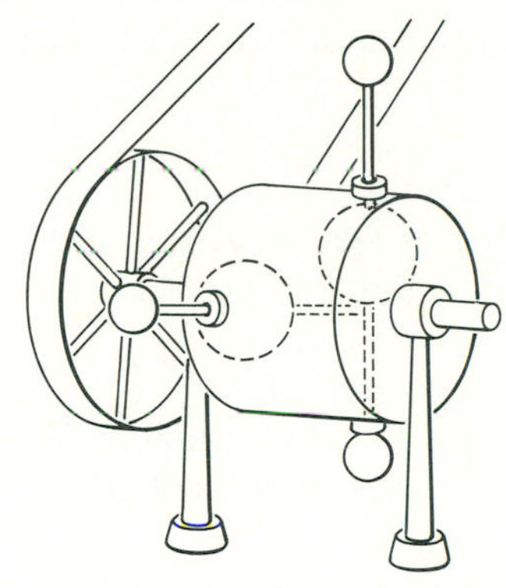

(b) A pair of hollow floats within a water-filled drum move weights so that the force of gravity (reputedly) produces a net torque about the axle.

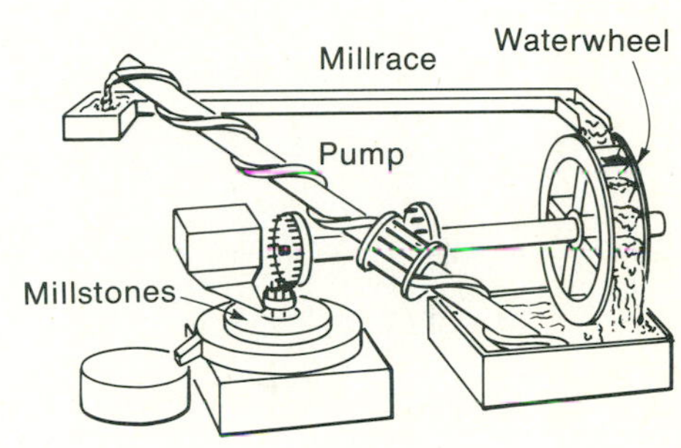

(c) This closed-cycle mill was proposed in 1618 as a source of perpetual power in areas that lacked streams. Archimedes first devised the hollow helical screw for raising water.

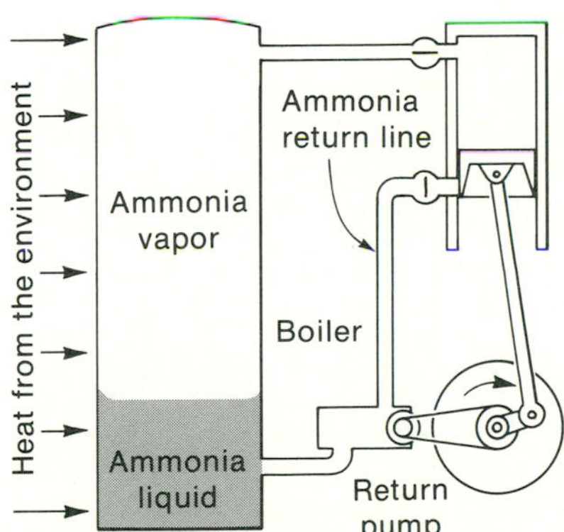

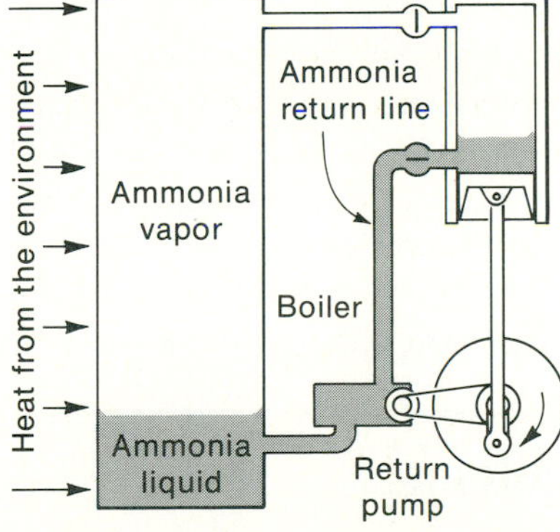

(d) This ammonia engine utilizes heat from the surroundings. The vapor pressure drives the piston. As the piston moves, the expanding vapor cools and thereby condenses the ammonia, returning the liquid to the reservoir. The second law is violated in this perpetual motion machine: There is no low-temperature reservoir for the rejection of heat during the cycle.

(e) This perpetual waterfall is by the Dutch artist M.C. Escher.

Summary

Because of the statistical tendency of macroscopic systems to proceed toward states of greater probability and greater disorder, all natural processes are irreversible. *Entropy* is a measure of the disorder of a system (and also a measure of the unavailability of energy for doing work). In all natural processes, entropy increases. A change in entropy ΔS is defined as

$$\Delta S = \int_1^2 \frac{dQ}{T} \qquad \text{(for any } \textit{reversible} \text{ path connecting states 1 and 2)}$$

Entropy is a state variable. Therefore, changes in entropy from one state to another depend only on the states themselves and not on the path that connects the states.

For a closed, *reversible* process, the entropy change is zero.[7]

$$\oint \frac{dQ}{T} = 0$$

For all real, *irreversible* processes, the entropy increases.

$$\Delta S > 0$$

The connection between entropy S and probability W is

$$S = k \ln W$$

where k is Boltzmann's constant and W is the probability of the system being in a particular state. There is a close relation between this equation and the defining equation for *information* I:

$$I = -\ln W$$

which points to a close mathematical analogy between equations involving entropy and the equations of information theory in general.

Questions

1. Analyze several different processes that occur in nature and explain why each is an *irreversible* change.
2. Can you think of some process in nature that occurs *reversibly*?
3. Rub your hands together vigorously until your palms begin to heat up. Explain why this process increases entropy.
4. Without refreshing your memory by referring to the discussion in the text, give *macroscopic* and *microscopic* definitions of entropy.
5. An ideal gas is compressed to one-half its original volume by two different processes: adiabatic and isothermal. In which case (if either) is the final temperature greater? Is there a change in entropy in either process?
6. At autumn, a leaf falls from a tree. Explain where the increase of entropy occurs.

Problems

In the following problems, assume all gases are ideal.

21A–1 Calculate the entropy change when 4 g of steam at 100°C condense to water at the same temperature. *Answer:* -24.2 J/K

[7] Recall that the circle on the integral sign specifies that the integration be carried out over a closed path.

21A–2 Find the change in entropy when 1 g of water at 100°C is changed to steam at 100°C. The process occurs at constant pressure.

21A–3 A glass of ice water contains 100 g of ice in 100 g of water, both at 0°C. Find the entropy change that occurs when all the ice melts to water at 0°C. (Hint: Where does the heat come from that melts the ice?) *Answer:* 123 J/K

21A–4 If 200 g of copper at 90°C are added to 100 g of water at 0°C in a thermally insulated container, find the total change in entropy that occurs when the two substances come to thermal equilibrium.

21A–5 A 200-g block of aluminum (specific heat = 0.215 cal/g·C°) at 100°C is added to 400 g of water at 20°C in a thermally insulated container. Calculate the change of entropy that occurs as the combination comes to thermal equilibrium. *Answer:* 5.27 J/K

21A–6 A student cleans her desk, putting papers and books into a highly ordered state (of low probability). Explain why this does not violate the second law of thermodynamics.

21B–1 If n mol of an ideal gas, initially at P_0, V_0, T_0, undergo an isobaric expansion that doubles its volume, show that the change in entropy of the gas is $nR[\gamma/(\gamma - 1)] \ln 2$.

21B–2 Sketch a Carnot cycle on a T versus S diagram, identifying the points A, B, C, and D in Figure 20–3.

21B–3 If 40 g of ice at 0°C are added to 100 g of water at 90°C in a thermally insulated container, find the total change in entropy as the mixture comes to thermal equilibrium. *Answer:* 12.6 J/K

21B–4 Complete the table below by placing the symbols +, 0, or − to indicate the sign of the change (or the fact that no change occurs) in the thermodynamic quantities associated with the cyclic process indicated in Figure 21–6.

Path	ΔQ	ΔU	ΔW	ΔP	ΔT	ΔV	ΔS
$a \to b$							
$b \to c$			−	0			
$c \to a$					+		

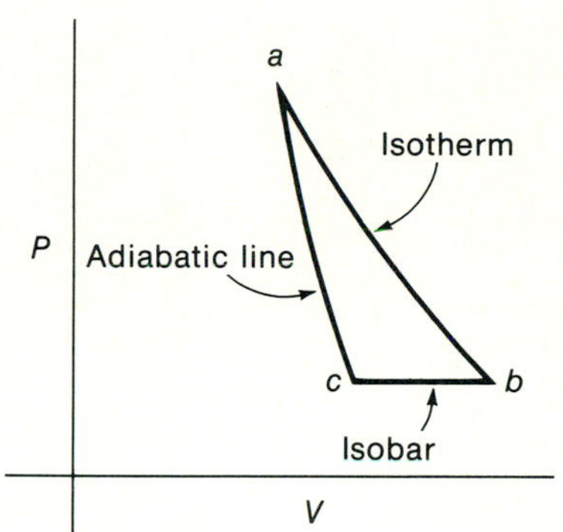

Figure 21–6
Problem 21B–4

21B–5 A meteorite punctures an unmanned spacecraft so that the air inside escapes in a free expansion process into the volume of our galaxy (about 10^{61} m^3). If the spacecraft originally contained 10 m^3 of air under standard conditions, what change in entropy occurred? *Answer:* $\sim 5 \times 10^5$ J/K

21B–6 On a cold day, 3000 cal leak through a window pane from a warm room maintained at 27°C to the outside air at a constant temperature of 4°C. Find the entropy change of the universe.

21B–7 A heat engine operates between two reservoirs at $T_2 = 600$ K and $T_1 = 350$ K. It absorbs 1000 J of heat from the higher temperature reservoir and performs 250 J of work.

(a) Find the entropy change of the universe $\Delta S_{\text{universe}}$ for this process.
(b) Find the work W that could have been done by an ideal Carnot engine.
(c) Show that the difference between the work done in parts (a) and (b) is $T_1 \, \Delta S_{\text{universe}}$.

Answers: (a) 0.476 J/K (b) 417 J

21B–8 Coins in the bottom of a box show 10 heads and 10 tails. The box is shaken. There are now 7 heads and 13 tails. Has the second law of thermodynamics been violated? Explain.

21C–1 Show that when an ideal monatomic gas undergoes an isobaric change from an initial temperature T_i to a final temperature T_f, the change in entropy is $\frac{5}{2}nR \ln (T_f/T_i)$.

21C–2 A gas is heated at constant pressure from T_1 to T_2. The molar specific heat at constant pressure over this range is given by $C_P = a + bT$. Calculate the change in entropy for n moles of the gas.

21C–3 Two identical objects, each of mass m and specific heat c, are at temperatures T_2 and T_1, respectively. A Carnot engine uses the objects as (varying) temperature reservoirs as they are brought to a common final temperature while the engine performs an amount of work W.

(a) Show that the final temperature T_f of the two objects when the engine shuts down is $\sqrt{T_1 T_2}$.

(b) What is the total work W done by the engine? (Hint: Since the process is reversible, find the total entropy change of the system.)

Answer: (b) $mc[(T_2 + T_1) - 2\sqrt{T_2 T_1}]$

21C–4 Consider the situation in the previous problem. If instead of using the Carnot engine, the two objects were merely put in thermal contact, they would have come to a final temperature of $(T_2 + T_1)/2$ (the *arithmetic* mean rather than the *geometric* mean, $\sqrt{T_1 T_2}$). Using thermodynamic reasoning as opposed to mathematical arguments, explain why the final temperature in this case is higher than the final temperature using the Carnot engine process.

21C–5 Show that when an ideal monatomic gas undergoes an isovolumic change from initial pressure P_i to final pressure P_f, the change in entropy is $\frac{3}{2}nR \ln (P_f/P_i)$.

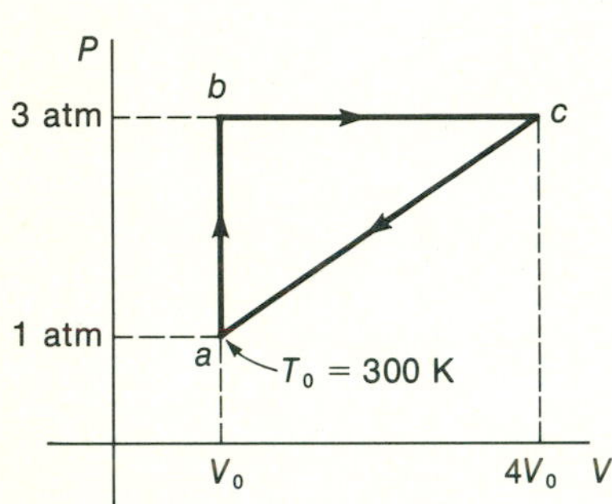

Figure 21–7
Problem 21C–6

21C–6 Assume that 2 mol of an ideal monatomic gas, initially at $P_0 = 1$ atm and $T_0 = 300$ K, are carried reversibly along the path $a \rightarrow b \rightarrow c$ in Figure 21–7. Process $a \rightarrow b$ is isovolumic, $b \rightarrow c$ is isobaric, and $c \rightarrow a$ is a straight line.

(a) Calculate the change in entropy for each step of the complete cycle: $a \rightarrow b \rightarrow c \rightarrow a$.

(b) Find the overall change in entropy for the complete cycle.

(c) Determine the net work done in one cycle.

21C–7 Show that when an ideal gas undergoes an isothermal process from P_1 and V_1 to P_2 and V_2, the change in entropy is $nR \ln (V_2/V_1) = nR \ln (P_1/P_2)$.

21C–8 A box of volume V contains n_1 molecules of an ideal gas (molecular mass $= M_1$). Another box of volume $2V$ contains n_2 molecules of an ideal gas (molecular mass $= M_2$). Both are at the same pressure. The boxes are brought into contact, and a hole is made in the wall between the boxes. Each gas then expands into the other volume without interacting with the other molecules, so that essentially each undergoes a free expansion. Calculate the entropy change that occurs when the gases mix into the combined volume $3V$.

21C–9 An evacuated box has a partition that separates its volume into two equal halves. If eight molecules are placed initially in one half, where they undergo random motions, find the change in entropy when the partition is removed and the molecules move around in the total volume V. *Answer:* 8k ln 2

21C–10 One gram of water is heated from 0°C to 100°C by bringing it into contact with a heat reservoir at 100°C. What is the entropy change (a) of the water and (b) of the heat reservoir? (c) What is the net entropy change of the universe? Suppose, instead, the water had been heated by first bringing it into thermal equilibrium with a reservoir at 50°C, then to equilibrium with a reservoir at 100°C. In this two-step procedure, what is the total change in entropy (d) for the water, (e) for the two heat reservoirs, and (f) for the universe? (g) Describe a method by which the water could be heated (in principle) without any change in the entropy of the universe.

21C–11 The temperature of a room is maintained constant at 300 K. Suppose 100 g of water at room temperature are converted to ice at 273 K by a Carnot engine operating between room temperature and the (varying) temperature of the water as it is cooled. What total amount of work is required to freeze the water? (Hint: Since it is done reversibly, the total entropy change of the water plus the room is zero. Calculate these separate entropy changes individually and sum to zero. Refer to Example 21–2.)

Answer: 385 J

21C–12 A paddle-wheel is immersed in 800 g of water in a thermally insulated container. A 20-kg mass, allowed to move vertically 3 m under gravity, turns the paddle-wheel as it descends, raising the temperature of the water, which was initially 300 K.

(a) Find the change in internal energy of the water.
(b) What heat is transferred to the water?
(c) What is the entropy change of the water?
(d) What is the entropy change of the universe?

21C–13 (a) Show that if the Kelvin-Planck statement of the second law were violated, the entropy of the universe would decrease. (Hint: Consider a "perfect" heat engine that, for each cycle, converts an amount of heat Q entirely into work W.) (b) Show that if the Clausius statement of the second law were violated, the entropy of the universe would decrease. (Hint: Consider a "perfect" refrigerator that extracts Q units of heat from a cold reservoir and transfers the same amount Q to a hot reservoir without any work being performed on the refrigerator.) Note that this problem shows that the entropy form of the second law is equivalent to the other forms of the second law.

Perspective

Let us pause again to reflect on the varied landscape we have traveled through. You should now begin to appreciate the power and beauty of Newton's great accomplishments. It is truly astonishing how a few profound ideas about force, energy, and momentum provide insight into such a wide variety of phenomena. We have seen numerous examples of how these ideas reduce a large array of happenings in nature to just a few fundamental principles in action.

The application of mechanical principles to thermal effects led to the laws of thermodynamics. They were not derived from previous principles; instead, they are generalizations made from our observations of nature's behavior. These laws, particularly the second law, have a power and generality that perhaps you will appreciate only if you study further topics in thermodynamics. They are truly universal.

The remaining chapters treat new areas of physics: electricity and magnetism, optics, and (briefly) the intriguing world of quantum phenomena. As you proceed, keep your eyes on the clever way physics has of explaining the myriad phenomena about us with just a few dominant themes. This characteristic of physics, which enables us to simplify and unify our view of nature, can be most satisfying.

Electricity is of two kinds, positive and negative. The difference is, I presume, that one comes a little more expensive, but is more durable; the other is a cheaper thing, but the moths get into it.

STEPHEN LEACOCK
[*Literary Lapses* (1910)]

chapter 22
Electrostatics

We are all familiar with the fact that after combing dry hair, the comb becomes "electrified" with the ability to attract bits of paper. If you stop to think about it, this phenomenon is baffling: somehow the bits of paper mysteriously sense the presence of the electrified comb without actually touching it. Magnets have similar powers of attracting iron and steel objects without touching them.

Such behavior has been observed for a long time. The ancient Greeks discovered that when amber was rubbed by any of a variety of materials, it became capable of attracting small objects. In fact, the word "electricity" comes from the Greek word for amber: *electron*, a fossilized resin that becomes electrified when rubbed. In describing atoms, the term *electron* is applied to the negative charges surrounding the nucleus of the atom. Let us trace the evolution of our understanding of electricity from the electrification of certain materials to the elegantly unified description of electric and magnetic phenomena known as Maxwell's equations (Chapter 28).

It took many intelligent investigators a long time to unravel the story. About 200 years elapsed between the publication of Newton's *Principia* (1687) and the comparable achievement by James Clerk Maxwell in his *Treatise on Electricity and Magnetism* (1873). Despite this relatively long gap in the progress of physics, many scientists were struggling to make sense of electromagnetic phenomena during this period, and there were numerous sparks of insight that helped to illuminate the separate pieces of the puzzle. But it required the genius of Maxwell to finally fit all the pieces together in a coherent and unified theory.

Perhaps much of the delay in progress was due also to the difference between mechanical and electrical phenomena. The study of mechanical phenomena benefited heavily from the everyday experiences of pushing and pulling objects and observing their motions. But there is no comparable sensory experience with electromagnetism (except on the superficial level of static electricity and magnets). So the subject is inherently more abstract and more obscure from everyday observations. Furthermore, quantitative experiments in electricity and magnetism are vastly more difficult to carry out than experiments in mechanics. The electric force is so large that just a slight unknown imbalance of electrical charge easily spoils the measurements. As Richard Feynman explains it, if you were standing at arm's length from someone and each of you had just 1% more electrons than protons, the repulsive force on you would be enough to lift a "weight" equal to that of the entire earth!

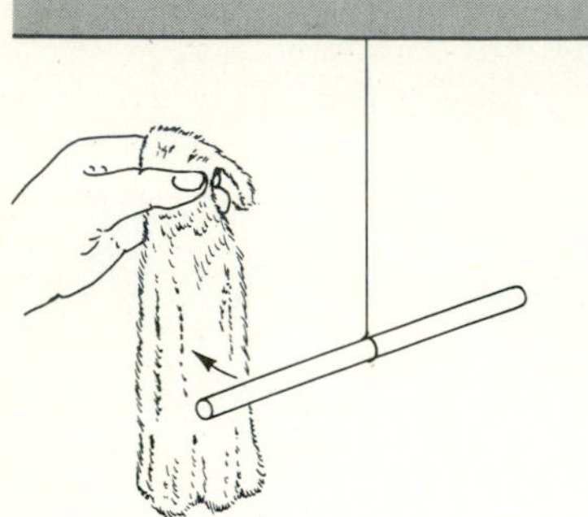

(**a**) When a fur is used to rub a hard rubber rod, that end of the rod is attracted to the fur.

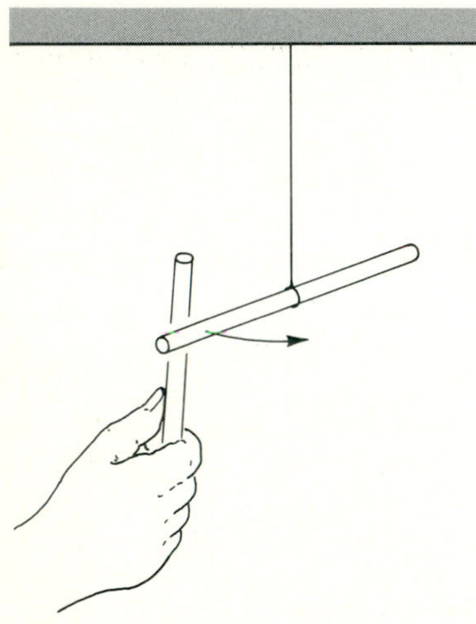

(**b**) When two such rods are rubbed by a fur, the rods repel each other.

Figure 22–1

Electrical forces may be either attractive or repulsive.

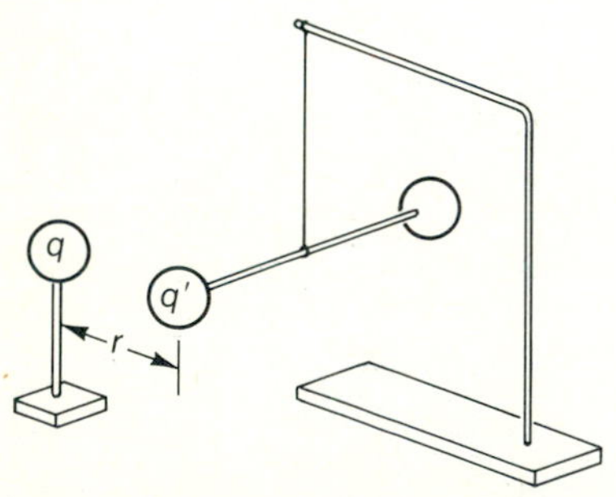

Figure 22–2

The force of interaction between the charges q and q' twist the fiber supporting q'.

Electrical forces are everywhere about us. All so-called "contact forces" are electrical in origin, such as the forces described by Newton's third law (equal and opposite forces), which occur between adjacent links in a chain, between your pencil lead and the paper, and between a tire and the roadway. All of these originate in forces of attraction or repulsion between electric charges. We shall begin our discussion of electricity and magnetism by investigating forces between electrified objects that are *at rest* with respect to each other. This branch of electrical phenomena is known as **electrostatics.**

22.1 Electrostatic Forces

If we rub an animal fur against a hard rubber rod, the rod acquires new characteristics. For example, it readily attracts bits of paper, and it can deflect a jet of water without actually touching it. In the process of being rubbed, the rod has changed. We say it has become *electrified*, or *charged*, without really knowing what these terms mean.

Let us sharpen our terminology and understanding of electrical forces by carrying out some simple experiments. First, suppose we suspend a hard rubber rod by a thread as shown in Figure 22–1. If a piece of fur is brought near the rod, there is no noticeable interaction. However, when the rod is rubbed with the fur, it is then attracted to the fur even at a distance. We call the attraction an **electrostatic force** and conclude:

> **Electrostatic forces (*like* gravitational forces) can be forces of attraction**

Suppose we now rub another hard rubber rod with fur. We find that the second rod repels the suspended rod that had been previously rubbed, and conclude:

> **Electrostatic forces (*unlike* gravitational forces) can also be forces of repulsion.**

Since the charged objects interact without touching, we further conclude:

> **Electrostatic forces (*like* gravitational forces) act through empty space.**

Thus, we would find that the results of this experiment would be the same if it were conducted in a vacuum.

With our knowledge of Newton's law of universal gravitation, we could estimate the force of gravity between the fur and the rod and at least qualitatively conclude:

> **Electrostatic forces are much stronger than gravitational forces.**

In order to focus our attention on the nature of the interaction between two charged objects, we refine the experimental apparatus to that shown in Figure 22–2. This arrangement is a form of *torsional balance*, which the English physicist Henry Cavendish (1731–1810) used to measure gravitational forces. The charged objects are small spheres that have an *electrical charge* on them, designated by the symbols q and q'. The numerical value of q and q' (to be specified later) indicates the amount of charge the objects have. Since the spheres are small, they approximate *point charges.*

The force of interaction can be determined by the amount of torque required to twist the supporting fiber. The distance between the charges is measured directly. After a series of measurements is made with differing separation of the spheres and with various amounts of charge on the spheres, we conclude:

Electrostatic forces (*like* gravitational forces) are inverse-square forces; that is, they decrease with distance r as $1/r^2$.

Electrostatic forces are mutual forces of interaction that obey Newton's third law.

Electrostatic forces are proportional to the product of the amount of charge on each of the interacting charges.

These results may be summarized into a single statement:

$$F = k\frac{qq'}{r^2} \tag{22–1}$$

where k is a constant of proportionality. This result was first published in 1785 by the French physicist Charles Coulomb (1736–1806).

We have referred to *charged* objects and *charges* without really knowing what constitutes the charge. During the 1740s, Benjamin Franklin proposed that the charge was a single fluid. When he rubbed glass with a silk cloth he noted that the glass became "electrified" and attracted bits of paper. He hypothesized that the glass had a surplus of an "electrical fluid," and therefore was *positively charged*, and that the fluid came from the cloth, which therefore had a *negative charge*. The terms "negative" and "positive" are thus derived from his description. Franklin further hypothesized that the fluid was *conserved*, that is, that the total amount of fluid in a *closed* system remains unchanged. In other words, when one object becomes charged positively, another becomes equally charged negatively. Even though Franklin's single-fluid concept was eventually shown to be incorrect, his *conservation of charge* remains as one of the fundamental laws of physics. We now know that the negative charge of a rubber rod that has been rubbed by a piece of fur results from a surplus of electrons stripped from the hairs of the fur. The corresponding deficiency of electrons in the fur causes it to be positively charged.

22.2 Coulomb's Law

Equation (22–1) describes the force between two point charges qualitatively. To make the equation a quantitative one, first we must define the unit of charge and then we must experimentally determine the constant of proportionality k.

The unit of charge is the *coulomb*. Rather than defining the coulomb through Coulomb's law, it is experimentally easier and more precise to define the **coulomb** (C) as the amount of charge per second passing through any cross section of a wire carrying a constant current of one *ampere*. In turn, the **ampere** (A) is defined as the amount of current in the same direction through each of two parallel wires separated by one meter that produces a certain force of attraction between the wires exactly 2×10^{-7} N on each one-meter length of wire. *In this manner, the definition of the coulomb is connected to the SI unit for force.*

In metal wires, the charge is carried by electrons. If the flow of electrons were in opposite directions in the two wires, the wires would repel each other. Note that the force of interaction between the wires is *not* a Coulomb force, since neither wire has a net charge; it is an *electromagnetic* force (to be discussed in Chapter 25). From this definition of a coulomb, the charge on a single electron (designated by the symbol e) has the magnitude:

MAGNITUDE OF THE ELECTRON CHARGE

$$e = 1.602 \times 10^{-19}\ \text{C}$$

This is the smallest electrical charge that has been found; it is equal in magnitude to the positive charge on a single proton.

Knowing the charge on each electron, one can calculate that the rate of flow of electrons for one ampere is 6.249×10^{18} electrons/s. Since the unit for charge q is now defined, the value of k in Coulomb's law can be determined experimentally so that the force F is in newtons. The value of k is found to be $8.99 \times 10^{9}\ \text{N}\cdot\text{m}^2/\text{C}^2$. A good approximation (and one that is easy to remember) is

$$k \approx 9 \times 10^{9}\ \frac{\text{N}\cdot\text{m}^2}{\text{C}^2}$$

In order to simplify the equations that will be developed later, it is convenient to express the constant of proportionality in another way:

$$k = \frac{1}{4\pi\varepsilon_0}$$

where ε_0, called the **permittivity of free space,** has the value

$$\varepsilon_0 = 8.854 \times 10^{-12}\ \frac{\text{C}^2}{\text{N}\cdot\text{m}^2} \tag{22–2}$$

Thus, Coulomb's law becomes

COULOMB'S LAW

$$F = \left(\frac{1}{4\pi\varepsilon_0}\right)\frac{qq'}{r^2} \tag{22–3}$$

or, in SI units:

$$F = \left(9 \times 10^{9}\ \frac{\text{N}\cdot\text{m}^2}{\text{C}^2}\right)\frac{qq'}{r^2} \tag{22–4}$$

The value of the coulomb is here defined using SI units, so F and r in Coulomb's law must also be in SI units. We will use these units in the remaining chapters, since they lead to the familiar electrical units of amperes and volts.

The Coulomb force between two charges is a *mutual* force described by Newton's third law: that is, the force on one charge is equal and opposite to the force on the other. It is therefore convenient to express Coulomb's law in vector form:

$$\mathbf{F} = \left(\frac{1}{4\pi\varepsilon_0}\right)\frac{qq'}{r^2}\,\hat{\mathbf{r}} \tag{22–5}$$

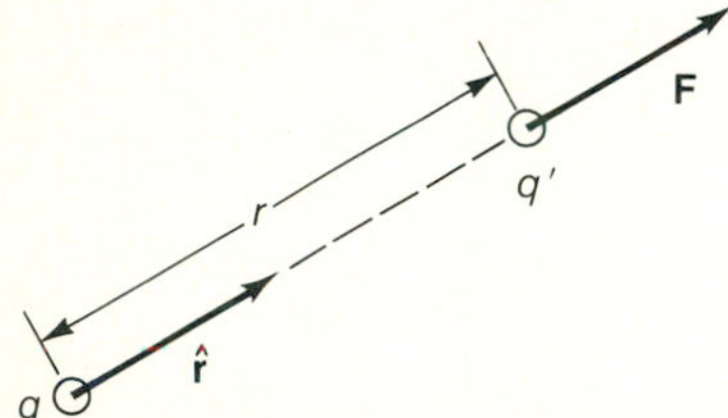

Figure 22–3

The force on the charge q' is in the direction of $\hat{\mathbf{r}}$ if the product qq' is positive. The situation illustrated here would be the case where both q and q' are positive charges or both are negative. The vector from q' to q is $\mathbf{r} = r\hat{\mathbf{r}}$. By Newton's third law, the force on q is equal in magnitude to $\mathbf{F}$ but opposite in direction.

where the force $\mathbf{F}$ is on the charge q' and $\hat{\mathbf{r}}$ is the unit vector (magnitude = 1) from q toward q', as shown in Figure 22–3. (We always use unit vectors that define force directions in this way: the unit vector is drawn *from the source of the force toward the object upon which the force acts.*) The third-law character of the force is part of this definition because *either* of the two charges may be identified as q'. Equation (22–5) further provides the correct direction of $\mathbf{F}$ if we use the following sign convention:

A *positive* charge is given the algebraic sign +.
A *negative* charge is given the algebraic sign −.

Thus, the force between dissimilar charges will be attractive and the force between similar charges will be repulsive.

□ □ □

Coulomb's law describes the electrostatic interaction between two point charges. We will now use Coulomb's law to describe the interaction of several point charges, as well as the interaction of a point charge with a distribution of charge. Such distributions may be along a line, throughout a volume, or over a surface.

As with the gravitational force between two point masses described by Newton's law of gravitation, we find experimentally that the electrostatic forces on a single charge due to the presence of many other charges may be *superimposed*, or added together as vectors. Called the **principle of superposition,** this is, from a practical standpoint, just as significant as the scalar statement of Coulomb's law given in Equation (22–3). Since Newton's law of gravitation and Coulomb's law have the same mathematical form, similar conclusions may be made for each. For example, we have shown that the gravitational attraction for two uniform, solid spheres is as though all the mass were concentrated at a point at the center of each sphere. Similarly, the electrostatic force between two spheres of charge is as though the total charge of each sphere were located at its center. However, as with the gravitational case, the density of charge within each sphere may vary only with the distance from the center of the sphere. That is, the density distribution must have *spherical symmetry.*

EXAMPLE 22–1

Two spheres of negligible size, each with a mass of 2 g, are suspended from a common support by massless strings 1 m long. When each sphere is given the same charge, the spheres diverge until they are 15 cm apart, as shown in Figure 22–4(a). (a) Find the charge on each sphere. (b) Is there more than one answer to this example?

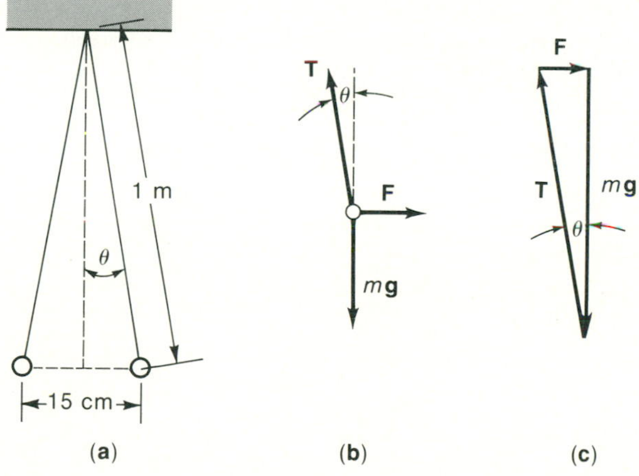

Figure 22–4
Example 22–1

SOLUTION

First we draw a free-body diagram for the sphere on the right, as shown in Figure 22–4(b). (Choosing the *right* or *left* sphere is arbitrary because of *symmetry*. That is, the spheres have identical masses and charges, and are suspended by strings of the same length. Therefore, the Coulomb and gravitational forces on one sphere are the same as on the other one, except for the directions of the Coulomb force.)

(a) The net force on the sphere is zero, so the three forces add to form a closed triangle, as indicated in Figure 22–4(c). Thus:

$$\tan\theta = \frac{F}{mg} = \frac{0.075\text{ m}}{1\text{ m}} = 0.075$$

Substituting Coulomb's law for the force F, we have

$$\tan\theta = \left(\frac{1}{4\pi\varepsilon_0}\right)\frac{qq'}{r^2mg}$$

We then solve for qq' (each of the same magnitude):

$$qq' = (4\pi\varepsilon_0)(\tan\theta)r^2mg$$

Substituting SI values gives

$$qq' = \left(\frac{1}{9\times10^9\,\frac{\text{N}\cdot\text{m}^2}{\text{C}^2}}\right)(0.075)(15\times10^{-2}\text{ m})^2(2\times10^{-3}\text{ kg})\left(9.80\,\frac{\text{m}}{\text{s}^2}\right)$$

$$qq' = 3.68\times10^{-15}\text{ C}^2$$

Since the charges are equal, we have

$$q = q' = \sqrt{3.68\times10^{-15}\text{ C}^2} = \boxed{\pm6.07\times10^{-8}\text{ C}}$$

(b) There are two possibilities: both could be positively charged or both could be negatively charged. Moreover, q and q' could have any value as long as their product is 3.68×10^{-15} C^2.

EXAMPLE 22–2

Three different point charges are located at the vertices of a 3-4-5 triangle whose longest side is 0.5 m. Point charges of 5 μC (1 microcoulomb = 10^{-6} C), 3 μC, and 1 μC are placed as shown in Figure 22–5(a). Find the magnitude and direction of the force on the 5-μC charge.

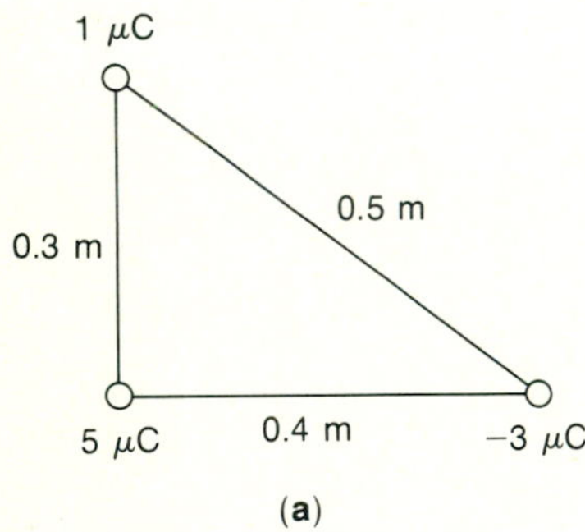

(a)

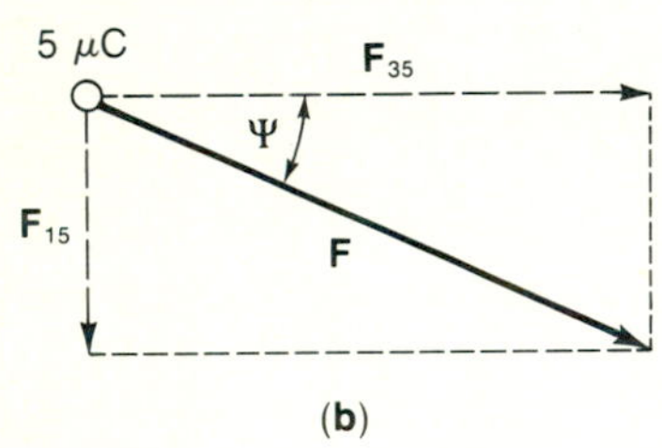

(b)

Figure 22–5
Example 22–2

SOLUTION

We draw a free-body diagram for the 5-μC charge, as shown in Figure 22–5(b). Then, using Coulomb's law, we determine the force each charge exerts on the 5-μC charge. (Note carefully the meaning of the double-subscript notation. It will be used throughout the rest of this text. Here, F_{15} means the force exerted *by* the 1-μC charge *on* the 5-μC charge.)

Due to the 1-μC charge

$$F = \left(\frac{1}{4\pi\varepsilon_0}\right)\frac{qq'}{r^2}$$

$$F_{15} = \left(9\times10^9\,\frac{\text{N}\cdot\text{m}^2}{\text{C}^2}\right)\times\frac{(1\times10^{-6}\text{ C})(5\times10^{-6}\text{ C})}{(0.3\text{ m})^2}$$

$$F_{15} = 0.5\text{ N}$$

Due to the −3-μC charge

$$F = \left(\frac{1}{4\pi\varepsilon_0}\right)\frac{qq'}{r^2}$$

$$F_{35} = \left(9\times10^9\,\frac{\text{N}\cdot\text{m}^2}{\text{C}^2}\right)\times\frac{(-3\times10^{-6}\text{ C})(5\times10^{-6}\text{ C})}{(0.4\text{ m})^2}$$

$$F_{35} = -0.844\text{ N}$$

The directions of the forces are determined by the fact that *like* charges repel and *unlike* charges attract.

The net force on the 5-μC charge is obtained by the superposition (vector addition) of the two forces. The *magnitude* of the net force is

$$F = \sqrt{(F_{15})^2 + (F_{35})^2}$$

$$F = \sqrt{(0.5\text{ N})^2 + (0.844\text{ N})^2}$$

$$\boxed{F = 0.981\text{ N}}$$

The *direction* of the force is

$$\Psi = \tan^{-1}\left(\frac{0.5}{0.844}\right)$$

$$\boxed{\Psi = 30.6^\circ}$$

where Ψ is the angle indicated in Figure 22–5(b).

The next example will illustrate how to find the force on a single point charge due to a *uniform* distribution of charges (such as a line of charge with a charge density λ, in units of charge per unit length). We first find the force between the point charge q and an *element of charge* dq', so it is simply the force between two *point* charges. Then we use integration to add up similar forces due to all the other elements of charge in the distribution.

EXAMPLE 22–3

Find the force on a 3-μC point charge due to a 5-μC charge distributed uniformly over a 1-m rod that is 0.4 m away, as in Figure 22–6(a).

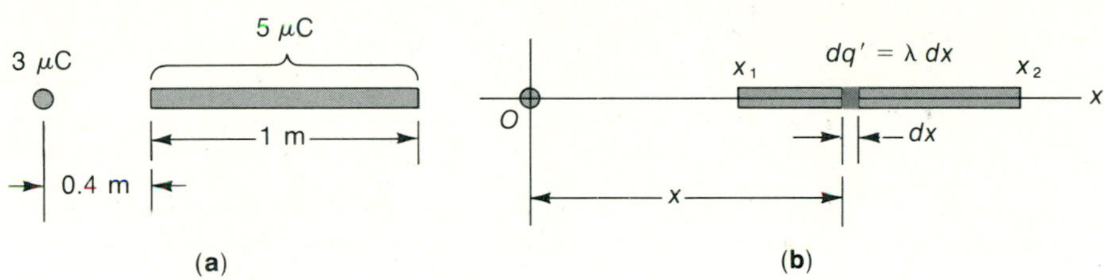

Figure 22–6
Example 22–3

SOLUTION

Since the charge is distributed uniformly along the rod, we find the total force by integrating all the infinitesimal forces dF on the elements of charge dq' along the length of the rod. We begin by placing the coordinate origin at the point charge. Using Coulomb's law in differential form, we have

$$dF = \left(\frac{1}{4\pi\varepsilon_0}\right)\frac{q\,dq'}{r^2}$$

In terms of the variable x in Figure 22–6(b), the element of charge dq' is equal to $\lambda\,dx$:

$$dF = \left(\frac{1}{4\pi\varepsilon_0}\right)\frac{q(\lambda\,dx)}{x^2}$$

where the *charge per unit length* λ is equal to q'/ℓ or 5 μC/1 m. Integrating, we have

$$F = \left(\frac{1}{4\pi\varepsilon_0}\right)q\lambda\int_{x_1}^{x_2}\frac{dx}{x^2} = -\frac{q\lambda}{4\pi\varepsilon_0}\left(\frac{1}{x}\right)\Bigg|_{x_1}^{x_2}$$

Substituting $x_1 = 0.4$ m and $x_2 = 1.4$ m, and other numerical values, gives

$$F = -\left(9\times10^9\,\frac{\text{N}\cdot\text{m}^2}{\text{C}^2}\right)(3\times10^{-6}\,\text{C})\left(5\times10^{-6}\,\frac{\text{C}}{\text{m}}\right)\left(\frac{1}{1.4\text{ m}}-\frac{1}{0.4\text{ m}}\right)$$

$$F = \boxed{0.241\text{ N}}$$

Because the rod and the point have *like* charges, the direction of the force on the point charge is *away from the rod.*

22.3 Electrostatic Fields

Think back for a moment to the concept of a *gravitational field* (Section 14.6). The field idea is useful because it enables us to avoid the conceptual difficulties of "action at a distance," which Newton's law of universal gravitation describes. For example, according to this law the earth exerts a force on a satellite in orbit even though the earth and satellite are separated by empty space. But the idea of a force operating through empty space was repugnant to Newton and to many later scientists: "action at a distance" just did not seem sensible. The concept of a field is a more modern view. This alternative way of describing the gravitational interaction is to imagine that the earth creates a gravitational field **g** in the surrounding space. Then, a satellite of mass m experiences a force $\mathbf{F} = m\mathbf{g}$ due to the *local* gravitational field **g** where the satellite is located. It is no longer a case of action at a distance.

Similar to a gravitational field **g**, an **electric field E** can be defined so that

$$\mathbf{F} = q\mathbf{E} \tag{22–6}$$

where **F** is the force on a charge q placed in an electric field **E**. Figures 22–7 and 22–8 illustrate the similarity in concept between the gravitational field and the electric field.

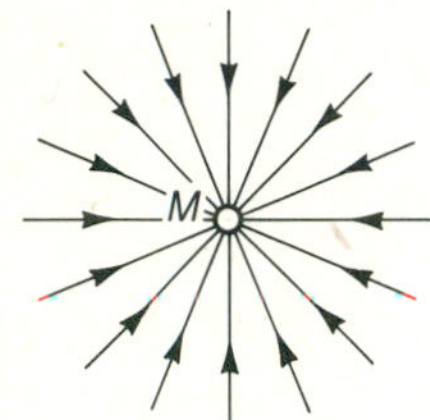

(**a**) The gravitational field **g** due to the point mass M.

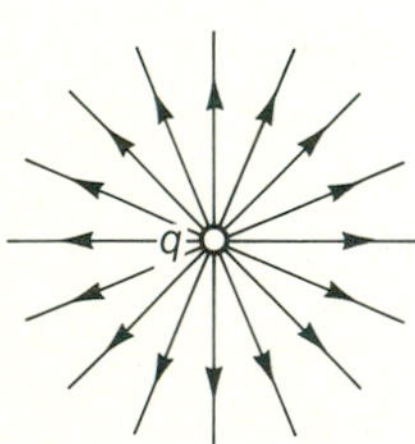

(**b**) The electric field **E** due to the point charge q. (For a negative charge $-q$, the field direction would be inward.)

Figure 22–7
The field concept can be defined for both gravitational and electric fields. Each is an inverse-square law, and each is a *vector* field. The gravitational field is always inward toward the mass M. The electric field may be in either direction, depending on the sign of the charge q. (For situations of this sort, which have spherical symmetry, use your imagination to extend the field-line pattern to fill three-dimensional space.)

Consider the field created by a single point charge q. We begin with Coulomb's law in vector form, as given in Equation (22–5):

$$\mathbf{F} = \left(\frac{1}{4\pi\varepsilon_0}\right)\frac{qq'}{r^2}\,\hat{\mathbf{r}}$$

The electric field **E** at a given point in space is defined to be the force per unit positive charge ($\mathbf{F}/q'$) when the charge q' is placed at that point. Dividing both sides of Equation (22–6) by q', we have $\mathbf{E} = \mathbf{F}/q'$, or

ELECTRIC FIELD E DUE TO A POINT CHARGE *q*

$$\mathbf{E} = \left(\frac{1}{4\pi\varepsilon_0}\right)\frac{q}{r^2}\,\hat{\mathbf{r}} \tag{22–7}$$

The units of the electric field are *newtons per coulomb* (N/C). As in Figure 22–3, the unit vector $\hat{\mathbf{r}}$ is always directed *away* from the *source* of the field (charge q). If the charge is positive, the field **E** is directed away from q; if it is negative, the field is toward q.

At a distance r from a point charge q, a field **E** exists and can be calculated using Equation (22–7) whether or not another charge q' is at that location. But if we consider a charge q' in the vicinity of q, it will experience a force as indicated by Equation (22–6):

FORCE ON A CHARGE *q'* IN THE PRESENCE OF A FIELD E

$$\mathbf{F} = q'\mathbf{E} \tag{22–8}$$

where **E** is the field created by the charge q. Note that the force on q' is in the same direction as **E** if q' is positive and in the opposite direction as **E** if q' is negative.

Since Coulomb forces due to a *collection of charges* add vectorially to produce a single force on another charge in their neighborhood, the electric fields produced by a collection of charges also add to produce a single electric

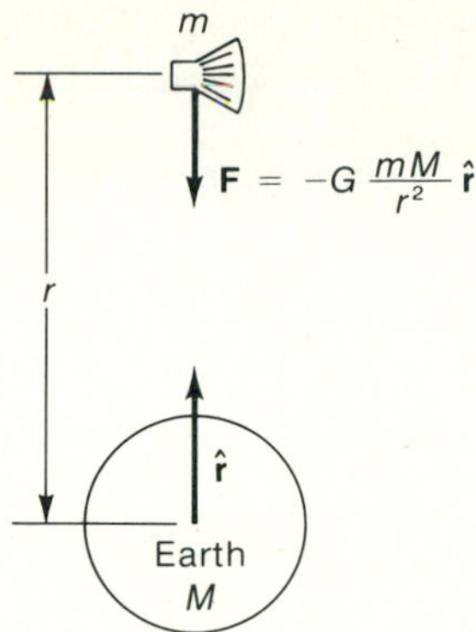

(a) "Action at a distance."

A gravitational field **g** exists in this region

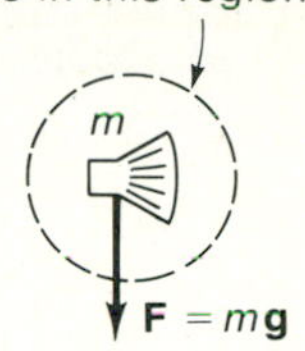

(b) The force due to the local gravitational field.

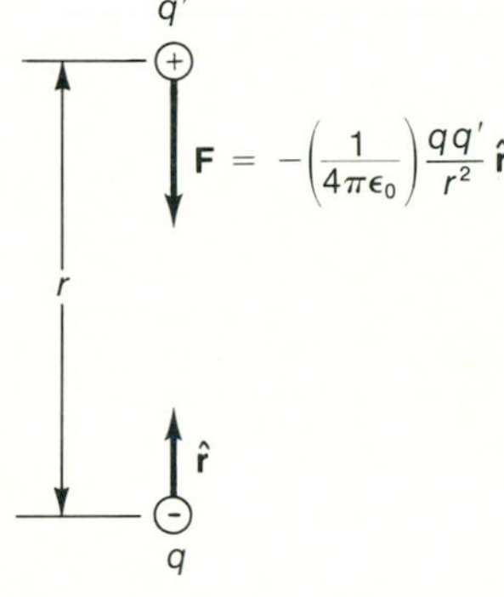

(c) "Action at a distance."

An electric field **E** exists in this region

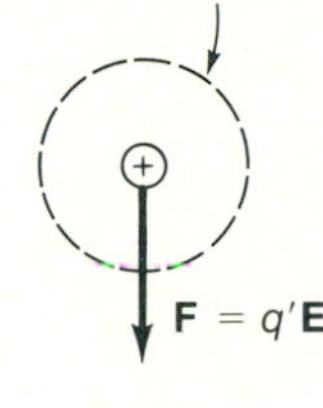

(d) The force due to the local electric field.

Figure 22–8
Gravitational and electric forces can each be considered as an interaction between two objects separated in space, as in (a) and (c), or as a force on one object due to the *local* field created by the other (distant) object, as in (b) and (d).

field. To illustrate this, we will reconsider Example 22–2, this time from the electric field point of view.

EXAMPLE 22–4

Consider a 3-4-5 right triangle with the longest side 0.5 m long. Point charges of 1 μC and -3 μC are placed as shown in Figure 22–9(a). (a) Find the magnitude and direction of the electric field at the right-angle corner of the triangle. (b) What force would be experienced by a point charge of 5 μC placed at the right angle of the triangle?

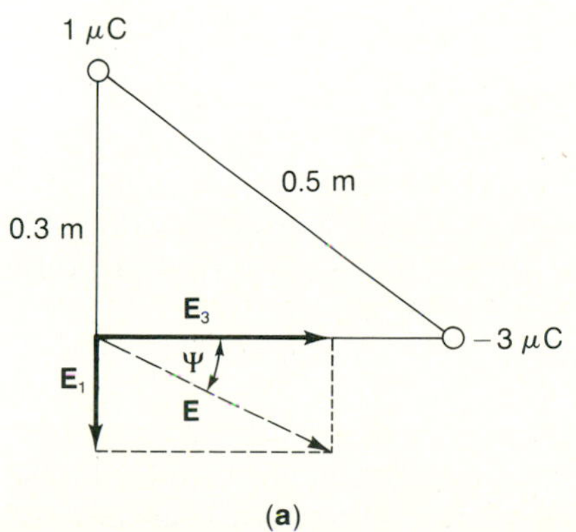

(a)

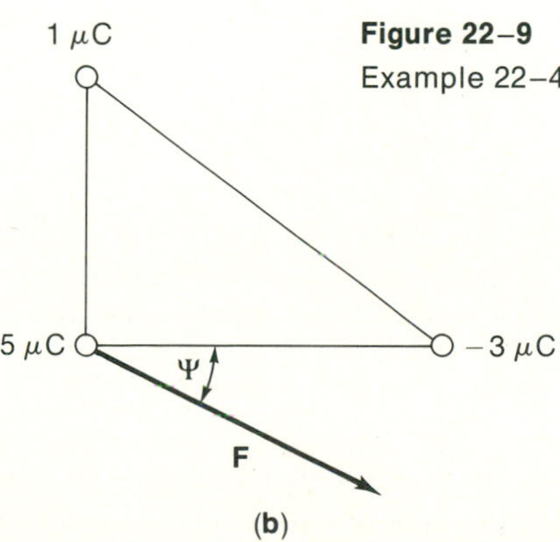

(b)

Figure 22–9
Example 22–4

SOLUTION

(a) The field at the right angle is the vector sum of the fields due to the two charges.

Due to the 1-μC charge

$$\mathbf{E} = \left(\frac{1}{4\pi\varepsilon_0}\right)\frac{q}{r^2}\,\hat{\mathbf{r}}$$

$$\mathbf{E}_1 = \left(9 \times 10^9\,\frac{\mathrm{N\cdot m^2}}{\mathrm{C^2}}\right)\frac{(1 \times 10^{-6}\ \mathrm{C})}{(0.3\ \mathrm{m})^2}\,\hat{\mathbf{r}}$$

$$\mathbf{E}_1 = (1.00 \times 10^5\,\hat{\mathbf{r}})\,\frac{\mathrm{N}}{\mathrm{C}}$$

$\mathbf{E}_1$ is *away* from the 1-μC charge because the charge is positive.

Due to the -3-μC charge

$$\mathbf{E} = \left(\frac{1}{4\pi\varepsilon_0}\right)\frac{q}{r^2}\,\hat{\mathbf{r}}$$

$$\mathbf{E}_3 = \left(9 \times 10^9\,\frac{\mathrm{N\cdot m^2}}{\mathrm{C^2}}\right)\frac{(-3 \times 10^{-6}\ \mathrm{C})}{(0.4\ \mathrm{m})^2}\,\hat{\mathbf{r}}$$

$$\mathbf{E}_3 = (-1.69 \times 10^5\,\hat{\mathbf{r}})\,\frac{\mathrm{N}}{\mathrm{C}}$$

$\mathbf{E}_3$ is *toward* the -3-μC charge because the charge is negative.

The total field $\mathbf{E} = \mathbf{E}_1 + \mathbf{E}_3$ has the magnitude

$$E = \sqrt{{E_1}^2 + {E_3}^2}$$

$$E = \sqrt{(1.00)^2 + (1.69)^2} \times 10^5\,\frac{\mathrm{N}}{\mathrm{C}}$$

$$E = \boxed{1.96 \times 10^5\,\frac{\mathrm{N}}{\mathrm{C}}}$$

The direction of $\mathbf{E}$ is given by the angle Ψ in Figure 22–9(a).

$$\Psi = \tan^{-1}\frac{1.00 \times 10^5\,\frac{\mathrm{N}}{\mathrm{C}}}{1.69 \times 10^5\,\frac{\mathrm{N}}{\mathrm{C}}}$$

$$\Psi = \tan^{-1} 0.592$$

$$\Psi = \boxed{30.6^\circ}$$

(b) The force on a 5-μC charge placed at the right-angle corner is found from Equation (22–8):

$$\mathbf{F} = q'\mathbf{E}$$

$$F = (5 \times 10^{-6}\ \mathrm{C})\left(1.96 \times 10^5\,\frac{\mathrm{N}}{\mathrm{C}}\right)$$

$$F = \boxed{0.980\ \mathrm{N}}$$

The direction of $\mathbf{F}$ is in the same direction as $\mathbf{E}$ because the 5-μC charge is positive, as shown in Figure 22–9(b).

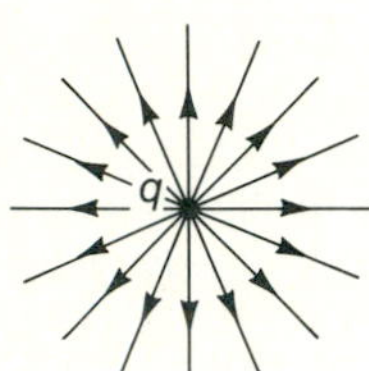

(**a**) A conventional way of depicting the electric field lines due to an isolated point charge q. (The diagram is an approximate cross-section of the field pattern. For a better illustration, mentally extend the pattern to three dimensions, somewhat like the quills on a porcupine.)

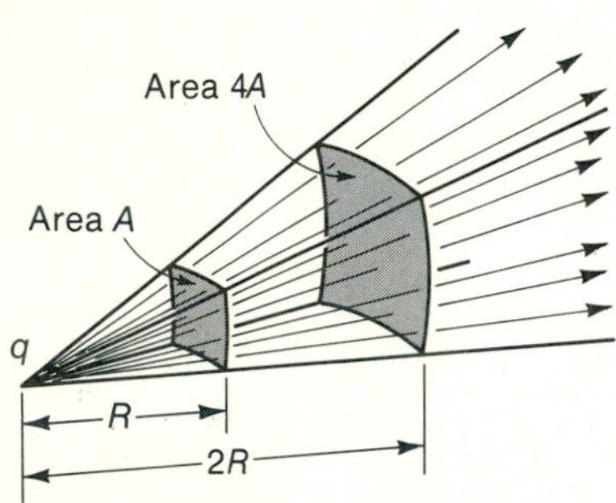

(**b**) A three-dimensional perspective sketch of the field lines diverging from a point charge q. The lines intersect portions of the surfaces of two concentric spheres (radii R and $2R$).

Figure 22–10
Electric field lines associated with an isolated point charge q.

22.4 Electric Field Lines

The purely mathematical concept of an electrostatic field can be made more tangible by introducing **field lines** (also called **lines of force**). Consider an isolated, positive point charge, as shown in Figure 22–10. The electric field at every point in space around the charge is defined by Equation (22–7):

$$\mathbf{E} = \left(\frac{1}{4\pi\varepsilon_0}\right)\frac{q}{r^2}\,\hat{\mathbf{r}}$$

The field is directly away from the positive charge and diminishes in magnitude in accordance with the inverse-square dependence. These two properties of the field can be represented by a number of equally spaced straight lines radiating from the point charge [as in Figure 22–10(a)]. We make the following identifications:

ELECTRIC FIELD LINES

(1) The direction of the lines is the *direction* of the electric field.

(2) The number of lines penetrating a unit area that is perpendicular to the lines represents the *intensity* of the electric field.

The second statement points up a particularly useful feature regarding field lines. Where they are crowded together, the field is stronger; where they are spread apart, the field is weaker. For an isolated point charge, the inverse-square-law behavior is obvious from geometric considerations. Imagine a series of spheres concentric with the point charge [Figure 22–10(b)]. Because the field lines extend radially (and symmetrically) from the source, the total number of lines penetrating each sphere is the same. But the area of each sphere increases with the square of the radius. Since $\mathbf{E}$ is proportional to the number of lines per unit area, the inverse-square relationship follows.

It is difficult to depict true three-dimensional fields in diagrams. Perhaps the best that can be done conveniently is as shown in Figure 22–10(a). One must always mentally extend such two-dimensional diagrams into three dimensions to grasp the true nature of the field.

The *number* of lines we imagine to emanate from a given charge is arbitrary. For example, a 1-μC charge may be associated with 100 field lines or with 1 million field lines. Thus we may choose any convenient "scale factor." But whatever convention we adopt, a 3-μC charge must have exactly three times as many lines emanating from it as a 1-μC charge. So the field lines are visualized as *proportional* to the strength of the field, and the lines themselves should not be taken literally. Keep in mind that field *lines* do not exist in nature; *they are just the mental image we use to help us think about the electric fields* (which *do* exist in the sense that they can be defined and experimentally measured).

The *operational definition* of an electric field is the force per unit positive charge ($\mathbf{E} = \mathbf{F}/q'$). An electric field at a given point in space could be determined simply by putting a known charge at that point, measuring the magnitude and direction of the force on the test charge, and applying the definition $\mathbf{E} = \mathbf{F}/q'$, where q' is the test charge. Operational definitions such as this are more theoretical than practical. For example, suppose we wish to measure the electric field of a charged spherical conductor, as shown in Figure 22–11(a). Since charges can move freely on a conductor, if we bring a test charge q' (of the same sign) near the sphere, the charges on the sphere will move toward the opposite side by Coulomb repulsion, as shown in Figure 22–11(b). The force on q' will then indicate an electric field strength that is *less* than the strength before the charge was introduced. That is, the *test charge disturbs the field it is measuring*. Such a difficulty is common in measurements. (If we wish to measure the temperature of a liquid, we alter the temperature by immersing a thermometer in the liquid.) We thus modify our definition from that given by Equation (22–8):

$$\mathbf{E} = \frac{\mathbf{F}}{q'}$$

to a better form:

OPERATIONAL DEFINITION OF AN ELECTRIC FIELD

$$\mathbf{E} = \lim_{q' \to 0} \left(\frac{\mathbf{F}}{q'} \right) \tag{22–9}$$

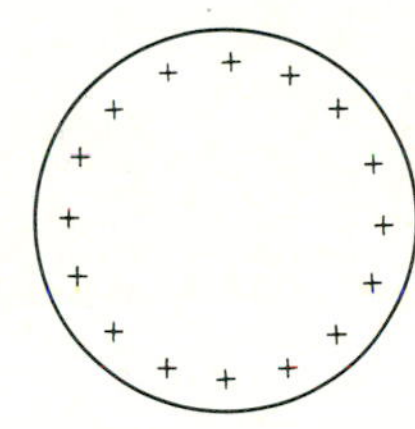

(**a**) When a charged conducting sphere is far from external charges, the distribution of charges on the sphere is symmetric.

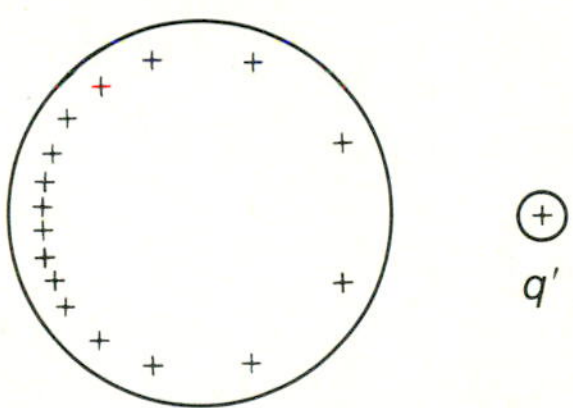

(**b**) When an external charge q' is brought near the charged conducting sphere, the distribution of charges on the sphere becomes asymmetric because of the repulsion of like charges.

Figure 22–11
A test charge used to determine an electric field may distort the very field to be determined.

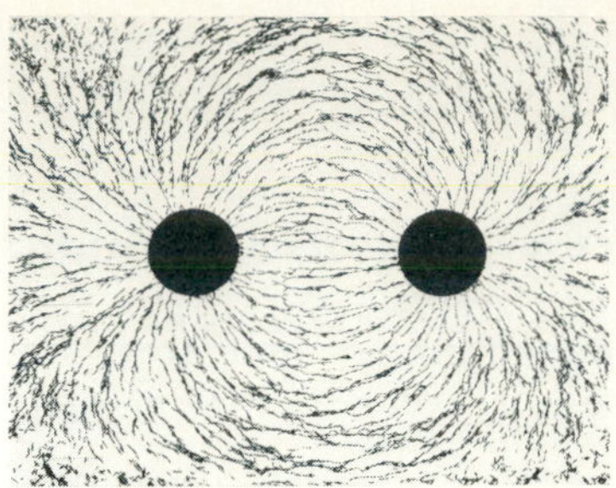

(**a**) The electric field near two parallel rods with opposite charges.

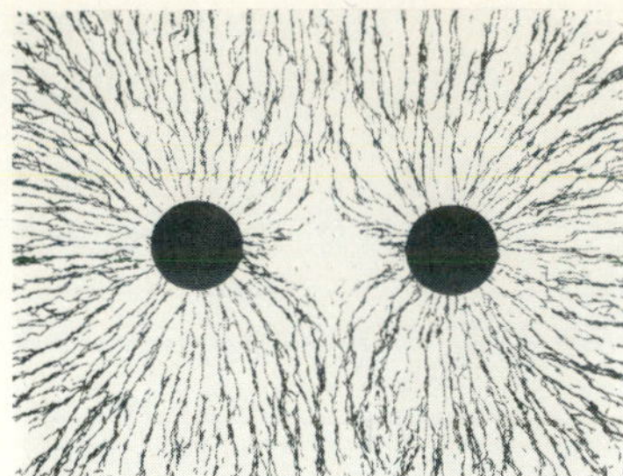

(**b**) The electric field near two parallel rods with the same charge.

Figure 22–12
The electric field can be visualized by sprinkling small, elongated, nonconducting particles on a glass plate. (Here, grass seed is used.) In the presence of a strong electric field, the particles align themselves in chains along the direction of the field.

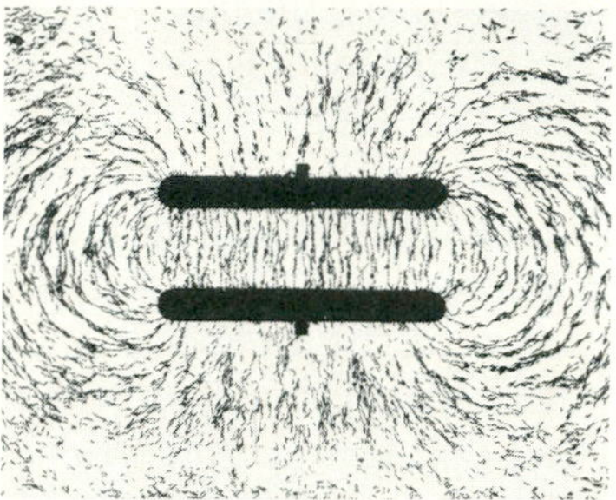

(**c**) The electric field near two parallel plates with opposite charges.

so that the influence of q' will become vanishingly small. Figure 22–12 shows an interesting method of illustrating electric fields.

22.5 The Electric Dipole

One particular configuration of electric charges has application in a great number of practical cases. This configuration is the **electric dipole:** two point charges, separated in space, with the same magnitude but with opposite signs. The many applications in atomic and molecular physics justify a rather thorough discussion of this topic. Figure 22–13(a) illustrates a dipole with electric field

Figure 22–13
Electric field patterns for two point charges. As with all diagrams representing three-dimensional fields, you should imagine the field lines filling three-dimensional space symmetrically. (In these cases, the pattern is symmetrical about the line joining the two charges.)

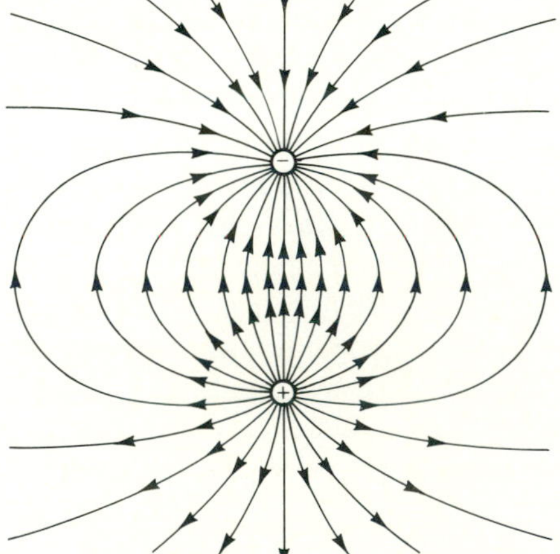

(**a**) The field of an *electric dipole*: point charges of equal magnitude but opposite sign.

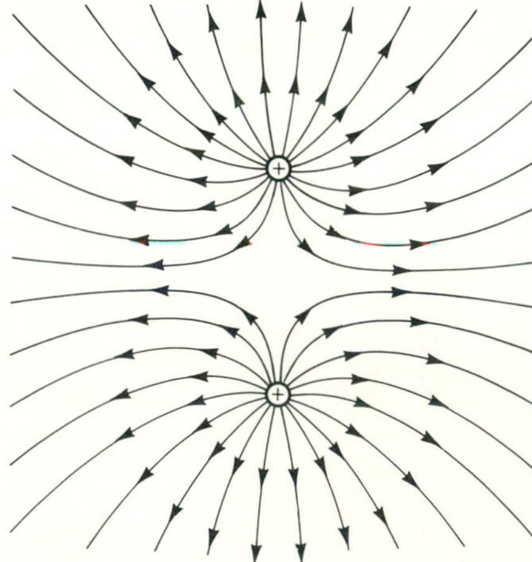

(**b**) The field of point charges of equal magnitude and the same sign (positive charges illustrated).

lines connecting the two charges. The *direction* of the field at any point is tangent to the field lines in the neighborhood. The *intensity* of the field is proportional to the spatial density of the lines.

EXAMPLE 22–5

Consider an electric dipole aligned along the y axis as in Figure 22–14. Find the magnitude and direction of the electric field at an arbitrary distance x along the x axis.

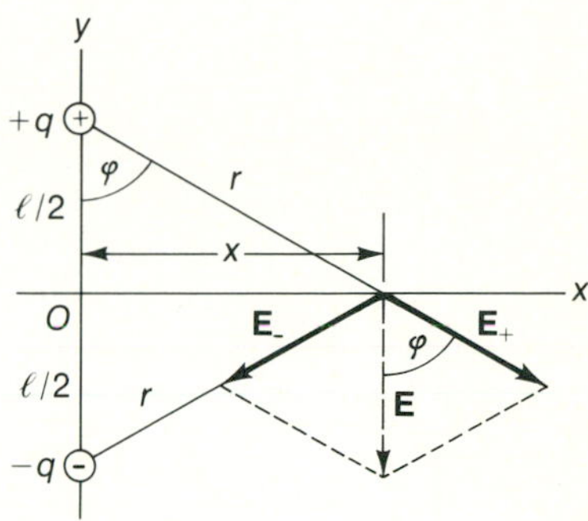

Figure 22–14
Example 22–5

SOLUTION

Let the separation of the charges be the distance ℓ, as shown in Figure 22–14. We will calculate the field $\mathbf{E}_+$ due to the positive charge and the field $\mathbf{E}_-$ due to the negative charge, and then add them vectorially. We start with Equation (22–7):

$$\mathbf{E} = \left(\frac{1}{4\pi\varepsilon_0}\right)\frac{q}{r^2}\,\hat{\mathbf{r}}$$

Recall that $\hat{\mathbf{r}}$ is the *unit vector* from q to the point in question and that r is the *distance* from q to the point. The value of r^2 is thus $(\ell/2)^2 + x^2$, and the magnitudes of E_+ and E_- are the same.

$$E_+ = E_- = \left(\frac{1}{4\pi\varepsilon_0}\right)\frac{q}{\left(\frac{\ell}{2}\right)^2 + x^2}$$

Adding $\mathbf{E}_+$ and $\mathbf{E}_-$ as vectors, the components along the x axis cancel, but the y components add together to yield

$$E = 2\left(\frac{1}{4\pi\varepsilon_0}\right)\frac{q}{\left(\frac{\ell}{2}\right)^2 + x^2}\cos\varphi$$

where $\cos\varphi$ may be written

$$\cos\varphi = \frac{\frac{\ell}{2}}{\left[\left(\frac{\ell}{2}\right)^2 + x^2\right]^{1/2}}$$

Thus the magnitude of the field becomes

$$E = \boxed{\left(\frac{1}{4\pi\varepsilon_0}\right)\frac{q\ell}{\left[\left(\frac{\ell}{2}\right)^2 + x^2\right]^{3/2}}}$$

The direction of $\mathbf{E}$ is parallel to the $(-)y$ axis and is therefore in the same direction as the line from $+q$ to $-q$.

For distances far from the dipole (compared with the distance between the charges), the field assumes a simple form known as *the far-field approximation* (Figure 22–15). When $x \gg \ell$, the following equation becomes valid:

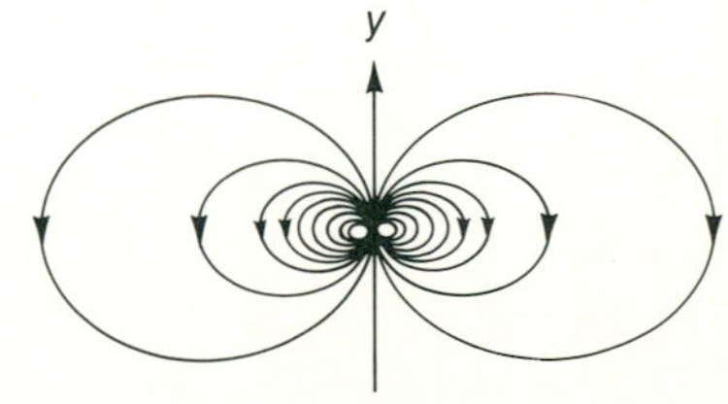

Figure 22–15
Lines of force for the dipole far-field approximation. The dipole itself is too small to be seen; it is aligned along the y axis as in Figure 22–14. Field lines near the center are not shown because they are so close together.

$$E \approx \left(\frac{1}{4\pi\varepsilon_0}\right)\frac{q\ell}{x^3}$$

Thus, for large distances, the field decreases with the *inverse cube of the distance*. Even along the line joining the charges, the field falls off with the inverse cube of the distance (as will be shown in a problem). In fact, it can be shown that for *all* directions away from the dipole, an inverse-cube behavior exists at large distances. We usually place the origin of the coordinate system at the center of the dipole (with the $-y$ direction along the direction of the field between the charges). Then, distances away from the origin are simply the vector **r**. With this convention in mind, we now rewrite the previous equation in the more general notation:

FAR-FIELD APPROXIMATION FOR THE ELECTRIC DIPOLE ($r \gg \ell$)

$$E \approx \left(\frac{1}{4\pi\varepsilon_0}\right)\frac{q\ell}{r^3} \tag{22–10}$$

An interesting feature about the far-field approximation is that if q were doubled and ℓ were halved, the field would still be the same. Indeed, *any* combination of q and ℓ whose product has the same numerical value leads to the same electric field. In other words, it is only the *product* $q\ell$ that determines the field at far distances. For this reason, the product is given a special name: the *electric dipole moment*, as discussed below.

□ □ □

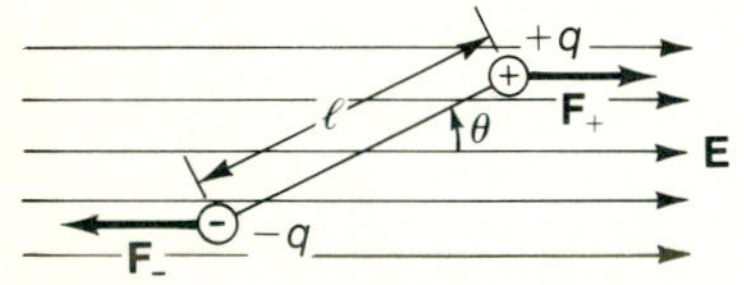

Figure 22–16
An electric dipole in a uniform external field **E**.

Of special interest is the behavior of an electric dipole placed in a uniform electric field **E**, as shown in Figure 22–16. Since the field is uniform, the force $\mathbf{F}_+$ on the $+q$ charge is equal in magnitude but opposite in direction to the force $\mathbf{F}_-$ on the $-q$ charge. The net force on the dipole is zero, so the torque on the dipole may be computed from any point. Let us choose the point at the negative charge $-q$. Recall from Chapter 9 that the torque τ about $-q$ is

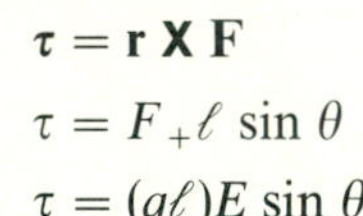

$$\boldsymbol{\tau} = \mathbf{r} \times \mathbf{F}$$

$$\tau = F_+\ell \sin\theta$$

or

$$\tau = (q\ell)E \sin\theta$$

which tends to rotate the dipole toward decreasing θ. The form of this equation suggests a vector notation:

$$\boldsymbol{\tau} = (q\boldsymbol{\ell}) \times \mathbf{E} \tag{22–11}$$

where $(q\boldsymbol{\ell})$ is the **electric dipole moment p** directed from the negative to the positive charge. Note that the direction of $\boldsymbol{\tau}$ also is specified by the cross-product.

ELECTRIC DIPOLE MOMENT p

$$\mathbf{p} = q\boldsymbol{\ell} \quad \text{(where } \boldsymbol{\ell} \text{ is directed from the negative to the positive charge)} \tag{22–12}$$

The dipole moment has units of coulomb·meters (C·m).

When the dipole is in an external electric field **E**, the torque is expressed in vector form as

TORQUE AN ELECTRIC FIELD E EXERTS ON AN ELECTRIC DIPOLE MOMENT p

$$\boldsymbol{\tau} = \mathbf{p} \times \mathbf{E} \tag{22–13}$$

Note that the torque tries to align the dipole *along* the field direction. We would have to do work against this torque to rotate the dipole away from the field-parallel direction. Thus, in the presence of the external field, the dipole possesses potential energy when not aligned along the field direction. The electric force is *conservative*, so the change in potential energy ΔU is the *negative* of the work done by the conservative force. For linear motion (Equation 6–3a), this change is

$$U_b - U_a = -\int_a^b \mathbf{F} \cdot d\mathbf{x}$$

For a torque $\boldsymbol{\tau}$ acting through an angle $d\boldsymbol{\theta}$, the relation is

$$U_\theta - U_{\theta_0} = -\int_{\theta_0}^{\theta} \boldsymbol{\tau} \cdot d\boldsymbol{\theta}$$

We note that the torque $\boldsymbol{\tau}$ is opposite to $d\boldsymbol{\theta}$, thus introducing a minus sign:

$$U_\theta - U_{\theta_0} = -\int_{\theta_0}^{\theta} (-pE \sin\theta)\, d\theta$$

$$U_\theta - U_{\theta_0} = -pE(\cos\theta - \cos\theta_0)$$

Choosing the zero reference level $U_{\theta_0} \equiv 0$ when $\theta_0 = 90°$, we have

$$U = -pE \cos\theta$$

which can be written as the scalar product:

POTENTIAL ENERGY *U* OF AN ELECTRIC DIPOLE IN AN ELECTRIC FIELD ($U \equiv 0$ when *p* and *E* are at an angle of 90°)

$$U = -(\mathbf{p} \cdot \mathbf{E}) \qquad (22\text{–}14)$$

Note that the potential energy of the dipole is a *maximum* when **p** is *antiparallel* to **E** and a *minimum* when **p** is *parallel* to **E**.

EXAMPLE 22–6

A certain electric dipole has opposite charges of 2×10^{-15} C separated a distance of 0.2 mm. It is placed in a uniform electric field of 10^3 N/C. (a) Find the torque the field exerts on the dipole when the dipole moment is at 60° with respect to the field. (b) Find the amount of work it would take to rotate the dipole to this position starting with the dipole aligned parallel to the field.

SOLUTION

(a) The magnitude of the dipole moment is

$$p = q\ell$$
$$p = (2 \times 10^{-15}\ \text{C})(2 \times 10^{-4}\ \text{m})$$
$$p = 4 \times 10^{-19}\ \text{C·m}$$

From Equation (22–13), we obtain

$$\boldsymbol{\tau} = \mathbf{p} \times \mathbf{E}$$
$$\tau = pE \sin\theta$$
$$\tau = (4 \times 10^{-19}\ \text{C·m})(10^3\ \text{N/C})(0.866)$$
$$\tau = \boxed{3.46 \times 10^{-16}\ \text{N·m}}$$

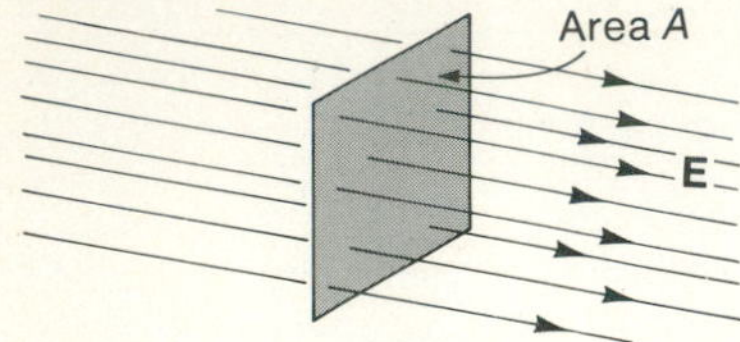

(a) When the area A is normal to the uniform field **E**, the flux Φ_E passing through the area is $\Phi_E = EA$.

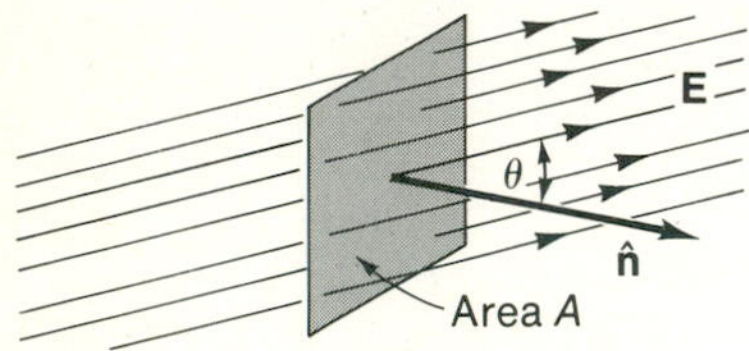

(b) When a uniform field **E** is at an angle θ with the normal $\hat{\mathbf{n}}$ to the area, the flux is $\Phi_E = (\mathbf{E} \cdot \hat{\mathbf{n}})A = EA \cos \theta$.

Figure 22–17

The electric flux Φ_E associated with a uniform field **E** and a plane area A.

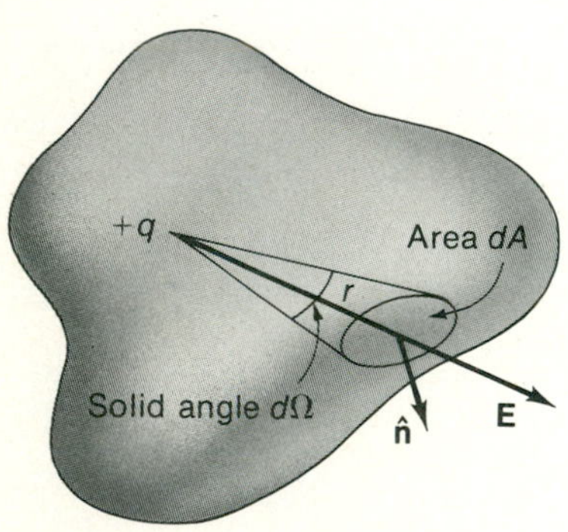

(a) A surface completely encloses the charge q.

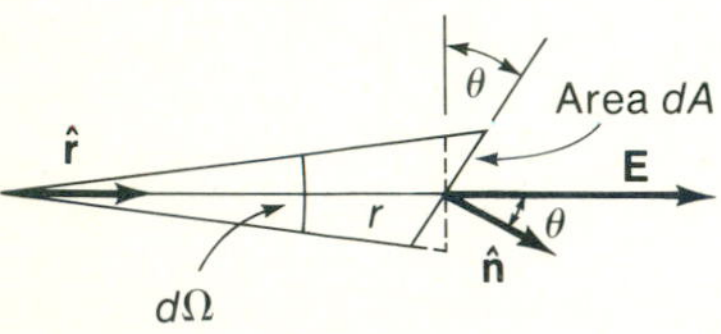

(b) An edge-on view of the surface element dA.

Figure 22–18

The electric field emanating from a positive charge passes through the surface surrounding the charge.

(b) The work an external torque must do on the system equals the change in potential energy of the system (the work-energy relation):

$$W = U_\theta - U_{\theta_0}$$
$$W = (-pE \cos \theta) - (-pE \cos \theta_0)$$
$$W = -pE(\cos 60^\circ - \cos 0^\circ)$$
$$W = -(4 \times 10^{-19}\ \text{C·m})(10^3\ \text{N/C})(0.500 - 1.00)$$
$$W = \boxed{2.00 \times 10^{-16}\ \text{J}}$$

In the above example, if the dipole were placed in a *nonuniform* electric field, the force on each charge would not be the same. Thus there would be a net force on the dipole as a whole (in addition to the torque), causing the dipole to move. Many molecules have an electric dipole. If a nonuniform electric field is established in a liquid suspension of such molecules, the molecules tend to drift toward the regions of the stronger field.

22.6 Gauss's Law

The application of Coulomb's law is often tedious in cases involving a large number of point charges or a continuous distribution of charge. Fortunately, because of its inverse-square form Coulomb's law can be cast into another powerful statement known as *Gauss's law.*

Consider a uniform electric field **E** whose field lines penetrate an area A as in Figure 22–17. We define the *electric flux* Φ_E as

$$\Phi_E = (\mathbf{E} \cdot \hat{\mathbf{n}})A \qquad \textbf{(22–15)}$$

For an element of area dA, the differential flux $d\Phi_E$ is

$$d\Phi_E = (\mathbf{E} \cdot \hat{\mathbf{n}})\, dA \qquad \textbf{(22–16)}$$

where $\hat{\mathbf{n}}$ is a unit vector perpendicular to the surface dA. In the case of a *closed* surface (which completely encloses a volume), it is customary always to define the vector $\hat{\mathbf{n}}$ to extend *outward* from the surface. It will be convenient to combine the symbols $\hat{\mathbf{n}}$ and dA into a new symbol $d\mathbf{A}$:

VECTOR ELEMENT OF AREA dA

$$d\mathbf{A} \equiv \hat{\mathbf{n}}\, dA \quad \text{(where } \hat{\mathbf{n}} \text{ is the outward normal vector at the surface)} \qquad \textbf{(22–17)}$$

Thus $d\mathbf{A}$ is a vector in the direction of the outward normal vector $\hat{\mathbf{n}}$ and has a magnitude of dA. Equation (22–16) then becomes the convenient form

$$d\Phi_E = \mathbf{E} \cdot d\mathbf{A} \qquad \textbf{(22–18)}$$

which, for finite areas, is defined[1] as

ELECTRIC FLUX Φ_E

$$\Phi_E = \int_A \mathbf{E} \cdot d\mathbf{A} \qquad \textbf{(22–19)}$$

Now imagine an isolated point charge q enclosed by an arbitrary closed surface, as shown in Figure 22–18(a). The surface may be a real surface that

---

[1] The integral may be over an arbitrary area, as written, or over a completely *closed* surface, in which case the symbol $\oint$ is used. (Note the similarity in notation with the integral over a closed *path*: $\oint d\boldsymbol{\ell}$.)

does not alter the electric field or any surface that we might imagine as long as it has a clearly defined inside and outside. We choose an element of area dA on this surface a distance r from the charge. There is a relation between the area element dA and the solid angle $d\Omega$ that it subtends (see Figure 22–19). Since the solid angle is defined in terms of an area perpendicular to the radius, we must find the projection of dA on a plane perpendicular to $\mathbf{r}$. It is $(dA \cos\theta)$, so the solid angle $d\Omega$ is

$$d\Omega = \frac{dA \cos\theta}{r^2}$$

Rearranging:

$$dA = \frac{r^2\, d\Omega}{\cos\theta}$$

Combining this expression with Equation (22–17) and the equation for the field $\mathbf{E}$ due to a point charge (Equation 22–7), we write Equation (22–18) as

$$d\Phi_E = \left(\frac{1}{4\pi\varepsilon_0}\right)\left(\frac{q}{r^2}\hat{\mathbf{r}}\right)\cdot\left(\frac{r^2\, d\Omega}{\cos\theta}\hat{\mathbf{n}}\right)$$

(Note how the $1/r^2$ nature of Coulomb's law permits cancellation of the r^2 term associated with the area element.)

$$d\Phi_E = \left(\frac{q}{4\pi\varepsilon_0}\right)\frac{\hat{\mathbf{r}}\cdot\hat{\mathbf{n}}}{\cos\theta}\, d\Omega$$

But from Figure 22–18(b), $\hat{\mathbf{r}}\cdot\hat{\mathbf{n}} = \cos\theta$, so that

$$d\Phi_E = \frac{q}{4\pi\varepsilon_0}\, d\Omega$$

Integrating over the whole surface that encloses the charge q is equivalent to integrating over the whole solid angle.

$$\Phi_E = \frac{q}{4\pi\varepsilon_0}\oint_A d\Omega$$

Since the whole solid angle is 4π in steradians, we have

$$\Phi_E = \frac{q}{\varepsilon_0}$$

Combining this result with Equation (22–19), we thus obtain **Gauss's law:**

GAUSS'S LAW

$$\oint_A \mathbf{E}\cdot d\mathbf{A} = \frac{q}{\varepsilon_0} \quad \text{(when a charge } q \text{ is inside the surface)} \qquad \textbf{(22–20)}$$

where A is *any* surface enclosing q. Such a surface is often referred to as a **Gaussian surface.**

Gauss's law states that the total electric flux emanating from an arbitrary surface enclosing a charge q *is independent of the location of that charge within the surface*. This is a direct consequence of the inverse-square form of Coulomb's law.[2] Since the superposition principle applies to electric fields, Equation (22–20) may be extended to include several charges within the surface or even a continuous distribution of charge. Therefore, in Equation (22–20), q may be

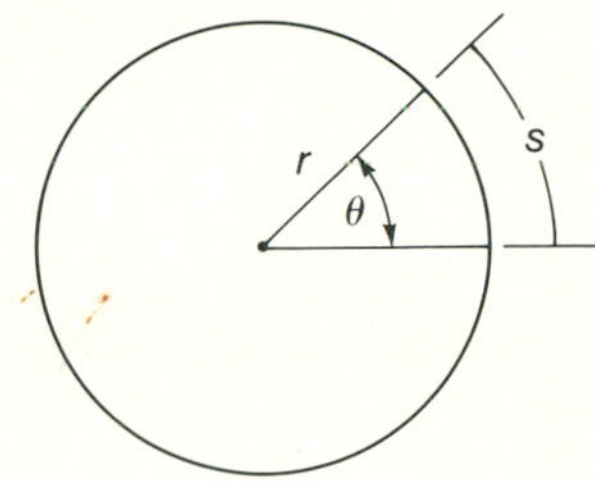

(a) A plane angle θ, measured in *radians*:

$$\theta = \frac{s}{r} \quad \text{(in radians)}$$

where s is the length of the arc subtended by θ. For the complete circle, the whole plane angle is 2π radians.

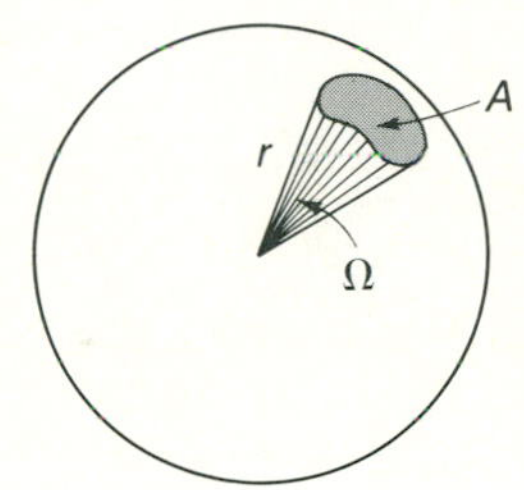

(b) A solid angle Ω, measured in *steradians*:

$$\Omega = \frac{A}{r^2} \quad \text{(in steradians)}$$

where A is the area subtended by Ω. The area may have any shape, but it must be everywhere perpendicular to the radius. The whole solid angle is 4π steradians.

Figure 22–19

The definition of a solid angle is analogous to the definition of a plane angle. Just as the arc length s is everywhere perpendicular to the radius, the area element A must be everywhere perpendicular to the radius.

[2] Newton's law of universal gravitation could also be cast into a form like Equation (22–20).

interpreted as the total net charge (the algebraic sum of the positive and negative charges) enclosed by the surface. If there is a charge density ρ within the volume V enclosed by the surface, the total charge q is equal to $\int_V \rho\, dV$. Thus:

GAUSS'S LAW

$$\oint_A \mathbf{E} \cdot d\mathbf{A} = \frac{1}{\varepsilon_0} \int_V \rho\, dV \quad \text{(when a charge density } \rho \text{ is inside the surface)} \tag{22–21}$$

It is important to note that if a surface contains zero net charge, it does not imply that the field is zero at every place on the surface; rather, Gauss's law states that the *net flux over the entire surface integrates to zero.* That is:

GAUSS'S LAW

$$\oint_A \mathbf{E} \cdot d\mathbf{A} = 0 \quad \text{(when zero net charge is inside the surface)} \tag{22–22}$$

which means *either* that $\mathbf{E}$ is zero everywhere on the surface, *or* that contributions of $\mathbf{E} \cdot d\mathbf{A}$ to the integral cancel when the integration is carried out over the entire surface.

To illustrate, suppose a dipole is located within a cubical Gaussian surface as in Figure 22–20. The electric field lines of the dipole will penetrate the walls of the surface, some entering and some leaving at a variety of angles, depending on the location of the dipole and the position on the surface. Gauss's law makes no statement about the field at any point on the surface, but instead states that the total electric flux through the surface integrates to zero if the net charge inside is zero. Note that in the integration, flux lines *leaving* the surface give a *positive* contribution, while lines *entering* the surface give a *negative* contribution.

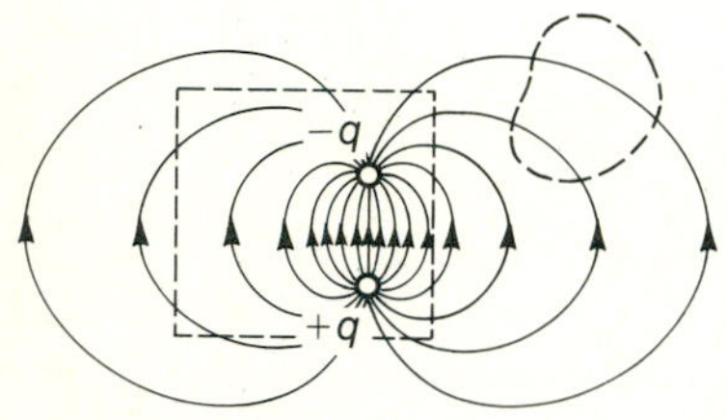

Figure 22–20
Each of the two Gaussian surfaces (shown dashed) encloses a volume of space. The rectangular surface surrounds a dipole, so the net charge inside the surface is zero. The other surface is in the field of the dipole and also has zero net charge inside. For each, $\oint \mathbf{E} \cdot d\mathbf{A} = 0$.

EXAMPLE 22–7

Using Gauss's law, calculate the electric field at a distance r from an isolated point charge $-Q$.

SOLUTION

Gauss's law enables us to calculate the electric field if we know at least some qualitative facts about the field. We know that because of symmetry, an isolated, negative point charge has a field radiating uniformly inward in all directions toward the negative charge. Therefore, we will choose a Gaussian surface best suited to this symmetry. Obviously, a sphere that is concentric with the charge (Figure 22–21) is best, because the field is everywhere perpendicular to the surface and has the same magnitude everywhere on the surface, thus making it easy to integrate $\oint \mathbf{E} \cdot d\mathbf{A}$. Applying Equation (22–20):

$$\oint \mathbf{E} \cdot d\mathbf{A} = \frac{q}{\varepsilon_0}$$

we obtain

$$\oint E \cos\theta\, dA = -\frac{Q}{\varepsilon_0}$$

Since E and θ are constant over the whole surface, we have

$$E \cos\theta \oint dA = -\frac{Q}{\varepsilon_0}$$

$$E \cos\theta\, 4\pi r^2 = -\frac{Q}{\varepsilon_0}$$

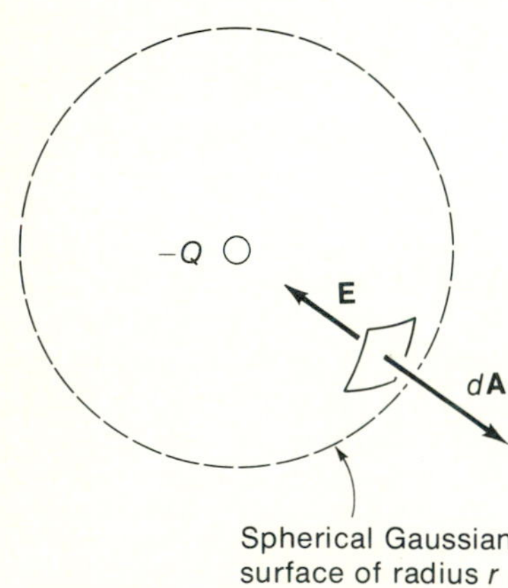

Figure 22–21
Example 22–7

The minus sign indicates that **E** and $d\mathbf{A}$ are antiparallel, that is, that $\theta = 180°$. Solving for E yields

$$\mathbf{E} = \boxed{-\left(\frac{1}{4\pi\varepsilon_0}\right)\frac{Q}{r^2}\hat{\mathbf{r}}} \qquad \text{(radially inward)}$$

EXAMPLE 22–8

Consider an isolated, charged spherical conductor (in which charges can move freely) such as that shown in Figure 22–22. Find the electric field (a) within the cavity of the sphere, (b) between the inner and outer walls of the sphere, and (c) outside the sphere.

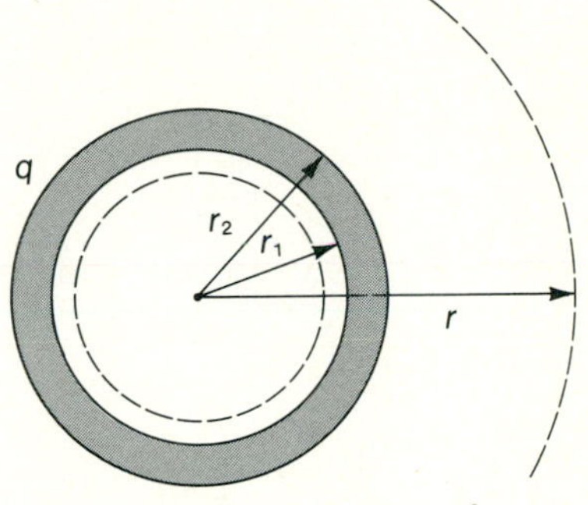

Figure 22–22
Example 22–8

SOLUTION

Since charges can move freely within the conductor, we utilize the symmetry of the sphere to conclude that the charge is *uniformly* distributed over the sphere.

(a) The field within the cavity can be determined by constructing a spherical Gaussian surface whose radius is just a bit less than r_1. The charge within this surface is zero. Therefore:

$$\oint \mathbf{E} \cdot d\mathbf{A} = 0$$

within the cavity. Because of the symmetry of charge distribution, there cannot be compensating values of $\mathbf{E} \cdot d\mathbf{A}$ as we integrate over the surface. Thus, **E** must be zero everywhere on the surface, and we conclude that **E** is zero everywhere inside the cavity. As we will see later, even for an *arbitrarily shaped* empty conducting surface, the electric field is always zero inside.

(b) We next show that the electric field within the conductor itself must be zero. If a field existed within the conductor, a force would be exerted on the electrons, which are free to move in the conductor. Indeed, they would move to locations such that there would be no field in the conductor. Therefore, in the case of electrostatics (which deals with *stationary* electric charges), no field can exist within a conductor.

(c) The electric field outside the conducting sphere can be determined by constructing a spherical Gaussian surface that is concentric with the spherical conductor. Let the radius of this sphere be r. As in Example 22–7, a *spherical* Gaussian surface was chosen to take advantage of the symmetry of the charge distribution. We apply Equation (22–20):

$$\oint \mathbf{E} \cdot d\mathbf{A} = \frac{q}{\varepsilon_0}$$

Since **E** is constant in magnitude over the Gaussian surface and perpendicular to the surface everywhere, the integral becomes

$$E\oint dA = \frac{q}{\varepsilon_0}$$

or

$$E(4\pi r^2) = \frac{q}{\varepsilon_0}$$

Solving for E gives

$$E = \boxed{\left(\frac{1}{4\pi\varepsilon_0}\right)\frac{q}{r^2}}$$

The field is directed away from the sphere and is the same as though all the charge were located at the center. Since the field is zero everywhere inside the outer surface of a spherical conductor, the charge must reside entirely on the outer surface of the sphere.

The results of Example 22–8 suggest an extremely sensitive test of the inverse-square form of Coulomb's law and consequently Gauss's law. The prediction of no electric field within a charged hollow sphere is a direct consequence of the inverse-square law. Because experiments seeking a null result are inherently more sensitive than those determining a nonzero value, the inverse-square nature of Coulomb's law is experimentally established to within one part in a billion. Unfortunately, the inverse-square nature of Newton's law of gravitation cannot be established in this way. Gravitational mass cannot flow in a hollow "conductor" in such a way as to establish zero gravitational field within the "conductor." Due to the influence of gravitational masses outside the sphere, a gravitational field will always exist inside a hollow, massive sphere. There is no "gravitational shield." As a result, the inverse-square behavior of Newton's law of universal gravitation is experimentally verified to no greater than one part in 10,000.

22.7 Electric Potential

The concept of energy has proved so useful in physics and engineering that we will next investigate the energy involved in the interaction of charges. A charge placed in an electric field has a force exerted on it, and consequently the field can do work on the charge. (This is completely analogous to the gravitational field **g** doing work on a mass as it falls, thereby decreasing the potential energy of the mass.)

Consider a charge q' as it moves from point P_1 to point P_2 under the influence of the field due to the charge q, as shown in Figure 22–23. The force on q' at any point along the path will be

$$\mathbf{F} = q'\mathbf{E}$$

where **E** is the field at that point. The incremental work dW done on q' in moving along the path a distance $d\ell$ is given by

$$dW = \mathbf{F} \cdot d\boldsymbol{\ell}$$

$$dW = q'\mathbf{E} \cdot d\boldsymbol{\ell} \tag{22–23}$$

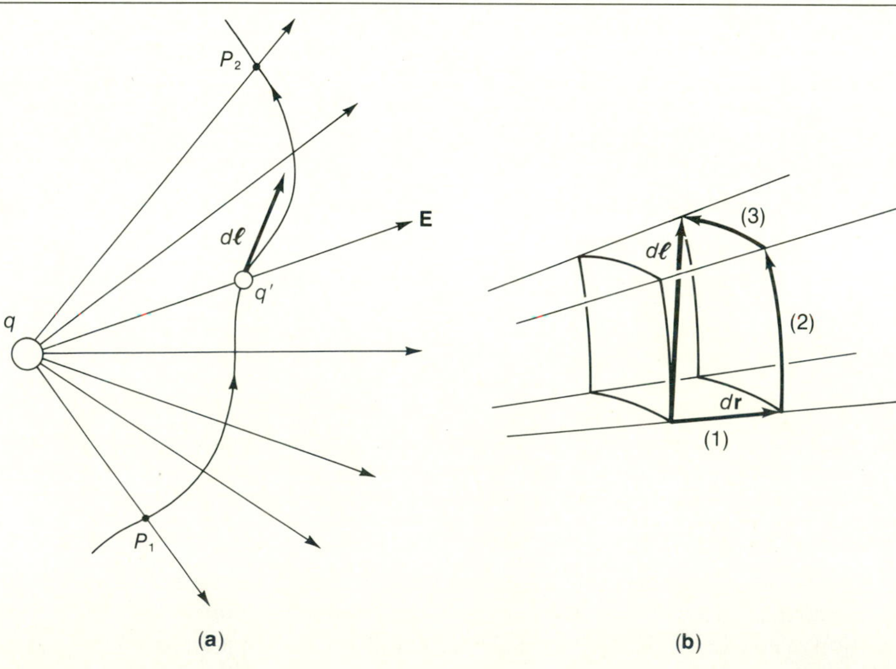

Figure 22–23
The work done by the electric field of q on q' depends only on the change in radial distance Δr between P_1 and P_2.

Evaluating the scalar product $\mathbf{E} \cdot d\boldsymbol{\ell}$, we observe (Figure 22–23a) that **E** has a component only in the radial direction, while in general $d\boldsymbol{\ell}$ not only has a component in the radial direction (1), but it also may have components in two directions perpendicular to the radial direction: (2) and (3). The dot product is zero for the components perpendicular to **r**, so we are left with the following:

$$dW = q'E\,dr$$

Simply stated, **E** does work only in a radial direction. The total work done by **E** in moving q' from P_1 to P_2 is therefore:

$$W = q' \int_{r_1}^{r_2} E\,dr$$

where r_1 and r_2 are the radial distances between q and q' at the points P_1 and P_2, respectively. Substituting the expression for E and evaluating the integral, we have

$$W = q' \int_{r_1}^{r_2} \left(\frac{1}{4\pi\varepsilon_0}\right)\frac{q}{r^2}\,dr$$

$$W = -\frac{qq'}{4\pi\varepsilon_0}\left(\frac{1}{r_2} - \frac{1}{r_1}\right) \qquad \textbf{(22–24)}$$

The work W done by *the field* **E** equals[3] the *negative* of the change in potential energy U of q':

$$-(U_{P_2} - U_{P_1}) = -\frac{qq'}{4\pi\varepsilon_0}\left(\frac{1}{r_2} - \frac{1}{r_1}\right)$$

Identifying corresponding terms on the left and right sides, we find

$$U_{P_1} = \frac{qq'}{4\pi\varepsilon_0 r_1} \qquad \text{and} \qquad U_{P_2} = \frac{qq'}{4\pi\varepsilon_0 r_2}$$

Or, more simply, for any point r:

POTENTIAL ENERGY *U* OF TWO CHARGES SEPARATED A DISTANCE *r* ($U \equiv 0$ for $r = \infty$)

$$U = \left(\frac{1}{4\pi\varepsilon_0}\right)\frac{qq'}{r} \qquad \textbf{(22–25)}$$

Note that from this definition, the zero reference level for the potential energy is when the two charges are infinitely far apart.

We defined the electric field due to q as the force per unit of positive charge exerted on a unit charge q' by q. Similarly, we will define the **electric potential *V*** due to q at a point as the potential energy per unit charge when q is placed at that point.

ELECTRIC POTENTIAL *V*

$$V \equiv \frac{U}{q'} \qquad \textbf{(22–26)}$$

or, for a point charge q:

ELECTRIC POTENTIAL *V* FOR A POINT CHARGE ($V \equiv 0$ for $r = \infty$)

$$V = \left(\frac{1}{4\pi\varepsilon_0}\right)\frac{q}{r} \qquad \textbf{(22–27)}$$

[3] Recall the analogous situation for gravity: The work done on a mass *by the gravitational field* equals the negative of the change in the potential energy of the mass.

Electric potential has the units of newton-meters/coulomb (N·m/C), which is called a **volt** (V).

□ □ □

We shall now develop a definition of potential in terms of the electric field for the general case (not just the point-charge situation). We begin with Equation (22–23), the work dW done by the force $q'\mathbf{E}$ as a field moves a unit charge q' a distance $d\boldsymbol{\ell}$.

$$dW = q'\mathbf{E} \cdot d\boldsymbol{\ell}$$

In this process, the charge q' is moved from a higher potential to a lower potential, thereby losing potential energy.

$$dU = -dW$$

Substituting for dW gives

$$dU = -q'\mathbf{E} \cdot d\boldsymbol{\ell}$$

Substituting $U = q'V$ and integrating, we obtain

RELATION BETWEEN *V* and E

$$V_2 - V_1 = -\int_1^2 \mathbf{E} \cdot d\boldsymbol{\ell} \qquad \textbf{(22–28)}$$

where the integral is a line integral from point 1 (at $\mathbf{r}_1$) to point 2 (at $\mathbf{r}_2$). Note that *the value of the integral depends only on the end points, not on the path taken.* This behavior shows that the coulomb force is a *conservative* force. Although we derived Equation (22–28) for a single point charge, it is a perfectly general result; a similar analysis shows that it is true for the electric field resulting from any distribution of charges.

One note of caution: Be sure you understand the difference between *electric potential energy U* and *electric potential V*. The situation is analogous to the electric field **E** and the force **F** a charge feels when placed in the field. These two variables differ by the factor q: $\mathbf{F} = q\mathbf{E}$. In a similar way, the potential energy U that a charge has differs from the potential V at that point by the factor q: $U = qV$.

The potential *difference* ΔV between two points is usually more important than just the potential V at a given point. Just as the definition of potential energy U implies a zero reference level, a reference level must also be chosen for the potential V. The choice is one of convenience. We may choose the point at infinity (as in Equation 22–27) or at some other location convenient for the problem.

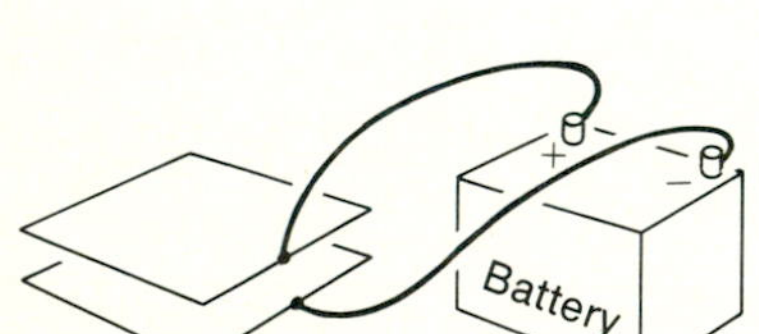

(**a**) A battery connected to two parallel metal plates.

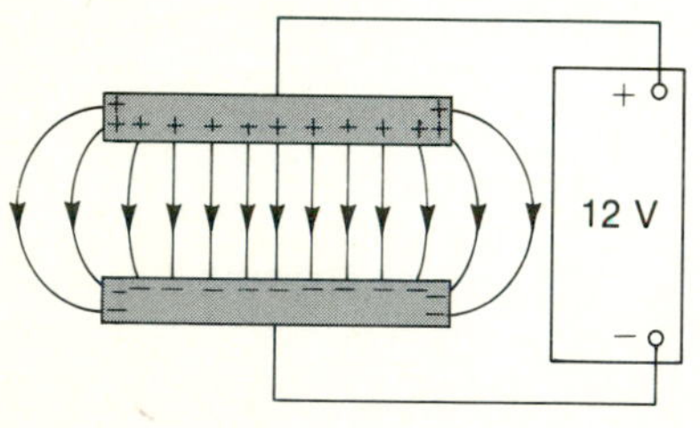

(**b**) A schematic diagram for (**a**). The charges are at the inner surfaces of the plates.

Figure 22–24
A battery connected to parallel metal plates creates an electric field between the plates.

□ □ □

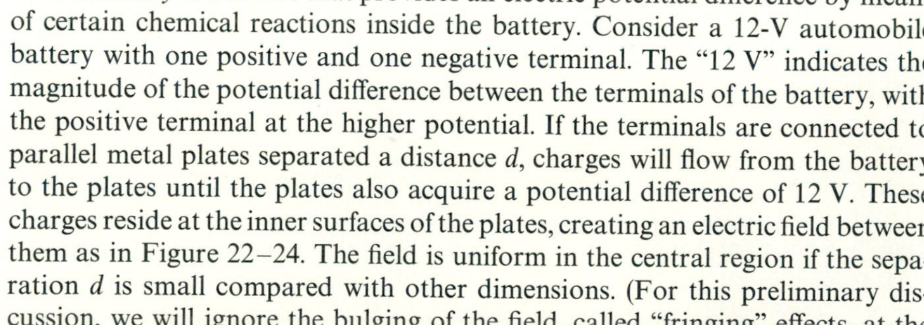

A *battery* is a device that provides an electric potential difference by means of certain chemical reactions inside the battery. Consider a 12-V automobile battery with one positive and one negative terminal. The "12 V" indicates the magnitude of the potential difference between the terminals of the battery, with the positive terminal at the higher potential. If the terminals are connected to parallel metal plates separated a distance d, charges will flow from the battery to the plates until the plates also acquire a potential difference of 12 V. These charges reside at the inner surfaces of the plates, creating an electric field between them as in Figure 22–24. The field is uniform in the central region if the separation d is small compared with other dimensions. (For this preliminary discussion, we will ignore the bulging of the field, called "fringing" effects, at the edges of the plates.) The next example makes use of this arrangement to further clarify the relation between **E** and V.

EXAMPLE 22–9

The terminals of a 12-V battery are connected to two parallel metal plates in a vacuum. An electron is released from rest at the negative plate and is accelerated by the electric field toward the positive plate. (Since the electron has a *negative* charge, the field exerts a force on it in a direction *opposite* to **E**.) What kinetic energy ΔK does the electron gain as it just reaches the positive plate?

SOLUTION

The potential difference through which the electron moves, $\Delta V = 12$ V, is the *negative* of the work done per unit charge by the force the electric field exerts as it moves the electron toward the positive plate. We begin with Equation (22–28):

$$V_2 - V_1 = -\int_1^2 \mathbf{E} \cdot d\ell$$

Letting ΔW represent the work done, we have (since $\Delta W = \mathbf{F} \cdot \Delta \ell$ and $\mathbf{F} = \mathbf{E}/q$):

$$\Delta V = -\frac{\Delta W}{q}$$

The force the field exerts is the only force acting on the electron.[4] The work done by this force accelerates the electron toward the positive plate, giving it kinetic energy ΔK equal to the work done ΔW. Thus:

$$\Delta V = -\frac{\Delta K}{q}$$

or

$$\Delta K = -\Delta V q$$

The charge q is the electron charge $e = -1.6 \times 10^{-19}$ C, so:

$$\Delta K = -(12 \text{ V})(-1.6 \times 10^{-19} \text{ C})$$

Recalling that one volt is one newton·meter/coulomb, or one joule/coulomb, the units are in joules. Since the electron starts at rest, its final kinetic energy K is equal to ΔK:

$$K = \boxed{1.92 \times 10^{-18} \text{ J}}$$

The prevalence of the electron charge in atomic and nuclear physics has led to defining a new energy unit, the *electron volt*. An **electron volt** (eV) is the amount of *energy* acquired by an object with a charge q equal in magnitude to the electronic charge when the object is accelerated through a potential difference of one volt.

$$\Delta W = q\Delta V$$

$$1 \text{ eV} = (1.602 \times 10^{-19} \text{ C})(1 \text{ J/C})$$

ELECTRON VOLT (an energy unit)

$$1 \text{ eV} = 1.602 \times 10^{-19} \text{ J}$$

In practice, the electron volt as an energy unit is often applied to nonelectrical situations. For example, an air molecule at room temperature is sometimes said to have a kinetic energy of about 1/40 eV.

[4] For all problems involving electrons, protons, charged nuclei, and so on, the changes in electrical potential energy are so much greater than the changes in gravitational potential energy that we ignore the latter when applying conservation of energy.

EXAMPLE 22–10

Consider two thin, conducting, spherical shells as in Figure 22–25. The inner shell has a radius $r_1 = 15$ cm and a charge of $+10$ nC (nC is the symbol for *nanocoulomb*, equal to 10^{-9} C). The outer shell has a radius $r_2 = 30$ cm and a charge of -15 nC. Find the electric field and the electric potential in these regions:

Region A: inside the inner shell ($r < r_1$)
Region B: between the shells ($r_1 < r < r_2$)
Region C: outside the outer shell ($r > r_2$)

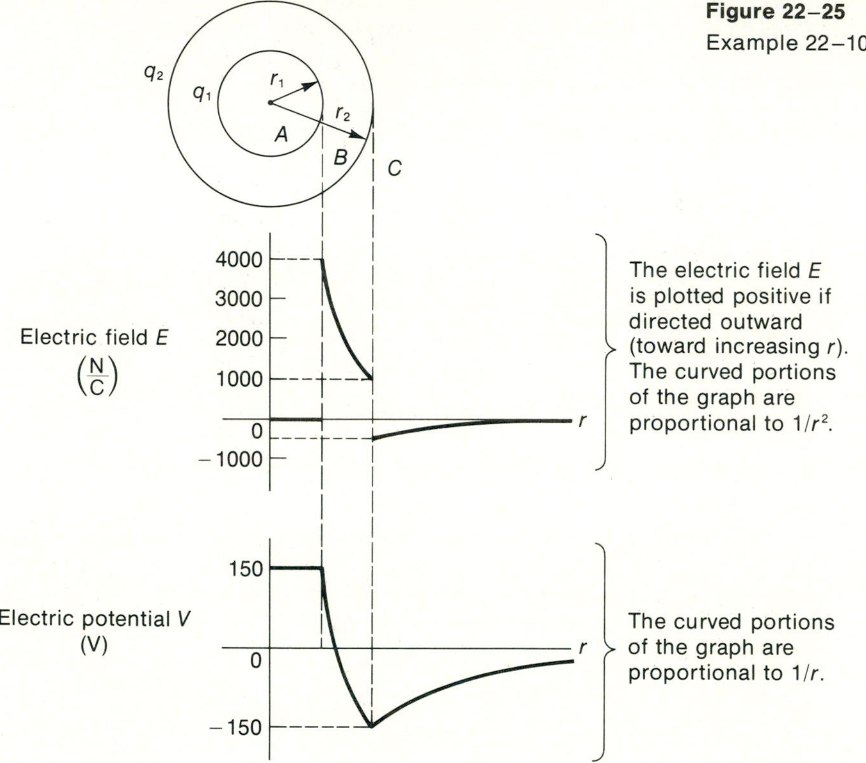

Figure 22–25
Example 22–10

SOLUTION

Because of the symmetry of the situation, the charges must distribute themselves symmetrically over the spheres. Also, because of this symmetry, it is easiest first to calculate the field using Gauss's law, then to obtain the potential from the relation between V and E (Equation 22–28). We will consider each region in turn, starting with the electric field.

(a) Region A: *inside the inner shell.* Noting the symmetry, we construct a Gaussian surface in the form of a sphere concentric with the center. We then apply Gauss's law:

$$\oint \mathbf{E} \cdot d\mathbf{A} = \frac{q}{\varepsilon_0}$$

Whatever magnitude E has at one point on this surface, by symmetry it must have the same value at all points. Since there is no charge inside the Gaussian surface, we conclude that the field is zero all over the surface. Furthermore, we could construct such a surface anywhere in Region A with an arbitrary radius ($0 < r < r_1$), so we conclude that *the field* **E** *is zero everywhere inside the inner shell.*

$$E_A = \boxed{0} \qquad \text{(inside the inner shell)}$$

(b) Region B: *between the shells.* Again, we recognize that the symmetry calls for a Gaussian surface in the form of a sphere of radius r (where $r_1 < r < r_2$) concentric with the center. Beginning with Gauss's law:

$$\oint \mathbf{E} \cdot d\mathbf{A} = \frac{q}{\varepsilon_0}$$

We recognize this problem is similar to Example 22–8, whose result is

$$\boxed{E_B = \left(\frac{1}{4\pi\varepsilon_0}\right)\frac{q}{r^2}} \qquad \text{(radially outward: for } r_1 < r < r_2\text{)}$$

Recalling that $1/(4\pi\varepsilon_0) = 9 \times 10^9 \text{ N·m}^2/\text{C}^2$, we calculate the value of E at the location just barely[5] outside the inner shell radius r_1.

$$E_{r_1} = \left(9 \times 10^9 \frac{\text{N·m}^2}{\text{C}^2}\right)\frac{(10 \times 10^{-9}\text{ C})}{(0.15\text{ m})^2}$$

$$E_{r_1} = \boxed{4000 \frac{\text{N}}{\text{C}}} \qquad \text{(just outside the inner shell)}$$

This field decreases as $1/r^2$ until just barely inside the outer shell, where its value is

$$E_{r_2} = \left(9 \times 10^9 \frac{\text{N·m}^2}{\text{C}^2}\right)\frac{(10 \times 10^{-9}\text{ C})}{(0.30\text{ m})^2}$$

$$E_{r_2} = \boxed{1000 \frac{\text{N}}{\text{C}}} \qquad \text{(just inside the outer shell)}$$

(c) Region C: *outside the outer shell.* Again, we construct a concentric Gaussian surface of radius r (where $r_2 < r$) and apply Gauss's law, recognizing that q is the *net* charge inside the surface.

$$\oint \mathbf{E} \cdot d\mathbf{A} = \frac{q}{\varepsilon_0}$$

The net charge is $q = q_1 + q_2$, or (10 nC) + (−15 nC) = −5 nC. As before:

$$\boxed{E_C = \left(\frac{1}{4\pi\varepsilon_0}\right)\frac{q}{r^2}}$$

The value just barely outside the outer shell (at $r = r_2$) is

$$E_{r_2} = \left(9 \times 10^9 \frac{\text{N·m}^2}{\text{C}^2}\right)\frac{(-5 \times 10^{-9}\text{ C})}{(0.30\text{ m})^2}$$

$$\boxed{E_{r_2} = -500 \frac{\text{N}}{\text{C}}} \qquad \text{(just outside the outer shell)}$$

The minus sign indicates that the field is directed inward (in the $-\mathbf{r}$ direction). It changes as $1/r^2$, approaching zero as $r \to \infty$.

Note that the electric field is *not* a continuous function of distance. As the Gaussian surface expands across one of the shells, it suddenly encloses a new layer of charge, causing the value of $\mathbf{E}$ to change discontinuously (at least in this idealized case, where we assume that the layer of point charges has zero thickness). As we approach a discontinuity from one side or the other, we thereby learn information about the way the values change at the discontinuity itself. Figure 22–25 summarizes these results.

[5] The phrase "just barely outside the inner shell radius r_1" recognizes that we really mean $r = r_1 + dr$, where we then let $dr \to 0$. This way, we approach the surface from outside, so we are sure to include the surface charge inside the Gaussian surface. We then speak of the field just barely outside the inner shell as if it were at the location $r = r_1$. You will spot other situations where we use this same sort of reasoning, approaching a discontinuity from one side (or the other), and then quoting values at the discontinuity itself.

Calculation of the potential V

Since we know the field **E** everywhere, we will use Equation (22–28):

$$V_2 - V_1 = -\int_1^2 \mathbf{E} \cdot d\ell$$

to calculate how the potential varies. First, we must choose a zero reference location for the potential. A convenient choice for this problem is to assign $V = 0$ at $r = \infty$. Then, we can start at $r = \infty$ and work our way into the center of the sphere, calculating the change of potential as we go. Our variable is $d\mathbf{r}$. Changing variables[6] and setting $V_1 = 0$, the above equation becomes

$$V_r - 0 = -\int_\infty^r \frac{q}{4\pi\varepsilon_0 r^2}\, dr \qquad \textbf{(22–29)}$$

Region C: *outside the outer shell.* Because of the spherical symmetry of the charge distribution, both the field and the potential outside the spheres are as though the net charge q (which equals $q_1 + q_2$) were concentrated at a point at the center. Integrating inward from ∞ to a point r (outside the spheres), we have

$$V_r = -\int_\infty^r \frac{q}{4\pi\varepsilon_0 r^2}\, dr = \left.\frac{q}{4\pi\varepsilon_0 r}\right|_\infty^r \qquad \textbf{(22–30)}$$

or

$$\boxed{V_C = \left(\frac{1}{4\pi\varepsilon_0}\right)\frac{q}{r}} \qquad \text{(Region } C\text{: } r \geqq r_2\text{)}$$

Note that we also obtained this result using different reasoning in Equation (22–27). The equation has the correct form: as r increases, the potential approaches zero as $1/r$, verifying our choice of $V = 0$ at infinity. Since the net charge q is -5 nC, the numerical value at $r = r_2$ is

$$V_{r_2} = \left(9 \times 10^9\, \frac{\text{N}\cdot\text{m}^2}{\text{C}^2}\right)\left(\frac{-5 \times 10^{-9}\ \text{C}}{0.30\ \text{m}}\right)$$

$$V_{r_2} = \boxed{-150\ \text{V}}$$

Region B: *between the shells.* As usual, the *change* of potential between r_2 and r (where $r_1 < r < r_2$) is given by

$$V_r - V_{r_2} = -\int_{r_2}^r \mathbf{E} \cdot d\ell$$

The magnitude of **E** is determined solely by the charge q_1 on the inner sphere (Gauss's law). Again changing variables, for the integration from r_2 to r we have

$$V_r - V_{r_2} = -\int_{r_2}^r \frac{q_1}{4\pi\varepsilon_0 r^2}\, dr$$

$$V_r - V_{r_2} = \frac{q_1}{4\pi\varepsilon_0}\left(\frac{1}{r} - \frac{1}{r_2}\right)$$

---

[6] Changing variables in this case is somewhat tricky, perhaps causing more errors from overlooked minus signs than any other single procedure in calculus. Since point 1 is at ∞ and point 2 is the location **r**, the path $d\ell$ is radially inward, opposite to the direction of **E**. Thus:

$$V_2 - V_1 = -\int_1^2 \mathbf{E} \cdot d\ell$$

becomes:

$$V_2 - V_1 = -\int_1^2 E \cos 180^\circ\, d\ell$$

However, since E is a function of r, the variable of integration must be r, not ℓ. As ℓ (the distance from infinity) *increases*, the variable r *decreases*. Thus, $d\ell = -dr$. Therefore:

$$V_2 - V_1 = -\int_1^2 \frac{q}{4\pi\varepsilon_0 r^2} \underbrace{\cos 180^\circ}_{(-1)} (-dr)$$

becomes:

$$V_2 - V_1 = -\int_1^2 \frac{q}{4\pi\varepsilon_0 r^2}\, dr$$

Since $V_{r_2} = -150$ V, the value within region B is

$$V_B = \boxed{-150\text{ V} + \frac{q_1}{4\pi\varepsilon_0}\left(\frac{1}{r} - \frac{1}{r_2}\right)} \qquad \text{(Region } B: r_1 < r < r_2) \qquad \textbf{(22–31)}$$

The numerical value for V_{r_1} at the inner shell is

$$V_{r_1} = (-150\text{ V}) + \left(9 \times 10^9 \frac{\text{N}\cdot\text{m}^2}{\text{C}^2}\right)(10 \times 10^{-9}\text{ C})\left(\frac{1}{0.15\text{ m}} - \frac{1}{0.30\text{ m}}\right)$$

$$V_{r_1} = -150\text{ V} + 300\text{ V}$$

$$V_{r_1} = \boxed{150\text{ V}}$$

It is worth noting that although **E** is discontinuous at the boundaries of the shells where the charges are located, *the potential V is continuous across these boundaries.* This is plausible when you recall that integrating $\int \mathbf{E} \cdot d\boldsymbol{\ell}$ may be interpreted as summing up the area under the curve for E as a function of distance (see Figure 22–25). Integrating across a discontinuity merely changes the *rate* at which area accumulates; the area itself does not change abruptly.

Region A: *inside the inner shell.* Again, we start with the same general relation:

$$V_2 - V_1 = -\int_1^2 \mathbf{E} \cdot d\boldsymbol{\ell}$$

But here **E** is zero everywhere inside the inner shell. So there is *no* change of potential as we move inward. Hence, the potential at r_1 (equal to 150 V) is the same (constant) value for all smaller values of r.

$$V_A = \boxed{150\text{ V}} \qquad \text{(Region } A: 0 \leqq r \leqq r_1)$$

It will be helpful to look at the previous problem from still one more viewpoint. The general relation we have used equates the change in potential to the *negative* of the work per unit charge done by the associated conservative force (in this case, the coulomb force) as a unit charge is moved from one point to another. But we may also think of some *external agent* exerting a force to move the charge at constant speed[7] against the coulomb forces. For example, at any point outside the outer shell, the unit charge would be attracted toward the shells. Thus the external agent would do *negative* work in bringing the charge from infinity to the outer shell. (That is, the external force would be in the opposite direction to the motion.) Since changes in kinetic energy are not involved, this work per unit charge would be -150 J/C, and the change in potential is -150 V. Once inside the outer shell, the external agent must force the unit charge against the field, doing 300 J/C of work to move the charge to the inner shell. The corresponding change of potential is 300 V. After passing through the inner shell, no further work would be necessary, so the potential does not change inside the inner shell. For some students, relating the change of potential directly to the work per unit charge that an *external agent* does may seem a bit more straightforward than the relation of Equation (22–28). Of course, either way of looking at it is correct. Just be aware of why there is a minus sign in Equation (22–28) and why there is no minus sign when one considers the work done by an external agent. (As always, whenever work is done, one must keep in mind *what system* is doing work on *what object.*)

The above discussion is analogous to the situation for gravitation. If we raise a mass vertically at constant speed, *we* do positive work per unit mass,

[7] The reason we assume constant speed is so that changes in kinetic energy are not involved. The only interchange of energy is between the work done by the forces and the change in potential energy. More precisely, the object may be imagined as moving infinitely slowly from one point to another—a *quasi-static* situation.

which directly equals the (positive) increase in the gravitational potential energy of the mass. On the other hand, the work per unit mass done by the downward conservative force F_{grav} as the mass is raised is *negative* and thus equals the negative of the change in potential energy.

The next few examples involve various extended charge distributions. Here is the customary notation:

Charge distribution	Notation
Along a line	*Linear* charge density λ $\left(\text{in } \frac{\text{C}}{\text{m}}\right)$
Over an area	*Surface* charge density σ $\left(\text{in } \frac{\text{C}}{\text{m}^2}\right)$
Throughout a volume	*Volume* charge density ρ $\left(\text{in } \frac{\text{C}}{\text{m}^3}\right)$

EXAMPLE 22–11

Consider a long line of charge with a linear (positive) charge density λ (in C/m). Find the electric field and the electric potential close to the line of charge and far from either end.

SOLUTION

"Far from either end" implies there are no end effects to worry about that would destroy the cylindrical symmetry of the line of charge. Because of this symmetry, we can easily calculate the electric field using Gauss's law. We construct a Gaussian surface that utilizes the symmetry: a cylinder concentric with the line of charge, as shown in Figure 22–26(a). According to Gauss's law:

$$\oint \mathbf{E} \cdot d\mathbf{A} = \frac{q}{\varepsilon_0}$$

Figure 22–26
Example 22–11

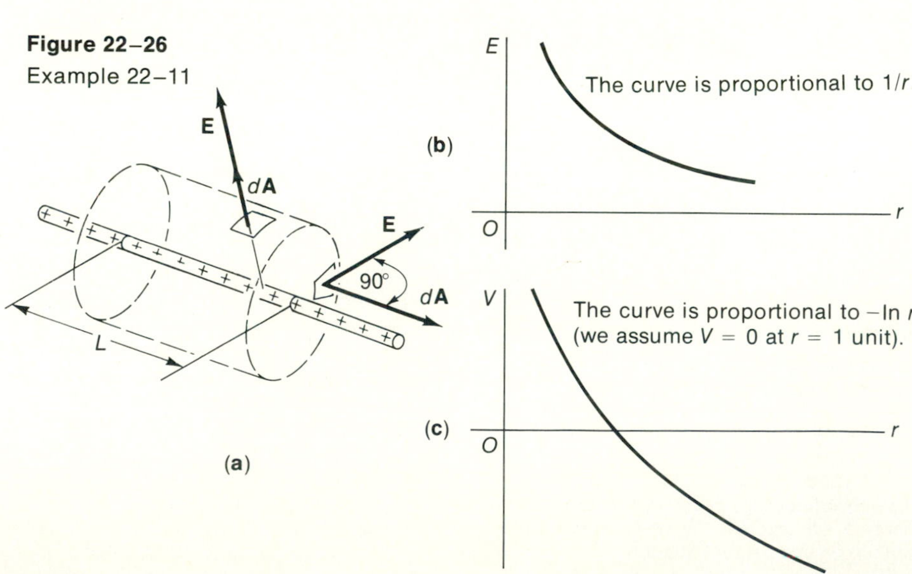
E
dA
E
90°
dA
L
(a)
(b)
E
The curve is proportional to 1/r.
O
r
V
The curve is proportional to −ln r
(we assume V = 0 at r = 1 unit).
(c)
O
r

the value of **E** is not the same over the entire Gaussian surface. At the ends of the cylinder, **E** varies for different distances from the line charge, but is everywhere at right angles to $d\mathbf{A}$. Therefore, the dot product gives zero for this portion of the integration. On the curved surface, **E** is parallel to $d\mathbf{A}$ and constant in magnitude. So the dot product gives $\cos 0° = 1$ for this integration, and we have

$$\int_{\text{ends}} \mathbf{E} \cdot d\mathbf{A} + \int_{\substack{\text{curved}\\\text{surface}}} \mathbf{E} \cdot d\mathbf{A} = \frac{\lambda L}{\varepsilon_0}$$

$$0 + EA \cos 0° = \frac{\lambda L}{\varepsilon_0}$$

where L is the length of the cylinder. Substituting $2\pi rL$ for A and solving for E yields

$$E(2\pi rL) = \frac{\lambda L}{\varepsilon_0}$$

$$E = \boxed{\frac{\lambda}{2\pi r\varepsilon_0}} \tag{22–32}$$

where **E** is outward since λ is positive.

We may calculate the potential from Equation (22–28):

$$V_2 - V_1 = -\int_1^2 \mathbf{E} \cdot d\boldsymbol{\ell}$$

Using an argument similar to the one used in calculating the potential of a point charge, the integral becomes

$$V_2 - V_1 = -\int_{r_1}^{r_2} E\, dr$$

Substituting the value for E given by Equation (22–32) gives

$$V_2 - V_1 = -\int_{r_1}^{r_2} \frac{\lambda}{2\pi r\varepsilon_0}\, dr$$

Integrating, we have

$$V_2 - V_1 = -\frac{\lambda}{2\pi\varepsilon_0}(\ln r_2 - \ln r_1)$$

Because of the logarithmic behavior, we cannot set $V_1 = 0$ at infinity, but must instead choose some finite point. For example, setting $V_1 = 0$ at $r_1 = 1$ unit, we have

$$V = \boxed{-\frac{\lambda}{2\pi\varepsilon_0} \ln r} \qquad \text{(for } V = 0 \text{ at } r = 1 \text{ unit)} \tag{22–33}$$

Graphs of E and V are shown in Figure 22–26.

EXAMPLE 22–12

A spherical volume of radius R contains a volume charge density ρ (in units of charge per unit volume). The charge density is not uniform, but varies with the distance r from the center according to

$$\rho = Ar^2$$

where A is a constant. (a) Find the SI units of A. (b) Find the total charge Q within the volume distribution.

SOLUTION

(a) Since the volume charge density ρ has SI units of coulombs per cubic meter, the units of A are coulombs/meter^5.

(b) Because the charge density varies with the distance r from the center, we must sum elements of charge dq contained within volume elements dV, where *all of each volume element* is the same distance r from the center. (This means that ρ will have the same value throughout the volume dV.) The volume element is thus a thin spherical shell, which has a radius r and a thickness dr. Its volume is

$$dV = 4\pi r^2\,dr$$

and the charge dq within dV is

$$dq \doteq \rho\,dV$$

$$dq = \rho 4\pi r^2\,dr$$

Since $\rho = Ar^2$, we have

$$dq = A4\pi r^4\,dr$$

The total charge Q is found by summing all the similar spherical shells from $r = 0$ to $r = R$.

$$\int_{\substack{\text{entire}\\\text{sphere}}} dq = \int_{\substack{\text{entire}\\\text{sphere}}} \rho\,dV$$

$$Q = \int_0^R A4\pi r^4\,dr$$

Removing constant terms from under the integral sign, we have

$$Q = A4\pi \int_0^R r^4\,dr$$

Integrating (Appendix G-II) gives

$$Q = A4\pi\left(\frac{r^5}{5}\right)\Bigg|_0^R$$

$$Q = \boxed{\tfrac{4}{5}A\pi R^5}$$

22.8 Relationships Between Fields and Potentials

There is a close relationship between electric fields and electric potentials. As we have seen, the potential at a point is obtained by calculating the negative of the work done per unit charge by the field as the unit charge moves from some reference location ($V \equiv 0$) to the point. In this section we will show that it is possible to work this in reverse. That is, given the potential we can derive the corresponding field. To illustrate the usefulness of this approach, consider the problems involved in finding the field of a collection of charges. The straightforward approach would be to calculate the field attributable to each charge separately and then to add the fields of all the charges. However, since fields are *vectors*, their addition may be cumbersome. The easier approach would be to find the potential of each charge and then add the potentials together. This procedure is relatively simple, since potentials are *scalars* and add algebraically. We could then derive the field from knowledge of the potential at every point.

A little thought will show that knowledge of the electric potential at a *single* point cannot yield a value of the electric field. For example, the potential 3 m from a 4-μC charge is the same as the potential 6 m from an 8-μC charge. Yet the electric field is not the same in both cases. We will see that to determine the electric field at a point, we must know the potential *in the region around the point.*

Consider a positive unit charge in a space where an electric field E exists. Suppose we move this charge in a direction *opposite* to the field direction (pushing against the electric force). We find that in moving the charge an infinitesimal distance from x to $x + dx$, we do an amount of work dW. We thus increase the electric potential energy of the unit charge by an amount dU. (That is, we have moved the charge to a place of higher potential.) Because the electric field is in the opposite direction, the field exerts a force F_x opposite to the direction of motion. Thus the work done *by the field* is the negative of the change in potential energy dU:

$$dU = -F_x\,dx$$

Since $U = qV$ and $F_x = qE_x$, we have

$$dV = -E_x\,dx$$

or

$$E_x = -\frac{dV}{dx} \tag{22–34}$$

where dV/dx is the rate of change in potential with respect to direction.

This relation shows that **E** may be expressed in units of *volts/meter* (as well as *newtons/coulomb*). The above equation relating E and V is expressed as a *derivative*. We have previously seen an *integral* expression for the same relationship; it is Equation (22–28), written here for the one-dimensional case:

$$V_2 - V_1 = -\int_1^2 \mathbf{E}_x \cdot \mathbf{dx}$$

In some situations, the integral form is more convenient; in others, the derivative form is more useful.

EXAMPLE 22–13

Terminals of a 6-V battery are connected to two large, parallel metal plates separated by a distance $d = 0.20$ cm. What is the electric field between the plates? (Ignore fringing effects near the edges.)

SOLUTION

As shown in Example 22–9 (and in Figure 22–24), except near the edges the field is uniform. Therefore, Equation (22–34) may be written

$$E = -\frac{\Delta V}{\Delta x}$$

or, for this case: $|E| = \frac{V}{d}$

$$|E| = \frac{6\text{ V}}{2 \times 10^{-3}\text{ m}}$$

$$|E| = \boxed{3000\text{ V/m}}$$

(The field extends from the positive plate toward the negative plate.)

22.9 The Gradient of *V*

If the field is nonuniform—that is, if it has changing values in all three coordinate directions—we may still write a relation between V and **E**. Consider, for example, the two-dimensional potential:

$$V = ax^2y$$

where a is a constant. We can find, separately, the x and y components of $\mathbf{E}$ by first holding y constant, so that the potential changes in x as follows:

$$E_x = -\frac{dV}{dx} \qquad \text{(holding } y \text{ constant)}$$

$$E_x = -2axy$$

A common mathematical notation for this procedure is the *partial derivative* symbol, $\partial V/\partial x$, which means "take the derivative of V with respect to x, holding all other variables (here, y) constant." In a similar way, the y component is

$$E_y = -\frac{\partial V}{\partial y}$$

Therefore, a more complete form of Equation (22–34) is

$$E_x = -\frac{\partial V}{\partial x}$$

$$E_y = -\frac{\partial V}{\partial y}$$

$$E_z = -\frac{\partial V}{\partial z}$$

The total field then becomes

$$\mathbf{E} = E_x\hat{\mathbf{x}} + E_y\hat{\mathbf{y}} + E_z\hat{\mathbf{z}}$$

or

$$\mathbf{E} = -\left(\frac{\partial V}{\partial x}\hat{\mathbf{x}} + \frac{\partial V}{\partial y}\hat{\mathbf{y}} + \frac{\partial V}{\partial z}\hat{\mathbf{z}}\right) \qquad \textbf{(22–35)}$$

The expression in the parentheses is called the *gradient* of V. In short, Equation (22–35) states that *the electric field is the negative gradient of the potential.* For convenience, the gradient of a function is represented by the symbol $\mathbf{\nabla}$ (called *del* or *grad*).

THE GRADIENT OF *V* (Cartesian coordinates)

$$\nabla V = \frac{\partial V}{\partial x}\hat{\mathbf{x}} + \frac{\partial V}{\partial y}\hat{\mathbf{y}} + \frac{\partial V}{\partial z}\hat{\mathbf{z}} \qquad \textbf{(22–36)}$$

In the case of spherical coordinates (see Figure 22–27), we define the unit vectors $\hat{\mathbf{r}}$, $\hat{\boldsymbol{\theta}}$, and $\hat{\boldsymbol{\phi}}$. Then, without presenting the proof, the gradient is

THE GRADIENT OF *V* (spherical coordinates)

$$\nabla V = \frac{\partial V}{\partial r}\hat{\mathbf{r}} + \frac{1}{r}\frac{\partial V}{\partial \theta}\hat{\boldsymbol{\theta}} + \frac{1}{r\sin\theta}\frac{\partial V}{\partial \phi}\hat{\boldsymbol{\phi}} \qquad \textbf{(22–37)}$$

Using the vector symbol $\mathbf{\nabla}$, we may write the following expression:

RELATION BETWEEN E and *V*

$$\mathbf{E} = -\nabla V \qquad \textbf{(22–38)}$$

Note the convenience of the vector notation; it is easy to write and it is true for *all* coordinate systems.

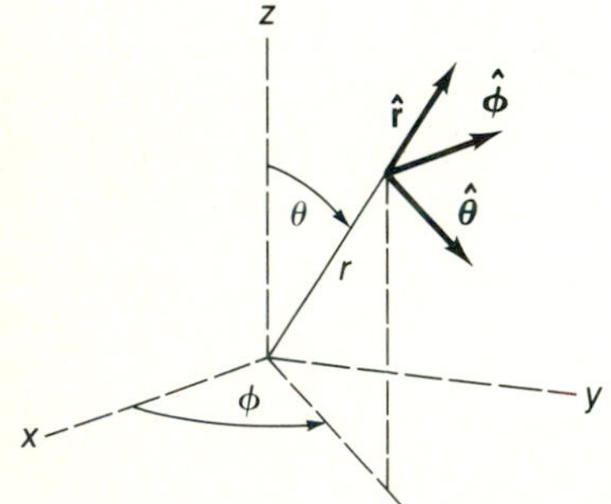

Figure 22–27
Spherical coordinates. The unit vectors $\hat{\mathbf{r}}$, $\hat{\boldsymbol{\theta}}$, and $\hat{\boldsymbol{\phi}}$ represent the direction of increase of the variables r, θ, and ϕ, respectively. They are mutually at right angles.

EXAMPLE 22–14

(a) Derive an expression for the potential of a dipole at distances that are large compared with the separation of the charge. (b) Using Equation (22–38), derive an expression for the field of a dipole.

SOLUTION

(a) The potential of a dipole is the sum of the potential of each of the two charges. We apply Equation (22–27) for the potential for a single charge:

$$V = \left(\frac{1}{4\pi\varepsilon_0}\right)\frac{q}{r}$$

For both charges, the potential at point P is

$$V = \frac{q}{4\pi\varepsilon_0}\left(\frac{1}{r_+} - \frac{1}{r_-}\right) \tag{22–39}$$

where r_+ and r_- are defined in Figure 22–28. The above equation may be written:

$$V = \frac{q}{4\pi\varepsilon_0}\left(\frac{r_- - r_+}{r_- r_+}\right) \tag{22–40}$$

For $r \gg \ell$:

$$r_+ r_- \approx r^2$$

and

$$r_- - r_+ \approx \ell \cos\theta$$

Equation (22–40) then becomes

$$V = \frac{q\ell \cos\theta}{4\pi\varepsilon_0 r^2}$$

or, in terms of the dipole moment $p = q\ell$:

$$\boxed{V = \left(\frac{1}{4\pi\varepsilon_0}\right)\frac{p\cos\theta}{r^2}} \tag{22–41}$$

(b) To obtain the electric field in the r, θ, and ϕ directions, we apply Equations (22–39) and (22–40):

$$\mathbf{E} = -\left(\frac{\partial V}{dr}\hat{\mathbf{r}} + \frac{1}{r}\frac{\partial V}{\partial\theta}\hat{\boldsymbol{\theta}} + \frac{1}{r\sin\theta}\frac{\partial V}{\partial\phi}\hat{\boldsymbol{\phi}}\right)$$

Evaluating the appropriate partial derivatives, we have

$$\frac{\partial V}{\partial r} = -\frac{2p\cos\theta}{4\pi\varepsilon_0 r^3}$$

$$\frac{\partial V}{\partial\theta} = \frac{-p\sin\theta}{4\pi\varepsilon_0 r^2}$$

$$\frac{\partial V}{\partial\phi} = 0$$

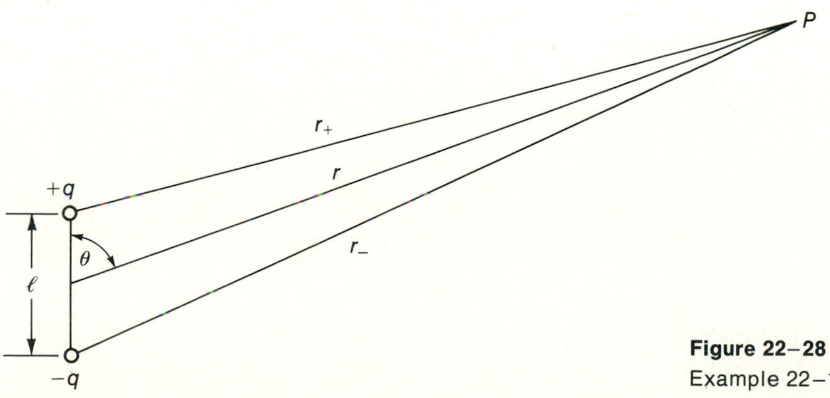

Figure 22–28
Example 22–14

Thus, the vector field **E** is

$$\mathbf{E} = \boxed{\left(\frac{2p\cos\theta}{4\pi\varepsilon_0 r^3}\right)\hat{\mathbf{r}} + \left(\frac{p\sin\theta}{4\pi\varepsilon_0 r^3}\right)\hat{\boldsymbol{\theta}}} \tag{22–42}$$

Note that this result agrees with that of Example 22–5 for $\theta = \pi/2$. The fact that there is no field in the $\hat{\boldsymbol{\phi}}$ direction may also be deduced from symmetry considerations and from the knowledge that electric field lines must terminate on charges. Equation (22–42) reveals an interesting feature of the dipole: At large distances from the dipole, the field along the axis of the dipole has twice the magnitude of the field perpendicular to the axis. Also note that at large distances, the dipole field is proportional to $1/r^3$.

22.10 Equipotential Surfaces

We have seen that diagrams of electric field lines are useful for understanding the nature of electric charges and their interactions. In a similar way, it is helpful to visualize electric potentials. Consider an imaginary surface that is

Figure 22–29
Lines of force (solid) and cross sections of equipotential surfaces (dashed). The lines of force are always perpendicular to the equipotential surfaces, a property called *orthogonality*. Lines representing equipotential surfaces are always spaced to indicate equal intervals of potential ΔV. Figures (a), (b), and (d) also represent, in a qualitative way, patterns for point charges.

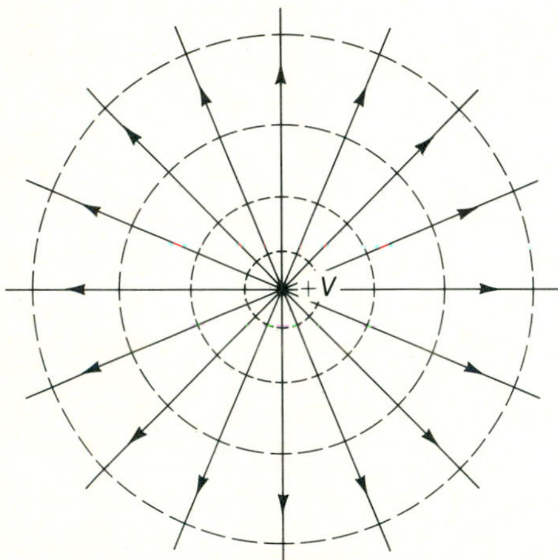

(**a**) A line (perpendicular to the paper) at positive potential. The field lines are imagined to extend to infinity, where they terminate on negative charges.

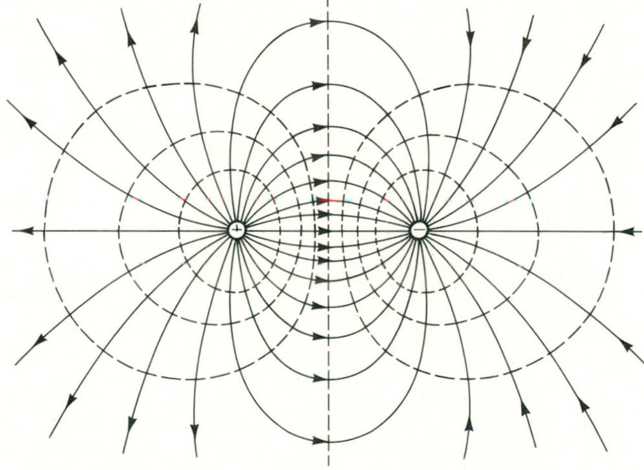

(**b**) Two parallel wires at opposite potentials.

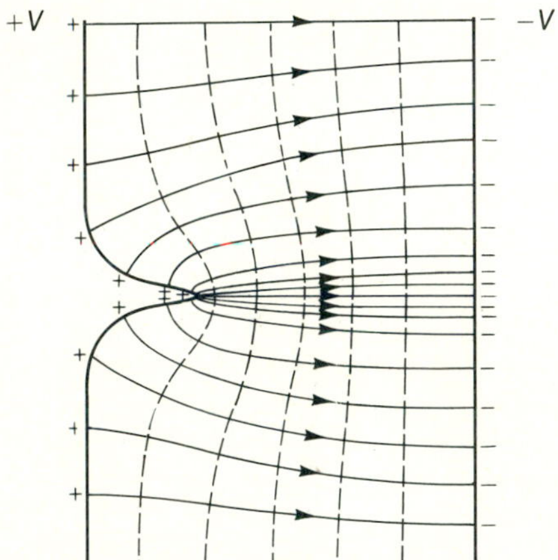

(**c**) Two conducting planes at opposite potential. One plane has a pointed ridge extending perpendicular to the paper.

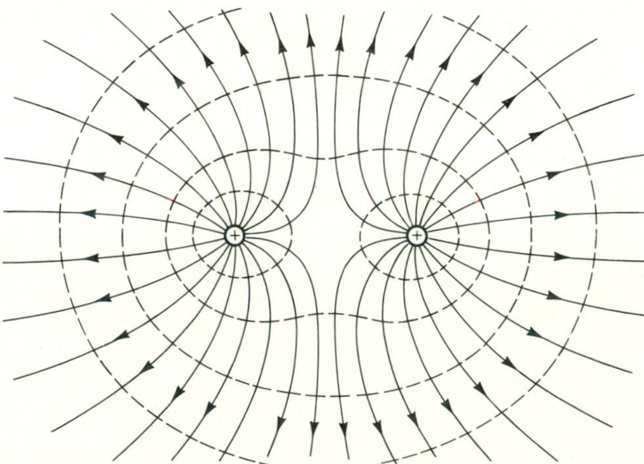

(**d**) Two parallel wires at the same positive potential. As in (**a**), the field lines are imagined to extend to infinity, where they terminate on negative charges.

everywhere perpendicular to the field lines. It would take no work to move a small test charge q' around on such a surface, since the force $\mathbf{F} = q'\mathbf{E}$ is always perpendicular to the motion. So (from Equation 22–28) the entire surface is at the same potential; it is an **equipotential surface.** A family of such surfaces, spaced apart at equal intervals of potential ΔV, gives one an intuitive "feel" for the physical situation. Figure 22–29 shows several examples. For a point charge, the equipotential surfaces are spheres concentric with the charge.

Equipotentials are easier to locate experimentally than field lines. For complicated two-dimensional geometries, the field pattern is most easily found experimentally by first determining a series of equipotentials *spaced at equal intervals of potential difference.* The correct field pattern can then be sketched by drawing field lines perpendicular to the equipotentials, such that the equipotential lines and the field lines form a grid of near-rectangles, with each of the rectangles as square as possible.

A perfect conductor is, of course, an equipotential surface. Therefore, *electric field lines must always intersect conductors at right angles.* (If they did not, there would be a component of **E** parallel to the surface, thus requiring work to move a test charge along the surface.) Furthermore, since field lines must terminate on charges, when a field line intersects a conductor there must be a net charge at that point on the surface of the conductor. These properties make possible some interesting assertions. For example, a hollow conducting sphere may be placed concentric to a point charge without altering the field outside the shell. Moreover, once the shell is in place, the charge inside may move about within the shell without changing the external field (see Figure 22–30).

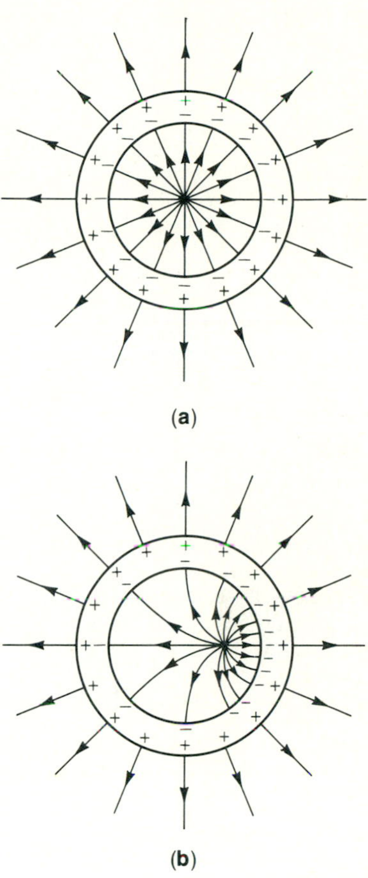

Figure 22–30
A point charge inside a hollow conducting sphere. The external field is the same in both diagrams.

The concept of equipotential surfaces and associated electric fields allows us to conclude that *no electric field exists within any empty, closed conductor, whether the conductor is charged or not.* We have already shown this to be the case for a hollow conducting *sphere* (Example 22–8). Consider now an *irregular* hollow conductor, such as that in Figure 22–31. We construct a Gaussian surface just within the surface and apply Gauss's law:

$$\oint \mathbf{E} \cdot d\mathbf{A} = \frac{q}{\varepsilon_0}$$

Since there is no charge inside the Gaussian surface:

$$\oint \mathbf{E} \cdot d\mathbf{A} = 0$$

But note that we cannot invoke symmetry arguments to assert that the field is zero. (There could be some field lines entering the surface and some leaving, so that the integral is zero.) Let us suppose a field line enters and leaves the Gaussian surface as shown in Figure 22–31. Then an electron at A could leave the conductor, work could be done on it by the field between A and B, and it could subsequently enter the conductor at B. The electron could then be moved through the conductor *without doing work* from B to A (since the conductor is an equipotential surface). The process could be repeated, giving still more energy to the electron. The energy of the system would increase without end; it would indicate a perpetual motion machine, which violates the first law of thermodynamics. *Therefore no field exists within an empty, hollow conductor.* Stated another way, a closed conductor is a perfect electrostatic shield.

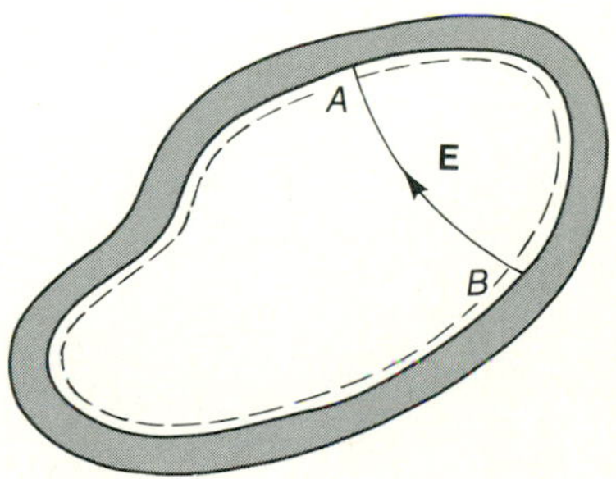

Figure 22–31
No electric field exists within any empty, closed conductor, whether the conductor is charged or not.

Another conclusion we may draw from the use of equipotentials and field lines is that *charges tend to accumulate on the points of conductors.* Consider two conducting spheres, one larger than the other, connected by a conducting wire (Figure 22–32). You may think of this assembly of conductors as approximating an irregular surface that has large as well as small radii of curvatures. The potential of a sphere with a charge q is given by

$$V = \left(\frac{1}{4\pi\varepsilon_0}\right)\frac{q}{r}$$

Since the two spheres are connected by a conductor, their potentials are equal. Therefore:

$$\left(\frac{1}{4\pi\varepsilon_0}\right)\frac{q_1}{r_1} = \left(\frac{1}{4\pi\varepsilon_0}\right)\frac{q_2}{r_2}$$

or

$$\frac{q_1}{r_1} = \frac{q_2}{r_2} \qquad \textbf{(22–43)}$$

The surface charge density σ on a sphere is

$$\sigma = \frac{q}{4\pi r^2} \qquad \textbf{(22–44)}$$

so Equation (22–43) becomes

$$r_1\sigma_1 = r_2\sigma_2 \qquad \textbf{(22–45)}$$

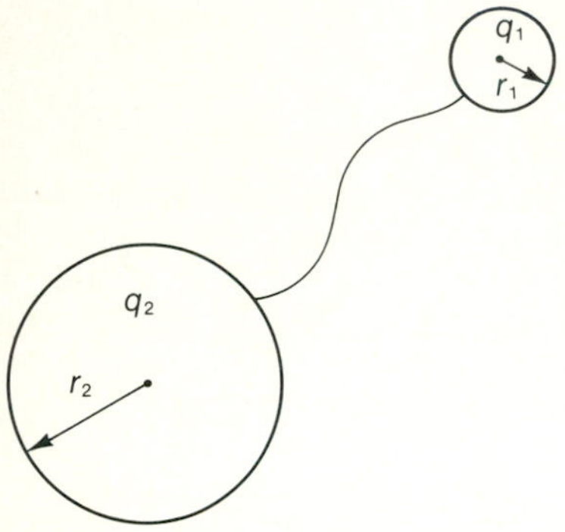

From this we can see that places on a conductor of irregular shape that have a relatively small radius of curvature have a relatively high surface charge density. As will be shown in a problem, the electric field just barely outside a charged conductor is

Figure 22–32
A charge placed on this assembly of conductors will distribute itself so that the small sphere has a larger surface charge density. (Because the spheres are connected by a conducting wire, they are at the same potential.)

ELECTRIC FIELD BARELY OUTSIDE A CONDUCTING SURFACE WITH SURFACE CHARGE σ

$$E = \frac{\sigma}{\varepsilon_0} \qquad \textbf{(22–46)}$$

Thus, if σ is high the electric field is also high. Charges tend to leak off conductors with sharp points because the air tends to ionize and become a conductor of charge if the electric field exceeds about 10^6 V/m.

EXAMPLE 22–15

On a clear, sunny day, there is a vertical electric field pointing up over flat ground (or water) of about 100 V/m. (The field can vary considerably in magnitude and may be reversed if clouds are overhead.) What is the surface charge density on the ground for these conditions?

SOLUTION
From Equation (22–46), we have

$$\sigma = \varepsilon_0 E$$

Recalling the value for ε_0 given in Equation (22–2) yields

$$\sigma = \left(8.85 \times 10^{-12}\,\frac{\text{C}^2}{\text{N}\cdot\text{m}^2}\right)\left(100\,\frac{\text{V}}{\text{m}}\right)$$

$$\sigma = \boxed{8.85 \times 10^{-10}\,\frac{\text{C}}{\text{m}^2}}$$

(**a**) A field ion micrograph of surface atoms in an iridium crystal needle point. Each spot corresponds to a single atom.

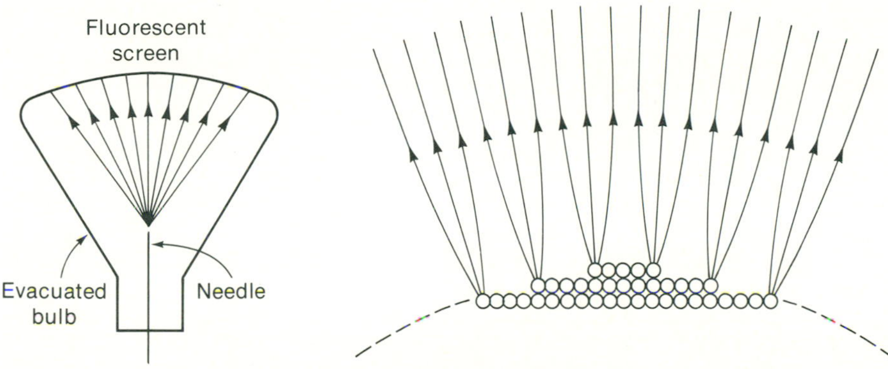

(**b**) A field ion microscope.

(**c**) A simplified sketch of the electric field near atoms on the surface of the needle point.

Figure 22–33

The field ion microscope developed by Erwin Müller at Pennsylvania State University gives a "picture" of individual atoms on the surface of a needle point. Its operation depends on the fact that the electric field is strongest at the sharp corners of a surface. In an evacuated glass bulb, a very sharp needle point with a tip radius of only a few tens of nanometers is held at a large positive voltage with respect to a fluorescent screen on the inside of the bulb wall. At the surface of the crystalline structure of the needle tip, atoms at the edge of a plane of atoms form sharp "corners," causing particularly strong fields just above them. Helium atoms are then introduced into the evacuated bulb. When the helium atoms encounter regions of extremely strong fields just above individual atoms, they lose an electron to the surface, become positive ions, and are accelerated along the field lines to the fluorescent screen, causing a bright spot on the screen. Each spot thus corresponds to the location of a particular atom on the needle surface.

Since the charge on each electron is $e = -1.6 \times 10^{-19}$ C, this implies a deficiency of more than 5 billion electrons per square meter on the surface of the ground.

Summary

The following concepts and equations were introduced in this chapter:

Coulomb's law (which defines the force between two point charges q and q' separated a distance r):

$$\mathbf{F} = \left(\frac{1}{4\pi\varepsilon_0}\right)\frac{qq'}{r^2}\hat{\mathbf{r}}$$

where
$$\left(\frac{1}{4\pi\varepsilon_0}\right) \approx 9 \times 10^9\ \frac{\mathrm{N{\cdot}m^2}}{\mathrm{C^2}}$$

Electron charge:

$$e = -1.602 \times 10^{-19}\ \mathrm{C}$$

Force on a charge in the presence of an *electric field* $\mathbf{E}$:

$$\mathbf{F} = q\mathbf{E}$$

Electric dipole moment $\mathbf{p}$ (for two opposite charges separated a distance ℓ):

$$\mathbf{p} = q\boldsymbol{\ell} \qquad \text{(where } \boldsymbol{\ell} \text{ is directed from } -q \text{ to } +q\text{)}$$

Torque τ *on a dipole* $\mathbf{p}$ in an electric field $\mathbf{E}$:

$$\boldsymbol{\tau} = \mathbf{p} \times \mathbf{E}$$

Electric potential energy U (for a dipole in an electric field):

$$U = -(\mathbf{p} \cdot \mathbf{E})$$

Electric flux Φ_E:

$$\Phi_E = \int_A \mathbf{E} \cdot d\mathbf{A}$$

Gauss's law:

$$\oint_A \mathbf{E} \cdot d\mathbf{A} = \frac{q}{\varepsilon_0} \qquad \text{(where } A \text{ is any surface enclosing the charge } q\text{)}$$

$$\oint_A \mathbf{E} \cdot d\mathbf{A} = \frac{1}{\varepsilon_0}\int_V \rho\, dV \qquad \text{(where the charge density } \rho \text{ is inside the surface)}$$

Electric potential energy U (for two point charges separated a distance r):

$$U = \left(\frac{1}{4\pi\varepsilon_0}\right)\frac{qq'}{r} \qquad \text{(in joules)}$$

Electric potential V (for a point charge q):

$$V = \left(\frac{1}{4\pi\varepsilon_0}\right)\frac{q}{r} \qquad \text{(in volts)}$$

Note the distinction between potential energy U (in joules) and potential V (in volts).

The *electron volt* (eV) is the amount of *energy* acquired by an object with a charge equal in magnitude to the electron charge accelerated through a potential difference of one volt.

$$1 \text{ eV} = 1.602 \times 10^{-19} \text{ J}$$

Relation between V and E (integral form):

$$V_2 - V_1 = -\int_1^2 \mathbf{E} \cdot d\boldsymbol{\ell}$$

Relation between V and E (differential form):

$$\mathbf{E} = -\nabla V$$

where the symbol ∇ (called *del* or *grad*) is the vector operator for the *gradient*, illustrated here:

Cartesian coordinates $\nabla V = \frac{\partial V}{\partial x}\hat{\mathbf{x}} + \frac{\partial V}{\partial y}\hat{\mathbf{y}} + \frac{\partial V}{\partial z}\hat{\mathbf{z}}$

Spherical coordinates $\nabla V = \frac{\partial V}{\partial r}\hat{\mathbf{r}} + \frac{1}{r}\frac{\partial V}{\partial \theta}\hat{\boldsymbol{\theta}} + \frac{1}{r \sin \theta}\frac{\partial V}{\partial \Phi}\hat{\boldsymbol{\phi}}$

Electric field just barely outside a conducting surface with *surface charge density* σ:

$$E = \frac{\sigma}{\varepsilon_0}$$

Questions

1. What is the distinction between electric potential energy difference and electric potential difference?
2. Does the attraction of a small positive charge toward a large metal sphere necessarily mean that the sphere is negatively charged?
3. Why can electric field lines not cross one another or form closed loops?
4. Why can equipotential lines not cross one another?
5. Can the electric potential be zero at a point where the electric field is not zero? If so, give an example.
6. Can the electric field be zero at a point where the electric potential is not zero? If so, give an example.
7. Can the movement of a charge within a hollow conducting sphere alter the electric field outside the sphere?
8. Why are uncharged objects attracted to charged objects?
9. Why do electric dipoles experience a net force as well as a torque in a nonuniform electric field?
10. Why, in general, is a charge within a hollow metal sphere attracted toward the walls of the sphere whether or not the sphere is charged?
11. How can the surface charge density on the outer surface of a hollow sphere be uniform while the surface charge density on the inner surface is not?
12. Answer Question 11 interchanging the words "outer" and "inner."
13. A charge is at the center of a hollow, uncharged metal sphere. If a charge is placed external to the sphere, does the charge within the sphere experience a net force?
14. A small mass is at the center of a hollow, massive sphere. If another mass is placed external to the sphere, does the mass within the sphere experience a net gravitational force due to the presence of the external mass?
15. Why is Gauss's law impractical in finding the electric field outside a charged metal cube?

16. A "Faraday cage" consists of a hollow box with sides constructed of metallic wire screen. A sensitive voltmeter is connected between the screen and a probe inside the box. How does this device detect a net charge within the box?

17. Why is it impossible for a potential function of a charge distribution to have a finite discontinuity?

18. An inflated toy balloon becomes slightly larger as it acquires an electrical charge. Why?

19. A charge is deposited on a hollow metal sphere floating in oil. As a consequence of becoming charged, does the sphere float higher, lower, or remain at the same level in the oil? Why?

Problems

22A–1 Consider three charges located in the x-y plane. A charge of $+3\ \mu$C is located at $x = 4$ cm, $y = 0$; a charge of $-2\ \mu$C is located at $x = 0$, $y = 5$ cm. Find the force on a $+6\ \mu$C charge at the origin. *Answer:* 110 N at 157° from the $+x$ axis

22A–2 In Problem 22A–1, find the electric field at the origin if the $+6\ \mu$C charge were absent. Verify your answer by using the answer to Problem 22A–1.

22A–3 Calculate the ratio of the electrostatic force to the gravitational force between the electron and proton in a hydrogen atom. *Answer:* 2.27×10^{39}

22A–4 Suppose two objects, each with a net positive charge of one coulomb, were separated by a distance equal to the distance between New York and San Francisco (about 4140 km). Calculate the mutual force of repulsion between these objects.

22A–5 Two small silver spheres, each with a mass of 100 g, are separated by a distance of 1 m. Calculate the fraction of the electrons in one sphere that must be transferred to the other in order to produce an attractive force of 10^4 N (about a ton) between the spheres. (The number of electrons per atom of silver is 47 and the number of atoms per gram is Avogadro's number divided by the atomic weight of silver, 107.9.)
Answer: 2.15×10^{-10} (or about 1 in 4 billion)

22A–6 Express the units for an electric field in terms of the SI base units of mass (kg), length (m), time (s), and electric current (A).

22A–7 Calculate the mass of electrons which, when shared between the earth and moon, would produce a force of repulsion equal to the gravitational force between the earth and moon. *Answer:* 649 kg

22A–8 Under normal atmospheric conditions, an upward electric field of about 100 N/C exists just above the surface of the earth. If a toy helium-filled balloon is barely capable of lifting a mass of 50 g, find the amount of electric charge that must be distributed over the balloon's surface so that the balloon will not rise when the mass is removed. (The amount of charge required would produce repulsive forces along the surface of the balloon that would be more than sufficient to tear the balloon.)

22A–9 Consider an isolated conducting sphere of very large radius that possesses a surface charge density of σ (in coulombs/meter^2).

(a) Derive an expression for the electric field close to the surface of the sphere.

(b) Derive an expression for the electric field close to a large, planar conducting sheet that has the same area and total charge as the sphere.

Answers: (a) σ/ε_0 (b) $\sigma/2\varepsilon_0$

22A–10 Consider a hollow metallic sphere with a charge of $+10\ \mu$C and a radius of 10 cm. The center of the sphere is at the origin of a Cartesian coordinate system. Within the sphere, at $x = 5$ cm, is a negative charge of $-3\ \mu$C. Find the electric field external to the sphere along the x axis. Make a qualitative sketch of the field lines inside and outside the sphere.

22A–11 Explain qualitatively why an electric dipole placed in a nonuniform electric field will move in the direction of the field toward a region of stronger field. Include a diagram.

22A–12 Many molecules possess an electric dipole moment because the center of distribution of the positive charge (protons) does not exactly coincide with that of the negative charge (electrons). The dipole moment of a water molecule in its gaseous state is 6.1×10^{-30} C·m. If a water molecule is placed in an electric field of 10^4 N/C, calculate the maximum torque the field can exert on the molecule and the range of potential energies the molecule may have.

22A–13 Suppose that at an altitude of 300 m above the ground the electric field is 100 N/C upward and that at 100 m it is 150 N/C upward. Calculate the average charge density in the region between these altitudes. Express your answer in terms of electron surplus or deficiency. *Answer:* 1.38×10^7 electrons/m^3, surplus

22A–14 Four equal positive charges q form the corners of a square with a side length a. Find the potential difference between a point at the center of the square and a point midway along one side of the square. Which point is at the higher potential?

22A–15 Show that, for two positively charged, concentric conducting shells, the inner shell is always at a higher potential than the outer shell, regardless of the amount of charge on either shell.

22A–16 The potential V (in volts) in a region is defined by $V = (3\ \text{V/m}^2)x^2 + (0.2\ \text{V/m})y$, where x and y are expressed in meters. Find the magnitude and direction of the force on an electron placed at $x = 10$ cm, $y = 15$ cm.

22B–1 Richard Feynman asserts that if two persons stood at arm's length from each other and each person had 1% more electrons than protons, the force of repulsion between the two people would be enough to lift a "weight" equal to that of the entire earth. Carry out an order-of-magnitude calculation to substantiate this assertion.

22B–2 If there were a slight imbalance between the number of protons and the number of electrons in matter, the gravitational attraction between astronomical objects could be overcome by the electrostatic repulsion between these objects. Calculate the minimum fraction by which one charge would have to exceed the other for this to occur. The approximate average number of proton-electron pairs per kilogram of matter is 3×10^{26}.

22B–3 Two point charges are located as follows: a $-3\ \mu$C charge at the origin and a $+2\ \mu$C charge at $x = 0.15$ m. Find the location where a positive point charge q' may be placed so that the net force on the charge q' is zero. *Answer:* at $x = 0.817$ m

22B–4 Consider three point charges in the x-y plane such that a charge $-2q$ is at the origin, a charge $+q$ is at $y = +\ell/2$, and a charge $+q$ is at $y = -\ell/2$. Such an arrangement of charges is called an *electric quadrupole.* Derive expressions for the electric field along (a) the x axis as a function of x and (b) the y axis as a function of y. (c) Determine the direction of E in each case. (Optional: Show that $E \propto 1/r^4$ in each case for x or y much greater than ℓ.)

22B–5 A uniform electric field is described in Cartesian coordinates by $\mathbf{E} = E_0\hat{\mathbf{y}}$, where E_0 is a constant. A charge with a mass m and charge $+q$ is injected at the origin into the electric field with an initial velocity $\mathbf{v} = v_0\hat{\mathbf{x}}$. Find the equation of the subsequent trajectory of the charge. *Answer:* $y = (qE_0/2mv_0{}^2)x^2$

22B–6 An electric dipole with opposite charges of 3×10^{-12} C separated a distance of 2 mm is aligned with an electric field of 10^3 N/C so that the positive charge is in the region of higher potential. Find the amount of work done by the electric field in rotating the dipole so as to reverse the positions of the charges.

22B–7 An electric dipole is made of two point charges, $+q$ and $-q$, each of mass m, separated a distance ℓ. The dipole is placed in a uniform electric field E oriented near its lowest potential energy state. Show that the dipole will undergo oscillatory rotations about its center of mass, and derive an expression for the approximate period of small-amplitude oscillations. *Answer:* $2\pi\sqrt{m\ell/2qE}$

22B–8 Two isolated conducting spheres, one with a radius R and the other with a radius $3R$, each carry an equal charge Q_0. The spheres are brought into contact and then separated again. Find the charge on each sphere.

22B–9 Consider two hollow, metallic, concentric spheres. The inner sphere has a radius of 30 cm and a charge of $-80\ \mu$C. The outer sphere has a radius of 50 cm and a charge of

40 μC. For the regions outside the spheres, between the spheres, and inside the inner sphere, find: (a) the electric field and (b) the potential. (c) Sketch qualitative graphs for E and V.

Answers: inside the spheres: $\mathbf{E} = 0$
$V = -(140/3\pi\varepsilon_0)\ \mu\text{V}$
between the spheres: $\mathbf{E} = (80/4\pi\varepsilon_0 r^2)\ \mu\text{N/C}$, inward
$V = -(1/4\pi\varepsilon_0)(80/r - 40/0.5)\ \mu\text{V}$
outside the spheres: $\mathbf{E} = (40/4\pi\varepsilon_0 r^2)\ \mu\text{N/C}$, inward
$V = -(40/4\pi\varepsilon_0 r)\ \mu\text{V}$

22B–10 Consider again the electric quadrupole in Problem 22B–4, which was an assembly of three charges: $-2q$ at the origin, $+q$ at $y = \ell/2$, and $+q$ at $y = -\ell/2$. Find the potential at points along the x axis and the y axis. Show that at large distances from the quadrupole (that is, x and y much larger than ℓ), the potential varies as the inverse cube of the distance.

22C–1 Show that two small objects a given distance apart and sharing a given total charge will have a maximum force of repulsion when the charge is shared equally between the objects.

22C–2 Two point charges, each of charge $+Q$, are held fixed a distance d apart. A third positive charge q is confined to move along the straight line joining the original two charges.

(a) Show that if the charge q is displaced a small distance x (where $x \ll d$) from its position of equilibrium, it will execute approximately simple harmonic motion.

(b) Find the "spring constant" k associated with this motion.

22C–3 Calculate the amount of work required to accumulate a charge q on a sphere of radius R. This is done by bringing infinitesimal charges dq from infinity up to the surface of the sphere until the total charge q is reached. *Answer:* $W = q^2/8\pi\varepsilon_0 R$

22C–4 An electric field is described by $E = 2000\hat{\mathbf{x}} + 3000\hat{\mathbf{y}}$ (in SI units). Find the potential difference $(V_B - V_A)$ between the points A at $x = 0$, $y = 3$ m, $z = 2$ m and B at $x = 2$ m, $y = 1$ m, $z = 0$. (Hint: Since E is a conservative field, $V_B - V_A$ may be calculated along any path between A and B.)

22C–5 A long, thin ribbon with a width b has a uniform surface charge density σ on both top and bottom surfaces. Find the electric field at a distance a from the ribbon on the plane that perpendicularly bisects the length of the ribbon. (Hint: Consider the ribbon as an assembly of charged wires.) *Answers:* $E = (2\sigma/\pi\varepsilon_0)\tan^{-1}(b/2a)$

22C–6 A charge of $+10\ \mu$C is uniformly distributed along a rod 80 cm long. Calculate the electric field at a point on the perpendicular bisector of the rod that is a distance of 4 cm from the rod. Compare this value with that obtained using Gauss's law, assuming the rod is infinitely long but has the same charge density.

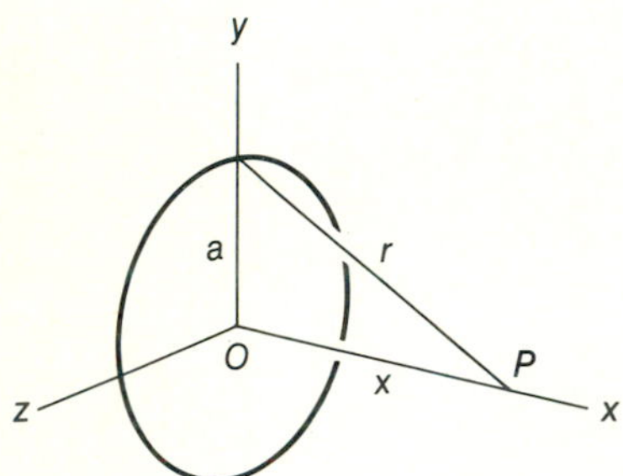

Figure 22–34
Problems 22C–7 and 22C–8

22C–7 In Figure 22–34, the circular hoop of radius a has a total charge Q distributed uniformly around the hoop. Derive an expression for the electric field E on the axis of the hoop at a distance x from the center of the hoop. Verify that for $x \gg a$, E varies as $1/x^2$.
Answer: $E = Qx/4\pi\varepsilon_0(a^2 + x^2)^{3/2}$

22C–8 The circular ring in Figure 22–34 has a charge per unit length λ (in coulombs/meter). Starting with an expression for the potential dV due to an element of charge dq, find the electric potential V at a point P along the axis a distance x from the ring's center. (Note: The parameter r should not appear in the answer.)

22C–9 A circular hoop of a nonconducting material with a uniform distribution of charge has zero electric field at the center of the hoop. Why? Consider such a hoop of radius R with a total charge $+Q$. A length ℓ along the circumference is cut from the hoop. Find an expression for the field at the center of curvature of the remaining segment.
Answer: $\mathbf{E} = (2/4\pi\varepsilon_0 R^2)\sin(\ell/2R)$, away from the remaining segment

22C–10 Consider a uniform spatial distribution of electrons within a sphere of radius R. If the volume charge density within the sphere is $-\rho$, derive an expression for the electric field and electric potential for (a) $r < R$ and (b) $r > R$.

22C–11 The interior of a sphere of radius R has a volume charge density ρ that is proportional to the distance r from the center:

$$\rho = Ar \qquad (\text{for } 0 < r < R)$$

where A is a constant.

(a) Find the SI units for A.
(b) Find the total charge Q inside the sphere in terms of A and R. (Hint: Following Example 22–12, sum the charges dq in shells of thickness dr.)
(c) Use Gauss's law to find the electric field E inside the sphere a distance r from the center.
(d) Setting $V = 0$ at $r = \infty$, find the potential V as a function of r both outside and inside the sphere.

Answers: (a) C/m^4
(b) $Q = A\pi R^4$
(c) $E = Ar^2/4\varepsilon_0$
(d) For $r \geq R$: $V = AR^4/4\varepsilon_0 r$
For $r \leq R$: $V = (A/12\varepsilon_0)(2R^3 + r^3)$

22C–12 Repeat the previous problem for a charge distribution $\rho = Ar^2$.

22C–13 As shown in Figure 22–35, a positive charge distribution exists within the volume of an infinitely long cylindrical shell between radii a and b. The charge density ρ is not uniform, but varies inversely as the radius r from the axis. That is, $\rho = \kappa/r$ for $a < r < b$, where κ is a constant in SI units. (a) Find the units of κ. (b) Find the total charge Q in a length L of the cylindrical shell. (c) Starting with Gauss's law, find the electric field E at a point r (for $a < r < b$).

Answers: (a) C/m^2
(b) $Q = \kappa 2\pi L(b - a)$
(c) $E = (\kappa/\varepsilon_0)(1 - a/r)$

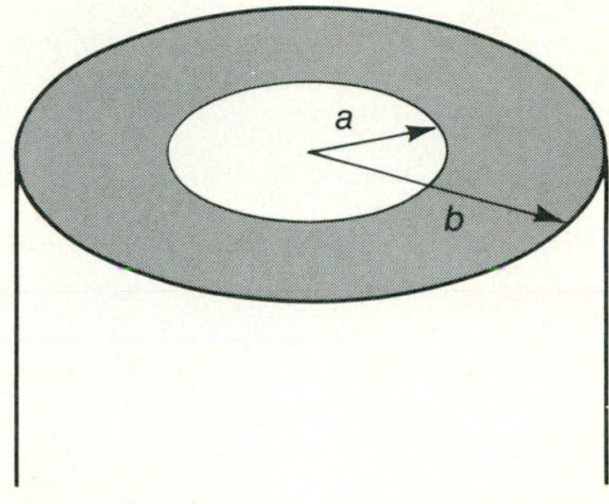

Figure 22–35
Problem 22–13

chapter 23 Capacitance and the Energy in Electric Fields

Penetrating so many secrets, we cease to believe in the unknowable. But there it sits nevertheless, calmly licking its chops.

H.L. MENCKEN
[*Minority Report* (1956)]

In this chapter, it will become clear why we have placed so much importance on the concept of an electric field. Compact configurations of conductors can be constructed so as to contain very intense electric fields. Such devices are called *capacitors*, a name derived from their capacity for storing positive and negative charges. We will show that the external work performed in establishing the separation of charge within the capacitor appears as energy stored in the electric field that is thereby created inside the capacitor. This chapter will lead us to the important conclusion that electric fields, wherever they exist, contain energy. Capacitors are widely used in electronic circuits.

23.1 Capacitance

The ability of a capacitor to maintain a separation of charge with a given potential difference is called *capacitance*. We begin by discussing a capacitor that has a simple geometry. Consider two concentric, conducting spherical shells (with very thin walls) such as those shown in Figure 23–1. Suppose we transfer some electrons from the inner to the outer shell, thereby leaving the inner shell with a charge $+q$ and the outer shell with a charge $-q$. We then can calculate the potential difference between the shells, using the method described in Example 22–10.

Applying Gauss's law, we find the value of the field between the shells:

$$E = \left(\frac{1}{4\pi\varepsilon_0}\right)\frac{q}{r^2} \qquad \text{(outward)} \qquad \textbf{(23–1)}$$

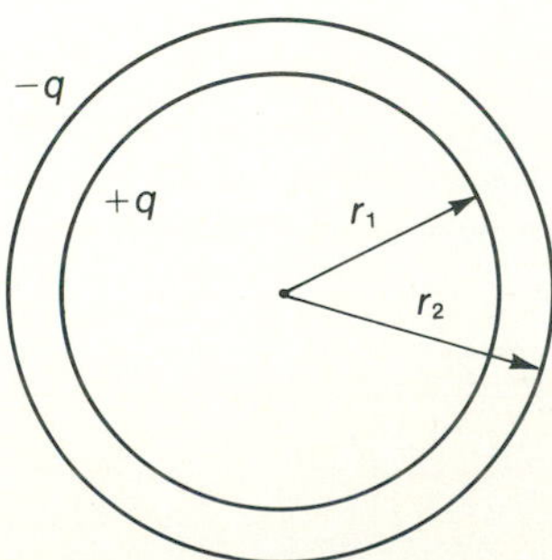

Figure 23–1
Two thin, concentric, conducting spheres have a potential difference proportional to the charge on the spheres.

Using Equation (22–28):

$$V_2 - V_1 = -\int_1^2 \mathbf{E} \cdot d\boldsymbol{\ell}$$

we obtain the potential difference between the spheres by integrating outward from r_1 to r_2. Changing variables from $d\ell$ to dr, we have

$$V_2 - V_1 = -\int_{r_1}^{r_2} \mathbf{E} \cdot d\mathbf{r}$$

Substituting the value for E from Equation (23–1) and noting that both $\mathbf{E}$ and $d\mathbf{r}$ are in the same (radially outward) direction, we have

$$V_2 - V_1 = -\int_{r_1}^{r_2} \left(\frac{q}{4\pi\varepsilon_0 r^2}\right) \cos 0° \; dr$$

$$V_2 - V_1 = -\left(\frac{q}{4\pi\varepsilon_0}\right)\left(-\frac{1}{r}\right)\Bigg|_{r_1}^{r_2}$$

$$V_2 - V_1 = \frac{q}{4\pi\varepsilon_0}\left(\frac{1}{r_2} - \frac{1}{r_1}\right)$$

Because $r_2 > r_1$, in this example V_2 is less than V_1. To make this potential difference a positive number, we change signs on both sides:

$$V_1 - V_2 = \frac{q}{4\pi\varepsilon_0}\left(\frac{1}{r_1} - \frac{1}{r_2}\right)$$

Solving this equation for q yields

$$q = \left(\frac{4\pi\varepsilon_0 r_1 r_2}{r_2 - r_1}\right)(V_1 - V_2) \tag{23–2}$$

Note that the magnitude of the charge on one of the shells is proportional to the potential difference between the shells. The constant of proportionality is given the symbol C:

$$\left(\frac{4\pi\varepsilon_0 r_1 r_2}{r_2 - r_1}\right) = C \tag{23–3}$$

It is dependent solely upon the geometry of the shells and indicates the capacity of the shells to hold a charge q with a potential difference $V_1 - V_2$. The constant of proportionality is called the **capacitance** C, measured in units of *farads* (F), which are equal to coulombs/volt (C/V). (In the context of its use, there is no confusion between the symbol C for capacitance and the symbol C for coulombs.)

Combining the previous two equations, we may write

CAPACITANCE *C* (in farads)

$$q = CV \tag{23–4}$$

where V is the potential difference between the shells. Any configuration of conductors that possesses capacitance is called a *capacitor.*

We commonly speak of "the charge q on a capacitor" meaning, of course, just the magnitude of the charge on *one* of the conductors. (The total net charge on both conductors is zero.) Similar usage also applies to the phrase "a charged capacitor," meaning that each conductor has a charge of magnitude q but of opposite sign. The following examples derive expressions for the capacitance for three additional configurations.

EXAMPLE 23–1

Find the capacitance of an isolated conducting sphere.

SOLUTION

Ordinarily capacitance is associated with two separated conductors. However, this need not be the case when the zero reference potential is associated with one conductor "at infinity." In this case we start with Equation (23–3):

$$C = \left(\frac{4\pi\varepsilon_0 r_1 r_2}{r_2 - r_1}\right)$$

As $r_2 \rightarrow \infty$, r_1 becomes insignificant in the denominator, and then the remaining r_2 essentially cancels the r_2 in the numerator, leaving (in the limit):

$$\lim_{r_2 \rightarrow \infty} C = 4\pi\varepsilon_0 r_1 \qquad \textbf{(23–5)}$$

or, for a sphere of radius R:

CAPACITANCE OF AN ISOLATED SPHERE

$$C = 4\pi\varepsilon_0 R \qquad \textbf{(23–6)}$$

The unit of capacitance, the *farad*, is an extremely large unit. For example, the capacitance of a conducting sphere the size of the earth ($R \approx 6.37 \times 10^6$ m) is only about 7×10^{-4} F. For this reason, the microfarad (μF $= 10^{-6}$ F), nanofarad (nF $= 10^{-9}$ F), and picofarad (pF $= 10^{-12}$ F) are commonly encountered.

EXAMPLE 23–2

Find the capacitance of two parallel, planar conductors, each with an area A. We will assume that the distance between them, d, is much smaller than the length of the edges of the plates, so we may ignore side effects.

SOLUTION

Method 1: We begin with two concentric spheres whose capacitance is given by Equation (23–3):

$$C = \left(\frac{4\pi\varepsilon_0 r_1 r_2}{r_2 - r_1}\right)$$

Suppose we let r_2 and r_1 be nearly the same value and let both become very large—so large that a small section of the concentric sphere will approximate parallel plates (see Figure 23–2a). We note that $r_1 \approx r_2 = R$ and $r_2 - r_1 = d$, so the equation becomes

$$C = \frac{4\pi\varepsilon_0 R^2}{d}$$

The area of the entire shell is $4\pi R^2$. For a small section of the shells of area A, we have

$$C = \frac{\varepsilon_0 A}{d} \quad \text{(for parallel plates)} \qquad \textbf{(23–7)}$$

The concentric-sphere capacitor has the property that the electric field is confined to the space between the shells. In extracting the small section of the large shells to approximate a parallel-plate capacitor, the field between the plates was not disturbed. Of course, for an actual parallel-plate capacitor, there will be some fringing field at the edges. But if

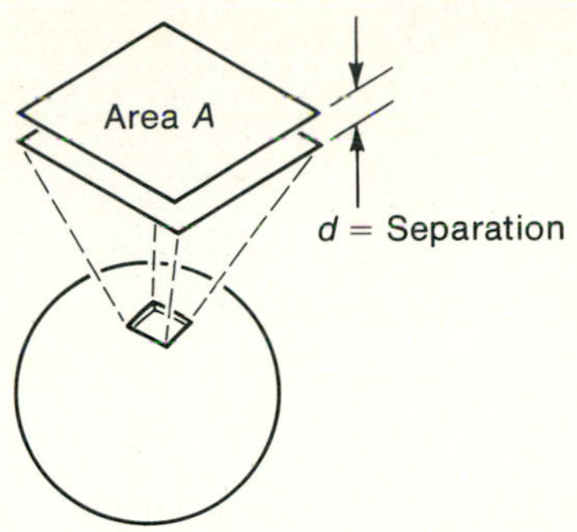

(a) A parallel-plate capacitor derived from a concentric-sphere capacitor.

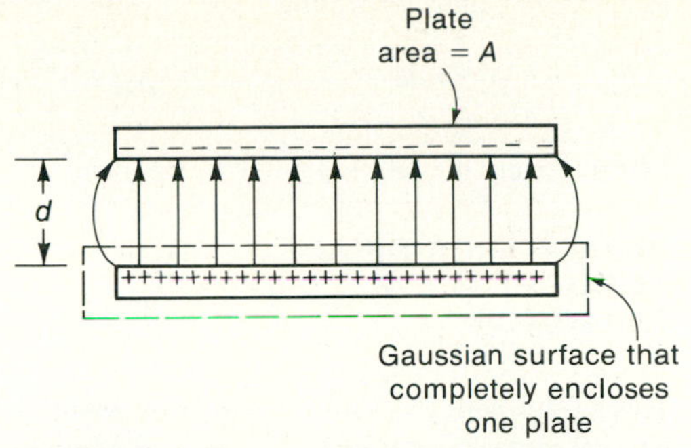

(b) Each plate of the parallel-plate capacitor has an area A with plate separation d.

Figure 23–2
Example 23–2

the plate separation is small compared with the dimensions of the capacitor, the fringing field is negligible and can usually be ignored.

Method 2: This method involves calculating the field between charged, parallel conducting plates and then determining the potential difference between the plates. Ignoring the fringing field at the edges, we assert that the charge on each plate resides only on the inner surfaces of the capacitor. This assertion is based on the previous argument that parallel plates are considered a section of very large, concentric shells. And since no field exists inside the inner shell and outside the outer shell, the charges must reside entirely on the adjacent surfaces of the shells.

To calculate the field, we will use Gauss's law:

$$\oint \mathbf{E} \cdot d\mathbf{A} = \frac{q}{\varepsilon_0}$$

and construct a Gaussian surface that completely encloses one of the plates, as shown in Figure 23–2(b). We evaluate the integral as follows:

$$\oint \mathbf{E} \cdot d\mathbf{A} = \begin{cases} \displaystyle\int_{\substack{\text{Top}\\ \text{surface}}} \mathbf{E} \cdot d\mathbf{A} = EA & \text{E is constant and parallel to } d\mathbf{A} \text{ on this surface} \\ + \\ \displaystyle\int_{\substack{\text{Bottom}\\ \text{surface}}} \mathbf{E} \cdot d\mathbf{A} = 0 & \\ + & E = 0 \text{ on these surfaces} \\ \displaystyle\int_{\substack{\text{Surface}\\ \text{around the edges}}} \mathbf{E} \cdot d\mathbf{A} = 0 & \end{cases}$$

Thus:

$$\oint \mathbf{E} \cdot d\mathbf{A} = EA$$

Since the integral equals q/ε_0, we have

$$E = \frac{q}{\varepsilon_0 A} \tag{23–8}$$

Using Equation (22–28), we now calculate the potential difference between the plates.

$$V_2 - V_1 = -\int_1^2 \mathbf{E} \cdot d\boldsymbol{\ell}$$

We evaluate the line integral by going from the negatively charged to the positively charged plate and noting that $d\boldsymbol{\ell}$ is opposite to $\mathbf{E}$.

$$V_2 - V_1 = -\int_0^d \frac{q}{\varepsilon_0 A} \cos 180^\circ \, d\ell$$

$$V_2 - V_1 = \frac{d}{\varepsilon_0 A} q$$

Rearranging and substituting V for $V_2 - V_1$, the potential difference:

$$q = \left(\frac{\varepsilon_0 A}{d}\right) V$$

Incorporating the relation $q = CV$, we have

CAPACITANCE OF A PARALLEL-PLATE CAPACITOR

$$C = \frac{\varepsilon_0 A}{d} \quad \text{(for parallel plates)} \qquad \textbf{(23–9)}$$

This agrees with the value obtained by Method 1. Note that this expression is only for parallel-plate capacitors; other formulas apply for other geometries.

EXAMPLE 23–3

Find the capacitance of two metal plates, each 2 m^2 in area, separated by 1 mm. Ignore fringing effects at the edges.

SOLUTION

For parallel plates:

$$C = \frac{\varepsilon_0 A}{d}$$

$$C = \left(8.85 \times 10^{-12}\,\frac{\text{F}}{\text{m}}\right)\left(\frac{2\ \text{m}^2}{10^{-3}\ \text{m}}\right) = \boxed{17.7\ \text{nF}}$$

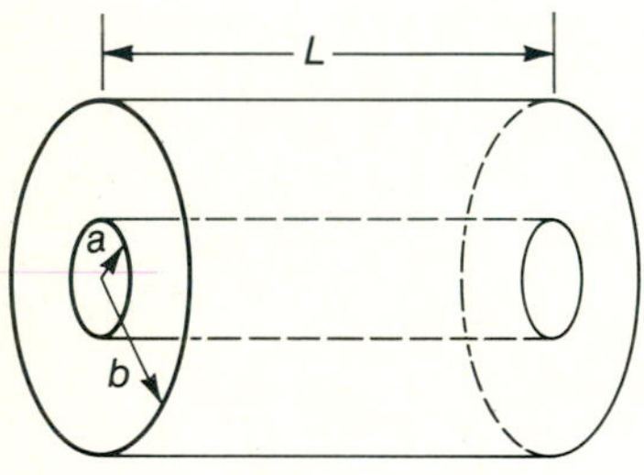

(**a**) A short segment of length L.

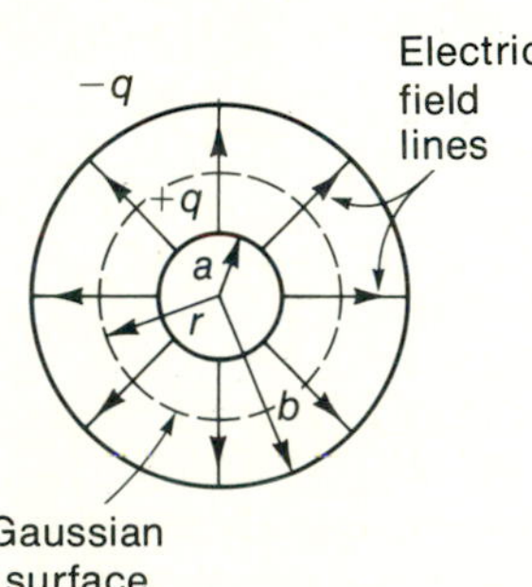

(**b**) End view.

Figure 23–3
In a cylindrical capacitor, the electric field **E** is radial.

The *cylindrical capacitor* consists of two concentric conducting cylinders (Figure 23–3). The radius of the inner cylinder is designated by a and that of the outer cylinder by b. We assume the total length L of the cylinders is much greater than a or b, so that the fringing fields at the ends may be ignored. In a cylinder length L, the total charge on the inner cylinder is $+q$, while the corresponding charge on the outer cylinder is $-q$, producing an electric field that is radially outward everywhere. This symmetry suggests a Gaussian surface in the shape of a coaxial cylinder of length L and radius r $(a < r < b)$ with closed ends. We apply Gauss's law:

$$\oint_A \mathbf{E} \cdot d\mathbf{A} = \frac{q}{\varepsilon_0}$$

Since the field **E** is radial everywhere, the integration over the ends of the cylinder is zero (**E** and $d\mathbf{A}$ are at right angles there, so we have $\cos 90^\circ = 0$). For the integration over the cylindrical surface, **E** is parallel to $d\mathbf{A}$. Thus:

$$\int_{\substack{\text{Cylindrical}\\ \text{surface}}} \mathbf{E} \cdot d\mathbf{A} = \frac{q}{\varepsilon_0}$$

$$E \cos 0^\circ\,(2\pi r L) = \frac{q}{\varepsilon_0}$$

$$E = \frac{q}{2\pi\varepsilon_0 r L}$$

This field is equivalent to that of a line charge along the axis.

The potential difference $(V_b - V_a)$ is found from

$$V_b - V_a = -\int_a^b \mathbf{E} \cdot d\ell$$

Since $d\ell$ is in the same direction as $d\mathbf{r}$ (which is in the same direction as $\mathbf{E}$), we have

$$V_b - V_a = -\frac{q}{2\pi\varepsilon_0 L} \int_a^b \frac{dr}{r}$$

Integrating (Appendix G-II) gives

$$V_b - V_a = -\frac{q}{2\pi\varepsilon_0 L} \ln r \Big|_a^b$$

$$V_b - V_a = -\frac{q}{2\pi\varepsilon_0 L} \ln\left(\frac{b}{a}\right)$$

If we designate the potential difference by V, its *magnitude* is

$$V = \frac{q}{2\pi\varepsilon_0 L} \ln\left(\frac{b}{a}\right)$$

The capacitance C is thus

$$C = \frac{q}{V}$$

CAPACITANCE OF A CYLINDRICAL CAPACITOR

$$C = \frac{2\pi\varepsilon_0 L}{\ln\left(\frac{b}{a}\right)} \qquad \textbf{(23–10)}$$

As with all expressions for capacitance, C depends only on *geometrical* factors (here, a, b, and L).

23.2 Combinations of Capacitors

In the construction of electronic circuits, it is often necessary to combine two or more capacitors. Combinations of capacitors consist of *parallel* and/or *series* connections, as shown in Figure 23–4(a) and (b), respectively. The electronic symbol ⊣⊢ for a capacitor is used in the figure. (The symbol actually implies a parallel plate capacitor, but it is used for any type of capacitor.)

In the *parallel* combination, the potential difference V is the same for all capacitors, but the charge on each may be different. The total charge on all capacitors is

$$q = q_1 + q_2 + q_3$$

Substituting:

$$q = C_1 V + C_2 V + C_3 V$$

$$q = (C_1 + C_2 + C_3)V$$

Therefore, the equivalent capacitance C of the combination is

$$C = C_1 + C_2 + C_3$$

Since the analysis could be extended to include any number of capacitors in parallel, we may write the general formula:

CAPACITORS IN PARALLEL

$$C = C_1 + C_2 + C_3 + \cdots \qquad \textbf{(23–11)}$$

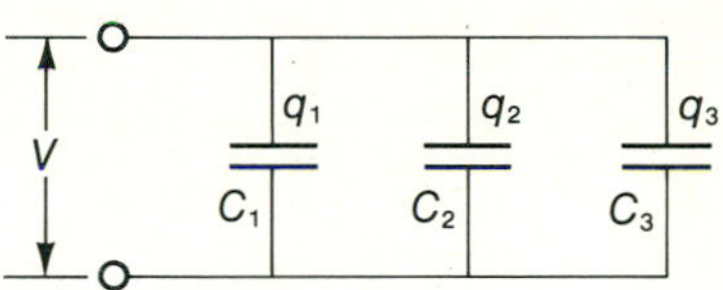

(a) A *parallel* combination of capacitors, whose equivalent capacitance is: $C = C_1 + C_2 + C_3$

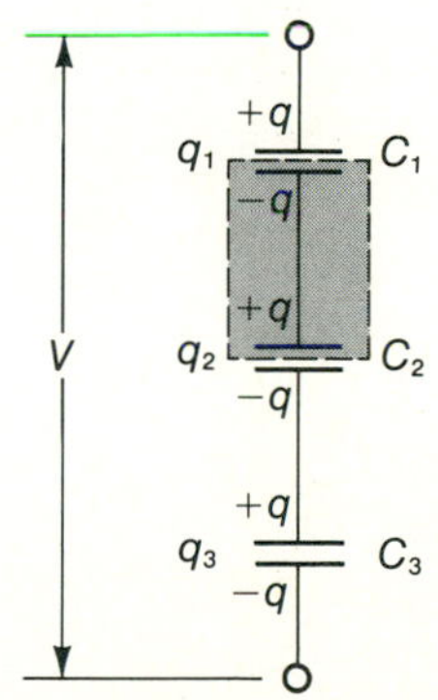

(b) A *series* combination of capacitors, whose equivalent capacitance is: $\frac{1}{C} = \frac{1}{C_1} + \frac{1}{C_2} + \frac{1}{C_3}$

Figure 23–4
Combinations of capacitors.

In the *series* combination, suppose the capacitors are originally uncharged, and we connect a battery of voltage V across the ends of the combination. A charge $+q$ will be established on the outer conductor of C_1 and a charge $-q$ will be established on the outer conductor of C_3. No charge will be transferred to the inner conductors, but the charges on these inner conductors will be polarized as shown in the shaded box in the figure. Electrons will be attracted toward the inner plate of C_1, leaving the other plate of C_2 positive. Similarly, charges will move in the other connected parts until there is a $+q$ or $-q$ charge on each plate. Thus, each capacitor acquires the same magnitude of charge q. The potential V across the combination is the sum of the potentials across each capacitor.

$$V = V_1 + V_2 + V_3$$

or

$$V = \frac{q}{C_1} + \frac{q}{C_2} + \frac{q}{C_3}$$

Solving for q:

$$q = \frac{V}{\dfrac{1}{C_1} + \dfrac{1}{C_2} + \dfrac{1}{C_3}}$$

Therefore, from $q = CV$, the equivalent capacitance C of the combination is

$$\frac{1}{C} = \frac{1}{C_1} + \frac{1}{C_2} + \frac{1}{C_3}$$

Since the analysis could be extended to include any number of capacitors in series, we may write the general formula:

CAPACITORS IN SERIES

$$\frac{1}{C} = \frac{1}{C_1} + \frac{1}{C_2} + \frac{1}{C_3} + \cdots \qquad \textbf{(23–12)}$$

EXAMPLE 23–4

Find the equivalent capacitance of the combination of capacitors shown in Figure 23–5. Assume that all of the capacitors have the same capacitance C.

SOLUTION

The equivalent capacitance C' of the *series* combination of the two capacitors in the upper or lower branch is obtained by applying Equation (23–12):

$$\frac{1}{C'} = \frac{1}{C} + \frac{1}{C}$$

$$C' = \frac{C}{2}$$

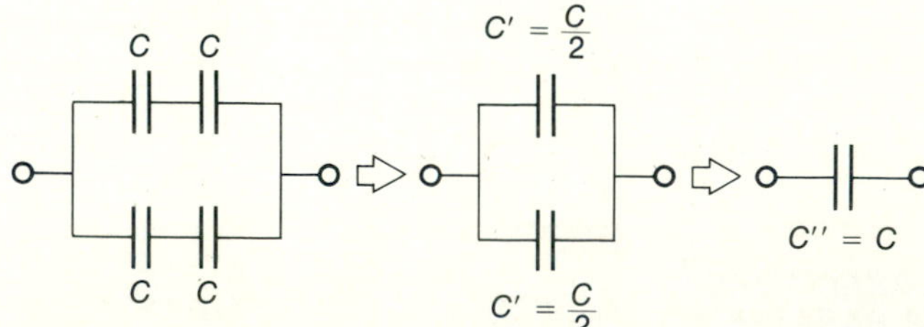

Figure 23–5
Example 23–4

The equivalent capacitance C'' of the *parallel* combination of the two branches, each C', is obtained by applying Equation (23–11):

$$C'' = C' + C'$$
$$C'' = 2C'$$

Substituting $C/2$ for C':

$$\boxed{C'' = C}$$

This result appears to make such a combination of capacitors senseless. However, in combination, each of the capacitors has half the potential difference it would have if it were used alone. This is sometimes important because there is a practical limit to the potential difference that may be applied to a capacitor, without causing a spark from one plate to the other.

23.3 Dielectrics

In our discussion of capacitors so far, we have assumed that the space between the conductors is a vacuum. However, it is usually impractical and even undesirable to construct vacuum capacitors. We will find that if the space between the conductors is occupied by certain insulating materials, the capacitance as well as the voltage that can be applied can be greatly increased.

Suppose we place a nonconducting material called a **dielectric** between isolated charged plates of a parallel-plate capacitor (see Figure 23–6). We will observe experimentally that the potential difference between the plates *decreases.* To understand why, we must examine the behavior of the dielectric material at the molecular level when it is placed in an electric field. Figure 23–7 shows two types of dielectric materials: *polar* and *nonpolar.* The molecules of

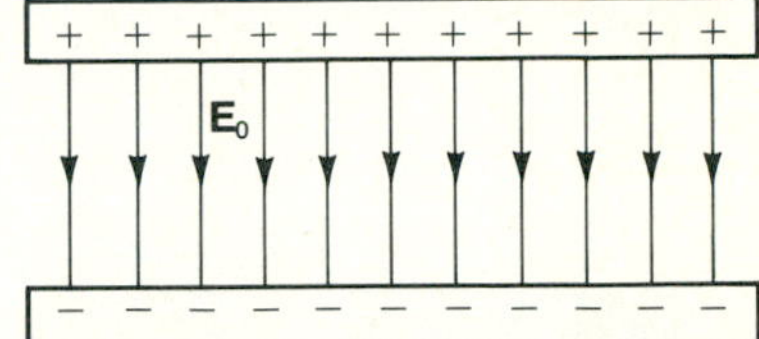

(**a**) The original field $\mathbf{E}_0$ due to the isolated charged plates.

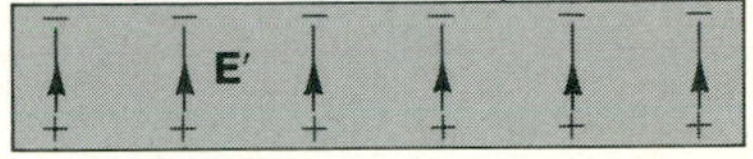

(**b**) The induced field $\mathbf{E}'$ due to the induced charges on the surface of the dielectric material when it is placed between the plates. (Plates are not shown.) Note the direction of the field.

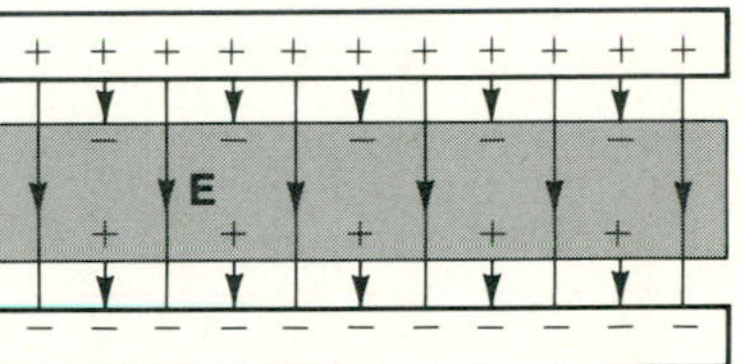

(**c**) The resultant field $\mathbf{E} = \mathbf{E}_0 + \mathbf{E}'$ is *less* than the original field $\mathbf{E}_0$.

Figure 23–6
A dielectric material placed between the charged (isolated) plates of a parallel-plate capacitor reduces the net field between the plates. As a result, the capacitance increases.

POLAR

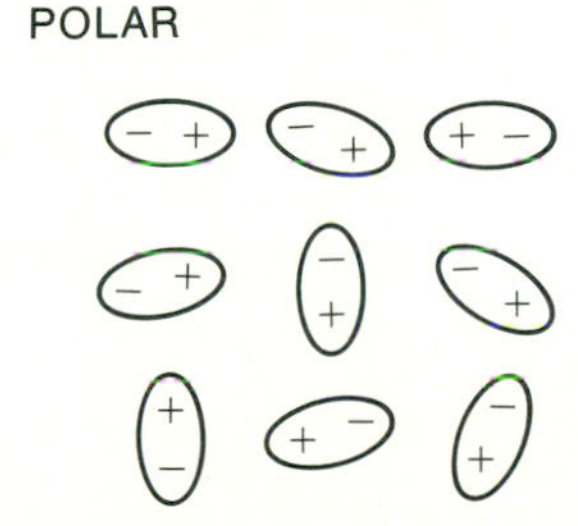

(**a**) Polar dielectric, $E_0 = 0$.

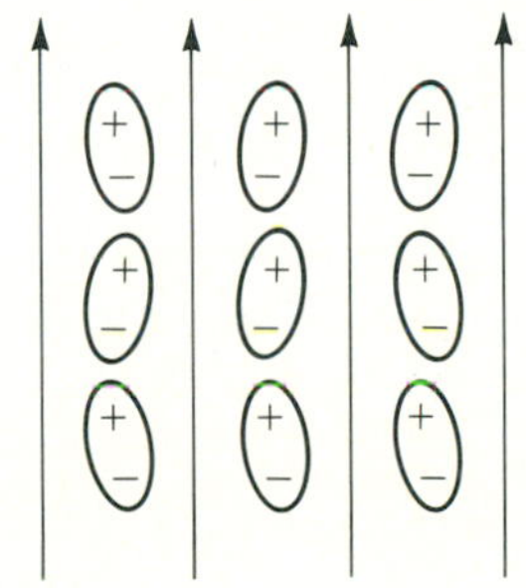

(**b**) Polar dielectric, with external field E_0.

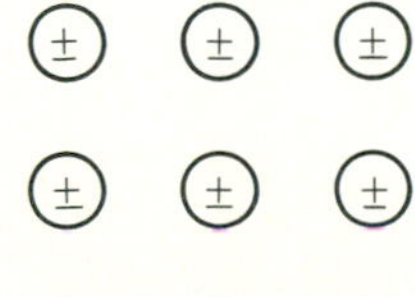

(**c**) Nonpolar dielectric, $E_0 = 0$.

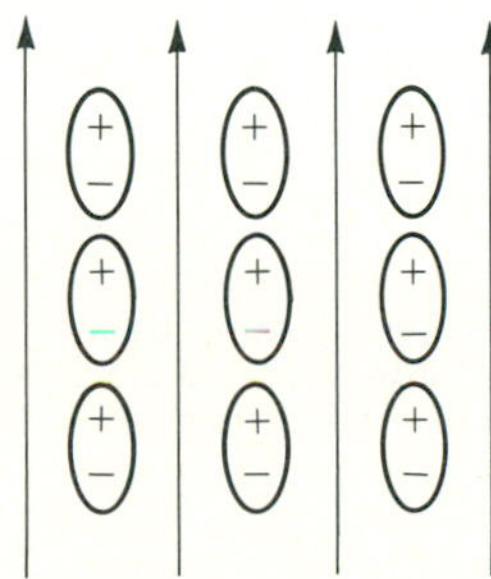

(**d**) Nonpolar dielectric, with external field E_0.

Figure 23–7
The effect of an external electric field on the molecules of a dielectric material.

a *polar* dielectric material have a permanent dipole moment; when placed in an electric field, they tend to align themselves in the direction of the field, as shown in Figure 23–7(b). The molecules of a *nonpolar* dielectric material have no inherent dipole moment, since the center of a positive charge within the molecule is coincident with the center of negative charge. However, when a nonpolar molecule is subjected to an external electric field, the centers of charge are drawn apart slightly to form dipole moments aligned in the direction of the field, as shown in Figure 23–7(d). In either material, the overall effect of dipole alignments with the electric field is that the surfaces of the dielectric perpendicular to the applied field become charged, as shown in Figure 23–6(b). This **induced surface charge** q' in turn produces an electric field $\mathbf{E}'$ within the dielectric *in a direction opposite to the applied field* $\mathbf{E}_0$. The net field is $\mathbf{E}_0 + \mathbf{E}'$, *smaller* than the original field $\mathbf{E}_0$. By Equation (22–28):

$$V_2 - V_1 = -\int_1^2 \mathbf{E} \cdot d\boldsymbol{\ell}$$

a reduced field between the plates for a given charge on the plates results in a lower potential difference between them.

Gauss's law may be applied to cases in which dielectric materials are present. Consider the situation illustrated in Figure 23–8. Suppose that the net field $\mathbf{E}$ is a fraction $1/\kappa$ of what the field would be without the dielectric $\mathbf{E}_0$. Then:

DIELECTRIC CONSTANT κ

$$E = \frac{E_0}{\kappa} \tag{23–13}$$

where κ (Greek letter *kappa*) is called the **dielectric constant.** Applying Gauss's law:

$$\oint \mathbf{E} \cdot d\mathbf{A} = \frac{q}{\varepsilon_0}$$

we obtain

$$\oint \mathbf{E} \cdot d\mathbf{A} = \frac{q - q'}{\varepsilon_0} \tag{23–14}$$

Evaluating over the surface gives

$$(E)(A) = \frac{q}{\varepsilon_0} - \frac{q'}{\varepsilon_0}$$

$$\frac{E_0}{\kappa} A = \frac{q}{\varepsilon_0} - \frac{q'}{\varepsilon_0}$$

Since:

$$E_0 A = \frac{q}{\varepsilon_0}$$

Gaussian surface enclosing both the charge on the plate and the induced surface charge on the dielectric

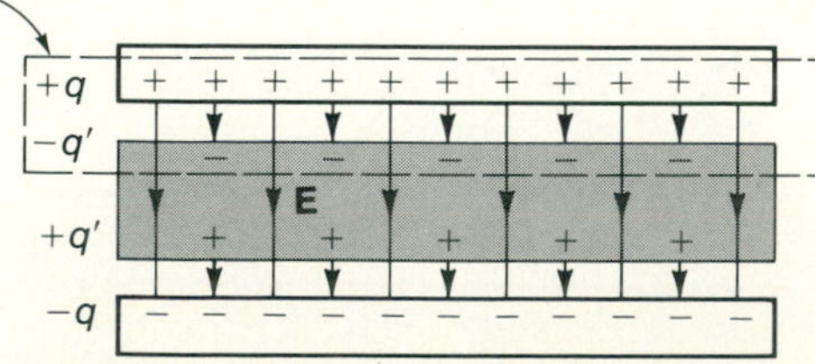

Figure 23–8

we have

$$\frac{q}{\kappa\varepsilon_0} = \frac{q}{\varepsilon_0} - \frac{q'}{\varepsilon_0}$$

Rearranging, the induced charge is

$$q' = \left(1 - \frac{1}{\kappa}\right)q \tag{23-15}$$

Substituting this expression for the induced charge into Equation (23–14), we obtain

GAUSS'S LAW FOR DIELECTRIC MATERIALS

$$\oint \mathbf{E} \cdot d\mathbf{A} = \frac{q}{\kappa\varepsilon_0} \tag{23-16}$$

Notice that only the *free* charge q appears and that the *induced* charge q' does not.

The various effects of introducing a dielectric material between plates with charges $+q$ and $-q$ may be summarized as follows:

(1) A reduced potential difference, $V = V_0/\kappa$
(2) A reduced electric field, $E = E_0/\kappa$
(3) An induced charge on the dielectric, $q' = (1 - 1/\kappa)q$
(4) As a consequence of the form of Gauss's law for dielectrics, our previous derivations for capacitances can be modified by simply replacing ε_0 by $\kappa\varepsilon_0$, thus increasing the capacitance:

$$C = \kappa C_0 \tag{23-17}$$

where V_0, E_0, and C_0 refer to values without the dielectric.

Table 23–1 shows a tabulation of approximate dielectric constants for a few materials, along with the approximate maximum field they can withstand

TABLE 23–1

Approximate Dielectric Constants and Dielectric Strengths

Material	Dielectric constant κ	Dielectric strength (V/m)
Vacuum	1	—
Air	1.000 59	1×10^6
Water	78	—
Glass	5	13×10^6
Castor oil	4.6	10×10^6
Polystyrene	2.5	15×10^6
Hard rubber	3	500×10^6
Mica	5	3000×10^6
Titanium dioxide	100	150×10^6

without being rendered conducting. This maximum electric field is called the **dielectric strength.** Because the dielectric constant for air is so close to unity, unless otherwise specified we will always assume $\kappa = 1$ for air in numerical problems.

EXAMPLE 23–5

Consider the parallel-plate capacitor shown in Figure 23–9(a), where the space between the plates is filled with rectangular slabs of different dielectric materials. The area of each plate is A. Find the capacitance.

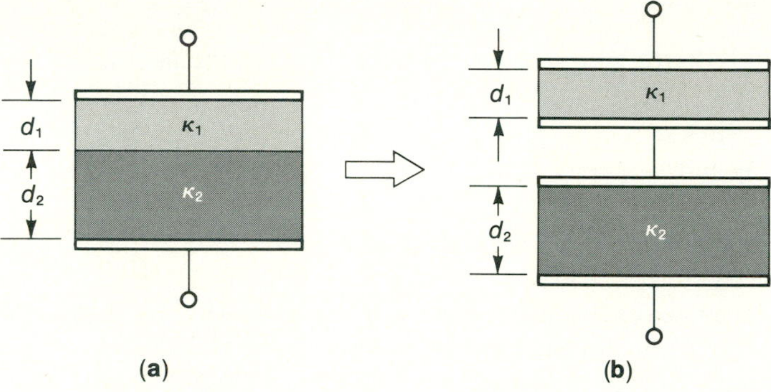

Figure 23–9
Example 23–5

SOLUTION

Because the field is perpendicular to the boundary between the dielectrics, the boundary is an equipotential surface. Therefore, a conducting sheet may be introduced at the boundary without altering the fields within the dielectrics. The conducting sheet may then be split as shown in Figure 23–9(b), forming two capacitors in series.

The capacitance $(C_1)_0$ of the upper capacitor without a dielectric is, by Equation (23–9);

$$(C_1)_0 = \frac{\varepsilon_0 A}{d_1}$$

By inserting the dielectric, the capacitance is increased by a factor κ_1:

$$C_1 = \kappa_1 (C_1)_0$$

$$C_1 = \frac{\kappa_1 \varepsilon_0 A}{d_1}$$

Similarly, the capacitance of the lower capacitor is

$$C_2 = \frac{\kappa_2 \varepsilon_0 A}{d_2}$$

The series combination of C_1 and C_2, by Equation (23–12), becomes

$$\frac{1}{C} = \frac{1}{C_1} + \frac{1}{C_2}$$

$$\frac{1}{C} = \frac{1}{\varepsilon_0 A}\left(\frac{d_1}{\kappa_1} + \frac{d_2}{\kappa_2}\right)$$

$$C = \boxed{\varepsilon_0 A\left[\frac{\kappa_1 \kappa_2}{\kappa_2 d_1 + \kappa_1 d_2}\right]}$$

23.4 Potential Energy of Charged Capacitors

As we have seen, configurations of charges have electric potential energy U. This potential energy implies the system could do work. For a charged parallel-plate capacitor, there are several ways this could be accomplished. For example, the force of attraction between the plates could do work if the plates were free to move toward each other. Or if the charges could move, work could be done by each charge as it moves through the potential difference.

The amount of potential energy can be determined by calculating the amount of work done by an external agent to charge the capacitor in the first place. The incremental work dW required to move a charge dq' from the plate at the lower potential to the plate at the higher potential is

$$dW = V\,dq' \tag{23–18}$$

where V is the potential difference between the plates. However, the potential difference depends upon the charge q' already deposited on the plates:

$$V = \frac{q'}{C} \tag{23–19}$$

where C is the capacitance of the capacitor. Substituting this relation into Equation (23–18), we have

$$dW = \frac{q'}{C}\,dq' \tag{23–20}$$

We obtain the total amount of work required to charge the capacitor to a final charge q by integrating.

$$W = \int_0^q \frac{q'}{C}\,dq'$$

$$W = \frac{1}{2}\left(\frac{q^2}{C}\right)$$

Since the work done by the external agent is the gain in electric potential energy U of the capacitor:

$$U = \frac{1}{2}\left(\frac{q^2}{C}\right) \tag{23–21}$$

It is usually easier to determine the potential difference V rather than the charge q. Since $q = CV$, we may write Equation (23–21) in terms of V and C:

ENERGY STORED IN A CHARGED CAPACITOR

$$U = \tfrac{1}{2}CV^2 \tag{23–22}$$

EXAMPLE 23–6

A 2-nF parallel-plate capacitor is charged to an initial potential difference $V_i = 100$ V and then isolated. The dielectric material between the plates is mica ($\kappa = 5$). (a) How much work is required to withdraw the mica sheet? (b) What is the potential difference of the capacitor after the mica is withdrawn?

SOLUTION

(a) Work must be done to withdraw the mica because the charges on the plates exert forces of attraction on the induced charges of the mica (see Figure 23–10). The work required will be the difference in potential energy between the capacitor without the dielectric and the capacitor with the dielectric. Since the charge q on the plates does not change when the dielectric is removed, we use Equation (23–21) to find the potential energy rather than Equation (23–22). (As we will see, the potential V changes as the dielectric is withdrawn.) The initial and final energies are

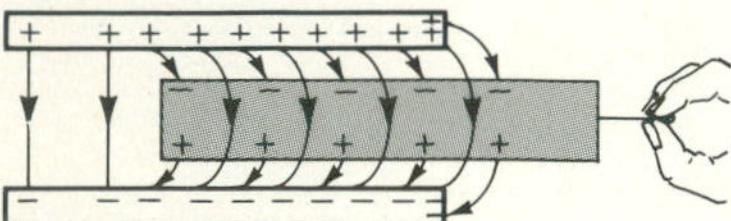

Figure 23–10
Example 23–6

$$U_i = \frac{1}{2}\left(\frac{q^2}{C_i}\right) \quad \text{and} \quad U_f = \frac{1}{2}\left(\frac{q^2}{C_f}\right)$$

But the initial capacitance (with the dielectric) is $C_i = \kappa C_f$. Therefore:

$$U_f = \tfrac{1}{2}\kappa\left(\frac{q^2}{C_i}\right)$$

Since the work done by the external force in removing the dielectric equals the change in potential energy, we have

$$W = U_f - U_i$$

$$W = \tfrac{1}{2}\kappa\left(\frac{q^2}{C_i}\right) - \frac{1}{2}\left(\frac{q^2}{C_i}\right)$$

$$W = \frac{1}{2}\left(\frac{q^2}{C_i}\right)(\kappa - 1)$$

To express this relation in terms of the potential V_i, we substitute $q = C_iV_i$.

$$W = \tfrac{1}{2}(C_iV_i^2)(\kappa - 1)$$

Evaluating, we obtain

$$W = \tfrac{1}{2}(2 \times 10^{-9}\ \text{F})(100\ \text{V})^2(5 - 1)$$

$$W = \boxed{4.00 \times 10^{-5}\ \text{J}}$$

The positive result confirms that the final energy of the capacitor is greater than the initial energy. The extra energy comes from the work done *on* the system by the external force that pulled out the dielectric.

(b) The final potential difference across the capacitor is given by

$$V_f = \frac{q}{C_f}$$

Substituting $C_f = C_i/\kappa$ and $q = C_iV_i$ gives

$$V_f = \kappa V_i$$

Evaluating, we obtain

$$V_f = (5)(100\ \text{V})$$

$$V_f = \boxed{500\ \text{V}}$$

Even though the capacitor is isolated and its charge remains constant, the potential difference across the plates does increase in this case.

Just as mechanical systems tend toward configurations that represent lower potential energies, charges distribute themselves so as to achieve the lowest possible potential energy.

EXAMPLE 23–7

Consider the capacitors shown in Figure 23–11(a). The 4-μF and 12-μF capacitors are connected in series across a potential difference of 50 V. After becoming charged, the capacitors are disconnected from the source of potential, separated, and then rejoined, with positive plates together and negative plates together, as shown in Figure 23–11(b). (a) Find the initial and final potential energies. (b) Find the final voltage across the two capacitors in parallel.

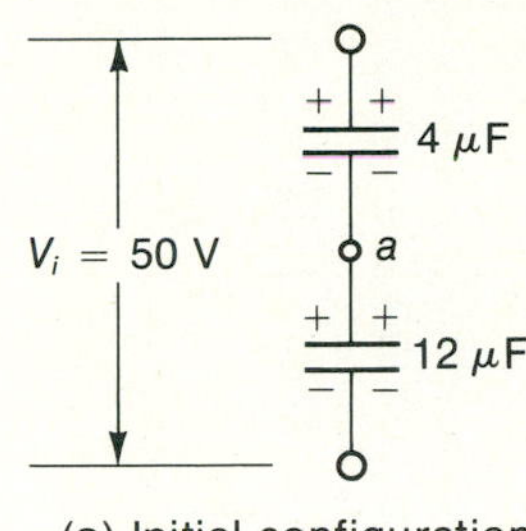

(a) Initial configuration

V_f 4 μF 12 μF

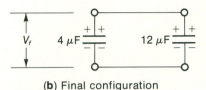

(b) Final configuration

Figure 23–11
Example 23–7

SOLUTION

(a) The value of the series combination of two capacitors is given by Equation (23–12):

$$\frac{1}{C} = \frac{1}{C_1} + \frac{1}{C_2}$$

Substituting:

$$\frac{1}{C_i} = \frac{1}{4\ \mu\text{F}} + \frac{1}{12\ \mu\text{F}}$$

$$C_i = 3\ \mu\text{F}$$

The initial potential energy of the system of capacitors is given by Equation (23–22):

$$U_i = \tfrac{1}{2}C_i V_i^2$$

Evaluating, we obtain

$$U_i = \tfrac{1}{2}(3 \times 10^{-6}\ \text{F})(50\ \text{V})^2$$

$$U_i = \boxed{3.75 \times 10^{-3}\ \text{J}}$$

As explained earlier, when capacitors in series are charged, *each* capacitor acquires the *same* magnitude of charge q. This is:

$$q = C_i V_i$$

Substituting:

$$q = (3 \times 10^{-6}\ \text{F})(50\ \text{V})$$

$$q = 1.50 \times 10^{-4}\ \text{C}$$

When connected in the new arrangement, the charge on the parallel combination of capacitors will be $2q$, or 3×10^{-4} C. The capacitance of the parallel combination is given by Equation (23–11):

$$C = C_1 + C_2$$

Evaluating:

$$C_f = 4\ \mu\text{F} + 12\ \mu\text{F}$$

$$C_f = 16\ \mu\text{F}$$

The final potential energy is given by Equation (23–21):

$$U = \frac{1}{2}\left(\frac{q^2}{C}\right)$$

Substituting the appropriate values, we obtain

$$U_f = \frac{1}{2}\frac{(3 \times 10^{-4}\ \text{C})^2}{(16 \times 10^{-6}\ \text{F})}$$

$$U_f = \boxed{2.81 \times 10^{-3}\ \text{J}}$$

Note that a *loss* in potential energy has occurred:

$$\Delta U = U_f - U_i$$
$$\Delta U = 2.81 \times 10^{-3}\ \text{J} - 3.75 \times 10^{-3}\ \text{J}$$
$$\Delta U = -9.4 \times 10^{-4}\ \text{J}$$

(b) The final potential difference V_f is obtained from the relation:

$$q = CV$$

In this case:

$$V_f = \frac{q_f}{C_f}$$
$$V_f = \frac{3 \times 10^{-4}\ \text{C}}{16 \times 10^{-6}\ \text{F}}$$
$$V_f = \boxed{18.8\ \text{V}}$$

23.5 Energy Stored in an Electric Field

The previous example raises a few questions: Since the final energy of the system is less than the initial energy, where does the "missing" energy go? Also, where does the potential energy of a charged capacitor (or, for that matter, a single charged particle) reside? To answer the first question, we must realize that the redistribution of charge causes a current flow through the wires connecting the capacitors. It can be shown that the resultant heating of the wires, no matter how small their resistance (excluding zero resistance), exactly accounts for the energy loss of the charged capacitors.

The other question, regarding *where* the potential energy resides, leads to an important new concept. Consider a charged capacitor. If an incremental charge dq is freed from the positive plate, it will be accelerated toward the negative plate by the electric field between the plates. The kinetic energy acquired by dq results in a corresponding reduction in the electric field (because the charge on the plates is now less). Therefore, it is reasonable to assume that *the potential energy of a charged capacitor resides in the electric field.*

We can derive an expression for the energy stored in an electric field by considering a parallel-plate capacitor, where the field is uniform. We have seen that for a parallel-plate capacitor, $C = \varepsilon_0 A/d$ (Equation 23–9) and $V = Ed$ (Example 22–13). Substituting these expressions for the energy of a charged capacitor:

$$U_C = \tfrac{1}{2}CV^2$$

we obtain

$$U_C = \frac{1}{2}\left(\frac{\varepsilon_0 A}{d}\right)(Ed)^2$$
$$U_C = \tfrac{1}{2}\varepsilon_0 E^2(Ad)$$

But (Ad) is the volume occupied by the electric field. We now define the *energy per unit volume* in the electric field as the **energy density** u_E (in joules/meter^3). Thus:

$$u_E = \frac{U}{Ad}$$

Substituting, we obtain

ENERGY DENSITY u_E IN AN ELECTRIC FIELD (in free space)

$$u_E = \tfrac{1}{2}\varepsilon_0 E^2 \qquad (23\text{–}23)$$

Had there been a dielectric present, the capacitance C would have been increased by the factor κ. The previous analysis[1] would then lead to:

ENERGY DENSITY u_E IN AN ELECTRIC FIELD (in the presence of a dielectric)

$$u_E = \tfrac{1}{2}\kappa\varepsilon_0 E^2 \qquad (23\text{–}24)$$

EXAMPLE 23–8

An isolated conducting sphere of radius R has a charge q. Show that the total energy stored in the surrounding electric field equals the expression for the potential energy of a charged capacitor: $U = \frac{1}{2}(q^2/C)$, Equation 23–21.

SOLUTION

The isolated sphere has a capacitance $C = 4\pi\varepsilon_0 R$ (Equation 23–6). Its potential energy is thus:

$$U = \frac{q^2}{8\pi\varepsilon_0 R} \qquad (23\text{–}25)$$

The electric field surrounding the charged sphere is given by

$$E = \left(\frac{1}{4\pi\varepsilon_0}\right)\frac{q}{r^2} \qquad (\text{for } r \geqq R)$$

At any point in this field, the energy density u_E is

$$u_E = \tfrac{1}{2}\varepsilon_0 E^2$$

Substituting for E:

$$u_E = \tfrac{1}{2}\varepsilon_0\left(\frac{q}{4\pi\varepsilon_0 r^2}\right)^2$$

$$u_E = \left(\frac{q^2}{32\pi^2\varepsilon_0}\right)\frac{1}{r^4}$$

To find the total energy U stored in the surrounding electric field, we first express the energy dU in the spherical shell element of radius r and thickness dr. This thin shell has a volume $4\pi r^2\,dr$. So the energy dU within this shell is

$$dU = 4\pi r^2\,dr\,u_E$$

Substituting for u_E gives

$$dU = 4\pi r^2\,dr\left(\frac{q^2}{32\pi^2\varepsilon_0}\right)\frac{1}{r^4}$$

[1] Certain dielectric materials can be given a permanent electric dipole moment by melting them and then allowing them to solidify in the presence of an electric field. The resulting *electret* has a permanent electric field analogous to the permanent magnetic field of a magnet.

Rearranging and integrating from $r = R$ to $r = \infty$, we have

$$U = \frac{q^2}{8\pi\varepsilon_0}\int_R^\infty \frac{1}{r^2}\,dr$$

$$U = -\frac{q^2}{8\pi\varepsilon_0 r}\Bigg|_R^\infty$$

$$U = \boxed{\frac{q^2}{8\pi\varepsilon_0 R}}$$

Note that this is indeed equal to Equation (23–25).

Summary

The following concepts and equations were introduced in this chapter:

Capacitance C (which is dependent on the geometry of an object):

General formula: $C = \dfrac{q}{V}$

For an isolated sphere: $C = 4\pi\varepsilon_0 R$

For parallel plates: $C = \dfrac{\varepsilon_0 A}{d}$

For a cylindrical capacitor: $C = \dfrac{2\pi\varepsilon_0 L}{\ln \dfrac{b}{a}}$

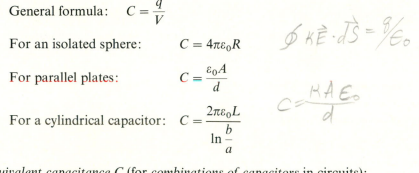

Equivalent capacitance C (for *combinations of capacitors* in circuits):

In parallel: $C = C_1 + C_2 + C_3 + \cdots$

In series: $\dfrac{1}{C} = \dfrac{1}{C_1} + \dfrac{1}{C_2} + \dfrac{1}{C_3} + \cdots$

Certain nonconductors become polarized in the presence of an electric field. The *dielectric constant* κ describes the effects of such materials. In a capacitor, the capacitance is increased when a dielectric is introduced:

$$C = \kappa C_0$$

Induced charge q' that appears on the surface (perpendicular to the field) of a dielectric:

$$q' = \left(1 - \frac{1}{\kappa}\right)q$$

Energy U stored in a charged capacitor:

$$U = \tfrac{1}{2}CV^2 \qquad \text{(in joules)}$$

Energy density u_E in an electric field:

$$u_E = \tfrac{1}{2}\varepsilon_0 E^2 \qquad \text{(in joules/meter}^3\text{)}$$

In the presence of a dielectric, the energy stored in the field is increased by a factor κ:

$$u_E = \kappa(u_E)_0$$

Note the difference between *energy* U (in joules) and *energy density* u_E (in joules/meter^3).

Questions

1. The pattern of electric field lines between opposite but equal charges is undisturbed by placing a thin metal sheet halfway between the two charges so that the plane of the sheet is perpendicular to the line joining the charges. Why?
2. In terms of basic concepts, why is the capacitance of an isolated spherical conductor proportional to its radius?
3. Does the fringing effect in a parallel-plate capacitor tend to increase or decrease its calculated capacitance? Why?
4. Is it possible for the plates of a capacitor to have different magnitudes of charge?
5. Why should air bubbles be avoided in oil-filled capacitors?
6. A dielectric slab is inserted between the plates of a charged parallel-plate capacitor. The capacitor is not connected to a battery. What happens to the energy of the capacitor? What happens to the potential difference across the plates of the capacitor?
7. Given three capacitors of different capacitances, how many capacitance values can be obtained using one or more of the capacitors?
8. In view of its high dielectric constant, why is water not commonly used as a dielectric material in capacitors?
9. Capacitors are often stored with a wire connected across their terminals. Why?
10. How does the size of a given type of capacitor depend on its maximum energy storage capacity?
11. The oil in an isolated (but charged) oil-filled capacitor leaks out. What happens to the potential differences between the terminals of the capacitor?
12. The edge of a parallel-plate capacitor is placed in a pool of oil. The oil rises between the plates due to capillary action. Will the height to which the oil rises depend on the potential differences between the plates? In what way?
13. Due to the normal potential gradient in the earth's atmosphere, an electric field exists there. What are the difficulties in extracting the energy associated with this field and applying the energy for useful purposes?

Problems

23A–1 Consider the parallel-plate capacitor configuration shown in Figure 23–12. Derive an expression for its capacitance. Ignore fringing effects.

Answer: $C = \varepsilon_0 A(a + b)/[b(a - b)]$

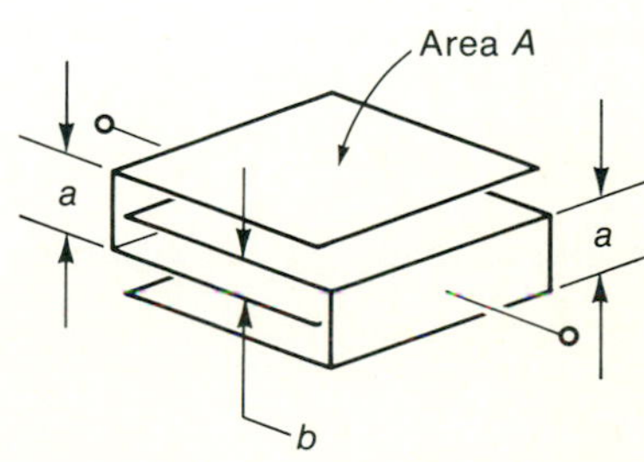

Figure 23–12
Problem 23A–1

23A–2 A potential difference of 200 V is applied to a series combination of a 2 μF capacitor and a 6 μF capacitor.

(a) For each individual capacitor, find the potential difference and the charge.
(b) The charged capacitors are isolated, then connected together with positive polarities joined and negative polarities joined. Find the new potential difference and the charge on each capacitor.
(c) If the above procedure were repeated, except that the capacitors were connected with opposing polarities, what would be the final potential difference and the charge on each capacitor?

23A–3 Determine the capacitance for each of the networks of capacitors shown in Figure 23–13. Each capacitor has the same capacitance C.

Answers: (a) $\frac{3}{5}C$ (b) $3C$ (c) C
(d) Capacitors are shorted by a conductor.

Figure 23–13
Problem 23A–3

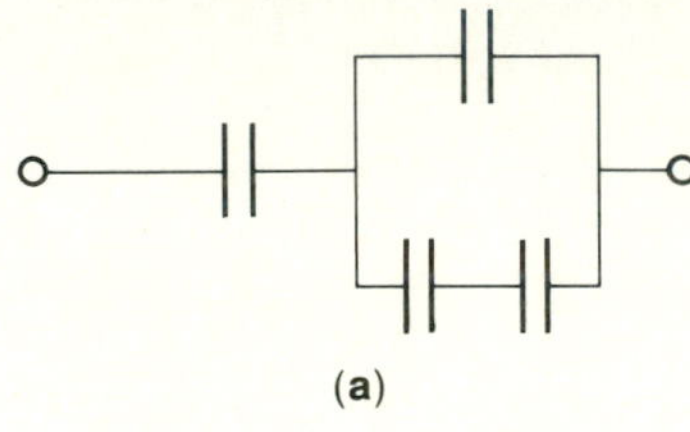
(a)

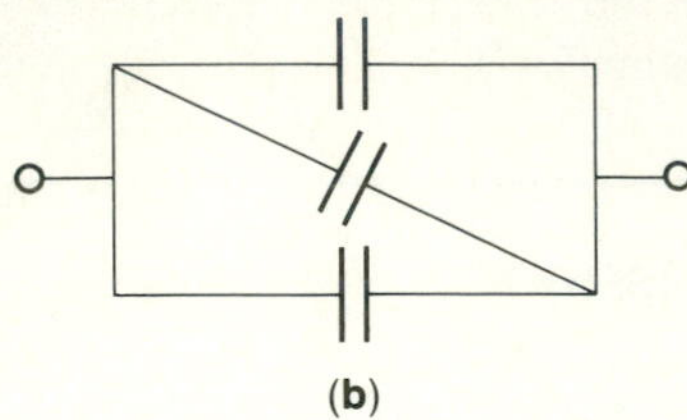
(b)

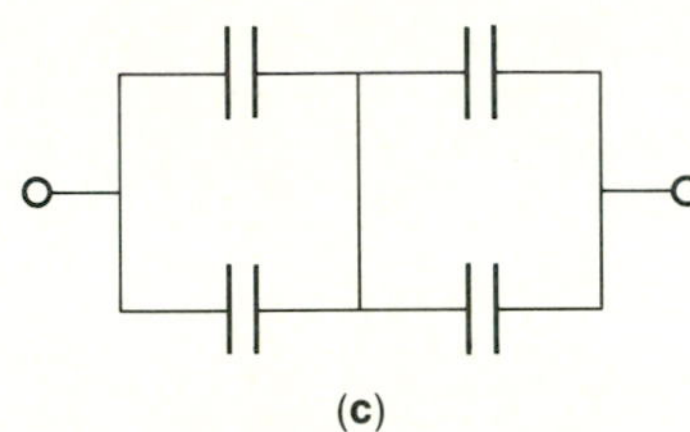
(c)

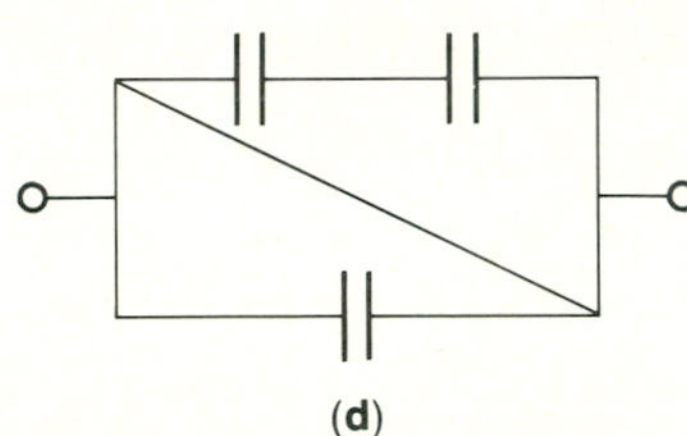
(d)

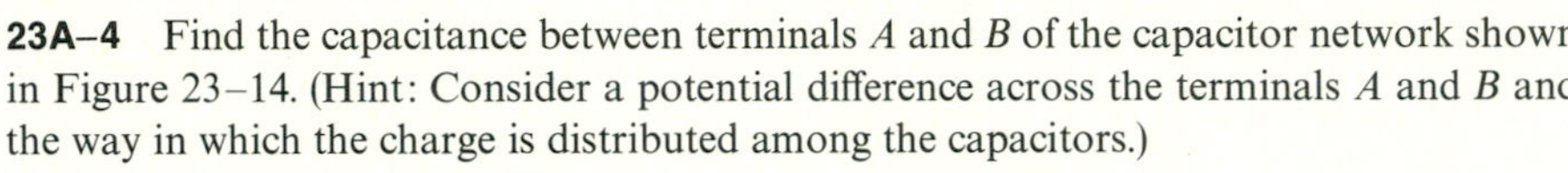

23A–4 Find the capacitance between terminals A and B of the capacitor network shown in Figure 23–14. (Hint: Consider a potential difference across the terminals A and B and the way in which the charge is distributed among the capacitors.)

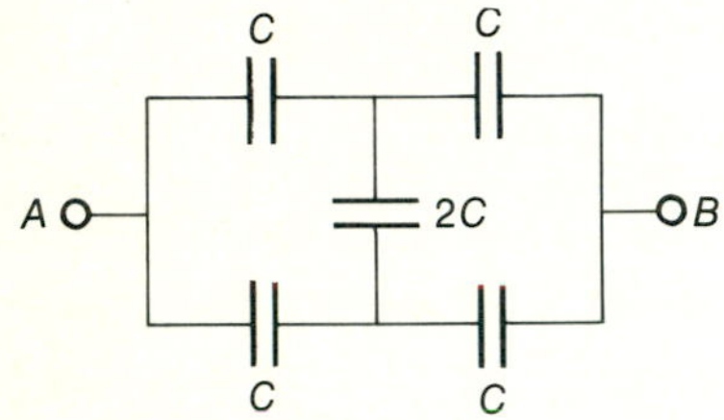

Figure 23–14
Problem 23A–4

23A–5 Show that the energy storage capability of a parallel-plate capacitor is proportional to the volume of the capacitor.

23A–6 Estimate the maximum voltage to which a smooth, metallic sphere 10 cm in diameter can be charged without exceeding the dielectric strength of the air around the sphere.

23A–7 A metal sphere 50 cm in diameter is charged to a potential of 10 kV. Determine the energy density just next to the outer surface of the sphere.

Answer: 7.08×10^{-3} J/m^3

23A–8 Einstein asserted that energy is associated with mass according to the famous relation $E = mc^2$. Estimate the radius of an electron, assuming that its charge is distributed uniformly over the surface of a sphere of radius r and that the rest mass energy of an electron is equal to the total electric field energy of the electron. The result is called the *classical radius* of the electron.

23B–1 A parallel-plate capacitor is constructed using a dielectric material whose dielectric constant is 3 and whose dielectric strength is 2×10^8 V/m. The desired capacitance is 0.25 μF, and the capacitor must withstand a maximum potential difference of 4000 V. Find the minimum area of the capacitor plates. *Answer:* 0.188 m^2

23B–2 A parallel-plate capacitor has a dielectric material between its plates with a dielectric constant κ and dielectric strength E_{max}. Derive an expression for the maximum voltage that may be applied if the capacitor has a capacitance C and volume $\mathscr{V}$.

23B–3 A parallel-plate capacitor with a polystyrene dielectric between the plates has a capacitance of 10 nF (equal to 10×10^{-9} F). While the capacitor is attached to a 100 V battery, the dielectric is withdrawn. Find (a) the change in the charge on one of the plates, (b) the change in energy stored in the capacitor, and (c) the amount of work required to remove the dielectric.

Answers: (a) 600 nC, decrease
(b) 30 μJ, decrease
(c) 30 μJ

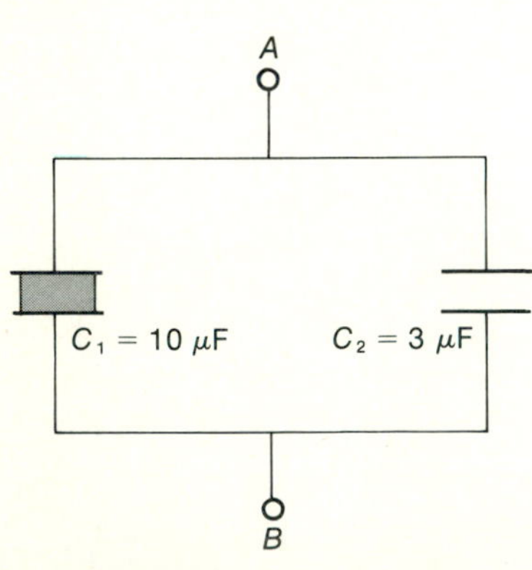

Figure 23–15
Problem 23B–4

23B–4 Consider two parallel-plate capacitors that are connected to each other as shown in Figure 23–15. The capacitors are identical except for the dielectric material in C_1. A potential difference of 150 V is applied across the terminals A and B and then removed.

(a) Find the charge on each capacitor.
(b) Find the total energy stored in the capacitors.

(c) If the dielectric material is removed from C_1, determine the total energy stored in the capacitors.

(d) Find the final voltage across the terminals A and B.

23B–5 A parallel-plate capacitor with air between its plates has a capacitance C. A slab of dielectric material with a dielectric constant κ and a thickness equal to a fraction f of the separation of the plates is inserted between the plates in contact with one plate. Find the capacitance in terms of f, κ, and C_0. Check your result by first letting f approach zero and then letting it approach one. *Answer:* $C = \kappa C_0/[f + (1 - f)\kappa]$

23B–6 A parallel-plate capacitor has plate area A and separation d. A slab of copper of the same area and thickness t ($t < d$) is inserted symmetrically between the plates.

(a) Find the new capacitance C of the capacitor.

(b) Suppose the copper slab is moved closer to one plate so that the separation from the plate is half the separation between the slab and the other plate. Find the capacitance for this geometry.

23B–7 Figure 23–16 shows a variable capacitor commonly used in the tuning circuit of radios. Alternate plates are connected together, with one group held fixed while the other group rotates together, resulting in a variable meshing of the plates. The area of each plate is A, with a spacing d between a plate of one group and the adjacent plate of the other group. The total number of plates is n. Ignoring fringing effects at the edges, show that the maximum capacitance is $C_{max} = (\varepsilon_0 A/d)(n - 1)$.

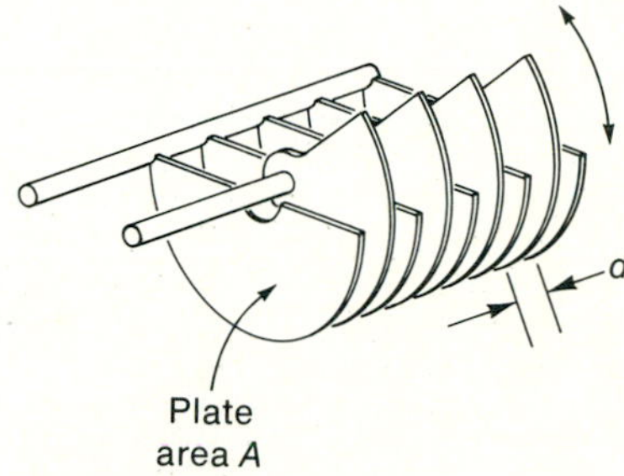

Figure 23–16
Problem 23B–7

23B–8 Consider a cylindrical capacitor with two layers of dielectric material between the inner and outer cylinder, as shown in Figure 23–17. Derive an expression for the capacitance of the capacitor in terms of the given parameters.

Figure 23–17
Problem 23B–8

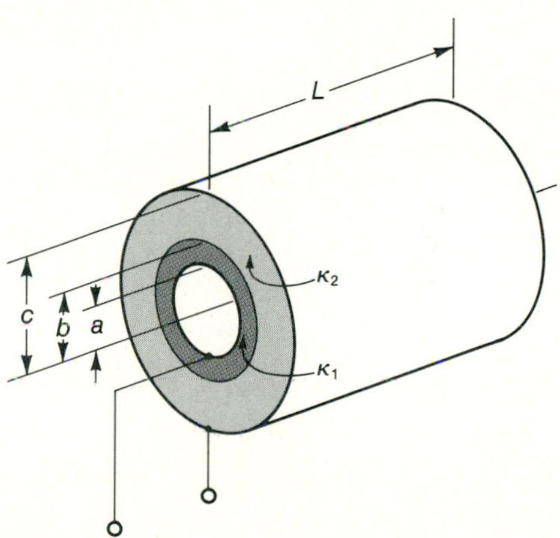

23B–9 Derive an expression for the force of attraction between the plates of a parallel-plate capacitor in terms of the capacitance C, the separation d of the plates, and the potential difference V between the plates. (Hint: Consider the difference in stored energy dU when the plate separation is changed from x to $x + dx$. This equals the work done $dW = F\,dx$.) *Answer:* $CV^2/2d$

23C–1 Coaxial cable consists of a central wire surrounded by a plastic insulator, which in turn is surrounded by a woven metallic cylindrical conductor. Let κ be the dielectric constant of the insulator, a the radius of the central wire, and b the outer radius. Derive an expression for the capacitance C per unit length L of the cable. *Answer:* $C/L = \kappa 2\pi\varepsilon_0/[\ln(b/a)]$

23C–2 A cylindrical capacitor is made of an inner conducting cylinder of radius a and a concentric outer conducting cylinder of radius b. The length L of the capacitor is sufficiently large that end effects may be ignored. The total charge on the inner cylinder is $+Q$, while the charge on the outer cylinder is an equal-magnitude negative charge $-Q$.

(a) Starting with Gauss's law, find the electric field E between the cylinders.

(b) Find the potential difference $V_b - V_a$ between the cylinders in terms of the given symbols.

(c) Find the capacitance C.

23C–3 A dielectric slab fills only half of the space between the plates of a parallel-plate capacitor, as shown in Figure 23–18. Derive an expression for the fraction of the total energy that is stored in the dielectric.

Answer: $\kappa/(1 - \kappa)$

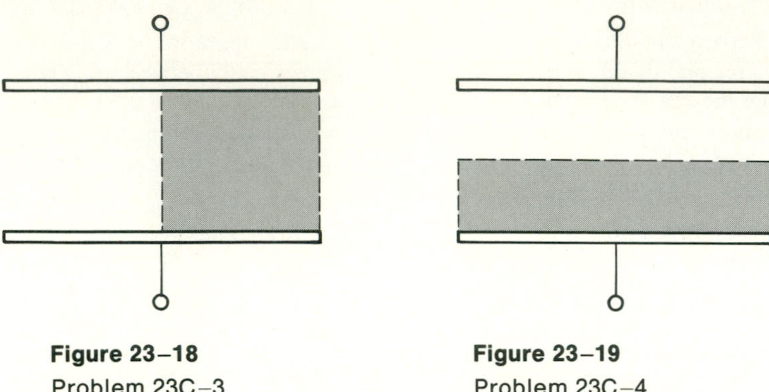

Figure 23–18
Problem 23C–3

Figure 23–19
Problem 23C–4

23C–4 Repeat Problem 23C–3 for the case shown in Figure 23–19.

Answer: $\kappa/(1 - \kappa)$

23C–5 Consider two concentric, conducting spherical shells with equal but opposite charges. Beginning with $u_E = \frac{1}{2}\varepsilon_0 E^2$, calculate the total energy contained in the field between the shells. Show that this agrees with the energy given by $\frac{1}{2}CV^2$.

Don't worry—
Lightning
Never strikes twice
In the same

BILLY BEE

chapter 24 Electrodynamics

In this chapter we will investigate the steady flow of charge (a *current*) through conductors as the charges move from a higher to a lower potential. Since the flow of charge is continuous, there is no net accumulation of charge at any one place. The conductor paths form closed loops, some of which contain energy sources to maintain the current. Such networks of conductors and energy sources are called *circuits*. We will show that the current through various parts of a circuit is determined by two conservation laws: the *conservation of charge*, which simply means that the charge carriers are neither created nor destroyed in a circuit, and the familiar *conservation of energy*.

24.1 Electromotive Force

A source of electrical energy is essential to any practical electrical circuit. Such an energy source is called a **seat of electromotive force.** (As we shall see, the use of the word *force* in this context is somewhat of a misnomer.) For convenience, we will use EMF as an abbreviation for electromotive force.

A SEAT OF ELECTROMOTIVE FORCE (EMF) — **A seat of electromotive force is any device that transforms one source of energy into a source of electrical energy.**

Examples of seats of EMF are:

(a) Any of a number of batteries, including flashlight cells and automobile batteries (a "battery" really means a series or battery of cells). These transform chemical energy into electrical energy.
(b) Generators such as a power station generator driven by a water turbine or an automobile generator (commonly called an *alternator*) driven by the automobile engine. These transform mechanical energy into electrical energy.
(c) Solar cells such as those that provide power for spacecraft. These transform radiant energy into electrical energy.

By whatever means (chemical, mechanical, or radiant), a seat of EMF does work on electrical charges. Specifically, a battery maintains a potential difference between its terminals, so that if an external electrical circuit is connected to the terminals, electric charge will be driven around the circuit.[1] When the charge returns to the battery at the lower-potential terminal, the seat of electromotive force does work on the charge, moving it through the seat to the higher-potential terminal, ready to be driven again around the external circuit. Even if no external circuit is connected to the battery, the potential difference between the terminals is maintained.

We define the unit of EMF in terms of the ability to do work on a charge:

ELECTROMOTIVE FORCE $\mathscr{E}$

$$\mathscr{E} = \frac{dW}{dq} \tag{24-1}$$

Thus, the electromotive force $\mathscr{E}$ is the work done per unit charge. The units of EMF are *joules/coulomb* (J/C), or simply *volts* (V), which are the same units used to measure potential difference. A 12 V battery does 12 J of work in transporting one coulomb of charge around the external circuit.

In our discussion of electrical circuits, we will restrict ourselves to batteries as a source of electrical energy. However, our analysis of the circuits would be valid for any source of electrical energy. The symbol for a battery, shown in Figure 24–1(a), is somewhat similar to the symbol for a capacitor, but in the context of a circuit they are seldom confused. By convention, the longer of the two vertical lines in Figure 24–1(a) is always at the higher potential, so the plus and minus signs are often omitted. A number of single symbols are sometimes connected end-to-end (in series), as shown in Figure 24–1(b), to designate a battery with a high potential difference between the terminals.

(a) (b)

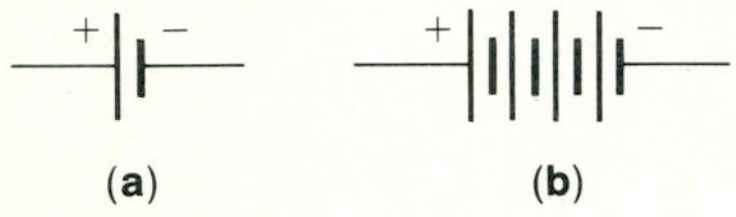

Figure 24–1
The electrical symbol for a battery.

An EMF is somewhat analogous to a pump in a circulating water system that raises water vertically, increasing its gravitational potential energy. If the water pipes form a closed loop, the pump drives water around the system (see Figure 24–2). A partial obstruction in the system (indicated by the portion of the pipe containing screens or gravel) will offer some mechanical resistance to the flow of water, somewhat reducing the rate of flow of the water. If the water pipe is blocked completely, so that no water can flow, the pump still exerts a pressure that will cause water to flow when the obstruction is removed. In the electrical case, a partial obstruction to the flow of charge is called an electrical *resistance* (symbolized by —ww—). If a switch (symbolized by o⁄ o) in the electrical circuit is opened, so that no complete electrical path exists from

[1] The first experiment to test the biological effects of electricity was perhaps made by Count Alessandro Volta (1745–1827), the Italian physicist who invented the voltaic cell. He connected 50 cells in series, put the ends of the wires in his ears, and reported it felt like a strong blow to the head, followed by sounds of boiling soup.

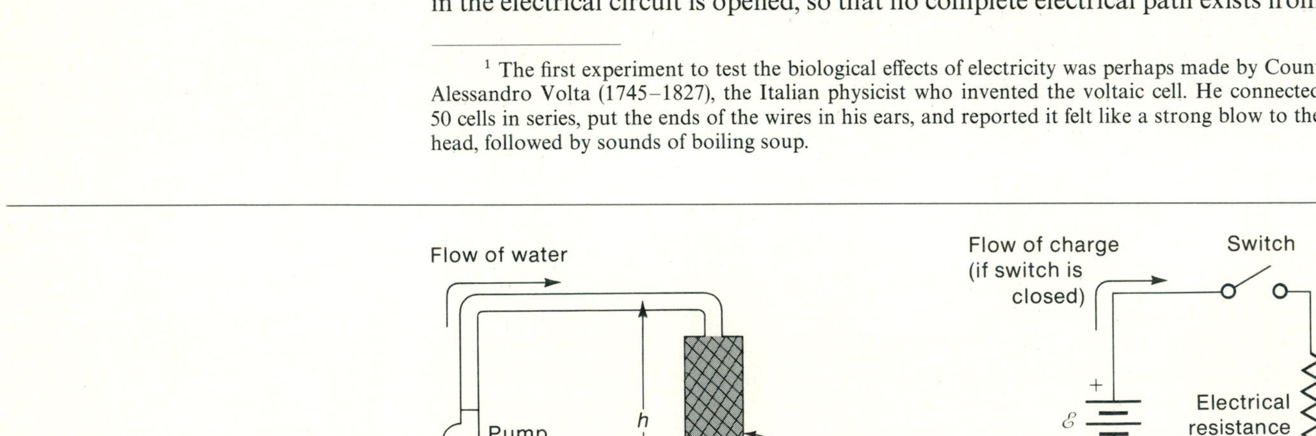

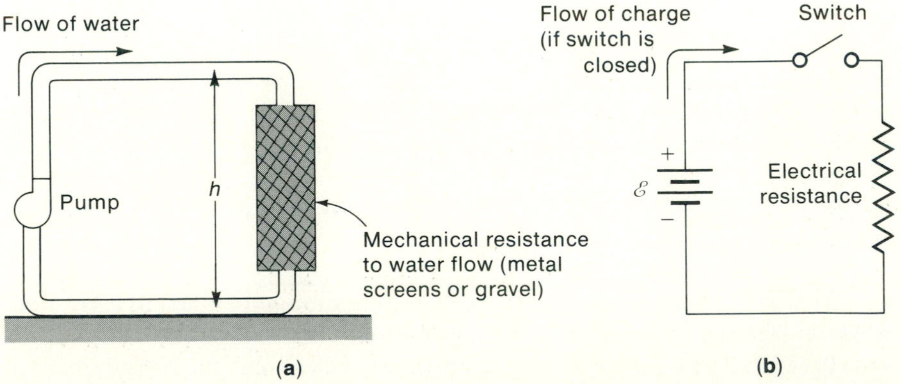

Figure 24–2
A water-pump analogy for a seat of electromotive force.

one terminal to the other, the seat of EMF still exerts an electromotive force that appears as an electrical potential V across the open switch terminals, ready to cause a flow of charge when the switch is closed.

24.2 Conductors and Resistors

Substances may be classified according to the ease with which electric charges can move within them. In a *conductor*, electric charges can move freely. Most metals are conductors because the outer electrons associated with each atom—the "conduction electrons"—can travel easily throughout the material; in certain conducting liquids, positive as well as negative charges can move. On the other hand, substances such as glass, wood, and plastics are classified as *nonconductors* or *insulators*, since electric charges are much less free to move within them. Substances that are intermediate between these two extremes are called *semiconductors*.

Suppose that the ends of a long, uniform metal wire are attached to the terminals of a seat of EMF. What happens to the conduction electrons in the wire? Because a difference of potential $(V_2 - V_1)$ exists at the terminals, an electric field $\mathbf{E}$ will exist within the wire, given by Equation (22–28):

$$V_2 - V_1 = -\int_1^2 \mathbf{E} \cdot d\ell$$

In this case, since the wire is uniform the field will be constant throughout the length of the wire. (This does not contradict the information in Chapter 22, which stated that in a perfect conductor, the electric field is zero. There, we were concerned with the *static* case, in which charges are at rest, and we assumed that no battery established a potential difference between two different points on the conductor. Here, we discuss *dynamic* situations, in which charges are in motion, because a battery *does* maintain two different points on a conductor at different electric potentials, thereby establishing an electric field within the conductor.)

The electric field the battery creates in the conductor will exert a force on charges within the wire. Charges free to move will then be forced along the wire; the ions (charged atoms) of the conductor are held in place in a "lattice" array by elastic forces between them due to their net charges. In metallic conductors, some of the "conduction electrons" are relatively free to move along the wire. In reality, even though there is a force applied to these free electrons, they will not (on the average) accelerate along the wire, but instead will drift along it at an average speed ricocheting through the atomic lattice of the metal. In these collisions, the electrons transfer some of their kinetic energy to the vibrational energy of the lattice, heating the metal. This behavior is similar to a stream of water descending a rapids: On the average, the water does not accelerate as it falls through a gravitational field.

Let us examine the drift of charges through the wire in a more quantitative way. Consider a segment of the wire shown in Figure 24–3. All of the electrons within the shaded volume will pass the plane perpendicular to the wire at P in a time Δt. The length $\Delta\ell$ of the shaded volume is

$$\Delta\ell = v_d\,\Delta t \tag{24–2}$$

where v_d is the average drift velocity of the electrons. The total charge within the volume $A\,\Delta\ell$ is

$$\Delta q = neA\,\Delta\ell \tag{24–3}$$

where n is the number of conduction electrons per unit volume moving along the wire, e is the charge on the electrons, and A is the cross-sectional area of

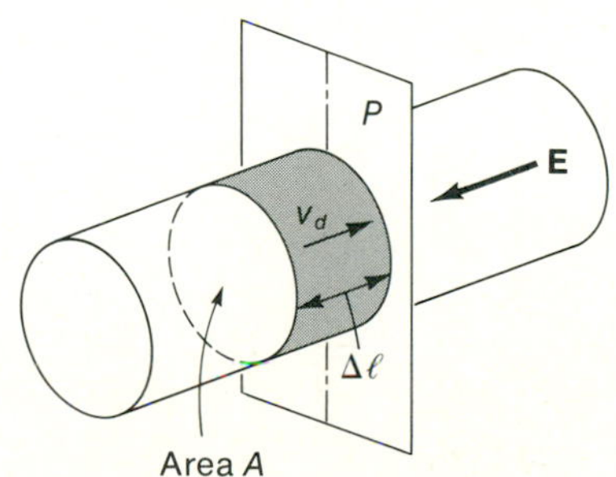

Figure 24–3
All the electrons in the shaded volume drift past the plane P in a time $\Delta\ell/v_d$.

the wire. Combining Equations (24–2) and (24–3), we obtain the amount of charge Δq passing a given point per unit time Δt:

$$\frac{\Delta q}{\Delta t} = nev_d A \qquad \textbf{(24–4)}$$

The amount of charge per second passing through a cross section of a conductor is called an electric *current*. The symbol used for a constant current is I. Upper-case (capital) letters are used for constant values of current or voltage, and lower-case letters signify varying values. The **electric current** I is defined as:

ELECTRIC CURRENT I

$$I = \frac{\Delta q}{\Delta t} \qquad \textbf{(24–5)}$$

The unit of current is *coulomb/second* (C/s), also called the *ampere* (A).[2] (*Milliamperes*, mA $= 10^{-3}$ A, and *microamperes*, μA $= 10^{-6}$ A, are also commonly used.) Combining Equations (24–4) and (24–5), we have

$$I = nev_d A \qquad \textbf{(24–6)}$$

(Be careful not to confuse the area A with the abbreviation for ampere: A.)

EXAMPLE 24–1

Calculate the average drift speed of electrons traveling through a copper wire with a cross-sectional area of 1 mm^2 when carrying a current of 1 A. It is known that about one electron per atom of copper contributes to the current flow. The atomic weight of copper is 63.54, and its density is 8.92 g/cm^3.

SOLUTION

We first calculate n, the number of current-carrying electrons per unit volume in copper.

$$n = \left(\frac{\text{electrons}}{\text{atom}}\right)\left(\frac{\text{atoms}}{\text{mol}}\right)\left(\frac{1}{\left(\frac{\text{mass}}{\text{mol}}\right)}\right)\left(\frac{\text{mass}}{\text{volume}}\right)$$

$$n = \left(1\,\frac{\text{electron}}{\cancel{\text{atom}}}\right)\left(6.02 \times 10^{23}\,\frac{\cancel{\text{atoms}}}{\cancel{\text{mol}}}\right)\left(\frac{1}{63.54\,\frac{\cancel{\text{g}}}{\cancel{\text{mol}}}}\right)\left(8.92\,\frac{\cancel{\text{g}}}{\cancel{\text{cm}^3}}\right)\left(\frac{10^6\,\cancel{\text{cm}^3}}{1\text{ m}^3}\right)$$

$$n = 8.45 \times 10^{28}\,\frac{\text{electrons}}{\text{m}^3}$$

From Equation (24–6), we obtain

$$v_d = \frac{I}{neA}$$

$$v_d = \frac{1\text{ A}}{\left(8.45 \times 10^{28}\,\frac{\text{electrons}}{\text{m}^3}\right)\left(1.602 \times 10^{-19}\,\frac{\text{C}}{\text{electron}}\right)(10^{-6}\text{ m}^2)}$$

$$v_d = \boxed{7.39 \times 10^{-5}\,\frac{\text{m}}{\text{s}}}$$

or less than 0.1 millimeter per second!

[2] This unit honors André Ampère (1775–1836), a French physicist who gained considerable knowledge of the magnetic effects of currents.

You were probably surprised at the slow drift speed calculated in the example. If electrons typically travel along a wire at such a slow speed, why is it that when we flip a wall switch, the light goes on almost instantaneously? The reason is that when a circuit is connected to a source of EMF, the electric field is established in all parts of the circuit at nearly the speed of light. So when the final connection is made, forming a complete, closed path with the battery, electrons start to flow more or less simultaneously in all parts of the circuit. Even though the average drift speed of each electron is slow, all parts of the circuit feel the effects of the current almost instantaneously. Also, even though the cross-sectional area may vary along the length of the wire (thus causing the drift speed to vary), the current has the same numerical value I throughout the circuit.

The conduction electrons also take part in another motion. These electrons behave somewhat like the molecules of a gas, with random thermal velocities between collisions, whose average velocity (for copper at room temperature) is about 1.6×10^6 m/s. So a typical current in a wire consists of random conduction-electron velocities of more than a million meters per second, upon which is superposed a slow drift speed of much less than a millimeter per second.

Let us examine each factor of Equation (24–6) to determine which are intrinsic properties of the current-carrying material and which are determined by the potential difference across the material. Equation (24–6) states that

$$I = nev_dA$$

where e is constant and A is a geometrical parameter of the material. Of the other factors, the number of current-carrying charges per unit volume n is certainly an intrinsic property. The factor v_d is, in part, also an intrinsic property, since it depends on the mobility of the charges through the material. However, it also depends on the force driving the charges through the material (that is, the electric field within the material).

It is convenient to write another form of Equation (24–6) by considering a definite length of conducting material with a constant cross-sectional area, across which a potential difference is applied, as shown in Figure 24–4. The potential difference $V = V_2 - V_1$ can be related to the field $\mathbf{E}$ of the material by

$$V = -\int_1^2 \mathbf{E} \cdot d\ell$$

Since the material is uniform, the integral reduces to $E\ell$, or

$$E = \frac{V}{\ell} \tag{24–7}$$

The drift speed v_d is proportional to E:

$$v_d \propto E \tag{24–8}$$

Rewriting Equation (24–6) using the above two equations gives

$$I \propto \frac{VA}{\ell}$$

The constant of proportionality depends on the intrinsic properties of the conducting material. Defining a constant of proportionality $1/\rho$, we have

$$I = \left(\frac{1}{\rho}\right)\left(\frac{A}{\ell}\right)V \tag{24–9}$$

where ρ is called the **resistivity** of the material. From Equation (24–9), we know that the units associated with ρ are (volts/ampere)·meter, or (V/A)·m. The unit

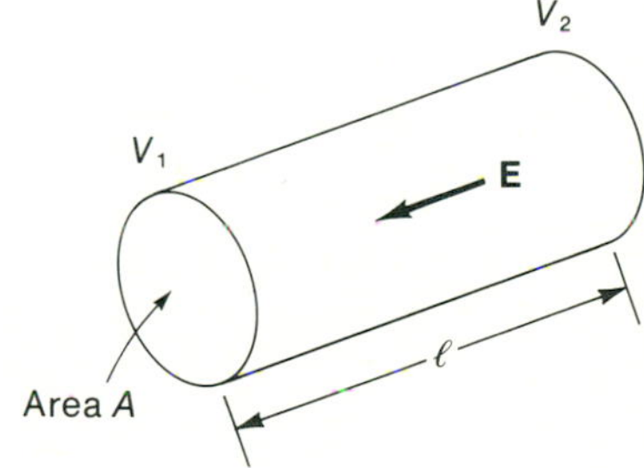

Figure 24–4
A uniform conducting material of constant cross-sectional area A and length ℓ.

volt/ampere (V/A) is called an **ohm** (Ω, the Greek capital letter *omega*). The unit of resistivity is thus the *ohm·meter* (Ω·m).

We have discussed resistivity in terms of a *constant* of proportionality. In reality, the resistivity of a given material depends on a number of factors, such as moisture content, pressure, crystalline structure, and temperature. Analytically, temperature dependence is most easily handled. It is known from experiment that the fractional change in resistivity is approximately proportional to the corresponding change in temperature. That is:

$$\frac{\rho - \rho_0}{\rho_0} = \alpha(T - T_0) \tag{24–10}$$

where α is the constant of proportionality called the **thermal coefficient of resistivity.** Equation (24–10) is often written in the more convenient form:

CHANGE OF RESISTIVITY WITH TEMPERATURE

$$\rho = \rho_0[1 + \alpha(T - T_0)] \tag{24–11}$$

Values of ρ and α are given in Table 24–1 for a few common substances.

Materials are often classified according to their resistivity, as shown in Table 24–1. The solid-state electronics industry has been developed largely through techniques of fabricating semiconductor devices. Such devices are rarely a pure element such as germanium or silicon, but are modified with traces of other elements to enhance their conductivity. (Not listed in Table 24–1 are a number of alloys and a few elements called *superconductors*, which exhibit zero resistivity at temperatures near absolute zero.)

TABLE 24–1

Resistivities and Thermal Coefficients of Resistivity

Material	Resistivity ρ at 20°C (Ω·m)	Thermal coefficient of resistivity α (1/°C)
Insulators		
Beeswax (white)	5×10^{12}	—
Mica (clear)	2×10^{15}	-50×10^{-3}
Sulfur	1×10^{15}	-80×10^{-3}
Glass (plate)	2×10^{11}	-70×10^{-3}
Semiconductors		
Carbon (graphite)	1.4×10^{-5}	-0.5×10^{-3}
Conductors		
Aluminum	2.8×10^{-8}	3.9×10^{-3}
Bronze	18×10^{-8}	0.5×10^{-3}
Copper	1.6×10^{-8}	3.9×10^{-3}
Gold	2.4×10^{-8}	3.4×10^{-3}
Mercury	96×10^{-8}	0.8×10^{-3}
Nichrome (trademark)	100×10^{-8}	0.4×10^{-3}
Iron	10×10^{-8}	5×10^{-3}
Silver	1.6×10^{-8}	3.8×10^{-3}
Manganin {84% Cu, 12% Mn, 4% Ni}	44×10^{-8}	$<0.0005 \times 10^{-3}$

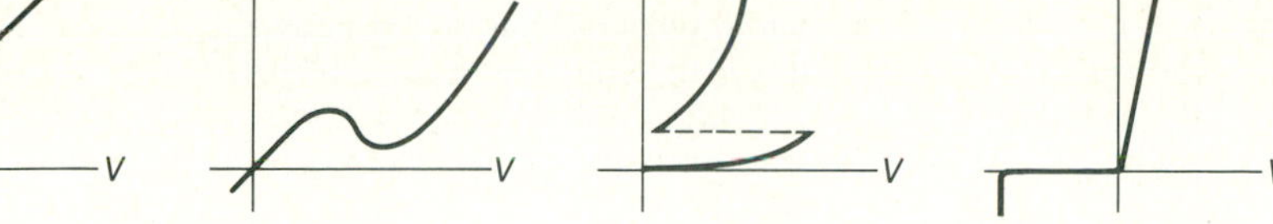

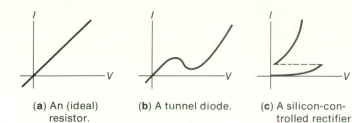

(**a**) An (ideal) resistor. (**b**) A tunnel diode. (**c**) A silicon-controlled rectifier. (**d**) A zener diode.

Figure 24–5
Relationships between I and V for some common resistive devices. Only the ideal resistor in (a) obeys Ohm's law.

24.3 Ohm's Law

Equation (24–9) implies that if a bar of resistive material were fitted with terminals at each end, the current through the bar would be proportional to the potential difference between the terminals. Surprisingly, this is nearly the case for a wide variety of substances. If the proportionality is exact, the substance conforms to **Ohm's law.** From Equation (24–9), we obtain

OHM'S LAW
$$V = IR \tag{24–12}$$

where R is a constant called the *resistance*, which is independent of the current I through the substance. Comparing Equations (24–9) and (24–12) leads to

$$R = \frac{\rho \ell}{A} \tag{24–13}$$

for a resistive substance of uniform cross-sectional area A and a length ℓ between the terminals. Based on the units already assigned to resistivity, the unit for resistance[3] is *ohms* (Ω).

The functional relationship between the potential difference across the terminals of a resistive device and the current through it is not necessarily a linear one. Figure 24–5 describes relationships between I and V for a number of devices. Of those shown, only the (ideal) resistor in (a) obeys Ohm's law, since there the value of R does not depend on the current. For all the others, R is still defined by the ratio of V to I, but it varies as the current changes. We will restrict our discussion to those resistive devices that obey Ohm's law.

EXAMPLE 24–2

A resistor is constructed of a carbon rod that has a cross-sectional area of 5 mm^2. When a potential difference of 15 V is applied across the ends of the rod, a current of 4×10^{-3} A passes through the rod. Find (a) the resistance of the rod and (b) the rod's length.

SOLUTION
(a) Applying Ohm's law (Equation 24–12), we find the resistance of the rod:

$$R = \frac{V}{I}$$

$$R = \frac{15 \text{ V}}{4 \times 10^{-3} \text{ A}} = 3750 \ \Omega$$

or
$$R = \boxed{3.75 \text{ k}\Omega}$$

[3] This unit is in recognition of Georg Ohm, a German physicist who in 1827 discovered the proportionality between current and potential difference.

where kΩ designates the *kilohm* (kΩ = 10^3 Ω). Similarly, MΩ is used for the *megohm* (MΩ = 10^6 Ω). Note that if R is written in units of kilohm and potential difference is written in terms of volts, then the current is in milliamperes (mA = 10^{-3} A), a more practical unit in modern electronics.

(b) The length of the rod may be determined from Equation (24–13):

$$R = \frac{\rho \ell}{A}$$

or

$$\ell = \frac{RA}{\rho}$$

Substituting numerical values for R, A, and the values of ρ given for carbon in Table 24–1, we obtain:

$$\ell = \frac{(3.75 \times 10^3\ \Omega)(5 \times 10^{-6}\ \text{m}^2)}{(1.4 \times 10^{-5}\ \Omega\cdot\text{m})}$$

$$\ell = \boxed{1.34 \times 10^3\ \text{m}}$$

Obviously, a resistor of 3.75 kΩ (a rather common value) could not be constructed of pure carbon and still be part of a miniaturized electronic circuit. Resistors are constructed of a mixture of materials that is engineered not only to have the desired resistance and size, but also to contribute to its physical strength and constancy of resistance value under a variety of environmental conditions.

24.4 Joule's Law

We have seen that the potential difference across a resistor forces electrons through the resistor, with the electrons emerging with the same drift velocity they had when they entered. Recall that the potential energy lost by the electrons appears as thermal energy within the resistor. (An analogous situation is that of of a boat driven at constant speed through the water by a motor; energy is dissipated in the water, heating it slightly.) For resistors, this effect is called *Joule heating.*

To calculate the Joule heating in a resistor, consider a simple circuit of a seat of EMF and a resistor, as shown in Figure 24–6. The symbol —ww— is used to describe resistors that obey Ohm's law and solid lines indicate resistanceless conductors of current. Consistent with the usage introduced by Benjamin Franklin, as well as that used today, *outside a seat of EMF, the current I always flows from a point of higher potential to one of lower potential.* This is sometimes called **conventional current.** Of course, conduction electrons in metals move in the opposite direction because of their negative charge. On the other hand, positive charges contribute to current flow in many substances, including liquids, gases, and certain solid-state devices. So the direction *positive* charges would flow (whether or not such charges are actually present) is the direction of the conventional current.

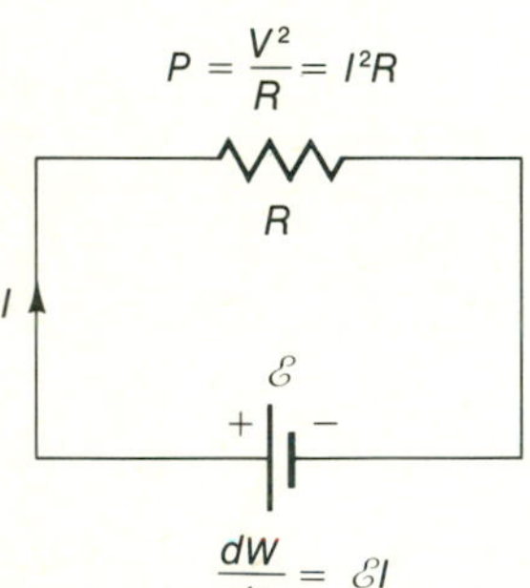

Figure 24–6
The work done by the seat of EMF appears as thermal energy in the resistor.

As we have done before, we will isolate the system consisting of the seat of EMF and the resistor, so that no energy enters or leaves the system. Because energy is conserved within the system, we know that the energy acquired by the charges through the work done on them as they move through the seat of EMF must be equal to the thermal energy developed in the resistor. The work done by the seat of EMF per unit charge is given in Equation (24–1):

$$\mathcal{E} = \frac{dW}{dq}$$

or

$$dW = \mathcal{E}\, dq$$

The time rate at which work is done by the seat of EMF is

$$\frac{dW}{dt} = \mathscr{E}\frac{dq}{dt}$$

or, by the definition of current (Equation 24–5), we have

$$\frac{dW}{dt} = \mathscr{E}I \tag{24–14}$$

It is important to notice that while $\mathscr{E}$ is the work done per unit charge by the seat of EMF, the charge acquires a corresponding increase in potential energy, that is, a potential energy per unit charge (V), while moving through the seat of EMF. The potential difference across the terminals of the seat of EMF is therefore V, the same as that across the resistor R. Conservation of energy requires that the rate at which work is done (dW/dt) by the seat of EMF must equal the rate at which thermal energy is developed in the resistor. (Because this thermal energy is usually radiated away and thus "disappears" from the circuit, one often uses the phrase "dissipated in the resistor" for this thermal energy.) Using the symbol P for the power dissipated in the resistor, we have

$$\frac{dW}{dt} = P$$

or, since $\mathscr{E} = V$:

$$P = VI \tag{24–15}$$

This equation may be stated in another way. Using Ohm's law:

$$V = IR$$

Equation (24–15) becomes

$$P = I^2R \tag{24–16}$$

Indeed, the power dissipated in resistors is often referred to as "I squared R" losses. Ohm's law also leads to $P = V^2/R$, so we have:

POWER *P* DISSIPATED IN RESISTIVE CIRCUIT ELEMENTS

$$P \begin{cases} = VI \\ = I^2R \\ = \dfrac{V^2}{R} \end{cases} \tag{24–17}$$

EXAMPLE 24–3

Referring to Figure 24–6, if $\mathscr{E} = 6$ V and $R = 12\ \Omega$, find (a) the rate at which the seat of EMF does work and (b) the power dissipated in the resistor.

SOLUTION

Conservation of energy requires that the answers to parts (a) and (b) be the same. In both cases, the current must be calculated first. Using Ohm's law (Equation 24–12) and realizing that $\mathscr{E}$ has the same value as the potential difference across the resistor:

$$\mathscr{E} = IR$$

$$I = \frac{\mathscr{E}}{R}$$

Substituting the numerical values, $\mathscr{E} = 6$ V and $R = 12\ \Omega$, we have

$$I = 0.5 \text{ A}$$

Superconductivity

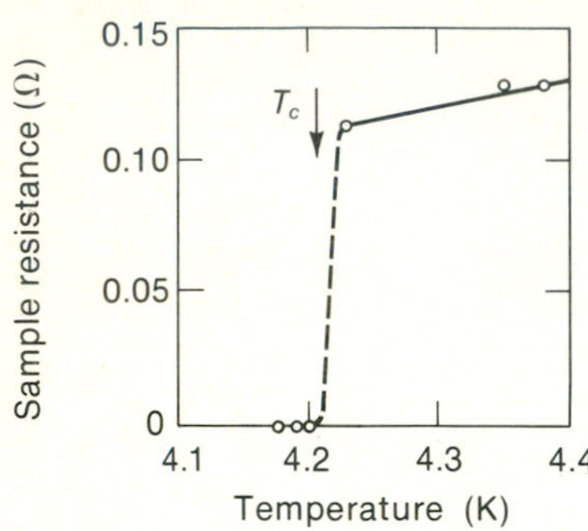

(**a**) The resistance of a sample of mercury versus temperature showing the sudden drop in resistance at the critical temperature T_c. Modern experiments indicate a measured resistivity for the superconducting state of no more than 10^{-25} Ω·m; it may well be truly zero.

(**b**) Current set up in a superconducting ring and ball produces magnetic forces which levitate the ball in space. The currents persist for many years without measurable change.

Figure 24–7
Examples of superconductivity.

In 1908, the Dutch physicist H. Kammerlingh Onnes succeeded in liquefying helium at 4.2 K *and began to investigate various properties of metals at low temperatures. Three years later he discovered the phenomenon of* superconductivity, *the astonishing behavior of some metals in which the electrical resistance drops abruptly to zero below a certain* critical temperature, *usually a few degrees above absolute zero. The consequences of zero resistance are surprising. For example, if a current is established around a superconducting ring, it continues to flow indefinitely without the production of heat (there are no* I^2R *losses) and with no driving* EMF *necessary! Circulating currents have been set up in superconducting rings that persist by themselves for years with no measurable loss. At least 24 metallic elements have been found to exhibit the phenomenon, as well as over 1000 alloys. One of the highest critical temperatures to date is that of a niobium compound*: 20.7 K.

Superconductivity is not fully understood, though it is known to be a quantum mechanical phenomenon. The search for compounds that become superconductors

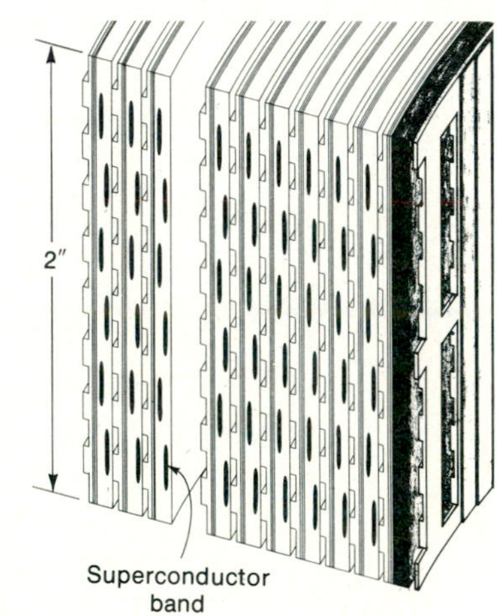

(**c**) The coil windings of this magnet at Brookhaven National Laboratory were one of the first large-scale applications of superconductors. Six thin bands of superconducting material were embedded within each copper conductor. At liquid helium temperature, the six bands carry 5880 A. Later magnet designs increase the flexibility of the windings by dividing the superconductor into many small filaments, with a single conductor carrying as much as 100 000 A in a magnetic field of 8 T.

The rate at which work is done by the seat of EMF is given by Equation (24–14):

$$\frac{dW}{dt} = \mathcal{E}I$$

Substituting the appropriate values gives

$$\frac{dW}{dt} = (6\ \text{V})(0.5\ \text{A}) = 3\ \frac{\text{J}}{\text{s}} = \boxed{3.00\ \text{W}}$$

The power dissipated in the resistor, although equal to dW/dt, may also be obtained from Equation (24–16):

$$P = I^2R$$

$$P = (0.5\ \text{A})^2(12\ \Omega) = \boxed{3.00\ \text{W}}$$

at still higher temperatures has far-reaching implications because about half the electrical power generated in the United States is lost as I^2R *losses in transmission lines. If substances with higher critical temperatures could be found, superconducting cables for power transmission could have tremendous economic advantages in spite of the extra refrigeration equipment necessary. Such cables could also be used as transmission lines where extremely small signals are involved: for example, between radio telescopes and amplifiers. At the present time, superconducting electromagnets are in use in many scientific laboratories. In some cases, the magnet windings carry many tens of thousands of amperes without troublesome heating effects.*

An interesting example of the use of superconductivity is the superconducting gravimeter (Figure 24–8) for remarkably precise measurements of the earth's gravitational field. An aluminum shell, 2.54 cm *in diameter, is plated with lead, which becomes superconducting at the temperature of liquid helium,* 4.2 K. *The sphere is supported in midair by establishing currents in two horizontal superconducting coils. As the coil currents build up, they produce an increasing magnetic field, which induces currents in the sphere. The resultant magnetic forces levitate the sphere in space without its physically touching any supports. The sphere is positioned symmetrically between six metal plates. If the value of* g *changes, the sphere will rise or fall vertically by a tiny amount. Its altered position changes the capacitance between the plates, which is sensed by external electrical circuits, thereby indicating a change in* g. *The entire instrument is surrounded by a superconducting shield in liquid helium to minimize spurious magnetic field changes, and the sensing apparatus itself is placed in a vacuum to control the temperature to within a few microkelvins. Figure 14–8 shows a graph of variations in the earth's field obtained with this instrument.*

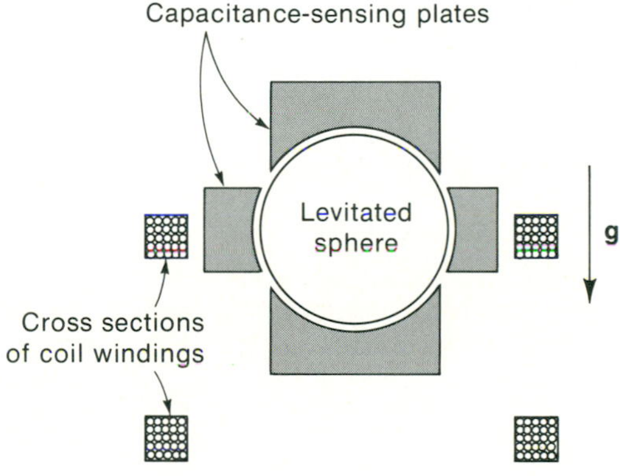

Figure 24–8
A superconducting gravimeter employs two horizontal current loops that levitate a sphere between capacitance-sensing plates. Two additional plates above and below the plane of the figure are not shown.

24.5 Resistors in Series and in Parallel

The combination of two or more resistors connected *in series*, as in Figure 24–9(a), is equivalent to a single resistance R whose value can be found from the following analysis. The potential difference V across the combination is the sum of the potential differences across each resistor:

$$V = V_1 + V_2$$

Each has the same current I, so from Ohm's law ($V = IR$), we have

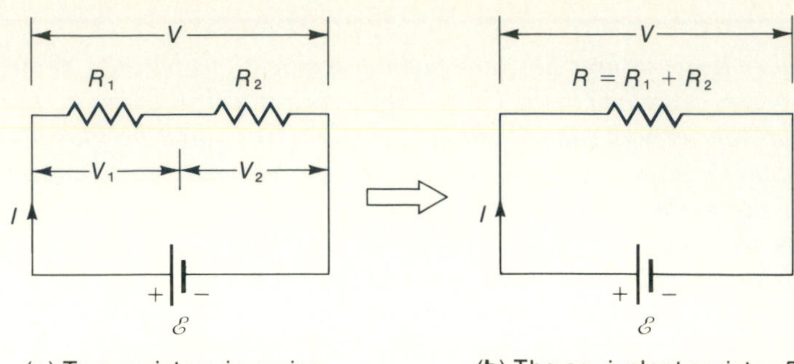

(a) Two resistors in series.
(b) The equivalent resistor R.

Figure 24–9
The combined resistance of two resistors in series is the sum of the two individual resistances.

$$IR = IR_1 + IR_2$$

or

$$R = R_1 + R_2$$

If more than two resistors are connected in series, a similar reasoning shows that the equivalent single resistance R is

RESISTORS IN SERIES

$$R = R_1 + R_2 + R_3 + \cdots \qquad \textbf{(24–18)}$$

The combination of two or more resistors connected *in parallel*, as in Figure 24–10(a), is equivalent to a single resistance R whose value can be found by recognizing that at point a, the current I splits into two parts: I_1 through R_1 and I_2 through R_2. From the **conservation of charge** we conclude that the rate $dq/dt = I$ at which charge enters point a equals the rate at which charge leaves (since no charge accumulates at point a as time goes on). Thus:

$$I = I_1 + I_2 \qquad \textbf{(24–19)}$$

From Ohm's law ($I = V/R$), we have

$$\frac{V}{R} = \frac{V_1}{R_1} + \frac{V_2}{R_2} \qquad \textbf{(24–20)}$$

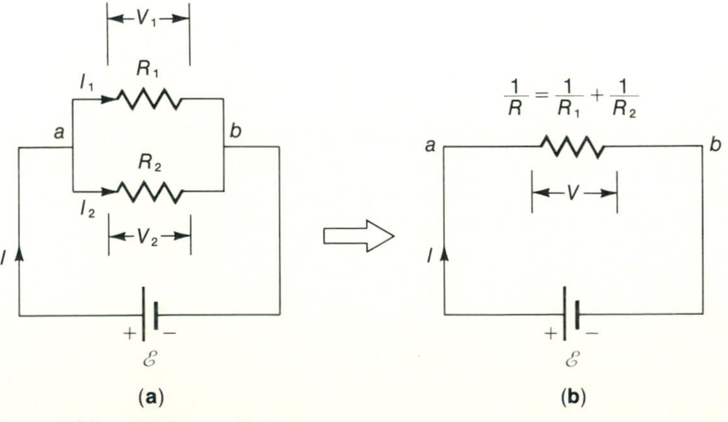

(a)
(b)

Figure 24–10
The resistance of two resistors in parallel.

Because both resistors are connected between the same two points, a and b, the potential difference across each of the resistors is the same value V:

$$V_1 = V_2 = V$$

Therefore, Equation (24–20) becomes

$$\frac{V}{R} = \frac{V}{R_1} + \frac{V}{R_2}$$

Dividing by V, we have

$$\frac{1}{R} = \frac{1}{R_1} + \frac{1}{R_2}$$

When more than two resistors are connected in parallel between the same two junction points, a similar analysis gives

RESISTORS IN PARALLEL

$$\frac{1}{R} = \frac{1}{R_1} + \frac{1}{R_2} + \frac{1}{R_3} + \cdots \qquad \textbf{(24–21)}$$

for the equivalent resistance R. Note that the equivalent resistance of a parallel combination is always *less* than any of the individual resistances alone. Also, it is helpful to remember that resistors add in parallel the way that capacitors add in series, and vice versa.

EXAMPLE 24–4

Find the equivalent resistance of the resistor network shown in Figure 24–11(a).

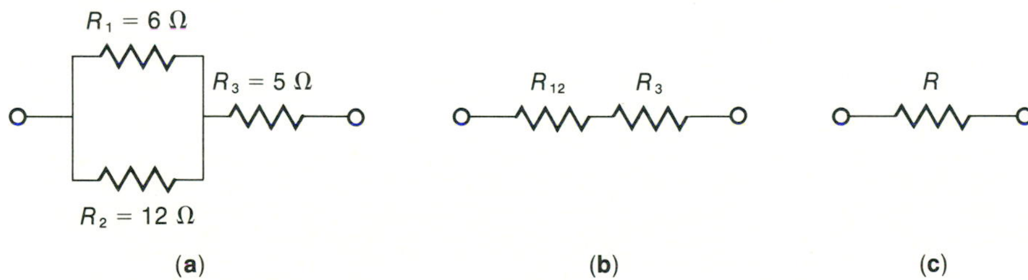

Figure 24–11
Example 24–4

SOLUTION

Usually the best procedure is to combine groups of parallel resistors to form a single equivalent resistor and groups of series resistors to form a single equivalent resistor. These combinations can then be combined further to reduce the entire network to a single equivalent resistor. In this example, we will combine R_1 and R_2 to form a single resistor R_{12}. Since they are in *parallel*, we utilize Equation (24–21):

$$\frac{1}{R_{12}} = \frac{1}{R_1} + \frac{1}{R_2}$$

Substituting the appropriate values gives

$$\frac{1}{R_{12}} = \frac{1}{6\ \Omega} + \frac{1}{12\ \Omega}$$

$$R_{12} = 4\ \Omega$$

We next combine R_{12} and R_3 as shown in Figure 24–11(b). Since these are in *series*:

$$R = R_{12} + R_3$$

$$R = 4\ \Omega + 5\ \Omega = \boxed{9.00\ \Omega}$$

EXAMPLE 24–5

Three 60 W, 120 V light bulbs are connected across a 120 V power source, as shown in Figure 24–12. Find (a) the total power dissipation in the three light bulbs and (b) the voltage across each of the bulbs. Assume that the resistance of each bulb conforms to Ohm's law (even though in reality the resistance increases markedly with current).

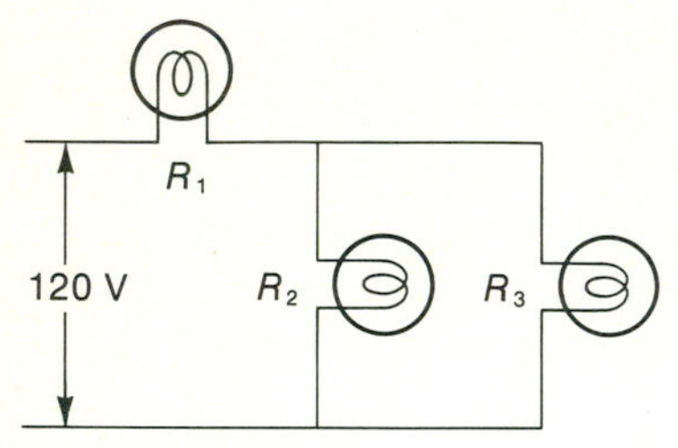

Figure 24–12
Example 24–5

SOLUTION

(a) The first step is to determine the resistance of each light bulb. From Equation (24–17):

$$P = \frac{V^2}{R}$$

$$R = \frac{(120\ \text{V})^2}{60\ \text{W}} = 240\ \Omega$$

The resistance of the network of light bulbs is obtained by applying Equations (24–18) and (24–21):

$$R = R_1 + \frac{1}{\left(\dfrac{1}{R_2} + \dfrac{1}{R_3}\right)}$$

$$R = 240\ \Omega + 120\ \Omega = 360\ \Omega$$

The total power dissipated in the equivalent resistance of 360 Ω is

$$P = \frac{V^2}{R}$$

$$P = \frac{(120\ \text{V})^2}{360\ \Omega} = \boxed{40.0\ \text{W}}$$

(b) The current through the network is given by Equation (24–16):

$$P = I^2R$$

Solving for I:

$$I = \sqrt{\frac{40\ \text{W}}{360\ \Omega}} = \tfrac{1}{3}\ \text{A}$$

The potential difference across R_1 is:

$$V_1 = IR_1$$

$$V_1 = (\tfrac{1}{3}\ \text{A})(240\ \Omega) = \boxed{80.0\ \text{V}}$$

The potential difference V_{23} across the parallel combination of R_2 and R_3 is:

$$V_{23} = IR_{23}$$

$$V_{23} = (\tfrac{1}{3}\ \text{A})\left(\frac{1}{\dfrac{1}{240\ \Omega} + \dfrac{1}{240\ \Omega}}\right) = \boxed{40.0\ \text{V}}$$

24.6 Multiloop Circuits and Kirchhoff's Rules

The most interesting as well as practical circuits are those that contain more than a single closed path or loop and that contain more than one seat of EMF. An example of such a circuit is shown in Figure 24–13. Ordinarily the circuit parameters are given; in this case, they are $\mathscr{E}_1$, $\mathscr{E}_2$, R_1, R_2, and R_3. The desired quantity to be calculated may be the potential difference across a particular portion of the circuit or the current through a part of the circuit. In most cases, the current must first be found in one or all of the circuit branches.

Utilizing the two basic conservation laws, conservation of energy and conservation of charge, we will formulate a procedure for finding the current in every branch of any circuit consisting of resistors and seats of EMF. The conservation laws are often formulated into what are known as *Kirchhoff's rules*:

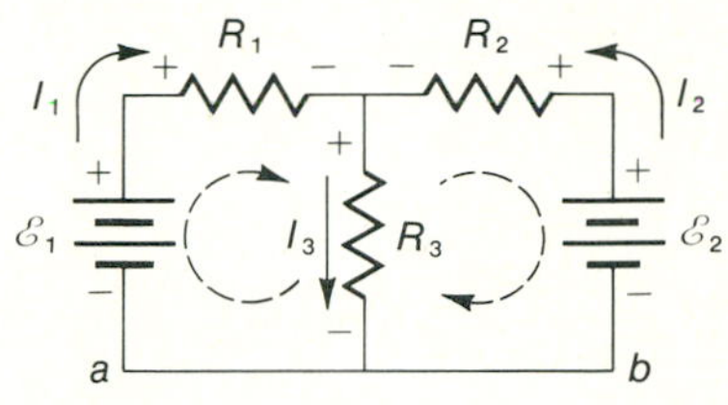

Figure 24–13
A multiloop circuit.

KIRCHHOFF'S RULES

(1) Conservation of energy (loop rule): **The sum of the voltage increases and decreases around any closed loop is zero.**

(2) Conservation of charge (branch rule): **The algebraic sum of all currents into and out of a junction is zero. (Currents entering a junction are positive and currents leaving are negative.)**

The application of these rules is most easily accomplished by following a rather formal procedure. We will illustrate by solving for the currents in the circuit of Figure 24–13. Here are the steps:

(1) *Label the polarity of each seat of EMF.* Notice that in the circuit shown, the seats of EMF in the outside loops oppose each other. (Thus, one EMF may be able to force current through the other EMF in the "backward" direction.)
(2) *Indicate a current direction in each branch of the circuit.* If you can guess ahead of time which direction is reasonable, choose it. If not, assign the current in some direction. (If a wrong guess is made, the numerical answer for that current will be a *negative* number, indicating that the actual current is in the opposite direction.)
(3) *According to the direction assumed for each current, label each resistor with a "+" at the end with the higher potential, and a "−" at the other end.* Note that current flows through a resistor from a higher to a lower potential. Thus the end at which the current *enters* has a "+" label.
(4) *Establish a direction for traveling around each individual loop.* In this example, we will traverse each loop in a *clockwise* direction, as indicated by the dashed circular arrows. (The directions are arbitrary: We could have chosen a counterclockwise direction for either or both loops.) There is a third loop path around the outer branches. But as we will see, having chosen the other two loops, this third path is redundant, so it need not be considered. The only criterion that must be met is that *every branch be traversed at least once by a loop path.*
(5) *Starting at any convenient point, travel around each independent loop in the direction chosen, keeping track of the potential increases and decreases. (Increases are positive; decreases are negative.) Equate the sum to zero.* For the circuit chosen, we start at point *a* in the left-hand loop and point *b* in

the right-hand loop, obtaining the following equations:

$$\mathscr{E}_1 - I_1R_1 - I_3R_3 = 0 \tag{24-22}$$

$$I_3R_3 + I_2R_2 - \mathscr{E}_2 = 0 \tag{24-23}$$

Notice that if we add these equations, we obtain the equation for the loop going clockwise around the outer branches. Therefore, traversing the outer loop adds no new information in the solution; it is redundant.

(6) *Equate the sum of the currents entering each independent junction to zero. (Currents that enter are positive and currents that leave are negative.)* For the circuit shown, at the upper junction we obtain:

$$I_1 + I_2 - I_3 = 0 \tag{24-24}$$

The lower junction would yield the same equation except for the sign. So it is not an *independent* junction.

We now have *three* equations: Equations (24–22), (24–23), and (24–24), and *three* unknowns: I_1, I_2, and I_3. To organize the solution, it is helpful to rewrite these equations in a "standard" format, aligning terms for each unknown in vertical columns (and adding zeros for missing terms):

$$-R_1I_1 + 0 - R_3I_3 = -\mathscr{E}_1 \tag{24-25}$$

$$0 + R_2I_2 + R_3I_3 = \mathscr{E}_2 \tag{24-26}$$

$$I_1 + I_2 - I_3 = 0 \tag{24-27}$$

From this point on, any of the usual methods for solving simultaneous equations may be used.[4] The solutions are:

$$I_1 = \frac{(R_2 + R_3)\mathscr{E}_1 - R_3\mathscr{E}_2}{-R_1R_2 + R_1R_3 + R_2R_3} \tag{24-28}$$

$$I_2 = \frac{(R_1 + R_3)\mathscr{E}_2 - R_3\mathscr{E}_1}{-R_1R_2 + R_1R_3 + R_2R_3} \tag{24-29}$$

$$I_3 = \frac{R_1\mathscr{E}_2 + R_2\mathscr{E}_1}{-R_1R_2 + R_1R_3 + R_2R_3} \tag{24-30}$$

If any of the currents is opposite to the direction we initially guessed, the value of I for that current will turn out to be negative when the numerical values are substituted for the network parameters in the equation.

24.7 The Superposition Principle

Provided circuit elements are *linear*—that is, resistors and EMFs maintain their values regardless of the amount of current flow—we may solve the circuit of Figure 24–13 by using the *principle of superposition*. This general procedure recognizes that the effect of just one EMF working alone is independent of the effects produced by other EMFs. So we may pretend that all EMFs are absent except one, replacing the absent ones by conductors whose resistance is zero, and then solving for the currents in the circuit due to the single remaining EMF. Repeating the procedure in turn for the other EMFs gives other sets of currents due to each of them working separately. The actual currents in the original

[4] The actual calculation is somewhat tedious. Perhaps the most convenient procedure is the *determinant method* (Cramer's rule) for solving linear algebraic equations. Consult a mathematics text for details.

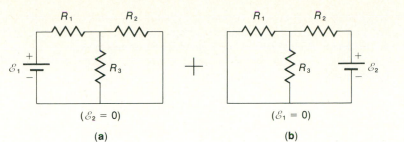

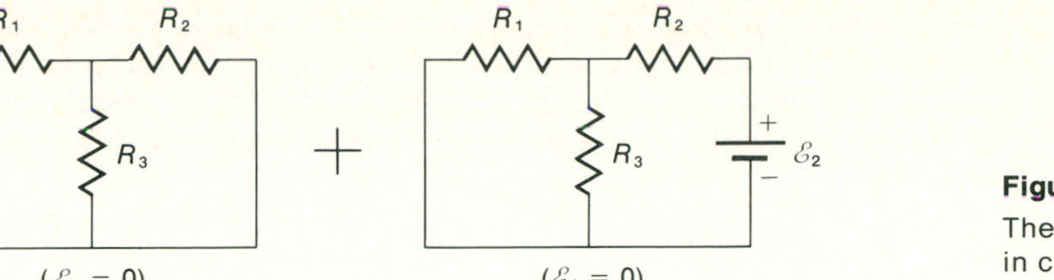

Figure 24–14
The superposition of the currents in circuits (a) and (b) gives the currents in the circuit shown on Figure 24–13.

network are the sum, or *superposition*, of these sets of partial currents. Figure 24–14 illustrates the procedure for a network with two EMFs. Often this procedure is simpler than the brute-force solving of simultaneous equations.

EXAMPLE 24–6

In Figure 24–15(a), find the current in each branch of the network.

SOLUTION

Method 1: Kirchhoff's rules

We choose currents in each branch as indicated in Figure 24–15(a). The polarities for the (assumed) potential differences across each resistor are labeled with plus and minus signs (remembering that a current *enters* a resistor at the positive end). Starting at the bottom junction, we travel around the loops, equating the sum of the potential increases and decreases to zero (Kirchhoff's first rule). Omitting the units for simplicity, we obtain:

Left loop (clockwise) $\quad 10 - 2I_1 - 4I_3 = 0$

Right loop (counterclockwise) $\quad 4 - 4I_3 = 0$

We next equate the sum of all currents entering the top junction to zero (Kirchhoff's second rule). Currents entering are positive, currents leaving are negative.

$$I_1 + I_2 - I_3 = 0$$

Rewriting the above equations in the standard format:

$$\begin{aligned} -2I_1 + 0 - 4I_3 &= -10 \\ 0 + 0 - 4I_3 &= -4 \\ I_1 + I_2 - I_3 &= 0 \end{aligned}$$

These simultaneous equations are simple to solve by direct algebraic substitution. Substituting the value for I_3 from the second equation into the other two, we obtain:

$$\boxed{I_1 = 3.00 \text{ A}}$$

$$\boxed{I_2 = -2.00 \text{ A}}$$

$$\boxed{I_3 = 1.00 \text{ A}}$$

The minus sign for I_2 signifies the current in that branch is actually opposite to the direction assumed.

Method 2: Principle of superposition

With this method, we successively replace each of the EMFs by a conductor with zero resistance and solve for the currents due to the other EMF, obtaining the two simplified circuits in Figures 24–15(b) and (c). To indicate that the currents in these circuits are only partial currents, we use single and double primes. We here assume currents in each branch

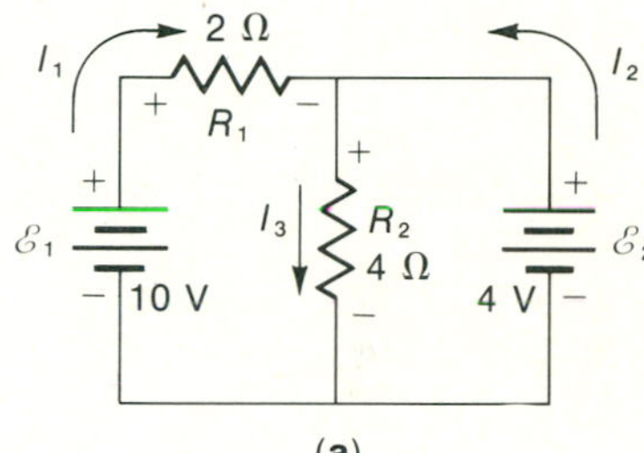

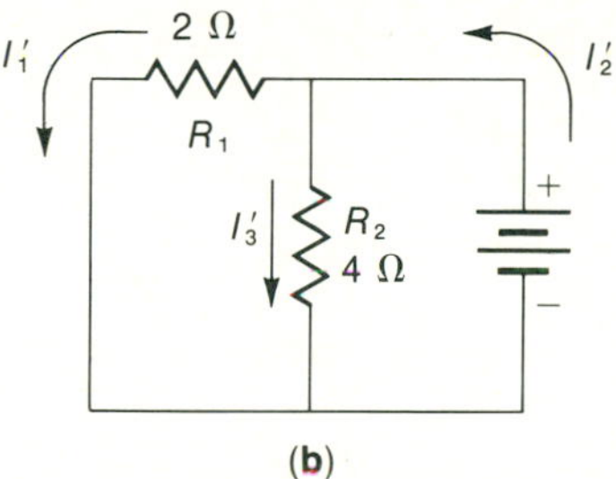

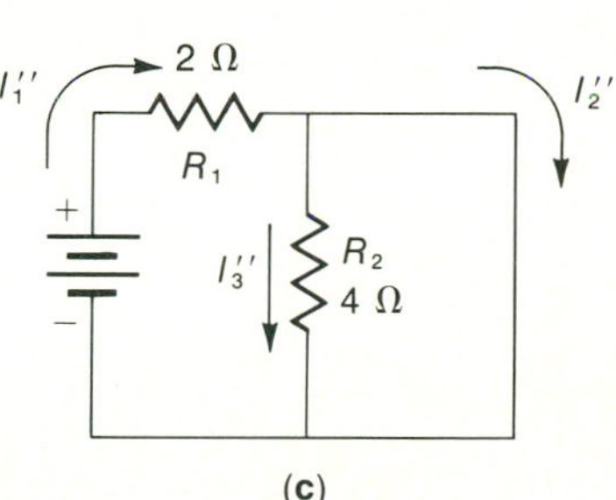

Figure 24–15
Example 24–6

that are plausible for the modified circuits. Thus, in (a), I'_1 would be the current if $\mathscr{E}_2$ were the only battery; similarly, I''_1 would be the current if $\mathscr{E}_1$ were the only battery. The actual current direction in R_1 will depend on which of these currents is larger.

Circuit (b) is simply two resistors in parallel across $\mathscr{E}_2$. From Ohm's law ($I = V/R$), the current in each is

$$I'_1 = \frac{4\ V}{2\ \Omega} = 2\ \text{A}$$

$$I'_3 = \frac{4\ V}{4\ \Omega} = 1\ \text{A}$$

From Kirchhoff's junction rule:

$$I'_2 = I'_1 + I'_3$$
$$I'_2 = 2\ \text{A} + 1\ \text{A} = 3\ \text{A}$$

In circuit (c), the conducting wire on the right-hand side has zero resistance, so all the current flows through this parallel branch and none through R_2. (We say that R_2 has been "shorted out.") Thus:

$$I''_3 = 0$$

and

$$I''_1 = I''_2 = \frac{10\ V}{2\ \Omega} = 5\ \text{A}$$

We now superpose these two sets of currents (noting their assumed directions) to find the actual currents in the original circuit. In resistor R_1, $I'_1 = 2$ A is toward the left while $I''_1 = 5$ A is toward the right. The actual current in that branch is therefore

$$I''_1 - I'_1 = 5\ \text{A} - 2\ \text{A} = \boxed{3.00\ \text{A} \quad \text{(toward the right)}}$$

In R_2, both I'_3 and I''_3 were assumed to be in the same direction. So

$$I'_3 + I''_3 = 1\ \text{A} + 0 = \boxed{1.00\ \text{A} \quad \text{(down)}}$$

In the right-hand branch, I'_2 and I''_2 were assumed to be in opposite directions, so the actual current is

$$I''_2 - I'_2 = 5\ \text{A} - 3\ \text{A} = \boxed{2.00\ \text{A} \quad \text{(down)}}$$

EXAMPLE 24–7

Calculate the potential difference between the points A and B for the circuit shown in Figure 24–16 and identify which point is at the higher potential.

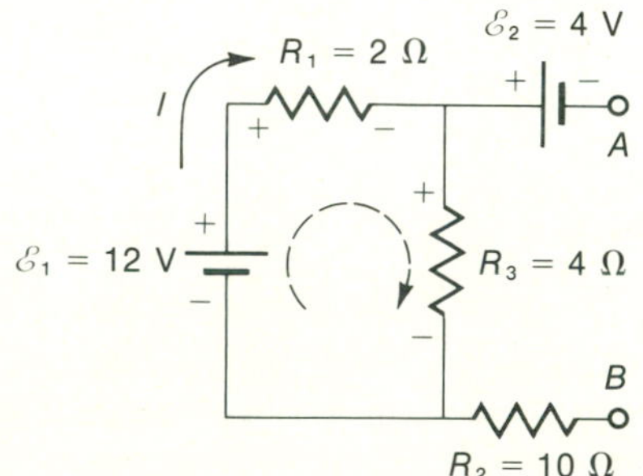

Figure 24–16
Example 24–7

SOLUTION

We identify the circuit as a *single*-loop circuit because points A and B are not connected. (No current can exist in the branches containing $\mathscr{E}_2$ and R_2. Consequently, there is no potential difference across R_2.) The only current is a clockwise one in the loop at the left. The potential difference between A and B is the sum of the potential differences across

$\mathscr{E}_2$, R_3, and R_2 (which is zero). We will first find the potential difference across R_3 by applying Kirchhoff's rules for the current in the loop. Assuming a clockwise direction, we sum the voltage increases and decreases around the loop:

$$\mathscr{E}_1 - IR_1 - IR_3 = 0$$

Solving for I and substituting numerical values:

$$I = \frac{\mathscr{E}_1}{R_1 + R_3}$$

$$I = \frac{12\text{ V}}{2\,\Omega + 4\,\Omega} = 2\text{ A}$$

The potential difference V_3 across R_3 is thus:

$$V_3 = IR_3$$

$$V_3 = (2\text{ A})(4\,\Omega) = 8\text{ V}$$

with the polarity indicated in Figure 24–16.

Starting at point B and moving along the network to point A, we find the potential V_{AB} of A with respect to B:

$$V_{AB} = R_2 + IR_3 - \mathscr{E}_2$$

$$V_{AB} = 0 + 8\text{ V} - 4\text{ V}$$

$$V_{AB} = \boxed{4.00\text{ V}}$$

The potential at point A is thus 4 V higher than the potential at point B.

24.8 Applications

A number of different measuring devices are used to measure the parameters of a circuit. They include the voltmeter, the ammeter, the Wheatstone bridge, and the potentiometer.

The Voltmeter

Potential differences across the components of a circuit are often measured with a *voltmeter*. A voltmeter usually has a sensitive current-measuring meter called a *galvanometer*, shown in Figure 24–17. The *sensitivity* of a galvanometer is the current that will cause a full-scale deflection of the needle, usually in the range of 10 μA to 1 mA. The meter movement itself has a resistance R_G. (In circuit diagrams, it is usually drawn as a separate resistor, though one should remember that this resistance is an internal part of the meter movement itself.) Usually the external voltages to be measured are much greater than that which will cause a full-scale deflection, so a resistance R is added *in series* to reduce the voltage that appears across the meter movement. Let us now calculate the resistance R for full-scale deflection when a potential difference V is applied to the terminals AB. If we let I_G be the current in the galvanometer that will produce a full-scale deflection, and R_G is the internal resistance of the galvanometer, then, from Ohm's law:

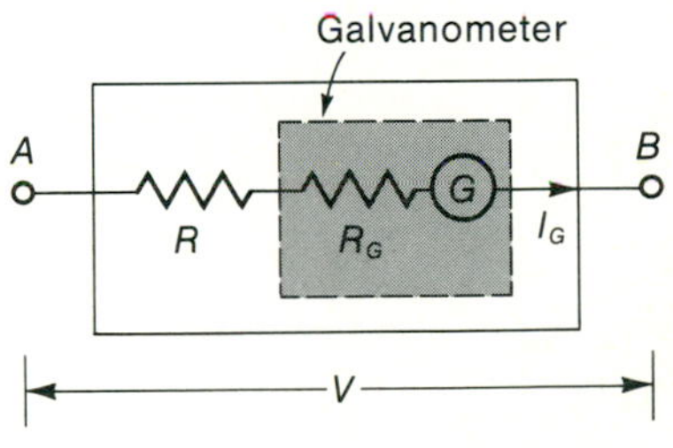

Figure 24–17
The voltmeter.

$$V = I_G(R + R_G) \qquad \textbf{(24–31)}$$

or

$$R = \frac{V}{I_G} - R_G \qquad \textbf{(24–32)}$$

In order to change the range of a voltmeter, it is necessary to change the value of the series resistor. In a multirange voltmeter, this is usually accomplished by a switching arrangement.

EXAMPLE 24–8

A galvanometer with a full-scale sensitivity of 1 mA requires a 900 Ω series resistor to make a voltmeter reading full scale when 1 V is across the terminals. What series resistor is required to make the same galvanometer into a 50 V (full-scale) voltmeter?

SOLUTION

We will use the values required for the 1 V voltmeter to obtain the internal resistance of the galvanometer. Applying Equation (24–31):

$$V = I_G(R + R_G)$$

we solve for R_G:

$$R_G = \frac{V}{I_G} - R$$

Substituting the appropriate values:

$$R_G = \frac{1\ \text{V}}{0.001\ \text{A}} - 900\ \Omega = 100\ \Omega$$

We then apply Equation (24–32) to obtain the series resistance required for the 50 V voltmeter:

$$R = \frac{V}{I_G} - R_G$$

$$R = \frac{50\ \text{V}}{0.001\ \text{A}} - 100\ \Omega = \boxed{49\,900\ \Omega}$$

Since a current I_G is required to operate a voltmeter, the introduction of a voltmeter into a circuit alters the currents in the circuit. Consequently, the voltmeter reading does not exactly represent the potential difference before the voltmeter was introduced. It is therefore desirable that a voltmeter have a very high internal resistance so that it does not draw much current from the circuit being measured. To compare various voltmeters, one calculates the *figure of merit*, or "quality," defined as the total resistance of the meter divided by the full-scale voltage reading. For the voltmeter described in Example 24–8, the quality is 1000 Ω/V. (This means that it is not a particularly good meter; a high-quality meter has a typical value of 20 000 Ω/V.) Analysis shows that a multirange meter that utilizes a given galvanometer movement will have the same figure of merit on all voltage scales. It may also be shown that the figure of merit is equal to the reciprocal of the current in the galvanometer that produces a full-scale deflection.

The Ammeter

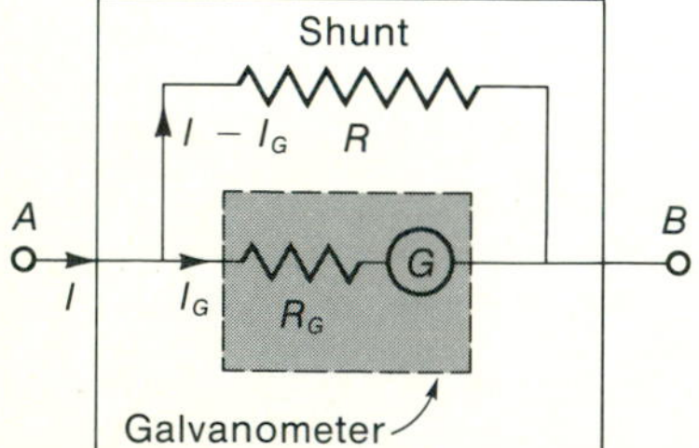

Figure 24–18
The ammeter.

A galvanometer measures very small currents. An *ammeter* measures larger currents by detouring, or *shunting*, some of the current around the galvanometer, as shown in Figure 24–18. Of the current I entering terminal A of the instrument, only a smaller portion I_G flows through the galvanometer movement. The voltage across R is the same as that across the galvanometer movement. Thus:

$$V_R = V_G$$

$$(I - I_G)R = I_G R_G$$

The value of the shunt resistor R is

$$R = \frac{I_G R_G}{I - I_G} \tag{24–33}$$

In a multirange ammeter, as in a multirange voltmeter, the value of R is usually changed by a switching arrangement.

EXAMPLE 24–9

An ammeter is constructed using a galvanometer that requires a potential difference of 50 mV across the meter movement and a current of 1 mA through the movement to cause a full-scale deflection. Find the shunt resistance R that will produce a full-scale deflection when a current of 5 A enters the ammeter.

SOLUTION

Direct application of Equation (24–33) requires a knowledge of R_G. However, R_G may be derived from the given quantities by using Ohm's law:

$$R_G = \frac{V_G}{I_G}$$

Substituting this into Equation (24–33):

$$R = \frac{I_G R_G}{I - I_G}$$

we obtain

$$R = \frac{V_G}{I - I_G}$$

Note that the resulting equation is simply Ohm's law applied to the shunt resistor alone, where $I - I_G$ is the current through the shunt. Substituting the appropriate values into the equation, we have

$$R = \frac{50 \times 10^{-3}\ \text{V}}{5\ \text{A} - 0.001\ \text{A}}$$

$$R = \boxed{0.010\ \Omega}$$

Note that the shunt resistance is very low for the measurement of currents much larger than the current requirements of the galvanometer. Just as in the construction of a voltmeter, a high-sensitivity galvanometer produces an ammeter that will introduce little change in a circuit when making a measurement.

A word of caution on the use of meters is in order. Of course, these instruments must not be connected to a circuit that will exceed their maximum range of values. But an additional hazard should be mentioned. Since an ammeter has an extremely low resistance, if it were mistakenly connected as a voltmeter *across* a source of voltage (instead of *in series* with other components), the resultant large current through the meter might easily destroy the ammeter. On the other hand, because a voltmeter has a large resistance, if it were mistakenly inserted in a circuit as an ammeter, probably no damage would result. Just remember: *Ammeters are connected in series in a line; voltmeters are connected in parallel across a potential difference.*

The Wheatstone Bridge

The primary use of a Wheatstone bridge is the measurement of resistance. Bridge-type circuits also have extensive application in electronics control circuits that detect small electrical imbalances.

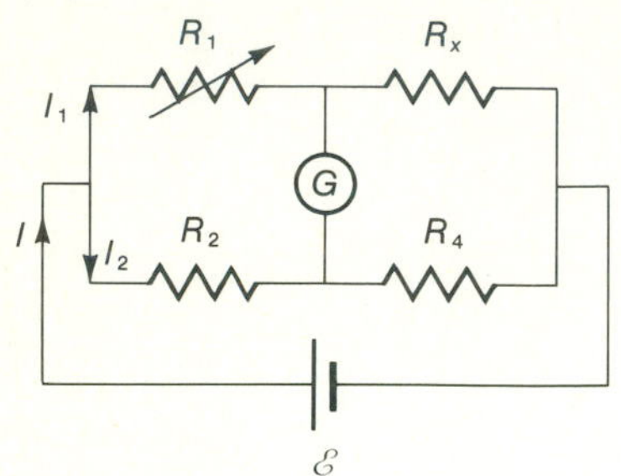

Figure 24–19
The Wheatstone bridge.

A Wheatstone bridge circuit is shown in Figure 24–19. When measuring an unknown resistance R_x, the procedure is to adjust R_1 (the symbol —⩘— indicates a *variable* resistor) until no measurable current passes through the galvanometer. This is known as the *null-balance condition.* What are the conditions in the circuit at null balance? If no current passes through the galvanometer, the potential differences across R_1 and R_2 must be the same

$$I_1R_1 = I_2R_2 \tag{24–34}$$

Moreover, the current through R_1 is the same as that through R_x. Similarly, the current through R_2 is the same as that through R_4. Therefore, the potential differences across R_x and R_4 are the same.

$$I_1R_x = I_2R_4 \tag{24–35}$$

Eliminating I_1 and I_2 between Equations (24–34) and (24–35), and solving for R_x, we have

$$R_x = \left(\frac{R_4}{R_2}\right)R_1 \tag{24–36}$$

In practice the ratio of R_4 to R_2 is known, as is the value of the adjustable resistance R_1, thereby yielding the value of the unknown resistance R_x. Note that the value of the seat of EMF need not be known. (However, the magnitude of the seat of EMF and the sensitivity of the galvanometer are important in the precision that the instrument can achieve since both contribute to the galvanometer deflection when the bridge is nearly balanced.)

The Potentiometer

A potentiometer is an extremely important laboratory instrument because it is capable of measuring potential differences in a circuit without drawing any current from the circuit. (This is in contrast to a voltmeter, which always requires some current for its operation.)

In Figure 24–20, the external battery causes a current in a long, uniform resistance wire called a *slide wire.* With the battery polarity shown, the potential

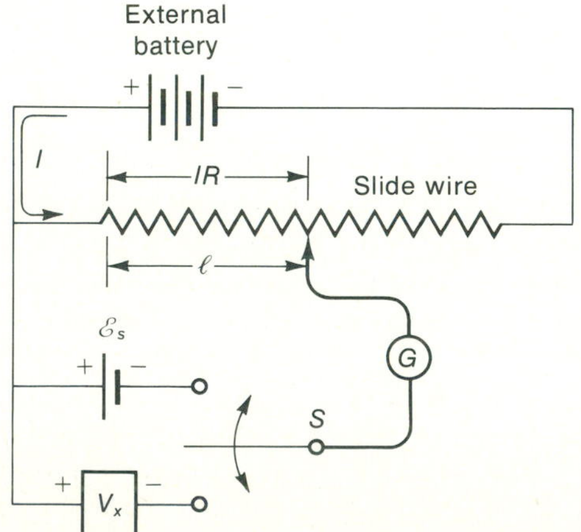

Figure 24–20
The basic potentiometer circuit. Note that the external battery, the standard cell $\mathcal{E}_s$, and the unknown potential V_x have the same polarity (here, positive) connected to the left end of the slide wire.

along the slide wire drops uniformly with distance as one proceeds from the left end toward the right. The symbol $\mathscr{E}_s$ denotes a *standard cell* whose potential difference is precisely known. When the switch S introduces the standard cell in the circuit, the sliding contact (the small arrow) is moved along the slide wire until it reaches the point ℓ_s where the IR voltage decrease along the wire equals $\mathscr{E}_s$. This condition is indicated by a lack of current passing through the galvanometer G, that is, a null-balance condition. Because the potential change along the slide wire is uniform, this procedure calibrates the potential at all points along the wire in terms of the distance ℓ from the left end. Thus, the voltage V along the slide wire is proportional to the distance ℓ.

$$\frac{\mathscr{E}_s}{\ell_s} = \frac{V}{\ell} \tag{24–37}$$

After calibrating the slide wire in this fashion, the switch is changed to replace $\mathscr{E}_s$ with the voltage V_x to be measured. The sliding contact is moved again to achieve the null condition. The new setting ℓ_x then gives sufficient information to determine V_x.

$$\frac{\mathscr{E}_s}{\ell_s} = \frac{V_x}{\ell_x} \tag{24–38}$$

Solving for V_x:

$$V_x = \left(\frac{\mathscr{E}_s}{\ell_s}\right)\ell_x \tag{24–39}$$

Standard cells are available whose EMF $\mathscr{E}_s$ has been calibrated by the National Bureau of Standards or other agencies. In practice, the ratio $\mathscr{E}_s/\ell_s$ is set to a convenient value by inserting a variable resistance (not shown) in series with the external battery, allowing control over the amount of current in the slide wire (and thus the magnitude of the IR decrease along the wire). Also, a protective resistance is sometimes added in series with the galvanometer to protect it from excessive currents in case the initial trial contact with the slide wire is far from the correct null-condition point. When the correct point is found (or closely approached), the protective resistance is shorted out, giving maximum sensitivity to the galvanometer reading.

The virtue of the potentiometer, *that it measures a potential difference without drawing any current*, makes it a valuable instrument for measuring potential differences when no disturbance of the circuit being measured can be tolerated.

EXAMPLE 24–10

Batteries always have an internal resistance. An automobile battery may have a resistance as low as 0.01 Ω, while that of an old flashlight battery may be as high as 50 Ω. Suppose the potential difference across the terminals of a particular battery is measured in two ways: first with a potentiometer, which indicates 1.50 V, and then with a voltmeter, which indicates 1.48 on a 2 V scale. The voltmeter is known to have a figure of merit of 1000 Ω/V. Find the internal resistance of the battery.

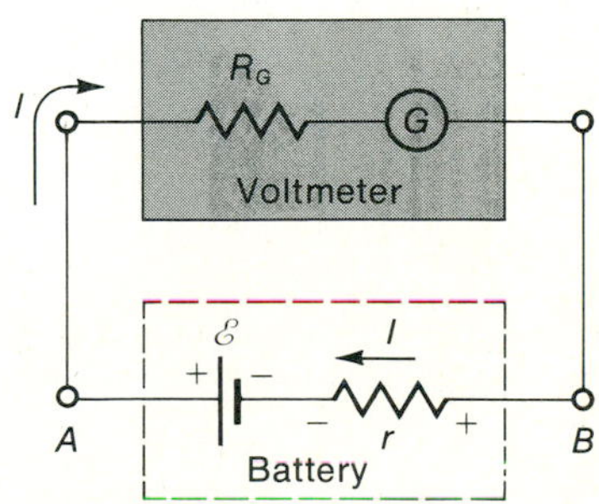

Figure 24–21
Example 24–10

SOLUTION

Although the internal resistance is distributed throughout the battery, for purposes of analysis we often draw circuit diagrams showing the EMF and the internal resistance r separately, as in Figure 24–21. In the first case, if the voltage across the battery terminals AB is measured by a potentiometer under null conditions, the potentiometer measures the true EMF of the battery (because no current passes through r, which would lower the potential difference across the terminals). On the other hand, when measured by a voltmeter, some current passes through the battery. As a result, the internal resistance r has a potential difference across it, which causes the terminal voltage to be less than the EMF. That is:

$$V = \mathscr{E} - Ir \tag{24–40}$$

where I is the current through not only the internal resistance r but also through the galvanometer. We can calculate I. We know that a figure of merit of 1000 Ω/V means that 1 mA produces a full-scale deflection of the voltmeter and that the meter was deflected (1.48/2.00) of the full scale. Thus:

$$I = \left(\frac{1.48}{2.00}\right)(1\ \text{mA}) = 0.740\ \text{mA}$$

Solving Equation (24–40) for r:

$$r = \frac{\mathscr{E} - V}{I}$$

and substituting the appropriate values:

$$r = \frac{1.50\ \text{V} - 1.48\ \text{V}}{0.74 \times 10^{-3}\ \text{A}} = \boxed{27.0\ \Omega}$$

24.9 *RC* Circuits

Until now we have discussed *steady state circuits*, that is, circuits in which the currents are constant in time. In this section, we will analyze time-varying currents in circuits that contain resistors, capacitors, and seats of EMF.

Consider the circuit shown in Figure 24–22(a). Whenever the switch is in the position shown, the seat of EMF will charge the capacitor with the polarity shown. The charging current is indicated by the symbol i. (We use lower-case letters for time-varying quantities.) Applying the Kirchhoff loop rule to the left loop of the circuit:

$$\mathscr{E} - iR_1 - \frac{q}{C} = 0 \tag{24–41}$$

where q/C is the potential difference across the capacitor. The above equation indicates that as the charge on the capacitor increases, the current must decrease. Suppose at $t = t_0$ the capacitor is uncharged and the switch is put in the charging position. Then the charge on the capacitor will increase with time, as shown in Figure 24–22(b). The mathematical form of this curve may be obtained by the solution of Equation (24–41), remembering that $i = dq/dt$. Rearranging this equation, we obtain the differential equation

$$\frac{dq}{dt} = \frac{\mathscr{E}}{R_1} + \frac{q}{R_1 C}$$

The following expression satisfies this differential equation.

CHARGING A CAPACITOR THROUGH A RESISTOR

$$q = \mathscr{E}C(1 - e^{-t/R_1 C}) \tag{24–42}$$

Note that as time progresses, q asymptotically approaches (but, in principle, never reaches) a value $\mathscr{E}C$. Note also that if $R_1 C$ is small, the capacitor acquires charge more rapidly.

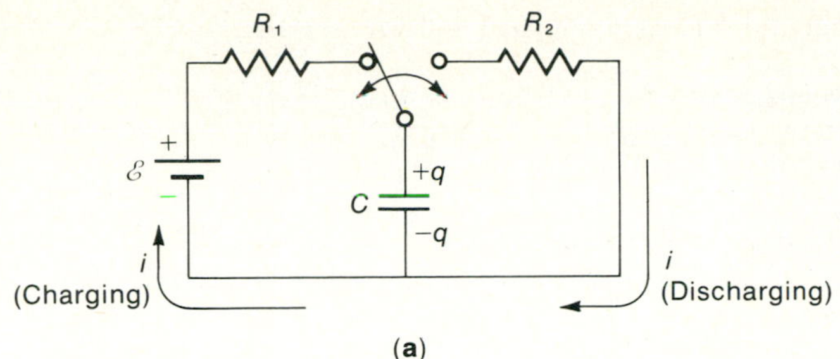

(**a**)

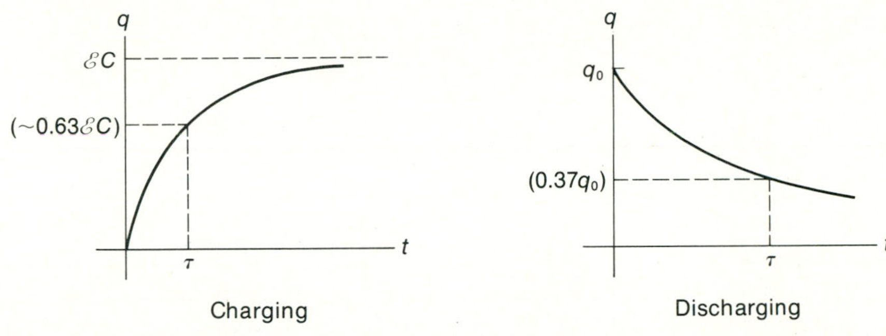

(**b**) The charge on the capacitor varies with time. Note that in this particular example, the charging rate is faster than the discharging rate because $R_1 < R_2$.

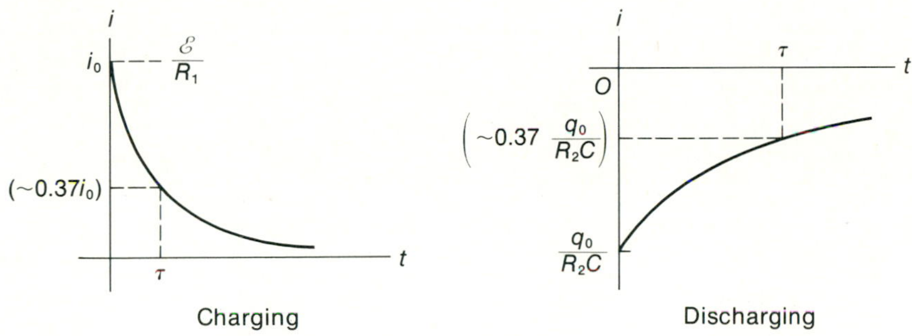

(**c**) The current in the branch containing the capacitor changes with time. Note that the charging current through C is opposite to the direction of the discharging current through C.

Figure 24–22
An *RC* circuit for charging and discharging a capacitor through resistors. Each of the curves varies exponentially.

Now suppose that when the capacitor has acquired a charge q_0, the switch in Figure 24–22(a) is moved to the position that connects the charged capacitor to the resistance R_2. The charge will drain off slowly through R_2. Applying Kirchhoff's loop rule to the right loop, we obtain

$$\frac{q}{C} - iR_2 = 0 \qquad \textbf{(24–43)}$$

where q/C is the potential difference across the capacitor. Since $i = -dq/dt$ (the minus sign results from the fact that q is decreasing with increasing time), we may write

$$\frac{q}{C} = -\frac{dq}{dt} R_2$$

or

$$\frac{dq}{q} = -\frac{dt}{R_2 C}$$

Integrating and setting $q = q_0$ at $t = 0$, we obtain

DISCHARGING A CAPACITOR THROUGH A RESISTOR

$$q = q_0 e^{-t/R_2C} \tag{24–44}$$

The charging and discharging currents as a function of time can be obtained by differentiating Equations (24–42) and (24–44), respectively. The charging current is

$$i = \left(\frac{\mathscr{E}}{R_1}\right) e^{-t/R_1C} \tag{24–45}$$

and the discharging current is

$$i = \left(-\frac{q_0}{R_2C}\right) e^{-t/R_2C} \tag{24–46}$$

where the minus sign indicates that the two currents are in opposite directions.

A useful parameter for describing RC circuits is the characteristic time $\tau = RC$ (the time at which the power of the exponential term is -1). This value is called the ***RC* time constant.** It is related to the speed with which the currents and voltages change. For example, in discharging the capacitor, in *one* time constant the charge falls to $1/e$ ($\approx 37\%$) of its original value. Similarly, in charging the capacitor, the charge rises to $1 - 1/e$ ($\approx 63\%$) of its maximum value. All of the curves in Figure 24–22 change in an exponential fashion, so it will be useful to remember the 37% and 63% values.

EXAMPLE 24–11

An initially-uncharged capacitor is charged by attaching it to a seat of EMF with wires of resistance R. When the capacitor is fully charged, the seat of EMF will have done an amount of work W given by

$$W = qV \tag{24–47}$$

where V is the terminal potential difference of the seat of EMF. The energy stored in the capacitor is

$$U_C = \tfrac{1}{2}CV^2$$

but since

$$C = \frac{q}{V}$$

we obtain

$$U_C = \tfrac{1}{2}qV \tag{24–48}$$

Thus, the work done by the seat of EMF is twice the energy stored in the capacitor. What has happened to the other half of the work done?

SOLUTION

By the conservation of energy principle, it seems reasonable that the "missing" energy must have appeared as I^2R heating of the resistor. Let us verify this supposition. The charging current is given by Equation (24–45):

$$i = \left(\frac{\mathscr{E}}{R}\right) e^{-t/RC}$$

Since $\mathscr{E}$ is equal to the terminal potential difference V:

$$i = \left(\frac{V}{R}\right) e^{-t/RC} \tag{24–49}$$

The instantaneous power dissipated in the resistance is:

$$P = i^2 R$$

If dE_R represents the thermal energy dissipated in R in a time dt, then

$$\frac{dE_R}{dt} = i^2 R$$

Substituting the value for i given by Equation (24–49), and integrating:

$$\int_0^{E_R} dE_R = R \int_0^{\infty} \left(\frac{V^2}{R^2}\right) e^{-2t/RC}\, dt$$

$$E_R = -\frac{V^2}{R}\left(\frac{RC}{2}\right) e^{-2t/RC}\Big|_0^{\infty}$$

$$E_R = \frac{CV^2}{2}$$

But since

$$C = \frac{q}{V}$$

we have

$$E_R = \tfrac{1}{2}qV \qquad \text{(half of the work } W \text{ done by the seat of EMF)}$$

Note that this result is true regardless of the value of R (as long as it is not zero).

When a seat of EMF charges a capacitor through any resistor, exactly *half* of the work done by the seat of EMF always appears as joule heating of the resistor.

Summary

The following terms and equations were introduced in this chapter:

Electromotive force $\mathscr{E}$ (the work done per unit charge by a seat of electromotive force in raising the potential energy of the charge):

$$\mathscr{E} = \frac{dW}{dq}$$

Electric current I (the amount of charge dq passing a given point in time dt):

$$I = \frac{dq}{dt}$$

Electric current I (for a conductor of cross-sectional area A, volume charge density ne, and average drift speed v_d of the charges):

$$I = nev_dA$$

Electrical resistivity ρ (for a conductor with cross-sectional area A and length ℓ):

$$I = \left(\frac{1}{\rho}\right)\left(\frac{A}{\ell}\right) V \qquad \text{(where } V \text{ is the potential difference across the conductor that carries } I\text{)}$$

Thermal coefficient of resistivity α (the proportionality factor between the fractional change in resistivity and the change in temperature):

$$\frac{\rho - \rho_0}{\rho_0} = \alpha(T - T_0)$$

or, more conveniently:

$$\rho = \rho_0[1 + \alpha(T - T_0)]$$

Resistance R (for a rod of cross-sectional area A and length ℓ):

$$R = \frac{\rho\ell}{A}$$

Ohm's law:

$$V = IR \qquad \text{(where } R \text{ is independent of } I\text{)}$$

Power P dissipated in a resistor:

$$P = VI = I^2R = \frac{V^2}{R}$$

Equivalent resistance R (for *combinations of resistors* in circuits):

In *series:* $\quad R = R_1 + R_2 + R_3 + \ldots$

In *parallel:* $\quad \frac{1}{R} = \frac{1}{R_1} + \frac{1}{R_2} + \frac{1}{R_3} + \ldots$

Kirchhoff's loop rule: The sum of the potential increases and decreases around any closed circuit is zero.

Kirchhoff's branch rule: The algebraic sum of all currents into and out of a junction is zero. (Currents entering are positive; currents leaving are negative.)

Superposition theorem: In a linear circuit containing more than one seat of EMF, the current in any branch is the superposition of all the currents contributed by each seat of EMF acting individually (with all other EMFs replaced by conducting wires of zero resistance).

RC circuit: In a series combination of a seat of EMF, resistor, and capacitor, the charge on the capacitor is

Charging: $\quad q = \mathscr{E}C(1 - e^{-t/RC})$

Discharging: $\quad q = q_0e^{-t/RC}$

In one *time constant*, $\tau = RC$, the charging exponential rises to $\approx 63\%$ of its final (maximum) value and the discharging exponential falls to $\approx 37\%$ of its original value.

Questions

1. Suppose you had a battery with unmarked terminals. How can the polarity of the terminals be determined? List as many ways as you can.
2. What are the merits, if any, in defining *conventional* current flow?
3. Why is the thermal coefficient of resistivity negative for insulators and semiconductors?
4. Since the drift speed of electrons in a conductor is generally very slow, why does a ceiling light bulb go on so soon after the wall switch is closed?
5. What is the principal reason that resistors do not conform to Ohm's law?
6. A solid copper wire has a resistance R_1. The wire is used to form a hollow tube of the same length as the wire, so that the inside diameter is half the outside diameter. If the resistance of the tube is R_2, what is the value of the ratio R_2/R_1?

7. Early Edison light bulbs had essentially a carbon filament. Why was it necessary to operate these light bulbs with an external series resistor?
8. In Chapter 22 we were careful to point out that $\mathbf{E} = 0$ inside a conductor and that $\mathbf{E}$ is often not zero outside a conductor. Why, in this chapter, do we assert just the opposite?
9. How can the terminal voltage of a battery exceed the EMF of the battery?
10. A 10 W, 110 V light bulb connected to a series of batteries may produce a brighter light than a 250 W, 110 V light bulb connected to the same batteries. Why?
11. Estimate the power consumption of a household in which all of the heating is done electrically. What is the cost per month based on a rate of seven cents per kilowatt-hour?
12. At one time automobiles utilized a 6 V electrical system. Why was a change made to the 12 V system, which is now used?
13. Of the two light bulbs designated by 25 W, 110 V and 100 W, 110 V, which has the higher filament resistance?
14. Consider a circular hoop of resistance wire with two terminals attached to different places on the hoop. How does the resistance between the terminals depend on their relative positions on the hoop?
15. Imagine a closed surface in the midst of a complicated electrical network, so that current-carrying conductors penetrate the surface and so that some of the circuit components such as resistors, batteries, and capacitors are within the surface. Is the net current through the surface zero? Does Gauss's law hold for this surface?
16. A potentiometer is often used to measure open-circuit voltages of batteries. How can the potentiometer also be used to measure current and resistance?
17. How does a run-down battery affect the operation of a Wheatstone bridge?
18. If the battery and galvanometer of a Wheatstone bridge were interchanged, the circuit is still that of a Wheatstone bridge. Suppose a Wheatstone bridge is balanced. Does interchanging the battery and galvanometer result in a balanced bridge?
19. A volt-ohm-meter is a single meter movement with circuits and switches that make it appropriate for use as an ammeter, a voltmeter, or an ohmeter. When not in use, why is it best to leave the switch of a volt-ohm-meter on a high-voltage scale rather than on a current scale or resistance scale?
20. Why is it more practical to specify the meter-current sensitivity of a voltmeter in ohms per volt rather than in amperes?
21. How can a voltmeter be used to measure capacitance?
22. In the slide-wire potentiometer of Figure 24–20, a variable resistor is usually added in series with the external battery in order to control the amount of current through the slide wire. Suppose this variable resistor were a combination of two variable resistors in parallel, one large and the other small, that act as coarse and fine controls of the current. Which resistor is the coarse control and which is the fine control?

Problems

24A–1 A current of 5 A flows through a conductor. How many electrons pass a given point per second? *Answer:* 3.12×10^{19} electrons/s

24A–2 Find the resistance of a nichrome wire 1 m long with a cross-sectional area of 0.1 mm^2 at (a) 20°C and (b) 1000°C.

24A–3 A wire with a resistance R is stretched to 1.25 times its original length. Find the resistance of the wire after it is stretched. *Answer:* $1.56R$

24A–4 A wire of constant diameter is composed of equal lengths of copper and iron wires joined end-to-end. If a potential difference of 12 V is applied across the ends of the combination, find the potential difference across the copper portion of the wire. Assume a temperature of 20°C.

24A–5 Find the temperature at which the resistance of a length of copper wire will be double its value at 20°C. *Answer:* 276°C

24A–6 Find the cost of electrically heating 100 L of water from 20°C to 90°C if the power company charges 8.4¢ per kW·h.

24A–7 A 1300 W electric heater is designed to operate from 120 V. Find (a) its resistance and (b) the current it draws. *Answers:* (a) 11.1 Ω (b) 10.8 A

24A–8 A 1000 Ω resistor is capable of dissipating a maximum power of 2 W. What is the maximum potential difference that should be applied to the resistor?

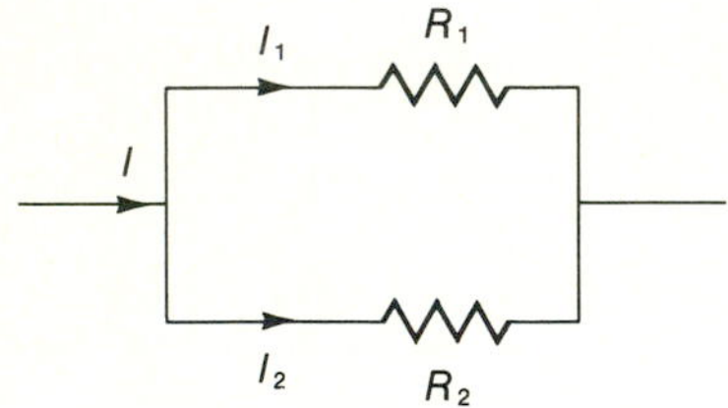

Figure 24–23
Problem 24A–9

24A–9 Consider a current I entering a circuit junction as shown in Figure 24–23. Show that the fraction I_1/I of I going through the branch that contains R_1 is given by $R_1/(R_1 + R_2)$.

24A–10 Using no more than three resistors, 2 Ω, 3 Ω, and 4 Ω, find all 14 resistance values that may be obtained by combining the three resistors.

24A–11 To achieve different values of power consumption, four 40 W, 120 V light bulbs are connected in a variety of ways across a 120 V power source. Sketch nine different ways and calculate the total power consumption in each case. Assume that the resistance of the light bulb is independent of the current through it (a poor assumption).
Answers: In watts: 10, 16, 24, 30, 40, $53\frac{1}{3}$, $66\frac{2}{3}$, 100, 160

24A–12 Three resistors, R, $2R$, and $3R$, are connected in parallel, producing an equivalent resistance of 20 Ω. Find their equivalent resistance when connected in series.

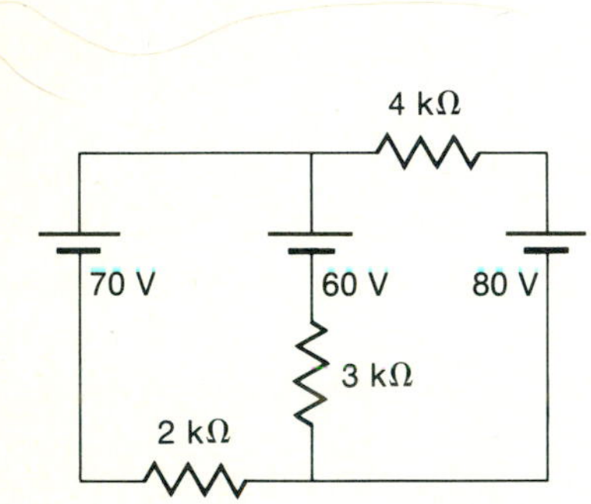

Figure 24–24
Problem 24A–14

24A–13 A 12 V car battery has an internal resistance of 0.02 Ω. Find the terminal voltage while the starter motor draws 140 A from the battery. *Answer:* 9.20 V

24A–14 Using Kirchhoff's rules, find the current in each of the resistors in the circuit shown in Figure 24–24.

24A–15 Consider the circuit shown in Figure 24–25. Show that the work done by the seat of EMF equals the energy dissipated in the resistors.

24A–16 A 12 V car battery is rated at 120 A·h (meaning its initial charge is 120 ampere·hours). While the car is parked, the two 80 W headlights are inadvertently left on. Assuming the terminal voltage remains constant, determine the number of hours that elapse before half the initial charge in the battery is used up.

24A–17 A generating station supplies power at 60 kV over transmission lines to a distant load.

(a) If the voltage can be raised to 100 kV without damage to the power lines, how much additional power (at the same current) can be transmitted?

(b) Will there be an additional transmission loss because of extra heating in the lines?

Answers: (a) 66.7% more power (b) No

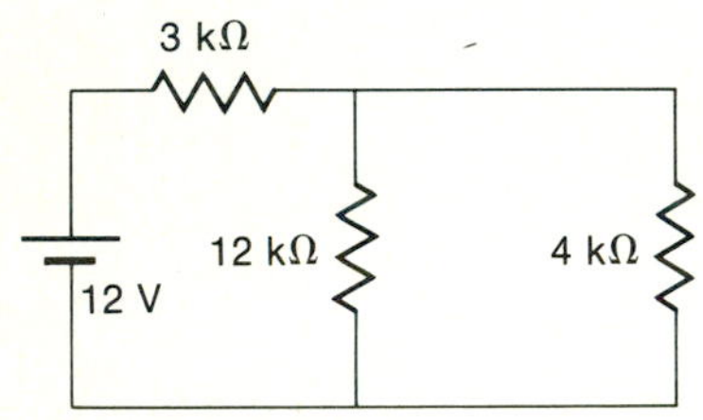

Figure 24–25
Problem 24A–15

24A–18 A certain meter movement has an internal resistance of 100 Ω and requires a current of 200 μA for full-scale deflection. Find the resistances that will convert the meter to (a) a 10 V voltmeter and (b) a 5 A ammeter. Include sketches showing how the resistance is connected in each case.

24A–19 The value of a resistance may be found using a circuit such as that shown in Figure 24–26.

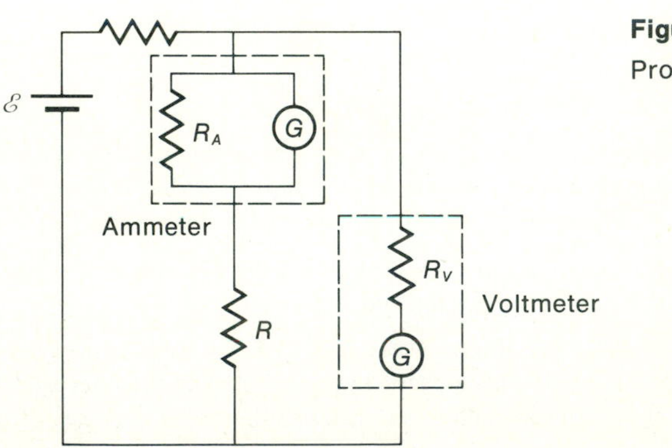

Figure 24–26
Problem 24A–19

(a) If the ammeter, which has a resistance of 50 Ω, reads 5 mA and the voltmeter reads 12.3 V, determine the value of R.
(b) If the ammeter had zero resistance, what would the value of R be?

Answers: (a) 2.41 kΩ (b) 2.46 kΩ

24A–20 A capacitance C discharges through a resistance R. How long does it take for the charge on the capacitor to reduce to $1/e^2$ of its initial value?

24A–21 Verify that the product RC has dimensions of time.

24A–22 A 20 000 Ω/V voltmeter set on a 100 V scale is connected to a charged capacitor. If the reading reduces to half its initial reading in 2 s, find the capacitance of the capacitor.

24A–23 A 10 μF capacitor is charged by a 10 V battery through a resistance R. The capacitor reaches a potential difference of 4 V in a period of 3 s after the charging began. Find the value of R. *Answer:* 0.587 MΩ

24B–1 A beam of high-energy alpha particles strike an absorbing target. (An alpha particle is a helium nucleus, which has a positive charge equal in magnitude to twice the electronic charge.) If the beam current is 0.3 μA and the energy of the particles is 20 MeV, find (a) the number of particles striking the target per second and (b) the power absorbed by the target. *Answers:* (a) 9.38×10^{11} particles/s (b) 3.00 W

24B–2 Consider a network of 12 resistors, all with a resistance R, joined so that each resistor forms the edge of a cube. Find the resistance between diametrically opposite vertices. (Hint: Apply a potential difference across these vertices and identify points of equal potential, which then may be joined by a resistanceless wire.)

24B–3 Find the resistance between terminals A and B of the resistor network shown in Figure 24–27. (Hint: Use the "delta-wye" transformation of Problem 24C–5.)

Answer: $R_{AB} = \frac{7}{5}R$

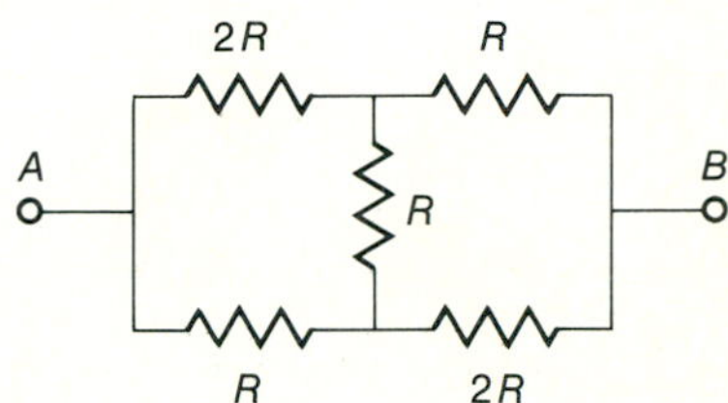

Figure 24–27
Problem 24B–3

24B–4 The value of a resistance is obtained by measuring the current through the resistor and the voltage across it, as shown in Figure 24–28. The voltmeter indicates 30 V on a 50 V scale and the ammeter indicates 150 mA on a 500 mA scale. What is the value of the resistance R if both meters have a galvanometer with a 1 mA full-scale sensitivity?

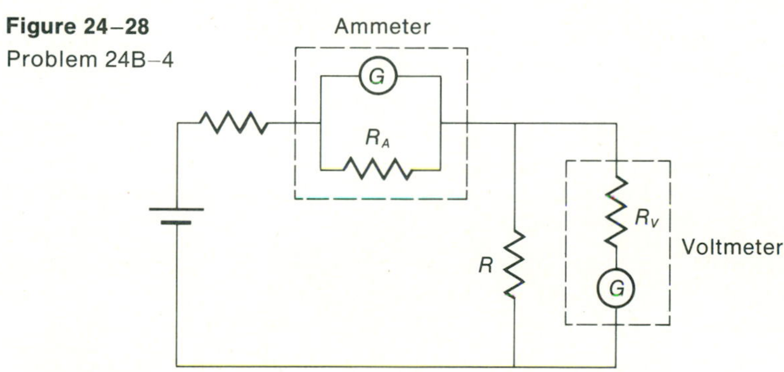

Figure 24–28
Problem 24B–4

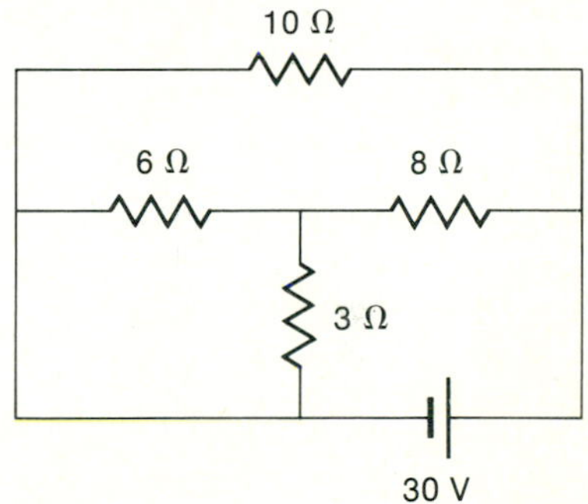

Figure 24–29
Problem 24B–6

24B–5 An electric utility company supplies a customer's house from the main power lines (120 V) with two copper wires, each 140 ft long and having a resistance of 0.108 Ω per 1000 ft. (a) Find the voltage at the customer's house for a load current of 110 A. For this load current, find (b) the power the customer is receiving and (c) the power dissipated in the copper wires. *Answers:* (a) 116.7 V (b) 12.8 kW (c) 366 W

24B–6 In Figure 24–29, calculate (a) the current in the 6 Ω resistor and (b) the current through the battery.

24B–7 Consider the circuit shown in Figure 24–30. Find the current in each of the resistors using (a) a direct application of Kirchhoff's rules and (b) the superposition theorem.

Answers: In the 6 Ω resistor: $2\frac{2}{3}$ mA
In the 4 Ω resistor: $2\frac{1}{2}$ mA
In the 12 Ω resistor: $\frac{1}{6}$ mA

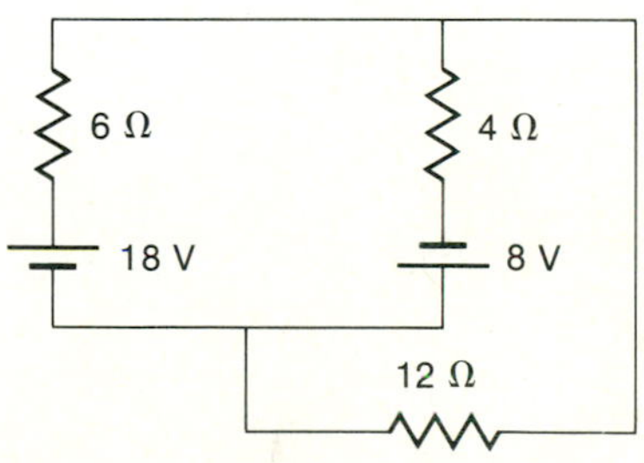

Figure 24–30
Problem 24B–7

24B–8 When two resistors, A and B, are connected in series, their total resistance is R_s. When connected in parallel, their equivalent resistance is R_p. Find R_A and R_B in terms of R_s and R_p.

24B–9 A certain run-down battery has an open-circuit voltage across its terminals of 7.22 V. While a battery charger is charging the battery with a current of 8.60 A, the terminal voltage is 7.96 V. Find the internal resistance of the battery. *Answer:* 0.0860 Ω

24B–10 An electric hoist operates at 240 V and uses a steady current of 9 A while lifting a 1700 lb load at the rate of 26 ft/min. Find (a) the power input to the hoist, (b) the power output (in horsepower), and (c) the efficiency of the system.

24B–11 A galvanometer is often made into a multirange ammeter through the use of an *Ayrton shunt* such as that shown in Figure 24–31. If the galvanometer has a resistance of

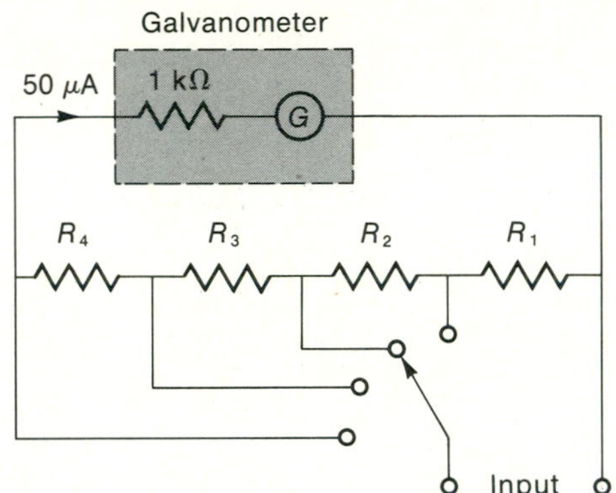

Figure 24–31
Problem 24B–11

1000 Ω and a full-scale sensitivity of 50 μA, find the values of R_1, R_2, R_3, and R_4 such that the meter will deflect full scale for 10 mA, 100 mA, 1 A, and 10 A.

Answers: $R_1 = 5.025 \times 10^{-3}\ \Omega$
$R_2 = 4.523 \times 10^{-2}\ \Omega$
$R_3 = 4.523 \times 10^{-1}\ \Omega$
$R_4 = 4.523\ \Omega$

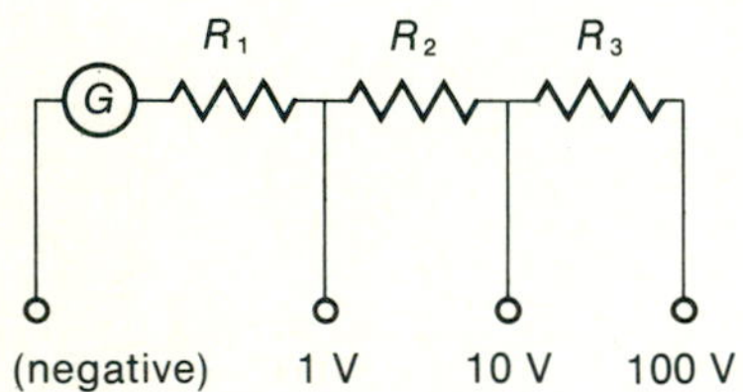

Figure 24–32
Problem 24B–12

24B–12 Figure 24–32 shows the series resistances inside a multirange voltmeter. The meter movement G has an internal resistance of 500 Ω and indicates full-scale deflection when a current of 0.500 mA is present. The markings on the terminals are as indicated. Find the values of R_1, R_2, and R_3.

24B–13 The *figure of merit* of a voltmeter is defined as the total resistance of the meter divided by the full-scale voltage reading. Prove that for a multirange voltmeter, the figure of merit is the same on all voltage scales.

24B–14 Refer to the previous problem and prove that the figure of merit is also equal to the reciprocal of the current in the galvanometer movement that produces a full-scale deflection.

24B–15 In the potentiometer circuit of Figure 24–20, the slide wire is 100 cm long. For an unknown EMF, the null position occurs when the sliding contact is 58 cm from the left end with an uncertainty in position of 0.30 mm.

(a) Find the percentage error in determining the unknown EMF, assuming the instrument is accurately calibrated.

(b) If the current in the slide wire were doubled, find the percentage error in the measurement, assuming the position uncertainty is still 0.30 mm.

Answers: (a) 0.517% (b) 0.103%

24B–16 How many time constants elapse while charging a capacitor in an RC circuit to within 2% of its maximum charge?

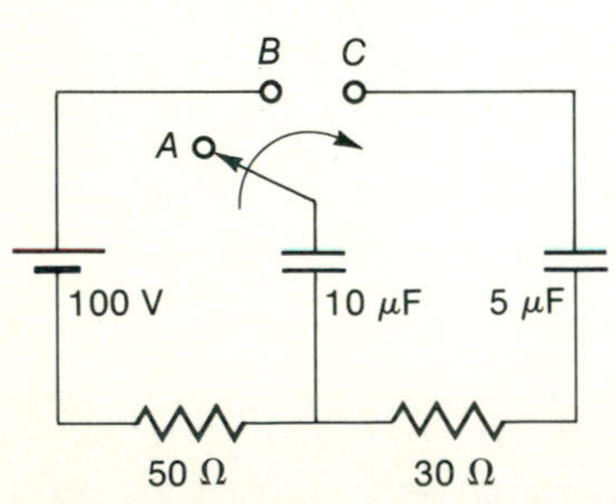

Figure 24–33
Problem 24B–17

24B–17 Consider the circuit shown in Figure 24–33. With the capacitors initially uncharged, the switch is moved from A to B, remaining there until the 10 μF capacitor is fully charged. The switch is then moved to position C. Calculate the total energy dissipated in each of the resistors while the final equilibrium is established.

Answers: 0.050 J in the 50 Ω resistor
0.0167 J in the 30 Ω resistor

24B–18 A 3 μF capacitor is initially charged to a potential of 200 V, then isolated. Because of leakage through the dielectric, 5 min later the potential has dropped to 185 V. Find the leakage resistance between the plates.

24C–1 The two resistors shown in Figure 24–34 are fabricated from the same resistive material. Show that they have the same resistance if the radius r of the cylindrical resistor is the geometrical mean $\sqrt{r_1 r_2}$ of the two radii associated with the resistor in the shape of a truncated cone. The ends of the resistors are plated with a conductor. Assume the current density is uniform over any cross section.

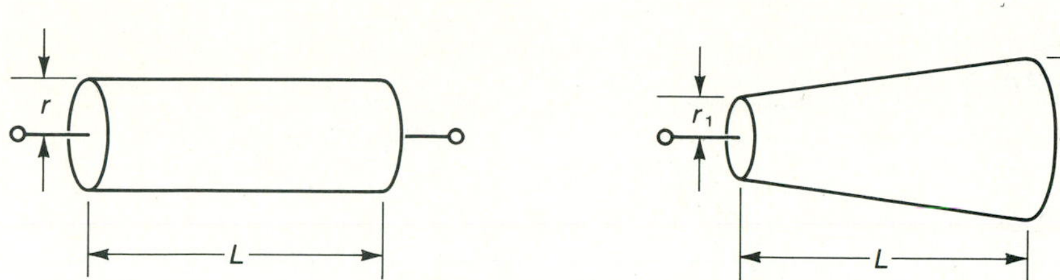

Figure 24–34
Problem 24C–1

24C–2 Consider the resistor network shown in Figure 24–35. All of the resistors have the same value R. Find the resistance between points A and B as the number of network elements becomes very large.

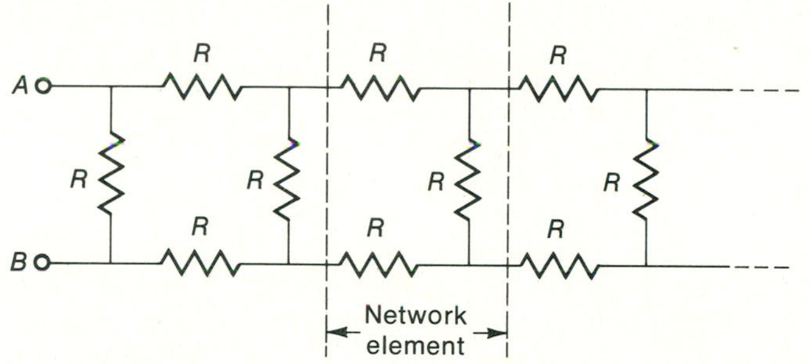

Figure 24–35
Problems 24C–2 and 24C–3

24C–3 A long, parallel pair of current-carrying wires may be represented by a network similar to that shown in Figure 24–35. The horizontal resistors represent the resistance per unit length L of the wires, $r_1 = 2R/L$, and the vertical resistors represent the resistance per unit length, $r_2 = R/L$, between the wires (that is, the resistance of the insulation separating the wires). Show that if $r_2 \gg r_1$, the resistance per unit length between A and B is $\sqrt{r_1 r_2}$.

24C–4 Consider an infinite network of resistors, as shown in Figure 24–36. If each resistor has the same value R, find the resistance between points A and B. (Hint: Connect

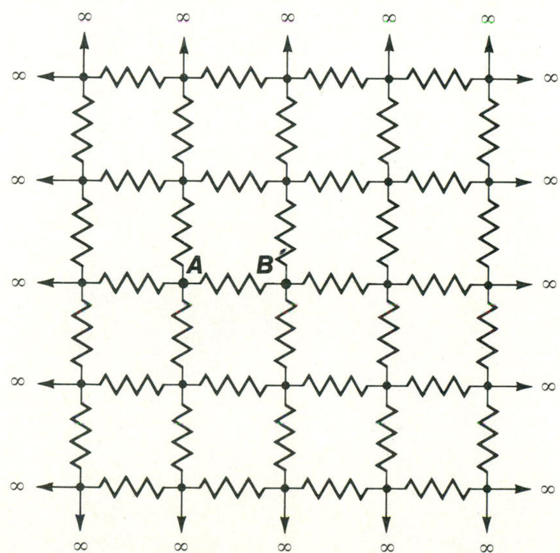

Figure 24–36
Problem 24C–4

a battery between point A and infinity, causing a current I into point A. Next connect another battery between point B and infinity, causing a current I out of point B. Then apply the superposition theorem.)

24C–5 The "delta" network of resistors shown in Figure 24–37(a) may be transformed into a "wye" network, shown in Figure 24–37(b), such that the resistance between corresponding terminals is equal. Find values for R_A, R_B, and R_C.

Answers: $R_A = R_1R_3/(R_1 + R_2 + R_3)$
$R_B = R_1R_2/(R_1 + R_2 + R_3)$
$R_C = R_2R_3/(R_1 + R_2 + R_3)$

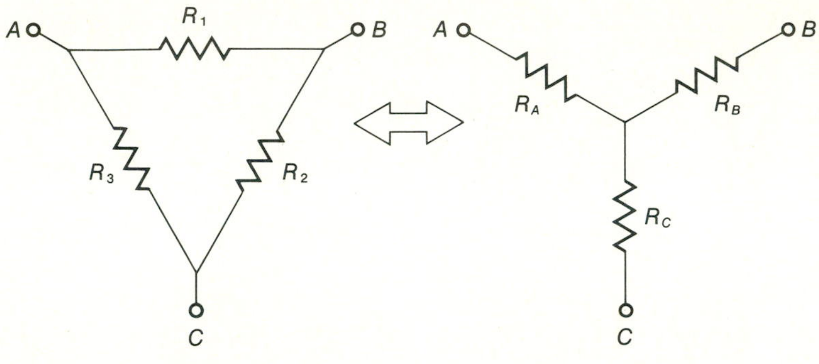

Figure 24–37
Problems 24C–5 and 24C–7

(**a**) A "delta" circuit. (**b**) A "wye" circuit.

24C–6 A power source consists of a seat of EMF and an internal series resistance R_s. The source delivers power to an external load resistor with resistance R_L. Show that the maximum fraction of the total work done by the seat of EMF (which appears as thermal energy in R_L) is one-half. Describe the relation between R_s and R_L for which this is achieved. This is known as the *maximum-power-transfer theorem.*

24C–7 Derive the equations that transform the "wye" configuration of resistors shown in Figure 24–37(b) into the "delta" configuration shown in Figure 24–37(a).

Answers: $R_1 = (R_AR_B + R_BR_C + R_CR_A)/R_C$
$R_2 = (R_AR_B + R_BR_C + R_CR_A)/R_A$
$R_3 = (R_AR_B + R_BR_C + R_CR_A)/R_B$

24C–8 A *DC* voltage source may be considered as a seat of EMF $\mathscr{E}$ in series with an internal resistance R_s. When measured by a 20 000 Ω/V multirange voltmeter, the terminal voltage is 95 V on the 100 V scale, and 120 V on the 200 V scale. Determine $\mathscr{E}$ and R_s. (The difference in the voltage readings is not a meter malfunction; the meter accurately reads the terminal voltage.)

24C–9 Verify that $q = \mathscr{E}C(1 - e^{-t/RC})$ satisfies $\mathscr{E} - iR - q/C = 0$.

24C–10 An ideal capacitor is charged by a seat of EMF through wires of very small resistance. The work done by the seat of EMF is $q\mathscr{E}$, where q is the final charge on the capacitor. The energy stored in the capacitor, however, is $\frac{1}{2}q\mathscr{E}$. Where is the remaining energy? (Hint: Assume a small, but nonzero, resistance R and evaluate $\int_0^\infty i^2R\,dt$.

chapter 25 Magneto-statics

The leading distinction of magnets is sex. . . . The kind that is found in Troas is black, and of the female sex, and consequently destitute of attractive power.

PLINY
c. 77 A.D.

In previous chapters we have discussed gravitational and Coulomb forces, both of which are inverse-square laws that do not depend on the relative motion of masses or charges. We now take up a type of force that does depend on the motion of charges. If two charges are both moving, in addition to Coulomb forces they exert a *magnetic* force on each other. The situation is a bit complicated, so we will separate the discussion into two parts. In one part we will show how a moving charge generates a magnetic field; in the other part, a second charge moves in the presence of this field and experiences a force. (We also followed this procedure in our discussion of Coulomb forces by considering one charge as the source of an electric field; the field, in turn, produces a force on another charge.) This chapter describes the effect a magnetic field has on a moving charge, and the next chapter will discuss the origin of the magnetic field.

25.1 Magnetic Fields

The earliest recorded observations of magnetism were those of the Greeks about 2500 years ago. The word *magnetism* comes from the Greek *magnetis lithos*, a certain type of stone containing iron oxide found in Magnesia, a district in northern Greece. This "lodestone" could exert forces on similar stones and on pieces of iron. It would also impart this magnetic property to a piece of iron it touched. The early Chinese were perhaps the first to discover that if a splinter of lodestone were suspended by a thread, it would align itself in a north-south direction. No doubt you have seen iron-filing patterns of the magnetic field surrounding a bar magnet (Figure 25–1). In the presence of a

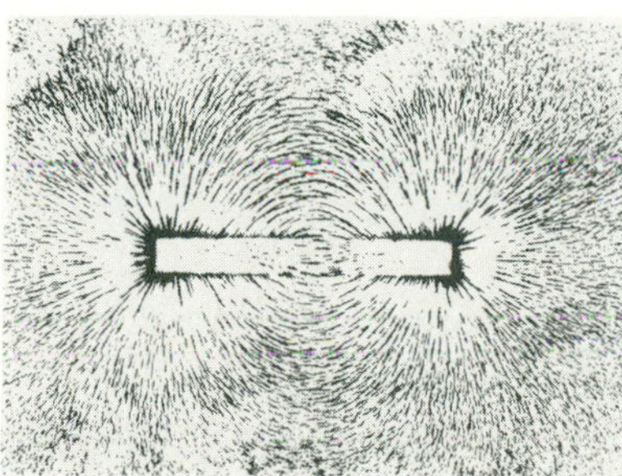

Figure 25–1
Iron filings sprinkled on a piece of paper covering a bar magnet arrange themselves in lines that suggest the magnetic field pattern.

magnetic field, iron filings themselves become small magnets, aligning along the field directions and attracting each other to form chains that suggest the pattern of the field.

Since a compass needle always points in a unique direction in a magnetic field, the field has *vector* properties. How do we determine the existence of a field in a given region of space? The formal operational definition of a magnetic field is as follows. We place a test charge in the space. If there is a force exerted on the charge when it is at rest, we conclude that an *electrostatic* field is present. If still another force arises when the charge is moving, we conclude that a *magnetic* field also exists in the space. As a result of such experiments, the following facts concerning magnetic fields emerge:

> *The magnitude of the force is proportional to the magnitude of the test charge.*
>
> *The direction of the force is always perpendicular to the direction of motion.*
>
> *When the charge is moving in a given direction, the force is proportional to the speed, but for a given speed the force varies with the direction of motion. (Thus the field must be a vector.)*

The fact that the force is always perpendicular to the velocity implies a vector *cross-product* definition for the magnetic field. The following equation, based on experiment, defines the **magnetic induction,** also called the **magnetic flux density, B.** We will follow the current widespread (although somewhat loose) usage and call it simply the **magnetic field.**[1] The force **F** on a charge q that has a velocity **v** in the presence of a magnetic field **B** is

MAGNETIC FIELD B

$$\mathbf{F} = q\mathbf{v} \times \mathbf{B} \tag{25–1}$$

The units for the magnetic field are newton·seconds/coulomb·meters (N·s/C·m), also called a *tesla*[2] (T). Because **F** is always perpendicular to the plane containing **v** and **B**, we will often need to depict three-dimensional situations.

A magnetic field is represented graphically in the same way we represent an electric field. Lines are drawn so that their density is proportional to the magnitude of the magnetic field, and the tangent to a field line at a given point represents the direction of the field at that point. As in the representation of electric fields, the number of lines used to represent a given magnitude of magnetic field is arbitrary. For example, we may associate 10 lines/meter^2 or 10^4 lines/meter^2 with a given field, depending on convenience. There is no such thing in reality as a field line; we sketch the lines merely to help us understand the properties of the magnetic field. The iron-filing patterns shown in Figure 25–1 accurately depict the field directions but do not give a good representation of the magnitude of the fields.

The end of a magnetized compass needle, which seeks the northerly direction, is called the *north pole* of the needle; the other end is the *south pole*. Consistent with Equation (25–1), the direction of the magnetic field created by a magnet is that the field lines *leave* the north pole and *enter* the south pole.

---

[1] Formally, the term "magnetic field strength" has been assigned to the vector $\mathbf{H} = \mathbf{B}/\mu$, where μ is the *permeability* of the space occupied by **B**.

[2] This unit honors Nikola Tesla (1856–1943), a Serbo-American engineer who devised many ingenious methods of electrical power generation and distribution. Among other accomplishments, he designed the Niagara Falls power system. A unit for **B** that is still used occasionally is the *gauss* (from the cgs system of units): 1 T = 10^4 gauss.

A convenient way of indicating magnetic fields *perpendicular* to the plane of a diagram is shown in Figure 25–2. In perspective sketches (refer to Figure 25–3b), idealized magnet poles are sometimes used to help establish the three-dimensionality of the diagram, with field lines emerging from the north pole and entering the south pole. The fringing fields are usually omitted in such sketches.

The spatial relationship among the force, velocity, and magnetic field vectors expressed in Equation (25–1) may be visualized by the usual right-hand rule shown in Figure 25–3(a). In this convention, the fingers of the right hand curl around in the sense of rotation established when **v** is rotated (through the smallest angle) into the direction of **B**. The extended thumb then points in the direction of **F**. An alternative convention useful in dealing with fields is shown in Figure 25–3(b). Here, the hand is held *flat* (with the thumb in the plane of the fingers). The fingers of the right hand point in the direction of the magnetic field. (This may be remembered by identifying the four fingers with field lines, which are spread through space.) The thumb points in the direction of the velocity of the charged particle. By the definition of a vector cross-product, the force is in the direction:

$$q\mathbf{v} \times \mathbf{B} = (qvB \sin \theta)\hat{\mathbf{n}}$$

where $\hat{\mathbf{n}}$ is a unit vector perpendicular to both **v** and **B** according to the right-hand rule. The angle θ in this expression is then the angle between the thumb and first finger. The magnetic force is outward from the palm of the hand and can be identified by the direction one would push.

When applying the right-hand rule, we will always consider q as a *positive* charge. If q is negative, we simply determine the direction of the force for a positive charge, then reverse the direction of the force. As an illustration, consider the magnetic force on a negative charge moving in an easterly direction near the equator, where the magnetic field of the earth is approximately horizontal in a northerly direction. Applying the right-hand rule, the fingers point north and the outstretched thumb points east. The palm is upward, indicating an upward force on a positive charge. However, since the charge is negative, the force is downward.

(**a**) *Out of the paper.* (The dots suggest the points of arrows coming *toward* the reader.)

(**b**) *Into the paper.* (The crosses suggest the tail feathers of arrows going *away* from the reader.)

Figure 25–2
Conventional ways of depicting uniform fields perpendicular to the plane of the diagram.

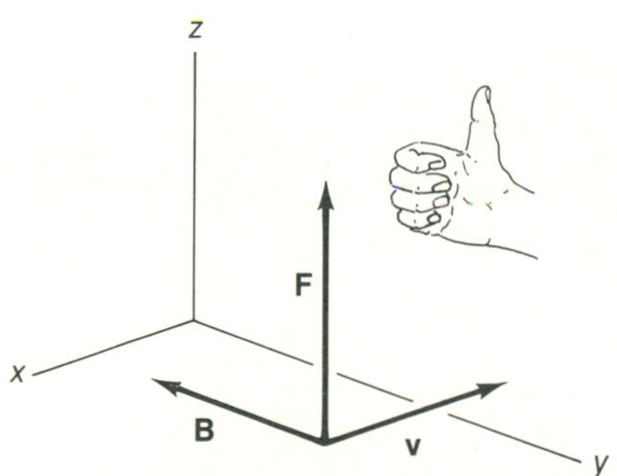

(**a**) The usual right-hand rule for the cross product **F** = q**v** × **B**.

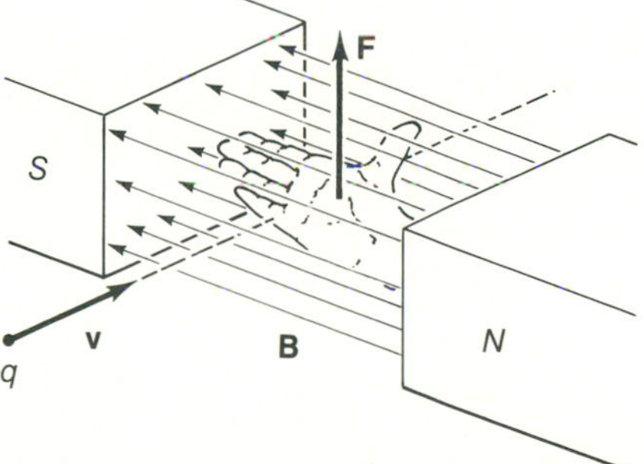

(**b**) Another way of thinking about the right-hand rule for cross products.

Figure 25–3
Two different ways of remembering the right-hand rule.

25.2 Motion of a Charged Particle in a Magnetic Field

An important feature of the motion of a charged particle in the presence of a magnetic field arises from the fact that the magnetic force is always at right angles to the velocity. Therefore, *the magnetic force does no work on the particle;*

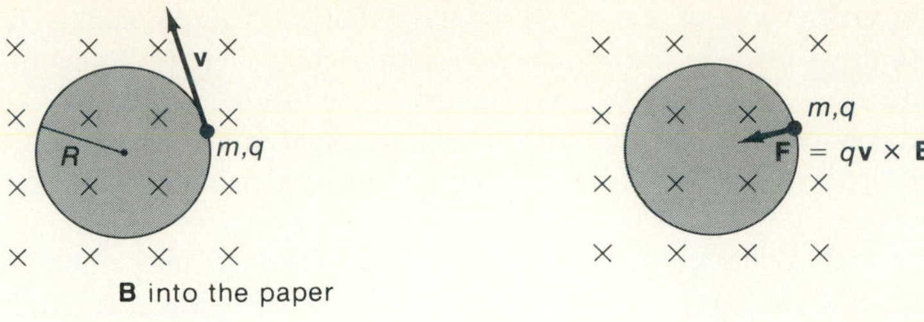

(a) The velocity **v**.

(b) The magnetic force **F**.

Figure 25–4
A charged particle traveling with velocity **v** at right angles to a uniform magnetic field **B** moves in a circle of radius R at constant speed. The frequency of revolution is called the *cyclotron frequency*.

the particle's speed remains constant, though its direction changes in response to the sideways deflecting force of the magnetic field.

If the charged particle is given a velocity **v** at right angles to **B**, the particle will travel in a circular path at constant speed, with the magnetic force providing the centripetal force necessary for the centripetal acceleration: v^2/R (see Figure 25–4). Since **v** and **B** are at 90°, the magnitude of the magnetic force is

$$F = q|\mathbf{v} \times \mathbf{B}|$$
$$F = qvB \sin 90^\circ$$
$$F = qvB$$

The radius R of the circular path may be found from Newton's second law:

$$\Sigma\mathbf{F} = m\mathbf{a}$$

$$qvB = m\left(\frac{v^2}{R}\right) \tag{25–2}$$

or

$$R = \frac{mv}{qB} \tag{25–3}$$

The momentum mv of the particle is related to its kinetic energy K by

$$mv = \sqrt{2mK} \qquad \text{(nonrelativistic)} \tag{25–4}$$

Combining these two equations yields:

$$R = \frac{\sqrt{2mK}}{qB} \qquad \text{(nonrelativistic)} \tag{25–5}$$

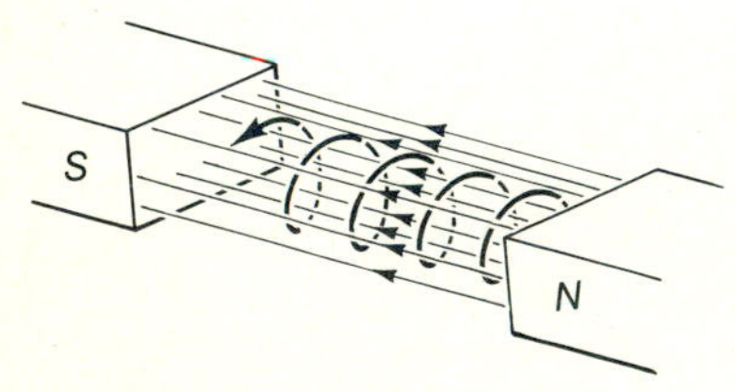

Figure 25–5
In a uniform magnetic field, a charged particle can travel in a helical path at constant speed. The path lies on an imaginary cylinder of constant radius.

EXAMPLE 25–1

An electron with an energy of 500 eV moves at right angles to a uniform magnetic field of 0.010 T. Find the radius of the circular motion.

SOLUTION

Since the kinetic energy of the electron is much less than its rest energy ($m_0c^2 =$ 0.511 MeV), we may use the nonrelativistic formula. After ensuring that all numerical values are in SI units,[3] we substitute them into Equation (25–5):

$$R = \frac{\sqrt{2mK}}{qB}$$

$$R = \frac{\left[(2)(9.11 \times 10^{-31}\text{ kg})(500\text{ eV})\left(\frac{1.602 \times 10^{-19}\text{ J}}{1\text{ eV}}\right)\right]^{1/2}}{(1.602 \times 10^{-19}\text{ C})(0.010\text{ T})}$$

$$R = \boxed{7.54 \times 10^{-3}\text{ m}}$$

Note that $1\text{ eV} = 1.602 \times 10^{-19}$ J was used to make the conversion to SI units.

The rotational frequency of the circular motion is called the *cyclotron frequency*. The name comes from the fact that motion of this type originates in a *cyclotron*, a type of machine that accelerates charged particles (see Figure 25–8). We may obtain the cyclotron frequency from Equation (25–2):

$$qvB = m\frac{v^2}{R}$$

For circular motion, $v = 2\pi fR$. Substituting this value and solving for f leads to

CYCLOTRON FREQUENCY

$$f = \frac{B}{2\pi}\left(\frac{q}{m}\right) \quad \text{(relativistic if } m \text{ is the relativistic mass)} \qquad (25\text{–}6)$$

where f is the rotational frequency of circular motion in units of revolutions per unit of time. This is the characteristic frequency of a particle of a given *charge-to-mass ratio* (q/m) in a uniform magnetic field. Note that the cyclotron frequency is independent of the speed and energy of the charged particle. Although this equation was derived using nonrelativistic physics, it is also valid for particles moving near the speed of light, provided that m is taken to be the relativistic mass: $m = m_0/\sqrt{1 - \beta^2}$.

If a charged particle moves parallel to the field **B**, there is no force on the particle because the cross product **v X B** is zero. For motion at an arbitrary angle (other than 90°) with respect to the field, its motion will be a *helix* rather than a circle (Figure 25–5). Since the velocity of the particle can be resolved into two components, *parallel* and *perpendicular* to the field, the cyclotron frequency is also the characteristic frequency for the helical motion.

The motion of charged particles in *non*uniform fields can be rather complicated. However, there is one simple example worth mentioning. Figure 25–6 depicts an axially-symmetric magnetic field that is stronger at the ends than in the middle. A charged particle approaching one end as it moves in a helical path will experience a magnetic force **F** having a horizontal component that "reflects" the particle back toward the middle. This configuration is called a *magnetic bottle* because it can trap charged particles within a confined region as they oscillate in helical paths back and forth between the ends of the bottle. In recent years, magnetic bottles have been used to confine plasmas in controlled fusion experiments. Unfortunately, the bottle "leaks" somewhat, since particles traveling *along* the magnetic field lines escape out the ends.

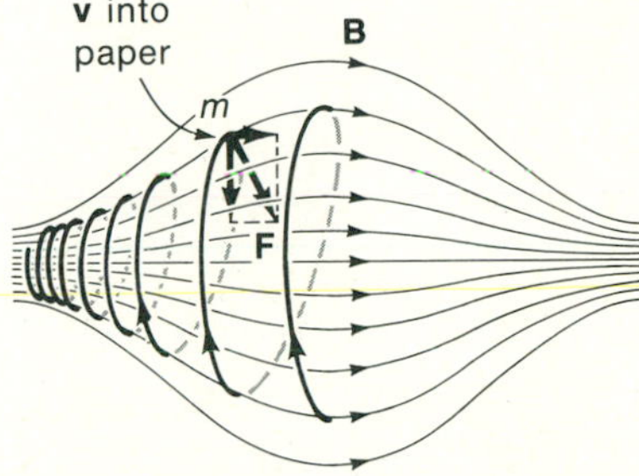

Figure 25–6
A *magnetic bottle* can trap charged particles by "reflecting" their helical motions at each end.

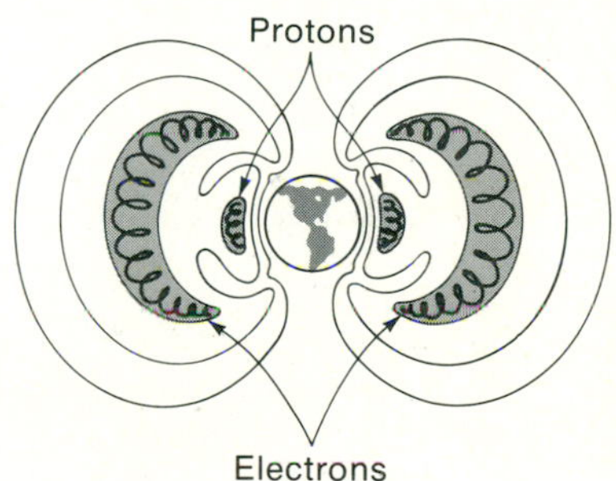

Figure 25–7
A cross section of the *Van Allen belts* that surround the earth. The earth's magnetic field acts as a magnetic bottle, trapping high-energy electrons and protons within two concentrated regions. The charged particles spiral between the north and south poles of the earth, with a typical round trip taking about one second. When some particles near the poles leak from the ends and penetrate the atmosphere, they excite various atoms and molecules, causing spectacular auroras.

---

[3] Up to this point, including units in our numerical substitutions allowed easy verification of the consistency of the units. However, because we will now be using more derived units such as the tesla in combination with basic units (the meter, kilogram, and second), such verification becomes difficult. For this reason, we will *ensure that all values are in SI units* and then assume with confidence that the answer is also in SI units.

The Cyclotron

Ernest O. Lawrence was awarded the Nobel Prize in 1939 for his development (with M.S. Livingston) of the cyclotron, a device that accelerates charged particles to high energies for use in nuclear experiments. Its basic components are a short cylindrical box made of copper sheet metal, divided into two sections called dees *(see Figure 25–8). The dees are in a vacuum chamber that is evacuated so the charged particles can move without colliding with air molecules. A magnetic field is established normal to the plane of the dees. A source of alternating voltage is connected to the dees, creating an electric field across the gap between the dees that reverses its direction every half cycle. Near the center of the dees, an ion source supplies charged particles such as protons, deuterons, or alpha particles, giving them a small velocity in the plane of the dees. Within the dees, the metal walls shield the ions from electric fields. However, the magnetic field is not shielded, causing the ions to move in a semicircle. Consider an ion that arrives at the gap between the dees just when the electric field between them is a maximum and in a direction to accelerate the ion across the gap. Subsequently, the ion will move in a larger semicircle because of its greater speed. If the frequency of the voltage reversals is correct, the ion arrives again at the gap just as the electric field reaches its maximum value in the opposite direction, again accelerating the ion. Each time it crosses the gap, the ion thus gains kinetic energy, traveling in larger and larger radii until it approaches the circumference of the cylinder, where a negatively charged* deflecting plate *pulls the ion from its circular path and allows it to pass out of the chamber through a thin window. The key to the operation of a cyclotron is that* the travel time for each semicircular path is the same. *As Equation (25–6) shows, the cyclotron frequency is independent of the speed or of the radius of the circle.*

There is an upper energy limit (about 22 MeV *for protons) because of relativity. As the particle increases its relativistic mass, its speed does not increase sufficiently to keep in step with the voltage reversals. The difficulty is overcome in the* synchrocyclotron, *whose frequency is reduced as a group of ions progresses outward. In another type of accelerator called the* synchrotron, *both the magnetic field and the frequency are varied, keeping the orbit radius essentially constant. This method has the economical advantage of requiring a magnetic field only in the region of the orbit, rather than in the entire area of the circle. The synchrotron at the Enrico Fermi National Accelerator Laboratory, Batavia, Illinois, shown in Figure 25-9, is designed to produce* 1000 GeV *protons that travel in a circular orbit* 2 km *in diameter.*

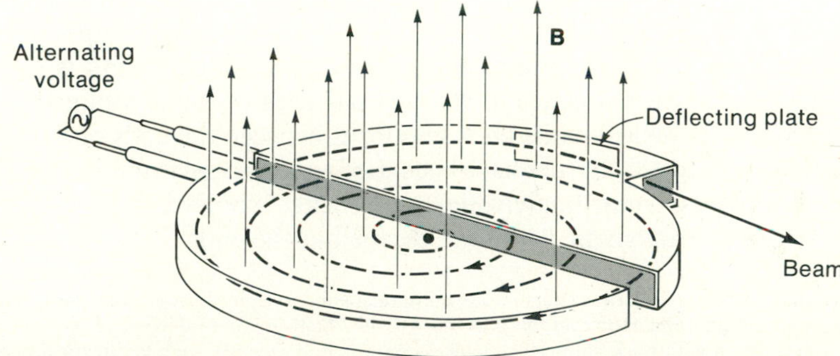

Figure 25–8 Under the influence of the magnetic field, the charges move in semicircular paths within the dees of the cyclotron.

Figure 25–9
The synchrotron at the Enrico Fermi National Accelerator Laboratory, Batavia, Illinois.

EXAMPLE 25–2

Find the cyclotron frequency of an electron moving in a uniform magnetic field of 0.020 T.

SOLUTION
From Equation (25–6), we obtain

$$f = \frac{B}{2\pi}\left(\frac{q}{m}\right)$$

$$f = \frac{(2 \times 10^{-2}\ \text{T})(1.602 \times 10^{-19}\ \text{C})}{2\pi(9.11 \times 10^{-31}\ \text{kg})}$$

$$f = \boxed{5.59 \times 10^{8}\ \text{Hz}}$$

25.3 The Lorentz Force Law

In general, a charged particle may simultaneously experience the effects of both an electric field and a magnetic field. Since the electric and magnetic forces resulting from these fields add as vectors, the net force on a charge may be written as:

LORENTZ FORCE

$$\mathbf{F} = q(\mathbf{E} + \mathbf{v} \times \mathbf{B}) \qquad \textbf{(25–7)}$$

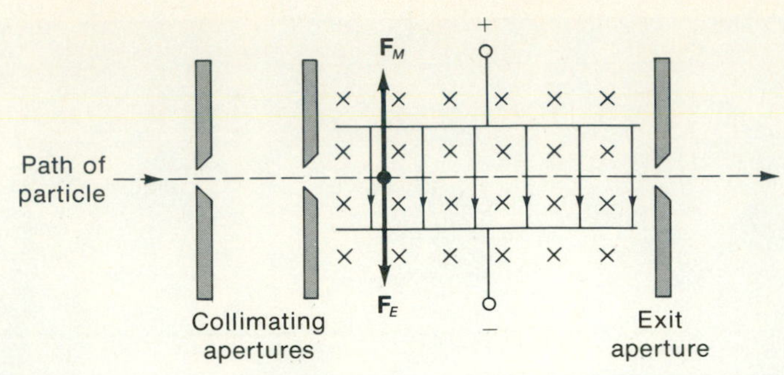

Figure 25–10
A charged-particle velocity filter. The magnetic force $\mathbf{F}_M$ balances the electric force $\mathbf{F}_E$.

where **F** is the net force on a charge q moving with a velocity **v** in the presence of an electric field **E** and magnetic field **B**. This equation is called the **Lorentz force law.**

A useful application of the Lorentz force law is a charged-particle *velocity filter*. Consider a particle of mass m and charge q moving with speed v along a straight path defined by collimating apertures as shown in Figure 25–10. The particle will pass through the exit aperture if there is no net force on the particle while it is in the region between the collimating and exit apertures. To accomplish this, magnetic and electric fields are established in the region so that the magnetic force on the particle is equal and opposite to the electric force. The directions of these forces, for positive charges, are as shown in Figure 25–10. Both forces are reversed in direction for negative charges. We apply Equation (25–7):

$$\mathbf{F} = q(\mathbf{E} + \mathbf{v} \times \mathbf{B})$$

For zero net force we have $\mathbf{E} + \mathbf{v} \times \mathbf{B} = 0$. In magnitude, this is

$$v = \frac{E}{B} \tag{25–8}$$

Only particles with this speed will travel in a straight line and emerge from the exit apertures, thus giving the device the name "velocity filter."

EXAMPLE 25–3

A stream of electrons passes through a velocity filter when the crossed magnetic and electric fields are 2×10^{-2} T and 5×10^4 V/m, respectively. Find the energy (in electron volts) of the electrons passing through the filter.

SOLUTION

We must first determine the speed of the particles to determine whether or not relativity need be applied. Substituting numerical values into Equation (25–8), we have

$$v = \frac{E}{B} = \frac{5 \times 10^4 \text{ V/m}}{2 \times 10^{-2} \text{ T}} = 2.50 \times 10^6 \frac{\text{m}}{\text{s}}$$

Since this is less than 1% of the speed of light, we may use the nonrelativistic expression:

$$K = \tfrac{1}{2}mv^2$$

Substituting numerical values yields

$$K = \tfrac{1}{2}(9.11 \times 10^{-31}\ \text{kg})\left(2.50 \times 10^{6}\ \frac{\text{m}}{\text{s}}\right)^2$$

$$K = 2.85 \times 10^{-18}\ \text{J}\underbrace{\left(\frac{1\ \text{eV}}{1.602 \times 10^{-19}\ \text{J}}\right)}_{\text{Conversion ratio}} = \boxed{17.8\ \text{eV}}$$

25.4 Magnetic Force on a Current-Carrying Conductor

In most applications, moving charges are confined to move through conductors. In the case of a metal wire, the charges are electrons moving with the drift velocity v_d. We shall now investigate the total force on all these moving charges when the conductor is in the presence of a magnetic field.

Equation (25–1) gives the force on one charge:

$$\mathbf{F} = q\mathbf{v} \times \mathbf{B}$$

The total number of moving charges in a wire of length ℓ is the number of conduction charges per unit volume n times the volume of the wire segment $A\ell$. Thus, the total force on the wire segment of length ℓ is

$$\mathbf{F} = q(\mathbf{v} \times \mathbf{B})nA\ell \qquad \textbf{(25–9)}$$

In the previous chapter (Equation 24–6), we found the current I to be

$$I = nev_dA$$

Combining these two equations, and identifying $\mathbf{v}$ with the drift velocity $\mathbf{v}_d$ of (positive) charges, we obtain

FORCE ON A CURRENT-CARRYING CONDUCTOR IN THE PRESENCE OF A MAGNETIC FIELD

$$\mathbf{F} = I\boldsymbol{\ell} \times \mathbf{B} \qquad \textbf{(25–10)}$$

Here, we maintain the vector form of the magnetic force by defining the length of the conductor as a vector $\boldsymbol{\ell}$ *in the direction of the conventional current* (the direction *positive* charges move).

Equation (25–10) assumes the wire segment is straight and the magnetic field is uniform. If the wire segment is of arbitrary shape and if the field varies, we recognize that the force $d\mathbf{F}$ on a small element $d\boldsymbol{\ell}$ of the wire is

$$d\mathbf{F} = I\, d\boldsymbol{\ell} \times \mathbf{B} \qquad \textbf{(25–11)}$$

Then, to find the total force, we integrate over the entire length of the wire using the value of $\mathbf{B}$ appropriate for each element $d\boldsymbol{\ell}$.

EXAMPLE 25–4

In Figure 25–11, a straight wire carries a current of 8 A in the presence of a uniform magnetic field of 3×10^{-3} T. The field is at an angle $\theta = 48°$ with respect to the wire. Find the force per unit length the field exerts on the current-carrying wire.

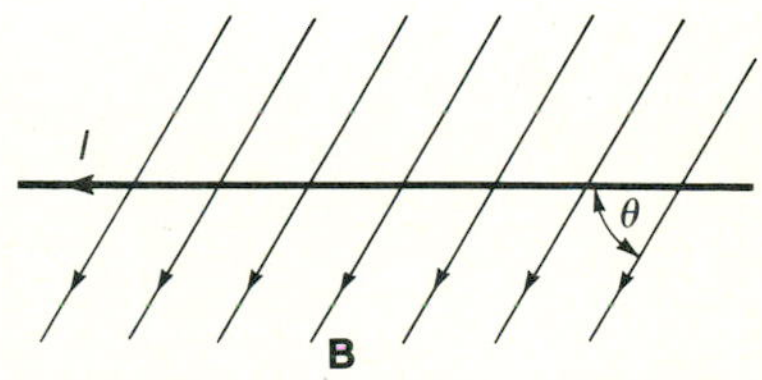

Figure 25–11
Example 25–4

SOLUTION

From Equation (25–10):

$$F = I|\boldsymbol{\ell} \times \mathbf{B}|$$

$$F = I\ell B \sin\theta$$

$$\frac{F}{\ell} = IB \sin\theta$$

$$\frac{F}{\ell} = (8\ \text{A})(3 \times 10^{-3}\ \text{T})(\sin 48^\circ)$$

$$\frac{F}{\ell} = \boxed{1.78 \times 10^{-2}\ \frac{\text{N}}{\text{m}}} \quad \text{(out of the paper)}$$

EXAMPLE 25–5

A current I is in a rigid semicircular loop of wire that has a radius R, as shown in Figure 25–12. A uniform magnetic field **B** is perpendicular to the loop. The current enters and leaves the loop by conductors (not shown) that are perpendicular to the plane of the paper and therefore parallel to the field. Find the net force on the semicircular wire.

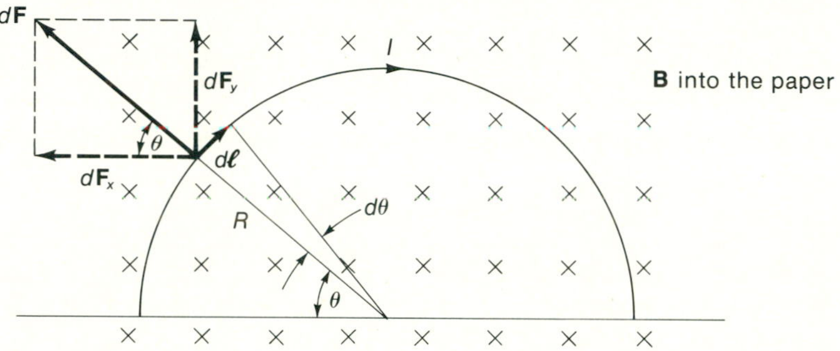

Figure 25–12
Example 25–5

SOLUTION

Since the conductors leading into and out of the wire are parallel to the magnetic field, the cross product in $\mathbf{F} = I\boldsymbol{\ell} \times \mathbf{B}$ involves $\sin 0^\circ = 0$. Thus the force on these conductors is zero.

The force $d\mathbf{F}$ on an incremental length $d\ell$ of the loop is given by Equation (25–11): $d\mathbf{F} = I\,d\boldsymbol{\ell} \times \mathbf{B}$. Since **B** is perpendicular to $d\ell$ over the entire length of the semicircular loop, the incremental force $d\mathbf{F}$ is directed radially outward and has a magnitude

$$dF = IB\,d\ell$$

We now make use of a symmetry argument. For every incremental force $d\mathbf{F}$ on the left side of the semicircular loop, there will be a corresponding increment symmetrically located on the right-hand side. The x components of these two forces are equal but in opposite directions, so they add to zero. However, the y components are in the same direction. Therefore, as we sum up the forces for the entire semicircle, we are left with only the sum of the y components dF_y:

$$dF_y = IBR \sin\theta$$

Integrating:

$$\int dF_y = IBR \int_0^\pi \sin\theta\, d\theta$$

$$F = IBR(-\cos\theta)\Big|_0^\pi$$

$$F = \boxed{2IBR} \quad \text{(in the } +y \text{ direction)}$$

Note that this would be the force on a straight conductor *along the diameter* of the circular loop. Actually, the shape of the loop is unimportant. As shown in a problem, the net force on any arbitrarily-shaped segment of wire that lies in a plane perpendicular to a field depends only on the *length of the gap* between the current input and the current output. This example leads to the conclusion that the net force on a *closed* current-carrying loop (of any shape) in a uniform magnetic field is zero. The net force is zero not because the force on each segment of the loop is zero, but because the *sum* of the forces on all segments is zero.

25.5 Magnetic Dipoles

Although a planar current-carrying loop in a uniform magnetic field experiences no net *force*, it may experience a *torque*. The behavior is analogous to that of an electric dipole in a uniform electric field. In fact, the analogy is so close that we will define a current-carrying loop as a *magnetic dipole*, in much the same way as we called a pair of charges of opposite sign an *electric dipole.*

We begin by defining a vector that is normal to a current-carrying loop. As shown in Figure 25–13, the direction of the normal is defined by curling the fingers of the right hand around the loop so that the fingers circle the loop in the direction of the conventional current. The extended thumb points in the direction of the desired normal. This is the direction of the vector $\boldsymbol{\mu}$ defined shortly.

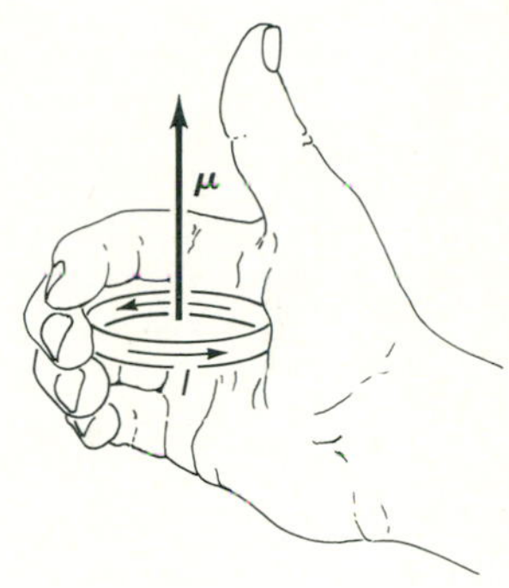

Figure 25–13
The normal to a current-carrying loop is determined by this right-hand rule: The fingers curl around in the current direction and the extended thumb points in the direction of $\boldsymbol{\mu}$.

Consider the rectangular loop in a magnetic field shown in Figure 25–14. Note how $\boldsymbol{\mu}$ is related to the direction of the current around the loop by the angle θ between $\boldsymbol{\mu}$ and **B**. The force on each of the sides of the rectangle is given by Equation (25–10):

$$\mathbf{F} = I\boldsymbol{\ell} \times \mathbf{B}$$

The forces $\mathbf{F}_1$ and $\mathbf{F}_2$ shown in Figure 25–15(a) are equal, but opposite in direction, so their contribution to the net *force* is zero. Also, since they are collinear, they produce zero net *torque*. The forces $\mathbf{F}_3$ and $\mathbf{F}_4$ are equal and opposite, so they, too, contribute zero net *force*. However, because they are not collinear, they form a *couple* (see Section 10.7) and produce a torque about an axis (☆) normal to the view of the loop in Figure 25–15(b). If θ is the angle between **B** and the normal to the loop (the vector $\boldsymbol{\mu}$), the magnitude of the torque about (☆) is

$$\tau = F_3\left(\frac{a}{2}\right)\sin\theta + F_4\left(\frac{a}{2}\right)\sin\theta$$

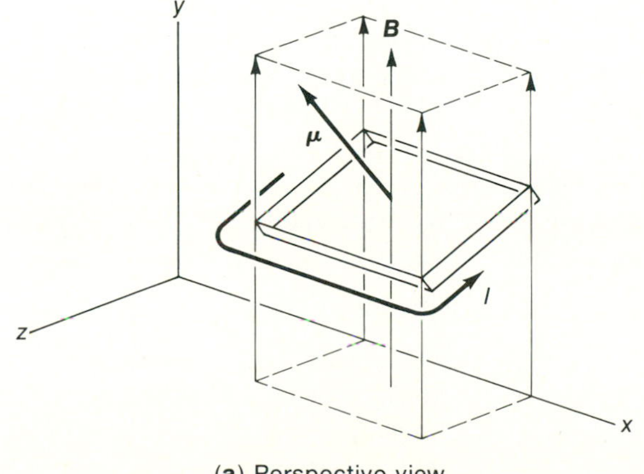

(**a**) Perspective view.

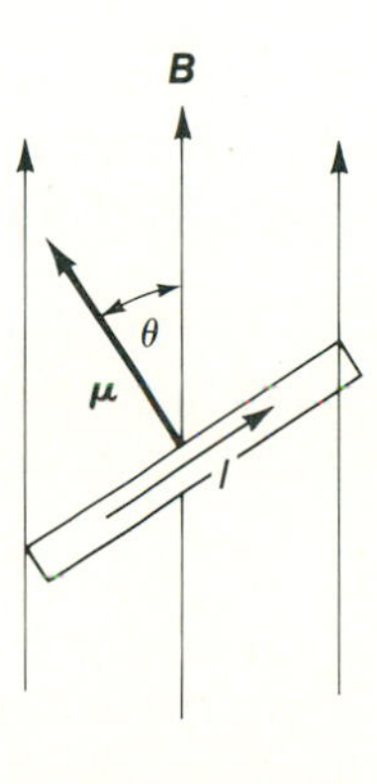

(**b**) Side view.

Figure 25–14
A rectangular current-carrying loop in a uniform magnetic field **B**.

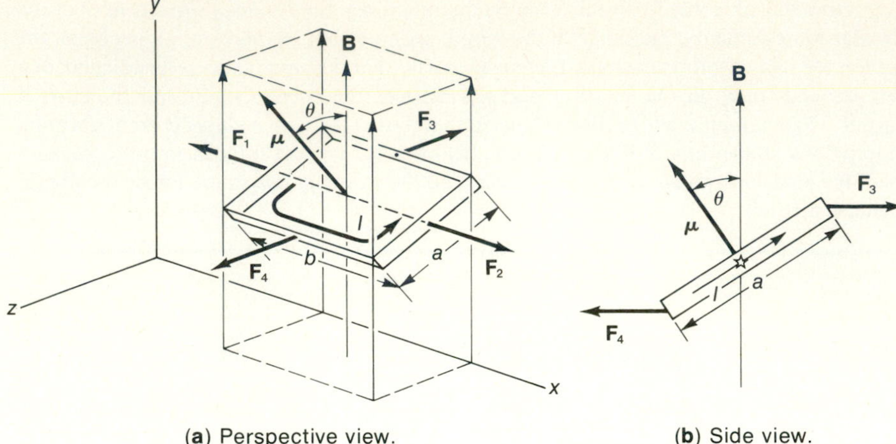

Figure 25–15
The forces on each side of a rectangular current-carrying loop in a uniform magnetic field **B**.

(**a**) Perspective view.

(**b**) Side view.

Since $F_3 = F_4$:

$$\tau = F_3 a \sin\theta \qquad (\text{or } F_4 a \sin\theta) \tag{25–12}$$

Since sides 3 and 4 are perpendicular to **B**, from $\mathbf{F} = I\boldsymbol{\ell} \times \mathbf{B}$ we have

$$F_3 = IbB$$

Substituting this value in Equation (25–12):

$$\tau = IabB \sin\theta \tag{25–13}$$

Letting A represent the area ab, the factor IA is called the magnitude of the **magnetic dipole moment** $\boldsymbol{\mu}$, whose direction is defined by the right-hand rule described previously. Using the notation **A** for the area normal *vector* (whose magnitude is the area A), we write

MAGNETIC DIPOLE MOMENT $\boldsymbol{\mu}$

$$\boldsymbol{\mu} = I\mathbf{A} \tag{25–14}$$

(the direction of $\boldsymbol{\mu}$ is normal to the plane of the loop of area A according to the right-hand rule)

The units of μ are ampere-meters² ($\text{A}\cdot\text{m}^2$).

We may write Equation (25–13) using vector notation as follows:

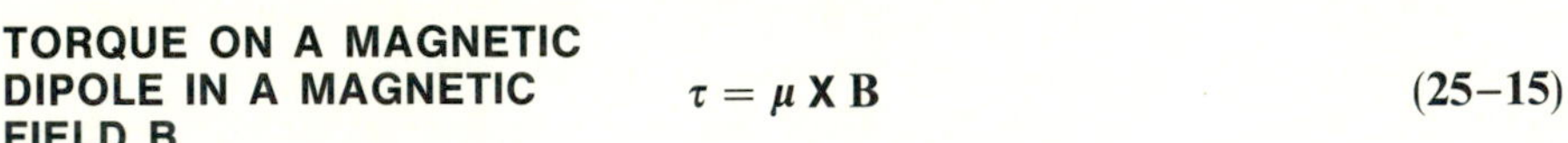

TORQUE ON A MAGNETIC DIPOLE IN A MAGNETIC FIELD B

$$\boldsymbol{\tau} = \boldsymbol{\mu} \times \mathbf{B} \tag{25–15}$$

Note the close similarity to the expression for torque on an electric dipole **p** in an electric field **E**:

$$\boldsymbol{\tau} = \mathbf{p} \times \mathbf{E}$$

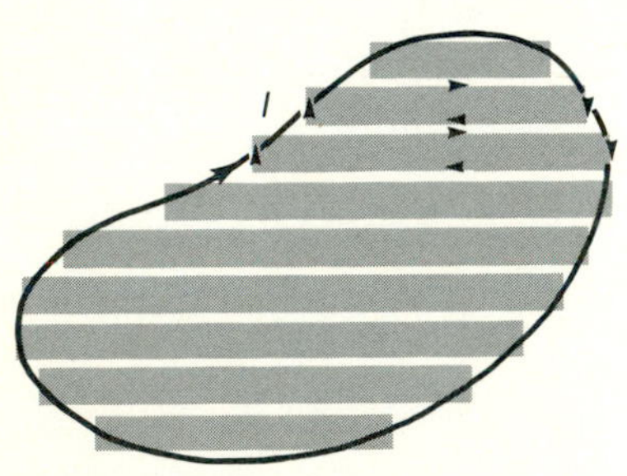

Figure 25–16
The clockwise currents around all the individual rectangles approximating the loop form the current I around the loop. All currents inside the loop cancel. (Rectangles are shown slightly separated for clarity.)

Although this derivation was based on a rectangular loop, it is valid for *any* planar-loop shape. That is:

$$\mu = (I)(\text{area of the loop}) \tag{25–16}$$

Figure 25–16 provides the basis for this conclusion. A current-carrying loop of arbitrary shape may be considered as a group of adjacent current-carrying rectangles. (The greater the number of rectangles, the better the approximation to the loop.) The currents in all the rectangles are clockwise. Thus, the currents of adjacent rectangles cancel out in the interior of the loop, leaving only the current around the perimeter. In this way, we generalize the derivation from that of a rectangular loop to a (planar) loop of any shape whatever.

□ □ □

A torque on a current-carrying loop in a magnetic field implies a potential energy associated with the orientation of the loop with respect to the field direction. Following the similar development of a potential energy associated with an electric dipole in an electric field, we start with the general definition of potential energy for rotation (see Chapter 22, Section 22.5).

$$U = -\int_{\theta_0}^{\theta} \boldsymbol{\tau} \cdot d\boldsymbol{\theta}$$

Since θ increases counterclockwise, as indicated in Figure 25–14(b), $\boldsymbol{\tau}$ and $d\boldsymbol{\theta}$ are antiparallel. Therefore, cos 180° = −1, and

$$U = \int_{\theta_0}^{\theta} \tau \, d\theta$$

Substituting the expression for τ given by Equations (25–13) and (25–14) and integrating, we have

$$U = \int_{\theta_0}^{\theta} \mu B \sin\theta \, d\theta$$

$$U = -\mu\beta(\cos\theta - \cos\theta_0) \qquad \textbf{(25–17)}$$

To simplify the expression, we define θ_0 as $\pi/2$, so $\cos\theta_0 = 0$. The form of the relationship between the vector quantities $\boldsymbol{\mu}$ and $\mathbf{B}$ indicated by Equation (25–17) suggests the vector dot product:

POTENTIAL ENERGY *U* OF A MAGNETIC DIPOLE IN A MAGNETIC FIELD ($U \equiv 0$ when μ and B are at 90°)

$$U = -(\boldsymbol{\mu} \cdot \mathbf{B}) \qquad \textbf{(25–18)}$$

Note the similarity with Equation (22–14) for the electric dipole case: $U = -(\mathbf{p} \cdot \mathbf{E})$.

The potential energy of the dipole is a maximum when the magnetic dipole $\boldsymbol{\mu}$ is *antiparallel* to the magnetic field **B**. Since physical systems tend to move toward positions of minimum potential energy, the magnetic dipole $\boldsymbol{\mu}$ tends to align itself in the direction of the magnetic field **B**.

EXAMPLE 25–6

A wire is formed into a circle with a diameter of 10 cm and placed in a uniform magnetic field of 3×10^{-3} T. A current of 5 A passes through the wire. Find (a) the maximum torque that can be experienced by the current-carrying loop and (b) the range of potential energy the loop possesses for different orientations.

SOLUTION

The magnetic dipole moment of the current-carrying loop of wire is given by Equation (25–16):

$$\mu = (I)(\text{area of the loop})$$

Substituting numerical values:

$$\mu = (5\ \text{A})(\pi)(0.05\ \text{m})^2$$
$$\mu = 3.93 \times 10^{-2}\ \text{A·m}^2$$

(a) The torque exerted on a magnetic dipole in a uniform magnetic field is given by Equation (25–15):

$$\tau = \boldsymbol{\mu} \times \mathbf{B}$$

which has a maximum value when the field and dipole moment are perpendicular, that is, when the plane of the wire loop is parallel to the magnetic field. Its maximum magnitude is

$$\tau = \mu B$$

Substituting:

$$\tau = (3.93 \times 10^{-2}\ \text{A·m}^2)(3 \times 10^{-3}\ \text{T})$$
$$\tau = \boxed{1.18 \times 10^{-4}\ \text{N·m}}$$

(b) The potential energy possessed by a magnetic dipole in a magnetic field is given by Equation (25–18):

$$U = -\boldsymbol{\mu} \cdot \mathbf{B}$$

The maximum potential energy occurs when the dipole moment is antiparallel to the field and a minimum occurs when the magnetic moment is parallel to the field. The range of potential energy is

$$\Delta U = U_{\text{max}} - U_{\text{min}}$$

or

$$\Delta U = -\mu B \cos \pi - (-\mu B \cos 0^\circ)$$
$$\Delta U = 2\mu B$$

Substituting the values for μ and B:

$$\Delta U = 2(3.93 \times 10^{-2}\ \text{A·m}^2)(3 \times 10^{-3}\ \text{T})$$
$$\Delta U = \boxed{2.36 \times 10^{-4}\ \text{J}}$$

25.6 Applications

Galvanometer

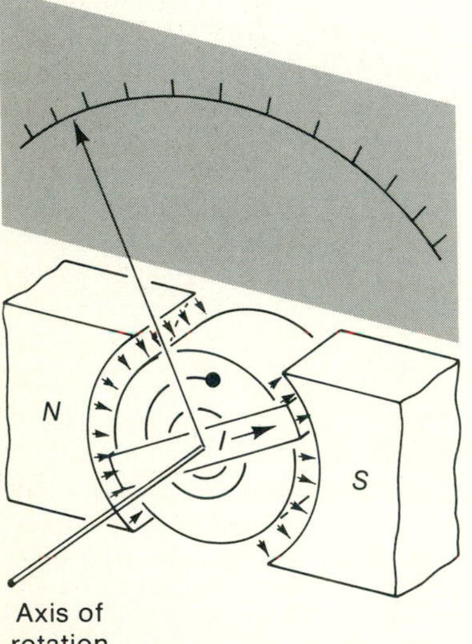

Figure 25–17
The basic meter movement of a galvanometer.

In Chapter 24 we discussed the construction of voltmeters and ammeters, both of which utilize a sensitive current-measuring device called a galvanometer. We shall now describe the basic principles underlying the operation of a galvanometer.

A galvanometer consists of a current-carrying coil in a magnetic field, as shown in Figure 25–17. Current is conducted to the coil of wire through the bearings that support the coil and allow rotation about a fixed axis. The connection from one bearing to the coil is through a spiral spring, which not only conducts the current, but also exerts a restoring torque when the coil is rotated from its equilibrium position. As the loop rotates, the sides of the loop move in a region of magnetic field **B**, which is constant in magnitude and always perpendicular to the $\boldsymbol{\mu}$ of the coil. This is achieved by specially-shaped pole faces of a permanent magnet and a (fixed) iron cylinder inside the loop. Thus the torque on the coil due to the current depends only on the current and is independent of the orientation of the loop. The coil is restrained by a spiral spring that conforms to Hooke's law:

$$\tau_{\text{spring}} = -\kappa\theta$$

The torque on the coil is given by Equation (25–15):

$$\tau_{\text{coil}} = \boldsymbol{\mu} \times \mathbf{B}$$

But since $\boldsymbol{\mu}$ is always perpendicular to $\mathbf{B}$:

$$\tau_{\text{coil}} = \mu B$$

When the coil is in static equilibrium:

$$\tau_{\text{spring}} = -\tau_{\text{coil}}$$

or:

$$\kappa\theta = \mu B$$

Expressing the angle of rotation from equilibrium θ in terms of the current through the coil, we have

$$\theta = \left(\frac{AB}{\kappa}\right) I \tag{25–19}$$

where A is the area of the coil. If the coil has n turns of wire, the total current I in the loop is

$$I = nI_0$$

where I_0 is the current through the wire. Then Equation (25–19) becomes

$$\theta = \left(\frac{nAB}{\kappa}\right) I_0 \tag{25–20}$$

The angle θ is measured by a pointer attached to the coil. *The angular deflection is directly proportional to the current in the coil*, so the scale along which the pointer moves is linear.

EXAMPLE 25–7

A typical galvanometer has the following specifications and parameters: coil area, 1 cm^2; number of turns of wire on the coil, 100 turns; spring constant of the spiral spring, 3×10^{-7} N·m/rad; and a current sensitivity of 50 μA for a coil rotation of $\pi/2$ rad (full-scale deflection). Find the magnitude of the magnetic field through which the coil moves.

SOLUTION

From Equation (25–20):

$$B = \frac{\kappa\theta}{nAI_0}$$

Substituting the appropriate values (all in SI units) yields

$$B = \frac{(3 \times 10^{-7}\ \text{N·m/rad})\left(\dfrac{\pi}{2}\ \text{rad}\right)}{(100\ \text{turns})(10^{-4}\ \text{m}^2)(50 \times 10^{-6}\ \text{A/turn})} = \boxed{0.942\ \text{T}}$$

Sensitive galvanometers such as this are very delicate because of the small torque exerted on a practical-size coil moving in a field that can be realistically achieved by a magnet. The bearings are often jewel bearings similar to those found in a watch, and the spiral spring consists of several turns of extremely fine spring wire.

Hall Effect

An effect used by E.H. Hall in 1879 to determine the sign of current carriers in conductors is now used extensively to measure current and magnetic fields. The *Hall effect* describes the potential difference that develops between the sides of a current-carrying conductor when placed in a magnetic field. In order to understand how such a potential difference develops, we will consider an idealized conductor in which the charge-carriers are free electrons.[4]

Consider an idealized conductor of rectangular cross section placed in a magnetic field **B**, as shown in Figure 25–18. The magnetic force $\mathbf{F}_M$ on a single electron is

$$\mathbf{F}_M = (-e)\mathbf{v}_d \times \mathbf{B} \tag{25–21}$$

where $-e$ is the charge on the electron and $\mathbf{v}_d$ is the drift velocity of the electron. Initially, the magnetic force will cause the electrons to drift toward the right-hand edge of the conductor. Eventually, however, the accumulation of charge produces an electric field **E** within the conductor, thus inhibiting further lateral drift of the charge. In equilibrium, the electric force $\mathbf{F}_E$ that results from the electric field will just balance the magnetic force:

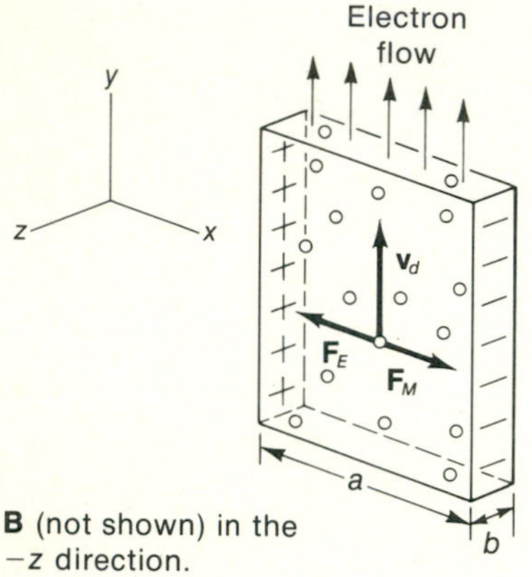

B (not shown) in the $-z$ direction.

Figure 25–18
The Hall effect. A magnetic field in the $-z$ direction forces the moving electrons to the right edge of the conductor, creating an electric field in the $+x$ direction. When equilibrium is attained, the magnetic force $\mathbf{F}_M$ on the moving electrons is equal and opposite to the electric force $\mathbf{F}_E$.

$$|\mathbf{F}_E| = |\mathbf{F}_M|$$

Applying the Lorentz force law, the forces balance if

$$eE = ev_dB$$

or

$$E = v_dB \tag{25–22}$$

The drift speed v_d can be expressed in terms of the current and the parameters of the conductor through the definition of current:

$$I = nev_dA \tag{25–23}$$

where n is the number of current-carriers per unit volume and A is the cross-sectional area of the conductor. In this instance:

$$A = ab$$

Substituting and solving for v_d, we have

$$v_d = \frac{I}{neab} \tag{25–24}$$

Further substitution into Equation (25–22) gives

$$Ea = \frac{IB}{neb} \tag{25–25}$$

The electric field E times the width of the conductor a is the potential difference across the width. This potential difference is referred to as the *Hall potential*, V_H.

$$V_H = \frac{BI}{neb} \tag{25–26}$$

Since the Hall potential depends on the product BI, knowledge of either one of the variables leads to determination of the other by measurement of the Hall potential. Note that if the current-carriers were positive charges, the po-

[4] Monovalent metals such as copper and silver behave as nearly-idealized current-carriers in the analysis of the Hall effect. The analysis of the Hall effect in magnetic current-carriers such as iron and in semiconductor current-carriers is complicated by quantum effects.

larity of the Hall potential would be reversed for the same direction of magnetic field and current. So the Hall effect can be used to determine the *number* and *sign* of the current-carriers.

EXAMPLE 25–8

Suppose the conductor shown in Figure 25–18 is copper and is carrying a current of 10 A in a magnetic field of 0.5 T. The width of the conductor d is 1 cm and the thickness is 1 mm. Find the Hall potential across the width of the conductor.

SOLUTION

The Hall potential is given by Equation (25–26):

$$V_H = \frac{BI}{neb}$$

Copper has a density ρ of 8.92×10^6 g/m^3, a molecular weight (MW) of 63.546 g/mol. Assuming that each copper atom contributes one electron to the current, the number n of conduction electrons per unit volume is

$$n = \frac{\rho N_A}{(\text{MW})}$$

where N_A is Avogadro's number: 6.022×10^{23} atoms/mol. Substituting the appropriate values:

$$n = \frac{\left(8.92 \times 10^6 \ \frac{\text{g}}{\text{m}^3}\right)\left(6.022 \times 10^{23} \ \frac{\text{electrons}}{\text{mol}}\right)}{63.546 \ \frac{\text{g}}{\text{mol}}}$$

$$n = 8.45 \times 10^{28} \ \frac{\text{electrons}}{\text{m}^3}$$

Substituting this and other values into Equation (25–26) yields

$$V_H = \frac{BI}{neb}$$

$$V_H = \frac{(0.5 \text{ T})(10 \text{ A})}{\left(8.45 \times 10^{28} \ \frac{\text{electrons}}{\text{m}^3}\right)\left(1.602 \times 10^{-19} \ \frac{\text{coulomb}}{\text{electron}}\right)\left(1 \times 10^{-3} \text{ m}\right)}$$

$$V_H = \boxed{3.69 \times 10^{-7} \text{ V}}$$

While this potential difference is very small for conductors, the corresponding potential difference for semiconductors is much greater. [See Equation (25–26): n is smaller for semiconductors than for ordinary conductors.] For this reason semiconductors are useful as probes in measuring magnetic fields by the Hall effect.

Analysis of the Hall effect gives us a clearer understanding of the nature of the force on a current-carrying conductor in a magnetic field. The force on the conductor is actually an electric force arising from the Hall field. Note that in Figure 24–17 the net sideways force on the moving charge is zero:

$$\mathbf{F}_E + \mathbf{F}_M = 0$$

The magnetic force $\mathbf{F}_M$ is produced by a field external to the conductor, whereas the electric force $\mathbf{F}_E$ arises within the conductor due to the Hall effect. The force on the conductor is (by Newton's third law) equal and opposite to the electric force on the charge-carriers. So we see that the magnetic force on a conductor is actually electrical in nature.

Linear Mass Spectrometer

Charged particles may be sorted according to their charge-to-mass ratio q/m by a device illustrated in Figure 25–19. The material to be analyzed is placed in the oven and heated to a temperature high enough to produce a gas of ionized particles. The particles leave the oven with a relatively low velocity and are accelerated by a potential difference between the oven and an aperture. The particles then leave the aperture with velocities that have essentially the same component in the x direction. Since the aperture does not collimate the charged particles perfectly, the particles may also have a component of velocity perpendicular to the x direction. After leaving the aperture, the particles enter a longitudinal magnetic field, causing them to execute a helical trajectory. Because the cyclotron frequency is the same for all particles having the same q/m ratio, after one turn all such particles will cross the axis at the same point (if their x components of velocity are the same).

Let us now solve for the charge-to-mass ratio in terms of the other parameters. The x component of the velocity v_x of a particle leaving the aperture is obtained by the energy relation:

$$qV = \tfrac{1}{2}mv_x^{\,2}$$

Solving for $v_x^{\,2}$:

$$v_x^{\,2} = 2V\left(\frac{q}{m}\right) \tag{25–27}$$

Another expression for v_x is

$$v_x = \frac{L}{T}$$

where T is the time for the particle to execute one turn of its helical trajectory. It is equal to the reciprocal of the cyclotron frequency f of the particle, given by Equation (25–6):

$$f = \frac{1}{2\pi} B\left(\frac{q}{m}\right)$$

Therefore:

$$v_x = \frac{1}{2\pi} BL\left(\frac{q}{m}\right)$$

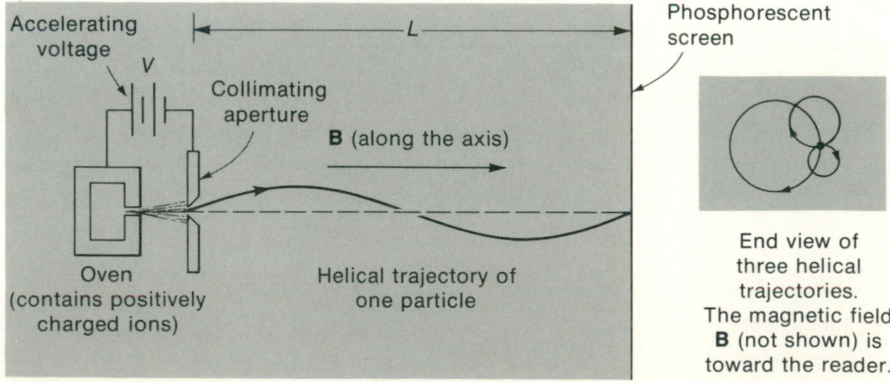

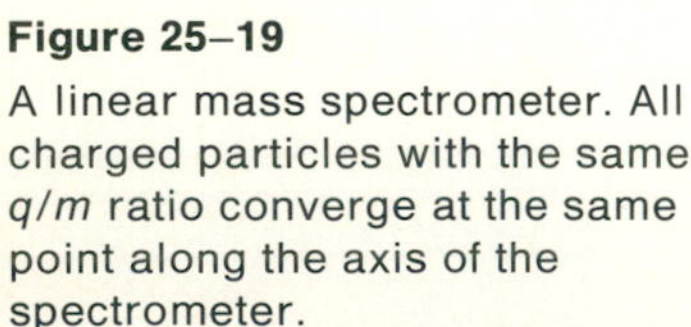

Figure 25–19

A linear mass spectrometer. All charged particles with the same q/m ratio converge at the same point along the axis of the spectrometer.

Substituting this expression for v_x into Equation (25–27) and solving for q/m yields

$$\frac{q}{m} = \frac{8\pi^2 V}{B^2 L^2} \tag{25–28}$$

In practice, a small fixed collector of charged particles is placed on the axis of the spectrometer. The potential V is adjusted until the collection of charges is a maximum, indicating the convergence of particles. The ratio q/m can then be calculated using Equation (25–28).

EXAMPLE 25–9

An electron microscope produces a magnified image on a photographic plate, utilizing an electron beam rather than light rays. The electron beam is "focused" by a magnetic field in the same way that a linear mass spectrometer converges charged particles. (a) Find the magnitude of the minimum magnetic field that will focus 10 keV electrons at a distance of 10 cm from the source of electrons. (b) Calculate another value of the magnetic field that will also produce a focusing of the electrons.

SOLUTION

(a) The focusing of an electron beam is identical to the operation of the linear mass spectrometer. Therefore Equation (25–28) is applicable:

$$\frac{q}{m} = \frac{8\pi^2 V}{B^2 L^2}$$

Solving for the magnetic field B:

$$B = \frac{\pi}{L}\left(\frac{8Vm}{q}\right)^{1/2}$$

Substituting the appropriate values:

$$B = \frac{\pi}{0.1\ \text{m}}\left[\frac{(8)(10^4\ \text{V})(9.11 \times 10^{-31}\ \text{kg})}{(1.602 \times 10^{-19}\ \text{C})}\right]^{1/2}$$

$$B = \boxed{2.12 \times 10^{-2}\ \text{T}}$$

(b) This is the minimum field required to produce one turn of the helical paths of the electrons. Equation (25–28) indicates that if the magnitude of B were doubled, *two* turns of the helical paths would be executed in the same distance L, again producing a focused spot. Therefore, focusing would also occur for:

$$B = \boxed{4.24 \times 10^{-2}\ \text{T}}$$

25.7 Magnetic Flux

When we discussed electric fields, we defined the *electric flux* Φ_E as:

$$\Phi_E = \int \mathbf{E} \cdot d\mathbf{A}$$

Corresponding to this definition of electric flux, the definition of **magnetic flux** Φ_B is:

MAGNETIC FLUX Φ_B

$$\Phi_B \equiv \int \mathbf{B} \cdot d\mathbf{A} \tag{25–29}$$

Magnetic flux is measured in units of tesla-meters² (T·m²), also called a *weber*[5] (Wb).

EXAMPLE 25–10

A uniform magnetic field $B = 2 \times 10^{-3}$ T is perpendicular to the plane of a wire loop of radius 3 cm. (a) Find the magnetic flux Φ_B that the loop encloses (called the *flux linkage*). (b) If the loop were tilted so its plane makes an angle of 30° with respect to the field direction, find the magnetic flux that now passes through the loop.

SOLUTION

(a) As shown in Figure 25–20, the area vector **A** is parallel to the uniform field **B**. Equation (25–29) reduces to

$$\Phi_B = \mathbf{B} \cdot \mathbf{A}$$

$$\Phi_B = BA \cos \theta$$

$$\Phi_B = (2 \times 10^{-3}\ \text{T})(\pi)(0.03\ \text{m})^2(1) = \boxed{5.65 \times 10^{-6}\ \text{Wb}}$$

(b) When the plane of the loop makes an angle of 30° with the field direction, the vector **A** (normal to the plane) makes an angle of 60° with **B**. Therefore:

$$\Phi_B = BA \cos \theta$$

$$\Phi_B = (2 \times 10^{-3}\ \text{T})(\pi)(0.03\ \text{m})^2(\cos 60°) = \boxed{2.83 \times 10^{-6}\ \text{Wb}}$$

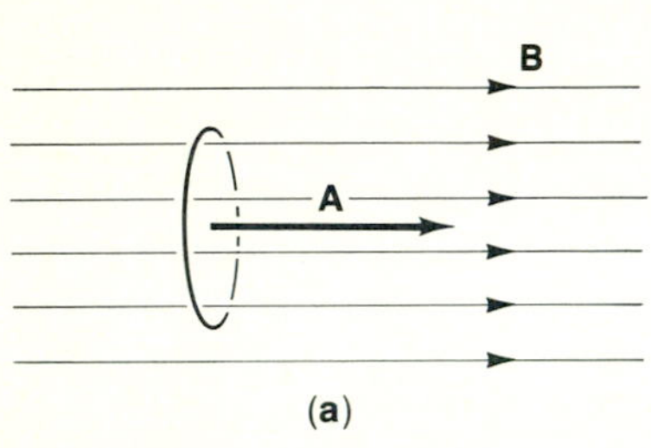

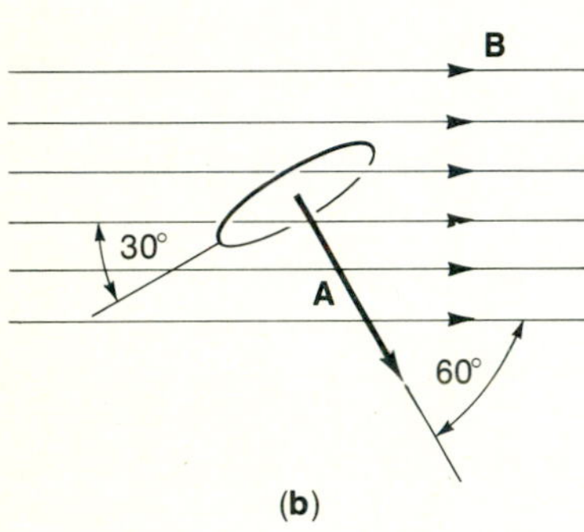

Figure 25–20
Example 25–10

25.8 Comments About Units

A difficulty arises in electricity and magnetism because many quantities are given special names in honor of the early investigators. This obscures the more fundamental units of meters, kilograms, seconds, and coulombs, making a check of the consistency of units in a given equation difficult. Furthermore, the same quantity may be expressed in a variety of ways, depending on the problem. For example, here is a partial list of the different units that electric and magnetic fields may have (the unit listed first is the most commonly used):

Electric Field E

$$\left[\frac{\text{V}}{\text{m}}\right] = \left[\frac{\text{N}}{\text{C}}\right] = \left[\frac{\text{T}\cdot\text{m}}{\text{s}}\right]$$

Magnetic Field B

$$[\text{T}] = \left[\frac{\text{Wb}}{\text{m}^2}\right] = \left[\frac{\text{N}\cdot\text{s}}{\text{C}\cdot\text{m}}\right] = \left[\frac{\text{V}\cdot\text{s}}{\text{m}^2}\right] = \left[\frac{\text{N}\cdot\text{m}^3}{\text{A}}\right] = \left[\frac{\text{H}\cdot\text{A}}{\text{m}^2}\right] = \left[\frac{\text{W}\cdot\text{H}}{\text{V}\cdot\text{m}^2}\right]$$

(The unit H, which will be defined in the next chapter, stands for *henries*.)

Because of this variety, we again stress the importance of making certain all numerical values are expressed in SI units before substituting them into equations. Then one may confidently write the answer in the most appropriate SI units, even though a consistency check is not carried out for each problem.

[5] This unit honors Wilhelm Weber (1814–1891), a German physicist who did theoretical and experimental work on magnetism. The unit is older than the *tesla*. Therefore, in many texts the magnetic field is referred to in units of *webers per square meter*, rather than in units of *tesla*.

Summary

The *magnetic induction* or *magnetic flux density* **B** (commonly called the *magnetic field*) is defined from this relation:

$$\mathbf{F} = q\mathbf{v} \times \mathbf{B}$$

where **F** is the force on a charge q moving in the field with velocity **v**. The unit is the *tesla* [T].

The *Lorentz force law* expresses the forces on a charge in the presence of both **E** and **B** fields:

$$\mathbf{F} = q(\mathbf{E} + \mathbf{v} \times \mathbf{B})$$

The *force on a current-carrying conductor* of length ℓ carrying current I in the presence of a magnetic field is

$$\mathbf{F} = I\boldsymbol{\ell} \times \mathbf{B}$$

The *magnetic dipole moment* $\boldsymbol{\mu}$ of a current loop is

$$\boldsymbol{\mu} = I\mathbf{A} \qquad \text{(in A·m}^2\text{)}$$

where the direction of $\boldsymbol{\mu}$ is given by the right-hand rule: The fingers curl around in the direction of the current and the extended thumb points in the direction of $\boldsymbol{\mu}$.

The *torque* $\boldsymbol{\tau}$ on a magnetic dipole in a magnetic field is

$$\boldsymbol{\tau} = \boldsymbol{\mu} \times \mathbf{B}$$

Note the similarity with the electric dipole case: $\boldsymbol{\tau} = \mathbf{p} \times \mathbf{E}$.

The *potential energy* U of a magnetic dipole in a magnetic field is

$$U = -(\boldsymbol{\mu} \cdot \mathbf{B}) \qquad \text{(where } U \equiv 0 \text{ for } \boldsymbol{\mu} \text{ and } \mathbf{B} \text{ at } 90^\circ\text{)}$$

Note the similarity with the electric case: $U = -(\mathbf{p} \cdot \mathbf{E})$.

The *magnetic flux* Φ_B is

$$\Phi_B = \int \mathbf{B} \cdot d\mathbf{A} \qquad \text{(in Wb)}$$

Note the similarity with the electric case: $\Phi_E = \int \mathbf{E} \cdot d\mathbf{A}$.

Questions

1. Which pairs of vectors in the equation $\mathbf{F} = q(\mathbf{v} \times \mathbf{B})$ are always perpendicular to each other and which are not necessarily so?
2. An oscilloscope has a cathode ray tube, which at one end produces a stream of electrons that travel the length of the tube and strike its face, forming a light spot. By observing the spot while orienting the tube in various directions, how is it possible to detect magnetic fields as well as electric fields? How can the fields be distinguished?
3. An electron, in passing between the poles of a magnet, experiences a change in momentum. Where is the source of the force required to produce such a change in momentum?
4. A *cloud chamber* consists of a chamber filled with supersaturated water vapor. A charged particle passing through the chamber leaves a trail of ions upon which small water droplets form, thus making the particle's path visible. A uniform magnetic field is often imposed upon the chamber, so that the *sign* of the charged particle as well as its energy can be determined. Electrons often produce spiral tracks rather than circular tracks. Why?

5. Using simple equipment, is it easier to deflect an electron beam by an electric field or by a magnetic field?

6. The speed of a charged particle moving in only an electric field may or may not change, while the speed of a charged particle moving in only a magnetic field does not change. Explain.

7. An electron with a kinetic energy greater than its rest energy has a circular orbit in a magnetic field. Is the radius of the orbit larger or smaller than that predicted using nonrelativistic formulas? Explain.

8. A current-carrying loop lies on the top of a table. Suddenly a vertical magnetic field penetrates the table top. What changes in external forces does the loop experience?

9. Conventional current in one direction through a conductor is equivalent to electron flow in the opposite direction. Is a magnetic force on the conductor the same whether we consider the current to be electron current, conventional current, or a mixture of both?

10. A magnetic dipole is aligned with a magnetic field so that it is in stable equilibrium with the field. The work required to turn the dipole end-for-end is $2\,\mu$B. Does the work required to do this depend on the initial orientation of the dipole?

11. The magnetic moment of a magnetic dipole is antiparallel to a magnetic field. Is there a torque on the dipole? Is the dipole in stable equilibrium, unstable equilibrium, or not in static equilibrium?

12. The precise measurement of an electric field involves the measurement of the force on a charge that is necessarily very small. For similar reasons, does the precise measurement of a magnetic field involve the measurement of the torque on a magnetic dipole that must have a small magnetic dipole moment?

13. In n-type semiconductors, electrons are the principal current-carriers, while in p-type semiconductors, the current is carried by deficiencies of electrons called *holes*, which behave like positive charges. How can the Hall effect be used to determine whether a semiconductor is n-type or p-type?

14. How would you design a magnetic compass without using iron or any other magnetic material?

15. Using a galvanometer movement as a start, how would you design an electric motor?

16. Why is it usually desirable to have a large number of turns of wire in the rotating coil of a galvanometer?

17. Why is a linear mass spectrometer designed for the analysis of ionized atoms unsuitable for electrons?

18. Why is the Hall potential greater for semiconductors than for conductors?

19. Can the drift velocity of charge carriers in a conductor be measured using the Hall effect? If so, how?

20. How is the description of magnetic flux using no more than a number of webers incomplete? Why is magnetic flux Φ_B not a vector quantity?

Problems

25A–1 At a certain location, the horizontal component of the earth's magnetic field is $30\ \mu$T in a northerly direction. An electron moving westward perpendicular to this field has enough speed so that the magnetic force on the electron balances its weight. Find the speed of the electron. (The answer reveals one difficulty in "weighing" a single electron.)

Answer: 1.86×10^{-6} m/s

25A–2 At the equator, near the surface of the earth, the magnetic field is approximately $50\ \mu$T northward, and the electric field is about 100 N/C downward. Find the gravitational, electric, and magnetic forces on a 100 eV electron moving eastward in this environment.

25A–3 An electron (mass $= m_e$ and charge $-e$), a proton (mass $= 1836m_e$ and charge $+e$), and an alpha particle (mass $= 4 \times 1836m_e$ and charge $+2e$) all have the same kinetic energy

(nonrelativistic) as they move in circular orbits in a uniform magnetic field. In terms of the radius R of the electron's path, find the radii of the paths of the proton and alpha particle.
Answer: $R_\alpha = R_p = 42.8R$

25A–4 A 0.15 MeV beta particle (electron) emitted during the radioactive decay of ^{14}C enters a magnetic field of 0.04 T in a direction perpendicular to the magnetic field. Find the radius of curvature of the particle's trajectory. The particle may be assumed to have a nonrelativistic speed.

25A–5 A proton moves in a circle perpendicular to a magnetic field. If the radius of the proton's path is 1.00 cm and the field is 0.5 T, find the kinetic energy of the proton in units of electron volts. *Answer:* 1.20 keV

25A–6 A 4.2 MeV alpha particle (a helium nucleus consisting of two protons and two neutrons) emitted during the radioactive decay of ^{238}U enters a magnetic field of 0.04 T in a direction perpendicular to the velocity of the particle. Find the radius of curvature of the particle's trajectory.

25A–7 One type of radar oscillator, a magnetron, utilizes the cyclotron frequency of electrons circulating in a magnetic field to determine the transmitting frequency. Find the magnitude of the magnetic field necessary to generate radar radiation with a 3 cm wavelength. *Answer:* 0.357 T

25A–8 A weighing scale supports a 12 V battery, to which a rigid rectangular wire hoop is attached, as shown in Figure 25–21. The lower portion of the hoop is in a magnetic field $B = 0.10$ T. If the total mass of the battery and wire hoop is 100 g, calculate the resistance of the wire necessary for the scales to indicate zero weight. Which pole of the battery is positive?

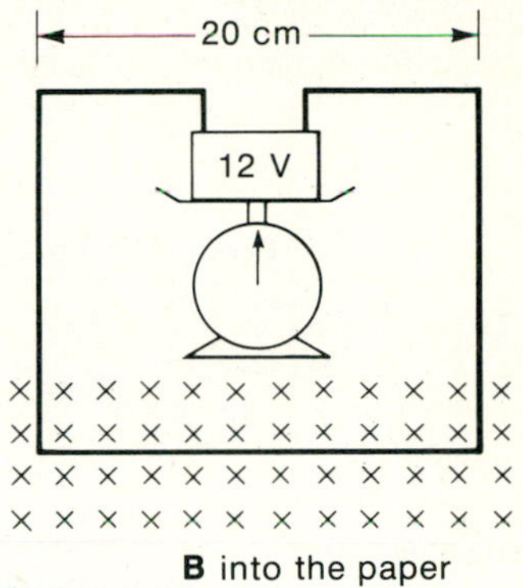

Figure 25–21
Problem 25A–8

25A–9 A rectangular loop of current-carrying wire is oriented as shown in Figure 25–22. A magnetic field $\mathbf{B} = 0.15\,\hat{\mathbf{x}}$ (in tesla) exerts a torque on the loop. If $a = 8$ cm, $b = 12$ cm, $\theta = 30°$, and $I = 2$ A, calculate the torque on the loop.
Answer: $\boldsymbol{\tau} = (-1.44 \times 10^{-3}\ \text{N·m})\,\hat{\mathbf{z}}$

25A–10 A silver ribbon 4 cm wide and 0.1 mm thick carries a current of 5 A. If the plane of the ribbon is perpendicular to a magnetic field of 0.15 T, calculate the Hall voltage across the ribbon. Assume that on the average, each silver atom contributes one electron to the current flow. The density of silver is 10.5 g/cm^3 and its atomic weight is 107.87 g/mol.

25A–11 A galvanometer has a full-scale sensitivity of 50 μA. By what factor must the spring constant of the galvanometer movement be changed in order to change the full-scale sensitivity to 10 μA? *Answer:* $\frac{1}{5}$

25A–12 A velocity filter consists of magnetic and electric fields described by $\mathbf{E} = E\,\hat{\mathbf{z}}$ and $\mathbf{B} = B\hat{\mathbf{y}}$. If $B = 0.015$ T, find the value of E such that a 750 eV electron moving outward along the x axis will be undeflected.

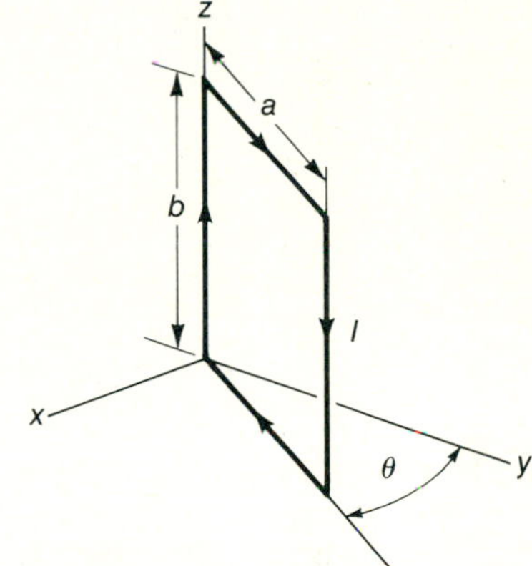

Figure 25–22
Problems 25A–9, 25B–3, and 25B–4

25A–13 Show that the units of magnetic dipole moment, ampere-meters^2, can also be expressed as joules/tesla.

25A–14 At a certain location in Michigan, the earth's magnetic field is 5.80×10^{-5} T and the *dip angle* is 74°. (The dip angle is between **B** and the horizontal.) Find the magnetic flux Φ_B that links a flat, horizontal coil of 10 cm diameter.

25B–1 An electron moves with a speed of 3×10^6 m/s outward along the x axis. Find the force on the electron if there is a magnetic field $\mathbf{B} = 0.4\,\hat{\mathbf{x}} + 0.7\,\hat{\mathbf{y}} + 0.3\,\hat{\mathbf{z}}$ (in tesla).
Answer: $\mathbf{F} = 1.44 \times 10^{-13}\,\hat{\mathbf{y}} - 3.36 \times 10^{-13}\,\hat{\mathbf{z}}$ (in newtons)

25B–2 A metal axle with conducting wheels rolls down a pair of inclined metal rails, as shown in Figure 25–23. A uniform magnetic field **B** is vertically upward between the rails. A battery is connected to the rails with the polarity indicated. The axle moving in the field **B** generates a current (see Sec. 26.3) and the system acts as a battery charger, forcing current into the + terminal. A constant speed v eventually results when $|dU_{grav}/dt|$ = the power input to the battery. Find v in terms of the symbols in the figure.

25B–3 Calculate the magnetic dipole moment $\boldsymbol{\mu}$ of the current loop shown in Figure 25–22. *Answer:* $\boldsymbol{\mu} = -Iab\cos\theta\,\hat{\mathbf{x}} + Iab\sin\theta\,\hat{\mathbf{y}}$

25B–4 Calculate the potential energy of the current loop shown in Figure 25–22. Use the numerical values given in Problem 25A–9.

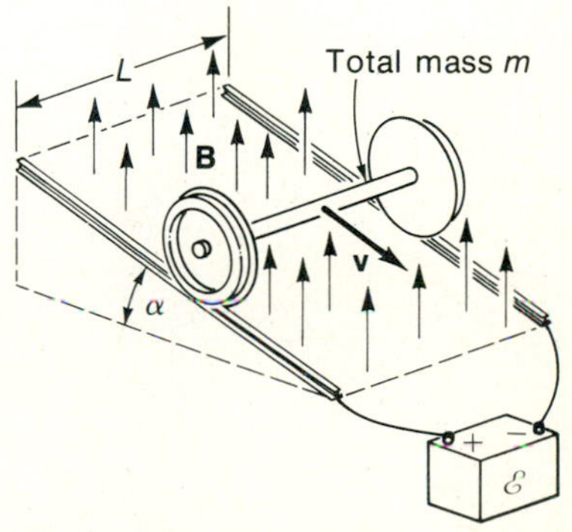

Figure 25–23
Problem 25B–2

25B–5 A circular current-carrying loop of wire experiences a maximum torque τ_0 when placed in a given magnetic field. If the same loop were re-formed to a smaller circular loop containing two turns of the wire, find the maximum torque on this loop in terms of τ_0.

Answer: $\tau = \tau_0/2$

25B–6 A circular hoop of wire with a radius R carries a current I. If the plane of the hoop is perpendicular to a uniform magnetic field B, the wire comes under tension T. Derive an expression for the tension in terms of R, I, and B. The leads that carry current to and from the hoop are parallel to the magnetic field.

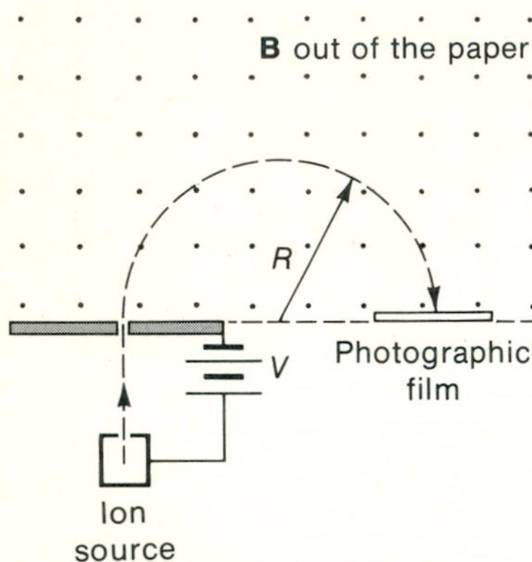

Figure 25–24
Problems 25B–7 and 25B–8

25B–7 As shown in Figure 25–24, in one type of mass spectrometer charged particles (mass m and charge q) are accelerated from rest by a potential difference V. They then enter a region of uniform magnetic field B perpendicular to the plane of the diagram. Starting with Newton's second law, derive an expression for the radius R of the particles' path in the field in terms of m, q, V, and B. *Answer:* $R = \sqrt{2mV/qB^2}$

25B–8 In the mass spectrometer shown in Figure 25–24, singly-charged lithium ions of mass 6 u and 7 u are accelerated by a potential difference of 900 V before they enter the uniform magnetic field $B = 0.040$ T. (One *unified mass unit* u $\equiv 1.66 \times 10^{-27}$ kg.) After traveling through a semicircle, they strike a photographic film, producing two spots on the film separated a distance x. Find x.

25B–9 A cyclotron at the University of California, Berkeley, has a diameter of 60 in (1.52 m) and operates with a magnetic field of 1.6 T. It can be used to accelerate deuterons, which have the same charge as a proton, but twice its mass.

(a) Find the frequency of the accelerating voltage applied to the dees when deuterons are accelerated.
(b) Find the kinetic energy (in MeV) of the emerging deuterons.
(c) Calculate (a) and (b) for protons.
(d) It is usually a major operation to change the frequency of a cyclotron, so for protons the field B is often reduced to a lower value without changing the original frequency. Find the final kinetic energy of protons following this procedure.
(e) Keeping the original frequency, find the magnetic field for alpha particles (mass = $4m_p$ and charge = $2e$).
(f) Again, keeping the original frequency, find the kinetic energy for alpha particles.
(g) If we raised the voltage applied to the dees, which of these answers would be different?

Answers: (a) 12.2 MHz (b) 35.4 MeV (c) 24.4 MHz, 70.8 MeV (d) 17.7 MeV (e) 1.60 T (f) 35.4 MeV (g) None

25C–1 The color purity of a color television set requires that the electron beam strike a given location on the face of the picture tube with an error of less than one millimeter. Show that a component of the earth's magnetic field perpendicular to the electron beam of about 10 μT may well deflect a 20 keV electron beam enough to affect color purity. (Note: The deflection corresponding to a circular trajectory may be approximated using the *sagitta formula*, explained in Appendix E.)

25C–2 An evacuated glass tube with a diameter of 8 cm has a uniform magnetic field $B = 5 \times 10^{-5}$ T throughout its volume, parallel to the axis of the tube. Electrons are injected into the tube at a point on the axis with a speed of 2×10^6 m/s.

(a) Find the largest angle the electron velocity may have with respect to the axis such that the subsequent spiral motion of the electrons will not strike the tube walls.
(b) How far along the tube does such an electron again cross the axis?

25C–3 An irregular open loop of current-carrying wire lies in the x-y plane. The current input to the loop is along the z axis, and the output is parallel to the z axis at $x = h$. A uniform magnetic field is described by $\mathbf{B} = B\,\hat{\mathbf{z}}$. Show that the net force on the loop is independent of the shape of the loop and that the force is given by $\mathbf{F} = -BhI\,\hat{\mathbf{y}}$.

25C–4 A rigid hoop of wire (mass m and radius r) rests on a horizontal surface in a region where there is a uniform magnetic field $\mathbf{B} = B_x\,\hat{\mathbf{x}} + B_y\,\hat{\mathbf{y}}$, where $\hat{\mathbf{y}}$ is vertically upward. Find the minimum current I that will barely cause one side of the hoop to lift off the surface.

25C–5 The maximum torque on a current-carrying rectangular loop of wire placed in a magnetic field depends on the shape of the loop. Show that for a given length of wire, the greatest maximum torque is achieved when the loop is a square.

25C–6 A wire of length ℓ is formed into a flat, circular coil of N turns.

(a) Show that for a given current I in the coil, the greatest magnetic dipole moment is for $N = 1$.

(b) Explain why a coil of any shape other than circular would have a smaller magnetic moment.

25C–7 In the Bohr model of the hydrogen atom, the electron moves in a circle about the proton, with the Coulomb force being the centripetal force necessary for circular motion. In the lowest energy state, the radius of the path is 52.9 pm (1 pm = 10^{-12} m).

(a) Find the equivalent current the moving electron generates.

(b) Find the magnetic dipole moment of this current loop (called the *Bohr magneton*).

Answers: (a) 1.05×10^{-3}A (b) 9.27×10^{-24} A · m^2

25C–8 A thin rod of length ℓ is made of a nonconducting material and carries a uniform charge per unit length λ. The rod is rotated with angular velocity ω about an axis through its center, perpendicular to the length of the rod. Show that the magnetic dipole moment is $\omega\lambda\ell^3/24$. (Hint: Consider the charge dq located within the element dx a distance x from the axis.)

25C–9 A disk of nonconducting material has a radius R, and on one side there is a surface charge density σ. The disk is rotated with angular velocity ω about its axis. Show that the magnetic dipole moment is $\omega\sigma\pi R^4/4$. (Hint: Consider the current loop formed by the motion of the charge within the annular ring of radius r and width dr.)

chapter 26
Magneto-dynamics

The rotating armatures of every generator and motor in this age of electricity are steadily proclaiming the truth of the relativity theory to all who have ears to hear.

LEIGH PAGE
[*American Journal of Physics*, *43*, 330 (1975)]

The last chapter described static magnetic fields and the forces they exert on moving charges. In this chapter we will discuss the origin of static magnetic fields. One interesting fact in electromagnetism is that a steady current of moving charges produces a static magnetic field. We will also show a satisfying symmetry between electric and magnetic fields. In particular, a *changing* magnetic field produces a *static* electric field, and a *changing* electric field produces a *static* magnetic field. The English physicist James Clerk Maxwell (1831–1879) put the finishing touch on the elegant electromagnetic theory, which expressed this symmetry between electricity and magnetism.

26.1 The Biot-Savart Law

In our study of electric fields, we found they had their origin in electrical charges. To find the field at a given point due to an arbitrary distribution of charges, we considered that each *charge element* dq produces a field $d\mathbf{E}$ at a distance r from the charge according to Coulomb's law for electric fields. It is an *inverse-square* law:

$$d\mathbf{E} = \left(\frac{1}{4\pi\varepsilon_0}\right)\frac{dq}{r^2}\,\hat{\mathbf{r}}$$

Here, the unit vector $\hat{\mathbf{r}}$ extends *from the source of the field* (the charge dq) *toward the point in question.* To find the total electric field $\mathbf{E}$, we sum over all the charge elements present.

We will now introduce a similar equation that describes how an element of current-carrying wire $I\,d\ell$ produces a magnetic field $d\mathbf{B}$ at a point a distance r from the element. Consider a current-carrying wire of arbitrary shape (Figure 26–1). Jean Baptiste Biot and Felix Savart first gave the expression for the field $d\mathbf{B}$ produced at a distance r from an element of the wire $d\ell$ carrying a current I. It is an inverse-square law known as the **Biot-Savart law:**

BIOT-SAVART LAW

$$d\mathbf{B} = \left(\frac{\mu_0}{4\pi}\right)\frac{I\,d\ell \times \hat{\mathbf{r}}}{r^2} \qquad \textbf{(26–1)}$$

The direction of the vector $d\ell$ is along the wire in the direction of the current I. The unit vector $\hat{\mathbf{r}}$ is *from the source of the field* (the current-carrying element $I\,d\ell$) *toward the point in question.* Thus, $\mathbf{r} = r\hat{\mathbf{r}}$. To find the total magnetic field $\mathbf{B}$ at the point, we sum over all the current-carrying elements present.

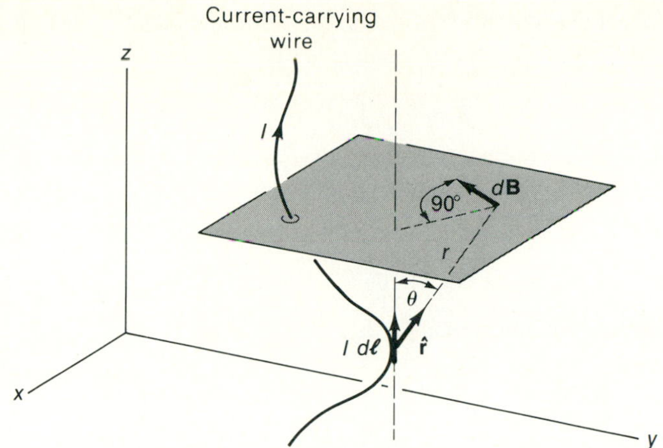

Figure 26–1
The Biot-Savart law. In the figure, the incremental magnetic field $d\mathbf{B}$ lies in the plane of the elevated square, perpendicular to both $\hat{\mathbf{r}}$ and $I\,d\ell$. The unit vector $\hat{\mathbf{r}}$ is directed from the current-carrying element $I\,d\ell$ toward the point where the field $d\mathbf{B}$ exists a distance r away. (*Note:* $\mathbf{r} = r\hat{\mathbf{r}}$.) The cross-product ($I\,d\ell \times \hat{\mathbf{r}}$) involves the angle θ.

The constant μ_0 is called the **permeability of free space:**

PERMEABILITY OF FREE SPACE

$$\mu_0 = 4\pi \times 10^{-7}\ \frac{\text{Wb}}{\text{A·m}}$$

This numerical value is chosen to be consistent with the definition of the unit of current, the *ampere* (A). (The constant μ_0 should not be confused with the symbol for the magnetic dipole moment $\boldsymbol{\mu}$.) The most significant feature of Equation (26–1) is that magnetic fields, like electric fields, are inverse-square fields. The calculation of the total field for all but very simple configurations of current-carrying conductors is quite cumbersome, so we will restrict our examples to simple, yet important, symmetrical configurations.

EXAMPLE 26–1

Calculate the magnetic field 10 cm from a very long, straight wire carrying a current of 10 A.

SOLUTION

We first develop a general expression for the field in the vicinity of a straight current-carrying conductor. In Figure 26–2(b), the incremental field $d\mathbf{B}$ due to the current element $I\,d\ell$ is directed into the plane of the figure at point P. Equation (26–1) states:

$$d\mathbf{B} = \left(\frac{\mu_0}{4\pi}\right)\frac{I\,d\ell \times \hat{\mathbf{r}}}{r^2}$$

The magnitude of $d\mathbf{B}$ is given by

$$dB = \left(\frac{\mu_0}{4\pi}\right)\frac{I d\ell}{r^2}\sin\theta$$

where θ is the angle between $I\,d\ell$ and $\hat{\mathbf{r}}$.

Introducing the perpendicular distance a from the point to the wire, we note that $r^2 = \ell^2 + a^2$ and that $\sin\theta = a/\sqrt{\ell^2 + a^2}$. Thus:

$$dB = \left(\frac{\mu_0}{4\pi}\right)\frac{Ia}{(\ell^2 + a^2)^{3/2}}\,d\ell$$

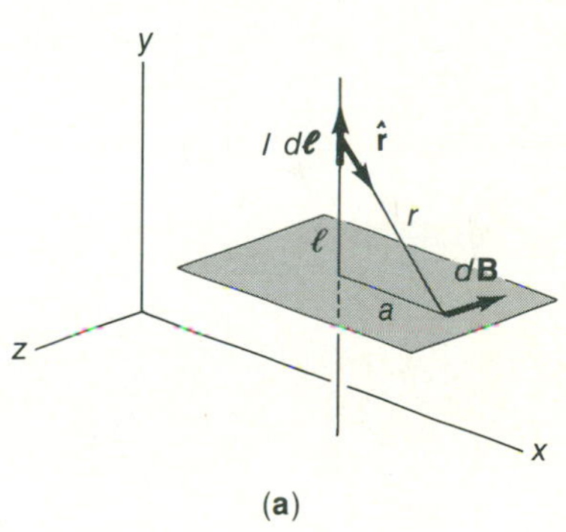

(a)

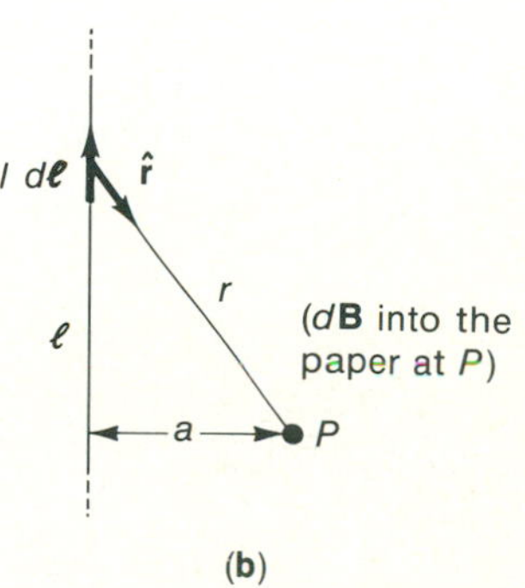

(b)

Figure 26–2
Example 26–1

Since each element produces a field $d\mathbf{B}$ in the same direction, the total field B is merely the scalar sum $\int dB$:

$$B = \frac{\mu_0 I a}{4\pi} \int_{-\infty}^{+\infty} \frac{d\ell}{(\ell^2 + a^2)^{3/2}}$$

Using the table of integrals in Appendix G, we obtain

$$B = \frac{\mu_0 I a}{4\pi} \left[\frac{\ell}{a^2(\ell^2 + a^2)^{1/2}} \right]_{-\infty}^{+\infty}$$

Substituting the limits of integration yields

MAGNETIC FIELD DUE TO A CURRENT IN A LONG, STRAIGHT WIRE

$$B = \frac{\mu_0 I}{2\pi a} \qquad \textbf{(26–2)}$$

Substituting numerical values, all of which must be in SI units:

$$B = \frac{\left(4\pi \times 10^{-7} \frac{\text{Wb}}{\text{A·m}}\right)(10\text{ A})}{2\pi(0.10\text{ m})} = \boxed{2.00 \times 10^{-5}\text{ T}}$$

The direction of **B** is found from the cross-product in the Biot-Savart law ($I\,d\ell \times \hat{\mathbf{r}}$). From symmetry considerations, the field lines form concentric circles surrounding the wire. Their direction is easily remembered using the **right-hand rule** as defined in Figure 26–3(c).

Figure 26–3
The magnetic field associated with a straight, current-carrying conductor.

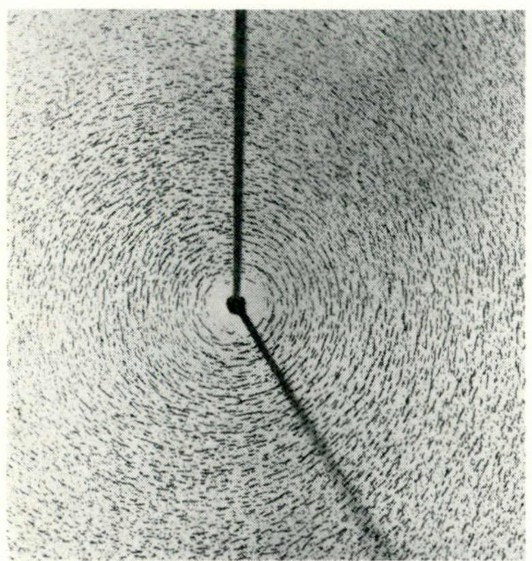

(**a**) If iron filings are sprinkled on a horizontal plane perpendicular to a straight, current-carrying wire, they form a pattern that suggests the magnetic field lines.

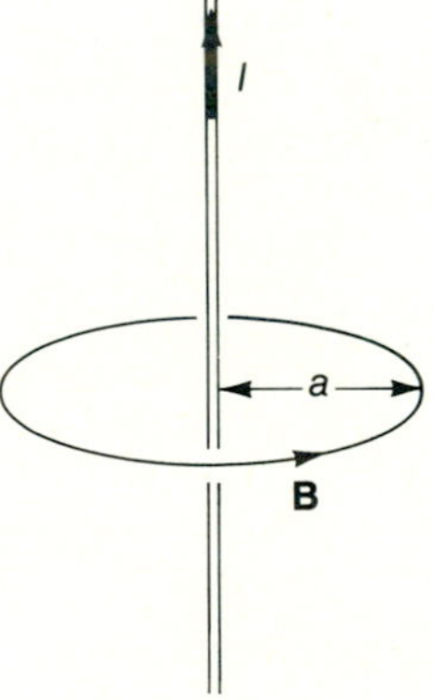

(**b**) One of the field lines that circle the wire symmetrically.

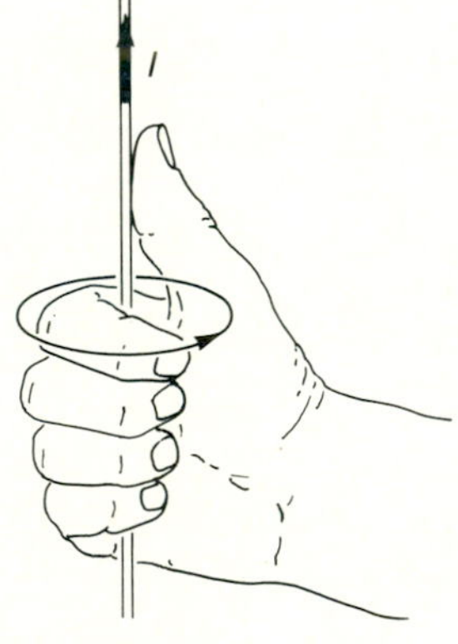

(**c**) The magnetic field lines circle the conductor in the direction of the fingers of the right hand when the extended thumb is in the direction of the current.

The previous example illustrates a very important characteristic of magnetic field lines: *Magnetic field lines are closed loops.* This closure of magnetic field lines is in contrast to electrostatic field lines, which always terminate on plus and minus charges.

Having developed an expression for the magnetic field around a long, straight wire enables us to define the *ampere* and thus the *coulomb*. Consider two parallel conductors, each carrying the same current I in the same direction, as shown in Figure 26–4. The field produced by wire A a distance a from the wire is given by Equation (26–2):

$$B = \frac{\mu_0 I}{2\pi a}$$

The direction of this field is straight down (in the $-z$ direction). The force $d\mathbf{F}$ exerted on an incremental length $d\ell$ of wire B is, by Equation (25–11):

$$d\mathbf{F} = I\,d\boldsymbol{\ell} \times \mathbf{B}$$

Since $I\,d\ell$ and $\mathbf{B}$ are perpendicular:

$$dF = IB\,d\ell$$

Substituting the expression for the field B and rearranging, we have

$$\frac{dF}{d\ell} = \frac{\mu_0 I^2}{2\pi a}$$

where $\mu_0 = 4\pi \times 10^{-7}$ Wb/A·m. Then:

$$\frac{dF}{d\ell} = \frac{2I^2}{a} \times 10^{-7}\,\frac{\text{Wb}}{\text{A·m}}$$

If the separation of the wires is one meter and the current in each wire is one ampere, the force of attraction per unit length of wire is

$$\frac{dF}{d\ell} = 2 \times 10^{-7}\,\frac{\text{Wb·A}}{\text{m}^2}$$

or

$$\frac{dF}{d\ell} = 2 \times 10^{-7}\,\frac{\text{N}}{\text{m}}$$

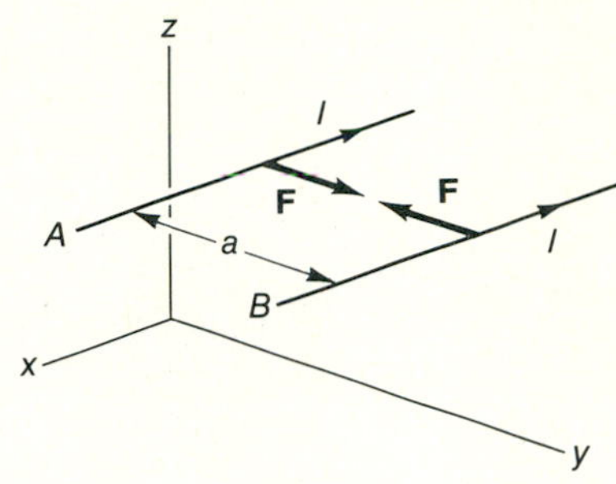

Figure 26–4
The definition of the ampere. Two parallel conductors one meter apart will experience a mutual force of attraction equal to exactly 2×10^{-7} N/m if each conductor carries a current of one ampere in the same direction.

DEFINITION OF THE AMPERE **If one ampere is in the same direction in each of two long, parallel conductors one meter apart, the conductors will be attracted to each other with a force of *exactly* 2×10^{-7} N per meter of length.**

This basic definition of the ampere is the crucial link between electrical quantities and mechanical quantities. It extends the SI system to include electrical units by defining the *ampere* in terms of the meter, kilogram, and second. As mentioned in Chapter 22, it also leads to the *coulomb*, since that unit is defined as the amount of charge per second passing a cross section of a conductor carrying a steady current of one ampere. Mechanical experiments that measure forces between current-carrying wires are much easier to carry out and give greater precision than experiments that measure the coulomb force between charges. Thus there are strong practical reasons for basing the fundamental electrical definition on the ampere rather than on the coulomb.

The two constants μ_0 and ε_0 are related, since the first arises when analyzing forces between current-carrying elements and the second arises from forces between charge elements. And, of course, currents and charges are intimately related. Formally (through the ampere), we *define* μ_0 as exactly $4\pi \times 10^{-7}$ Wb/A·m, so ε_0 becomes a constant of nature that must be determined *experimentally*.

EXAMPLE 26–2

A wire bent into a loop with a radius of 10 cm carries a current of 10 A (Figure 26–5). Find the magnetic field at the center of the loop.

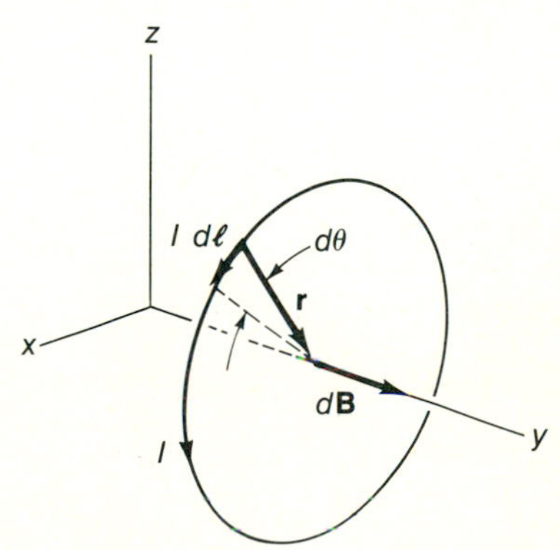

Figure 26–5
Example 26–2

SOLUTION

Each of the current elements $I d\ell$ and the unit vector **r** are perpendicular. Equation (26–1) therefore becomes

$$dB = \left(\frac{\mu_0}{4\pi}\right)\frac{I\,d\ell}{r^2}$$

We change to a more convenient variable of integration by expressing the element $d\ell$ as

$$d\ell = r\,d\theta$$

where $d\theta$ is the angle subtended by $d\ell$ from the center of the loop. Then:

$$dB = \left(\frac{\mu_0}{4\pi}\right)\frac{I}{r}\,d\theta$$

At the center of the loop, each field increment $d\mathbf{B}$ is in the same direction. So the total field B is merely the sum $\int dB$ integrated around the entire loop from $\theta = 0$ to $\theta = 2\pi$ rad. All the terms are constant except $d\theta$, so we have

$$B = \left(\frac{\mu_0}{4\pi}\right)\frac{I}{r}\int_0^{2\pi} d\theta$$

Integrating, and substituting limits:

$$B = \left(\frac{\mu_0}{4\pi}\right)\frac{I}{r}(2\pi - 0)$$

MAGNETIC FIELD AT THE CENTER OF A CURRENT-CARRYING LOOP

$$B = \frac{\mu_0 I}{2r} \tag{26–3}$$

Substituting numerical values:

$$B = \frac{\left(4\pi \times 10^{-7}\,\frac{\text{Wb}}{\text{A}\cdot\text{m}}\right)(10\text{ A})}{(2)(0.10\text{ m})}$$

$$B = \boxed{6.28 \times 10^{-5}\text{ T}}$$

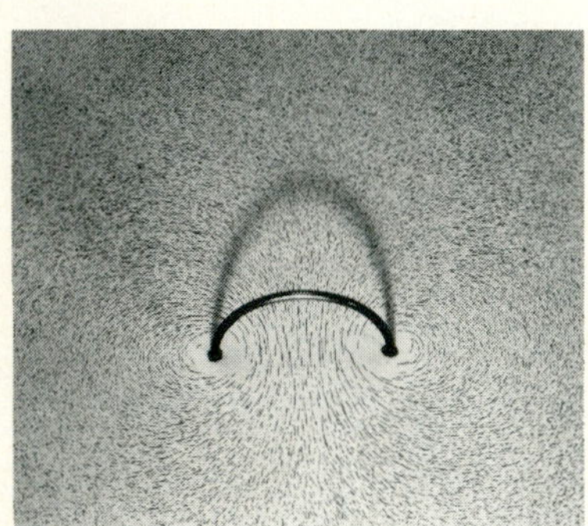

(**a**) A piece of paper placed horizontally is perpendicular to the plane of a current-carrying loop. If iron filings are sprinkled on the paper, they form a pattern of lines similar to the magnetic field in the plane.

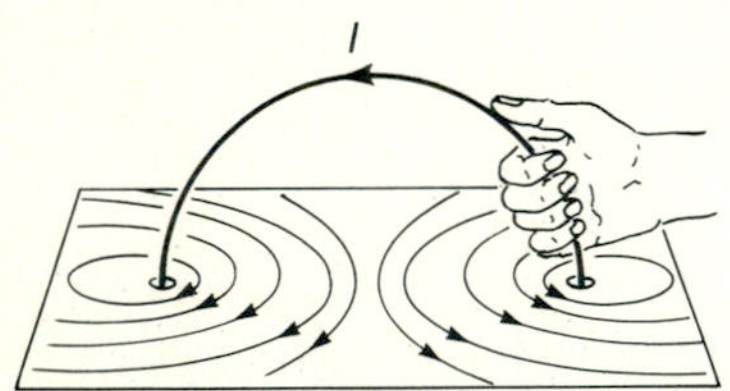

(**b**) The right-hand rule for determining the direction of magnetic field lines.

Figure 26–6
The magnetic field produced by a current-carrying loop.

Most practical devices used for the production of magnetic fields are constructed of loops or coils of wire. The usual right-hand rule determines the field direction. If the current-carrying wire is grasped with the fingers of the right hand so that the extended thumb is in the direction of the current, the curled fingers indicate the magnetic field direction. *Inside* the loop, the field at the center is along the axis of the loop. The field lines elsewhere are shown in Figure 26–6.

26.2 Ampère's Law

If the configuration of a current-carrying conductor is simple, an equivalent and simpler form of the Biot-Savart law, known as Ampère's law, may be used. The basic idea involves a *closed path of integration*, sometimes called an *ampere loop*. **Ampère's law** states that the integral of $\mathbf{B} \cdot d\ell$ around *any* closed path is $\mu_0 I$, where I is the current crossing any surface bounded by the path of integration.

AMPÈRE'S LAW $$\oint \mathbf{B} \cdot d\ell = \mu_0 I \tag{26–4}$$

We will now illustrate the use of Ampère's law for three important configurations.

(1) *The field of a long, straight, current-carrying wire* Although Ampère's law is true for *any* closed path, noting the *symmetry* of **B** we choose a path that coincides with a field line (for example, refer to Figure 26–3b). This is a wise choice because B is *constant in magnitude* along this path and therefore may be brought outside the integral sign. The line element $d\ell$ is chosen to be in the same direction as **B**, so the dot product gives $\cos 0^\circ = 1$. Thus, the integral $\oint d\ell$ is simply around a circle of radius a, which, of course, just equals the circumference $2\pi a$. The current passing through the circular area bounded by the path is I. Therefore:

$$B \oint d\ell = \mu_0 I$$

$$B(2\pi a) = \mu_0 I$$

MAGNETIC FIELD DUE TO A CURRENT IN A LONG, STRAIGHT WIRE $$B = \frac{\mu_0 I}{2\pi a}$$

This is the same expression obtained using the Biot-Savart law, Equation (26–2). Ampère's law and the Biot-Savart law are completely equivalent. The context of a problem determines which form is easier to use.

(2) *The field of a toroid* Our next example in the use of Ampère's law is that of a wire wound around a *toroid:* a donut-shaped coil, as illustrated in Figure 26–7. To find the field inside the windings of a toroidal coil, symmetry suggests that the appropriate path of integration is a circle of radius R in the plane of the toroid along the axis of the windings. The reason for this choice is that because of symmetry, **B** is constant in magnitude along such a path and is parallel to $d\ell$ everywhere on the circle. Thus B may be brought outside the integral sign. The total current through the integration loop is

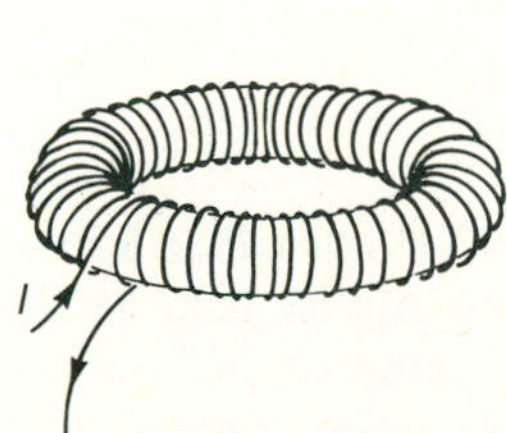

(a) A toroidal coil is formed by winding a current-carrying conductor around a toroid.

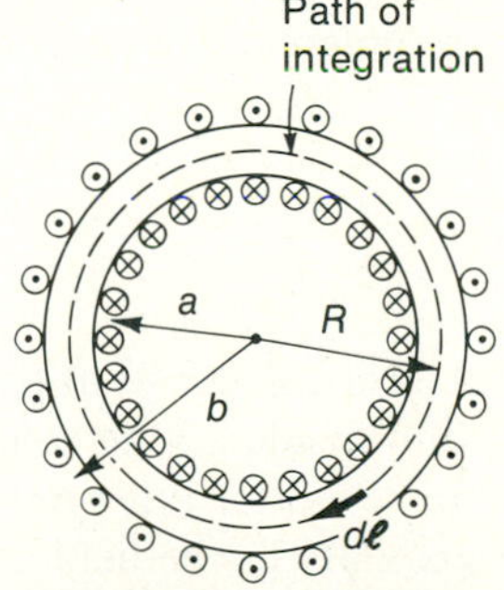

(b) In the inside edge of the coil windings, the current is into the plane of the paper; at the outside edge of the windings, the current is out of the paper.

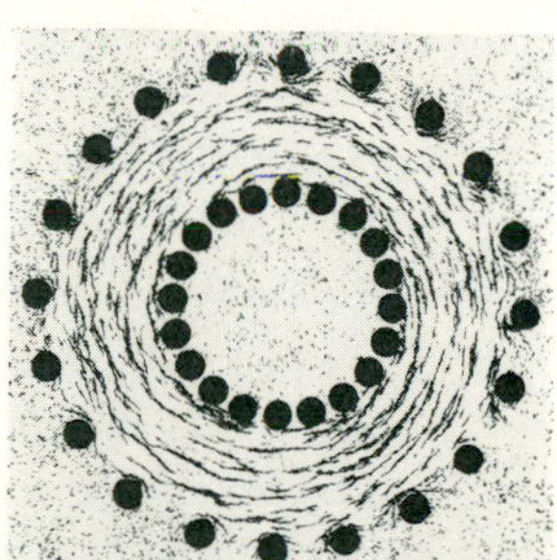

(c) Iron filings reveal the pattern of the magnetic field. Even for this loosely-wound toroid, the field is confined almost wholly within the windings.

Figure 26–7
A toroidal coil. From the right-hand rule, the magnetic field inside the windings in (b) is clockwise.

NI, where N is the total number of turns of wire around the toroid and I is the current through the wire. Applying Ampère's law:

$$\oint \mathbf{B} \cdot d\ell = \mu_0 I$$

we obtain:

$$B(2\pi R) = \mu_0 NI$$

MAGNETIC FIELD INSIDE THE WINDINGS OF A TOROIDAL COIL (average circumferential length of the toroid: $2\pi R$)

$$B = \frac{\mu_0 NI}{2\pi R} \tag{26-5}$$

Note that the field varies slightly within the windings, being somewhat stronger near the inner radius of the toroid. The shape of the toroid itself acts like a single large loop of wire of radius R carrying a current I. So the field outside the windıngs is relatively small. For this reason, a toroid is useful in electronic circuits whenever a magnetic field must be confined.

(3) *The field of a long solenoid* A solenoid is a straight coil of wire, as shown in Figure 26–8(a). Because of the relative ease of its fabrication, it is the most common configuration used to produce a magnetic field electrically. Calculation of the magnetic field is complicated for a loosely-wound solenoid that is short compared with its diameter. However, the field at the center of the solenoid can be closely approximated by considering an *ideal* solenoid: one that is long compared with the diameter, with the turns of wire close together, as in Figure 26–8(b). Just as we considered a parallel-plate capacitor to be a section of large concentric spheres, we may consider a solenoid to be a short section of a toroid with a very large outside diameter and nearly as large an inside diameter. The field will then be essentially uniform within the solenoid and confined to the solenoid's interior. The direction of the field lines is into one end of the section and out of the other end. Rewriting Equation (26–5), we have

$$B = \left(\frac{N}{2\pi R}\right)\mu_0 I$$

For a large value of R that does not change appreciably from the inner to outer diameter of the toroid, the quantity within the parentheses is the number of turns n per circumferential length of the toroid. Then:

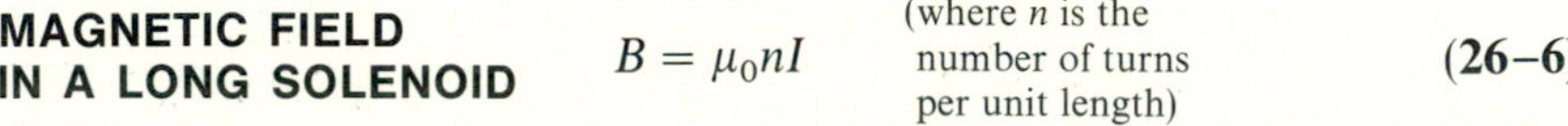

MAGNETIC FIELD IN A LONG SOLENOID

$$B = \mu_0 nI \quad \text{(where } n \text{ is the number of turns per unit length)} \tag{26-6}$$

Just a short bit of reasoning will lead us to the field at *one end* of a long solenoid. Consider the point inside a long solenoid equally far from either end. (The above equation is valid for this point.) By symmetry, each half of the long coil contributes equally to the field at this midpoint. Therefore, if we remove one-half of the solenoid, the field at the (newly created) open end is just half that of Equation (26–6):

MAGNETIC FIELD AT ONE END OF A LONG SOLENOID

$$B = \frac{\mu_0 nI}{2} \tag{26-7}$$

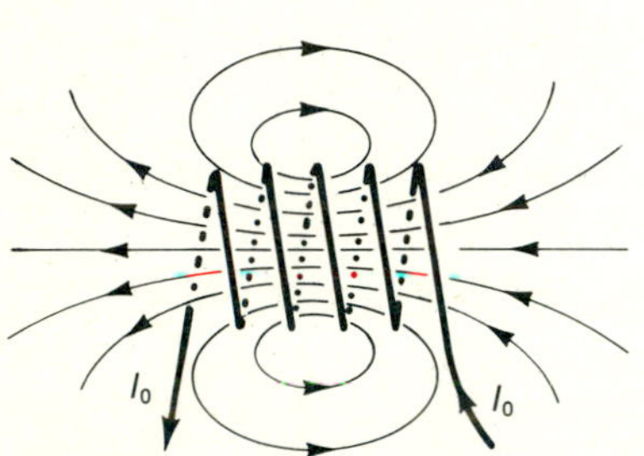

(**a**) A loosely-wound, short solenoid.

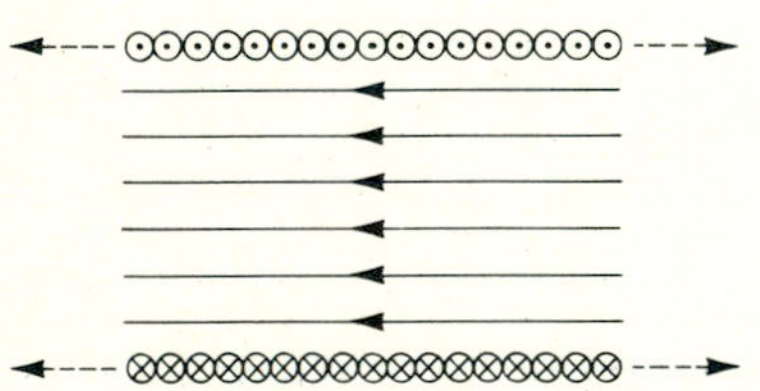

(**b**) An *ideal* solenoid has closely-wound windings that extend to infinity in both directions, confining the field wholly within the solenoid.

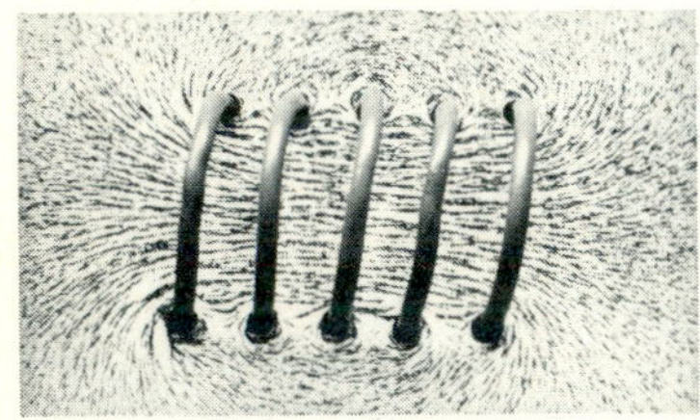

(**c**) Iron filings reveal the magnetic field pattern.

Figure 26–8
The magnetic field of a solenoid.

EXAMPLE 26–3

A utility permanent magnet similar to the one shown in Figure 26–9(a) has a magnetic field of 0.4 T in the air gap between its pole pieces. In order to produce the same magnetic field within a solenoid of comparable size, how much current would have to pass through its windings? Assume the solenoid is 30 cm long with a small cross section and is wound with copper wire so that the total number of turns is 2000.

SOLUTION

With the assumption that the solenoid is ideal, we apply Equation (26–6):

$$B = \mu_0 n I$$

$$I = \frac{B}{\mu_0 n}$$

Substituting numerical values:

$$I = \frac{0.4\ \text{T}}{\left(4\pi \times 10^{-7}\ \dfrac{\text{Wb}}{\text{A}\cdot\text{m}}\right)\left(\dfrac{2000\ \text{turns}}{0.30\ \text{m}}\right)}$$

$$I = \boxed{47.7\ \text{A}}$$

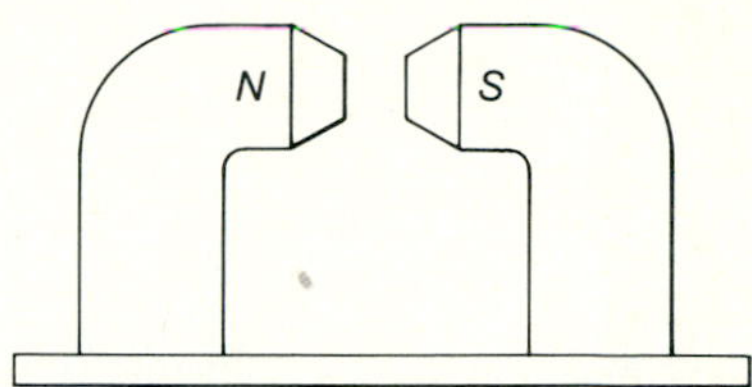

(a) A permanent magnet.

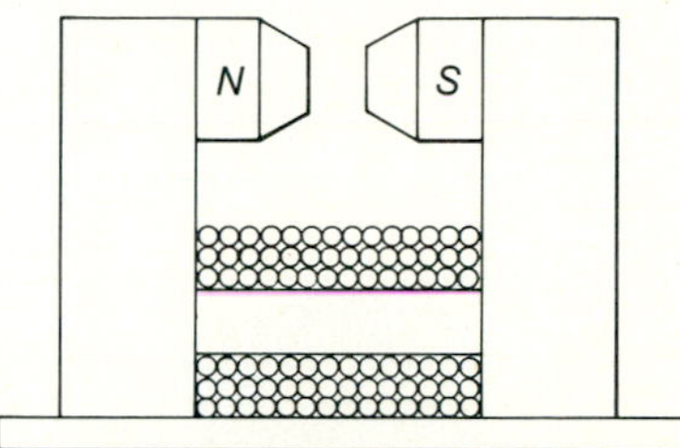

(b) An electromagnet (cross-sectional view of the windings).

Figure 26–9
Laboratory magnets

To see whether or not the result in Example 26–3 is consistent with a practical laboratory device, suppose the average length of each turn of wire is 10 cm and that the cross-sectional area of the wire is 1.0 mm^2. The resistance of the wire is given by:

$$R = \rho\left(\frac{\ell}{A}\right)$$

where ρ is the resistivity of copper (1.8×10^{-8} ohm·m), ℓ is the length of the wire, and A is the cross-sectional area. Substituting the appropriate values in SI units:

$$R = (1.8 \times 10^{-8}\ \Omega\cdot\text{m})\,\frac{(2000\ \text{turns})(0.1\ \text{m})}{(1.0 \times 10^{-6}\ \text{m}^2)}$$

$$R = 3.60\ \Omega$$

The rate at which heat would be generated due to Joule heating of the copper is:

$$P = I^2 R$$

$$P = (47.7\ \text{A})^2(3.6\ \Omega)$$

$$P = 8.19\ \text{kW}$$

Clearly, the cooling requirements of such a solenoid make it impractical as a source of magnetic field. However, the presence of an iron core in a solenoid greatly increases the resultant magnetic field (as will be discussed later in the chapter). So, in practice, the solenoid would be constructed similarly to that in Figure 26–9(b), greatly reducing the current requirements.

EXAMPLE 26–4

A long, hollow conducting wire carries a current I that is uniformly distributed over the cross-sectional area of the wire between radii a and b, as shown: Find the magnetic field $\mathbf{B}$ for the regions

Region 1:	$r < a$
Region 2:	$a < r < b$
Region 3:	$r > b$

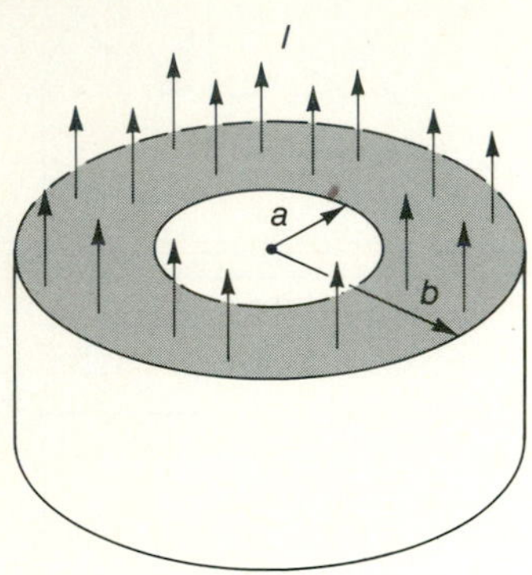

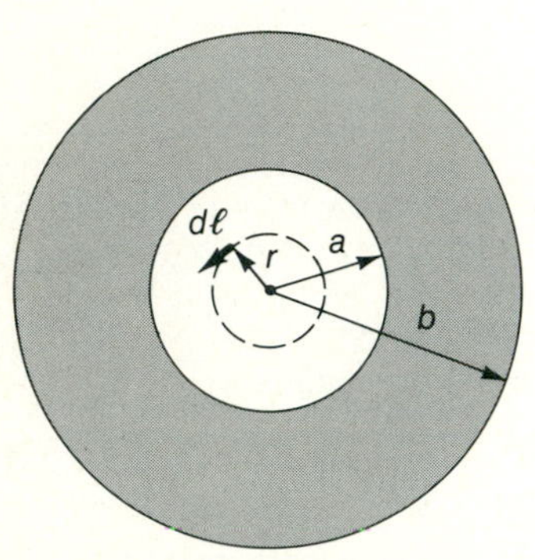

I out of the paper (shaded area)

SOLUTION

From the symmetry of the situation, the only directions the magnetic field lines can have are concentric circles about the axis of the wire (right-hand rule). Furthermore, by symmetry the magnitude B must be constant everywhere along such a line. We purposely match this symmetry by choosing paths of integration for $\oint \mathbf{B} \cdot d\ell$ that are concentric circles about the axis, in the direction of $\mathbf{B}$.

In region 1 ($r < a$):

$$\oint \mathbf{B} \cdot d\ell = \mu_0 I$$

The dot product gives $\cos 0° = 1$. Because B is constant along the path, it may be brought out from under the integral sign. Since $\oint d\ell = 2\pi r$, and the value of I enclosed by the integration path is zero, we have

$$B_1(2\pi r) = \mu_0(0)$$

$$B_1 = \boxed{0}$$

In region 2 ($a < r < b$): Again, by symmetry, we choose the integration path to be a concentric circle. However, we now need to know the fraction of the total current I that is enclosed by the path of integration. Since the current is distributed uniformly over the cross-sectional area, it is the fraction

$$\left(\frac{\text{Area inside } r}{\text{Total area}}\right) I = \left[\frac{\pi(r^2 - a^2)}{\pi(b^2 - a^2)}\right] I$$

Therefore:

$$\oint \mathbf{B} \cdot d\ell = \mu_0 I$$

$$B_2(2\pi r) = \mu_0 \left(\frac{r^2 - a^2}{b^2 - a^2}\right) I$$

$$B_2 = \boxed{\frac{\mu_0}{2\pi r}\left(\frac{r^2 - a^2}{b^2 - a^2}\right) I}$$

The direction of $\mathbf{B}$ is counterclockwise in the figure (right-hand rule).

In region 3 ($r > b$): Here, the path of integration encloses the entire current I.

$$\oint \mathbf{B} \cdot d\ell = \mu_0 I$$

$$B_3(2\pi r) = \mu_0 I$$

$$B_3 = \boxed{\frac{\mu_0 I}{2\pi r}}$$

The direction of $\mathbf{B}$ is counterclockwise in the figure (right-hand rule).

TABLE 26–1

Similarities Between Electric and Magnetic Fields

	Magnetic field of a long, straight current-carrying conductor	Electric field of a long line of charge
1. General equations (both equations are inverse square)	$d\mathbf{B} = \left(\frac{\mu_0}{4\pi}\right)\frac{I\, d\ell \times \hat{\mathbf{r}}}{r^2}$	$d\mathbf{E} = \left(\frac{1}{4\pi\varepsilon_0}\right)\frac{\lambda\, d\ell}{r^2}\hat{\mathbf{r}}$ where λ is the linear charge density
2. Alternative general equations	$\oint \mathbf{B} \cdot d\ell = \mu_0 I$ (line integral) Ampère's law	$\oint \mathbf{E} \cdot d\mathbf{A} = \frac{q}{\varepsilon_0}$ (surface integral) Gauss's law
3. Field equations for a long line a distance a away from the line (both equations are inverse first power)	$B = \frac{\mu_0 I}{2\pi a}$ where B circles the line	$E = \frac{\lambda}{\varepsilon_0 2\pi a}$ where E is directed away from a positively charged line

One of the most aesthetically pleasing aspects of electricity and magnetism is the similarity of form among the equations describing both phenomena. As an illustration, compare the equations in Table 26–1, which describe the magnetic field of a long, straight, current-carrying conductor and the electric field of a long line of charge.

26.3 Faraday's Law

We began our discussion of electrical circuits by describing batteries as devices that were capable of doing work on charges. We defined the amount of work dW per unit charge dq done by such a device as the electromotive force (EMF) $\mathscr{E}$ of the device:

$$\mathscr{E} = \frac{dW}{dq} \tag{26–8}$$

The EMF is thus capable of moving charges from one point to another (that is, it can produce a current).

The discovery that a current produces a magnetic field was first made in 1820 by the Danish schoolteacher Hans Oersted (1777–1851). It was felt that the connection between electricity and magnetism could not be in one direction only, so many investigators tried to find an "inverse" effect. Namely, they asked if a magnetic field could produce a current. The answer is *yes*, though it did not become obvious until it was discovered that *moving* charges produce the field. Thus, they reasoned, perhaps a *changing* field could produce a current. The discovery was made in 1831 by the English experimenter Michael Faraday and shortly thereafter (independently) in the United States by Joseph Henry (1797–1878). The effect is called **electromagnetic induction.**

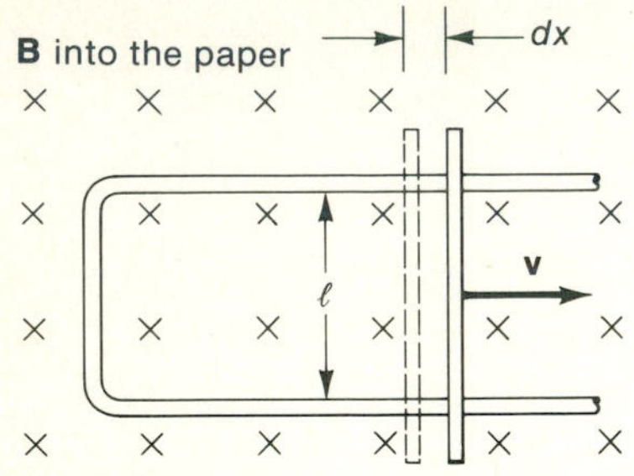

Figure 26–10
As the area of the rectangular loop of wire increases, an EMF is induced in the moving bar, causing the top end of the bar to become charged positively and the bottom end to become negatively charged.

Consider a rectangular loop of conductor formed by a stationary U-shaped wire with a sliding bar for one edge, as illustrated in Figure 26–10. As the bar slides, it maintains electrical contact with the wire. The loop is oriented in a magnetic field so that the plane of the loop is perpendicular to the field. Suppose some external force moves the bar with a velocity **v** as in the figure. A charge dq in the bar will experience a force $d\mathbf{F}$ along the bar given by the Lorentz force law:

$$d\mathbf{F} = (dq)\,\mathbf{v} \times \mathbf{B} \tag{26–9}$$

The work done by this force in moving the charge dq from one end of the bar to the other is:

$$dW = \ell\, dF \tag{26–10}$$

where ℓ is the length of the bar.

Substituting the expression for dF given by Equation (26–9):

$$dW = (dq)\, vB\ell$$

Rearranging, and noting that $v = dx/dt$:

$$\frac{dW}{dq} = \frac{dx}{dt}\ell B$$

or

$$\frac{dW}{dq} = B\frac{dA}{dt} \tag{26–11}$$

where dA/dt is the rate of change of area of the loop. Further, since the magnetic flux Φ_B through the loop is BA, we note that $B\,dA = d\Phi_B$:

$$\frac{dW}{dq} = \frac{d\Phi_B}{dt} \tag{26–12}$$

where $d\Phi_B/dt$ is the rate of change of magnetic flux through the loop (often called the *change in flux linkage* through the loop). Comparing Equation (26–12) with Equation (26–8), we conclude that

$$\mathscr{E} = \frac{d\Phi_B}{dt}$$

This and similar results from a variety of experiments have led to the formulation of **Faraday's law of induction:**

FARADAY'S LAW OF INDUCTION

$$\mathscr{E} = -\frac{d\Phi_B}{dt} \tag{26–13}$$

The minus sign indicates the direction of the EMF (to be discussed in the next section). The moving bar is thus a seat of EMF with the top end of the bar positive, forcing current *out* of the top end of the bar into any external circuit connected to the ends of the bar. Unlike a localized seat of EMF such as a battery, the EMF is distributed along the entire length of the moving bar.

EXAMPLE 26–5

Referring to Figure 26–10, suppose that the bar is 10 cm long and that the magnetic field within the loop is 2×10^{-2} T. How rapidly must the bar move in order to produce a potential difference of one volt between the ends of the bar?

SOLUTION

The magnetic flux Φ_B within the loop is changing at the rate

$$\frac{d\Phi_B}{dt} = B\frac{dA}{dt}$$

where

$$\frac{dA}{dt} = \ell v$$

Then, by Equation (26–13), the magnitude of the EMF is

$$|\mathscr{E}| = \frac{d\Phi_B}{dt}$$

where

$$\mathscr{E} = B\ell v$$

Solving for the speed of the bar v:

$$v = \frac{\mathscr{E}}{B\ell}$$

$$v = \frac{(1\ \text{V})}{(2 \times 10^{-2}\ \text{T})(0.10\ \text{m})}$$

$$v = \boxed{500\ \text{m/s}}$$

The previous discussion shows that the central feature of Faraday's law of induction is the **change of flux linkages** $d\Phi_B/dt$ through the loop. There are several ways we may interpret Equation (26–13):

$$\mathscr{E} = -\frac{d\Phi_B}{dt}$$

(1) If the field is constant, but the area of the loop is changing:

$$\mathscr{E} = -B\frac{dA}{dt} \qquad \textbf{(26–14)}$$

(2) If the area of the loop is constant, but the field is changing:

$$\mathscr{E} = -A\frac{dB}{dt} \qquad \textbf{(26–15)}$$

(3) If both the area and the field are changing:

$$\mathscr{E} = -\left(B\frac{dA}{dt} + A\frac{dB}{dt}\right) \qquad \textbf{(26–16)}$$

All of these interpretations boil down to just one fact: *The rate of change of flux linkages $d\Phi_B/dt$ determines the induced EMF.* Figure 26–11 illustrates some of the methods of changing flux linkages.

Figure 26–11
Methods of inducing an EMF in the conducting loop A by causing a *change in the flux linkage* through that loop. The current direction in R due to the induced EMF is shown by the arrow.

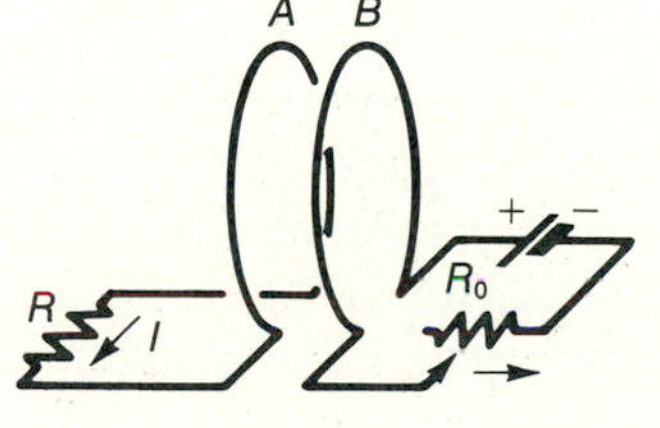

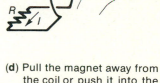

(**a**) Change the current in loop B. (Induced current in A is shown for the case of an *increasing* current in B.)

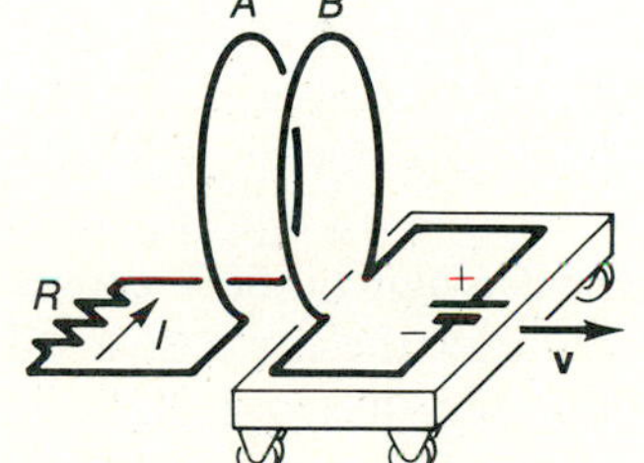

(**b**) Move the coils farther apart.

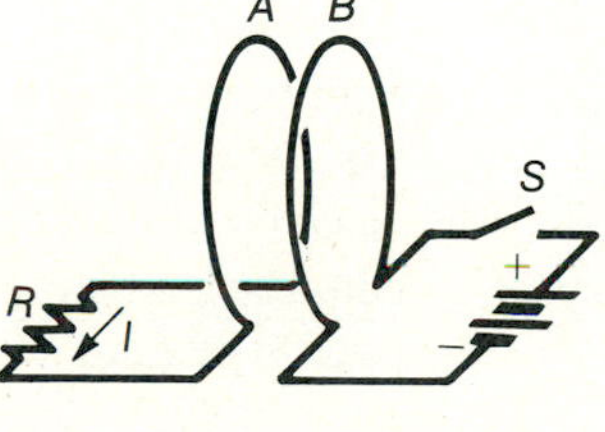

(**c**) Close the switch S.

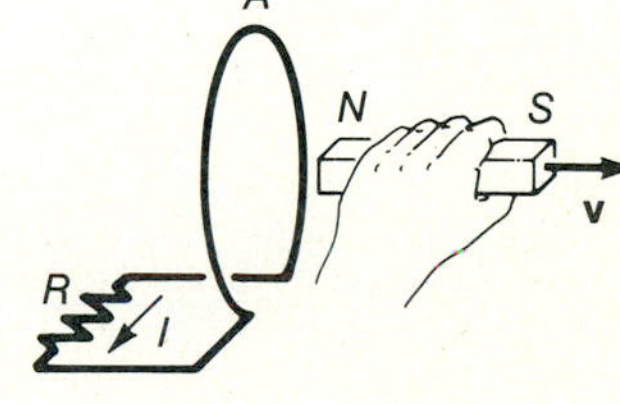

(**d**) Pull the magnet away from the coil or push it into the coil. (Induced current in A is shown for pulling the magnet *away*.)

EXAMPLE 26–6

A solenoid with 100 turns and a cross-sectional area of 40 cm^2 has a magnetic field of 0.450 T along its axis. Assuming that the field is confined within the solenoid, find the induced EMF at the terminals of the solenoid if the field is changing at a rate of 0.500 T/s.

SOLUTION

The magnitude of the field is not relevant in determining the EMF; only the *change* of field is involved. In this case, the area of the loop is constant while the field changes. So the *change in flux linkage* $d\Phi_B/dt$ is $A\,dB/dt$, and we have for the magnitude of the EMF [Equation (26–15)]:

$$|\mathscr{E}_1| = A\frac{dB}{dt}$$

for *each* turn of the solenoid. Substituting values:

$$\mathscr{E}_1 = (40 \times 10^{-4}\ \text{m}^2)\left(5 \times 10^{-1}\ \frac{\text{T}}{\text{s}}\right)$$

$$\mathscr{E}_1 = 2 \times 10^{-3}\ \text{V} \qquad \text{(for one turn)}$$

Since all of the turns are in series, the induced EMFs are additive. So, for 100 turns or loops:

$$\mathscr{E} = 100\,\mathscr{E}_1$$

$$\mathscr{E} = \boxed{0.200\ \text{V}}$$

Faraday's law may be generalized to include cases in which the field within a loop is not uniform or perpendicular to the plane of the loop. In general, as in the previous chapter:

$$\Phi_B = \int \mathbf{B} \cdot d\mathbf{A}$$

and

$$\mathscr{E} = \oint \mathbf{E} \cdot d\boldsymbol{\ell}$$

where the EMF is the work per unit charge done by the electric field **E** in moving the unit charge around the loop ℓ. Faraday's law may thus be written

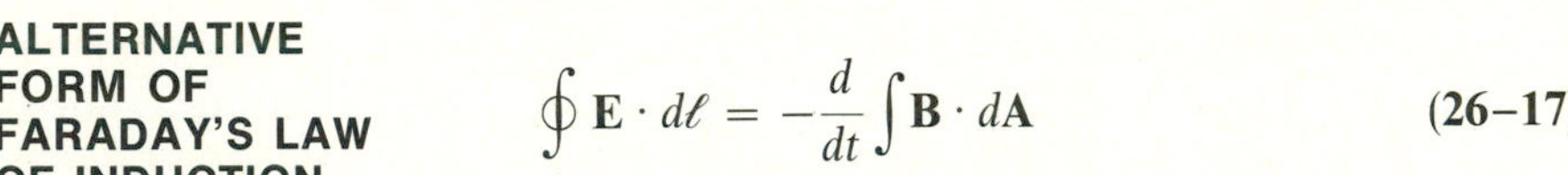

ALTERNATIVE FORM OF FARADAY'S LAW OF INDUCTION

$$\oint \mathbf{E} \cdot d\boldsymbol{\ell} = -\frac{d}{dt}\int \mathbf{B} \cdot d\mathbf{A} \qquad \textbf{(26–17)}$$

To illustrate the use of this form of Faraday's law, we will analyze the operation of a *betatron*, a type of accelerator for electrons. A betatron such as that shown in Figure 26–12 can accelerate electrons to energies of 100 million electron volts (MeV). Electrons are injected into a horizontal toroidal vacuum chamber, where a vertical magnetic field causes them to travel in a circular path. The magnetic field inside the circle is varied, inducing an EMF along the path, thereby accelerating the electrons. The conditions must be such that as the electrons speed up, the magnetic field at the orbit must also increase to keep the electrons traveling in a circle of constant radius. There are two regions of the magnetic field we must investigate: the field *at* the orbit (which we designate B_1) and the average field over *the area within the orbit* (which we

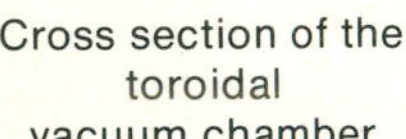

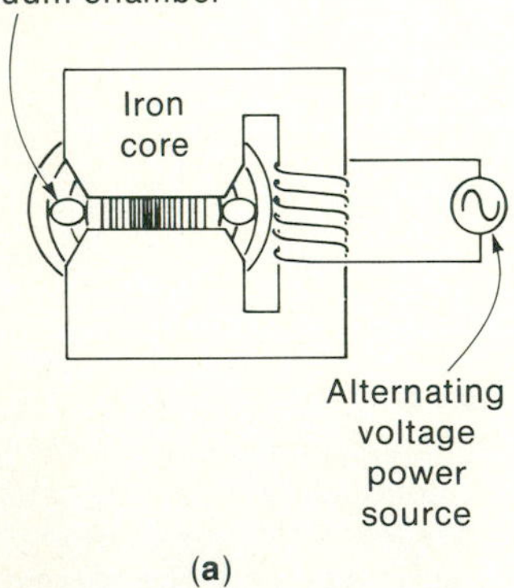

(a)

(b)

Figure 26–12
The betatron

designate B_2). We now determine the conditions necessary to maintain an orbit of constant radius as the electron speed increases.

The electric field E along the orbit path that accelerates the electrons is due to the induced EMF along that path, given by Equation (26–17):

$$\oint \mathbf{E} \cdot d\ell = -\frac{d}{dt}\int \mathbf{B} \cdot d\mathbf{A}$$

If R is the radius of the orbit, and B_2 is the average magnetic field inside the orbit (perpendicular to the plane of the orbit), the integrals become

$$2\pi RE = \pi R^2 \frac{dB_2}{dt}$$

or

$$E = \tfrac{1}{2}R\frac{dB_2}{dt} \qquad \textbf{(26–18)}$$

The magnetic field B_1 *at* the orbit necessary to keep the accelerating electrons traveling in a circle of constant radius can be determined using the magnetic force law:

$$\mathbf{F} = q\mathbf{v} \times \mathbf{B}$$

where **F** is the centripetal force required. Since **v** and **B** are at right angles, $\sin 90^\circ = 1$. Thus:

$$\Sigma F = ma$$

$$evB_1 = m\left(\frac{v^2}{R}\right)$$

Rearranging and differentiating gives

$$B_1 = \left(\frac{1}{eR}\right)mv$$

$$\frac{dB_1}{dt} = \left(\frac{1}{eR}\right)\frac{d(mv)}{dt} \qquad \textbf{(26–19)}$$

Now $d(mv)/dt$ is the force accelerating the electron *along* the orbit. This is eE, where E is the accelerating field along the orbit. Therefore:

$$eE = \frac{d(mv)}{dt} \qquad \textbf{(26–20)}$$

Substituting this into Equation (26–19) and rearranging:

$$E = R\frac{dB_1}{dt}$$

Comparing this with Equation (26–18), we see that

$$\frac{dB_2}{dt} = 2\frac{dB_1}{dt} \qquad \textbf{(26–21)}$$

or, by integration:

BETATRON CONDITION

$$B_2 = 2B_1 \qquad \textbf{(26–22)}$$

Thus, the pole pieces of a betatron must be designed so that the average magnetic field B_2 within the orbit must always be exactly twice that of B_1 at the orbit. This condition is sometimes known as the *betatron "one-two" relation.* As the field B_2 *within* the orbit increases to induce the EMF along the orbit, the field B_1 *at* the orbit must also change, keeping the one-two relation at all times. This is accomplished by carefully designing the contours of the pole faces of the iron core. Although our development was nonrelativistic, the result is the same when treated relativistically.

26.4 Lenz's Law

The information in Faraday's law was originally expressed by Faraday in a rather cumbersome form that involved several relations. Later investigators revised and reduced these relations to the succinct equation we have today. An important clarification was made by Heinrich Lenz (1804–1865), who contributed the minus sign. This minus sign has an importance greater than it might seem at first glance, since an understanding of its meaning gives the direction of the induced EMF. Lenz's law is the interpretation we give to the minus sign. Let us develop the law using a specific example.

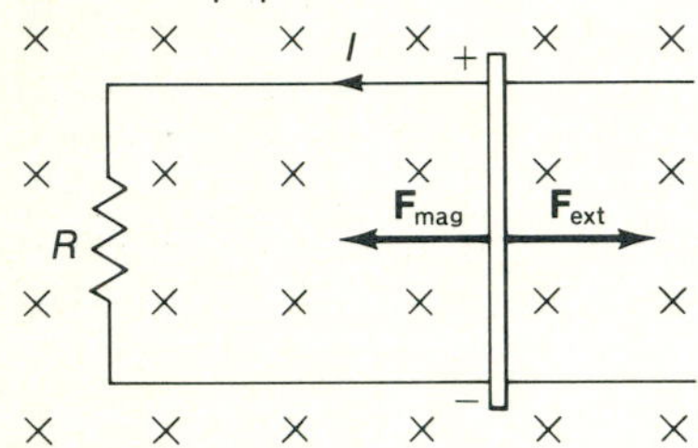

Figure 26–13
The external force $\mathbf{F}_{ext}$ moves the bar at constant speed. It is equal and opposite to the magnetic force $\mathbf{F}_{mag}$.

Consider the situation illustrated in Figure 26–13. The movable bar maintains electrical contact with the rest of the wires as the external force $\mathbf{F}_{ext}$ moves the bar toward the right at constant speed v. From the analysis of Figure 26–10 we have seen that charges will move in the bar so that the top end of the bar becomes positive with respect to the bottom end. This induced EMF produces a current I around the loop in the direction shown. But because of the presence of the external field $\mathbf{B}$, the induced current I in the bar produces a magnetic force on the bar toward the left given by

$$F_{mag} = B\ell I \tag{26–23}$$

Since the bar is not accelerating, this force has the same magnitude as the external force moving the bar. The power expended in moving the bar with a speed v is therefore

$$P = F_{ext}v$$

From Equation (26–23) and $F_{ext} = F_{mag}$, we have

$$P = B\ell Iv$$

but

$$\ell v = \frac{dA}{dt}$$

which is the change in area of the loop. Thus:

$$P = IB\frac{dA}{dt}$$

Since the induced EMF $\mathscr{E}$ is $B\,dA/dt$:

$$P = I\mathscr{E} \tag{26–24}$$

The current through the loop is given by Ohm's law:

$$\mathscr{E} = IR$$

Substituting into Equation (26–24):

$$P = I^2R \tag{26–25}$$

The work done by the force moving the bar appears as Joule heating of the resistance R. The most important thing to note in this development is *the direction of the current flow in the loop.* The counterclockwise current shown in Figure 26–13 produces a magnetic field *out* of the plane of the illustration, which is *opposite* to the *change* of the flux linkages. In all cases it is found that:

LENZ'S LAW **The induced EMF (and the induced current) always has a polarity such as to *oppose the change* of flux linkages that produces it.**

It should be emphasized again that the induced EMF opposes the *change* of flux, not the flux itself. Overlooking this distinction is a common error.

Let us examine the changing flux for each of the situations shown in Figure 26–11.

(a) Initially, the current in loop B produces a field toward the right, so flux lines initially thread through loop A toward the right. If the current in B were increased, this would cause *more* flux lines to thread through A toward the right. Since the induced EMF in A will be in such a direction as to *oppose* this *change* of flux linkages, the induced EMF tries to make a current in A that will produce flux lines toward the left. The direction of this current is as shown in the diagram. (Conversely, if the current in loop B were reduced, the induced EMF in A would be in the opposite direction, making a current that would produce flux lines toward the right.)
(b) Initially, the magnetic flux lines are toward the right in both loops. Moving B toward the right causes *fewer* lines to thread through loop A. Since the induced EMF in A will be in such a direction as to *oppose* this *change*, the induced EMF tries to make a current flow in A that will produce flux toward the right.
(c) By closing the switch in the right loop, the magnetic flux would increase toward the right in loop A. Therefore, the induced current in loop A is such that it produces flux lines toward the left within the loop.
(d) Since flux lines come out of the north pole of a magnet, initially flux lines thread through loop A toward the left. If the magnet is pulled away toward the right, the flux through A *decreases*, so the induced EMF produces a current as shown to produce flux lines toward the left in loop A.

As all these examples show, the induced EMF always tends to produce an effect that maintains the existing flux-line situation. In other words, it tends to *maintain the status quo*. It does not oppose flux linkages; it opposes any *change* in the flux linkages.

An induced current is actually present in these examples. If there is a gap in the circuit that prevents a current from being established, we can pretend the circuit is closed and analyze the situation to determine the direction in which the induced current would be. In this way, we can determine the polarity of the induced EMF across the gap, remembering that if an external wire were connected across the gap, the current would be from plus to minus.

EXAMPLE 26–7

In Figure 26–14(a), the wire loop with gap AB is held fixed, while the permanent magnet is withdrawn as indicated. Find the polarity of the induced EMF across the gap while the magnet is being withdrawn.

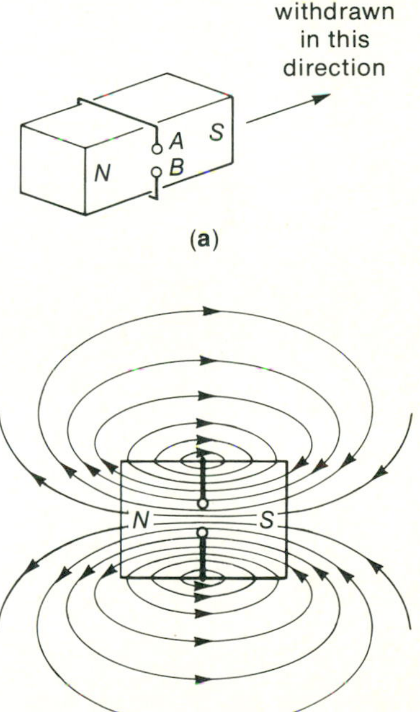

(a)

(b) Field lines of a permanent magnet form complete, closed loops.

Figure 26–14
Example 26–7

SOLUTION

Magnetic field lines always form complete, closed loops. Therefore, due to the field lines *inside* the magnet, the net flux linkages through the wire loop are initially toward the left (Figure 26–14b). If an external wire were connected across the gap, the induced current in this wire would be from B to A as the magnet is withdrawn (so the induced current in the loop creates a magnetic field that opposes the change of flux linkages through the loop). Thus point B is at a *higher* potential than point A. The points AB are, in effect, the terminals of a source of EMF that will cause current from B to A in an external wire connected between these terminals. The EMF is present across the gap whether or not an external wire is connected across points AB.

26.5 Self-Inductance

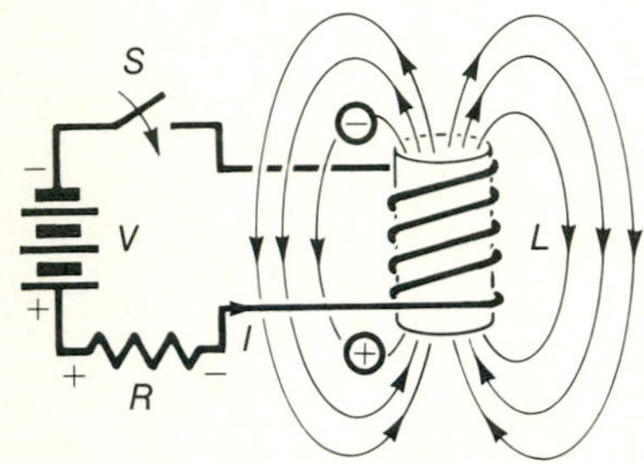

Figure 26–15
An inductance in series with a battery and a resistor. As the current builds up, note the polarities of the potential differences across R and L.

Just as a resistor in a direct-current circuit restricts the amount of current, a loop or coil of wire in a circuit restricts any *change* of current in a circuit. This may be understood by considering the circuit shown in Figure 26–15. When the switch S is closed, current will be established in the direction indicated, which will produce a growing magnetic field in the coil, as shown. By Faraday's and Lenz's laws, these changing flux lines induce an EMF in the coil itself, a process called *self-induction.* Its polarity tends to cause an induced current in the coil (opposite to the original current), which generates flux lines *downward* to oppose the growing upward flux linkages. In the figure, the induced EMF has a polarity such that the bottom wire leading into the coil is *positive* with respect to the top wire. (This polarity is shown in the small circles.) If no iron or similar magnetic materials are nearby, the magnitude of this *self-induced EMF*, or "*back-EMF*," depends on the physical dimensions of the coil and the rate of change of current dI/dt in the coil (since these are the factors that determine the rate of change of flux linkages). Designating the factor of proportionality due to the physical dimensions as L, this relationship may be expressed by the equation

SELF-INDUCTANCE *L*

$$\mathscr{E} = -L\frac{dI}{dt} \tag{26–26}$$

where L is called the **self-inductance** of the coil (symbol: -ꝏꝏ-). The unit[1] of self-inductance is the *henry* (H). Circuit elements that have inductance are called **inductors.** (Note the correspondence to capacitance and capacitors, and to resistance and resistors.) The minus sign reflects Lenz's law: The polarity of the induced EMF opposes the *change* in current through the coil.

Ordinarily, the value of L for a given physical configuration of a coil is somewhat difficult to calculate. However, to illustrate the procedure, we will evaluate L for the simple case of a long solenoid. For each single turn, the induced EMF is given by

$$\mathscr{E} = -\frac{d\Phi_B}{dt}$$

Each turn of a long solenoid encircles the same magnetic flux, so the induced EMF for N turns is

$$\mathscr{E} = -N\frac{d\Phi_B}{dt} \tag{26–27}$$

[1] This unit honors the American physicist Joseph Henry (1797–1878), who discovered electromagnetic induction independently of Faraday's discoveries in England.

Comparing this with Equation (26–26), we have

$$N\frac{d\Phi_B}{dt} = L\frac{dI}{dt} \qquad (26\text{–}28)$$

or

$$N\,d\Phi_B = L\,dI$$

Integrating both sides of this equation and noting that $\Phi_B = 0$ when $I = 0$, we obtain

$$\int_0^{\Phi_B} N\,d\Phi_B = \int_0^I L\,dI$$

$$N\Phi_B = LI$$

Solving for the self-inductance L:

$$L = \frac{N\Phi_B}{I} \qquad (26\text{–}29)$$

From this relation, we see that alternative units for self-inductance are weber·turns/ampere as well as the henry. Thus, we may think of self-inductance as the number of *flux linkages per ampere* produced as a result of physical dimensions of the coil (its radius, the number of turns, and so on). Occasionally, L is called merely the *inductance* of the coil.

Earlier in this chapter we showed that the magnetic field of a long solenoid is uniform within the solenoid and is given by

$$B = \mu_0 nI$$

where n is the number of turns per unit length and I is the current through each turn of wire. If the cross-sectional area of the solenoid is A, the total flux Φ_B through the coil is BA, or

$$\Phi_B = \mu_0 nIA$$

For a solenoid of length ℓ and a total number of turns N, so that $n = N/\ell$, this equation becomes

$$\Phi_B = \frac{\mu_0 ANI}{\ell} \qquad (26\text{–}30)$$

Substituting this expression for Φ_B into Equation (26–29), we obtain

SELF-INDUCTANCE OF A LONG SOLENOID (ignoring end effects)

$$L = \frac{\mu_0 N^2 A}{\ell} \qquad (26\text{–}31)$$

EXAMPLE 26–8

(a) Find the self-inductance of a solenoid that has a cross-sectional area of 1 cm^2, a length of 10 cm, and 1000 turns of wire. (b) If the current through the inductor is increasing at the rate of 15 A/s, find the magnitude of the induced EMF.

SOLUTION

(a) The length of the solenoid is large compared with the cross-sectional radius, and the turns of wire are closely wound. So we treat it as a long solenoid. Substituting the appropriate values in SI units into Equation (26–31) yields

$$L = \frac{\left(4\pi \times 10^{-7}\,\frac{\text{H}}{\text{m}}\right)(1000\ \text{turns})^2(10^{-4}\ \text{m}^2)}{(0.10\ \text{m})}$$

$$L = \boxed{1.26\ \text{mH}}$$

(b) From Equation (26–26), we obtain

$$\mathscr{E} = -L\frac{dI}{dt}$$

$$|\mathscr{E}| = (1.26 \times 10^{-3}\ \text{H})\left(15\ \frac{\text{A}}{\text{s}}\right)$$

$$|\mathscr{E}| = \boxed{18.9\ \text{mV}}$$

26.6 Mutual Inductance

In the previous section, we defined the *self-inductance* L of a coil that involves the back-EMF generated in a coil due to a changing current in the coil itself. Similar effects occur between two coils that are close enough together so that flux lines generated in one coil can link the other coil. Then, an EMF will be induced in either coil due to current changes in the other coil (see Figure 26–16). This process is known as *mutual induction*, defined in the following way. The EMF generated in coil 1 due to a changing current dI_2/dt in coil 2 is

$$\mathscr{E}_1 = -M_{12}\frac{dI_2}{dt} \tag{26–32}$$

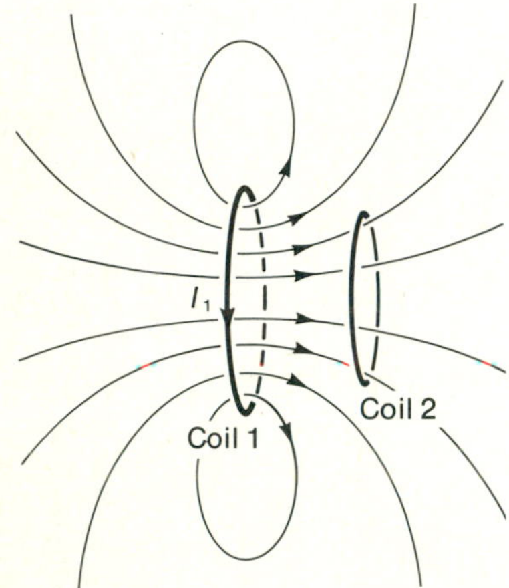

Figure 26–16
Mutual inductance between two coils occurs when they are close enough together so that a current in one coil causes flux linkages in the other. (Here, current in coil 1 creates flux linkages in coil 2.) It is a mutual effect: A *changing* current in either coil will induce EMFs in the other coil.

where M_{12} is the **mutual inductance** of coil 1 with respect to coil 2. Mutual inductance has the same unit as self-inductance: the *henry* (H). Similarly, the EMF generated in coil 2 due to a changing current dI_1/dt in coil 1 is

$$\mathscr{E}_2 = -M_{21}\frac{dI_1}{dt} \tag{26–33}$$

where M_{21} is the mutual inductance of coil 2 with respect to coil 1. We omit the proof and merely state that it can be shown that

$$M_{12} = M_{21}$$

Thus the symbol M (without subscripts) may be used for mutual inductance:

$$\mathscr{E}_1 = -M\frac{dI_2}{dt} \quad \text{and} \quad \mathscr{E}_2 = -M\frac{dI_1}{dt} \tag{26–34}$$

Provided no iron or similar material is nearby, the value of M depends only on geometrical factors such as how close together the two coils are and their orientations. Except when the two coils are wound together so that *all* the flux from one coil links the other coil, the calculation of M may be quite complicated.

As a practical example of mutual inductance, telephone lines in a cable sometimes suffer from "cross-talk" when current changes in one line generate EMFs in adjacent lines. (Capacitive effects can similarly cause trouble.) Another example is the "hum" in audio amplifiers. This occurs, for example, because alternating currents in the power supply can induce alternating EMFs in sensitive circuits unless these circuits are properly shielded from the magnetic fields or are placed sufficiently far away.

26.7 *RL* Circuits

Since the wire used to form a coil has some electrical resistance,[2] it is inevitable that each inductor will also have some finite resistance—how much depends on the resistivity of the wire material, its cross-sectional area, and its length.

[2] Superconductors (discussed in Chapter 24) are an exception. The resistivity of these materials becomes zero as the temperature approaches absolute zero.

For many applications, this resistance is negligible compared with other resistances in the circuit. Nevertheless, it is important to investigate the behavior of a "pure" inductance L in series with a resistance R. Such "*RL* circuits" are common in electrical networks.

Consider the series *RL* circuit shown in Figure 26–17. We assume the inductance is "pure," so that all the resistance of the circuit is assumed to be in R. The moment switch S_1 is closed, the battery tries to make a current flow through the coil. But because of self-inductance, a back-EMF is generated, which opposes the increase of current. So just after closing the switch, no appreciable current has yet begun to flow; therefore, there is no appreciable voltage drop ($V_R = IR$) across the resistor. But the *rate of change* of current is at its maximum value. In fact, it is just that value necessary to make the voltage across the inductor ($V_L = -L\,dI/dt$) be the value demanded by Kirchhoff's rule. The polarity is as shown in the figure. As time goes on, the current increases (thus V_R increases), but the rate of change of I decreases (so V_L decreases). Finally, after steady state conditions are reached ($dI/dt = 0$), $V_L = 0$, $I = V/R$, and $V_R = V$. At this point, the inductance no longer impedes the current flow, but a magnetic field has been established in the vicinity of the inductance due to the steady current in the coil.

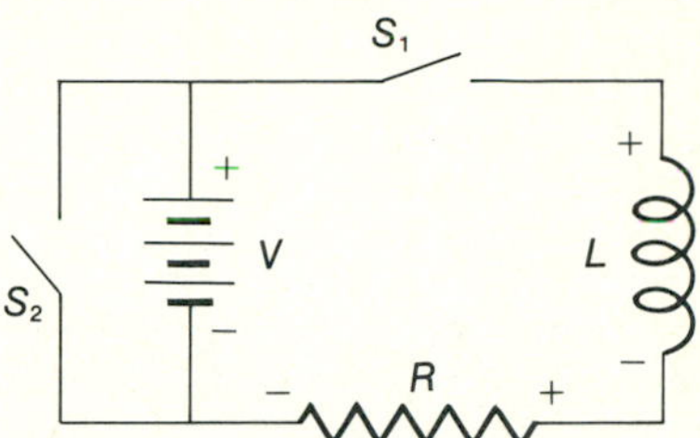

Figure 26–17
An *RL* series circuit for investigating current increases and decreases through an inductor. (Polarities shown are for the current buildup after S_1 is closed.)

We can derive an equation that expresses this scenario. At all times after closing the switch, Kirchhoff's loop rule must hold true:

$$\Sigma V = 0$$

$$V - L\frac{dI}{dt} - IR = 0 \tag{26–35}$$

The solution[3] of this differential equation for the current I as a function of time is the following:

$$I = \frac{V}{R}\left(1 - e^{-(R/L)t}\right) \tag{26–36}$$

[3] This equation may be solved by the mathematical technique called *separation of variables.* A rearrangement of the terms in Equation (26–35) produces

$$L\frac{dI}{dt} = V - IR$$

or

$$\frac{dI}{V - IR} = \frac{dt}{L}$$

thereby separating the variables I and t by the equal sign. Both sides of the equation may be integrated so that

$$\int \frac{dI}{V - IR} = \int \frac{dt}{L}$$

Using the table in Appendix GII, we obtain

$$-\frac{1}{R}\ln\,(V - IR) = \frac{t}{L} + c$$

The constant of integration c may be obtained from the initial condition, when $t = 0$ and $I = 0$:

$$c = -\frac{1}{R}\ln V$$

Substituting this value of c into the previous equation, we have

$$\ln\left(\frac{V - IR}{V}\right) = -\frac{R}{L}t$$

Since $e^{\ln x} = x$, we have

$$\left(\frac{V - IR}{V}\right) = e^{-(R/L)t}$$

or

$$I = \frac{V}{R}\left(1 - e^{-(R/L)t}\right)$$

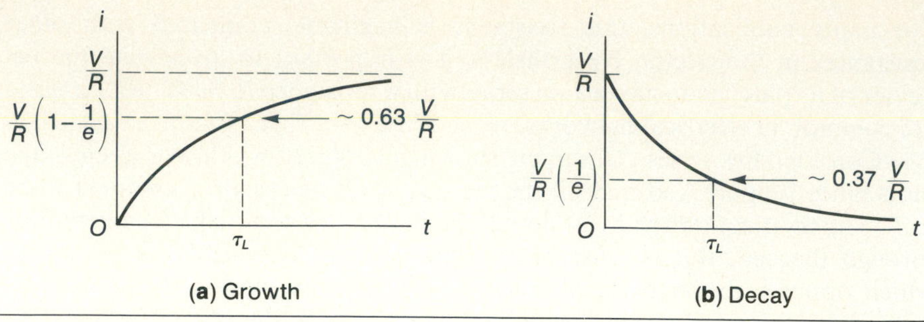

(a) Growth (b) Decay

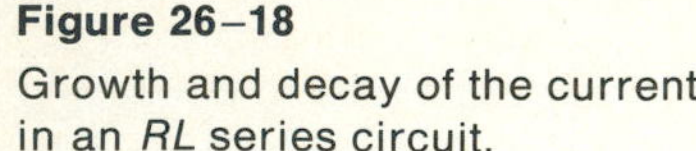

Figure 26–18
Growth and decay of the current in an *RL* series circuit.

A graph of this solution is shown in Figure 26–18(a). Because of the exponential factor, the current increases asymptotically toward the value V/R. The rate of increase depends on the ratio L/R; the larger this ratio, the more slowly the current increases. The ratio L/R is called the **time constant** τ_L of the circuit. In a period of time equal to *one* time constant after the switch is closed, the current will rise to $(1 - 1/e)$ of its maximum value. This is $\sim 63\%$ of the maximum value.

After the steady state condition (constant $I = V/R$) is reached, what happens if we close switch S_2, effectively shorting out the battery?[4] Because the battery is no longer present, the current starts to decrease. The rate of change of the current is exactly that which, as the magnetic field collapses, induces a back-EMF across the inductance equal in magnitude to the voltage across the resistance (in accordance with Kirchhoff's loop rule). The current falls toward zero exponentially, such that the magnitudes V_R and V_L are equal at all times.

We can derive an equation that describes this exponential decay. From Kirchhoff's loop rule:

$$\Sigma V = 0$$

$$IR + L\frac{dI}{dt} = 0 \tag{26-37}$$

This equation may be solved in the manner used for the solution of Equation (26–35):

$$I = \frac{V}{R}e^{-(R/L)t} \tag{26-38}$$

In this case, the current will drop to $1/e$ ($\sim 37\%$) of its initial value in the time constant $\tau_L = L/R$. A graph of the solution is shown in Figure 26–18(b).

□ □ □

The exponential growth and decay of the current in *RL* circuits is similar to the exponential changes occurring in *RC* circuits. It will be helpful if you review the discussion of *RC* circuits (Section 24.9) to clarify these similarities. It is always easier to remember facts if one can relate them to similar behavior in other situations.

---

[4] Some batteries would be severely damaged by a direct "short circuit," even for a second or so. However, once the switch is closed, the battery could be immediately removed without affecting the rest of the circuit. In this theoretical discussion, we ignore these troublesome realities. But they should not be ignored in the laboratory.

A final comment of a practical nature. In order to obtain the decay described by Equation (26–38), it was necessary to reduce V to zero *without interrupting the current loop*. Opening the switch S_1 after the current had reached its maximum value would tend to interrupt the current suddenly, causing a high value of dI/dt. Consequently, a very high EMF would be induced, creating an electric spark or arc across the switch contacts. In the next section we will show that the energy dissipated in the spark (and in the I^2R losses) is derived from the energy stored in the magnetic field of the inductance. In electrical machinery and power-transmission design, a great deal of care must be taken to avoid arcing at switch contacts in circuits that contain inductances.

EXAMPLE 26–9

Reconsider the inductance-resistance circuit shown in Figure 26–17. Let $V = 10$ V, $R = 2\ \Omega$, and $L = 5$ H. The switch S_1 is closed and the current is allowed to reach essentially its maximum value. The switch S_1 is then suddenly opened, causing the current to drop to zero in 0.20 s. Find the approximate EMF developed across the inductance as the switch is opened.

SOLUTION

We will make the assumption that I decreases linearly with time. (This assumption yields a minimum value of dI/dt.) The initial (steady state) value of the current is

$$I_0 = \frac{V}{R} = \frac{10\text{ V}}{2\ \Omega} = 5\text{ A}$$

The rate of change of current is

$$\frac{\Delta I}{\Delta t} = \frac{I - I_0}{\Delta t} = \frac{0 - 5\text{ A}}{0.20\text{ s}} = -25\,\frac{\text{A}}{\text{s}}$$

So the induced EMF is

$$\mathscr{E} = -L\frac{dI}{dt}$$

$$\mathscr{E} = -(5\text{ H})\left(-25\,\frac{\text{A}}{\text{s}}\right) = \boxed{125\text{ V}}$$

The polarity is such as to oppose the *change* of current; therefore, the top end of the coil is positive with respect to the bottom end.

26.8 Energy in Inductors

In the last section we referred to energy stored within an inductor. In order to determine the amount of energy stored, we will start with an energy-balance equation applied to the circuit shown in Figure 26–19.

Applying Kirchhoff's loop rule, multiplying by I, and rearranging, we have

$$V - IR - L\frac{dI}{dt} = 0$$

$$\Sigma V = 0$$

$$VI = I^2R + LI\frac{dI}{dt} \tag{26–39}$$

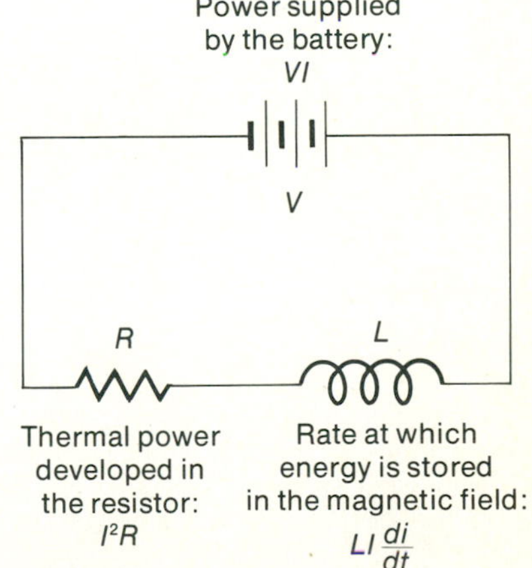

Figure 26–19
As the current rises toward its steady state value, part of the power supplied by the battery appears as energy stored in the inductor.

where

$$VI = \text{the power supplied by the battery}$$
$$I^2R = \text{the thermal power dissipated in the resistor}$$
$$LI\frac{dI}{dt} = \text{the rate at which energy is stored in the inductor}$$

Letting U_L represent the energy stored in the inductor:

$$\frac{dU_L}{dt} = LI\frac{dI}{dt}$$

or

$$dU_L = LI\,dI$$

Since $U_L = 0$ when $I = 0$, we integrate this equation to obtain

ENERGY STORED IN AN INDUCTOR

$$U_L = \tfrac{1}{2}LI^2 \tag{26–40}$$

Note the similarity in form between this expression and the energy-storage equation for a capacitor:

$$U_C = \tfrac{1}{2}CV^2 \tag{26–41}$$

where C is the capacitance and V is the potential difference across the capacitor. An interesting difference between energy storage in a capacitor and in an inductor is that a charged capacitor may be removed from the circuit retaining its stored energy, whereas an inductor can retain its stored energy only by maintaining a current through it.

Recall that starting with the expression for the energy stored in a capacitor, $U_C = \frac{1}{2}CV^2$, we obtained an expression for the energy density u_E in an electric field:

$$u_E = \tfrac{1}{2}\varepsilon_0 E^2 \tag{26–42}$$

We now follow a similar procedure for magnetic fields. We will start with Equation (26–40) for the energy stored in an inductor and develop an expression for the energy density u_B associated with a magnetic field. The inductor we will consider is a large toroid, since all the magnetic field is confined to its interior and thus is easy to calculate. The toroid field is given by Equation (26–5):

$$B = \frac{\mu_0 NI}{2\pi R}$$

As the radius of the toroid R becomes larger while the cross-sectional area remains constant, B becomes uniform over the cross section, yielding

$$B = \frac{\mu_0 NI}{\ell} \tag{26–43}$$

where ℓ is the circumferential length of the toroid.

From the same reasoning we used to find the self-inductance of a long solenoid, the self-inductance L of a toroid with a very large radius R has exactly the same form [Equation (26–31)]:

$$L = \frac{\mu_0 N^2 A}{\ell}$$

where ℓ is the circumference, $2\pi R$, of the toroid.

Equation (26–40) gives the total energy U_L stored in an inductor:

$$U_L = \tfrac{1}{2}LI^2$$

Since this energy is stored in the magnetic field, we may determine the energy U_B contained in the magnetic field by using Equations (26–43) and (26–31) and substituting for I and L, respectively:

$$U_B = \frac{1}{2}\left(\frac{\mu_0 N^2 A}{\ell}\right)\left(\frac{B\ell}{\mu_0 N}\right)^2$$

or

$$U_B = \frac{1}{2}\left(\frac{B^2}{\mu_0}\right)A\ell$$

Since $A\ell$ is the volume of the large toroid, the **energy density** u_B is

$$u_B = \frac{U_B}{A\ell}$$

or

ENERGY DENSITY u_B IN A MAGNETIC FIELD

$$u_B = \frac{1}{2}\left(\frac{B^2}{\mu_0}\right) \qquad \textbf{(26–44)}$$

The units of energy density are *joules/meter*^{3}. Though for ease of calculation we used a particular configuration for the inductor, the result is perfectly general and applies to any magnetic field B.

EXAMPLE 26–10

Find the energy stored in the gap of a permanent magnet such as the one illustrated previously in Figure 26–9(a). Assume that the field is uniform within the gap and equal to 0.5 T and that the volume of the gap is 2.0 cm^3.

SOLUTION

Since Equation (26–44) applies to any magnetic field:

$$u_B = \frac{1}{2}\left(\frac{B^2}{\mu_0}\right)$$

Substituting:

$$u_B = \frac{(1)(0.5\ \text{T})^2}{(2)\left(4\pi \times 10^{-7}\ \dfrac{\text{H}}{\text{m}}\right)}$$

$$u_B = 9.95 \times 10^4\ \frac{\text{J}}{\text{m}^3}$$

The total energy stored is

$$U_B = u_B V$$

where V is the volume of the gap.

$$U_B = \left(9.95 \times 10^4\ \frac{\text{J}}{\text{m}^3}\right)(2.0 \times 10^{-6}\ \text{m}^3)$$

$$U_B = \boxed{0.199\ \text{J}}$$

While it is possible to extract the energy associated with an electric field, no practical method has yet been devised to extract the considerable energy associated with the magnetic field of a permanent magnet. It is interesting to speculate concerning the vast reservoir of magnetic-field energy associated with the earth's magnetic field.

26.9 Magnetic Properties of Materials

The origin of the magnetic properties of materials is within their atomic structures. A detailed explanation therefore involves quantum mechanics. However, a brief discussion of the phenomena in classical terms provides a valuable insight without unduly compromising the accuracy of the description.

For our purposes, we may consider an atom to be made up of a positively charged nucleus with electrons circulating in orbits about the nucleus. In addition to these orbital motions, each electron also spins about its own axis, similar to a spinning top. As a result, each electron produces a magnetic dipole moment μ [Equation (25–14)] partly due to the orbital motion and partly due to the spin. There is a tendency for all the individual dipole moments within a single atom to combine in pairs, with opposite orientations, so that the net magnetic dipole moment for the atom as a whole can be zero. In other cases, however, the dipole moments do not exactly cancel, so the atoms of some materials have a finite magnetic dipole moment.

Paramagnetism

Thermal motions of atoms that have net dipole moments randomly orient their dipoles so that the bulk material has zero net dipole moment. However, as discussed earlier in this chapter, in the presence of an external magnetic field the dipoles experience a torque that tends to align them parallel to the field. Depending on the field strength and the temperature, some of the atoms will align their magnetic dipoles in the field direction. In this way the bulk material exhibits a *net dipole moment* in the presence of a magnetic field. When the material is removed from the field, thermal motions again randomize the orientations of individual atomic dipoles and the material no longer possesses a net dipole moment. Substances that exhibit this property are called **paramagnetic.** A permanent magnet will exert an *attractive* force on such materials.

Diamagnetism

A few elements are *repelled* by a permanent magnet. Such materials are called **diamagnetic.** Michael Faraday noticed this effect in bismuth; silver is also noticeably diamagnetic. The effect is quite weak. Elements whose atoms have zero net magnetic dipole moments are diamagnetic. Atoms that have a permanent dipole moment (and that are not *ferro*magnetic) may be either diamagnetic or paramagnetic, depending on which effect is stronger. To understand the phenomenon, consider two electrons of the same atom circulating in opposite directions, as shown in Figure 26–20(a). (For clarity, the centers of rotation have been separated.) The centripetal force that keeps the electrons in their circular orbits is the electrostatic attraction of the positively charged nucleus. Since the electrons circulate in opposite directions, their combined dipole moment is zero. When an external magnetic field is applied, the circulating electrons

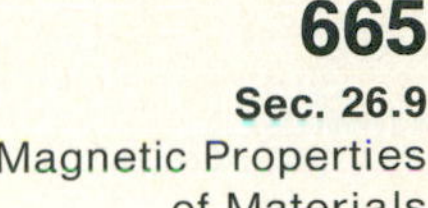

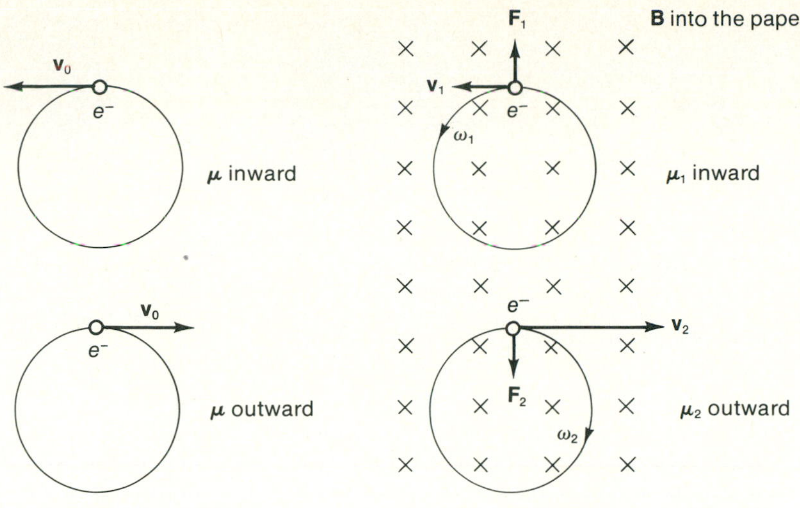

(**a**) With no external magnetic field, the net dipole moment is zero, because both dipole moments are equal in magnitude.

(**b**) The external magnetic field produces the additional forces $\mathbf{F}_1$ and $\mathbf{F}_2$, so that $\omega_2 > \omega_1$. As a result, $\mu_2 > \mu_1$, making the net dipole moment opposite to the external field.

Figure 26–20
Diamagnetism

experience an additional radial force given by the Lorentz force law:

$$\mathbf{F} = q(\mathbf{v} \times \mathbf{B})$$

In one case the force *aids* the electrostatic force, and in the other case it *opposes* the electrostatic force. Since the centripetal force is

$$F = m\omega^2 r$$

the greater the inward force, the greater the angular velocity ω of the electron. (It can be shown that the radius of the orbit remains unchanged.) Thus the lower illustration in Figure 26–20(b) has the greater dipole moment, resulting in a net dipole moment *opposite* to the applied field (rather than *in* the field direction as in paramagnetism). Because the induced dipole moment is opposite to the field direction, a permanent magnet exerts a *repulsive* force on diamagnetic materials.

Ferromagnetism

The third class of magnetic materials contains five elements: iron, cobalt, nickel, gadolinium, and dysprosium, as well as some alloys made from them. These are the **ferromagnetic** materials whose magnetic effects are orders of magnitude greater than those of paramagnetic or diamagnetic substances. The basic distinction is that because of quantum mechanical effects, the dipole moments of these ferromagnetic atoms exert forces on their neighbors, causing all the dipoles to align parallel to one another within a region called a *domain.* Magnetic domains may have volumes from about 10^{-6} cm^3 to 10^{-2} cm^3, so each may contain from 10^{17} to 10^{21} atoms. Boundaries between regions in which domains have different orientations are called *domain walls.* Although each domain is magnetized as strongly as it can be, neighboring domains may be aligned along different directions. In fact, even within a single crystal there may be several different *directions of easy magnetization* for domain orientation. In unmagnetized bulk material, the domains have sufficiently random orientations that

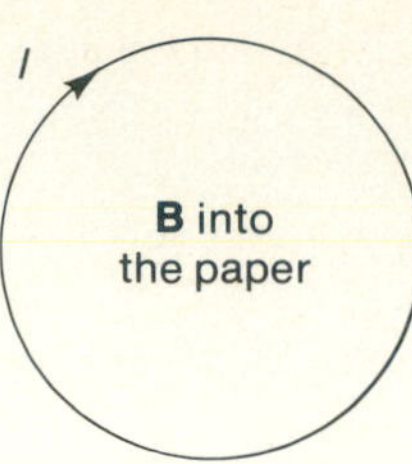

(a) A solenoid with an air core.

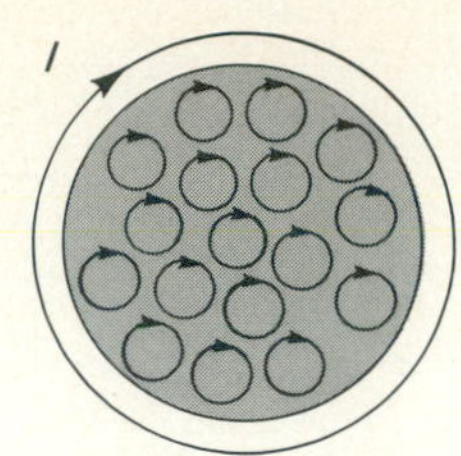

(b) A solenoid with a paramagnetic material in the core.

(c) The additional current I' is due to the paramagnetic material.

Figure 26–21
The cross section of a solenoid, showing the effect of a paramagnetic material within its windings.

there is no net magnetic moment. But if an external field is applied, domains oriented *in* the field direction will grow in size (at the expense of adjacent, less-favorably-oriented domains) by displacement of the domain boundaries. Also, dipoles within an isolated domain may suddenly "flip around" together to align themselves in the field direction.[5] When the external field is removed, the dipole moment of the bulk material remains, because domains are not easily dislodged into random orientation by thermal agitation. Thus "magnetized" materials retain a permanent magnetic moment. Ferromagnetic materials are used to fabricate permanent magnets. Permalloy and Alnico are trade names of well-known examples of certain aluminum-nickel-cobalt-iron alloys that retain a high degree of permanent magnetic moment.

Each ferromagnetic material has a critical temperature, called the *Curie temperature*, above which the energies of thermal motions are great enough to upset the alignment of magnetic moments in a domain. Above the Curie temperature, the material becomes paramagnetic.

□ □ □

Having described three basic types of magnetic materials: *para*magnetic, *dia*magnetic, and *ferro*magnetic, we now investigate the effect of placing a paramagnetic material inside the windings of a solenoid. Consider a long solenoid whose interior magnetic field (without the paramagnetic material) is described by Equation (26–6). Letting $n = N/\ell$, the number of turns per unit length, the field B inside is

$$B = \mu_0 \frac{NI}{\ell}$$

This equation is for a vacuum (and, to a close approximation, an "air core") within the windings. In Figure 26–21(a), the windings carry a current I around the cross section as shown, producing a magnetic field inside that is directed *into* the plane of the figure. In Figure 26–21(b), a paramagnetic material has been introduced into the core of the solenoid. The field has aligned some of the dipole moments of the material so that all the current loops corresponding

[5] As a domain near the wall boundary reorients its direction of magnetization, the domain as a whole does not rotate as a unit. Rather, as a result of quantum mechanical forces, almost simultaneously each *atom* within the domain reorients its magnetic moment in the new direction. If a coil of wire is wound around the material and connected to a sensitive amplifier and loudspeaker, the sudden, slight changes of flux in the coil as each domain flips are detected as tiny "ticks" in the amplifier output. This is known as the *Barkhausen effect*, after the experimenter who first discovered it in 1919.

to these dipoles circulate in the same direction. (Unaligned current loops are not shown.) At the boundaries between adjacent current loops, the currents are in opposite directions and therefore cancel. In effect, this cancellation occurs *throughout the cross section* of the paramagnetic material, leaving only one large loop current I' around the outside perimeter. Since I' is in the *same* direction as the current I in the windings, the net magnetic field inside the solenoid is increased.

The additional magnetic field B' due to the current loops I' is proportional to the same quantities that determined the original magnetic field in the vacuum. Namely:

$$B' = \chi \frac{NI}{\ell} \tag{26–45}$$

where the proportionality factor χ is called the **magnetic susceptibility** of the material in units of henrys/meter (H/m). The total field B inside the solenoid is due to both the actual current in the windings and the induced current loops:

$$B = \mu_0 \frac{NI}{\ell} + \chi \frac{NI}{\ell}$$

or, simply:

$$B = (\mu_0 + \chi)H \tag{26–46}$$

where:

MAGNETIC FIELD STRENGTH *H*

$$H = \frac{NI}{\ell} \tag{26–47}$$

The symbol H is called[6] the **magnetic field strength** and is expressed in units of ampere·turns/meter (A·turns/m). Equation (26–46) may be rewritten in the form:

RELATION BETWEEN *B* AND *H*

$$B = \mu H$$

where

$$\mu \equiv \mu_0 + \chi$$

The symbol μ is called the **permeability** of the magnetic material. It includes the *permeability of free space*, μ_0, as well as the additional magnetic effects of various materials (described by χ). (Do not confuse μ with the vector quantity $\boldsymbol{\mu}$, the symbol for a magnetic dipole moment.) Representative values of χ are given in Table 26–2.

In summary, the symbol H represents the contribution to the magnetic field B (through $\mu_0 H$) produced *only* by currents in coil windings. The symbol B includes not only the field $\mu_0 H$, but also additional effects produced by the alignment of the magnetic dipole moments of the atoms of any material that is placed inside the coil windings.

[6] At last our rather loose terminology catches up with us. In calling B the *magnetic field* (rather than by its precise name, *magnetic induction* or *magnetic flux density*), we are following a very widespread practice that is still in vogue. Originally, H was defined as the "magnetic field strength". It has been proposed that these quantities be redefined to conform more to usage, but as yet the change has not occurred. In the meantime, be careful to keep B and H distinct: They *are* different.

TABLE 26–2

Magnetic Susceptibilities (at 20°C unless otherwise noted)

Material	χ (H/m) (negative values indicate diamagnetic materials)
Aluminum	1.03×10^{-11}
Bismuth	-2.12×10^{-11}
Cerium	$24. \times 10^{-11}$
Copper	-0.14×10^{-11}
Ferric chloride	$135. \times 10^{-11}$
Mercury	-0.30×10^{-11}
Silver	-0.32×10^{-11}
Liquid oxygen (−219°C)	3.9×10^{-4}
Iron (for $H = 20$ A·turns/m)	$5. \times 10^{-4}$
Iron (for $H = 60$ A·turns/m)	$16. \times 10^{-4}$

EXAMPLE 26–11

If the air within the windings of a long solenoid were replaced by iron, by what factor would the total magnetic field B increase? Assume that the magnetic field strength H due to current in the windings remains constant at 20 A·turns/m.

SOLUTION

By Equation (26–47), the magnetic field B_0 inside the solenoid without the iron is

$$B_0 = \mu_0 H$$

With the iron, it is

$$B = (\mu_0 + \chi)H$$

The ratio of B/B_0 is

$$\frac{B}{B_0} = \frac{\mu_0 + \chi}{\mu_0}$$

Substituting the appropriate values (noting from Table 26–2 that $\chi = 5 \times 10^{-4}$ for $H =$ 20 A·turns/m):

$$\frac{B}{B_0} = \frac{4\pi \times 10^{-7} + 5 \times 10^{-4}}{4\pi \times 10^{-7}}$$

$$\frac{B}{B_0} = \boxed{399}$$

Thus, the use of the iron core increases the magnetic flux density inside the solenoid by a factor of about 400.

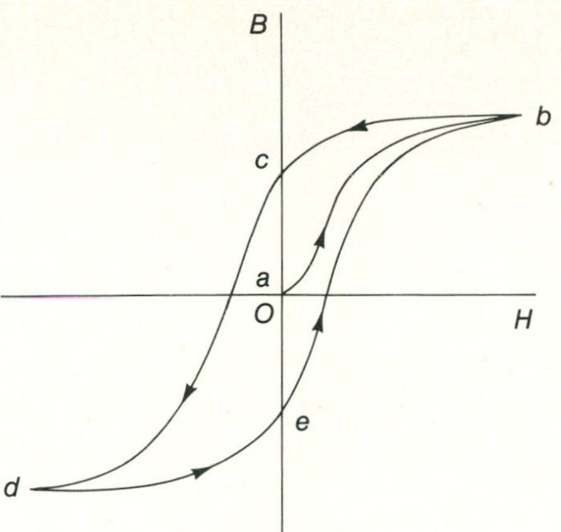

(**a**) Starting with an unmagnetized sample at *O*, the curve $a \rightarrow b$ is the *magnetization curve*. Repeated reversals of the solenoid current then trace out the outer portion (*bcdeb*), called a *hysteresis loop*.

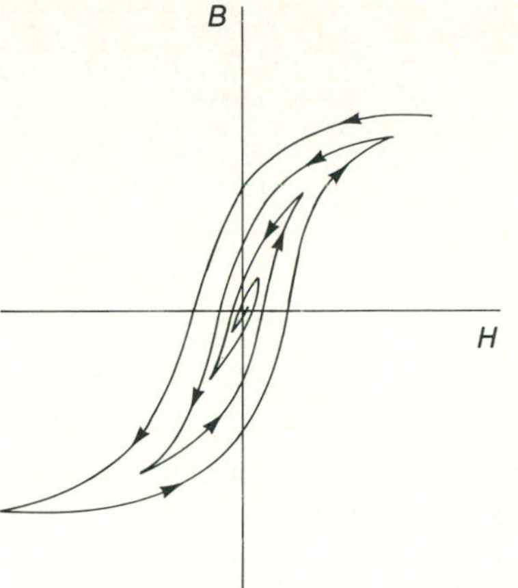

(**b**) Demagnetizing a ferromagnetic material involves traveling around successive hysteresis loops, gradually decreasing the magnitude of *H* with each cycle.

Figure 26–22
A graph of the magnetic field *B* in an iron core inside a solenoid versus the magnetic field strength *H* produced by current in the coil windings.

The previous example calculated *B* in a sample of iron for just one value of the magnetic field strength *H*. If the current in the solenoid is varied (changing the value of *H*), the resultant values of *B* change in an interesting way because of a phenomenon called *hysteresis*. If we measure *B* inside a ferromagnetic material as the current in the surrounding solenoid is increased, the *magnetization curve* of Figure 26–22 is obtained. Starting with unmagnetized material at point *a*, the field *B* increases to point *b* as the magnetizing current in the solenoid increases. The curve begins to level off near *b* because of *saturation* as the majority of the domains become oriented in the "proper" direction. If the saturation is 100%, a further increase of *H* increases *B* only to the extent of the $\mu_0 H$ term in Equation (26–46). If the current is then reduced to zero, some residual magnetic field remains (point *c*) because domain boundaries do not all return to their original states. The material is now a "permanent" magnet. The fact that the material does not retrace the original magnetizing curve is called **hysteresis,** a term meaning "to lag behind." Increasing the magnetizing current in the opposite direction and back again produces the characteristic curve called a *hysteresis loop* (*bcdeb*). The area within the loop is proportional to the amount of electrical energy converted to thermal energy within the material during one cycle (because the magnetization process is not reversible). To demagnetize a sample, the material is carried around successive hysteresis loops by repeatedly reversing the current in the solenoid, decreasing the magnitude of the reversals during each cycle.

26.10 Transformers

One of the most universally useful electrical devices is a *transformer*. It is capable of raising or lowering the amplitude of a varying voltage without appreciable loss of power. In order to transmit power over great distances, the sinusoidally varying voltage at the source is often raised by a transformer to a very high value. Since the total power (VI) remains the same, raising the voltage means the current is lower. Consequently, the I^2R losses in the transmission lines are

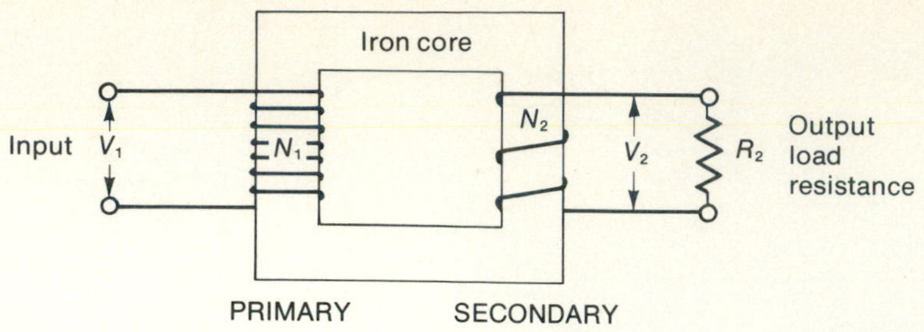

Figure 26–23
A transformer

reduced. At the consumer end, another transformer lowers the voltage to a safe and practical value.

Figure 26–23 is a conventional way of indicating a transformer. One side of a transformer is designated the **primary** winding, where the *input* voltage is applied; the other side is the **secondary** winding, which supplies the *output* voltage. An iron core usually connects the primary and secondary windings so that the magnetic flux generated by the primary winding is large. Because of the magnetic effects of the iron core, the flux lines are almost entirely confined within the core. (That is, there is very little "leakage." In an *ideal* transformer, there is assumed to be no leakage and no thermal losses in the core or windings.) So both windings surround essentially the same varying magnetic flux. Therefore, the EMF $\mathscr{E}_1$ per turn N_1 in the primary is the same as the EMF $\mathscr{E}_2$ per turn N_2 in the secondary. Mathematically, this is

$$\frac{\mathscr{E}_1}{N_1} = \frac{\mathscr{E}_2}{N_2} \tag{26–48}$$

The primary may be considered an inductor in which the EMF $\mathscr{E}_1$ is equal to the applied voltage V_1. The EMF $\mathscr{E}_2$ across the secondary appears as the output voltage V_2. Then, the above equation may be written as

$$\frac{N_2}{N_1} = \frac{V_2}{V_1} \tag{26–49}$$

showing that the *turns ratio* (N_2/N_1) for an ideal transformer is the same as the *voltage ratio* (V_2/V_1). If the output voltage is larger than the input voltage, the transformer is called a *step-up* transformer; if the output voltage is lower, it is a *step-down* transformer.

When a resistive load R_2 is attached to the secondary, current I_2 will flow in the secondary winding. Ideally, with no I^2R losses in the windings or core, the instantaneous power input equals the instantaneous power output. That is:

$$V_1 I_1 = V_2 I_2 \tag{26–50}$$

Using Equation (26–49), we see that the turns ratio is the *inverse* of the current ratio:

$$\frac{N_2}{N_1} = \frac{I_1}{I_2} \tag{26–51}$$

Thus a transformer may be used to transform a varying current as well as a varying voltage. (Note that a step-up transformer steps down the current.)

EXAMPLE 26–12

Consider the ideal transformer shown in Figure 26–23. What is the ratio V_1/I_1 in terms of N_1, N_2, and R_2? (This is an important ratio because it is the equivalent *input resistance* that the source "sees" when a resistive load R_2 is placed across the secondary.)

SOLUTION
In the secondary circuit, at any instant:

$$V_2 = R_2 I_2 \qquad \textbf{(26–52)}$$

We begin by solving for V_2 in Equation (26–49):

$$V_2 = \left(\frac{N_2}{N_1}\right) V_1$$

and for I_2 in Equation (26–51):

$$I_2 = \left(\frac{N_1}{N_2}\right) I_1$$

Substituting these into Equation (26–52) and rearranging, we obtain

$$\left(\frac{N_2}{N_1}\right) V_1 = R_2 \left(\frac{N_1}{N_2}\right) I_1$$

$$\frac{V_1}{I_1} = \left(\frac{N_1}{N_2}\right)^2 R_2$$

Notice that $(N_1/N_2)^2 R_2$ is the *effective resistance of the load* (R_2) input from the point of view of the primary. This technique of changing the effective resistance using a transformer is very important in power transfer from one part of an electronic circuit to another. The next chapter will discuss this subject in more detail.

Summary

The following concepts and equations were introduced in this chapter:

Biot-Savart law:

$$d\mathbf{B} = \left(\frac{\mu_0}{4\pi}\right) \frac{I\, d\ell \times \hat{\mathbf{r}}}{r^2}$$

where μ_0 is the *permeability of free space:*

$$\mu_0 = 4\pi \times 10^{-7} \frac{\text{Wb}}{\text{A}\cdot\text{m}}$$

Ampere: A mutual force per unit length of exactly 2×10^{-7} N/m exists between two parallel conductors one meter apart, each carrying a current of one ampere.

Ampère's law:

$$\oint \mathbf{B}\cdot d\ell = \mu_0 I$$

where I is the total current through the area enclosed by the path of integration. Ampère's law is of practical use when symmetry indicates that $\mathbf{B}$ has a constant magnitude and a constant angle with respect to $d\ell$ along the path of integration.

Magnetic field B:
Due to a current in a long, straight wire:

$$B = \frac{\mu_0 I}{2\pi a}$$

Center of a current-carrying circular loop:

$$B = \frac{\mu_0 I}{2r}$$

Inside the windings of a toroidal coil (N turns, with an average circumference of $2\pi R$):

$$B = \frac{\mu_0 N I}{2\pi R}$$

Inside a long solenoid (n turns per unit length):

$$B = \mu_0 n I$$

Faraday's law:

$$\mathscr{E} = -\frac{d\Phi_B}{dt}$$

or

$$\oint \mathbf{E} \cdot d\boldsymbol{\ell} = -\frac{d}{dt}\int \mathbf{B} \cdot d\mathbf{A}$$

Lenz's law: The induced EMF always has a polarity such as to *oppose the change* in magnetic flux that produces the EMF.

Inductance L (also called *self-inductance*):

$$\mathscr{E} = -L\frac{dI}{dt}$$

The inductance of a *long solenoid* with a length ℓ or a *large toroid* with a circumference ℓ:

$$L = \frac{\mu_0 N^2 A}{\ell} \qquad (N = \text{total number of turns})$$

Mutual inductance M:

$$\mathscr{E}_1 = -M\frac{dI_2}{dt} \qquad \text{and} \qquad \mathscr{E}_2 = -M\frac{dI_1}{dt}$$

RL circuit: In a series combination of a seat of EMF with terminal voltage V, a resistor R, and an inductor L, the current is:

Growth: $I = \frac{V}{R}(1 - e^{-(R/L)t})$

Decay: $I = \frac{V}{R}e^{-(R/L)t}$

In one *time constant*, $\tau_L = L/R$, the growing exponential rises to $\sim 63\%$ of its final (maximum) value and the decaying exponential falls to $\sim 37\%$ of its original value.

Energy stored in a current-carrying inductor:

$$U_L = \tfrac{1}{2}LI^2 \qquad \text{(in joules)}$$

Energy density in a magnetic field:

$$u_B = \frac{1}{2}\left(\frac{B^2}{\mu_0}\right) \qquad \text{(in joules/meter}^3\text{)}$$

Magnetic materials:

Paramagnetic materials temporarily become magnetically polarized *in* the direction of an applied magnetic field.

Diamagnetic materials temporarily become magnetically polarized *opposite* to the direction of an applied magnetic field.

Ferromagnetic materials can acquire residual magnetism by orientation of magnetic domains.

Magnetic field strength H of a long solenoid:

$$H = \frac{NI}{\ell}$$

Magnetic field B (more precisely called the *magnetic induction*) is related to the magnetic field strength H:

$$B = \mu H$$

where μ is the *permeability* of the magnetic material.

Magnetic susceptibility χ represents the change in permeability due to the presence of a magnetic material:

$$\mu \equiv \mu_0 + \chi$$

Transformers: Letting the subscript 1 refer to the primary and the subscript 2 refer to the secondary, in an *ideal* transformer (no I^2R losses and no flux leakage) the input power equals the output power:

$$V_1 I_1 = V_2 I_2$$

The *turns ratio* is:

For the voltage

$$\frac{N_2}{N_1} = \frac{V_2}{V_1}$$

For the current

$$\frac{N_2}{N_1} = \frac{I_1}{I_2}$$

In a *step-up* transformer, $V_2 > V_1$ (while $I_2 < I_1$); in a *step-down* transformer, $V_2 < V_1$ (while $I_2 > I_1$).

The *effective resistance* $(R_2)_{\text{input}}$ of the load resistor R_2 viewed from the primary is

$$(R_2)_{\text{input}} = \left(\frac{N_1}{N_2}\right)^2 R_2$$

Questions

1. In what way is the Biot-Savart law similar to Coulomb's law? In what way are these two laws dissimilar?
2. If current were to pass through the helical turns of a stretched coil spring, would the force the spring exerts increase, decrease, or remain the same? Explain.
3. For what kind of situation is it more appropriate to use Ampère's law rather than the Biot-Savart law for computing the magnetic field?
4. Is there a magnetic field inside copper tubing that is carrying a current? If not, why not?
5. Parallel current-carrying conductors interact with each other. How do current-carrying conductors perpendicular to each other interact?
6. An airplane with a metal propeller is flying along the direction of the magnetic field lines of the earth. (a) Is there a potential difference between the tips of the propeller blades? (b) Is there a potential difference between the propeller hub and the propeller tips?

7. Where and in which direction would an airplane fly so that the earth's magnetic field would produce the greatest potential difference between the wing tips, with the right tip positive with respect to the left tip?

8. Can electric field lines form closed loops as well as originate on charges? Explain.

9. What is the connection between Lenz's law and the conservation of energy?

10. Is the net magnetic flux through a closed surface surrounding the north pole of a magnet zero? Is the net electric field flux through a surface surrounding the positive charge of an electric dipole zero?

11. A toroid inductor has essentially no external magnetic field. However, there is a small, unavoidable external field. Describe this field and its origin.

12. Pairs of wires carrying current to and from electrical devices are often twisted together. Why?

13. The sensing element of most metal detectors is a coil of wire. Considering mutual inductance phenomena, how does such a metal detector work?

14. Why does increased resistance increase the time constant of an *RC* circuit, while it decreases the time constant of an *RL* circuit?

15. An unmagnetized iron placed halfway into a solenoid will be suddenly drawn into the solenoid as soon as current begins to pass through the windings of the solenoid. What is the explanation of this phenomenon on the basis of the microscopic changes within the iron and on the basis of energy considerations?

16. Consider two iron bars that are identical except that only one of them is magnetized. Using only the two bars, how can the magnetized bar be identified?

17. Wrenches and screwdrivers sometimes become magnetized even though they are not used in electrical work. Why?

18. In order to shield a device from a magnetic field, the device is often enclosed in a box made of iron or a special metal with a high permeability (called *mu-metal*). How does such a shield work?

19. An isolated long, straight wire has inductance. How could its inductance be calculated?

20. How could a transformer be used as a variable inductance? Your answer should indicate why the technique is not used.

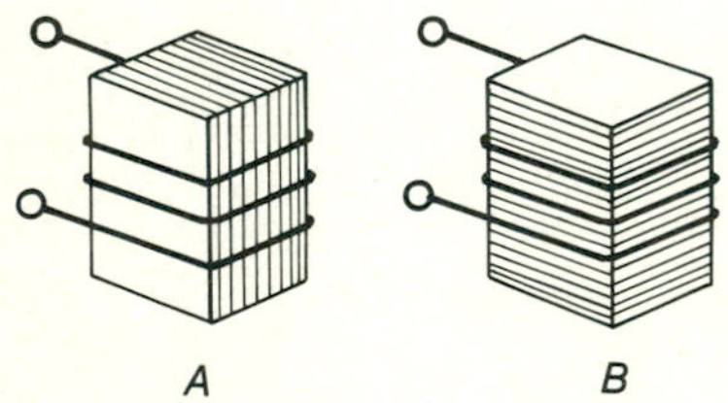

Figure 26–24
Question 22

21. Two identical, rectangular loops of wire are situated so that the plane of each loop is perpendicular to a uniform magnetic field **B**. Loop A is then rotated with angular velocity ω about a central axis parallel to the longer side of the loop, while loop B is rotated with the same angular velocity about a central axis parallel to the shorter side. Is the peak value of the induced EMF in loop A greater than, equal to, or less than the peak EMF induced in loop B? Is the answer the same if the axes are coincident with the sides of the loops (rather than passing through their centers)?

22. Figure 26–24 shows coils of wires with laminated iron cores (thin slabs electrically insulated from each other). When an *AC* voltage is applied to the windings, the changing magnetic flux induces alternating currents in the iron, called *eddy currents*. To reduce I^2R heating in the core, should the laminations be oriented as in A or as in B?

Problems

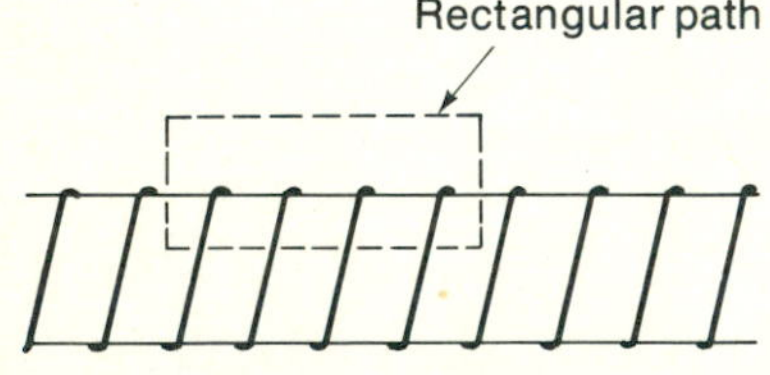

Figure 26–25
Problem 26A–2

26A–1 A straight, solid-metal wire of circular cross section with a radius R carries a current I. Derive an expression for the magnetic field B inside the wire. (For a steady current, the current density inside the wire is uniform.) *Answer:* $B = \mu_0 Ir/2\pi R^2$

26A–2 Derive an expression for the magnetic field B inside an ideal solenoid by applying Ampère's law to the rectangular path shown dashed in Figure 26–25. Assume that B is uniform inside the solenoid and negligible outside.

26A–3 A hiker observes a pocket compass while standing 40 m directly below a single power line that carries a steady current of 150 A. If the horizontal component of the earth's magnetic field is 3×10^{-5} T at the hiker's location, calculate the maximum possible error in the compass reading due to the power line. *Answer:* 1.43°

26A–4 An air-core toroid has individual windings that form loops 2 cm in diameter. The effective circumference of the toroid is 50 cm. Find the number of turns required to produce a magnetic field of 0.07 T within the windings when the current is 5 A. Compare your result with the answer to Problem 26A–5.

26A–5 A magnetic field B of 0.07 T is required within a solenoid 50 cm long and 2 cm in diameter.

(a) Calculate the magnetic flux within the solenoid.

(b) Calculate the necessary number of turns of wire if the current is to be 5 A.

Answers: (a) 2.20×10^{-5} Wb (b) 5570 turns

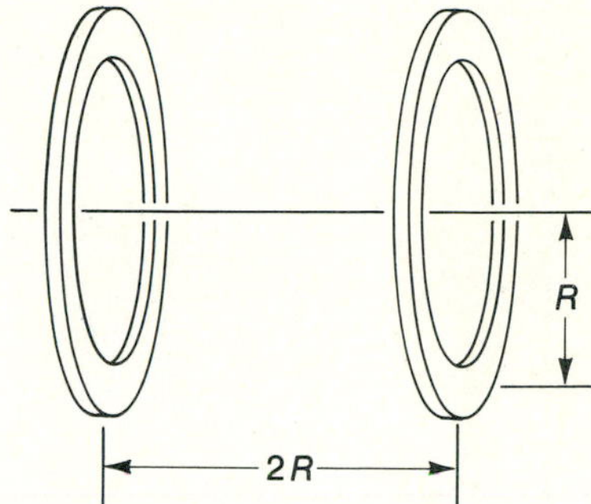

Figure 26–26
Problem 26A–6

26A–6 Two circular coils, each containing N turns of wire, have a radius R and are separated by a distance $2R$, as shown in Figure 26–26. Find the magnetic field at a point on the axis of the coils midway between them. Assume that the coils are in series (so that the circulation of the current I is in the same direction in both coils) and that the cross section of the coils is small compared with R^2.

26A–7 A circular hoop is linked to a toroidal inductor as shown in Figure 26–27. The switch is closed to produce a surge of current through the inductor. Calculate the induced EMF in the hoop when the magnetic flux within the toroid is changing at the rate of 30 Wb/s. Determine the direction of the resulting current flow through the hoop.

Answer: 30 V, clockwise

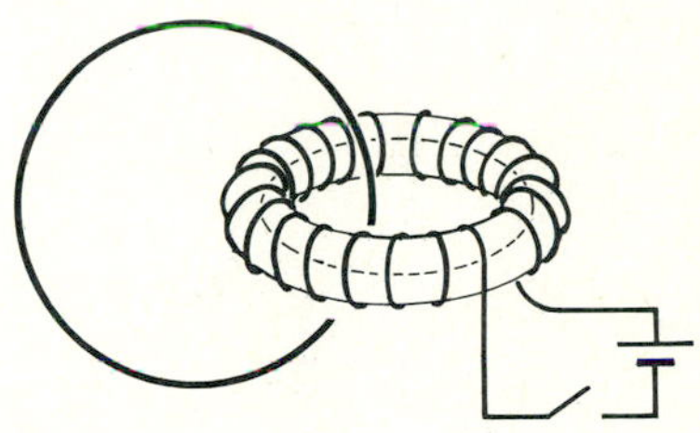
Figure 26–27
Problem 26A–7

26A–8 An airplane with a wingspan of 70 m is flying horizontally at 1000 km/h toward the north magnetic pole of the earth. If the vertical component of the earth's magnetic field at the airplane's position is 2×10^{-5} T, calculate the potential difference between the wing tips. Which wing tip is at the higher potential? Explain why this potential difference cannot be used as a source of power.

26A–9 A wire loop of radius r and resistance R lies in a plane perpendicular to a magnetic field **B**. The loop is rapidly turned over in a time Δt. Find the average EMF $\mathscr{E}$ induced in the loop during the time Δt.

Answer: $\mathscr{E} = \dfrac{2B\pi r^2}{\Delta t}$

26A–10 The current in a 12 mH inductor that has negligible resistance varies with time according to the sawtooth waveform shown in Figure 26–28. Make a graph (with numerical values) of the voltage across the inductor as a function of time.

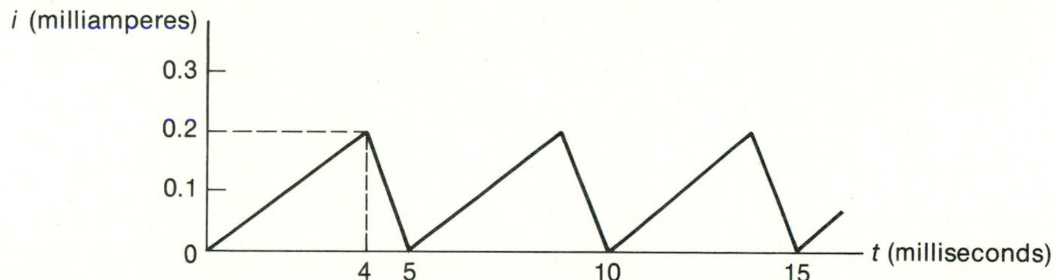

Figure 26–28
Problem 26A–10

26A–11 Beginning with the basic definitions of inductance L and resistance R, show that L/R has the dimensions of time.

26A–12 A 10 V battery, a 5 Ω resistor, and a 10 H inductor are connected in series. Assuming the current in the circuit has reached its maximum value, calculate (a) the power supplied to the circuit by the battery, (b) the power dissipated in the resistor, (c) the power dissipated in the inductor, and (d) the energy stored in the magnetic field of the inductor.

Answers: (a) 20 W (b) 20 W (c) 0 (d) 20 J

26A–13 A seat of EMF $\mathcal{E} = 10$ V is in a series circuit with a switch S, a resistance $R = 50\,\Omega$, and an inductance $L = 5$ H. Find the time after the switch is closed for the current to reach (a) half its final value and (b) 99% of its final value.

Answers: (a) 69.3 ms (b) 461 ms

26A–14 A step-down transformer has 1000 turns in its primary winding and 50 turns in its secondary winding. At a particular instant, a voltage of 10 V appears across a 25 Ω load resistor in the secondary circuit of the transformer. Calculate for the same instant (a) the voltage across the primary of the transformer and (b) the current through the primary winding. Assume that no energy is lost in the transformer.

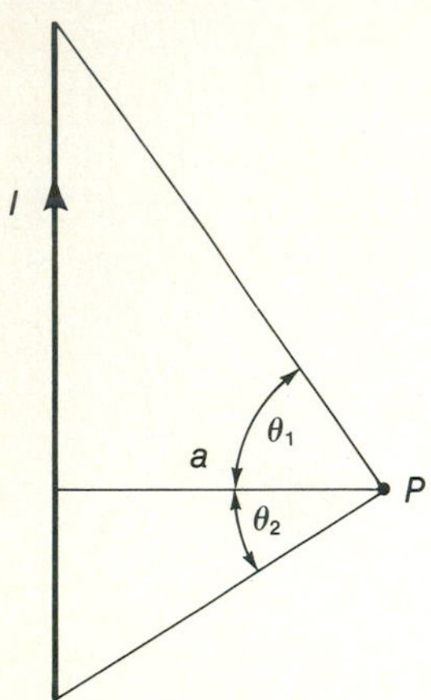

Figure 26–29
Problem 26B–1

26B–1 Refer to Figure 26–29. Starting with the Biot-Savart law, show that the magnetic field B at point P near the straight segment of current-carrying wire is given by $B = (\mu_0 I/4\pi a)(\sin\theta_1 + \sin\theta_2)$.

26B–2 As shown in Figure 26–30, a circular loop of radius R carries a current I. Show that the magnetic field on the axis of the loop a distance x from the plane of the loop is

$$\mathbf{B} = \left(\frac{\mu_0 I}{2}\right)\frac{R^2}{(x^2 + R^2)^{3/2}}\,\hat{\mathbf{x}}$$

(Hint: As you sum the fields $d\mathbf{B}$ due to the current elements $I\,d\boldsymbol{\ell}$ around the loop, what happens to the field components $d\mathbf{B}_\perp$ perpendicular to the x direction?)

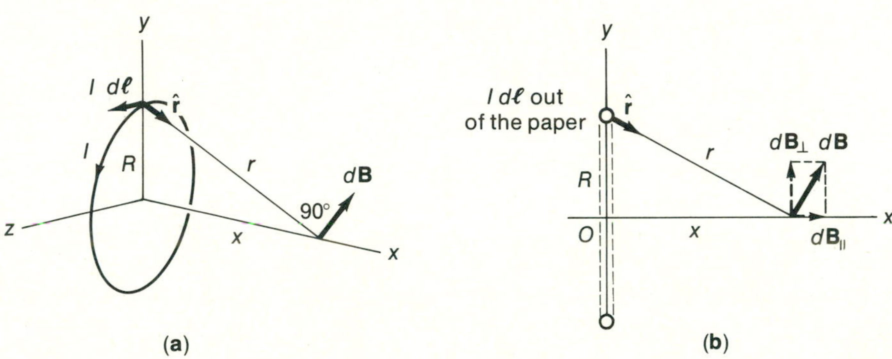

Figure 26–30
Problem 26B–2

26B–3 The steady magnetic field between the pole pieces of a magnet cannot end abruptly at the edges of the pole pieces, as shown in Figure 26–31(a). Instead, the field must fringe outward, as in Figure 26–31(b). Prove this by applying Ampère's law to the region at the edge of the field, as in Figure 26–31(a).

Figure 26–31
Problem 26B–3

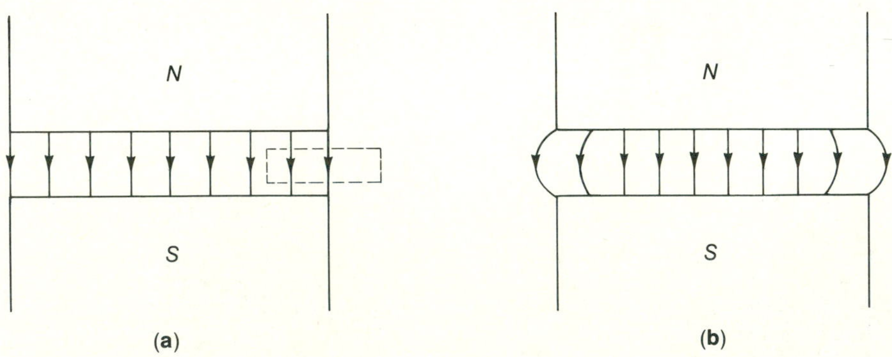

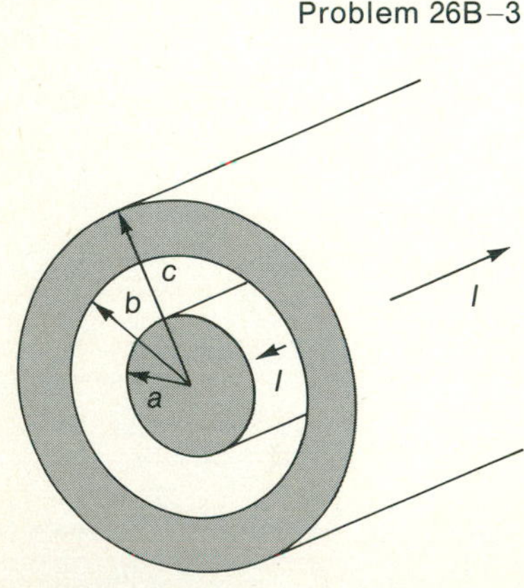

Figure 26–32
Problem 26B–4

26B–4 Consider the long, straight coaxial cable shown in Figure 26–32. A current I is in one direction in the inner conductor and in the opposite direction in the outer conductor. Find expressions for the magnetic field B in the following regions: (a) $r < a$, (b) $a < r < b$, (c) $b < r < c$, and (d) $r > c$. (e) Make a qualitative graph of the magnetic field as a function of distance r from the center of the cable.

26B–5 A rectangular wire loop of mass m, total resistance R, and dimensions as shown in Figure 26–33 is falling freely under gravity as it emerges from a region of uniform, horizontal magnetic field B. The plane of the loop is perpendicular to B.

(a) Is the induced current in the loop clockwise or counterclockwise?

(b) At a certain speed v, the loop falls without acceleration while emerging from the field. Show that this speed is $v = mgR/B^2a^2$.

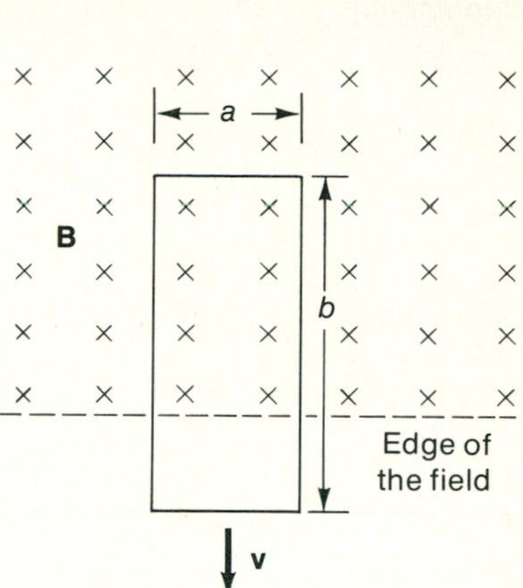

Figure 26–33
Problem 26B–5

26B–6 A 30-turn coil of wire is placed at the end of a long solenoid wound with 4000 turns/m. The coil and solenoid have the same radius, $R = 5$ cm, and their axes are coincident. Find the rate of change of current in the solenoid if there is an induced EMF of 2 mV in the coil.

26B–7 In Figure 26–34, the rectangular wire loop and the long, straight conductor lie in the same plane. The total electrical resistance of the wire loop is 2 Ω.

(a) For a steady current I in the straight conductor, find the total magnetic flux Φ_B that passes through the loop. (Hint: Choose an element of area $dA = \ell\,dr$ and find the flux $d\Phi_B$ through this area. Then integrate to find the total flux.)

(b) If the current I in the conductor decreases uniformly from 10 A to 2 A in 2 s, find the induced current I' in the loop for the case $\ell = 30$ cm.

Answers: (a) $\Phi_B = \mu_0 I\ell(\ln 3)/2\pi$ (b) 1.32×10^{-7} A

Figure 26–34
Problem 26B–7

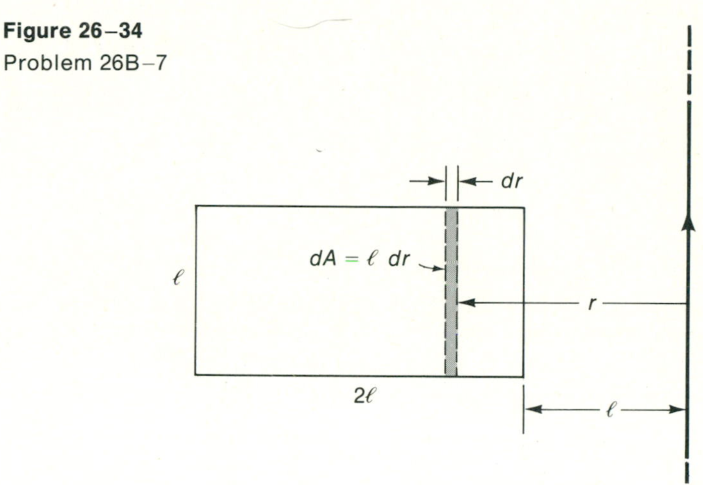

26B–8 A time-varying current I is applied to an inductance of 5 H, as shown in Figure 26–35. Make a quantitative graph of the potential of point a relative to that at point b. The current arrow should indicate the direction of positive current flow.

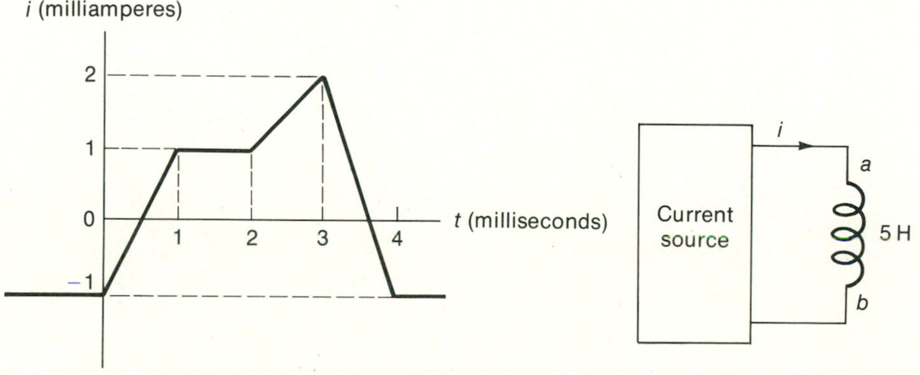

Figure 26–35
Problem 26B–8

26B–9 A long solenoid of length ℓ and cross-sectional area A contains a total of N_1 turns. A second coil of N_2 turns is closely wound around the center of the solenoid (keeping the two coils electrically insulated from each other). Find the mutual inductance M between the coil and solenoid. *Answer:* $M = \mu_0 A N_1 N_2/\ell$

26B–10 A source of EMF, $\mathscr{E} = 500$ V, is applied to a coil that has an inductance of 0.80 H and a resistance of 30 Ω.

(a) Find the energy stored in the magnetic field after the current reaches half its maximum value.

(b) How long after the EMF is connected does it take for the current to reach this value?

26B–11 Verify by direct substitution that $I = (V/R)(1 - e^{-(R/L)t})$ is a solution of the differential equation $V - IR - L\,dI/dt = 0$.

26B–12 A battery is in series with a switch and a 2 H inductor whose windings have a resistance R. After closing the switch, the current rises to 80% of its final value in 0.4 s. Find the value of R.

26B–13 A circuit contains a series combination of a voltage source, an inductor, and a resistor. If the output voltage of the voltage source were to suddenly drop to zero, the current would begin to decrease. If the initial rate of current decrease were to continue at a constant rate (rather than to decrease exponentially), calculate the number of time constants required for the current to reach zero. *Answer:* One

26B–14 Find the energy stored in the magnetic field within the toroid windings of Problem 26A–4.

26B–15 A long iron-core solenoid with 2000 turns/m carries a current of 10 mA.

(a) Calculate the magnetic field B within the solenoid.

(b) With the iron core removed, calculate the current necessary to produce the magnetic field obtained in (a). *Answers:* (a) $B = 0.0100$ T (b) $I = 3.98$ A

26B–16 A toroid with an effective circumference of 50 cm has 1000 turns that carry a current of 200 mA. The core has a permeability of 3×10^{-3} T·m/A·turns.

(a) Calculate the magnetic field B in the core.

(b) Calculate the magnetic field strength H in the toroid.

(c) Calculate the fraction of the magnetic field B due to the current in the windings.

26C–1 Consider the magnetic field B at a point P near a long, straight current-carrying wire. Find the fraction of the field B that is due to the nearest segment of the wire that subtends an angle of $\pi/2$ rad from that point. *Answer:* $1/\sqrt{2}$

26C–2 A uniform, thin, plastic disk of radius R has a uniform surface charge density σ over both its top and bottom surfaces. Calculate the magnetic field at the center of the disk when the disk is rotating about its axis of symmetry with an angular velocity ω. (Hint: Consider the current produced by the charge contained within an annular ring of radius r and width dr.)

26C–3 A thin, uniform, plastic disk of mass m and radius R has a charge Q distributed uniformly over its surface. When the disk is rotating about its axis with angular velocity ω, show (a) that the magnetic field at its center is $B = \mu_0 Q\omega/2\pi R$ and (b) that its magnetic dipole moment is $\mu = Q\omega R^2/4$. (Hint: Consider the current loop due to the moving charge within the annular ring of radius r and width dr.) (c) Show that the ratio of the magnetic moment of the disk to its angular momentum (called the *gyromagnetic ratio*) is $Q/2m$.

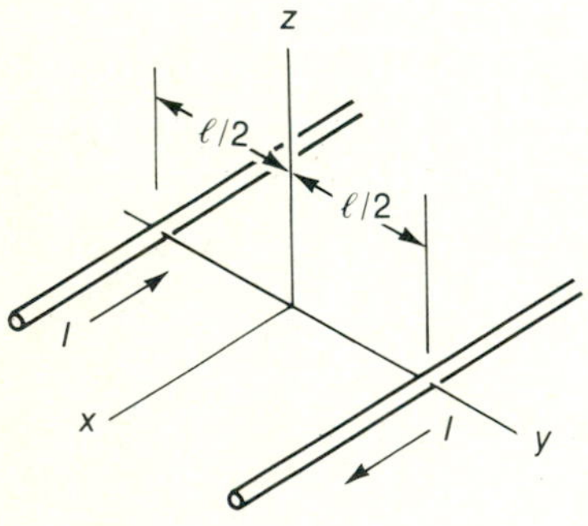

Figure 26–36
Problems 26C–4 and 26C–5

26C–4 Consider the straight, parallel conductors in the x-y plane carrying current in opposite directions, as shown in Figure 26–36. Find the direction and magnitude of the magnetic field **B** on the y axis for (a) $0 < y < \ell/2$ and (b) $y > \ell/2$. (c) Show that the field diminishes as the inverse square for $y \gg \ell$.

26C–5 Two straight, parallel wires in the x-y plane carry current I in opposite directions (Figure 26–36).

(a) Find the direction and magnitude of the magnetic field on the z axis as a function of z.

(b) Show that the field diminishes as the inverse square for z much greater than the separation of the wires. *Answers:* (a) $\mathbf{B} = -(\mu_0 I\ell/2\pi[z^2 + (\ell/2)^2])\,\hat{\mathbf{z}}$

(b) $\lim_{z \gg \ell} \mathbf{B} = -(\mu_0 I\ell/2\pi z^2)\,\hat{\mathbf{z}}$

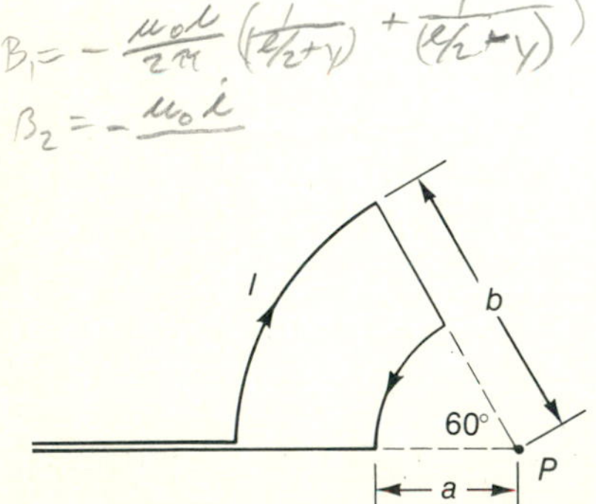

Figure 26–37
Problem 26C–6

26C–6 Consider the current-carrying loop shown in Figure 26–37, formed of radial lines and segments of circles whose centers are at point P. Find the magnitude and direction of the magnetic field **B** at P.

26C–7 A square loop of wire, with side length b, carries a current I. Find the magnetic field in the plane of the square at its center. Ignore the magnetic field due to lead-in wires.

Answer: $\mu_0 I 2\sqrt{2}/\pi b$

26C–8 An electron is moving at 3×10^6 m/s parallel to and at a distance of 1.0 cm from a long, straight wire. Suddenly a current of 10 A passes through the wire in a direction parallel to the velocity of the electron.

(a) Find the magnitude and direction of the initial acceleration of the electron.
(b) Describe qualitatively the subsequent motion of the electron.

26C–9 A pair of *Helmholtz coils* is often used to produce a uniform magnetic field over a small region of space. The pair consists of two flat, circular coils separated by the radius of the coils, as in Figure 26–38. The current flows in the same direction in both coils. Show that for a separation equal to the radius of the coils, the field on the axis halfway between the coils is such that dB/dx and d^2B/dx^2 are both zero, where x is the distance along the axis.

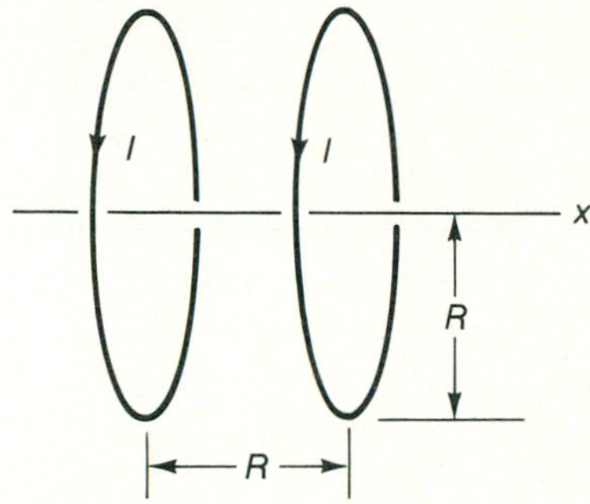

Figure 26–38
Problem 26C–9

26C–10 A circular loop of wire of area A and resistance R is held fixed with its plane normal to a magnetic field **B**. The field is then reduced from an initial value of B_0 so that it changes as a function of time according to $B = B_0 e^{-\alpha t}$, where α is a constant.

(a) Sketch the loop, showing the magnetic field directed *into* the paper, and indicate on the diagram the direction of the induced current.
(b) Do the electromagnetic forces associated with the induced current tend to make the loop expand, contract, or neither?
(c) Derive an expression in terms of B_0, A, and R for the total quantity of charge Q that flows past a point in the loop during the time the field is reduced from B_0 to zero.
(d) Derive an expression in terms of B_0, A, R, and α for the amount of thermal energy dissipated in the loop while the field is reduced from B_0 to zero.

26C–11 A thin metal rod of length 0.8 m falls from rest under the action of gravity. It remains horizontal with its length oriented along the magnetic east-west direction. At this location, the earth's magnetic field **B** has a magnitude of 5×10^{-5} T and a downward direction at 70° below the horizontal (the "dip" angle).

(a) Find the induced EMF in the rod after it falls 8 m.
(b) Which end of the rod has the higher potential?

Answers: (a) 0.171 mV (b) East

26C–12 In Figure 26–39, the rolling axle, 1.5 m long, is pushed along the horizontal rails at a constant speed $v = 3$ m/s. A resistor $R = 0.4\ \Omega$ is connected to the rails at points A and B, directly opposite each other. (The wheels make good electrical contact with the rails, so the axle, rails, and R form a complete, closed-loop circuit. The only significant resistance in the circuit is R.) There is a uniform magnetic field $\mathbf{B} = 0.08$ T vertically downward.

(a) Find the induced current I in the resistor.
(b) What horizontal force F is required to keep the axle rolling at constant speed?
(c) Which end of the resistor, A or B, is at the higher electric potential?
(d) After the axle rolls past the resistor, does the current in R reverse direction?

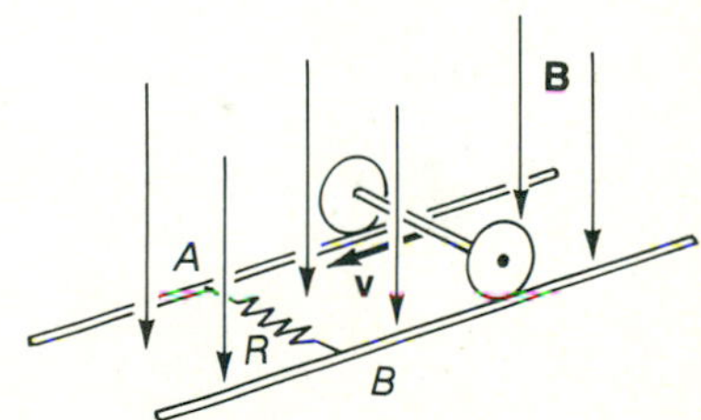

Figure 26–39
Problem 26C–12

26C–13 A flat coil of wire has an inductance of 2 H and a resistance of 40 Ω. At $t = 0$, a battery of EMF, $\mathscr{E} = 60$ V, is connected to the coil. Consider the state of affairs one time constant later. At this instant, find (a) the power delivered by the battery, (b) the joule power developed in the resistance of the windings, and (c) the instantaneous rate at which energy is being stored in the magnetic field.

Answers: (a) 56.9 W (b) 36.0 W (c) 20.9 W

chapter 27
AC Circuit Theory

The Buddha, the Godhead, resides quite as comfortably in the circuits of a digital computer or the gears of a cycle transmission as he does at the top of a mountain or in the petals of a flower.

ROBERT PIRSIG
(*Zen and the Art of Motorcycle Maintenance*)

We have discussed the response of a series *RC* circuit when a battery voltage is applied. The current initially is large, limited only by the resistance, and decreases exponentially to zero. In a similar fashion, when a battery voltage is applied to an *RL* circuit the current grows exponentially from zero to a value limited only by the resistance. In both instances, the response is transient; that is, the varying part lasts only momentarily, until steady-state conditions have been achieved. The *time constant* of the circuit determines how steep the exponential curves are.

In this chapter we will investigate the response of a circuit to a constantly changing applied voltage.[1] Most electromechanical generators of electricity produce a *sinusoidally* varying voltage, resulting in *alternating current* (*AC*). The resultant voltages and currents are called *AC* voltage and *AC* current. (The latter is firmly entrenched in common usage, so we shall go along with it despite the redundancy.) Let us consider alternating voltages and currents of the following type:

$$v = V \sin(\omega t + \alpha) \qquad \text{and} \qquad i = I \sin(\omega t + \beta) \qquad \textbf{(27–1)}$$

The *amplitudes*, or *peak values*, of the voltage and current are represented by capital letters (V and I, respectively). Small letters represent voltage and current values that change in time (v and i, respectively). At any given instant, the sinusoidally varying voltage has a particular *phase angle* given by $\omega t + \alpha$, where α is the *phase constant*. Similarly, $\omega t + \beta$ is the phase of the current and β is a phase constant. As defined previously, $\omega = 2\pi f = 2\pi/T$, where f is the frequency and T is the period of variation.

Limiting the discussion to just one frequency is justifiable even though many situations, such as in hi-fi amplifiers for music and speech reproduction, involve numerous frequencies simultaneously. The reason is that any complicated waveshape that is *periodic* (that is, repeats itself again and again) may be replaced by a combination of sinusoidal voltage sources involving a fundamental frequency (f_0) and multiples ($2f_0, 3f_0, 4f_0, \ldots$). The mathematical

[1] Historically, formulation of the laws governing the flow of direct current was relatively simple compared with those describing alternating current. It wasn't until just before the end of the last century that the brilliant mathematician-engineer Charles Proteus Steinmetz developed the laws that describe alternating current. The initial publication of his work consisted of three volumes of detailed and complicated mathematical development of alternating-current circuit theory.

method is known as Fourier analysis (see Appendix F). Thus, more complicated (periodic) waveshapes are understandable in terms of the simple sine and cosine waves we will examine in this chapter.

We will discover that the current through a series combination of resistors, inductors, and capacitors will also vary sinusoidally, but the voltage across these elements will not necessarily have the same phase as the current through them. How much current is present depends not only on the value of the circuit components, but also on the frequency of the applied voltage.

27.1 Circuits with Resistance Only

Consider the circuit shown in Figure 27–1. We choose the applied voltage to be $v = V \sin \omega t$. By setting $\alpha = 0$ in Equation (27–1), we choose the particular value of the voltage variation at which we begin to measure time ($v = 0$, going positive, at $t = 0$). In circuit diagrams, the symbol for an *AC* voltage source is —⊙—. To find the current i in Figure 27–1(a), we use the fact that conservation-of-charge and conservation-of-energy relationships hold just as they do in direct-current (*DC*) circuits. Thus, *at every instant*, the sum of the potential increases and decreases around a current loop must equal zero. Using minus signs for potential drops, we have

$$\Sigma v = 0$$

$$v - iR = 0$$

Substituting for v from Equation (27–1) and rearranging, we obtain an expression similar to Ohm's law:

$$V \sin \omega t = iR$$

$$i = \frac{V}{R} \sin \omega t \tag{27–2}$$

Notice that the current has the *same phase* as the applied voltage, as shown in Figure 27–1(b).

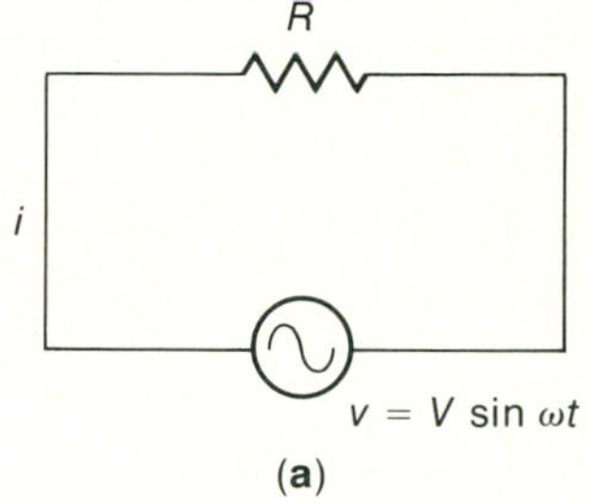

(a)

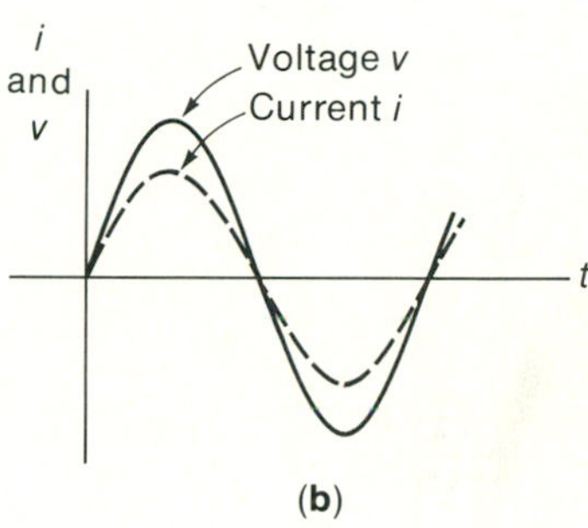

(b)

Figure 27–1
A purely resistive *AC* circuit.

27.2 Circuits with Capacitance Only

Now consider the circuit shown in Figure 27–2(a). As always, the sum of the potential differences around a current loop must be zero at every instant. Thus:

$$\Sigma v = 0$$

$$V \sin \omega t - \frac{q}{C} = 0 \tag{27–3}$$

where q is the charge on the capacitor at time t. The current through the circuit is the rate at which the charge on the plates of the capacitor is changing. Thus:

$$i = \frac{dq}{dt} \tag{27–4}$$

Differentiating Equation (27–3) and solving for dq/dt, we have

$$\frac{dq}{dt} = V\omega C \cos \omega t$$

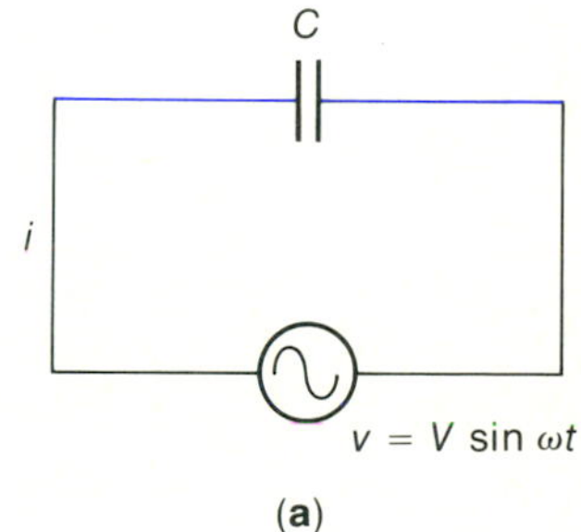

(a)

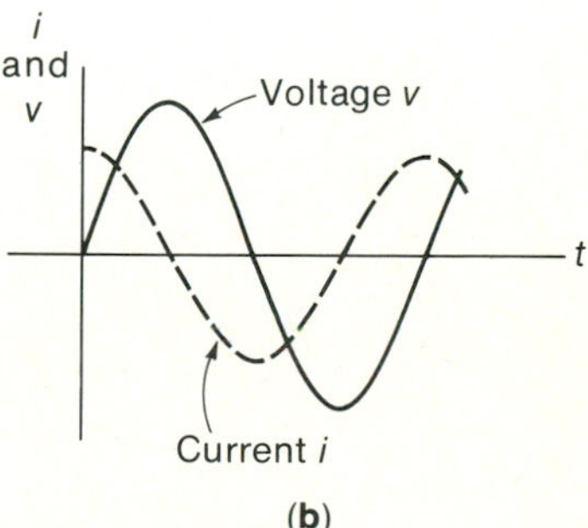

(b)

Figure 27–2
A purely capacitive *AC* circuit.

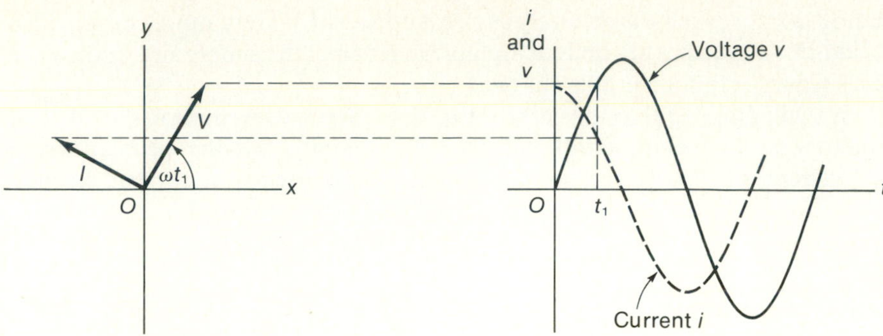

Figure 27–3
Voltage and current relations for the capacitive circuit of Figure 27–2 are represented conveniently by phasors 90° apart.

(**a**) A phasor diagram. The phasors rotate counterclockwise with an angular frequency ω.

(**b**) A graph of the voltage and current as a function of time.

Using Equation (27–4), we can write this expression in a form similar to Ohm's law:

$$i = V\omega C \cos \omega t$$

$$i = \frac{V}{X_C} \cos \omega t$$

where we introduce the new concept

CAPACITIVE REACTANCE

$$X_C = \frac{1}{\omega C} \tag{27–5}$$

The symbol X_C is called the **capacitive reactance,** measured in *ohms* (Ω). It limits the amplitude of the current flow in the way that resistance limits the flow in a purely resistive circuit. (It is left as an exercise to show that capacitive reactance does have dimensions of ohms.) Notice that the current leads the applied voltage by $\pi/2$ rad (or 90°), as shown in Figure 27–2(b). The phrase "leading the applied voltage" means that as time progresses (that is, as we move along the t axis), the current reaches its peak value *before* the applied voltage reaches its peak value.

A useful way of portraying the relationship between the applied voltage and the resulting current is by using a **phasor diagram.** The phasor diagram for a purely capacitive circuit is shown in Figure 27–3. In this diagram, the voltage and current are represented by vectorlike arrows, called **phasors,**[2] that rotate counterclockwise with an angular frequency ω, maintaining their relative angular separations as they rotate. The lengths of the phasors are the amplitudes of the time-varying voltage and current. Their angular separation represents the **phase constant** between the voltage v and the current i. The projection of the phasors on a vertical axis is then expressed by the equations

$$v = V \sin \omega t$$

and

$$i = I \cos \omega t$$

or

$$i = I \sin\left(\omega t + \frac{\pi}{2}\right) \tag{27–6}$$

[2] Although voltage and current are not vectors in the usual sense, they do follow the rules for vector addition. Their representation on a phasor diagram is a useful mathematical technique.

thus representing the time variation of voltage and current. The phase constant, $\pi/2$ rad (equal to 90°), indicates that for a purely capacitive reactance, the current leads the voltage by just one-quarter cycle of the sinusoidal variation. As the phasor diagram rotates around, the current phasor **I** is ahead of the voltage phasor **V** (that is, the current leads the voltage). The phasor diagram thus helps us visualize the phase relationship between the applied voltage and the current.

27.3 Circuits with Inductance Only

Consider the circuit shown in Figure 27–4(a). We assume that L represents a "pure" inductance (that is, the resistance of the windings is negligible). As always, the instantaneous sum of the potential increases and decreases around the circuit loop must be zero. Recall that the voltage v_L across the inductor due to the changing current through it is

$$v_L = -L\frac{di}{dt}$$

where the minus sign indicates opposition to the applied voltage. Then for $\Sigma v = 0$, we have

$$v - L\frac{di}{dt} = 0 \qquad \textbf{(27–7)}$$

Substituting $v = V \sin \omega t$ and rearranging, Equation (27–7) becomes

$$L\frac{di}{dt} = V \sin \omega t$$

We solve this equation by separating the variables i and t, so that they appear on opposite sides of the equal sign, and integrating (see Appendix G):

$$\int di = \frac{V}{L}\int \sin \omega t \, dt$$

$$i = -\frac{V}{\omega L}\cos \omega t + c$$

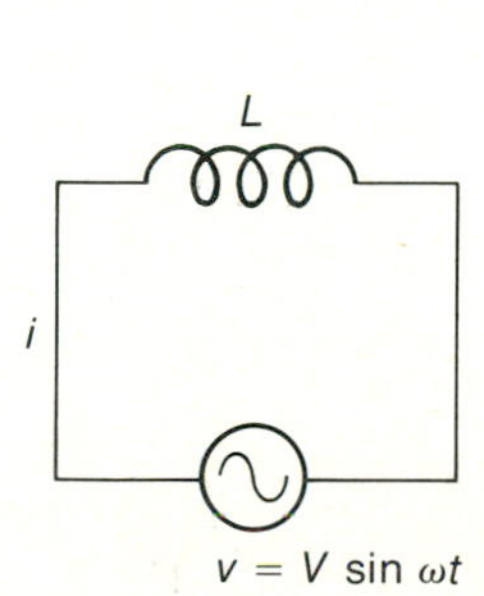

(a) A purely inductive *AC* circuit.

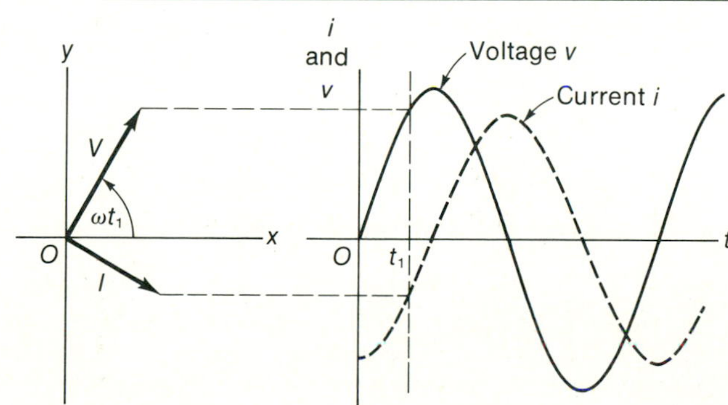

(b) A phasor diagram, which shows the 90° phase relationship between **V** and **I**.

(c) A graph of the voltage and current as a function of time.

Figure 27–4
Voltage and current relations for a purely inductive *AC* circuit.

Setting[3] $c = 0$ and using the fact that $-\cos \omega t = \sin(\omega t - \pi/2)$, we write this in a form similar to Ohm's law:

$$i = \frac{V}{X_L} \sin\left(\omega t - \frac{\pi}{2}\right) \tag{27–8}$$

where X_L is defined as the **inductive reactance** which, like capacitive reactance, is measured in ohms (Ω).

INDUCTIVE REACTANCE X_L

$$X_L = \omega L \tag{27–9}$$

The inductive reactance limits the amplitude of the current just as resistance limits the current in a purely resistive circuit.

The phasor diagram is shown in Figure 27–4(b). For a pure inductance, the phase constant is $-\pi/2$ (or $-90°$). This means that the current *lags* the applied voltage by one-quarter of the sinusoidal variation. The phrase "lags the applied voltage" means that as time progresses, the current reaches its peak value *after* the voltage reaches its peak value. As the phasor diagram rotates, the current phasor **I** lags behind the voltage phasor **V**.

EXAMPLE 27–1

An *AC* voltage with an amplitude of 15 V and frequency of 60 Hz is applied across an inductor whose inductance is 30 mH. Find the resulting *AC* current.

SOLUTION

From Equation (27–8), the amplitude of the current is

$$I = \frac{V}{X_L}$$

Since $X_L = \omega L = 2\pi f L$, we have

$$I = \frac{V}{2\pi f L}$$

$$I = \frac{(15\ \text{V})}{(2\pi)(60\ \text{s}^{-1})(3 \times 10^{-2}\ \text{H})} = 1.33\ \text{A}$$

We know that in a pure inductance, the current lags the applied voltage by $\pi/2$ rad. The frequency of the current is the same as that of the applied voltage, 60 Hz. Thus:

$$i = I \sin(\omega t + \beta)$$

$$i = \boxed{1.33 \sin\left(120\pi t - \frac{\pi}{2}\right) \text{A}} \qquad \text{(where } t \text{ is in seconds)}$$

27.4 Series *RLC* Circuits

Rather than investigating various pairs of resistance, capacitance, and inductance in series, we will present the case in which all three are present. Consider the circuit shown in Figure 27–5. By Kirchhoff's loop rule, at any instant the

[3] The constant of integration c represents a constant *DC* current, which could have been established only if a *DC* source of voltage were in the circuit initially.

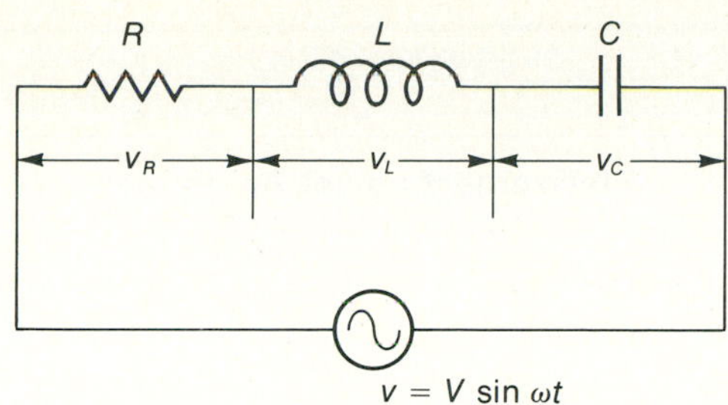

Figure 27–5
A series *RLC* circuit. At any instant, $v = v_R + v_L + v_C$.

applied voltage $v = V \sin \omega t$ must equal the sum of the back-EMF across the inductor $v_L = L\,di/dt$, the voltage drop across the resistor $v_R = iR$, and the voltage across the capacitor due to the charge on the capacitor $v_C = q/C$.

$$\Sigma V = 0$$

$$V \sin \omega t - L\frac{di}{dt} - iR - \frac{q}{C} = 0$$

or

$$L\frac{di}{dt} + Ri + \frac{q}{C} = V \sin \omega t \tag{27–10}$$

In order to understand the physical significance of each term in Equation (27–10), as well as to perform the initial step in the solution of the equation, it is necessary to express the current i (and its derivatives) as derivatives of the charge q:

$$L\frac{d^2q}{dt^2} + R\frac{dq}{dt} + \frac{q}{C} = V \sin \omega t \tag{27–11}$$

This equation is identical in form to the equation that describes a forced mechanical oscillator with viscous damping [Chapter 16, Equation (16–35)]:

$$m\frac{d^2x}{dt^2} + b\frac{dx}{dt} + kx = F_0 \sin \omega t \tag{27–12}$$

A term-by-term comparison of Equations (27–11) and (27–12) reveals: (a) An inductance resists the surge of charge through an electrical circuit in a way that is analogous to mass resisting acceleration in a mechanical system. (b) Resistance in an electrical circuit is analogous to viscosity in a mechanical system, each being responsible for energy loss in the system. (c) The reciprocal of capacitance provides the "resilience" to an electrical circuit in the way the spring constant in a mechanical system determines the restoring force. These (plus other analogies) are summarized in Table 27–1.

Equation (27–11) may be solved for q as a function of time, then differentiated with respect to time to yield the current. The solution of this equation requires a mathematical technique beyond the scope of this text. The result is

$$i = \frac{V}{\sqrt{R^2 + (X_L - X_C)^2}} \sin(\omega t - \phi) + i_0(t) \tag{27–13}$$

As before, $X_L = \omega L$ and $X_C = 1/\omega C$. The term $i_0(t)$ is called the *transient term*. It describes the current variations that occur immediately after the voltage is first applied. In most circuits, it becomes essentially zero soon after the voltage

TABLE 27–1

Electromechanical Analogues

Mechanical system	Electrical circuit
Mass M (resists change of velocity)	Inductance L (resists change of current)
Viscosity constant b (dissipates energy into thermal form)	Resistance R (dissipates energy into thermal form)
Spring constant k (determines restoring force and "elasticity" of mechanical motion)	Reciprocal of capacitance $1/C$ (provides "resilience" to an electrical current)
Displacement x	Charge q
Velocity $v = dx/dt$	Current $i = dq/dt$
Force F	Voltage V

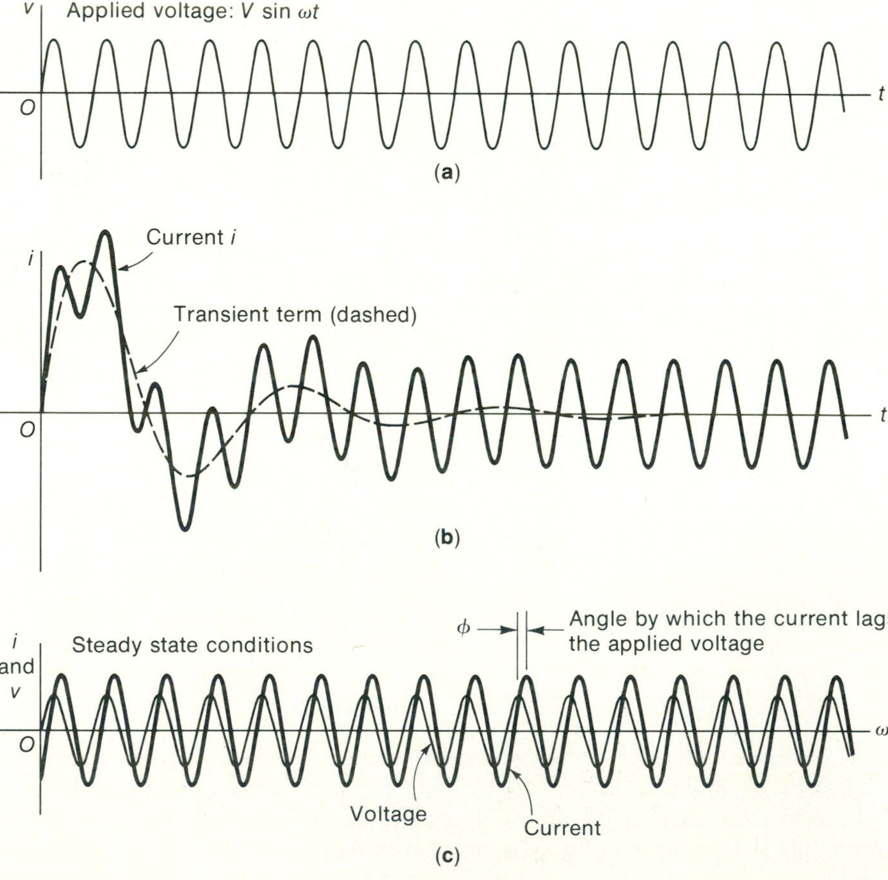

Figure 27–6
Current in an *RLC* circuit that is predominantly an inductive reactance.

is applied.[4] A typical example is shown in Figure 27–6, wherein the transient effects die out rapidly as the *AC* current settles down to its steadystate condition. For our purposes, we will not analyze these transient effects, but instead will concentrate on steady state conditions:

$$i = I \sin(\omega t - \phi) \tag{27–14}$$

where the **phase constant** ϕ is given by

PHASE CONSTANT ϕ

$$\phi = \tan^{-1}\left(\frac{X_L - X_C}{R}\right) \tag{27–15}$$

It may have any value between $-\pi/2$ rad and $+\pi/2$ rad, depending on the relative magnitudes of X_L and X_C. In the next section we show that ϕ is the phase constant between the voltage v applied to the circuit and the current i that flows in the circuit.

There is an easy way to keep track of the various phase relationships in a series *RLC* circuit. Because at any instant the current is the *same* in all components, *we use the current as a reference*, measuring all other phase angles with respect to the current. In Figure 27–7(a) we develop a *voltage phasor diagram* that depicts voltages and current in their correct phase relationships. We represent each as a phasor: **V** or **I**. By custom, the phasor for the reference current **I** is usually drawn horizontally toward the right. Because the voltage across the resistor is *in phase* with the current, both $\mathbf{V}_R$ and **I** are in the same direction. In an inductor, the current *lags* the voltage across the inductor by $\pi/2$ rad, so $\mathbf{V}_L$ is shown as a phasor 90° ahead of the current phasor **I**. In a capacitor, the current *leads* the voltage across the capacitor by $\pi/2$ rad, so $\mathbf{V}_C$ is shown 90° behind the current phasor **I**. From Kirchhoff's law, the voltage phasors for the individual circuit components add *vectorially* to give the applied voltage **V** (Figure 27–7b).

$$\mathbf{V} = \mathbf{V}_R + \mathbf{V}_L + \mathbf{V}_C \tag{27–16}$$

The voltage phasor diagram portrays the various phase relationships in a series *AC* circuit. As the phasor diagram rotates counterclockwise with angular frequency ω, the projections of the phasors on the vertical axis give the instantaneous values of all voltages and currents.

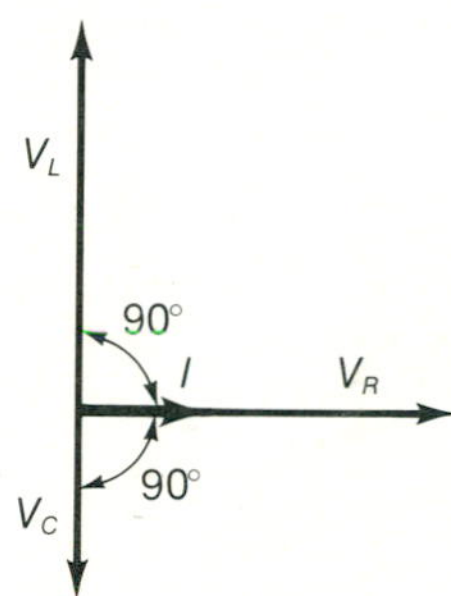

(**a**) A voltage phasor diagram with the current **I** as a reference.

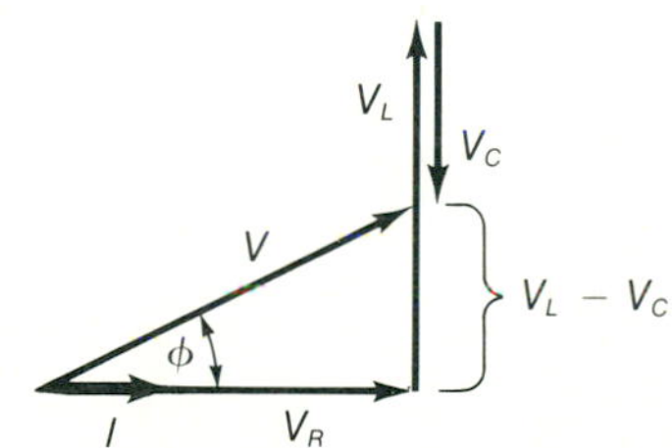

(**b**) The applied voltage **V** is the vector sum of the voltage phasors for individual circuit elements: $\mathbf{V} = \mathbf{V}_R + \mathbf{V}_L + \mathbf{V}_C$.

Figure 27–7
Phase relationships between voltages and current in a series *RLC* circuit. (In this illustration, the net reactance for the circuit as a whole is *inductive*, so the current **I** *lags* the applied voltage **V** by the phase constant ϕ.)

27.5 Impedance in Series *RLC* Circuits

The alternating current through a series *RLC* circuit is impeded by an amount dependent upon the value of the components as well as the frequency. The amplitude I of the current is, from Equation (27–13):

$$I = \frac{V}{\sqrt{R^2 + (X_L - X_C)^2}}$$

where
V = amplitude of the applied voltage
R = resistance
$X_L = \omega L$, the inductive reactance
$X_C = 1/\omega C$, the capacitive reactance

[4] The size of the transient effect depends on the initial conditions. For example: What is the phase of the *AC* voltage at the instant it is applied? Are capacitors initially charged or uncharged? With suitable adjustment of the initial conditions, the transient can be eliminated entirely. Unfortunately, under certain adverse circumstances, the transient can cause extreme surges of current that damage circuit components.

The combination of resistance and reactances is defined as the **impedance** Z measured in ohms (Ω).

IMPEDANCE Z IN A SERIES *RLC* CIRCUIT

$$Z = \sqrt{R^2 + (X_L - X_C)^2} \qquad \textbf{(26–17)}$$

Thus, the amplitude of the current is related to the amplitude of the applied voltage by the simple relation $I = V/Z$, or

OHM'S LAW FOR *AC*

$$V = IZ \qquad \textbf{(27–18)}$$

For *AC*, the impedance Z plays a role similar to resistance in *DC* circuits.

The mathematical form of Equation (27–17) suggests the Pythagorean Theorem, in which R and $X_L - X_C$ are lengths of the legs of a right triangle, with Z forming the hypotenuse, as illustrated in Figure 27–8(a). The angle between Z and R is defined by

$$\phi = \tan^{-1}\left(\frac{X_L - X_C}{R}\right)$$

(**a**) A right triangle formed by R, $X_L - X_C$, and Z.

which is the phase constant ϕ between the applied voltage and the current in the series circuit.[5] The triangle is related to the *impedance diagram* of Figure 27–8(b). To sketch an impedance diagram, we draw the resistance as a vectorlike arrow $\mathbf{R}$ along the $+x$ axis, we draw $\mathbf{X}_L$ as a vectorlike arrow along the $+y$ axis, and we draw $\mathbf{X}_C$ along the $-y$ axis. The vector sum of these three arrows is the total impedance $\mathbf{Z}$.

The impedance diagram is closely related to the voltage phasor diagram of Figure 27–7, since the voltages in that diagram are merely the scalar I times the corresponding resistance, reactance, and impedance of Figure 27–8(b). Thus, the two representations differ only by the scale factor I. (The impedance diagram does not rotate, however.)

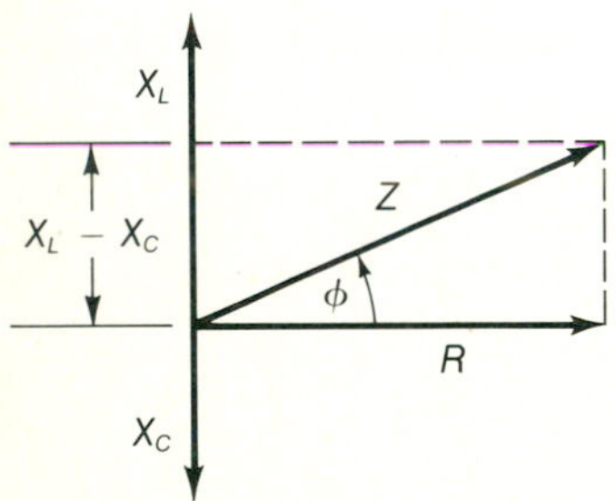

(**b**) An impedance diagram.

Figure 27–8
An *impedance diagram* for a series *RLC* circuit. The angle ϕ is the phase constant between the applied voltage and the current in the circuit. (For this example, the net reactance is *inductive*; that is, $X_L > X_C$.)

EXAMPLE 27–2

Consider the series *RLC* circuit of Figure 27–9(a) with the following circuit parameters:

$$R = 200\ \Omega$$
$$L = 663\ \text{mH}$$
$$C = 26.5\ \mu\text{F}$$

The applied voltage has an amplitude of 50 V and a frequency of 60 Hz. Find the following:

(a) The current i, including its phase constant ϕ relative to the applied voltage v.
(b) The voltage V_R across the resistor and its phase relative to the current.
(c) The voltage V_C across the capacitor and its phase relative to the current.
(d) The voltage V_L across the inductor and its phase relative to the current.

[5] The current is always *in phase* with $\mathbf{V}_R$. A *positive* phase constant ϕ means the current *lags* the applied voltage $\mathbf{V}$, and vice versa. We should mention that this sign association is different from that for β in Equation (27–1) because of the minus sign for ϕ in the general solution [Equation (27–13)]. Fortunately, the leading or lagging relationship is easy to determine by inspecting a voltage phasor diagram.

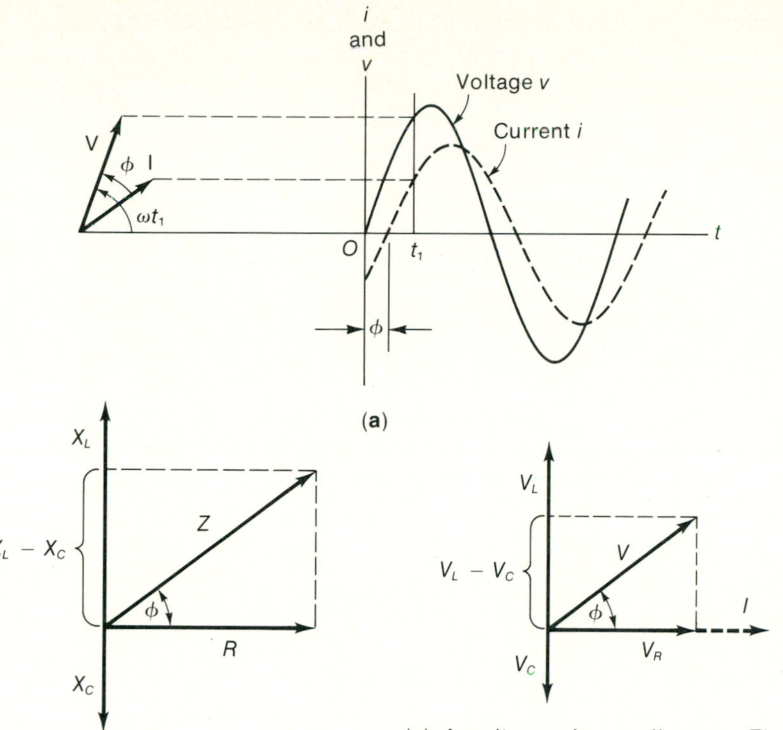

(**b**) An impedance diagram.

(**c**) A voltage phasor diagram. The current **I** is in phase with $\mathbf{V}_R$.

Figure 27–9
Example 27–2

SOLUTION

In general, the initial step is to calculate reactances and impedances and then apply Ohm's law for *AC* circuits.

$$X_C = \frac{1}{\omega C}$$

$$X_C = \frac{1}{(2\pi f)(C)} = \frac{1}{(2\pi)(60\ \text{s}^{-1})(26.5 \times 10^{-6}\ \text{F})} = 100\ \Omega$$

$$X_L = \omega L$$

$$X_L = (2\pi f)(L) = (2\pi)(60\ \text{s}^{-1})(663 \times 10^{-3}\ \text{H}) = 250\ \Omega$$

$$Z = \sqrt{R^2 + (X_L - X_C)^2}$$

$$Z = [(200)^2 + (250 - 100)^2]^{\frac{1}{2}}\ \Omega = 250\ \Omega$$

Figure 27–9(b) is an impedance diagram for this circuit. Because $X_L > X_C$, it has a net inductive reactance, so the current will *lag* the applied voltage.

(a) Applying Ohm's law for *AC* circuits, the magnitude of the current I is

$$I = \frac{V}{Z}$$

$$I = \frac{50\ \text{V}}{250\ \Omega} = 0.200\ \text{A}$$

It has the same frequency, $f = 60$ Hz, as the applied voltage. The phase constant ϕ between the current and the applied voltage is found from Equation (27–15):

$$\phi = \tan^{-1}\left(\frac{X_L - X_C}{R}\right)$$

$$\phi = \tan^{-1}\left(\frac{250\ \Omega - 100\ \Omega}{200\ \Omega}\right) = 36.9^\circ$$

(The net reactance is inductive, so the current *lags* the applied voltage.)

Incorporating these values in the general expression [Equation (27–14)]:

$$i = I \sin(\omega t - \phi)$$

we have

$$i = 0.200 \sin[(2\pi)(60\ \text{s}^{-1}) - 36.9°]\ \text{A}$$

$$i = \boxed{0.200 \sin(120\pi t - 36.9°)\ \text{A}} \quad \text{(where } t \text{ is in seconds)}$$

The current is expressed relative to the applied voltage, $v = 50 \sin(120\pi t)$ V, and *lags* the voltage by 36.9°.

(b) The voltage V_R across the resistor is

$$V_R = IR$$

$$V_R = (0.200\ \text{A})(200\ \Omega) = \boxed{40.0\ \text{V}}$$

The instantaneous voltage across a resistor is *in phase* with the current through it, so $\phi = 0°$.

(c) The voltage V_C across the capacitor is

$$V_C = IX_C$$

$$V_C = (0.200\ \text{A})(100\ \Omega) = \boxed{20.0\ \text{V}}$$

The instantaneous current through a pure capacitor always *leads* the voltage across it by $\pi/2$ rad.

(d) The voltage V_L across the inductor is

$$V_L = IX_L$$

$$V_L = (0.200\ \text{A})(250\ \Omega) = \boxed{50.0\ \text{V}}$$

The instantaneous current through a pure inductor always *lags* the voltage across it by $\pi/2$ rad.

Figure 27–10 is a phasor diagram showing all voltages. Note the way in which the *AC* voltages combine. In particular, the *algebraic* sum of their magnitudes is *not* the applied voltage V.

$$V \neq V_R + V_L + V_C$$

(This sum is 40 V + 50 V + 20 V = 110 V, instead of the correct value of 50 V.) On the other hand, the algebraic sum of the *instantaneous* voltages across the circuit elements always equals the applied voltage:

$$v = v_R + v_L + v_C$$

These instantaneous voltages are the *projections* on the vertical axis of the voltage phasors in Figure 27–10. The fact that the projections of the phasors add algebraically implies that the phasors themselves add *vectorially:*

$$\mathbf{V} = \mathbf{V}_R + \mathbf{V}_L + \mathbf{V}_C$$

From the vector diagram, we have

$$V^2 = V_R^2 + (V_L - V_C)^2$$

$$V^2 = (40\ \text{V})^2 + (50\ \text{V} - 20\ \text{V})^2$$

$$V = 50\ \text{V}$$

which is the correct value of the applied voltage.

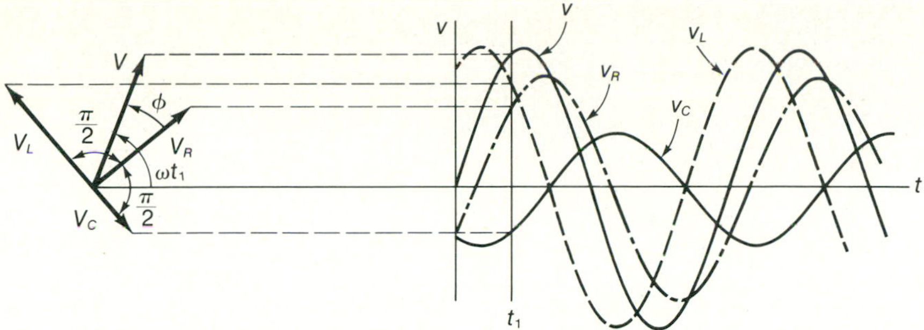

(**a**) A phasor diagram. The current is in phase with the phasor V_R.

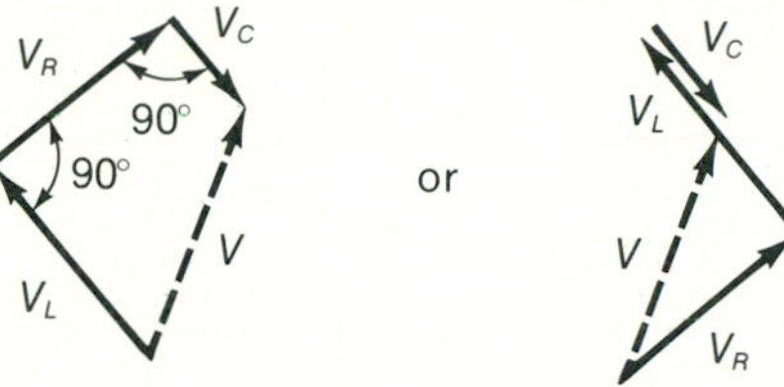

(**b**) Vector addition for the voltage phasors: $\mathbf{V} = \mathbf{V}_R + \mathbf{V}_L + \mathbf{V}_C$.

Figure 27–10
The phasors $\mathbf{V}_R$, $\mathbf{V}_L$, and $\mathbf{V}_C$ add *vectorially* to equal the applied voltage phasor $\mathbf{V}$. At any instant, the projections of the phasors add algebraically so that $v = v_R + v_L + v_C$.

EXAMPLE 27–3

Consider the circuit shown in Figure 27–11(a). Find (a) the impedance, (b) the amplitude of the current in the circuit, and (c) the phase constant between the applied voltage and current.

SOLUTION

(a) The impedance is

$$Z = \sqrt{R^2 + (X_L - X_C)^2}$$

$$Z = [(1200\ \Omega)^2 + (300\ \Omega - 800\ \Omega)^2]^{\frac{1}{2}} = \boxed{1300\ \Omega}$$

Figure 27–11(b) depicts the impedance diagram.

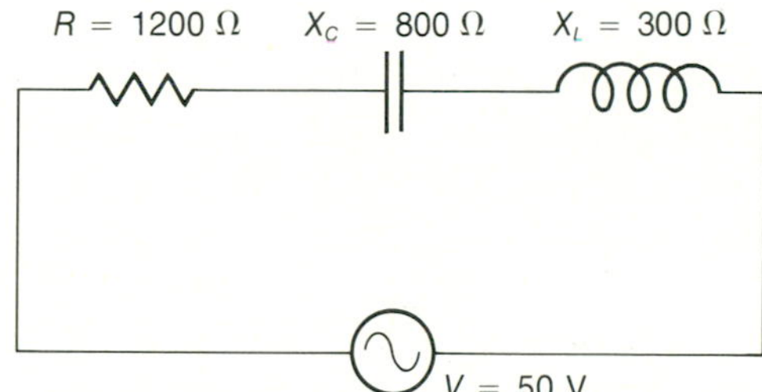

(**a**) A series *RLC* circuit with an applied *AC* voltage of 50 V amplitude.

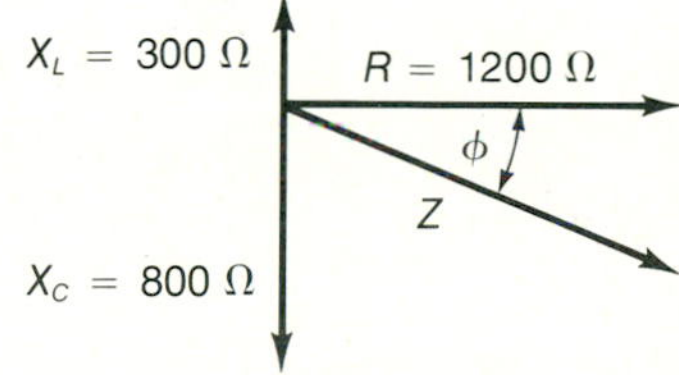

(**b**) The impedance diagram for the circuit shown in (**a**).

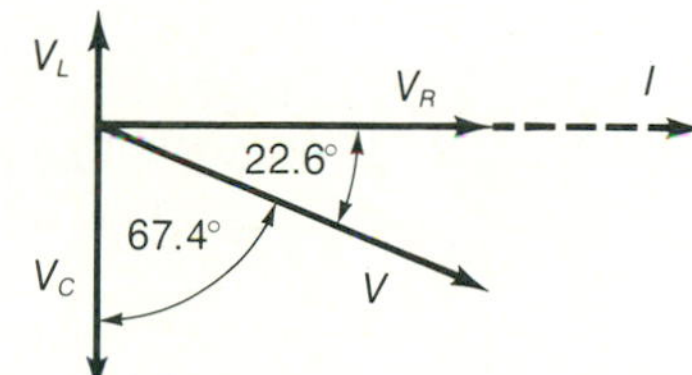

(**c**) The voltage phasor diagram.

Figure 27–11
Example 27–3

(b) From Ohm's law for AC:

$$I = \frac{V}{Z}$$

$$I = \frac{50\text{ V}}{1300\ \Omega} = \boxed{0.0385\text{ A}}$$

(c) The phase constant is

$$\phi = \tan^{-1}\left(\frac{X_L - X_C}{R}\right)$$

$$\phi = \tan^{-1}\left(\frac{300\ \Omega - 800\ \Omega}{1200\ \Omega}\right) = \boxed{-22.6^\circ}$$

(The current *leads* the applied voltage.)

The negative phase angle implies that the current leads the applied voltage. This agrees with the fact that since $X_C > X_L$, the net reactance is capacitive.

EXAMPLE 27–4

In Example 27–3, find the voltage across the capacitor and its phase relative to the applied voltage.

SOLUTION

The amplitude of the voltage across the capacitor is

$$V_C = IX_C$$

$$V_C = (0.0385\text{ A})(800\ \Omega) = \boxed{30.8\text{ V}}$$

Multiplying the impedance diagram by I, we obtain the *voltage phasor diagram*, Figure 27–11(c). The voltage across the capacitor V_C *lags* the applied voltage by $90^\circ - \phi = 90^\circ - 22.6^\circ = \boxed{67.4^\circ}$.

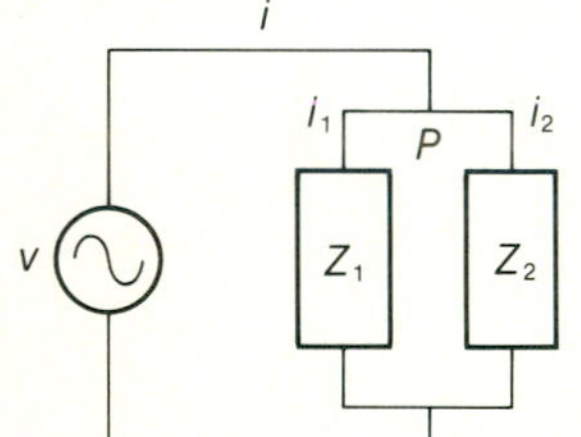

(a) The voltage v is across the parallel combination of Z_1 and Z_2.

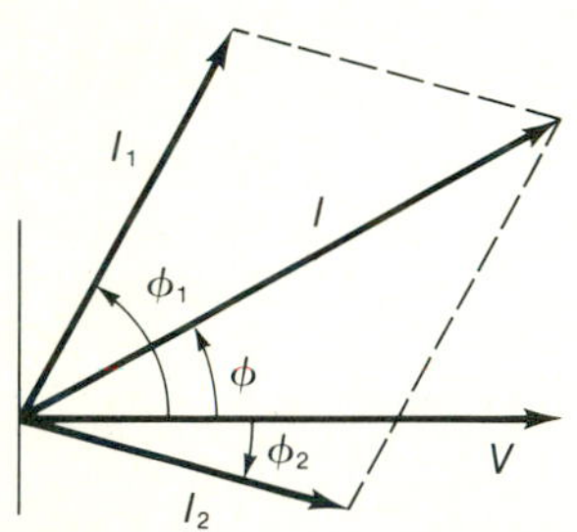

(b) A current phasor diagram with the voltage V across the parallel branches as a reference for phase relations.

Figure 27–12
A circuit with two impedances in parallel. In this illustration, the resultant impedance of the parallel combination is capacitive, so the current i *leads* the applied voltage v by the phase constant ϕ.

27.6 Impedance in Parallel *RLC* Circuits

Consider the circuit shown in Figure 27–12(a), with two impedances in parallel across a voltage v. The analysis of a parallel circuit differs in one important feature from the way we analyzed a series circuit. In a series circuit, the current is common to all components, so we used the current as a reference for the phases of various voltages. In a parallel combination, the *voltage* across the combination is common to both branches, so *we use the voltage* V *as a reference for phase relations.* The method is to first find the currents i_1 and i_2 in each branch, then add the currents together *vectorially* (to preserve their phase relations) to obtain the current i. This reasoning is based on the Kirchhoff junction rule: $\Sigma i = 0$. That is, the instantaneous current i entering a junction must equal the sum of the instantaneous currents leaving the junction. Just as we used a voltage phasor diagram to add voltages vectorially, we construct a current phasor diagram to add currents vectorially. The method is illustrated in the following example.

EXAMPLE 27–5

Consider the circuit in Figure 27–13(a). Find the total current i in both amplitude and phase relative to the applied voltage v.

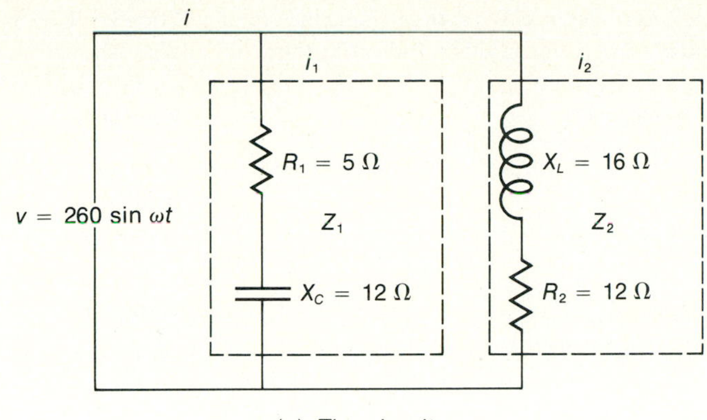

(**a**) The circuit.

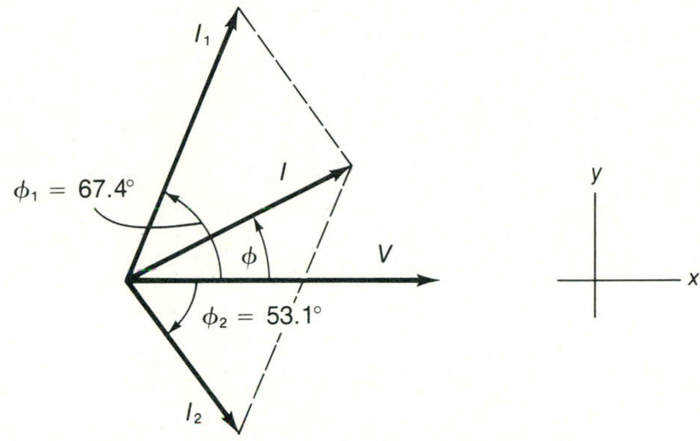

(**b**) The current phasor diagram using the voltage V, which is common to both branches as a reference.

Figure 27–13
Example 27–5

SOLUTION

The solution involves the following steps:

(1) Calculate the impedance of each branch.
(2) For each branch, find the current amplitude and its phase relative to the applied voltage.
(3) Construct a current phasor diagram and add the branch currents *vectorially* to find the total current *i*.

Branch 1	**Branch 2**
Step 1:	
$Z_1 = \sqrt{R_1^{\,2} + (X_{L_1} - X_{C_1})^2}$	$Z_2 = \sqrt{R_2^{\,2} + (X_{L_2} - X_{C_2})^2}$
$Z_1 = \sqrt{(5\ \Omega)^2 + (-12\ \Omega)^2} = 13\ \Omega$	$Z_2 = \sqrt{(12\ \Omega)^2 + (16\ \Omega)^2} = 20\ \Omega$
Step 2:	
$I_1 = \dfrac{V}{Z_1}$	$I_2 = \dfrac{V}{Z_2}$
$I_1 = \dfrac{260\ \text{V}}{13\ \Omega} = 20\ \text{A}$	$I_2 = \dfrac{260\ \text{V}}{20\ \Omega} = 13\ \text{A}$
$\phi_1 = \tan^{-1}\left(\dfrac{X_{L_1} - X_{C_1}}{R_1}\right)$	$\phi_2 = \tan^{-1}\left(\dfrac{X_{L_2} - X_{C_2}}{R_2}\right)$
$\phi_1 = \tan^{-1}\left(\dfrac{-12\ \Omega}{5\ \Omega}\right) = -67.4°$	$\phi_2 = \tan^{-1}\left(\dfrac{16\ \Omega}{12\ \Omega}\right) = 53.1°$
The current $\mathbf{I}_1$ *leads* the voltage **V** by 67.4°.	The current $\mathbf{I}_2$ *lags* the voltage **V** by 53.1°.

Step 3: We plot the currents as phasors in a current phasor diagram with the applied voltage **V** as a reference. We then calculate the vector addition $\mathbf{I} = \mathbf{I}_1 + \mathbf{I}_2$, using the method of component addition. Indicating the x and y axes as shown, we have:

	***x* component**	***y* component**
I_1:		
	$I_{1x} = I_1 \cos \phi_1$	$I_{1y} = I_1 \sin \phi_1$
	$I_{1x} = (20\text{ A})(\frac{5}{13}) = 7.69\text{ A}$	$I_{1y} = (20\text{ A})(\frac{12}{13}) = 18.5\text{ A}$
I_2:		
	$I_{2x} = I_2 \cos \phi_2$	$I_{2y} = I_2 \sin \phi_2$
	$I_{2x} = (13\text{ A})(\frac{3}{5}) = 7.80\text{ A}$	$I_{2y} = (13\text{ A})(-\frac{4}{5}) = -10.4\text{ A}$
I:		
	$I_x = I_{1x} + I_{2x}$	$I_y = I_{1y} + I_{2y}$
	$I_x = 7.69\text{ A} + 7.80\text{ A} = 15.5\text{ A}$	$I_y = 18.5\text{ A} - 10.4\text{ A} = 8.10\text{ A}$

Combining I_x and I_y, we obtain

$$I = \sqrt{I_x^2 + I_y^2}$$

$$I = \sqrt{(15.5\text{ A})^2 + (8.10\text{ A})^2} = 17.5\text{ A}$$

Finally:

$$\phi = -\tan^{-1}\left(\frac{I_y}{I_x}\right)$$

(Note that this expression for ϕ in terms of current-phasor components has a minus sign. Recall that a negative value of ϕ means that the current *leads* the voltage.)

$$\phi = -\tan^{-1}\left(\frac{8.10\text{ A}}{15.5\text{ A}}\right) = -27.6°$$

So the total current i supplied by the source is

$$i = \boxed{17.5 \sin(\omega t + 27.6°)\text{ A}}$$

with the current leading the voltage by 27.6°.

Note that in the previous example the parallel network as a whole behaves as a *series RC* combination. Yet the branch containing the capacitance has the lower impedance. This is similar to the situation in *DC* parallel resistive circuits: The *lower* resistance branch dominates in determining the total resistance (whereas in series combinations the largest resistance dominates in determining the total resistance).

27.7 Resonance

Even though most of us are aware of natural resonances in mechanical systems such as springboards, tuning forks, and springs, we do not usually view them as frequency-selection mechanisms. Yet this is exactly how they behave. If the driving frequency coincides with one of the natural frequencies of the system, large-amplitude oscillations occur. Electrical circuits composed of inductors, capacitors, and resistors behave in a similar way. That is, if an *AC* voltage is applied to such a circuit, the response of the circuit is greatest when the frequency of the voltage coincides with the resonant frequency of the circuit. We will discover that the **resonant frequency** of a circuit is the frequency at which *the current through the circuit is in phase with the driving voltage.* The most practical way to view electrical resonance is as a *frequency-selection* phenomenon. Whenever we tune to a particular radio or television broadcast, we utilize this selection capability of resonant circuits.

Series Resonance

Consider a series *RLC* combination, as shown in Figure 27–14(a). In order to examine the behavior of the circuit as the angular frequency ω changes, we will construct a series of impedance diagrams. We begin by constructing the diagram shown in Figure 27–14(b), for which $X_L = X_C$; thus the impedance Z is just the resistance R. The angular frequency ω_0 corresponding to this condition is found as follows:

$$X_L = X_C$$

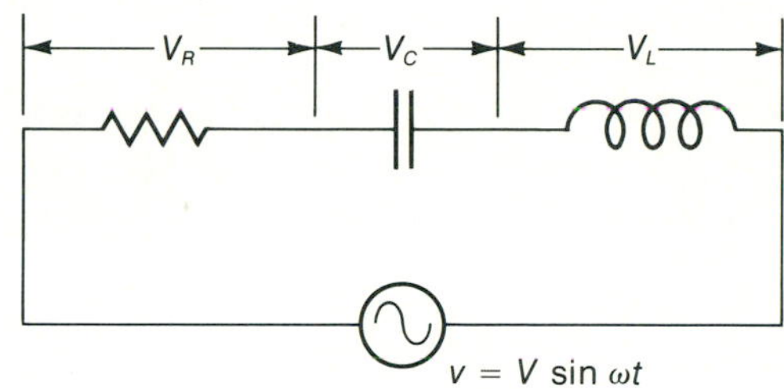

(**a**) A series *RLC* circuit.

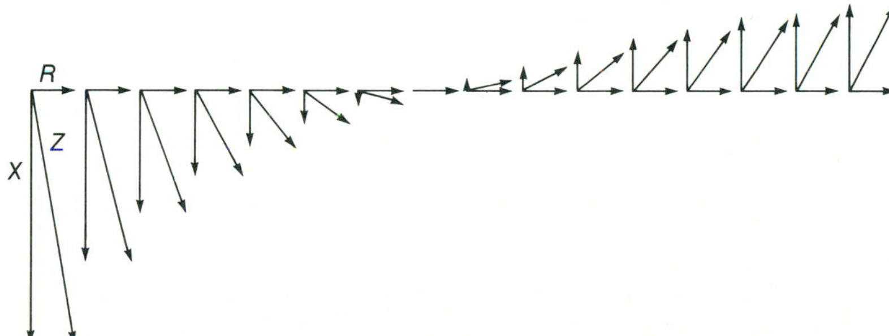

(**b**) Impedance diagrams for various frequencies.

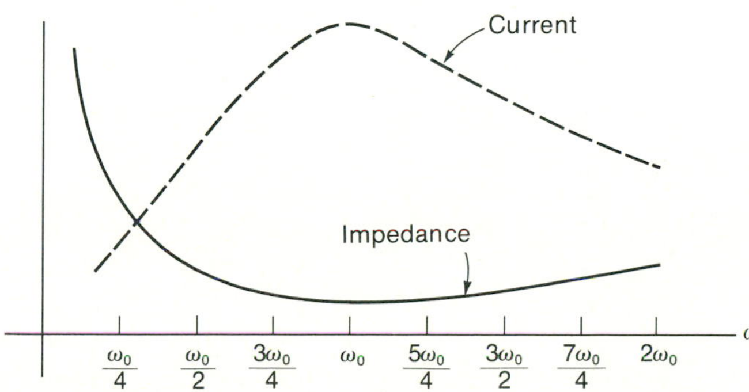

(**c**) Impedance and current as a function of frequency ω.

Figure 27–14
Series resonance.

Substituting for X_L and X_C and solving for ω_0:

$$\omega_0 L = \frac{1}{\omega_0 C}$$

RESONANT ANGULAR FREQUENCY ω_0 FOR A SERIES *RLC* CIRCUIT

$$\omega_0 = \frac{1}{\sqrt{LC}} \tag{27–19}$$

We then construct impedance diagrams for $\frac{1}{8}\omega_0$, $\frac{2}{8}\omega_0$, $\frac{3}{8}\omega_0$, . . . all the way to $\frac{16}{8}\omega_0$. The arrow representing **Z** has two reactive components: $\mathbf{X}_L$, pointing upward, and $\mathbf{X}_C$, pointing downward. As the frequency increases, $\mathbf{X}_L$ increases linearly and $\mathbf{X}_C$ decreases hyperbolically. The resistive component **R** remains constant.

The magnitudes of the impedance from each of the impedance diagrams are plotted as a function of frequency, generating the impedance curve shown in Figure 27–14(c). The corresponding values of the current I (equal to V/Z) are plotted and represented by the dashed curve. The current versus frequency curve is called the *resonance curve* and reveals the following important features of the series resonant circuit:

(1) The *sharpness* of the resonance curve increases as the value of the resistance decreases relative to the inductive or capacitive reactance. The sharpness is described by the Q (or *quality*) of the circuit. By definition:

SHARPNESS Q OF A RESONANT CIRCUIT

$$Q \equiv \frac{\omega_0 L}{R} \tag{27–20}$$

Since Q is a ratio of ohms over ohms, it is dimensionless. Typical low-frequency resonant circuits may have a Q of less than 10, while a very high-frequency resonant circuit may have a Q of several thousand (see Figure 27–15).

(2) The resonance curve is not symmetrical in shape when plotted linearly with frequency. However, if the current is plotted versus the logarithm of the frequency, the resonance curve is (essentially) symmetrical.

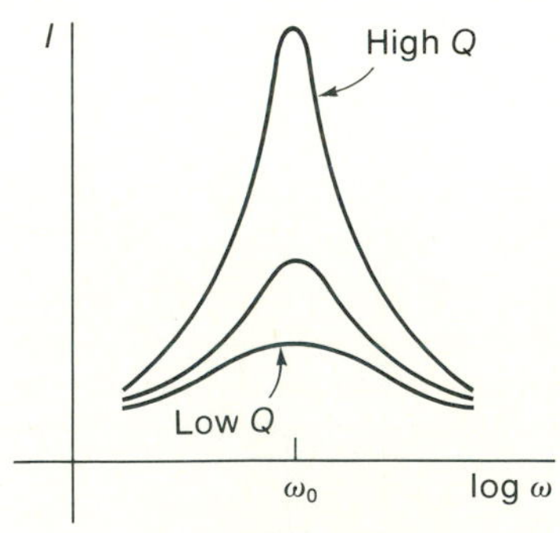

Resonant frequency

(a) Resonance curves for *RLC* circuits having different sharpness Q.

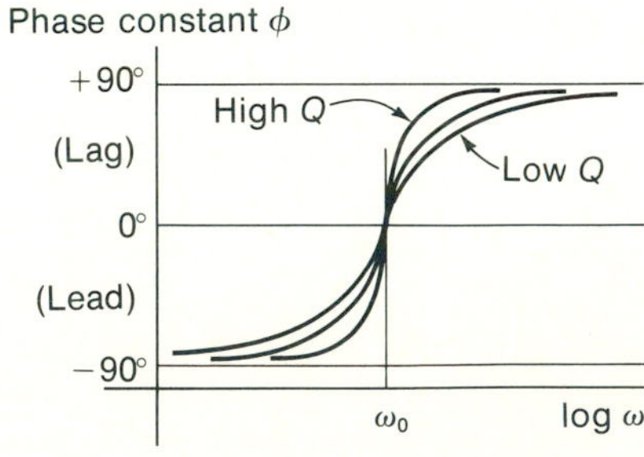

(b) The phase constant by which the current leads or lags the applied voltage in a series *RLC* circuit.

Figure 27–15
Series *RLC* resonance.

(3) At the resonant frequency, the current becomes very large, limited only by the value of R. At resonance:

$$I = \frac{V}{R} \tag{27–21}$$

where V is the magnitude of the applied voltage.

(4) At resonance, the magnitude of the voltage across the inductor equals that across the capacitor. However, the voltage across one is 180° out of phase with the voltage across the other, so they add vectorially to zero. *But each, by itself, may be a very large value*; in high-Q circuits, the voltage across a reactance may be thousands of times larger than the applied voltage.

EXAMPLE 27–6

A series RLC circuit has the following values:

$$L = 20 \text{ mH}$$
$$C = 100 \text{ nF}$$
$$R = 20\ \Omega$$
$$V = 100 \text{ V} \qquad (\text{where } v = V \sin \omega t)$$

Find (a) the resonant frequency, (b) the magnitude of the current at the resonant frequency, (c) the Q of the circuit, and (d) the magnitude of the voltage across the inductor at resonance.

SOLUTION

(a) The resonant frequency is obtained from Equation (27–19):

$$\omega_0 = \frac{1}{\sqrt{LC}}$$

Substituting numerical values:

$$\omega_0 = [(20 \times 10^{-3} \text{ H})(100 \times 10^{-9} \text{ F})]^{-\frac{1}{2}} = 2.24 \times 10^4 \frac{\text{rad}}{\text{s}}$$

$$f_0 = \left(2.24 \times 10^4 \frac{\text{rad}}{\text{s}}\right)\underbrace{\left(\frac{1 \text{ cycle}}{2\pi \text{ rad}}\right)}_{\text{conversion ratio}} = \boxed{3.56 \text{ kHz}}$$

(b) At resonance, the magnitude of the current is simply the magnitude of the applied voltage divided by the resistance:

$$I = \frac{V}{R}$$

$$I = \frac{100 \text{ V}}{20\ \Omega} = \boxed{5.00 \text{ A}}$$

(c) The Q of the circuit is obtained from Equation (27–20):

$$Q = \frac{\omega_0 L}{R}$$

$$Q = \frac{\left(2.24 \times 10^4 \frac{\text{rad}}{\text{s}}\right)\left(20 \times 10^{-3} \text{ H}\right)}{20\ \Omega} = \boxed{22.4}$$

Note that Q is dimensionless.

(d) The magnitude of the voltage V_L across the inductor is given by

$$V_L = X_L I$$

where X_L is the inductive reactance at the resonant frequency and I is the magnitude of the current at resonance:

$$V_L = (\omega_0 L)(I)$$

$$V_L = \left(2.24 \times 10^4 \frac{\text{rad}}{\text{s}}\right)\left(20 \times 10^{-3}\ \text{H}\right)(5\ \text{A}) = \boxed{2240\ \text{V}}$$

Note that this voltage is considerably higher than the applied voltage of 100 V.

Parallel Resonance

One of the most common forms of a resonant circuit is a parallel combination of a capacitor and an inductor, such as that illustrated in Figure 27–16(a). A resistor is shown in the branch containing the inductor to represent the resistance of the windings inherent to all inductors.[6] This circuit is analyzed in a manner similar to that used for the series resonant circuit. Such an analysis reveals a *current minimum* at the resonant frequency, corresponding to the situation of the *current being in phase with the applied voltage.* A current phasor diagram at this frequency is shown in Figure 27–16(b). The total current represented by the phasor $\mathbf{I}$ is the vector sum of $\mathbf{I}_1$, the current through the capacitor, and $\mathbf{I}_2$, the current through the series combination of the inductor and resistor. The current $\mathbf{I}_1$ leads the applied voltage $\mathbf{V}$ by $\pi/2$ rad, while the current through the RL branch lags the applied voltage by ϕ, where

$$\phi = \tan^{-1}\left(\frac{\omega_0 L}{R}\right) \tag{27–22}$$

[6] The reason we do not show a resistance in the capacitive branch is the following. If the dielectric material of a capacitor "leaks," allowing some current under *DC* conditions, this electrical resistance would be represented as a resistance *in parallel* with the capacitor. (Thus, for *DC*, some current would flow.) Since we usually try to design high-Q circuits, the *DC* resistance of capacitors can be made so high that it can be neglected in circuit analyses.

Figure 27–16
Resonance in a parallel circuit.

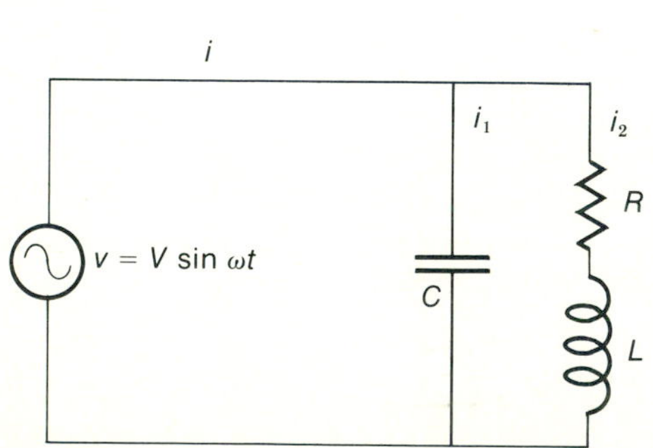

(a) A parallel resonant circuit.

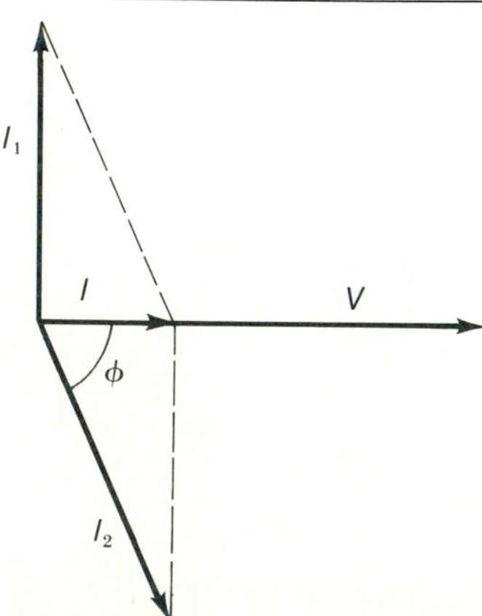

(b) The current phasor diagram at resonance.

The magnitudes of $\mathbf{I}_1$ and $\mathbf{I}_2$ are

$$I_1 = \frac{V}{X_C} \tag{27–23}$$

and

$$I_2 = \frac{V}{\sqrt{X_L^2 + R^2}} \tag{27–24}$$

At resonance, the vertical components of $\mathbf{I}_1$ and $\mathbf{I}_2$ in the current phasor diagram must be equal. Thus:

$$\frac{V}{X_C} = \frac{V}{\sqrt{X_L^2 + R^2}} \sin\phi \tag{27–25}$$

where X_L, X_C, and ϕ are the values at resonance. Using Equation (27–22):

$$\tan\phi = \frac{\omega_0 L}{R}$$

Therefore:

$$\sin\phi = \frac{\omega_0 L}{\sqrt{(\omega_0 L)^2 + R^2}}$$

Substituting the appropriate quantities into Equation (27–25) and solving for ω_0 yields:

$$\omega_0 C = \left(\frac{1}{\sqrt{(\omega_0 L)^2 + R^2}}\right)\left(\frac{\omega_0 L}{\sqrt{(\omega_0 L)^2 + R^2}}\right)$$

RESONANT ANGULAR FREQUENCY ω_0 FOR THE PARALLEL *RLC* CIRCUIT OF FIGURE 27–16

$$\omega_0 = \sqrt{\frac{1}{LC} - \frac{R^2}{L^2}} \tag{27–26}$$

Note that if R is small compared with L (corresponding to a high Q), the condition for parallel resonance is the same as that for series resonance.

27.8 Power in *AC* Circuits

When considering *DC* circuits, the energy balance is quite simple. The rate at which the seats of EMF supply energy to the circuit equals the rate at which energy is lost through the joule heating of the resistors. For *AC* circuits, the energy balance seems complicated at first glance. At any instant, the rate at which energy is supplied by the *AC* source must be balanced not only by the rate of joule heating of resistors but also by the rate at which energy associated with magnetic and electric fields is stored or released in the inductors and capacitors. Fortunately, from a practical standpoint, power considerations in *AC* circuits are reasonably straightforward. At any instant, the incremental work dW done by a source of varying voltage v in changing the potential of an incremental charge dq is expressed as

$$dW = v\,dq$$

The rate at which work is being done at that instant is the instantaneous power p supplied by the source of the circuit:

$$p = \frac{dW}{dt}$$

Combining these two equations:

$$p = v\frac{dq}{dt}$$

or

$$p = vi \tag{27–27}$$

Since

$$v = V \sin \omega t$$

and

$$i = I \sin (\omega t - \phi)$$

the instantaneous power becomes

$$p = VI \sin \omega t \sin (\omega t - \phi) \tag{27–28}$$

The power supplied to the circuit thus varies in time. However, we are most often concerned about the *average power* P_{ave} supplied to the circuit. From the mathematical definition for the average over time (that is, the *time-weighted* average):

$$P_{\text{ave}} = \frac{1}{T}\int_0^T p\,dt$$

where T is one period of power variation. (Note the similarity to the *mass-weighted* average used in the determination of center of mass.)

Substituting from Equation (27–28), we have

$$P_{\text{ave}} = \frac{1}{T}\int_0^T VI \sin \omega t \sin (\omega t - \phi)\,dt \tag{27–29}$$

From Appendix D, $\sin (\omega t - \phi) = (\sin \omega t \cos \phi - \cos \omega t \sin \phi)$. Therefore, Equation (27–29) becomes

$$P_{\text{ave}} = \frac{VI \cos \phi}{T}\int_0^T \sin^2 \omega t\,dt - \frac{VI \sin \phi}{T}\int_0^T \sin \omega t \cos \omega t\,dt$$

Using Appendix GII to evaluate the integrals, we have

$$P_{\text{ave}} = \frac{VI \cos \phi}{T}\left(\frac{t}{2} - \frac{\sin 2\omega t}{4}\right)\Bigg|_0^T - \frac{VI \sin \phi}{T}\left(\frac{\sin^2 \omega t}{2}\right)\Bigg|_0^T$$

Substituting the limits and using the relation $T = 2\pi/\omega$, we have

$$P_{\text{ave}} = \frac{VI}{2}\cos \phi \tag{27–30}$$

where ϕ is the phase angle between the voltage v and current i. (Note that the integral of either $\sin^2 \omega t$ or $\cos^2 \omega t$ over a period T is equal to $\frac{1}{2}$.)

The fact that the average power supplied to the circuit depends on the cosine of the phase angle has important implications concerning how the power is dissipated in the components of the circuit. In a purely inductive circuit, the phase angle $\phi = \pi/2$. Since the cosine of $\pi/2$ is zero, the *average* power dissipated in the inductor is zero. We may interpret this physically by realizing that the work done by the source of current in building the magnetic field of the inductor is returned to the source when the field collapses. Similarly, since for a purely capacitive circuit $\phi = -\pi/2$, the *average* power dissipated in a capacitor is also zero. The work done in creating the electric field in the capacitor is returned to the source when the field collapses. If you follow the buildup and reduction of these fields, you will discover that the processes occur exactly 180° out of phase; while the electric field is building up, the magnetic field is collapsing, and vice versa. In effect, the inductance and capacitance merely exchange energy back and forth between themselves. If the reactances are pure (if there is no resistance associated with them), there is no average energy loss in the reactances. *The only energy dissipation occurs in the resistance* R *by joule heating.* That is, all of the power loss in an alternating-current circuit occurs in the resistor by joule heating. Since the voltage v_R across the resistor is in phase ($\phi = 0$) with the current i through it, Equation (27–30) becomes

$$P_{\text{ave}} = \frac{V_R I}{2} \tag{27–31}$$

This equation may also be deduced from Figure 27–9(c), where

$$V_R = V \cos \phi \tag{27–32}$$

Using this fact, Equation (27–31) can also be written as

$$P_{\text{ave}} = \frac{VI}{2} \cos \phi \tag{27–33}$$

The cosine term is called the **power factor.** From Figure 27–10, it equals

POWER FACTOR

$$\cos \phi = \frac{R}{Z} \tag{27–34}$$

We may also express the average power in terms of the *root-mean-square*[7] (rms) values of V and I:

$$V_{\text{rms}} = \frac{V}{\sqrt{2}} \qquad I_{\text{rms}} = \frac{I}{\sqrt{2}} \tag{27–35}$$

leading to:

$$P_{\text{ave}} = V_{\text{rms}} I_{\text{rms}} \cos \phi \tag{27–36}$$

Using Equation (27–32), we may express P_{ave} in still another form:

$$P_{\text{ave}} = (V_R)_{\text{rms}} I_{\text{rms}} \tag{27–37}$$

The several forms for P_{ave} are listed together for easy reference:

AVERAGE POWER DISSIPATED IN AN *RLC* CIRCUIT

$$P_{\text{ave}} \begin{cases} = \dfrac{VI}{2} \cos \phi & (27–33) \\ = V_{\text{rms}} I_{\text{rms}} \cos \phi & (27–36) \\ = (V_R)_{\text{rms}} I_{\text{rms}} & (27–37) \end{cases}$$

[7] The root-mean-square value of any quantity is the square *root* of the average (or *mean*) value of the *square* of the quantity. Thus, in the case of a sinusoidally varying voltage:

$$V_{\text{rms}} = \left(\int_0^T V^2 \sin^2 \omega t \, dt \right)^{\frac{1}{2}}$$

where T is the period of the variation. Since $\int_0^T \sin^2 \omega t \, dt = \frac{1}{2}$,

$$V_{\text{rms}} = \frac{V}{\sqrt{2}}$$

The form of Equation (27–37) is similar to the expression for *DC* circuits in which

$$P = V_R I \tag{27–38}$$

where P is the constant power dissipated in a resistor that has a constant potential difference V_R across its terminals, resulting in a constant current I through the resistor. The similarity between Equations (27–37) and (27–38) is the basis for describing rms values as *effective* values. The rms values of current and voltage *produce the same joule heating in a resistor* as *DC* current and voltage of the same magnitudes; they are just as "effective" in producing I^2R losses in a resistor.

It is important to note that effective values are defined for *sinusoidal* currents and voltages. Power-line currents and voltages are always quoted in rms values, even though the subscript is commonly omitted. For example, an electrical outlet supplying a voltage of 110 V, 60 Hz, has a *peak* value of $(\sqrt{2})(110\text{ V})$, or 156 V. Such a line voltage would be expressed analytically in SI units as

$$110\text{ V } 60\text{ Hz} \Rightarrow 156 \sin(120\pi t)\text{ V}$$

EXAMPLE 27–7

A voltage of the form (in SI units):

$$v = 100 \sin(1000t) \qquad \text{(in volts if } t \text{ is in seconds)}$$

is applied to a series *RLC* circuit. If $R = 400\ \Omega$, $C = 5.0\ \mu\text{F}$, and $L = 0.50$ H, find the average power dissipated in the circuit.

SOLUTION

All three expressions for the average power involve the current, so we first solve for the current in the circuit. From Equation (27–17), the impedance Z is

$$Z = \sqrt{R^2 + (X_L - X_C)^2}$$

where

$$X_L = \omega L$$

$$X_L = \left(1000\ \frac{\text{rad}}{\text{s}}\right)(0.50\text{ H}) = 500\ \Omega$$

and

$$X_C = \frac{1}{\omega C}$$

$$X_C = \frac{1}{\left(1000\ \frac{\text{rad}}{\text{s}}\right)(5.0 \times 10^{-6}\text{ F})} = 200\ \Omega$$

Substituting:

$$Z = \sqrt{(400\ \Omega)^2 + (500\ \Omega - 200\ \Omega)^2} = 500\ \Omega$$

The amplitude of the current is

$$I = \frac{V}{Z}$$

$$I = \frac{100\text{ V}}{500\ \Omega} = 0.200\text{ A}$$

Knowing the current, we may use any one of the expressions for the average power. We will illustrate the use of all three.

Using Equation (27–33):

$$P_{ave} = \frac{VI}{2}\cos\phi$$

where $$\cos\phi = \frac{R}{Z}$$

$$\cos\phi = \frac{400\ \Omega}{500\ \Omega} = 0.800$$

Substituting:

$$P_{ave} = \frac{(100\ \text{V})(0.200\ \text{A})}{2}(0.800) = \boxed{8.00\ \text{W}}$$

Using Equation (27–36):

$$P_{ave} = V_{rms}I_{rms}\cos\phi$$

$$P_{ave} = \left(\frac{100\ \text{V}}{\sqrt{2}}\right)\left(\frac{0.200\ \text{A}}{\sqrt{2}}\right)(0.800) = \boxed{8.00\ \text{W}}$$

Using Equation (27–37):

$$P_{ave} = (V_R)_{rms}I_{rms}$$

where $$(V_R)_{rms} = (IR)_{rms}$$

$$(V_R)_{rms} = \frac{(0.200\ \text{A})(400\ \Omega)}{\sqrt{2}} = \frac{80}{\sqrt{2}}\ \text{V}$$

Substituting:

$$P_{ave} = \left(\frac{80}{\sqrt{2}}\ \text{V}\right)\frac{(0.200\ \text{A})}{\sqrt{2}} = \boxed{8.00\ \text{W}}$$

Summary

The following concepts and equations were introduced in this chapter:

Phase between voltage and current in *AC* circuit components:

Resistance: The current is *in phase* with the voltage.
Capacitance: The current *leads* the voltage by $\pi/2$ rad.
Inductance: The current *lags* the voltage by $\pi/2$ rad.

Reactance

Capacitive: $X_C = 1/\omega C$
Inductive: $X_L = \omega L$
Total reactance: $X = X_L - X_C$

Series RLC circuits

$$i = \frac{V}{\sqrt{R^2 + (X_L - X_C)^2}}\sin(\omega t - \phi) + i_0(t)$$

where $$\phi = \tan^{-1}\left(\frac{X_L - X_C}{R}\right)$$

Impedance Z

$$Z = \sqrt{R^2 + (X_L - X_C)^2}$$

In *phasor diagrams*, the amplitudes of voltages and currents are represented by vectorlike arrows called *phasors*, drawn to depict their phase relationships. The phasor diagram rotates counterclockwise, with angular frequency ω. The projections of the phasors on a vertical axis give the instantaneous values of v and i.

An *impedance diagram* depicts $\mathbf{R}$ along the $+x$ axis, $\mathbf{X}_L$ along the $+y$ axis, and $\mathbf{X}_C$ along the $-y$ axis. The arrows add vectorially to give the impedance $\mathbf{Z}$, with the phase angle ϕ between $\mathbf{R}$ and $\mathbf{Z}$.

Resonance

Series:

$$X_L = X_C \Rightarrow \omega_0 = \frac{1}{\sqrt{LC}}$$

Parallel: For a capacitor in parallel with an inductor-resistor series combination:

$$\omega_0 = \sqrt{\frac{1}{LC} - \frac{R^2}{L^2}}$$

Sharpness of resonance Q:

$$Q = \frac{\omega_0 L}{R}$$

The *effective* (or rms) value of a sinusoidally-varying current is that *DC* current which produces the same heating effect in a resistor. It is related to the peak value I of the *AC* current:

$$I_{\text{eff}} = I_{\text{rms}} = \frac{I}{\sqrt{2}}$$

Average power in AC circuits: All of the power developed in *AC* circuits is in the resistive components. If a power supply delivers to a circuit a current I at a voltage V (peak values), then

$$P_{\text{ave}} \begin{cases} = \dfrac{VI}{2}\cos\phi \\ = V_{\text{rms}} I_{\text{rms}} \cos\phi \\ = (V_R)_{\text{rms}} I_{\text{rms}} \end{cases}$$

The *power factor* in an *AC* circuit is $\cos\phi$.

Questions

1. Using nonmathematical reasoning, can you explain why the current through a capacitor leads the voltage across the capacitor and why the current through an inductor lags the voltage across the inductor?
2. A square-wave voltage is applied to a series combination of a resistor and an inductor. If the resistance is large compared with the inductive reactance (corresponding to the lowest Fourier component of the square-wave), what is the voltage waveform across the inductor?
3. If the secondary winding of a transformer is open circuit, why does a small current still pass through the primary winding?
4. An *AC* voltage is applied to a series *RLC* circuit. How does the phase constant change as the frequency of the applied voltage changes from zero to a very high value?

5. In what ways do Kirchhoff's junction and loop rules for *DC* circuits have to be modified to apply to *AC* circuits?
6. As the frequency of an *AC* power source varies from zero to a very high value, how does the behavior of a series combination of a capacitor and inductor compare with that of a parallel combination of a capacitor and inductor?
7. In a parallel circuit, one branch has a capacitive reactance X_C, while the other branch has an inductive reactance X_L, where $X_L > X_C$. Is the parallel combination capacitive or inductive?
8. Why is it often "hazardous to your health" to experiment with high-Q resonant circuits (unless you take careful precautions)?
9. In a series *RLC* circuit, how should the frequency be adjusted so as to dissipate the maximum amount of power in the resistor?
10. Is the rms current through a series *RLC* circuit at $1/N$ times the resonant frequency equal to the rms current at N times the resonant frequency, where N is any number?
11. Is it possible to have resonance in a power transmission line? If so, and if such resonance presents a serious problem to power transmission, how could the problem be avoided?
12. An *RLC* circuit is analogous to a driven mechanical oscillator. What are the analogies between the two systems?
13. An *AC* voltage is applied to a series *RLC* circuit. In what ways could you determine whether the circuit is above or below resonance? Repeat for a parallel *RLC* circuit (capacitance in one branch and inductance in the other).
14. The resonant power circuit of a radio transmitter has an inductor made of very heavy wire mounted on large insulators. Why?
15. Why is it inadvisable to interchange the input and output terminals of a step-down transformer in order to make it a step-up transformer? (Hint: What limits the primary current with an open-circuit secondary?)
16. Is the power dissipation in an *RLC* circuit continuous or pulsating?
17. In order to reduce household electrical power consumption, why not decrease the power factor rather than decrease the rms current?
18. A resistor is connected to the secondary winding of a transformer while a square-wave voltage is applied to the primary winding. What is the voltage waveform across the secondary?
19. The average power dissipated in an ideal inductor or capacitor is zero. How does the instantaneous power input to these devices vary with time? How does this variation lead to the conclusion that the average power input is zero?
20. Edison proposed that power distribution systems should be direct current. What are the advantages and disadvantages of such a system?
21. Why is the engineer in a commercial power station concerned about the power factor of the load the station supplies? (Hint: Consider power losses in transmission lines.)
22. An *AC* voltage source whose frequency can be varied is applied to a series *RLC* circuit. As the frequency is raised from ω_1 to ω_2, the current gradually decreases. Suppose a capacitor is now added in series with the circuit. Will this increase or decrease the original impedance in this range of frequencies?

Problems

27A–1 The voltage at a household electrical outlet is often described as 110 volt, 60 cycle. The 110 volts is the rms value of the voltage and the 60 cycle represents a frequency of 60 Hz. Describe this voltage in the form $v = V \sin \omega t$, including numerical values.

Answer: $v = 156 \sin (377t)$ V

27A–2 The voltage $v = 240 \sin 500t$ (in SI units) is applied across a parallel combination of a 600 Ω resistor and a 2.5 μF capacitor. Express the current from the voltage source in the form $i = I \sin (500t - \phi)$, including numerical values for I and ϕ.

27A–3 The voltage $v = 30 \sin 5000t$ (in SI units) is applied across a parallel combination of a 75 Ω resistor and a 30 mH inductor. Express the current from the voltage source in the form $i = I \sin (5000t + \phi)$, including numerical values for I and ϕ.

Answer: $i = 0.477 \sin (5000t - 26.6°)$ A

274–4 A voltage $v = 100 \sin 2500t$ (in SI units) is applied to a series combination of a 30 Ω resistor and a 10 μF capacitor.

(a) Make impedance and phasor diagrams for the circuit.
(b) Calculate the maximum energy stored in the electric field of the capacitor.

27A–5 A voltage $v = 200 \sin 2000t$ is applied across a series combination of a 2500 Ω resistor and a 1.5 H inductor. (a) Sketch an impedance diagram and a phasor diagram for this circuit. Calculate the rms values of (b) the applied voltage, (c) the current, (d) the voltage across the inductor, and (e) the voltage across the resistor.

Answers: (b) 141 V (c) 36.2 mA (d) 109 V (e) 90.5 V

27A–6 A sinusoidally-varying voltage with an amplitude of 100 V is connected across a series combination of a 10 Ω resistor, a 100 mH inductor, and a 0.1 μF capacitor. Calculate the amplitude of the voltage across the capacitor at (a) the resonant frequency and (b) $\frac{1}{10}$ the resonant frequency.

27A–7 The tuning circuit of an AM radio is a parallel LC combination that has negligible resistance. The inductance is 0.2 mH and the capacitor is variable, so that the circuit can resonate at frequencies between 550 kHz and 1650 kHz. Find the range of C.

Answer: 46.5 pF to 419 pF

27A–8 A voltage $v = 100 \sin 5000t$ (in SI units) is applied across a series combination of a 700 Ω resistor and a 100 mH inductor.

(a) Sketch impedance and phasor diagrams for the circuit.
(b) Calculate the rms current in the circuit.
(c) Find the power dissipated in the resistor.
(d) Calculate the power supplied to the circuit by the voltage source.

27A–9 Beginning with the definitions of capacitance and inductance, show that (a) capacitive reactance and inductive reactance have the dimensions of ohms and (b) that $(LC)^{\frac{1}{2}}$ has the dimensions of time.

27A–10 The windings of a 150 mH inductor have 30 Ω resistance. A 20 V (rms), 60 Hz voltage is applied to the inductor. Assuming the equivalent circuit is a resistance in series with a pure inductance, find (a) the power factor and (b) the power developed in the windings. (c) Suppose the frequency of the applied voltage were changed to 50 Hz (with the same rms value). Find the power developed in the windings.

27A–11 A sinusoidal voltage with an amplitude of 156 V is connected to a heater with a resistance of 100 Ω. Calculate the power dissipated in the heater. *Answer:* 122 W

27B–1 Show that $i = (V/\omega L) \sin (\omega t - 90°)$ is a solution to the differential equation $V \sin \omega t - L\,di/dt = 0$.

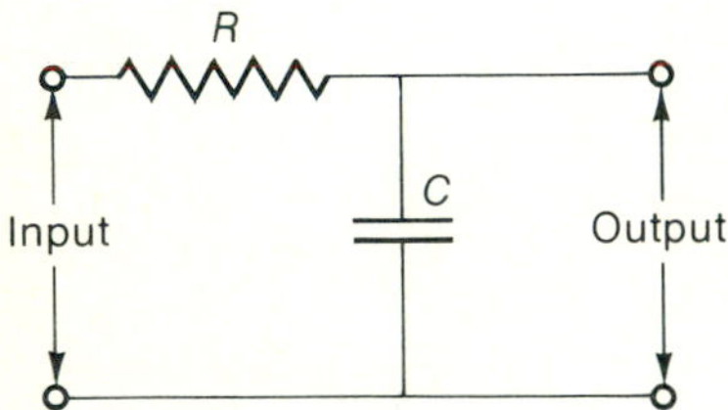

Figure 27–17
Problem 27B–2

27B–2 The circuit shown in Figure 27–17 is called a "low-pass" filter. (The impedance of the capacitor becomes less at higher frequencies, so the output voltage for higher frequencies is reduced. Such a filter is often used in *DC* power supplies to reduce unwanted *AC* components.) The *half-power frequency* is defined as the frequency above which the amplitude of the output voltage is smaller than $1/\sqrt{2}$ times the input voltage.

(a) Derive the expression for the half-power frequency.
(b) Find the phase of the output voltage relative to the input voltage at this frequency.

27B–3 A voltage $v = 10 \sin 1000t$ (in SI units) is applied across a 1.0 μF capacitor in series with a 1.5 kΩ resistor.

(a) Draw a phasor diagram showing the input voltage, the voltages across the resistor and the capacitor, and the current.
(b) Describe the voltage across the resistor in a functional form similar to that describing the input voltage. Include the phase constant.

Answer: (b) $v = 8.32 \times 10^{-3} \sin (1000t + 33.7°)$ (in SI units)

27B–4 Consider the "phase-shifter" circuit shown in Figure 27–18. The input voltage is described by $v = 10 \sin 200t$ (in SI units). If $L = 500$ mH, (a) find the value of R such that the output voltage v_0 lags the input voltage by 30° and (b) find the amplitude of the output voltage.

27B–5 Consider the circuit shown in Figure 27–18. The input voltage is a time-varying voltage (not necessarily sinusoidal). Show that the output voltage is approximately the integral of the input voltage if $R \ll \omega L$, where ω represents the frequencies of the input voltage Fourier components.

27B–6 Consider the circuit shown in Figure 27–19. The input voltage is a time-varying voltage (not necessarily sinusoidal). Show that the output voltage is approximately the derivative of the input voltage if $R \ll 1/\omega C$, where ω represents the angular frequencies of the input voltage Fourier components.

27B–7 Figure 27–20 shows a simple AC generator. As the wire loop rotates in the presence of a uniform magnetic field, the induced EMF in the loop is of the form $v = V \sin \omega t$. Consider a rectangular loop of sides $a = 0.2$ m and $b = 0.4$ m, rotating at 3600 rpm in the presence of a uniform field $B = 0.8$ T.

(a) Write an equation for the induced EMF, including numerical values for V and ω.

(b) Describe the orientation of the loop relative to $\mathbf{B}$ at the instant $t = 0$.

Answers: (a) 24.1 V (b) Plane of the loop is perpendicular to **B**

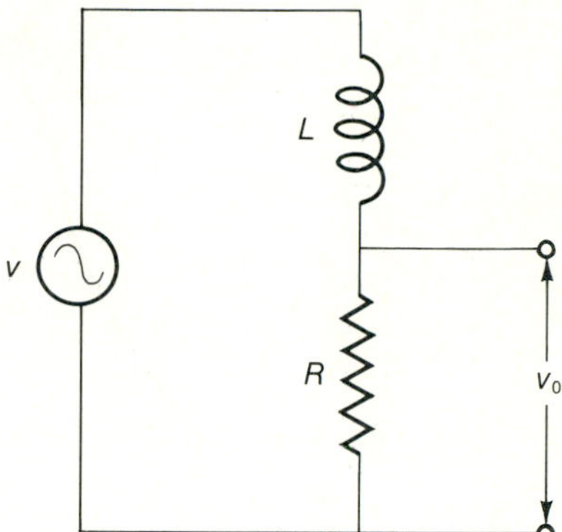

Figure 27–18
Problems 27B–4 and 27B–5

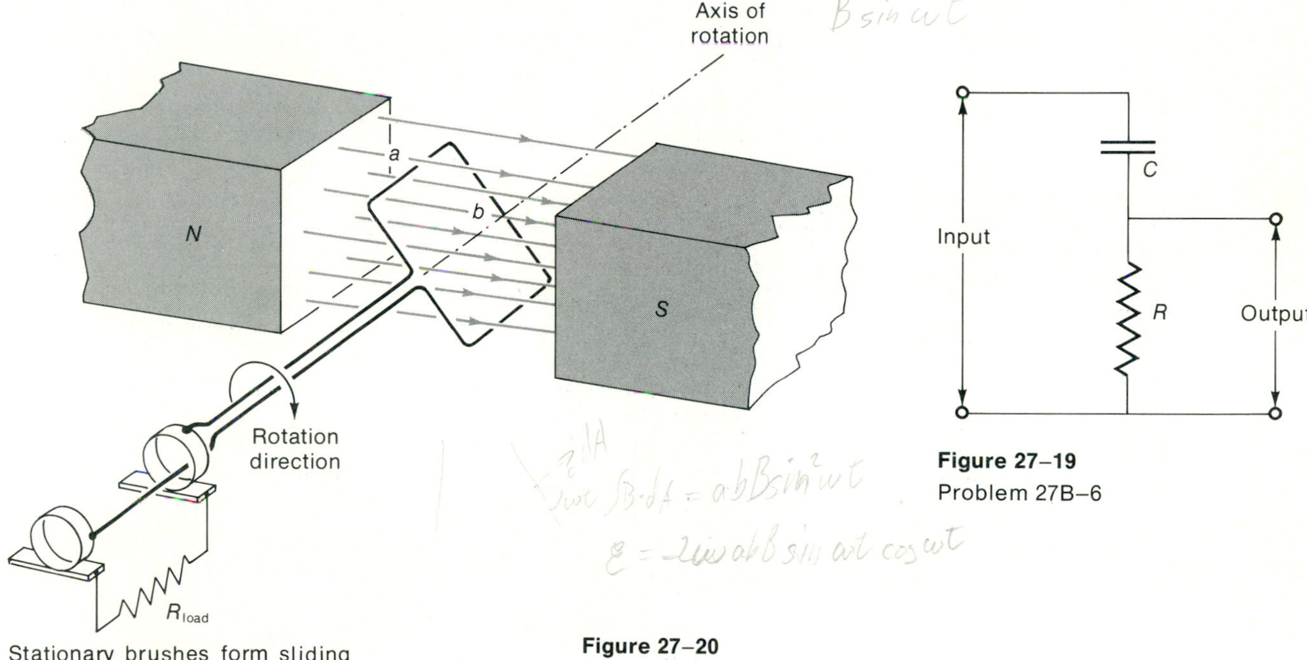

Figure 27–19
Problem 27B–6

Figure 27–20
Problems 27B–7 and 27B–8

27B–8 Refer to the AC voltage generator shown in Figure 27–20.

(a) Show that the torque required to turn the generator is given by $\tau = [\omega(abB)^2/R] \sin^2 \omega t$.

(b) Describe the orientation of the loop relative to **B** at the instant when the torque is a maximum.

27B–9 A series RLC circuit has the following values: $R = 20\ \Omega$ and $X_L = 10\ \Omega$. The applied voltage is 50 V (rms) at $\omega = 400$ rad/s, and the value of the capacitance is unknown. The power factor is 0.800 and the current of 2 A (rms) leads the applied voltage.

(a) Find the value of the capacitor.

(b) There are several ways to bring the circuit into resonance. To what value should the angular frequency be changed to make resonance occur?

(c) At this new resonant frequency, what power is developed in the circuit?

(d) At resonance, what is the rms voltage across the inductor?

(e) Suppose we kept the original frequency of 400 rad/s and instead changed the value of C to achieve resonance. Find the value of a single capacitor that could be *added* to the circuit to bring it into resonance. Would it be added in series or in parallel with the original capacitor?

Answers: (a) 100 μF (b) 632 rad/s (c) 125 W (d) 39.5 V (e) 150 μF, in parallel

27B–10 Consider a series combination of a 10 mH inductor, a 100 μF capacitor, and a 10 Ω resistor. A 50 V (rms) sinusoidal voltage is applied to the combination. Calculate the rms current for (a) the resonant frequency, (b) half the resonant frequency, and (c) double the resonant frequency.

27B–11 An inductor is in series with an 80 Ω resistor across a 110 V (rms), 60 Hz power source. If the resistor dissipates 50 W of power, find the inductance of the inductor.

Answer: 239 mH

27B–12 A 60 Hz, sinusoidally-varying voltage with an amplitude of 156 V is applied to a 0.15 H inductor that has a resistance of 50 Ω. Calculate the rate at which heat is produced in the inductor when (a) the resistance of the inductor is considered to be a resistance in series with a resistanceless inductance and (b) when the resistance of the inductor is considered to be a resistance in parallel with a resistanceless inductance.

27B–13 The power delivered by a 110 V (rms), 60 Hz source is 480 W. The power factor is 0.70 and current lags the voltage.

(a) Find the value of the capacitor C added in series that will change the power factor to unity.

(b) Find the power delivered by the source under these new conditions.

Answers: (a) 211 μF (b) 979 W

27C–1 A 30 mH inductor and a 40 kΩ resistor are connected across a voltage source described by $v = 100 \sin 10^6 t$ (in SI units). Find the maximum rate at which the current is changing in the circuit. *Answer:* 2000 A/s

27C–2 A voltage $v = 100 \sin 2000t$ (in SI units) is applied across a series combination of a 2500 Ω resistor and a 1.5 H inductor. (a) Sketch an impedance diagram for this circuit. For the time $t = 1$ ms, calculate the instantaneous value of (b) the applied voltage, (c) the current, (d) the voltage across the resistor, and (e) the voltage across the inductor. The sum of your answers to (d) and (e) should equal the answer to (b).

27C–3 A voltage $v = 100 \sin 1000t$ (in SI units) is applied across a series combination of a 1000 Ω resistor, a 0.5 μF capacitor, and a 1.5 H inductor. (a) Sketch an impedance diagram for the circuit. At $t = 0.7$ ms, calculate the instantaneous voltage across (b) the resistor, (c) the capacitor, and (d) the inductor. (e) Calculate the algebraic sum of these voltages and compare with the applied voltage at that instant. (f) On a sketch of the circuit, indicate the instantaneous polarities of these voltages.

Answers: (b) 82.1 V (c) −70.8 V (d) 53.1 V (e) 64.4 V

27C–4 Using a method similar to that used to demonstrate resonance in a series inductance-capacitance-resistance circuit, plot a resonance curve for a capacitor in parallel with a series combination of a resistance and an inductance.

27C–5 A nonideal inductor whose windings have appreciable resistance is connected in series with a 4 μF capacitor across a 120 V (rms), 60 Hz power source. The rms voltage across the capacitor is 180 V and the rms voltage across the inductor is 75 V. If the nonideal inductor is assumed to be equivalent to a resistor in series with an ideal inductor, find (a) the inductance of such an ideal inductor and (b) the resistance of the series resistor.

Answers: (a) 641 mH (b) 134 Ω

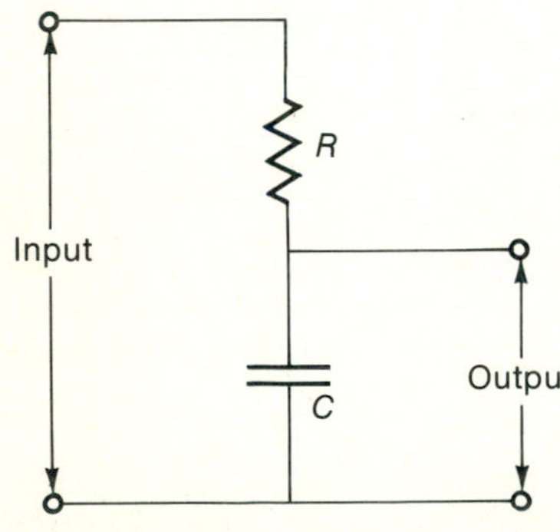

Figure 27–21
Problem 27C–7

27C–6 "Phase-shifters" with only a resistor and a capacitor *or* an inductor can only produce phase shifts of less than 90°. Greater phase shifts can be achieved by using a series combination of a resistor, a capacitor, and an inductor. Consider such a circuit containing an 80 mH inductor, a 10 μF capacitor, and resistance R.

(a) Determine the value of R and the location of the output terminals to produce an output voltage $v_0 = V_0 \sin(1000t + 120°)$, where the input voltage $v_i = 10 \sin 1000t$.

(b) Find the value of V_0.

27C–7 Consider the circuit shown in Figure 27–21. The input voltage is time varying (but not necessarily sinusoidal). Show that the output voltage is approximately the integral of the input voltage if $1/\omega C \ll R$, where ω represents the frequencies of the input voltage Fourier components.

27C–8 A series resonant circuit consists of an ideal inductor and a capacitor that "leaks," as indicated in Figure 27–22. Sketch a qualitative phasor diagram at the resonant frequency. Indicate the phasor representing the current through each component and the voltage across each component.

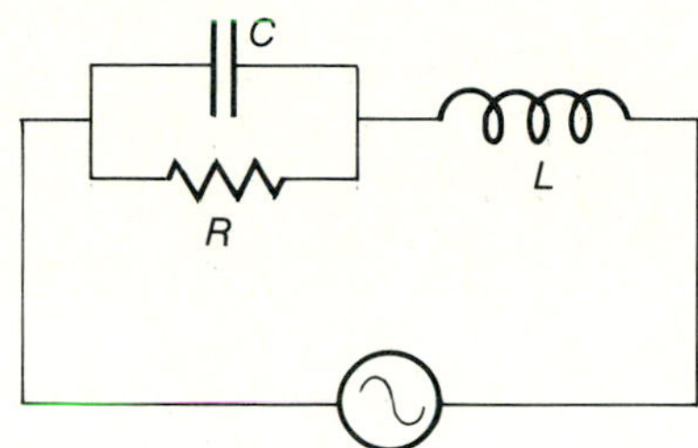

Figure 27–22
Problem 27C–8

chapter 28
Electromagnetic Radiation

I have also a paper afloat, with an electromagnetic theory of light, which, till I am convinced to the contrary, I hold to be great guns.

J.C. MAXWELL, in a letter to C.H. Cay, 5 January 1865
[*American Journal of Physics, 44, 676* (1976)]

In this chapter we will draw together the laws of electricity and magnetism into a single theory of *electromagnetism.* The possibility of forming such a theory should not be surprising: In previous chapters we have frequently noted a parallelism between the equations of electricity and those of magnetism. To further dramatize the intimate relations between electricity and magnetism, we will consider a phenomenon that can be analyzed as a problem in magnetism *or* in electricity, depending on the frame of reference. Although the situation we will describe may seem artificial, it will clarify some interesting relations between relativity and electromagnetism.

Suppose a single electron e is moving parallel to a current-carrying wire, as shown in Figure 28–1(a). For the sake of clarity, the electrons in the wire are shown separated from the positive ions and moving in straight-line motion with the drift velocity $\mathbf{v}$. We will assign the same velocity to the electron outside the wire, and assume that the wire has no net charge. The current in the wire produces a magnetic field out of the plane of the diagram at the moving (negative) electron, resulting in a magnetic force $\mathbf{F}_B$ toward the wire. The electron is accelerated toward the wire.

Now consider the same situation viewed from the moving charge, as shown in Figure 28–1(b). Here the wire moves to the left and the electron e is stationary. The wire is still observed to carry a current because, although the electrons are at rest, the positive charges are now moving toward the left. But since the electron is not moving, there is no magnetic force $\mathbf{F}_m = q\mathbf{v} \times \mathbf{B}$ on the electron. Obviously, if the electron accelerates toward the wire in one frame of reference, it must also do so when viewed from any other frame. What (if not a *magnetic* force) is the origin of a force that could produce such an acceleration?

The theory of special relativity provides the answer.[1] Viewed in frame S (Figure 28–1a), the positive ions are at rest and the electrons in the wire are moving with a velocity $\mathbf{v}$ to the right. The electrons appear closer together along the wire than their "proper" separation by the Lorentz length-contraction

[1] The only fact we need from special relativity is the *Lorentz contraction effect*. This states that objects moving past us with speed v are contracted along the direction of motion by the factor $\sqrt{1 - v^2/c^2}$, where c is the speed of light. Specifically, if L_0 is the length when at rest, the length L when moving is $L = L_0\sqrt{1 - v^2/c^2}$.

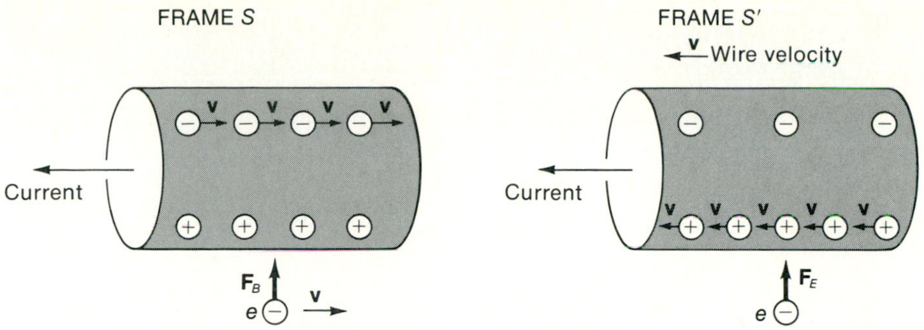

Conditions:

(1) The ⊕ charges are stationary.

(2) The ⊖ charges are moving to the right with a speed v. Therefore the separation of the ⊖ charges is Lorentz contracted.

(3) The net linear charge density on the wire is *zero* (because the Lorentz-contracted distance between the moving ⊖ charges is the same as the distance between the ⊕ charges at rest).

(4) **The force on the electron** e **is entirely magnetic**.

(**a**)

Conditions

(1) The ⊖ charges are stationary.

(2) The ⊕ charges are moving to the left with a speed v. Therefore the separation of the ⊕ charges is Lorentz contracted.

(3) The net linear charge density on the wire is *positive* (because the distance between the ⊖ charges at rest is greater than the Lorentz-contracted distance between the moving ⊕ charges).

(4) **The force on the electron** e **is entirely electrostatic.**

(**b**)

Figure 28–1
Whether the force on the electron is magnetic, electrostatic, or a combination of both depends on the frame of reference. In FRAME S, the electron e moves to the right with a speed v, and the wire is stationary. *The force is entirely magnetic.* In FRAME S', the wire moves to the left with a speed v and the charge is stationary. *The force is entirely electrostatic.*

factor $\sqrt{1 - v^2/c^2}$. However, the contracted separation is just equal to the proper separation of the positive ions *because the wire has no net charge*. Viewed from frame S' (which is moving with the charge e), the situation is quite different: The electrons in the wire are at rest and thus are more widely separated than there were in frame S. At the same time, the positive ions now appear closer together by the Lorentz contraction factor $\sqrt{1 - v^2/c^2}$. The net effect is that the *wire now has a net positive charge*. Therefore, the electron e viewed in S' is attracted to the wire by an *electrostatic* force. A detailed analysis shows that *the magnetic force viewed in the* S *frame is exactly equivalent to the electrostatic force viewed in the* S′ *frame*.

The validity of this analysis is based on the supposition that the magnitude of the electronic charge does not vary with relative motion between a charge and the observer. A variety of experiments indicates that this is true. For example, when a block of metal is heated, the thermal motion of the electrons increases much more than that of the positive ions. Yet the net charge on the block does not change.

A *historical note:* This discussion shows that electric and magnetic fields are viewed differently in frames of reference that have relative motion. Actually, this was the main subject of Einstein's original paper on special relativity, as evidenced by the title, "On the Electrodynamics of Moving Bodies," *Annalen der Physiks*, 17, 1905. All of the startling ideas about space and time for which special relativity is famous emerged, in fact, rather unexpectedly from delving into a question about charged objects in motion.

TABLE 28–1

The Laws of Electricity and Magnetism

Law	Phenomenon	Equation
Coulomb's law	The electrostatic force between charges	$\mathbf{F} = \left(\frac{1}{4\pi\varepsilon_0}\right)\frac{q_1 q_2}{r^2}\,\hat{\mathbf{r}}$
Gauss's law	A mathematical consequence of the inverse-square form of Coulomb's law	$\oint \mathbf{E} \cdot d\mathbf{A} = \frac{q}{\varepsilon_0}$
Lorentz force law	The magnetic force on a moving charge (the definition of a magnetic field)	$\mathbf{F} = q\mathbf{v} \times \mathbf{B}$
	The electric force on a stationary charge (the definition of an electric field)	$\mathbf{F} = q\mathbf{E}$
Biot-Savart law	The magnetic field of a current-carrying conductor	$d\mathbf{B} = \left(\frac{\mu_0}{4\pi}\right)\frac{I\,d\boldsymbol{\ell} \times \hat{\mathbf{r}}}{r^2}$
Ampère's law	A mathematical consequence of the Biot-Savart law	$\oint \mathbf{B} \cdot d\boldsymbol{\ell} = \mu_0 I$
Faraday's law	An electric field produced by a changing magnetic flux	$\oint \mathbf{E} \cdot d\boldsymbol{\ell} = -\frac{d\Phi_B}{dt}$

28.1 Maxwell's Equations

Certainly one of the greatest achievements in the physics of the nineteenth century was the complete unification of the laws of electricity and magnetism. James Clerk Maxwell[2] drew together the great discoveries of Coulomb, Faraday, Oersted, Ampère, and others into a single theory of electromagnetism.

The work of Maxwell has a profound effect on the philosophical foundations of physics. The laws of physics began to assume a unity that was not apparent previously. Maxwell established the basis for Einstein's work in relativity and showed the feasibility of communication by electromagnetic waves (soon achieved experimentally by Hertz, and later exploited by Marconi).

We begin the development of Maxwell's equations by summarizing the laws of electricity and magnetism. Table 28–1 is a list of the important laws we have thus far discussed. Faraday's law is the only one that involves both a magnetic field and an electric field without reference to current or charges. The implication of Faraday's law is that, even in empty space, an electric field may

[2] James Clerk Maxwell (1831–1879) was born in Edinburgh in the year Faraday discovered electromagnetic induction. Maxwell's theory of electromagnetism rivals Newton's laws of mechanics for their elegance and wide applicability. In spite of their brevity, Maxwell's four equations include all that is known concerning macroscopic effects of electricity, magnetism, and electromagnetic waves (light, radio waves, and so on). True, on an atomic scale, quantum mechanics and relativity must be introduced. But these modern theories were purposely developed so as to reduce to the classical expressions of Maxwell and Newton in the limit of low velocities and macroscopic dimensions. Maxwell also made contributions in numerous other fields of physics before his death at age 48.

arise without the presence of charges; only a *varying* magnetic flux is necessary. If, by Faraday's law, a changing magnetic flux produces an electric field, can a changing electric flux produce a magnetic field? The answer, which is yes, may be demonstrated in the following way.

Consider the situation illustrated in Figure 28–2. The capacitor in the circuit is a parallel-plate capacitor in a vacuum, with plate area A and plate separation d. The capacitance is given by

$$C = \varepsilon_0 \frac{A}{d} \tag{28–1}$$

After the switch S is closed, current begins to flow, charging the capacitor. Ampère's law describes the resulting magnetic field around the current-carrying conductor. Experiment indicates that a magnetic field is also produced in the space between the plates of the capacitor, as though it, too, were a conductor. Yet this space is a complete vacuum: There are no charges anywhere! From this experiment, we conclude that Ampère's law, as stated in Table 28–1, is incomplete. A magnetic field *can* exist in the absence of moving charges. Indeed, *the changing electric field between the plates of the capacitor can assume the role of current in Ampère's law.*

To see how this arises, consider Figure 28–2. The instantaneous potential difference v_c across the capacitor is given by

$$v_c = \frac{q}{C} \tag{28–2}$$

where q is the instantaneous charge on the capacitor. The electric field resulting from this charge is

$$E = \frac{v_c}{d} \tag{28–3}$$

Substituting for v_c from Equation (28–2):

$$E = \frac{q}{Cd}$$

and substituting for C from Equation (28–1), we solve for q:

$$q = \varepsilon_0 AE \tag{28–4}$$

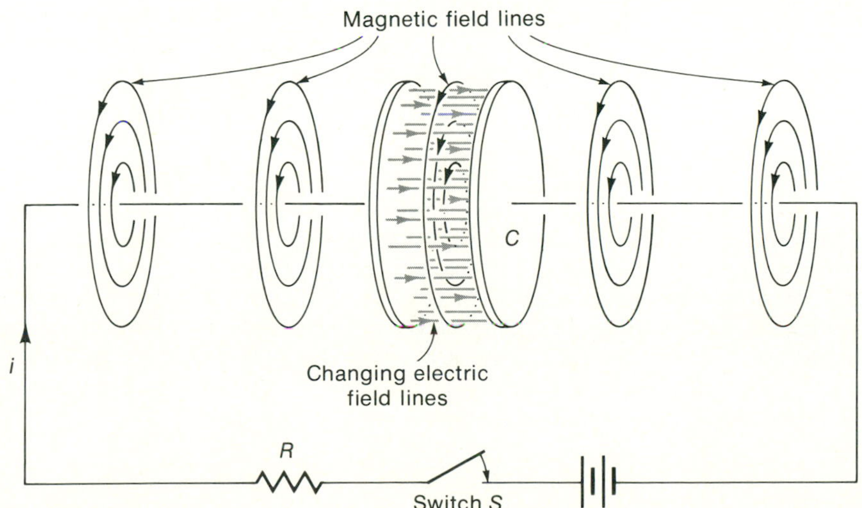

Figure 28–2
After the switch is closed, the current begins to flow, charging the capacitor. The current i produces a magnetic field around the wire *and also between the capacitor plates.*

The current i_d corresponding to the *changing* charge on the capacitor is

$$i_d = \frac{dq}{dt}$$

By Equation (28–4):

$$i_d = \varepsilon_0 A \frac{dE}{dt}$$

Since A is constant, we use the relation $E = \Phi_E/A$ to obtain:

DISPLACEMENT CURRENT i_d

$$i_d = \varepsilon_0 \frac{d\Phi_E}{dt} \tag{28–5}$$

where Φ_E is the **electric field flux** between the plates of the capacitor. Ampère's law should thus be modified to include displacement current.

NEW FORM OF AMPÈRE'S LAW (MODIFIED BY MAXWELL)

$$\oint \mathbf{B} \cdot d\ell = \mu_0\left(I + \varepsilon_0 \frac{d\Phi_E}{dt}\right) \tag{28–6}$$

In the process of modifying Ampère's law, we used the specific case of a parallel-plate capacitor. Nonetheless, the result is perfectly general. Displacement current occurs whenever the electric field changes, even in free space where there are no actual charges flowing.

EXAMPLE 28–1

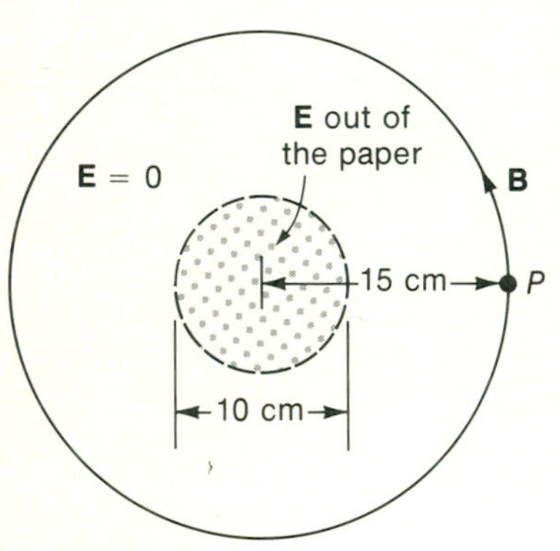

Figure 28–3
Example 28–1

Consider the situation illustrated in Figure 28–3. An electric field of 300 V/m is confined to a circular area 10 cm in diameter and directed outward from the plane of the figure. If the field is increasing at a rate of 20 V/m·s, what is the direction and magnitude of the magnetic field at the point P, 15 cm from the center of the circle?

SOLUTION

We use the modified form of Ampère's law, Equation (28–6):

$$\oint \mathbf{B} \cdot d\ell = \mu_0\left(I + \varepsilon_0 \frac{d\Phi_E}{dt}\right)$$

Since no moving charges are present:

$$I = 0$$

Then:

$$\oint \mathbf{B} \cdot d\ell = \mu_0 \varepsilon_0 \frac{d\Phi_E}{dt} \tag{28–7}$$

In order to evaluate the integral, we make use of the symmetry of the situation. Symmetry requires that no particular direction from the center can be any different from any other direction. Therefore, there must be *circular symmetry* about the central axis. From the experiment of Figure 28–2, we know the magnetic field lines are circles about the axis. Therefore, as we travel around such a magnetic field circle, the magnetic field remains constant in magnitude. Setting aside until later the determination of the *direction* of **B**, we integrate $\int \mathbf{B} \cdot d\ell$ around the circle at $R = 0.15$ m to obtain:

$$\oint \mathbf{B} \cdot d\ell = 2\pi R B$$

Differentiating the expression $\Phi_E = AE$, we have

$$\frac{d\Phi_E}{dt} = \left(\frac{\pi d^2}{4}\right)\frac{dE}{dt}$$

Equation (28–7) becomes

$$2\pi RB = \mu_0\varepsilon_0\left(\frac{\pi d^2}{4}\right)\frac{dE}{dt}$$

and

$$B = \frac{\mu_0\varepsilon_0}{2\pi R}\left(\frac{\pi d^2}{4}\right)\frac{dE}{dt}$$

Substituting the numerical values:

$$B = \frac{\left(4\pi \times 10^{-7}\,\frac{\text{H}}{\text{m}}\right)\left(8.85 \times 10^{-12}\,\frac{\text{F}}{\text{m}}\right)(\pi)(0.10\text{ m})^2\left(20\,\frac{\text{V}}{\text{m}\cdot\text{s}}\right)}{(2\pi)(0.15\text{ m})(4)}$$

$$B = \boxed{1.85 \times 10^{-18}\text{ T}}$$

Equation (28–6) determines the field direction, because it states that an *increasing* electric field produces a magnetic field in the same manner as a current I. In Figure 28–3, the direction of the *increase* of electric field is out of the plane of the paper. By the right-hand rule, this implies that the direction of **B** is *counterclockwise.* Note that the magnitude of the electric field is irrelevant; only the *change* in the electric field determines the magnetic field.

Table 28–1 has a notable omission. It does not include the fact that to our knowledge magnetic monopoles do not exist. The concept of a magnetic *monopole* has its origin in comparing a magnetic dipole with an electric dipole. Since electric dipoles are made up of two distinct electric charges, $+q$ and $-q$, it seems tempting to visualize a magnetic dipole similarly as a pair of *monopoles*, $+p$ and $-p$. The north pole of a magnet would contain a $+p$ monopole (from which field lines would emanate) and the south pole a $-p$ monopole (toward which field lines would converge). Of course, this way of thinking about it is contrary to the model of a magnetic dipole, which consists of a *current loop.* For a current loop, it seems inconceivable that a monopole could exist by itself: the current loop inherently generates both "poles" together, so that a magnetic *di*pole is the most fundamental magnetic structure. In any event, monopoles have not yet been experimentally found in nature. Breaking a long bar magnet in half produces two separate *dipoles* (not monopoles). Presumably the fragmenting process could be continued until just a single atom was left, with its inherent "loop current" and electron spin, also creating a dipole. Needless to say, if a monopole were discovered in nature, it would precipitate a profound change in electromagnetic theory.

Despite this disclaimer, it will be helpful to use the concept of a monopole for a short discussion. Figure 28–4 shows a comparison between charges and magnetic poles. Figure 28–4(a) demonstrates Gauss's law for electric fields

GAUSS'S LAW FOR ELECTRIC FIELDS

$$\oint \mathbf{E}\cdot d\mathbf{A} = \frac{q}{\varepsilon_0} \qquad \textbf{(28–8)}$$

where the total electric flux emanating from the surface S is not zero. Figure 28–4(b) shows the north-seeking pole of a long magnet surrounded by a surface

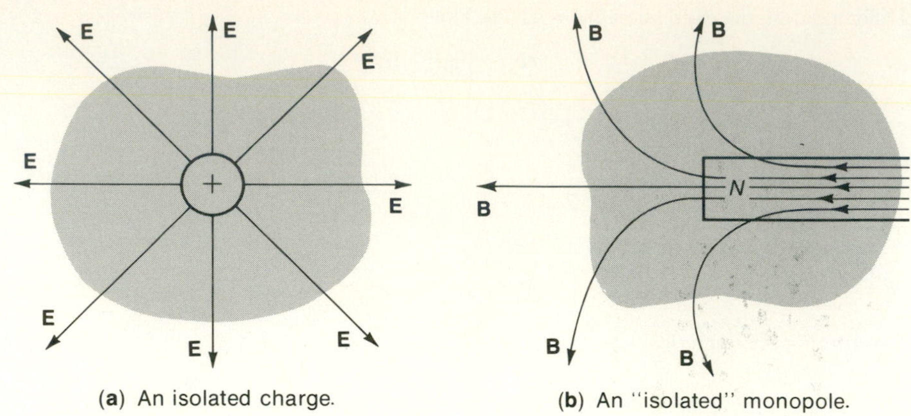

(a) An isolated charge.

(b) An "isolated" monopole.

Figure 28–4
The net electric field flux through a closed surface may not be zero, but the net magnetic field flux through a closed surface is always zero.

S. Since isolated magnetic monopoles do not exist, the north pole of the magnet is always paired with a south pole. Thus, the magnetic flux emanating from any closed surface must equal that entering from the paired south pole. This fact may be formulated as

GAUSS'S LAW FOR MAGNETIC FIELDS

$$\oint \mathbf{B} \cdot d\mathbf{A} = 0 \tag{28–9}$$

Table 28–2 is a revised version of Table 28–1. Coulomb's law and the Biot-Savart law have been deleted because they are represented, respectively, by Gauss's law for electric fields and Ampère's law. The Lorentz force law has been deleted because it is essentially a definition of electric and magnetic fields. Ampère's law has been extended to include magnetic fields arising from changing electric field flux, and Gauss's law for magnetic fields has been added. The resulting Table 28–2 is a collection of equations known as **Maxwell's electromagnetic field equations.**[3] The presence of a dielectric material may be incorporated by simply replacing ε_0, the permittivity of free space (a vacuum),

[3] Maxwell's equations are often expressed as *differentials.* But the use of the differential form leads to mathematical procedures best postponed to a more advanced course in electromagnetism.

TABLE 28–2

Maxwell's Equations

Gauss's law for electric fields	$\oint \mathbf{E} \cdot d\mathbf{A} = \dfrac{q}{\varepsilon_0}$
Gauss's law for magnetic fields	$\oint \mathbf{B} \cdot d\mathbf{A} = 0$
Ampère's law (extended form)	$\oint \mathbf{B} \cdot d\boldsymbol{\ell} = \mu_0\left(I + \varepsilon_0 \dfrac{d\Phi_E}{dt}\right)$
Faraday's law	$\oint \mathbf{E} \cdot d\boldsymbol{\ell} = -\dfrac{d\Phi_B}{dt}$

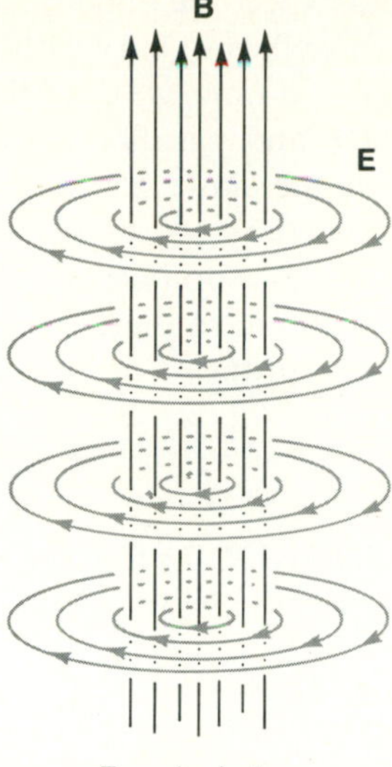

Faraday's law

$$\oint \mathbf{E} \cdot d\boldsymbol{\ell} = -\frac{d\Phi_B}{dt}$$

(**a**) If Φ_B increases uniformly, a constant **E** field is generated.

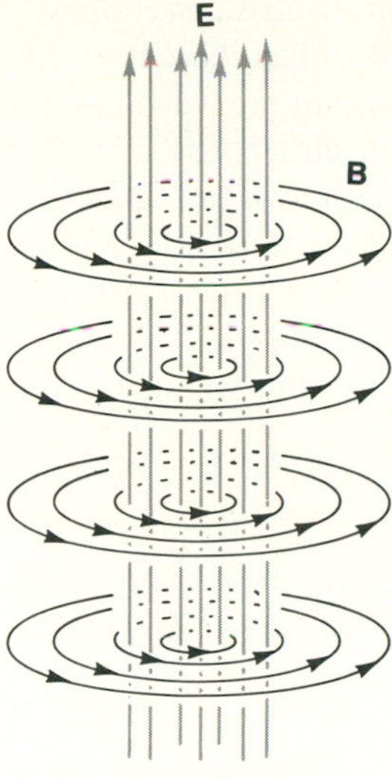

Ampere's law (as extended by Maxwell)

$$\oint \mathbf{B} \cdot d\boldsymbol{\ell} = \mu_0\varepsilon_0\frac{d\Phi_E}{dt}$$

(for current-free regions)

(**b**) If Φ_E increases uniformly, a constant **B** field is generated.

Figure 28–5
The symmetry of **E** and **B** in the absence of moving charges. Note that the symmetry is not quite exact, however, since the circular fields are in opposite directions. (One equation has a minus sign.)

with ε, the permittivity of the dielectric material. Similarly, for magnetic materials, μ_0, the permeability of free space, is replaced by μ, the permeability of the magnetic material.

28.2 Electromagnetic Waves

The pinnacle of Maxwell's achievements was the formulation of the theory of electromagnetic radiation. One of the consequences of Maxwell's equations is that a time-varying combination of electric and magnetic fields propagates through completely empty space—a vacuum—with the speed of light, c.

It takes some mathematical manipulation to start with Maxwell's equations and derive electromagnetic waves. So we will not present the complete, step-by-step story. However, if we start with a simple combination of **E** and **B** fields, we will show that electromagnetic *waves* follow and that the equations for these waves do agree with Maxwell's equations.

The starting point is a combination of "crossed" **E** and **B** fields[4] in a vacuum, as shown in Figure 28–6(a). The fields are *uniform*, extending without change to plus or minus infinity in the y-z plane. (We have sketched only a small segment of the field pattern.) As we will demonstrate in a later chapter, it is easy to verify experimentally that a traveling wave has **E** and **B** fields *perpendicular* to the direction of propagation. Therefore, we suggest that this particular distribution is applicable to *plane wave propagation along the x axis.*

---

[4] This arrangement of crossed fields is a simpler version of the fields associated with displacement current: If you examine the space between the capacitor plates of Figure 28–2, you will see that E and B are always at right angles.

For a wave traveling along the x axis, we would expect the magnitudes of **E** and **B** to be different at different points along x, as well as to vary with time. To ferret out the ways these fields vary in both space and time, we examine the fields on either side of a thin slab of space Δx thick and parallel to the y-z plane, as shown in Figure 28–6(b). Both E_y and B_z on the plane at x may differ from the corresponding fields on the plane at $x + \Delta x$.

We now apply Ampère's law, Equation (28–6), to the top face of the slab [Figure 28–6(c)]. There are no actual charges in a vacuum, so there can be no current I. Thus we have only the displacement-current term.

$$\oint \mathbf{B} \cdot d\ell = \mu_0 \varepsilon_0 \frac{d\Phi_E}{dt} \tag{28–10}$$

The edges of the slab are L. Beginning at the corner marked P, the four segments of the closed-path integration give

$$\oint \mathbf{B} \cdot d\ell = \underbrace{B_z(x)L}_{①} + \underbrace{0}_{②} - \underbrace{B_z(x + \Delta x)L}_{③} + \underbrace{0}_{④} \tag{28–11}$$

For paths ② and ④, **B** is perpendicular to $d\ell$, so the dot product is zero for these segments.

The right-hand side of Equation (28–10) may be written in terms of **E** using the fact that $\Phi_E = AE_y$, and therefore $d\Phi_E/dt = A\,dE_y/dt$. The area A enclosed by the path is $L\,\Delta x$. Since E_y varies in both space and time, we use *partial* derivative symbols[5] to indicate that all other variables are to be held constant as we take the derivative indicated. Thus the right-hand side is

$$\mu_0 \varepsilon_0 \frac{\partial \Phi_E}{\partial t} = \mu_0 \varepsilon_0 L\,\Delta x \frac{\partial E_y}{\partial t} \tag{28–12}$$

Combining the previous two equations, we have

$$B_z(x)L - B_z(x + \Delta x)L = \mu_0 \varepsilon_0 L\,\Delta x \frac{\partial E_y}{\partial t} \tag{28–13}$$

Canceling L from both sides and allowing the thickness of the slab to become infinitesimally small:

$$\lim_{\Delta x \to 0} \frac{B_z(x + \Delta x) - B_z(x)}{\Delta x} = -\mu_0 \varepsilon_0 \frac{\partial E_y}{\partial t} \tag{28–14}$$

The left-hand side is just the definition of the derivative dB_z/dx, so

$$\frac{\partial B_z}{\partial x} = -\mu_0 \varepsilon_0 \frac{\partial E_y}{\partial t} \tag{28–15}$$

(Again, the *partial* derivative $\partial B_z/\partial x$ acknowledges that B_z may also vary in time.) In a similar fashion, we apply Faraday's law

$$\oint \mathbf{E} \cdot d\ell = -\frac{d\Phi_B}{dt} \tag{28–16}$$

to the face perpendicular to the z direction and obtain

$$\frac{\partial E_y}{\partial x} = -\frac{\partial B_z}{\partial t} \tag{28–17}$$

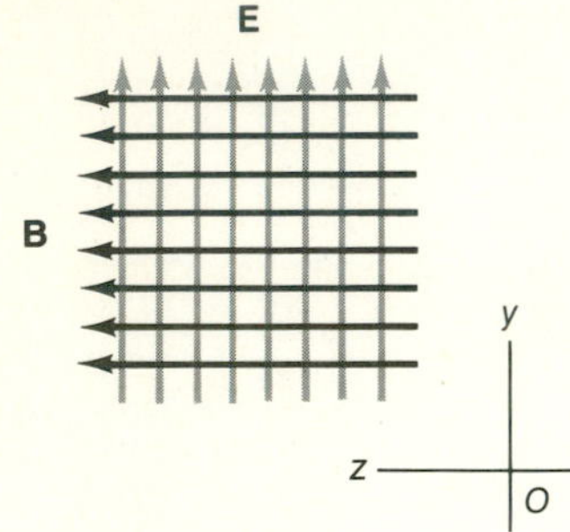

(a) The **E** and **B** fields are at right angles. (They extend like this, uniformly to infinity, all over the y-z plane.)

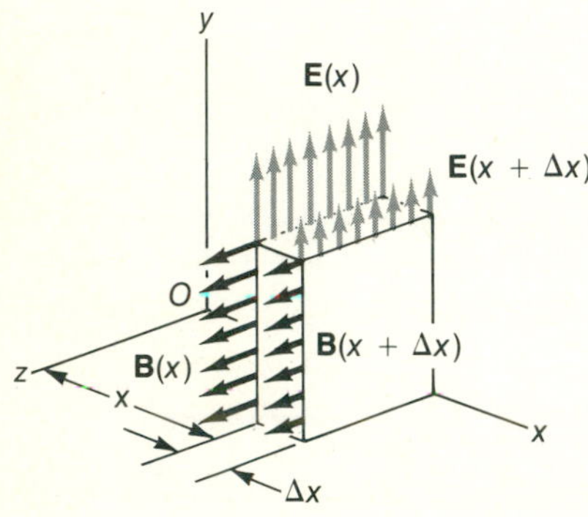

(b) The **E** and **B** fields at two different y-z planes, spaced a distance Δx apart. The fields at x may differ from the fields at $x + \Delta x$.

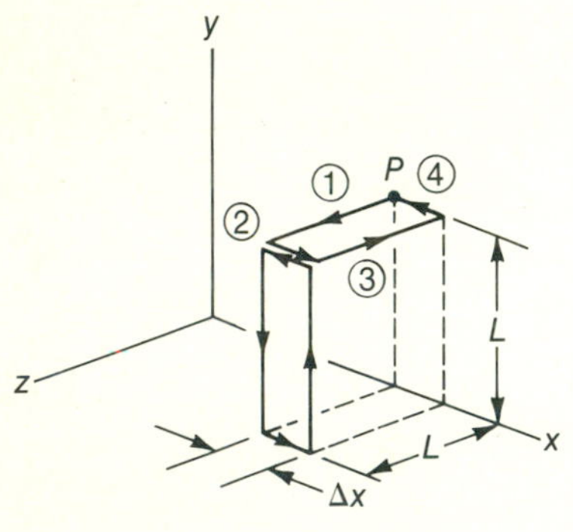

(c) The paths of integration along the edges of the slab depicted in (**b**).

Figure 28–6
A plane electromagnetic wave traveling in the $+x$ direction has this pattern of "crossed" **E** and **B** fields.

[5] For a comment on partial derivatives, see Chapter 16, page 386.

Equations (28–15) and (28–17) are now solved simultaneously to obtain two equations: one involving only the electric field E_y and the other involving only the magnetic field B_z. The procedure is not difficult. We first differentiate both sides of Equation (28–15) with respect to x and obtain

$$\frac{\partial^2 B_z}{\partial x^2} = -\mu_0 \varepsilon_0 \frac{\partial^2 E_y}{\partial t\, \partial x} \tag{28–18}$$

We next differentiate Equation (28–17) with respect to t and obtain

$$\frac{\partial^2 E_y}{\partial x\, \partial t} = -\frac{\partial^2 B_z}{\partial t^2} \tag{28–19}$$

Substituting this value for the mixed derivative $\partial^2 B_z/\partial x\, \partial t$ into Equation (28–18), we obtain an expression involving B alone:

WAVE EQUATION FOR B_z

$$\frac{\partial^2 B_z}{\partial x^2} = \mu_0 \varepsilon_0 \frac{\partial^2 B_z}{\partial t^2} \tag{28–20}$$

By a similar process, Equations (28–15) and (28–17) may be combined to obtain an expression involving E alone:

WAVE EQUATION FOR E_y

$$\frac{\partial^2 E_y}{\partial x^2} = \mu_0 \varepsilon_0 \frac{\partial^2 E_y}{\partial t^2} \tag{28–21}$$

The previous two equations have the same form as the *wave equation* we developed in Chapter 16 [Equation (16–8)] for the propagation of transverse waves on a rope. A solution to the wave equation is

$$A = A_0 \sin (kx - \omega t) \tag{28–22}$$

where A_0 = amplitude of the wave

$$k = \frac{2\pi}{\lambda} \qquad (\lambda = \text{wavelength})$$

$$\omega = \frac{2\pi}{T} \qquad \left(T = \text{period} = \frac{1}{f}\right)$$

$$\frac{\omega}{k} = v \qquad (\text{speed of propagation of the wave} = \lambda f)$$

The electric field E_y thus varies in space and time according to

$$E_y = E_{y0} \sin (kx - \omega t) \tag{28–23}$$

A graphical representation of E is shown in Figure 28–7.

Evaluating the derivatives $\partial^2 E_y/\partial x^2$ and $\partial^2 E_y/\partial t^2$, and substituting into Equation (28–21), we obtain

$$\frac{\omega}{k} = \frac{1}{\sqrt{\mu_0 \varepsilon_0}} \tag{28–24}$$

Since $\omega/k = v$, the variation in the electric field propagates with a speed $(\mu_0 \varepsilon_0)^{-\frac{1}{2}}$ in a direction perpendicular to the orientation of $\mathbf{E}$.

Because the magnetic field B_z satisfies an equation identical to that of E_y, we also have:

$$B_z = B_{z0} \sin (kx - \omega t) \tag{28–25}$$

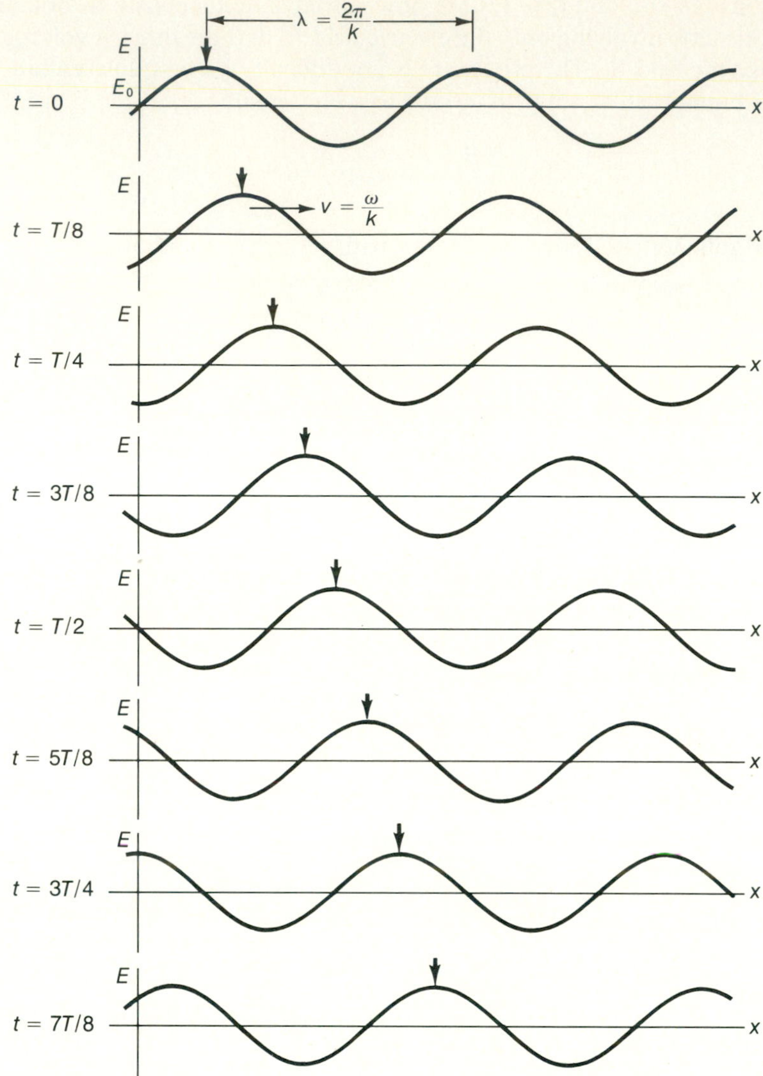

Figure 28–7
A series of "snapshots," taken at intervals of $T/8$ seconds (where $T = 2\pi/\omega$), showing the electric field variation moving in the $+x$ direction with a speed v.

where, again:

$$\frac{\omega}{k} = \frac{1}{\sqrt{\mu_0\varepsilon_0}}$$

In order to emphasize the implications of our wave-equation development, let us enumerate them:

(1) **The E and B fields are perpendicular to each other.** This was assumed initially but proved to be entirely consistent with Maxwell's equations (and, indeed, can be derived from Maxwell's equations).

(2) **The E and B fields are transverse waves (perpendicular to the direction of propagation)** and are in phase with each other. This is ensured by the identical form of the wave equations for **E** and **B** (see Figure 28–8a). The sine-wave curves in (a) and (b) represent the magnitudes of the **E** and **B** fields. The diagram should *not* be interpreted as vibrations of something like a string or water waves. Instead, the sine curves are the *envelope* of the tips of the field vectors, where the *length of the vector represents the*

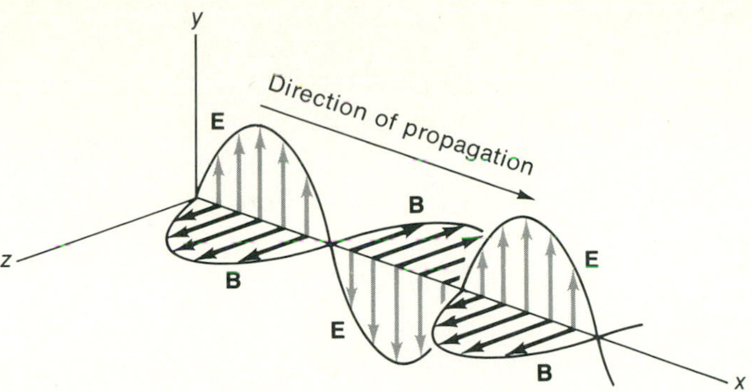

(a) A "snapshot" of the spatial variation of a plane electromagnetic wave moving in the $+x$ direction. The *length of the vectors* corresponds to the *field strength*. The pattern moves along the $+x$ direction with a speed c.

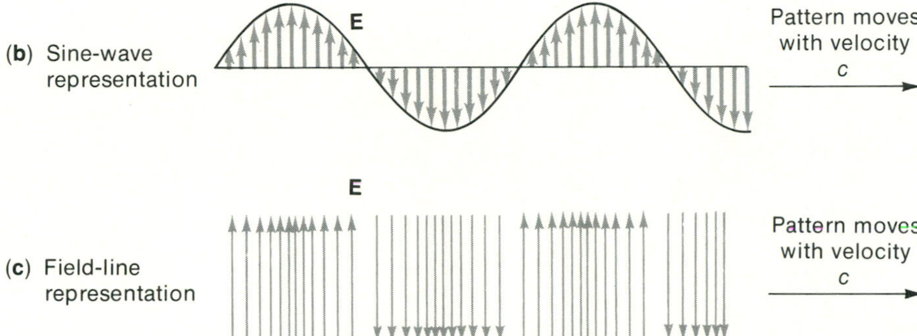

(b) and (c) In the sine-wave representation (b) for the electric field, the vectors themselves are often omitted. This curve implies the field lines are crowded together where the field is stronger and farther apart where the field is weaker, as shown in (c). Since it is a *plane* wave (that is, uniform over the y-z plane), *the field lines should be mentally extended to infinity in the $\pm y$ direction, and the pattern should be duplicated in and out of the paper to fill all space in the $\pm z$ direction.* The *wavefronts* are planes in the y-z direction; they move, of course, in the $+x$ direction with the speed c.

Figure 28–8
Representations of a plane electromagnetic wave.

strength of the field. Another representation (c) of the spatial distribution makes use of the convention that the *density* of field lines corresponds to the field strength. Although the sketch in (c) is more cumbersome to draw, perhaps it gives the best impression of the actual field distribution in a plane wave. Careful study of Figure 28–8 will help you avoid misconceptions regarding the nature of plane waves.

(3) **The speed of propagation of the wave is $(\mu_0\varepsilon_0)^{-1/2}$**, a term whose numerical value constitutes one of the most remarkable aspects of the electromagnetic wave. Before evaluating this combination of μ_0 and ε_0, let us review the origin of these constants.

The constant ε_0 appears in Coulomb's law:

$$F = \left(\frac{1}{4\pi\varepsilon_0}\right)\frac{q_1 q_2}{r^2} \qquad \textbf{(28–26)}$$

In the modern definition, the quantity of charge q is determined in terms of the ampere. The ampere I (a coulomb per second) is defined in terms of the force per unit length F/ℓ between parallel current-carrying conductors a distance d apart:

$$\frac{F}{\ell} = \frac{\mu_0 I^2}{2\pi d} \tag{28–27}$$

On the other hand, μ_0 is an *assigned* number that fixes the value of F/ℓ in Equation (28–27) to be *exactly* $2\pi \times 10^{-7}$ N/m for $d = 1$ m. Thus:

$$\varepsilon_0 = 8.8542 \times 10^{-12} \frac{\text{C}^2}{\text{N}\cdot\text{m}^2} \qquad \text{(experimentally determined)}$$

$$\mu_0 = 4\pi \times 10^{-7} \frac{\text{N}\cdot\text{s}^2}{\text{C}^2} \qquad \text{(defined exactly)}$$

Substituting these values into the expression for the speed of propagation, we obtain

$$(\mu_0\varepsilon_0)^{-\frac{1}{2}} = \left[\left(4\pi \times 10^{-7} \frac{\cancel{\text{N}}\cdot\text{s}^2}{\cancel{\text{C}^2}}\right)\left(8.85 \times 10^{-12} \frac{\cancel{\text{C}^2}}{\cancel{\text{N}}\cdot\text{m}^2}\right)\right]^{-\frac{1}{2}}$$

$$(\mu_0\varepsilon_0)^{-\frac{1}{2}} = 3.00 \times 10^8 \, \frac{\text{m}}{\text{s}}$$

which is *the speed of light in a vacuum.* The fact that the speed of propagation of an electromagnetic wave appeared to be the speed of light led to the realization that light *is* electromagnetic in nature. Therefore[6]

SPEED OF LIGHT c

$$c = \frac{1}{\sqrt{\mu_0\varepsilon_0}} \tag{28–28}$$

Thus it became accepted that light is an electromagnetic wave, making up just a small portion of the electromagnetic spectrum, which includes radio waves, x-rays, and so on—all described by Maxwell's four equations.

From Equation (28–28), the speed of light can be determined by purely electrical methods (rather than by a direct velocity measurement). For example, consider a parallel-plate capacitor charged by a battery. By measuring both the force between the wires leading to the capacitor and the force between the plates of the charged capacitor, the value of c can be experimentally determined. In 1906, E.B. Rosa and N.E. Dorsey of the National Bureau of Standards performed a beautifully precise experiment that determined the speed of light by electrical methods. It was the most accurate determination of c at that time. The value they obtained was $c = 299\,784 \pm 15$ km/s, which is in agreement with the current best value (1980):

SPEED OF LIGHT c

$$c = 299\,792\,458 \pm 1.2 \text{ m/s} \tag{28–29}$$

(4) **The magnitudes of E and B are related.** Solutions of the wave equation are Equations (28–23) and (28–25):

$$E_y = E_{y0} \sin(kx - \omega t)$$

and

$$B_z = B_{z0} \sin(kx - \omega t)$$

[6] Since the speed of light is easier to determine experimentally than ε_0, the value of ε_0 is now defined in terms of the experimentally determined speed of light: $\varepsilon_0 \equiv 1/\mu_0 c^2$.

These solutions must satisfy Equation (28–17):

$$\frac{\partial E_y}{\partial x} = -\frac{\partial B_z}{\partial t}$$

obtained in the first part of the wave-equation development. Evaluating the derivatives indicated in Equation (28–17), we have

$$kE_{y0} \sin (kx - \omega t) = \omega B_{z0} \sin (kx - \omega t)$$

or

$$kE_y = \omega B_z$$

$$\frac{E_y}{B_z} = \frac{\omega}{k}$$

But since by Equation (28–24):

$$\frac{\omega}{k} = \frac{1}{\sqrt{\mu_0 \varepsilon_0}}$$

we have

RELATION BETWEEN E_y AND B_z IN ELECTROMAGNETIC WAVES

$$\frac{E_y}{B_z} = c \qquad (28\text{–}30)$$

EXAMPLE 28–2

The electric field in an electromagnetic wave is described by the equation

$$E_y = 100 \sin (10^7 x - \omega t) \qquad \text{(in SI units)}$$

Find (a) the amplitude of the corresponding magnetic wave, (b) the wavelength λ, and (c) the value of ω.

SOLUTION

(a) The magnitudes of **E** and **B** are related by Equation (28–30):

$$\frac{E_y}{B_z} = c$$

Solving for B_z and substituting numerical values:

$$B_z = \frac{E_y}{c}$$

$$B_z = \frac{100 \dfrac{\text{V}}{\text{m}}}{3 \times 10^8 \dfrac{\text{m}}{\text{s}}} = \boxed{3.33 \times 10^{-7}\ \text{T}}$$

(b) To find the wavelength λ and the angular frequency $\omega = 2\pi f$, we recall the following relations from Chapter 16:

$$\lambda = \frac{2\pi}{k} \qquad (28\text{–}31)$$

$$f = \frac{\omega}{2\pi} \qquad (28\text{–}32)$$

and

$$v = \lambda f \qquad (28\text{–}33)$$

where λ is the wavelength, f is the frequency, and v is the speed of the wave. From the given data, $k = 10^7\ \text{m}^{-1}$. Therefore:

$$\lambda = \frac{2\pi}{k}$$

$$\lambda = \frac{2\pi}{10^7\ \text{m}^{-1}} = 6.28 \times 10^{-7}\ \text{m}$$

From Figure 28–9, this value corresponds to the *visible* part of the electromagnetic spectrum, whose wavelengths are usually expressed in *nanometers* (nm), where 1 nm $\equiv 10^{-9}$ m. Thus:

$$\lambda = 6.28 \times 10^{-7}\ \text{m}\underbrace{\left(\frac{10^9\ \text{nm}}{1\ \text{m}}\right)}_{\text{Conversion ratio}} = \boxed{628\ \text{nm}}$$

The color of this light is red-orange.

(c) To find the value of ω, we use

$$\frac{\omega}{k} = v \qquad (=c)$$

Rearranging and substituting:

$$\omega = ck$$

$$\omega = \left(3 \times 10^8\ \frac{\text{m}}{\text{s}}\right)(10^7\ \text{m}^{-1})$$

$$\omega = 3 \times 10^{15}\ \frac{\text{rad}}{\text{s}}\underbrace{\left(\frac{1\ \text{cycle}}{2\pi\ \text{rad}}\right)}_{\text{Conversion ratio}} = \boxed{4.77 \times 10^{14}\ \text{Hz}}$$

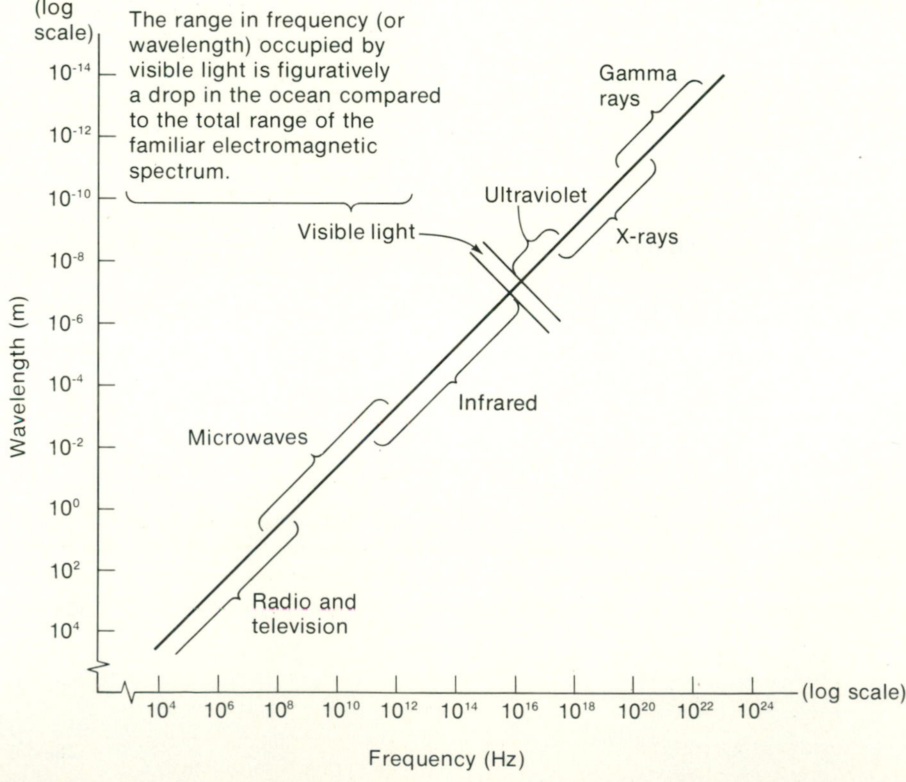

Figure 28–9
The electromagnetic spectrum. The slope of the line is the speed of light c. Overlapping regions imply two modes of production; for example, microwaves produced electrically are identical to infrared radiation of atomic origin.

28.3 Radiation

There are many ways of generating radiation. All of them rely on the phenomenon that *accelerated charges radiate electromagnetic waves.* Figure 28–10 explains the origin of this radiation. In Figure 28–10(c), the charge is originally at rest at O. At $t = 0$, it accelerates for a very short time Δt to acquire a speed $v = 0.2c$ at O'. From there, it travels at constant speed to reach point P at time t. Now the "kink" in the field lines introduced by the acceleration travels outward with the speed c. It is just this kink which carries the information that the charge has accelerated. For distances away from O larger than t/c, news of the sudden acceleration has not yet arrived, so the field lines farther away point to the original location O. For distances from O' smaller than t/c, the field lines center on the charge at its present location P [similar to the field pattern in Figure 28–10(b)]. Note an important feature of the kink: It contains a

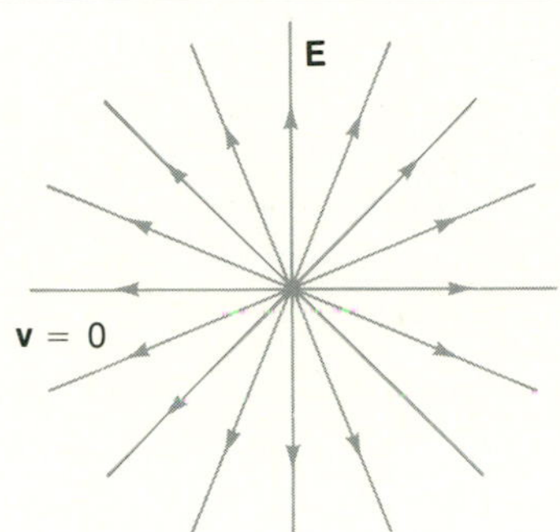

(**a**) Field lines for a charge at rest.

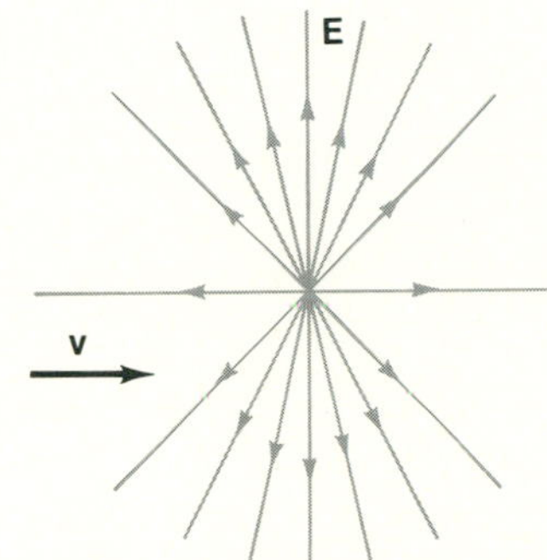

(**b**) Field lines for a charge moving with constant velocity. Because of relativity, the pattern of field lines is "squashed together" by the factor $\sqrt{1 - v^2/c^2}$ along the direction of motion. As a result, the field lines are not quite so close together along the direction of motion as in (**a**) and are closer together perpendicular to that direction.

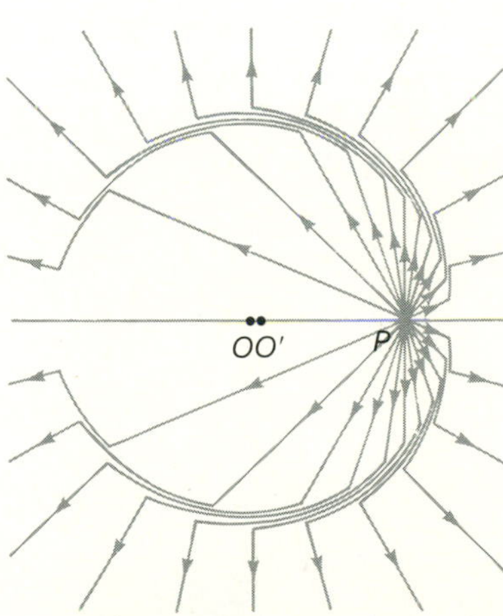

(**c**) Field lines for a charge that undergoes a very brief acceleration from rest from O to O' and then travels at constant speed to point P. The "kink" in the field lines produced by the acceleration travels outward with speed c.

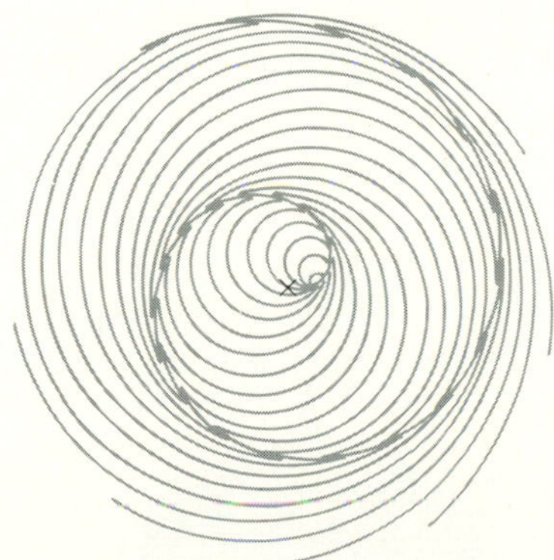

(**d**) Field lines around an isolated charge moving clockwise at constant speed $v = 0.9c$ in a circle centered on the ×. The kink in the spiral pattern travels outward with speed c.

Figure 28–10
Radiation from accelerated charges.

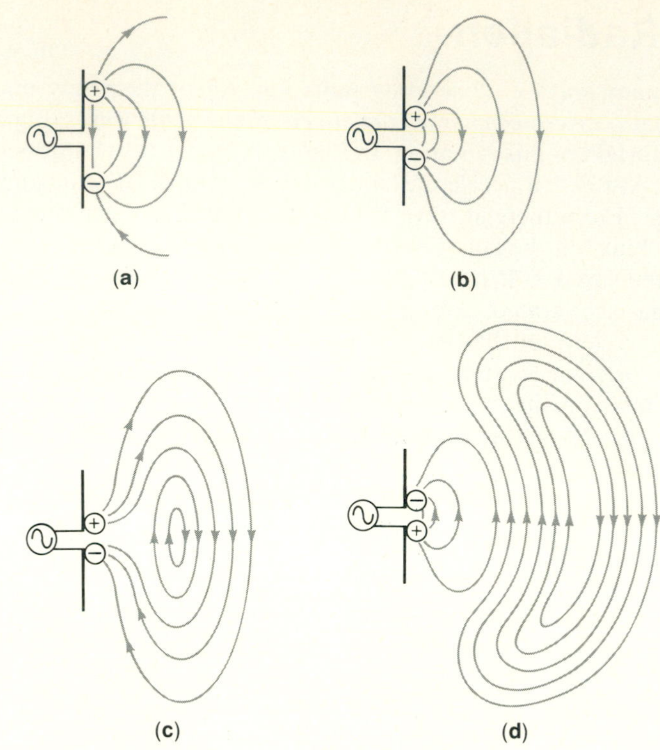

Figure 28–11
The generation of an electromagnetic wave by the accelerating charges in a dipole antenna. (Only the electric field is shown; the associated magnetic field is omitted for clarity.) The complete field pattern forms a figure-of-revolution about the axis of the dipole wires.

transverse component of the electric field. This is the origin of the transverse **E** field in the traveling wave.

A common example of radiation is the *dipole antenna* illustrated in Figure 28–11, composed of two wires that are connected to an *AC* voltage source. Electrons are accelerated first in one direction, and then in the other, making one wire positive and the other negative, and then vice versa. These oscillations produce a growing electric field pattern, as shown in Figure 28–11(a). At the instant when the potential reverses, there is no net charge on the dipole, so no field lines can terminate there. Consequently, the loops of electric field are "pinched off" and propagate away from the antenna with the speed c. At any given point in space, the electric field changes in time, so according to Maxwell's equations, there is also a changing magnetic field (not shown). At very large distances from the antenna, the waves become approximately *plane waves*, as described previously in Figure 28–8.

Figure 28–12
In three dimensions, dipole radiation in various directions (far from the dipole) may be depicted as a doughnut-shaped pattern, where the energy radiated along a particular direction is proportional to the length of a vector drawn from the center of the dipole to the surface of the figure. The radiation is a maximum at right angles to the dipole, with no radiation occurring along the axis of the dipole.

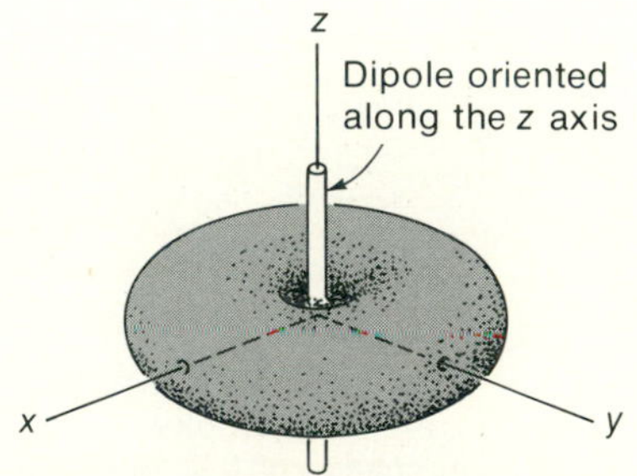

(a) Doughnut-shaped radiation pattern.

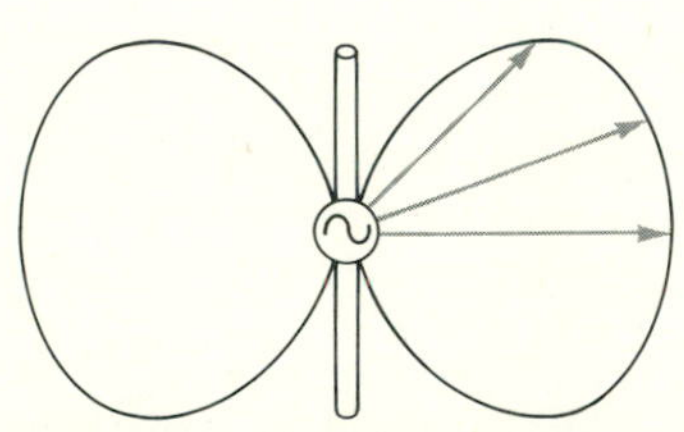

(b) Cross section of the radiation pattern.

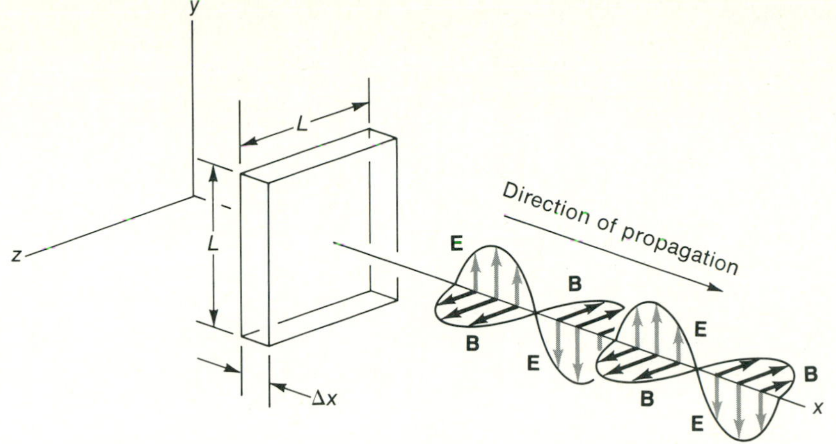

Figure 28–13
A plane wave carrying electromagnetic energy through the thin slab with a speed *c*. The electric field varies in the $\pm y$ direction and the magnetic field varies in the $\pm z$ direction.

28.4 Energy in Electromagnetic Waves

Electromagnetic waves from the sun bring to the earth about 174 trillion kilowatts of power striking the top of the earth's atmosphere. This inflow of energy undoubtedly was essential to the origin of life and to the storage of immense reserves of fossil fuels. It continues to be important in driving the earth's winds and ocean currents, in the evaporation of water to produce rain which replenishes fresh-water supplies, and in other energy-transfer processes that are so important in sustaining living systems.[7] The flow of energy to the earth appears to be in balance with enough energy radiated from the earth to maintain thermal equilibrium. Although living matter relies directly on only a few hundredths of one percent of this incoming radiant energy, life could not continue very long without this constant flow of energy from the sun.

In this section, we will explain how electromagnetic waves transport energy along the direction of propagation. As shown in previous chapters, the *energy density* of electric and magnetic fields is

ENERGY DENSITY IN FIELDS

$$\text{Electric:} \quad u_E = \tfrac{1}{2}\varepsilon_0 E^2 \qquad \textbf{(28–34)}$$

$$\text{Magnetic:} \quad u_B = \frac{1}{2}\left(\frac{B^2}{\mu_0}\right) \qquad \textbf{(28–35)}$$

To see how this energy is carried along by the wave, we apply these equations to a thin volume of space, as illustrated in Figure 28–13.

At a given instant, the volume contains a total energy ΔU that consists of electric field energy ΔU_E and magnetic field energy ΔU_B:

$$\Delta U = \Delta U_E + \Delta U_B \qquad \textbf{(28–36)}$$

Since the volume of the slab is $L^2\,\Delta x$, the energy ΔU in the slab is $L^2\,\Delta x(u_E + u_B)$. Using Equations (28–34) and (28–35), we have

$$\Delta U = \tfrac{1}{2}L^2\,\Delta x\left(\varepsilon_0 E_y{}^2 + \frac{1}{\mu_0}B_z{}^2\right) \qquad \textbf{(28–37)}$$

[7] See an interesting discussion by A.H. Oort, "The Energy Cycle of the Earth," *Scientific American 223*, 54, Sept. 1970. For an analysis of some of the problems of energy and man's interaction with the environment, see *Energy and the Environment*, John M. Fowler, McGraw-Hill, 1975. Also see *Energy*, G.M. Crawley, Macmillan, 1975.

But using $$c = (\mu_0 \varepsilon_0)^{-\frac{1}{2}}$$
and $$E_y = cB_z$$

Equation (28–37) may be written so that each term contains the product EB:

$$\Delta U = \tfrac{1}{2}L^2\,\Delta x\left(\varepsilon_0 c E_y B_z + \frac{1}{\mu_0 c}\,E_y B_z\right)$$

or

$$\Delta U = \tfrac{1}{2}L^2 E_y B_z\,\frac{\Delta x}{c}\left(\varepsilon_0 c^2 + \frac{1}{\mu_0}\right)$$

Since $c^2 = 1/\varepsilon_0 \mu_0$, we have

$$\Delta U = L^2 E_y B_z\,\frac{\Delta x}{c}\left(\frac{1}{\mu_0}\right) \tag{28–38}$$

The time Δt required for the energy ΔU to pass through the face of the volume is

$$\Delta t = \frac{\Delta x}{c}$$

Designating the amount of energy per unit time that flows through a unit area as S:

$$S = \frac{(\text{Energy})}{(\text{Area})(\text{Time})}$$

$$S = \frac{\Delta U}{L^2\,\Delta t}$$

and substituting the previous expressions, we have:

$$S = \frac{1}{\mu_0}\,E_y B_z \tag{28–39}$$

Because **E** and **B** are both vectors perpendicular to the direction of propagation (Figure 28–13), we know that **E X B** is *along the direction of propagation.* Therefore, we may write the previous equation as

THE POYNTING VECTOR[8] **(instantaneous value)**

$$\mathbf{S} = \frac{1}{\mu_0}\,\mathbf{E} \times \mathbf{B} \tag{28–40}$$

The vector **S** is called the **Poynting vector** in honor of its originator, John Henry Poynting (1852–1914). Its direction is that of the *energy flow per unit area per unit time*, measured in units of (J/s·m^2), or *power per unit area* (W/m^2).

The Poynting vector gives the *instantaneous* energy flow in terms of E and B. For the waves we consider, these quantities vary sinusoidally, so the *instantaneous* power oscillates between zero and some maximum value. Of more practical importance is the *average* energy flow during a period of time that includes numerous cycles of the variation. This energy flow is easy to calculate. Substituting the basic sine-wave expressions for E_y and B_z into the

---

[8] As mentioned in Chapter 26, Equation (26–47), in a vacuum the *magnetic field* **H** is related to the *magnetic induction* **B**: $\mathbf{H} = (1/\mu_0)\mathbf{B}$. So Equation (28–40) is sometimes written as:

$$\mathbf{S} = \mathbf{E} \times \mathbf{H} \tag{28–41}$$

Poynting vector:

$$S = \frac{1}{\mu_0} E_{y0} B_{z0} \sin^2 (kx - \omega t) \tag{28–42}$$

This implies the energy received at a given point varies in time as the sine squared, which repeats itself every half cycle of the basic period T. Thus, to find the average energy flow, we calculate

$$S_{\text{ave}} = \frac{E_{y0} B_{z0}}{\mu_0 \left(\frac{T}{2}\right)} \int_0^{T/2} \sin^2 (kx - \omega t)\, dt \tag{28–43}$$

Evaluating the integral yields

AVERAGE ENERGY FLOW (power per unit area)

$$S_{\text{ave}} = \frac{1}{2\mu_0} E_{y0} B_{z0} \tag{28–44}$$

EXAMPLE 28–3

Consider a lamp that emits essentially monochromatic green light uniformly in all directions. If the lamp is 3% efficient and consumes 100 W of power, find the amplitude of the electric field associated with the electromagnetic radiation at a distance of 10 m from the lamp.

SOLUTION

Since the lamp is 3% efficient, it emits 3.0 W of electromagnetic power, which is spread uniformly over a sphere of radius 10 m. Thus, the average power per unit area is

$$S_{\text{ave}} = \frac{P}{4\pi R^2}$$

Substituting values for P and R:

$$S_{\text{ave}} = \frac{3.0\ \text{W}}{4\pi(10\ \text{m})^2} = \frac{0.030}{4\pi} \left(\frac{\text{W}}{\text{m}^2}\right)$$

Since the light is essentially of only one color, we can assume a single electromagnetic wave of wavelength λ and use Equation (28–44):

$$S_{\text{ave}} = \frac{1}{2\mu_0} (E_{y0} B_{z0})$$

and

$$E_{y0} = cB_{z0}$$

to give

$$S_{\text{ave}} = \frac{1}{2\mu_0} \left(\frac{E_{y0}^2}{c}\right)$$

Solving for E_{y0}

$$E_{y0} = (2\mu_0 c S_{\text{ave}})^{\frac{1}{2}}$$

Substituting values for μ_0 and S_{ave}:

$$E_{y0} = \left[2\left(\cancel{4\pi} \times 10^{-7} \frac{\text{H}}{\text{m}}\right)\left(3 \times 10^8 \frac{\text{m}}{\text{s}}\right) \frac{0.030 \frac{\text{W}}{\text{m}^2}}{\cancel{4\pi}} \right]^{\frac{1}{2}} = \boxed{1.34 \frac{\text{V}}{\text{m}}}$$

28.5 Momentum of Electromagnetic Waves

We have shown that energy is transported in an electromagnetic wave. We will now show that the wave also possesses momentum. We shall begin by demonstrating that the electromagnetic wave exerts a force on a charged particle in the direction of the wave propagation. This is in spite of the fact that the electric field **E** and the magnetic field **B** are entirely *transverse* (perpendicular to the direction of propagation). Thus, if an electromagnetic wave interacts with matter (which, of course, contains electrons), it will give some momentum to the matter.

Consider an electromagnetic wave traveling in the $+x$ direction, as in Figure 28–14, and striking an electron that is free to move in a sheet of resistive material. Let us suppose that at the sheet, the electric field oscillates in the x-y plane and the magnetic field oscillates in the x-z plane. The field can then be described as

$$\mathbf{E} = (E_0 \sin \omega t)\,\hat{\mathbf{y}} \tag{28–45}$$

and

$$\mathbf{B} = (B_0 \sin \omega t)\,\hat{\mathbf{z}} \tag{28–46}$$

The electric field will force the negatively charged electron $-e$ to move through the resistive material. For our purposes, we will assume that the electron moves with a drift velocity $\mathbf{v}_d$ as though it were in a viscous medium, where the electron is essentially always at its terminal velocity. Thus:

$$\mathbf{F}_E = b\mathbf{v}_d$$

where b is a constant and F_E is the force produced by the electric field:

$$\mathbf{F}_E = (-eE_0 \sin \omega t)\,\hat{\mathbf{y}}$$

Combining the previous two equations, we have

$$\mathbf{v}_d = -\left(\frac{eE_0}{b} \sin \omega t\right)\hat{\mathbf{y}} \tag{28–47}$$

Note that the oscillating velocity of the electron is exactly 180° out of phase with the electric field oscillation. The moving electron also experiences a magnetic force $\mathbf{F}_B$, where

$$\mathbf{F}_B = -e(\mathbf{v}_d \mathbf{X}\, \mathbf{B})$$

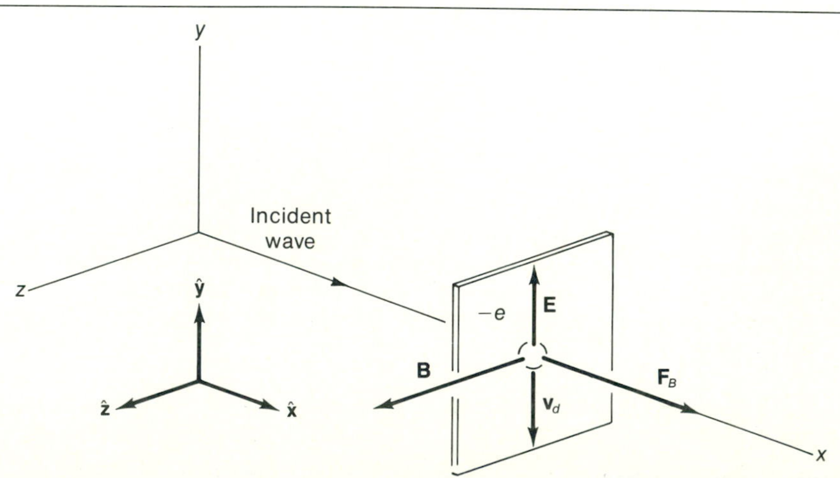

Figure 28–14
An electromagnetic wave exerts forces on electrons residing within a sheet of resistive material. The net average force is in the direction of the wave propagation.

Substituting the expressions for $\mathbf{v}_d$ and $\mathbf{B}$, we have

$$\mathbf{F}_B = -e\left(\frac{-eE_0}{b}\sin\omega t\right)\hat{\mathbf{y}} \times (B_0 \sin\omega t)\,\hat{\mathbf{z}}$$

which becomes

$$\mathbf{F}_B = \left(\frac{e^2E_0B_0}{b}\sin^2\omega t\right)\hat{\mathbf{x}} \qquad \textbf{(28–48)}$$

Because the $\sin^2 \omega t$ factor is always positive, the force is always in the $+x$ direction—the direction the electromagnetic wave travels. Of course, the sheet of resistive material experiences the sum of all the forces on all of the electrons in the sheet.

In being forced through the resistive material, the electron absorbs energy from the electromagnetic wave to overcome the viscous force on the electron. The power or rate that energy is given to the electron by the electromagnetic wave is[9]

$$\frac{dU}{dt} = \mathbf{F}_E \cdot \mathbf{v}_d$$

where U is the energy absorbed by the electron. Substituting expressions for $\mathbf{F}_E$ and $\mathbf{v}_d$ in this equation, we have

$$\frac{dU}{dt} = (-eE_0 \sin\omega t)\,\hat{\mathbf{y}} \cdot \left(\frac{-eE_0}{b}\sin\omega t\right)\hat{\mathbf{y}}$$

$$\frac{dU}{dt} = \frac{e^2{E_0}^2}{b}\sin^2\omega t \qquad \textbf{(28–49)}$$

Since $E = cB$, the rate of energy absorption can be written as

$$\frac{dU}{dt} = c\left(\frac{e^2E_0B_0}{b}\sin^2\omega t\right) \qquad \textbf{(28–50)}$$

Comparing this with Equation (28–48), we have

$$\frac{dU}{dt} = cF_B$$

Since F_B is the rate of momentum change acquired by the electron in the $+x$ direction, this equation becomes

$$\frac{dU}{dt} = c\,\frac{dp}{dt} \qquad \textbf{(28–51)}$$

where p is the momentum of the electron. This equation states that the rate of energy absorption by the electron equals the speed of light times the rate of momentum change of the electron. Since both the energy absorbed and momentum acquired by the electron were extracted from the electromagnetic wave, we may apply the conservation of energy and momentum principles, integrate Equation (28–51) with respect to time, and obtain

MOMENTUM *p* CARRIED BY A WAVE OF ENERGY *U*

$$U = cp \qquad \textbf{(28–52)}$$

[9] Recall that a *magnetic* field does no work because $\mathbf{F}_B$ is always perpendicular to $\mathbf{v}$.

Thus, if an electron absorbs energy U from an electromagnetic wave, it also receives momentum p from the wave. If the wave strikes matter, it interacts with all the electrons in the matter to transfer momentum to the object as a whole.

Because an electromagnetic wave possesses momentum, we may assume that a wave exerts a force on any surface it strikes, analogous to the force a stream of water exerts on a surface. Since $F = dp/dt$, from Equation (28–51) we have

$$F = \frac{1}{c}\frac{dU}{dt} \tag{28–53}$$

Considering radiation absorbed per unit area, the force per unit area exerted on an object equals the rate of energy loss per unit area. This force per unit area is called the **radiation pressure,** or *light pressure.*[10] It may be expressed in terms of the Poynting vector. Since:

$$S_{\text{ave}} = \frac{dU}{dt}$$

we have

PRESSURE *P* EXERTED BY A WAVE

$$P = \frac{S_{\text{ave}}}{c} \quad \text{(total absorption)} \tag{28–54}$$

If the electromagnetic wave strikes a perfectly reflecting mirror, the change in momentum would be twice as great, or

$$P = \frac{2S_{\text{ave}}}{c} \quad \text{(total reflection)} \tag{28–55}$$

EXAMPLE 28–4

What is the approximate force exerted on a small pocket mirror, 40 cm^2, held perpendicular to the light from a 100 W bulb at a distance of 10 m from the bulb?

SOLUTION

The calculations performed in Example 28–3 are relevant here. The average magnitude of the Poynting vector 10 m from a 100 W bulb was found to be

$$S_{\text{ave}} = \frac{0.030}{4\pi}\left(\frac{\text{W}}{\text{m}^2}\right)$$

Assuming that at 10 m the wave is essentially a plane wave over the face of the mirror, and that the waves are totally reflected, the pressure on the mirror given by Equation (28–55) is

$$P = \frac{2S_{\text{ave}}}{c} \quad \text{(for total reflection)}$$

The force F is the pressure on the mirror times the area A of the mirror. Then:

$$F = \left(\frac{2S_{\text{ave}}}{c}\right)(A)$$

[10] Pressure, momentum, and power all have the symbol p or P. Since all three concepts are involved here, be careful not to confuse them. For an interesting discussion of radiation pressure, see G.E. Henry, "Radiation Pressure," *Scientific American*, June 1957, p. 99.

Substituting numerical values:

$$F = \frac{2\left(\dfrac{0.030\ \dfrac{\text{W}}{\text{m}^2}}{4\pi}\right)(40 \times 10^{-4}\,\text{m}^2)}{3 \times 10^8\ \dfrac{\text{m}}{\text{s}}} = \boxed{6.37 \times 10^{-14}\ \text{N}}$$

This is an extremely small force, comparable to the gravitational attraction between two cups of coffee separated a distance of 10 m.

You may be familiar with a toy called a *radiometer*, illustrated in Figure 28–15(a). This device consists of vanes blackened on one side and silvered on the other. The vanes are mounted on a vertical axle in a glass bulb from which most of the air is removed. When the radiometer is exposed to moderately strong light (or even the infrared radiation from a flatiron), the vanes rotate about the axle, with the blackened faces trailing in the motion. Being aware of radiation pressure, a person may hastily conclude that the motion is due to radiation pressure. But this conclusion is incorrect for three reasons:

(1) If the torque producing this rotation of the radiometer vanes is attributable to light pressure, the vanes are rotating in the wrong direction. (We have shown that the force exerted on the silvered side of the vane is twice the magnitude of that on the blackened side. Therefore, the silvered side should trail the rotation.)
(2) The force exerted by the electromagnetic wave is far too small to account for the rapid angular acceleration of the vanes when the radiometer is suddenly exposed to light. (Example 28–4 indicated how small the force would be on the radiometer vanes even if the radiometer were to be placed close to a light bulb.)
(3) If the radiometer bulb is evacuated to an extremely low pressure, the vanes will not rotate. (The torque on the vanes due to light pressure is too small to overcome the friction on the bearings of the vane support.)[11]

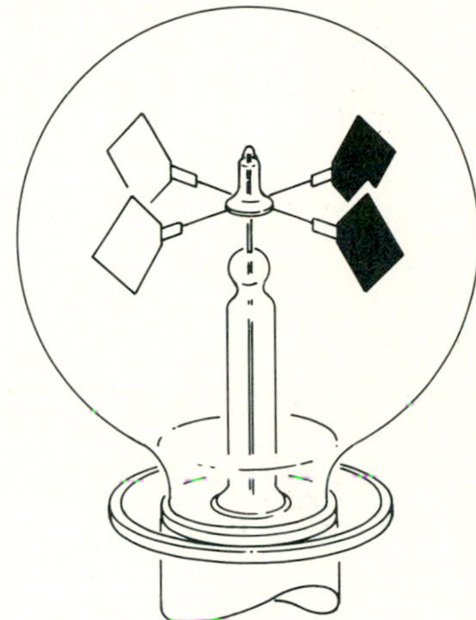

(a) A radiometer

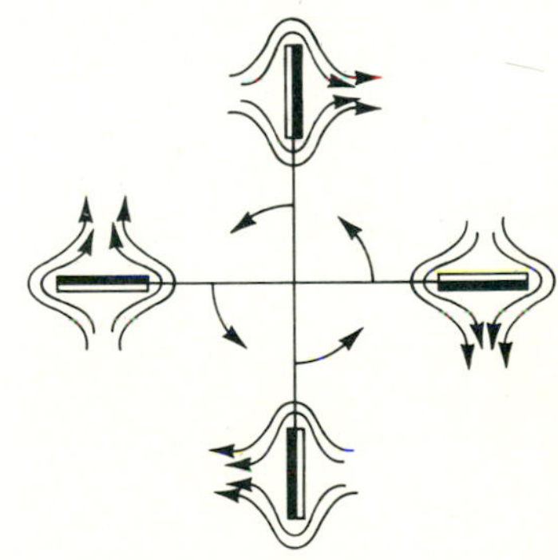

(b) The thermal creep of air around the edges of the vanes (view from above).

Figure 28–15
A radiometer turns on its axis when exposed to a moderately strong light. The torque causing it to turn is *not* produced by the pressure of the light.

The explanation of the moving vanes in a radiometer was first suggested by Maxwell in 1879. The explanation is based on the fact that air moves along the surface of an unevenly heated object toward regions of higher temperature. This phenomenon is known as *thermal creep*.[12] In the case of a radiometer vane, air flows over the edge of the vane toward the warmer blackened side. The resulting increase in air pressure on the blackened side produces the rotation of the vanes. In a typical radiometer, the air-pressure effect is about 10 000 times greater than the radiation pressure.

In spite of the relative smallness of the radiation pressure, in certain situations it can become a large effect. For example, sunlight exerts a force on the earth of about 6×10^8 N (over 60 000 tons). Sunlight falling on balloon satellites circling the earth (such as the Echo satellite launched in the 1960s) produces noticeable alterations of the orbit. Spacecraft that have extended vanes of solar

[11] If the vanes are suspended by a thin quartz fiber, and if the air pressure is extremely low, then the true radiation-pressure effect can be demonstrated. If just one of the vanes is illuminated, the vanes can be turned through an angle in opposition to the restoring torque of the fiber.

[12] Experiments establishing the thermal-creep explanation of radiometers are described in E.H. Kennard, *Kinetic Theory of Gases*, McGraw-Hill, 1938.

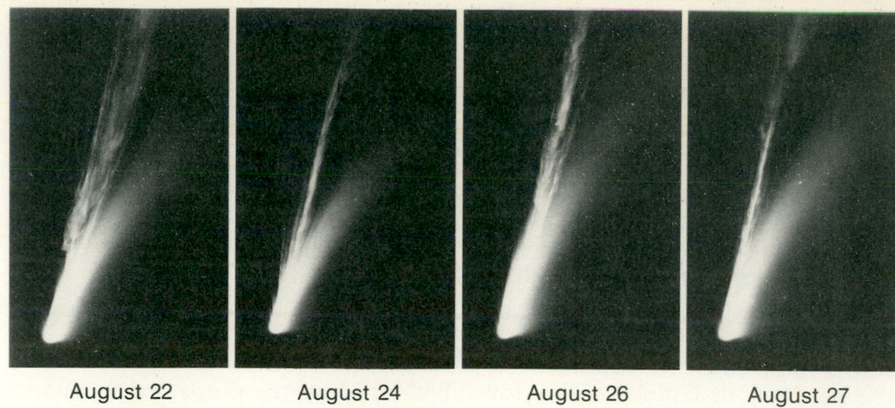

Figure 28–16

The comet Mrkos, photographed in 1957, is traveling toward the left in these pictures. The curving, diffuse tail, which extends almost at right angles to the path, is formed from dust particles "blown away" by *radiation pressure* from the sun. The straight, ragged tail is composed of ionized atoms and molecules pushed away with greater speeds by the *solar wind,* a stream of charged particles ejected by the sun.

---

cells to capture sunlight will experience net torques due to radiation pressure if the center of mass of the satellite does not coincide with the center of zero torque due to radiation pressures.

Many comets have two tails: one composed of ionized atoms and molecules and the other made of dust particles (Figure 28–16). The dust tails are produced by radiation pressure from the sun, which "blows away" dust grains ejected from the comet's head to form a curved, diffuse tail. The atoms and molecules in the ion tails are generally accelerated to faster speeds (up to 100 km/s) and are formed by the *solar wind*—a stream of mostly electrons and protons ejected from the sun and flowing throughout the solar system.

Calculations show that with sufficiently large "sails," space vehicles might feasibly be propelled away from the sun through interplanetary space by radiation pressure from the sun. (The method does not work for *interstellar* journeys, however, because the spacecraft moves too far away from the source of radiation.)

Summary

The following concepts and equations were introduced in this chapter:

Displacement current:

$$i_d = \varepsilon_0 \frac{d\Phi_E}{dt}$$

Maxwell's equations (for a vacuum):

$\oint \mathbf{E} \cdot d\mathbf{A} = \dfrac{q}{\varepsilon_0}$	$\oint \mathbf{B} \cdot d\mathbf{A} = 0$
$\oint \mathbf{B} \cdot d\boldsymbol{\ell} = \mu_0\left(I + \varepsilon_0 \dfrac{d\Phi_E}{dt}\right)$	$\oint \mathbf{E} \cdot d\boldsymbol{\ell} = -\dfrac{d\Phi_B}{dt}$

Wave equation (for a plane wave traveling in the $+x$ direction):

$$\frac{\partial^2 E_y}{\partial x^2} = \mu_0 \varepsilon_0 \frac{\partial^2 E_y}{\partial t^2} \quad \text{and} \quad \frac{\partial^2 B_z}{\partial x^2} = \mu_0 \varepsilon_0 \frac{\partial^2 B_z}{\partial t^2}$$

For a sinusoidal *electromagnetic wave:*

$$E_y = E_{y0} \sin \omega t$$
$$B_z = B_{z0} \sin \omega t$$

where **E** and **B** are perpendicular to each other, so that **E X B** is the direction of the wave velocity.

Speed of light in a vacuum:

$$c = \frac{1}{\sqrt{\mu_0 \varepsilon_0}}$$

E *and* **B** *fields in electromagnetic waves:*

$$E_y = cB_z$$

Energy density in **E** *and* **B** *fields:*

Electric: $$u_E = \tfrac{1}{2}\varepsilon_0 E^2$$

Magnetic: $$u_B = \frac{1}{2\mu_0} B^2$$

Energy flow in electromagnetic waves:

Instantaneous: $$\mathbf{S} = \frac{1}{\mu_0}(\mathbf{E} \times \mathbf{B})$$

Average: $$S_{ave} = \frac{1}{2\mu_0} E_{y0} B_{z0}$$

where **S**, the *Poynting vector*, has units of watts/meter^2 and is in the direction of the electromagnetic wave propagation.

Momentum carried by electromagnetic waves: An object acquires a momentum p in the absorption of electromagnetic energy U according to

$$U = cp$$

Light pressure: A pressure P is exerted on an object absorbing the radiant energy flux S_{ave}

$$P = \frac{S_{ave}}{c} \quad \text{(total absorption)}$$

If the light is reflected by the object:

$$P = \frac{2S_{ave}}{c} \quad \text{(total reflection)}$$

Questions

1. What kind of simple apparatus would be needed to demonstrate that a changing magnetic field produces an electric field? Similarly, what simple apparatus would be required to show that a changing electric field produces a magnetic field?

2. A physicist discovers that in his laboratory the magnetic field is directed upward and increasing. When he directs a beam of electrons upward (along the direction of **B**), the beam is deflected in a certain direction. What causes the deflection? What information about the extent of the magnetic field does this provide?
3. A parallel-plate capacitor in series with a resistor is charged by a battery. How would the displacement current between the plates of the capacitor depend on the dielectric material?
4. Does the magnitude or direction of an electric field that is induced by a changing magnetic field give any information about the instantaneous direction or magnitude of the magnetic field?
5. The behavior of magnetic dipoles and quadrupoles is consistent with Maxwell's equations. Is it possible to construct a magnetic tripole (two north poles and one south pole, for example) that also has properties consistent with Maxwell's equations?
6. At a given point in space, there is an instant when both the electric and magnetic fields associated with an electromagnetic wave are zero. How can the wave propagate from that point if no fields exist there?
7. Straight-wire radio receiving antennae are designed to detect the electric field variation of an electromagnetic wave rather than the magnetic field variation. Explain.
8. A directional radio receiving antenna is in the form of a circular coil of wire. Is such an antenna sensitive to the magnetic field variation of the transmitted electromagnetic wave or to the electric field variation? How should this antenna be oriented with respect to a straight vertical radio transmitter antenna?
9. Design an electrical apparatus by which, in principle, the speed of light could be determined through the measurement of time-varying forces alone.
10. Since the measured values of ε_0 and c are related by the defined constant μ_0, what form would Maxwell's equations take if μ_0 or ε_0 did not appear explicitly?
11. In what ways does the radiation from a light bulb differ from the radiation from a radio transmitter antenna?
12. Identify what is wrong with the following statement: "The electric field associated with the electromagnetic wave is much greater than the magnetic field because $E = cB$."
13. An electromagnetic wave transports energy in its electric and magnetic fields. Which, if either, of the fields contains the greater amount of energy?
14. Does a detector of a monochromatic electromagnetic wave experience a continuous or pulsating flow of momentum and energy? If pulsating, what is the frequency of the pulses?
15. Explain what is inappropriate about the way the following question is worded: "What fraction of the total electromagnetic spectrum does visible light represent?" What would be a better way to ask the question?
16. In what ways is an electromagnetic wave similar to a stream of particles?

Problems

28A–1 Find the distance in centimeters that light travels in one nanosecond.

Answer: 300 cm

28A–2 A 0.5 μF parallel-plate capacitor is being charged through a resistance of 100 Ω by a 9 V battery. Calculate the displacement current in the capacitor 50 μs after the charging is initiated.

28A–3 Show that the displacement current i_d between the plates of a parallel-plate capacitor may be expressed by $i_d = C\,dV/dt$, where C is the capacitance of the capacitor and dV/dt is the rate of voltage change across the capacitor.

28A–4 Consider the region between the plates of a charging circular parallel-plate capacitor. Make a qualitative plot of the magnitude of the magnetic field as a function of the distance from the axis of the capacitor. Include the region beyond the edge of the plates. Neglect the fringing of the electric field at the edge of the plates.

28A–5 Standard wire tables indicate that 12 gauge copper wire has a diameter of 0.080 81 in and a resistance of 1.588 Ω/1000 ft. Calculate the values of **E**, **B**, and **S** on the surface of the wire when it carries a direct current of 20 A.

Answers: **E** = 0.104 V/m, in the direction of the current
B = 3.90×10^{-3} T, clockwise around the wire, looking in the direction of the current
S = 323 W/m^2, toward the wire

28A–6 The electric field associated with an electromagnetic wave traveling in the $+x$ direction is described in SI units by $\mathbf{E} = 6 \sin(kx - 10^{16}t)\,\hat{\mathbf{y}}$.

(a) Write the corresponding expression for the magnetic field.
(b) Calculate the wavelength of the radiation.
(c) Calculate the average energy density in the radiation.

28A–7 Show that the equation $E = cB$ balances dimensionally in SI units.

28A–8 The electric field oscillations received at an FM radio antenna have an amplitude of 5×10^{-5} V/m.

(a) Calculate the amplitude of the associated magnetic field oscillations.
(b) Calculate the average rate of energy flow of the radiation in watts per square meter.

28A–9 A monochromatic light source emits 100 W of electromagnetic power uniformly in all directions.

(a) Calculate the average electric field energy density one meter from the source.
(b) Calculate the average magnetic field energy density at the same distance from the source.

Answer: $u_E = u_B = 1.33 \times 10^{-8}$ J/m^3

28B–1 Show that the displacement current defined by $i_d = \varepsilon_0\, d\Phi_E/dt$ has the units of amperes.

28B–2 A parallel-plate capacitor consists of circular plates 10 cm in diameter and separated by 1 mm. Calculate the magnitude of the magnetic field between the plates at their outer edge while the potential difference on the capacitor is changing at the rate of 1000 V/s. (Neglect fringing of the electric field.)

28B–3 A parallel-plate capacitor with circular plates of radius R has a capacitance C. The potential across the capacitor is increasing at the constant rate dV/dt. Assuming there is no fringing of the electric field, show that the expressions for the magnetic field at distances radially away from the center of the capacitor are (in SI units):

Answers: $B = (2rC/R^2)\,dV/dt \times 10^{-7}$ (for $r < R$)
$B = (2C/r)dV/dt \times 10^{-7}$ (for $r > R$)

28B–4 Show that for a sinusoidal electromagnetic wave, the average value of the Poynting vector $|\mathbf{S}_{\text{ave}}|$ is related to the root-mean-square value of the electric field by $E_{\text{rms}} = \sqrt{\mu_0 c S_{\text{ave}}}$.

28B–5 Using the value of S obtained in Problem 28A–5, verify numerically that $\oint \mathbf{S} \cdot d\mathbf{A} = I^2R$ for a 1000 ft length of 12 gauge copper wire.

28B–6 A copper sphere has a net positive charge. It is placed at the origin of a coordinate system in a region where there is a uniform, constant magnetic field B_z in the $+z$ direction. Show qualitatively that for these fields, which do not change with time, Poynting's vector implies that electromagnetic energy is flowing around the sphere.

28B–7 A pulsed laser produces a flash of light 4 ns in duration, with a total energy of 2 J, in a beam 3 mm in diameter.

(a) Find the spatial length of the traveling pulse of light.
(b) Find the energy density in joules/meter^3 within the pulse.
(c) Find the amplitude E_0 of the electric field in the wave.

Answers: (a) 1.20 m (b) $u_E = 2.36 \times 10^5$ J/m^3 (c) $E_0 = 2.31 \times 10^8$ V/m

28B–8 A 100 mW laser beam is reflected back upon itself by a mirror. Calculate the force on the mirror.

28B–9 An inflated mylar balloon 50 m in diameter orbits the earth at an altitude of approximately 1000 km. Calculate the force on the balloon due to the direct electromagnetic radiation from the sun, assuming the radiation is totally absorbed.

28B–10 Assuming the earth absorbs all the sunlight incident upon it, find the force the sun exerts on the earth due to radiation pressure. Compare this value with the sun's gravitational attraction.

28B–11 Derive the relationship between the radiation pressure on a nonreflecting surface with the energy density associated with the radiation just outside the surface. Explain why the relationship does not depend on whether or not the surface is nonreflecting.

Answer: they are numerically equal

28C–1 Show that the function $E = E_0 e^{k(x-ct)}$ satisfies the wave equation $\partial^2 E/\partial x^2 = (1/c^2)\,\partial^2 E/\partial t^2$. (Any function of the form $f(x \pm ct)$ satisfies the wave equation.)

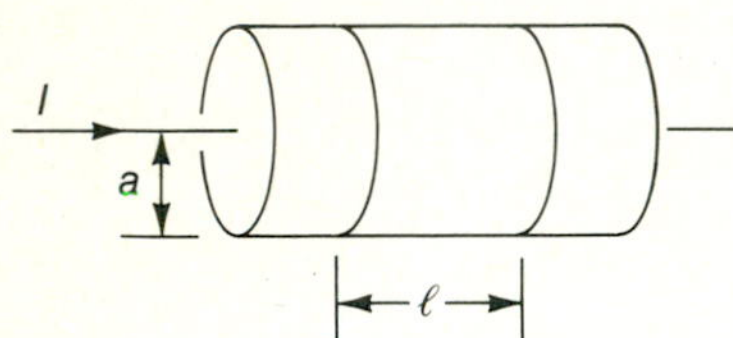

Figure 28–17
Problem 28C–2

28C–2 Figure 28–17 shows a long, cylindrical resistor of radius a and resistivity ρ, carrying a current I.

(a) Show that the Poynting vector **S** is radially inward everywhere on the surface of the resistor.

(b) Integrate **S** over the surface for a length ℓ of the resistor to show that it equals the I^2R losses within that length: that is, $\int \mathbf{S} \cdot d\mathbf{A} = I^2R$. Note that at the surface, E is parallel to the axis of the resistor. (This calculation implies that the thermal energy developed in the resistor does not originate inside the resistor that carries the current, but from the space surrounding the resistor.)

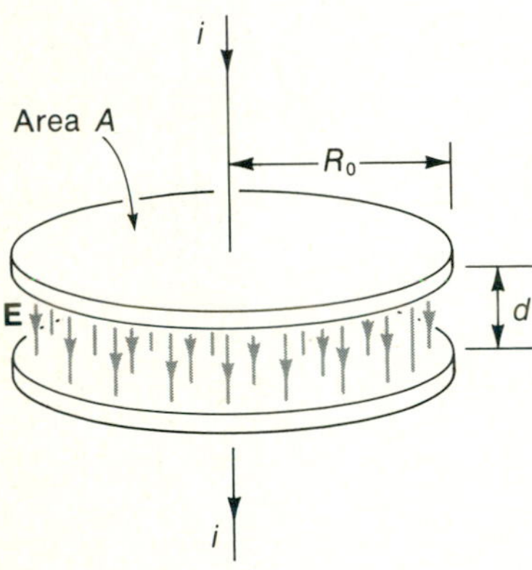

Figure 28–18
Problem 28C–3

28C–3 Figure 28–18 shows the charging of a parallel-plate capacitor by a current i. As the electric field is increasing, (a) show that the Poynting vector **S** is toward the axis everywhere throughout the volume between the plates. (Ignore fringing of the electric field.) (b) The integral of the Poynting vector over the cylindrical surface surrounding the volume between the plates represents the energy flow into the volume. Show that this energy flow equals the rate of increase of energy stored in the electric field between the plates. (In this view, the energy stored in a capacitor does not come through the wires carrying the current, but flows in from the surrounding space.)

28C–4 A dust particle in outer space is attracted toward the sun by gravity and repelled by the radiation from the sun. Suppose that a particle is spherical, with a radius R and density $\rho = 2\ \text{g/cm}^3$, and that it absorbs all the radiation falling on its surface. Determine the value of R such that the gravitational and radiation forces are equal. Obtain the necessary constants from the appendices. Explain why the distance from the sun is irrelevant.

292 nm

28C–5 An astronaut, stranded in space "at rest" 10 m from his spacecraft, has a mass (including equipment) of 110 kg. He has a 100 W light source that forms a directed beam, so he decides to use the beam of light as a photon rocket to propel himself continuously toward the spacecraft.

(a) Calculate how long it will take him to reach the spacecraft by this method.

(b) Suppose, instead, he decides to throw the light source away in a direction opposite to the spacecraft. If the mass of the light source is 3 kg and after being thrown it moves with a speed of 12 m/s relative to the recoiling astronaut, how long will it take the astronaut to reach the spacecraft?

Answers: (a) 22.6 h (b) 30.5 s

Three brothers bought a cattle ranch and named it "Focus." When their father asked why they chose that name, they replied: "It's the place where the sons raise meat."

Triple pun attributed
to Professor W.B. Pietenpol, Physics Department,
University of Colorado, Boulder, Colorado

chapter 29 Geometrical Optics

In this chapter we will discuss the interaction of visible light with smooth reflecting surfaces and transparent materials. Since we will be concerned with the *result* of the interaction rather than with the physical details of *how* the interaction takes place, the chapter will contain relatively less new physics but more geometry than previous chapters. We will limit our discussion to electromagnetic radiation in the visible-light region, where all frequencies behave in a similar way. If we were to go very far beyond the visible portion of the spectrum, the interaction would change. For example, a thin sheet of aluminum foil, which reflects visible light, is essentially transparent to x-rays and gamma rays.

Visible wavelengths extend through the full range of the colors we see in a spectrum from deep violet to dark red, corresponding to wavelengths from about 400 nm to 700 nm. The **nanometer** (nm), a unit of length where 1 nm $\equiv 10^{-9}$ m, is the customary unit for specifying wavelengths in the visible region. A person with normal eyesight can barely distinguish[1] two colors with a wavelength difference of 1 nm. Another unit of length commonly used by spectroscopists (but gradually being replaced by the nanometer) is the *angstrom* (Å), where 1 Å $\equiv 10^{-8}$ m. Table 29–1 correlates colors of the visible spectrum with their approximate wavelengths.

29.1 Wavefronts and Rays

As discussed in the previous chapter, visible light is a small portion of the electromagnetic spectrum of traveling waves of electric and magnetic fields. Glowing hot bodies emit waves of a variety of wavelengths. For this introductory discussion, however, we will consider a *point source* of light that emits radiation of a *single* wavelength λ. A cross section of the spherical waves moving away from this point source is analogous to circular water waves moving away from a small object that is moved up and down on the surface of a pond. We may

[1] The human ear, in many ways, is much more discerning than the eye. While the visible spectrum covers less than one octave (a factor of two in frequency), the audible range of sounds is about 10 octaves, with the smallest discernible change in pitch of about one *cent*, where one octave contains 1200 cents.

TABLE 29–1

Wavelength and Color

Approximate wavelength (nm)	Color
420	Violet
470	Blue
520	Green
570	Yellow
620	Orange
670	Red

identify the *electric* field variations in the electromagnetic waves with the crests and troughs of the water waves, as shown in Figure 29–1.

The similarity between water waves and electromagnetic waves is more than just geometrical. They also share other properties. For example, electromagnetic waves bend around obstacles just as ocean waves bend around the end of a breakwater. If the obstacle has an opening, or *aperture*, in it, the waves will spread out as they pass through the opening. The amount of bending depends on the size of the aperture compared with the wavelength of the waves. (The closer this dimension is to the wavelength, the more the bending.) But let us postpone these *wave optics* phenomena to later chapters and treat here only those cases where the obstacle or aperture size is very large compared with the wavelength, and where the bending and spreading effects can thus be ignored. So in this chapter *we assume that the light always travels in straight lines, or rays.* This approximation defines the subject called *geometrical optics.*

The spreading of water waves across the surface of a pond is essentially a two-dimensional situation. The line along the crest of each wave is called a *wavefront*. For light waves emitted from a point source, the spreading waves move in three dimensions, so the wavefronts are *surfaces* that form concentric spheres around the source. If we associate wavefronts with the peak values

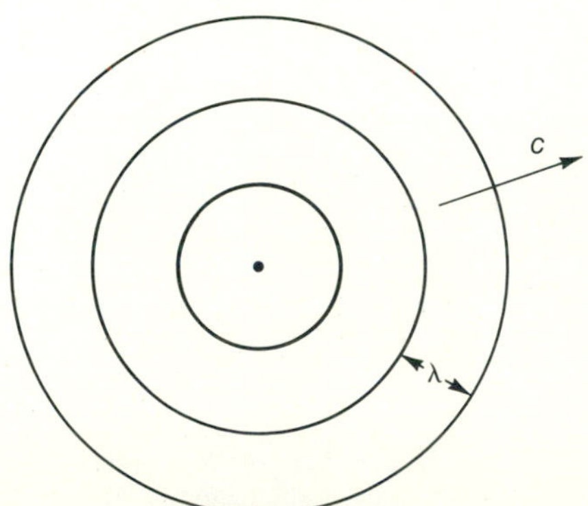

(**a**) Representation of light waves traveling outward from a point source.

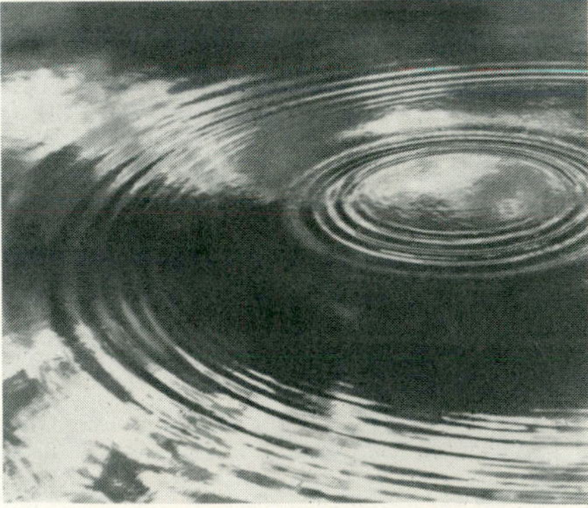

(**b**) Circular water waves.

Figure 29–1
A cross section of the spherical waves emanating from a point source of light is geometrically similar to water waves moving outward from a localized disturbance on the surface of the water.

of the electromagnetic wave, the wavefronts are spaced one wavelength λ apart and move away from the source with the speed of light c in a vacuum. A more general definition of a **wavefront** is *the locus of neighboring points where the waves have the same phase.* In sketching diagrams of light waves spreading from a point source such as Figure 29–2(a), we draw the wavefronts as lines. However, keep in mind that in the case of electromagnetic waves, the wavefronts are *surfaces.*

The direction a wavefront moves is always perpendicular to the wavefront itself. Any line drawn perpendicular to a wavefront is called a **ray;** an arrow on the ray indicates the direction of motion of the wavefront. Figure 29–2(b) illustrates a set of rays associated with the spherical wavefronts emerging from a point source of light. Figure 29–2(c) shows plane waves, where the rays and wavefronts are a very great distance from the source, such that the spherical wavefronts have negligible curvature. While the spacing of wavefronts is usually considered to be one wavelength apart, the spacing of the rays has no significance. As we will show, often just two rays from a source are adequate for the analysis of an optical system.

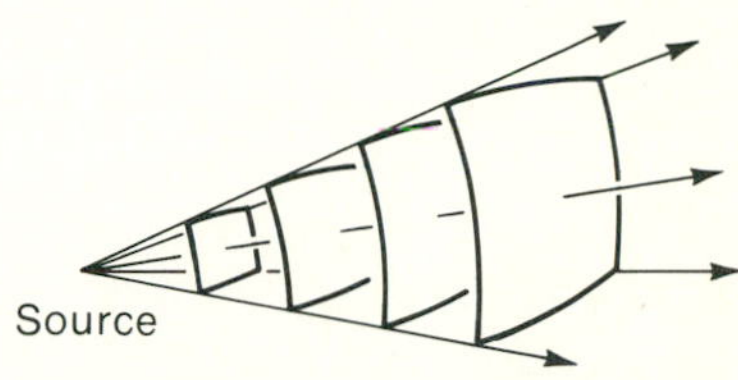

(**a**) A portion of the spherical wavefronts emerging from a point source.

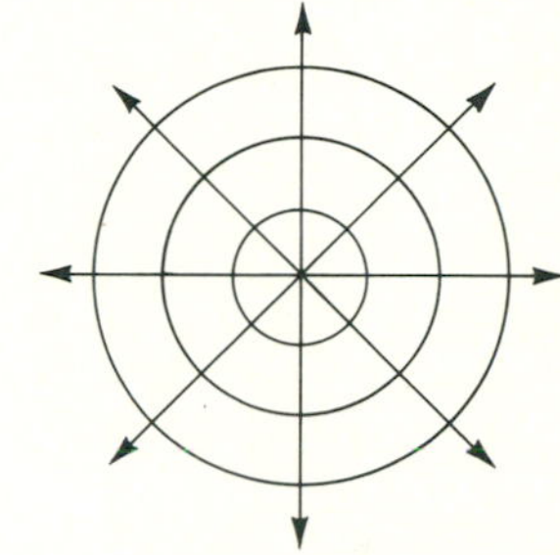

(**b**) The usual way of depicting spherical wavefronts from a point source.

(**c**) A plane wave traveling toward the right.

Figure 29–2
Rays are perpendicular to wavefronts. The arrows on the rays indicate the direction of wavefront motion.

29.2 Huygens' Principle

An extremely useful technique for the analysis of optical systems was devised by the Dutch physicist and astronomer Christian Huygens (1629–1695). He proposed that each successive wavefront could be generated from the preceding one by considering every point on a wavefront to be a secondary point source of waves. After traveling outward for a time equal to one period of the vibration, each secondary wavelet will be a spherical wavefront centered about its own point source. According to Huygens, the new wavefront is the *envelope,* or *tangent surface,* of all these secondary wavelets.[2] This method of analysis is called **Huygens' principle.** Figure 29–3 illustrates the procedure. Each point along a wavefront AA' is considered to be a series of point sources, each radiating

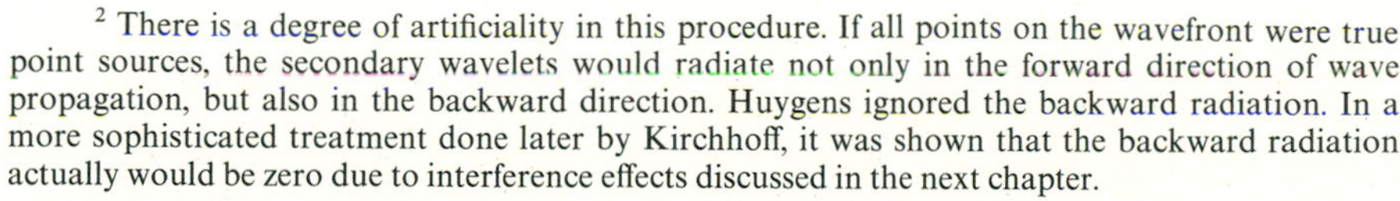
[2] There is a degree of artificiality in this procedure. If all points on the wavefront were true point sources, the secondary wavelets would radiate not only in the forward direction of wave propagation, but also in the backward direction. Huygens ignored the backward radiation. In a more sophisticated treatment done later by Kirchhoff, it was shown that the backward radiation actually would be zero due to interference effects discussed in the next chapter.

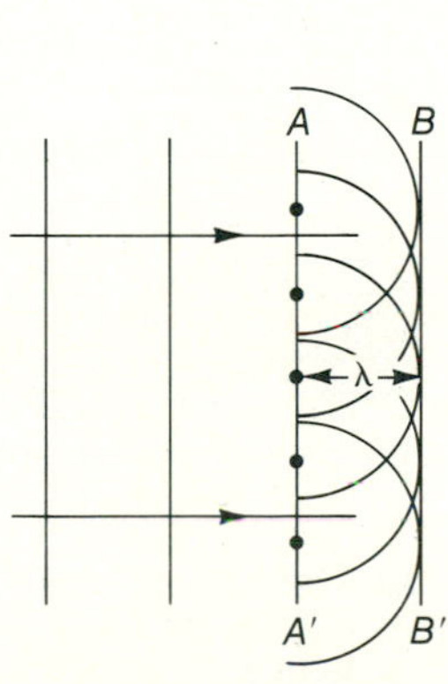

(**a**) Plane wavefronts

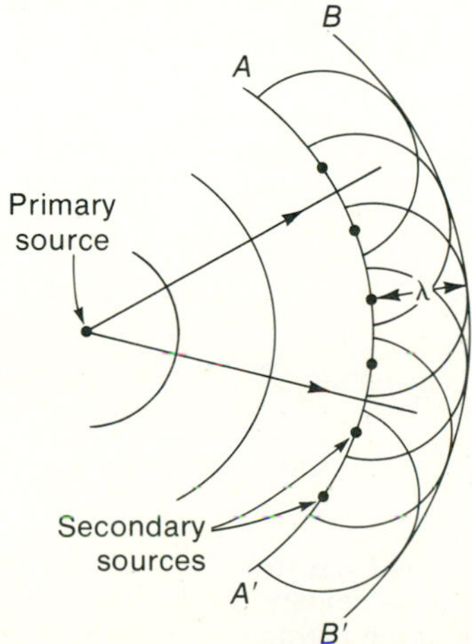

(**b**) Spherical wavefronts

Figure 29–3
Huygens' principle. A wavefront AA' is considered to be a series of point sources for secondary wavelets. After a time t, the secondary wavelets travel a distance λ. The envelope of these secondary wavelets forms the new wavefront BB'. The arrows on the rays indicate the direction of wave propagation.

secondary wavelets. At a later time t, the envelope of these wavelets forms the new wavefront BB'. The method works for a wavefront of any arbitrary shape, not just the plane and spherical wavefronts illustrated.

29.3 Reflection by a Plane Mirror

Laws describing reflection of light by mirrors were probably known as early as the time of Plato in the fourth century B.C. We now deduce these laws by two different methods, each illustrating an important principle in physics.

Using Huygens' Principle

We often speak of looking "into" a mirror in the same sense as looking into a room. We see images that certainly appear to be on the other side of the mirror, and every child has wondered what it would be like to pass through the looking glass into that other world, whose contents have a one-to-one relationship with objects in the real world. How far behind the mirror is the image of a given object? Consider a plane wave approaching a mirror as in Figure 29–4(a). The rays associated with incoming wavefront AB form an angle α_1 with the surface of the mirror. As each portion of the incoming electromagnetic wave strikes the mirror, electrons in the surface of the mirror are set into oscillations. These oscillating electrons reradiate electromagnetic waves, so each becomes a source of secondary wavelets.[3]

Let us look more closely at the reflected wavefronts shown in Figure 29–4(a). As the point A on the wavefront AB strikes the mirror, a circular wavelet originating at A will proceed to a point C on the reflected wavefront CD. Meanwhile, a wavelet originating at B will proceed toward the point D on the mirror. If the time required for a wavelet to travel from A to C equals the time required for a wavelet to travel from B to D, the points C and D will be in phase, thus constituting parts of a reflected wavefront. (Of course, all wavelets originating from points between A and B will be reflected to reach

[3] The idea of Huygens' wavelets originating from every point on a wavefront in *free space* does seem to be merely a "trick" that gives the right answer. However, when a material medium is present, with oscillating electrons acting as sources of reradiated waves, the idea becomes plausible and, indeed, correctly describes the mechanism of electromagnetic waves interacting with matter. In Huygens' time, it was believed that the medium that transmitted light waves—the ether, as it was called—was present everywhere, even in a vacuum, so it is easy to see how Huygens' principle arose. Of course, following Einstein, present-day theory makes no use of the ether concept.

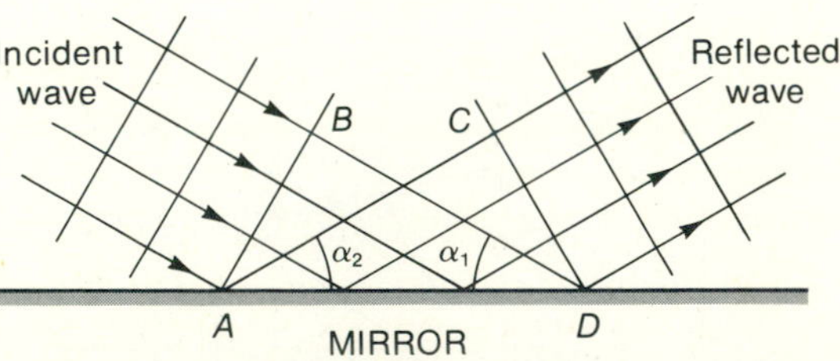

(a) The reflected wavefront is formed by the envelope of secondary waves originating at the surface of the mirror.

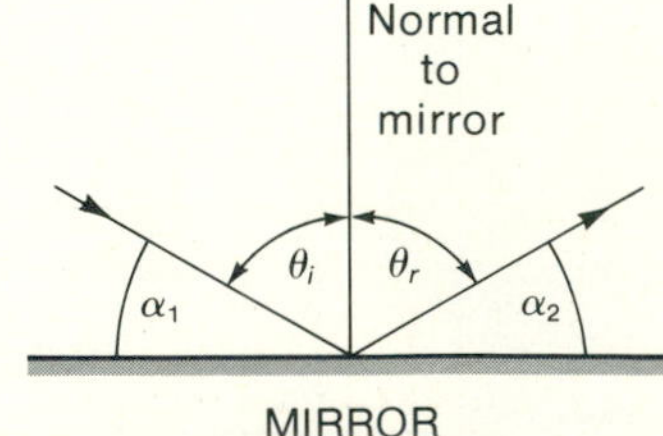

(b) The angle of incidence equals the angle of reflection: $\theta_i = \theta_r$.

Figure 29–4
A plane wave reflected by a plane mirror.

corresponding points between C and D.) Therefore, the distances AC and BD are equal, and the right triangle ABD is congruent to the right triangle ACD. (They have the common hypotenuse AD and equal sides.) Thus angles α_1 and α_2 are equal. It follows that their complements, θ_i and θ_r, are also equal. In optics it is customary to measure angles of rays with respect to the *normal* to a surface. Therefore, in Figure 29–4(b) we see that the *angle of incidence* θ_i is equal to the *angle of reflection* θ_r. Moreover, if we carry out the analysis in three dimensions, it can be shown that the incident ray, the normal to the mirror, and the reflected ray lie in the same plane.

LAWS OF REFLECTION

(1) The angle of incidence equals the angle of reflection: $\theta_i = \theta_r$.

(2) The incident ray, the normal to the mirror, and the reflected ray lie in the same plane.

If the surface is rough as in Figure 29–5, a bundle of parallel rays will be reflected at various angles. This type of reflection, called *diffuse* reflection, is illustrated by the surface of the page you are now reading. Even though the illumination on the page is essentially parallel rays from a single study lamp, you can observe the page from any angle. Most nonluminous objects you see are observed by diffuse reflection. The difference between diffuse and *specular* (mirrorlike) reflection depends on the size of surface irregularities compared with the wavelength of the illumination. If such irregularities are small compared with the wavelength of light, specular reflection occurs. On the other hand, if such irregularities are of the order of a wavelength or larger, the reflection is diffuse. Thus the roughened surface of a piece of aluminum that has been sanded would cause diffuse reflection of visible light, but specular reflection of radar waves of 5 cm wavelength.

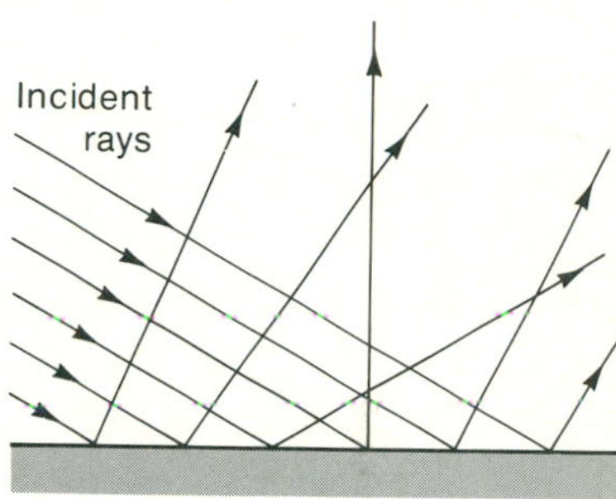

Figure 29–5
In the case of diffuse reflection, parallel rays are reflected in various directions.

Using Fermat's Principle

The laws of reflection may also be deduced from Fermat's[4] principle, another important relation of physics.

FERMAT'S PRINCIPLE

In going from one point to another, a light ray travels a path that requires equal or less time in transit than the time required for neighboring paths.

To illustrate this principle, we will apply it to the situation shown in Figure 29–6. The source and observer lie in a plane perpendicular to the surface of the mirror. An arbitrary path of a ray is shown as a dashed line. Clearly this path is not the shortest from source to observer, so it will not be traveled in the least time. While it seems obvious that the shortest path (the solid line) lies in the plane containing the normal to the mirror, it is also true that the angle of incidence equals the angle of reflection. Proof of the latter by Fermat's principle is left as a problem.

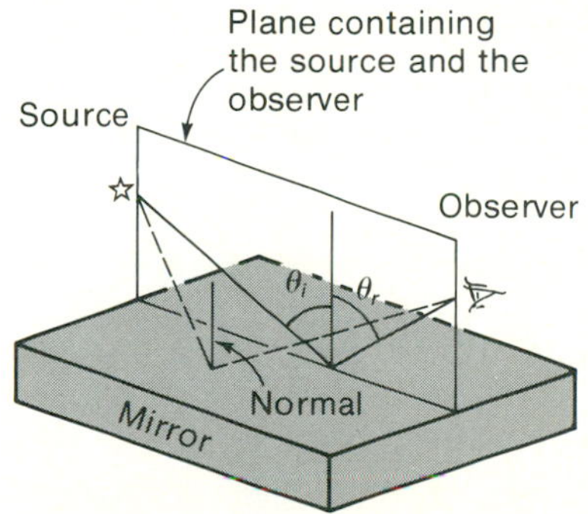

Figure 29–6
Fermat's principle: A light ray will be reflected in such a way that the total time in transit from the source to the observer is a minimum.

[4] Pierre de Fermat (1601–1665), a French nobleman, founded modern probability theory as a result of his interest in calculating gambling odds. In addition to Fermat's principle, he is also famous for "Fermat's last theorem," a tantalizing puzzle that still frustrates mathematicians. In a note (discovered posthumously) written on the margin of a book page, he claimed to have proved there are no nontrivial integral solutions of $x^n + y^n = z^n$ for $n > 2$. To date, no one else has been able to prove or disprove it.

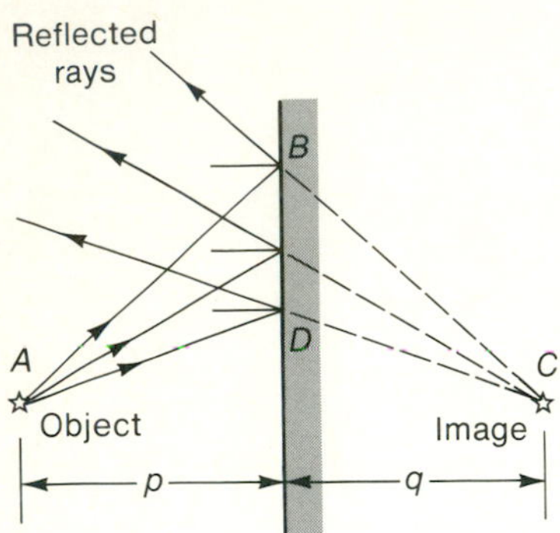

Figure 29–7
The image formed by a plane mirror lies behind the mirror at a distance equal to the distance the object is from the mirror.

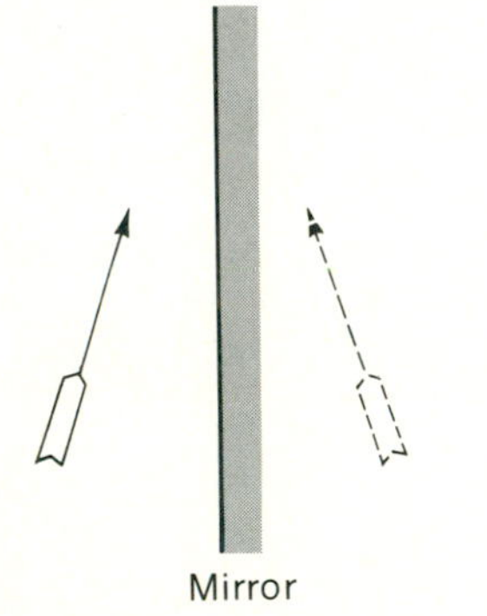

Figure 29–8
The image of an object formed by a plane mirror is the same distance behind the mirror as the object is in front. The image and object are of equal size.

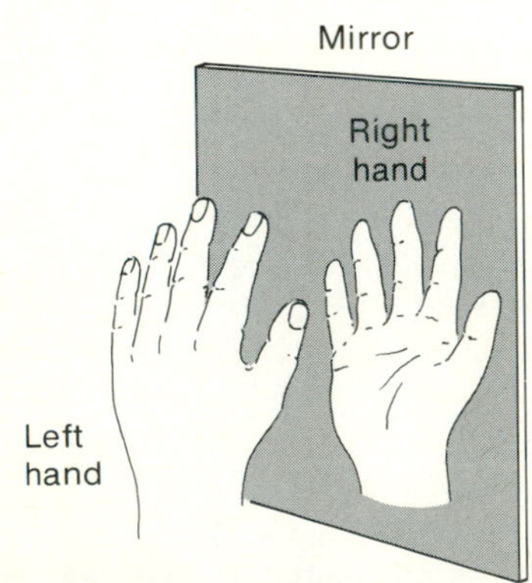

Figure 29–9
The images in a plane mirror have left and right interchanged.

Let us now return to the question asked at the beginning of this section: How far behind the mirror is the image of an object? To find the answer, we trace the paths of representative rays in accordance with the laws of reflection. Figure 29–7 is a ray diagram that shows rays leaving the source (☆) at A and being reflected by the mirror. The directions along which the reflected rays travel make them appear to come from the single point C behind the mirror, a point that is the *image* of the source. We now introduce a notation that will simplify the discussion. The *object distance* p is the perpendicular distance from the object to the mirror, and the *image distance* q is the perpendicular distance from the image to the mirror. The second law of reflection ensures that the rays shown lie in the plane of the figure. The first law of reflection leads to the conclusion that the triangle ABD is congruent to the triangle CBD. (They have a common side BD, and the other two sides of the triangles form equal angles with BD.) Thus:

IMAGE LOCATION IN PLANE MIRRORS	**The image distance q equals the object distance p.**

This conclusion is based on a point source.

An object of finite size may be thought of as a distribution of point sources, each with its own image. Thus there is a point-to-point correspondence between an object and its image in the mirror. Because $p = q$ for each point, the object and image are located symmetrically on opposite sides of the mirror and are the same size, as shown in Figure 29–8.

An interesting feature of mirror images is the fact that left and right are interchanged (see Figure 29–9). Also, a right-handed coordinate system has a mirror image that is a left-handed coordinate system. Thus plane mirror images, while having a one-to-one correspondence with their objects, do have an important characteristic that is different: Left and right are interchanged. Such images are called *perverted*.

29.4 Reflection by Spherical Mirrors

Much to the distress of some of us, we are greeted in the morning by a larger-than-life-size image of ourselves as we look into a shaving or makeup mirror. In most of our encounters with mirrors, the image is behind the mirror, though we will see that under certain circumstances, images can also be formed in front of the mirror. Mirrors with curved surfaces—*spherical mirrors*—may be concave or convex, depending on the type of surface curvature that the incident light rays encounter. The surface of a spherical shell approached from inside the shell is *concave*; when approached from outside the shell, the surface is *convex*.

To locate and describe an image produced by a spherical mirror, we use the technique of *ray-tracing*. A line called the *axis* is sketched symmetrically through the center of the mirror, perpendicular to the mirror surface. We then consider a point (☆) on the axis and investigate how the mirror affects light rays that leave the object. After reflection, the rays may either converge to form a *real* image, as shown in Figure 29–10(a), or diverge to form a *virtual* image, as shown in Figure 29–10(b). The word "real" signifies that light rays actually converge at the image location to form an image. If a screen were placed there (without interfering with the passage of the rays), an image would appear on the screen. The word "virtual" signifies that light rays do not actually reach the image location, and if a screen were placed there, no image would

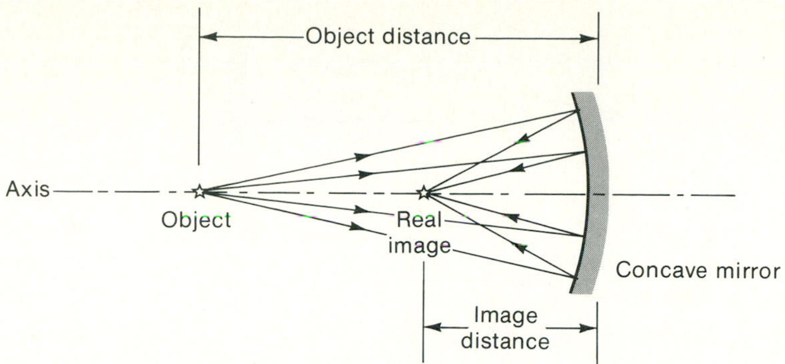

(**a**) All rays from this point object (☆) are reflected by the concave mirror and converge to form a *real* image.

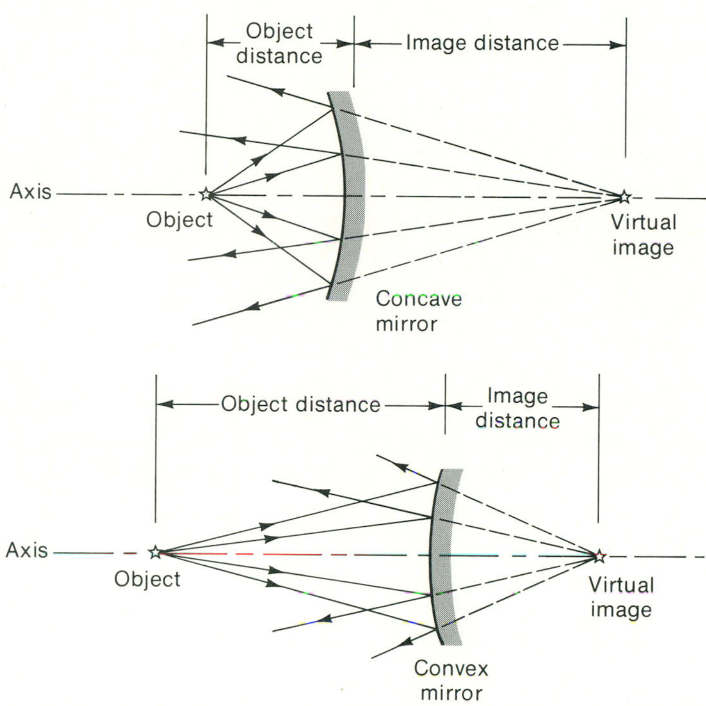

(**b**) All rays from these point objects (☆) are reflected by a concave or a convex mirror, so they diverge to form a *virtual* image.

Figure 29–10
Image formation by spherical mirrors.

appear on the screen. In either case, if our eyes were in a position to intercept the rays after they leave the mirror, we would visually see an image at that location. Without other clues we could not know whether the image was real or virtual: Both types of images have the same visual appearance.

To find the location of the image, we trace two rays whose paths we can easily determine. All reflected rays pass through the same point, so determining the paths of just two rays is sufficient to locate the image. We use the following notation, as shown in Figure 29–11: The center of curvature is at C, the object is at O, the image is at I, and the position of the mirror is at M. We restrict our considerations to those cases for which the angles involved are small enough that the tangent of the angle is approximately equal to the angle itself in radian measure. Accepting this approximation, the results of the ray-tracing analysis are valid for mirrors whose diameters are much smaller than

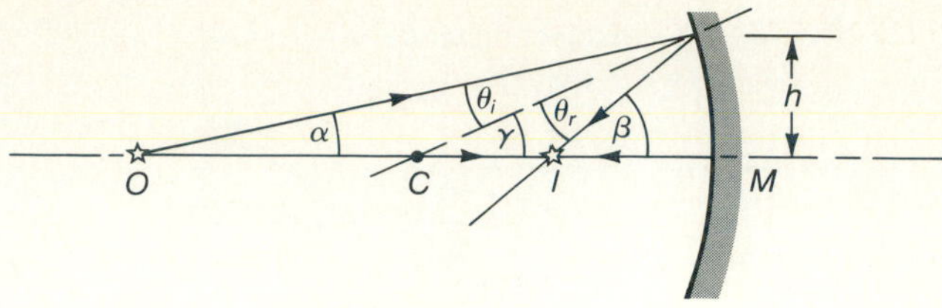

(a) Case 1. Concave mirror: real image

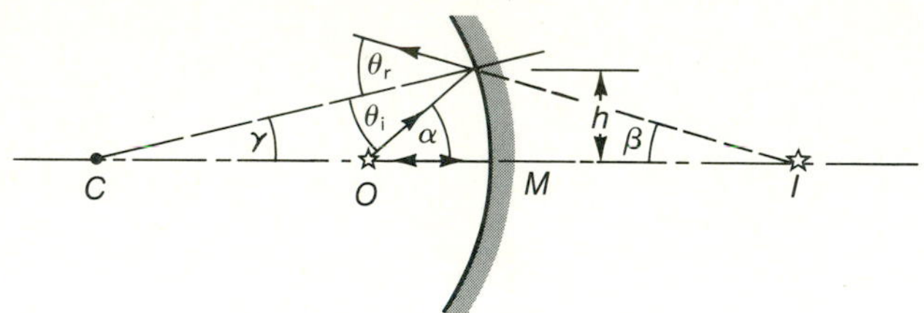

(b) Case 2. Concave mirror: virtual image

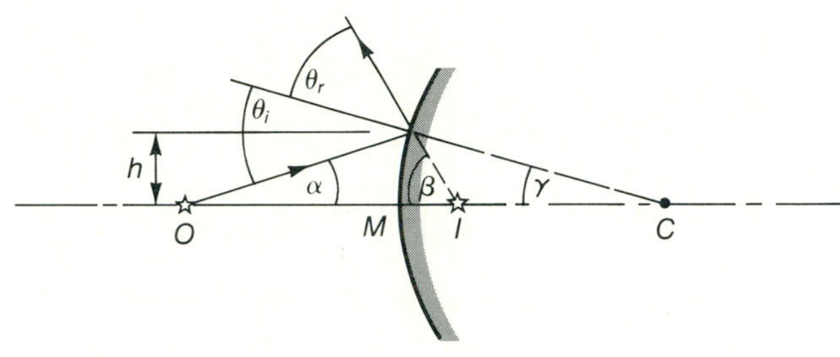

(c) Case 3. Convex mirror: virtual image

Figure 29–11
Ray-tracing analysis for spherical mirrors. One ray, from the object O, travels along the axis and is reflected backwards along the axis. The other ray travels at an angle α to the axis and is reflected along a direction at an angle β to the axis. The image is located at the intersection of the two reflected rays.

the radius of curvature.[5] In all cases, we will apply the laws of reflection, and for simplicity, we will drop the subscripts, so that

$$\theta_i = \theta_r = \theta$$

Let us now analyze three different cases of image formation by spherical mirrors and summarize the results in a single, convenient equation known as *the mirror equation.*

Case 1. Concave Mirror: Real Image

In Figure 29–11(a), we trace these two rays: One ray travels from the object *O along the axis* and (since it strikes the mirror perpendicularly) is reflected back along the axis. The other ray travels at *an angle α to the axis* and is reflected along a direction at an angle β to the axis. The point where these two reflected rays intersect is the image location. Because the exterior angle of a triangle is equal to the sum of the opposite interior angles, we have for one triangle

$$\beta = \gamma + \theta$$

[5] Often this criterion is expressed by the phrase "a small-aperture mirror." That is, the aperture (diameter) of the mirror is small compared with its radius of curvature. Since this is a relative matter, an astronomical mirror 3 m in diameter may be classified as a small-aperture mirror, while a mirror 5 cm in diameter may not be.

and for another triangle

$$\beta = \alpha + 2\theta$$

Eliminating θ, we obtain

$$\alpha + \beta = 2\gamma \tag{29–1}$$

Using the small-angle approximation:

$$\alpha \approx \frac{h}{OM}$$

$$\beta \approx \frac{h}{IM}$$

$$\gamma \approx \frac{h}{CM}$$

Treating these expressions as equalities and substituting into Equation (29–1) gives

$$\frac{1}{OM} + \frac{1}{IM} = \frac{2}{CM} \tag{29–2}$$

Note that h and θ do not appear in the expression. This implies that *all* rays emanating from the object and reflected by the mirror will converge to the image point (at least, within the validity of the small-angle approximations used in the derivation).

Case 2. Concave Mirror: Virtual Image

Referring to Figure 29–11(b) and proceeding as in Case 1, we have

$$\theta = \beta + \gamma$$

and

$$\alpha = \theta + \gamma$$

Eliminating θ gives

$$\alpha - \beta = 2\gamma$$

Identifying these angles with their tangents, we have

$$\frac{1}{OM} - \frac{1}{IM} = \frac{2}{CM} \tag{29–3}$$

Case 3. Convex Mirror: Virtual Image

Referring to Figure 29–11(c), again we have

$$\theta = \alpha + \gamma$$

and

$$2\theta = \alpha + \beta$$

Eliminating θ, we obtain

$$\frac{1}{OM} - \frac{1}{IM} = -\frac{2}{CM} \tag{29–4}$$

Note that unlike a concave mirror, which may produce either a real or a virtual image, a convex mirror *always* produces a virtual image of an object.

□ □ □

The ray-tracing analysis of image formation by spherical mirrors produced similar results in all three cases. Equations (29–2), (29–3), and (29–4) are identical in form, varying only in the signs of some terms. It is convenient to summarize the results by deducing a *single* equation that is valid for all cases. We do this by establishing a **sign convention** to determine the sign of the numerical values to be used in that equation. Observe that in Figure 29–11, *OM* is the object distance *p*, *IM* is the image distance *q*, and *CM* is the radius of curvature *R* of the mirror. All object and image distances are measured along the axis to the center (*M*) of the mirror. Equations (29–2), (29–3), and (29–4) may then be combined in a single equation:

MIRROR EQUATION

$$\frac{1}{p}+\frac{1}{q}=\frac{2}{R} \tag{29–5}$$

where p = object distance

q = image distance

R = radius of curvature of the mirror

To use this equation, we must adopt the following sign convention:

SIGN CONVENTION FOR MIRRORS

(1) The numerical value of *p* is positive if the rays *approaching* the surface are *divergent*. Otherwise *p* is negative.

(2) The numerical value of *q* is positive if the rays *leaving* the surface are *convergent*. Otherwise *q* is negative.

(3) The numerical value of *R* is positive if the mirror is concave, and it is negative if the mirror is convex.

You should memorize the sign convention because problems in this chapter cannot be solved without following these rules for signs. It is important to remember that the mirror equation is always written as in Equation (29–5). *Minus signs are introduced only when substituting numerical values.* This same procedure is followed with all general equations in physics.

In certain cases of multiple-mirror systems, the object distance *p* can be negative. For example, in Figure 29–12 the first mirror, acting alone, would produce a real image at I_1. In a sense, this image becomes the object for mirror 2 (with an object distance p_2). However, since mirror 2 intercepts the rays before they form the image, the rays that strike mirror 2 are *converging.*

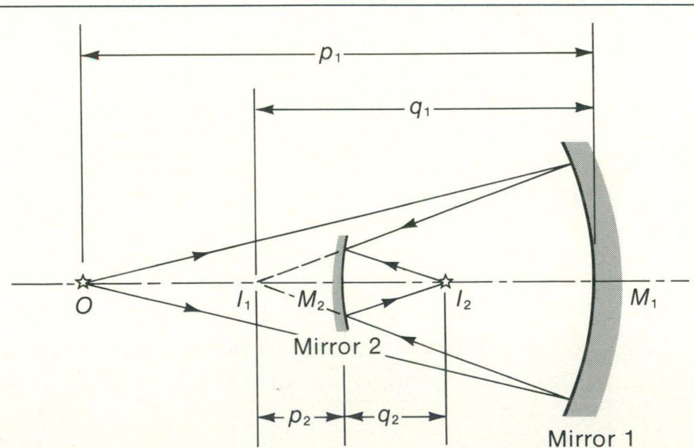

Figure 29–12
A multiple-mirror system in which the object distance p_2 is negative ($p_2 < 0$) according to the sign convention. This is because the rays approaching mirror 2 are converging.

According to the sign convention, the numerical value of p_2 would therefore be negative. In such cases, the object is called a *virtual object.*

A common term applied to mirrors (and lenses) is the **focal length** f (see Figure 29–13). A group of rays parallel to the axis will be reflected by a *concave* mirror so that they converge to a point a focal length f in front of the mirror. The point at which they focus is the **focal point**[6] F. If parallel rays are incident on a *convex* mirror, they reflect along divergent lines that meet at a focal-length distance behind the mirror. Parallel incoming rays imply that the object is at infinity, or $p = \infty$. Substituting this value into the mirror equation:

$$\frac{1}{p} + \frac{1}{q} = \frac{2}{R}$$

we obtain

$$\frac{1}{\infty} + \frac{1}{q} = \frac{2}{R}$$

where q becomes equal to the focal length f.

Substituting f for q, we have

$$f = \frac{R}{2} \qquad \textbf{(29–6)}$$

Since the numerical value of the focal length is *positive* for *concave* mirrors and *negative* for *convex* mirrors, Equation (29–5) becomes

MIRROR EQUATION (alternative form)

$$\frac{1}{p} + \frac{1}{q} = \frac{1}{f} \qquad \textbf{(29–7)}$$

In this chapter it is easy to become confused in the discussions of numerous cases of mirrors and lenses in a variety of situations. However, the major content of the chapter is the single equation $1/p + 1/q = 1/f$, which is the starting point for locating images formed by both mirrors and lenses. *Knowing the sign convention is essential.* One easy way to remember the sign convention is the following: For the "standard setup" of an object situated farther from a converging mirror than the focal-length distance, the symbols p, q, f, and R each have *positive* numerical values. If any of these distances are on the opposite side of the mirror (compared with their locations in this standard setup), they are negative. The following two examples will illustrate the use of the mirror equation and the sign convention.

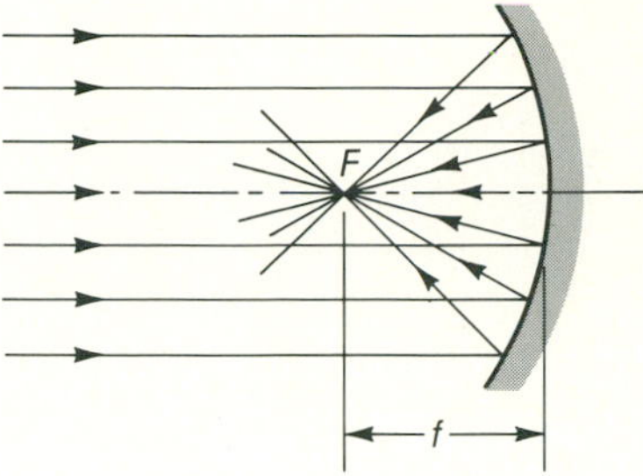

(a) Concave mirror

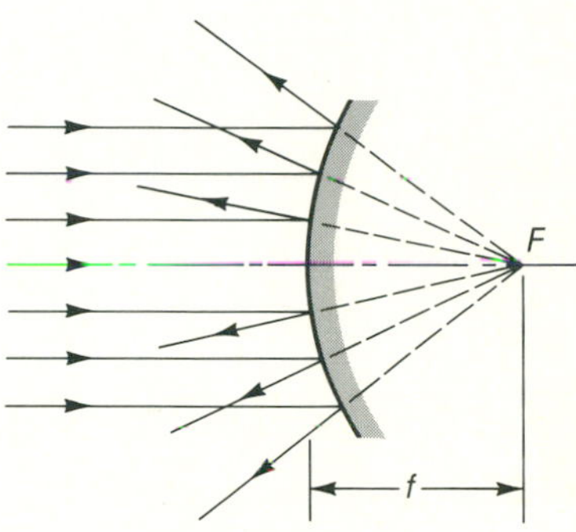

(b) Convex mirror

Figure 29–13
When light parallel to the axis is incident on a mirror, the image distance is called the *focal length* f of the mirror. The point F is the *focal point.* For concave mirrors, f is positive; for convex mirrors, it is negative.

EXAMPLE 29–1

While holding his shaving mirror near a window, a man is able to produce the image of the sun on the wall next to the window. The window is 50 cm from the wall. When shaving, his chin is 20 cm in front of the mirror. Find the image of his chin.

SOLUTION

Light rays from the sun are essentially parallel, so that the image of the sun is produced at the focal point of the mirror. Thus, $f = +50$ cm. (We know that f is positive from the fact that only concave mirrors are capable of producing a *real* image of an object, and

[6] Calling F the focal point of the mirror does *not* mean that all images are formed at that location. Only for the single case of incident light parallel to the axis is this true; in all other cases, the image is elsewhere. It is helpful to think of F as a point that "belongs" to the mirror and that we find useful in constructing ray diagrams.

according to the sign convention, concave mirrors have positive focal lengths.) Light from a point on the man's chin is diverging as it approaches the mirror, so according to the sign convention, $p = +20$ cm. Starting with the mirror equation:

$$\frac{1}{p} + \frac{1}{q} = \frac{1}{f}$$

and then substituting:

$$\frac{1}{+20 \text{ cm}} + \frac{1}{q} = \frac{1}{+50 \text{ cm}}$$

$$q = \boxed{-33.3 \text{ cm}}$$

According to the sign convention, the minus sign indicates that the light diverges from the surface as if it came from a virtual image *behind* the mirror. (Again, "virtual" implies that no rays are actually present at the image location.) So the image is 33 cm behind the mirror (see Figure 29–14).

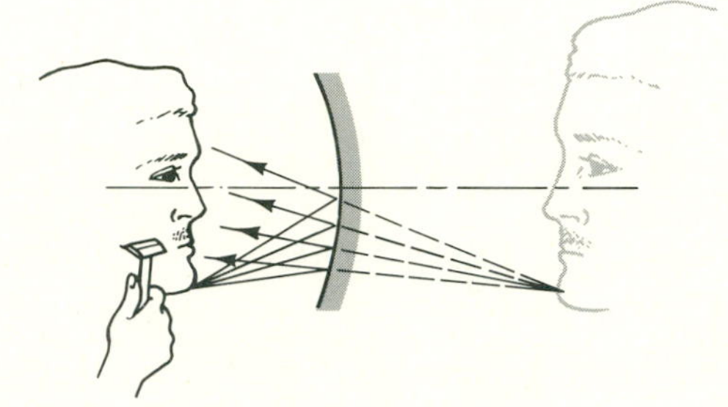

Figure 29–14
Example 29–1

EXAMPLE 29–2

Consider the system of mirrors shown in Figure 29–15. Locate the final image of the object. Is the image real or virtual?

SOLUTION

The procedure in a multiple-mirror system is to find the image formed by each mirror acting alone in the order in which the rays are reflected. In this case mirror 1 is first. Starting with the mirror equation:

$$\frac{1}{p} + \frac{1}{q} = \frac{1}{f}$$

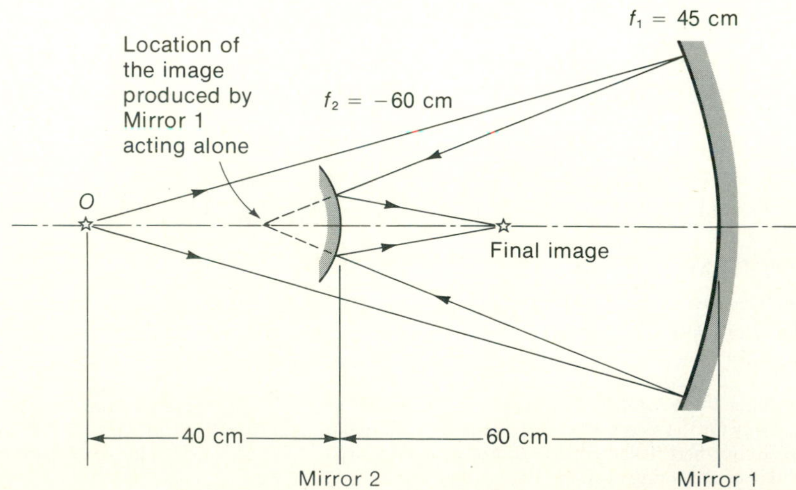

Figure 29–15
Example 29–2

and then substituting:

$$\frac{1}{+(40\text{ cm} + 60\text{ cm})} + \frac{1}{q_1} = \frac{1}{45\text{ cm}}$$

$$q_1 = +82\text{ cm}$$

According to the sign convention, the positive value signifies that light rays are converging as they leave the first mirror. If this mirror were the only one present, a real image would be formed 82 cm in front of mirror 1. However, mirror 2 intercepts these rays 22 cm before that image can be formed. Nevertheless, we consider that hypothetical "image" to be the object for mirror 2 to work upon. Because the rays impinging on mirror 2 are converging, the object distance p_2 is negative ($p_2 = -22$ cm, according to the sign convention). The object is a virtual object because no light rays are actually present at the object. Substituting appropriate numerical values into the mirror equation gives

$$\frac{1}{p} + \frac{1}{q} = \frac{1}{f}$$

$$\frac{1}{-22\text{ cm}} + \frac{1}{q_2} = \frac{1}{-60\text{ cm}}$$

$$q_2 = \boxed{34.7\text{ cm}}$$

The positive sign indicates that the rays converge upon leaving the second mirror. The final image is thus *real* and is located 34.7 cm to the right of mirror 2, as shown in Figure 29–15. Unless otherwise intercepted, convergent rays always produce real images.

When virtual objects are involved, it is usually not easy to construct significant ray diagrams for the multiple reflections. After all, no light actually travels from the location of the virtual object to the next mirror. So in these cases, just a preliminary ray diagram may be sketched to verify the location of the first image, then the mirror equation is used to find the final image produced by the second mirror.

29.5 Ray Diagrams: Linear Magnification

The primary function of a shaving or makeup mirror is to produce an enlarged image. In this section we will extend the ray-tracing technique to discover how an object point that is *off the axis* of the mirror is imaged. In sketches, an object that extends off the axis is usually indicated by an arrow labeled O, as in Figure 29–16. All the rays that leave the tip and strike the mirror are brought to a

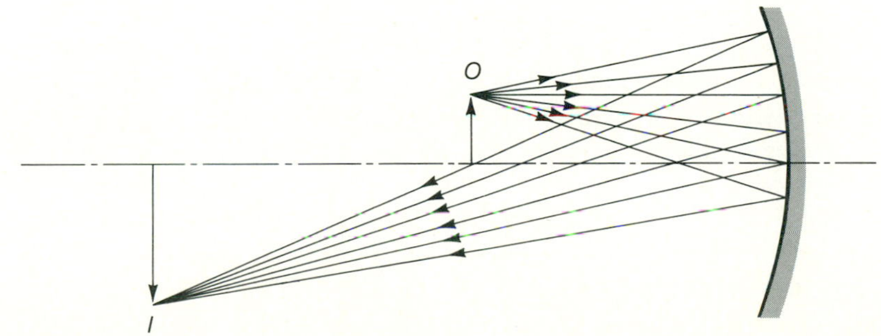

Figure 29–16
All rays emerging from the arrow tip that strike the mirror are brought to a focus at the image of the arrow tip.

focus at the tip of the image I (at least, this is true for small-aperture mirrors). If we can identify just two rays that intersect at I, it will be sufficient to locate the image by ray-tracing. Although other choices are possible, we choose the following two rays for such diagrams because it is easy to sketch what happens to them after they strike the mirror:

RAYS USED IN RAY-TRACING DIAGRAMS

(1) A ray striking the center of the mirror is reflected symmetrically. (The angle of incidence equals the angle of reflection.)

(2) A ray parallel to the axis passes through the focal point *F*.

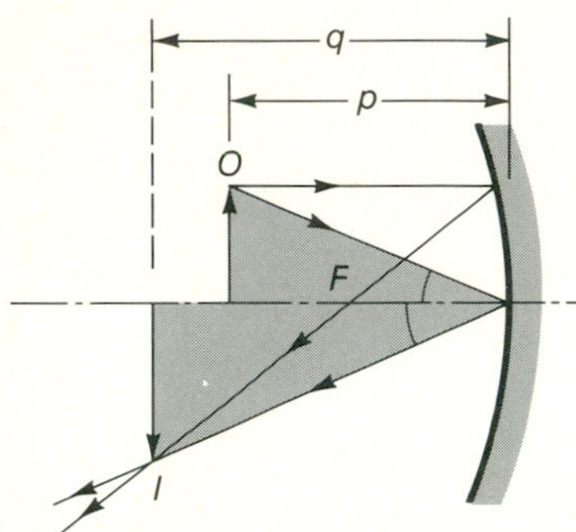

(a) Case 1. Concave mirror: real image

From these two reflected rays we locate the image; other portions of the arrow are similarly imaged on a point-to-point basis.

As in the last section, we will treat each of the three possible cases separately. Remember that *concave* mirrors have positive focal lengths and positive radii of curvature, both located in front of the mirror. In contrast, *convex* mirrors have negative focal lengths and negative radii of curvature, both located behind the mirror. For all mirrors, $|f| = R/2$.

Case 1. Concave Mirror: Real Image

Referring to Figure 29–17(a), the object O is the tip of the arrow located a distance p from the mirror. Two rays are drawn from the tip. One ray strikes the center of the mirror and is reflected symmetrically (the angle of incidence equals the angle of reflection). The other ray approaches the mirror parallel to its axis and is reflected through the focal point F. The intersection of these two rays locates the image I of the arrow tip.

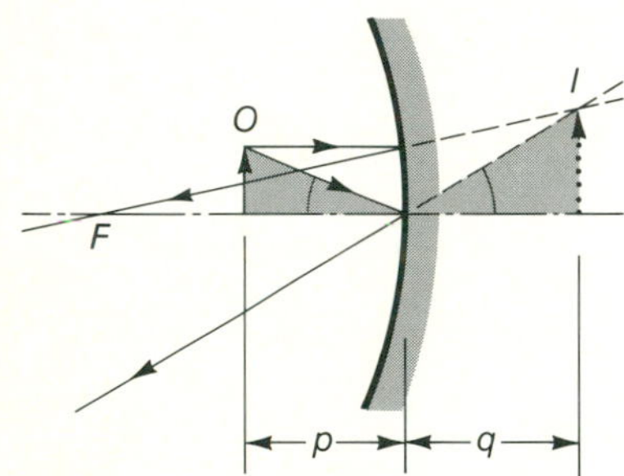

(b) Case 2. Concave mirror: virtual image

The size of the images formed is a significant feature of optical systems. They may be larger or smaller than the object. In a ray-tracing diagram, the triangles formed by the axis, the object, and the image lead to a simple expression of the **linear magnification** M:

$$M \equiv \frac{\text{Image size}}{\text{Object size}} \qquad \textbf{(29–8)}$$

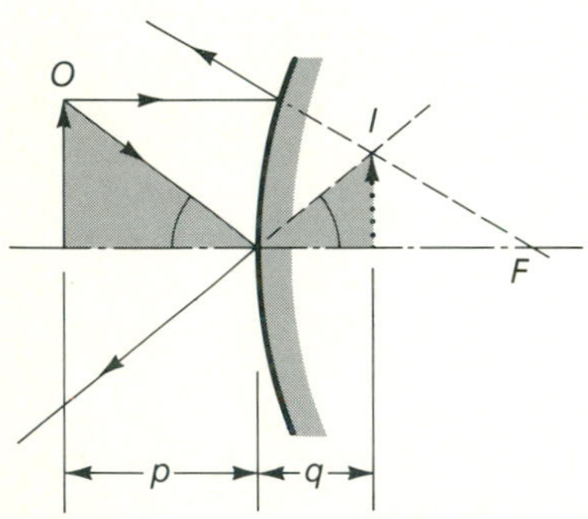

(c) Case 3. Convex mirror: virtual image

Figure 29–17

Magnification by spherical mirrors. Extensions of rays behind the mirrors are represented by dashed lines. Virtual images are indicated by dotted lines.

Note that the shaded triangles in Figure 29–17(a) are similar right triangles with corresponding sides having the same ratio. Then:

LINEAR MAGNIFICATION

$$M = -\frac{q}{p} \qquad \textbf{(29–9)}$$

The minus sign is introduced so that a *negative* value of M indicates an *inverted* image and a *positive* value of M indicates an *erect* image. This same sign convention holds true for all cases of mirrors and lenses.

As may be verified by sketching ray diagrams, if a *real* image is created by a concave mirror, it is always inverted and may be larger than, the same size as, or smaller than the object. It is not necessary to memorize such details for various cases, since the information is contained inherently in the sign convention and in the definition for the linear magnification.

Case 2. Concave Mirror: Virtual Image

If the object is placed closer than the focal point to a concave mirror, the image is virtual, as shown in Figure 29–17(b). As in the first case, we use two rays: One leaves the tip of the arrow *parallel to the* axis and is then reflected through the focal point F; the other ray is *reflected symmetrically at the center of the mirror.* Unlike the first case, the rays diverge after reflection, seemingly from a point behind the mirror. This point is the image of the arrow tip. We locate it by extending the two reflected rays backward along their directions until they intersect. The point of intersection is the image point. But because no actual light rays travel along these extended lines, we draw them dashed. Also, because no actual light rays form the image, it is a *virtual* image, which we sketch with dotted lines.

The shaded triangles are again similar, so that (as before) the linear magnification is

$$M = -\frac{q}{p}$$

As may be verified by sketching ray diagrams, if a concave mirror forms a virtual image, it is always erect and always larger than the object. A shaving or makeup mirror is concave and, when held the proper distance from the face, produces an erect, virtual image behind the mirror.

The two cases just discussed differ in important ways. In Case 1, the object is *farther* than a focal-length distance from the mirror and produces an inverted, real image. In Case 2, the object is *closer* than a focal-length distance from the mirror and produces an erect, virtual image.

Case 3. Convex Mirror: Virtual Image

Referring to Figure 29–17(c), one ray from the tip of the arrow is reflected symmetrically at the center of the mirror. The other ray approaches the mirror parallel to the axis and is reflected in a direction *away* from the focal point. (Remember that the center of curvature as well as the focal point of a convex mirror lie behind the mirror.) Again, the linear magnification is

$$M = -\frac{q}{p}$$

Convex mirrors always form virtual images (of real objects) and are always smaller than the object. Images seen in polished balls are of this type.

To describe an image, we specify the following:

real or virtual
erect or inverted
magnification

Again, do not try to memorize rules for all the types of imaging that result when objects are at various distances from converging and diverging mirrors (and lenses). Instead, gain skill in rapidly sketching ray diagrams, which reveal the nature of the image. This approach is much simpler and enables you to deal with new situations you have not seen before. For numerical calculations, knowing the sign convention is essential.

EXAMPLE 29–3

A concave mirror rests face up on a table 0.50 m below a desk lamp bulb, as in Figure 29–18. An inverted image of the bulb appears on the ceiling in clear focus and is five times the size of the bulb in the lamp. (a) How high is the ceiling above the table top? (b) Find the focal length of the mirror.

Figure 29–18
Example 29–3

SOLUTION

(a) The situation described in this example is highly unlikely (except by chance). Ordinarily a clear image would not be produced on the ceiling because, for a given focal length, a definite relationship between the object and image distances must exist. This relationship is the mirror equation:

$$\frac{1}{p} + \frac{1}{q} = \frac{1}{f}$$

Since, in this example, we are given only the object distance p, we still have two unknowns: q and f. An additional relationship between p and the image distance q is needed. With the linear magnification M known, Equation (29–9) is appropriate:

$$M = -\frac{q}{p}$$

Using the numerical values of $M = -5$ (it is negative because the image is inverted) and $p = 0.5$ m (it is positive because the rays from the bulb diverge when striking the mirror):

$$-5 = -\frac{q}{0.5\text{ m}}$$

or

$$q = \boxed{2.5\text{ m}}$$

The distance from the table top to the ceiling is 2.5 m. Because q has a positive numerical value, we know that the rays leaving the mirror are converging (as they must to form a *real* image on the ceiling).

(b) Now that we know two of the three unknowns, we can apply the mirror equation:

$$\frac{1}{p} + \frac{1}{q} = \frac{1}{f}$$

Substituting numerical values, we obtain

$$\frac{1}{+0.5\text{ m}} + \frac{1}{+2.5\text{ m}} = \frac{1}{f}$$

Solving for the focal length f, we obtain

$$f = \boxed{0.417\text{ m}}$$

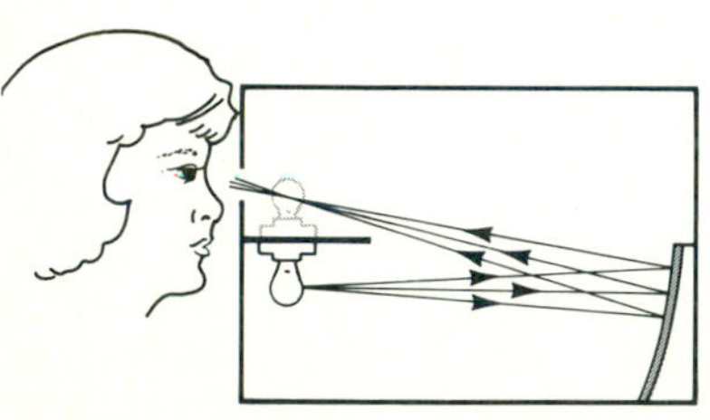

Figure 29–19
The observer views a real image of the lighted bulb in the empty socket. With a good-quality mirror, the illusion is strikingly realistic.

Because concave mirrors can produce real images of almost any magnification, they are particularly useful in optical instruments. An unusual example of the varied applications of a concave mirror is shown in Figure 29–19. A light bulb in a socket is mounted upside down beneath a flat surface and positioned so the bulb is at the center of curvature of a concave mirror at the back of the box. An empty socket is mounted on top of the surface, also at the center of curvature, in clear view of the observer. When the bulb is lighted, a real image of the bulb appears in the empty socket. This is one of the many fascinating illusions that may be produced with mirrors.

29.6 Refraction at a Plane Surface

The speed of light we have referred to so often has been the speed in a vacuum or in air. Actually, the speed of light in air v_{air} is slightly less than that in a vacuum:

$$\frac{c}{v_{air}} = 1.000\,29 \qquad \text{(at 0°C and 1 atm)}$$

The ratio is defined as the **refractive index** n of the air. For any medium:

REFRACTIVE INDEX

$$n = \frac{c \text{ (speed of light in a vacuum)}}{v \text{ (speed of light in a medium)}} \tag{29–10}$$

The refractive index (also called the *index of refraction*) of air is thus 1.000 29. Although for most purposes, the speed of light in air may be approximated as its speed in a vacuum, the speed of light in solid, transparent materials is considerably less than its speed in a vacuum. Table 29–2 illustrates how widely the index of refraction varies from one material to another.

Notice that the speed of light in water is only about three-fourths its speed in a vacuum. The brilliance of a cut diamond exists primarily because the speed of light in a diamond is only about four-tenths its speed in a vacuum. In general, we speak of materials with a high index of refraction as being *optically dense* materials. If we compare different transparent materials, we find that sometimes those with the greater (physical) density also have the greater optical density. But there are many exceptions. For example, most oils, which float on water, have a greater index of refraction than water. And many transparent

TABLE 29–2

The Refractive Index of Some Representative Materials

Substance	Refractive index n (for $\lambda \approx 550$ nm)
Air (0°C, 1 atm)	1.000 29
Hydrogen (0°C, 1 atm)	1.000 13
Ice	1.31
Water	1.333 ($= \frac{4}{3}$)
Crown glass	1.52
Polystyrene	1.59
Flint glass	1.66
Diamond	2.42

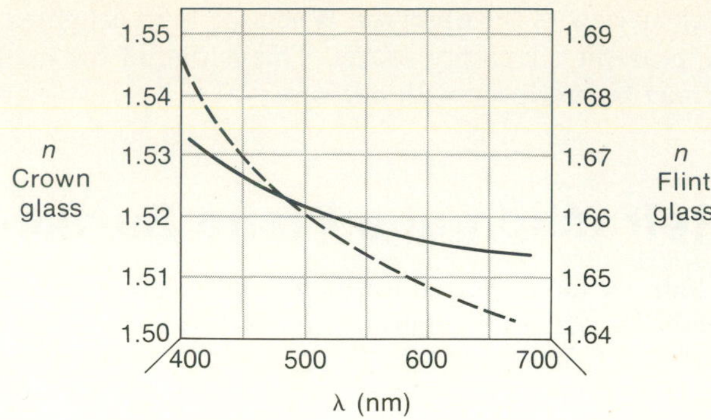

Figure 29–20
Dispersion curves. The refractive index for crown glass is on the left-hand vertical axis, with the curve shown as a solid line. The refractive index for dense flint glass is shown on the right, with the curve indicated as a dashed line.

plastics have greater indices of refraction than ordinary crown glass, which is denser. So there is no general rule that relates physical density with optical density.

One of the complicating factors in the design of optical instruments is that the glass used for making lenses does not have a constant index of refraction. The index of refraction not only depends on temperature and pressure but also on the wavelength. Typically the refraction index will vary about 2% over the visible spectrum, decreasing with increasing wavelength. A dense flint glass has a refractive index of 1.69 at 410 nm, decreasing to 1.64 at 670 nm, as indicated in Figure 29–20. Curves displaying the refraction index as a function of wavelength are called *dispersion curves.* The property of **dispersion** can be useful or troublesome, as we will see in later discussions.

A direct consequence of different refractive indices for different materials is that the direction of a light ray changes when it encounters a change in refractive index. The bending of a light ray in this manner is called **refraction.** Consider the situation shown in Figure 29–21. A plane wave traveling in a less optically dense material (one with a lower refractive index n_1) encounters a plane interface and goes into a more optically dense material (one with a higher refractive index n_2). In the material with a refractive index n_1, the wave moves with a speed v_1, where (by the definition of a refractive index):

$$v_1 = \frac{c}{n_1} \qquad \textbf{(29–11)}$$

Similarly, for the material with refractive index v_2:

$$v_2 = \frac{c}{n_2} \qquad \textbf{(29–12)}$$

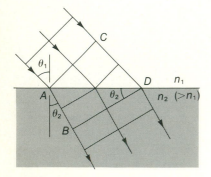

Figure 29–21
Refraction of a plane wave by a plane interface between two materials.

Applying Huygens' principle to the wavefront AC, we note that in the time t required for a secondary wavelet to move from C to D, another secondary wavelet will move from A to B. That is:

$$t = \frac{CD}{v_1}$$

and

$$t = \frac{AB}{v_2}$$

Therefore: $$\frac{CD}{v_1} = \frac{AB}{v_2}$$

Utilizing Equations (29–11) and (29–12), we obtain

$$n_1 CD = n_2 AB \tag{29–13}$$

Since CD is a side of the right triangle ACD and AB is a side of the right triangle ABD, and both triangles have the common side AD:

$$CD = AD \sin \theta_1$$

and $$AB = AD \sin \theta_2$$

Substituting these relations into Equation (29–13), we obtain:

SNELL'S LAW FOR REFRACTION

$$n_1 \sin \theta_1 = n_2 \sin \theta_2 \tag{29–14}$$

This equation expresses **Snell's law,**[7] where θ_1 is the angle between the incident ray and the *normal* to the interface between the materials and θ_2 is the angle between the refracted ray and the same normal. Just as in the case of reflection, the incident ray, the normal to the interface, and the refracted ray lie in the same plane.

EXAMPLE 29–4

A tin can 14 cm high and 12 cm in diameter is filled with an unknown liquid. An observer looking along a direction 25° above the horizontal (see Figure 29–22) can barely see the inside bottom edge of the can. Find the index of refraction of the liquid.

SOLUTION

The light ray from the bottom edge incident within the liquid on the top surface has an angle of incidence $\theta_1 = \tan^{-1}(\frac{12}{14}) = 40.6°$. The index of refraction of air is $n_2 = 1.000$ (to four significant figures). We apply Snell's law to the refraction of the ray as it emerges into the air with an angle of refraction $\theta_2 = 65°$.

$$n_1 \sin \theta_1 = n_2 \sin \theta_2$$

$$n_1 \sin 40.6° = (1.00) \sin 65°$$

$$n_1 = \frac{\sin 65°}{\sin 40.6°}$$

$$n_1 = \frac{0.906}{0.651}$$

$$n_1 = \boxed{1.39}$$

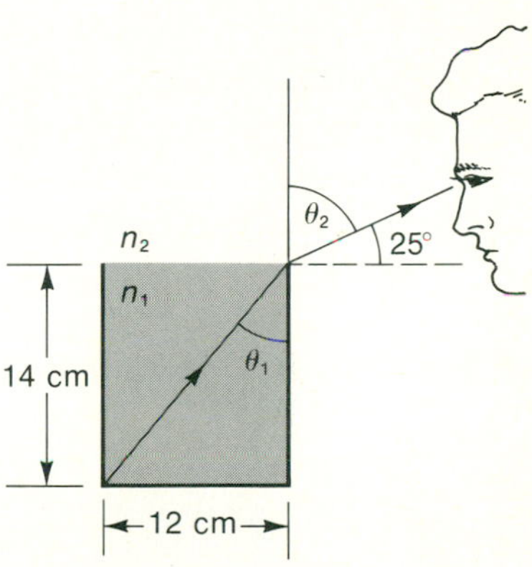

Figure 29–22
Example 29–4

When looking straight down into a pail of water resting on the floor, the bottom of the pail appears to be noticeably above the floor level. How do we visually judge distance? Human depth perception involves a variety of mechanisms. One clue is the comparison we make between the known size of an object and its perceived size. For distant landscapes, atmospheric haze provides additional helpful information. (In the absence of such haze, one can be fooled into

[7] Snell's law was named after its discoverer, the Dutch physicist Willebrand Snell (1591–1626).

greatly underestimating the distance of "nearby" mountains.) For objects close to us, an aid is the parallax effect that occurs when we move our head slightly. Also, we need to "aim" each eye along slightly different directions in order to match the divergence of light rays as they leave a nearby object: Our minds, through experience, relate this "aiming" effect to distance estimation. The next example utilizes this last method of judging distance.

EXAMPLE 29–5

An observer looks straight down into the same tin can of fluid that was described in Example 29–4. What is the apparent depth of the fluid?

SOLUTION

Referring to Figure 29–23(a), the two rays shown coming from a point on the bottom of the can diverge as they approach the top of the fluid. As they proceed into the air, they diverge even more due to refraction. To the observer, the rays will appear to originate from a point at a depth *d* below the surface. To emphasize the refraction at the surface water, an exaggerated view is shown in Figure 29–23(b). The angles involved in this example are so small that we may use the small-angle approximations:

$$\sin\theta_1 \approx \tan\theta_1 \approx \theta_1 \qquad \textbf{(29–15)}$$

and

$$\sin\theta_2 \approx \tan\theta_2 \approx \theta_2$$

(For simplicity of notation, we replace the approximately equal sign with the equal sign in the discussion that follows.) From trigonometry, we have

$$x = d\tan\theta_2 \qquad \textbf{(29–16)}$$

and

$$x = H\tan\theta_1$$

Eliminating x between these equations and using the small-angle approximations, we obtain

$$\theta_1 H = \theta_2 d \qquad \textbf{(29–17)}$$

Snell's law relates θ_1 and θ_2:

$$n_1 \sin\theta_1 = n_2 \sin\theta_2$$

Again applying the small-angle approximations:

$$n_1\theta_1 = n_2\theta_2 \qquad \textbf{(29–18)}$$

We combine Equations (29–17) and (29–18) to obtain

$$d = H\left(\frac{n_2}{n_1}\right) \qquad \textbf{(29–19)}$$

Substituting numerical values yields

$$d = 14\text{ cm}\left(\frac{1.00}{1.39}\right)$$

$$d = \boxed{10.1\text{ cm}}$$

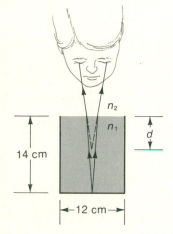

(a) Looking straight down into a can of fluid.

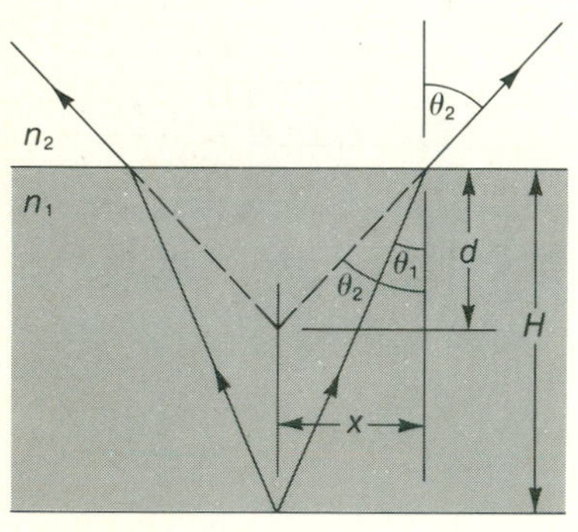

(b) An exaggerated sketch of the refraction that occurs at the liquid surface.

Figure 29–23
Example 29–5

Fermat's principle applies to refracted rays just as it applies to reflected rays. That is, of all the rays emanating from a point on one side of a plane interface between two media, only one will pass through a given point on the other side of the interface. *That ray takes the path that requires the least time in transit.* Refer to Figure 29–24 for a discussion of this principle.

29.7 Total Internal Reflection

Whenever a light ray traveling in a medium of high refractive index strikes an interface or a medium with lower refractive index, there is always some reflection. However, under certain conditions the reflection is 100% and no light is transmitted through the interface. To see this, we start with Snell's law:

$$n_1 \sin\theta_1 = n_2 \sin\theta_2$$

In Figure 29–25, as the angle of incidence increases, the angle of refraction approaches 90°. Since $\sin 90° = 1$, we have

$$\sin\theta_1 = \left(\frac{n_2}{n_1}\right)1$$

For angles of incidence larger than this **critical angle** θ_c, *total internal reflection* occurs and there is no refracted ray.

CRITICAL ANGLE θ_c FOR TOTAL INTERNAL REFLECTION

$$\sin\theta_c = \frac{n_2}{n_1} \quad (\text{for } n_2 < n_1) \qquad \textbf{(29–20)}$$

(There is, of course, no critical angle of light traveling from a medium with a lower refractive index into a medium with a higher refractive index.)

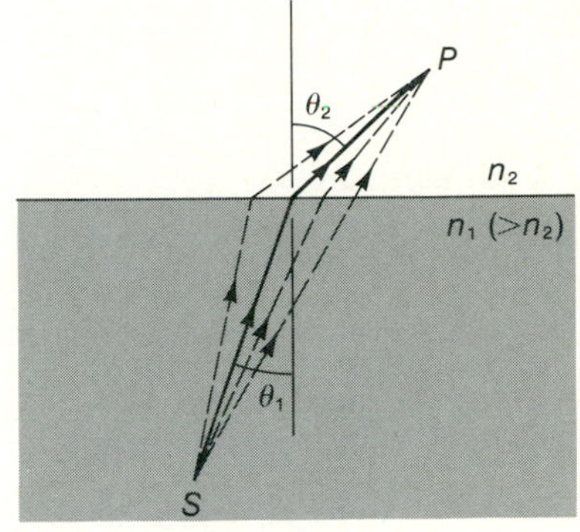

Figure 29–24
Fermat's principle applied to refraction. Of all the rays emanating from a source S, only one will pass through the point P. This ray lies in a plane, satisfies Snell's law ($n_1 \sin\theta_1 = n_2 \sin\theta_2$), *and requires the least time of transit from S to P.* Even though some of the alternative paths indicated by the dashed lines are shorter in distance, they are longer in *travel time,* so that no rays of light take these other routes in traveling from S to P.

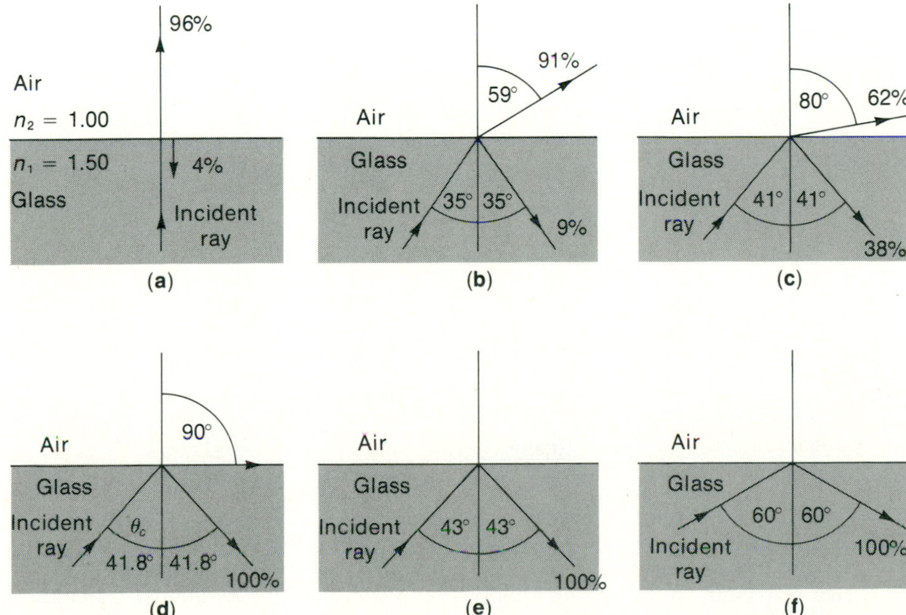

Figure 29–25
Reflection of light at an interface. *Total internal reflection* occurs for all angles of incidence greater than the critical angle θ_c. The percentages of the incident light energy present in the reflected and refracted rays are indicated in each case.

Total internal reflection is used in a variety of practical applications. For example, in binoculars (Figure 29–26), the image is made erect by several reflections at 45°. Since 45° is greater than the critical angle for glass, a 45° prism is used rather than a mirror with a silver coating (which might become tarnished with time); the reflection is 100% from the interior glass interfaces. Similarly, "corner" reflectors made of solid glass, whose three faces meet mutually at 90°, are used to reverse the direction of any light ray incident on the reflector. Arrays of these corner reflectors have been placed on the moon to reflect laser pulses sent from the earth. Precise timing of the round trip of a pulse enables earth-moon distances to be determined within a few centimeters, aiding studies of continental drifts, effects of tides, and numerous other investigations.

Because of total internal reflection, light can be conducted along a flexible, transparent tube called a *light pipe* provided the angle of bending is small enough to ensure that all angles of incidence are greater than θ_c. Such light pipes only a few hundredths of a millimeter in diameter are used for long-distance transmission of a laser beam whose light is modulated to carry simultaneously many radio and TV programs or several hundred telephone channels. The theoretical limit is far above these figures, and rapid progress

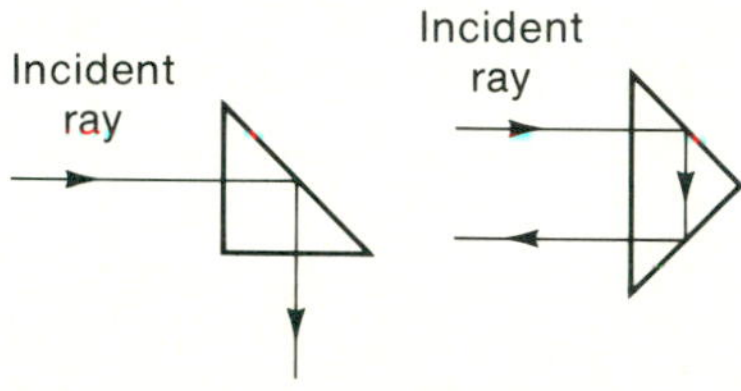

(**a**) Glass prisms (45°–45°–90°) reflect light rays by total internal reflection.

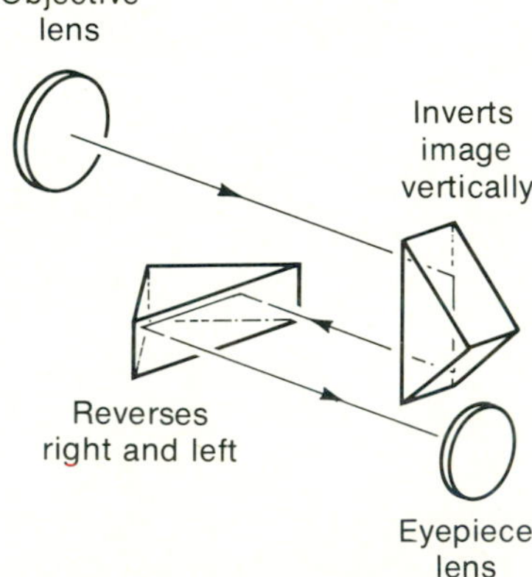

(**b**) Optical path for one eye of prism binoculars. The use of two 45° prisms oriented at right angles causes the final image to be upright and not reversed right-for-left. Because the magnification is proportional to the focal length of the objective lens, the use of prisms "folds" the long optical path into a shorter, more convenient length.

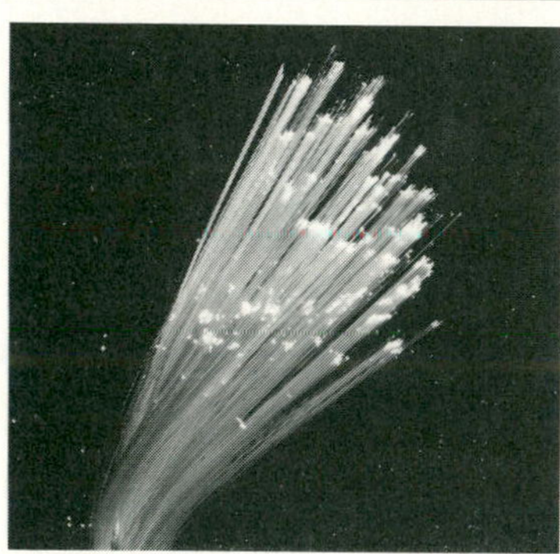

(**c**) Light is transmitted through glass fibers by total internal reflection.

(**d**) Thousands of transparent fibers are held parallel forming a *light pipe*, so that they transmit a true image along the length of the pipe.

Figure 29–26
Examples of total internal reflection.

is being made in increasing the rate of information transmission by this method. Glass fibers a few thousandths of a millimeter are sufficiently flexible that bundles of them can be used as probes, enabling physicians to see internal parts of the body or technicians to view inaccessible parts of a mechanism. Nature has used this principle of *fiber optics* for millions of years: Certain insects and crustaceans have visual sensors that consist of bundles of crystalline light pipes, which transmit light between an array of outer corneal lenses and light-sensing cells deeper within the insect body. Finally a clever coding device can be constructed of a short length of thousands of parallel fibers that are fixed at each end, *A* and *B*. The fibers in the middle of the cable are twisted in a random fashion and held fast with epoxy resin as the cable is cut in two. The message to be coded is held against end *A*. The cut end is then photographed, forming a picture of random light and dark dots. The coding is unbreakable. Only when the other matched half of the cut cable is held against the photograph can the random pattern be unscrambled, as the original message emerges at end *B*.

EXAMPLE 29–6

Find the critical angle for transparent plastic of refractive index 2.14 immersed in oil of refractive index 1.63.

SOLUTION

The critical angle is given by Equation (29–20):

$$\sin \theta_c = \frac{n_2}{n_1}$$

$$\sin \theta_c = \frac{1.63}{2.14} = 0.762$$

$$\theta_c = \boxed{49.6^\circ}$$

29.8 Refraction at a Spherical Surface

Most familiar optical instruments utilize lenses rather than mirrors because of their durability and ease of combination with other elements of an optical system. As a first step in the study of lenses, we will investigate how light is refracted when it is incident on a glass surface that has a spherical curvature. This approach introduces a technique for studying lenses and also has useful applications in itself. Consider a point object on the axis of a spherical interface between two media, as in Figure 29–27(a). We first take the case where all the rays are refracted sufficiently to intersect the axis inside the medium. We will show that within the small-angle approximation we have been making, all rays converge to form a real image at the point *I*.

Tracing a single ray, as in Figure 29–27(b), we find that it intersects the axis at the image distance *CI*. Using the fact that the exterior angle of a triangle is equal to the sum of the opposite interior angles, we have for one triangle

$$\theta_1 = \alpha + \gamma \tag{29–21}$$

and for another triangle

$$\gamma = \theta_2 + \beta \tag{29–22}$$

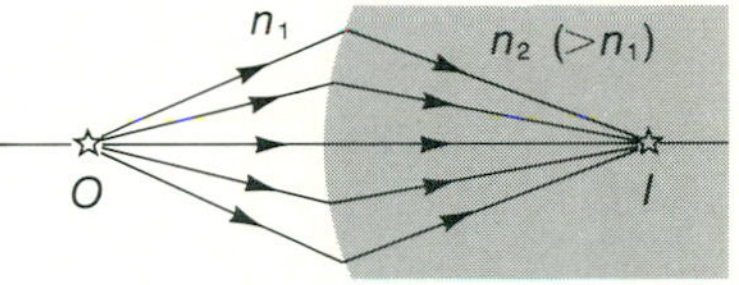

(a) A real image is formed by the convergence of rays refracted at the spherical interface.

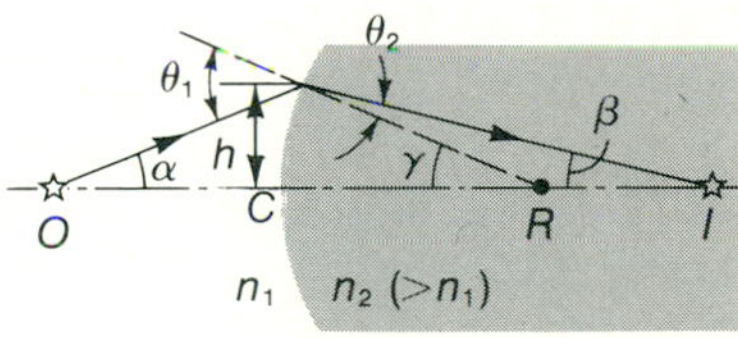

(b) Tracing a single ray from the object *O* to the image *I*.

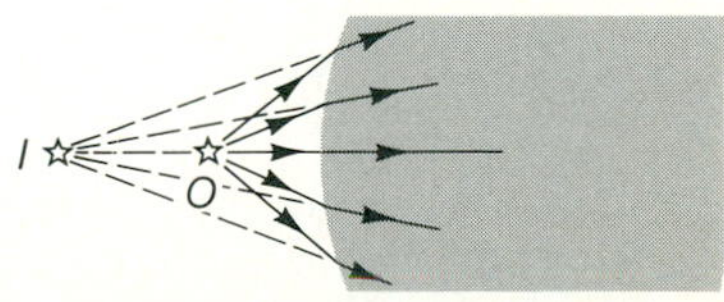

(c) If the refraction is insufficient to produce converging rays, a virtual image is formed outside the medium.

Figure 29–27
Refraction at a spherical interface between two media.

We may eliminate θ_1 and θ_2 by multiplying these equations by the appropriate refractive indices:

$$n_1\theta_1 = n_1\alpha + n_1\gamma$$
$$n_2\theta_2 = n_2\gamma - n_2\beta$$

and combining them with the small-angle approximation of Snell's law (Equation 29–18):

$$n_1\theta_1 = n_2\theta_2$$

to obtain

$$n_1\alpha + n_2\beta = (n_2 - n_1)\gamma \qquad \textbf{(29–23)}$$

Since α, β, and γ are small:

$$\alpha = \tan\alpha = \frac{h}{OC}$$

$$\beta = \tan\beta = \frac{h}{IC}$$

$$\gamma = \tan\gamma = \frac{h}{RC}$$

Equation (29–23) then becomes

$$\frac{n_1}{OC} + \frac{n_2}{IC} = \frac{n_2 - n_1}{RC} \qquad \textbf{(29–24)}$$

In terms of the object distance p, the small distance q, and the radius of curvature R, we have

REFRACTION AT A SINGLE SPHERICAL INTERFACE

$$\frac{n_1}{p} + \frac{n_2}{q} = \frac{n_2 - n_1}{R} \qquad \textbf{(29–25)}$$

The usual sign convention applies for p and q, with R being positive for convex surfaces (that is, the center of curvature is inside the medium). Since h and θ do not appear in this expression, we know that *all* rays refracted by the interface will converge to the same image point I (at least within the validity of the small-angle approximations we have made). If the rays are not bent sufficiently to converge inside the medium, their diverging directions may be traced backward to a point of intersection to the left of the interface, forming a virtual image I as in Figure 29–27(c). The following example illustrates this type of situation.

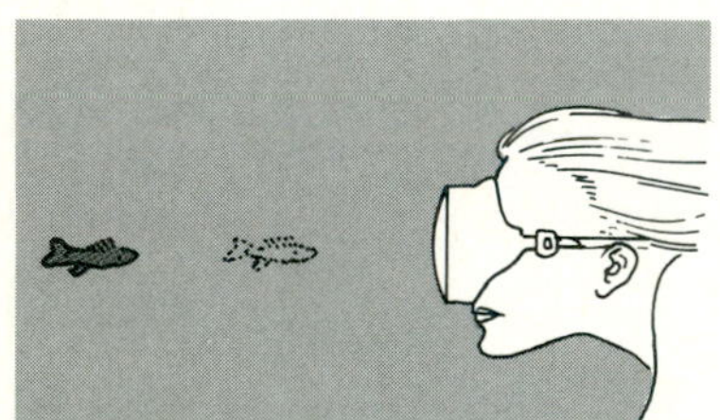

(**a**) A swimmer wearing a diving mask with a face plate that bulges outward. The image of a fish is much closer than the fish itself.

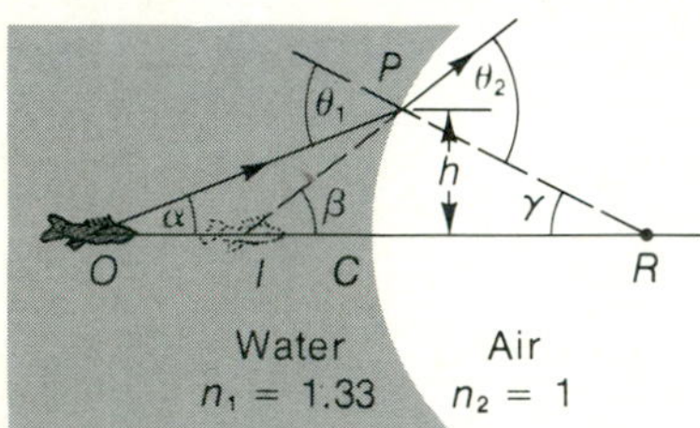

(**b**) The divergence of the light ray from the fish has been exaggerated to show more clearly the refraction at the interface between the water and air inside the diving mask.

Figure 29–28
Example 29–7

EXAMPLE 29–7

A swimmer views a small fish through a face plate on her diving mask, as shown in Figure 29–28(a). The face plate bulges outward, forming an outer convex surface with a radius of curvature of 0.40 m. If the actual distance to the fish is 3.0 m, find the apparent distance to the fish as viewed by the swimmer.

SOLUTION

Ignoring the thickness of the face plate itself, the plate forms an interface between the water ($n_1 = 1.33$) and the air within the mask ($n_2 = 1$). Tracing a single ray from the fish as in Figure 29–28(b), two triangles are formed in which

$$\theta_1 = \alpha + \gamma$$

and

$$\theta_2 = \beta + \gamma$$

Using Snell's law:

$$n_1\theta_1 = n_2\theta_2$$

we proceed as before and obtain

$$\frac{n_1}{OC} - \frac{n_2}{IC} = \frac{n_2 - n_1}{RC}$$

Substituting the appropriate numerical values, we have

$$\frac{1.33}{3.00\text{ m}} - \frac{1}{IC} = \frac{1.00 - 1.33}{0.40\text{ m}}$$

$$IC = \boxed{0.788\text{ m}}$$

Instead of 3 m, the apparent distance is only 0.788 m. Obviously, a convex face plate produces large distortion of actual distances. As indicated in Example 29–5, a flat face plate would produce an image of the fish at 2.3 m, much closer to the actual location of the fish.

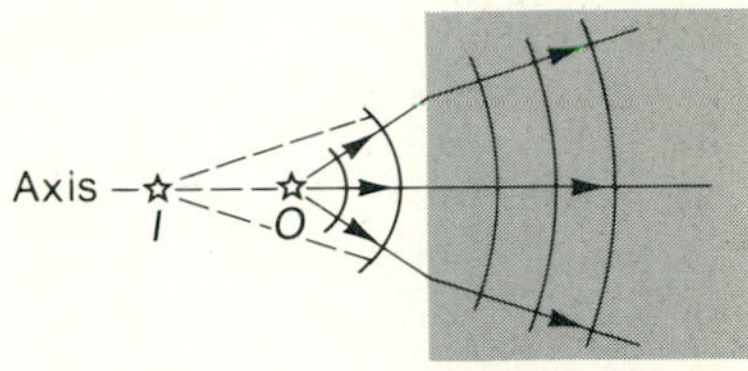

(a) Plane

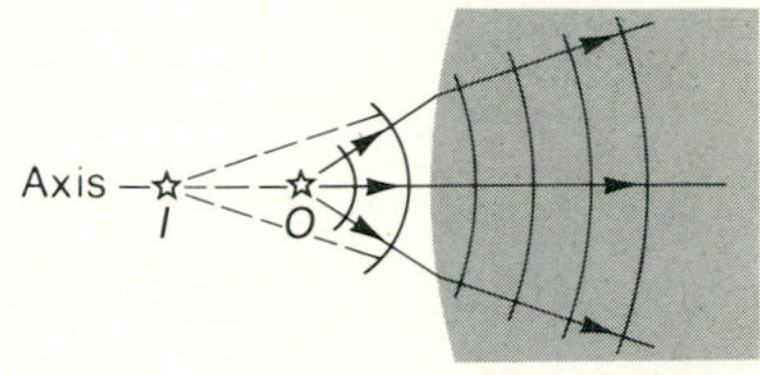

(b) Convex

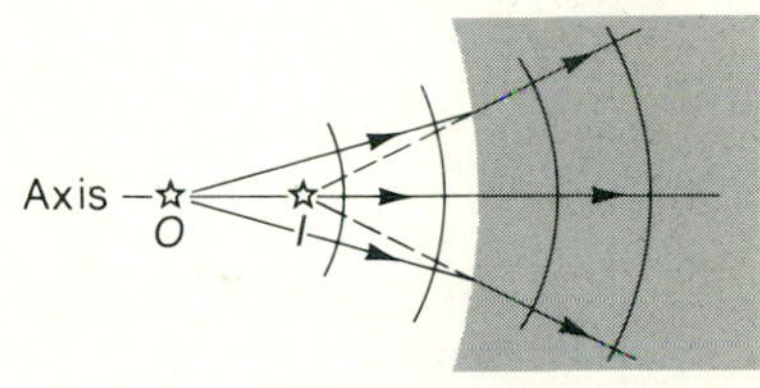

(c) Concave

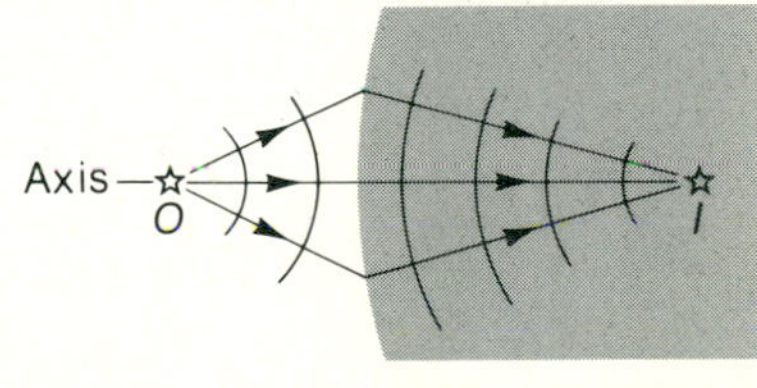

(d) Convex

Figure 29–29
Refraction at a single surface. Shaded regions indicate a higher refractive index. Dashed lines are extensions of the refracted rays and form virtual images.

Figure 29–29 shows examples of how rays from an object on the axis are refracted at an interface between two media. A convex surface may create either a real or a virtual image.

29.9 Thin Lenses

Most lenses have spherical surfaces, with each surface contributing some refraction. Thus, unless a ray strikes a surface at normal incidence, the ray will bend as it enters the lens and also as it leaves the lens (see Figure 29–30a). We will

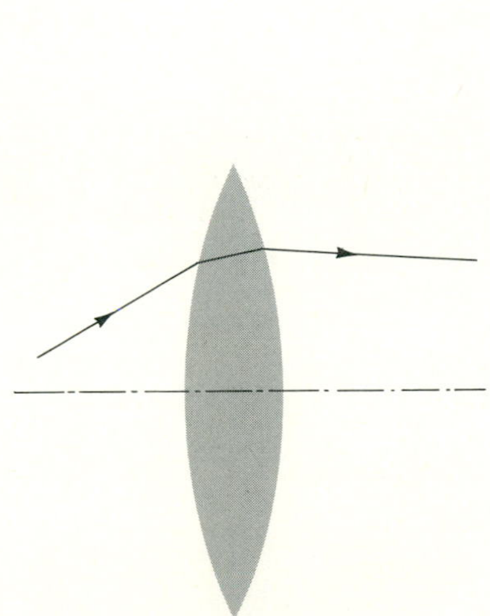

(a) In actual lenses, a ray is refracted at both surfaces (unless it happens to strike a surface at normal incidence).

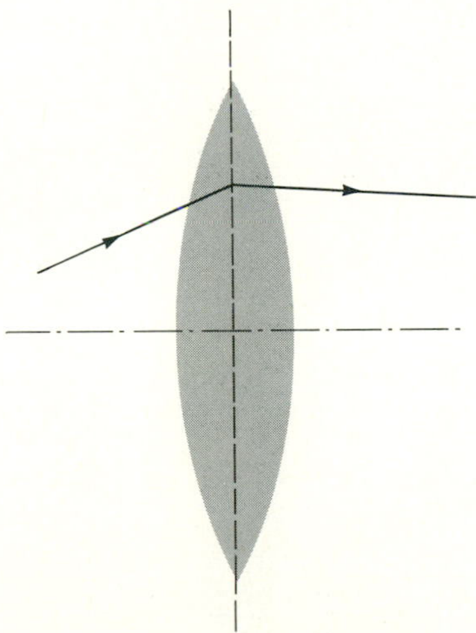

(b) In ray-tracing diagrams, we assume that all bending of a ray occurs at a plane passing through the center of the lens. The distances *p*, *q*, and *f* are also measured from this plane.

Figure 29–30
In the *thin-lens approximation,* the physical thickness of a lens is ignored and all refraction is assumed to occur at a plane passing through the center of the lens.

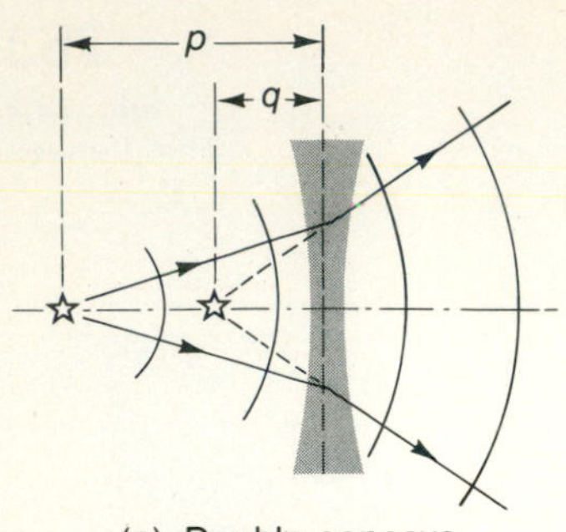

(**a**) Double concave.

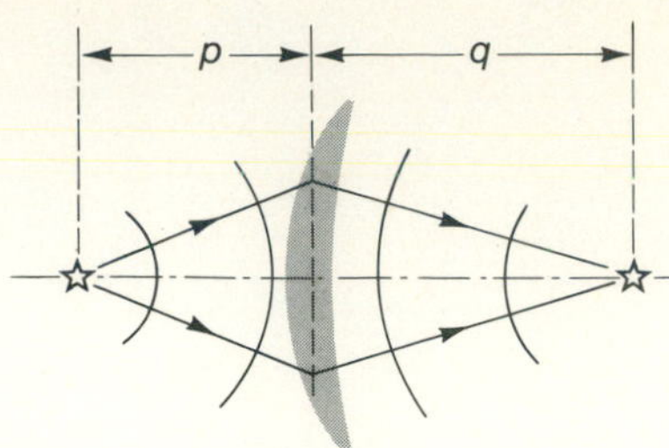

(**b**) Concave-convex, forming a real image.

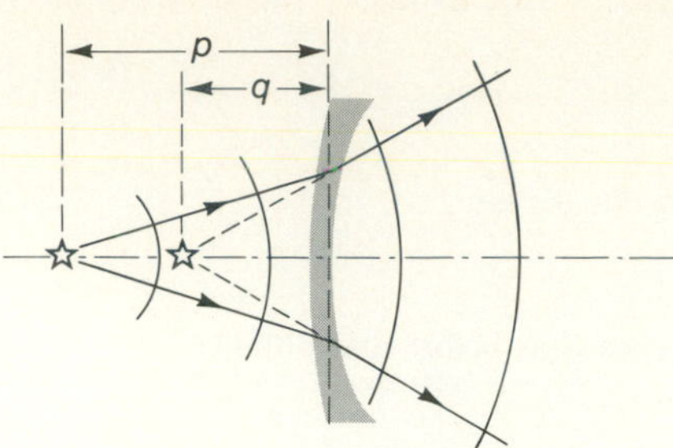

(**c**) Concave-convex, forming a virtual image.

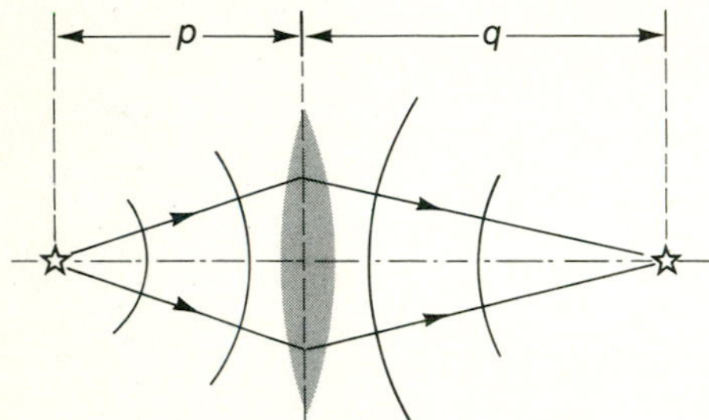

(**d**) Double convex, forming a real image.

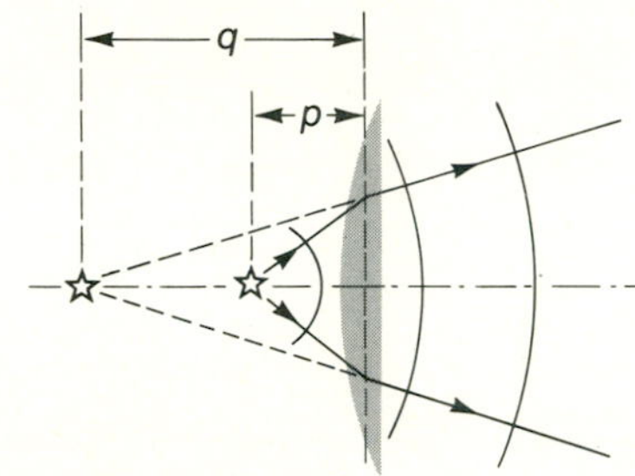

(**e**) Plano-convex, forming a virtual image.

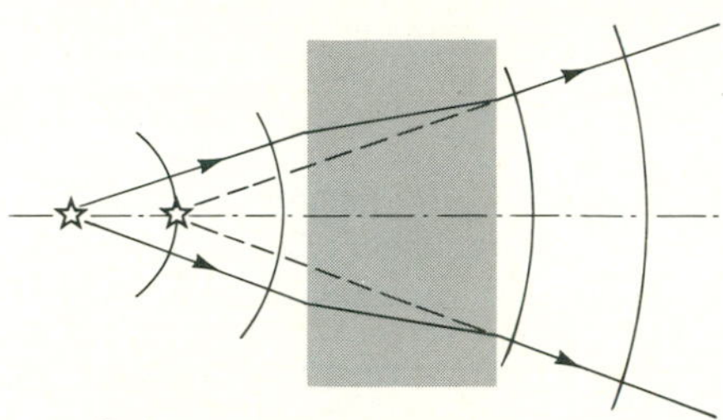

(**f**) Thick glass plate with parallel surfaces.

Figure 29–31

Examples of lenses, named according to the types of surfaces they have. Dashed lines are extensions of rays to locate virtual images.

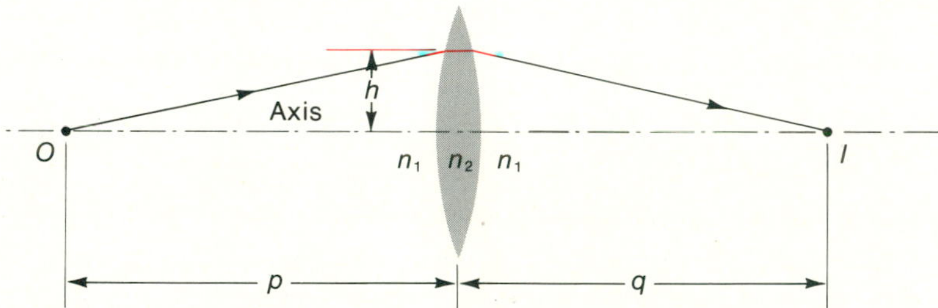

(**a**) An image I of an object O is formed by a lens. The refractive index of the lens is n_2, where $n_2 > n_1$.

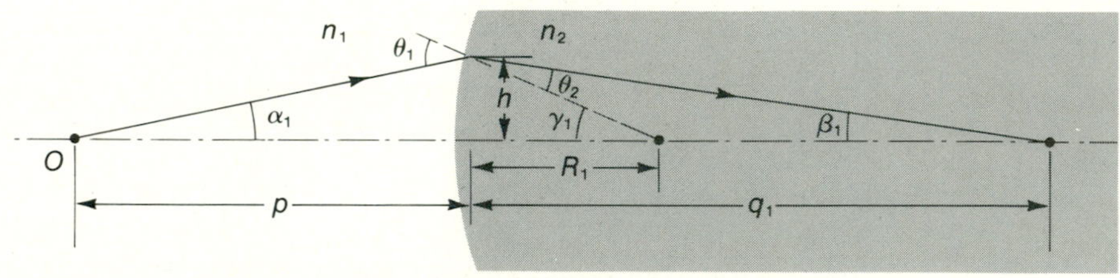

(**b**) The ray from the object O is bent by the first surface alone. The dashed line is the normal to the surface.

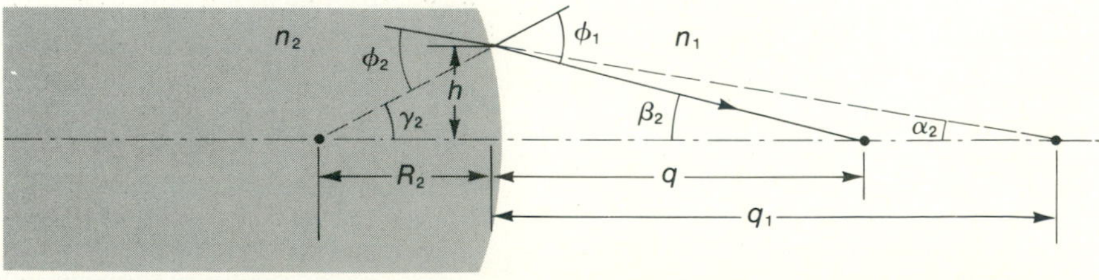

(**c**) The ray is bent as it emerges from inside the lens to intersect the axis at the final image distance q from the surface.

Figure 29–32

To determine the net effect of a lens, we must consider that in general a light ray is bent twice by the lens, once at the first surface and again at the second surface.

limit our discussion to the thin-lens case, in which the thickness of the lens is negligible compared with other dimensions. This *thin-lens approximation* means that it makes no difference whether object and image distances are measured from the front surface or the back surface. To simplify ray-tracing diagrams, we assume that all bending of a ray occurs at a plane passing through the center of the lens and that all distances are measured from this plane. We use the same notation as for mirrors:

p = object distance

q = image distance

f = focal length

R = radius of curvature of a surface

By restricting our discussion to "thin" lenses, we can use the small-angle approximations, greatly simplifying the analysis.

Figure 29–31 illustrates various types of lenses. Those which are thicker in the center than on the edge are called *convergent* or *positive* lenses and have positive values of f. Those with a center that is thinner than the edge are called *divergent* or *negative* lenses and have negative values of f. Centers of curvature all lie on the axis.

The analysis of refraction by a lens is a three-step process: (a) calculating the refraction of a light ray by the lens surface first encountered by the ray, (b) calculating the refraction of the ray as it emerges from the second surface, and (c) combining the results of (a) and (b) to obtain a general formula relating object distance p, image distance q, and the lens parameters. Fortunately, the thin-lens approximation makes the final result a simple expression.

The First Surface

Consider the case shown in Figure 29–32(a). The first surface the light ray encounters is sketched by itself in Figure 29–32(b), where p is the object distance and q_1 is the image distance that would exist if the second surface were absent. This step of the analysis is the same as that done previously for a single refractive surface [Equation (29–25)]. In the notation of this case, it is

$$\frac{n_1}{p} + \frac{n_2}{q_1} = \frac{n_2 - n_1}{R_1} \qquad \textbf{(29–26)}$$

The Second Surface

The ray is further refracted as it emerges from the second surface into the air, as in Figure 29–32(c). Rather than a real image being formed at q_1, it is formed at q. Since the exterior angle of a triangle is equal to the sum of the opposite interior angles, we have

$$\phi_1 = \gamma_2 + \beta_2 \qquad \textbf{(29–27)}$$

and

$$\phi_2 = \gamma_2 + \alpha_2 \qquad \textbf{(29–28)}$$

Snell's law for small angles is

$$n_1\phi_1 = n_2\phi_2 \qquad \textbf{(29–29)}$$

Using this equation to eliminate ϕ_1 and ϕ_2 in Equations (29–27) and (29–28), we obtain

$$n_1\beta_2 - n_2\alpha_2 = (n_2 - n_1)\gamma_2 \qquad \textbf{(29–30)}$$

Using the small-angle approximations:

$$\alpha_2 \approx \tan \alpha_2 = \frac{h}{q_1}$$

$$\beta_2 \approx \tan \beta_2 = \frac{h}{q}$$

$$\gamma_2 \approx \tan \gamma_2 = \frac{h}{R_2}$$

we obtain

$$\frac{n_1}{q} - \frac{n_2}{q_1} = \frac{n_2 - n_1}{R_2} \tag{29–31}$$

The Combined Result

For a double-convex lens forming a real image, we have obtained

$$\frac{n_1}{p} + \frac{n_2}{q_1} = \frac{n_2 - n_1}{R_1} \tag{29–26}$$

and

$$\frac{n_1}{q} - \frac{n_2}{q_1} = \frac{n_2 - n_1}{R_2} \tag{29–31}$$

Adding these equations to eliminate n_2/q_1 gives

$$\frac{n_1}{p} + \frac{n_1}{q} = (n_2 - n_1)\left(\frac{1}{R_1} + \frac{1}{R_2}\right) \tag{29–32}$$

which may be simplified by introducing the *relative refractive index* n of the lens material (relative to the surrounding medium, n_1):

$$n \equiv \frac{n_2}{n_1} \tag{29–33}$$

the result is the *lens-maker's formula:*

LENS-MAKER'S FORMULA

$$\frac{1}{p} + \frac{1}{q} = (n - 1)\left(\frac{1}{R_1} + \frac{1}{R_2}\right) \tag{29–34}$$

As in the case of mirrors, the *focal length* f of a lens is defined as the image distance of parallel light incident upon the lens ($p = \infty$). Substituting this value in Equation (29–34), the focal length of the lens is

$$f = \left[(n - 1)\left(\frac{1}{R_1} + \frac{1}{R_2}\right)\right]^{-1} \tag{29–35}$$

Finally, combining Equations (29–34) and (29–35), we obtain the *lens equation:*

THE THIN-LENS EQUATION

$$\frac{1}{p} + \frac{1}{q} = \frac{1}{f} \tag{29–36}$$

Our development of the lens equation was based on the analysis of a real image produced by a double-convex lens. If we analyze the other cases shown in Figure 29–31, we obtain equations similar to the lens-maker's formula, but with various changes in the signs of the terms. However, the lens-maker's formula is valid for all cases, with the following sign convention:

SIGN CONVENTION FOR THIN LENSES

(1) The numerical value of p is positive if the rays *approaching* the lens are *divergent.* Otherwise p is negative.

(2) The numerical value of q is positive if the rays *leaving* the lens are *convergent*. Otherwise q is negative.

(3) The radius of curvature of a lens surface is positive if it is convex, and negative if it is concave (assuming the refractive index of the lens is greater than that of the surrounding medium).

Note that the first two of these sign-convention rules are identical to those used for mirrors. Moreover, the lens formula and mirror formula have the same form. The sign convention may be remembered from the fact that for the "standard setup" of an object situated farther than the focal-length distance from a converging lens, all symbols are positive; if any distances are on the opposite side of the lens (compared with their locations in this standard setup), the symbols are negative.

Let us examine further the definition for the focal length of a lens, as described in Equation (29–35):

$$\frac{1}{f} = (n - 1)\left(\frac{1}{R_1} + \frac{1}{R_2}\right) \tag{29–37}$$

The smaller radius of curvature is dominant in determining the focal length. Thus, by the sign convention, any lens that is thicker in the center than at the edge, such as those shown in Figures 29–31(b), (d), and (e), would have a *positive* focal length. The lenses shown in Figures 29–31(a) and (c) have *negative* focal lengths. The *relative* refractive index n of a lens (relative to the surrounding medium) may be less than one (for example, this is true for a lens formed by air trapped between two plastic sheets submerged under water). In such a case, a lens thicker at the center than on the edge would have a negative focal length. To summarize, an easy rule for determining the signs of focal lengths that is *applicable to both mirrors and lenses* is: If parallel incoming rays are bent so they *converge*, f is *positive*; if they *diverge*, f is *negative*.

29.10 Diopter Power

The *strength* of a lens is a measure of its ability to alter the direction of light rays. This strength is measured in **diopters,** defined as the reciprocal of the focal length measured in meters:

$$\text{Strength (in diopters)} = \frac{1}{f\text{ (in meters)}}$$

EXAMPLE 29–8

A lens is constructed of crown glass with a refractive index of 1.50 and has surfaces with radii of curvature of 0.10 m and 0.075 m. Both centers of curvature are on the same side of the lens, as illustrated in Figure 29–33. Find (a) the focal length and (b) the strength of the lens. (c) Describe the image produced by the lens of an object placed 0.080 m from the lens.

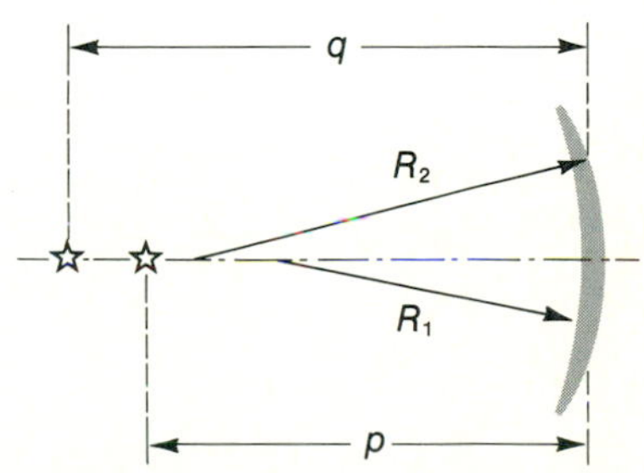

Figure 29–33
Example 29–8

SOLUTION

(a) The focal length may be determined by direct substitution into Equation (29–37):

$$\frac{1}{f} = (n - 1)\left(\frac{1}{R_1} + \frac{1}{R_2}\right)$$

where we assume that the lens is in air, so that the relative refractive index is just that of the crown glass. Substituting numerical values:

$$\frac{1}{f} = (1.50 - 1)\left(-\frac{1}{0.10\text{ m}} + \frac{1}{0.075\text{ m}}\right)$$

and solving for f:

$$f = \boxed{+0.600\text{ m}}$$

(b) The strength (in diopters) is

$$\text{Strength} = \frac{1}{f\text{ (in meters)}}$$

$$\text{Strength} = \frac{1}{+0.60}$$

$$\text{Strength} = \boxed{+1.67\text{ diopters}}$$

Note the radii have opposite signs. Since R_1 is for a concave surface, it has a negative numerical value; since R_2 is for a convex surface, it is positive. The positive sign for the focal length f indicates a converging lens, thicker at the center than at the edge.

(c) Just as in the application of any general equation in physics, we do not anticipate the sign of unknown quantities. Possible minus signs are introduced only when we substitute numerical values in accordance with the sign convention. Starting with the lens equation:

$$\frac{1}{p} + \frac{1}{q} = \frac{1}{f}$$

we have

$$\frac{1}{0.080\text{ m}} + \frac{1}{q} = \frac{1}{0.60\text{ m}}$$

where the sign of the numerical value of p is positive because rays from the object are divergent as they strike the lens. Solving for q, we obtain

$$q = \boxed{-0.0923\text{ m}}$$

By the sign convention, the negative sign of the image distance indicates that the rays are still divergent as they leave the lens. The image is therefore virtual. That is, if we were to view the object through the lens, it would appear to be 0.092 m from the lens, on the same side of the lens as the object.

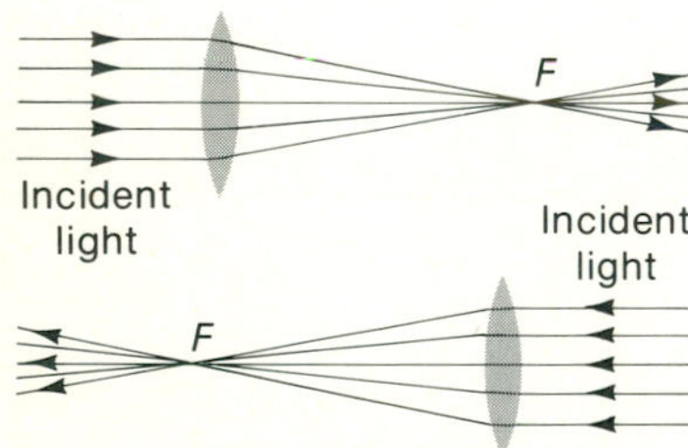

(a) Parallel light incident on opposite sides of a lens. If the lens is thin, the focal distance f is the same for each case.

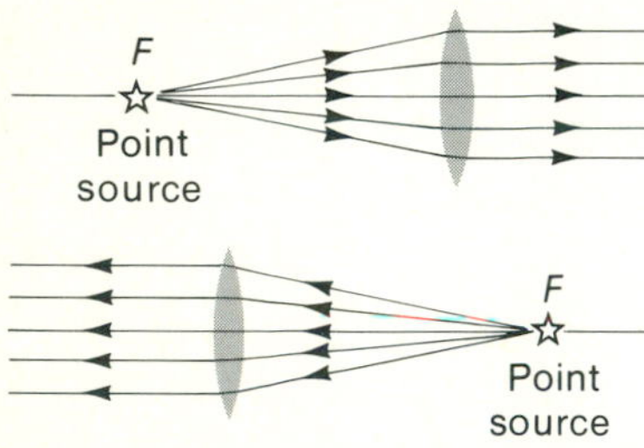

(b) Light rays from a point source at either focal point F emerge from the lens in directions parallel to the axis.

Figure 29–34

A thin lens affects light the same way regardless of which side of the lens the incident light strikes. (This is not true for thick lenses.)

29.11 Thin Lens Ray-Tracing and Image Size

We shall now describe ray-tracing techniques used for locating images formed by thin lenses. They are similar to the ray-tracing methods used for mirrors. We will sometimes indicate two focal points F located equidistant on either side of the lens, called *principal foci.* This recognizes that for thin lenses, parallel light incident on a converging lens *from either side* is brought to a focus at a focal-length distance on the other side of the lens (for divergent lenses, it is brought to a focus on the *same* side). Furthermore, rays from a point source at either focal point of a converging lens emerge from the lens parallel to the axis. As shown in Figure 29–34, the situation is symmetrical regarding the direction the light passes through the lens. (This is not true for thick lenses.)

As for mirrors, a focal point F is *not* the location of the image, except in the one case of incident parallel light. It is helpful to think of the focal points as points "belonging" to a lens, which we find useful in constructing ray diagrams. As before, we concern ourselves only with two particular rays that can be easily traced. These rays and their characteristics are:

RAYS USED IN RAY-TRACING FOR THIN LENSES

(1) A ray passing through the center of the lens is undeviated. (It is undeviated because, near the center, the lens acts like a thin piece of glass with parallel sides.)
(2) A ray parallel to the axis is refracted so that it passes directly through the focal point F of the lens. (For a convergent lens, the ray passes directly through F on the other side of the lens. For a divergent lens, if the ray is extended backwards, this line passes through F on the same side of the lens.)

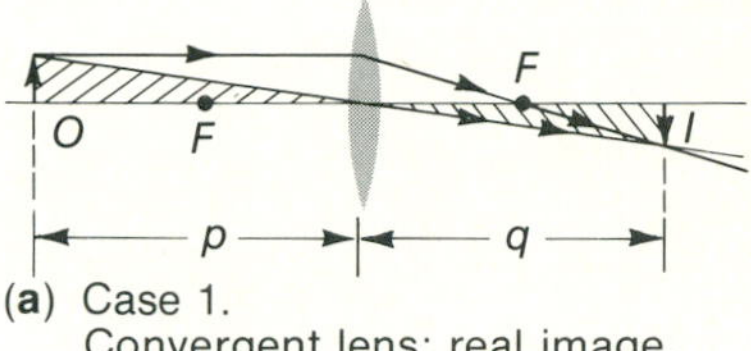

(a) Case 1.
Convergent lens: real image.

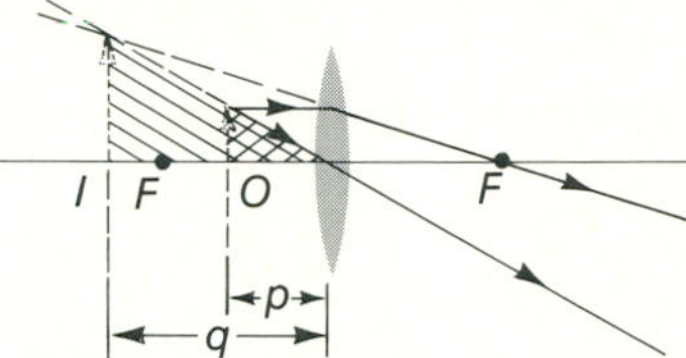

(b) Case 2.
Convergent lens: virtual image

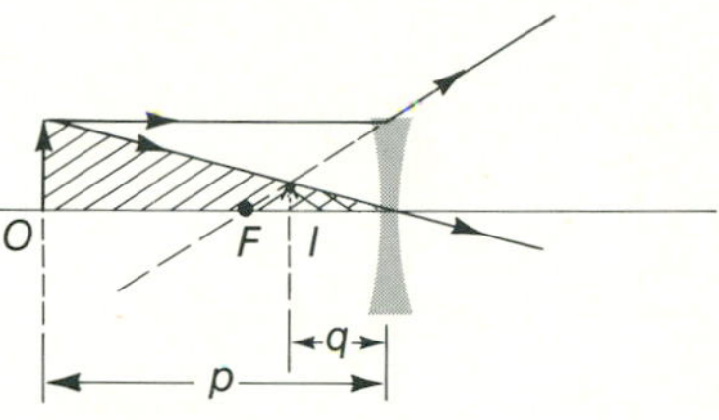

(c) Case 3.
Divergent lens: virtual image

Figure 29–35
Image formation by ray-tracing. If the rays leaving the lens do not intersect, the virtual image is located by extending dashed lines backwards from the directions of the actual rays.

The image is located at the intersection of these two rays. Since all other rays from a point on the object also pass through this image point, it is sufficient to trace just two rays. Figure 29–35 illustrates the three possible cases of refraction by a thin lens with a refractive index greater than that of the surrounding medium.

Case 1. Converging Lens: Real Image

In Figure 29–35(a), the object is the tip of the arrow at O, located more than a focal-length distance from the lens. (We investigate an object that extends only *above* the axis, recognizing that because of symmetry about the axis, an object that extends below the axis would produce the same result.) We trace two rays from the tip whose directions we can easily determine. One ray passes through the center of the lens undeviated, and the other approaches the lens parallel to the axis and is refracted so that it converges toward the focal point F of the lens. The intersection of these two rays locates the image at I. (Other parts of the arrow may similarly be imaged to form the image of the complete arrow.) Since light rays actually converge to form the image, it is *real*; a screen placed at that location would have an image formed on it. The image is *inverted*, as revealed by the ray diagram (and also by the fact that the numerical value of q is negative).

The **linear magnification** M is defined as the ratio of the image size to the object size:

$$M = \frac{\text{Image size}}{\text{Object size}}$$

In Figure 29–35(a), the image and object are corresponding parts of the similar right triangles shown shaded. Therefore, corresponding sides have the same ratio, and we have

LINEAR MAGNIFICATION

$$M = -\frac{q}{p} \tag{29–38}$$

The minus sign is introduced so that a *negative* value of M indicates an inverted image and a *positive* value of M indicates an *erect* image. This same sign convention holds true for all cases of lenses as well as mirrors. The term magnification is somewhat of a misnomer, since the image can be smaller than the object, in which case M is less than one.

Case 2. Converging Lens: Virtual Image

In Figure 29–35(b), the object is located closer than a focal-length distance from the lens. Again we trace two rays from the tip of the arrow. One ray passes through the center of the lens undeviated, and the other approaches the lens parallel to the axis and is refracted so that it passes through the focal point of the lens. The intersection of these two rays is determined by extending them backwards along their directions until they intersect at the image location I. Because no actual light rays travel along these extended lines, we draw them dashed. Since no actual light rays form the image, it is *virtual*. (A screen placed at that location would *not* have an image formed on it.) The image is *erect*, as revealed by the ray diagram (and also by the fact that the numerical value of M is positive).

The linear magnification is the ratio of image size to object size. By comparing similar triangles as in Case 1, we conclude that the linear magnification is

$$M = -\frac{q}{p}$$

You may verify that for this case, the image is always larger than the object, always virtual, and always erect.

Case 3. Divergent Lens: Virtual Image

Referring to Figure 29–35(c), the rays from the arrow tip always diverge after passing through a divergent lens, no matter how far the object is from the lens. As a consequence, a virtual image is formed that is always smaller than the object, and always erect. The linear magnification is, as before:

$$M = -\frac{q}{p}$$

□ □ □

TABLE 29–3

Characteristics of Images Produced by Thin Lenses

	Image characteristics		
Lens	**Linear magnification**	**Real or virtual**	**Erect or inverted**
Convergent, $p < f$	$M > 1$	Virtual	Erect
Convergent, $p > f$	$M \gtreqless 1$	Real	Inverted
Divergent	$M < 1$	Virtual	Erect

The results of ray-tracing through thin lenses are summarized in Table 29–3. It is not worth memorizing these relations, since all information is contained in the sign convention and the relation $M = -q/p$. Furthermore, these characteristics are clearly revealed in a ray-tracing diagram. Note the similarity between lenses and mirrors. With regard to their image-forming characteristics, convergent lenses correspond to concave mirrors and divergent lenses correspond to convex mirrors.

29.12 Combinations of Lenses

Most optical instruments contain a system of several lenses. In many cases, the use of multiple lenses helps to correct certain image defects. In other instances, if the final image is formed in a series of steps, the overall length of the instrument is much shorter than it would be if just a single lens were used. In this text, we will limit the discussion to simple two-lens combinations.

We start with two thin lenses in contact, a common situation in many optical instruments. Figure 29–36 depicts two such lenses, which have positive focal lengths f_1 and f_2. Parallel light from the left strikes lens ① and would focus at F_1 if the second lens were absent. After leaving lens ①, the rays are convergent as they strike lens ②. Therefore, the object distance p_2 for the second lens is $-f_1$. Applying the lens equation:

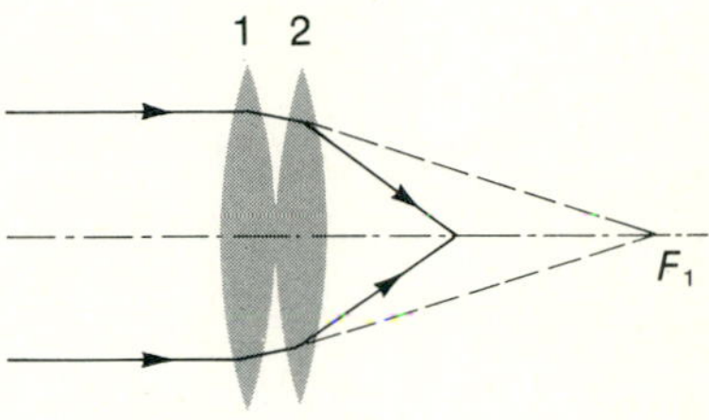

Figure 29–36
Two thin, converging lenses in contact.

$$\frac{1}{p} + \frac{1}{q} = \frac{1}{f}$$

and noting that q_2 is the resultant focal length f for the combination, we have

$$\frac{1}{-f_1} + \frac{1}{f} = \frac{1}{f_2}$$

Rearranging, we obtain

THIN-LENS COMBINATIONS (lenses in contact)

$$\frac{1}{f} = \frac{1}{f_1} + \frac{1}{f_2} \qquad \textbf{(29–39)}$$

or, expressed as diopters:

$$\begin{pmatrix}\text{Resultant}\\ \text{diopter}\\ \text{power}\end{pmatrix} = \begin{pmatrix}\text{Diopter}\\ \text{power of}\\ \text{lens 1}\end{pmatrix} + \begin{pmatrix}\text{Diopter}\\ \text{power of}\\ \text{lens 2}\end{pmatrix} \qquad \textbf{(29–40)}$$

Diopter notation is particularly convenient for lens combinations because the strength of two thin lenses in contact is merely the sum of the strengths of the individual lenses. These are general relations, valid for any combination of positive and negative lenses.

EXAMPLE 29–9

Two thin lenses of focal lengths $f_1 = 20$ cm and $f_2 = 60$ cm are placed in contact. (a) Find the focal length f' of the combination. (b) Find the focal length f_3 of a third lens placed in contact with these two that would result in an overall focal length $f'' = -40$ cm.

SOLUTION

(a) Substituting numerical values in the lens-combination formula:

Using focal lengths

$$\frac{1}{f} = \frac{1}{f_1} + \frac{1}{f_2}$$

$$\frac{1}{f'} = \frac{1}{20\text{ cm}} + \frac{1}{60\text{ cm}}$$

$$f' = \boxed{15.0\text{ cm}}$$

Using diopters (D)

$$D_1 = \frac{1}{0.2\text{ m}} = 5\text{ diopters}$$

$$D_2 = \frac{1}{0.6\text{ m}} = 1.67\text{ diopters}$$

$$D' = D_1 + D_2$$

$$D' = 5\text{ diopters} + 1.67\text{ diopter}$$

$$D' = \boxed{6.67\text{ diopters}}$$

(b) Adding one more lens in contact, we repeat the same analyses.

$$\frac{1}{f''} = \frac{1}{f'} + \frac{1}{f_3}$$

$$\frac{1}{-40\text{ cm}} = \frac{1}{15\text{ cm}} + \frac{1}{f_3}$$

$$f_3 = \boxed{-10.9\text{ cm}}$$

$$D'' = \frac{1}{-0.4\text{ m}} = -2.5\text{ diopters}$$

$$D'' = D' + D_3$$

$$-2.5\text{ diopters} = 6.67\text{ diopters} + D_3$$

$$D_3 = \boxed{-9.17\text{ diopters}}$$

Combining positive and negative lenses together helps to correct certain image defects, as we will discuss later.

EXAMPLE 29–10

Consider the combination of lenses shown in Figure 29–37. An object is placed 0.15 m from a convergent lens with a focal length f_1 of 0.10 m. A divergent lens with a focal length f_2 of -0.20 m is placed 0.15 m from the convergent lens. Locate and describe the image formed by the two lenses.

Figure 29–37
Example 29–10

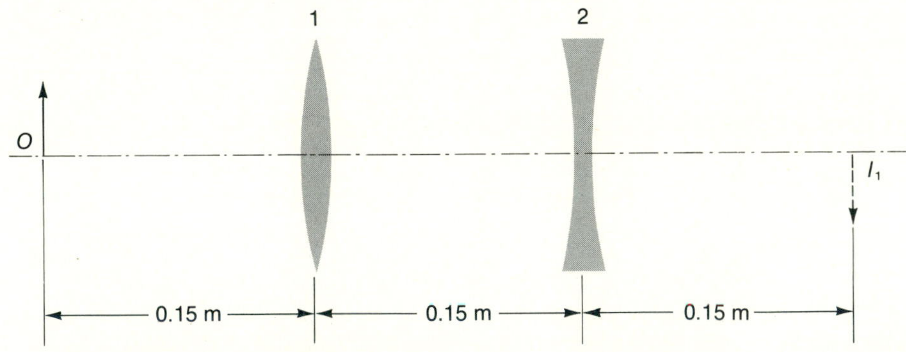

SOLUTION

Because these two lenses are *not* in contact, we cannot use the lens-combination formula. Instead, we investigate the focusing properties of each individual lens by itself. The first step is to locate the image formed by the first lens, pretending that the second lens is absent. Applying the lens equation:

$$\frac{1}{p} + \frac{1}{q} = \frac{1}{f}$$

$$\frac{1}{+0.15\text{ m}} + \frac{1}{q_1} = \frac{1}{0.10\text{ m}}$$

Solving for q_1:

$$q_1 = +0.300 \text{ m}$$

Thus the image would fall 0.30 m to the right of the first lens if the second lens were absent. Because the second lens is 0.15 m to the right of the first lens, the rays are still convergent when they strike the second lens. Since the rays are *converging* to a point 0.15 m from the second lens, the object distance p_2 for the second lens is -0.15 m. Thus:

$$\frac{1}{p} + \frac{1}{q} = \frac{1}{f}$$

Substituting numerical values and solving for q_2 gives

$$\frac{1}{-0.15 \text{ m}} + \frac{1}{q_2} = \frac{1}{-0.20 \text{ m}}$$

$$q_2 = \boxed{0.600 \text{ m}}$$

The final image is 0.60 m to the right of the second lens. The image is inverted because the first lens produced a real, inverted image, and the second lens, being divergent, cannot itself produce a further inversion. The first lens produced a linear magnification of

$$M_1 = -\frac{q_1}{p_1}$$

and the second produced a magnification of

$$M_2 = -\frac{q_2}{p_2}$$

The total magnification is the product:

$$M = M_1 M_2$$

or

$$M = \left(-\frac{q_1}{p_1}\right)\left(-\frac{q_2}{p_2}\right)$$

$$M = \left(-\frac{0.30 \text{ m}}{0.15 \text{ m}}\right)\left(-\frac{0.60 \text{ m}}{-0.15 \text{ m}}\right)$$

$$M = \boxed{-8.00}$$

The final image is 8 times the size of the object, *real* (because q_2 is positive), and *inverted* (because M is negative).

A word of caution: As this example illustrates, in solving multiple-lens systems, care must be taken to apply the sign convention correctly each step of the way. It will often be helpful to verify each step by sketching a ray diagram. In this example, although the second lens causes the rays to diverge, they are still converging as they finally emerge, resulting in an extension of the image an additional 45 cm to the right.

29.13 Optical Instruments

The Simple Magnifier

Probably the simplest optical instrument is the single-lens magnifier or reading glass. A convergent lens is placed in front of fine print or a small object and moved closer or farther away until the best magnified image appears. Because the object is closer than a focal-length distance from the lens, the image is erect, as shown in Figure 29–38(a). How much is the object magnified and how does that depend on the focal length of the lens? The answer is complicated because the image distance may be fairly short, or even infinitely far away, depending on the preference of the observer regarding the most comfortable viewing distance. Also, the observer's eye may be at various distances from the lens, so the angular size of the image the eye sees may vary. To reduce the number of possibilities, we will discuss only cases where the observer's eye is close to the lens.

When we use a magnifier, we are interested in how much larger the image appears *with the magnifier* compared to viewing the object *with the unaided eye.* A person with so-called normal eyesight can see clearly objects located anywhere from infinity to about 25 cm from the eye. The largest angular size of an object will be when it is held as close to the eye as possible. *By definition*, the "closest distance for comfortable viewing" is taken to be 25 cm. Of course, some persons can see objects closer than this, while others cannot see objects that close; the 25 cm figure is chosen as an average value.

We define the **angular magnification** m as the ratio:

ANGULAR MAGNIFICATION

$$m = \frac{\begin{bmatrix}\text{Angle subtended by the image} \\ \text{when using the magnifier}\end{bmatrix}}{\begin{bmatrix}\text{Angle subtended by the object} \\ \text{when viewed from 25 cm by} \\ \text{the unaided eye}\end{bmatrix}}$$

Let us first consider using the magnifier so that it forms an image at infinity. This is achieved by placing the object at a focal point F, Figure 29–38(b). If the height of the object is small compared with the focal distance (as in the case of printed material), the small-angle approximation applies, so that

$$\alpha \approx \frac{h}{f}$$

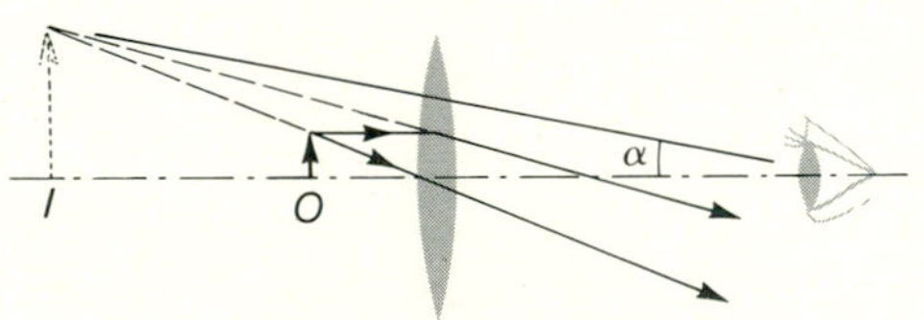

(**a**) Viewing with a magnifier. The angle α the image subtends at the eye depends on how close the eye is to the lens.

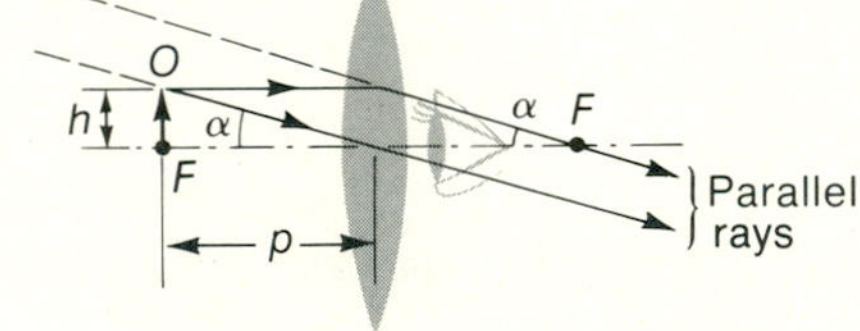

(**b**) When the object is placed at one of the focal points F, the image is at infinity.

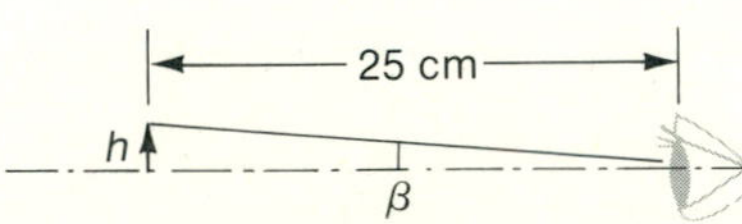

(**c**) Viewing the object with the unaided eye at the closest distance for comfortable viewing, 25 cm. The object subtends an angle β at the eye.

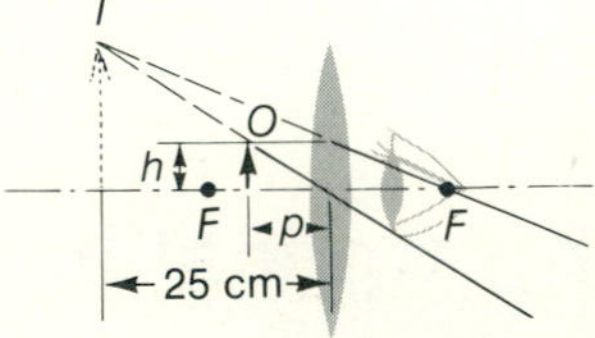

(**d**) The greatest angular magnification is achieved when the eye is close to the magnifying lens and the image is at the closest distance for comfortable viewing: that is, $q = -25$ cm.

Figure 29–38
The single-lens magnifier.

The angle β the object subtends when viewed with the unaided eye (Figure 29–38c) is

$$\beta \approx \frac{h}{25\text{ cm}}$$

Using the definition of angular magnification:

$$m = \frac{\alpha}{\beta}$$

we obtain

ANGULAR MAGNIFICATION OF A MAGNIFIER (image at infinity)

$$m = \frac{25\text{ cm}}{f} \quad \text{(where } f \text{ is in centimeters)} \qquad \textbf{(29–41)}$$

Now consider the case where the magnifier is used so that the image is at the closest distance for comfortable viewing, Figure 29–38(d). This method gives slightly greater magnification than with the object at infinity. Assuming that the eye is close to the lens, the image distance q will be approximately -25 cm. (Remember that the image is *virtual*, so q has a negative value.) The image subtends an angle α of approximately

$$\alpha \approx \frac{h}{|p|}$$

Substituting p obtained from the lens equation:

$$\frac{1}{p} + \frac{1}{q} = \frac{1}{f}$$

With $q = -25$ cm, we have

$$\alpha = h\left(\frac{1}{f} + \frac{1}{25\text{ cm}}\right)$$

The angle β subtended by the object at 25 cm when viewed with the unaided eye is

$$\beta = \frac{h}{25\text{ cm}}$$

where h is in centimeters. Using the definition of angular magnification, we obtain

$$m = \frac{\alpha}{\beta}$$

ANGULAR MAGNIFICATION OF A MAGNIFIER (with image at 25 cm and the eye placed close to the lens)

$$m = \frac{25\text{ cm}}{f} + 1 \quad \text{(where } f \text{ is in centimeters)} \qquad \textbf{(29–42)}$$

If we compare this result with that for the image at infinity, we see that when the focal length is small compared to 25 cm, the magnification of a single-lens magnifier is essentially independent of the image distance. Common magnifiers have focal lengths of about 5 cm.

The Eye

The eye, with its linkage to that master computer—the brain—is surely one of the most remarkable of human organs. The overwhelming majority of information input comes to us via our eyes, and the way we analyze and sort out the ever-changing pattern we see is astonishing. Basically, the eye is similar to a camera with a lens that forms images on a photosensitive surface, but there are many unique features that no camera can duplicate (see Figure 29–39). The focusing of light rays occurs primarily at the outer surface of the* cornea, *where the change in refractive index from air to the cornea is greatest. (Its power is about 40 to 45 diopters in the average person.) On the other hand, the* lens *is surrounded by fluids whose refractive indices are not too different from that of the lens material, so relatively less refraction occurs at these surfaces. (The major reason you cannot see clearly under water is that the refractive index for water,* $n_{\text{water}} = 1.33$, *is too close to that of the cornea,* $n_c = 1.367$, *to cause sufficient refraction. A face plate corrects the problem by maintaining an air contact with the eye.)*

The lens is somewhat flexible, enabling the ciliary muscles *to adjust its power from about 20 to 24 diopters. In this way, even though the lens is a fixed distance from the photosensitive surface, sharp images can be formed for varying object distances. The overall power of the cornea plus the lens is about 60 to 65 diopters. The ability of the eye to change its focal length is called* accommodation. *With age, the lens material gradually hardens, so the degree of accommodation becomes less as we grow older. The closest object distance for which sharp images can be formed is called the* near point. *For an average 10-year-old, it is around* 7 cm, *increasing to about* 22 cm *in middle age, and to about* 100 cm *at age 60, often requiring "reading" glasses to assist vision at closer distances.*

The retina *consists of roughly 125 million photoreceptor cells called* rods *and* cones. *An elaborate network of neurons and nerve fibers connects them to the*

* *Scientific American* has many interesting articles on vision. Among them are the following:

"The Visual Cortex of the Brain," David Hubel, November 1963.
"Attitude and Pupil Size," E. Hess, April 1965.
"Retinal Processing of Visual Images," Charles Michael, May 1969.
"The Neurophysiology of Binocular Vision," John Pettigrew, August 1972.
"Visual Pigments and Color Blindness," W. Rushton, March 1975.
"The Resources of Binocular Perception," John Ross, March 1976.

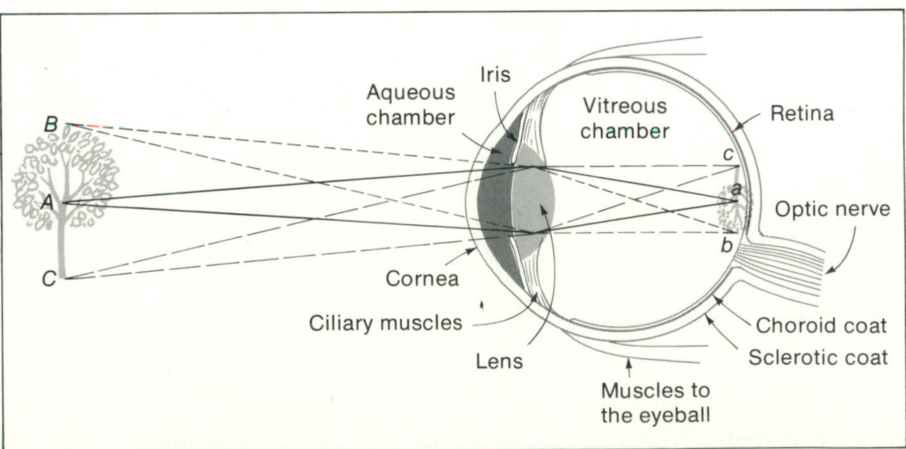

Figure 29–39
A simplified diagram of the human eye. The optic nerve is actually to one side of a vertical plane through the eye.

brain via the optic nerve (Figure 29–40). As you read these words, your eye jumps abruptly from point to point, so that the center of your field of view falls on the fovea, *a small area about* 0.3 mm *in diameter containing only cones packed closely together. (To get some idea of the field of view that covers the fovea, the moon's image on the retina is about* 0.2 mm *in diameter.) The eye's ability to detect detail* (resolution) *is greatest in the fovea. Only the cones are sensitive to colors. Away from the fovea, the rods become relatively more numerous, and though they have no color sensitivity, they can detect very dim light. You can test the rods' sensitivity to low light levels by trying to observe a faint star. The star may not be seen if you look directly at it, but shifting the direction of vision to one side a bit so the image falls on the rods makes the star's presence detectable. It is believed that some data analysis of the image occurs at the retina, particularly in certain animals with photoreceptors that send signals to the brain only for specific orientations of light-dark edges or for motions in certain directions.*

The iris *is an adjustable diaphragm that controls the amount of light passing into the eye. The size of the iris opening, the* pupil, *is affected not only by the amount of incident light but also by drugs and by our emotions. If something pleases us, our pupils tend to enlarge; if we are displeased, they tend to contract. Clever poker players aware of this effect claim they can sometimes discern the value of their opponents' hands by watching changes in the sizes of their pupils. Although the iris controls light intensity only by a factor of 16 or so, the retina itself has an enormously larger range of sensitivity. Light causes chemical changes in the rods and cones, reducing their sensitivity; after about half an hour in the dark, the eye becomes "dark adapted" and the greatest sensitivity is achieved. There is no completely adequate theory of color vision that explains all phenomena, though it is reasonably certain that the cones are of three types whose color sensitivities overlap somewhat; one type is most responsive to blue light, one to green light, and the third to yellow light (not red, as previously thought). About 8% of males and 1% of females have some defects of color vision, a hereditary malady that is recessive and sex linked.*

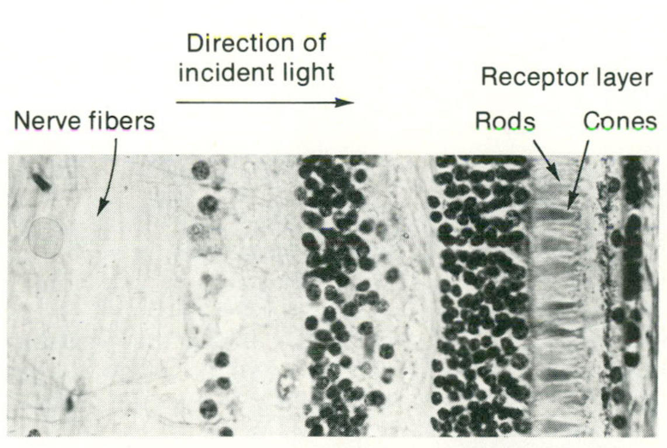

(**a**) A cross section of the human retina shows the elaborate layer of nerve fibers, blood vessels, and other tissues through which the incident light must pass to reach the rods and cones. You can see this overlying layer by closing your eye and placing a tiny source of light near the lid. With practice, you can see a pattern of shadows cast by the blood vessels on the rods and cones and even the shadows of blood cells coursing through small blood vessels with each heartbeat.

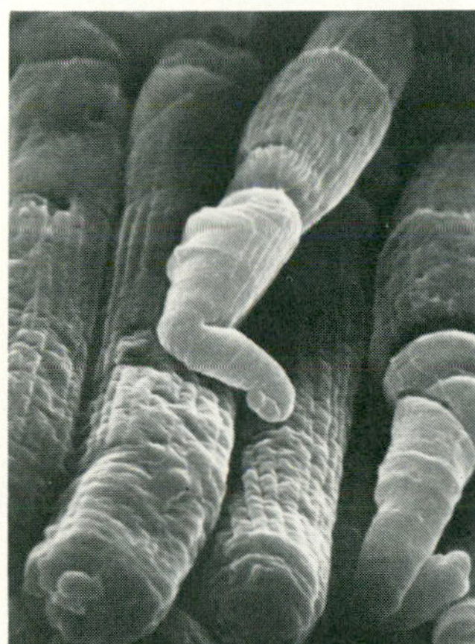

(**b**) Rods and cones magnified 1600 times with a scanning electron microscope.

Figure 29–40

(a) Cross section of a human retina; (b) rods and cones from the retina of a salamander.

The Eye (*Continued*)

In the fovea, where resolution is greatest, each cone has a separate path to the optic nerve, but near the edge of the retina several receptors may be connected to the same nerve path. The region where the optic nerve leaves the retina produces a blind spot *in the field of vision* (*Figure 29–41*). *A portion of the nerve pathways from each eye cross over and lead to the opposite half of the* visual cortex *in the brain, a feature of the "wiring diagram" believed to be involved in depth perception.*

The eye-plus-brain combination is a surprisingly effective visual system that enables us to rapidly scan a scene, investigating interesting portions with the high-resolution fovea, sorting out the varying images, and picking up significant information on intensity, form, motion, and color to store temporarily in our memory, thereby building up a single, three-dimensional concept of our surroundings.

Figure 29–41
Diagram for revealing the blind spot. Close the left eye and look at the circle as you move the book closer to your eye. When about 20 cm away, the star will disappear. (The same effect occurs when closing the right eye and looking at the star.) The brain tends to "fill in" the missing portion of the field of view with a pattern similar to its surroundings. For example, if a row of X's is observed, the blank space of the missing symbol seems filled with an X.

Eyeglasses

A familiar optical device is the eyeglass. Actually, eyeglasses are not optical instruments in themselves, but become part of the highly sophisticated human viewing system. The lens of a normal eye can slightly change its focal length (*accommodation*), so that even though the retina is a fixed distance from the

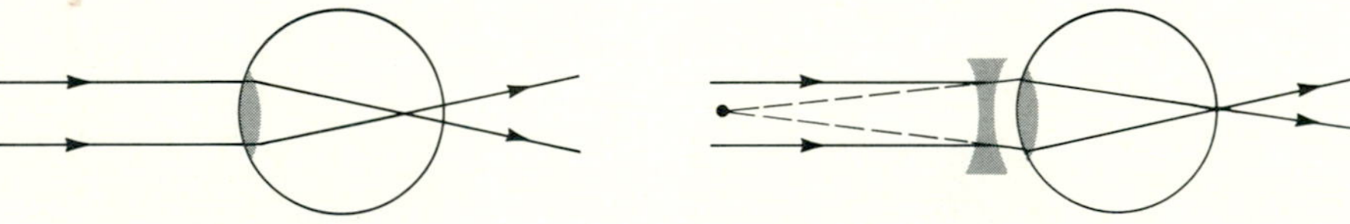

(**a**) In *nearsightedness* (*myopia*), a person cannot see distant objects because the eye converges incoming light too strongly and the image is thus formed in front of the retina.

(**b**) A negative lens corrects nearsightedness by forming a virtual image of a distant object closer to the eye. The eye itself can then form an image of it on the retina.

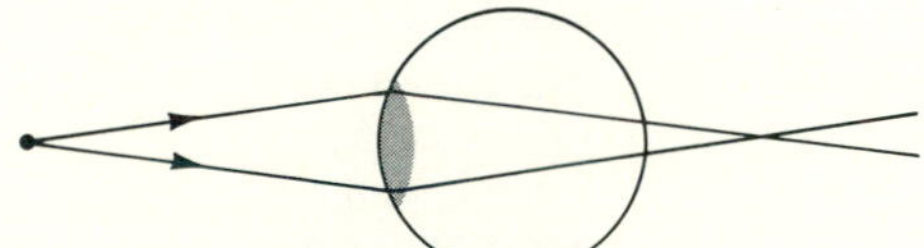

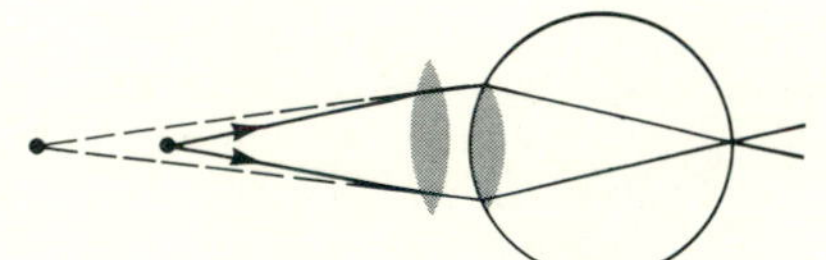

(**c**) In *farsightedness* (*hyperopia*), a person cannot see nearby objects because the eye does not refract the light sufficiently and the image distance is thus behind the retina.

(**d**) A positive lens corrects farsightedness by forming a virtual image of a nearby object farther from the eye. The eye itself then can form an image of it on the retina.

Figure 29–42
Eyeglasses can correct the visual defects caused by the eye's inability to form images on the retina.

lens, it produces sharp images on the retina for both near and distant objects. Sometimes the lens of the eye is not symmetrical and tends to produce elongated images of point sources of light, an abnormality called *astigmatism.* Both lack of accommodation and astigmatism can be corrected by eyeglasses.

The inability to see sharp images of near or distant objects arises because the objects lie outside the range of distances the eye can accommodate (see Figure 29–42). For example, if the unaided eye is *nearsighted* and cannot form sharp images of distant objects, an eyeglass can produce an image of the object at a sufficiently close distance so that the eye can look at this image and, with its own lens, focus it on the retina. Conversely, if the eye is *farsighted* and nearby objects seem blurred, an eyeglass can form an image of the object farther away, so that the eye can look at it and form a sharp image on the retina. In each case, it is easiest to think of the eyeglass as first forming an image in front of the observer within the distance the eye can accommodate. Then the eye itself looks at this image and properly focuses it on the retina. The following example illustrates these procedures.

EXAMPLE 29–11

A nearsighted person (Figure 29–43) can see easily and comfortably objects within the range of 15 cm to 100 cm. (a) Describe the eyeglasses that will provide a normal range of 25 cm to infinity. (b) Find the image distance these eyeglasses would produce of an object held at the convenient reading and working distance of 25 cm.

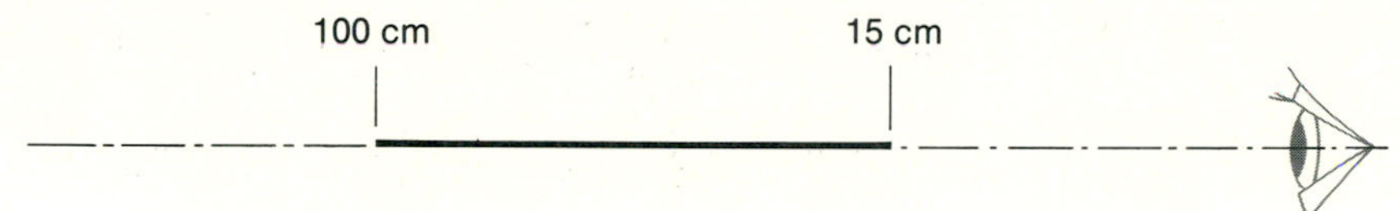

(**a**) A nearsighted person's range of clear vision without eyeglasses.

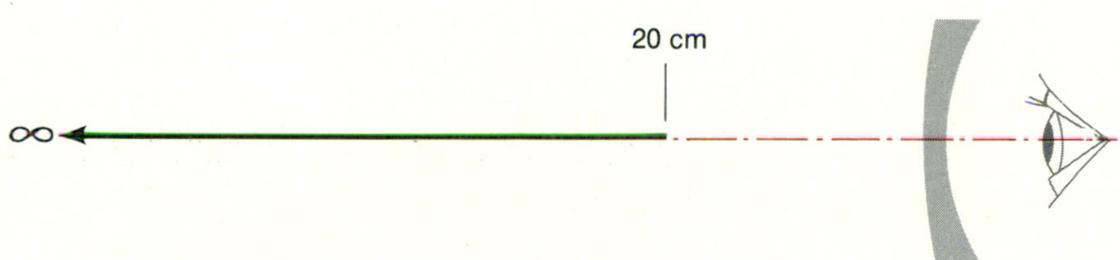

(**b**) The person's range of clear vision with eyeglasses.

Figure 29–43
Example 29–11

SOLUTION

(a) While this person can read easily without glasses, distant objects are out of focus. The glasses should therefore produce images of distant objects that fall within the range of 15 to 100 cm. In practice, objects at infinity should have images at the most distant point of clear vision, so that reasonably close vision is not impaired with the eyeglasses on. Thus, for this person, an object at infinity should have its image at $q = -100$ cm. Starting with the lens equation:

$$\frac{1}{p} + \frac{1}{q} = \frac{1}{f}$$

we have

$$\frac{1}{\infty} + \frac{1}{-100\text{ cm}} = \frac{1}{f}$$

where the signs are in accordance with the sign convention. Solving for the focal length of the eyeglass, we obtain

$$f = \boxed{-100 \text{ cm}} \quad \text{or} \quad \boxed{-1.00 \text{ diopter}}$$

(b) Again, using the lens equation:

$$\frac{1}{p} + \frac{1}{q} = \frac{1}{f}$$

we have

$$\frac{1}{25 \text{ cm}} + \frac{1}{q} = \frac{1}{-100 \text{ cm}}$$

Solving for the image distance q:

$$q = \boxed{-20.0 \text{ cm}}$$

which is well within the clear viewing range of the person. The eyeglasses would expand the range of clear vision as shown in Figure 29–43.

Toward middle age, people often lose accommodation, so that their range of clear vision becomes smaller. (This condition is called *presbyopia.*) A person with a narrow range of clear vision near 100 cm would need a convergent portion of the eyeglass for viewing nearby objects and a divergent portion for distance objects. Such eyeglasses are *bifocals.* Trifocal eyeglasses are sometimes needed to accommodate intermediate distances also.

The Astronomical Telescope

The simplest form of the astronomical telescope consists of only two lenses: the objective lens (focal length f_o) and the eyepiece lens (focal length f_e). As shown in Figure 29–44(a), the objective lens creates a real, inverted image I_1 at its focal-length distance because the object's distance is essentially infinite. In turn, the eyepiece forms a virtual image of I_1. In practice, most viewers focus the eyepiece so that the final image is at infinity. (In doing so, the eye may be shifted from the eyepiece to the object without eye accommodation.) For a final image at infinity, the image I_1 must be at the focal point of the eyepiece.

The angular magnification m of an astronomical telescope is the ratio of the angle subtended by the image formed by the eyepiece to that formed by the object. Thus:

$$m = \frac{\alpha}{\beta} \tag{29–43}$$

where α and β are as indicated in Figure 29–44(a). Ordinarily, the image formed by the objective is small compared with either the objective focal length f_o or the eyepiece focal length f_e. Then:

$$\alpha = \tan^{-1}\left(\frac{h}{f_e}\right)$$

or, since h is much smaller than f_e:

$$\alpha \approx \frac{h}{f_e}$$

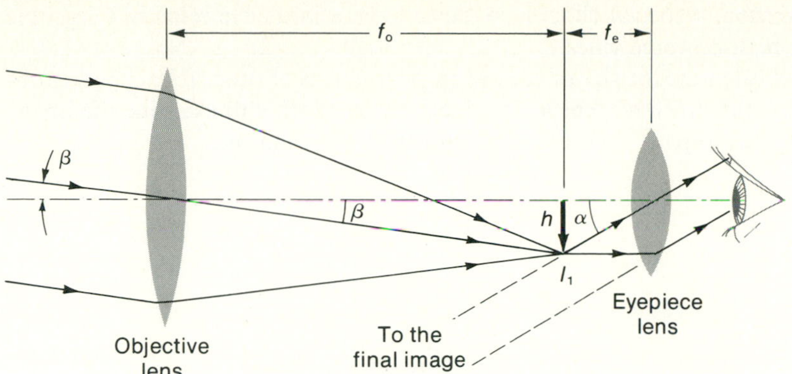

(a) The objective lens forms a real, inverted image I_1 of a distant object. The eyepiece lens forms a distant virtual image of I_1.

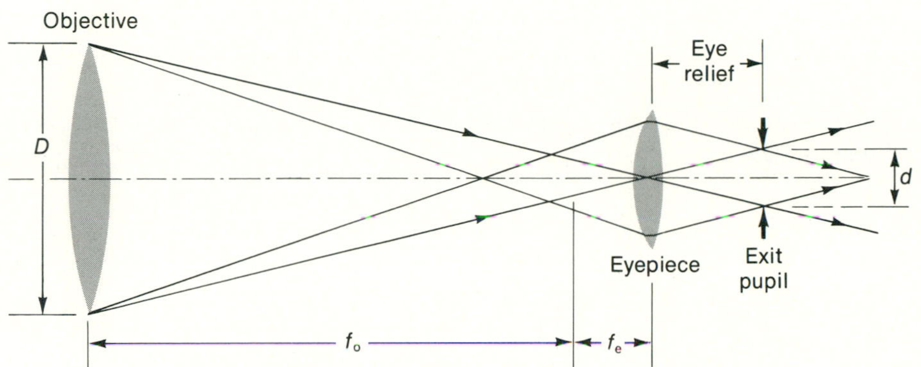

(b) The eyepiece lens forms an image of the objective lens, called the *exit pupil*. All the light that passes through the telescope goes out through the exit pupil.

Figure 29–44
The astronomical telescope.

Similarly:

$$\beta \approx \frac{h}{f_o}$$

Then Equation (29–43) becomes

ANGULAR MAGNIFICATION OF AN ASTRONOMICAL TELESCOPE

$$m = \frac{f_o}{f_e} \qquad \textbf{(29–44)}$$

An important characteristic of a telescope is its light-gathering ability. Ideally, when using the telescope visually, all of the light that enters the objective lens should ultimately enter the pupil of the eye. Unless special care is taken with the design, some light may be lost because the light emerging from the eyepiece covers a larger area than the pupil of the observer's eye. As shown in Figure 29–44(b), the bundle of light rays emerging from the eyepiece becomes constricted, then spreads out. The area of constriction, called the *eye ring* or *exit pupil*, is actually the image of the objective lens formed by the eyepiece lens. If the exit pupil diameter is smaller than the observer's eye pupil, the eye can capture all the light passing through the telescope. The distance from the eyepiece to the exit pupil is called the *eye relief*; it should be large enough so that the viewer may wear glasses and still view easily. For an obvious reason, a special effort is made to have a large eye relief in telescopic sights for high-powered rifles.

Let us calculate the size of the exit pupil in terms of other telescope parameters. Using the objective lens as an object, we need only to examine the image formed by the eyepiece lens. Starting with the lens equation:

$$\frac{1}{p} + \frac{1}{q} = \frac{1}{f}$$

we have

$$\frac{1}{f_o + f_e} + \frac{1}{q} = \frac{1}{f_e}$$

where the sum of the focal length f_o of the objective and the eyepiece focal length f_e is the object distance p. Solving for the image distance (the eye relief), we obtain

$$q = \frac{f_e(f_o + f_e)}{f_o} \tag{29–45}$$

The size of the exit pupil is obtained from the linear magnification relation, Equation (29–38):

$$M = -\frac{q}{p}$$

The diameter of the exit pupil d is then

$$d = MD$$

where D is the diameter of the objective lens. Substituting appropriate values:

$$d = \frac{q}{p} D$$

or

$$d = \frac{f_e(f_o + f_e)}{f_o(f_o + f_e)} D$$

$$d = \frac{f_e}{f_0} D$$

In terms of the angular magnification m of the telescope:

$$d = \frac{D}{m} \tag{29–46}$$

Consideration of the exit pupil diameter is particularly important when purchasing either a telescope or a pair of binoculars. For example, should a person buy a pair of binoculars with the designation 7 × 50 for viewing sporting events? The designation 7 × 50 means that the angular magnification is 7 and the objective lens diameter is 50 mm. Substituting numerical values into Equation (29–46), we find that the exit pupil is $d = 7.1$ mm. Since even on a cloudy day, the pupil size is only about 3 mm, the eye could gather only about one-fifth of the light entering the objective lens. Obviously, 7 × 50 binoculars have an exit pupil far too large for daytime viewing. The extra ease in finding the exit pupil usually does not offset the fact that 7 × 50 binoculars are heavy and rather cumbersome to use. A more reasonable choice for sports viewing and general daytime use is an 8 × 35 pair of binoculars.

The Simple Microscope

Practical microscopes are extremely complicated optical systems. As we will see, the objective lens must have a very short focal length, which requires elaborate measures to correct its various defects. In addition, adequate illumination of the object and efficient light-gathering impose further requirements on the optical system.

In the simple two-lens microscope in Figure 29–45, the object is placed just outside the focal point of the *objective* lens (whose focal length f_o is very short), producing a greatly enlarged real image I_1. This image is magnified by the *ocular* or *eyepiece* lens (focal length f_e) in the way we discussed for a simple magnifier with the final virtual image at the closest distance for comfortable viewing. The eyepiece lens has an angular magnification m [Equation (29–42)] of approximately 25 cm/f_e. (We drop the 1 in the formula because f_e is usually much shorter than 25 cm.) The objective lens has a linear magnification M, so that the total *magnifying power* of the microscope is Mm, or:

$$\text{Magnifying power} = M\left(\frac{25\text{ cm}}{f_e}\right) \qquad \textbf{(29–47)}$$

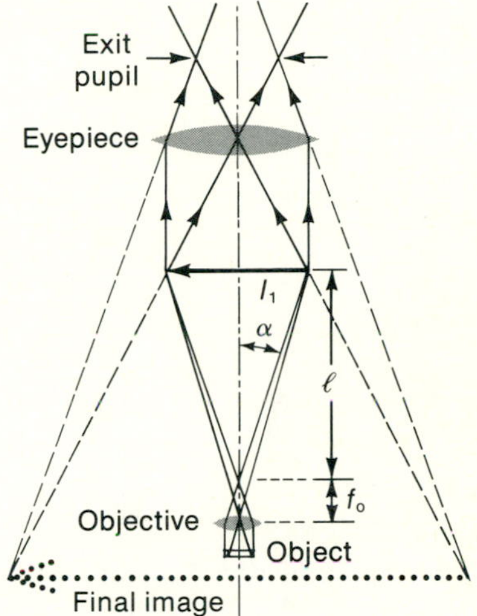

Figure 29–45
A simple microscope. The objective lens forms a greatly enlarged real image I_1 of the object. The eyepiece is a magnifier that creates a virtual image of I_1. The final image is usually chosen at the closest distance for comfortable viewing.

The linear magnification of the objective lens is ℓ/f_o, where ℓ is called the *tube length* of the microscope. A typical value of ℓ is 18 cm. Substituting these values, we have

$$\text{Magnifying power} = \left(\frac{\ell}{f_o}\right)\left(\frac{25\text{ cm}}{f_e}\right) \qquad \textbf{(29–48)}$$

$$\text{Magnifying power} = \frac{450\text{ cm}^2}{f_o f_e} \qquad \text{(where } f_o \text{ and } f_e \text{ are in centimeters)}$$

In order to achieve good exit pupil size and eye relief, f_e must be of the order of 1 cm. Thus, by Equation (29–48) the effective focal length of the objective must be very short. For a magnifying power of 2000, f_o is of the order of 2 mm.

Microscopes often have interchangeable objective and eyepiece lenses. Each of them is marked with its individual magnification. The total magnifying power of the microscope is simply the product of the objective and eyepiece magnification. For example, the objective may have an $M = 200$ (indicated by the notation × 200), which when combined with an eyepiece of $m = 10$ (indicated by × 10) produces a magnifying power of 2000.

As with using a telescope, an awareness of the location of the exit pupil is important since, to perceive the maximum field of view, the viewer's eye should be placed at the exit pupil.

The Slide Projector

We have indicated the importance of maximum light transmission through the optical systems of telescopes and microscopes. Such consideration is particularly important in slide projectors. The function of a slide projector is to project the brightest possible image on a screen without overheating the slide.

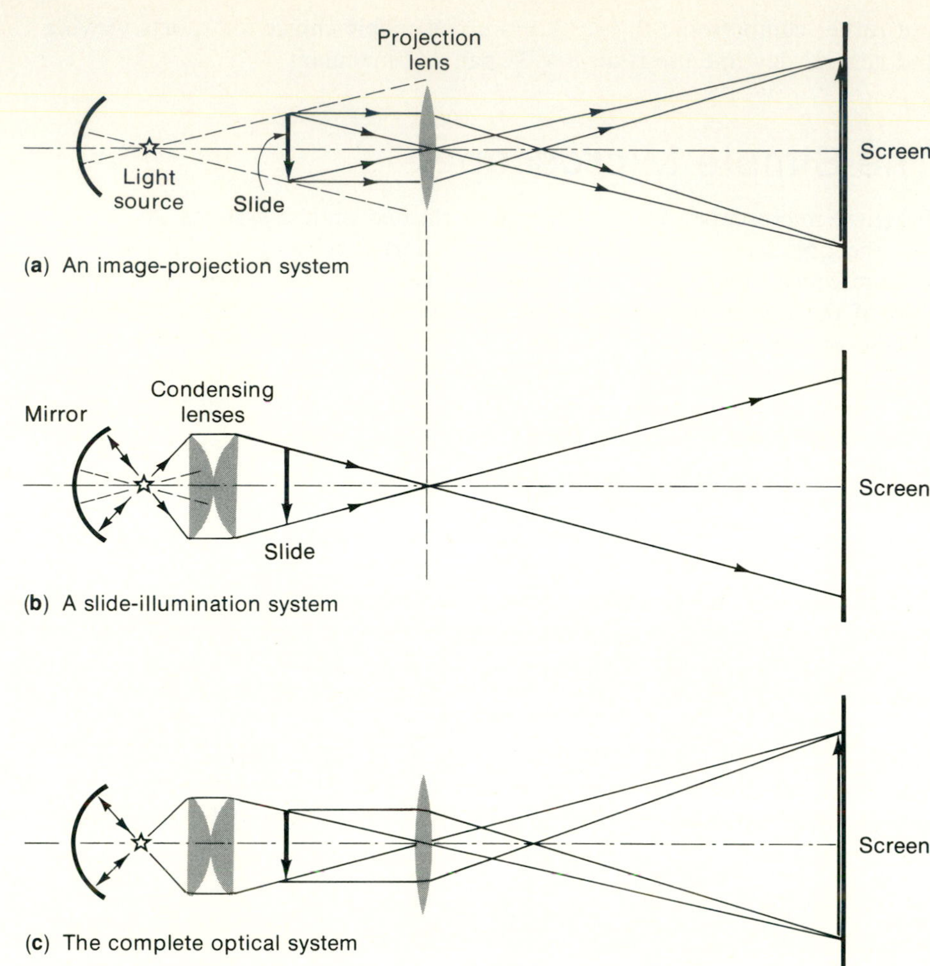

Figure 29–46
The slide projector. The optical system of a slide projector consists of two parts: an *image-projection system* and a *slide-illumination system.*

Consider the simple optical system shown in Figure 29–46(a). The slide is illuminated by a light source, which usually is backed by a concave mirror. The illuminated slide is then projected by the projection lens. Not only is the light reaching the slide divergent, but generally only a small portion of the light emitted by the source is intercepted by the slide and only part of the light passing through the slide passes through the projection lens to reach the screen.

A great improvement can be made by the insertion of *condensing lenses*, shown in Figure 29–46(b). The condensing lenses form an image of the light source at the position of the projection lens. Note that (except for reflection losses) all of the light from the source that is intercepted by the condensing lenses falls on the screen occupied by the image of the slide (Figure 29–46c). Condensing lenses need not be of high quality, since imperfections in the image they produce are irrelevant. Only the projection lens determines the quality of the slide image.

29.14 Aberrations

The lens-maker's formula [Equation (29–34)] implies that all a lens maker must do is apply the formula to obtain a lens capable of accurate images. Such is not the case. Remember, in developing the formula we used several assumptions: the small-angle and thin-lens assumptions and the assumption that the object was a point on the axis of the lens. We also neglected the fact that the index of refraction is not the same for all colors of light. As a consequence, every lens produces defects, or *aberrations*, in the image.[8]

Spherical Aberration

The effect of the small-aperture assumption in mirrors and lenses is most evident when comparing the focal distances of rays entering the edge of the mirror or lens with those entering near the axis. As indicated in Figure 29–47, those near the edge have a different focal distance than those near the axis. Short-focal-length lenses of large diameter (that is, "thick" lenses) produce a great deal of spherical aberration. The net effect of the aberration is to produce an image of a point object on the axis that is diffuse (that appears "out of focus" everywhere.)

Figure 29–47
Spherical aberration.

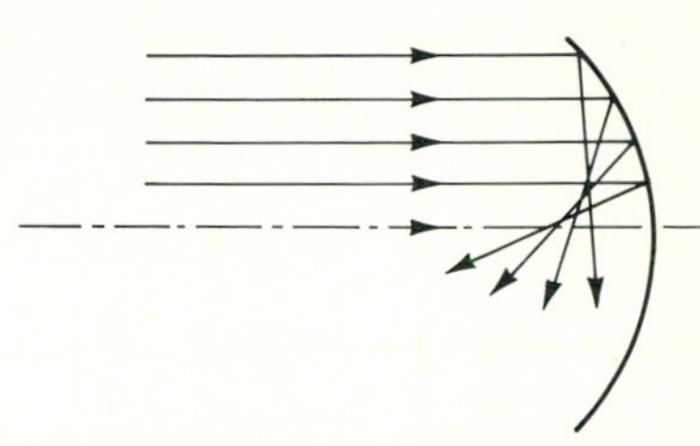

(**a**) Rays farther from the axis of a *spherical* mirror are brought to focus closer to the mirror than are rays near the axis.

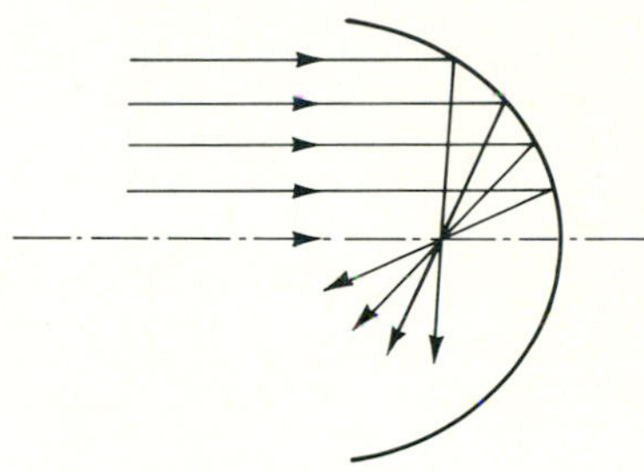

(**b**) A *parabolic* mirror focuses all parallel rays at the same point: thus it has no spherical aberration.

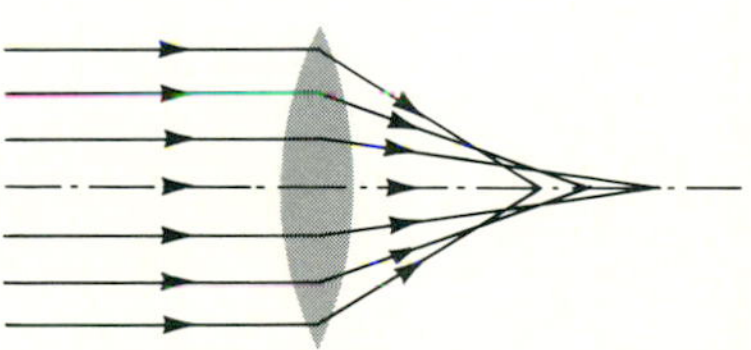

(**c**) Parallel rays near the edge of a spherical lens have a shorter focal length than those rays near the axis of the lens.

Astigmatism

A symmetrically spherical lens becomes nonsymmetric when tilted with respect to the direction of the oncoming light. Equivalently, a symmetrical lens exhibits *astigmatism*[9] whenever the incident light is not parallel to the axis of the lens, as shown in Figure 29–48(a). The figure depicts two views of the same situation. Light from a distant point source enters the lens obliquely. Viewed from the top, the rays seem parallel to the axis of the lens and converge to a focal point

[8] A good discussion of aberrations in camera lenses can be found in the article: William H. Price, "The Photographic Lens," *Scientific American*, August 1976.

[9] The word is from the Greek: *a*, meaning "no," and *stigma*, meaning "point."

Figure 29–48
Astigmatism.

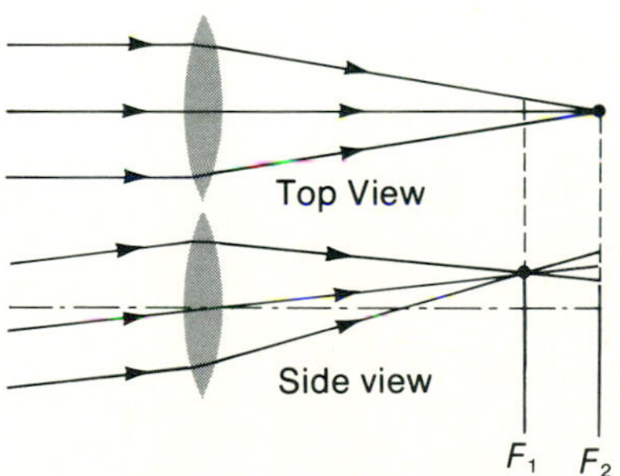

(**a**) Parallel rays entering a lens obliquely have a shorter focal distance than those entering parallel to the axis of the lens.

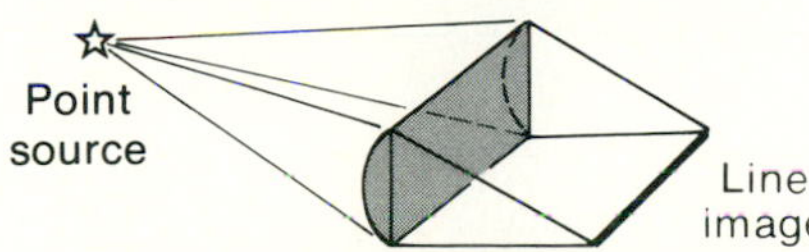

(**b**) A cylindrical lens forms a line image of a point source.

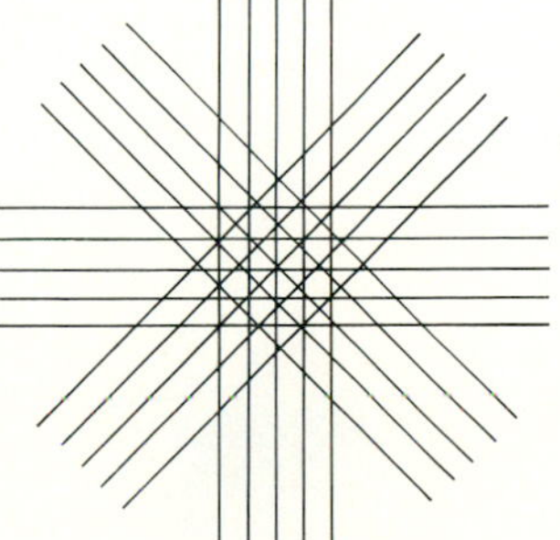

(**c**) A test for astigmatism in human vision. If all lines are not equally sharp, the eye lens may be astigmatic.

F_2. However, if viewed from the side, the light approaches what appears to be a thicker lens and converges to a shorter focal point F_1. If a screen were placed at F_1, the image of a point source would be a horizontal line. If the screen were moved farther from the lens to F_2, the image would become a vertical line. The net effect of astigmatism is generally an indistinct, nonsymmetrical image of a symmetrical object.

Off-axis astigmatism is usually not a problem in human vision because we can shift our line-of-sight by moving our eyeballs. However, if the curvature of the human eye lens is not precisely symmetrical, it may produce an astigmatic image even for a point on the axis. This defect is produced by a slightly cylindrical property of the lens, as illustrated in Figure 29–48(b). As a consequence, images of lines at various angles may be focused on the retina with varying degrees of sharpness (Figure 29–48c). This sort of astigmatism may be corrected by adding a degree of cylindricality to eyeglass lenses.

Curvature of Field

In most optical instruments, it is essential that the image formed by a lens be in focus on a *plane* area. The inability of a lens to do this is termed *curvature of field* (see Figure 29–49).

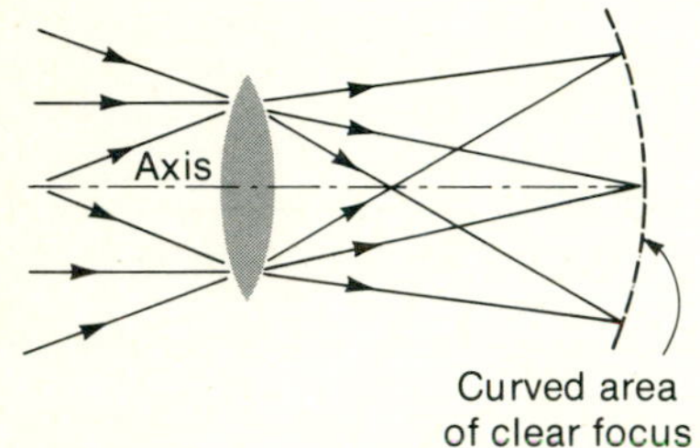

Figure 29–49
Curvature-of-field aberration.

Distortion

The primary function of any optical system is to produce an image that is an accurate representation of the object. Photographs taken with a toy camera often exhibit distorted images. For example, the edges of buildings may appear curved. *Distortion* arises from the fact the linear magnification of an object depends upon its distance from the axis of the lens. (Those parts of the object situated at an angle far from the lens axis will be magnified differently from those parts situated close to the axis.) Artistic effects in photography are produced through the use of wide-angle lenses, which produce a large amount of distortion.

Chromatic Aberration of Lenses

The focal length of a lens depends on the wavelength of light passing through it, resulting in an image with colored edges. This effect is called *chromatic aberration.*

A converging lens is like a prism in its ability to disperse white light into its component wavelengths, as shown in Figure 29–50. If the incoming light has a range of wavelengths, different colors will be brought to a focus at different

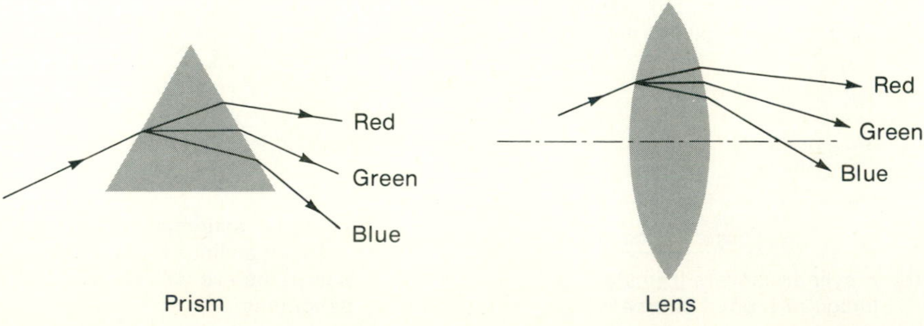

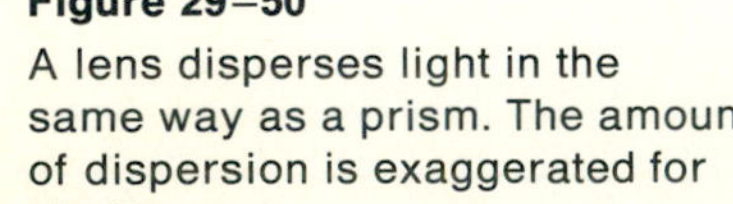

Figure 29–50
A lens disperses light in the same way as a prism. The amount of dispersion is exaggerated for clarity.

points, as in Figure 29–51(a). This effect is called *dispersion.* A screen placed at F_B would display an image with a blue center and a red halo. An image at F_R would have a red center with a blue halo. Fortunately, the dispersion of one lens can be partially undone by a second lens (Figure 29–51b). Of the so-called *acromat* pair of lenses shown, the converging lens is crown glass and the diverging lens is flint glass. The relative characteristics of these glasses were illustrated in Figure 29–20. If the divergent lens has the greater dispersion, it can bring the colors together at the focus, with the combination still having a net convergent effect.

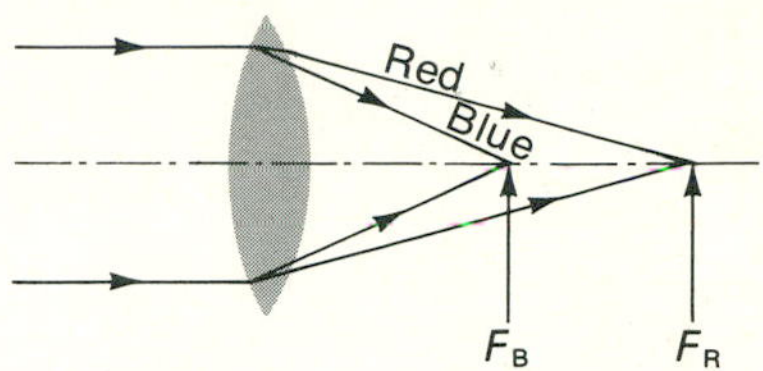

(a) A single convergent lens has a longer focal length for red light than for blue light.

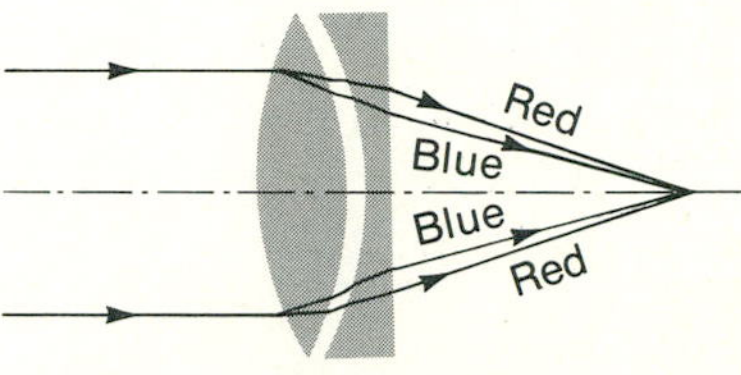

(b) The dispersion of two colors by a converging lens can be "undone" by a divergent lens. The divergent lens must have greater dispersion than the convergent lens.

Figure 29–51
Chromatic aberration. (The amount of dispersion shown is exaggerated.)

With all of their aberrations, how is it possible to make telescopes, microscopes, and cameras that produce sharp, accurate images? The answer lies in multiple-element lens systems, such as that shown in Figure 29–52. Because of the ease of fabrication, almost all lenses have spherical surfaces with low-reflection coatings. The design of near-perfect complicated lens systems is feasible only through the development of new optical-glass materials with advantageous index-of-refraction characteristics. Also, extensive computer programs have made possible the use of nonspherical lens surfaces by carrying out complex ray-tracing calculation for multiple-lens systems.

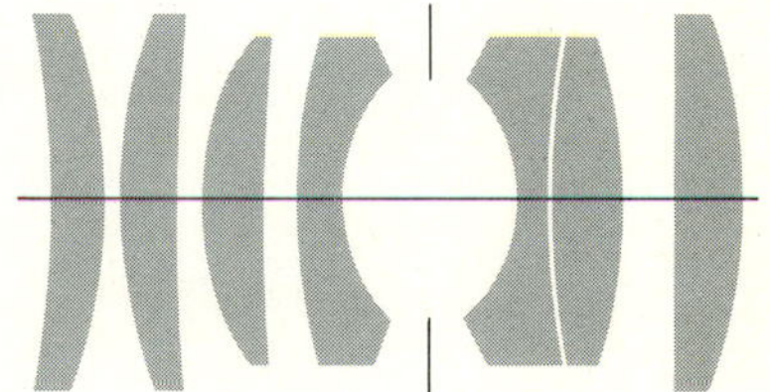

Figure 29–52
A system of several lenses is used to almost completely remove lens aberration. This is an example of a high-quality camera lens system.

Summary

The following concepts and equations were introduced in this chapter:

Propagation of light (which is characterized by rays and wavefronts):

Rays: Imaginary lines in the direction of propagation.
Wavefronts: Imaginary surfaces perpendicular to rays, moving in the direction of propagation; each point on a wavefront has the same phase.

Huygens' principle:

Wavefronts can be generated from the preceding one by considering every point on a wavefront to be a secondary source. The envelope of wavelets emanating from each of the secondary sources forms the next wavefront.

Mirror-lens equation:

$$\frac{1}{p}+\frac{1}{q}=\frac{1}{f}$$

where p is the object distance
q is the image distance
f is the focal length

The equation is to be used with the *sign convention:*

(1) The value of p is *positive* if the rays that impinge on the mirror or lens are *divergent.*
(2) The value of q is *positive* if the rays leaving the mirror or lens are *convergent.*
(3) *For mirrors,* the sign of the focal distance is determined by:

$$f=\frac{R}{2}$$

where R is the radius of curvature of the mirror surface. It is *positive* if the surface is *concave* and *negative* if the surface is *convex.*

(4) *For lenses*, the focal length f for a *converging* lens is *positive*; for a *diverging* lens, it is *negative*. In more formal terms:

$$\frac{1}{f} = (n - 1)\left(\frac{1}{R_1} + \frac{1}{R_2}\right)$$

where n is the refractive index of the lens relative to the surrounding medium and R_1 and R_2 are the radii of curvature of the lens surfaces. The values of R_1 and R_2 are *positive* if the corresponding surfaces are *convex*, and *negative* if they are *concave*.

The sign convention may be remembered from the fact that for the "standard setup" of an object situated farther than a focal-length distance from a converging mirror or lens, all the symbols in the corresponding equation are positive; if any of the distances are on the opposite side of the mirror or lens, they are negative.

Law of reflection:

$$\theta_i = \theta_r$$

Ray diagrams are constructed by tracing the following two rays:

(1) A ray striking the center is reflected symmetrically (or passes through a lens undeviated).
(2) A ray parallel to the axis passes through the focal point F.

Images are real or virtual, erect or inverted, with linear magnification according to the following:

Linear magnification:

$$M = -\frac{q}{p}$$

Index of refraction:

$$n = \frac{c}{v}$$

Snell's law:

$$n_1 \sin \theta_1 = n_2 \sin \theta_2$$

Total internal reflection: The critical angle θ_c is given by

$$\sin \theta_c = \frac{n_2}{n_1} \qquad (n_2 < n_1)$$

Lens-maker's formula:

$$\frac{1}{p} + \frac{1}{q} = (n - 1)\left(\frac{1}{R_1} + \frac{1}{R_2}\right)$$

Thin lenses in contact:

$$\frac{1}{f} = \frac{1}{f_1} + \frac{1}{f_2}$$

Separated lenses: The image produced by the first lens (acting alone) is the object for the second lens.

Angular magnification:

Simple magnifier: $m = \frac{25\text{ cm}}{f} + 1$ (image at 25 cm)

where f is in centimeters.

Telescope: $m = \frac{f_o}{f_e}$ (image at infinity)

Microscope: Magnifying power $= \left(\frac{\ell}{f_o}\right)\left(\frac{25\text{ cm}}{f_e}\right)$

where ℓ is the tube length, usually 18 cm, and focal lengths are in centimeters.

Common aberrations:

Spherical aberration
Astigmatism
Curvature of field
Distortion
Chromatic aberration (of lenses)

Questions

1. A plane mirror produces an image that is reversed right and left. Why does a plane mirror not produce an upside-down image?

2. Will convergent rays reflected by a plane mirror produce a real or virtual image?

3. A sign painted on a store window is reversed when viewed from inside the store. By viewing the reversed sign in a mirror, does the image of the sign appear as the sign would appear from outside the store?

4. Devise a system of plane mirrors that will produce an image that is not right-left reversed as it is with a single plane mirror.

5. At one corner of a room, the ceiling and the two walls are plane mirrors. As you look into the corner, how many images of yourself can you see?

6. Is the focal length of a spherical mirror altered by immersing the mirror in water?

7. Under what conditions will a convex mirror produce a real image?

8. Why does the apparent depth of a swimming pool depend on the observer's distance from the edge of the pool?

9. Why does a straight pole penetrating the surface of a pond often appear to be bent at the point where the pole enters the water?

10. If a fisherman can see the eye of a fish in a still pond, can the fish always see the eye of the fisherman? That is, are there situations for which total internal reflection prevents either from seeing the other?

11. What does a swimmer see as she looks upward toward the smooth surface of a swimming pool? Include considerations of total internal reflection.

12. What are the optical properties of an air bubble in glass?

When measuring the angle between the late afternoon sun and the horizon with a sextant, a navigator must apply a correction to the observed angle. Why is a correction necessary and what is the sign of the correction?

14. Is it possible for a lens to be convergent in air and divergent in water?

15. How does the focal distance of a converging lens depend on the color of light? Is the dependence the same for a diverging lens?

16. The two focal points of a thin lens are the same distance from the lens. Can you show by sketching ray diagrams that the two focal points of a thick lens may not be the same distance from the center of the lens?

17. What is a procedure for determining the focal lengths of (a) a diverging lens and (b) a convex mirror?

18. A person's eyes appear to be smaller when he wears his glasses. Is he nearsighted or farsighted?

19. While swimming under water without a diving mask, does the swimmer become more nearsighted or more farsighted? Can this be corrected by wearing eyeglasses? If so, what kind of eyeglasses?

20. A simple two-lens astronomical telescope is used to view a distant sign. Is the image simply inverted or is the lettering on the sign reversed, as in a plane mirror image?

21. Why does a person with normal vision often adjust the eyepiece of an astronomical telescope so that the image is at infinity?

22. Without asking the wearer (but being allowed to experiment with the lens), how would you determine if an eyeglass lens includes a correction for astigmatism?

Problems

29A–1 A woman whose eyes are 1.59 m from the floor stands before a mirror.
(a) If the top of her hat is 14 cm above her eyes, find the minimum vertical dimension of a wall mirror that would enable her to see an entire image of herself (hat included).
(b) How far from the floor is the bottom edge of the mirror?

Answers: (a) 86.5 cm (b) 79.5 cm

29A–2 A light beam strikes a mirror and is reflected. Show that if the mirror is rotated through an angle α about an axis in the plane of the mirror, the reflected beam moves through an angle 2α.

29A–3 An object placed 5 cm from a concave mirror produces a real image four times as large as the object. Find the radius of curvature of the mirror. *Answer:* 8 cm

29A–4 Complete the following table for *mirrors*. In every case assume that the diameter of the mirror is small compared with the radius of curvature of its surface. All numerical values are expressed in centimeters. Indicate the appropriate sign of the values in accordance with the sign convention.

Type of mirror	Radius of curvature	Focal distance	Object distance	Image Distance	Image Real?	Image Inverted?	Magnification
Convex	−120	−60	+30	−20	No	No	$+\frac{2}{3}$
Plane			+30				
		+10		−20			
	−100		+5				
				+100			−2
Convex				−20			$\frac{1}{4}$ (sign?)
Concave	20 (sign?)			+100			

29A–5 A microscope may be used to measure the refractive index of a plane sheet of glass. The top surface of the glass is brought into focus by the microscope. The objective of the microscope is then lowered 2.50 mm to bring the lower surface into focus. The measured thickness of the glass is 3.80 mm. Calculate the refractive index of the glass.
Answer: $n = 1.52$

29A–6 A tin can 20 cm high and 15 cm in diameter has one end removed, and a small hole is punched in the center of the other end. When peering through the small hole, you have a cone of vision that is limited by the edge of the other end of the can.

(a) Calculate the approximate solid angle of this cone.

(b) Calculate the semivertical angle of the cone (between the axis of the cone and the surface of the cone).

(c) Calculate the semivertical angle of vision if the can is filled with a clear plastic that has a refractive index of 1.65.

29A–7 A small air bubble is at the center of a large glass sphere that has a refractive index n and radius R. Determine how far the air bubble appears to be from the surface of the sphere.
Answer: R

29A–8 A lens made of glass ($n = 1.62$) has a concave surface with a radius of curvature of 100 cm and a convex surface with a radius of curvature of 40 cm. Calculate the focal distance of the lens.

29A–9 A pair of eyeglasses is to have a power of 1.25 diopter, they must be made of glass with a refractive index of 1.50, and the surface next to the eye must be concave with a radius of curvature of 80 cm. Calculate the radius of curvature of the outer surface of the eyeglass lenses.
Answer: 26.7 cm

29A–10 Complete the following table for *lenses*. In every case assume that the diameter of the lens is small compared with the radii of curvature of its surfaces. All numerical values are expressed in centimeters. Indicate the appropriate sign of the values in accordance with the sign criteria.

Type of lens	Focal distance	Object distance	Image			Magnification
			Distance	Real?	Inverted?	
Converging	+60	+20	−30	No	No	$+\frac{3}{2}$
	−40	+120				
		+50				−4
			200 (sign?)		No	+5
		+40				−5
Diverging	120 (sign?)		−30			
		+50	−30			

29A–11 A thin converging lens forms a real image a distance x' from the focal point when the object is a distance x from the other focal point. Show that $xx' = f^2$, where f is the focal distance of the lens.

29A–12 A 6-diopter magnifying glass is held 10 cm from a printed page. Find the image size of a letter 4 mm high.

29A–13 A simple camera has a single thin lens with a focal length of 50 mm. Determine how far and in which direction the lens must be moved relative to the film in order to change the focus from very distant to 75 cm. *Answer:* 3.57 mm, outward

29A–14 One way to determine the focal length of a thin divergent lens is to place the lens in contact with a convergent lens strong enough that the combination produces a real image of a very distant object. Suppose a divergent lens of unknown focal length is combined with a 2-diopter converging lens to produce a real image of a distant object on a screen 75 cm away from the lenses. Calculate the focal length of the divergent lens.

29A–15 A farsighted person can comfortably view objects no nearer than 2 m but can see very distant objects clearly.

(a) Calculate the power of eyeglasses necessary for the person to read a book held 25 cm away.

(b) Find the farthest object that the person could see comfortably while wearing these glasses, assuming the eye cannot make more accommodation for distant vision than when unaided. *Answers:* (a) +3.50 diopters (b) 28.6 cm

29B–1 The edge of a plane mirror is in contact with the edge of another plane mirror, and the angle between the reflecting surfaces is 90°. Let the mirrors be oriented so that one mirror surface is along the $+x$ axis and the other is along the $+y$ axis. (The joined edges are along the z axis.) By drawing ray diagrams, locate all three images of an object located at the position (x, y), where $x = 30$ and $y = 40$. (All distances are in centimeters.) *Answer:* $(30, -40)$, $(-30, 40)$, $(-30, -40)$

29B–2 The size of a real image produced by a concave mirror is doubled if the object distance is decreased from 80 cm to 50 cm. Find the radius of curvature of the mirror.

29B–3 Sketch ray diagrams for each of the cases given in Problem 29A–4.

29B–4 An optical system consists of a large concave mirror and a small mirror, as shown in Figure 29–53. Light from a very distant object strikes the large mirror and is reflected by the small mirror to form an image on the surface of the large mirror. If the radius of curvature of the large mirror is 5 m and the two mirrors are separated by 2 m, (a) determine whether the small mirror should be concave or convex and (b) find the radius of curvature of the small mirror.

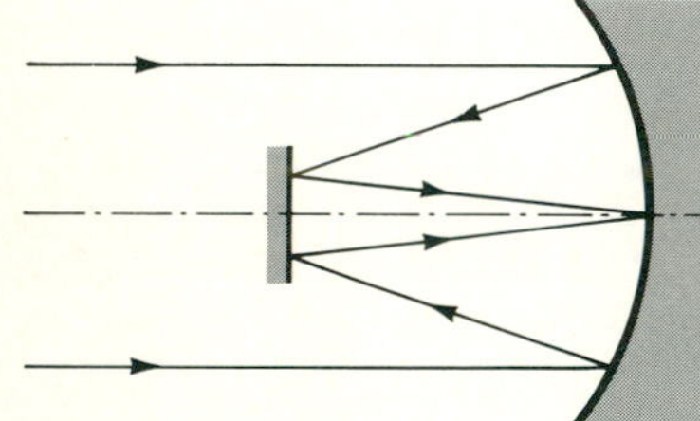

Figure 29–53
Problem 29B–4

29B–5 A beam of light strikes a plane slab of glass at an angle of 40° with the surface of the glass. The glass is 1.5 cm thick and has a refractive index of 1.60. The beam emerging from the other side of the slab will be parallel to the incident beam but displaced laterally. Calculate the distance the emerging beam is displaced laterally from the incident beam by the glass. *Answer:* 0.624 cm

29B–6 A ray of light enters a 45° prism (refer to Figure 29–26a).

(a) Find the minimum refractive index of the prism to produce total internal reflection, as shown.

(b) If the prism were immersed in water, calculate the minimum index of refraction to produce the same result.

29B–7 The time required for a light signal to travel from the bottom to the top of an empty vessel is t_0. Show that when the same vessel is filled with a liquid $(n > 1)$, the time required for a signal to travel in the liquid the distance of the apparent depth of the vessel is also t_0.

29B–8 A can 12 cm deep is filled with a layer of water $(n = 1.33)$ 5 cm deep and a layer of oil $(n = 1.48)$ 7 cm deep. Calculate the apparent depth of the can when viewed from a point directly above the can. (Hint: Use the result of Problem 29B–7.)

29B–9 A fish at the bottom of a still pond sees the entire region above the surface of the water in a circular field of view centered on a vertical line above the fish. Calculate the solid angle (in steradians) that the circular area subtends at the fish's eye. *Answer:* 1.125 steradians

29B–10 Sketch ray diagrams for each of the cases given in Problem 29A–10.

29B–11 A small-diameter parallel light beam is directed toward the center of a large solid sphere made of transparent plastic. The beam is brought to a focus on the opposite side of the sphere. Find the refractive index of the plastic. *Answer:* $n = 2$

29B–12 A lens made of polystyrene ($n = 1.59$) has a power of 2 diopters. The radius of curvature of a convex surface is 50 cm. Calculate the radius of curvature of the other surface. Is it concave or convex?

29B–13 A nearsighted person wearing eyeglasses with a power of -1.5 diopters can see clearly objects as close as 25 cm as well as very distant objects. Determine the person's range of vision without eyeglasses, assuming no further accommodation for distant vision is possible. *Answer:* from 18.2 cm to 66.7 cm

29B–14 An object is located at the origin of the x axis. Two converging lenses of focal lengths 10 cm and 20 cm are placed, respectively, at $x = 15$ cm and $x = 35$ cm.

(a) Locate and describe the final image.

(b) Sketch a ray diagram.

29C–1 An observer views a point source of light reflected in a mirror as in Figure 29–6. Show by application of Fermat's principle that the angle of incidence θ_i equals the angle of reflection θ_r. Assume that the incident ray and reflected ray lie in the same plane. However, do not assume that the source and observer are the same distance above the mirror. (Hint: Choose the variable, upon which both θ_i and θ_r depend, to be the distance between the point on the mirror directly below the source and the point of reflection.)

29C–2 Two plane mirrors have their reflecting surfaces facing one another, with the edge of one mirror in contact with an edge of the other, so that the angle between the mirrors is α. When an object is placed between the mirrors, a number of images are formed. In general, if the angle α between the two mirrors is such that $n\alpha = 360°$, where n is an integer, the number of images formed is $n - 1$. Graphically, find all of the image positions for the case $n = 6$.

29C–3 A concave mirror with a focal length of 25 cm produces an image 200 cm away from the object. Find the *two* object distances that produce such an object-to-image separation. Describe the image in each case.

Answers: for $p = 228$ cm, the image is inverted, real, and $M = -0.123$
for $p = 21.9$ cm, the image is erect, virtual, and $M = 10.1$

29C–4 A man discovers that he can focus clearly on objects no closer than 70 cm when he does not wear glasses.

(a) When using a shaving mirror with a focal length of $+75$ cm, he prefers to view the image of his face at a distance of 80 cm. Determine how far his face should be from the mirror.

(b) Calculate the linear magnification.

29C–5 A "floating coin" illusion consists of two parabolic mirrors, each with a focal length of 7.5 cm, facing each other so that their centers are 7.5 cm apart (Figure 29–54). If a few coins are placed in the lower mirror, an image is formed at the small opening at the center of the top mirror. Show that the final image is formed at that location and describe its characteristics: real or virtual, erect or inverted, and the overall linear magnification. *Answer:* Real, erect, and unit magnification

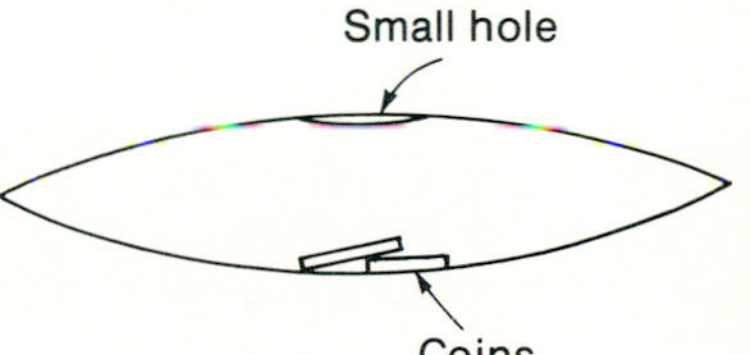

Figure 29–54
Problem 29C–5

29C–6 Derive Snell's law of refraction using Fermat's principle. Use assumptions similar to those in Problem 29C–1.

29C–7 A luminous object and a screen are a distance L apart. A converging lens with a focal length f placed at either of two positions between the object and screen will produce an image of the object on the screen. Derive an expression for the separation of those positions. *Answer:* $(L^2 - 4fL)^{\frac{1}{2}}$

29C–8 Consider the combination of lenses shown in Figure 29–37 but with the convergent ($f_1 = 0.1$ m) and the divergent ($f_2 = -0.2$ m) lenses *interchanged.* Locate and describe the final image. Compare your results with the results of Example 29–10.

Answer: real, inverted image 0.174 m beyond the convergent lens;
the net magnification is $M = -0.42$

29C–9 A small change in object distance Δp corresponds to the thickness of a thin object. Show that the image of the object produced by either a lens or mirror has an apparent thickness equal to $M^2\Delta p$, where M is the linear magnification of the lens or mirror.

29C–10 Find (in terms of f) the minimum object-to-real-image distance for a converging lens with a focal length f.

chapter 30
Interference

Mirrors have one limitation: You can't
either by hook or by crook
Use them to see how you look when you aren't
looking to see how you look.

PIET HEIN
(*Grooks 4*)

Maxwell's electromagnetic theory describes light as a changing pattern of electric and magnetic fields that travel through space as waves. For the next few chapters we will discuss phenomena that demonstrate these wavelike properties. However, we must also mention that visible light, as well as all other forms of electromagnetic radiation, possesses a dual, apparently contradictory, nature. Light in transit seems to behave as a wave, but as we will show in Chapter 33, when radiation is absorbed by matter it behaves as particles. This dual nature of light—explainable as a wave in certain instances, but as particles in other cases—will be our main concern in the remaining chapters.

We have developed the laws of reflection and refraction of light using a wave model. Interestingly, these laws can be derived just as easily with a particle model. In fact, Newton was the first to work out a particle model in some detail, explaining reflection and refraction on that basis. However, as we now discuss the interference, diffraction, and polarization of light, the particle model is clearly unworkable. Only *waves* seem to make sense.

30.1 Double-Slit Interference

In 1802 and 1803, Thomas Young[1] presented papers before the Royal Society signaling the downfall of the particle theory of light, which had been so firmly established by Newton. By passing light from a single small source through two small adjacent openings in an opaque screen, Young observed a series of light and dark fringes on a viewing screen. The double-slit arrangement he used was similar to that shown in Figure 30–1. (The figure is not to scale. In a typical experiment, the slit separation is about 0.2 mm and the screen is about 1 m from the slits, producing fringes about 2 mm apart.)

Unlike the phenomena of geometrical optics, the fringe pattern cannot be explained using a particle theory of light. The reason is the following. If we cover one of the slits, the result is a general illumination of the screen, as shown in Figure 30–2(a). It does not matter which slit we cover, the screen illumination is the same broad pattern. Now suppose we uncover the slit, so that both slits

[1] Thomas Young (1773–1829) was a brilliant scientist who not only contributed to the wave theory of light and the three-color theory of light perception, but also to Egyptology. It was largely through his efforts that the Rosetta stone, the key to Egyptian hieroglyphics, was deciphered.

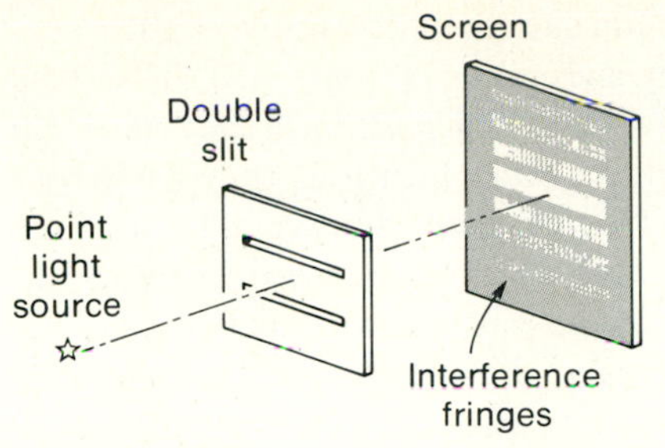

(**a**) A pictorial view showing the arrangement of the point light source, the double slit, and the screen.

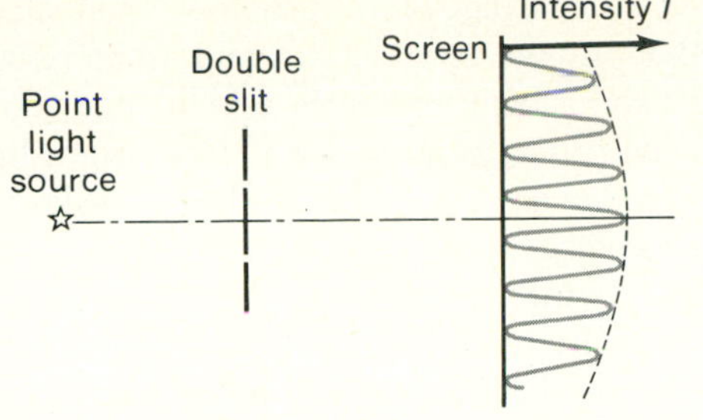

(**b**) A schematic sketch that includes a graph of the intensity versus position on the screen.

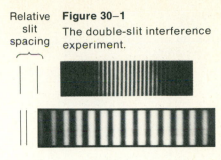

Relative slit spacing

(**c**) As the slit separation *decreases*, the distance between fringes *increases*.

Figure 30–1
The double-slit interference experiment.

are open. If light were a stream of particles, uncovering the slit should merely add the two individual patterns together to produce an overall intensity of about twice the original value. Instead, the pattern of light and dark fringes is produced. Furthermore, the intensity on the screen at the central axis is now four times the intensity with just one slit open, so obviously the light passing through the two individual slits is not simply additive.

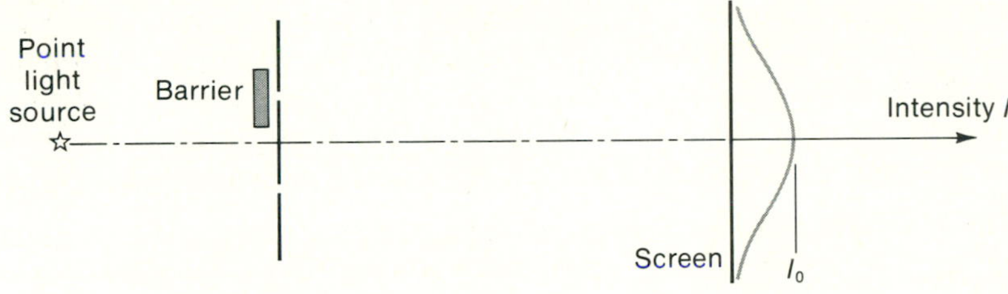

(**a**) With only *one slit* open, the illumination on the screen is diffuse, diminishing gradually in intensity for distances away from the center.

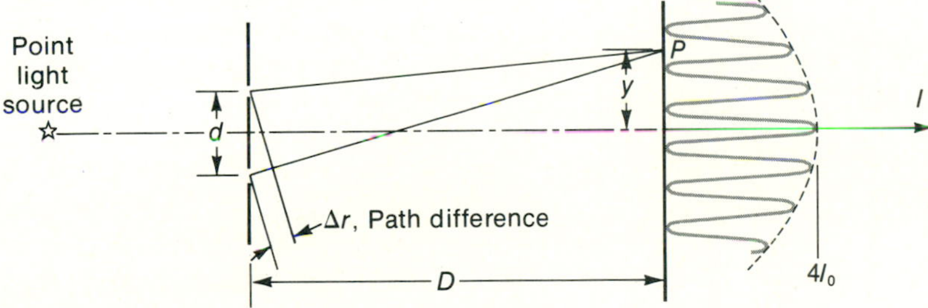

(**b**) With *both slits* open, the pattern on the screen is equally spaced bright and dark fringes.

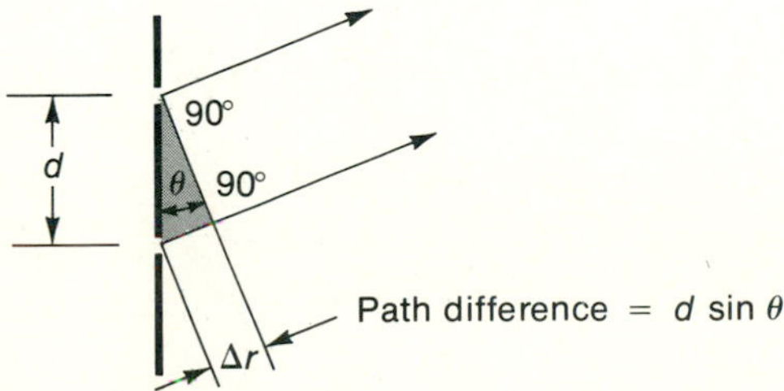

(**c**) If the distance D is very much larger than y and d, we may consider the two rays as essentially parallel. With this approximation, the shaded triangle is a right triangle and the path difference Δr is equal to $d \sin\theta$.

Figure 30–2
A screen illuminated by coherent light from two slits produces an interference pattern.

The fringe pattern can be explained as the superposition of two light *waves* that emerge from the slits and interfere with each other as they reach the screen. At some locations on the screen, the two waves arrive *in phase* and they reinforce each other, producing an extra bright light. At other locations they arrive *out of phase* and cancel each other (see Figure 30–3). The *interference* of light waves produces the array of light and dark fringes called an *interference pattern.*

Before we examine the details of the fringe pattern, we must discuss the necessity of a single point source of light. If each slit had its own separate light source, the pattern of interference would not be produced, for the following reason. To produce, say, a bright fringe at a particular location on the screen, all of the light reaching that location from the two slits must arrive there in phase *and remain in phase as time goes on.* At other locations on the screen, the phase difference ϕ between the two waves will be a different value, but whatever the value, it must remain constant in time to produce the proper type of interference for that part of the pattern. *Two light waves with a phase difference that remains constant in time* are said to be **coherent.** Obviously, only waves of the same wavelength (and frequency) can be coherent. With a separate light source for each slit, different atoms would radiate light through each slit; since each atom radiates independently of other atoms, the two light waves would not be coherent and there would be no interference pattern.

There is another important feature worth mentioning. As we will see in Chapter 33, light is emitted as the result of an energy transition within an atom. Each transition produces a single *wave train* of finite length (Figure 30–3b). When a single wave train illuminates both slits, the radiation passing through the slits acts as two coherent sources, since the two waves will be in phase and of equal amplitude. Because each atom emits a wave train independently of other atoms, the various wave trains from different atoms are *not* coherent with

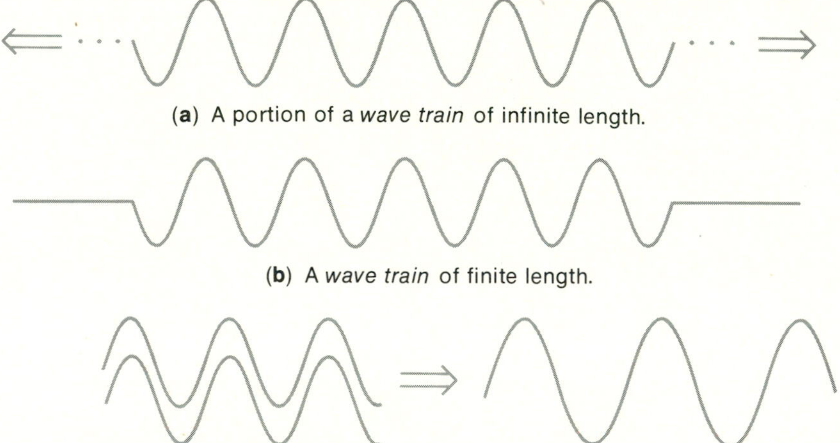

(**a**) A portion of a *wave train* of infinite length.

(**b**) A *wave train* of finite length.

(**c**) Two light waves *in phase* superpose to produce an extra bright light (increased intensity).

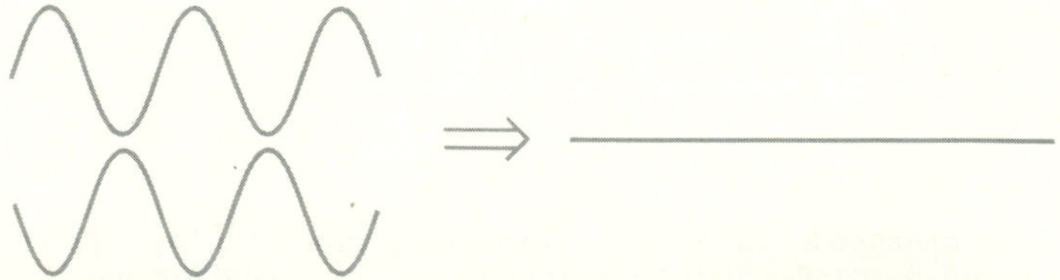

(**d**) Two light waves *180° out of phase* (with equal amplitudes) superpose to produce darkness (zero intensity).

Figure 30–3
The result of the *superposition* of two coherent light waves depends strongly on the phase relation between the two waves.

one another. The interference pattern develops because a wave train *interferes only with itself*, not with wave trains from other atoms. The pattern is the result of an extremely large number of individual wave trains, each interfering only with itself. This interpretation is verified by experiments in which, on the average, only one wave train at a time passes through the slits. After waiting long enough to accumulate the same total illumination on a photographic film, the pattern is the same as in an experiment when many wave trains pass through more or less simultaneously. Thus, *each wave train from a single atom interferes only with itself as it passes through the apparatus.*

Here are the criteria necessary for producing a stationary pattern of wave interference:

CRITERIA FOR INTERFERENCE OF TRANSVERSE WAVES

(1) The waves emerging from the two slits must be *coherent*; that is, the waves must have the same frequency and *a phase difference that remains constant in time.*

(2) The waves must oscillate in the same plane. In the case of light waves, the electric field oscillations must be in the same direction.

A single point source behind the slits meets these criteria (Figure 30–4). If the source is equidistant from each slit, the light passing through the two slits will

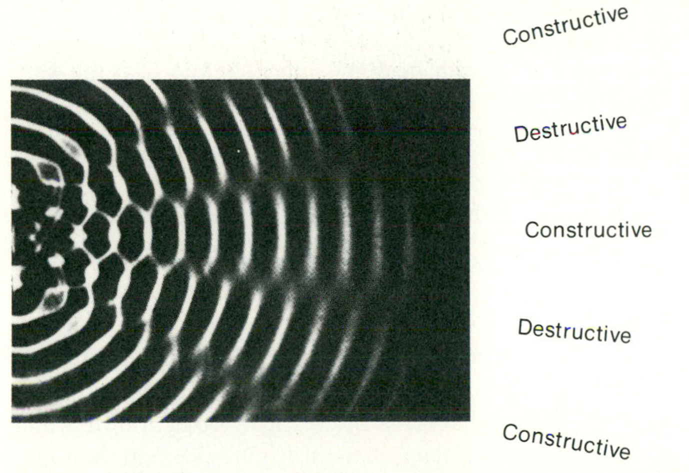

(**a**) Water waves spreading from two coherent sources produce a stationary pattern of constructive and destructive interference where the waves overlap.

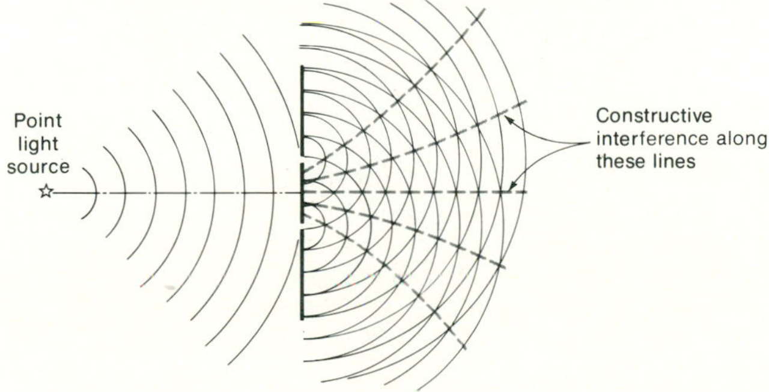

(**b**) According to Huygens' principle, light waves emerging from the two slits spread out in all directions, causing interference in the region where they overlap. If the slits are located symmetrically from the source, the waves emerging from them have equal amplitudes, are in phase with one another, and are coherent.

Figure 30–4
Waves from coherent sources produce stationary interference patterns.

be *in phase* and of *equal amplitude*. In the discussion to follow, we will always assume this to be the case.

We now investigate the details of the interference. Consider the rays leaving each of the slits shown in Figure 30–2(b) and arriving on the screen at the point P, a distance y from the center of the fringe pattern. The light from the lower slit will be out of phase with that from the upper slit because it travels a greater distance. As shown in Figure 30–2(c), the extra path distance Δr is essentially

Path difference $$\Delta r = d \sin \theta \qquad \textbf{(30–1)}$$

where d is the slit separation (center-to-center). This introduces a phase difference ϕ between the two waves when they arrive at the screen. The phase difference will be 2π rad for each wavelength λ in the distance Δr. That is:

Phase difference (due to the extra path length Δr) $$\phi = 2\pi\left(\frac{\Delta r}{\lambda}\right) \quad \text{(in radians)} \qquad \textbf{(30–2)}$$

Note that ϕ is greater than 2π if Δr is greater than λ. The two light waves arriving at the point P on the screen may be represented by corresponding electric-field amplitudes:

$$E_1 = E_0 \sin \omega t \qquad \textbf{(30–3)}$$

and

$$E_2 = E_0 \sin (\omega t + \phi) \qquad \textbf{(30–4)}$$

where E_1 is the wave amplitude from the upper slit and E_2 is the wave amplitude from the lower slit. Note that E_1 and E_2 will be in phase for

$$\phi = m2\pi \qquad \textbf{(30–5)}$$

where $m = 0, 1, 2, 3, \ldots$, thus producing a resultant wave E.

In-phase condition $$E = 2E_0 \sin \omega t \qquad \textbf{(30–6)}$$

For intermediate values of the phase difference ϕ, the resultant wave is best obtained using the mathematical technique of *phasors* (which we employed in Chapter 27 in the addition of alternating currents at an *AC* circuit junction). Equation (30–3) suggests that E_1 is the vertical projection of a phasor $\mathbf{E}_0$ that is rotating at an angular velocity ω, as in Figure 30–5(a). Similarly, E_2 is the vertical projection of $\mathbf{E}_0$ that is rotating at the same angular velocity but is leading the phasor $\mathbf{E}_0$ of (a) by the angle ϕ. This is shown in Figure 30–5(b). The sum E_3 of the *vertical projections*, E_1 and E_2, is then the sum of the waves:

$$E_3 = E_1 + E_2$$

or

$$E_3 = E_0 \sin \omega t + E_0 \sin (\omega t + \phi) \qquad \textbf{(30–7)}$$

Figure 30–5(c) shows that the vector sum of the two rotating phasors $\mathbf{E}_0$ shown in (a) and (b) produces a projection equal to E_3. The application of simple trigonometry for the sum gives

$$E_3 = \underbrace{2E_0 \cos \frac{\phi}{2}}_{\text{Amplitude}} \sin\left(\omega t + \frac{\phi}{2}\right) \qquad \textbf{(30–8)}$$

where $2E_0 \cos (\phi/2)$ is the projection of both phasors $\mathbf{E}_0$ in the direction of $\mathbf{E}$ and $\omega t + \phi/2$ is the angle that E makes with the horizontal axis.

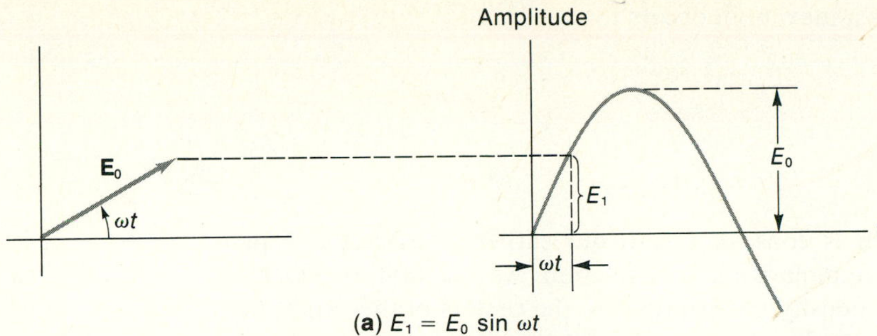

(a) $E_1 = E_0 \sin \omega t$

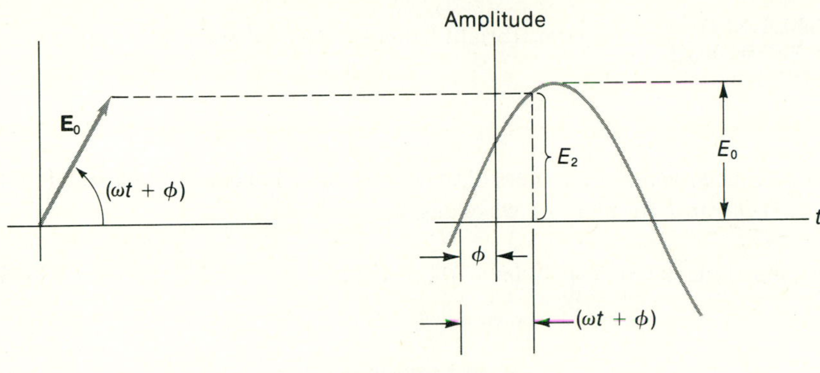

(b) $E_2 = E_0 \sin (\omega t + \phi)$

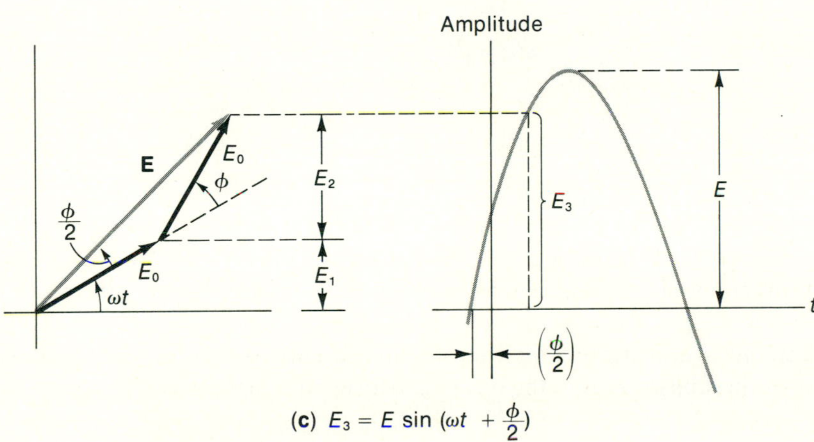

(c) $E_3 = E \sin (\omega t + \frac{\phi}{2})$

Figure 30–5
Phasor diagrams for two waves, E_1 and E_2, and their sum, $E_3 = E_1 + E_2$.

The intensity of the light is proportional to the square of the amplitude of the resultant wave.

$$I \propto \left(2E_0 \cos \frac{\phi}{2}\right)^2$$

or

$$I \propto 4E_0{}^2 \cos^2 \frac{\phi}{2} \qquad \textbf{(30–9)}$$

Expressed in terms of the intensity I_0 at the central maximum ($\phi = 0$), the intensity at other locations is

$$I = I_0 \cos^2 \frac{\phi}{2} \qquad \textbf{(30–10)}$$

Thus a maximum occurs for

$$\frac{\phi}{2} = m\pi$$

or $$\phi = m2\pi$$

which is consistent with our earlier observation, Equation (30–5).

Summarizing this relation, we see that the location of the *maxima* in the intensity pattern (that is, the centers of the bright fringes) will occur when the path difference Δr is an *integral* number of wavelengths:

DOUBLE-SLIT INTERFERENCE PATTERN

Maxima (bright fringes) $$m\lambda = d \sin \theta \qquad \textbf{(30–11)}$$

(where $m = 0, 1, 2, 3, \ldots$)

Similarly, the *minima* (the centers of the dark fringes) occur for a path difference of a *half-integral* number of wavelengths:

Minima (dark fringes) $$(m + \tfrac{1}{2})\lambda = d \sin \theta \qquad \textbf{(30–12)}$$

(where $m = 0, 1, 2, 3, \ldots$)

In practice, the distance $D \gg y$, so that $\sin \theta \approx \tan \theta = \theta$. Using the tangent approximation, we may write the above two equations as

Maxima $$m\lambda = d\left(\frac{y}{D}\right) \qquad \textbf{(30–13)}$$

Minima $$\left(m + \frac{1}{2}\right)\lambda = d\left(\frac{y}{D}\right) \qquad \textbf{(30–14)}$$

(small-angle approximation)

(where $m = 0, 1, 2, 3, \ldots$)

The central bright fringe is called the *zero-order* fringe ($m = 0$). Moving away on either side of the central maximum, successive bright fringes are the *first-order* fringes ($m = \pm 1$), the *second-order* fringes ($m = \pm 2$), and so on. A characteristic feature of double-slit interference is that as the separation between the slits *decreases*, the distance between the fringes *increases*.

EXAMPLE 30–1

In a double-slit experiment using light of wavelength 486 nm, the slit spacing is 0.60 mm and the screen is 2 m from the slits. Find the distance along the screen between adjacent bright fringes.

SOLUTION

Assuming the small-angle approximation, Equation (30–13) gives the location y of the mth maximum:

$$m\lambda = d\left(\frac{y}{D}\right)$$

The separation between adjacent maxima is then

$$y_{m+1} - y_m = \frac{\lambda D}{d}[(m+1) - m]$$

$$y_{m+1} - y_m = \frac{(486 \times 10^{-9}\ \text{m})(2\ \text{m})}{(0.60 \times 10^{-3}\ \text{m})}[1]$$

$$y_{m+1} - y_m = 1.62 \times 10^{-3}\ \text{m} = \boxed{1.62\ \text{mm}}$$

Because this is such a small distance relative to the slit-to-screen distance, the small-angle approximation is justified.

EXAMPLE 30–2

Consider the situation shown in Figure 30–6. The source illuminates the slits with green light from a mercury lamp ($\lambda = 546$ nm). The screen is $D = 1$ m from the slits, and the slit separation d is 0.30 mm. (a) Find the intensity I of the light at a distance $y = 1$ cm from the center of the pattern relative to the intensity of the central fringe maximum I_0. (b) Find the number of bright fringes between the central fringe and the point y.

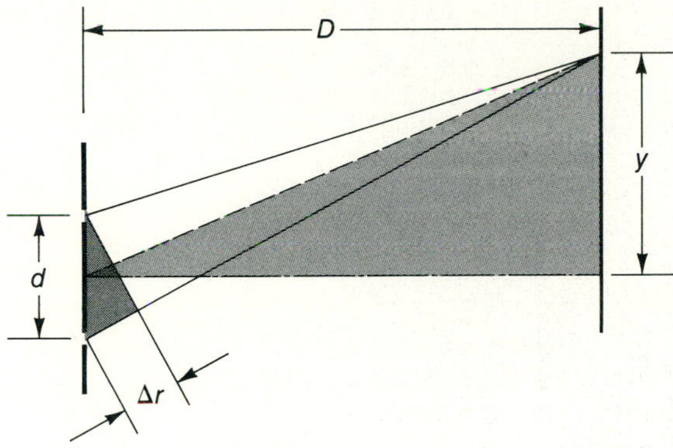

(**a**) For cases where $D >> y$ and d, the two shaded triangles are similar.

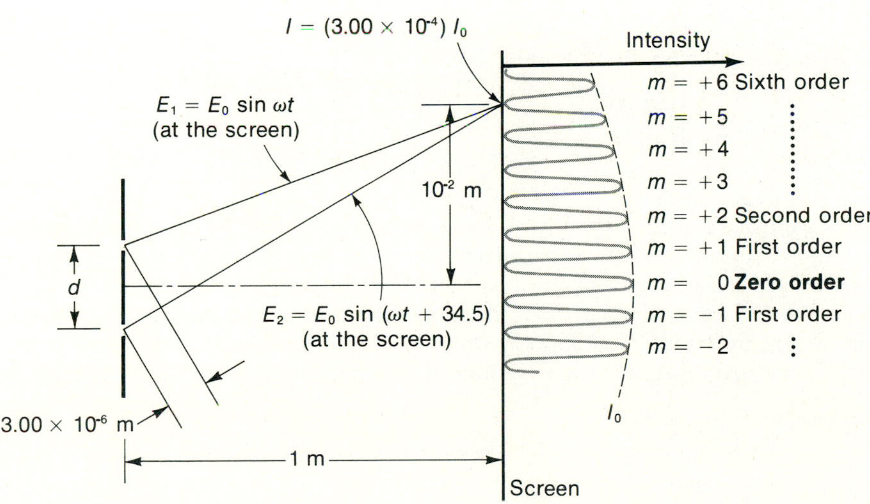

(**b**) The numerical values. (The sizes of the slit separation and fringe pattern are greatly exaggerated for clarity.)

Figure 30–6
Example 30–2

SOLUTION

(a) To find the difference in phase between the waves originating at the upper and lower slits shown in Figure 30–6, we first find the path difference Δr. Within the approximation of Figure 30–2(c) and recognizing that $\sin\theta \approx \tan\theta$, the shaded triangles in the figure are similar. Therefore, corresponding sides of the triangles are proportional:

$$\frac{\Delta r}{y} = \frac{d}{D} \tag{30–15}$$

or

$$\Delta r = \frac{d}{D}y$$

$$\Delta r = \left(\frac{0.30 \times 10^{-3}\ \text{m}}{1\ \text{m}}\right)(1 \times 10^{-2}\ \text{m})$$

$$\Delta r = 3.00 \times 10^{-6}\ \text{m}$$

Equation (30–2) yields the phase angle in terms of the wavelength and the distance Δr:

$$\phi = 2\pi\left(\frac{\Delta r}{\lambda}\right)$$

$$\phi = 2\pi\left(\frac{3 \times 10^{-6}\ \text{m}}{5.46 \times 10^{-7}\ \text{m}}\right)$$

$$\phi = 34.523\ \text{rad}$$

Applying Equation (30–10):

$$I = I_0 \cos^2 \frac{\phi}{2}$$

$$I = \boxed{(2.98 \times 10^{-4})I_0}$$

This answer indicates that the point y lies very near a point of minimum intensity.

(b) As we move along the screen away from the central fringe, the path difference Δr increases. As Δr increases by one full wavelength, the two waves from the slits are again in phase, corresponding to moving from the central bright fringe to the adjacent fringe, and so on. How many wavelengths are there in the total path difference $\Delta r = 3.00 \times 10^{-6}$ m?

$$\frac{\Delta r}{\lambda} = \frac{3.00 \times 10^{-6}\ \text{m}}{5.46 \times 10^{-7}\ \text{m}}$$

$$\frac{\Delta r}{\lambda} = 5.49\ \text{wavelengths}$$

Thus 5 bright fringes will exist between the central maximum and the point y. The remaining 0.49 wavelength indicates that the waves from the upper and lower slits are nearly π rad out of phase, which is consistent with the answer in part (a).

An alternative approach is to determine the number of times the phase angle ϕ is divisible by 2π. As we move away from the central fringe, each increase of 2π in the phase angle corresponds to moving from one bright fringe to the next. Thus, for $\phi = 34.5$ rad:

$$\frac{\phi}{2\pi} = \frac{34.5\ \text{rad}}{2\pi}$$

$$\frac{\phi}{2\pi} = 5.49\ \text{multiples of}\ 2\pi$$

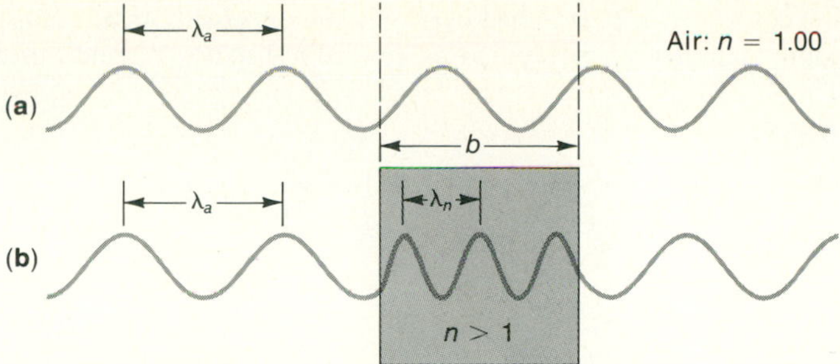

Figure 30–7
A phase difference may be produced by inserting a material with refractive index n in the path of a light wave.

Thus, we conclude that 5 bright fringes appear between the central bright fringe and the point $y = 1$ cm.

The solution to this example is summarized in Figure 30–6(b).

A phase difference between the light waves emitted from a double slit also may be produced by introducing a transparent material with a different refractive index into the path of one of the waves (see Figure 30–7). Inserting a refractive material of thickness b and refractive index n increases the number of wavelengths in that path. If the wavelength in air is λ_a, a distance b (in air) contains b/λ_a wavelengths. In a material of refractive index n, the wavelength is shorter: $\lambda_n = \lambda_a/n$. Thus the same distance b contains b/λ_n wavelengths, or

$$\frac{b}{\left(\dfrac{\lambda_a}{n}\right)} = n\left(\frac{b}{\lambda_a}\right) \text{ wavelengths}$$

The *increase* in number of wavelengths is therefore

$$\frac{nb}{\lambda_a} - \frac{b}{\lambda_a} = \frac{b}{\lambda_a}(n - 1) \tag{30–16}$$

Since a phase difference of 2π corresponds to each full wavelength increase, the phase difference ϕ is

Phase difference (due to inserting in one path a material of thickness b and refractive index n)

$$\phi = 2\pi\left(\frac{b}{\lambda_a}\right)(n - 1) \tag{30–17}$$

(where λ_a is the wavelength in air)

The following example illustrates the sensitivity of an interference pattern to small changes in the refractive index associated with one of the light paths.

EXAMPLE 30–3

Consider the double-slit arrangement shown in Figure 30–8, where the separation d of the slits is 0.30 mm and the distance D to the screen is 1 m. A very thin sheet of transparent plastic, with a thickness of $b = 0.050$ mm (about the thickness of this page) and a refractive index of $n = 1.50$, is placed over only the upper slit. As a result, the central maximum of the interference pattern moves upward a distance y'. Find this distance.

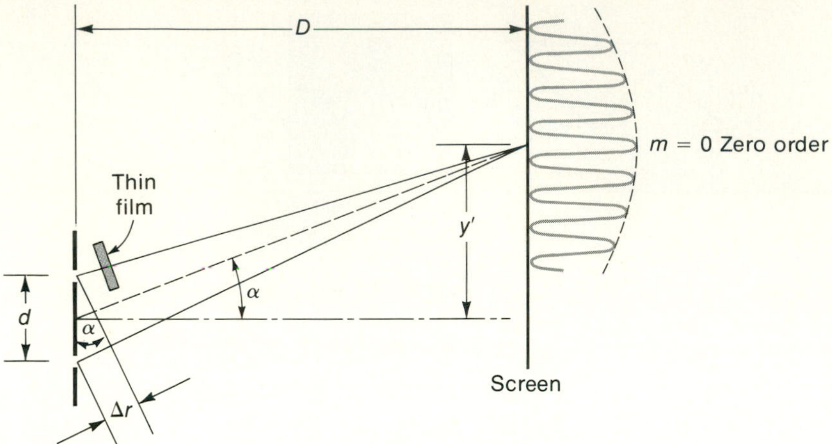

Figure 30–8
Example 30–3

SOLUTION

The central maximum corresponds to zero phase difference. Thus the added distance Δr traveled by the light from the lower slit must introduce a phase difference equal to that introduced by the plastic film. The *phase difference* ϕ is given by Equation (30–17):

$$\phi = 2\pi\left(\frac{b}{\lambda_a}\right)(n-1)$$

The corresponding difference in *path length* Δr is, from Equation (30–2):

$$\phi = 2\pi\left(\frac{\Delta r}{\lambda_a}\right)$$

$$\Delta r = \phi\left(\frac{\lambda_a}{2\pi}\right)$$

Substituting for ϕ, we obtain

$$\Delta r = 2\pi\left(\frac{b}{\lambda_a}\right)(n-1)\left(\frac{\lambda_a}{2\pi}\right)$$

$$\Delta r = b(n-1) \tag{30–18}$$

Note that the wavelength of the light does not appear in this equation. In Figure 30–8 the two rays from the slits are essentially parallel, so the angle α may be expressed as

$$\tan\alpha = \frac{\Delta r}{d}$$

or as

$$\tan\alpha = \frac{y'}{D}$$

Equating these expressions and solving for y' gives

$$y' = \Delta r\left(\frac{D}{d}\right)$$

Inserting the value of Δr from Equation (30–18):

$$y' = \frac{b(n-1)D}{d}$$

and substituting numerical values:

$$y' = \frac{(5 \times 10^{-5}\ \text{m})(1.50 - 1)(1\ \text{m})}{(3 \times 10^{-4}\ \text{m})}$$

$$y' = 0.0833\ \text{m}$$

$$y' = \boxed{8.33\ \text{cm}}$$

30.2 Multiple-Slit Interference

The mathematical technique of phasors developed in the last section for a double-slit interference pattern is easily extended to include interference of light from three or more slits. The physical arrangement of the slits is similar to that for a double-slit pattern. However, as shown in Figure 30–9, the pattern is quite different, composed of alternating *major* and *minor* maxima. (They are also called *primary* and *secondary* maxima.) The major maxima are spaced in the *same* way as the double-slit maxima if adjacent slits have the same separation distance d.

The development of the triple-slit interference pattern by the use of phasors is shown in Figure 30–10. The central maximum corresponds to the addition of three electric phasors, all in phase, as in Figure 30–10(a). As the distance from the central maximum increases, the phase angle ϕ between the electric phasors from adjacent slits also increases, in increments of $\pi/3$. Note that one *minor* maximum (d) occurs between each of the *major* maxima [(a) and (g)].

As the number of slits increases, the number of minor maxima between major maxima also increases, as illustrated in Figure 30–11. The number of these minor peaks is always *two less* than the total number of slits in the array. Furthermore, *as the number of slits increases, these minor peaks are suppressed in intensity, while the major maxima become much more intense and narrower.* Since the *positions* of the *major* maxima depend only on the slit

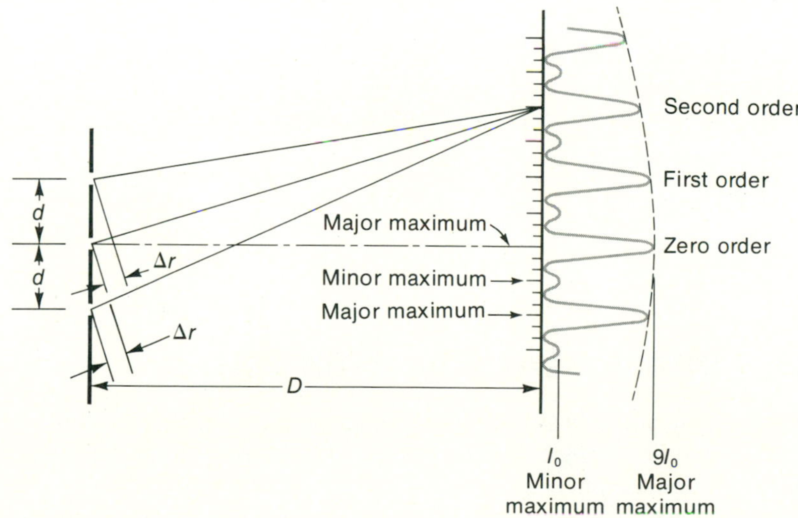

Figure 30–9
Triple-slit interference. On the screen, each marked interval from the center of the pattern corresponds to a phase difference between waves from adjacent slits of $\pi/3$ rad. (The slit separations and fringe pattern are greatly enlarged for clarity.)

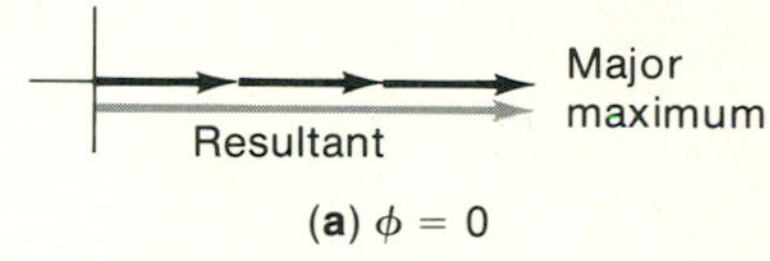

(a) $\phi = 0$

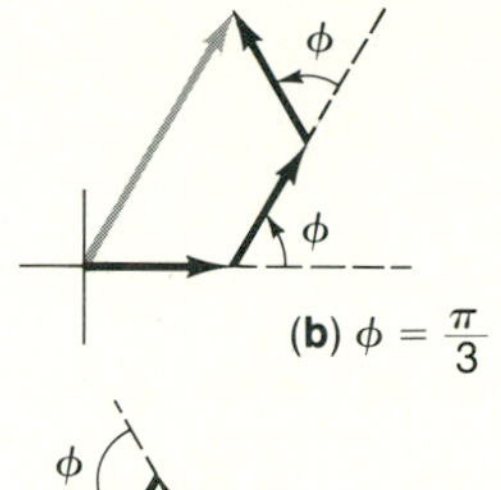

(b) $\phi = \frac{\pi}{3}$

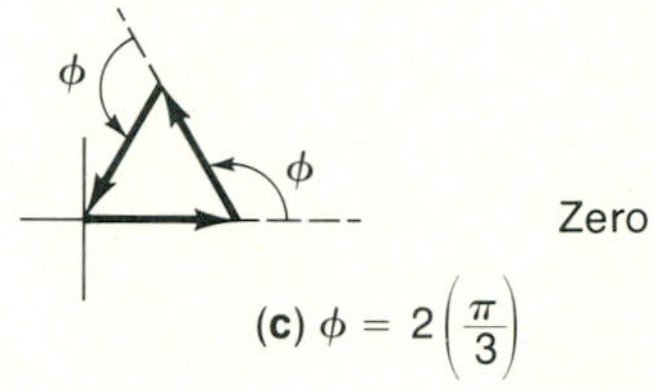

(c) $\phi = 2\left(\frac{\pi}{3}\right)$

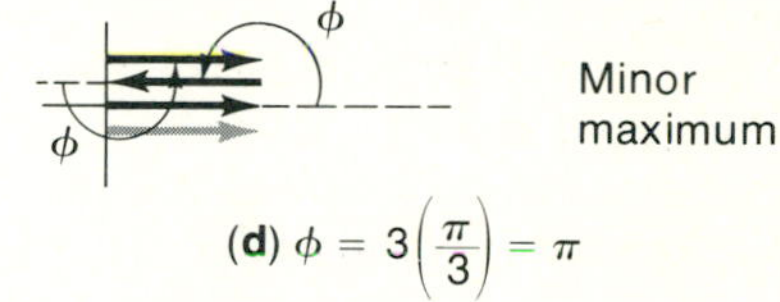

(d) $\phi = 3\left(\frac{\pi}{3}\right) = \pi$

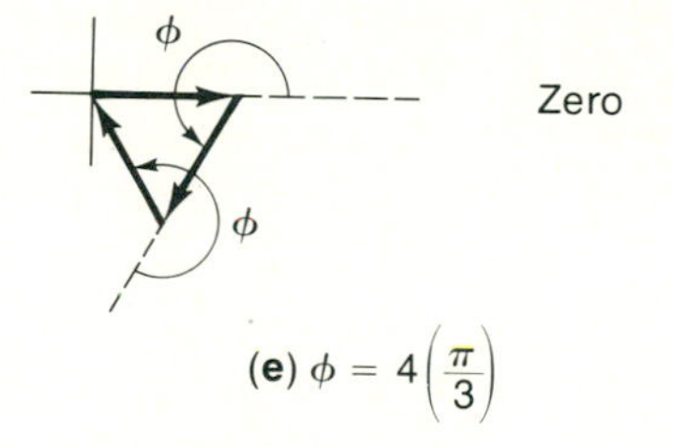

(e) $\phi = 4\left(\frac{\pi}{3}\right)$

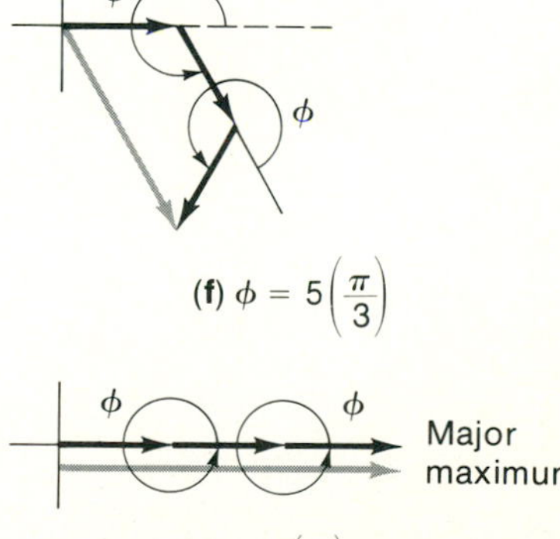

(f) $\phi = 5\left(\frac{\pi}{3}\right)$

(g) $\phi = 6\left(\frac{\pi}{3}\right) = 2\pi$

Figure 30–10
A series of phasor diagrams for triple-slit interference. Each successive diagram represents an additional phase delay of $\phi = \pi/3$ rad between *adjacent* phasor components.

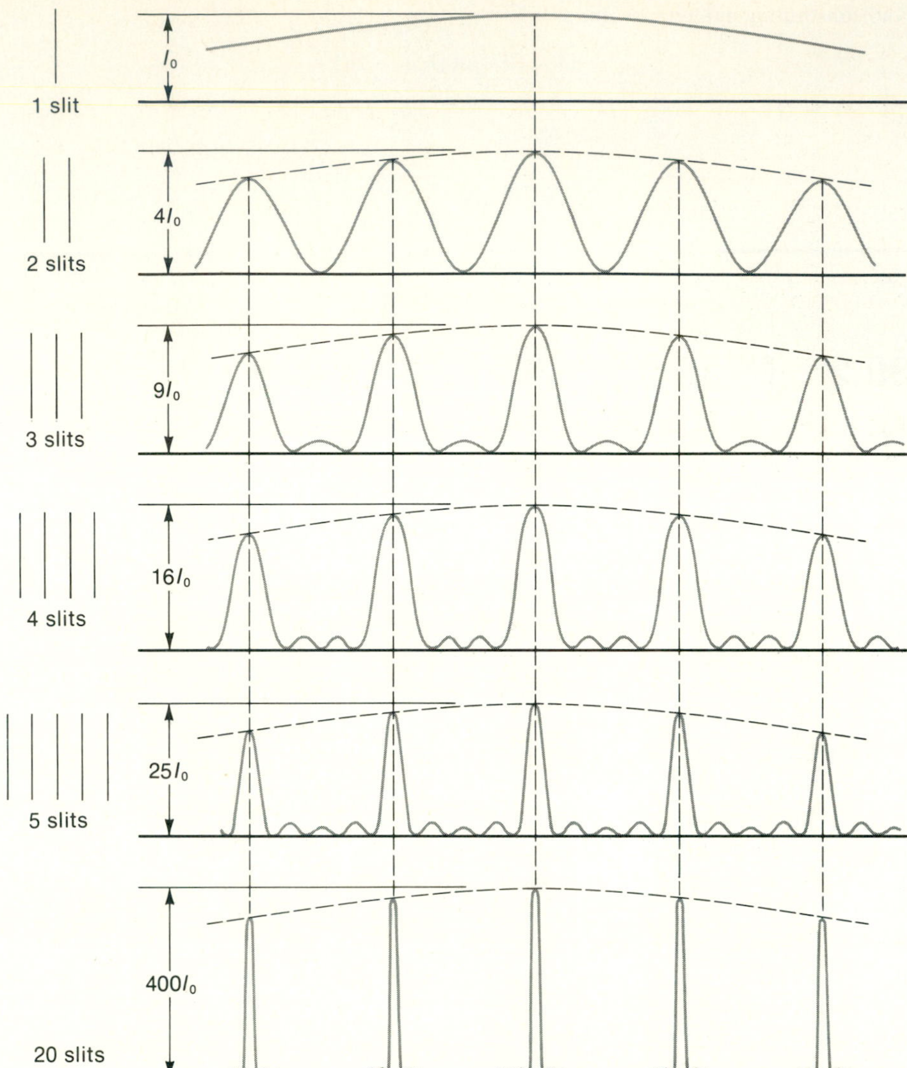

Figure 30–11
Multiple-slit interference patterns. As the number of slits is increased (keeping the slit separation constant), the major maxima remain fixed in position as they become narrower and more intense. The intensity of the sharp peaks increases with the square of the number of slits. (Note the changes in vertical scale.)

separation d (and not on the number of slits), Equation (30–11) expresses the location of the major maxima for any number of slits.

MULTIPLE-SLIT INTERFERENCE PATTERN (major maxima)

$$d \sin \theta = m\lambda \qquad \text{(where } m = 0, 1, 2, \ldots\text{)} \tag{30–19}$$

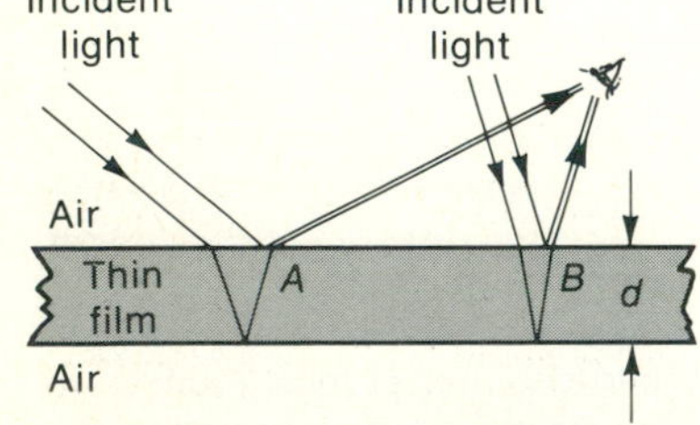

Figure 30–12
Interference of light reflected from a thin film.

30.3 Interference by Thin Films

We have all enjoyed a beautiful display of colors from a thin oil film on the surface of a puddle, or the colored reflection of light from the surface of a soap bubble. Both of these phenomena result from the interference of light.

Consider a thin film of refractive material such as glass. Figure 30–12 illustrates the observation of white light reflected from two different places, A and B, on the film. At both places, the light reaching the eye of the observer is a combination of light reflected from the top surface and from the lower surface. In each case, these two light waves interfere with each other, reinforcing certain wavelengths and canceling others, depending on the particular path difference between them. For example, suppose that at A, wavelengths in the red portion of the spectrum undergo destructive interference. Then the observer will see a predominantly blue-green reflection at that location. On the other hand, if at B, where the path difference is shorter, blue wavelengths experience destructive interference, the observer will see a predominantly reddish reflection. In this manner, an entire rainbow of colors is often reflected from various portions of the film.

Light reflected from a very thin soap film also shows another interesting feature. Generally, a freshly blown soap bubble displays a swirl of reflected colors when viewed against a dark background. This is partially due to the nonuniform refractive index of the soap solution as well as the varying thickness of the film. However, if we continue to watch a soap film supported vertically, as in Figure 30–13, the various colors gradually sort themselves into horizontal rainbow stripes, slowly compressing together toward the bottom. This happens because the action of gravity drains fluid from the upper portion of the film, causing it to be thinner at the top than at the bottom. But now a surprising effect occurs. As the top part of the film becomes much thinner than a wavelength of visible light, *no light at all is reflected from the film.* It has become invisible! The reason is that light reflected from the front and back surfaces interferes *destructively* because of a phase change of π rad (180°) that occurs at one surface and not the other. A detailed analysis of the reflection of light from refractive materials shows that:

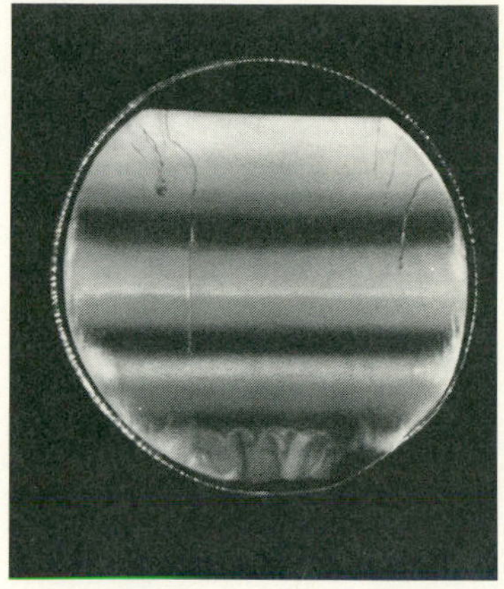

Figure 30–13
Interference of white light reflected from a thin vertical film of soap solution. Gravity pulls the fluid downward, causing the film to become very thin near the top. As the path difference between reflections from the front and rear surfaces approaches zero, the 180° phase change for the front reflection causes destructive interference for all wavelengths of visible light, making the top segment of the film invisible.

PHASE CHANGE UPON REFLECTION

(1) When light traveling in a given medium reflects from another medium of *higher* refractive index, it undergoes *a phase change of π rad (180°)*.

(2) When light traveling in a given medium reflects from another medium of *lower* refractive index, *no phase change* occurs.

Reflections from the front and back surfaces of the soap film are of these two different types, so the reflections alone introduce a 180° phase difference. Thus, as the film thickness shrinks toward zero, making the path differences negligible, the two reflected rays become 180° out of phase and undergo destructive interference. If you observe reflections from a soap bubble against a dark background and watch carefully as the bubble ages, you will see the color contrasts diminish. Then, just before the film breaks, no light is reflected from the spot where the break originates.

EXAMPLE 30–4

An important application of interference produced by thin films is the use of *nonreflecting coatings* for camera lenses. These coatings reduce the loss of light at various surfaces of multiple-lens systems, as well as prevent internal reflections, which might mar the image. Find the minimum thickness of a layer of magnesium fluoride ($n' = 1.38$) on flint glass ($n = 1.80$) that will cause *destructive* interference of reflected light of wavelength $\lambda = 550$ nm near the middle of the visible spectrum. Consider normal incidence on the coating.

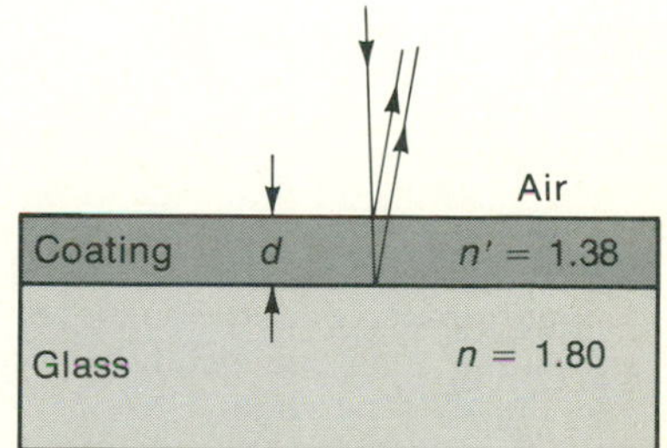

Figure 30–14
Example 30–4

SOLUTION

In Figure 30–14, both rays reflect from a medium of higher refractive index than the medium they are traveling in, so *both* undergo a phase shift of π rad upon reflection. Therefore, the only factor contributing to a net phase shift is the extra path length of one ray.

For destructive interference, the (minimum) round trip distance $2d$ should be $\lambda_{n'}/2$, where $\lambda_{n'} = \lambda_a/n'$ is the wavelength *in the coating*. Thus:

$$2d = \frac{\lambda_a}{2n'}$$

Solving for d:

$$d = \frac{\lambda_a}{4n'}$$

$$d = \frac{(5.50 \times 10^{-7}\ \text{m})}{4(1.38)}$$

$$d = \boxed{99.6\ \text{nm}}$$

Though such coatings are very thin (approximately a hundred atomic diameters thick), they are easily applied by evaporating the magnesium fluoride and allowing it to condense on the glass surface. For complete destruction, the *amplitudes* of the reflected rays must be equal. It can be shown this is true only if the refractive indices of the coating (n') are the geometric mean between the refractive indices of the materials on either side of the coating. For air, $n_a = 1$, and we have

$$n' = \sqrt{nn_a}$$

or

$$n' = \sqrt{n}$$

30.4 Interference Produced by Thin Wedges

Consider two glass plates that are in contact at one edge and separated slightly at the opposite edge by inserting a hair or small wire between the plates. A side view of such an arrangement is shown in Figure 30–15. Parallel, mono-

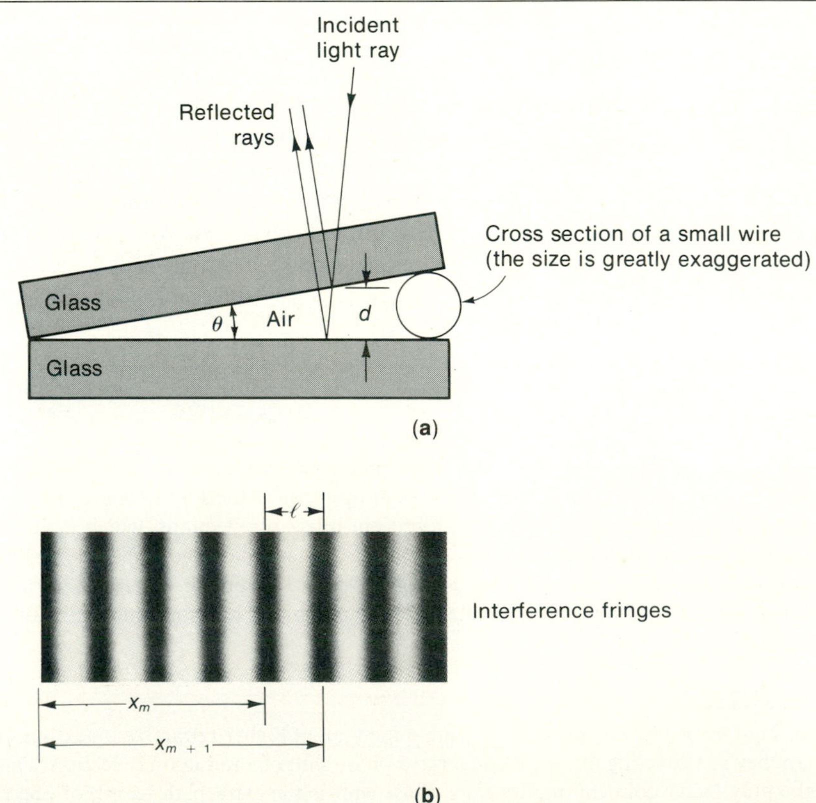

Figure 30–15
The interference pattern produced by a wedge of air between two glass plates. (The angle θ between the plates is greatly exaggerated to emphasize the variation in thickness of the air wedge.)

chromatic light rays incident downward are reflected back to an observer above the plates. (The rays reflected from the top of the upper plate and from the bottom of the lower plate need not be considered if opposite sides of the plate are not quite parallel.) The reflected light is thus composed of a combination of the light rays reflected from the lower surface of the top plate (no phase shift) and the light rays reflected from the upper surface of the lower plate (phase shift π). Destructive interference thus occurs if the extra distance traveled by ray B ($2d$ for the round trip) is an *integral* number of wavelengths λ. That is:

$$2d = m\lambda \qquad \text{(where } m = 0, 1, 2, 3, \ldots \text{ and } d = \text{plate separation)} \tag{30–20}$$

Note that this equation is the condition for *destructive* interference and includes the phase shift that occurs in one of the reflections.

If the glass plates have plane surfaces, the interference pattern is a series of equally spaced bright and dark fringes. As we proceed from one dark fringe to the next, the air wedge increases in thickness by $\lambda/2$ (making the *round trip* path increase by λ). The separation ℓ of adjacent dark fringes is found as follows. In traveling along the plate a distance ℓ, the wedge increases by $\lambda/2$. Therefore, the tangent of the wedge angle θ is

$$\tan\theta = \frac{\left(\frac{\lambda}{2}\right)}{\ell}$$

Since θ is ordinarily a very small angle, we may substitute $\tan\theta = \theta$ to obtain for ℓ:

$$\ell = \frac{\lambda}{2\theta} \tag{30–21}$$

The flatness of a glass surface is often determined by the interference pattern produced when it is placed in contact with an *optical flat*, a surface known to be flat to within a small fraction of a wavelength of light (see Figure 30–16). A *dark* fringe is located at the region where the two surfaces touch because of the 180° phase shift that occurs for (only) one of the two reflections.

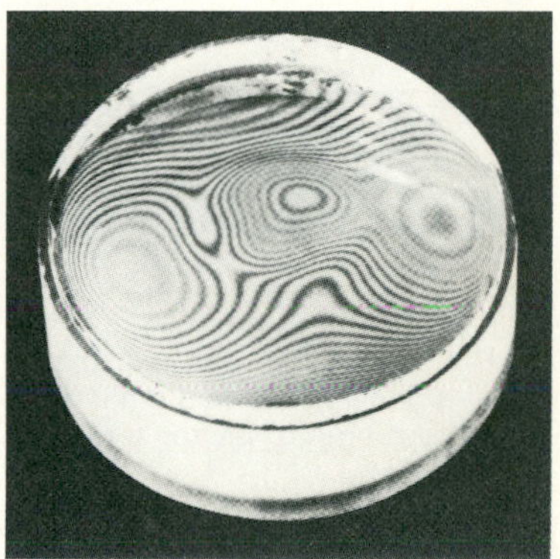

(a) A wavy fringe pattern indicates an uneven surface. Three "high" spots are revealed by the regions of circular fringes.

(b) The surfaces are in contact at one edge and separated a small amount at the opposite edge. The regularly spaced bright and dark fringes indicate that the surface is uniformly flat.

Figure 30–16
The flatness of a glass surface is tested by placing it in contact with an *optical flat* and observing the interference pattern of reflected monochromatic light.

EXAMPLE 30–5

Suppose two flat glass plates 30 cm long are in contact along one end and separated by a human hair at the other end, as indicated in Figure 30–15. If the diameter of the hair is 50 μm, find the separation of the interference fringes when the plates are illuminated by green light, $\lambda = 546$ nm.

SOLUTION

The angle of the air wedge between the plates is given by

$$\theta = \frac{D}{L} \text{ radians}$$

where D is the diameter of the hair and L is the length of the plates. Substituting this expression for θ into Equation (30–21), we obtain:

$$\ell = \frac{\lambda}{2\theta}$$

$$\ell = \frac{L\lambda}{2D}$$

Substituting numerical values gives:

$$\ell = \frac{(0.3\text{ m})(5.46 \times 10^{-7}\text{ m})}{2(5.0 \times 10^{-5}\text{ m})} = 1.64 \times 10^{-3}\text{ m}$$

$$\ell = \boxed{1.64\text{ mm}}$$

A plano-convex lens placed on an optical flat produces a circular interference pattern known as *Newton's rings* (Figure 30–17). The thickness d of the air wedge between the lens and the flat glass plate is related to the radius of curvature R of the lens surface and the distance r from the center of the pattern. Applying the Pythagorean Theorem, we obtain

$$R^2 = r^2 + (R - d)^2$$
$$R^2 = r^2 + R^2 - 2Rd + d^2$$

Since the radius of curvature R of the lens is much greater than the thickness d of the air wedge, we ignore the d^2 term and obtain

$$2d \approx \frac{r^2}{R}$$

The condition for *destructive* interference exists when the extra (round trip) path $2d$ for the ray reflected from the bottom surface is an *integral* number of wavelengths (because of the 180° phase change for one of the reflections):

$$2d = m\lambda \qquad (\text{where } m = 0, 1, 2, 3, \ldots)$$

Equating these two values for $2d$, we obtain an expression for r_m, the radius of the mth ring:

RADII OF NEWTON'S RINGS

$$r_m = \sqrt{Rm\lambda} \tag{30–22}$$

As we proceed outward from one dark ring to the next, the radii r_m increase in size with $\sqrt{m}$, becoming closer together. One of the most interesting aspects of this interference pattern is that the area between each of the successive circles is a constant.

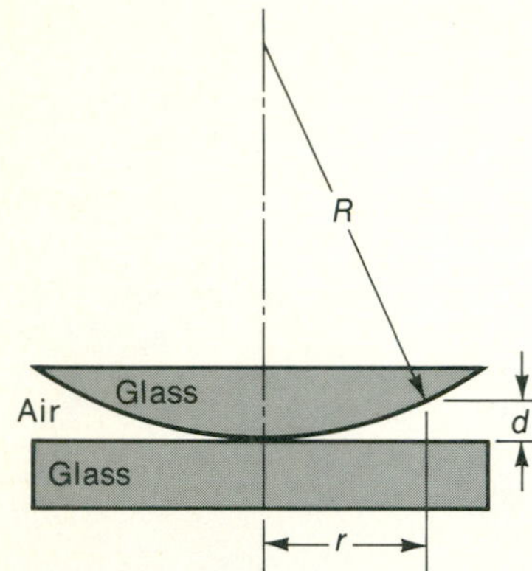

(a) Reflections between the surface of a convex lens and a flat glass plate produce Newton's rings.

(b) Photograph of Newton's rings obtained with monochromatic light.

Figure 30–17
Newton's rings. [Note: The faint patterns in (b) are spurious.]

30.5 The Michelson Interferometer

The Michelson[2] interferometer is an ingenious device that utilizes the interference of light to measure distances, or *changes* of distance, with great accuracy. The basic components, shown in Figure 30–18, include an *extended* light source, such as a ground-glass screen illuminated uniformly from behind with monochromatic light. (The reason a *point* source is unsatisfactory will be evident

---

[2] Albert Michelson (1852–1931) was the son of Polish immigrants who were somewhat poor, and his prospects for education beyond high school were not promising. However, when his application to the U.S. Naval Academy was turned down, he shrewdly arranged to meet President Grant "by chance" while walking his dog on the White House grounds. He so highly impressed the President with his determination that a special appointment to Annapolis was granted. After graduating in 1873, Michelson became a physics and chemistry instructor at the Academy, where he began a lifelong interest in precision measurements of the speed of light. He then became a professor of physics at Case Institute of Technology, where he improved earlier interferometer experiments on the ether drift, this time with a collaborator, Edward Morley, a chemist at nearby Western Reserve. Michelson was keenly disappointed in the null result: He would much have preferred to report a finite velocity through the ether and felt that the absence of a positive value was somehow due to an unknown defect in his method. In 1907, Michelson became the first American Nobel prize winner for his work on light.

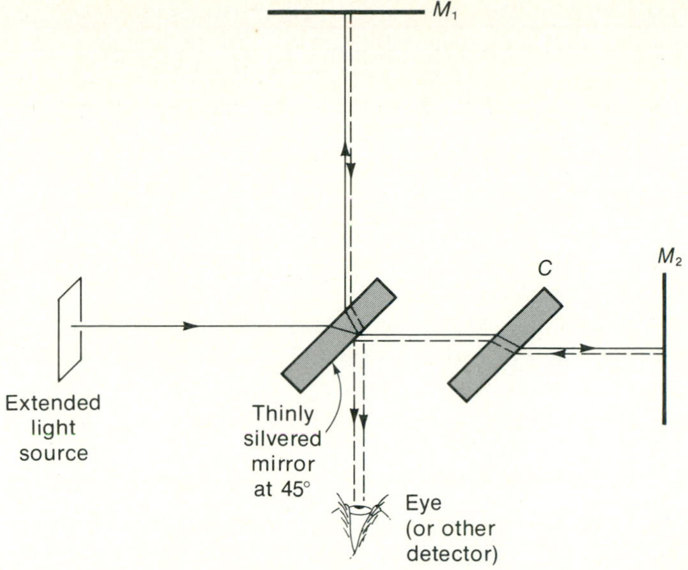

Figure 30–18
The basic components of a Michelson interferometer. The clear glass slab C is called a *compensating plate*. It has the same dimensions and orientation as the 45° mirror in order to make the light paths in glass equal along the two arms, a condition necessary when a white-light source is used.

after we discuss the origin of the interference pattern.) Light from the source falls on a thinly silvered, semitransparent mirror at 45°, an angle that reflects approximately half the light to mirror M_1 and transmits half to mirror M_2. Light reflected from M_1 and M_2 eventually merges together at the eye or other detector (minus, of course, that part further diverted by the 45° mirror). If we straighten out the several right-angle deflections caused by the 45° mirror, the situation is essentially as shown in Figure 30–19. The extended source is reflected by the two mirrors, forming two images, I_1 and I_2. The mirrors can be aligned so that the two images are parallel. If the distance between M_1 and M_2 is d, the images are separated by a distance of $2d$.

The significant feature of these images is that *light waves from corresponding points in the images are coherent*. These waves come from a wave train emitted from a *single atom* in the source at point P. Thus, the light waves that enter the eye from the image points P_1 and P_2 are coherent and they will interfere. The phase relation between the two rays depends on their path difference, $2d \cos \theta$. The *in-phase* condition for *bright* fringes is

$$2d \cos \theta = m\lambda$$

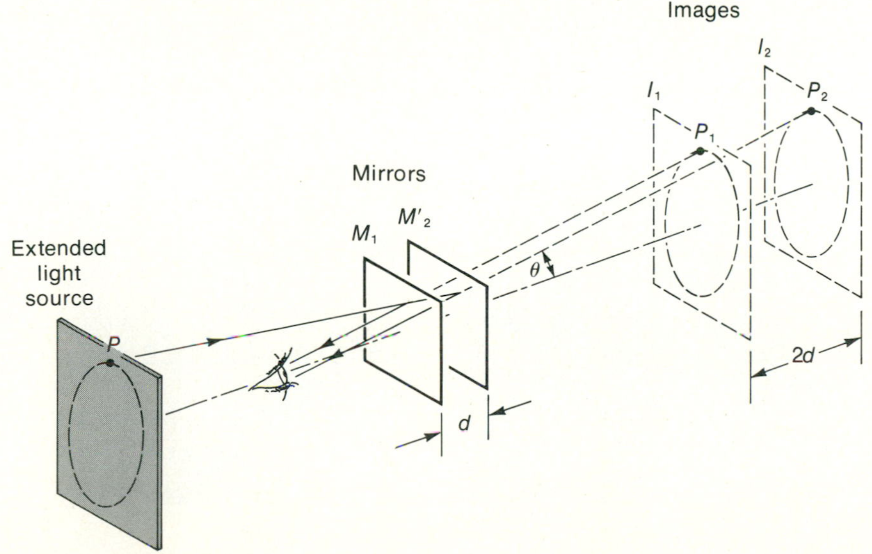

Figure 30–19
The origin of the circular fringes in a Michelson interferometer. In this figure, the right-angle deflections produced by the thinly silvered mirror at 45° have been straightened out. Mirror M_1 is observed directly (through the 45° mirror), while M'_2 is the virtual image of M_2 produced by reflection in the 45° mirror. These mirrors form two images, I_1 and I_2, of the extended light source. Light waves from corresponding points in these images are *coherent*.

When the two image planes are parallel, all corresponding points on a circle surrounding the central axis have the same phase relationship, producing an overall fringe pattern of concentric circles similar to Newton's rings. If one mirror is moved by $\lambda/2$, the path difference changes by λ and we are again at an in-phase condition: Each fringe has moved to the position previously occupied by the adjacent fringe. This shifting of the fringe pattern enables tiny motions to be observed. (For example, if one of the interferometer arms is arranged vertically and a small mirror is attached to a mushroom, the growth rate of the mushroom can be accurately observed as fringes sweep past, usually at the rate of about one per second!) Slowly moving one mirror continuously in the same direction causes the circular fringes to shrink in size and vanish at the center (or, for the other direction, to expand from the center). As d approaches zero, the path differences for *all points* approach zero and the entire field of view thus becomes bright (or dark), depending on the *net* phase change due to reflections at the various glass surfaces. If one mirror is tilted slightly, the separation of the image planes becomes a thin wedge. In effect, this moves the center of the fringe pattern off to one side, so we now see an array of slightly curved, almost parallel bright and dark fringes that are part of the ring pattern far away from the center (Figure 30–20).

With monochromatic light, the fringe pattern remains sharp for path differences of 10 cm or more. However, an interferometer may also be used with *white* light, provided the path difference $2d$ is no more than a few wavelengths and the field of view is near the center of the pattern. With a range of wavelengths between 400 nm and 700 nm, the spacing between fringes varies for different colors; hence each bright ring for the monochromatic case becomes a spread-out rainbow of colors. Beyond about a dozen fringes from the center, the patterns overlap so much they fade out to produce essentially white illumination. The interference is still taking place for each individual color, however, as can be verified by viewing the pattern with a filter that allows only one color to pass through. In some applications, it is necessary to have a reference position that can be found again, even though the mirror has been moved in the meantime. This can be accomplished with white-light fringes, because there is a unique, *color-free*, all-bright (or all-dark) field of view when d is exactly zero. It thus serves as a fixed reference position that can be repeatedly reached at will to within a fraction of a wavelength of visible light.

An important early use Michelson made of the interferometer was to determine the length of the then-standard meter bar in Paris in terms of the wavelengths of certain spectral lines of cadmium, counting the number of fringes that swept by as one mirror was moved along the meter bar. The present standard of length is defined in terms of wavelengths of light, a much more

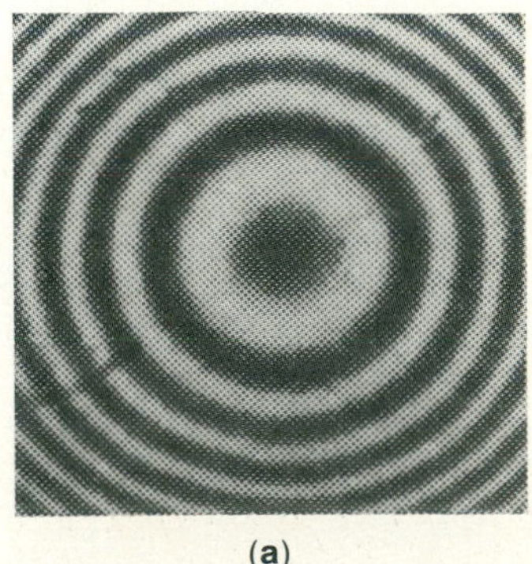

(a)

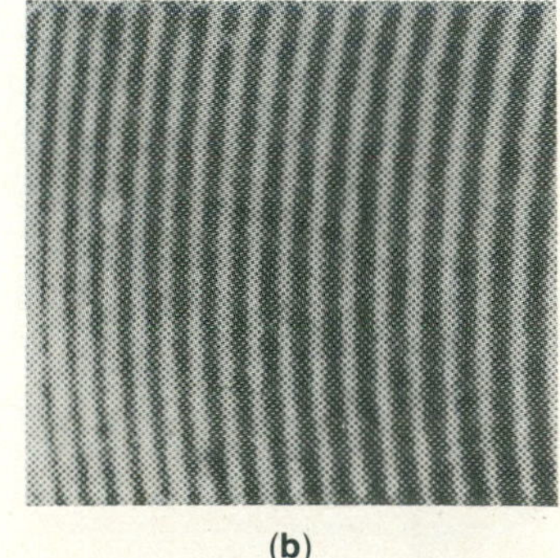

(b)

Figure 30–20
The interference patterns seen in a Michelson interferometer are similar to Newton's rings. In (a), the image planes are parallel (Figure 30–19) and the pattern is a series of concentric circles, whose overall size depends on the separation of the image planes. In (b), the image planes are not parallel and the pattern is a series of curved, almost parallel lines.

accessible and easily reproducible standard than a particular metal bar located in France.

1 meter ≡ 1 650 763.73 wavelengths of the orange light of krypton-86

There is an almost endless list of applications using a Michelson interferometer or modifications thereof, particularly when laser light is used as a source. Instruments using microwaves or other portions of the electromagnetic spectrum have also been constructed. The interferometer has proved to be a versatile and extremely precise measuring instrument, helpful in all areas of science and technology.

Summary

Coherent light waves have a phase difference that remains constant in time. When two different portions of a wave train (emitted from a single atom) are combined, they are coherent and they *interfere*. The sum of two waves:

$$E_1 = E_0 \sin \omega t \qquad \text{and} \qquad E_2 = E_0 \sin(\omega t + \phi)$$

is

$$E_1 + E_2 = 2E_0 \cos\frac{\phi}{2} \sin\left(\omega t + \frac{\phi}{2}\right)$$

The *phase difference* ϕ may result from three effects:

(1) A difference in *path length* Δr of the waves:

$$\phi = 2\pi\left(\frac{\Delta r}{\lambda}\right) \qquad \text{(in radians)}$$

(2) Placement of a material with refractive index n and thickness b in the path of one of the waves:

$$\phi = \frac{2\pi b}{\lambda_a}(n - 1) \qquad \text{(in radians)}$$

(3) Reflections the two waves may undergo:
 (a) A phase change of π rad (180°) occurs for a wave traveling in one medium when reflected from a medium of *higher* refractive index.
 (b) No phase change occurs on reflection from a medium of *lower* refractive index.

For double-slit interference:

d = slit separation (center-to-center)
D = slit-to-screen distance
y = distance along the screen from the central maximum
m = order

The *maxima* are given by

$$m\lambda = d \sin\theta \qquad \text{(where } m = 0, 1, 2, 3, \ldots)$$

For small angles:

$$m\lambda = d\left(\frac{y}{D}\right)$$

The *minima* are spaced halfway between the bright fringes:

$$(m + \tfrac{1}{2})\lambda = d \sin \theta \qquad \text{(where } m = 0, 1, 2, 3, \ldots)$$

For small angles:

$$(m + \tfrac{1}{2})\lambda = d\left(\frac{y}{D}\right)$$

Multiple-slit interference: For the same slit separation, the major maxima are the same as for the double-slit case.

The *maxima* are given by

$$m\lambda = d \sin \theta \qquad \text{(where } m = 0, 1, 2, 3, \ldots)$$

For small angles:

$$m\lambda = d\left(\frac{y}{D}\right)$$

The major characteristics of multiple-slit interference may be summarized as follows:

(1) The angular separation of major maxima depends on the phase difference of waves from adjacent slits, not on the number of slits.
(2) The number of minor maxima between major maxima is two less than the number of slits.
(3) The sharpness and intensity of major maxima increases as the number of slits increases.

Interference patterns produced by thin film and wedges depend on the phase difference (upon recombination) of waves reflected from the two surfaces. Phase differences are due to the extra path length (round trip) for one of the waves and the different types of reflections at the two surfaces.

The *Michelson interferometer* is an ingenious and versatile instrument capable of measuring distances to within a small fraction of a wavelength of light.

Questions

1. Why is it impossible for all the fringes of a double-slit interference pattern to be of exactly the same intensity?

2. Would longitudinal waves such as sound waves produce double-slit interference effects?

3. Two closely-spaced parallel fluorescent light tubes, both covered with a green filter, illuminate a distant wall. Is an interference pattern produced?

4. Our discussion of double-slit interference was based on a plane light wave falling with normal incidence upon a screen containing two slits. What changes in the interference pattern would be observed if the screen containing the slits were tilted relative to the incident light? Consider tilting about an axis parallel to the slits and about an axis perpendicular to the slits.

5. Describe the interference pattern produced by two closely-spaced pinholes.

6. If a pure tone is sounded in a room, a listener experiences large changes in intensity by moving his head from side to side. Is this an interference phenomenon? Why is the effect less pronounced when music is heard?

7. Suppose a double-slit experiment is immersed under water. What changes, if any, occur in the pattern of fringes on the screen?

8. In a Young's double-slit experiment, the lower halves of the two vertical slits are covered with a blue filter and the upper halves are covered with a red filter. (a) What is the appearance of the resultant interference pattern observed on a screen? (b) Suppose, instead, one slit is covered with a blue filter and the other slit is covered with a red filter. Describe the pattern and explain the reasoning behind your conclusions.

9. How would a triple-slit interference pattern be altered if the center slit were covered by a gray filter to reduce the intensity of the light emanating from that slit?

10. An oil slick on water seems brightest where the oil film is much thinner than a wavelength of visible light. Is the refractive index of the oil greater or less than that of water?

11. A Newton's rings interference pattern produced by a convex lens and plane-glass plate is usually observed by reflection, and the center of the pattern is dark. What is the appearance of the pattern transmitted through the lens and plate?

12. A lens is coated to reduce reflection. What happens to the light energy that had previously been reflected?

13. When looking at the light reflected from a windowpane, why is an interference pattern not observed? After all, light is reflected by both the front and rear surfaces of the glass.

14. Why do coated camera lenses look purplish when observed by reflected light?

15. Suppose reflected white light is used to observe a thin, transparent coating on glass as the coating material is gradually being deposited by evaporation in a vacuum. Describe possible color changes that occur during the process of building up the thickness of the coating.

16. A person in a dark room looking out of a window can easily see a person outside in the sunlight. However, the person outside is often unable to see the person inside the room. Is this an interference phenomenon? If so, explain. If not, what is the nature of the phenomenon?

17. Consider two glass plates in contact at one edge and separated slightly at the opposite edge. In analyzing the visual appearance of the interference pattern produced by reflections from the "air wedge" between the plates, why can we ignore interference between waves reflected from the top surface of the top plate and bottom surface of the bottom plate, even if the plates have perfectly parallel surfaces?

18. What change, if any, would occur in the pattern of Newton's rings if the space between the lens and plate were filled with water?

19. Could an acoustical Michelson interferometer be used to measure the wavelength of ultrasonic sound waves? If so, how would such an interferometer be constructed and what procedure would be used in the measurement?

20. The fringe pattern in a Michelson interferometer is sharpest when the two arms of the interferometer are of approximately equal length. In fact, if the arms are of considerably different lengths so the two paths differ by about a meter or more, the fringes blur together. Explain.

Problems

30A–1 Light of wavelength 600 nm illuminates a double slit with a slit separation of 0.30 mm. An interference pattern is produced on a screen 2.5 m from the slits. Calculate the separation of the interference fringes on the screen near the central maximum.

Answer: 5.00 mm

30A–2 Design a double-slit system that will produce fringes 2 mm apart on a screen 3 m away using light of 550 nm.

30A–3 In a double-slit experiment, sodium light ($\lambda = 589$ nm) produces fringes spaced 1.8 mm apart on a screen. Find the fringe spacing when mercury light ($\lambda = 436$ nm) is used.

Answer: 1.33 mm

30A–4 Find the thickness of the thinnest soap film ($n = 1.33$) that will reflect blue light of wavelength 400 nm.

30A–5 An air wedge is formed between two glass plates separated at one edge by a very fine wire, as was shown in Figure 30–15. When the wedge is illuminated from above by light with a wavelength of 600 nm, 30 dark fringes are observed. Calculate the radius of the wire.
Answer: 4.50 μm

30A–6 In Example 30–4, the minimum thickness of a nonreflecting coating was found to be 99.6 nm. Calculate the next thicker coating that will produce the same effect.

30A–7 A lens is made of glass with a refractive index of 1.70 at a wavelength of 550 nm. Find (a) the thickness and (b) the refractive index of a nonreflecting coating for use at this wavelength. *Answers:* (a) 105 nm (b) 1.30

30A–8 As the mirror M_1 of the Michelson interferometer shown in Figure 30–18 is moved through a distance of 0.163 mm, 500 bright fringes move across the field of view. Calculate the wavelength of the light illuminating the mirrors of the interferometer.

30B–1 Using light of wavelength 500 nm, a double-slit interference pattern is produced on a screen 1.5 m from a pair of vertical slits separated by 0.50 mm. Find the number of interference maxima that lie between the central maximum and 1.00 cm to the left of the central maximum. *Answer:* Six

30B–2 Light composed of two wavelengths illuminates a double slit, forming two interference patterns that are superimposed on a screen. The fifth-order maximum of one color falls exactly at the location of the third-order maximum of the other color. Calculate the ratio of the two wavelengths.

30B–3 A double slit with a separation of 0.45 mm is illuminated by light of wavelength λ_1 and produces an interference pattern on a screen 3 m away. The tenth-order interference maximum is 4 cm from the central maximum. When light of another wavelength λ_2 also illuminates the slits, the combination of fringes overlaps such that the tenth fringe remains distinct while neighboring fringes become less clear. (a) Calculate λ_1 and (b) find the two closest values for λ_2. *Answers:* (a) 600 nm (b) 545 nm and 667 nm

30B–4 A double slit is illuminated by light of wavelength 600 nm and produces an interference pattern on a screen. A very thin slab of flint glass ($n = 1.65$) is placed over only one of the slits. As a consequence, the central maximum of the pattern moves to the position originally occupied by the tenth order maximum. Find the thickness of the glass slab.

30B–5 Consider the radii r_m in a Newton's rings pattern. Show that for $m \gg 1$, the area between successive rings is approximately equal to the constant value $\pi R\lambda$, where R is the radius of curvature of the plano-convex lens and λ is the wavelength of light.

30B–6 When a liquid is introduced into the air space between the lens and plate in a Newton's rings apparatus, the diameter of the tenth ring changes from 1.50 to 1.31 cm. Find the index of refraction of the liquid.

30B–7 Repeat the construction shown in Figure 30–10 for a four-slit interference pattern. Show phasor combinations corresponding to major maxima, minima, and near-minor maxima.

30B–8 A glass plate ($n = 1.62$) is coated with a thin, transparent film ($n = 1.27$). Light reflected at normal incidence is observed as the wavelength is varied continuously. Constructive interference occurs for light at 680 nm, while destructive interference occurs at 544 nm (with no other such instances between these wavelengths). Find the thickness of the film.

30B–9 A glass plate 0.4 mm thick, with a refractive index of 1.50, is placed in a light beam ($\lambda = 580$ nm) such that the plane of the plate is perpendicular to the beam.

(a) Calculate to eight significant figures the number of wavelengths of light within the glass plate.

(b) Find the net phase shift in the light beam resulting from the introduction of the glass plate into the beam.

Answers: (a) 1034.4827 wavelengths
(b) 62.1°, lagging the uninterrupted beam

30B–10 The beam from a helium-neon laser ($\lambda = 633$ nm) is directed toward a screen. Find the number of additional wavelengths of light in the optical path from the laser to the screen when a thin slab of glass, with a thickness of 0.110 mm and a refractive index of 1.55, is inserted into the beam. The surface of the slab is perpendicular to the beam.

30B–11 An oil film ($n = 1.45$) on water is illuminated by white light at normal incidence. If the film is 280 nm thick, find the dominant color in the reflected light.

Answer: green ($\lambda = 541$ nm)

30B–12 A film of soap solution is illuminated by white light at normal incidence and reflects bands of color, as was shown in Figure 30–13. Calculate the thickness of the film at the first green band ($\lambda = 530$ nm) below the nonreflecting portion of the film. The soap solution has a refractive index of 1.33.

30B–13 An air wedge is formed between two glass plates in contact along one edge and slightly separated at the opposite edge. When illuminated with monochromatic light from above, the reflected light reveals a total of 85 dark fringes. Calculate the number of dark fringes that would appear if water ($n = 1.33$) were to replace the air between the plates.

Answer: 113

30B–14 Interference fringes can be produced using a *Lloyd's mirror* arrangement with a single source S_0, as in Figure 30–21. The image S' of the source formed by the mirror acts as a second coherent source that interferes with S_0. If fringes spaced 1.2 mm apart are formed on a screen 2 m from the source S_0 (which emits light of wavelength 606 nm), find the vertical distance h of the source above the plane of the reflecting surface.

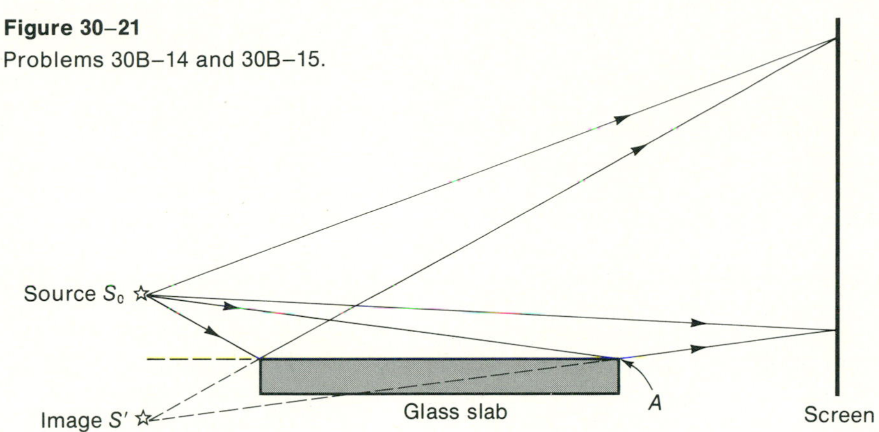

Figure 30–21
Problems 30B–14 and 30B–15.

30B–15 Referring to the Lloyd's mirror setup of the previous problem, light waves are interfering in space wherever the two sets of waves pass through each other. Suppose we use a lens of high magnification to examine the interference in the vertical plane just above the edge A of the mirror. Assuming the mirror reflection is from glass, will the fringe nearest the edge of the mirror be light or dark? Explain. *Answer:* Dark

30C–1 Show that the dashed curves representing lines of constant phase difference shown previously in Figure 30–4(b) are hyperbolas. The general form for a hyperbola in rectangular coordinates is $y^2/a^2 - x^2/b^2 = 1$.

30C–2 Yellow light from the mercury spectrum ($\lambda = 579$ nm) illuminates a pair of vertical slits separated by 0.20 mm. An interference pattern is produced on a screen 2.5 m from the slits. Find the intensity of the light at a distance of 1.5 cm to the right of the central maximum relative to the intensity at the central maximum.

30C–3 Show that the half-width of a double-slit interference maximum (defined by the width where $I = I_0/2$) subtends an angle $\theta = \lambda/2d$. Assume that $\theta \approx \sin\theta$.

30C–4 In terms of the slit separation d and wavelength λ, derive an expression for the total angular width $\Delta\theta$ of the central maximum for (a) a three-slit interference pattern, (b) a four-slit interference pattern, and (c) an N-slit interference pattern.

30C–5 The expression for the radius of Newton's rings, $r_m = (Rm\lambda)^{\frac{1}{2}}$, is the result of an approximation. Show that an exact expression is $r_m = (Rm\lambda - m^2\lambda^2/4)^{\frac{1}{2}}$.

30C–6 An air-tight tube with parallel end windows 6.0 cm apart is placed in one arm of a Michelson interferometer so that light with a wavelength of 570 nm passes through the tube, is reflected by the mirror, and again passes through the tube. When the air is withdrawn from the tube by a vacuum pump, 63 fringes pass the field of view. Calculate the refractive index of air.

30C–7 Figure 30–22 shows a *Fresnel biprism* for producing interference fringes using a single source S_0. The biprism is two identical glass prisms joined at their bases with very small vertex angles α. The prisms form two coherent virtual images, S_1 and S_2, separated a distance d. If the glass has an index of refraction n, show that $d = 2x(n - 1)\alpha$.

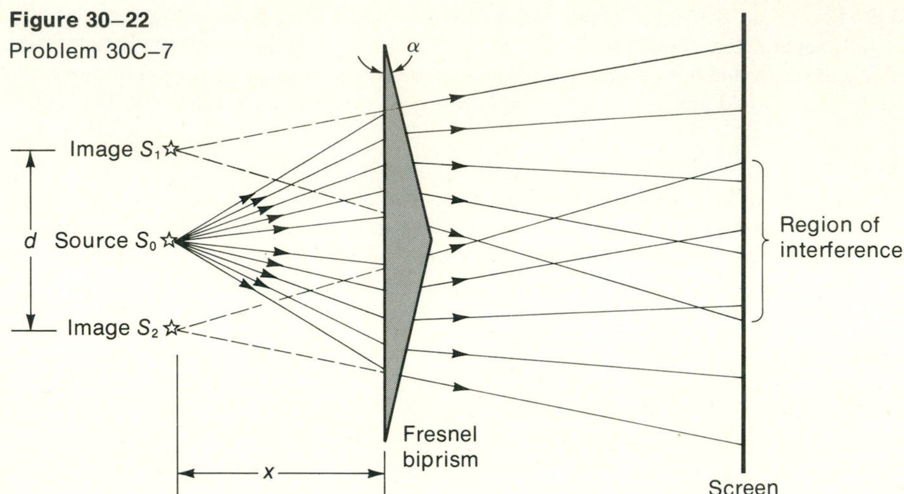

Figure 30–22
Problem 30C–7

30C–8 The yellow light emitted by a sodium source has two wavelengths at 589.0 nm and 589.6 nm. Consider a Michelson interferometer used with this light. When the mirror at the end of one arm is moved continuously in one direction, the observed fringes "wash out," then reappear sharply, then wash out, and so on.

(a) Explain this effect.

(b) Calculate the distance between two successive positions of the mirror when the fringes are sharp.

Where the telescope ends, the microscope begins.
Which of the two has the grander view?

VICTOR HUGO
(Saint Dennis)

chapter 31
Diffraction

As we proceed into this chapter entitled "Diffraction," you will see that the discussion is really one involving interference. There is no physical difference between interference and diffraction. In both cases, light waves interfere to produce regions of extra brightness or darkness. It has become customary to use the term *interference* for situations involving a finite number of point or line sources (such as multiple slits) and the term *diffraction* for the interference of waves from an area source (essentially an infinity of neighboring point sources).

If light traveled only in straight lines, the shadows of opaque objects would have sharp edges, changing abruptly from bright to dark. The fact is, however, that light does bend somewhat around the edge of an object into the shadow region, often producing bright and dark fringes as a result of the *interference* of light waves. *The bending of light away from straight-line paths as it passes near an object* is an example of the diffraction of light. Figure 31–1 shows a magnified view of the shadow of a sharp knife-edge, illustrating these diffraction effects.

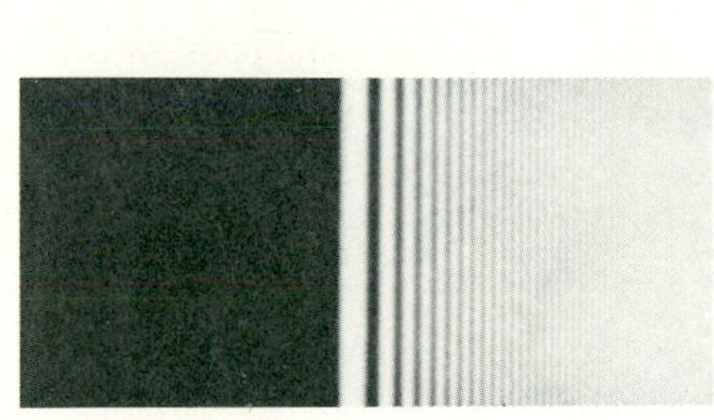

(**a**) A magnified view of the transition from a dark shadow on the left to the bright region on the right.

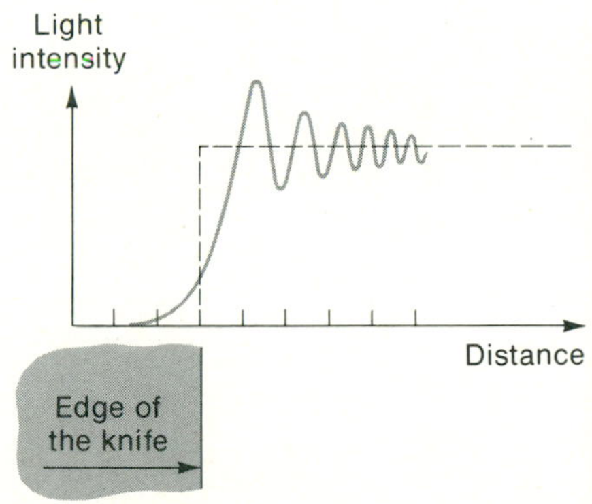

(**b**) A plot of the light intensity versus distance for the knife-edge diffraction pattern. If there were no diffraction, the intensity would change abruptly from dark to light as shown by the dashed line.

Figure 31–1
The diffraction pattern produced by a sharp knife-edge.

One way to observe diffraction easily is to place your hand over your eye so that you can see light from a *point* source penetrating the cracks between your fingers. (A *line* source such as a straight neon tube far away may also be satisfactory if it is aligned in the same direction as the crack between your fingers.) If the crack is narrow enough, you will observe a pattern of bright and dark fringes. They are particularly pronounced when viewing a distant mercury-vapor street light because of the dominance of only a few different wavelengths of light emitted from such a source. For things we look at in our everyday experience, diffraction effects are usually quite small and therefore overlooked. Another consideration is that most light sources have an *extended area*, so that diffraction patterns from one part of the source overlap with patterns from another part of the source, making them difficult to distinguish. Furthermore, each wavelength of light produces its own distinct pattern, so when many wavelengths are present, as in white light, the various patterns again overlap. It is important to understand diffraction effects because they place inescapable upper limits on the sharpness of images formed by all optical instruments. They also limit the accuracy of certain measurements.

Diffraction effects were known to both Newton and Huygens, but it was not until the nineteenth century that an explanation was proposed by Augustin J. Fresnel (1788–1827), a brilliant French physicist. His work, coupled with that of the British physicist Thomas Young (1773–1829), firmly established the *wave* theory of light.

The general situation that produces diffraction effects is either an *aperture* or an *obstacle* placed between a light source and a screen, as pictured in Figure 31–2. To find out what happens, we adopt Huygens' approach and imagine that *each point on the wavefront acts as a new point source of radiation.* Thus, the light falling on any given location on the screen (for example, P_1 in the directly illuminated part of the screen or P_2 in the geometrical shadow region) contains contributions from *all parts* of the wavefront passing through the aperture. The case shown in this figure is complicated for two reasons:

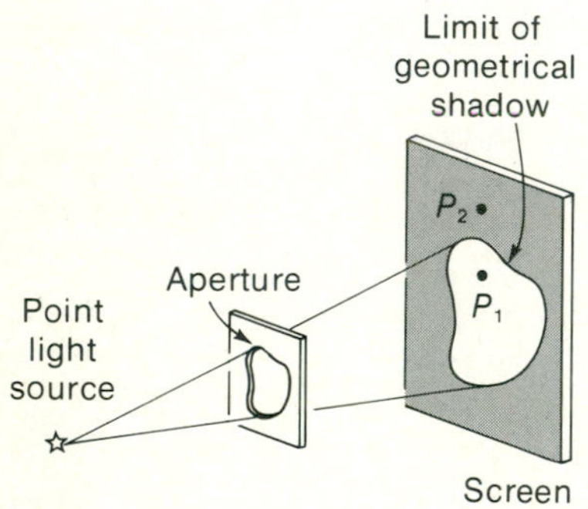

Figure 31–2
Diffraction: a general case. Light reaching the screen is composed of waves emanating from all parts of the wavefront as it emerges from the aperture.

(1) The wavefront at the aperture is *divergent* rather than *plane*. This means that as we consider different points on the wavefront, the *angle* between the normal to the wavefront and the direction to a given point P varies for different points on the wavefront.
(2) The *distances* from various points on the wavefront to a given point P are all different.

Another representation of this same situation is shown in Figure 31–3(a). The light diverges from a nearby point source as it moves toward the aperture The light reaching point P on the screen is made up of Huygens wavelets that emanated from all parts of the wavefront as it emerges from the aperture. This general case is known as **Fresnel diffraction** and is quite complicated to analyze. We will consider only a few such cases, at the end of the chapter.

Figure 31–3(b) illustrates a situation that is easier to analyze. Rays approaching the aperture are *parallel* (with a *plane* wavefront), and rays leaving the aperture that reach a given point P on the screen are *parallel* (or essentially parallel), because the screen is so far away. This case is known as **Fraunhofer diffraction.** If large distances for the source and screen are not available, we can achieve this condition experimentally by using lenses with a nearby source and screen, as in Figure 31–3(c).[1] Fraunhofer diffraction is easy to analyze

[1] In Figure 31–3 and later figures, we draw only those rays from the wavefront at the aperture that *reach the given point P*. Of course, simultaneously there are other rays at other angles, which travel to other points on the screen.

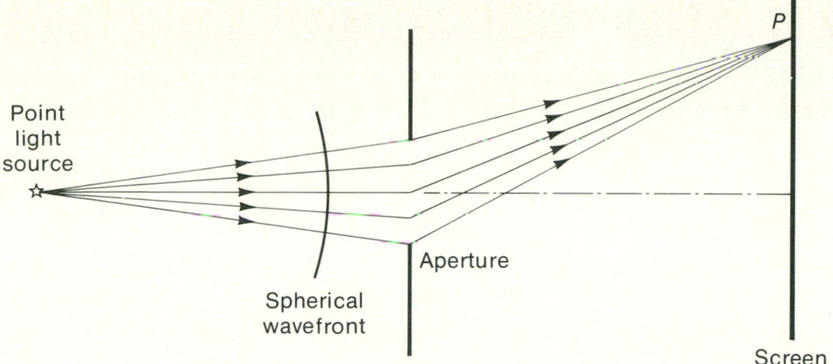

(a) Fresnel diffraction. The source and screen are both near the aperture. Rays from the source and rays to the screen cannot be considered parallel.

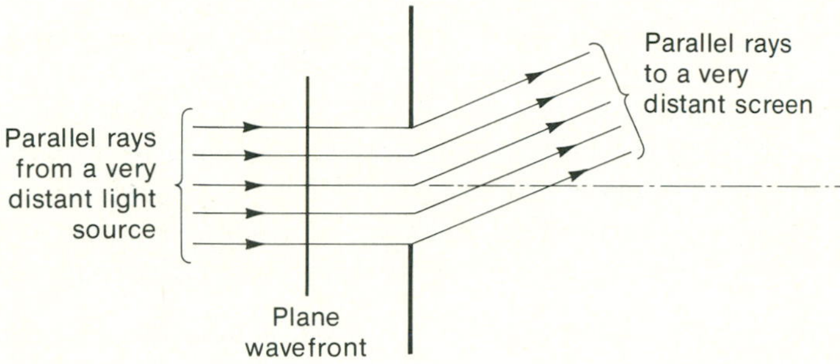

(b) Fraunhofer diffraction. The light source and the screen are both very far from the aperture. Rays incident on the aperture are parallel, and rays leaving the aperture toward the screen are parallel.

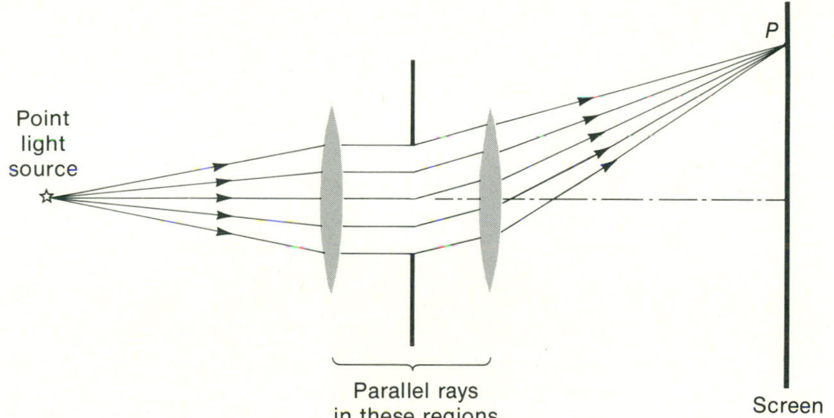

(c) With the use of 2 lenses, conditions for *Fraunhofer diffraction* can be produced using a nearby light source and screen.

Figure 31–3
The distinction between Fresnel and Fraunhofer diffraction. In Fraunhofer diffraction, the light rays striking the aperture are parallel, and the light rays leaving the aperture are parallel.

because we do not have to deal with the varying angles and distances characteristic of Fresnel diffraction.

The distinction between Fresnel and Fraunhofer diffraction patterns sometimes cannot be sharply defined. For example, if we start with a nearby source and screen and gradually move them farther away from the aperture, the Fresnel diffraction pattern gradually changes over into the Fraunhofer pattern. Thus, Fraunhofer diffraction is really just a *limiting case* of the more general Fresnel diffraction.

31.1 Single-Slit Diffraction

We will discuss two approaches to single-slit Fraunhofer diffraction. The first is a simple but useful technique of *half-wave zones* that yields the criterion for constructive and destructive interference. The second is a more detailed approach utilizing *phasors*, which yields a quantitative expression for the intensity distribution within a diffraction pattern.

Half-Wave Zones

Consider the Fraunhofer diffraction apparatus illustrated in Figure 31–4. To restrict the problem to two dimensions, we analyze a slit of width a aligned perpendicular to the plane of the figure. The slit is divided into *zones* such that the path length of a ray emanating from one edge of a zone is one-half wavelength longer than that from the corresponding edge of the adjacent zone. Such zones are called *half-wave zones.* In Figure 31–4 the aperture is wide enough to contain exactly four such zones.

We will now consider what happens to the rays from two adjacent zones. The rays coming from two corresponding points, such as P_1 and P_2, will differ in path length by one-half wavelength as they reach the screen. Combining similar pairs of rays for the two zones, we conclude that the light from one zone will interfere *destructively* with that from the neighboring zone.

HALF-WAVE ZONE CRITERION FOR SINGLE-SLIT DIFFRACTION MINIMA

A *minimum* in the diffraction pattern occurs if the slit viewed from that point on the screen contains exactly an *even* number of half-wave zones.

In reference to Figure 31–4, for point A on the screen the slit contains *four* half-wave zones, while for point B the slit contains *two* half-wave zones.

An alternative criterion for a minimum in a single-slit diffraction pattern may be based upon the total width of the slit.

ALTERNATIVE CRITERION FOR SINGLE-SLIT MINIMA

A *minimum* in the diffraction pattern occurs if the path for a ray of light arriving at that point from one edge of the slit is an *integral number of wavelengths longer* than the path of a ray from the opposite edge of the slit.

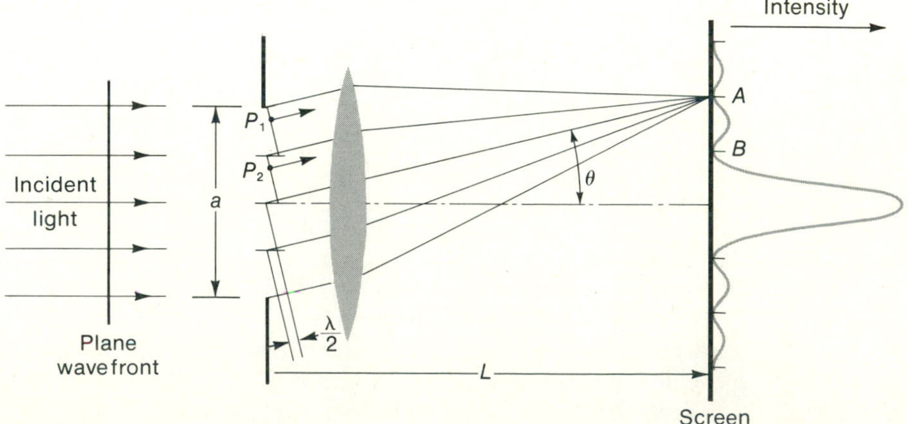

Figure 31–4
Fraunhofer diffraction. The criterion for destructive interference at a point on the screen may be determined by dividing the slit into *half-wave zones.* (For clarity, the width of the slit is greatly exaggerated relative to the screen distance *L*.) The incident light rays are parallel, forming a *plane* wavefront at the aperture.

Thus, in Figure 31–4 a minimum in the diffraction pattern occurs when

SINGLE-SLIT FRAUNHOFER DIFFRACTION PATTERN MINIMA

$$a \sin \theta = m\lambda \qquad (\textit{minima} \text{ for } m = 1, 2, 3, \ldots) \tag{31–1}$$

Note that the central *maximum* corresponds to $m = 0$, with all other values of m designating *minima.* In most situations, the angle θ is small enough to justify the small-angle approximation: $\sin \theta \approx \tan \theta \approx \theta$. When this is true, the *central maximum and all the other minima are equally spaced from one another.* Thus, the full width of the central maximum is *twice* the separation of adjacent minima.

Do not confuse this relation with Equation (30–11):

$$d \sin \theta = m\lambda \qquad (\textit{maxima} \text{ for } m = 0,1,2, \ldots)$$

The equations have the same form, but Equation (31–1) is for the *single-slit* diffraction pattern *minima,* while the *double-slit* relation [Equation (30–11)] is for the interference pattern *maxima.*

Phasors

The use of phasors to determine the intensity distribution in a single-slit Fraunhofer diffraction pattern is an extension of the technique used in multiple-slit interference patterns. We consider the slit to be divided into small incremental zones, Δy wide, as illustrated in Figure 31–5. Each of these zones, or strips, may be considered a source of radiation contributing an incremental electric field amplitude ΔE at the point P on the screen. The total field amplitude E_θ at the point P will be the sum of such increments from all of the zones. However, depending on the angle θ, the incremental field amplitudes will be slightly out of phase with one another. Since:

$$\frac{\text{Path difference}}{\lambda} = \frac{\text{Phase difference}}{2\pi}$$

we have:

$$\frac{\Delta y \sin \theta}{\lambda} = \frac{\Delta \phi}{2\pi}$$

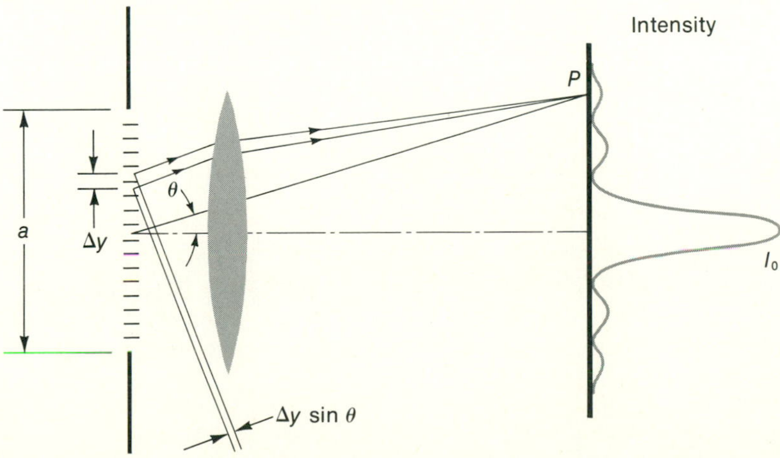

Figure 31–5
Fraunhofer diffraction. The electric field at P is the sum of incremental fields emanating from incremental zones Δy wide at the aperture.

where λ is the wavelength and $\Delta\phi$ is the phase difference of the electric field increments from adjacent zones. Rearranging, we have

$$\Delta\phi = \left(\frac{2\pi}{\lambda}\right)\Delta y \sin\theta \tag{31–2}$$

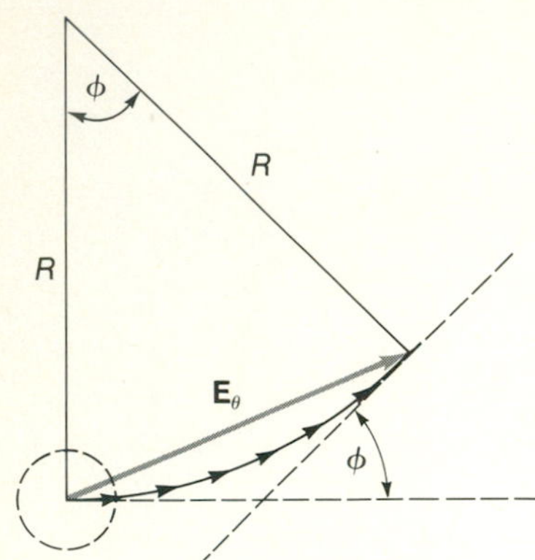

(a) The phasor addition of incremental electric fields $\Delta\mathbf{E}_n$ to produce the total field $\mathbf{E}_\theta$.

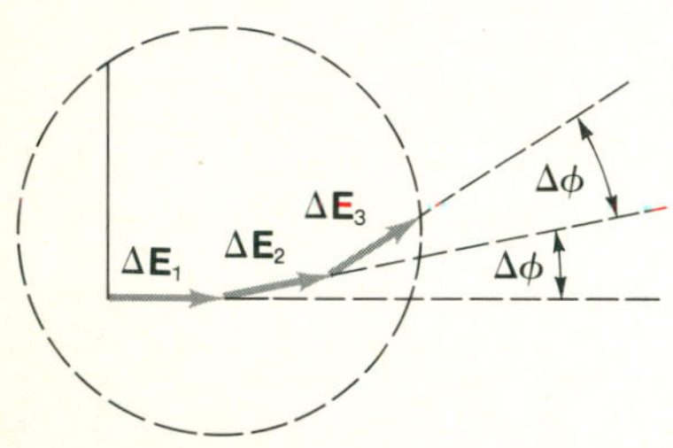

(b) A magnified view of the first three increments.

Figure 31–6

Phasor addition to determine the total electric field amplitude E_θ at a point *P* on the screen.

Figure 31–6(b) shows the difference in phase between electric field increments from three adjacent zones at the top of the slit shown in Figure 31–5. If θ is small, all of the incremental field elements may be considered equal in amplitude. The angle ϕ between the last incremental zone at the bottom of the slit and the first zone at the top is shown in Figure 31–6(a). The sum of all the incremental phasors is then $\mathbf{E}_\theta$, the base of the isosceles triangle with equal sides R. From trigonometry:

$$E_\theta = 2R\sin\left(\frac{\phi}{2}\right) \tag{31–3}$$

where, from Equation (31–2):

$$\phi = \left(\frac{2\pi}{\lambda}\right)a\sin\theta \tag{31–4}$$

The value of R may be obtained by letting the incremental phasor amplitude approach zero as the number of increments approaches infinity. In this limit, the sum of increments forms the arc of a circle with radius R. The length of the arc is simply the incremental phasor sum when all of the increments are in phase. This occurs for light rays parallel to the axis ($\theta = 0$) forming the central peak of the diffraction pattern. Thus the amplitude E_0 of the central maximum is

$$E_0 = R\phi \tag{31–5}$$

Combining Equations (31–3) and (31–5), we have

$$E_\theta = \frac{2E_0\sin\left(\frac{\phi}{2}\right)}{\phi}$$

or, more simply:

$$E_\theta = E_0\left(\frac{\sin\alpha}{\alpha}\right) \tag{31–6}$$

where

$$\alpha \equiv \frac{\phi}{2}$$

From Equation (31–4):

$$\alpha = \left(\frac{\pi}{\lambda}\right)a\sin\theta \tag{31–7}$$

Mathematically, ($\sin\alpha/\alpha$) approaches unity as α approaches zero. Therefore, E_θ approaches E_0 as θ approaches zero. Recall that the intensity I of light is proportional to the *square* of the amplitude of the electric field strength E.

$$I \propto E^2$$

We summarize the single-slit diffraction equations as follows:

$$I_\theta = I_0\left(\frac{\sin\alpha}{\alpha}\right)^2 \qquad \textbf{(31–8)}$$

where

$$\alpha = \left(\frac{\pi}{\lambda}\right) a \sin\theta$$

Minima occur when

$$\alpha = m\pi \qquad \text{(where } m = 1, 2, 3, \ldots) \qquad \textbf{(31–9)}$$

Combining Equations (31–7) and (31–9), we have

$$\left(\frac{\pi}{\lambda}\right) a \sin\theta = m\pi$$

or

SINGLE-SLIT FRAUNHOFER DIFFRACTION MINIMA

$$a \sin\theta = m\lambda \qquad (\textit{minima} \text{ for } m = 1, 2, 3, \ldots) \qquad \textbf{(31–10)}$$

which is the same equation derived using half-wave zones.

The mathematical form of Equation (31–8) makes it difficult to determine the exact relative amplitude of diffraction *maxima* and their locations. However, approximate relative amplitudes may be obtained by assuming that the maxima lie halfway between the minima. That is, from Equation (31–9) an approximate maximum occurs when

$$\alpha = (m + \tfrac{1}{2})\pi \qquad \text{(where } m = 1, 2, 3, \ldots) \qquad \textbf{(31–11)}$$

Substituting this into Equation (31–8), we obtain

$$\frac{I_\theta}{I_0} = \left[\frac{\sin(m + \frac{1}{2})\pi}{(m + \frac{1}{2})\pi}\right]^2 \qquad \text{(where } m = 1, 2, 3, \ldots)$$

or

$$\frac{I_\theta}{I_0} = \frac{1}{(m + \frac{1}{2})^2\pi^2} \qquad \text{(approximate maxima for } m = 1, 2, 3, \ldots)$$

Thus, if I_0 is the intensity at the central peak, for $m = 1$, $I_\theta = 0.045I_0$; for $m = 2$, $I_\theta = 0.016I_0$; and for $m = 3$, $I_\theta = 0.0083I_0$. Clearly, almost all of the light in a diffraction pattern falls within the central maximum peak.

A graphical approach to diffraction pattern intensities is often enlightening. In Figure 31–7(a), at the central maximum the sum of the incremental electric-field phasors ΔE_i forms a straight electric field phasor E_0. As we move away from the central maximum (that is, as θ increases), the sum of incremental phasors forms an ever-tightening arc that closes around on itself to form successive minima, with maxima occurring between closures. In the limit of negligibly small increments, the length of the arc remains constant as it winds up. In Figure 31–7(c) and (e) we see that the lengths of the resultant phasor are a maximum just *slightly* before $\phi = 3\pi$ and 5π because, as the arc tightens, the diameter of the circle becomes smaller. However, this difference is very small, justifying the approximation used in deriving Equation (31–10).

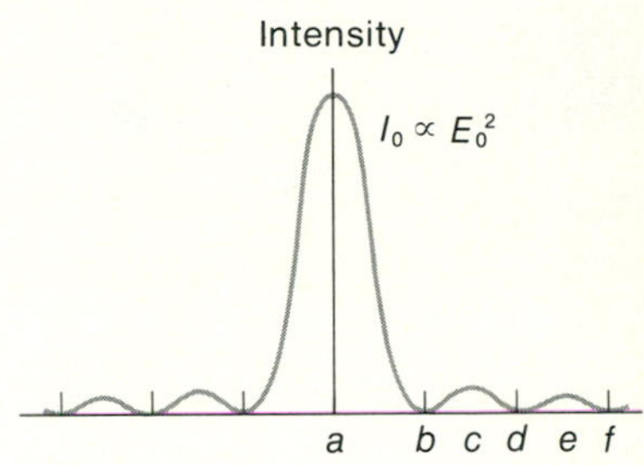

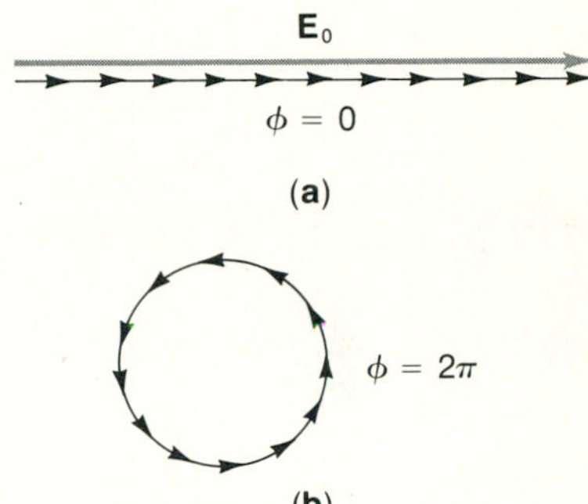

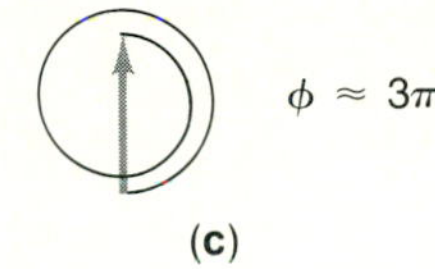

$\phi = 4\pi$

(d)

$\phi \approx 5\pi$

(e)

$\phi = 6\pi$

(f)

Figure 31–7
Phasor-addition diagrams corresponding to the maxima and minima of a single-slit diffraction pattern. For clarity, the arcs shown in (c) through (f) are drawn as spirals instead of circles.

As you look at Figures 31–8 and 31–9, it will be helpful to remember the following general characteristics of a single-slit diffraction pattern (when θ is small):

(1) The minima are *equally spaced* from one another.
(2) The full width of the central peak is *twice* the spacing between all other minima.
(3) The maxima of other peaks are relatively faint and approximately midway between the minima. (Actually, they are displaced slightly toward the central peak.)
(4) As the width of the slit is made smaller, the diffraction pattern becomes larger.
(5) As the wavelength is made smaller, the diffraction pattern becomes smaller.

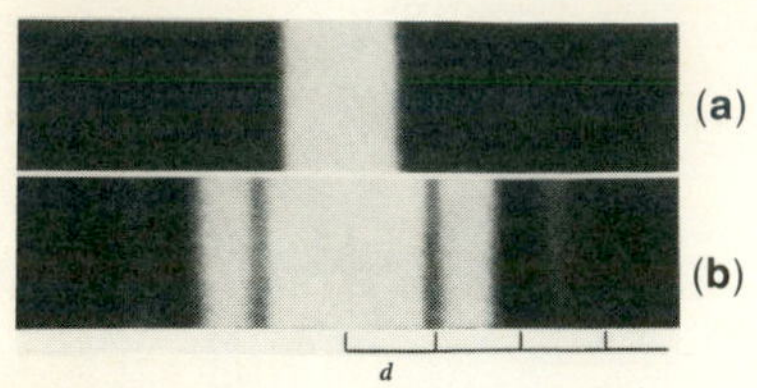

Figure 31–8
Two photographs of the same single-slit diffraction pattern. Ninety percent of the light passing through the slit falls in the central peak. In (b), the exposure time has been increased greatly, overexposing the central peak to bring out some of the faint maxima on either side.

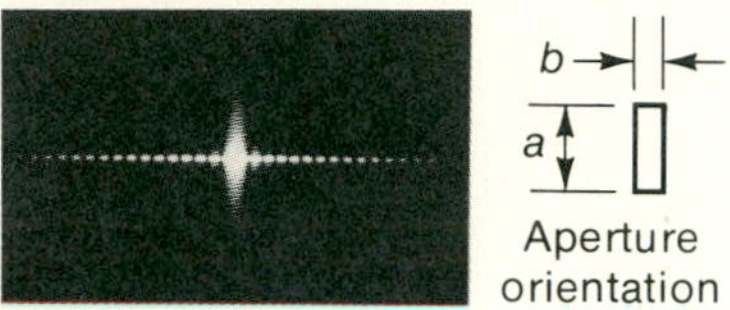

Figure 31–9
The diffraction pattern of a rectangular aperture. Along the horizontal axis, the minima are spaced *farther apart* than along the vertical axis because the aperture width is *narrower* along the horizontal direction.

EXAMPLE 31–1

The width of the central maximum in the diffraction pattern is of particular interest. Suppose that a slit 3×10^{-4} m wide is illuminated by a yellow-green light ($\lambda = 500$ nm). Find the total width of the central maximum on a screen 2 m from the slit.

SOLUTION

We will define the total width of the central maximum as the distance between the first minima on either side of the peak. The value of θ shown in Figure 31–10 is obtained by using Equation (31–1):

$$a \sin \theta = m\lambda$$

where $m = 1$ for the first minimum. Substituting values for a and λ, we obtain

$$\sin \theta = \frac{5.00 \times 10^{-7} \text{ m}}{3 \times 10^{-4} \text{ m}}$$

$$\sin \theta = (\tfrac{5}{3}) \times 10^{-3}$$

The distance y is half the width of the central maximum:

$$y = L \tan \theta$$

Since $\sin \theta \approx \tan \theta$ for small angles, we have

$$y = (2 \text{ m})(\tfrac{5}{3} \times 10^{-3})$$

$$y = 3.33 \times 10^{-3} \text{ m}$$

Thus the total width $2y$ of the central maximum is

$$2y = \boxed{6.67 \text{ mm}}$$

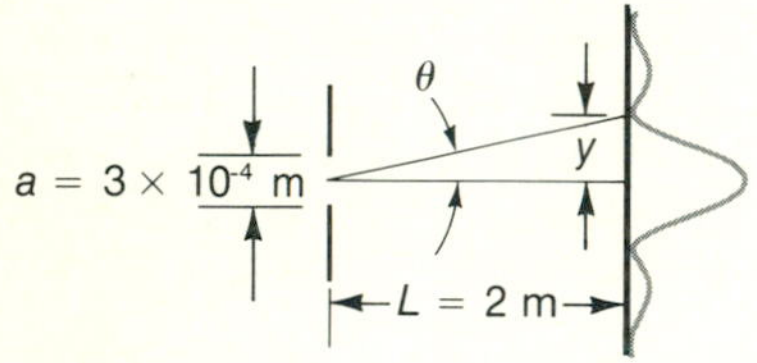

Figure 31–10
Example 31–1

EXAMPLE 31–2

(a) Find the angular width $\Delta\theta$ of the half-maximum intensity within the central maximum for the situation described in Example 31–1. (b) Find the width on the screen of the central peak at half maximum (see Figure 31–11).

SOLUTION

(a) The intensity distribution is given by Equation (31–8):

$$\frac{I_\theta}{I_0} = \left(\frac{\sin \alpha}{\alpha}\right)^2$$

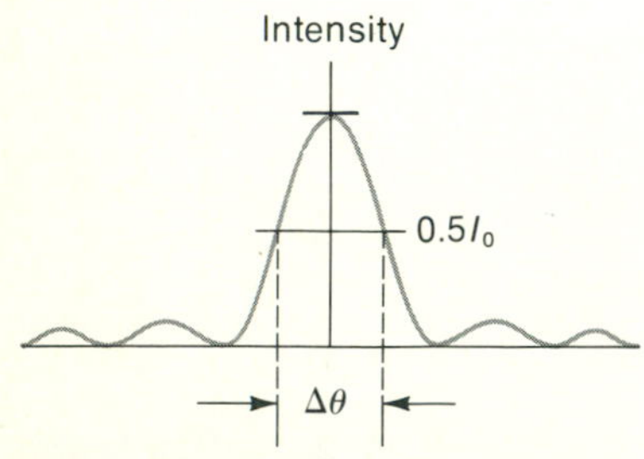

Figure 31–11
Example 31–2

For $I_\theta/I_0 = 0.5$, we have

$$\left(\frac{\sin \alpha}{\alpha}\right)^2 = 0.5$$

Because this equation cannot be solved algebraically, advanced optics texts contain tables of values for $(\sin \alpha)/\alpha$ versus α. However, we may use successive approximations to find α. A good first approximation is that $\alpha = \pi/2$ (equal to 90°), since the first minimum is $\alpha = \pi$. We find that

$$\left(\frac{\sin \frac{\pi}{2}}{\frac{\pi}{2}}\right)^2 = 0.405$$

As we decrease α, $[(\sin \alpha)/\alpha]^2$ increases. Thus, by decreasing α we find that for $\alpha = 80° \approx 1.40$ rad:

$$\left(\frac{\sin 80°}{1.40}\right)^2 \approx 0.500$$

The angle θ may be obtained from α using Equation (31–7):

$$\alpha = \left(\frac{\pi}{\lambda}\right) a \sin \theta$$

or

$$\sin \theta = \left(\frac{\lambda}{a\pi}\right) \alpha$$

Substituting numerical values:

$$\sin \theta = \frac{(5 \times 10^{-7}\ \text{m})(1.40)}{(3 \times 10^{-4}\ \text{m})\pi}$$

$$\sin \theta = 7.43 \times 10^{-4}$$

Then, the full angular width $\Delta\theta$ is

$$\Delta\theta = 2 \sin^{-1}(7.43 \times 10^{-4})$$

or

$$\Delta\theta = \boxed{1.49 \times 10^{-3}\ \text{rad}}$$

(b) On the screen, the full width of the central peak at half maximum is

$$\text{Width} = L\,\Delta\theta$$

$$\text{Width} = (2\ \text{m})(1.49 \times 10^{-3}\ \text{rad}) = 2.98 \times 10^{-3}\ \text{m}$$

$$\text{Width} = \boxed{2.98\ \text{mm}}$$

31.2 The Diffraction Grating

A *diffraction grating* is essentially a multiple-slit device in which the slits are extremely narrow and very closely spaced. The first gratings constructed by Fraunhofer were simply arrays of closely spaced, fine, parallel wires or threads. A typical modern grating is formed by making parallel scratches on glass or metal, and often has the equivalent of more than five or ten thousand slits per centimeter. Because it is desirable to have as many slits per centimeter as possible, the width of each slit is very small, producing wide-angle diffraction effects. Thus the diffraction grating involves a combination of two phenomena: *multiple-slit interference* and *single-slit diffraction.* As we shall now explain,

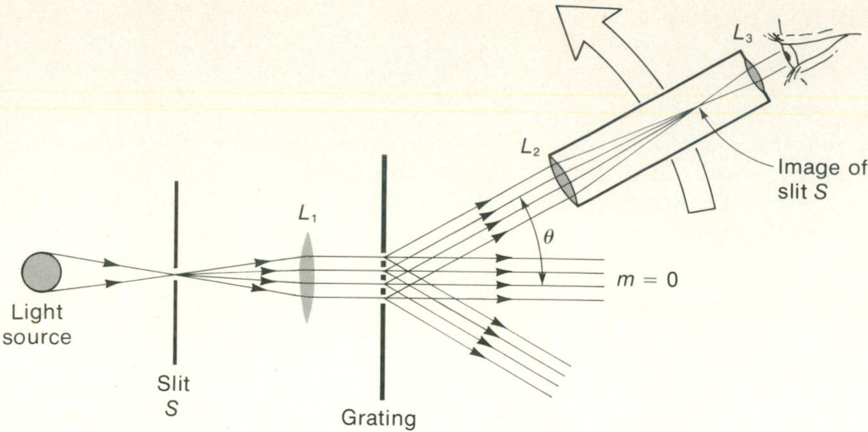

Figure 31–12
A simple grating spectroscope.

diffraction gratings are used to make very accurate measurements of wavelengths of light.

Look again at Figure 30–11 in the previous chapter. As the number of slits in an array increases, the major maxima become much more narrow and intense. Indeed, for several thousand slits, the major peaks will be extremely narrow, and the intensities of the minor maxima in between become truly negligible. Figure 31–12 shows a source of monochromatic light passing through a slit S (aligned parallel to the grating slits). A *collimating* lens L_1 makes the light parallel as it strikes the diffraction grating (to achieve the Fraunhofer condition). The different orders of diffracted light ($m = 0, \pm 1, \pm 2, \ldots$) are emitted at various angles θ_m, where they are collected by a movable telescope that may be rotated around to the appropriate angles. Lens L_2 brings the parallel light to a line focus (really just an image of the slit S), where it is further magnified by a lens L_3 for examination by the eye. Once the device is calibrated and the grating space d is known, the angle θ permits a determination of the wavelength λ.

If the source emits several different discrete wavelengths, then instead of a single line at each order position, there will be a cluster of lines spread out at various angles, one for each wavelength present. The greater the number of lines per centimeter in the grating, the more this cluster will be spread out, allowing very precise measurements of wavelengths. If a source emits a continuous spectrum, the full distribution of wavelengths is displayed over a range of angles. Unfortunately, sometimes the spectrum of one order will overlap a portion of the spectrum of an adjacent order, a possible source of confusion that must be taken into account. Many instruments record the spectrum photographically or analyze the light with a sensitive photoelectric cell. Such devices are called *grating spectrographs* (see Figure 31–13). They have advantages over *prism spectrographs* in that they spread out the wavelengths over a wider angle than does a prism, particularly in the higher orders. Also, gratings that use the *reflection* of light rather than transmission through the grating can be curved to achieve focusing properties, omitting the lens system to allow studies of infrared and ultraviolet spectra that would otherwise be absorbed by glass or quartz lenses. The grating spectroscope does have the disadvantage that it wastes light over several different orders, so it is not suitable for faint sources (though the grooves of reflection gratings can be shaped to throw extra light in the direction of one particular order). Prism spectrographs thus have advantages for investigating weak sources, while grating spectrographs achieve the highest precision in separating two (nearly identical) wavelengths.

Gratings are made by scratching a series of parallel lines with a diamond stylus on a clear glass plate (forming a *transmission* grating) or on a flat metal plate (forming a *reflection* grating, in which the interference effects are viewed by reflected light).[2] Because a good grating is so difficult to manufacture, most gratings in use are replicas formed by pouring a thin layer of a transparent collodion solution on the grating, allowing it to harden, and peeling it off, producing a transmission grating. The collodion sheet is then mounted on glass or supported in a rigid frame. This transparent plastic replica contains a series of ridges where the scratches were, separated by undisturbed clear strips. In an overly simplified picture, we may think of such a transmission grating as allowing light to transmit through the clear strips, which therefore act as slits, while the somewhat irregular ridges scatter the light in all directions and are thus effectively opaque.

We begin a discussion of the theory of gratings by analyzing a transmission grating with just four slits, as shown in Figure 31–14. Parallel light is incident so that as the plane wave front passes through the grating, the slits act as a series of coherent light sources. Unlike the double-slit interference discussion in Chapter 30, where we ignored the diffraction occurring at each slit, here we take diffraction into account. Note that the slit widths a are comparable to the slit separation d. The parallel light rays that leave the slit at an angle θ are brought to a focus on the screen as a line image perpendicular to the plane of the diagram. (Of course, diffraction causes light rays to leave the slit at other angles, too, which are similarly brought to a focus at other points on the screen; we show just one particular angle θ on the diagram.) The lens enables us to use the Fraunhofer single-slit diffraction theory we developed in the previous section.

Before proceeding further, it is important to note that because of the lens, the diffraction patterns produced by all of the slits *exactly* superimpose on the

[2] Making a series of parallel scratches sounds like a simple task. However, in practice, the procedure is full of unexpected difficulties. An interesting discussion of one of the most precise mechanical devices ever invented can be found in "Ruling Engines," A.G. Ingalls, *Scientific American*, June 1952. Most modern gratings are made on a thin layer of aluminum evaporated on a glass plate "optically flat" to within a fraction of a wavelength of light.

Figure 31–13
Fraunhofer was the first person to investigate the spectrum of sunlight with a diffraction grating and in so doing observed thousands of dark lines (the "Fraunhofer lines"). He noted that some lines fell in the same positions as known bright lines in the spectra of certain elements he had studied in the laboratory, but he was unable to explain the mechanism that produced the dark lines. More than half a century later, Kirchhoff gave the correct explanation that the cool atmosphere of gas atoms above the sun's glowing surface *absorbed* the characteristic wavelengths of those atoms from the continuous spectrum of the sun. Helium, a Greek word meaning "the sun," was first discovered in the Fraunhofer lines of sunlight, as were several other elements first discovered outside the earth.

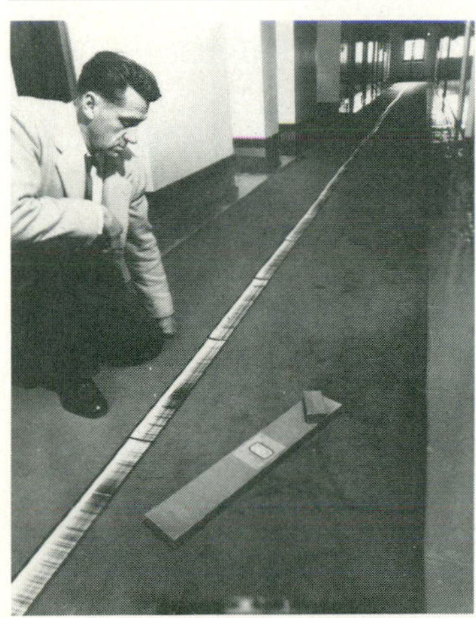

(a) Astronomer R.S. Richardson displays a 40-ft record of the sun's spectrum obtained by a grating spectrograph.

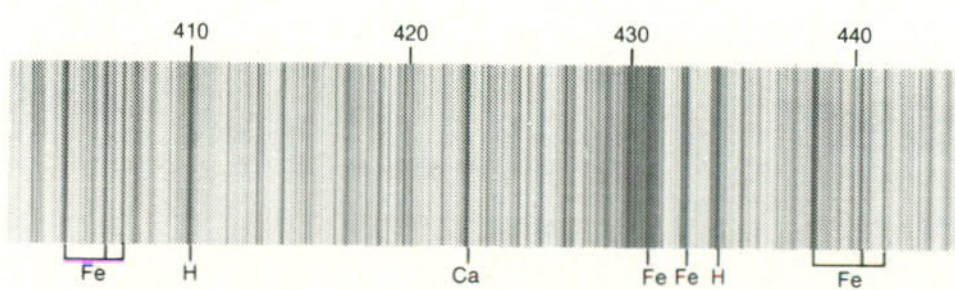

(b) A segment of the visible spectrum of the sun. Wavelengths (in nanometers) are listed above the spectrum, while elements responsible for certain strong lines are shown below.

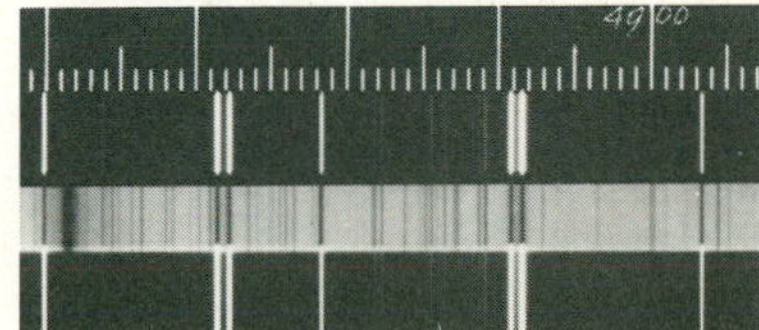

(c) A laboratory spectrum of iron compared with the Fraunhofer absorption lines of the sun's spectrum indicates the presence of iron in the sun.

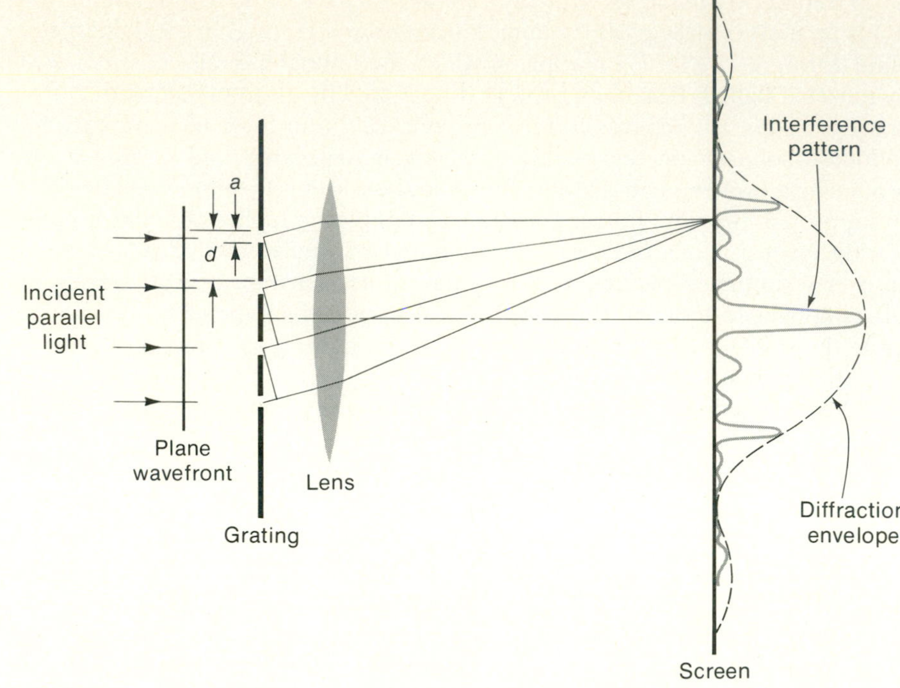

Figure 31–14
A four-slit grating. As the slits become narrower and the number of slits increases to that in a typical diffraction grating, the interference major maxima become sharper and more intense, while the diffraction envelope becomes broader.

screen. That is, the diffraction maximum formed by one slit falls in precisely the same place as that formed by another slit. (All parallel rays entering a lens converge to the same focal point.)

In Section 30.3 we showed that the following equation describes the condition for the major maxima in a multiple-slit interference pattern:

MULTIPLE-SLIT INTERFERENCE $\quad d \sin\theta = m\lambda \quad$ (major maxima at $m = 0, 1, 2, \ldots$) **(30–19)**

where d is the slit separation, θ is the angle between the central maximum ($m = 0$) and other major maxima, and λ is the wavelength. When we derived this equation, diffraction effects were ignored. However, if we now take into account the slit width a, we can interpret the distance d to be the *distance between corresponding points within adjacent slits.*

This equation specifies the angular location of the major interference peaks. But the overall intensity of the pattern is reduced by the diffraction effects of Equation (31–8):

$$I_\theta = I_0\left(\frac{\sin\alpha}{\alpha}\right)^2$$

where
$$\alpha = \left(\frac{\pi}{\lambda}\right)a \sin\theta$$

and
$$a = \text{Slit width}$$

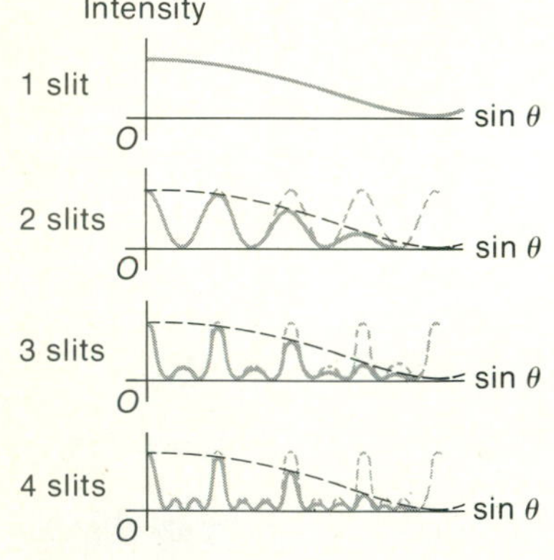

Figure 31–15
The diffraction and interference patterns for 1, 2, 3, and 4 slits. The resultant intensity is shown with the solid line.

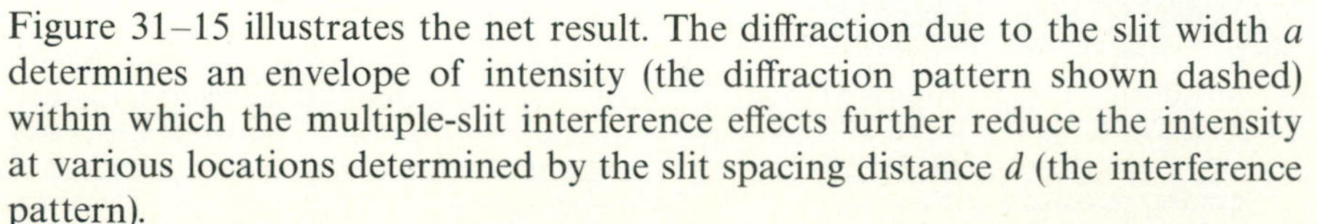

Figure 31–15 illustrates the net result. The diffraction due to the slit width a determines an envelope of intensity (the diffraction pattern shown dashed) within which the multiple-slit interference effects further reduce the intensity at various locations determined by the slit spacing distance d (the interference pattern).

EXAMPLE 31–3

For the multiple-slit pattern depicted in Figure 31–15, find the ratio of the slit width a to the slit separation d.

SOLUTION

Note that the *first* diffraction minimum falls at the *fourth* interference maximum. From Equation (30–11), the *interference maxima* are given by

$$d \sin \theta = m\lambda$$

Rearranging and substituting $m = 4$:

$$\sin \theta = \frac{4\lambda}{d} \qquad \textbf{(31–12)}$$

Equation (31–10) gives the diffraction minima:

$$a \sin \theta = m\lambda$$

Rearranging and substituting $m = 1$:

$$\sin \theta = \frac{\lambda}{a} \qquad \textbf{(31–13)}$$

Combining Equations (31–12) and (31–13), we obtain

$$\frac{4\lambda}{d} = \frac{\lambda}{a}$$

or

$$\boxed{\frac{a}{d} = \frac{1}{4}}$$

An interesting corollary to this example is the pattern resulting from a grating in which the space between the slits is equal to the slit widths. In this case:

$$d = 2a$$

Substituting this value into Equation (30–11):

$$d \sin \theta = m\lambda$$

we obtain

$$a = \frac{m\lambda}{2 \sin \theta}$$

Using this expression for a in Equation (31–7):

$$\alpha = \left(\frac{\pi}{\lambda}\right) a \sin \theta$$

we obtain

$$\alpha = \frac{m\pi}{2}$$

for all wavelengths. Since from Equation (31–8):

$$I_\theta = I_0\left(\frac{\sin\alpha}{\alpha}\right)^2$$

we have

$$I_\theta = \frac{I_0}{\left(\frac{m\pi}{2}\right)^2}\sin^2\left(\frac{m\pi}{2}\right)$$

Thus, the sine-squared factor causes the intensity to be zero for all *even* values of m. That is, a diffraction minimum occurs at every alternate interference major maximum, beginning with the second maximum from the center of the pattern. By noting the "missing" orders in an interference pattern, one can thus determine the ratio of slit width to slit separation.

Until now we have concerned ourselves with a single wavelength λ. However, when observing light that includes several wavelengths, the spectrum is dispersed through a range of angles for each value of m. This spectral pattern is repeated for other orders of diffraction. That is, $m = 1$ corresponds to the *first order*, $m = 2$ corresponds to the *second order*, and so on.

EXAMPLE 31–4

A diffraction grating disperses white light so that the red wavelength $\lambda = 650$ nm appears in the second-order pattern at $\theta = 20°$. (a) Find the so-called *grating constant*, that is, the number of slits/centimeter. (b) Determine whether or not visible light of the third-order pattern appears at $\theta = 20°$.

SOLUTION

(a) Equation (30–11), which gives the multiple-slit interference maxima, is applicable.

$$d\sin\theta = m\lambda$$

Rearranging, we have

$$d = \frac{m\lambda}{\sin\theta}$$

Substituting the appropriate values:

$$d = \frac{(2)(650\text{ nm})}{(\sin 20°)}$$

$$d = 3800\text{ nm}$$

The number of slits/centimeter ($\mathcal{N}$) becomes

$$\mathcal{N} = \frac{1}{d}$$

$$\mathcal{N} = \frac{1\text{ slit}}{3800\times10^{-9}\text{ m}}\underbrace{\left(\frac{1\text{ m}}{100\text{ cm}}\right)}_{\text{Conversion ratio}}$$

$$\mathcal{N} = \boxed{2630\text{ slits/cm}}$$

(b) Again, Equation (30–11) is appropriate:

$$d\sin\theta = m\lambda$$

However, in this case we solve for λ and substitute the given values.

$$\lambda = \frac{d \sin \theta}{m}$$

$$\lambda = \frac{(3800 \text{ nm})(\sin 20^\circ)}{3}$$

$$\lambda = \boxed{433 \text{ nm}}$$

A wavelength of 433 nm is a faintly visible violet. Thus, for this grating visible portions of the second and third order do overlap.

Diffraction gratings are often used rather than prisms in the analysis of spectra, because gratings are capable of spreading the spectrum over a wider range of angles, enabling more precise measurements of λ to be made. The dispersion D expresses the ability of a grating or prism to spread a range of wavelengths $d\lambda$ over an angular spread of $d\theta$.

DISPERSION

$$D \equiv \frac{d\theta}{d\lambda} \tag{31–14}$$

The greater the dispersion, the farther apart will be the angular separation of two wavelengths that are close together. As shown in Table 31–1, the dispersion is greater than that produced by a prism and greater for larger values of the order m.

Since the relationship between θ and λ for a diffraction grating is $d \sin \theta = m\lambda$, we have

$$\sin \theta = \left(\frac{m}{d}\right)\lambda$$

The dispersion D is obtained by differentiating θ with respect to λ:

$$\cos \theta \frac{d\theta}{d\lambda} = \frac{m}{d}$$

and rearranging:

$$\frac{d\theta}{d\lambda} = \frac{m}{d \cos \theta} \tag{31–15}$$

TABLE 31–1

Comparison of the Dispersion of a Prism and a Diffraction Grating

			4500 rulings/cm grating			
λ (nm)	60° flint-glass prism		$m = 1$		$m = 3$	
670.8	50.51°		17.57°		64.90°	
656.3	50.61°		17.18°		62.38°	
589.3	51.17°		15.38°		52.71°	
546.1	51.64°	$\Delta\theta = 4.32°$	14.26°	$\Delta\theta = 7.08°$	47.50°	$\Delta\theta = 31.78°$
486.1	52.58°		12.64°		41.01°	
404.7	54.83°		10.49°		33.12°	

DISPERSION OF A GRATING

$$D = \frac{m}{d \cos \theta} \tag{31–16}$$

The units of D are radians/meter (or, more commonly, degrees/nanometer).

While dispersion is an important consideration in diffraction-grating design, the ability of a grating to perceptibly separate, or *resolve*, two spectral lines of nearly the same wavelength is also important. The **resolving power** R of a diffraction grating or prism is defined as:

RESOLVING POWER

$$R \equiv \frac{\lambda}{\Delta\lambda} \tag{31–17}$$

where λ is the average wavelength of two spectral lines with a wavelength difference of $\Delta\lambda$. Referring to Figures 31–15 or 31–16, we see that the principal maxima become sharper as the total number of slits N increases. The relationship between the sharpness of a principle maximum (other than the central maximum) and the total number of slits N can be shown to be

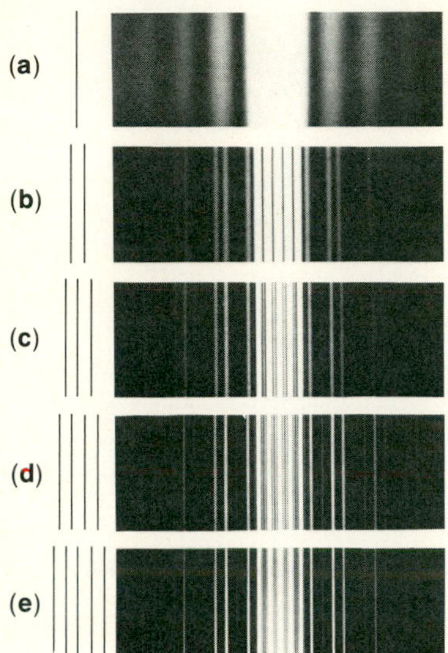

Figure 31–16
(a) The Fraunhofer diffraction patterns for a single slit. Multiple-slit patterns are shown in (b) through (e) for the slit systems shown at left.

$$\theta_R = \frac{\lambda}{Nd \cos \theta} \tag{31–18}$$

The "sharpness" of the central peak is measured by θ_R, the angular separation between the center of the peak and the *first* minimum. The symbol d is the separation of the slits, and θ is the diffraction angle of the peak. The notation θ_R we adopt for this angle comes from its use in the criterion proposed by Lord Rayleigh for the minimum resolvable angular separation for two overlapping diffraction patterns. According to **Rayleigh's criterion,**[3] two closely-spaced patterns are acceptably "resolved" (that is, one can decide they are definitely due to *two* point sources instead of *one*) if:

RAYLEIGH'S CRITERION FOR MINIMUM RESOLUTION

The peak of one diffraction pattern is located at the *first* minimum of the other pattern.

Figure 31–17 illustrates the criterion. For a diffraction grating, the angle θ_R is exceedingly small, corresponding to a wavelength difference $\Delta\lambda$, as given by Equation (31–15):

$$\theta_R = \frac{m}{d \cos \theta} \Delta\lambda \tag{31–19}$$

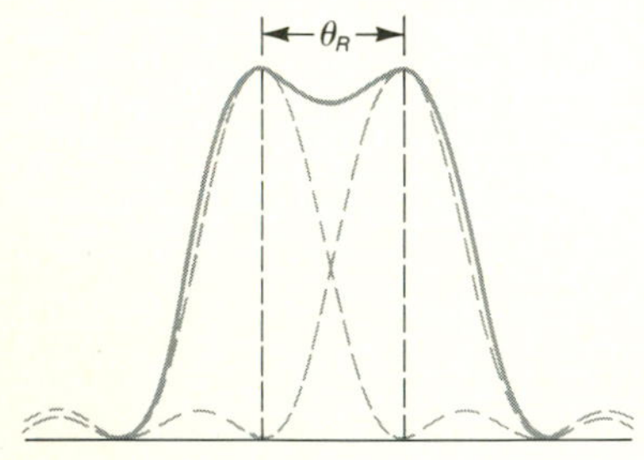

Figure 31–17
The Rayleigh criterion for the resolution of two diffraction maxima. The peak of one pattern falls on the first minimum of the other pattern.

Combining Equations (31–18) and (31–19), we have

$$\frac{m}{d \cos \theta} \Delta\lambda = \frac{\lambda}{Nd \cos \theta}$$

from which we obtain for the *resolving power of a grating* $R = \lambda/\Delta\lambda$:

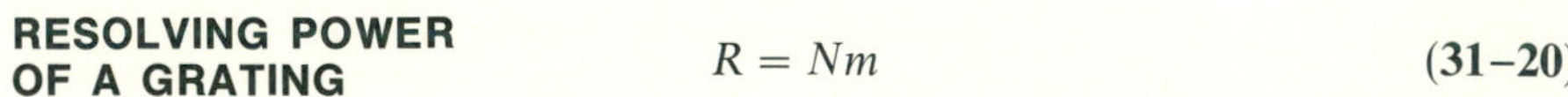

RESOLVING POWER OF A GRATING

$$R = Nm \tag{31–20}$$

where N is the total number of slits in the grating and m is the order.

[3] Rayleigh's criterion is arbitrary. Sometimes other criteria are defined in slightly different form.

TABLE 31–2

The First-Order Spectrum ($m = 1$) for Light near Wavelength $\lambda = 550$ nm

Grating	N_t	d (nm)	θ	R	D (10^{-2} degrees/nm)
A	10 000	2500	12.7°	10 000	2.35
B	20 000	2500	12.7°	20 000	2.35
C	10 000	1500	21.5°	10 000	4.11

The distinction between *dispersion* and *resolving power* becomes obvious from Table 31–2, which compares data for three different gratings. As illustrated in Figure 31–18, gratings A and B have the same dispersion D (they separate two given wavelengths by the same angular distance), while gratings A and C have the same resolving power R (the ability to distinguish two wavelengths very close together, limited only by the *width* of each diffraction peak). Note that grating B has the highest resolving power, while grating C has the highest dispersion.

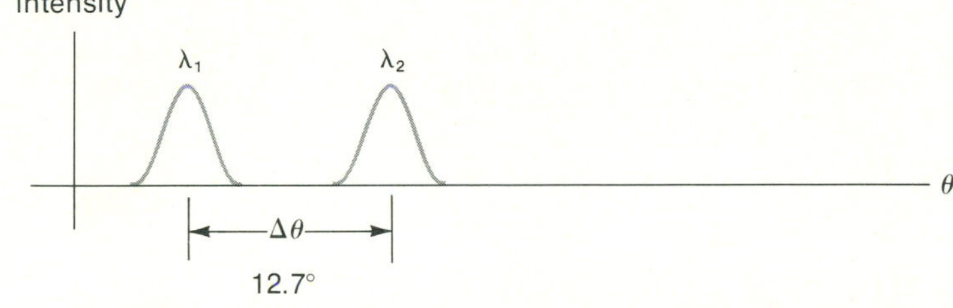

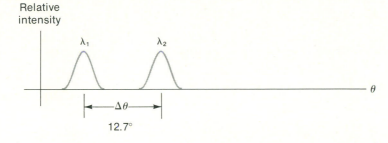

Grating B

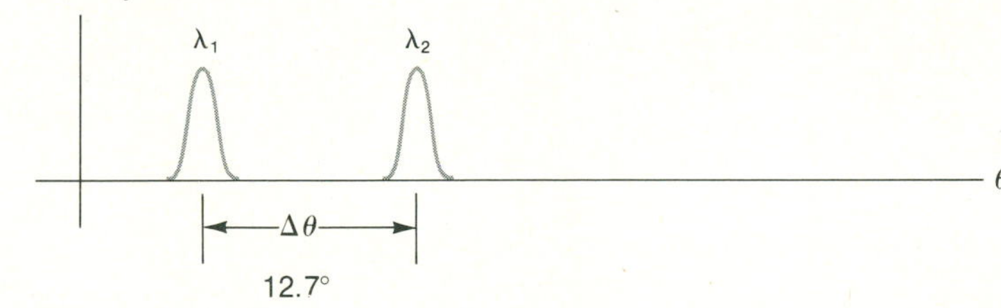

Grating C

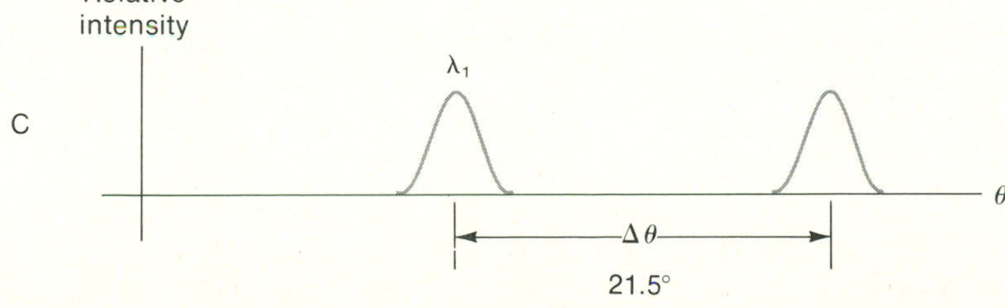

Figure 31–18
The relative intensity patterns produced by the gratings of Table 31–2 for two wavelengths, λ_1 and λ_2, near 550 nm.

EXAMPLE 31–5

A sodium vapor lamp emits a yellow light corresponding to two wavelengths, 589.00 nm and 589.59 nm. How many rulings must a grating have so that it will barely resolve this sodium doublet in the first order?

SOLUTION

The required resolving power is given by Equation (31–17):

$$R \equiv \frac{\lambda}{\Delta\lambda}$$

where

$$\lambda = \frac{589.00\text{ nm} + 589.59\text{ nm}}{2}$$

$$\lambda = 589.30\text{ nm}$$

and

$$\Delta\lambda = 589.59\text{ nm} - 589.00\text{ nm}$$

$$\Delta\lambda = 0.59\text{ nm}$$

Thus:

$$R = \frac{589.3\text{ nm}}{0.59\text{ nm}}$$

$$R = 1000$$

The resolving power for a diffraction grating is [Equation (31–20)]:

$$R = Nm$$

For the first order ($m = 1$), the number of rulings is equal to the resolving power:

$$N = \boxed{1000\text{ rulings}}$$

Since a typical diffraction grating has approximately 5000 rulings per centimeter, the sodium doublet can be easily resolved without resorting to either very fine rulings or large gratings.

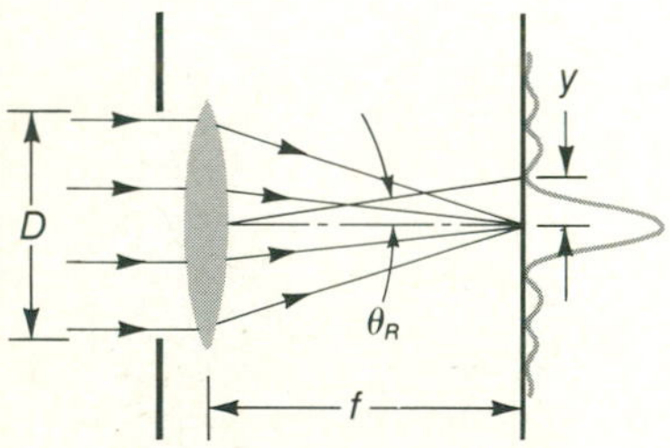

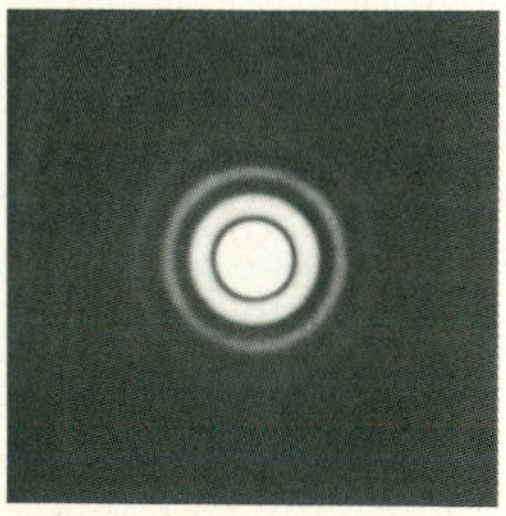

31.3 Fraunhofer Diffraction by a Circular Aperture

Diffraction effects impose a serious limitation on the resolving power of all optical instruments, such as telescopes and microscopes, and for all portions of the electromagnetic spectrum. Whether observing a star system through a telescope or a living cell through a microscope, the primary aim is to see fine details. That is, light emitted from two adjacent parts of the object should, after passing through an optical system, produce sharp, distinct images of those parts. Since the light from the two parts passes through the aperture of the optical system, it is always diffracted, producing overlapping diffraction patterns that tend to obscure the image.

In this section we will investigate the very practical problem of the diffraction produced by a telescope objective lens. (Similar considerations will apply to microscopes.) The light rays entering the aperture of the telescope are parallel and focus onto a photographic plate (or a focal plane that is viewed through an eyepiece). Thus, a distant point source of light produces a Fraunhofer diffraction pattern at the focus of the telescope objective.

Figure 31–19 illustrates the pattern produced by a point source. The mathematical technique used to calculate the radial intensity distribution of

Figure 31–19
The Fraunhofer diffraction pattern of a distant point source produced by a circular aperture. The size of the pattern is always larger than the diameter D of the hole. Also, the smaller the hole, the larger the pattern. The location of the first diffraction minimum determines the minimum angle of resolution θ_R.

the circular pattern is similar to that used for the single-slit pattern. Although the cylindrical geometry makes the derivation more complicated, the result is also similar to that for the minima in a single-slit pattern ($a \sin\theta = m\lambda$). For a *circular* aperture of diameter D, the minima are located by

CIRCULAR APERTURE FRAUNHOFER DIFFRACTION MINIMA

$$D \sin\theta = p_m\lambda \tag{31–21}$$

where the subscript m indicates the number of the minimum from the center of the pattern. Specifically, $p_1 = 1.220$, $p_2 = 2.233$, $p_3 = 3.238$, $p_4 = 4.241$, and $p_5 = 5.243$. With a very bright source, 20 or 30 circular rings can be observed, though photographically they are difficult to reproduce without greatly over-exposing the center of the pattern. The central spot is called the *Airy disk* (after Sir George Airy who first analyzed the diffraction by a circular aperture in 1835). The Airy disk contains 84% of the light passing through the aperture, while 91% is contained within the central spot plus the first diffraction ring.

The first minimum ($m = 1$) is at an angle θ away from the center; this angle is usually so small we can substitute $\sin\theta \approx \theta$. Thus, when two closely-spaced point sources are observed as in Figure 31–20(b), *Rayleigh's criterion* for barely resolving the image as that of *two* point sources (instead of one) is met when the peak of one diffraction pattern falls on the first minimum of the other pattern. Under these conditions, the minimum angular separation θ_R of the two sources is [from Equation (31–21)]:

MINIMUM ANGLE OF RESOLUTION θ_R FOR A CIRCULAR APERTURE (Rayleigh's criterion)

$$\theta_R = \frac{(1.22)\lambda}{D} \tag{31–22}$$

where D is the diameter of the circular lens or aperture and λ is the wavelength.

Although the Rayleigh criterion is arbitrary, it does represent a simple and precisely defined rule. Applied to the unaided human eye, which has an average iris opening of a few millimeters, this rule predicts a resolving power of roughly 20 seconds of arc. In practice, however, the resolving power of the average eye is somewhat worse than this, due to the finite size of the receptors in the retina. On the other hand, careful analysis of photographic images can routinely do somewhat better than the Rayleigh limit.

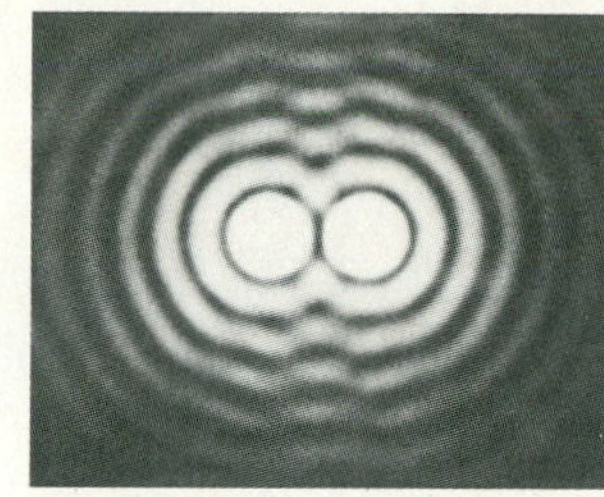

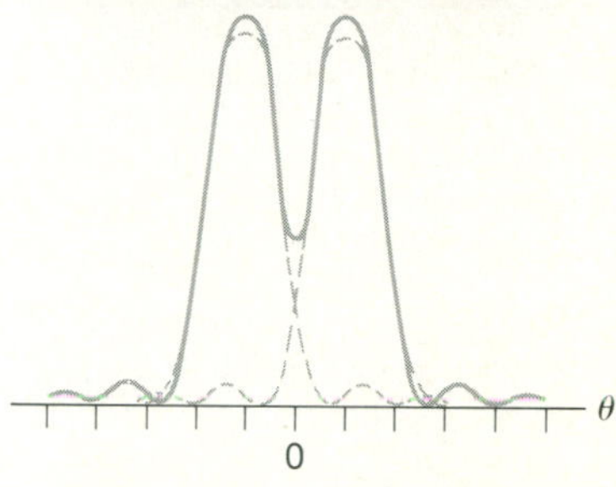

(a) The angular separation of the two patterns is clearly large enough to reveal two sources.

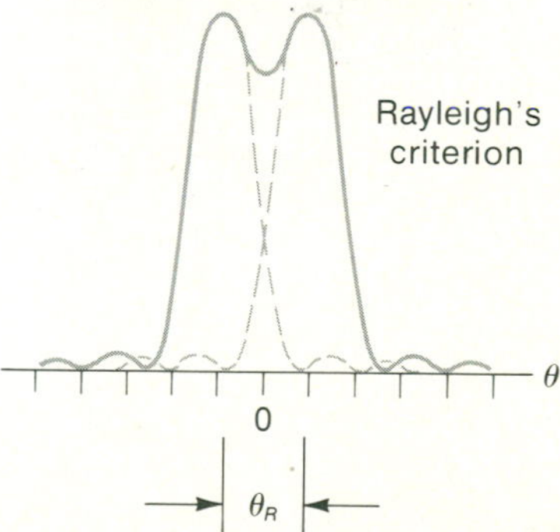

(b) The patterns overlap according to the Rayleigh criterion. The resulting pattern is barely discernible as two overlapping diffraction patterns.

Figure 31–20
Superimposed Fraunhofer diffraction patterns associated with the images of two distant, incoherent point sources.

EXAMPLE 31–6

The world's largest operating refracting telescope is the University of Chicago's Yerkes telescope. The objective lens of the telescope is 1.02 m (40 in) in diameter and has a focal length of about 18.9 m. The image of a distant star produced at the focal point of the objective lens is a Fraunhofer diffraction pattern. Find the total width of the central peak (Airy disk) of the diffraction pattern. Assume an average wavelength of 500 nm for the light emitted by the star.

SOLUTION

The total angular width of the central diffraction peak is $2\theta_R$, where θ_R is given by Equation (31–22):

$$\theta_R = \frac{(1.22)\lambda}{D}$$

Substituting numerical values:

$$\theta_R = \frac{(1.22)(500 \times 10^{-9}\ \text{m})}{(1.02\ \text{m})}$$

$$\theta_R = 5.98 \times 10^{-7}\ \text{rad}$$

The total width is twice this value, or

$$2\theta_R = 1.20 \times 10^{-6}\ \text{rad}$$

The linear width y corresponding to this angular width is given by

$$y = (f)(2\theta_R)$$

where f is the focal length of the objective lens (Figure 31–19). Substituting numerical values:

$$y = (18.9\ \text{m})(1.20 \times 10^{-6}\ \text{rad})$$

$$y = \boxed{2.27 \times 10^{-5}\ \text{m}}$$

(a) The world's largest radio telescope, prior to 1980, was the 305-m-diameter, fixed-dish reflector at Arecibo, Puerto Rico. Its movable overhead antenna near the focus can collect signals within ±20° from the vertical. Its 20-acre surface was greater than the combined area of all other telescopes ever built.

(b) The Very Large Array (VLA) system in New Mexico became operational in 1981, employing 27 steerable dishes, each 26 m in diameter, arranged in a movable array in the shape of a "Y" extending over 27-km baseline. The signals will be simultaneously analyzed with an interferometric technique known as *aperture synthesis* by a large computer at the center of the "Y." The angular resolution depends on the baseline distance and will be comparable to the 1-arcsecond resolution of visible-light observations from large telescopes.

Figure 31–21

Radio telescopes. The Arecibo Observatory (a) is part of the National Astronomy and Ionosphere Center which is operated by Cornell University under contract with the National Science Foundation.

While the value obtained for the width in Example 31–6 is small, the central maximum of a pattern associated with a neighboring star may overlap this pattern enough to make the two images indistinguishable. The minimum angular separation of two stars resolvable by the Yerkes telescope is thus about 6×10^{-7} rad. Because of the difficulty of fabricating larger lenses and supporting them at their edges without distortion due to gravity, the Yerkes 40-in-diameter lens will probably remain the world's largest *refracting* telescope.

Larger aperture *reflecting* telescopes can be supported from the rear of the mirror at various points to prevent sagging under gravity; they achieve greater resolution and create brighter images because of their greater light-gathering ability. Currently, the largest reflector for visible light is the 6-m reflector atop Mt. Semirodriki in the Caucasus, near Zelenchukskaya, USSR. Unfortunately, the limiting angular resolution is not the only consideration in operating visible-light telescopes. The blurring of the image by atmospheric turbulence is also a problem, particularly with larger telescopes, so a good site with steady air is of crucial importance.

Radio telescopes are not affected by air turbulence and can operate day and night (see Figure 31–21). Signals from several radio telescopes many kilometers apart can be combined with interferometric methods. The signal from an astronomical source will arrive at each antenna with slightly different phase shifts and time delays, allowing an accurate determination of the direction of the source. Pairs of radio telescopes 10 000 km apart in California and Australia have been used in this fashion, achieving an angular resolution of a few ten-thousandths of a second of arc, far surpassing the resolution of optical telescopes. The method relies on very accurate international time standards to correlate the signals.

EXAMPLE 31–7

(a) Find the resolving power of a reflecting telescope with a diameter of 6 m. Assume a wavelength of 500 nm. (b) Find the diameter of a radio telescope with the same resolving power when receiving radio waves with a 21 cm wavelength.

SOLUTION

(a) The resolving power defined in accordance with the Rayleigh criterion is given by Equation 31–22:

$$\theta_R = \frac{(1.22)\lambda}{D}$$

Substituting the appropriate values for the reflecting telescope, we obtain

$$\theta_R = \frac{(1.22)(500 \times 10^{-9}\ \text{m})}{(6\ \text{m})}$$

$$\theta_R = \boxed{1.02 \times 10^{-7}\ \text{rad}}$$

(b) Under ideal conditions, this telescope will resolve two objects with an angular operation of only about 10^{-7} rad, or 0.02 s of arc.

Using the same criteria for a radio telescope:

$$\theta_R = \frac{(1.22)\lambda}{D}$$

we find that for equal resolving power, the ratio λ/D must be the same for both telescopes:

$$\frac{\lambda_1}{D_1} = \frac{\lambda_2}{D_2}$$

where the subscript 1 corresponds to the radio telescope and 2 corresponds to the optical telescope. Rearranging:

$$D_1 = \left(\frac{\lambda_1}{\lambda_2}\right) D_2$$

Using the given values, we obtain

$$D_1 = \left(\frac{0.21\ \text{m}}{500 \times 10^{-9}\ \text{m}}\right)(6.00\ \text{m})$$

$$D_1 = \boxed{2.52 \times 10^{6}\ \text{m}}$$

Obviously a single reflecting telescope of this size is out of the question. However, as mentioned earlier, by using two separate telescopes as an interferometer with a baseline this long, the equivalent resolution is achieved (but with far less light-gathering ability).

31.4 X-Ray Diffraction

In 1912, the German physicist Max von Laue (1879–1960) first suggested that a crystalline array of atoms might act as a *three-dimensional* "diffraction grating" for x-rays of wavelengths comparable to the atomic spacing in the crystal ($\sim$0.1 nm). The incoming radiation would be absorbed by electrons, and according to the Huygens theory, each electron would reradiate expanding wavelets in a process called *scattering*. Thus, just as the slits in a diffraction grating act as coherent sources of radiation, the three-dimensional array of scattering centers would act as coherent sources. Depending on the location of the scattering centers, there would be fixed phase differences between the wavelets coming from the various centers. (Because electrons are concentrated near the atoms, each *atom* is effectively a scattering center.) In certain directions, the scattered waves will be in phase, producing a high intensity of scattered radiation in that direction. For certain other directions, the waves will be out of phase, resulting in destructive interference and no scattering.

Consider the line of scattering centers in Figure 31–22. For radiation incident at an angle θ_1 as shown, the scattered waves will be in phase if the two distances *AB* and *CD* are equal. By symmetry, this happens when the scattering

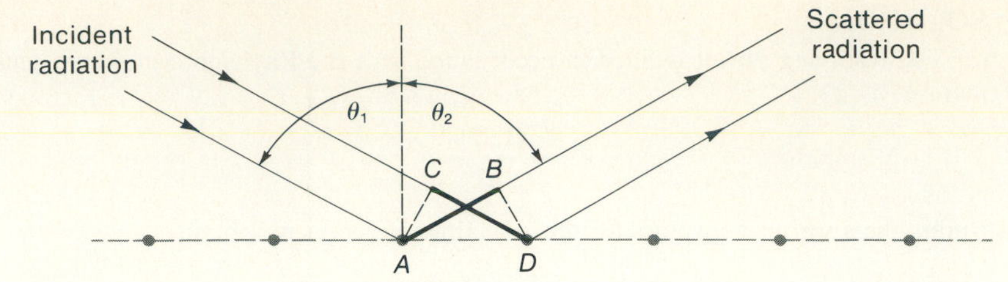

Figure 31–22
A line of equally spaced scattering centers.

angle θ_2 equals the incident angle θ_1. Sir William Bragg[4] noted the similarity to optical reflection ("the angle of incidence equals the angle of reflection"), so he proposed an alternative explanation involving "Bragg reflection" from atomic planes. Though this "reflection" is an incorrect picture of the scattering process, it is a simple and useful way of thinking about the phenomenon. Adopting this simpler view, look at Figure 31–23, wherein the incident radiation strikes a cubical array of atoms, which form atomic planes spaced a distance d apart. Consider rays ① and ②. The lines aA and aC are drawn perpendicular to the incident and reflected ray ②, so that the distance ABC is the extra path length traveled by ray ②. Thus, for incident radiation at an angle ϕ with respect to the atomic planes (not to the *normal to the plane*, as in optical reflection), the extra path length is

$$\text{Path length difference} = 2(d \sin \phi) \qquad \textbf{(31–23)}$$

When this path difference is an *integral* number of wavelengths $m\lambda$, the scattered rays will be *in phase*. (Similar relations also apply to other rays, such as ③, scattered from deeper regions.) The relation for constructive interference is called the *Bragg scattering condition.*

BRAGG SCATTERING CONDITION

$$m\lambda = 2d \sin \phi \qquad \textbf{(31–24)}$$

where $m = 1, 2, 3, \ldots$ (the *order* of scattering)
$\phi =$ the *glancing* angle between the incident ray and the *plane* (not between the ray and the normal, as in optical reflection)
$d =$ atomic plane spacing

The scattered radiation is very sharply "peaked" at these angles. As for a plane diffraction grating with a great many slits, the three-dimensional "grating"

[4] The British father–son team W.H. and W.L. Bragg received the Nobel Prize in 1915 for their studies of crystal structures by x-ray diffraction; this was just a year after von Laue received the Nobel Prize for his basic discovery of the diffraction of x-rays.

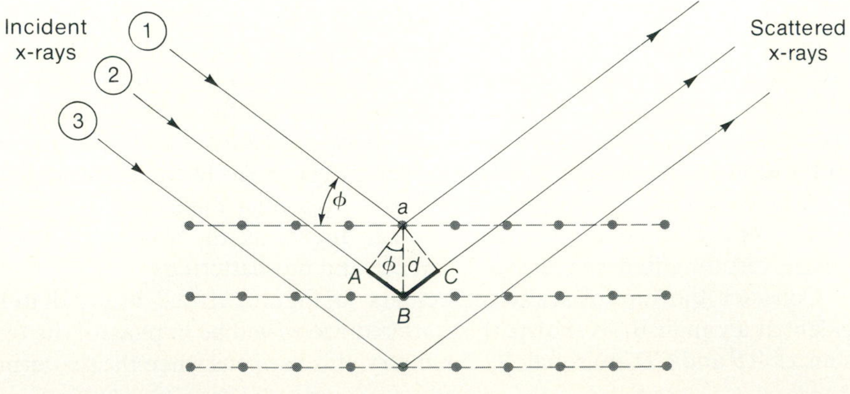

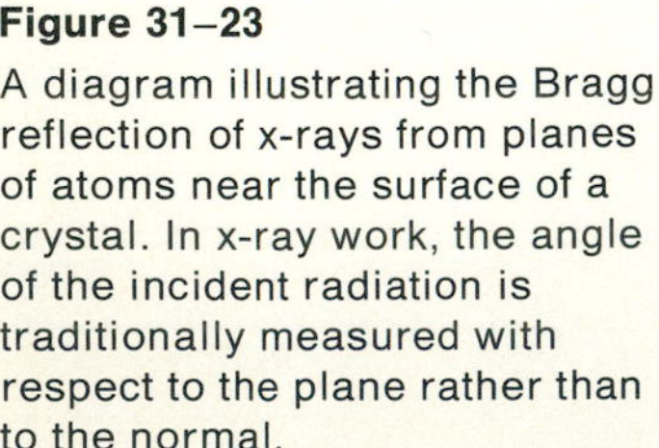
Figure 31–23
A diagram illustrating the Bragg reflection of x-rays from planes of atoms near the surface of a crystal. In x-ray work, the angle of the incident radiation is traditionally measured with respect to the plane rather than to the normal.

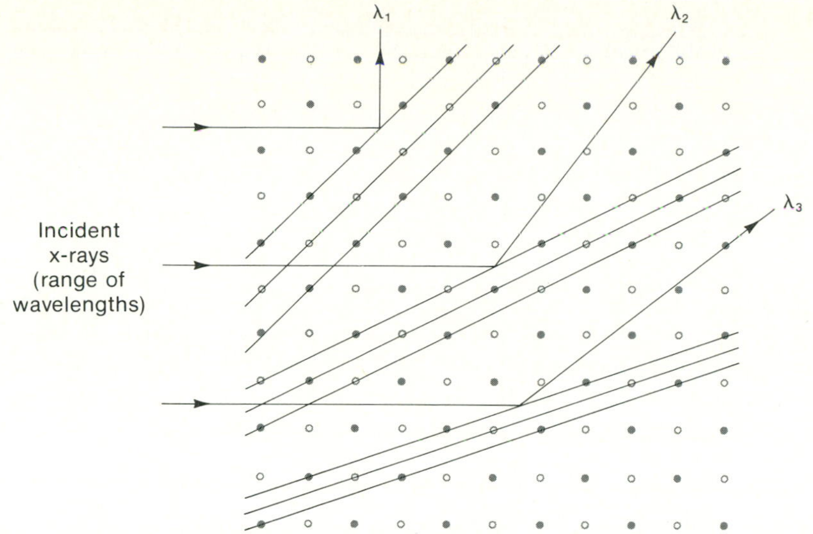

Figure 31–24
The lattice of atoms in a crystal may be grouped into parallel planes at various angles, each with its own spacing. The wavelength that matches the Bragg condition will be reflected from its corresponding set of planes.

has an enormous number of scattering centers, which causes the major maxima to become extremely narrow and intense, while suppressing all minor maxima. If a *continuous spectrum* of radiation containing all wavelengths (called "white" radiation) is incident, we may also consider the simultaneous Bragg reflections from other planes in the crystal (Figure 31–24). The various sets of parallel planes will have different spacings between them, depending on the geometric positions of the atoms. Since the incident radiation contains all wavelengths, there will be some radiation at the correct values of $m\lambda$ that match all the various Bragg conditions. A photograph of the scattered spots (Figure 31–25) is called a

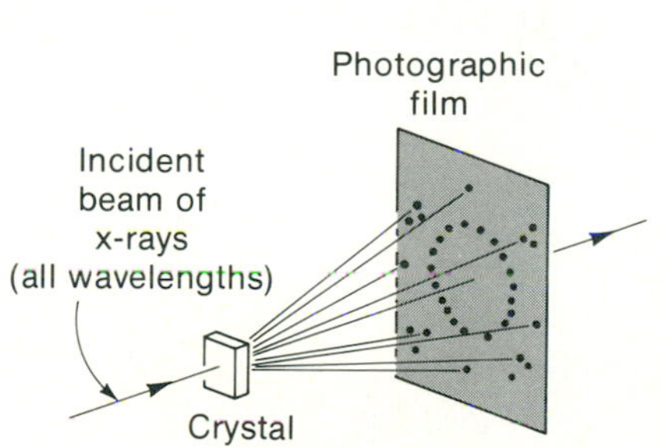

(a) Experimental arrangement for making a Laue-spot diffraction pattern.

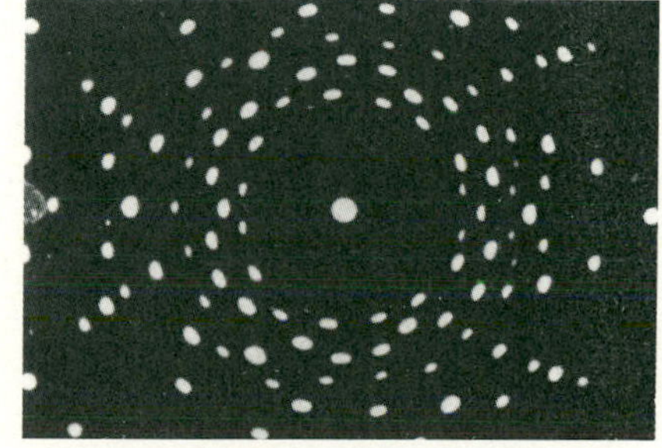

(b) A Laue diffraction pattern for a single quartz crystal.

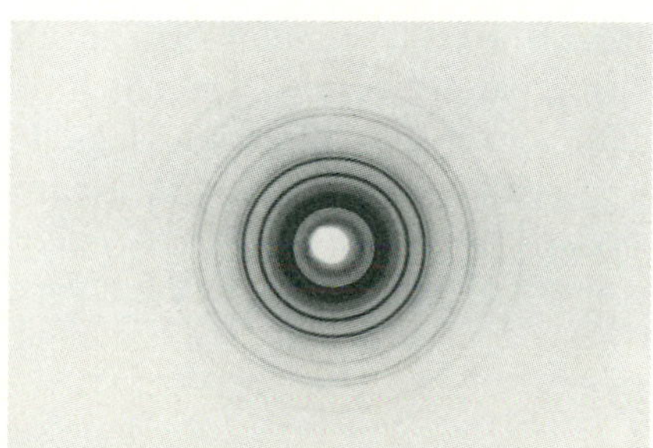

(c) A *powder pattern* of the diffraction of x-rays from polycrystalline aluminum.

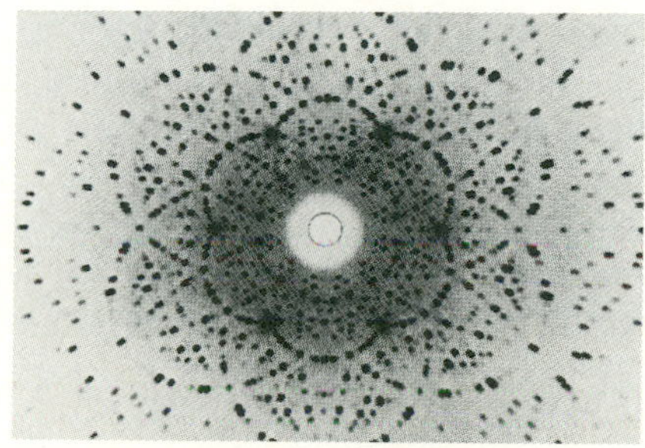

(d) Complex cyrstals produce strikingly beautiful Laue-spot patterns.

Figure 31–25
X-ray diffraction patterns.

Laue diffraction pattern. The positions of such spots correlate with the configuration of the atoms in the specimen. So for a crystal whose atomic structure is unknown, we can work backwards from the Laue pattern to figure out the exact atomic configuration. If single crystals large enough to use are not available, a *polycrystalline* target with microscopic crystals at all orientations can be used with *monochromatic* x-rays. Here, by chance, some crystals will be oriented properly to reflect the monochromatic x-rays from important sets of planes, forming a pattern of concentric circles known as a *Debye-Scherrer diffraction pattern* (also called a *powder pattern*), shown in Figure 31–25(c).

X-ray diffraction studies are a most powerful method of determining atomic arrangements in crystals as well as in large complicated molecules of interest in biology. For example, in 1962 James Watson, Francis Crick, and Maurice Wilkins[5] received the Nobel Prize in biology for discovering the double-helix structure of DNA using x-ray diffraction methods.

31.5 Fresnel Diffraction

Certain diffraction effects were known as early as 1650, when an experimenter named Professor Grimaldi, at Bologna, Italy, observed that upon using a point source of light, a thin rod cast an unexpected shadow that had light and dark bands. However, the importance of this observation was not appreciated at the time and did not lead to any speculation of a possible wave nature for light. Robert Hooke later suggested a wave theory in an attempt to explain the colored patterns reflected from thin films, but he did not hit upon the idea of interference. At the same time, Newton, at age twenty-three, began his famous prism experiments and concluded that light was a mixture of colored particles, or corpuscles, which produced the different colors. Huygens proposed a more complete wave theory, and the battle lines of the wave-particle controversy became clearly drawn. Newton's great prestige seemed to stifle belief in waves, and it was not until the beginning of the nineteenth century that Thomas Young proposed a carefully-worked-out wave theory to the Royal Society. This theory included a new idea: the concept that waves could *interfere*. Young was sharply attacked by Newton's supporters, and the wave-particle argument became especially impassioned. Unaware of Young's ideas, thirteen years earlier the French physicist Augustin Fresnel (1788–1827) revived interest in a wave theory by combining the new (to him) ideas of interference with the Huygens' wavelets approach. Two cases studied theoretically by Fresnel were the diffraction pattern of a circular opening and that of a circular obstacle.

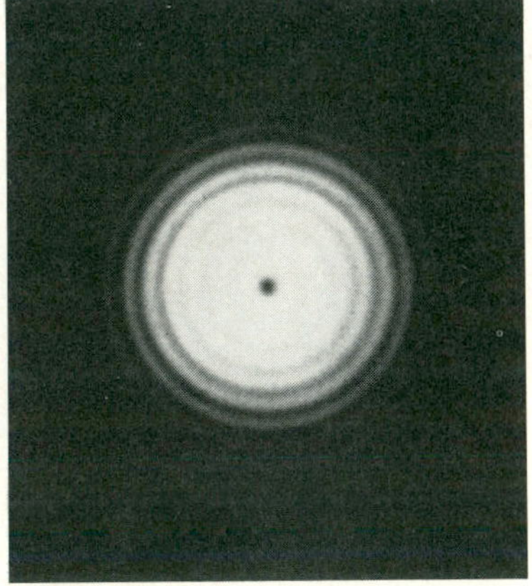

Figure 31–26
A Fresnel diffraction pattern of a circular aperture.

Circular Aperture

If parallel, monochromatic light passes through a small circular hole in an opaque plate and falls on a screen, a surprising pattern results. The patch of light will be larger than the hole and contain concentric diffraction rings. It may even have a dark spot in the center of the patch, as illustrated in Figure 31–26. A dark spot in the center is unexpected, since one would anticipate that the straight-through direction from the center of the opening to the screen would be bright. As the screen is gradually moved farther away, the center of

[5] For a fascinating story of a scientific quest, see James D. Watson, *The Double Helix*, Atheneum Press, New York, 1968. Also see Horace Judson, *The Eighth Day of Creation*, Simon & Schuster, 1979.

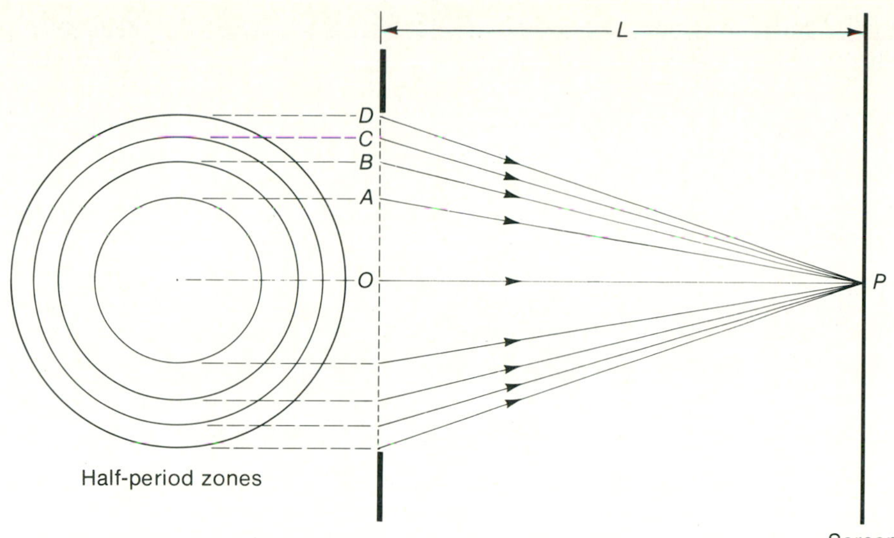

Figure 31–27
Half-period zones of a circular aperture. The diameter of the aperture relative to the screen distance L is greatly exaggerated.

the pattern changes back and forth from dark to bright. We can understand this surprising behavior on the basis of *half-period zones.*

In Figure 31–27, parallel light (plane wavefronts) passes through the circular aperture and proceeds to the screen. Let us examine the situation at the center of the light patch, point P on the screen, which (keeping Huygens' wavelets in mind) receives light from all parts of the wavefront in the aperture. To add the contributions from various parts of the wavefront, we divide the wavefront into circular zones, as shown. The radius $\overline{OA}$ of the central circular zone is such that

$$\overline{AP} = \overline{OP} + \frac{\lambda}{2}$$

The next zone includes the ring area between the central zone and the circle of radius $\overline{OB}$ such that

$$\overline{BP} = \overline{AP} + \frac{\lambda}{2}$$

A similar procedure is followed for the remaining ring zones. It is left as a problem to show that *all the zones have approximately the same area,* $(\pi\lambda L)$, where L is the aperture-to-screen distance.

The reason we chose these particular zones is that the contributions from any two adjacent zones are (on the average) *one-half wavelength out of phase* and therefore tend to cancel each other by destructive interference. If the resultant electric vector for the light from the first zone is $+\mathbf{E}_1$, the light from the second zone will be $-\mathbf{E}_2$, that from the third zone will be $+\mathbf{E}_3$, and so on, where each of the vectors has approximately the same amplitude. For our example, the geometry results in just four zones filling the aperture. Thus, when added together as in Figure 31–28, the net electric vector is zero, causing a dark spot at P. If the screen is slightly moved either toward or away from the opening so that an *odd* number of zones fills the aperture, then point P will become bright. Continuing to move still further in the same direction, the spot changes alternately from bright to dark repeatedly. The rest of the pattern off the axis is a series of concentric rings that increase (or decrease) in number as the screen is moved, collapsing inward toward the center or growing outward. An exact

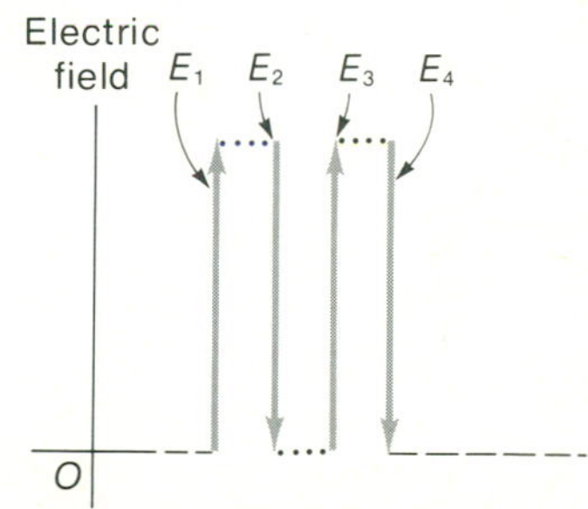

Figure 31–28
The average electric field vectors for the light from each of the four zones. They add together to zero. (For clarity, the vectors have been displaced sideways from one another.)

mathematical treatment for the off-axis pattern is complicated, but the result agrees with the observed patterns.

As we include more zones farther and farther from the center, the distance to P increases, causing the magnitude of the electric vector terms to decrease (the inverse-square law). An obliquity factor also enters in. If we take these into account, the modification is as shown in Figure 31–29. For an extremely large number of zones, the intensity at P approaches half that due to the first zone acting alone.

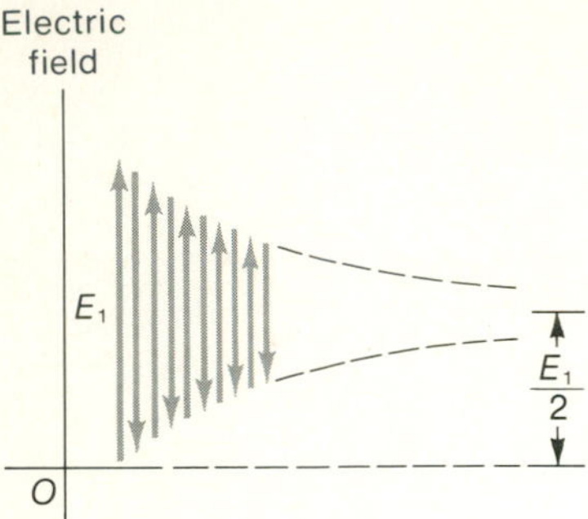

Figure 31–29
Adding together the electric field vectors for a large number of zones, the resultant amplitude approaches *half* that due to the first zone acting alone.

Circular Obstacle

One of the most startling of all Fresnel diffraction phenomena is the small bright spot in the center of the shadow of a small circular obstacle. Moreover, the bright spot exists for all wavelengths of light and does not depend on the obstacle-to-screen distance (see Figure 31–30). An explanation is simple if the ideas of half-period zones are used.

Replacing the circular aperture by an opaque circular obstacle, such as a small ball bearing (with free space around it), we see that now a few of the half-period zones near the center of the wavefront are obscured by the obstacle, while light from *all* the other zones reaches the point P. As shown in Figure 31–31, the vector sum of the average electric vectors from all the outer zones (working inward) will always be finite if we stop short of including some of the zones near the center. Thus, there will always be a bright (or fairly bright) spot at the center of the shadow.

There is an interesting anecdote about the spot. In response to a Prize Essay competition, Fresnel submitted his basic diffraction theories to the French Academy in 1818. Poisson, a member of the judging committee and a firm believer in the corpuscular theory of light, strongly ridiculed Fresnel's theories. To clinch his objections and hoping to deal a death blow to the wave theory of light, Poisson told a committee member, Arago, that (as Fresnel had not realized) the theory unrealistically predicted the existence of a bright spot at the center of the shadow of a circular obstruction—clearly an absurd prediction! Arago immediately tried the experiment and rediscovered the bright spot (which actually had been found 85 years earlier by Miraldi, but had been long forgotten). The bright spot's existence gave a big boost to Fresnel's wave theory, though Poisson stubbornly clung to the Newtonian particle model for light until his death 22 years later. Ironically, today the spot is sometimes called *Poisson's* bright spot, ignoring the true heroes in the story: Miraldi, Fresnel, and Arago. Fresnel ultimately did win first prize for his essay.

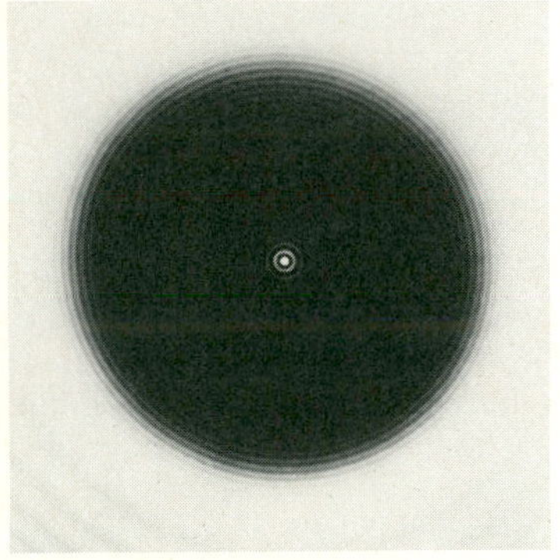

Figure 31–30
A diffraction pattern due to a penny reveals the Poisson bright spot at the center. To produce this pattern, a penny was placed mid-way between a monochromatic point source of light and a screen 40 m away.

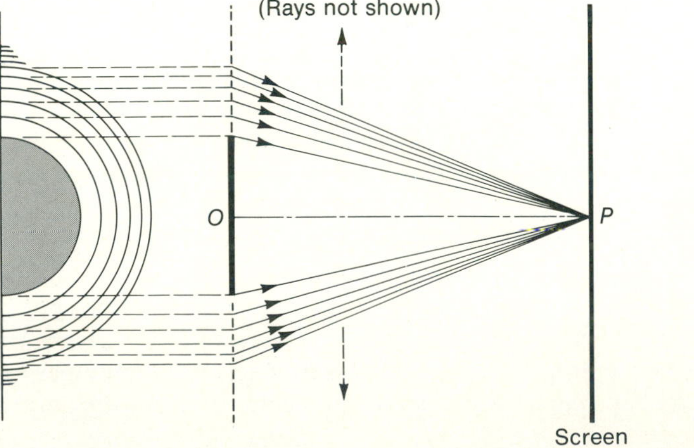

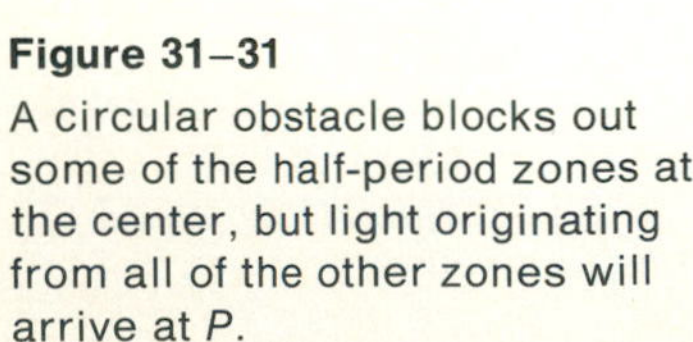
Figure 31–31
A circular obstacle blocks out some of the half-period zones at the center, but light originating from all of the other zones will arrive at P.

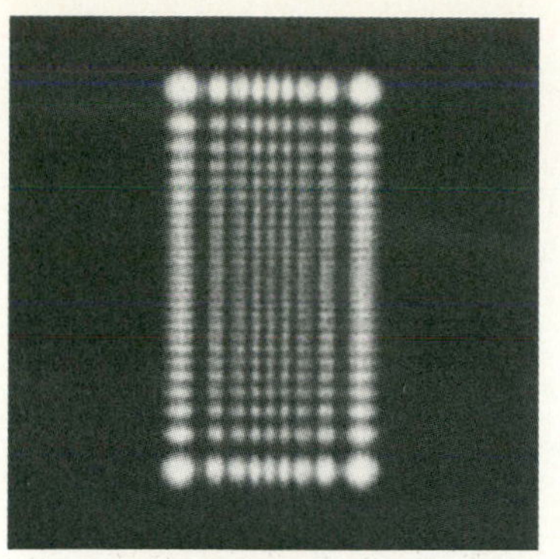

(**a**) A rectangular aperture.

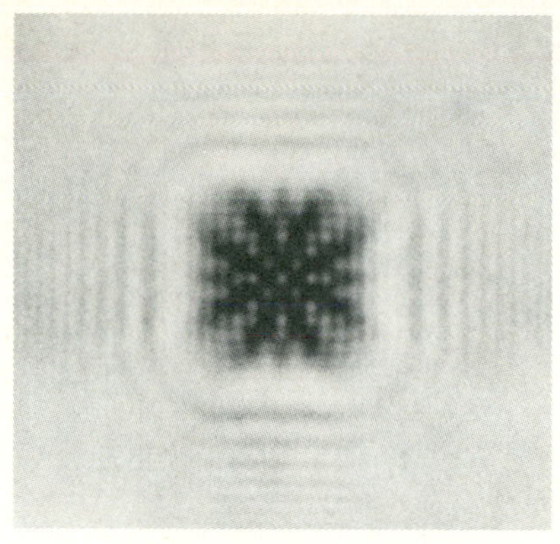

(**b**) An opaque square.

(**c**) The teeth of a saw.

(**d**) The diffraction pattern of an opaque ball bearing with a source consisting of an illuminated negative of a portrait of Woodrow Wilson. The opaque disk acts as a sort of lens, since for every point in the source there is a Poisson bright spot in the image.

(**e**) A triangular aperture.

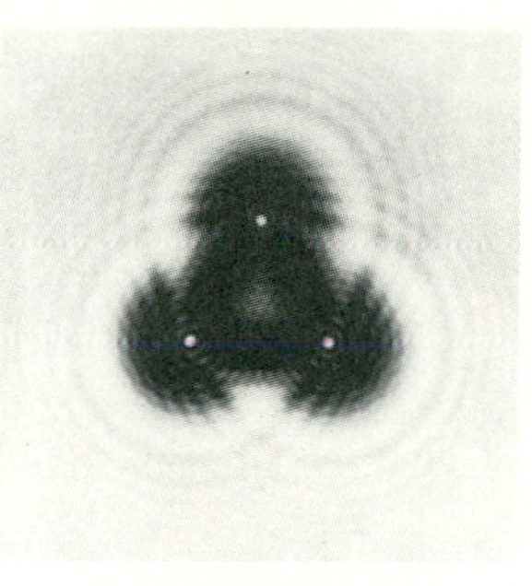

(**f**) Three opaque circular disks.

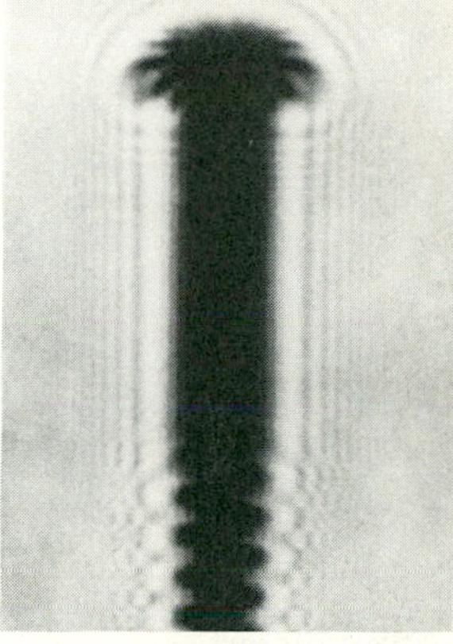

(**g**) The shadow of a screw.

Figure 31–32
Fresnel diffraction patterns.

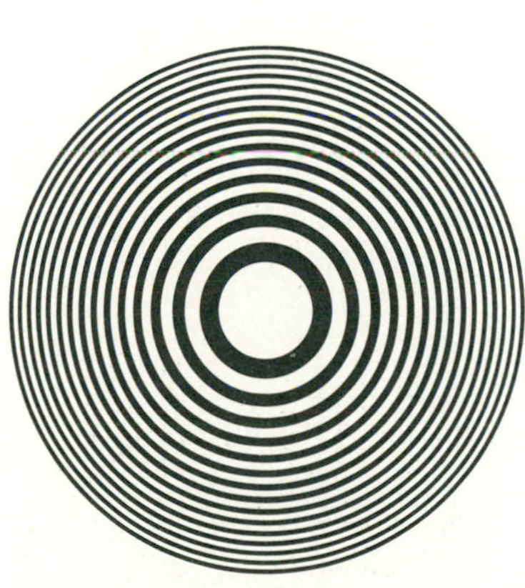

Figure 31–33
Alternate half-period zones are made opaque to form a *Fresnel zone plate*. The negative of this pattern (with the central zone opaque) is also a Fresnel zone plate. In most cases, the area of each zone ($\pi\lambda L$) is quite small. For example, for $L = 1$ m and $\lambda = 500$ nm, each area would be about 1.6 mm^2.

31.6 The Fresnel Zone Plate

As we pointed out in the previous section, light from two adjacent half-period zones tends to destructively interfere. This suggests that if we blocked out every other zone (either the odd ones or the even ones), there should be a great increase in the light that reaches point P on the screen: All the light from exposed zones would be essentially in phase. A transparent film in which alternate half-period zones have been blocked out is called a **Fresnel zone plate** (see Figure 31–33).

Isn't it interesting that by making half the area of an aperture opaque, the light transmitted to the center of the pattern on a screen increases? If, for example, an open aperture included 30 zones, the electric vector magnitude at point P would be approximately $(E_1/2)$. However, with a zone plate put into the aperture, blocking out half the area, the field at P is about $30E_1$, an increase in intensity of 3600. Thus, the zone plate acts as a sort of lens, diffracting incident parallel light so it converges to a real point image a distance L from the zone plate.

In addition to the real image formed on the screen, the diffraction effects also form a virtual image on the opposite side of the zone plate (on the side of the incident light). This virtual image can be seen by an observer located so as to receive the diffracted light (Figure 31–34).

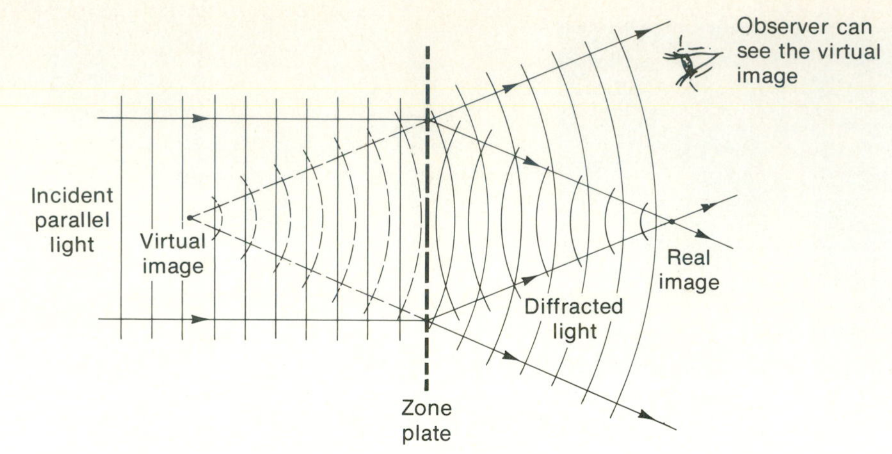

Figure 31–34
When parallel light is incident on a zone plate, real and virtual point images are formed on opposite sides of the plate. (Additional point images [not shown] for other orders of diffraction are also located along the central axis.)

In this brief discussion, we have omitted many of the details of zone plates. For example, in addition to the *first-order* (or *primary*) focal length, there are other point images formed at other locations along the central axis. However, the primary focal-length image is the brightest.

31.7 Holography

Everyone is by now familiar with holographic images, those fascinating ghost-like images that have full three-dimensional properties, formed without the use of lenses by passing coherent light through a flat sheet of film. In the old-fashioned stereoscopic image, a pair of almost identical pictures are viewed separately by each eye, producing the mental impression of a three-dimensional image *as viewed from a fixed perspective.* In contrast, when viewing a holographic image with the unaided eye by looking through the hologram as through a window, the image has a true three-dimensional property and one can easily see behind an object in the foreground by merely changing the viewer's position. In fact, 360° holograms in the form of a cylinder have been made, allowing the viewer to move completely around the image, seeing all sides.

The principles of *holography* (Greek, meaning "whole writing") were first presented in 1948 by Dennis Gabor, who was awarded the Nobel Prize in 1971 for his theories. The basic principle of holography can be simply explained using the idea of a zone plate. Consider Figure 31–35, in which two sets of

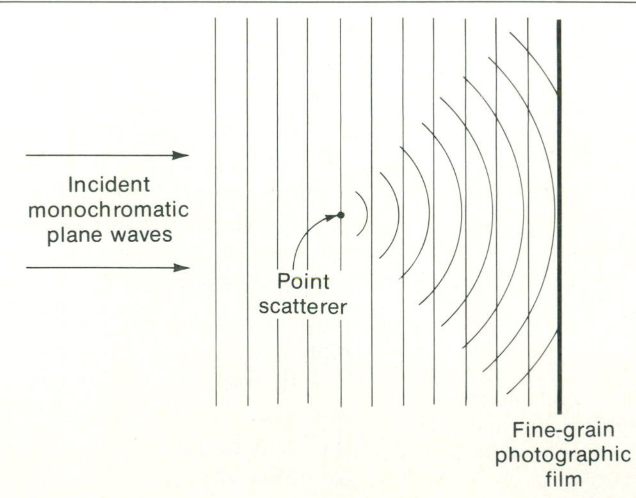

Figure 31–35
Plane monochromatic waves and the waves scattered coherently from the point object produce an interference pattern on the photographic film. When the film is developed, the resulting hologram is a series of light and dark concentric circles similar to a Fresnel zone plate.

monochromatic coherent waves impinge on a photographic film. One set, the *reference beam*, consists of plane waves. The other set is the light scattered from a point object. At the film, the interference of these two sets of coherent waves produces a pattern of light and dark rings. Upon development, the film will have opaque and clear regions, forming a *Gabor zone plate*, similar to a Fresnel-zone-plate pattern. If the developed film (called a *hologram*) is then illuminated with coherent monochromatic light, an observer located properly to receive the diffracted light coming through the hologram will see a virtual point image, as in Figure 31–34. (The real image is also present on the viewer's side of the hologram.)

Suppose that instead of a point object, a small *extended* object is used. Then each point of the object forms its own zone-plate pattern, which superimposes with patterns for all the other points. The resulting hologram is a very complicated array of fringes (Figure 31–36) that contains the full information about the zone-plate pattern *for each point on the object*. When the hologram is illuminated with coherent monochromatic light, the diffracted light reconstructs a full virtual image of the object.

In practice, to make about equal intensities for the two sets of waves and thus achieve maximum contrast in the interference pattern, the arrangement diagrammed in Figure 31–37 is often used.[6] Since no lenses are used at any stage

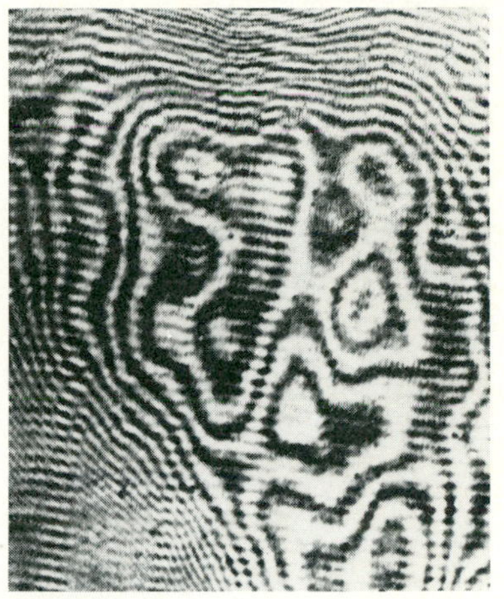

Figure 31–36
A magnified portion of a hologram, showing the complicated pattern of interference fringes.

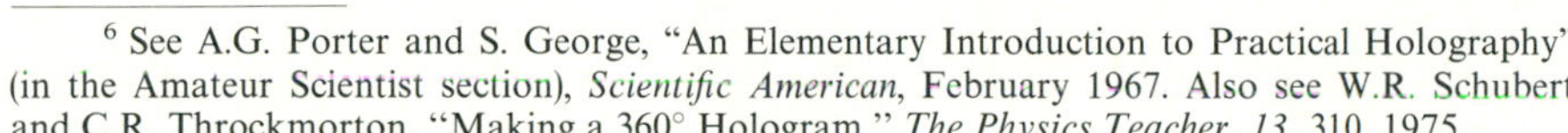

[6] See A.G. Porter and S. George, "An Elementary Introduction to Practical Holography" (in the Amateur Scientist section), *Scientific American*, February 1967. Also see W.R. Schubert and C.R. Throckmorton, "Making a 360° Hologram," *The Physics Teacher*, *13*, 310, 1975.

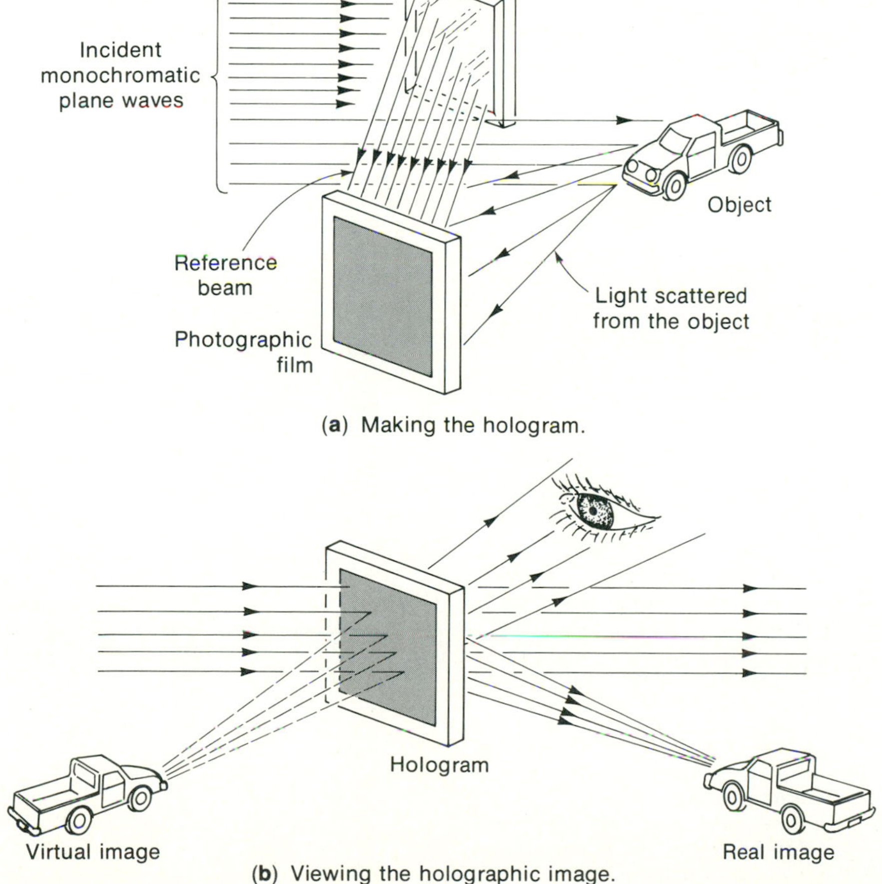

Figure 31–37
A common arrangement for making and viewing a hologram.

of the procedure, the troublesome Rayleigh limit of resolution is avoided. The detail in the reconstructed image can actually be better than that produced by any conventional photography using lenses: It is limited only by the grain size of the hologram (which, incidentally, does not have the same consequence as grain size in a photographic negative). Each small fragment of a hologram contains information about the entire object (as seen from that vantage point) and will reconstruct the entire image.

The applications of holography are impressive (see Figure 31–38). There are no depth-of-field limitations, so a flash picture of a cloud of mosquitoes reconstructs sharp images of mosquitoes close to the film as well as those far in the background. In terms of information storage capabilities, use of thick emulsions makes it possible to store multiple images in the same hologram by slightly changing the reference beam angle for each exposure. The pages of an entire book, for example, may be stored in a single film so that when viewed, the pages are "turned" by slightly shifting the illuminating beam angle. Holographic motion pictures have been produced, and full-color reconstructions are feasible by making a (black and white) hologram with a reference beam containing the three basic colors (red, green, and blue) to be viewed with (noncoherent) reflected white light. Acoustical holography using sound waves instead of light reproduces three-dimensional images of the internal body organs of a living person. Even computer-generated holograms are possible, resulting in a three-dimensional image of an object that never existed. Can holographic billboards be far behind?

Figure 31–38
A few applications of holography.

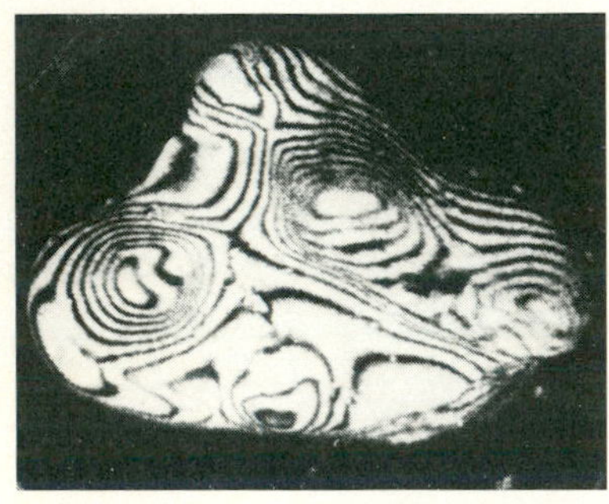

(**a**) A holographic contour map of a fossil badger tooth (8 mm long). Two holograms of the specimen were recorded on the same photographic plate using two slightly different wavelengths. When the resulting hologram is reconstructed using one wavelength, the interference between the two images creates fringes in the form of height contours.

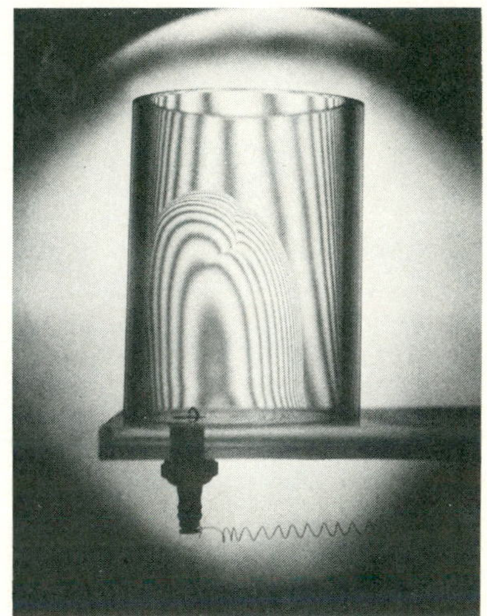

(**b**) Details of the spark detonation of acetylene gas inside a transparent cylinder are visualized by making a double exposure using a pulsed ruby laser that illuminates the scene from the rear through a ground-glass diffuser. The first exposure was made prior to ignition. The second exposure was recorded 10 ms after ignition. Upon reconstruction, three-dimensional patterns are formed by the interference of the two holographic images.

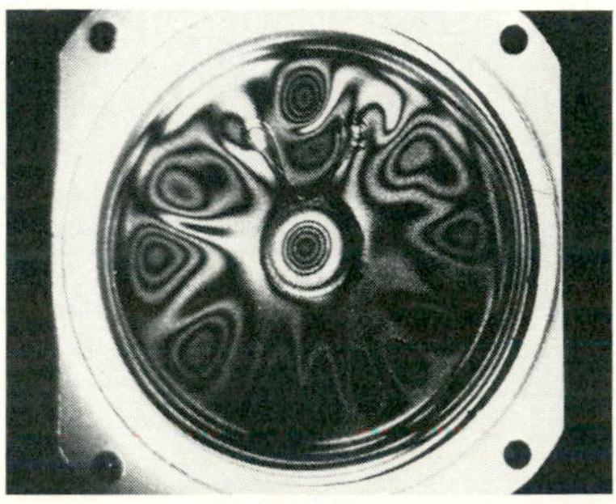

(**c**) A photograph of a "time-averaged holographic interferogram" of a loudspeaker vibrating at 3000 Hz. The hologram was a time exposure over several thousand cycles. Only the nodal lines and the stationary portions of the scene reconstruct brightly.

Summary

In Fraunhofer single-slit diffraction:

a = slit width
θ = angle measured from the center line
λ = wavelength
m = order

A diffraction *minimum* occurs when

$$m\lambda = a \sin\theta \qquad \text{(where } m = 1, 2, 3, \ldots)$$

The intensity distribution I_θ is given by

$$I_\theta = I_0\left(\frac{\sin\alpha}{\alpha}\right)^2 \qquad \left(\text{where } \alpha = \left(\frac{\pi}{\lambda}\right) a \sin\theta\right)$$

The *maxima* of the diffraction pattern fall *approximately* halfway between the minima.

In the diffraction grating:

d = separation of the slits (center-to-center)
θ = angle measured from the center line
λ = wavelength
m = order

A diffraction *maximum* occurs when

$$m\lambda = d \sin\theta \qquad \text{(where } m = 0, 1, 2, 3, \ldots)$$

The *dispersion* D of a diffraction grating is defined as

$$D \equiv \frac{d\theta}{d\lambda}$$

and may be calculated for different orders m from

$$D = \frac{m}{d\cos\theta}$$

The *resolving power* R of a diffraction grating is defined as

$$R \equiv \frac{\lambda}{\Delta\lambda}$$

and may be calculated for different orders m from

$$R = Nm$$

where N is the total number of rulings in the grating.

In Fraunhofer diffraction by a circular aperture, the angle θ_r between the center of the diffraction pattern and the *first* minimum (measured from the center of the aperture) is

$$\sin\theta_R = \frac{(1.22)\lambda}{D}$$

where D is the diameter of the aperture and λ is the wavelength.

The Rayleigh criterion for the minimum angle of resolution θ_R is that two adjacent point sources are distinguishable if the *central peak* of one diffraction pattern falls on the *first minimum* of the diffraction pattern of the other. Thus, for a telescope aperture of diameter D (or other optical instrument with a circular aperture):

$$\theta_R = \frac{(1.22)\lambda}{D}$$

In x-ray diffraction:

d = atomic plane spacing
ϕ = the *glancing* angle between the incident ray and the plane (not between the ray and the normal, as in optical reflection)
λ = wavelength
m = order

For the *Bragg scattering condition*, a diffraction *maximum* occurs at

$$m\lambda = 2d \sin \phi \qquad \text{(where } m = 1, 2, 3, \ldots\text{)}$$

Fresnel diffraction occurs when either the source of light or the observing screen (or both) lie at a *finite* distance from the diffracting aperture or obstacle. Fresnel's method of analysis employs *half-period zones*, for which the average light from any two adjacent zones is out of phase by one-half wavelength.

A *Fresnel zone plate* is a special screen in which alternate half-period zones are made opaque. The zone plate has lenslike focusing properties (with multiple focal lengths).

Holography is a two-step process in which an object illuminated by coherent light produces a complicated diffraction pattern on a photographic film. When the developed film (a *hologram*) is illuminated with coherent light, diffraction effects produce a three-dimensional image in which true differences in perspective occur if the viewer's position is changed. Since no lenses are used, the conventional Rayleigh resolution limits are avoided (though other limits eventually are present).

Questions

1. A small hole illuminated by monochromatic light produces a diffraction pattern on a screen in which the edges of the hole are poorly defined. If a lens is properly placed between the hole and the screen, the diffraction effects seem to disappear and the edges of the hole are well defined. Explain.
2. What happens to a Fraunhofer single-slit diffraction displayed on a screen if water replaces air in the space between the slit and the screen?
3. Since interference and diffraction effects depend on the addition of the electric fields associated with electromagnetic waves, why isn't it necessary to have a light source in which all electric field variations are in the same plane?
4. What are the limitations to the half-wave-zone analysis of single-slit Fraunhofer diffraction?
5. Suppose rather than a long narrow slit producing a diffraction pattern, a "slit" only twice as long as it is wide is used to produce the pattern. Qualitatively, what is the appearance of the pattern?
6. The central maximum of a double-slit interference pattern is the same width and approximate intensity as adjoining maxima. Why is this not true for a single-slit Fraunhofer diffraction pattern?
7. Two diffraction gratings, one larger than the other, are of the same quality and have the same number of rulings per centimeter. What are the advantages of using the larger of the two gratings?

8. Light from a slit is collimated by a lens, then passes through a diffraction grating whose rulings are parallel to the slit. What happens to the diffraction pattern on a screen as the grating is tilted about an axis parallel to its rulings?
9. Suppose a grating or prism is used in the spectral analysis of light containing a mixture of wavelengths. Under what circumstances is resolving power more important than dispersion and vice versa?
10. Describe the diffraction pattern produced by two crossed diffraction gratings.
11. What are the advantages, if any, of diffraction gratings over a prism in displaying the spectral components of a light source? What are the disadvantages, if any?
12. At night it is easier to read distant road signs if they are painted in green and white rather than red and white. Why?
13. A diffraction grating produces a continuous spectrum when illuminated by white light. How does a crystal produce a discontinuous array of dots ("Laue spots") when illuminated by "white x-rays"?
14. What are the similarities and differences between Fraunhofer and Fresnel diffraction?
15. The shadows of objects cast by the sun seem to have a fuzzy edge. Is this a diffraction phenomenon in which fringes are not evident because of the mixture of wavelengths in sunlight? If this is not a diffraction phenomenon, what is the cause of the fuzziness?
16. If you peer through very small cracks between your fingers at a distant light source, you will see light and dark fringes. Is this a diffraction phenomenon? If so, is it an example of Fraunhofer or Fresnel diffraction?
17. Why is it necessary to have a very nearly circular obstacle in order to observe Poisson's bright spot?
18. In describing the diffraction of sound waves, how would our development of light-diffraction analysis have to be modified? Remember that sound waves are longitudinal pressure waves.
19. In what way is a Fresnel zone plate like a converging lens? In what ways is it dissimilar?

Problems

31A–1 Light of wavelength 550 nm passes through a single slit and forms a diffraction pattern on a screen 3 m away. The distance between the third minima on opposite sides of the central maximum is 25 mm. Find the width of the slit. *Answer:* 0.396 mm

31A–2 A single slit is illuminated by light of wavelength 550 nm and produces a diffraction pattern on a screen 3 m from the slit. Find the total width of the central maximum for a slit width of (a) 0.2 mm and (b) 0.4 mm.

31A–3 When illuminated with monochromatic light, a certain diffraction grating produces a pattern in which the third, sixth, ninth, etc., orders are missing. Determine the ratio of slit width to slit separation for this grating. *Answer:* $\frac{1}{3}$

31A–4 In a Young's double-slit experiment, green light (520 nm) produces a pattern of bright fringes spaced 1.5 mm apart on a screen 1.8 m away.

(a) Find the distance between the slits.

(b) Every sixth fringe is missing. Calculate the width of each slit.

31A–5 A diffraction grating is 2.5 cm square and has a grating constant of 5000 rulings/cm. Calculate (a) the dispersion and (b) the resolving power of this grating in the second order for a wavelength of 600 nm.

Answers: (a) 7.16×10^{-2} deg/nm (b) 25 000

31A–6 Calculate the diameter of a reflecting-telescope mirror that by the Rayleigh criterion can resolve two point sources whose angular separation is $\frac{1}{4}$ second. Assume a wavelength of 550 nm.

31A–7 A person observing the taillights of an automobile as it recedes in the distance at night can barely distinguish them as separate sources of light. Assuming the lights are 1.5 m apart and they emit at an average wavelength of 640 nm, estimate the distance between the observer and the automobile. The pupil size of the observer's eye is 6 mm in diameter. *Answer:* 11.5 km

31A–8 A parabolic microwave antenna has a diameter of 1.5 m and is designed to receive x-band microwave signals ($\lambda = 3$ cm). Calculate the minimum angular separation in degrees of two microwave sources that can be resolved by this antenna.

31A–9 A television picture is composed of about 500 horizontal lines of varying light intensity. Assume that your ability to resolve the lines is limited only by the Rayleigh criterion and that the pupils of your eyes are 5 mm in diameter. Calculate the ratio of minimum viewing distance to the vertical dimension of the picture such that you will not be able to resolve the lines. Assume the average wavelength of the light coming from the screen is 550 nm. *Answer:* 14.9

31A–10 Monochromatic x-rays incident upon a crystal produce first-order Bragg reflection at a glancing angle of 20°. Calculate the expected angle for the second-order reflection.

31A–11 X-rays of wavelength 0.30 nm produce a first-order reflection from a crystal of NaCl when the glancing angle of incidence is 30°. Calculate the lattice spacing that corresponds to this reflection. *Answer:* 0.300 nm

31B–1 A single slit 0.20 mm wide is illuminated by monochromatic light of wavelength 600 nm, producing a diffraction pattern on a screen 1.5 m away. Find the distance between the first and fifth diffraction minima. *Answer:* 18.0 mm

31B–2 A vertical single slit 0.25 mm wide is illuminated by light with a wavelength of 600 nm, and a pattern is produced on a screen 2.5 m from the slit.

(a) Find the intensity relative to the central maximum intensity at a point 2.0 cm left of the central maximum position.

(b) Describe the position in terms of the nearest minimum location.

31B–3 A single slit is illuminated by light composed of two wavelengths, λ_1 and λ_2. The diffraction patterns produced overlap such that the first minimum created by the light of wavelength λ_1 falls at the second minimum of the pattern produced by light of wavelength λ_2.

(a) Calculate the ratio λ_1/λ_2.

(b) At what other places in the combined diffraction pattern will the minima coincide? *Answer:* (a) $\lambda_1/\lambda_2 = 2$

31B–4 A double-slit diffraction pattern is produced by slits that are one-third as wide as the separation of their centers. Calculate the ratio of the intensity of the first maximum relative to the central maximum.

31B–5 A certain grating has 20 000 slits spread over 5.5 cm. Find the wavelength of light for which the angle between the two second-order maxima is 60°. *Answer:* 688 nm

31B–6 A diffraction grating with 2500 rulings/cm is used to examine the sodium spectrum. Calculate the angular separation of the sodium yellow doublet lines (588.995 nm and 589.592 nm) in each of the first three orders.

31B–7 A certain diffraction grating has a dispersion of 2.5×10^{-2} deg/nm and a resolving power of 10^4 in the first order. Calculate the angular separation of two spectral lines near 550 nm that can barely be resolved in accordance with the Rayleigh criterion. *Answer:* 1.375×10^{-3} deg

31B–8 The neoimpressionist painter Georges Seurat perfected a technique known as "pointillism," whereby paintings were composed of small, closely spaced dots of pure color, each about 2 mm in diameter. The illusion of color mixing is produced in the eye of the viewer. Calculate the distance away a viewer should be in order to achieve a blending of the color dots into a smooth variation of color. Assume that the viewer has normal vision and that the ambient lighting causes the viewer to have a pupil diameter of about 3 mm.

31B–9 Show that the area of each zone in a zone plate is approximately $\pi\lambda L$, where L is the primary focal length of the zone plate.

31C–1 Show that the condition for diffraction *maxima* is $\tan\alpha = \alpha$, where $\alpha = (\pi/\lambda)a \sin\theta$.

31C–2 The condition for the angular position θ of single-slit diffraction maxima is given by $\tan\alpha = \alpha$ (again, where $\alpha = (\pi/\lambda)a \sin\theta$, a being the slit width). This equation is most easily solved by successive approximations using a pocket calculator. Assuming $a = 20\lambda$, (a) show that the angular position of the first-order maximum does not lie exactly midway between the first and second minima and (b) find its value. (Hint: As a first approximation for α, use an angle θ_{ave} that is the average value of θ_1 and θ_2, the first two minima. Then try slightly smaller values of α until you "zero in" on the correct value.)

Answer: (b) 0.0716 rad

31C–3 Using the information in the previous problem, find the angular position (in radians) of the second-order diffraction maximum when $a = 20\lambda$. *Answer:* 0.1233 rad

31C–4 As shown in Figure 31–39, a beam of parallel monochromatic light passes through a large, circular hole to produce a spot of light on the screen. Consider a point *P*, away from the straight-through beam, located where no light strikes the screen. Now suppose an opaque disk with an arbitrary opening in it (object *A*) is placed in the hole, causing some diffracted light to reach *P*. Next, suppose we replace object *A* with its "complement," object *B*, which is opaque where object *A* is transparent, and vice versa. (If both objects were present at the same time, the combination would be completely opaque.) According to Babinet's principle, *the diffracted light reaching P is exactly the same in the two cases.* Prove the theorem using superposition concepts.

Figure 31–39
Problem 31C–4

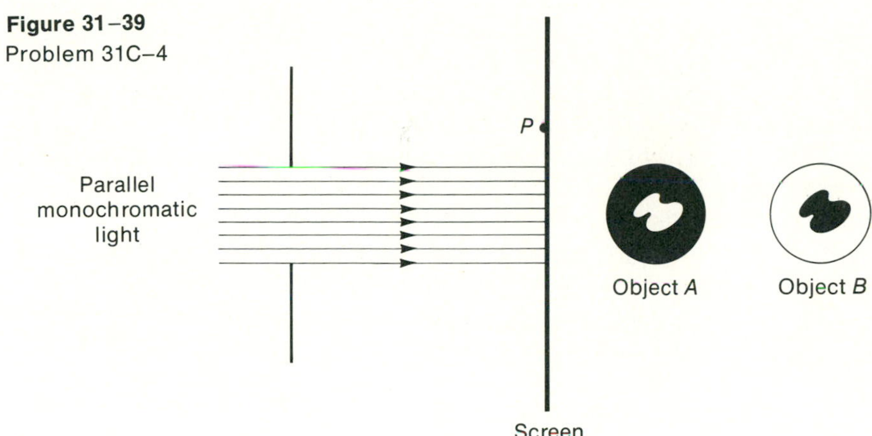

31C–5 Consider the Bragg planes indicated in the two-dimensional lattice shown in Figure 31–24.

(a) Show that the spacing of these planes can be represented by $d = a(n^2 + 1)^{-\frac{1}{2}}$, where $n = 1, 2, 3, \ldots$, and a is the atomic spacing in the lattice.

(b) In general, $d = a(n^2 + m^2)^{-\frac{1}{2}}$, where both n and m are integers. Make a sketch similar to Figure 31–24, showing the planes separated by $d = a(13)^{-\frac{1}{2}}$.

chapter 32
Polarization

We can scarcely avoid the inference that light consists in the transverse undulation of the medium which is the cause of electric and magnetic phenomena.

JAMES CLERK MAXWELL (1856)

Maxwell's equations describe electromagnetic radiation as a transverse wave of oscillating electric and magnetic fields. It is called a *transverse* wave because the **E** and **B** fields are represented by vectors that lie in a plane perpendicular to the direction of propagation. By convention, the direction of the *electric* vibration is usually called the **direction of polarization** of the **linearly polarized wave.** Figure 32–1 shows methods of depicting linearly polarized light rays in diagrams. The arrays of short arrows indicate the direction of polarization; if you wish, you may instead think of them as oscillating electric field vectors. Other words are also used to describe polarized waves. The terms *plane of polarization* and *plane-polarized* waves are common, but these may have possible ambiguities. For example, in Figure 32–2 the two rays *A* and *B* have the same plane of polarization (the plane of the paper), but their directions of polarization are different.

Radio waves and microwaves emitted from antennae are polarized in directions related to the direction of the accelerated charges in the antenna wires (Figure 32–2). A receiving dipole oriented parallel to the direction of polarization will absorb energy from the waves because the alternating electric field causes electrons in the receiving dipole to accelerate back and forth along the wires, producing an oscillating potential difference between the dipole halves. However, if the receiving dipole is oriented perpendicular to the direction of polarization, the two halves of the dipole remain at the same potential and the waves are not detected by the receiver.

The fact that electromagnetic waves can be polarized is conclusive evidence that they are *transverse* waves. Interference and diffraction give evidence of their wave nature, but these effects do not differentiate between longitudinal and transverse waves. (Sound waves, for example, are longitudinal and do show interference, but they cannot be polarized. Only transverse waves can be polarized.)

Visible light emitted by ordinary sources, such as light bulbs and glowing hot objects, has its origin in excited atoms and molecules. Classically, each atom or molecule emits a short burst of electromagnetic waves lasting about 10^{-8} s and containing a few million vibrations, thereby sending out a *wave train* that extends about 3 m along the direction of propagation. Because the atoms emit light *independently* of one another, the resultant light is a superposition of many wave trains whose electric vectors are oriented randomly in all possible directions perpendicular to the direction of propagation. We call (**a**) Linearly polarized light rays approaching the viewer (that is, the rays are perpendicular to the plane of the paper).

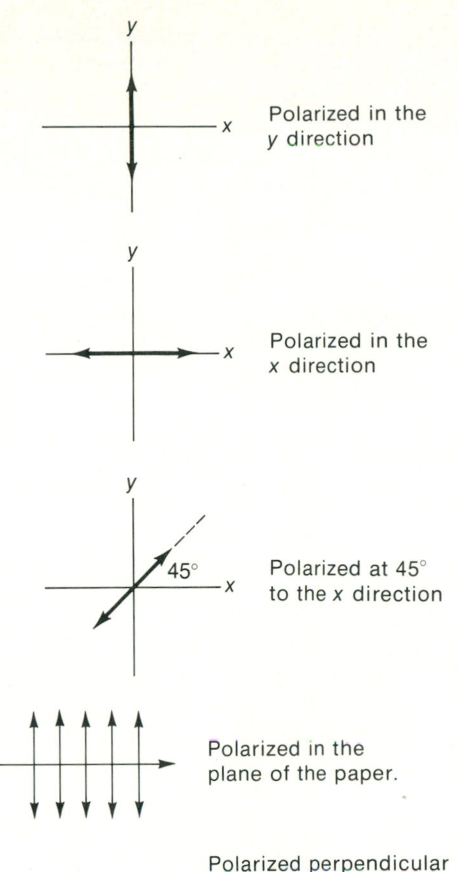

(**b**) Linearly polarized light rays traveling in the plane of the paper. The direction of polarization is indicated by an array of short arrows or a series of dots.

Figure 32–1
Ways of indicating the direction of polarization of the electric field for *linearly polarized* light rays.

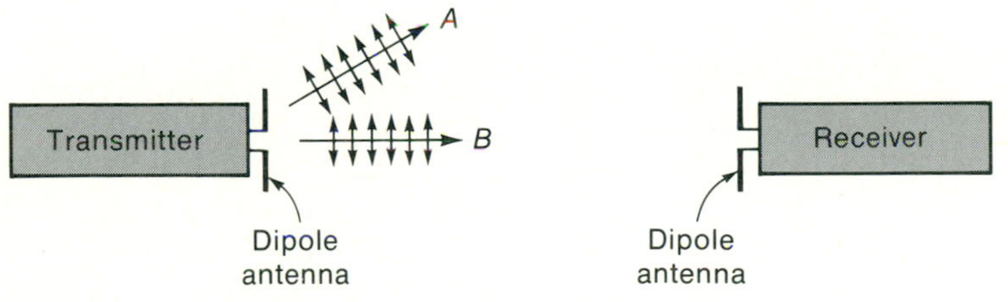

Figure 32–2
Radiation from a dipole antenna is polarized. The direction of polarization is perpendicular to the direction of propagation and lies in the plane containing the dipole antenna.

such light **unpolarized.** As shown in Figure 32–3, there are two customary ways of depicting unpolarized light in diagrams. In (a), the light ray is approaching the viewer, and the array of arrows represents the superposition of many wave trains plane-polarized with random orientations. Since an electric field at any arbitrary direction in the x-y plane may be resolved into components along

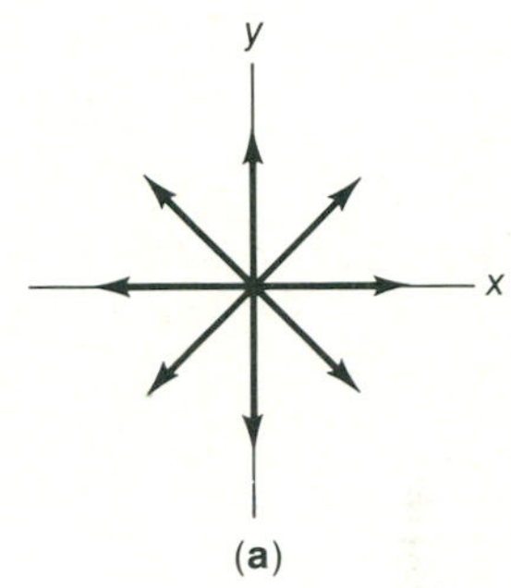

(**a**)

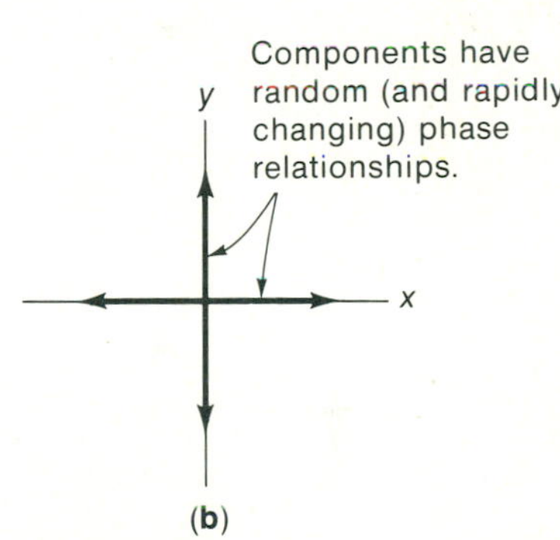

(**b**)

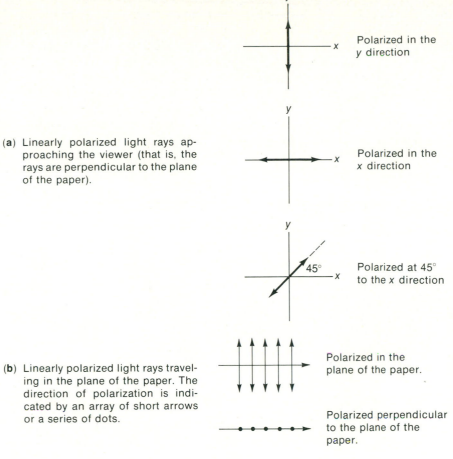

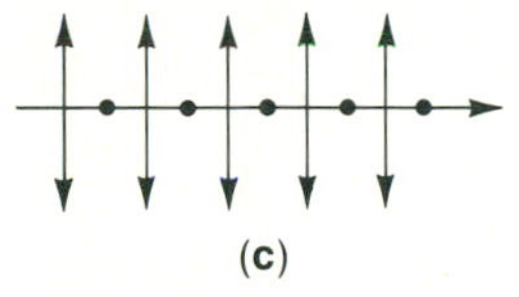

(**c**)

Figure 32–3
(a) and (b) Two ways of depicting *unpolarized* transverse waves approaching the viewer. (c) An unpolarized ray traveling in the plane of the paper; the arrays of short arrows and dots are separated in space to emphasize that there is no fixed phase relationship between the two components, which vibrate incoherently.

the x and y axes, an equivalent representation is shown in (b). Here, the electric field of each individual wave train has been resolved separately; when summed along the x and y axes, the two net components are equal in (average) magnitude. One important characteristic should be noted. Since the *phases* of the wave trains are completely random (the light from the various atoms is *incoherent*), there is no fixed phase relationship in the net components. In fact, the components have a random and rapidly changing phase relationship. However, their *time average* is the same in each direction. Consequently, our choice for the orientation of the x and y axes about the direction of propagation makes no difference for unpolarized light: In each case the (average) components at right angles are equal.

32.1 Polaroid

The human eye is not very sensitive to the direction of polarization.[1] However, polarized light can be produced and analyzed easily with a commercial material called Polaroid, first invented in 1928 by Edwin H. Land while he was a 19-year-old undergraduate at Harvard. The modern version of Polaroid is made by heating and stretching a plastic sheet that contains long-chain molecules of polyvinyl alcohol. The stretching process aligns the molecules parallel to one another. The sheet is then dipped into an iodine solution, which causes iodine atoms to attach themselves to the alcohol molecules, forming chains of their own that apparently act as microscopic conducting wires. An incident electromagnetic wave that has a component of **E** parallel to the chains will drive conduction electrons along them, absorbing essentially all the energy of that component of the wave. On the other hand, if the E field is perpendicular to the chains, only a small absorption takes place and most of this component passes through, a property called *selective absorption*. The *transmission axis* of a sheet of Polaroid is thus perpendicular to the direction the film was stretched. Sheets as large as 1 m by 30 m (or longer) are available. For protection and strength, the material is usually laminated between thin sheets of cellulose or glass. The way a sheet of Polaroid affects light has its macroscopic counterpart in a grid of parallel conducting wires. The grid affects an unpolarized beam of radio waves or microwaves in exactly the same fashion, *transmitting* only the component whose electric vector is *perpendicular* to the direction of the wires, provided the separation between wires is somewhat less than the wavelengths of the radiation.

Figure 32–4
When polarizing sheets are *crossed*, their transmission axes are at right angles. Each individual sheet appears gray because it absorbs approximately half of the incident unpolarized light.

Ideally, a "perfect" polarizing sheet would transmit 50% of an incident unpolarized beam and absorb 50%. However, in practice the transmission is about 40% or less because of reflection at surfaces and some unwanted absorption. As shown in Figure 32–4, if a pair of Polaroid sheets are "crossed" so that their transmission axes are at an angle of 90°, approximately 90% of the light is absorbed. It has been proposed that all automobile headlights and windshield visors be equipped with sheets of Polaroid whose optic axes are parallel and aligned the same way at 45° with respect to the vertical. Thus, at night a motorist could see objects illuminated by his own headlights, but light from the headlights of approaching cars would be polarized at right angles to the transmission axis of the motorist's visor and thus be almost totally absorbed.

---

[1] The unaided eye can sometimes detect the direction of polarization through a faint pattern known as *Haidinger's brush*, which some, but not all, people can observe. For a description of this effect and other interesting features of polarized light, see the Science Series paperback: W.A. Shurcliff and S.S. Ballard, *Polarized Light*, D. Van Nostrand Co., 1965.

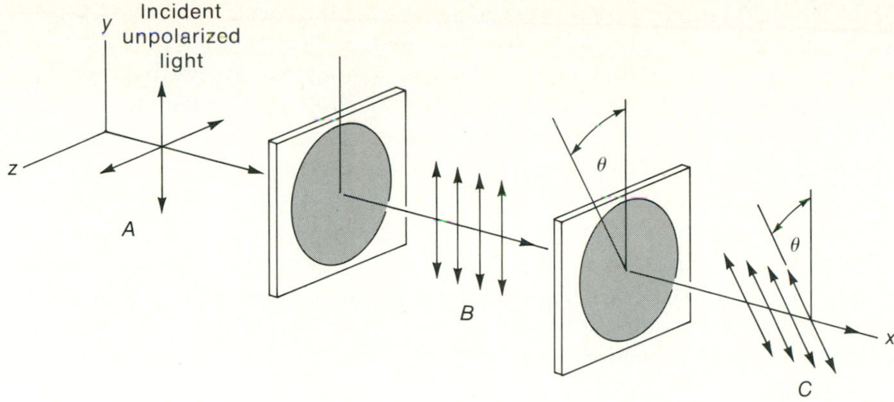

Figure 32–5
Two parallel polarizing sheets, one rotated so that its transmission axis is at an angle θ with respect to the other. Unpolarized light traveling along the x axis at A becomes linearly polarized in the y direction at B and polarized at an angle θ with respect to the y axis at C. If the intensity at B is I_0, the intensity at C is $I_0 \cos^2 \theta$ (for "ideal" polarizing materials).

When two polarizing sheets are used together, the first is often called the *polarizer* and the second, which may be used to determine the direction of polarization of light coming from the first, is called the *analyzer*. Consider two polarizing sheets whose transmission axes are at angle θ with respect to each other, as in Figure 32–5. If light coming from the polarizer has an electric field amplitude E_0, the analyzer (assumed "ideal") will transmit only the component $E_0 \cos \theta$ parallel to its transmission axis. Since the intensity I of an electromagnetic wave is proportional to the *square* of the amplitude, the transmitted intensity varies with the angle θ as

$$I = I_0 \cos^2 \theta \tag{32–1}$$

where I_0 is the intensity of incident polarized light whose direction of polarization is at an angle θ with the transmission axis of a sheet of polarizing material. Equation (32–1) is called the *law of Malus* after its discoverer, Captain Etienne Malus, a military engineering officer in Napoleon's army. When several polarizing sheets at various angles are used in series, Equation (32–1) is applied to each successive sheet.

EXAMPLE 32–1

When two (ideal) polarizing sheets have their transmission axes parallel, light passing through them emerges with an intensity I_0. If one sheet is rotated through an angle of 45°, what is the intensity of the emerging light?

SOLUTION
From Equation (32–1):

$$I = I_0 \cos^2 \theta$$
$$I = I_0 (\cos 45°)^2$$
$$I = I_0 \left(\frac{1}{\sqrt{2}}\right)^2$$
$$I = \boxed{\frac{I_0}{2}}$$

32.2 Birefringence

In a few crystalline substances, the atoms are arranged in arrays that have high degrees of symmetry. As a result, they have just a single index of refraction, which is independent of the polarization direction of the incident light. Most gases, liquids, and amorphous solids, such as unstressed glass or plastic, also behave this way. They are called *optically isotropic*. However, many crystalline substances and stressed amorphous materials have considerable asymmetries in their basic atomic structures. As a result, they have *two* indices of refraction, depending on the direction of polarization of the incident light. These *doubly refractive*, or *birefringent*, substances are *optically anisotropic*. The reason for the two indices of refraction is straightforward. If the crystal lattice of atoms is not symmetrical, the binding force on the electrons is also not symmetrical. That is, electrons displaced from their equilibrium positions along one direction have a greater effective "spring constant" than along another direction. Because the propagation of electromagnetic waves through materials is a process of causing electrons to vibrate and then reradiate this energy, the fact that electrons respond differently along one direction than along another causes the waves to be transmitted with different speeds in different directions.

Calcite, quartz, sugar, mica, and ice are examples of birefringent materials. Figure 32–6 shows that an unpolarized ray incident on calcite splits into two components: an *ordinary* ray (called the "*o*-ray"), which obeys Snell's law of refraction, and an *extraordinary* ray (the "*e*-ray"), which does not. Within the

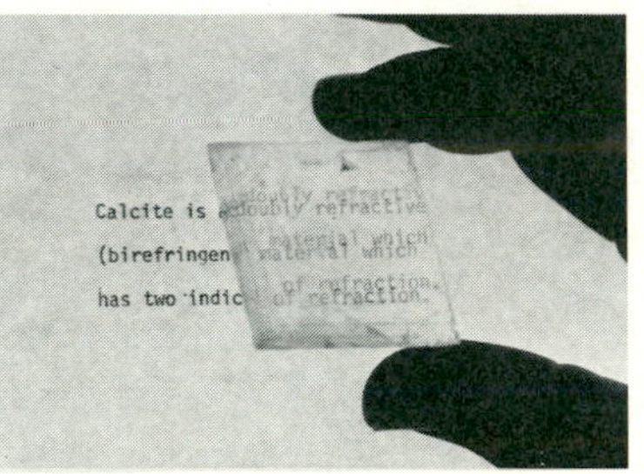

(**a**) A calcite crystal forms a double image.

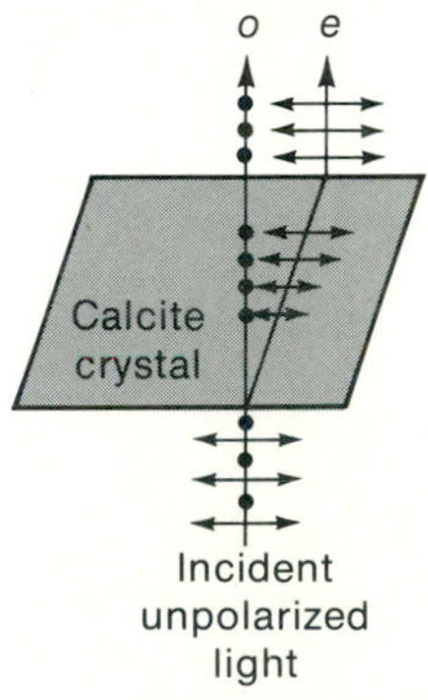

(**b**) An unpolarized ray incident perpendicularly on the face of a calcite crystal splits into two polarized rays. The *o*-ray continues in the same straight line; the *e*-ray is at an angle inside the crystal, emerging parallel to the *o*-ray but displaced to one side.

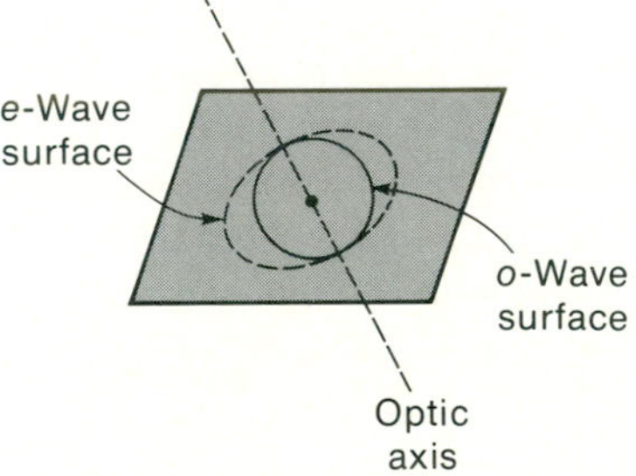

(**c**) A point source inside a calcite crystal generates two different Huygens' wave surfaces. The *o*-wave surface (solid line) is a sphere. The *e*-wave surface (dashed line) is an ellipsoid of revolution formed by rotating an ellipse about the axis that passes through the two points where the circle and ellipse shown in the figure are in contact. This axis is called the *optic axis* and is the direction in which both the *o*-wave and *e*-wave propagate with the same speed. (Note that the optic axis is a direction, not a line.)

Figure 32–6
Some optical properties of calcite, a birefringent crystal.

crystal, the extraordinary ray generally does not propagate in the same direction as the incident ray. To observe this effect, place a crystal of calcite on a piece of paper with a black dot on it; two images of the dot can be seen. Rotating the crystal causes one image to remain stationary, while the other image revolves around it. Furthermore, the two images are linearly polarized with their directions of polarization at right angles. Since the two rays emerging from the calcite are linearly polarized at right angles, if one of the rays could be eliminated the crystal would make a very efficient polarizer. A device called the *Nicol prism*, named after the Scottish physicist William Nicol (1768–1851), accomplishes this goal (Figure 32–7). The main drawbacks to Nicol prisms are their cost, small size, and limited field of view ($\sim 28°$).

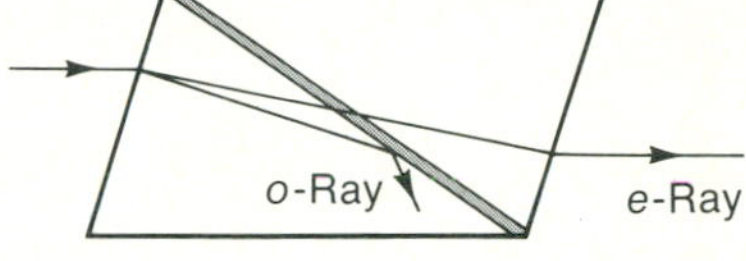

Figure 32–7
The Nicol prism. A calcite crystal is cut along a special plane and then cemented together with Canada balsam. The refractive index of Canada balsam is about midway between those of the *o*-ray and the *e*-ray in calcite. As a consequence, the *o*-ray is totally internally reflected at the interface (and absorbed in black paint on the side), while the *e*-ray continues on to emerge from the prism.

A few crystalline substances are natural polarizers in that they absorb one component of polarization while being transparent to the other component. Tourmaline, a semiprecious stone often used in jewelry, is an example. This property of *selective absorption* is called *dichroism* (from the Greek *di*, meaning "two," and *chros*, meaning "skin" or "color"), because when viewed by transmitted light along two different directions these crystals usually exhibit two different colors. Unfortunately, dichroic crystals are generally very small.

32.3 Polarization by Reflection and Scattering

In 1808, Malus happened to be looking through a calcite crystal at reflections of the setting sun from a window of Luxembourg Palace in Paris. To his surprise, he observed that the two images were extinguished in turn as he rotated the crystal, thus implying that the reflected light was polarized. Sir David Brewster, a Scottish physicist, investigated the phenomenon in 1812 and found the reflected wave was 100% polarized when the refracted and reflected waves at the surface of the glass were at right angles. This relation becomes plausible when we think of the incident unpolarized light as made up of two (incoherent) *E*-field components at right angles (Figure 32–8). As the light is refracted into the material, it causes electrons to vibrate along these right-angle directions. However, since accelerating electrons cannot radiate energy along the direction of acceleration, the electron vibration component in the plane of the diagram in the material cannot reradiate in the direction of the reflected beam. Only the

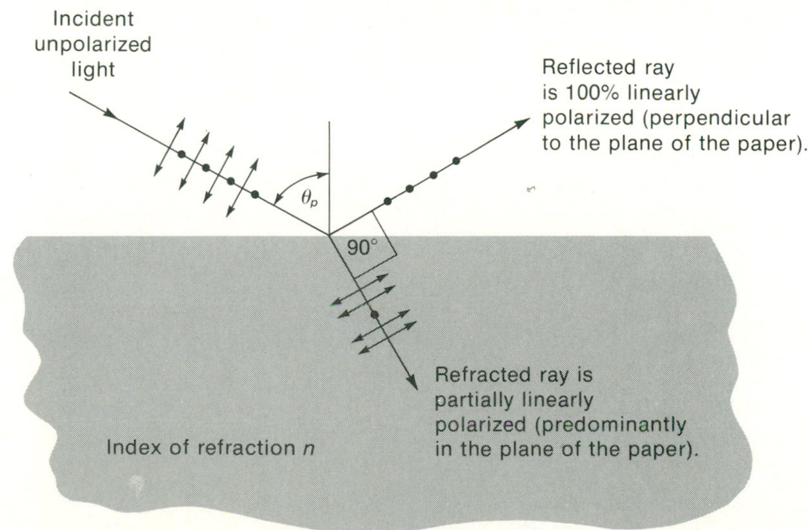

Figure 32–8
When light is incident at the *polarizing angle* θ_p, the reflected and refracted rays are at right angles.

vibration component perpendicular to the plane of the paper radiates in that direction, producing a reflected beam that is 100% polarized, as shown.

Letting θ_p be the *polarizing angle* of incidence that produces this right-angle condition, we have

$$\theta_p + \theta_2 = 90^\circ \tag{32-2}$$

Combining this equation with Snell's law for refraction:

$$n_1 \sin \theta_p = n_2 \sin \theta_2 \tag{32-3}$$

we can derive the following relation, found in 1812 by Sir David Brewster:

BREWSTER'S LAW (for 100% polarization of light by reflection from dielectric materials)

$$\tan \theta_p = n \tag{32-4}$$

where $n = n_2/n_1$, the index of refraction of the material relative to that of the surrounding medium. The phenomenon works only for *dielectric* materials. (The process of reflection by *conducting* surfaces is more complex, and we will not take up those cases. In general, metallic surfaces reflect all components of polarization with varying degrees of effectiveness, depending on the angle of incidence.)

EXAMPLE 32-2

What is the polarizing angle for light incident on water (index of refraction = 1.33)?

SOLUTION

From Equation (32-4):

$$\tan \theta_p = 1.33$$

$$\theta_p = \tan^{-1}(1.33) = \boxed{53.1^\circ}$$

Sunglasses made of polarizing sheets make use of the fact that glare reflections from water surfaces, roadways, and other horizontal surfaces are

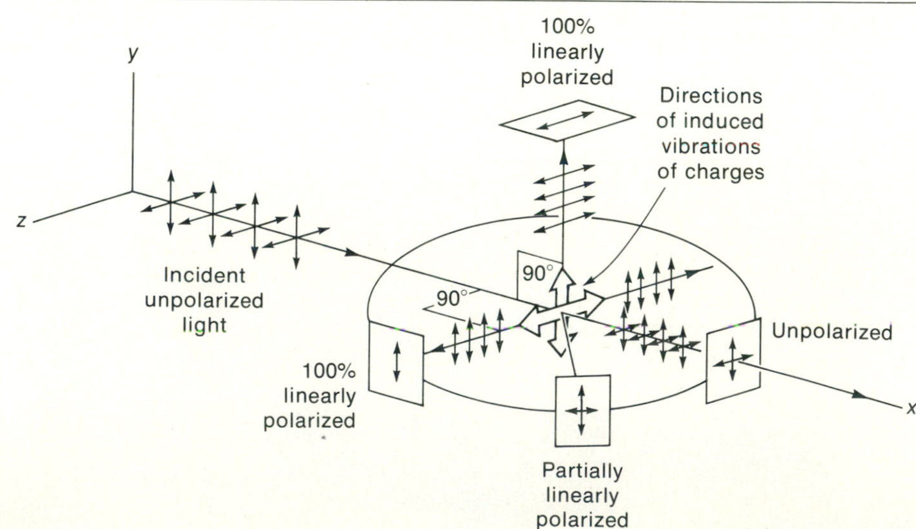

Figure 32-9
Scattering of unpolarized light by molecules. The incident unpolarized light is represented by components vibrating (incoherently) at right angles. The oscillating electric fields set charges into vibration at all directions in the *y-z* plane (shown here resolved into *y* and *z* components). The waves reradiated by these oscillating charges at 90° to the incident direction are 100% linearly polarized. At other angles they are partially polarized, while scattering in the forward direction is unpolarized.

(at least partially) linearly polarized and hence can easily be reduced if the transmission axis of the sunglasses is oriented correctly. (What direction is correct?) Scattered sunlight from the clear sky is partially polarized because the incident (unpolarized) sunlight sets molecules in the air into oscillations and the vibrating charges in these molecules reradiate the light, with planes of polarization related to the direction of accelerations of the charges (Figure 32–9). Look overhead at the sky through Polaroid sunglasses during sunrise or sunset, when the sun's rays are at right angles to the overhead sky. The scattered light is partially polarized because the induced vibrations along the line-of-sight direction do not reradiate energy in that direction: Only the oscillations at right angles to that direction contribute to scattering. The light is only partially polarized because some of it is scattered more than once before it reaches the eye. Honeybees, ants, and certain other insects have polarizing lenses in their eyes and are believed by some biologists to use the polarization of skylight as an aid in navigation.

The clear daylight sky is blue because of two effects. For scattering particles that are small compared with the wavelength of the radiation (O_2 and N_2 molecules are only about one-thousandth the wavelength of visible light), the scattering efficiency is proportional to the fourth power of the frequency. So blue light is scattered more than red light. Also, molecules of air have natural resonance frequencies in the violet and ultraviolet, so they are driven into larger amplitude oscillations by blue light than by red, causing preferential reradiation of wavelengths near the blue end of the spectrum. When the sun is close to the horizon, we see it through an extra thick layer of air that scatters blue and violet light sideways, leaving mostly reds and yellows along the line of sight, to produce flame-colored sunsets and sunrises. The volcano Krakatoa, west of Java, emphasized this effect when it blew up in 1883, injecting into the atmosphere a great amount of fine volcanic dust. Especially spectacular sunrises and sunsets were observed worldwide for several years thereafter as the dust slowly settled. In the absence of an atmosphere such as on the moon or in space, the sky is jet black (Figure 32–10).

Figure 32–10
On the moon, the earth appears in a jet-black sky because of the lack of an atmosphere to scatter sunlight.

Smoke from a cigarette appears blue when observed against a dark background because smoke particles are smaller than wavelengths of light and therefore preferentially scatter blue light. On the other hand, smoke exhaled from the lungs, which usually has larger water droplets mixed with it, appears whitish because the larger droplets cause the light to be refracted and reflected many times, processes that are not particularly frequency sensitive. The whiteness of clouds in the sky, and of sugar, salt, snow, fog, ground glass, and so on, is similarly due to multiple refractions and reflections.

32.4 Wave Plates and Circular Polarization

As mentioned previously, a birefringent material has two indices of refraction, one each for the *o*-ray and the *e*-ray. This means that light travels with two different speeds through the material, depending on the direction of polarization of the incident light. (In calcite, the *e*-ray is faster; in some other materials, the *o*-ray is faster.) Suppose we cut a piece of calcite into a thin slab[2] such that for

---

[2] In general, a ray of light incident on a birefringent material is split into two distinct beams traveling *at an angle* to each other. However, there are certain directions in which the rays travel *along the same direction* at different speeds. The slab is constructed with this direction perpendicular to the front and back surfaces of the plate so that the two rays do not get out of alignment in traveling through the slab.

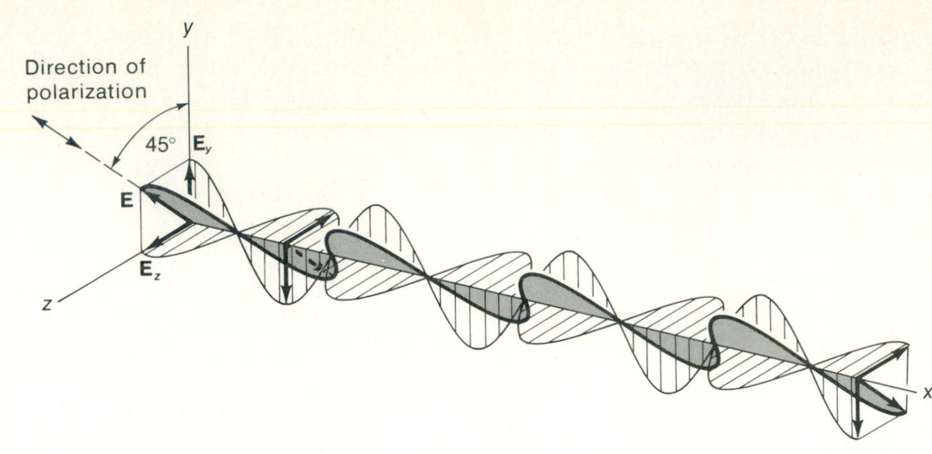

Figure 32–11
A linearly polarized light wave travels along the x axis, with its direction of polarization at 45° with respect to the y axis. The electric field **E** is resolved into two equal-amplitude components along the y and z directions, respectively.

a given wavelength of light the o-ray emerges from the slab just half a wavelength behind the e-ray. The two rays are thus out of phase by 180°. Such a slab is called a *half-wave plate*. It has interesting properties.

In Figure 32–11, a beam of light traveling along the x axis is linearly polarized at 45° with respect to the y axis. We can represent its electric field as two electric field components along the y and z axes that *vibrate in phase with each other*. (Do not confuse this representation with that of Figure 32–3(b), where components of *unpolarized* light have random and changing phase relationships.) Now allow this polarized light to enter a half-wave plate oriented so that these two components ecome the o- and e-rays in the plate. As they pass through the plate, the o-ray is retarded slightly relative to the e-ray. When they emerge, the components will be exactly 180° out of phase, shifting the direction of polarization 90°, as shown in Figure 32–12.

Figure 32–13 shows the effect of a *quarter-wave plate* (that is, the slow and fast rays become out of phase by 90°). The two components of the electric field emerge 90° out of phase, producing **circularly polarized** light. If you look toward the source of such a wave as it approaches you, its electric field vector **E** will rotate at an angular frequency $\omega = 2\pi f$ (where f = light frequency). Depending on which component lags behind, the direction of rotation will be *clockwise* or *counterclockwise*, corresponding to the two possible states of

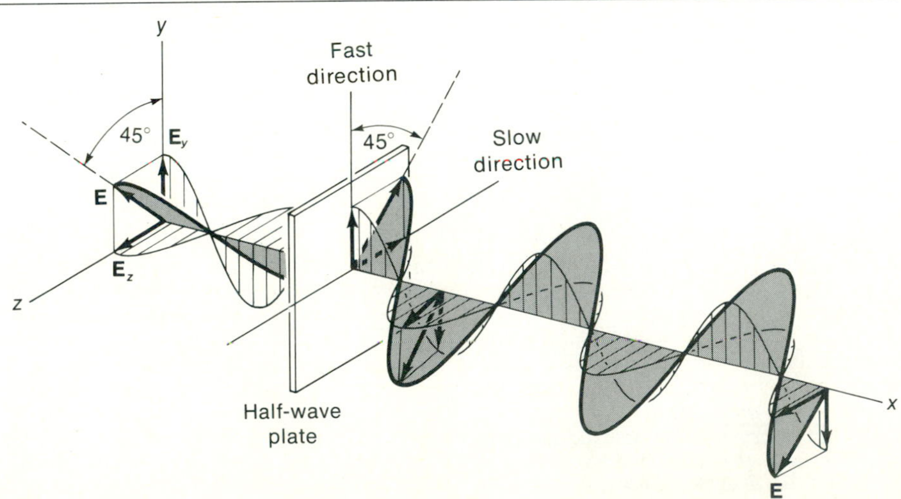

Figure 32–12
The half-wave plate is oriented so that $\mathbf{E}_y$ is along the "fast" direction and $\mathbf{E}_z$ is along the "slow" direction. As they emerge from the half-wave plate, the $\mathbf{E}_z$ vibrations have been retarded half a wavelength (180°) behind the $\mathbf{E}_y$ vibrations, shifting the direction of polarization by 90°.

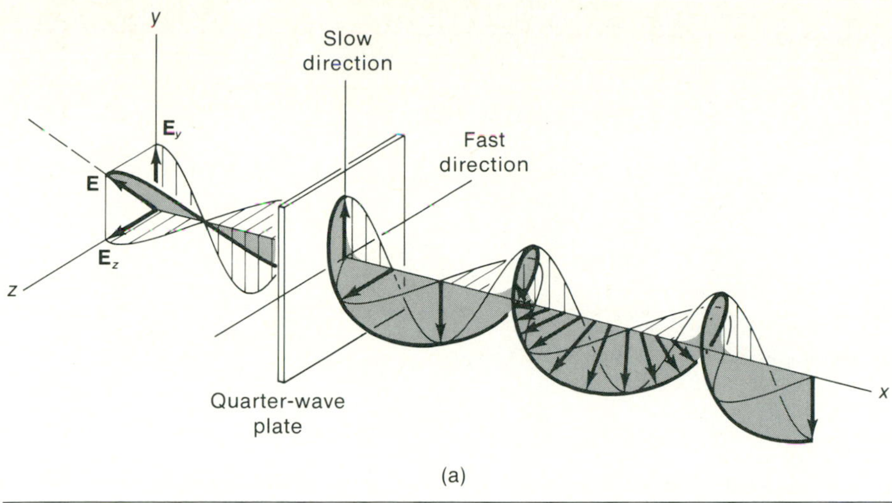

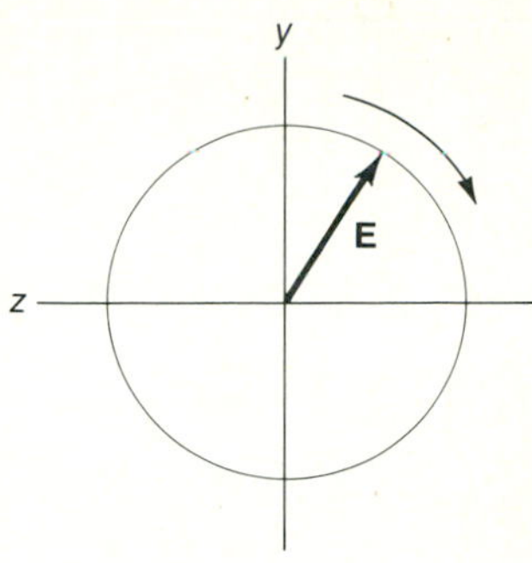

(**b**) Looking along the negative *x* direction, the **E** vector of the approaching wave rotates clockwise.

Figure 32–13
The quarter-wave plate is oriented so that $\mathbf{E}_y$ is along the "fast" direction and $\mathbf{E}_z$ is along the "slow" direction. As they emerge from the quarter-wave plate, the $\mathbf{E}_z$ vibrations have been retarded a quarter-wavelength (90°) behind the $\mathbf{E}_y$ vibrations, producing a circularly polarized wave.

circularly polarized light.[3] If the direction of polarization is at an angle other than 90° with the fast and slow axes, the y and z components of the electric field are unequal (but still 90° out of phase), producing **elliptically polarized** light (Figure 32–14).

The general name for half-wave plates, quarter-wave plates, and so on, is *retardation plates.* You can make fairly good ones for amateur experimentation from certain brands of sticky (glossy) cellophane tape, or from cellophane wrappers on cigarette packages. In the manufacturing process of some of these types of films, a sheet of organic polymers is stretched, causing it to become birefringent. Transparent plastic food-wrap films, which are not birefringent, may become so if physically stretched somewhat.[4] With a little experimenting, one or more layers (keeping their stretch axes parallel) will form retardation plates. If sheets of mica are available, a sheet about 0.018 mm thick will form a quarter-wave plate for yellow light.

EXAMPLE 32–3

What minimum thickness of calcite will make a half-wave plate for yellow light of wavelength $\lambda = 589.3$ nm? The indices of refraction for the *o*- and *e*-rays are $n_o = 1.6584$ and $n_e = 1.4864$, respectively.

SOLUTION

The times t_o and t_e for the *o*- and *e*-rays to travel through a plate of thickness d are

$$t_o = \frac{d}{v_o} \quad \text{and} \quad t_e = \frac{d}{v_e} \tag{32–5}$$

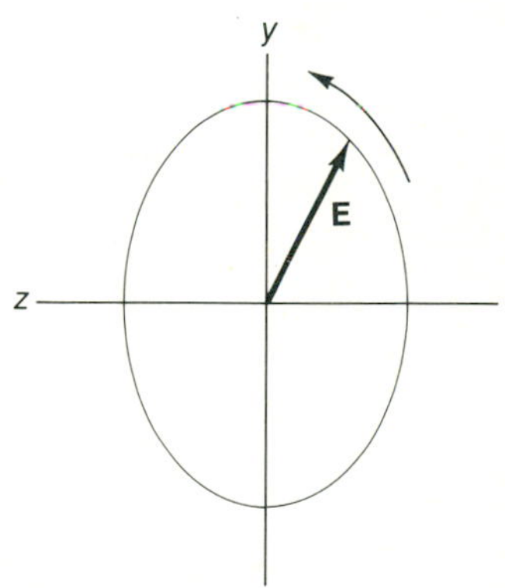

Figure 32–14
If a quarter-wave plate is oriented at some arbitrary angle with respect to the direction of polarization of the incident light, the electric field components are unequal in magnitude (but 90° out of phase) and the emerging light is *elliptically polarized.*

[3] As we will discuss in the next chapter, a light beam carries *linear* momentum, so that when it strikes an absorber it imparts a force against the absorber. It can be verified experimentally that in addition to this force, a *circularly* polarized light beam has *angular* momentum and therefore also exerts a torque on the absorber. It is interesting that in the particle, or *photon*, model for light, every individual photon is circularly polarized and carries one unit of angular momentum. The conservation of angular momentum requires that an atom that emits a photon must therefore itself undergo a change of angular momentum by one unit in a sense of rotation opposite to that of the photon. Plane-polarized light is actually an equal mixture of photons, with clockwise and counter-clockwise senses of rotation.

[4] For interesting experiments you can perform yourself, see The Amateur Scientist section of *Scientific American*, December 1977.

where the respective velocities in the plate are $v_o = c/n_o$ and $v_e = c/n_e$. The time difference, $\Delta t = (t_o - t_e)$, is thus

$$\Delta t = \frac{d}{c}(n_o - n_e) \tag{32–6}$$

To form a half-wave plate, we want the emerging e-ray to travel (in air) a half-wavelength before the o-ray finally emerges from the plate. The time required to do this is therefore

$$\Delta t = \frac{\left(\dfrac{\lambda}{2}\right)}{c} \tag{32–7}$$

Combining Equations (32–6 and 32–7) gives

$$\frac{\lambda}{2} = d(n_o - n_e) \tag{32–8}$$

Solving for the plate thickness d and substituting numerical values yields

$$d = \frac{\lambda}{2}\left(\frac{1}{n_o - n_e}\right) \tag{32–9}$$

$$d = \frac{(5.893 \times 10^{-7}\text{ m})}{2(1.6584 - 1.4864)}$$

$$d = \boxed{1.713 \times 10^{-6}\text{ m}}$$

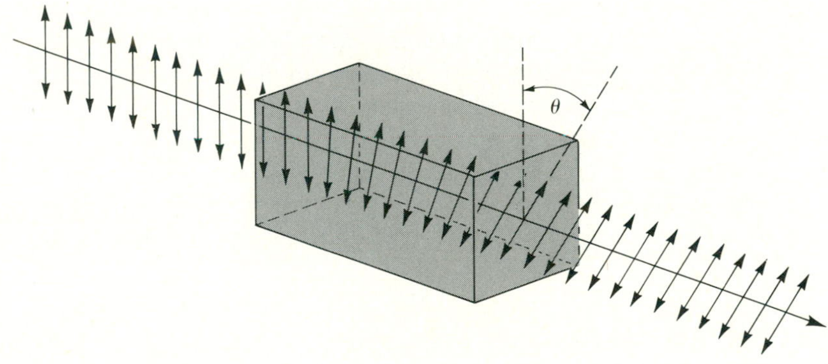

(a) As a linearly polarized wave travels through an *optically active* substance, the plane of polarization gradually rotates about the direction of propagation.

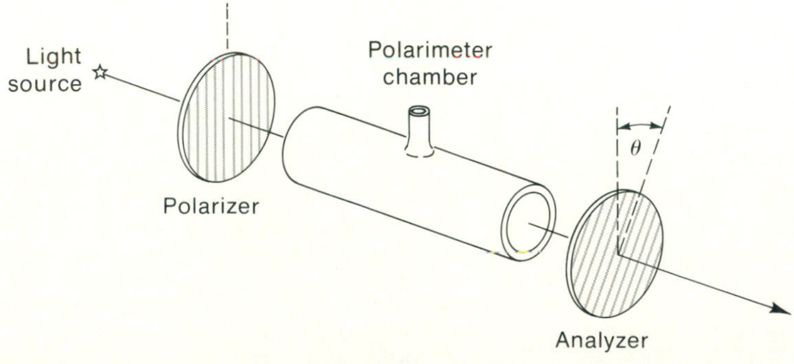

(b) A *polarimeter* measures the angle through which the direction of polarization is rotated.

Figure 32–15

Optically active substances cause a shift in the direction of linearly polarized light.

32.5 Optical Activity

Just as certain materials transmit linearly polarized light with two different speeds, some substances transmit circularly polarized light with two different speeds, depending on the sense of rotation of the electric vector. This has the interesting effect of causing a shift in the direction of *linearly* polarized light. For example, if a sugar solution is placed between a polarizer and an analyzer, the solution will rotate the direction of polarization, as shown in Figure 32–15. Such substances are called *optically active.* The amount of rotation is proportional to the distance traveled; in solutions, it is also proportional to the concentration of the optically active substance.

The mechanism causing the rotation is explained by recognizing that linearly polarized light may be considered as the sum of two circular polarizations rotating in opposite directions (Figure 32–16). In optically active substances, one of the rotating components of light travels faster through the material than the other, causing a continual change in phase between the two components. Consequently, the direction of linear polarization gradually changes as the light travels through the substance. The shift may be in a clockwise or counterclockwise sense, depending on the arrangement of atoms in the molecules. Sugar, for example, comes in two different forms that have the same chemical formula but whose atoms are arranged in mirror-image configurations. Such pairs of molecules are called *stereoisomers* (Figure 32–17). One form of sugar, *dextrose* (from the Latin *dextro*, meaning "right"), causes the direction of vibration to revolve clockwise as seen by an observer toward whom the light is moving, a sense of rotation called *right-handed.* Another form of sugar, *levulose* (from the Latin *levo*, meaning "left"), causes a counterclockwise or *left-handed* rotation. Often different wavelengths are rotated by different amounts, causing color changes as the analyzer is rotated. You can easily observe optical activity in ordinary transparent corn syrup (dextrose), which causes about 12°/cm rotation. A *saccharimeter* uses this effect to measure the sugar concentration in commercial syrups, wines, and so on, and in urine samples to test for suspected diabetes.

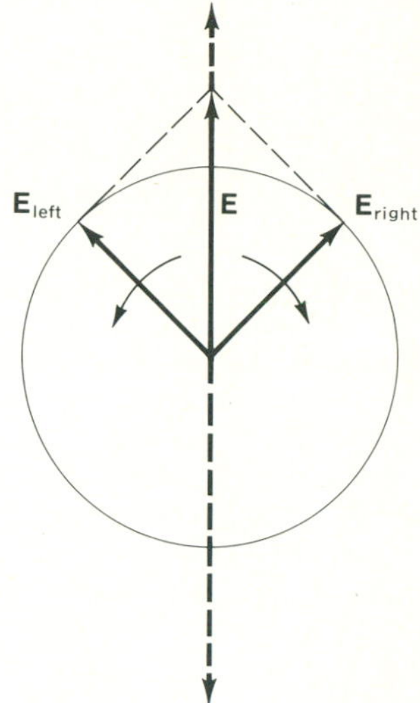

Figure 32–16
The oscillating electric field vector **E** in linearly polarized light may be considered as the sum of two circularly polarized components, $\mathbf{E}_{left}$ and $\mathbf{E}_{right}$, rotating in opposite directions.

32.6 Interference Colors and Photoelasticity

If a sheet of birefringent cellophane is folded randomly several times and placed between polarizing sheets, the transmitted light shows a pattern of vividly colored areas. These colors arise because certain layers of cellophane may act as a half-wave plate for red light, while also acting as a quarter-wave plate for blue light, and so forth. Thus the direction of polarization of the light striking the cellophane may be rotated different amounts for different wavelengths, allowing some wavelengths to pass the second polarizer while others are blocked. The emerging light is therefore deficient in certain portions of the spectrum, producing striking color effects. Rotating the cellophane or either one of the polarizers produces changing colors that are beautiful to see.

This aesthetically pleasing effect has practical uses. Transparent scale models of mechanical structures such as I-beams and arches are made from special *photoelastic* plastics. When the models are placed between a polarizer and an analyzer and "loaded" by applying forces to them, the plastic becomes birefringent in amounts proportional to the applied stress. The resultant patterns of light and dark (and colors) give a map of the regions of mechanical

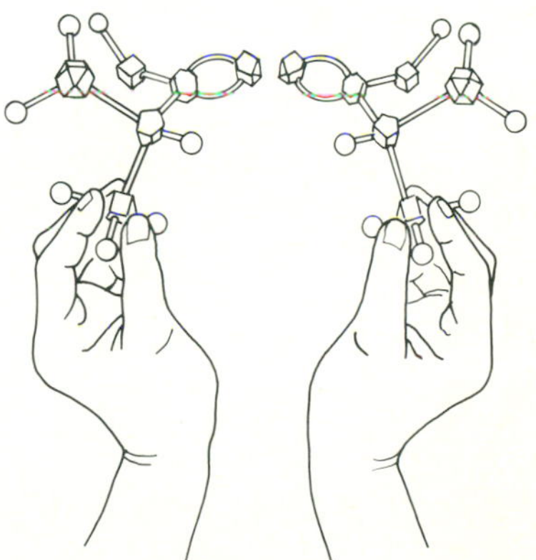

Figure 32–17
Stereoisomers have the same chemical composition, but the physical configurations of their atoms are mirror images.

(a)

(b)

(d)

(c)

(e)

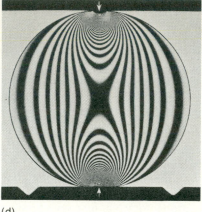

Figure 32–18
Models of mechanical structures are made of special photoelastic plastic and placed between polarizing sheets. When forces are applied to the models, they become birefringent, producing patterns that indicate the stress distribution within the models.

stress within the model (see Figure 32–18). Similar photoelastic colors can be observed if you use polarizing glasses to view light reflected from plastic boxes, plastic T-squares used in drafting, and other transparent objects. (Some plastics are not strongly birefringent, so you may have to search to find those that show marked effects.) Rear windows of automobiles are often purposely heat-stressed nonuniformly so that in case of breakage, the sudden release of stresses causes the entire window to crumble into tiny gravel-sized fragments rather than to shatter into large shards as ordinary glass does. The stresses make the glass birefringent. You can observe this stress pattern when wearing polarized glasses or, with the unaided eye, by noting the reflections of polarized sky light from the window.

Polarized light is useful in numerous other applications. For example, atoms in the presence of magnetic fields emit polarized light (the Zeeman effect); this polarization is used in the investigation of magnetic fields near sunspots and in distant stars. Also, magnetic fields in far regions of our galaxy cause elongated dust grains present in interstellar gas and dust clouds to align parallel to one another. Light from nearby stars scattered by these clouds is partially polarized, so by analyzing the percentage and direction of polarization of this scattered starlight, information about these distant magnetic fields can be obtained. Much information about crystal structure, biological specimens, and other materials is obtained by analysis with polarized light.

Summary

Transverse waves are *linearly polarized* if all the vibrations associated with the waves are parallel to the direction of a fixed line in space. The *direction of polarization* of an electromagnetic wave is the direction of the *electric* field vector.

When a single (free) atom undergoes a transition from a higher to a lower energy state, it emits a *wave train* of radiation, which for visible light is of the order of 1 to 3 m long in the direction of propagation. *Unpolarized* light is the superposition of many wave trains whose electric vectors are oriented randomly in all possible directions.

Certain transparent materials such as Polaroid selectively absorb some directions of polarization more than other directions, so they transmit electromagnetic waves that are partially or completely linearly polarized. If the direction of polarization of incident polarized light (intensity I_0) has an angle θ with respect to the *transmission axis* of an ideal polarizer, the transmitted intensity I (proportional to E^2) is given by

$$I = I_0 \cos^2 \theta$$

Birefringent substances have two indices of refraction, depending on the direction of polarization of the incident light. *Retardation plates* (or *wave plates*) are constructed of birefringent materials so that the *ordinary* (o) and *extraordinary* (e) waves emerge out of phase. When a polarized wave passes through a quarter-wave plate, one component is shifted 90° relative to the other component; when a polarized wave passes through a half-wave plate, one component is shifted 180° relative to the other. *Circularly polarized* light is composed of o and e components of equal amplitude that are out of phase by 90°. *Optically active substances* (sugar solution, for example) transmit circularly polarized light with two different speeds, causing a shift in the direction of *linearly* polarized light.

Unpolarized light becomes 100% polarized when reflected from dielectric materials at an angle of incidence called *Brewster's angle* θ_p, given by

$$\tan \theta_p = n$$

where n is the index of refraction of the material relative to that of the surrounding medium. (At other angles, the light will be partially polarized.)

Interference colors are produced from white light when various thicknesses of birefringent films are placed between polarizing sheets (a *polarizer* and an *analyzer*). The colors arise because certain layers of the film act as a half-wave plate for, say, red light, but as a quarter-wave plate for blue light, thus shifting the direction of polarization more for certain wavelengths than for others and allowing some wavelengths to pass the second polarizer while others are blocked. Therefore, the emerging light has certain portions of the spectrum missing; the remaining portions produce the color effects.

Mechanical structures may be analyzed by constructing transparent models from *photoelastic* materials. When a model is placed between polarizing sheets and mechanically stressed, the material becomes birefringent, producing fringe patterns that reveal the stress conditions within the structure.

Questions

1. Can longitudinal waves such as sound waves be polarized? If so, how?
2. Which phenomenon, polarization or interference, provides the most convincing evidence for the wave nature of light?
3. What aspect of the wave nature of light do polarization phenomena reveal that interference does not?
4. A radio-telephone transmitter in an automobile uses an antenna that is straight and vertical. Is the electromagnetic radiation from such an antenna vertically or horizontally polarized? Why?
5. A grid of closely spaced vertical wires is opaque to vertically polarized microwaves. Why?

6. Light is not transmitted through crossed polarizers. However, if a third polarizer is placed between the crossed polarizers, some light may be transmitted. Explain.

7. How can a stack of polarizing sheets be used to rotate the plane of polarization of polarized light?

8. One form of a variable-density light filter consists of two polarizing sheets placed together such that the orientations of their transmission axes may be rotated relative to each other. Does a small rotation produce a greater change in transmitted intensity when the axes are nearly aligned, nearly crossed, or at some angle in between?

9. One sheet of polarizing material is removed from a stack of randomly oriented polarizing sheets. As a result, the light transmitted through the stack decreases. How could this happen?

10. An ideal polarizing sheet transmits only half of the incident unpolarized light. What happens to the other half?

11. How can the direction of the optic axis of a birefringent crystal be found?

12. When looking through a calcite prism resting on a printed page, the two images of printing do not seem to be in the same plane. Explain.

13. Many fishermen use polarized sunglasses while fishing. Why?

14. Can light be polarized by reflection at an interface between two transparent media if the light is traveling toward the interface from the region of higher refractive index?

15. How would you determine whether a beam of light is unpolarized, plane polarized, or circularly polarized?

16. In some situations a photographer uses a polarizing filter over the lens of his or her camera. What would be some of these situations?

17. A beam of plane-polarized light may be represented by the superposition of two circularly polarized beams of opposite rotation. What is the effect of changing the relative phase of the two beams?

18. A fascinating device consists of a pair of polarizing sheets, each of which has a quarter-wave plate laminated to it. Light is transmitted when one of the pair is placed over the other, but is not transmitted when the order of the pair is interchanged. What are the details of their construction and why do they behave as they do?

19. If one slit of a double-slit interference apparatus were covered by a polarizing sheet with its axis perpendicular to the slit, while the other slit were covered by a polarizing sheet with its axis parallel to the slit, would an interference pattern be produced? Explain.

20. Photoelastic plastic models of mechanical structures placed between polarizers exhibit stress by producing colored bands, as shown in Figure 32–18. How can the spacing of the bands be interpreted?

Problems

32A–1 Unpolarized light passes through two polarizing sheets. If the angle between the transmission axes of the sheets is 60°, determine the fraction of the incident light energy absorbed by the sheets. Assume that the polarizing sheets are ideal; that is, each of the sheets alone transmits 50% of unpolarized incident light energy. *Answer:* $\frac{7}{8}$

32A–2 Two ideal polarizing sheets are placed together so that there is an angle θ between their transmission axes. Find the angle θ such that the sheets transmit 45% of the incident unpolarized light.

32A–3 Derive Brewster's law for polarization by reflection.

32A–4 An unpolarized light beam reflected from the surface of water is plane polarized for an incident angle of 53°.

(a) Calculate the index of refraction for the water.

(b) Show that the angle the refracted beam makes with the normal to the surface is the complement of 53°.

32A–5 A beam of unpolarized light is incident upon a sheet of glass at the polarizing angle of 58°. Find the angle of the refracted beam inside the glass. *Answer:* 32°

32A–6 For a particular wavelength, the index of refraction is 1.50 for a sample of glass. Calculate the Brewster angle θ_p for this refractive index. In general, does the Brewster angle increase or decrease as the wavelength of incident light increases?

32A–7 The critical angle for total internal reflection in a dielectric material is θ_c. Derive an expression for Brewster's angle θ_p in terms of θ_c for the material.

Answer: $\tan\theta_p = \csc\theta_c$

32B–1 Two polarizing sheets are placed together with their transmission axes crossed so that no light is transmitted. A third sheet is inserted between them with its transmission axis at an angle of 45° with respect to each of the other axes. Find the fraction of incident unpolarized light that will be transmitted by the combination of the three sheets. (Assume each polarizing sheet is ideal.) *Answer:* $\frac{1}{8}$

32B–2 Unpolarized light falls upon a stack of ideal polarizing sheets. The transmission axis of the second sheet is rotated 30° with respect to that of the first sheet, and the transmission axis of the third sheet is rotated 30° with respect to that of the second sheet. Calculate the fraction of the incident light transmitted by the stack of sheets.

32B–3 Brewster's angle for a plate of glass is 57° when the plate is in air. Calculate Brewster's angle for the glass plate when under water ($n = 1.33$). *Answer:* 49.2°

32B–4 A half-wave plate is inserted between two polarizing sheets whose directions of polarization are parallel. The half-wave plate is oriented with respect to the first sheet as in Figure 32–12.

(a) Explain why *no* light passes through the combination.

(b) If, instead, the two polarizing sheets are crossed at 90°, explain why *all* the light transmitted by the first sheet passes through the second sheet.

(c) In part (a), explain qualitatively the nature of the light emerging from the combination as the half-wave plate is slowly rotated through 360°.

32B–5 A beam of circularly polarized light is incident upon a polarizing sheet. Explain why the light is transmitted equally well for all orientations of the sheet.

32B–6 Quartz is birefringent, with indices of refraction of 1.553 and 1.544 for incident light of wavelength 590 nm. Find the minimum thickness of quartz that acts as a quarter-wave plate at this wavelength.

32B–7 Show that when a beam of circularly polarized light is incident upon a quarter-wave plate, the emerging light is plane polarized. Also show that if the rotation sense of the circularly polarized light is reversed, the direction of polarization is changed by 90°.

32C–1 A variable transmission filter is composed of two polarizing sheets, one of which can be rotated relative to the other. Determine the angle between the transmission axes where an incremental change in rotation produces the greatest change in light transmission.

Answer: 0° and 90°

32C–2 A stack of polarizing sheets may be used to rotate the plane of polarization of plane-polarized light. If a stack of 10 sheets were used to produce a 90° rotation, determine the maximum percentage of the incident plane-polarized light that will be transmitted. Assume that each sheet alone transmits 50% of the incident unpolarized light.

32C–3 The minimum thickness of the half-wave plate described in Example 32–3 is too thin to be practical. Determine an exact thickness near 0.1 mm that will produce the same effect as the minimum thickness. *Answer:* 0.085 65 mm or 0.1199 mm

chapter 33

The Quantum Nature of Radiation

Physics is very muddled again at the moment; it is much too hard for me anyway, and I wish I were a movie comedian or something like that and had never heard anything about physics!

WOLFGANG PAULI
(in a letter to R. Kronig, 25 May 1925)
[*American Journal of Physics*, *43*, 208 (1975)]

I do not like it, and I am sorry I ever had anything to do with it.

E. SCHRÖDINGER
(on quantum mechanics)

Toward the end of the nineteenth century, our understanding of what is now called *classical physics* had reached an impressive stage. It was believed that almost everything was known about the physical world and its interactions—at least, this was the opinion expressed by several well-known scientists at that time. *A more embarrassing misconception can hardly be imagined.* Yet, considering the widespread success of Newtonian mechanics in explaining the motion of all kinds of objects from baseballs to the solar system, and the fact these same ideas also brought all heat phenomena under the rules of mechanics, it seemed reasonable that we had, at last, found a great unifying theory that explained all phenomena. There were also radio waves, light, and so forth, which were obviously apart from mechanics, but these, too, were brought together in another unifying theory, that of Maxwell's electromagnetism. Together these two theories seemed to complete our understanding of all natural phenomena in terms of *particles* and *waves.*

However, a few surprises began to surface. In 1895 Wilhelm Konrad Roentgen discovered x-rays; the next year Antoine Becquerel discovered nuclear radioactivity; and the year after that, J.J. Thomson's measurements of e/m for electrons showed that they were a fundamental component of all atoms, so the model of an atom needed revision. In addition, there were a few well-known phenomena that still remained a mystery. For example, the spectral distribution of wavelengths emitted by hot, glowing bodies had no satisfactory theoretical explanation. And the fact that ultraviolet light could eject electrons from metals had some very puzzling aspects. But most scientists felt that these were merely a few isolated instances that sooner or later would also be explained by the two "complete" theories of the day, Newton's mechanics and Maxwell's electromagnetism. If this had been true, the future activity for physicists would have been quite dull; that is, it would have consisted of applying these theories to the few remaining puzzles and of determining the next decimal places in the fundamental constants of nature (the charge on the electron, the speed of light, Avogadro's constant, and so on).

We now tell the story of how the few minor cracks in the foundations of physics widened and brought the smug complacency of the nineteenth century tumbling down. In the process, physics itself expanded rapidly and became greatly strengthened. The revolution that occurred—the *quantum revolution*—was even more troubling and difficult to accept than Einstein's theory of relativity was a few years later. In a sense, relativity is considered part of classical physics (prequantum, that is) because the fundamental concepts of mass, momentum, energy, and the way systems interchange energy remain essentially unchanged. Einstein's revolution was to completely revise the structure of space and time within which measurements are made and to extend classical concepts so that physical laws would be correct for high velocities. The quantum revolution revised classical concepts so that they were correct for very small distances. The new physics of both relativity and quantum mechanics includes classical physics as special cases. But the quantum revolution was perhaps the more revolutionary because it altered our most basic concepts of *particles* and of *electromagnetic waves*—the only "stuff" physicists in those days believed the universe was made of. The new quantum physics demonstrated that these classical ideas were inadequate and often led to profound contradictions, both in disagreeing with experiment and in challenging basic philosophical issues about the nature of matter and our perception of it.

33.1 The Spectrum of Cavity Radiation

One outstanding unsolved puzzle in physics in the late nineteenth century was the spectral distribution of so-called *cavity radiation*, also referred to as *blackbody radiation*. It was shown by Kirchhoff that the most efficient *radiator* of electromagnetic waves was also the most efficient *absorber*. A "perfect" absorber would be one that absorbs all incident radiation; since no light would be reflected, it is called a *blackbody*.

To investigate the nature of radiation, it seems best to construct the most efficient radiator of all. How does one make a blackbody? The nearest practical approach to an ideal blackbody is a tiny hole in a cavity with rough walls (Figure 33–1). Any radiation that enters the hole has negligible chance of being reflected out through the opening: It is essentially 100% absorbed. As the walls of the cavity absorb this incoming radiation, their temperature rises and they begin to radiate. They continue to radiate until *thermal equilibrium* is reached, at which time they radiate electromagnetic energy at the same rate they absorb it. The radiation inside is then called blackbody radiation, or cavity radiation, and the tiny amount that manages to leak out the hole can be studied. *The hole itself is the blackbody.*

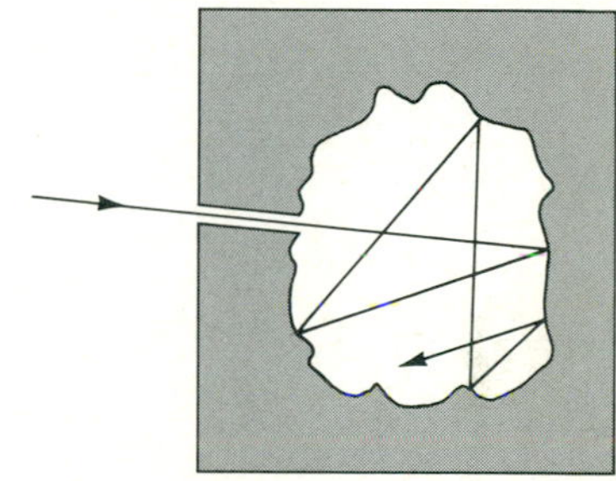

Figure 33–1
A practical approximation of an ideal blackbody is a hole that leads to a cavity with rough walls. The hole is the blackbody, since essentially all of the radiation incident on the hole is absorbed.

In 1879, the Austrian physicist J. Stefan first measured the total amount of radiation emitted by a blackbody at all wavelengths and found it varied as the fourth power of the absolute temperature. This was later explained through a theoretical derivation by L. Boltzmann, so the result became known as the *Stefan-Boltzmann radiation law.*

STEFAN-BOLTZMANN RADIATION LAW

$$R = \sigma T^4 \qquad \textbf{(33–1)}$$

where the **total emittance** R is the total energy of all wavelengths emitted per unit time and per unit area of the blackbody, T is the kelvin temperature, and σ is the *Stefan-Boltzmann constant*, equal to $5.672 \times 10^{-8}\ \text{W/m}^2\cdot\text{K}^4$. (It should be noted that the total emittance for an *outside* surface is always somewhat less than this and is different for different materials.)

In examining the spectral distribution of cavity radiation (the amount of energy radiated at various wavelengths), a startling discovery was made. *The spectral distribution does not depend on the material of the cavity, but only on the absolute temperature T.* No matter what the cavity is made of, the spectral distribution is the same for a given temperature. Whenever physicists discover a phenomenon that is independent of the material involved, they feel there is a strong probability that the effect involves a very basic interaction. So it is important to them to understand the effect thoroughly.

33.2 Attempts to Explain Cavity Radiation

Many capable physicists tried to develop a theory based on classical ideas that could predict the spectral distribution of cavity radiation. The goal was to derive the **spectral energy density** (in joules/meter^3) for the cavity radiation between wavelengths λ and $\lambda + d\lambda$. This is defined in terms of a mathematical function, $f(\lambda,T)$, that depends on both the wavelength λ and the absolute temperature T. Figure 33–2 shows experimental curves for three different temperatures. Note that as the temperature increases, the wavelength at the peak of each curve is displaced toward shorter wavelengths. The German physicist W. Wien derived a relationship for this feature, known as *Wien's displacement law.*

WIEN'S DISPLACEMENT LAW

$$\lambda_m T = \text{constant} \tag{33–2}$$

where λ_m is the wavelength at the maximum of the spectral distribution, T is the absolute temperature, and the constant is experimentally found to be 2.898×10^{-3} m·K.

The total energy density at all wavelengths is the area under the curve:

$$\begin{matrix}\text{Total energy density} \\ \text{(all wavelengths)}\end{matrix} = \int_0^\infty f(\lambda,T)\,d\lambda \tag{33–3}$$

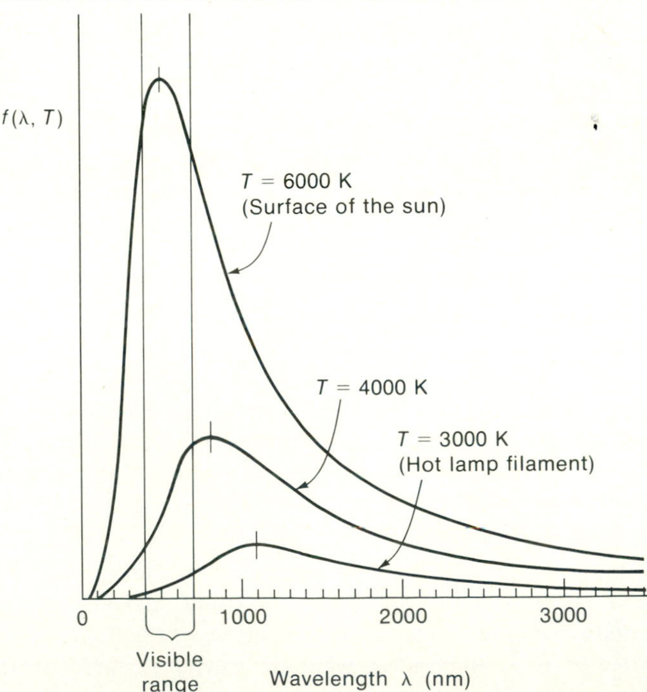

Figure 33–2
Spectral distribution curves for cavity radiation at three different temperatures. The function $f(\lambda,T)$ is the energy density per wavelength interval of the cavity radiation between wavelengths λ and $\lambda + d\lambda$. The small vertical lines on the curves show that as the temperature becomes hotter, the wavelength at the peak becomes shorter, according to Wien's displacement law.

According to the Stefan-Boltzmann law, the total emittance is proportional to the fourth power of T, so the area under the curve for $T = 6000$ K is 16 times that for $T = 3000$ K.

Also note that the fraction of the radiation that falls within the *visible* range[1] is not uniform. At low temperatures, there is relatively more energy radiated at long wavelengths (red) than at shorter wavelengths (blue). As the temperature increases, this changes to relatively more radiation in the blue, explaining the common observation of color changes that occur as a solid is heated: It first begins to glow with a dull red, progressing through orange, yellow-white, and finally, at very high temperatures, blue-white.

EXAMPLE 33–1

The wavelength at the peak of the spectral distribution for a blackbody at 4300 K is 674 nm (red). At what temperature would the peak be 420 nm (violet)?

SOLUTION

From the Wien displacement law, we have (for the wavelengths at the maximum):

$$\lambda_1 T_1 = \lambda_2 T_2$$

Substituting numerical values:

$$(674 \times 10^{-9}\ \text{m})(4300\ \text{K}) = (420 \times 10^{-9}\ \text{m})(T_2)$$

$$\boxed{6900\ \text{K}} = T_2$$

Wien's Theory

The search for a theoretical basis for the radiation formula is one of the most fascinating chapters in the history of physics. We will relate just a few high points here. In 1884, Boltzmann used a thermodynamic approach to the problem, assuming the cavity radiation was in a cylinder with a movable piston and calculating the results of a Carnot-cycle process. (The radiation exerts a force on the walls, so ideas of work came into the analysis.) In 1893, Wien expanded on this result and, from considerations of the Doppler shift upon reflection from the moving piston, derived the fact that some function of the product (λT) was involved. Making certain assumptions about the emission and absorption of radiation, he derived the following for $f(\lambda,T)$, the *spectral energy density* (in J/m^3) for cavity radiation between wavelengths λ and $\lambda + d\lambda$:

WIEN'S RADIATION LAW

$$f(\lambda,T) = \frac{c_1\lambda^{-5}}{e^{(c_2/\lambda T)}} \qquad (33\text{–}4)$$

The unknown constants, c_1 and c_2, are to be determined by making the "best fit" to the experimental data. The curve fit the data well at short wavelengths,

[1] It is interesting that the visual sensitivity of our eyes centers on the peak of the sun's radiation distribution. If there are sensing beings on planets around a star of different temperature, perhaps, through evolution, their "eyes" evolved to respond to a different portion of the electromagnetic spectrum.

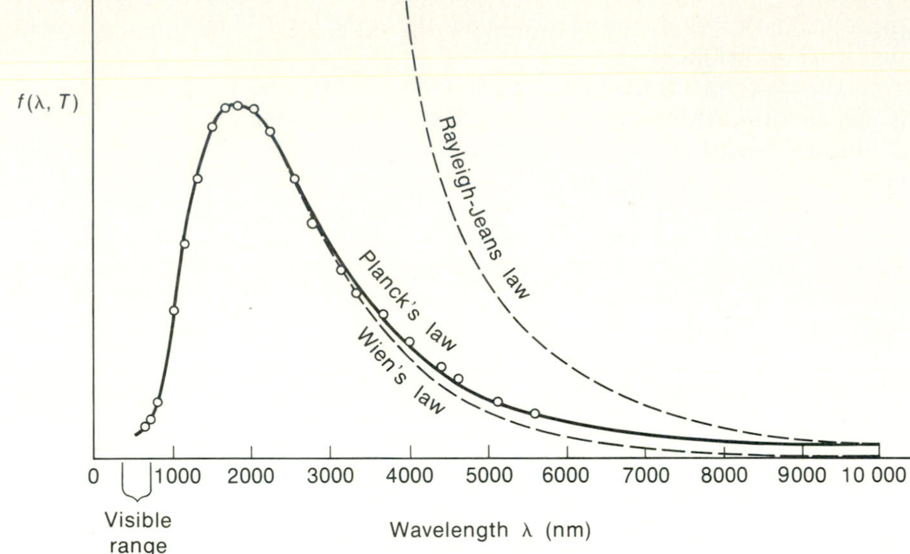

Figure 33–3
The circles are experimental points for cavity radiation at 1600 K. Curves for three different theories are shown for comparison.

but as more and more experimental data at long wavelengths were obtained, the disagreement became obvious. Figure 33–3 shows how the Wien curve falls below the experimental points at long wavelengths.

The Rayleigh-Jeans Theory

The thermodynamic derivation of Boltzmann and Wien was a helpful step in revealing that probably some function of λ and T was involved. However, since thermodynamics is based on very general principles that apply to *all* systems, often thermodynamic arguments do not give insight into the particular processes involved in a given system. Perhaps more success would come if one focused on the source of the cavity radiation, the actual process of radiation, and the absorption by the walls.

Rayleigh (1900) approached the problem from this viewpoint. He considered a rectangular cavity with metallic walls and assumed that the electric charges in the walls were the source of the radiation. They behaved as simple harmonic oscillators and could radiate as well as absorb radiation, each with its "characteristic" natural frequency of oscillation. For any sufficiently large enclosure, there were such an extremely great number of oscillators that the resulting negligible differences between adjacent frequencies caused the radiation to appear continuous over all wavelengths. At a given temperature T, constant operation of the oscillators means that *standing waves* would be set up in the enclosure. With perfectly conducting walls, the standing waves must have nodes at each wall. The total number[2] of such standing waves (per unit volume) turned out to be $8\pi\lambda^{-4}$.

The *equipartition theorem* (Chapter 30) states that, on the average, $\frac{1}{2}kT$ of energy is associated with each variable required to specify the energy of a system

[2] Rayleigh made a trivial error of a factor of 8 in the derivation. After the result was published, Sir James Jeans pointed out the obvious mistake, so the corrected formula became known as the Rayleigh-Jeans law. In this instance, considerable fame resulted from a rather minor contribution. Of course, Jeans also made a great many other contributions to physics.

in thermal equilibrium at absolute temperature T. For electromagnetic waves,[3] there are two variables (the two directions of polarization), so the total energy associated with each is

AVERAGE ENERGY OF A CLASSICAL SHM OSCILLATOR (in a system at thermal equilibrium)

$$E_{\text{ave}} = kT \tag{33-5}$$

where the Boltzmann constant k is equal to 1.381×10^{-23} J/K. Multiplying the number of standing waves by the average energy of each gives the Rayleigh-Jeans law:

RAYLEIGH-JEANS RADIATION LAW

$$f(\lambda,T) = 8\pi kT\lambda^{-4} \tag{33-6}$$

where k is the Boltzmann constant.

The theory fits the data at extremely long wavelengths, but as shown in Figure 33–3, it was in drastic error everywhere else. The curve never "bent over": As the wavelength approached zero, the curve continued to increase toward infinity. Since the discrepancy was greatest at short wavelengths, it became known as the *ultraviolet catastrophe*. And a catastrophe for classical physics it was. The Rayleigh-Jeans derivation was based on classical concepts of thermodynamics and statistical mechanics, which had been completely successful in every other application. Each step of the derivation seemed so plausible that it was extremely disturbing to find the result so inaccurate. Where was the error in thinking?

33.3 Planck's Theory

In 1900, the German physicist Max Planck stumbled upon a solution to the difficulties. He first found it by some purely mathematical reasoning, then tried to figure out the physical implications of the mathematical trick he employed. Even though he obtained a radiation law that agreed with the experimental data, thereby avoiding the ultraviolet catastrophe, the physical implications were so startling that for many years Planck himself did not want to accept them as describing what the "real world" was like. The quantum ideas were just too radical.

Planck's stratagem was the following. In the Rayleigh-Jeans derivation, an important step in the procedure was to find the average energy of a SHM oscillator by integrating over all possible energies the oscillator might have. Classically, such an oscillator (as, for example, a mass on a spring) could vibrate with any amplitude from zero on up. Since the energy is equal to the square of the amplitude, the oscillator could have any of a *continuum* of energy states, a range of values that varied smoothly from 0 to ∞. The trouble was that *integrating* over a *continuous* range of energies from 0 to ∞ made the function become infinite as $\lambda \to 0$. Planck was a good enough mathematician to realize that if, instead, he made a *summation* over a *discrete* range of energies from 0 to ∞, the result was a function that "turned over" and approached zero as $\lambda \to 0$, just like the experimental radiation curves. As it turned out, the curve Planck

[3] One could also apply the equipartition theorem to the SHM oscillators in the walls. In SHM, two variables are required: one for the kinetic energy and one for the potential energy. Therefore, each SHM oscillator has an average energy of kT.

obtained matched the experimental points exactly. This put Planck in a position similar to that of a student who has looked in the back of the book to find the right answer to a problem, but is then faced with finding out how to get there from the given facts. What was it about nature that made a summation of *discrete* energy states the proper approach?

Planck decided on a bold step. Although it disagreed with all classical theories, he assumed that a SHM oscillator with a natural frequency f was "allowed" to have only one of a discrete series of energies: 0, hf, $2hf$, $3hf$, . . . , where h is a constant.

ALLOWED ENERGIES FOR A QUANTIZED SHM OSCILLATOR

$$E_n = nhf \qquad \text{(where } n = 0, 1, 2, 3, \ldots) \tag{33–7}$$

Planck first determined the constant by fitting experimental data to the expression for $f(\lambda,T)$ that evolved from his theory. He obtained a value very close to the currently accepted value:

PLANCK'S CONSTANT $\quad h = 6.626 \times 10^{-34}\ \text{J·s}$

Figure 33–4 compares *energy-level* diagrams for the classical and the quantum cases.

As a second assumption, Planck proposed that the only amount of energy ΔE an oscillator could emit or absorb was a **quantum**[4] of energy:

$$\Delta E = hf \tag{33–8}$$

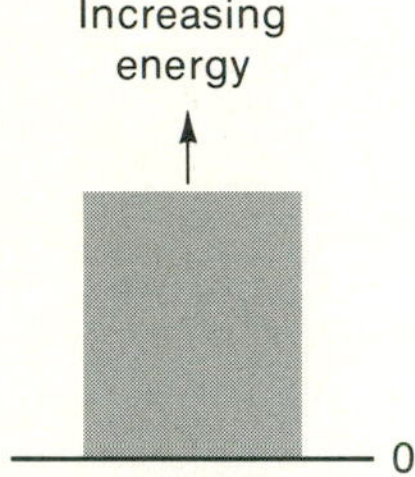

(**a**) According to classical mechanics, the possible energy states form a *continuous* distribution.

With these assumptions, Planck found the average energy for a collection of oscillators in thermal equilibrium at absolute temperature T to be

$$E_{\text{ave}} = \frac{hf}{(e^{hf/kT} - 1)} \tag{33–9}$$

Transforming the variable from frequency f to wavelength λ through $f\lambda = c$, we have

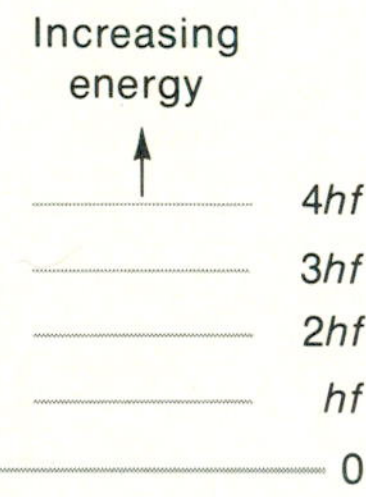

(**b**) According to quantum mechanics, the possible energy states form a *discrete* distribution.

Figure 33–4
Energy-level diagrams for a SHM oscillator of natural frequency f.

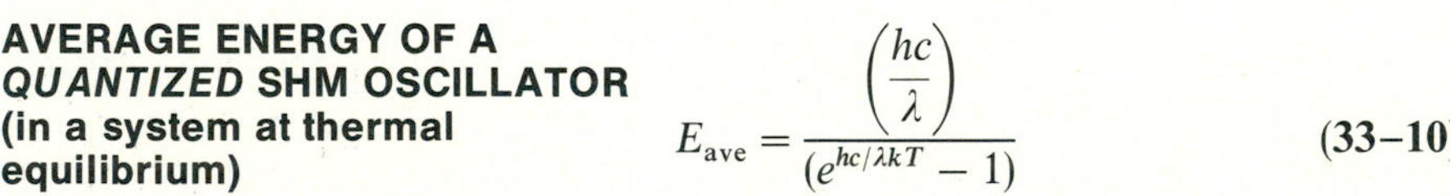

AVERAGE ENERGY OF A *QUANTIZED* SHM OSCILLATOR (in a system at thermal equilibrium)

$$E_{\text{ave}} = \frac{\left(\dfrac{hc}{\lambda}\right)}{(e^{hc/\lambda kT} - 1)} \tag{33–10}$$

This is quite different from the classical value of $E_{\text{ave}} = kT$ [Equation (33–5)].

If the oscillators had this average energy, then it must also be the average energy of the waves in the cavity (because everything was at thermal equilibrium). Multiplying this average energy by the Rayleigh-Jeans calculation for the number of standing waves, $8\pi\lambda^{-4}$, Planck obtained his expression for $f(\lambda,T)$, the *spectral energy density* (in joules/meter^3) for cavity radiation between wavelengths λ and $\lambda + d\lambda$:

PLANCK'S RADIATION LAW

$$f(\lambda,T) = \frac{8\pi hc\lambda^{-5}}{(e^{hc/\lambda kT} - 1)} \tag{33–11}$$

[4] The word *quantum* comes from the Latin word *quantus*, meaning "how much." Planck originally proposed that quanta could have integral multiples of hf, but Einstein and others later showed that only single units of hf were permissible.

As you can see from Figure 33–3, the Planck theory fits the experimental points beautifully. For short wavelengths, the Planck equation approaches the Wien expression, which was correct in that region. For long wavelengths, the Planck equation approaches the Rayleigh-Jeans law, correct for long wavelengths. Planck effectively built a bridge between the two classical radiation theories. However, to do so, he had to make a radical break with all previous ideas about the energy a system could possess. If nature really behaved this way and all systems had quantized energy states, why wasn't it discovered long ago? The following example will explain why.

EXAMPLE 33–2

A 5 g mass is hung from a string 10 cm long and set into motion so the string at extreme positions makes an angle of 0.1 rad with the vertical. Because of friction with the air, the amplitude gradually decreases. Can we detect the quantum jumps in energy as the amplitude decreases?

SOLUTION

The frequency of oscillation is given by Equation (12–8):

$$f = \frac{1}{2\pi}\sqrt{\frac{g}{\ell}}$$

$$f = \frac{1}{2\pi}\sqrt{\frac{9.8\ \frac{\text{m}}{\text{s}^2}}{0.1\ \text{m}}}$$

$$f = 1.58\ \text{s}^{-1}$$

The energy of the pendulum is equal to the gravitational potential energy at an extremity.

$$E = mgh$$

$$E = mg\ell(1 - \cos\theta)$$

$$E = (0.005\ \text{kg})\left(9.8\ \frac{\text{m}}{\text{s}^2}\right)(0.1\ \text{m})(1 - \cos 0.1)$$

$$E = 2.45 \times 10^{-5}\ \text{J}$$

The quantum jumps in energy would be

$$\Delta E = hf$$

$$\Delta E = (6.63 \times 10^{-34}\ \text{J·s})(1.58\ \text{s}^{-1})$$

$$\Delta E = 1.05 \times 10^{-33}\ \text{J}$$

The ratio is $\Delta E/E = 4.28 \times 10^{-29}$. Therefore, in order to detect the quantized nature of the energy states, we would have to measure energy to better than 4 parts in 10^{-29}, a sensitivity far beyond the capability of any experimental technique.

As the example shows, the quantization of energy states is undetectable for *macroscopic* mechanical systems. The "graininess" of energy transfers is usually not noticed in everyday phenomena because of the smallness of h. If h were bigger, we would see quantum effects all around us. Quantum effects are always present, but they become noticeable only for *microscopic* systems on an atomic scale, that is, for cases where ΔE is of the order of E. This condition is what makes blackbody radiation (at high frequencies) behave in an unusual way, traceable to quantum effects. It is interesting that if, in all quantum equations, we let $h \rightarrow 0$, the equations turn into the corresponding classical expressions. Thus the new quantum mechanics is a more general theory, which contains classical mechanics as a special case.

33.4 The Photoelectric Effect

Today it is hard to realize the magnitude of the break with classical thinking that Planck initiated. Planck himself, who strongly resisted giving up the continuity of possible energy states, spent much effort in trying (unsuccessfully) to find an alternative solution to the ultraviolet catastrophe within the framework of classical physics. Though he grudgingly came to accept the idea that oscillators could have only quantized energy states and emit or absorb radiation in units of hf, he held to the classical view of radiation: Electromagnetic waves were not quantized. But soon even this link to classical physics was broken.

Heinrich Hertz was the first (in 1887) to experimentally produce the electromagnetic waves predicted by Maxwell's equations. Using an *induction coil* (a step-up transformer with a great many turns on the secondary) attached to two metal spheres as shown in Figure 33–5, he initiated an *oscillating*[5] spark across gap A. A nearby metal ring with a gap B would respond by sparking across its gap, verifying that electromagnetic waves had traveled from A to B. Quite by accident, Hertz discovered that the spark at B could be initiated much more easily if the gap were illuminated by ultraviolet light. Ten years later Thomson discovered the electron, and it was then verified that the ultraviolet light ejected electrons from the gap electrodes, making the spark easier to form. The phenomenon of electron ejection by light was called the *photoelectric effect*.

Figure 33–6 shows an experimental apparatus for investigating the effect. At any one time, light of essentially just one wavelength (*monochromatic* light)

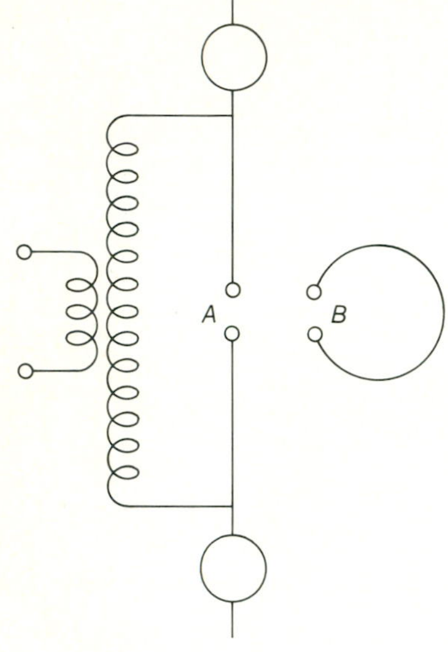

Figure 33–5
The experimental apparatus used by Hertz to detect electromagnetic waves.

[5] The frequency of the oscillation was determined by the capacitance of the spheres and the inductance of the induction coil.

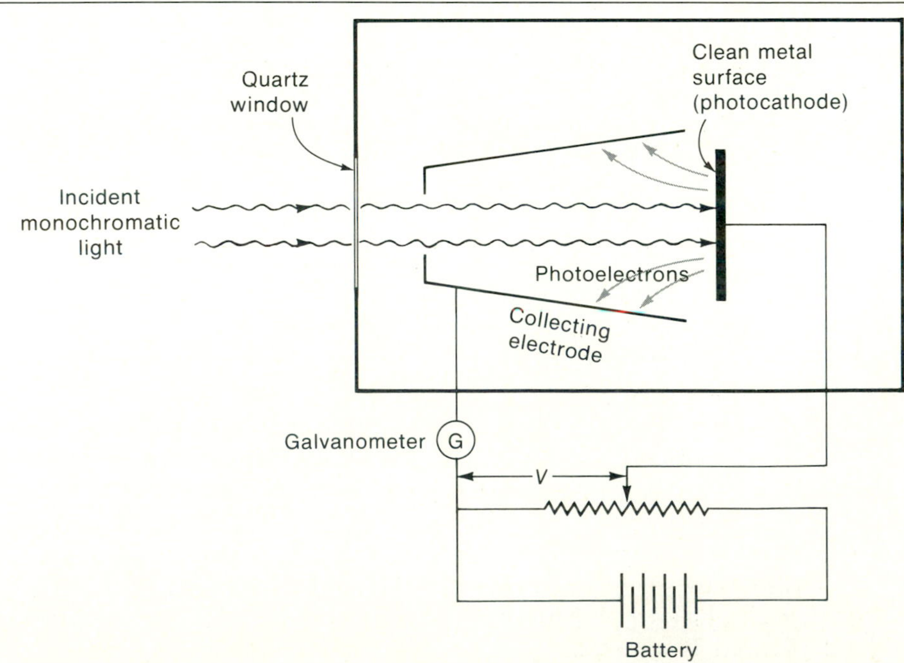

Figure 33–6
An experimental arrangement for investigating the photoelectric effect. The quartz window passes wavelengths in the ultraviolet that would be stopped by ordinary glass. The battery voltage can be reversed by a switching circuit (not shown).

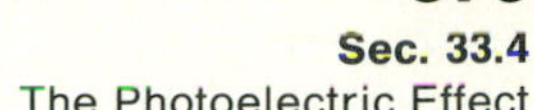

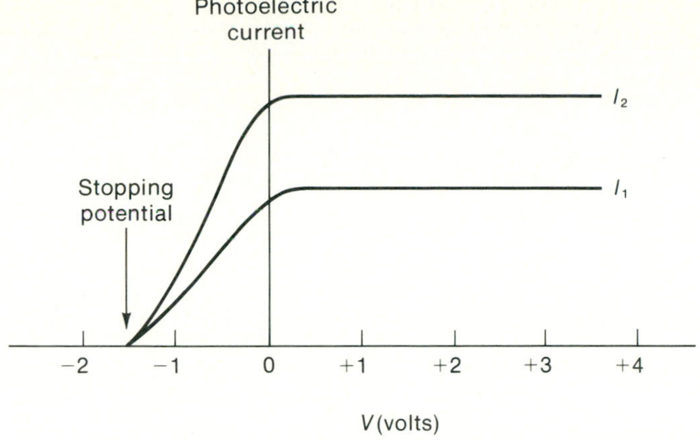

Figure 33–7
The photoelectric current versus the potential V of the collecting electrode with respect to the photocathode. Curves for monochromatic light of two different intensities are shown. Both have the same stopping potential.

was used. According to classical wave theory, the electric field of the incident light could transfer some of its energy to electrons in the surface of the metal, allowing them to acquire sufficient energy to escape. If the intensity of the light were increased, the ejected *photoelectrons* should acquire greater kinetic energy because of the stronger electric field of the light. The frequency of the light, however, should not make any difference at all. *Both of these deductions from classical theory disagree with experimental data.*

Figure 33–7 shows experimental curves for the *photocurrent* resulting when light of (essentially) a single wavelength is incident. The *stopping potential* is the negative voltage V_0 applied to the collecting electrode such that the kinetic energy of the most energetic electrons will be converted to potential energy at the collector. That is, the voltage V_0 barely stops the most energetic photoelectrons from reaching the collector. The relation is

$$eV_0 = \tfrac{1}{2}mv_{max}^2 \qquad \textbf{(33–12)}$$

The two just-mentioned features of the photoelectric effect that are contrary to predictions of classical theory, along with a third feature, may be summarized as follows:

Classical prediction	**Experimental fact**
(1) As the intensity of light is increased, the electric field E becomes larger. Since the force on an electron is eE, increasing the intensity should increase the kinetic energy acquired by the electrons.	(1) Figure 33–7 shows that even though the light intensity is increased, the maximum[6] kinetic energy of the photoelectrons remains the same.
(2) The *frequency* of light should not affect the kinetic energy of the ejected photoelectrons. Only the *amplitude* of the electric field should change their energies.	(2) As the frequency of the light is reduced, a *threshold frequency* is reached, below which *no* photoelectrons are produced, regardless of the light intensity (see Figure 33–8).

[6] The photoelectrons emerge with a spectrum of energies from zero to the maximum value indicated in Equation (33–12). Presumably many come from varying depths, just within the surface, where they must expend some energy to make their way through the lattice of metal atoms as well as overcome the attractive forces at the surface.

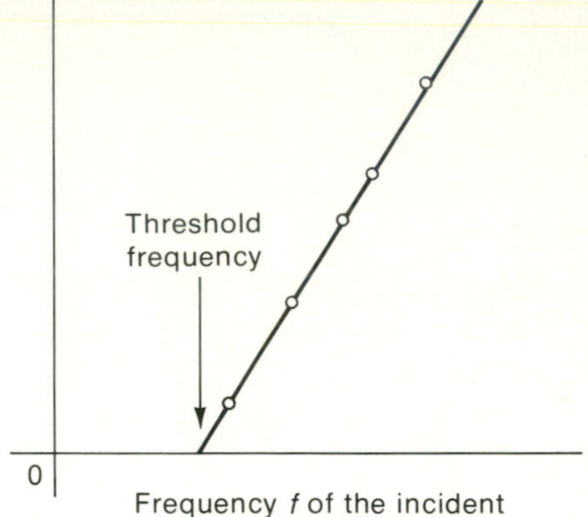

Figure 33–8
The frequency of the incident light determines the maximum kinetic energy of the photoelectrons. Below a certain "cut-off" value called the *threshold frequency*, no photoelectrons are ejected regardless of the intensity of the incident light.

(3) Assuming a single electron in the metal surface could absorb energy over an "effective target area" about the size of an atom, very dim light should require a longer time before the electron absorbs sufficient energy to escape.

(3) No appreciable time delay has ever been observed (though *fewer* electrons are produced by the dimmer light). The upper limit on measurements of the time delay is $<10^{-9}$ s; the actual time delay may be much less than that.

Clearly, classical ideas just do not come up with the correct predictions. The following example illustrates one of the discrepancies.

EXAMPLE 33–3

A cesium surface is 2 m from a point light source of 1 μW power that emits light uniformly in all directions. Assume that a single electron can absorb energy over a circular area of one atom (radius $\approx 10^{-10}$ m). The minimum energy required to extract an electron from the surface is 1.96 eV. Estimate how much time is required for the electron to absorb this amount of energy according to classical theory.

SOLUTION

The effective target area is $\pi r^2 \approx 3 \times 10^{-20}$ m^2. According to classical theory, the energy from the point source is spread uniformly over a spherical wave front of radius $R = 2$ m. Therefore, the power/meter^2 at the cesium surface is

$$\frac{P}{4\pi R^2} = \frac{10^{-6}\ \text{W}}{16\pi\ \text{m}^2} \approx 2 \times 10^{-8}\ \frac{\text{W}}{\text{m}^2}$$

For power that falls on the area of one atom, we obtain

$$\left(2 \times 10^{-8}\ \frac{\text{W}}{\text{m}^2}\right)(3 \times 10^{-20}\ \text{m}^2) = 6 \times 10^{-28}\ \frac{\text{J}}{\text{s}}$$

The minimum energy needed to escape the surface is

$$(1.96\ \text{eV})\left(\frac{1.6 \times 10^{-19}\ \text{J}}{1\ \text{eV}}\right) \approx 3 \times 10^{-19}\ \text{J}$$

and the time required to absorb this much energy is then

$$\frac{3 \times 10^{-19}\ \text{J}}{6 \times 10^{-28}\ \frac{\text{J}}{\text{s}}} = \boxed{5 \times 10^{8}\ \text{s}}$$

This is equal to about 16 years! Yet, experimentally, the upper limit to any possible time delay is less than 10^{-9} s—a discrepancy of a factor of $\sim 10^{17}$.

A plausible argument from electromagnetic theory can be made that an electron might absorb energy over a larger target area of the order of λ^2, where λ is the wavelength of the incident radiation. For visible light ($\lambda \approx 500$ nm), this improves the situation by only a factor of $\sim 10^8$, still leaving a factor of $\sim 10^9$ unaccounted for. There are not many experiments that disagree with theory so drastically!

□ □ □

In 1905, Einstein[7] proposed a solution to the photoelectric dilemma. Though Planck was reluctant to accept the possibility that electromagnetic waves were quantized, Einstein saw that if one assumed that radiation was actually well-localized "bundles" or quanta (later called **photons**), then the photoelectric effect could be simply explained. Einstein proposed the following:

EINSTEIN'S ASSUMPTION OF THE QUANTIZATION OF RADIATION

The emission and absorption of radiation of frequency *f* always occur in quanta (or *photons*) of energy: $E = hf$. The photon remains localized in space as it moves away from the source with a velocity *c*.

If photons remain well localized, then, Einstein reasoned, in the photoelectric process the photon could be completely absorbed by a single electron. After gaining an energy hf, the electron would use part of this energy in escaping from the surface, and its remaining energy would appear as kinetic energy of the electron. The *minimum* energy required to barely escape from a surface is called the **work function** w_0. (Typical values for metals are about 2 to 6 eV. Visible photons have energies of around 2 eV in the red to somewhat above 3 eV for blue. For this reason, some materials exhibit a photoelectric effect only for the more energetic photons of ultraviolet light.) Applying conservation of energy to the process, Einstein proposed that the maximum kinetic energy of the electrons would be related to the photon energy hf according to

EINSTEIN'S PHOTOELECTRIC EQUATION

$$hf = K_{\text{max}} + w_0 \tag{33–13}$$

[7] It was an incredible year for 26-year-old Einstein. Volume 17 of *Annalen der Physik* (1905) included his revolutionary paper on special relativity, a treatise on Brownian motion that enabled Perrin to determine Avogadro's number, and his article on the photoelectric effect. It was this article that led to Einstein's Nobel Prize in 1921. (See the chronology of quantum theory development at the end of Chapter 34.)

This simple idea immediately explained the three baffling features of the photoelectric effect mentioned previously:

(1) Since K_{max} depends only on the frequency of the light, and not on its intensity, dim light as well as bright light has the same stopping potential (Figure 33–7).
(2) For certain materials, the photon energy at a given wavelength may be less than the work function. Therefore, there is a threshold frequency, below which no photoelectrons would be produced.
(3) Since the photon energy is localized in space (rather than spread uniformly over a wave front), its total energy can be transferred to an electron in a single step, ejecting the electron with negligible time delay no matter how dim the illumination. (Of course, the number of photoelectrons depends on the light intensity.)

This close agreement with experiment in another area, distinct from blackbody radiation, seemed to force acceptance of the photon's existence. However, as we will discuss shortly, it was a large pill to swallow.

Photoelectric experiments yield a great deal of important information. For example, combining Equations (33–12 and 33–13) and rearranging, we have

$$V_0 = \left(\frac{h}{e}\right)f - \left(\frac{w_0}{e}\right) \qquad \textbf{(33–14)}$$

This is a straight-line function for the stopping potential V_0 as a function of frequency f (Figure 33–8). The slope of the line is h/e, furnishing another experimental method of determining Planck's constant h. (These values agree with those found previously from the completely different phenomenon of blackbody radiation. It is reassuring that separate pieces of evidence lock together like this to form an overall coherent picture.) Another feature of Equation (33–14) is shown in Figure 33–9. The intercept of the straight line on the horizontal axis is the threshold frequency, and the intercept with the vertical axis is the work function.

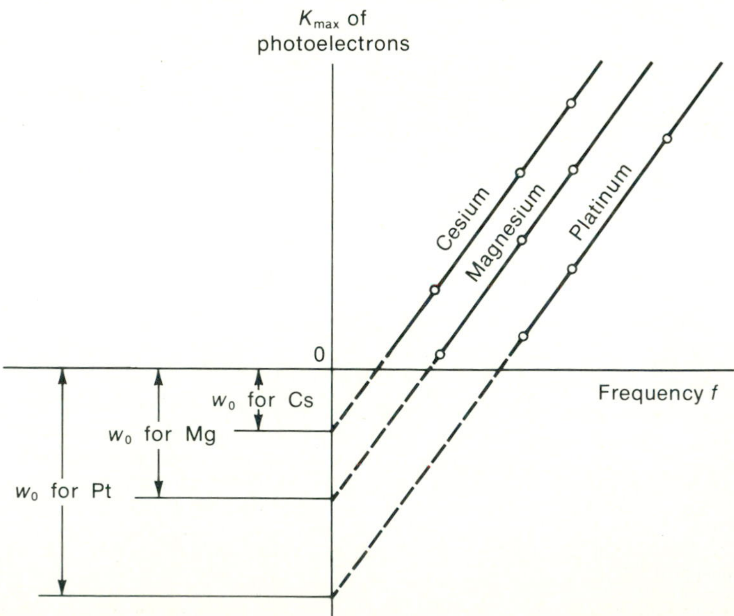

Figure 33–9
Photoelectric data for various substances produce straight lines whose slope is h/e. The lines intersect the horizontal axis at the threshold frequencies and the vertical axis at the respective work functions.

EXAMPLE 33–4

If the 1 μW light source of Example 33–3 emitted only monochromatic light of wavelength $\lambda = 550$ nm, (a) find the number of photons per second striking a circular target area 1 cm in diameter, and located $R = 2$ m from the source. (b) Find the maximum kinetic energy (in electron volts) of the photoelectrons. (c) Find the threshold frequency for cesium.

SOLUTION

(a) The fraction of the energy output of the source that falls on a circular target area ($r = 5$ mm) that is 2 m away is

$$\frac{\pi r^2}{4\pi R^2} = \frac{\pi(0.005\text{ m})^2}{4\pi(2\text{ m})^2} = 1.56 \times 10^{-6}$$

The power incident on the target is therefore

$$\left(1 \times 10^{-6}\,\frac{\text{J}}{\text{s}}\right)(1.56 \times 10^{-6}) = 1.56 \times 10^{-12}\,\frac{\text{J}}{\text{s}}$$

Each photon has an energy of

$$hf = h\left(\frac{c}{\lambda}\right) = \frac{(6.63 \times 10^{-34}\text{ J}\cdot\text{s})(3 \times 10^{8}\text{ m}\cdot\text{s}^{-1})}{550 \times 10^{-9}\text{ m}}$$

$$hf = 3.62 \times 10^{-19}\text{ J}$$

The number of photons per second striking the target is therefore

$$\frac{\left(1.56 \times 10^{-12}\,\frac{\text{J}}{\text{s}}\right)}{3.62 \times 10^{-19}\text{ J}} = \boxed{4.33 \times 10^{6}\,\frac{\text{photons}}{\text{second}}}$$

(b) The photon energy in electron volts is

$$(3.62 \times 10^{-19}\text{ J})\underbrace{\left(\frac{1\text{ eV}}{1.6 \times 10^{-19}\text{ J}}\right)}_{\text{Conversion ratio}} = 2.26\text{ eV}$$

The work function w_0 for cesium is (from Example 33–3) 1.96 eV. The maximum kinetic energy, K_{max}, of the photoelectrons is given by Equation (33–13):

$$hf = K_{max} + w_0$$

Solving for K_{max}:

$$K_{max} = hf - w_0$$

$$K_{max} = 2.26\text{ eV} - 1.96\text{ eV} = \boxed{0.300\text{ eV}}$$

(c) At the threshold frequency, f_{th}, the photon energy hf_{th} $(= hc/\lambda_{th})$ equals the work function w_0. Solving for λ_{th}, and substituting numerical values, we obtain

$$\lambda_{th} = \frac{hc}{w_0}$$

$$\lambda_{th} = \frac{(6.63 \times 10^{-34}\text{ J}\cdot\text{s})\left(3 \times 10^{8}\,\frac{\text{m}}{\text{s}}\right)}{(1.96\text{ eV})\underbrace{\left(\frac{1.6 \times 10^{-19}\text{ J}}{1\text{ eV}}\right)}_{\text{Conversion ratio}}} = 6.34 \times 10^{-7}\text{ m} = \boxed{634\text{ nm}}$$

This is in the red portion of the spectrum, so almost all wavelengths of visible light will eject photoelectrons from cesium.

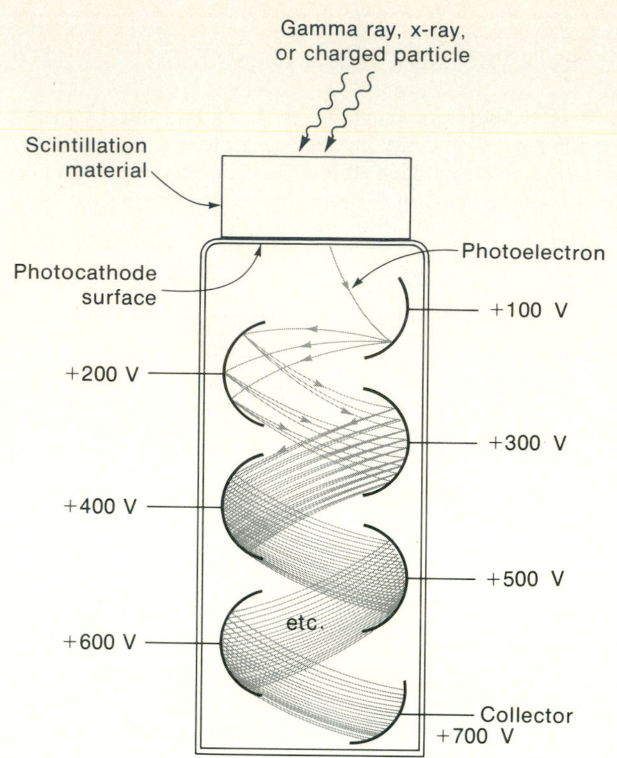

Figure 33–10 A scintillation counter uses a scintillation material with a photomultiplier tube to produce a large electrical pulse at the collector when a gamma ray, x-ray, or charged particle is absorbed in the scintillator.

The photoelectric effect has many practical applications. Most light meters for determining proper exposures in photography use the photocurrent produced by incident light for operating the meter. *Photocells* are the "electric eye" that opens doors when a beam of light is interrupted. They are also used in computers to detect holes in punched cards or paper tape. An instrument widely used in nuclear physics experiments is the *scintillation counter*, shown in Figure 33–10. A typical detector uses certain materials that emit tiny flashes of light or scintillations, when energy is absorbed from photons or charged particles. This light, in turn, falls on a photocathode surface, ejecting photoelectrons that subsequently strike a series of *dynodes*. If the impact velocity is high enough, a single electron striking a dynode will eject one or more additional electrons in a process called *secondary emission*. Typical multiplication factors are from 2 to 5 or more. Assuming a *photomultiplier* with 10 dynodes and a multiplication factor of 4 at each impact, a single photoelectron that starts down the chain produces 4^{10} ($\approx 10^6$) electrons at the collector. Many photomultipliers have gains as high as 10^9 or more.

33.5 The Compton Effect

An additional piece of evidence for the existence of photons was presented by A.H. Compton in 1923. He directed a monochromatic beam of x-rays at a thin slab of low-element material, such as carbon. He observed that the x-rays that

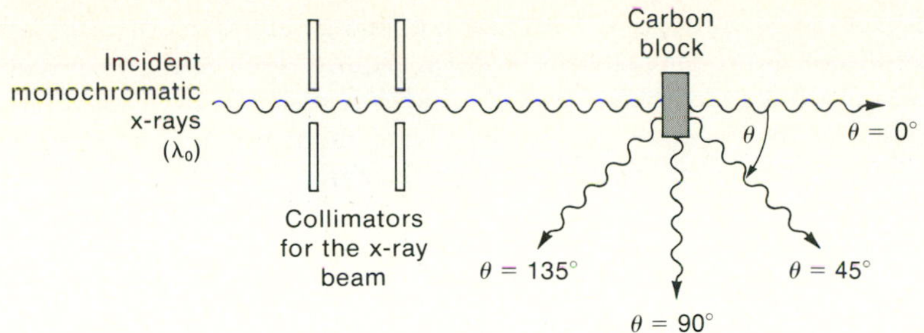

Figure 33–11
X-rays scattered at various angles have longer wavelengths λ' than the incident wavelength λ_0.

were scattered from the carbon at various angles had a *longer* wavelength λ' than the incident wavelength λ_0. Figure 33–11 shows the experimental arrangement, and Figure 33–12 shows the experimental data. The amount of wavelength shift, $\Delta\lambda = \lambda' - \lambda_0$, was the same regardless of the target material, implying it is an effect involving electrons rather than the atom as a whole. *Classical wave theory cannot explain this result.*

According to classical theory, the oscillating electric field of the incoming wave would set electrons in the target material into oscillations. These vibrating electrons would then reradiate electromagnetic waves, *but necessarily at the same frequency of the incident wave*, contrary to what was observed.

Compton invoked the photon model to explain the results in a simple way. From Einstein, the energy of a photon is

$$E = hf \tag{33–15}$$

According to relativity, energy and mass are related by

$$E = mc^2 \tag{33–16}$$

Combining these equations gives

$$hf = mc^2 \tag{33–17}$$

If photons travel with a speed c, their momentum is $p = mc$, which, from Equation (33–17), becomes

$$p = \frac{hf}{c}$$

or

MOMENTUM p OF A PHOTON

$$p = \frac{h}{\lambda} \tag{33–18}$$

It should be noted that even though photons have momentum, *they have zero rest mass.* This is seen from the relativistic relation [Chapter 13, Equation (13–10)] between energy E, momentum p, and rest mass m_0:

$$E^2 = c^2p^2 + (m_0c^2)^2 \tag{33–19}$$

Since the momentum of a photon is $p = hf/c = E/c$, it becomes clear that the rest-mass term in Equation (33–19) must be zero.

Compton viewed the interaction as a billiard-ball type of "collision" between the incoming photon and an (essentially) "free" electron[8] at rest. Figure

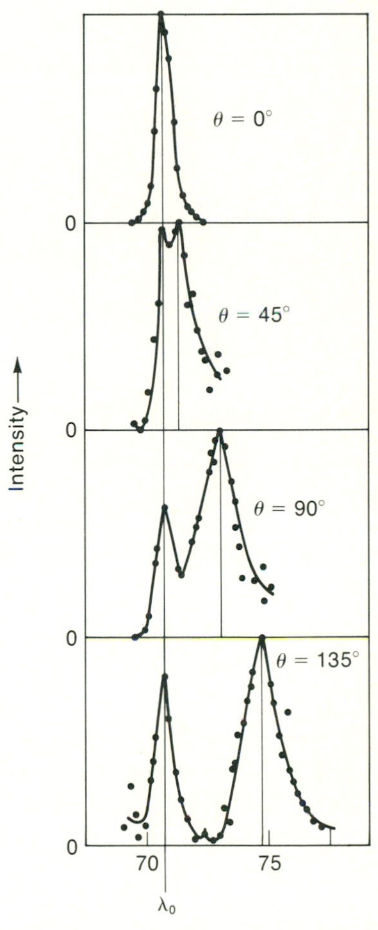

Figure 33–12
Experimental data for Compton scattering. The intensity of the x-rays scattered at various angles is plotted versus the wavelength.

[8] The bonds that hold outer electrons to atoms have energies of only a few electron volts. The x-rays Compton used had energies thousands of times greater, so the outer electrons were essentially "free" in their interactions with the incoming photons.

33–13 sketches the process. Conservation of energy and of momentum applies in the collision. Since the scattered electron acquires some energy, the scattered photon must have *less* energy than the incident photon. Applying relativistic equations for the conservation of energy and momentum, Compton derived the following expression for the shift in wavelength:

COMPTON SHIFT
$$\lambda' - \lambda_0 = \frac{h}{m_0 c}(1 - \cos\theta) \tag{33-20}$$

where θ is the angle through which the photon is scattered, and m_0 is the rest mass of the electron. The quantity h/m_0c is known as the *Compton wavelength* and has a value of $\lambda_C = 0.002\ 43$ nm. Since Compton shifts are of this order, the effect is noticeable only for photons of comparably short wavelengths (x-rays and gamma rays).

This equation agrees with the experimental data of Figure 33–12. (The presence of the *unshifted* line at λ_0 is the result of scattering from inner-shell electrons, which are firmly bound to the atom, so that the atom as a whole recoils. Because of its relatively great mass, the atom acquires negligible energy in the collision.) The success of the photon model in explaining Compton scattering further reinforced belief in the particle-like nature of radiation.

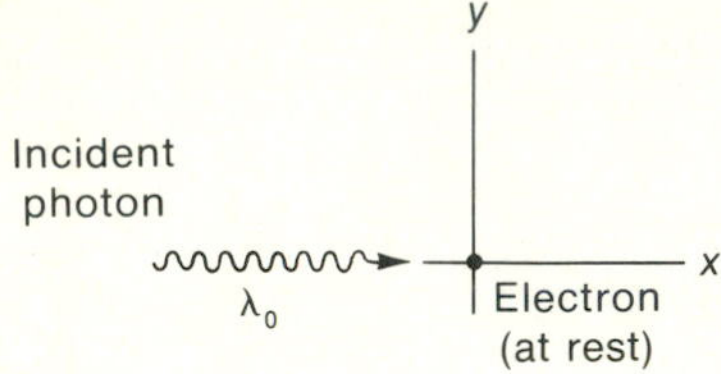

(a) Before

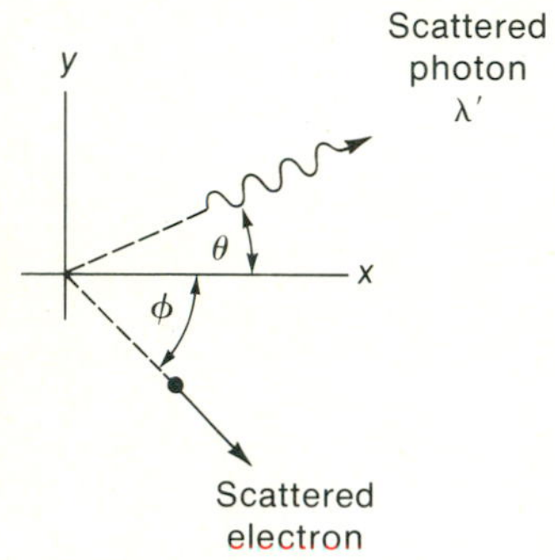

(b) After

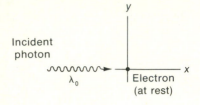

Figure 33–13
In a Compton scattering process, a photon of wavelength λ_0 undergoes a particle-like collision with an electron initially at rest.

EXAMPLE 33–5

A 2 W, helium–neon laser beam ($\lambda = 632$ nm) is completely absorbed as it strikes a target. From considerations of the momentum carried by the photon, find the force the light beam exerts on the target.

SOLUTION

The energy of each photon is

$$E = hf = \frac{hc}{\lambda} = \frac{(6.63 \times 10^{-34}\ \text{J}\cdot\text{s})\left(3 \times 10^8\ \frac{\text{m}}{\text{s}}\right)}{6.32 \times 10^{-7}\ \text{m}}$$

$$E = 3.15 \times 10^{-19}\ \text{J}$$

For the number of photons that strike the target in one second, we have

$$\frac{2\ \frac{\text{J}}{\text{s}}}{3.15 \times 10^{-19}\ \text{J}} = 6.35 \times 10^{18}\ \frac{\text{photons}}{\text{second}}$$

The momentum of each photon is

$$p = \frac{h}{\lambda} = \frac{6.63 \times 10^{-34}\ \text{J}\cdot\text{s}}{6.32 \times 10^{-7}\ \text{m}} = 1.05 \times 10^{-27}\ \frac{\text{kg}\cdot\text{m}}{\text{s}}$$

The force on the target is therefore

$$\text{F} = \frac{\Delta p}{\Delta t}$$

$$F = \left(1.05 \times 10^{-27}\ \frac{\text{kg}\cdot\text{m}}{\text{s}}\right)\left(6.35 \times 10^{18}\ \frac{\text{photons}}{\text{s}}\right)$$

$$F = \boxed{6.67 \times 10^{-9}\ \text{N}}$$

33.6 The Dual Nature of Electromagnetic Radiation

Up to this point, we have reviewed some of the experimental evidence for the particle-like behavior of radiation. No doubt the photon model now appears logical and straightforward. However, its acceptance was a slow and painful process for most physicists. Robert Millikan, the noted American physicist, expressed his reluctance thus (in 1916):

I spent ten years of my life testing the 1905 equation of Einstein's and contrary to all my expectations, I was compelled in 1915 to assert its unambiguous experimental verification in spite of its unreasonableness since it seemed to violate everything that we knew about the interference of light.

The reasons for the reluctance are as follows. All interference and diffraction phenomena seem to furnish ample evidence that radiation is a *wave*. If we accept the photon model, can we interpret an effect such as double-slit interference on the basis of *photons*? You will recall we explained the light and dark fringes as an interference between two coherent *waves* that spread out as they emerge from the slits. What happens if we assume the incident light is a stream of *photons*?

First, we can clearly associate the light intensity pattern on the screen with the varying numbers of photons that arrive at different locations. Each individual photon arrival is a localized "point event," perhaps knocking an electron off a silver-halide molecule in a photographic emulsion, causing the molecule to deposit a silver grain during the development process. If a photon is, indeed, a localized particle small enough to interact with a single electron, it certainly should go through just one of the slits at a time. Therefore, it should not make any difference if we close one of the slits for half the exposure time, then open it and close the other slit for the other half of the exposure time. *Yet if we do that experiment, we do not obtain the double-slit pattern.* As shown in Figure 33–14, the light pattern is just a superposition of two *single*-slit patterns, due to each slit acting alone. Apparently the photon, even though it is a well-localized particle, "knows" whether or not the other slit is open.

How do photons cause interference effects? Could one photon pass through one slit and interfere with another photon going through the other slit? No. Experiments have been performed using extremely dim light, which guarantees

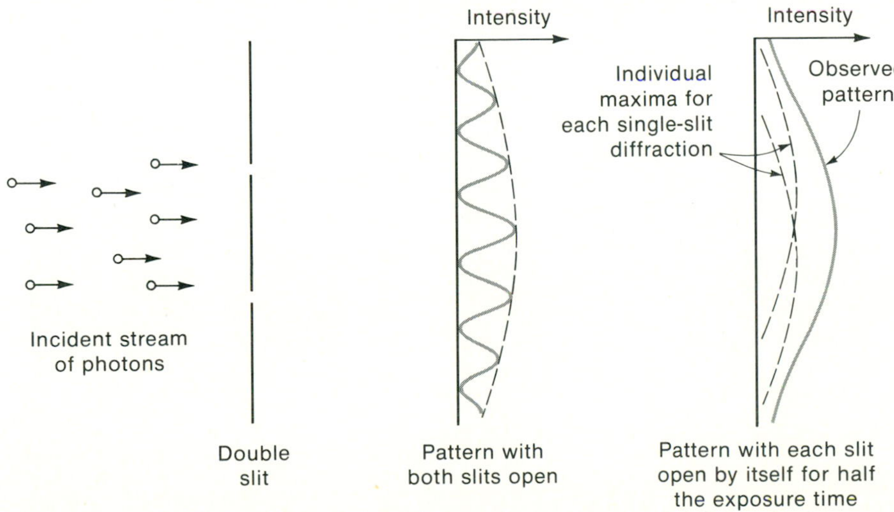

Figure 33–14
An attempt to interpret a double-slit interference experiment in terms of photons.

that (on the average) only one photon at a time passes from the source to the screen. In one such case, the experimenter started the exposure in an interference experiment, then went sailing for a few months. Upon his return, he developed the photographic film and found the usual fringe pattern, even though only one photon at a time had passed through the apparatus. **Each photon interferes only with itself.**[9]

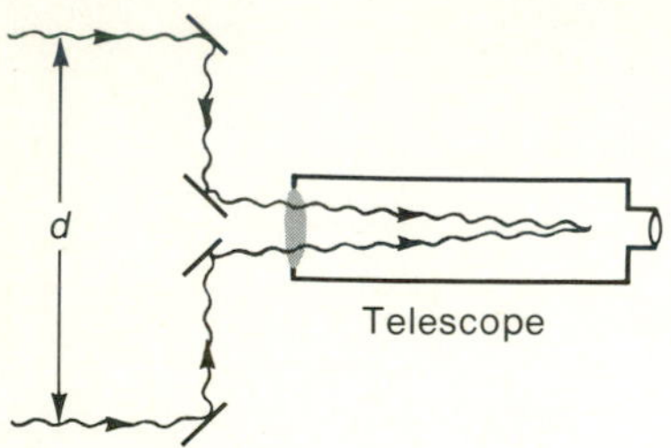

Figure 33–15
A stellar interferometer. Mirrors at 45° reflect light from a distant star into a telescope, causing certain interference effects in the image. Essentially the stellar interferometer is a double-slit apparatus, where the slit separation d may be as large as 10 m.

Does this imply the photon is "smeared out" so that part of it went through each slit? This is hard to imagine when we consider an instrument known as the *stellar interferometer*, Figure 33–15. Basically this instrument is a double-slit apparatus with the two slits separated by up to 10 m. Both slits must be open simultaneously to obtain the correct interference effect. But if we try to imagine a photon as spread out so much that parts of it can go through both slits simultaneously, we must keep in mind that the photon must also be capable of giving up all its energy to a single electron should the photon, instead, just happen to undergo a photoelectric process. Such a scenario is certainly inconsistent. We do run into serious difficulties if we try to imagine photons as spread out in space. Figure 33–16 shows another experiment that cannot be explained using a photon model.

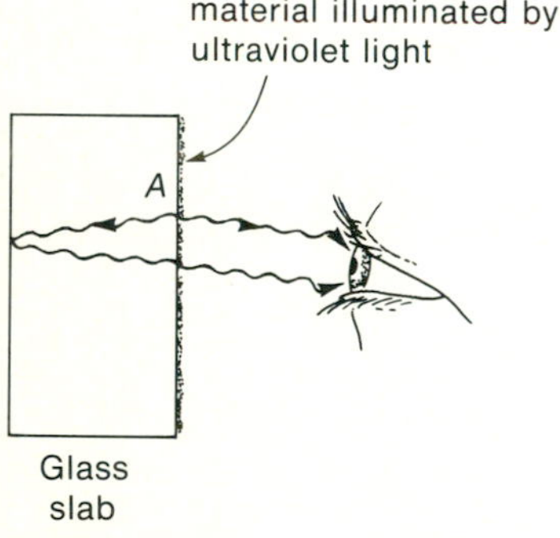

Figure 33–16
A single atom at A emits visible light. The light that reflects from the rear surface of the glass slab interferes with the light traveling directly to the eye, and the observer sees a pattern of light and dark rings similar to Newton's rings (Figure 30–17). The effect is easily understandable in terms of spherical *wavefronts* that expand outward from the atom. But if the atom emits a single *photon*, does the photon start to travel outward simultaneously in two opposite directions? Because of coherence requirements for forming an interference pattern, light from a single atom interferes only with itself, not with light from other atoms.

Well, then, how should we interpret light phenomena? One has to become an expert in "double-think." For some experiments, a wave model for light gives us insight into what is occurring; for another class of experiments, only a particle model makes sense. Are there any hints we can find for choosing a model? One clue is the following. If the dimensions of the apparatus (slit widths, apertures, and so on) are of the order of λ, then the *wave* nature of the radiation is usually most important because of interference and diffraction. On the other hand, when significant dimensions are $\gg \lambda$ (such as in Chapter 29, "Geometrical Optics"), we are usually not interested in the wave characteristics, so we can assume light rays do not bend around edges but travel in straight lines (as particles, if we wish). Another clue is that if the energy and momentum of a photon are comparable with other energies and momenta in the system, then we must treat the photon as a *particle* (as in the photoelectric effect and Compton scattering). However, all of these clues are only rule-of-thumb considerations. We must use care: For example, in a stellar interferometer we think of *waves* passing through the apparatus, but *photons* arriving at the astronomer's eye or the photographic plate.

At this stage in the development of physics (the early 1900s), light seemed to develop a split personality. Even today, for most applications we still think of light in terms of *waves* or *particles* (though in the next chapter we will discuss a more sophisticated view). But one significant fact should be noted. *Whenever we detect light experimentally, it always involves a particle-like interaction, not a wave-like one.* We need the wave model to understand such effects as interference and diffraction, *yet we never physically detect light in those regions where we think of it as waves.* If light interacts with matter, we must use a particle model. The formation of an interference pattern is the result of a very large number of photons that *statistically* sort themselves out to gradually form the pattern of light and dark fringes. This statistical behavior of photons is present in all image formation (see Figure 33–17). It is a bit ironic that we need the wave model to understand the propagation of light *only through that part of the system where it leaves no trace.*

Perhaps the moral of the story is that we should not take either the particle or the wave models too seriously. They are useful, but inherently contradictory:

[9] Photon-photon interactions do take place under certain circumstances, but they are very rare and of no consequence here.

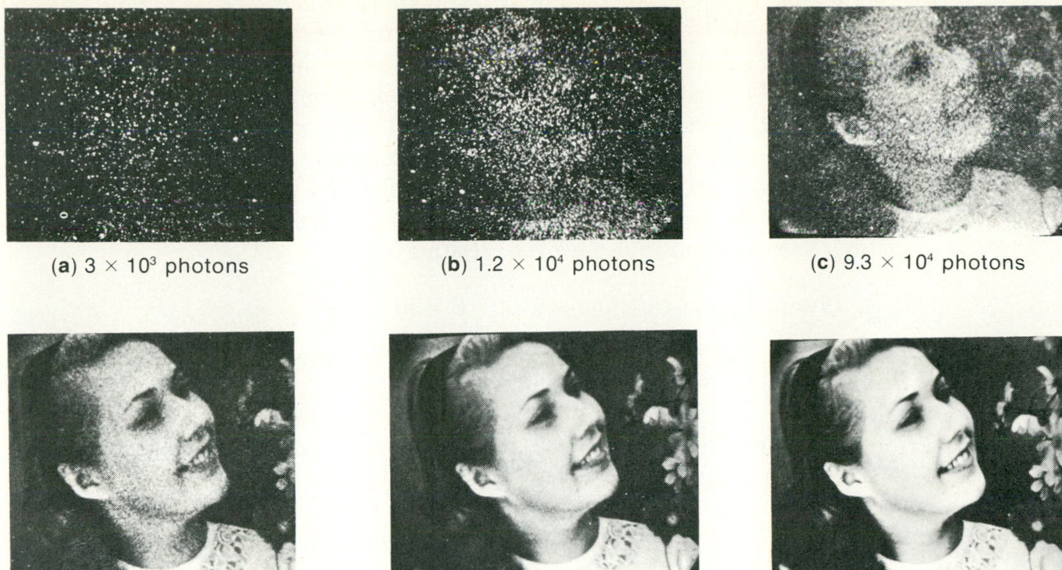
(a) 3 × 10³ photons (b) 1.2 × 10⁴ photons (c) 9.3 × 10⁴ photons (d) 7.6 × 10⁵ photons (e) 3.6 × 10⁶ photons (f) 2.8 × 10⁷ photons

Figure 33–17
A great many photons are needed to form a complete image. The number of photons involved is indicated below each picture.

Particles are *localized*, waves are *spread out*. Conceptually, we cannot melt them together. The modern resolution to this paradoxical duality is revealed in the next chapter.

Summary

One of the characteristics of *blackbody* (or *cavity*) radiation is that the *total emittance* R at all wavelengths (in watts/meter^2) is proportional to the fourth power of the Kelvin temperature T.

Stefan-Boltzmann radiation law $$R = \sigma T^4$$

where the *Stefan-Boltzmann constant* is $\sigma = 5.672 \times 10^{-8}\ \mathrm{W/(m^2 \cdot K^4)}$. Another characteristic is that as the Kelvin temperature increases, the wavelength λ_m at the maximum of the spectral distribution becomes shorter according to:

Wien displacement law $$\lambda_m T = \text{constant}$$

The classical radiation laws of Wien and Rayleigh-Jeans are approximations that are correct only for short and long wavelengths, respectively. Planck derived the correct expression by assuming:

Planck assumption of quantization: SHM oscillators (with a natural frequency f) can exist only in quantized energy states:

$$E_n = nhf$$

where $n = 0, 1, 2, 3, \ldots$ and $h = 6.626 \times 10^{-34}$ J·s, known as *Planck's constant*. The oscillators emit or absorb only quanta of energy: $\Delta E = hf$. (Planck originally proposed nhf, but it was shown later that only $n = 1$ occurs.)

Planck's radiation law $$f(\lambda, T) = \frac{8\pi hc\lambda^{-5}}{(e^{hc/\lambda kT} - 1)}$$

where $f(\lambda,T)\,d\lambda$ is the energy per unit volume from wavelength λ to $\lambda + d\lambda$ and the *Boltzmann constant* k is equal to 1.381×10^{-23} J/K. As $h \to 0$, quantum mechanical expressions approach the corresponding classical expressions.

Einstein explained the *photoelectric effect* by assuming that radiation is quantized as *photons* that remain localized in space as they travel with speed c (in a vacuum) and that have zero rest mass.

Photon energy $$E = hf$$

Photon momentum $$p = \frac{h}{\lambda}$$

Einstein's photoelectric equation $$hf = K_{\text{max}} + w_0$$

where f is the frequency of the illumination, K_{max} is the maximum kinetic energy of the photoelectrons, and w_0 is the *work function* of the surface.

The *Compton shift* for the scattering of photons by free electrons is:

$$\lambda' - \lambda_0 = \frac{h}{m_0 c}(1 - \cos\theta)$$

where m_0 is the rest mass of the electron. The quantity h/m_0c is the *Compton wavelength* $\lambda_C = 0.002\,43$ nm.

As a result of the dual nature of electromagnetic radiation, we use both *wave* and *particle* models. The wave model enables us to predict interference and diffraction effects, but all interactions of radiation with matter that are experimentally detected are particle-like interactions.

Questions

1. As the power input to an incandescent bulb is reduced, the brightness decreases. Why does the color of the emitted light also change?
2. Metal objects put into a heat-treating furnace are heated to incandescence. Upon peering through a small hole in the oven door, the objects seem to have almost disappeared. Why?
3. What is meant by the adjective "black" in "blackbody radiation"?
4. Materials that are heated radiate energy in accordance with a modified form of the Stefan-Boltzmann radiation law and may be written as $R = e\sigma T^4$, where e (called the *emissivity*) is equal to one for a blackbody and less than one for other materials. Two different metals at the same temperature may glow at different intensities. On a thermodynamic basis, how is the emissivity of a material related to the material's ability to absorb radiation?
5. At a temperature that causes metals to become incandescent, glass does not even glow. Why not?
6. Doesn't an inconsistency exist in ascribing an energy $E = hf$ to a *photon*, when f refers to the frequency of a *wave*?
7. Could a faint star be visible to the eye if the light from the star were not corpuscular in nature?

8. In order to observe quantum effects in simple mechanical oscillators, about how large would Planck's constant have to be?

9. Why is the maximum kinetic energy with which photoelectrons leave the surface of a metal independent of the intensity of the light falling on the surface?

10. In a photocell, electrons emanating from a photosensitive cathode are drawn to an anode that is normally at a higher potential than the cathode. How does the electron current through the photocell depend on the intensity of the light falling on the cathode and upon the potential difference between the cathode and anode?

11. An isolated sheet of zinc exposed to ultraviolet light will emit photoelectrons when first exposed to the light, then seems to stop. Why?

12. Why does it seem reasonable that the Compton shift in wavelength of scattered photons is independent of the scattering material?

13. Why is the Compton effect not readily observable for visible light?

14. In what way does the Compton effect reinforce the photoelectric effect in substantiating the quantum theory of radiation?

15. What is wrong with the following explanation of the Compton effect? Electromagnetic radiation is only a wave phenomenon. The wave interacts with electrons, causing the electrons to recoil due to the momentum carried by the wave as well as causing the electrons to oscillate at the frequency of the incoming electric wave. The frequency shift observed is simply a Doppler shift of radiation produced by the oscillating electrons, which are also moving under the recoil.

Problems

33A–1 Find the wavelength at the maximum of the blackbody radiation curve for a room temperature of 27°C. *Answer:* 9660 nm

33A–2 The universe is filled with blackbody radiation believed to be the remnant of the original "big bang," which occurred about 15 to 20 billion years ago. The radiation is in thermal equilibrium at 3 K. Calculate the wavelength of the radiation with maximum intensity.

33A–3 A cavity radiator is maintained at a temperature of 6000 K (approximately the surface temperature of the sun). Calculate the wavelength at which the radiation per unit wavelength is most intense. *Answer:* 483 nm

33A–4 Radiation from the sun falls perpendicularly just outside the earth's atmosphere at the rate of 1.34×10^3 W/m^2. Assuming an average wavelength of 500 nm, find how many photons per square centimeter strike the earth at normal incidence each minute.

33A–5 The photoelectric work function for sodium is 2.3 eV. Find the longest wavelength of light that will eject photoelectrons from the surface of sodium. *Answer:* 540 nm

33A–6 Monochromatic light with a wavelength of 550 nm falling on a cesium surface produces photoelectrons that may be prevented from leaving the surface with a stopping potential as low as 0.29 V. Find the photoelectric work function for cesium.

33A–7 A photon undergoes a 180° Compton scattering from a free electron that is initially at rest. Calculate the resulting change in the wavelength of the photon.
Answer: 4.85×10^{-12} m

33B–1 A hot oven with an interior temperature of 1000 K has a small peephole out of which energy flows. Calculate the increase in temperature required to double the flow of radiant energy emanating from the hole. *Answer:* $\Delta T = 189$ K

33B–2 Radiant energy emerges through a hole 4.0 mm in diameter in a heated cavity. The spectral distribution of the radiation is shown in Figure 33–18. Estimate the rate of energy flow in the wavelength range from 1000 nm to 2000 nm. (Hint: Use a graphical approach.)

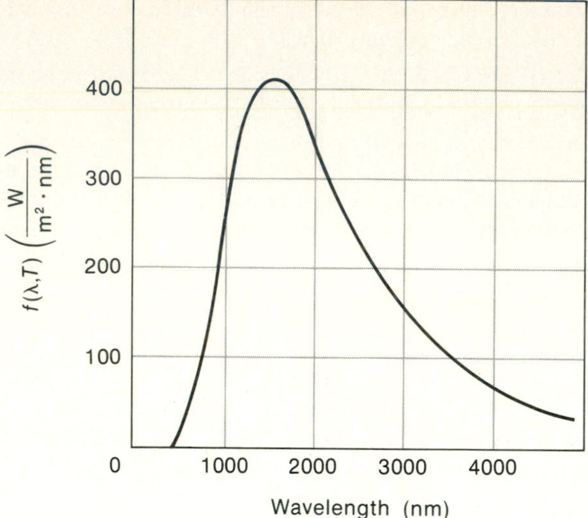

Figure 33–18
Problems 33B–2 and 33C–1

33B–3 Experiments indicate that a dark-adapted human eye can detect a single photon of visible light. Consider a point source of monochromatic light of wavelength 555 nm that emits 2 W of luminous power uniformly in all directions. How far away would this source have to be so that, on the average, one photon per second enters an eye whose pupil is 6 mm in diameter?

Answer: 3.54×10^6 m (about the distance between New York and London)

33B–4 Find the wavelength of a photon that has an energy equal to the rest energy of an electron (0.511 MeV).

33B–5 A very useful relation between the energy of a photon and its wavelength is $E\lambda = 1.240 \times 10^{-3}$ MeV·nm. Derive this expression.

33B–6 A 10 g mass oscillates with an amplitude of 3.0 cm under the influence of a restoring force provided by a spring with a spring constant of 0.01 N/m. Find the decrease in amplitude of oscillation corresponding to the loss of a single quantum of energy.

33B–7 Show that for low energies ($hf \ll kT$), Planck's radiation law reduces to the Rayleigh-Jeans radiation law.

33B–8 Show that Planck's radiation law reduces to Wien's radiation law for short wavelengths and low temperatures.

33B–9 A gamma-ray photon with an energy equal to the rest energy of an electron (511 keV) collides with an electron that is initially at rest. Calculate the kinetic energy acquired by the electron if the photon is scattered 30° from its original line of approach.

Answer: 60.6 keV

33B–10 The nucleus of a radioactive isotope of chlorine (^{38m}Cl) decays by the emission of a 660 keV photon. (The symbol m indicates a *metastable* state. Instead of decaying immediately, the nucleus exists in this excited state a relatively long time.) If the nucleus is initially at rest, determine the ratio of the kinetic energy acquired by the nucleus to the energy of the emitted photon. The rest mass of the ^{38m}Cl nucleus is 35.4 GeV.

33C–1 Referring to Figure 33–18, which shows a blackbody spectral energy density $f(\lambda,T)$ as a function of wavelength, (a) estimate the total emittance of the blackbody. (b) On the basis of the result obtained in part (a), determine the temperature of the blackbody. Your answer should be consistent with the data shown in Figure 33–2.

Answers: (a) $R \approx 9 \times 10^5$ W/m² (b) $T = 2000$ K

33C–2 A square meter of a perfectly reflecting metal sheet is held so that its surface is perpendicular to the sun's rays at the surface of the earth where the solar constant is 840 W/m². Assuming an average wavelength of 500 nm for photons from the sun, calculate (a) the number of photons per second that strike the sheet. (b) Find the force on the sheet as a consequence of the momentum carried by the incoming photons. (c) Verify the answer to

(b) using the analysis in Chapter 28 (Section 28.5) describing the momentum of electromagnetic waves.

33C–3 In an experiment to study the photoelectric effect, the maximum kinetic energy of photoelectrons is determined for each of several different wavelengths of illumination:

Wavelength (in nm)	Maximum kinetic energy of photoelectrons (in eV)
588	0.67
505	0.98
445	1.35
399	1.63

Using these data, make an appropriate graph that plots as a straight line and determine (a) an experimental value for Planck's constant (in joules·second) and (b) the work function (in electron volts). *Answers:* (a) 6.5×10^{-34} J·s (b) about 1.4 eV

33C–4 An electron initially at rest recoils from a head-on collision with a photon. Show that the kinetic energy acquired by the electron is given by

$$hf\left(\frac{2\alpha}{1 + 2\alpha}\right)$$

where α is the ratio of the photon's initial energy to the rest energy of the electron.

33C–5 Show that the photoelectric effect is impossible for free electrons; that is, a photon cannot transfer all of its energy to a free electron. (Hint: Analyze the collision using relativistic mechanics.)

33C–6 A photon strikes a free *proton* initially at rest in a Compton type of collision. Find the minimum energy of the photon that will give the proton a kinetic energy of 4 MeV.

33C–7 A metal target is placed in a beam of 662 keV gamma rays emitted by a radioactive isotope of cesium (^{137}Cs). Find the energy of those photons that are scattered through an angle of 90°. The electrons in the metal target may be considered to be essentially free electrons. *Answer:* 288 keV

33C–8 A low-energy photon ($E \ll$ electron rest-mass energy) collides head-on with a free electron initially at rest. The photon is scattered backwards along the line of approach. Show that the ratio of the scattered photon energy to the kinetic energy acquired by the electron is approximately c/v, where v is the speed of the electron. (Hint: This is a nonrelativistic Compton scattering problem.)

chapter 34 Wave Nature of Particles

Heisenberg may have been here. . .

A variation of the usual "Kilroy Was Here" graffito

Those who are not shocked when they first come across quantum mechanics cannot possibly have understood it.

NIELS BOHR

The discovery of the dual nature of radiation was a fascinating revelation in its own right, but in the 1920s, an equally startling development occurred when particles of matter were found to exhibit wave-like behavior. This rounded out the physicist's "picture" of nature in a particularly symmetrical and satisfying way. Radiation *and* matter exhibit particle-like characteristics as well as wave-like characteristics. To place this discovery in context, we will describe some related developments that set the stage for this important step.

34.1 Models of an Atom

At the turn of the century it was believed atoms were made of just two components: protons and electrons. But how were these components put together so they formed stable atoms? What configuration of charged particles could produce the extraordinary complexity of atomic spectral lines observed when a gas is excited by passing an electrical current through it (Figure 34–1)? These spectra had been studied and catalogued carefully, and many attempts were made to discover some mathematical relationship between wavelengths that might reveal a clue to the atom's structure. Also, the cyclic variation in chemical properties of atoms in the periodic table was another clue to the puzzle. As a

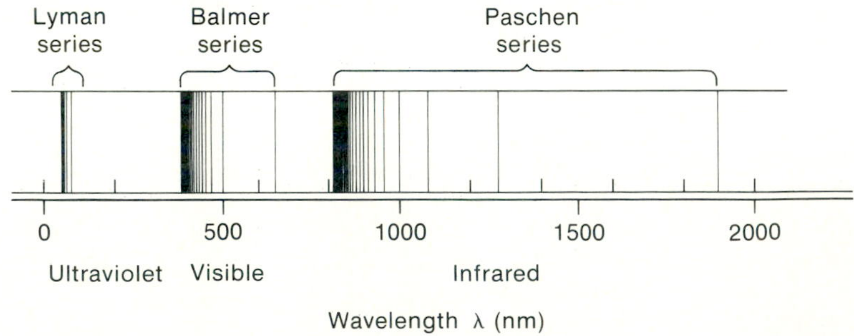

Figure 34–1
When hydrogen gas is heated by passing an electrical current through it, the gas emits light consisting of a series of spectral lines called a **bright-line spectrum**, or an **emission spectrum** (indicated here by dark lines).

starting point, atoms were assumed to be spherical with radii $\sim 10^{-10}$ m. (This could be calculated from the density, the atomic number, and Avogadro's number.)

The Thomson Model

One notable attempt to devise an atomic model was that of the British physicist J.J. Thomson at Cambridge University. He suggested a kind of fluid of positive charge Ze (where Z is the atomic number) that contained most of the mass of the atom. The electrons were embedded within this positive fluid somewhat like plums in a plum pudding (Figure 34–2a). Supposedly, the electrons could then vibrate in various modes of oscillation and thereby (according to classical theory) emit radiation at these natural frequencies of oscillation. Unfortunately, quantitative agreement with observed spectral frequencies was lacking.

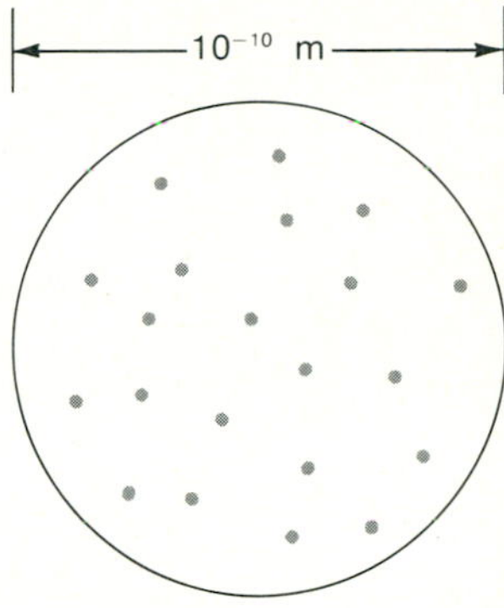

(**a**) Thomson's "plum pudding" model, with electrons embedded in a sphere of positively charged fluid.

The Rutherford Model

Before 1910, many attempts had been made to discover the secrets of atomic structure by observing how incident particles and radiation were scattered from atoms. X-rays, electrons, and alpha particles were the main projectiles. A former student of Thomson's, Professor Ernest Rutherford,[1] was conducting experiments at the University of Manchester in England on the scattering of alpha particles by matter. An alpha particle was known to have a positive charge twice the magnitude of the electronic charge and a mass about four times that of hydrogen. Alpha particles were a convenient projectile since they were emitted with several million electron volts of energy by certain naturally radioactive elements. Rutherford wanted a very thin target because he hoped to observe the scattering by just a single atom rather than the multiple scattering by many atoms; multiple encounters would tend to obscure the characteristics of the single collision he wished to investigate. Although several different elements were investigated, gold was a particularly convenient target substance because it could be hammered to extremely thin foils, only a few hundred atoms thick. As shown in Figure 34–3, the scattered alpha particles struck a small screen coated with zinc sulfide, causing tiny flashes of light that were observed by watching the screen with a microscope. It was tedious work,

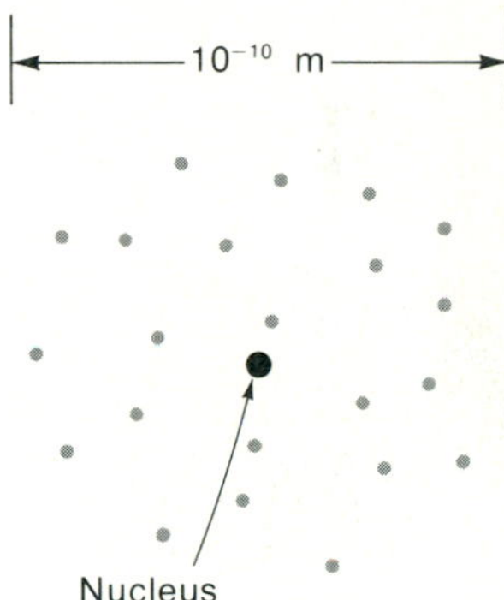

(**b**) Rutherford's nuclear model, with all the positive charge (and most of the mass) concentrated in a very small region at the center. Electrons surround the nucleus in an unknown way.

Figure 34–2
Classical models of the atom.

[1] Rutherford received the Nobel prize in *chemistry* in 1908 for discovering that the radiation from uranium consisted of at least two types he called *alpha* and *beta* radiation. He later showed that alpha "radiation" actually consisted of particles, being nuclei of helium atoms.

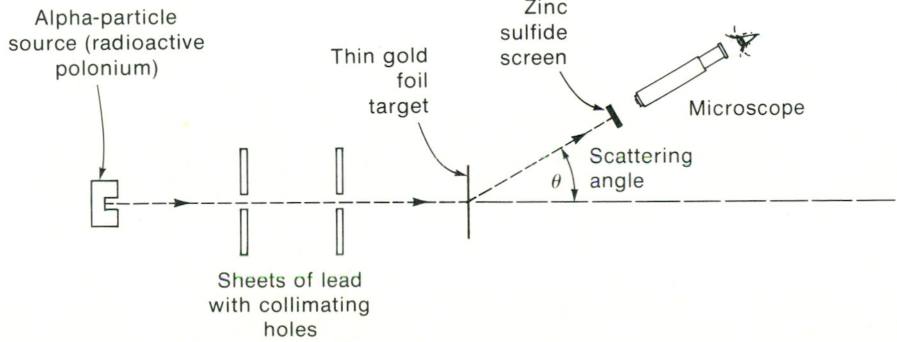

Figure 34–3
The Rutherford alpha-scattering experiment. The zinc sulfide detector can be moved to record scattering at various angles. The entire apparatus is placed within an evacuated chamber.

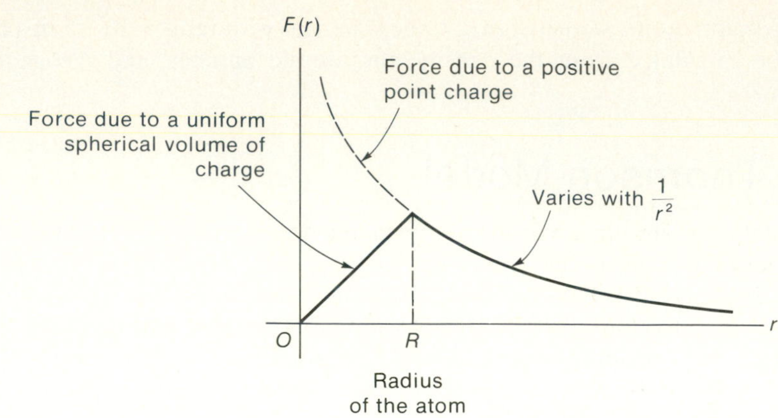

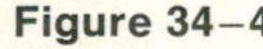

Figure 34–4
The force on an alpha particle due to a positive charge in two different configurations: a point charge and a uniform spherical volume of charge.

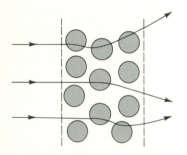

requiring well-dark-adapted eyes. Rutherford's assistants were Dr. Hans Geiger[2] and an undergraduate student, Ernest Marsden.

Early data for small-angle scattering of 1° or 2° seemed to confirm the Thomson model. Wishing to start Marsden on a research project of his own, Rutherford suggested he look for scatterings in the backward direction ($>90°$), though Rutherford personally felt that the chance of a fast alpha particle being scattered backward by a Thomson atom was truly negligible. Much to everyone's amazement, many alphas were back-scattered. The reason for surprise is clear from the estimates of the scattering probabilities. The mass of an alpha is about 8000 times the mass of an electron, so electrons have negligible effect on the scattering: All the scattering occurs from the massive positive charge. In a Thomson atom, the positive charge is spread uniformly throughout a spherical volume, so the maximum force a single atom could exert on an alpha particle was limited (Figure 34–4), causing a deflection of just a few hundredths of a degree at most. Thus, thousands of scatterings would have to take place, *with a majority adding up in the same direction*, to cause a net deflection of 90° or more. The chance of a backward scattering by Thomson atoms in the foil used in one experiment was calculated to be incredibly small—about 1 in 10^{3500}. Yet Geiger and Marsden found roughly 1 in 10^4! Undoubtedly this discrepancy of a factor of $\sim 10^{3496}$ takes the all-time prize for the greatest disagreement between theory and experimental results ever encountered. Rutherford later wrote of his reaction:

It was quite the most incredible event that has ever happened to me in my life. It was almost as incredible as if you fired a 15-inch shell at a piece of tissue paper and it came back and hit you.

Recognizing that a single scattering at large angles could occur only if the forces were extremely strong, in 1911 Rutherford proposed his nuclear model of an atom. In it, the massive positive charge was concentrated in a region he called the **nucleus**, no bigger than 10^{-14} m, since to experience the force required for backward scattering an alpha particle would have to approach that close to a point charge Ze. Figure 34–5 compares the two situations, and Figure 34–6 shows experimental points for one experiment.

(**a**) According to the Thomson model, multiple scattering could occur if the alpha particle penetrates more than one atom. (The scattering is greatly exaggerated.)

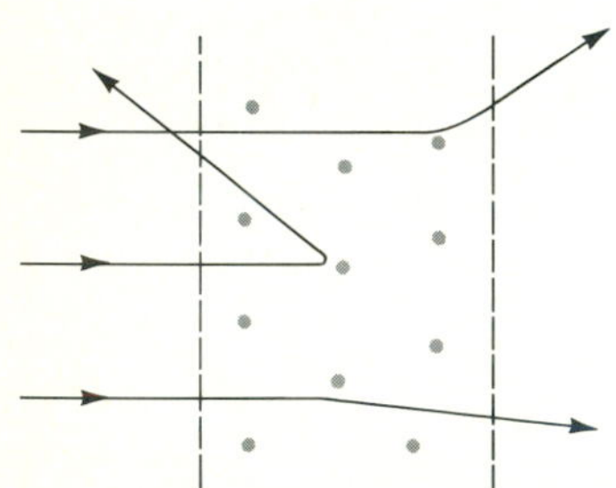

(**b**) According to the Rutherford model, a single close encounter with a nucleus could produce a large-angle scattering.

Figure 34–5
Scattering of alpha particles by a thin foil. The target foil is typically several hundred atoms thick.

[2] To avoid the painstaking and boring method of data taking, Geiger later invented an electronic gadget for detecting charged particles: the "Geiger" counter, widely used today.

Though the Rutherford model was clearly superior to the Thomson model, there were still some troublesome aspects. For example, what held the positive charges in the nucleus together? And what held the negatively charged electrons away from the positively charged nucleus? They could not rotate around the nucleus in a "solar system" motion because Maxwell's equations predicted that accelerated charges radiate electromagnetic waves. Indeed, such radiation was observed in every instance in which electrons were accelerated. According to classical physics, if you started electrons moving in circular orbits, they would radiate energy and spiral into the nucleus in less than 10^{-8} seconds. Obviously atoms did not do this, so what was wrong?

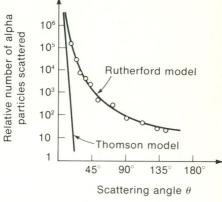

Figure 34–6
Typical data by Geiger and Marsden for the scattering of alpha particles by gold foils. The solid lines are theoretical curves based on the Thomson and Rutherford models.

The Bohr Model

As shown in Figure 34–7, the spectrum of a hydrogen atom—the simplest atom of all—had a baffling complexity and regularity. How could just a proton and an electron interact to produce this series of spectral lines? A Swiss high school teacher of descriptive geometry, J. Balmer, had found an *empirical* formula by trial and error that agreed with the observed wavelengths.

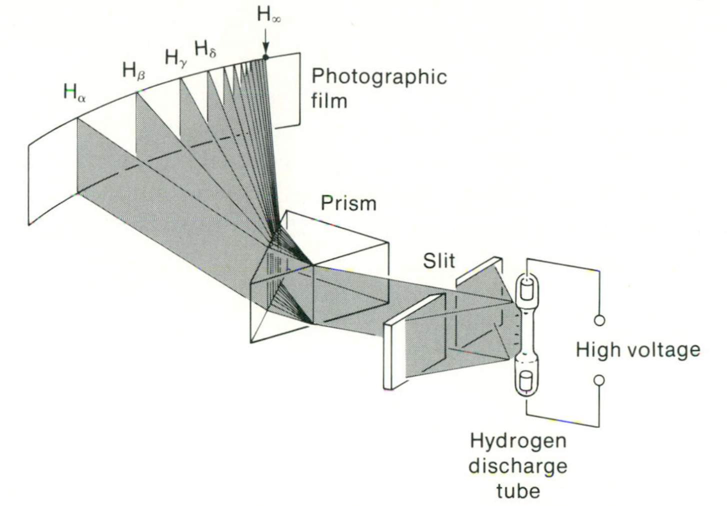

(**a**) A prism spectrometer. Light from the hydrogen discharge tube is refracted by the prism to form the line spectrum on the photographic film.

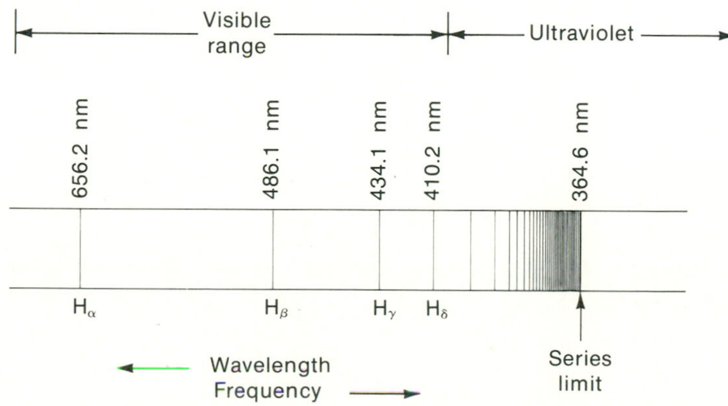

(**b**) The Balmer series is a group of an infinite number of spectral lines whose spacings regularly converge toward the short-wavelength limit of 364.6 nm.

Figure 34–7
The Balmer series emission spectrum of hydrogen.

THE BALMER SERIES IN HYDROGEN

$$\lambda = (364.6 \text{ nm})\left(\frac{n^2}{n^2 - 4}\right) \qquad \text{(where } n = 3, 4, 5, \ldots) \qquad \textbf{(34–1)}$$

But how the hydrogen atom produced this mathematically simple series of lines remained a nagging puzzle.

In 1913, the Danish physicist Niels Bohr proposed his famous model of the hydrogen atom. Bohr was young (age 28) and fearless. His theory contained radical ideas that were clearly contrary to classical physics, but his model predicted all observed lines almost exactly. It was based on the following assumptions:

(1) The electron travels in circular orbits around the proton, obeying the classical laws of mechanics. (The coulomb force of attraction is the centripetal force.)
(2) Contrary to classical theory, the electron can move in certain *allowed* orbits of radius r_n *without radiating*. Since the energy E_n is constant in such orbits, the electron is said to be in a *stationary state*.
(3) The allowed orbits are those for which the angular momentum mvr of the electron is an integral multiple of Planck's constant divided by 2π (notation: $\hbar \equiv h/2\pi$).

$$mvr = n\hbar \qquad \begin{array}{l}\text{(where } n = 1, 2, 3, 4, \ldots \\ \text{and } \hbar = 1.0546 \times 10^{-34}\,\text{J}\cdot\text{s)}\end{array} \qquad \textbf{(34–2)}$$

(4) *Transitions* between stationary states are possible when the electron somehow "jumps" from one allowed orbit to another. Electromagnetic radiation is emitted or absorbed by the atom, and the difference in the two energy states is the energy hf of the radiation emitted or absorbed.

$$hf = E_{\text{final}} - E_{\text{initial}} \qquad \textbf{(34–3)}$$

Bohr's proposal was a peculiar mixture of classical and quantum physics. Thanks to the classical coulomb force, the electron moved in circular orbits according to classical mechanics. Contrary to classical physics, it did not radiate. Also, Planck's quantum constant entered the picture in two ways: in the energy hf associated with the radiation and in an entirely new way by quantizing the angular momentum, a parameter that had previously been nonquantized.

The allowed radii and energy states can be calculated as follows. Applying Newton's second law to the circular motion of the electron of charge e and mass m about a nucleus[3] of charge Ze (Figure 34–8), we have

$$\Sigma F = ma$$

$$\left(\frac{1}{4\pi\varepsilon_0}\right)\frac{(Ze)(e)}{r^2} = m\left(\frac{v^2}{r}\right) \qquad \textbf{(34–4)}$$

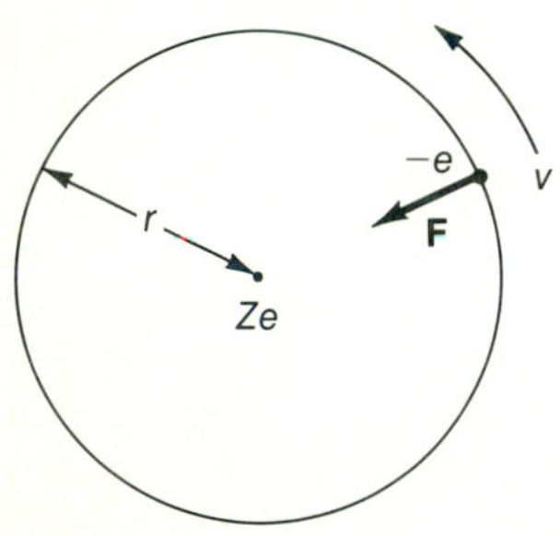

Figure 34–8
The Bohr model for a one-electron atom. The electron of charge $-e$ travels in a circular orbit around a fixed nucleus of charge Ze. The coulomb force **F** is the centripetal force on the electron.

The quantum restriction on the angular momentum is

$$mvr_n = n\hbar \qquad \textbf{(34–5)}$$

Combining the two equations to eliminate v, we obtain the radii r_n for the allowed orbits:

$$r_n = \frac{\varepsilon_0 h^2 n^2}{\pi m Z e^2} \qquad \textbf{(34–6)}$$

[3] If we consider a charge Ze in the nucleus (Z = atomic number), the analysis also applies to a singly-ionized helium, doubly-ionized lithium, and so on. The equations obtained predict the observed spectra for all these cases very well.

Substituting numerical values gives

RADII OF BOHR ORBITS FOR HYDROGEN

$$r_n = (0.0529 \text{ nm})n^2 \qquad (\text{where } n = 1, 2, 3, 4, \ldots) \tag{34-7}$$

The allowed radii are thus proportional to n^2.

The energy state E of the atom is found from

$$E = K + U$$

Defining the zero reference for U to exist when the electron is infinitely far from the nucleus, we have

$$U = -\left(\frac{1}{4\pi\varepsilon_0}\right)\frac{(Ze)(e)}{r}$$

Therefore:

$$E = \tfrac{1}{2}mv^2 - \left(\frac{1}{4\pi\varepsilon_0}\right)\frac{(Ze)(e)}{r} \tag{34-8}$$

Substituting values of v and r from Equations (34–5) and (34–6), we obtain

$$E_n = -\frac{mZ^2e^4}{8\varepsilon_0{}^2h^2n^2} \tag{34-9}$$

Substituting numerical values (for hydrogen, $Z = 1$):

ENERGY STATES OF THE BOHR HYDROGEN ATOM

$$E_n = -\frac{13.6 \text{ eV}}{n^2} \qquad (n = 1, 2, 3, 4, \ldots) \tag{34-10}$$

The allowed energy states are thus *negative* and proportional to $1/n^2$ (see Figure 34–9). Each series of spectral lines is characterized by the common *final* state involved in the transition.

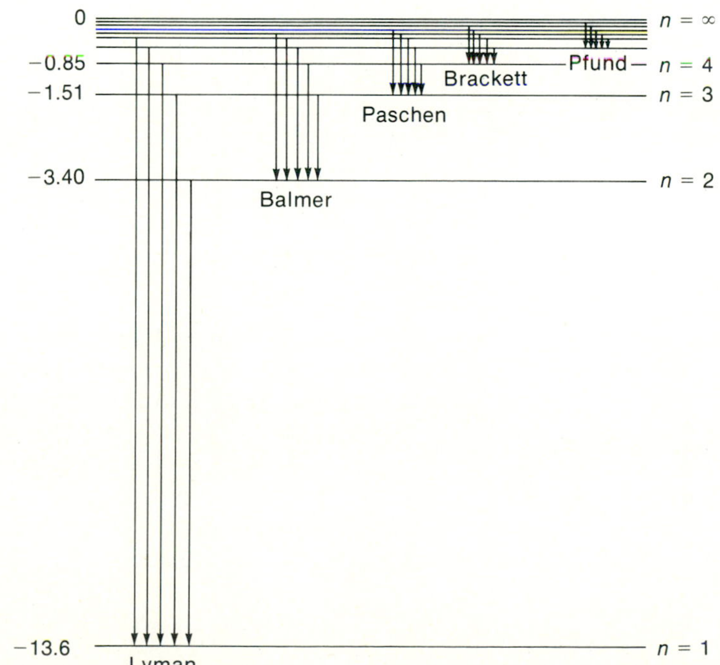

Figure 34–9
Energy states of a hydrogen atom. Between the levels $n = 4$ and $n = \infty$, there are an infinite number of energy levels. Transitions from higher to lower energy states result in emission of radiation of energy hf. The names of the experimenters who investigated the different spectral series are shown. Only a portion of the Balmer series is in the visible range of wavelengths.

34.2 The Correspondence Principle

Every new revolution in physics introduces concepts radically different from the older, established theories. For example, in relativity the equations appropriate at high speeds are quite different from those of Newtonian mechanics. Similarly, the quantum ideas of radiation are radically different from the classical Maxwell equations. Yet, physically, the transition between cases where classical equations apply and where the newer ideas must be used cannot be an abrupt one; there must be a smooth transition in the "overlap" region from one theory to the other.

In quantum physics, the relation between the new and old theories was pointed out by Bohr in a statement he called the *correspondence principle.* According to classical electromagnetic theory, the frequency emitted by an electron traveling in a circular orbit is just the orbital frequency of revolution f_0. From Equations (34–5) and (34–6), we obtain for this orbital frequency for hydrogen ($Z = 1$):

$$f_0 = \frac{me^4}{4\varepsilon_0 h^3 n^3} \qquad \textbf{(34–11)}$$

In the newer Bohr theory, the frequency f emitted in a transition between adjacent energy states is intermediate between the two orbital frequencies, given by Equation (34–3).

$$hf = E_{\text{final}} - E_{\text{initial}}$$

From Equation (34–9):

$$hf = \frac{me^4}{8\varepsilon_0{}^2 h^2}\left[\frac{1}{n^2} - \frac{1}{(n+1)^2}\right] \qquad \textbf{(34–12)}$$

The factor in brackets may be written as

$$\left[\frac{1}{n^2} - \frac{1}{(n+1)^2}\right] = \left[\frac{\cancel{n^2} + 2n + 1 - \cancel{n^2}}{n^2(n+1)^2}\right]$$

When n becomes very large, we have

$$\lim_{n \gg 1}\left[\frac{2n+1}{n^2(n+1)^2}\right] = \frac{2}{n^3} \qquad \textbf{(34–13)}$$

So for large n, the frequency of emission is

$$f \approx \frac{me^4}{4\varepsilon_0{}^2 h^3 n^3} \qquad \textbf{(34–14)}$$

Comparing this equation with Equation (34–11), we see that in the limit of large n, the quantum expression agrees with the classical expression. This illustrates Bohr's **correspondence principle.**

BOHR'S CORRESPONDENCE PRINCIPLE	**Any new theory must reduce to the classical theory to which it corresponds when applied to situations appropriate to the classical theory.**

This means that the new theory must contain the old theory as a special case. For example, in the limit of large n, the predictions of quantum physics must agree with the corresponding predictions of classical physics. The correspondence principle provides a valuable check on the validity of new theoretical developments.

□ □ □

Bohr's model for the hydrogen atom was a great triumph. It agreed very closely with wavelengths of the Balmer series, and it correctly predicted the spectrum of other series outside the visible range. Yet small but unmistakable discrepancies were still present. The reason for part of these discrepancies originated in the fact that energies were calculated on the basis of a *fixed* nucleus (which is equivalent to assuming the proton has an infinitely large mass compared with the electron mass). Agreement with experimental data was improved by considering the proton's motion about the *CM* of the rotating proton–electron system. Still further improvements were made by A. Sommerfeld, who considered elliptical as well as circular orbits and included relativistic effects for the electron's motion.

In some respects, this improved theory was still not completely satisfactory. What was the reason for the strange quantum restriction on angular momentum? It implied, for example, that a top could spin only with certain discrete values of angular velocity ω instead of with any arbitrary value among a smooth continuum of possible velocities. As experiments continued, more puzzles were uncovered. Some individual spectral lines are apparently multiple lines at the same frequency, because subjecting the gas to an electric or magnetic field "split" the lines into a cluster of two or more lines spaced closely together. Other elements exhibited this splitting to an even greater extent, implying considerable hidden complexity. One spectral line of dysprosium, for example, splits into 137 closely spaced lines.

34.3 De Broglie Waves

A crucial step toward understanding these mysteries was made by a graduate student in physics at the University of Paris, Prince Louis-Victor de Broglie.[4] While studying for his doctor's degree in physics, de Broglie began to think that perhaps the wave-particle duality applied not only to radiation but also to particles of matter. It would, indeed, form a grand sort of symmetry in nature if particles showed wave-like characteristics just as waves have particle-like characteristics. In his doctoral thesis (1924), de Broglie proposed the following ideas (somewhat simplified here).[5] Since photons of electromagnetic radiation have momentum p according to

Photons $$p = \frac{h}{\lambda} \tag{34–15}$$

de Broglie proposed that a wavelength λ is also associated with any particle having momentum mv according to:

Particles $$mv = \frac{h}{\lambda} \tag{34–16}$$

Just as for electromagnetic radiation, the question as to *what* it is that is "waving" (if anything) requires a long explanation. (It definitely is *not* electromagnetic waves.) De Broglie called them *matter waves*, or *phase waves*, since he

---

[4] The de Broglies are an old, noble French family that pronounces its name *de Brolïe*, to rhyme approximately with the English word *troy*. Louis-Victor de Broglie originally majored in medieval history at the Sorbonne, specializing in Gothic cathedrals. However, he became interested in physics, switched majors, and received his first degree in physics in 1913. During World War I he served at a military radio installation in the Eiffel Tower, returning to graduate work after the war at the University of Paris. While a graduate student, he published several scientific papers.

[5] For a summary of the development of de Broglie's ideas, see H. Medicus, "Fifty Years of Matter Waves," *Physics Today*, February 1947.

believed there might be interference between the phase of the waves as there is for light waves.

EXAMPLE 34–1

A 1 g mass moves at a speed of 1 mm/s. Calculate the de Broglie wavelength associated with this particle.

SOLUTION

The momentum p of the particle is

$$p = mv$$

$$p = (10^{-3}\ \text{kg})\left(10^{-3}\ \frac{\text{m}}{\text{s}}\right) = 10^{-6}\ \frac{\text{kg·m}}{\text{s}}$$

The associated de Broglie wavelength is

DE BROGLIE WAVELENGTH

$$\lambda = \frac{h}{p}$$

$$\lambda = \frac{6.63 \times 10^{-34}\ \text{J·s}}{10^{-6}\ \dfrac{\text{kg·m}}{\text{s}}} = \boxed{6.63 \times 10^{-28}\ \text{m}}$$

This is an impossibly small wavelength to measure, since the size of a single proton is about 10^{13} times larger. Indeed, de Broglie waves are of little consequence for macroscopic particles. However, for microscopic particles such as electrons, neutrons, and atoms, interference effects due to these waves are clearly evident and lead to some surprising effects.

In 1923, de Broglie showed that if one assumes there are matter waves for electrons, there is a reasonable explanation for the Bohr quantum condition on angular momentum that originally seemed so baffling. According to de Broglie, it is simply a case of a *standing-wave pattern* for the electron's motion. This requires that for stationary states, only an integral number of wavelengths can fit around the circular orbit, as in Figure 34–10.

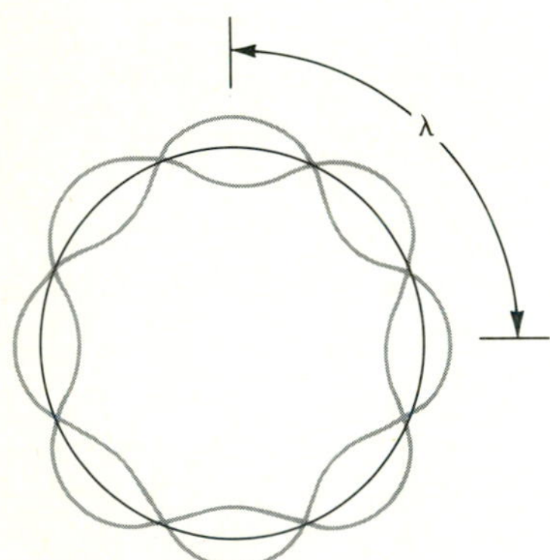

Figure 34–10
De Broglie waves for the orbiting electron in the Bohr model for hydrogen form a standing-wave pattern. The distance between adjacent nodes is $\lambda/2$. This illustration is for the energy state $n = 4$.

$$n\lambda = 2\pi r$$

Substituting the de Broglie relation for the wavelength, $\lambda = h/mv$:

$$n\left(\frac{h}{mv}\right) = 2\pi r$$

and rearranging, we obtain

$$mvr = n\left(\frac{h}{2\pi}\right)$$

This is just the Bohr condition for allowed orbits. Thus the *arbitrary* assumption Bohr made for no reason other than that it led to the right answer could be derived in a plausible way by assuming that only the motions of the electron that lead to *standing-wave patterns* represent *stationary states* of the atom. Note that what is involved is interference between different parts of the de Broglie wave associated with a *single* electron. (This is similar to the case of light, where interference between different parts of the electric field wave of a *single* photon is significant, not interference between waves of one photon and waves of another.)

De Broglie's proposal did not win immediate acceptance. While it was recognized as a worthy exercise in theoretical physics, it was treated more as a curious hypothesis that might turn out to have some validity, but on the other hand, might not. During the oral examination for his doctoral degree, de Broglie was asked how one might detect these waves. He suggested that perhaps a beam of electrons impinging on a crystal would exhibit interference effects, since the crystal lattice of atoms would provide the necessary close spacing of the order of λ that was required to bring out the interference behavior of the waves. The first experiment to detect de Broglie waves did not succeed because of a variety of experimental difficulties. But three years after de Broglie presented his thesis, a dramatic confirmation of matter waves occurred in the United States.

34.4 The Davisson-Germer Experiments

The experiments that first verified de Broglie began in 1921, when an American physicist, Clinton Davisson, was investigating the reflection of electrons by metal surfaces for the Western Electric Company (now the Bell Telephone Laboratories).[6] Some of the results he obtained were puzzling. Instead of being scattered uniformly at all angles, the electrons seemed to be scattered at certain angles more than at others. Davisson published the results, but could give no satisfactory explanation for the unusual scattering. The experiments continued with an assistant, Lester Germer.

In 1925, while Davisson was using a target of pure nickel metal in the usual metallic form (innumerable microcrystals with random orientations), a small, accidental explosion in the laboratory shattered the glass enclosure that kept the apparatus in a vacuum. The exposure to air oxidized the surface of the nickel, making it unusable for the experiment. To remove the layer of oxide, Davisson and Germer rebuilt the vacuum enclosure and then heated the target, inadvertently heating it so much that the nickel melted and recrystallized into just a few large crystals at the spot the electrons struck the target. When they resumed the experiment, the data showed unmistakable peaks in the scattering distribution when the velocity of the electrons was adjusted to certain values. They traced the difference to the fact that the target now consisted of just a few large crystals rather than being in a polycrystalline state. However, unaware of de Broglie's ideas, they proposed an incorrect origin for the peculiar scattering: They felt that the crystal lattice planes somehow "channeled" electrons in certain directions. In 1927, after Davisson attended a physics meeting at Oxford University and learned that matter waves might be responsible, he checked de Broglie's theory with the data and found an excellent agreement. Figure 34–11 shows results from an experiment using a single large crystal.

The Davisson-Germer experiment is most easily understood by analogy with the scattering of x-rays from crystals, a field of study that had been going on for over a decade. The effect, called *Bragg scattering*, was discussed in Section 31–4. In an overly simplified view, the incident x-rays are "reflected" from

[6] The General Electric Company had brought a patent suit against Western Electric over a vacuum-tube design. This experimental work on the scattering of electrons was undertaken to obtain evidence in fighting the suit. Western Electric won.

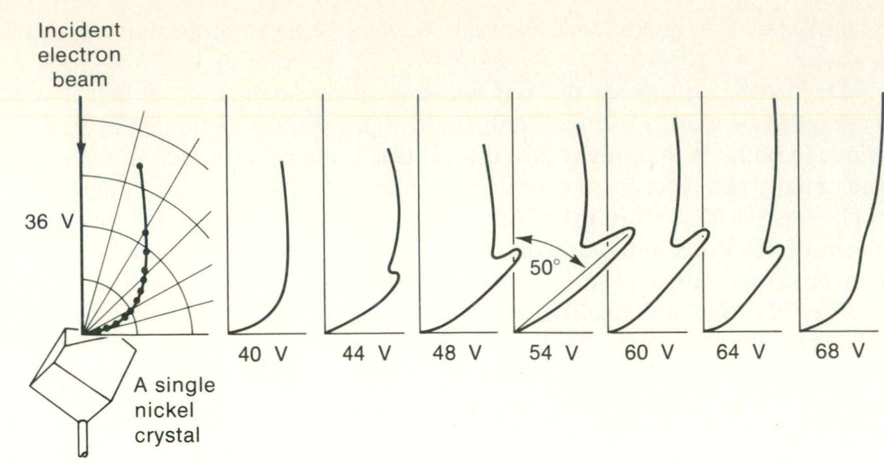

Figure 34–11

The diffraction of electrons in a Davisson-Germer experiment. Each plot is a polar graph for the number of scattered electrons as a function of angle. Several different values of the accelerating voltage are shown. When the de Broglie wavelength of the incident electrons matches the Bragg condition for the spacing between atomic planes, the diffracted beam is a maximum.

planes of atoms spaced a distance d apart (see Figure 34–12). Constructive interference occurs for the reflected waves whenever the *Bragg scattering condition* is met [Equation (31–24)]:

$$m\lambda = 2d \sin \phi \tag{34–17}$$

where $m = 1, 2, 3, \ldots$ (the *order* of the scattering)
$\phi =$ the *glancing* angle between the incident radiation and the *plane*
$d =$ atomic plane spacing

If the de Broglie wavelength of the electrons is comparable to the spacing between the planes (~ 0.1 nm), they, like x-rays, should exhibit diffraction effects. In the Davisson-Germer experiment, electrons (charge e and mass m) are accelerated through a potential difference V of the order of 100 V or less. Since 100 eV $\ll m_e c^2$, the velocity v of the electrons is nonrelativistic. Therefore, from the conservation of energy, we have

$$eV = \tfrac{1}{2}mv^2 \tag{34–18}$$

Solving for v:

$$v = \sqrt{\frac{2eV}{m}}$$

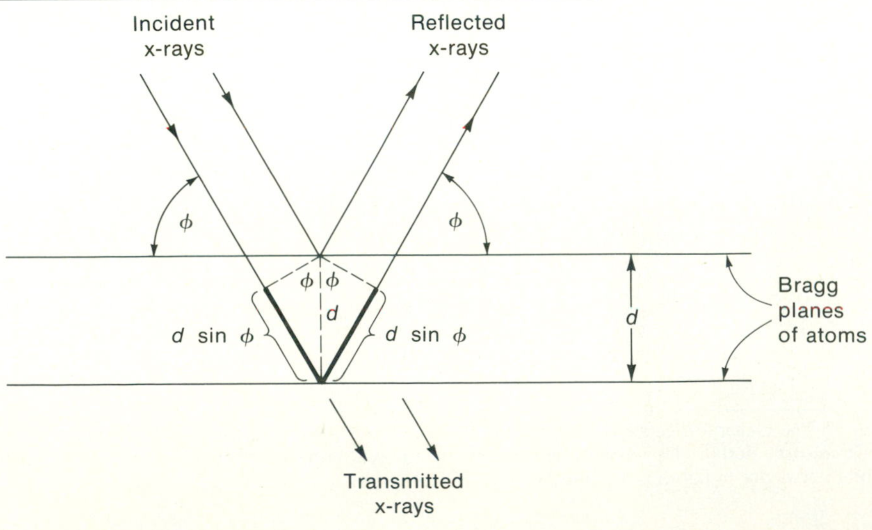

Figure 34–12

The scattering, or reflection, of x-rays from two adjacent Bragg planes. The path-length difference for the two rays shown is $2d \sin \phi$.

The momentum $p = mv$ is thus

$$p = m\sqrt{\frac{2eV}{m}} = \sqrt{2meV} \tag{34-19}$$

and the de Broglie wavelength $\lambda = h/p$ is

$$\lambda = \left(\frac{h}{\sqrt{2me}}\right)\frac{1}{\sqrt{V}} \tag{34-20}$$

Substituting numerical values, we obtain the useful relation:

DE BROGLIE WAVELENGTH FOR ELECTRONS (nonrelativistic)

$$\lambda = \frac{1.226 \text{ nm}}{\sqrt{V}} \quad \text{(where } V \text{ is in volts)} \tag{34-21}$$

In Figure 34–11, the most prominent peak at 50° occurred for 54 eV electrons with a de Broglie wavelength of

$$\lambda = \frac{1.226 \text{ nm}}{\sqrt{54}} = 0.167 \text{ nm}$$

The 50° angle in the diagram (130° scattering from the incident direction) corresponds to a grazing angle of incidence $\phi = 65°$. The effective spacing between atomic planes is known (from x-ray scattering data) to be $d = 0.091$ nm. Substituting these numbers into the Bragg condition (assuming first-order diffraction, $m = 1$):

$$\lambda = 2d \sin \phi$$
$$\lambda = (2)(0.091 \text{ nm})(\sin 65°)$$
$$\lambda = 0.165 \text{ nm}$$

This calculated value for the wavelength that would produce the observed scattering is in excellent agreement with the de Broglie wavelength for 54 eV electrons.

The experiments of Davisson and Germer in 1925–1927 (and similar studies by G.P. Thomson in Scotland) were the first experimental confirmation of the de Broglie wave properties of particles.[7] Essentially all of the interference and diffraction effects of electromagnetic waves were later duplicated with particles (see Figure 34–13).

[7] Davisson and Thomson shared the Nobel Prize in 1937 for demonstrating the *wave* properties of electrons. Thirty-one years earlier, Thomson's father, J.J. Thomson, received the Nobel Prize for investigating the conduction of electricity by gases, a phenomenon involving the *particle* properties of electrons.

Figure 34–13
Diffraction patterns produced by electromagnetic waves and particles. (a) and (b): Laue-spot patterns demonstrate the wave nature of photons and of neutrons. (c), (d), and (e): Diffraction rings produced by scattering from polycrystalline metal samples.

(**a**) X-rays on NaCl.

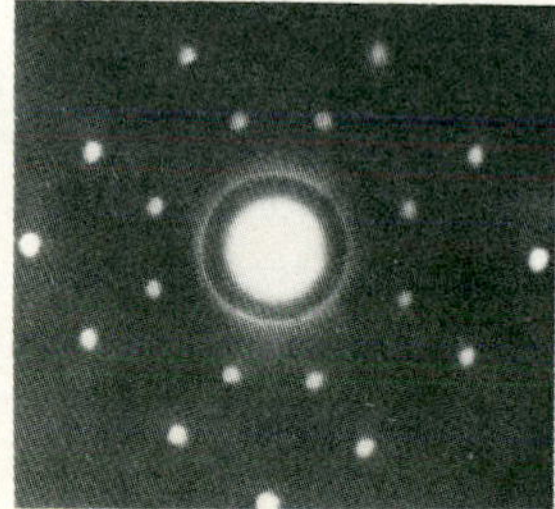
(**b**) Neutrons on NaCl.

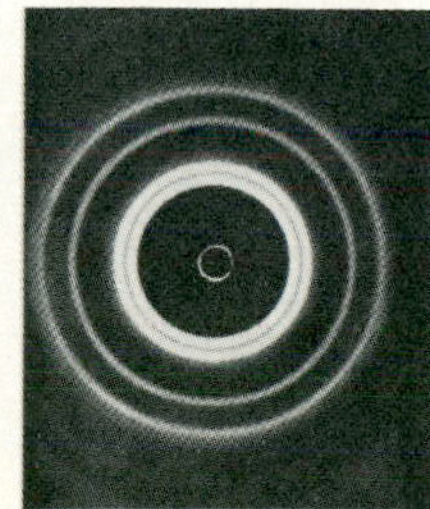
(**c**) 0.071 nm x-rays.

(**d**) 600 eV electrons.

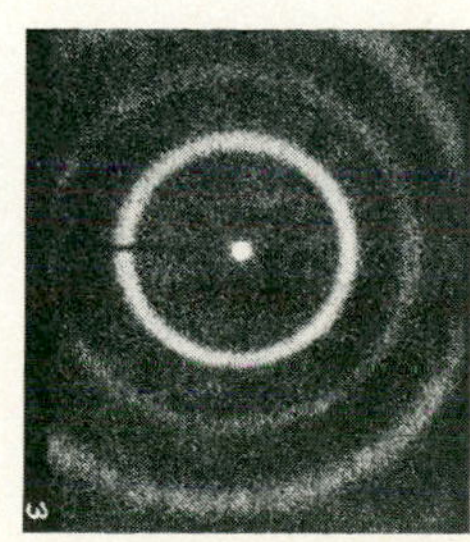
(**e**) 0.057 eV neutrons.

(**a**) Visible light

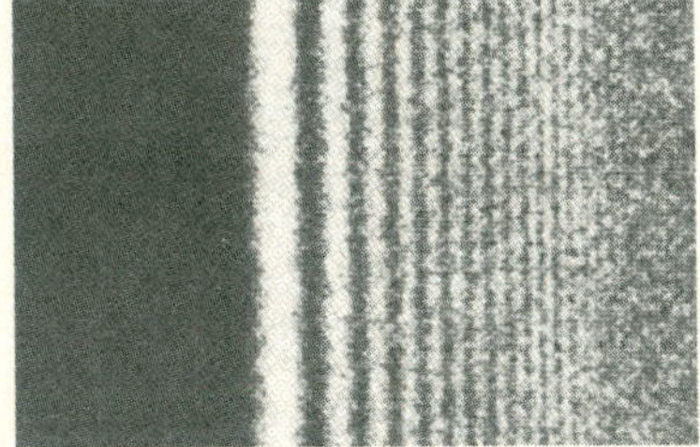

(**b**) Electrons

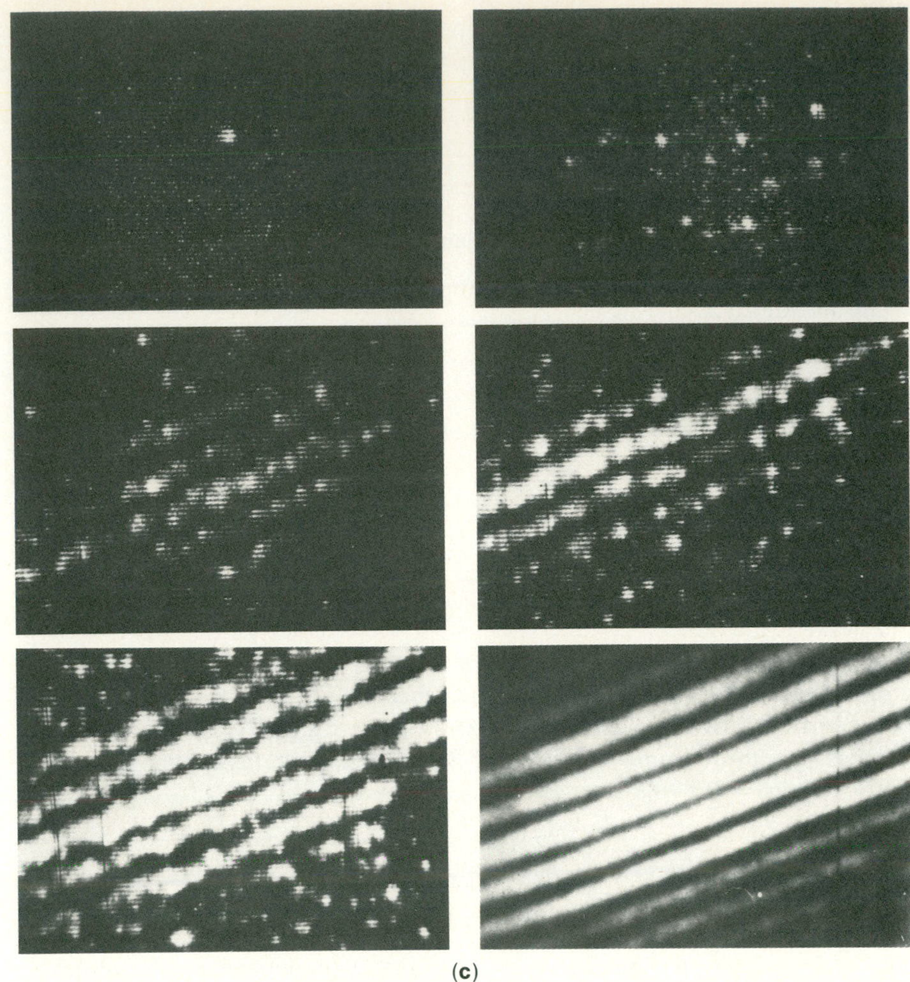

(**c**)

Figure 34–14
Fringes formed in the shadow of a straightedge by visible light and by electrons. In (b), the fringes were recorded with the aid of an electron microscope. (c) An interference pattern produced by electrons is the sum of many independent events. As the number of events increases, the pattern becomes more distinct.

34.5 Wave Mechanics

Before matter waves were experimentally verified, two physicists used de Broglie's ideas in 1925–1926 to develop a theory called *wave mechanics*, or *quantum mechanics*, which describes what happens when a force acts on a de Broglie wave. The two theories are vastly different in mathematical form. The German physicist Werner Heisenberg used sophisticated matrix methods, while the Austrian physicist Erwin Schrödinger devised a differential equation approach.[8] Shortly after the theories were published, it was discovered that they were entirely equivalent; either could be derived from the other. Since matrix methods are usually treated in more advanced mathematics courses, we will discuss only the Schrödinger theory here.

For all but the simplest cases, the theory is mathematically difficult to apply. Perhaps the most troublesome aspect of the theory is that its concepts are foreign to our everyday experience and common sense. Yet it has proved to be the only correct way of analyzing the microphysical world. In fact, in its complete relativistic form, known as *quantum electrodynamics* ("Q.E.D."), there is *no* discrepancy with experimental data (at least, to the date of this publication). It

[8] Nobel prizes were awarded to Heisenberg in 1932 and to Schrödinger (along with P.A.M. Dirac) in 1933 for their accomplishments in developing quantum mechanics.

is a "perfect" theory. To the experienced physicist, quantum mechanics is the only plausible and workable way to describe the interaction between the electromagnetic field and the electronic structure of atoms and molecules. It is the most all-inclusive theory we have, since in the macroscopic limit the equations become the familiar Newtonian mechanics as a special case and in the limit of low velocities the relativistic aspects fade out to become the well-known classical equations. In addition, the wave-particle duality is built in, so the theory correctly deals with particles and radiation under all circumstances. There is a limit, however, to quantum mechanics because it applies primarily to the world of atoms and molecules. There are still plenty of unanswered questions. For example, the physics of the nucleus and of fundamental particles themselves still offers ample mysteries to keep physicists interested for a long time to come. While some progress has been made, we are not certain of the appropriate potential energy function for nuclear forces, or even whether the Schrödinger equation needs modification to be applicable in these contexts. Perhaps some entirely new approach will be necessary before we can understand the nucleus adequately.

The central idea of quantum mechanics is contained in a differential equation called "the Schrödinger equation." (Its counterpart in classical mechanics is the differential equation of Newton's second law: $m\,d^2x/dt^2 = F$.) A rigorous derivation would lead us too far astray, so we will give just a plausibility argument here for its origin.

The (nonrelativistic) kinetic energy K of a particle may be written in terms of the momentum p as

$$K = \tfrac{1}{2}mv^2 = \frac{p^2}{2m} \qquad \textbf{(34–22)}$$

If the potential energy is U, the total energy E becomes

$$E = K + U$$

$$E = \frac{p^2}{2m} + U$$

Solving for p:

$$p = \sqrt{2m(E - U)} \qquad \textbf{(34–23)}$$

If we put this value into the de Broglie relation $\lambda = h/p$, we obtain

$$\lambda = \frac{h}{\sqrt{2m(E - U)}} \qquad \textbf{(34–24)}$$

As developed in Chapter 16, a solution to the classical wave equation for a wave traveling in the $+x$ direction is

$$y = A \sin(kx - \omega t) \qquad \left(\text{where } k = \frac{2\pi}{\lambda} \text{ and } \omega = \frac{2\pi}{T}\right) \qquad \textbf{(16–16)}$$

If we take partial derivatives with respect to x, we can obtain the following:

$$\frac{\partial^2 y}{\partial x^2} + \left(\frac{2\pi}{\lambda}\right)^2 y = 0 \qquad \textbf{(34–25)}$$

where y is the amplitude of the wave motion. *Any* type of mechanical wave—sound waves, waves on a stretched rope, and so on—satisfies this relation.

Schrödinger put into this equation the value for λ from Equation (34–24) to obtain the time-independent *Schrödinger wave equation:*

SCHRÖDINGER'S TIME-INDEPENDENT WAVE EQUATION (one-dimension)

$$\frac{\partial^2 \psi}{\partial x^2} + \left(\frac{2m(E - U)}{\hbar^2}\right)\psi = 0 \qquad \textbf{(34–26)}$$

where ψ represents the amplitude of the matter wave.

The Schrödinger equation is used in the following way. To find the effect of applying a force to a particle, we substitute the potential energy function U that is associated with the force into the Schrödinger equation. Solutions to the differential equation then express the behavior of the matter wave for the particle. For example, if we put the coulomb potential $U(r) = -(1/4\pi\varepsilon_0)(qq'/r)$ into the Schrödinger equation (for three dimensions), we obtain the ψ functions that represent the matter waves for the stationary states of the electron in a hydrogen atom.

But what does ψ itself represent? We have called it a "matter wave," but naming it does not give us much insight. Since waves are inherently spread out in space, does this mean, for example, that an electron in a hydrogen atom is somehow "smeared out" in space in a way described by the amplitude of ψ? Schrödinger originally proposed this interpretation, but it did not gain much support. The difficulties arose in the complete time-dependent theory, where the *wave packet* representing a free electron gradually spreads out in space as time goes on. Interpreting this to mean that the charge and mass of an electron in free space similarly spreads out seemed impossible for most physicists to accept.

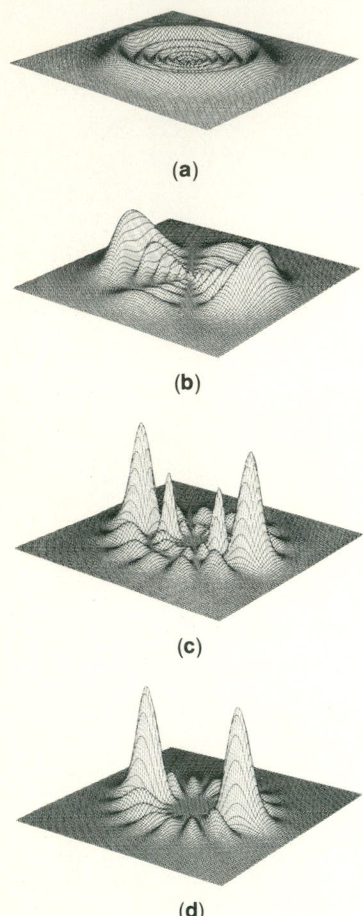

Figure 34–15
Representations of the probability-density distributions for highly excited states ($n = 8$) of the hydrogen atom that have different values of angular momentum. The nodal lines are either concentric circles or straight lines passing through the nucleus. The true three-dimensional distributions may be visualized by imagining that the graph is rotated about a horizontal line passing through the nucleus, forming nodal surfaces that are spherical shells or cones. In these excited states, the hydrogen atom is much larger than it is in its lowest energy state. The distance from the nucleus to the edge of these graphs is 380 times the Bohr radius for $n = 1$.

In 1926, a more reasonable interpretation for ψ was proposed by Max Born, a professor at the University of Göttingen.[9] Born noted that Einstein had put forth a new interpretation of the amplitude of the electric field E for electromagnetic waves. Since the square of the amplitude is proportional to the intensity of the wave, Einstein suggested that E^2 is proportional to the *probability* of finding a photon near that location. Thus the light and dark fringes on a photographic film (which can be predicted from wave interference) may be interpreted as the probability of a photon arriving near that particular location on the film. Born extended this idea to the wave function ψ. He proposed that ψ^2 represents the probability that the particle is located near that region of space. This interpretation gave back to the electron its status as a particle rather than a smeared-out entity. *Only our ability to predict the electron's location becomes spread out.*

Generally ψ is a *complex* mathematical function (that is, it involves $\sqrt{-1}$). Because only mathematically real numbers correlate with physically real objects, Born removed the complex characteristics of ψ by suggesting that the *square of the absolute value* of ψ be used. In particular:

BORN'S PROBABILITY INTERPRETATION OF ψ

$|\psi|^2\,\Delta x =$ the probability the particle will be found within the region Δx

Thus the electron itself does not become smeared out in space; only the *mathematical statement regarding the probability of finding it in a given region* is spread out in space.

If the potential U forms a potential well that can trap electrons in a certain finite region (for example, the coulomb potential for the hydrogen atom), then the Schrödinger solutions are *standing-wave patterns* in three dimensions such that $|\psi|^2\,\Delta V$ is the probability of finding the electron within a volume ΔV. Figure 34–15 illustrates pictorially the probability distributions for various energy states of the hydrogen atom. Where the "cloud" is most dense, the mathematical

[9] In the 1920s, the most exciting and productive centers in the world for physics were at Göttingen and Copenhagen. However, in the 1930s the rise of Nazism brought about purges of non-Aryan professors, and the influence and prestige of Göttingen was greatly diminished.

probability is highest. These probability distributions are vastly different from the simple electron orbits of the Bohr model.

We have given only the briefest introduction to quantum mechanics. Numerous innovations and additions to the theory were made by many physicists, most notably the British physicist P. Dirac (1928), who developed the *relativistic* wave equation (which accounts for the splitting of spectral lines in the presence of a magnetic field and predicts the existence of *antimatter*).[10]

34.6 The Uncertainty Principle

Wave mechanics replaces the precise trajectories of particles with a "cloud" of probability estimates spread out in space. This is a profound change in the way we deal with nature. The most all-inclusive theory we have—quantum mechanics—is not based upon the kind of *physical* models that all previous theories were. It does not tell us exactly where the electron is or how it moves, but only how to *estimate the probability of finding it near a certain region traveling within a certain range of velocities.* But the nagging question remains: The electron must be *somewhere.* Can't we improve our measuring technique to pin down its location exactly and find out precisely how it moves from one place to another?

In 1927, Heisenberg pointed out that there is a *fundamental* limit, inherent in *all* measurements, that prevents us from doing this. No amount of cleverness or refinement of our measuring apparatus will get around this basic obstacle, because *the limitation is a consequence of the wave-particle duality of nature,* and we cannot change that.

The uncertainty principle can be illustrated in the following way. Suppose we wish to determine the position of an electron along the x axis with a very powerful microscope (Figure 34–16). Because of diffraction effects due to the lens diameter D, the image of a (point) electron will be a *diffraction pattern* whose central peak has an angular size θ_R according to Equation (31–22):

$$\theta_R = \frac{(1.22)\lambda}{D}$$

where λ is the wavelength of light used and D is the lens diameter. This minimum angle of resolution θ_R may also be written as $\Delta x/d$. It implies the electron's position is known only within an uncertainty $\pm\Delta x$.

$$\frac{\Delta x}{d} = \frac{(1.22)\lambda}{D}$$

or, rearranging:

$$\Delta x = \frac{(1.22)\lambda}{\left(\dfrac{D}{d}\right)}$$

If 2α is the angle of the cone of light from the object the lens gathers, then $\tan\alpha = (D/2)/d = \frac{1}{2}(D/d)$. For an order-of-magnitude estimate, we may replace

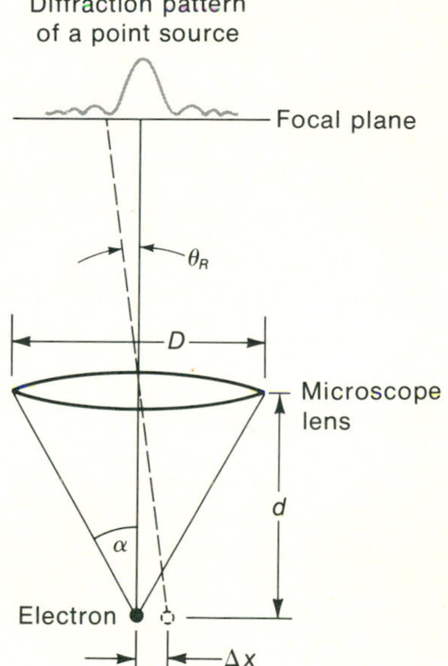

Figure 34–16
Observing the position of an electron with a microscope. The central peak of the diffraction pattern is within $\pm\theta_R$ of the axis.

[10] Antimatter—antielectrons, antiprotons, antineutrons, and so forth—is another form of matter, created in high-energy interactions of photons and particles. If an antimatter particle comes in contact with a particle of matter, they mutually annihilate, forming an equivalent amount of energy (mc^2) in photons. Our own galaxy is perhaps composed only of matter. Since matter and antimatter are always experimentally formed in equal amounts, it may well be that other galaxies are composed entirely of antimatter, to maintain an overall balance in the universe.

tan α with the approximation sin α (not an overwhelmingly good approximation, but it is still in the ballpark).

$$\Delta x \approx \frac{(1.22)\lambda}{2 \sin \alpha}$$

Finally, in the same spirit of estimation, we drop the factor 1.22/2 to obtain

$$\Delta x \approx \frac{\lambda}{\sin \alpha} \tag{34–27}$$

This is the inherent uncertainty in determining the x coordinate of the position of the electron. It is due to the fact that we used a lens of diameter D. If we used a lens with a smaller diameter, the uncertainty would be greater (because sin α would be smaller).

Perhaps we could try to improve matters by using light of shorter wavelength, say, in the x-ray region. But, unfortunately, a photon of shorter wavelength has a greater momentum $p = h/\lambda$ and would give the electron a harder "kick" as it scatters off the electron into the microscope lens. The scattered photon can enter the lens *anywhere* within an angle 2α. We do not know the exact direction because we do not detect the photon until after it travels through the lens to reach the image location. All we know is that it went through the lens at *some* point. As the photon scatters off the electron in a Compton interaction, its x component of momentum can vary anywhere from $+(p_x \sin \alpha)$ to $-(p_x \sin \alpha)$. And, by the conservation of momentum, this uncertain amount is transferred to the electron. So the uncertainty in the x component of the electron's momentum becomes

$$\Delta p_x \approx 2p \sin \alpha$$

or

$$\Delta p_x \approx 2\left(\frac{h}{\lambda}\right) \sin \alpha \tag{34–28}$$

Combining these uncertainties in position and momentum, we have

$$\Delta x\, \Delta p_x \approx \frac{\lambda}{\sin \alpha}\, 2\left(\frac{h}{\lambda}\right) \sin \alpha$$

$$\Delta x\, \Delta p_x \approx 2h \tag{34–29}$$

As the uncertainty in position is reduced, inevitably the uncertainty in momentum increases, and vice versa.

Of course, there are other ways of measuring the position and momentum of an electron. Certain schemes might lead to an improvement by a factor of 2 or π or some small number such as that. But the basic limit of this order of magnitude always remains. Note that it is not due in any way to lack of refinement in our measuring instruments. Even with the most ideal apparatus imaginable, the fundamental limitation still remains; this limitation is traceable to the wave-particle aspects of both matter and radiation.

A more rigorous statement of the **Heisenberg uncertainty relation** is:

HEISENBERG UNCERTAINTY RELATION

$$\Delta x\, \Delta p_x \gtrsim \hbar \tag{34–30}$$

In a simultaneous measurement of the position and momentum of a particle, the product of the uncertainties is equal to or greater than a number of the order of $\hbar$.

No amount of ingenuity or improvement in measurement techniques can outwit this limitation. Because of the wave-particle aspects of matter and radiation, the very act of measurement itself inevitably disturbs the system under investigation in an *unknown* way that cannot be avoided. This is a consequence of the wave-particle duality of matter and radiation and is therefore a built-in limitation in nature. In essence, the uncertainty principle underscores the fact that classical models of atomic phenomena are bound to be misleading.

Note, however, that there is no limit on determining the position (only) of a particle to any desired degree of accuracy, or the momentum (only). But as we narrow down the uncertainty in position ($\Delta x \to 0$), *inevitably* the uncertainty in the simultaneous determination of the momentum of the particle becomes larger and larger ($\Delta p_x \to \infty$), and vice versa. The precise relation between Δx and Δp_x depends on how one defines the limits of uncertainty in a particular case. The product may vary somewhat in the range of $2h$ down to about $\hbar$. Similar relations also apply in the y and z directions.

$$\Delta y \, \Delta p_y \gtrsim \hbar \tag{34–31}$$

$$\Delta z \, \Delta p_z \gtrsim \hbar \tag{34–32}$$

Different sets of variables are also related in the same way. It can be shown that

$$\Delta E \, \Delta t \gtrsim \hbar \tag{34–33}$$

where ΔE is the uncertainty in the measurement of energy E and Δt is the time interval for determining the energy. The principle also applies to angular measurements. For example, if we wish to determine where the electron is located in the orbit of a Bohr-model hydrogen atom, the uncertainty in the angle ϕ measurement is related to the uncertainty in the angular momentum L_ϕ:

$$\Delta \phi \, \Delta L_\phi \gtrsim \hbar \tag{34–34}$$

This form of the uncertainty principle essentially leads to the destruction of the planetary view of the Bohr model, in which the electron occupies a well-defined position in an orbit. Consider the following example:

EXAMPLE 34–2

Estimate the uncertainty in the angular position $\Delta\phi$ of the electron in a Bohr orbit.

SOLUTION

The value $\Delta\phi$ is related to the electron's uncertainty in angular momentum ΔL_ϕ by Equation 34–34:

$$\Delta \phi \, \Delta L_\phi \gtrsim \hbar$$

Since L_ϕ is quantized according to one of the Bohr-model postulates, it has discrete values only, with no uncertainty in any of the Bohr orbits:

$$\Delta L_\phi = 0$$

Equation (34–34) then states that $\Delta\phi$ must have no finite value, which is equivalent to stating that ϕ is *completely uncertain.* The electron is equally likely to be anywhere in the orbit all of the time. Thus, it is meaningless to speak of the electron as moving from point to point along its orbit.

The uncertainty principle has profound philosophical consequences. Just as Einstein showed that absolute space, absolute time, and absolute simultaneity are inherently unmeasurable and therefore meaningless concepts that should be eliminated from our theories, Heisenberg pointed out that precise knowledge of the position and momentum of an electron at a given instant is inherently limited.

This is in contrast to the situation in classical physics, where *any* measurement could, in principle, be made with increasing precision without limit. Possible disturbance to the object being measured could be eliminated or accounted for. And if the position and velocity of a particle could be precisely known at a given instant (and the forces acting on the particle are also known), classical physics predicts the future behavior of the particle exactly.

The uncertainty principle denies this. It points out the impossibility of making a measurement without disturbing the object *by an unknown amount*, thereby reducing our knowledge of some related quantity. This is true even with "perfect" measuring instruments that have no technical imperfections because the uncertainties do not originate in defects in the equipment or in the measuring techniques: They originate in the wave-particle duality of matter and radiation. Since we can never determine the exact behavior of particles at the atomic level, we should not speak of their motions in classical terms.

It now becomes clear why the paradoxes arose in analyzing the double-slit interference effect in terms of classical trajectories for photons (or particles) as they pass through the slit system. In an experiment in which a beam of electrons incident on two slits whose spacing is of the same order of magnitude as the de Broglie wavelength, the usual two-slit interference pattern results. Even if we send only one electron at a time at the slit, the interference fringes are still formed (statistically) if enough electrons are used. However, if one slit at a time is covered alternatively during the exposure time, we do not get the two-slit fringe pattern, but just the single-slit diffraction pattern. Thus we must conclude that with both slits open, each electron somehow interacts simultaneously with both slits in spite of our classical model of an electron as a well-defined particle that could go through only one slit at a time. As far as we can experimentally verify, electrons are not classical particles with well-defined trajectories, so we should not talk as if they were. This is the essence of the *positivist* philosophy that gained a strong foothold in physics, first through Einstein's relativity (which rejected the idea of an ether because it was unmeasurable) and later through quantum mechanics (which rejected precise classical descriptions of atomic phenomena as unmeasurable). In its place, quantum mechanics offers only probability estimates. If a series of identical measurements are made of a property of a system, quantum mechanics can predict *precisely* the average value of these measurements, yet it can give only a *probability estimate* for any single measurement.

This probabilistic interpretation of quantum mechanics is associated with the *Copenhagen* school of thought, so-named because of its main architect, the Danish physicist Niels Bohr. The majority of physicists today accept this interpretation. However, there are some notable exceptions. Einstein, for example, never accepted the abandonment of the strict causality on which classical physics is based. "God does not play dice with the universe," he said, and felt there must be some underlying causal relations that produce the statistical behavior we observe. He had faith that some future theory could reveal a strict causality at a deeper level. A few good theorists have devoted years in attempts to devise such a "hidden parameter" theory. None has succeeded to date.

EXAMPLE 34–3

An electron ($m = 9.11 \times 10^{-31}$ kg) and a bullet ($m = 0.02$ kg) each have speeds of 500 m/s, accurate to within 0.01%. Within what limits could we determine the position of the objects?

SOLUTION

(a) The electron's momentum is

$$p = mv = (9.11 \times 10^{-31}\ \text{kg})\left(500\ \frac{\text{m}}{\text{s}}\right) = 4.56 \times 10^{-28}\ \frac{\text{kg·m}}{\text{s}}$$

The uncertainty Δp_x in this momentum measurement is given as 0.01%. Thus:

$$\Delta p_x = \left(4.56 \times 10^{-28}\ \frac{\text{kg·m}}{\text{s}}\right)(0.0001) = 4.56 \times 10^{-32}\ \frac{\text{kg·m}}{\text{s}}$$

From the Heisenberg uncertainty relation [Equation (34–30)], the uncertainty Δx in position is of the order of

$$\Delta x \approx \frac{\hbar}{\Delta p_x} = \frac{h}{(2\pi)\,\Delta p_x} = \frac{6.63 \times 10^{-34}\ \text{J·s}}{(2\pi)\left(4.56 \times 10^{-32}\ \frac{\text{kg·m}}{\text{s}}\right)}$$

$$\Delta x \approx 0.002\ 31\ \text{m} \quad \text{or} \quad \boxed{2.31\ \text{mm}}$$

This is an unbeatable lower limit on the uncertainty with which we could determine the electron's position. A model of an electron as a small point mass is not valid for this situation.

(b) The bullet's momentum is

$$p = mv = (0.02\ \text{kg})\left(500\ \frac{\text{m}}{\text{s}}\right) = 10.0\ \frac{\text{kg·m}}{\text{s}}$$

The uncertainty Δp_x in this momentum measurement is given as 0.01%, or:

$$\Delta p_x = \left(10.0\ \frac{\text{kg·m}}{\text{s}}\right)(0.0001) = 10^{-3}\ \frac{\text{kg·m}}{\text{s}}$$

From the Heisenberg uncertainty relation [Equation (34–30)], the uncertainty Δx in position is of the order of

$$\Delta x \approx \frac{\hbar}{\Delta p_x} = \frac{h}{(2\pi)(\Delta p_x)} = \frac{6.63 \times 10^{-34}\ \text{J·s}}{(2\pi)\left(10^{-3}\ \frac{\text{kg·m}}{\text{s}}\right)}$$

$$\Delta x \approx \boxed{1.00 \times 10^{-31}\ \text{m}}$$

This uncertainty in position is far below any conceivable possibility of measurement (an atomic nucleus is about 10^{-15} m in size), so for macroscopic objects under everyday circumstances, we may confidently treat them as *classical* particles.

34.7 The Complementarity Principle

We have described how physicists came to believe in a certain symmetry in nature involving particles and waves. But this new unity came at the price of new conceptual difficulties. The best theory we have—quantum electrodynamics—does not allow us to picture the motions and interactions of objects as we did in classical physics. Objects are neither particles nor waves, yet on occasion they show more strongly one or the other of these attributes. An experiment designed to bring out the *wave* aspects (such as double-slit interference) cannot be dealt with in terms of particles. Similarly, an experiment that brings out *particle* aspects (such as Compton scattering) cannot be visualized in terms of waves. Bohr (1928) recognized this essential characteristic of nature by suggesting a **principle of complementarity** at the atomic level.

BOHR'S COMPLEMENTARITY PRINCIPLE

In the quantum domain, wave and particle aspects complement each other. Though the choice of one description precludes the simultaneous choice of the other, *both* are required for a complete understanding.

In explaining this principle, Bohr suggested an analogy: Both sides of a coin must be included for a complete description of the coin, yet we cannot see both sides simultaneously. As with the Copenhagen interpretation of quantum mechanics, a few physicists and philosophers still seek an alternative view. Nevertheless, Bohr's principle of complementarity does seem to express in general terms why we find ourselves in a dilemma when we try to cling to classical ideas at the atomic level.

Our concepts, our modes of thought and language—indeed, what we call common sense—all originate in our experiences. Classical physics is the crowning achievement of this common sense. In the 1920s, however, our experiences in the microworld and in the relativistic domain began to include observations that violated classical ideas, so our common sense had to be enlarged and changed to include these new types of experiences. Nature continues to challenge us with a myriad of mysteries. What concepts will we need to accept in the future in order to unravel them?

34.8 A Brief Chronology of Quantum Theory Development

1900 Explanation of blackbody radiation by energy quantization.
Max Planck (*Nobel Prize 1918*).

1900 Discovery that the energy of electrons emitted by the photoelectric effect was independent of the light intensity.
Philip von Lenard (*Nobel Prize 1905*).

1905 Explanation of the photoelectric effect.
Albert Einstein (*Nobel Prize 1921*).

1905 The theory of special relativity.
Albert Einstein.

1907 Explanation of the specific heats of solids by energy quantization.
Albert Einstein.

1913 First quantized model of the hydrogen atom.
Niels Bohr (*Nobel Prize 1922*).

1916 Experimental studies of the photoelectric effect.
Robert Millikan (*Nobel Prize 1923*).

1923 Discovery *and* explanation of the collisions between light quanta and electrons.
Arthur Compton (*Nobel Prize, with C.T. Wilson, 1927*).

1924 Proposal that electrons have an associated wavelength $\lambda = h/p$.
Prince Louis-Victor de Broglie (*Nobel Prize 1929*).

1925 Mathematical theory of wave mechanics.
Erwin Schrödinger (*Nobel Prize, with P. Dirac, 1933*).

1925 Mathematical theory of matrix mechanics.
Werner Heisenberg (*Nobel Prize 1932*).

1926 Statistical interpretation of the wave function.
Max Born (*Nobel Prize 1954*).

1927 The Uncertainty Principle.
Werner Heisenberg.

1927 Observation of electron-wave diffraction by crystals.
Clinton Davisson (Nobel Prize, with G.P. Thompson, 1937).

1928 Relativistic theory of quantum mechanics and the prediction of the positron.
Paul Dirac (Nobel Prize, with E. Schrödinger, 1933).

Summary

The *Bohr model for hydrogen* assumes:

(1) The electron travels in circular orbits about the proton. The coulomb force is the centripetal force.
(2) There exist allowed energy states E_n for which the electron moves without radiating.
(3) The allowed energy states are those for which

$$mvr = n\hbar$$

(4) Transitions between allowed energy states involve the emission or absorption of photons of energy hf, where

$$hf = E_{\text{final}} - E_{\text{initial}}$$

The orbital radii and the energy of allowed energy states in the Bohr model are:

$$r_n = \frac{\varepsilon_0 h^2 n^2}{\pi m Z e^2} = (0.0529 \text{ nm})n^2$$

$$E_n = -\frac{mZ^2e^4}{8\varepsilon_0{}^2h^2n^2} = -\frac{13.6 \text{ eV}}{n^2} \qquad (n = 1, 2, 3 \ldots)$$

Bohr's *correspondence principle* is:

Any new theory must reduce to the corresponding classical theory when applied to situations appropriate to the classical theory.

Under certain circumstances, particles exhibit wave characteristics *with a de Broglie wavelength:*

$$\lambda = \frac{h}{p}$$

where p is the momentum of the particle. For electrons accelerated from rest through a potential difference V:

$$\lambda = \frac{1.226 \text{ nm}}{\sqrt{V}} \qquad \text{(where } V \text{ is in volts)}$$

Wave mechanics, or *quantum mechanics*, is a theory developed by Erwin Schrödinger (and independently by Heisenberg in a different format) that includes the wave and particle characteristics for both matter and radiation. It is a differential equation for an amplitude ψ. In Born's interpretation, it became

$$|\psi|^2 \,\Delta x = \textit{the probability the particle will be found within the region } \Delta x$$

Heisenberg's uncertainty relation places a fundamental limit on the accuracy with which certain pairs of variables can be measured simultaneously. The product of the uncertainties is $\gtrsim \hbar$. Following is a partial list of these variables:

Position and momentum:	$\Delta x \, \Delta p_x \gtrsim \hbar$
Energy and time:	$\Delta E \, \Delta t \gtrsim \hbar$
Angular position and angular momentum:	$\Delta\phi \Delta L_\phi \gtrsim \hbar$

Questions

1. How does the correspondence principle apply to Einstein's theory of special relativity?
2. Insofar as the wave nature of particles is concerned, what would be the observable consequences of Planck's constant being of the order of 0.1 J·s?
3. What are the similarities between particle waves and electromagnetic waves? What are the dissimilarities?
4. In what ways are high-energy electrons and photons similar? In what ways are they dissimilar?
5. Do the wave-like properties of particles imply that a baseball pitched through an open door will be deflected?
6. In what ways does the wave-like concept of particles contradict Bohr's model of the hydrogen atom?
7. Attempt to clarify the statement that if a beam of electrons were used to produce a double-slit interference pattern, each of the electrons would have to pass through both slits.
8. In what way is the uncertainty principle a direct consequence of the wave-like nature of particles?
9. How is the de Broglie concept of an orbital standing wave for the electron in the hydrogen atom inconsistent with the uncertainty principle?
10. What is the role of the complementarity principle in an experiment that demonstrates electron diffraction?

Problems

34A–1 Calculate the series limit (shortest wavelength) of the Lyman series in the hydrogen spectrum. (The final energy state for the Lyman series corresponds to $n = 1$.)

Answer: $\lambda = 91.2$ nm

34A–2 Suppose that the angular momentum of the moon revolving about the earth is quantized in accordance with the relation $mvr = nH/2\pi$, where H is a constant.

(a) Assuming the moon now occupies its lowest energy state, calculate the value of H in SI units.

(b) Calculate the distance between the earth and moon in the moon's next higher quantized orbit.

34A–3 The Paschen series of spectral lines in the hydrogen spectrum corresponds to a final energy quantum number n of 3. Calculate the *shortest* wavelength in the Paschen series. Your answer should indicate that the Paschen series is entirely in the infrared.

Answer: 821 nm

34A–4 The Lyman series of spectral lines in the hydrogen spectrum corresponds to a final atomic energy level of -13.6 eV ($n = 1$). Calculate the *longest* wavelength in the Lyman series. Your answer should indicate that the Lyman series is entirely in the ultraviolet.

34A–5 Show that the orbital speed of the electron in the ground state of the Bohr hydrogen atom is $c/137$. (The number 1/137.036 is called the *fine structure constant*, a prominent constant in atomic theories.)

34A–6 Determine the de Broglie wavelength of an electron in the $n = 3$ energy state of the Bohr hydrogen atom.

34A–7 Show that a rifle bullet with a mass of 15 g and a speed of 1000 m/s does not exhibit measurable wave-like properties. *Answer:* $\lambda = 4.42 \times 10^{-35}$ m

34A–8 Electrons emerge from the Stanford linear accelerator with a kinetic energy of 20 GeV.

(a) Find the de Broglie wavelength of these electrons.

(b) Find the de Broglie wavelength of 500 GeV protons from the accelerator at Fermilab, Batavia, Ill. (Note: For extreme relativistic energies, $E \approx pc$.)

34A–9 A certain meson (a particle produced in high-energy nuclear reactions) has a lifetime of about 10^{-23} s. To what uncertainty can its rest-mass energy be determined? *Answer:* ~65.8 MeV

34B–1 A prominent blue spectral line in the Balmer series of hydrogen corresponds to a wavelength of 486.1 nm. Find the energy (in electron volts) corresponding to the initial and final energy states of the atom. *Answer:* $E_i = -0.85$ eV $E_f = -3.4$ eV

34B–2 Assuming that the angular momentum of the moon revolving about the earth is quantized in accordance with the relation $mvr = n\hbar$, calculate the approximate change in distance between the earth and moon corresponding to a single quantum jump.

34B–3 Neutrons in a nuclear reactor are most likely to be absorbed by metals (such as cadmium) when the neutrons have a kinetic energy of about 0.025 eV. Calculate the de Broglie wavelength of such neutrons. *Answer:* $\lambda = 0.181$ nm

34B–4 Calculate the de Broglie wavelength of an H_2 molecule that has the rms speed for gas molecules at 17°C.

34B–5 Raindrops 1 mm in diameter traveling at 60 m/s fall through an opening (0.5 cm wide) between two roof boards. Because of their wave-like character, the raindrops are diffracted as they pass through the opening. Calculate the angular separation of the resulting diffraction fringes. Comment on the applicability of the correspondence principle. *Answer:* 4.22×10^{-27} rad

34B–6 Find the de Broglie wavelength of an electron that has been accelerated through a potential difference of one million volts. Since the kinetic energy of the electron is comparable to its rest energy, $E^2 = p^2c^2 + m_0{}^2c^4$ applies. (The rest mass of an electron is 0.511 MeV.)

34B–7 Show that particles with an energy much greater than their rest energy have a de Broglie wavelength approximately equal to the wavelength of photons of the same energy.

34B–8 An electron microscope uses a beam of electrons (instead of a light beam), which is focused by electric and magnetic fields (instead of lenses). The theoretical limit of resolution for such a microscope is of the order of the de Broglie wavelength of the electrons.

(a) Find the limit of resolution for an electron microscope that uses electrons accelerated through 60 kV.

(b) What is the limit of resolution for an ordinary microscope using visible light?

34B–9 Many physical properties of atomic nuclei are examined by the way in which they scatter an incident beam of high-energy electrons. However, to be effective the electrons must have a de Broglie wavelength smaller than the size of the nucleus. Calculate the energy (in electron volts) of electrons with a de Broglie wavelength of 10^{-15} m. (Assume that the electrons have nearly the speed of light and show that your answer justifies this assumption.) *Answer:* $E = 1.24$ GeV

34B–10 Suppose the location of a particle is determined to within an uncertainty of one nanometer. Find the uncertainty in velocity of the particle if it is (a) an electron and (b) a proton.

34C–1 Beginning with Equation (34–10), $E_n = -13.6\ \text{eV}/n^2$, derive Equation (34–1), $\lambda = (364.6\ \text{nm})(n^2/[n^2 - 4])$, where $n = 3, 4, 5, \ldots$.

34C–2 Show that for a Bohr hydrogen atom, an energy level transition from $n = 171$ to $n = 170$ results in the emission of radiation with a frequency within 1% of the orbital frequency of the electron. (This is an example of the Bohr correspondence principle.)

34C–3 The wave characteristics of an electron confined to move in one dimension between two rigid walls gives rise to quantized energy levels, because only certain wavelengths produce standing waves between the walls. Show that if the walls are separated by a distance L, the allowable energy levels are given by $E_n = (n^2)(h^2/8mL^2)$, where $n = 1, 2, 3, \ldots$ and m is the mass of the electron.

34C–4 Using the results of Problem 34C–3, calculate the two lowest energies (in electron volts) that an electron may have as it moves between two rigid walls that are separated by 2×10^{-9} m, which is about 10 atomic diameters.

34C–5 Referring again to Problem 34C–3, the possible standing waves may be represented as a wave function $\psi = A \sin(n\pi x/L)$, where $n = 1, 2, 3, \ldots$, L is the distance between the walls, and A is the amplitude of the wave function.

(a) Calculate the ratio of the probability of the electron in its lowest energy state being in a small region Δx one-third the distance from one wall to the probability of the electron being in a region Δx halfway between the walls.

(b) Repeat the calculation for the third energy level.

Answers: (a) $\frac{3}{4}$ (b) 0

34C–6 Complete the following table:

	Wavelength (nm)	Momentum (kg·m/s)	Energy (J)	Speed (m/s)
Photon	1.00			
Electron	1.00			
Proton	1.00			

34C–7 A photon of frequency f undergoes a Compton scattering at 90° from a free electron initially at rest. The scattered photon has frequency f'. Show that the de Broglie wavelength of the recoil electron is given by $\lambda = c(f^2 - f'^2)^{-\frac{1}{2}}$.

34C–8 An electron in the hydrogen atom resides in the energy state corresponding to $n = 2$ for about 10^{-8} s before making the transition to the ground state.

(a) Estimate the uncertainty of the energy in the $n = 2$ state.

(b) Compare this uncertainty with the Bohr prediction of -3.40 eV as the energy in the $n = 2$ state.

34C–9 If a particle is confined somewhere within a region Δx, the uncertainty principle implies that its momentum (and hence its kinetic energy) must also be uncertain. Estimate this uncertainty in kinetic energy for an electron confined within a typical nucleus of diameter 10^{-14} m. (Since electrons emitted from nuclei in a process called *beta decay* have maximum energies much less than this—for example, 0.16 MeV from a ^{14}C nucleus—this fact is considered evidence that electrons do not exist, as electrons, in the nucleus.)

Answer: $\Delta K \approx 20$ MeV

There was a young lady named Bright,
Whose speed was far faster than light.
She left one day
In a relative way,
And returned home the previous night!

ANONYMOUS

Supplemental Topic Derivations of Special Relativity Conclusions*

ST.1 Setting Clocks in Synchronism

Einstein points out that to make measurements of events, a "local" observer situated where the event occurs determines the (x,y,z) coordinates by comparison with the meter-stick framework and the time (t) by comparison with the observer's local clock, which has been synchronized with all other clocks in the frame of reference. In principle, *all measurements are to be made in this fashion.* We now discuss the procedure for synchronizing a system of clocks that are stationed at various points throughout the frame of reference. This matter of synchronization is the source of many of the so-called "paradoxes" of relativity, so be sure to understand the reasoning in this procedure.

We cannot synchronize clocks when they are together, then move them to their respective positions. Because of time dilation, transporting them in this fashion would move them out of synchronization. Instead, Einstein proposed the method illustrated in Figure 1. A flashbulb situated at the point midway between two clocks is set off, sending light pulses in opposite directions. *If the clocks are set so they indicate the same times at the arrivals of the pulses, they will be properly synchronized.* In principle, all other clocks in the frame of reference could similarly be synchronized with these two. Together, the entire array of clocks establishes a time scale by which the simultaneity of events separated in space is judged in that frame. This procedure is based on the

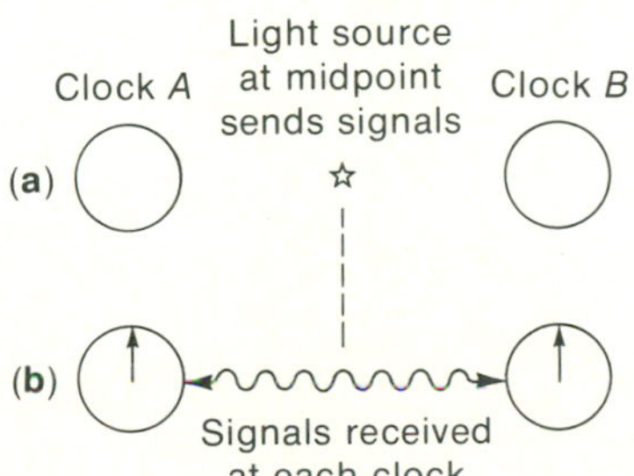

Figure 1
One method of synchronizing two clocks that are separated in space. A flashbulb at the midpoint sends light signals to each clock. If the clocks are set to read the same times when the signals arrive at each clock, they are in synchronism.

* This supplement outlines the derivation of some conclusions of special relativity already presented in Chapter 13, "Special Relativity." It assumes the reader is familiar with the material in Chapter 13.

second postulate: the constancy of the speed of light. It gives to the speed of light c a more fundamental significance than just one of the constants of nature: in particular, it is intimately related to our concepts of time and of simultaneity.

ST.2 The Galilean Transformation

In classical physics there is a straightforward way to relate measurements of an event made in the two frames:

A point event	(x,y,z,t)	in the S frame
	(x',y',z',t')	in the S' frame

The set of equations that do this is called the Galilean transformation. These equations are easily obtained from the geometrical considerations shown in Figure 2.

THE GALILEAN TRANSFORMATION

$$\begin{aligned} x &= x' + Vt' & x' &= x - Vt \\ y &= y' & y' &= y \\ z &= z' & z' &= z \\ t &= t' & t' &= t \end{aligned} \tag{1}$$

The transformation equations are a sort of dictionary that translates the description of an event as measured in one frame into the description of the same event as measured in the other frame. Note that in the left-hand set, primed quantities appear only on the right side, while in the other set they appear only on the left side. Many of our basic assumptions about the nature of space and time are contained in the transformation equations. For example, the fact that we write $t = t'$ implies a universal (or absolute) time scale that is valid for all frames of reference. Similarly, the equations imply that the space in which events happen is the same in both frames. The difference in the x-coordinate descriptions clearly has its origin in the relative motion of the frames; it does not imply that *space itself* is different in the two frames. These classical ideas about space and time are so strongly ingrained in experience that it seems impossible to imagine they are not correct. Indeed, for centuries philosophers have accepted them without question. This makes all the more remarkable the great revolution Einstein brought about when his relativity theory showed these classical ideas to be wrong.

To illustrate the use of the transformation equations, we consider a stick in the S' frame and determine its length as measured in both the S' and the

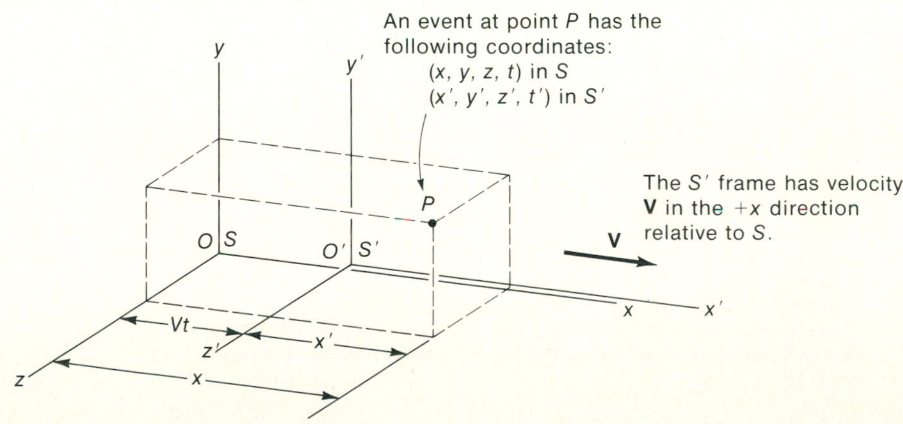

Figure 2
The coordinate systems S and S'. The frames are coincident at the time $t = t' = 0$. At a later time, the origins are a distance Vt apart.

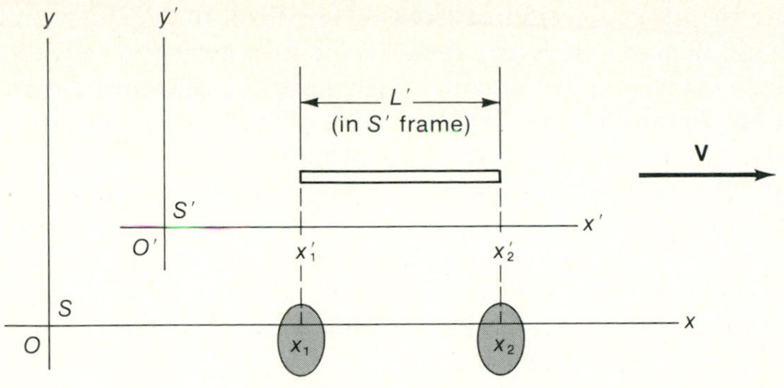

Figure 3
In the S frame, the two events that determine the location of the ends of the moving stick at x_1 and x_2 are *simultaneous* events.

S frame. The stick is oriented along the x' axis as shown in Figure 3. We define the length of the stick to be $L' = x'_2 - x'_1$. *Since the locations of the ends do not change with time*, we measure these values in the usual way by placing a ruler next to the stick and locate the ends in terms of two point events:

Event 1, locating the left end: (x'_1, y'_1, z'_1, t'_1)

Event 2, locating the right end: (x'_2, y'_2, z'_2, t'_2)

The length L' of the stick depends only on x'_2 and x'_1. In particular, the times t'_2 and t'_1 are not involved.

However, in the S frame, the stick is moving. Let us investigate the quantity $x'_2 - x'_1$ as expressed in terms of measurements in S. Applying the Galilean transformation, Equation (1), we have

$$x'_2 - x'_1 = (x_2 - Vt_2) - (x_1 - Vt_1)$$

Or, rearranging the right-hand side:

$$x'_2 - x'_1 = (x_2 - x_1) - V(t_2 - t_1) \tag{2}$$

The quantity $x_2 - x_1$ is the length L of the stick as measured in the S frame. Obviously, it would not make much sense to locate the left end of the stick at t_1 and later, after the stick has moved, to locate the other end at a different time t_2. So, in the S frame we adopt the reasonable procedure of *determining the locations of the ends simultaneously*, when $t_2 = t_1$. Thus, Equation (2) becomes

$$x'_2 - x'_1 = x_2 - x_1$$

or

$$L' = L$$

By locating the ends of the moving object *simultaneously*, the length as measured in S (where the object is moving) corresponds to the length as measured in S' (where the object is at rest). In Galilean relativity, these two length measurements give the same answer. Einstein showed that this conclusion is incorrect.

ST.3 The Lorentz Transformation

Special relativity is based upon two postulates:

BASIC POSTULATES OF SPECIAL RELATIVITY

(1) All the laws of physics have the same form in all inertial frames (the Principle of Relativity).

(2) The speed of light in a vacuum has the same value c in all inertial frames (the Principle of the Constancy of the Speed of Light).

The entire theory is derived just from these two postulates. Adopting these assumptions, Einstein derived a new set of transformation equations that replaced the Galilean transformation. They have the same mathematical form as an earlier transformation developed by Lorentz (though Einstein derived them using reasoning different from that of Lorentz, and the interpretation of the equations is vastly different from the meaning Lorentz attached to them). The derivation is based upon the second postulate of relativity and certain assumptions about the homogeneity of space and time: for example, that as far as physical experiments are concerned, all points in space are equivalent.[1] To simplify the mathematical form, we define $\beta \equiv V/c$, where V is the relative speed of the two frames of reference.

THE LORENTZ TRANSFORMATION (where $\beta \equiv V/c$)

$$x = \frac{x' + Vt'}{\sqrt{1-\beta^2}} \qquad x' = \frac{x - Vt}{\sqrt{1-\beta^2}}$$
$$y = y' \qquad y' = y$$
$$z = z' \qquad z' = z \tag{3}$$
$$t = \frac{t' + Vx'/c^2}{\sqrt{1-\beta^2}} \qquad t' = \frac{t - Vx/c^2}{\sqrt{1-\beta^2}}$$

Note that the two sets of equations "turn the crank" in opposite directions. Either set may be obtained from the other by interchanging primed and unprimed quantities and changing the signs of V and β. (To observers in S, the other moving frame has a velocity $+V$; but in S', the other frame has a velocity $-V$. Hence there is a sign change.)

The Lorentz transformation has an interesting characteristic. If the velocity of the moving frame is much smaller than c, then the factor β approaches zero and the Lorentz transformation becomes identical to the Galilean transformation. So classical relativity is just a special case contained within Einstein's more comprehensive special relativity theory.

We now discuss specific details. Note that the only novel features of the derivations are the use of Einstein's two postulates (as expressed by the Lorentz transformation). All the surprising conclusions that follow are contained implicitly in these two assumptions. Their justification rests on the tremendous successes that special relativity has had in explaining physical phenomena.

ST.4 Comparison of Clock Rates

How do we determine the rate at which a moving clock runs? We cannot just compare a moving clock with a stationary clock at a single instant of time; the procedure necessarily involves measuring a time interval between two events. Figure 4 illustrates how it is done. *Three* clocks are involved. In the first event, clock B is at rest in our frame, coincident at time t_1 with the moving clock A'. In the second event, clock C is at rest in our frame, coincident with the moving clock at a later time t_2. Local observers in each frame record the readings on the clocks at each event. As measured in S', the time interval between the events is $T' = t_2' - t_1'$, while as measured in S the time interval is $T = t_2 - t_1$. Using the Lorentz transformation, Equation (3), we obtain the relation between

[1] For the derivation see, for example, Robert Resnick, *Basic Concepts in Relativity and Early Quantum Mechanics*, Wiley (1972).

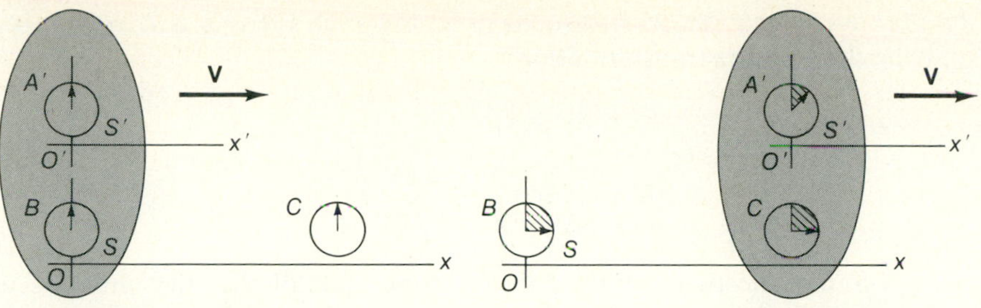

(**a**) *The first event.* The moving clock A' is coincident with the stationary clock B in the S frame. (For convenience, we set all clocks to read zero at this instant.)

(**b**) *The second event.* The moving clock A' is coincident with the stationary clock C in the S frame.

Figure 4
The two events, *measured in the* S *frame*, by which we compare the rate of a moving clock A' with synchronized clocks B and C at rest in S.

these two time intervals:

$$T = t_2 - t_1$$

$$T = \frac{\left(t'_2 + \dfrac{Vx'_2}{c^2}\right)}{\sqrt{1-\beta^2}} - \frac{\left(t'_1 + \dfrac{Vx'_1}{c^2}\right)}{\sqrt{1-\beta^2}}$$

In the S' frame, the two events occur at the same location, ($x'_2 = x'_1$), so the above expression becomes

$$T = \frac{t'_2 - t'_1}{\sqrt{1-\beta^2}}$$

The time interval $T' = t'_2 - t'_1$ is measured by a *single* clock in S' (in contrast to the time interval T, which is measured by *two different* clocks in S). As we will point out in Section 7, this has special significance. Since single-clock readings may occur in *either* frame, instead of primes we will use a zero subscript to signify this type of measurement.

TIME DILATION

$$T = \frac{T_0}{\sqrt{1-\beta^2}} \qquad \text{(where } T_0 \text{ must be a time interval measured by a } \textit{single} \text{ clock)} \tag{4}$$

Thus, as measured in S, there is a *longer* time interval T between the events than the time interval T_0 measured in S'. That is, *moving clocks run slower than stationary clocks.* All clocks read the correct times at their respective locations. *Time itself* is different in the two frames. Also, note that observers in *each* frame find that clocks in the other frame moving past them run slower than clocks at rest in their own frame. This is no paradox since the two sets of measurements are made in different frames of reference, and each frame has its own scale of space and time.

ST.5 Comparison of Length Measurements Parallel to the Direction of Motion

Using classical ideas, in Section 2 we compared the length L' of a meter stick at rest in the S' frame with the length measurement L in the S frame (in which the stick is moving). Instead of the Galilean transformation, here we use the

Lorentz transformation. As measured in S', the stick's length is $L' = x'_2 - x'_1$. Applying the Lorentz transformation:

$$L' = x'_2 - x'_1$$

$$L' = \frac{x_2 - Vt_2}{\sqrt{1-\beta^2}} - \frac{x_1 - Vt_1}{\sqrt{1-\beta^2}}$$

In the S frame, the two events of locating the ends of the stick must occur *simultaneously* ($t_2 = t_1$), so the above expression becomes

$$L' = \frac{x_2 - x_1}{\sqrt{1-\beta^2}}$$

The length L' is made in the S' frame, in which the object is *at rest* (in contrast to measuring L in the S frame, in which the object is moving). Again this has special significance. Because measurements of an object at rest may occur in either frame, instead of primes we will use a zero subscript to signify this type of measurement.

LENGTH CONTRACTION $\quad L = L_0\sqrt{1-\beta^2}$ $\quad$ (where L_0 must be a measurement in a frame in which the object is at rest) $\quad$ **(5)**

The length of a moving object is shorter along the direction of motion than its length measured at rest. As with time dilation, this is a symmetrical effect. Observers in *each* frame measure the other meter stick to be shorter than theirs. This is also no paradox, since the two sets of measurements are made in different frames. The length of an object is not some attribute possessed by that object; rather, it is the result of a measurement. Because there is no absolute space and no absolute time, measurements in one frame do not necessarily agree with those made in another frame. Nevertheless, measurements made in each frame are equally valid. As Martin Gardner[2] points out, if two people stand on opposite sides of a huge reducing lens, each sees the other as smaller. But that is not the same as making the paradoxical statement that each person *is* smaller than the other. It only makes the statement that in each person's frame of reference, the other person is smaller.

ST.6 Relativistic Momentum

To obtain the relativistic expression for momentum, we analyze an elastic collision between two identical particles and require that momentum conservation hold true in all frames of reference (Einstein's first postulate). Consider two railroad flatcars (in the S and S' frames) approaching each other on parallel tracks with equal speeds in opposite directions. Figure 5(a) shows the situation in the earth's frame of reference. Observers in both frames have balls whose rest masses m_0 are equal. Each observer throws the ball perpendicular to the direction of motion with equal speeds as measured in their respective frames, launching the balls so that each travels the same distance y (perpendicular to the motion of the two cars) before colliding. In an elastic collision, the y components of velocities simply reverse directions (while the x component of

[2] Martin Gardner, *The Relativity Explosion*, Vintage Books (1976), is an excellent non-mathematical introduction to relativity, black holes, and related topics.

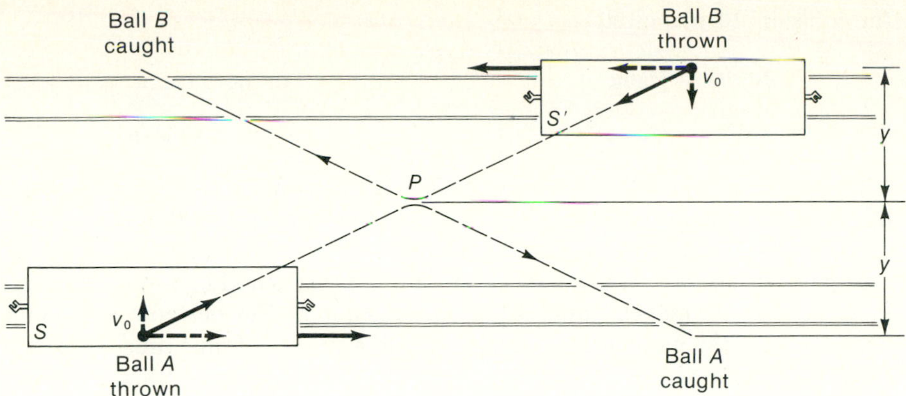

(a) In the earth's frame, the elastic collision is completely symmetrical.

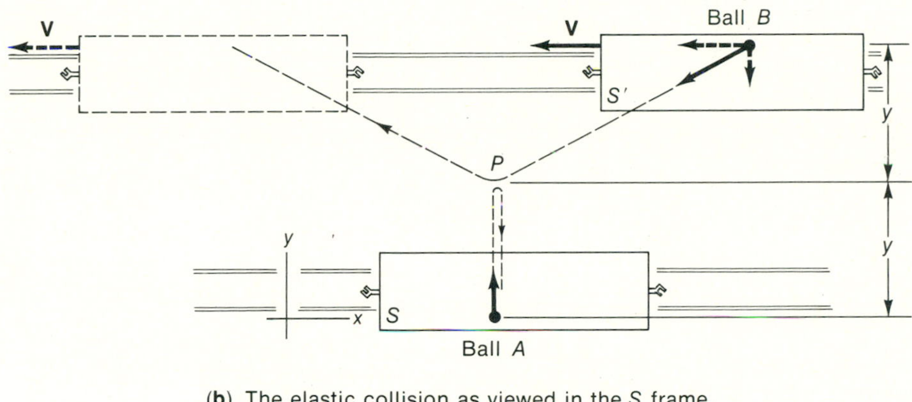

(b) The elastic collision as viewed in the S frame.

Figure 5
A hypothetical experiment in the earth's frame involving a symmetrical collision between identical balls.

velocity remains unchanged), so after traveling the distance y again at the same speed, each ball is caught by its respective observer.

Now let us analyze the collision in the S frame of reference [Figure 5(b)]. Because distances perpendicular to the motion are not contracted, the distance y each ball travels remains the same. However, the time intervals between throwing and catching the balls are not the same because of time dilation. This causes a change in the y component of velocity for Ball B as measured in the S frame. The speed each ball travels along the y direction is the total distance $2y$ (the same for each) divided by the time interval between throwing and catching each ball. In S', this time interval is between two events that occur at the *same location* and is thus measured by a *single* clock. The notation we use for such a time interval is T_0. In S, however, the same two events occur at *different locations*, so the time interval T as measured in the S frame is related to T_0 according to the time dilation formula, Equation (4): $T = T_0/\sqrt{1-\beta^2}$. Thus we arrive at two different expressions for the y-component speed of Ball B:

In the S' frame	**In the S frame**
$v'_{yB} = \dfrac{2y'}{T_0}$	$v_{yB} = \dfrac{2y}{T}$

Since the y components of velocities simply reverse their directions, the momentum change for Ball B along y is *twice* the momentum along y before (or after) the collision. Its magnitude is

In the S' frame	**In the S frame**
$\Delta p'_{yB} = 2m'_B v'_{yB}$	$\Delta p_{yB} = 2m_B v_{yB}$

where m'_B and m_B are the masses of Ball B in the respective S' and S frames. (Note that we are allowing for the possibility that the mass of Ball B, as determined in the two frames, may be different.)

Now, by symmetry, the mass m_A of Ball A as measured in S must have the same value as the mass m'_B of Ball B as measured in S'. Furthermore, symmetry demands that the momentum change of Ball A in the S frame must be of the same magnitude as the momentum change of Ball B in the S' frame. We will use these facts in the following step. Requiring that the conservation of momentum hold true in the S frame means that the y momentum lost by Ball A has the same magnitude as the y momentum gained by Ball B.

In the S frame

$$\Delta p_{yA} = \Delta p_{yB}$$

$$2m_A\left(\frac{2y}{T_0}\right) = 2m_B\left(\frac{2y}{T}\right)$$

(Note the appropriate time T_0 for the round trip of Ball A in the S frame.) Rearranging:

$$m_B = m_A\left(\frac{T}{T_0}\right)$$

$$m_B = \frac{m_A}{\sqrt{1-\beta^2}}$$

In the limit, as launching speeds (that is, perpendicular speeds) approach zero, m_B becomes the *relativistic mass m* (having only the relative speed V of the frames of reference), while m_A becomes the *rest mass* m_0.

RELATIVISTIC MASS *m*

$$m = \frac{m_0}{\sqrt{1-\beta^2}} \tag{6}$$

Note that we have associated the square root factor with the mass, obscuring the fact that it really came into the derivation in connection with a velocity measurement. Nevertheless, relativistic mass is a convenient concept, since it allows relativistic energy to be written as $E = mc^2$ and momentum as

RELATIVISTIC MOMENTUM

$$\mathbf{p} = m\mathbf{v} = \frac{m_0\mathbf{v}}{\sqrt{1-\beta^2}} \tag{7}$$

ST.7 Proper Measurements

Since observers in different frames of reference may find different answers for measurements of lengths, time intervals, and masses, it is customary to use a special name for measurements made as follows:

Proper length A length determination made in a frame of reference in which the object is at rest.

Proper time interval A time interval between two events when measured in a frame of reference for which the events occur at the same location in space. Proper time can only be measured by a single clock. Each clock indicates proper time at its respective location.

Proper mass (or rest mass) The mass of an object measured in a frame of reference in which the mass is at rest.

The use of the word "proper" does not imply that other measurements are somehow improper or incorrect. The adjective is used in the sense of "naturally belonging to" or "characteristic of." Although one can always find the proper length of an object, there are situations in which the concept of a proper time interval does not apply. Note that proper time is measured only by a single clock. If two events occur so far apart in space, but so close together in time, that a single clock cannot move fast enough (without exceeding the speed of light c) to be located where both events occur, then the concept of a proper time interval between the events does not apply. It is important to remember that the formulas for time dilation, length contraction, and relativistic mass all involve *proper* measurements, regardless of which particular frame of reference is designated the primed frame.

EXAMPLE 1

This rather facetious example illustrates a situation in which the concept of proper time does not apply. Imagine a spaceship of proper length 300 m traveling past the earth with a speed $V = 0.80c$. Spacemen send a light signal from the tail to the nose of the ship. Find the time that elapses between the two events of *sending* and *receiving* the signal in (a) the spaceship frame and (b) the earth frame (see Figure 6).

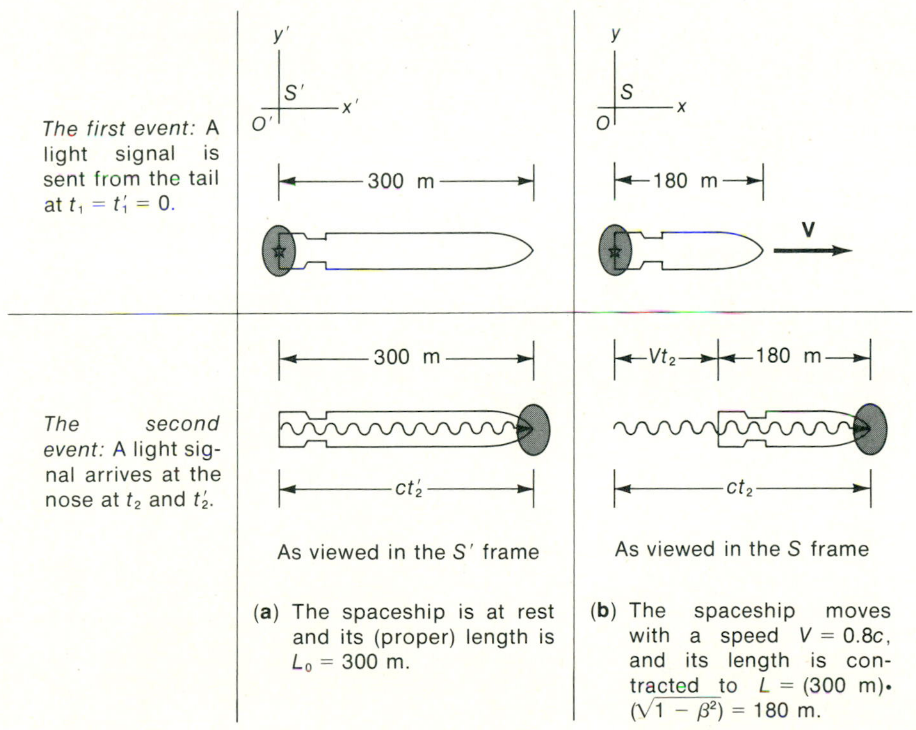
The first event: A light signal is sent from the tail at $t_1 = t_1' = 0$.
The second event: A light signal arrives at the nose at t_2 and t_2'.
y'
S'
x'
O'
300 m
ct_2'
y
S
x
O
180 m
V
Vt_2
ct_2
As viewed in the S' frame
As viewed in the S frame
(a) The spaceship is at rest and its (proper) length is $L_0 = 300$ m.
(b) The spaceship moves with a speed $V = 0.8c$, and its length is contracted to $L = (300 \text{ m})\cdot(\sqrt{1-\beta^2}) = 180$ m.

Figure 6
The sending and receiving of the light signals as viewed in the S' and S frames of reference.

SOLUTION

In this case, the concept of proper time does not apply (since no *single* clock can travel fast enough to be located where the two events occur). Therefore, the time dilation formula, Equation (4), cannot be used. Instead, we use either of the methods illustrated below.

Method 1

(a) In the spaceship frame (S'), the light signal travels 300 m at the speed c. Thus, the time $\Delta t' = t'_2 - t'_1 = t'_2 - 0$ is

$$t'_2 = \frac{\text{distance}}{\text{velocity}} = \boxed{\frac{300\text{ m}}{c}}$$

(b) In the earth frame (S), the moving ship has a contracted length of $L = L_0\sqrt{1-\beta^2} = (300\text{ m})(\sqrt{1-0.80^2}) = (300\text{ m})(0.60) = 180$ m. While the spaceship moves a distance Vt_2, the light travels a distance ct_2. Thus, from Figure 6(b), we have

$$ct_2 = Vt_2 + L$$

$$ct_2 = (0.80c)t_2 + 180\text{ m}$$

$$(0.20c)t_2 = 180\text{ m}$$

$$t_2 = \boxed{\frac{900\text{ m}}{c}}$$

Method 2 [for part (b)]

An alternative solution is to use the Lorentz transformation for the events. In S', the events are (0,0,0,0) and $(x'_2,0,0,t'_2)$, where $x'_2 = 300$ m and $t'_2 = 300$ m/c. We seek $\Delta t = t_2 - t_1 = t_2 - 0 = t_2$. From the Lorentz transformation, we have

$$t_2 = \frac{t'_2 + Vx'_2/c^2}{\sqrt{1-\beta^2}} = \frac{\dfrac{300\text{ m}}{c} + \dfrac{(0.80c)(300\text{ m})}{c^2}}{\sqrt{1-0.80^2}} = \frac{1.8\left(\dfrac{300\text{ m}}{c}\right)}{0.60} = \boxed{\frac{900\text{ m}}{c}}$$

Because classical ideas of space and time are so deeply ingrained in our thought processes, it is surprisingly easy to be led astray in solving relativity problems. For this reason, it is prudent always to think in terms of *point events* and to make careful sketches of these events as seen in a particular frame of reference.

ST.8 Relativistic Velocity Addition

At the instant the S and S' frames coincide (at $t = t' = 0$), assume that a particle passes the origin moving along the $+x$ (and $+x'$) direction with a constant speed v' as measured in S' (see Figure 7). We obtain the velocity addition relation by considering the following two events:

	In S'	In S
First event (particle at origin) (x_1,y_1,z_1,t_1)	(0,0,0,0)	(0,0,0,0)
Second event (particle at x at later time t) (x_2,y_2,z_2,t_2)	$(x',0,0,t')$	$(x,0,0,t)$

In the S' frame, during the time t' the particle moves a distance x' with constant speed v', so $x' = v't'$. The S' frame itself moves along the $+x$ direction with constant speed V relative to the S frame. This motion is the second velocity V we will add to the particle velocity v' to give the velocity v of the particle as measured in the S frame. In S, the particle's speed is $v = xt$, so we make use

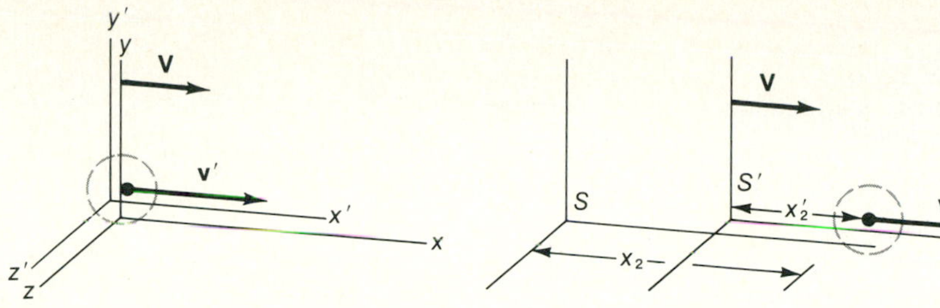

(**a**) *The first event:* The particle passes the origin ($x_1 = x_1' = 0$) at the instant ($t_1 = t_1' = 0$) when the two frames are coincident.

(**b**) *The second event:* At the times t_2 and t_2', the particle is at the location x_2 as measured in S and at x_2' as measured in S'.

Figure 7
The velocity of a moving particle as measured in two different frames of reference.

of the Lorentz transformation (Equation 3) and substitute for x and t as follows:

$$x = \frac{x' + Vt'}{\sqrt{1-\beta^2}} = \frac{t'(v' + V)}{\sqrt{1-\beta^2}}$$

and

$$t = \frac{t' + Vx'/c^2}{\sqrt{1-\beta^2}} = \frac{t'(1 + v'V/c^2)}{\sqrt{1-\beta^2}}$$

Thus the speed v of the particle in the S frame is

$$v = \frac{x}{t} = \frac{\left[\dfrac{t'(v' + V)}{\sqrt{1-\beta^2}}\right]}{\left[\dfrac{t'(1 + v'V/c^2)}{\sqrt{1-\beta^2}}\right]}$$

which reduces to

RELATIVISTIC VELOCITY ADDITION (for velocities along the $\pm x$ direction)

$$v = \frac{v' + V}{\left(1 + \dfrac{v'V}{c^2}\right)} \qquad (8)$$

Here, v' is the speed of the particle in the x' direction as measured in the S' frame; V is the speed of S' along the $+x$ direction relative to S; and v is the speed of the particle in the $+x$ direction as measured in the S frame.

ST.9 The Nonsynchronism of Moving Clocks

A system of clocks, properly synchronized in the moving S' frame, will appear to have been improperly synchronized when viewed from the S frame of reference. This effect is in addition to the time dilation phenomenon and is perhaps relativity's greatest jolt to our common-sense ideas. The effect is the source of most of the so-called "paradoxes" of special relativity.

Recall the procedure for synchronizing two clocks, A' and B', at rest in the S' frame. As seen in the S' frame, a light flash at the midpoint between the clocks sends light signals in opposite directions. When a pulse arrives at a clock, that clock is set to indicate $t' = 0$. In this manner, A' and B' are correctly synchronized in the S' frame.

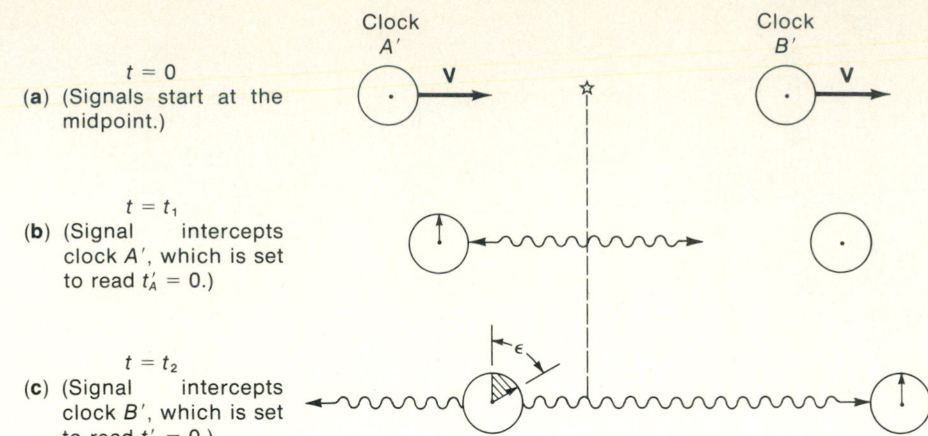

Figure 8
As viewed in the S frame, the moving clocks are set out of synchronism by an amount ε. (Of course, in the S' frame, they are properly synchronized according to time in that frame.)

Now let us view this procedure from the S frame of reference (Figure 8). Since clock A' is moving toward its light signal, it will intercept the light pulse first and be set to read $t'_A = 0$. At some *later* time (as seen in the S frame), the other light signal reaches clock B', which has been moving away from its light signal, and clock B' is set to read $t'_B = 0$. Thus, according to observers in the S frame, the "chasing" clock is set to read a later time than the "leading" clock. As observed at a given instant in the S frame, the clocks are not in synchronism.

As usual, the situation is a symmetrical one. Observers in S' similarly find that clocks in S are not properly set in synchronism. Yet, the synchronizing of clocks establishes a time scale *by which the simultaneity of events is judged in that particular frame of reference.* There is no reason, however, to prefer one sense of simultaneity over another. Thus, events (separated in space) that appear simultaneous in one frame are not necessarily simultaneous in another frame. The amount of nonsynchronism is directly related to the (Vx'/c^2) term in the Lorentz transformation for time. Interestingly, the time t' depends not only on t and V, but also on the space coordinate x. Space and time are truly interdependent in relativity. It can be shown that the discrepancy ε between two moving clocks is:

NONSYNCHRONISM OF MOVING CLOCK SYSTEMS

Two clocks, separated a distance $\Delta x'$ and correctly synchronized in the S' frame, are incorrectly synchronized to observers in the S frame by an amount ε:

$$|\varepsilon| = \frac{V\,\Delta x'}{c^2} \tag{9}$$

The "chasing" clock indicates a *later* time than the "leading" clock.

Only clocks spaced along the $\pm x'$ direction are out of synchronism when viewed in the S frame; moving clocks located along lines in the y' or z' directions are correctly synchronized in both frames.

Another feature of nonsynchronism is illustrated if we consider *three* frames of reference, each with a line of several clocks along the direction of motion (Figure 9). We consider the S frame to be at rest, the S' frame moving in the $+x$ direction, and the S'' frame moving in the $-x$ direction. For simplicity, we assume all of the center clocks read zero at the instant depicted in the S

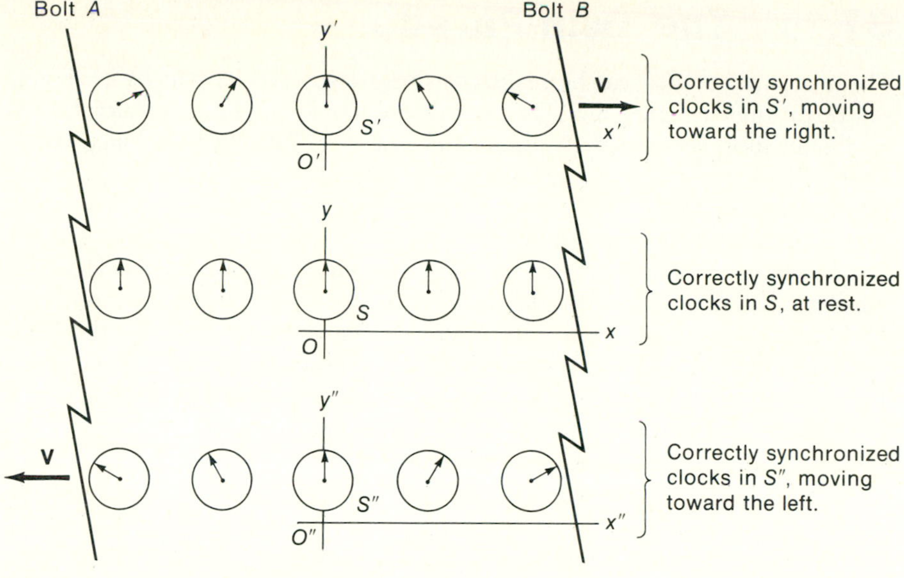

Figure 9
A comparison of clocks in three frames of reference *as measured at a given instant in the* S *frame*. Each set of clocks is correctly synchronized in its own frame. To simplify the comparison, we suppose that the clocks located at the respective origins read zero at the instant the origins coincide.

frame. For a line of moving clocks, each individual clock reads a *later* time than its predecessor. Now suppose that at the instant sketched, two lightning bolts, A and B, strike the left and right groups of clocks, respectively. These two events would be judged simultaneous in the S frame because clocks in that frame indicate the same time. However, as measured in S', the clock readings indicate that B occurs before A, and in the S'' frame, that A occurs before B. There is no such thing as absolute simultaneity.

Does this reversal of the time sequence of events imply that in some frame of reference an "effect" might occur before a "cause"? Could the arrow hit the target before the bowstring is released? No. A careful analysis reveals that only those events that could *not* conceivably be related in any way can occur in a reversed time sequence in some frame of reference. So the important principle of *cause and effect* is still preserved in relativity theory.

It should be emphasized that this lack of agreement regarding the time sequence of certain events is *not* due to the fact that light signals from a distant event take a finite time to reach an observer (and thus the observer may visually see one event after the other). Even after all corrections for finite transit times of light signals are made, the same peculiarities of simultaneity (or the lack thereof) still remain. Of course, within any given frame, the concept of simultaneity is clearly defined; it just does not agree with the scale of simultaneity in other frames. Ultimately, all the so-called "paradoxes" of special relativity are traceable to the lack of absolute simultaneity.

One possible misunderstanding about relativity should be clarified. The message of relativity is not that "everything is relative." True, we must discard absolute space, absolute time, and a few other "absolute" concepts. But the major significance of Einstein's theory (aside from being the theory that agrees best with experimental facts) is that *it is possible to express the laws of nature in such a way that they are the* same *for all observers*. What a chaotic situation it would be if each frame had its own fundamental laws of nature, which would not agree with laws valid in other frames. (This is actually the situation if one clings to Newtonian concepts.) By devising a model for the universe in which nature behaves exactly the same way for all frames of reference, Einstein made a great unifying simplification to our understanding of the universe.

ST.10 The Twin Paradox

The twin paradox effect can be analyzed using just special relativity by imagining a straight-line trip in which the turnaround time involving acceleration is negligibly short compared to other time intervals. The acceleration times in starting and stopping are also assumed to be negligible.[3] Consider a trip to the star Alpha Centauri, 4 light-years away. One twin, in the S' frame, travels at a constant velocity $V = \frac{4}{5}c$ to the destination, turns around in negligibly short time, and returns to the earth at the same constant speed. His twin brother remains on the earth, the S frame. The journey starts on January 1st. To keep track of the elapsed time, each twin agrees to send the other a New Year's message via radio waves on January 1st of each year during the journey. These signals travel with the speed of light c and are emitted at a frequency f_0 of 1 per year according to the sender's local time scale. Figure 10 shows a diagram of the journey as drawn in the *earth* frame of reference. Here, we plot distance (in units of light-years) on the horizontal axis and time (in years) on the vertical axis.

We now make use of a well-verified effect known as the *relativistic Doppler shift* for light (similar to the Doppler shift for sound discussed in Section 16.10). The effect describes how light signals (or any electromagnetic wave) received from a moving source are shifted in frequency f. (Remember, however, the *speed* of light received from a moving source is always c.) When the light source is *receding* along the line of sight with a speed $V = \beta c$, the received frequency f is *lower* than the frequency f_0 emitted by the source. When the source is *approaching* along the line of sight, the received frequency is *higher* than f_0.

RELATIVISTIC DOPPLER SHIFT FOR LIGHT[4]

Light source moving away	Light source approaching	
$f = f_0\sqrt{\frac{1-\beta}{1+\beta}}$	$f = f_0\sqrt{\frac{1+\beta}{1-\beta}}$	**(10)**

For the twin paradox example, the radio signals are sent with a frequency f_0 of 1 pulse per year. The speed of the source is $\beta = \frac{4}{5}$. Putting these values into the Doppler shift formulas, we calculate the rate of signals received for the two cases:

When separating: $$f = f_0\sqrt{\frac{1-\beta}{1+\beta}} = f_0\sqrt{\frac{1-\frac{4}{5}}{1+\frac{4}{5}}} = \tfrac{1}{3}f_0$$

When approaching: $$f = f_0\sqrt{\frac{1+\beta}{1-\beta}} = f_0\sqrt{\frac{1+\frac{4}{5}}{1-\frac{4}{5}}} = 3f_0$$

Each twin agrees that in the earth frame, the destination is 4 light-years away, requiring, in S, a total of 10 years for the round trip at a speed of $\frac{4}{5}c$. Therefore, 10 signals are sent from the earth. *Thus, both twins conclude that the elapsed time in* S *is* 10 *years.* Both twins also know the Doppler effect formula,

---

[3] It can be experimentally verified (by utilizing the Mössbauer effect) that accelerations up to the order of $10^{16}g$ produce no effect in clock rates. Only relative *velocities* alter clock rates. See C.W. Sherwin, *Physical Review, 120*, 17 (1960).

[4] If there is an angle θ between the line of sight and the velocity of the source, the equation is

$$f = f_0\left(\frac{\sqrt{1-\beta^2}}{1+\beta\cos\theta}\right)$$

For $\theta = 90°$, the shift in frequency is just the time dilation effect.

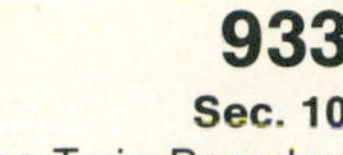

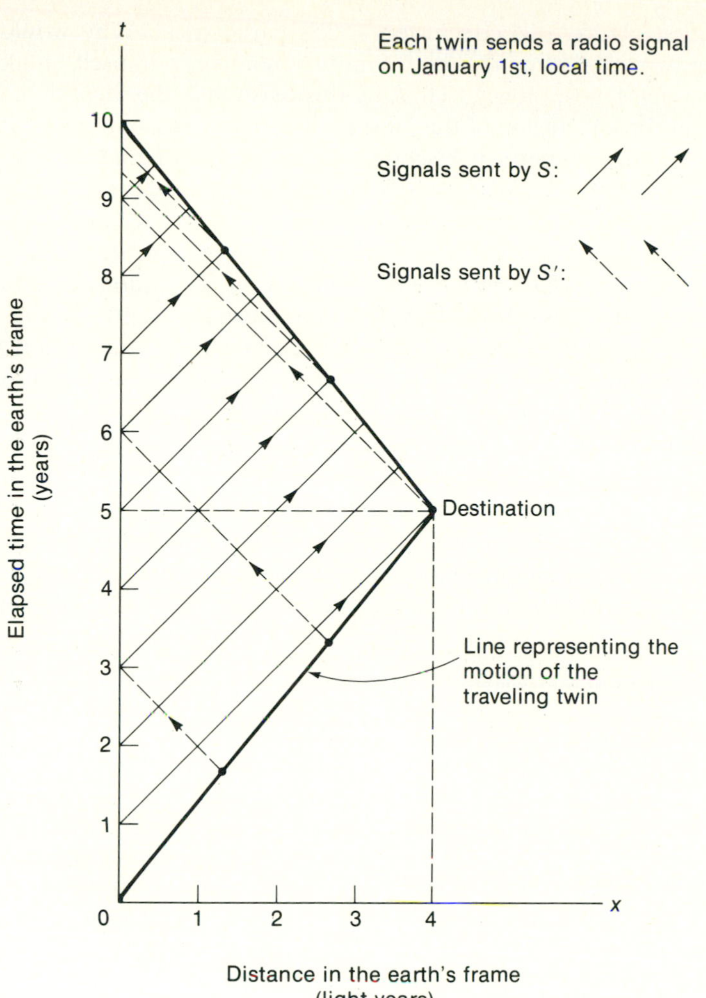

Figure 10
A diagram that illustrates the so-called "twin paradox" as drawn by the twin who remains on earth. The traveling twin moves in a straight line at constant speed $v = (\frac{4}{5})c$. The time intervals involved in starting, stopping, and turnaround are assumed to be negligibly short.

so they can calculate the elapsed time that occurs in the traveling twin's frame:

Calculated in S': The twin in S' receives signals at the rate of $\frac{1}{3}$ per year for half the journey and 3 per year for half the journey. The average rate of receiving signals during the entire journey is thus:

$$(\tfrac{1}{2})(\tfrac{1}{3}) + (\tfrac{1}{2})(3) = \tfrac{5}{3} \text{ per year}$$

Ten signals were sent altogether, so the total time for S' is $10/(\frac{5}{3}) = 6$ years.[5]

Calculated in S: The twin in S receives signals at the rate of $\frac{1}{3}$ per year for 9 years and 3 per year for 1 year. The total number of signals received by S is thus $(\frac{1}{3})(9) + (3)(1) = 6$, signifying 6 years has elapsed in S'.

Thus, both twins conclude that the elapsed time in S′ *is* 6 *years.* Altough both twins have aged during the trip, after they are reunited the space traveler is 4 years younger than the twin who remained on earth.

[5] Here is another way of looking at it. In the moving frame of reference, the 4 light-years distance is contracted to only $4\sqrt{1-\beta^2} = 2.4$ light-years. Writing this as 2.4 c·yr, the traveler finds the time required to traverse this distance at a relative speed $\frac{4}{5}c$ to be distance/velocity = 2.4 c·yr/$\frac{4}{5}c$ = 3 yr. Thus, the round trip is 6 years.

At first glance, this conclusion seems paradoxical. Why would not the traveling twin, observing the earth's motion relative to himself, think that the earth twin ages more slowly? Thus, by this reasoning, the *earth* twin would be younger at the conclusion of the journey. The faulty reasoning here is that the situation is *not* symmetrical because the traveling frame accelerated and the earth frame did not. And the *acceleration* of a frame of reference relative to any inertial frame is an absolute phenomenon, not a relative one. The crucial feature is that *the acceleration of a frame of reference radically changes the sense of simultaneity for that frame* (refer to Figure 9). To figure out what occurs, sketch arrays of clocks in the two frames at various instants during the turn-around, being careful to depict only *point events* as seen at a given instant in a frame of reference. Remember that events that are simultaneous in one frame are not necessarily simultaneous in another frame.[6]

To summarize, the absolute acceleration of the traveling twin at turn-around provides the asymmetry to the situation, causing a change in the judgment of simultaneity for that frame of reference. The twin paradox effect has been experimentally verified. *There is no such thing as absolute time. There is no such thing as absolute simultaneity.* All the apparent paradoxes of relativity are traceable ultimately to the lack of an absolute scale of simultaneity for all frames.

Problems

Note: Until you become accustomed to relativity, you will find some of these problems difficult to solve. However, if you always think in terms of point events and remember that each frame has its own unique time scale by which simultaneity is judged, these so-called paradoxes can be solved with just the information in this supplement.

1 A spaceship has a proper length of 100 m. It travels close to the earth's surface with a constant speed of $0.8c$. Earth observers decide to measure the length of the ship by erecting two towers that coincide with the ends of the ship simultaneously (in the earth's frame) as it passes by.

(a) How far apart do the earthmen build the towers?

(b) How long do the earthmen say it takes for the nose of the ship to travel from tower A to tower B?

(c) How long, according to measurements in the spaceship frame, does it take for the nose of the ship to travel from tower A to tower B?

(d) As measured by spacemen, how far apart are the towers?

(e) Find the proper time interval between the event 1, in which the nose of the ship coincides with tower A, and event 2, in which the nose of the ship coincides with tower B. *Answers:* (a) 60 m (b) 75 m/c (c) 45 m/c (d) 36 m (e) 45 m/c

2 Refer to the previous problem.

(a) In the spaceship frame, how long does it take a beam of light to travel from the front to the rear end of the spaceship?

(b) How long, according to earthmen, is required for a beam of light to travel from the front to the rear end of the moving spaceship?

(c) A projectile is fired from the rear of the spaceship toward the front end with a speed of $0.6c$ as measured by the spacemen. Find the speed of the projectile in the earth frame of reference.

---

[6] The consequences of this are explained in an article by E. S. Lowry, *American Journal of Physics, 31*, 59 (1963). Good discussions of the twin paradox will also be found in articles by G. David Scott, *American Journal of Physics, 27*, 580 (1959), and by A. Schild, *American Mathematical Monthly, 66*, 1 (1959).

(d) What would be the earth speed of the projectile if it had been fired in the opposite direction with the same speed relative to the spaceship?

3 Referring to Problem 1, as the spaceship passes the towers, the following two events are simultaneous in the earth frame:

Event 1: Coincidence of tower *A* with the tail of the ship.
Event 2: Coincidence ot tower *B* with the nose of the ship.

Make sketches to show how the same two events look in the spaceship frame of reference. Find the time interval (if any) between the events as measured in the spaceship frame.

Answer: 80 m/*c*

4 The "pole-in-the-barn" paradox is one of the classic puzzles of special relativity. Consider a 20-ft pole carried along so fast that it is only 10 ft long as measured in the earth's frame (see Figure 11). The pole is carried through a barn that has doors *C* and *D* on opposite walls. The barn is 12 ft long (2 ft longer than the moving pole), so both doors could be simultaneously shut for a brief period, trapping the pole inside the barn. (Door *D* would then be opened to permit the pole to travel on through.) On the other hand, to the runner carrying the pole, the barn is only 6 ft long because of length contraction. Here is the apparent paradox: "To the runner, how can his 20-ft pole fit inside the 6-ft barn with both doors closed?" Obviously, there is an inconsistency somewhere. Can you resolve this apparent paradox? (Hint: As with most paradoxes in relativity, the root of the problem lies in the fact that two events that are simultaneous in one frame of reference are not necessarily simultaneous in another frame.)

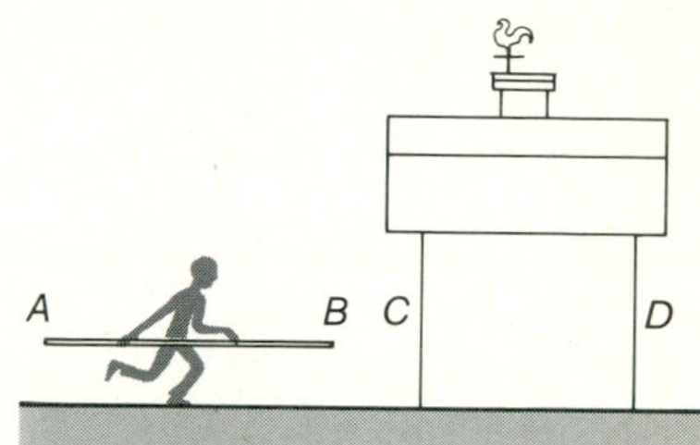

Figure 11
Problem 4

5 Two clocks are located in the nose and tail of a spaceship whose proper length is 300 m. They are correctly synchronized in the spaceship frame of reference. If the spaceship moves past the earth with a speed $V = 0.90c$, (a) find the difference in the readings of the two clocks as measured simultaneously in the earth's frame. (b) Which clock reads the earlier time?

Answer: Clock in nose earlier by 270 m/*c* or 9.00×10^{-7} s.

6 In a space war, two identical rocketships pass close to each other traveling in opposite directions at speeds close to that of light. When the tail of ship *B* is adjacent to the nose of ship *A*, a mortar shell is fired sideways (perpendicular to the relative motion) from a gun barrel located near the tail of *A* in an attempt to hit ship *B* (see Figure 12). Obviously, one of the statements and the diagram are wrong. Ship *B* is either hit or it is not. Find the ambiguities in the statements of this problem, and discuss briefly what really happens and why there is no paradox. Assume the ships pass very close to each other and that the shell's speed is very great, so that the transit time of the shell itself is not a factor in the analysis.

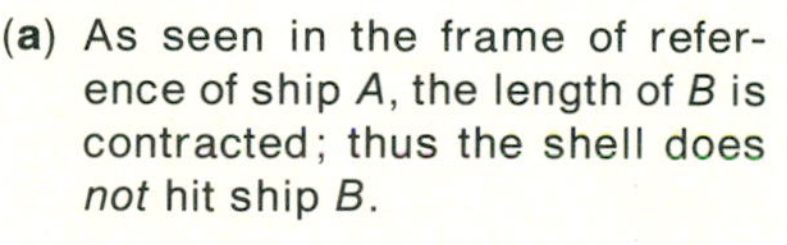

(a) As seen in the frame of reference of ship A, the length of B is contracted; thus the shell does not hit ship B.

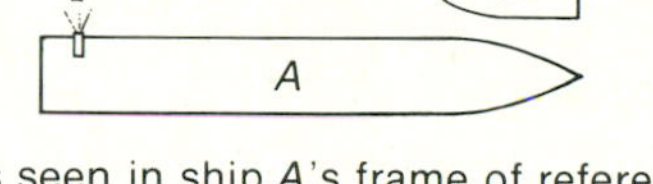

As seen in ship *A*'s frame of reference

(b) As seen in the frame of reference of ship *B*, the length of *A* is contracted; thus the shell *does* hit ship *B*.

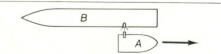

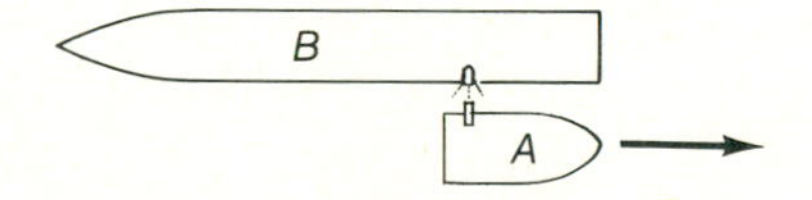

As seen in ship *B*'s frame of reference

Figure 12
Problem 6

7 The "stick-in-the-hole" paradox is one of the most puzzling paradoxes in special relativity. A 100-cm stick is moving horizontally with relativistic speed such that its length is contracted to 50 cm as seen in the earth's frame (Figure 13). An observer in the earth's frame has a thin board with a circular hole, 70 cm in diameter, cut out of it. As the pole passes by, the observer quickly lifts the board vertically (keeping its plane horizontal), allowing the stick to pass through the hole. Thus, at some instant, the (contracted) stick fits entirely inside the horizontal hole. Here is the paradox: "In the stick's frame of reference,

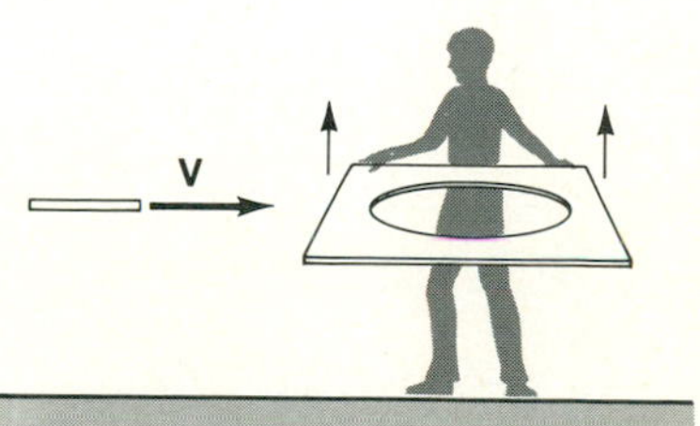

Figure 13
Problem 7

the stick is 100 cm long, and the hole is only 35 cm across. How can the 35-cm opening engulf the 100-cm stick?" (Hint: Focus your attention on *point events.* For example, consider four points equally spaced along the stick's length. At some instant in the earth's frame, these four points simultaneously lie in the plane of the board. What do these four events look like in the pole's frame?)

8 Bandits try to stop a train (which is moving forward) by setting off explosive charges at (a) the engine and (b) the caboose. The two explosions are simultaneous in the earth's frame of reference. In the train's frame, which explosion, if either, occurred first according to relativity? Does it make any difference whether the train is traveling in the $+x$ or $-x$ direction? Justify your answers.

9 A certain galaxy moves away from the earth so fast that the spectral lines in its light emission are Doppler-shifted to one-half their frequencies here on the earth. Find the galaxy's speed. *Answer:* $c/3$

10 Consider the twin paradox. In the traveling twin's frame the earth clocks move away and come back, so as measured in that frame the moving earth clocks run more slowly than clocks at rest in that frame. Therefore, why is it that upon his return, the traveling twin finds that *more* time has elapsed on the earth than in the traveling frame of reference?

Appendices

A. Prefixes for Decimal Multiples and Submultiples of Ten 938
B. Mathematical Symbols 939
C. Conversion Factors 940
D. Trigonometric Formulas 943
E. Mathematical Approximations, Formulas, and Conversions 945
F. Fourier Analysis 947
G. Calculus Formulas 949
H. Finite Rotations 952
I. Trigonometric Functions 953
J. Periodic Table of the Elements 954
K. Constants and Standards 955
L. Terrestrial and Astronomical Data 956
M. SI Units 957

Appendix A
Prefixes for Decimal Multiples and Submultiples of Ten

Multiple	Prefix		Symbol
10^{12}	tera	(tĕr′ȧ)	T
10^{9}	giga	(jĭ′gȧ)	G
10^{6}	mega	(mĕg′ȧ)	M
10^{3}	kilo	(kĭl′ō)	k
10^{2}	hecto	(hĕc′tō̍)	h
10^{1}	deka	(dĕk′ȧ)	da
10^{-1}	deci	(dĕs′ĭ)	d
10^{-2}	centi	(sĕn′tĭ)	c
10^{-3}	milli	(mĭl′ĭ)	m
10^{-6}	micro	(mī′krō̍)	μ
10^{-9}	nano	(năn′ō)	n
10^{-12}	pico	(pē′cō̍)	p
10^{-15}	femto	(fĕm′tō̍)	f
10^{-18}	atto	(ăt′tō̍)	a

Appendix B
Mathematical Symbols

Symbols	
$=$	is equal to
$\neq$	is not equal to
$\equiv$	is identical to or by definition
$a > b$	a is greater than b
$a \gg b$	a is much greater than b
$a < b$	a is less than b
$a \ll b$	a is much less than b
$a \geqq b$	a is equal to or greater than b
$a \leqq b$	a is equal to or less than b
$\pm$	plus or minus (for example, $\sqrt{4} = \pm 2$)
$\propto$	is proportional to
$\approx$	is approximately equal to
$r \rightarrow \infty$	r approaches infinity
$\Rightarrow$	implies
$\sum$	the sum of
$\| \|$	the absolute value of
$\|\mathbf{A}\|$ or A	the magnitude of the vector **A**
$\oint$	a line integral around a closed path or a surface integral over a closed surface
$\cdot$	multiplication symbol
$\cdot$	(as in $\mathbf{A} \cdot \mathbf{B}$) dot product
X	(as in **A X B**) cross product
$\times$	(as in 3.2×10^4) multiplication symbol in scientific notation

The Greek Alphabet		
Alpha	A	α
Beta	B	β
Gamma	Γ	γ
Delta	Δ	δ
Epsilon	E	ε
Zeta	Z	ζ
Eta	H	η
Theta	Θ	θ
Iota	I	ι
Kappa	K	κ
Lambda	Λ	λ
Mu	M	μ
Nu	N	ν
Xi	Ξ	ξ
Omicron	O	o
Pi	Π	π
Rho	P	ρ
Sigma	Σ	σ
Tau	T	τ
Upsilon	Υ	υ
Phi	Φ	ϕ
Chi	X	χ
Psi	Ψ	ψ
Omega	Ω	ω

Appendix C
Conversion Factors

Use of Conversion Factors

The ratio of any pair of quantities listed in a given table of conversion factors is dimensionless, having the value of 1. To illustrate, consider the expression 1 mi = 5280 ft. Dividing both sides by 5280 ft, we obtain the ratio:

$$\left(\frac{1 \text{ mi}}{5280 \text{ ft}}\right) = 1$$

Ratios that equal unity are useful in converting a quantity from one system of units to another.

EXAMPLE C–1

To express 44 ft/s in units of miles per hour, we make use of two conversion factors.

$$1 \text{ mi} = 5280 \text{ ft} \quad \text{implies} \quad \left(\frac{1 \text{ mi}}{5280 \text{ ft}}\right) = 1$$

$$1 \text{ h} = 3600 \text{ s} \quad \text{implies} \quad \left(\frac{3600 \text{ s}}{1 \text{ h}}\right) = 1$$

Multiplying and canceling units:

$$44\,\frac{\text{ft}}{\text{s}} = \left(\frac{44\,\cancel{\text{ft}}}{\cancel{\text{s}}}\right)\underbrace{\left(\frac{3600\,\cancel{\text{s}}}{1 \text{ h}}\right)}_{=1}\underbrace{\left(\frac{1 \text{ mi}}{5280\,\cancel{\text{ft}}}\right)}_{=1} = 30\,\frac{\text{mi}}{\text{h}}$$

Thus:

$$44\,\frac{\text{ft}}{\text{s}} = 30\,\frac{\text{mi}}{\text{h}} \qquad \text{(exactly)}$$

Length

1 **m** = 39.3701 **in** = 3.230 39 **ft**
1 **km** = 0.621 37 **mi** = 0.539 96 **international nautical mile**
1 **in** = 2.54 **cm** (exactly)
1 **mi** = 5280 **ft**
1 astronomical unit[1] (**AU**) = 1.4960×10^{11} **m**
 = 4.8481×10^{-6} parsec (**pc**) = 1.5812×10^{-5} **light-year**
1 parsec (**pc**) = 3.262 **light-year** = 3.084×10^{16} **m**
1 angstrom (**A**) = 10^{-10} **m** = 10^{-4} **micrometer** (μ**m** or μ)

Mass

1 **kg** = 0.068 522 **slug**

1-kg mass weighs[2] 2.205 **lb** (where $g = 9.807\ \text{m/s}^2$)

Time

1 **min** = 60 **s**

1 **hour** = 3600 **s**

1 **day** = 86 400 **s** (mean solar day)

1 **week** = 604 800 **s**

1 **year** (tropical) = $3.155\,69 \times 10^7$ **s** = 365.242 **days**

Plane Angle

1 **rad** = 57.3° = 0.1592 **rev**

1 **rev** = 2π **rad** = 360°

$$1^\circ = 60\ \textbf{min} = 3600\ \textbf{sec} = 0.017\,45\ \textbf{rad} = \frac{1}{360}\ \textbf{rev}$$

$$1\ \frac{\textbf{rev}}{\textbf{min}} = 0.1047\ \frac{\textbf{rad}}{\textbf{s}}$$

Solid Angle

1 **sphere** = 4π **steradian**

Area

1 $\mathbf{m^2}$ = 10.76 $\mathbf{ft^2}$ = 1550 $\mathbf{in^2}$

1 $\mathbf{km^2}$ = 0.3861 $\mathbf{mi^2}$

Volume

1 $\mathbf{m^3}$ = 10^3 **L** = 35.31 $\mathbf{ft^3}$ = 6.102×10^4 $\mathbf{in^3}$

1 $\mathbf{km^3}$ = 0.2399 $\mathbf{mi^3}$

Density

1 $\mathbf{kg/m^3}$ = 1.940 $\mathbf{slug/ft^3}$

1 $\mathbf{ft^3}$ of water weighs[3] 62.4 **lb**

Velocity

1 **m/s** = 2.2369 **mi/h**

30 **mi/h** = 44 **ft/s** (exactly)

Force

1 **N** = 10^5 **dynes** = 0.224 81 **lb**

---

[1] One **astronomical unit** (AU) is defined as the mean radius of the earth's orbit.

One **parsec** (pc) is the distance at which one astronomical unit subtends an angle of one second of arc. Its name comes from the fact that as seen from the earth, a star at this distance has an annual *par*allax of one *sec*ond. Note the convenient relation (distance in parsecs) = (parallax in seconds)$^{-1}$.

One **light-year** ($= 9.46 \times 10^{15}$ m) is the *distance* light travels in one year in a vacuum. In numerical calculations in which the speed of light is represented by c, the unit may be conveniently written as $c\cdot$yr, so that the symbol c may cancel in the calculation as other units do.

[2] It is incorrect to state that 1 kg = 2.205 lb, since there are units of *mass* on one side of the equal sign and units of *force* on the other. Furthermore, the value of the force depends on the local value of g, varying from point to point on the earth. However, with care, the fact that a mass of 1 kg weighs 2.205 lb (for standard g) can be used to change the mass of an object as expressed in one system to its weight in another system. It is generally safest to make this conversion *prior* to solving a numerical problem.

[3] From this fact, one can obtain the *weight density* (in pounds/foot^3), which is dimensionally different from the *mass density* (in slugs/foot^3).

Work and Energy

1 **J** = 10^7 **erg** = 0.738 **ft·lb** = 2.389 × 10^{-2} **cal** = 9.48 × 10^{-4} **Btu**
= 2.778 × 10^{-7} **kW·h** = 3.725 × 10^{-7} **horsepower·h**
= 9.87 × 10^{-3} **L·atm**

1 **eV** = 1.602 10 × 10^{-19} **J**

1 **L·atm** = 101.3 **J**

1 **cal** = 4.184 **J** (exactly)

Power

1 **W** = 1 **J/s** = 0.738 **ft·lb/s** = 1.341 × 10^{-5} **horsepower** = 3.413 **Btu/h**

1 **horsepower** = 550 **ft·lb/s** (exactly) ≈ 746 **W** (approximately)

Pressure

1 **atm** = 1.013 × 10^5 **N/m²** (or **Pa**) = 1.013 × 10^6 **dyne/cm²**
= 1.013 **bar** = 14.70 **lb/in²** (or **psi**) = 2116 **lb/ft²**

1 **atm** (at 0°C) = 76.00 **cm Hg** = 760.0 **mm Hg** (or **torr**)
= 29.92 **in Hg** = 33.90 **ft of water**

Appendix D
Trigonometric Formulas

Right Triangles (see Figure D–1)

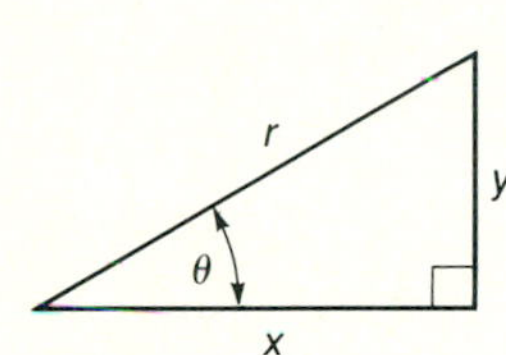

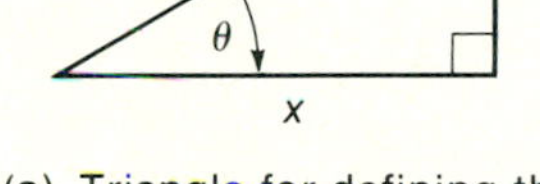

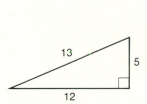

(**a**) Triangle for defining the trigonometric functions.

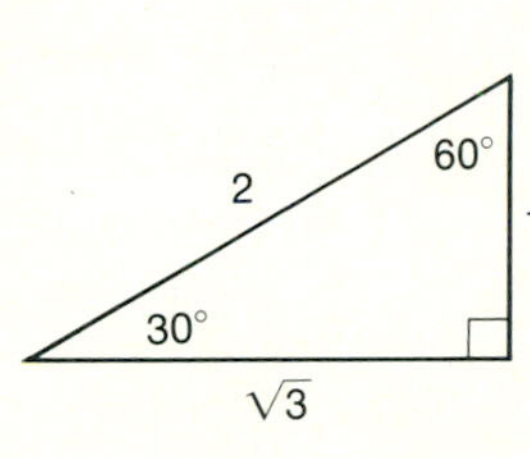

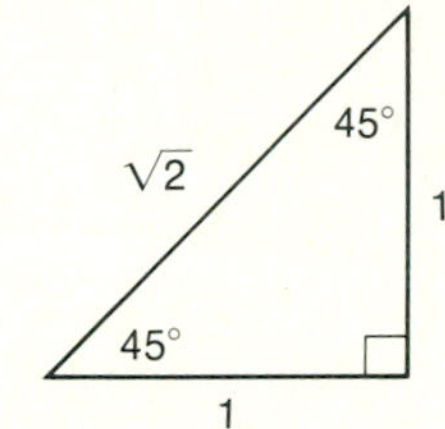

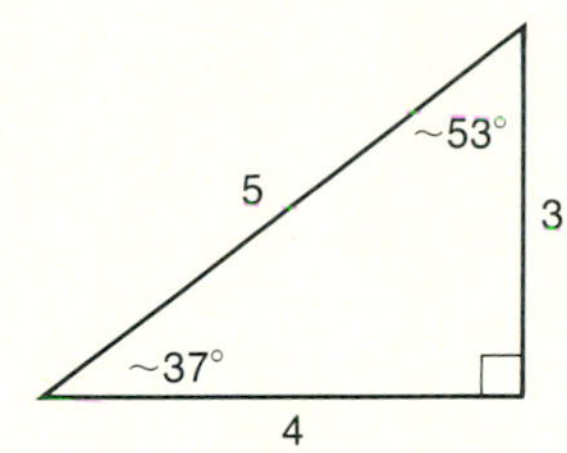

13
5
12

(**b**) Some common right triangles.

Figure D–1
Right triangles.

$$\sin\theta = \frac{y}{r} \qquad \csc\theta = \frac{r}{y}$$

$$\cos\theta = \frac{x}{r} \qquad \sec\theta = \frac{r}{x}$$

$$\tan\theta = \frac{y}{x} \qquad \cot\theta = \frac{x}{y}$$

$$\sin^2\theta + \cos^2\theta = 1$$
$$\sec^2\theta - \tan^2\theta = 1$$
$$\csc^2\theta - \cot^2\theta = 1$$

All Plane Triangles (see Figure D–2)

Law of sines: $\dfrac{a}{\sin\alpha} = \dfrac{b}{\sin\beta} = \dfrac{c}{\sin\gamma} = D$

(where D is the diameter of the circumscribed circle)

Law of cosines: $c^2 = a^2 + b^2 - 2ab\cos\gamma$

Law of tangents: $$\frac{(a+b)}{(a-b)} = \frac{\tan\left(\dfrac{\alpha+\beta}{2}\right)}{\tan\left(\dfrac{\alpha-\beta}{2}\right)}$$

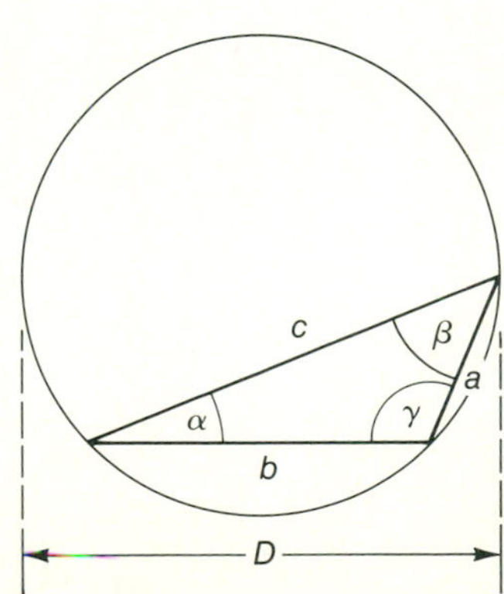

Figure D–2
Triangle for defining the relationships among the parameters of all triangles.

Solution of All Oblique Triangles (see Figure D–2)

I. **Given any two sides a and b and the included angle γ:**

$$\text{Area} = \tfrac{1}{2}ab \sin \gamma$$

$$\tan \frac{\alpha - \beta}{2} = \frac{a - b}{a + b} \cot \frac{\gamma}{2} \quad \text{and} \quad \frac{\alpha + \beta}{2} = \frac{\pi - \gamma}{2}$$

Obtaining $(\alpha - \beta)$ and $(\alpha + \beta)$ from the above equations, find α and β.

Then
$$c = \frac{a \sin \gamma}{\sin \alpha}$$

II. **Given any two sides a and b and an angle opposite one of them, say α:**

$$\sin \beta = \frac{b \sin \alpha}{a} \begin{cases} \text{If } a \geq b, \text{ then } \beta < \dfrac{\pi}{2} \text{ and has only one value.} \\ \text{If } a < b, \text{ then} \\ \quad (1)\ \beta \text{ has two values for } b \sin \alpha < a \text{ such that} \\ \qquad \beta_1 + \beta_2 = \pi. \\ \quad (2)\ \beta = \dfrac{\pi}{2} \text{ for } b \sin \alpha = a. \\ \quad (3)\ \text{The triangle does not exist for } b \sin \alpha > a. \end{cases}$$

Knowing β, then $\gamma = \pi - (\alpha + \beta)$ and

$$c = \frac{a \sin \gamma}{\sin \alpha}, \qquad \text{Area} = \tfrac{1}{2}ab \sin \gamma$$

III. **Given any two angles α and β, and any side a:**

$$\gamma = \pi - (\alpha + \beta), \qquad b = a \frac{\sin \beta}{\sin \alpha}, \qquad c = a \frac{\sin \gamma}{\sin \alpha}$$

IV. **Given three sides a, b, c:**

$$\text{Area} = [s(s - a)(s - b)(s - c)]^{1/2} \qquad \text{where } s = \tfrac{1}{2}(a + b + c)$$

$$\tan \frac{\alpha}{2} = \frac{\text{area}}{s(s - a)}, \qquad \tan \frac{\beta}{2} = \frac{\text{area}}{s(s - b)}, \qquad \tan \frac{\gamma}{2} = \frac{\text{area}}{s(s - c)}$$

Miscellaneous Formulas

$$\sin(\alpha \pm \beta) = \sin \alpha \cos \beta \pm \cos \alpha \sin \beta$$
$$\cos(\alpha \pm \beta) = \cos \alpha \cos \beta \mp \sin \alpha \sin \beta$$
$$\tan(\alpha \pm \beta) = \frac{\tan \alpha \pm \tan \beta}{1 \mp \tan \alpha \tan \beta}$$
$$\sin 2\alpha = 2 \sin \alpha \cos \alpha$$
$$\cos 2\alpha = 1 - 2 \sin^2 \alpha = 2 \cos^2 \alpha - 1 = \cos^2 \alpha - \sin^2 \alpha$$
$$\sin \alpha \cos \beta = \tfrac{1}{2}[\sin(\alpha - \beta) + \sin(\alpha + \beta)]$$
$$\sin \alpha \sin \beta = \tfrac{1}{2}[\cos(\alpha - \beta) - \cos(\alpha + \beta)]$$
$$\cos \alpha \cos \beta = \tfrac{1}{2}[\cos(\alpha - \beta) + \cos(\alpha + \beta)]$$

Appendix E

Mathematical Approximations, Formulas, and Conversions

Approximations

For $x \ll 1$:

$$\frac{1}{1 \pm x} \approx 1 \mp x$$

$$\sqrt{1 \pm x} \approx 1 \pm \frac{x}{2}$$

$$\frac{1}{(1 \pm x^2)^{3/2}} \approx 1 \mp \frac{3x^2}{2}$$

$$\frac{1}{\sqrt{1 \mp x^2}} \approx 1 \pm \frac{x^2}{2}$$

For $x \approx 1$:

$$(1 - x^2) \approx 2(1 - x)$$

(The last two: Often useful for solving problems in special relativity)

For small angles θ (in radians) ($<1\%$ discrepancy for $\theta < 10°$)

$$\begin{cases} \sin \theta \approx \theta \\ \cos \theta \approx 1 - \dfrac{\theta^2}{2} \\ \tan \theta \approx \theta \end{cases}$$

Quadratic Formula

If

$$ax^2 + bx + c = 0,$$

then

$$x = \frac{-b \pm \sqrt{b^2 - 4ac}}{2a}$$

Binomial Formula

$$(x + y)^n = x^n + nx^{n-1}y + \frac{n(n-1)}{2!}x^{n-2}y^2 + \frac{n(n-1)(n-2)}{3!}x^{n-3}y^3 + \cdots + y^n$$

Taylor's Expansion

$$y(x) = y(a) + \left.\frac{dy}{dx}\right|_{x=a}(x-a) + \frac{1}{2!}\left.\frac{d^2y}{dx^2}\right|_{x=a}(x-a)^2 + \frac{1}{3!}\left.\frac{d^3y}{dx^3}\right|_{x=a}(x-a)^3 + \cdots + \frac{1}{n!}\left.\frac{d^ny}{dx^n}\right|_{x=a}(x-a)^n + \cdots$$

Stirling's Approximation for Factorials

$n! \approx \sqrt{2\pi n}\, n^n e^{-n}$, for large n ($<1\%$ discrepancy for $n > 10$)

Exponentials

$$e = 2.718\ 28 \qquad e^0 = 1 \qquad \frac{1}{e} = 0.367\ 879$$

If $y = e^x$, then $x = \ln y$

$e^{\ln x} = x$

$$e^x = 1 + x + \frac{x^2}{2!} + \frac{x^3}{3!} + \cdots$$

Logarithms

If $\log x = y$, then $x = 10^y$
If $\ln x = y$, then $x = e^y$
If $\log_b x = y$, then $x = b^y$

Change of base: $\ln x = (\ln 10)(\log x) = 2.3026 \log x$
$\log x = (\log e)(\ln x) = 0.434\ 29 \ln x$

$\ln e = 1 \qquad \ln e^x = x \qquad \ln 0 = -\infty$
$\ln 1 = 0 \qquad \ln a^x = x \ln a \qquad \log 0 = -\infty$
$\ln xy = \ln x + \ln y \qquad \ln a^x = x \ln a$

$$\ln \frac{x}{y} = \ln x - \ln y \qquad \ln \sqrt[b]{a} = \frac{\ln a}{b}$$

$$\ln(1 + x) = x - \frac{x^2}{2} + \frac{x^3}{3} - \cdots \qquad (\text{for } -1 < x \leq 1)$$

Sagitta Formula (see Figure E–1)

$$R \approx \frac{b^2}{2a} \qquad \left(\text{for } \frac{a}{b} \ll 1\right)$$

where R is the radius of the arc, a is the arc-to-chord distance, and b is the half-chord length.

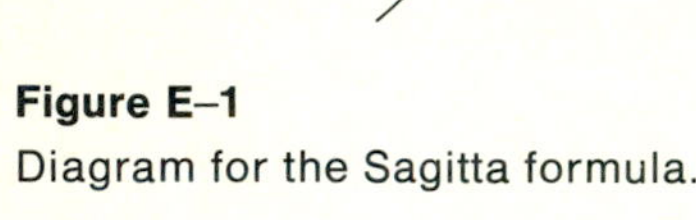

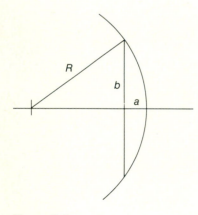

Figure E–1
Diagram for the Sagitta formula.

Vector Relationships

Let α be the angle between the forward directions of the vectors **A** and **B**.

Scalar (or Dot) Product

$$\mathbf{A} \cdot \mathbf{B} = \mathbf{B} \cdot \mathbf{A} = |\mathbf{A}|\,|\mathbf{B}| \cos \alpha = AB \cos \alpha$$

$$= A_x B_x + A_y B_y + A_z B_z$$

(where α is the smaller of the two angles between the forward directions of **A** and **B**)

Vector (or Cross) Product

$$\mathbf{A} \times \mathbf{B} = -\mathbf{B} \times \mathbf{A} = \begin{vmatrix} \hat{\mathbf{x}} & \hat{\mathbf{y}} & \hat{\mathbf{z}} \\ A_x & A_y & A_z \\ B_x & B_y & B_z \end{vmatrix}$$

$$= (A_y B_z - B_y A_z)\hat{\mathbf{x}} + (A_z B_x - B_z A_x)\hat{\mathbf{y}} + (A_x B_y - B_x A_y)\hat{\mathbf{z}}$$

$$|\mathbf{A} \times \mathbf{B}| = |\mathbf{A}|\,|\mathbf{B}| \sin \alpha = AB \sin \alpha$$

Appendix F
Fourier Analysis

The French mathematician François Fourier (1772–1836) showed that almost any periodic function,[1] such as that shown in Figure F–1, may be expressed as an infinite sum of sine and cosine functions and possibly a constant term. Such a sum is called a *Fourier series* with the general form:

$$f(t) = a_0 + \sum_{n=1}^{\infty} (a_n \cos n\omega t + b_n \sin n\omega t)$$

where
$$\omega = \frac{2\pi}{T}$$

$$a_0 = \frac{1}{T}\int_0^T f(t)\,dt \qquad \text{[the average value of } f(t)\text{]}$$

$$a_n = \frac{2}{T}\int_0^T f(t) \cos n\omega t\,dt$$

$$b_n = \frac{2}{T}\int_0^T f(t) \sin n\omega t\,dt$$

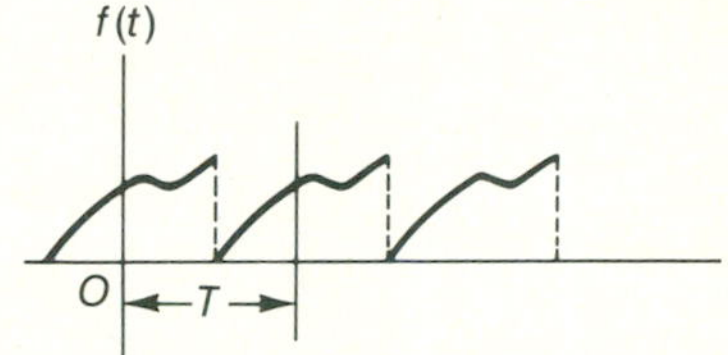

Figure F–1
A periodic function with a period T that may be expressed by an infinite sum of sine and cosine functions and a constant.

Note that for even functions $[f(t) = f(-t)]$, all the *b*s are zero, and for odd functions $[f(-t) = -f(t)]$, all the *a*s (except possibly a_0) are zero. Very often a function may be made even or odd by a shift of the origin, as shown in Figure F–2.

EXAMPLE F–1

Example F–1 (Figure F–3)

$$f(t) = \begin{cases} A, & 0 < t < \dfrac{T}{2} \\ -A, & \dfrac{T}{2} < t < T \end{cases}$$

$$f(t) = \frac{4A}{\pi}\left(\frac{\sin t}{1} + \frac{\sin 3t}{3} + \frac{\sin 5t}{5} + \cdots\right)$$

Figure F–3
Example F–1

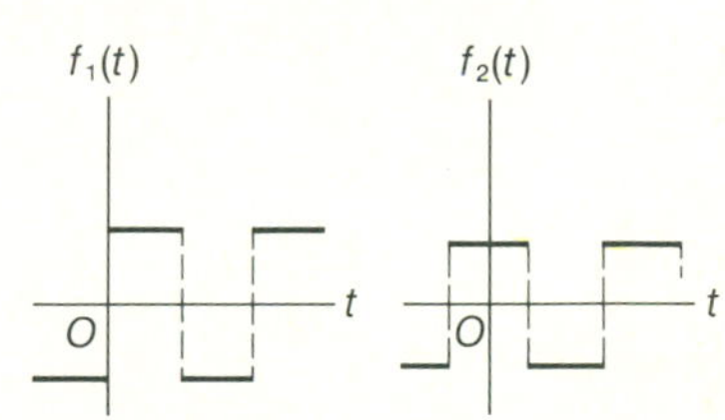

Figure F–2
A square wave may be expressed either as a sum of only sine terms or as a sum of only cosine terms. The Fourier series of $f_1(t)$ contains only sine terms (since it is an odd function), and $f_2(t)$ contains only cosine terms (since it is an even function).

[1] Certain mathematical criteria must be met. The function representing the motion must be single-valued and continuous except for a finite number of finite discontinuities, and must not have an infinite number of maxima or minima in the neighborhood of any given point. However, all motions of real physical objects, electrical currents, and so forth, meet these criteria, so we may always use this method for analyzing physical phenomena.

EXAMPLE F–2

Example F–2 (Figure F–4)

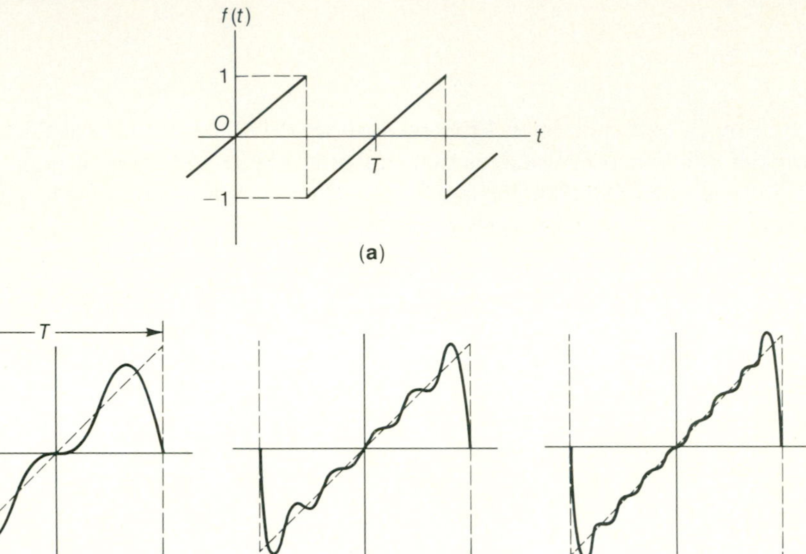

Three terms   Six terms   Nine terms

(**b**) These illustrations show the approximation of a sawtooth waveshape of period T by combining. respectively, the first three, six, and nine terms of the Fourier series.

Figure F–4
Example F–2

$$f(t) = t, \qquad -\frac{T}{2} < t < \frac{T}{2}$$

$$f(t) = 2\left(\frac{\sin t}{1} - \frac{\sin 2t}{2} + \frac{\sin 3t}{3} - \cdots\right)$$

EXAMPLE F–3

Example F–3 (Figure F–5)

$$f(t) = A \sin t, \qquad -\frac{T}{2} < t < \frac{T}{2}$$

$$f(t) = \frac{2A}{\pi} - \frac{4A}{\pi}\left(\frac{\cos 2t}{1 \cdot 3} + \frac{\cos 4t}{3 \cdot 5} + \frac{\cos 6t}{5 \cdot 7} + \cdots\right)$$

f(t)
A
O
T
t

Figure F–5
Example F–3

Appendix G
Calculus Formulas

In the following, a, b, c, and n are constants; u and v are functions of x; and x and y are functions of t. Logarithmic expressions are to the base $e = 2.718\,28\ldots$ All angles are measured in radians.

G–I Derivatives

1. $$\frac{d}{dx}[cu] = c\frac{du}{dx}$$
2. $$\frac{d}{dx}[u+v] = \frac{du}{dx} + \frac{dv}{dx}$$
3. $$\frac{d}{dx}\left[\frac{u}{v}\right] = \frac{v\dfrac{du}{dx} - u\dfrac{dv}{dx}}{v^2}$$
4. $$\frac{d}{dx}[uv] = u\frac{dv}{dx} + \frac{du}{dx}v$$
5. $$\frac{d}{dx}[u]^n = nu^{n-1}\frac{du}{dx}$$
6. $$\frac{d}{dx}[a^u] = (a^u \ln a)\frac{du}{dx}$$
7. $$\frac{d}{dx}[\sin ax] = a\cos ax$$
8. $$\frac{d}{dx}[\cos ax] = -a\sin ax$$
9. $$\frac{d}{dx}[\tan ax] = a\sec^2 ax$$
10. $$\frac{d}{dx}[\ln u] = \frac{1}{u}\frac{du}{dx}$$
11. $$\frac{du}{dt} = \frac{du}{dx}\frac{dx}{dt}$$ (the "chain rule")
12. $$\frac{du}{dv} = \frac{\left(\dfrac{du}{dx}\right)}{\dfrac{dv}{dx}}$$

G–II Integrals

1. $$\int a\,dx = ax + c$$
2. $$\int [u+v]\,dx = \int u\,dx + \int v\,dx + c$$
3. $$\int_a^b u\,dv = (uv)\Big|_a^b - \int_a^b v\,du$$
4. $$\int \frac{dx}{ax+b} = \frac{1}{a}\ln(ax+b) + c$$
5. $$\int x^n\,dx = \frac{x^{n+1}}{n+1} + c \qquad (n \neq -1)$$
6. $$\int \sin ax\,dx = -\frac{1}{a}\cos ax + c$$
7. $$\int \cos ax\,dx = \frac{1}{a}\sin ax + c$$
8. $$\int \tan ax\,dx = -\frac{1}{a}\ln(\cos ax) + c$$
9. $$\int \sin^2 ax\,dx = \frac{x}{2} - \frac{\sin 2ax}{4a} + c$$
10. $$\int \cos^2 ax\,dx = \frac{x}{2} + \frac{\sin 2ax}{4a} + c$$
11. $$\int (\sin ax)(\cos ax)\,dx = \frac{\sin^2 ax}{2a} + c$$
12. $$\int \frac{dx}{\sqrt{a^2 - x^2}} = \begin{cases} \sin^{-1}\left(\dfrac{x}{|a|}\right) + c \\ -\cos^{-1}\left(\dfrac{x}{|a|}\right) + c \end{cases}$$
13. $$\int \frac{dx}{a^2 + x^2} = \frac{1}{a}\tan^{-1}\left(\frac{x}{a}\right) + c$$
14. $$\int \frac{dx}{(a^2 + x^2)^{3/2}} = \frac{x}{a^2\sqrt{a^2 + x^2}} + c$$

G–III The Definite Integral

Most of the use of integration in this text involves the *definite integral*, that is, the integration between two specific values of the variable. The procedure may be illustrated by the following example.

Consider a one-dimensional force $F(x)$, which varies as a function of distance x, as shown in Figure G–1. Let us find the work done by this force as it moves through a displacement from x_1 to x_2. We divide the total displacement into a large number N of small intervals, Δx_1, Δx_2, $\Delta x_3, \ldots, \Delta x_i, \ldots, \Delta x_n$. Since $F(x)$ is nearly constant during each small displacement, we may assume it has the average value $F(x_i)$ during the displacement Δx_i. Thus, the work ΔW_1 accomplished during the first interval Δx_1 is approximately

$$\Delta W_1 \approx F(x_1)\,\Delta x_1$$

and so on for the rest of the intervals. The total work done in moving from x_1 to x_2 is therefore

$$W_{12} \approx \sum_{i=1}^{N} F(x_i)\,\Delta x_i$$

To make a better approximation, we divide the displacement into an even greater number of intervals, so that each Δx_i becomes smaller and the total number of intervals N becomes larger. Continuing to improve the approximation, we let the intervals become smaller and smaller as the total number of intervals becomes larger and larger. The *exact* value for the work done is obtained as Δx shrinks to zero and N goes to infinity. This defines the *definite integral of* F(x) *with respect to* x *from* x_1 *to* x_2. The notation is:

THE DEFINITE INTEGRAL

$$\lim_{N\to\infty} \sum_{i=1}^{N} F(x_i)\,\Delta x_i = \int_{x_1}^{x_2} F(x)\,dx \qquad \textbf{(G–1)}$$

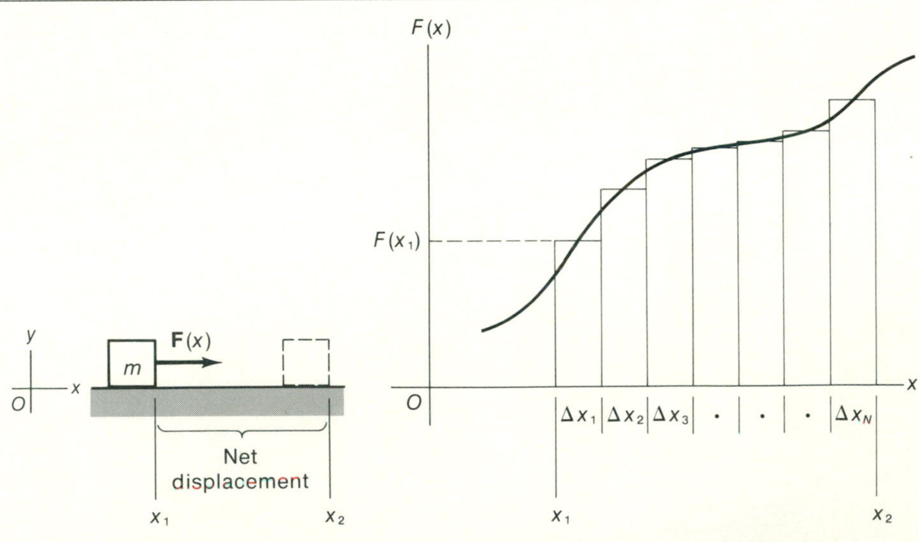

(**a**) A mass m is moved through a displacement $x_2 - x_1$ by the *variable* force $F(x)$.

(**b**) The area under the curve of $F(x)$ versus x between the limits x_1 and x_2 equals the work done by the force $F(x)$.

Figure G–1
Illustration of the definite integral.

The definite integral is equal to the area under the curve of $F(x)$ versus x between the limits x_1 and x_2.

There is a close connection between the *definite integral*, such as Equation (G–1), and the *indefinite integral*, such as $\int F(x)\,dx$ and those integrals listed in G–II. The connection is known as *the fundamental theorem of calculus*, which we state here without proof:

$$\int_{x_1}^{x_2} F(x)\,dx = \underbrace{\int F(x)\,dx}_{\substack{\text{Evaluated} \\ \text{at } x = x_2}} - \underbrace{\int F(x)\,dx}_{\substack{\text{Evaluated} \\ \text{at } x = x_1}} \qquad \textbf{(G–2)}$$

Thus, in calculating a definite integral between two limits, one merely evaluates the integral at the upper limit of x_2 and subtracts its value at the lower limit x_1. In the process, the constant of integration c is eliminated.

Appendix H
Finite Rotations

Finite rotations of an object cannot be represented by vectors. Consider a book that is to be turned through two rotations, one about a vertical axis and the other about a horizontal axis. Suppose a 90° rotation about the vertical axis is represented by the axial vector **V** and a 90° rotation about the horizontal axis by the axial vector **H**. The sum **V** + **H** is the vertical rotation followed by the horizontal rotation, as shown in Figure H–1(a).

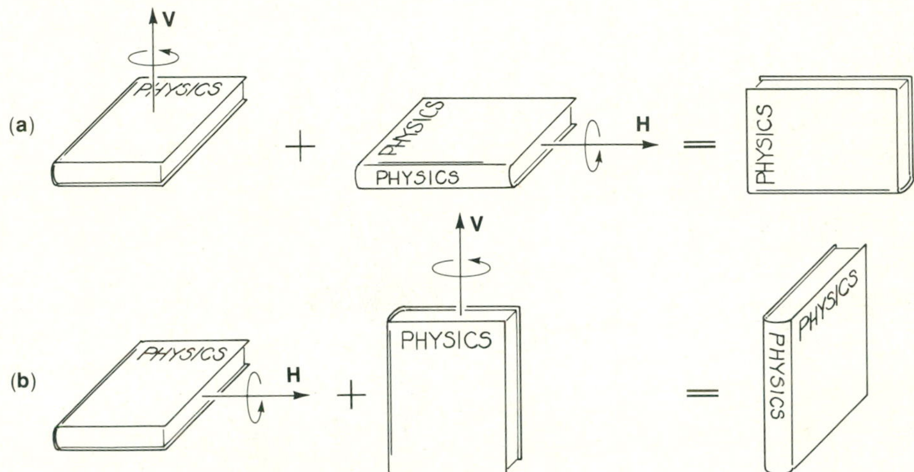

Figure H–1
Successive 90° rotations of a book.

The sum **H** + **V** is represented by the horizontal rotation followed by the vertical rotation, as shown in Figure H–1(b). The final orientation of the book depends on which rotation is first. Mathematically, $\mathbf{V} + \mathbf{H} \neq \mathbf{H} + \mathbf{V}$. Therefore, finite rotations cannot be represented as axial vectors because vectors must have the property of commuting in addition.

However, as the angular displacements become smaller and smaller, the final orientation depends less and less upon the order of the rotations. In the limit of *infinitesimal* rotations, axial vectors may be used to describe such rotations because they commute in addition. Thus, while $\Delta\theta$ is not an axial vector, $d\boldsymbol{\theta}$ is. For this reason, angular velocity $\omega = \lim_{\Delta t \to 0} (\Delta\theta/\Delta t) = d\theta/dt$ may be represented by the axial vector $\boldsymbol{\omega}$ according to the right-hand rule, as shown in Figure H–2. Angular acceleration $\boldsymbol{\alpha} = d\boldsymbol{\omega}/dt$ may similarly be defined.

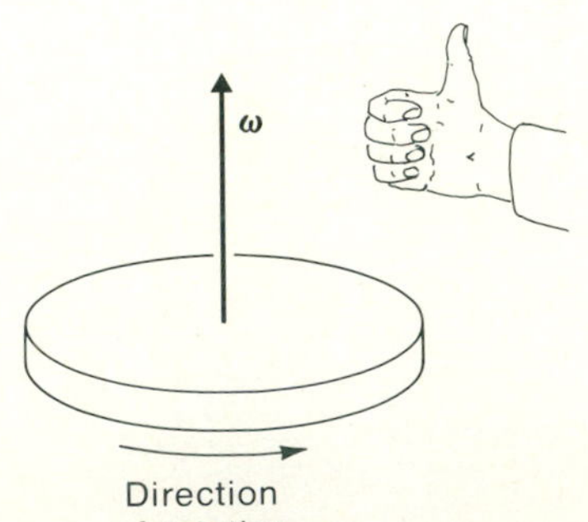

Figure H–2
Right-hand rule: If the fingers of the *right* hand are curled around in the rotational sense, the extended thumb points in the direction of $\boldsymbol{\omega}$. The axial vector $\boldsymbol{\omega}$ represents the angular velocity of rotation of the disk.

Appendix I
Trigonometric Functions

Degree-radian conversions: $1^\circ = \left(\frac{\pi}{180}\right)$ rad $= 0.017\,45$ rad; 1 rad $= \left(\frac{180}{\pi}\right)$ deg $= 57.2958^\circ$

Angle θ					Angle θ				
Degree	Radian	sin θ	cos θ	tan θ	Degree	Radian	sin θ	cos θ	tan θ
0	0.0000	0.0000	1.0000	0.0000					
1	0.0175	0.0175	0.9998	0.0175	46	0.8029	0.7193	0.6947	1.0355
2	0.0349	0.0349	0.9994	0.0349	47	0.8203	0.7314	0.6820	1.0724
3	0.0524	0.0523	0.9986	0.0524	48	0.8378	0.7431	0.6691	1.1106
4	0.0698	0.0698	0.9976	0.0699	49	0.8552	0.7547	0.6561	1.1504
5	0.0873	0.0872	0.9962	0.0875	50	0.8727	0.7660	0.6428	1.1918
6	0.1047	0.1045	0.9945	0.1051	51	0.8901	0.7771	0.6293	1.2349
7	0.1222	0.1219	0.9925	0.1228	52	0.9076	0.7880	0.6157	1.2799
8	0.1396	0.1392	0.9903	0.1405	53	0.9250	0.7986	0.6018	1.3270
9	0.1571	0.1564	0.9877	0.1584	54	0.9425	0.8090	0.5878	1.3764
10	0.1745	0.1736	0.9848	0.1763	55	0.9599	0.8192	0.5736	1.4281
11	0.1920	0.1908	0.9816	0.1944	56	0.9774	0.8290	0.5592	1.4826
12	0.2094	0.2079	0.9781	0.2126	57	0.9948	0.8387	0.5446	1.5399
13	0.2269	0.2250	0.9744	0.2309	58	1.0123	0.8480	0.5299	1.6003
14	0.2443	0.2419	0.9703	0.2493	59	1.0297	0.8572	0.5150	1.6643
15	0.2618	0.2588	0.9659	0.2679	60	1.0472	0.8660	0.5000	1.7321
16	0.2793	0.2756	0.9613	0.2867	61	1.0647	0.8746	0.4848	1.8040
17	0.2967	0.2924	0.9563	0.3057	62	1.0821	0.8829	0.4695	1.8807
18	0.3142	0.3090	0.9511	0.3249	63	1.0996	0.8910	0.4540	1.9626
19	0.3316	0.3256	0.9455	0.3443	64	1.1170	0.8988	0.4384	2.0503
20	0.3491	0.3420	0.9397	0.3640	65	1.1345	0.9063	0.4226	2.1445
21	0.3665	0.3584	0.9336	0.3839	66	1.1519	0.9135	0.4067	2.2460
22	0.3840	0.3746	0.9272	0.4040	67	1.1694	0.9205	0.3907	2.3559
23	0.4014	0.3907	0.9205	0.4245	68	1.1868	0.9272	0.3746	2.4751
24	0.4189	0.4067	0.9135	0.4452	69	1.2043	0.9336	0.3584	2.6051
25	0.4363	0.4226	0.9063	0.4663	70	1.2217	0.9397	0.3420	2.7475
26	0.4538	0.4384	0.8988	0.4877	71	1.2392	0.9455	0.3256	2.9042
27	0.4712	0.4540	0.8910	0.5095	72	1.2566	0.9511	0.3090	3.0777
28	0.4887	0.4695	0.8829	0.5317	73	1.2741	0.9563	0.2924	3.2709
29	0.5061	0.4848	0.8746	0.5543	74	1.2915	0.9613	0.2756	3.4874
30	0.5236	0.5000	0.8660	0.5774	75	1.3090	0.9659	0.2588	3.7321
31	0.5411	0.5150	0.8572	0.6009	76	1.3265	0.9703	0.2419	4.0108
32	0.5585	0.5299	0.8480	0.6249	77	1.3439	0.9744	0.2250	4.3315
33	0.5760	0.5446	0.8387	0.6494	78	1.3614	0.9781	0.2079	4.7046
34	0.5934	0.5592	0.8290	0.6745	79	1.3788	0.9816	0.1908	5.1446
35	0.6109	0.5736	0.8192	0.7002	80	1.3963	0.9848	0.1736	5.6713
36	0.6283	0.5878	0.8090	0.7265	81	1.4137	0.9877	0.1564	6.314
37	0.6458	0.6018	0.7986	0.7536	82	1.4312	0.9903	0.1392	7.115
38	0.6632	0.6157	0.7880	0.7813	83	1.4486	0.9925	0.1219	8.144
39	0.6807	0.6293	0.7771	0.8098	84	1.4661	0.9945	0.1045	9.514
40	0.6981	0.6428	0.7660	0.8391	85	1.4835	0.9962	0.0872	11.430
41	0.7156	0.6561	0.7547	0.8693	86	1.5010	0.9976	0.0698	14.301
42	0.7330	0.6691	0.7431	0.9004	87	1.5184	0.9986	0.0523	19.081
43	0.7505	0.6820	0.7314	0.9325	88	1.5359	0.9994	0.0349	28.636
44	0.7679	0.6947	0.7193	0.9657	89	1.5533	0.9998	0.0175	57.290
45	0.7854	0.7071	0.7071	1.0000	90	1.5708	1.0000	0.0000	∞

Appendix J

Periodic Table of the Elements[1]

1 **H** 1.0080	← Atomic number ← Atomic mass (u)

1 **H** 1.0080																	2 **He** 4.0026
3 **Li** 6.941	4 **Be** 9.0122											5 **B** 10.81	6 **C** 12.011	7 **N** 14.0067	8 **O** 15.9994	9 **F** 18.9984	10 **Ne** 20.179
11 **Na** 22.9898	12 **Mg** 24.305											13 **Al** 26.9815	14 **Si** 28.086	15 **P** 30.9738	16 **S** 32.06	17 **Cl** 35.453	18 **Ar** 39.948
19 **K** 39.102	20 **Ca** 40.08	21 **Sc** 44.956	22 **Ti** 47.90	23 **V** 50.941	24 **Cr** 51.996	25 **Mn** 54.9380	26 **Fe** 55.847	27 **Co** 58.9332	28 **Ni** 58.71	29 **Cu** 63.54	30 **Zn** 65.37	31 **Ga** 69.72	32 **Ge** 72.59	33 **As** 74.9216	34 **Se** 78.96	35 **Br** 79.909	36 **Kr** 83.80
37 **Rb** 85.467	38 **Sr** 87.62	39 **Y** 88.906	40 **Zr** 91.22	41 **Nb** 92.906	42 **Mo** 95.94	43 **Tc** (99)	44 **Ru** 101.07	45 **Rh** 102.906	46 **Pd** 106.4	47 **Ag** 107.870	48 **Cd** 112.40	49 **In** 114.82	50 **Sn** 118.69	51 **Sb** 121.75	52 **Te** 127.60	53 **I** 126.9045	54 **Xe** 131.30
55 **Cs** 132.906	56 **Ba** 137.34	57 **La** 138.906	72 **Hf** 178.49	73 **Ta** 180.948	74 **W** 183.85	75 **Re** 186.2	76 **Os** 190.2	77 **Ir** 192.2	78 **Pt** 195.09	79 **Au** 196.967	80 **Hg** 200.59	81 **Tl** 204.37	82 **Pb** 207.2	83 **Bi** 208.981	84 **Po** (210)	85 **At** (210)	86 **Rn** (222)
87 **Fr** (223)	88 **Ra** (226)	89 **Ac** (227)	104 **Rf** (261)	105 **Ha** (262)													

Lanthanide Series

58 **Ce** 140.12	59 **Pr** 140.908	60 **Nd** 144.24	61 **Pm** (147)	62 **Sm** 150.4	63 **Eu** 151.96	64 **Gd** 157.25	65 **Tb** 158.925	66 **Dy** 162.50	67 **Ho** 164.930	68 **Er** 167.26	69 **Tm** 168.934	70 **Yb** 173.04	71 **Lu** 174.97

Actinide Series

90 **Th** (232)	91 **Pa** (231)	92 **U** (238)	93 **Np** (237)	94 **Pu** (242)	95 **Am** (243)	96 **Cm** (248)	97 **Bk** (249)	98 **Cf** (249)	99 **Es** (254)	100 **Fm** (257)	101 **Md** (258)	102 **No** (259)	103 **Lr** (260)

[1] Atomic masses (in atomic mass units u) are based on $^{12}C = 12$ u exactly. Values for stable elements are those adopted in 1969 by the International Union of Pure and Applied Chemistry. For elements having no stable isotope, the mass number of the "most stable" well-investigated isotope is given in parentheses.

Appendix K
Constants and Standards

Constants and Standards

[Based on values in *Rev. Mod. Phys.*, Vol. 52, No. 2, Part II, April 1980, and *CODATA Bulletin* No. 11 (Dec 1973).]

Name	Symbol	Value[1]
Speed of light	c	2.998×10^{8} m/s
Electronic charge	e	1.602×10^{-19} C
Avogadro's constant[2]	N_A	6.022×10^{23} mol^{-1}
Unified atomic mass unit	u	1.661×10^{-27} kg ($=\frac{1}{12}$ mass of ^{12}C atom $= 931.5$ MeV/c^2)
Electron mass	m_e	9.110×10^{-31} kg ($= 0.5110$ MeV/c^2)
Proton mass	m_p	1.673×10^{-27} kg ($= 938.3$ MeV/c^2)
Neutron mass	m_n	1.675×10^{-27} kg ($= 939.6$ MeV/c^2)
Ratio of proton mass to electron mass	m_p/m_e	1836
Electron (charge/mass) ratio	e/m_e	1.759×10^{11} C/kg
Permittivity of free space	ε_0	8.854×10^{-12} C^2/N·m^2 (or F/m)
Permeability of free space	μ_0	$4\pi \times 10^{-7}$ N/A^2 (or H/m)
Planck's constant	h	6.626×10^{-34} J·s ($= 4.136 \times 10^{-15}$ eV·s)
Universal gas constant[2]	R	8.314 J/mol·K $\left(\begin{array}{l} = 8.205 \times 10^{-2} \dfrac{\text{L·atm}}{\text{mol·K}} \\ = 1.986 \text{ cal/mol·K} \end{array}\right)$
Molar[2] volume of an ideal gas at 1 atm and 0°C	—	2.241×10^{-2} m^3/mol
Boltzmann constant	$k\left(= \dfrac{R}{N_A}\right)$	1.381×10^{-23} J/K
Gravitational constant	G	6.672×10^{-11} N·m^2/kg^2
Standard gravitational acceleration	g	9.806 m/s^2 ($\doteq$ 32.17 ft/s^2)
Standard atmospheric pressure	1 atm	1.013×10^{5} Pa (or N/m^2) $\left(\begin{array}{l} = 14.70 \text{ lb/in}^2 = 2116 \text{ lb/ft}^2 \\ = 76.00 \text{ cm of Hg} \\ = 760 \text{ torr} \end{array}\right)$
Standard temperature	—	273.15 K

[1] Accurate to four significant figures (rounded values).
[2] In this text, "mol" (mole) means "gram-molecular weight" ($= 10^{-3}$ kg-molecular weight).

Appendix L
Terrestrial and Astronomical Data

Terrestrial and Astronomical Data

Terrestrial	
Equatorial radius	6.378×10^6 m
Polar radius	6.357×10^6 m
Radius of a sphere having the earth's volume	6.371×10^6 m
Volume	1.087×10^{21} m^3
Mean orbital speed	2.977×10^4 m/s
Rotational velocity (sidereal)	7.29×10^{-5} rad/s
Tangential velocity of rotation of the equator	4.65×10^2 m/s
Solar constant (solar energy incident perpendicularly on a unit area per unit time):	
At the top of the earth's atmosphere	1.34×10^3 W/m^2
At the earth's surface	0.84×10^3 W/m^2 = 1.13 hp/m^2
Astronomical	
Solar energy output	3.86×10^{26} J/s
Solar surface temperature	5780 K
Number of stars in our galaxy	$\sim 1.6 \times 10^{11}$
Diameter of our galaxy	$\sim 10^{21}$ m
Total mass of our galaxy	$\sim 1 \times 10^{41}$ kg
Number of galaxies in the observable universe	$\sim 10^{11}$

Data of the Solar System

Body	Mass (kg)	Radius of object (m)	Mean density (kg/m^3)	Acceleration due to gravity at surface (m/s^2)	Period of revolution about the sun (days)	Distance from the sun (m)
Sun	1.971×10^{30}	6.960×10^8	1.42×10^3	274.4	—	—
Mercury	3.284×10^{23}	2.439×10^6	5.61×10^3	3.92	87.97	5.8×10^{10}
Venus	4.830×10^{24}	6.052×10^6	5.16×10^3	8.82	2.247×10^2	1.08×10^{11}
Earth	5.983×10^{24}	6.378×10^6	5.52×10^3	9.80	3.653×10^2	1.49×10^{11}
Mars	6.372×10^{23}	3.397×10^6	3.95×10^3	3.92	6.870×10^2	2.28×10^{11}
Jupiter	1.882×10^{27}	7.140×10^7	1.34×10^3	26.46	4.333×10^3	7.78×10^{11}
Saturn	5.628×10^{26}	6.033×10^7	6.90×10^2	11.76	1.076×10^4	1.43×10^{12}
Uranus	8.616×10^{25}	2.54×10^7	1.36×10^3	9.80	3.069×10^4	2.87×10^{12}
Neptune	1.000×10^{26}	2.43×10^7	1.30×10^3	9.80	6.019×10^4	4.50×10^{12}
Pluto	$1. \times 10^{22}$ (?)	1.7×10^6 (?)	(?)	(?)	9.089×10^4	5.90×10^{12}
Moon	7.347×10^{22}	1.738×10^6	3.36×10^3	1.67	27.32*	3.84×10^{8}†

* Revolution about the earth.
† Distance from the earth.

Appendix M
SI Units

The General Conference on Weights and Measures has developed *Le Système International d'Unités*, abbreviated SI, a system of units that has been adopted by almost all industrial nations of the world. It is an outgrowth of the MKSA (*m*eter-*k*ilogram-*s*econd-*a*mpere) system. SI units are divided into three classes: *base units, derived units,* and *supplementary units.* Although such a division is not logically essential, it does have certain practical advantages. The General Conference meets from time to time, occasionally revising or adding to the list of official standards. The following information is from the second revision (1973) of the publication NASA SP-7012, available from the Superintendent of Documents, U.S. Government Printing Office, Washington, D.C. 20402.

Names and Symbols of SI Units

Quantity	Name of Unit	Symbol	
	SI Base Units		
length	meter	m	
mass	kilogram	kg	
time	second	s	
electric current	ampere	A	
thermodynamic temperature	kelvin	K	
luminous intensity	candela	cd	
amount of substance	mole	mol	
	SI Derived Units		
area	square meter	m^2	
volume	cubic meter	m^3	
frequency	hertz	Hz	s^{-1}
mass density (density)	kilogram per cubic meter	kg/m^3	
speed, velocity	meter per second	m/s	
angular velocity	radian per second	rad/s	
acceleration	meter per second squared	m/s^2	
angular acceleration	radian per second squared	rad/s^2	
force	newton	N	$kg{\cdot}m/s^2$
pressure (mechanical stress)	pascal	Pa	N/m^2
kinematic viscosity	square meter per second	m^2/s	
dynamic viscosity	newton-second per square meter	$N{\cdot}s/m^2$	
work, energy, quantity of heat	joule	J	N·m
power	watt	W	J/s
quantity of electricity	coulomb	C	A·s
potential difference, electromotive force	volt	V	W/A
electric field strength	volt per meter	V/m	
electric resistance	ohm	Ω	V/A
capacitance	farad	F	A·s/V
magnetic flux	weber	Wb	V·s
inductance	henry	H	V·s/A

Names and Symbols of SI Units (*continued*)

Quantity	Name of Unit	Symbol	
magnetic flux density	tesla	T	Wb/m^2
magnetic field strength	ampere per meter	A/m	
magnetomotive force	ampere	A	
luminous flux	lumen	lm	cd·sr
luminance	candela per square meter	cd/m^2	
illuminance	lux	lx	lm/m^2
wave number	1 per meter	m^{-1}	
entropy	joule per kelvin	J/K	
specific heat capacity	joule per kilogram kelvin	J/(kg·K)	
thermal conductivity	watt per meter kelvin	W/(m·K)	
radiant intensity	watt per steradian	W/sr	
activity (of a radioactive source)	1 per second	s^{-1}	
	SI Supplementary Units		
plane angle	radian	rad	
solid angle	steradian	sr	

Definitions of SI Units

meter (m)
The *meter* is the length equal to 1 650 763 73 wavelengths in vacuum of the radiation corresponding to the transition between the levels 2 p_{10} and 5 d_5 of the krypton-86 atom.

kilogram (kg)
The *kilogram* is the unit of mass; it is equal to the mass of the international prototype of the kilogram. (The international prototype of the kilogram is a particular cylinder of platinum-iridium alloy which is preserved in a vault at Sèvres, France, by the International Bureau of Weights and Measures.)

second (s)
The *second* is the duration of 9 192 631 770 periods of the radiation corresponding to the transition between the two hyperfine levels of the ground state of the cesium-133 atom.

ampere (A)
The *ampere* is that constant current which, if maintained in two straight parallel conductors of infinite length, of negligible circular cross section, and placed 1 meter apart in vacuum, would produce between these conductors a force equal to 2×10^{-7} newton per meter of length.

kelvin (K)
The *kelvin*, unit of thermodynamic temperature, is the fraction 1/273.16 of the thermodynamic temperature of the triple point of water.

candela (cd)
The *candela* is the luminous intensity, in the perpendicular direction, of a surface of 1/600 000 square meter of a blackbody at the temperature of freezing platinum under a pressure of 101 325 newtons per square meter.

mole (mol)
The *mole* is the amount of substance of a system which contains as many elementary entities as there are carbon atoms in 0.012 kg of carbon-12. The elementary entities must be specified and may be atoms, molecules, ions, electrons, other particles, or specified groups of such particles.

newton (N)
The *newton* is that force which gives to a mass of 1 kilogram an acceleration of 1 meter per second per second.

joule (J)
The *joule* is the work done when the point of application of 1 newton is displaced a distance of 1 meter in the direction of the force.

watt (W)
The *watt* is the power which gives rise to the production of energy at the rate of 1 joule per second.

volt (V)
The *volt* is the difference of electric potential between two points of a conducting wire carrying a constant current of 1 ampere, when the power dissipated between these points is equal to 1 watt.

ohm (Ω)
The *ohm* is the electric resistance between two points of a conductor when a constant difference of potential of 1 volt, applied between these two points, produces in this conductor a current of 1 ampere, this conductor not being the source of any electromotive force.

coulomb (C)
The *coulomb* is the quantity of electricity transported in 1 second by a current of 1 ampere.

farad (F)
The *farad* is the capacitance of a capacitor between the plates of which there appears a difference of potential of 1 volt when it is charged by a quantity of electricity equal to 1 coulomb.

henry (H)
The *henry* is the inductance of a closed circuit in which an electromotive force of 1 volt is produced when the electric current in the circuit varies uniformly at a rate of 1 ampere per second.

weber (Wb)
The *weber* is the magnetic flux which, linking a circuit of one turn, produces in it an electromotive force of 1 volt as it is reduced to zero at a uniform rate in 1 second.

lumen (lm)
The *lumen* is the luminous flux emitted in a solid angle of 1 steradian by a uniform point source having an intensity of 1 candela.

radian (rad)
The *radian* is the plane angle between two radii of a circle which cut off on the circumference an arc equal in length to the radius.

steradian (sr)
The *steradian* is the solid angle which, having its vertex in the center of a sphere, cuts off an area of the surface of the sphere equal to that of a square with sides of length equal to the radius of the sphere.

Units Outside the International System

Though not official SI units, certain other units are in widespread use with the International System of Units.

Units in Use with the International System		
Name	Symbol	Value in SI unit
minute	min	1 min = 60 s
hour	h	1 h = 60 min = 3600 s
day	d	1 d = 24 h = 86 400 s
degree	°	$1^\circ = (\pi/180)$ rad
minute	′	$1' = (1/60)^\circ = (\pi/10\,800)$ rad
second	″	$1'' = (1/60)' = (\pi/648\,000)$ rad
liter	L	$1\ \text{L} = 1\ \text{dm}^3 = 10^{-3}\ \text{m}^3$
tonne	t	$1\ \text{t} = 10^3$ kg

Units Used with the International System Whose Values in SI Units Are Obtained Experimentally

One **electron volt** (eV) is the kinetic energy acquired by an electron in passing through a potential difference of 1 volt in a vacuum.

The **unified atomic mass unit** (u) is equal to the fraction $\frac{1}{12}$ of the mass of an atom of the nuclide ^{12}C (carbon-12).

The **astronomical unit** (AU in English) is the mean distance of the earth from the sun: 1 AU = $1.495\,978\,92 \times 10^{11}$ m (with an uncertainty of about 5 km).

The **parsec** (pc) is the distance at which 1 astronomical unit subtends an angle of 1 second of arc: 1 pc = 2.063×10^5 AU = 3.262 light-year.

Photograph and Illustration Sources

Chapter 2
Fig. 2-1: Adapted from "Variations in the Rate of the Rotation of the Earth," R.A. Challinor, *Science*, Vol. 172, pp. 1022–1024 (figure 1), 4 June, 1971; copyright 1971 by the American Association for the Advancement of Science. **Fig. 2-30:** *PSSC Physics*, 5th Ed.; copyright 1981, D.C. Heath and Company.

Chapter 4
Fig. 4-18: General Motors Corporation. **Fig. 4-19:** Six Flags Magic Mountain.

Chapter 5
Unn. 5B Fig.: Adapted from *Men, Ants and Elephants* by Peter K. Weyl and illustrated by Anthony Rabielli; copyright 1959 by Peter K. Weyl; adapted by permission of Viking Penguin, Inc. **Fig. 5-17:** Adapted from "The Conversion of Energy," Claude Summers; copyright 1971, Scientific American, Inc.; all rights reserved.

Chapter 7
Fig. 7-4: Dr. Harold Edgerton, MIT, Cambridge, Massachusetts.

Chapter 8
Fig. 8-3(a): From *PSSC Physics*, 2nd Ed., 1965 (figure 24-6), D.C. Heath and Company and Education Development Center, Newton, Massachusetts. **Fig. 8-3(b):** Lawrence Berkeley Laboratory, University of California.

Chapter 10
Fig. 10-8(a): Adapted from "Birds as Flying Machines" by Carl Welty; copyright 1955, Scientific American, Inc.; all rights reserved. **Fig. 10-15(a):** From *PSSC Physics*, 2nd Ed., 1965, D.C. Heath and Company and Education Development Center, Newton, Massachusetts. **Fig. 10-15(c):** Illustration adapted from *Physics* by Robert Resnick and David Halliday, John Wiley & Sons, Inc., 1966. **Fig. 10-33(ab):** Adapted from Tricker and Tricker, *The Science of Movement*, American Elsevier Publishing Co., 1967, and Mills & Boon Limited.

Chapter 11
Fig. 11-15: *VECTORS*, a publication of Hughes Aircraft Company. **Fig. 11-16(a):** Annan Photo Features. **Fig. 11-16(b):** Adapted from *Science for the Airplane Passenger* by Elizabeth Wood; copyright 1968 by Elizabeth Wood; reprinted by permission of Houghton Mifflin Company.

Chapter 12
Fig. 12-16: Adapted from Sears, *Mechanics, Wave Motion and Heat*; copyright 1958, Addison-Wesley Publishing Company, Inc. [figure appears on p. 318 (adapted)]; reprinted with permission. **Fig. 12-17:** Adapted from Sears, *Mechanics, Wave Motion and Heat*; copyright 1958, Addison-Wesley Publishing Company, Inc. [figure appears on p. 318 (adapted)]; reprinted with permission. **Fig. 12-19:** Adapted from "The Physics of Brasses" by Arthur H. Benade; copyright 1973, Scientific American, Inc.; all rights reserved. **Fig. 12-20:** United Press International.

Chapter 13
Fig. 13-3: Adapted from W. Bertoozi, *American Journal of Physics*, Vol. 32, p. 555, 1964, with permission from the American Journal of Physics. **Fig. 13-4:** Courtesy of Stanford Linear Accelerator Center. **Unn. 13A Fig.:** AIP Niels Bohr Library.

Chapter 14
Fig. 14-8: Adapted material from Warburton and Goodkind (1978). **Fig. 14-10:** From *PSSC Physics*, 2nd Ed., 1965, D.C. Heath and Company and Education Development Center, Newton, Massachusetts.

Chapter 15

Fig. 15-3: Deutsches Museum Lichtbildstelle München. **Fig. 15-8:** United Press International. **Fig. 15-16(a):** Courtesy of Education Development Center, Newton, Massachusetts. **Fig. 15-16(b):** From the film "Flow Visualization" by S.J. Kline, Stanford University; produced by the National Committee for Fluid Mechanics Films, Educational Services, Inc. **Fig. 15-16(c):** Graduate Aeronautical Laboratories, California Institute of Technology.

Chapter 16

Fig. 16-8: Adapted from H. Fletcher, *Reviews of Modern Physics*, January 1940. **Fig. 16-9:** Adapted from photographs of Wave Demonstration Machine by The Bell Telephone Laboratories. **Fig. 16-17:** Richards, Sears, Wehr, and Zemansky, *Modern University Physics*, copyright 1960, Addison-Wesley Publishing Company, Inc., page 266; courtesy of Dr. Harvey Fletcher. **Fig. 16-18(a):** From the film "Vibrations of a Drum"—Kalmia Company. **Fig. 16-18(b):** From Mary Waller, *Chladni Figures: A Study in Symmetry*, Bell and Hyman, London, 1961. **Fig. 16-20(c):** Courtesy United States Army Ballistic Research Laboratories, Aberdeen Proving Ground, Maryland.

Chapter 17

Fig. 17-3: Wide World Photos. **Fig. 17-9:** Adapted from Richards, Sears, Wehr, and Zemansky, *Modern University Physics*, copyright 1960, Addison-Wesley Publishing Company, Inc., Chap. 15, page 293 [figure 15.4 (adapted)]; reprinted with permission. **Fig. 17-10:** Illustration adapted from *Physics* by Robert Resnick and David Halliday, John Wiley & Sons, Vol. I, page 531, 1966.

Chapter 18

Fig. 18-7: Adapted from F.H. Crawford, *Heat Thermodynamics and Statistical Physics*, 1963, Harcourt Brace Jovanovich, Inc.

Chapter 19

Fig. 19-13: Adapted from Ford, *Classical and Modern Physics*, Xerox College Publishing, 1973.

Chapter 20

Fig. 20-8: From Knut Schmidt-Nielsen, *How Animals Work*, Cambridge University Press, 1972.

Chapter 21

Fig. 21-3: Photography by Lotte Jacobi, processed by Case Western Reserve University, Cleveland, Ohio. **Fig. 21-5(abcd):** Adapted from "Perpetual Motion Machines," Stanley W. Angrist; copyright 1968, Scientific American, Inc.; all rights reserved. **Fig. 21-5(e):** M.C. Escher, "Perpetual Waterfall"; copyright Beeldrecht Amsterdam/Vaga, New York, 1982, Collection Haags Grameente Museum.

Chapter 22

Fig. 22-12: From O.D. Jefimenko, *Electricity and Magnetism*, Appleton-Century-Crofts, 1966. **Fig. 22-13:** Adapted from *Physics* by Robert Resnick and David Halliday, John Wiley & Sons, Inc., 1966. **Fig. 22-33(a):** Professor O. Nishikawa, The Graduate School of Nagatsuta, Tokyo Institute of Technology, 4259 Nagatsuta, Midon-ku, Yokohama 227 Japan.

Chapter 24

Fig. 24-7(a): Graph adapted from J.S. Blakemore, *Solid State Physics*, W.B. Saunders Company, 1969. **Fig. 24-7(b):** Photograph courtesy of CTI Cryogenics. **Fig. 24-7(c):** Courtesy of Brookhaven National Laboratory. **Fig. 24-8:** Adapted from Prothero and Goodkind, *Review of Scientific Instruments*, American Institute of Physics, Vol. 39, p. 1247, 1968.

Chapter 25

Fig. 25-1: From *PSSC Physics*, 2nd Ed., 1965, D.C. Heath and Company and Education Development Center, Newton, Massachusetts. **Fig. 25-6:** Courtesy Fermi National Accelerator Laboratory.

Chapter 26

Fig. 26-3(a): *PSSC Physics*, 5th Ed., 1981, D.C. Heath and Company and Education Development Center, Newton, Massachusetts. **Fig. 26-6(a):** *PSSC Physics*, 5th Ed., 1981, D.C. Heath and Company and Education Development Center, Newton, Massachusetts. **Fig. 26-7(c):** From O.D. Jefimenko, *Electricity and Magnetism*, Appleton-Century-Crofts, 1966. **Fig. 26-8(c):** Copyright Kodansha. **Fig. 26-12(b):** University of Illinois.

Chapter 28

Fig. 28-5: Hecht/Zajac, *Optics*; copyright 1974, Addison-Wesley, Reading, Massachusetts (figures 3.2 and 3.7); reprinted with permission. **Fig. 28-10(abc):** Adapted from E.M. Purcell, *Electricity and Magnetism*, Berkeley Physics Course, Volume 3, McGraw-Hill, 1965. **Fig. 28-10(d):** By permission of the American Journal of Physics [*American Journal of Physics*, Vol. 40, p. 46 (1972)]. **Fig. 28-16:** Palomar Observatory, California Institute of Technology.

Chapter 29

Fig. 29-1(b): Helen Faye. **Fig. 29-26(cd):** Courtesy of American Optical Corporation. **Fig. 29-40(a):** Manfred Kage; copyright Peter Arnold, Inc. **Fig. 29-40(b):** Courtesy of E.R. Lewis, F.S. Werblin, Y.Y. Zeevi. **Fig. 29-52:** From "The Photographic Lens" by William H. Price; copyright August 1976 by Scientific American, Inc.; all rights reserved.

Chapter 30

Fig. 30-1(c—top): From *Fundamentals of Optics* by Jenkins and White; copyright 1976, McGraw-Hill; used with the permission of McGraw-Hill Book Company. **Fig. 30-1(c—bottom):** From *Atlas of Optical Phenomena* by M. Cagnet, M. Francon, and J. Thrien; copyright 1962, Springer-Verlag. **Fig. 30-4(a):** Miller, *College Physics*, 4th Ed.; copyright 1977, Harcourt Brace Jovanovich, Inc., E. Leybolds Nachfolder, courtesy J. Klinger. **Fig. 30-13:** Francis W. Sears, *Optics*, 3rd Ed., 1949, Addison-Wesley, Reading, Massachusetts; frontispiece copyright holder Mildred C. Sears, Norwich, Vermont; reprinted with permission from the copyright holder. **Fig. 30-15(b):** From *Atlas of Optical Phenomena*, M. Cagnet, M. Francon, and J. Thrien; copyright 1962, Springer-Verlag. **Fig. 30-16:** Miller, *College Physics*, 4th Ed.; copyright 1977, Harcourt Brace Jovanovich, Inc.; courtesy Bausch & Lomb. **Fig. 30-17(b):** Courtesy Bausch & Lomb. **Fig. 30-19:** Adapted from *Fundamentals of Optics* by Jenkins andWhite; copyright 1976, McGraw-Hill; used with the permission of McGraw-Hill Book Company. **Fig. 30-20:** From *Fundamentals of Optics* by Jenkins and White; copyright 1976, McGraw-Hill; used with permission of McGraw-Hill Book Company.

Chapter 31

Fig. 31-1(a): From *Atlas of Optical Phenomena* by M. Cagnet, M. Francon, and J. Thrien; copyright 1962, Springer-Verlag. **Fig. 31-8:** From *Fundamentals of Optics* by Jenkins and White; copyright 1976, McGraw-Hill; used with permission of McGraw-Hill Book Company. **Fig. 31-9:** From *Atlas of Optical Phenomena* by M. Cagnet, M. Francon, and J. Thrien; copyright 1962, Springer-Verlag. **Fig. 31-13(a):** J.R. Eyerman, *Life Magazine*; copyright Time, Inc. **Fig. 31-13(bc):** Copyright by Palomar Observatory, California Institute of Technology; reproduced by permission. **Fig. 31-15:** Adapted from Francis W. Sears, *Optics*, 3rd Ed.; copyright 1949, Addison-Wesley, Reading, Massachusetts (figure 9-20); reprinted with permission of the author. **Fig. 31-16:** Francis W. Sears, *Optics*, 3rd Ed.; copyright 1949, Addison-Wesley, Reading, Massachusetts. **Fig. 31-19:** From *Atlas of Optical Phenomena* by M. Cagnet, M. Francon, and J. Thrien; copyright 1962, Springer-Verlag. **Fig. 31-20:** From *Atlas of Optical Phenomena* by M. Cagnet, M. Francon, and J. Thrien; copyright 1962, Springer-Verlag. **Fig. 31-21(a):** The Arecibo Observatory is part of the National Astronomy and Ionosphere Center which is operated by Cornell University under contract with the National Science Foundation; reprinted with permission. **Fig. 31-21(b):** The National Radio Astronomy Observatory VLA Program is operated by Associated Universities, Inc., under contract with the National Science Foundation. **Fig. 31-25(b):** From B. Rossi, *Optics*, Addison-Wesley, 1957. **Fig. 31-25(c):** Courtesy of Bell Laboratories, Mrs. M.H. Read. **Fig. 31-25(d):** Permission requested from M.H.F. Wilkins, Kings College, London. **Fig. 31-26:** From *Atlas of Optical Phenomena* by M. Cagnet, M. Francon, J. Thrien; copyright 1962, Springer-Verlag,

p. 21. **Fig. 31-30:** From P.M. Rinard, p. 70, *American Journal of Physics*, Vol. 44, No. 1, January 1976. **Fig. 31-32 (abcefg):** From *Atlas of Optical Phenomena* by M. Cagnet, M. Francon, J. Thrien, copyright 1962, Springer-Verlag. **Fig. 31-32(d):** Courtesy M.E. Hufford. **Fig. 31-36:** Conductron Corporation. **Fig. 31-37:** Adapted from Hecht/Zajac, *Optics*; copyright 1974, Addison-Wesley, Reading, Massachusetts (figure 14.36); reprinted with permission. **Fig. 31-38(a):** Courtesy Professor Stu Elliott, Occidental College. **Fig. 31-38(b):** Courtesy Dr. Ralph Wverker, TRW. **Fig. 31-38(c):** Courtesy Dr. Ralph Wverker, TRW.

Chapter 32

Fig. 32-4: Alvin Hudson and Rex Nelson. **Fig. 32-6(a):** Alvin Hudson and Rex Nelson. **Fig. 32-10:** NASA. **Fig. 32-18(abce):** Courtesy of the Measurements Group, Raleigh, North Carolina, USA. **Fig. 32-18(d):** Frocht, *Photo Elasticity*, Vol. I, John Wiley & Sons, 1948.

Chapter 33

Fig. 33-3: Adapted from K. Richtmyer, E. Kernard, and J. Cooper, *Introduction to Modern Physics*, 6th Ed., McGraw-Hill, 1969. **Fig. 33-9:** R.A. Millikan, *Physical Review, American Institute of Physics*. **Fig. 33-12:** R.A. Millikan, *Physical Review, American Institute of Physics*. **Fig. 33-17:** Miller, *College Physics*, 4th Ed., Harcourt Brace Jovanovich, 1977; courtesy Dr. Albert Rose.

Chapter 34

Fig. 34-1: Adapted from Borowitz/Beiser, *Essentials of Physics*, 2nd Ed.; copyright 1971, Addison-Wesley, Reading, Massachusetts (figure 32-4); reprinted with permission. **Fig. 34-6:** Adapted from *The Nature of Physics* by Peter Broncazio, Macmillan Publishing Company, Inc., copyright 1975. **Fig. 34-7:** B.J. Graham and Virgilo Acosta. **Fig. 34-13(ab):** R.E. Lapp. **Fig. 34-13(cd):** Courtesy of Education Development Center, Newton, Massachusetts. **Fig. 34-13(e):** C.G. Shull. **Fig. 34-14(a):** From J. Valasek *Introduction to Theoretical and Experimental Optics*, Wiley, New York, 1949. **Fig. 34-14(b):** H. Raether "Elektron Interferenzen," *Handbuch Der Physik*, XXXII, Springer, Berlin, 1957. **Fig. 34-14(c):** From P.G. Merli *et al.*, p. 306, *American Journal of Physics*, Vol. 44, No. 3, January 1976. **Fig. 34-15:** William P. Spencer, MIT.

Index

A

Aberrations, lens, 784
Absolute temperature scale, 490
Absolute zero, 427, 428, 491
Acceleration, 19
  angular, 198
  average, 19
  centripetal, 48, 199
  due to gravity, 24, 35
  instantaneous, 19
  in simple harmonic motion, 199
  tangential, 48
Accelerometer, 291
AC generator, 707
Achromat, 787
Action at a distance, 347, 524
Adhesion, 371
Adiabatic, 460
Aerodynamics, 363
Airy disk, 837
Alternating current, 680
  effective value, 702
  peak value, 702
Ammeter, 600
Ampere
  definition, 643
  electron flow, 520
  unit, 519
Ampère's law, 644
  modified by Maxwell, 714
Amplitude, 296
  simple harmonic motion, 141
amu (atomic mass unit), 64, *440*
Analyzer, 857
Angstrom, 9
Angular
  acceleration, 198
  displacement, 198, 952
  frequency, 296
  momentum, 204
    conservation of, 245
    quantization of, 898, 902
  motion
    conservation of energy, 246
      table, 248
    dynamics, 238
    tangential motion, 236
  position, 48, 197
  speed, 198
Antimatter, 909
Antinodes, 400
Antireflection coating, 399
Apogee, 354
Apparent depth, 758
Approximations, mathematical, 945
Archimedes' principle, 368
Area expansion, thermal, 419
Astigmatism, 785
Astronomical and terrestrial data (Appendix L), 956
Astronomical telescope, 780
  angular magnification of, 781
Astronomical unit, 941, 959
Atmosphere, Jupiter, 352
Atmospheric pressure, standard, 366
Atom, models of, 894
  Bohr, 897
  Rutherford, 895
  Thomson, 895
Atomic clock, 9
Atomic mass unit, unified, 64, *440*
Avogadro's number, 439
Axes of rotation, moving, 243
Axial vectors, 240
Ayrton shunt, 612

B

Babinet's principle, 853
Balmer series (hydrogen), 897
Banked roadway, 82
Bar, 366
Barkhausen effect, 666
Barometer, 366
Beats, 404
Bernoulli's principle, 374
  equation, 376
Betatron, 653
  "one-two" relation, 653, 654
Binding energy, 332
  gravitation, 352
Binomial formula, 945
Biot-Savart law, 640
Birefringence, 858
Black hole, 356
Blackbody radiation, 871
  Planck's law, 876
  Rayleigh-Jeans law, 875
  Wien's law, 873
Bohr magneton, 639
Bohr model of atom, 897
  assumptions, 898
  energy states of, 899
  radii of orbits, 899
Bohr's complementarity principle, 913
Bohr's correspondence principle, 900
Boltzmann's constant, 446
Born, M., 908
Born's probability, interpretation of, 908
Boyle's law, 439
Bragg scattering condition, 840, 904
Branch rule, Kirchhoff's, 595
Brewster's law, 860
British engineering units, 68
Btu (British thermal unit), 422, 942
Bubble chamber, 175
Bulk modulus, 254
Buoyant force, 368

C

Calcite, optical properties, 858
Calculus formulas (Appendix G), 949
Calorie, definition, 421
Capacitance, 560, 561
  charging of, 606
  combinations of, 565
    in parallel, 565
    in series, 566
  cylindrical capacitor, 565
  effect of dielectric, 567
  isolated sphere, 562
  parallel-plate, 564
  spherical capacitor, 561, 562
Capacitive reactance, 682
Capacitor
  charging of, 606
  cylindrical, 564
  effect of dielectric, 567
  energy stored in, 571
  parallel-plate, 564
  spherical, 561, 562
Capillary tubes, 371
Carnot, S., 480
Carnot cycle, 481

Carnot efficiency, 485
Carnot's theorem, 489
Cavendish experiment, 350
Cavity radiation, 871
  Planck's law, 876
  Rayleigh-Jeans law, 875
  Wien's law, 873
Celsius temperature scale, 415
Center of buoyancy, 386
Center of gravity, 215
  definition of, 217
Center of mass, 217
  collisions, 184
  definition, 218
  extended object, 217
  motion of, *177*, 182
  two particles, 177
Center of oscillation, 305
Center of percussion, 305
Centigrade temperature, 415
Central force, 280
Centrifugal force, 280
Centripetal acceleration, *48*, 199
  definition, 49
Centripetal force, 288
Cerenkov radiation, 404
Changing units, 28
Charge
  negative, 519
  positive, 519
Charge density, 544
  electric field and surface charges, 552
  line charge, 525
Charles' law, 439
Chladni figures, 401
Chromatic aberration, 786
Circle of reference, simple harmonic motion, 300
Circuits
  alternating current, 680
    C only, 681
    LC resonance, 694, 696
    L only, 683
    RLC, parallel, 692
    RLC, series, 684, 687
  direct current, multiloop, 595
  Kirchhoff's rules for, 595
  RC (with battery), 604
  RL (with battery), 658
Circular motion, 47
  derivation of acceleration using unit vectors, 51
  general curvilinear motion, 53
Clocks
  atomic, 7
  comparing rates of, 922
  nonsynchronism of moving, 929
  setting in synchronism, 919
Cloud chamber, 635
Coherent light, 796
Cohesion, 371
Collision, *171*, 184
  center-of-mass frame, 184
  elastic, 172, 182
  inelastic, 172, 183, 184
  momentum vector diagram, 186
Combinations of capacitors
  parallel, 565
  series, 566
Combinations of resistors
  parallel, 593
  series, 592
Complementarity principle, 913
Compton effect, 884
  Compton shift, equation, 886
  Compton wavelength, 886
Concave mirror, 744
  image, 746, 752
Conduction, heat, 425
  electrons, 520
    drift speed, 583, 584
Conductor, 583
  current, 584
  drift speed of charges, 583
Conjugate points, 305
Conservation of
  angular momentum, 207, 245
  charge, 592
  energy, 138, 145
  linear momentum, 154
  mechanical energy, 138
  rotational energy, 246
Conservative force, 133
  definition, 135
  and potential energy, 136
Conservative system, 135
Constant, gravitational, 350
Constants and standards (Appendix K), 955
Continuity equation, 374
Conventional current, 588
Converging lens, 769, 770
Conversion factors (Appendix C), 940
Conversion of units, 28
Convex mirror, 744
  image, 747, 753
Conveyor belt, 162
Coordinate systems, 10
  cartesian, 10, 16
  cylindrical, 16
  polar, 47
  spherical, 16
  three dimensions, 16
Coriolis force, 280
Correspondence principle, 900
Coulomb, C., 519
Coulomb (unit), 519
Coulomb's law, 520
Couple, 230
Covariance, principle of, 355, 357
Critical angle
  light pipe, 760
  for total internal reflection, 759
Critical temperature, 441
Cross product, 203
Current, 584
  alternating, 680
  conventional, 588
  displacement, 714
  electron flow, 520, *583*, 584
  induced, 654, 655
Current loop
  dipole moment of, 625
  energy in magnetic field, 627
  magnetic field of, 644
Curvature of field, 786
Cyclone, 286
Cyclotron, 620
  at Berkeley, 638
Cyclotron frequency, 619

D

Damped harmonic motion, 310
Damped oscillations, 310
Davisson-Germer experiment, 903
De Broglie, L., 901
De Broglie waves, 901
  wavelength equation, 905
Decibel, 394
Dees, 621
Definite integral (Appendix G-III), 950
Degrees of freedom (energy variables), 469
Delta-wye transformation, 614
Density, 364
  tables, 364
Derivative, 17
  partial, 386
  second, 387
  table of (Appendix G), 949
Derivatives (Appendix G-I), 949
Diamagnetism, 664
Diatomic molecule, 144
Dichroism, 859
Dielectric, 567
  constant, 568
    table, 569
  nonpolar, 567
  polar, 567
  strength, 570
    table, 569
Dielectric constants, table, 569
Diesel engine, 488

Differentiation formulas (Appendix G-I), 949
Diffraction, 819
  circular aperture, 836, 837
  circular obstacle, 844
  four slits, 830
  Fraunhofer, 821
  Fresnel, 820, 821, *842*
  Fresnel zone plate, 845
  grating, 827
  half-period zones, 843
  half-wave zones, 822
  multiple slits, 830
  Poisson's bright spot, 844
  rectangular aperture, 826
  resolving power, 834, 837
  single slit, 822
  single-slit pattern, 823, 825
  X-ray diffraction, 839
Dimensions, 28
Diopter power, 767
Dip angle, 637
Dipole
  electric, 528
  magnetic, 625
Dipole antenna, 726
Dispersion, *395*, 787, 833
  of grating, 834
Displacement, 12
  angular, 198, 952
  in simple harmonic motion, 297
Displacement current, 714
Distances, measuring methods and comparison table, 8
Distortion, lens, 786
Divergent lens, 770
DNA, 509
Domains, magnetic, 665
Doppler effect, 402
  for light, 932
Dot product, 97
Double-slit interference, 794
  double-slit pattern, 800
Drag racing, physics of, 92
Drift speed of electrons, 583, 584
  Hall effect, 630
Duality of electromagnetic radiation, 887
Dumbbell model for diatomic gas,
Dynamics, 59
  fluid, 372, 376
  particle, 59
  rigid bodies, 215
  rotational, 238

E

Earthquake waves, 392
Eddy currents, 674
Efficiency, 123
  Carnot, 485
  diesel engine, 488
  engine, 484
  machines, 123
  Otto cycle, 488
  performance coefficient, refrigerator, 486
  Stirling engine, 489
  table, 124
Einstein, A., 324
Einstein's photoelectric equation, 881
Einstein's quantization of radiation, 881
Elastic collisions, 172
Elastic properties of matter, 253
  moduli, 254
Electric dipole, 528
  far-field approximation, 529
  moment, 530
  potential energy of, 531
  torque on, 530
Electric field, 524
  of charged sphere, 535
  in dielectrics, 568
  of dipole, 528, 548
  energy stored in, 575
  intensity, 527
  of line charge, 544
  lines, 526
  operational definition of, 527
  of point charge, 524
  relation to electric potential, 546, 548
  near sharp points, 552
  similarities with magnetic field, 649
  of surface charge, 552
Electric flux, 532
Electric potential, 536
  of a dipole, 548
  gradient of, 547
  of line charge, 544
  of point charge, 537
  relation to electric field, 546, 548
Electrodynamics, 581
Electromagnetic radiation, 710
  dual nature of, 887
  energy in, 727
Electromagnetic spectrum, 724
Electromagnetic waves, 717
  blackbody radiation, 871
  energy density, 727
  energy flow, 728, 729
  energy in, 727
  momentum of, 730, 731
  plane waves, 721
  polarization, 854
  Poynting vector, 728, 729
  pressure exerted by, 732
  radiation of, 725
  relation between E_y and B_z, 723
  spectrum, 724
  transverse, 720, 726
  wave equation, 719
Electromagnetism, 710
Electromechanical analogues, 686
Electromotive force, *581*, 582
  seat of, 581
Electron charge, 520
Electron microscope, 633
Electron-volt, 539
Electrostatic
  field, 524
  force, 518
Electrostatics, 517
Elliptic polarization, 863
EMF, 581
Emittance, 871
Energy, 95
  binding, 332, 352
  comparison table, 2
  conservation of mechanical, 138
  definition, 113
  density, 575
  electron-volt, 539
  equipartition of, 470
  for the future, 120
  gravitational, 350
  gravitational potential energy, 106
  internal, 457
  kilowatt-hour, 122
  kinetic, 103
  kinetic, relativistic, 330
  relativistic, 330
  rest-mass, 331
  rotational, 247, 248
  sources, tables, 480
  spring potential energy, 112
  stored in charged capacitor, 574
  stored in electric field, 574
  theorem, 138
  thermal, 113
  wave, 396
  work and mechanical energy, 108
  work-energy relation, 115
  (*also see names of kinds of energy*)
Energy density
  in E and B fields, 727
  in electric field, 575
  in magnetic field, 663
Energy flow in electromagnetic waves, 728, 729
Energy variables, 469
Engine efficiency, 484
  Carnot, 484, 485

Otto cycle, 488
performance coefficient, refrigerator, 486
Enrico Fermi National Accelerator Laboratory, 620
Entropy, 497
information and, 510
macroscopic view, 497
microscopic view, 501
probability and, 502
second law and, 503
unavailable energy and, 506
Entropy change, free expansion, 499
Equation of state, 456
ideal gas, 440
Equilibrium, 220
conditions, 222
dynamic, 221
rotational, 222
stable, unstable, neutral, 221
static, 221
thermodynamic, 456
Equipartition theorem, 469
of energy, 470
Equipotential surface, 550
Equivalence, principle of, 355, 357
Erg (unit), 944
Escape velocity, 351
Event, point, 321
Exit pupil, 781
Expansion
adiabatic, 464
isobaric, 463
thermal, 416, 417, 419
Extraneous roots, 31
Eye, 776
closest distance of comfortable viewing, 774
defects of, 779
retina, 777

F

Farad (unit), 561
Faraday's law of induction, 649, 650
Farenheit temperature scale, 415
Fermat's last theorem, 743
Fermat's principle, 743
Ferromagnetism, 665
Fiber optics, 761
Field
electrostatic, 524
gravitational, 347
lines, 526
magnetic, 615
near conductor, 552
Field, electric, 524
of charged conducting surface, 552
of dipole, 528, 548
intensity, 527
of line charge, 544
lines, 526
operation definition, 527
of point charge, 524
relation to electric potential, 546, 548
Field ion microscope, 553
Figure of merit, 600, 612
Fluids, 363
Bernoulli's principle, 374
pressure in, 365
thermal expansion of, 419
Flux
electric, 532
linkage, 651, 657
magnetic, 633
Finite rotations (Appendix H), 952
First law
Newton's, 60
of thermodynamics, 457
Fizeau, H., 337
Focal length, 749
Focal point, 749
Foot-pound (unit), 96
Force
buoyant, 368
central, 207
centrifugal, 280
centripetal, 288
conservative, 133, 135
constant, 102, 296
contact, 65
Coriolis, 280
electrostatic, 518
gravitational, 66, 67, *342*
Hooke's law, 296
inertial, 271, 272
Lorentz, *621*, 623
magnetic, on current-carrying wire, 623
moment of, 201
nonconservative, 137
normal, 74
real, 271
restoring, 296
spring, 101
third-law pair of, 79
time-average, 160
Forced oscillation, 311
steady-state, 312
transient therm, 312
Foucault pendulum, 285
Fourier analysis (Appendix F), 947
Frames of reference, 10
center-of-mass, 182
coordinate systems, 10
inertial, 273, 288
moving, 272
rotating, 278
Franklin, B., 519
Fraunhofer diffraction, 821
circular aperture, 836, 837
four slits, 830
multiple slits, 830
single slit, 823, 825
Free-body diagram, 69
Free expansion, 499
Free-fall, 30, *35*
Frequency
angular, 296
fundamental, 400
harmonic, 400
of simple harmonic motion, 296
Fresnel biprism, 818
Fresnel diffraction, 820, 821, *842*
circular obstacle, 844
half-period zones, 843
Fresnel zone plate, 845
Friction, 74
coefficient of, 75
kinetic, 75
rolling, 249
static, 75
thermal energy and, 113
Fundamental frequency, 400
Fusion, latent heat of, 422

G

g, acceleration due to gravity, 24
variations in, 348
Galilean relativity principle, 322
Galilean transformation, 920
Galvanometer, 599, 628
sensitivity of, 599
Gas
constant *R*, 440
ideal, 440
thermometer, 426, *429*
work done by, 460, 461
Gauge pressure, 366
Gauss (unit), 616
Gaussian surface, 533
Gauss's law, 532
Gauss's law for dielectric, 569
Gauss's law for magnetic fields, 716
Gay-Lussac's law, 439
General relativity, 355
Generator, *AC*, 707
Geometrical optics, 739
Glass, nonreflecting, 807
Gradient, 547
Grating, diffraction, 827
dispersion, 833, 834
four slits, 830
multiple slits, 830
reflection, 829

resolving power, 834
transmission, 829
Gravitation, 340
action-at-a-distance, 347
field, 347
field, nonuniform, 218
potential energy, 350
universal constant of, 350
universal law of, 342
Gravitational constant, 350
Gravitational force, 66, 67, *342*
Gravitational mass, 354
Gravitational potential energy, 106, 108
definition, 108
Gravity
acceleration due to, 24
action-at-a-distance, 347
center of, 215
Great American Revolution, 85
Greek alphabet (Appendix B), 939
Gyroscope, 250

H

Half-power frequency, 706
Half-wave plate, 862
Hall effect, 630
Harmonic frequencies, 400
Harmonic motion
damped, 310
forced, 311
simple, 296
Hearing sensitivity, 394
Heat, 420
conduction, 423
definition, 113
engine, 481
latent, 422
mechanical equivalent of, 421
phase change, 422
pump, 487
and work, 421
Heat capacity (specific heat), 421
Heat death, 507
Heisenberg's uncertainty principle, 909
relation, 910
Helmholz coils, 679
Henry (unit), 656
Hologram, 847
Holography, 846
Hooke, R., 101
Hooke's law, 101, *296*
Horsepower, 119, 122
Huygens' principle, 741
reflection by plane mirror, 742
and refraction, 756
Hydrodynamics, 363
Hydrogen
Bohr model for, 897
de Broglie waves for, 902
Hysteresis, 669

I

Ice point, 415
Ideal gas, 438
law, 440
model of, 443
speed of sound in, 392
thermometer, 427, *429*
Image
characteristics, 770
erect and inverted, 752
in plane mirror, 744
real, 744
size, 768
in spherical mirrors, 744, 745
virtual, 744
Image distance, 744
in plane mirror, 744
Image size, 768
angular magnification, 775
Impedance, 687
Impedance diagram, 688
parallel *RLC*, 692
series *RLC*, 688
Impulse, 159
Indeterminacy principle, 909
Index of refraction, 755
birefringence, 858
relative, 766
Induced EMF
Faraday's law, 650
Lenz's law, 655
Induced surface charge, 568
Inductance
energy in, 661
long solenoid, 657
mutual, 658
self, 656
Induction, Faraday's law of, 650
Inductive reactance, 684
Inductors, 656
energy in, 661, 662
Inelastic collision, 172
Inertia, moment of, 202
Inertial force, 272
Inertial frame, 60
Inertial mass, 63
and gravitational mass, 354
Information
bits, 508
entropy and, 510
Insulator, 583
Integration formulas (Appendix G-II), 949
Intensity wave, 397
Interference, 794
colors by, 865
constructive, 399
criteria, 797
destructive, 399
by double slit, 794
double-slit pattern, 800
by multiple slits, 805
multiple-slit pattern, 806
Newton's rings, 810
by thin films, 806
by thin wedges, 808
Interferometer, Michelson, 810
Internal energy, *420*, 457
Internal work, 118
Irreversible process, 458
Irrotational flow, 374
Isobaric, 460
Isochoric, 460
Isolation diagram, 69
Isotherm, 441
Isothermal, *460*, 461
Isovolumic, 460

J

Jet engine, 165
Joule (unit), 96
Joule, J., 96
Joule energy, 588
Joule heating, 588
Joule's law, 588
Junction rule, 595

K

Kelvin temperature scale, 490
Kepler's law, 341
Kilogram (standard), 63
Kilowatt-hour, 122
Kinematics, 6
graphical relations, 26
kinematic equations
derivation using calculus, 25
linear, 23
rotational, 234, 236
three dimensions, 36
linear motion, 21
rotational motion, 234, 236
Kinetic energy, *103*, 106
relativistic, 330
rotational, 248
in SHM, 141
units, 104
variable force, 106
and work, 115
Kinetic friction, 75
Kinetic theory, 438

Kirchhoff's rules, 595
Krypton-86, 9

L

Laminar flow, 372
Latent heat, 422
Laue diffraction, 841
LC resonance, 694, 696
LCR Circuit (*AC*), 684, 687
Length contraction, 326, 924
Length, proper, 926
Length, standard, 9
Lens
  aberrations, 784
  combinations, 771
  divergent, 765
  eyepiece, 781
  images in, 769, 770
  negative, 765
  objective, 781
  positive, 765
  principal foci, 768
  thin, 763
  thin-lens approximation, 763, 765
  thin-lens equation, 766
Lens-maker's formula, 766
Lenz's law, 654, *655*
Light
  momentum delivered by, 730, 731
  moving source of, 932
  phase change on reflection, 807
  photon, theory of, 881
    momentum of photon, 885
  polarization of, 854
  pressure of, 732
  speed of, 722
  transverse waves, 854
  unpolarized, 855
  wavelength and color, 740
Light pipe, 760
Light-year (unit), 941
Lilliputians and Brobdingnagians, 110
Limiting process, 16
Linear charge density, 544
Linear expansion, thermal, 417
Linear momentum, 154 (*also see* momentum)
Lines of force
  electric, 526
  gravitational, 349
Lloyd's mirror, 817
Long solenoid
  magnetic field of, 646
  self-inductance of, 657
Longitudinal waves, 384
Loop rule, Kirchhoff's, 595
Lorentz force, 621
Lorentz transformation, 921
  equations, 922
LR (with battery), 658
LRC circuit (*AC*), 684, 687

M

Magnetic bottle, 621
Magnetic domain, 665
Magnetic dipole, 625
  of Bohr magneton, 639
  potential energy of, in magnetic field, 627
  torque on, 626
Magnetic dipole moment, 626
  torque on, 626
Magnetic field, *B*, 615, 616
  charged particles, motion in, 617
  due to current in straight wire, 642, 645
  due to current loop, 644
  cyclotron frequency, 618
  dip angle, earth's, 637
  in electromagnetic waves, 717, 718, 721
  energy density in, 663
  flux of, 616, 633
  induction, 650
  lines of, 617, 633
  similarities with electric field, 649
  in solenoid, 646
  strength, *H*, 667
  inside toroidal coil, 646
  units of, 616
  vector **B**, 616
Magnetic field strength, *H*, 667
Magnetic flux, 633
Magnetic flux density, *B*, 616, *633*
Magnetic force
  on current-carrying wire, 623
  on moving charge, 616
  between parallel wires, 643
  special relativity and, 710
Magnetic induction, *B*, 616
  (*also see* magnetic field, *B*)
Magnetic monopole, 715
Magnetic properties of materials, 664
  diamagnetism, 664
  ferromagnetism, 665
  hysteresis, 669
  paramagnetism, 664
Magnetic susceptibility, 667
Magnetization curve, 669
Magnetostatics, 615
Magnification, 751, *752*
  angular, 774, 775, 781
  image size, 768
  linear, 752
  of microscope, 783
  of telescope, 781
Magnifier, 773
  angular magnification, 775
Malus, law of, 857
Mass-energy, 330
Mass, 63, 64
  atomic unit of, 440
  center of, 217
  comparison, table, 64
  equivalence principle, 355
  gravitational, 354
  inertial, 63, 354
  relativistic, 327, 926
  rest, 327, 926
  rest-mass energy, 330
  standard, 63
  units, 67, 68
  and weight, 66
Mass spectrometer, 632
Mathematical formulas (Appendix E), 945
Mathematical symbols (Appendix B), 939
Matter wave, 901, 905
  probability interpretation of, 908
Maxwell, J., 712
Maxwell-Boltzmann, 447
Maxwell's equations, 712
  table, 716
Mechanical advantage, 125
  actual, 125
  ideal, 126
Mechanical energy, 108
  conservation of, 138
  work and, 108
Mechanical equivalent of heat, 421
Meniscus, 371
Meter, standard bar, 9
  atomic standard, 9
Michelson, A., 810
Michelson interferometer, 810
Micrometer (micron), 9
Microscope, 783
Mirror equation, 748, 749
  sign convention for, 748
Mirrors, images in, 752, 753
Molar specific heat, 462
  table, gases, 472
Mole, 439
Molecular weight, 440
Moment arm, 201
Moment of inertia, 202, 231
  point mass about *O*, 202
  solid object, 231
  (*also see* rotational inertia)

Momentum, 64
  angular, 204
  of electromagnetic waves, 730, 731
  linear, 64, 154
  of photon, 885
  relativistic, 327, 924, 926
Monopole, magnetic, 715
Motion
  angular, 197
  of center of mass, 177
  with constant acceleration, 21, 36
  linear, 21
  Newton's laws of,
    first, 60
    second, 64
    third, 77
  oscillatory, 295, 296
  periodic, 295
  planetary, 353
  rotational, 197
  satellite, 353
  simple harmonic, 296
  in three dimensions, 35
Mrkos comet, 734
Multiple-slit interference, 805
  multiple-slit pattern, 806
Mutual inductance, 658

N

Neutrinos, 509
Newton, I., 59, 342
Newton's first law, 60
Newton's law of gravitation, 342
Newton's rings, 810
Newton's second law, 64
  accelerated frame, 274
  angular motion, 202
  definition, 65
  rotational motion, 207, 239
Newton's third law, 77
Nodal lines, surfaces, 401
Nodes, 400
Nonconductor, 583
Nonconservative forces, 137
  and energy conservation, 145
Nonpolar dielectric, 567
Nonreflecting coatings, 807
Normal force, 74
North pole of magnet, 616
Nucleus, 896

O

Ohm, 586
Ohm's law, 587
  for *AC*, 688
"One-two" relation (betatron), 654
Operators, mathematical, 279
Optic axis, 858
Optical activity, 865
Optical flat, 809
Optical instruments, 773
  astronomical telescope, 780
  eyeglasses, 778
  microscope, 783
  simple magnifier, 773
  slide projector, 783
Optical path difference, 798
Organ pipes, 400
Oscillations
  damped, 310
  forced, 311
  LC resonance, 696
  simple harmonic, 296
  two-mass system, 306
Otto cycle, 488

P

Parallel-axis theorem, 234
Parallel combinations of
  capacitors, 565
  resistors, 593
Paramagnetism, 664
Parsec (unit), 941, 959
Partial derivative, 386
Particle dynamics, 59
Particle kinematics, 6
  linear, 26
  rotational, 234, 236
Pascal's principle, 368
Pascal (unit), 366
Path difference (optical), 798
Pendulum, 302
  conical, 73
  Foucault, 285
  physical, 304
  simple, 302
  torsional, 303
Performance coefficient, refrigerator, 486
Perigee, 354
Periodic motion, 295
  Fourier analysis of, 947
  simple harmonic, 296
Periodic table of the elements (Appendix J), 954
Permeability
  of free space, μ_0, 641
  of magnetic materials, μ, 667
Permittivity, free space, 520
Perpendicular axis theorem, 268
Perpetual motion, 510
Perspectives, 94, 196, 413, 516
Perverted image, 744
Phase
  change, 421
  velocity, 389
Phase angle, 296
  AC circuits, 680
Phase change on reflection, 807
Phase constant:
  AC circuits, 680
  RLC circuits, 687
  in *RLC* resonance, 696
Phasors, 798, 823
Phasor diagrams, 682
  C only, 682
  L only, 683
  RLC (parallel), 692
  RLC (series), 687
Photoelectric effect, 878
  Einstein's equation, 881
  quantization of radiation, 881
  work function, 881
Photomultiplier, 884
Photon, 881
  "interference" of, 888
  momentum of, 885
  quantized energy of, 881, 885
Physical constants (Appendix K), 955
Physical pendulum, 304
  center of oscillation, 305
  center of percussion, 305
  conjugate points, 305
Planck's constant, 876
Planck's quantum hypothesis, 473
Planck's quantum theory, 875
Planck's radiation law, 876
Plane mirror, 742
Plane wave, *395*, 717, 721
Planetary motion, 353
Plasma, 363
Point event, 321
Poisson's bright spot, 844
Polar dielectric, 567
Polarization, 854
  analyzer, 857
  Brewster's law, 860
  circular, 862, 863
  direction of, 854
  elliptic, 863
  indicating direction of, 855
  polarizing angle (reflection), 860
  Polaroid, 856
  by reflection, 859, 860
  retardation plates, 861
  by scattering, 860, 861
  wave plates, 861
Polarizer, 857
Pole-in-the-barn paradox, 935

Position, angular, 197
Position vector, 11
Postulates of special relativity, 322, 921
of general relativity, 355
Potential, electric, 536
of dipole, 548
of line charge, 544
of point charge, 537
relation to electric field, 546, 548
Potential energy
and conservative forces, 136
of electric dipole in field, 531
gravitational, 350
in simple harmonic motion, 141
of two charges, 537
Potential energy, spring, 112
Potentiometer, 602
Pound, 68
Power, 119
in *AC* circuits, 699
average (*AC*), 700, *701*
in *DC* circuits, 589
definition, 119
half-power frequency, 706
horsepower, 120
transmitted by waves, 396
units of, 119
Power factor, 701
Poynting vector, 728
average, 729
instantaneous, 728
Precession, 251
Prefixes (Appendix A), 938
Prefixes, metric, 9
Pressure, 365
absolute, 366
of electromagnetic waves, 732
in fluids, 365
gauge, 366
standard atmosphere, units, 366
Principal foci, 768
Principle of complementarity, 913
Principle of covariance, 355
Principle of equivalence, 355
Principle of superposition, *399*, 521, 596
Probability
and entropy, 502
in quantum mechanics, 908, 912
Process
irreversible, 458
reversible, 458
Processes, thermodynamic, 460
adiabatic, 460, *464*
irreversible, 458
isobaric, 460, *463*
isochoric, 460
isothermal, 460, *461*
isovolumic, 460, *462*
quasi-static, 459
reversible, 458
Projectile motion, 35
Proper measurements, 926
Pulley, 125, 126

Q

Q (resonance), 696
Quadratic, roots of, 945
Quantization
of angular momentum, 898, 902
of radiation, 876, *881*
of SHM oscillators, 876
Quantum, 876
Quantum hypothesis, 473
Quantum mechanics, 906
chronology, 914
probability interpretation of, 908
Quarter-wave plate, 862
Quasi-static process, 459, 543

R

Radial acceleration, 199
Radian, definition, 197
Radiometer, 733
Radius of gyration, 233
Ray, 739
Ray-tracing, 744, 751, 768
and linear magnification, 751
mirrors, 752
spherical mirrors, 744, 746
thin lens, 768
Rayleigh-Jeans theory, 874
Rayleigh's criterion, 834
circular aperture, 837
slits, 834
RC circuits, 604
RC time constant, 606
RCL circuits, 684, 687
Reactance
capacitive, 682
inductive, 684
Real image, 744
Reduced mass, 307
Reference frame
center-of-mass, 182
inertial, 273, 288
moving, 272
rotating, 278
Reflection, 742
Fermat's principle, 743
Huygens' principle, 742
laws of, 743
phase change in, 807
by plane mirror, 742
by spherical mirror, 744
by thin films, 806
total internal, 759
of waves, 397
Refraction, 755
Huygens' principle, 756
index of, 755
plane surface, 755
Snell's law, 757
spherical surface, 761, 762
thin lenses, 766
Refractive index, 755
birefringence, 858
relative, 766
Refrigerator, performance coefficient, 486
Relative refractive index, 766
Relative velocity, 180
in relativity, 329, 928
Relativity, general theory of, 355
curvature of space-time, 356
principle of covariance, 355
principle of equivalence, 355
Relativity, special theory of, 321
clock synchronism, 919
Doppler effect, 932
and electromagnetic force, 710
energy, relativistic, 330
Galilean relativity principle, 322
length contraction, 326
mass, 327
mass-energy, 330
momentum, 327, 924
point event, 321
pole-in-the-barn paradox, 935
postulates, 322
principle of
relativity, 322
constancy of speed of light, 322
special theory, 321
stick-in-the-hole paradox, 935
time dilation, 323
transformation equations
Galilean, 920
Lorentz, 921
twin paradox, 333, 932
velocity addition, 329
Resistance, 587
equivalent, 592, 593
Resistivity, 585
table, 586
thermal coefficient of, 586
units of, 586
Resistors, 583
in parallel, 593
power in, 589, 699
in series, 592

thermal energy in, 589
Resolving power, 834
of circular aperture, 837
of grating, 834
Resonance, 309
AC, 694
parallel *RLC*, 698
series *RLC*, 695
sharpness, *Q*, 696
Resonant frequency, 310, 695
RLC parallel circuit, 699
RLC series circuit, 696
Rest mass, 327, 926
Rest-mass energy, 330
table, 331
Restoring force, 101
Reversible process, 458
entropy, change in, 504
Right-hand rule
for magnetic field lines, 642
for torques, 204
for vector product, 203
Rigid-body dynamics, 215
RL circuit (with battery), 658
rms speed, 447
Rocket, 164
Rolling bodies
kinematics, 243
friction, 249
Rosa, E., and Dorsey, N., 722
Rotating frame of reference, 278
Rotational dynamics, 238
conservation of angular momentum, 245
conservation of energy, 246
of point mass, 197
of rigid bodies, 238
(*also see* Rotational motion)
Rotational flow, 374
Rotational inertia, 231
parallel-axis theorem, 234
radius of gyration, 233
various shapes (table), 233
Rotational kinetic energy, 248
Rotational motion, 197, 228
conservation of angular momentum, 245
conservation of energy, 246
dynamics, 197, 238
kinematic equations, 236
kinetic energy of, 248
large objects, 228
point mass, 197
rigid bodies, 238
and tangential motion, 236
Rumford, Count (B. Thompson), 420
Rutherford model of atom, 895

S

Sagitta formula, 946
Satellite, motion of, 208, 353
Scalar, 11
Scalar product, 97
Schrödinger, E., 906
Schrödinger's wave equation, 907
Scintillation counter, 884
Second (time) unit, 9
Second law of thermodynamics, 481
Clausius's statement, 481
entropy and, 503
Kelvin-Planck statement, 481
Various statements of, 507
Self-inductance, 656
long solenoid, 657
Semicircular canals, 286
Semiconductor, 583
Separation of variables, 659
Series combinations of
capacitors, 566
resistors, 592
Sharpness, *Q*, 696
Shear modulus, 254
SHM, 140 (*also see* Simple harmonic motion)
Shock waves, 403
SI units, 9, 67
SI units (Appendix M), 957
Sign convention
for lenses, 766
for mirrors, 748
Significant figures, 29
Simple harmonic motion, 296
acceleration, 297
amplitude, 141, 296
angular frequency, 296
circle of reference, 300
displacement, 297
energy graphs, 140
frequency, 297
initial phase angle, 296
period, 297
velocity, 297
Simple pendulum, 302
Size, comparison table, 2
Slide projector, 783
Slug, 68
Snell's law, 757
Solar system data (Appendix L), 956
Solar wind, 734
Solenoid
magnetic field of, 646
self-inductance of, 657
Solid angle, 533
Sound
decibels, 394
in gas, 392
intensity, table, 394
in solids, 392
speed of, 391, 393
Sound waves
antinodes, 400
nodes, 400
speed, 392
standing, 399
South pole, of magnet, 616
Space, homogeneous and isotropic, 204
Special theory of relativity, 321
(*also see* Relativity, special theory of)
Specific gravity, 364
Specific heat, 421
table, solids, 422
table, gases, 472
Specific heats, molar, 462
table, gases, 472
Spectral energy density, 872
Spectrograph, 828, 829
Spectrum, electromagnetic, 724
Speed, 17
angular, 198
molecular, 447
wave, 391
Speed of light, 722
moving source, 932
Spherical aberration, 785
Spherical mirror, 744
concave, 744
convex, 744
mirror equation, 748, 749
sign convention for, 748
Spring, 101
constant, 102
force law, 101
potential energy, 112
work to stretch, 101
Standard
length, 9
mass, 63
time interval, 9
Standard cell, 603
Standard conditions (STP), 439
Standing waves, 399
Stanford two-mile accelerator, 328
State variables, 456, *498*
Static equilibrium, 222
Static friction, 75
Steady-state SHM, 313
Steam point, 415
Stefan-Boltzmann law, 871
Step-up/down transformer, 670
Stereoisomers, 865
Stirling engine, 489
Strain, 253

Streamline flow, 372
Strength (of lenses), 767
Stress, 253
Superconducting gravimeter, 591
Superconductivity, 590
Superconductors, 586
Superposition, electrostatics, 521
Superposition principle, *399*, 521, 596
Surface charge density, 544
  and electric field, 552
  induced, 567, 568
Surface charge density, 544
Surface charge, induced, 568
Surface tension, 370
Symmetry of E and B, 717
Synchrocyclotron, 621
Synchronism, clocks, 919
Synchrotron, 621
System, thermodynamic, 455

T

Tacoma bridge collapse, 314
Tangential acceleration, 48
Tangential motion, rotational, 236
Telescope, 780
  angular magnification of, 781
  radio, 838
Temperature, 414, *420*
  Celsius, 415
  critical, 441
  Fahrenheit, 415
  ideal gas, scale of, 426
  Kelvin, 416, 427
  and kinetic theory, 446, 447
  significant temperatures, table, 432
  triple point, 429
Tension, 71
  surface, 370
Terminal velocity, 363
Terrestrial and astronomical data (Appendix L), 956
Tesla (unit), 616
Thermal
  conductivity, 423
    table, 424
  equilibrium, 456
  expansion, 417
  expansion, table, 418
Thermal coefficient of resistivity, table, 586
Thermal creep of air, 733
Thermal energy, *113*, 420
  in resistor, 588, 589
Thermal equilibrium, 456
Thermal expansion, 416
  area, 419
  linear, 416
  table, coefficient of, 418
  volume, 418
Thermodynamics
  equilibrium, 456
  first law, 457
  heat death, 507
  second law, 481
    entropy and, 503
  temperature scale, 427, 428
  third law, 491
  zeroth law, 456
Thermometer, gas, 426, *429*
  liquid column, 416
Thin films, reflection from, 806
Thin lens, 763
  equation, 766
  ray-tracing, 768
  sign convention, 766
Thin-lens approximation, 763, 765
Third law of thermodynamics, 491
Thompson, B. (Count Rumford), 420
Thomson model of atom, 895
Timbre, 401
Time, standard, 9
Time, proper, 926
Time constant
  capacitive, 606
  inductive, 660
Time dilation, 323, 923
Time intervals, comparison table, 2
Torque, 200
  and angular acceleration, 239
  and angular momentum, 204
Torsion balance, 518
Torsion constant, 303
Torsional pendulum, 303
Toroidal coil, magnetic field of, 646
Torque, 200
  on electric dipole, 530
  on magnetic dipole, 626
  on a particle, 204
  on rigid bodies, 238
  vector, 204
Torricelli's law, 376
Total internal reflection, 759
  critical angle for, 759
  fiber optics, 761
  light pipe, 760
Trajectory motion, 35
Transformation
  Galilean, 920
  Lorentz, 921
Transformer, 669
  eddy currents, 674
  step-up/down, 670
  turns ratio, 674
Transient term, 685
Translational motion, 226
Transverse waves, 384
  electromagnetic, 720, 726
Trigonometric formulas (Appendix D), 943
Trigonometric functions, table (Appendix I), 953
Triple point, water, 429
Twin paradox, 333, 932
Two-mass system, 306

U

Unavailable energy, 506
Uncertainty principle, 909
Unified atomic mass unit, u, 64, *440*
Unit vector, 12
Units, electric and magnetic, comment on, 634
Universal gas constant, R, 440

V

Vaporization, heat of, 422
Variable of state, 498
Vector, 11, 12
  addition of, 14
    commutative law, 14
    polygon method, 14
  axial, 240
  components of, 13
  displacement, 12
  element of area, 532
  minus signs, 15
  multiplication
    scalar product, 97
    vector product, 203
  position, 11
  resolving of, 11
  subtraction of, 14
  three dimensions, 15
  unit, 12
Vector product, 203
Velikovsky problem, 269
Velocity, 16
  average, 16
  of efflux, 376
  escape, 351
  Galilean addition, 180
  instantaneous, 17
  phase, 389
  relative, 180
  in simple harmonic motion, 297
  tangential, 48
Velocity addition
  Galilean, 180
  relativistic, 329, 928
Velocity filter, 622
Venturi effect, 377
Virtual image, 744

Virtual object, 749
Viscosity, 363
Volt, 538
Voltmeter, 599
  figure of merit of, 600
Volume charge density, 544
Volume expansion, thermal, 418

W

Water waves, 395, 396
Watt, J., 119
Wave equation
  electromagnetic, 719
  mechanical, 385
  solutions to, 387, 389, 719
Wave function, 907
  meaning of, 908
Wave interference, criteria, 797
Wave mechanics, 906
  probability, interpretation of, 908
Wave nature of particles, 894
Wave number, 390
Wave-particle duality, 794, *887*, 913
Wave plates (optical), 862
Wave train, 796
Wavefront, *395*, 739, 741
Wavelength
  de Broglie, 901
  of electrons, 905
  of light, 740
Waves
  beats, 404
  coherent, 796
  dispersion, 395
  electromagnetic, 710
    energy in, 727
    equation, 719
    radiation of, 725
  energy of, 396
  equation, 385, 719
    solution, 387, 389
  front, 395
  intensity of, 397
  interference, criteria, 797
  light, 796
  longitudinal, 384
  and Maxwell's equations, 718
  mechanical, 384
  number, 390
  phase, 389
  plane, 395, 717, 721
  polarized, 854 (*also see* Polarization)
  power transmitted, 396
  pulse, 384
  reflection, 397
  shock, 403
  speed, 391
  standing, 399
  torsional, 395
  train, 385
  transverse, 384, 854
  traveling, 389
    amplitude, 390
    frequency, 390
    period, 390
    wavelength, 390
  water, 395, 396
Waveshape, periodic, 680
Weber (unit), 634
Weight, 66
  and mass, 66
  units, 67, 68
Wheatstone bridge, 602
Wien's displacement law, 872
Wien's radiation law, 873
Wire, current-carrying, magnetic field of, 642, 645
Work, *96*
  area under F versus x graph, 99
  area under P–V diagram, 460
  constant force, 96
  internal work, 118
  and kinetic energy, 115
  scalar product, 97
  in stretching spring, 101
  varying force, 9
Work-energy relation, 115
Work function, 881
Wye-delta transformation, 614
Wye network, 614

X, Y, Z

X-ray diffraction, 839
  Bragg condition, 840
Young, T., 794
Young's double-slit experiment, 794
Young's modulus, 254
Zero-momentum frame, 183
Zeroth law of thermodynamics, 456

C 3
D 4
E 5
F 6
G 7
H 8
I 9
J 0